CLASSICAL MECHANICS

Gregory's *Classical Mechanics* is a major new textbook for undergraduates in mathematics and physics. It is a thorough, self-contained and highly readable account of a subject many students find difficult. The author's clear and systematic style promotes a good understanding of the subject: each concept is motivated and illustrated by worked examples, while problem sets provide plenty of practice for understanding and technique. Computer assisted problems, some suitable for projects, are also included. The book is structured to make learning the subject easy; there is a natural progression from core topics to more advanced ones and hard topics are treated with particular care. A theme of the book is the importance of conservation principles. These appear first in vectorial mechanics where they are proved and applied to problem solving. They reappear in analytical mechanics, where they are shown to be related to symmetries of the Lagrangian, culminating in Noether's theorem.

- Suitable for a wide range of undergraduate mechanics courses given in mathematics and physics departments: no prior knowledge of the subject is assumed

- Profusely illustrated and thoroughly class-tested, with a clear direct style that makes the subject easy to understand: all concepts are motivated and illustrated by the many worked examples included

- Good, accurately-set problems, with answers in the book: computer assisted problems and projects are also provided. Model solutions for problems available to teachers from www.cambridge.org/Gregory

The author

Douglas Gregory is Professor of Mathematics at the University of Manchester. He is a researcher of international standing in the field of elasticity, and has held visiting positions at New York University, the University of British Columbia, and the University of Washington. He is highly regarded as a teacher of applied mathematics: this, his first book, is the product of many years of teaching experience.

Bloody instructions, which, being taught,
Return to plague th' inventor.

SHAKESPEARE, *Macbeth*, act I, sc. 7

Front Cover The photograph on the front cover shows Mimas, one of the many moons of Saturn; the huge crater was formed by an impact. Mimas takes 22 hours 37 minutes to orbit Saturn, the radius of its orbit being 185,500 kilometres. After reading Chapter 7, you will be able to estimate the mass of Saturn from this data!

CLASSICAL MECHANICS

AN UNDERGRADUATE TEXT

R. DOUGLAS GREGORY

University of Manchester

CAMBRIDGE
UNIVERSITY PRESS

CAMBRIDGE
UNIVERSITY PRESS

University Printing House, Cambridge CB2 8BS, United Kingdom

One Liberty Plaza, 20th Floor, New York, NY 10006, USA

477 Williamstown Road, Port Melbourne, VIC 3207, Australia

314-321, 3rd Floor, Plot 3, Splendor Forum, Jasola District Centre, New Delhi - 110025, India

79 Anson Road, #06-04/06, Singapore 079906

Cambridge University Press is part of the University of Cambridge.

It furthers the University's mission by disseminating knowledge in the pursuit of education, learning and research at the highest international levels of excellence.

www.cambridge.org
Information on this title: www.cambridge.org/9780521534093

First published 2006
16th printing 2018

A catalogue record for this publication is available from the British Library

ISBN 978-0-521-82678-5 Hardback
ISBN 978-0-521-53409-3 Paperback

Contents

Preface

Information for readers

What is this book about and who is it for?

This is a book on **classical mechanics** for **university undergraduates**. It aims to cover all the material normally taught in classical mechanics courses from Newton's laws to Hamilton's equations. If you are attending such a course, you will be unlucky not to find the course material in this book.

What prerequisites are needed to read this book?

It is expected that the reader will have attended an elementary **calculus** course and an elementary course on **differential equations** (ODEs). A previous course in mechanics is helpful but not essential. *This book is self-contained in the sense that it starts from the beginning and assumes no prior knowledge of mechanics.* However, in a general text such as this, the early material is presented at a brisker pace than in books that are specifically aimed at the beginner.

What is the style of the book?

The book is written in a crisp, no nonsense style; in short, there is no waffle! The object is to get the reader to the important points as quickly and easily as possible, consistent with good understanding.

Are there plenty of examples with full solutions?

Yes there are. Every new concept and technique is reinforced by **fully worked examples**. The author's advice is that the reader should think how he or she would do each worked example *before* reading the solution; much more will be learned this way!

Are there plenty of problems with answers?

Yes there are. At the end of each chapter there is a large collection of problems. For convenience, these are arranged by topic and trickier problems are marked with a star. **Answers are provided to all of the problems**. A feature of the book is the inclusion of computer assisted problems. These are interesting physical problems that cannot be solved analytically, but can be solved easily with computer assistance.

Where can I find more information?

More information about this book can be found on the book's homepage

`http://www.cambridge.org/Gregory`

All feedback from readers is welcomed. Please e-mail your comments, corrections and good ideas by clicking on the comments button on the book's homepage.

Information for lecturers

Scope of the book and prerequisites

This book aims to cover all the material normally taught in undergraduate mechanics courses from Newton's laws to Hamilton's equations. It assumes that the students have attended an elementary calculus course and an elementary course on ODEs, but no more. The book is self contained and, in principle, it is not essential that the students should have studied mechanics before. However, their lives will be made easier if they have!

Inspection copy and Solutions Manual

Any lecturer who is giving an undergraduate course on classical mechanics can request an **inspection copy** of this book. Simply go to the book's homepage

`http://www.cambridge.org/Gregory`

and follow the links.

Lecturers who adopt this book for their course may receive the **Solutions Manual**. This has a **complete set of detailed solutions** to the problems at the end of the chapters. To obtain the Solutions Manual, just send an e-mail giving your name, affiliation, and details of the course to `solutions@cambridge.org`

Feedback

All feedback from instructors and lecturers is welcomed. Please e-mail your comments via the link on the book's homepage

Acknowledgements

I am very grateful to many friends and colleagues for their helpful comments and suggestions while this book was in preparation. But most of all I thank my wife Win for her unstinting support and encouragement, without which the book could not have been written at all.

Part One

NEWTONIAN MECHANICS OF A SINGLE PARTICLE

CHAPTERS IN PART ONE

Chapter One

The algebra and calculus of vectors

KEY FEATURES

The key features of this chapter are the **rules of vector algebra** and **differentiation of vector functions** of a scalar variable.

This chapter begins with a review of the rules and applications of **vector algebra.** Almost every student taking a mechanics course will already have attended a course on vector algebra, and so, instead of covering the subject in full detail, we present, for easy reference, a summary of vector operations and their important properties, together with a selection of worked examples.

The chapter closes with an account of the **differentiation of vector functions** of a scalar variable. Unlike the vector algebra sections, this is treated in full detail. Applications include the **tangent vector** and **normal vector** to a curve. These will be needed in the next chapter in order to interpret the velocity and acceleration vectors.

1.1 VECTORS AND VECTOR QUANTITIES

Most physical quantities can be classified as being **scalar quantities** or **vector quantities**. The temperature in a room is an example of a scalar quantity. It is so called because its *value* is a scalar, which, in the present context, means a *real number*. Other examples of scalar quantities are the volume of a can, the density of iron, and the pressure of air in a tyre. Vector quantities are defined as follows:

Definition 1.1 ***Vector quantity*** *If a quantity Q has a* ***magnitude*** *and a* ***direction*** *associated with it, then Q is said to be a* ***vector quantity***. [Here, magnitude means a positive real number and direction is specified relative to some underlying reference frame* that we regard as fixed.]

The **displacement** of a particle† is an example of a vector quantity. Suppose the particle starts from the point A and, after moving in a general manner, ends up at the

* See section 2.2 for an explanation of the term 'reference frame'.

† A particle is an idealised body that occupies only a single point of space.

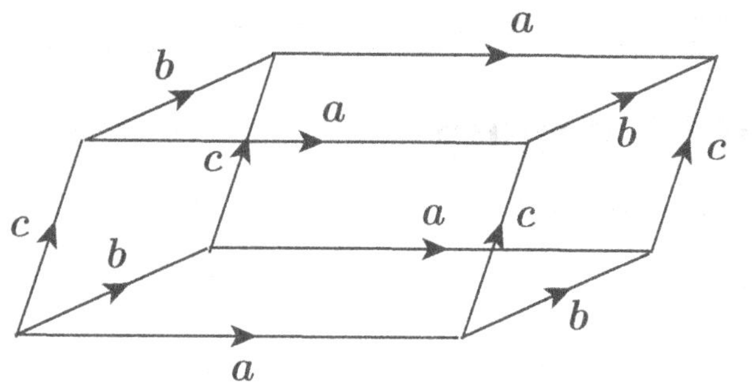

FIGURE 1.1 Four different representations of each of the vectors ***a***, ***b*** ***c*** form the twelve edges of the parallelepiped box.

point B. The *magnitude* of the displacement is the distance AB and the *direction* of the displacement is the direction of the straight line joining A to B (in that order). Another example is the **force** applied to a body by a rope. In this case, the *magnitude* is the strength of the force (a real positive quantity) and the *direction* is the direction of the rope (away from the body). Other examples of vector quantities are the velocity of a body and the value of the electric (or magnetic) field. In order to manipulate all such quantities without regard to their physical origin, we introduce the concept of a vector as an *abstract quantity*.

Definition 1.2 ***Vector*** *A* ***vector*** *is an* ***abstract*** *quantity characterised by the two properties* ***magnitude*** *and* ***direction***. *Thus two vectors are equal if they have the same magnitude and the same direction.**

Notation. Vectors are written in bold type, for example $\boldsymbol{a}$, $\boldsymbol{b}$, $\boldsymbol{r}$ or $\boldsymbol{F}$. The **magnitude** of the vector $\boldsymbol{a}$, which is a real positive number, is written $|\boldsymbol{a}|$, or sometimes† simply a.

It is convenient to define operations involving abstract vectors by reference to some simple, easily visualised vector quantity. The standard choice is the set of directed **line segments**. Each straight line joining two points (P and Q say, in that order) is a vector quantity, where the magnitude is the distance PQ and the direction is the direction of Q relative to P. We call this the line segment $\overrightarrow{PQ}$ and we say that it *represents* some abstract vector $\boldsymbol{a}$.‡ Note that each vector $\boldsymbol{a}$ is represented by infinitely many different line segments, as indicated in Figure 1.1.

* In order that our set of vectors should have a standard algebra, we also include a special vector whose magnitude is zero and whose direction is not defined. This is called the **zero vector** and written **0**. The zero vector is not the same thing as the number zero!

† It is often useful to denote the magnitudes of the vectors $\boldsymbol{a}$, $\boldsymbol{b}$, $\boldsymbol{c}$, ... by a, b, c, ..., but this does risk confusion. Take care!

‡ The zero vector is represented by line segments whose end point and starting point are coincident.

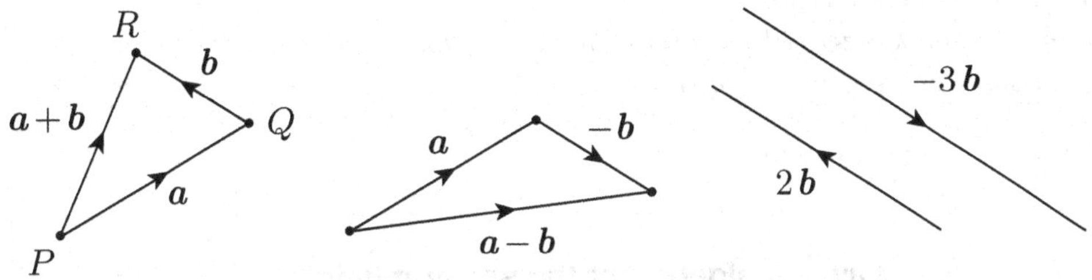

FIGURE 1.2 Addition, subtraction and scalar multiplication of vectors.

1.2 LINEAR OPERATIONS: $\boldsymbol{a} + \boldsymbol{b}$ AND $\lambda \boldsymbol{a}$

Since vectors are abstract quantities, we can define sums and products of vectors in any way we like. However, in order to be of any use, the definitions must create some coherent algebra and represent something of interest when applied to a range of vector quantities. Also, our definitions must be independent of the particular representations used to construct them. The definitions that follow satisfy all these requirements.

The vector sum $\boldsymbol{a} + \boldsymbol{b}$

Definition 1.3 ***Sum of vectors*** *Let* $\boldsymbol{a}$ *and* $\boldsymbol{b}$ *be any two vectors. Take any representation* $\overrightarrow{PQ}$ *of* $\boldsymbol{a}$ *and suppose the line segment* $\overrightarrow{QR}$ *represents* $\boldsymbol{b}$*. Then the* ***sum*** $\boldsymbol{a} + \boldsymbol{b}$ *of* $\boldsymbol{a}$ *and* $\boldsymbol{b}$ *is the* ***vector*** *represented by the line segment* $\overrightarrow{PR}$*, as shown in Figure 1.2 (left).*

Laws of algebra for the vector sum

(i) $\boldsymbol{b} + \boldsymbol{a} = \boldsymbol{a} + \boldsymbol{b}$ (commutative law)

(ii) $\boldsymbol{a} + (\boldsymbol{b} + \boldsymbol{c}) = (\boldsymbol{a} + \boldsymbol{b}) + \boldsymbol{c}$ (associative law)

Definition 1.4 ***Negative of a vector*** *Let* $\boldsymbol{b}$ *be any vector. Then the vector with the same magnitude as* $\boldsymbol{b}$ *and the* ***opposite*** *direction is called the* ***negative*** *of* $\boldsymbol{b}$ *and is written* $-\boldsymbol{b}$*.* ***Subtraction*** *by* $\boldsymbol{b}$ *is then defined by*

$$\boldsymbol{a} - \boldsymbol{b} = \boldsymbol{a} + (-\boldsymbol{b}).$$

[That is, to subtract $\boldsymbol{b}$ just add $-\boldsymbol{b}$, as shown in Figure 1.2 (centre).]

The scalar multiple $\lambda \boldsymbol{a}$

Definition 1.5 ***Scalar multiple*** *Let* $\boldsymbol{a}$ *be a vector and* λ *be a scalar (a real number). Then the* ***scalar multiple*** $\lambda \boldsymbol{a}$ *is the vector whose magnitude is* $|\lambda||\boldsymbol{a}|$ *and whose direction is*

(i) *the same as* $\boldsymbol{a}$ *if* λ *is positive,*
(ii) *undefined if* λ *is zero (the answer is the zero vector),*
(iii) *the same as* $-\boldsymbol{a}$ *if* λ *is negative.*

It follows that $-(\lambda\boldsymbol{a}) = (-\lambda)\boldsymbol{a}$.

Laws of algebra for the scalar multiple

(i) $\lambda(\mu\boldsymbol{a}) = (\lambda\mu)\boldsymbol{a}$ (associative law)

(ii) $\lambda(\boldsymbol{a}+\boldsymbol{b}) = \lambda\boldsymbol{a}+\lambda\boldsymbol{b}$ and $(\lambda+\mu)\boldsymbol{a} = \lambda\boldsymbol{a}+\mu\boldsymbol{a}$ (distributive laws)

The effect of the above laws is that **linear combinations** of vectors can be manipulated just *as if* the vectors were symbols representing real or complex numbers.

Example 1.1 *Laws for vector sum and scalar multiple*

Simplify the expression $3(2\boldsymbol{a}-4\boldsymbol{b}) - 2(2\boldsymbol{a}-\boldsymbol{b})$.

Solution

On this one occasion we will do the simplification by strict application of the laws. It is instructive to decide which laws are being used at each step!

$$\begin{aligned}3(2\boldsymbol{a}-4\boldsymbol{b}) - 2(2\boldsymbol{a}-\boldsymbol{b}) &= 3\Big(2\boldsymbol{a}+(-4)\boldsymbol{b}\Big) + (-2)\Big(2\boldsymbol{a}+(-1)\boldsymbol{b}\Big)\\ &= \Big(6\boldsymbol{a}+(-12)\boldsymbol{b}\Big) + \Big((-4)\boldsymbol{a}+2\boldsymbol{b}\Big)\\ &= \Big(6\boldsymbol{a}+(-4)\boldsymbol{a}\Big) + \Big((-12)\boldsymbol{b}+2\boldsymbol{b}\Big)\\ &= 2\boldsymbol{a}+(-10)\boldsymbol{b} = 2\boldsymbol{a}-10\boldsymbol{b}. \blacksquare\end{aligned}$$

Unit vectors

A vector of **unit magnitude** is called a **unit vector**. If any vector $\boldsymbol{a}$ is divided by its own magnitude, the result is a *unit vector* having the *same direction* as $\boldsymbol{a}$. This new vector is denoted by $\widehat{\boldsymbol{a}}$ so that

$$\widehat{\boldsymbol{a}} = \boldsymbol{a}/|\boldsymbol{a}|.$$

Basis sets

Suppose $\boldsymbol{a}$ and $\boldsymbol{b}$ are two non-zero vectors, with the direction of $\boldsymbol{b}$ neither the same nor opposite to that of $\boldsymbol{a}$. Let $\overrightarrow{OA}$, $\overrightarrow{OB}$ be representations of $\boldsymbol{a}$, $\boldsymbol{b}$ and let $\mathcal{P}$ be the plane

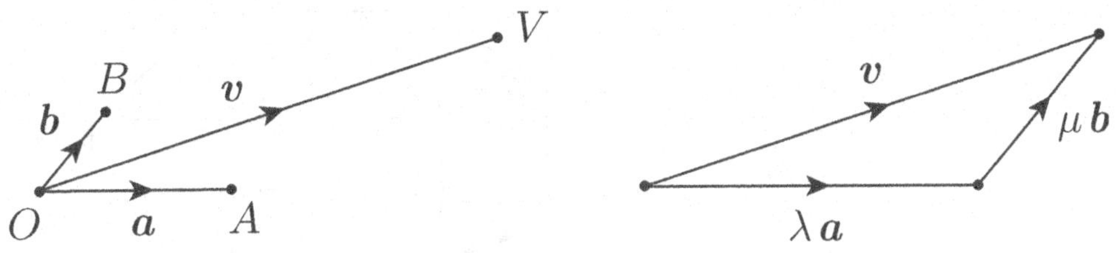

FIGURE 1.3 The set $\{a, b\}$ is a basis for all vectors lying in the plane OAB.

containing the triangle OAB. Then (see Figure 1.3) any vector $\boldsymbol{v}$ whose representation $\overrightarrow{OV}$ lies in the plane $\mathcal{P}$ can be written in the form

$$\boldsymbol{v} = \lambda \boldsymbol{a} + \mu \boldsymbol{b}, \tag{1.1}$$

where the coefficients λ, μ are unique. Vectors that have their directions parallel to the same plane are said to be **coplanar**. Thus we have shown that *any vector coplanar with* $\boldsymbol{a}$ *and* $\boldsymbol{b}$ *can be expanded uniquely in the form* (1.1). It is also apparent that this expansion set cannot be reduced in number (in this case to a single vector). For these reasons the pair of vectors $\{\boldsymbol{a}, \boldsymbol{b}\}$ is said to be a **basis set** for vectors lying* in the plane $\mathcal{P}$.

Suppose now that $\{\boldsymbol{a}, \boldsymbol{b}, \boldsymbol{c}\}$ is a set of three non-coplanar vectors. Then any vector $\boldsymbol{v}$, *without restriction*, can be written in the form

$$\boldsymbol{v} = \lambda \boldsymbol{a} + \mu \boldsymbol{b} + \nu \boldsymbol{c}, \tag{1.2}$$

where the coefficients λ, μ, ν are unique. In this case we say that the set $\{\boldsymbol{a}, \boldsymbol{b}, \boldsymbol{c}\}$ is a **basis set** for all three-dimensional vectors. Although *any* set of three non-coplanar vectors forms a basis, it is most convenient to take the basis vectors to be *orthogonal unit vectors*. In this case the basis set† is usually denoted by $\{\boldsymbol{i}, \boldsymbol{j}, \boldsymbol{k}\}$ and is said to be an **orthonormal basis**. The representation of a general vector $\boldsymbol{v}$ in the form

$$\boldsymbol{v} = \lambda \boldsymbol{i} + \mu \boldsymbol{j} + \nu \boldsymbol{k}$$

is common in problem solving.

In applications involving the cross product of vectors, the distinction between **right-** and **left-handed** basis sets actually matters. There is no experiment in classical mechanics or electromagnetism that can distinguish between right- and left-handed sets. The difference can only be exhibited by a model or some familiar object that exhibits 'handedness', such as a corkscrew.‡ Figure 1.4 shows a **right-handed orthonormal basis set** attached to a well known object.

---

* Strictly speaking vectors are abstract quantities that do not *lie* anywhere. This phrase should be taken to mean '*vectors whose directions are parallel to the plane* $\mathcal{P}$'.

† It should be remembered that there are *infinitely* many basis sets made up of orthogonal unit vectors.

‡ Suppose that the non-coplanar vectors $\{\boldsymbol{a}, \boldsymbol{b}, \boldsymbol{c}\}$ have representations $\overrightarrow{OA}$, $\overrightarrow{OB}$, $\overrightarrow{OC}$ respectively. Place an ordinary corkscrew with the screw lying along the line through O perpendicular to the plane OAB,

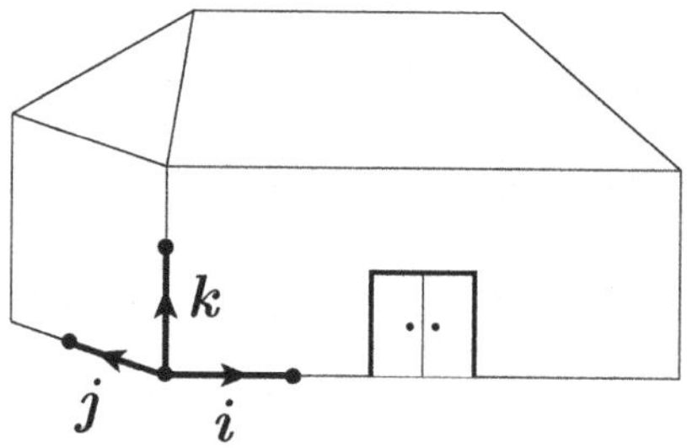

FIGURE 1.4 A **standard basis** set $\{\boldsymbol{i}, \boldsymbol{j}, \boldsymbol{k}\}$ is both **orthonormal** and **right-handed**.

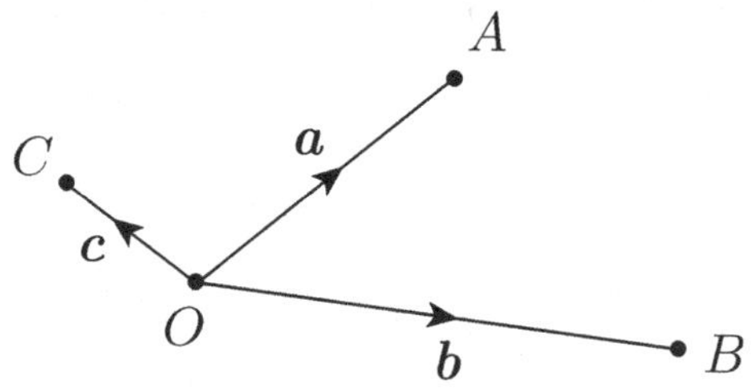

FIGURE 1.5 The points A, B, C have position vectors $\boldsymbol{a}, \boldsymbol{b}, \boldsymbol{c}$ relative to the origin O.

Definition 1.6 ***Standard basis set*** *If an **orthonormal basis** $\{\boldsymbol{i}, \boldsymbol{j}, \boldsymbol{k}\}$ is also **right-handed** (as shown in Figure 1.4), we will call it a **standard basis**.*

Position vectors and vector geometry

Suppose that O is a fixed point of space. Then relative to the origin O (and relative to the underlying reference frame), any point of space, such as A, has an associated line segment, $\overrightarrow{OA}$, which represents some vector $\boldsymbol{a}$. Conversely, the vector $\boldsymbol{a}$ is sufficient to specify the position of the point A.

Definition 1.7 ***Position vector*** *The vector **a** is called the **position vector** of the point A relative to the origin O, [It is standard practice, and very convenient, to denote the position vectors of the points $A, B, C, \ldots$ by **a**, **b**, **c**, and so on, as shown in Figure 1.5.]*

Since vectors can be used to specify the positions of points in space, we can now use the laws of vector algebra to prove* results in Euclidean geometry. This is not just an academic exercise. Familiarity with geometrical concepts is an important part of mechanics. We begin with the following useful result:

---

and the handle parallel to OA. Now turn the corkscrew until the handle is parallel to OB and note the direction in which the corkscrew *would* move if it were 'in action'. (The direction of the turn must be such that the angle turned through is at most 180°.) If OC makes an *acute angle* with this direction, the set $\{\boldsymbol{a}, \boldsymbol{b}, \boldsymbol{c}\}$ (in that order) is *right-handed*; if OC makes an *obtuse* angle with this direction then the set is *left-handed*.

* Some properties of Euclidean geometry have been used to prove the laws of vector algebra. However, this does not prevent us from giving valid proofs of other results.

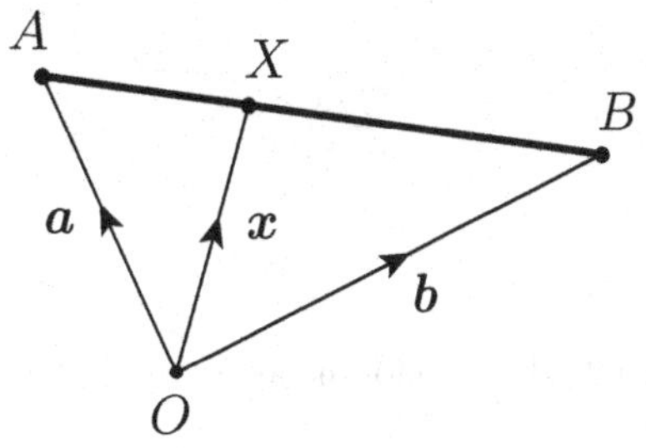

FIGURE 1.6 The point X divides the line AB in the ratio $\lambda : \mu$.

Example 1.2 *Point dividing a line in a given ratio*

The points A and B have position vectors $\boldsymbol{a}$ and $\boldsymbol{b}$ relative to an origin O. Find the position vector $\boldsymbol{x}$ of the point X that divides the line AB in the ratio $\lambda : \mu$ (that is $AX/XB = \lambda/\mu$).

Solution

It follows from Figure 1.6 that $\boldsymbol{x}$ is given by*

$$\boldsymbol{x} = \boldsymbol{a} + \overrightarrow{AX} = \boldsymbol{a} + \left(\frac{\lambda}{\lambda+\mu}\right)\overrightarrow{AB}$$
$$= \boldsymbol{a} + \left(\frac{\lambda}{\lambda+\mu}\right)(\boldsymbol{b}-\boldsymbol{a}) = \frac{\mu\boldsymbol{a}+\lambda\boldsymbol{b}}{\lambda+\mu}.$$

In particular, the **mid-point** of the line AB has position vector $\frac{1}{2}(\boldsymbol{a}+\boldsymbol{b})$. ■

Example 1.3 *Centroid of a triangle*

Show that the three medians of any triangle meet in a point (the centroid) which divides each of them in the ratio 2:1.

Solution

Let the triangle be ABC where the points A, B, C have position vectors $\boldsymbol{a}$, $\boldsymbol{b}$, $\boldsymbol{c}$ relative to some origin O. Then the mid-point P of the side BC has position vector $\boldsymbol{p} = \frac{1}{2}(\boldsymbol{b}+\boldsymbol{c})$. The point X that divides the median AP in the ratio 2:1 therefore has position vector

$$\boldsymbol{x} = \frac{\boldsymbol{a}+2\boldsymbol{p}}{2+1} = \frac{\boldsymbol{a}+\boldsymbol{b}+\boldsymbol{c}}{3}.$$

The position vectors of the corresponding points on the other two medians can be found by cyclic permutation of the vectors $\boldsymbol{a}$, $\boldsymbol{b}$, $\boldsymbol{c}$ and clearly give the same value. Hence all three points are coincident and so the three medians meet there. ■

---

* Strictly speaking we should not write expressions like $\boldsymbol{a} + \overrightarrow{AX}$ since the sum we defined was the sum of two *vectors*, not a vector and a line segment. What we really mean is '*the sum of* $\boldsymbol{a}$ *and the vector represented by the line segment* $\overrightarrow{AX}$'. Pure mathematicians would not approve but this notation is so convenient we will use it anyway. It's all part of living dangerously!

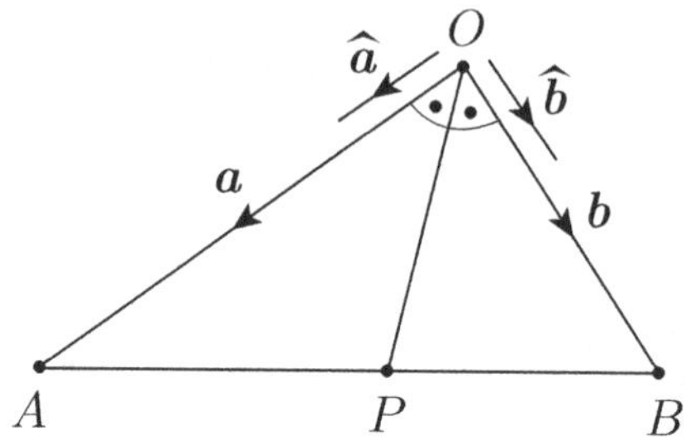

FIGURE 1.7 The bisector theorem: $AP/PB = OA/OB$.

Example 1.4 *The bisector theorem*

In a triangle OAB, the bisector of the angle $A\widehat{O}B$ meets the line AB at the point P. Show that $AP/PB = OA/OB$.

Solution

Let the vertex O be the origin of vectors* and let the position vectors of the vertices A, B relative to O be $\boldsymbol{a}$, $\boldsymbol{b}$ as shown in Figure 1.7. The point with position vector $\boldsymbol{a} + \boldsymbol{b}$ does *not* lie the bisector OP in general since the vectors $\boldsymbol{a}$ and $\boldsymbol{b}$ have different magnitudes a and b. However, by symmetry, the point with position vector $\widehat{\boldsymbol{a}} + \widehat{\boldsymbol{b}}$ *does* lie on the bisector and a general point X on the bisector has a position vector $\boldsymbol{x}$ of the form

$$\boldsymbol{x} = \lambda\left(\widehat{\boldsymbol{a}} + \widehat{\boldsymbol{b}}\right) = \lambda\left(\frac{\boldsymbol{a}}{a} + \frac{\boldsymbol{b}}{b}\right) = \lambda\left(\frac{b\,\boldsymbol{a} + a\,\boldsymbol{b}}{ab}\right) = \left(\frac{b\,\boldsymbol{a} + a\,\boldsymbol{b}}{K}\right),$$

where $K = ab/\lambda$ is a new constant. Now X will lie on the line AB if its position vector has the form $(\mu\boldsymbol{a} + \lambda\boldsymbol{b})/(\lambda + \mu)$, that is, if $K = a + b$. Hence the position vector $\boldsymbol{p}$ of P is

$$\boldsymbol{p} = \frac{b\,\boldsymbol{a} + a\,\boldsymbol{b}}{a + b}.$$

Moreover we see that P divides that line AB in the ratio $a : b$, that is, $AP/PB = OA/OB$ as required. ■

1.3 THE SCALAR PRODUCT $\boldsymbol{a} \cdot \boldsymbol{b}$

Definition 1.8 ***Scalar product*** *Suppose the vectors* $\boldsymbol{a}$ *and* $\boldsymbol{b}$ *have representations* $\overrightarrow{OA}$ *and* $\overrightarrow{OB}$. *Then the* ***scalar product*** $\boldsymbol{a} \cdot \boldsymbol{b}$ *of* $\boldsymbol{a}$ *and* $\boldsymbol{b}$ *is defined by*

$$\boldsymbol{a} \cdot \boldsymbol{b} = |\boldsymbol{a}||\boldsymbol{b}|\cos\theta, \tag{1.3}$$

where θ *is the angle between* OA *and* OB. [Note that $\boldsymbol{a} \cdot \boldsymbol{b}$ is a **scalar** quantity.]

---

* One can always take a special point of the figure as origin. The penalty is that the symmetry of the labelling is lost.

Laws of algebra for the scalar product

(i) $b \cdot a = a \cdot b$ (commutative law)

(ii) $a \cdot (b + c) = a \cdot b + a \cdot c$ (distributive law)

(iii) $(\lambda a) \cdot b = \lambda(a \cdot b)$ (associative with scalar multiplication)

Properties of the scalar product

(i) $a \cdot a = |a|^2$.

(ii) The scalar product $a \cdot b = 0$ if (and only if) a and b are perpendicular (or one of them is zero).

(iii) If $\{i, j, k\}$ is an orthonormal basis then

$$i \cdot i = j \cdot j = k \cdot k = 1, \qquad i \cdot j = j \cdot k = k \cdot i = 0.$$

(iv) If $a_1 = \lambda_1 i + \mu_1 j + \nu_1 k$ and $a_2 = \lambda_2 i + \mu_2 j + \nu_2 k$ then

$$a_1 \cdot a_2 = \lambda_1\lambda_2 + \mu_1\mu_2 + \nu_1\nu_2.$$

Example 1.5 *Numerical example on the scalar product*

If $a = 2i - j + 2k$ and $b = 4i - 3k$, find the magnitudes of a and b and the angle between them.

Solution

$|a|^2 = a \cdot a = (2i - j + 2k) \cdot (2i - j + 2k) = 2^2 + (-1)^2 + 2^2 = 9$. Hence $|a| = 3$. Similarly $|b|^2 = 4^2 + 0^2 + (-3)^2 = 25$ so that $|b| = 5$. Also $a \cdot b = 8 + 0 + (-6) = 2$. Since $a \cdot b = |a||b| \cos\theta$, it follows that $2 = 3 \times 5 \times \cos\theta$ so that $\cos\theta = 2/15$.

Hence the magnitudes of a and b are 3 and 5, and the angle between them is $\cos^{-1}(2/15)$. ■

Example 1.6 *Apollonius's theorem*

In the triangle OAB, M is the mid-point of AB. Show that $(OA)^2 + (OB)^2 = 2(OM)^2 + 2(AM)^2$.

Solution

Let the vertex O be the origin of vectors and let the position vectors of A and B be a and b. Then the position vector of M is $\frac{1}{2}(a + b)$. Then

$$\begin{aligned} 4(OM)^2 &= |a + b|^2 = (a + b) \cdot (a + b) \\ &= a \cdot a + b \cdot b + 2a \cdot b = |a|^2 + |b|^2 + 2a \cdot b \end{aligned}$$

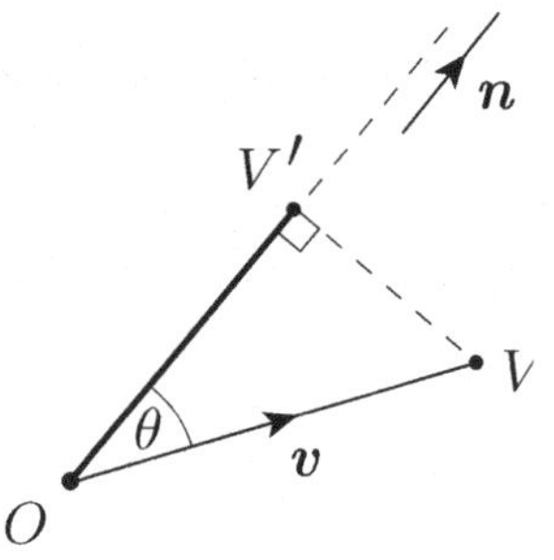

FIGURE 1.8 The component of $\boldsymbol{v}$ in the direction of the unit vector $\boldsymbol{n}$ is equal to OV', the **projection** of OV onto the line through O parallel to $\boldsymbol{n}$.

and

$$\begin{aligned}4(AM)^2 = (AB)^2 = |\boldsymbol{a}-\boldsymbol{b}|^2 &= (\boldsymbol{a}-\boldsymbol{b})\cdot(\boldsymbol{a}-\boldsymbol{b})\\ &= \boldsymbol{a}\cdot\boldsymbol{a}+\boldsymbol{b}\cdot\boldsymbol{b}-2\boldsymbol{a}\cdot\boldsymbol{b} = |\boldsymbol{a}|^2+|\boldsymbol{b}|^2-2\boldsymbol{a}\cdot\boldsymbol{b}.\end{aligned}$$

Hence

$$2(OM)^2+2(AM)^2 = |\boldsymbol{a}|^2+|\boldsymbol{b}|^2 = (OA)^2+(OB)^2$$

as required. ■

Components of a vector

Definition 1.9 *Components of a vector* *Let $\boldsymbol{n}$ be a unit vector. Then the* ***component*** *of the vector $\boldsymbol{v}$ in the direction of $\boldsymbol{n}$ is defined to be $\boldsymbol{v}\cdot\boldsymbol{n}$. The component of $\boldsymbol{v}$ in the direction of a general vector $\boldsymbol{a}$ is therefore $\boldsymbol{v}\cdot\widehat{\boldsymbol{a}}$.*

Properties of components

(i) The component $\boldsymbol{v}\cdot\boldsymbol{n}$ has a simple geometrical significance. Let $\overrightarrow{OV}$ be a representation of $\boldsymbol{v}$ as shown in Figure 1.8. Then

$$\boldsymbol{v}\cdot\boldsymbol{n} = |\boldsymbol{v}||\boldsymbol{n}|\cos\theta = OV\cos\theta = OV',$$

where OV' is the **projection** of OV onto the line through O parallel to $\boldsymbol{n}$.

(ii) Suppose that $\boldsymbol{v}$ is a sum of vectors, $\boldsymbol{v} = \boldsymbol{v}_1+\boldsymbol{v}_2+\boldsymbol{v}_3$ say. Then the component of $\boldsymbol{v}$ in the direction of $\boldsymbol{n}$ is

$$\boldsymbol{v}\cdot\boldsymbol{n} = (\boldsymbol{v}_1+\boldsymbol{v}_2+\boldsymbol{v}_3)\cdot\boldsymbol{n} = (\boldsymbol{v}_1\cdot\boldsymbol{n})+(\boldsymbol{v}_2\cdot\boldsymbol{n})+(\boldsymbol{v}_3\cdot\boldsymbol{n}),$$

by the distributive law for the scalar product. Thus, the *component of the sum of a number of vectors in a given direction is equal to the sum of the components of the individual vectors* in that direction.

(iii) If a vector $\boldsymbol{v}$ is expanded in terms of a general basis set $\{\boldsymbol{a}, \boldsymbol{b}, \boldsymbol{c}\}$ in the form $\boldsymbol{v} = \lambda\boldsymbol{a}+\mu\boldsymbol{b}+\nu\boldsymbol{c}$, the coefficients λ, μ, ν are *not* the components of the vector $\boldsymbol{v}$ in the

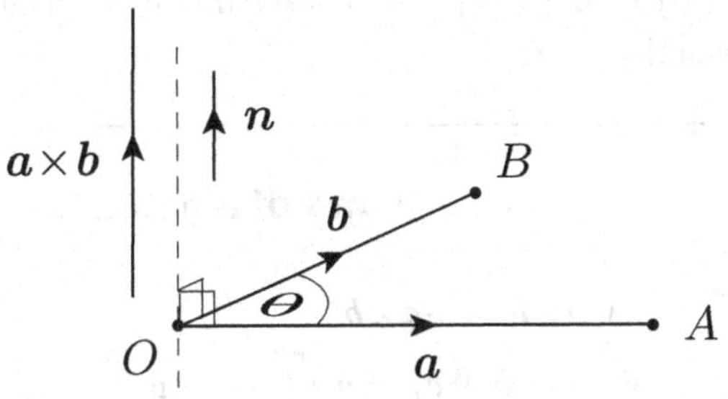

FIGURE 1.9 The vector product $\boldsymbol{a}\times\boldsymbol{b} = (|\boldsymbol{a}||\boldsymbol{b}|\sin\theta)\,\boldsymbol{n}$.

directions of $\boldsymbol{a}$, $\boldsymbol{b}$, $\boldsymbol{c}$. However if $\boldsymbol{v}$ is expanded in terms of an *orthonormal* basis set $\{\boldsymbol{i}, \boldsymbol{j}, \boldsymbol{k}\}$ in the form $\boldsymbol{v} = \lambda\boldsymbol{i} + \mu\boldsymbol{j} + \nu\boldsymbol{k}$, then the component of $\boldsymbol{v}$ in the $\boldsymbol{i}$-direction is

$$\begin{aligned}\boldsymbol{v}\cdot\boldsymbol{i} &= (\lambda\boldsymbol{i} + \mu\boldsymbol{j} + \nu\boldsymbol{k})\cdot\boldsymbol{i} = \lambda(\boldsymbol{i}\cdot\boldsymbol{i}) + \mu(\boldsymbol{j}\cdot\boldsymbol{i}) + \nu(\boldsymbol{k}\cdot\boldsymbol{i})\\ &= \lambda + 0 + 0 = \lambda.\end{aligned}$$

Similarly μ and ν are the components of $\boldsymbol{v}$ in the $\boldsymbol{j}$- and $\boldsymbol{k}$-directions. Hence *when a vector $\boldsymbol{v}$ is expanded in terms of an orthonormal basis set $\{\boldsymbol{i}, \boldsymbol{j}, \boldsymbol{k}\}$ in the form $\boldsymbol{v} = \lambda\boldsymbol{i} + \mu\boldsymbol{j} + \nu\boldsymbol{k}$, the coefficients λ, μ, ν are the components of $\boldsymbol{v}$ in the* **i**- **j**- *and* **k**-*directions.*

Example 1.7 *Numerical example on components*

If $\boldsymbol{v} = 6\boldsymbol{i} - 3\boldsymbol{j} + 15\boldsymbol{k}$ and $\boldsymbol{a} = 2\boldsymbol{i} - \boldsymbol{j} - 2\boldsymbol{k}$, find the component of $\boldsymbol{v}$ in the direction of $\boldsymbol{a}$.

Solution

$|\boldsymbol{a}|^2 = \boldsymbol{a}\cdot\boldsymbol{a} = 2^2 + (-1)^2 + (-2)^2 = 9$. Hence $|\boldsymbol{a}| = 3$ and

$$\widehat{\boldsymbol{a}} = \frac{\boldsymbol{a}}{|\boldsymbol{a}|} = \frac{2\boldsymbol{i} - \boldsymbol{j} - 2\boldsymbol{k}}{3}.$$

The required component of $\boldsymbol{v}$ is therefore

$$\boldsymbol{v}\cdot\widehat{\boldsymbol{a}} = (6\boldsymbol{i} - 3\boldsymbol{j} + 15\boldsymbol{k})\cdot\left(\frac{2\boldsymbol{i} - \boldsymbol{j} - 2\boldsymbol{k}}{3}\right) = \frac{12 + 3 - 30}{3} = -5. \blacksquare$$

1.4 THE VECTOR PRODUCT $a \times b$

Definition 1.10 ***Vector product*** *Suppose the vectors* **a** *and* **b** *have representations $\overrightarrow{OA}$ and $\overrightarrow{OB}$ and let* **n** *be the unit vector perpendicular to the plane OAB and such that $\{\boldsymbol{a}, \boldsymbol{b}, \boldsymbol{n}\}$ is a* **right-handed** *set. Then the* **vector product** $\boldsymbol{a}\times\boldsymbol{b}$ *of* **a** *and* **b** *is defined by*

$$\boldsymbol{a}\times\boldsymbol{b} = (|\boldsymbol{a}||\boldsymbol{b}|\sin\theta)\,\boldsymbol{n}, \qquad (1.4)$$

where θ $(0 \le \theta \le 180°)$ *is the angle between* OA *and* OB. [Note that $\boldsymbol{a}\times\boldsymbol{b}$ is a **vector** quantity.]

Laws of algebra for the vector product

(i) $\boldsymbol{b}\times\boldsymbol{a} = -\boldsymbol{a}\times\boldsymbol{b}$ (**anti**-commutative law)

(ii) $\boldsymbol{a}\times(\boldsymbol{b}+\boldsymbol{c}) = \boldsymbol{a}\times\boldsymbol{b}+\boldsymbol{a}\times\boldsymbol{c}$ (distributive law)

(iii) $(\lambda\boldsymbol{a})\times\boldsymbol{b} = \lambda(\boldsymbol{a}\times\boldsymbol{b})$ (associative with scalar multiplication)

Since the vector product is anti-commutative, the *order of the terms in vector products must be preserved.* The vector product is not associative.

Properties of the vector product

(i) $\boldsymbol{a}\times\boldsymbol{a} = \boldsymbol{0}$.

(ii) The vector product $\boldsymbol{a}\times\boldsymbol{b} = \boldsymbol{0}$ if (and only if) $\boldsymbol{a}$ and $\boldsymbol{b}$ are parallel (or one of them is zero).

(iii) If $\{\boldsymbol{i}, \boldsymbol{j}, \boldsymbol{k}\}$ is a standard basis then

$$\boldsymbol{i}\times\boldsymbol{j} = \boldsymbol{k}, \quad \boldsymbol{k}\times\boldsymbol{i} = \boldsymbol{j}, \quad \boldsymbol{j}\times\boldsymbol{k} = \boldsymbol{i}, \qquad \boldsymbol{i}\times\boldsymbol{i} = \boldsymbol{j}\times\boldsymbol{j} = \boldsymbol{k}\times\boldsymbol{k} = \boldsymbol{0}.$$

(iv) If $\boldsymbol{a}_1 = \lambda_1\boldsymbol{i} + \mu_1\boldsymbol{j} + \nu_1\boldsymbol{k}$ and $\boldsymbol{a}_2 = \lambda_2\boldsymbol{i} + \mu_2\boldsymbol{j} + \nu_2\boldsymbol{k}$ then

$$\boldsymbol{a}_1\times\boldsymbol{a}_2 = \begin{vmatrix} \boldsymbol{i} & \boldsymbol{j} & \boldsymbol{k} \\ \lambda_1 & \mu_1 & \nu_1 \\ \lambda_2 & \mu_2 & \nu_2 \end{vmatrix}$$

where the determinant is to be evaluated by the first row.

Example 1.8 *Numerical example on vector product*

If $\boldsymbol{a} = 2\boldsymbol{i} - \boldsymbol{j} + 2\boldsymbol{k}$ and $\boldsymbol{b} = -\boldsymbol{i} - 3\boldsymbol{k}$, find a unit vector perpendicular to both $\boldsymbol{a}$ and $\boldsymbol{b}$.

Solution

The vector $\boldsymbol{a}\times\boldsymbol{b}$ is perpendicular to both $\boldsymbol{a}$ and $\boldsymbol{b}$. Now

$$\begin{aligned}\boldsymbol{a}\times\boldsymbol{b} &= \begin{vmatrix} \boldsymbol{i} & \boldsymbol{j} & \boldsymbol{k} \\ 2 & -1 & 2 \\ -1 & 0 & -3 \end{vmatrix} \\ &= (3-0)\,\boldsymbol{i} - ((-6)-(-2))\,\boldsymbol{j} + (0-1)\,\boldsymbol{k} \\ &= 3\boldsymbol{i} + 4\boldsymbol{j} - \boldsymbol{k}.\end{aligned}$$

The magnitude of this vector is $\left(3^2 + 4^2 + (-1)^2\right)^{1/2} = (26)^{1/2}$. Hence the required unit vector can be either of $\pm\,(3\boldsymbol{i} + 4\boldsymbol{j} - \boldsymbol{k})\,/(26)^{1/2}$. ■

1.5 TRIPLE PRODUCTS

Triple products are not new operations but are simply one product followed by another. There are two kinds of triple product whose values are scalar and vector respectively.

Triple scalar product

An expression of the form $\boldsymbol{a} \cdot (\boldsymbol{b} \times \boldsymbol{c})$ is called a **triple scalar product**; its value is a scalar.

Properties of the triple scalar product

(i)

$$\boldsymbol{a} \cdot (\boldsymbol{b} \times \boldsymbol{c}) = \boldsymbol{c} \cdot (\boldsymbol{a} \times \boldsymbol{b}) = \boldsymbol{b} \cdot (\boldsymbol{c} \times \boldsymbol{a}), \tag{1.5}$$

that is, *cyclic permutation of the vectors* $\boldsymbol{a}, \boldsymbol{b}, \boldsymbol{c}$ *in a triple scalar product leaves its value unchanged.* [Interchanging two vectors reverses the sign.]
This formula can alternatively be written

$$\boldsymbol{a} \cdot (\boldsymbol{b} \times \boldsymbol{c}) = (\boldsymbol{a} \times \boldsymbol{b}) \cdot \boldsymbol{c}, \tag{1.6}$$

that is, *interchanging the positions of the 'dot' and the 'cross' in a triple scalar product leaves its value unchanged.*
Because of this symmetry, the triple scalar product can be denoted unambiguously by $[\boldsymbol{a}, \boldsymbol{b}, \boldsymbol{c}]$.

(ii) The triple scalar product $[\boldsymbol{a}, \boldsymbol{b}, \boldsymbol{c}] = 0$ if (and only if) $\boldsymbol{a}, \boldsymbol{b}, \boldsymbol{c}$ are **coplanar** (or one of them is zero). In particular *a triple scalar product is zero if two of its vectors are the same.*

(iii) If $[\boldsymbol{a}, \boldsymbol{b}, \boldsymbol{c}] > 0$ then the set $\{\boldsymbol{a}, \boldsymbol{b}, \boldsymbol{c}\}$ is *right-handed.* If $[\boldsymbol{a}, \boldsymbol{b}, \boldsymbol{c}] < 0$ then the set $\{\boldsymbol{a}, \boldsymbol{b}, \boldsymbol{c}\}$ is *left-handed.*

(iv) If $\boldsymbol{a}_1 = \lambda_1 \boldsymbol{i} + \mu_1 \boldsymbol{j} + \nu_1 \boldsymbol{k}$, $\boldsymbol{a}_2 = \lambda_2 \boldsymbol{i} + \mu_2 \boldsymbol{j} + \nu_2 \boldsymbol{k}$, $\boldsymbol{a}_3 = \lambda_3 \boldsymbol{i} + \mu_3 \boldsymbol{j} + \nu_3 \boldsymbol{k}$, where $\{\boldsymbol{i}, \boldsymbol{j}, \boldsymbol{k}\}$ is a standard basis, then

$$[\boldsymbol{a}_1, \boldsymbol{a}_2, \boldsymbol{a}_3] = \begin{vmatrix} \lambda_1 & \mu_1 & \nu_1 \\ \lambda_2 & \mu_2 & \nu_2 \\ \lambda_3 & \mu_3 & \nu_3 \end{vmatrix}. \tag{1.7}$$

Triple vector product

An expression of the form $\boldsymbol{a} \times (\boldsymbol{b} \times \boldsymbol{c})$ is called a **triple vector product**; its value is a vector.

Property of the triple vector product

Since $\boldsymbol{b} \times \boldsymbol{c}$ is perpendicular to both $\boldsymbol{b}$ and $\boldsymbol{c}$, it follows that $\boldsymbol{a} \times (\boldsymbol{b} \times \boldsymbol{c})$ must lie in the same plane as $\boldsymbol{b}$ and $\boldsymbol{c}$. It can therefore be expanded in the form $\mu \boldsymbol{b} + \nu \boldsymbol{c}$. The actual formula is

$$\boldsymbol{a}\times(\boldsymbol{b}\times\boldsymbol{c}) = (\boldsymbol{a}\cdot\boldsymbol{c})\,\boldsymbol{b} - (\boldsymbol{a}\cdot\boldsymbol{b})\,\boldsymbol{c}. \tag{1.8}$$

Since the vector product is anti-commutative and non-associative, it is wise to use this formula *exactly* as it stands.

Example 1.9 ***Using triple products***

Expand the expression $(\boldsymbol{a}\times\boldsymbol{b})\cdot(\boldsymbol{c}\times\boldsymbol{d})$ in terms of scalar products.

Solution

Use the triple scalar product formula (1.6) to interchange the first 'dot' and 'cross', and then expand the resulting triple vector product by the formula (1.8), as follows:

$$\begin{aligned}(\boldsymbol{a}\times\boldsymbol{b})\cdot(\boldsymbol{c}\times\boldsymbol{d}) &= \boldsymbol{a}\cdot[\,\boldsymbol{b}\times(\boldsymbol{c}\times\boldsymbol{d})\,] = \boldsymbol{a}\cdot[\,(\boldsymbol{b}\cdot\boldsymbol{d})\,\boldsymbol{c} - (\boldsymbol{b}\cdot\boldsymbol{c})\,\boldsymbol{d}\,]\\ &= (\boldsymbol{a}\cdot\boldsymbol{c})(\boldsymbol{b}\cdot\boldsymbol{d}) - (\boldsymbol{a}\cdot\boldsymbol{d})(\boldsymbol{b}\cdot\boldsymbol{c})\ \blacksquare\end{aligned}$$

1.6 VECTOR FUNCTIONS OF A SCALAR VARIABLE

In practice, the value of a vector quantity often depends on a scalar variable such as the time t. For example, if A is the label of a particle moving through space, then its position vector $\boldsymbol{a}$ (relative to a fixed origin O) will vary with time, that is, $\boldsymbol{a} = \boldsymbol{a}(t)$. The vector $\boldsymbol{a}$ is therefore a function of the scalar variable t.

The time dependence of a vector need not involve motion. The value of the electric or magnetic field at a *fixed point** of space will generally vary with time so that $\boldsymbol{E} = \boldsymbol{E}(t)$ and $\boldsymbol{B} = \boldsymbol{B}(t)$. More generally, the scalar variable need not be the time. Consider the space curve $\mathcal{C}$ shown in Figure 1.10, whose points are parametrised by the parameter α. Each point of the curve has a unique tangent line whose *direction* can be characterised by the unit vector $\boldsymbol{t}$. This is called the **unit tangent vector** to $\mathcal{C}$ and it depends on α, that is, $\boldsymbol{t} = \boldsymbol{t}(\alpha)$. In this case the independent variable is the scalar α and (just to confuse matters) the dependent variable is the vector $\boldsymbol{t}$.

Differentiation

The most important operation that can be carried out on a vector function of a scalar variable is **differentiation.**

Definition 1.11 ***Differentiation of vectors*** *Suppose that the vector $\boldsymbol{v}$ is a function of the scalar variable α, that is, $\boldsymbol{v} = \boldsymbol{v}(\alpha)$. Then the **derivative** of the function $\boldsymbol{v}(\alpha)$ with respect to α is defined by the limit*†

$$\frac{d\boldsymbol{v}}{d\alpha} = \lim_{\Delta\alpha\to 0}\left(\frac{\boldsymbol{v}(\alpha+\Delta\alpha) - \boldsymbol{v}(\alpha)}{\Delta\alpha}.\right) \tag{1.9}$$

* We will not be concerned here with vector functions of *position*. These are called vector *fields*.

† *Mathematical note*: The statement $\boldsymbol{u}(\alpha) \to \boldsymbol{U}$ as $\alpha \to A$ means that $|\,\boldsymbol{u}(\alpha) - \boldsymbol{U}\,| \to 0$ as $\alpha \to A$.

This looks identical to the definition of the derivative of an ordinary real function, but there is a difference. When α changes to $\alpha+\Delta\alpha$, the function $\boldsymbol{v}$ changes from $\boldsymbol{v}(\alpha)$ to $\boldsymbol{v}(\alpha+\Delta\alpha)$, a difference of $\boldsymbol{v}(\alpha+\Delta\alpha)-\boldsymbol{v}(\alpha)$. However, this 'difference' now means *vector* subtraction and its value is a vector; it remains a vector after dividing by the scalar increment $\Delta\alpha$. Hence $d\boldsymbol{v}/d\alpha$, the limit of this quotient as $\Delta\alpha \to 0$, is a **vector**. Furthermore, since $d\boldsymbol{v}/d\alpha$ depends on α, it is itself a vector function of the scalar variable α. The rules for differentiating combinations of vector functions are similar to those for ordinary scalar functions.

Differentiation rules for vector functions

Let $\boldsymbol{u}(\alpha)$ and $\boldsymbol{v}(\alpha)$ be vector functions of the scalar variable α, and let $\lambda(\alpha)$ be a scalar function. Then:

(i) $\dfrac{d}{d\alpha}(\boldsymbol{u}+\boldsymbol{v}) = \dot{\boldsymbol{u}}+\dot{\boldsymbol{v}}$ (ii) $\dfrac{d}{d\alpha}(\lambda\,\boldsymbol{u}) = \dot{\lambda}\,\boldsymbol{u}+\lambda\dot{\boldsymbol{u}}$

(iii) $\dfrac{d}{d\alpha}(\boldsymbol{u}\cdot\boldsymbol{v}) = \dot{\boldsymbol{u}}\cdot\boldsymbol{v}+\boldsymbol{u}\cdot\dot{\boldsymbol{v}}$ (iv) $\dfrac{d}{d\alpha}(\boldsymbol{u}\times\boldsymbol{v}) = \dot{\boldsymbol{u}}\times\boldsymbol{v}+\boldsymbol{u}\times\dot{\boldsymbol{v}}$

where $\dot{\boldsymbol{u}}$ means $d\boldsymbol{u}/d\alpha$ and so on. Note that the order of the terms in the vector product formula must be preserved.

Example 1.10 ***Differentiating vector functions***

(i) The position vector of a particle P at time t is given by

$$\boldsymbol{r} = (2t^2-5t)\,\boldsymbol{i} + (4t+2)\,\boldsymbol{j} + t^3\,\boldsymbol{k},$$

where $\{\boldsymbol{i},\boldsymbol{j},\boldsymbol{k}\}$ is a *constant* basis set. Find $d\boldsymbol{r}/dt$ and $d^2\boldsymbol{r}/dt^2$. (These are the velocity and acceleration vectors of P at time t.)

(ii) If $\boldsymbol{a} = \boldsymbol{a}(t)$ and $\boldsymbol{b}$ is a constant vector, show that

$$\frac{d}{dt}[\boldsymbol{a}\cdot(\dot{\boldsymbol{a}}\times\boldsymbol{b})] = \boldsymbol{a}\cdot(\ddot{\boldsymbol{a}}\times\boldsymbol{b}).$$

Solution

(i) Since $\boldsymbol{i},\boldsymbol{j},\boldsymbol{k}$ are constant vectors, it follows from the differentiation rules that

$$\frac{d\boldsymbol{r}}{dt} = (4t-5)\,\boldsymbol{i} + 4\,\boldsymbol{j} + 3t^2\,\boldsymbol{k}, \qquad \frac{d^2\boldsymbol{r}}{dt^2} = 4\boldsymbol{i} + 6t\,\boldsymbol{k}.$$

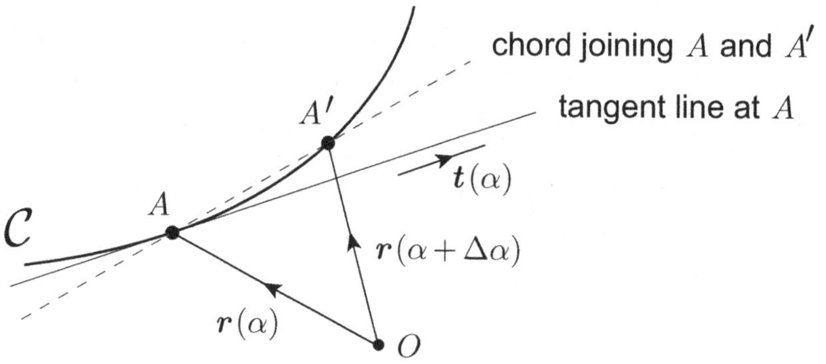

FIGURE 1.10 The unit tangent vector $\boldsymbol{t}(\alpha)$ at a typical point A on the curve $\mathcal{C}$, defined parametrically by $\boldsymbol{r} = \boldsymbol{r}(\alpha)$.

(ii)

$$\frac{d}{dt}[\boldsymbol{a}\cdot(\dot{\boldsymbol{a}}\times\boldsymbol{b})] = \dot{\boldsymbol{a}}\cdot\left(\dot{\boldsymbol{a}}\times\boldsymbol{b}\right)+\boldsymbol{a}\cdot\left(\frac{d}{dt}(\dot{\boldsymbol{a}}\times\boldsymbol{b})\right) = \boldsymbol{0}+\boldsymbol{a}\cdot\left(\ddot{\boldsymbol{a}}\times\boldsymbol{b}+\dot{\boldsymbol{a}}\times\dot{\boldsymbol{b}}\right)$$
$$= \boldsymbol{a}\cdot\left(\ddot{\boldsymbol{a}}\times\boldsymbol{b}+\dot{\boldsymbol{a}}\times\boldsymbol{0}\right) = \boldsymbol{a}\cdot(\ddot{\boldsymbol{a}}\times\boldsymbol{b}),$$

as required. ■

1.7 TANGENT AND NORMAL VECTORS TO A CURVE

In the next chapter we will define the velocity and acceleration of a particle moving in a space of three dimensions. In order to be able to interpret these definitions, we need to know a little about the differential geometry of curves. In particular, it is useful to know what the **unit tangent** and **unit normal** vectors of a curve are.

Unit tangent vector

Consider the curve $\mathcal{C}$ shown in Figure 1.10 which is defined by the parametric equation $\boldsymbol{r} = \boldsymbol{r}(\alpha)$. In general this can be a curve in *three-dimensional* space. Let A be a typical point of $\mathcal{C}$ corresponding to the parameter α and A' a nearby point corresponding to the parameter $\alpha + \Delta\alpha$. The chord $\overrightarrow{AA'}$ represents the vector

$$\Delta\boldsymbol{r} = \boldsymbol{r}(\alpha+\Delta\alpha) - \boldsymbol{r}(\alpha)$$

and so $\Delta\boldsymbol{r}/|\Delta\boldsymbol{r}|$ is a *unit* vector parallel to the chord $\overrightarrow{AA'}$. The **unit tangent vector $\boldsymbol{t}(\alpha)$** at the point A is defined to be the limit of this expression as $A' \to A$, that is

$$\boldsymbol{t}(\alpha) = \lim_{\Delta\alpha\to 0}\frac{\Delta\boldsymbol{r}}{|\Delta\boldsymbol{r}|}.$$

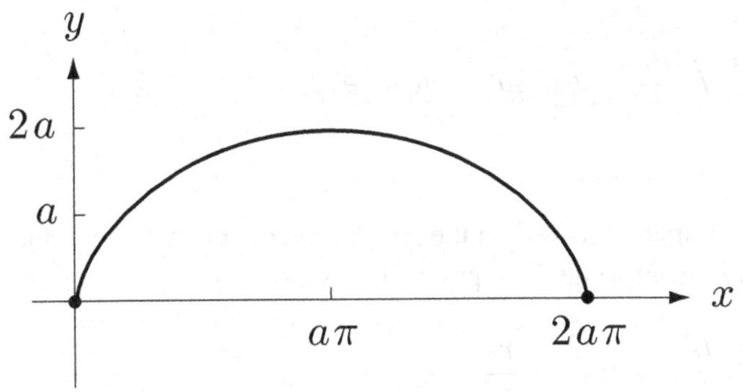

FIGURE 1.11 The cycloid $x = a(\theta - \sin\theta)$, $y = a(1 - \cos\theta)$, $z = 0$, where $0 < \theta < 2\pi$.

The tangent vector $\boldsymbol{t}$ is related to the derivative $d\boldsymbol{r}/d\alpha$ since

$$\frac{d\boldsymbol{r}}{d\alpha} = \lim_{\Delta\alpha\to 0}\frac{\Delta\boldsymbol{r}}{\Delta\alpha} = \lim_{\Delta\alpha\to 0}\frac{\Delta\boldsymbol{r}}{|\Delta\boldsymbol{r}|}\times\lim_{\Delta\alpha\to 0}\frac{|\Delta\boldsymbol{r}|}{\Delta\alpha}$$
$$= \boldsymbol{t}(\alpha)\times\left|\lim_{\Delta\alpha\to 0}\frac{\Delta\boldsymbol{r}}{\Delta\alpha}\right| = \boldsymbol{t}(\alpha)\times\left|\frac{d\boldsymbol{r}}{d\alpha}\right|,$$

that is,

$$\boxed{\frac{d\boldsymbol{r}}{d\alpha} = \left|\frac{d\boldsymbol{r}}{d\alpha}\right|\boldsymbol{t}(\alpha).} \tag{1.10}$$

Example 1.11 *Finding the unit tangent vector*

Figure 1.11 shows the cycloid $x = a(\theta - \sin\theta)$, $y = a(1 - \cos\theta)$, $z = 0$, where $0 < \theta < 2\pi$. Find the unit tangent vector to the cycloid at the point with parameter θ.

Solution

Let $\boldsymbol{i}$, $\boldsymbol{j}$ be unit vectors in the directions Ox, Oy respectively. Then the vector form of the equation for the cycloid is

$$\boldsymbol{r} = a(\theta - \sin\theta)\,\boldsymbol{i} + a(1 - \cos\theta)\,\boldsymbol{j}.$$

Then

$$\frac{d\boldsymbol{r}}{d\theta} = a(1 - \cos\theta)\,\boldsymbol{i} + (a\sin\theta)\,\boldsymbol{j}$$

and

$$\left|\frac{d\boldsymbol{r}}{d\theta}\right| = a\,(2 - 2\cos\theta)^{1/2} = 2a\sin\tfrac{1}{2}\theta.$$

Hence the **unit tangent vector** to the cycloid is

$$\boldsymbol{t}(\theta) = \frac{d\boldsymbol{r}}{d\theta} \Big/ \left|\frac{d\boldsymbol{r}}{d\theta}\right| = (\sin\tfrac{1}{2}\theta)\,\boldsymbol{i} + (\cos\tfrac{1}{2}\theta)\,\boldsymbol{j},$$

after simplification. ■

The formula (1.10) takes its simplest form when the parameter α is taken to be s, the *distance along the curve* measured from some fixed point. In this case,

$$\left|\frac{d\boldsymbol{r}}{ds}\right| = \lim_{\Delta s\to 0} \frac{|\Delta \boldsymbol{r}|}{\Delta s} = 1$$

so that $\boldsymbol{t}$ (pointing in the direction of increasing s) is given by the simple formula

$$\boxed{\boldsymbol{t} = \frac{d\boldsymbol{r}}{ds}.} \qquad (1.11)$$

This is the most convenient formula for *theoretical* purposes.

Unit normal vector

Let $\boldsymbol{t}(s)$ be the unit tangent vector to the curve C, where the parameter s represents distance along the curve. Then, since $\boldsymbol{t}$ is a vector function of the scalar variable s, it has a derivative $d\boldsymbol{t}/ds$ which is another vector function of s.

Since $\boldsymbol{t}$ is a unit vector it follows that $\boldsymbol{t}(s)\cdot\boldsymbol{t}(s) = 1$ and if we differentiate this identity with respect to s, we obtain

$$\begin{aligned} 0 &= \frac{d}{ds}(\boldsymbol{t}\cdot\boldsymbol{t}) = \frac{d\boldsymbol{t}}{ds}\cdot\boldsymbol{t} + \boldsymbol{t}\cdot\frac{d\boldsymbol{t}}{ds} \\ &= 2\left(\frac{d\boldsymbol{t}}{ds}\cdot\boldsymbol{t}\right). \end{aligned}$$

It follows that $d\boldsymbol{t}/ds$ is always perpendicular to $\boldsymbol{t}$. It is usual to write $d\boldsymbol{t}/ds$ in the form

$$\boxed{\frac{d\boldsymbol{t}}{ds} = \kappa\,\boldsymbol{n}} \qquad (1.12)$$

where $\kappa = |d\boldsymbol{t}/ds|$, a positive scalar called the **curvature**, and $\boldsymbol{n}$ is a *unit* vector called the (principal) **unit normal vector**. At each point of the curve, the unit vectors $\boldsymbol{t}(s)$ and $\boldsymbol{n}(s)$ are mutually perpendicular.

The quantities $\boldsymbol{n}$ and κ have a nice geometrical interpretation. Let A be any point on the curve and suppose that the distance parameter s is measured from A. Then, by Taylor's theorem, the form of the curve C near A is given approximately by

$$\boldsymbol{r}(s) = \boldsymbol{r}(0) + s\left[\frac{d\boldsymbol{r}}{ds}\right]_{s=0} + \tfrac{1}{2}s^2\left[\frac{d^2\boldsymbol{r}}{ds^2}\right]_{s=0} + O\left(s^3\right),$$

that is,

$$\boldsymbol{r}(s) = \boldsymbol{a} + s\,\boldsymbol{t} + \left(\tfrac{1}{2}\kappa s^2\right)\boldsymbol{n} + O\left(s^3\right), \tag{1.13}$$

where $\boldsymbol{a}$ is the position vector of the point A, and $\boldsymbol{t}$, κ and $\boldsymbol{n}$ are evaluated at the point A. Thus, near A, *the curve C lies* in the plane through A parallel to the vectors $\boldsymbol{t}$ and $\boldsymbol{n}$.* We can also see from equation (1.13) that, near A, the curve C is approximately a parabola. To the *same order of approximation*, it is equally true that, near A, the curve C is given by

$$\boldsymbol{r}(s) = \boldsymbol{a} + \kappa^{-1}(\sin\kappa s)\,\boldsymbol{t} + \kappa^{-1}(1 - \cos\kappa s)\,\boldsymbol{n} + O\left(s^3\right). \tag{1.14}$$

Thus, near A, the curve C is approximately a **circle** of radius κ^{-1}; the vector $\boldsymbol{t}$ is tangential to this circle and the vector $\boldsymbol{n}$ points towards its centre. The radius κ^{-1} is called the **radius of curvature** of C at the point A.

Example 1.12 *Finding the unit normal vector and curvature*

Find the unit normal vector and curvature of the cycloid $x = a(\theta - \sin\theta)$, $y = a(1 - \cos\theta)$, $z = 0$, where $0 < \theta < 2\pi$.

Solution

The **tangent vector** to the cycloid has already been found to be

$$\boldsymbol{t}(\theta) = \frac{d\boldsymbol{r}}{d\theta}\bigg/\left|\frac{d\boldsymbol{r}}{d\theta}\right| = (\sin\tfrac{1}{2}\theta)\,\boldsymbol{i} + (\cos\tfrac{1}{2}\theta)\,\boldsymbol{j}.$$

Hence, by the chain rule,

$$\frac{d\boldsymbol{t}}{ds} = \frac{d\boldsymbol{t}/d\theta}{ds/d\theta} = \frac{d\boldsymbol{t}/d\theta}{|d\boldsymbol{r}/d\theta|} = \frac{\frac{1}{2}(\cos\frac{1}{2}\theta)\,\boldsymbol{i} - \frac{1}{2}(\sin\frac{1}{2}\theta)\,\boldsymbol{j}}{2a\sin\frac{1}{2}\theta}$$

$$= \left(4a\sin\tfrac{1}{2}\theta\right)^{-1}\left((\cos\tfrac{1}{2}\theta)\,\boldsymbol{i} - (\sin\tfrac{1}{2}\theta)\,\boldsymbol{j}\right).$$

Hence the **unit normal vector** and **curvature** of the cycloid are given by

$$\boldsymbol{n}(\theta) = (\cos\tfrac{1}{2}\theta)\,\boldsymbol{i} - (\sin\tfrac{1}{2}\theta)\,\boldsymbol{j}, \qquad \kappa(\theta) = \left(4a\sin\tfrac{1}{2}\theta\right)^{-1}.$$

The **radius of curvature** of the cycloid is therefore $4a\sin\frac{1}{2}\theta$. ■

* More precisely, this plane makes *three point contact* with the curve C at the point A.

Problems on Chapter 1

Answers and comments are at the end of the book.

Harder problems carry a star ($*$).

1.1 In terms of the standard basis set $\{\boldsymbol{i}, \boldsymbol{j}, \boldsymbol{k}\}$, $\boldsymbol{a} = 2\boldsymbol{i} - \boldsymbol{j} - 2\boldsymbol{k}$, $\boldsymbol{b} = 3\boldsymbol{i} - 4\boldsymbol{k}$ and $\boldsymbol{c} = \boldsymbol{i} - 5\boldsymbol{j} + 3\boldsymbol{k}$.

(i) Find $3\boldsymbol{a} + 2\boldsymbol{b} - 4\boldsymbol{c}$ and $|\boldsymbol{a} - \boldsymbol{b}|^2$.
(ii) Find $|\boldsymbol{a}|$, $|\boldsymbol{b}|$ and $\boldsymbol{a} \cdot \boldsymbol{b}$. Deduce the angle between $\boldsymbol{a}$ and $\boldsymbol{b}$.
(iii) Find the component of $\boldsymbol{c}$ in the direction of $\boldsymbol{a}$ and in the direction of $\boldsymbol{b}$.
(iv) Find $\boldsymbol{a} \times \boldsymbol{b}$, $\boldsymbol{b} \times \boldsymbol{c}$ and $(\boldsymbol{a} \times \boldsymbol{b}) \times (\boldsymbol{b} \times \boldsymbol{c})$.
(v) Find $\boldsymbol{a} \cdot (\boldsymbol{b} \times \boldsymbol{c})$ and $(\boldsymbol{a} \times \boldsymbol{b}) \cdot \boldsymbol{c}$ and verify that they are equal. Is the set $\{\boldsymbol{a}, \boldsymbol{b}, \boldsymbol{c}\}$ right- or left-handed?
(vi) By evaluating each side, verify the identity $\boldsymbol{a} \times (\boldsymbol{b} \times \boldsymbol{c}) = (\boldsymbol{a} \cdot \boldsymbol{c})\,\boldsymbol{b} - (\boldsymbol{a} \cdot \boldsymbol{b})\,\boldsymbol{c}$.

Vector geometry

1.2 Find the angle between any two diagonals of a cube.

1.3 $ABCDEF$ is a regular hexagon with centre O which is also the origin of position vectors. Find the position vectors of the vertices C, D, E, F in terms of the position vectors $\boldsymbol{a}$, $\boldsymbol{b}$ of A and B.

1.4 Let $ABCD$ be a general (skew) quadrilateral and let P, Q, R, S be the mid-points of the sides AB, BC, CD, DA respectively. Show that $PQRS$ is a parallelogram.

1.5 In a general tetrahedron, lines are drawn connecting the mid-point of each side with the mid-point of the side opposite. Show that these three lines meet in a point that bisects each of them.

1.6 Let $ABCD$ be a general tetrahedron and let P, Q, R, S be the median centres of the faces opposite to the vertices A, B, C, D respectively. Show that the lines AP, BQ, CR, DS all meet in a point (called the *centroid* of the tetrahedron), which divides each line in the ratio 3:1.

1.7 A number of particles with masses $m_1, m_2, m_3, \ldots$ are situated at the points with position vectors $\boldsymbol{r}_1, \boldsymbol{r}_2, \boldsymbol{r}_3, \ldots$ relative to an origin O. The centre of mass G of the particles is defined to be the point of space with position vector

$$\boldsymbol{R} = \frac{m_1\boldsymbol{r}_1 + m_2\boldsymbol{r}_2 + m_3\boldsymbol{r}_3 + \cdots}{m_1 + m_2 + m_3 + \cdots}$$

Show that if a different origin O' were used, this definition would still place G at the same point of space.

1.8 Prove that the three perpendiculars of a triangle are concurrent.
[Construct the two perpendiculars from A and B and take their intersection point as O, the origin of position vectors. Then prove that OC must be perpendicular to AB.]

Vector algebra

1.9 If $a_1 = \lambda_1 i + \mu_1 j + \nu_1 k$, $a_2 = \lambda_2 i + \mu_2 j + \nu_2 k$, $a_3 = \lambda_3 i + \mu_3 j + \nu_3 k$, where $\{i, j, k\}$ is a standard basis, show that

$$a_1 \cdot (a_2 \times a_3) = \begin{vmatrix} \lambda_1 & \mu_1 & \nu_1 \\ \lambda_2 & \mu_2 & \nu_2 \\ \lambda_3 & \mu_3 & \nu_3 \end{vmatrix}.$$

Deduce that cyclic rotation of the vectors in a triple scalar product leaves the value of the product unchanged.

1.10 By expressing the vectors a, b, c in terms of a suitable standard basis, prove the identity $a \times (b \times c) = (a \cdot c)\, b - (a \cdot b)\, c$.

1.11 Prove the identities

(i) $(a \times b) \cdot (c \times d) = (a \cdot c)(b \cdot d) - (a \cdot d)(b \cdot c)$
(ii) $(a \times b) \times (c \times d) = [a, b, d]\, c - [a, b, c]\, d$
(iii) $a \times (b \times c) + c \times (a \times b) + b \times (c \times a) = 0$ (Jacobi's identity)

1.12 ***Reciprocal basis*** Let $\{a, b, c\}$ be any basis set. Then the corresponding **reciprocal basis** $\{a^*, b^*, c^*\}$ is defined by

$$a^* = \frac{b \times c}{[a, b, c]}, \qquad b^* = \frac{c \times a}{[a, b, c]}, \qquad c^* = \frac{a \times b}{[a, b, c]}.$$

(i) If $\{i, j, k\}$ is a standard basis, show that $\{i^*, j^*, k^*\} = \{i, j, k\}$.
(ii) Show that $[a^*, b^*, c^*] = 1/[a, b, c]$. Deduce that if $\{a, b, c\}$ is a right handed set then so is $\{a^*, b^*, c^*\}$.
(iii) Show that $\{(a^*)^*, (b^*)^*, (c^*)^*\} = \{a, b, c\}$.
(iv) If a vector v is expanded in terms of the basis set $\{a, b, c\}$ in the form

$$v = \lambda a + \mu b + \nu c,$$

show that the coefficients λ, μ, ν are given by $\lambda = v \cdot a^*$, $\mu = v \cdot b^*$, $\nu = v \cdot c^*$.

1.13 *Lamé's equations* The directions in which X-rays are strongly scattered by a crystal are determined from the solutions x of Lamé's equations, namely

$$x \cdot a = L, \qquad x \cdot b = M, \qquad x \cdot c = N,$$

where $\{a, b, c\}$ are the basis vectors of the crystal lattice, and L, M, N are *any* integers. Show that the solutions of Lamé's equations are

$$x = L\, a^* + M\, b^* + N\, c^*,$$

where $\{a^*, b^*, c^*\}$ is the reciprocal basis to $\{a, b, c\}$.

Differentiation of vectors

1.14 If $\boldsymbol{r}(t) = (3t^2 - 4)\,\boldsymbol{i} + t^3\,\boldsymbol{j} + (t+3)\,\boldsymbol{k}$, where $\{\boldsymbol{i}, \boldsymbol{j}, \boldsymbol{k}\}$ is a constant standard basis, find $\dot{\boldsymbol{r}}$ and $\ddot{\boldsymbol{r}}$. Deduce the time derivative of $\boldsymbol{r} \times \dot{\boldsymbol{r}}$.

1.15 The vector $\boldsymbol{v}$ is a function of the time t and $\boldsymbol{k}$ is a constant vector. Find the time derivatives of (i) $|\,\boldsymbol{v}\,|^2$, (ii) $(\boldsymbol{v} \cdot \boldsymbol{k})\,\boldsymbol{v}$, (iii) $\left[\boldsymbol{v}, \dot{\boldsymbol{v}}, \boldsymbol{k}\right]$.

1.16 Find the unit tangent vector, the unit normal vector and the curvature of the circle $x = a\cos\theta$, $y = a\sin\theta$, $z = 0$ at the point with parameter θ.

1.17 Find the unit tangent vector, the unit normal vector and the curvature of the helix $x = a\cos\theta$, $y = a\sin\theta$, $z = b\theta$ at the point with parameter θ.

1.18 Find the unit tangent vector, the unit normal vector and the curvature of the parabola $x = ap^2$, $y = 2ap$, $z = 0$ at the point with parameter p.

Chapter Two

Velocity, acceleration and scalar angular velocity

KEY FEATURES

The key concepts in this chapter are the **velocity** and **acceleration** of a particle and the **angular velocity** of a rigid body in planar motion.

Kinematics is the study of the **motion of material bodies** without regard to the forces that cause their motion. The subject does not seek to answer the question of *why* bodies move as they do; that is the province of dynamics. It merely provides a geometrical description of the possible motions. The basic building block for bodies in mechanics is the **particle**, an idealised body that occupies only a single point of space. The important kinematical quantities in the motion of a particle are its **velocity** and **acceleration**. We begin with the simple case of straight line particle motion, where velocity and acceleration are scalars, and then progress to three-dimensional motion, where velocity and acceleration are vectors.

The other important idealisation that we consider is the **rigid body**, which we regard as a collection of particles linked by a light rigid framework. The important kinematical quantity in the motion of a rigid body is its **angular velocity**. In this chapter, we consider only those rigid body motions that are essentially two-dimensional, so that angular velocity is a scalar quantity. The general three-dimensional case is treated in Chapter 16.

2.1 STRAIGHT LINE MOTION OF A PARTICLE

Consider a particle P moving along the x-axis so that its displacement x from the origin O is a known function of the time t. Then the **mean velocity** of P over the time

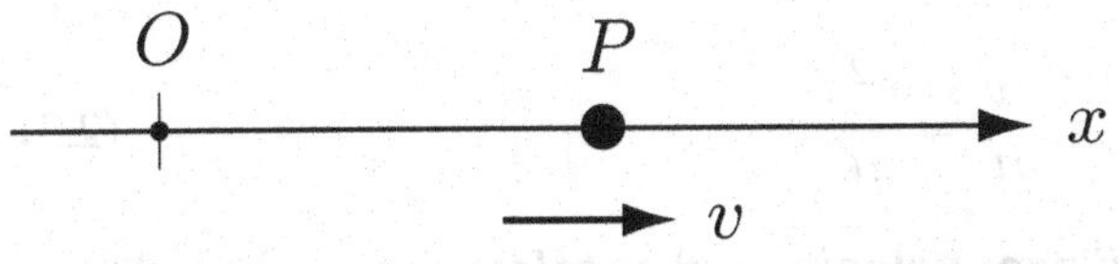

FIGURE 2.1 The particle P moves in a straight line and has displacement x and velocity v at time t.

interval $t_1 \le t \le t_2$ is defined to be the increase in the displacement of P divided by the time taken, that is,

$$\frac{x(t_2) - x(t_1)}{t_2 - t_1}. \tag{2.1}$$

Example 2.1 *Mean velocity*

Suppose the displacement of P from O at time t is given by $x = t^2 - 6t$, where x is measured in metres and t in seconds. Find the mean velocity of P over the time interval $1 \le t \le 3$.

Solution

In this case, $x(1) = -5$ and $x(3) = -9$ so that the mean velocity of P is $((-9) - (-5))/(3 - 1) = -2\ \mathrm{m\,s^{-1}}$. ■

The mean velocity of a particle is less important to us than its *instantaneous* velocity, that is, its velocity at a given instant in time. We cannot find the instantaneous velocity of P at time t_1 merely by letting $t_2 = t_1$ in the formula (2.1), since the quotient would then be undefined. However, we can define the instantaneous velocity as the *limit* of the mean velocity as the time interval *tends* to zero, that is, as $t_2 \to t_1$. Thus $v(t_1)$, the instantaneous velocity of P at time t_1 can be defined by

$$v(t_1) = \lim_{t_2 \to t_1} \left(\frac{x(t_2) - x(t_1)}{t_2 - t_1} \right).$$

But this is precisely the definition of dx/dt, the derivative of x with respect to t, evaluated at $t = t_1$. This leads us to the official definition:

Definition 2.1 ***1-D velocity*** *The (instantaneous)* ***velocity*** v *of* P*, in the positive* x*-direction, is defined by*

$$v = \frac{dx}{dt}. \tag{2.2}$$

The ***speed*** *of* P *is defined to be the rate of increase of the total distance travelled and is therefore equal to* $|v|$.

Similarly, the acceleration of P, the rate of increase of v, is defined as follows:

Definition 2.2 ***1-D acceleration*** *The (instantaneous)* ***acceleration*** a *of* P*, in the positive* x*-direction, is defined by*

$$a = \frac{dv}{dt} = \frac{d^2x}{dt^2}. \tag{2.3}$$

Example 2.2 *Finding rectilinear velocity and acceleration*

Suppose the displacement of P from O at time t is given by $x = t^3 - 6t^2 + 4$, where x is measured in metres and t in seconds. Find the velocity and acceleration of P at time t. Deduce that P comes to rest twice and find the position and acceleration of P at the later of these two times.

Solution

Since $v = dx/dt$ and $a = dv/dt$, we obtain

$$v = 3t^2 - 12t \qquad \text{and} \qquad a = 6t - 12$$

as the velocity and acceleration of P at time t.

P comes to rest when its velocity v is zero, that is, when

$$3t^2 - 12t = 0.$$

This is a quadratic equation for t having the solutions $t = 0, 4$. Thus P is at rest when $t = 0$ s and $t = 4$ s.

When $t = 4$ s, $x = -28$ m and $a = 12$ m s^{-2}. Note that merely because $v = 0$ at some instant it does not follow that $a = 0$ also. ■

Example 2.3 *Reversing the process*

A particle P moves along the x-axis with its acceleration a at time t given by

$$a = 12t^2 - 6t + 6 \text{ m s}^{-2}.$$

Initially P is at the point $x = 4$ m and is moving with speed 8 m s^{-1} in the negative x-direction. Find the velocity and displacement of P at time t.

Solution

Since $a = dv/dt$ we have

$$\frac{dv}{dt} = 12t^2 - 6t + 6,$$

and integrating with respect to t gives

$$v = 4t^3 - 3t^2 + 6t + C,$$

where C is a constant of integration. This constant can be determined by using the given initial condition on v, namely, $v = -8$ when $t = 0$. This gives $C = -8$ so that the velocity of P at time t is

$$v = 4t^3 - 3t^2 + 6t - 8 \text{ m s}^{-1}.$$

By writing $v = dx/dt$ and integrating again, we obtain

$$x = t^4 - t^3 + 3t^2 - 8t + D,$$

where D is a second constant of integration. D can now be determined by using the given initial condition on x, namely, $x = 4$ when $t = 0$. This gives $D = 4$ so that the displacement of P at time t is

$$x = t^4 - t^3 + 3t^2 - 8t + 4 \text{ m.} ■$$

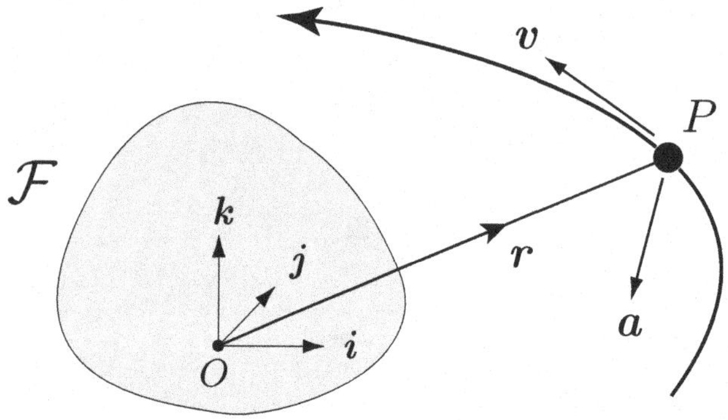

FIGURE 2.2 The particle P moves in three-dimensional space and, relative to the reference frame $\mathcal{F}$ and origin O, has position vector $\boldsymbol{r}$ at time t.

2.2 GENERAL MOTION OF A PARTICLE

When a particle P moves in two or three-dimensional space, its position can be described by its vector displacement $\boldsymbol{r}$ from an origin O that is fixed in a rigid **reference frame** $\mathcal{F}$. Whether $\mathcal{F}$ is moving or not is irrelevant here; the position vector $\boldsymbol{r}$ is simply measured *relative* to $\mathcal{F}$. Figure 2.2 shows a particle P moving in three-dimensional space with position vector $\boldsymbol{r}$ (relative to the reference frame $\mathcal{F}$) at time t.

Question *Reference frames*

What is a reference frame and why do we need one?

Answer

A rigid reference frame $\mathcal{F}$ is essentially a **rigid body** whose particles can be labelled to create reference points. The most familiar such body is the Earth. Relative to a single particle, the only thing that can be specified is distance from that particle. However, relative to a rigid body, one can specify both distance and direction. Thus the value of any vector quantity can be specified relative to $\mathcal{F}$. In particular, if we label some particle O of the body as origin, we can specify the position of any point of space by its position vector relative to the frame $\mathcal{F}$ and the origin O.

The specification of vectors relative to a reference frame is much simplified if we introduce a Cartesian coordinate system. This can be done in infinitely many different ways. Imagine that $\mathcal{F}$ is extended by a set of three mutually orthogonal planes that are *rigidly embedded* in it. The coordinates x, y, z of a point P are then the distances of P from these three planes. Let O be the origin of this coordinate system, and $\{\boldsymbol{i}, \boldsymbol{j}, \boldsymbol{k}\}$ its unit vectors. We can then conveniently refer to the frame $\mathcal{F}$, together with the embedded coordinate system $Oxyz$, by the notation $\mathcal{F}\{O\,;\,\boldsymbol{i}, \boldsymbol{j}, \boldsymbol{k}\}$. ■

In general motion, the velocity and acceleration of a particle are *vector quantities* and are defined by:

Definition 2.3 ***3-D velocity and acceleration*** *The velocity* $\boldsymbol{v}$ *and acceleration* $\boldsymbol{a}$ *of* P *are defined by*

$$\boldsymbol{v} = \frac{d\boldsymbol{r}}{dt} \quad \text{and} \quad \boldsymbol{a} = \frac{d\boldsymbol{v}}{dt}. \tag{2.4}$$

Connection with the rectilinear case

The scalar velocity and acceleration defined in section 2.1 for the case of straight line motion are simply related to the corresponding vector quantities defined above. It would be possible to use the vector formalism in all cases but, for the case of straight line motion along the x-axis, $\boldsymbol{r}$, $\boldsymbol{v}$, and $\boldsymbol{a}$ would have the form

$$\boldsymbol{r} = x\,\boldsymbol{i}, \qquad \boldsymbol{v} = v\,\boldsymbol{i}, \qquad \boldsymbol{a} = a\,\boldsymbol{i},$$

where $v = dx/dt$ and $a = dv/dt$. It is therefore sufficient to work with the scalar quantities x, v and a; use of the vector formalism would be clumsy and unnecessary.

Example 2.4 *Finding 3-D velocity and acceleration*

Relative to the reference frame $\mathcal{F}\{O\,;\,\boldsymbol{i},\,\boldsymbol{j},\,\boldsymbol{k}\}$, the position vector of a particle P at time t is given by

$$\boldsymbol{r} = (2t^2 - 3)\,\boldsymbol{i} + (4t + 4)\,\boldsymbol{j} + (t^3 + 2t^2)\,\boldsymbol{k}.$$

Find (i) the distance OP when $t = 0$, (ii) the velocity of P when $t = 1$, (iii) the acceleration of P when $t = 2$.

Solution

In this solution we will make use of the rules for differentiation of sums and products involving vector functions of the time. These rules are listed in section 1.6.

(i) When $t = 0$, $\boldsymbol{r} = -3\boldsymbol{i} + 4\boldsymbol{j}$ so that $OP = |\boldsymbol{r}| = 5$.

(ii) Relative to the reference frame $\mathcal{F}$, the unit vectors $\{\boldsymbol{i},\,\boldsymbol{j},\,\boldsymbol{k}\}$ are *constant* and so their time derivatives are zero. The velocity $\boldsymbol{v}$ of P is therefore

$$\boldsymbol{v} = d\boldsymbol{r}/dt = 4t\,\boldsymbol{i} + 4\,\boldsymbol{j} + (3t^2 + 4t)\,\boldsymbol{k}.$$

When $t = 1$, $\boldsymbol{v} = 4\boldsymbol{i} + 4\boldsymbol{j} + 7\boldsymbol{k}$.

(iii) Relative to the reference frame $\mathcal{F}$, the acceleration $\boldsymbol{a}$ of P is

$$\boldsymbol{a} = d\boldsymbol{v}/dt = 4\boldsymbol{i} + (6t + 4)\,\boldsymbol{k}.$$

When $t = 2$, $\boldsymbol{a} = 4\boldsymbol{i} + 16\boldsymbol{k}$. ■

Interpretation of the vectors $\boldsymbol{v}$ and $\boldsymbol{a}$

The velocity vector $\boldsymbol{v}$ has a simple interpretation. Suppose that s is the arc-length travelled by P, measured from some fixed point of its path, and that s is increasing with time.*

* The arguments that follow assume a familiarity with the unit tangent and normal vectors to a general curve, as described in section 1.7

Then, by the chain rule,

$$\begin{aligned} \boldsymbol{v} &= \frac{d\boldsymbol{r}}{dt} = \frac{d\boldsymbol{r}}{ds} \times \frac{ds}{dt} \\ &= v\,\boldsymbol{t} \end{aligned}$$

where $\boldsymbol{t}$ is the **unit tangent vector** to the path and v $(= ds/dt)$ is the **speed*** of P. Thus, at each instant, *the direction of the velocity vector* $\boldsymbol{v}$ *is along the tangent to its path, and* $|\boldsymbol{v}|$ *is the speed of* P.

The acceleration vector $\boldsymbol{a}$ is harder to picture. This is partly because we are too accustomed to the special case of straight line motion. However, in general,

$$\begin{aligned} \boldsymbol{a} &= \frac{d\boldsymbol{v}}{dt} = \frac{d(v\boldsymbol{t})}{dt} = \frac{dv}{dt}\boldsymbol{t} + v\frac{d\boldsymbol{t}}{dt} = \left(\frac{dv}{dt}\right)\boldsymbol{t} + v\left(\frac{d\boldsymbol{t}}{ds} \times \frac{ds}{dt}\right) \\ &= \left(\frac{dv}{dt}\right)\boldsymbol{t} + \left(\frac{v^2}{\rho}\right)\boldsymbol{n}, \end{aligned} \tag{2.5}$$

where $\boldsymbol{n}$ is the unit normal vector to the path of P and ρ $(= \kappa^{-1})$ is its radius of curvature. Hence, *the acceleration vector* $\boldsymbol{a}$ *has a component* dv/dt *tangential to the path and a component* v^2/ρ *normal to the path.*

This formula is surprising. Since each small segment of the path is ‘approximately straight’ one might be tempted to conclude that only the first term $(dv/dt)\boldsymbol{t}$ should be present. However, what we have shown is that the acceleration vector of P does not generally point along the path but has a component *perpendicular* to the local path direction. The full meaning of formula (2.5) will become clear when we have treated particle motion in polar coordinates.

Uniform circular motion

The simplest example of non-rectilinear motion is motion in a circle. Circular motion is important in practical applications such as rotating machinery. Here we consider the special case of *uniform* circular motion, that is, circular motion with *constant speed.*

Consider a particle P moving with *constant speed* u in the anti-clockwise direction around a circle centre O and radius b, as shown in Figure 2.3. At time $t = 0$, P is at the point $B(b, 0)$. What are its velocity and acceleration vectors at time t?

The first step is to find the position vector of P at time t. Since P moves with constant speed u, the *arc* length BP travelled in time t must be ut. It follows that the angle θ shown in Figure 2.3 is given by $\theta = ut/b$. The position vector of P at time t is therefore

$$\begin{aligned} \boldsymbol{r} &= b\cos\theta\,\boldsymbol{i} + b\sin\theta\,\boldsymbol{j}, \\ &= b\cos(ut/b)\,\boldsymbol{i} + b\sin(ut/b)\,\boldsymbol{j}. \end{aligned}$$

* As in the rectilinear case, *speed* means the rate of increase of the total distance travelled, which, in the present context, is ds/dt, the rate of increase of arc length along the path of P.

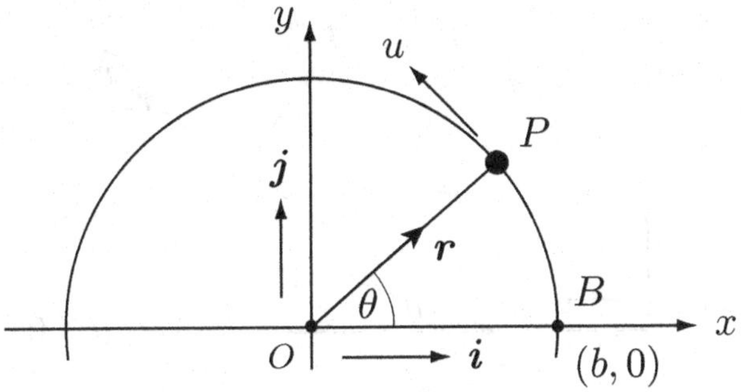

FIGURE 2.3 Particle P moves with constant speed u around a circle of radius b.

It follows that the **velocity** and **acceleration** of P at time t are given by

$$\boldsymbol{v} = \frac{d\boldsymbol{r}}{dt} = -u \sin(ut/b)\, \boldsymbol{i} + u \cos(ut/b)\, \boldsymbol{j},$$

$$\boldsymbol{a} = \frac{d\boldsymbol{v}}{dt} = -\frac{u^2}{b} \cos(ut/b)\, \boldsymbol{i} - \frac{u^2}{b} \sin(ut/b)\, \boldsymbol{j}.$$

We note that the speed of P, calculated from $\boldsymbol{v}$, is

$$|\,\boldsymbol{v}\,| = \left(u^2 \cos^2(ut/b) + u^2 \sin^2(ut/b)\right)^{1/2} = u,$$

which is what it was specified to be.

The *magnitude* of the acceleration $\boldsymbol{a}$ is given by

$$|\,\boldsymbol{a}\,| = \left(\left(\frac{u^2}{b}\right)^2 \cos^2(ut/b) + \left(\frac{u^2}{b}\right)^2 \sin^2(ut/b)\right)^{1/2} = \frac{u^2}{b}$$

and, since $\boldsymbol{a} = -(u^2/b^2)\boldsymbol{r}$, the *direction* of $\boldsymbol{a}$ is opposite to that of $\boldsymbol{r}$. This proves the following important result:

Uniform circular motion

When a particle P moves with constant speed u around a fixed circle with centre O and radius b, its acceleration vector is in the direction $\overrightarrow{PO}$ and has constant magnitude u^2/b.

This result is consistent with the general formula (2.5). In this special case, we have $v = u$ and $\rho = b$ so that $dv/dt = 0$ and $\boldsymbol{a} = (u^2/b)\boldsymbol{n}$.

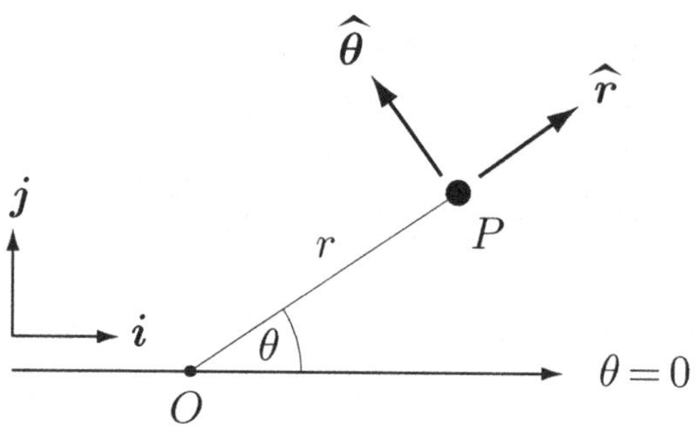

FIGURE 2.4 The plane polar co-ordinates r, θ of the point P and the polar unit vectors $\widehat{\boldsymbol{r}}$ and $\widehat{\boldsymbol{\theta}}$ at P.

Example 2.5 *Uniform circular motion*

A body is being whirled round at 10 m s^{-1} on the end of a rope. If the body moves on a circular path of 2 m radius, find the magnitude and direction of its acceleration.

Solution

The acceleration is directed towards the centre of the circle and its magnitude is $10^2/2 = 50$ m s^{-1}, five times the acceleration due to Earth's gravity! ■

2.3 PARTICLE MOTION IN POLAR CO-ORDINATES

When a particle is moving in a plane, it is sometimes very convenient to use polar co-ordinates r, θ in the analysis of its motion; the case of circular motion is an obvious example. Less obviously, polar co-ordinates are used in the analysis of the orbits of the planets. This famous problem stimulated Newton to devise his laws of mechanics.

Figure 2.4 shows the polar co-ordinates r, θ of a point P and the **polar unit vectors** $\widehat{\boldsymbol{r}}$, $\widehat{\boldsymbol{\theta}}$ at P. The directions of the vectors $\widehat{\boldsymbol{r}}$ and $\widehat{\boldsymbol{\theta}}$ are called the **radial** and **transverse** directions respectively at the point P. As P moves around, the polar unit vectors do not remain constant. They have constant magnitude (unity) but their directions depend on the θ co-ordinate of P; they are however independent of the r co-ordinate.* In other words, $\widehat{\boldsymbol{r}}$, $\widehat{\boldsymbol{\theta}}$ are *vector functions of the scalar variable* θ.

We will now evaluate the two derivatives $d\widehat{\boldsymbol{r}}/d\theta$, $d\widehat{\boldsymbol{\theta}}/d\theta$. These will be needed when we derive the formulae for the velocity and acceleration of P in polar co-ordinates. First we expand† $\widehat{\boldsymbol{r}}$, $\widehat{\boldsymbol{\theta}}$ in terms of the Cartesian basis vectors $\{\boldsymbol{i}, \boldsymbol{j}\}$. This gives

$$\widehat{\boldsymbol{r}} = \cos\theta\, \boldsymbol{i} + \sin\theta\, \boldsymbol{j}, \tag{2.6}$$

$$\widehat{\boldsymbol{\theta}} = -\sin\theta\, \boldsymbol{i} + \cos\theta\, \boldsymbol{j}. \tag{2.7}$$

Since $\widehat{\boldsymbol{r}}$, $\widehat{\boldsymbol{\theta}}$ are now expressed in terms of the *constant* vectors $\boldsymbol{i}$, $\boldsymbol{j}$, the differentiations with respect to θ are simple and give

* If this is not clear, sketch the directions of the polar unit vectors for P in a few different positions.

† Recall that *any* vector $\boldsymbol{V}$ lying in the plane of $\boldsymbol{i}$, $\boldsymbol{j}$ can be expanded in the form $\boldsymbol{V} = \alpha\, \boldsymbol{i} + \beta\, \boldsymbol{j}$, where the coefficients α, β are the components of $\boldsymbol{V}$ in the $\boldsymbol{i}$- and $\boldsymbol{j}$-directions respectively.

$$\boxed{\frac{d\widehat{\boldsymbol{r}}}{d\theta} = \widehat{\boldsymbol{\theta}} \qquad \frac{d\widehat{\boldsymbol{\theta}}}{d\theta} = -\widehat{\boldsymbol{r}}} \tag{2.8}$$

Suppose now that P is a moving particle with polar co-ordinates r, θ that are functions of the time t. The position vector of P relative to O has magnitude $OP = r$ and direction $\widehat{\boldsymbol{r}}$ and can therefore be written

$$\boldsymbol{r} = r\widehat{\boldsymbol{r}}. \tag{2.9}$$

In what follows, one must distinguish carefully between the position vector $\boldsymbol{r}$, which is the *vector* $\overrightarrow{OP}$, the co-ordinate r, which is the *distance* OP, and the polar unit vector $\widehat{\boldsymbol{r}}$.

To obtain the polar formula for the velocity of P, we differentiate formula (2.9) with respect to t. This gives

$$\boldsymbol{v} = \frac{d\boldsymbol{r}}{dt} = \frac{d}{dt}(r\widehat{\boldsymbol{r}}) = \left(\frac{dr}{dt}\right)\widehat{\boldsymbol{r}} + r\left(\frac{d\widehat{\boldsymbol{r}}}{dt}\right) \tag{2.10}$$

$$= \dot{r}\widehat{\boldsymbol{r}} + r\left(\frac{d\widehat{\boldsymbol{r}}}{dt}\right) \tag{2.11}$$

We will use the **dot notation** for time derivatives throughout this section; $\dot{r}$ means dr/dt, $\dot{\theta}$ means $d\theta/dt$, $\ddot{r}$ means d^2r/dt^2 and $\ddot{\theta}$ means $d^2\theta/dt^2$.

Now $\widehat{\boldsymbol{r}}$ is a function of θ which is, in its turn, a function of t. Hence, by the chain rule and formula (2.8),

$$\frac{d\widehat{\boldsymbol{r}}}{dt} = \frac{d\widehat{\boldsymbol{r}}}{d\theta} \times \frac{d\theta}{dt} = \widehat{\boldsymbol{\theta}} \times \dot{\theta} = \dot{\theta}\widehat{\boldsymbol{\theta}}.$$

If we now substitute this formula into equation (2.11) we obtain

$$\boldsymbol{v} = \dot{r}\widehat{\boldsymbol{r}} + \left(r\dot{\theta}\right)\widehat{\boldsymbol{\theta}}, \tag{2.12}$$

which is the polar formula for the **velocity** of P.

To obtain the polar formula for acceleration, we differentiate the velocity formula (2.12) with respect to t. This gives*

$$\begin{aligned}
\boldsymbol{a} = \frac{d\boldsymbol{v}}{dt} &= \frac{d}{dt}(\dot{r}\widehat{\boldsymbol{r}}) + \frac{d}{dt}\left((r\dot{\theta})\widehat{\boldsymbol{\theta}}\right)\\
&= \ddot{r}\widehat{\boldsymbol{r}} + \dot{r}\,\frac{d\widehat{\boldsymbol{r}}}{dt} + \left(\dot{r}\dot{\theta} + r\ddot{\theta}\right)\widehat{\boldsymbol{\theta}} + \left(r\dot{\theta}\right)\frac{d\widehat{\boldsymbol{\theta}}}{dt}\\
&= \ddot{r}\widehat{\boldsymbol{r}} + \dot{r}\left(\frac{d\widehat{\boldsymbol{r}}}{d\theta} \times \frac{d\theta}{dt}\right) + \left(\dot{r}\dot{\theta} + r\ddot{\theta}\right)\widehat{\boldsymbol{\theta}} + \left(r\dot{\theta}\right)\left(\frac{d\widehat{\boldsymbol{\theta}}}{d\theta} \times \frac{d\theta}{dt}\right)\\
&= \ddot{r}\widehat{\boldsymbol{r}} + \left(\dot{r}\dot{\theta}\right)\widehat{\boldsymbol{\theta}} + \left(\dot{r}\dot{\theta} + r\ddot{\theta}\right)\widehat{\boldsymbol{\theta}} - \left(r\dot{\theta}^2\right)\widehat{\boldsymbol{r}}\\
&= \left(\ddot{r} - r\dot{\theta}^2\right)\widehat{\boldsymbol{r}} + \left(r\ddot{\theta} + 2\dot{r}\dot{\theta}\right)\widehat{\boldsymbol{\theta}},
\end{aligned}$$

* Be a hero. Obtain this formula yourself without looking at the text.

which is the polar formula for the **acceleration** of P. These results are summarised below:

Polar formulae for velocity and acceleration

If a particle is moving in a plane and has polar coordinates r, θ at time t, then its velocity and acceleration vectors are given by

$$\boldsymbol{v} = \dot{r}\,\widehat{\boldsymbol{r}} + \left(r\dot{\theta}\right)\widehat{\boldsymbol{\theta}}, \tag{2.13}$$

$$\boldsymbol{a} = \left(\ddot{r} - r\dot{\theta}^2\right)\widehat{\boldsymbol{r}} + \left(r\ddot{\theta} + 2\dot{r}\dot{\theta}\right)\widehat{\boldsymbol{\theta}}. \tag{2.14}$$

The formula (2.13) shows that the velocity of P is the vector sum of an outward radial velocity $\dot{r}$ and a transverse velocity $r\dot{\theta}$; in other words $\boldsymbol{v}$ is just the sum of the velocities that P *would* have if r and θ varied separately. This is *not* true for the acceleration as it will be observed that adding together the separate accelerations would not yield the term $2\dot{r}\dot{\theta}\,\widehat{\boldsymbol{\theta}}$. This 'Coriolis term' is certainly present however, but is difficult to interpret intuitively.

Example 2.6 *Velocity and acceleration in polar coordinates*

A particle sliding along a radial groove in a rotating turntable has polar coordinates at time t given by

$$r = ct \qquad \theta = \Omega t,$$

where c and Ω are positive constants. Find the velocity and acceleration vectors of the particle at time t and find the speed of the particle at time t.

Deduce that, for $t > 0$, the angle between the velocity and acceleration vectors is always acute.

Solution

From the polar formulae (2.13), (2.14) for velocity and acceleration, we obtain

$$\boldsymbol{v} = c\,\widehat{\boldsymbol{r}} + (ct)\Omega\,\widehat{\boldsymbol{\theta}} = c\left(\widehat{\boldsymbol{r}} + \Omega t\,\widehat{\boldsymbol{\theta}}\right)$$

and

$$\boldsymbol{a} = \left(0 - (ct)\Omega^2\right)\widehat{\boldsymbol{r}} + (0 + 2c\Omega)\,\widehat{\boldsymbol{\theta}} = c\Omega\left(-\Omega t\,\widehat{\boldsymbol{r}} + 2\widehat{\boldsymbol{\theta}}\right).$$

The *speed* of the particle at time t is thus given by $|\boldsymbol{v}| = c\left(1 + \Omega^2 t^2\right)^{1/2}$.

To find the angle between $\boldsymbol{v}$ and $\boldsymbol{a}$, consider

$$\begin{aligned}\boldsymbol{v}\cdot\boldsymbol{a} &= c^2\Omega(-\Omega t + 2\Omega t) = c^2\Omega^2 t\\ &> 0\end{aligned}$$

for $t > 0$. Hence, for $t > 0$, the angle between $\boldsymbol{v}$ and $\boldsymbol{a}$ is acute. ■

General circular motion

An important application of polar coordinates is to circular motion. We have already considered the special case of *uniform* circular motion, but now we suppose that P moves in any manner (not necessarily with constant speed) around a circle with centre O and radius b. If we take O to be the origin of polar coordinates, the condition $r = b$ implies that $\dot{r} = \ddot{r} = 0$ and the formula (2.13) for the **velocity** of P reduces to

$$\boldsymbol{v} = \left(b\dot{\theta}\right)\widehat{\boldsymbol{\theta}}. \tag{2.15}$$

This result is depicted in Figure 2.5. The transverse velocity component $b\dot{\theta}$ (which is not necessarily the speed of P since $\dot{\theta}$ may be negative) is called the **circumferential velocity** of P. Circumferential velocity will be important when we study the motion of a rigid body rotating about a fixed axis; in this case, each particle of the rigid body moves on a circular path.

The corresponding formula for the acceleration of P is

$$\begin{aligned} \boldsymbol{a} &= \left(0 - b\dot{\theta}^2\right)\widehat{\boldsymbol{r}} + + \left((b\ddot{\theta} + 0\right)\widehat{\boldsymbol{\theta}} \\ &= -\left(b\dot{\theta}^2\right)\widehat{\boldsymbol{r}} + + \left((b\ddot{\theta}\right)\widehat{\boldsymbol{\theta}} \\ &= -\left(\frac{v^2}{b}\right)\widehat{\boldsymbol{r}} + + \dot{v}\widehat{\boldsymbol{\theta}} \end{aligned}$$

where v is the circumferential velocity $b\dot{\theta}$. These results are summarised below:

General circular motion

Suppose a particle P moves in any manner around the circle $r = b$, where r, θ are plane polar coordinates. Then the velocity and acceleration vectors of P are given by

$$\boldsymbol{v} = v\widehat{\boldsymbol{\theta}}, \tag{2.16}$$

$$\boldsymbol{a} = -\left(\frac{v^2}{b}\right)\widehat{\boldsymbol{r}} + \dot{v}\widehat{\boldsymbol{\theta}}, \tag{2.17}$$

where v $(= b\dot{\theta})$ is the circumferential velocity of P.

The formula (2.17) shows that, in *general* circular motion, the acceleration of P is the (vector) sum of an inward radial acceleration v^2/b and a transverse acceleration $\dot{v}$. This is consistent with the general formula (2.5). Indeed, what the formula (2.5) says is that, when P moves along a completely general path, its acceleration vector is the same *as if* it were moving on the circle of curvature at each point of its path.

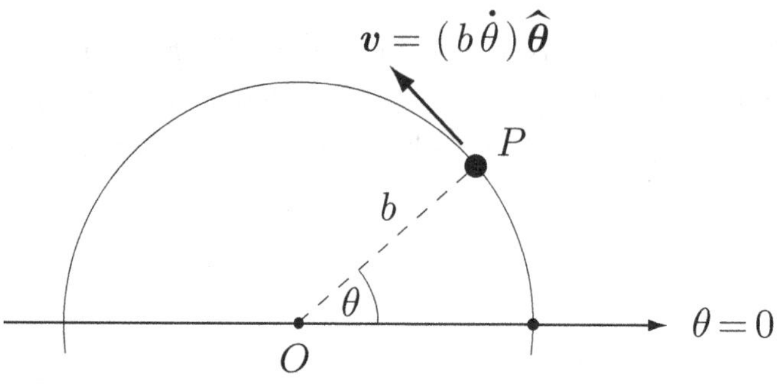

FIGURE 2.5 The particle P moves on the circle with centre O and radius b. At time t its angular displacement is θ and its circumferential velocity is $b\dot{\theta}$.

Example 2.7 *Pendulum motion*

The bob of a certain pendulum moves on a vertical circle of radius b and, when the string makes an angle θ with the downward vertical, the circumferential velocity v of the bob is given by

$$v^2 = 2gb\cos\theta,$$

where g is a positive constant. Find the acceleration of the bob when the string makes angle θ with the downward vertical.

Solution

From the acceleration formula (2.17), we have

$$\boldsymbol{a} = -\left(\frac{v^2}{b}\right)\widehat{\boldsymbol{r}} + \dot{v}\widehat{\boldsymbol{\theta}} = -(2g\cos\theta)\widehat{\boldsymbol{r}} + \dot{v}\widehat{\boldsymbol{\theta}}.$$

It remains to express $\dot{v}$ in terms of θ. On differentiating the formula $v^2 = 2gb\cos\theta$ with respect to t, we obtain

$$2v\dot{v} = -(2gb\sin\theta)\dot{\theta},$$

and, since $b\dot{\theta} = v$, we find that

$$\dot{v} = -g\sin\theta.$$

Hence the **acceleration** of the bob when the string makes angle θ with the downward vertical is

$$\boldsymbol{a} = -(2g\cos\theta)\widehat{\boldsymbol{r}} - (g\sin\theta)\widehat{\boldsymbol{\theta}}. \blacksquare$$

2.4 RIGID BODY ROTATING ABOUT A FIXED AXIS

Some objects that we find in everyday life, such as a brick or a thick steel rod, are so difficult to deform that their shape is virtually unchangeable. We model such an

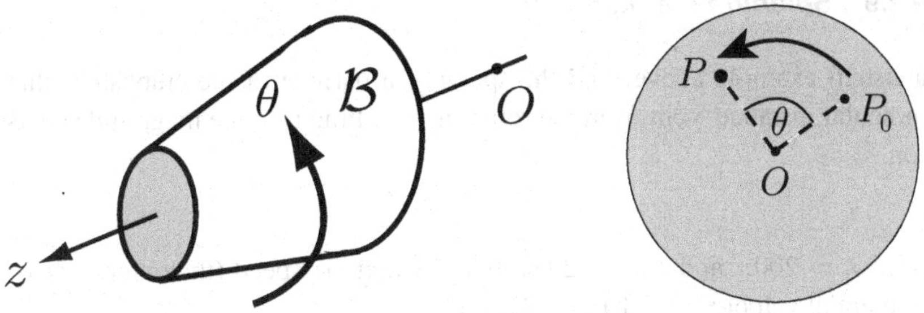

FIGURE 2.6 The rigid body $\mathcal{B}$ rotates about the fixed axis Oz and has angular displacement θ at time t. Each particle P of $\mathcal{S}$ moves on a circular path; the point P_0 is the reference position of P.

object by a **rigid body**, a collection of particles forming a perfectly rigid framework. Any motion of the rigid body must maintain this framework.

An important type of rigid body motion is **rotation about a fixed axis**; a spinning fan, a door opening on its hinges and a playground roundabout are among the many examples of this type of motion. Suppose $\mathcal{B}$ is a rigid body which is constrained to rotate about the fixed axis Oz as shown in Figure 2.6. (This means that the particles of $\mathcal{B}$ that lie on Oz are held fixed. Rotation about Oz is then the only motion of $\mathcal{B}$ consistent with rigidity.) At time t, $\mathcal{B}$ has **angular displacement** θ measured from some reference position. The angular displacement θ is the rotational counterpart of the Cartesian displacement x of a particle in straight line motion. By analogy with the rectilinear case, we make the following definitions:

Definition 2.4 ***Angular velocity*** *The* ***angular velocity*** ω *of* $\mathcal{B}$ *is defined to be* $\omega = d\theta/dt$ *and the absolute value of* ω *is called the* ***angular speed*** *of* $\mathcal{B}$.

Units. Angular velocity (and angular speed) are measured in radians per second (rad s^{-1}).

Example 2.8 *Spinning crankshaft 1*

The crankshaft of a motorcycle engine is spinning at 6000 revolutions per minute. What is its angular speed in S.I. units?

Solution

6000 revolutions per minute is 100 revolutions per second which is 200π radians per second. This is the angular speed in S.I. units. ■

Particle velocities in a rotating rigid body

In rotational motion about a fixed axis, each particle P of $\mathcal{B}$ moves on a circle of some radius ρ, where ρ is the (fixed) perpendicular distance of P from the rotation axis. It then follows from (2.16) that the **circumferential velocity** v of P is given by $\rho\dot{\theta}$, that is

$$\boxed{v = \omega\,\rho} \qquad (2.18)$$

Example 2.9 ***Spinning crankshaft 2***

In the crankshaft example above, find the speed of a particle of the crankshaft that has perpendicular distance 5 cm from the rotation axis. Find also the magnitude of its acceleration.

Solution

In this case, $|\omega| = 200\pi$ and $\rho = 1/20$ so that the particle speed (the magnitude of the circumferential velocity v) is $10\pi \approx 31.4 \text{ m s}^{-1}$.

Since the circumferential velocity is constant, $|\boldsymbol{a}| = v^2/\rho = (10\pi)^2/0.05 \approx 2000 \text{ m s}^{-2}$, which is two hundred times the value of the Earth's gravitational acceleration! ■

2.5 RIGID BODY IN PLANAR MOTION

We now consider a more general form of rigid body motion called **planar motion**.

Definition 2.5 ***Planar motion*** *A rigid body $\mathcal{B}$ is said to be in* ***planar motion*** *if each particle of $\mathcal{B}$ moves in a fixed plane and all these planes are parallel to each other.*

Planar motion is quite common. For instance, any flat-bottomed rigid body sliding on a flat table is in planar motion. Another example is a circular cylinder rolling on a rough flat table.

The **particle velocities** in planar motion can be calculated by the following method; the proof is given in Chapter 16. First select some particle C of the body as the reference particle. The velocity of a general particle P of the body is then the vector sum of

(i) a **translational contribution** equal to the velocity of C (as if the body did not rotate) and

(ii) a **rotational contribution** (as if C were fixed and the body were rotating with angular velocity ω about a fixed axis through C).

This result is illustrated in Figure 2.7, where the body is a rectangular plate and the reference particle C is at a corner of the plate. The velocity $\boldsymbol{v}$ of P is given by $\boldsymbol{v} = \boldsymbol{v}^C + \boldsymbol{v}^R$, where the translational contribution $\boldsymbol{v}^C$ is the velocity of C and the rotational contribution $\boldsymbol{v}^R$ is caused by the angular velocity ω about C. Although the reference particle can be any particle of the body, it is usually taken to be the centre of mass or centre of symmetry of the body.

Example 2.10 ***The rolling wheel***

A circular wheel of radius b rolls in a straight line with speed u on a fixed horizontal table. Find the velocities of its particles.

Solution

This is an instance of planar motion and so the particle velocities can be found by the method above. Let the position of the wheel at some instant be that shown in

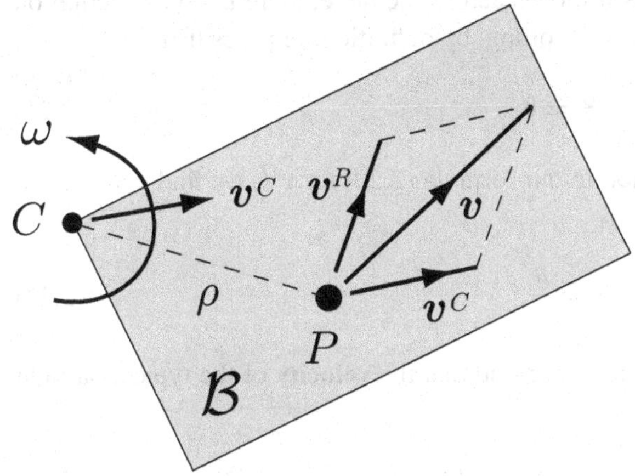

FIGURE 2.7 The velocity of the particle P belonging to the rigid body $\mathcal{B}$ is the sum of the translational contribution $\boldsymbol{v}^C$ and the rotational contribution $\boldsymbol{v}^R$. The reference particle C can be *any* particle of the body.

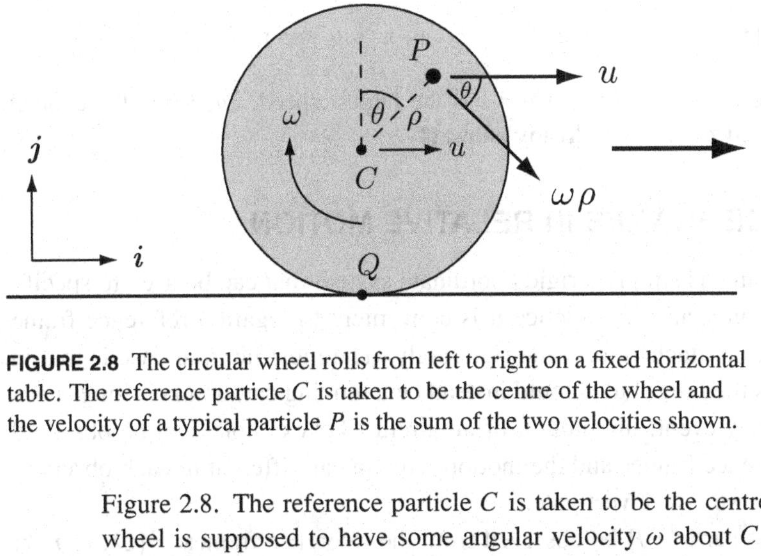

FIGURE 2.8 The circular wheel rolls from left to right on a fixed horizontal table. The reference particle C is taken to be the centre of the wheel and the velocity of a typical particle P is the sum of the two velocities shown.

Figure 2.8. The reference particle C is taken to be the centre of the wheel, and the wheel is supposed to have some angular velocity ω about C. The velocity $\boldsymbol{v}^P$ of a typical particle P is then the sum of the two velocities shown. In terms of the vectors $\{\boldsymbol{i}, \boldsymbol{j}\}$

$$\begin{aligned} \boldsymbol{v}^P &= u\,\boldsymbol{i} + \omega\rho\,(\cos\theta\,\boldsymbol{i} - \sin\theta\,\boldsymbol{j}) \\ &= (u + \omega\rho\cos\theta)\,\boldsymbol{i} - (\omega\rho\sin\theta)\,\boldsymbol{j}. \end{aligned} \tag{2.19}$$

In particular, on taking $\rho = b$ and $\theta = \pi$, the velocity $\boldsymbol{v}^Q$ of the contact particle Q is given by

$$\boldsymbol{v}^Q = (u - \omega b)\,\boldsymbol{i}. \tag{2.20}$$

If the wheel is allowed to slip as it moves across the table, there is no restriction on $\boldsymbol{v}^Q$ so that u and ω are unrelated. But rolling, by definition, requires that

$$\boldsymbol{v}^Q = \mathbf{0}. \tag{2.21}$$

On applying this **rolling condition** to our formula (2.20) for $\boldsymbol{v}^Q$, we find that ω must be related to u by

$$\omega = \frac{u}{b}, \tag{2.22}$$

and on using this value of ω in (2.19) we find that the **velocity** of the typical particle P is given by

$$\boldsymbol{v}^P = u\left(1 + \frac{\rho}{b}\cos\theta\right)\boldsymbol{i} - u\left(\frac{\rho}{b}\sin\theta\right)\boldsymbol{j}. \tag{2.23}$$

When P lies on the circumference of the wheel, this formula simplifies to

$$\boldsymbol{v}^P = u\,(1 + \cos\theta)\,\boldsymbol{i} - u\sin\theta\,\boldsymbol{j}, \tag{2.24}$$

in which case the *speed* of P is given by

$$|\,\boldsymbol{v}^P\,| = 2u\cos(\theta/2), \qquad (-\pi \le \theta \le \pi).$$

Thus the highest particle of the wheel has the largest speed, $2u$, while the contact particle has speed zero, as we already know. ■

2.6 REFERENCE FRAMES IN RELATIVE MOTION

A reference frame is simply a rigid coordinate system that can be used to specify the positions of points in space. In practice it is convenient to regard a reference frame as being embedded in, or attached to, some rigid body. The most familiar case is that in which the rigid body is the Earth but it could instead be a moving car, or an orbiting space station. In principle, any event, the motion of an aircraft for example, can be observed from any of these reference frames and the motion will appear different to each observer. It is this difference that we now investigate.

Let the motion of a particle P be observed from the reference frames $\mathcal{F}\{O\,;\boldsymbol{i},\boldsymbol{j},\boldsymbol{k}\}$ and $\mathcal{F}'\{O'\,;\boldsymbol{i},\boldsymbol{j},\boldsymbol{k}\}$ as shown in Figure 2.9. Here we are supposing that the frame $\mathcal{F}'$ *does not rotate* relative to $\mathcal{F}$. This is why, without losing generality, we can suppose that $\mathcal{F}$ and $\mathcal{F}'$ have the same set of unit vectors $\{\boldsymbol{i},\boldsymbol{j},\boldsymbol{k}\}$. For example, P could be an aircraft, $\mathcal{F}$ could be attached to the Earth, and $\mathcal{F}'$ could be attached to a car driving along a *straight* road.

Then, $\boldsymbol{r}$, $\boldsymbol{r}'$, the position vectors of P relative to $\mathcal{F}$, $\mathcal{F}'$ are connected by

$$\boldsymbol{r} = \boldsymbol{r}' + \boldsymbol{D}, \tag{2.25}$$

where $\boldsymbol{D}$ is the position vector of O' relative to $\mathcal{F}$.

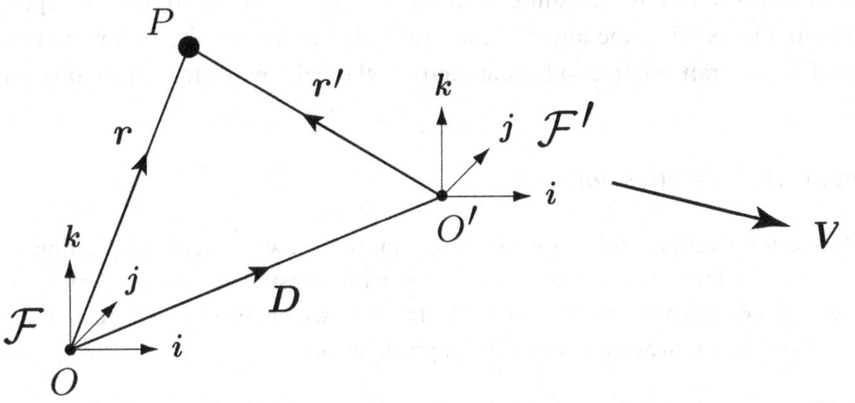

FIGURE 2.9 The particle P is observed from the two reference frames $\mathcal{F}$ and $\mathcal{F}'$.

We now differentiate this equation with respect to t, a step that requires some care. Let us consider the rates of change of the vectors in equation (2.25), as observed from the frame $\mathcal{F}$. Then

$$\boldsymbol{v} = \left(\frac{d\boldsymbol{r}'}{dt}\right)_{\mathcal{F}} + \boldsymbol{V}, \tag{2.26}$$

where $\boldsymbol{v}$ is the velocity of P observed in $\mathcal{F}$ and $\boldsymbol{V}$ is the velocity of $\mathcal{F}'$ relative to $\mathcal{F}$.

Now when two different reference frames are used to observe the same vector, the observed *rates of change* of that vector will generally be different. In particular, it is *not* generally true that

$$\left(\frac{d\boldsymbol{r}'}{dt}\right)_{\mathcal{F}} = \left(\frac{d\boldsymbol{r}'}{dt}\right)_{\mathcal{F}'}.$$

However, as we will show in Chapter 17, these two rates of change *are* equal if the frame $\mathcal{F}'$ *does not rotate* relative to $\mathcal{F}$. Hence, in our case, we do have

$$\left(\frac{d\boldsymbol{r}'}{dt}\right)_{\mathcal{F}} = \left(\frac{d\boldsymbol{r}'}{dt}\right)_{\mathcal{F}'} = \boldsymbol{v}',$$

where $\boldsymbol{v}'$ is the velocity of P observed in $\mathcal{F}'$.

Equation (2.26) can then be written

$$\boxed{\boldsymbol{v} = \boldsymbol{v}' + \boldsymbol{V}} \tag{2.27}$$

Thus *the velocity of P observed in $\mathcal{F}$ is the sum of the velocity of P observed in $\mathcal{F}'$ and the velocity of the frame $\mathcal{F}'$ relative to $\mathcal{F}$*. This result applies only when $\mathcal{F}'$ *does not rotate* relative to $\mathcal{F}$.

This is the well known rule for handling 'relative velocities'. In the aircraft example, it means that the true velocity of the aircraft (relative to the ground) is the vector sum of (i) the velocity of the aircraft relative to the car, and (ii) the velocity of the car relative to the road.

Example 2.11 *Relative velocity*

The Mississippi river is a mile wide and has a uniform flow. A steamboat sailing at full speed takes 12 minutes to cover a mile when sailing upstream, but only 3 minutes when sailing downstream. What is the shortest time in which the steamboat can *cross* the Mississippi to the nearest point on the opposite bank?

Solution

The way to handle this problem is to view the motion of the boat from a reference frame $\mathcal{F}'$ moving with the river. In this reference frame the water is at rest and the boat sails with the same speed *in all directions*. The relative velocity formula (2.27) then gives us the true picture of the motion of the boat relative to the river bank, which is the reference frame $\mathcal{F}$.

Let u^B be the speed of the boat in still water and u^R be the speed of the river, both measured in miles per hour. The upstream and downstream times are just a sneaky way of telling us the values of u^B and u^R. When the boat sails downstream, (2.27) implies that its speed relative to the bank is $u^B + u^R$. But this speed is stated to be 1/3 mile per minute (or 20 miles per hour). Hence

$$u^B + u^R = 20.$$

Similarly the upstream speed is $u^B - u^R$ and is stated to be 1/12 mile per minute (or 5 miles per hour). Hence

$$u^B - u^R = 5.$$

Solving these equations yields

$$u^B = 12.5 \text{ mph}, \qquad u^R = 7.5 \text{ mph}.$$

Now the boat must cross the river. In order to cross by a straight line path to the nearest point on the opposite bank, the boat's velocity (relative to the water) must be directed at some angle α to the required path (as shown in Figure 2.10) so that its *resultant* velocity is perpendicular to the stream. For this to be true, α must satisfy

$$u^B \sin\alpha = u^R,$$

which gives $\sin\alpha = 3/5$. The resultant speed of the boat when crossing the river is therefore $u^B \cos\alpha = 12.5 \times (4/5) = 10$ mph. Since the river is one mile wide, the time taken for the crossing is $1/10$ hour $= 6$ minutes. ∎

The relative velocity formula (2.27) can be differentiated again with respect to t to give a similar connection between accelerations. The result is that

$$\boldsymbol{a} = \boldsymbol{a}' + \boldsymbol{A}, \tag{2.28}$$

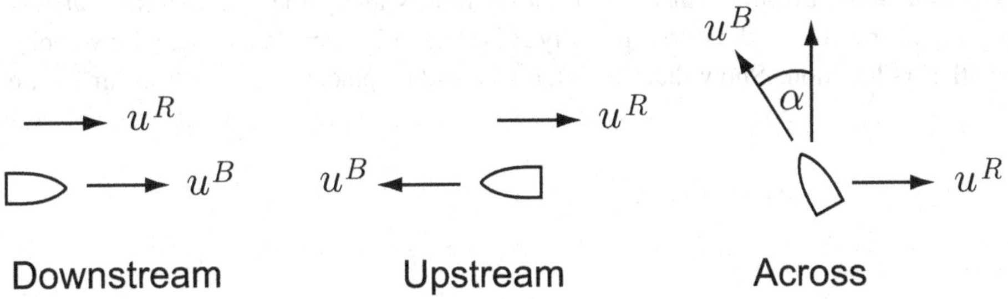

FIGURE 2.10 The river flows from left to right with speed u^R and the boat sails with speed u^B *relative to the river*. In each case the velocity of the boat relative to the bank is the vector sum of the two velocities shown.

where $\boldsymbol{a}$ and $\boldsymbol{a}'$ are the accelerations of P relative to the frames $\mathcal{F}$ and $\mathcal{F}'$ respectively, and $\boldsymbol{A}$ is the acceleration of the frame $\mathcal{F}'$ relative to the frame $\mathcal{F}$. Once again, this result applies only when $\mathcal{F}'$ does not rotate relative to $\mathcal{F}$.

Mutually unaccelerated frames

An important special case of equation (2.28) occurs when the frame $\mathcal{F}'$ is moving with *constant velocity* (and no rotation) relative to $\mathcal{F}$. We will then say that $\mathcal{F}$ and $\mathcal{F}'$ are **mutually unaccelerated** frames. In this case $\boldsymbol{A} = \boldsymbol{0}$ and (2.28) becomes

$$\boldsymbol{a} = \boldsymbol{a}'. \tag{2.29}$$

This means that *when mutually unaccelerated frames are used to observe the motion of a particle P, the observed acceleration of P is the same in each frame.*

This result will be vital in our discussion of *inertial frames* in Chapter 3.

Problems on Chapter 2

Answers and comments are at the end of the book.

Harder problems carry a star (∗).

Rectilinear particle motion

2.1 A particle P moves along the x-axis with its displacement at time t given by $x = 6t^2 - t^3 + 1$, where x is measured in metres and t in seconds. Find the velocity and acceleration of P at time t. Find the times at which P is at rest and find its position at these times.

2.2 A particle P moves along the x-axis with its acceleration a at time t given by

$$a = 6t - 4 \text{ m s}^{-2}.$$

Initially P is at the point $x = 20$ m and is moving with speed 15 m s^{-1} in the negative x-direction. Find the velocity and displacement of P at time t. Find when P comes to rest and its displacement at this time.

2.3 *Constant acceleration formulae* A particle P moves along the x-axis with *constant* acceleration a in the positive x-direction. Initially P is at the origin and is moving with velocity u in the positive x-direction. Show that the velocity v and displacement x of P at time t are given by*

$$v = u + at, \qquad x = ut + \tfrac{1}{2}at^2,$$

and deduce that

$$v^2 = u^2 + 2ax.$$

In a standing quarter mile test, the Suzuki Bandit 1200 motorcycle covered the quarter mile (from rest) in 11.4 seconds and crossed the finish line doing 116 miles per hour. Are these figures consistent with the assumption of constant acceleration?

General particle motion

2.4 The trajectory of a charged particle moving in a magnetic field is given by

$$\boldsymbol{r} = b\cos\Omega t\,\boldsymbol{i} + b\sin\Omega t\,\boldsymbol{j} + ct\,\boldsymbol{k},$$

where b, Ω and c are positive constants. Show that the particle moves with constant speed and find the magnitude of its acceleration.

2.5 *Acceleration due to rotation and orbit of the Earth* A body is at rest at a location on the Earth's equator. Find its acceleration due to the Earth's rotation. [Take the Earth's radius at the equator to be 6400 km.]

Find also the acceleration of the Earth in its orbit around the Sun. [Take the Sun to be fixed and regard the Earth as a particle following a circular path with centre the Sun and radius 15×10^{10} m.

2.6 An insect flies on a spiral trajectory such that its polar coordinates at time t are given by

$$r = be^{\Omega t}, \qquad \theta = \Omega t,$$

where b and Ω are positive constants. Find the velocity and acceleration vectors of the insect at time t, and show that the angle between these vectors is always $\pi/4$.

2.7 A racing car moves on a circular track of radius b. The car starts from rest and its *speed* increases at a constant rate α. Find the angle between its velocity and acceleration vectors at time t.

---

* These are the famous **constant acceleration formulae**. Although they are a mainstay of school mechanics, we will make little use of them since, in most of the problems that we treat, the acceleration is *not* constant. *It is a serious offence to use these formulae in non-constant acceleration problems.*

2.8 A particle P moves on a circle with centre O and radius b. At a certain instant the speed of P is v and its acceleration vector makes an angle α with PO. Find the magnitude of the acceleration vector at this instant.

2.9* A bee flies on a trajectory such that its polar coordinates at time t are given by

$$r = \frac{bt}{\tau^2}(2\tau - t) \qquad \theta = \frac{t}{\tau} \qquad (0 \le t \le 2\tau),$$

where b and τ are positive constants. Find the velocity vector of the bee at time t.

Show that the least speed achieved by the bee is b/τ. Find the acceleration of the bee at this instant.

2.10* ***A pursuit problem: Daniel and the Lion*** The luckless Daniel (D) is thrown into a circular arena of radius a containing a lion (L). Initially the lion is at the centre O of the arena while Daniel is at the perimeter. Daniel's strategy is to run with his maximum speed u around the perimeter. The lion responds by running at its maximum speed U in such a way that it remains on the (moving) radius OD. Show that r, the distance of L from O, satisfies the differential equation

$$\dot{r}^2 = \frac{u^2}{a^2}\left(\frac{U^2a^2}{u^2} - r^2\right).$$

Find r as a function of t. If $U \ge u$, show that Daniel will be caught, and find how long this will take.

Show that the path taken by the lion is an arc of a circle. For the special case in which $U = u$, sketch the path taken by the lion and find the point of capture.

2.11 ***General motion with constant speed*** A particle moves along any path in three-dimensional space with *constant speed.* Show that its velocity and acceleration vectors must always be perpendicular to each other. [*Hint.* Differentiate the formula $\boldsymbol{v} \cdot \boldsymbol{v} = v^2$ with respect to t.]

2.12 A particle P moves so that its position vector $\boldsymbol{r}$ satisfies the differential equation

$$\dot{\boldsymbol{r}} = \boldsymbol{c} \times \boldsymbol{r},$$

where $\boldsymbol{c}$ is a constant vector. Show that P moves with constant speed on a circular path. [*Hint.* Take the dot product of the equation first with $\boldsymbol{c}$ and then with $\boldsymbol{r}$.]

Angular velocity

2.13 A large truck with double rear wheels has a brick jammed between two of its tyres which are 4 ft in diameter. If the truck is travelling at 60 mph, find the maximum speed of the brick and the magnitude of its acceleration. [Express the acceleration as a multiple of $g = 32$ ft s^{-2}.]

2.14 A particle is sliding along a smooth radial grove in a circular turntable which is rotating with constant angular speed Ω. The distance of the particle from the rotation axis at time t is

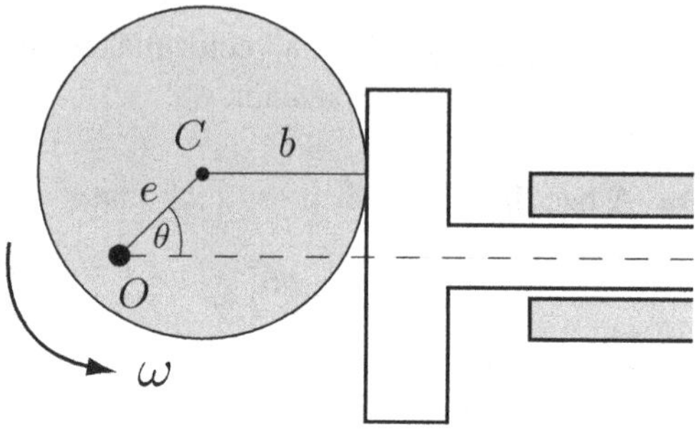

FIGURE 2.11 Cam and valve mechanism

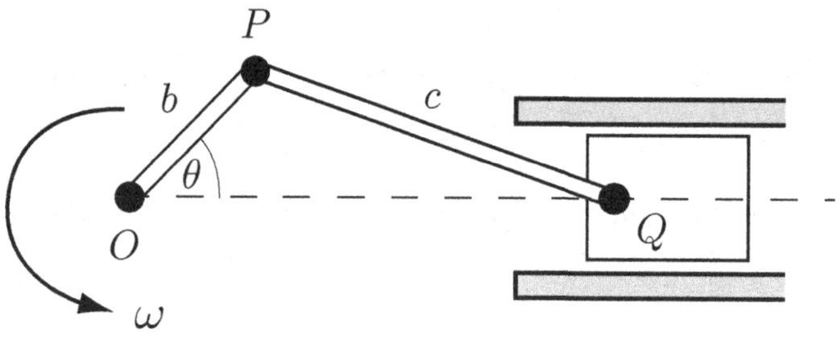

FIGURE 2.12 Crank and piston mechanism

observed to be

$$r = b \cosh \Omega t$$

for $t \geq 0$, where b is a positive constant. Find the speed of the particle (relative to a fixed reference frame) at time t, and find the magnitude and direction of the acceleration.

2.15 Figure 2.11 shows an eccentric circular cam of radius b rotating with constant angular velocity ω about a fixed pivot O which is a distance e from the centre C. The cam drives a valve which slides in a straight guide. Find the maximum speed and maximum acceleration of the valve.

2.16 Figure 2.12 shows a piston driving a crank OP pivoted at the end O. The piston slides in a straight cylinder and the crank is made to rotate with constant angular velocity ω. Find the distance OQ in terms of the lengths b, c and the angle θ. Show that, when b/c is small, OQ is given approximately by

$$OQ = c + b\cos\theta - \frac{b^2}{2c}\sin^2\theta,$$

on neglecting $(b/c)^4$ and higher powers. Using this approximation, find the maximum acceleration of the piston.

2.17 Figure 2.13 shows an epicyclic gear arrangement in which the 'sun' gear $\mathcal{G}_1$ of radius b_1 and the 'ring' gear $\mathcal{G}_2$ of inner radius b_2 rotate with angular velocities ω_1, ω_2 respectively

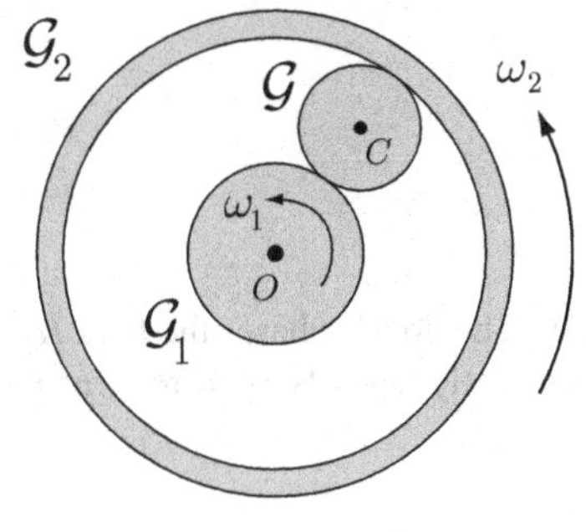

FIGURE 2.13 Epicyclic gear mechanism

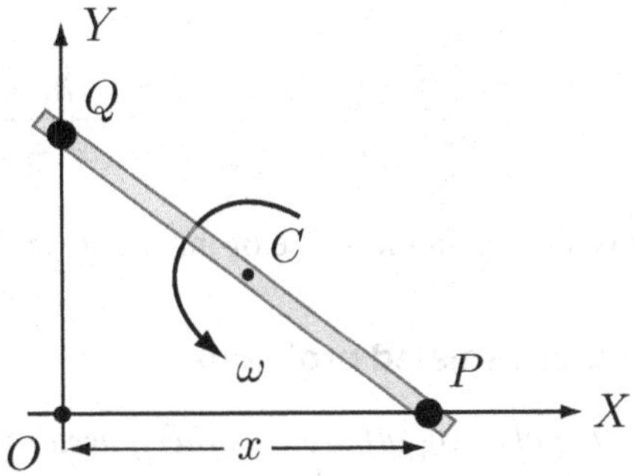

FIGURE 2.14 The pins P and Q at the ends of a rigid link move along the axes OX, OY respectively.

about their fixed common centre O. Between them they grip the 'planet' gear $\mathcal{G}$, whose centre C moves on a circle centre O. Find the circumferential velocity of C and the angular velocity of the planet gear $\mathcal{G}$. If O and C were connected by an arm pivoted at O, what would be the angular velocity of the arm?

2.18 Figure 2.14 shows a straight rigid link of length a whose ends contain pins P, Q that are constrained to move along the axes OX, OY. The displacement x of the pin P at time t is prescribed to be $x = b \sin \Omega t$, where b and Ω are positive constants with $b < a$. Find the angular velocity ω and the speed of the centre C of the link at time t.

Relative velocity

2.19 An aircraft is to fly from a point A to an airfield B 600 km due north of A. If a steady wind of 90 km/h is blowing from the north-west, find the direction the plane should be pointing and the time taken to reach B if the cruising speed of the aircraft in still air is 200 km/h.

2.20 An aircraft takes off from a horizontal runway with constant speed U, climbing at a constant angle α to the horizontal. A car is moving on the runway with constant speed u directly towards the front of the aircraft. The car is distance a from the aircraft at the instant of take-off. Find the distance of closest approach of the car and aircraft. [Don't try this one at home.]

2.21* An aircraft has cruising speed v and a flying range (out and back) of R_0 in still air. Show that, in a north wind of speed u ($u < v$) its range in a direction whose true bearing from

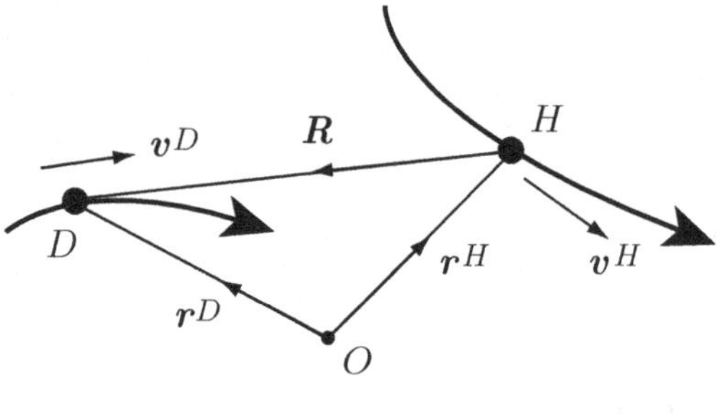

FIGURE 2.15 The dog D chases the hare H by running directly towards the hare's current position.

north is θ is given by

$$\frac{R_0(v^2 - u^2)}{v(v^2 - u^2 \sin^2\theta)^{1/2}}.$$

What is the maximum value of this range and in what directions is it attained?

Computer assisted problems

2.22 *Dog chasing a hare; another pursuit problem.* Figure 2.15 shows a dog with position vector $\boldsymbol{r}^D$ and velocity $\boldsymbol{v}^D$ chasing a hare with position vector $\boldsymbol{r}^H$ and velocity $\boldsymbol{v}^H$. The dog's strategy is to run directly towards the current position of the hare. Given the motion of the hare and the *speed* of the dog, what path does the dog follow?

Since the dog runs directly towards the hare, its velocity $\boldsymbol{v}^D$ must satisfy

$$\frac{\boldsymbol{v}^D}{v^D} = \frac{\boldsymbol{r}^H - \boldsymbol{r}^D}{|\boldsymbol{r}^H - \boldsymbol{r}^D|}.$$

In terms of the position vector of the dog *relative to the hare*, given by $\boldsymbol{R} = \boldsymbol{r}^D - \boldsymbol{r}^H$, this equation becomes

$$\dot{\boldsymbol{R}} = -\frac{\boldsymbol{R}}{R}v^D - \boldsymbol{v}^H.$$

Given the velocity $\boldsymbol{v}^H$ of the hare and the *speed* v^D of the dog as functions of time, this differential equation determines the trajectory of the dog relative to the hare; capture occurs when $\boldsymbol{R} = \boldsymbol{0}$. The actual trajectory of the dog is given by $\boldsymbol{r}^D = \boldsymbol{R} + \boldsymbol{r}^H$.

If the motion takes place in a plane with $\boldsymbol{R} = X\,\boldsymbol{i} + Y\,\boldsymbol{j}$ then X and Y satisfy the coupled differential equations

$$\dot{X} = -\frac{v^D X}{(X^2 + Y^2)^{1/2}} - v_x^H, \qquad \dot{Y} = -\frac{v^D Y}{(X^2 + Y^2)^{1/2}} - v_y^H,$$

together with initial conditions of the form $X(0) = x_0$ and $Y(0) = y_0$. Such equations cannot usually be solved analytically but are extremely easy to solve with computer assistance. Two interesting cases to consider are as follows. In each case the *speeds* of the dog and the hare are constants.

(i) Initially the hare is at the origin with the dog at some point (x_0, y_0). The hare then runs along the positive x-axis and is chased by the dog. Show that the hare gets caught if $v^D > v^H$, but when $v^D = v^H$ the dog always misses (unless he starts directly in the path of the hare). This remarkable result can be proved analytically.

(ii) The hare runs in a circle (like the lion problem). In this case, with $v^D = v^H$, the dog seems to miss no matter where he starts.

Try some examples of your own and see if you can find interesting paths taken by the dog.

2.23 Consider further the piston problem described in Problem 2.16. Use computer assistance to calculate the exact and approximate accelerations of the piston as functions of θ. Compare the exact and approximate formulae (non-dimensionalised by $\omega^2 b$) by plotting both on the same graph against θ. Show that, when $b/c < 0.5$, the two graphs are close, but when b/c gets close to unity, large errors occur.

Chapter Three

Newton's laws of motion and the law of gravitation

KEY FEATURES

The key features of this chapter are **Newton's laws** of motion, the definitions of **mass** and **force**, the **law of gravitation**, the **principle of equivalence**, and **gravitation by spheres**.

This chapter is concerned with the **foundations of dynamics** and **gravitation**. Kinematics is concerned purely with geometry of motion, but dynamics seeks to answer the question as to what motion will actually occur when specified forces act on a body. The rules that allow one to make this connection are **Newton's laws of motion**. These are laws of physics that are founded upon experimental evidence and stand or fall according to the accuracy of their predictions. In fact, Newton's formulation of mechanics has been astonishingly successful in its accuracy and breadth of application, and has survived, essentially intact, for more than three centuries. The same is true for Newton's universal **law of gravitation** which specifies the forces that all masses exert upon each other.

Taken together, these laws represent virtually the entire foundation of classical mechanics and provide an accurate explanation for a vast range of motions from large molecules to entire galaxies.

3.1 NEWTON'S LAWS OF MOTION

Isaac Newton's* three famous laws of motion were laid down in *Principia*, written in Latin and published in 1687. These laws set out the founding principles of mechanics and have survived, essentially unchanged, to the present day. Even when translated into English, Newton's original words are hard to understand, mainly because the terminology

* Sir Isaac Newton (1643–1727) is arguably the greatest scientific genius of all time. His father was completely uneducated and Isaac himself had no contact with advanced mathematics before the age of twenty. However, by the age of twenty seven, he had been appointed to the Lucasian chair at Cambridge and was one of the foremost scientists in Europe. His greatest achievements were his discovery of the calculus, his laws of motion, and his theory of universal gravitation. On the urging of Halley (the Astronomer Royal), Newton wrote up an account of his new physics and its application to astronomy. *Philosophiae Naturalis Principia Mathematica* was published in 1687 and is generally recognised as the greatest scientific book ever written.

of the seventeenth century is now archaic. Also, the laws are now formulated as applying to particles, a concept never used by Newton. A **particle** is an idealised body that occupies only a **single point of space** and has no internal structure. True particles do not exist† in nature, but it is convenient to regard realistic bodies as being made up of particles. Using modern terminology, Newton's laws may be stated as follows:

Newton's laws of motion

First Law When all external influences on a particle are removed, the particle moves with constant velocity. [This velocity may be zero in which case the particle remains at rest.]

Second Law When a force $\boldsymbol{F}$ acts on a particle of mass m, the particle moves with instantaneous acceleration $\boldsymbol{a}$ given by the formula

$$\boldsymbol{F} = m\boldsymbol{a},$$

where the unit of force is implied by the units of mass and acceleration.

Third Law When two particles exert forces upon each other, these forces are (i) equal in magnitude, (ii) opposite in direction, and (iii) parallel to the straight line joining the two particles.

Units

Any consistent system of units can be used. The standard scientific units are **SI units** in which the unit of mass is the **kilogram**, the unit of length is the **metre**, and the unit of time is the **second**. The unit of force implied by the Second Law is called the **newton**, and written N. An excellent description of the SI system of units can be found on

`http://www.physics.nist.gov/PhysRefData`

the website of the US National Institute of Standards & Technology.

In the Imperial system of units, the unit of mass is the pound, the unit of length is the foot, and the unit of time is the second. The unit of force implied by the Second Law is called the poundal. These units are still used in some industries in the US, a fact which causes frequent confusion.

Interpreting Newton's laws

Newton's laws are clear enough in themselves but they leave some important questions unanswered, namely:

(i) In what **frame of reference** are the laws true?

† The nearest thing to a particle is the electron, which, unlike other elementary particles, does seem to be a *point* mass. The electron does however have an internal structure, having spin and angular momentum.

(ii) What are the definitions of **mass** and **force**?

These questions are answered in the sections that follow. What we do is to set aside Newton's laws for the time being and go back to simple experiments with particles. These are 'thought experiments' in the sense that, although they are perfectly meaningful, they are unlikely to be performed in practice. The supposed 'results' of these experiments are taken to be the *primitive governing laws* of mechanics on which we base our definitions of mass and force. Finally, these laws and definitions are shown to be equivalent to Newton's laws as stated above. This process could be said to provide an *interpretation* of Newton's laws. The interpretation below is quite sophisticated and is probably only suitable for those who have already seen a simpler account, such as that given by French [3].

3.2 INERTIAL FRAMES AND THE LAW OF INERTIA

The first law states that, when a particle is unaffected by external influences, it moves with constant velocity, that is, it moves in a straight line with constant speed. Thus, contrary to Aristotle's view, the particle needs no agency of any kind to maintain its motion.* Since the influence of the Earth's gravity rules out any verification of the First Law by an experiment conducted on Earth, Newton showed remarkable insight in proposing a law he could not possibly verify. In order to verify the First Law, all *external influences* must be removed, which means that we must carry out our thought experiment in a place as remote as possible from any material bodies, such as the almost empty space between the galaxies. In our minds then we go to such a place armed with a selection of test particles† which we release in various ways and observe their motion. According to the First Law, each of these particles should move with constant velocity.

Inertial reference frames

So far we have ignored the awkward question as to what reference frame we should use to observe the motion of our test particles. When confronted with this question for the first time, one's probable response is that the reference frame should be 'fixed'. But fixed to what? The Earth rotates and is in orbital motion around the Sun. Our entire solar system is part of a galaxy that rotates about its centre. The galaxies themselves move relative to each other. The fact is that everything in the universe is moving relative to everything else and nothing can properly be described as fixed. From this it might be concluded that any reference frame is as good as any other, but this is not so, for, if the First Law is true at all, it can only be true in certain special reference frames. Suppose for instance that the First Law has been found to be true in the reference frame $\mathcal{F}$. Then it is also true in any other frame $\mathcal{F}'$ that is mutually unaccelerated relative to $\mathcal{F}$ (see section 2.6). This follows because, if the test particles have constant velocities in $\mathcal{F}$, then they have

* Such a law was proposed prior to Newton by Galileo but, curiously, Galileo did not accept the consequences of his own statement.

† Since true particles do not exist, we will have to make do with uniform rigid spheres of various kinds.

zero accelerations in $\mathcal{F}$. But, since $\mathcal{F}$ and $\mathcal{F}'$ are mutually unaccelerated frames, the test particles must have zero accelerations in $\mathcal{F}'$ and thus have constant velocities in $\mathcal{F}'$. Moreover, the First Law does *not* hold in any other reference frame.

Definition 3.1 ***Inertial frame*** *A reference frame in which the First Law is true is said to be an* ***inertial frame****.*

It follows that, *if* there exists one inertial frame, then there exist infinitely many, with each frame moving with constant velocity (and no rotation) relative to any other.

It may appear that the First Law is without physical content since we are saying that it is true in those reference frames in which it is true. However, this is not so since *inertial frames need not have existed at all, and the fact that they do is the real physical content of the First Law. Why* there should exist this special class of reference frames in which the laws of physics take simple forms is a very deep and interesting question that we do not have to answer here!

Our discussion is summarised by the following statement which we take to be a law of physics:

The law of inertia *There exists in nature a unique class of mutually unaccelerated reference frames (the inertial frames) in which the First Law is true.*

Practical inertial frames

The preceding discussion gives no clue as to how to set up an inertial reference frame and, in practice, *exact* inertial frames are not available. Practical reference frames have to be tied to real objects that are actually available. The most common practical reference frame is **the Earth**. Such a frame is sufficiently close to being inertial for the purpose of observing most Earth-bound phenomena. The orbital acceleration of the Earth is insignificant and the effect of the Earth's rotation is normally a small correction. For example, when considering the motion of a football, a pendulum or a spinning top, the Earth may be assumed to be an inertial frame.

However, the Earth is not a suitable reference frame from which to observe the motion of an orbiting satellite. In this case, the **geocentric frame** (which has its origin at the centre of mass of the Earth and has no rotation relative to distant stars) would be appropriate. Similarly, the **heliocentric frame** (which has its origin at the centre of mass of the solar system and has no rotation relative to distant stars) is appropriate when observing the motion of the planets.

Example 3.1 *Inertial frames*

Suppose that a reference frame fixed to the Earth is *exactly* inertial. Which of the following are then inertial frames?

A frame fixed to a motor car which is

(i) moving with constant speed around a flat race track,

(ii) moving with constant speed along a straight undulating road,

(iii) moving with constant speed up a constant gradient,

(iv) freewheeling down a hill.

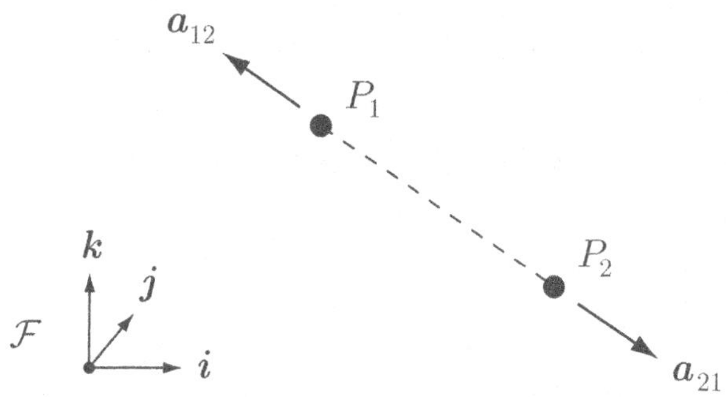

FIGURE 3.1 The particles P_1 and P_2 move under their mutual interaction and, relative to an inertial reference frame $\mathcal{F}$, have accelerations $\boldsymbol{a}_{12}$ and $\boldsymbol{a}_{21}$ respectively. These accelerations are found to satisfy the **law of mutual interaction**.

Solution

Only (iii) is inertial. In the other cases, the frame is accelerating or rotating relative to the Earth.

3.3 THE LAW OF MUTUAL INTERACTION; MASS AND FORCE

We first dispose of the question of what frame of reference should be used to observe the particle motions mentioned in the Second Law. The answer is that any **inertial reference frame** can be used and we will always assume this to be so, unless stated otherwise. As stated earlier, the problem in understanding the Second and Third Laws is that the concepts of mass and force are not defined, which is obviously unsatisfactory.

Our second thought experiment is concerned with the motion of a pair of **mutually interacting particles**. The nature of their mutual interaction can be of any kind* and all other influences are removed. Since each particle is influenced by the other, the First Law does not apply. The particles will, in general, have accelerations, these being independent of the inertial frame in which they are measured. Our second law of physics is concerned with the 'observed' values of these mutually induced accelerations.

The law of mutual interaction *Suppose that two particles P_1 and P_2 interact with each other and that P_2 induces an instantaneous acceleration $\boldsymbol{a}_{12}$ in P_1, while P_1 induces an instantaneous acceleration $\boldsymbol{a}_{21}$ in P_2. Then*

(i) *these accelerations are opposite in direction and parallel to the straight line joining P_1 and P_2,*

* The mutual interaction might be, for example, (i) mutual gravitation, (ii) electrostatic interaction, caused by the particles being electrically charged, or (iii) the particles being connected by a fine elastic cord.

(ii) *the ratio of the magnitudes of these accelerations,* $|\boldsymbol{a}_{21}|/|\boldsymbol{a}_{12}|$ *is a constant independent of the nature of the mutual interaction between* P_1 *and* P_2*, and independent of the positions and velocities*[†] *of* P_1 *and* P_2.

Moreover, suppose that when P_2 *interacts with a third particle* P_3 *the induced accelerations are* $\boldsymbol{a}_{23}$ *and* $\boldsymbol{a}_{32}$*, and when* P_1 *interacts with* P_3 *the induced accelerations are* $\boldsymbol{a}_{13}$ *and* $\boldsymbol{a}_{31}$*. Then the magnitudes of these accelerations satisfy the consistency relation*[*]

$$\frac{|\boldsymbol{a}_{21}|}{|\boldsymbol{a}_{12}|} \times \frac{|\boldsymbol{a}_{32}|}{|\boldsymbol{a}_{23}|} \times \frac{|\boldsymbol{a}_{13}|}{|\boldsymbol{a}_{31}|} = 1. \tag{3.1}$$

Definition of inertial mass

The law of mutual interaction leads us to our definitions of mass and force. *The qualitative definition of the (inertial) mass of a particle is that it is a numerical measure of the reluctance of the particle to being accelerated.* Thus, when particles P_1 and P_2 interact, we attribute the fact that the induced accelerations $\boldsymbol{a}_{12}$ and $\boldsymbol{a}_{21}$ have different magnitudes to the particles having different masses. This point of view is supported by the fact that the ratio $|\boldsymbol{a}_{21}|/|\boldsymbol{a}_{12}|$ depends only upon the particles themselves, and not on the interaction, or where the particles are, or how they are moving. We define the mass ratio m_1/m_2 of the particles P_1, P_2 to be the inverse ratio of the magnitudes of their mutually induced accelerations, as follows:

Definition 3.2 ***Inertial mass*** *The* **mass ratio** m_1/m_2 *of the particles* P_1, P_2 *is defined to be*

$$\frac{m_1}{m_2} = \frac{|\boldsymbol{a}_{21}|}{|\boldsymbol{a}_{12}|}. \tag{3.2}$$

There is however a possible inconsistency in this definition of mass ratio. Suppose that we introduce an third particle P_3. Then, by performing three experiments, we could *independently* determine the three mass ratios m_1/m_2, m_2/m_3 and m_3/m_1 and there is no guarantee that the product of these three ratios would be unity. However, the **consistency relation** (3.1) assures us that it would be found to be unity, and this means that the above definition defines the mass ratios of particles unambiguously.

In order to have a numerical measure of mass, we simply choose some particle A as the reference mass (having mass one unit), in which case the mass of any other particle can be expressed as a number of 'A-units'. If we were to use a different particle B as the reference mass, we would obtain a second measure of mass in B-units, but this second measure would just be proportional to the first, differing only by a multiplied constant. In SI units, the reference body (having mass one kilogram) is a cylinder of platinum iridium alloy kept under carefully controlled conditions in Paris.

† This is true when relativistic effects are negligible.
* The significance of the consistency relation will be explained shortly.

Example 3.2 ***A strange definition of mass***

Suppose the mass ratio m_2/m_1 were defined in some other way, such as

$$\frac{m_1}{m_2} = \left(\frac{|a_{21}|}{|a_{12}|}\right)^{1/2}.$$

Is this just as good as the standard definition?

Solution

For some purposes it would be just as good. It would lead to the non-standard form

$$\boldsymbol{F} = m^2 \boldsymbol{a}$$

for the second law, and, for the motion of a single particle, the theory would be essentially unaffected. We will see later however that, if mass were defined in this way, then the mass of a multi-particle system would *not* be equal to the sum of the masses of its constituent particles! This is not contradictory, but it is a very undesirable feature and explains why the standard definition is used. ■

Definition of force

We now turn to the definition of force. *Qualitatively, the presence of a force is the reason we give for the acceleration of a particle.* Thus, when interacting particles cause each other to accelerate, we say it is because they *exert forces upon each other*. How do we know that these forces are present? Because the particles are accelerating! These statements are obviously circular and without real content. Force is therefore a quantity of our own invention, but a very useful one nonetheless and an essential part of the Newtonian formulation of classical mechanics. It should be noted though that the concept of force is not an *essential* part of the Lagrangian or Hamiltonian formulations of classical mechanics.*

In mutual interactions, the forces that the particles exert upon each other are defined as follows:

Definition 3.3 ***Force*** *Suppose that the particles P_1 and P_2 are in mutual interaction and have accelerations $\boldsymbol{a}_{12}$ and $\boldsymbol{a}_{21}$ respectively. Then the force $\boldsymbol{F}_{12}$ that P_2 exerts on P_1, and the force $\boldsymbol{F}_{21}$ that P_1 exerts on P_2 are defined to be*

$$\boldsymbol{F}_{12} = m_1 \boldsymbol{a}_{12}, \qquad \boldsymbol{F}_{21} = m_2 \boldsymbol{a}_{21}, \tag{3.3}$$

where the unit of force is implied by the units of mass and acceleration.

It follows that, in the case of two-particle interactions, the Second Law is true by the definition of force. Also, since $\boldsymbol{a}_{12}$ and $\boldsymbol{a}_{21}$ are opposite in direction and are parallel

* This fact is important when making connections between classical mechanics and other theories, such as general relativity or quantum mechanics. The concept of force does not appear in either of these theories.

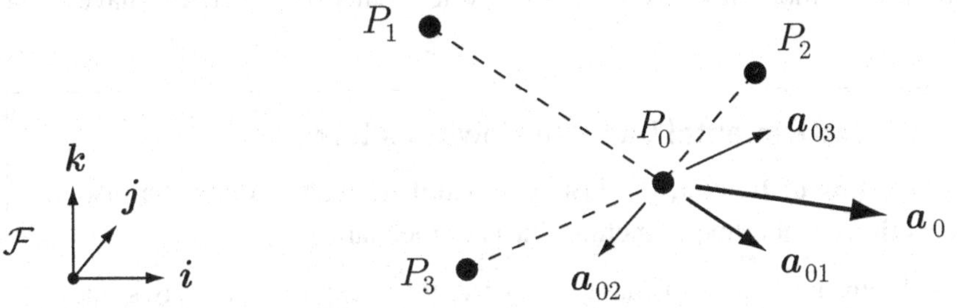

FIGURE 3.2 The **law of multiple interactions.** In the presence of interactions from the particles P_1, P_2, P_3, the acceleration $\boldsymbol{a}_0$ of particle P_0 is given by $\boldsymbol{a}_0 = \boldsymbol{a}_{01} + \boldsymbol{a}_{02} + \boldsymbol{a}_{03}$.

to the line P_1P_2, so then are $\boldsymbol{F}_{12}$ and $\boldsymbol{F}_{21}$; thus parts (ii) and (iii) of the Third Law are automatically true. Furthermore

$$|\boldsymbol{F}_{12}| = m_1|\boldsymbol{a}_{12}| = m_2|\boldsymbol{a}_{21}| = |\boldsymbol{F}_{21}|,$$

on using the definition (3.2) of the mass ratio m_1/m_2. Thus part (i) of the Third Law is also true. Hence the *law of mutual interaction, together with our definitions (3.2), (3.3) of mass and force, implies the truth of the Second and Third Laws.*

3.4 THE LAW OF MULTIPLE INTERACTIONS

Our third and final thought experiment is concerned with what happens when a particle is subject to more than one interaction.

The law of multiple interactions *Suppose the particles P_0, P_1, ... P_n are interacting with each other and that all other influences are removed. Then the acceleration $\boldsymbol{a}_0$ induced in P_0 can be expressed as*

$$\boldsymbol{a}_0 = \boldsymbol{a}_{01} + \boldsymbol{a}_{02} + \cdots + \boldsymbol{a}_{0N}, \tag{3.4}$$

where $\boldsymbol{a}_{01}$, $\boldsymbol{a}_{02}$ ... are the accelerations that P_0 would have if the particles P_1, P_2, ... were individually interacting with P_0.

This result is sometimes expressed by saying that **interaction forces act independently** of each other. It follows that

$$\begin{aligned} m_0\boldsymbol{a}_0 &= m_0(\boldsymbol{a}_{01} + \boldsymbol{a}_{02} + \cdots + \boldsymbol{a}_{0N}) \\ &= \boldsymbol{F}_{01} + \boldsymbol{F}_{02} + \cdots + \boldsymbol{F}_{0N}, \end{aligned}$$

on using the definition (3.3) of mutual interaction forces. Thus the *Second Law remains true for multiple interactions provided that the 'effective force' $\boldsymbol{F}_0$ acting on P_0 is understood to mean the (vector) resultant of the individual interaction forces acting on P_0*, that is

$$\boldsymbol{F}_0 = \boldsymbol{F}_{01} + \boldsymbol{F}_{02} + \cdots + \boldsymbol{F}_{0N}.$$

This result is not always thought of as a law of physics, but it is.* It *could* have been otherwise!

Experimental basis of Newton's Laws

1. We accept the **law of inertia**, the **law of mutual interaction** and the **law of multiple interactions** as the 'experimental' basis of mechanics.

2. Together with our definitions of mass and force, these experimental laws imply that **Newton's laws are true in any inertial reference frame.**

3.5 CENTRE OF MASS

We can now introduce the notion of the centre of mass of a collection of particles. Suppose we have a system of particles $P_1, P_2, \ldots, P_N$ with masses $m_1, m_2, \ldots, m_N$, and position vectors $\boldsymbol{r}_1, \boldsymbol{r}_2, \ldots, \boldsymbol{r}_N$ respectively. Then:

Definition 3.4 ***Centre of mass*** *The* ***centre of mass*** *of this system of particles is the* ***point of space*** *whose position vector* $\boldsymbol{R}$ *is defined by*

$$\boldsymbol{R} = \frac{m_1\boldsymbol{r}_1 + m_2\boldsymbol{r}_2 + \cdots + m_N\boldsymbol{r}_N}{m_1 + m_2 + \cdots + m_N} = \frac{\sum_{i=1}^N m_i\boldsymbol{r}_i}{\sum_{i=1}^N m_i} = \frac{\sum_{i=1}^N m_i\boldsymbol{r}_i}{M}, \qquad (3.5)$$

where M *is the sum of the separate masses.*

The centre of mass of a system of particles is simply a 'weighted' mean of the position vectors of the particles, where the 'weights' are the particle masses. Centre of mass is an important concept in the mechanics of multi-particle systems. Unfortunately, there is a widespread belief that the centre of mass has a magical ability to describe the behaviour of the system in all circumstances. This is simply not true. For instance, we will show in the next section that it is *not* generally true that the total gravitational force that a system of masses exerts on a test mass is equal to the force that would be exerted by a particle of mass M situated at the centre of mass.

Example 3.3 *Finding centres of mass*

Find the centre of mass of (i) a pair of particles of different masses, (ii) three identical particles.

Solution

(i) For a pair of particles P_1, P_2, the position vector of the centre of mass is given by

$$\boldsymbol{R} = \frac{m_1\boldsymbol{r}_1 + m_2\boldsymbol{r}_2}{m_1 + m_2}.$$

* It certainly does not follow from the observation that 'forces are vector quantities'!

It follows that the centre of mass lies on the line P_1P_2 and divides this line in the ratio $m_2 : m_1$.

(ii) For three identical particles P_1, P_2, P_3, the position vector of the centre of mass is given by

$$\boldsymbol{R} = \frac{m\boldsymbol{r}_1 + m\boldsymbol{r}_2 + m\boldsymbol{r}_3}{m + m + m} = \frac{\boldsymbol{r}_1 + \boldsymbol{r}_2 + \boldsymbol{r}_3}{3}.$$

It follows that the centre of mass lies at the centroid of the triangle $P_1P_2P_3$. ■

The centres of mass of most of the systems we meet in mechanics are easily determined by symmetry considerations. However, when symmetry is lacking, the position of the centre of mass has to be worked out from first principles by using the definition (3.5), or its counterpart for continuous mass distributions. The Appendix at the end of the book contains more details and examples.

3.6 THE LAW OF GRAVITATION

Physicists recognise only four distinct kinds of interaction forces that exist in nature. These are gravitational forces, electromagnetic forces and weak/strong nuclear forces. The nuclear forces are important only within the atomic nucleus and will not concern us at all. The electromagnetic forces include electrostatic attraction and repulsion, but we will encounter them mainly as 'forces of contact' between material bodies. Since such forces are intermolecular, they are ultimately electromagnetic although we will make no use of this fact! The present section however is concerned with **gravitation**.

It is an observed fact that any object with mass attracts any other object with mass with a force called gravitation. When gravitational interaction occurs between particles, the Third Law implies that the interaction forces must be equal in magnitude, opposite in direction and parallel to the straight line joining the particles. The *magnitude* of the gravitational interaction forces is given by:

The law of gravitation

The gravitational forces that two particles exert upon each other each have magnitude

$$\frac{m_1 m_2 G}{R^2}, \qquad (3.6)$$

where m_1, m_2 are the particle masses, R is the distance between the particles, and G, the *constant of gravitation*, is a universal constant. Since G is not dimensionless, its numerical value depends on the units of mass, length and force.

This is the famous **inverse square law of gravitation** originally suggested by Robert Hooke,* a scientific contemporary (and adversary) of Newton. In SI units, the constant of gravitation is given approximately by

$$G = 6.67 \times 10^{-11} \quad \mathrm{N\,m^2\,kg^{-2}}, \tag{3.7}$$

this value being determined by observation and experiment. There is presently no theory (general relativity included) that is able to predict the value of G. Indeed, the theory of general relativity does not exclude *repulsion* between masses!

To give some idea of the magnitudes of the forces involved, suppose we have two uniform spheres of lead, each with mass 5000 kg (five metric tons). Their common radius is about 47 cm which means that they can be placed with their centres 1 m apart. What gravitational force do they exert upon each other when they are in this position? We will show later that the gravitational force between uniform spheres of matter is exactly the same *as if* all the mass of each sphere were concentrated at its centre. Given that this result is true, we can find the force that each sphere exerts on the other simply by substituting $m_1 = m_2 = 5000$ and $R = 1$ into equation (3.6). This gives $F = 0.00167$ N approximately, the weight of a few grains of salt! Such forces seem insignificant, but gravitation is the force that keeps the Moon in orbit around the Earth, and the Earth in orbit around the Sun. The reason for this disparity is that the masses involved are so much larger than those of the lead spheres in our example. For instance, the mass of the Sun is about 2×10^{30} kg.

3.7 GRAVITATION BY A DISTRIBUTION OF MASS

It is important to be able to calculate the gravitational force exerted on a particle by a *distribution* of mass, such as a disc or sphere. The Earth, for example, is an approximately spherical mass distribution. We first treat an introductory problem of gravitational attraction by a pair of *particles* and then progress to *continuous distributions* of matter. In all cases, the law of multiple interactions means that the effective force exerted on a particle is the resultant of the individual forces of interaction exerted on that particle.

Example 3.4 *Attraction by a pair of particles*

A particle C, of mass m, and two particles A and B, each of mass M, are placed as shown in Figure 3.3. Find the gravitational force exerted on the particle C.

Solution

By the law of gravitation, each of the particles A and B attracts C with a force of magnitude F' where

$$F' = \frac{mMG}{R^2},$$

* It was Newton however who proved that Kepler's laws of planetary motion follow from the inverse square law.

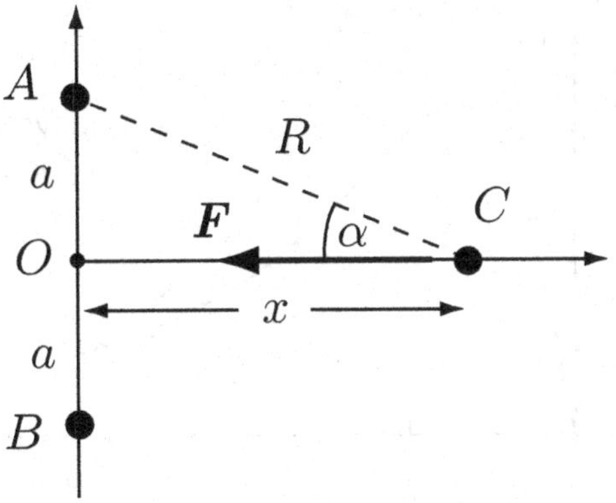

FIGURE 3.3 The particle C, of mass m, is attracted by the particles A and B, each of mass M. The resultant force on C points towards O.

where $R = (a^2 + x^2)^{1/2}$ is the distance AC (= BC). By symmetry, the resultant force $\boldsymbol{F}$ points in the direction CO and so its magnitude F can be found by summing the components of the contributing forces in this direction. Hence

$$F = \frac{2mMG\cos\alpha}{R^2} = 2mMG\left(\frac{R\cos\alpha}{R^3}\right) = 2mMG\left(\frac{x}{(a^2+x^2)^{3/2}}\right)$$

for $x \geq 0$. [The angle α is shown in Figure 3.3.] Thus the resultant force exerted on C looks nothing like the force exerted by a *single* gravitating particle. In particular, it is *not* equal to the force that would be exerted by a mass $2M$ placed at O. However, on writing F in the form

$$F = \frac{2mMG}{x^2}\left(1 + \frac{a^2}{x^2}\right)^{-3/2},$$

we see that

$$F \sim \frac{m(2M)G}{x^2}$$

when x/a is large. Thus, when C is very distant from A and B, the gravitational force exerted on C is *approximately* the same as that of a single particle of mass $2M$ situated at O.

The graph of the *exact* value of F as a function of x is shown in Figure 3.4. Dimensionless variables are used. $F = 0$ when $x = 0$, and rises to a maximum when $x = a/\sqrt{2}$ where $F = 4mMG/3\sqrt{3}a^2$. Thereafter, F decreases, becoming ever closer to its asymptotic form $m(2M)G/x^2$. ■

General asymptotic form of F as $r \to \infty$

The asymptotic result in the last example is true for attraction by any bounded* distribution of mass. The general result can be stated as follows:

* This excludes mass distributions that extend to infinity, such as an infinite straight wire.

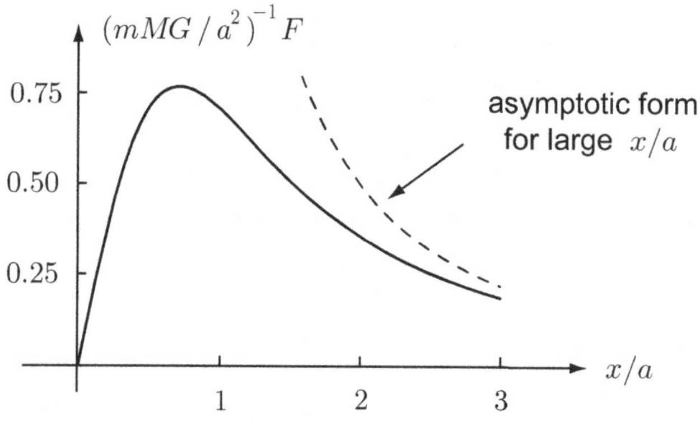

FIGURE 3.4 The dimensionless resultant force $(mMG/a^2)^{-1}F$ plotted against x/a.

Let S be any bounded system of masses with total mass M. Then the force $\boldsymbol{F}$ exerted by S on a particle P, of mass m and position vector $\boldsymbol{r}$, has the asymptotic form*

$$\boldsymbol{F} \sim -\frac{mMG}{r^2}\widehat{\boldsymbol{r}},$$

as $r \to \infty$, where $r = |\boldsymbol{r}|$ and $\widehat{\boldsymbol{r}} = \boldsymbol{r}/r$.

In other words, the force exerted by S on a distant particle is *approximately* the same as that exerted by a particle of mass equal to the total mass of S, situated at the centre of mass of S.

Example 3.5 *Gravitation by a uniform rod*

A particle P, of mass m, and a uniform rod, of length $2a$ and mass M, are placed as shown in Figure 3.5. Find the gravitational force that the rod exerts on the particle.

Solution

Consider the element $[x, x+dx]$ of the rod which has mass $M\,dx/2a$ and exerts an attractive force of magnitude

$$\frac{m(M\,dx/2a)G}{R^2}$$

on P, where R is the distance shown in Figure 3.5. By symmetry, the resultant force acts towards the centre O of the rod and can be found by summing the components of the contributing forces in the direction PO. Since the rod is a *continuous* distribution

* The result, as stated, is true for any choice of the origin O of position vectors. However, the asymptotic error is least if O is located at the centre of mass of S. In this case the *relative* error is of order $(a/r)^2$, where a is the maximum 'radius' of the mass distribution about O.

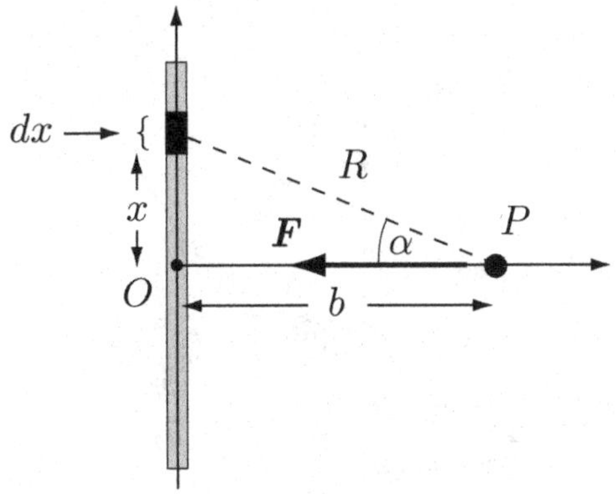

FIGURE 3.5 A particle P, of mass m, is attracted by a uniform rod of length $2a$ and mass M. The resultant force $\boldsymbol{F}$ on P points towards the centre O of the rod.

of mass, this sum becomes an integral. The resultant force exerted by the rod thus has magnitude F given by

$$F = \frac{mMG}{2a}\int_{-a}^{a}\frac{\cos\alpha}{R^2}\,dx = \frac{mMG}{2a}\int_{-a}^{a}\frac{R\cos\alpha}{R^3}\,dx$$
$$= \frac{mMG}{2a}\int_{-a}^{a}\frac{b\,dx}{(x^2+b^2)^{3/2}},$$

where b is the distance of P from the centre of the rod. This integral can be evaluated by making the substitution $x = b\tan\theta$, the limits on θ being $\theta = \pm\beta$, where $\tan\beta = a/b$. This gives

$$F = \frac{mMG}{2a}\left(\frac{2\sin\beta}{b}\right) = \frac{mMG}{2a}\left(\frac{2a}{b(b^2+a^2)^{1/2}}\right)$$
$$= \frac{mMG}{b(b^2+a^2)^{1/2}}. \blacksquare$$

Example 3.6 *Gravitation by a uniform disk*

A particle P, of mass m, is situated on the axis of a uniform disk, of mass M and radius a, as shown in Figure 3.6. Find the gravitational force that the disk exerts on the particle.

Solution

Consider the element of area dA of the disk which has mass $M\,dA/\pi a^2$ and attracts P with a force of magnitude

$$\frac{m(M\,dA/\pi a^2)G}{R^2},$$

where R is the distance shown in Figure 3.6. By symmetry, the resultant force acts towards the centre O of the disk and can be found by summing the components of the

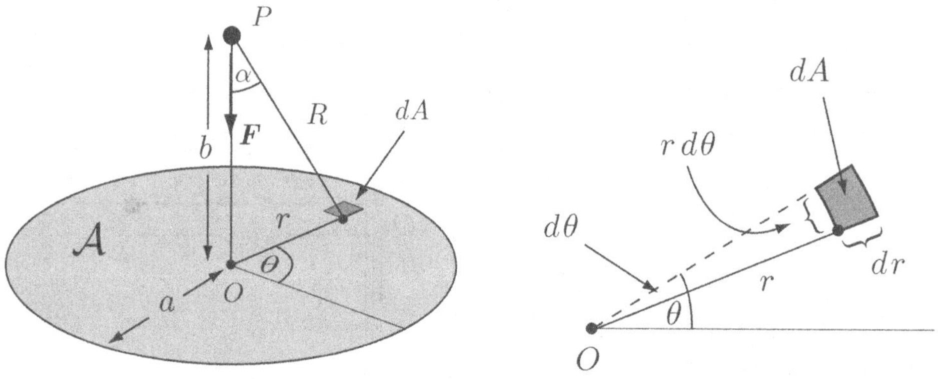

FIGURE 3.6 **Left**: A particle P, of mass m, is attracted by a uniform disk of mass M and radius a. The resultant force $\boldsymbol{F}$ on P points towards the centre O of the disk. **Right**: The element of area dA in polar coordinates r, θ.

contributing forces in the direction PO. The resultant force exerted by the disk thus has magnitude F given by

$$F = \frac{mMG}{\pi a^2} \int_{\mathcal{A}} \frac{\cos\alpha}{R^2}\, dA,$$

where the integral is to be taken over the region $\mathcal{A}$ occupied by the disk. This integral is most easily evaluated using polar coordinates. In this case $dA = (dr)(r\,d\theta) = r\,dr\,d\theta$, and the integrand becomes

$$\frac{\cos\alpha}{R^2} = \frac{R\cos\alpha}{R^3} = \frac{b}{(r^2+b^2)^{3/2}},$$

where b is the distance of P from the centre of the disk. The ranges of integration for r, θ are $0 \le r \le a$ and $0 \le \theta \le 2\pi$. We thus obtain

$$F = \frac{mMG}{\pi a^2} \int_{r=0}^{r=a} \int_{\theta=0}^{\theta=2\pi} \left(\frac{b}{(r^2+b^2)^{3/2}} \right) r\,dr\,d\theta.$$

Since the integrand is independent of θ, the θ-integration is trivial leaving

$$\begin{aligned} F &= \frac{mMG}{\pi a^2} \int_{r=0}^{r=a} \frac{2\pi b r\,dr}{(r^2+b^2)^{3/2}} = \frac{2mMG}{a^2} \left[-b(r^2+b^2)^{-1/2} \right]_{r=0}^{r=a} \\ &= \frac{2mMG}{a^2} \left[1 - \frac{b}{(a^2+b^2)^{1/2}} \right]. \blacksquare \end{aligned}$$

Gravitation by spheres

Because of its applications to astronomy and space travel, and because we live on a nearly spherical body, gravitation by a spherical mass distribution is easily the most important case. We suppose that the mass distribution occupies a spherical volume and is also **spherically symmetric** so that the *mass density depends only on distance from the centre of the*

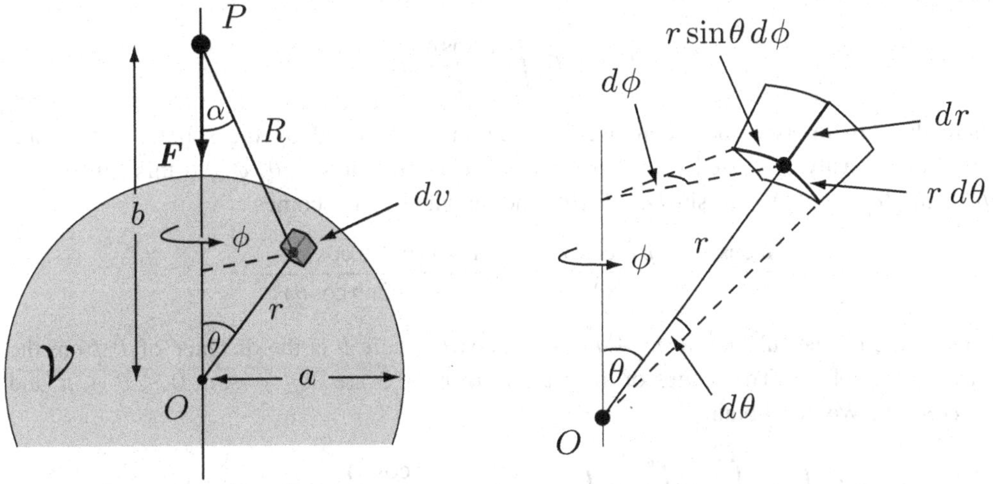

FIGURE 3.7 **Left**: A particle P, of mass m, is attracted by a symmetric sphere of radius a and total mass M. **Right**: The element of volume dV in spherical polar coordinates r, θ, ϕ.

sphere. We call such a body a **symmetric sphere**. The fact that we do not require the density to be uniform is very important in practical applications. The Earth, for instance, has a density of about 3,000 kg m^{-3} near its surface, but its density at the centre is about 16, 500kg m^{-3}. Similar remarks apply to the Sun. Thus, if our results were restricted to spheres of uniform density, they would not apply to the Earth or the Sun, the two most important cases.

The fundamental result concerning gravitation by a symmetric sphere was proved by Newton himself and confirmed his universal theory of gravitation. It is presented here as a theorem.

Theorem 3.1 *The gravitational force exerted by a symmetric sphere of mass M on a particle external to itself is* ***exactly*** *the same as if the sphere were replaced by a particle of mass M located at the centre.*

Proof. Figure 3.7 shows a symmetric sphere with centre O and radius a, and a particle P, of mass m, exterior to the sphere. We wish to calculate the force exerted by the sphere on the particle. The calculation is similar to that in the 'disk' example, but this time the integration must be carried out over the spherical volume occupied by the mass distribution.

Consider the element of volume dv of the sphere which has mass $\rho\, dv$ and attracts P with a force of magnitude

$$\frac{m(\rho\, dv)G}{R^2},$$

where R is the distance shown in Figure 3.7. By symmetry, the resultant force acts towards the centre O of the sphere and can be found by summing the components of the contributing forces in the direction PO. The resultant force exerted by the sphere thus has magnitude F

given by

$$F = mG \int_{\mathcal{V}} \frac{\rho \cos\alpha}{R^2}\, dv,$$

where the integral is to be taken over the region $\mathcal{V}$ occupied by the sphere. This integral is most easily evaluated using spherical polar coordinates r, θ, ϕ. In this case $dv = (dr)(r\,d\theta)(r\sin\theta\,d\phi) = r^2\sin\theta\,dr\,d\theta\,d\phi$, and the integrand becomes

$$\frac{\rho\cos\alpha}{R^2} = \frac{\rho R\cos\alpha}{R^3} = \frac{\rho(r)\,(b - r\cos\theta)}{(r^2 + b^2 - 2rb\cos\theta)^{3/2}},$$

on using the cosine rule $R^2 = r^2 + b^2 - 2rb\cos\theta$, where b is the distance of P from the centre of the sphere. The ranges of integration for r, θ, ϕ are $0 \le r \le a$, $0 \le \theta \le \pi$ and $0 \le \phi \le 2\pi$. We thus obtain

$$F = mG \int_{r=0}^{r=a} \int_{\theta=0}^{\theta=\pi} \int_{\phi=0}^{\phi=2\pi} \left(\frac{\rho(r)\,(b - r\cos\theta)}{(r^2 + b^2 - 2rb\cos\theta)^{3/2}} \right) r^2 \sin\theta\, dr\, d\theta\, d\phi.$$

This time the ϕ-integration is trivial leaving

$$\begin{aligned} F &= mG \int_{r=0}^{r=a} \int_{\theta=0}^{\theta=\pi} \frac{2\pi\rho(r)\,(b - r\cos\theta)}{(r^2 + b^2 - 2rb\cos\theta)^{3/2}}\, r^2 \sin\theta\, dr\, d\theta \\ &= 2\pi mG \int_{r=0}^{r=a} r^2\rho(r) \left\{ \int_{\theta=0}^{\theta=\pi} \frac{(b - r\cos\theta)\sin\theta\, d\theta}{(r^2 + b^2 - 2rb\cos\theta)^{3/2}} \right\} dr, \end{aligned}$$

on taking the θ-integration first and the r-integration second.

The θ-integration is tricky if done directly, but it comes out nicely on making the change of variable from θ to R given by

$$R^2 = r^2 + b^2 - 2rb\cos\theta, \qquad (R > 0).$$

(In this change of variable, r has the status of a constant.) The range of integration for R is $b - r \le R \le b + r$. Then

$$2R\,dR = 2rb\sin\theta\,d\theta,$$

$$b - r\cos\theta = \frac{2b^2 - 2rb\cos\theta}{2b} = \frac{R^2 + (b^2 - r^2)}{2b},$$

and the θ-integral becomes

$$\int_{b-r}^{b+r} \left(\frac{R^2 + (b^2 - r^2)}{2bR^3} \right) \frac{R\,dR}{rb} = \frac{1}{2rb^2} \int_{b-r}^{b+r} \left(1 + \frac{b^2 - r^2}{R^2} \right) dR = \frac{2}{b^2},$$

on performing the now elementary integration.

Hence

$$F = \frac{mG}{b^2} \left(4\pi \int_{r=0}^{r=a} r^2 \rho(r)\, dr \right)$$

and this is as far as we can go without knowing the density function $\rho(r)$. The answer that we are looking for is that $F = mMG/b^2$, where M is the total mass of the sphere. Now M can also be calculated as a volume integral. Since the mass of the volume element dv is $\rho\, dv$, the total mass M is given by

$$M = \int_{\mathcal{V}} \rho\, dv = \int_{r=0}^{r=a} \int_{\theta=0}^{\theta=\pi} \int_{\phi=0}^{\phi=2\pi} r^2 \rho(r) \sin\theta\, dr\, d\theta\, d\phi$$
$$= 4\pi \int_{r=0}^{r=a} r^2 \rho(r)\, dr,$$

on performing the θ- and ϕ-integrations.

Hence, we finally obtain

$$F = \frac{mMG}{b^2},$$

which is the required result.

Since there is no reason why the density $\rho(r)$ should not be zero over part of its range, this result also applies to the case of a particle external to a hollow sphere. The case of a particle *inside* a hollow sphere is different (see Problem 3.5). ■

Spheres attracted by other spheres

Since any element of mass is attracted by a symmetric sphere as if the sphere were a particle, it follows that the force that a symmetric sphere exerts on *any other mass distribution* can be calculated by replacing the sphere by a particle of equal mass located at its centre. In particular then, the force that two symmetric spheres exert upon each other is the same as if *each* sphere were replaced by its equivalent particle. Thus, as far as the forces of gravitational attraction are concerned, *symmetric spheres behave exactly as if they were particles.*

3.8 THE PRINCIPLE OF EQUIVALENCE AND g

Although we have so far not mentioned it, the law of gravitation hides a deep and very surprising fact, namely, that *the force between gravitating particles is proportional to each of their inertial masses.* Now **inertial mass**, as defined by equation (3.2), has no necessary connection with gravitation. It is a measure of the reluctance of that particle to being accelerated and can be determined by non-gravitational means, for instance, by using electrostatic interactions between the particles. It is a matter of extreme surprise then that a quantity that seems to have no necessary connection with gravitation actually *determines* the force of gravitation between particles. What we would have expected was that each particle would have a second property m^*, called **gravitational mass** (not the same as m), which appears in the law of gravitation (3.6) and determines the gravitational force. For example, suppose that we have three uniform spheres of gold, silver and bronze and that the silver and bronze spheres have equal inertial mass. Then the law of gravitation states that, when separated by equal distances, the gold sphere will attract the silver and bronze spheres with *equal forces*, whereas we would have expected these forces to be different.

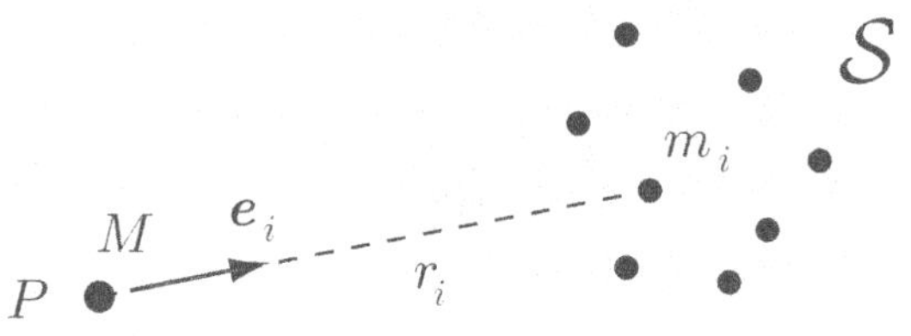

FIGURE 3.8 A particle of mass M is attracted by the gravitation of the system $\mathcal{S}$ which consists of N particles with masses $\{m_i\}$ $(1 \le i \le N)$.

The question arises then as to whether m and m^* are actually equal or just nearly equal so that the difference is difficult to detect. Newton himself did experiments with pendulums made of differing materials, but could not detect any difference in the period. Newton's experiment could have detected a difference of about one part in 10^3. However, the classic experiment of Eötvös (1890) and its later refinements have now shown that any difference between m and m^* is less than one part in 10^{11}. This leads us to believe that m and m^* really are equal and that the law of gravitation means exactly what it says.

The proposition that inertial and gravitational mass are *exactly* equal is called the **principle of equivalence**. Although we accept the principle of equivalence as being true, we still have no explanation why this is so! In this context, it is worth remarking that Einstein made the principle of equivalence into a fundamental *assumption* of the theory of general relativity.

The gravitational acceleration g

Suppose a particle P of mass M is under the gravitational attraction of the system $\mathcal{S}$, as shown in Figure 3.8. Then, by the law of gravitation, the resultant force $\boldsymbol{F}$ that $\mathcal{S}$ exerts upon P is given by

$$\begin{aligned}\boldsymbol{F} &= \frac{Mm_1G}{r_1^2}\boldsymbol{e}_1 + \frac{Mm_2G}{r_2^2}\boldsymbol{e}_2 + \cdots + \frac{Mm_NG}{r_N^2}\boldsymbol{e}_N = M\left(\sum_{i=1}^{N}\frac{m_iG}{r_i^2}\boldsymbol{e}_i\right)\\ &= M\boldsymbol{g},\end{aligned}$$

where the vector $\boldsymbol{g}$, defined by

$$\boldsymbol{g} = \sum_{i=1}^{N}\frac{m_iG}{r_i^2}\boldsymbol{e}_i,$$

is *independent of* M. Then, by the Second Law, the induced acceleration $\boldsymbol{a}$ of particle P is determined by the equation

$$M\boldsymbol{g} = M\boldsymbol{a},$$

that is,

$$\boldsymbol{a} = \boldsymbol{g}.$$

Thus *the induced acceleration* $\mathbf{g}$ *is the same for any particle* situated at that point. This rather remarkable fact is a direct consequence of the principle of equivalence. Tradition has it that, prior to Newton, Galileo did experiments in which he released different masses from the top of the Tower of Pisa and found that they reached the ground at the same time. Galileo's result is thus a colourful but rather inaccurate verification of the principle of equivalence!

Gravitation by the Earth (rotation neglected)

In the present treatment, the rotation of the Earth is neglected and we regard the Earth as an inertial frame of reference. A more accurate treatment which takes the Earth's rotation into account is given in Chapter 17.

When the system $\mathcal{S}$ is the Earth (or some other celestial body) it is convenient to introduce the notion of the local **vertical** direction. The unit vector $\boldsymbol{k}$, which has the *opposite* direction to $\mathbf{g}$, is called the **vertically upwards unit vector** relative to the Earth. In terms of $\boldsymbol{k}$, the force exerted by the Earth on a particle of mass M is given by

$$\boldsymbol{F} = -Mg\boldsymbol{k},$$

where the **gravitational acceleration** g is the *magnitude* of the gravitational acceleration vector $\mathbf{g}$. Both g and $\boldsymbol{k}$ are functions of position on the Earth.

Weight

The positive quantity Mg (which is a function of position) is called the **weight** of the particle P. It is the *magnitude of the gravity force* exerted on P by the Earth. Thus the same body will have different weights depending upon where it is situated. However, at a fixed point of space, the weights of bodies are proportional to their masses. This fact, which is a consequence of the principle of equivalence, enables *masses* to be compared simply by comparing their *weights* at the same location (by using a balance, for instance).

The approximation of uniform gravity

It is easy to see that the Earth's gravitational acceleration g and the vertical direction $\boldsymbol{k}$ depend upon position. The Earth is approximately a symmetric sphere which exerts its gravitational force as if all its mass were at its centre. Thus, if the value of g at a point on the Earth's surface is g_1, then the value of g at a height of 6,400 km (the Earth's radius) must be $g_1/4$ approximately. On the other hand, the vertical vector $\boldsymbol{k}$ changes from point to point on the Earth's surface. These changes will be significant for motions whose extent is significant compared to the Earth's radius; this is true for a ballistic missile, for instance. However, most motions taking place on Earth have an extent that is insignificant compared to the Earth's radius and for which the variations of g and $\boldsymbol{k}$ are negligible. Simple examples include the motion of a tennis ball, a javelin or a bullet.

The approximation in which g and $\boldsymbol{k}$ are assumed to be constants is called **uniform gravity**. Uniform gravity is the most common force field in mechanics. Many of the problems solved in this book make this simplifying (and accurate) approximation.

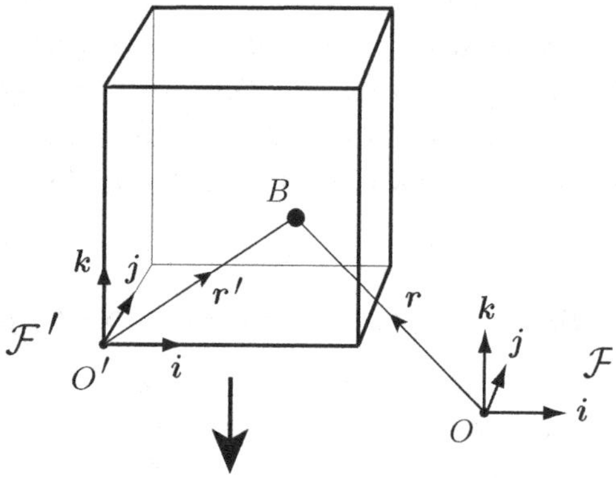

FIGURE 3.9 An elevator contains a ball B and both are freely falling under uniform gravity. $\mathcal{F}$ is an inertial reference frame and $\mathcal{F}'$ is a reference frame attached to the falling elevator.

Numerical values of g

The value of g at any location on the Earth can be measured experimentally (by using a pendulum for instance). The value of g is not quite constant over the Earth's surface since the Earth does not quite have spherical symmetry and different locations have differing altitudes. At sea level on **Earth**, $g = 9.8\ \mathrm{m\,s^{-2}}$ approximately, and a rough value of $10\ \mathrm{m\,s^{-2}}$ is often assumed. The corresponding value for the **Moon** is $1.6\ \mathrm{m\,s^{-2}}$, roughly a sixth of the Earth's value.

Example 3.7 *Particle inside a falling elevator*

An elevator cable has snapped and the elevator and its contents are falling under uniform gravity. One of the passengers takes a ball from his pocket and throws it to another passenger.* What is the motion of the ball *relative to the elevator*?

Solution

Suppose that the ball has mass m and that the local (vector) gravitational acceleration is $-g\boldsymbol{k}$. Then the motion of the ball relative to an *inertial* reference frame $\mathcal{F}$ (fixed to the ground, say) is determined by the Second Law, namely,

$$m\boldsymbol{a} = -mg\boldsymbol{k},$$

where $\boldsymbol{a}$ is the acceleration of the ball measured in $\mathcal{F}$.

Let the frame $\mathcal{F}'$ be attached to the elevator, as shown in Figure 3.9. Then the acceleration $\boldsymbol{a}'$ of the ball measured in $\mathcal{F}'$ is given (see section 2.6) by

$$\boldsymbol{a} = \boldsymbol{a}' + \boldsymbol{A},$$

where $\boldsymbol{A}$ is the acceleration of the frame $\mathcal{F}'$ relative to $\mathcal{F}$. But the elevator, to which the frame $\mathcal{F}'$ is attached, is also moving under uniform gravity and its acceleration $\boldsymbol{A}$ is therefore, by the principle of equivalence, the same as that of the ball, namely,

$$\boldsymbol{A} = -g\boldsymbol{k}.$$

* People do react oddly when put under pressure.

Hence $\boldsymbol{a} = \boldsymbol{a}' - g\boldsymbol{k}$ and so

$$\boldsymbol{a}' = \boldsymbol{0}.$$

Thus, relative to the elevator, the **ball moves with constant velocity**. To observers resident in the frame $\mathcal{F}'$, gravity appears to be absent and $\mathcal{F}'$ appears to be an inertial frame. This provides a practical method for simulating conditions of weightlessness. Fortunately for those wishing to experience weightlessness, there is no need to use an elevator; the same acceleration can be achieved by an aircraft in a vertical dive!

This result is of considerable importance in the theory of general relativity. It shows that, locally at least, a gravitational field can be 'transformed away' by observing the motion of bodies from a freely falling reference frame. ■

Problems on Chapter 3

Answers and comments are at the end of the book.

Harder problems carry a star (∗).

Gravitation

3.1 Four particles, each of mass m, are situated at the vertices of a regular tetrahedron of side a. Find the gravitational force exerted on any one of the particles by the other three.

Three uniform rigid spheres of mass M and radius a are placed on a horizontal table and are pressed together so that their centres are at the vertices of an equilateral triangle. A fourth uniform rigid sphere of mass M and radius a is placed on top of the other three so that all four spheres are in contact with each other. Find the gravitational force exerted on the upper sphere by the three lower ones.

3.2 Eight particles, each of mass m, are situated at the corners of a cube of side a. Find the gravitational force exerted on any one of the particles by the other seven.

Deduce the total gravitational force exerted on the four particles lying on one face of the cube by the four particles lying on the opposite face.

3.3 A uniform rod of mass M and length $2a$ lies along the interval $[-a, a]$ of the x-axis and a particle of mass m is situated at the point $x = x'$. Find the gravitational force exerted by the rod on the particle.

Two uniform rods, each of mass M and length $2a$, lie along the intervals $[-a, a]$ and $[b - a, b + a]$ of the x-axis, so that their centres are a distance b apart ($b > 2a$). Find the gravitational forces that the rods exert upon each other.

3.4 A uniform rigid disk has mass M and radius a, and a uniform rigid rod has mass M' and length b. The rod is placed along the axis of symmetry of the disk with one end in contact with the disk. Find the forces necessary to pull the disk and rod apart. [*Hint.* Make use of the solution in the 'disk' example.]

3.5 Show that the gravitational force exerted on a particle *inside* a hollow symmetric sphere is zero. [*Hint.* The proof is the same as for a particle *outside* a symmetric sphere, except in one detail.]

3.6 A narrow hole is drilled through the centre of a *uniform* sphere of mass M and radius a. Find the gravitational force exerted on a particle of mass m which is inside the hole at a distance r from the centre.

3.7 A symmetric sphere, of radius a and mass M, has its centre a distance b ($b > a$) from an infinite plane containing a uniform distribution of mass σ per unit area. Find the gravitational force exerted on the sphere.

3.8* Two uniform rigid hemispheres, each of mass M and radius a are placed in contact with each other so as to form a complete sphere. Find the forces necessary to pull the hemispheres apart.

Computer assisted problem

3.9 A uniform wire of mass M has the form of a circle of radius a and a particle of mass m lies in the plane of the wire at a distance b ($b < a$) from the centre O. Show that the gravitational force exerted by the wire on the particle (in the direction OP) is given by

$$F = \frac{mMG}{2\pi a^2} \int_0^{2\pi} \frac{(\cos\theta - \xi)d\theta}{\{1 + \xi^2 - 2\xi\cos\theta\}^{3/2}},$$

where the dimensionless distance $\xi = b/a$.

Use computer assistance to plot the graph of (dimensionless) F against ξ for $0 \le \xi \le 0.8$ and confirm that F is *positive* for $\xi > 0$. Is the position of equilibrium at the centre of the circle stable? Could the rings of Saturn be solid?

Chapter Four

Problems in particle dynamics

KEY FEATURES

The key features in this chapter are (i) the **vector equation of motion** and its reduction to **scalar equations**, (ii) motion in a **force field**, (iii) geometrical constraints and **forces of constraint**, and (iv) linear and quadratic **resistance forces**.

Particle dynamics is concerned with the problem of calculating the motion of a particle that is acted upon by specified forces. Our starting point is Newton's laws. However, since the First Law merely tells us that we should observe the motion from an inertial frame, and the Third Law will never be used (since there is only one particle), the entirety of particle dynamics is based on the Second Law

$$m\boldsymbol{a} = \boldsymbol{F}_1 + \boldsymbol{F}_2 + \cdots + \boldsymbol{F}_N,$$

where $\boldsymbol{F}_1, \boldsymbol{F}_2, \ldots, \boldsymbol{F}_N$ are the various forces that are acting on the particle. The typical method of solution is to write the Second Law in the form

$$m\frac{d\boldsymbol{v}}{dt} = \boldsymbol{F}_1 + \boldsymbol{F}_2 + \cdots + \boldsymbol{F}_N, \tag{4.1}$$

which is a first order ODE for the unknown velocity function $\boldsymbol{v}(t)$ and is called the **equation of motion** of the particle. If the initial value of $\boldsymbol{v}$ is given, then equation (4.1) can often be solved to yield $\boldsymbol{v}$ as a function of the time t. Once $\boldsymbol{v}$ is determined (and if the initial position of the particle is given), the position vector $\boldsymbol{r}$ of the particle at time t can be found by solving the first order ODE $d\boldsymbol{r}/dt = \boldsymbol{v}$. The sections that follow contain many examples of the implementation of this method. Indeed, it is remarkable how many interesting problems can be solved in this way.

Question ***When can real bodies be modelled as particles?***

Newton's laws apply to particles, but real bodies are not particles. When can real bodies, such as a tennis ball, a spacecraft, or the Earth, be treated as if they were particles?

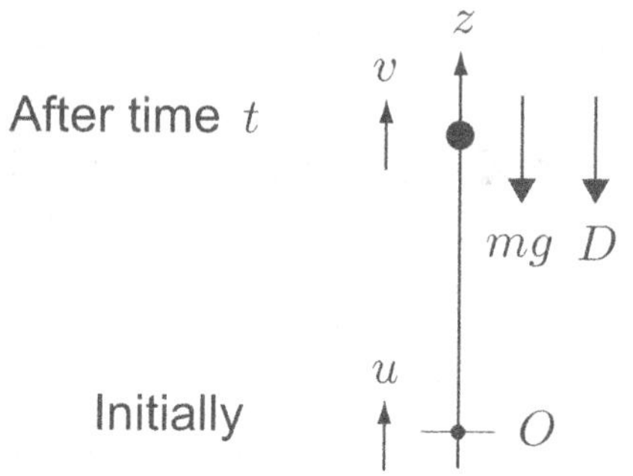

FIGURE 4.1 The particle is initially at the origin and is projected vertically upwards with speed u. The particle moves in a vertical straight line (the axis Oz) under the uniform gravity force mg and possibly a resistance (or drag) force D. At time t the particle has upward velocity v.

Answer

This is quite a tricky question which is not fully discussed until Chapter 10. What we will show is that the **centre of mass** of any body moves *as if* it were a particle of mass equal to the total mass, and all the forces on the body acted upon it. In particular, a rigid body, moving without rotation, can be treated *exactly* as if it were a particle. For example, a block sliding without rotation on a table can be treated exactly as if it were a particle. In other cases we gain only partial information about the motion. If the body is a brick thrown through the air, then particle dynamics can tell us exactly where its centre of mass will go, but not which point of the brick will hit the ground first. ■

4.1 RECTILINEAR MOTION IN A FORCE FIELD

Our first group of problems is concerned with the straight line motion of a particle moving in a force field. A force $\boldsymbol{F}$ is said to be a **field** if it depends only on the **position** of the particle, and not, for instance, on its velocity or the time. For example, the gravitational attraction of any *fixed* mass distribution is a field of force, but resistance forces, which are usually velocity dependent, are not.

If the rectilinear motion takes place along the z-axis, the equation of motion (4.1) reduces to the scalar equation

$$m\frac{dv}{dt} = F(z), \tag{4.2}$$

where v is the (one-dimensional) velocity of the particle and $F(z)$ is the (one-dimensional) force field, both measured in the *positive* z-direction.

First we consider the problem of **vertical motion** of a particle under **uniform gravity** with **no air resistance**. This is fine on the Moon (which has no atmosphere) but, on Earth, the motion of a body is resisted by its passage through the atmosphere and this will introduce errors. The effect of resistance forces is investigated in section 4.3.

Example 4.1 *Vertical motion under uniform gravity*

A particle is projected vertically upwards with speed u and moves in a vertical straight line under uniform gravity with no air resistance. Find the maximum height achieved by the particle and the time taken for it to return to its starting point.

Solution

Let v be the upwards velocity of the particle after time t, as shown in Figure 4.1. Then the scalar equation of motion (4.2) takes the form

$$m\frac{dv}{dt} = -mg,$$

since the drag force D is absent. A simple integration gives

$$v = -gt + C,$$

where C is the integration constant, and, on applying the initial condition $v = u$ when $t = 0$, we obtain $C = u$. Hence the **velocity** v at time t is given by

$$v = u - gt.$$

To find the upward displacement z at time t, write

$$\frac{dz}{dt} = v = u - gt.$$

A second simple integration gives

$$z = ut - \tfrac{1}{2}gt^2 + D,$$

where D a second integration constant, and, on applying the initial condition $z = 0$ when $t = 0$, we obtain $D = 0$. Hence the upward **displacement** of the body at time t is given by

$$z = ut - \tfrac{1}{2}gt^2.$$

The **maximum height** $z_{\max}$ is achieved when $dz/dt = 0$, that is, when $v = 0$. Thus $z_{\max}$ is achieved when $t = u/g$ and is given by

$$z_{\max} = u\left(\frac{u}{g}\right) - \tfrac{1}{2}g\left(\frac{u}{g}\right)^2 = \frac{u^2}{2g}.$$

The particle returns to O when $z = 0$, that is, when

$$t\left(u - \tfrac{1}{2}gt\right) = 0.$$

Thus the **particle returns** after a time $2u/g$.

For example, if we throw a body vertically upwards with speed 10 m s^{-1}, it will rise to a height of 5 m and return after 2 s. [Here we are neglecting atmospheric resistance and taking $g = 10$ m s^{-2}.] ■

Question *Saving oneself in a falling elevator*

An elevator cable has snapped and the elevator is heading for the ground. Can the occupants save themselves by leaping into the air just before impact in order to avoid the crash?

Answer

Suppose that the elevator is at rest at a height H when the cable snaps. The elevator will fall and reach the ground with speed $(2gH)^{1/2}$. In order to save themselves, the occupants must leap upwards (relative to the elevator) with this same speed so that their speed relative to the ground is zero. If they were able do this, then they would indeed be saved. However, if they were able to project themselves upwards with this speed, they would also be able to stand outside the building and leap up to the same height H that the elevator fell from! Even athletes cannot jump much more than a metre off the ground, so the answer is that escape is possible in principle but not in practice. ■

Uniform gravity is the simplest force field because it is constant. In the next example we show how to handle a **non-constant force field**.

Example 4.2 *Rectilinear motion in the inverse square field*

A particle P of mass m moves under the gravitational attraction of a mass M fixed at the origin O. Initially P is at a distance a from O when it is projected with speed u directly away from O. Find the condition that P will 'escape' to infinity.

Solution

By symmetry, the motion of P takes place in a straight line through O. By the law of gravitation, the scalar equation of motion is

$$m\frac{dv}{dt} = -\frac{mMG}{r^2},$$

where r is the distance OP and $v = \dot{r}$. Equations like this can always be integrated once by first eliminating the time. Since

$$\frac{dv}{dt} = \frac{dv}{dr} \times \frac{dr}{dt} = v\frac{dv}{dr},$$

the equation of motion can be written as

$$v\frac{dv}{dr} = -\frac{MG}{r^2},$$

a first order ODE for v as a function of r. This is to be solved with the initial condition $v = u$ when $r = a$. The equation separates to give

$$\int v\,dv = -MG \int \frac{dr}{r^2},$$

and so

$$\tfrac{1}{2}v^2 = \frac{MG}{r} + C,$$

where C is the integration constant. On applying the initial condition $v = u$ when $r = a$, we find that $C = (u^2/2) - (MG/a)$ so that

$$v^2 = \left(u^2 - \frac{2MG}{a}\right) + \frac{2MG}{r}.$$

This determines the outward **velocity** v as a function of r.

Whether the particle escapes to infinity, or not, depends on the *sign* of the bracketed constant term.

(i) Suppose first that this term is *positive* so that

$$u^2 - \frac{2MG}{a} = V^2,$$

where V is a positive constant. Then, since the term $2MG/r$ is positive, it follows that $v > V$ at all times. It further follows that $r > a + Vt$ for all t and so the *particle escapes to infinity*.

(ii) On the other hand, if $u^2 - (2MG/a)$ is *negative*, then v becomes zero when

$$r = \frac{a}{1 - (u^2 a/2MG)},$$

after which the particle falls back towards O and *does not escape*.

(iii) The critical case, in which $u^2 = 2MG/a$ is treated in Problem 4.10; the result is that the *particle escapes*.

Hence the **particle escapes** if (and only if)

$$u^2 \geq \frac{2MG}{a}. \blacksquare$$

Question ***Given u, find r_{max} and the time taken to get there***

For the particular case in which $u^2 = MG/a$, find the maximum distance from O achieved by P and the time taken to reach this position.

Answer

For this value of u, the equation connecting v and r becomes

$$v^2 = MG\left(\frac{2}{r} - \frac{1}{a}\right).$$

Since $r = r_{\text{max}}$ when $v = 0$, it follows that the **maximum distance** from O achieved by P is $2a$.

To find the time taken, we write $v = dr/dt$ and solve the ODE

$$\left(\frac{dr}{dt}\right)^2 = MG\left(\frac{2}{r} - \frac{1}{a}\right)$$

with the initial condition $r = a$ when $t = 0$. After taking the positive square root of each side ($dr/dt \geq 0$ in this motion), the equation separates to give

$$\int_a^{2a} \left(\frac{ar}{2a - r}\right)^{1/2} dr = (MG)^{1/2} \int_0^{\tau} dt.$$

(Here we have introduced the initial and final conditions directly into the limits of integration; τ is the elapsed time.) On simplifying, we obtain

$$\tau = (MG)^{-1/2} \int_a^{2a} \left(\frac{ar}{2a - r}\right)^{1/2} dr.$$

This integral can be evaluated by making the substitution $r = 2a \sin^2 \theta$; the details are unimportant. The result is that the **time taken** for P to progress from $r = a$ to $r = 2a$ is

$$\tau = \left(\frac{a^3}{MG}\right)^{1/2} \left(1 + \tfrac{1}{2}\pi\right). \blacksquare$$

Question ***Speed of escape from the Moon***

A body is projected vertically upwards from the surface of the Moon. What projection speed is necessary for the body to escape the Moon?

Answer

We regard the Moon as a fixed symmetric sphere of mass M and radius R. In this case, the gravitational force exerted by the Moon is the same as that of a particle of mass M situated at the centre. Thus the preceding theory applies with the distance a replaced by the radius R. The **escape speed** is therefore $(2MG/R)^{1/2}$, which evaluates to about 2.4 km s^{-1}. [For the Moon, $M = 7.35 \times 10^{22}$ kg and $R = 1740$ km.] ■

4.2 CONSTRAINED RECTILINEAR MOTION

Figure 4.2 shows a uniform rigid rectangular block of mass m sliding down the inclined surface of a *fixed* rigid wedge of angle α. The initial conditions are supposed to be such that the block slides, without rotation, down the line of steepest slope of the wedge. The block is subject to uniform gravity, but it is clear that there must be other forces as well. If there were no other forces and the block were released from rest, then the block would move vertically downwards. However, solid bodies cannot pass through each other like ghosts, and interpenetration is prevented by (equal and opposite) forces that they exert upon each other. These are **material contact forces** which come into play only when bodies are in physical contact. They are examples of **forces of constraint**, which are not prescribed beforehand but are sufficient to enforce a specified **geometrical constraint**. Tradition has it that the constraint force that the wedge exerts on the block is

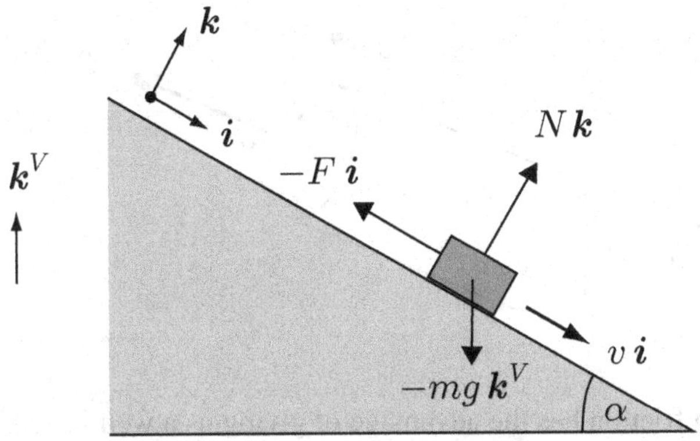

FIGURE 4.2 A rigid rectangular block slides down the inclined surface of a fixed rigid wedge of angle α. Note that $\boldsymbol{k}^V$ is the vertically upwards unit vector, while $\boldsymbol{i}$ and $\boldsymbol{k}$ are parallel to and perpendicular to the inclined surface of the wedge.

called the total **reaction force** $\boldsymbol{R}$. It is convenient to write this force in the form

$$\boldsymbol{R} = -F\boldsymbol{i} + N\boldsymbol{k},$$

where the unit vectors $\boldsymbol{i}$ and $\boldsymbol{k}$ are parallel to and normal to the slope of the wedge. The scalar N is called the **normal reaction** component and the scalar F is called the **frictional** component.*

The equation of motion of the block is the vector equation (4.1) which becomes

$$m\frac{d(v\boldsymbol{i})}{dt} = -mg\boldsymbol{k}^V - F\boldsymbol{i} + N\boldsymbol{k},$$

where $\boldsymbol{k}^V$ is the vertically upwards unit vector. The easiest way of proceeding is to take components of this vector equation in the $\boldsymbol{i}$- and $\boldsymbol{k}$-directions (the $\boldsymbol{j}$-component gives nothing). On noting that $\boldsymbol{k}^V = -\sin\alpha\,\boldsymbol{i} + \cos\alpha\,\boldsymbol{k}$, this gives

$$m\frac{dv}{dt} = mg\sin\alpha - F \qquad \text{and} \qquad 0 = N - mg\cos\alpha.$$

The second of these equations determines the normal reaction $N = mg\cos\alpha$. However, in the first equation, both v and F are unknown and this prevents any further progress in the solution of this problem.* One can proceed by proposing some empirical 'law of friction', but such laws hold only very roughly. It is not surprising then that, in much of mechanics, frictional forces are neglected. In this case, the total reaction force exerted by the surface is in the normal direction and we describe such surfaces as **smooth**, meaning

* The minus sign is introduced so that F will be positive when the scalar velocity v is positive.
* This reflects the fact that we have said nothing about the roughness of the surface of the wedge!

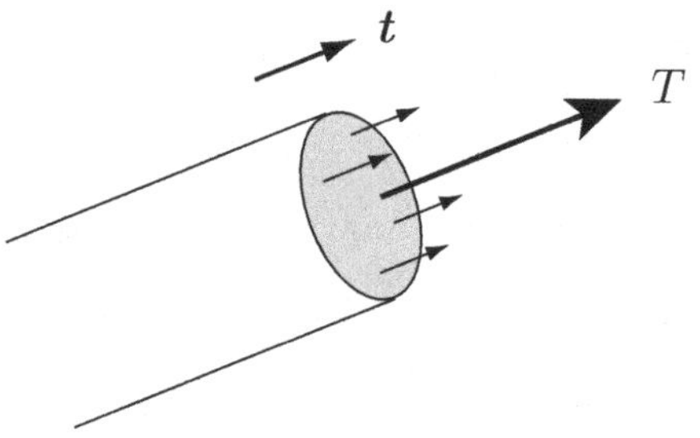

FIGURE 4.3 The idealised string is depicted here as having a small circular cross-section. At each cross-section only tensile stresses exist and their resultant is the tension T in the string at that point.

'perfectly smooth'. Doing away with friction has the advantage of giving us a well-posed problem that we can solve; however the solution will then apply only approximately to real surfaces.

If we now suppose that the inclined surface of the **wedge** is **smooth**, then $F = 0$ and the first equation reduces to

$$\frac{dv}{dt} = g \sin\alpha.$$

Thus, in the absence of friction, the block slides down the plane with the *constant acceleration* $g \sin\alpha$.

Inextensible strings

Another agency that can cause a geometrical constraint is the **inextensible string**. If a particle P of a system is connected to a fixed point O by an inextensible string of length a then, *if the string is taut*, P is constrained to move so that the distance $OP = a$. This geometrical constraint is enforced by the (unknown) constraint force that the string applies to particle P. Our 'string' is an idealisation of real cords and ropes in that it is *infinitely thin*, has *no bending stiffness*, and is *inextensible*. The only force that one part of the string exerts on another is the **tension** T in the string, which acts parallel to the tangent vector $\boldsymbol{t}$ to the string at each point (see Figure 4.3).

It is evident that, in general, T varies from point to point along the string. Suppose for example that a uniform string of mass ρ per unit length is suspended vertically under uniform gravity. Then, since the tension at the lower end is zero, the string will not be in equilibrium unless the tension at a height z above the lowest point is given by $T = \rho g z$; the tension thus rises linearly with height.

The situation is simpler when the mass of the string is negligible; this is the case of the **light inextensible string.*** In this case, it is obvious that the tension is constant when the string is straight. In fact, the tension also remains constant when the string slides over a *smooth* body. This is proved in Chapter 10. The tension in a light string is also constant when the string passes over a *light, smoothly pivoted* pulley wheel.

* In this context, 'light' means 'of zero mass'.

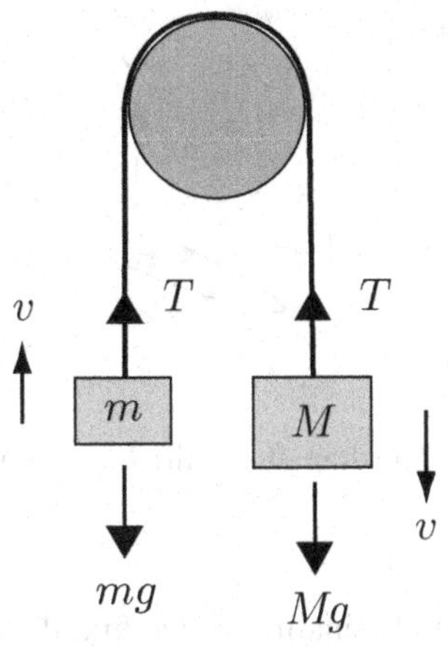

FIGURE 4.4 Attwood's machine: two bodies of masses m and M are connected by a light inextensible string which passes over a smooth rail.

Example 4.3 *Attwood's machine*

Two bodies with masses m, M are connected by a light inextensible string which passes over a smooth horizontal rail. The system moves in a vertical plane with the bodies moving in vertical straight lines. Find the upward acceleration of the mass m and the tension in the string.

Solution

The system is shown in Figure 4.4. Let v be the *upward* velocity of the mass m. Then, since the string is inextensible, v must also be the *downward* velocity of the mass M. Also, since the string is light and the rail is *smooth*, the string has constant tension T. The scalar equations of motion for the two masses are therefore

$$m\frac{dv}{dt} = T - mg, \qquad M\frac{dv}{dt} = Mg - T.$$

It follows that

$$\frac{dv}{dt} = \left(\frac{M-m}{M+m}\right)g \quad \text{and} \quad T = \left(\frac{2Mm}{M+m}\right)g. \blacksquare$$

Question *The monkey puzzle*

Suppose that, in the last example, both bodies have the same mass M and one of them is a monkey which begins to climb up the rope. What happens to the other mass?

Answer

Suppose that the monkey climbs with velocity V *relative to the rope*. Then its upward velocity relative to the ground is $V - v$. The equations of upward motion for the mass and the monkey are therefore

$$M\frac{dv}{dt} = T - Mg, \qquad M\frac{d(V-v)}{dt} = T - Mg.$$

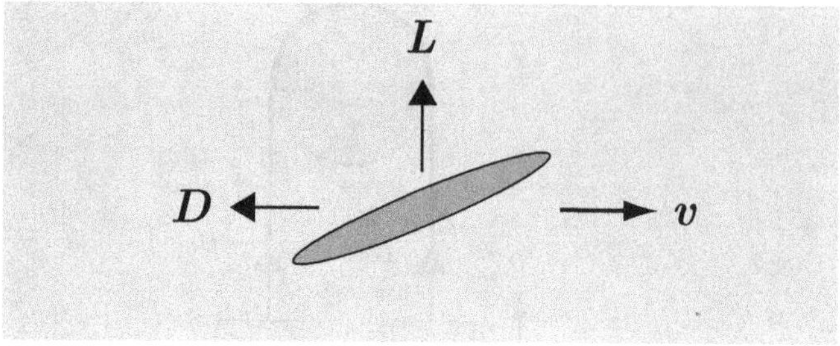

FIGURE 4.5 The drag $\boldsymbol{D}$ and lift $\boldsymbol{L}$ on a body moving through a fluid.

On eliminating T, we find that

$$\frac{dv}{dt} = \frac{1}{2}\frac{dV}{dt},$$

so that, if the whole system starts from rest,

$$v = \tfrac{1}{2}V, \qquad \text{and} \qquad V - v = \tfrac{1}{2}V.$$

Thus the monkey and the mass rise (relative to the ground) with the same velocity; the monkey cannot avoid hauling up the mass as well as itself! ■

4.3 MOTION THROUGH A RESISTING MEDIUM

The physics of fluid drag

When a body moves through a fluid such as air or water, the fluid exerts forces on the surface of the body. This is because the body must push the fluid out of the way, and to do this the body must exert forces on the fluid. By the Third Law, the fluid must then exert equal and opposite forces on the body. A person wading through water or riding a motorcycle is well aware of the existence of such forces, which fall into the general category of **material contact forces**. We are interested in the *resultant* force that the fluid exerts on the body and it is convenient to write this resultant in the form

$$\boldsymbol{F} = \boldsymbol{D} + \boldsymbol{L},$$

where the vector **drag force** $\boldsymbol{D}$ has the opposite direction to the velocity of the body, and the vector **lift force** $\boldsymbol{L}$ is at right angles to this velocity. The existence of lift makes air travel possible and is obviously very important. However, we will be concerned only with drag since we will restrict our attention to those cases in which the body is a rigid body of revolution which moves (without rotation) in the direction of its axis of symmetry. In this case, the lift is zero, by symmetry. We are then left with the scalar drag D, acting in the opposite direction to the velocity of the body.

The theoretical determination of lift and drag forces is one of the great unsolved problems of hydrodynamics and most of the available data has been obtained by experiment. Even for the case of a rigid sphere moving with constant velocity through an incompressible* fluid, a *general* theoretical solution for the drag is not available. In this problem, the drag depends on the radius a and speed V of the sphere, and the density ρ and viscosity μ of the fluid. Straightforward dimensional analysis shows that D must have the form

$$D = \rho a^2 V^2 F\left(\frac{\rho V a}{\mu}\right),$$

where F is a function of a single variable.

Definition 4.1 *Reynold's number* *The dimensionless quantity $R = \rho V a/\mu$ is called the* ***Reynolds number.***† *It is more commonly written $R = Va/\nu$, where the quantity $\nu = \mu/\rho$ is called the* ***kinematic viscosity*** *of the fluid.*

The function $F(R)$ has never been calculated theoretically, and experimental data must be used. It is a surprising fact that, for a wide range of values of R (about $1000 < R < 100,000$), the *function F is found to be roughly constant.* Subject to this approximation, the formula for the drag becomes

$$D = C\rho a^2 V^2,$$

where the dimensionless constant C is called the **drag coefficient**‡ for the sphere; its value is about 0.8.

A similar formula holds (with a different value of C) for any body of revolution moving parallel to its axis. In this case a is the radius of the maximum cross sectional area of the body perpendicular to the direction of motion. For example, the drag coefficient for a circular disk moving at right angles to its own plane is about 1.7. We thus obtain the result that (subject to the conditions mentioned above) the *drag is proportional to the square of the speed of the body through the fluid.* This is the **quadratic law of resistance.**

This result does not hold for low Reynolds numbers. This was shown theoretically by Stokes§ in his analysis of the creeping flow of a fluid past a sphere. Stokes proved that, as $R \to 0$, the function $F(R) \sim 6\pi/R$ so that the drag formula becomes

$$D \sim 6\pi a\mu V.$$

On dimensional grounds, a similar formula (with a different coefficient) should hold for other bodies of revolution. Thus *at low Reynolds numbers*¶ *the drag is proportional to speed of the body through the fluid.* This is the **linear law of resistance.**

Which (if either) of these laws is appropriate in any particular case depends on the Reynolds number. However, it is quickly apparent that the low Reynolds number condition requires quite special physical conditions, as the following example shows.

---

* In this treatment, the effects of fluid compressibility are neglected. In practice, this means that the speed of the body must be well below the speed of sound in the fluid.

† After the great English hydrodynamicist Osborne Reynolds 1842–1912. At the age of twenty six he was appointed to the University of Manchester's first professorship of engineering.

‡ The drag coefficient C_D used by aerodynamicists is $2C/\pi$.

§ George Gabriel Stokes 1819–1903, a major figure in British applied mathematics.

¶ *Low* means R less than about 0.5.

Table 1 Some fluid properties relevant to drag calculations (Kaye & Laby [14]).

	Density ρ (kg m^{-3})	Kinematic viscosity ν (m^2s^{-1})	Sound speed (m s^{-1})
Air (20°C, 1 atm.)	1.20	1.50×10^{-5}	343
Water (20°C)	998	1.00×10^{-6}	1480
Castor oil (20°C)	950	1.04×10^{-3}	1420

Example 4.4 *Which law of resistance?*

A stainless steel ball bearing of radius 1 mm is falling vertically with constant speed in air. Find the speed of the ball bearing. [The density of stainless steel is 7800 kg m^{-3}.]

If the medium were castor oil, what then would be the speed of the ball bearing?

Solution

Suppose the ball bearing is falling with constant speed V. (We will later call V the *terminal speed* of the ball bearing.) Then, since its acceleration is zero, the total of the forces acting upon it must also be zero. Thus

$$mg + D = m'g,$$

where m' is the mass of the ball bearing, m is the mass of the displaced fluid, and D is the drag. The term $m'g$ is the gravity force acting downwards and the term mg is the (Archimedes) buoyancy force acting upwards.* (In air, the buoyancy force is negligible.)

Hence, if the **linear** resistance law holds, then

$$6\pi a\rho\nu V = \tfrac{4}{3}\pi a^3 \left(\rho' - \rho\right) g,$$

where a is the radius of the ball bearing, and ρ', ρ are the densities of the ball bearing and air respectively. This gives

$$V = \frac{2a^2 g}{9\nu}\left(\frac{\rho'}{\rho} - 1\right).$$

On using the numerical values given in Table 1 we obtain $V = 940\ \text{m s}^{-1}$ with the corresponding Reynolds number $R = 63,000$. Quite apart from the fact that the calculated speed is nearly three times the speed of sound, this solution is disqualified on the grounds that the Reynolds number is 100,000 times too large for the low Reynolds number approximation to hold!

On the other hand, if the **quadratic** law of resistance holds then

$$C\rho a^2 V^2 = \tfrac{4}{3}\pi a^3 \left(\rho' - \rho\right) g,$$

* It is not entirely obvious that the total force exerted by the fluid on the sphere is the sum of the drag and buoyancy forces, but it is true for an incompressible fluid.

where C is the drag coefficient for a sphere which we will take to be 0.8. In this case

$$V^2 = \frac{4\pi a g}{3C}\left(\frac{\rho'}{\rho} - 1\right).$$

This gives the value $V = 19\ \mathrm{m\,s^{-1}}$ with the corresponding Reynolds number $R = 1250$. This Reynolds number is nicely within the range in which the quadratic resistance law is applicable, and so provides a consistent solution. Thus the answer is that, **in air, the ball bearing falls with a speed of** $19\ \mathrm{m\,s^{-1}}$.

When the medium is **castor oil**, a similar calculation shows that it is the **linear resistance** law which provides the consistent solution. The answer is that, **in castor oil, the ball bearing falls with a speed of** $1.5\ \mathrm{cm\,s^{-1}}$, the Reynolds number being 0.015.

This example illustrates the conditions needed for low Reynolds number flow: slow motion of a small body through a sticky fluid. Perhaps the most celebrated application of the low Reynolds number drag formula is **Millikan**'s oil drop method of determining the electronic charge (see Problem 4.20 at the end of the chapter). ■

Example 4.5 *Vertical motion under gravity with linear resistance*

A body is projected vertically upwards with speed u in a medium that exerts a drag force $-mK\boldsymbol{v}$, where K is a positive constant and $\boldsymbol{v}$ is the velocity of the body.* Find the maximum height achieved by the body, the time taken to reach that height, and the terminal speed.

Solution

On including the linear resistance force, the scalar equation of motion becomes

$$m\frac{dv}{dt} = -mg - mKv,$$

with the initial condition $v = u$ when $t = 0$ (see Figure 4.1). This first order ODE for v separates in the form

$$\int \frac{dv}{g + Kv} = -\int dt,$$

and, on integration, gives

$$\frac{1}{K}\ln(g + Kv) = -t + C,$$

where C is the integration constant. On applying the initial condition $v = u$ when $t = 0$, we obtain $C = K^{-1}\ln(g + Ku)$ and so

$$t = \frac{1}{K}\ln\left(\frac{g + Ku}{g + Kv}\right).$$

* This is the vector drag force acting on the body; hence the minus sign. The coefficient is taken in the form mK for algebraic convenience.

This expression gives t in terms of v, which is what we need for finding the time taken to reach the maximum height. The maximum height is achieved when $v = 0$ so that τ, **the time taken to reach the maximum height**, is given by

$$\tau = \frac{1}{K} \ln\left(1 + \frac{Ku}{g}\right).$$

The expression for t in terms of v can be inverted to give

$$v = ue^{-Kt} - \frac{g}{K}\left(1 - e^{-Kt}\right)$$

for the upward **velocity** of the body at time t.

The terminal speed of the body is the limit of $|v|$ as $t \to \infty$. In this limit, the exponential terms tend to zero and

$$v \to -\frac{g}{K}.$$

Thus, in contrast to motion with no resistance, the speed of the body does not increase without limit as it falls, but tends to the finite value g/K. Thus the **terminal speed** of the body is g/K.

The terminal speed can also be deduced directly from the equation of motion. If the body is falling with the terminal speed, then $dv/dt = 0$ and the equation of motion implies that $0 = -mg - mKv$. It follows that the (upward) terminal velocity is $-g/K$.

The **maximum height** z_{max} can now be found by integrating the equation $dz/dt = v$ and then putting $t = \tau$. However we can also obtain z_{max} by starting again with a modified equation of motion. For some laws of resistance, this trick is essential. If we write

$$\frac{dv}{dt} = \frac{dv}{dz} \times \frac{dz}{dt} = v\frac{dv}{dz},$$

the equation of motion becomes

$$v\frac{dv}{dz} = -g - Kv,$$

with the initial condition $v = u$ when $z = 0$. This equation also separates to give

$$\begin{aligned} -\int dz &= \int \frac{v\,dv}{g + Kv} = \frac{1}{K}\int\left(1 - \frac{g}{g + Kv}\right) dv \\ &= \frac{v}{K} - \frac{g}{K^2}\ln(g + Kv) + D, \end{aligned}$$

where D is the integration constant. On applying the initial condition $v = u$ when $z = 0$, we obtain

$$\begin{aligned} z &= -\frac{v}{K} + \frac{g}{K^2}\ln(g + Kv) + \frac{u}{K} - \frac{g}{K^2}\ln(g + Ku) \\ &= \frac{1}{K}(u - v) - \frac{g}{K^2}\ln\left(\frac{g + Ku}{g + Kv}\right). \end{aligned}$$

This expression cannot be inverted to give v as a function of z, but it is exactly what we need to find z_{max}. Since z_{max} is achieved when $v = 0$, we find that the **maximum height** achieved by the body is given by

$$z_{max} = \frac{u}{K} - \frac{g}{K^2}\ln\left(1 + \frac{Ku}{g}\right). \blacksquare$$

Question *Approximate form of z_{max} for small Ku/g*

Find an approximate expression for z_{max} when Ku/g is small.

Answer

When Ku/g is small, the log term can be expanded as a power series. This gives

$$\begin{aligned} z_{max} &= \frac{u}{K} - \frac{g}{K^2}\left[\frac{Ku}{g} - \frac{1}{2}\left(\frac{Ku}{g}\right)^2 + \frac{1}{3}\left(\frac{Ku}{g}\right)^3 + \cdots\right] \\ &= \frac{u^2}{2g}\left[1 - \frac{2}{3}\left(\frac{Ku}{g}\right) + \cdots\right]. \end{aligned}$$

In this expression, the leading term $u^2/2g$ is just the value of z_{max} in the absence of resistance. The first correction term has a negative sign which means that z_{max} is reduced by the presence of resistance, as would be expected. ■

Question *Ball bearing released in castor oil*

The ball bearing in Example 4.4 is released from rest in castor oil. How long does it take for the ball bearing to achieve 99% of its terminal speed?

Answer

Recall that the linear law of resistance *is* appropriate for this motion. Since the motion is entirely downwards, it is more convenient to measure v downwards in this problem, in which case the solution for v becomes

$$v = \frac{g}{K}\left(1 - e^{-Kt}\right) = V\left(1 - e^{-gt/V}\right),$$

where V is the terminal velocity. When $v = 0.99V$, $e^{-gt/V} = 0.01$ and so the **time required** is

$$t = \ln(100)V/g,$$

which evaluates to about 7 milliseconds on using the value for V calculated in Example 4.4. ■

Note on the sign of resistance forces In the last example we used the same scalar equation of motion whether the body was rising or falling. This is correct in the case of linear resistance since, when the sign of v is reversed, so is the sign of Kv. In the case of quadratic resistance however, when the sign of v is reversed, the sign of Kv^2 remains unchanged and so the correct sign must be inserted manually. Thus, for *quadratic resistance, the scalar equations of motion for ascent and descent are different.* The same is true when the drag is proportional to *any even power* of v.

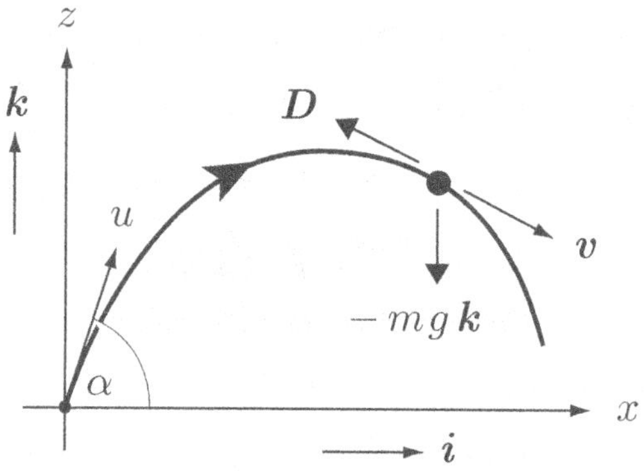

FIGURE 4.6 A particle, initially at the origin, is projected with speed u in a direction making an angle α with the horizontal. The particle moves under the uniform gravity force $-mg\boldsymbol{k}$ and the resistance (drag) force $\boldsymbol{D}$.

4.4 PROJECTILES

A body that moves freely under uniform gravity, and possibly air resistance, is called a **projectile**. Projectile motion is very common. In ball games, the ball is a projectile, and controlling its trajectory is a large part of the skill of the game. On a larger scale, artillery shells are projectiles, but guided missiles, which have rocket propulsion, are not.

The projectile problem differs from the problems considered in section 4.3 in that projectile motion is not restricted to take place in a vertical straight line. However, we will continue to assume that the effect of the air is to exert a drag force opposing the current velocity of the projectile.* It is then evident by symmetry that each **projectile motion takes place in a vertical plane**; this vertical plane contains the initial position of the projectile and is parallel to its initial velocity.

Projectiles without resistance

The first (and easiest) problem is that of a projectile moving without air resistance. This is fine on the Moon, but will be only an approximation to projectile motion on Earth. The effect of air resistance can be very significant, as our later examples will show.

Example 4.6 *Projectile without air resistance*

A particle which is subject solely to uniform gravity is projected with speed u in a direction making an angle α with the horizontal. Find the subsequent motion.

Solution

Suppose that the motion takes place in the (x, z)-plane as shown in Figure 4.6. In the absence of the drag force, the vector equation of motion becomes

$$m\frac{d\boldsymbol{v}}{dt} = -mg\boldsymbol{k},$$

with the initial condition $\boldsymbol{v} = (u\cos\alpha)\boldsymbol{i} + (u\sin\alpha)\boldsymbol{k}$ when $t = 0$. If we now write $\boldsymbol{v} = v_x\boldsymbol{i} + v_z\boldsymbol{k}$ and take components of this equation (and initial condition) in the

* This will be true if the projectile is a rigid sphere moving without rotation.

$\boldsymbol{i}$- and $\boldsymbol{k}$-directions, we obtain the two scalar equations of motion

$$\frac{dv_x}{dt} = 0, \qquad \frac{dv_z}{dt} = -g,$$

with the respective initial conditions $v_x = u\cos\alpha$ and $v_z = u\sin\alpha$ when $t = 0$. Simple integrations then give the components of the particle **velocity** to be

$$v_x = u\cos\alpha, \qquad v_z = u\sin\alpha - gt.$$

The position of the particle at time t can now be found by integrating the expressions for v_x, v_z and applying the initial conditions $x = 0$ and $z = 0$ when $t = 0$. This gives

$$x = (u\cos\alpha)\,t, \qquad z = (u\sin\alpha)\,t - \tfrac{1}{2}gt^2,$$

the solution for the **trajectory** of the particle. ■

Question ***Form of the path***

Show that the path taken by the particle is an inverted parabola.

Answer

To find the path, eliminate t from the trajectory equations. This gives

$$z = (\tan\alpha)\,x - \left(\frac{g}{2u^2\cos^2\alpha}\right)x^2,$$

which is indeed an inverted parabola. ■

Question ***Time of flight and the range***

Find the time of flight and the range of the projectile on level ground.

Answer

On level ground, the motion will terminate when $z = 0$ again. From the second trajectory equation, this happens when $(u\sin\alpha)\,t - \frac{1}{2}gt^2 = 0$. Hence the time of flight τ is given by $\tau = 2u\sin\alpha/g$. The horizontal range R is then obtained by putting $t = \tau$ in the first trajectory equation, which gives

$$R = \frac{u^2\sin 2\alpha}{g}. \ \blacksquare$$

Question ***Maximum range***

Find the value of α that gives the maximum range on level ground when u is fixed.

Answer

R is a maximum when $\sin 2\alpha = 1$, that is when $\alpha = \pi/4$ in which case $R_{max} = u^2/g$. Thus, if an artillery shell is to be projected over a horizontal range of 4 km, then the gun must have a muzzle speed of at least 200 m s^{-1}. ■

There is a myriad of problems that can be found on the projectile with no air resistance, and some interesting examples are included at the end of the chapter. It should be noted though that all these problems are *dynamically* equivalent to the problem solved above. Any difficulties lie in the geometry!

Projectiles with resistance

We now proceed to include the effect of air resistance. From our earlier discussion of fluid drag, it is evident that in most practical instances of projectile motion through the Earth's atmosphere, it is the **quadratic law** of resistance that is appropriate. On the other hand, only the **linear law** of resistance gives rise to linear equations of motion and simple analytical solutions. This explains why mechanics textbooks contain extensive coverage of the linear case, even though this case is almost never appropriate in practice; the case that *is* appropriate cannot be solved! In the following example, we treat the linear resistance case.

Example 4.7 *Projectile with linear resistance*

A particle is subject to uniform gravity and the linear resistance force $-mK\boldsymbol{v}$, where K is a positive constant and $\boldsymbol{v}$ is the velocity of the particle. Initially the particle is projected with speed u in a direction making an angle α with the horizontal. Find the subsequent motion.

Solution

With the linear resistance term included, the vector equation of motion becomes

$$m\frac{d\boldsymbol{v}}{dt} = -mK\boldsymbol{v} - mg\boldsymbol{k},$$

with the initial condition $\boldsymbol{v} = (u\cos\alpha)\boldsymbol{i} + (u\sin\alpha)\boldsymbol{k}$ when $t = 0$. As in the last example, this equation resolves into the two scalar equations of motion

$$\frac{dv_x}{dt} + Kv_x = 0, \qquad \frac{dv_z}{dt} + Kv_z = -g,$$

with the respective initial conditions $v_x = u\cos\alpha$ and $v_z = u\sin\alpha$ when $t = 0$. These first order ODEs are both separable and linear and can be solved by either method; if they are regarded as linear, the integrating factor is e^{Kt}. The equations integrate to give the components of the particle **velocity** to be

$$v_x = (u\cos\alpha)e^{-Kt}, \qquad v_z = (u\sin\alpha)e^{-Kt} - \frac{g}{K}\left(1 - e^{-Kt}\right).$$

The position of the particle at time t can now be found by integrating the expressions for v_x, v_z and applying the initial conditions $x = 0$ and $z = 0$ when $t = 0$. This gives

$$x = \frac{u\cos\alpha}{K}\left(1 - e^{-Kt}\right), \qquad z = \frac{Ku\sin\alpha + g}{K^2}\left(1 - e^{-Kt}\right) - \frac{g}{K}t, \tag{4.3}$$

the solution for the **trajectory of the particle**. Figure 4.7 shows typical paths taken by the particle for the same initial conditions and three different values of the dimensionless resistance parameter λ $(= Ku/g)$. (The case $\lambda = 0$ corresponds to zero resistance so that the path is a parabola.) It is apparent that resistance can have a dramatic effect on the motion. ■

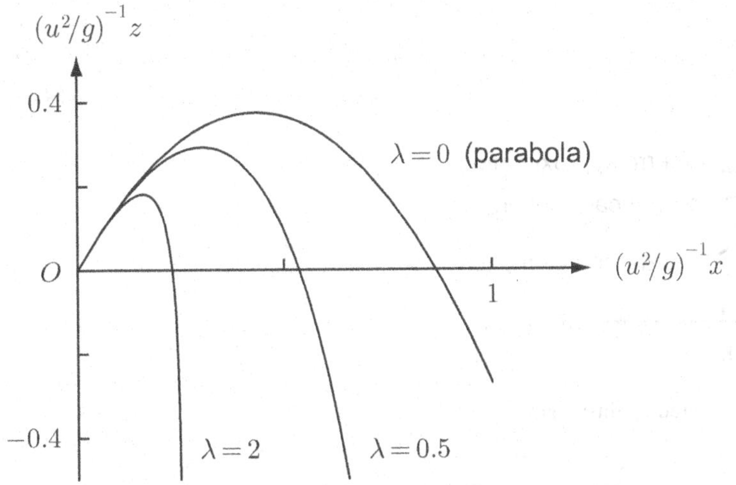

FIGURE 4.7 Projectile motion under uniform gravity and **linear resistance**. The graphs show the paths of the particle for $\alpha = \pi/3$ and three different values of the dimensionless resistance parameter λ $(= Ku/g)$. Except when $\lambda = 0$, the paths have vertical asymptotes.

Question ***Vertical asymptote of the path***

Show that the path has a vertical asymptote.

Answer

Since e^{-Kt} decreases and tends to zero as $t \to \infty$, it follows from equations (4.3) that the horizontal displacement x *increases* and tends to the value $u\cos\alpha/K$ as $t \to \infty$, while the vertical displacement z tends to negative infinity. Thus the vertical line $x = u\cos\alpha/K$ is an asymptote to the path. In terms of the dimensionless variables used in Figure 4.7, this is the line $(u^2/g)^{-1}x = \cos\alpha/\lambda$. ■

Question ***Approximate formula for the range when λ is small***

Find an approximate formula for the range on level ground when the resistance parameter λ is small.

Answer

Since the particle returns to Earth again when $z = 0$, it follows from the second of equations (4.3) that the flight time τ satisfies the equation

$$(Ku\sin\alpha + g)\left(1 - e^{-K\tau}\right) - Kg\tau = 0,$$

which can be written in the form

$$(\lambda\sin\alpha + 1)\left(1 - e^{-K\tau}\right) - K\tau = 0, \qquad (4.4)$$

where $\lambda(= Ku/g)$ is the dimensionless resistance parameter. Unfortunately, this equation cannot be solved explicitly for τ, and hence the need for an approximate solution. We know from the last example that, in the absence of resistance, the flight time τ is given by $\tau = 2u\sin\alpha/g$. It is

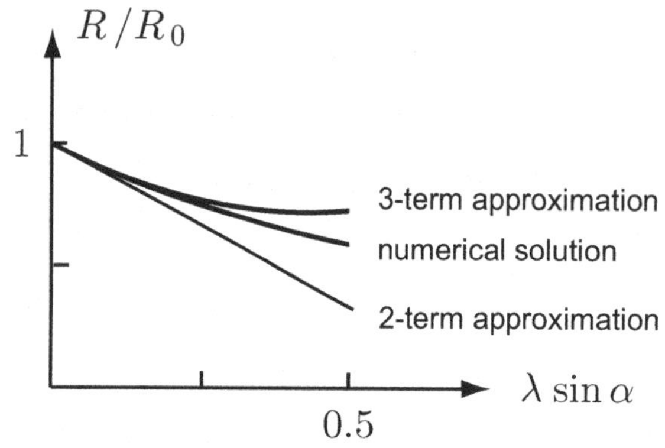

FIGURE 4.8 The ratio R/R_0 plotted against $\lambda \sin\alpha$.

reasonable then, when λ is small, to seek a solution for τ in the form

$$\tau = \frac{2u\sin\alpha}{g}\left[1 + b_1\lambda + b_2\lambda^2 + \cdots\right], \tag{4.5}$$

where the coefficients $b_1, b_2, \ldots$ are to be determined. To find the expansion coefficients we substitute the expansion (4.5) (truncated after the required number of terms) into the left side of equation (4.4), re-expand in powers of λ, and then set the coefficients in this expansion equal to zero. The corresponding formula for the range can then be found by substituting this approximate formula for τ into the first equation of (4.3) and re-expanding in powers of λ. The details are tedious and, in fact, such operations are best done with computer assistance. The completion of this solution is the subject of computer assisted Problem 4.34 at the end of this chapter. The answer (to three terms) is that the range R on level ground is given by

$$\frac{R}{R_0} = 1 - \left(\frac{4\sin\alpha}{3}\right)\lambda + \left(\frac{14\sin^2\alpha}{9}\right)\lambda^2 + O(\lambda^3),$$

where R_0 is the range when resistance is absent. Figure 4.8 compares two different approximations to R with the 'exact' value obtained by numerical solution of equation (4.4). As would be expected, the three term approximation is closer to the exact value. ■

4.5 CIRCULAR MOTION

In this section we examine some important problems in which a body moves on a circular path. Our first problem is concerned with a body executing a circular orbit under the gravitational attraction of a fixed mass. This is a fairly accurate model of the motion of the planets* around the Sun.

Example 4.8 *Circular orbit in the inverse square field*

A particle of mass m moves under the gravitational attraction of a fixed mass M situated at the origin. Show that circular orbits with centre O and any radius are

* The orbits of Mercury, Mars and Pluto are the most elliptical with eccentricities of 0.206, 0.093 and 0.249 respectively. The eccentricity of Earth's orbit is 0.017.

possible, and find the speed of the particle in such an orbit. Deduce the period of the orbit.

Solution

Note that we are not required to find the *general* orbit; we may assume from the start that the orbit is a circle. Suppose then that the particle is executing a circular orbit with centre O and radius R. We need to confirm that the vector equation of motion can be satisfied. Take polar coordinates r, θ with centre at O. Then the acceleration $\boldsymbol{a}$ of the particle is given in terms of the usual polar unit vectors by the formula (2.14), that is,

$$\begin{aligned}\boldsymbol{a} &= \left(\ddot{r} - r\dot{\theta}^2\right)\widehat{\boldsymbol{r}} + \left(r\ddot{\theta} + 2\dot{r}\dot{\theta}\right)\widehat{\boldsymbol{\theta}} \\ &= -\frac{v^2}{R}\widehat{\boldsymbol{r}} + \dot{v}\,\widehat{\boldsymbol{\theta}}\end{aligned}$$

for motion on the circle $r = R$, where the circumferential velocity $v = R\dot{\theta}$. The equation of motion for the particle is therefore

$$m\left[-\frac{v^2}{R}\widehat{\boldsymbol{r}} + \dot{v}\widehat{\boldsymbol{\theta}}\right] = -\frac{mMG}{R^2}\widehat{\boldsymbol{r}},$$

which, on taking components in the radial and transverse directions, gives

$$\frac{v^2}{R} = \frac{MG}{R^2} \qquad \text{and} \qquad \dot{v} = 0.$$

Hence the equation of motion is satisfied if v is a constant given by

$$v^2 = \frac{MG}{R}.$$

Thus a circular orbit of radius R is possible provided that the particle has **constant speed** $(MG/R)^{1/2}$.

The **period** τ of the orbit is the time taken for one circuit and is given by

$$\tau = \frac{2\pi R}{v} = \left(\frac{4\pi^2 R^3}{MG}\right)^{1/2}.$$

Thus the square of the period of a circular orbit is proportional to the cube of its radius. This is a special case of Kepler's third law of planetary motion (see Chapter 7). ■

A particle may move on a circular path because it is *constrained* to do so. The simplest and most important example of this is the simple pendulum, a mass suspended from a fixed point by a string.

Example 4.9 *The simple pendulum*

A particle P is suspended from a fixed point O by a light inextensible string of length b. P is subject to uniform gravity and moves in a vertical plane through O with the string taut. Find the equation of motion.

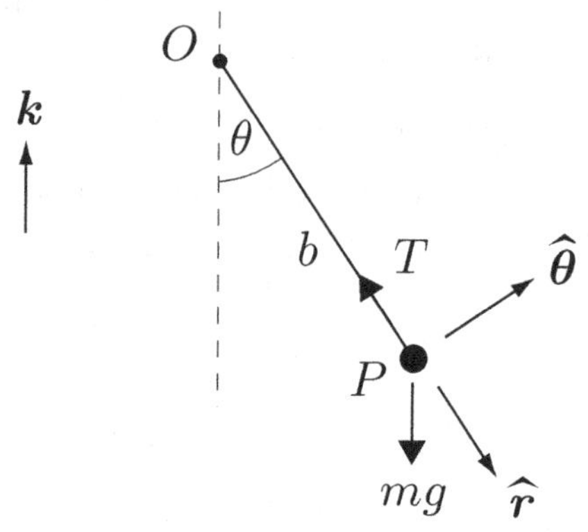

FIGURE 4.9 The simple pendulum

Solution

The system is shown in Figure 4.9. Since the string is of fixed length b, the position of P is determined by the angle θ shown. The acceleration of P can be expressed in the polar form

$$\boldsymbol{a} = -\left(b\dot{\theta}^2\right)\widehat{\boldsymbol{r}} + \left(b\ddot{\theta}\right)\widehat{\boldsymbol{\theta}},$$

where $\widehat{\boldsymbol{r}}$ and $\widehat{\boldsymbol{\theta}}$ are the polar unit vectors shown in Figure 4.9.

P moves under the uniform gravity force $-mg\boldsymbol{k}$ and the tension T in the string which acts in the direction $-\widehat{\boldsymbol{r}}$. It should be noted that the tension T is a force of constraint and not known beforehand. The equation of motion is therefore

$$m\left[-\left(b\dot{\theta}^2\right)\widehat{\boldsymbol{r}} + \left(b\ddot{\theta}\right)\widehat{\boldsymbol{\theta}}\right] = -mg\boldsymbol{k} - T\widehat{\boldsymbol{r}}.$$

If we now take components of this equation in the radial and transverse directions we obtain

$$-mb\dot{\theta}^2 = mg\cos\theta - T, \qquad mb\ddot{\theta} = -mg\sin\theta.$$

The second of these equations is the effective equation of motion in terms of the 'coordinate' θ, namely,

$$\ddot{\theta} + \left(\frac{g}{b}\right)\sin\theta = 0, \tag{4.6}$$

while the first equation determines the unknown tension T once $\theta(t)$ is known.

Equation (4.6) is the **exact equation of motion** for the simple pendulum. Because of the presence of the term in $\sin\theta$, this second order ODE is **non-linear** and cannot be solved by using the standard technique for linear ODEs with constant coefficients. ■

Question *The linear theory for small amplitude oscillations*

Find an approximate linear equation for the case in which the pendulum undergoes oscillations of small amplitude.

Answer

If θ is always small then $\sin\theta$ can be approximated by θ in which case the equation of motion becomes

$$\ddot{\theta} + \left(\frac{g}{b}\right)\theta = 0. \tag{4.7}$$

This is the **linearised equation** for the simple pendulum, which holds approximately for oscillations of small amplitude. Although we do not cover linear oscillations until Chapter 5, many readers will recognise equation (4.7) as the simple harmonic motion equation and will know that the period τ of the oscillations is given by $\tau = 2\pi(b/g)^{1/2}$, independent of the (small) amplitude. ■

Question ***Period of large oscillations***

Find the period of the pendulum when the (angular) amplitude of its oscillations is α, where α may not be small.

Answer

This requires that we integrate the exact equation of motion (4.6). We start with a familiar trick. If we write $\Omega = \dot{\theta}$, then

$$\ddot{\theta} = \frac{d\Omega}{dt} = \frac{d\Omega}{d\theta} \times \frac{d\theta}{dt} = \Omega\frac{d\Omega}{d\theta}$$

and the equation of motion becomes

$$\Omega\frac{d\Omega}{d\theta} = -\left(\frac{g}{b}\right)\sin\theta.$$

This is a separable first order ODE for Ω which integrates to give

$$\tfrac{1}{2}\Omega^2 = \left(\frac{g}{b}\right)\cos\theta + C,$$

where C is the constant of integration. On applying the initial condition $\Omega = 0$ when $\theta = \alpha$, we find that $C = -(g/b)^{1/2}\cos\alpha$ and the integrated equation can be written

$$\left(\frac{d\theta}{dt}\right)^2 = \left(\frac{2g}{b}\right)(\cos\theta - \cos\alpha), \tag{4.8}$$

where we have now replaced Ω by $d\theta/dt$.

Since the pendulum comes to rest only when $d\theta/dt = 0$ (that is, when $\theta = \pm\alpha$) it follows that θ must oscillate in the range $-\alpha \le \theta \le \alpha$. The period τ is the time taken for one complete oscillation but, by the symmetry of equation (4.8) under the transformation $\theta \to -\theta$, it follows that the time taken for the pendulum to swing from $\theta = 0$ to $\theta = +\alpha$ is $\tau/4$. To evaluate this time we take the positive square root of each side of equation (4.8) and integrate over the time interval $0 \le t \le \tau/4$. This gives

$$\int_0^{\alpha} \frac{d\theta}{(\cos\theta - \cos\alpha)^{1/2}} = \left(\frac{2g}{b}\right)^{1/2} \int_0^{\tau/4} dt,$$

so that

$$\tau = \left(\frac{8b}{g}\right)^{1/2} \int_0^{\alpha} \frac{d\theta}{(\cos\theta - \cos\alpha)^{1/2}}. \tag{4.9}$$

This is the **exact period** of the pendulum when the amplitude of its oscillations is α. It is not possible to perform this integration in terms of standard functions* and so the integral must either be evaluated numerically or be approximated.

Numerical evaluation shows that the *exact period is longer than that predicted by the linearised theory*. When $\alpha = \pi/6$, the period is 1.7% longer, and when $\alpha = \pi/3$ it is 7.3% longer. The period can also be approximated by expanding the integral in equation (4.9) as a power series in α. This is the subject of Problem 4.35 at the end of the chapter. The answer is that, expanded to two terms,

$$\tau = 2\pi \left(\frac{b}{g}\right)^{1/2} \left[1 + \frac{\alpha^2}{16} + O(\alpha^4)\right]. \tag{4.10}$$

This two term approximation predicts an increase in the period of 1.7% when $\alpha = \pi/6$. Note that there is no term in this expansion proportional to α and that the term in α^2 has the small coefficient 1/16. This explains why the prediction of the linearised theory is rather accurate even when α is not so small! ■

In our final example, we solve the important problem of an **electrically charged particle moving in a uniform magnetic field**. It turns out that plane motions are circular, but the most general motion is helical. The solution in this case differs from the previous examples in that we use Cartesian coordinates instead of polars. This is because we do not know beforehand where the centre of the circle (or the axis of the helix) is, which means that we do not know on which point (or axis) to centre the polar coordinates.

Example 4.10 *Charged particle in a magnetic field*

A particle of mass m and charge e moves in a uniform magnetic field of strength B_0. Show that the most general motion is helical with the axis of the helix parallel to the direction of the magnetic field.

Solution

The total force $\boldsymbol{F}$ that an electric field $\boldsymbol{E}$ and magnetic field $\boldsymbol{B}$ exert on a charge e is given by the **Lorentz force formula**†

$$\boldsymbol{F} = e\boldsymbol{E} + e\boldsymbol{v}\times\boldsymbol{B},$$

where $\boldsymbol{v}$ is the velocity of the charge. In our problem, there is no electric field and the magnetic field is uniform. If the direction of $\boldsymbol{B}$ is the z-direction of Cartesian coordinates, then $\boldsymbol{B} = B_0\boldsymbol{k}$. The equation of motion of the particle is then

$$m\frac{d\boldsymbol{v}}{dt} = eB_0\boldsymbol{v}\times\boldsymbol{k}.$$

If we now write $\boldsymbol{v}$ in the component form $\boldsymbol{v} = v_x\boldsymbol{i} + v_y\boldsymbol{j} + v_z\boldsymbol{k}$, the vector equation of motion resolves into the three scalar equations

$$\frac{dv_x}{dt} = \Omega v_y, \qquad \frac{dv_y}{dt} = -\Omega v_x, \qquad \frac{dv_z}{dt} = 0, \tag{4.11}$$

* The integral is related to a special function called the *complete elliptic integral of the first kind.*

† This form is correct in SI units.

where

$$\Omega = eB_0/m. \tag{4.12}$$

The last of these equations shows that $v_z = V$, a constant, so that the component of $\boldsymbol{v}$ parallel to the magnetic field is a constant. The first two equations are **first order coupled ODEs** but they are easy to uncouple. If we differentiate the first equation with respect to t and use the second equation, we find that v_x satisfies the equation

$$\frac{d^2 v_x}{dt^2} + \Omega^2 v_x = 0.$$

This equation is a second order ODE with constant coefficients and can be solved in the standard way. However, many readers will recognise this as the SHM equation whose general solution can be written in the form

$$v_x = A \sin(\Omega t + \alpha),$$

where A and α are arbitrary constants. It is more convenient if we introduce a new arbitrary constant R defined by $A = -\Omega R$, so that

$$v_x = -\Omega R \sin(\Omega t + \alpha).$$

If we now substitute this formula for v_x into the first equation of (4.11), we obtain

$$v_y = -\Omega R \cos(\Omega t + \alpha).$$

Having obtained the solution for the three components of $\boldsymbol{v}$, we can now find the trajectory simply by integrating with respect to t. This gives

$$x = R\cos(\Omega t + \alpha) + a, \qquad y = -R\sin(\Omega t + \alpha) + b, \qquad z = Vt + c,$$

where a, b, and c are constants of integration. These constants may be removed by a shift of the origin of coordinates to the point (a, b, c), and the constant α may be removed by a shift in the origin of t. Also, the constant R may be assumed positive; if it is not, make a shift in the origin of t by π/Ω. With these simplifications, the final form for the trajectory is

$$x = R\cos\Omega t, \qquad y = -R\sin\Omega t, \qquad z = Vt, \tag{4.13}$$

where R is a positive constant and $\Omega = eB_0/m$. This is the most **general trajectory** for a charged particle moving in a uniform magnetic field.

To identify this trajectory as a helix, suppose first that $V = 0$ so that the motion takes place in the (x, y)-plane. Then the first two equations of (4.13) imply that the path is a circle of radius R traversed with constant speed $R|\Omega|$ and with period $2\pi/|\Omega|$. When $V \neq 0$, this circular motion is supplemented by a uniform velocity V in the z-direction. The result is a helical path of radius R, with its axis parallel to the magnetic field, which is traversed with constant speed $(V^2 + R^2\Omega^2)^{1/2}$. ■

The above problem has important applications to the **cyclotron** particle accelerator and the **mass spectrograph**. The cyclotron depends for its operation on thefact that the constant

Ω, known as the **cyclotron frequency**, is independent of the velocity of the charged particles. The mass spectrograph is the subject of Problem 4.32 at the end of the chapter.

Problems on Chapter 4

Answers and comments are at the end of the book.
Harder problems carry a star (∗).

Introductory problems

4.1 Two identical blocks each of mass M are connected by a light inextensible string and can move on the surface of a *rough* horizontal table. The blocks are being towed at constant speed in a straight line by a rope attached to one of them. The tension in the tow rope is T_0. What is the tension in the connecting string? The tension in the tow rope is suddenly increased to $4T_0$. What is the instantaneous acceleration of the blocks and what is the instantaneous tension in the connecting string?

4.2 A body of mass M is suspended from a fixed point O by an inextensible uniform rope of mass m and length b. Find the tension in the rope at a distance z below O. The point of support now begins to rise with acceleration $2g$. What now is the tension in the rope?

4.3 Two uniform lead spheres each have mass 5000 kg and radius 47 cm. They are released from rest with their centres 1 m apart and move under their mutual gravitation. Without doing any integration show that they will collide in *less* than 425 s. [$G = 6.67 \times 10^{-11}$ N m^2 kg^{-2}.]

4.4 The block in Figure 4.2 is sliding down the inclined surface of a fixed wedge. This time the frictional force F exerted on the block is given by $F = \mu N$, where N is the normal reaction and μ is a positive constant. Find the acceleration of the block. How do the cases $\mu < \tan\alpha$ and $\mu > \tan\alpha$ differ?

4.5 A stuntwoman is to be fired from a cannon and projected a distance of 40 m over level ground. What is the least projection speed that can be used? If the barrel of the cannon is 5 m long, show that she will experience an acceleration of at least $4g$ in the barrel. [Take $g = 10$ m s^{-2}.]

4.6 In an air show, a pilot is to execute a circular loop at the speed of sound (340 m s^{-1}). The pilot may black out if his acceleration exceeds $8g$. Find the radius of the smallest circle he can use. [Take $g = 10$ m s^{-2}.]

4.7 A body has terminal speed V when falling in still air. What is its terminal velocity (relative to the ground) when falling in a steady horizontal wind with speed U?

4.8 ***Cathode ray tube*** A particle of mass m and charge e is moving along the x-axis with speed u when it passes between two charged parallel plates. The plates generate a uniform electric field $E_0\boldsymbol{j}$ in the region $0 \le x \le b$ and no field elsewhere.* Find the angle through

* This is only approximately true.

which the particle is deflected by its passage between the plates. [The cathode ray tube uses this arrangement to deflect the electron beam.]

Straight line motion in a force field

4.9 An object is dropped from the top of a building and is in view for time τ while passing a window of height h some distance lower down. How high is the top of the building above the top of the window?

4.10 A particle P of mass m moves under the gravitational attraction of a mass M fixed at the origin O. Initially P is at a distance a from O when it is projected with the *critical* escape speed $(2MG/a)^{1/2}$ directly away from O. Find the distance of P from O at time t, and confirm that P escapes to infinity.

4.11 A particle P of mass m is attracted towards a fixed origin O by a force of magnitude $m\gamma/r^3$, where r is the distance of P from O and γ is a positive constant. [It's gravity Jim, but not as we know it.] Initially, P is at a distance a from O, and is projected with speed u directly away from O. Show that P will escape to infinity if $u^2 > \gamma/a^2$.

For the case in which $u^2 = \gamma/(2a^2)$, show that the maximum distance from O achieved by P in the subsequent motion is $\sqrt{2}a$, and find the time taken to reach this distance.

4.12 If the Earth were suddenly stopped in its orbit, how long would it take for it to collide with the Sun? [Regard the Sun as a *fixed* point mass. You may make use of the formula for the period of the Earth's orbit.]

Constrained motion

4.13 A particle P of mass m slides on a smooth horizontal table. P is connected to a second particle Q of mass M by a light inextensible string which passes through a small smooth hole O in the table, so that Q hangs below the table while P moves on top. Investigate motions of this system in which Q remains at rest vertically below O, while P describes a circle with centre O and radius b. Show that this is possible provided that P moves with constant speed u, where $u^2 = Mgb/m$.

4.14 A light pulley can rotate freely about its axis of symmetry which is fixed in a horizontal position. A light inextensible string passes over the pulley. At one end the string carries a mass $4m$, while the other end supports a second light pulley. A second string passes over this pulley and carries masses m and $4m$ at its ends. The whole system undergoes planar motion with the masses moving vertically. Find the acceleration of each of the masses.

4.15 A particle P of mass m can slide along a *smooth* rigid straight wire. The wire has one of its points fixed at the origin O, and is made to rotate in the (x, y)-plane with angular speed Ω. By using the vector equation of motion of P in polar co-ordinates, show that r, the distance of P from O, satisfies the equation

$$\ddot{r} - \Omega^2 r = 0,$$

and find a second equation involving N, where $N\widehat{\boldsymbol{\theta}}$ is the force the wire exerts on P. [Ignore gravity in this question.]

Initially, P is at rest (relative to the wire) at a distance a from O. Find r as a function of t in the subsequent motion, and deduce the corresponding formula for N.

Resisted motion

4.16 A body of mass m is projected with speed u in a medium that exerts a resistance force of magnitude (i) $mk|\boldsymbol{v}|$, or (ii) $mK|\boldsymbol{v}|^2$, where k and K are positive constants and $\boldsymbol{v}$ is the velocity of the body. Gravity can be ignored. Determine the subsequent motion in each case. Verify that the motion is bounded in case (i), but not in case (ii).

4.17 A body is projected vertically upwards with speed u and moves under uniform gravity in a medium that exerts a resistance force proportional to the square of its speed and in which the body's terminal speed is V. Find the maximum height above the starting point attained by the body and the time taken to reach that height.

Show also that the speed of the body when it returns to its starting point is $uV/(V^2 + u^2)^{1/2}$. [*Hint.* The equations of motion for ascent and descent are different. See the note at the end of section 4.3.]

4.18* A body is released from rest and moves under uniform gravity in a medium that exerts a resistance force proportional to the square of its speed and in which the body's terminal speed is V. Show that the time taken for the body to fall a distance h is

$$\frac{V}{g}\cosh^{-1}\left(e^{gh/V^2}\right).$$

In his famous (but probably apocryphal) experiment, Galileo dropped different objects from the top of the tower of Pisa and timed how long they took to reach the ground. If Galileo had dropped two iron balls, of 5 mm and 5 cm radius respectively, from a height of 25 m, what would the descent times have been? Is it likely that this difference could have been detected? [Use the quadratic law of resistance with $C = 0.8$. The density of iron is 7500 kg m^{-3}.]

4.19 A body is projected vertically upwards with speed u and moves under uniform gravity in a medium that exerts a resistance force proportional to the fourth power its speed and in which the body's terminal speed is V. Find the maximum height above the starting point attained by the body.

Deduce that, however large u may be, this maximum height is always less than $\pi V^2/4g$.

4.20 ***Millikan's experiment*** A microscopic spherical oil droplet, of density ρ and unknown radius, carries an unknown electric charge. The droplet is observed to have terminal speed v_1 when falling vertically in air of viscosity μ. When a uniform electric field E_0 is applied in the vertically upwards direction, the same droplet was observed to move *upwards* with terminal speed v_2. Find the charge on the droplet. [Use the low Reynolds number approximation for the drag.]

Projectiles

4.21 A mortar gun, with a maximum range of 40 m on level ground, is placed on the edge of a vertical cliff of height 20 m overlooking a horizontal plain. Show that the horizontal range R of the mortar gun is given by

$$R = 40\left\{\sin\alpha + \left(1 + \sin^2\alpha\right)^{\frac{1}{2}}\right\}\cos\alpha,$$

where α is the angle of elevation of the mortar above the horizontal. [Take $g = 10\ \mathrm{m\,s^{-2}}$.]

Evaluate R (to the nearest metre) when $\alpha = 45°$ and $35°$ and confirm that $\alpha = 45°$ does not yield the maximum range. [Do not try to find the optimum projection angle this way. See Problem 4.22 below.]

4.22 It is required to project a body from a point on level ground in such a way as to clear a thin vertical barrier of height h placed at distance a from the point of projection. Show that the body will just skim the top of the barrier if

$$\left(\frac{ga^2}{2u^2}\right)\tan^2\alpha - a\tan\alpha + \left(\frac{ga^2}{2u^2} + h\right) = 0,$$

where u is the speed of projection and α is the angle of projection above the horizontal.

Deduce that, if the above trajectory is to exist for some α, then u must satisfy

$$u^4 - 2ghu^2 - g^2a^2 \geq 0.$$

Find the least value of u that satisfies this inequality.

For the special case in which $a = \sqrt{3}h$, show that the minimum projection speed necessary to clear the barrier is $(3gh)^{\frac{1}{2}}$, and find the projection angle that must be used.

4.23 A particle is projected from the origin with speed u in a direction making an angle α with the horizontal. The motion takes place in the (x, z)-plane, where Oz points vertically upwards. If the projection speed u is fixed, show that the particle can be made to pass through the point (a, b) for some choice of α if (a, b) lies *below* the parabola

$$z = \frac{u^2}{2g}\left(1 - \frac{g^2x^2}{u^4}\right).$$

This is called the **parabola of safety**. Points *above* the parabola are 'safe' from the projectile.

An artillery shell explodes on the ground throwing shrapnel in all directions with speeds of up to $30\ \mathrm{m\,s^{-1}}$. A man is standing at an open window 20 m above the ground in a building 60 m from the blast. Is he safe? [Take $g = 10\ \mathrm{m\,s^{-2}}$.]

4.24 A projectile is fired from the top of a conical mound of height h and base radius a. What is the least projection speed that will allow the projectile to clear the mound? [*Hint*. Make use of the parabola of safety.]

A mortar gun is placed on the summit of a conical hill of height 60 m and base diameter 160 m. If the gun has a muzzle speed of $25\ \mathrm{m\,s^{-1}}$, can it shell anywhere on the hill? [Take $g = 10\ \mathrm{m\,s^{-2}}$.]

4.25 An artillery gun is located on a plane surface inclined at an angle β to the horizontal. The gun is aligned with the line of steepest slope of the plane. The gun fires a shell with speed u in the direction making an angle α with the (upward) line of steepest slope. Find where the shell lands.

Deduce the maximum ranges R^U, R^D, up and down the plane, and show that

$$\frac{R^U}{R^D} = \frac{1 - \sin\beta}{1 + \sin\beta}.$$

4.26 Show that, when a particle is projected from the origin in a medium that exerts *linear* resistance, its position vector at time t has the general form

$$\boldsymbol{r} = -\alpha(t)\boldsymbol{k} + \beta(t)\boldsymbol{u},$$

where $\boldsymbol{k}$ is the vertically upwards unit vector and $\boldsymbol{u}$ is the *velocity* of projection. Deduce the following results:

(i) A number of particles are projected simultaneously from the same point, with the same speed, but in *different directions*. Show that, at each later time, the particles all lie on the surface of a sphere.

(ii) A number of particles are projected simultaneously from the same point, in the same direction, but with *different speeds*. Show that, at each later time, the particles all lie on a straight line.

(iii) Three particles are projected simultaneously in a completely general manner. Show that the plane containing the three particles remains parallel to some fixed plane.

4.27 A body is projected in a steady horizontal wind and moves under uniform gravity and *linear* air resistance. Show that the influence of the wind is the same as if the magnitude and direction of gravity were altered. Deduce that it is possible for the body to return to its starting point. What is the shape of the path in this case?

Circular motion and charged particles

4.28 The radius of the Moon's approximately circular orbit is 384,000 km and its period is 27.3 days. Estimate the mass of the Earth. [$G = 6.67 \times 10^{-11}$ N m^2 kg^{-2}.] The actual mass is 5.97×10^{24} kg. What is the main reason for the error in your estimate?

An artificial satellite is to be placed in a circular orbit around the Earth so as to be 'geo-stationary'. What must the radius of its orbit be? [The period of the Earth's rotation is 23 h 56 m, *not* 24 h. Why?]

4.29 ***Conical pendulum*** A particle is suspended from a fixed point by a light inextensible string of length a. Investigate 'conical motions' of this pendulum in which the string maintains a constant angle α with the downward vertical. Show that, for any acute angle α, a conical motion exists and that the particle speed u is given by $u^2 = ag \sin\alpha \tan\alpha$.

4.30 A particle of mass m is attached to the highest point of a *smooth* rigid sphere of radius a by a light inextensible string of length $\pi a/4$. The particle moves in contact with the outer surface of the sphere, with the string taut, and describes a horizontal circle with constant speed u. Find the reaction of the sphere on the particle and the tension in the string. Deduce the maximum value of u for which such a motion could take place. What will happen if u exceeds this value?

4.31 A particle of mass m can move on a *rough* horizontal table and is attached to a fixed point on the table by a light inextensible string of length b. The resistance force exerted on the particle is $-mK\boldsymbol{v}$, where $\boldsymbol{v}$ is the velocity of the particle. Initially the string is taut and the particle is projected horizontally, at right angles to the string, with speed u. Find the angle turned through by the string before the particle comes to rest. Find also the tension in the string at time t.

4.32 ***Mass spectrograph*** A stream of particles of various masses, all carrying the same charge e, is moving along the x-axis in the positive x-direction. When the particles reach the origin they encounter an electronic 'gate' which allows only those particles with a specified speed V to pass. These particles then move in a uniform magnetic field B_0 acting in the z-direction. Show that each particle will execute a semicircle before meeting the y-axis at a point which depends upon its mass. [This provides a method for determining the masses of the particles.]

4.33 ***The magnetron*** An electron of mass m and charge $-e$ is moving under the combined influence of a uniform electric field $E_0\boldsymbol{j}$ and a uniform magnetic field $B_0\boldsymbol{k}$. Initially the electron is at the origin and is moving with velocity $u\,\boldsymbol{i}$. Show that the trajectory of the electron is given by

$$x = a(\Omega t) + b\sin\Omega t, \qquad y = b(1 - \cos\Omega t), \qquad z = 0,$$

where $\Omega = eB_0/m$, $a = E_0/\Omega B_0$ and $b = (uB_0 - E_0)/\Omega B_0$. Use computer assistance to plot typical paths of the electron for the cases $a < b$, $a = b$ and $a > b$. [The general path is called a *trochoid*, which becomes a *cycloid* in the special case $a = b$. Cycloidal motion of electrons is used in the **magnetron** vacuum tube, which generates the microwaves in a microwave oven.]

Computer assisted problems

4.34 Complete Example 4.7 on the projectile with linear resistance by obtaining the quoted asymptotic formula for the range of the projectile.

4.35 Find a series approximation for the period of the simple pendulum, in powers of the angular amplitude α. Proceed as follows:

The exact period τ of the pendulum was found in Example 4.9 and is given by the integral (4.9). This integral is not suitable for expansion as it stands. However, if we write $\cos\theta - \cos\alpha = 2(\sin^2(\alpha/2) - \sin^2(\theta/2))$ and make the sneaky substitution $\sin(\theta/2) = \sin(\alpha/2)\sin\phi$, the formula for τ becomes

$$\tau = 4\left(\frac{b}{g}\right)^{1/2}\int_0^{\pi/2}\left(1 - \epsilon^2\sin^2\phi\right)^{-1/2} d\phi$$

where $\epsilon = \sin(\alpha/2)$. This new integrand is easy to expand as a power series in the variable ϵ and the limits of integration are now constants. Use computer assistance to expand the integrand to the required number of terms and then integrate term by term over the interval $[0, \pi/2]$. Finally re-expand as a power series in the variable α. The answer to two terms is given by equation (4.10), but it is just as easy to obtain any number of terms.

4.36 ***Baseball trajectory*** A baseball is struck with an initial speed of 45 m s^{-1} (just over 100 mph) at an elevation angle of 40°. Find its path and compare this with the corresponding path when air resistance is neglected. [A baseball has mass 0.30 kg and radius 3.5 cm. Assume the quadratic law of resistance.]

Show that the equation of motion can be written in the form

$$\frac{d\boldsymbol{v}}{dt} = -g\left(\boldsymbol{k} + \frac{\boldsymbol{v}|\boldsymbol{v}|}{V^2}\right),$$

where V is the terminal speed. Resolve this vector equation into two (coupled) scalar equations for v_x and v_z and perform a numerical solution. In this example, air resistance reduces the range by about 35%. It really is easier to hit a home run in Mile High stadium!

Chapter Five

Linear oscillations

and normal modes

KEY FEATURES

The key features of this chapter are the properties of **free undamped** oscillations, **free damped** oscillations, **driven** oscillations, and **coupled** oscillations.

Oscillations are a particularly important part of mechanics and indeed of physics as a whole. This is because of their widespread occurrence and the practical importance of oscillation problems. In this chapter we study the classical **linear theory** of oscillations, which is important for two reasons: (i) the linear theory usually gives a good approximation to the motion when the amplitude of the oscillations is small, and (ii) in the linear theory, most problems can be solved explicitly in closed form. The importance of this last fact should not be underestimated! We develop the theory in the context of the oscillations of a body attached to a spring, but the same equations apply to many different problems in mechanics and throughout physics.

In the course of this chapter we will need to solve linear second order ODEs with constant coefficients. For a description of the standard method of solution see Boyce & DiPrima [8].

5.1 BODY ON A SPRING

Suppose a body of mass m is attached to one end of a light spring. The other end of the spring is attached to a fixed point A on a smooth horizontal table, and the body slides on

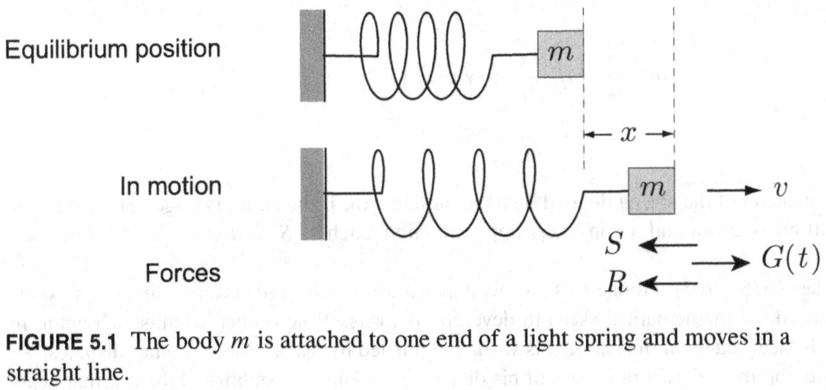

FIGURE 5.1 The body m is attached to one end of a light spring and moves in a straight line.

the table in a straight line through A. Let x be the displacement and v the velocity of the body at time t, as shown in Figure 5.1; note that x is measured from the *equilibrium position* of the body.

Consider now the forces acting on the body. When the spring is extended, it exerts a **restoring force** S in the opposite direction to the extension. Also, the body may encounter a **resistance force** R acting in the opposite direction to its velocity. Finally, there may be an external **driving force** $G(t)$ that is a specified function of the time. The equation of motion for the body is then

$$m\frac{dv}{dt} = -S - R + G(t). \tag{5.1}$$

The **restoring force** S is determined by the design of the spring and the extension x. For sufficiently *small strains*,* the relationship between S and x is approximately *linear*, that is,

$$S = \alpha x, \tag{5.2}$$

where α is a positive constant called the **spring constant** (or **strength**) of the spring. A powerful spring, such as those used in automobile suspensions, has a large value of α; the spring behind a doorbell has a small value of α. The formula (5.2) is called **Hooke's law**† and a spring that obeys Hooke's law exactly is called a **linear** spring.

The **resistance force** R depends on the physical process that is causing the resistance. For fluid resistance, the linear or quadratic resistance laws considered in Chapter 4 may be appropriate. However, neither of these laws represents the frictional force exerted by a rough table. In this chapter we assume the law of **linear resistance**

$$R = \beta v, \tag{5.3}$$

where β is a positive constant called the **resistance constant**; it is a measure of the strength of the resistance. There is no point in disguising the fact that our major reason for assuming linear resistance is that (together with Hooke's law) it leads to a linear equation of motion that can be solved explicitly. However, it does give insight into the general effect of all resistances, and actually is appropriate when the resistance arises from slow viscous flow (automobile shock absorbers, for instance); it is also appropriate in the electric circuit analogue, where it is equivalent to Ohm's law.

With **Hooke's law** and **linear resistance**, the equation of motion (5.1) for the body becomes

$$m\frac{d^2x}{dt^2} + \beta\frac{dx}{dt} + \alpha x = G(t),$$

---

* The strain is the extension of the spring divided by its natural length. If the strain is large, then the linear approximation will break down and a non-linear approximation, such as $S = ax + bx^3$ must be used instead.

† After Robert Hooke (1635–1703). Hooke was an excellent scientist, full of ideas and a first class experimenter, but he lacked the mathematical skills to develop his ideas. When other scientists (Newton in particular) did so, he accused them of stealing his work and this led to a succession of bitter disputes. So that his rivals could not immediately make use of his discovery, Hooke first published the law that bears his name as an anagram on the Latin phrase '*ut tensio, sic vis*' (as the extension, so the force).

where α is the spring constant, β is the resistance constant and $G(t)$ is the prescribed driving force. *This is a second order, linear ODE with constant coefficients for the unknown displacement* $x(t)$. We could go ahead with the solution of this equation as it stands, but the algebra is made much easier by introducing two new constants Ω and K (instead of α and β) defined by the relations

$$\alpha = m\Omega^2 \qquad\qquad \beta = 2mK.$$

The equation of motion for the body then becomes

$$\boxed{\frac{d^2x}{dt^2} + 2K\frac{dx}{dt} + \Omega^2 x = F(t)} \tag{5.4}$$

where $F(t) = G(t)/m$, the *driving force per unit mass*. This is the standard form of the **equation of motion** for the body. Any system that leads to an equation of this form is called a **damped*** **linear oscillator**. When the force $F(t)$ is absent, the oscillations are said to be **free**; when it is present, the oscillations are said to be **driven**.

5.2 CLASSICAL SIMPLE HARMONIC MOTION

A linear oscillator that is both **undamped** and **undriven** is called a **classical linear oscillator**. This is the simplest case, but arguably the most important system in physics! The equation (5.4) reduces to

$$\frac{d^2x}{dt^2} + \Omega^2 x = 0, \tag{5.5}$$

which, because of the solutions we are about to obtain, is called the **SHM equation**.

Solution procedure

Seek solutions of the form $x = e^{\lambda t}$. Then λ must satisfy the equation

$$\lambda^2 + \Omega^2 = 0,$$

which gives $\lambda = \pm i\Omega$. We have thus found the pair of complex solutions

$$x = e^{\pm i\Omega t},$$

which form a basis for the space of complex solutions. The real and imaginary parts of the first complex solution are

$$x = \begin{cases} \cos\Omega t \\ \sin\Omega t \end{cases}$$

* Damping is another term for resistance. Indeed, automobile shock absorbers are sometimes called dampers.

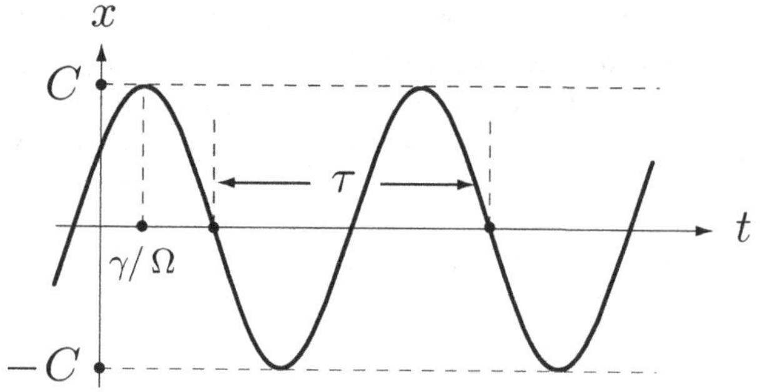

FIGURE 5.2 Classical simple harmonic motion $x = C\cos(\Omega t - \gamma)$.

and these functions form a basis for the space of real solutions. The **general real solution** of the SHM equation is therefore

$$x = A\cos\Omega t + B\sin\Omega t, \tag{5.6}$$

where A and B are real arbitrary constants. This general solution can be written in the alternative form*

$$x = C\cos(\Omega t - \gamma), \tag{5.7}$$

where C and γ are real arbitrary constants with $C > 0$.

General form of the motion

The general form of the motion is most easily deduced from the form (5.7) and is shown in Figure 5.2. This is called **simple harmonic motion (SHM)**. The body makes infinitely many oscillations of constant **amplitude** C; the constant γ is simply a 'phase factor' which shifts the whole graph by γ/Ω in the t-direction. Since the cosine function repeats itself when the argument Ωt increases by 2π, it follows that the **period** of the oscillations is given by

$$\tau = \frac{2\pi}{\Omega}. \tag{5.8}$$

The quantity Ω, which is related to the frequency ν by $\Omega = 2\pi\nu$, is called the **angular frequency** of the oscillations.

Example 5.1 *An initial value problem for classical SHM*

A body of mass m is suspended from a fixed point by a light spring and can move under uniform gravity. In equilibrium, the spring is found to be extended by a distance b. Find the period of vertical oscillations of the body about this equilibrium position. [Assume small strains.]

* This transformation is based on the result from trigonometry that $a\cos\theta + b\sin\theta$ can always be written in the form $c\cos(\theta - \gamma)$, where $c = (a^2 + b^2)^{1/2}$ and $\tan\gamma = b/a$.

The body is hanging in its equilibrium position when it receives a sudden blow which projects it upwards with speed u. Find the subsequent motion.

Solution

When the spring is subjected to a constant force of magnitude mg, the extension is b. Hence α, the strength of the spring, is given by $\alpha = mg/b$.

Let z be the *downwards* displacement of the body from its equilibrium position. Then the extension of the spring is $b + z$ and the restoring force is $\alpha(b + z) = g(b + z)/b$. The equation of motion for the body is therefore

$$m\frac{d^2z}{dt^2} = mg - \frac{mg(b+z)}{b}$$

that is

$$\frac{d^2z}{dt^2} + \left(\frac{g}{b}\right) z = 0.$$

This is the **SHM equation** with $\Omega^2 = g/b$. It follows that the **period** τ of vertical oscillations about the equilibrium position is given by

$$\tau = \frac{2\pi}{\Omega} = 2\pi\left(\frac{b}{g}\right)^{1/2}.$$

In the **initial value problem**, the subsequent motion must have the form

$$x = A\cos\Omega t + B\sin\Omega t,$$

where $\Omega = (g/b)^{1/2}$. The initial condition $x = 0$ when $t = 0$ shows that $A = 0$ and the initial condition $\dot{x} = -u$ when $t = 0$ then gives $\Omega B = -u$, that is, $B = -u/\Omega$. The subsequent motion is therefore

$$x = -\frac{u}{\Omega}\sin\Omega t,$$

where $\Omega = (g/b)^{1/2}$. ■

5.3 DAMPED SIMPLE HARMONIC MOTION

When **damping** is present but there is no external force, the general equation (5.4) reduces to

$$\frac{d^2x}{dt^2} + 2K\frac{dx}{dt} + \Omega^2 x = 0, \tag{5.9}$$

the **damped SHM equation.**

The solution procedure is the same as in the last section. Seek solutions of the form $x = e^{\lambda t}$. Then λ must satisfy the equation

$$\lambda^2 + 2K\lambda + \Omega^2 = 0,$$

that is

$$(\lambda + K)^2 = K^2 - \Omega^2.$$

We see that *different cases arise depending on whether* $K < \Omega$, $K = \Omega$ *or* $K > \Omega$. These cases give rise to different kinds of solution and must be treated separately.

Under-damping (sub-critical damping): $K < \Omega$

In this case, we write the equation for λ in the form

$$(\lambda + K)^2 = -\Omega_D^2,$$

where $\Omega_D = (\Omega^2 - K^2)^{1/2}$, a *positive real number*. The λ values are then $\lambda = -K \pm i\Omega_D$. We have thus found the pair of complex solutions

$$x = e^{-Kt}e^{\pm i\Omega_D t},$$

which form a basis for the space of complex solutions. The real and imaginary parts of the first complex solution are

$$x = \begin{cases} e^{-Kt}\cos\Omega_D t \\ e^{-Kt}\sin\Omega_D t \end{cases}$$

and these functions form a basis for the space of real solutions. The **general real solution** of the damped SHM equation in this case is therefore

$$x = e^{-Kt}\left(A\cos\Omega_D t + B\sin\Omega_D t\right), \tag{5.10}$$

where A and B are real arbitrary constants. This general solution can be written in the alternative form

$$x = Ce^{-Kt}\cos(\Omega_D t - \gamma), \tag{5.11}$$

where C and γ are real arbitrary constants with $C > 0$.

General form of the motion

The general form of the motion is most easily deduced from the form (5.11) and is shown in Figure 5.3. This is called **under-damped SHM**. The body still executes infinitely many oscillations, but now they have *exponentially decaying amplitude* Ce^{-Kt}. Suppose the **period** τ of the oscillations is defined as shown in Figure 5.3.* The introduction of damping decreases the angular frequency of the oscillations from Ω to Ω_D, which *increases* the period of the oscillations from $2\pi/\Omega$ to

$$\tau = \frac{2\pi}{\Omega_D} = \frac{2\pi}{(\Omega^2 - K^2)^{1/2}}. \tag{5.12}$$

* The period might also be defined as the time interval between successive maxima of the function $x(t)$. Since these maxima do *not* occur at the points at which $x(t)$ touches the bounding curves, it is not obvious that this time interval is even a constant. However, it *is* a constant and has the same value as (5.12) (see Problem 5.5)

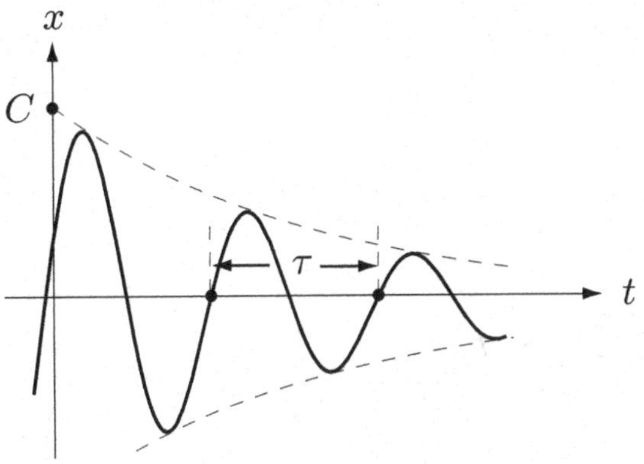

FIGURE 5.3 Under-damped simple harmonic motion $x = Ce^{-Kt}\cos(\Omega_D t - \gamma)$.

Over-damping (super-critical damping): $K > \Omega$

In this case, we write the equation for λ in the form

$$(\lambda + K)^2 = \delta^2,$$

where $\delta = (K^2 - \Omega^2)^{1/2}$, a *positive real number*. The λ values are then $\lambda = -k \pm \delta$, which are now real. We have thus found the pair of real solutions

$$x = e^{-Kt}e^{\pm\delta t},$$

which form a basis for the space of real solutions. The **general real solution** of the damped SHM equation in this case is therefore

$$x = e^{-Kt}\left(Ae^{\delta t} + Be^{-\delta t}\right), \tag{5.13}$$

where A and B are real arbitrary constants.

General form of the motion

Three typical forms for the motion are shown in Figure 5.4. This is called **over-damped SHM**. Somewhat surprisingly, *the body does not oscillate at all*. For example, if the body is released from rest, then it simply drifts back towards the equilibrium position. On the other hand, if the body is projected towards the equilibrium position with sufficient speed, then it passes the equilibrium position once and then drifts back towards it from the other side.

Critical damping: $K = \Omega$

The case of critical damping is solved in Problem 5.6. Qualitatively, the motions look like those in Figure 5.4.

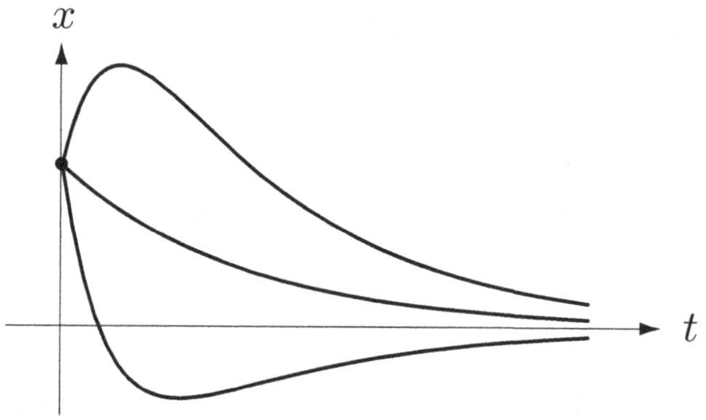

FIGURE 5.4 Three typical cases of over-damped simple harmonic motion.

5.4 DRIVEN (FORCED) MOTION

We now include the effect of an external **driving force** $G(t)$ which we suppose to be a *given* function of the time. In the case of a body suspended by a spring, we could apply such a force directly, but, in practice, the external 'force' often arises indirectly by virtue of the suspension point being made to oscillate in some prescribed way. The seismograph described in the next section is an instance of this.

Whatever the origin of the driving force, the **governing equation** for driven motion is (5.4), namely

$$\frac{d^2x}{dt^2} + 2K\frac{dx}{dt} + \Omega^2 x = F(t), \tag{5.14}$$

where $2mK$ is the damping constant, $m\Omega^2$ is the spring constant and $mF(t)$ is the driving force. Since this equation is linear and inhomogeneous, its general solution is the sum of (i) the general solution of the corresponding homogeneous equation (5.9) (the complementary function) and (ii) *any* particular solution of the inhomogeneous equation (5.14) (the particular integral). The complementary function has already been found in the last section, and it remains to find the particular integral for interesting choices of $F(t)$. Actually there is a (rather complicated) formula for a particular integral of this equation for *any choice* of the driving force $mF(t)$. However, the most important case by far is that of **time harmonic** forcing and, in this case, it is easier to find a particular integral directly. Time harmonic forcing is the case in which

$$F(t) = F_0 \cos pt, \tag{5.15}$$

where F_0 and p are positive constants; mF_0 is the amplitude of the applied force and p is its angular frequency.

Solution procedure

We first replace the forcing term $F_0 \cos pt$ by its complex counterpart $F_0 e^{ipt}$. This gives the complex equation

$$\frac{d^2x}{dt^2} + 2K\frac{dx}{dt} + \Omega^2 x = F_0 e^{ipt}. \tag{5.16}$$

We then seek a particular integral of this complex equation in the form

$$x = ce^{ipt}, \tag{5.17}$$

where c is a complex constant called the **complex amplitude**. On substituting (5.17) into equation (5.16) we find that

$$c = \frac{F_0}{\Omega^2 - p^2 + 2iKp}, \tag{5.18}$$

so that the complex function

$$\frac{F_0 e^{ipt}}{\Omega^2 - p^2 + 2iKp} \tag{5.19}$$

is a particular integral of the complex equation (5.16). A particular integral of the real equation (5.14) is then given by the real part of the complex expression (5.19). It follows that a **particular integral** of equation (5.14) is given by

$$x^D = a\cos(pt - \gamma),$$

where $a = |c|$ and $\gamma = -\arg c$. This particular integral, which is also time harmonic with the same frequency as the applied force, is called the **driven response** of the oscillator to the force $mF_0 \cos pt$; a is the **amplitude** of the driven response and γ $(0 < \gamma \le \pi)$ is the **phase angle** by which the response lags behind the force. From the expression (5.18) for c, it follows that

$$a = \frac{F_0}{\left((\Omega^2 - p^2)^2 + 4K^2p^2\right)^{1/2}}, \qquad \tan\gamma = \frac{2Kp}{\Omega^2 - p^2}. \tag{5.20}$$

The **general solution** of equation (5.14) therefore has the form

$$x = a\cos(pt - \gamma) + x^{CF}, \tag{5.21}$$

where x^{CF} is the complementary function, that is, the general solution of the corresponding *undriven* problem.

The undriven problem has already been solved in the last section. The solution took three different forms depending on whether the damping was supercritical, critical or subcritical. However, all these forms have one feature in common, that is, *they all decay to zero with increasing time*. For this reason, the complementary function for this equation is often called the **transient response** of the oscillator. Any solution of equation (5.21) is therefore the sum of the driven response x^D (which persists) and a transient response x^{CF} (which dies away). Thus, *no matter what the initial conditions, after a sufficiently long time we are left with just the driven response*. In many problems, the transient response can be disregarded, but it must be included if initial conditions are to be satisfied.

Example 5.2 ***An initial value problem for driven motion***

The equation of motion of a certain driven damped oscillator is

$$\frac{d^2x}{dt^2} + 3\frac{dx}{dt} + 2x = 10\cos t$$

and initially the particle is at rest at the origin. Find the subsequent motion.

Solution

First we find the **driven response** x^D. The complex counterpart of the equation of motion is

$$\frac{d^2x}{dt^2} + 3\frac{dx}{dt} + 2x = 10e^{it}$$

and we seek a solution of this equation of the form $x = ce^{it}$. On substituting in, we find that

$$c = \frac{10}{1+3i} = 1 - 3i.$$

It follows that the **driven response** x^D is given by

$$x^D = \Re\left[(1-3i)e^{it}\right] = \cos t + 3\sin t.$$

Now for the **complementary function** x^{CF}. This is the general solution of the corresponding undriven equation

$$\frac{d^2x}{dt^2} + 3\frac{dx}{dt} + 2x = 0,$$

which is easily found to be

$$x = Ae^{-t} + Be^{-2t},$$

where A and B are arbitrary constants. The **general solution** of the equation of motion is therefore

$$x = \cos t + 3\sin t + Ae^{-t} + Be^{-2t}.$$

It now remains to choose A and B so that the **initial conditions** are satisfied. The condition $x = 0$ when $t = 0$ implies that

$$0 = 1 + A + B,$$

and the condition $\dot{x} = 0$ when $t = 0$ implies that

$$0 = 3 - A - 2B.$$

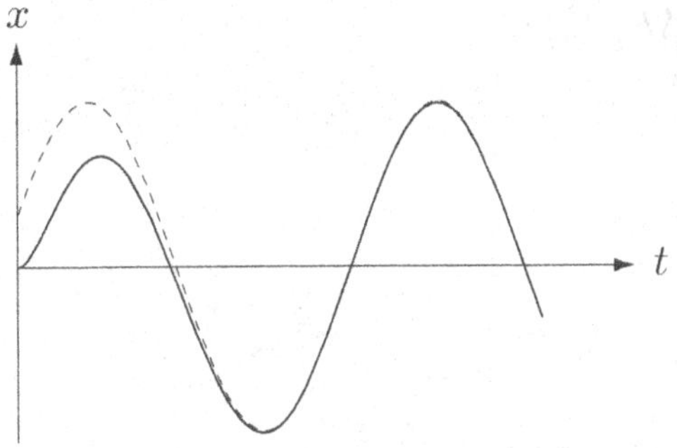

FIGURE 5.5 The solid curve is the actual response and the dashed curve the driven response only.

Solving these simultaneous equations gives $A = -5$ and $B = 4$. The **subsequent motion** of the oscillator is therefore given by

$$x = \underbrace{\cos t + 3\sin t}_{\text{driven response}} \underbrace{-5e^{-t} + 4e^{-2t}}_{\text{transient response}}.$$

This solution is shown in Figure 5.5 together with the driven response only. In this case, the transient response is insignificant after less than one cycle of the driving force. The **amplitude** of the driven response is $(1^2+3^2)^{1/2} = \sqrt{10}$ and the **phase lag** is $\tan^{-1}(3/1) \approx 72°$. ■

Resonance of an oscillating system

Consider the general formula

$$a = \frac{F_0}{\left((\Omega^2 - p^2)^2 + 4K^2p^2\right)^{1/2}}$$

for the amplitude a of the driven response to the force $mF_0 \cos pt$ (see equation (5.20)). Suppose that the amplitude of the applied force, the spring constant, and the resistance constant are held fixed and that the *angular frequency* p of the applied force is varied. Then a is a function of p only. Which value of p produces the largest driven response? Let

$$f(q) = (\Omega^2 - q)^2 + 4K^2 q.$$

Then, since $a = F_0/\sqrt{f(p^2)}$, we need only find the minimum point of the function $f(q)$ lying in $q > 0$. Now

$$f'(q) = -2(\Omega^2 - q) + 4K^2 = 2\left(q - (\Omega^2 - 2K^2)\right)$$

so that $f(q)$ decreases for $q < \Omega^2 - 2K^2$ and increases for $q > \Omega^2 - 2K^2$. Hence $f(q)$ has a unique minimum point at $q = \Omega^2 - 2K^2$. Two cases arise depending on whether this value is positive or not.

Case **1**. When $\Omega^2 > 2K^2$, the minimum point $q = \Omega^2 - 2K^2$ is positive and a has its maximum value when $p = p^R$, where

$$p^R = (\Omega^2 - 2K^2)^{1/2}.$$

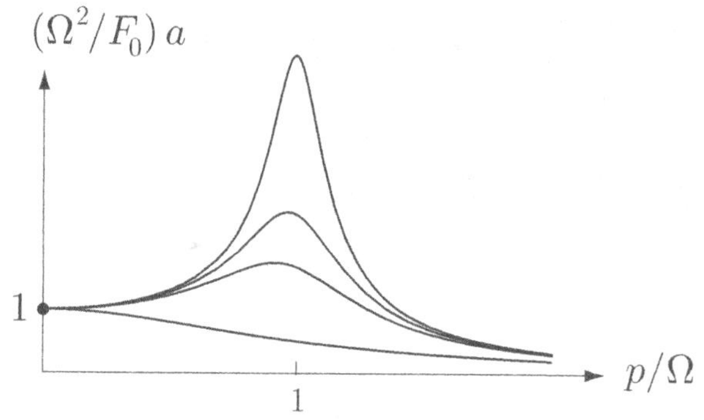

FIGURE 5.6 The dimensionless amplitude $(F_0/\Omega^2)a$ against the dimensionless driving frequency p/Ω for (from the top) $K/\Omega = 0.1, 0.2, 0.3, 1$.

The angular frequency p^R is called the **resonant frequency** of the oscillator. The value of a at the resonant frequency is

$$a_{\max} = \frac{F_0}{2K(\Omega^2 - K^2)^{1/2}}.$$

Case **2**. When $\Omega^2 \le 2K^2$, a is a decreasing function of p for $p > 0$ so that a has *no maximum point*.

These results are illustrated in Figure 5.6. They are an example of the general physical phenomenon known as **resonance**, which can be loosely stated as follows:

The phenomenon of resonance

Suppose that, in the absence of damping, a physical system can perform free oscillations with angular frequency Ω. Then a driving force with angular frequency p will induce a large response in the system when p is close to Ω, providing that the damping is not too large.

This principle does not just apply to the mechanical systems we study here. It is a general physical principle that also applies, for example, to the oscillations of electric currents in circuits and to the quantum mechanical oscillations of atoms.

Note that the resonant frequency p^R is always less than Ω, but is *close* to Ω when K/Ω is small. The height of the resonance peak, $a_{\max}$, is given approximately by

$$a_{\max} \sim \frac{F_0}{2\Omega^2}\left(\frac{\Omega}{K}\right)^{-1}$$

in the limit in which K/Ω is small; $a_{\max}$ therefore tends to infinity in this limit. In the same limit, the width of the resonance peak is directly proportional to K/Ω and consequently tends to zero.

General periodic driving force

The method we have developed for the time harmonic driving force can be extended to any periodic driving force $mF(t)$. A function $f(t)$ is said to be **periodic** with period τ if the values taken by f in any interval of length τ are then repeated in the next interval of length τ. An example is the 'square wave' function shown in Figure 5.7. The solution method requires that $F(t)$ be expanded as a **Fourier series.*** A textbook on mechanics is not the place to develop the theory of Fourier series. Instead we will simply quote the essential results and then give an example of how the method works. To keep the algebra as simple as possible, we will suppose that the driving force has period 2π.†

Fourier's Theorem

Fourier's theorem states that any function $f(t)$ that is periodic with period 2π can be expanded as a **Fourier series** in the form

$$f(t) = \tfrac{1}{2}a_0 + \sum_{n=1}^{\infty} a_n \cos nt + b_n \sin nt, \tag{5.22}$$

where the **Fourier coefficients** $\{a_n\}$ and $\{b_n\}$ are given by the formulae

$$a_n = \frac{1}{\pi}\int_{-\pi}^{\pi} f(t)\cos nt\, dt, \qquad b_n = \frac{1}{\pi}\int_{-\pi}^{\pi} f(t)\sin nt\, dt. \tag{5.23}$$

What this means is that *any* function $f(t)$ with period 2π can be expressed as a sum of *time harmonic* terms, each of which has period 2π. In order to find the driven response of the oscillator when the force $mF(t)$ is applied, we first expand $F(t)$ in a Fourier series. We then find the driven response that would be induced by each of the terms of this Fourier series applied separately, and then simply add these responses together. The method depends on the equation of motion being *linear*.

Example 5.3 *Periodic non-harmonic driving force*

Find the driven response of the damped linear oscillator

$$\frac{d^2x}{dt^2} + 2K\frac{dx}{dt} + \Omega^2 x = F(t)$$

for the case in which $F(t)$ is periodic with period 2π and takes the values

$$F(t) = \begin{cases} F_0 & (0 < t < \pi), \\ -F_0 & (\pi < t < 2\pi), \end{cases}$$

* After Jean Baptiste Joseph Fourier 1768–1830. The memoir in which he developed the theory of trigonometric series '*On the Propagation of Heat in Solid Bodies*' was submitted for the mathematics prize of the Paris Institute in 1811; the judges included such luminaries as Lagrange, Laplace and Legendre. They awarded Fourier the prize but griped about his lack of mathematical rigour.

† The general case can be reduced to this one by a scaling of the unit of time.

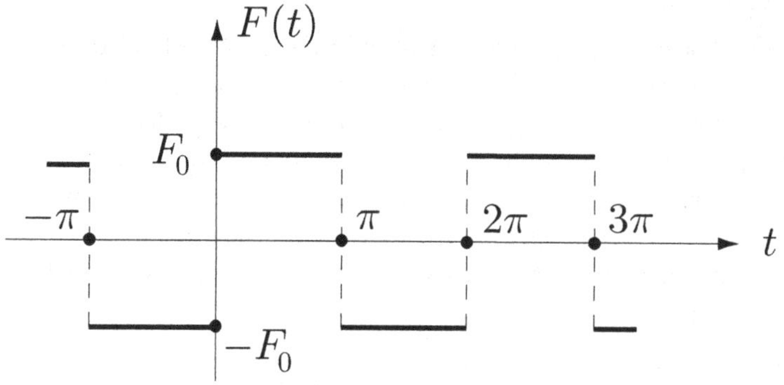

FIGURE 5.7 The 'square wave' input function $F(t)$ is periodic with period 2π. Its value alternates between $\pm F_0$.

in the interval $0 < t < 2\pi$. This function* is shown in Figure 5.7.

Solution

The first step is to find the Fourier series of the function $F(t)$. From the formula (5.23), the coefficient a_n is given by

$$\begin{aligned} a_n &= \frac{1}{\pi}\int_{-\pi}^{\pi} F(t)\cos nt\,dt = \frac{1}{\pi}\int_{-\pi}^{0} (-F_0)\cos nt\,dt + \frac{1}{\pi}\int_{0}^{\pi} (+F_0)\cos nt\,dt \\ &= 0, \end{aligned}$$

since both integrals are zero for $n \ge 1$ and are equal and opposite when $n = 0$. In the same way,

$$\begin{aligned} b_n &= \frac{1}{\pi}\int_{-\pi}^{\pi} F(t)\sin nt\,dt = \frac{1}{\pi}\int_{-\pi}^{0} (-F_0)\sin nt\,dt + \frac{1}{\pi}\int_{0}^{\pi} (+F_0)\sin nt\,dt \\ &= \frac{2F_0}{\pi}\int_0^{\pi} \sin nt\,dt, \end{aligned}$$

since this time the two integrals are equal. Hence

$$\begin{aligned} b_n &= \frac{2F_0}{\pi}\left[\frac{-\cos nt}{n}\right]_0^{\pi} = \frac{2F_0}{\pi}\left(\frac{1-\cos n\pi}{n}\right) \\ &= \frac{2F_0}{\pi}\left(\frac{1-(-1)^n}{n}\right). \end{aligned}$$

Hence the **Fourier series** of the function $F(t)$ is

$$F(t) = \sum_{n=1}^{\infty} \frac{2F_0}{\pi}\left(\frac{1-(-1)^n}{n}\right)\sin nt.$$

* This function is the mechanical equivalent of a 'square wave input' in electric circuit theory.

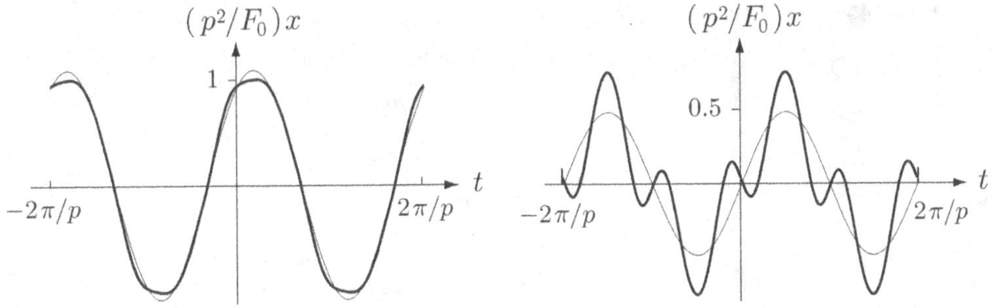

FIGURE 5.8 Driven response of a damped oscillator to the alternating constant force $\pm mF_0$ with angular frequency p: **Left** $\Omega/p = 1.5$, $K/p = 1$. **Right** $\Omega/p = 2.5$, $K/p = 0.1$. The light graphs show the first term of the expansion series.

The next step is to find the driven response of the oscillator to the force $m(b_n \sin nt)$, that is, the particular integral of the equation

$$\frac{d^2x}{dt^2} + 2K\frac{dx}{dt} + \Omega^2 x = b_n \sin nt. \tag{5.24}$$

The complex counterpart of this equation is

$$\frac{d^2x}{dt^2} + 2K\frac{dx}{dt} + \Omega^2 x = b_n e^{int}$$

for which the particular integral is ce^{int}, where the complex amplitude c is given by

$$c = \frac{b_n}{\Omega^2 - n^2 + 2iK}.$$

The particular integral of the real equation (5.24) is then given by

$$\Im\left(\frac{b_n e^{int}}{\Omega^2 - n^2 + 2iKn}\right) = b_n\left(\frac{(\Omega^2 - n^2)\sin nt + 2Kn\cos nt}{(\Omega^2 - n^2)^2 + 4K^2n^2}\right).$$

Finally we add together these separate responses to find the **driven response** of the oscillator to the force $mF(t)$. On inserting the value of the coefficient b_n, this gives

$$x = \frac{2F_0}{\pi}\sum_{n=1}^{\infty}\left(\frac{1-(-1)^n}{n}\right)\left(\frac{(\Omega^2 - n^2)\sin nt + 2Kn\cos nt}{(\Omega^2 - n^2)^2 + 4K^2n^2}\right). \tag{5.25}$$

In order to deduce anything from this complicated formula, we must either sum the series numerically or approximate the formula in some way. When Ω and K are both small compared to the forcing frequency p, the series (5.25) converges quite quickly and can be approximated (to within a few percent) by the first term. Even when $\Omega/p = 1.5$ and $K/p = 1$, this is still a reasonable approximation (see Figure 5.8 (left)). However, for larger values of Ω/p, the higher harmonics in the Fourier expansion of $F(t)$ that have frequencies close to Ω produce large contributions (see Figure 5.8 (right)). In this case, the series (5.25) must be summed numerically. ■

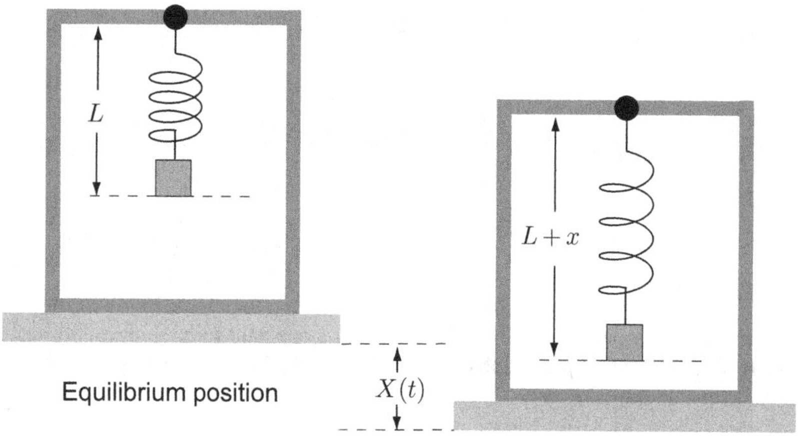

FIGURE 5.9 A simple seismograph for measuring vertical ground motion.

5.5 A SIMPLE SEISMOGRAPH

The seismograph is an instrument that measures the motion of the ground on which it stands. In real earthquakes, the ground motion will generally have both vertical and horizontal components, but, for simplicity, we describe here a device for measuring **vertical motion** only.

Our simple seismograph (see Figure 5.9) consists of a mass which is suspended from a rigid support by a spring; the motion of the mass relative to the support is resisted by a damper. The support is attached to the ground so that the suspension point has the same motion as the ground below it. This motion sets the suspended mass moving and the resulting spring extension is measured as a function of the time. Can we deduce what the ground motion was?

Suppose the ground (and therefore the support) has downward displacement $X(t)$ at time t and that the extension $x(t)$ of the spring is measured from its equilibrium length. Then the *displacement* of the mass is $x + X$, relative to an inertial frame. The equation of motion (5.9) is therefore modified to become

$$m\frac{d^2(x+X)}{dt^2} = -(2mK)\frac{dx}{dt} - (m\Omega^2)x,$$

that is,

$$\frac{d^2x}{dt^2} + 2K\frac{dx}{dt} + \Omega^2 x = -\frac{d^2X}{dt^2}.$$

This means *that the motion of the body relative to the moving support is the same as if the support were fixed and the external driving force* $-m(d^2X/dt^2)$ *were applied to the body.*

First consider the driven response of our seismograph to a train of harmonic waves with amplitude A and angular frequency p, that is,

$$X = A\cos pt.$$

The equation of motion for the spring extension x is then

$$\frac{d^2x}{dt^2} + 2K\frac{dx}{dt} + \Omega^2 x = Ap^2\cos pt.$$

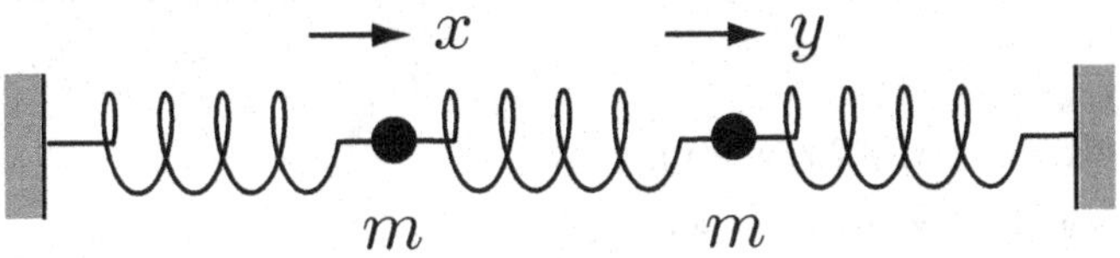

FIGURE 5.10 Two particles are connected between three springs and perform longitudinal oscillations.

The complex amplitude of the driven motion is

$$c = \frac{p^2 A}{-p^2 + 2iKp + \Omega^2},$$

and the real driven motion is

$$x = a\cos(pt - \gamma),$$

where

$$a = |c| = \frac{A}{\left|-1 + 2i(K/p) + (\Omega/p)^2\right|}. \tag{5.26}$$

Thus, providing that the spring and resistance constants are accurately known, the angular frequency p and amplitude A of the incident wave train can be deduced.

In practice, things may not be so simple. In particular, the incident wave train may be a mixture of harmonic waves with different amplitudes and frequencies, and these are not easily disentangled. However, if K and Ω are chosen so that K/p and Ω/p are small compared with unity (for all likely values of p), then $c = -A$ and $X = -x$ approximately. Thus, in this case, the record for $x(t)$ is simply the negative of the ground motion $X(t)$.* Since this result is independent of the incident frequency, it should also apply to complicated inputs such as a pulse of waves.

5.6 COUPLED OSCILLATIONS AND NORMAL MODES

Interesting new effects occur when two or more oscillators are coupled together. Figure 5.10 shows a typical case in which two bodies are connected between three springs and the motion takes place in a straight line. We restrict ourselves here to the classical theory in which the *restoring forces are linear and damping is absent*. If the springs are non-linear, then the displacements of the particles must be small enough so that the linear approximation is adequate.

Let x and y be the displacements of the two bodies from their respective equilibrium positions at time t; because two coordinates are needed to specify the configuration, the system is said to have two *degrees of freedom*. Then, at time t, the extensions of the three springs are x, $y - x$ and $-y$ respectively. Suppose that the strengths of the three springs are α, 2α and

* What is actually happening is that the mass is hardly moving at all (relative to an inertial frame).

4α respectively. Then the three restoring forces are αx, $2\alpha(y-x)$, $-4\alpha y$ and the equations of motion for the two bodies are

$$m\ddot{x} = -\alpha x + 2\alpha(y-x),$$
$$m\ddot{y} = -2\alpha(y-x) - 4\alpha y,$$

which can be written in the form

$$\begin{aligned} \ddot{x} + 3n^2 x - 2n^2 y &= 0, \\ \ddot{y} - 2n^2 x + 6n^2 y &= 0, \end{aligned} \tag{5.27}$$

where the positive constant n is defined by $n^2 = \alpha/m$. These are the **governing equations** for the motion. They are a *pair of simultaneous second order homogeneous linear ODEs* with constant coefficients. The equations are **coupled** in the sense that both unknown functions appear in each equation; thus neither equation can be solved on its own.

The solution procedure: normal modes

The solution procedure is simply an extension of the usual method for finding the complementary function for a single homogeneous linear ODE with constant coefficients. However, rather than seek solutions in exponential form, it is simpler to seek solutions directly in the trigonometric form

$$\begin{aligned} x &= A\cos(\omega t - \gamma), \\ y &= B\cos(\omega t - \gamma), \end{aligned} \tag{5.28}$$

where A, B, ω and γ are constants. A solution of the governing equations (5.27) that has the form (5.28) is called a **normal mode** of the oscillating system. In a normal mode, all the coordinates that specify the configuration of the system vary harmonically in time with the *same frequency* and the *same phase*; however, they generally have *different amplitudes*. On substituting the normal mode form (5.28) into the governing equations (5.27), we obtain

$$-\omega^2 A\cos(\omega t-\gamma) + 3n^2 A\cos(\omega t-\gamma) - 2n^2 B\cos(\omega t-\gamma) = 0,$$
$$-\omega^2 B\cos(\omega t-\gamma) - 2n^2 A\cos(\omega t-\gamma) + 6n^2 B\cos(\omega t-\gamma) = 0,$$

which simplifies to give

$$\begin{aligned} (3n^2-\omega^2)A - 2n^2 B &= 0, \\ -2n^2 A + (6n^2-\omega^2)B &= 0, \end{aligned} \tag{5.29}$$

a pair of *simultaneous linear algebraic equations* for the amplitudes A and B. Thus a normal mode will exist if we can find constants A, B and ω so that the equations (5.29) are satisfied. Since the equations are homogeneous, they always have the *trivial solution* $A = B = 0$, whatever the value of ω. However, the trivial solution corresponds to the *equilibrium solution* $x = y = 0$ of the governing equations (5.27), which is not a motion at all. We therefore require the equations (5.29) to have a **non-trivial solution** for A, B. There is a simple condition that this should be so, namely that the determinant of the system of equations should be zero, that is,

$$\det\begin{pmatrix} 3n^2-\omega^2 & -2n^2 \\ -2n^2 & 6n^2-\omega^2 \end{pmatrix} = 0. \tag{5.30}$$

On simplification, this gives the condition

$$\omega^4 - 9n^2\omega^2 + 14n^4 = 0, \tag{5.31}$$

a quadratic equation in the variable ω^2. If this equation has *real positive* roots ω_1^2, ω_2^2, then, for *each* of these values, the linear equations (5.29) will have a non-trivial solution for the amplitudes A, B. In the present case, the equation (5.31) factorises and the roots are found to be

$$\omega_1^2 = 2n^2, \qquad \omega_2^2 = 7n^2. \tag{5.32}$$

Hence there are **two normal modes** with (angular) frequencies $\sqrt{2}n$ and $\sqrt{7}n$ respectively. These frequencies are known as the **normal frequencies** of the oscillating system.

Slow mode: In the slow mode we have $\omega^2 = 2n^2$ so that the linear equations (5.29) become

$$\begin{aligned} n^2A - 2n^2B &= 0, \\ -2n^2A + 4n^2B &= 0. \end{aligned}$$

These two equations are each equivalent to the single equation $A = 2B$. This is to be expected since, if the equations were linearly independent, then there would be no non-trivial solution for A and B. We have thus found a family of non-trivial solutions $A = 2\delta$, $B = \delta$, where δ can take any (non-zero) value. Thus the *amplitude of the normal mode is not uniquely determined*; this happens because the governing ODEs are linear and homogeneous. The **slow normal mode** therefore has the form

$$\begin{aligned} x &= 2\delta\cos(\sqrt{2}\,nt - \gamma), \\ y &= \delta\cos(\sqrt{2}\,nt - \gamma), \end{aligned} \tag{5.33}$$

where the amplitude factor δ and phase factor γ can take any values. We see that, in the slow mode, the two bodies always move in the *same* direction with the body on the left having twice the amplitude as the body on the right.

Fast mode: In the fast mode we have $\omega^2 = 7n^2$ and, by following the same procedure, we find that the form of the **fast normal mode** is

$$\begin{aligned} x &= \delta\cos(\sqrt{7}\,nt - \gamma), \\ y &= -2\delta\cos(\sqrt{7}\,nt - \gamma), \end{aligned} \tag{5.34}$$

where the amplitude factor δ and phase factor γ can take any values. We see that, in the fast mode, the two bodies always move in *opposite* directions with the body on the right having twice the amplitude as the body on the left.

The general motion

Since the governing equations (5.27) are linear and homogeneous, a sum of normal mode solutions is also a solution. Indeed, the general solution can be written as a sum of normal

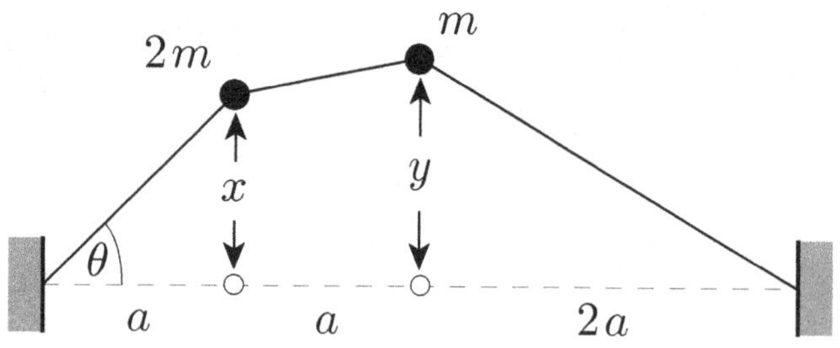

FIGURE 5.11 The two particles are attached to a light stretched string and perform *small* transverse oscillations. The displacements are shown to be large for clarity.

modes. Consider the expression

$$\begin{aligned} x &= 2\delta_1 \cos(\sqrt{2}\,nt - \gamma_1) + \delta_2 \cos(\sqrt{7}\,nt - \gamma_2), \\ y &= \delta_1 \cos(\sqrt{2}\,nt - \gamma_1) - 2\delta_2 \cos(\sqrt{7}\,nt - \gamma_2). \end{aligned} \tag{5.35}$$

This is simply a sum of the first normal mode (with amplitude factor δ_1 and phase factor γ_1) and the second normal mode (with amplitude factor δ_2 and phase factor γ_2). Since it is possible to choose these *four* arbitrary constants so that x, y, $\dot{x}$, $\dot{y}$ take any set of assigned values when $t = 0$, this must be the **general solution** of the governing equations (5.27).

Question *Periodicity of the general motion*

Is the general motion periodic?

Answer

The general motion is a sum of normal mode motions with periods τ_1, τ_2 respectively. This sum will be periodic with period τ if (and only if) τ is an integer multiple of both τ_1 and τ_2, that is, if τ_1/τ_2 is a *rational* number. (In this case, the periods are said to be *commensurate*.) This in turn requires that ω_1/ω_2 is a rational number. In the present case, $\omega_1/\omega_2 = (2/7)^{1/2}$, which is irrational. The general motion is therefore **not periodic** in this case. ■

We conclude by solving another typical normal mode problem.

Example 5.4 *Small transverse oscillations*

Two particles P and Q, of masses $2m$ and m, are secured to a light string that is stretched to tension T_0 between two fixed supports, as shown in Figure 5.11. The particles undergo *small transverse oscillations* perpendicular to the equilibrium line of the string. Find the normal frequencies, the forms of the normal modes, and the general motion of this system. Is the general motion periodic?

Solution

First we need to make some simplifying assumptions.* We will assume that the transverse displacements x, y of the two particles are small compared with a; the three sections of the string then make small angles with the equilibrium line. We will also neglect any change in the tensions of the three sections of string.

The left section of string then has constant tension T_0. When the particle P is displaced, this tension force has the transverse component $-T_0 \sin\theta$, which acts as a restoring force on P; since θ is small, this component is approximately $-T_0 x/a$. Similar remarks apply to the other sections of string. The **equations of transverse motion** for P and Q are therefore

$$2m\ddot{x} = -\frac{T_0 x}{a} + \frac{T_0(y-x)}{a},$$
$$m\ddot{y} = -\frac{T_0(y-x)}{a} - \frac{T_0 y}{2a}.$$

which can be written in the form

$$2\ddot{x} + 2n^2 x - n^2 y = 0, \tag{5.36}$$
$$2\ddot{y} - 2n^2 x + 3n^2 y = 0, \tag{5.37}$$

where the positive constant n is defined by $n^2 = T_0/ma$.

These equations will have **normal mode** solutions of the form

$$x = A\cos(\omega t - \gamma),$$
$$y = B\cos(\omega t - \gamma),$$

when the simultaneous linear equations

$$\begin{aligned}(2n^2 - 2\omega^2)A - n^2 B \quad &= 0,\\ -2n^2 A + (3n^2 - 2\omega^2)B &= 0,\end{aligned} \tag{5.38}$$

have a non-trivial solution for the amplitudes A, B. The condition for this is

$$\det\begin{pmatrix} 2n^2 - 2\omega^2 & -n^2 \\ -n^2 & 3n^2 - 2\omega^2 \end{pmatrix} = 0. \tag{5.39}$$

On simplification, this gives

$$2\omega^4 - 5n^2\omega^2 + 2n^4 = 0, \tag{5.40}$$

a quadratic equation in the variable ω^2. This equation factorises and the roots are found to be

$$\omega_1^2 = \tfrac{1}{2}n^2, \qquad \omega_2^2 = 2n^2. \tag{5.41}$$

Hence there are **two normal modes** with **normal frequencies** $n/\sqrt{2}$ and $\sqrt{2}n$ respectively.

* These assumptions are consistent with the more complete treatment given in Chapter 15.

Slow mode: In the slow mode we have $\omega^2 = n^2/2$ so that the linear equations (5.38) become

$$n^2 A - n^2 B = 0,$$
$$-2n^2 A + 2n^2 B = 0.$$

These two equations are each equivalent to the single equation $A = B$ so that we have the family of non-trivial solutions $A = \delta$, $B = \delta$, where δ can take any (non-zero) value. The **slow normal mode** therefore has the form

$$\begin{aligned} x &= \delta \cos(nt/\sqrt{2} - \gamma), \\ y &= \delta \cos(nt/\sqrt{2} - \gamma), \end{aligned} \tag{5.42}$$

where the amplitude factor δ and phase factor γ can take any values. We see that, in the slow mode, the two particles always have the *same displacement*.

Fast mode: In the fast mode we have $\omega^2 = 2n^2$ and, by following the same procedure, we find that the form of the **fast normal mode** is

$$\begin{aligned} x &= \delta \cos(\sqrt{2}\,nt - \gamma), \\ y &= -2\delta \cos(\sqrt{2}\,nt - \gamma), \end{aligned} \tag{5.43}$$

where the amplitude factor δ and phase factor γ can take any values. We see that, in the fast mode, the two particles always move in *opposite* directions with Q having twice the amplitude of P.

The **general motion** is now the sum of the first normal mode (with amplitude factor δ_1 and phase factor γ_1) and the second normal mode (with amplitude factor δ_2 and phase factor γ_2). This gives

$$\begin{aligned} x &= \delta_1 \cos(nt/\sqrt{2} - \gamma_1) + \delta_2 \cos(\sqrt{2}\,nt - \gamma_2), \\ y &= \delta_1 \cos(nt/\sqrt{2} - \gamma_1) - 2\delta_2 \cos(\sqrt{2}\,nt - \gamma_2). \end{aligned} \tag{5.44}$$

For this system $\tau_1/\tau_2 = \omega_2/\omega_1 = 2$ so that the general motion is **periodic** with period $\tau_1 = 2\sqrt{2}\pi/n$. ■

Problems on Chapter 5

Answers and comments are at the end of the book.

Harder problems carry a star (∗).

Free linear oscillations

5.1 A certain oscillator satisfies the equation

$$\ddot{x} + 4x = 0.$$

Initially the particle is at the point $x = \sqrt{3}$ when it is projected towards the origin with speed 2. Show that, in the subsequent motion,

$$x = \sqrt{3}\cos 2t - \sin 2t.$$

Deduce the amplitude of the oscillations. How long does it take for the particle to first reach the origin?

5.2 When a body is suspended from a fixed point by a certain linear spring, the angular frequency of its vertical oscillations is found to be Ω_1. When a different linear spring is used, the oscillations have angular frequency Ω_2. Find the angular frequency of vertical oscillations when the two springs are used together (i) in parallel, and (ii) in series. Show that the first of these frequencies is at least twice the second.

5.3 A particle of mass m moves along the x-axis and is acted upon by the restoring force $-m(n^2 + k^2)x$ and the resistance force $-2mk\dot{x}$, where n, k are positive constants. If the particle is released from rest at $x = a$, show that, in the subsequent motion,

$$x = \frac{a}{n}e^{-kt}(n\cos nt + k\sin nt).$$

Find how far the particle travels before it next comes to rest.

5.4 An overdamped harmonic oscillator satisfies the equation

$$\ddot{x} + 10\dot{x} + 16x = 0.$$

At time $t = 0$ the particle is projected from the point $x = 1$ towards the origin with speed u. Find x in the subsequent motion.

Show that the particle will reach the origin at some later time t if

$$\frac{u - 2}{u - 8} = e^{6t}.$$

How large must u be so that the particle will pass through the origin?

5.5 A damped oscillator satisfies the equation

$$\ddot{x} + 2K\dot{x} + \Omega^2 x = 0$$

where K and Ω are positive constants with $K < \Omega$ (under-damping). At time $t = 0$ the particle is released from rest at the point $x = a$. Show that the subsequent motion is given by

$$x = ae^{-Kt}\left(\cos\Omega_D t + \frac{K}{\Omega_D}\sin\Omega_D t\right),$$

where $\Omega_D = (\Omega^2 - K^2)^{1/2}$.

Find all the turning points of the function $x(t)$ and show that the ratio of successive maximum values of x is $e^{-2\pi K/\Omega_D}$.

A certain damped oscillator has mass 10 kg, period 5 s and successive maximum values of its displacement are in the ratio 3 : 1. Find the values of the spring and damping constants α and β.

5.6 ***Critical damping*** Find the general solution of the damped SHM equation (5.9) for the special case of critical damping, that is, when $K = \Omega$. Show that, if the particle is initially released from rest at $x = a$, then the subsequent motion is given by

$$x = ae^{-\Omega t}(1 + \Omega t).$$

Sketch the graph of x against t.

5.7* ***Fastest decay*** The oscillations of a galvanometer satisfy the equation

$$\ddot{x} + 2K\dot{x} + \Omega^2 x = 0.$$

The galvanometer is released from rest with $x = a$ and we wish to bring the reading permanently within the interval $-\epsilon a \le x \le \epsilon a$ as quickly as possible, where ϵ is a small positive constant. What value of K should be chosen? One possibility is to choose a sub-critical value of K such that the first minimum point of $x(t)$ occurs when $x = -\epsilon a$. [Sketch the graph of $x(t)$ in this case.] Show that this can be achieved by setting the value of K to be

$$K = \Omega\left[1 + \left(\frac{\pi}{\ln(1/\epsilon)}\right)^2\right]^{-1/2}.$$

If K has this value, show that the time taken for x to reach its first minimum is approximately $\Omega^{-1}\ln(1/\epsilon)$ when ϵ is small.

5.8 A block of mass M is connected to a second block of mass m by a linear spring of natural length $8a$. When the system is in equilibrium with the first block on the floor, and with the spring and second block vertically above it, the length of the spring is $7a$. The upper block is then pressed down until the spring has half its natural length and is then released from rest. Show that the lower block will leave the floor if $M < 2m$. For the case in which $M = 3m/2$, find when the lower block leaves the floor.

Driven linear oscillations

5.9 A block of mass 2 kg is suspended from a fixed support by a spring of strength $2000\ \text{N m}^{-1}$. The block is subject to the vertical driving force $36\cos pt$ N. Given that the spring will yield if its extension exceeds 4 cm, find the range of frequencies that can safely be applied.

5.10 A driven oscillator satisfies the equation

$$\ddot{x} + \Omega^2 x = F_0 \cos[\Omega(1+\epsilon)t],$$

where ϵ is a positive constant. Show that the solution that satisfies the initial conditions $x = 0$ and $\dot{x} = 0$ when $t = 0$ is

$$x = \frac{F_0}{\epsilon(1+\frac{1}{2}\epsilon)\Omega^2}\sin\tfrac{1}{2}\epsilon\Omega t\,\sin\Omega(1+\tfrac{1}{2}\epsilon)t.$$

Sketch the graph of this solution for the case in which ϵ is small.

5.11 Figure 5.12 shows a simple model of a car moving with constant speed c along a gently undulating road with profile $h(x)$, where $h'(x)$ is small. The car is represented by a chassis

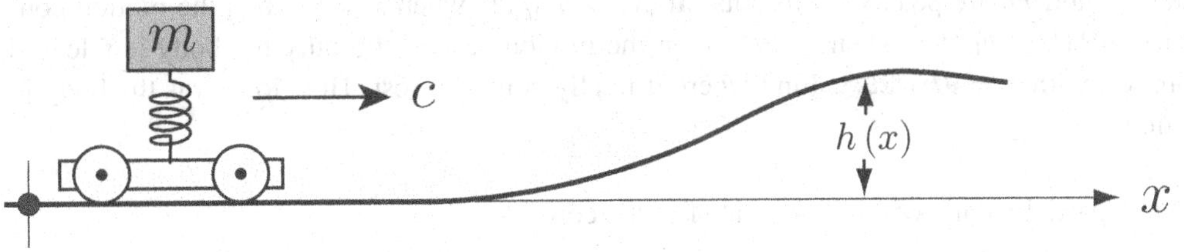

FIGURE 5.12 The car moves along a gently undulating road.

which keeps contact with the road, connected to an upper mass m by a spring and a damper. At time t the upper mass has displacement $y(t)$ above its equilibrium level. Show that, under suitable assumptions, y satisfies a differential equation of the form

$$\ddot{y} + 2K\dot{y} + \Omega^2 y = 2Kch'(ct) + \Omega^2 h(ct)$$

where K and Ω are positive constants.

Suppose that the profile of the road surface is given by $h(x) = h_0 \cos(px/c)$, where h_0 and p are positive constants. Find the amplitude a of the *driven* oscillations of the upper mass.

The vehicle designer adjusts the damper so that $K = \Omega$. Show that

$$a \le \frac{2}{\sqrt{3}} h_0,$$

whatever the values of the constants Ω and p.

5.12 ***Solution by Fourier series*** A driven oscillator satisfies the equation

$$\ddot{x} + 2K\dot{x} + \Omega^2 x = F(t),$$

where K and Ω are positive constants. Find the driven response of the oscillator to the saw tooth' input, that is, when $F(t)$ is given by

$$F(t) = F_0 t \qquad (-\pi < t < \pi)$$

and $F(t)$ is periodic with period 2π. [It is a good idea to sketch the graph of the function $F(t)$.]

Non-linear oscillations that are piecewise linear

5.13 A particle of mass m is connected to a fixed point O on a smooth horizontal table by a linear elastic string of natural length $2a$ and strength $m\Omega^2$. Initially the particle is released from rest at a point on the table whose distance from O is $3a$. Find the period of the resulting oscillations.

5.14 ***Coulomb friction*** The displacement x of a spring mounted body under the action of Coulomb friction satisfies the equation

$$\ddot{x} + \Omega^2 x = \begin{cases} -F_0 & \dot{x} > 0 \\ F_0 & \dot{x} < 0 \end{cases}$$

where Ω and F_0 are positive constants. If $|x| > F_0/\Omega^2$ when $\dot{x} = 0$, then the motion continues; if $|x| \le F_0/\Omega^2$ when $\dot{x} = 0$, then the motion ceases. Initially the body is released from rest with $x = 9F_0/2\Omega^2$. Find where it finally comes to rest. How long was the body in motion?

5.15 A partially damped oscillator satisfies the equation

$$\ddot{x} + 2\kappa\,\dot{x} + \Omega^2 x = 0,$$

where Ω is a positive constant and κ is given by

$$\kappa = \begin{cases} 0 & x < 0 \\ K & x > 0 \end{cases}$$

where K is a positive constant such that $K < \Omega$. Find the period of the oscillator and the ratio of successive maximum values of x.

Normal modes

5.16 A particle P of mass $3m$ is suspended from a fixed point O by a light linear spring with strength α. A second particle Q of mass $2m$ is in turn suspended from P by a second spring of the same strength. The system moves in the vertical straight line through O. Find the normal frequencies and the form of the normal modes for this system. Write down the form of the general motion.

5.17 Two particles P and Q, each of mass m, are secured at the points of trisection of a light string that is stretched to tension T_0 between two fixed supports a distance $3a$ apart. The particles undergo small *transverse* oscillations perpendicular to the equilibrium line of the string. Find the normal frequencies, the forms of the normal modes, and the general motion of this system. [Note that the forms of the modes could have been deduced from the symmetry of the system.] Is the general motion periodic?

5.18 A particle P of mass $3m$ is suspended from a fixed point O by a light inextensible string of length a. A second particle Q of mass m is in turn suspended from P by a second string of length a. The system moves in a vertical plane through O. Show that the linearised equations of motion for *small* oscillations near the downward vertical are

$$3\ddot{\theta} + 4n^2\theta - n^2\phi = 0,$$
$$\ddot{\theta} + \ddot{\phi} + n^2\phi = 0,$$

where θ and ϕ are the angles that the two strings make with the downward vertical, and $n^2 = g/a$. Find the normal frequencies and the forms of the normal modes for this system.

Chapter Six

Energy conservation

KEY FEATURES

The key features of this chapter are the **energy principle** for a particle, **conservative fields** of force, **potential energies** and **energy conservation**.

In this Chapter, we introduce the notion of **mechanical energy** and its **conservation**. Although energy methods are never indispensable* for the solution of problems, they do give a greater insight and allow many problems to be solved in a quick and elegant manner. Energy has a fundamental rôle in the Lagrangian and Hamiltonian formulations of mechanics. More generally, the notion of energy has been so widely extended that energy conservation has become the most pervasive and important principle in the whole of physics.

6.1 THE ENERGY PRINCIPLE

Suppose a particle P of mass m moves under the influence of a force $\boldsymbol{F}$. Then its equation of motion is

$$m\frac{d\boldsymbol{v}}{dt} = \boldsymbol{F}, \qquad (6.1)$$

where $\boldsymbol{v}$ is the velocity of P at time t. At this stage we place no restrictions on the force $\boldsymbol{F}$. It may depend on the position of P, the velocity of P, the time, or anything else; if more than one force is acting on P, then $\boldsymbol{F}$ means the *vector resultant* of these forces. On taking the scalar product of both sides of equation (6.1) with $\boldsymbol{v}$, we obtain the scalar equation

$$m\boldsymbol{v}\cdot\frac{d\boldsymbol{v}}{dt} = \boldsymbol{F}\cdot\boldsymbol{v}$$

and, since

$$m\boldsymbol{v}\cdot\frac{d\boldsymbol{v}}{dt} = \frac{d}{dt}\left(\tfrac{1}{2}m\boldsymbol{v}\cdot\boldsymbol{v}\right),$$

* Energy is never mentioned in the work of Newton!

this can be written in the form

$$\frac{dT}{dt} = \boldsymbol{F} \cdot \boldsymbol{v}, \tag{6.2}$$

where $T = \frac{1}{2}m\boldsymbol{v} \cdot \boldsymbol{v}$.

Definition 6.1 ***Kinetic energy*** *The scalar quantity* $T = \frac{1}{2}m\boldsymbol{v} \cdot \boldsymbol{v} = \frac{1}{2}m|\boldsymbol{v}|^2$ *is called the* ***kinetic energy*** *of the particle* P.

If we now integrate equation (6.2) over the time interval $[t_1, t_2]$, we obtain

$$\boxed{T_2 - T_1 = \int_{t_1}^{t_2} \boldsymbol{F} \cdot \boldsymbol{v}\, dt} \tag{6.3}$$

where T_1 and T_2 are the kinetic energies of P at times t_1 and t_2 respectively. This is the **energy principle** for a particle moving under a force $\boldsymbol{F}$.

Definition 6.2 ***Work done*** *The scalar quantity*

$$W = \int_{t_1}^{t_2} \boldsymbol{F} \cdot \boldsymbol{v}\, dt \tag{6.4}$$

is called the ***work done*** *by the force* $\boldsymbol{F}$ *during the time interval* $[t_1, t_2]$. *The* ***rate of working*** *of* $\boldsymbol{F}$ *at time* t *is thus* $\boldsymbol{F} \cdot \boldsymbol{v}$.
[The SI unit of work is the joule (J) and one joule per second is one watt (W).]

Our result can now be stated as follows:

Energy principle for a particle

In any motion of a particle, the increase in the kinetic energy of the particle in a given time interval is equal to the total work done by the applied forces during this time interval.

The energy principle is a *scalar* equality which is derived by integrating the *vector* equation of motion (6.1). Thus the energy principle will generally contain less information than the equation of motion, so that we have no right to expect the motion of P to be determined from the energy principle *alone*. The situation is simpler when P has **one degree of freedom**, which means that the position of P can be specified by a single scalar variable. In this case the equation of motion and the energy principle are equivalent and the energy principle alone *is* sufficient to determine the motion.

Example 6.1 *Verify the energy principle*

A man of mass 100 kg can pull on a rope with a maximum force equal to two fifths of his own weight. [Take $g = 10\ \mathrm{m\,s^{-2}}$.] In a competition, he must pull a block of mass 1600 kg across a smooth horizontal floor, the block being initially at rest. He is able to apply his maximum force horizontally for 12 seconds before falling exhausted. Find the total work done by the man and confirm that the energy principle is true in this case.

Solution

In this problem, the block is subjected to three forces: the force exerted by the man, uniform gravity, and the vertical reaction of the smooth floor. However, since the last two of these are equal and opposite, they can be ignored.

The man has weight 1000 N so that the force he applies to the block is a constant 200 N. The Second Law then implies that, while the man is pulling on the rope, the block must have constant rectilinear acceleration $200/1600 = 1/8 \text{ m s}^{-2}$. Since the block is initially at rest, its velocity v at time t is therefore $v = t/8 \text{ m s}^{-1}$. The total work W done by the man is then given by the formula (6.4) to be

$$W = \int_0^{12} \boldsymbol{F} \cdot \boldsymbol{v}\, dt = \int_0^{12} 200 \left(\frac{t}{8}\right) dt = 1800 \text{ J}.$$

When $t = 12$ s, the block has velocity $v = 12/8 = 3/2 \text{ m s}^{-1}$, so that the final kinetic energy of the block is $\frac{1}{2}(1600)(3/2)^2 = 1800$ J. Since the initial kinetic energy of the block is zero, the kinetic energy of the block increases by 1800 J, the same as the work done by the man. This confirms the truth of the energy principle. ■

6.2 ENERGY CONSERVATION IN RECTILINEAR MOTION

The energy principle is not normally used in the general form (6.3). When possible, it is transformed into a conservation principle. This is most easily illustrated by the special case of rectilinear motion.

Suppose that the particle P moves along the x-axis under the force F acting in the positive x-direction. In this case, the 'work done' integral (6.4) reduces to

$$W = \int_{t_1}^{t_2} F v\, dt,$$

where $v = \dot{x}$ is the velocity of P in the positive x-direction. For the case in which F is a **force field** (so that $F = F(x)$), the formula for W becomes

$$W = \int_{t_1}^{t_2} F v\, dt = \int_{t_1}^{t_2} F(x) \frac{dx}{dt}\, dt = \int_{x_1}^{x_2} F(x)\, dx,$$

where $x_1 = x(t_1)$ and $x_2 = x(t_2)$. Thus, when P moves over the interval $[x_1, x_2]$ of the x-axis, the work done by the field F is given by

$$\boxed{W = \int_{x_1}^{x_2} F(x)\, dx} \qquad (6.5)$$

(This is a common definition of the work done by a force F. It can be used when $F = F(x)$, but not in general.) It follows that the energy principle for a particle moving in a rectilinear force field can be written

$$T_2 - T_1 = \int_{x_1}^{x_2} F(x)\,dx.$$

Now let $V(x)$ be the indefinite integral of $-F(x)$, so that

$$F = -\frac{dV}{dx} \qquad \text{and} \qquad \int_{x_1}^{x_2} F(x)\,dx = V(x_1) - V(x_2). \tag{6.6}$$

Such a V is called the **potential energy*** function of the force field F. In terms of V, the energy principle in rectilinear motion can be written

$$T_2 + V(x_2) = T_1 + V(x_1),$$

which is equivalent to the **energy conservation** formula

$$\boxed{T + V = E} \tag{6.7}$$

where E is a constant called the **total energy** of the particle. This result can be stated as follows:

> **Energy conservation in rectilinear motion**
>
> When a particle undergoes rectilinear motion in a force field, the sum of its kinetic and potential energies remains constant in the motion.

Example 6.2 *Finding potential energies*

Find the potential energies of (i) the (one-dimensional) SHM force field, (ii) the (one-dimensional) attractive inverse square force field.

Solution

(i) The one-dimensional SHM force field is $F = -\alpha x$, where α is a positive constant. The corresponding V is given by

$$V = -\int_a^x F(x)\,dx = \alpha \int_a^x x\,dx,$$

where a, the lower limit of integration, can be arbitrarily chosen. (This corresponds to the arbitrary choice of the constant of integration.) Note that, by beginning the

* The potential energy corresponding to a given F is uniquely determined apart from a constant of integration; this constant has no physical significance.

integration at $x = a$, we make $V(a) = 0$. In the present case it is conventional to take $a = 0$ so that $V = 0$ at $x = 0$. With this choice, the potential energy is $V = \frac{1}{2}\alpha x^2$.

(ii) The one-dimensional attractive inverse square force field is $F = -K/x^2$, where $x > 0$ and K is a positive constant. The corresponding V is given by

$$V = -\int_a^x F(x)\,dx = K\int_a^x \frac{1}{x^2}\,dx.$$

This time it is not possible to take $a = 0$ (the integral would then be meaningless) and it is conventional to take $a = +\infty$; this makes $V = 0$ when $x = +\infty$. With this choice, the potential energy is $V = -K/x$ $(x > 0)$. ■

Example 6.3 *Rectilinear motion under uniform gravity*

A particle P is projected vertically upwards with speed u and moves under uniform gravity. Find the maximum height achieved and the speed of P when it returns to its starting point.

Solution

Suppose that P is projected from the origin and moves along the z-axis, where Oz points vertically upwards. The force F exerted by uniform gravity is $F = -mg$ and the corresponding potential energy V is given by

$$V = -\int_0^z (-mg)\,dz = mgz.$$

Energy conservation then implies that

$$\tfrac{1}{2}mv^2 + mgz = E,$$

where $v = \dot{z}$, and the constant E is determined from the initial condition $v = u$ when $z = 0$. This gives $E = \frac{1}{2}mu^2$ so that the energy conservation equation for the motion is

$$\tfrac{1}{2}mv^2 + mgz = \tfrac{1}{2}mu^2.$$

Since $v = 0$ when $z = z_{\text{max}}$, it follows that $z_{\text{max}} = u^2/(2g)$. This result was obtained from the Second Law in Chapter 4. When P returns to O, $z = 0$ and so $|v| = u$. Thus P returns to O with speed u, the projection speed. ■

Example 6.4 *Simple harmonic motion*

A particle of mass m is projected from the point $x = a$ with speed u and moves along the x-axis under the SHM force field $F = -m\omega^2 x$. Find the maximum distance from O and the maximum speed achieved by the particle in the subsequent motion.

Solution

The potential energy correspondingto the force field $F = -m\omega^2 x$ is $V = \frac{1}{2}m\omega^2 x^2$.

Energy conservation then implies that

$$\tfrac{1}{2}mv^2 + \tfrac{1}{2}m\omega^2x^2 = E,$$

where $v = \dot{x}$, and the constant E is determined from the initial condition $|v| = u$ when $x = a$. This gives $E = \frac{1}{2}m(u^2+\omega^2a^2)$ so that the energy conservation equation for the motion becomes

$$v^2 + \omega^2x^2 = u^2 + \omega^2a^2.$$

Since $v = 0$ when $|x|$ takes its maximum value, it follows that

$$|x|_{\max} = \left(\frac{u^2}{\omega^2} + a^2\right)^{1/2}.$$

Also, since the left side of the energy conservation equation is the sum of two positive terms, it follows that $|v|$ takes its maximum value when $x = 0$. Hence

$$|v|_{\max} = \left(u^2 + \omega^2a^2\right)^{1/2}.$$

These results could also be obtained (less quickly) by using the methods described in Chapter 5. ■

6.3 GENERAL FEATURES OF RECTILINEAR MOTION

The energy conservation equation

$$\tfrac{1}{2}mv^2 + V(x) = E \tag{6.8}$$

enables us to deduce the general features of rectilinear motion in a force field. Since $T \geq 0$ (and is equal to zero only when $v = 0$) it follows that the position of the particle is restricted to those values of x that satisfy

$$V(x) \leq E,$$

and that equality will occur only when $v = 0$. Suppose that $V(x)$ has the form shown in Figure 6.1 and that E has the value shown. Then the motion of P must take place either (i) in the bounded interval $a \leq x \leq b$, or (ii) in the unbounded interval $c \leq x \leq \infty$. Thus, if the particle was situated in the interval $[a, b]$ initially, this is the interval in which the motion will take place.

Bounded motions

Suppose that the motion is started with P in the interval $[a, b]$ and with v positive, so that P is moving to the right. Then, since v can only be zero at $x = a$ and $x = b$, v will remain positive until P reaches the point $x = b$, where it comes to rest*. From equation (6.8), it follows that

* Strictly speaking, we should exclude the possibility that P might approach the point $x = b$ asymptotically as $t \to \infty$, and never actually get there. This can happen, but only in the case in which the line $V = E$ is a *tangent* to the graph of $V(x)$ at $x = b$. In the general case depicted in Figure 6.1, P does arrive at $x = b$ in a finite time.

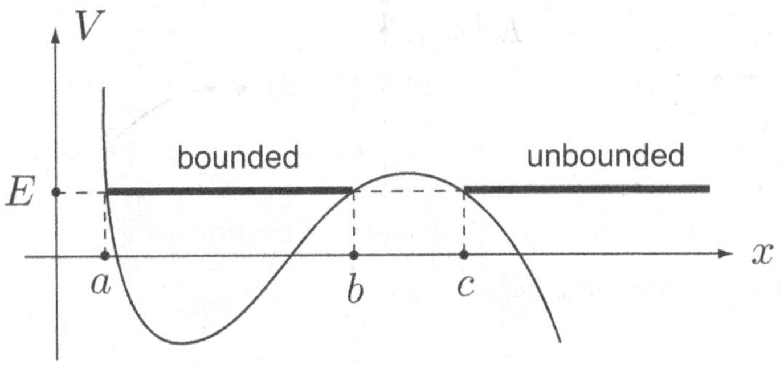

FIGURE 6.1 Bounded and unbounded motions in a rectilinear force field.

the ODE that governs this 'right' part of the motion is

$$\frac{dx}{dt} = +[2(E - V(x))/m]^{1/2}.$$

At the point $x = b$, $V' > 0$ which implies that $F < 0$. P therefore moves to the left and does not stop until it reaches the point $x = a$. The ODE that governs this 'left' part of the motion is

$$\frac{dx}{dt} = -[2(E - V(x))/m]^{1/2}.$$

At the point $x = a$, $V' < 0$, which implies that $F > 0$ and that P moves to the right once again. The result is that P performs **periodic oscillations** between the extreme points $x = a$ and $x = b$. Since the 'left' and 'right' parts of the motion take equal times, the period τ of these oscillations can be found by integrating either equation over the interval $a \le x \le b$. Each equation is a separable ODE and integration gives

$$\tau = 2\int_a^b \frac{dx}{[2(E - V(x))/m]^{1/2}}.$$

It should be noted that these oscillations are generally *not* simple harmonic. In particular, their *period is amplitude dependent.*

Example 6.5 *Periodic oscillations*

A particle P of mass 2 moves on the positive x-axis under the force field $F = (4/x^2) - 1$. Initially P is released from rest at the point $x = 4$. Find the extreme points and the period of the motion.

Solution

The force field F has potential energy $V = (4/x) + x$, so that the energy conservation equation for P is

$$\tfrac{1}{2}(2)v^2 + (4/x) + x = E,$$

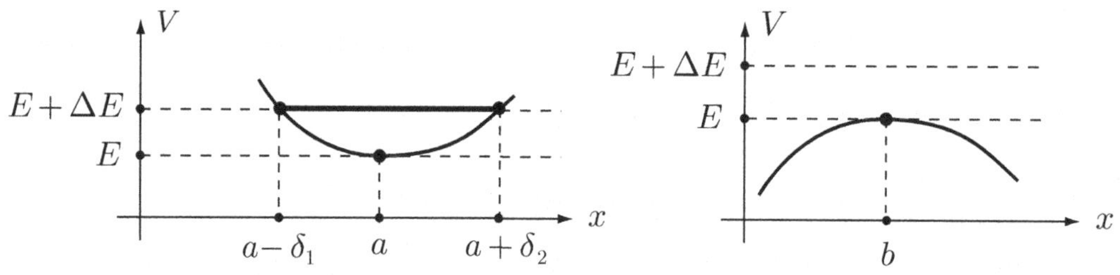

FIGURE 6.2 Positions of stable and unstable equilibrium.

where $v = \dot{x}$ and E is the total energy. The initial condition $v = 0$ when $x = 4$ gives $E = 5$ so that

$$v^2 = 5 - (4/x) - x.$$

The extreme points of the motion occur when $v = 0$, that is, when $x = 1$ and $x = 4$. To find the period τ of the oscillations, write $v = dx/dt$ in the last equation and take square roots. This gives the separable ODEs

$$\frac{dx}{dt} = \pm\left[\frac{(x-1)(4-x)}{x}\right]^{1/2},$$

where the plus and minus signs refer to the motion of P in the positive and negative x-directions respectively. Integration of either equation gives

$$\tau = 2\int_1^4 \left[\frac{x}{(x-1)(4-x)}\right]^{1/2} dx \approx 9.69. \blacksquare$$

Unbounded motions

Suppose now that the motion is started with P in the interval $[c, \infty)$ and with v negative, so that P is moving to the left. Then, since v can only be zero at $x = c$, v will remain negative until P reaches the point $x = c$, where it comes to rest. At the point $x = c$, $V' < 0$ which implies that $F > 0$. P therefore moves to the right and continues to do so indefinitely.

Stable equilibrium and small oscillations

First, we define what we mean by an equilibrium position.

Definition 6.3 ***Equilibrium*** *The point A is said to be an* ***equilibrium position*** *of P if, when P is released from rest at A, P remains at A.*

In the case of rectilinear motion under a force field $F(x)$, the point $x = a$ will be an equilibrium position of P if (and only if) $F(a) = 0$, that is, if $V'(a) = 0$. It follows that *the equilibrium positions of P are the stationary points of the potential energy function $V(x)$.* Consider the equilibrium positions shown in Figure 6.2. These occur at stationary points of V that are a minimum and a maximum respectively. Suppose that P is at rest at the minimum point $x = a$ when it receives an impulse of magnitude J which gives it kinetic energy ΔE ($= J^2/2m$). The total energy of P is now $E + \Delta E$, and so P will oscillate in the interval $[a - \delta_1, a + \delta_2]$ shown. It is clear from Figure 6.2 that, as the magnitude of J (and therefore

ΔE) tends to zero, the 'amplitude' δ of the resulting motion (the larger of δ_1 and δ_2) also tends to zero. This is the definition of **stable equilibrium**.

Definition 6.4 ***Stable equilibrium*** *Suppose that a particle P is in equilibrium at the point A when it receives an impulse of magnitude J; let δ be the amplitude of the subsequent motion. If $\delta \to 0$ as $J \to 0$, then the point A is said to be a position of* **stable equilibrium** *of P.*

On the other hand, if P is at rest at the *maximum* point $x = b$ when it receives an impulse of magnitude J, it is clear that the amplitude of the resulting motion does not tend to zero as J tends to zero, so that a maximum point of $V(x)$ is *not* a position of stable equilibrium of P. The same applies to stationary inflection points.

> **Equilibrium positions of a particle**
>
> The stationary points of the potential energy $V(x)$ are the equilibrium points of P and the minimum points of $V(x)$ are the positions of stable equilibrium. If A is a position of stable equilibrium, then P can execute small-amplitude oscillations about A.

Approximate equation of motion for small oscillations

Suppose that the point $x = a$ is a minimum point of the potential energy $V(x)$. Then, when x is sufficiently close to a, we may approximate $V(x)$ by the first three terms of its Taylor series in powers of the variable $(x - a)$, as follows:

$$\begin{aligned} V(x) &= V(a) + (x-a)V'(a) + \tfrac{1}{2}(x-a)^2 V''(a) \\ &= V(a) + \tfrac{1}{2}(x-a)^2 V''(a), \end{aligned} \tag{6.9}$$

since $V'(a) = 0$. Thus, for small amplitude oscillations about $x = a$, the energy conservation equation is approximately

$$\tfrac{1}{2}mv^2 + V(a) + \tfrac{1}{2}(x-a)^2 V''(a) = E.$$

If we now differentiate this equation with respect to t (and divide by v), we obtain the approximate (linearised) equation of motion

$$m\frac{d^2x}{dt^2} + V''(a)\,(x-a) = 0.$$

Provided that $V''(a) > 0$, this is the equation for **simple harmonic oscillations** with angular frequency $(V''(a)/m)^{1/2}$ about the point $x = a$. The small oscillations of P about $x = a$ are therefore **approximately simple harmonic** with **approximate period** $\tau = 2\pi(m/V''(a))^{1/2}$.

Example 6.6 *Finding the period of small oscillations*

A particle P of mass 8 moves on the x-axis under the force field whose potential energy is

$$V = \frac{x(x-3)^2}{3}.$$

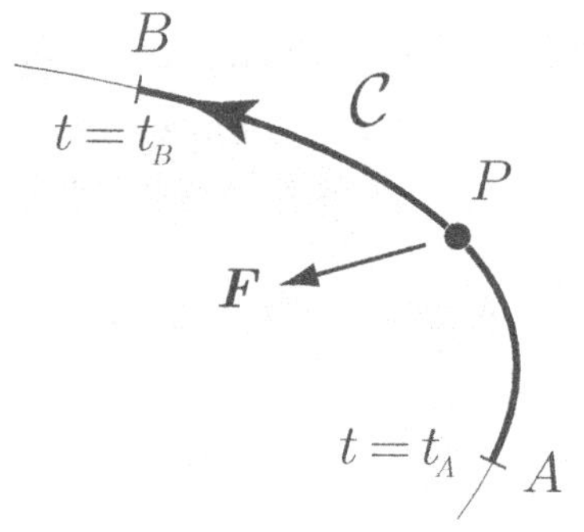

FIGURE 6.3 Particle P is in general motion under the force $\boldsymbol{F}$. The arc C is the path taken by P between the points A and B.

Show that there is a single position of stable equilibrium and find the approximate period of small oscillations about this point.

Solution

For this V, $V' = x^2 - 4x + 3$ and $V'' = 2x - 4$. The equilibrium positions occur when $V' = 0$, that is when $x = 1$ and $x = 3$. Since $V''(1) = -2$ and $V''(3) = 2$, we deduce that the only position of *stable* equilibrium is at $x = 3$. The approximate period τ of small oscillations about this point is therefore given by $\tau = 2\pi(8/V''(3))^{1/2} = 4\pi$. ■

6.4 ENERGY CONSERVATION IN A CONSERVATIVE FIELD

Suppose now that the particle P is in **general three-dimensional motion** under the force $\boldsymbol{F}$ and that, in the time interval $[t_A, t_B]$, P moves from the point A to the point B along the path C, as shown in Figure 6.3. Then, by the energy principle (6.3),

$$T_B - T_A = \int_{t_A}^{t_B} \boldsymbol{F} \cdot \boldsymbol{v}\, dt, \tag{6.10}$$

where T_A and T_B are the kinetic energies of P when $t = t_A$ and $t = t_B$ respectively. When $\boldsymbol{F}$ is a **force field** $\boldsymbol{F}(\boldsymbol{r})$, the 'work done' integral on the right side of equation (6.10) can be written in the form

$$\int_{t_A}^{t_A} \boldsymbol{F} \cdot \boldsymbol{v}\, dt = \int_{t_A}^{t_B} \boldsymbol{F}(\boldsymbol{r}) \cdot \frac{d\boldsymbol{r}}{dt}\, dt = \int_C \boldsymbol{F}(\boldsymbol{r}) \cdot d\boldsymbol{r},$$

where C is the path taken by P in the time interval $[t_A, t_B]$. It follows that the energy principle for a particle moving in a 3D force field can be written

$$T_B - T_A = \int_C \boldsymbol{F}(\boldsymbol{r}) \cdot d\boldsymbol{r}. \tag{6.11}$$

Integrals like that on the right side of equation (6.11) are called **line integrals**. They differ from ordinary integrals in that the range of integration is not an interval of the x-axis, but a *path* in three-dimensional space. Line integrals are treated in detail in texts on vector field theory (see for example Schey [11]), but their *physical* meaning in the present context is clear enough. The quantity $\boldsymbol{F} \cdot d\boldsymbol{r}$ is the infinitesimal work done by $\boldsymbol{F}$ when P traverses the element $d\,\boldsymbol{r}$ of the path C. The line integral sums these contributions to give the *total* work done by $\boldsymbol{F}$.

The line integral of $\boldsymbol{F}$ along C is taken to be the *definition* of the work done by the force field $\boldsymbol{F}(\boldsymbol{r})$ when its point of application moves along *any* path C that connects A and B.

Definition 6.5 ***3-D work done*** *The expression*

$$W[\,A \to B\,;\,C\,] = \int_C \boldsymbol{F}(\boldsymbol{r}) \cdot d\boldsymbol{r} \tag{6.12}$$

is called the ***work done*** *by the force field $\boldsymbol{F}(\boldsymbol{r})$ when its point of application moves from A to B along any path C.*

The above definition is more than just an alternative definition of the work done by a force field acting on a particle. It defines the quantity $W[\,A \to B\,;\,C\,]$ *whether or not C is an actual path traversed by the particle P*. In this wider sense, the concept of 'work done' is purely notional, as is the concept of the 'point of application of the force'. In this sense, W exists for *all* paths joining the points A and B, but W should be regarded as the real work done by $\boldsymbol{F}$ only when C is an actual path of a particle moving under the field $\boldsymbol{F}(\boldsymbol{r})$.

Energy conservation

In order to develop an energy conservation principle for the general three-dimensional case, we need the right side of equation (6.11) to be expressible in the form

$$\int_C \boldsymbol{F}(\boldsymbol{r}) \cdot d\boldsymbol{r} = V_A - V_B, \tag{6.13}$$

for *some* scalar function of position $V(\boldsymbol{r})$, where V_A and V_B are the values of V at the points A and B. In the rectilinear case, there was no difficulty in finding such a V (it was the indefinite integral of $-F(x)$). In the general case however, it is far from clear that any such V should exist. For, if there does exist a function $V(\boldsymbol{r})$ satisfying equation (6.13), this must mean that the line integral $W[\,A \to B\,;\,C\,]$ has the same value for *all* paths C that connect the points A and B. There is no reason why this should be true and, in general, it is *not* true. There is however an important class of fields $\boldsymbol{F}(\boldsymbol{r})$ for which $V(\boldsymbol{r})$ does exist, and it is these fields that we shall consider from now on.

Definition 6.6 ***Conservative field*** *If the field $\boldsymbol{F}(\boldsymbol{r})$ can be expressed in the form**

$$\boldsymbol{F} = -\operatorname{grad} V, \tag{6.14}$$

where $V(\boldsymbol{r})$ is a scalar function of position, then $\boldsymbol{F}$ is said to be a ***conservative*** *field and the function V is said to be the* ***potential energy*** *function† for $\boldsymbol{F}$.*

* If $\psi(\boldsymbol{r})$ is a *scalar* field then grad ψ is the *vector* field defined by

$$\operatorname{grad}\psi = \frac{\partial\psi}{\partial x}\,\boldsymbol{i} + \frac{\partial\psi}{\partial y}\,\boldsymbol{j} + \frac{\partial\psi}{\partial z}\,\boldsymbol{k}.$$

Thus if $\psi = xy^3z^5$, then $\operatorname{grad}\psi = y^3z^5\boldsymbol{i} + 3xy^2z^5\boldsymbol{j} + 5xy^3z^4\boldsymbol{k}$. We could omit the minus sign in the definition (6.14), but the potential energy of $\boldsymbol{F}$ would then be $-V$ instead of V.

† If V exists, then it is unique apart from a constant of integration. As in the rectilinear case, this constant has no physical significance.

Example 6.7 ***Conservative or not conservative?***

Show that the field $\boldsymbol{F}_1 = -2x\boldsymbol{i} - 2y\boldsymbol{j} - 2z\boldsymbol{k}$ is conservative but that the field $\boldsymbol{F}_2 = y\boldsymbol{i} - x\boldsymbol{j}$ is not.

Solution

(i) If $\boldsymbol{F}_1$ is conservative, then its potential energy V must satisfy

$$-\frac{\partial V}{\partial x} = -2x, \qquad -\frac{\partial V}{\partial y} = -2y, \qquad -\frac{\partial V}{\partial z} = -2z,$$

and these equations integrate to give

$$V = x^2 + p(y, z), \qquad V = y^2 + q(x, z), \qquad V = z^2 + r(x, y),$$

where p, q and r are 'constants' of integration, which, in this case, are functions of the other variables. If V really exists, then these three representations of V can be made identical by making special choices of the functions p, q and r. In this example it is clear that this can be achieved by taking $p = y^2 + z^2$, $q = x^2 + z^2$, and $r = x^2 + y^2$. Hence $\boldsymbol{F}_1 = -\operatorname{grad}(x^2 + y^2 + z^2)$ and so $\boldsymbol{F}_1$ is conservative.

(ii) If $\boldsymbol{F}_2$ is conservative, then its potential energy V must satisfy

$$-\frac{\partial V}{\partial x} = y, \qquad -\frac{\partial V}{\partial y} = -x, \qquad -\frac{\partial V}{\partial z} = 0.$$

There is no V that satisfies these equations simultaneously. The easiest way to show this is to observe that, from the first equation, $\partial^2 V/\partial y \partial x = -1$ while, from the second equation, $\partial^2 V/\partial x \partial y = +1$. Since these mixed partial derivatives of V should be equal, we have a contradiction. The conclusion is that no such V exists and that $\boldsymbol{F}_2$ is not conservative. ■

Suppose now that the field $\boldsymbol{F}(\boldsymbol{r})$ is conservative with potential energy $V(\boldsymbol{r})$ and let $\mathcal{C}$ be *any* path connecting the points A and B. Then

$$\begin{aligned}
W[A \to B\,;\,\mathcal{C}] &= \int_{\mathcal{C}} \boldsymbol{F}(\boldsymbol{r}) \cdot d\boldsymbol{r} = \int_{\mathcal{C}} (-\operatorname{grad} V) \cdot d\boldsymbol{r} \\
&= -\int_{\mathcal{C}} \left(\frac{\partial V}{\partial x} \boldsymbol{i} + \frac{\partial V}{\partial y} \boldsymbol{j} + \frac{\partial V}{\partial z} \boldsymbol{k} \right) \cdot (dx\,\boldsymbol{i} + dy\,\boldsymbol{j} + dz\,\boldsymbol{k}) \\
&= -\int_{\mathcal{C}} \frac{\partial V}{\partial x} dx + \frac{\partial V}{\partial y} dy + \frac{\partial V}{\partial z} dz \\
&= -\int_{\mathcal{C}} dV = V_A - V_B. \qquad (6.15)
\end{aligned}$$

Thus, when $\boldsymbol{F}$ is conservative with potential energy V,

$$\int_{\mathcal{C}} \boldsymbol{F}(\boldsymbol{r}) \cdot d\boldsymbol{r} = V_A - V_B,$$

for any path C connecting the points A and B. The energy principle (6.11) can therefore be written

$$T_B + V_B = T_A + V_A$$

which is equivalent to the **energy conservation** formula

$$\boxed{T + V = E} \tag{6.16}$$

Our result can be summarised as follows:

> **Energy conservation in 3-D motion**
>
> When a particle moves in a **conservative** force field, the sum of its kinetic and potential energies remains constant in the motion.

The condition that $\boldsymbol{F}$ be conservative seems restrictive, but most force fields encountered in mechanics actually are conservative!

Example 6.8 *Finding 3-D potential energies*

(a) Show that the uniform gravity field $\boldsymbol{F} = -mg\boldsymbol{k}$ is conservative with potential energy $V = mgz$.

(b) Show that any force field of the form

$$\boldsymbol{F} = h(r)\,\widehat{\boldsymbol{r}}$$

(a central field) is conservative with potential energy $V = -H(r)$, where $H(r)$ is the indefinite integral of $h(r)$. Use this result to find the potential energies of (i) the 3-D SHM field $\boldsymbol{F} = -\alpha r\,\widehat{\boldsymbol{r}}$, and (ii) the attractive inverse square field $\boldsymbol{F} = -(K/r^2)\,\widehat{\boldsymbol{r}}$, where α and K are positive constants.

Solution

Since the potential energies are given, it is sufficient to evaluate $-\operatorname{grad} V$ in each case and show that this gives the appropriate $\boldsymbol{F}$. Case (a) is immediate. In case (b),

$$\frac{\partial H(r)}{\partial x} = \frac{dH}{dr}\frac{\partial r}{\partial x} = H'(r)\frac{x}{r} = h(r)\frac{x}{r},$$

since $r = \left(x^2 + y^2 + z^2\right)^{1/2}$ and $H'(r) = h(r)$. Thus

$$-\operatorname{grad}\left[-H(r)\right] = h(r)\left(\frac{x}{r}\,\boldsymbol{i} + \frac{y}{r}\,\boldsymbol{j} + \frac{z}{r}\,\boldsymbol{k}\right) = h(r)\frac{\boldsymbol{r}}{r} = h(r)\,\widehat{\boldsymbol{r}},$$

as required.

In particular then, the potential energy of the SHM field $\boldsymbol{F} = -\alpha r\,\widehat{\boldsymbol{r}}$ is $V = \frac{1}{2}\alpha r^2$, and the potential energy of the attractive inverse square field $\boldsymbol{F} = -(K/r^2)\,\widehat{\boldsymbol{r}}$ is $V = -K/r$. ■

Example 6.9 *Projectile motion*

A body is projected from the ground with speed u and lands on the flat roof of a building of height h. Find the speed with which the projectile lands. [Assume uniform gravity and no air resistance.]

Solution

Since uniform gravity is a conservative field with potential energy mgz, energy conservation applies in the form

$$\tfrac{1}{2}m|\boldsymbol{v}|^2 + mgz = E,$$

where O is the initial position of the projectile and Oz points vertically upwards. From the initial conditions, $E = \frac{1}{2}mu^2$. Hence, when the body lands,

$$\tfrac{1}{2}m|\boldsymbol{v}^L|^2 + mgh = \tfrac{1}{2}mu^2,$$

where $\boldsymbol{v}^L$ is the landing velocity. The **landing speed** is therefore

$$|\boldsymbol{v}^L| = \left(u^2 - 2gh\right)^{1/2}.$$

Thus, energy conservation determines the *speed* of the body on landing, but not its *velocity*. ■

Example 6.10 *Escape from the Moon*

A body is projected from the surface of the Moon with speed u *in any direction*. Show that the body cannot escape from the Moon if $u^2 < 2MG/R$, where M and R are the mass and radius of the Moon. [Assume that the Moon is spherically symmetric.]

Solution

If the Moon is spherically symmetric, then the force $\boldsymbol{F}$ that it exerts on the body is given by $\boldsymbol{F} = -(mMG/r^2)\widehat{\boldsymbol{r}}$, where m is the mass of the body, and $\boldsymbol{r}$ is the position vector of the body relative to an origin at the centre of the Moon. This force is a conservative field with potential energy $V = -mMG/r$. Hence energy conservation applies in the form

$$\tfrac{1}{2}m|\boldsymbol{v}|^2 - \frac{mMG}{r} = E,$$

and, from the initial conditions, $E = \frac{1}{2}mu^2 - (mMG)/R$. Thus the energy conservation equation is

$$|\boldsymbol{v}|^2 = u^2 + 2MG\left(\frac{1}{r} - \frac{1}{R}\right).$$

Since the left side of the above equation is *positive*, the values of r that occur in the motion must satisfy the inequality

$$u^2 + 2MG\left(\frac{1}{r} - \frac{1}{R}\right) \geq 0.$$

If the body is to escape, this inequality must hold for *arbitrarily large* r. This means that the condition

$$u^2 - \frac{2MG}{R} \geq 0$$

is necessary for escape. Hence if $u^2 < 2MG/R$, the body cannot escape. The interesting feature here is that the 'escape speed' is the same for *all directions* of projection from the surface of the Moon. (The special case in which the body is projected vertically upwards was solved in Chapter 4.) ■

Example 6.11 ***Stability of equilibrium in a 3-D conservative field***

A particle P of mass m can move under the gravitational attraction of two particles, of equal mass M, fixed at the points $(0, 0, \pm a)$. Show that the origin O is a position of equilibrium, but that it is not stable. [This illustrates the general result that *no* free-space static gravitational field can provide a position of stable equilibrium.]

Solution

When P is at O, the fixed particles exert equal and opposite forces so that the total force on P is zero. The origin is therefore an equilibrium position for P.

Just as in rectilinear motion, O will be a position of *stable* equilibrium if the potential energy function $V(x, y, z)$ has a minimum at O. This means that the value of V at O must be less than its values at *all* nearby points. But at points on the z-axis between $z = -a$ and $z = a$

$$V(0, 0, z) = -\frac{mMG}{a - z} - \frac{mMG}{a + z} = -\frac{2amMG}{a^2 - z^2},$$

which has a *maximum* at $z = 0$. Hence the equilibrium at O is unstable to disturbances in the z-direction. ■

6.5 ENERGY CONSERVATION IN CONSTRAINED MOTION

Some of the most useful applications of energy conservation occur when the moving particle is subject to geometrical constraints, such as being connected to a fixed point by a light inextensible string, or being required to remain in contact with a fixed rigid surface (see section 4.2). Since constraint forces are not known beforehand one may wonder how to find the work that they do. The answer is that, in the idealised problems that we study, the *work done by the constraint forces is often zero*. In these cases the constraint forces make no contribution to the energy principle and they can be disregarded. Situations in which constraint forces do no work include:

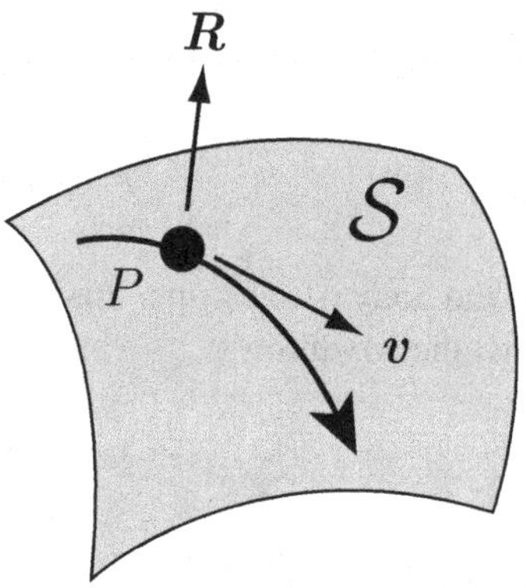

FIGURE 6.4 The particle P slides over the fixed smooth surface $\mathcal{S}$. The reaction $\boldsymbol{R}$ is normal to $\mathcal{S}$ and hence perpendicular to $\boldsymbol{v}$, the velocity of P.

Some constraint forces that do no work

- A particle connected to a fixed point by a light inextensible string; here the *string tension does no work.*
- A particle sliding along a smooth fixed wire; here the *reaction of the wire does no work.*
- A particle sliding over a smooth fixed surface; here the *reaction of the surface does no work.*

Consider for example the case of a particle P sliding over a smooth fixed surface $\mathcal{S}$ as shown in Figure 6.4. Because $\mathcal{S}$ is smooth, any reaction force $\boldsymbol{R}$ that it exerts must always be normal to $\mathcal{S}$. But, since P remains on $\mathcal{S}$, its velocity $\boldsymbol{v}$ must always be tangential to $\mathcal{S}$. Hence $\boldsymbol{R}$ is always perpendicular to $\boldsymbol{v}$ so that $\boldsymbol{R} \cdot \boldsymbol{v} = 0$. Thus the *rate of working* of $\boldsymbol{R}$ is zero and so $\boldsymbol{R}$ makes no contribution to the energy principle. Very similar arguments apply to the other two cases.

We may now extend the use of conservation of energy as follows:

Energy conservation in constrained motion

When a particle moves in a **conservative** force field and is subject to **constraint forces that do no work**, the sum of its kinetic and potential energies remains constant in the motion.

Example 6.12 *The snowboarder*

A snowboarder starts from rest and descends a slope, losing 320 m of altitude in the process. What is her speed at the bottom? [Neglect all forms of resistance and take $g = 10\ \mathrm{m\,s^{-2}}$.]

Solution

The snowboarder moves under uniform gravity and the reaction force of the smooth hillside. Since this reaction force does no work, energy conservation applies in the form

$$\tfrac{1}{2}m|\boldsymbol{v}|^2 + mgz = E,$$

where m and $\boldsymbol{v}$ are the mass and velocity of the snowboarder, and z is the *altitude* of the snowboarder relative to the bottom of the hill. If the snowboarder starts from rest at altitude h, then $E = 0 + mgh$. Hence, at the bottom of the hill where $z = 0$, her speed is

$$|\boldsymbol{v}| = (2gh)^{1/2},$$

just as if she had fallen down a vertical hole! This speed evaluates to 80 m s^{-1}, about 180 mph. [At such speeds, air resistance would have an important influence.] ■

Our next example concerns a particle constrained to move on a vertical circle. This is one of the classical applications of the energy conservation method. There are two distinct cases: (i) where the particle is constrained always to remain on the circle, or (ii) where the particle is constrained to remain on the circle only while the constraint force has a particular sign.

Example 6.13 *Motion in a vertical circle*

A fixed hollow sphere has centre O and a smooth inner surface of radius b. A particle P, which is inside the sphere, is projected horizontally with speed u from the lowest interior point (see Figure 6.5). Show that, in the subsequent motion,

$$v^2 = u^2 - 2gb(1 - \cos\theta),$$

provided that P remains in contact with the sphere.

Solution

While P remains in contact with the sphere, the motion is as shown in Figure 6.5. The forces acting on P are uniform gravity mg and the constraint force N, which is the normal reaction of the smooth sphere. Since N is always perpendicular to v (the circumferential velocity of P), it follows that N does no work. Hence energy conservation applies in the form

$$\tfrac{1}{2}mv^2 - mgb\cos\theta = E,$$

where m is the mass of P, and the zero level of the potential energy is the horizontal plane through O. Since $v = u$ when $\theta = 0$, it follows that $E = \tfrac{1}{2}mu^2 - mgb$ and the energy conservation equation becomes

$$v^2 = u^2 - 2gb(1 - \cos\theta), \qquad (6.17)$$

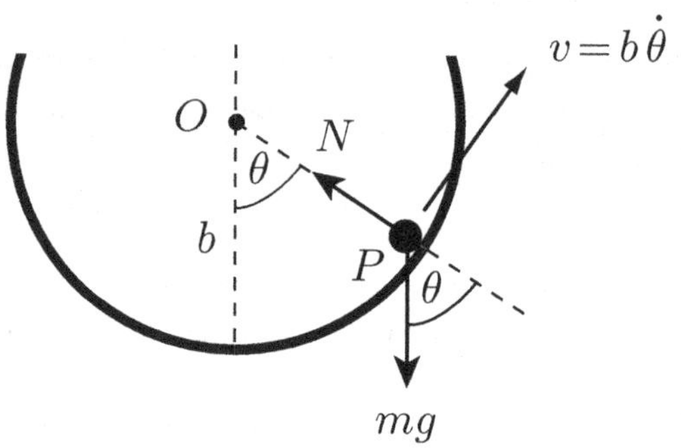

FIGURE 6.5 Particle P slides on the smooth inner surface of a fixed sphere. The motion takes place in a vertical plane through the centre O.

as required. This gives the value of v as a function of θ while P remains in contact with the sphere. ■

Question *The reaction force*

Find the reaction force N as a function of θ.

Answer

Once the motion is determined (for example by equation (6.17)), the unknown constraint forces may be found by using the Second Law in reverse. In the present case, consider the component of the Second Law $\boldsymbol{F} = m\boldsymbol{a}$ in the direction $\overrightarrow{PO}$. This gives

$$N - mg\cos\theta = mv^2/b,$$

where we have made use of the formula (2.17) for the acceleration of a particle in general circular motion. On using the formula for v^2 from equation (6.17), we obtain

$$N = \frac{mu^2}{b} + mg(3\cos\theta - 2). \tag{6.18}$$

This gives the value of N as a function of θ while P remains in contact with the sphere. ■

Question *Does P leave the surface of the sphere?*

For the particular case in which $u = (3gb)^{1/2}$, show that P will leave the surface of the sphere, and find the value of θ at which it does so.

Answer

When $u = (3gb)^{1/2}$, the formulae (6.17), (6.18) for v^2 and N become

$$v^2 = gb(1 + 2\cos\theta), \qquad N = mg(1 + 3\cos\theta).$$

If P remains in contact with the sphere, then it comes to rest when $v = 0$, that is, when $\cos\theta = -1/2$. This first happens when $\theta = 120°$ (a point on the upper half of the sphere, higher than O). If P were threaded on a circular *wire* (from which it could not fall off) this is exactly what would happen; P would perform periodic oscillations in the range $-120° \leq \theta \leq 120°$. However, in the present case, the reaction N is restricted to be *positive* and this condition will be violated when $\theta > \cos^{-1}(-1/3) \approx 109°$. Since this angle is less than 120°, the conclusion is that P loses contact with the sphere when $\theta = \cos^{-1}(-1/3)$; at this instant, the speed of P is $(gb/3)^{1/2}$. P then moves as a free projectile until it strikes the sphere. ■

Question ***Complete circles***

How large must the initial speed be for P to perform complete circles?

Answer

For complete circles to be executed, it is necessary (and sufficient) that $v > 0$ and $N \geq 0$ at all times, that is,

$$u^2 > 2gb(1 - \cos\theta) \qquad \text{and} \qquad u^2 \geq gb(2 - 3\cos\theta)$$

for all values of θ. For these inequalities to hold for *all* θ, the speed u must satisfy $u^2 > 4gb$ and $u^2 \geq 5gb$ respectively. Since the second of these conditions implies the first, it follows that P will execute complete circles if $u^2 \geq 5gb$. ■

Example 6.14 *Small oscillations in constrained motion*

A particle P of mass m can slide freely along a long straight wire. P is connected to a fixed point A, which is at a distance $4a$ from the wire, by a light elastic cord of natural length $3a$ and strength α. Find the approximate period of small oscillations of P about its equilibrium position.

Solution

Suppose P has displacement x from its equilibrium position. In this position, the length of the cord is $(16a^2 + x^2)^{1/2}$ and its potential energy V is

$$\begin{aligned} V &= \tfrac{1}{2}\alpha\left[\left(16a^2 + x^2\right)^{1/2} - 3a\right]^2 \\ &= \tfrac{1}{2}\alpha\left[25a^2 + x^2 - 6a\left(16a^2 + x^2\right)^{1/2}\right]. \end{aligned}$$

The energy conservation equation for P is therefore

$$\tfrac{1}{2}m\dot{x}^2 + \tfrac{1}{2}\alpha\left[25a^2 + x^2 - 6a\left(16a^2 + x^2\right)^{1/2}\right] = E,$$

which, on neglecting powers of x higher than the second, becomes

$$\tfrac{1}{2}m\dot{x}^2 + \tfrac{1}{2}\alpha\left[a^2 + \frac{x^2}{4}\right] = E.$$

On differentiating this equation with respect to t, we obtain the approximate **linearised equation of motion**

$$m\ddot{x} + \frac{\alpha}{4}x = 0.$$

This is the SHM equation with $\omega^2 = \alpha/4m$. It follows that the approximate **period** of small oscillations about $x = 0$ is $4\pi(m/\alpha)^{1/2}$. ■

Energy conservation from a physical viewpoint

Suppose that a particle P of mass m can move on the x-axis and is connected to a fixed post at $x = -a$ by a light elastic spring of natural length a and strength α. Then the force $F(x)$ exerted on P by the spring is given by $F = -\alpha x$, where x is the displacement of P in the positive x-direction. This force field has potential energy $V = \frac{1}{2}\alpha x^2$ and the energy conservation equation for P takes the form

$$\tfrac{1}{2}mv^2 + \tfrac{1}{2}\alpha x^2 = E, \tag{6.19}$$

where $v = \dot{x}$. Here, the spring is regarded merely as an agency that supplies a force field with potential energy $V = \frac{1}{2}\alpha x^2$. However, there is a much more satisfying interpretation of the energy conservation principle (6.19) that can be made.

To see this we consider the spring as described above, but now with no particle attached to its free end. Suppose that the spring is in equilibrium (with its free end at $x = 0$) when an external force $G(t)$ is applied there. This force is initially zero and increases so that, at any time t, the spring has extension $X = G(t)/\alpha$. Suppose that this process continues until the spring has extension Δ. Then the total work done by the force $G(t)$ in producing this extension is given by

$$\int_0^\tau G(t)\dot{X}\,dt = \int_0^\tau \alpha X \frac{dX}{dt}dt = \int_0^\Delta \alpha X\,dX = \tfrac{1}{2}\alpha\Delta^2.$$

Since the force exerted by the fixed post does no work, the *total* work done by the external forces in producing the extension Δ is $\frac{1}{2}\alpha\Delta^2$. Suppose now that the spring is 'frozen' in its extended state (by being propped open, for example) while the particle P is connected to the free end. The system is then released from rest. The energy conservation equation for P is given by equation (6.19), where, from the initial condition $v = 0$ when $x = \Delta$, the total energy $E = \frac{1}{2}\alpha\Delta^2$. This gives

$$\tfrac{1}{2}mv^2 + \tfrac{1}{2}\alpha x^2 = \tfrac{1}{2}\alpha\Delta^2. \tag{6.20}$$

Thus, the total energy in the subsequent motion is equal to the original work done in stretching the spring. The natural physical interpretation of this is that the spring is able to *store* the work that is done upon it as **internal energy**. Then, when the particle is connected and the system released, this stored energy is available to be transferred to the particle in the form of kinetic energy. Equation (6.20) can thus be interpreted as an **energy conservation** principle for the **particle** and **spring** together, as follows: *In any motion of the particle and spring, the sum of the kinetic energy of the particle and the internal energy of the spring remains constant.* In this interpretation, the particle has no potential energy; instead, the spring has internal energy.

In the above example, the particle and the spring can pass energy to each other, but the total of the two energies is conserved. This is the essential nature of energy. It is an entity that can appear in different forms but whose **total is always conserved**. Energy is probably the most important notion in the whole of physics. However, it should be remembered that, in the context of mechanics, it is not usual to take account of forms of energy such as heat or light.

As a result, we will find situations (inelastic collisions, for example) in which energy seems to disappear. There is no contradiction in this; the energy has simply been transferred into forms that we choose not to recognise.

Problems on Chapter 6

Answers and comments are at the end of the book.

Harder problems carry a star (∗).

Unconstrained motion

6.1 A particle P of mass 4 kg moves under the action of the force $\boldsymbol{F} = 4\boldsymbol{i} + 12t^2\,\boldsymbol{j}$ N, where t is the time in seconds. The initial velocity of the particle is $2\boldsymbol{i} + \boldsymbol{j} + 2\boldsymbol{k}$ m s^{-1}. Find the work done by $\boldsymbol{F}$, and the increase in kinetic energy of P, during the time interval $0 \le t \le 1$. What principle does this illustrate?

6.2 In a competition, a man pushes a block of mass 50 kg with constant speed 2 m s^{-1} up a smooth plane inclined at 30° to the horizontal. Find the rate of working of the man. [Take $g = 10$ m s^{-2}.]

6.3 An athlete putts a shot of mass 7 kg a distance of 20 m. Show that the athlete must do to at least 700 J of work to achieve this. [Ignore the height of the athlete and take $g = 10$ m s^{-2}.]

6.4 Find the work that must be done to lift a satellite of mass 200 kg to a height of 2000 km above the Earth's surface. [Take the Earth to be spherically symmetric and of radius 6400 km. Take the surface value of g to be 9.8 m s^{-2}.]

6.5 A particle P of unit mass moves on the positive x-axis under the force field

$$F = \frac{36}{x^3} - \frac{9}{x^2} \qquad (x > 0).$$

Show that each motion of P consists of either (i) a periodic oscillation between two extreme points, or (ii) an unbounded motion with one extreme point, depending upon the value of the total energy. Initially P is projected from the point $x = 4$ with speed 0.5. Show that P oscillates between two extreme points and find the period of the motion. [You may make use of the formula

$$\int_a^b \frac{x\,dx}{[(x-a)(b-x)]^{1/2}} = \frac{\pi(a+b)}{2}.\ ]$$

Show that there is a single equilibrium position for P and that it is stable. Find the period of *small* oscillations about this point.

6.6 A particle P of mass m moves on the x-axis under the force field with potential energy $V = V_0(x/b)^4$, where V_0 and b are positive constants. Show that any motion of P consists of a periodic oscillation with centre at the origin. Show further that, when the oscillation has amplitude a, the period τ is given by

$$\tau = 2\sqrt{2}\left(\frac{m}{V_0}\right)^{1/2}\frac{b^2}{a}\int_0^1 \frac{d\xi}{(1-\xi^4)^{1/2}}.$$

[Thus, the larger the amplitude, the shorter the period!]

6.7 A particle P of mass m, which is on the negative x-axis, is moving towards the origin with constant speed u. When P reaches the origin, it experiences the force $F = -Kx^2$, where K is a positive constant. How far does P get along the positive x-axis?

6.8 A particle P of mass m moves on the x-axis under the combined gravitational attraction of two particles, each of mass M, fixed at the points $(0, \pm a, 0)$ respectively (see Figure 3.3). Example 3.4 shows that the force field $F(x)$ acting on P is given by

$$F = -\frac{2mMGx}{(a^2 + x^2)^{3/2}}.$$

Find the corresponding potential energy $V(x)$.

Initially P is released from rest at the point $x = 3a/4$. Find the maximum speed achieved by P in the subsequent motion.

6.9 A particle P of mass m moves on the axis Oz under the gravitational attraction of a uniform circular disk of mass M and radius a as shown in Figure 3.6. Example 3.6 shows that the force field $F(z)$ acting on P is given by

$$F = -\frac{2mMG}{a^2}\left[1 - \frac{z}{(a^2 + z^2)^{1/2}}\right] \qquad (z > 0).$$

Find the corresponding potential energy $V(z)$ for $z > 0$.

Initially P is released from rest at the point $z = 4a/3$. Find the speed of P when it hits the disk.

6.10 A catapult is made by connecting a light elastic cord of natural length $2a$ and strength α between two fixed supports, which are distance $2a$ apart. A stone of mass m is placed at the center of the cord, which is pulled back a distance $3a/4$ and then released from rest. Find the speed with which the stone is projected by the catapult.

6.11 A light spring of natural length a is placed on a horizontal floor in the upright position. When a block of mass M is resting in equilibrium on top of the spring, the compression of the spring is $a/15$. The block is now lifted so that its underside is at height $3a/2$ above the floor and released from rest. Find the compression of the spring when the block first comes to rest.

6.12 A particle P carries a charge e and moves under the influence of the static magnetic field $\boldsymbol{B}(\boldsymbol{r})$ which exerts the force $\boldsymbol{F} = e\boldsymbol{v}\times\boldsymbol{B}$ on P, where $\boldsymbol{v}$ is the velocity of P. Show that P travels with constant *speed*.

6.13* A mortar shell is to be fired from level ground so as to clear a flat topped building of height h and width a. The mortar gun can be placed anywhere on the ground and can have any angle of elevation. What is the least projection speed that will allow the shell to clear the building? [*Hint* How is the required minimum projection speed changed if the mortar is raised to rooftop level?]

For the special case in which $h = \frac{1}{2}a$, find the optimum position for the mortar and the optimum elevation angle to clear the building.

If you are a star at electrostatics, try the following two problems:

6.14* An *earthed* conducting sphere of radius a is fixed in space, and a particle P, of mass m and charge q, can move freely outside the sphere. Initially P is a distance b ($> a$) from the centre O of the sphere when it is projected directly away from O. What must the projection speed be for P to escape to infinity? [Ignore electro*dynamic* effects. Use the method of images to solve the electro*static* problem.]

6.15* An *uncharged* conducting sphere of radius a is fixed in space and a particle P, of mass m and charge q, can move freely outside the sphere. Initially P is a distance b ($> a$) from the centre O of the sphere when it is projected directly away from O. What must the projection speed be for P to escape to infinity? [Ignore electro*dynamic* effects. Use the method of images to solve the electro*static* problem.]

Constrained motion

6.16 A bead of mass m can slide on a smooth circular wire of radius a, which is fixed in a vertical plane. The bead is connected to the highest point of the wire by a light spring of natural length $3a/2$ and strength α. Determine the stability of the equilibrium position at the lowest point of the wire in the cases (i) $\alpha = 2mg/a$, and (ii) $\alpha = 5mg/a$.

6.17 A smooth wire has the form of the helix $x = a\cos\theta$, $y = a\sin\theta$, $z = b\theta$, where θ is a real parameter, and a, b are positive constants. The wire is fixed with the axis Oz pointing vertically upwards. A particle P, which can slide freely on the wire, is released from rest at the point $(a, 0, 2\pi b)$. Find the speed of P when it reaches the point $(a, 0, 0)$ and the time taken for it to do so.

6.18 A smooth wire has the form of the parabola $z = x^2/2b$, $y = 0$, where b is a positive constant. The wire is fixed with the axis Oz pointing vertically upwards. A particle P, which can slide freely on the wire, is performing oscillations with x in the range $-a \le x \le a$. Show that the period τ of these oscillations is given by

$$\tau = \frac{4}{(gb)^{1/2}} \int_0^a \left(\frac{b^2 + x^2}{a^2 - x^2}\right)^{1/2} dx.$$

By making the substitution $x = a\sin\psi$ in the above integral, obtain a new formula for τ. Use this formula to find a two-term approximation to τ, valid when the ratio a/b is small.

6.19* A smooth wire has the form of the cycloid $x = c(\theta + \sin\theta)$, $y = 0$, $z = c(1 - \cos\theta)$, where c is a positive constant and the parameter θ lies in the range $-\pi \le \theta \le \pi$. The wire is fixed with the axis Oz pointing vertically upwards. [Make a sketch of the wire.] A particle

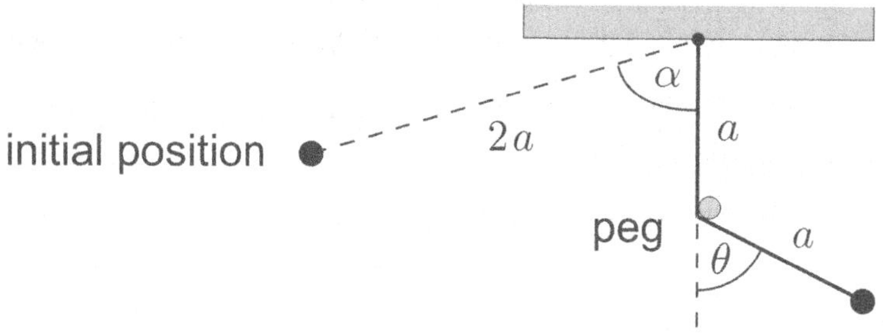

FIGURE 6.6 The swing of the pendulum is obstructed by a fixed peg.

can slide freely on the wire. Show that the energy conservation equation is

$$(1 + \cos\theta)\,\dot{\theta}^2 + \frac{g}{c}(1 - \cos\theta) = \text{constant}.$$

A new parameter u is defined by $u = \sin\frac{1}{2}\theta$. Show that, in terms of u, the equation of motion for the particle is

$$\ddot{u} + \left(\frac{g}{4c}\right)u = 0.$$

Deduce that the particle performs oscillations with period $4\pi(c/g)^{1/2}$, independent of the amplitude!

6.20 A smooth horizontal table has a vertical post fixed to it which has the form of a circular cylinder of radius a. A light inextensible string is wound around the base of the post (so that it does not slip) and its free end is attached to a particle that can slide on the table. Initially the unwound part of the string is taut and of length b. The particle is then projected with speed u at right angles to the string so that the string winds itself *on* to the post. How long does it take for the particle to hit the post?

6.21 A heavy ball is suspended from a fixed point by a light inextensible string of length b. The ball is at rest in the equilibrium position when it is projected horizontally with speed $(7gb/2)^{1/2}$. Find the angle that the string makes with the upward vertical when the ball begins to leave its circular path. Show that, in the subsequent projectile motion, the ball returns to its starting point.

6.22* A new *avant garde* mathematics building has a highly polished outer surface in the shape of a huge hemisphere of radius 40 m. The Head of Department, Prof. Oldfart, has his student, Vita Youngblood, hauled to the summit (to be photographed for publicity purposes) but a small gust of wind causes Vita to begin to slide down. Oldfart's displeasure is increased when Vita lands on (and severely damages) his car which is parked nearby. How far from the outer edge of the building did Oldfart park his car? Did he get what he deserved? (Happily, Vita escaped injury and found a new supervisor.)

6.23* * A heavy ball is attached to a fixed point O by a light inextensible string of length $2a$. The ball is drawn back until the string makes an acute angle α with the downward vertical and is then released from rest. A thin peg is fixed a distance a vertically below O in the path of the string, as shown in Figure 6.6. In a game of skill, the contestant chooses the value of α and wins a prize if the ball strikes the peg. Show that the winning value of α is approximately 86°.

Chapter Seven

Orbits in a central field

including Rutherford scattering

KEY FEATURES

For motion in *general* central force fields, the key results are the **radial motion equation** and the **path equation**. For motion in the *inverse square* force field, the key formulae are the **E-formula**, the **L-formula** and the **period formula**.

The theory of orbits has a special place in classical mechanics for it was the desire to understand why the planets move as they do which provided the major stimulus in the development of mechanics as a scientific discipline. Early in the seventeenth century, Johannes Kepler * published his 'laws of planetary motion', which he deduced by analysing the accurate experimental observations made by the astronomer Tycho Brahe.†

---

* The German mathematician and astronomer Johannes Kepler (1571–1630) was a firm believer in the Copernican (heliocentric) model of the solar system. In 1596 he became mathematical assistant to Tycho Brahe, the foremost observational astronomer of the day, and began working on the intractable problem of the orbit of Mars. This work continued after Tycho's death in 1601 and, after much labour, Kepler showed that Tycho's observations of Mars corresponded very precisely to an elliptic orbit with the Sun at a focus. This result, together with the 'law of areas' (the second law) was published in 1609. Kepler then found similar orbits for other planets and his third law was published in 1619.

† Tycho Brahe (1546–1601) was a Danish nobleman. He had a lifelong interest in observational astronomy and developed a succession of new and more accurate instruments. The King of Denmark gave him money to create an observatory and also the island of Hven on which to build it. It was here that Tycho made his accurate observations of the planets from which Kepler was able to deduce his laws of planetary motion. Tycho's other claim to fame is that he had a metal nose. When the original was cut off in a duel, he had an artificial nose made from an alloy of silver and gold. Tycho is perhaps better remembered for his nose job than he is for a lifetime of observations.

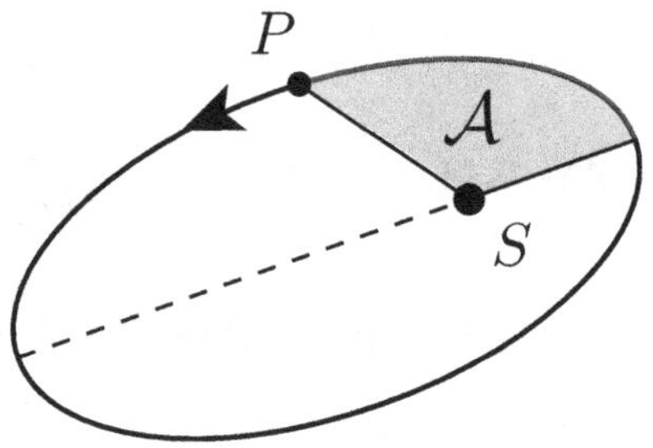

FIGURE 7.1 Each planet P moves on an elliptical path with the Sun S at one focus. The area $\mathcal{A}$ is that referred to in Kepler's second law.

Kepler's laws of planetary motion

First law Each of the planets moves on an elliptical path with the Sun at one focus of the ellipse.

Second law For each of the planets, the straight line connecting the planet to the Sun sweeps out equal areas in equal times.

Third law The squares of the periods of the planets are proportional to the cubes of the major axes of their orbits.

The problem of determining the law of force that causes the motions described by Kepler (and *proving* that it does so) was the most important scientific problem of the seventeenth century. In what must be the finest achievement in the whole history of science, Newton's publication of *Principia* in 1687 not only proved that the inverse square law of gravitation implies Kepler's laws, but also laid down the entire framework of the science of mechanics. Orbit theory is just as important today, the principal fields of application being astronomy, particle scattering and space travel.

In this chapter, we treat the problem of a particle moving in a central force field with a *fixed centre*; this is called the **one-body problem**. The assumption that the centre of force is fixed is an accurate approximation in the context of planetary orbits. The combined mass of all the planets, moons and asteroids is less than 0.2% of the mass of the Sun. We therefore expect the motion of the Sun to be comparatively small, as are inter-planetary influences.* However, we do not confine our interest to motion under the attractive inverse square field. At first, we consider motion in *any* central force field with a fixed centre. This part of the theory will then apply not only to gravitating bodies, but also (for example) to the scattering of neutrons. The important cases of inverse square attraction and repulsion are then examined in greater detail.

* The more general **two-body problem** is treated in Chapter 10. The two-body theory must be used to analyse problems in which the *masses of the two interacting bodies are comparable*, as they are in binary stars.

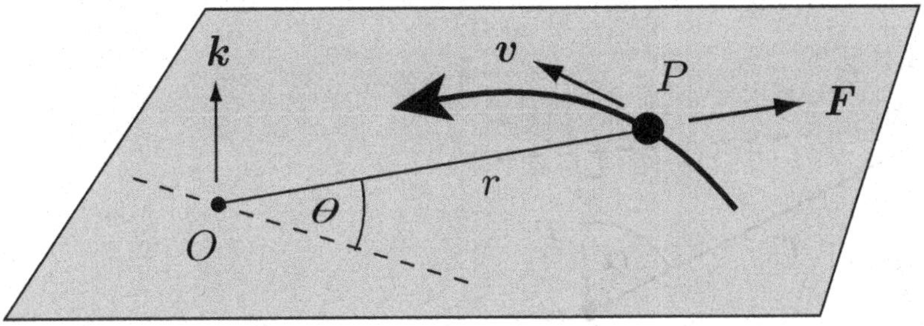

FIGURE 7.2 Each orbit of a particle P in a central force field with centre O takes place in a plane through O. The position of P in the plane of motion is specified by polar coordinates r, θ with centre at O.

7.1 THE ONE-BODY PROBLEM – NEWTON'S EQUATIONS

First we define what we mean by a central force field.

Definition 7.1 ***Central field*** *A force field $\boldsymbol{F}(\boldsymbol{r})$ is said to be a* ***central field*** *with centre O if it has the form*

$$\boldsymbol{F}(\boldsymbol{r}) = F(r)\,\widehat{\boldsymbol{r}},$$

where $r = |\boldsymbol{r}|$ and $\widehat{\boldsymbol{r}} = \boldsymbol{r}/r$. A central field is thus ***spherically symmetric*** *about its centre.*

A good example of a central force is the gravitational force exerted by a *fixed* point mass. Suppose P has mass m and moves under the gravitational attraction of a point mass M fixed at the origin. In this case, the force acting on P is given by the law of gravitation to be

$$\boldsymbol{F}(\boldsymbol{r}) = -\frac{mMG}{r^2}\,\widehat{\boldsymbol{r}},$$

where G is the constant of gravitation. This is a central field with

$$F(r) = -\frac{mMG}{r^2}.$$

Each orbit lies in a plane through the centre of force

The first thing to observe is that, when a particle P moves in a central field with centre O, *each orbit of P takes place in a plane through O*, as shown in Figure 7.2. This is the plane that contains O and the initial position and velocity of P. One may give a vectorial proof of this, but it is quite clear on symmetry grounds that P will never leave this plane. Each motion is therefore two-dimensional and we take polar coordinates r, θ (centred on O) to specify the position of P in the plane of motion. On using the formulae (2.14) for the components of acceleration in polar coordinates, the **Newton equations of motion** for

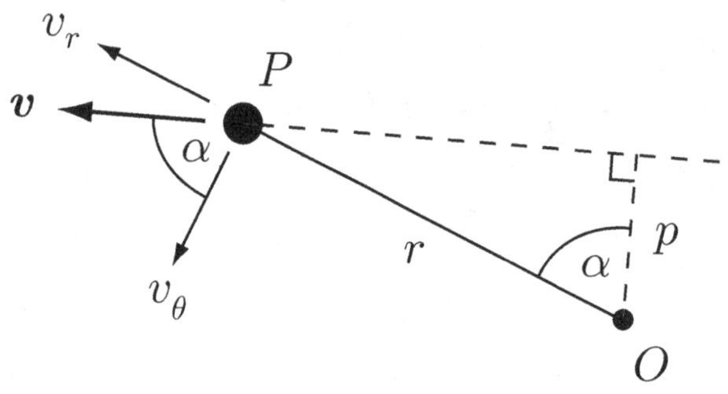

FIGURE 7.3 The angular momentum $mr^2\dot{\theta} = mp\,v$, where $v = |\boldsymbol{v}|$.

P become

$$m\left(\ddot{r} - r\dot{\theta}^2\right) = F(r), \tag{7.1}$$

$$m\left(r\ddot{\theta} + 2\dot{r}\dot{\theta}\right) = 0. \tag{7.2}$$

Angular momentum conservation

Equation (7.2) can be written in the form

$$\frac{1}{r}\frac{d}{dt}\left(mr^2\dot{\theta}\right) = 0,$$

which can be integrated with respect to t to give

$$mr^2\dot{\theta} = \text{constant}.$$

The quantity $mr^2\dot{\theta}$, which is a constant of the motion, is called the **angular momentum*** of P. The general theory of angular momentum (and its conservation) is described in Chapter 11, but for now it is sufficient to regard 'angular momentum' simply as a *name* that we give to the conserved quantity $mr^2\dot{\theta}$. This angular momentum has a simple kinematical interpretation. From Figure 7.3 it follows that

$$\begin{aligned} mr^2\dot{\theta} &= mr(r\dot{\theta}) = mr\,v_\theta = m(r\cos\alpha)\left(\frac{v_\theta}{\cos\alpha}\right) \\ &= mp\,v, \end{aligned}$$

where p is the perpendicular distance of O from the tangent to the path of P, and $v = |\boldsymbol{v}|$. This formula provides the usual way of calculating the constant value the angular momentum from the initial conditions.

* More precisely, it is the angular momentum of the particle about the *axis* $\{O, \boldsymbol{k}\}$, where the unit vector $\boldsymbol{k}$ is perpendicular to the plane of motion (see Figure 7.2). The angular momentum of P about the *point* O is the *vector* quantity $m\boldsymbol{r}\times\boldsymbol{v}$, but the axial angular momentum used in the present chapter is the component of this vector in the $\boldsymbol{k}$-direction.

Newton equations in specific form

It is usual and convenient to eliminate the mass m from the theory. If we write

$$F(r) = mf(r),$$

where $f(r)$ is the outward force *per unit mass*, and let L $(= r^2\dot{\theta})$ be the angular momentum *per unit mass* then the Newton equations (7.1), (7.2) reduce to the **specific form**

$$\ddot{r} - r\dot{\theta}^2 = f(r), \tag{7.3}$$
$$r^2\dot{\theta} = L, \tag{7.4}$$

where L is a constant.* Note that these equations apply to orbits in *any central field.* The second of these equations appears throughout this chapter and we will call it the *angular momentum equation.*

Angular momentum equation

$$r^2\dot{\theta} = L \tag{7.5}$$

Kepler's second law

Angular momentum conservation is equivalent to Kepler's second law. The area $\mathcal{A}$ shown in Figure 7.1 can be expressed (with an obvious choice of initial line) as

$$\mathcal{A} = \tfrac{1}{2}\int_0^\theta r^2\,d\theta.$$

Then, by the chain rule,

$$\frac{d\mathcal{A}}{dt} = \frac{d\mathcal{A}}{d\theta} \times \frac{d\theta}{dt} = \tfrac{1}{2}r^2\dot{\theta} = \tfrac{1}{2}L,$$

where L is the constant value of the angular momentum. Thus $\mathcal{A}$ increases at a constant rate, which is what Kepler's second law says. Thus Kepler's second law holds for *all* central force fields, not just the inverse square law.

7.2 GENERAL NATURE OF ORBITAL MOTION

In our first method of solution, we take as our starting point the principles of conservation of **angular momentum** and **energy**.

---

* Without losing generality, we will take L to be positive, that is, we suppose θ is *increasing* with time. (The special case in which $L = 0$ corresponds to rectilinear motion through O.)

Energy conservation

Every central field $\boldsymbol{F} = mf(r)\widehat{\boldsymbol{r}}$ is **conservative** with potential energy $mV(r)$, where

$$f(r) = -\frac{dV}{dr}. \tag{7.6}$$

Energy conservation then implies that

$$T + V = E,$$

where T is the specific kinetic energy, V is the specific potential energy, and the constant E is the specific total energy. On replacing T by its expression in polar coordinates, we obtain

Energy equation

$$\tfrac{1}{2}\left(\dot{r}^2 + (r\dot{\theta})^2\right) + V(r) = E \tag{7.7}$$

as the **energy conservation** equation. The conservation equations (7.5), (7.7) are equivalent to the Newton equations (7.1), (7.2) and are a convenient starting point for investigating the *general* nature of orbital motion.

The radial motion equation

From the angular momentum conservation equation (7.5), we have

$$\dot{\theta} = L/r^2$$

and, on eliminating $\dot{\theta}$ from the energy conservation equation (7.7), we obtain

$$\tfrac{1}{2}\dot{r}^2 + V(r) + \frac{L^2}{2r^2} = E, \tag{7.8}$$

an ODE for the radial distance $r(t)$. We call this the **radial motion equation** for the particle P. Equation (7.8) (together with the initial conditions) is sufficient to determine the variation of r with t, and the angular momentum equation (7.5) then determines the variation of θ with t. Unfortunately, for most laws of force, this procedure cannot be carried through analytically. However, it is still possible to make important deductions about the general nature of the motion.

Equation (7.8) can be written in the form

$$\tfrac{1}{2}\dot{r}^2 + V^*(r) = E, \tag{7.9}$$

where

$$V^*(r) = V(r) + \frac{L^2}{2r^2}. \tag{7.10}$$

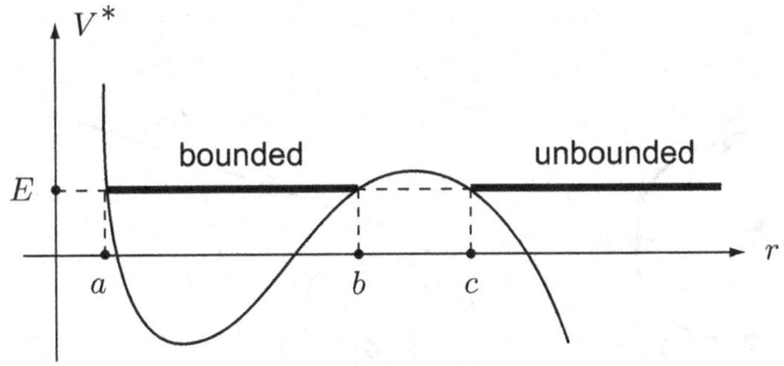

FIGURE 7.4 The effective potential V^* shown admits bounded and unbounded orbits, depending on the initial conditions.

The function $V^*(r)$ is called the **effective potential** of the radial motion and its use reduces the radial motion of P to a rectilinear problem. It must be emphasised though that the whole motion is *two*-dimensional since θ is increasing in accordance with (7.5).

Because r satisfies the radial motion equation (7.9), the variation of r with t can be analysed by using the same methods as were used in Chapter 6 for rectilinear particle motion. In particular, the general nature of the motion depends on the shape of the graph of V^* (which depends on L) and the value of E. The values of the constants L and E depend on the initial conditions.

Suppose for example that the law of force and the initial conditions are such that V^* has the form shown in Figure 7.4 and that E has the value shown. Then, since $T \geq 0$, it follows that the motion is restricted to those values r that satisfy the inequality

$$V^*(r) \leq E,$$

with equality holding when $\dot{r} = 0$. There are two possible motions, in each of which the variation of r with t is governed by the radial motion equation (7.8).

(i) a **bounded motion** in which r oscillates in the range $[a, b]$. In this motion, $r(t)$ is a periodic function.*

(ii) an **unbounded motion** in which r lies in the interval $[c, \infty)$. In this motion r is not periodic but decreases until the minimum value $r = c$ is achieved and then increases without limit.

The bounded orbit. A typical bounded orbit is shown in Figure 7.5 (left). The orbit alternately touches the inner and outer circles $r = a$ and $r = b$, which corresponds to the radial coordinate r oscillating in the interval $[a, b]$. Without losing generality, suppose that P is at the point B_1 when $t = 0$ and that OB_1 is the line $\theta = 0$. Consider the part of

* The fact that $r(t)$ is periodic does *not* mean that the whole motion must be periodic.

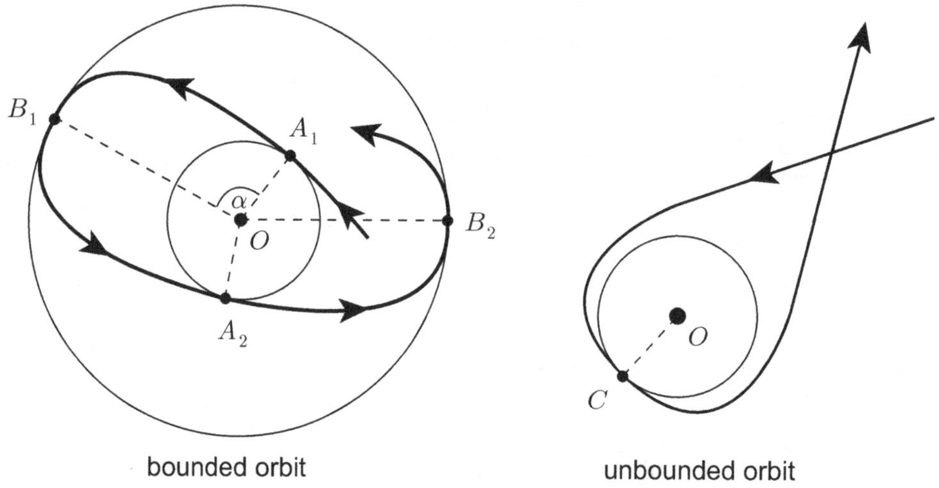

FIGURE 7.5 Typical bounded and unbounded orbits.

the orbit between A_1 and A_2. It follows from the governing equations (7.8), (7.5) that r is an *even* function of t while θ is an *odd* function of t. This means that the segment B_1A_2 of the orbit is just the reflection of the segment A_1B_1 in the line OB_1. This argument can be repeated to show that the segment A_2B_2 is the reflection of the segment B_1A_2 in the line OA_2, and so on. Thus the whole orbit can be constructed from a knowledge of a single segment such as A_1B_1.

It follows from what has been said that the angles $A_1\widehat{O}B_1$, $B_1\widehat{O}A_2$, $A_2\widehat{O}B_2$ (and so on) are all equal. Let α be the common value of these angles. Then the orbit will eventually close itself if some integer multiple of α is equal to some whole number of complete revolutions, that is, if α/π is a *rational number*. There is no reason to expect this condition to hold and, in general, it does not. It follows that these bounded orbits are *not generally closed*. The closed orbits associated with the attractive inverse square field are therefore exceptional, rather than typical!

The unbounded orbit. In the unbounded case there are just two segments both of which are semi-infinite (see Figure 7.5 (right)). The segment in which P recedes from O is the reflection of the segment in which P approaches O in the line OC.

Apses and apsidal distances

The points at which an orbit touches its bounding circles are important and are given a special name:

Definition 7.2 ***Apse, apsidal distance, apsidal angle*** *A point of an orbit at which the distance OP achieves its maximum or minimum value is called an* **apse** *of the orbit. These maximum and minimum distances are called the* **apsidal distances** *and the angular displacement between successive apses (the angle α in Figure 7.5 (left)) is called the* **apsidal angle**.

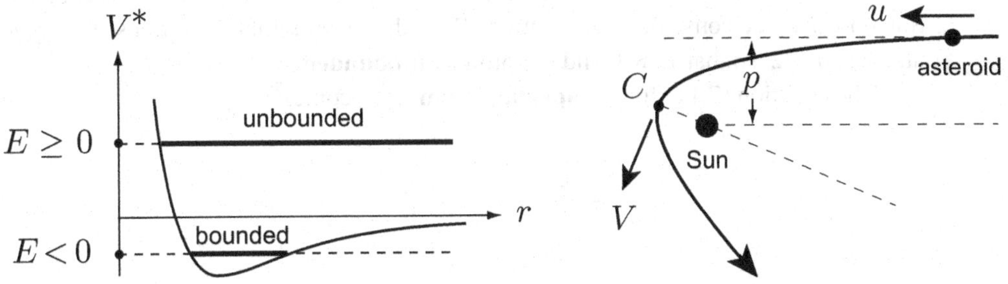

FIGURE 7.6 Left: The effective potential V^* for the attractive inverse square force. **Right:** The path of the asteroid around the Sun. C is the point of closest approach.

In the special case of orbits around the Sun, the point of closest approach is called the **perihelion** and the point of maximum distance the **aphelion**. The corresponding terms for orbits around the Earth are **perigee** and **apogee**.

The **apsidal distances**, the maximum and minimum distances of P from O, are easily found from the radial motion equation (7.8). At an apse, $\dot{r} = 0$ and so r must satisfy

$$V(r) + \frac{L^2}{2r^2} = E. \tag{7.11}$$

The positive roots of this equation are the apsidal distances.

Example 7.1 *Asteroid deflected by the Sun*

A particle P of mass m moves in the central force field $-(m\gamma/r^2)\widehat{\boldsymbol{r}}$, where γ is a positive constant. Show that bounded and unbounded orbits are possible depending on the value of E.

An asteroid is approaching the Sun from a great distance. At this time it has constant speed u and is moving in a straight line whose perpendicular distance from the Sun is p. Find the equation satisfied by the apsidal distances of the subsequent orbit. For the special case in which $u^2 = 4M_\odot G/3p$ (where $M_\odot$ is the mass of the Sun), find (i) the distance of closest approach of the asteroid to the Sun, and (ii) the speed of the asteroid at the time of closest approach.

Solution

For this law of force, $V = -\gamma/r$ and the effective potential V^* is

$$V^* = -\frac{\gamma}{r} + \frac{L^2}{2r^2}.$$

This V^* has the form shown in Figure 7.6 (left), from which it is clear that the orbit will be

(i) **bounded** if $E < 0$,

(ii) **unbounded** if $E \geq 0$,

whatever the value of L.

In the asteroid example, the constant $\gamma = M_{\odot}G$, where $M_{\odot}$ is the mass of the Sun and G is the constant of gravitation. With the given initial conditions, $L = p\,u$ and $E = u^2/2$, so that $E > 0$ and the orbit is **unbounded**.

The equation (7.11) for the **apsidal distances** becomes

$$-\frac{\gamma}{r} + \frac{p^2u^2}{2r^2} = \tfrac{1}{2}u^2,$$

that is,

$$u^2r^2 + 2\gamma r - p^2u^2 = 0,$$

where $\gamma = M_{\odot}G$.

For the special case in which $u^2 = 4M_{\odot}G/3p$, this equation simplifies to

$$2r^2 + 3pr - 2p^2 = 0.$$

The **distance** of closest approach of the asteroid is the *positive* root of this quadratic equation, namely $r = p/2$.

The **speed** V of the asteroid at closest approach is easily deduced from angular momentum conservation. Initially, $L = pu$ and, at closest approach, $L = (p/2)V$. It follows that $V = 2u$. ■

7.3 THE PATH EQUATION

In principle, the method of the last section allows us to determine the complete motion of the orbiting body as a function of the time. However, the procedure is usually too difficult to be carried through analytically. We can make the problem easier (and make more progress) by seeking just the **equation of the path** taken by the body, and not enquiring where the body is on this path at any particular time.

We start from the Newton equation (7.3) and try to eliminate the time by using the angular momentum equation (7.5). In doing this it is helpful to introduce the new dependent variable u, given by

$$u = 1/r. \tag{7.12}$$

This transformation has a magically simplifying effect. We begin by transforming $\dot{r}$ and $\ddot{r}$. By the chain rule,

$$\dot{r} = \frac{d}{dt}\left(\frac{1}{u}\right) = -\frac{1}{u^2} \times \frac{du}{d\theta} \times \frac{d\theta}{dt} = -\left(r^2\dot{\theta}\right)\frac{du}{d\theta}$$

which, on using the angular momentum equation (7.5), gives

$$\dot{r} = -L\frac{du}{d\theta}. \tag{7.13}$$

A second differentiation with respect to t then gives

$$\ddot{r} = -L\frac{d}{dt}\left(\frac{du}{d\theta}\right) = -L\frac{d^2u}{d\theta^2} \times \frac{d\theta}{dt} = -L^2u^2\frac{d^2u}{d\theta^2}, \qquad (7.14)$$

on using the angular momentum equation again.

The term $r\dot{\theta}^2 = L^2u^3$ so that the Newton equation (7.3) is transformed into

$$-L^2u^2\frac{d^2u}{d\theta^2} - L^2u^3 = f(1/u),$$

that is,

The path equation

$$\frac{d^2u}{d\theta^2} + u = -\frac{f(1/u)}{L^2u^2} \qquad (7.15)$$

This is the **path equation**. Its solutions are the polar equations of the paths that the body can take when it moves under the force field $\boldsymbol{F} = mf(r)\widehat{\boldsymbol{r}}$.

Despite the appearance of the left side of equation (7.15), the path equation is **not linear** in general. This is because the right side is a function of u, the *dependent* variable. Only for the **inverse square** and **inverse cube** laws does the path equation become linear. It is a remarkable piece of good luck that the inverse square law (the most important case by far) is one of only two cases that can be solved easily.

Initial conditions for the path equation

Suitable initial conditions for the path equation are provided by specifying the values of u and $du/d\theta$ when $\theta = \alpha$, say. Since $u = 1/r$, the initial value of u is given directly by the initial data. The value of $du/d\theta$ is not given directly but can be deduced from equation (7.13) in the form

$$\frac{du}{d\theta} = -\frac{\dot{r}}{L}, \qquad (7.16)$$

where $\dot{r}$ and L are obtained from the initial data.

Example 7.2 *Path equation for the inverse cube law*

The engines of the starship Enterprise have failed and the ship is moving in a straight line with speed V. The crew calculate that their present course will miss the planet B–Zar by a distance p. However, B–Zar is known to exert the force

$$\boldsymbol{F} = -\frac{m\gamma}{r^3}\widehat{\boldsymbol{r}}$$

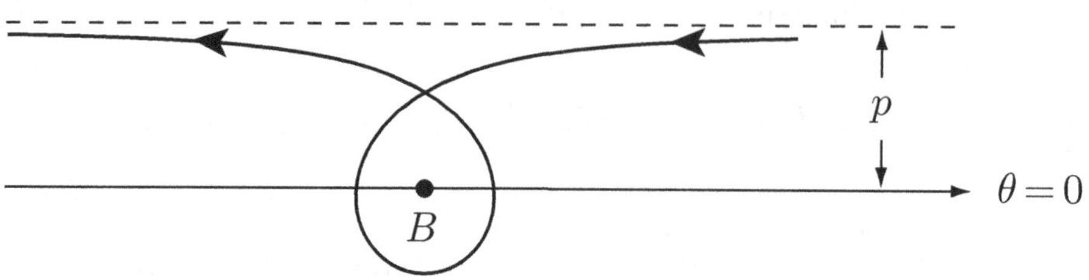

FIGURE 7.7 The path of the Enterprise around the planet B – Zar (B).

on any mass m in its vicinity. A measurement of the constant γ reveals that

$$\gamma = \frac{8p^2V^2}{9}.$$

Show that the crew of the Enterprise will get a free tour around B –Zar before continuing along their *original* path. What is the distance of closest approach and what is the speed of the Enterprise at that instant?

Solution

For the given law of force, $f(r) = -\gamma/r^3$ so that $f(1/u) = -\gamma u^3$. Also, from the initial conditions, $L = pV$. The path equation is therefore

$$\frac{d^2u}{d\theta^2} + u = \frac{\gamma u^3}{p^2V^2u^2},$$

which simplifies to

$$\frac{d^2u}{d\theta^2} + \frac{u}{9} = 0,$$

on using the stated value of γ. The general solution of this equation is

$$u = A\cos(\theta/3) + B\sin(\theta/3).$$

The constants A and B can now be determined from the initial conditions. Take the initial line $\theta = 0$ as shown in Figure 7.7. Then:

(i) The initial condition $r = \infty$ when $\theta = 0$ implies that $u = 0$ when $\theta = 0$. It follows that $A = 0$.

(ii) The initial condition on $du/d\theta$ is given by (7.16) to be

$$\frac{du}{d\theta} = -\frac{\dot{r}}{L} = -\left(\frac{-V}{pV}\right) = \frac{1}{p}$$

when $\theta = 0$. It follows that $B = 3/p$.

The required solution is therefore

$$u = \frac{3}{p}\sin(\theta/3),$$

that is

$$r = \frac{p}{3\sin(\theta/3)}.$$

This is the polar equation of the **path** of the Enterprise, as shown in Figure 7.7. The Enterprise recedes to infinity when $\sin(\theta/3) = 0$ again, that is when $\theta = 3\pi$. Thus the Enterprise makes one circuit of B–Zar before continuing on as before.

The distance of **closest approach** is $p/3$ and is achieved when $\theta = 3\pi/2$. By angular momentum conservation, the **speed** of the Enterprise at that instant is $3V$. ■

7.4 NEARLY CIRCULAR ORBITS

Although the path equation cannot be solved exactly for most laws of force, it is possible to obtain approximate solutions when the body is slightly perturbed from a *known* orbit. In particular, this can always be done when the unperturbed orbit is a circle with centre O.

Suppose that a particle P moves in a circular orbit of radius a under the force $f(r)$ per unit mass. This is only possible if $f(a)$ is *negative* and the particle speed v satisfies $v^2/a = -f(a)$; the angular momentum L is then given by $L^2 = -a^3 f(a)$. Suppose that P is now slightly disturbed by a small *radial* impulse. The angular momentum is unchanged but P now moves along some new path

$$u = \frac{1}{a}(1 + \xi(\theta)),$$

where ξ is a **small perturbation**. In terms of ξ, the path equation becomes

$$\frac{d^2\xi}{d\theta^2} + 1 + \xi = -\frac{(1+\xi)^{-2}}{f(a)} f\left(\frac{a}{1+\xi}\right).$$

This exact equation for ξ is non-linear, but we will now approximate it by expanding the right side in powers of ξ. On expanding the function $f(r)$ in a Taylor series about $r = a$ we obtain

$$\begin{aligned} f\left(\frac{a}{1+\xi}\right) &= f\left(a - \frac{a\xi}{1+\xi}\right) \\ &= f(a) - \left(\frac{a\xi}{1+\xi}\right) f'(a) + O\left(\frac{\xi}{1+\xi}\right)^2 \\ &= f(a) - af'(a)\,\xi + O\left(\xi^2\right), \end{aligned}$$

and a simple binomial expansion gives

$$(1+\xi)^{-2} = 1 - 2\xi + O\left(\xi^2\right).$$

On combining these results together, the constant terms cancel and we obtain

$$\frac{d^2\xi}{d\theta^2} + \left(3 + \frac{af'(a)}{f(a)}\right)\xi = 0, \tag{7.17}$$

on neglecting terms of order $O(\xi^2)$. This is the approximate **linearised equation** satisfied by the perturbation $\xi(\theta)$.

The general behaviour of the solutions of equation (7.17) depends on the *sign* of the coefficient of ξ.

(i) If

$$3 + \frac{af'(a)}{f(a)} < 0, \tag{7.18}$$

then the solutions are linear combinations of *real* exponentials, one of which has a positive exponent. In this case, the solution for ξ will not remain small, contrary to assumption. The conclusion is that the original circular orbit is **unstable**.

(ii) Alternatively, if

$$\Omega^2 \equiv 3 + \frac{af'(a)}{f(a)} > 0, \tag{7.19}$$

then the solutions are linear combinations of real cosines and sines, which remain bounded. The conclusion is that the original circular orbit is **stable** (at least to small radial impulses).

Closure of the perturbed orbits

From now on we will assume that the stability condition (7.19) is satisfied. The general solution of equation (7.17) then has the form

$$\xi = A\cos\Omega\theta + B\sin\Omega\theta.$$

We see that the perturbed orbit will **close** itself after one revolution if Ω is a **positive integer**. When the law of force is the **power law**

$$f(r) = kr^{\nu},$$

the perturbed orbit is stable for $\nu > -3$ and will close itself after one revolution if

$$\nu = m^2 - 3,$$

where m is a positive integer. The case $m = 1$ corresponds to inverse square attraction and $m = 2$ corresponds to simple harmonic attraction. The exponents $\nu = 6, 13, \ldots$ are also predicted to give closed orbits. It should be remembered though that these are only the predictions of the approximate linearised theory.* It is possible (but not pretty) to improve on the linear approximation by including quadratic terms in ξ as well as linear ones. The result of this refined theory is that the powers $\nu = -2$ and $\nu = 1$ still give

* It makes no sense to say that an orbit *approximately* closes itself!

closed orbits, but the powers $\nu = 6,\ 13,\ \ldots$ do not. This shows that the power laws with $\nu = 6,\ 13,\ \ldots$ do *not* give perturbed orbits that close after one revolution, but the cases $\nu = -2$ and $\nu = 1$ are still not finally decided. Mercifully, there is no need to carry the approximation procedure any further because all the paths corresponding to both inverse square and simple harmonic attraction can be calculated exactly. It is found that, for these two laws of force, *all bounded orbits close after one revolution.** There remains the possibility that the perturbed orbits might close themselves after more than one revolution, but a similar analysis shows that this does not happen. We have therefore shown that *the only power laws for which all bounded orbits are closed are the simple harmonic and inverse square laws*. This result is actually true for all central fields (not just power laws) and is known as **Bertrand's theorem**.

Precession of the perihelion of Mercury

The fact that the inverse square law leads to closed orbits, whilst very similar laws do not, provides an extremely sensitive test of the law of gravitation. Suppose for instance that the attractive force experienced by a planet were

$$f(r) = -\frac{\gamma}{r^{2+\epsilon}}$$

(per unit mass), where $\gamma > 0$ and $|\epsilon|$ is small. Then the value of Ω for a nearly circular orbit is

$$\Omega = (1-\epsilon)^{1/2} = 1 - \tfrac{1}{2}\epsilon + O(\epsilon^2).$$

This perturbed orbit does not close but has **apsidal angle** α, where

$$\alpha = \frac{\pi}{\Omega} = \frac{\pi}{1 - \frac{1}{2}\epsilon + O(\epsilon^2)} = \pi(1 + \tfrac{1}{2}\epsilon) + O(\epsilon^2).$$

Hence successive perihelions of the planet will not occur at the same point, but the **perihelion will advance** 'annually' by the small angle $\pi\epsilon$. The position of the perihelion of a planet can be measured with great accuracy. For the planet Mercury it is found (after all known perturbations have been subtracted out) that the perihelion advances by 43 (± 0.5) seconds of arc per century, or 5×10^{-7} radians per revolution. This corresponds to $\epsilon = 1.6 \times 10^{-7}$ and a power of -2.00000016 instead of -2. Miniscule though this discrepancy from the inverse square law seems, it is considerably greater than the error in the observations and for a considerable time was something of a puzzle.

This puzzle was resolved in a striking fashion by the theory of **general relativity**, published by Einstein in 1915. Einstein showed that one consequence of his theory was that planetary orbits *should* precess slightly and that, in the case of Mercury, the rate of precession should be 43 seconds of arc per century!

---

* In the inverse square case, the bounded orbits are ellipses with a *focus* at O, and, in the simple harmonic case, they are ellipses with the *centre* at O.

7.5 THE ATTRACTIVE INVERSE SQUARE FIELD

Because of its many applications to **astronomy**, the attractive inverse square field is the most important force field in the theory of orbits. The same field occurs in particle scattering when the two particles carry unlike electric charges. Because of these important applications, we will treat the inverse square field in more detail than other fields. In particular, we will obtain formulae that enable inverse square problems to be solved quickly and easily without referring to the equations of motion at all.

The paths

Suppose that $f(r) = -\gamma/r^2$ where $\gamma > 0$. Then $f(1/u) = -\gamma u^2$ and the path equation becomes

$$\frac{d^2u}{d\theta^2} + u = \frac{\gamma}{L^2},$$

where L is the angular momentum of the orbit. This has the form of the SHM equation with a constant on the right. The general solution is

$$u = A\cos\theta + B\sin\theta + \frac{\gamma}{L^2},$$

which can be written in the form

$$\frac{1}{r} = \frac{\gamma}{L^2}\Big(1 + e\cos(\theta - \alpha)\Big), \tag{7.20}$$

where e, α are constants with $e \geq 0$. This is the **polar equation of a conic** of eccentricity e and with one focus at O; α is the angle between the major axis of the conic and the initial line $\theta = 0$. If $e < 1$, then the conic is an **ellipse**; if $e = 1$ then the conic is a **parabola**; and when $e > 1$ the conic is the *near* branch of a **hyperbola**. The necessary geometry of the ellipse and hyperbola is summarised in Appendix A at the end of the chapter; the special case of the parabolic orbit is of marginal interest and we will make little mention of it.

Kepler's first law

It follows from the above that the only bounded orbits in the attractive inverse square field are **ellipses** with one **focus at the centre of force**. This is Kepler's first law, which is therefore a consequence of inverse square law attraction by the Sun. It would not be true for other laws of force.

The L-formula and the E-formula

By comparing the path formula (7.20) with the standard polar forms given in Appendix A, we see that the angular momentum L of the orbit is related to the conic parameters a,

b by the formula

$$\frac{\gamma}{L^2} = \frac{a}{b^2},$$

that is,

The L-formula

$$L^2 = \gamma b^2/a \tag{7.21}$$

We will call this result the **L-formula**. It applies to both elliptic and hyperbolic orbits. It is the first of two important formulae that relate L, E, the dynamical constants of the motion, to the conic parameters of the resulting orbit.

The second such formula involves the energy E. At the point of closest approach $r = c$,

$$E = \tfrac{1}{2}V^2 - \frac{\gamma}{c},$$

where V is the speed of P when $r = c$. Since P is moving transversely at the point of closest approach, it follows that $cV = L$, so that E may be written

$$E = \frac{L^2}{2c^2} - \frac{\gamma}{c} = \frac{\gamma b^2}{ac^2} - \frac{\gamma}{c}$$

on using the L-formula.

From this point on, the different types of conic must be treated separately. When the orbit is an ellipse, $c = a(1 - e)$, where e is the eccentricity, and a, b and e are related by the formula

$$e^2 = 1 - \frac{b^2}{a^2}.$$

Then E can be written

$$\begin{aligned} E &= \frac{\gamma a^2(1 - e^2)}{2a^3(1 - e)^2} - \frac{\gamma}{a(1 - e)} \\ &= -\frac{\gamma}{2a}. \end{aligned}$$

Thus the total energy E in the orbit is directly connected to a, the semi-major axis of the elliptical orbit. The parabolic and hyperbolic orbits are treated similarly and the full result, which we will call the **E-formula**, is

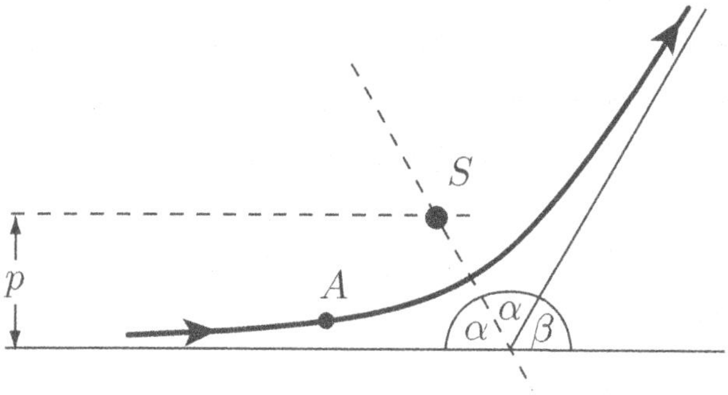

FIGURE 7.8 The asteroid A moves on a hyperbolic orbit around the Sun S as a focus and is deflected through the angle β.

The E-formula

Ellipse:	$E < 0$	$E = -\dfrac{\gamma}{2a}$	
Parabola:	$E = 0$		(7.22)
Hyperbola:	$E > 0$	$E = +\dfrac{\gamma}{2a}$	

Note that the type of orbit is determined solely by the *sign* of the total energy E. It follows that the **escape condition** (the condition that the body should eventually go off to infinity) is simply that $E \geq 0$. This useful result is true only for the inverse square law.

Example 7.3 *Asteroid deflected by the Sun*

An asteroid approaches the Sun with speed V along a line whose perpendicular distance from the Sun is p. Find the angle through which the asteroid is deflected by the Sun.

Solution

In this case we have the attractive inverse square field with $\gamma = M_\odot G$, where $M_\odot$ is the mass of the Sun. This problem can be solved from first principles by using the path equation, but here we make short work of it by using the L- and E-formulae.

From the initial conditions, $L = pV$ and $E = \frac{1}{2}V^2$. Since $E > 0$, the orbit is the near branch of a **hyperbola** and the L- and E-formulae give

$$p^2V^2 = \frac{M_\odot G b^2}{a} \qquad \text{and} \qquad \tfrac{1}{2}V^2 = +\frac{M_\odot G}{2a}.$$

It follows that

$$a = \frac{M_\odot G}{V^2}, \qquad b = p.$$

The semi-angle α between the asymptotes of the hyperbola is then given (see Appendix A) by

$$\tan\alpha = \frac{b}{a} = \frac{pV^2}{M_\odot G}.$$

Let β be the angle through which the asteroid is deflected. Then (see Figure 7.8) $\beta = \pi - 2\alpha$ and

$$\tan(\beta/2) = \tan(\pi/2 - \alpha) = \cot\alpha = \frac{M_\odot G}{pV^2}. \blacksquare$$

Period of the elliptic orbit

Whatever the law of force, once the path of P has been found, the progress of P along that path can be deduced from the angular momentum equation

$$r^2\dot{\theta} = L.$$

If we take $\theta = 0$ when $t = 0$, then the time t taken for P to progress to the point of the orbit with polar coordinates r, θ is given by

$$t = \frac{1}{L}\int_0^\theta r^2\, d\theta, \tag{7.23}$$

where $r = r(\theta)$ is the equation of the path. In particular then, the period τ of the elliptic orbit is given by

$$\tau = \frac{1}{L}\int_0^{2\pi} r^2\, d\theta,$$

where the path $r = r(\theta)$ is given by

$$\frac{1}{r} = \frac{a}{b^2}\left(1 + e\cos\theta\right). \tag{7.24}$$

Fortunately there is no need to evaluate the above integral since, for any path that closes itself after one circuit,

$$\tfrac{1}{2}\int_0^{2\pi} r^2\, d\theta = A,$$

where A is the area enclosed by the path. For the elliptical path, $A = \pi ab$ so that

$$\tau = \frac{2\pi ab}{L},$$

and on using the L-formula, the period of the elliptic orbit is given by:

The period formula

$$\tau = 2\pi \left(\frac{a^3}{\gamma} \right)^{1/2} \tag{7.25}$$

Kepler's third law

In the case of the planetary orbits, $\gamma = M_{\odot} G$, where $M_{\odot}$ is the mass of the Sun. Equation (7.25) can then be written

$$\tau^2 = \left(\frac{4\pi^2}{M_{\odot} G} \right) a^3. \tag{7.26}$$

This is Kepler's third law, which is therefore a consequence of inverse square law attraction by the Sun and would not be true for other laws of force.

Masses of celestial bodies

Once the constant of gravitation G is known, the formula (7.26) provides an accurate way to find the mass of the Sun. The same method applies to *any celestial body that has a satellite*. All that is needed is to measure the major axis $2a$ and the period τ of the satellite's orbit.*

Question ***Finding the mass of Jupiter***

The Moon moves in a nearly circular orbit of radius 384,000 km and period 27.32 days. Callisto, the fourth moon of the planet Jupiter, moves in a nearly circular orbit of radius 1,883,000 km and period 16.69 days. Estimate the mass of Jupiter as a multiple of the mass of the Earth.

Answer

$M_J = 316 M_E$.

Astronomical units

For astronomical problems, it is useful to write the period equation (7.26) in **astronomical units**. In these units, the unit of mass is the mass of the Sun ($M_{\odot}$), the unit of length (the AU) is the semi-major axis of the Earth's orbit, and the unit of time is the (Earth) year. On

* It should be noted that here we are neglecting the motion of the centre of force. We will see later that, when this is taken into account, formula (7.26) actually gives the *sum* of the masses of the body and its satellite. Usually, the satellite has a much smaller mass than the body and its contribution can be disregarded.

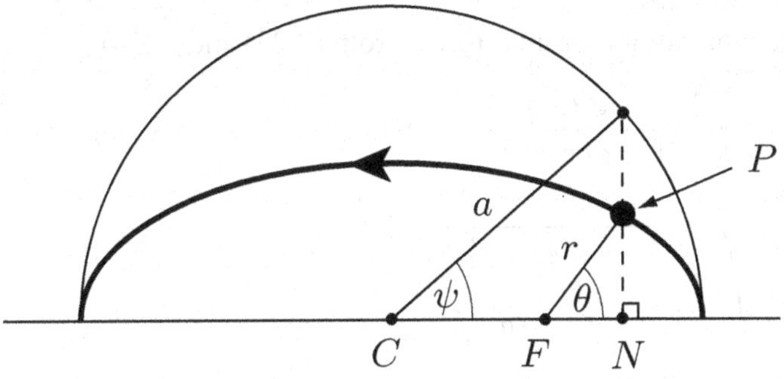

FIGURE 7.9 The eccentric angle ψ corresponding to the polar angle θ.

substituting the data for the Earth and Sun into equation (7.26), we find that $G = 4\pi^2$ in astronomical units. Hence, in **astronomical units** the period formula becomes

$$\tau^2 = \frac{a^3}{M}.$$

Question ***The major axis of the orbit of Pluto***

The period of Pluto is 248 years. What is the semi-major axis of its orbit?

Answer

39.5 AU. ■

Time dependence of the motion – Kepler's equation

The formula (7.23) can be used to find how long it takes for P to progress to a general point of the orbit. However, although the integration with respect to θ can be done in closed form, it is a *very* complicated expression. In order to obtain a manageable formula, we make a cunning change of variable, replacing the polar angle θ by the **eccentric angle** ψ. The relationship between these two angles is shown in Figure 7.9. Since $CN = CF + FN$, it follows that

$$a\cos\psi = ae + r\cos\theta,$$

and, on using the polar equation for the ellipse (7.24) together with the formula $b^2 = a^2(1 - e^2)$, the relation between ψ and θ can be written in the symmetrical form

$$(1 - e\cos\psi)(1 + e\cos\theta) = \frac{b^2}{a^2}. \tag{7.27}$$

Implicit differentiation of equation (7.27) with respect to ψ then gives

$$\frac{d\theta}{d\psi} = \frac{b}{a\,(1 - e\cos\psi)}, \tag{7.28}$$

after more manipulation.

We can now make the change of variable from θ to ψ. From (7.23) and (7.24)

$$\begin{aligned} t &= \frac{b^4}{a^2 L}\int_0^{\theta} \frac{d\theta}{(1+e\cos\theta)^2} \\ &= \frac{b^4}{a^2 L}\int_0^{\psi} \frac{1}{(1+e\cos\theta)^2}\left(\frac{d\theta}{d\psi}\right) d\psi \\ &= \frac{ab}{L}\int_0^{\psi} (1-e\cos\psi)\, d\psi, \\ &= \frac{ab}{L}\left(\psi - e\sin\psi\right), \end{aligned}$$

on using (7.27), (7.28). Finally, on making use of the L-formula $L^2 = \gamma b^2/a$, we obtain

Kepler's equation

$$t = \frac{\tau}{2\pi}\left(\psi - e\sin\psi\right) \qquad (7.29)$$

where τ (given by (7.25)) is the period of the orbit. This is **Kepler's equation** which gives the time as a function of position on the elliptical orbit.

If one needs to calculate the position of the orbiting body after a *given time*, then equation (7.29) must be solved numerically for the eccentric angle ψ. The corresponding value of θ is then given by equation (7.27) and the r value by equation (7.24) which, in view of (7.27), can be written in the form

$$r = a\left(1 - e\cos\psi\right). \qquad (7.30)$$

The need to solve Kepler's equation for the unknown ψ was a major stimulus in the development of approximate numerical methods for finding roots of equations.

Example 7.4 *Kepler's equation*

A body moving in an inverse square attractive field traverses an elliptical orbit with eccentricity e and period τ. Find the time taken for the body to traverse the half of the orbit that is nearer the centre of force.

Solution

The half of the orbit nearer the centre of force corresponds to the range $-\pi/2 \le \psi \le \pi/2$. The time taken is therefore

$$\frac{\tau}{\pi}\left(\frac{\pi}{2} - e\right) = \tau\left(\frac{1}{2} - \frac{e}{\pi}\right).$$

For example, Halley's comet moves on an elliptic orbit whose eccentricity is almost unity. It therefore spends only about 18% of its time on the half of its orbit that is nearer the Sun.

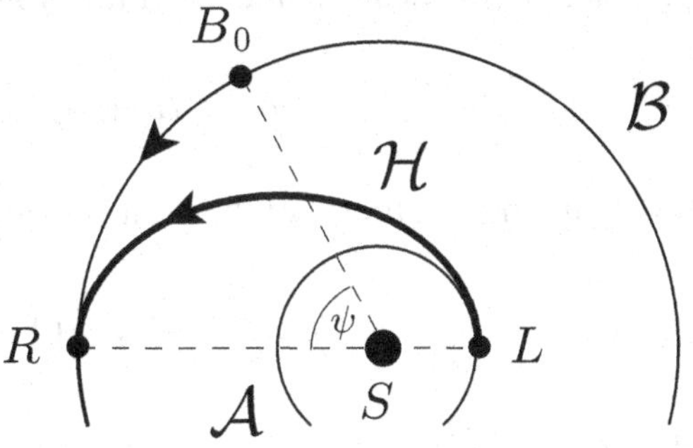

FIGURE 7.10 Two planets move on the circular orbits $\mathcal{A}$ and $\mathcal{B}$. A spacecraft is required to depart from one planet and rendezvous with the other planet at some point of its orbit. The Hohmann orbit $\mathcal{H}$ achieves this with the least expenditure of fuel.

7.6 SPACE TRAVEL – HOHMANN TRANSFER ORBITS

An important problem in space travel, and one that nicely illustrates the preceding theory, is that of transferring a spacecraft from one planet to another (from Earth to Jupiter say). In order to simplify the analysis, we will assume that both of the planetary orbits are circular. We will also suppose that the spacecraft has already effectively been removed from Earth's gravity, but is still in the vicinity of the Earth and is orbiting the Sun on the same orbit as the Earth. The object is to use the rocket motors to transfer the spacecraft to the vicinity of Jupiter, orbiting the Sun on the same orbit as Jupiter. Like everything else on board a spacecraft, fuel has to be transported from Earth at huge cost, so the transfer from Earth to Jupiter must be achieved using the *least mass of fuel*. In our analysis we will neglect the time during which the rocket engines are firing so that the engines are regarded as delivering an impulse to the spacecraft, resulting in a sudden change of velocity. After the initial firing impulse, the spacecraft is assumed to move freely under the Sun's gravitation until it reaches the orbit of Jupiter, when a second firing impulse is required to circularise the orbit. This is called a **two-impulse transfer**.

If the two firings produce velocity changes of $\Delta\boldsymbol{v}^A$ and $\Delta\boldsymbol{v}^B$ respectively, then the quantity Q that must be minimised if the least fuel is to be used is

$$Q = |\Delta\boldsymbol{v}^A| + |\Delta\boldsymbol{v}^B|.$$

The orbit that connects the two planetary orbits and minimises Q is called the **Hohmann transfer orbit*** and is shown in Figure 7.10. It has its perihelion at the lift-off point L and its aphelion at the rendezvous point R. It is not at all obvious that this is the optimal orbit; a proof is given in Appendix B at the end of the chapter. However, it is quite easy to find its properties.

Since the perihelial and aphelial distances in the Hohmann orbit are A and B (the radii of the orbits of Earth and Jupiter), it follows that

$$A = a(1 - e), \qquad B = a(1 + e),$$

* After Walter Hohmann, the German space research pioneer.

so that the geometrical parameters of the orbit are given by

$$a = \tfrac{1}{2}(B + A), \qquad e = \frac{B - A}{B + A}.$$

The angular momentum L of the orbit is then given by the L-formula to be

$$L^2 = \frac{\gamma b^2}{a} = \gamma\left(1 - e^2\right) a = \frac{\gamma B A}{B + A},$$

where $\gamma = M_{\odot} G$.

From L we can find the **speed** V^L of the spacecraft just after the lift-off firing, and the **speed** V^R at the rendezvous point just before the second firing. These are

$$V^L = \left(\frac{2\gamma B}{A(B + A)}\right)^{1/2}, \qquad V^R = \left(\frac{2\gamma A}{B(B + A)}\right)^{1/2}.$$

The **travel time** T, which is half the period of the Hohmann orbit, is given by

$$T^2 = \frac{\pi^2 a^3}{\gamma} = \frac{\pi^2 (B + A)^3}{8\gamma}.$$

Finally, in order to rendezvous with Jupiter, the lift-off must take place when Earth and Jupiter have the correct relative positions, so that Jupiter arrives at the meeting point at the right time. Since the speed of Jupiter is $(\gamma/B)^{1/2}$ and the travel time is now known, the angle ψ in Figure 7.10 must be

$$\psi = \pi\left(\frac{B + A}{2B}\right)^{3/2}.$$

Numerical results for the Earth–Jupiter transfer

In astronomical units, $G = 4\pi^2$, $A = 1$ AU and, for Jupiter, $B = 5.2$ AU. A speed of 1 AU per year is 4.74 km per second. Simple calculations then give:

(i) The travel time is 2.73 years, or 997 days.

(ii) V^L is 8.14 AU per year, which is 38.6 km per second. This is the speed the spacecraft must have after the lift-off firing.

(iii) V^R is 1.56 AU per year, which is 7.4 km per second. This is the speed with which the spacecraft arrives at Jupiter before the second firing.

(iv) The angle ψ at lift-off must be 83°.

The speeds V^L and V^R should be compared with the speeds of Earth and Jupiter in their orbits. These are 29.8 km/sec and 13.1 km/sec respectively. Thus the first firing must boost the speed of the spacecraft from 29.8 to 38.6 km/sec, and the second firing must boost the speed from 7.4 to 13.1 km/sec. The sum of these speed increments, 14.5 km/sec, is greater than the speed increment needed (12.4 km/sec) to escape from the Earth's orbit to infinity. Thus it takes more fuel to transfer a spacecraft from Earth's orbit to Jupiter's orbit than it does to escape from the solar system altogether!

7.7 THE REPULSIVE INVERSE SQUARE FIELD

The force field with $f(r) = +\gamma/r^2$, ($\gamma > 0$), is the **repulsive inverse square field**. It occurs in the interaction of charged particles carrying *like* charges and is required for the analysis of Rutherford scattering. Below we summarise the important properties of orbits in a repulsive inverse square field. These results are obtained in exactly the same way as for the attractive case.

The paths

The path equation is

$$\frac{d^2u}{d\theta^2} + u = -\frac{\gamma}{L^2},$$

where L is the angular momentum of the orbit. Its general solution can be written in the form

$$\frac{1}{r} = \frac{\gamma}{L^2}\left[-1 + e\cos(\theta - \alpha)\right],$$

where e, α are constants with $e \geq 0$. By comparing this path with the standard polar forms of conics given in Appendix A, we see that the path can only be the *far* branch of a **hyperbola** with focus at the centre O.

The L- and E-formulae

The formulae relating L, E, the dynamical constants of the orbit, to the hyperbola parameters are

$$L^2 = \gamma b^2/a, \tag{7.31}$$

$$E = +\gamma/2a. \tag{7.32}$$

7.8 RUTHERFORD SCATTERING

The most celebrated application of orbits in a repulsive inverse square field is Rutherford's* famous experiment in which a beam of alpha particles was scattered by gold nuclei in a sheet of gold leaf. We will analyse Rutherford's experiment in detail, beginning with the basic problem of a single alpha particle being deflected by a single fixed gold nucleus.

Alpha particle deflected by a heavy nucleus

An alpha particle A of mass m and charge q approaches a gold nucleus B of charge Q (see Figure 7.11). B is initially at rest and A is moving with speed V along a line whose

* Ernest Rutherford (1871–1937), a New Zealander, was one of the greatest physicists of the twentieth century. His landmark work on the structure of the nucleus in 1911 (and with Geiger and Marsden in 1913) was conducted at the University of Manchester, England.

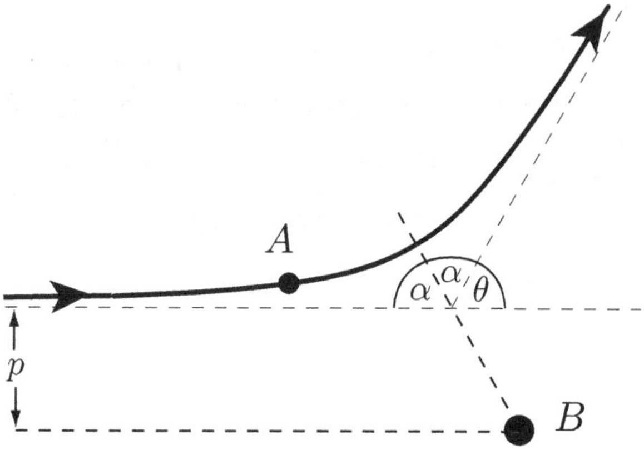

FIGURE 7.11 The alpha particle A of mass m and charge q is repelled by the fixed nucleus B of charge Q and moves on a hyperbolic orbit with the nucleus at the far focus. The alpha particle is deflected through the angle θ.

perpendicular distance from B is p. In the present treatment, we neglect the motion of the gold nucleus. This is justified since the mass of the gold nucleus is about fifty times larger than that of the alpha particle. A then moves in the electrostatic field due to B, which we now suppose to be fixed at the origin O. The force exerted on A is then

$$\boldsymbol{F} = +\frac{qQ}{r^2}\,\widehat{\boldsymbol{r}}$$

in cgs units. This is the **repulsive inverse square** field with $\gamma = qQ/m$.

We wish to find θ, the angle through which the alpha particle is deflected. This is obtained in exactly the same way as that of the asteroid in Example 7.1. From the initial conditions, $L = pV$ and $E = \frac{1}{2}V^2$. The L-formula (7.31) and the E-formula (7.32) then give

$$p^2V^2 = \frac{\gamma b^2}{a}, \qquad \tfrac{1}{2}V^2 = +\frac{\gamma}{2a}.$$

It follows that

$$a = \frac{\gamma}{V^2}, \qquad b = p.$$

The semi-angle α between the asymptotes of the hyperbola is then given (see Appendix A) by

$$\tan\alpha = \frac{b}{a} = \frac{pV^2}{\gamma}.$$

Hence, θ, the angle through which the asteroid is deflected, is given by

$$\tan(\theta/2) = \tan(\pi/2 - \alpha) = \cot\alpha = \frac{\gamma}{pV^2}.$$

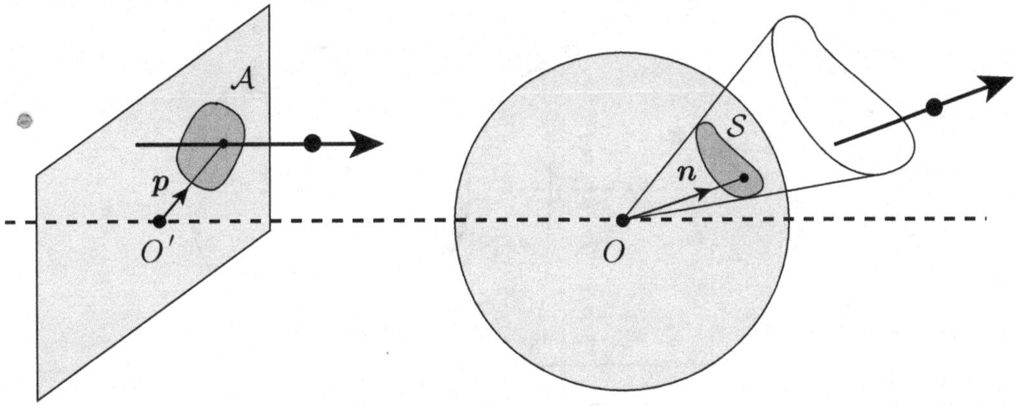

FIGURE 7.12 General scattering. A typical particle crosses the reference plane at the point $\boldsymbol{p}$ and finally emerges in the direction of the unit vector $\boldsymbol{n}$. Particles that cross the reference plane within the region $\mathcal{A}$ emerge within the (generalised) cone shown.

On writing $\gamma = qQ/m$, we obtain

$$\tan(\theta/2) = \frac{qQ}{mpV^2}. \tag{7.33}$$

as the formula for the **deflection angle** of the alpha particle. The quantity p, the distance by which the incident particle would miss the scatterer if there were no interaction, is called the **impact parameter** of the particle.

The deflection formula (7.33) cannot be confirmed directly by experiment since this would require the observation of a single alpha particle, a single nucleus, and a knowledge of the impact parameter p. What is actually done is to irradiate a gold target by a uniform beam of alpha particles of the same energy. Thus the target consists of many gold nuclei together with their associated electrons. However, the electrons have masses that are very small compared to that of an alpha particle and so their influence can be disregarded. Also, the gold target is taken in the form of thin foil to minimise the chance of multiple collisions. If multiple collisions are eliminated, then the gold nuclei act as *independent* scatterers and the problem reduces to that of a *single* fixed gold nucleus irradiated by a *uniform beam* of alpha particles. In this problem the alpha particles come in with different values of the impact parameter p and are scattered through different angles in accordance with formula (7.33). What *can* be measured is the **angular distribution** of the scattered alpha particles.

Differential scattering cross-section

The angular distribution of scattered particles is expressed by a function $\sigma(\boldsymbol{n})$, called the **differential scattering cross section**, where the unit vector $\boldsymbol{n}$ specifies the final direction of emergence of a particle from the scatterer O. One may imagine the values of $\boldsymbol{n}$ corresponding to points on the surface of a sphere with centre O and unit radius, as shown in Figure 7.12. Then values of $\boldsymbol{n}$ that lie in the shaded patch $\mathcal{S}$ correspond to particles whose final direction of emergence lies inside the (generalised) cone shown.

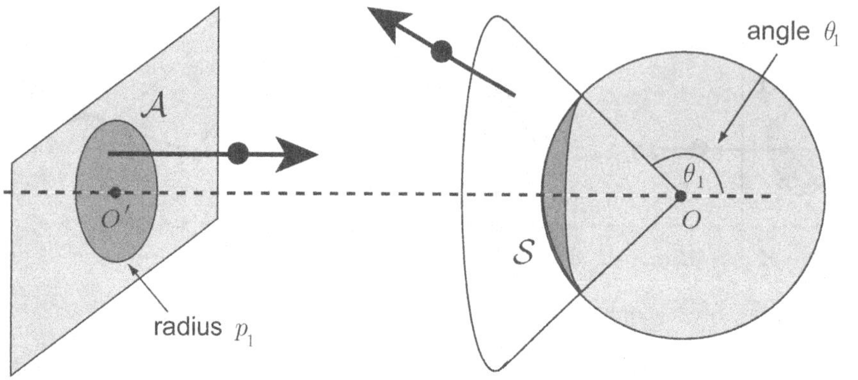

FIGURE 7.13 Axisymmetric scattering. Particles crossing the reference plane within the shaded circular disk are scattered and emerge in directions within the circular cone.

Take a reference plane far to the left of the scatterer and perpendicular to the incident beam, as shown in Figure 7.12. Suppose that there is a uniform flux of incoming particles crossing the reference plane such that N particles cross any unit area of the reference plane in unit time. When these particles have been scattered, they will emerge in different directions and some of the particles will emerge with directions lying within the (generalised) cone shown in Figure 7.12. The **differential scattering cross section** is defined to be that function $\sigma(\boldsymbol{n})$ such that the flux of particles that emerge with directions lying within the cone is given by the surface integral

$$N \int_{\mathcal{S}} \sigma(\boldsymbol{n})\, dS. \tag{7.34}$$

It is helpful to regard $\sigma(\boldsymbol{n})$ as a **scattering density**, analogous to a probability density, that must be integrated to give the flux of particles scattered within any given solid angle.

The particles that finally emerge within the cone must have crossed the reference plane within some region $\mathcal{A}$ as shown in Figure 7.12. A typical particle crosses the reference plane at the point $\boldsymbol{p}$ (relative to O') and eventually emerges in the direction $\boldsymbol{n}$ lying within the cone. However because the incoming beam is uniform, the flux of these particles across $\mathcal{A}$ is just $N|\mathcal{A}|$, where $|\mathcal{A}|$ is the **area** of the region $\mathcal{A}$. On equating the incoming and outgoing fluxes, we obtain the relation

$$\int_{\mathcal{S}} \sigma(\boldsymbol{n})\, dS = |\mathcal{A}|. \tag{7.35}$$

This is the general relation that *any* differential scattering cross section must satisfy; it simply expresses the equality of incoming and outgoing fluxes of particles. However, Rutherford scattering is axisymmetric and this provides a major simplification.

Axisymmetric scattering and Rutherford's formula

Rutherford scattering is simpler than the general case outlined above in that the problem is **axisymmetric** about the axis $O'O$. Thus σ depends on θ (the angle between $\boldsymbol{n}$ and the axis $O'O$), but is independent of ϕ (the azimuthal angle measured around the axis). In this case $\sigma(\theta)$ can be determined by using the axisymmetric regions shown in Figure 7.13. Particles that cross the reference plane within the *circle* centre O' and radius p_1 emerge within the *circular* cone $\theta_1 \le \theta \le \pi$, where p_1 and θ_1 are related by the deflection formula for a single particle, in our case formula (7.33). On applying equation (7.35) to the present case, we obtain

$$\int_S \sigma(\theta)\, dS = \pi p_1^2.$$

We evaluate the surface integral using θ, ϕ coordinates. The element of surface area on the unit sphere is given by $dS = \sin\theta\, d\theta d\phi$ so that

$$\begin{aligned}\int_S \sigma(\theta)\, dS &= \int_{\theta_1}^{\pi} \left\{ \int_0^{2\pi} \sigma(\theta) \sin\theta\, d\phi \right\} d\theta \\ &= 2\pi \int_{\theta_1}^{\pi} \sigma(\theta) \sin\theta\, d\theta.\end{aligned}$$

Hence

$$\begin{aligned}2\pi \int_{\theta_1}^{\pi} \sigma(\theta) \sin\theta\, d\theta = \pi p_1^2 & \\ &= 2\pi \int_0^{p_1} p\, dp \\ &= -2\pi \int_{\theta_1}^{\pi} p \frac{dp}{d\theta}\, d\theta,\end{aligned}$$

on changing the integration variable from p to θ. Here the impact parameter p is regarded as a function of the scattering angle θ. Now the above equality holds for all choices of the integration limit θ_1 and this can only be true if the two *integrands* are equal. Hence:

Axisymmetric scattering cross section

$$\sigma(\theta) = -\left(\frac{p}{\sin\theta}\right) \frac{dp}{d\theta} \tag{7.36}$$

This is the formula for the differential scattering cross section σ in any problem of **axisymmetric scattering**. All that is needed to evaluate it in any particular case is the expression for the impact parameter p in terms of the scattering angle θ.

In the case of **Rutherford scattering**, the expression for p in terms of θ is provided by solving equation (7.33) for p, which gives

$$p = \frac{qQ}{mV^2} \tan(\theta/2).$$

On substituting this function into the formula (7.36), we obtain

Rutherford's scattering cross-section

$$\sigma(\theta) = \frac{q^2 Q^2}{16E^2}\left(\frac{1}{\sin^4(\theta/2)}\right) \tag{7.37}$$

where $E(= \frac{1}{2}mV^2)$ is the energy of the incident alpha particles. This is **Rutherford's formula** for the angular distribution of the scattered alpha particles.

Significance of Rutherford's experiment

In the above description we have used the term 'nucleus' for convenience. What we really mean is '*the positively charged part of the atom that carries most of the mass*'. If this positive charge is distributed in a spherically symmetric manner, then the above results still hold, irrespective of the radius of the charge, provided that the alpha particles *do not penetrate* into the charge itself. What Rutherford found was that, when using alpha particles from a radium source, the formula (7.37) held even for particles that were scattered through angles close to π. These are the particles that get closest to the nucleus, the distance of closest approach being qQ/E. This meant that the nuclear radius of gold must be smaller than this distance, which was about 10^{-12} cm in Rutherford's experiment. The radius of an atom of gold is about 10^{-8} cm. This result completely contradicted the Thompson model, in which the positive charge was distributed over the whole volume of the atom, by showing that the nucleus (as it became known) must be a very small and very dense core at the centre of the atom.

Note on two-body scattering problems

Throughout this section we have neglected the motion of the target nucleus. This will introduce only small errors when the target nucleus is much heavier than the incident particles, as it was in Rutherford's experiment. However, if lighter nuclei are used as the target, then the motion of the nucleus cannot be neglected and we have a **two-body scattering problem**. Such problems are treated in Chapter 10.

Appendix A The geometry of conics

Ellipse

(i) In **Cartesian coordinates**, the standard ellipse with **semi-major axis** a and **semi-minor axis** b ($b \le a$) has the equation

$$\frac{x^2}{a^2} + \frac{y^2}{b^2} = 1.$$

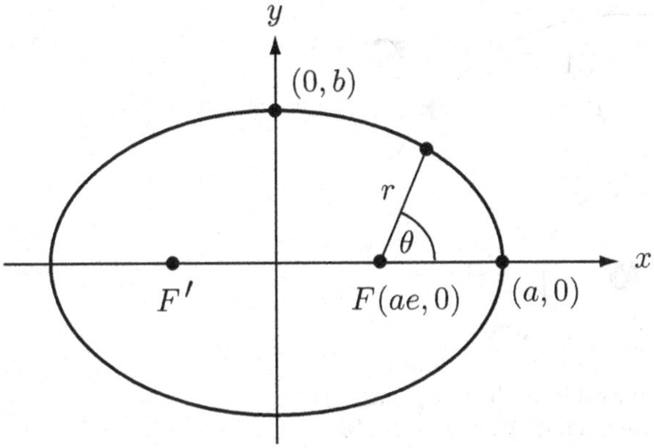

FIGURE 7.14 The standard ellipse $x^2/a^2 + y^2/b^2 = 1$.

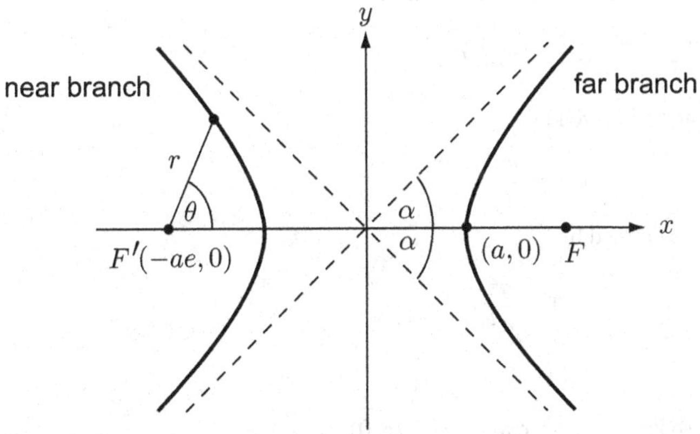

FIGURE 7.15 The standard hyperbola $x^2/a^2 - y^2/b^2 = 1$. The near and far branches are relative to the focus F', which is the origin of polar coordinates.

(ii) The **eccentricity** e of the ellipse is defined by

$$e^2 = 1 - \frac{b^2}{a^2}$$

and lies in the range $0 \le e < 1$. When $e = 0$, $b = a$ and the ellipse is a circle.

(iii) The **focal points** F, F' of the ellipse lie on the major axis at $(\pm ae, 0)$.

(iv) In **polar coordinates** with origin at the focus F and with initial line in the positive x-direction, the equation of the ellipse is

$$\frac{1}{r} = \frac{a}{b^2}(1 + e\cos\theta).$$

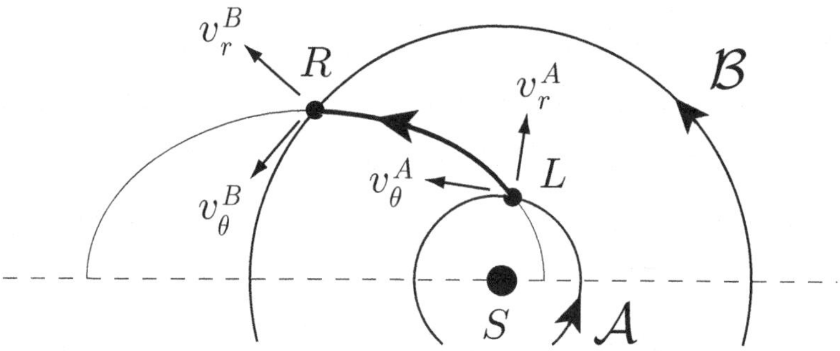

FIGURE 7.16 The circular orbits $\mathcal{A}$ and $\mathcal{B}$ are the orbits of the two planets. The elliptical orbit shown is a possible path for the spacecraft, which travels along the arc LR. The velocities shown are those *after* the first firing at L and *before* the second firing at R.

Hyperbola

(i) In **Cartesian coordinates**, the standard hyperbola has the equation

$$\frac{x^2}{a^2} - \frac{y^2}{b^2} = 1 \qquad (a, b > 0)$$

so that the angle 2α between the asymptotes is given by

$$\tan\alpha = \frac{b}{a}.$$

(ii) The **eccentricity** e of the hyperbola is defined by

$$e^2 = 1 + \frac{b^2}{a^2}$$

and lies in the range $e > 1$.

(iii) The **focal points** F, F' of the hyperbola lie on the x-axis at $(\pm ae, 0)$.

(iv) In **polar coordinates** with origin at the focus F' and with initial line in the positive x-direction, the equations of the near and far branches of the hyperbola are

$$\frac{1}{r} = \frac{a}{b^2}(1 + e\cos\theta), \qquad \frac{1}{r} = \frac{a}{b^2}(-1 + e\cos\theta),$$

respectively.

Appendix B The Hohmann orbit is optimal

The result that the Hohmann orbit is the connecting orbit that minimises Q is not at all obvious and correct proofs are rare.* Hopefully, the proof given below *is* correct!

* It is sometimes stated that the optimality requirement is to minimise the *energy* of the connecting orbit, which is not true. In any case, the Hohmann orbit is *not* the connecting orbit of minimum energy!

Proof of optimality Consider the general two-impulse transfer orbit LR shown in Figure 7.16, where the orbit is regarded as being generated by the velocity components v_θ^A, v_r^A of the spacecraft after the first impulse. Then, by angular momentum and energy conservation,

$$Av_\theta^A = Bv_\theta^B,$$

$$\left(v_r^A\right)^2 + \left(v_\theta^A\right)^2 - \frac{2\gamma}{A} = \left(v_r^B\right)^2 + \left(v_\theta^B\right)^2 - \frac{2\gamma}{B},$$

where A, B are the radii of the circular orbits of Earth and Jupiter and $\gamma = M_\odot G$. Thus

$$v_\theta^B = \frac{A}{B} v_\theta^A,$$

$$\left(v_r^B\right)^2 = \left(1 - \frac{A^2}{B^2}\right)\left(v_\theta^A\right)^2 + \left(v_r^A\right)^2 - 2\gamma\left(\frac{1}{A} - \frac{1}{B}\right).$$

Since the orbital speeds of Earth and Jupiter are $(\gamma/A)^{1/2}$ and $(\gamma/B)^{1/2}$, it follows that the velocity changes $\Delta \boldsymbol{v}^A$, $\Delta \boldsymbol{v}^B$ required at L and R have magnitudes given by

$$|\Delta \boldsymbol{v}^A|^2 = \left(v_\theta^A - \left(\frac{\gamma}{A}\right)^{1/2}\right)^2 + \left(v_r^A\right)^2,$$

$$\begin{aligned}|\Delta \boldsymbol{v}^B|^2 &= \left(\left(\frac{\gamma}{B}\right)^{1/2} - v_\theta^B\right)^2 + \left(v_r^B\right)^2\\ &= \left(\left(\frac{\gamma}{B}\right)^{1/2} - \frac{A}{B}v_\theta^A\right)^2 + \left(1 - \frac{A^2}{B^2}\right)\left(v_\theta^A\right)^2 + \left(v_r^A\right)^2 - 2\gamma\left(\frac{1}{A} - \frac{1}{B}\right)\\ &= \left(v_\theta^A - \frac{\gamma^{1/2}A}{B^{3/2}}\right)^2 + \left(v_r^A\right)^2 + \gamma\left(\frac{3}{B} - \frac{2}{A} - \frac{A^2}{B^3}\right).\end{aligned}$$

It is evident that, with v_θ^A fixed, both $|\Delta \boldsymbol{v}^A|$ and $|\Delta \boldsymbol{v}^B|$ are *increasing* functions of v_r^A. Thus Q may be reduced by reducing v_r^A provided that the resulting orbit still meets the circle $r = B$. Q can be thus reduced until either

(i) v_r^A is reduced to zero, or
(ii) the orbit shrinks until it touches the circle $r = B$ and any further reduction in v_r^A would mean that the orbit would not meet $r = B$.

In the first case, L becomes the perihelion of the orbit and, in the second case, R becomes the aphelion of the orbit. We will proceed assuming the first case, the second case being treated in a similar manner and with the same result.

Suppose then that L is the perihelion of the connecting orbit. Then $v_r^A = 0$ and, from now on, we will simply write v instead of v_θ^A. The velocity v must be such that the orbit reaches the circle $r = B$, which now means that the major axis of the orbit must not be less than $A + B$. On using the E-formula, this implies that v must satisfy

$$v^2 \geq \frac{2\gamma B}{A(A+B)}.$$

The formulae for $|\Delta \boldsymbol{v}^A|$ and $|\Delta \boldsymbol{v}^B|$ now simplify to

$$|\Delta \boldsymbol{v}^A|^2 = \left(v - \left(\frac{\gamma}{A}\right)^{1/2}\right)^2,$$

$$|\Delta \boldsymbol{v}^B|^2 = \left(v - \frac{\gamma^{1/2}A}{B^{3/2}}\right)^2 + \gamma\left(\frac{3}{B} - \frac{2}{A} - \frac{A^2}{B^3}\right)$$

from which it is evident that, for v in the permitted range, both of $|\Delta \boldsymbol{v}^A|$ and $|\Delta \boldsymbol{v}^B|$ are *increasing* functions v. Hence the minimum value of Q is achieved when v takes its smallest permitted value, namely

$$v = \left(\frac{2\gamma B}{A(A+B)}\right)^{1/2}.$$

With this value of v, the orbit touches the circle $r = B$ and so has its aphelion at R. Hence *the optimum orbit has its perihelion at L and its aphelion at R*. This is precisely the **Hohmann orbit.** ■

Problems on Chapter 7

Answers and comments are at the end of the book.

Harder problems carry a star (∗).

Radial motion equation, apses

7.1 A particle P of mass m moves under the repulsive inverse cube field $\boldsymbol{F} = (m\gamma/r^3)\widehat{\boldsymbol{r}}$. Initially P is at a great distance from O and is moving with speed V towards O along a straight line whose perpendicular distance from O is p. Find the equation satisfied by the apsidal distances. What is the distance of closest approach of P to O?

7.2 A particle P of mass m moves under the attractive inverse square field $\boldsymbol{F} = -(m\gamma/r^2)\widehat{\boldsymbol{r}}$. Initially P is at a point C, a distance c from O, when it is projected with speed $(\gamma/c)^{1/2}$ in a direction making an acute angle α with the line OC. Find the apsidal distances in the resulting orbit.

Given that the orbit is an ellipse with O at a focus, find the semi-major and semi-minor axes of this ellipse.

7.3 A particle of mass m moves under the attractive inverse square field $\boldsymbol{F} = -(m\gamma/r^2)\widehat{\boldsymbol{r}}$. Show that the equation satisfied by the apsidal distances is

$$2Er^2 + 2\gamma r - L^2 = 0,$$

where E and L are the specific total energy and angular momentum of the particle. When $E < 0$, the orbit is known to be an ellipse with O as a focus. By considering the sum and product of the roots of the above equation, establish the elliptic orbit formulae

$$L^2 = \gamma b^2/a, \qquad E = -\gamma/2a.$$

7.4 A particle P of mass m moves under the simple harmonic field $\boldsymbol{F} = -(m\Omega^2 r)\widehat{\boldsymbol{r}}$, where Ω is a positive constant. Obtain the radial motion equation and show that all orbits of P are bounded.

Initially P is at a point C, a distance c from O, when it is projected with speed Ωc in a direction making an acute angle α with OC. Find the equation satisfied by the apsidal distances. Given that the orbit of P is an ellipse with centre O, find the semi-major and semi-minor axes of this ellipse.

Path equation

7.5 A particle P moves under the attractive inverse square field $\boldsymbol{F} = -(m\gamma/r^2)\widehat{\boldsymbol{r}}$. Initially P is at the point C, a distance c from O, and is projected with speed $(3\gamma/c)^{1/2}$ perpendicular to OC. Find the polar equation of the path make a sketch of it. Deduce the angle between OC and the final direction of departure of P.

7.6 A comet moves under the gravitational attraction of the Sun. Initially the comet is at a great distance from the Sun and is moving towards it with speed V along a straight line whose perpendicular distance from the Sun is p. By using the path equation, find the angle through which the comet is deflected and the distance of closest approach.

7.7 A particle P of mass m moves under the attractive inverse cube field $\boldsymbol{F} = -(m\gamma^2/r^3)\widehat{\boldsymbol{r}}$, where γ is a positive constant. Initially P is at a great distance from O and is projected towards O with speed V along a line whose perpendicular distance from O is p. Obtain the path equation for P.

For the case in which

$$V = \frac{15\gamma}{\sqrt{209}\, p},$$

find the polar equation of the path of P and make a sketch of it. Deduce the distance of closest approach to O, and the final direction of departure.

7.8* A particle P of mass m moves under the central field $\boldsymbol{F} = -(m\gamma^2/r^5)\widehat{\boldsymbol{r}}$, where γ is a positive constant. Initially P is at a great distance from O and is projected towards O with speed $\sqrt{2}\gamma/p^2$ along a line whose perpendicular distance from O is p. Show that the polar equation of the path of P is given by

$$r = \frac{p}{\sqrt{2}} \coth\left(\frac{\theta}{\sqrt{2}}\right).$$

Make a sketch of the path.

7.9* A particle of mass m moves under the central field

$$\boldsymbol{F} = -m\gamma^2\left(\frac{4}{r^3} + \frac{a^2}{r^5}\right)\widehat{\boldsymbol{r}},$$

where γ and a are positive constants. Initially the particle is at a distance a from the centre of force and is projected at right angles to the radius vector with speed $3\gamma/\sqrt{2}a$. Find the polar equation of the resulting path and make a sketch of it.

Find the time taken for the particle to reach the centre of force.

Nearly circular orbits

7.10 A particle of mass m moves under the central field

$$\boldsymbol{F} = -m\left(\frac{\gamma e^{-\epsilon r/a}}{r^2}\right)\widehat{\boldsymbol{r}},$$

where γ, a and ϵ are positive constants. Find the apsidal angle for a nearly circular orbit of radius a. When ϵ is small, show that the perihelion of the orbit advances by approximately $\pi\epsilon$ on each revolution.

7.11 ***Solar oblateness*** A planet of mass m moves in the equatorial plane of a star that is a uniform oblate spheroid. The planet experiences a force field of the form

$$\boldsymbol{F} = -\frac{m\gamma}{r^2}\left(1 + \frac{\epsilon\, a^2}{r^2}\right)\widehat{\boldsymbol{r}},$$

approximately, where γ, a and ϵ are positive constants and ϵ is small. If the planet moves in a nearly circular orbit of radius a, find an approximation to the 'annual' advance of the perihelion. [It has been suggested that oblateness of the Sun might contribute significantly to the precession of the planets, thus undermining the success of general relativity. This point has yet to be resolved conclusively.]

7.12 Suppose the solar system is embedded in a gravitating dust cloud of uniform density ρ. Find an approximation to the 'annual' advance of the perihelion of a planet moving in a nearly circular orbit of radius a. (For convenience, let $\rho = \epsilon M/a^3$, where M is the solar mass and ϵ is small.)

7.13 ***Orbits in general relativity*** In the theory of general relativity, the path equation for a planet moving in the gravitational field of the Sun is, in the standard notation,

$$\frac{d^2u}{d\theta^2} + u = \frac{MG}{L^2} + \left(\frac{3MG}{c^2}\right)u^2,$$

where c is the speed of light. Find an approximation to the 'annual' advance of the perihelion of a planet moving in a nearly circular orbit of radius a.

Scattering

7.14 A uniform flux of particles is incident upon a fixed hard sphere of radius a. The particles that strike the sphere are reflected elastically. Find the differential scattering cross section.

7.15 A uniform flux of particles, each of mass m and speed V, is incident upon a fixed scatterer that exerts the repulsive radial force $\boldsymbol{F} = (m\gamma^2/r^3)\widehat{\boldsymbol{r}}$. Find the impact parameter p as a function of the scattering angle θ, and deduce the differential scattering cross section. Find the total back-scattering cross-section.

Assorted inverse square problems

Some useful **data**:

The radius R of the Earth is 6380 km. To obtain the value of MG, where M is the mass of the Earth, use the formula $MG = R^2 g$, where $g = 9.80$ m s^{-2}.

1 AU per year is 4.74 km per second. In astronomical units, $G = 4\pi^2$.

7.16 In Yuri Gagarin's first manned space flight in 1961, the perigee and apogee were 181 km and 327 km above the Earth. Find the period of his orbit and his maximum speed in the orbit.

7.17 An Earth satellite has a speed of 8.60 km per second at its perigee 200 km above the Earth's surface. Find the apogee distance above the Earth, its speed at the apogee, and the period of its orbit.

7.18 A spacecraft is orbiting the Earth in a circular orbit of radius c when the motors are fired so as to multiply the speed of the spacecraft by a factor k ($k > 1$), its direction of motion being unaffected. [You may neglect the time taken for this operation.] Find the range of k for which the spacecraft will escape from the Earth, and the eccentricity of the escape orbit.

7.19 A spacecraft travelling with speed V approaches a planet of mass M along a straight line whose perpendicular distance from the centre of the planet is p. When the spacecraft is at a distance c from the planet, it fires its engines so as to multiply its current speed by a factor k ($0 < k < 1$), its direction of motion being unaffected. [You may neglect the time taken for this operation.] Find the condition that the spacecraft should go into orbit around the planet.

7.20 A body moving in an inverse square attractive field traverses an elliptical orbit with major axis $2a$. Show that the time average of the potential energy $V = -\gamma/r$ is $-\gamma/a$. [Transform the time integral to an integral with respect to the eccentric angle ψ.]

Deduce the time average of the kinetic energy in the same orbit.

7.21 A body moving in an inverse square attractive field traverses an elliptical orbit with eccentricity e and major axis $2a$. Show that the time average of the distance r of the body from the centre of force is $a(1 + \frac{1}{2}e^2)$. [Transform the time integral to an integral with respect to the eccentric angle ψ.]

7.22 A spacecraft is 'parked' in a circular orbit 200 km above the Earth's surface. The spacecraft is to be sent to the Moon's orbit by Hohmann transfer. Find the velocity changes Δv^E and Δv^M that are required at the Earth and Moon respectively. How long does the journey take? [The radius of the Moon's orbit is 384,000 km. Neglect the gravitation of the Moon.]

7.23* A spacecraft is 'parked' in an *elliptic* orbit around the Earth. What is the most fuel efficient method of escaping from the Earth by using a single impulse?

7.24 A satellite already in the Earth's heliocentric orbit can fire its engines only once. What is the most fuel efficient method of sending the satellite on a 'flyby' visit to another planet? The satellite can visit either Mars or Venus. Which trip would use less fuel? Which trip would take the shorter time? [The orbits of Mars and Venus have radii 1.524 AU and 0.723 AU respectively.]

7.25 A satellite is 'parked' in a circular orbit 250 km above the Earth's surface. What is the most fuel efficient method of transferring the satellite to an (elliptical) synchronous orbit by using a single impulse? [A synchronous orbit has a period of 23 hr 56 m.] Find the value of $\Delta \boldsymbol{v}$ and apogee distance.

Effect of resistance

7.26 A satellite of mass m moves under the attractive inverse square field $-(m\gamma/r^2)\widehat{\boldsymbol{r}}$ and is also subject to the linear resistance force $-mK\boldsymbol{v}$, where K is a positive constant. Show that the governing equations of motion can be reduced to the form

$$\ddot{r} + K\dot{r} + \frac{\gamma}{r^2} - \frac{L_0^2 e^{-2Kt}}{r^3} = 0, \qquad r^2\dot{\theta} = L_0 e^{-Kt},$$

where L_0 is a constant which will be assumed to be positive.

Suppose now that the effect of resistance is slight and that the satellite is executing a 'circular' orbit of slowly changing radius. By neglecting the terms in $\dot{r}$ and $\ddot{r}$, find an approximate solution for the time variation of r and θ in such an orbit. Deduce that small resistance causes the circular orbit to contract slowly, but that the satellite speeds up!

7.27 Repeat the last problem for the case in which the particle moves under the simple harmonic attractive field $-(m\Omega^2 r)\widehat{\boldsymbol{r}}$ with the same law of resistance. Show that, in this case, the body slows down as the orbit contracts. [This problem can be solved exactly in Cartesian coordinates, but do not do it this way.]

Computer assisted problems

7.28 ***See the advance of the perihelion of Mercury*** It is possible to 'see' the advance of the perihelion of Mercury predicted by general relativity by direct numerical solution. Take Einstein's path equation (see Problem 7.13) in the dimensionless form

$$\frac{d^2 v}{d\theta^2} + v = \frac{1}{1-e^2} + \eta v^2,$$

where $v = au$. Here a and e are the semi-major axis and eccentricity of the non-relativistic elliptic orbit and $\eta = 3MG/ac^2$ is a small dimensionless parameter. For the orbit of Mercury, $\eta = 2.3 \times 10^{-7}$ approximately.

Solve this equation numerically with the initial conditions $r = a(1+e)$ and $\dot{r} = 0$ when $\theta = 0$; this makes $\theta = 0$ an aphelion of the orbit. To make the precession easy to see, use a fairly eccentric ellipse and take η to be about 0.005, which speeds up the precession by a factor of more than 10^4!

7.29 ***Orbit with linear resistance*** Confirm the approximate solution for small resistance obtained in Problem 7.26 by numerical solution of the governing simultaneous ODEs. First write the governing equations in dimensionless form. Suppose that, in the absence of resistance, a circular orbit with $r = a$ and $\dot{\theta} = \Omega$ is possible; then $\gamma = a^3\Omega$ and $L_0 = a^2\Omega$. On taking dimensionless variables ρ, τ defined by $\rho = r/a$ and $\tau = \Omega t$, and taking $L_0 = a^2\Omega$, the governing equations become

$$\frac{d^2\rho}{d\tau^2} + \epsilon\frac{d\rho}{d\tau} + \frac{1}{\rho^2} - \frac{e^{-2\epsilon\tau}}{\rho^3} = 0, \qquad \rho^2\frac{d\theta}{d\tau} = e^{-2\epsilon\tau},$$

where $\epsilon = K/\Omega$ is the dimensionless resistance parameter. Solve these equations with the initial conditions $\rho = 1$, $d\rho/d\tau = 0$ and $\theta = 0$ when $\tau = 0$. Choose some small value for ϵ and plot a polar graph of the path.

Chapter Eight

Non-linear oscillations and phase space

KEY FEATURES

The key features of this chapter are the use of **perturbation theory** to solve weakly non-linear problems, the notion of **phase space**, the **Poincaré–Bendixson** theorem, and **limit cycles**.

In reality, most oscillating mechanical systems are governed by **non-linear equations**. The linear oscillation theory developed in Chapter 5 is generally an approximation which is accurate only when the amplitude of the oscillations is small. Unfortunately, non-linear oscillation equations do not have nice exact solutions as their linear counterparts do, and this makes the non-linear theory difficult to investigate analytically.

In this chapter we describe two different analytical approaches, each of which is successful in its own way. The first is to use **perturbation theory** to find successive corrections to the linear theory. This gives a more accurate solution than the linear theory when the non-linear terms in the equation are small. However, because the solution is close to that predicted by the linear theory, new phenomena associated with non-linearity are unlikely to be discovered by perturbation theory! The second approach involves the use of geometrical arguments in **phase space**. This has the advantage that the non-linear effects can be large, but the conclusions are likely to be qualitative rather than quantitative. A particular triumph of this approach is the **Poincaré–Bendixson** theorem, which can be used to prove the existence of **limit cycles**, a new phenomenon that exists only in the non-linear theory.

8.1 PERIODIC NON-LINEAR OSCILLATIONS

Most oscillating mechanical systems are not exactly linear but are approximately linear when the oscillation amplitude is small. In the case of a body on a spring, the restoring force might actually have the form

$$S = m\Omega^2 x + m\Lambda x^3, \qquad (8.1)$$

which is approximated by the linear formula $S = m\Omega^2 x$ when the displacement x is small. The new constant Λ is a measure of the strength of the non-linear effect. If $\Lambda < 0$, then

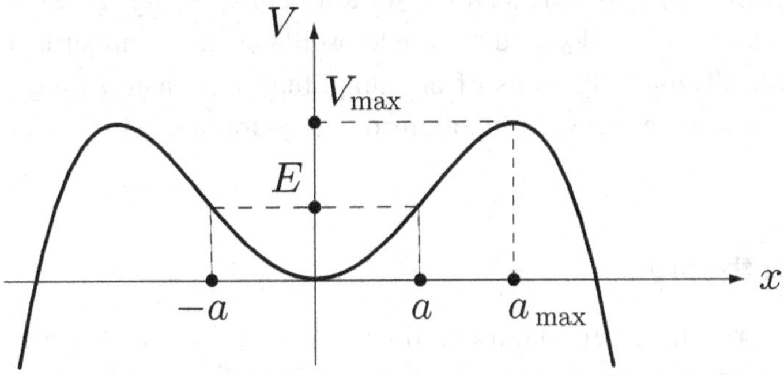

FIGURE 8.1 Existence of periodic oscillations for the quartic potential energy $V = \frac{1}{2}m\Omega^2x^2 + \frac{1}{4}m\Lambda x^4$ with $\Lambda < 0$.

S is less than its linear approximation and the spring is said to be **softening** as x increases. Conversely, if $\Lambda > 0$, then the spring is **hardening** as x increases. The formula (8.1) is typical of non-linear restoring forces that are *symmetrical* about $x = 0$. If the restoring force is unsymmetrical about $x = 0$, the leading correction to the linear case will be a term in x^2.

Existence of non-linear periodic oscillations

Consider the **free undamped oscillations** of a body sliding on a smooth horizontal table* and connected to a fixed point of the table by a spring whose restoring force is given by the cubic formula (8.1). In rectilinear motion, the **governing equation** is then

$$\frac{d^2x}{dt^2} + \Omega^2 x + \Lambda x^3 = 0, \qquad (8.2)$$

which is Duffing's equation with no forcing term (see section 8.5). The existence of periodic oscillations can be proved by the energy method described in Chapter 6. The restoring force has potential energy

$$V = \tfrac{1}{2}m\Omega^2x^2 + \tfrac{1}{4}m\Lambda x^4,$$

so that the energy conservation equation is

$$\tfrac{1}{2}mv^2 + \tfrac{1}{2}m\Omega^2x^2 + \tfrac{1}{4}m\Lambda x^4 = E,$$

where $v = \dot{x}$. The motion is therefore restricted to those values of x that satisfy

$$\tfrac{1}{2}m\Omega^2x^2 + \tfrac{1}{4}m\Lambda x^4 \le E,$$

* Would the motion be the same (relative to the equilibrium position) if the body were suspended vertically by the same spring?

with equality when $v = 0$. Figure 8.1 shows a sketch of V for a *softening* spring ($\Lambda < 0$). For each value of E in the range $0 < E < V_{\text{max}}$, the particle oscillates in a symmetrical range $-a \le x \le a$ as shown. Thus oscillations of any amplitude less than a_{max} ($= \Omega/|\Lambda|^{1/2}$) are possible. For a *hardening* spring, oscillations of any amplitude whatsoever are possible.

Solution by perturbation theory

Suppose then that the body is performing periodic oscillations with amplitude a. In order to reduce the number of parameters, we non-dimensionalise equation (8.2). Let the *dimensionless displacement* X be defined by $x = aX$. Then X satisfies the equation

$$\frac{1}{\Omega^2}\frac{d^2X}{dt^2} + X + \epsilon X^3 = 0, \tag{8.3}$$

with the initial conditions $X = 1$ and $dX/dt = 0$ when $t = 0$. The dimensionless parameter ϵ, defined by

$$\epsilon = \frac{a^2\Lambda}{\Omega^2}, \tag{8.4}$$

is a measure of the strength of the non-linearity. Equation (8.3) contains ϵ as a parameter and hence so does the solution. A major feature of interest is how the period τ of the motion varies with ϵ.

The non-linear equation of motion (8.3) cannot be solved explicitly but it reduces to a simple linear equation when the parameter ϵ is zero. In these circumstances, one can often find an *approximate* solution to the non-linear equation *valid when ϵ is small.* Equations in which the non-linear terms are small are said to be **weakly non-linear** and the solution technique is called **perturbation theory**. There is a well established theory of such perturbations. The simplest case is as follows:

Regular perturbation expansion

If the parameter ϵ appears as the coefficient of any term of an ODE that is *not* the highest derivative in that equation, then, when ϵ is small, the solution corresponding to fixed initial conditions can be expanded as a **power series** in ϵ.

This is called a **regular perturbation expansion*** and it applies to the equation (8.3). It follows that the solution $X(t, \epsilon)$ can be expanded in the **regular perturbation series**

$$X(t, \epsilon) = X_0(t) + \epsilon X_1(t) + \epsilon^2 X_2(t) + \cdots . \tag{8.5}$$

* The case in which the small parameter multiplies the *highest* derivative in the equation is called a **singular perturbation**. For experts only!

The standard method is to substitute this series into the equation (8.3) and then to try to determine the functions $X_0(t)$, $X_1(t)$, $X_2(t)$, ... In the present case however, this leads to an unsatisfactory result because the functions $X_1(t)$, $X_2(t)$, ... , turn out to be *non-periodic* (and unbounded) even though the exact solution $X(t, \epsilon)$ is periodic!* Also, it is not clear how to find approximations to τ from such a series.

This difficulty can be overcome by replacing t by a new variable s so that the solution $X(s, \epsilon)$ has period 2π in s *whatever the value of* ϵ. Every term of the perturbation series will then also be periodic with period 2π. This trick is known as *Lindstedt's method.*

Lindstedt's method

Let $\omega(\epsilon)$ $(= 2\pi/\tau(\epsilon))$ be the angular frequency of the required solution of equation (8.3). Now introduce a new independent variable s (the *dimensionless time*) by the equation $s = \omega(\epsilon)t$. Then $X(s, \epsilon)$ satisfies the equation

$$\left(\frac{\omega(\epsilon)}{\Omega}\right)^2 X'' + X + \epsilon X^3 = 0 \qquad (8.6)$$

with the initial conditions $X = 1$ and $X' = 0$ when $s = 0$. (Here $'$ means d/ds.) We now seek a solution of this equation in the form of the perturbation series

$$X(s, \epsilon) = X_0(s) + \epsilon X_1(s) + \epsilon^2 X_2(s) + \cdots . \qquad (8.7)$$

which is possible when ϵ is small. By construction, this solution must have period 2π *for all* ϵ from which it follows that each of the functions $X_0(s)$, $X_1(s)$, $X_2(s)$, ... must also have period 2π. However we have paid a price for this simplification since the *unknown* angular frequency $\omega(\epsilon)$ now appears in the equation (8.6); indeed, the function $\omega(\epsilon)$ is part of the *answer* to this problem! We must therefore also expand $\omega(\epsilon)$ as a perturbation series in ϵ. From equation (8.3), it follows that $\omega(0) = \Omega$ so we may write

$$\frac{\omega(\epsilon)}{\Omega} = 1 + \omega_1\epsilon + \omega_2\epsilon^2 + \cdots , \qquad (8.8)$$

where $\omega_1, \omega_2, \ldots$ are unknown constants that must be determined along with the functions $X_0(s)$, $X_1(s)$, $X_2(s)$, ...

On substituting the expansions (8.7) and (8.8) into the governing equation (8.6) and its initial conditions, we obtain:

$$(1 + \omega_1\epsilon + \omega_2\epsilon^2 + \cdots)^2(X_0'' + \epsilon X_1'' + \epsilon^2 X_2'' + \cdots) +$$

$$(X_0 + \epsilon X_1 + \epsilon^2 X_2 + \cdots) + \epsilon(X_0 + \epsilon X_1 + \epsilon^2 X_2 + \cdots)^3 = 0,$$

* This 'paradox' causes great bafflement when first encountered, but it is inevitable when the period τ of the motion depends on ϵ, as it does in this case. To have a series of non-periodic terms is not *wrong*, as is sometimes stated. However, it is certainly unsatisfactory to have a non-periodic approximation to a periodic function.

with

$$
\begin{aligned}
X_0 + \epsilon X_1 + \epsilon^2 X_2 + \cdots &= 1, \\
X_0' + \epsilon X_1' + \epsilon^2 X_2' + \cdots &= 0,
\end{aligned}
$$

when $s = 0$. If we now equate coefficients of powers of ϵ in these equalities, we obtain a succession of ODEs and initial conditions, the first two of which are as follows:

From coefficients of ϵ^0, we obtain the **zero order** equation

$$X_0'' + X_0 = 0, \tag{8.9}$$

with $X_0 = 1$ and $X_0' = 0$ when $s = 0$.
From coefficients of ϵ^1, we obtain the **first order** equation

$$X_1'' + X_1 = -2\omega_1 X_0'' - X_0^3, \tag{8.10}$$

with $X_1 = 0$ and $X_1' = 0$ when $s = 0$.

This procedure can be extended to any number of terms but the equations rapidly become very complicated. The method now is to solve these equations in order; the only sticking point is how to determine the unknown constants $\omega_1, \omega_2, \ldots$ that appear on the right sides of the equations. The solution of the **zero order** equation and initial conditions is

$$X_0 = \cos s \tag{8.11}$$

and this can now be substituted into the first order equation (8.10) to give

$$
\begin{aligned}
X_1'' + X_1 &= 2\omega_1 \cos s - \cos^3 s \\
&= \tfrac{1}{4}(8\omega_1 - 3)\cos s - \tfrac{1}{4}\cos 3s,
\end{aligned} \tag{8.12}
$$

on using the trigonometric identity $\cos 3s = 4\cos^3 s - 3\cos s$. This equation can now be solved by standard methods. The particular integral corresponding to the $\cos 3s$ on the right is $-(1/8)\cos 3s$, but the particular integral corresponding to the $\cos s$ on the right is $(1/2)s \sin s$, since $\cos s$ is a solution of the equation $X'' + X = 0$. The general solution of the first order equation is therefore

$$X_1 = \left(\omega_1 - \tfrac{3}{8}\right) s \sin s - \tfrac{1}{32}\cos 3s + A\cos s + B\sin s,$$

where A and B are arbitrary constants. Observe that the functions $\cos s$, $\sin s$ and $\cos 3s$ are all periodic with period 2π, but the term $s \sin s$ is *not periodic*. Thus, *the coefficient of $s \sin s$ must be zero, for otherwise $X_1(s)$ would not be periodic, which we know it must be.* Hence

$$\omega_1 = \frac{3}{8}, \tag{8.13}$$

which determines the first unknown coefficient in the expansion (8.8) of $\omega(\epsilon)$. The solution of the **first order** equation and initial conditions is then

$$X_1 = \frac{1}{32}\left(\cos 3s - \cos s\right). \tag{8.14}$$

We have thus shown that, when ϵ is small,

$$\frac{\omega}{\Omega} = 1 + \frac{3}{8}\epsilon + O\left(\epsilon^2\right),$$

and

$$X = \cos s + \frac{\epsilon}{32}\left(\cos 3s - \cos s\right) + O\left(\epsilon^2\right),$$

where $s = \left(1 + \frac{3}{8}\epsilon + O\left(\epsilon^2\right)\right)\Omega t$.

Results

When ϵ $(= a^2\Lambda/\Omega^2)$ is small, the **period** τ of the oscillation of equation (8.2) with amplitude a is given by

$$\tau = \frac{2\pi}{\omega} = \frac{2\pi}{\Omega}\left(1 + \frac{3}{8}\epsilon + O\left(\epsilon^2\right)\right)^{-1} = \frac{2\pi}{\Omega}\left(1 - \frac{3}{8}\epsilon + O\left(\epsilon^2\right)\right) \tag{8.15}$$

and the corresponding displacement $x(t)$ is given by

$$x = a\left[\cos s + \frac{\epsilon}{32}\left(\cos 3s - \cos s\right) + O\left(\epsilon^2\right)\right], \tag{8.16}$$

where $s = \left(1 + \frac{3}{8}\epsilon + O\left(\epsilon^2\right)\right)\Omega t$.

This is the *approximate solution correct to the first order in the small parameter* ϵ. More terms can be obtained in a similar way but the effort needed increases exponentially and this is best done with computer assistance (see Problem 8.15).

These formulae apply only when ϵ is small, that is, when the *non- linearity in the equation has a small effect.* Thus we have laboured through a sizeable chunk of mathematics to produce an answer that is only slightly different from the linear case. This sad fact is true of all *regular* perturbation problems. However, in non-linear mechanics, one must be thankful for even modest successes.

8.2 THE PHASE PLANE ((x_1, x_2)–plane)

The second approach that we will describe could not be more different from perturbation theory. It makes use of qualitative geometrical arguments in the phase space of the system.

Systems of first order ODEs

The notion of **phase space** springs from the theory of **systems of first order ODEs**. Such systems are very common and need have no connection with classical mechanics. A standard example is the predator-prey system of equations

$$\dot{x}_1 = ax_1 - bx_1x_2,$$
$$\dot{x}_2 = bx_1x_2 - cx_2,$$

which govern the population density $x_1(t)$ of a prey and the population density $x_2(t)$ of its predator. In the general case there are n unknown functions satisfying n first order ODEs, but here we will only make use of *two* unknown functions $x_1(t)$, $x_2(t)$ that satisfy a *pair* of first order ODEs of the form

$$\begin{aligned} \dot{x}_1 &= F_1(x_1, x_2, t), \\ \dot{x}_2 &= F_2(x_1, x_2, t). \end{aligned} \tag{8.17}$$

Just to confuse matters, a system of ODEs like (8.17) is called a **dynamical system**, whether it has any connection with classical mechanics or not! In the predator-prey dynamical system, the function $F_1 = ax_1 - bx_1x_2$ and the function $F_2 = bx_1x_2 - cx_2$. In this case F_1 and F_2 have no explicit time dependence. Such systems are said to be *autonomous*; as we shall see, more can be said about the behaviour of autonomous systems.

Definition 8.1 ***Autonomous system*** *A system of equations of the form*

$$\begin{aligned} \dot{x}_1 &= F_1(x_1, x_2), \\ \dot{x}_2 &= F_2(x_1, x_2), \end{aligned} \tag{8.18}$$

is said to be ***autonomous****.*

The phase plane

The values of the variables x_1, x_2 at any instant can be represented by a point in the (x_1, x_2)-plane. This plane is called the **phase plane*** of the system. A solution of the system of equations (8.17) is then represented by a point moving in the phase plane. The path traced out by such a point is called a **phase path**† of the system and the set of all phase paths is called the **phase diagram**. In the predator–prey problem, the variables x_1, x_2 are positive quantities and so the physically relevant phase paths lie in the first quadrant of the phase plane. It can be shown that they are all closed curves! (See Problem 8.10).

Phase paths of autonomous systems

The problem of finding the phase paths is much easier when the system is **autonomous**. The method is as follows:

* In the general case with n unknowns, the phase space is n-dimensional.

† Also called an *orbit* of the system.

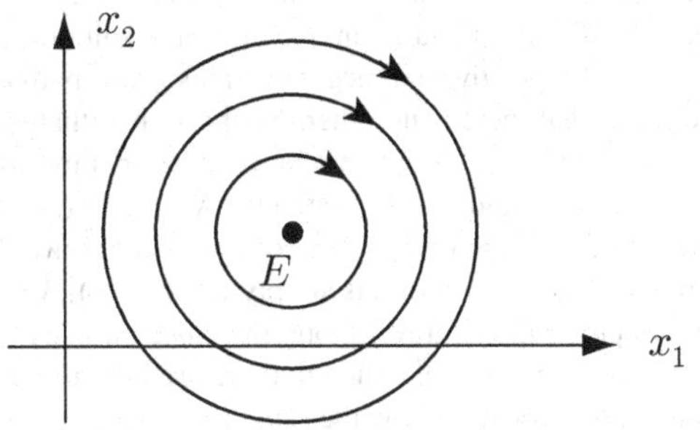

FIGURE 8.2 Phase diagram for the system $dx_1/dt = x_2 - 1$, $dx_2/dt = -x_1 + 2$. The point $E(2, 1)$ is an equilibrium point of the system.

Example 8.1 ***Finding phase paths for an autonomous system***

Sketch the phase diagram for the autonomous system of equations

$$\frac{dx_1}{dt} = x_2 - 1,$$

$$\frac{dx_2}{dt} = -x_1 + 2.$$

Solution

The phase paths of an *autonomous* system can be found by eliminating the time derivatives. The path gradient is given by

$$\frac{dx_2}{dx_1} = \frac{dx_2/dt}{dx_1/dt} = -\frac{x_1 - 2}{x_2 - 1}$$

and this is a first order separable ODE satisfied by the phase paths. The general solution of this equation is

$$(x_1 - 2)^2 + (x_2 - 1)^2 = C$$

and each (positive) choice for the constant of integration C corresponds to a phase path. The **phase paths** are therefore circles with centre (2, 1); the **phase diagram** is shown in Figure 8.2.

The direction in which the phase point progresses along a path can be deduced by examining the *signs* of the right sides in equations (8.18). This gives the signs of $\dot{x}_1$ and $\dot{x}_2$ and hence the direction of motion of the phase point. ■

When the system is autonomous, one can say quite a lot about the *general nature* of the phase paths without finding them. The basic result is as follows:

Theorem 8.1 ***Autonomous systems: a basic result*** *Each point of the phase space of an autonomous system has exactly one phase path passing through it.*

Proof. Let (a, b) be any point of the phase space. Suppose that the motion of the phase point (x_1, x_2) satisfies the equations (8.18) and that the phase point is at (a, b) when $t = 0$. The general theory of ODEs then tells us that a solution of the equations (8.18), that satisfies the initial conditions $x_1 = a$, $x_2 = b$ when $t = 0$, exists and is unique. Let this solution be $\{X_1(t), X_2(t)\}$, which we will suppose is defined for all t, both positive and negative. This phase path certainly passes through the point (a, b) and we must now show that there is no other. Suppose then that there is another solution of the equations in which the phase point is at (a, b) when $t = \tau$, say. This motion also exists and is uniquely determined and, in the general case, would not be related to $\{X_1(t), X_2(t)\}$. However, for autonomous systems, the right sides of equations (8.18) are independent of t so that *the two motions differ only by a shift in the origin of time*. To be precise, the new motion is simply $\{X_1(t - \tau), X_2(t - \tau)\}$. Thus, although the two motions are distinct, the two phase points travel along the *same path* with the second point delayed relative to the first by the constant time τ. Hence, although there are infinitely many *motions* of the phase point that pass through the point (a, b), they all follow the same path. This proves the theorem. ■

Some important deductions follow from this basic result.

Phase paths of autonomous systems

- Distinct phase paths of an autonomous system **do not cross** or touch each other.
- **Periodic motions** of an autonomous system correspond to phase paths that are simple* **closed loops**.

Figure 8.2 showsthe phase paths of an autonomous system. For this system, *all* of the phase paths are simple closed loops and so every motion is periodic. An exception occurs if the phase point is started from the point (2, 1). In this case the system has the constant solution $x_1 = 2$, $x_2 = 1$ so that the phase point never moves; for this reason, the point (2, 1) is called an **equilibrium point** of the system. In this case, the 'path' of the phase point consists of the *single point* (2, 1). However, this still qualifies as a path and the above theory still applies. Consequently no 'real' path may pass through an equilibrium point of an autonomous system.†

8.3 THE PHASE PLANE IN DYNAMICS ((x, v)–plane)

The above theory seems unconnected to classical mechanics since dynamical equations of motion are *second order* ODEs. However, *any second order ODE can be expressed as a pair of first order ODEs*. For example, consider the general linear oscillator equation

$$\frac{d^2x}{dt^2} + 2K\frac{dx}{dt} + \Omega^2 x = F(t). \tag{8.19}$$

If we introduce the new variable $v = dx/dt$, then

$$\frac{dv}{dt} + 2kv + \Omega^2 x = F(t).$$

* A simple curve is one that does not cross (or touch) itself (except possibly to close).

† It may appear from diagrams that phase paths *can* pass through equilibrium points. This is not so. Such a path approaches *arbitrarily close* to the equilibrium point in question, but never reaches it!

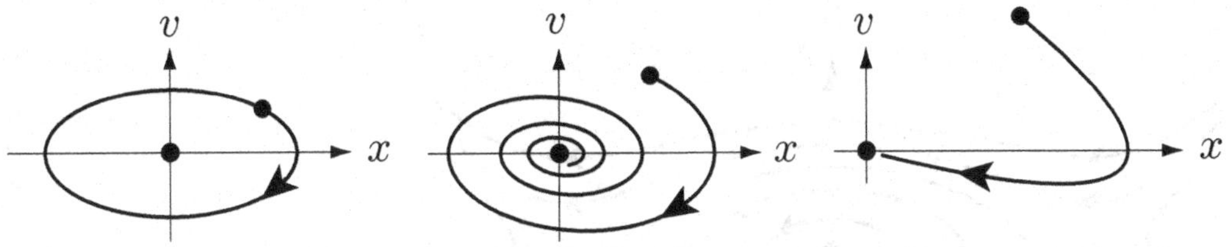

FIGURE 8.3 Typical phase paths for the simple harmonic oscillator equation. **Left**: No damping. **Centre**: Sub-critical damping. **Right**: Super-critical damping.

It follows that the second order equation (8.19) is equivalent to the pair of first order equations

$$\frac{dx}{dt} = v,$$
$$\frac{dv}{dt} = F(t) - 2kv - \Omega^2 x.$$

We may now apply the theory we have developed to this system of first order ODEs, where the **phase plane** is now the (x, v)-plane. It is clear that **driven motion** leads to a **non-autonomous system** because of the presence of the explicit time dependence of $F(t)$; **undriven motion** (in which $F(t) = 0$) leads to an **autonomous** system. It is also clear that equilibrium points in the (x, v)-plane lie on the x-axis and correspond to the ordinary equilibrium positions of the particle.

The form of the phase paths for the *undriven* SHO equation

$$\frac{d^2x}{dt^2} + 2K\frac{dx}{dt} + \Omega^2 x = 0$$

depends on the parameters K and Ω. We could find these paths by the method used in Example 8.1, but there is no point in doing so since we have already solved the equation explicitly in Chapter 5. For instance, when $K = 0$, the general solution is given by

$$x = C\cos(\Omega t - \gamma),$$

from which it follows that

$$v = \frac{dx}{dt} = -C\Omega\sin(\Omega t - \gamma).$$

The phase paths in the (x, v)-plane are therefore similar ellipses centred on the origin, which is an equilibrium point. This, and two typical cases of damped motion, are shown in Figure 8.3. In the presence of damping, the phase point *tends* to the equilibrium point at the origin as $t \to \infty$. Although the equilibrium point is never actually reached, it is convenient to say that these paths 'terminate' at the origin.

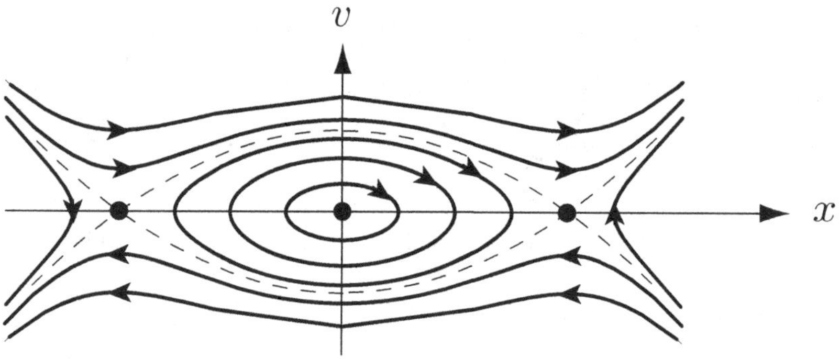

FIGURE 8.4 The phase diagram for the undamped Duffing equation with a softening spring.

Example 8.2 *Phase diagram for equation* $d^2x/dt^2 + \Omega^2 x + \Lambda x^3 = 0$

Sketch the phase diagram for the non-linear oscillation equation

$$d^2x/dt^2 + \Omega^2 x + \Lambda x^3 = 0,$$

when $\Lambda < 0$ (the softening spring).

Solution

This equation is equivalent to the pair of first order equations

$$\frac{dx}{dt} = v,$$
$$\frac{dv}{dt} = -\Omega^2 x - \Lambda x^3,$$

which is an **autonomous** system. The **phase paths** satisfy the equation

$$\frac{dv}{dx} = -\frac{\Omega^2 x + \Lambda x^3}{v},$$

which is a first order separable ODE whose general solution is

$$v^2 = C - \Omega^2 x^2 - \tfrac{1}{2}\Lambda x^4,$$

where C is a constant of integration. Each *positive* value of C corresponds to a phase path. The phase diagram for the case $\Lambda < 0$ is shown in Figure 8.4. There are three **equilibrium points** at $(0, 0)$, $(\pm\Omega/|\Lambda|^{1/2}, 0)$. The closed loops around the origin correspond to **periodic oscillations** of the particle about $x = 0$. Such oscillations can therefore exist for any amplitude less than $\Omega/|\Lambda|^{1/2}$; this confirms the prediction of the energy argument used earlier. Outside this region of closed loops, the paths are unbounded and correspond to unbounded motions of the particle. These two regions of differing behaviour are separated by the dashed paths (known as *separatrices*) that 'terminate' at the equilibrium points $(\pm\Omega/|\Lambda|^{1/2}, 0)$. ■

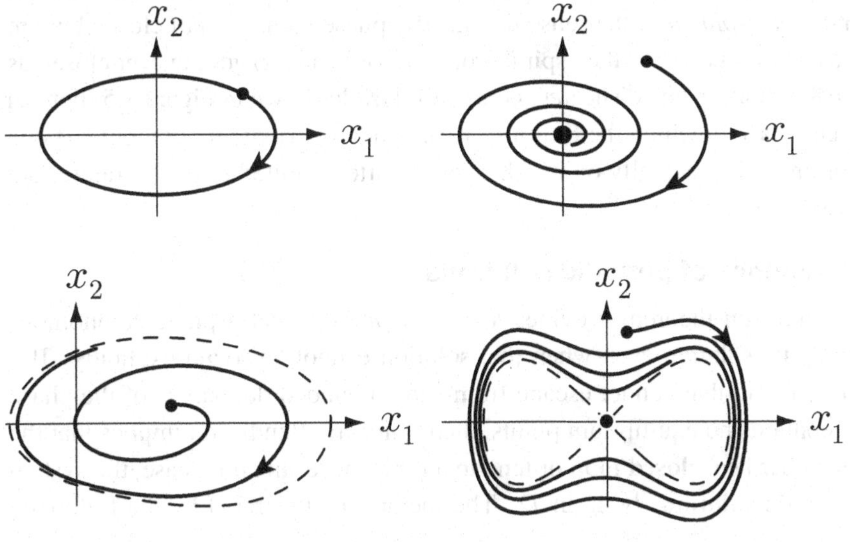

FIGURE 8.5 The Poincaré–Bendixson theorem. Any *bounded* phase path of a plane autonomous system must either close itself (**top left**), terminate at an equilibrium point (**top right**), or tend to a limit cycle (normal case **bottom left**, degenerate case **bottom right**).

8.4 POINCARÉ–BENDIXSON THEOREM: LIMIT CYCLES

In the autonomous systems we have studied so far, those phase paths that are *bounded* either (i) form a closed loop (corresponding to periodic motion), or (ii) 'terminate' at an equilibrium point (so that the motion dies away). Figure 8.3 shows examples of this. The famous Poincaré–Bendixson theorem* which is stated below, says that there is just *one* further possibility.

> **Poincaré–Bendixson theorem**
>
> Suppose that a phase path of a **plane autonomous system** lies in a **bounded** domain of the phase plane for $t > 0$. Then the path must either
>
> - **close** itself, or
> - **terminate** at an equilibrium point as $t \to \infty$, or
> - tend to a **limit cycle** (or a degenerate limit cycle) as $t \to \infty$.

A proper proof of the theorem is long and difficult (see Coddington & Levinson [9]).

* After Jules Henri Poincaré (1854–1912) and Ivar Otto Bendixson (1861–1935). The theorem was first proved by Poincaré but a more rigorous proof was given later by Bendixson.

The third possibility is new and needs explanation. A **limit cycle** is a **periodic motion** of a special kind. It is *isolated* in the sense that nearby phase paths are *not* closed but are attracted towards the limit cycle* ; they spiral around it (or inside it) getting ever closer, as shown in Figure 8.5 (bottom left). The *degenerate* limit cycle shown in Figure 8.5 (bottom right) is an obscure case in which the limiting curve is not a periodic motion but has one or more equilibrium points actually on it. This case is often omitted in the literature, but it definitely exists!

Proving the existence of periodic solutions

The Poincaré–Bendixson theorem provides a way of *proving* that a plane autonomous system has a periodic solution even when that solution cannot be found explicitly. If a phase path can be found that cannot escape from some bounded domain $\mathcal{D}$ of the phase plane, and if $\mathcal{D}$ contains no equilibrium points, then Poincaré–Bendixson implies that the phase path must either be a **closed loop** or tend to a **limit cycle**. In either case, the system must have a **periodic solution** lying in $\mathcal{D}$. The method is illustrated by the following examples.

Example 8.3 *Proving existence of a limit cycle*

Prove that the autonomous system of ODEs

$$\dot{x} = x - y - (x^2 + y^2)x,$$
$$\dot{y} = x + y - (x^2 + y^2)y,$$

has a limit cycle.

Solution

This system clearly has an equilibrium point at the origin $x = y = 0$, and a little algebra shows that there are no others. Although we have not proved this result, it is true that any periodic solution (simple closed loop) in the phase plane must have an equilibrium point lying *inside* it. In the present case, it follows that, if a periodic solution exists, then it *must* enclose the origin. This suggests taking the domain $\mathcal{D}$ to be the annular region between two circles centred on the origin.

It is convenient to express the system of equations in **polar coordinates** r, θ. The transformed equations are (see Problem 8.5)

$$\dot{r} = \frac{x_1\dot{x}_1 + x_2\dot{x}_2}{r}, \qquad \dot{\theta} = \frac{x_1\dot{x}_2 - x_2\dot{x}_1}{r^2},$$

where $x_1 = r\cos\theta$ and $x_2 = r\sin\theta$. In the present case, the polar equations take the simple form

$$\dot{r} = r(1 - r^2), \qquad \dot{\theta} = 1.$$

* This actually describes a *stable* limit cycle, which is the only kind likely to be observed.

These equations can actually be solved explicitly, but, in order to illustrate the method, we will make no use of this fact. Let $\mathcal{D}$ be the annular domain $a < r < b$, where $0 < a < 1$ and $b > 1$. On the circle $r = b$, $\dot{r} = b(1 - b^2) < 0$. Thus a phase point that starts anywhere on the outer boundary $r = b$ *enters* the domain $\mathcal{D}$. Similarly, on the circle $r = a$, $\dot{r} = a(1 - a^2) > 0$ and so a phase point that starts anywhere on the inner boundary $r = a$ also *enters* the domain $\mathcal{D}$. It follows that *any phase path that starts in the annular domain $\mathcal{D}$ can never leave*. Since $\mathcal{D}$ is a *bounded* domain with *no equilibrium points* within it or on its boundaries, it follows from Poincaré–Bendixson that any such path must either be a simple closed loop or tend to a limit cycle. In either case, the system must have a **periodic solution** lying in the annulus $a < r < b$.

We can say more. Phase paths that begin on either *boundary* of $\mathcal{D}$ enter $\mathcal{D}$ and can never leave. These phase paths cannot close themselves (that would mean leaving $\mathcal{D}$) and so can only tend to a limit cycle. It follows that the system must have (at least one) **limit cycle** lying in the domain $\mathcal{D}$. [The explicit solution shows that the circle $r = 1$ is a limit cycle and that there are no other periodic solutions.] ■

Not all examples are as straightforward as the last one. Often, considerable ingenuity has to be used to find a suitable domain $\mathcal{D}$. In particular, the boundary of $\mathcal{D}$ cannot always be composed of circles. Most readers will find our second example rather difficult!

Example 8.4 *Rayleigh's equation has a limit cycle*

Show that **Rayleigh's equation**

$$\ddot{x} + \epsilon\dot{x}\left(\dot{x}^2 - 1\right) + x = 0,$$

has a limit cycle for any *positive* value of the parameter ϵ.

Solution

Rayleigh's equation arose in his theory of the bowing of a violin string. In the context of particle oscillations however, it corresponds to a simple harmonic oscillator with a strange damping term. When $|\dot{x}| > 1$, we have ordinary (positive) damping and the motion decays. However, when $|\dot{x}| < 1$, we have *negative damping* and the motion grows. The possibility arises then of a periodic motion that is positively damped on some parts of its cycle and negatively damped on others. Somewhat surprisingly, this actually exists. The proof, which is rather subtle, is as follows.

Rayleigh's equation is equivalent to the autonomous system of ODEs

$$\begin{aligned} \dot{x} &= v, \\ \dot{v} &= -x - \epsilon v(v^2 - 1), \end{aligned} \tag{8.20}$$

for which the only equilibrium position is at $x = v = 0$. It follows that, if there is a periodic solution, then it must enclose the origin. At first, we proceed as in Example 8.3. By writing the equations (8.20) in polar form we can quickly deduce that any circle with centre at the origin and radius less than unity provides a suitable *inner* boundary for the domain $\mathcal{D}$.

Sadly, one cannot simply take a large circle to be the outer boundary of $\mathcal{D}$ since $\dot{r}$ has the wrong sign on those segments of the circle that lie in the strip $-1 < v < 1$. This allows any number of phase paths to escape and so invalidates our argument. However, this does not prevent us from choosing a boundary of a different shape. A suitable outer boundary for $\mathcal{D}$ can be constructed as shown in Figure 8.6. This contour is made up from six segments! The first segment

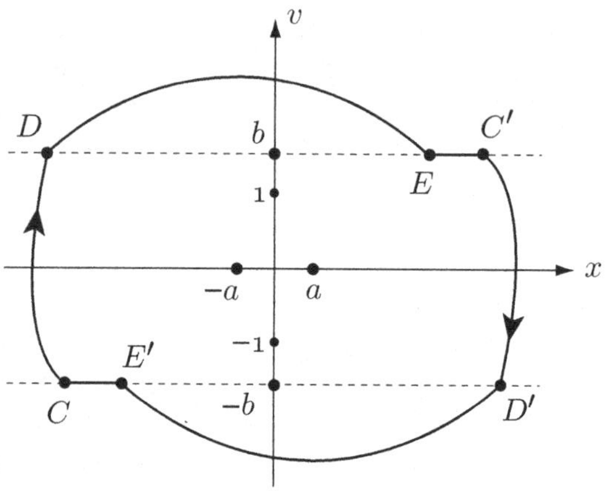

FIGURE 8.6 A suitable outer boundary to show that Rayleigh's equation has a limit cycle.

CD is part of an *actual phase path* of the system which starts at the point $C(-c, -b)$ on the line $v = -b$ and meets the line $v = b$ at the point $D(-d, b)$. Here, b and c are positive constants; the value d is determined from equations (8.20) once b and c have been chosen. Similarly, the segment $C'D'$ is part of a second actual phase path that begins at $C'(c, b)$. Because of the symmetry of the equations (8.20) under the transformation $x \to -x$, $v \to -v$, this segment is just the reflection of the segment CD in the origin; the point D' is therefore $(d, -b)$. The segment DE is an arc of a circle whose centre is at the point $(-a, 0)$, where a is a positive constant. Similarly, the segment $D'E'$ is an arc of a circle whose centre is at the point $(a, 0)$. (The reason why we must take the centres of these circles offset from the origin will only become apparent at the end of the proof.) Finally, the contour is closed by inserting the straight line segments EC' and $E'C$.

We will now show that the positive constants a, b and c can be chosen so that this is a suitable outer boundary for the domain $\mathcal{D}$. Consider first the segment CD. Since this *is* a phase path, no other phase path may cross it; the same applies to the segment $C'D'$. Now consider the circular arc DE. Let R be the distance of the point (x, v) from the point $(-a, 0)$, that is,

$$R^2 = (x + a)^2 + v^2.$$

Then if $(x(t), v(t))$ is the trajectory of a phase point,

$$\begin{aligned} R\dot{R} &= (x + a)\dot{x} + v\dot{v} = (x + a)v - v\left(x + \epsilon v(v^2 - 1)\right) \\ &= v\left(a - \epsilon v(v^2 - 1)\right). \end{aligned}$$

Hence $\dot{R}$ will be *negative* on the arc DE if

$$a < \epsilon v(v^2 - 1)$$

for all $v \geq b$. This will be true if the constant $b > 1$ and the constant a is chosen so that

$$a < \epsilon b(b^2 - 1).$$

With these choices, phase points that begin on the arc DE *enter* the domain $\mathcal{D}$. Similar remarks apply to the circular arc $D'E'$. Finally, consider the straight segment EC'. When $v = b$, $\dot{v} = -x - \epsilon b(b^2 - 1) < 0$ for $x > 0$ so that phase points that begin on the segment EC' move *downwards*. Similarly, phase points that begin on the segment $E'C$ move *upwards*.

It might appear that we have succeeded in finding a suitable outer boundary for $\mathcal{D}$ but there is a flaw in the argument that could be fatal. We have assumed that the point E lies to the *left* of C' as shown in Figure 8.6. If it actually lies to the right, then phase points that begin on the straight line EC' still move downwards but now this takes them *away* from $\mathcal{D}$. This will happen if D, the end point of the phase path CD, lies so far to the left that no permissible choice of a can make E lie to the left of C'. Thus we must somehow estimate where the phase path that starts at

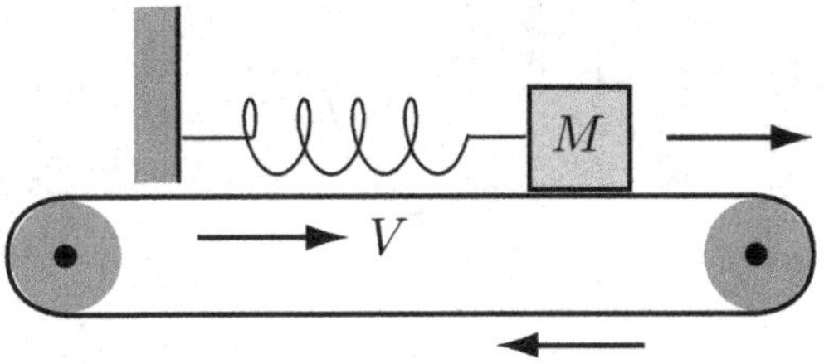

FIGURE 8.7 The body is supported by a rough moving belt and is attached to a fixed post by a light spring.

C actually goes. Since phase paths satisfy the ODE

$$\frac{dx}{dv} = -\frac{v}{x + \epsilon v(v^2 - 1)},$$

it follows that, for sufficiently large $|x|$,

$$\left|\frac{dx}{dv}\right| \le \frac{|v|}{|x| - \epsilon\left|v(v^2 - 1)\right|} \le \frac{b}{|x| - \epsilon M} \tag{8.21}$$

for $|v| \le b$, where M $(= M(b))$ is the maximum value achieved by the function $\left|v(v^2 - 1)\right|$ for v in the range $-b \le v \le b$. This bound for dx/dv tends to zero as $|x|$ tends to infinity. Hence, by taking c sufficiently large, the phase path CD can be made almost parallel to the v-axis so that d *differs only slightly* from c. If we take the constant c sufficiently large so that $|d - c| < 2a$, then the point E *does* lie to the left of C'. Note that we could not achieve this with certainty if the arc DE had its centre at the origin; this is why the centre must be offset. Similar remarks apply to the phase path $C'D'$.

We have thus found a suitable *outer* boundary for the domain $\mathcal{D}$. It follows that *any phase path that starts in the domain $\mathcal{D}$ can never leave.* Since $\mathcal{D}$ is a *bounded* domain with *no equilibrium points* within it or on its boundaries, it follows from Poincaré–Bendixson that any such path must either be a simple closed loop or tend to a limit cycle. In either case, Rayleigh's equation must have a **periodic solution** lying in $\mathcal{D}$. We can say more. Phase paths that begin on any of the straight or circular segments of the outer boundary enter $\mathcal{D}$ and never leave. These phase paths cannot close themselves (that would mean leaving $\mathcal{D}$) and so can only tend to a limit cycle. It follows that Rayleigh's equation must have (at least one) **limit cycle** lying in the domain $\mathcal{D}$. [There is in fact only one.] ■

A realistic mechanical system with a limit cycle

Finding *realistic* mechanical systems that exhibit limit cycles is not easy. Driven oscillations are eliminated by the requirement that the system be autonomous. Undamped oscillators have bounded *periodic* motions, and the introduction of damping causes the motions to die away to zero, not to a limit cycle. In order to keep the motion going, the system needs to be *negatively damped* for part of the time. This is an unphysical requirement, but it can be simulated in a physically realistic system as follows.

Consider the system shown in Figure 8.7. A block of mass M is supported by a rough horizontal belt and is attached to a fixed post by a light linear spring. The belt is made to move with constant speed V. Suppose that the motion takes place in a straight line and that $x(t)$ is the extension of the spring beyond its natural length at time t. Then the

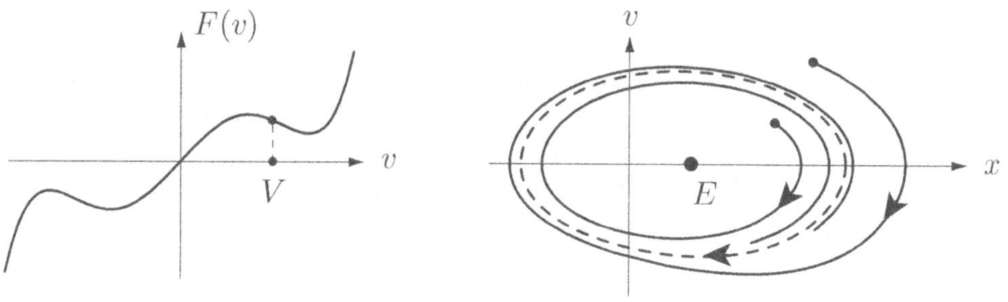

FIGURE 8.8 Left: The form of the frictional resistance function $G(v)$. **Right**: The limit cycle in the phase plane; E is the unstable equilibrium point.

equation of motion of the block is

$$M\frac{dv}{dt} = -M\Omega^2 x - F(v - V),$$

where $v = dx/dt$, $M\Omega^2$ is the spring constant, and $F(v)$ is the frictional force that the belt *would* exert on the block if the block had velocity v and the belt were *at rest*; in the actual situation, the argument v is replaced by the relative velocity $v - V$. The function $F(v)$ is supposed to have the form shown in Figure 8.8 (left). Although this choice is unusual ($F(v)$ is not an increasing function of v for all v), it is *not* unphysical!

Under the above conditions, the block has an equilibrium position at $x = F(V)/(M\Omega^2)$. The linearised equation for small motions near this equilibrium position is given by

$$M\frac{d^2x'}{dt^2} = -M\Omega^2 x' - F'(V)\frac{dx'}{dt},$$

where x' is the displacement of the block from the equilibrium position. If we select the belt velocity V so that $F'(V)$ is negative (as shown in Figure 8.8 (left)), then the *effective* damping is negative and small motions will grow. The equilibrium position is therefore *unstable*; oscillations of the block about the equilibrium position then do not die out, but instead tend to a **limit cycle**. This limit cycle is shown in Figure 8.8 (right). The formal proof that such a limit cycle exists is similar to that for Rayleigh's equation. Indeed, this system is essentially Rayleigh's model for the bowing of a violin string, where the belt is the bow, and the block is the string.

Chaotic motions

Another important conclusion from Poincaré–Bendixson is that *no bounded motion of a plane autonomous system can exhibit chaos*. The phase point cannot just wander about in a bounded region of the phase plane for ever. It must either close itself, terminate at an equilibrium point, or tend to a limit cycle and none of these motions is chaotic. In particular, no bounded motion of an undriven non-linear oscillator can be chaotic. As we will see in the next section however, the *driven* non-linear oscillator (a non-autonomous system) *can* exhibit bounded chaotic motions.

It should be remembered that Poincaré–Bendixson applies only to the bounded motion of *plane* autonomous systems. If the phase space has dimension three or more, then other motions, including chaos, are possible.

8.5 DRIVEN NON-LINEAR OSCILLATIONS

Suppose that we now introduce damping and a **harmonic driving force** into equation (8.2). This gives

$$\frac{d^2x}{dt^2} + k\frac{dx}{dt} + \Omega^2 x + \Lambda x^3 = F_0 \cos pt, \tag{8.22}$$

which is known as **Duffing's equation.**

The presence of the driving force $F_0 \cos pt$ makes this system *non-autonomous*. The behaviour of non-autonomous systems is considerably more complex than that of autonomous systems. Phase space is still a useful aid in *depicting* the motion of the system, but little can be said about the general behaviour of the phase paths. In particular, phase paths can cross each other any number of times, and Poincaré–Bendixson does not apply. Our treatment of driven non-linear oscillations is therefore restricted to perturbation theory.

In view of the large number of parameters, it is sensible to non-dimensionalise equation (8.22). The *dimensionless displacement* X is defined by $x = (F_0/p^2)X$ and the *dimensionless time* s by $s = pt$. The function $X(s)$ then satisfies the dimensionless equation

$$X'' + \left(\frac{k}{p}\right) X' + \left(\frac{\Omega}{p}\right)^2 X + \epsilon X^3 = \cos s, \tag{8.23}$$

where the dimensionless parameter ϵ is defined by

$$\epsilon = \frac{F_0^2 \Lambda}{\Omega^6}. \tag{8.24}$$

When $\epsilon = 0$, equation (8.23) reduces to the linear problem. This suggests that, when ϵ is small, we may be able to find approximate solutions by perturbation theory. The linear problem always has a periodic solution for X (the driven motion) that is harmonic with period 2π. Proving the existence of **periodic solutions** of Duffing's equation is an interesting and difficult problem. Here we address this problem for the case in which ϵ is small, a regular perturbation on the linear problem. To simplify the working we will suppose that damping is absent; the general features of the solution remain the same. The governing equation (8.23) then simplifies to

$$X'' + \left(\frac{\Omega}{p}\right)^2 X + \epsilon X^3 = \cos s. \tag{8.25}$$

Initial conditions do not come into this problem. We are simply seeking a family of solutions $X(s, \epsilon)$, parametrised by ϵ, that are (i) periodic, and (ii) reduce to the linear solution when $\epsilon = 0$. We need to consider first the **periodicity** of this family of solutions. In the non-linear problem, we have no right to suppose that the angular frequency of the driven motion is equal to that of the driving force, as it is in the linear problem; it could depend on ϵ. However, suppose that the driving force has minimum period τ_0 and that a family of solutions $X(s, \epsilon))$ of equation (8.25) exists with minimum period τ $(= \tau(\epsilon))$. Then, since the derivatives and powers of X also have period τ, it follows that the left side of equation (8.25) must have period τ. The right side however has period τ_0 and this is known to be the minimum period. It follows that *τ must be an integer multiple of τ_0*; note that τ is not compelled to be *equal* to τ_0.* However, in the present case, the period $\tau(\epsilon)$ is supposed to be a *continuous* function of ϵ with $\tau = \tau_0$ when $\epsilon = 0$. It follows that the only possibility is that $\tau = \tau_0$ for all ϵ. Thus *the period of the driven motion is independent of ϵ and is equal to the period of the driving force.* This argument leaves open the possibility that other driven motions may exist that have periods that are integer multiples of τ_0. However, even if they exist, they cannot occur in our perturbation scheme.

We therefore expand $X(s, \epsilon)$ in the **perturbation series**

$$X(s, \epsilon) = X_0(s) + \epsilon X_1(s) + \epsilon^2 X_2(s) + \cdots , \tag{8.26}$$

and seek a solution of equation (8.25) that has period 2π. It follows that the expansion functions $X_0(s)$, $X_1(s)$, $X_2(s)$, ... must also have period 2π. If we now substitute this series into the equation (8.25) and equate coefficients of powers of ϵ, we obtain a succession of ODEs the first two of which are as follows:

From coefficients of ϵ^0:

$$X_0'' + \left(\frac{\Omega}{p}\right)^2 X_0 = \cos s. \tag{8.27}$$

From coefficients of ϵ^1:

$$X_1'' + \left(\frac{\Omega}{p}\right)^2 X_1 = -X_0^3. \tag{8.28}$$

For $p \neq \Omega$, the general solution of the zero order equation (8.27) is

$$X_0 = \left(\frac{p^2}{\Omega^2 - p^2}\right) \cos s + A \cos(\Omega s/p) + B \sin(\Omega s/p),$$

where A and B are arbitrary constants. Since X_0 is known to have period 2π, it follows that A and B must be zero unless Ω is an integer multiple of p; we will assume this is *not*

---

* The fact that τ is the minimum period of X does not *necessarily* make it the minimum period of the left side of equation (8.25).

the case. Then the required solution of the **zero order** equation is

$$X_0 = \left(\frac{p^2}{\Omega^2 - p^2}\right)\cos s. \tag{8.29}$$

The **first order** equation (8.28) can now be written

$$\begin{aligned} X_1'' + \left(\frac{\Omega}{p}\right)^2 X_1 &= -\left(\frac{p^2}{\Omega^2 - p^2}\right)^3 \cos^3 s \\ &= -\left(\frac{p^6}{4(\Omega^2 - p^2)^3}\right)(3\cos s + \cos 3s), \end{aligned} \tag{8.30}$$

on using the trigonometric identity $\cos 3s = 4\cos^3 s - 3\cos s$. Since Ω/p is not an integer, the only solution of this equation that has period 2π is

$$X_1 = -\left(\frac{p^8}{4(\Omega^2 - p^2)^3}\right)\left(\frac{3\cos s}{\Omega^2 - p^2} + \frac{\cos 3s}{\Omega^2 - 9p^2}\right). \tag{8.31}$$

Results

When ϵ $(= F_0^3\Lambda/p^6)$ is small, the **driven response** of the Duffing equation (8.22) (with $k = 0$) is given by

$$x = \frac{F_0}{\Omega^2 - p^2}\left[\cos pt - \left(\frac{3p^6 \cos pt}{(\Omega^2 - p^2)^3} + \frac{p^6 \cos 3pt}{(\Omega^2 - p^2)^2(\Omega^2 - 9p^2)}\right)\epsilon + O\left(\epsilon^2\right)\right]. \tag{8.32}$$

This is the *approximate solution correct to the first order in the small parameter* ϵ. More terms can be obtained in a similar way but this is best done with computer assistance.

The most interesting feature of this formula is the behaviour of the first order correction term when Ω is close to $3p$, which suggests the existence of a *super-harmonic resonance* with frequency $3p$. Similar 'resonances' occur in the higher terms at the frequencies $5p, 7p, \ldots$, and are caused by the presence of the non-linear term Λx^3. It should not however be concluded that large amplitude responses occur at these frequencies.* The critical case in which $\Omega = 3p$ is solved in Problem 8.14 and reveals no infinities in the response.

Sub-harmonic responses and chaos

We have so far left open the interesting question of whether a driving force with minimum period τ can excite a **subharmonic response**, that is, a response whose minimum period is

* This is a subtle point. Like all power series, perturbation series have a certain 'radius of convergence'. When *all* the terms of the perturbation series are included, ϵ is restricted to some range of values $-\epsilon_0 < \epsilon < \epsilon_0$. What seems to happen when Ω approaches $3p$ is that ϵ_0 approaches zero so that the first order correction term never actually gets large.

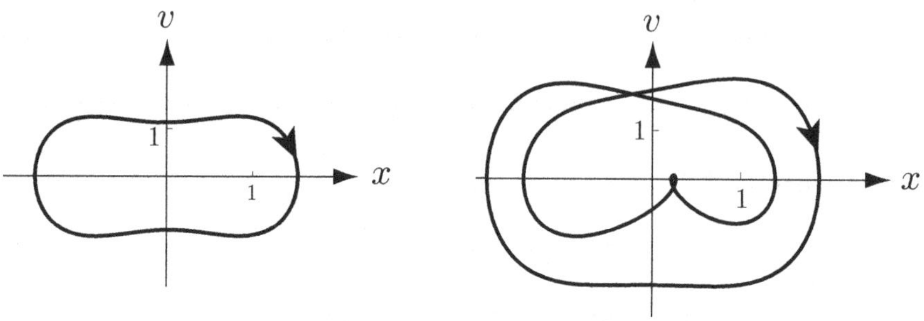

FIGURE 8.9 Two different periodic responses to the same driving force. **Left**: A response of period 2π, **Right**: A sub-harmonic response of period 4π.

an *integer multiple* of τ. This is certainly not possible in the linear case, where the driving force and the induced response always have the same period. One way of investigating this problem would be to expand the (unknown) response $x(t)$ as a Fourier series and to substitute this into the left side of Duffing's equation. One would then require all the odd numbered terms to magically cancel out leaving a function with period 2τ. Unlikely though this may seem, it can happen! *There are ranges of the parameters in Duffing's equation that permit a sub-harmonic response.* Indeed, it is possible for the *same* set of parameters to allow more than one periodic response. Figure 8.9 shows two different periodic responses of the equation $d^2x/dt^2 + kdx/dt + x^3 = A\cos t$, each corresponding to $k = 0.04$, $A = 0.9$. One response has period 2π while the other is a **subharmonic response** with period 4π. Which of these is the steady state response depends on the initial conditions. It is also possible for the motion to be **chaotic** with no steady state ever being reached, even though damping is present.

Problems on Chapter 8

Answers and comments are at the end of the book.

Harder problems carry a star (∗).

Periodic oscillations: Lindstedt's method

8.1 A non-linear oscillator satisfies the equation

$$\left(1+\epsilon x^2\right)\ddot{x} + x = 0,$$

where ϵ is a small parameter. Use Linstedt's method to obtain a two-term approximation to the oscillation frequency when the oscillation has unit amplitude. Find also the corresponding two-term approximation to $x(t)$. [You will need the identity $4\cos^3 s = 3\cos s + \cos 3s$.]

8.2 A non-linear oscillator satisfies the equation

$$\ddot{x} + x + \epsilon x^5 = 0,$$

where ϵ is a small parameter. Use Linstedt's method to obtain a two-term approximation to the oscillation frequency when the oscillation has unit amplitude. [You will need the identity $16\cos^5 s = 10\cos s + 5\cos 3s + \cos 5s$.]

8.3 ***Unsymmetrical oscillations*** A non-linear oscillator satisfies the equation

$$\ddot{x} + x + \epsilon x^2 = 0,$$

where ϵ is a small parameter. Explain why the oscillations are unsymmetrical about $x = 0$ in this problem.

Use Linstedt's method to obtain a two-term approximation to $x(t)$ for the oscillation in which the *maximum* value of x is unity. Deduce a two-term approximation to the *minimum* value achieved by $x(t)$ in this oscillation.

8.4* ***A limit cycle by perturbation theory*** Use perturbation theory to investigate the limit cycle of **Rayleigh's equation**, taken here in the form

$$\ddot{x} + \epsilon\left(\tfrac{1}{3}\dot{x}^2 - 1\right)\dot{x} + x = 0,$$

where ϵ is a small positive parameter. Show that the zero order approximation to the limit cycle is a circle and determine its centre and radius. Find the frequency of the limit cycle correct to order ϵ^2, and find the function $x(t)$ correct to order ϵ.

Phase paths

8.5 ***Phase paths in polar form*** Show that the system of equations

$$\dot{x}_1 = F_1(x_1, x_2, t), \qquad \dot{x}_2 = F_2(x_1, x_2, t)$$

can be written in polar coordinates in the form

$$\dot{r} = \frac{x_1F_1 + x_2F_2}{r}, \qquad \dot{\theta} = \frac{x_1F_2 - x_2F_1}{r^2},$$

where $x_1 = r\cos\theta$ and $x_2 = r\sin\theta$.

A dynamical system satisfies the equations

$$\dot{x} = -x + y,$$
$$\dot{y} = -x - y.$$

Convert this system into polar form and find the polar equations of the phase paths. Show that every phase path encircles the origin infinitely many times in the clockwise direction. Show further that every phase path terminates at the origin. Sketch the phase diagram.

8.6 A dynamical system satisfies the equations

$$\dot{x} = x - y - (x^2 + y^2)x,$$
$$\dot{y} = x + y - (x^2 + y^2)y.$$

Convert this system into polar form and find the polar equations of the phase paths that begin in the domain $0 < r < 1$. Show that all these phase paths spiral anti-clockwise and tend to the limit cycle $r = 1$. Show also that the same is true for phase paths that begin in the domain $r > 1$. Sketch the phase diagram.

8.7 A damped linear oscillator satisfies the equation

$$\ddot{x} + \dot{x} + x = 0.$$

Show that the polar equations for the motion of the phase points are

$$\dot{r} = -r \sin^2 \theta, \qquad \dot{\theta} = -\left(1 + \tfrac{1}{2} \sin 2\theta\right).$$

Show that every phase path encircles the origin infinitely many times in the clockwise direction. Show further that these phase paths terminate at the origin.

8.8 A non-linear oscillator satisfies the equation

$$\ddot{x} + \dot{x}^3 + x = 0.$$

Find the polar equations for the motion of the phase points. Show that phase paths that begin within the circle $r < 1$ encircle the origin infinitely many times in the clockwise direction. Show further that these phase paths terminate at the origin.

8.9 A non-linear oscillator satisfies the equation

$$\ddot{x} + \left(x^2 + \dot{x}^2 - 1\right) \dot{x} + x = 0.$$

Find the polar equations for the motion of the phase points. Show that any phase path that starts in the domain $1 < r < \sqrt{3}$ spirals clockwise and tends to the limit cycle $r = 1$. [The same is true of phase paths that start in the domain $0 < r < 1$.] What is the period of the limit cycle?

8.10 ***Predator–prey*** Consider the symmetrical predator–prey equations

$$\dot{x} = x - xy, \qquad \dot{y} = xy - y,$$

where $x(t)$ and $y(t)$ are positive functions. Show that the phase paths satisfy the equation

$$\left(xe^{-x}\right)\left(ye^{-y}\right) = A,$$

where A is a constant whose value determines the particular phase path. By considering the shape of the surface

$$z = \left(xe^{-x}\right)\left(ye^{-y}\right),$$

deduce that each phase path is a simple closed curve that encircles the equilibrium point at $(1, 1)$. Hence *every solution* of the equations is periodic! [This prediction can be confirmed by solving the original equations numerically.]

Poincaré–Bendixson

8.11 Use Poincaré–Bendixson to show that the system

$$\dot{x} = x - y - (x^2 + 4y^2)x,$$
$$\dot{y} = x + y - (x^2 + 4y^2)y,$$

has a limit cycle lying in the annulus $\frac{1}{2} < r < 1$.

8.12 ***Van der Pol's equation*** Show that Van der Pol's equation*

$$\ddot{x} + \epsilon\dot{x}\left(x^2 - 1\right) + x = 0$$

is equivalent to the system of first order equations

$$\dot{x} = u - \epsilon x\left(\tfrac{1}{3}x^2 - 1\right),$$
$$\dot{u} = -x,$$

and, by making appropriate changes of variable, that this system is in turn equivalent to the system

$$\dot{X} = V,$$
$$\dot{V} = -X - \epsilon V\left(V^2 - 1\right).$$

By comparing this last system with the system (8.20) discussed in Example 8.4, deduce that Van der Pol's equation has a limit cycle for any positive value of the parameter ϵ.

Driven oscillations

8.13 A driven non-linear oscillator satisfies the equation

$$\ddot{x} + \epsilon\dot{x}^3 + x = \cos pt,$$

where ϵ, p are positive constants. Use perturbation theory to find a two-term approximation to the driven response when ϵ is small. Are there any restrictions on the value of p?

Computer assisted problems

8.14 ***Super-harmonic resonance*** A driven non-linear oscillator satisfies the equation

$$\ddot{x} + 9x + \epsilon x^3 = \cos t,$$

---

* After the extravagantly named Dutch physicist Balthasar Van der Pol (1889–1959). The equation arose in connection with the current in an electronic circuit. In 1927 Van der Pol observed what is now called *deterministic chaos*, but did not investigate it further.

where ϵ is a small parameter. Use perturbation theory to investigate the possible existence of a superharmonic resonance. Show that the zero order solution is

$$x_0 = \frac{1}{8}\left(\cos t + a_0 \cos 3t + b_0 \sin 3t\right),$$

where a_0, b_0 are constants that are not known at the zero order stage.

By proceeding to the first order stage, show that $b_0 = 0$ and that a_0 is the unique real root of the cubic equation

$$3a_0^3 + 6a_0 + 1 = 0,$$

which is about -0.164. Thus, when driving the oscillator at this sub-harmonic frequency, the non-linear correction appears in the *zero order* solution. However, there are no infinities to be found in the perturbation scheme at this (or any other) stage.

Plot the graph of $x_0(t)$ and the path of the phase point $(x_0(t), x_0'(t))$.

8.15 ***Linstedt's method*** Use computer assistance to implement Lindstedt's method for the equation

$$\ddot{x} + x + \epsilon x^3 = 0.$$

Obtain a three-term approximation to the oscillation frequency when the oscillation has unit amplitude. Find also the corresponding three-term approximation to $x(t)$.

8.16 ***Van der Pol's equation*** A classic non-linear oscillation equation that has a limit cycle is Van der Pol's equation

$$\ddot{x} + \epsilon\left(x^2 - 1\right)\dot{x} + x = 0,$$

where ϵ is a positive parameter. Solve the equation numerically with $\epsilon = 2$ (say) and plot the motion of a few of the phase points in the (x, v)-plane. All the phase paths tend to the limit cycle. One can see the same effect in a different way by plotting the solution function $x(t)$ against t.

8.17 ***Sub-harmonic and chaotic responses*** Investigate the *steady state* responses of the equation

$$\ddot{x} + k\dot{x} + x^3 = A\cos t$$

for various choices of the parameters k and A and various initial conditions. First obtain the responses shown in Figure 8.9 and then go on to try other choices of the parameters. Some very exotic results can be obtained! For various chaotic responses try $K = 0.1$ and $A = 7$.

Part Two

MULTI-PARTICLE SYSTEMS AND CONSERVATION PRINCIPLES

CHAPTERS IN PART TWO

Chapter Nine

The energy principle

and energy conservation

KEY FEATURES

The key features of this chapter are the **energy principle** for a multi-particle system, the **potential energies** arising from **external** and **internal** forces, and **energy conservation**.

This is the first of three chapters in which we study the mechanics of **multi-particle systems**. This is an important development which greatly increases the range of problems that we can solve. In particular, multi-particle mechanics is needed to solve problems involving the rotation of rigid bodies.

The chapter begins by obtaining the **energy principle** for a multi-particle system. This is the first of the three great principles of multi-particle mechanics* that apply to *every* mechanical system without restriction. We then show that, under appropriate conditions, the total energy of the system is conserved. We apply this **energy conservation** principle to a wide variety of systems. When the system has just one degree of freedom, the energy conservation equation is sufficient to determine the whole motion.

9.1 CONFIGURATIONS AND DEGREES OF FREEDOM

A **multi-particle system** $\mathcal{S}$ may consist of any number of particles $P_1, P_2, \ldots,$ P_N, with masses $m_1, m_2 \ldots, m_N$ respectively.† A possible 'position' of the system is called a **configuration**. More precisely, if the particles $P_1, P_2, \ldots, P_N$ of a system have position vectors $\boldsymbol{r}_1, \boldsymbol{r}_2, \ldots, \boldsymbol{r}_N$, then any *geometrically possible* set of values for the position vectors $\{\boldsymbol{r}_i\}$ is a configuration of the system.

If the system is **unconstrained**, then each particle can take up any position in space (independently of the others) and all choices of the $\{\boldsymbol{r}_i\}$ are possible. This would be the case, for instance, if the particles of $\mathcal{S}$ were moving freely under their mutual gravitation. On the other hand, when **constraints** are present, the $\{\boldsymbol{r}_i\}$ are restricted. Suppose for instance that the particles P_1 and P_2 are connected by a light rigid rod of length a.

* The other two are the *linear momentum* and *angular momentum* principles.

† To save space, we will usually express this by saying that $\mathcal{S}$ is the system of particles $\{P_i\}$ with masses $\{m_i\}$, the range of the index number i being understood to be $1 \le i \le N$.

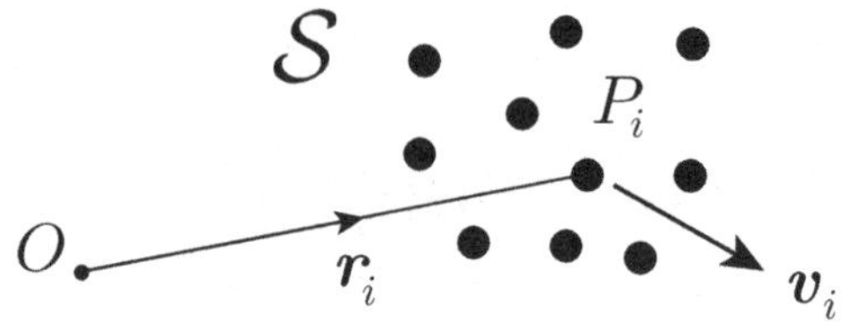

FIGURE 9.1 The multi-particle system $\mathcal{S}$ consists of N particles $P_1, P_2, \ldots, P_N$, of which the typical particle P_i is labelled. The particle P_i has mass m_i, position vector $\boldsymbol{r}_i$, and velocity $\boldsymbol{v}_i$.

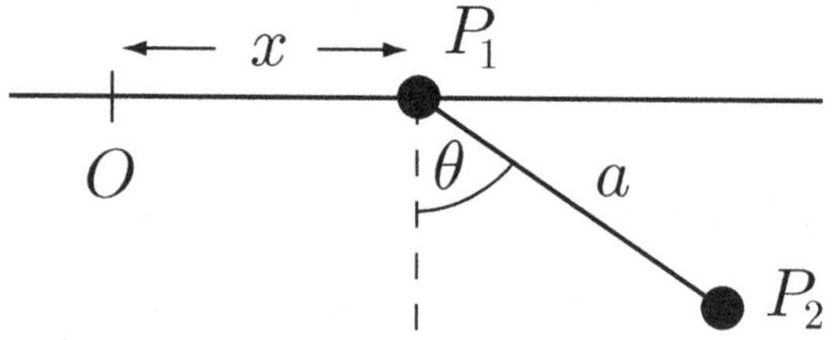

FIGURE 9.2 The generalised coordinates x and θ are sufficient to specify the configuration of this two-particle system in planar motion.

This imposes the geometrical restriction $|\boldsymbol{r}_1 - \boldsymbol{r}_2| = a$ so that not all choices of the $\{\boldsymbol{r}_i\}$ are then possible. This difference is reflected in the number of scalar variables needed to specify the configuration of $\mathcal{S}$. In the unconstrained case, all of the position vectors $\{\boldsymbol{r}_i\}$ must be specified separately. Since each of these vectors may be specified by three Cartesian coordinates, it follows that a total of $3N$ *scalar variables are needed to specify the configuration of an unconstrained N-particle system.* When constraints are present, this number is reduced, often dramatically so.

For example, consider the system shown in Figure 9.2, which consists of two particles P_1 and P_2 connected by a light rigid rod of length a. The particle P_1 is also constrained to move along a fixed horizontal rail and the whole system moves in the vertical plane through the rail. The *two* scalar variables x and θ shown are sufficient to specify the configuration of this system. This contrasts with the *six* scalar variables that would be needed if the two particles were in unconstrained motion. The variables x and θ are said to be a set of **generalised coordinates** for this system.* Other choices for the generalised coordinates could be made, but the *number* of generalised coordinates needed is always the same.

Definition 9.1 ***Degrees of freedom*** *The* ***number*** *of generalised coordinates needed to specify the configuration of a system* $\mathcal{S}$ *is called the number of* ***degrees of freedom*** *of* $\mathcal{S}$.

Importance of degrees of freedom

The number of degrees of freedom of a system is important because it is equal to the *number of equations that are needed to determine the motion of the system.* For example,

* Besides being sufficient to specify the configuration of the system, the generalised coordinates are also required to be *independent*, that is, there must be no functional relation between them. The coordinates x, θ in Figure 9.2 are certainly independent variables. If the coordinates were connected by a functional relation, they would not all be needed and one of them could be discarded.

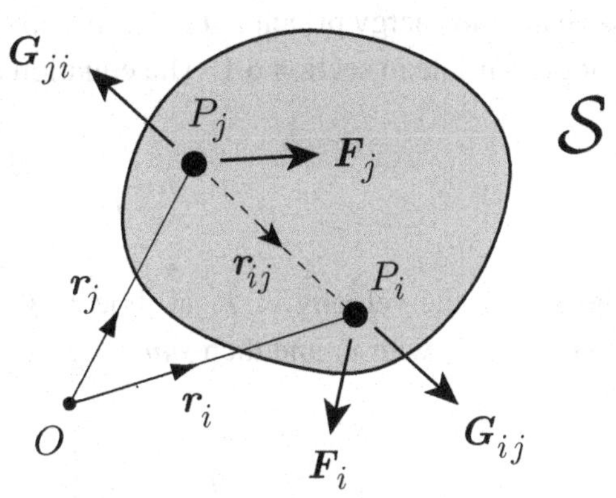

FIGURE 9.3 The multi-particle $\mathcal{S}$ consists of N particles $P_1, P_2, \ldots, P_N$, of which the typical particles P_i and P_j are shown explicitly. The force $\boldsymbol{F}_i$ is the *external* force acting on P_i and the force $\boldsymbol{G}_{ij}$ is the *internal* force exerted on P_i by the particle P_j.

the system shown in Figure 9.2 has *two* degrees of freedom and so needs *two* equations to determine the motion completely.

Example 9.1 ***Degrees of freedom***

Find the number of degrees of freedom of the following mechanical systems: (i) the simple pendulum (moving in a vertical plane), (ii) a door swinging on its hinges, (iii) a bar of soap (a particle) sliding on the inside of a hemispherical basin, (iv) a rigid rod sliding on a flat table, (v) four rigid rods flexibly jointed to form a quadrilateral which can slide on a flat table.

Solution

(i) 1 (ii) 1 (iii) 2 (iv) 3 (v) 4. ■

9.2 THE ENERGY PRINCIPLE FOR A SYSTEM

Let $\mathcal{S}$ be a system of N particles $\{P_i\}$, as shown in Figure 9.3. We classify the forces acting on the particles of $\mathcal{S}$ as being external or internal. **External forces** are those originating from *outside* $\mathcal{S}$. (In the case of a single particle, these are the only forces that act.) Uniform gravity is an example of an external force. However, in multi-particle systems, the particles are also subject to their own *mutual interactions*, that is, the forces that they exert upon each other. These mutual interactions are called the **internal forces** acting on $\mathcal{S}$. The situation is shown in Figure 9.3. $\boldsymbol{F}_i$ is the external force acting on the particle P_i, while $\boldsymbol{G}_{ij}$ is the internal force exerted on P_i by the particle P_j. By the Third Law, the force $\boldsymbol{G}_{ji}$ that P_i exerts on P_j must be equal and opposite to the force $\boldsymbol{G}_{ij}$, and both forces must be parallel to the straight line joining P_i and P_j. In short, the $\{\boldsymbol{G}_{ij}\}$ must satisfy

$$\boldsymbol{G}_{ji} = -\boldsymbol{G}_{ij}, \qquad \text{and} \qquad \boldsymbol{G}_{ij} \parallel (\boldsymbol{r}_i - \boldsymbol{r}_j). \tag{9.1}$$

To obtain the energy principle for the system $\mathcal{S}$, we proceed in the same way as we did for a single particle in section 6.1. The equation of motion for the particle P_i is*

$$m_i \frac{d\boldsymbol{v}_i}{dt} = \boldsymbol{F}_i + \sum_{j=1}^{N} \boldsymbol{G}_{ij}, \tag{9.2}$$

where $\boldsymbol{v}_i$ is the velocity of P_i at time t. On taking the scalar product of both sides of equation (9.2) with $\boldsymbol{v}_i$ and then *summing* the result over all the particles ($1 \le i \le N$), we obtain

$$\frac{dT}{dt} = \sum_{i=1}^{N} \left\{ \boldsymbol{F}_i + \sum_{j=1}^{N} \boldsymbol{G}_{ij} \right\} \cdot \boldsymbol{v}_i, \tag{9.3}$$

where

$$T = \sum_{i=1}^{N} \tfrac{1}{2} m_i |\boldsymbol{v}_i|^2,$$

the **total kinetic energy** of the whole system $\mathcal{S}$. Suppose that, in the time interval $[t_{\mathcal{A}}, t_{\mathcal{B}}]$, the system $\mathcal{S}$ moves from configuration $\mathcal{A}$ to configuration $\mathcal{B}$. On integrating equation (9.3) with respect to t over the time interval $[t_{\mathcal{A}}, t_{\mathcal{B}}]$ we obtain

$$T_{\mathcal{B}} - T_{\mathcal{A}} = \sum_{i=1}^{N} \int_{t_{\mathcal{A}}}^{t_{\mathcal{B}}} \boldsymbol{F}_i \cdot \boldsymbol{v}_i \, dt + \sum_{i=1}^{N} \sum_{j=1}^{N} \int_{t_{\mathcal{A}}}^{t_{\mathcal{B}}} \boldsymbol{G}_{ij} \cdot \boldsymbol{v}_i \, dt \tag{9.4}$$

where $T_{\mathcal{A}}$ and $T_{\mathcal{B}}$ are the kinetic energies of the system $\mathcal{S}$ at times $t_{\mathcal{A}}$ and $t_{\mathcal{B}}$ respectively. This is the **energy principle** for a multi-particle system moving under the external forces $\{\boldsymbol{F}_i\}$ and internal forces $\{\boldsymbol{G}_{ij}\}$. This impressive looking result can be stated quite simply as follows:

Energy principle for a multi-particle system

In any motion of a system, the increase in the total kinetic energy of the system in a given time interval is equal to the total work done by all the external and internal forces during this time interval.

* The summation over j in equation (9.2) contains the term $\boldsymbol{G}_{ii}$ which corresponds to the force that the particle P_i exerts upon *itself*. Since such a force is not actually present, we should really say that the summation is over the range $1 \le j \le N$ with $j \ne i$. Since this would make the formulae look messy, we adopt the device of regarding the terms $\boldsymbol{G}_{11}, \boldsymbol{G}_{22}, \ldots, \boldsymbol{G}_{NN}$ (which do not actually exist) as being zero.

9.3 ENERGY CONSERVATION FOR A SYSTEM

In order to develop an energy conservation principle, we need to write the right side of the energy principle (9.4) in the form $V(\mathcal{A}) - V(\mathcal{B})$, where V is the potential energy function for the *whole system*. We first consider *unconstrained* systems.

Unconstrained systems

When the system is unconstrained, all the forces that act on the system are specified directly. We will assume that the **external forces** $\boldsymbol{F}_i$ are *conservative fields*. In this case $\boldsymbol{F}_i = -\operatorname{grad}\phi_i$, where ϕ_i is the potential energy function of the field $\boldsymbol{F}_i$. Then the total work done by the external forces can be written

$$\sum_{i=1}^{N}\int_{t_\mathcal{A}}^{t_\mathcal{B}} \boldsymbol{F}_i \cdot \boldsymbol{v}_i \, dt = \sum_{i=1}^{N} \left(\phi_i(\boldsymbol{r}_\mathcal{A}) - \phi_i(\boldsymbol{r}_\mathcal{B})\right) = \Phi(\mathcal{A}) - \Phi(\mathcal{B}),$$

where

$$\Phi(\boldsymbol{r}_1, \boldsymbol{r}_2, \ldots, \boldsymbol{r}_N) = \phi_1(\boldsymbol{r}_1) + \phi_2(\boldsymbol{r}_2) + \cdots + \phi_N(\boldsymbol{r}_N)$$

is the potential energy of $\mathcal{S}$ arising from the **external** forces.

Example 9.2 *Potential energy under uniform gravity*

Find the potential energy Φ when the external forces on $\mathcal{S}$ arise from uniform gravity.

Solution

Under uniform gravity, the force $\boldsymbol{F}_i$ exerted on particle P_i is $\boldsymbol{F}_i = -m_i g \boldsymbol{k}$, where the unit vector $\boldsymbol{k}$ points vertically upwards. This conservative field has potential energy $\phi_i = m_i g z_i$, where z_i is the z-coordinate of P_i. The total potential energy of $\mathcal{S}$ due to uniform gravity is therefore

$$\Phi = m_1 g z_1 + m_2 g z_2 + \cdots + m_N g z_N.$$

On using the definition of centre of mass given in section 3.5, this can be written in the alternative form

$$\Phi = MgZ,$$

where M is the total mass of $\mathcal{S}$, and Z is the z-coordinate of the centre of mass of $\mathcal{S}$. Thus *the potential energy of any system due to uniform gravity is the same as if all its mass were concentrated at its centre of mass.* ∎

We now need to make a similar transformation to show that the work done by the **internal forces** can be written in the form $\Psi(\mathcal{A}) - \Psi(\mathcal{B})$, where Ψ is the internal potential energy. The argument is as follows:

We know from the Third Law that the $\{\boldsymbol{G}_{ij}\}$ satisfy the conditions (9.1), but a little more must be assumed. We further assume that the *magnitude* of $\boldsymbol{G}_{ij}$ depends only on r_{ij}, the distance between P_i and

P_j.* Internal forces that satisfy this conditions will be called **conservative**; mutual gravitation forces are a typical example. Hence, when the internal forces are conservative, $\boldsymbol{G}_{ij}$ must have the form

$$\boldsymbol{G}_{ij} = h_{ij}(r_{ij})\widehat{\boldsymbol{r}}_{ij} \tag{9.5}$$

where (see Figure 9.3)

$$\boldsymbol{r}_{ij} = \boldsymbol{r}_i - \boldsymbol{r}_j \qquad r_{ij} = |\boldsymbol{r}_{ij}| \qquad \widehat{\boldsymbol{r}}_{ij} = \boldsymbol{r}_{ij}/r_{ij}. \tag{9.6}$$

Note that h_{ij} is the *repulsive* force that the particles P_i and P_j exert upon each other.

Consider now the rate of working of the *pair* of forces $\boldsymbol{G}_{ij}$ and $\boldsymbol{G}_{ji}$. This is

$$\begin{aligned}\boldsymbol{G}_{ij}\cdot\boldsymbol{v}_i + \boldsymbol{G}_{ji}\cdot\boldsymbol{v}_j = \boldsymbol{G}_{ij}\cdot\left(\boldsymbol{v}_i - \boldsymbol{v}_j\right) = h_{ij}(r_{ij})\widehat{\boldsymbol{r}}_{ij}\cdot\frac{d\boldsymbol{r}_{ij}}{dt} = \left(\frac{h_{ij}(r_{ij})}{r_{ij}}\right)\boldsymbol{r}_{ij}\cdot\frac{d\boldsymbol{r}_{ij}}{dt} \\ = h_{ij}(r_{ij})\frac{dr_{ij}}{dt},\end{aligned}$$

on using equations (9.1), (9.6) and the identity $\boldsymbol{r}_{ij}\cdot\dot{\boldsymbol{r}}_{ij} = r_{ij}\dot{r}_{ij}$. The total work done by the forces $\boldsymbol{G}_{ij}$ and $\boldsymbol{G}_{ji}$ during the time interval $[t_{\mathcal{A}}, t_{\mathcal{B}}]$ is therefore

$$\int_{t_{\mathcal{A}}}^{t_{\mathcal{B}}} h_{ij}(r_{ij})\frac{dr_{ij}}{dt}\,dt = \int_{r_{ij}(\mathcal{A})}^{r_{ij}(\mathcal{B})} h_{ij}(r_{ij})\,dr_{ij} = H_{ij}(r_{ij}(\mathcal{A})) - H_{ij}(r_{ij}(\mathcal{B})),$$

where H_{ij} is the indefinite integral of $-h_{ij}$. The function $H_{ij}(r_{ij})$ is called the **mutual potential energy** of the particles P_i and P_j.

It follows that the total work done by all the internal forces in the time interval $[t_{\mathcal{A}}, t_{\mathcal{B}}]$ can be written in the form

$$\sum_{i=1}^{N}\sum_{j=1}^{N}\int_{t_{\mathcal{A}}}^{t_{\mathcal{B}}} \boldsymbol{G}_{ij}\cdot\boldsymbol{v}_i\,dt = \Psi(\mathcal{A}) - \Psi(\mathcal{B}),$$

where

$$\Psi(\boldsymbol{r}_1, \boldsymbol{r}_2, \ldots, \boldsymbol{r}_N) = \sum_{i=1}^{N}\sum_{j=1}^{i-1} H_{ij}(r_{ij})$$

is the **potential energy** of $\mathcal{S}$ arising from the **internal** forces. This potential energy is just the sum of the mutual potential energies of all pairs of particles.

Example 9.3 *Internal energy of three charged particles*

Three particles P_1, P_2, P_3 carry electric charges e_1, e_2, e_3 respectively. Find the internal potential energy Ψ.

Solution

In cgs/electrostatic units, the particles P_1 and P_2 repel each other with the force $h_{12}(r_{12}) = e_1e_2/(r_{12})^2$, where r_{12} is the distance between P_1 and P_2. Their mutual potential energy is therefore

$$H_{12} = -\int h_{12}(r_{12})\,dr_{12} = -\int \frac{e_1e_2}{(r_{12})^2}\,dr_{12} = \frac{e_1e_2}{r_{12}}.$$

* This is equivalent to the very reasonable assumptions that the magnitude of $\boldsymbol{G}_{ij}$ is invariant under spatial translations and rotations of each pair of particles P_i and P_j, and is independent of the time.

The **internal potential energy** of the whole system is therefore

$$\Psi = \frac{e_1 e_2}{r_{12}} + \frac{e_1 e_3}{r_{13}} + \frac{e_2 e_3}{r_{23}}. \blacksquare$$

On combining the above results, the energy principle (9.4) can be written

$$T_{\mathcal{B}} - T_{\mathcal{A}} = V(\mathcal{A}) - V(\mathcal{B}),$$

where $V = \Phi + \Psi$ is the **total potential energy** of the system $\mathcal{S}$. This is equivalent to the **energy conservation** formula

$$\boxed{T + V = E} \tag{9.7}$$

where E is the total energy of the system.This result can be summarised as follows:

> **Energy conservation for an unconstrained system**
>
> When both the external and internal forces acting on a system are *conservative*, the sum of its kinetic and potential energies* remains constant in the motion.

Example 9.4 *A star with two planets*

A star of very large mass M is orbited by two planets P_1 and P_2 of masses m_1 and m_2. Find the energy conservation equation for this system.

Solution

Since the mass of the star is supposed to be very much larger than the planetary masses, we will neglect its motion and suppose that it is fixed at the origin O. We then have a *two-particle* problem in which the planets move under the (external) gravitational attraction of the star and their (internal) mutual gravitational interaction. This is an unconstrained system.

The total potential energy arising from **external forces** is then

$$\Phi = -\frac{Mm_1G}{r_1} - \frac{Mm_2G}{r_2},$$

where r_1,r_2 are the distances OP_1, OP_2.

The particles P_1 and P_2 repel each other with the force $h_{12}(r_{12}) = -m_1m_2G/(r_{12})^2$, where r_{12} is the distance between P_1 and P_2. Their **mutual potential energy** is therefore

$$H_{12} = -\int h_{12}(r_{12})\,dr_{12} = \int \frac{m_1m_2G}{(r_{12})^2}\,dr_{12} = -\frac{m_1m_2G}{r_{12}},$$

and this is the only contribution to the internal potential energy Ψ.

* The potential energy is the total of the potential energies arising from both the external and internal forces.

Since the system is unconstrained and the external and internal forces are conservative, energy conservation applies. The **energy conservation equation** for the system is

$$\tfrac{1}{2}m_1 |\boldsymbol{v}_1|^2 + \tfrac{1}{2}m_2 |\boldsymbol{v}_2|^2 - MG\left(\frac{m_1}{r_1} + \frac{m_2}{r_2}\right) - \frac{m_1 m_2 G}{r_{12}} = E,$$

where $\boldsymbol{v}_1$, $\boldsymbol{v}_2$ are the velocities of the planets P_1, P_2, and E is the constant total energy. The value of E is determined from the initial conditions.

Since this system has six degrees of freedom (four if the motions are confined to a plane through O), the energy conservation equation is by no means sufficient to determine the motion! ■

Question *Can a planet escape?*

If the initial conditions are such that $E < 0$, is it possible for a planet to escape to infinity?

Answer

If $E < 0$, then it is certainly not possible for *both* planets to escape to infinity, since the total energy would then be positive. However, the escape of *one* planet is not prohibited by energy conservation. This does not mean however that such an escape will actually happen.

Constrained systems

When a system is subject to constraints, not all the forces that act on the system are specified. This is because constraints are enforced by **constraint forces** that are not part of the specification of the problem; all we know is that their *effect* is to enforce the given constraints. The work done by constraint forces cannot generally be calculated (or expressed in terms of a potential energy) and we are restricted to those *systems for which the total work done by the constraint forces happens to be zero.**

The constraint forces acting on the system may be **external** (for example, when a particle of the system is constrained to remain at rest), or **internal** (for example, when two particles of the system are constrained to remain the same distance apart).

A The list of **external** constraint forces that do no work is the same as that given in Section 6.5 for single particle motion.

B The most important result regarding **internal** constraint forces that do no work is this: *The total work done by any pair of mutual interaction forces is zero when the particles on which they act are constrained to remain a fixed distance apart.* The proof is as follows:

Suppose two particles P_i and P_j are constrained to remain a fixed distance apart and that their mutual interaction forces are $\boldsymbol{G}_{ij}$ and $\boldsymbol{G}_{ji}$ (see Figure 9.3). Since the distance between P_i and P_j is constant, it follows that $(\boldsymbol{r}_i - \boldsymbol{r}_j) \cdot (\boldsymbol{r}_i - \boldsymbol{r}_j)$ is constant, which, on differentiating with respect

* *Individual* constraint forces may do work.

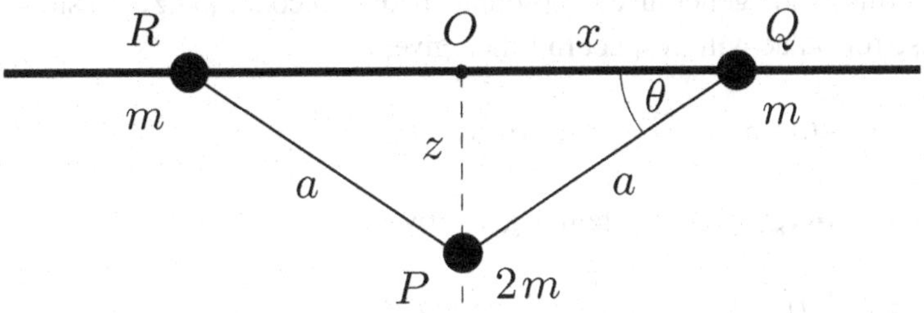

FIGURE 9.4 The particles Q and R slide along a smooth horizontal rail while the particle P moves vertically.

to t, gives

$$(\boldsymbol{r}_i - \boldsymbol{r}_j) \cdot (\boldsymbol{v}_i - \boldsymbol{v}_j) = 0.$$

Thus the vector $(\boldsymbol{v}_i - \boldsymbol{v}_j)$ must be *perpendicular* to the straight line joining P_i and P_j. Hence, the rate of working of the *two* forces $\boldsymbol{G}_{ij}$ and $\boldsymbol{G}_{ji}$ is

$$\boldsymbol{G}_{ij} \cdot \boldsymbol{v}_i + \boldsymbol{G}_{ji} \cdot \boldsymbol{v}_j = \boldsymbol{G}_{ij} \cdot \left(\boldsymbol{v}_i - \cdot \boldsymbol{v}_j\right) = 0,$$

since $\boldsymbol{G}_{ij}$ is known to be *parallel* to the straight line joining P_i and P_j. Thus the internal constraint forces $\boldsymbol{G}_{ij}$ and $\boldsymbol{G}_{ji}$ do no work *in total.*

It follows, for example, that the two tension forces exerted by a light inextensible string do no work in total. It further follows that the *internal forces that enforce rigidity in a rigid body do no work in total.* This important result allows us to solve rigid body problems by energy methods.

Our result for constrained systems can be summarised as follows:

Energy conservation for a constrained system

When the specified external and internal forces acting on a system are *conservative*, and the constraint forces *do no work in total*, the sum of the kinetic and potential energies of the system remains constant in the motion.

Example 9.5 *A constrained three-particle system*

Figure 9.4 shows a ball P of mass $2m$ suspended by light inextensible strings of length a from two sliders Q and R, each of mass m, which can move on a smooth horizontal rail. The system moves symmetrically so that O, the mid-point of Q and R, remains fixed and P moves on the downward vertical through O. Initially the system is released from rest with the three particles in a straight line and with the strings taut. Find the energy conservation equation for the system.

Solution

This is a system with one degree of freedom and we take the angle θ as the generalised coordinate. Let z and x be the displacements of the particles P and Q from the fixed point O. Then, in terms of the generalised coordinate θ, $x = a\cos\theta$ and $z = a\sin\theta$. Differentiating these formulae with respect to t then gives

$$\dot{x} = -(a\sin\theta)\dot{\theta}, \qquad \dot{z} = (a\cos\theta)\dot{\theta}.$$

Hence the total **kinetic energy** of the system is given by

$$T = \tfrac{1}{2}(2m)\dot{z}^2 + \tfrac{1}{2}m\dot{x}^2 + \tfrac{1}{2}m\dot{x}^2 = ma^2\dot{\theta}^2.$$

The only contribution to the **potential energy** comes from uniform gravity, so that

$$V = -(2m)gz + 0 + 0 = -2mga\sin\theta,$$

where we have taken the zero level of potential energy to be at the rail.

We must now show that the constraint forces do no work. The reactions exerted by the smooth rail on the particles Q and R are perpendicular to the rail and therefore perpendicular to the velocities of Q and R; these reactions therefore do no work. Also, the tension forces exerted by the inextensible strings do no work in total. Hence, the **constraint forces** do no work in total.

Energy conservation therefore applies in the form

$$ma^2\dot{\theta}^2 - 2mga\sin\theta = E.$$

From the initial conditions $\theta = \dot{\theta} = 0$ when $t = 0$, it follows that $E = 0$. The **energy conservation equation** for the system is therefore

$$\dot{\theta}^2 - \frac{2g}{a}\sin\theta = 0. \blacksquare$$

Question *When do the sliders collide?*

Find the time that elapses before the sliders collide.

Answer

Since this system has only one degree of freedom, the motion can be found from energy conservation alone. From the energy conservation equation, it follows that

$$\frac{d\theta}{dt} = \pm\left(\frac{2g}{a}\right)^{1/2}(\sin\theta)^{1/2},$$

and, since θ is an *increasing* function of t, we take the *positive* sign. This equation is a first order separable ODE.

Since the sliders collide when $\theta = \pi/2$, the time τ that elapses is given by

$$\tau = \left(\frac{a}{2g}\right)^{1/2}\int_0^{\pi/2}\frac{d\theta}{(\sin\theta)^{1/2}} \approx 1.85\left(\frac{a}{g}\right)^{1/2}. \blacksquare$$

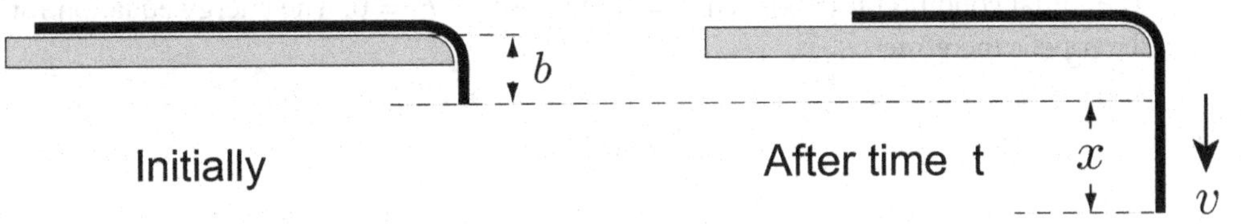

FIGURE 9.5 A uniform rope is released from rest hanging over the edge of a smooth table (left). After time t it has displacement x (right).

Example 9.6 *Rope sliding off a table*

A uniform inextensible rope of mass M and length a is released from rest hanging over the edge of a smooth horizontal table, as shown in Figure 9.5. Find the speed of the rope when it has the displacement x shown.

Solution

A rope is a continuous distribution of mass, unlike the discrete masses that appear in our theory. We regard the rope as being represented by a *light* inextensible string of length a with N particles, each of mass M/N, attached to the string at equally spaced intervals along its length. When N is very large, we expect this discrete set of masses to approximate the behaviour of the rope.

Since each particle of the rope has the same *speed* $v\,(=\dot{x})$, the total **kinetic energy** of the rope is simply

$$T = \tfrac{1}{2}Mv^2.$$

The only contribution to the **potential energy** comes from uniform gravity. If we take the reference state for V to be the initial configuration (Figure 9.5 (left)), then the potential energy in the displaced configuration (right) is the same *as if* a length x of the rope lying on the table were cut off and this piece were then suspended from the hanging end. In the continuous limit (that is, as $N \to \infty$), this piece of rope has mass Mx/a and its centre of mass is lowered a distance $b + (x/2)$ by this operation. The potential energy of the rope in the displaced configuration is therefore

$$V = -\left(\frac{Mx}{a}\right) g \left(b + \tfrac{1}{2}x\right).$$

We must now show that the constraint forces do no work. The reactions exerted by the smooth table on the particles of the rope are always perpendicular to the velocities of these particles; these reactions therefore do no work. Also, the tension forces exerted by each segment of the inextensible string (connecting adjacent particles of the rope) do no work in total. Hence, the **constraint forces** do no work in total.

Energy conservation therefore applies in the form

$$\tfrac{1}{2}Mv^2 - \left(\frac{Mx}{a}\right) g \left(b + \tfrac{1}{2}x\right) = E.$$

The initial condition $v = 0$ when $x = 0$ implies that $E = 0$. The energy equation for the rope is therefore

$$v^2 = \frac{g}{a}\,x(x + 2b).$$

This gives the **speed** of the rope when it has displacement x. This formula holds while there is still some rope left on the table *top*. ■

Note. In the above solution we have assumed that the rope follows the contour of the table edge and then falls vertically. However, it can be shown that this *cannot* be true when the rope is close to leaving the table. What actually happens is that the end of the rope overshoots the table edge. This is a tricky point which we will not investigate further.

Question *Displacement at time t*

Find the displacement of the rope at time t.

Answer

Since this system has only one degree of freedom, the motion can be found from energy conservation alone. From the energy conservation equation, it follows that

$$\frac{dx}{dt} = \pm n\, x^{1/2}(x + 2b)^{1/2},$$

where $n^2 = g/a$. Since x is an *increasing* function of t, we take the *positive* sign. This equation is a first order separable ODE.

It follows that

$$\begin{aligned} nt &= \int \frac{dx}{x^{1/2}(x + 2b)^{1/2}} \\ &= 2\sinh^{-1}\left(\frac{x}{2b}\right)^{1/2} + C, \end{aligned}$$

on using the substitution $x = 2b\sinh^2 w$. The initial condition $x = 0$ when $t = 0$ implies that $C = 0$ and, after some simplification, we obtain

$$x = b(\cosh nt - 1)$$

as the **displacement** of the rope after time t. As before, this formula holds while there is still some rope left on the table top. ■

Example 9.7 *Stability of a plank on a log*

A uniform thin rigid plank is placed on top of a rough circular log and can roll without slipping. Show that the equilibrium position, in which the plank rests symmetrically on top of the log, is stable.

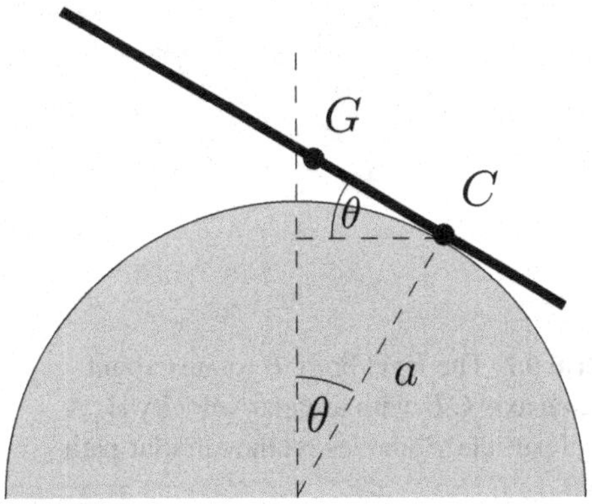

FIGURE 9.6 A thin uniform plank is placed symmetrically on top of a fixed rough circular log. Is the equilibrium position of the plank stable?

Solution

Suppose that the plank is disturbed from its equilibrium position and is tilted by an angle θ as shown in Figure 9.6. The plank is known to *roll* on the log, which means that the distance GC from the centre G of the plank to the contact point C must always be equal to the arc length of the log that has been traversed. If the radius of the log is a, then this arc length is $a\theta$.

We are not yet able to calculate the **kinetic energy** of the plank in terms of the coordinate θ. This is done in the next section. However, we do not need it to investigate stability.

The only contribution to the **potential energy** of the plank comes from uniform gravity. This is given by $V = MgZ$, where Z is the vertical displacement of the centre of mass G of the plank. Elementary trigonometry (see Figure 9.6) shows that $Z = a\cos\theta + a\theta\sin\theta - a$, so that

$$V = Mga(\cos\theta + \theta\sin\theta - 1).$$

We must now show that the constraint forces do no work. The rate of working of the constraint force $\boldsymbol{R}$ that the log exerts on the plank is $\boldsymbol{R} \cdot \boldsymbol{v}^C$, where $\boldsymbol{v}^C$ is the velocity of the particle C of the plank that is *instantaneously* in contact with the log. But, since the plank rolls on the log, $\boldsymbol{v}^C = \boldsymbol{0}$ so that the rate of working of $\boldsymbol{R}$ is zero. Also, the internal constraint forces that enforce the rigidity of the plank do no work in total. Hence, the **constraint forces** do no work in total.

Energy conservation therefore applies in the form

$$T + Mga(\cos\theta + \theta\sin\theta - 1) = E.$$

It follows that the equilibrium position (with the plank on top of the log) will be stable if V has a *minimum* at $\theta = 0$. Now $V' = Mga\theta\cos\theta$ and $V'' = Mga(\cos\theta - \theta\sin\theta)$ so that, when $\theta = 0$, $V' = 0$ and $V'' = 1$. Hence V has a minimum at $\theta = 0$ and so the **equilibrium position** is stable. ■

9.4 KINETIC ENERGY OF A RIGID BODY

The general theory we have presented applies to *any* multi-particle system; in particular, it applies to the rigid array of particles that we call a **rigid body**. However, in

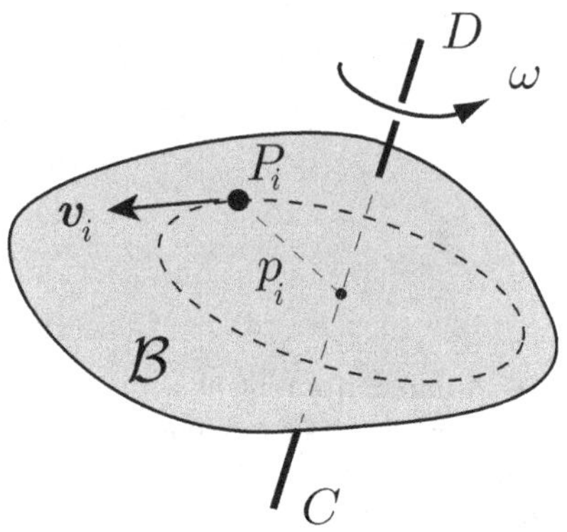

FIGURE 9.7 The rigid body $\mathcal{B}$ rotates about the *fixed* axis CD with angular velocity ω. A typical particle P_i moves on the circular path shown.

order to make use of energy conservation in rigid body dynamics, we need to be able to express the **kinetic energy** T of the body in terms of the generalised coordinates.

Rigid body with a fixed axis

Figure 9.7 shows a rigid body $\mathcal{B}$ which is rotating about the *fixed* axis CD. (Imagine that the body is penetrated by a thin light spindle, which is smoothly pivoted in a fixed position.) A typical particle P_i of the body can move on the circular path shown. This circle has radius p_i, where p_i is the perpendicular distance of P_i from the axis CD. Suppose that, at some instant, the angular velocity of $\mathcal{B}$ about the axis CD is ω. Then the *speed* of particle P_i at this instant is $|\omega| p_i$, and its kinetic energy is $\frac{1}{2}m_i (\omega p_i)^2$. The total kinetic energy of $\mathcal{B}$ is therefore

$$T = \sum_{i=1}^{N} \left(\tfrac{1}{2}m_i (\omega p_i)^2\right) = \tfrac{1}{2}\left(\sum_{i=1}^{N} m_i p_i^2\right)\omega^2.$$

Definition 9.2 ***Moment of inertia*** *The quantity*

$$I_{CD} = \sum_{i=1}^{N} m_i p_i^2 \tag{9.8}$$

where p_i is the perpendicular distance of the mass m_i from the axis CD, is called the ***moment of inertia*** *of the body $\mathcal{B}$ about the axis CD.*

The **moment of inertia**, as defined above, does not depend on the motion of the body $\mathcal{B}$. It is a purely *geometrical* quantity (like centre of mass), which describes how the mass in $\mathcal{B}$ is distributed relative to the axis CD. The further the mass in $\mathcal{B}$ lies from the axis, the larger is the moment of inertia of $\mathcal{B}$ about that axis. In the theory of rotating rigid bodies, the moment of inertia plays a similar rôle to that played by mass in the translational motion of a particle.

Our result may be summarised as follows:

Kinetic energy of a rigid body with a fixed axis

Suppose the rigid body $\mathcal{B}$ is rotating about the fixed axis CD with angular velocity ω. Then the kinetic energy of $\mathcal{B}$ is given by

$$T = \tfrac{1}{2} I_{CD}\, \omega^2, \qquad (9.9)$$

where I_{CD} is the moment of inertia of $\mathcal{B}$ about the axis CD.

Example 9.8 *Moment of inertia of a hoop*

Find the moment of inertia of a uniform hoop of mass M and radius a about its axis of rotational symmetry.

Solution

This is the easiest case to treat since each particle of the hoop has perpendicular distance a from the specified axis. The required moment of inertia is therefore

$$I = \sum_{i=1}^{N} m_i\, a^2 = \left(\sum_{i=1}^{N} m_i \right) a^2 = M a^2,$$

where M is the mass of the whole hoop. ■

It is evident that, in order to solve problems that include rotating rigid bodies, we need to know their moments of inertia. These can be worked out from the definition (9.8), or its counterpart for continuous mass distributions. The Appendix at the end of the book contains examples of how to do this and also contains a table of common moments of inertia, including those for the uniform **rod, hoop, disk** and **sphere**. Most readers will find it convenient to remember the moments of inertia in these four cases.

Example 9.9 *Rotational kinetic energy of the Earth*

Estimate the rotational kinetic energy of the Earth, regarded as a rigid uniform sphere rotating about a *fixed* axis through its centre.

Solution

From the Appendix, we find that I, the moment of inertia of a uniform sphere about an axis through its centre is given by $I = 2MR^2/5$, where M is the mass of the sphere and R its radius. The kinetic energy of the Earth is therefore given by

$$T = \tfrac{1}{2} I \omega^2 = \tfrac{1}{5} M R^2 \omega^2,$$

where M is the mass the Earth, R is its radius, and ω is its angular velocity.

On inserting the values $M = 6.0 \times 10^{24}$ kg, $R = 6400$ km and $\omega = 7.3 \times 10^{-5}$ radians per second, $T = 2.6 \times 10^{29}$ J approximately. ■

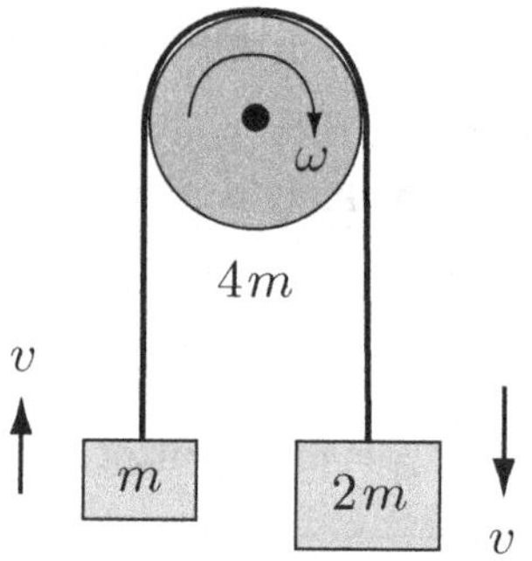

FIGURE 9.8 Two blocks of masses m and $2m$ are connected by a light inextensible string which passes over a circular pulley of mass $4m$ and radius a.

Example 9.10 *Attwood's machine*

Two blocks of masses m and $2m$ are connected by a light inextensible string which passes over a uniform circular pulley of radius a and mass $4m$. Find the upward acceleration of the mass m.

Solution

The system is shown in Figure 9.8. We suppose that the string does not slip on the pulley and that the pulley is smoothly pivoted about its axis of symmetry.

Let z be the upward displacement of the mass m (from some reference configuration) and $v\ (= \dot{z})$ its upward velocity at time t. Then, since the string is *inextensible*, the mass $2m$ must have the same displacement and velocity, but measured downwards. The angular velocity ω of the pulley is determined from the condition that the string does not slip. In this case, the velocity of the rim of the pulley and the velocity of the string must be the same at each point where they are in contact, that is, $a\omega = v$. Hence $\omega = v/a$. Also, from the table in the Appendix, the moment of inertia of a uniform circular disk of mass M and radius a about its axis of symmetry is $\frac{1}{2}Ma^2$. Hence, the total **kinetic energy** of the system is

$$T = \tfrac{1}{2}mv^2 + \tfrac{1}{2}(2m)v^2 + \tfrac{1}{2}\left(\tfrac{1}{2}(4m)a^2\right)\left(\frac{v}{a}\right)^2 = \tfrac{5}{2}mv^2.$$

The gravitational **potential energy** of the system (relative to the reference configuration) is

$$V = mgz - (2m)gz = -mgz.$$

We must now dispose of the **constraint forces**. (i) At the smooth pivot that supports the pulley, the reactions are perpendicular to the velocities of the particles on which they act. Hence these reactions do no work. (ii) Since there is no slippage between the string and the three material bodies of the system, the total work done by the string on the bodies must be equal and opposite to the total work done by the bodies on the string.* (iii) The internal forces that keep the pulley rigid do no work in total. Hence the constraint forces do no work in total.

* Since this string is massless and inextensible, it can have neither kinetic nor potential energy so that the total work done on the string must actually be zero.

Energy conservation therefore applies in the form

$$\tfrac{5}{2}mv^2 - mgz = E,$$

where E is the total energy. If we now differentiate this equation with respect to t (and cancel by mv), we obtain

$$\frac{dv}{dt} = \tfrac{1}{5}g$$

which is the **equation of motion** of the system. Thus the upward **acceleration** of the mass m is $g/5$. (If the pulley were massless, the result would be $g/3$.) ∎

Rigid body in general motion

We now go on to find the kinetic energy of a rigid body that has translational as well as rotational motion. The method depends on the following theorem.

Theorem 9.1 *Suppose a general system of particles $\mathcal{S}$ has total mass M and that its centre of mass G has velocity $\boldsymbol{V}$. Then the total kinetic energy of $\mathcal{S}$ can be written in the form*

$$T = \tfrac{1}{2}MV^2 + T^G, \tag{9.10}$$

where $V = |\boldsymbol{V}|$ and T^G is the kinetic energy of $\mathcal{S}$ in its motion ***relative to*** *G.*

Proof. By definition,

$$\begin{aligned} T^G &= \frac{1}{2}\sum_{i=1}^{N} \tfrac{1}{2}m_i\,(\boldsymbol{v}_i - \boldsymbol{V})\cdot(\boldsymbol{v}_i - \boldsymbol{V}) \\ &= \frac{1}{2}\sum_{i=1}^{N} m_i\boldsymbol{v}_i\cdot\boldsymbol{v}_i - \frac{1}{2}\left(\sum_{i=1}^{N} m_i\boldsymbol{v}_i\right)\cdot\boldsymbol{V} - \frac{1}{2}\boldsymbol{V}\cdot\left(\sum_{i=1}^{N} m_i\boldsymbol{v}_i\right) + \frac{1}{2}\left(\sum_{i=1}^{N} m_i\right)\boldsymbol{V}\cdot\boldsymbol{V} \\ &= T - \tfrac{1}{2}(M\boldsymbol{V})\cdot\boldsymbol{V} - \tfrac{1}{2}\boldsymbol{V}\cdot(M\boldsymbol{V}) + \tfrac{1}{2}M(\boldsymbol{V}\cdot\boldsymbol{V}) \\ &= T - \tfrac{1}{2}MV^2, \end{aligned}$$

as required. ∎

The term $\frac{1}{2}MV^2$ can be regarded as the **translational** contribution to T. When the system $\mathcal{S}$ is a **rigid body**, T^G also has a nice physical interpretation. In this case, the motion of $\mathcal{S}$ relative to G is an angular velocity ω about an axis CD passing through G, as shown in Figure 9.9. It then follows from equation (9.9) that $T^G = \frac{1}{2}I_{CD}\,\omega^2$. This can be regarded as the **rotational** contribution to T. We therefore have the result:

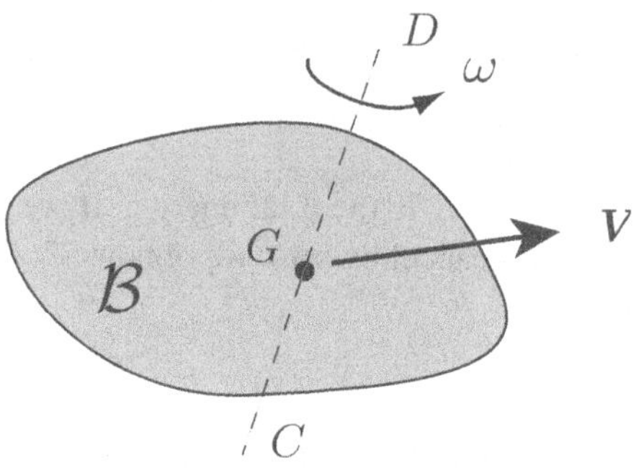

FIGURE 9.9 A rigid body $\mathcal{B}$ in general motion. The centre of mass G has velocity $\boldsymbol{V}$ and $\mathcal{B}$ is also rotating with angular velocity ω about an axis through G.

Kinetic energy of a rigid body in general motion

Let $\mathcal{B}$ be a rigid body of mass M and let G be its centre of mass. Suppose that G has velocity $\boldsymbol{V}$ and that the body is also rotating with angular velocity ω about an axis CD passing through G. Then the **kinetic energy** of $\mathcal{B}$ is given by

$$T = \tfrac{1}{2}MV^2 + \tfrac{1}{2}I_{CD}\,\omega^2, \qquad (9.11)$$

where $V = |\boldsymbol{V}|$ and I_{CD} is the moment of inertia of $\mathcal{B}$ about the axis CD. The term $\frac{1}{2}MV^2$ is called the **translational** kinetic energy and the term $\frac{1}{2}I_{CD}\,\omega^2$ the **rotational** kinetic energy of $\mathcal{B}$.

Example 9.11 *Kinetic energy of a rolling wheel*

Find the kinetic energy of the rolling wheel shown in Figure 2.8.

Solution

Assume the wheel to be uniform with mass M and radius b. Then its centre of mass C has speed u so that the **translational** kinetic energy is $\frac{1}{2}Mu^2$. Because of the rolling condition, the angular velocity of the wheel is given by $\omega = u/b$ so that the **rotational** kinetic energy is $\frac{1}{2}I(u/b)^2$, where $I = \frac{1}{2}Mb^2$. The **total** kinetic energy of the wheel is therefore given by

$$T = \tfrac{1}{2}Mu^2 + \tfrac{1}{2}\left(\tfrac{1}{2}Mb^2\right)\left(\frac{u}{b}\right)^2 = \frac{3Mu^2}{4}. \blacksquare$$

Example 9.12 *Cylinder rolling down a plane*

A uniform hollow circular cylinder is rolling down a *rough* plane inclined at an angle α to the horizontal. Find the acceleration of the cylinder.

Solution

Suppose that, at time t, the cylinder has displacement x down the plane (from some reference configuration) and that the centre of mass G of the cylinder has velocity

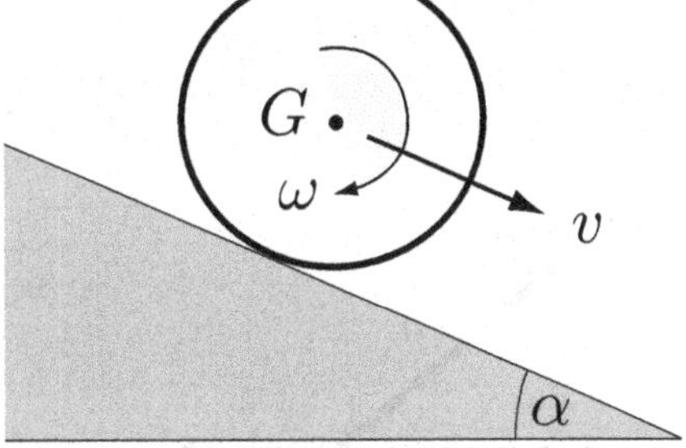

FIGURE 9.10 A hollow circular cylinder rolls down a plane inclined at angle α to the horizontal.

$v\ (= \dot{x})$ down the plane. The angular velocity ω of the cylinder is then determined by the rolling condition to be $\omega = v/b$. The kinetic energy of the cylinder is therefore

$$T = \tfrac{1}{2}Mv^2 + \tfrac{1}{2}I\omega^2 = \tfrac{1}{2}Mv^2 + \tfrac{1}{2}I\left(\frac{v}{b}\right)^2$$

where M is the mass of the cylinder, and I is its moment of inertia about its axis of symmetry. From the Appendix, we find that $I = Mb^2$ so that the **kinetic energy** of the cylinder is given by $T = Mv^2$.

The gravitational **potential energy** of the cylinder is given by $V = -Mgx \sin\alpha$.

We must now dispose of the constraint forces. The reaction forces that the inclined plane exerts on the cylinder act on particles of the cylinder which, because of the rolling condition, have zero velocity. These reaction forces therefore do no work. Also the internal forces that keep the cylinder rigid do no work in total. Hence the **constraint forces** do no work in total.

Conservation of energy therefore applies in the form

$$Mv^2 - Mgx\sin\alpha = E,$$

where E is the total energy. If we now differentiate this equation with respect to t (and cancel by Mv), we obtain

$$\frac{dv}{dt} = \tfrac{1}{2}g\sin\alpha,$$

which is the **equation of motion** of the cylinder. Thus the **acceleration** of the cylinder down the plane is $\frac{1}{2}g\sin\alpha$. (A block sliding down a *smooth* plane would have acceleration $g\sin\alpha$.) ■

Example 9.13 *The sliding ladder*

A uniform ladder of length $2a$ is supported by a smooth horizontal floor and leans against a smooth vertical wall.* The ladder is released from rest in a position making an angle of 60° with the downward vertical. Find the energy conservation equation for the ladder.

* Don't try this at home!

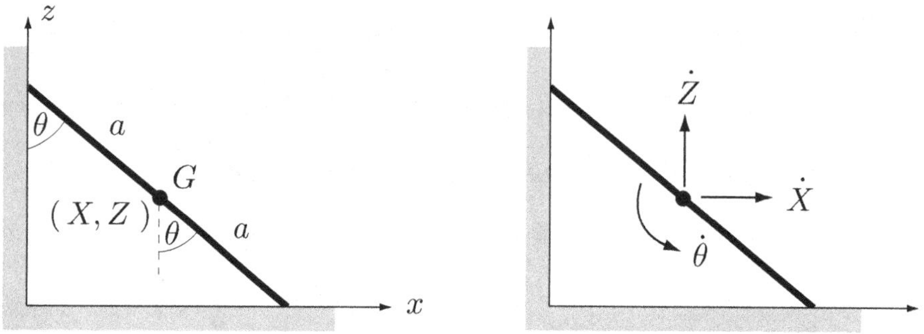

FIGURE 9.11 A uniform ladder of mass M and length $2a$ is supported by a smooth horizontal floor and leans against a smooth vertical wall. At time t, its centre of mass G has (x, z)-coordinates (X, Z) and the ladder makes an angle θ with the downward vertical.

Solution

Let θ be the angle that the ladder makes with the downward vertical after time t. The (x, z)-coordinates of the centre of mass G are then given by

$$X = a\sin\theta, \qquad Z = a\cos\theta,$$

and the corresponding velocity components by

$$\dot{X} = (a\cos\theta)\dot{\theta}, \qquad \dot{Z} = -(a\sin\theta)\dot{\theta}.$$

The angular velocity of the ladder at time t is simply $\dot{\theta}$ (see Figure 9.11). The **kinetic energy** of the ladder is therefore given by

$$T = \tfrac{1}{2}M\left(\dot{X}^2 + \dot{Y}^2\right) + \tfrac{1}{2}I\dot{\theta}^2 = \tfrac{1}{2}Ma^2\dot{\theta}^2 + \tfrac{1}{2}I\dot{\theta}^2,$$

where M is the mass of the ladder and I is its moment of inertia about the horizontal axis through G. From the Appendix, we find that $I = Ma^2/3$ so that the **kinetic energy** of the ladder is given by $T = (2Ma^2/3)\dot{\theta}^2$.

The gravitational **potential energy** of the ladder is given by $V = MgZ = Mga\cos\theta$.

We must now dispose of the constraint forces. The reaction forces that the smooth floor and wall exert on the ladder are both perpendicular to the particles of the ladder on which they act. These reaction forces therefore do no work. Also, the internal forces that keep the ladder rigid do no work in total. Hence the **constraint forces** do no work in total.

Conservation of energy therefore applies in the form

$$\tfrac{2}{3}Ma^2\dot{\theta}^2 + Mga\cos\theta = E,$$

where E is the total energy. From the initial conditions $\dot{\theta} = 0$ and $\theta = \pi/3$ when $t = 0$, it follows that $E = \frac{1}{2}Mga$. The **energy conservation equation** for the ladder

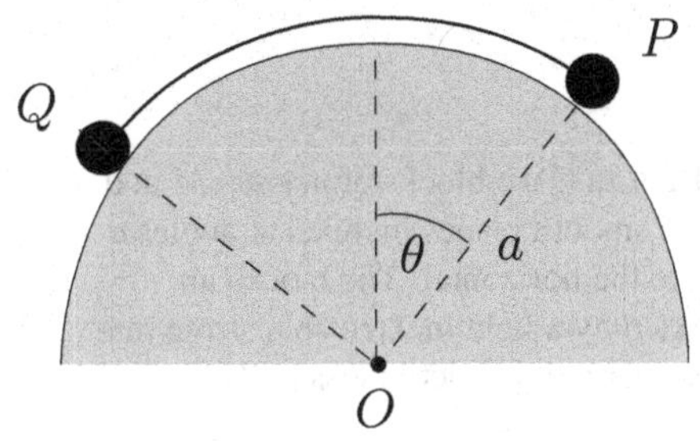

FIGURE 9.12 Two particles P and Q are connected by a light inextensible string and can move, with the string taut, on the surface of a smooth horizontal cylinder.

is therefore

$$\dot{\theta}^2 = \frac{3g}{4a}(1 - 2\cos\theta).$$

Since the system has only one degree of freedom, this equation is sufficient to determine the motion.

A curious feature of this problem (not proved here) is that the ladder does not maintain contact with the wall all the way down, but leaves the wall when θ becomes equal to $\cos^{-1}(1/3) \approx 71°$. ■

Problems on Chapter 9

Answers and comments are at the end of the book.
Harder problems carry a star (∗).

Potential energy and stability

9.1 Figure 9.12 shows two particles P and Q, of masses M and m, that can move on the smooth outer surface of a fixed horizontal cylinder. The particles are connected by a light inextensible string of length $\pi a/2$. Find the equilibrium configuration and show that it is unstable.

9.2 A uniform rod of mass m and of length $2a$ has one end smoothly pivoted at a fixed point O. The other end is connected to a fixed point A, which is a distance $2a$ vertically above O, by a light elastic spring of natural length a and modulus $\frac{1}{2}mg$. The rod moves in a vertical plane through O. Show that there are two equilibrium positions for the rod, and determine their stability. [The vertically upwards position for the rod would compress the spring to zero length and is excluded.]

9.3 The internal potential energy function for a diatomic molecule is approximated by the **Morse potential**

$$V(r) = V_0\left(1 - e^{-(r-a)/b}\right)^2 - V_0,$$

where r is the distance of separation of the two atoms, and V_0, a, b are positive constants. Make a sketch of the Morse potential.

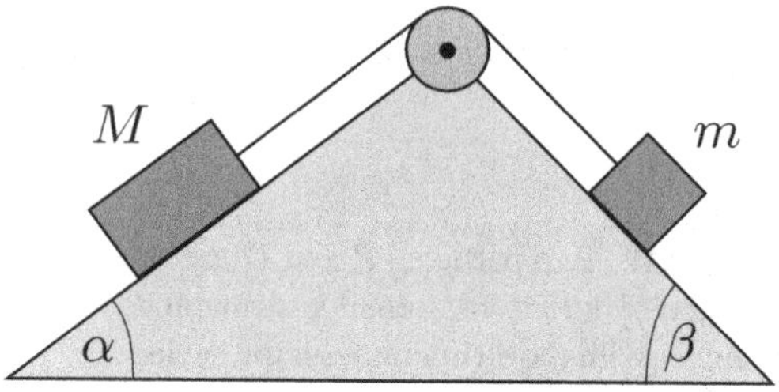

FIGURE 9.13 Two blocks of masses M and m slide on smooth planes inclined at angles α and β to the horizontal. The blocks are connected by a light inextensible string that passes over a light frictionless pulley.

Suppose the molecule is restricted to *vibrational* motion in which the centre of mass G of the molecule is fixed, and the atoms move on a fixed straight line through G. Show that there is a single equilibrium configuration for the molecule and that it is stable. If the atoms each have mass m, find the angular frequency of small vibrational oscillations of the molecule.

9.4* The internal gravitational potential energy of a system of masses is sometimes called the **self energy** of the system. (The reference configuration is taken to be one in which the particles are all a great distance from each other.) Show that the self energy of a uniform sphere of mass M and radius R is $-3M^2G/5R$. [Imagine that the sphere is built up by the addition of successive thin layers of matter brought in from infinity.]

Particles only

9.5 Figure 9.13 shows two blocks of masses M and m that slide on smooth planes inclined at angles α and β to the horizontal. The blocks are connected by a light inextensible string that passes over a light frictionless pulley. Find the acceleration of the block of mass m up the plane, and deduce the tension in the string.

9.6 Consider the system shown in Figure 9.12 for the special case in which the particles P, Q have masses $2m$, m respectively. The system is released from rest in a symmetrical position with θ, the angle between OP and the upward vertical, equal to $\pi/4$. Find the energy conservation equation for the subsequent motion in terms of the coordinate θ.

* Find the normal reactions of the cylinder on each of the particles. Show that P is first to leave the cylinder and that this happens when $\theta = 70°$ approximately.

Ropes

9.7 A heavy uniform rope of length $2a$ is draped symmetrically over a *thin* smooth horizontal peg. The rope is then disturbed slightly and begins to slide off the peg. Find the speed of the rope when it finally leaves the peg.

9.8 A uniform heavy rope of length a is held at rest with its two ends close together and the rope hanging symmetrically below. (In this position, the rope has two long vertical segments connected by a small curved segment at the bottom.) One of the ends is then released. Find the velocity of the free end when it has descended by a distance x.

Deduce a similar formula for the acceleration of the free end and show that it always *exceeds* g. Find how far the free end has fallen when its acceleration has risen to $5g$.

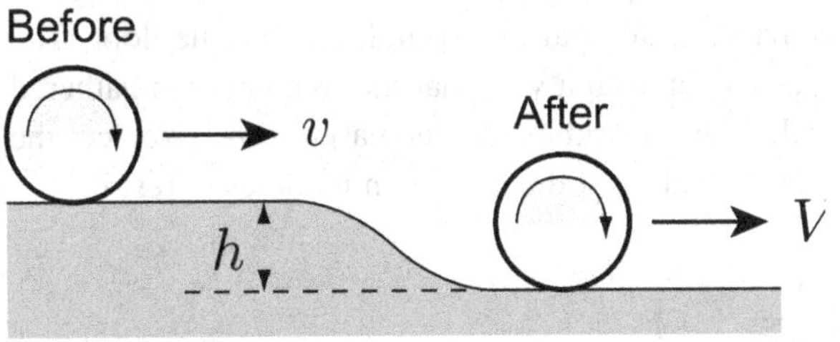

FIGURE 9.14 The circular hoop *rolls* down the slope from one level to another.

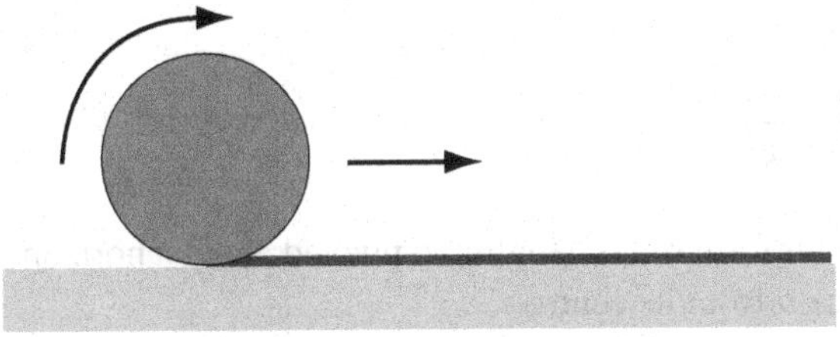

FIGURE 9.15 The roll of paper moves to the right and the free paper is gathered on to the roll.

9.9 A heavy uniform rope of mass M and length $4a$ has one end connected to a fixed point on a smooth horizontal table by light elastic spring of natural length a and modulus $\frac{1}{2}Mg$, while the other end hangs down over the edge of the table. When the spring has its natural length, the free end of the rope hangs a distance a vertically below the level of the table top. The system is released from rest in this position. Show that the free end of the rope executes simple harmonic motion, and find its period and amplitude.

Rigid bodies

9.10 A circular hoop is rolling with speed v along level ground when it encounters a slope leading to more level ground, as shown in Figure 9.14. If the hoop loses altitude h in the process, find its final speed.

9.11 A uniform ball is rolling in a straight line down a *rough* plane inclined at an angle α to the horizontal. Assuming the ball to be in planar motion, find the energy conservation equation for the ball. Deduce the acceleration of the ball.

9.12 A uniform circular cylinder (a yo-yo) has a light inextensible string wrapped around it so that it does not slip. The free end of the string is secured to a fixed point and the yo-yo descends in a vertical straight line with the straight part of the string also vertical. Explain why the string does no work on the yo-yo. Find the energy conservation equation for the yo-yo and deduce its acceleration.

9.13 Figure 9.15 shows a partially unrolled roll of paper on a horizontal floor. Initially the paper on the roll has radius a and the free paper is laid out in a straight line on the floor. The roll is then projected horizontally with speed V in such a way that the free paper is gathered up on to the roll. Find the speed of the roll when its radius has increased to b. [Neglect the bending stiffness of the paper.] Deduce that the radius of the roll when it comes to rest is

$$a\left(\frac{3V^2}{4ga}+1\right)^{1/3}.$$

9.14 A rigid body of general shape has mass M and can rotate freely about a fixed horizontal axis. The centre of mass of the body is distance h from the rotation axis, and the moment of inertia of the body about the rotation axis is I. Show that the period of small oscillations of the body about the downward equilibrium position is

$$2\pi\left(\frac{I}{Mgh}\right)^{1/2}.$$

Deduce the period of small oscillations of a uniform rod of length $2a$, pivoted about a horizontal axis perpendicular to the rod and distance b from its centre.

9.15 A uniform ball of radius a can roll without slipping on the *outside* surface of a fixed sphere of (outer) radius b and centre O. Initially the ball is at rest at the highest point of the sphere when it is slightly disturbed. Find the speed of the centre G of the ball in terms of the variable θ, the angle between the line OG and the upward vertical. [Assume planar motion.]

9.16 A uniform ball of radius a and centre G can roll without slipping on the *inside* surface of a fixed hollow sphere of (inner) radius b and centre O. The ball undergoes planar motion in a vertical plane through O. Find the energy conservation equation for the ball in terms of the variable θ, the angle between the line OG and the downward vertical. Deduce the period of small oscillations of the ball about the equilibrium position.

9.17* Figure 9.6 shows a uniform thin rigid plank of length $2b$ which can roll without slipping on top of a rough circular log of radius a. The plank is initially in equilibrium, resting symmetrically on top of the log, when it is slightly disturbed. Find the period of small oscillations of the plank.

Chapter Ten

The linear momentum principle

and linear momentum conservation

KEY FEATURES

The key features of this chapter are the **linear momentum principle**; its equivalent form, the **centre of mass equation**; and **conservation of linear momentum**. These principles are applied to **rocket propulsion**, **collision theory**, the **two-body problem** and **two-body scattering**.

This chapter is essentially based on the **linear momentum principle** and its consequences. The linear momentum principle is the second of the three great principles of multi-particle mechanics* that apply to *every* mechanical system without restriction. Under appropriate conditions, the linear momentum of a system (or one of its components) is **conserved**. Important applications include rocket propulsion, collision theory, the two-body problem and two-body scattering.

10.1 LINEAR MOMENTUM

We begin with the definition of linear momentum for a single particle and for a system of particles.

Definition 10.1 ***Linear momentum*** *If a particle has mass m and velocity $\boldsymbol{v}$, then $\boldsymbol{p}$, its* ***linear momentum****, is defined to be*

$$\boldsymbol{p} = m\boldsymbol{v}. \tag{10.1}$$

For a multi-particle system $\mathcal{S}$ consisting of particles $P_1, P_2, \dots, P_N$, with masses $m_1, m_2, \dots, m_N$ and velocities $\boldsymbol{v}_1, \boldsymbol{v}_2, \dots, \boldsymbol{v}_N$ (see Figure 9.1), $\boldsymbol{P}$, the ***linear momentum*** *of $\mathcal{S}$, is defined to be the* ***vector sum*** *of the linear momenta of the individual particles, that is,*

$$\boldsymbol{P} = \sum_{i=1}^{N} \boldsymbol{p}_i = \sum_{i=1}^{N} m_i \boldsymbol{v}_i. \tag{10.2}$$

* The other two are the energy and angular momentum principles.

Newton's Second Law can be written in terms of linear momentum in the form

$$\frac{d\boldsymbol{p}}{dt} = \boldsymbol{F}.$$

Although this offers no advantage in the mechanics of a single particle, we will find that this type of formulation is very useful in multi-particle mechanics. The expression (10.2) can be written simply in terms of the motion of G, the centre of mass of $\mathcal{S}$. Since the position vector $\boldsymbol{R}$ of G is given by

$$\boldsymbol{R} = \frac{\sum_{i=1}^{N} m_i \boldsymbol{r}_i}{\sum_{i=1}^{N} m_i},$$

where $\boldsymbol{r}_i$ is the position vector of the particle P_i, it follows that $\boldsymbol{V}$, the velocity of G is given by

$$\boldsymbol{V} = \frac{\sum_{i=1}^{N} m_i \boldsymbol{v}_i}{\sum_{i=1}^{N} m_i} = \frac{\boldsymbol{P}}{M},$$

where M $(= \sum m_i)$ is the total mass of the system $\mathcal{S}$. Hence

$$\boldsymbol{P} = M\boldsymbol{V}. \tag{10.3}$$

Thus the *linear momentum of any system is the same as if all its mass were concentrated at its centre of mass.*

Although true for all systems, this result is most useful when finding the linear momentum of a moving **rigid body**. Note that the rotational motion of the rigid body does not contribute to its linear momentum; this contrasts with the corresponding calculation of the kinetic energy of a rigid body (see Chapter 9).

10.2 THE LINEAR MOMENTUM PRINCIPLE

We now derive the fundamental result which relates the linear momentum of any system to the external forces that act upon it: **the linear momentum principle**.

Suppose that the system $\mathcal{S}$ is acted upon by the **external** forces $\{\boldsymbol{F}_i\}$ and **internal** forces $\{\boldsymbol{G}_{ij}\}$, as shown in Figure 9.3. Then the equation of motion for the particle P_i is

$$m_i \frac{d\boldsymbol{v}_i}{dt} = \boldsymbol{F}_i + \sum_{j=1}^{N} \boldsymbol{G}_{ij}, \tag{10.4}$$

where, as in Chapter 9, we take $\boldsymbol{G}_{ij} = \boldsymbol{0}$ when $i = j$. Then the rate of increase of the linear momentum of the system $\mathcal{S}$ can be written

$$\frac{d\boldsymbol{P}}{dt} = \frac{d}{dt}\left(\sum_{i=1}^{N} m_i \boldsymbol{v}_i\right) = \sum_{i=1}^{N} m_i \frac{d\boldsymbol{v}_i}{dt}, \tag{10.5}$$

which, on using the equation of motion (10.4), gives

$$\frac{d\boldsymbol{P}}{dt} = \sum_{i=1}^{N}\left\{\boldsymbol{F}_i + \sum_{j=1}^{N}\boldsymbol{G}_{ij}\right\} = \sum_{i=1}^{N}\boldsymbol{F}_i + \sum_{i=1}^{N}\sum_{j=1}^{N}\boldsymbol{G}_{ij}$$

$$= \sum_{i=1}^{N}\boldsymbol{F}_i + \sum_{i=1}^{N}\left(\sum_{j=1}^{i-1}(\boldsymbol{G}_{ij}+\boldsymbol{G}_{ji})\right),$$

where the terms of the double sum have been grouped in pairs and those terms known to be zero have been omitted. Now the internal forces $\{\boldsymbol{G}_{ij}\}$ satisfy the Third Law, so that $\boldsymbol{G}_{ji} = -\boldsymbol{G}_{ij}$. Hence, each term of the double sum in equation (10.5) is zero and we obtain

Linear momentum principle

$$\frac{d\boldsymbol{P}}{dt} = \boldsymbol{F} \qquad (10.6)$$

where $\boldsymbol{F}$ is the **total external force** acting on $\mathcal{S}$. This is the **linear momentum principle**. This fundamental principle can be expressed as follows:

Linear momentum principle

In any motion of a system, the rate of increase of its linear momentum is equal to the total *external* force acting upon it.

It should be noted that only the external forces appear in the linear momentum principle so that the *internal forces need not be known.* It is this fact which gives the linear momentum principle its power.

10.3 MOTION OF THE CENTRE OF MASS

The linear momentum principle can be written in an alternative form called the centre of mass equation, which is more useful for some purposes. If we substitute the expression (10.3) for $\boldsymbol{P}$ into the linear momentum principle (10.6) we obtain

Centre of mass equation

$$M\frac{d\boldsymbol{V}}{dt} = \boldsymbol{F} \qquad (10.7)$$

which is called the **centre of mass equation**. It has the form of an equation of motion for a *fictitious* particle of mass M situated at the centre of mass, which moves under the total of the external forces acting on the system $\mathcal{S}$. This important result can be simply expressed as follows:

Motion of the centre of mass

The centre of mass of any system moves as if it were a particle of mass the total mass, and all the *external* forces acted upon it.

Example 10.1 *Jumping cat*

A cat leaps off a table and lands on the floor. Show that, while the cat is in the air, its centre of mass moves on a parabolic path.

Solution

While the cat is in the air, the total external force on its body is due to uniform gravity, that is, $\boldsymbol{F} = -Mg\boldsymbol{k}$. The centre of mass equation for the cat is therefore

$$M\frac{d\boldsymbol{V}}{dt} = -Mg\boldsymbol{k},$$

which is precisely the equation of projectile motion for a single particle. The path of the centre of mass of the cat is therefore the same as if it were a particle of mass M moving freely under uniform gravity. This path is known (see Chapter 4) to be a parabola. ■

In previous examples, we have often used the Second Law to find an unknown constraint force acting on a particle, once the motion of a system has been found by other means (see, for instance, Example 6.13). The centre of mass equation allows us to do the same thing when the unknown constraint force acts on a rigid body. The following examples illustrate the method.

Example 10.2 *Cylinder rolling down an inclined plane*

Consider again a hollow cylinder of mass M rolling down a rough inclined plane as shown in Figure 9.10. In Example 9.12, energy conservation was used to show that the acceleration of the cylinder down the plane is $\frac{1}{2}g \sin\alpha$. Deduce the reaction force exerted by the plane on the cylinder.

Solution

Suppose that the component of the reaction force normal to the plane is N, while the component of the reaction force up the plane is F. (The plane is rough so both components are present.) The cylinder is therefore subject to these 'two' external forces together with uniform gravity. The **centre of mass equation** for the cylinder (when resolved into components tangential and normal to the plane) is given by

$$M\frac{dv}{dt} = Mg\sin\alpha - F, \qquad 0 = N - Mg\cos\alpha,$$

where $dv/dt = \frac{1}{2}g\sin\alpha$. It follows that the required **reactions** are given by

$$F = \tfrac{1}{2}Mg\sin\alpha, \qquad N = Mg\cos\alpha.$$

Thus, if F and N are restricted by the 'law of friction' $F/N < \mu$, then the supposed rolling motion of the cylinder cannot take place if $\tan\alpha > 2\mu$. ■

Example 10.3 *Sliding ladder*

Consider again the uniform ladder of length $2a$ supported by a smooth horizontal floor and leaning against a smooth vertical wall, as shown in Figure 9.11. The ladder is released from rest with θ, the angle between the ladder and the downward vertical, equal to 60°. In Example 9.13, we used energy conservation to show that, in the subsequent motion, θ satisfies the differential equation

$$\dot{\theta}^2 = \frac{3g}{4a}(1 - 2\cos\theta),$$

provided that the ladder maintains contact with the wall. Deduce that the ladder loses contact with the wall when $\theta = \cos^{-1}(1/3)$.

Solution

Let the normal reactions exerted on the ladder by the smooth floor and wall be N^F and N^W respectively. Then the centre of mass equation for the ladder, resolved into horizontal and vertical components, is given by

$$M\ddot{X} = N^W, \qquad M\ddot{Z} = N^F - Mg,$$

where (X, Z) are the coordinates of the centre of mass of the ladder (see Figure 9.11). Hence

$$N^F = M\ddot{Z} + Mg, \qquad N^W = M\ddot{X}.$$

Now, in terms of the angle θ, $X = a\sin\theta$ and $Z = a\cos\theta$. On differentiating twice with respect to t, we obtain the corresponding acceleration components

$$\ddot{X} = -a(\sin\theta)\,\dot{\theta}^2 + a(\cos\theta)\,\ddot{\theta},$$

$$\ddot{Z} = -a(\cos\theta)\,\dot{\theta}^2 - a(\sin\theta)\,\ddot{\theta}.$$

Hence

$$N^F = -Ma\left((\cos\theta)\,\dot{\theta}^2 - (\sin\theta)\,\ddot{\theta}\right) + Mg,$$
$$N^W = Ma\left(-(\sin\theta)\,\dot{\theta}^2 + (\cos\theta)\,\ddot{\theta}\right).$$

In order to express these reactions in terms of θ alone, we need to know $\dot{\theta}^2$ and $\ddot{\theta}$ as functions of θ. From the previously derived equation of motion, we already have

$$\dot{\theta}^2 = \frac{3g}{4a}(1 - 2\cos\theta)$$

and, if we differentiate this equation with respect to t (and cancel by $\dot{\theta}$), we obtain

$$\ddot{\theta} = \frac{3g}{4a}\sin\theta.$$

On making use of the above expressions for $\dot{\theta}^2$ and $\ddot{\theta}$, the required **reactions** are found to be

$$N^F = \frac{Mg}{4}(1 - 3\cos\theta + 9\cos^2\theta), \qquad N^W = \frac{3Mg}{4}\sin\theta(3\cos\theta - 1).$$

We observe that the predicted value of N^W becomes zero when $\theta = \cos^{-1}(1/3)$ and is *negative* thereafter. Since negative values of N^W cannot occur (the wall can only *push*), we conclude that the condition that the ladder maintains contact with the wall is violated when $\theta > \cos^{-1}(1/3)$. Therefore, the **ladder leaves the wall** when $\theta = \cos^{-1}(1/3)$. ■

10.4 CONSERVATION OF LINEAR MOMENTUM

Suppose that $\mathcal{S}$ is an **isolated** system, meaning that **no external force** acts on any of its particles. Then $\boldsymbol{F}$, the total external force acting on $\mathcal{S}$, is obviously zero. The linear momentum principle (10.6) for $\mathcal{S}$ then takes the form $d\boldsymbol{P}/dt = \boldsymbol{0}$, which implies that $\boldsymbol{P}$ must remain constant. This simple but important result can be stated as follows:

Conservation of linear momentum

In any motion of an isolated system, the total linear momentum is conserved.

It follows from equation (10.3) that the above result can also be stated in the alternative form '*In any motion of an isolated system, the centre of mass of the system moves with constant velocity*'. Clearly the same result applies to any system for which the *total* external force is zero, whether isolated or not.

It is also possible for a particular component of $\boldsymbol{P}$ to be conserved while other components are not. Let $\boldsymbol{n}$ be a *constant* unit vector and suppose that $\boldsymbol{F} \cdot \boldsymbol{n} = 0$ at all times. Then

$$\frac{d}{dt}(\boldsymbol{P} \cdot \boldsymbol{n}) = \frac{d\boldsymbol{P}}{dt} \cdot \boldsymbol{n} + \boldsymbol{P} \cdot \frac{d\boldsymbol{n}}{dt} = \frac{d\boldsymbol{P}}{dt} \cdot \boldsymbol{n} = \boldsymbol{F} \cdot \boldsymbol{n} = 0.$$

Hence the component $\boldsymbol{P} \cdot \boldsymbol{n}$ is conserved. This result can be stated as follows:

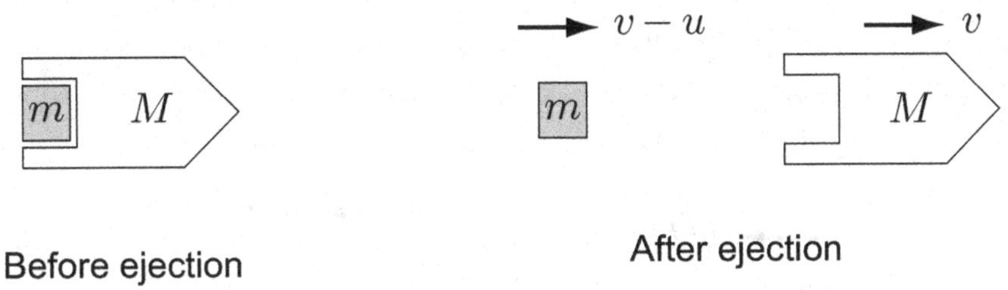

FIGURE 10.1 A rigid body of mass M (the rocket) contains a removable rigid block of mass m (the fuel). An internal source of energy causes the fuel block to be ejected backwards with speed u *relative to the rocket* and the rocket is projected forwards.

> **Conservation of a component of linear momentum**
>
> If the total force acting on a system has zero component in a *fixed* direction, then, in any motion of the system, the component of the total linear momentum in that direction is conserved.

Conservation of linear momentum is an important property of a system and the sections that follow rely heavily upon it. Two examples of momentum conservation are as follows:

- The **solar system** is an example of an *isolated* system, being extremely remote from any other masses. It follows that the total linear momentum of the solar system is conserved. Thus the centre of mass of the solar system moves with constant velocity.
- On the other hand, a **grasshopper** trying to move on a perfectly smooth horizontal table is *not isolated*, being subject to gravity and the vertical reaction of the table. However, since the grasshopper is not subject to any external *horizontal* force, it follows that, whatever the grasshopper tries to do, his component of total linear momentum in any *horizontal* direction is conserved. His vertical component of linear momentum is not conserved; he can leap into the air if he wishes.

10.5 ROCKET MOTION

An important application of linear momentum conservation is **rocket propulsion**. Figure 10.1 shows a rigid body of mass M (the rocket) which contains a removable rigid block of mass m (the fuel). The system is at rest when an internal source of energy causes the fuel block to be ejected backwards with speed u *relative to the rocket*. If the system is isolated, then its **total linear momentum is conserved**, which implies that

$$Mv + m(v - u) = 0$$

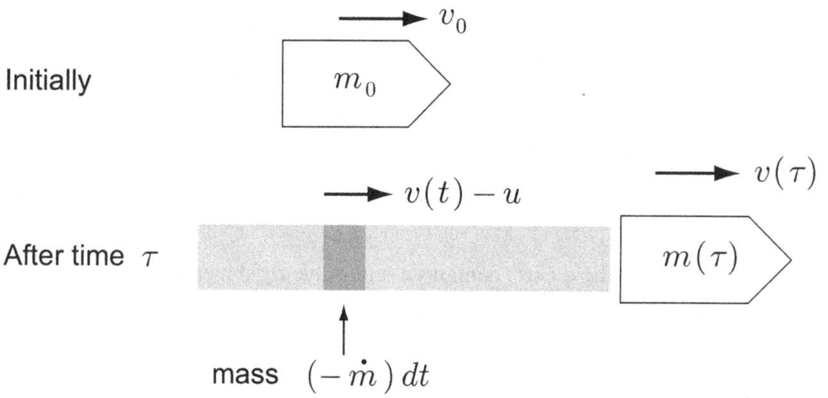

FIGURE 10.2 The rocket and its fuel at times $t = 0$ and $t = \tau$. The element of fuel ejected in the time interval $[t, t + dt]$ has mass $(-\dot{m})dt$ and (forward) velocity $v(t) - u$.

where v is the forward velocity of the rocket after the ejection of the fuel. As a result of this process the rocket acquires the forward velocity

$$v = \left(\frac{m}{M + m}\right) u.$$

This is the basic principle of **rocket propulsion**. The only mechanically significant difference between the simple example above and real rocket propulsion is that, in the case of the real rocket, the fuel mass is ejected continuously over a period of time and not in a single lump. In practice, the fuel is burned continuously and the combustion products eject themselves due to their rapid expansion.

Rocket motion in free space

Figure 10.2 shows a more realistic situation. Initially the rocket and its fuel have combined mass m_0 and are moving with constant velocity v_0. At time $t = 0$ the motors are started and fuel products are ejected backwards with speed u *relative to the rocket.* The fuel 'burn' continues for a time T, at the end of which the rocket and unburned fuel have mass m_1. Let $m = m(t)$ be the mass of the rocket and its unburned fuel after time t. Then m is a decreasing function of t and the rate of ejection of mass at time t is $-\dot{m}$. Let the system $\mathcal{S}$ consist of the rocket together with its fuel at time $t = 0$. After some time τ into the burn, the mass of $\mathcal{S}$ is distributed as shown in Figure 10.2. The rocket and unburned fuel have mass $m(\tau)$ and the remaining mass has been ejected as expended fuel. We will suppose that, once an element of fuel is ejected, it continues to move with the velocity it had at the instant of ejection.* Since we are assuming $\mathcal{S}$ to be an isolated system, its *total linear momentum is conserved.* The initial linear momentum in this one-dimensional problem is $m_0 v_0$ and the final linear momentum of the rocket and *unburned* fuel is $m(\tau)v(\tau)$,

* This assumption simplifies our derivation but, as we will see, it is not essential.

where v (= $v(t)$) is the velocity of the rocket at time t. It remains to take account of the linear momentum of the ejected fuel. Consider the element of fuel that was ejected in the time interval $[t, t + dt]$. This has mass $(-\dot{m}(t))\,dt$ and its forward velocity at the instant of ejection was $v(t) - u$. The linear momentum of this fuel element is therefore $(-\dot{m})(v - u)\,dt$ and the total linear momentum of the fuel expended in the time interval $[0, \tau]$ is

$$-\int_0^\tau \dot{m}(v - u)\,dt.$$

Linear momentum conservation for the system $\mathcal{S}$ therefore requires that

$$m_0 v_0 = m(\tau)v(\tau) - \int_0^\tau \dot{m}(v - u)\,dt,$$

which can be written in the form

$$\int_0^\tau \left[\frac{d}{dt}(mv) - \dot{m}(v - u)\right] dt = 0.$$

Since this equality must hold for *any choice* of τ during the burn, it follows that the integrand must be zero, that is

$$\frac{d}{dt}(mv) - \dot{m}(v - u) = 0$$

for $0 \le t \le T$. This simplifies to give

Rocket equation in free space

$$m\frac{dv}{dt} = (-\dot{m})u \tag{10.8}$$

the **rocket equation**, which holds for $0 < t < T$. The rocket equation can be interpreted physically as the Second Law applied to a *system of variable mass** $m(t)$, namely the rocket and its *unburned* fuel. In this interpretation, the term on the right, $-\dot{m}u$, plays the rôle of force and is called the **thrust** supplied by the motors.

Note. In our derivation, we assumed that, once an element fuel is ejected, it continues to move with the velocity it had at the instant of ejection. This is equivalent to assuming that each element of ejected

* This terminology is undesirable since, in classical mechanics, a 'system' means a fixed set of masses (or, at the very least, fixed total mass). No standard mechanical principle applies to a 'system' whose total mass is changing with time.

fuel is isolated from other fuel and from the rocket. It clearly makes no difference to the momentum of the ejected fuel if momentum is exchanged between *elements of itself* so that this assumption is actually unnecessary. However we must retain the assumption that ejected fuel has no further interaction *with the rocket*. This seems likely to be true in free space, but whether it is true just after take off from solid ground is questionable.

Providing that the ejection speed u is constant, the rocket equation (10.8) can easily be solved for any mass ejection rate. On dividing through by m and integrating with respect to t, we obtain

$$\int dv = \int \frac{(-\dot{m})u}{m} dt = -u \int \frac{dm}{m} = -u \ln m + \text{constant}$$

and, on applying the initial condition $v = v_0$ when $t = 0$, we obtain

$$v(t) = v_0 + u \ln\left(\frac{m_0}{m(t)}\right).$$

This gives the **rocket velocity** at time t. In particular, at the end of the fuel burn, the rocket velocity has increased by

$$\Delta v = v_1 - v_0 = u \ln\left(\frac{m_0}{m_1}\right), \tag{10.9}$$

where m_1 and v_1 are the final mass and velocity of the rocket. One can make some interesting deductions from this solution.

(i) Δv, the increase in the rocket velocity, is directly proportional to u, the fuel ejection speed. Thus it pays to make u as large as possible. Chemical processes can produce values of u as high as 5000 m s^{-1}.

(ii) If the fuel were all ejected in a single lump, Δv would never exceed the ejection speed u. But when the fuel is ejected over a period of time, it is possible for the rocket to attain any velocity by making the mass ratio m_0/m_1 large enough. For example, if we wish to make $\Delta v = 3u$, then we need $m_0/m_1 = e^3 \approx 20$. This means that 19 kg of fuel would be required for every kilogram of payload. The amount of fuel needed to achieve higher velocities quickly makes the process impractical. To achieve $\Delta v = 10u$ takes 22 metric tons of fuel for every kilogram of payload!

Rocket motion under gravity

Suppose now that the rocket is moving vertically under gravity. If we regard the governing equation as the equation of motion for the variable mass $m(t)$, then, when gravity is introduced, the equation of motion becomes

Rocket equation including gravity

$$m\frac{dv}{dt} = (-\dot{m})u - mg \tag{10.10}$$

where v is measured vertically upwards, and the weight force mg means $m(t)g$. In this case, the effective force on the right is the sum of the **thrust** $(-\dot{m})u$ acting upwards and the **weight force** mg acting downwards.

When the gravity is uniform and the ejection speed u is constant, the new rocket equation (10.10) can also be solved easily for any mass ejection rate. On dividing through by m and integrating with respect to t, we obtain

$$\int dv = \int \left[\frac{(-\dot{m})u}{m} - g \right] dt = -u \ln m + gt + \text{constant}$$

and, on applying the initial condition $v = v_0$ when $t = 0$, we obtain

$$v(t) = v_0 + u \ln \left(\frac{m_0}{m(t)} \right) - gt.$$

This gives the **rocket velocity** at time t. In particular, at the end of the fuel burn, the rocket velocity has increased by

$$\Delta v = v_1 - v_0 = u \ln \left(\frac{m_0}{m_1} \right) - gT, \tag{10.11}$$

where m_1 and v_1 are the final mass and velocity of the rocket and T is the time taken to burn all the fuel.

It will be noticed that, if T is too large, then Δv will be negative, which is hardly possible for a rocket standing on the ground. The reason for this paradox is that, if the fuel is burned too slowly then the thrust will be less than the initial weight of the rocket, which will not take off until its weight has become less than the thrust. We will therefore assume that $(-\dot{m})u > mg$ at all times during the burn so that the rocket has positive upward acceleration and achieves its maximum speed when $t = T$. If the rocket starts from rest, it then follows that the **maximum speed** achieved is

$$v_{\max} = u \ln \left(\frac{m_0}{m_1} \right) - gT.$$

In this and the zero gravity case, the *distance* travelled during the burn depends on the functional form of $m(t)$.

10.6 COLLISION THEORY

Another important application of linear momentum conservation occurs when we have an isolated system of **two particles**, and one particle is in **collision** with the other.

Collision processes

It is important to understand the meaning of the term 'collision'. Suppose that the mutual interaction between the two particles tends to zero as the distance between them tends to

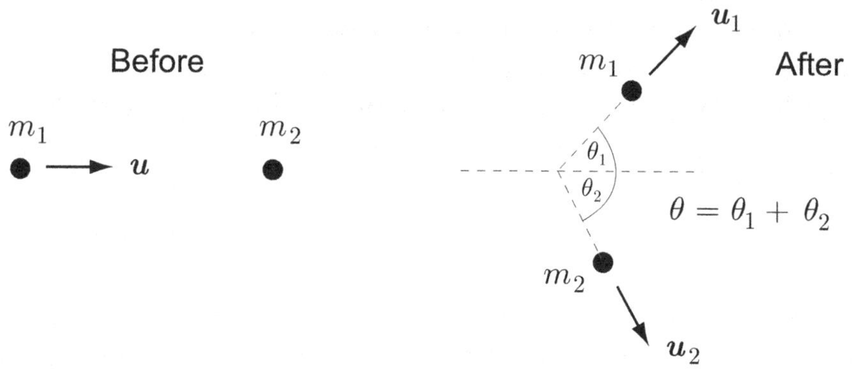

FIGURE 10.3 A collision between two particles viewed from the **laboratory frame**. A particle of mass m_1 and initial velocity $\boldsymbol{u}$ collides with a 'target' particle of mass m_2, which is initially at rest. After the collision, the particles have velocities $\boldsymbol{u}_1$ and $\boldsymbol{u}_2$ respectively. θ_1 is the **scattering angle** of the mass m_1, θ_2 is the **recoil angle** of the mass m_2, and θ $(= \theta_1 + \theta_2)$ is the **opening angle** between the emerging paths.

infinity, so that, if the particles are initially a great distance apart, each must be moving with constant velocity. If the particles approach each other, then there follows a period during which their mutual interaction causes their straight line motions to be disturbed. If the particles finally retreat to a great distance from each other, then they will move with constant velocities again, and these final velocities will generally be different to the initial velocities. This is what we mean by a **collision process**. Note that a collision process is not restricted to those cases in which the particles make physical contact with each other. This can of course happen, as in the 'real' collision of two pool balls. However, the deflection suffered by an alpha particle in passing close to a nucleus is also a 'collision', even though the alpha particle and the nucleus never made contact. Collision processes are particularly important in **nuclear** and **particle physics**, where they are the major source of experimental information.

General collisions

Consider the collision shown in Figure 10.3. A particle of mass m_1 and initial velocity $\boldsymbol{u}$ is incident upon a 'target' particle of mass m_2 which is initially at rest.* This is typical of the collisions observed in nuclear physics. After the collision we will suppose the particles retain their identities (and therefore their masses) and emerge with velocities $\boldsymbol{u}_1$ and $\boldsymbol{u}_2$ respectively. How are the final and initial motions of the particles related? Clearly we cannot 'solve the problem' since we have not even said what the mutual interaction between the particles is. However, it is surprising how much can be deduced simply from conservation laws without any detailed knowledge of the interaction.

Since the two particles form an isolated system, their total **linear momentum** is **conserved**, that is,

* This means 'at rest in the *laboratory reference frame*'.

$$\boxed{m_1 \boldsymbol{u} = m_1 \boldsymbol{u}_1 + m_2 \boldsymbol{u}_2} \tag{10.12}$$

This linear relation between the vectors $\boldsymbol{u}$, $\boldsymbol{u}_1$ and $\boldsymbol{u}_2$ implies that these three velocities must lie in the same plane so that scattering processes are *two-dimensional*.

Generally, collisions are **not energy preserving**. The energy principle for the collision has the form

$$\tfrac{1}{2}m_1 u^2 + Q = \tfrac{1}{2}m_1 u_1^2 + \tfrac{1}{2}m_2 u_2^2,$$

where $u = |\boldsymbol{u}|$, $u_1 = |\boldsymbol{u}_1|$, $u_2 = |\boldsymbol{u}_2|$, and Q is the energy gained in the collision. In 'real' collisions between large bodies, energy is usually lost in the form of heat, so that Q is negative. However, in nuclear collisions in which the particles change their identities, it is perfectly possible for energy to be gained.

Example 10.4 *Making Kraptons*

A little known particle physicist has proposed the existence of a new particle, with charge +2 and mass 2, which he has named the Krapton. He has calculated that this can be produced by the collision of two protons in the reaction*

$$p^+ + p^+ + 10\text{ MeV} \to K^{++}$$

Having failed to obtain funding to verify his theory, he has built his own equipment with which he accelerates protons to an energy of 16 MeV and uses them to bombard a stationary target of hydrogen. Could he succeed in making a Krapton?

Solution

Suppose a proton with kinetic energy E collides with proton at rest. Then this system has initial linear momentum $(2mE)^{1/2}$, where m is the mass of a proton. This *linear momentum is preserved* by the collision so that, if a Krapton of mass $2m$ *were* produced, it would have linear momentum $(2mE)^{1/2}$ and therefore kinetic energy $E/2$. Hence, only 8 MeV of the initial energy is available for Krapton building and, according to the physicist's own calculation, this is not enough. (On the other hand, a head-on collision between two 5 MeV protons *would* be enough. Why?) ■

Elastic collisions

The linear momentum equation (10.12) holds whether the collision is between pool balls, protons or peaches. Much more can be said if the collision is also **energy preserving**.

Definition 10.2 ***Elastic collision*** *A collision between particles is said to be* ***elastic*** *if the* ***total kinetic energy*** *of the particles is* ***conserved*** *in the collision.*

---

* The electron volt (eV) is a unit of energy equal to 1.6×10^{-19} J approximately.

Frame invariance In order that the above definition be physically meaningful, it is necessary that a collision observed to be elastic in one inertial frame should also be elastic when observed from any other. This is not obviously true, since kinetic energy is not a linear quantity. However, since the total kinetic energy of the system can be written in the form $T = T^{CM} + T^G$ (see Theorem 9.10), where T^{CM} is preserved in the collision and T^G is frame independent, it follows that *any gain or loss of kinetic energy in the collision is independent of the inertial reference frame used to observe the event.*

Elastic collisions are very common and extremely important. For example, *any collision in which the mutual interaction force is conservative is elastic.* In particular, the collisions that occur in Rutherford scattering are elastic. In elastic collisions, we have **energy conservation** in the form

$$\boxed{\tfrac{1}{2}m_1 u^2 = \tfrac{1}{2}m_1 u_1^2 + \tfrac{1}{2}m_2 u_2^2} \tag{10.13}$$

and, together with linear momentum conservation (10.12), we can make some interesting deductions. If we take the scalar product of each side of the linear momentum equation (10.12) with itself, we obtain

$$m_1^2 u^2 = m_1^2 u_1^2 + 2m_1 m_2 \boldsymbol{u}_1 \cdot \boldsymbol{u}_2 + m_2^2 u_2^2,$$

and, if we now eliminate the term in u^2 between this equation and the energy conservation equation (10.13), we obtain, after simplification,

$$2m_1 \boldsymbol{u}_1 \cdot \boldsymbol{u}_2 = (m_1 - m_2)\, u_2^2. \tag{10.14}$$

Since $\boldsymbol{u}_1 \cdot \boldsymbol{u}_2 = u_1 u_2 \cos\theta$, where θ is the **opening angle** between the paths of the emerging particles, the formula (10.14) can also be written

$$\cos\theta = \frac{(m_1 - m_2)\, u_2}{2m_1 u_1}, \tag{10.15}$$

provided that $u_1 \neq 0$, that is, provided that the incident particle is not brought to rest by the collision.* This formula holds for **all elastic collisions**, whatever the nature of the particles and the interaction. It therefore applies equally well to pool balls† and protons, but not peaches. Given the mass ratio of the two particles, formula (10.15) relates the speeds of the particles and the opening angle between their paths after the collision.

* The incident particle can be brought to rest in a head-on collision with a particle of equal mass.

† Collisions between pool balls are very nearly elastic. However, in the present treatment, we are disregarding the rotation of the balls.

Example 10.5 *Finding the final energies*

A ball of mass m and (kinetic) energy E is in an *elastic* collision with a second ball of mass $4m$ that is initially at rest. The two balls depart in directions making an angle of 120° with each other. What are the final energies of the two balls?

Solution

On substituting the given data into the formula (10.15), we find that $u_1/u_2 = 3$. It follows that

$$\frac{E_1}{E_2} = \frac{\frac{1}{2}mu_1^2}{\frac{1}{2}(4m)u_2^2} = \frac{1}{4}\left(\frac{u_1}{u_2}\right)^2 = \frac{9}{4}.$$

Hence $E_1 = \frac{9}{13}E$ and $E_2 = \frac{4}{13}E$. ■

An important special case occurs when the two particles have equal masses. In this case, formula (10.15) shows that the opening angle must always be a right angle. Thus, *in an elastic collision between particles of equal mass, the particles depart in directions at right angles*. Note that this result applies only when the target particle is initially at rest.

Example 10.6 *Elastic collision between two electrons*

In an elastic collision between an electron with kinetic energy E and an electron at rest, the incoming electron is observed to be deflected through an angle of 30°. What are the energies of the two electrons after the collision?

Solution

Since the collision is elastic and the electrons have equal mass, the opening angle between the emerging paths must be 90°. The target electron must therefore recoil at an angle of 60° to the initial direction of the incoming electron. Let the speed of the incoming electron be u and speeds of the electrons after the collision be u_1 and u_2 respectively. Then conservation of linear momentum implies that

$$mu = mu_1 \cos 30° + mu_2 \cos 60°,$$
$$0 = mu_1 \sin 30° - mu_2 \sin 60°,$$

which gives $u_1 = \frac{1}{2}\sqrt{3}\,u$ and $u_2 = \frac{1}{2}u$. Hence, after the collision, the electrons have energies $\frac{3}{4}E$ and $\frac{1}{4}E$ respectively. ■

10.7 COLLISION PROCESSES IN THE ZERO-MOMENTUM FRAME

We have so far supposed that the inertial reference frame from which the scattering process is observed is the one occupied by the experimental observer. This is called the **laboratory frame** (or lab frame) since it is the frame in which measurements (of scattering angles, for instance) are actually taken. In the lab frame, the target particle is initially at rest.

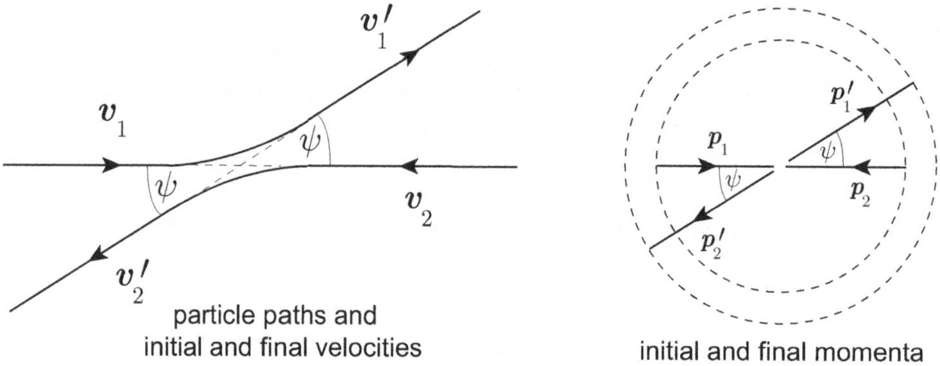

FIGURE 10.4 A collision between two particles viewed from the **zero-momentum frame**. The **initial momenta** $\boldsymbol{p}_1$, $\boldsymbol{p}_2$ are equal and opposite, as are the **final momenta** $\boldsymbol{p}'_1$, $\boldsymbol{p}'_2$. The angle ψ is the angle through which each of the masses is scattered.

However, it is very convenient to 'view' the scattering process from a different inertial frame. Since the two particles form an *isolated* system, their centre of mass G moves with constant velocity and so the frame* in which G is at rest is inertial. In this frame, the total linear momentum of the two particles is zero and, for this reason, we call it the **zero-momentum frame**† or **ZM frame**.

Consider, for example, the scattering problem which, in the lab frame, is shown in Figure 10.3. Then the total linear momentum $\boldsymbol{P} = m_1\boldsymbol{u}$ and the velocity $\boldsymbol{V}$ of the centre of mass of the two particles is

$$\boxed{\boldsymbol{V} = \frac{m_1\boldsymbol{u}}{m_1 + m_2}.} \tag{10.16}$$

This therefore is the **velocity of the ZM frame** relative to the lab frame for this collision process.

Collisions viewed from the ZM frame

Two-particle collisions look simple when viewed from the ZM frame. This is because, since the total linear momentum is now zero, the initial linear momenta $\boldsymbol{p}_1$, $\boldsymbol{p}_2$ of the two particles and the final momenta $\boldsymbol{p}'_1$, $\boldsymbol{p}'_2$ of the two particles must satisfy

$$\boldsymbol{p}_1 + \boldsymbol{p}_2 = \boldsymbol{0}, \qquad \boldsymbol{p}'_1 + \boldsymbol{p}'_2 = \boldsymbol{0}. \tag{10.17}$$

Thus, when a two-particle collision is **viewed from the ZM frame, the initial momenta are equal and opposite and so are the final momenta**. Figure 10.4 shows what a two-particle collision looks like when viewed from the ZM frame. Because of the relations

* This frame has the same velocity as G, and no rotation, relative to the lab frame.

† The term 'centre of mass frame' is also used. However, 'zero-momentum frame' is preferable since this notion holds good in relativistic mechanics.

(10.17), the particles both arrive and depart in opposite directions, so that **each particle is deflected through that same angle ψ**. All this follows solely from conservation of linear momentum. We can say more if we also have an energy principle of the form

$$\tfrac{1}{2}m_1|\boldsymbol{v}_1|^2 + \tfrac{1}{2}m_2|\boldsymbol{v}_2|^2 + Q = \tfrac{1}{2}m_1|\boldsymbol{v}_1'|^2 + \tfrac{1}{2}m_2|\boldsymbol{v}_2'|^2,$$

where $\boldsymbol{v}_1$, $\boldsymbol{v}_2$, $\boldsymbol{v}_1'$, $\boldsymbol{v}_2'$ are the initial and final velocities of the particles (as shown in Figure 10.4), and Q is the kinetic energy gained as a result of the collision.* Let p be the common *magnitude* of the initial momenta $\boldsymbol{p}_1$, $\boldsymbol{p}_2$, and p' be the common *magnitude* of the final momenta $\boldsymbol{p}_1'$, $\boldsymbol{p}_2'$. Then the energy balance equation can be re-written in the form

$$\frac{p^2}{2m_1} + \frac{p^2}{2m_2} + Q = \frac{p'^2}{2m_1} + \frac{p'^2}{2m_2},$$

that is,

$$p'^2 = p^2 + \left(\frac{2Qm_1m_2}{m_1+m_2}\right). \tag{10.18}$$

Thus the magnitudes of the initial and final momenta are related through Q, the energy gained in the collision. This is depicted in the momentum diagram in Figure 10.4. The magnitudes of the initial and final momenta (p and p') are represented by the radii of the two dashed circles. The diagram shows the case in which $p' > p$, which corresponds to $Q > 0$. For an **elastic collision**, the circles are coincident and **all four momenta have equal magnitudes.**

In a typical scattering problem, the masses m_1, m_2 and the initial momenta $\boldsymbol{p}_1$, $\boldsymbol{p}_2$ are known. For the scattering problem shown (in the lab frame) in Figure 10.3, $\boldsymbol{v}_1 = \boldsymbol{u} - \boldsymbol{V}$ and $\boldsymbol{v}_2 = -\boldsymbol{V}$, where $\boldsymbol{V}$ is given by equation (10.16). It follows that the initial momentum magnitude in the ZM frame are given by

$$p = \frac{m_1m_2u}{m_1+m_2}, \tag{10.19}$$

where $u = |\boldsymbol{u}|$. The scattering process is now entirely determined by the parameters Q (the energy gain) and ψ (the ZM scattering angle). Given p and Q, p' is determined from equation (10.18). Together with ψ, this determines the final momenta $\boldsymbol{p}_1$ and $\boldsymbol{p}_2$. The parameters Q and ψ depend on the physics of the actual collision. For instance, the collision may be known to be elastic, in which case $Q = 0$. The question of how the scattering angle ψ is related to the actual interaction and initial conditions is addressed in section 10.9.

Returning to the lab frame (elastic collisions only)

Although the scattering process looks simpler in the ZM frame, we usually need to know the details of the scattering actually observed by the experimenter in the lab frame. This

* As remarked earlier, Q is frame independent and so is the same as the energy gain measured in the lab frame.

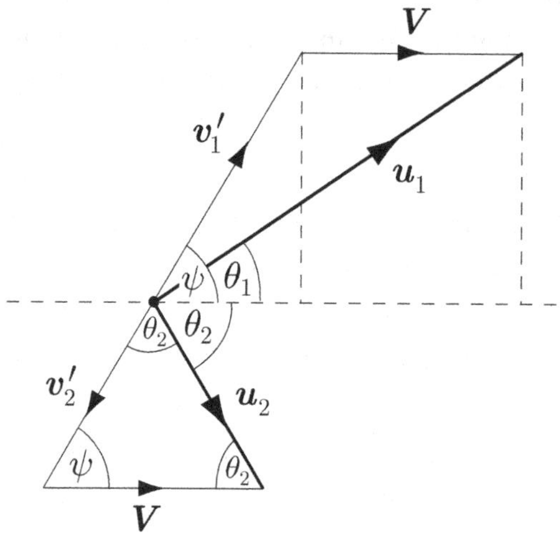

FIGURE 10.5 The final particle velocities $\boldsymbol{u}_1$, $\boldsymbol{u}_2$ in the lab frame are obtained from the final velocities $\boldsymbol{v}_1'$, $\boldsymbol{v}_2'$ in the ZM frame by the relations $\boldsymbol{u}_1 = \boldsymbol{v}_1' + \boldsymbol{V}$, $\boldsymbol{u}_2 = \boldsymbol{v}_2' + \boldsymbol{V}$. The diagram shows the elastic case, in which the velocity triangle for $\boldsymbol{u}_2$ is isosceles.

entails transforming the properties of the final state (velocities, momenta and kinetic energies) from the ZM frame back to the lab frame. Since the ZM frame has velocity $\boldsymbol{V}$ (given by (10.16)) relative to the lab frame, the final velocities $\boldsymbol{u}_1$, $\boldsymbol{u}_2$ observed in the lab frame are related to the final velocities $\boldsymbol{v}_1'$, $\boldsymbol{v}_2'$ in the ZM frame by

$$\boldsymbol{u}_1 = \boldsymbol{v}_1' + \boldsymbol{V}, \qquad \boldsymbol{u}_2 = \boldsymbol{v}_2' + \boldsymbol{V}. \tag{10.20}$$

Any other properties can then be found from $\boldsymbol{u}_1$, $\boldsymbol{u}_2$. The transformations (10.20) are depicted geometrically in Figure 10.5. The transformation formulae become rather complicated in the general case, but simplify nicely when the collision is elastic. From now on we will restrict ourselves to **elastic collisions only**. In this case, $Q = 0$ and the collisions are parametrised by ψ alone. The energy equation (10.18) then implies that $p' = p$ so that the four momentum magnitudes are equal, and given by equation (10.19). The final *speeds* of the particles in the ZM frame are therefore given by

$$v_1' = \frac{m_2 u}{m_1 + m_2}, \qquad v_2' = \frac{m_1 u}{m_1 + m_2} = V, \tag{10.21}$$

where $V = |\boldsymbol{V}|$. We may now deduce the required information from Figure 10.5. The lab **scattering angle** θ_1 can be expressed in terms of the parameter angle ψ by

$$\tan\theta_1 = \frac{v_1' \sin\psi}{v_1' \cos\psi + V} = \frac{\sin\psi}{\cos\psi + (V/v_1')} = \frac{\sin\psi}{\cos\psi + (m_1/m_2)},$$

on using equations (10.21). The lab **recoil angle** θ_2 is easily found since, in an elastic collision, the velocity triangle for $\boldsymbol{u}_2$ is isosceles with angles ψ, θ_2 and θ_2, as shown in Figure 10.5. It follows that

$$\theta_2 = \tfrac{1}{2}(\pi - \psi).$$

The expression for the lab **opening angle** θ $(= \theta_1 + \theta_2)$ is therefore given by

$$\tan\theta = \tan(\theta_1 + \theta_2) = \frac{\tan\theta_1 + \tan\theta_2}{1 + \tan\theta_1 \tan\theta_2} = \left(\frac{m_1 + m_2}{m_1 - m_2}\right)\cot(\tfrac{1}{2}\psi),$$

after some simplification.

To find the final energies, we observe that

$$u_2 = 2V\,\sin(\tfrac{1}{2}\psi)$$

so that E_2, the final lab **energy** of the mass m_2 is given by

$$\frac{E_2}{E_0} = \frac{\frac{1}{2}m_2\left(2V\,\sin(\frac{1}{2}\psi)\right)^2}{\frac{1}{2}m_1 u^2} = \frac{4m_1 m_2}{(m_1 + m_2)^2}\sin^2(\tfrac{1}{2}\psi),$$

where E_0 $(= \frac{1}{2}m_1 u^2)$ is the lab energy of the incident mass m_1. Since the collision is elastic, the final lab energy of the mass m_1 is simply deduced from the energy conservation formula $E_1 + E_2 = E_0$.

The above formulae give the properties of the final state following an elastic two-particle collision in terms of the ZM scattering angle ψ. We will call them the **elastic collision formulae** and they are summarised below:

Elastic collision formulae

A. $\tan\theta_1 = \dfrac{\sin\psi}{\cos\psi + \gamma}$ **B.** $\theta_2 = \frac{1}{2}(\pi - \psi)$

C. $\tan\theta = \left(\dfrac{\gamma + 1}{\gamma - 1}\right)\cot(\tfrac{1}{2}\psi)$ **D.** $\dfrac{E_2}{E_0} = \dfrac{4\gamma}{(\gamma + 1)^2}\sin^2(\tfrac{1}{2}\psi)$

(10.22)

ψ is the scattering angle in the ZM frame, and $\gamma = m_1/m_2$, the mass ratio of the two particles.

Using the elastic collision formulae A word of advice about the use of these formulae may be helpful. Most questions on this topic tell you some property of the scattering in the lab frame and ask you to find another property of the scattering in the lab frame; the ZM frame is never mentioned. *It is inadvisable to start manipulating the elastic scattering formulae.* This is almost guaranteed to cause errors. The simplest method is as follows: (i) Use the given data to find ψ by using the appropriate formula 'backwards', and then (ii) use this value of ψ to find the required scattering property. In short, the advice is '*go via* ψ'.

Example 10.7 ***Using the elastic scattering formulae***

In an experiment, particles of mass m and energy E are used to bombard stationary target particles of mass $2m$.

Q. The experimenters wish to select particles that, after scattering, have energy $E/3$. At what scattering angle will they find such particles?

A. If $E_1/E_0 = 1/3$, then by energy conservation $E_2/E_0 = 2/3$. First use formula **D** to find ψ. Since the mass ratio $\gamma = 1/2$, this gives

$$\frac{2}{3} = \frac{8}{9}\sin^2(\tfrac{1}{2}\psi),$$

so that $\psi = 120°$. Now use formula **A** to find the scattering angle θ_1. This gives $\tan\theta_1 = \infty$ so that $\theta_1 = 90°$. Particles scattered with energy $E/3$ will therefore be found emerging at right angles to the incident beam.

Q. In one collision, the opening angle was measured to be 45°. What were the individual scattering and recoil angles?

A. First use formula **C** to find ψ. This gives

$$\cot(\tfrac{1}{2}\psi) = \frac{1}{3},$$

so that $\frac{1}{2}\psi = 72°$, to the nearest degree. Now use formula **B** to find the recoil angle θ_2. This gives $\theta_2 = 90° - 72° = 18°$. The scattering angle θ_1 must therefore be $\theta_1 = \theta - \theta_2 = 45° - 18° = 27°$.

Q. In another collision, the scattering angle was measured to be 45°. What was the recoil angle?

A. First use formula **A** to find ψ. This shows that ψ satisfies the equation

$$2\cos\psi - 2\sin\psi = 1,$$

which can be written in the form*

$$\sqrt{8}\cos\left(\psi + 45°\right) = 1.$$

This gives $\psi = 24°$, to the nearest degree. Formula **B** now gives the recoil angle θ_2 to be $\theta_2 = 78°$, to the nearest degree.

10.8 THE TWO-BODY PROBLEM

The problem of determining the motion of two particles, moving solely under their mutual interaction, is called the **two-body problem**. Strictly speaking, all of the orbit

* Recall that equations of the form $a\cos\psi + b\sin\psi = c$ are solved by writing the left side in the 'polar form' $R\cos(\psi - \alpha)$, where $R^2 = a^2 + b^2$ and $\tan\alpha = b/a$.

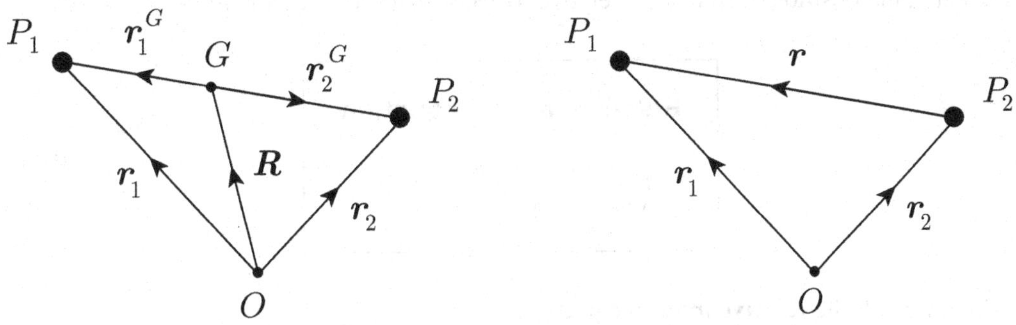

FIGURE 10.6 The motion of P_1 and P_2 relative to their centre of mass (left), and the motion of P_1 relative to P_2 (right).

problems considered in Chapter 7 should have been treated as two-body problems since centres of force are never actually fixed. The one-body theory is a good approximation when one particle is much more massive than the other. When the two particles have similar masses, the problem must be treated by two-body theory, in which neither particle is assumed to be fixed.

Let P_1 and P_2 be two particles moving under their mutual interaction. By the Third Law, the forces that they exert on each other are equal in magnitude, opposite in direction, and act along the line joining them. We will further suppose that the magnitude of these interaction forces depends only on r, the distance separating P_1 from P_2. The forces $\boldsymbol{F}_1$, $\boldsymbol{F}_2$, acting on P_1, P_2, then have the form

$$\boldsymbol{F}_1 = F(r)\,\widehat{\boldsymbol{r}}, \qquad \boldsymbol{F}_2 = -F(r)\,\widehat{\boldsymbol{r}},$$

where $\boldsymbol{r} = \boldsymbol{r}_1 - \boldsymbol{r}_2$, $r = |\boldsymbol{r}_1 - \boldsymbol{r}_2|$ and $\widehat{\boldsymbol{r}} = \boldsymbol{r}/r$ (see Figure 10.6). The **equations of motion** for P_1, P_2 are therefore

$$m_1\ddot{\boldsymbol{r}}_1 = F(r)\,\widehat{\boldsymbol{r}}, \qquad m_2\ddot{\boldsymbol{r}}_2 = -F(r)\,\widehat{\boldsymbol{r}}. \tag{10.23}$$

This is a generalisation of central force motion in which each particle moves under a force centred upon the other particle. Although this problem appears to be complicated, it can be quickly reduced to an **equivalent one-body problem**.

We first observe that the two particles form an isolated system so that their total linear momentum is conserved, or (equivalently) their centre of mass G moves with constant velocity. The motion of G is therefore determined from the initial conditions and it remains to find the motion of each particle *relative* to G, that is, their motions in the ZM frame. It turns out however that it is easier to find the motion of one particle *relative to the other*. The motion of each particle relative to G can then be easily deduced.

The equation of relative motion

It follows from the equations of motion (10.23) that

$$\ddot{\boldsymbol{r}}_1 - \ddot{\boldsymbol{r}}_2 = \frac{F(r)\,\widehat{\boldsymbol{r}}}{m_1} + \frac{F(r)\,\widehat{\boldsymbol{r}}}{m_2} = \left(\frac{m_1 + m_2}{m_1 m_2}\right) F(r)\,\widehat{\boldsymbol{r}},$$

so that $\boldsymbol{r}$, the position vector of P_1 relative to P_2, satisfies the equation

Relative motion equation

$$\left(\frac{m_1 m_2}{m_1 + m_2}\right)\ddot{\boldsymbol{r}} = F(r)\,\widehat{\boldsymbol{r}}, \qquad (10.24)$$

which we call the **relative motion equation.**

Definition 10.3 ***Reduced mass*** *The quantity μ, defined by*

$$\mu = \frac{m_1 m_2}{m_1 + m_2}. \qquad (10.25)$$

*is called the **reduced mass**.*

Our result can be expressed as follows:

Two-body problem – the relative motion

In the two-body problem, the motion of P_1 relative to P_2 is the same *as if* P_2 were held fixed and P_1 had the reduced mass μ instead of its actual mass m_1.

This rule* allows us to replace the problem of the motion of P_1 relative to P_2 by an **equivalent one-body problem** in which P_2 is fixed. The solution of such problems is fully described in Chapter 7.

Example 10.8 *Escape from a free gravitating body*

Two particles P_1 and P_2, with masses m_1 and m_2, can move freely under their mutual gravitation. Initially both particles are at rest and separated by a distance c. With what speed must P_1 be projected so as to escape from P_2?

Solution

Since this is a mutual gravitation problem, we take our rule in the form: *The motion of P_1 relative to P_2 is the same as if P_2 were held fixed and the constant of gravitation G replaced by G', where*

$$G' = \left(\frac{m_1 + m_2}{m_2}\right) G.$$

* The rule is ambiguous when the force F also depends on m_1, as in mutual gravitation. Do you also replace *this* m_1 by μ? The answer is no, but the easiest way to avoid this glitch in the mutual gravitation problem is to make the transformation $G \to (m_1 + m_2)G/m_2$ instead. This has the correct effect and is not ambiguous.

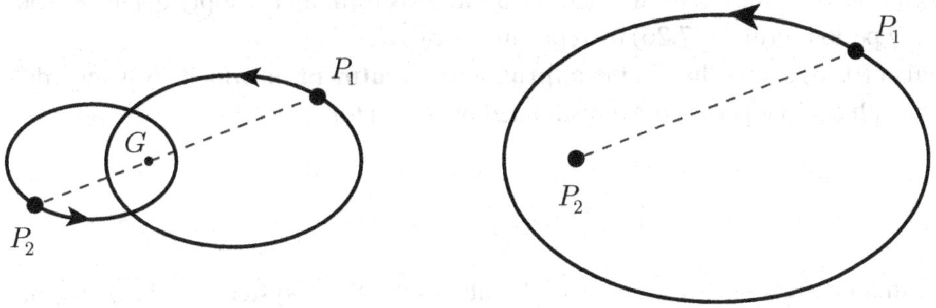

FIGURE 10.7 Particles P_1 and P_2 move under their mutual gravitation. In the zero momentum frame, the orbits are similar conics, each with a focus at G (left). The orbit of P_1 relative to P_2 is a third similar conic with P_2 at a focus (right).

From the one-body theory in Chapter 7, we know that P_1 will escape from a *fixed* P_2 if it has positive energy, that is if

$$\tfrac{1}{2}m_1V^2 - \frac{m_1m_2G}{c} \geq 0.$$

Hence, when P_2 is not fixed, P_1 will escape if

$$\tfrac{1}{2}m_1V^2 - \frac{m_1m_2G'}{c} \geq 0,$$

that is, if

$$V^2 \geq \frac{2(m_1+m_2)G}{c}.$$

This is the required **escape condition.** ■

Once the *relative* motion of the particles has been found, one may easily deduce the motion of each particle in the ZM frame since (see Figure 10.6)

$$\boldsymbol{r}_1^G = \left(\frac{m_2}{m_1+m_2}\right)\boldsymbol{r}, \qquad \boldsymbol{r}_2^G = -\left(\frac{m_1}{m_1+m_2}\right)\boldsymbol{r}.$$

It follows that the orbits of P_1, P_2 in the ZM frame are **geometrically similar** to the orbit in the relative motion. For instance, suppose that the mutual interaction of P_1 and P_2 is gravitational attraction, and that the orbit of P_1 relative to P_2 has been found to be an ellipse. Then the orbits of P_1 and P_2 in the ZM frame are *similar ellipses*, as shown in Figure 10.7. The ratio of the major axes of these orbits is $m_2 : m_1$, and the sum of their major axes is equal to the major axis of the orbit of P_1 relative to P_2. All three orbits have the same period τ given by

$$\tau^2 = \frac{4\pi^2a^3}{G\,(m_1+m_2)}, \qquad (10.26)$$

where a is the semi-major axis of the *relative* orbit. This formula is simply obtained from the one-body period formula (7.26) by replacing G by G'.

Formula (10.26) shows that, in the **approximate treatment** in which P_2 is regarded as fixed, the value of the period is overestimated by the factor

$$\left(1 + \frac{m_1}{m_2}\right)^{1/2},$$

which is a small correction when m_1/m_2 is small. In the Solar system, the largest value of m_1/m_2 for a planetary orbit is that for Jupiter, which is about 1/1000.

Binary stars

It is probable that over half of the 'stars' in our galaxy are not single stars, like the Sun, but occur in pairs* that move under their mutual gravitation. Such a pair is called a **binary star**.

Binary stars are important in astronomy and also provide a nice application of our two-body theory. In particular, the two components of the binary must orbit their centre of mass on similar ellipses, as shown in Figure 10.7; the orbit of either component relative to the other is a third similar ellipse; and the period of all three motions is given by formula (10.26), where a is the semi-major axis of the *relative* orbit.

One reason why binary stars are important in astronomy is that the masses of their component stars can be found by direct measurement; indeed they are the only stars for which this can be done. Suppose that the star is an *optical* binary, which means that both components are visible through a suitably large telescope. Then the period of the binary can be measured by direct observation. It is also possible to measure the major axis of the relative orbit. Once τ and a are known, formula (10.26) tells us the **sum of the masses** of the two components of the binary.

Example 10.9 *Sirius A and B*

A typical example of a binary is **Sirius** in the constellation *Canis Major*, the brightest star in the night sky. The large bright component is called Sirius A and its small dim companion Sirius B. The period of their mutual orbital motion is 50 years and the value of a is 20 AU. (This is about the distance from the Sun to the planet Uranus.) Find the sum of the masses of the two components of Sirius.

Solution

In terms of astronomical units, in which $G = 4\pi^2$, formula (10.26) gives

$$M_A + M_B = \frac{20^3}{50^2} = 3.2\ M_\odot \blacksquare$$

* Groups of three or more also occur.

In order to determine the **individual masses** of the components by optical means, it is necessary to find the a values for one of the individual components in its motion *relative to the centre of mass*. The procedure is essentially the same as before, but much more difficult observationally since the motion of the chosen component must be measured absolutely, that is, relative to background stars. In the case of Sirius, it is found that $M_A = 2.1\ M_\odot$ and $M_B = 1.1\ M_\odot$.

10.9 TWO-BODY SCATTERING

An important application of two-body motion is the **two-body scattering** problem. In our treatment of collision theory, we considered the whole class of possible collisions between two particles that were consistent with momentum and energy conservation. These collisions were parametrised by the ZM scattering angle ψ. We now consider the problem in more detail. Given the interaction between the particles and the impact parameter p, what *is* the resulting ZM scattering angle? This question can be answered by using two-body theory. We break up the process into a number of steps:

1. Find the $\{p, \theta\}$–relation for the one-body problem

First consider the **one-body problem** in which the particle P_2 is held *fixed*, and work out (or look up) the relation between the impact parameter p and the scattering angle θ. For example, the $\{p, \theta\}$–relation for **Rutherford scattering** was derived in Chapter 7 and was found to be

$$\tan \tfrac{1}{2}\theta = \frac{q_1 q_2}{m_1 p u^2}. \tag{10.27}$$

2. Find the $\{p, \phi\}$–relation for the relative motion problem

The next step is to find the relation between the impact parameter p and the scattering angle ϕ observed in the **relative motion problem**. This is easily obtained from the one-body formula (10.27) by replacing m_1 by μ (the reduced mass) and replacing θ by ϕ. This gives

$$\tan \tfrac{1}{2}\phi = \frac{q_1 q_2 (1+\gamma)}{m_1 p u^2}, \tag{10.28}$$

where γ $(= m_1/m_2)$ is the ratio of the two masses.

3. Find the $\{p, \psi\}$–relation observed in the ZM frame

The angle ϕ that appears in the formula (10.28) is the scattering angle in the motion of m_1 *relative to* m_2. However, by an amazing stroke of good fortune, it is actually the same angle as the ZM scattering angle ψ that we used in collision theory.* Hence, the

* The reason is as follows: The relative motion in the lab frame must be the same as the relative motion in the ZM frame. In this frame, the initial relative velocity of P_1 is equal to $(\boldsymbol{p}_1/m_1) - (\boldsymbol{p}_2/m_2)$, which has the same *direction* as $\boldsymbol{p}_1$. Likewise, the final relative velocity of P_1 is equal to $(\boldsymbol{p}_1'/m_1) - (\boldsymbol{p}_2'/m_2)$, which has the same *direction* as $\boldsymbol{p}_1'$. Hence the scattering angle in the relative motion is the same as that in the ZM frame.

$\{p, \psi\}$–relation when **two-body scattering** is observed from the **ZM frame** is obtained by simply replacing ϕ in formula (10.28) by ψ, that is,

$$\tan \tfrac{1}{2}\psi = \frac{q_1 q_2(1+\gamma)}{m_1 p u^2}. \qquad (10.29)$$

As always, u means the speed of the incident particle observed in the lab frame.

4. Find θ_1 and θ_2 in terms of p from the elastic collision formulae

Since the $\{p, \psi\}$–relation (10.29) gives the ZM scattering angle ψ in terms of p, this expression for ψ can now be substituted into the elastic scattering formulae (10.22**A**) and (10.22**B**) to give expressions for the **two-body scattering angle** θ_1, and **recoil angle** θ_2 in terms of p. For Rutherford scattering, this gives, after some simplification,

Two-body Rutherford scattering formulae

$$\tan\theta_1 = \frac{4q_1q_2pE}{4p^2E^2 - (1-\gamma^2)q_1^2q_2^2} \qquad \tan\theta_2 = \frac{2pE}{q_1q_2(1+\gamma)} \qquad (10.30)$$

where $E\ (= \tfrac{1}{2}m_1u^2)$ is the energy of the incident particle and $\gamma = m_1/m_2$.

These formulae simplify further when the particles have **equal masses**. In this special case, $\gamma = 1$ and the scattering and recoil angles are given by

$$\tan\theta_1 = \frac{q_1q_2}{pE}, \qquad \tan\theta_2 = \frac{pE}{q_1q_2}. \qquad (10.31)$$

(As expected, $\theta_1 + \theta_2 = \tfrac{1}{2}\pi$.) These formulae would apply, for example, to the scattering of alpha particles by helium nuclei.

Two-body scattering cross section

Having found the $\{p, \theta_1\}$–relation for the two-body scattering problem, the **two-body scattering cross section** σ^{TB} is given, in principle, by the formula

$$\sigma^{TB}(\theta_1) = -\frac{p}{\sin\theta_1}\frac{dp}{d\theta_1}.$$

However, this requires that the $\{p, \theta_1\}$–relation be solved to give p as a function of θ_1 and the resulting algebra is formidable.

The following method has the advantage that σ^{TB} is determined directly from the corresponding one-body scattering cross section. The trick is to introduce the ZM scattering angle ψ. By the chain rule,

$$\frac{dp}{d\theta_1} = \frac{dp}{d\psi} \times \frac{d\psi}{d\theta_1},$$

and so σ^{TB} can be written

$$\begin{aligned}\sigma^{TB} &= -\frac{p}{\sin\theta_1}\frac{dp}{d\theta_1} = -\frac{p}{\sin\theta_1}\left(\frac{dp}{d\psi}\times\frac{d\psi}{d\theta_1}\right)\\ &= \left(\frac{\sin\psi}{\sin\theta_1}\right)\left(\frac{d\psi}{d\theta_1}\right)\left(-\frac{p}{\sin\psi}\frac{dp}{d\psi}\right)\\ &= \left(\frac{\sin\psi}{\sin\theta_1}\right)\left(\frac{d\psi}{d\theta_1}\right)\sigma^{ZM}(\psi),\end{aligned}$$

where σ^{ZM} is defined by

$$\sigma^{ZM}(\psi) = -\frac{p}{\sin\psi}\frac{dp}{d\psi}. \qquad (10.32)$$

Now $\sigma^{ZM}(\psi)$ is easily obtained from the one-body cross-section $\sigma(\theta)$ by replacing m_1 by μ and θ by ψ. The two-body cross section is then given by

Two-body scattering cross section

$$\sigma^{TB}(\theta_1) = \left(\frac{\sin\psi}{\sin\theta_1}\right)\left(\frac{d\psi}{d\theta_1}\right)\sigma^{ZM}(\psi) \qquad (10.33)$$

In this formula, we have yet to replace ψ by its expression in terms of θ_1. To do this, we must invert the formula (10.22 **A**) to obtain ψ as a function of θ_1. Formula (10.22 **A**) can be rearranged in the form

$$\sin(\psi - \theta_1) = \gamma\sin\theta_1,$$

from which we obtain

$$\psi = \theta_1 + \sin^{-1}(\gamma\sin\theta_1) \qquad (10.34)$$

and, by differentiation,

$$\frac{d\psi}{d\theta_1} = 1 + \frac{\gamma\cos\theta_1}{(1-\gamma^2\sin^2\theta_1)^{1/2}}. \qquad (10.35)$$

These expressions for ψ and $d\psi/d\theta_1$ in terms of θ_1 must now be substituted into equation (10.33) to obtain the final formula for the **two-body scattering cross section** $\sigma^{TB}(\theta_1)$. These operations can be done with computer assistance.

For example, in **Rutherford scattering**, we first obtain $\sigma^{ZM}(\psi)$ by replacing m_1 by μ (and θ by ψ) in the one-body cross section formula (7.37) obtained in Chapter 7. This gives

$$\sigma^{ZM}(\psi) = \frac{q_1^2 q_2^2(1+\gamma)^2}{4m_1^2u^4}\left(\frac{1}{\sin^4\frac{1}{2}\psi}\right). \qquad (10.36)$$

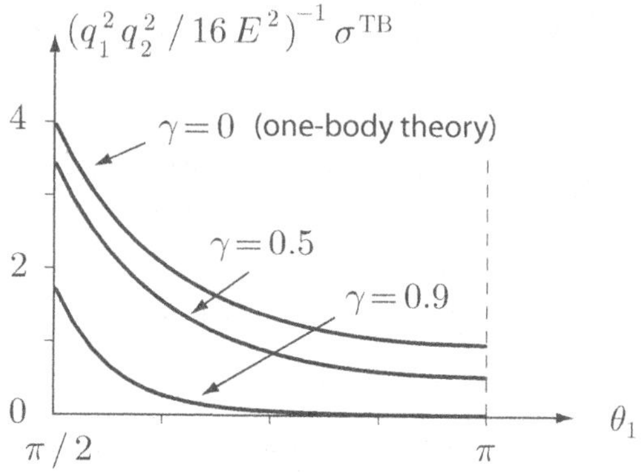

FIGURE 10.8 The **Rutherford two-body scattering cross section** σ^{TB} plotted against the scattering angle θ_1 ($\pi/2 \le \theta_1 \le \pi$) for various values of the mass ratio γ ($= m_1/m_2$). E is the kinetic energy of the incident particles.

The **two-body Rutherford scattering cross section** is now obtained by substituting the expression (10.36) into the general formula (10.33) and then replacing ψ and $d\psi/d\theta_1$ by the expressions (10.34), (10.35). After much manipulation, the answer is found to be

$$\sigma^{TB} = \frac{q_1^2 q_2^2}{16E^2}\left(\frac{4(1+\gamma)^2(\gamma\cos\theta_1 + S)^2}{S(1+\gamma\sin^2\theta_1 - \cos\theta_1 S)^2}\right), \qquad (10.37)$$

where

$$S = \left(1 - \gamma^2 \sin^2\theta_1\right)^{1/2}$$

and E ($= \frac{1}{2}m_1 u^2$) is the energy of the incident particle.

Figure 10.8 shows graphs of $\sigma^{TB}(\theta_1)$ in Rutherford scattering for various choices of the mass ratio γ. In Rutherford's actual experiment with alpha particles and gold nuclei, the value of γ was about 0.02 and the error in the scattering cross section caused by using the one-body theory was less than 0.1%. However, as the graphs show, larger values of γ can give rise to a substantial deviation from the one-body theory.

When the mass ratio γ ($= m_1/m_2$) is *small*, the formula (10.37) is approximated by

$$\sigma^{TB} = \frac{q_1^2 q_2^2}{16E^2}\left(\frac{1}{\sin^4(\theta_1/2)} - 2\gamma^2 + O\left(\gamma^4\right)\right).$$

Thus, when γ is small, the leading correction to the one-body approximation is a constant.

Equal masses

The whole process of finding σ^{TB} simplifies wonderfully when the two particles have equal masses. In this case, $\psi = 2\theta_1$, $d\psi/d\theta_1 = 2$, and the general formula (10.33) becomes

$$\sigma^{TB}(\theta_1) = 4\cos\theta_1\,\sigma^{ZM}(2\theta_1) \qquad (0 < \theta_1 \le \pi/2).$$

For example, in Rutherford scattering where the particles have equal masses, σ^{TB} has the simple form

$$\sigma^{TB}(\theta_1) = \frac{q_1^2 q_2^2}{E^2}\left(\frac{\cos\theta_1}{\sin^4\theta_1}\right) \qquad (0 \le \theta_1 \le \pi/2).$$

This formula would apply, for example, to the scattering of protons by protons.

10.10 INTEGRABLE MECHANICAL SYSTEMS

A mechanical system is said to be **integrable** if *its equations of motion are soluble in the sense that they can be reduced to integrations.** The most important class of integrable systems are those that satisfy as many conservation principles as they have degrees of freedom. Suppose that a mechanical system $\mathcal{S}$ has n degrees of freedom and that it satisfies n conservation principles. Then it is certainly true that the n conservation equations are sufficient to *determine* the motion of the system, in the sense that no more equations are needed. More importantly though, it can be shown† that *these equations can always be reduced to integrations*. The system $\mathcal{S}$ is therefore **integrable**.

Before we can apply this method to particular systems, there is a **kinematical problem** to be overcome, namely: how does one find the velocities (and angular velocities) of the elements‡ of $\mathcal{S}$ when there are two or more generalised coordinates which vary simultaneously? The answer is by drawing a **velocity diagram** for $\mathcal{S}$ as described below:

Drawing a velocity diagram

- Draw the system in general position and select a set of generalised coordinates.
- Let the first generalised coordinate vary (with the other coordinates held constant) and mark in the velocity of each element.
- Now let the second generalised coordinate vary (with the other coordinates held constant) and, on the same diagram, mark in the velocity of each element. Continue in this way through all the generalised coordinates.
- Then, when all the generalised coordinates are varying simultaneously, the **velocity** of each element of $\mathcal{S}$ is the vector sum of the velocities given to that element when the coordinates vary individually.

In the above, 'velocity' means 'velocity and/or angular velocity'.

---

* The system is still said to be integrable even when the integrals cannot be evaluated in terms of standard functions!

† This is Liouville's theorem on integrable systems (see Problem 14.15)

‡ The elements of $\mathcal{S}$ are the particles and/or rigid bodies of which $\mathcal{S}$ is made up. One needs to find (i) the velocity of each particle, (ii) the velocity of the centre of mass of each rigid body, and (iii) the angular velocity of each rigid body, in each case in terms of the chosen coordinates and their time derivatives.

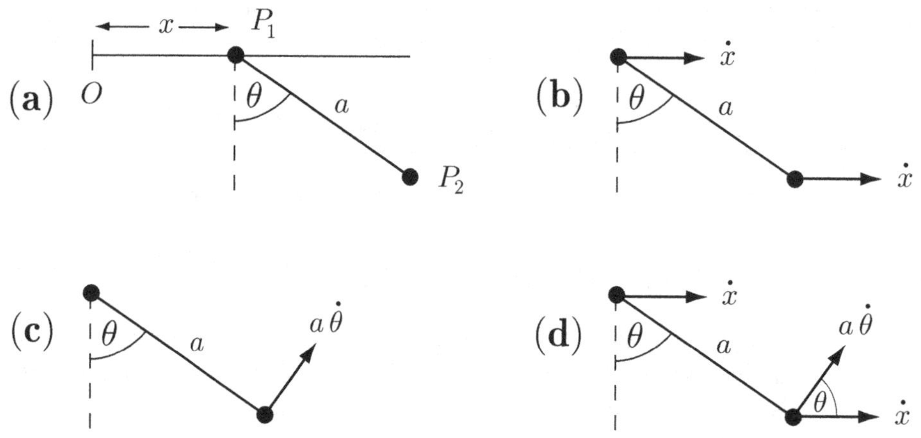

FIGURE 10.9 Constructing a **velocity diagram**. Figure (a) shows the system and the coordinates x and θ. Figure (b) shows the velocities generated when x varies with θ held constant. Figure (c) shows the velocities generated when θ varies with x held constant. Figure (d) is the **velocity diagram** which is formed by superposing the velocities in diagrams (b) and (c). Note that the velocity of P_2 is the *vector* sum of the two contributions shown.

Example 10.10 *Drawing a velocity diagram 1*

The system shown in Figure 10.9 consists of two particles P_1 and P_2 connected by a light inextensible string of length a. The particle P_1 is also constrained to move along a fixed horizontal rail and the whole system moves in the vertical plane through the rail. Take the variables x and θ shown as generalised coordinates and draw the velocity diagram.

Solution

The construction of the velocity diagram is shown in Figure 10.9. ■

Example 10.11 *Drawing a velocity diagram 2*

Two rigid rods CD and DE, of lengths $2a$ and $2b$, are flexibly jointed at D and can move freely on a horizontal table. Choose generalised coordinates and draw a velocity diagram for this system.

Solution

Let Oxy be a system of Cartesian coordinates in the plane of the table. Let (X, Y) be the Cartesian coordinates of the centre of the rod CD, and let θ and ϕ be the angles that the two rods make with positive x-axis. Then X, Y, θ, ϕ are a set of generalised coordinates for this system. These coordinates, and the corresponding velocity diagram are shown in Figure 10.10. There are *four* contributions to the velocity of the centre of the rod DE. Also, each rod has an angular velocity. ■

We will now solve the system shown in Figure 10.9 by using conservation principles.

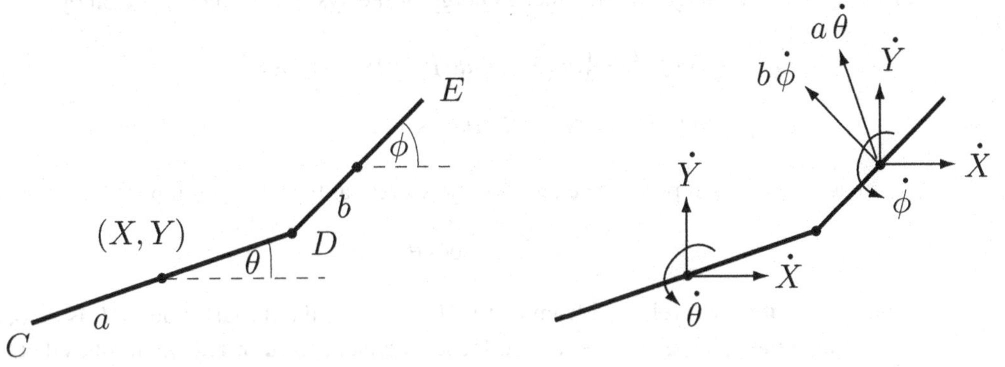

FIGURE 10.10 The **velocity diagram** for a system with four degrees of freedom. The figure on the left shows the system and the generalised coordinates X, Y, θ, ϕ. The figure on the right is the completed **velocity diagram**.

Example 10.12 *Solving an integrable system*

Consider the system shown in Figure 10.9 for the case in which P_1 and P_2 have masses $3m$ and m, the rail is smooth, and the system moves under uniform gravity. Initially, the system is released from rest with the string making an angle of $\pi/3$ with the downward vertical. Use conservation principles to obtain two equations for the subsequent motion.

Solution

Let $\boldsymbol{i}$ be the unit vector parallel to the rail (in the direction of increasing x). Since the rail is smooth, all the *external* forces on the system are vertical which means that $\boldsymbol{F} \cdot \boldsymbol{i} = 0$. This implies that $\boldsymbol{P} \cdot \boldsymbol{i}$, the *horizontal component of the total linear momentum, is conserved.* From the velocity diagram, the value of $\boldsymbol{P} \cdot \boldsymbol{i}$ at time t is given by

$$\boldsymbol{P} \cdot \boldsymbol{i} = 3m\dot{x} + m\left(\dot{x} + (a\dot{\theta})\cos\theta\right) = 4m\dot{x} + ma\dot{\theta}\cos\theta.$$

Also, since the motion is started from rest, $\boldsymbol{P} \cdot \boldsymbol{i} = 0$ initially. Hence, **conservation** of $\boldsymbol{P} \cdot \boldsymbol{i}$ implies that

$$4\dot{x} + a\dot{\theta}\cos\theta = 0, \tag{10.38}$$

on cancelling by m. This is our first equation for the subsequent motion.

Since the rail is smooth, the constraint force exerted by the rail does no work and the tensions in the inextensible string do no total work. Hence **energy is conserved.**

From the velocity diagram, the kinetic energy of the system at time t is given by*

$$
\begin{aligned}
T &= \tfrac{1}{2}(3m)\dot{x}^2 + \tfrac{1}{2}m\left(\dot{x}^2 + (a\dot{\theta})^2 + 2\dot{x}(a\dot{\theta})\cos\theta\right)\\
&= \tfrac{1}{2}m\left(4\dot{x}^2 + a^2\dot{\theta}^2 + 2a\dot{x}\dot{\theta}\cos\theta\right).
\end{aligned}
$$

The gravitational potential energy of the system at time t is given by

$$V = 0 - mga\cos\theta.$$

Since the system was released from rest with $\theta = 60°$, the initial value of T is zero, while the initial value of $V = -\tfrac{1}{2}mga$. Hence, **conservation of energy** implies that

$$\tfrac{1}{2}m\left(4\dot{x}^2 + a^2\dot{\theta}^2 + 2a\dot{x}\dot{\theta}\cos\theta\right) - mga\cos\theta = -\tfrac{1}{2}mga,$$

which simplifies to give

$$4\dot{x}^2 + a^2\dot{\theta}^2 + 2a\dot{x}\dot{\theta}\cos\theta = ga\,(2\cos\theta - 1)\,. \tag{10.39}$$

This is our second equation for the subsequent motion.

Since this system has *two* degrees of freedom and satisfies *two* conservation principles, it must be **integrable**. Hence, the conservation equations (10.38), (10.39) must be soluble in the sense described above. ■

Question *Equation for* θ

Deduce an equation satisfied by θ alone and find the speeds of P_1 and P_2 when the string becomes vertical.

Answer

From the linear momentum equation (10.38),

$$\dot{x} = -\tfrac{1}{4}a\dot{\theta}\cos\theta$$

and, if we now eliminate $\dot{x}$ from the energy equation (10.39), we obtain, after simplification,

$$\dot{\theta}^2 = \frac{4g}{a}\left(\frac{2\cos\theta - 1}{4 - \cos^2\theta}\right), \tag{10.40}$$

which is an equation for θ alone.

It follows from this equation that, when the string becomes vertical (that is, when $\theta = 0$), $\dot{\theta}^2 = 4g/3a$. Hence, at this instant, $\dot{\theta} = -(4g/3a)^{1/2}$ and (from the momentum conservation equation) $\dot{x} = +(g/12a)^{1/2}$. Hence, the **speed** of P_1 is $(ag/12)^{1/2}$ and the **speed** of P_2 is $|\dot{x} + a\dot{\theta}| = (3ag/4)^{1/2}$. ■

---

* Suppose a velocity $\boldsymbol{V}$ is the sum of two contributions, $\boldsymbol{v}_1$ and $\boldsymbol{v}_2$, so that $\boldsymbol{V} = \boldsymbol{v}_1 + \boldsymbol{v}_2$. Then

$$|\boldsymbol{V}|^2 = \boldsymbol{V}\cdot\boldsymbol{V} = (\boldsymbol{v}_1 + \boldsymbol{v}_2)\cdot(\boldsymbol{v}_1 + \boldsymbol{v}_2) = \boldsymbol{v}_1\cdot\boldsymbol{v}_1 + \boldsymbol{v}_2\cdot\boldsymbol{v}_2 + 2\boldsymbol{v}_1\cdot\boldsymbol{v}_2 = |\boldsymbol{v}_1|^2 + |\boldsymbol{v}_2|^2 + 2\boldsymbol{v}_1\cdot\boldsymbol{v}_2.$$

This formula was used to find the kinetic energy of particle P_2.

Question ***Period of oscillation***

Find the period of oscillation of the system.

Answer

From the equation (10.40), it follows that the motion is restricted to those values of θ that make the right side *positive*, and that $\dot{\theta} = 0$ when the right side is zero. Hence, θ oscillates periodically in the range $-\pi/3 < \theta < \pi/3$. Consider the first half-oscillation. In this part of the motion, $\dot{\theta} < 0$ and so θ satisfies the equation

$$\dot{\theta} = -\left(\frac{4g}{a}\right)^{1/2}\left(\frac{2\cos\theta - 1}{4 - \cos^2\theta}\right)^{1/2},$$

a first order separable ODE. On separating, we find that τ, the **period** of a full oscillation, is given by

$$\tau = \left(\frac{a}{g}\right)^{1/2}\int_{-\pi/3}^{\pi/3}\left(\frac{4 - \cos^2\theta}{2\cos\theta - 1}\right)^{1/2} \approx 6.23\left(\frac{a}{g}\right)^{1/2}.$$

Thus the determination of $\theta(t)$ has been reduced to an integration, and, with $\theta(t)$ 'known', the equation(10.38) can be solved to give $x(t)$ as an integral. This confirms that the system is **integrable.** ■

Appendix A Modelling bodies by particles

When can a large body, such as a tennis ball, a spacecraft, or the Earth, be modelled by a particle?

The answer commonly given is that '*a body may be modelled by a particle if its size is small compared with the extent of its motion*'. For example, since the radius of the Earth is small compared with the radius of its solar orbit, it is argued that the Earth may be modelled by a particle, at least in respect of its translational motion. This argument sounds reasonable enough, but it is derived only from intuition and, although it often gives the correct answer, it is not the correct condition at all!

We can make some more definite statements on this quite tricky question by using the **centre of mass equation**. This states that '*the centre of mass of any system moves as if it were a particle of mass the total mass, and all the external forces acted upon it*'. It might appear that this principle enables us to predict the motion of the centre of mass of any system, but this is not so. The reason is that, in general, the *total external force acting on a system does not depend solely on the motion of its centre of mass*; it may depend on the positions of the *individual* particles and also other factors such as the particle velocities. Suppose, for example, that the system is a rigid body of *general shape* moving under the gravitational attraction of a fixed mass. Then the total gravitational force acting on the body is only *approximately* given by supposing all the mass to be concentrated at the centre of mass G. The exact force depends on the *orientation* of the body as well as the position of G. The centre of mass equation tells us nothing about this orientation and so the total force on the body is not known and the motion of G cannot be determined.

There are however some important exceptions:

- Consider a **rigid body** moving **without rotation**. In this case the motion of G determines the motion of every particle of the body. Then the total external force on the body is known and the motion of G can be determined. This, in turn, determines the motion of the whole body. For example, the problem of a block sliding without rotation on a table can be completely solved by particle mechanics.
- Consider any system moving solely under **uniform gravity**. In this case, the total external force on the system is a known constant and the motion of G can be determined. This does not however determine the motion of the individual particles. For example, if the system were a brick thrown through the air, then particle mechanics can calculate exactly where its centre of mass will go, but not which particle of the brick will hit the ground first.

In the general case however, we must use approximations. For example, suppose that the particles of the system move in the force fields $\boldsymbol{F}_i(\boldsymbol{r})$ so that the total force on the system is

$$\sum_{i=1}^{N} \boldsymbol{F}_i(\boldsymbol{r}_i).$$

In general, this is not equal to $\sum \boldsymbol{F}_i(\boldsymbol{R})$. We can however *approximate* $\sum \boldsymbol{F}_i(\boldsymbol{r}_i)$ by $\sum \boldsymbol{F}_i(\boldsymbol{R})$, in which case we are assuming that the ratio

$$\frac{|\boldsymbol{F}_i(\boldsymbol{r}_i) - \boldsymbol{F}_i(\boldsymbol{R})|}{|\boldsymbol{F}_i(\boldsymbol{R})|} \ll 1 \qquad (10.41)$$

for all i. In the following argument we investigate when this condition can be expected to be hold.

Let $\boldsymbol{\delta}_i$ be the position vector of the particle P_i of the system *relative to* G. Then

$$\boldsymbol{F}_i(\boldsymbol{R} + \boldsymbol{\delta}_i) - \boldsymbol{F}_i(\boldsymbol{R}) = \left(\left.\frac{d\boldsymbol{F}_i}{ds}\right|_{\boldsymbol{r}=\boldsymbol{R}} \right) |\boldsymbol{\delta}_i| + O\left(|\boldsymbol{\delta}_i|^2\right),$$

where $d\boldsymbol{F}_i/ds$ means the *directional derivative* of $\boldsymbol{F}_i$ in the direction of the displacement $\boldsymbol{\delta}_i$. The condition (10.41) therefore requires that

$$\frac{\left(\left.\frac{d\boldsymbol{F}_i}{ds}\right|_{\boldsymbol{r}=\boldsymbol{R}} \right) |\boldsymbol{\delta}_i|}{|\boldsymbol{F}_i(\boldsymbol{R})|} \ll 1$$

for all i and for all values of $\boldsymbol{R}$ that are attained in the motion of the system. This will hold if

$$\Delta \ll \frac{|\boldsymbol{F}_i(\boldsymbol{R})|}{\max \left(\left.\frac{d\boldsymbol{F}_i}{ds}\right|_{\boldsymbol{r}=\boldsymbol{R}} \right)} \qquad (10.42)$$

for all i and for all points on the path of the centre of mass. Here 'max' means the maximum over all directions, and Δ is the 'radius' of the system (the maximum distance of any particle of the system from the centre of mass). Thus the radius of the system is required to be small compared with the quantities above, not the lateral extent of the motion.

Although the condition (10.42) looks formidable, its physical meaning is quite simple: *the radius of the system is required to be small compared with a length scale over which any of the force fields vary significantly.*

Consider for example a body moving under the gravitational attraction of a mass M_0 which is fixed at the origin O. In this field, the particle P_i of the system moves under the force field

$$\boldsymbol{F}_i(\boldsymbol{r}) = -\frac{m_i M_0 G}{r_i^2}\widehat{\boldsymbol{r}}_i.$$

For this field, the right side of the condition (10.42) evaluates to give $R/2$, where R $(= |\boldsymbol{R}|)$ is the distance of the centre of mass of the body from O. Therefore the total gravitational force on the body will be accurately approximated by the force

$$\boldsymbol{F}(\boldsymbol{R}) = -\frac{M M_0 G}{R^2}\widehat{\boldsymbol{R}}$$

(where M is the total mass of the body), if $\Delta \ll R$ at each point on the path of the centre of mass. This means that the *radius of the body must be small compared with its distance of closest approach to the centre O*. This condition has no direct connection with the 'extent of the motion'. Indeed, on a hyperbolic orbit, the path is infinite, but the condition (10.42) will not hold if the path passes too close to the centre of force. Similar remarks apply to motion of a body in any central field governed by a *power* law. If however the field were that corresponding to the Yukawa potential

$$V = -k\frac{e^{-r/a}}{r},$$

where k, a are positive constants, then Δ is required to be small compared with the length scale a as well as the distance of closest approach to O.

Problems on Chapter 10

Answers and comments are at the end of the book.
Harder problems carry a star (∗).

Linear momentum principle & centre of mass equation

10.1 Show that, if a system moves from one state of rest to another over a certain time interval, then the average of the total external force over this time interval must be zero.

An hourglass of mass M stands on a fixed platform which also measures the apparent weight of the hourglass. The sand is at rest in the upper chamber when, at time $t = 0$, a tiny disturbance causes the sand to start running through. The sand comes to rest in the lower chamber after a time $t = \tau$. Find the time average of the apparent weight of the hourglass over the time interval $[0, \tau]$. [The apparent weight of the hourglass is however *not constant* in time. One can advance an argument that, when the sand is steadily running through, the apparent weight of the hourglass *exceeds* the real weight!]

10.2 Show that, if a system moves periodically, then the average of the total external force over a period of the motion must be zero.

A juggler can juggle heavy balls in a periodic manner. The juggler wishes to cross a shaky bridge that cannot support the combined weight of the juggler and his balls. Would it help if he juggles his balls while he crosses?

10.3* A boat of mass M is at rest in still water and a man of mass m is sitting at the bow. The man stands up, walks to the stern of the boat and then sits down again. If the water offers a resistance to the motion of the boat proportional to the velocity of the boat, show that the boat will *eventually* come to rest at its original position. [This remarkable result is independent of the resistance constant and the details of the man's motion.]

10.4 A uniform rope of mass M and length a is held at rest with its two ends close together and the rope hanging symmetrically below. (In this position, the rope has two long vertical segments connected by a small curved segment at the bottom.) One of the ends is then released. It can be shown by energy conservation (see Problem 9.8) that the velocity of the free end when it has descended by a distance x is given by

$$v^2 = \left(\frac{x(2a - x)}{a - x}\right) g.$$

Find the reaction R exerted by the support at the *fixed* end when the free end has descended a distance x. The support will collapse if R exceeds $\frac{3}{2}Mg$. Find how far the free end will fall before this happens.

10.5 A fine uniform chain of mass M and length a is held at rest hanging vertically downwards with its lower end just touching a fixed horizontal table. The chain is then released. Show that, while the chain is falling, the force that the chain exerts on the table is always *three times* the weight of chain actually lying on the table. [Assume that, before hitting the table, the chain falls freely under gravity.]

* When all the chain has landed on the table, the loose end is pulled upwards with the constant force $\frac{1}{3}Mg$. Find the height to which the chain will first rise. [This time, assume that the force exerted on the chain by the table is *equal* to the weight of chain lying on the table.]

10.6 A uniform ball of mass M and radius a can roll without slipping on the rough outer surface of a fixed sphere of radius b and centre O. Initially the ball is at rest at the highest point of the sphere when it is slightly disturbed. Find the speed of the centre G of the ball in terms of the variable θ, the angle between the line OG and the upward vertical. [Assume planar motion.] Show that the ball will leave the sphere when $\cos\theta = \frac{10}{17}$.

Rocket motion

10.7 A rocket of initial mass M, of which $M - m$ is fuel, burns its fuel at a constant rate in time τ and ejects the exhaust gases with constant speed u. The rocket starts from rest and moves vertically under uniform gravity. Show that the maximum speed achieved by the rocket is $u \ln\gamma - g\tau$ and that its height at burnout is

$$u\tau\left(1 - \frac{\ln\gamma}{\gamma - 1}\right) - \tfrac{1}{2}g\tau^2,$$

where $\gamma = M/m$. [Assume that the thrust is such that the rocket takes off immediately.]

10.8 ***Saturn V rocket*** In first stage of the Saturn V rocket, the initial mass was 2.8×10^6 kg, of which 2.1×10^6 kg was fuel. The fuel was burned at a constant rate over 150 s and the exhaust speed was 2, 600 m s^{-1}. Use the results of the last problem to find the speed and height of the Saturn V at first stage burnout. [Take g to be constant at 9.8 m s^{-2} and neglect air resistance.]

10.9 ***Rocket in resisting medium*** A rocket of initial mass M, of which $M - m$ is fuel, burns its fuel at a constant rate k and ejects the exhaust gases with constant speed u. The rocket starts from rest and moves through a medium that exerts the resistance force $-\epsilon k v$, where v is the forward velocity of the rocket, and ϵ is a small positive constant. Gravity is absent. Find the maximum speed V achieved by the rocket. Deduce a two term approximation for V, valid when ϵ is small.

10.10 ***Two-stage rocket*** A two-stage rocket has a first stage of initial mass M_1, of which $(1 - \eta)M_1$ is fuel, a second stage of initial mass M_2, of which $(1 - \eta)M_2$ is fuel, and an inert payload of mass m_0. In each stage, the exhaust gases are ejected with the same speed u. The rocket is initially at rest in free space. The first stage is fired and, on completion, the first stage carcass (of mass ηM_1) is discarded. The second stage is then fired. Find an expression for the final speed V of the rocket and deduce that V will be maximised when the mass ratio $\alpha = M_2/(M_1 + M_2)$ satisfies the equation

$$\alpha^2 + 2\beta\alpha - \beta = 0,$$

where $\beta = m_0/(M_1 + M_2)$. [Messy algebra.]

Show that, when β is small, the optimum value of α is approximately $\beta^{1/2}$ and the maximum velocity reached is approximately $2u \ln \gamma$, where $\gamma = 1/\eta$.

10.11* A raindrop falls vertically through stationary mist, collecting mass as it falls. The raindrop remains spherical and the rate of mass accretion is proportional to its speed and the square of its radius. Show that, if the drop starts from rest with a negligible radius, then it has constant acceleration $g/7$. [Tricky ODE.]

Collisions

10.12 A body of mass $4m$ is at rest when it explodes into *three* fragments of masses $2m$, m and m. After the explosion the two fragments of mass m are observed to be moving with the same speed in directions making 120° with each other. Find the proportion of the total kinetic energy carried by each fragment.

10.13 Show that, in an elastic head-on collision between two spheres, the relative velocity of the spheres after impact is the negative of the relative velocity before impact.

A tube is fixed in the vertical position with its lower end on a horizontal floor. A ball of mass M is released from rest at the top of the tube followed closely by a second ball of mass m. The first ball bounces off the floor and immediately collides with the second ball coming down. Assuming that both collisions are elastic, show that, when m/M is small, the second ball will be projected upwards to a height nearly nine times the length of the tube.

10.14 Two particles with masses m_1, m_2 and velocities $\boldsymbol{v}_1$, $\boldsymbol{v}_2$ collide and stick together. Find the velocity of this composite particle and show that the loss in kinetic energy due to the collision is

$$\frac{m_1 m_2}{2(m_1+m_2)}\,|\boldsymbol{v}_1-\boldsymbol{v}_2|^2 .$$

10.15 In an elastic collision between a proton and a helium nucleus at rest, the proton was scattered through an angle of 45°. What proportion of its initial energy did it lose? What was the recoil angle of the helium nucleus?

10.16 In an elastic collision between an alpha particle and an unknown nucleus at rest, the alpha particle was deflected through a right angle and lost 40% of its energy. Identify the mystery nucleus.

10.17 ***Some inequalities in elastic collisions*** Use the elastic scattering formulae to show the following inequalities:

(i) When $m_1 > m_2$, the scattering angle θ_1 is restricted to the range $0 \le \theta_1 \le \sin^{-1}(m_2/m_1)$.

(ii) If $m_1 < m_2$, the opening angle is obtuse, while, if $m_1 > m_2$, the opening angle is acute.

(iii)

$$\frac{E_1}{E_0} \ge \left(\frac{m_1-m_2}{m_1+m_2}\right)^2, \qquad \frac{E_2}{E_0} \le \frac{4m_1m_2}{(m_1+m_2)^2}.$$

10.18 ***Equal masses*** Show that, when the particles are of equal mass, the elastic scattering formulae take the simple form

$$\theta_1 = \tfrac{1}{2}\psi \qquad \theta_2 = \tfrac{1}{2}\pi - \tfrac{1}{2}\psi \qquad \theta = \tfrac{1}{2}\pi \qquad \frac{E_1}{E_0} = \cos^2 \tfrac{1}{2}\psi \qquad \frac{E_2}{E_0} = \sin^2 \tfrac{1}{2}\psi$$

where ψ is the scattering angle in the ZM frame.

In the scattering of neutrons of energy E by neutrons at rest, in what directions should the experimenter look to find neutrons of energy $\frac{1}{4}E$?

10.19 Use the elastic scattering formulae to express the energy of the scattered particle as a function of the scattering angle, and the energy of the recoiling particle as a function of the recoil angle, as follows:

$$\frac{E_1}{E_0} = \frac{1+\gamma^2\cos 2\theta_1 + 2\gamma\cos\theta_1\left(1-\gamma^2\sin^2\theta_1\right)^{1/2}}{(\gamma+1)^2}, \qquad \frac{E_2}{E_0} = \frac{4\gamma}{(\gamma+1)^2}\cos^2\theta_2 .$$

Make polar plots of E_1/E_0 as a function of θ_1 for the case of neutrons scattered by the nuclei of hydrogen, deuterium, helium and carbon.

Two-body problem and two-body scattering

10.20 ***Binary star*** The observed period of the binary star Cygnus X-1 (of which only one component is visible) is 5.6 days, and the semi-major axis of the orbit of the visible component is about 0.09 AU. The mass of the visible component is believed to be about $20M_\odot$. Estimate the mass of its dark companion. [Requires the numerical solution of a cubic equation.]

10.21 In two-body elastic scattering, show that the angular distribution of the *recoiling* particles is given by

$$4\cos\theta_2\,\sigma^{ZM}(\pi - 2\theta_2),$$

where $\sigma^{ZM}(\psi)$ is defined by equation (10.32).

In a Rutherford scattering experiment, alpha particles of energy E were scattered by a target of ionised helium. Find the angular distribution of the emerging particles.

10.22* Consider two-body elastic scattering in which the incident particles have energy E_0. Show that the energies of the *recoiling* particles lie in the interval $0 \le E \le E_{\max}$, where $E_{\max} = 4\gamma E_0/(1+\gamma)^2$. Show further that the energies of the recoiling particles are distributed over the interval $0 \le E \le E_{\max}$ by the frequency distribution

$$f(E) = \left(\frac{4\pi}{E_{\max}}\right)\sigma^{ZM}(\psi),$$

where σ^{ZM} is defined by equation (10.32), and

$$\psi = 2\sin^{-1}\left(\frac{E}{E_{\max}}\right)^{1/2}.$$

In the elastic scattering of neutrons of energy E_0 by protons at rest, the energies of the recoiling protons were found to be uniformly distributed over the interval $0 \le E \le E_0$, the total cross section being A. Find the *angular* distribution of the recoiling protons and the scattering cross section of the incident neutrons.

Integrable systems

10.23 A particle Q has mass $2m$ and two other particles P, R, each of mass m, are connected to Q by light inextensible strings of length a. The system is free to move on a smooth horizontal table. Initially P, Q R are at the points $(0, a)$, $(0, 0)$, $(0, -a)$ respectively so that they lie in a straight line with the strings taut. Q is then projected in the positive x-direction with speed u. Express the conservation of linear momentum and energy for this system in terms of the coordinates x (the displacement of Q) and θ (the angle turned by each of the strings).

Show that θ satisfies the equation

$$\dot\theta^2 = \frac{u^2}{a^2}\left(\frac{1}{2-\cos^2\theta}\right)$$

and deduce that P and R will collide after a time

$$\frac{a}{u}\int_0^{\pi/2}\left[2-\cos^2\theta\right]^{\frac{1}{2}}\,d\theta.$$

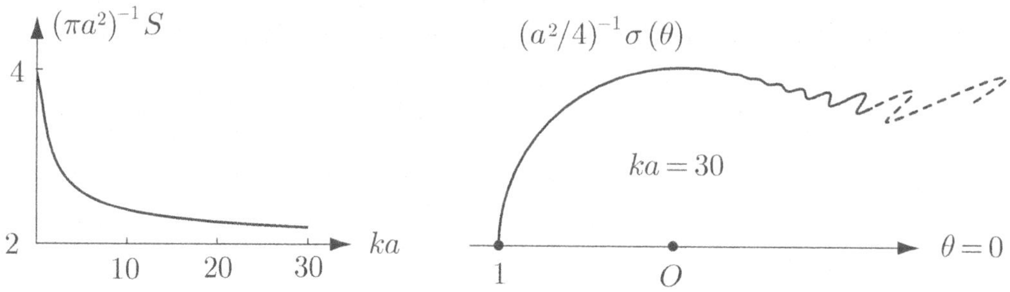

FIGURE 10.11 The quantum mechanical solution of the problem in which a uniform beam of particles, each with momentum $\hbar k$, is scattered by an impenetrable sphere of radius a. **Left**: The (dimensionless) total cross section $(\pi a^2)^{-1} S$ against ka. **Right**: A polar graph of the (dimensionless) scattering cross section $(a^2/4)^{-1}\sigma(\theta)$ against θ when $ka = 30$.

10.24 A uniform rod of length $2a$ has its lower end in contact with a smooth horizontal table. Initially the rod is released from rest in a position making an angle of 60° with the upward vertical. Express the conservation of linear momentum and energy for this system in terms of the coordinates x (the horizontal displacement of the centre of mass of the rod) and θ (the angle between the rod and the upward vertical). Deduce that the centre of mass of the rod moves in a vertical straight line, and that θ satisfies the equation

$$\dot{\theta}^2 = \frac{3g}{a}\left(\frac{1 - 2\cos\theta}{4 - 3\cos^2\theta}\right).$$

Find how long it takes for the rod to hit the table.

Computer assisted problems

10.25 ***Two-body Rutherford scattering*** Calculate the two-body scattering cross section σ^{TB} for Rutherford scattering and obtain the graphs shown in Figure 10.8. Obtain also an approximate formula for σ^{TB} valid for small γ $(= m_1/m_2)$, and correct to order $O(\gamma^2)$.

10.26 ***Comparison with quantum scattering*** A uniform flux of particles is incident upon a fixed hard sphere of radius a. The particles that strike the sphere are reflected elastically. Show that the differential scattering cross section is $\sigma(\theta) = a^2/4$ and that the total cross section is $S = \pi a^2$.

The solution of the same problem given by quantum mechanics is

$$\sigma(\theta) = \frac{a^2}{(ka)^2}\left|\sum_{l=0}^{\infty}\frac{(2l+1)\,j_l(ka)P_l(\cos\theta)}{h_l(ka)}\right|^2, \qquad S = \frac{4\pi a^2}{(ka)^2}\sum_{l=0}^{\infty}\left|\frac{(2l+1)\,j_l(ka)}{h_l(ka)}\right|^2,$$

where $P_l(z)$ is the Legendre polynomial of degree l, and $j_l(z)$, $h_l(z)$ are spherical Bessel functions order l. (Stay cool: these special functions should be available on your computer package.) The parameter k is related to the particle momentum p by the formula $p = \hbar k$, where $\hbar$ is the modified Planck constant. When ka is large, one would expect the quantum mechanical values for $\sigma(\theta)$ and S to approach the classical values. Calculate the quantum mechanical values numerically for ka up to about 30 (the calculation becomes increasingly difficult as ka increases), using about 100 terms of the series.

The author's results are shown in Figure 10.11. The quantum mechanical value for $\sigma(\theta)$ does approach the classical value for larger scattering angles, but behaves very erratically for small scattering angles. Also, the value of S tends to *twice* the value expected! Your physics lecturer will be pleased to explain these interesting anomalies.

Chapter Eleven

The angular momentum principle

and angular momentum conservation

KEY FEATURES

The key features of this chapter are the **angular momentum principle** and **conservation of angular momentum**. Together, the linear and angular momentum principles provide the governing equations of **rigid body motion**.

This chapter is essentially based on the **angular momentum principle** and its consequences. The angular momentum principle is the last of the three great principles of multiparticle mechanics* that apply to *every* mechanical system without restriction. Under appropriate conditions, the angular momentum of a system (or one of its components) is **conserved**, and we use this conservation principle to solve a variety of problems.

Together, the linear and angular momentum principles provide the governing equations of **rigid body motion**; the linear momentum principle determines the *translational* motion of the centre of mass, while the angular momentum principle determines the *rotational* motion of the body relative to the centre of mass. In this chapter, we restrict our attention to the special case of **planar rigid body motion**. Three-dimensional motion of rigid bodies is considered in Chapter 19.

11.1 THE MOMENT OF A FORCE

We begin with the definition of the moment of a force about a *point*, which is a vector quantity. The moment of a force about an *axis*, a scalar quantity, is the component along the axis of the corresponding vector moment.

Definition 11.1 ***Moment of a force about a point*** *Suppose a force* $\boldsymbol{F}$ *acts on a particle* P *with position vector* $\boldsymbol{r}$ *relative to an origin* O. *Then* $\boldsymbol{k}_O$, *the* ***moment***† *of the force* $\boldsymbol{F}$ *about the point* O *is defined to be*

$$\boldsymbol{k}_O = \boldsymbol{r} \times \boldsymbol{F}, \tag{11.1}$$

* The other two are the energy and linear momentum principles.

† Also called **torque**, especially in the engineering literature.

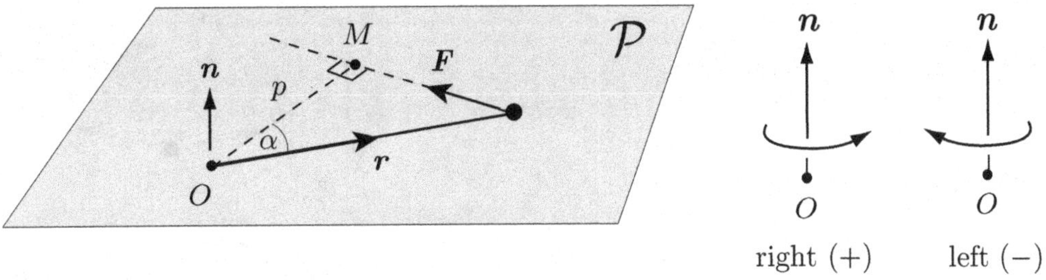

FIGURE 11.1 **Left**: Geometrical interpretation of the vector moment $\boldsymbol{k}_O = \boldsymbol{r}\times\boldsymbol{F}$. **Right**: The right- and left-handed senses around the 'axis' $\{O, \boldsymbol{n}\}$.

a ***vector*** *quantity. If the system of particles $P_1, P_2, \dots, P_N$, with position vectors $\boldsymbol{r}_1, \boldsymbol{r}_2, \dots, \boldsymbol{r}_N$ are acted upon by the system of forces $\boldsymbol{F}_1, \boldsymbol{F}_2, \dots, \boldsymbol{F}_N$ respectively, then $\boldsymbol{K}_O$, the* ***total moment*** *of the system of forces about O is defined to be the* ***vector sum*** *of the moments of the individual forces, that is,*

$$\boldsymbol{K}_O = \sum_{i=1}^{N} \boldsymbol{r}_i \times \boldsymbol{F}_i. \tag{11.2}$$

Since any fixed point can be taken to be the origin O, there is no loss of generality in the above definitions. However, there are occasions on which it is convenient to take moments about a general point A whose position vector is $\boldsymbol{a}$. To find $\boldsymbol{K}_A$, we simply replace $\boldsymbol{r}_i$ in the above definitions by the position vector of P_i *relative to* A, namely, $\boldsymbol{r}_i - \boldsymbol{a}$. This gives

$$\boldsymbol{K}_A = \sum_{i=1}^{N} (\boldsymbol{r}_i - \boldsymbol{a}) \times \boldsymbol{F}_i. \tag{11.3}$$

It follows that $\boldsymbol{K}_A$ and $\boldsymbol{K}_O$ are simply related by

$$\boldsymbol{K}_A = \boldsymbol{K}_O - \boldsymbol{a} \times \boldsymbol{F},$$

where $\boldsymbol{F}$ is the resultant force. Hence, if $\boldsymbol{F}$ is zero, the total moment of the forces $\{\,\boldsymbol{F}_i\,\}$ is the *same about every point*. Such a force system is said to be a **couple** with moment $\boldsymbol{K}$.

Geometrical interpretation of vector moment

The formula (11.1) has a nice geometrical interpretation. Let $\mathcal{P}$ be the plane that contains the origin and the force $\boldsymbol{F}$, as shown in Figure 11.1. Let $\boldsymbol{n}$ be a unit vector normal to $\mathcal{P}$, and suppose that $\boldsymbol{F}$ acts in the right-handed (or positive) sense around the 'axis' $\{O, \boldsymbol{n}\}$. (This is the case shown in Figure 11.1.) Then, from the definition (1.4) of the vector product,

$$\boldsymbol{K}_O = \boldsymbol{r}\times\boldsymbol{F} = \Big(|\boldsymbol{r}|\,|\boldsymbol{F}|\sin(\tfrac{1}{2}\pi + \alpha)\Big)\,\boldsymbol{n} = F(r\cos\alpha)\,\boldsymbol{n} = (F \times p)\,\boldsymbol{n},$$

where F is the magnitude of $\boldsymbol{F}$ and $p\,(= OM)$ is the perpendicular distance of O from the 'line of action' of $\boldsymbol{F}$. Thus, $\boldsymbol{K}_O$ has magnitude $F \times p$ and points in the $\boldsymbol{n}$-direction. If $\boldsymbol{F}$ has the left-handed (or negative) sense around $\{O, \boldsymbol{n}\}$, then $\boldsymbol{K}_O = -(F \times p)\,\boldsymbol{n}$.

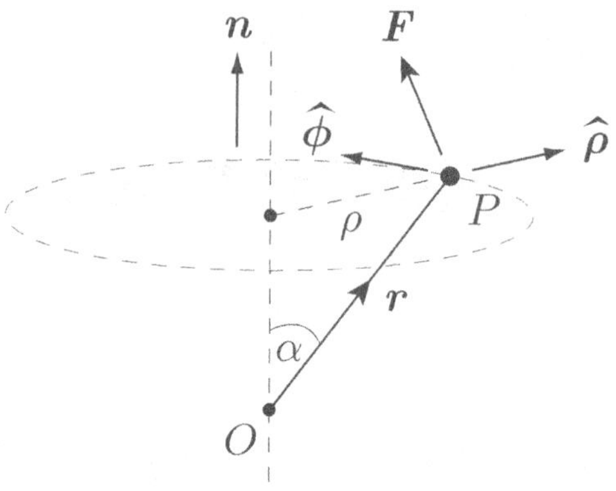

FIGURE 11.2 The moment of the force $\boldsymbol{F}$ about the axis $\{O, \boldsymbol{n}\}$ is $\rho \times (\boldsymbol{F} \cdot \widehat{\boldsymbol{\phi}})$.

Motion in a plane

Suppose we have a **system** of particles that lie in a plane, and the forces acting on the particles also lie in this plane. Such a system is said to be **two-dimensional**. Then the **total moment** $\boldsymbol{K}_O$ of these forces about a point O of the plane is given by

$$\boldsymbol{K}_O = \sum_{i=1}^{N} \pm F_i\, p_i\, \boldsymbol{n} = \left(\sum_{i=1}^{N} \pm F_i\, p_i \right) \boldsymbol{n},$$

where the plus (or minus) sign is taken when the sense of $\boldsymbol{F}_i$ around the axis $\{O, \boldsymbol{n}\}$ is right- (or left-) handed. This formula explains why, in *two-dimensional* mechanics, the moment of a force can be represented by the *scalar* quantity $\pm F \times p$. In the two-dimensional case, the directions of all the moments are parallel, so that they add like scalars. However, in three dimensional mechanics, the moments have general directions and must be summed as vectors.

Moments about an axis

Definition 11.2 ***Moment of a force about an axis*** *The component of the moment* $\boldsymbol{K}_O$ *in the direction of a unit vector* $\boldsymbol{n}$ *is called the* ***moment*** *of* $\boldsymbol{F}$ *about the* ***axis**** $\{O, \boldsymbol{n}\}$; *it is the scalar quantity* $\boldsymbol{K}_O \cdot \boldsymbol{n}$.

This axial moment can be written (see Figure 11.2)

$$\begin{aligned} \boldsymbol{K}_O \cdot \boldsymbol{n} &= (\boldsymbol{r} \times \boldsymbol{F}) \cdot \boldsymbol{n} = (\boldsymbol{n} \times \boldsymbol{r}) \cdot \boldsymbol{F} = \big((r \sin\alpha)\, \widehat{\boldsymbol{\phi}}\big) \cdot \boldsymbol{F} \\ &= \rho\, \big(\boldsymbol{F} \cdot \widehat{\boldsymbol{\phi}}\big), \end{aligned}$$

where ρ is the distance of P from the axis $\{O, \boldsymbol{n}\}$ and ϕ is measured around the axis. The direction of the unit vector $\widehat{\boldsymbol{\phi}}$ is called the **azimuthal** direction around the axis $\{O, \boldsymbol{n}\}$. Thus $\boldsymbol{F} \cdot \widehat{\boldsymbol{\phi}}$ is the **azimuthal component** of $\boldsymbol{F}$.

* This 'axis' is merely a directed line in space. It does not necessarily correspond to the rotation of any rigid body.

Example 11.1 ***Finding moments (numerical example)***

A force $\boldsymbol{F} = 2\boldsymbol{i} - \boldsymbol{j} - 2\boldsymbol{k}$ acts on a particle located at the point $P(0, 3, -1)$. Find the moment of $\boldsymbol{F}$ about the origin O and about the point $A(-2, 4, -3)$. Find also the moment of $\boldsymbol{F}$ about the axis through O in the direction of the vector $3\boldsymbol{i} - 4\boldsymbol{k}$.

Solution

The moment $\boldsymbol{K}_O$ is given by

$$\boldsymbol{K}_O = \boldsymbol{r} \times \boldsymbol{F} = (3\boldsymbol{j} - \boldsymbol{k}) \times (2\boldsymbol{i} - \boldsymbol{j} - 2\boldsymbol{k}) = -7\boldsymbol{i} - 2\boldsymbol{j} - 6\boldsymbol{k}.$$

Similarly,

$$\boldsymbol{K}_A = (\boldsymbol{r} - \boldsymbol{a}) \times \boldsymbol{F} = (2\boldsymbol{i} - \boldsymbol{j} + 2\boldsymbol{k}) \times (2\boldsymbol{i} - \boldsymbol{j} - 2\boldsymbol{k}) = 4\boldsymbol{i} + 8\boldsymbol{j}.$$

The required axial moment is $\boldsymbol{K}_O \cdot \boldsymbol{n}$, where $\boldsymbol{n}$ is the unit vector in the direction of $3\boldsymbol{i} - 4\boldsymbol{k}$, namely

$$\boldsymbol{n} = \frac{3\boldsymbol{i} - 4\boldsymbol{k}}{|3\boldsymbol{i} - 4\boldsymbol{k}|} = \frac{3\boldsymbol{i} - 4\boldsymbol{k}}{5}.$$

Hence

$$\boldsymbol{K}_O \cdot \boldsymbol{n} = (-7\boldsymbol{i} - 2\boldsymbol{j} - 6\boldsymbol{k}) \cdot \left(\frac{3\boldsymbol{i} - 4\boldsymbol{k}}{5}\right) = \frac{3}{5}. \blacksquare$$

Example 11.2 ***Total moment of gravity forces***

A system $\mathcal{S}$ moves under uniform gravity. Show that the total moment of the gravity forces about any point is the same as if all the mass of $\mathcal{S}$ were concentrated at its centre of mass.

Solution

Without losing generality, let the point about which moments are taken be the origin O. Under uniform gravity, $\boldsymbol{F}_i = -m_i g\boldsymbol{k}$, where the unit vector $\boldsymbol{k}$ points vertically upwards, so that

$$\begin{aligned} \boldsymbol{K}_O &= \sum_{i=1}^{N} \boldsymbol{r}_i \times (-m_i g\boldsymbol{k}) = \left(\sum_{i=1}^{N} m_i \boldsymbol{r}_i\right) \times (-g\boldsymbol{k}) = (M\boldsymbol{R}) \times (-g\boldsymbol{k}) \\ &= \boldsymbol{R} \times (-Mg\boldsymbol{k}), \end{aligned}$$

where M is the total mass of $\mathcal{S}$ and $\boldsymbol{R}$ is the position vector of its centre of mass. This is the required result. Note that it is only true for *uniform* gravity. ■

11.2 ANGULAR MOMENTUM

We begin with the definition of the angular momentum of a particle about a fixed point. The old name for angular momentum is 'moment of momentum' and that is exactly what it is - the moment of the linear momentum of the particle about the chosen point.

Definition 11.3 ***Angular momentum about a point*** *Suppose a particle P of mass m has position vector $\boldsymbol{r}$ and velocity $\boldsymbol{v}$. Then $\boldsymbol{l}_O$, the* ***angular momentum*** *of P about O is defined to be*

$$\boldsymbol{l}_O = \boldsymbol{r} \times (m\boldsymbol{v}), \tag{11.4}$$

a ***vector*** *quantity. If the system of particles $P_1, P_2, \dots, P_N$, with masses $m_1, m_2, \dots, m_N$, have position vectors $\boldsymbol{r}_1, \boldsymbol{r}_2, \dots, \boldsymbol{r}_N$ and velocities $\boldsymbol{v}_1, \boldsymbol{v}_2, \dots, \boldsymbol{v}_N$ respectively, then $\boldsymbol{L}_O$, the* ***angular momentum*** *of the system about O, is defined to be the* ***vector sum*** *of the angular momenta of the individual particles, that is,*

$$\boldsymbol{L}_O = \sum_{i=1}^{N} \boldsymbol{r}_i \times (m_i \boldsymbol{v}_i). \tag{11.5}$$

The corresponding formula for angular momentum about a general point A is therefore

$$\boldsymbol{L}_A = \sum_{i=1}^{N} (\boldsymbol{r}_i - \boldsymbol{a}) \times (m_i \boldsymbol{v}_i),$$

from which it follows that $\boldsymbol{L}_A$ and $\boldsymbol{L}_O$ are simply related by

$$\boldsymbol{L}_A = \boldsymbol{L}_O - \boldsymbol{a} \times \boldsymbol{P},$$

where $\boldsymbol{P}$ is the total *linear* momentum of the system.

The geometrical interpretation of the angular momentum of a particle is similar to that of moment of a force (see Figure 11.1). Let $\mathcal{P}$ be the plane that contains O, P and the velocity $\boldsymbol{v}$, and let $\boldsymbol{n}$ be a unit vector normal to $\mathcal{P}$. Then

$$\boldsymbol{L}_O = \pm(mv \times p)\,\boldsymbol{n},$$

where v is the magnitude of $\boldsymbol{v}$ and p is the perpendicular distance of O from the line through P parallel to $\boldsymbol{v}$. The $\pm$ sign is decided by the sense of $\boldsymbol{v}$ around the axis $\{O, \boldsymbol{n}\}$, as shown in Figure 11.1.

Example 11.3 *Calculating the angular momentum of a particle*

The position of a particle P of mass m at time t is given by $x = a\theta^2$, $y = 2a\theta$, $z = 0$, where $\theta = \theta(t)$. Find the angular momentum of P about the point $B(a, 0, 0)$ at time t.

Solution

The position vector of the particle relative to B at time t is

$$\boldsymbol{r} - \boldsymbol{b} = \left(a\theta^2\boldsymbol{i} + 2a\theta\,\boldsymbol{j}\right) - a\boldsymbol{i} = a\left[(\theta^2 - 1)\boldsymbol{i} + 2\theta\,\boldsymbol{j}\right]$$

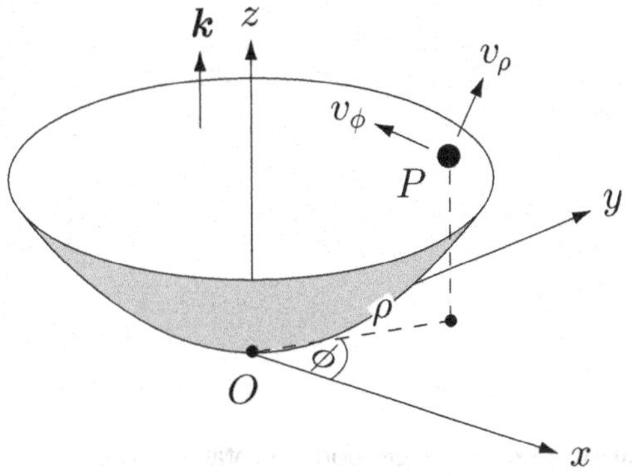

FIGURE 11.3 The particle P slides on the inside surface of the axially symmetric bowl $z = f(\rho)$.

and the velocity of the particle at time t is

$$\boldsymbol{v} = \frac{d\boldsymbol{r}}{dt} = \frac{d\boldsymbol{r}}{d\theta} \times \frac{d\theta}{dt} = 2a(\theta\boldsymbol{i} + \boldsymbol{j})\dot{\theta}.$$

The angular momentum of the particle about B at time t is therefore

$$\begin{aligned} \boldsymbol{L}_B &= (\boldsymbol{r} - \boldsymbol{b}) \times (m\boldsymbol{v}) = 2ma^2\dot{\theta}\left[(\theta^2 - 1)\boldsymbol{i} + 2\theta\,\boldsymbol{j}\right] \times [\theta\boldsymbol{i} + \boldsymbol{j}] \\ &= -2ma^2(\theta^2 + 1)\dot{\theta}\,\boldsymbol{k}. \blacksquare \end{aligned}$$

Angular momentum about an axis

Definition 11.4 ***Angular momentum about an axis*** *The component of the angular momentum* $\boldsymbol{L}_O$ *in the direction of a unit vector* $\boldsymbol{n}$ *is called the* ***angular momentum*** *of* P *about the* ***axis*** $\{O, \boldsymbol{n}\}$*; it is the scalar quantity* $\boldsymbol{L}_O \cdot \boldsymbol{n}$.

By that same argument as was used for moments about an axis, the angular momentum of a particle of mass m and velocity $\boldsymbol{v}$ about the axis $\{O, \boldsymbol{n}\}$ can be written in the form

$$\boldsymbol{L}_O \cdot \boldsymbol{n} = m\rho\left(\boldsymbol{v} \cdot \widehat{\boldsymbol{\phi}}\right), \tag{11.6}$$

where ρ is the perpendicular distance of the particle from the axis and $\boldsymbol{v} \cdot \widehat{\boldsymbol{\phi}}$ is the azimuthal component of $\boldsymbol{v}$ around the axis.

Example 11.4 *Particle sliding inside a bowl*

A particle P of mass m slides on the inside surface of an axially symmetric bowl. Find its angular momentum about its axis of symmetry in terms of the coordinates ρ, ϕ shown in Figure 11.3.

Solution

In order to express $\boldsymbol{L}_O \cdot \boldsymbol{k}$ in terms of coordinates, we draw a velocity diagram for the system as explained in section 10.10. The velocities v_ρ and v_ϕ, corresponding to the coordinates ρ and ϕ, have the directions shown in Figure 11.3. These two velocities are perpendicular, with the v_ϕ contribution in the *azimuthal* direction around the

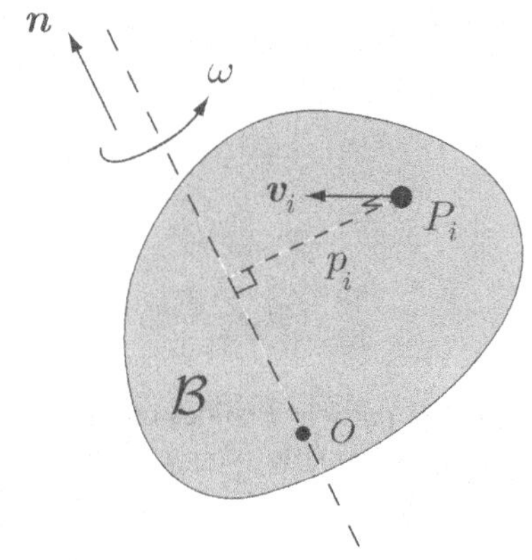

FIGURE 11.4 The rigid body $\mathcal{B}$ rotates about the fixed axis $\{O, \boldsymbol{n}\}$ with angular velocity ω.

vertical axis $\{O, \boldsymbol{k}\}$. It follows that $\boldsymbol{v} \cdot \widehat{\boldsymbol{\phi}} = v_\phi = \rho\dot{\phi}$. The required axial angular momentum is therefore

$$\boldsymbol{L}_O \cdot \boldsymbol{k} = m\rho(\boldsymbol{v} \cdot \widehat{\boldsymbol{\phi}}) = m\rho(\rho\dot{\phi}) = m\rho^2\dot{\phi}.$$

Just for the record, the velocity v_ρ is the (vector) sum of $\dot{\rho}$ *radially outwards* and $\dot{z}$ *vertically upwards*. Note that $\dot{z}$ is not an independent quantity. If the equation of the bowl is $z = f(\rho)$, then $\dot{z} = f'(\rho)\dot{\rho}$. In particular then, the kinetic energy of P is given by

$$T = \tfrac{1}{2}m\left[\dot{\rho}^2 + \left(f'(\rho)\dot{\rho}\right)^2 + \left(\rho\dot{\phi}\right)^2\right].$$

and its potential energy by $V = mgf(\rho)$. ■

11.3 ANGULAR MOMENTUM OF A RIGID BODY

The problem of finding the angular momentum of a moving rigid body in the general three-dimensional case is tricky and is deferred until Chapter 19. In the present chapter we essentially restrict ourselves to the case of planar rigid body motion, for which it is sufficient to find the angular momentum of the body *about its axis of rotation*. This axis may be fixed (as in the armature of a motor) or, more generally, may be the instantaneous rotation axis through the centre of mass of the body (as in the case of a rolling penny). In this section we consider only the case of rotation about a fixed axis; the case of planar motion is treated in section 11.6.

Consider a rigid body $\mathcal{B}$ rotating with angular velocity ω about the *fixed* axis $\{O, \boldsymbol{n}\}$, as shown in Figure 11.4. Then the angular momentum of the body about this axis is

$$\boldsymbol{L}_O \cdot \boldsymbol{n} = \left(\sum_{i=1}^{N} \boldsymbol{l}_O^{(i)}\right) \cdot \boldsymbol{n} = \sum_{i=1}^{N} (\boldsymbol{l}_O^{(i)} \cdot \boldsymbol{n}) = \sum_{i=1}^{N} m_i p_i (\boldsymbol{v}_i \cdot \widehat{\boldsymbol{\phi}}),$$

where p_i is the perpendicular distance of m_i from the axis, and $\boldsymbol{v}_i \cdot \widehat{\boldsymbol{\phi}}$ is the azimuthal component of $\boldsymbol{v}_i$ around the axis (see formula (11.6)). But, since the body is rigid, the velocity of m_i is entirely azimuthal and is equal to ωp_i. Hence

$$\boldsymbol{L}_O \cdot \boldsymbol{n} = \left(\sum_{i=1}^{N} m_i p_i^2 \right) \omega = I\,\omega, \tag{11.7}$$

where I is the moment of inertia of $\mathcal{B}$ about the rotation axis $\{O, \boldsymbol{n}\}$. We have thus proved that:

Angular momentum of a rigid body about its rotation axis

If a rigid body is rotating with angular velocity ω about the fixed axis $\{A, \boldsymbol{n}\}$, then the angular momentum of the body about this axis is given by

$$\boldsymbol{L}_A \cdot \boldsymbol{n} = I\,\omega, \tag{11.8}$$

where I is the moment of inertia of the body about the axis $\{O, \boldsymbol{n}\}$.

It should be remembered that, if a rigid body of general shape is rotating about the fixed axis $\{O, \boldsymbol{n}\}$, then $\boldsymbol{L}_O$, the angular momentum of the body about O, is not generally parallel to the rotation axis. If the rotation axis happens to be an axis of **rotational symmetry** of the body, then $\boldsymbol{L}_O$ *will* be parallel to the rotation axis and $\boldsymbol{L}_O$ is simply given by

$$\boldsymbol{L}_O = (I\,\omega)\boldsymbol{n}. \tag{11.9}$$

Example 11.5 *Axial angular momentum of a hollow sphere*

A hollow sphere of inner radius a and outer radius b is made of material of uniform density ρ. The sphere is spinning with angular velocity Ω about a fixed axis through its centre. Find the angular momentum of the sphere about its rotation axis.

Solution

From equation (SysA:L=Iomega), the angular momentum of the sphere about its rotation axis is given by $L = I\,\omega$, where I is its moment of inertia and ω is its angular velocity about this axis. In the present case,

$$I = \tfrac{2}{5}Mb^2 - \tfrac{2}{5}ma^2,$$

where $M = 4\rho b^3/3$ and $m = 4\rho a^3/3$, giving

$$I = \frac{8\rho}{15}\left(b^5 - a^5\right).$$

The angular momentum of the sphere about its rotation axis is therefore

$$L = \frac{8\rho}{15}\left(b^5 - a^5\right)\Omega. \blacksquare$$

11.4 THE ANGULAR MOMENTUM PRINCIPLE

We now derive the fundamental result which relates the angular momentum of any system to the external forces that act upon it – **the angular momentum principle.**

Consider the general multi-particle system $\mathcal{S}$ which consists of particles $P_1, P_2, \dots, P_N$, with masses $m_1, m_2, \dots, m_N$ and velocities $\boldsymbol{v}_1, \boldsymbol{v}_2, \dots, \boldsymbol{v}_N$, as shown in Figure 9.1. Suppose that $\mathcal{S}$ is acted upon by **external** forces $\boldsymbol{F}_i$ and **internal** forces $\boldsymbol{G}_{ij}$, as shown in Figure 9.3. Then the equation of motion for the particle P_i is

$$m_i \frac{d\boldsymbol{v}_i}{dt} = \boldsymbol{F}_i + \sum_{j=1}^{N} \boldsymbol{G}_{ij}, \tag{11.10}$$

where, as in Chapter 9, we take $\boldsymbol{G}_{ij} = \boldsymbol{0}$ when $i = j$. Then the rate of increase of the angular momentum of the system $\mathcal{S}$ about the origin O can be written

$$\begin{aligned}\frac{d\boldsymbol{L}_O}{dt} &= \frac{d}{dt}\left(\sum_{i=1}^{N} \boldsymbol{r}_i \times (m_i \boldsymbol{v}_i)\right) = \sum_{i=1}^{N} \left\{\boldsymbol{r}_i \times \left(m_i \frac{d\boldsymbol{v}_i}{dt}\right) + \dot{\boldsymbol{r}}_i \times (m_i \boldsymbol{v}_i)\right\}\\ &= \sum_{i=1}^{N} \boldsymbol{r}_i \times \left(m_i \frac{d\boldsymbol{v}_i}{dt}\right),\end{aligned}$$

since $\dot{\boldsymbol{r}}_i \times (m_i \boldsymbol{v}_i) = m_i \boldsymbol{v}_i \times \boldsymbol{v}_i = \boldsymbol{0}$. On using the equation of motion (11.10), we obtain

$$\begin{aligned}\frac{d\boldsymbol{L}_O}{dt} &= \sum_{i=1}^{N} \boldsymbol{r}_i \times \left\{\boldsymbol{F}_i + \sum_{j=1}^{N} \boldsymbol{G}_{ij}\right\} = \sum_{i=1}^{N} \boldsymbol{r}_i \times \boldsymbol{F}_i + \sum_{i=1}^{N}\sum_{j=1}^{N} \boldsymbol{r}_i \times \boldsymbol{G}_{ij}\\ &= \boldsymbol{K}_O + \sum_{i=2}^{N}\left(\sum_{j=1}^{i-1} \left(\boldsymbol{r}_i \times \boldsymbol{G}_{ij} + \boldsymbol{r}_j \times \boldsymbol{G}_{ji}\right)\right),\end{aligned} \tag{11.11}$$

where $\boldsymbol{K}_O$ is the **total moment** about O of the *external* forces. We have also grouped the terms of the double sum in pairs and omitted those terms known to be zero. Now the internal forces $\{\boldsymbol{G}_{ij}\}$ satisfy the Third Law, which means that $\boldsymbol{G}_{ij}$ must be equal and opposite to $\boldsymbol{G}_{ji}$, and that $\boldsymbol{G}_{ij}$ must be parallel to the line $P_i P_j$. It follows that

$$\boldsymbol{r}_i \times \boldsymbol{G}_{ij} + \boldsymbol{r}_j \times \boldsymbol{G}_{ji} = \boldsymbol{r}_i \times \boldsymbol{G}_{ij} - \boldsymbol{r}_j \times \boldsymbol{G}_{ij} = (\boldsymbol{r}_i - \boldsymbol{r}_j) \times \boldsymbol{G}_{ij} = \boldsymbol{0},$$

since $\boldsymbol{G}_{ij}$ is parallel to the vector $\boldsymbol{r}_i - \boldsymbol{r}_j$. Thus each pair of terms of the double sum in equation (11.11) is zero and we obtain

$$\frac{d\boldsymbol{L}_O}{dt} = \boldsymbol{K}_O,$$

which is the **angular momentum principle**. Since *any fixed point* can be taken to be the origin, this proves that:

Angular momentum principle about fixed points

$$\frac{d\boldsymbol{L}_A}{dt} = \boldsymbol{K}_A \tag{11.12}$$

for *any fixed point* A. This fundamental principle can be stated as follows:

Angular momentum principle about a fixed point

In any motion of a system $\mathcal{S}$, the rate of increase of the angular momentum of $\mathcal{S}$ about any fixed point is equal to the total moment about that point of the external forces acting on $\mathcal{S}$.

It should be noted that only the external forces appear in the angular momentum principle so that the **internal forces need not be known**. It is this fact which gives the principle its power.

Question ***Overusing the angular momentum principle***

The angular momentum principle can be applied about any point. Are all the resulting equations independent of each other?

Answer

The short answer is obviously no. The long answer is as follows: From the definitions of $\boldsymbol{K}$ and $\boldsymbol{L}$, we have already shown that

$$\boldsymbol{K}_A = \boldsymbol{K}_O - \boldsymbol{a} \times \boldsymbol{F},$$

and

$$\boldsymbol{L}_A = \boldsymbol{L}_O - \boldsymbol{a} \times \boldsymbol{P},$$

where $\boldsymbol{F}$ is the total force acting on the system $\mathcal{S}$, and $\boldsymbol{P}$ is its linear momentum. It follows that, for any fixed point A,

$$\boldsymbol{K}_A - \dot{\boldsymbol{L}}_A = \left(\boldsymbol{K}_O - \dot{\boldsymbol{L}}_O\right) - \boldsymbol{a} \times \left(\boldsymbol{F} - \dot{\boldsymbol{P}}\right).$$

Hence, if the linear momentum principle $\dot{\boldsymbol{P}} = \boldsymbol{F}$ and the angular momentum principle $\dot{\boldsymbol{L}}_O = \boldsymbol{K}_O$ have already been used, then nothing new is obtained by applying the angular momentum principle about another point A. ■

Angular momentum principle about the centre of mass

The angular momentum principle in the form (11.12) does not generally apply if A is a *moving* point. However, the standard form does apply when moments and angular momenta are taken about the **centre of mass** G, even though G may be accelerating. This follows from the theorem below. The corresponding result for kinetic energy appeared in Chapter 9.

Theorem 11.1 *Suppose a general system of particles $\mathcal{S}$ has total mass M and that its centre of mass G has position vector $\boldsymbol{R}$ and velocity $\boldsymbol{V}$. Then the angular momentum of $\mathcal{S}$ about O can be written in the form*

$$\boldsymbol{L}_O = \boldsymbol{R}\times(M\boldsymbol{V}) + \boldsymbol{L}_G, \tag{11.13}$$

where $\boldsymbol{L}_G$ is the angular momentum of $\mathcal{S}$ about G in its motion **relative** *to G.*

Proof. By definition,

$$\begin{aligned}
\boldsymbol{L}_G &= \sum_{i=1}^{N} m_i\,(\boldsymbol{r}_i - \boldsymbol{R})\times(\boldsymbol{v}_i - \boldsymbol{V})\\
&= \sum_{i=1}^{N} m_i\boldsymbol{r}_i\times\boldsymbol{v}_i - \left(\sum_{i=1}^{N} m_i\boldsymbol{r}_i\right)\times\boldsymbol{V} - \boldsymbol{R}\times\left(\sum_{i=1}^{N} m_i\boldsymbol{v}_i\right) + \left(\sum_{i=1}^{N} m_i\right)\boldsymbol{R}\times\boldsymbol{V}\\
&= \boldsymbol{L}_O - (M\boldsymbol{R})\times\boldsymbol{V} - \boldsymbol{R}\times(M\boldsymbol{V}) + M(\boldsymbol{R}\times\boldsymbol{V})\\
&= \boldsymbol{L}_O - \boldsymbol{R}\times(M\boldsymbol{V}),
\end{aligned}$$

as required. ∎

The two terms on the right of equation (11.13) have a nice physical interpretation. The term $\boldsymbol{R}\times(M\boldsymbol{V})$ is the **translational** contribution to $\boldsymbol{L}_O$ while the term $\boldsymbol{L}_G$ is the contribution from the motion of $\mathcal{S}$ **relative** to G. If $\mathcal{S}$ is a **rigid body**, then the motion of $\mathcal{S}$ relative to G is an angular velocity about some axis through G, and the term $\boldsymbol{L}_G$ then represents the **rotational contribution** to $\boldsymbol{L}_O$.

The angular momentum principle for $\mathcal{S}$ about O can therefore be written

$$\begin{aligned}
\boldsymbol{K}_O &= \frac{d}{dt}(M\boldsymbol{R}\times\boldsymbol{V}) + \frac{d\boldsymbol{L}_G}{dt}\\
&= M\boldsymbol{R}\times\dot{\boldsymbol{V}} + \frac{d\boldsymbol{L}_G}{dt}.
\end{aligned}$$

Furthermore, since $\boldsymbol{K}_O = \boldsymbol{K}_G + \boldsymbol{R}\times\boldsymbol{F}$, it follows that

$$\begin{aligned}
\boldsymbol{K}_G &= \frac{d\boldsymbol{L}_G}{dt} + \boldsymbol{R}\times\left(M\dot{\boldsymbol{V}} - \boldsymbol{F}\right)\\
&= \frac{d\boldsymbol{L}_G}{dt},
\end{aligned}$$

on using the *linear* momentum principle. We therefore obtain:

Angular momentum principle about G

$$\frac{d\boldsymbol{L}_G}{dt} = \boldsymbol{K}_G \tag{11.14}$$

Thus *the standard form of the angular momentum principle applies to the motion of* $\mathcal{S}$ *relative to the centre of mass* G.

The rigid body equations

The linear and angular momentum principles provide sufficient equations to determine the motion of a **single rigid body** moving under **known forces**. The standard form of the **rigid body equations** is

Rigid body equations

$$M\frac{d\boldsymbol{V}}{dt} = \boldsymbol{F} \qquad \frac{d\boldsymbol{L}_G}{dt} = \boldsymbol{K}_G \tag{11.15}$$

in which we have taken both the linear and angular momentum principles in their centre of mass form. The linear momentum principle thus determines the **translational motion** of G (as if it were a particle), and the angular momentum principle determines the **rotational motion** of the body about G.

We will use a subset of these equations later in this chapter to solve problems of **planar** rigid body motion. The delights of general **three-dimensional** rigid body motion* are revealed in Chapter 19.

Example 11.6 *Rigid body moving under uniform gravity*

A rigid body is moving in any manner under uniform gravity. Show that its motion *relative to its centre of mass* is the same as if gravity were absent.

Solution

Under uniform gravity, the total moment of the gravity forces about any point is the same as if they all acted at G, the centre of mass of the body (see Example 11.2). It follows that $\boldsymbol{K}_G = \boldsymbol{0}$.

The rigid body equations (11.15) therefore take the form

$$M\frac{d\boldsymbol{V}}{dt} = -Mg\boldsymbol{k}, \qquad \frac{d\boldsymbol{L}_G}{dt} = \boldsymbol{0}.$$

Hence, when a rigid body moves under uniform gravity, G undergoes projectile motion (which we already knew), and the equation for the motion of the body relative to G is the same as if the body were moving in free space. ■

* The difficulty in the three-dimensional case is the calculation of $\boldsymbol{L}$.

11.5 CONSERVATION OF ANGULAR MOMENTUM

Isolated systems

Suppose that $\mathcal{S}$ is an **isolated** system, and let A be any fixed point. Then $\boldsymbol{L}_A$, the total moment about A of the external forces acting on $\mathcal{S}$, is obviously zero. The angular momentum principle (11.12) then implies that $d\boldsymbol{L}_A/dt = \boldsymbol{0}$, which implies that $\boldsymbol{L}_A$ remains constant. The same argument holds for $\boldsymbol{L}_G$. This simple but important result can be stated as follows:

> **Conservation of angular momentum about a point**
>
> In any motion of an isolated system, the angular momentum of the system about any fixed point is conserved. The angular momentum of the system about its centre of mass is also conserved.

For example, the angular momentum of the **solar system** about any fixed point (or about its centre of mass) is conserved. The same is true for an astronaut floating freely in space (irrespective of how he moves his body). The angular momentum of a system about its centre of mass may still be conserved even when external forces are present. For any system moving under **uniform gravity** (a falling cat trying to land on its feet, say) $\boldsymbol{K}_G = \boldsymbol{0}$ which implies that $\boldsymbol{L}_G$ is conserved.

Angular momentum in central field orbits

In the case of a particle P moving in a **central field** with centre O,

$$\boldsymbol{K}_O = \boldsymbol{r}\times\boldsymbol{F} = \boldsymbol{0},$$

since $\boldsymbol{r}$ and $\boldsymbol{F}$ are parallel. This implies that $\boldsymbol{L}_O$ is conserved. (Angular momentum about other points is not conserved.) By symmetry, each possible motion of P must take place in a plane through O and we may take polar coordinates r, θ (centred on O) to specify the position of P in the plane of motion. In terms of these coordinates,

$$\boldsymbol{L}_O = \boldsymbol{r}\times(m\boldsymbol{v}) = m(r\,\widehat{\boldsymbol{r}})\times\left(\dot{r}\,\widehat{\boldsymbol{r}} + (r\dot{\theta})\,\widehat{\boldsymbol{\theta}}\right) = mr^2\dot{\theta}\,\boldsymbol{n},$$

where the constant unit vector $\boldsymbol{n}$ $(=\widehat{\boldsymbol{r}}\times\widehat{\boldsymbol{\theta}})$ is perpendicular to the plane of motion. Hence, in this case, conservation of $\boldsymbol{L}_O$ is equivalent to conservation of $\boldsymbol{L}_O \cdot \boldsymbol{n}$, the angular momentum of P about the axis $\{O, \boldsymbol{n}\}$. The conclusion is then that the quantity

$$\boldsymbol{L}_O \cdot \boldsymbol{n} = mr^2\dot{\theta} = L,$$

where L is a constant. This important result was obtained in Chapter 7 by integrating the azimuthal equation of motion for P. We now see that it is a consequence of angular momentum conservation and that the constant L is the angular momentum* of P about the axis $\{O, \boldsymbol{n}\}$.

---

* The constant L used in Chapter 7 was actually the angular momentum *per unit mass*.

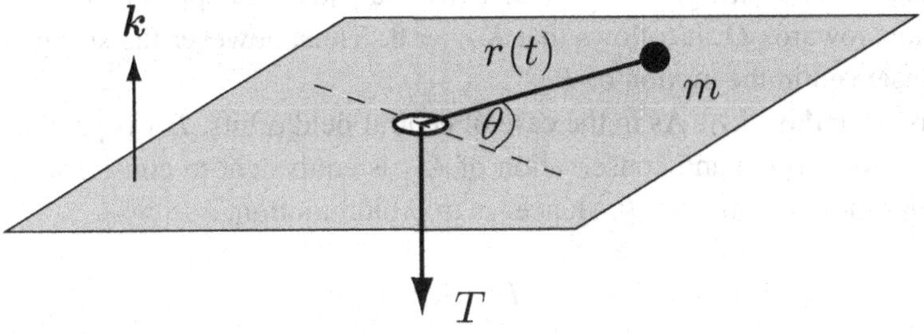

FIGURE 11.5 The particle slides on the table while the string is pulled down through the hole.

Conservation of angular momentum about an axis

Even when $\boldsymbol{K}_A \neq \boldsymbol{0}$ it is still possible for angular momentum to be conserved about a *particular axis* through A. Let $\boldsymbol{n}$ be a fixed unit vector and A a fixed point so that $\{A, \boldsymbol{n}\}$ is a *fixed axis* through the point A. Then

$$\frac{d}{dt}(\boldsymbol{L}_A \cdot \boldsymbol{n}) = \frac{d\boldsymbol{L}_A}{dt} \cdot \boldsymbol{n} + \boldsymbol{L}_A \cdot \frac{d\boldsymbol{n}}{dt} = \frac{d\boldsymbol{L}_A}{dt} \cdot \boldsymbol{n} = \boldsymbol{K}_A \cdot \boldsymbol{n}$$

Hence, if $\boldsymbol{K}_A \cdot \boldsymbol{n} = 0$ at all times, it follows that $\boldsymbol{L}_A \cdot \boldsymbol{n}$ is conserved. This result can be stated as follows:

Conservation of angular momentum about an axis

If the external forces acting on a system have no total moment about a fixed axis, then the angular momentum of the system about that axis is conserved. The same applies for a moving axis which passes through G and maintains a constant direction.

In our first example, angular momentum conservation is sufficient to determine the entire motion.

Example 11.7 *Pulling a particle through a hole*

A particle P of mass m can slide on a smooth horizontal table. P is connected to a light inextensible string which passes through a small smooth hole O in the table, so that the lower end of the string hangs vertically below the table while P moves on top with the string taut (see figure 11.5). Initially the lower end of the string is held fixed with P moving with speed u on a circle of radius a. The string is now pulled down from below in such a way that the string above the table has the length $r(t)$ at time t. Find the velocity of P and the tension in the string at time t.

Solution

We must first establish that some component of angular momentum is conserved in this motion. The forces acting on P are gravity, the normal reaction of the smooth table, and the tension in the string. Since the first two are equal and opposite and the tension force points towards O, it follows that $\boldsymbol{K}_O = \boldsymbol{0}$. Thus, however the string is pulled, $\boldsymbol{L}_O$ is **conserved** in the motion of P.

Now we must **calculate** $\boldsymbol{L}_O$. As in the case of central field orbits, $\boldsymbol{L}_O$ is perpendicular to the plane of motion and conservation of $\boldsymbol{L}_O$ is equivalent to conservation of the axial angular momentum $\boldsymbol{L}_O \cdot \boldsymbol{k}$. Hence, as in orbital motion,

$$\boldsymbol{L}_O \cdot \boldsymbol{k} = mr^2\dot{\theta} = L,$$

where the constant L is given by the initial conditions to be $L = mau$. Hence, in the motion of P, the **conservation equation**

$$mr^2\dot{\theta} = mau$$

is satisfied. Since $r(t)$ is given, this equation is sufficient to determine the motion of P. In particular, the velocity of P at time t is given by

$$\boldsymbol{v} = \dot{r}\,\widehat{\boldsymbol{r}} + (r\dot{\theta})\,\widehat{\boldsymbol{\theta}} = \dot{r}\,\widehat{\boldsymbol{r}} + \left(\frac{au}{r}\right)\widehat{\boldsymbol{\theta}},$$

from which we see that the transverse velocity of P tends to infinity as r tends to zero.

The **string tension** T can be found from the radial equation of motion for P, namely,

$$m\left(\ddot{r} - r\dot{\theta}^2\right) = -T,$$

which gives

$$T = m\left(r\dot{\theta}^2 - \ddot{r}\right) = m\left(\frac{a^2u^2}{r^3} - \ddot{r}\right).$$

For example, in order to pull the string down with constant speed, the applied tension must be

$$T = \frac{ma^2u^2}{r^3}.$$

This tends rapidly to infinity as r tends to zero, making it impossible to pull the particle through the hole! ■

Our second example belongs to a class of problems that could be called '*before* and *after* problems'. We have encountered the same notion before. In elastic collision problems, the linear momentum and energy of the system are conserved and these conservation laws are used to relate the initial state of the system (*before*) to the final state (*after*). This provides information about the final state that is independent of the nature of the particle interaction. Conservation of **angular momentum** can be exploited in the same way. In the following example, angular momentum conservation is sufficient to determine the final state uniquely.

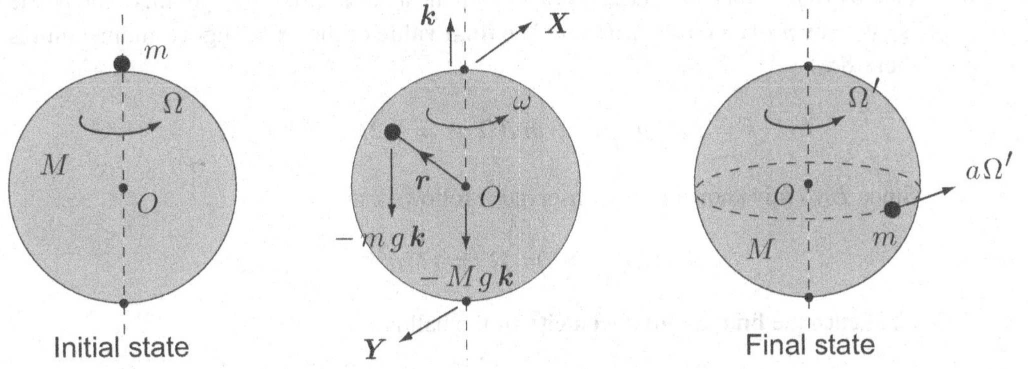

FIGURE 11.6 The beetle and the ball: the ball is smoothly pivoted about a vertical diameter and the beetle crawls on the surface of the ball.

Example 11.8 *The beetle and the ball*

A uniform ball of mass M and radius a is pivoted so that it can turn freely about one of its diameters which is fixed in a vertical position. A beetle of mass m can crawl on the surface of the ball. Initially the ball is rotating with angular speed Ω with the beetle at the 'North pole' (see Figure 11.6 (left)). The beetle then walks (in any manner) to the 'equator' of the ball and sits down. What is the angular speed of the ball now?

Solution

We must first establish that some component of angular momentum is conserved. The external forces acting on the system of 'beetle and ball' are shown in Figure 11.6 (centre). The forces $\boldsymbol{X}$ and $\boldsymbol{Y}$ are the constraint forces exerted by the pivots. The total moment of the external forces about O is therefore

$$\boldsymbol{K}_O = \boldsymbol{0}\times(-Mg\boldsymbol{k}) + \boldsymbol{r}\times(-mg\boldsymbol{k}) + (a\boldsymbol{k})\times\boldsymbol{X} + (-a\boldsymbol{k})\times\boldsymbol{Y}.$$

It follows that

$$\boldsymbol{K}_O \cdot \boldsymbol{k} = 0,$$

since all the resulting triple scalar products contain two $\boldsymbol{k}$'s. Hence $\boldsymbol{L}_O \cdot \boldsymbol{k}$, the angular momentum of the system about the rotation axis, is conserved, irrespective of the wanderings of the beetle. It follows that this axial angular momentum is the same *after* as it was *before*.

In the **initial** state, the angular momentum of the ball about its rotation axis is given by $I\,\Omega$, where $I = 2Ma^2/5$. Initially the beetle has zero velocity so its angular momentum is zero. Hence the **initial value** of the axial angular momentum is

$$\boldsymbol{L}_O \cdot \boldsymbol{k} = \left(\tfrac{2}{5}Ma^2\right)\Omega.$$

In the **final** state the ball has an unknown angular velocity Ω' and axial angular momentum $2Ma^2\Omega'/5$. The velocity of the beetle is entirely azimuthal and is equal to $\Omega' a$. Hence, on using the formula (11.6), the axial angular momentum of the beetle is given by $m\rho(\boldsymbol{v} \cdot \widehat{\boldsymbol{\phi}}) = ma(\Omega a)$. The **final value** of the axial angular momentum is therefore

$$\boldsymbol{L}_O \cdot \boldsymbol{k} = \left(\tfrac{2}{5}Ma^2\right)\Omega' + ma(\Omega' a) = \tfrac{1}{5}(2M + 5m)a^2\Omega'.$$

Since $\boldsymbol{L}_O \cdot \boldsymbol{k}$ is known to be conserved it follows that

$$\tfrac{1}{5}(2M + 5m)a^2\Omega' = \tfrac{2}{5}Ma^2\Omega,$$

and hence the **final angular velocity** of the ball is

$$\Omega' = \left(\frac{2M}{2M + 5m}\right)\Omega. \blacksquare$$

Question ***Change in kinetic energy***

Find the change in kinetic energy of the system caused by the beetle's journey.

Answer

The initial and final kinetic energies of the system are

$$\tfrac{1}{2}\left(\tfrac{2}{5}Ma^2\right)\Omega^2 \quad \text{and} \quad \tfrac{1}{2}\left(\tfrac{2}{5}Ma^2\right)\Omega'^2 + \tfrac{1}{2}m(a\Omega')^2$$

respectively. On using the value of Ω' found above and simplifying, the **kinetic energy** of the system is found to *decrease* by

$$\frac{mMa^2\Omega^2}{2M + 5m}. \blacksquare$$

Question ***Red hot beetle***

Does this loss of energy mean that the beetle arrives in a red-hot condition?

Answer

Your mechanics lecturer will be pleased to answer this question. ■

Our last example, the spherical pendulum, is a system with two degrees of freedom. By using both angular momentum and energy conservation, a complete solution can be found.

Example 11.9 *The spherical pendulum: an integrable system*

A particle P of mass m is suspended from a fixed point O by a light inextensible string of length a and moves with the string taut in three-dimensional space (the spherical pendulum). Show that angular momentum about the vertical axis through O is conserved and express this conservation law in terms of the generalised coordinates θ, ϕ, as shown in Figure 11.7. Obtain also the corresponding equation for conservation of energy.

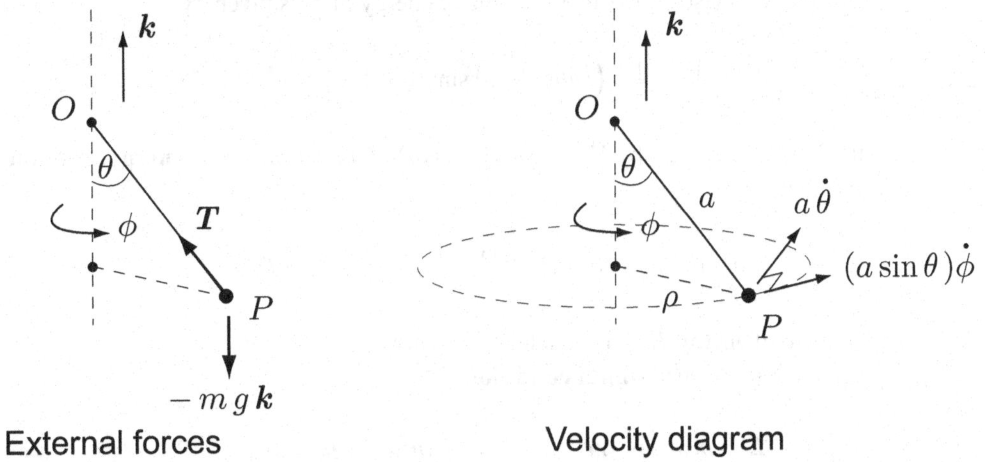

FIGURE 11.7 The spherical pendulum with generalised coordinates θ and ϕ. **Left**: the external forces. **Right**: the velocity diagram.

Initially the string makes an acute angle α with the downward vertical and the particle is projected with speed u in a horizontal direction at right angles to the string. Determine the constants of the motion, and deduce an equation satisfied by $\theta(t)$ in the subsequent motion.

Solution

The external forces on the particle are gravity and the tension in the string (see Figure 11.7 (left)). Hence,

$$\boldsymbol{K}_O = \boldsymbol{r}\times(-mg\boldsymbol{k}) + \boldsymbol{r}\times\boldsymbol{T} = -mg\boldsymbol{r}\times\boldsymbol{k},$$

the second term being zero since $\boldsymbol{r}$ and $\boldsymbol{T}$ are parallel. It follows that

$$\boldsymbol{K}_O \cdot \boldsymbol{k} = -mg(\boldsymbol{r}\times\boldsymbol{k})\cdot\boldsymbol{k} = 0,$$

since the triple scalar product has two $\boldsymbol{k}$'s. Hence $\boldsymbol{L}_O \cdot \boldsymbol{k}$ is **conserved**.

In order to express this conservation law in terms of coordinates, we draw a velocity diagram for the system as explained in section 10.10. The velocities corresponding to the coordinates θ and ϕ are $a\dot{\theta}$ and $\rho\dot{\phi}$ $(= (a\sin\theta)\dot{\phi})$ respectively in the directions shown in Figure 11.7 (right). These two velocities are perpendicular, with the $(a\sin\theta)\dot{\phi}$ contribution in the *azimuthal* direction around the vertical axis $\{O, \boldsymbol{k}\}$. It follows that $\boldsymbol{v}\cdot\widehat{\boldsymbol{\phi}} = (a\sin\theta)\dot{\phi}$. The required axial angular momentum is therefore

$$\boldsymbol{L}_O \cdot \boldsymbol{k} = m\rho(\boldsymbol{v}\cdot\widehat{\boldsymbol{\phi}}) = m(a\sin\theta)(a\sin\theta\,\dot{\phi}) = ma^2\sin^2\theta\,\dot{\phi}$$

and **conservation** of $\boldsymbol{L}_O \cdot \boldsymbol{k}$ is expressed by

$$ma^2\sin^2\theta\,\dot{\phi} = L,$$

where the axial angular momentum L is a constant of the motion.

The diagram also shows that the kinetic energy of P is given by

$$\tfrac{1}{2}m\left((a\dot{\theta})^2 + (a\sin\theta\,\dot{\phi})^2\right)$$

and the potential energy by $V = -mg(a\cos\theta)$. **Conservation of energy** therefore requires that

$$\tfrac{1}{2}m\left((a\dot{\theta})^2 + (a\sin\theta\,\dot{\phi})^2\right) - mg(a\cos\theta) = E,$$

where the total energy E is a constant of the motion.

From the prescribed **initial conditions,**

$$L = m(a\sin\alpha)u, \qquad E = \tfrac{1}{2}mu^2 - mga\cos\alpha,$$

so that the subsequent motion of P satisfies the **conservation equations**

$$ma^2\sin^2\theta\,\dot{\phi} = ma\sin\alpha\,u, \tag{11.16}$$

$$\tfrac{1}{2}m\left(a^2\dot{\theta}^2 + a^2\sin^2\theta\,\dot{\phi}^2\right) - mga\cos\theta = \tfrac{1}{2}mu^2 - mga\cos\alpha. \tag{11.17}$$

Since the spherical pendulum has two degrees of freedom, these two conservation equations are **sufficient to determine the motion**. Moreover, the system is **integrable** (see section (10.10) so that it must be possible to reduce the solution of the problem to integrations.

The equations (11.16), (11.17) are a pair of *coupled* first order ODEs for the unknown functions $\theta(t)$, $\phi(t)$. However, because the coordinate ϕ only appears as $\dot{\phi}$ in both equations, ϕ can be eliminated (θ can not!). From equation (11.16) we have

$$\dot{\phi} = \frac{u\sin\alpha}{a\sin^2\theta} \tag{11.18}$$

and this can now be substituted into equation (11.17) to obtain an equation for $\theta(t)$ alone. After some algebra we find that

$$\dot{\theta}^2 = \frac{u^2}{a^2}(\cos\alpha - \cos\theta)\left(\frac{\cos\alpha + \cos\theta}{\sin^2\theta} - \frac{2ag}{u^2}\right), \tag{11.19}$$

which is the **required equation** satisfied by $\theta(t)$. On taking square roots, this equation becomes a first order *separable* ODE whose solution can be written as an integral. Now that $\theta(t)$ is 'known', $\phi(t)$ can be found (as another integral) from equation (11.18). Thus, as predicted, the solution has thus been reduced to integrations. ■

Question *Form of the motion*

This is all very well, but what does the motion actually look like?

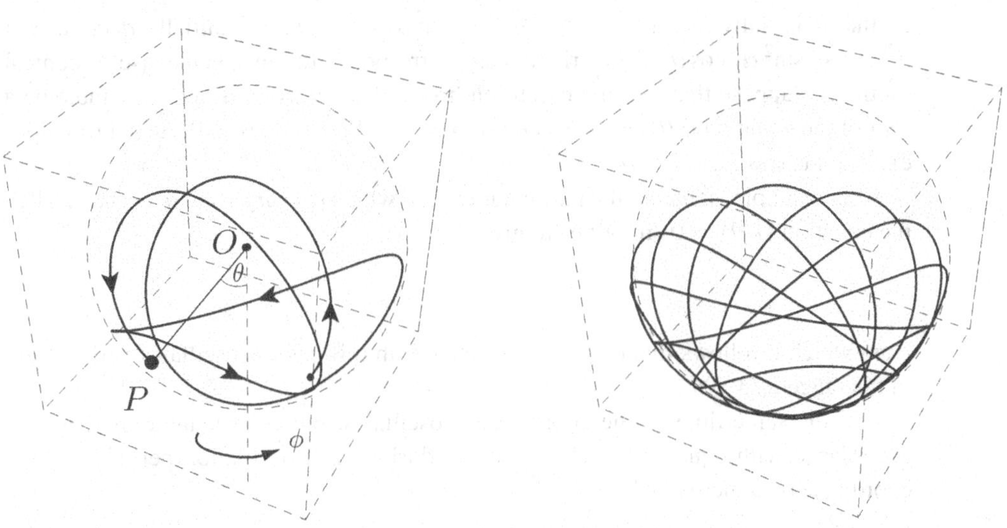

FIGURE 11.8 The calculated path of the spherical pendulum for the case $\alpha = \pi/6$ and $u^2/ag = 1.9$. **Left**: After four oscillations of θ. **Right**: After ten oscillations of θ. The surrounding boxes show the perspective.

Answer

Despite the problem being called integrable, the integrals arising from the separation procedure cannot be evaluated and no explicit solution is possible. However, equation (11.19) has the form of an energy equation for a system with one degree of freedom. We have met this situation before with the radial motion equation in orbit theory and the deductions we can make are the same. Because the left side of (11.19) is positive, it follows that the motion is restricted to those values of θ that make the function

$$F = (\cos\alpha - \cos\theta)\left(\frac{\cos\alpha + \cos\theta}{\sin^2\theta} - \frac{2ag}{u^2}\right)$$

positive. Moreover, maximum and minimum values of θ can only occur when $F(\theta) = 0$.

Since $F(\alpha) = 0$, $\theta = \alpha$ is one extremum* and any other extremum must be a root of the equation $G(\theta) = 0$, where

$$G = \frac{\cos\alpha + \cos\theta}{\sin^2\theta} - \frac{2ag}{u^2}.$$

Whether α is a maximum or minimum point of θ depends on the value of the initial projection speed u. On differentiating equation (11.19) with respect to t, we find that the initial value of $\ddot{\theta}$ is given by

$$\ddot{\theta}\big|_{\theta=\alpha} = \frac{u^2}{a^2}\left(\frac{\cos\alpha}{\sin^2\alpha} - \frac{ag}{u^2}\right),$$

* This is because of the form of the initial conditions.

so that θ initially *increases* if $u^2/ag > \sin^2\alpha/\cos\alpha$, and θ initially *decreases* if $u^2/ag < \sin^2\alpha/\cos\alpha$. (The critical case corresponds to the special case of conical motion.) Suppose that the first condition holds. Then $\theta_{min} = \alpha$ and θ_{max} must be a root of the equation $G(\theta) = 0$. Since $G(\alpha) > 0$ and $G(\pi - \alpha) < 0$, such a root does exist and is less than $\pi - \alpha$.

For example, consider the particular case in which $\alpha = \pi/3$ and $u^2 = 4ag$. Then the equation $G(\theta) = 0$ simplifies to give

$$\cos\theta(\cos\theta + 2) = 0,$$

from which it follows that $\theta_{max} = \pi/2$. Hence, in this case, θ oscillates periodically in the range $\pi/3 \leq \theta \leq \pi/2$.

At the same time as the coordinate θ oscillates, the coordinate ϕ increases in accordance with equation (11.18). Hence, during each oscillation period τ of the coordinate θ, ϕ increases by

$$\Phi = \frac{u\sin\alpha}{a}\int_0^\tau \frac{dt}{\sin^2\theta}.$$

This *pattern* of motion repeats itself with period τ, but the motion is only truly periodic if it eventually links up with itself; this occurs only when the initial conditions are such that Φ/π is a *rational number*.

Figure 11.8 shows an actual path of the spherical pendulum, corresponding to the initial conditions $\alpha = \pi/6$ and $u^2/g = 1.9$. The results are entirely consistent with the theory above. ■

11.6 PLANAR RIGID BODY MOTION

What is planar motion?

Planar motion is a generalisation of two-dimensional motion in which two-dimensional methods are still valid. The complications of the full three-dimensional theory (represented by the two *vector* equations (11.15)) melt away to leave three *scalar* equations, which often have a very simple form. This enables a variety of fascinating problems to be solved in simple closed form. Planar rigid body motion is good value for money!

A system is said to be in **planar motion** if each of its particles moves in a plane and all of these planes are parallel to a fixed plane $\mathcal{P}$ called the **plane of motion**. For example, any *purely translational* motion of a rigid body in which G moves in the plane $\mathcal{P}$ is a planar motion, as is any *purely rotational* motion about an axis through G that is perpendicular to $\mathcal{P}$. The same is true when both of these motions are present together. For example, a cylinder (of any cross-section) rolling down a rough inclined plane is in planar motion. It is not necessary for the bodies that make up our system to be cylinders, nor even to be bodies of revolution. We merely suppose that *each constituent of the system should have reflective symmetry in the plane of motion*,* as shown in Figure 11.9.

* The reason for this symmetry restriction is that, if a body of completely general shape (a potato, say) were started in planar motion and moved under realistic forces (uniform gravity, say), then *the motion would*

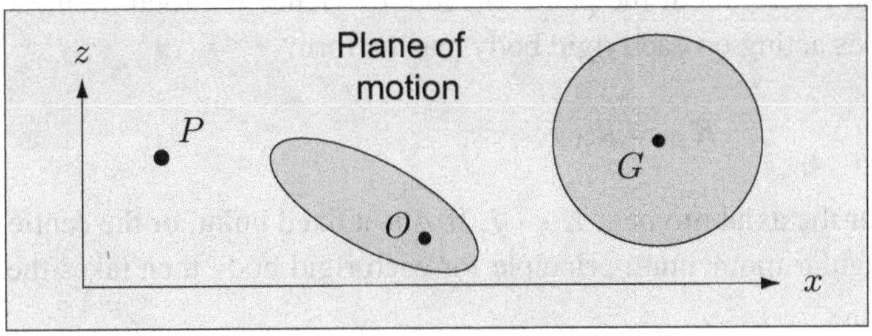

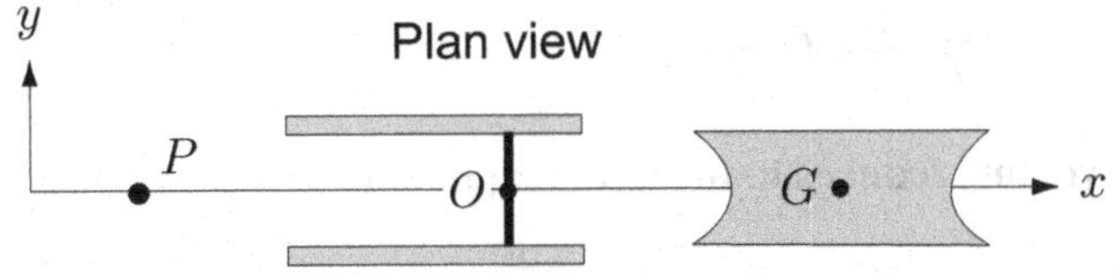

FIGURE 11.9 Three typical elements of a system in planar motion. The particle P moves in the plane of motion $y = 0$; the elliptical crank rotates about the fixed axis $\{O, \boldsymbol{j}\}$; and the circular pulley is in general planar motion. In the last case, G moves in the plane of motion, and the pulley also rotates about the axis $\{G, \boldsymbol{j}\}$.

Question ***Bodies in planar motion***

Decide whether the following rigid bodies are in planar motion: (i) a cotton reel rolling on a table, (ii) the Earth, and (iii) a snooker ball rolling after being struck with 'side'. [If you don't know what this means, get a player to show you.]

Answer

(i) Yes. (ii) No, because the Earth's rotation axis is not perpendicular to the plane of its orbit. (iii) No, because the rotation axis of the ball is not horizontal when the ball is struck with 'side'. ■

Angular momentum in planar motion

The fact that makes planar motion so special is that the total **angular momentum** of each rigid body in the system has a **constant direction** normal to the plane of motion. This follows from the reflective symmetry that each body has in the plane of motion. In other words, if A is any point lying in the plane of motion, then the angular momentum of each rigid body about A has the simple form

$$\boldsymbol{L}_A = L_A \, \boldsymbol{j},$$

where, as shown in Figure 11.9, the plane of motion has been taken to be $y = 0$, and L_A is a short form for the axial angular momentum $\boldsymbol{L}_A \cdot \boldsymbol{j}$. Similar remarks apply to the total moment about A of the external forces acting on each rigid body. This follows from the

not remain planar. However, if the system and the external forces have reflective symmetry in the plane of motion, then a motion that is initially planar will remain planar.

supposed symmetry of these forces about the plane of motion. Hence the total moment about A of the external forces acting on each rigid body has the form

$$\boldsymbol{K}_A = K_A\,\boldsymbol{j},$$

where K_A is a short form for the axial moment $\boldsymbol{K}_A \cdot \boldsymbol{j}$. If A is a fixed point, or the centre of mass of the body, the angular momentum principle for each rigid body then takes the form

$$\frac{d}{dt}(L_A\,\boldsymbol{j}) = K_A\,\boldsymbol{j}$$

which, **since $\boldsymbol{j}$ is a constant vector**, reduces to the scalar equation

$$\frac{dL_A}{dt} = K_A.$$

Since the axis of rotation of the body in its motion relative to A is also normal to the plane of motion, it follows from equation (11.8) that

$$L_A = \boldsymbol{L}_A \cdot \boldsymbol{j} = I_A\,\omega,$$

where ω is the angular velocity of the body, and I_A its moment of inertia, about the axis $\{A, \boldsymbol{j}\}$. We therefore obtain the planar angular momentum principle in the form

$$\frac{d}{dt}(I_A\,\omega) = K_A, \qquad (11.20)$$

where A is either some fixed point in the plane of motion, or the centre of mass of the body. In applications, the moment of inertia I_A is usually constant.

Planar rigid body equations

We are now in a position to reduce the full rigid body equations (11.15) to planar form. Since there is no motion in the $\boldsymbol{j}$-direction, only the $\boldsymbol{i}$- and $\boldsymbol{k}$-components of the linear momentum principle survive, and, as we showed in the last section, only the $\boldsymbol{j}$-component of the angular momentum principle survives. Thus, each rigid body in planar motion satisfies the *three scalar equations of motion*:

Planar rigid body equations

$$M\frac{dV_x}{dt} = F_x \qquad M\frac{dV_z}{dt} = F_z \qquad I_G\frac{d\omega}{dt} = K_G \qquad (11.21)$$

where we have taken the angular momentum principle about the centre of mass G; I_G is then constant. These are the **planar rigid body equations**.

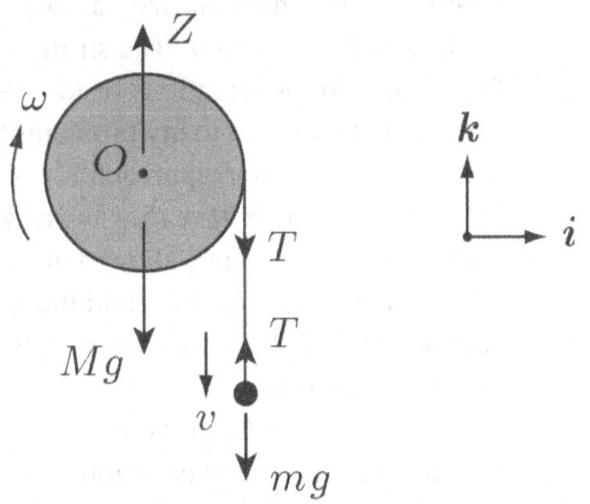

FIGURE 11.10 The pulley rotates about the fixed horizontal axis $\{O, \boldsymbol{j}\}$ and the suspended particle moves vertically.

A special case arises when the body is rotating about a *fixed* axis like the elliptical crank shown in Figure 11.9. Although the equations (11.21) could still be used, this is not the quickest way. Let the fixed axis be $\{O, \boldsymbol{j}\}$, where O lies in the plane of motion (see Figure 11.9). If the angular momentum principle is now applied about O (instead of G), then the unknown reactions exerted by the pivots make no contribution to K_O and *do not appear* in the third equation of (11.21). The first two equations in (11.21) serve only to determine these reactions, once the motion has been calculated. Hence, unless the pivot reactions are actually required, it is sufficient to use the single equation:

Rigid body equation – fixed axis

$$I_O \frac{d\omega}{dt} = K_O \tag{11.22}$$

In this case also, the moment of inertia I_O is constant.

Example 11.10 *Planar motion: mass hanging from a pulley*

A circular pulley of mass M and radius a is smoothly pivoted about the axis $\{O, \boldsymbol{j}\}$, as shown in Figure 11.10. A light inextensible string is wrapped round the pulley so that it does not slip, and a particle of mass m is suspended from the free end. The system undergoes planar motion with the particle moving vertically. Find the downward acceleration of the particle.

Solution

This problem is most easily solved by using energy conservation, but it is instructive to solve it as a planar motion problem to illustrate the difference between the two approaches. In the energy conservation approach, the *whole system* is considered to be a single entity, and in this case, the string tensions do no *total* work and need not be considered. In the setting of planar motion however, the system consists of **two elements**, (i) a **particle** moving in a vertical straight line, and (ii) a **rigid pulley** rotating about a fixed horizontal axis. For each constituent, the tension force exerted by the string is **external**. The string tensions (which are equal in this problem) therefore appear in the equations of motion. For this reason, the conservation method is simpler, but there are many problems where there is no useful conservation principle and the planar motion approach is essential.

Let the particle have downward vertical velocity v, and the pulley have angular velocity ω (in the sense shown) at time t. Since the vector $\boldsymbol{j}$ points *into* the page, this is the positive sense around the axis $\{O, \boldsymbol{j}\}$. Thus, in the notation used here, *the* ***positive sense*** *for angular velocity is* ***clockwise****. The same applies to moments and angular momenta.*

First consider the motion of the **particle**. Since the motion is in a vertical straight line, the only surviving equation is

$$m\frac{dv}{dt} = mg - T,$$

where T is the string tension at time t.

Now consider the motion of the **pulley**. Since this is rotating about a *fixed* axis, the equation of motion is $I_O d\omega/dt = K_O$, that is,

$$I_O\frac{d\omega}{dt} = aT,$$

since the weight force Mg and the pivot reaction Z have zero moment about O.

Hence, the unknown tension T can be eliminated to give

$$\left(ma\frac{dv}{dt} + I_O\frac{d\omega}{dt}\right) = mga.$$

This equation applies whether or not the string slips on the pulley. However, since we are given that the string does not slip, v and ω must be related by the **no-slip condition** $v = \omega a$. On using this condition in the last equation, we obtain

$$\frac{dv}{dt} = \left(\frac{ma^2}{ma^2 + I_O}\right) g.$$

This is the downward **acceleration** of the particle. If the moment of inertia of the pulley is $\frac{1}{2}Ma^2$, then the value of this acceleration is $[2m/(2m + M)]g$. ■

Our next example, the cotton reel problem, is a famous problem in planar mechanics. The mathematics is elementary, but the solution needs to be interpreted carefully.

Example 11.11 *The cotton reel problem*

A cotton reel is at rest on a rough horizontal table when the free end of the thread is pulled horizontally with a constant force T, as shown in Figure 11.11. Given that the reel undergoes planar motion,* how does it move?

* In practice, it is impossible to maintain planar motion in the problem as described (try it). However, the problem is the same if the thread is replaced by a broad flat tape for which planar motion is easier to achieve.

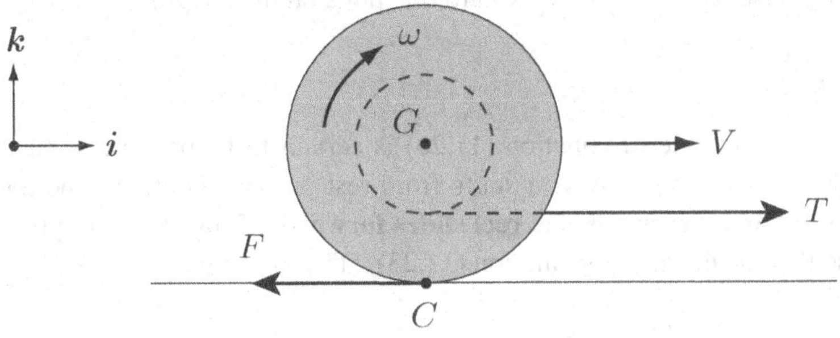

FIGURE 11.11 The reel is initially at rest on a rough horizontal table, when the free end of the thread is pulled with a constant force.

Solution

Suppose the ends of the reel have radius a, the axle wound with thread has radius b (with $b < a$), and the whole reel together with its thread has mass M. (We will neglect the mass of any thread pulled from the reel.) Let the reel have horizontal velocity V and angular velocity ω in the directions shown at time t. Note that we are not presuming the variables V, ω (or F) must take positive values. The signs of these variables will be deduced in the course of solving the problem.

The external forces on the reel are the string tension T and the friction force F at the table. (The weight force and the normal reaction at the table cancel.) Hence, the equations of motion for the reel are

$$M\frac{dV}{dt} = T - F, \tag{11.23}$$

$$Mk^2\frac{d\omega}{dt} = aF - bT, \tag{11.24}$$

where Mk^2 is the moment of inertia of the reel about $\{G, \boldsymbol{j}\}$. We are not making any prior assumption about whether the reel slides or rolls. We will simply presume that the friction force F is bounded in magnitude by some maximum $F^{\max}$, that is,

$$-F^{\max} \le F \le F^{\max}, \tag{11.25}$$

and that $F = +F^{\max}$ (or $-F^{\max}$) when the reel is sliding forwards (or backwards).

Whether the reel slides or rolls depends on v^C, the velocity of the contact particle C. Since $v^C = V - \omega a$, it follows by manipulating the equations (11.23), (11.24) that v^C satisfies the equation

$$\left(\frac{Mk^2}{k^2 + ab}\right)\frac{dv^C}{dt} = T - \gamma F, \tag{11.26}$$

where the constant γ is given by

$$\gamma = \frac{k^2 + a^2}{k^2 + ab}. \tag{11.27}$$

Different cases arise depending on how hard one pulls on the thread.

Strong pull $T > \gamma F^{\max}$

In this case, the right side of equation (11.26) is certain to be positive so that $dv^C/dt > 0$ for all t. Since the system starts from rest, $v^C = 0$ initially and so $v^C > 0$ for all $t > 0$. In other words, the **reel slides forwards**. This in turn implies that $F = F^{\max}$ so that the equations of motion (11.23), (11.24) become

$$\frac{dV}{dt} = \frac{T - F^{\max}}{M},$$
$$\frac{d\omega}{dt} = \frac{aF^{\max} - bT}{Mk^2}.$$

These equations imply that the reel slides forwards with **constant acceleration** and **constant angular acceleration**. Note that ω is positive for $\gamma F^{\max} < T < (a/b)F^{\max}$ and negative for $T > (a/b)F^{\max}$.

Gentle pull $T < \gamma F^{\max}$

In this case, the reel must **roll**. The proof of this is by contradiction, as follows.

Suppose that the reel were to *slide forwards* at any time in the subsequent motion. Then there must be a time τ, at which v^C and dv^C/dt are both positive. The condition $v^C > 0$ implies that $F = F^{\max}$ when $t = \tau$, and the condition $dv^C/dt > 0$ then implies that $T > \gamma F^{\max}$ when $t = \tau$. This is contrary to assumption and so forward sliding can never take place. A similar argument excludes backward sliding and so the only possibility is that the reel must roll.

The reel must therefore satisfy the rolling condition $V = \omega a$ and this, together with the equations of motion (11.23), (11.24), implies that the reel must **roll forwards** with constant acceleration

$$\frac{dV}{dt} = \frac{a(a-b)T}{M(k^2 + a^2)}. \blacksquare$$

Example 11.12 *A circus trick*

In a circus trick, a performer of mass m causes a large ball of mass M and radius a to accelerate to the right (see Figure 11.12) by running to the left on the upper surface of the ball. The man does not fall off the ball because he maintains this motion in such a way that the angle α shown remains constant. Find the conditions necessary for such a motion to take place.

Solution

Suppose the motion is planar and that, at time t, the ball has velocity V in the $\boldsymbol{i}$-direction and angular velocity ω $(= V/a)$ around the axis $\{O, \boldsymbol{j}\}$. If the man is to maintain his position on the ball, then he must run up the surface of the ball (towards the highest point) with velocity V. This maintains his vertical height and his acceleration is then the same as that of the ball, namely $(dV/dt)\boldsymbol{i}$.

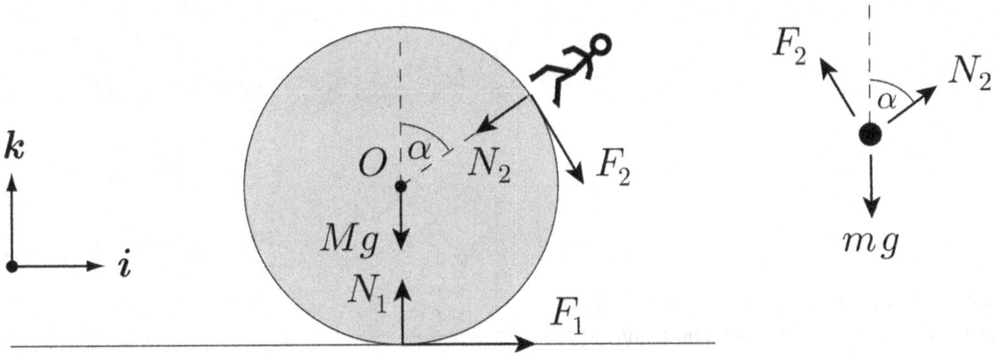

FIGURE 11.12 The circus trick: forces on the ball and the performer.

The equations of motion for the **man** are therefore

$$m\frac{dV}{dt} = N_2 \sin\alpha - F_2 \cos\alpha$$
$$0 = N_2 \cos\alpha + F_2 \sin\alpha - mg$$

and the equations of motion for the **ball** are

$$M\frac{dV}{dt} = F_1 - N_2 \sin\alpha + F_2 \cos\alpha$$
$$0 = N_1 - N_2 \cos\alpha - F_2 \sin\alpha - Mg$$
$$I_O\frac{d\omega}{dt} = aF_2 - aF_1$$

where I_O is the moment of inertia of the ball about $\{O, j\}$.

These five equations, together with the rolling condition $V = \omega a$, are sufficient to determine the six unknowns dV/dt, $d\omega/dt$, F_1, N_1, F_2 and N_2. After some algebra, the solution for the forward **acceleration** of the ball turns out to be

$$\frac{dV}{dt} = \frac{mg\sin\alpha}{M + (I_O/a^2) + m(1 + \cos\alpha)}.$$

Hence the motion is possible for any acute angle α provided that the performer can accelerate relative to the ball with this acceleration. For the case in which the ball is hollow, the masses of the man and the ball are equal, and $\alpha = 45°$, the acceleration required is approximately $0.21\ g$. ■

11.7 RIGID BODY STATICS IN THREE DIMENSIONS

Although we are not yet able to attempt problems in which a rigid body undergoes three-dimensional motion, we *are* able to solve problems in which a rigid body is in *equilibrium* under a three-dimensional system of forces. In equilibrium, the linear and angular momentum of the body are known to be zero, so that the rigid body equations become

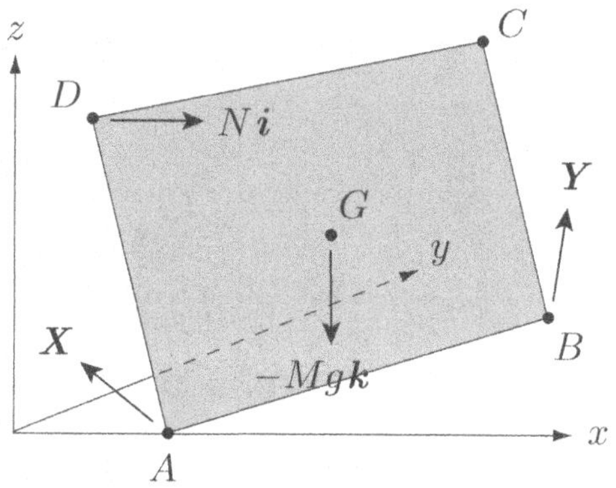

FIGURE 11.13 The rectangular panel $ABCD$ rests on the rough floor $z = 0$ and leans against the smooth wall $x = 0$.

Equations of rigid body statics

$$\boldsymbol{F} = \boldsymbol{0} \qquad \boldsymbol{K}_A = \boldsymbol{0} \tag{11.28}$$

where A is any fixed point of space.* In other words, *when a system is in equilibrium, the resultant force and the resultant moment of the external forces must be zero.*

Since there is no motion, one may wonder what there is left to calculate in statical problems. However, the body is usually supported or restrained in some prescribed way, and it is the unknown constraint forces that are to be determined. If these constraint forces can be determined solely from the equilibrium equations (11.28), then the problem is said to be **statically determinate.**†

Example 11.13 *Leaning panel*

A rectangular panel $ABCD$ of mass M is (rather carelessly) placed with its edge AB on the rough horizontal floor $z = 0$ and with the vertex D resting against the smooth wall $x = 0$, as shown in Figure 11.13. The four vertices of the panel are at the points $A(2, 0, 0)$, $B(6, 4, 0)$, $C(4, 6, 6)$ and $D(0, 2, 6)$ respectively. Given that the panel does not slip on the floor, find the reaction force exerted by the wall.

Solution

The external forces acting on the panel are the normal reaction of the wall $P\boldsymbol{i}$, the weight force $-Mg\boldsymbol{k}$, and the reaction of the floor on the edge AB. Now the reaction of the floor is distributed along the edge AB and, although we could treat it as such,

* It is unnecessary and inconvenient to restrict moments to be taken about G.

† Not all problems are statically determinate by any means. In the two-dimensional theory, if a heavy rigid plank is resting on three or more supports, then the individual reactions at the supports cannot be found from the equilibrium equations. What this means is that modelling the plank as a *rigid* body is not appropriate in such a problem. One should instead model the plank as a *deformable* body, solve the problem using the theory of elasticity, and then pass to the rigid limit.

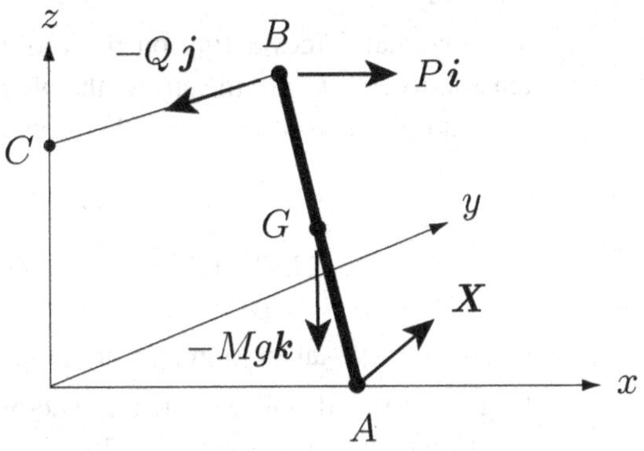

FIGURE 11.14 The rod AB is in equilibrium with A on a rough floor and B resting against a smooth wall. The rod is prevented from falling by the string BC.

this is an irrelevant complication. We will therefore suppose that (in order to avoid damage to the floor) the panel has been supported on two small pads beneath the corners A and B, in which case the reaction of the floor consists of the forces $\boldsymbol{X}$ and $\boldsymbol{Y}$ as shown.

We now apply the equilibrium conditions. The condition $\boldsymbol{F} = \boldsymbol{0}$ yields

$$N\boldsymbol{i} + \boldsymbol{X} + \boldsymbol{Y} - Mg\boldsymbol{k} = \boldsymbol{0}. \qquad (11.29)$$

If we take moments about the corner A, the reaction $\boldsymbol{X}$ makes no contribution and the condition $\boldsymbol{K}_A = \boldsymbol{0}$ becomes

$$\overrightarrow{AD} \times (N\boldsymbol{i}) + \overrightarrow{AB} \times \boldsymbol{Y} + \overrightarrow{AG} \times (-Mg\boldsymbol{k}) = \boldsymbol{0}.$$

On inserting the given numbers, this equation becomes

$$N(6\boldsymbol{j} - 2\boldsymbol{k}) + (4\boldsymbol{i} + 4\boldsymbol{j}) \times \boldsymbol{Y} - Mg\,(3\boldsymbol{i} - \boldsymbol{j}) = \boldsymbol{0}. \qquad (11.30)$$

The six scalar equations in (11.29), (11.30) contain the seven scalar unknowns $\boldsymbol{X}$, $\boldsymbol{Y}, N$ which means that the problem is *not statically determinate*. However, this does not stop us from finding N, since $\boldsymbol{Y}$ can be eliminated from equation (11.30) by taking the scalar product with the vector $4\boldsymbol{i}+4\boldsymbol{j}$. (This is equivalent to taking moments about the *axis* AB so that both $\boldsymbol{X}$ and $\boldsymbol{Y}$ disappear to leave N as the only unknown.) This gives the **reaction exerted** by the wall to be $N = Mg/3$. ■

Example 11.14 *Leaning rod*

A rough floor lies in the horizontal plane $z = 0$ and the plane $x = 0$ is occupied by a smooth vertical wall. A uniform rod of mass M has its lower end on the floor at the point $(a, 0, 0)$ and its upper end rests in contact with the wall at the point $(0, b, c)$. The rod is prevented from falling by having its upper end connected to the point $(0, 0, c)$ by a light inextensible string. Given that the rod does not slip, find the tension in the string and the reaction exerted by the wall.

Solution

The external forces acting on the rod are the normal reaction $N\boldsymbol{i}$ of the wall, the tension force $-Q\boldsymbol{j}$ in the string, the weight force $-Mg\boldsymbol{k}$, and reaction $\boldsymbol{X}$ of the floor. The equilibrium equations are therefore

$$P\boldsymbol{i} - Q\boldsymbol{j} - Mg\boldsymbol{k} + \boldsymbol{X} = \boldsymbol{0}, \tag{11.31}$$

$$(2L\boldsymbol{n})\times(P\boldsymbol{i} - Q\boldsymbol{j}) + (L\boldsymbol{n})\times(-Mg\boldsymbol{k}) = \boldsymbol{0}, \tag{11.32}$$

where we have taken moments about A to eliminate the reaction $\boldsymbol{X}$. Here, $2L$ is the length of the rod, and $\boldsymbol{n}$ is the unit vector in the direction $\overrightarrow{AB}$.

Equation (11.31) serves only to determine the reaction $\boldsymbol{X}$ once P and Q are known. To extract P and Q from the vector equation (11.32), we take components in any two directions other than the $\boldsymbol{n}$-direction; the easiest choices are the $\boldsymbol{i}$- and $\boldsymbol{j}$-directions. On taking the scalar product of equation (11.32) with $\boldsymbol{i}$, we obtain

$$\begin{aligned}0 &= 2\,[\boldsymbol{n}, P\boldsymbol{i} - Q\boldsymbol{j}, \boldsymbol{i}] + [\boldsymbol{n}, -Mg\boldsymbol{k}, \boldsymbol{i}]\\ &= 2P\,[\boldsymbol{n}, \boldsymbol{i}, \boldsymbol{i}] - 2Q\,[\boldsymbol{n}, \boldsymbol{j}, \boldsymbol{i}] - Mg\,[\boldsymbol{n}, \boldsymbol{k}, \boldsymbol{i}]\\ &= 0 - 2Q\,\boldsymbol{n}\cdot(\boldsymbol{j}\times\boldsymbol{i}) - Mg\,\boldsymbol{n}\cdot(\boldsymbol{k}\times\boldsymbol{i})\\ &= 2Q\,(\boldsymbol{n}\cdot\boldsymbol{k}) - Mg\,(\boldsymbol{n}\cdot\boldsymbol{j}),\end{aligned}$$

where we have used the notation $[\boldsymbol{u}, \boldsymbol{v}, \boldsymbol{w}]$ to mean the triple scalar product of the vectors $\boldsymbol{u}$, $\boldsymbol{v}$ and $\boldsymbol{w}$. Hence

$$Q = \frac{Mg\,(\boldsymbol{n}\cdot\boldsymbol{j})}{2\,(\boldsymbol{n}\cdot\boldsymbol{k})},$$

and, by taking the scalar product of equation (11.32) with $\boldsymbol{j}$ and proceed in the same way, we obtain

$$P = -\frac{Mg\,(\boldsymbol{n}\cdot\boldsymbol{i})}{2\,(\boldsymbol{n}\cdot\boldsymbol{k})}.$$

Finally, we need to express these answers in terms of the data given in the question. Since the unit vector $\boldsymbol{n}$ is given by

$$\boldsymbol{n} = \frac{-a\,\boldsymbol{i} + b\,\boldsymbol{j} + c\,\boldsymbol{k}}{2L},$$

it follows that the **reaction** exerted by the wall, and the **tension** in the string, are

$$P = \frac{Mga}{2c}, \qquad Q = \frac{Mgb}{2c}. \blacksquare$$

Problems on Chapter 11

Answers and comments are at the end of the book.

Harder problems carry a star (∗).

11.1 ***Non-standard angular momentum principle*** If A is a generally moving point of space and $\boldsymbol{L}_A$ is the angular momentum of a system $\mathcal{S}$ about A *in its motion relative to* A, show that the angular momentum principle for $\mathcal{S}$ about A takes the non-standard form

$$\frac{d\boldsymbol{L}_A}{dt} = \boldsymbol{K}_A - M(\boldsymbol{R} - \boldsymbol{a}) \times \frac{d^2\boldsymbol{a}}{dt^2}.$$

[Begin by expanding the expression for $\boldsymbol{L}_A$.]

When does this formula reduce to the standard form? [This non-standard version of the angular momentum principle is rarely needed. However, see Problem 11.9.]

Problems soluble by conservation principles

11.2 A fairground target consists of a uniform circular disk of mass M and radius a that can turn freely about a diameter which is fixed in a vertical position. Initially the target is at rest. A bullet of mass m is moving with speed u along a horizontal straight line at right angles to the target. The bullet embeds itself in the target at a point distance b from the rotation axis. Find the final angular speed of the target. [The moment of inertia of the disk about its rotation axis is $Ma^2/4$.]

Show also that the energy lost in the impact is

$$\frac{1}{2}mu^2\left(\frac{Ma^2}{Ma^2 + 4mb^2}\right).$$

11.3 A uniform circular cylinder of mass M and radius a can rotate freely about its axis of symmetry which is fixed in a vertical position. A light string is wound around the cylinder so that it does not slip and a particle of mass m is attached to the free end. Initially the system is at rest with the free string taut, horizontal and of length b. The particle is then projected horizontally with speed u at right angles to the string. The string winds itself around the cylinder and eventually the particle strikes the cylinder and sticks to it. Find the final angular speed of the cylinder.

11.4 ***Rotating gas cloud*** A cloud of interstellar gas of total mass M can move freely in space. Initially the cloud has the form of a uniform sphere of radius a rotating with angular speed Ω about an axis through its centre. Later, the cloud is observed to have changed its form to that of a thin uniform circular disk of radius b which is rotating about an axis through its centre and perpendicular to its plane. Find the angular speed of the disk and the increase in the kinetic energy of the cloud.

11.5 ***Conical pendulum with shortening string*** A particle is suspended from a support by a light inextensible string which passes through a small fixed ring vertically below the support. Initially the particle is performing a conical motion of angle 60°, with the moving part of the string of a. The support is now made to move slowly upwards so that the motion remains nearly conical. Find the angle of this conical motion when the support has been raised by a distance $a/2$. [Requires the numerical solution of a trigonometric equation.]

11.6 ***Baseball bat*** A baseball bat has mass M and moment of inertia Mk^2 about any axis through its centre of mass G that is perpendicular to the axis of symmetry. The bat is at rest when a ball of mass m, moving with speed u, is normally incident along a straight line through the axis of symmetry at a distance b from G. Show that, whether the impact is elastic or not, there is a point on the axis of symmetry of the bat that is instantaneously at rest after the impact and that the distance c of this point from G is given by $bc = k^2$. In the elastic case, find the speed of the ball after the impact. [Gravity (and the batter!) should be ignored throughout this question.]

11.7 ***Hoop mounting a step*** A uniform hoop of mass M and radius a is rolling with speed V along level ground when it meets a step of height h ($h < a$). The particle C of the hoop that makes contact with the step is suddenly brought to rest. Find the instantaneous speed of the centre of mass, and the instantaneous angular velocity of the hoop, immediately after the impact. Deduce that the particle C cannot remain at rest on the edge of the step if

$$V^2 > (a-h)g\left(1-\frac{h}{2a}\right)^{-2}.$$

Suppose that the particle C *does* remain on the edge of the step. Show that the hoop will go on to mount the step if

$$V^2 > hg\left(1-\frac{h}{2a}\right)^{-2}.$$

Deduce that the hoop cannot mount the step in the manner described if $h > a/2$.

11.8 ***Particle sliding on a cone*** A particle P slides on the smooth inner surface of a circular cone of semi-angle α. The axis of symmetry of the cone is vertical with the vertex O pointing downwards. Show that the vertical component of angular momentum about O is conserved in the motion. State a second dynamical quantity that is conserved.

Initially P is a distance a from O when it is projected horizontally along the inside surface of the cone with speed u. Show that, in the subsequent motion, the distance r of P from O satisfies the equation

$$\dot{r}^2 = (r-a)\left[\frac{u^2(r+a)}{r^2} - 2g\cos\alpha\right].$$

Case A For the case in which gravity is absent, find r and the azimuthal angle ϕ explicitly as functions of t. Make a sketch of the path of P (as seen from 'above') when $\alpha = \pi/6$.

Case B For the case in which $\alpha = \pi/3$, find the value of u such that r oscillates between a and $2a$ in the subsequent motion. With this value of u, show that r will first return to the value $r = a$ after a time

$$2\sqrt{3}\left(\frac{a}{g}\right)^{1/2}\int_1^2 \frac{\xi\, d\xi}{[(\xi-1)(2-\xi)(2+3\xi)]^{1/2}}.$$

11.9* ***Bug running on a hoop*** A uniform circular hoop of mass M can slide freely on a smooth horizontal table, and a bug of mass m can run on the hoop. The system is at rest when the bug starts to run. What is the angle turned through by the hoop when the bug has completed one lap of the hoop? [This is a classic problem, but difficult. Apply the angular momentum principle about the centre of the hoop, using the *non-standard* version given in Problem 11.1]

Planar rigid body motion

11.10 ***General rigid pendulum*** A rigid body of general shape has mass M and can rotate freely about a fixed horizontal axis. The centre of mass of the body is distance h from the rotation axis, and the moment of inertia of the body about the rotation axis is I. Show that the period of small oscillations of the body about the downward equilibrium position is

$$2\pi\left(\frac{I}{Mgh}\right)^{1/2}.$$

Deduce the period of small oscillations of a uniform rod of length $2a$, pivoted about a horizontal axis perpendicular to the rod and distance b from its centre.

11.11 ***From sliding to rolling*** A snooker ball is at rest on the table when it is projected forward with speed V and no angular velocity. Find the speed of the ball when it eventually begins to roll. What proportion of the original kinetic energy is lost in the process?

11.12 ***Rolling or sliding?*** A uniform ball is released from rest on a rough plane inclined at angle α to the horizontal. The coefficient of friction between the ball and the plane is μ. Will the ball roll or slide down the plane? Find the acceleration of the ball in each case.

11.13 A circular disk of mass M and radius a is smoothly pivoted about its axis of symmetry which is fixed in a horizontal position. A bug of mass m runs with constant speed u around the rim of the disk. Initially the disk is held at rest and is released when the bug reaches its lowest point. What is the condition that the bug will reach the highest point of the disk?

11.14 ***Yo-yo with moving support*** A uniform circular cylinder (a yo-yo) has a light inextensible string wrapped around it so that it does not slip. The free end of the string is fastened to a support and the yo-yo moves in a vertical straight line with the straight part of the string also vertical. At the same time the support is made to move vertically having upward displacement $Z(t)$ at time t. Find the acceleration of the yo-yo. What happens if the system starts from rest and the support moves upwards with acceleration $2g$?

11.15 ***Supermarket belt*** A circular cylinder, which is axially symmetric but not uniform, has mass M and moment of inertia Mk^2 about its axis of symmetry. The cylinder is placed on a rough horizontal belt at right angles to the direction in which the belt can move. Initially the cylinder and the belt are both at rest when the belt begins to move with velocity $V(t)$. Given that there is no slipping, find the velocity of the cylinder at time t.

Explain why drinks bottles tend to spin on a supermarket belt (instead of moving forwards) if they are placed at right-angles to the belt.

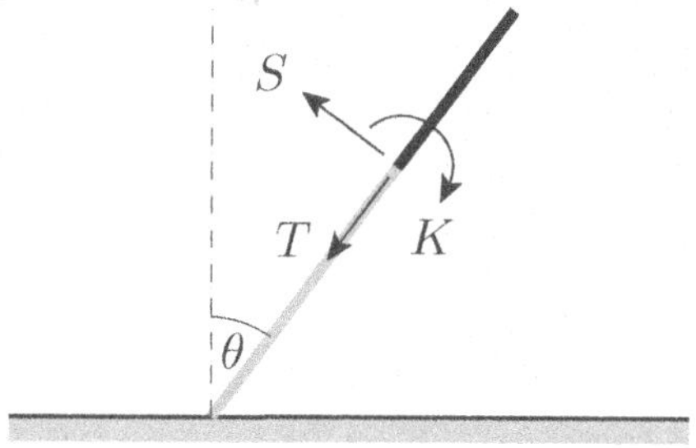

FIGURE 11.15 The tension force T, the shear force S and the couple K exerted on the upper part of the rod (black) by the lower part (grey).

11.16* ***Falling chimney*** A uniform rod of length $2a$ has one end on a rough table and is balanced in the vertically upwards position. The rod is then slightly disturbed. Given that its lower end does not slip, show that, in the subsequent motion, the angle θ that the rod makes with the upward vertical satisfies the equation

$$2a\dot{\theta}^2 = 3g(1 - \cos\theta).$$

Consider now the *upper part* of the rod of length $2\gamma a$, as shown in Figure 11.15. Let T, S and K be the tension force, the shear force and the couple exerted on the upper part of the rod by the lower part. By considering the upper part of the rod to be a rigid body in planar motion, find expressions for S and K in terms of θ.

If a tall thin chimney begins to fall, at what point along its length would you expect it to break first?

Rigid body statics

11.17 ***Leaning triangular panel*** A rough floor lies in the horizontal plane $z = 0$ and the planes $x = 0$, $y = 0$ are occupied by smooth vertical walls. A rigid uniform triangular panel ABC has mass m. The vertex A of the panel is placed on the floor at the point $(2, 2, 0)$ and the vertices B, C rest in contact with the walls at the points $(0, 1, 6)$, $(1, 0, 6)$ respectively. Given that the vertex A does not slip, find the reactions exerted by the walls. Deduce the reaction exerted by the floor.

11.18 ***Triangular coffee table*** A trendy swedish coffee table has an unsymmetrical triangular glass top supported by a leg at each vertex. Show that, whatever the shape of the triangular top, each leg bears one third of its weight.

11.19 ***Pile of balls*** Three identical balls are placed in contact with each other on a horizontal table and a fourth identical ball is placed on top of the first three. Show that the four balls cannot be in equilibrium unless (i) the coefficient of friction between the balls is at least $\sqrt{3} - \sqrt{2}$, and (ii) the coefficient of friction between each ball and the table is at least $\frac{1}{4}(\sqrt{3} - \sqrt{2})$.

Part Three

ANALYTICAL MECHANICS

CHAPTERS IN PART THREE

Chapter Twelve

Lagrange's equations and conservation principles

KEY FEATURES

The key features of this chapter are **generalised coordinates** and **configuration space**, the derivation and use of **Lagrange's equations**, the **Lagrangian**, and the connection between **symmetry** of the Lagrangian and **conservation principles**.

Lagrange's equations mark a change in direction in our development of mechanics. Building on the work of d'Alembert, Lagrange* devised a general method for obtaining the **equations of motion** for a very wide class of mechanical systems. In earlier chapters we have used conservation principles for this purpose, but there is no guarantee that enough conservation principles exist. In contrast, Lagrange's method is completely general and is not restricted to problems soluble by conservation principles. The method is so simple to apply that it is quite possible to solve complex mechanical problems whilst knowing very little about mechanics! However, the supporting theory has its subtleties.

Lagrange's equations also mark the beginning of **analytical mechanics** in which general principles, such as the connection between symmetry and conservation principles, begin to take over from actual problem solving.

12.1 CONSTRAINTS AND CONSTRAINT FORCES

A **general mechanical system** $\mathcal{S}$ consists of any number of particles P_1, P_2, ..., P_N. The particles of $\mathcal{S}$ may have interconnections of various kinds (light strings, springs and so on) and also be subject to external connections and constraints. These could include features such as a particle being forced to remain on a fixed surface or suspended from a

* Joseph-Louis Lagrange (Giuseppe Lodovico Lagrangia), (1736–1813). Although Lagrange is often considered to be French, he was in fact born in Turin, Italy and did not move to Paris until 1787. Lagrange had a long career in Turin and Berlin during which time he made major contributions to mechanics, fluid mechanics and the calculus of variations. His famous book *Mécanique Analitique*, published in Paris in 1788, is a definitive account of his contributions to mechanics. This work transformed mechanics into a branch of mathematical analysis. Perhaps to emphasise this, there is not a single diagram in the whole book!

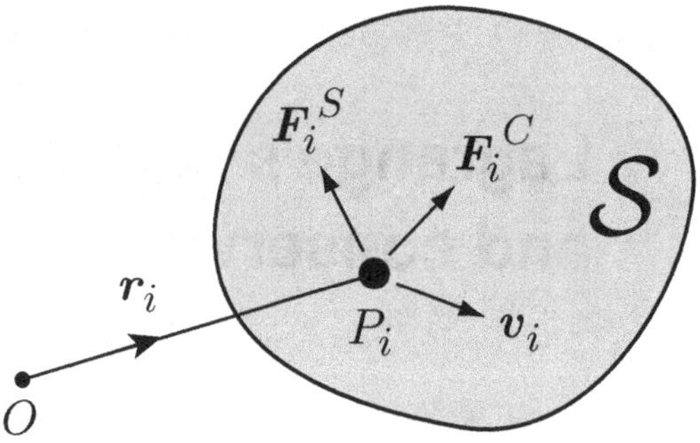

FIGURE 12.1 The general mechanical system $\mathcal{S}$ consists of any number of particles $\{P_i\}$ $(i = 1 \dots, N)$. The typical particle P_i has mass m_i, position vector $\boldsymbol{r}_i$ and velocity $\boldsymbol{v}_i$. $\boldsymbol{F}_i^S$ is the specified force and $\boldsymbol{F}_i^C$ the constraint force acting on P_i.

fixed point by a light inextensible string. The pendulum, the spinning top, the bicycle and the solar system are examples of mechanical systems.

Unconstrained systems

If the particles of $\mathcal{S}$ are free to move anywhere in space *independently of each other* then $\mathcal{S}$ is said to be an **unconstrained system**. In this special case, the equations of motion for $\mathcal{S}$ are simply Newton's equations for the N individual particles. Suppose that the typical particle P_i has mass m_i, position vector $\boldsymbol{r}_i$ and velocity $\boldsymbol{v}_i$. Then the **equations of motion** for the system $\mathcal{S}$ are

$$m_i \dot{\boldsymbol{v}}_i = \boldsymbol{F}_i \qquad (i = 1 \dots, N),$$

where $\boldsymbol{F}_i$ is the force acting on the particle P_i.

Example 12.1 *Two-body problem*

Write down the Newton equations for the two-body gravitation problem.

Solution

In this problem $\mathcal{S}$ consists of two particles moving solely under their mutual gravitational attraction. There are no constraints. The motion of the system is therefore governed by the two Newton equations

$$m_1 \dot{\boldsymbol{v}}_1 = m_1 m_2 G \frac{\boldsymbol{r}_2 - \boldsymbol{r}_1}{|\boldsymbol{r}_1 - \boldsymbol{r}_2|^3}, \qquad m_2 \dot{\boldsymbol{v}}_2 = m_1 m_2 G \frac{\boldsymbol{r}_1 - \boldsymbol{r}_2}{|\boldsymbol{r}_1 - \boldsymbol{r}_2|^3},$$

where G is the constant of gravitation. These equations, together with the initial conditions, are sufficient to determine the motion of the two particles. ■

Constrained systems

Unconstrained mechanical systems are relatively rare. Indeed many of the problems solved in earlier chapters involve mechanical systems that are subject to **geometrical** or **kinematical constraints**. Geometrical constraints are those that involve only the position vectors $\{\boldsymbol{r}_i\}$; kinematical constraints involve the $\{\boldsymbol{v}_i\}$ as well. Some **typical constraints** are as follows:

- The bob of a pendulum *must* remain a fixed distance from the point of support.
- The particles of a rigid body *must* maintain fixed distances from each other.
- A particle sliding on a wire *must not* leave the wire.
- The contact particle of a body rolling on a fixed surface *must* be at rest.

The rolling condition is a *kinematical constraint* since it involves the *velocity* of a particle. All the other constraints are geometrical.

These, and all other constraints, are enforced by **constraint forces**. Constraint forces are not part of the specification of a system and are therefore *unknown*. For example, when a particle is constrained to slide on a wire, it is prevented from leaving the wire by the force that the wire exerts upon it. This constraint force (which would commonly be called the reaction of the wire on the particle) is unknown; we know only that it is sufficient to keep the particle on the wire.

For **constrained systems** the straightforward approach of using the Newton equations runs into the following difficulties:

A The equations of motion do not incorporate the constraints
The Newton equations (in Cartesian coordinates) do not incorporate the constraints. These must therefore be included in the form of additional conditions to be *solved simultaneously* with the dynamical equations.

B The constraint forces are unknown
For constrained systems, the Newton equations have the form

$$m_i \dot{\boldsymbol{v}}_i = \boldsymbol{F}_i^S + \boldsymbol{F}_i^C \qquad (1 \le i \le N), \tag{12.1}$$

where $\boldsymbol{F}_i^S$ is the **specified force** and $\boldsymbol{F}_i^C$ is the **constraint force** acting on the particle P_i. The $\boldsymbol{F}_i^S$ are known but the $\boldsymbol{F}_i^C$ are not.

Because of these two difficulties, only the simplest problems of constrained motion are tackled this way. In the following sections we show how these difficulties can be overcome. The first difficulty is overcome by using a new (reduced) set of coordinates called **generalised coordinates**, while the second is overcome by using **Lagrange's equations** instead of Newton's.

12.2 GENERALISED COORDINATES

Suppose that the system is subject to **geometrical constraints** only. Then the position vectors $\{\boldsymbol{r}_i\}$ of its particles are not independent variables, but are related to each other by these constraints. A possible 'position' of such a system is called a **configuration.** More precisely, a set of values for the position vectors $\{\boldsymbol{r}_i\}$ that is *consistent with the geometrical constraints* is a configuration of the system.

The trick is to select new 'coordinates' that *are* independent of each other but are still sufficient to specify the configuration of the system. These new coordinates are called **generalised coordinates** and their official definition is as follows:

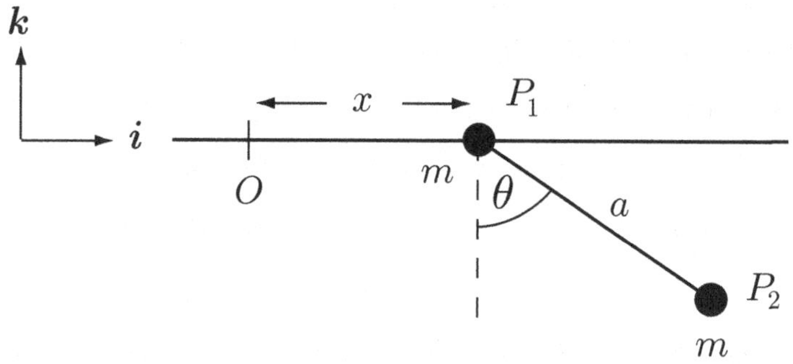

FIGURE 12.2 The variables x and θ are a set of generalised coordinates for this system.

Definition 12.1 ***Generalised coordinates*** *If the configuration of a system $\mathcal{S}$ is determined by the values of a set of independent variables $q_1, \ldots, q_n$, then $\{q_1, \ldots, q_n\}$ is said to be a set of* ***generalised coordinates*** *for $\mathcal{S}$.*

This definition deserves some explanation.

(i) When we say the generalised coordinates must be **independent variables**, we mean that there must be *no functional relation connecting them.* If there were, one of the coordinates could be removed and the remaining $n - 1$ coordinates would still determine the configuration of the system. The set of generalised coordinates must not be reducible in this way.

(ii) When we say the generalised coordinates $q_1, \ldots, q_n$ **determine the configuration** of the system $\mathcal{S}$, we mean that, when the values of the coordinates $q_1, \ldots, q_n$ are given, the position of every particle of $\mathcal{S}$ is determined. In other words, the position vectors $\{\boldsymbol{r}_i\}$ of the particles must be known functions of the independent variables $q_1, \ldots, q_n$, that is,

$$\boldsymbol{r}_i = \boldsymbol{r}_i(q_1, \ldots, q_n) \qquad (i = 1, \ldots, N). \tag{12.2}$$

Abstract though this concept may seem, generalised coordinates are remarkably easy to use. In practice, *they are chosen to be displacements or angles that appear naturally in the problem.* This is illustrated by the following examples.

Example 12.2 *Choosing generalised coordinates*

Let $\mathcal{S}$ be the system shown in Figure 12.2 which consists of two particles P_1 and P_2 connected by a light rigid rod of length a. The particle P_1 is constrained to move along a fixed horizontal rail and the system moves in the vertical plane through the rail. Select generalised coordinates for this system and obtain expressions for the position vectors $\boldsymbol{r}_1$, $\boldsymbol{r}_2$ in terms of these coordinates.

Solution

Consider the variables x, θ shown. These are certainly independent variables (they are not connected by any functional relation) and, when they are given, the configuration of $\mathcal{S}$ is determined. Thus $\{x, \theta\}$ is a set of **generalised coordinates** for the system $\mathcal{S}$.

In terms of the coordinates x and θ, the positions of the particles P_1 and P_2 are given by

$$r_1 = x\,i,$$
$$r_2 = (x + a\sin\theta)\,i - (a\cos\theta)\,k,$$

which are the expressions (12.2) for this system and this choice of coordinates. ■

Example 12.3 *Choosing more generalised coordinates*

Choose generalised coordinates for the system consisting of three particles P_1, P_2, P_3 where P_1, P_2 are connected by a light rigid rod of length a and P_2, P_3 are connected by a light rigid rod of length b. The system slides on a horizontal table. [Make a sketch of the system.]

Solution

Many choices of generalised coordinates are possible. Let $Oxyz$ be a system of rectangular coordinates with O on the table and Oz pointing vertically upwards. One set of generalised coordinates consists of (i) the x and y coordinates of the particle P_1, (ii) the angle θ between the line P_1P_2 and the x-axis, (iii) the angle ϕ between the line P_2P_3 and the x-axis. A second set of generalised coordinates consists of (i) the x and y coordinates of the particle P_2, (ii) the angle θ between the line P_1P_2 and the y-axis, (iii) the angle ϕ between the line P_1P_2 and the line P_2P_3. ■

Degrees of freedom

It is evident from the above example that the configuration of a system can be specified by many different sets of generalised coordinates. However the *number of coordinates needed is always the same*. In the last example, the number of generalised coordinates needed is *always* three.

Definition 12.2 *Degrees of freedom* *Let $\mathcal{S}$ be a mechanical system subject to geometrical constraints. Then the* **number** *of generalised coordinates needed to specify the configuration of $\mathcal{S}$ is called the number of* **degrees of freedom** *of $\mathcal{S}$.*

The number of degrees of freedom is an important property of a mechanical system. Suppose, for example, that we have a system with *three* degrees of freedom and generalised coordinates q_1, q_2, q_3. Suppose also that the system is in some given configuration when it is started into motion in some given way. This means that we know the initial values of the coordinates q_1, q_2, q_3, and their time derivatives $\dot{q}_1, \dot{q}_2, \dot{q}_3$. How many equations of motion (second order ODEs) do we need to determine the functions $q_1(t)$, $q_2(t)$, $q_3(t)$ that describe the subsequent motion of the system? The answer is provided by the general theory of ODEs. If the three functions $q_1(t)$, $q_2(t)$, $q_3(t)$ satisfy *three* (independent) second order ODEs, then the general theory guarantees that there is precisely one solution that satisfies the prescribed initial conditions. If there are fewer equations, the solution is not uniquely determined; if there are more, the equations are not independent. This gives us the following important result:

> **Degrees of freedom and equations of motion**
>
> The number of degrees of freedom of a system is equal to the number of equations of motion (second order ODEs) that are needed to determine the motion of the system.

Example 12.4 *Degrees of freedom*

State the number of degrees of freedom of the following mechanical systems: (i) the simple pendulum, (ii) the spherical pendulum, (iii) a door swinging on its hinges, (iv) a bar of soap (a particle) sliding on the inside of a basin, (v) four rigid rods flexibly jointed to form a quadrilateral which can move on a flat table, (vi) a ball rolling on a rough table.

Solution

(i) 1 (ii) 2 (iii) 1 (iv) 2 (v) 4 (vi) Not defined! This system has a *kinematical* constraint, namely, the rolling condition at the contact point. ■

Kinematical constraints

So far we have not discussed kinematical constraints such as the rolling condition. We can now handle geometrical constraints since they are automatically taken into account by using generalised coordinates. But kinematical constraints involve the particle velocities which in turn depend not only on the coordinates $q_1, \ldots, q_n$, but also their time derivatives $\dot{q}_1, \ldots, \dot{q}_n$. *In general, kinematical constraints cannot be incorporated by selecting some new set of generalised coordinates.* As a result, such constraints have to remain as additional ODEs that must be solved along with the equations of motion.

All is not lost however since, in some special but important cases, the ODE representing the kinematical constraint can be immediately integrated to yield an *equivalent geometrical constraint.* Such a constraint is said to be **integrable**.

Example 12.5 *An integrable kinematical constraint*

A circular cylinder rolls down a rough inclined plane. Show that, in this problem, the rolling condition is an integrable constraint.

Solution

In the absence of the rolling condition, this system has two degrees of freedom; take as generalised coordinates x (the displacement of the cylinder axis down the plane) and θ (the rotation angle of the cylinder). The **rolling condition** is then given by the first order ODE

$$\dot{x} = a\dot{\theta}, \tag{12.3}$$

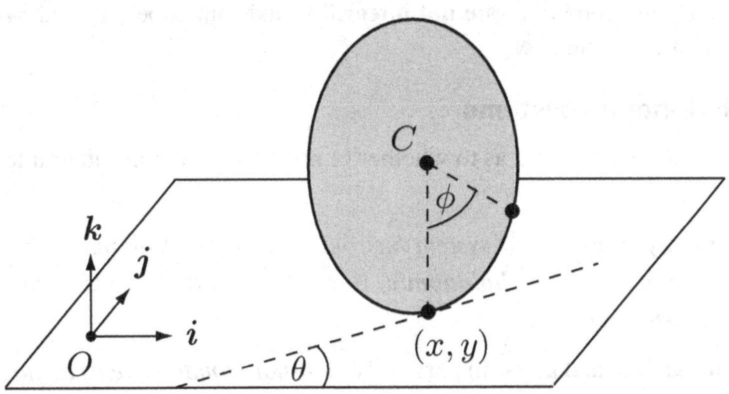

FIGURE 12.3 For the generally rolling wheel, the rolling conditions are non-integrable.

where a is the radius of the cylinder. But this constraint can be integrated (without solving the problem!) to give

$$x = a\theta, \tag{12.4}$$

on taking $x = \theta = 0$ in the reference configuration. Thus the kinematical constraint (12.3) is equivalent to the geometrical constraint (12.4). This geometrical constraint can now be incorporated by selecting a new (reduced) set of generalised coordinates. In this example, only one generalised coordinate is finally required (either x or θ) so that the rolling cylinder has **one degree of freedom.** ■

Example 12.6 *A non-integrable kinematical constraint*

Figure 12.3 shows a circular disk of radius a which is constrained to roll on a horizontal floor with its plane vertical. Show that, in this problem, the rolling conditions are not integrable.

Solution

In the absence of the rolling condition, this system has four degrees of freedom. Let $Oxyz$ be a fixed system of rectangular coordinates with O on the floor and Oz pointing vertically upwards. Then a set of generalised coordinates is given by

(i) the x and y coordinates of the centre C of the disk,

(ii) the angle θ between the plane of the disc and the x-axis,

(iii) the angle ϕ that the disk has rotated about its axis (relative to some reference position).

Now we impose the **rolling condition**, namely, that the contact particle should have zero velocity. In terms of the chosen coordinates, this gives

$$\dot{x} + a\dot{\phi}\cos\theta = 0, \qquad \dot{y} + a\dot{\phi}\sin\theta = 0,$$

a *pair* of first order ODEs. These equations cannot be integrated since θ is an unknown function of the time and $\dot{\theta}$ is absent from both equations. It follows that, in this problem, the rolling conditions are **not integrable** and cannot be replaced by equivalent geometrical constraints. ■

Holonomic and non-holonomic systems

Mechanical systems are classified according as to whether or not they have non-integrable kinematical constraints.

Definition 12.3 ***Holonomic systems*** *If a system has only geometrical or integrable kinematical constraints, then it is said to be* ***holonomic****. If it has non-integrable kinematical constraints, then it is* ***non-holonomic****.*

Non-holonomic systems are the bad guys. In particular, *non-holonomic systems do not satisfy Lagrange's equations* (as presented later in this chapter). It is beyond the scope of this book to proceed any further with the *analytical* mechanics of such systems and, from now on, we will deal only with holonomic systems. (A way of extending the Lagrange method to non-holonomic systems is described by Goldstein [4].) Such systems can still be treated by standard Newtonian methods however. The problem of the rolling wheel is solved in this way in Chapter 19.

12.3 CONFIGURATION SPACE (q–space)

Let $\mathcal{S}$ be a **holonomic mechanical system** with generalised coordinates $q_1 \ldots, q_n$. It is convenient to regard the list of values $q_1, \ldots, q_n$ as the coordinates of a 'point' $\boldsymbol{q}$ in a space of n dimensions, that is,

$$\boldsymbol{q} = (q_1, \ldots, q_n). \tag{12.5}$$

Mathematicians call such a space $\mathbb{E}_n$ (the Euclidean space of n dimensions), but we will denote it by $\mathcal{Q}$ (the space to which $\boldsymbol{q}$ belongs) and call it **configuration space**. Since the values of $q_1, \ldots, q_n$ determine the configuration of the system $\mathcal{S}$, it follows that *the configuration of $\mathcal{S}$ is determined by the 'position' of the point $\boldsymbol{q}$ in configuration space*, that is,

$$\boldsymbol{r}_i = \boldsymbol{r}_i(\boldsymbol{q}) \qquad (i = 1, \ldots, N).$$

This abstract view becomes much clearer when applied to a particular example. Let $\mathcal{S}$ be the two-particle system shown in Figure 12.4. This system has two degrees of freedom and generalised coordinates x, θ. In this case the configuration space $\mathcal{Q}$ is the (x, θ)-plane. Each point $\boldsymbol{q} = (x, \theta)$ lying in $\mathcal{Q}$ corresponds to a configuration of the mechanical system $\mathcal{S}$. Moreover, as the configuration of the system changes with time, the point $\boldsymbol{q}$ moves through the configuration space as shown.

Generalised velocities

When the configuration of $\mathcal{S}$ changes with time, the point $\boldsymbol{q}$ moves through the configuration space $\mathcal{Q}$ so that $\boldsymbol{q} = \boldsymbol{q}(t)$. This leads to the notion of **generalised velocities**.

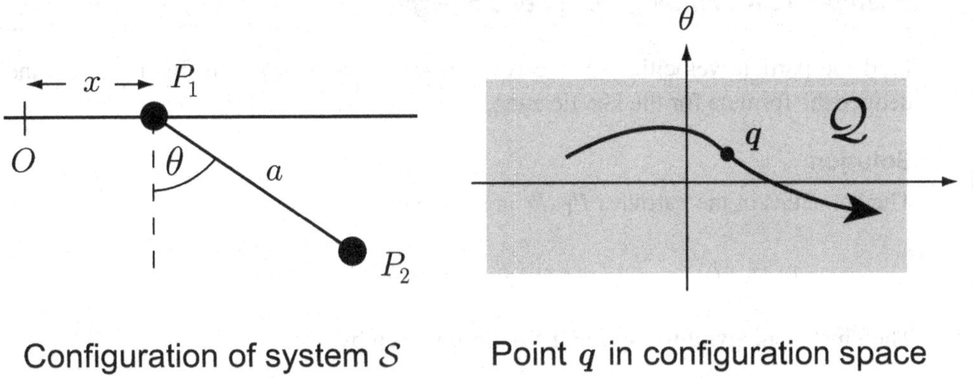

FIGURE 12.4 The configuration of the system $\mathcal{S}$ is represented by the point $\boldsymbol{q} = (x, \theta)$ in the configuration space $\mathcal{Q}$. As the configuration of $\mathcal{S}$ changes with time, the point $\boldsymbol{q}$ moves on a path lying in the configuration space $\mathcal{Q}$.

Definition 12.4 ***Generalised velocities*** *The time derivatives $\dot{q}_1, \dots, \dot{q}_n$ of the generalised coordinates $q_1 \dots, q_n$ are called the* **generalised velocities** *of the system $\mathcal{S}$.*

The n-dimensional vector $(\dot{q}_1, \dots, \dot{q}_n)$, formed from the $\{\dot{q}_j\}$, is just the time derivative of the vector $\boldsymbol{q}$, that is,

$$\dot{\boldsymbol{q}} = (\dot{q}_1, \dots, \dot{q}_n). \tag{12.6}$$

The vector $\dot{\boldsymbol{q}}$ can be regarded as the 'velocity' of the point $\boldsymbol{q}$ as it moves through the configuration space $\mathcal{Q}$.

Particle velocities

The values of $\boldsymbol{q}$ and $\dot{\boldsymbol{q}}$ determine the position and velocity of every particle of the system $\mathcal{S}$. For, since $\boldsymbol{r}_i = \boldsymbol{r}_i(\boldsymbol{q})$ and $\boldsymbol{q} = \boldsymbol{q}(t)$, it follows from the chain rule that

$$\boldsymbol{v}_i = \frac{\partial \boldsymbol{r}_i}{\partial q_1}\dot{q}_1 + \cdots + \frac{\partial \boldsymbol{r}_i}{\partial q_n}\dot{q}_n = \sum_{j=1}^{n} \frac{\partial \boldsymbol{r}_i}{\partial q_j}\dot{q}_j. \tag{12.7}$$

This expression for $\boldsymbol{v}_i$ is **linear** in the variables $\dot{q}_1, \dots, \dot{q}_n$ with coefficients that depend on $\boldsymbol{q}$.

Example 12.7 *Rule for finding particle velocities*

What is the connection between the formula (12.7) and the rule we have often used to find particle velocities?

Solution

Formula (12.7) says that $\boldsymbol{v}_i$ is the vector sum of n contributions, each one arising from the variation of a particular q_j. This therefore *justifies* our rule for finding particle velocities. ■

Example 12.8 *Finding the kinetic energy*

Find the particle velocities for the two-particle system shown in Figure 12.2, and deduce the formula for the kinetic energy.

Solution

The velocities of the particles P_1, P_2 are given by

$$\boldsymbol{v}_1 = \dot{x}\,\boldsymbol{i}, \qquad \boldsymbol{v}_2 = \dot{x}\,\boldsymbol{i} + (a\cos\theta\,\boldsymbol{i} + a\sin\theta\,\boldsymbol{k})\,\dot{\theta}.$$

The kinetic energy of the system is therefore given by

$$\begin{aligned} T &= \tfrac{1}{2}m\,(\boldsymbol{v}_1\cdot\boldsymbol{v}_1) + \tfrac{1}{2}m\,(\boldsymbol{v}_2\cdot\boldsymbol{v}_2) \\ &= \tfrac{1}{2}m\dot{x}^2 + \tfrac{1}{2}m\left(\dot{x}^2 + (a\dot{\theta})^2 + 2\dot{x}\,(a\dot{\theta})\cos\theta\right) \\ &= m\,\dot{x}^2 + (\tfrac{1}{2}ma^2)\,\dot{\theta}^2 + (ma\cos\theta)\,\dot{x}\dot{\theta}. \blacksquare \end{aligned}$$

Example 12.9 *General form of the kinetic energy*

Show that the kinetic energy of *any holonomic mechanical system* has the form

$$T = \sum_{j=1}^{n}\sum_{k=1}^{n} a_{jk}(\boldsymbol{q})\,\dot{q}_j\dot{q}_k$$

that is, a **homogeneous quadratic form** in the variables $\dot{q}_1, \dots, \dot{q}_n$, with coefficients depending on $\boldsymbol{q}$.

Solution

Let P be a typical particle of $\mathcal{S}$ with position vector $\boldsymbol{r}$ and velocity $\boldsymbol{v}$. Then

$$\boldsymbol{v} = \frac{\partial \boldsymbol{r}}{\partial q_1}\dot{q}_1 + \cdots + \frac{\partial \boldsymbol{r}}{\partial q_n}\dot{q}_n = \sum_{j=1}^{n}\frac{\partial \boldsymbol{r}}{\partial q_j}\dot{q}_j$$

and so

$$\begin{aligned} \boldsymbol{v}\cdot\boldsymbol{v} &= \left(\frac{\partial \boldsymbol{r}}{\partial q_1}\dot{q}_1 + \cdots + \frac{\partial \boldsymbol{r}}{\partial q_n}\dot{q}_n\right)\cdot\left(\frac{\partial \boldsymbol{r}}{\partial q_1}\dot{q}_1 + \cdots + \frac{\partial \boldsymbol{r}}{\partial q_n}\dot{q}_n\right) \\ &= \left(\sum_{j=1}^{n}\frac{\partial \boldsymbol{r}}{\partial q_j}\dot{q}_j\right)\cdot\left(\sum_{k=1}^{n}\frac{\partial \boldsymbol{r}}{\partial q_k}\dot{q}_k\right) = \sum_{j=1}^{n}\sum_{k=1}^{n}\left(\frac{\partial \boldsymbol{r}}{\partial q_j}\cdot\frac{\partial \boldsymbol{r}}{\partial q_k}\right)\dot{q}_j\dot{q}_k, \end{aligned}$$

which is a homogeneous quadratic form in the variables $\dot{q}_1, \dots, \dot{q}_n$, with coefficients depending on $\boldsymbol{q}$. The kinetic energy of $\mathcal{S}$ is then given by

$$T = \tfrac{1}{2}\sum_{i=1}^{N} m_i\,(\boldsymbol{v}_i\cdot\boldsymbol{v}_i) = \sum_{j=1}^{n}\sum_{k=1}^{n} a_{jk}(\boldsymbol{q})\,\dot{q}_j\dot{q}_k,$$

where

$$a_{jk}(\boldsymbol{q}) = \tfrac{1}{2}\sum_{i=1}^{N} m_i \left(\frac{\partial \boldsymbol{r}_i}{\partial q_j} \cdot \frac{\partial \boldsymbol{r}_i}{\partial q_k}\right).$$

It follows that T is also a **homogeneous quadratic form** in the variables $\dot{q}_1, \dots, \dot{q}_n$, with coefficients depending on $\boldsymbol{q}$. ■

12.4 D'ALEMBERT'S PRINCIPLE

For a holonomic system, we can overcome the problem that the position vectors $\{\boldsymbol{r}_i\}$ are not independent variables by using generalised coordinates. We must now overcome the problem that the **constraint forces are unknown.**

The Newton equations of motion for the general mechanical system $\mathcal{S}$ are

$$m_i \dot{\boldsymbol{v}}_i = \boldsymbol{F}_i^S + \boldsymbol{F}_i^C \qquad (1 \le i \le N), \tag{12.8}$$

where $\boldsymbol{F}_i^S$ is the **specified force** and $\boldsymbol{F}_i^C$ is the **constraint force** acting on the particle P_i. The $\{\boldsymbol{F}_i^S\}$ are known while the $\{\boldsymbol{F}_i^C\}$ are unknown. The trick is to construct linear combinations of the equations (12.8) so as to eliminate the $\{\boldsymbol{F}_i^C\}$.

Let $\boldsymbol{a}_1(t)$, $\boldsymbol{a}_2(t)$, ..., $\boldsymbol{a}_N(t)$ be *any* vector functions of the time. Then, by taking the scalar product of equation (12.8) with $\boldsymbol{a}_i$ and summing over i, we obtain the scalar equation

$$\sum_{i=1}^{N} m_i \dot{\boldsymbol{v}}_i \cdot \boldsymbol{a}_i = \sum_{i=1}^{N} \boldsymbol{F}_i^S \cdot \boldsymbol{a}_i + \sum_{i=1}^{N} \boldsymbol{F}_i^C \cdot \boldsymbol{a}_i. \tag{12.9}$$

The question now is whether we can make

$$\sum_{i=1}^{N} \boldsymbol{F}_i^C \cdot \boldsymbol{a}_i = 0 \tag{12.10}$$

by a cunning choice of the functions $\{\boldsymbol{a}_i\}$. More precisely, since $\mathcal{S}$ has n degrees of freedom, we need n linearly independent choices of the $\{\boldsymbol{a}_i\}$ that make the equation (12.10) true.

Actually, we already know one choice of the $\{\boldsymbol{a}_i\}$ that makes the equation (12.10) true. Suppose that the total rate of working of the constraint forces is zero, which is true for many constraints (see Chapter 6). This condition can be written

$$\sum_{i=1}^{N} \boldsymbol{F}_i^C \cdot \boldsymbol{v}_i = 0,$$

where $\boldsymbol{v}_i$ is the velocity of the particle P_i at time t. Thus the condition (12.10) certainly holds for such a system if the $\{\boldsymbol{a}_i\}$ are chosen to be the particle velocities $\{\boldsymbol{v}_i\}$. With

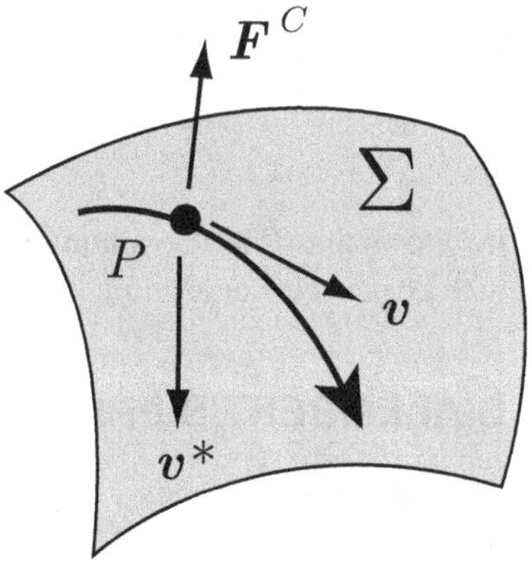

FIGURE 12.5 The particle P belonging to the system $\mathcal{S}$ is constrained to slide on the smooth fixed surface Σ.

this choice, the $\{F_i^C\}$ are eliminated from equation (12.9). The result of this operation is actually well known to us; it leads to the energy principle for the system! This is *not* quite what we are looking for, but it does suggest what the correct choices of the $\{a_i\}$ might be.

Now comes the clever bit. For all the usual constraints that do no work, it is also true that the stronger condition

$$\sum_{i=1}^{N} F_i^C \cdot v_i^* = 0 \tag{12.11}$$

holds, where the $\{v_i^*\}$ are *any kinematically possible set of particle velocities at time t*. The $\{v_i^*\}$ need not be the *actual* particle velocities at time t. For example, suppose a particle P of $\mathcal{S}$ is constrained to move on a *smooth* fixed surface Σ. Let v be the actual velocity of P as shown in Figure 12.5. Since Σ is a smooth surface, the constraint force F^C that it exerts must be *normal* to Σ. Moreover, any kinematically possible motion of $\mathcal{S}$ at time t gives P a velocity v^* that is *tangential* to Σ. It follows that

$$F^C \cdot v^* = 0,$$

for any choice of v^* that is kinematically possible. Although there is no *theorem* to this effect, a similar conclusion can be drawn for all the usual constraint forces that do no work.

A set of velocities $\{v_i^*\}$ that is kinematically possible at time t is called a **virtual motion** of the system. The condition (12.11) is therefore equivalent to the statement that the total rate of working of the constraint forces is zero in all virtual motions, or, more briefly, that the constraint forces do **no virtual work**. We have therefore obtained the following result, known as **d' Alembert's principle.***

* After Jean le Rond d'Alembert (1717–1783). He was baptised Jean le Rond after being found abandoned on the steps of a Paris church of that name. His principle was published in 1743 in his *Traite de Dynamique*.

D'Alembert's principle

If the constraint forces on a system do **no virtual work**, then

$$\sum_{i=1}^{N} m_i \dot{\boldsymbol{v}}_i \cdot \boldsymbol{v}_i^* = \sum_{i=1}^{N} \boldsymbol{F}_i^S \cdot \boldsymbol{v}_i^*, \qquad (12.12)$$

where $\{\boldsymbol{v}_i^*\}$ is any virtual motion of the system at time t.

Differential form of d'Alembert's principle

D'Alembert's principle is often quoted in the equivalent differential form

$$\sum_i m_i \dot{\boldsymbol{v}}_i \cdot d\boldsymbol{r}_i = \sum_i \boldsymbol{F}_i^S \cdot d\boldsymbol{r}_i,$$

where the $\{d\boldsymbol{r}_i\}$ are any *kinematically possible set of infinitesimal displacements* of the particles $\{P_i\}$ at time t. This form is also known as the **principle of virtual work**.

D'Alembert's principle is not often applied directly, except in statical problems. We have obtained it because it leads to Lagrange's equations.

12.5 LAGRANGE'S EQUATIONS

From now on, we will suppose that our mechanical system is holonomic *and* that its constraint forces do no virtual work.

Definition 12.5 ***Standard system*** *If a mechanical system is holonomic and its constraint forces do no virtual work, we will call it a* ***standard system****.*

Consider then a standard mechanical system with n degrees of freedom and generalised coordinates $\boldsymbol{q} = (q_1, q_2, \ldots, q_n)$. Consider first the virtual motion $\{\boldsymbol{v}_i^*\}$ generated by *prescribing* the generalised velocities at time t to be

$$\dot{q}_1 = 1, \; \dot{q}_2 = \cdots = \dot{q}_n = 0.$$

From (12.7) it follows that the corresponding particle velocities are given by

$$\boldsymbol{v}_i^* = \frac{\partial \boldsymbol{r}_i}{\partial q_1} \qquad (1, \ldots, N).$$

Since we are assuming our system to be *holonomic*, these $\{\boldsymbol{v}_i^*\}$ *are* a kinematically possible set of velocities.* Furthermore, since we are assuming that the constraint forces do *no*

* This is the only point in the derivation of Lagrange's equations where it is *essential* that the system be holonomic.

virtual work, d'Alembert's principle holds. It therefore follows that

$$\sum_{i=1}^{N} m_i \dot{\boldsymbol{v}}_i \cdot \frac{\partial \boldsymbol{r}_i}{\partial q_1} = \sum_{i=1}^{N} \boldsymbol{F}_i^S \cdot \frac{\partial \boldsymbol{r}_i}{\partial q_1}.$$

A similar argument holds when the $\{\dot{q}_j\}$ are prescribed to be $\dot{q}_1 = 0,\ \dot{q}_2 = 1, \dot{q}_3 = \cdots = \dot{q}_n = 0$, and so on. We thus obtain the system of equations

$$\sum_{i=1}^{N} m_i \dot{\boldsymbol{v}}_i \cdot \frac{\partial \boldsymbol{r}_i}{\partial q_j} = \sum_{i=1}^{N} \boldsymbol{F}_i^S \cdot \frac{\partial \boldsymbol{r}_i}{\partial q_j} \qquad (j = 1, \dots, n). \tag{12.13}$$

These are essentially Lagrange's equations. It remains only to put them into a form that is easy to use. In fact, the left sides of equations (12.13) can be constructed simply from the **kinetic energy** of the system. The result is as follows:

Suppose the holonomic system S has kinetic energy $T(\boldsymbol{q}, \dot{\boldsymbol{q}})$. Then the left sides of equations (12.13) can be written in the form

$$\sum_{i} m_i \dot{\boldsymbol{v}}_i \cdot \frac{\partial \boldsymbol{r}_i}{\partial q_j} = \frac{d}{dt}\left(\frac{\partial T}{\partial \dot{q}_j}\right) - \frac{\partial T}{\partial q_j} \tag{12.14}$$

$(1 \le j \le n)$*, where, for the purpose of calculating the partial derivatives, T is considered to be a function of the $2n$ independent variables $q_1, \dots, q_n, \dot{q}_1, \dots \dot{q}_n$.*

Lagrange partial derivatives

The partial derivatives of T that appear in equations (12.14) are peculiar to Lagrange's equations. In the expression for T, the coordinate velocities $\dot{q}_1, \dots \dot{q}_n$ are considered to be *independent variables* in addition to the coordinates $q_1, \dots, q_n$. Consider, for example, the two particle system in Example 12.2. For this system

$$T = m\dot{x}^2 + (\tfrac{1}{2}ma^2)\dot{\theta}^2 + (ma\cos\theta)\,\dot{x}\dot{\theta}$$

and this expression is considered to be a function of the *four* independent variables $x, \theta, \dot{x}, \dot{\theta}$ (x is absent). The Lagrange partial derivatives of T are therefore

$$\frac{\partial T}{\partial x} = 0, \quad \frac{\partial T}{\partial \dot{x}} = 2m\dot{x} + (ma\cos\theta)\dot{\theta}, \quad \frac{\partial T}{\partial \theta} = -(ma\sin\theta)\dot{x}\dot{\theta}, \quad \frac{\partial T}{\partial \dot{\theta}} = ma^2\dot{\theta} + (ma\cos\theta)\dot{x}.$$

The proof of the formula (12.14) is straightforward (once Lagrange had found the answer!) but a bit messy because of the many suffices.

Proof of the formula (12.14)

Since

$$\boldsymbol{v}_i = \frac{\partial \boldsymbol{r}_i}{\partial q_1}\dot{q}_1 + \cdots + \frac{\partial \boldsymbol{r}_i}{\partial q_n}\dot{q}_n,$$

it follows that

$$\frac{\partial}{\partial \dot{q}_j}\left(\tfrac{1}{2}\boldsymbol{v}_i\cdot\boldsymbol{v}_i\right)=\boldsymbol{v}_i\cdot\frac{\partial \boldsymbol{v}_i}{\partial \dot{q}_j}=\boldsymbol{v}_i\cdot\frac{\partial \boldsymbol{r}_i}{\partial q_j}.$$

The last step follows since, in the formula for $\boldsymbol{v}_i$, $\boldsymbol{q}$ and $\dot{\boldsymbol{q}}$ are regarded as independent variables*. Then

$$\begin{aligned}\frac{d}{dt}\left[\frac{\partial}{\partial \dot{q}_j}\left(\tfrac{1}{2}\boldsymbol{v}_i\cdot\boldsymbol{v}_i\right)\right]&=\dot{\boldsymbol{v}}_i\cdot\frac{\partial \boldsymbol{r}_i}{\partial q_j}+\boldsymbol{v}_i\cdot\frac{d}{dt}\left(\frac{\partial \boldsymbol{r}_i}{\partial q_j}\right)\\&=\dot{\boldsymbol{v}}_i\cdot\frac{\partial \boldsymbol{r}_i}{\partial q_j}+\boldsymbol{v}_i\cdot\sum_{k=1}^{n}\frac{\partial^2 \boldsymbol{r}_i}{\partial q_k\partial q_j}\dot{q}_k,\end{aligned}$$

after a further application of the chain rule. In a similar way,

$$\frac{\partial}{\partial q_j}\left(\tfrac{1}{2}\boldsymbol{v}_i\cdot\boldsymbol{v}_i\right)=\boldsymbol{v}_i\cdot\frac{\partial \boldsymbol{v}_i}{\partial q_j}=\boldsymbol{v}_i\cdot\frac{\partial}{\partial q_j}\left(\sum_{k=1}^{n}\frac{\partial \boldsymbol{r}_i}{\partial q_k}\dot{q}_k\right)=\boldsymbol{v}_i\cdot\sum_{k=1}^{n}\frac{\partial^2 \boldsymbol{r}_i}{\partial q_j\partial q_k}\dot{q}_k,$$

where, in the formula for $\partial\boldsymbol{v}_i/\partial q_j$, we have regarded $\boldsymbol{q}$ and $\dot{\boldsymbol{q}}$ as independent variables. Combining these two results gives

$$\frac{d}{dt}\left[\frac{\partial}{\partial \dot{q}_j}\left(\tfrac{1}{2}\boldsymbol{v}_i\cdot\boldsymbol{v}_i\right)\right]-\frac{\partial}{\partial q_j}\left(\tfrac{1}{2}\boldsymbol{v}_i\cdot\boldsymbol{v}_i\right)=\dot{\boldsymbol{v}}_i\cdot\frac{\partial \boldsymbol{r}_i}{\partial q_j}.$$

If we now multiply by m_i and sum over i we obtain

$$\frac{d}{dt}\left(\frac{\partial T}{\partial \dot{q}_j}\right)-\frac{\partial T}{\partial q_j}=\sum_i m_i\dot{\boldsymbol{v}}_i\cdot\frac{\partial \boldsymbol{r}_i}{\partial q_j},$$

which is the required result. ■

For **general specified forces** $\{\boldsymbol{F}_i\}$ there is no simplification for the right sides of the equations (12.13), but we do give them names:

Definition 12.6 ***Generalised force*** *The quantity Q_j, defined by*

$$Q_j=\sum_i \boldsymbol{F}_i^S\cdot\frac{\partial \boldsymbol{r}_i}{\partial q_j}$$

is called the ***generalised force*** *corresponding to the coordinate q_j.*

We have therefore proved that:

---

* The formula

$$\frac{\partial \dot{\boldsymbol{r}}_i}{\partial \dot{q}_j}=\frac{\partial \boldsymbol{r}_i}{\partial q_j}$$

is sometimes facetiously referred to as 'cancelling the dots'. Only mathematicians find this amusing.

Lagrange's equations for a general standard system

Let $\mathcal{S}$ be a **standard system** with generalised coordinates $\boldsymbol{q}$, kinetic energy $T(\boldsymbol{q}, \dot{\boldsymbol{q}})$ and generalised forces $\{Q_j\}$. Then, in any motion of $\mathcal{S}$, the coordinates $\boldsymbol{q}(t)$ must satisfy the system of equations

$$\frac{d}{dt}\left(\frac{\partial T}{\partial \dot{q}_j}\right) - \frac{\partial T}{\partial q_j} = Q_j \qquad (1 \le j \le n). \tag{12.15}$$

This is the **form of Lagrange's equations** that applies to **any standard system**.

Conservative systems

When the standard system is also **conservative**, the generalised forces $\{Q_j\}$ can be written in terms of the **potential energy** $V(\boldsymbol{q})$ as

$$Q_j = -\frac{\partial V}{\partial q_j}. \tag{12.16}$$

[This result is simply a generalisation of the formula $\boldsymbol{F} = -\operatorname{grad} V$.]

Proof of the formula (12.16)
Let $\boldsymbol{q}^A$, $\boldsymbol{q}^B$ be any two points of configuration space that can be joined by a straight line parallel to the q_j-axis. Then

$$\begin{aligned}\int_{\boldsymbol{q}^A}^{\boldsymbol{q}^B} Q_j\, dq_j &= \int_{\boldsymbol{q}^A}^{\boldsymbol{q}^B} \left(\sum_i \boldsymbol{F}_i^S \cdot \frac{\partial \boldsymbol{r}_i}{\partial q_j}\right) dq_j = \sum_i \int_{C_i} \boldsymbol{F}_i^S \cdot d\boldsymbol{r}\\ &= V(\boldsymbol{q}^A) - V(\boldsymbol{q}^B) = -\int_{\boldsymbol{q}^A}^{\boldsymbol{q}^B} \frac{\partial V}{\partial q_j}\, dq_j.\end{aligned}$$

This equality holds for all $\boldsymbol{q}^A$, $\boldsymbol{q}^B$ chosen as described, which implies that the two integrands must be equal. Hence

$$Q_j = -\frac{\partial V}{\partial q_j},$$

as required. ■

We have therefore proved that:

Lagrange's equations for a conservative standard system

Let $\mathcal{S}$ be a **conservative standard system** with generalised coordinates $\boldsymbol{q}$, kinetic energy $T(\boldsymbol{q}, \dot{\boldsymbol{q}})$ and potential energy $V(\boldsymbol{q})$. Then, in any motion of $\mathcal{S}$, the coordinates $\boldsymbol{q}(t)$ must satisfy the system of equations

$$\frac{d}{dt}\left(\frac{\partial T}{\partial \dot{q}_j}\right) - \frac{\partial T}{\partial q_j} = -\frac{\partial V}{\partial q_j} \qquad (1 \le j \le n). \qquad (12.17)$$

These are **Lagrange's equations** for a **conservative standard system**. This is by far the most important case; most of analytical mechanics deals with conservative systems. It is remarkable that *all one needs to obtain the equations of motion for a conservative system are the expressions for the* ***kinetic*** *and* ***potential energies***.

Sufficiency of the Lagrange equations

We have shown that if $\mathcal{S}$ is a conservative standard system then Lagrange's equations (12.17) must hold. Thus Lagrange's equations are *necessary* conditions for $\boldsymbol{q}(t)$ to be a motion of $\mathcal{S}$. It does not seem possible to reverse this argument to show that the Lagrange equations are also *sufficient*. (Where does the reverse argument break down?) However, from the general theory of ODEs, we are assured that there is a unique solution of the Lagrange equations corresponding to each set of initial values for $\boldsymbol{q}$, $\dot{\boldsymbol{q}}$. Thus the Lagrange equations actually *are* sufficient to determine the motion of $\mathcal{S}$.

The Lagrange method for finding the equations of motion of a conservative system is summarised below:

Lagrange's method for conservative systems

- Confirm that the system is standard and that the specified forces are conservative.
- Select generalised coordinates.
- Evaluate the expressions for T and V in terms of the chosen coordinates*.
- Substitute these expressions into the Lagrange equations (12.17) and turn the handle. It's a piece of cake!

Example 12.10 ***Using Lagrange's equations: I***

Consider a block of mass m sliding on a smooth wedge of mass M and angle α which itself slides on a smooth horizontal floor, as shown in Figure 12.6. The whole motion

* See Chapter 9 for the details of how to find T.

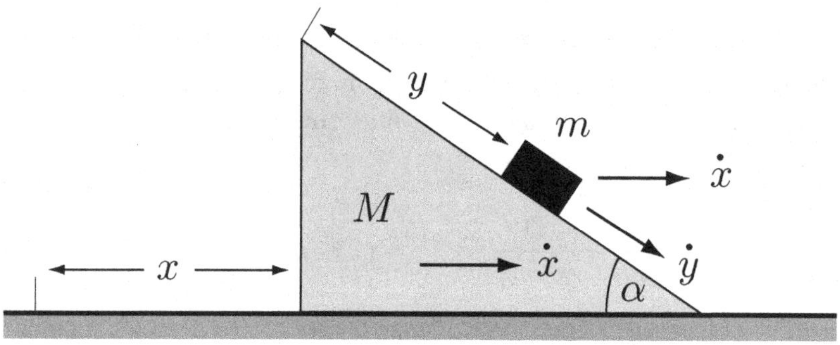

FIGURE 12.6 The block slides on the smooth surface of the wedge which slides on a smooth horizontal floor.

is planar. Find Lagrange's equations for this system and deduce (i) the acceleration of the wedge, and (ii) the acceleration of the block relative to the wedge.

Solution

This is a standard conservative system with two degrees of freedom. Take as generalised coordinates x, the displacement of the wedge from a fixed point on the floor, and y, the displacement of the block from a fixed point on the wedge. The calculation of the **kinetic** and **potential energies** in terms of x, y is performed exactly as in Chapter 9 and gives

$$T = \tfrac{1}{2}M\dot{x}^2 + \tfrac{1}{2}m\left(\dot{x}^2 + \dot{y}^2 + 2\dot{x}\dot{y}\cos\alpha\right),$$
$$V = -mgy\sin\alpha.$$

The required partial derivatives of T and V are then given by

$$\frac{\partial T}{\partial x} = 0, \qquad \frac{\partial T}{\partial \dot{x}} = (M+m)\dot{x} + (m\cos\alpha)\dot{y}, \qquad \frac{\partial V}{\partial x} = 0.$$

$$\frac{\partial T}{\partial y} = 0, \qquad \frac{\partial T}{\partial \dot{y}} = (m\cos\alpha)\dot{x} + m\dot{y}, \qquad \frac{\partial V}{\partial y} = -mg\sin\alpha.$$

We can now form up the **Lagrange equations**. The equation corresponding to the coordinate x is

$$\frac{d}{dt}\left[(M+m)\dot{x} + (m\cos\alpha)\dot{y}\right] - 0 = 0, \tag{12.18}$$

and the equation corresponding to the coordinate y is

$$\frac{d}{dt}\left[(m\cos\alpha)\dot{x} + m\dot{y}\right] - 0 = mg\sin\alpha. \tag{12.19}$$

If we now perform the time derivatives in equations (12.18), (12.19) and solve for the unknowns $\ddot{x}$, $\ddot{y}$ we obtain

$$\ddot{x} = -\frac{mg\sin\alpha\cos\alpha}{M + m\sin^2\alpha}, \qquad \ddot{y} = \frac{(M+m)g\sin\alpha}{M + m\sin^2\alpha},$$

which are the required **accelerations**. They are both constant. ■

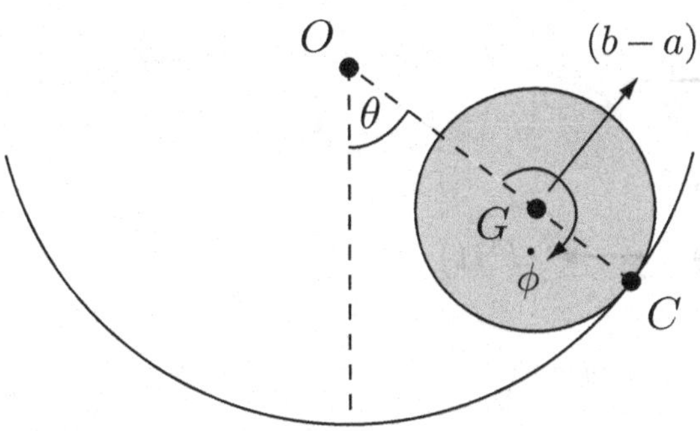

FIGURE 12.7 The small solid cylinder rolls on the inside surface of the large fixed cylinder.

These results can of course be obtained by more elementary means. For instance we could solve this problem by appealing to conservation of horizontal linear momentum and energy. However the Lagrange method does have the advantage that less physical insight is needed to solve the problem. If the system is a standard one and T and V can be calculated, then turning the handle produces the equations of motion.

Example 12.11 *Using Lagrange's equations: II*

Figure 12.7 shows a solid cylinder with centre G and radius a rolling on the rough inside surface of a *fixed* cylinder with centre O and radius $b > a$. Find the Lagrange equation of motion and deduce the period of small oscillations about the equilibrium position.

Solution

If the cylinder were not obliged to roll, the system would have two degrees of freedom with generalised coordinates θ (the angle between OG and the downward vertical) and ϕ (the rotation angle of the cylinder measured from some reference position). The **rolling condition** imposes the kinematical constraint

$$(b-a)\dot{\theta} - a\dot{\phi} = 0.$$

This constraint is **integrable** and is equivalent to the geometrical constraint

$$(b-a)\theta - a\phi = 0$$

on taking $\phi = 0$ when $\theta = 0$. Thus the rolling cylinder is a standard conservative system with *one* degree of freedom.

Take θ as the generalised coordinate. Then the **kinetic energy** is given by

$$\begin{aligned}
T &= \tfrac{1}{2}m\left((b-a)\dot{\theta}\right)^2 + \tfrac{1}{2}\left(\tfrac{1}{2}ma^2\right)\dot{\phi}^2 \\
&= \tfrac{1}{2}m\left((b-a)\dot{\theta}\right)^2 + \tfrac{1}{2}\left(\tfrac{1}{2}ma^2\right)\left(\frac{b-a}{a}\right)^2\dot{\theta}^2 \\
&= \tfrac{3}{4}m(b-a)^2\dot{\theta}^2
\end{aligned}$$

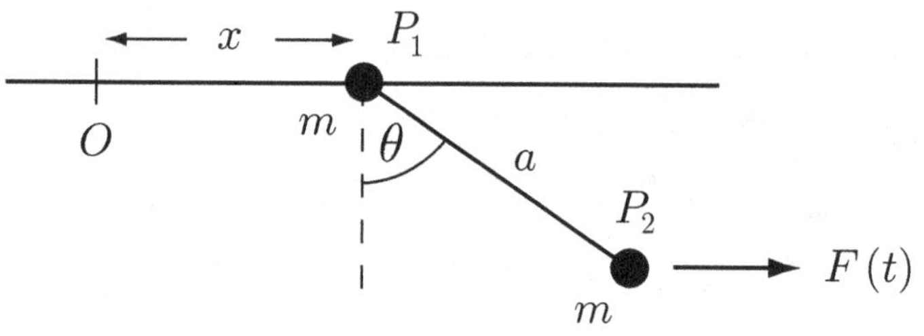

FIGURE 12.8 The system moves under the prescribed force $F(t)$.

and the **potential energy** by

$$V = -mg(b-a)\cos\theta.$$

There is only one **Lagrange equation**, namely

$$\frac{d}{dt}\left[\tfrac{3}{2}m(b-a)^2\dot{\theta}\right] - 0 \;=\; -mg(b-a)\sin\theta$$

which simplifies to give

$$\ddot{\theta} + \frac{2g}{3(b-a)}\sin\theta = 0.$$

Interestingly, this equation is identical to the exact equation for the oscillations of a simple pendulum of length $3(b-a)/2$ as obtained in Chapter 6.

The linearised equation governing small oscillations of the cylinder about $\theta = 0$ is

$$\ddot{\theta} + \frac{2g}{3(b-a)}\theta = 0$$

so that the **period** τ of small oscillations is given by

$$\tau = 2\pi\left(\frac{3(b-a)}{2g}\right)^{1/2}. \blacksquare$$

Example 12.12 *Using Lagrange's equations: III*

Let $\mathcal{S}$ be the system shown in Figure 12.8. The rail is smooth and the prescribed force $F(t)$ acts on the particle P_2 as shown. Gravity is absent. Find the Lagrange equations for $\mathcal{S}$.

Solution

$\mathcal{S}$ is a standard system with two degrees of freedom. The new feature is the prescribed external force $F(t)$ acting on P_2. This time dependent force cannot be represented by a potential energy and so the generalised forces $\{Q_j\}$ must be evaluated direct from the definition (12.16).

Take generalised coordinates x, θ as shown and let the corresponding generalised forces be called Q_x, Q_θ. Then, since $\mathcal{S}$ has just two particles,

$$Q_x = \boldsymbol{F}_1^S \cdot \frac{\partial \boldsymbol{r}_1}{\partial x} + \boldsymbol{F}_2^S \cdot \frac{\partial \boldsymbol{r}_2}{\partial x},$$

$$Q_\theta = \boldsymbol{F}_1^S \cdot \frac{\partial \boldsymbol{r}_1}{\partial \theta} + \boldsymbol{F}_2^S \cdot \frac{\partial \boldsymbol{r}_2}{\partial \theta},$$

where

$$\boldsymbol{F}_1^S = \boldsymbol{0}, \qquad \boldsymbol{F}_2^S = F(t)\,\boldsymbol{i},$$

and

$$\boldsymbol{r}_1 = x\,\boldsymbol{i}, \qquad \boldsymbol{r}_2 = (x + a\sin\theta)\,\boldsymbol{i} - (a\cos\theta)\,\boldsymbol{k}.$$

The **generalised forces** Q_x, Q_θ are therefore given by

$$Q_x = 0 + (F(t)\,\boldsymbol{i}) \cdot \boldsymbol{i} = F(t)$$

and

$$Q_\theta = 0 + (F(t)\,\boldsymbol{i}) \cdot (a\cos\theta\,\boldsymbol{i} + a\sin\theta\,\boldsymbol{k}) = (a\cos\theta)\,F(t).$$

The **kinetic energy** is given by

$$T = m\dot{x}^2 + (ma\cos\theta)\dot{x}\dot{\theta} + \tfrac{1}{2}m\dot{\theta}^2$$

and so the **Lagrange equations** are

$$\frac{d}{dt}\left[2m\dot{x} + (ma\cos\theta)\dot{\theta}\right] = F(t),$$

$$\frac{d}{dt}\left[(ma\cos\theta)\dot{x} + m\dot{\theta}\right] - \left[-(ma\sin\theta)\dot{x}\dot{\theta}\right] = (a\cos\theta)F(t). \blacksquare$$

Question ***Incorporating extra forces***

How would you incorporate gravity into the last example?

Answer

Since the expression (12.16) for the $\{Q_j\}$ is linear in the $\{\boldsymbol{F}_i\}$, the extra forces are incorporated by just *adding* in their respective contributions to the $\{Q_j\}$. Thus when gravity is present in the last example, Q_x, Q_θ become

$$Q_x = F(t) + 0, \qquad Q_\theta = (a\cos\theta)F(t) - mga\sin\theta. \blacksquare$$

Many more examples of the use of Lagrange's equations are given in the problems at the end of the chapter.

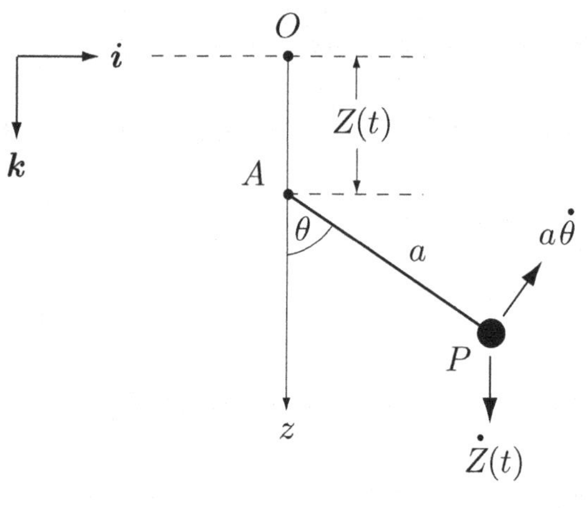

FIGURE 12.9 The pendulum with a moving support.

12.6 SYSTEMS WITH MOVING CONSTRAINTS

The theory of Lagrange's equations can be extended to include a fascinating class of problems in which the constraints are time dependent. Consider the system shown in Figure 12.9 which is a simple pendulum in which the support point A is made to move vertically so that its downward displacement from the *fixed* origin O at time t is some *specified* function $Z(t)$. For example it could be made to oscillate so that $Z(t) = Z_0 \cos pt$. With this constraint, the coordinate θ is no longer sufficient to specify the position of the particle P. In fact, relative to the origin O, the position vector of P at time t is given by

$$\boldsymbol{r} = (a \sin\theta)\,\boldsymbol{i} + (Z(t) + a\cos\theta)\,\boldsymbol{k},$$

so that $\boldsymbol{r}$ is a function of θ and t, not just θ. Constraints which cause the $\{\boldsymbol{r}_i\}$ to depend on $\boldsymbol{q}$ and t (and not just $\boldsymbol{q}$) are called **time dependent constraints**, or simply **moving constraints**. Systems that have moving constraints include:

Systems with moving constraints

- Systems in which particles are forced to move *in a prescribed manner*.
- Systems in which particles are forced to remain on boundaries that move *in a prescribed manner*.
- Systems in which the motion is viewed from a frame of reference that is accelerating or rotating *in a prescribed manner*.
- Systems in which beetles, mice (or lions!) move around *in a prescribed manner*. (These creatures are highly trained!)

We assume our systems are such that they *would* be standard *if* the constraints were fixed. We will refer to such systems as **standard systems with moving constraints**.

Kinematics of systems with moving constraints

The configuration $\{\boldsymbol{r}_i\}$ of a system with moving constraints is specified by

$$\boldsymbol{r}_i = \boldsymbol{r}_i(\boldsymbol{q}, t) \qquad (1 \le i \le N). \tag{12.20}$$

Here, the time t has the rôle of an 'additional coordinate'. However, it is not a true coordinate and we will still regard the system as being holonomic with n degrees of freedom. The corresponding particle velocities are given by

$$\boldsymbol{v}_i = \frac{\partial \boldsymbol{r}_i}{\partial q_1}\dot{q}_1 + \cdots + \frac{\partial \boldsymbol{r}_i}{\partial q_n}\dot{q}_n + \frac{\partial \boldsymbol{r}_i}{\partial t}. \tag{12.21}$$

This expression is still a linear form in the variables $\{\dot{q}_j\}$ but it is not homogeneous; there is now a 'constant' term (a function of $\boldsymbol{q}$ and t).

Question ***Form of the kinetic energy***

What is the form of the kinetic energy when moving constraints are present?

Answer

It follows from the above expression for the particle velocities that T has the form

$$T(\boldsymbol{q}, \dot{\boldsymbol{q}}, t) = \sum_{j=1}^{n}\sum_{k=1}^{n} a_{jk}(\boldsymbol{q}, t)\dot{q}_j\dot{q}_k + \sum_{j=1}^{n} b_j(\boldsymbol{q}, t)\dot{q}_j + c(\boldsymbol{q}, t),$$

which is still a quadratic form in the variables $\{\dot{q}_j\}$, but it is not homogeneous; there are now linear terms and a constant term. ■

Energy not conserved with moving constraints

Another feature of systems with moving constraints is that the **constraint forces do work**. This is quite obvious from the driven pendulum example above. The constraint force that causes the specified displacement of the support point A will generally have a vertical component and, since A is *moving* vertically, this force will do work. So, even when the *specified* forces are conservative (as gravity is in the driven pendulum example), the total energy $T + V$ is not a constant because the constraint force does work. Hence, systems with moving constraints are generally **not conservative**.

Lagrange's equations with moving constraints

There are good reasons to expect that systems with moving constraints do not satisfy Lagrange's equations. In general, constraint forces that enforce moving constraints do work. Since virtual motions include the special case of real motion, surely such constraints must also do *virtual* work; then d'Alemberts principle and Lagrange's equations will not hold. Compelling though this argument seems, it is false. Systems with moving constraints *do* satisfy Lagrange's equations! To see why this is so, one must identify the crucial steps in the derivation of Lagrange's equations. There are actually only three:

- Are the equations $\sum_{i=1}^{N} \boldsymbol{F}_i^C \cdot (\partial \boldsymbol{r}_i/\partial q_j) = 0$ still true?
  This question could be posed in the form '*do the constraint forces do virtual work?*' and we have presented a plausible argument that they do. Consider however the meaning of the partial derivatives $\partial \boldsymbol{r}_i/\partial q_j$. Since we now have $\boldsymbol{r}_i = \boldsymbol{r}_i(\boldsymbol{q}, t)$, $\partial \boldsymbol{r}_i/\partial q_1$ means the derivative of $\boldsymbol{r}_i$ with respect to q_1 keeping $q_2, q_3, \dots, q_n$ *and the time t* constant. Thus these derivatives are calculated at constant t. It follows that the virtual motion defined by the $\{\partial \boldsymbol{r}_i/\partial q_1\}$ is kinematically consistent with constraints that are *fixed* at time t, not with the actual moving constraints. Hence, since $\sum_{i=1}^{N} \boldsymbol{F}_i^C \cdot (\partial \boldsymbol{r}_i/\partial q_j)$ would be zero if the constraints were fixed, it is still zero when the constraints are moving!*
- Is the formula (12.14) still true?
  The point here is that the formula (12.21) for the particle velocities now has the extra term $\partial \boldsymbol{r}_i/\partial t$ which might upset the formula (12.14). However, it does not. The proof of this is left as an exercise.
- When the specified forces are conservative, is the formula $\sum_{i=1}^{N} \boldsymbol{F}_i^S \cdot (\partial \boldsymbol{r}_i/\partial q_j) = \partial V/\partial q_j$ still true?
  The potential energy V is a function of the configuration of the system, but, since the configuration is now specified by $\boldsymbol{q}$ and t, $V = V(\boldsymbol{q}, t)$. However, since the partial derivatives $\partial \boldsymbol{r}_i/\partial q_j$ and $\partial V/\partial q_j$ are evaluated at constant t, the proof of the formula is unchanged.

We have therefore obtained the following result:

Lagrange's equations with moving constraints

Lagrange's equations still hold when moving constraints are present provided that, in the expressions for T and V, the time t is regarded as an *independent* variable.

Example 12.13 *Pendulum with an oscillating support*

Find the Lagrange equation for the driven pendulum for the case in which the displacement function $Z(t) = Z_0 \cos pt$. [Assume that the 'string' is a light rigid rod that cannot go slack.]

Solution

This system has one degree of freedom and a moving constraint at A. Take θ as the generalised coordinate. It follows from Figure 12.9 that the **kinetic energy** T is given by

$$T = \tfrac{1}{2}m\left(a^2\dot{\theta}^2 + \dot{Z}^2 - 2a\dot{\theta}\dot{Z}\sin\theta\right)$$

* The situation is sometimes loosely expressed by the mysterious statement that '*moving constraints do real work but no virtual work*'.

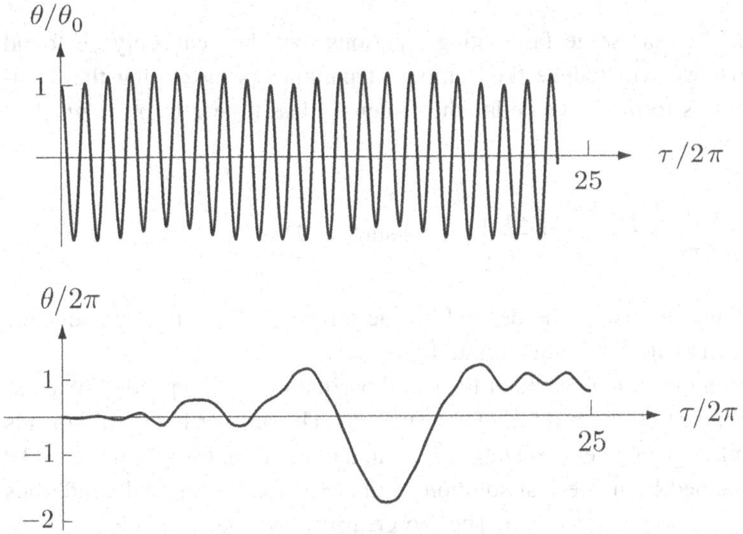

FIGURE 12.10 Motions of the driven pendulum. **Top:** $p/\Omega = 1.1$, $Z_0/a = 0.2$ and $\theta_0 = 0.1$. **Bottom:** $p/\Omega = 1.9$, $Z_0/a = 0.2$ and $\theta_0 = 0.1$.

and the **potential energy** V by

$$V = -mg(Z + a\cos\theta).$$

The required partial derivatives are therefore

$$\frac{\partial T}{\partial\dot\theta} = m\left(a^2\dot\theta - a\dot Z\sin\theta\right), \qquad \frac{\partial T}{\partial\theta} = -ma\,\dot\theta\dot Z\cos\theta, \qquad \frac{\partial V}{\partial\theta} = mga\sin\theta.$$

The Lagrange equation corresponding to the coordinate θ is therefore

$$\frac{d}{dt}ma\left(a\dot\theta + \dot Z\sin\theta\right) - ma\dot\theta\dot Z\cos\theta = -mga\sin\theta,$$

which simplifies to give

$$\ddot\theta + (\Omega^2 - a^{-1}\ddot Z)\sin\theta = 0,$$

where $\Omega^2 = g/a$. Hence, for the case in which $Z = Z_0\cos pt$, the **Lagrange equation** is

$$\ddot\theta + \left(\Omega^2 + \frac{Z_0p^2}{a}\cos pt\right)\sin\theta = 0. \blacksquare \tag{12.22}$$

Question *Motions of the driven pendulum*

What do the pendulum motions look like?

Answer

The equation (12.22) has some fascinating solutions, but they can only be found numerically. First we will reduce the number of parameters by putting the equation in dimensionless form. If we define the dimensionless time τ by $\tau = pt$, then the equation becomes

$$\frac{d^2\theta}{d\tau^2} + \left(\frac{\Omega^2}{p^2} + \left(\frac{Z_0}{a} \right) \cos\tau \right) \sin\theta = 0.$$

We can now see that the solutions depend on the dimensionless driving frequency p/Ω and the dimensionless driving amplitude Z_0/a.

One interesting question is whether the small oscillations of the pendulum about $\theta = 0$ are destabilised by the motion of the support. The answer is that it depends on the dimensionless parameters p/Ω and Z_0/a in a complicated way. Figure 12.10 shows results obtained by numerical solution of the equation with initial conditions of the form $\theta = \theta_0, \dot{\theta} = 0$ when $t = 0$. The **top graph** shows the motion for the case $p/\Omega = 1.1$, $Z_0/a = 0.2$ and $\theta_0 = 0.1$. In this graph, θ/θ_0 is plotted against $\tau/2\pi$ (the number of oscillations of the support). The motion turns out to be **stable** with the amplitude of the oscillations remaining close to their initial value. The **bottom graph** shows the motion for the case $p/\Omega = 1.9$, $Z_0/a = 0.2$ and $\theta_0 = 0.1$. In this graph, $\theta/2\pi$ (the number of *revolutions* of the pendulum) is plotted against $\tau/2\pi$ (the number of oscillations of the support). This motion turns out to be **unstable**. The amplitude of the oscillations grows until the pendulum performs complete circles; it then stops and goes the opposite way. Numerical results suggest that this chaotic motion continues indefinitely. ■

12.7 THE LAGRANGIAN

The Lagrange equations of motion (12.17) for a conservative standard system are expressed in terms of the kinetic energy $T = T(\boldsymbol{q}, \dot{\boldsymbol{q}})$ and the potential energy $V = V(\boldsymbol{q})$. They can however be written in terms of the single function $T - V$. Since $\partial V/\partial \dot{q}_j = 0$, the equations can be written

$$\frac{d}{dt}\left(\frac{\partial T}{\partial \dot{q}_j} \right) - \frac{\partial T}{\partial q_j} = \frac{d}{dt}\left(\frac{\partial V}{\partial \dot{q}_j} \right) - \frac{\partial V}{\partial q_j} \qquad (1 \le j \le n),$$

that is,

$$\frac{d}{dt}\left(\frac{\partial L}{\partial \dot{q}_j} \right) - \frac{\partial L}{\partial q_j} = 0 \qquad (1 \le j \le n),$$

where $L(\boldsymbol{q}, \dot{\boldsymbol{q}}) = T(\boldsymbol{q}, \dot{\boldsymbol{q}}) - V(\boldsymbol{q})$ is called the **Lagrangian** of the system. The same operation can be applied to systems with moving constraints whose specified forces are conservative. The only difference is that, in this case, $L = L(\boldsymbol{q}, \dot{\boldsymbol{q}}, t)$.

Writing the Lagrange equations in this form makes no difference whatever to problem solving. However, any system of equations that can be written in this way has special properties. In particular, it is equivalent to a **stationary principle** (see Chapter 13), and can also be written in **Hamiltonian form** (see Chapter 14). This is the form most suitable for advanced developments and for making the transition to quantum mechanics. There is therefore a strong interest in *any* physical system whose equations can be written in Lagrangian form.

Definition 12.7 ***Lagrangian form*** *If the equations of motion of a holonomic system with generalised coordinates* $\boldsymbol{q}$ *can be written in the form*

$$\frac{d}{dt}\left(\frac{\partial L}{\partial \dot{q}_j}\right) - \frac{\partial L}{\partial q_j} = 0 \qquad (1 \le j \le n), \qquad (12.23)$$

for some function $L = L(\boldsymbol{q}, \dot{\boldsymbol{q}}, t)$*, then* L *is called the* ***Lagrangian*** *of the system and the equations are said to have* ***Lagrangian form****.*

For example, the Lagrangian for the driven pendulum is

$$L(\theta, \dot{\theta}, t) = \tfrac{1}{2}m\left(a^2\dot{\theta}^2 + \dot{Z}^2 - 2a\dot{\theta}\dot{Z}\sin\theta\right) + mg(Z + a\cos\theta),$$

where $Z = Z(t)$ is the displacement of the support point.

Velocity dependent potential

There are systems whose specified forces are **not conservative** (so that V does not exist), but their equations of motion can still be written in Lagrangian form. Any standard system with generalised forces $\{Q_j\}$ satisfies the Lagrange equations (12.15). If it happens that the generalised forces can be written in the form

$$Q_j = \frac{d}{dt}\left(\frac{\partial U}{\partial \dot{q}_j}\right) - \frac{\partial U}{\partial q_j} \qquad (1 \le j \le n), \qquad (12.24)$$

for some function $U(\boldsymbol{q}, \dot{\boldsymbol{q}}, t)$, then clearly the equations (12.15) can be written in Lagrangian form by taking

$$L(\boldsymbol{q}, \dot{\boldsymbol{q}}, t) = T(\boldsymbol{q}, \dot{\boldsymbol{q}}, t) - U(\boldsymbol{q}, \dot{\boldsymbol{q}}.t).$$

The function $U(\boldsymbol{q}, \dot{\boldsymbol{q}}, t)$ is called the **velocity dependent potential** of the system.

This seems to be a mathematical artifice that has no importance in practice. It is true that there is only *one* important case in which a velocity dependent potential exists, but that case is very important; it is the case of a charged particle moving in electromagnetic fields. The following example proves this to be so for *static* fields. The corresponding result for electro*dynamic* fields is the subject of Problem 12.15.

Example 12.14 *Charged particle in static EM fields*

A particle P of mass m and charge e can move freely in the static electric field $\boldsymbol{E} = \boldsymbol{E}(\boldsymbol{r})$ and the static magnetic field $\boldsymbol{B} = \boldsymbol{B}(\boldsymbol{r})$. The electric and magnetic fields exert a force on P given by the **Lorentz force** formula

$$\boldsymbol{F} = e\,\boldsymbol{E} + e\,\boldsymbol{v}\times\boldsymbol{B},$$

where $\boldsymbol{v}$ is the velocity of P. Show that this force can be represented by a velocity dependent potential $U(\boldsymbol{r}, \dot{\boldsymbol{r}})$ and find the Lagrangian of the system.

Solution

In the static case, Maxwell's equations for the electromagnetic field reduce to

$$\text{div}\,\boldsymbol{D} = \rho, \qquad \text{curl}\,\boldsymbol{E} = \boldsymbol{0}, \qquad \text{curl}\,\boldsymbol{H} = \boldsymbol{j}, \qquad \text{div}\,\boldsymbol{B} = 0.$$

In particular, the equation curl $\boldsymbol{E} = \boldsymbol{0}$ implies that $\boldsymbol{E}(\boldsymbol{r})$ is a conservative field and can be written in the form

$$\boldsymbol{E} = -\,\text{grad}\,\phi$$

where $\phi = \phi(\boldsymbol{r})$ is the **electrostatic potential**. The equation div $\boldsymbol{B} = 0$ implies that $\boldsymbol{B}(\boldsymbol{r})$ can be written in the form

$$\boldsymbol{B} = \text{curl}\,\boldsymbol{A},$$

where $\boldsymbol{A} = \boldsymbol{A}(\boldsymbol{r})$ is the **magnetic vector potential.***

Take the generalised coordinates to be the Cartesian coordinates x, y z of the particle and, from now on, let $\boldsymbol{r}$ mean (x, y, z). What we are looking for is a velocity dependent potential $U(\boldsymbol{r}, \dot{\boldsymbol{r}})$ that yields the correct generalised forces when substituted into the equations (12.24). In the present case the generalised forces Q_x, Q_y, Q_z are simply the x- y- and z-components of the Lorentz force $\boldsymbol{F}$. The **electric** part of the force, $e\boldsymbol{E}$ is easily dealt with since it is conservative and can be represented by the ordinary potential energy $V = e\,\phi(\boldsymbol{r})$. It is the **magnetic** part of the force that needs the velocity dependent potential. One wonders how anyone found the correct U, but they did, and it turns out to be $-e\,\dot{\boldsymbol{r}} \cdot \boldsymbol{A}(\boldsymbol{r})$. All *we* need to do is to check that this potential is correct.

Consider therefore the potential

$$U = \dot{\boldsymbol{r}} \cdot \boldsymbol{A}(\boldsymbol{r}) = \dot{x}A_x(\boldsymbol{r}) + \dot{y}A_y(\boldsymbol{r}) + \dot{z}A_z(\boldsymbol{r}).$$

* The potential ϕ is unique to within an added constant, but, for any fixed $\boldsymbol{B}$, there are many possibilities for $\boldsymbol{A}$. Adding the grad of any scalar function to $\boldsymbol{A}$ does not change the value of $\boldsymbol{B}$. This ambiguity in $\boldsymbol{A}$ makes no difference in the present context; *any* choice of $\boldsymbol{A}$ such that $\boldsymbol{B} = \text{curl}\,\boldsymbol{A}$ will do. The actual determination of vector potentials is described in textbooks on vector field theory (see Schey [11]) for example.

Then

$$\frac{\partial U}{\partial \dot{x}} = A_x, \qquad \frac{\partial U}{\partial x} = \dot{x}\frac{\partial A_x}{\partial x} + \dot{y}\frac{\partial A_y}{\partial x} + \dot{z}\frac{\partial A_z}{\partial x}$$

and so

$$\begin{aligned}
\frac{d}{dt}\left(\frac{\partial U}{\partial \dot{x}}\right) - \frac{\partial U}{\partial x} &= \frac{d}{dt}(A_x) - \dot{x}\frac{\partial A_x}{\partial x} - \dot{y}\frac{\partial A_y}{\partial x} - \dot{z}\frac{\partial A_z}{\partial x} \\
&= \left(\frac{\partial A_x}{\partial x}\dot{x} + \frac{\partial A_x}{\partial y}\dot{y} + \frac{\partial A_x}{\partial z}\dot{z}\right) - \dot{x}\frac{\partial A_x}{\partial x} - \dot{y}\frac{\partial A_y}{\partial x} - \dot{z}\frac{\partial A_z}{\partial x} \\
&= -\dot{y}\left(\frac{\partial A_y}{\partial x} - \frac{\partial A_x}{\partial y}\right) + \dot{z}\left(\frac{\partial A_x}{\partial z} - \frac{\partial A_z}{\partial x}\right) \\
&= -\dot{y}\,(\text{curl}\,\boldsymbol{A})_z + \dot{z}\,(\text{curl}\,\boldsymbol{A})_y = -\dot{y}B_z + \dot{z}B_y \\
&= -\left(\dot{\boldsymbol{r}}\times\boldsymbol{B}\right)_x.
\end{aligned}$$

When multiplied by $-e$ this is Q_x for the magnetic part of the force. The values of Q_y and Q_z are confirmed in the same way.

We have therefore proved that the **Lorentz force** is derivable from the **velocity dependent potential**

$$U = e\,\phi(\boldsymbol{r}) - e\,\dot{\boldsymbol{r}}\cdot\boldsymbol{A}(\boldsymbol{r}) \tag{12.25}$$

and the **Lagrangian** of the particle is therefore

$$\boxed{L = \tfrac{1}{2}m\,\dot{\boldsymbol{r}}\cdot\dot{\boldsymbol{r}} - e\,\phi(\boldsymbol{r}) + e\,\dot{\boldsymbol{r}}\cdot\boldsymbol{A}(\boldsymbol{r})} \tag{12.26}$$

Question ***Why bother?***

Why should we find the Lagrangian for this system when we already know that the equation of motion is

$$m\frac{d\boldsymbol{v}}{dt} = e\,\boldsymbol{E} + e\,\boldsymbol{v}\times\boldsymbol{B}\ ?$$

Answer

The interest in this Lagrangian is that, from it, one can find the **Hamiltonian**, and this is what is needed to formulate the corresponding problem in **quantum mechanics**. This problem has important applications to the spectra of atoms in magnetic fields. ■

12.8 THE ENERGY FUNCTION h

Let $\mathcal{S}$ be any holonomic mechanical system with Lagrangian $L(\boldsymbol{q}, \dot{\boldsymbol{q}}, t)$. Then the equations of motion for $\mathcal{S}$ take the form (12.23). On multiplying the j-th equation by $\dot{q}_j$

and summing over j we obtain

$$
\begin{aligned}
0 &= \sum_{j=1}^{n} \left[\frac{d}{dt}\left(\frac{\partial L}{\partial \dot{q}_j}\right) - \frac{\partial L}{\partial q_j} \right] \dot{q}_j \\
&= \sum_{j=1}^{n} \left[\frac{d}{dt}\left(\frac{\partial L}{\partial \dot{q}_j}\dot{q}_j\right) - \frac{\partial L}{\partial q_j}\dot{q}_j - \frac{\partial L}{\partial \dot{q}_j}\ddot{q}_j \right] \\
&= \frac{d}{dt}\left[\sum_{j=1}^{n} \left(\frac{\partial L}{\partial \dot{q}_j}\dot{q}_j\right) - L \right] + \frac{\partial L}{\partial t}.
\end{aligned}
$$

Note that $\partial L/\partial t$ means the partial derivative of $L(\boldsymbol{q}, \dot{\boldsymbol{q}}, t)$ with respect to its final argument, holding $\boldsymbol{q}$ and $\dot{\boldsymbol{q}}$ constant. Thus

$$
\boxed{\frac{dh}{dt} + \frac{\partial L}{\partial t} = 0} \tag{12.27}
$$

where

$$
\boxed{h = \sum_{j=1}^{n} \left(\frac{\partial L}{\partial \dot{q}_j}\dot{q}_j\right) - L} \tag{12.28}
$$

Definition 12.8 ***Energy function*** *The function h defined by equation (12.28) is called the* ***energy function*** *of the system $\mathcal{S}$.*

The energy function h is a *generalisation of the notion of energy*. For conservative systems, we will show that it is identical with the total energy $E = T + V$. However, for non-conservative systems, V may not exist and, even if it does, h and E are not generally equal. There are three typical cases:

Case A If $L = L(\boldsymbol{q}, \dot{\boldsymbol{q}}, t)$, then $\partial L/\partial t \neq 0$ and h is **not conserved**.

Example 12.15 *h for the driven pendulum*

Find the energy function h for the driven pendulum problem.

Solution

In the driven pendulum problem,

$$
L = \tfrac{1}{2}m\left(a^2\dot{\theta}^2 + \dot{Z}^2 - 2a\dot{\theta}\dot{Z}\sin\theta\right) + mg(Z + a\cos\theta),
$$

and so

$$
h = \dot{\theta}\frac{\partial T}{\partial \dot{\theta}} - L = \tfrac{1}{2}m\left(a^2\dot{\theta}^2 - \dot{Z}^2\right) - mg(Z + a\cos\theta).
$$

This is not the same as the total energy

$$T+V=\tfrac{1}{2}m\left(a^2\dot{\theta}^2+\dot{Z}^2-2a\dot{\theta}\dot{Z}\sin\theta\right)-mg(Z+a\cos\theta),$$

and neither quantity is conserved. ■

Case B If $L = L(\boldsymbol{q}, \dot{\boldsymbol{q}})$ then $\partial L/\partial t = 0$ so that h is a **constant**. The conservation formula

$$\boxed{\sum_{j=1}^{n}\left(\frac{\partial L}{\partial \dot{q}_j}\dot{q}_j\right)-L=\text{constant}} \tag{12.29}$$

is called the **energy integral** of the system $\mathcal{S}$.

Systems for which $L = L(\boldsymbol{q}, \dot{\boldsymbol{q}})$ are said to be **autonomous**. The above result can therefore be expressed in the form:

Autonomous systems conserve h

In any motion of an autonomous system, the energy function $h(\boldsymbol{q}, \dot{\boldsymbol{q}})$ is conserved.

Example 12.16 *A charge moving in a magnetic field*

Find the energy integral for a particle of mass m and charge e moving in the *static* magnetic field $\boldsymbol{B}(\boldsymbol{r})$.

Solution

For this problem

$$L=\tfrac{1}{2}m\,\dot{\boldsymbol{r}}\cdot\dot{\boldsymbol{r}}+e\,\dot{\boldsymbol{r}}\cdot\boldsymbol{A}(\boldsymbol{r}),$$

where $\boldsymbol{A}$ is the magnetic vector potential. Since $\partial L/\partial t = 0$, the energy integral exists and has the form

$$\dot{x}\,\frac{\partial L}{\partial \dot{x}}+\dot{y}\,\frac{\partial L}{\partial \dot{y}}+\dot{z}\,\frac{\partial L}{\partial \dot{z}}-L=h,$$

where h is a constant. On using the formula for L, this becomes

$$m\left(\dot{x}^2+\dot{y}^2+\dot{z}^2\right)+e\,\dot{\boldsymbol{r}}\cdot\boldsymbol{A}-\tfrac{1}{2}m\left(\dot{x}^2+\dot{y}^2+\dot{z}^2\right)-e\,\dot{\boldsymbol{r}}\cdot\boldsymbol{A}=h,$$

that is

$$\tfrac{1}{2}m\left(\dot{x}^2+\dot{y}^2+\dot{z}^2\right)=h.$$

This is the required **energy integral**. In this case, the constant h is the **kinetic energy** of the particle.

This result is well known. When a charged particle moves in a magnetic field, the force is perpendicular to the velocity of the charge. Thus no work is done by the magnetic field and so the *kinetic energy of the particle is conserved*. For this system, V does not exist since the force exerted by the magnetic field is velocity dependent; the total energy E is therefore not defined. ■

Case C If $\mathcal{S}$ is a **conservative standard system**, then $\mathcal{S}$ is autonomous and so h is **conserved**. In addition, the energy integral can be written in a more familiar form. In this case, $L = T - V$, where T has the form

$$T = \sum_{j=1}^{n} \sum_{k=1}^{n} a_{jk}(\boldsymbol{q})\,\dot{q}_j \dot{q}_k$$

(see Example 12.9), and $V = V(\boldsymbol{q})$. Hence

$$\frac{\partial L}{\partial \dot{q}_j} = \frac{\partial T}{\partial \dot{q}_j} - 0 = 2\sum_{k=1}^{n} a_{jk}(\boldsymbol{q})\,\dot{q}_k$$

and so

$$\sum_{j=1}^{n} \frac{\partial L}{\partial \dot{q}_j}\dot{q}_j = 2\sum_{j=1}^{n}\sum_{k=1}^{n} a_{jk}(\boldsymbol{q})\,\dot{q}_j\dot{q}_k = 2T.$$

The **energy integral** therefore becomes

$$2T - (T - V) = \text{constant},$$

that is

$$\boxed{T + V = \text{constant}} \tag{12.30}$$

which is the classical form of **conservation of energy**. In this case, the constant is the **total energy** E of the system.

12.9 GENERALISED MOMENTA

The generalised momenta of a mechanical system are defined in a different way to conventional linear and angular momentum.

Definition 12.9 ***Generalised momenta*** *Consider a holonomic mechanical system with Lagrangian $L = L(\boldsymbol{q}, \dot{\boldsymbol{q}}, t)$. Then the scalar quantity p_j defined by*

$$p_j = \frac{\partial L}{\partial \dot{q}_j}$$

*is called the **generalised momentum** corresponding to the coordinate q_j. It is also called the momentum **conjugate** to q_j.*

Example 12.17 ***Finding generalised momenta***

Consider the problem in Example 12.10 whose Lagrangian is

$$L = \tfrac{1}{2}M\dot{x}^2 + \tfrac{1}{2}m\left(\dot{x}^2 + \dot{y}^2 + 2\dot{x}\dot{y}\cos\alpha\right) + mgy\sin\alpha.$$

Find the generalised momenta.

Solution

With this Lagrangian, the momenta p_x and p_y are given by

$$p_x = \frac{\partial L}{\partial \dot{x}} = M\dot{x} + m\left(\dot{x} + \dot{y}\cos\alpha\right),$$
$$p_y = \frac{\partial L}{\partial \dot{y}} = m\left(\dot{y} + \dot{x}\cos\alpha\right). \blacksquare$$

Generalised momenta are often recognisable as components of linear or angular momentum of the system. In the above example, p_x is the horizontal component of the linear momentum of $\mathcal{S}$, but p_y is *not* a component of linear momentum.

Conservation of generalised momenta

In terms of the generalised momentum p_j, the j-th Lagrange equation can be written

$$\frac{dp_j}{dt} = \frac{\partial L}{\partial q_j}.$$

It follows that *if $\partial L/\partial q_j = 0$ (that is, if the coordinate q_j is absent from the Lagrangian), then the generalised momentum p_j is constant in any motion.* Such 'absent' coordinates are said to be **cyclic**. We have therefore shown that:

Conservation of momentum

If q_j is a cyclic coordinate (in the sense that it does not appear in the Lagrangian), then p_j, the generalised momentum conjugate to q_j, is constant in any motion.

In the last example, the coordinate x is cyclic but y is not. It follows that p_x is conserved but p_y is not.

Example 12.18 ***A cyclic coordinate for the spherical pendulum***

Consider the spherical pendulum shown in Figure 11.7. The Lagrangian L is given by

$$L = \tfrac{1}{2}ma^2\left[\dot{\theta}^2 + (\sin\theta\,\dot{\phi})^2\right] + mga\cos\theta,$$

where θ, ϕ are the polar angles shown. Verify that ϕ is a cyclic coordinate and find the corresponding conserved momentum.

Solution

Since $\partial L/\partial \phi = 0$, the coordinate ϕ is **cyclic**. It follows that the conjugate momentum p_ϕ is conserved, where

$$p_\phi = \frac{\partial L}{\partial \dot{\phi}} = ma^2 \sin^2\theta\,\dot{\phi}.$$

This generalised momentum is actually the angular momentum of the pendulum about the polar axis. ■

12.10 SYMMETRY AND CONSERVATION PRINCIPLES

The existence of a cyclic coordinate is not the only reason why a generalised momentum (or momentum-like quantity) may be conserved. Indeed, whether a cyclic coordinate is present depends not only on the system, but also on *which coordinates are chosen*; if the 'wrong' coordinates are chosen then the conserved quantity will be missed. The existence of conserved quantities of the form $F(\boldsymbol{q}, \dot{\boldsymbol{q}})$ is in fact closely linked with **symmetries of the system**. We illustrate this by the following two results, which are the most important of such cases.

Theorem 12.1 ***Invariance of*** V ***under translation*** *Let $\mathcal{S}$ be a conservative standard system with potential energy V. Then if $\mathcal{S}$ can be translated (as if rigid) parallel to a constant vector $\boldsymbol{n}$ without violating any constraints, and if V is unchanged by this translation, then, in any motion of $\mathcal{S}$, the component of* ***linear momentum*** *in the $\boldsymbol{n}$-direction is* ***conserved****.*

Proof. Let $\{\boldsymbol{r}_i\}$ be any configuration of $\mathcal{S}$ and let the corresponding point in configuration space be $\boldsymbol{q}$. Then a (rigid) displacement λ in the $\boldsymbol{n}$-direction will have the effect

$$\boldsymbol{r}_i \to \boldsymbol{r}_i^\lambda,$$

where

$$\boldsymbol{r}_i^\lambda = \boldsymbol{r}_i + \lambda \boldsymbol{n}.$$

Since this displacement is consistent with the system constraints, $\{\boldsymbol{r}_i^\lambda\}$ is also a configuration of $\mathcal{S}$ and corresponds to some point $\boldsymbol{q}^\lambda$ in configuration space. Thus, in configuration space, the displacement has the effect

$$\boldsymbol{q} \to \boldsymbol{q}^\lambda,$$

where

$$\boldsymbol{r}_i^\lambda = \boldsymbol{r}_i(\boldsymbol{q}^\lambda).$$

Note that $\lambda = 0$ corresponds to the undisplaced state so that $\boldsymbol{r}_i^\lambda = \boldsymbol{r}_i$ and $\boldsymbol{q}^\lambda = \boldsymbol{q}$ when $\lambda = 0$.

Suppose now that $\boldsymbol{q}(t)$ is a motion of $\mathcal{S}$ under the potential $V(\boldsymbol{q})$. Then $\boldsymbol{q}$ satisfies Lagrange's equations which we choose to take in the form

$$\sum_i m_i \dot{\boldsymbol{v}}_i \cdot \frac{\partial \boldsymbol{r}_i}{\partial q_j} = -\frac{\partial V}{\partial q_j} \qquad (j = 1, \ldots, n).$$

On multiplying the j-th Lagrange equation by

$$\left[\frac{\partial q_j^\lambda}{\partial \lambda}\right]_{\lambda=0}$$

and summing over j we obtain

$$\sum_i m_i \dot{\boldsymbol{v}}_i \cdot \left(\sum_{j=1}^{n} \frac{\partial \boldsymbol{r}_i}{\partial q_j}\left[\frac{\partial q_j^\lambda}{\partial \lambda}\right]_{\lambda=0}\right) = -\sum_{j=1}^{n} \frac{\partial V}{\partial q_j}\left[\frac{\partial q_j^\lambda}{\partial \lambda}\right]_{\lambda=0}.$$

Now

$$\sum_{j=1}^{n} \frac{\partial \boldsymbol{r}_i}{\partial q_j}\left[\frac{\partial q_j^\lambda}{\partial \lambda}\right]_{\lambda=0} = \sum_{j=1}^{n}\left[\left(\frac{\partial}{\partial q_j^\lambda}\boldsymbol{r}_i(\boldsymbol{q}^\lambda)\right)\frac{\partial q_j^\lambda}{\partial \lambda}\right]_{\lambda=0} = \left[\frac{d}{d\lambda}\boldsymbol{r}_i(\boldsymbol{q}^\lambda)\right]_{\lambda=0}, \qquad (12.31)$$

by the chain rule. Furthermore

$$\frac{d}{d\lambda}\boldsymbol{r}_i(\boldsymbol{q}^\lambda) = \frac{\partial \boldsymbol{r}_i^\lambda}{\partial \lambda} = \boldsymbol{n},$$

since $\boldsymbol{r}_i^\lambda = \boldsymbol{r}_i(\boldsymbol{q}) + \lambda \boldsymbol{n}$ in the given displacement.

In the same way,

$$\sum_{j=1}^{n} \frac{\partial V}{\partial q_j}\left[\frac{\partial q_j^\lambda}{\partial \lambda}\right]_{\lambda=0} = \sum_{j=1}^{n}\left[\left(\frac{\partial}{\partial q_j^\lambda}V(\boldsymbol{q}^\lambda)\right)\frac{\partial q_j^\lambda}{\partial \lambda}\right]_{\lambda=0} = \left[\frac{d}{d\lambda}V(\boldsymbol{q}^\lambda)\right]_{\lambda=0} = 0,$$

since V is unchanged by the displacement, that is, $V(\boldsymbol{q}^\lambda) = V(\boldsymbol{q})$. On combining these results together, we obtain

$$\sum_i m_i \dot{\boldsymbol{v}}_i \cdot \boldsymbol{n} = 0.$$

Finally, since $\boldsymbol{n}$ is a constant vector, we may integrate with respect to t to obtain

$$\left(\sum_i m_i \boldsymbol{v}_i\right) \cdot \boldsymbol{n} = C,$$

where C is a constant. Thus the component of **linear momentum in the $\boldsymbol{n}$-direction is conserved.** ■

For example, this theorem applies to the system shown in Figure 12.6 (the wedge and block). This system can be translated in the x-direction without violating any constraints, and this translation leaves the potential energy unchanged. The conserved quantity is the component of linear momentum in the x-direction.

Theorem 12.2 ***Invariance of*** V ***under rotation*** *Let S be a conservative standard system with potential energy V. Then if S can be rotated (as if rigid) about the fixed axis $\{O, \boldsymbol{k}\}$ without violating any constraints, and if V is unchanged by this rotation, then, in any motion of S, the* ***angular momentum*** *about the axis $\{O, \boldsymbol{k}\}$ is* ***conserved.***

Proof. The proof closely follows that in the last theorem. Let λ be the angle turned in a rotation of S about the fixed axis $\{O, \boldsymbol{k}\}$, where O is also the origin of position vectors. Then by following the same steps, we obtain, as before,

$$\sum_i m_i \dot{\boldsymbol{v}}_i \cdot \left[\frac{\partial \boldsymbol{r}_i^\lambda}{\partial \lambda}\right]_{\lambda=0} = 0.$$

This time the λ-derivative means the rate of change with respect to the *rotation* angle λ so that

$$\frac{\partial \boldsymbol{r}_i^\lambda}{\partial \lambda} = \boldsymbol{k} \times \boldsymbol{r}_i^\lambda.$$

Since $\boldsymbol{r}_i^\lambda = \boldsymbol{r}_i$ when $\lambda = 0$, it follows that

$$\sum_i m_i \dot{\boldsymbol{v}}_i \cdot (\boldsymbol{k} \times \boldsymbol{r}_i) = 0.$$

Finally, since $\boldsymbol{k}$ is a constant vector, we may integrate with respect to t to obtain

$$\left(\sum_i m_i \boldsymbol{r}_i \times \boldsymbol{v}_i\right) \cdot \boldsymbol{k} = C,$$

where C is a constant. Thus the **angular momentum about the axis $\{O, \boldsymbol{k}\}$ is conserved.** ■

For example, this theorem applies to the spherical pendulum. The pendulum can be rotated about the axis $\{O, \boldsymbol{k}\}$ (where O is the support and $\boldsymbol{k}$ points vertically upwards) without violating any constraints, and this rotation leaves the potential energy unchanged. The conserved quantity is the angular momentum of the pendulum about the vertical axis through O.

These two theorems are powerful tools for identifying conserved components of linear or angular momentum even when the system is very complex. For example, any conservative standard system whose potential energy is invariant under *all* translations and rotations conserves all three components of linear and angular momentum, as well as the total energy, making seven conserved quantities in all.

Noether's theorem

The two theorems above are particular instances of an abstract result known as **Noether's theorem.*** In each of the above cases, there is a one-parameter family of mappings $\{\mathfrak{M}^\lambda\}$,

* After the German mathematician Emmy Amalie Noether (1882–1935). Despite the obstacles placed in the way of women academics at the time, she made fundamental contributions to pure mathematics in the areas of invariance theory and abstract algebra. The result now known as Noether's theorem was published in 1918.

parametrised by a real variable λ, that act on the configuration space $\mathcal{Q}$, that is,

$$\boldsymbol{q} \xrightarrow{\mathfrak{M}^\lambda} \boldsymbol{q}^\lambda. \tag{12.32}$$

In each case $\lambda = 0$ corresponds to the identity mapping (that is, $\boldsymbol{q} \to \boldsymbol{q}$), and in each case the potential energy $V(\boldsymbol{q})$ is invariant under $\{\mathfrak{M}^\lambda\}$, that is,

$$V(\boldsymbol{q}^\lambda) = V(\boldsymbol{q}).$$

From these facts, we were able to prove that, in each case, a certain momentum component was a constant of the motion.

This idea was generalised by Noether to apply to *any* Lagrangian system and *any* family of mappings $\{\mathfrak{M}^\lambda\}$, provided that the **Lagrangian** L is **invariant** under $\{\mathfrak{M}^\lambda\}$ in the sense that

$$L(\boldsymbol{q}^\lambda, \dot{\boldsymbol{q}}^\lambda, t) = L(\boldsymbol{q}, \dot{\boldsymbol{q}}, t)$$

for all λ. In this formula, $\boldsymbol{q}^\lambda$ is a known function of the variables $\boldsymbol{q}$ and λ (as defined by the mapping $\mathfrak{M}^\lambda$), but we have not yet said what we mean by $\dot{\boldsymbol{q}}^\lambda$. This is however defined in the following commonsense way: let λ be fixed and let $\boldsymbol{q}^\lambda$ be the image point of a typical point $\boldsymbol{q}$. Suppose now that the point $\boldsymbol{q}$ has velocity $\dot{\boldsymbol{q}}$ in the configuration space $\mathcal{Q}$. This motion of $\boldsymbol{q}$ *imparts* a velocity to the image point $\boldsymbol{q}^\lambda$, and it is this velocity that we call $\dot{\boldsymbol{q}}^\lambda$. This definition is expressed by the formula

$$\dot{\boldsymbol{q}}^\lambda = \sum_{j=1}^{n} \frac{\partial \boldsymbol{q}^\lambda}{\partial q_j} \dot{q}_j \tag{12.33}$$

from which we see that $\dot{\boldsymbol{q}}^\lambda$ is a known function of the variables $\boldsymbol{q}$, $\dot{\boldsymbol{q}}$ and λ.

The formal statement and proof of Noether's theorem are as follows:

Theorem 12.3 *Noether's theorem* *Let $\mathcal{S}$ be a holonomic mechanical system with* **Lagrangian** *$L(\boldsymbol{q}, \dot{\boldsymbol{q}}, t)$ and let $\{\mathfrak{M}^\lambda\}$ be a one-parameter family of* **mappings** *that have the action*

$$\boldsymbol{q} \xrightarrow{\mathfrak{M}^\lambda} \boldsymbol{q}^\lambda \tag{12.34}$$

where $\boldsymbol{q}^\lambda = \boldsymbol{q}$ when $\lambda = 0$. If the mappings $\{\mathfrak{M}^\lambda\}$ leave L ***invariant*** *in the sense that*

$$L(\boldsymbol{q}^\lambda, \dot{\boldsymbol{q}}^\lambda, t) = L(\boldsymbol{q}, \dot{\boldsymbol{q}}, t) \tag{12.35}$$

for all λ, then the quantity

$$\sum_{j=1}^{n} p_j \left[\frac{\partial q_j^\lambda}{\partial \lambda}\right]_{\lambda=0} \tag{12.36}$$

*is **conserved** in any motion of* $\mathcal{S}$. [Note that the conserved quantity is not generally one of the momenta $\{p_j\}$ but a linear combination of all of them with coefficients depending on $\boldsymbol{q}$.]

Proof. Let $\boldsymbol{q}(t)$ be any physical motion of the system $\mathcal{S}$, that is, a solution of the Lagrange equations

$$\frac{d}{dt}\left(\frac{\partial L}{\partial \dot{q}_j}\right) - \frac{\partial L}{\partial q_j} = 0 \qquad (1 \le j \le n),$$

where $L(\boldsymbol{q}, \dot{\boldsymbol{q}}, t)$ is the Lagrangian of the system $\mathcal{S}$. Now consider the expression

$$\begin{aligned}\frac{d}{dt}\left(p_j \left[\frac{\partial q_j^\lambda}{\partial \lambda}\right]_{\lambda=0}\right) &= \frac{d}{dt}\left(\frac{\partial L}{\partial \dot{q}_j}\left[\frac{\partial q_j^\lambda}{\partial \lambda}\right]_{\lambda=0}\right) \\ &= \frac{d}{dt}\left(\frac{\partial L}{\partial \dot{q}_j}\right)\left[\frac{\partial q_j^\lambda}{\partial \lambda}\right]_{\lambda=0} + \frac{\partial L}{\partial \dot{q}_j}\left(\frac{d}{dt}\left[\frac{\partial q_j^\lambda}{\partial \lambda}\right]_{\lambda=0}\right) \\ &= \frac{\partial L}{\partial q_j}\left[\frac{\partial q_j^\lambda}{\partial \lambda}\right]_{\lambda=0} + \frac{\partial L}{\partial \dot{q}_j}\left(\frac{d}{dt}\left[\frac{\partial q_j^\lambda}{\partial \lambda}\right]_{\lambda=0}\right)\end{aligned}$$

on using the j-th Lagrange equation. Now, by the chain rule,

$$\frac{d}{dt}\left(\frac{\partial q_j^\lambda}{\partial \lambda}\right) = \sum_{k=1}^{n} \frac{\partial}{\partial q_k}\left(\frac{\partial q_j^\lambda}{\partial \lambda}\right)\dot{q}_k = \frac{\partial}{\partial \lambda}\left(\sum_{k=1}^{n} \frac{\partial q_j^\lambda}{\partial q_k}\dot{q}_k\right) = \frac{\partial \dot{q}_j^\lambda}{\partial \lambda}$$

by the definition (12.33) of $\dot{\boldsymbol{q}}^\lambda$. It follows that

$$\begin{aligned}\frac{d}{dt}\left(p_j \left[\frac{\partial q_j^\lambda}{\partial \lambda}\right]_{\lambda=0}\right) &= \frac{\partial L}{\partial q_j}\left[\frac{\partial q_j^\lambda}{\partial \lambda}\right]_{\lambda=0} + \frac{\partial L}{\partial \dot{q}_j}\left[\frac{\partial \dot{q}_j^\lambda}{\partial \lambda}\right]_{\lambda=0} \\ &= \left[\frac{\partial}{\partial q_j^\lambda} L(\boldsymbol{q}^\lambda, \dot{\boldsymbol{q}}^\lambda, t)\,\frac{\partial q_j^\lambda}{\partial \lambda} + \frac{\partial}{\partial \dot{q}_j^\lambda} L(\boldsymbol{q}^\lambda, \dot{\boldsymbol{q}}^\lambda, t)\,\frac{\partial \dot{q}_j^\lambda}{\partial \lambda}\right]_{\lambda=0}\end{aligned}$$

since $\boldsymbol{q}^\lambda = \boldsymbol{q}$ and $\dot{\boldsymbol{q}}^\lambda = \dot{\boldsymbol{q}}$ when $\lambda = 0$. On summing this result over j, we obtain

$$\begin{aligned}\frac{d}{dt}\left(\sum_{j=1}^{n} p_j \left[\frac{\partial q_j^\lambda}{\partial \lambda}\right]_{\lambda=0}\right) &= \left[\sum_{j=1}^{n} \frac{\partial}{\partial q_j^\lambda} L(\boldsymbol{q}^\lambda, \dot{\boldsymbol{q}}^\lambda, t)\,\frac{\partial q_j^\lambda}{\partial \lambda} + \sum_{j=1}^{n} \frac{\partial}{\partial \dot{q}_j^\lambda} L(\boldsymbol{q}^\lambda, \dot{\boldsymbol{q}}^\lambda, t)\,\frac{\partial \dot{q}_j^\lambda}{\partial \lambda}\right]_{\lambda=0} \\ &= \left[\frac{d}{d\lambda} L(\boldsymbol{q}^\lambda, \dot{\boldsymbol{q}}^\lambda, t)\right]_{\lambda=0}\end{aligned}$$

by the chain rule. Finally, we appeal to the invariance of L under the mappings $\{\mathfrak{M}^\lambda\}$. In this case, $L(\boldsymbol{q}^\lambda, \dot{\boldsymbol{q}}^\lambda, t) = L(\boldsymbol{q}, \dot{\boldsymbol{q}}, t)$ and so

$$\frac{d}{d\lambda} L(\boldsymbol{q}^\lambda, \dot{\boldsymbol{q}}^\lambda, t) = \frac{d}{d\lambda} L(\boldsymbol{q}, \dot{\boldsymbol{q}}, t) = 0.$$

It follows that

$$\frac{d}{dt}\left(\sum_{j=1}^{n} p_j \left[\frac{\partial q_j^{\lambda}}{\partial \lambda}\right]_{\lambda=0}\right) = 0$$

and this proves the theorem. ∎

The importance of Noether's theorem lies in the general notion that **an invariance of the Lagrangian gives rise to a constant of the motion**. Such invariance properties are of great importance when the Lagrangian formalism is extended to continuous systems and fields. For more details, see Goldstein [4] who will tell you more about Noether's theorem than you wish to know!

Problems on Chapter 12

Answers and comments are at the end of the book.
Harder problems carry a star (∗).

Conservative systems

12.1 A bicycle chain consists of N freely jointed links forming a closed loop. The chain can slide freely on a smooth horizontal table. How many degrees of freedom has the chain? How many conserved quantities are there in the motion? What is the maximum number of links the chain can have for its motion to be determined by conservation principles alone?

12.2 ***Attwood's machine*** A uniform circular pulley of mass $2m$ can rotate freely about its axis of symmetry which is fixed in a horizontal position. Two masses m, $3m$ are connected by a light inextensible string which passes over the pulley without slipping. The whole system undergoes planar motion with the masses moving vertically. Take the rotation angle of the pulley as generalised coordinate and obtain Lagrange's equation for the motion. Deduce the upward acceleration of the mass m.

12.3 ***Double Attwood machine*** A light pulley can rotate freely about its axis of symmetry which is fixed in a horizontal position. A light inextensible string passes over the pulley. At one end the string carries a mass $4m$, while the other end supports a second light pulley. A second string passes over this pulley and carries masses m and $4m$ at its ends. The whole system undergoes planar motion with the masses moving vertically. Find Lagrange's equations and deduce the acceleration of each of the masses.

12.4 ***The swinging door*** A uniform rectangular door of width $2a$ can swing freely on its hinges. The door is misaligned and the line of the hinges makes an angle α with the upward vertical. Take the rotation angle of the door from its equilibrium position as generalised coordinate and obtain Lagrange's equation for the motion. Deduce the period of small oscillations of the door about the equilibrium position.

12.5 A uniform solid cylinder C with mass m and radius a rolls on the rough outer surface of a fixed horizontal cylinder of radius b. In the motion, the axes of the two cylinders remain parallel to each other. Let θ be the angle between the plane containing the cylinder axes and the upward vertical. Taking θ as generalised coordinate, obtain Lagrange's equation and verify that it is equivalent to the energy conservation equation.

Initially the cylinder C is at rest on top of the fixed cylinder when it is given a very small disturbance. Find, as a function of θ, the normal component of the reaction force exerted on C. Deduce that C will leave the fixed cylinder when $\theta = \cos^{-1}(4/7)$. Is the assumption that rolling persists up to this moment realistic?

12.6 A uniform disk of mass M and radius a can roll along a rough horizontal rail. A particle of mass m is suspended from the centre C of the disk by a light inextensible string of length b. The whole system moves in the vertical plane through the rail. Take as generalised coordinates x, the horizontal displacement of C, and θ, the angle between the string and the downward vertical. Obtain Lagrange's equations. Show that x is a cyclic coordinate and find the corresponding conserved momentum p_x. Is p_x the horizontal linear momentum of the system?

Given that θ remains small in the motion, find the period of small oscillations of the particle.

12.7 A uniform ball of mass m rolls down a rough wedge of mass M and angle α, which itself can slide on a smooth horizontal table. The whole system undergoes planar motion. How many degrees of freedom has this system? Obtain Lagrange's equations. For the special case in which $M = 3m/2$, find (i) the acceleration of the wedge, and (ii) the acceleration of the ball relative to the wedge.

12.8 A rigid rod of length $2a$ has its lower end in contact with a smooth horizontal floor. Initially the rod is at an angle α to the upward vertical when it is released from rest. The subsequent motion takes place in a vertical plane. Take as generalised coordinates x, the horizontal displacement of the centre of the rod, and θ, the angle between the rod and the upward vertical. Obtain Lagrange's equations. Show that x remains constant in the motion and verify that the θ-equation is equivalent to the energy conservation equation.

* Find, in terms of the angle θ, the reaction exerted on the rod by the floor.

Moving constraints

12.9 A particle P is connected to one end of a light inextensible string which passes through a small hole O in a smooth horizontal table and extends below the table in a vertical straight line. P slides on the upper surface of the table while the string is pulled downwards from below in a prescribed manner. (Suppose that the length of the horizontal part of the string is $R(t)$ at time t.) Take θ, the angle between OP and some fixed reference line in the table, as generalised coordinate and obtain Lagrange's equation. Show that θ is a cyclic coordinate and find (and identify) the corresponding conserved momentum p_θ. Why is the kinetic energy not conserved?

If the constant value of p_θ is mL, find the tension in the string at time t.

12.10 A particle P of mass m can slide along a smooth rigid straight wire. The wire has one of its points fixed at the origin O, and is made to rotate in the (x, y)-plane with angular speed Ω. Take r, the distance of P from O, as generalised coordinate and obtain Lagrange's equation.

Initially the particle is a distance a from O and is at rest relative to the wire. Find its position at time t. Find also the energy function h and show that it is conserved even though there is a time dependent constraint.

12.11 ***Yo-yo with moving support*** A uniform circular cylinder of mass m (a yo-yo) has a light inextensible string wrapped around it so that it does not slip. The free end of the string is fastened to a support and the yo-yo moves in a vertical straight line with the straight part of the string also vertical. At the same time the support is made to move vertically having upward displacement $Z(t)$ at time t. Take the rotation angle of the yo-yo as generalised coordinate and obtain Lagrange's equation. Find the acceleration of the yo-yo. What upwards acceleration must the support have so that the centre of the yo-yo can remain at rest?

Suppose the whole system starts from rest. Find an expression for the total energy $E = T + V$ at time t.

12.12 ***Pendulum with a shortening string*** A particle is suspended from a support by a light inextensible string which passes through a small fixed ring vertically below the support. The particle moves in a vertical plane with the string taut. At the same time the support is made to move vertically having an upward displacement $Z(t)$ at time t. The effect is that the particle oscillates like a simple pendulum whose string length at time t is $a - Z(t)$, where a is a positive constant. Take the angle between the string and the downward vertical as generalised coordinate and obtain Lagrange's equation. Find the energy function h and the total energy E and show that $h = E - m\dot{Z}^2$. Is either quantity conserved?

12.13* ***Bug on a hoop*** A uniform circular hoop of mass M can slide freely on a smooth horizontal table, and a bug of mass m can run on the hoop. The system is at rest when the bug starts to run. What is the angle turned through by the hoop when the bug has completed one lap of the hoop?

Velocity dependent potentials and Lagrangians

12.14 Suppose a particle is subjected to a time dependent force of the form $\boldsymbol{F} = f(t)\,\mathrm{grad}\,W(\boldsymbol{r})$. Show that this force can be represented by the time dependent potential $U = -f(t)W(\boldsymbol{r})$. What is the value of U when $\boldsymbol{F} = f(t)\,\boldsymbol{i}$?

12.15 ***Charged particle in an electrodynamic field*** Show that the velocity dependent potential

$$U = e\,\phi(\boldsymbol{r}, t) - e\,\dot{\boldsymbol{r}} \cdot \boldsymbol{A}(\boldsymbol{r}, t)$$

represents the Lorentz force $\boldsymbol{F} = e\,\boldsymbol{E} + e\,\boldsymbol{v} \times \boldsymbol{B}$ that acts on a charge e moving with velocity $\boldsymbol{v}$ in the general *electrodynamic* field $\{\boldsymbol{E}(\boldsymbol{r}, t), \boldsymbol{B}(\boldsymbol{r}, t)\}$. Here $\{\phi, \boldsymbol{A}\}$ are the *electrodynamic potentials* that generate the field $\{\boldsymbol{E}, \boldsymbol{B}\}$ by the formulae

$$\boldsymbol{E} = -\operatorname{grad}\phi - \frac{\partial \boldsymbol{A}}{\partial t}, \qquad \boldsymbol{B} = \operatorname{curl}\boldsymbol{A}.$$

Show that the potentials $\phi = 0$, $\boldsymbol{A} = tz\,\boldsymbol{i}$ generate a field $\{\boldsymbol{E}, \boldsymbol{B}\}$ that satisfies all four Maxwell equations in free space. A particle of mass m and charge e moves in this field. Find the Lagrangian of the particle in terms of Cartesian coordinates. Show that x and y are cyclic coordinates and find the conserved momenta p_x, p_y.

12.16* ***Relativistic Lagrangian*** The relativistic Lagrangian for a particle of rest mass m_0 moving along the x-axis under the simple harmonic potential field $V = \frac{1}{2}m_0\Omega^2 x^2$ is given by

$$L = m_0 c^2\left(1 - \left(1 - \frac{\dot{x}^2}{c^2}\right)^{1/2}\right) - \tfrac{1}{2}m_0\Omega^2 x^2.$$

Obtain the energy integral for this system and show that the period of oscillations of amplitude a is given by

$$\tau = \frac{4}{\Omega}\int_0^{\pi/2} \frac{1 + \frac{1}{2}\epsilon^2\cos^2\theta}{\left(1 + \frac{1}{4}\epsilon^2\cos^2\theta\right)^{1/2}}\,d\theta,$$

where the dimensionless parameter $\epsilon = \Omega a/c$.

Deduce that

$$\tau = \frac{2\pi}{\Omega}\left[1 + \tfrac{3}{16}\epsilon^2 + O\left(\epsilon^4\right)\right],$$

when ϵ is small.

Conservation principles and symmetry

12.17 A particle of mass m moves under the gravitational attraction of a fixed mass M situated at the origin. Take polar coordinates r, θ as generalised coordinates and obtain Lagrange's equations. Show that θ is a cyclic coordinate and find (and identify) the conserved momentum p_θ.

12.18 A particle P of mass m slides on the smooth inner surface of a circular cone of semi-angle α. The axis of symmetry of the cone is vertical with the vertex O pointing downwards. Take as generalised coordinates r, the distance OP, and ϕ, the azimuthal angle about the vertical through O. Obtain Lagrange's equations. Show that ϕ is a cyclic coordinate and find (and identify) the conserved momentum p_ϕ.

12.19 A particle of mass m and charge e moves in the magnetic field produced by a current I flowing in an infinite straight wire that lies along the z-axis. The vector potential $\boldsymbol{A}$ of the induced magnetic field is given by

$$A_r = A_\theta = 0, \qquad A_z = -\left(\frac{\mu_0 I}{2\pi}\right)\ln r,$$

where r, θ, z are cylindrical polar coordinates. Find the Lagrangian of the particle. Show that θ and z are cyclic coordinates and find the corresponding conserved momenta.

12.20 A particle moves freely in the gravitational field of a fixed mass distribution. Find the conservation principles that correspond to the symmetries of the following fixed mass distributions: (i) a uniform sphere, (ii) a uniform half plane, (iii) two particles, (iv) a uniform right circular cone, (v) an infinite uniform circular cylinder.

12.21* ***Helical symmetry*** A particle moves in a conservative field whose potential energy V has *helical symmetry*. This means that V is invariant under the *simultaneous* operations (i) a rotation through any angle α about the axis Oz, and (ii) a translation $c\alpha$ in the z-direction. What conservation principle corresponds to this symmetry?

Computer assisted problem

12.22 ***Upside-down pendulum*** A particle P is attached to a support S by a light rigid rod of length a, which is freely pivoted at S. P moves in a vertical plane through S and at the same time the support S is made to oscillate vertically having upward displacement $Z = \epsilon a \cos pt$ at time t. Take θ, the angle between SP and the *upward* vertical, as generalised coordinate and show that Lagrange's equation is

$$\ddot{\theta} - \left(\Omega^2 + \epsilon p^2 \cos pt\right) \sin\theta = 0,$$

where $\Omega^2 = g/a$. The object is to show that, for suitable choices of the parameters, the pendulum is stable in the vertically *upwards* position!

First write the equation in dimensionless form by introducing the dimensionless time $\tau = pt$. Then $\theta(\tau)$ satisfies

$$\frac{d^2\theta}{d\tau^2} - \left(\frac{\Omega^2}{p^2} + \epsilon \cos\tau\right) \sin\theta = 0.$$

Solve this equation numerically with initial conditions in which the pendulum starts from rest near the upward vertical. Plot the solution $\theta(\tau)$ as a function of τ for about twenty oscillations of the support. Try $\epsilon = 0.3$ with increasing values of the parameter p/Ω in the range $1 \leq p/\Omega \leq 10$. You will know that the upside-down pendulum is stable when θ remains small in the subsequent motion.

Even more surprisingly, it is possible to stabilise the double pendulum (or any multiple pendulum) in the upside-down position by vibrating the support. See Acheson [1] for photographs of a triple pendulum (and even a length of floppy wire) stabilised in the upside-down position by vibrating the support. However, the famous but elusive 'Indian Rope Trick', in which a small boy climbs up a self-supporting vertical rope, has yet to be demonstrated!

Chapter Thirteen

The calculus of variations and Hamilton's principle

KEY FEATURES

The key features of this chapter are **integral functionals** and the functions that make them **stationary**, the **Euler–Lagrange** equation and **extremals**, and the importance of **variational principles**.

The notion that physical processes are governed by **minimum principles** is older than most of science. It is based on the long held belief that nature arranges itself in the most 'economical' way. Actually, many 'minimum' principles have, on closer inspection, turned out to make their designated quantity *stationary*, but not necessarily a minimum. As a result, they are now known to be **variational principles**, but they are no less important because of this. A good example of a variational principle is **Fermat's principle** of geometrical optics, which was proposed in 1657 as *Fermat's principle of least time* in the form:

Of all the possible paths that a light ray might take between two fixed points, the actual path is the one that minimises the travel time of the ray.

Fermat showed that the laws of reflection and refraction could be derived from his principle, and proposed that the principle was true in general. Not only did Fermat's principle 'explain' the known laws of optics, it was simple and elegant, and was capable of extending the laws of optics far beyond the results that led to its conception. This example explains why variational principles continue to be sought; it is because of their innate simplicity and elegance, and the generality of their application.

The variational principle on which it is possible to base the whole of classical mechanics was discovered by Hamilton* and is known as **Hamilton's principle.**† In its original form, it stated that:

Of all the kinematically possible motions that take a mechanical system from one given configuration to another within a given time interval, the actual motion is the one that minimises the time integral of the Lagrangian of the system.

Lagrange's equations of motion can be derived from Hamilton's principle, which can therefore be taken as the basic postulate of classical mechanics, instead of Newton's laws. More importantly however, Hamilton's principle has had a far reaching influence on many areas of physics, where apparently non-mechanical systems (fields, for example) can be described in the language of classical mechanics, and their behaviour characterised by 'Lagrangians'. Hamilton's principle is generally regarded as one of the most elegant and far reaching principles in physics.

In order to get concrete results from a variational principle, it is usually necessary to convert it to a differential equation. This can be done by using the **calculus of variations**, which is concerned with minimising or maximising the value of an integral functional. The calculus of variations is a large subject and we develop only those aspects most relevant to interesting physical problems, and to the understanding and use of variational principles.

13.1 SOME TYPICAL MINIMISATION PROBLEMS

The **calculus of variations** arose from attempts to solve minimisation and maximisation problems that occur naturally in physics and mathematics, but the scope of applications has since widened greatly. We begin by describing three minimisation problems taken from geometry, physics and economics respectively. Maximisation problems also occur, but these can be converted into minimisation problems merely by reversing the sign of the quantity to be maximised. Thus we lose no generality by presenting the theory for minimisation problems only.

1. Shortest paths – geodesics A basic problem of the calculus of variations is that of finding the **path of shortest length** that connects two given points A and B. If the path has no constraints to satisfy, such as having to go round obstacles or lie on a given curved

* Sir William Rowan Hamilton (1805–1865), was a great genius but an unhappy man. He was appointed Professor of Astronomy at Trinity College Dublin at the age of twenty one, whilst still an undergraduate. Much of his early work is on optics where he introduced the notion of the characteristic function. His paper *On a General Method in Dynamics*, which contains what is now called Hamilton's principle, was presented to the Royal Irish Academy in 1834. He was knighted in 1835. However, his personal life was as chaotic as his academic achievements were brilliant. He was frustrated in love, frequently depressed and a heavy drinker; this culminated in his making an exhibition of himself at a meeting of the Irish Geological Society. He spent the later years of his life working on the theory of quaternions, but they were never the great discovery he had hoped for.

† Hamilton's principle is sometimes called the *principle of least action*. The terminology in this area is confusing, since another variational principle of mechanics, Maupertuis's principle, is also referred to as the principle of least action.

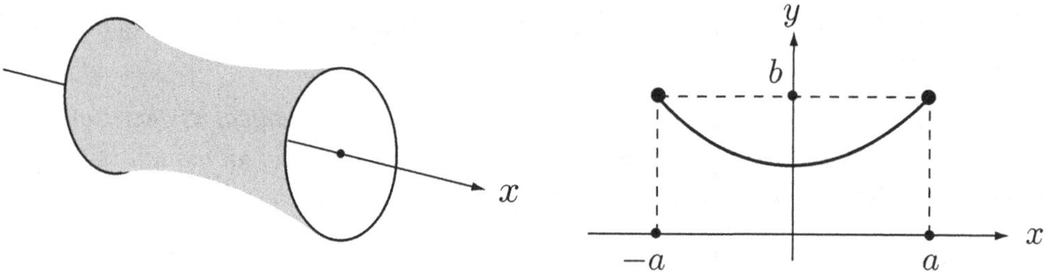

FIGURE 13.1 A soap film is stretched between two circular wires. It has the form of a surface of revolution generated by rotating the curve $y = y(x)$ about the x-axis.

surface, the answer is well known; it is the straight line joining A and B. However, it is still instructive to formulate this problem and to check that the calculus of variations does yield the expected result.

Suppose that $A = (0, 0)$, $B = (1, 0)$ and that the general path in the (x, y)-plane connecting these points is $y = y(x)$. Then the total length of the path is

$$L[y] = \int_0^1 \left[1 + \left(\frac{dy}{dx}\right)^2\right]^{1/2} dx. \tag{13.1}$$

The **problem** is to find the function $y(x)$, satisfying the end conditions $y(0) = 0$, $y(1) = 0$, that minimises the length L. This is the subject of Problem 13.3; the answer is indeed the straight line $y = 0$.

In general, paths of shortest length are called **geodesics**. For example, geodesics on the surface of a sphere are great circles. Some surfaces (such as the cylinder and cone) are *developable*, which means that they can be rolled out flat without changing any lengths. The geodesic can be drawn while the surface is flat (it is now a straight line) and the surface can then be rolled back up again. In general however, surfaces are not developable and geodesics have to be found by the calculus of variations.

2. The soap film problem Two rigid circular wires each of radius b have the same axis of symmetry and are fixed at a distance $2a$ from each other. A soap film is created which spans the two wires as shown in Figure 13.1. The soap film has the form of a surface of revolution with the two circular ends open. What is the shape of the soap film?

Fortunately, we can formulate this problem without resorting to the theory of thin membranes! Since the air pressure is the same on either side of the film, and since the effect of gravity is negligible, surface tension is the dominant effect. The condition that the total energy be a minimum in equilibrium is therefore equivalent to the condition that the *area of the film be a minimum*.

Let the film be the surface generated by rotating the curve $y = y(x)$ about the x-axis. Then the surface area A of the film is

$$A[y] = 2\pi \int_{-a}^{a} y \left\{1 + \dot{y}^2\right\}^{1/2} dx, \tag{13.2}$$

where $\dot{y}$ means dy/dx. The **problem** is to find the function $y(x)$, satisfying the end conditions $y(-a) = b$, $y(a) = b$, that minimises the area A. This is the subject of Problem 13.7.

3. A minimum cost strategy A manufacturer must produce a volume X of a product in time T. Let $x = x(t)$ be the volume produced after time t and suppose that there is a production cost $\alpha + \beta\dot{x}$ per unit volume of product and a storage cost γx per unit time, where α, β and γ are positive constants. The term $\beta\dot{x}$ is a simple model of the increased costs associated with faster production. Then the total cost C of the production run is

$$C[x] = \int_0^X (\alpha + \beta\dot{x})\, dx + \int_0^T \gamma x\, dt,$$

which can be written in the form

$$C[x] = \int_0^T \{(\alpha + \beta\dot{x})\dot{x} + \gamma x\}\, dt. \tag{13.3}$$

The **problem** is to find the function $x(t)$, satisfying the end conditions $x(0) = 0$, $x(T) = X$, that minimises the cost C. This is the subject of Problem 13.6; it is found that producing the goods at a uniform rate is *not* the best strategy.

Integral functionals

The expressions $L[y]$, $A[y]$ and $C[x]$ are examples of **integral functionals**. Functionals differ from ordinary functions in that the independent variable is a *function*, not a number; however, the dependent variable is a number, as usual. *The calculus of variations is concerned with minimising or maximising integral functionals.*

13.2 THE EULER–LAGRANGE EQUATION

Before we begin the theory proper, it is useful to recall the procedure for finding the value of x that minimises an *ordinary function* $f(x)$ on the interval $a \le x \le b$. The procedure is as follows:

1. First find the values of x (in the range $a < x < b$) that satisfy the equation $f'(x) = 0$. These are the **stationary points** of $f(x)$. They are so called because, if x^* is a stationary point, then

$$f(x^* + h) - f(x^*) = O\left(h^2\right) \tag{13.4}$$

for all sufficiently small h. (That is, at a stationary point, the change in f due to a small change h in x is of order h^2.)

2. Now determine the nature of each stationary point, that is, whether it is a minimum point, a maximum point, or neither. This can usually be done by examining the sign of $f''(x^*)$. The minimum points are the **local minima** of f. They are so called because

$$f(x^* + h) \ge f(x^*) \tag{13.5}$$

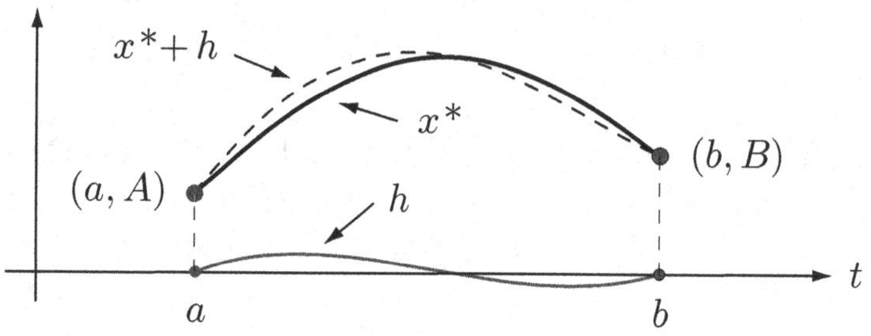

FIGURE 13.2 The minimising function x^* is perturbed by the admissible variation h.

for *sufficiently small* h, but not necessarily for all h. (In other words, the inequality $f(x) \geq f(x^*)$ is true when x is close enough to x^*.)

3. Determine the values of f at the extreme points $x = a$, $x = b$.
4. The **global minimum** of f is then the least of the local minima of f and the extreme values of f.

Each of these steps has its counterpart in the calculus of variations. However, since this material is large enough to fill a book by itself, we will mainly be concerned with the first step. This will still be enough to narrow down the search for the minimising function $x^*(t)$ to a finite number of possibilities and often one ends up with only *one* possibility. Thus, if it is 'known' (rigorously or otherwise!) that a minimising function does exist, then the problem is solved.

The **general problem** in the calculus of variations is that of finding a function $x^*(t)$ that minimises an integral functional of the form

$$J[x] = \int_a^b F(x, \dot{x}, t)\, dt, \tag{13.6}$$

where F is a given function of *three independent variables.** Suppose that the function $x^*(t)$ minimises the functional $J[x]$. This means that

$$J[x] \geq J[x^*] \tag{13.7}$$

for all *admissible* functions $x(t)$. Here **admissible** means that x must satisfy whatever end conditions are prescribed at $t = a$ and $t = b$. We will always assume that these

* This means that, despite the fact that x, $\dot{x}$ and t are clearly *not* independent of each other (x is a function of t and $\dot{x}$ is the derivative of x), the *partial derivatives* of F are evaluated as if x, $\dot{x}$ and t were three independent variables. For example, in the case of the cost functional C given by equation (13.3), $F = (\alpha + \beta\dot{x})\dot{x} + \gamma x$ and the partial derivatives of F are

$$\frac{\partial F}{\partial x} = \gamma \qquad \frac{\partial F}{\partial \dot{x}} = \alpha + 2\beta\dot{x} \qquad \frac{\partial F}{\partial t} = 0.$$

conditions have the form $x(a) = A$ and $x(b) = B$, where A, B are given. It is convenient to regard the function $x(t)$ that appears in (13.7) as being composed of $x^*(t)$ together with a **variation*** $h(t)$ so that we may alternatively write

$$J[x^* + h] \geq J[x^*] \tag{13.8}$$

for all admissible variations $h(t)$. Since x must satisfy the same end conditions as x^*, the **admissible variations** are those for which $h(a) = h(b) = 0$ (see Figure 13.2).

Most readers will find the theory that follows quite difficult. To aid understanding, the argument is broken down into three separate steps.

The variation in J and the meaning of 'stationary'

The first step is to give a meaning to the statement that a function $x(t)$ makes the functional $J[x]$ stationary.

Let $x^*(t)$ be *any* admissible function and $h(t)$ an admissible variation. When h is a *small* variation, we can estimate the corresponding variation in $J[x]$ by ordinary calculus, as follows:

Let t have any fixed value. Then x and $\dot{x}$ are just real numbers and the variation in F due to the variation h in x is†

$$F(x^* + h, \dot{x}^* + \dot{h}, t) - F(x^*, \dot{x}^*, t) = h\frac{\partial F}{\partial x}(x^*, \dot{x}^*, t) + \dot{h}\frac{\partial F}{\partial \dot{x}}(x^*, \dot{x}^*, t) + O\left(h^2 + \dot{h}^2\right),$$

when h and $\dot{h}$ are both small. On integrating both sides of this equation with respect to t over the interval $[a, b]$, the corresponding variation in J is given by

$$J[x^* + h] - J[x^*] = \int_a^b \left[h\frac{\partial F}{\partial x}(x^*, \dot{x}^*, t) + \dot{h}\frac{\partial F}{\partial \dot{x}}(x^*, \dot{x}^*, t)\right] dt + O\left(||h||^2\right), \tag{13.9}$$

for small $||h||$, where $||h||$ is defined by

$$||h|| = \max_{a \leq t \leq b} |h(t)| + \max_{a \leq t \leq b} |\dot{h}(t)|$$

and is called the **norm**‡ of h. (When $||h||$ is small, both $|h(t)|$ and $|\dot{h}(t)|$ are small throughout the interval $[a, b]$.) The second term in the integrand of equation (13.9) can be integrated by parts to give

$$\int_a^b \dot{h}\frac{\partial F}{\partial \dot{x}}(x, x^*, t)\, dt = \left[h\frac{\partial F}{\partial \dot{x}}(x^*, \dot{x}^*, t)\right]_{t=a}^{t=b} - \int_a^b h\frac{d}{dt}\left(\frac{\partial F}{\partial \dot{x}}(x^*, \dot{x}^*, t)\right) dt$$

* These are the variations that give the 'calculus of variations' its name.

† The formula we are using is that, if $G(u, v)$ is a function of the independent variables u and v, then the variation in G caused by the variations $u_0 \to u_0 + h$ and $v_0 \to v_0 + k$ is given by

$$G(u_0 + h, v_0 + k) - G(u_0, v_0) = h\frac{\partial G}{\partial u}(u_0, v_0) + k\frac{\partial G}{\partial v}(u_0, v_0) + O\left(h^2 + k^2\right),$$

for small h, k. Note that the partial derivatives are evaluated at the 'starting point' (u_0, v_0).

‡ Do not be too concerned over the exact definition of $||h||$. It must be written in some such way for mathematical correctness, but the only property that we will need is that $||h||$ is *proportional* to h in the sense that, if h is multiplied by a constant λ, then so is $||h||$.

and, since h is an *admissible* variation satisfying $h(a) = h(b) = 0$, the integrated term evaluates to zero. We thus obtain

$$J[x^* + h] - J[x^*] = \int_a^b \left[\frac{\partial F}{\partial x}(x^*, \dot{x}^*, t) - \frac{d}{dt}\left(\frac{\partial F}{\partial \dot{x}}(x^*, \dot{x}^*, t) \right) \right] h \, dt + O\left(||h||^2\right). \tag{13.10}$$

This is the variation in J caused by the admissible variation h in x when $||h||$ is small. The variation in J is therefore linear in to h with an error term of order $||h||^2$. By analogy with the case of ordinary functions, we say that x^* makes $J[x]$ stationary if the linear term is zero, leaving only the error term.

Definition 13.1 ***Stationary*** J *The function $x^*(t)$ is said to make the functional $J[x]$* ***stationary*** *if*

$$J[x^* + h] - J[x^*] = O\left(||h||^2\right) \tag{13.11}$$

when $||h||$ is small.

It follows that the condition that x^* makes $J[x]$ stationary is equivalent to the condition that

$$\int_a^b \left[\frac{\partial F}{\partial x}(x^*, \dot{x}^*, t) - \frac{d}{dt}\left(\frac{\partial F}{\partial \dot{x}}(x^*, \dot{x}^*, t) \right) \right] h \, dt = 0 \tag{13.12}$$

for all admissible variations h.

Minimising functions make J stationary

Suppose now that $x^*(t)$ provides a **local minimum** for $J[x]$ in the sense that

$$J[x^* + h] \geq J[x^*] \tag{13.13}$$

when $||h||$ is small. We will now show that such an x^* makes $J[x]$ stationary.

If we substitute equation (13.10) into the inequality (13.13), we obtain

$$\int_a^b \left[\frac{\partial F}{\partial x}(x^*, \dot{x}^*, t) - \frac{d}{dt}\left(\frac{\partial F}{\partial \dot{x}}(x^*, \dot{x}^*, t) \right) \right] h \, dt + O\left(||h||^2\right) \geq 0 \tag{13.14}$$

for small $||h||$. It follows from this inequality that the integral term must be zero. The proof is as follows:

In the inequality (13.14), let h be replaced by λh, where λ is a positive constant. On dividing through by λ, this gives

$$\int_a^b \left[\frac{\partial F}{\partial x}(x^*, \dot{x}^*, t) - \frac{d}{dt}\left(\frac{\partial F}{\partial \dot{x}}(x^*, \dot{x}^*, t) \right) \right] h \, dt + \lambda O\left(||h||^2\right) \geq 0.$$

On letting $\lambda \to 0$, we find that

$$\int_a^b \left[\frac{\partial F}{\partial x}(x^*, \dot{x}^*, t) - \frac{d}{dt}\left(\frac{\partial F}{\partial \dot{x}}(x^*, \dot{x}^*, t) \right) \right] h \, dt \geq 0$$

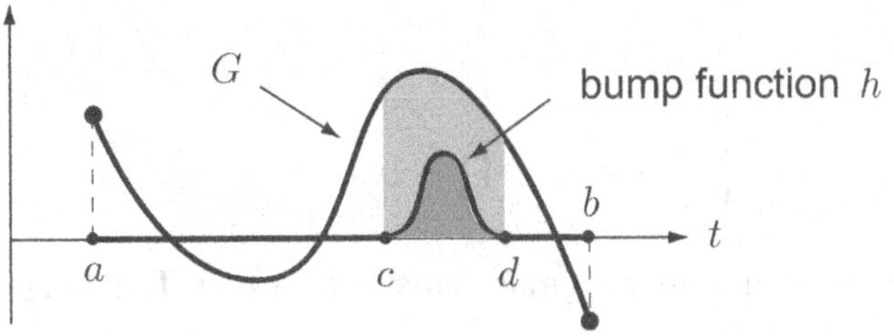

FIGURE 13.3 An interval (c, d) in which $G(t) > 0$ and a corresponding bump function $h(t)$. The integral of $G \times h$ must be positive.

for *all* admissible variations h. In particular, this inequality must remain true if h is replaced by $-h$, but this is only possible if the integral has the value zero.

Equation (13.10) therefore reduces to

$$J[x^* + h] - J[x^*] = O\left(||h||^2\right) \tag{13.15}$$

for small $||h||$. By definition, this means that x^* makes $J[x]$ stationary. The same result applies to functions that provide a local maximum for $J[x]$. Our result is summarised as follows:

Functions that minimise or maximise J make J stationary

If the function x^* provides a local minimum for the integral functional $J[x]$, then x^* makes J stationary. The same applies to functions that provide a local maximum for J.

Euler–Lagrange equation

We will now obtain the differential equation that must be satisfied by a function $x^*(t)$ that makes $J[x]$ stationary. This is the counterpart of the elementary condition $f'(x) = 0$. To find this, we return to equation (13.12). In the integrand, the function inside the square brackets looks complicated but it is just a function of t and, if we denote it by $G(t)$, then

$$\int_a^b G(t)\, h(t)\, dt = 0 \tag{13.16}$$

for *all* admissible variations h. In fact, the only function $G(t)$ for which this is possible is the *zero function*, that is, $G(t) = 0$ for $a < t < b$. The proof is as follows:

Suppose that $G(t)$ is *not* the zero function. Then there must exist some interval (c, d), lying inside the interval (a, b) in which $G(t) \neq 0$ and thus has *constant sign* (positive, say). Take $h(t)$ to be a 'bump function' (as shown in Figure 13.3), which is zero outside the interval (c, d) and positive inside. For such a choice of h,

$$\int_a^b G(t)\, h(t)\, dt = \int_c^d G(t)\, h(t)\, dt > 0,$$

since the integral of a positive function must be positive. This contradicts equation (13.16) and so G must be the zero function.

We have therefore shown that

$$\frac{\partial F}{\partial x}(x^*, \dot{x}^*, t) - \frac{d}{dt}\left(\frac{\partial F}{\partial \dot{x}}(x^*, \dot{x}^*, t)\right) = 0$$

for $a < t < b$, which is the same as saying that x^* must satisfy the **Euler–Lagrange** differential equation

$$\frac{\partial F}{\partial x} - \frac{d}{dt}\left(\frac{\partial F}{\partial \dot{x}}\right) = 0.$$

The above argument is reversible and so the converse result is also true. Our result is summarised as follows:

Euler–Lagrange equation

If the function x^* makes the integral functional

$$J[x] = \int_a^b F(x, \dot{x}, t)\, dt, \tag{13.17}$$

stationary, then x^* must satisfy the **Euler–Lagrange** differential equation

$$\frac{d}{dt}\left(\frac{\partial F}{\partial \dot{x}}\right) - \frac{\partial F}{\partial x} = 0. \tag{13.18}$$

The converse result is also true.

It is very convenient to give solutions of the Euler–Lagrange equation a special name.

Definition 13.2 ***Extremals*** *Any solution of the Euler–Lagrange equation is called an* ***extremal**** *of the functional* $J[x]$.

Key results of the calculus of variations

- If the function x^* minimises or maximises the functional J, then x^* makes J stationary and so x^* must be an extremal of J.
- If x^* is an extremal of J, then x^* makes J stationary, but it may not minimise or maximise J.

* The term *extremal* should not be confused with *extremum* (plural: *extrema*). Extremum means maximum or minimum. Thus a function that provides an extremum of J must make J stationary and so must be an extremal of J (that is, it must satisfy the Euler–Lagrange equation). The converse is not true. An extremal may not provide a minimum or maximum of J. (Try reading this again!)

Fortunately, solving problems with the E–L equation is much easier than the preceding theory! The E–L equation is a second order *non-linear* ODE, and so one needs a measure of luck when it comes to finding solutions. Nevertheless, many interesting cases can be solved in closed form.

Example 13.1 ***Finding extremals 1***

Find the extremal of the functional

$$J[x] = \int_1^2 \frac{\dot{x}^2}{4t}\,dt$$

that satisfies the end conditions $x(1) = 5$ and $x(2) = 11$.

Solution

By definition, extremals are solutions of the E–L equation. In the present case, $F = \dot{x}^2/4t$ so that

$$\frac{\partial F}{\partial x} = 0, \qquad \frac{\partial F}{\partial \dot{x}} = \frac{\dot{x}}{2t},$$

and the E–L equation takes the form

$$\frac{d}{dt}\left(\frac{\dot{x}}{2t}\right) - 0 = 0.$$

On integrating, we obtain

$$x = ct^2 + d,$$

where c and d are constants of integration. The **extremals** of J are therefore a family of parabolas in the (t, x)-plane. The *admissible* extremals are those that satisfy the prescribed end conditions $x(1) = 5$ and $x(2) = 11$. On applying these conditions, we find that $c = 2$ and $d = 3$ so that the only **admissible extremal** of $J[x]$ is given by

$$\widehat{x} = 2t^2 + 3. \blacksquare$$

Question ***Maximum, minimum, or neither?***

Does the extremal $\widehat{x} = 2t^2 + 3$ maximise or minimise J?

Answer

The admissible extremal $\widehat{x}$ is known to make J *stationary*. It may minimise J, maximise J, or do neither. With the theory that we have at our disposal, we cannot generally decide what happens. However, in a few simple cases (including this one), we can decide very easily.

Let h be *any* admissible variation (not necessarily small) and consider the variation in J that it produces, namely,

$$\begin{aligned} J[\widehat{x}+h]-J[\widehat{x}] &= \int_1^2 \frac{(4t+\dot{h})^2}{4t}\,dt - \int_1^2 \frac{(4t)^2}{4t}\,dt \\ &= \int_1^2 \left(4t+2\dot{h}+\frac{\dot{h}^2}{4t}\right)dt - \int_1^2 4t\,dt \\ &= 2\Big[\,h\,\Big]_{t=1}^{t=2} + \int_1^2 \frac{\dot{h}^2}{4t}\,dt \\ &= \int_1^2 \frac{\dot{h}^2}{4t}\,dt \end{aligned}$$

since h is an admissible extremal satisfying $h(1) = h(2) = 0$. Hence

$$J[\widehat{x}+h]-J[\widehat{x}] = \int_1^2 \frac{\dot{h}^2}{4t}\,dt \geq 0,$$

since the integral of a positive function must be positive. Thus $\widehat{x}$ actually provides the **global minimum** of $J[x]$. The global minimum *value* of J is therefore $J[2t^2+3] = 6$. ■

A useful integral of the Euler–Lagrange equation

Not all examples are as easy as the last one and the E–L equation often has a complicated form. However, for the case in which the function $F(x,\dot{x},t)$ has no *explicit* dependence on t (that is, $F = F(x,\dot{x})$), the second order E–L equation can always be integrated once to yield a first order ODE. This offers a great simplification in many important problems.

Suppose that $F = F(x,\dot{x})$. Then it follows from the product rule and the chain rule that

$$\begin{aligned} \frac{d}{dt}\left(\dot{x}\frac{\partial F}{\partial \dot{x}} - F\right) &= \ddot{x}\frac{\partial F}{\partial \dot{x}} + \dot{x}\frac{d}{dt}\left(\frac{\partial F}{\partial \dot{x}}\right) - \left(\ddot{x}\frac{\partial F}{\partial \dot{x}} + \dot{x}\frac{\partial F}{\partial x}\right) \\ &= \dot{x}\left(\frac{d}{dt}\left(\frac{\partial F}{\partial \dot{x}}\right) - \frac{\partial F}{\partial x}\right). \end{aligned} \tag{13.19}$$

Thus, if x satisfies the E–L equation

$$\frac{d}{dt}\left(\frac{\partial F}{\partial \dot{x}}\right) - \frac{\partial F}{\partial x} = 0,$$

it follows that x satisfies the first order equation

$$\dot{x}\frac{\partial F}{\partial \dot{x}} - F = \text{constant}. \tag{13.20}$$

for some choice of the constant c. Conversely, if x is any *non-constant* solution of equation (13.20) then it satisfies the E–L equation.

It should be noted that equation (13.20) always has solutions of the form x = constant, but these solutions usually do *not* satisfy the corresponding E–L equation. They occur because of the factor $\dot{x}$ that appears on the right in equation (13.19). When overlooked, this glitch can give baffling results. *Do not believe that constant solutions of equation (13.20) satisfy the E–L equation unless you have checked it directly!*

Our result is summarised as follows:

A first integral of the E–L equation

Suppose that $F = F(x, \dot{x})$. Then any function that satisfies the E–L equation

$$\frac{d}{dt}\left(\frac{\partial F}{\partial \dot{x}}\right) - \frac{\partial F}{\partial x} = 0$$

also satisfies the first order differential equation

$$\dot{x}\frac{\partial F}{\partial \dot{x}} - F = c, \tag{13.21}$$

for some value of the constant c. Conversely, any *non-constant* solution of equation (13.21) satisfies the E–L equation. Constant solutions of equation (13.21) may or may not satisfy the E–L equation.

Example 13.2 *Finding extremals 2*

Find the extremal of the functional

$$J[x] = \int_0^7 \frac{(1+\dot{x}^2)^{1/2}}{x}\, dt$$

that lies in $x > 0$ and satisfies the end conditions $x(0) = 4$ and $x(7) = 3$. [The restriction $x > 0$ ensures that the integrand does not become singular.]

Solution

By definition, extremals are solutions of the E–L equation and, since t is not explicitly present in this functional, we can use the integrated form (13.21). On substituting $F = (1+\dot{x}^2)^{1/2}/x$ into (13.21) and simplifying, we obtain

$$x(1+\dot{x}^2)^{1/2} = C,$$

where C is a constant; since x is assumed positive, C must be positive. This equation can be rearranged in the form

$$\dot{x} = \pm\frac{(C^2 - x^2)^{1/2}}{x},$$

a pair of first order separable ODEs.

The solutions are*

$$\pm\left(C^2 - x^2\right)^{1/2} = t + D,$$

where D is a constant of integration. Hence the **extremals** of J are (the upper halves of) the family of circles

$$x^2 + (t + D)^2 = C^2$$

in the (t, x)-plane. On applying the given end conditions, we find that $C = 5$ and $D = -3$, so that the only **admissible extremal** is an arc of the circle with centre $(3, 0)$ and radius 5, namely

$$\widehat{x} = +\sqrt{16 + 6t - t^2} \qquad (0 \le t \le 7).$$

Since there is only one admissible extremal, it follows that, if it were *known* that a minimising (or maximising) function existed, then this must be it. However, we have no such knowledge and no means of deciding whether $\widehat{x}$ provides a minimum or maximum for J, or neither. (It actually provides the global minimum of J.) ■

Our final example is the famous brachistochrone problem.†

Example 13.3 *The brachistochrone (shortest time) problem*

Two fixed points P and Q are connected by a smooth wire lying in the vertical plane that contains P and Q. A particle is released from rest at P and slides, under uniform gravity, along the wire to Q. What shape should the wire be so that the transfer is completed in the shortest time?

Solution

Suppose that the wire lies in (x, z)-plane, with Oz pointing vertically *downwards*, with P at the origin, and Q at the point (a, b). Let the shape of the wire be given by the curve $z = z(x)$. Then, since the particle is released from rest when $z = 0$, energy conservation implies that the speed of the particle when its downward displacement is z is $(2gz)^{1/2}$. The time T taken for the particle to complete the transfer is therefore

$$T[z] = (2g)^{-1/2}\int_0^a \frac{\left\{1 + \dot{z}^2\right\}^{1/2}}{z^{1/2}}\, dx, \qquad (13.22)$$

* It is evident that these equations also admit the constant solution $x = C$. However, it may be verified that the E–L equation for this problem has no constant solutions.

† This famous minimisation problem was posed in 1696 by Johann Bernoulli (who had already found the solution) as a not-so-friendly challenge to his mathematical contemporaries. Solutions were found by Jacob Bernoulli, de l'Hôpital, Leibnitz, and Newton, who (according to his publicity manager) had the answer within a day. Newton published his solution anonymously, but Johann Bernoulli identified Newton as the author declaring '*one can recognise the lion by the marks of his claw*'. The last word goes to Newton. He complained '*I do not love to be pestered and teased by foreigners about mathematical things* ...'.

where $\dot{z}$ means dz/dx. The problem is to find the function $z(x)$, satisfying the end conditions $z(0) = 0$, $z(a) = b$, that minimises T.

If x^* minimises T, then it must make T stationary and so be an extremal of T. Since x is not explicitly present in this functional, we can use the integrated form (13.21) of the E–L equation. On substituting in $F = (1+\dot{z}^2)^{1/2}/z^{1/2}$ and simplifying, we obtain

$$z\left(1 + \dot{z}^2\right) = 2C,$$

where C is a positive constant. (The constant is called $2C$ for later convenience.) This equation can be arranged in the form

$$\dot{z} = \pm\left(\frac{2C - z}{z}\right)^{1/2},$$

a pair of first order separable ODEs. (The constant solution $z = 2C$ is not an extremal of J and can be disregarded.) Integration gives

$$x = \pm\int\left(\frac{z}{2C - z}\right)^{1/2} dz.$$

To perform the integral, we make the substitution* $z = C(1 - \cos\psi)$, in which case

$$\begin{aligned} x &= \pm C\int\left(\frac{1 - \cos\psi}{1 + \cos\psi}\right)^{1/2} \sin\psi\, d\psi \\ &= \pm C\int 2\sin^2 \tfrac{1}{2}\psi\, d\psi \\ &= \pm C(\psi - \sin\psi) + D, \end{aligned}$$

where D is a constant of integration. Thus the **extremals** of J have the *parametric* form

$$x = \pm C(\psi - \sin\psi) + D, \qquad z = C(1 - \cos\psi),$$

with ψ as parameter. Since the two choices of sign correspond only to a change of sign of the parameter, we may assume the positive choice. These curves are a family of cycloids† with 'radius' C and shift D in the x-direction.

Now we find the **admissible extremals**. The condition that $z = 0$ when $x = 0$ implies that the shift constant $D = 0$. The radius C of the cycloid is then determined from the second end condition $z = b$ when $x = a$. In general, C must be determined numerically but, in special cases, C may be found analytically. For example, if Q is the point $(a, 0)$ (so that P and Q are on the same horizontal level) it is found that $C = a/2\pi$. A more typical case is shown in Figure 13.4.

* It's easy to spot the smart substitution when you already know the answer!

† The cycloid is the path traced out by a point on the rim of a disk rolling on a plane; the 'radius' referred to is the radius of this disk. Since the E–L equation cannot be satisfied at a cusp, each extremal must lie on a single loop of the cycloid.

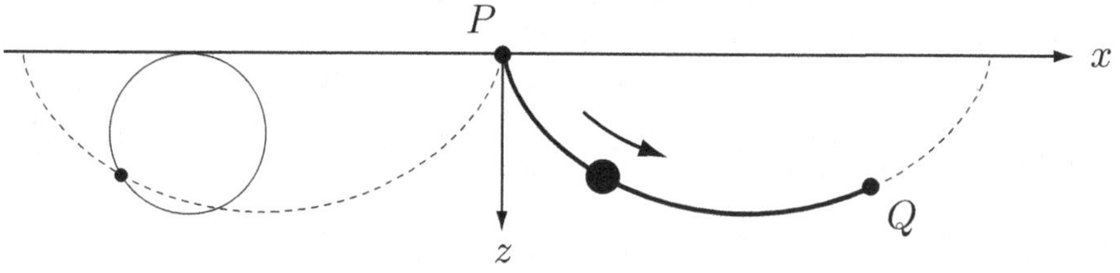

FIGURE 13.4 The curve that minimises $T[z]$ is an arc of a cycloid.

Since there is only one admissible extremal, it must be the **minimising curve** for $T[z]$, *provided that a minimising curve exists at all.* In view of the physical origin of the problem, most of us are happy to take this for granted, but purists may sleep more soundly in the knowledge that this can also be proved mathematically. ■

13.3 VARIATIONAL PRINCIPLES

The laws of physics are usually formulated in terms of variables or fields that satisfy differential equations. Thus, the generalised coordinates $\boldsymbol{q}(t)$ of a mechanical system satisfy Lagrange's equations (a system of ODEs), the electromagnetic field satisfies Maxwell's equations (a system of PDEs), the wave function of quantum mechanics satisfies Schrödinger's wave equation, and so on. But there is an alternative way of expressing these laws in terms of **variational principles**. *In the variational approach, the actual physical behaviour of the system is distinguished by the fact that it makes a certain integral functional stationary.* Thus all of the physics is somehow contained in the integrand of this functional!

Expressing physical laws in variational form does not make it any easier to solve problems. Indeed, problems will continue to be solved by using differential equations. The virtue of the variational formulation is that it is much easier to extend existing theory to new situations. For example, the theory of fields can be developed in the language of classical mechanics by using the variational formulation.

Fermat's principle

These ideas are nicely illustrated by an example most readers will be familiar with: the paths of **light rays** in geometrical optics. When travelling through a **homogeneous medium**, light rays travel in straight lines. But, on meeting a plane interface between two different homogeneous media, the ray is either reflected or suffers a sharp change of direction called refraction, as shown in Figure 13.5. This change of direction is governed by **Snell's law** of refraction $n_1 \sin\theta_1 = n_2 \sin\theta_2$, where n_1 and n_2 are the refractive indices of the two media, and θ_1 and θ_2 are the angles that the ray makes with the *normal* to the interface. In terms of the angle ψ used in Figure 13.5, Snell's law takes the form

$$n_1 \cos\psi_1 = n_2 \cos\psi_2 = n_3 \cos\psi_3.$$

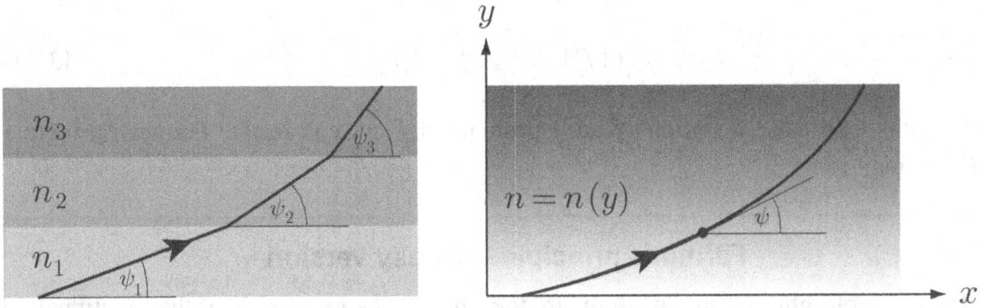

FIGURE 13.5 When a light ray passes between homogeneous media (left), it satisfies Snell's law $n_1 \cos\psi_1 = n_2 \cos\psi_2 = n_3 \cos\psi_3$. In the continuous case with $n = n(y)$ (right), Snell's law becomes $n \cos\psi$ = constant.

In the more general case of an **inhomogeneous medium** in which n varies continuously in the y-direction, one would then expect curved rays that satisfy Snell's law in the form

$$n \cos\psi = \text{constant}.$$

A variational principle consistent with these rules was proposed by Fermat in 1657 and became known as **Fermat's principle** of least time. This stated that:

Of all the possible paths that a light ray might take between two fixed points, the actual path is the one that minimises the travel time of the ray.

Fermat showed that his principle implied the truth of the laws of reflection and refraction (as well as predicting straight rays in a homogeneous medium). Fermat's original principle is a beautifully simple and general statement about the paths taken by light rays but, sadly, it is not quite correct. The correct version is as follows:

Fermat's principle

The actual path taken by a light ray between two fixed points makes the travel time of the ray **stationary**.

The difference between the original and correct versions is that the path taken by the ray does not necessarily make the travel time a minimum, but it does make the travel time stationary.* In practice, the travel time is usually a minimum, but there are exceptional cases where it is not.

If the free-space speed of light is c, then the speed of light at a point of a medium where the refractive index is n is c/n. A (hypothetical) path $\mathcal{P}$ in the medium would

* Surprisingly, incorrect statements about Fermat's principle abound in the literature. It is often claimed that *'the path of a light ray makes the travel time a minimum or (occasionally) a maximum'*. This is untrue. The path of a ray can *never* make T a maximum. It usually makes T a minimum, occasionally it provides neither a minimum nor a maximum, but it *never* provides a maximum.

therefore be traversed in time T given by the line integral

$$T[\mathcal{P}] = c^{-1} \int_{\mathcal{P}} n\, ds. \tag{13.23}$$

Since paths that make T stationary are extremals of T we can restate Fermat's principle in the elegant form:

Fermat's principle – Classy version

The paths of light rays in a medium are the same as the **extremals** of the functional T for that medium.

Suppose that the refractive index n in the medium depends only on y (as in Figure 13.5) and consider rays that lie in the (x, y)-plane. Then a ray that connects the points (x_0, y_0) and (x_1, y_1) must be an extremal of the functional T which, in Cartesian coordinates, takes the form

$$T[y] = c^{-1} \int_{x_0}^{x_1} n \left(1 + \dot{y}^2\right)^{1/2} dx, \tag{13.24}$$

where $\dot{y}$ means dy/dx, and $n = n(y)$. Since n does not depend upon x, we may use the integrated form (13.21) of the E–L equation, which gives

$$\frac{n}{(1 + \dot{y}^2)^{1/2}} = \text{constant}.$$

If we write $\dot{y} = \tan\psi$, where ψ is the angle between the tangent to the ray and the x-axis (see Figure 13.5), then this equation becomes

$$n \cos\psi = \text{constant},$$

exactly as anticipated from Snell's law for layered media.

Question ***A puzzle***

When $n = n(y)$, it is easy to verify that the straight lines $y = \text{constant}$ are *not* extremals of $T[y]$ and are therefore *not* rays (although they do satisfy Snell's law!). But since such a 'ray' would experience a constant value of n, how does the ray know that it must bend?

Answer

Your physics lecturer will be pleased to answer this question. ■

The variational approach really comes into its own however when we extend our theory to other inhomogeneous media, where the correct generalisation of Snell's law is difficult to spot. There is no such difficulty with the variational approach; Fermat's principle still holds. The case of a light ray propagating in an axially symmetric medium is solved in Problem 13.9. From Fermat's principle, this is quite straightforward, but, starting from Snell's law, one would probably guess the wrong formula!

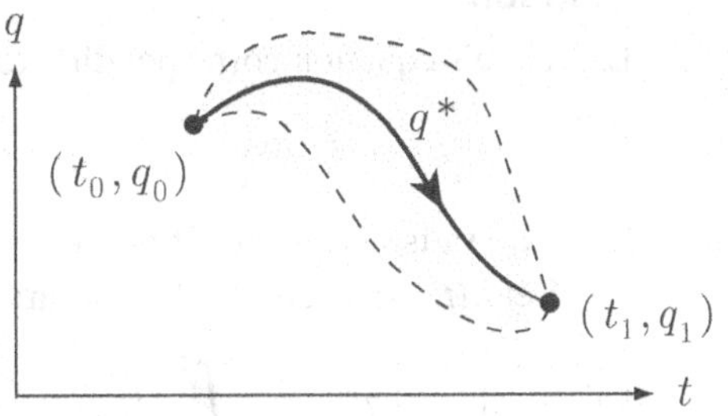

FIGURE 13.6 Hamilton's principle Of all the kinematically possible trajectories of a system that connect the configurations $q = q_1$ and $q = q_2$ in the time interval $[t_1, t_2]$, the actual motion $q^*(t)$ makes the action functional of the system stationary.

13.4 HAMILTON'S PRINCIPLE

Hamilton's principle is the variational principle that is equivalent to Lagrange's equations of motion. The comparison with geometrical optics is that Hamilton's principle corresponds to Lagrange's equations as Fermat's principle corresponds to Snell's law. We consider first the special case of systems with one degree of freedom.

Systems with one degree of freedom

Consider a Lagrangian system with a single generalised coordinate q and Lagrangian $L(q, \dot{q}, t)$. Then the trajectory $q^*(t)$ is an actual motion of the system if, and only if, it satisfies the Lagrange equation

$$\frac{d}{dt}\left(\frac{\partial L}{\partial \dot{q}}\right) - \frac{\partial L}{\partial q} = 0. \qquad (13.25)$$

It is impossible not to notice that equation (13.25) is the **Euler–Lagrange** equation one would get by making stationary the functional $S[q]$ defined by

$$S[q] = \int_{t_0}^{t_1} L(q, \dot{q}, t)\, dt. \qquad (13.26)$$

The scalar quantity S is called the **action** and the functional $S[q]$ is called the **action functional** corresponding to the Lagrangian L (for the time interval $[t_0, t_1]$).

From this simple observation, it follows that $q^*(t)$ *is an actual motion of the system if, and only if, it makes the action functional* $S[q]$ *stationary*. The situation is as shown in Figure 13.6. This is **Hamilton's principle** for a mechanical system with one degree of freedom.

Example 13.4 *Hamilton's principle*

A certain oscillator with generalised coordinate q has Lagrangian

$$L = \tfrac{1}{2}\dot{q}^2 - \tfrac{1}{2}q^2.$$

Verify that $q^* = \sin t$ is a motion of the oscillator, and show directly that it makes the action functional $S[q]$ stationary in any time interval $[0, \tau]$.

Solution

Lagrange's equation corresponding to the Lagrangian $L = \frac{1}{2}\dot{q}^2 - \frac{1}{2}q^2$ is

$$\ddot{q} + q = 0.$$

Since q^* satisfies this equation, it is a motion of the oscillator.

Let $h(t)$ be an admissible variation. Then

$$\begin{aligned}
S[q^* + h] - S[q^*] &= \tfrac{1}{2}\int_0^\tau \left[\left(\cos t + \dot{h}\right)^2 - (\sin t + h)^2 - \cos^2 t + \sin^2 t\right] dt \\
&= \tfrac{1}{2}\int_0^\tau \left(2\dot{h}\cos t + \dot{h}^2 - 2h\sin t - h^2\right) dt \\
&= \Big[h\cos t\Big]_0^\tau + \tfrac{1}{2}\int_0^\tau \left(\dot{h}^2 - h^2\right) dt \\
&= \tfrac{1}{2}\int_0^\tau \left(\dot{h}^2 - h^2\right) dt,
\end{aligned}$$

since $h(0) = h(\tau) = 0$. It follows that

$$\begin{aligned}
|\,S[q^* + h] - S[q^*]\,| &\le \tfrac{1}{2}\tau\left(\max_{0\le t\le\tau} |h(t)|^2 + \max_{0\le t\le\tau} |\dot{h}(t)|^2\right) \\
&\le \tfrac{1}{2}\tau\left(\max_{0\le t\le\tau} |h(t)| + \max_{0\le t\le\tau} |\dot{h}(t)|\right)^2 \\
&= \tfrac{1}{2}\tau\,||h||^2.
\end{aligned}$$

Hence

$$S[q^* + h] - S[q^*] = O\left(||h||^2\right),$$

which, by definition, means that q^* makes the action functional $S[q]$ **stationary**. [It may *not* make $S[q]$ a minimum.] ■

Systems with many degrees of freedom

Hamilton's principle can be extended to systems with any number of degrees of freedom. In this more general case, the system has generalised coordinates $\boldsymbol{q} = (q_1, q_2, \ldots, q_n)$, the Lagrangian has the form $L = L(\boldsymbol{q}, \dot{\boldsymbol{q}}, t)$, and Lagrange's equations of motion are the *n simultaneous equations*

$$\frac{d}{dt}\left(\frac{\partial L}{\partial \dot{q}_j}\right) - \frac{\partial L}{\partial q_j} = 0 \qquad (1 \le j \le n). \tag{13.27}$$

The action functional is now defined to be:

Definition 13.3 ***Action functional*** *The functional*

$$S[\boldsymbol{q}] = \int_{t_0}^{t_1} L(\boldsymbol{q}, \dot{\boldsymbol{q}}, t)\, dt \tag{13.28}$$

is called the ***action functional*** *corresponding to the Lagrangian $L(\boldsymbol{q}, \dot{\boldsymbol{q}}, t)$ (for the time interval $[t_0, t_1]$).*

The notation $S[\boldsymbol{q}]$ is really a shorthand form for $S[q_1, q_2, \ldots, q_n]$ so that now there are n functions that can be varied by the n independent variations $h_1, h_2, \ldots, h_n$ respectively. In the vector notation, such a variation is denoted by $\boldsymbol{h}$, where $\boldsymbol{h} = (h_1, h_2, \ldots, h_n)$. The theory that we have developed does not cover the case where the functional has more than one 'independent variable', but it can be extended to do so. An outline of this extension is as follows:

Consider the general situation in which

$$J[\boldsymbol{x}] = \int_a^b F(\boldsymbol{x}, \dot{\boldsymbol{x}}, t)\, dt,$$

where the vector function $\boldsymbol{x}(t) = (x_1(t), x_2(t), \ldots, x_n(t))$. By using the same argument as before, the variation in J caused by the admissible* variation $\boldsymbol{h}$ in $\boldsymbol{x}^*$ is found to be

$$J[\boldsymbol{x}^* + \boldsymbol{h}] - J[\boldsymbol{x}^*] = \sum_{j=1}^{n} \int_a^b \left[\frac{\partial F}{\partial x_j}(\boldsymbol{x}^*, \dot{\boldsymbol{x}}^*, t) - \frac{d}{dt}\left(\frac{\partial F}{\partial \dot{x}_j}(\boldsymbol{x}^*, \dot{\boldsymbol{x}}^*, t) \right) \right] h_j \, dt + O\left(||\boldsymbol{h}||^2 \right),$$

where $||\boldsymbol{h}||^2 = ||h_1||^2 + ||h_2||^2 + \cdots + ||h_n||^2$. This variation is linear in $\boldsymbol{h}$ with an error term of order $||\boldsymbol{h}||^2$. As before we say that $\boldsymbol{x}^*$ makes $J[\boldsymbol{x}]$ stationary if the linear term is zero, leaving only the error term.

Definition 13.4 ***Stationary*** J *The vector function* $\boldsymbol{x}^*(t)$ *is said to make the functional* $J[\boldsymbol{x}]$ ***stationary*** *if*

$$J[\boldsymbol{x}^* + \boldsymbol{h}] - J[\boldsymbol{x}^*] = O\left(||\boldsymbol{h}||^2 \right)$$

when $||\boldsymbol{h}||$ *is small.*

If $\boldsymbol{x}^*$ makes the functional $J[\boldsymbol{x}]$ stationary then

$$\sum_{j=1}^{n} \int_a^b \left[\frac{\partial F}{\partial x_j}(\boldsymbol{x}^*, \dot{\boldsymbol{x}}^*, t) - \frac{d}{dt}\left(\frac{\partial F}{\partial \dot{x}_j}(\boldsymbol{x}^*, \dot{\boldsymbol{x}}^*, t) \right) \right] h_j \, dt = 0,$$

for all admissible variations $\boldsymbol{h}$. By allowing each of the $\{x_j\}$ to vary separately (while the others remain constant), the 'bump function' argument can be applied exactly as before to show that

$$\frac{\partial F}{\partial x_j}(\boldsymbol{x}^*, \dot{\boldsymbol{x}}^*, t) - \frac{d}{dt}\left(\frac{\partial F}{\partial \dot{x}_j}(\boldsymbol{x}^*, \dot{\boldsymbol{x}}^*, t) \right) = 0 \qquad (1 \le j \le n).$$

This is the same as saying that $\boldsymbol{x}^*$ must satisfy the *simultaneous* **Euler–Lagrange** equations

$$\frac{\partial F}{\partial x_j} - \frac{d}{dt}\left(\frac{\partial F}{\partial \dot{x}_j} \right) = 0 \qquad (1 \le j \le n).$$

Our result is summarised as follows:

* The vector variation $\boldsymbol{h}$ is admissible if $\boldsymbol{h}(a) = \boldsymbol{h}(b) = \boldsymbol{0}$, that is, if $h_1, h_2, \ldots, h_n$ are all admissible.

Euler–Lagrange equations with many variables

The vector function $\boldsymbol{x}^*$ makes the integral functional

$$J[\boldsymbol{x}] = \int_a^b F\left(\boldsymbol{x}, \dot{\boldsymbol{x}}, t\right) dt$$

stationary if, and only if, $\boldsymbol{x}^*$ satisfies the simultaneous **Euler–Lagrange** differential equations

$$\frac{d}{dt}\left(\frac{\partial F}{\partial \dot{x}_j}\right) - \frac{\partial F}{\partial x_j} = 0 \qquad (1 \le j \le n).$$

This is a natural generalisation of the single variable theory and corresponds to the elementary result that a function of n variables $f(x_1, x_2, \ldots, x_n)$ has a stationary point if, and only if, all its first partial derivatives vanish at that point.

The statement of Hamilton's principle for systems with many degrees of freedom is therefore:

Hamilton's principle

The trajectory $\boldsymbol{q}^*(t)$ is an actual motion of a mechanical system if, and only if, $\boldsymbol{q}^*$ makes the action functional of the system stationary.

The only essential difference between this correct version of Hamilton's principle and the original version (quoted at the beginning of the chapter) is that an actual motion of the system does not necessarily make the action functional a *minimum*, but it always makes the action functional *stationary*.* In practice, the action functional is usually minimised, but there are exceptional cases where it is not (see Problem 13.11).

As with Fermat's principle, there is a classy version of Hamilton's principle, which is less wordy and more satisfactory generally. It makes use of the concept of the **extremals** of J, which are simply solutions of the n simultaneous Euler–Lagrange equations.

Hamilton's principle – Classy version

The actual motions of a mechanical system are the same as the extremals of its action functional.

* Incorrect statements about Hamilton's principle also abound in the literature. It is often claimed that *'an actual motion the system makes the action functional a minimum or (occasionally) a maximum'*. This is untrue. It is not possible to make S a maximum. The actual motion of the system usually makes S a minimum, occasionally it provides neither a minimum nor a maximum, but it *never* provides a maximum.

Significance of Hamilton's principle

Since Hamilton's principle is equivalent to Lagrange's equations, it can be regarded as the fundamental postulate of classical mechanics, instead of Newton's laws,* for any mechanical system that has a Lagrangian. It should be emphasised that this is *not* a new theory – the Newtonian theory is correct – but an alternative route to the same results. Thus we can derive Lagrange's equations of motion from the Newtonian theory (as we did) or, more directly, from Hamilton's principle. Because Hamilton's principle can be extended to apply to a wide range of physical phenomena while the Newtonian theory can not, Hamilton's principle is regarded as the more fundamental.

The problem with taking Hamilton's principle as the fundamental postulate of classical mechanics is that, had one not been exposed to the traditional treatment, one would have no idea what the Lagrangian ought to be for any particular system. To convince oneself of the difficulties involved, it is instructive to read Landau's [6] 'derivation' of the Lagrangian for the simplest system imaginable – a single particle moving in free space. Indeed, it seems difficult to introduce the concept of mass convincingly at all. Nevertheless, this is the route that must be followed when Hamilton's principle is extended, for example, to particle physics. The Lagrangian has to be found by intelligent guesswork, and, in particular, by taking account of all the symmetries that are known to exist.

Within classical mechanics itself, it may appear that Hamilton's principle has told us nothing new. It says that the motions of a mechanical system are the same as the extremals of the action functional, that is, the motions satisfy Lagrange's equations; this we already knew. However, because the equations of motion have the special form associated with variational principles, they can be shown to possess important properties that would be very difficult to prove directly. One example of this is the effect on the equations of motion of choosing a new set of generalised coordinates $\boldsymbol{q}' = (q_1', q_2', \dots, q_n')$. The $\boldsymbol{q}'$ are known functions of the old generalised coordinates $\boldsymbol{q}$ and *vice versa*. The direct approach would be to subject the Lagrange equations (13.27) to this general transformation of the coordinates and see what happens; the result would be a complicated mess. However, in the variational approach, one simply expresses the Lagrangian L as a function of the new variables, that is, $L = L(\boldsymbol{q}', \dot{\boldsymbol{q}}', t)$. Although L has a different *functional form* in terms of the coordinates $\boldsymbol{q}'$, its *values* are the same as before, so that the new action functional

$$S[\boldsymbol{q}'] = \int_{t_0}^{t_1} L(\boldsymbol{q}', \dot{\boldsymbol{q}}', t)\, dt,$$

takes the *same values* as the old, provided that $\boldsymbol{q}'(t)$ and $\boldsymbol{q}(t)$ refer to the same trajectory of the mechanical system. It follows that, if the trajectory $\boldsymbol{q}(t)$ makes $S[\boldsymbol{q}]$ stationary, then the corresponding trajectory $\boldsymbol{q}'(t)$ makes $S[\boldsymbol{q}']$ stationary. Hence the extremals of $S[\boldsymbol{q}]$ map into the extremals of $S[\boldsymbol{q}']$, and *vice versa*. It follows that the transformed equations of motion are just the same as the old ones with $\boldsymbol{q}$ replaced by $\boldsymbol{q}'$. This fact is expressed

* Actually, instead of the Second and Third Laws. The First Law is needed to ensure that the motion is observed from an inertial reference frame.

by saying that the Lagrange equations of motion are **invariant under transformations** of the generalised coordinates. This remarkable result clearly applies to *any* system of equations derived from a variational principle. This provides a general way of ensuring that any proposed set of governing equations should be invariant under a particular group of transformations (the Lorentz transformations, for instance). This will be so if the equations are derivable from a variational principle whose 'Lagrangian' is invariant under the same group of transformations.

Problems on Chapter 13

Answers and comments are at the end of the book.
Harder problems carry a star (∗).

Euler–Lagrange equation

13.1 Find the extremal of the functional

$$J[x] = \int_1^2 \frac{\dot{x}^2}{t^3}\, dt$$

that satisfies $x(1) = 3$ and $x(2) = 18$. Show that this extremal provides the global minimum of J.

13.2 Find the extremal of the functional

$$J[x] = \int_0^{\pi} (2x \sin t - \dot{x}^2)\, dt$$

that satisfies $x(0) = x(\pi) = 0$. Show that this extremal provides the global maximum of J.

13.3 Find the extremal of the path length functional

$$L[y] = \int_0^1 \left[1 + \left(\frac{dy}{dx}\right)^2\right]^{1/2} dx$$

that satisfies $y(0) = y(1) = 0$ and show that it does provide the global minimum for L.

13.4 An aircraft flies in the (x, z)-plane from the point $(-a, 0)$ to the point $(a, 0)$. ($z = 0$ is ground level and the z-axis points vertically upwards.) The cost of flying the aircraft at height z is $\exp(-kz)$ per unit *distance* of flight, where k is a positive constant. Find the extremal for the problem of minimising the total cost of the journey. [Assume that $ka < \pi/2$.]

13.5* ***Geodesics on a cone*** Solve the problem of finding a shortest path over the surface of a cone of semi-angle α by the calculus of variations. Take the equation of the path in the form $\rho = \rho(\theta)$, where ρ is distance from the vertex O and θ is the cylindrical polar angle measured around the axis of the cone. Obtain the general expression for the path length and find the extremal that satisfies the end conditions $\rho(-\pi/2) = \rho(\pi/2) = a$.

Verify that this extremal is the same as the shortest path that would be obtained by developing the cone on to a plane.

13.6 ***Cost functional*** A manufacturer wishes to minimise the cost functional

$$C[x] = \int_0^4 \left((3 + \dot{x})\dot{x} + 2x \right) dt$$

subject to the conditions $x(0) = 0$ and $x(4) = X$, where X is volume of goods to be produced. Find the extremal of C that satisfies the given conditions and prove that this function provides the global minimum of C.

Why is this solution not applicable when $X < 8$?

13.7 ***Soap film problem*** Consider the soap film problem for which it is required to minimise

$$J[y] = \int_{-a}^{a} y \left(1 + \dot{y}^2\right)^{\frac{1}{2}} dx$$

with $y(-a) = y(a) = b$. Show that the extremals of J have the form

$$y = c \cosh\left(\frac{x}{c} + d\right),$$

where c, d are constants, and that the end conditions are satisfied if (and only if) $d = 0$ and

$$\cosh \lambda = \left(\frac{b}{a}\right) \lambda,$$

where $\lambda = a/c$. Show that there are *two* admissible extremals provided that the aspect ratio b/a exceeds a certain critical value and *none* if b/a is less than this critical value. Sketch a graph showing how this critical value is determined.

The remainder of this question requires computer assistance. Show that the critical value of the aspect ratio b/a is about 1.51. Choose a value of b/a larger than the critical value ($b/a = 2$ is suitable) and find the two values of λ. Plot the two admissible extremals on the same graph. Which one looks like the actual shape of the soap film? Check your guess by perturbing each extremal by small admissible variations and finding the change in the value of the functional $J[y]$.

Fermat's principle

13.8 A sugar solution has a refractive index n that increases with the depth z according to the formula

$$n = n_0 \left(1 + \frac{z}{a}\right)^{1/2},$$

where n_0 and a are positive constants. A particular ray is horizontal when it passes through the origin of coordinates. Show that the path of the ray is not the straight line $z = 0$ but the parabola $z = x^2/4a$.

13.9 Consider the propagation of light rays in an axially symmetric medium, where, in a system of cylindrical polar co-ordinates (r, θ, z), the refractive index $n = n(r)$ and the rays lie in the plane $z = 0$. Show that Fermat's time functional has the form

$$T[r] = c^{-1} \int_{\theta_0}^{\theta_1} n \left(r^2 + \dot{r}^2\right)^{1/2} d\theta,$$

where $r = r(\theta)$ is the equation of the path, and $\dot{r}$ means $dr/d\theta$.

(i) Show that the extremals of T satisfy the ODE

$$\frac{n r^2}{(r^2 + \dot{r}^2)^{1/2}} = \text{constant}.$$

Show further that, if we write $\dot{r} = r \tan \psi$, where ψ is the angle between the tangent to the ray and the local cylindrical surface $r =$ constant, this equation becomes

$$r\, n \cos \psi = \text{constant},$$

which is the form of Snell's law for this case. Deduce that circular rays with centre at the origin exist only when the refractive index $n = a/r$, where a is a positive constant.

Hamilton's principle

13.10 A particle of mass 2 kg moves under uniform gravity along the z-axis, which points vertically downwards. Show that (in SI units) the action functional for the time interval $[0, 2]$ is

$$S[z] = \int_0^2 \left(\dot{z}^2 + 20z\right) dt,$$

where g has been taken to be 10 m s^{-2}.

Show directly that, of all the functions $z(t)$ that satisfy the end conditions $z(0) = 0$ and $z(2) = 20$, the actual motion $z = 5t^2$ provides the *least* value of S.

13.11 A certain oscillator with generalised coordinate q has Lagrangian

$$L = \dot{q}^2 - 4q^2.$$

Verify that $q^* = \sin 2t$ is a motion of the oscillator, and show directly that it makes the action functional $S[q]$ stationary in any time interval $[0, \tau]$.

For the time interval $0 \le t \le \pi$, find the variation in the action functional corresponding to the variations (i) $h = \epsilon \sin 4t$, (ii) $h = \epsilon \sin t$, where ϵ is a small parameter. Deduce that the motion $q^* = \sin 2t$ does not make S a minimum or a maximum.

13.12 A particle is constrained to move over a smooth fixed surface under *no forces* other than the force of constraint. By using Hamilton's principle and energy conservation, show that the

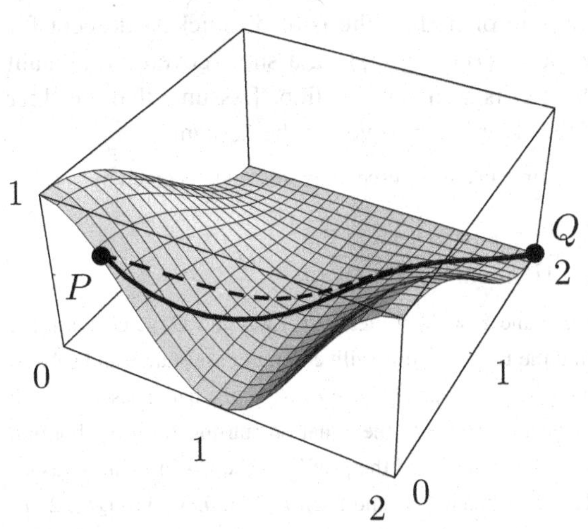

FIGURE 13.7 The path of quickest descent from P to Q in Cosine Valley. Those who lose their nerve at the summit can walk down by the *shortest* path (shown dashed).

path of the particle must be a geodesic of the surface. (The term geodesic has been extended here to mean those paths that make the length functional *stationary*).

This result has a counterpart in the theory of general relativity, where the concept of force does not exist and particles move along the geodesics of a curved space-time.

13.13 By using Hamilton's principle, show that, if the Lagrangian $L(\boldsymbol{q}, \dot{\boldsymbol{q}}, t)$ is modified to L' by any transformation of the form

$$L' = L + \frac{d}{dt}\, g(\boldsymbol{q}, t),$$

then the equations of motion still have the same family of solutions.

Computer assisted problems

13.14 *Geodesics on a paraboloid* Solve the problem of finding the shortest path between two points $P(0, 1, 1)$ and $Q(0, -1, 1)$ on the surface of the paraboloid $z = x^2 + y^2$.

Let C be a path lying in the surface that connects P and Q. Show that the length of C is given by

$$L[r] = \int_{-\pi/2}^{\pi/2} \left[r^2 + \left(1 + 4r^2\right)\dot{r}^2\right]^{1/2} d\theta,$$

where $r = r(\theta)$ is the *polar equation* of the projection of C on to the plane $z = 0$, and $\dot{r}$ means $dr/d\theta$. Now find the function $r(\theta)$ that minimises L. It is easier to work directly with the second order E–L-equation (which can be found with computer assistance). Solve the E–L-equation numerically with the initial conditions $r(0) = \lambda$, $\dot{r}(0) = 0$ and choose λ so that the path passes through P (and, by symmetry, Q). Plot the shortest path using 3D graphics.

13.15* ***The downhill skier*** Solve the problem of finding the path of quickest descent for a skier from the point $P(x_0, y_0, z_0)$ to the point $Q(x_1, y_1, z_1)$ on a snow covered mountain whose profile is given by $z = G(x, y)$, where G is a known function. [Assume that the skier starts from rest and that the total energy of the skier is conserved in the descent.]

Let C be a path connecting P and Q. Show that the time taken to descend by this route is given by

$$T[y] = (2g)^{-1/2} \int_{x_0}^{x_1} \left(\frac{1 + \dot{y}^2 + \left(G_{,x} + G_{,y}\dot{y}\right)^2}{G(x_0, y_0) - G(x, y)} \right)^{1/2} dx$$

where $y = y(x)$ is the projection of C on to the plane $z = 0$, $\dot{y}$ means dy/dx, and $G_{,x}$, $G_{,y}$ are the partial derivatives of G with respect to x, y. Obtain the E–L equation with computer assistance and solve it numerically with the initial conditions $y(x_1) = y_1$, $y'(x_1) = \lambda$ and choose λ so that the path passes through the P. [The numerical ODE integrator finds it easier to integrate the equation starting from the bottom. Why?] Plot the quickest route using 3D graphics. The author used the profile of the Cosine Valley resort, for which $G(x, y) = \cos^2(\pi x/2)\cos^2(\pi y/4)$. The skier had to descend from $P(1/3, 0, 3/4)$ to $Q(2, 2, 0)$. The computed quickest route down the valley is shown in Figure 13.7. Those who lose their nerve at the summit can walk down by the *shortest* route (shown dashed). You may make up your own mountain profile, but keep it simple.

Chapter Fourteen

Hamilton's equations and phase space

KEY FEATURES

The key features of this chapter are the equivalence of Lagrange's equations and **Hamilton's equations**, Hamiltonian **phase space, Liouville's theorem** and **recurrence**.

In this chapter we show how Lagrange's equations can be reformulated as a set of *first order* differential equations known as **Hamilton's equations**. Nothing new is added to the physics and the Hamilton formulation is not superior to that of Lagrange when it comes to problem solving. The value of Hamilton's supremely elegant formulation lies in providing a foundation for theoretical extensions both within and outside classical mechanics. Within classical mechanics, it is the basis for most further developments such as the Hamilton-Jacobi theory and chaos. Elsewhere, Hamiltonian mechanics provides the best route to statistical mechanics, and the notion of the Hamiltonian is at the heart of quantum mechanics. As applications of the Hamiltonian formulation, we prove Liouville's theorem and Poincaré's recurrence theorem and explore some of their remarkable consequences.

14.1 SYSTEMS OF FIRST ORDER ODES

The standard form for a system of first order ODEs in the n unknown functions $x_1(t), x_2(t), \ldots, x_n(t)$ is

$$\begin{aligned}\dot{x}_1 &= F_1(x_1, x_2, \ldots, x_n, t),\\ \dot{x}_2 &= F_2(x_1, x_2, \ldots, x_n, t),\\ &\vdots\\ \dot{x}_n &= F_n(x_1, x_2, \ldots, x_n, t),\end{aligned} \tag{14.1}$$

where $F_1, F_2, \ldots, F_n$ are given functions of the variables $x_1, x_2, \ldots, x_n, t$. This can also be written in the compact vector form

$$\dot{\boldsymbol{x}} = \boldsymbol{F}(\boldsymbol{x}, t), \tag{14.2}$$

where $\boldsymbol{x}$ and $\boldsymbol{F}$ are the n-dimensional vectors $\boldsymbol{x} = (x_1, x_2, \ldots, x_n)$ and $\boldsymbol{F} = (F_1, F_2, \ldots, F_n)$. If the value of $\boldsymbol{x}$ is given when $t = t_0$, the equations $\dot{\boldsymbol{x}} = \boldsymbol{F}(\boldsymbol{x}, t)$ determine the unknowns $\boldsymbol{x}(t)$ at all subsequent times.

A typical example is the **predator-prey** system of equations

$$\dot{x}_1 = ax_1 - bx_1x_2,$$
$$\dot{x}_2 = bx_1x_2 - cx_2,$$

which govern the population density $x_1(t)$ of the prey and the population density $x_2(t)$ of the predator. In this case, $F_1 = ax_1 - bx_1x_2$, and $F_2 = bx_1x_2 - cx_2$.

Converting higher order equations to first order

Higher order ODEs can always be converted into equivalent systems of first order ODEs. For example, consider the **damped oscillator** equation

$$\ddot{x} + 3\dot{x} + 4x = 0. \tag{14.3}$$

If we introduce the new variable v, defined by $v = \dot{x}$, then the second order equation (14.3) for $x(t)$ can be converted into the pair of first order equations

$$\dot{x} = v$$
$$\dot{v} = -3x - 4v,$$

in the unknowns $\{x(t), v(t)\}$. Since this step is reversible, this pair of first order equations is *equivalent* to the original second order equation (14.3). More generally:

Any system of n second order ODEs in n unknowns can be converted into an equivalent system of 2*n first order ODEs in* 2*n unknowns.*

Consider, for example, the **orbit equations** for a particle of mass m attracted by the gravitation of a mass M fixed at the origin O. In terms of polar coordinates centred on O, the Lagrangian is

$$L = \tfrac{1}{2}m\left(\dot{r}^2 + r^2\dot{\theta}^2\right) + \frac{mMG}{r}.$$

and the corresponding Lagrange equations are

$$\ddot{r} - r\dot{\theta}^2 = -\frac{MG}{r^2}, \qquad 2r\dot{r}\dot{\theta} + r^2\ddot{\theta} = 0,$$

a pair of second order ODEs in the unknowns $\{r(t), \theta(t)\}$. If we now introduce the new variables v_r and v_θ, defined by

$$v_r = \dot{r}, \qquad v_\theta = \dot{\theta},$$

the *two second order* Lagrange equations can be converted into

$$\dot{r} = v_r, \qquad \dot{v}_r = rv_\theta^2 - MG/r^2,$$
$$\dot{\theta} = v_\theta, \qquad \dot{v}_\theta = -2v_r v_\theta/r,$$

an equivalent system of *four first order* ODEs in the four unknowns $\{r, \theta, v_r, v_\theta\}$.

Hamilton form

In the above examples, we have performed the conversion to first order equations by introducing the **coordinate velocities** as new variables. This is not the only choice however. When transforming a system of Lagrange equations to a first order system, we could instead take the **conjugate momenta**, defined by

$$p_j = \frac{\partial L}{\partial \dot{q}_j}, \tag{14.4}$$

as the new variables. This seems quite attractive since the Lagrange equations already have the form

$$\dot{p}_j = \frac{\partial L}{\partial q_j} \qquad (1 \le j \le n).$$

The downside is that the right sides $\partial L/\partial q_j$ are functions of $\boldsymbol{q}, \dot{\boldsymbol{q}}, t$, and must be transformed to functions of $\boldsymbol{q}, \boldsymbol{p}, t$. This is achieved by inverting the equations (14.4) to express the $\dot{\boldsymbol{q}}$ as functions of $\boldsymbol{q}, \boldsymbol{p}, t$.

For the Lagrangian in the orbit problem, the conjugate momenta are

$$p_r = \frac{\partial L}{\partial \dot{r}} = m\dot{r}, \qquad p_\theta = \frac{\partial L}{\partial \dot{\theta}} = mr^2\dot{\theta},$$

and these equations are easily inverted* to give

$$\dot{r} = \frac{p_r}{m}, \qquad \dot{\theta} = \frac{p_\theta}{mr^2}.$$

The *two second order* Lagrange equations for the orbit problem are therefore equivalent to the system of *four first order* ODEs

$$\dot{r} = \frac{p_r}{m}, \qquad \dot{p}_r = \frac{p_\theta^2}{mr^3} - \frac{mMG}{r^2},$$

$$\dot{\theta} = \frac{p_\theta}{mr^2}, \qquad \dot{p}_\theta = 0,$$

in the four unknowns $\{r, \theta, p_r, p_\theta\}$. This is the **Hamilton form** of Lagrange's equations for the orbit problem.

Why bother?

The reader is probably wondering what is the point of converting Lagrange's equations into a system of first order ODEs. For the purpose of finding solutions to particular problems, like the orbit problem above, there is no point. Indeed, the new system of first order

* This step is difficult in the general case where there are *n coupled* linear equations to solve for $\dot{q}_1, \ldots, \dot{q}_n$.

equations may be harder to solve than the original second order equations. The real interest lies in the structure of the general theory. When Lagrange's equations are expressed in Hamilton form* *in a general manner*, the result is a system of first order equations of great simplicity and elegance, now known as **Hamilton's equations**. These equations are the foundation of further developments in analytical mechanics, such as the Hamilton-Jacobi theory and chaos. Also, the **Hamiltonian function**, which appears in Hamilton's equations, is at the heart of quantum mechanics.

14.2 LEGENDRE TRANSFORMS

The general problem of converting Lagrange's equations into Hamilton form hinges on the inversion of the equations that define $\boldsymbol{p}$, namely,

$$p_j = \frac{\partial}{\partial \dot{q}_j} L(\boldsymbol{q}, \dot{\boldsymbol{q}}, t) \qquad (1 \le j \le n), \tag{14.5}$$

so as to express $\dot{\boldsymbol{q}}$ in terms of $\boldsymbol{q}$, $\boldsymbol{p}$, t. This inversion is made easier by the fact that the $\{p_j\}$ are not general functions of $\boldsymbol{q}$, $\dot{\boldsymbol{q}}$ and t, but are the *first partial derivatives of a scalar function*, the Lagrangian $L(\boldsymbol{q}, \dot{\boldsymbol{q}}, t)$. It is a remarkable consequence that the inverse formulae can be written in a similar way.† The details of the argument follow below and the results are summarised at the end of the section.

The two-variable case

We develop the transformation theory for the case of functions of two variables. This has all the important features of the general case but is much easier to follow. Suppose that v_1 and v_2 are defined as functions of the variables u_1 and u_2 by the formulae

$$v_1 = \frac{\partial F}{\partial u_1}, \qquad v_2 = \frac{\partial F}{\partial u_2}, \tag{14.6}$$

where $F(u_1, u_2)$ is a given function of u_1 and u_2. Is it possible to write the inverse formulae‡ in the form

$$u_1 = \frac{\partial G}{\partial v_1}, \qquad u_2 = \frac{\partial G}{\partial v_2}, \tag{14.7}$$

* This is the form in which the new variables are the conjugate momenta $p_1, \ldots, p_n$. The form in which the new variables are the generalised velocities $\dot{q}_1, \ldots, \dot{q}_n$ does not lead to an elegant theory, and is therefore not used. It is often claimed that it is not *possible* to take the generalised velocities as new variables because 'they are the time derivatives of the generalised coordinates and therefore cannot be independent variables'. This objection is baseless, as the previous examples show. Indeed, if this objection had any substance, the conjugate momenta would be disqualified as well!

† There is a neat way of seeing that this must be true, which may appeal to mathematicians. If $\boldsymbol{v} = \text{grad}_{\boldsymbol{u}} F(\boldsymbol{u})$, then the Jacobian matrix of the transformation from $\boldsymbol{u}$ to $\boldsymbol{v}$ is symmetric. It follows that the Jacobian matrix of the inverse transformation must also be symmetric, which is precisely the condition that the inverse transformation has the form $\boldsymbol{u} = \text{grad}_{\boldsymbol{v}} G(\boldsymbol{v})$.

‡ We will always suppose that the inverse transformation does exist.

for some function $G(v_1, v_2)$? In the simplest cases one can answer the question by grinding through the details directly. For example, suppose $F = 2u_1^2 + 3u_1u_2 + u_2^2$. Then

$$v_1 = 4u_1 + 3u_2,$$
$$v_2 = 3u_1 + 2u_2.$$

The inverse formulae are easily obtained by solving these equations for u_1, u_2, which gives

$$u_1 = -2v_1 + 3v_2,$$
$$u_2 = 3v_1 - 4v_2.$$

There is no prior reason to expect that these formulae for u_1, u_2 can be expressed in terms of a single function $G(v_1, v_2)$ in the form (14.7), but it *is* true because the right sides of these equations happen to satisfy the necessary **consistency condition.*** Simple integration then shows that (to within a constant)

$$G = -v_1^2 + 3v_1v_2 - 2v_2^2.$$

This result is not a coincidence. Let $F(u_1, u_2)$ now be *any* function of the variables u_1, u_2, and suppose that a function $G(v_1, v_2)$ satisfying equation (14.7) *does* exist. Consider the expression

$$X = F(u_1, u_2) + G(v_1, v_2) - (u_1v_1 + u_2v_2),$$

which, as it stands, is a function of the four independent variables u_1, u_2, v_1, v_2. Suppose now that, in this formula, we imagine† that v_1 and v_2 are replaced by their expressions in terms of u_1 and u_2. Then X becomes a function of the variables u_1 and u_2 only. Its partial derivative with respect to u_1, holding u_2 constant, is then given by

$$\begin{aligned}\frac{\partial X}{\partial u_1} &= \frac{\partial F}{\partial u_1} + \left(\frac{\partial G}{\partial v_1} \times \frac{\partial v_1}{\partial u_1} + \frac{\partial G}{\partial v_2} \times \frac{\partial v_2}{\partial u_1}\right) - \left(v_1 + u_1\frac{\partial v_1}{\partial u_1} + u_2\frac{\partial v_2}{\partial u_1}\right)\\ &= \left(\frac{\partial F}{\partial u_1} - v_1\right) + \left(\frac{\partial G}{\partial v_1} - u_1\right)\frac{\partial v_1}{\partial u_1} + \left(\frac{\partial G}{\partial v_2} - u_2\right)\frac{\partial v_2}{\partial u_1}\\ &= 0 + 0 + 0 = 0,\end{aligned}$$

* If $u_1 = f_1(v_1, v_2)$ and $u_2 = f_2(v_1, v_2)$ then it is possible to express u_1, u_2 in the form (14.7) only if the functions f_1 and f_2 are related by the formula

$$\frac{\partial f_1}{\partial v_2} = \frac{\partial f_2}{\partial v_1},$$

which is called the **consistency condition.**

† Pure mathematicians strongly object to such feats of imagination. Unfortunately, the alternative is to introduce a welter of functional notation which obscures the essential simplicity of the argument. We will make frequent use of such 'imagined' substitutions.

on using first the chain rule and then the formulae (14.6) and (14.7). Hence X is independent of the variable u_1. In exactly the same way we may show that $\partial X/\partial u_2 = 0$ so that X is also independent of u_2. It follows that X must be a constant! This constant can be absorbed into the function G without disturbing the formulae (14.7), in which case $X = 0$. We have therefore shown that if F and G are related by the equations (14.6) and (14.7), then they must* satisfy the relation

$$F(u_1, u_2) + G(v_1, v_2) = u_1 v_1 + u_2 v_2. \tag{14.8}$$

The above argument is reversible so the converse result is also true. We have therefore shown that:

The required function $G(v_1, v_2)$ ***always exists*** *and can be generated from the function $F(u_1, u_2)$ by the formula*

$$\boxed{G(v_1, v_2) = (u_1 v_1 + u_2 v_2) - F(u_1, u_2)} \tag{14.9}$$

where u_1 and u_2 are to be replaced by their expressions in terms of v_1 and v_2.

It is evident that the relationship between the functions F and G is a symmetrical one. Each function is said to be the **Legendre transform** of the other.

Example 14.1 *Finding a Legendre transform*

Find the Legendre transform of the function $F(u_1, u_2) = 2u_1^2 + 3u_1 u_2 + u_2^2$ by using the formula (14.9).

Solution

For this F, $v_1 = \partial F/\partial u_1 = 4u_1 + 3u_2$ and $v_2 = \partial F/\partial u_2 = 3u_1 + 2u_2$. The inverse formulae are $u_1 = -2v_1 + 3v_2$ and $u_2 = 3v_1 - 4v_2$. From equation (14.9), the function G is given by

$$\begin{aligned}
G &= u_1 v_1 + u_2 v_2 - F(u_1, u_2) \\
&= (-2v_1 + 3v_2)v_1 + (3v_1 - 4v_2)v_2 - F(-2v_1 + 3v_2, 3v_1 - 4v_2) \\
&= -2v_1^2 + 6v_1 v_2 - 4v_2^2 - \\
&\qquad \left(2(-2v_1 + 3v_2)^2 + 3(-2v_1 + 3v_2)(3v_1 - 4v_2) + (3v_1 - 4v_2)^2\right) \\
&= -v_1^2 + 3v_1 v_2 - 2v_2^2,
\end{aligned}$$

the same as was obtained directly. This is the **Legendre transform** of the given function F. ■

* As we have seen, this may require a constant to be added to the function G.

Active and passive variables

The variables $\boldsymbol{u} = (u_1, u_2)$ and $\boldsymbol{v} = (v_1, v_2)$ are called **active variables** because they are the ones that are actually transformed. However, the functions F and G may also depend on additional variables that are not part of the transformation as such, but have the status of parameters. These are called **passive variables**. In the dynamical problem, $\dot{\boldsymbol{q}}$ and $\boldsymbol{p}$ are the active variables and $\boldsymbol{q}$ is the passive variable. We need to find how partial derivatives of F and G with respect to the passive variables are related.

Suppose then that $F = F(u_1, u_2, w)$ and $G = G(v_1, v_2, w)$ satisfy the formulae (14.6) and (14.7), where w is a passive variable. Then (14.6) defines v_1, v_2 as functions of u_1, u_2 and w, and (14.7) defines u_1, u_2 as functions of v_1, v_2 and w. The argument leading to the formula (14.8) still holds so that

$$F(u_1, u_2, w) + G(v_1, v_2, w) = u_1 v_1 + u_2 v_2. \tag{14.10}$$

In this formula, imagine that v_1 and v_2 are replaced by their expressions in terms of u_1, u_2 and w; then differentiate the resulting identity with respect to w, holding u_1 and u_2 constant. On using the chain rule, this gives

$$\frac{\partial F}{\partial w} + \left(\frac{\partial G}{\partial v_1} \times \frac{\partial v_1}{\partial w} + \frac{\partial G}{\partial v_2} \times \frac{\partial v_2}{\partial w} + \frac{\partial G}{\partial w} \right) = u_1 \frac{\partial v_1}{\partial w} + u_2 \frac{\partial v_2}{\partial w},$$

which can be written

$$\frac{\partial F}{\partial w} + \frac{\partial G}{\partial w} = \left(u_1 - \frac{\partial G}{\partial v_1} \right) \frac{\partial v_1}{\partial w} + \left(u_2 - \frac{\partial G}{\partial v_2} \right) \frac{\partial v_2}{\partial w} = 0 + 0 = 0,$$

on using the relations (14.7). Hence the partial derivatives of $F(u_1, u_2, w)$ and $G(v_1, v_2, w)$ with respect to w are related by

$$\boxed{\frac{\partial F}{\partial w} = -\frac{\partial G}{\partial w}} \tag{14.11}$$

This is the required result; it holds for *each* passive variable w.

The general case with many variables

The preceding theory can be extended to any number of variables. The results are exactly what one would expect and are summarised in the box below. This summary is presented in a compact vector form using the n-dimensional 'grad'.* It is a good idea to write these results out in expanded form.

* If $F = F(\boldsymbol{u}, \boldsymbol{w})$, where $\boldsymbol{u} = (u_1, u_2, \ldots, u_n)$ and $\boldsymbol{w} = (w_1, w_2, \ldots, w_m)$, then $\text{grad}_{\boldsymbol{u}} F$ and $\text{grad}_{\boldsymbol{w}} F$ mean

$$\text{grad}_{\boldsymbol{u}} F(\boldsymbol{u}, \boldsymbol{w}) = \left(\frac{\partial F}{\partial u_1}, \frac{\partial F}{\partial u_2}, \cdots, \frac{\partial F}{\partial u_n} \right), \qquad \text{grad}_{\boldsymbol{w}} F(\boldsymbol{u}, \boldsymbol{w}) = \left(\frac{\partial F}{\partial w_1}, \frac{\partial F}{\partial w_2}, \cdots, \frac{\partial F}{\partial w_m} \right).$$

Legendre transforms

Suppose that the variables $\boldsymbol{v} = (v_1, v_2, \dots, v_n)$ are defined as functions of the active variables $\boldsymbol{u} = (u_1, u_2, \dots, u_n)$ and passive variables $\boldsymbol{w} = (w_1, w_2, \dots, w_m)$ by the formula

$$\boldsymbol{v} = \operatorname{grad}_{\boldsymbol{u}} F(\boldsymbol{u}, \boldsymbol{w}), \tag{14.12}$$

where F is a given function of $\boldsymbol{u}$ and $\boldsymbol{w}$. Then the inverse formula can always be written in the form

$$\boldsymbol{u} = \operatorname{grad}_{\boldsymbol{v}} G(\boldsymbol{v}, \boldsymbol{w}), \tag{14.13}$$

where the function $G(\boldsymbol{v}, \boldsymbol{w})$ is related to the function $F(\boldsymbol{u}, \boldsymbol{w})$ by the formula

$$G(\boldsymbol{v}, \boldsymbol{w}) = \boldsymbol{u} \cdot \boldsymbol{v} - F(\boldsymbol{u}, \boldsymbol{w}), \tag{14.14}$$

where $\boldsymbol{u} \cdot \boldsymbol{v} = u_1 v_1 + u_2 v_2 + \cdots + u_n v_n$.

Furthermore, the derivatives of F and G with respect to the passive variables $\{w_j\}$ are related by

$$\operatorname{grad}_{\boldsymbol{w}} F(\boldsymbol{u}, \boldsymbol{w}) = -\operatorname{grad}_{\boldsymbol{w}} G(\boldsymbol{v}, \boldsymbol{w}). \tag{14.15}$$

The relationship between the functions F and G is symmetrical and each is said to be the **Legendre transform** of the other.

14.3 HAMILTON'S EQUATIONS

Let $\mathcal{S}$ be a Lagrangian mechanical system with n degrees of freedom and generalised **coordinates** $\boldsymbol{q} = (q_1, q_1, \dots, q_n)$. Then the Lagrange equations of motion for $\mathcal{S}$ are

$$\frac{d}{dt}\left(\frac{\partial L}{\partial \dot{q}_j}\right) - \frac{\partial L}{\partial q_j} = 0 \qquad (1 \le j \le n), \tag{14.16}$$

where $L = L(\boldsymbol{q}, \dot{\boldsymbol{q}}, t)$ is the Lagrangian of $\mathcal{S}$. This is a set of n second order ODEs in the unknowns $\boldsymbol{q}(t) = (q_1(t), q_2(t), \dots, q_n(t))$. We now wish to convert these equations into **Hamilton form**, that is, an equivalent set of $2n$ first order ODEs in the $2n$ unknowns $\boldsymbol{q}(t)$, $\boldsymbol{p}(t)$, where $\boldsymbol{p}(t) = (p_1(t), p_2(t), \dots, p_n(t))$, where the $\{p_j\}$ are the generalised **momenta** of $\mathcal{S}$. The $\{p_j\}$ are defined by

$$p_j = \frac{\partial L}{\partial \dot{q}_j} \qquad (1 \le j \le n), \tag{14.17}$$

which can be written in the vector form

$$\boldsymbol{p} = \text{grad}_{\dot{\boldsymbol{q}}}\, L(\boldsymbol{q}, \dot{\boldsymbol{q}}, t). \tag{14.18}$$

The first step is to eliminate the coordinate velocities $\dot{\boldsymbol{q}}$ from the Lagrange equations in favour of the momenta $\boldsymbol{p}$. This in turn requires that the formula (14.18) must be inverted so as to express $\dot{\boldsymbol{q}}$ in terms of $\boldsymbol{q}$, $\boldsymbol{p}$ and t. *This is precisely what Legendre transforms do.* It follows from the theory of the last section that the inverse formula to (14.18) can be written in the form

$$\dot{\boldsymbol{q}} = \text{grad}_{\boldsymbol{p}}\, H(\boldsymbol{q}, \boldsymbol{p}, t), \tag{14.19}$$

where the function $H(\boldsymbol{q}, \boldsymbol{p}, t)$ is the **Legendre transform** of the Lagrangian function $L(\boldsymbol{q}, \dot{\boldsymbol{q}}, t)$. Here, $\dot{\boldsymbol{q}}$ and $\boldsymbol{p}$ are the active variables and $\boldsymbol{q}$ is the passive variable.

Definition 14.1 ***Hamiltonian function*** *The function $H(\boldsymbol{q}, \boldsymbol{p}, t)$, which is the Legendre transform of the Lagrangian function $L(\boldsymbol{q}, \dot{\boldsymbol{q}}, t)$, is called the **Hamiltonian function** of the system $\mathcal{S}$.*

Since the functions H and L are Legendre transforms of each other, they satisfy the relations

$$\boxed{H(\boldsymbol{q}, \boldsymbol{p}, t) = \dot{\boldsymbol{q}} \cdot \boldsymbol{p} - L(\boldsymbol{q}, \dot{\boldsymbol{q}}, t)} \tag{14.20}$$

which can be used to generate H from L, and

$$\text{grad}_{\boldsymbol{q}}\, L(\boldsymbol{q}, \dot{\boldsymbol{q}}, t) = -\,\text{grad}_{\boldsymbol{q}}\, H(\boldsymbol{q}, \boldsymbol{p}, t), \tag{14.21}$$

which connects the derivatives of L and H with respect to the passive variables.

It is now quite easy to perform the transformation of Lagrange's equations. The Lagrange equations (14.16) can be written in terms of the generalised momenta $\{p_j\}$ in the form

$$\dot{p}_j = \frac{\partial L}{\partial q_j} \qquad (1 \le j \le n),$$

which is equivalent to the vector form

$$\dot{\boldsymbol{p}} = \text{grad}_{\boldsymbol{q}}\, L(\boldsymbol{q}, \dot{\boldsymbol{q}}, t). \tag{14.22}$$

The right sides of these equations still involve $\dot{\boldsymbol{q}}$, but, on using the formula (14.21), we obtain

$$\dot{\boldsymbol{p}} = -\,\text{grad}_{\boldsymbol{q}}\, H(\boldsymbol{q}, \boldsymbol{p}, t). \tag{14.23}$$

These are the transformed Lagrange equations! *The **Hamilton form** of the Lagrange equations therefore consists of equations (14.23) together with equations (14.19), which effectively define the generalised momentum $\boldsymbol{p}$.*

All of the above argument is reversible and so the Hamilton form and the Lagrange form are equivalent. Our results are summarised below:

Hamilton's equations

The n Lagrange equations (14.16) are equivalent to the system of $2n$ first order ODEs

$$\dot{\boldsymbol{q}} = \text{grad}_{\boldsymbol{p}} H(\boldsymbol{q}, \boldsymbol{p}, t), \qquad \dot{\boldsymbol{p}} = -\text{grad}_{\boldsymbol{q}} H(\boldsymbol{q}, \boldsymbol{p}, t), \tag{14.24}$$

where the **Hamiltonian function** $H(\boldsymbol{q}, \boldsymbol{p}, t)$ is the Legendre transform of the Lagrangian $L(\boldsymbol{q}, \dot{\boldsymbol{q}}, t)$ and is generated by the formula (14.20). This is the vector form of **Hamilton's equations**.* The expanded form is

$$\dot{q}_j = \frac{\partial H}{\partial p_j}, \qquad \dot{p}_j = -\frac{\partial H}{\partial q_j} \qquad (1 \le j \le n).$$

We have shownthat the n second order Lagrange equations in the n unknowns $\boldsymbol{q}(t)$ are mathematically equivalent to the $2n$ first order Hamilton equations in the $2n$ unknowns $\boldsymbol{q}(t)$, $\boldsymbol{p}(t)$. In each of these formulations of mechanics, the motion of the system is determined by the form of a single function, the **Lagrangian** $L(\boldsymbol{q}, \dot{\boldsymbol{q}}, t)$ in the Lagrange formulation, and the **Hamiltonian** $H(\boldsymbol{q}, \boldsymbol{p}, t)$ in the Hamilton formulation. Hamilton's equations are a particularly elegant first order system in which the functions $F_1, F_2, \dots$ that appear on the right are simply the first partial derivatives of a *single* function, the Hamiltonian H. Moreover these right hand sides also satisfy the special condition† div $\boldsymbol{F} = 0$, which allows Liouville's theorem to be applied to Hamiltonian mechanics.‡

Explicit time dependence

One final note. When the Lagrangian has an *explicit* time dependence, this t has the status of an extra passive variable. It follows that we then have the additional relation

$$\frac{\partial L}{\partial t} = -\frac{\partial H}{\partial t}, \tag{14.25}$$

* After Sir William Rowan Hamilton, whose paper *Second Essay on a General Method in Dynamics* was published in 1835. Hamilton's equations are sometimes called the *canonical equations*; no one seems to know the reason why.

† The scalar quantity div $\boldsymbol{F}$ is defined by

$$\text{div}\,\boldsymbol{F} = \frac{\partial F_1}{\partial x_1} + \frac{\partial F_2}{\partial x_2} + \cdots + \frac{\partial F_n}{\partial x_n}.$$

‡ None of these statements is true if Lagrange's equations are expressed as a first order system by taking the coordinate *velocities* $\dot{\boldsymbol{q}}$ as the new variables.

which shows that if either of L or H has an explicit time dependence, then so does the other.

Example 14.2 *Finding a Hamiltonian and Hamilton's equations*

Find the Hamiltonian and Hamilton's equations for the simple pendulum.

Solution

The **Lagrangian** for the simple pendulum is

$$L = \tfrac{1}{2}ma^2\dot{\theta}^2 + mga\cos\theta,$$

where θ is the angle between the string and the downward vertical, m is the mass of the bob, and a is the string length. The momentum p_θ conjugate to the coordinate θ is given by

$$p_\theta = \frac{\partial L}{\partial \dot{\theta}} = ma^2\dot{\theta}$$

and this formula is easily inverted to give

$$\dot{\theta} = \frac{p_\theta}{ma^2}. \tag{14.26}$$

The Hamiltonian H is then given by

$$H = \dot{\theta}\, p_\theta - L,$$

where $\dot{\theta}$ is given by equation (14.26). This gives

$$\begin{aligned} H &= \left(\frac{p_\theta}{ma^2}\right) p_\theta - \tfrac{1}{2}ma^2\left(\frac{p_\theta}{ma^2}\right)^2 - mga\cos\theta \\ &= \frac{p_\theta^2}{2ma^2} - mga\cos\theta. \end{aligned}$$

This is the **Hamiltonian** for the simple pendulum. From H we can find Hamilton's equations. They are

$$\begin{aligned} \dot{\theta} &= \frac{\partial H}{\partial p_\theta} = \frac{p_\theta}{ma^2}, \\ \dot{p}_\theta &= -\frac{\partial H}{\partial \theta} = -mga\sin\theta. \end{aligned}$$

These are **Hamilton's equations** for the simple pendulum.

This simple example illustrates clearly why Lagrange's equations are preferred over Hamilton's equations for the *practical* solution of problems. To solve Hamilton's equations in this case, we would differentiate the first equation with respect to t and then use the second equation to eliminate the unknown p_θ. This gives

$$\ddot{\theta} + \frac{g}{a}\sin\theta = 0,$$

which is precisely the Lagrange equation for the system! ■

Properties of the Hamiltonian H

The Hamiltonian function $H(\boldsymbol{q}, \boldsymbol{p}, t)$ has been defined as the Legendre transform of $L(\boldsymbol{q}, \dot{\boldsymbol{q}}, t)$ and, as such, it can be generated by the formula (14.20). We have met this expression before. It is identical to the **energy function** h

$$h = \sum_{j=1}^{n} \dot{q}_j p_j - L(\boldsymbol{q}, \dot{\boldsymbol{q}}, t), \tag{14.27}$$

defined in section 12.8. The only difference between h and H is that the functional form of H is vital. H *must* be expressed in terms of the variables $\boldsymbol{q}$, $\boldsymbol{p}$, t. On the other hand, since only the *values* taken by h are significant, its functional form is unimportant and it may be expressed in terms of any variables. However, since the values taken by H and h are the same, the results that we obtained in section 12.8 concerning h must also be true for the Hamiltonian H. In particular, when H has **no explicit time dependence**, H is a **constant of the motion**.* This result can also be proved independently, as follows.

Suppose that $H = H(\boldsymbol{q}, \boldsymbol{p})$ and that $\{\boldsymbol{q}(t), \boldsymbol{p}(t)\}$ is a motion of the system. Then, in this motion,

$$\begin{aligned}\frac{dH}{dt} &= \sum_{j=1}^{n} \frac{\partial H}{\partial q_j}\dot{q}_j + \sum_{j=1}^{n} \frac{\partial H}{\partial p_j}\dot{p}_j \\ &= \sum_{j=1}^{n} \frac{\partial H}{\partial q_j}\left(\frac{\partial H}{\partial p_j}\right) + \sum_{j=1}^{n} \frac{\partial H}{\partial p_j}\left(-\frac{\partial H}{\partial q_j}\right) \\ &= 0,\end{aligned}$$

where the first step follows from the chain rule and the second from Hamilton's equations. Hence H remains constant in the motion.

Systems for which $H = H(\boldsymbol{q}, \boldsymbol{p})$ are said to be **autonomous**. (This term was previously applied to systems for which $L = L(\boldsymbol{q}, \dot{\boldsymbol{q}})$, but equation (14.25) shows that these two classes of systems are the same.) The above result can therefore be expressed in the form:

Autonomous systems conserve H

In any motion of an autonomous system, the Hamiltonian $H(\boldsymbol{q}, \boldsymbol{p})$ is conserved.

In addition, when $\mathcal{S}$ is a **conservative standard system**, the Hamiltonian H can be expressed in the simpler form

$$\boxed{H(\boldsymbol{q}, \boldsymbol{p}) = T(\boldsymbol{q}, \boldsymbol{p}) + V(\boldsymbol{q})} \tag{14.28}$$

* As we remarked earlier, H has an explicit time dependence when L does; the circumstances under which this occurs are listed in section 12.6.

where $T(\boldsymbol{q}, \boldsymbol{p})$ is the kinetic energy of the system expressed in terms of the variables $\boldsymbol{q}, \boldsymbol{p}$. *In this case, H is simply the **total energy** of the system, expressed in terms of the variables* $\boldsymbol{q}, \boldsymbol{p}$. This is the quickest way of finding H when the system is conservative.

Example 14.3 *Finding a Hamiltonian 2*

Find the Hamiltonian for the inverse square orbit problem considered earlier and deduce Hamilton's equations for this system.

Solution

This is a **conservative** system so that $H = T + V$. With the polar coordinates r and θ as generalised coordinates, T and V are given by

$$T = \tfrac{1}{2}m\left(\dot{r}^2 + r^2\dot{\theta}^2\right) \qquad V = -\frac{mMG}{r}.$$

and the generalised momenta are given by

$$p_r = \frac{\partial L}{\partial \dot{r}} = m\dot{r}, \qquad p_\theta = \frac{\partial L}{\partial \dot{\theta}} = mr^2\dot{\theta}.$$

These equations are easily inverted to give

$$\dot{r} = \frac{p_r}{m}, \qquad \dot{\theta} = \frac{p_\theta}{mr^2}$$

so that the Hamiltonian is given by

$$\begin{aligned} H = T + V &= \tfrac{1}{2}m\left(\dot{r}^2 + r^2\dot{\theta}^2\right) - \frac{mMG}{r} \\ &= \frac{p_r^2}{2m} + \frac{p_\theta^2}{2mr^2} - \frac{mMG}{r}. \end{aligned}$$

This is the required **Hamiltonian**. Hamilton's equations are now found by using this Hamiltonian in the general equations (14.25). The partial derivatives of H are

$$\frac{\partial H}{\partial p_r} = \frac{p_r}{m}, \qquad \frac{\partial H}{\partial p_\theta} = \frac{p_\theta}{mr^2}, \qquad \frac{\partial H}{\partial r} = -\frac{p_\theta^2}{mr^3}, \qquad \frac{\partial H}{\partial \theta} = 0$$

and **Hamilton's equations** for the orbit problem are therefore

$$\dot{r} = \frac{p_r}{m}, \qquad \dot{p}_r = \frac{p_\theta^2}{mr^3} - \frac{mMG}{r^2},$$

$$\dot{\theta} = \frac{p_\theta}{mr^2}, \qquad \dot{p}_\theta = 0.$$

Naturally, these are the same equations as were obtained earlier by 'manual' transformation of Lagrange's equations. As in the last example, solution of the Hamilton equations by eliminating the momenta simply leads back to Lagrange's equations. ■

Momentum conservation

From the Hamilton equation $\dot{p}_j = -\partial H/\partial q_j$, it follows that:

If $\partial H/\partial q_j = 0$ (that is, if the coordinate q_j is absent from the Hamiltonian), then the generalised momentum p_j is constant in any motion.

The corresponding result in the Lagrangian formulation is that:

If $\partial L/\partial q_j = 0$ (that is, if the coordinate q_j is absent from the Lagrangian), then the generalised momentum p_j is constant in any motion.

These two results seem slightly different, but they *are* equivalent, since, from equation (14.21), $\partial H/\partial q_j = -\partial L/\partial q_j$. This means that the term **cyclic coordinate**, by which we previously meant a coordinate that did not appear in the *Lagrangian*, can be applied without ambiguity to mean that the coordinate does not appear in the *Hamiltonian*. Our result is then:

> **Conservation of momentum**
>
> If q_j is a cyclic coordinate (in the sense that it does not appear in the Hamiltonian), then p_j, the generalised momentum conjugate to q_j, is constant in any motion.

14.4 HAMILTONIAN PHASE SPACE ((q, p)–space)

Suppose the mechanical system $\mathcal{S}$ has generalised coordinates $\boldsymbol{q}$, conjugate momenta $\boldsymbol{p}$, and Hamiltonian $H(\boldsymbol{q}, \boldsymbol{p}, t)$. If the initial values of $\boldsymbol{q}$ and $\boldsymbol{p}$ are known,* then the subsequent motion of $\mathcal{S}$, described by the functions $\{\boldsymbol{q}(t), \boldsymbol{p}(t)\}$, is uniquely determined by Hamilton's equations. This motion can be represented geometrically by the motion of a 'point' (called a **phase point**) in Hamiltonian **phase space**. Hamiltonian phase space is a real space of $2n$ dimensions in which a 'point' is a set of values $(q_1, q_2, \ldots, q_n, p_1, p_2, \ldots, p_n)$ of the independent variables $\{\boldsymbol{q}, \boldsymbol{p}\}$. (Note that a point in Hamiltonian phase space represents not only the configuration of the system $\mathcal{S}$ but also its instantaneous momenta. This is the distinction between Hamiltonian phase space, which has $2n$ dimensions, and Lagrangian configuration space, which has n dimensions.) Each motion of the system $\mathcal{S}$ then corresponds to the motion of a phase point through the phase space.†

The only case in which we can actually draw the phase space is when $\mathcal{S}$ has just one degree of freedom. Then the phase space is two-dimensional and can be drawn on paper.

* We are more familiar with initial conditions in which $\boldsymbol{q}$ and $\dot{\boldsymbol{q}}$ are prescribed. However, these conditions are equivalent to those in which the initial values of $\boldsymbol{q}$ and $\boldsymbol{p}$ are prescribed.

† It should be noted that *Hamiltonian* phase space is generally not the same as the phase space introduced in Chapter 8, which, in the present notation, is $(\boldsymbol{q}, \dot{\boldsymbol{q}})$-space. In particular, our next result (Liouville's theorem) does not apply in $(\boldsymbol{q}, \dot{\boldsymbol{q}})$-space.

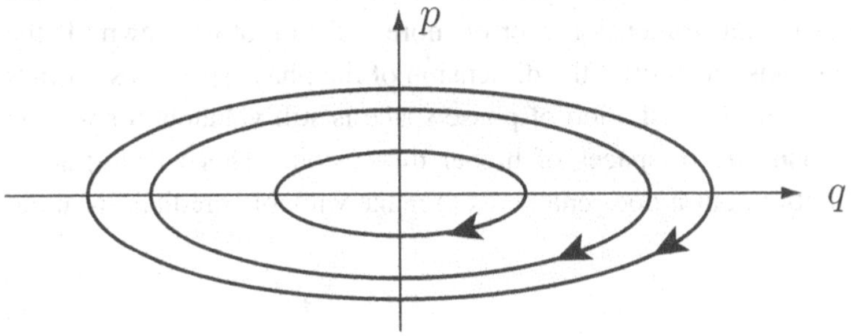

FIGURE 14.1 Typical paths in the phase space (q, p) corresponding to motions of a system $\mathcal{S}$ with Hamiltonian $H = p^2 + q^2/9$. The arrows show the direction that the phase point moves along each path as t increases.

Example 14.4 *Paths in phase space*

Suppose that $\mathcal{S}$ has the single coordinate q and Lagrangian

$$L = \frac{\dot{q}^2}{4} - \frac{q^2}{9}.$$

Find the paths in Hamiltonian phase space that correspond to the motions of $\mathcal{S}$.

Solution

The conjugate momentum $p = \partial L/\partial \dot{q} = \frac{1}{2}\dot{q}$, and the Hamiltonian is

$$H = \dot{q}p - L = (2p)\,p - \frac{1}{4}(2p)^2 + \frac{q^2}{9} = p^2 + \frac{q^2}{9}.$$

Hamilton's equations for $\mathcal{S}$ are therefore

$$\dot{q} = 2p, \qquad \dot{p} = -2q/9.$$

On eliminating p, we find that q satisfies the SHM equation

$$\ddot{q} + (4q/9) = 0.$$

The general solution of the Hamilton equations for $\mathcal{S}$ is therefore

$$q = 3A\cos((2t/3) + \alpha), \qquad p = -A\sin((2t/3) + \alpha),$$

where A and α are arbitrary constants. These are the parametric equations of the **paths** in phase space, the parameter being the time t; each path corresponds to a possible motion of the system $\mathcal{S}$. Some typical paths are shown in Figure 14.1. For this system , every motion is periodic so that the paths are *closed* curves in the (q, p)-plane. (They are actually concentric similar ellipses.) The arrows show the direction that the phase point moves along each path as t increases. ■

Of course, most mechanical systems have more than one degree of freedom so that the corresponding phase space has dimension four or more and cannot be drawn. If the system consists of a mole of gas molecules, the dimension of the phase space is six times Avogadro's number! Nevertheless, the notion of phase space is still valuable for we can still apply **geometrical reasoning** to spaces of higher dimension. This will not *solve* Hamilton's equations of motion, but it does enable us to make valuable predictions about the nature of the motion.

The phase fluid

The paths of phase points have a simpler structure when the system is **autonomous**, that is, $H = H(\boldsymbol{q}, \boldsymbol{p})$. In this case, H is a constant of the motion, so that each phase path must lie on a 'surface'* of constant energy† within the phase space. Thus the phase space is filled with the non-intersecting level surfaces of H, like layers in a multi-dimensional onion, and each phase path is restricted to one of these level surfaces.

For autonomous systems, *there can only be one phase path passing through any point of the phase space.* The reason is as follows: suppose that one phase point is at the point $(\boldsymbol{q}_0, \boldsymbol{p}_0)$ at time t_1, and another phase point is at $(\boldsymbol{q}_0, \boldsymbol{p}_0)$ at time t_2. Then, since H is independent of t, the second motion can be obtained from the first by simply making the substitution $t \to t + t_1 - t_2$, a shift in the origin of time. Therefore the two phase points travel along the *same path* with the second point delayed relative to the first by the constant time $t_2 - t_1$. Hence **phase paths cannot intersect**. This means that the phase space is filled with non-intersecting phase paths like the **streamlines** of a fluid in steady flow. Each motion of the system $\mathcal{S}$ corresponds to a phase point moving along one of these paths, just as the real particles of a fluid move along the fluid streamlines. The $2n$-dimensional vector quantity $\boldsymbol{u} = (\dot{\boldsymbol{q}}, \dot{\boldsymbol{p}})$ has the rôle of the fluid velocity field $\boldsymbol{u}(\boldsymbol{r})$‡ and Hamilton's equations serve to specify what this velocity is at the point $(\boldsymbol{q}, \boldsymbol{p})$ of the phase space. Because of this analogy with fluid mechanics, the motion of phase points in phase space is called the **phase flow**.

14.5 LIOUVILLE'S THEOREM AND RECURRENCE

Consider those phase points that, at some instant, occupy the region $\mathcal{R}_0$ of the phase space, as shown in Figure 14.2. As t increases, these points move along their various phase paths in accordance with Hamilton's equations and, after time t, will occupy some new region $\mathcal{R}_t$ of the phase space. This new region will have a different shape§

* If the phase space has dimension six, then a 'surface' of constant H has dimension five. This is therefore a generalisation of the notion of surface, which normally has dimension two.

† This 'energy' is the generalised energy $H(\boldsymbol{q}, \boldsymbol{p})$. For a conservative system, it is equal to the actual total energy $T + V$.

‡ $\boldsymbol{u}(\boldsymbol{r})$ is the velocity of the fluid particle instantaneously at the point with position vector $\boldsymbol{r}$.

§ Since the motion of many systems is sensitive to the initial conditions, the shape of $\mathcal{R}_t$ can become very weird indeed!

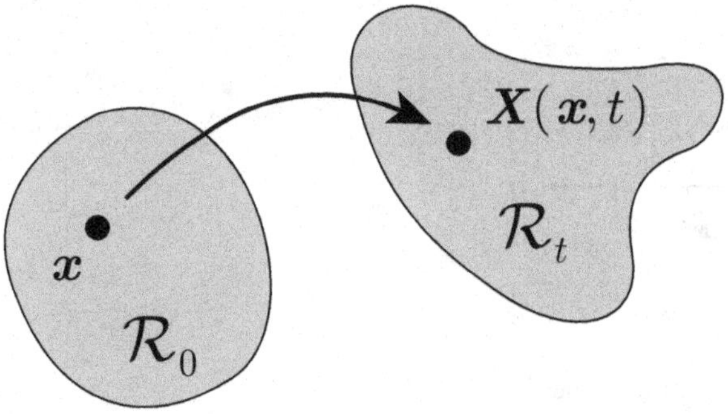

FIGURE 14.2 Liouville's theorem: the Hamiltonian phase flow preserves volume.

to $\mathcal{R}_0$, but **Liouville's theorem*** states that the volumes† of the two regions are equal. This remarkable result is expressed by saying that the *Hamiltonian phase flow preserves volume*. The theorem is easy to apply, but the proof is rather difficult.

Proof of Liouville's theorem

The proof is easier to follow if we use $x_1, x_2, \ldots, x_{2n}$ as the names of the variables (instead of $\boldsymbol{q}$, $\boldsymbol{p}$), and also call the right sides of Hamilton's equations $F_1, F_2, \ldots, F_{2n}$. Then, in vector notation, the equations of motion are $\dot{\boldsymbol{x}} = \boldsymbol{F}(\boldsymbol{x}, t)$. We will give the details for the case when the phase space is two-dimensional; the method in the general case is the same but uglier.

Consider a set of phase points moving in the (x_1, x_2)-plane, which, at some instant in time, occupies the region $\mathcal{R}_0$, as shown in Figure 14.2. Without losing generality, we may suppose that this occurs at time $t = 0$. After time t, a typical point $\boldsymbol{x}$ of $\mathcal{R}_0$ has moved on to position $\boldsymbol{X} = \boldsymbol{X}(\boldsymbol{x}, t)$ and the set as a whole now occupies the region $\mathcal{R}_t$. In this two-dimensional case, the 'volume' $v(t)$ of $\mathcal{R}_t$ is the *area* of this region in the (x_1, x_2)-plane. Now

$$v(t) = \int_{\mathcal{R}_t} dX_1 dX_2 = \int_{\mathcal{R}_0} J\, dx_1 dx_2,$$

where J is the Jacobian of the transformation $\boldsymbol{X} = \boldsymbol{X}(\boldsymbol{x}, t)$, that is,

$$J = \begin{vmatrix} \partial X_1/\partial x_1 & \partial X_1/\partial x_2 \\ \partial X_2/\partial x_1 & \partial X_2/\partial x_2 \end{vmatrix}. \tag{14.29}$$

Now, for small t, $\boldsymbol{X}$ may be approximated by

$$\begin{aligned} \boldsymbol{X}(\boldsymbol{x}, t) &= \boldsymbol{X}(\boldsymbol{x}, 0) + t\frac{\partial \boldsymbol{X}}{\partial t}(\boldsymbol{x}, 0) + O(t^2) \\ &= \boldsymbol{x} + t\,\boldsymbol{F}(\boldsymbol{x}, 0) + O(t^2), \end{aligned}$$

on using the equation of motion $\dot{\boldsymbol{x}} = \boldsymbol{F}(\boldsymbol{x}, t)$. The corresponding approximation for J is

$$J = 1 + t\left[\frac{\partial F_1}{\partial x_1} + \frac{\partial F_2}{\partial x_2}\right]_{t=0} + O(t^2) = 1 + t \operatorname{div} \boldsymbol{F}(\boldsymbol{x}, 0) + O(t^2).$$

* After the French mathematician Joseph Liouville (1809–1882).

† Since the dimension of the phase space can be any (even) number, this is a generalisation of the notion of volume.

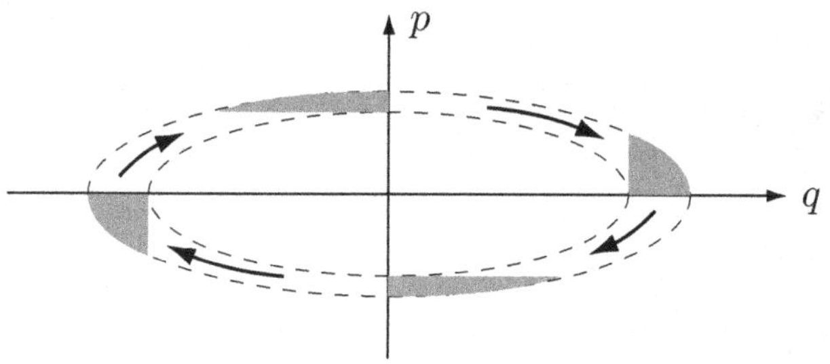

FIGURE 14.3 An instance of Liouville's theorem with the Hamiltonian $H = p^2 + q^2/9$. The shaded region moves through the phase space. Its shape changes but its area remains the same.

Hence the volume of $\mathcal{R}_t$ is approximated by

$$v(t) = \int_{\mathcal{R}_0} \Big(1 + t \operatorname{div} \boldsymbol{F}(\boldsymbol{x}, 0)\Big)\, dx_1 dx_2 + O(t^2)$$

when t is small. It follows that

$$\left.\frac{dv}{dt}\right|_{t=0} = \lim_{t\to 0}\left(\frac{v(t) - v(0)}{t}\right) = \int_{\mathcal{R}_0} \operatorname{div} \boldsymbol{F}(\boldsymbol{x}, 0)\; dx_1 dx_2.$$

Finally, since the initial instant $t = 0$ was arbitrarily chosen, this result must apply for any t, that is,

$$\frac{dv}{dt} = \int_{\mathcal{R}_t} \operatorname{div} \boldsymbol{F}(\boldsymbol{x}, t)\; dx_1 dx_2$$

at any time t. We see that, for *general* systems of equations, the phase flow does *not* preserve volume. However, if $\operatorname{div} \boldsymbol{F}(\boldsymbol{x}, t) = 0$, then volume *is* preserved. For the case of Hamilton's equations with one degree of freedom,

$$\begin{aligned}\operatorname{div} \boldsymbol{F} &= \frac{\partial F_1}{\partial x_1} + \frac{\partial F_2}{\partial x_2}\\ &= \frac{\partial}{\partial q}\left(\frac{\partial H}{\partial p}\right) + \frac{\partial}{\partial p}\left(-\frac{\partial H}{\partial q}\right) = 0.\end{aligned}$$

Hence the Hamiltonian phase flow satisfies the condition $\operatorname{div} \boldsymbol{F} = 0$ and so preserves volume. This completes the proof. ■

Liouville's theorem

The motions of a Hamiltonian system preserve volume in $(\boldsymbol{q}, \boldsymbol{p})$-space.

A particular instance of Liouville's theorem is shown in Figure 14.3. The phase paths of the Hamiltonian $H = p^2 + q^2/9$, shown in Figure 14.1, are concentric similar ellipses. Figure 14.3 shows the progress of a region of the phase space lying between two such elliptical paths. The region changes shape but its area remains the same.

Liouville's theorem has many applications and is particularly important in statistical mechanics. The following is a simple example.

Example 14.5 *No limit cycles in Hamiltonian mechanics*

In the theory of dynamical systems, a periodic solution is said to be an *asymptotically stable limit cycle* if it 'attracts' points in nearby volumes of the phase space (see Chapter 8). Show that limit cycles cannot occur in the dynamics of Hamiltonian systems.

Solution

Suppose there were a closed path $\mathcal{C}$ in the phase space that attracts points in a nearby region $\mathcal{R}$. Then eventually the points that lay in $\mathcal{R}$ must lie in a narrow 'tube' of *arbitrarily small* 'radius' enclosing the path $\mathcal{C}$. The volume of this tube tends to zero with increasing time so that the original volume of $\mathcal{R}$ cannot be preserved. This is contrary to Liouville's theorem and so asymptotically stable limit cycles cannot exist. ■

Poincaré's theorem and recurrence

Many Hamiltonian systems have the property that each path is confined to some **bounded region** within the phase space. Typically this is a consequence of **energy conservation**, where the energy surfaces happen to be bounded. Liouville's theorem has startling implications concerning the motion of such systems. First, we need to prove a result known as **Poincaré's recurrence theorem.** Poincaré's theorem is actually a result from ergodic theory and has many applications outside classical mechanics. However, since we are going to apply it to phase space, we will prove it in that context.

Theorem 14.1 ***Poincaré's recurrence theorem*** *Let $\mathcal{S}$ be an* ***autonomous*** *Hamiltonian system and consider the motion of the phase points that initially lie in a bounded region $\mathcal{R}_0$ of the phase space. If the paths of all of these points lie within a fixed* ***bounded*** *region of phase space for all time, then some of the points must eventually return to $\mathcal{R}_0$.*

Proof. Let $\mathcal{R}_1$ be the region occupied by the points after time τ. (We will suppose that $\mathcal{R}_1$ does not overlap $\mathcal{R}_0$ so that all the points that lay in $\mathcal{R}_0$ at time $t = 0$ have left $\mathcal{R}_0$ at time $t = \tau$.) We must show that some of them eventually return to $\mathcal{R}_0$. Let $\mathcal{R}_2, \mathcal{R}_3, \ldots, \mathcal{R}_n$ be the regions occupied by the same points after times $2\tau, 3\tau, \ldots, n\tau$. By Liouville's theorem, *all of these regions have the same volume*. Therefore, if they never overlap, their total volume will increase without limit. But, by assumption, all these regions lie within some *finite* volume, so that eventually one of them must overlap a previous one. This much is obvious, but we must now show that an overlap takes place with the *original* region $\mathcal{R}_0$.

Suppose it is $\mathcal{R}_m$ that overlaps $\mathcal{R}_k$ $(0 \le k < m)$. Each point of this overlap region corresponds to an intersection of the paths of two phase points that started out at some points $\boldsymbol{x}_1, \boldsymbol{x}_2$ of $\mathcal{R}_0$ at time $t = 0$. In the same notation used in the proof of Liouville's theorem, it means that $\boldsymbol{X}(\boldsymbol{x}_1, m\tau) = \boldsymbol{X}(\boldsymbol{x}_2, k\tau)$. But, since the system is **autonomous**, the two solutions $\boldsymbol{X}(\boldsymbol{x}_1, t)$ and $\boldsymbol{X}(\boldsymbol{x}_2, t)$ must therefore differ only by a shift $(m - k)\tau$ in the origin of time. It follows that $\boldsymbol{X}(\boldsymbol{x}_1, (m - k)\tau) = \boldsymbol{X}(\boldsymbol{x}_2, 0) = \boldsymbol{x}_2$. Thus the phase point that was at $\boldsymbol{x}_1$ when $t = 0$ is at $\boldsymbol{x}_2$ when $t = (m - k)\tau$. This phase point has therefore returned to $\mathcal{R}_0$ after time $(m - k)\tau$ and this completes the proof. ■

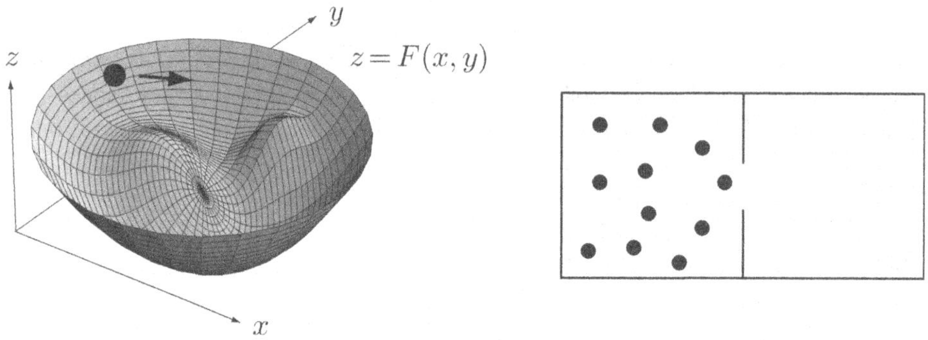

FIGURE 14.4 Consequences of Poincaré's recurrence theorem. **Left**: the particle sliding inside the smooth irregular bowl will eventually almost reassume its initial state. **Right**: The mole of gas molecules, initially all in the left compartment will eventually all be found there again.

Since the recurrence theorem holds for any sub-region of $\mathcal{R}_0$, it follows that, throughout $\mathcal{R}_0$, there are phase points that pass *arbitrarily close* to their original positions. Thus if the system $\mathcal{S}$ has one of these points as its initial state, then $\mathcal{S}$ will eventually become arbitrarily close to reassuming that state. Actually, such points are typical rather than exceptional. To show this requires a stronger version of Poincaré's theorem* than we have proved here, namely: *The path of* ***almost every***† *point in* $\mathcal{R}_0$ *passes* ***arbitrarily close*** *to its starting point.* This implies the remarkable result that *for almost every choice of the initial conditions, the system* $\mathcal{S}$ *becomes arbitrarily close to reassuming those conditions at later times.*

An example of this phenomenon is the motion of a single particle P sliding under gravity on the smooth inner surface of a bowl of some irregular shape $z = F(x, y)$, as shown in Figure 14.4. This is an autonomous Hamiltonian system with two degrees of freedom. Take the Cartesian coordinates (x, y) of P to be generalised coordinates. Then the Lagrangian is given by

$$L = \tfrac{1}{2}m\left(\dot{x}^2 + \dot{y}^2 + \dot{z}^2\right) - mgz$$
$$= \tfrac{1}{2}m\left(\dot{x}^2 + \dot{y}^2 + (F,_x\dot{x} + F,_y\dot{y})^2\right) - mgF,$$

where $F,_x = \partial F/\partial x$ and $F,_y = \partial F/\partial y$. The conjugate momenta are

$$p_x = m\left(\dot{x} + (F,_x\dot{x} + F,_y\dot{y})F,_x\right), \qquad p_y = m\left(\dot{y} + (F,_x\dot{x} + F,_y\dot{y})F,_y\right).$$

Just because energy is conserved, it does not necessarily mean that Poincaré's theorem applies. We must show that the energy surfaces in phase space are bounded. The proof of this is as follows:

* See Walters [12].

† This means that the set of exceptional points has measure zero. For example, in a two-dimensional phase space, a curve has zero measure.

In this case, $\mathcal{R}_0$ is some bounded region of the phase space (x, y, p_x, p_y). Suppose that z_0 is the maximum value of z and that T_0 is the maximum kinetic energy associated with points of $\mathcal{R}_0$. Then, by energy conservation, the maximum value of z in the subsequent motions cannot exceed $z^{\max} = z_0 + (T_0/mg)$. Hence, providing the bowl rises to at least this height, the motions are confined to values of (x, y) that satisfy $F(x, y) \le z^{\max}$. It follows that both x and y are bounded in the subsequent motions. Also, if the lowest point of the bowl is at $z = 0$, the value of T in the subsequent motions cannot exceed $T^{\max} = T_0 + mgz_0$. It follows that $\dot{x}$ and $\dot{y}$ are bounded in the subsequent motions and this implies that the same is true for p_x and p_y. Hence x, y, p_x, and p_y are all bounded in the subsequent motions. This means that the paths of the phase points that lie in $\mathcal{R}_0$ when $t = 0$ are confined to a bounded region of the phase space for all time. **Poincaré's theorem therefore applies.**

Hence, if the particle P is released from rest (say) at some point A on the surface of the bowl, then, whatever the shape of the bowl, P will become arbitrarily close to being at rest at A at later times.

In the same way, it follows that if a compartment containing a mole of gas molecules is separated from an empty compartment by a partition, and the partition is suddenly punctured, then at (infinitely many!) later times the molecules will *all be found in the first compartment again*. This remarkable result, which seems to be in contradiction to the second law of thermodynamics, appears less paradoxical when one realises that 'later times' may mean 10^{20} years later!

Question ***Exceptional points***

How do you know that the initial conditions you have chosen do not correspond to an 'exceptional point' for which Poincaré's theorem does not hold?

Answer

You don't know, but you would be very unlucky if this happened! ■

Problems on Chapter 14

Answers and comments are at the end of the book.

Harder problems carry a star (∗).

Finding Hamiltonians

14.1 Find the Legendre transform $G(v_1, v_2, w)$ of the function

$$F(u_1, u_2, w) = 2u_1^2 - 3u_1u_2 + u_2^2 + 3wu_1,$$

where w is a passive variable. Verify that $\partial F/\partial w = -\partial G/\partial w$.

14.2 A smooth wire has the form of the helix $x = a\cos\theta$, $y = a\sin\theta$, $z = b\theta$, where θ is a real parameter, and a, b are positive constants. The wire is fixed with the axis Oz pointing vertically upwards. A particle P of mass m can slide freely on the wire. Taking θ as generalised coordinate, find the Hamiltonian and obtain Hamilton's equations for this system.

14.3 ***Projectile*** Using Cartesian coordinates, find the Hamiltonian for a projectile of mass m moving under uniform gravity. Obtain Hamilton's equations and identify any cyclic coordinates.

14.4 ***Spherical pendulum*** The spherical pendulum is a particle of mass m attached to a fixed point by a light inextensible string of length a and moving under uniform gravity. It differs from the simple pendulum in that the motion is not restricted to lie in a vertical plane. Show that the Lagrangian is

$$L = \tfrac{1}{2}ma^2\left(\dot{\theta}^2 + \sin^2\theta\,\dot{\phi}^2\right) + mga\cos\theta,$$

where the polar angles θ, ϕ are shown in Figure 11.7. Find the Hamiltonian and obtain Hamilton's equations. Identify any cyclic coordinates.

14.5 The system shown in Figure 10.9 consists of two particles P_1 and P_2 connected by a light inextensible string of length a. The particle P_1 is constrained to move along a fixed smooth horizontal rail, and the whole system moves under uniform gravity in the vertical plane through the rail. For the case in which the particles are of equal mass m, show that the Lagrangian is

$$L = \tfrac{1}{2}m\left(2\dot{x}^2 + 2a\dot{x}\dot{\theta} + a^2\dot{\theta}^2\right) + mga\cos\theta,$$

where x and θ are the coordinates shown in Figure 10.9.

Find the Hamiltonian and verify that it satisfies the equations $\dot{x} = \partial H/\partial p_x$ and $\dot{\theta} = \partial H/\partial p_\theta$. [Messy algebra.]

14.6 ***Pendulum with a shortening string*** A particle is suspended from a support by a light inextensible string which passes through a small fixed ring vertically below the support. The particle moves in a vertical plane with the string taut. At the same time, the support is made to move vertically having an upward displacement $Z(t)$ at time t. The effect is that the particle oscillates like a simple pendulum whose string length at time t is $a - Z(t)$, where a is a positive constant. Show that the Lagrangian is

$$L = \tfrac{1}{2}m\left((a - Z)^2\dot{\theta}^2 + \dot{Z}^2\right) + mg(a - Z)\cos\theta,$$

where θ is the angle between the string and the downward vertical.

Find the Hamiltonian and obtain Hamilton's equations. Is H conserved?

14.7 ***Charged particle in an electrodynamic field*** The Lagrangian for a particle with mass m and charge e moving in the general electrodynamic field $\{\boldsymbol{E}(\boldsymbol{r}, t), \boldsymbol{B}(\boldsymbol{r}, t)\}$ is given in Cartesian coordinates by

$$L(\boldsymbol{r}, \dot{\boldsymbol{r}}, t) = \tfrac{1}{2}m\,\dot{\boldsymbol{r}}\cdot\dot{\boldsymbol{r}} - e\,\phi(\boldsymbol{r}, t) + e\,\dot{\boldsymbol{r}}\cdot\boldsymbol{A}(\boldsymbol{r}, t),$$

where $\boldsymbol{r} = (x, y, z)$ and $\{\phi, \boldsymbol{A}\}$ are the electrodynamic potentials of field $\{\boldsymbol{E}, \boldsymbol{B}\}$. Show that the corresponding Hamiltonian is given by

$$H(\boldsymbol{r}, \boldsymbol{p}, t) = \frac{(\boldsymbol{p} - e\boldsymbol{A})\cdot(\boldsymbol{p} - e\boldsymbol{A})}{2m} + e\,\phi,$$

where $\boldsymbol{p} = (p_x, p_y, p_x)$ are the generalised momenta conjugate to the coordinates (x, y, z). [Note that $\boldsymbol{p}$ is *not* the ordinary linear momentum of the particle.] Under what circumstances is H conserved?

14.8 ***Relativistic Hamiltonian*** The relativistic Lagrangian for a particle of rest mass m_0 moving along the x-axis under the potential field $V(x)$ is given by

$$L = m_0 c^2 \left(1 - \left(1 - \frac{\dot{x}^2}{c^2}\right)^{1/2}\right) - V(x).$$

Show that the corresponding Hamiltonian is given by

$$H = m_0 c^2 \left(1 + \left(\frac{p_x}{m_0 c}\right)^2\right)^{1/2} - m_0 c^2 + V(x),$$

where p_x is the generalised momentum conjugate to x.

14.9 ***A variational principle for Hamilton's equations*** Consider the functional

$$J[\boldsymbol{q}(t), \boldsymbol{p}(t)] = \int_{t_0}^{t_1} \Big(H(\boldsymbol{q}, \boldsymbol{p}, t) - \dot{\boldsymbol{q}} \cdot \boldsymbol{p}\Big)\, dt$$

of the $2n$ independent functions $q_1(t), \dots, q_n(t),\ p_1(t), \dots, p_n(t)$. Show that the extremals of J satisfy Hamilton's equations with Hamiltonian H.

Liouville's theorem and recurrence

14.10 In the theory of dynamical systems, a point is said to be an *asymptotically stable equilibrium point* if it 'attracts' points in a nearby volume of the phase space. Show that such points cannot occur in Hamiltonian dynamics.

14.11 A one dimensional damped oscillator with coordinate q satisfies the equation $\ddot{q} + 4\dot{q} + 3q = 0$, which is equivalent to the first order system

$$\dot{q} = v, \qquad \dot{v} = -3q - 4v.$$

Show that the area $a(t)$ of any region of points moving in (q, v)-space has the time variation

$$a(t) = a(0)\, e^{-4t}.$$

Does this result contradict Liouville's theorem?

14.12 ***Ensembles in statistical mechanics*** In statistical mechanics, a macroscopic property of a system $\mathcal{S}$ is calculated by averaging that property over a set, or *ensemble*, of points moving in the phase space of $\mathcal{S}$. The number of ensemble points in any volume of phase space is represented by a *density function* $\rho(\boldsymbol{q}, \boldsymbol{p}, t)$. If the system is autonomous and in *statistical equilibrium*, it is required that, even though the ensemble points are moving (in accordance with

Hamilton's equations), their density function should remain the same, that is, $\rho = \rho(\boldsymbol{q}, \boldsymbol{p})$. This places a restriction on possible choices for $\rho(\boldsymbol{q}, \boldsymbol{p})$. Let $\mathcal{R}_0$ be any region of the phase space and suppose that, after time t, the points of $\mathcal{R}_0$ occupy the region $\mathcal{R}_t$. Explain why statistical equilibrium requires that

$$\int_{\mathcal{R}_0} \rho(\boldsymbol{q}, \boldsymbol{p})\, dv = \int_{\mathcal{R}_t} \rho(\boldsymbol{q}, \boldsymbol{p})\, dv$$

and show that the *uniform* density function $\rho(\boldsymbol{q}, \boldsymbol{p}) = \rho_0$ satisfies this condition. [It can be proved that the above condition is also satisfied by any density function that is constant along the streamlines of the phase flow.]

14.13 Decide if the energy surfaces in phase space are bounded in the following cases:

(i) The two-body gravitation problem with $E < 0$.

(ii) The two-body gravitation problem viewed from the zero momentum frame and with $E < 0$.

(iii) The three-body gravitation problem viewed from the zero momentum frame and with $E < 0$. Does the solar system have the recurrence property?

Poisson brackets

14.14 ***Poisson brackets*** Suppose that $u(\boldsymbol{q}, \boldsymbol{p})$ and $v(\boldsymbol{q}, \boldsymbol{p})$ are any two functions of position in the phase space $(\boldsymbol{q}, \boldsymbol{p})$ of a mechanical system $\mathcal{S}$. Then the **Poisson bracket** $[u, v]$ of u and v is defined by

$$[u, v] = \text{grad}_{\boldsymbol{q}}\, u \cdot \text{grad}_{\boldsymbol{p}}\, v - \text{grad}_{\boldsymbol{p}}\, u \cdot \text{grad}_{\boldsymbol{q}}\, v = \sum_{j=1}^{n} \left(\frac{\partial u}{\partial q_j}\frac{\partial v}{\partial p_j} - \frac{\partial u}{\partial p_j}\frac{\partial v}{\partial q_j} \right).$$

The *algebraic* behaviour of the Poisson bracket of two functions resembles that of the cross product $\boldsymbol{U} \times \boldsymbol{V}$ of two vectors or the commutator $\boldsymbol{UV} - \boldsymbol{VU}$ of two matrices. The Poisson bracket of two functions is closely related to the commutator of the corresponding operators in quantum mechanics.*

Prove the following properties of Poisson brackets.

Algebraic properties

$$[u, u] = 0, \qquad [v, u] = -[u, v], \qquad [\lambda_1 u_1 + \lambda_2 u_2, v] = \lambda_1 [u_1, v] + \lambda_2 [u_2, v]$$

$$[[u, v], w] + [[w, u], v] + [[v, w], u] = 0.$$

This last formula is called *Jacobi's identity*. It is quite important, but there seems to be no way of proving it apart from crashing it out, which is very tedious. Unless you can invent a smart method, leave this one alone.

* The commutator $[\mathbf{U}, \mathbf{V}]$ of two quantum mechanical operators $\mathbf{U}, \mathbf{V}$ corresponds to $i\hbar[u, v]$, where $\hbar$ is the modified Planck constant, and $[u, v]$ is the Poisson bracket of the corresponding classical variables u, v.

Fundamental Poisson brackets

$$[q_j, q_k] = 0, \qquad [p_j, p_k] = 0, \qquad [q_j, p_k] = \delta_{jk},$$

where δ_{jk} is the Kroneker delta.

Hamilton's equations

Show that Hamilton's equations for $\mathcal{S}$ can be written in the form

$$\dot{q}_j = [q_j, H], \qquad \dot{p}_j = [p_j, H], \qquad (1 \le j \le n).$$

Constants of the motion

(i) Show that the *total* time derivative of $u(\boldsymbol{q}, \boldsymbol{p})$ is given by

$$\frac{du}{dt} = [u, H]$$

and deduce that u is a constant of the motion of $\mathcal{S}$ if, and only if, $[u, H] = 0$.

(ii) If u and v are constants of the motion of $\mathcal{S}$, show that the Poisson bracket $[u, v]$ is another constant of the motion. [Use Jacobi's identity.] Does this mean that you can keep on finding more and more constants of the motion ?

14.15 ***Integrable systems and chaos*** A mechanical system is said to be **integrable** if its equations of motion are soluble in the sense that they can be reduced to integrations. (You do not need to be able to evaluate the integrals in terms of standard functions.) A theorem due to Liouville states that *any Hamiltonian system with n degrees of freedom is integrable if it has n independent constants of the motion, and all these quantities commute in the sense that all their mutual Poisson brackets are zero.** The qualitative behaviour of integrable Hamiltonian systems is well investigated (see Goldstein [4]). In particular, *no integrable Hamiltonian system can exhibit chaos.*

Use Liouville's theorem to show that any autonomous system with n degrees of freedom and $n - 1$ cyclic coordinates must be integrable.

Computer assisted problem

14.16 ***The three body problem***

There is no general solution to the problem of determining the motion of three or more bodies moving under their mutual gravitation. Here we consider a restricted case of the three-body problem in which the mass of one of the bodies, P, is *much smaller* than that of the other two masses, which are called the primaries. In this case we neglect the effect of P on the primaries which therefore move in known fixed orbits. The body P moves in the time dependent, gravitational field of the primaries.

---

* This result is really very surprising. A *general* system of first order ODEs in $2n$ variables needs $2n$ integrals in order to be integrable in the Liouville sense. Hamiltonian systems need only half that number. The theorem does not rule out the possibility that there could be other classes of integrable systems. However, according to Arnold [2], every system that has ever been integrated is of the Liouville kind!

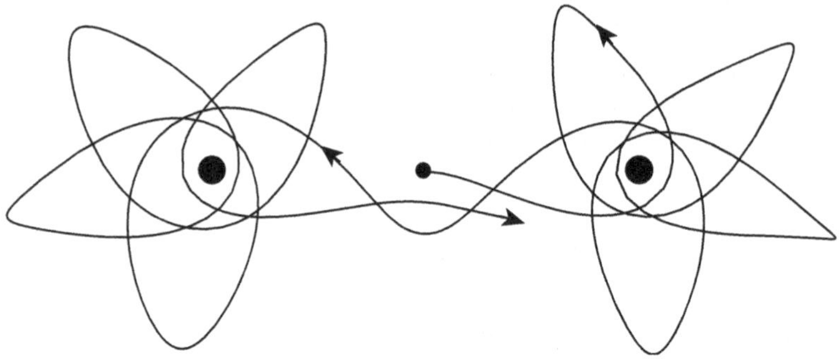

FIGURE 14.5 There is no such thing as a typical orbit in the three-body problem. The orbit shown corresponds to the initial conditions $x = 0$, $y = 0$, $p_x = 1.03$, $p_y = 0$ and is viewed from axes *rotating with the primaries*.

Suppose the primaries each have mass M and move under their mutual gravitation around a fixed circle of radius a, being at the opposite ends of a rotating diameter. The body P moves under the gravitational attraction of the primaries in the same plane as their circular orbit. Using Cartesian coordinates, write code to set up Hamilton's equations for this system and solve them with general initial conditions. [Take M as the unit of mass, a as the unit of length, and take the unit of time so that the speed of the primaries is unity. With this choice of units, the gravitational constant $G = 4$.]

By experimenting with different initial conditions, some very weird orbits can be found for P. It is interesting to plot these relative to fixed axes and also relative to axes rotating with the primaries, as in Figure 14.5. Some fascinating cases are shown by Acheson [1] and you should be able to reproduce these. [Acheson used a different normalisation however and his initial data needs to be doubled to be used in your code.]

Part Four

FURTHER TOPICS

CHAPTERS IN PART FOUR

Chapter Fifteen

The general theory of small oscillations

KEY FEATURES

The key features of this chapter are the existence of **small oscillations** near a position of stable equilibrium and the matrix theory of **normal modes**. A simpler account of the basic principles is given in Chapter 5.

Any mechanical system can perform oscillations in the neighbourhood of a position of stable equilibrium. These oscillations are an extremely important feature of the system whether they are intended to occur (as in a pendulum clock), or whether they are undesirable (as in a suspension bridge!). Analogous oscillations occur in continuum mechanics and in quantum mechanics. Here we present the theory of such oscillations for **conservative systems** under the assumption that the amplitude of the oscillations is small enough so that the **linear approximation** is adequate. A simpler account of the theory is given in Chapter 5. This treatment is restricted to systems with two degrees of freedom and does not make use of Lagrange's equations. Although the material in the present chapter is self-contained, it is helpful to have solved a few simple normal mode problems before.

The best way to develop the theory of small oscillations is to use Lagrange's equations. We will show that it is possible to approximate the expressions for T and V from the start so that the linearised equations of motion are obtained immediately. The theory is presented in an elegant matrix form which enables us to make use of concepts from linear algebra, such as eigenvalues and eigenvectors. We prove that fundamental result that a system with n degrees of freedom always has n harmonic motions known as **normal modes**, whose frequencies are generally different. These **normal frequencies** are the most important characteristic of the oscillating system. One important application of the theory is to the internal vibrations of molecules. Although this should really be treated by quantum mechanics, the classical model is extremely valuable in making qualitative predictions and classifying the vibrational modes of the molecule.

15.1 STABLE EQUILIBRIUM AND SMALL OSCILLATIONS

Let $\mathcal{S}$ be a standard mechanical system with n degrees of freedom and with generalised coordinates $\boldsymbol{q} = (q_1, q_2, \ldots, q_n)$. Suppose also that $\mathcal{S}$ is **conservative**. Then the motion of $\mathcal{S}$ is determined by the classical Lagrange equations of motion

$$\frac{d}{dt}\left(\frac{\partial T}{\partial \dot{q}_j}\right) - \frac{\partial T}{\partial q_j} = -\frac{\partial V}{\partial q_j} \qquad (1 \le j \le n), \tag{15.1}$$

where $T(\boldsymbol{q}, \dot{\boldsymbol{q}})$ and $V(\boldsymbol{q})$ are the kinetic and potential energies of $\mathcal{S}$. In particular, these equations determine the **equilibrium positions** of $\mathcal{S}$. The point $\boldsymbol{q}^{(0)}$ in configuration space is an equilibrium position of $\mathcal{S}$ if (and only if) the constant function $\boldsymbol{q} = \boldsymbol{q}^{(0)}$ satisfies the equations (15.1). For a standard system, T has the form

$$T = \sum_{j=1}^{n}\sum_{k=1}^{n} t_{jk}(\boldsymbol{q})\, \dot{q}_j \dot{q}_k, \tag{15.2}$$

that is, a homogeneous quadratic form in the variables $\dot{q}_1, \dot{q}_2, \ldots, \dot{q}_n$, with coefficients depending on $\boldsymbol{q}$. It follows that the left side of the j-th Lagrange equation has the form

$$2\sum_{k=1}^{n}\left(\frac{dt_{jk}}{dt}\,\dot{q}_k + t_{jk}\,\ddot{q}_k\right) - \sum_{j=1}^{n}\sum_{k=1}^{n}\left(\frac{\partial t_{jk}}{\partial q_j}\,\dot{q}_j\dot{q}_k\right),$$

which takes the value zero when the constant function $\boldsymbol{q} = \boldsymbol{q}^{(0)}$ is substituted in. It follows that $\boldsymbol{q} = \boldsymbol{q}^{(0)}$ will satisfy the equations (15.1) if (and only if)

$$\boxed{\frac{\partial V}{\partial q_j} = 0 \qquad (1 \le j \le n)} \tag{15.3}$$

when $\boldsymbol{q} = \boldsymbol{q}^{(0)}$. In other words, we have the result:

Stationary points of V

The **equilibrium positions** of a conservative system are the **stationary points** of its potential energy function $V(\boldsymbol{q})$.

Stable equilibrium

The stability of equilibrium is most easily understood in terms of the motion of the phase point of $\mathcal{S}$ in the Hamilton phase space $(\boldsymbol{q}, \boldsymbol{p})$ (see Chapter 14). If $\boldsymbol{q}^{(0)}$ is an equilibrium point of $\mathcal{S}$ in *configuration* space, then $(\boldsymbol{q}^{(0)}, \boldsymbol{0})$ is the corresponding equilibrium point of $\mathcal{S}$ in *phase* space; if the phase point starts at $(\boldsymbol{q}^{(0)}, \boldsymbol{0})$, then it remains there, its trajectory consisting of the single point $(\boldsymbol{q}^{(0)}, \boldsymbol{0})$.

Now consider phase paths that begin at points that lie inside the sphere S_δ in phase space which has centre $(\boldsymbol{q}^{(0)}, \boldsymbol{0})$ and radius δ. When δ is small, this corresponds to starting

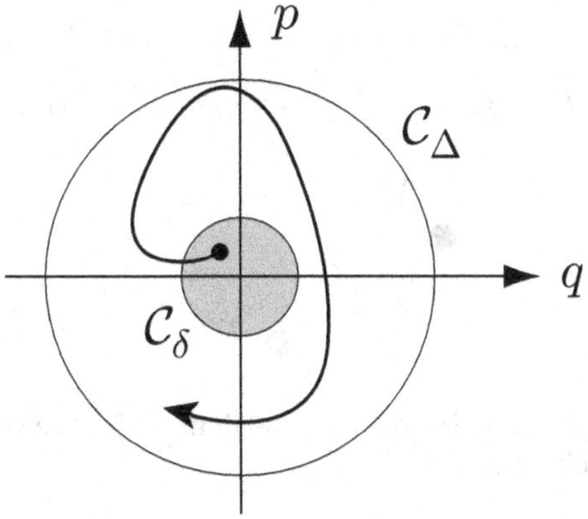

FIGURE 15.1 The circle $\mathcal{C}_\delta$ must be chosen small enough so that all the phase paths that start within it remain within the given circle $\mathcal{C}_\Delta$.

the system from a configuration close to the equilibrium configuration with a small kinetic energy. Let S_Δ be the smallest sphere in phase space with centre $(\boldsymbol{q}^{(0)}, \mathbf{0})$ *that contains all the phase paths that begin within* S_δ.

Definition 15.1 ***Stable equilibrium*** *If the radius* Δ *tends to zero as the radius* δ *tends to zero, then the equilibrium point at* $(\boldsymbol{q}^{(0)}, \mathbf{0})$ *is said to be* ***stable***.*

This means that, if $\mathcal{S}$ is given a small nudge from a configuration close to a position of *stable* equilibrium, then the subsequent motion of $\mathcal{S}$ (in configuration space) is restricted to a small neighbourhood of the equilibrium point. Thus *any mechanical system can perform small motions near a position of stable equilibrium.* These motions are generally called **small oscillations.**

We know that the equilibrium positions of $\mathcal{S}$ correspond to the stationary points of the potential energy $V(\boldsymbol{q})$, but we have yet to identify which of these points correspond to *stable* equilibrium. In fact it is quite easy to prove the following important result:

Minimum points of V

The **minimum points** of the potential energy function $V(\boldsymbol{q})$ are positions of **stable equilibrium** of the system $\mathcal{S}$.

Proof. Without loss of generality, suppose that the minimum point of the function $V(\boldsymbol{q})$ is at $\boldsymbol{q} = \mathbf{0}$ and that $V(\mathbf{0}) = \mathbf{0}$. Take any $\Delta > 0$. Then we must show that we can find a sphere S_δ in phase space such that all the paths that begin within it remain inside the sphere S_Δ. This is illustrated in Figure 15.1 for the only case that can be drawn, namely, when the phase space is two-dimensional; in this case, the 'spheres' are circles.

The result follows from **energy conservation.** Let $T(\boldsymbol{q}, \boldsymbol{p})$ be the kinetic energy of the system. The total energy is then

$$E(\boldsymbol{q}, \boldsymbol{p}) = T(\boldsymbol{q}, \boldsymbol{p}) + V(\boldsymbol{q}).$$

* In the dynamical systems literature, this is known as **Liapunov stability**.

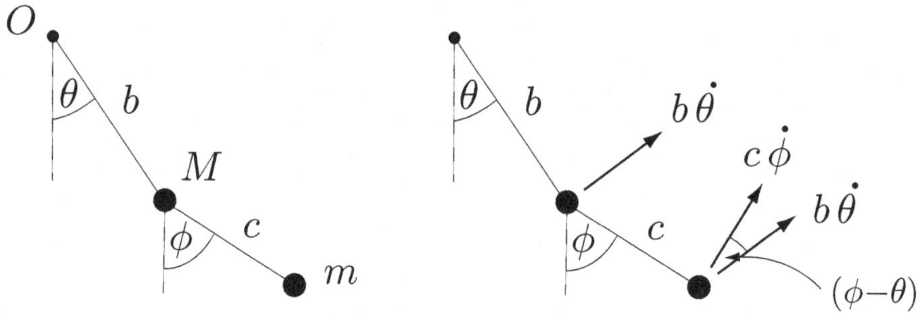

FIGURE 15.2 The double pendulum. **Left**: The generalised coordinates θ, ϕ. **Right**: The velocity diagram.

Since $\boldsymbol{q} = \boldsymbol{0}$ is a minimum point of $V(\boldsymbol{q})$ and $T(\boldsymbol{q}, \boldsymbol{p})$ is positive for $\boldsymbol{p} \neq \boldsymbol{0}$, it follows that $(\boldsymbol{0}, \boldsymbol{0})$, the origin of phase space, is a minimum point of the function $E(\boldsymbol{q}, \boldsymbol{p})$; the value of E at the minimum point is zero. The value of E on the sphere S_Δ must therefore be greater than some *positive* number E_Δ. On the other hand, by making the radius δ small enough, it follows by continuity that the value of E within the sphere S_δ can be made as close to $E(\boldsymbol{0}, \boldsymbol{0})$ $(= 0)$ as we wish; we can certainly make it less than E_Δ. Consider now any phase path starting within the circle C_δ. Then, by energy conservation, E is constant along this path and is *less* than E_Δ. Such a path cannot cross the sphere S_Δ, for, if it did, the value of E at the crossing point would be *greater* than E_Δ, which is not true. Hence, any phase path starting within the sphere S_δ must remain within the sphere S_Δ and this completes the proof. ■

Example 15.1 *Stability of equilibrium*

Consider the double pendulum shown in Figure 15.2 which moves under uniform gravity. Show that the vertically downwards configuration is a position of stable equilibrium.

Solution

The system is assumed to move in a vertical plane through the suspension point O. In terms of the generalised coordinates θ, ϕ shown, the vertically downwards configuration corresponds to the point $\theta = \phi = 0$ in configuration space. The gravitational potential energy is given by

$$V = (M + m)gb(1 - \cos\theta) + mgc(1 - \cos\phi)$$

so that $\partial V/\partial\theta = (M+m)gb\sin\theta = 0$ when $\theta = \phi = 0$. The same is true for $\partial V/\partial\phi$ and so the point $\theta = \phi = 0$ is a **stationary point** of the function $V(\theta, \phi)$. It follows that the downwards configuration is a position of **equilibrium**.

We could determine the nature of this stationary point by looking at the second derivatives of $V(\theta, \phi)$, but there is no need because it is evident that $V(\theta, \phi) > V(0, 0)$ unless $\theta = \phi = 0$. Thus $(0, 0)$ is a minimum point of V and so the downwards configuration is a position of **stable equilibrium.** ■

15.2 THE APPROXIMATE FORMS OF T AND V

Now that we know small oscillations can take place about any minimum point of V, we can go on to find approximate equations that govern such motions. The obvious (but not the best!) way of doing this is as follows: Take the example of the double pendulum. In this case, T and V are given (see Figure 15.2) by

$$T = \tfrac{1}{2}M(b\dot{\theta})^2 + \tfrac{1}{2}m\left((b\dot{\theta})^2 + (c\dot{\phi})^2 + 2(b\dot{\theta})(c\dot{\phi})\cos(\theta - \phi)\right), \tag{15.4}$$

$$V = (M + m)gb(1 - \cos\theta) + mgc(1 - \cos\phi). \tag{15.5}$$

If these expressions are substituted into the Lagrange's equations, we obtain (after some simplification) the **exact equations of motion**

$$(M + m)b\ddot{\theta} + mc\cos(\theta - \phi)\ddot{\phi} + mc\sin(\theta - \phi)\dot{\phi}^2 + (M + m)g\sin\theta = 0,$$

$$b\cos(\theta - \phi)\ddot{\theta} + c\ddot{\phi} - b\sin(\theta - \phi)\dot{\theta}^2 + g\sin\phi = 0.$$

This formidable pair of *coupled, second order, non-linear ODEs* govern the large oscillations of the double pendulum. However, for small oscillations about $\theta = \phi = 0$, these equations can be approximated by neglecting everything except linear terms in θ, ϕ and their time derivatives. On carrying out this approximation, the equations simplify dramatically to give

$$(M + m)b\,\ddot{\theta} + mc\,\ddot{\phi} + (M + m)g\,\theta = 0, \tag{15.6}$$

$$b\,\ddot{\theta} + c\,\ddot{\phi} + g\,\phi = 0. \tag{15.7}$$

These are the **linearised equations** governing small oscillations of the double pendulum about the downward vertical. They are a pair of *coupled, second order, linear ODEs with constant coefficients*. An explicit solution is therefore possible.

While the above method of finding the linearised equations of motion is perfectly correct, it is wasteful of effort and is also unsuitable when presenting the general theory. What we did was to obtain the *exact* expressions for T and V, derive the *exact* equations of motion, and then linearise. In the linearisation process, many of the terms we took pains to find were discarded. It makes far better sense to approximate the expressions for T and V from the start so that, when these approximations are used in Lagrange's equations, the linearised equations of motion are produced immediately. The saving in labour is considerable and this is also a nice way to present the general theory.

Consider the double pendulum for example. The exact expression for V is given by equation (15.5) and when θ, ϕ are small, this is given approximately by

$$V = \tfrac{1}{2}(M + m)gb\,\theta^2 + \tfrac{1}{2}mgc\,\phi^2 + \cdots,$$

where the neglected terms have power four or higher. Similarly, when θ, ϕ and their time derivatives are small, T is given approximately by

$$\begin{aligned} T &= \tfrac{1}{2}M(b\dot{\theta})^2 + \tfrac{1}{2}m\left((b\dot{\theta})^2 + (c\dot{\phi})^2 + 2(b\dot{\theta})(c\dot{\phi})(1 + \cdots)\right), \\ &= \tfrac{1}{2}(M + m)b^2\dot{\theta}^2 + mbc\,\dot{\theta}\dot{\phi} + \tfrac{1}{2}mc^2\dot{\phi}^2 + \cdots, \end{aligned}$$

where the neglected terms have power four (or higher) in small quantities. If these approximate forms for T and V are now substituted into Lagrange's equations, the linearised equations of motion (15.6), (15.7) are obtained immediately. [Check this.] This is clearly superior to our original method.

The general approximate form of V

In the general case, suppose that the potential energy $V(\boldsymbol{q})$ of the system $\mathcal{S}$ has a minimum at $\boldsymbol{q} = \boldsymbol{0}$ and that $V(\boldsymbol{0}) = 0$. (If the minimum point of V is not at $\boldsymbol{q} = \boldsymbol{0}$, it can always be made so by a simple change of coordinates.) Then, for $\boldsymbol{q}$ near $\boldsymbol{0}$, $V(\boldsymbol{q})$ can be expanded as an (n-dimensional) Taylor series in the variables $q_1, q_2, \ldots, q_n$. For the special case when $\mathcal{S}$ has two degrees of freedom, this series has the form

$$V(q_1, q_2) = V(0,0) + \left(\frac{\partial V}{\partial q_1} q_1 + \frac{\partial V}{\partial q_1} q_2\right) + \left(\frac{\partial^2 V}{\partial q_1^2} q_1^2 + 2\frac{\partial^2 V}{\partial q_1 \partial q_2} q_1 q_2 + \frac{\partial^2 V}{\partial q_2^2} q_2^2\right) + \cdots,$$

where all partial derivatives of V are evaluated at the expansion point $q_1 = q_1 = 0$. Now V has been selected so that $V(0,0) = 0$. Also, since $(0,0)$ is a stationary point of $V(q_1, q_2)$, it follows that $\partial V/\partial q_1 = \partial V/\partial q_2 = 0$ there. Thus the constant and linear terms are absent from the Taylor expansion of V. It follows that V can be approximated by

$$V^{\text{app}}(q_1, q_2) = v_{11} q_1^2 + 2v_{12} q_1 q_2 + v_{22} q_2^2,$$

where v_{11}, v_{12}, v_{22} are constants given by

$$v_{11} = \frac{\partial^2 V}{\partial q_1^2}(0,0) \quad v_{12} = \frac{\partial^2 V}{\partial q_1 \partial q_2}(0,0) \quad v_{22} = \frac{\partial^2 V}{\partial q_2^2}(0,0)$$

and the neglected terms have power three (or higher) in the small quantities q_1, q_2. The corresponding approximation to $V(\boldsymbol{q})$ in the case when $\mathcal{S}$ has n-degrees of freedom is

$$\boxed{V^{\text{app}}(\boldsymbol{q}) = \sum_{j=1}^{n} \sum_{k=1}^{n} v_{jk}\, q_j q_k} \tag{15.8}$$

where the $\{v_{jk}\}$ are constants given by

$$v_{jk} = v_{kj} = \left.\frac{\partial^2 V}{\partial q_j \partial q_k}\right|_{\boldsymbol{q}=\boldsymbol{0}},$$

and the neglected terms have power three (or higher) in the small quantities $q_1, q_2, \ldots,$ q_n. This is the general form of the **approximate potential energy** $V^{\text{app}}(\boldsymbol{q})$. It is a homogeneous quadratic form in the variables $q_1, q_2, \ldots, q_n$.

In the theory that follows, we will always assume that $\boldsymbol{q} = \boldsymbol{0}$ is also a minimum point of the *approximate* potential energy $V^{\text{app}}(\boldsymbol{q})$.* This condition is equivalent to requiring that the quadratic form (15.8) should be **positive definite**. This simply means that it takes positive values except when $\boldsymbol{q} = \boldsymbol{0}$.

The general approximate form of T

For any standard mechanical system with generalised coordinates $\boldsymbol{q}$, the kinetic energy T has the form

$$T(\boldsymbol{q}, \dot{\boldsymbol{q}}) = \sum_{j=1}^{n} \sum_{k=1}^{n} t_{jk}(\boldsymbol{q})\, \dot{q}_j \dot{q}_k,$$

a quadratic form in the variables $\dot{q}_1, \dot{q}_2, \ldots, \dot{q}_n$ with coefficients that depend on $\boldsymbol{q}$. If we expand each of these coefficients as a Taylor series about $\boldsymbol{q} = \boldsymbol{0}$, the constant term is simply $t_{jk}(\boldsymbol{0})$ and

$$T = \sum_{j=1}^{n} \sum_{k=1}^{n} t_{jk}(\boldsymbol{0})\, \dot{q}_j \dot{q}_k + \cdots .$$

It follows that T can be approximated by

$$\boxed{T^{\text{app}} = \sum_{j=1}^{n} \sum_{k=1}^{n} t_{jk}\, \dot{q}_j \dot{q}_k} \qquad (15.9)$$

where the constants $\{t_{jk}\}$ are what we previously called $\{t_{jk}(\boldsymbol{0})\}$, and the neglected terms have power three (or higher) in the small quantities $q_1, q_2, \ldots, q_n, \dot{q}_1, \dot{q}_2, \ldots, \dot{q}_n$. This is the general form of the **approximate kinetic energy** $V^{\text{app}}(\boldsymbol{q})$. It is a homogeneous quadratic form in the variables $\dot{q}_1, \dot{q}_2, \ldots, \dot{q}_n$. Since $T(\boldsymbol{q}, \dot{\boldsymbol{q}}) > 0$ except when $\dot{\boldsymbol{q}} = \boldsymbol{0}$, it follows that the quadratic form (15.9) must also be **positive definite**.

Useful tip: Since $T^{\text{app}}(\dot{\boldsymbol{q}}) = T(\boldsymbol{0}, \dot{\boldsymbol{q}})$, it follows that T^{app} can be found directly by calculating T *when the system is passing through the equilibrium position*; the general formula for T need never be found!

---

* It might appear that this follows the fact that $\boldsymbol{q} = \boldsymbol{0}$ is known to be a minimum point of the *exact* V, but this is not necessarily so. For example, if $V(q_1, q_2) = q_1^2 + q_2^4$, then $V^{\text{app}} = q_1^2$, which does not have a *strict* minimum at $q_1 = q_2 = 0$. The general theory of small oscillations does not apply to such cases, and we exclude them.

The *V*-matrix and the *T*-matrix

In order to express the general theory concisely, we introduce the $n \times n$ matrices **V** and **T** as follows:

Definition 15.2 ***The V-matrix and the T-matrix*** *The symmetric $n \times n$ matrix* **V** *whose elements are the coefficients $\{v_{jk}\}$ that appear in the formula (15.8) is called the V* ***-matrix****. The symmetric $n \times n$ matrix* **T** *whose elements are the coefficients $\{t_{jk}\}$ that appear in the formula (15.9) is called the T* ***-matrix****.*

In terms of **V** and **T**, the approximate potential and kinetic energies of $\mathcal{S}$ can be written in compact matrix notation:*

Quadratic forms for V^{app} and T^{app}

$$\begin{aligned} V^{\text{app}} &= \sum_{j=1}^{n}\sum_{k=1}^{n} v_{jk}\, q_j q_k = \mathbf{q}' \cdot \mathbf{V} \cdot \mathbf{q} \\ T^{\text{app}} &= \sum_{j=1}^{n}\sum_{k=1}^{n} t_{jk}\, \dot{q}_j \dot{q}_k = \dot{\mathbf{q}}' \cdot \mathbf{T} \cdot \dot{\mathbf{q}} \end{aligned} \tag{15.10}$$

where $\mathbf{q}$ is the column vector with elements $\{q_j\}$, and $\dot{\mathbf{q}}$ is the column vector with elements $\{\dot{q}_j\}$.

Example 15.2 *Finding* V *and* T *for the double pendulum*

Find the matrices **V** and **T** for the double pendulum.

Solution

For the double pendulum, V^{app} is given by

$$\begin{aligned} V^{\text{app}} &= \tfrac{1}{2}(M+m)gb\,\theta^2 + \tfrac{1}{2}mgc\,\phi^2, \\ &= \begin{pmatrix} \theta & \phi \end{pmatrix} \begin{pmatrix} \tfrac{1}{2}(M+m)gb & 0 \\ 0 & \tfrac{1}{2}mgc \end{pmatrix} \begin{pmatrix} \theta \\ \phi \end{pmatrix} \end{aligned}$$

and T^{app} is given by

$$\begin{aligned} T^{\text{app}} &= \tfrac{1}{2}(M+m)b^2\dot{\theta}^2 + mbc\,\dot{\theta}\dot{\phi} + \tfrac{1}{2}mc^2\dot{\phi}^2 \\ &= \begin{pmatrix} \dot{\theta} & \dot{\phi} \end{pmatrix} \begin{pmatrix} \tfrac{1}{2}(M+m)b^2 & \tfrac{1}{2}mbc \\ \tfrac{1}{2}mbc & \tfrac{1}{2}mc^2 \end{pmatrix} \begin{pmatrix} \dot{\theta} \\ \dot{\phi} \end{pmatrix}. \end{aligned}$$

* The notation $\mathbf{x}'$ means the *transpose* of the column vector $\mathbf{x}$. The alternative notation $\mathbf{x}^T$ would cause confusion here.

Hence, **V** and **T** are the 2×2 matrices

$$\mathbf{V} = \begin{pmatrix} \frac{1}{2}(M+m)gb & 0 \\ 0 & \frac{1}{2}mgc \end{pmatrix}, \qquad \mathbf{T} = \begin{pmatrix} \frac{1}{2}(M+m)b^2 & \frac{1}{2}mbc \\ \frac{1}{2}mbc & \frac{1}{2}mc^2 \end{pmatrix}. \blacksquare$$

15.3 THE GENERAL THEORY OF NORMAL MODES

In this section, we develop the general theory of normal modes for any oscillating system. This extends the method described in Chapter 5, which was restricted to two degrees of freedom.

The small oscillation equations

The first step is to obtain the general form of the small oscillation equations. This is done by substituting the approximate potential and kinetic energies V^{app} and T^{app} into Lagrange's equations. Now

$$\frac{\partial T^{\text{app}}}{\partial \dot{q}_j} = 2\sum_{k=1}^{n} t_{jk}\,\dot{q}_k, \qquad \frac{\partial T^{\text{app}}}{\partial q_j} = 0, \qquad \frac{\partial V^{\text{app}}}{\partial q_j} = 2\sum_{k=1}^{n} v_{jk}\,q_k,$$

so that Lagrange's equations become

Small oscillation equations

Expanded form:

$$\sum_{k=1}^{n} \left(t_{jk}\,\ddot{q}_k + v_{jk}\,q_k\right) = 0 \qquad (1 \le j \le n) \qquad (15.11)$$

Matrix form:

$$\mathbf{T}\cdot\ddot{\mathbf{q}} + \mathbf{V}\cdot\mathbf{q} = \mathbf{0}$$

in the expanded and matrix forms respectively. These are the **linearised equations** for the **small oscillations** of $\mathcal{S}$ about the point $\boldsymbol{q} = \mathbf{0}$. They are a set of *n coupled second order linear ODEs* satisfied by the unknown functions $q_1(t), q_2(t), \ldots, q_n(t)$.

Normal modes

The next step is to find a special class of solutions of the small oscillation equations known as **normal modes**. We will show later that the general solution of the small oscillation equations can be expressed as a sum of normal modes.

Definition 15.3 ***Normal mode*** *A solution of the small oscillation equations that has the special form*

Expanded form:	$q_j = a_j \cos(\omega t - \gamma)$ $(1 \le j \le n)$	(15.12)
Matrix form:	$\mathbf{q} = \mathbf{a}\cos(\omega t - \gamma)$	

where the $\{a_j\}$, ω *and* γ *are constants, is called a* **normal mode** *of the system* $\mathcal{S}$.

Notes. In a normal mode, the coordinates $q_1, q_2, \dots, q_n$ all vary harmonically in time with the *same frequency* ω and the *same phase* γ; however, they generally have *different amplitudes* $a_1, a_2, \dots, a_n$. The n-dimensional quantity $\boldsymbol{a} = (a_1, a_2, \dots, a_n)$ is called the **amplitude vector** of the mode and, when considered to be a column vector, will be written $\mathbf{a}$. Without losing generality, the angular frequency ω can be assumed to be *positive.*

On substituting the normal mode form (15.12) into the small oscillation equations (15.11), we obtain, on cancelling by the common factor $\cos(\omega t - \gamma)$,

Equations for the amplitude vector

Expanded form:

$$\sum_{k=1}^{n}\left(v_{jk} - \omega^2 t_{jk}\right) a_k = 0 \qquad (1 \le j \le n) \tag{15.13}$$

Matrix form:

$$\left(\mathbf{V} - \omega^2\,\mathbf{T}\right)\cdot\mathbf{a} = \mathbf{0}$$

This is an $n \times n$ system *simultaneous linear algebraic equations* for the coordinate amplitudes $\{a_k\}$. A normal mode will exist if we can find constants $\{a_k\}$, ω so that the equations (15.13) are satisfied. Since the equations are homogeneous, they always have the *trivial solution* $a_1 = a_2 = \cdots = a_n = 0$, whatever the value of ω. However, the trivial solution corresponds to the *equilibrium solution* $q_1 = q_2 = \cdots = q_n = 0$ of the governing equations (15.11), which is not a motion at all. We therefore need the equations (15.13) to have a **non-trivial solution** for the $\{a_k\}$. There is a simple condition that this should be so, namely that the **determinant** of the system of equations should be **zero**, that is,

Determinantal equation for ω

$$\det\left(\mathbf{V} - \omega^2\,\mathbf{T}\right) = 0 \tag{15.14}$$

This is the equation satisfied by the angular frequency ω in any normal mode of the system $\mathcal{S}$. When expanded, this is a polynomial equation of degree n in the variable ω^2. *If this equation has any* **real positive roots** $\omega_1^2, \omega_2^2, \ldots$, *then, for each of these values of* ω, *the linear equations (15.13) will have a non-trivial solution for the amplitudes* $\{a_k\}$ *and a normal mode will exist.*

Definition 15.4 ***Normal frequencies*** *The angular frequencies* $\omega_1, \omega_2, \ldots$ *of the normal modes are called the* **normal frequencies** *of the system* $\mathcal{S}$.

The normal frequencies are a very important characteristic of an oscillating system. They are found by solving the determinantal equation (15.14) for ω. In the example that follows, we find the normal frequencies of the double pendulum, and three further worked examples are given in section 15.5.

Example 15.3 *Normal frequencies of the double pendulum*

Find the normal frequencies of the double pendulum for the case in which $M = 3m$ and $c = b$.

Solution

With these special values, the matrices **V** and **T** become

$$\mathbf{V} = \tfrac{1}{2}mgb\begin{pmatrix} 4 & 0 \\ 0 & 1 \end{pmatrix}, \qquad \mathbf{T} = \tfrac{1}{2}mb^2\begin{pmatrix} 4 & 1 \\ 1 & 1 \end{pmatrix}.$$

The **determinantal equation** for ω is therefore

$$\det\left[\tfrac{1}{2}mgb\begin{pmatrix} 4 & 0 \\ 0 & 1 \end{pmatrix} - \tfrac{1}{2}mb^2\omega^2\begin{pmatrix} 4 & 1 \\ 1 & 1 \end{pmatrix}\right] = 0,$$

which can be simplified into the form

$$\begin{vmatrix} 4n^2 - 4\omega^2 & -\omega^2 \\ -\omega^2 & n^2 - \omega^2 \end{vmatrix} = 0,$$

where $n^2 = g/b$. On expanding this determinant, we obtain

$$3\omega^4 - 8n^2\omega^2 + 4n^4 = 0,$$

which is a quadratic equation in the variable ω^2. This is the equation satisfied by the normal frequencies. This quadratic factorises and has **two real positive roots** ω_1^2, ω_2^2 for ω^2, where

$$\omega_1^2 = \frac{2n^2}{3}, \qquad \omega_2^2 = 2n^2,$$

where $n^2 = g/b$. The double pendulum therefore has the **two normal frequencies** $(2g/3b)^{1/2}$ and $(2g/b)^{1/2}$. ∎

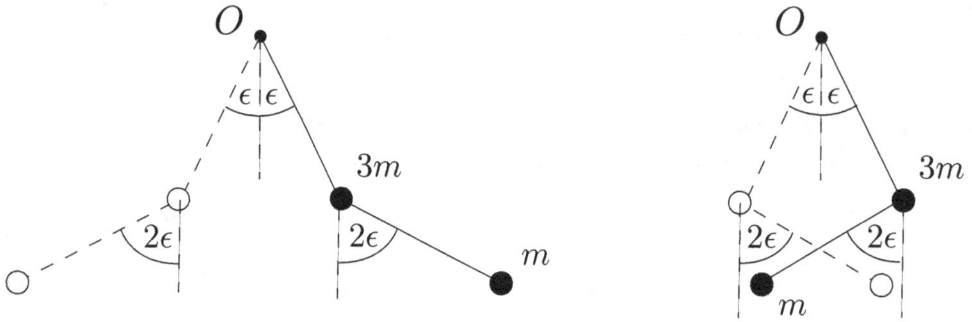

FIGURE 15.3 Normal modes of the double pendulum. **Left:** The slow mode. **Right:** The fast mode. (The angle ϵ, which should be small, is made large for clarity.)

Question *Form of the normal modes*

What do the normal mode motions of the double pendulum look like?

Answer

To answer this we need to find the coordinate amplitudes in each of the normal modes. If the amplitudes of θ and ϕ are a_1 and a_2 respectively, then these amplitudes satisfy the linear equations

$$\begin{pmatrix} 4n^2 - 4\omega^2 & -\omega^2 \\ -\omega^2 & n^2 - \omega^2 \end{pmatrix} \begin{pmatrix} a_1 \\ a_2 \end{pmatrix} = 0.$$

Slow mode: When $\omega^2 = \omega_1^2 = 2n^2/3$, the equations for the amplitudes a_1, a_2 become, after simplification,

$$\begin{pmatrix} 4 & -2 \\ -2 & 1 \end{pmatrix} \begin{pmatrix} a_1 \\ a_2 \end{pmatrix} = 0.$$

Each of these equations is equivalent to the single equation $2a_1 = a_2$ so that we have the family of non-trivial solutions $a_1 = \epsilon$, $a_2 = 2\epsilon$, where ϵ can take any (non-zero) value. There is therefore just **one slow normal mode**. It has the form

$$\begin{aligned} \theta &= \epsilon \cos(\sqrt{2/3}\, nt - \gamma), \\ \phi &= 2\epsilon \cos(\sqrt{2/3}\, nt - \gamma), \end{aligned} \tag{15.15}$$

where the amplitude factor ϵ and phase factor γ can take any values.* This mode is shown in Figure 15.3 (left).

Fast mode: In the fast mode, we have $\omega^2 = 2n^2$ and, by following the same procedure, we find that there is also **one fast normal mode**. It has the form

$$\begin{aligned} \theta &= \epsilon \cos(\sqrt{2}\, nt - \gamma), \\ \phi &= -2\epsilon \cos(\sqrt{2}\, nt - \gamma), \end{aligned} \tag{15.16}$$

where the amplitude factor ϵ and phase factor γ can take any values. This mode is shown in Figure 15.3 (right). ■

* However, the linearised theory is a good approximation only when ϵ is small.

15.4 EXISTENCE THEORY FOR NORMAL MODES

So far we have not said anything general about the *number* of normal mode motions that a system possesses. This is related to the number of *real positive* roots of the equation

$$\det(\mathsf{V} - \lambda\, \mathsf{T}) = 0. \tag{15.17}$$

When expanded, this is a polynomial equation of degree n in the variable λ, where n is the number of degrees of freedom of the system $\mathcal{S}$. Such an equation always has n roots in the *complex* plane, but there seems to be no reason why any of them should be real, let alone positive. In fact, *all of the roots of equation (15.17) are real and positive.* This follows from **generalised eigenvalue theory**, which we will now develop.

Definition 15.5 ***Generalised eigenvalues and eigenvectors*** *Let* K *and* L *be real* $n \times n$ *matrices. If there exists a number* λ *and a (non-zero) column vector* x *such that*

$$\mathsf{K} \cdot \mathsf{x} = \lambda\, \mathsf{L} \cdot \mathsf{x}, \tag{15.18}$$

then λ *is said to be a* ***generalised eigenvalue*** *of the matrix* K *(with respect to the matrix* L*) and* x *is a corresponding* ***generalised eigenvector***.*

The defining equation (15.18) can also be written

$$(\mathsf{K} - \lambda\, \mathsf{L}) \cdot \mathsf{x} = \mathsf{0}, \tag{15.19}$$

which has a non-zero solution for x only when

$$\det(\mathsf{K} - \lambda\, \mathsf{L}) = 0. \tag{15.20}$$

This is the equation satisfied by the eigenvalues. Provided that L is a non-singular matrix, the eigenvalue equation is a polynomial equation of degree n in λ, which has n roots in the complex plane. The more one knows about the matrices K and L, the more one can say about their eigenvalues and eigenvectors.

Theorem 15.1 ***Eigenvalues of symmetric, positive definite matrices*** *If* K *and* L *are* ***real symmetric*** *matrices and* L *is* ***positive definite***,† *then all the eigenvalues are* ***real***. *If the matrix* K *is also* ***positive definite***, *then all the eigenvalues are* ***positive***.

Proof. Let x be *any* complex column vector and consider the scalar quantity $\overline{\mathsf{x}}' \cdot \mathsf{K} \cdot \mathsf{x}$, where $\overline{\mathsf{x}}$ is the complex conjugate of x. Then, since K is real and symmetric,

$$\begin{aligned}\overline{\overline{\mathsf{x}}' \cdot \mathsf{K} \cdot \mathsf{x}} &= \mathsf{x}' \cdot \overline{\mathsf{K}} \cdot \overline{\mathsf{x}} = \mathsf{x}' \cdot \mathsf{K} \cdot \overline{\mathsf{x}} = (\mathsf{x}' \cdot \mathsf{K} \cdot \overline{\mathsf{x}})' = \overline{\mathsf{x}}' \cdot \mathsf{K}' \cdot \mathsf{x} \\ &= \overline{\mathsf{x}}' \cdot \mathsf{K} \cdot \mathsf{x}.\end{aligned}$$

* Ordinary eigenvalues and eigenvectors correspond to the special case when $\mathsf{L} = \mathsf{1}$, the identity matrix.

† A *matrix* A is called **positive definite** if its associated quadratic form $\mathsf{x}' \cdot \mathsf{A} \cdot \mathsf{x}$ is positive definite. Since this condition is known to hold for V and T, both these matrices must be positive definite.

Hence $\overline{\mathbf{x}}' \cdot \mathbf{K} \cdot \mathbf{x}$ must be real, and, by a similar argument, $\overline{\mathbf{x}}' \cdot \mathbf{L} \cdot \mathbf{x}$ must also be real.

Also, if we write $\mathbf{x}$ in terms of its real and imaginary parts in the form $\mathbf{x} = \mathbf{u} + i\,\mathbf{v}$, then

$$
\begin{aligned}
\overline{\mathbf{x}}' \cdot \mathbf{L} \cdot \mathbf{x} &= (\mathbf{u} - i\,\mathbf{v})' \cdot \mathbf{L} \cdot (\mathbf{u} + i\,\mathbf{v}) \\
&= \mathbf{u}' \cdot \mathbf{L} \cdot \mathbf{u} + \mathbf{v}' \cdot \mathbf{L} \cdot \mathbf{v} + i\,(\mathbf{u}' \cdot \mathbf{L} \cdot \mathbf{v} - \mathbf{v}' \cdot \mathbf{L} \cdot \mathbf{u}) \\
&= \mathbf{u}' \cdot \mathbf{L} \cdot \mathbf{u} + \mathbf{v}' \cdot \mathbf{L} \cdot \mathbf{v}
\end{aligned}
$$

since $\overline{\mathbf{x}}' \cdot \mathbf{L} \cdot \mathbf{x}$ is known to be real. Since $\mathbf{L}$ is a positive definite matrix, it follows that $\mathbf{u}' \cdot \mathbf{L} \cdot \mathbf{u}$ is positive except when $\mathbf{u} = \mathbf{0}$, and $\mathbf{v}' \cdot \mathbf{L} \cdot \mathbf{v}$ is positive except when $\mathbf{v} = \mathbf{0}$. Hence $\overline{\mathbf{x}}' \cdot \mathbf{L} \cdot \mathbf{x}$ is positive except when $\mathbf{x} = \mathbf{0}$.

Now suppose that $\mathbf{x}$ is a complex eigenvector corresponding to the complex eigenvalue λ. Then

$$
\begin{aligned}
\overline{\mathbf{x}}' \cdot \mathbf{K} \cdot \mathbf{x} &= \overline{\mathbf{x}}' \cdot (\mathbf{K} \cdot \mathbf{x}) = \overline{\mathbf{x}}' \cdot (\lambda\, \mathbf{L} \cdot \mathbf{x}) \\
&= \lambda\left(\overline{\mathbf{x}}' \cdot \mathbf{L} \cdot \mathbf{x}\right).
\end{aligned}
$$

But $\overline{\mathbf{x}}' \cdot \mathbf{K} \cdot \mathbf{x}$ is known to be real and $\overline{\mathbf{x}}' \cdot \mathbf{L} \cdot \mathbf{x}$ is known to be real and positive (since the complex eigenvector $\mathbf{x}$ is not zero). Hence the eigenvalue λ must be real.

The eigenvalues are now known to be real, and we may therefore restrict the eigenvectors to be real too. Suppose that the real eigenvalue λ has real eigenvector $\mathbf{x}$ and that the matrix $\mathbf{K}$ is now also positive definite. Then

$$
\mathbf{x}' \cdot \mathbf{K} \cdot \mathbf{x} = \lambda\left(\mathbf{x}' \cdot \mathbf{L} \cdot \mathbf{x}\right).
$$

But, since $\mathbf{K}$ and $\mathbf{L}$ are both positive definite matrices, the quantities $\mathbf{x}' \cdot \mathbf{K} \cdot \mathbf{x}$ and $\mathbf{x}' \cdot \mathbf{L} \cdot \mathbf{x}$ are both *positive*. It follows that λ must also be positive. ■

Since the matrices $\mathbf{V}$ and $\mathbf{T}$ are both symmetric and positive definite, the above theorem applies to normal mode theory. It follows that the roots of the determinantal equation (15.17) are all real and positive. If these roots are distinct (the most common case), then there are n distinct **normal frequencies** $\omega_1, \omega_2, \ldots, \omega_n$. It is however possible for the determinantal equation (15.14) to have repeated roots, so that there are fewer than n distinct normal frequencies. This usually happens when the system has symmetry; the spherical pendulum oscillating about the downward vertical is a simple example.

The number of normal modes associated with a particular normal frequency, ω_1 (say), depends on whether ω_1^2 is a simple or repeated root of the eigenvalue equation (15.17). It can be proved that, if ω_1^2 is a *simple root*, then the equations (15.13) for the amplitude vector $\boldsymbol{a}$ have a non-trivial solution that is *unique* to within a multiplied constant. There is therefore only *one normal mode* associated with the normal frequency ω_1. More generally, it can be proved that, if the root is repeated k times, then the equations (15.13) for the amplitude vector $\boldsymbol{a}$ have k linearly independent solutions.* The normal frequency then has k normal modes associated with it instead of one. It follows that, in all cases, we have the **fundamental result** that *the total number of normal modes is always equal to n, the number of degrees of freedom of the system.*

* It follows from the orthogonality relations (see section 15.6) that the amplitude vectors of the normal modes must be linearly independent and therefore cannot exceed n in number. Hence, when there are n *distinct* normal frequencies, each frequency must have exactly *one normal mode* associated with it. The corresponding result in the degenerate case is not easy to prove and is beyond the scope of a mechanics text. (See Anton [7] and Lang [10].)

Suppose for example that the oscillating system has six degrees of freedom and that the determinantal equation (15.14) is

$$(\Omega^2 - \omega^2)^2 (4\Omega^2 - \omega^2)^3 (25\Omega^2 - \omega^2) = 0,$$

after factorisation, where Ω is a positive constant. The normal frequencies are then $\omega_1 = \Omega$ (double root), $\omega_2 = 2\Omega$ (triple root) and $\omega_3 = 5\Omega$ (simple root). There are therefore two normal modes associated with the normal frequency ω_1, three normal modes associated with the normal frequency ω_2, and one normal mode associated with the normal frequency ω_3. The total number of normal modes is six, which is equal to the number of degrees of the system.

Definition 15.6 ***Degenerate frequencies*** *If a normal frequency has more than one normal mode associated with it, then that frequency is said to be* ***degenerate****.*

In the example above, the normal frequencies ω_1 and ω_2 are degenerate, but ω_3 is not. The notion of degeneracy is important in **quantum mechanics**, where normal frequencies correspond to the energies of stationary states. An unperturbed atom may have an energy level E that is (say) five-fold degenerate. When the atom is perturbed (by a magnetic field, for example) the energies of the five states may be changed by differing amounts so that the energy level is 'split' into five nearly equal levels. This is an important effect in the theory of atomic spectra.

Existence of normal modes

- For any oscillating system, the roots of the eigenvalue equation

$$\det\left(\mathbf{V} - \lambda\,\mathbf{T}\right) = 0$$

are all real and positive and their values are the *squares* of the normal frequencies $\{\omega_j\}$ of the system.

- If ω_1^2 is a *simple* root of the eigenvalue equation, then the equations

$$\left(\mathbf{V} - \omega_1^2\,\mathbf{T}\right)\cdot \mathbf{a} = \mathbf{0}$$

for the amplitude vector $\boldsymbol{a}$ have a non-trivial solution that is *unique* to within a multiplied constant. There is therefore only *one normal mode* associated with the normal frequency ω_1.

- More generally, if the root ω_1^2 is *repeated* k times, then the equations for the amplitude vector $\boldsymbol{a}$ have k linearly independent solutions. The normal frequency ω_1 is then degenerate with k normal modes associated with it.
- In all cases, **an oscillating system with n degrees of freedom has a total of n normal modes.**

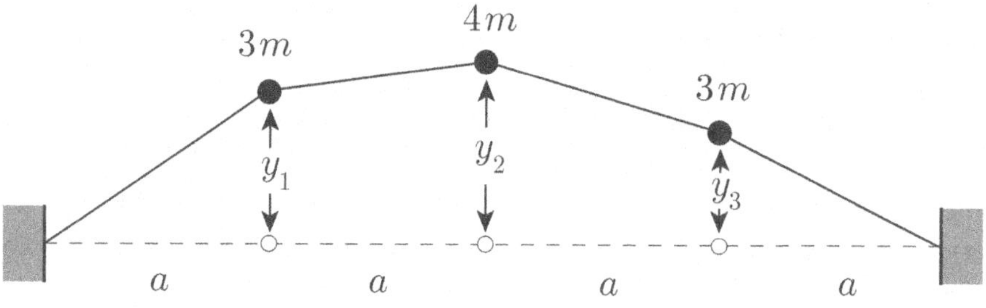

FIGURE 15.4 Transverse oscillations of three particles attached to a stretched string. (The particle displacements, which should be small, are shown large for clarity.)

15.5 SOME TYPICAL NORMAL MODE PROBLEMS

The determination of normal modes of vibration is an important subject with applications in physics, chemistry and mechanical engineering. The following three problems are typical. The first involves the transverse oscillations of a loaded stretched string; such problems make popular examination questions!

Example 15.4 *Transverse oscillations of a loaded stretched string*

A light string is stretched to a tension T_0 between two fixed points a distance $4a$ apart and particles of masses $3m$, $4m$ and $3m$ are attached to the string at equal intervals, as shown in Figure 15.4. The system performs small plane oscillations in which the particles move *transversely*, that is, at right angles to the equilibrium line of the string. Find the frequencies and forms of the normal modes.

Solution

Although it is clear by symmetry that *purely longitudinal* modes exist, it is not obvious that *purely transverse* modes exist. This question is investigated in Problem 15.5, where it is shown that the longitudinal and transverse modes uncouple in the linear theory. In this setting, purely transverse modes do exist and can be found by setting the longitudinal displacements equal to zero, which is what we will do here.

Let the transverse displacements of the three particles be y_1, y_2, y_3, as shown. Then the extension Δ_1 of the first section of string is given by

$$\begin{aligned}\Delta_1 &= \left(a^2 + y_1^2\right)^{1/2} - a = a\left(1 + \frac{y_1^2}{a^2}\right)^{1/2} - a \\ &= \frac{y_1^2}{2a} + \cdots,\end{aligned}$$

where the neglected terms have power four (or higher) in the small quantity y_1. Consider now the potential energy V_1 of this section of string. If the string had no initial tension then V_1 would be given by $V_1 = \frac{1}{2}\alpha\Delta_1^2$, where α is the 'spring constant' of the first section of the string. However, since there is an initial tension T_0, the formula is modified to

$$V_1 = T_0\Delta_1 + \tfrac{1}{2}\alpha\Delta_1^2 = T_0\left(\frac{y_1^2}{2a} + \cdots\right) + \tfrac{1}{2}\alpha\left(\frac{y_1^2}{2a} + \cdots\right)^2$$
$$= \frac{T_0 y_1^2}{2a} + \cdots,$$

where the neglected terms have power four (or higher) in the small quantity y_1. Note that, in the quadratic approximation, the spring constant α does not appear so that the increase in tension of the string is negligible.

In the same way, the potential energies of the other three sections of string are given by

$$V_2 = \frac{T_0(y_2 - y_1)^2}{2a} + \cdots, \qquad V_3 = \frac{T_0(y_3 - y_2)^2}{2a} + \cdots, \qquad V_4 = \frac{T_0 y_3^2}{2a} + \cdots,$$

and the total **approximate potential energy** is given by

$$V^{\text{app}} = \frac{T_0 y_1^2}{2a} + \frac{T_0(y_2 - y_1)^2}{2a} + \frac{T_0(y_3 - y_2)^2}{2a} + \frac{T_0 y_3^2}{2a}$$
$$= \frac{T_0}{2a}\left(2y_1^2 + 2y_2^2 + 2y_3^2 - 2y_1y_2 - 2y_2y_3\right).$$

The V-matrix is therefore

$$\mathbf{V} = \frac{T_0}{2a}\begin{pmatrix} 2 & -1 & 0 \\ -1 & 2 & -1 \\ 0 & -1 & 2 \end{pmatrix}.$$

In this problem, the exact and approximate **kinetic energies** are the same, namely

$$T = T^{\text{app}} = \tfrac{1}{2}(3m)\dot{y}_1^2 + \tfrac{1}{2}(4m)\dot{y}_2^2 + \tfrac{1}{2}(3m)\dot{y}_3^2,$$

so that the T-matrix is

$$\mathbf{T} = \tfrac{1}{2}m\begin{pmatrix} 3 & 0 & 0 \\ 0 & 4 & 0 \\ 0 & 0 & 3 \end{pmatrix}.$$

The **eigenvalue equation** $\det(\mathbf{V} - \lambda\mathbf{T}) = 0$ can therefore be written

$$\begin{vmatrix} 2 - 3\mu & -1 & 0 \\ -1 & 2 - 4\mu & -1 \\ 0 & -1 & 2 - 3\mu \end{vmatrix} = 0,$$

where $\mu = ma\omega^2/T_0$. When expanded, this is the **cubic equation**

$$18\mu^3 - 33\mu^2 + 17\mu - 2 = 0$$

for the parameter μ. Such an equation would have to be solved numerically in general, but problems that appear in textbooks (and in examinations!) are usually contrived so that an *exact factorisation* is possible; this is true in the present problem. Sometimes a factor can be spotted while the cubic is still in determinant form. In the present case, one can see that, by subtracting the third row of the determinant from the first, the cubic has the factor $2 - 3\mu$. If no factor can be spotted in this way, then one must try to spot that the expanded cubic equation has a (hopefully small) integer root. In the present case, one would have to spot that $\mu = 1$ is a root of the expanded cubic. By using either method, our cubic equation factorises into

$$(6\mu - 1)(3\mu - 2)(\mu - 1) = 0,$$

and its roots are $\mu_1 = 1/6$, $\mu_2 = 2/3$, $\mu_3 = 1$. Since $\mu = ma\omega^2/T_0$, the **normal frequencies** are given by

$$\omega_1^2 = \frac{T_0}{6ma}, \qquad \omega_2^2 = \frac{2T_0}{3ma}, \qquad \omega_3^2 = \frac{T_0}{ma}.$$

Since the normal frequencies are non-degenerate, the corresponding amplitude vectors are unique to within multiplied constants. In the **slow mode**, $\mu = 1/6$ and the equations $(\mathbf{V} - \lambda\mathbf{T}) \cdot \mathbf{a} = \mathbf{0}$ for the amplitude vector $\mathbf{a}$ become

$$\begin{pmatrix} 9 & -6 & 0 \\ -6 & 8 & -6 \\ 0 & -6 & 9 \end{pmatrix} \cdot \begin{pmatrix} a_1 \\ a_2 \\ a_3 \end{pmatrix} = \begin{pmatrix} 0 \\ 0 \\ 0 \end{pmatrix},$$

on clearing fractions. It is evident that $a_1 = 2, a_2 = 3, a_3 = 2$ is a solution so that the amplitude vector for the mode with frequency ω_1 is $\boldsymbol{a}_1 = (2,\ 3,\ 2)$. The other modes are treated in a similar way and the amplitude vectors are given by

$$\mathbf{a}_1 = \begin{pmatrix} 2 \\ 3 \\ 2 \end{pmatrix}, \qquad \mathbf{a}_2 = \begin{pmatrix} 1 \\ 0 \\ -1 \end{pmatrix}, \qquad \mathbf{a}_3 = \begin{pmatrix} 1 \\ -1 \\ 1 \end{pmatrix}.$$

These are the **forms** of the three normal modes. (It is a good idea to sketch the shapes of the three modes.) ■

Our second example is concerned with the **internal vibrations of molecules**, an important subject in physical chemistry. Although such problems should really be treated using quantum mechanics, the classical theory of normal modes gives much valuable information with far less effort. It would also be very difficult to understand the quantum treatment of molecular vibrations without first having studied the classical theory. The simplest case in which there is more than one frequency is the **linear triatomic molecule**. In this case the three atoms lie in a straight line and can perform *rectilinear* oscillations. The classic example of a linear triatomic molecule is carbon dioxide.

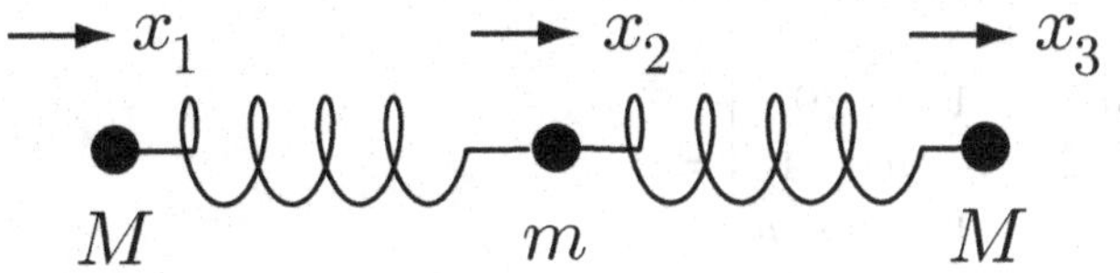

FIGURE 15.5 A simple classical model of a linear symmetric molecule.

Example 15.5 *The linear triatomic molecule*

A symmetric linear triatomic molecule is modelled by three particles connected by two springs, arranged as shown in Figure 15.5. Find the frequencies of rectilinear vibration of the molecule and the forms of the normal modes.

Solution

Let the centre atom have mass m, the outer atoms have mass M, and the springs have constant α. In this context, the spring constant is a measure of the 'strength' of the chemical bond between the two atoms. (We will suppose that the interaction between the two outer atoms is negligible.) Let the displacements of the three atoms from their equilibrium positions be x_1, x_2, x_3 as shown in Figure 15.5. Then the kinetic and potential energies of the molecule are given by

$$T = T^{\text{app}} = \tfrac{1}{2}M\dot{x}_1^2 + \tfrac{1}{2}m\dot{x}_2^2 + \tfrac{1}{2}M\dot{x}_3^2,$$

$$\begin{aligned} V = V^{\text{app}} &= \tfrac{1}{2}\alpha(x_2 - x_1)^2 + \tfrac{1}{2}\alpha(x_3 - x_2)^2 \\ &= \tfrac{1}{2}\alpha\left(x_1^2 + 2x_2^2 + x_3^2 - 2x_1x_2 - 2x_2x_3.\right) \end{aligned}$$

The T- and V- matrices are therefore

$$\mathbf{T} = \frac{1}{2}\begin{pmatrix} M & 0 & 0 \\ 0 & m & 0 \\ 0 & 0 & M \end{pmatrix}, \qquad \mathbf{V} = \frac{1}{2}\alpha\begin{pmatrix} 1 & -1 & 0 \\ -1 & 2 & -1 \\ 0 & -1 & 1 \end{pmatrix}.$$

Although everything looks normal, this problem has a non-standard feature in that the potential energy of the molecule does not have a true minimum* at $x_1 = x_2 = x_3 = 0$. This is actually clear from the start since the potential energy is unchanged if the whole molecule is translated to the right or left. Strictly speaking then, our theory does *not* apply to this problem, but, fortunately, only minor modifications are needed. In cases like this, it turns out that one or more of the normal frequencies is zero. These *zero frequencies do not correspond to true normal modes*. They correspond to uniform translational (or rotational) motions of the molecule as a whole. In the present case, the only uniform motion allowed is translational motion along the line of the molecule, so that we expect just *one* of the normal frequencies to be zero.

* As a result, the V-matrix is not positive definite.

The **eigenvalue equation** $\det(\mathbf{V} - \lambda\mathbf{T}) = 0$ can be written

$$\begin{vmatrix} 1-\mu & -1 & 0 \\ -1 & 2-\gamma^{-1}\mu & -1 \\ 0 & -1 & 1-\mu \end{vmatrix} = 0,$$

where $\mu = M\omega^2/\alpha$ and $\gamma = M/m$. The roots of this cubic equation are easily found to be $\mu = 0$, $\mu = 1$ and $\mu = 1+2\gamma$. The zero root corresponds to uniform translation and the other two are genuine oscillatory modes with **vibrational frequencies** given by

$$\omega_1^2 = \frac{\alpha}{M}, \qquad \omega_2^2 = \frac{(1+2\gamma)\alpha}{M}.$$

The **amplitude vectors** corresponding to the vibrational modes are

$$\mathbf{a}_1 = \begin{pmatrix} -1 \\ 0 \\ 1 \end{pmatrix}, \qquad \mathbf{a}_2 = \begin{pmatrix} 1 \\ -2\gamma \\ 1 \end{pmatrix},$$

respectively. We see that, in the slow ω_1-mode, the centre particle remains at rest while the outer pair oscillate symmetrically about the centre. This is called the **symmetric stretching mode** of the molecule. The fast ω_2-mode, in which all three particles move with the outer atoms remaining a constant distance apart, is called the **antisymmetric stretching mode**.

Comparison with experiment

Since the value of the constant α is unspecified, it can be chosen to fit any observed frequency. However, the *frequency ratio* ω_2/ω_1 $\left(= (1+2\gamma)^{1/2}\right)$ is independent of α and therefore affords a check on the theory.

The vibrational frequencies of real molecules can be measured with great accuracy by infrared and Raman spectroscopy. Spectroscopists measure the wavelength λ of radiation that excites each vibrational mode. The mode frequency is proportional to the reciprocal wavelength λ^{-1}, the standard units being cm^{-1}. Table 2 compares the observed and theoretical results for **carbon dioxide** and **carbon disulphide**,* both of which have linear symmetric molecules.

The theoretical values of the frequency ratio ω_2/ω_1 are within about 8% of those measured experimentally, which, considering the simplicity of the theory, is very good agreement.

Some more examples on vibrating molecules are given in the problems at the end of the chapter. The standard reference on this subject is the monumental work of Herzberg [13], Volume II. ■

Our third example involves a **rigid body suspended by three strings**. Problems of this type tend to be difficult because the constraints make it difficult to calculate the

* The atomic weights of carbon, oxygen and sulphur are C = 12, O = 16, S = 32.

Molecule	λ_1^{-1} (cm^{-1})	λ_2^{-1} (cm^{-1})	$\lambda_2^{-1}/\lambda_1^{-1}$	$(1+2\gamma)^{1/2}$
O–C–O	1337	2349	1.76	1.91
S–C–S	657	1532	2.33	2.52

Table 2 Vibrational frequencies of linear triatomic molecules; comparison of theory and experiment.

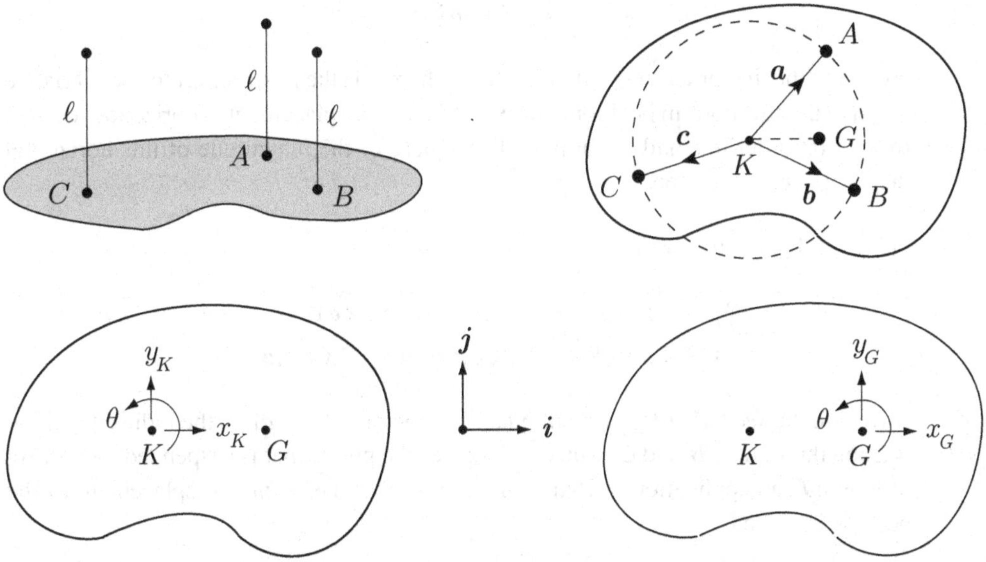

FIGURE 15.6 A flat plate of general shape is suspended by three equal strings.

potential energy. It is all the more surprising then that the following problem has a simple exact solution.

Example 15.6 *Plate supported by three strings*

A flat plate of general shape and general mass distribution is suspended in a horizontal position by three vertical strings of equal length. Find the normal frequencies of small oscillation.

Solution

The system is shown in its equilibrium position in Figure 15.6, top left. The three strings are of length ℓ and are attached to the points A, B, C of the plate. The point K (see Figure 15.6, top right) is the centre of the circle that passes through A, B, C. Our initial choice of generalised coordinates is shown in Figure 15.6, bottom left. X_K, Y_K are the horizontal Cartesian* displacement components of the point K from

* These can be any Cartesian axes K_0xyz with K_0z pointing vertically upwards; $\{\boldsymbol{i}, \boldsymbol{j}, \boldsymbol{k}\}$ are the corresponding unit vectors.

its equilibrium position K_0, and θ is the rotation angle of the plate about the vertical axis through K_0.* Three coordinates are sufficient since the three string constraints reduce the number of degrees of freedom of the plate from six to three.

We will now calculate the **potential energy** of the plate in terms of the coordinates X_K, Y_K and θ. This is the tricky step since the plate does *not* remain horizontal and this complicates the geometry. However, the vertical displacement of any point of the plate is *quadratic* in the small quantities X_K, Y_K, θ and, providing care is taken, this enables us to use approximations. Consider first $\Delta \boldsymbol{a}$, the displacement of the point A. This is given approximately by

$$\Delta \boldsymbol{a} = X_K \boldsymbol{i} + Y_K \boldsymbol{j} + (\theta \boldsymbol{k}) \times \boldsymbol{a} + \cdots ,$$

correct to the *first* order in small quantities, where $\boldsymbol{a}$ is the position vector of A relative to K_0 in the equilibrium position. As expected, this displacement is horizontal, correct to the first order in small quantities. The square of the magnitude of this horizontal displacement is therefore

$$\begin{aligned}
&(X_K \boldsymbol{i} + Y_K \boldsymbol{j} + \theta \boldsymbol{k} \times \boldsymbol{a})^2 \\
&\qquad = X_K^2 + Y_K^2 + \theta^2 |\boldsymbol{a}|^2 + 2X_K\theta\, (\boldsymbol{i} \cdot (\boldsymbol{k} \times \boldsymbol{a})) + 2Y_K\theta\, (\boldsymbol{j} \cdot (\boldsymbol{k} \times \boldsymbol{a})) \\
&\qquad = X_K^2 + Y_K^2 + R^2\theta^2 + 2Y_K\theta(\boldsymbol{a} \cdot \boldsymbol{i}) - 2X_K\theta(\boldsymbol{a} \cdot \boldsymbol{j}),
\end{aligned}$$

correct to the *second* order in small quantities, where R ($= |\boldsymbol{a}|$) is the radius the circle passing through A, B and C. Since A is one of the points that is suspended by a string of length ℓ, an application of Pythagoras shows that the *vertical* displacement of the point A is given by

$$z_A = \left(\frac{X_K^2 + Y_K^2 + R^2\theta^2}{2\ell} \right) + \left(\frac{2Y_K\theta}{2\ell} \right) (\boldsymbol{a} \cdot \boldsymbol{i}) - \left(\frac{2X_K\theta}{2\ell} \right) (\boldsymbol{a} \cdot \boldsymbol{j}),$$

correct to the *second* order in small quantities. Similar expressions exist for z_B and z_C with $\boldsymbol{a}$ replaced by $\boldsymbol{b}$ and $\boldsymbol{c}$ respectively. Now for the clever bit. Since the plate is flat, it follows that a *general* point of the plate with position vector $\boldsymbol{r}$ relative to K_0 in the equilibrium position must have vertical displacement†

$$z = \left(\frac{X_K^2 + Y_K^2 + R^2\theta^2}{2\ell} \right) + \left(\frac{2Y_K\theta}{2\ell} \right) (\boldsymbol{r} \cdot \boldsymbol{i}) - \left(\frac{2X_K\theta}{2\ell} \right) (\boldsymbol{r} \cdot \boldsymbol{j}) \qquad (15.21)$$

in the displaced position, correct to the second order in small quantities. This expression confirms that the plate is not generally horizontal in the displaced position.

---

* As we will see, the plate does *not* remain horizontal and so the angle θ ought to be defined more carefully. Let KP be any line fixed in the plate and let the projection of this line on to the equilibrium plane of the plate be $K'P'$. The angle θ can be properly defined as the angle turned through by the line $K'P'$. This is not quite the same as the angle turned through by the projection of some other line lying in the plate, but the differences are quadratic in the small quantities X_K, Y_K and θ and will turn out to be immaterial.

† This is because the expression (15.21) is linear in $\boldsymbol{r}$ and gives the correct z-values at A, B, C.

The purpose of all this is to calculate the potential energy $V = Mgz_G$, where M is the mass of the plate and z_G is the vertical displacement of the centre of mass G. With no loss of generality we may take the axis K_0x to point in the direction $\overrightarrow{K_0G_0}$ so that $\boldsymbol{r}_G = D\boldsymbol{i}$, where D is the distance KG. The general formula (15.21) then shows that

$$\begin{aligned} z_G &= \left(\frac{X_K^2 + Y_K^2 + R^2\theta^2}{2\ell}\right) + \left(\frac{2Y_K\theta}{2\ell}\right) D \\ &= \frac{X_K^2 + (Y_K + D\,\theta)^2 + (R^2 - D^2)\theta^2}{2\ell}, \end{aligned}$$

correct to the second order in small quantities. Hence the approximate **potential energy** is given by

$$V^{\text{app}} = \frac{Mg}{2\ell}\left[X_K^2 + (Y_K + D\,\theta)^2 + (R^2 - D^2)\theta^2\right].$$

This formula can be simplified further by a change of generalised coordinates. Let X_G, Y_G be the horizontal Cartesian displacement components of the centre of mass G from its equilibrium position G_0, and θ_G be the rotation angle of the plate about the vertical axis through G_0. Then, correct to the *first* order in small quantities,

$$X_G = X_K, \qquad Y_G = Y_K + D\,\theta, \qquad \theta_G = \theta.$$

In terms of the generalised coordinates X_G, Y_G, θ (we will drop the subscript from θ_G from now on) the expression for V^{app} is

$$V^{\text{app}} = \frac{Mg}{2\ell}\left(X_G^2 + Y_G^2 + (R^2 - D^2)\,\theta^2\right).$$

and the V-matrix is

$$\mathbf{V} = \frac{Mg}{2\ell}\begin{pmatrix} 1 & 0 & 0 \\ 0 & 1 & 0 \\ 0 & 0 & R^2 - D^2 \end{pmatrix},$$

a remarkably simple result in the end.

The approximate **kinetic energy** is calculated when the plate is passing through its equilibrium position. This is simply

$$T^{\text{app}} = \tfrac{1}{2}M\dot{X}_G^2 + \tfrac{1}{2}M\dot{Y}_G^2 + \tfrac{1}{2}I\,\dot{\theta}^2,$$

where I is the moment of inertia of the plate about the axis through G perpendicular to its plane. If we write $I = Mk^2$, then the T-matrix becomes

$$\mathbf{T} = \tfrac{1}{2}M\begin{pmatrix} 1 & 0 & 0 \\ 0 & 1 & 0 \\ 0 & 0 & k^2 \end{pmatrix}.$$

The **normal frequencies** are therefore given by

$$\omega_1^2 = \frac{g}{\ell} \text{ (doubly degenerate)}, \qquad \omega_2^2 = \left(\frac{R^2 - D^2}{k^2}\right)\frac{g}{\ell},$$

which correspond to two translational modes and one rotational mode.

In particular, if the lamina is a uniform circular **ring** of radius a, then $R = a$, $D = 0$ and $k = a$. Then $\omega_2 = \omega_1$ and the system has the single triply degenerate normal frequency $(g/\ell)^{1/2}$. In this case, any small motion of the system is periodic with period $2\pi(\ell/g)^{1/2}$. ■

15.6 ORTHOGONALITY OF NORMAL MODES

In this section we will show that the n **amplitude vectors** of the normal modes of an oscillating system are mutually orthogonal, in a sense that we will make clear. This is an important theoretical result, but it is not needed in the practical solution of normal mode problems. We will make use of orthogonality in our treatment of normal coordinates in section 15.8.

The basic theorem on orthogonality of eigenvectors is as follows:

Theorem 15.2 *Orthogonality of eigenvectors* *Suppose* K *and* L *are* ***symmetric*** $n \times n$ *matrices and that* $\mathbf{x}_1$ *and* $\mathbf{x}_2$ *are generalised eigenvectors of* K *(with respect to the matrix* L*) belonging to* ***distinct eigenvalues****. Then* $\mathbf{x}_1$ *and* $\mathbf{x}_2$ *are* ***mutually orthogonal*** *(with respect to the matrix* L*) in the sense that they satisfy the relation*

$$\mathbf{x}_1' \cdot \mathsf{L} \cdot \mathbf{x}_2 = 0.$$

Proof. Suppose that λ_1 and λ_2 are distinct eigenvalues with the corresponding eigenvectors $\mathbf{x}_1$ and $\mathbf{x}_2$ respectively. Consider the scalar quantity

$$\begin{aligned}\mathbf{x}_1' \cdot \mathsf{K} \cdot \mathbf{x}_2 &= \mathbf{x}_1' \cdot (\mathsf{K} \cdot \mathbf{x}_2) = \mathbf{x}_1' \cdot (\lambda_2 \mathsf{L} \cdot \mathbf{x}_2)\\ &= \lambda_2\,(\mathbf{x}_1' \cdot \mathsf{L} \cdot \mathbf{x}_2).\end{aligned}$$

However, the same quantity can also be written

$$\begin{aligned}\mathbf{x}_1' \cdot \mathsf{K} \cdot \mathbf{x}_2 &= (\mathsf{K}' \cdot \mathbf{x}_1)' \cdot \mathbf{x}_2 = (\mathsf{K} \cdot \mathbf{x}_1)' \cdot \mathbf{x}_2 = (\lambda_1 \mathsf{L} \cdot \mathbf{x}_1)' \cdot \mathbf{x}_2 = \lambda_1(\mathbf{x}_1' \cdot \mathsf{L}' \cdot \mathbf{x}_2)\\ &= \lambda_1\,(\mathbf{x}_1' \cdot \mathsf{L} \cdot \mathbf{x}_2),\end{aligned}$$

since K and L are both symmetric. It follows that

$$(\lambda_1 - \lambda_2)\left(\mathbf{x}_1' \cdot \mathsf{L} \cdot \mathbf{x}_2\right) = 0,$$

and, since $\lambda_1 \neq \lambda_2$, that

$$\mathbf{x}_1' \cdot \mathsf{L} \cdot \mathbf{x}_2 = 0. \blacksquare$$

Since the matrices V and T of an oscillating system are both real and symmetric, the above theorem applies to normal mode theory. It follows that if $\boldsymbol{a}_1$ and $\boldsymbol{a}_2$ are the amplitude vectors of two normal modes of an oscillating system with *distinct frequencies*, then

$$\mathbf{a}_1' \cdot \mathsf{T} \cdot \mathbf{a}_2 = 0.$$

This result is not necessarily true for amplitude vectors that belong to the same (degenerate) frequency, but they can always be *chosen* to do so. If this has been done, then the full set of amplitude vectors $\boldsymbol{a}_1, \boldsymbol{a}_2, \ldots \boldsymbol{a}_n$ are mutually orthogonal.

Orthogonality of normal modes

The amplitude vectors $\boldsymbol{a}_1, \boldsymbol{a}_2, \ldots \boldsymbol{a}_n$ of the normal modes of an oscillating system satisfy (or can be chosen to satisfy) the **orthogonality relations**

$$\mathbf{a}'_j \cdot \mathbf{T} \cdot \mathbf{a}_k = 0 \qquad (j \neq k). \tag{15.22}$$

For theoretical purposes, it is also convenient to *normalise* the amplitude vectors. Since $\mathbf{T}$ is a positive definite matrix, the quantities $\mathbf{a}'_1 \cdot \mathbf{T} \cdot \mathbf{a}_1, \mathbf{a}'_2 \cdot \mathbf{T} \cdot \mathbf{a}_2, \ldots, \mathbf{a}'_n \cdot \mathbf{T} \cdot \mathbf{a}_n$ are all *positive*. It follows that the amplitude vectors can be scaled so that

$$\mathbf{a}'_1 \cdot \mathbf{T} \cdot \mathbf{a}_1 = \mathbf{a}'_2 \cdot \mathbf{T} \cdot \mathbf{a}_2 = \cdots = \mathbf{a}'_n \cdot \mathbf{T} \cdot \mathbf{a}_n = 1, \tag{15.23}$$

in which case they are said to be **normalised**. The orthogonality and normalisation relations (15.23), (15.23) can then be combined into the single set of relations

$$\mathbf{a}'_j \cdot \mathbf{T} \cdot \mathbf{a}_k = \begin{cases} 0 & (j \neq k) \\ 1 & (j = k) \end{cases} \tag{15.24}$$

called the **orthonormality relations**.

Rayleigh's minimum principle

As an application of the orthogonality relations, we will now prove a far reaching result known as Rayleigh's minimum principle. Suppose an oscillating system $\mathcal{S}$ with n degrees of freedom has potential and kinetic energy matrices $\mathbf{V}$ and $\mathbf{T}$. Consider the function

$$F(\mathbf{x}) = \frac{\mathbf{x}' \cdot \mathbf{V} \cdot \mathbf{x}}{\mathbf{x}' \cdot \mathbf{T} \cdot \mathbf{x}} \tag{15.25}$$

where $\mathbf{x}$ is *any* non-zero column vector of dimension n. The function $F(\mathbf{x})$ is called **Rayleigh's function** for the system $\mathcal{S}$ and it has some interesting properties. To keep things simple we will suppose that $\mathcal{S}$ has *no degenerate normal frequencies*.

Theorem 15.3 ***Rayleigh's minimum principle*** *Suppose that an oscillating system $\mathcal{S}$ has Rayleigh function $F(\mathbf{x})$. Then*

$$F(\mathbf{x}) \geq \omega_1^2 \tag{15.26}$$

for all non-zero column vectors $\mathbf{x}$, *where* ω_1 *is the fundamental frequency* of* $\mathcal{S}$. *The minimum value is achieved when* $\mathbf{x}$ *is a multiple of the amplitude vector the fundamental mode.*

Proof. Let the n normal frequencies be ordered so that $\omega_1 < \omega_2 < \cdots < \omega_n$ and let the corresponding amplitude vectors be $\boldsymbol{a}_1, \boldsymbol{a}_2, \ldots, \boldsymbol{a}_n$. We will suppose that the amplitude vectors have been normalised so that they satisfy the orthonormality relations (15.24).

Now let $\mathbf{x}$ be any column vector. Since the n amplitude vectors form a basis set,† $\mathbf{x}$ can be expanded in the form $\mathbf{x} = \alpha_1\mathbf{a}_1 + \alpha_2\mathbf{a}_2 + \cdots + \alpha_n\mathbf{a}_n$. Then

$$\begin{aligned}
\mathbf{x}' \cdot \mathbf{V} \cdot \mathbf{x} &= \mathbf{x}' \cdot \mathbf{V} \cdot (\alpha_1\mathbf{a}_1 + \alpha_2\mathbf{a}_2 + \cdots + \alpha_n\mathbf{a}_n) \\
&= \alpha_1 \left(\mathbf{x}' \cdot \mathbf{V} \cdot \mathbf{a}_1\right) + \alpha_2 \left(\mathbf{x}' \cdot \mathbf{V} \cdot \mathbf{a}_2\right) + \cdots + \alpha_n \left(\mathbf{x}' \cdot \mathbf{V} \cdot \mathbf{a}_n\right) \\
&= \alpha_1\omega_1^2 \left(\mathbf{x}' \cdot \mathbf{T} \cdot \mathbf{a}_1\right) + \alpha_2\omega_2^2 \left(\mathbf{x}' \cdot \mathbf{T} \cdot \mathbf{a}_2\right) + \cdots + \alpha_n\omega_n^2 \left(\mathbf{x}' \cdot \mathbf{T} \cdot \mathbf{a}_n\right).
\end{aligned}$$

But

$$\begin{aligned}
\mathbf{x}' \cdot \mathbf{T} \cdot \mathbf{a}_k &= (\alpha_1\mathbf{a}_1 + \alpha_2\mathbf{a}_2 + \cdots + \alpha_n\mathbf{a}_n) \cdot \mathbf{T} \cdot \mathbf{a}_k \\
&= \alpha_1 (\mathbf{a}_1 \cdot \mathbf{T} \cdot \mathbf{a}_k) + \alpha_2 (\mathbf{a}_2 \cdot \mathbf{T} \cdot \mathbf{a}_k) + \cdots + \alpha_n (\mathbf{a}_n \cdot \mathbf{T} \cdot \mathbf{a}_k) \\
&= \alpha_k
\end{aligned}$$

on using the orthonormality relations. Hence

$$\mathbf{x}' \cdot \mathbf{V} \cdot \mathbf{x} = \alpha_1^2\,\omega_1^2 + \alpha_2^2\,\omega_2^2 + \cdots + \alpha_n^2\,\omega_n^2$$

and, by a similar argument,

$$\mathbf{x}' \cdot \mathbf{T} \cdot \mathbf{x} = \alpha_1^2 + \alpha_2^2 + \cdots + \alpha_n^2.$$

Hence

$$\begin{aligned}
F(\mathbf{x}) &= \frac{\alpha_1^2\,\omega_1^2 + \alpha_2^2\,\omega_2^2 + \cdots + \alpha_n^2\,\omega_n^2}{\alpha_1^2 + \alpha_2^2 + \cdots + \alpha_n^2} \ge \frac{\alpha_1^2\,\omega_1^2 + \alpha_2^2\,\omega_1^2 + \cdots + \alpha_n^2\,\omega_1^2}{\alpha_1^2 + \alpha_2^2 + \cdots + \alpha_n^2} \\
&= \omega_1^2
\end{aligned}$$

which is the required result. It is also evident that equality can only occur when $\alpha_2 = \alpha_3 = \cdots = \alpha_n = 0$, that is, when $\mathbf{x} = \alpha_1\mathbf{a}_1$. ■

This result means that $F(\mathbf{x})$ is an **upper bound** for ω_1^2, for *any* choice of the column vector $\mathbf{x}$; this upper bound has been obtained *without solving the oscillation problem.* Moreover, if we could substitute every value of $\mathbf{x}$ into the function $F(\mathbf{x})$, then the vectors that yield the least value of F must be multiples of the amplitude vector $\mathbf{a}_1$.

---

* The **fundamental frequency** is the lowest of the normal frequencies and the corresponding normal mode is the **fundamental mode.**

† This follows because the column vectors $\mathbf{a}_1, \mathbf{a}_2, \ldots, \mathbf{a}_n$ satisfy the orthogonality relations (15.22). A set of mutually orthogonal vectors must be linearly independent, and, since there are n of them, they form a basis for the space of column vectors of dimension n.

In normal mode theory, this result is of little consequence since the normal frequencies are simply the roots of a polynomial equation which can always be solved numerically. However, Rayleigh's principle has extensions to many areas of applied mathematics and physics such as continuum mechanics and quantum mechanics. In these subjects, the oscillation problems often cannot be solved, even numerically, and Rayleigh's principle is one of the few ways in which information can be gained about the fundamental mode. For example, in **quantum mechanics** Rayleigh's minimum principle takes the form:

Suppose $\mathcal{S}$ is a quantum mechanical system with Hamiltonian **H** *and ground state energy E_1. Then*

$$\frac{\langle \mathsf{x} \,|\, \mathsf{H} \,|\, \mathsf{x} \rangle}{\langle \mathsf{x} \,|\, \mathsf{x} \rangle} \geq E_1,$$

for any choice of the quantum state **x**.

15.7 GENERAL SMALL OSCILLATIONS

Normal modes are special small motions but, from them, we can generate the **general solution** of the small oscillation equations. The result is as follows:

General solution of small oscillation equations

The general solution of the small oscillation equations can be expressed as a linear combination of normal modes in the form

$$\boldsymbol{q}(t) = C_1 \boldsymbol{a}_1 \cos(\omega_1 t - \gamma_1) + C_2 \boldsymbol{a}_2 \cos(\omega_2 t - \gamma_2) + \cdots + C_n \boldsymbol{a}_n \cos(\omega_n t - \gamma_n)$$

where the amplitude factors $\{C_j\}$ and phase factors $\{\gamma_j\}$ are arbitrary constants.

Proof. Suppose that $\boldsymbol{q}(t)$ is *any* solution of the small oscillation equations (15.11). We will now show that we can construct a linear combination of normal modes that satisfies the small oscillation equations and also satisfies the same initial conditions as $\boldsymbol{q}(t)$. To do this, take a general linear combination of normal modes in the form

$$\begin{aligned}\boldsymbol{q}^*(t) &= C_1 \boldsymbol{a}_1 \cos(\omega_1 t - \gamma_1) + C_2 \boldsymbol{a}_2 \cos(\omega_2 t - \gamma_2) + \cdots + C_n \boldsymbol{a}_n \cos(\omega_n t - \gamma_n)\\ &= \boldsymbol{a}_1 (A_1 \cos\omega_1 t + B_1 \sin\omega_2 t) + \boldsymbol{a}_2 (A_2 \cos\omega_2 t + B_2 \sin\omega_2 t) + \\ &\qquad \cdots + \boldsymbol{a}_n (A_n \cos\omega_n t + B_n \sin\omega_n t),\end{aligned}$$

on writing $C_j \cos(\omega_j t - \gamma_j) = A_j \cos\omega_j t + B_j \sin\omega_j t$. Since the small oscillation equations are linear and homogeneous, $\boldsymbol{q}^*(t)$ is a solution for all choices of the coefficients $\{A_j\}$, $\{B_j\}$. We now need to choose the coefficients $\{A_j\}$, $\{B_j\}$ so that $\boldsymbol{q}^*(0) = \boldsymbol{q}(0)$ and $\dot{\boldsymbol{q}}^*(0) = \dot{\boldsymbol{q}}(0)$. This requires that the coefficients $\{A_j\}$ be chosen so that

$$A_1 \boldsymbol{a}_1 + A_2 \boldsymbol{a}_2 + \cdots + A_n \boldsymbol{a}_n = \boldsymbol{q}(0),$$

and that the coefficients $\{B_j\}$ be chosen so that

$$(B_1\omega_1)\, \boldsymbol{a}_1 + (B_2\omega_2)\, \boldsymbol{a}_2 + \cdots + (B_n\omega_n)\, \boldsymbol{a}_n = \dot{\boldsymbol{q}}(0).$$

This is *always possible* because the n amplitude vectors $\boldsymbol{a}_1, \boldsymbol{a}_2, \dots, \boldsymbol{a}_n$ form a basis for the space of vectors of dimension n. The vectors $\boldsymbol{q}(t)$ and $\dot{\boldsymbol{q}}(t)$ *can* therefore be expanded in the required forms.

We have thus constructed a solution $\boldsymbol{q}^*(t)$ of the small oscillation equations that satisfies the same initial conditions as the solution $\boldsymbol{q}(t)$. But ODE theory tells us that there can be only *one* such solution and so $\boldsymbol{q} = \boldsymbol{q}^*$. Since $\boldsymbol{q}$ can be any solution and $\boldsymbol{q}^*$ is a linear combination of normal modes, it follows that *any solution of the small oscillation equations can be expressed as a linear combination of normal modes.* ■

Example 15.7 *General small motion of the double pendulum*

Find the general solution of the small oscillation equations for the double pendulum problem.

Solution

The normal modes for the double pendulum problem have been found to be

$$\left.\begin{aligned} \theta &= \epsilon_1 \cos(\sqrt{2/3}\,nt - \gamma_1) \\ \phi &= 2\epsilon_1 \cos(\sqrt{2/3}\,nt - \gamma_1) \end{aligned}\right\}, \qquad \left.\begin{aligned} \theta &= \epsilon_2 \cos(\sqrt{2}\,nt - \gamma_2) \\ \phi &= -2\epsilon_2 \cos(\sqrt{2}\,nt - \gamma_2) \end{aligned}\right\}.$$

The general small motion is therefore

$$\begin{aligned} \theta &= \epsilon_1 \cos(\sqrt{2/3}\,nt - \gamma_1) + \epsilon_2 \cos(\sqrt{2}\,nt - \gamma_2), \\ \phi &= 2\epsilon_1 \cos(\sqrt{2/3}\,nt - \gamma_1) - 2\epsilon_2 \cos(\sqrt{2}\,nt - \gamma_2), \end{aligned}$$

where $\epsilon_1, \epsilon_2, \gamma_1, \gamma_2$ are arbitrary constants. ■

General small motion not usually periodic

The general small motion is a sum of periodic motions, but it is *not* usually periodic itself. Periodicity will occur only if there is some time interval τ that is an integer multiple of each of the periods $\tau_1, \tau_2, \dots, \tau_n$ of the normal modes. This only happens when the *ratios* of the normal mode periods are all *rational numbers*. In the double pendulum example, $\tau_1/\tau_2 = \omega_2/\omega_1 = \sqrt{3}$, which is irrational. The general small motion is therefore *not* periodic.

15.8 NORMAL COORDINATES

The preceding theory applies for any choice of the generalised coordinates $\{q_j\}$. Changing the generalised coordinates will change the V- and T-matrices, but the normal frequencies and the physical forms of the normal modes will be the same. This suggests that it might be possible to make a clever choice of coordinates so that the V- and T-matrices have a simple form leading to a much simplified theory. In particular, it would be very advantageous if T and V had **diagonal** form.

Definition 15.7 ***Normal coordinates*** *A set of generalised coordinates in terms of which the T- and V-matrices have diagonal form are called* ***normal coordinates***.

Actually, every oscillating system has normal coordinates, as we will now show. Let $\boldsymbol{q}$ be the original choice of coordinates with corresponding matrices V and T. Then

$$T^{\text{app}} = \dot{\mathsf{q}}' \cdot \mathsf{T} \cdot \dot{\mathsf{q}}, \qquad V^{\text{app}} = \mathsf{q}' \cdot \mathsf{V} \cdot \mathsf{q}. \qquad (15.27)$$

Now consider a change of coordinates from $\boldsymbol{q}$ to $\boldsymbol{\eta}$ defined by the linear transformation*

$$\mathsf{q} = \mathsf{P} \cdot \boldsymbol{\eta} \quad \Longleftrightarrow \quad \boldsymbol{\eta} = \mathsf{P}^{-1} \cdot \mathsf{q} \tag{15.28}$$

where P can be any non-singular matrix. On substituting the transformation (15.28) into the expressions (15.27), we obtain

$$T^{\text{app}} = (\mathsf{P} \cdot \dot{\boldsymbol{\eta}})' \cdot \mathsf{T} \cdot (\mathsf{P} \cdot \dot{\boldsymbol{\eta}}) = \dot{\boldsymbol{\eta}}' \cdot (\mathsf{P}' \cdot \mathsf{T} \cdot \mathsf{P}) \cdot \dot{\boldsymbol{\eta}},$$
$$V^{\text{app}} = (\mathsf{P} \cdot \boldsymbol{\eta})' \cdot \mathsf{V} \cdot (\mathsf{P} \cdot \boldsymbol{\eta}) = \boldsymbol{\eta}' \cdot (\mathsf{P}' \cdot \mathsf{V} \cdot \mathsf{P}) \cdot \boldsymbol{\eta},$$

from which we see that this transformation of coordinates causes V and T to be transformed as

$$\mathsf{T} \to \mathsf{P}' \cdot \mathsf{T} \cdot \mathsf{P}, \qquad \mathsf{V} \to \mathsf{P}' \cdot \mathsf{V} \cdot \mathsf{P}. \tag{15.29}$$

Can we now choose the transformation matrix P so that the new T- and V-matrices are diagonal?

Let $\boldsymbol{a}_1, \boldsymbol{a}_2, \ldots, \boldsymbol{a}_n$ be the amplitude vectors of the normal modes when they are expressed in terms of the coordinates $\boldsymbol{q}$ and let $\omega_1, \omega_2, \ldots, \omega_n$ be the corresponding normal frequencies. We will suppose that these amplitude vectors have been chosen so that they satisfy the **orthonormality relations** (15.24), that is

$$\mathsf{a}_j' \cdot \mathsf{T} \cdot \mathsf{a}_k = \begin{cases} 0 & (j \neq k), \\ 1 & (j = k). \end{cases} \tag{15.30}$$

Now consider the matrix P whose columns are the amplitude vectors $\{\mathsf{a}_j\}$, that is,

$$\mathsf{P} = (\mathsf{a}_1 \,|\, \mathsf{a}_2 \,|\, \cdots \,|\, \mathsf{a}_n)\,. \tag{15.31}$$

Since the amplitude vectors are known to be linearly independent, P has linearly independent columns and is therefore a non-singular matrix. Let us now try this P as the transformation matrix. Then

$$\mathsf{P}' \cdot \mathsf{T} \cdot \mathsf{P} = \begin{pmatrix} \mathsf{a}_1' \\ \hline \mathsf{a}_2' \\ \hline \vdots \\ \hline \mathsf{a}_n' \end{pmatrix} \cdot \mathsf{T} \cdot (\mathsf{a}_1 \,|\, \mathsf{a}_2 \,|\, \cdots \,|\, \mathsf{a}_n)\,.$$

* For example, in the case of two degrees of freedom, this transformation has the form

$$q_1 = p_{11}\,\eta_1 + p_{12}\,\eta_2,$$
$$q_2 = p_{21}\,\eta_1 + p_{22}\,\eta_2.$$

The jk-th element of this matrix is given by

$$\mathbf{a}_j' \cdot \mathbf{T} \cdot \mathbf{a}_k = \begin{cases} 0 & (j \neq k) \\ 1 & (j = k) \end{cases}$$

by the orthonormality relations. Hence, with this choice of $\mathbf{P}$,

$$\mathbf{P}' \cdot \mathbf{T} \cdot \mathbf{P} = \mathbf{1},$$

where $\mathbf{1}$ is the **identity matrix**. In the same way,

$$\mathbf{P}' \cdot \mathbf{V} \cdot \mathbf{P} = \begin{pmatrix} \mathbf{a}_1' \\ \hline \mathbf{a}_2' \\ \hline \vdots \\ \hline \mathbf{a}_n' \end{pmatrix} \cdot \mathbf{V} \cdot (\mathbf{a}_1 \,|\, \mathbf{a}_2 \,|\, \cdots \,|\, \mathbf{a}_n).$$

The jk-th element of this matrix is given by

$$\begin{aligned} \mathbf{a}_j' \cdot \mathbf{V} \cdot \mathbf{a}_k &= \mathbf{a}_j' \cdot (\mathbf{V} \cdot \mathbf{a}_k) = \mathbf{a}_j' \cdot \left(\omega_k^2 \mathbf{T} \cdot \mathbf{a}_k\right) = \omega_k^2 \left(\mathbf{a}_j' \cdot \mathbf{T} \cdot \mathbf{a}_k\right) \\ &= \begin{cases} 0 & (j \neq k), \\ \omega_j^2 & (j = k). \end{cases} \end{aligned}$$

Hence, with this choice of $\mathbf{P}$,

$$\mathbf{P}' \cdot \mathbf{V} \cdot \mathbf{P} = \mathbf{\Omega}^2,$$

where $\mathbf{\Omega}$ is the diagonal matrix whose diagonal elements are the normal frequencies, that is,

$$\mathbf{\Omega} = \begin{pmatrix} \omega_1 & 0 & \cdots & 0 \\ 0 & \omega_2 & \cdots & 0 \\ \vdots & \vdots & \ddots & \vdots \\ 0 & 0 & \cdots & \omega_n \end{pmatrix}.$$

We have thus succeeded in reducing both $\mathbf{V}$ and $\mathbf{T}$ to diagonal form. Hence the coordinates $\{\eta_j\}$ defined by (15.28) with $\mathbf{P} = (\mathbf{a}_1 \,|\, \mathbf{a}_2 \,|\, \cdots \,|\, \mathbf{a}_n)$ are a set of **normal coordinates**. They are given explicitly by

$$\boldsymbol{\eta} = \mathbf{P}^{-1} \cdot \mathbf{q} = \left(\mathbf{P}' \cdot \mathbf{T}\right) \cdot \mathbf{q},$$

on using the formula $\mathbf{P}' \cdot \mathbf{T} \cdot \mathbf{P} = \mathbf{1}$. This can also be written in the semi-expanded form

$$\eta_j = \left(\mathbf{a}_j' \cdot \mathbf{T}\right) \cdot \mathbf{q} \qquad (1 \leq j \leq n). \tag{15.32}$$

From this last formula, we can see that, if the amplitude vectors $\{\mathbf{a}_j\}$ are *not* normalised, then the coordinates $\{\eta_j\}$ are simply multiplied by constants. They are therefore *still normal coordinates*. The corresponding V- and T-matrices are still diagonal, but T is no longer reduced to the identity.

Our results are summarised as follows:

Finding normal coordinates

Let $\boldsymbol{a}_1, \boldsymbol{a}_2, \ldots, \boldsymbol{a}_n$ be the amplitude vectors of the normal modes when expressed in terms of the coordinates $\{q_j\}$. Then the coordinates $\{\eta_j\}$ defined by

$$\eta_j = \left(\mathbf{a}_j' \cdot \mathsf{T}\right) \cdot \mathbf{q} \qquad (1 \le j \le n)$$

are a set of **normal coordinates**, as are any constant multiples of them. (The amplitude vectors only need to be normalised if it is required to reduce the matrix T to the identity.)

When expressed in terms of normal coordinates, the **small oscillation equations** become

$$\ddot{\boldsymbol{\eta}} + \boldsymbol{\Omega}^2 \cdot \boldsymbol{\eta} = \mathbf{0}.$$

In expanded form, this is

$$\ddot{\eta}_j + \omega_j^2 \eta_j = 0 \qquad (1 \le j \le n),$$

a system of n *uncoupled* SHM equations. The solution $\eta_1 = C_1 \cos(\omega_1 t - \gamma_1)$, $\eta_2 = \eta_3 = \cdots = \eta_n = 0$ is the first normal mode, the solution $\eta_2 = C_2 \cos(\omega_2 t - \gamma_2)$, $\eta_1 = \eta_3 = \cdots = \eta_n = 0$ is the second normal mode, and so on.

Note. Using normal coordinates is *not* a practical way of solving normal mode problems. Indeed the problem has to be solved before the normal coordinates can be found! Normal coordinates are important because they simplify further developments of the general theory.

Example 15.8 *Finding normal coordinates*

Find a set of normal coordinates for the double pendulum problem.

Solution

For the double pendulum problem, we have already found that

$$\mathsf{T} = \tfrac{1}{2} m b^2 \begin{pmatrix} 4 & 1 \\ 1 & 1 \end{pmatrix}, \qquad \mathbf{a}_1 = \begin{pmatrix} 1 \\ 2 \end{pmatrix}, \qquad \mathbf{a}_2 = \begin{pmatrix} 1 \\ -2 \end{pmatrix}.$$

Hence, on dropping the inessential constant factor $\frac{1}{2}mb^2$, a set of normal coordinates is given by

$$\eta_1 = (1\ \ 2)\cdot\begin{pmatrix}4 & 1\\ 1 & 1\end{pmatrix}\cdot\begin{pmatrix}\theta\\ \phi\end{pmatrix} = 6\theta + 3\phi$$

and

$$\eta_2 = (1\ -2)\cdot\begin{pmatrix}4 & 1\\ 1 & 1\end{pmatrix}\cdot\begin{pmatrix}\theta\\ \phi\end{pmatrix} = 2\theta - \phi.$$

Since normal coordinates may always be scaled, we can equally well take

$$\eta_1 = 2\theta + \phi,$$
$$\eta_2 = 2\theta - \phi,$$

as our **normal coordinates.** ■

Problems on Chapter 15

Answers and comments are at the end of the book.

Harder problems carry a star (∗).

Two degrees of freedom

15.1 A particle P of mass $3m$ is connected to a particle Q of mass $8m$ by a light elastic spring of natural length a and strength α. Two similar springs are used to connect P and Q to the fixed points A and B respectively, which are a distance $3a$ apart on a smooth horizontal table. The particles can perform longitudinal oscillations along the straight line AB. Find the normal frequencies and the forms of the normal modes.

The system is in equilibrium when the particle P receives a blow that gives it a speed u in the direction $\overrightarrow{AB}$. Find the displacement of each particle at time t in the subsequent motion.

15.2 A particle A of mass $3m$ is suspended from a fixed point O by a spring of strength α and a second particle B of mass $2m$ is suspended from A by a second identical spring. The system performs small oscillations in the vertical straight line through O. Find the normal frequencies, the forms of the normal modes, and a set of normal coordinates.

15.3 ***Rod pendulum*** A uniform rod of length $2a$ is suspended from a fixed point O by a light inextensible string of length b attached to one of its ends. The system moves in a vertical plane through O. Take as coordinates the angles θ, ϕ between the string and the rod respectively and the downward vertical. Show that the equations governing small oscillations of the system about $\theta = \phi = 0$ are

$$b\ddot{\theta} + a\ddot{\phi} = -g\theta,$$

$$b\ddot{\theta} + \tfrac{4}{3}a\ddot{\phi} = -g\phi.$$

For the special case in which $b = 4a/5$, find the normal frequencies and the forms of the normal modes. Is the general motion periodic?

Three or more degrees of freedom

15.4 ***Triple pendulum*** A triple pendulum has three strings of equal length a and the three particles (starting from the top) have masses $6m$, $2m$, m respectively. The pendulum performs small oscillations in a vertical plane. Show that the normal frequencies satisfy the equation

$$4\mu^3 - 20\mu^2 + 27\mu - 9 = 0,$$

where $\mu = a\omega^2/g$. Find the normal frequencies, the forms of the normal modes, and a set of normal coordinates. [$\mu = 3$ is a root of the equation.]

15.5 A light *elastic* string is stretched to tension T_0 between two fixed points A and B a distance $3a$ apart, and two particles of mass m are attached to the string at equally spaced intervals. The strength of *each* of the three sections of the string is α. The system performs small oscillations in a plane through AB. Without making any prior assumptions, prove that the particles oscillate longitudinally in two of the normal modes and transversely in the other two. Find the four normal frequencies.

15.6 A rod of mass M and length L is suspended from two fixed points at the same horizontal level and a distance L apart by two equal strings of length b attached to its ends. From each end of the rod a particle of mass m is suspended by a string of length a. The system of the rod and two particles performs small oscillations in a vertical plane. Find $\mathbf{V}$ and $\mathbf{T}$ for this system. For the special case in which $b = 3a/2$ and $M = 6m/5$, find the normal frequencies. Show that the general small motion is periodic and find the period.

15.7 A uniform rod of length $2a$ is suspended in a horizontal position by *unequal* vertical strings of lengths b, c attached to its ends. Show that the frequency of the in-plane swinging mode is $((b+c)g/2bc)^{1/2}$, and that the frequencies of the other modes satisfy the equation

$$bc\mu^2 - 2a(b+c)\mu + 3a^2 = 0,$$

where $\mu = a\omega^2/g$. Find the normal frequencies for the particular case in which $b = 3a$ and $c = 8a$.

15.8* A uniform rod BC has mass M and length $2a$. The end B of the rod is connected to a fixed point A on a smooth horizontal table by an elastic string of strength α_1, and the end C is connected to a second fixed point D on the table by a second elastic string of strength α_2. In equilibrium, the rod lies along the line AD with the strings having tension T_0 and lengths b, c respectively. Show that the frequency of the longitudinal mode is $((\alpha_1 + \alpha_2)/M)^{1/2}$ and that the frequencies of the transverse modes satisfy the equation

$$bc\mu^2 - 2(2ab + 3bc + 2ac)\mu + 6a(2a + b + c) = 0,$$

where $\mu = Ma\omega^2/T_0$. [The calculation of V^{app} is very tricky.]

Find the frequencies of the transverse modes for the particular case in which $a = 3c$ and $b = 5c$.

15.9* A light *elastic* string is stretched between two fixed points A and B a distance $(n+1)a$ apart, and n particles of mass m are attached to the string at equally spaced intervals. The strength of *each* of the $n+1$ sections of the string is α. The system performs small *longitudinal* oscillations along the line AB. Show that the normal frequencies satisfy the determinantal equation

$$\Delta_n \equiv \begin{vmatrix} 2\cos\theta & -1 & 0 & \cdots & 0 & 0 \\ -1 & 2\cos\theta & -1 & \cdots & 0 & 0 \\ \vdots & \vdots & \vdots & \ddots & \vdots & \vdots \\ 0 & 0 & 0 & \cdots & 2\cos\theta & -1 \\ 0 & 0 & 0 & \cdots & -1 & 2\cos\theta \end{vmatrix} = 0,$$

where $\cos\theta = 1 - (m\omega^2/2\alpha)$.

By expanding the determinant by the top row, show that Δ_n satisfies the recurrence relation

$$\Delta_n = 2\cos\theta\,\Delta_{n-1} - \Delta_{n-2},$$

for $n \geq 3$. Hence, show by induction that

$$\Delta_n = \sin(n+1)\theta / \sin\theta.$$

Deduce the normal frequencies of the system.

15.10 A light string is stretched to a tension T_0 between two fixed points A and B a distance $(n+1)a$ apart, and n particles of mass m are attached to the string at equally spaced intervals. The system performs small plane *transverse* oscillations. Show that the normal frequencies satisfy the same determinantal equation as in the previous question, except that now $\cos\theta = 1 - (ma\omega^2/2T_0)$. Find the normal frequencies of the system.

Vibrating molecules

15.11 ***Unsymmetrical linear molecule*** A general linear triatomic molecule has atoms A_1, A_2, A_3 with masses m_1, m_2, m_3. The chemical bond between A_1 and A_2 is represented by a spring of strength α_{12} and the bond between A_2 and A_3 is represented by a spring of strength α_{23}. Show that the vibrational frequencies of the molecule satisfy the equation

$$m_1 m_2 m_3\,\omega^4 - [\alpha_{12} m_3(m_1+m_2) + \alpha_{23} m_1(m_2+m_3)]\,\omega^2$$

$$+\,\alpha_{12}\alpha_{23}(m_1+m_2+m_3) = 0.$$

Find the vibrational frequencies for the special case in which $m_1 = 3m$, $m_2 = m$, $m_3 = 2m$ and $\alpha_{12} = 3\alpha$, $\alpha_{23} = 2\alpha$.

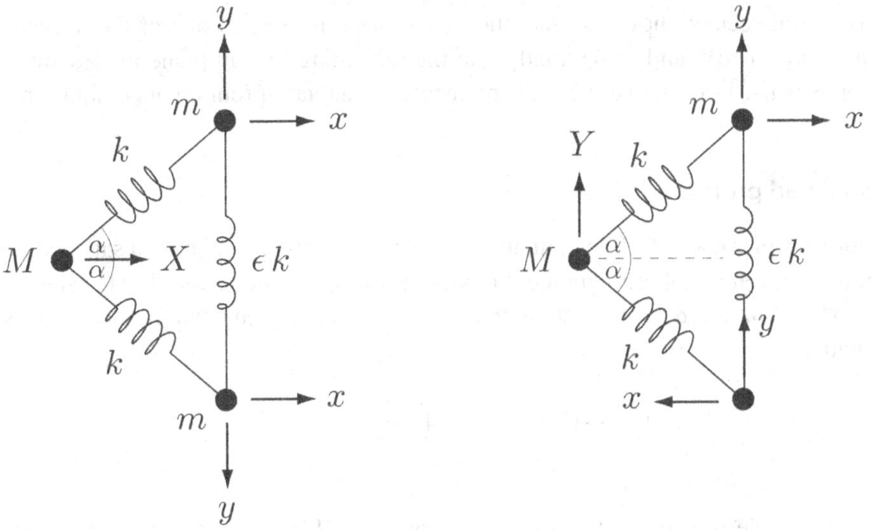

FIGURE 15.7 Vibrations of a symmetric V-shaped molecule. **Left:** a symmetric motion, **Right:** an antisymmetric motion.

The molecule O–C–S (carbon oxysulphide) is known to be linear. Use the λ_1^{-1} values given in Table 2 of the book (p. 441) to estimate its vibrational frequencies. [The experimentally measured values are 2174 cm^{-1} and 874 cm^{-1}.]

15.12* ***Symmetric V-shaped molecule*** Figure 15.7 shows the symmetric V-shaped triatomic molecule XY_2; the X–Y bonds are represented by springs of strength k, while the Y–Y bond is represented by a spring of strength $\epsilon\, k$. Common examples of such molecules include water, hydrogen sulphide, sulphur dioxide and nitrogen dioxide; the apex angle 2α is typically between 90° and 120°. In planar motion, the molecule has six degrees of freedom of which three are rigid body motions; there are therefore *three* vibrational modes. It is best to exploit the reflective symmetry of the molecule and solve separately for the symmetric and antisymmetric modes. Figure 15.7 (left) shows a symmetric motion while (right) shows an antisymmetric motion; the displacements X, Y, x, y are measured *from the equilibrium position.* Show that there is one antisymmetric mode whose frequency ω_3 is given by

$$\omega_3^2 = \frac{k}{mM}(M + 2m\sin^2\alpha),$$

and show that the frequencies of the symmetric modes satisfy the equation

$$\mu^2 - \left(1 + 2\gamma\cos^2\alpha + 2\epsilon\right)\mu + 2\epsilon\cos^2\alpha(1+2\gamma) = 0,$$

where $\mu = m\omega^2/k$ and $\gamma = m/M$.

Find the three vibrational frequencies for the special case in which $M = 2m$, $\alpha = 60°$ and $\epsilon = 1/2$.

15.13 ***Plane triangular molecule*** The molecule BCl_3 (boron trichloride) is plane and symmetrical. In equilibrium, the Cl atoms are at the vertices of an equilateral triangle with the

B atom at the centroid. Show that the molecule has six vibrational modes of which five are in the plane of the molecule; show also that the out-of-plane mode and one of the in-plane modes have axial symmetry; and show finally that the remaining four in-plane modes are in doubly degenerate pairs. Deduce that the BCl_3 molecule has a total of four distinct vibrational frequencies.

Computer assisted problem

15.14 ***Sulphur dioxide molecule*** Use computer assistance to obtain an equation satisfied by the squares of the frequencies of the symmetric modes of a V-shaped molecule. For the special case in which $M = 2m$ and $\alpha = 60°$, show that the frequencies of the symmetrical modes satisfy the equation

$$4\mu^2 - (5 + 8\epsilon)\mu + 4\epsilon = 0,$$

where $\mu = m\omega^2/k$.

The sulphur dioxide molecule O–S–O has mass ratio $M/m = 2$ and an apex angle very close to 120°. Its infrared absorption wave numbers are found to be $\lambda_1^{-1} = 1151\ \text{cm}^{-1}$, $\lambda_2^{-1} = 525\ \text{cm}^{-1}$, $\lambda_3^{-1} = 1336\ \text{cm}^{-1}$. Show that there is no value of ϵ that fits this data with reasonable accuracy. This is a deficiency of our simple (central force) model of interatomic forces, which gives poor results for V-shaped molecules (see Herzberg [13]).

Chapter Sixteen

Vector angular velocity and rigid body kinematics

KEY FEATURES

The key features in this chapter are **vector angular velocity** and the kinematics of **rigid bodies in general motion**.

This chapter is concerned with the kinematics of **rigid bodies in general motion**. In Chapter 2 we considered only those rigid body motions that were essentially two-dimensional, and angular velocity appeared there as a *scalar* quantity. In general three-dimensional rigid body motion, this approach is no longer adequate and angular velocity must be introduced in its proper rôle as a **vector** quantity. The principal result of the Chapter is that any motion of a rigid body can be represented as a sum of **translational** and **rotational** contributions.

16.1 ROTATION ABOUT A FIXED AXIS

In this chapter we adopt a more rigorous approach to rigid body rotation than we did in Chapter 2. We begin with a proper definition of rigidity.

Definition 16.1 ***Rigidity*** *A body $\mathcal{B}$ is said to be a **rigid body** if the distance between any pair of its particles remains constant. That is, if P_i and P_j are typical particles of $\mathcal{B}$ with position vectors $\boldsymbol{r}_i(t)$ and $\boldsymbol{r}_j(t)$ at time t, then*

$$|\boldsymbol{r}_i(t) - \boldsymbol{r}_j(t)| = c_{ij}, \tag{16.1}$$

where the c_{ij} are constants.

Suppose a rigid body $\mathcal{B}$ is rotating about a fixed axis with angular speed ω. This motion certainly satisfies the rigidity conditions (16.1). Let $\widehat{\boldsymbol{n}}$ be a *unit* vector parallel to the rotation axis. Then the **vector angular velocity** of $\mathcal{B}$ is defined as follows:

Definition 16.2 ***Vector angular velocity*** *The **angular velocity vector** of the body $\mathcal{B}$ is defined to be*

$$\boldsymbol{\omega} = \pm\,\omega\,\widehat{\boldsymbol{n}}, \tag{16.2}$$

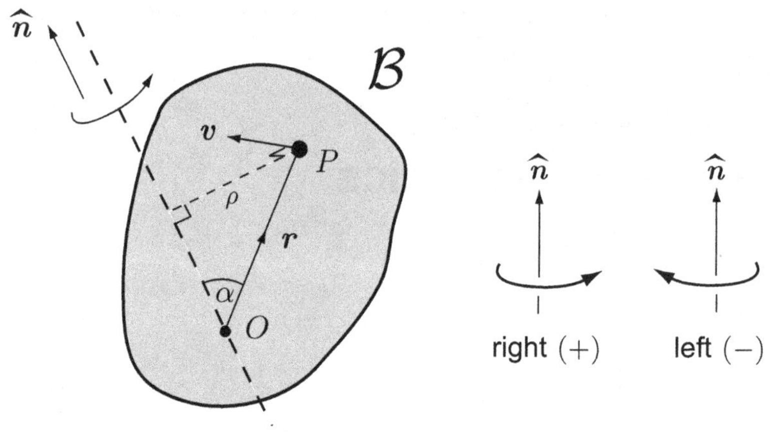

FIGURE 16.1 The rigid body $\mathcal{B}$ rotates with angular speed ω about a fixed axis parallel to the unit vector $\widehat{\boldsymbol{n}}$. Its angular velocity vector is defined by $\boldsymbol{\omega} = \pm\,\omega\,\widehat{\boldsymbol{n}}$, where the sign is determined by the sense of the rotation.

where the sign is taken to be plus or minus depending on whether the sense of the rotation (relative to the vector $\widehat{\boldsymbol{n}}$) is right- or left-handed. These senses are shown in Figure 16.1.

Note. From the vector $\boldsymbol{\omega}$ we can deduce the angular speed, the axis direction, and the rotation sense about the axis. It tells us nothing about the position of the axis either in space or in the body.

Example 16.1 *Calculation of* $\boldsymbol{\omega}$

A rigid body $\mathcal{B}$ is rotating with angular speed 7 radians per second about a fixed axis through the points $A(2, 3, -1)$, $B(-4, 0, 1)$. The rotation is in the left-handed sense relative to $\overrightarrow{AB}$. Find the angular velocity of $\mathcal{B}$.

Solution

The position vectors of the points A and B are $\boldsymbol{a} = 2\boldsymbol{i} + 3\boldsymbol{j} - \boldsymbol{k}$ and $\boldsymbol{b} = -4\boldsymbol{i} + \boldsymbol{k}$ so that

$$\overrightarrow{AB} = \boldsymbol{b} - \boldsymbol{a} = -6\boldsymbol{i} - 3\boldsymbol{j} + 2\boldsymbol{k}.$$

The vector $\widehat{\boldsymbol{n}}$ is then

$$\widehat{\boldsymbol{n}} = \frac{-6\boldsymbol{i} - 3\boldsymbol{j} + 2\boldsymbol{k}}{|-6\boldsymbol{i} - 3\boldsymbol{j} + 2\boldsymbol{k}|} = \frac{-6\boldsymbol{i} - 3\boldsymbol{j} + 2\boldsymbol{k}}{7}.$$

The rotation sense is left-handed relative to the direction of $\boldsymbol{n}$ so that the angular velocity of $\mathcal{B}$ is

$$\begin{aligned}\boldsymbol{\omega} &= -7\widehat{\boldsymbol{n}}\\ &= -7\left(\frac{-6\boldsymbol{i} - 3\boldsymbol{j} + 2\boldsymbol{k}}{7}\right)\\ &= 6\boldsymbol{i} + 3\boldsymbol{j} - 2\boldsymbol{k} \quad \text{radians per second.} \blacksquare\end{aligned}$$

Particle velocities and accelerations

The **particle velocities** can be conveniently calculated in terms of the vector $\boldsymbol{\omega}$. Let P be a particle of $\mathcal{B}$ with position vector $\boldsymbol{r}$ relative to an origin O located **on the rotation axis** (see Figure 16.1). Then the velocity of P has the same direction as the vector $\boldsymbol{\omega}\times\boldsymbol{r}$. Hence $\boldsymbol{v}$ can be written in the form

$$\boldsymbol{v} = \lambda\, \boldsymbol{\omega}\times\boldsymbol{r},$$

where λ is a positive scalar. To determine λ, consider the magnitude of each side. The magnitude of $\boldsymbol{v}$ is the circumferential speed $\omega\rho$ (see Figure 16.5) and so

$$\begin{aligned}\omega\rho &= \lambda|\boldsymbol{\omega}\times\boldsymbol{r}| = \lambda|\boldsymbol{\omega}||\boldsymbol{r}|\sin\alpha = \lambda\omega\,(OP\sin\alpha)\\ &= \lambda\omega\rho.\end{aligned}$$

Hence $\lambda = 1$ and the formula for $\boldsymbol{v}$ is

$$\boldsymbol{v} = \boldsymbol{\omega}\times\boldsymbol{r}. \qquad (16.3)$$

This formula applies only when the origin of vectors lies on the rotation axis, but a more general result is easy to obtain. Let B be *any* fixed point on the rotation axis. Then the velocity formula (16.3) still holds if the position vector $\boldsymbol{r}$ is replaced by $\overrightarrow{BP} = \boldsymbol{r} - \boldsymbol{b}$, where $\boldsymbol{b}$ is the position vector of the point B. Hence the general formula for the velocity of P is given by

$$\boxed{\boldsymbol{v} = \boldsymbol{\omega}\times(\boldsymbol{r} - \boldsymbol{b})} \qquad (16.4)$$

where B is any point on the rotation axis.

Example 16.2 *Finding particle velocities and accelerations*

A rigid body is rotating with constant angular speed 7 radians per second about a fixed axis through the points $A(2, 3, -1)$, $B(-4, 0, 1)$, distances being measured in centimetres. The rotation is in the left-handed sense relative to $\overrightarrow{AB}$. Find the instantaneous velocity, speed, and acceleration of the particle P of the body at the point $(-3, 3, 5)$.

Solution

The angular velocity of this body has been determined in the last example to be

$$\boldsymbol{\omega} = 6\boldsymbol{i} + 3\boldsymbol{j} - 2\boldsymbol{k} \quad \text{radians per second.}$$

The velocity of P can now be found using (16.4) with

$$\boldsymbol{r} = -3\boldsymbol{i} + 3\boldsymbol{j} + 5\boldsymbol{k} \quad \text{and} \quad \boldsymbol{b} = -4\boldsymbol{i} + \boldsymbol{k}.$$

This gives

$$\begin{aligned} \boldsymbol{v} &= (6\boldsymbol{i} + 3\boldsymbol{j} - 2\boldsymbol{k}) \times (\boldsymbol{i} + 3\boldsymbol{j} + 4\boldsymbol{k}) \\ &= \begin{vmatrix} \boldsymbol{i} & \boldsymbol{j} & \boldsymbol{k} \\ 6 & 3 & -2 \\ 1 & 3 & 4 \end{vmatrix} \\ &= 18\boldsymbol{i} - 26\boldsymbol{j} + 15\boldsymbol{k} \quad \text{cm s}^{-1}. \end{aligned}$$

This is the instantaneous **velocity** of P. The **speed** is therefore $|\boldsymbol{v}| = \left(18^2 + (-26)^2 + 15^2\right)^{1/2} = 35 \text{ cm s}^{-1}$.

The **acceleration** of P can be found by differentiating the formula (16.4) with respect to t. This gives

$$\boldsymbol{a} = \dot{\boldsymbol{\omega}} \times (\boldsymbol{r} - \boldsymbol{b}) + \boldsymbol{\omega} \times (\dot{\boldsymbol{r}} - \dot{\boldsymbol{b}}).$$

But $\boldsymbol{\omega}$ is known to have constant direction and magnitude and so $\dot{\boldsymbol{\omega}} = \boldsymbol{0}$. Also $\dot{\boldsymbol{b}} = \boldsymbol{0}$ since B is a fixed particle. This leaves

$$\begin{aligned} \boldsymbol{a} &= \boldsymbol{\omega} \times \dot{\boldsymbol{r}} \\ &= \boldsymbol{\omega} \times \boldsymbol{v} \\ &= (6\boldsymbol{i} + 3\boldsymbol{j} - 2\boldsymbol{k}) \times (18\boldsymbol{i} - 26\boldsymbol{j} + 15\boldsymbol{k}) \\ &= \begin{vmatrix} \boldsymbol{i} & \boldsymbol{j} & \boldsymbol{k} \\ 6 & 3 & -2 \\ 18 & -26 & 15 \end{vmatrix} \\ &= -7\boldsymbol{i} - 126\boldsymbol{j} - 210\boldsymbol{k} \quad \text{cm s}^{-2}. \blacksquare \end{aligned}$$

16.2 GENERAL RIGID BODY KINEMATICS

We now move on to the more general case in which the rigid body does not have a fixed rotation axis. We first consider a rigid body that has *one particle* O that does not move, and we take O to be the origin of position vectors. Now the rigidity conditions (16.1) are equivalent to

$$(\boldsymbol{r}_i - \boldsymbol{r}_j) \cdot (\boldsymbol{r}_i - \boldsymbol{r}_j) = c_{ij}^2 \tag{16.5}$$

and because O is a particle $\mathcal{B}$ it follows in particular that

$$\boldsymbol{r}_i \cdot \boldsymbol{r}_i = d_i, \tag{16.6}$$

where the d_i are constants. On expanding the dot product in (16.5) and using (16.6) we obtain

$$\boldsymbol{r}_i \cdot \boldsymbol{r}_j = e_{ij}, \tag{16.7}$$

where the e_{ij} are constants. If we now differentiate (16.7) with respect to t we obtain

$$\dot{\boldsymbol{r}}_i \cdot \boldsymbol{r}_j + \boldsymbol{r}_i \cdot \dot{\boldsymbol{r}}_j = 0 \qquad \text{for all} \quad i, j, \tag{16.8}$$

which is our preferred form of the **rigidity conditions**.

We now prove the **fundamental theorems** of rigid body kinematics. The details of the proofs are mainly of interest to mathematics students.

Theorem 16.1 ***Existence of angular velocity I*** *Let a rigid body $\mathcal{B}$ be in motion with one of its particles O fixed. Then there exists a unique vector $\boldsymbol{\omega}(t)$ such that the velocity of any particle P of $\mathcal{B}$ is given by the formula*

$$\boldsymbol{v} = \boldsymbol{\omega} \times \boldsymbol{r}, \tag{16.9}$$

where $\boldsymbol{r}$ is the position vector of P relative to O.

This result means that, at each instant, $\mathcal{B}$ is rotating about an *instantaneous* axis through O. This axis is not fixed in space or in the body.

Proof. Suppose that there exist particles E_1, E_2, E_3 of $\mathcal{B}$ such that their position vectors $\{\boldsymbol{e}_1, \boldsymbol{e}_2, \boldsymbol{e}_3\}$ relative to O form a standard basis set. Then if there does exist an $\boldsymbol{\omega}$ satisfying (16.9), it must in particular satisfy

$$\dot{\boldsymbol{e}}_k = \boldsymbol{\omega} \times \boldsymbol{e}_k \tag{16.10}$$

for $k = 1, 2, 3$. Taking the cross product of this equation with $\boldsymbol{e}_k$ gives

$$\begin{aligned}\boldsymbol{e}_k \times \dot{\boldsymbol{e}}_k &= \boldsymbol{e}_k \times (\boldsymbol{\omega} \times \boldsymbol{e}_k) = (\boldsymbol{e}_k \cdot \boldsymbol{e}_k)\, \boldsymbol{\omega} - (\boldsymbol{\omega} \cdot \boldsymbol{e}_k)\, \boldsymbol{e}_k \\ &= \boldsymbol{\omega} - (\boldsymbol{\omega} \cdot \boldsymbol{e}_k)\, \boldsymbol{e}_k\end{aligned}$$

since $\boldsymbol{e}_k$ is a unit vector. Summing these equations over $1 \le k \le 3$ gives

$$\sum_k \boldsymbol{e}_k \times \dot{\boldsymbol{e}}_k = 3\boldsymbol{\omega} - \sum_k (\boldsymbol{\omega} \cdot \boldsymbol{e}_k)\, \boldsymbol{e}_k = 3\boldsymbol{\omega} - \boldsymbol{\omega}$$

since the sum on the right is just the expansion of $\boldsymbol{\omega}$ with respect to the basis set $\{\boldsymbol{e}_1, \boldsymbol{e}_2, \boldsymbol{e}_3\}$. Hence

$$\boldsymbol{\omega} = \frac{1}{2} \sum_k \boldsymbol{e}_k \times \dot{\boldsymbol{e}}_k. \tag{16.11}$$

Thus if $\boldsymbol{\omega}$ does exist, it must be given by the formula (16.11), which shows that $\boldsymbol{\omega}$ is *unique*. We must now show that this $\boldsymbol{\omega}$ satisfies (16.9) for *all* the particles of the body. It is a simple exercise to verify that this is true for the particles E_1, E_2, E_3 by substituting (16.11) into (16.10) and using the rigidity conditions (16.8). Now let P be any other particle of the body and expand its position vector $\boldsymbol{r}$ with respect to the basis set $\{\boldsymbol{e}_1, \boldsymbol{e}_2, \boldsymbol{e}_3\}$ in the form

$$\boldsymbol{r} = \sum_k (\boldsymbol{r} \cdot \boldsymbol{e}_k)\, \boldsymbol{e}_k.$$

The velocity of P is then given by

$$\boldsymbol{v} = \dot{\boldsymbol{r}} = \sum_k (\dot{\boldsymbol{r}} \cdot \boldsymbol{e}_k + \boldsymbol{r} \cdot \dot{\boldsymbol{e}}_k)\, \boldsymbol{e}_k + \sum_k (\boldsymbol{r} \cdot \boldsymbol{e}_k)\, \dot{\boldsymbol{e}}_k.$$

Now $\dot{\boldsymbol{r}} \cdot \boldsymbol{e}_k + \boldsymbol{r} \cdot \dot{\boldsymbol{e}}_k = 0$ by the rigidity conditions (16.8), and we have directly verified that $\dot{\boldsymbol{e}}_k = \boldsymbol{\omega} \times \boldsymbol{e}_k$. Hence

$$\begin{aligned} \boldsymbol{v} &= \sum_k (\boldsymbol{r} \cdot \boldsymbol{e}_k)\,(\boldsymbol{\omega} \times \boldsymbol{e}_k) = \boldsymbol{\omega} \times \sum_k (\boldsymbol{r} \cdot \boldsymbol{e}_k)\,\boldsymbol{e}_k \\ &= \boldsymbol{\omega} \times \boldsymbol{r} \end{aligned}$$

as required. ■

The above proof cannot even begin if any of the particles E_1, E_2, E_3 are not actually present (for example the body could be a lamina). In such a case, suppose that the body has at least *two* particles A, B in addition to O, and that O, A, B are not collinear. Then define the standard basis set $\{\boldsymbol{e}_1, \boldsymbol{e}_2, \boldsymbol{e}_3\}$ by

$$\boldsymbol{e}_1 = \frac{\boldsymbol{a}}{|\boldsymbol{a}|}, \qquad \boldsymbol{e}_2 = \frac{\boldsymbol{b} - (\boldsymbol{a} \cdot \boldsymbol{b})\,\boldsymbol{a}}{|\boldsymbol{b} - (\boldsymbol{a} \cdot \boldsymbol{b})\,\boldsymbol{a}|}, \qquad \boldsymbol{e}_3 = \boldsymbol{e}_1 \times \boldsymbol{e}_2.$$

It can be shown that the points of space E_1, E_2, E_3 that have the position vectors $\boldsymbol{e}_1, \boldsymbol{e}_2, \boldsymbol{e}_3$ satisfy the *same rigidity conditions* as the real particles of the body. They can therefore be regarded as real particles and the proof given above then holds.

We now extend the result in Theorem 16.1 to the case of *completely general* motion in which no particle of the body is fixed.

Theorem 16.2 ***Existence of angular velocity II*** *Suppose a rigid body is in completely general motion and let B be any one of its particles. Then there exists a unique vector $\boldsymbol{\omega}(t)$ such that the velocity of any particle P of the body is given by the formula*

$$\boldsymbol{v} = \boldsymbol{v}^B + \boldsymbol{\omega} \times (\boldsymbol{r} - \boldsymbol{b}), \tag{16.12}$$

where $\boldsymbol{r}$ and $\boldsymbol{b}$ are the position vectors of P and B, and $\boldsymbol{v}^B$ is the velocity of B.

Proof. We view the motion of $\mathcal{B}$ from a reference frame $\mathcal{F}'$ with origin at B and moving without rotation relative to the original frame $\mathcal{F}$. Then (see section 1.4) the position vector $\boldsymbol{r}'$ and velocity $\boldsymbol{v}'$ of a particle P relative to $\mathcal{F}'$ are related to the original $\boldsymbol{r}$ and $\boldsymbol{v}$ by

$$\boldsymbol{r} = \boldsymbol{r}' + \boldsymbol{b}, \qquad \boldsymbol{v} = \boldsymbol{v}' + \boldsymbol{v}^B. \tag{16.13}$$

It follows that if P_i and P_j are particles of $\mathcal{B}$

$$|\boldsymbol{r}_i' - \boldsymbol{r}_j'| = |\boldsymbol{r}_i - \boldsymbol{r}_j| = c_{ij},$$

where the c_{ij} are constants, so that the rigidity conditions are also satisfied in $\mathcal{F}'$. But in $\mathcal{F}'$ the particle B is fixed (at the origin) and so Theorem 16.1 applies. Hence there exists a unique vector $\boldsymbol{\omega}(t)$ such that, for any particle of $\mathcal{B}$,

$$\boldsymbol{v}' = \boldsymbol{\omega} \times \boldsymbol{r}'. \tag{16.14}$$

On using (16.13) into (16.14) we obtain

$$\boldsymbol{v} - \boldsymbol{v}^B = \boldsymbol{\omega} \times (\boldsymbol{r} - \boldsymbol{b}),$$

that is

$$\boldsymbol{v} = \boldsymbol{v}^B + \boldsymbol{\omega} \times (\boldsymbol{r} - \boldsymbol{b}), \tag{16.15}$$

as required. ■

What the last theorem shows is that **any rigid body motion** can be resolved into a **translation** with velocity $\boldsymbol{v}^B$ and a **rotation** with some angular velocity $\boldsymbol{\omega}$ about an axis

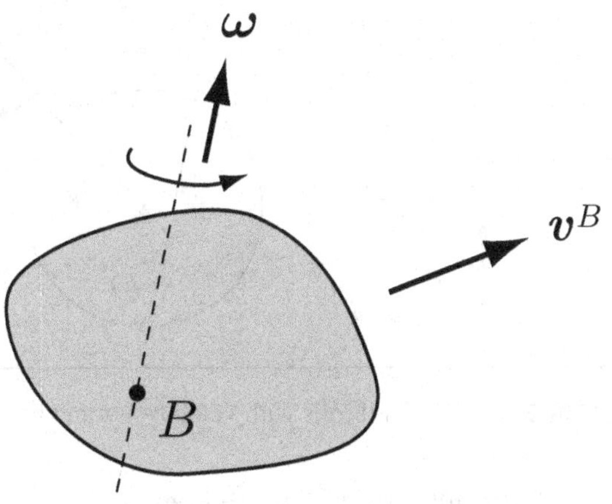

FIGURE 16.2 A rigid body in general motion. The velocities of its particles are the sum of those due to a **translation** with velocity $\boldsymbol{v}^B$ and a **rotation** with angular velocity $\boldsymbol{\omega}$ about an axis through B.

through B, where the particle B can be any particle of the body. This is shown in Figure 16.2.

Now suppose we choose a new reference particle C. Then the translational velocity would become $\boldsymbol{v}^C$, but what happens to the angular velocity $\boldsymbol{\omega}$? The answer is nothing; the angular velocity $\boldsymbol{\omega}$ is *independent* of the choice of reference particle. This means that we can refer to *the* angular velocity of a rigid body without specifying the reference particle. The proof of this is as follows:

Proof. Suppose that, with reference particles B, C, the body has angular velocities $\boldsymbol{\omega}^B$, $\boldsymbol{\omega}^C$ respectively. Then the velocity $\boldsymbol{v}$ of any particle P of the body is given by either of the two formulae

$$\boldsymbol{v} = \boldsymbol{v}^B + \boldsymbol{\omega}^B \times (\boldsymbol{r} - \boldsymbol{b}),$$
$$\boldsymbol{v} = \boldsymbol{v}^C + \boldsymbol{\omega}^C \times (\boldsymbol{r} - \boldsymbol{c}),$$

where $\boldsymbol{r}$ is the position vector of P. It follows that

$$\boldsymbol{v}^B + \boldsymbol{\omega}^B \times (\boldsymbol{r} - \boldsymbol{b}) = \boldsymbol{v}^C + \boldsymbol{\omega}^C \times (\boldsymbol{r} - \boldsymbol{c}) \tag{16.16}$$

for any $\boldsymbol{r}$ that is the position vector of a particle of the body. In particular, since B and C are particles of the body, it follows that

$$\boldsymbol{v}^B = \boldsymbol{v}^C + \boldsymbol{\omega}^C \times (\boldsymbol{b} - \boldsymbol{c}),$$

$$\boldsymbol{v}^B + \boldsymbol{\omega}^B \times (\boldsymbol{c} - \boldsymbol{b}) = \boldsymbol{v}^C,$$

and if we now subtract each of these formulae from the equality (16.16), we obtain

$$\boldsymbol{x} \times (\boldsymbol{r} - \boldsymbol{b}) = \boldsymbol{0},$$
$$\boldsymbol{x} \times (\boldsymbol{r} - \boldsymbol{c}) = \boldsymbol{0},$$

where $\boldsymbol{x} = \boldsymbol{\omega}^B - \boldsymbol{\omega}^C$. Now let P be any particle of the body not collinear with B and C and suppose that $\boldsymbol{x} \neq \boldsymbol{0}$. Then $\boldsymbol{x}$ must be parallel to both of the vectors $\boldsymbol{r} - \boldsymbol{b}$ and $\boldsymbol{r} - \boldsymbol{c}$, which are *not* parallel to each other. This is impossible and so $\boldsymbol{x} = \boldsymbol{0}$, which means that $\boldsymbol{\omega}^B = \boldsymbol{\omega}^C$. Hence the angular velocity of the body is the same, irrespective of the choice of reference particle. ■

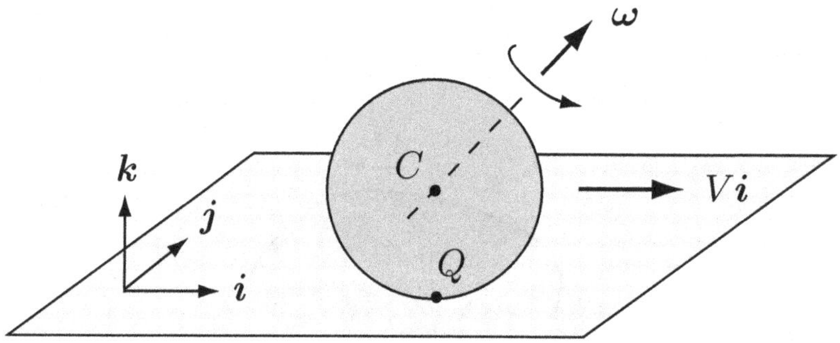

FIGURE 16.3 The ball rolls with velocity $V\boldsymbol{i}$ and has angular velocity $\boldsymbol{\omega}$.

Our results are summarised below:

> **Particle velocities in a rigid body**
>
> Suppose a rigid body is in general motion and that B is one of its particles. Then the velocity of any particle P of the body is given by the formula
>
> $$\boldsymbol{v} = \boldsymbol{v}^B + \boldsymbol{\omega}\times(\boldsymbol{r} - \boldsymbol{b}), \tag{16.17}$$
>
> where $\boldsymbol{v}^B$ is the velocity of the reference particle B and the angular velocity $\boldsymbol{\omega}$ is independent of the choice of reference particle.

Example 16.3 *Rolling snooker ball*

A rigid ball of radius b rolls without slipping on a horizontal table. Find the most general form of $\boldsymbol{\omega}$ consistent with the rolling condition.

Solution

Suppose the ball is rolling with velocity $V\boldsymbol{i}$ (where $\boldsymbol{i}$ is a horizontal unit vector), and has an unknown angular velocity $\boldsymbol{\omega}$. Then, on taking the centre C of the ball as the reference particle, the velocity of any particle P is given by (16.15) to be

$$\boldsymbol{v} = V\boldsymbol{i} + \boldsymbol{\omega}\times(\boldsymbol{r} - \boldsymbol{c}).$$

In particular, the velocity of the contact particle Q is given by

$$\boldsymbol{v}^Q = V\boldsymbol{i} + \boldsymbol{\omega}\times(-b\boldsymbol{k}),$$

where the unit vector $\boldsymbol{k}$ points vertically upwards. Since the rolling condition requires that $\boldsymbol{v}^Q = \boldsymbol{0}$, it follows that $\boldsymbol{\omega}$ must satisfy the condition

$$V\boldsymbol{i} + b\,\boldsymbol{k}\times\boldsymbol{\omega} = \boldsymbol{0}. \tag{16.18}$$

On taking the cross product of this equation with $\boldsymbol{k}$, we obtain

$$V\boldsymbol{k}\times\boldsymbol{i} + b\,\boldsymbol{k}\times(\boldsymbol{k}\times\boldsymbol{\omega}) = \boldsymbol{0},$$

that is

$$V\boldsymbol{j} + b\big((\boldsymbol{\omega}\cdot\boldsymbol{k})\boldsymbol{k} - (\boldsymbol{k}\cdot\boldsymbol{k})\boldsymbol{\omega}\big) = \boldsymbol{0}.$$

Since $\boldsymbol{k}$ is a unit vector, $\boldsymbol{k}\cdot\boldsymbol{k} = 1$ and we obtain

$$\boldsymbol{\omega} = \frac{V}{b}\,\boldsymbol{j} + (\boldsymbol{\omega}\cdot\boldsymbol{k})\boldsymbol{k}.$$

It follows that any $\boldsymbol{\omega}$ consistent with the rolling condition must have the form

$$\boldsymbol{\omega} = \frac{V}{b}\,\boldsymbol{j} + \lambda\boldsymbol{k}, \tag{16.19}$$

where λ is a scalar function of the time. Conversely, it is easy to verify that the formula (16.19) for $\boldsymbol{\omega}$ satisfies the rolling condition (16.18) for *any* choice of the scalar λ. This is therefore the most **general form** of $\boldsymbol{\omega}$ consistent with rolling.

This result is surprising at first. If the motion were planar, the value of ω would be V/b, the corresponding $\boldsymbol{\omega}$ being $(V/b)\boldsymbol{j}$. This is the special case $\lambda = 0$. But in general three dimensional rolling, $\lambda \neq 0$ and the rotation axis is not horizontal. This effect is well known to pool and snooker players and is achieved by striking the ball to the right (or left) of centre, thereby giving λ a positive (or negative) value. Players call this putting 'side' on the ball. It makes no difference to the rolling but affects the bounce when the ball strikes a cushion. ■

Time to relax Find a pool table and experiment by striking a ball slowly but firmly well to the right of centre. The marking on the ball should enable you to 'see' the rotation axis (a ball with spots is best). Check that giving the ball 'right hand side' produces a positive value of $\boldsymbol{\omega}\cdot\boldsymbol{k}$.

Example 16.4 *Wheel rolling around a circular path*

A circular wheel of radius b has its plane vertical and rolls with constant speed V around a circular path of radius R marked on a horizontal floor. Find the angular velocity of the wheel and the acceleration of the contact particle.

Solution

Suppose we view the motion of the wheel from the *rotating* reference frame $\{O;\ \boldsymbol{i}, \boldsymbol{j}, \boldsymbol{k}\}$ shown in Figure 16.4. This frame rotates about the axis $\{O, \boldsymbol{k}\}$ with angular velocity $\boldsymbol{\Omega}$ given by

$$\boldsymbol{\Omega} = \dot{\theta}\boldsymbol{k} = \frac{V}{R}\,\boldsymbol{k}. \tag{16.20}$$

Viewed from the rotating frame, the wheel is rotating about a *fixed* axis parallel to the vector $\boldsymbol{i}$. On applying the rolling condition, the angular velocity $\boldsymbol{\omega}'$ of the wheel, viewed from the rotating frame, is given by

$$\boldsymbol{\omega}' = -\frac{V}{b}\,\boldsymbol{i}. \tag{16.21}$$

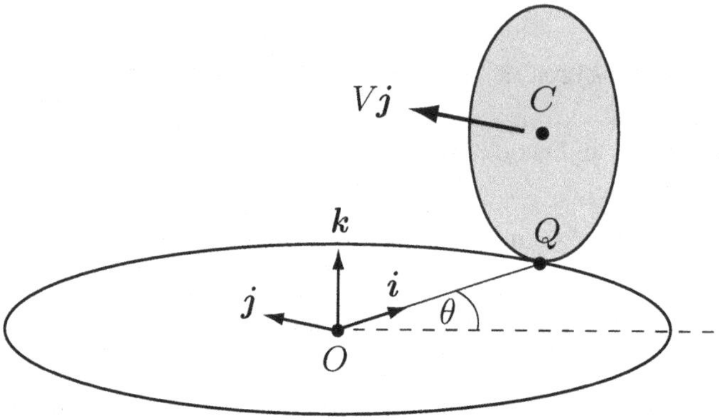

FIGURE 16.4 A wheel of radius b rolls around a circle of radius R. The unit vectors $\boldsymbol{i}(t)$ and $\boldsymbol{j}(t)$ follow the wheel as it moves around the circle; $\boldsymbol{k}$ is a constant unit vector.

The true angular velocity $\boldsymbol{\omega}$ of the wheel is then given by the sum

$$\boldsymbol{\omega} = \boldsymbol{\omega}' + \boldsymbol{\Omega}. \tag{16.22}$$

Here we are using a result from Chapter 17, the *angular velocity addition theorem*; it is the rotational counterpart of the addition theorem for linear velocities that we obtained in Chapter 2 . Although we have yet to prove this result, it is clear enough what it means and we will use it anyway! On substituting 16.20) and (16.21) into (16.22) we find the **angular velocity** of the wheel to be

$$\boldsymbol{\omega} = -\frac{V}{b}\boldsymbol{i} + \frac{V}{R}\boldsymbol{k}. \tag{16.23}$$

Now for the particle accelerations. These can be found by differentiating the velocity formula

$$\boldsymbol{v} = V\boldsymbol{j} + \boldsymbol{\omega}\times(\boldsymbol{r} - \boldsymbol{c}), \tag{16.24}$$

with respect to t, where we have taken C, the centre of the wheel, as the reference particle. This gives

$$\begin{aligned}\boldsymbol{a} &= V\frac{d\boldsymbol{j}}{dt} + \dot{\boldsymbol{\omega}}\times(\boldsymbol{r} - \boldsymbol{c}) + \boldsymbol{\omega}\times(\dot{\boldsymbol{r}} - \dot{\boldsymbol{c}})\\ &= V\frac{d\boldsymbol{j}}{dt} + \dot{\boldsymbol{\omega}}\times(\boldsymbol{r} - \boldsymbol{c}) + \boldsymbol{\omega}\times(\boldsymbol{v} - V\boldsymbol{j}),\end{aligned}$$

since $\dot{\boldsymbol{r}} = \boldsymbol{v}$ and $\dot{\boldsymbol{c}} = V\boldsymbol{j}$. In particular, since $\boldsymbol{r}^Q = \boldsymbol{c} - b\boldsymbol{k}$ and $\boldsymbol{v}^Q = \boldsymbol{0}$, the acceleration of the contact particle Q is given by

$$\begin{aligned}\boldsymbol{a}^Q &= V\frac{d\boldsymbol{j}}{dt} + \dot{\boldsymbol{\omega}}\times(-b\boldsymbol{k}) + \boldsymbol{\omega}\times(-V\boldsymbol{j})\\ &= V\frac{d\boldsymbol{j}}{dt} + V\frac{d\boldsymbol{i}}{dt}\times\boldsymbol{k} + \frac{V^2}{b}\boldsymbol{k} + \frac{V^2}{R}\boldsymbol{i},\end{aligned} \tag{16.25}$$

on using the formula (16.23) to replace $\boldsymbol{\omega}$ and $\dot{\boldsymbol{\omega}}$.

The only unknown quantities left are $d\boldsymbol{i}/dt$ and $d\boldsymbol{j}/dt$. However, the vectors $\boldsymbol{i}$, $\boldsymbol{j}$ correspond precisely to the polar unit vectors $\widehat{\boldsymbol{r}}, \widehat{\boldsymbol{\theta}}$ treated in section 2.3, from which we deduce that

$$\frac{d\boldsymbol{i}}{dt} = \dot{\theta}\,\boldsymbol{j} = \frac{V}{R}\,\boldsymbol{j} \qquad \text{and} \qquad \frac{d\boldsymbol{j}}{dt} = -\dot{\theta}\,\boldsymbol{i} = -\frac{V}{R}\,\boldsymbol{i}.$$

On substituting these formulae into equation (16.25) we obtain

$$\boldsymbol{a}^Q = \frac{V^2}{R}\,\boldsymbol{i} + \frac{V^2}{b}\,\boldsymbol{k}$$

as the **acceleration** of the contact particle Q. ■

Problems on Chapter 16

Answers and comments are at the end of the book.

Harder problems carry a star (∗).

16.1 A rigid body is rotating in the right-handed sense about the axis Oz with a constant angular speed of 2 radians per second. Write down the angular velocity vector of the body, and find the instantaneous velocity, speed and acceleration of the particle of the body at the point $(4, -3, 7)$, where distances are measured in metres.

16.2 A rigid body is rotating with constant angular speed 3 radians per second about a fixed axis through the points $A(4, 1, 1)$, $B(2, -1, 0)$, distances being measured in centimetres. The rotation is in the left-handed sense relative to the direction $\overrightarrow{AB}$. Find the instantaneous velocity and acceleration of the particle P of the body at the point $(4, 4, 4)$.

16.3 A spinning top (a rigid body of revolution) is in general motion with its vertex (a particle on the axis of symmetry) fixed at the origin O. Let $\boldsymbol{a}(t)$ be the unit vector pointing along the axis of symmetry and let $\boldsymbol{\omega}(t)$ be the angular velocity of the top. (In general, $\boldsymbol{\omega}$ does *not* point along the axis of symmetry.) By considering the velocities of particles of the top that lie on the axis of symmetry, show that $\boldsymbol{a}$ satisfies the equation

$$\dot{\boldsymbol{a}} = \boldsymbol{\omega} \times \boldsymbol{a}.$$

Deduce that the most general form $\boldsymbol{\omega}$ can have is

$$\boldsymbol{\omega} = \boldsymbol{a} \times \dot{\boldsymbol{a}} + \lambda\,\boldsymbol{a},$$

where λ is a scalar function of the time. [This formula is needed in the theory of the spinning top.]

16.4 A penny of radius a rolls without slipping on a rough horizontal table. The penny rolls in such a way that its centre G remains fixed (see Figure 16.5). The plane of the penny makes a constant angle α with the table and the point of contact C traces out a circle with centre O and

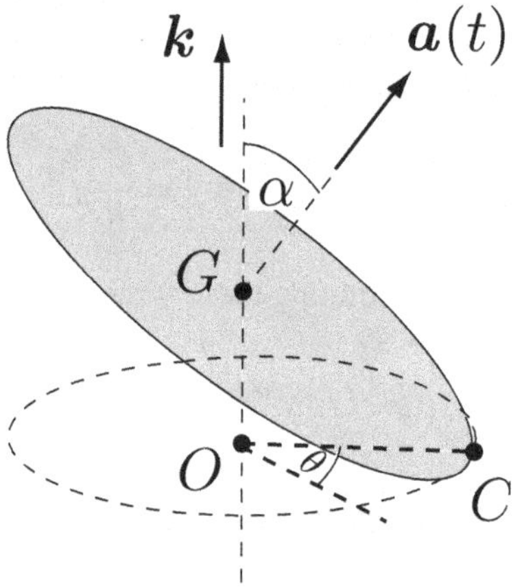

FIGURE 16.5 A penny of radius b rolls on a rough horizontal table in such a way that its centre G remains fixed.

radius $a\cos\alpha$, as shown. At time t, the angle between the radius OC and some fixed radius is θ. Find the angular velocity vector of the penny in terms of the unit vectors $\boldsymbol{a}(t)$, $\boldsymbol{k}$ shown.

Find the velocity of the highest particle of the penny.

16.5 A rigid circular cone with altitude h and semi-angle α rolls without slipping on a rough horizontal table. Explain why the vertex O of the cone never moves. Let $\theta(t)$ be the angle between OC, the line of the cone that is in contact with the table, and some fixed horizontal reference line OA. Show that the angular velocity $\boldsymbol{\omega}$ of the cone is given by

$$\boldsymbol{\omega} = -\left(\dot{\theta}\cot\alpha\right)\boldsymbol{i},$$

where $\boldsymbol{i}(t)$ is the unit vector pointing in the direction $\overrightarrow{OC}$. [First identify the *direction* of $\boldsymbol{\omega}$, and then consider the velocities of those particles of the cone that lie on the axis of symmetry.]

Identify the particle of the cone that has the maximum speed and find this speed.

16.6* Two rigid plastic panels lie in the planes $z = -b$ and $z = b$ respectively. A rigid ball of radius b can move in the space between the panels and is gripped by them so that it does not slip. The panels are made to rotate with angular velocities $\omega_1\boldsymbol{k}$, $\omega_2\boldsymbol{k}$ about fixed vertical axes that are a distance $2c$ apart. Show that, *with a suitable choice of origin*, the position vector $\boldsymbol{R}$ of the centre of the ball satisfies the equation

$$\dot{\boldsymbol{R}} = \boldsymbol{\Omega}\times\boldsymbol{R},$$

where $\boldsymbol{\Omega} = \frac{1}{2}(\boldsymbol{\omega}_1 + \boldsymbol{\omega}_2)$. Deduce that the ball must move in a circle and find the position of the centre of this circle. [This arrangement is sometimes seen as a shop window display. The panels are transparent and the ball seems to be executing a circle in mid-air.]

16.7 Two hollow spheres have radii a and b ($b > a$), and their common centre O is fixed. A rigid ball of radius $\frac{1}{2}(b-a)$ can move in the annular space between the spheres and is gripped by them so that it does not slip. The spheres are made to rotate with constant angular velocities $\boldsymbol{\omega}_1$, $\boldsymbol{\omega}_2$ respectively. Show that the ball must move in a circle whose plane is perpendicular to the vector $a\,\boldsymbol{\omega}_1 + b\,\boldsymbol{\omega}_2$.

Chapter Seventeen

Rotating reference frames

KEY FEATURES

The key features of this chapter are the **transformation** of **velocity** and **acceleration** between frames in general relative motion, and the dynamical **effects of the Earth's rotation**.

So far we have viewed the motion of mechanical systems from an inertial reference frame. The reason for this is simple; the Second Law, in its standard form, applies only in inertial frames. However, circumstances arise in which it is convenient to view the motion from a **non-inertial frame**. The most important instance of this occurs when the motion takes place near the surface of the Earth. Previously we have argued that the dynamical effects of the Earth's rotation are small enough to be neglected. While this is usually true, there are circumstances in which it has a significant effect. In long range artillery, the Earth's rotation gives rise to an important correction, and, in the hydrodynamics of the atmosphere and oceans, the Earth's rotation can have a dominant effect. If we wish to calculate such effects (as seen by an observer on the Earth), we must take our reference frame fixed to the Earth, thus making it a *non-inertial* frame. The downside of this choice is that the Second Law does not hold and must be replaced by a considerably more complicated equation.

In addition to applications involving the Earth's rotation, there are instances where the motion of a system looks much simpler when viewed from a suitably chosen rotating frame. The **Larmor precession** of a charged particle moving in a uniform magnetic field is one example; a second is the motion of a rigid body relative to its own principal axes of inertia, which leads to **Euler's equations**.

17.1 TRANSFORMATION FORMULAE

In this section we derive the transformation formulae that link the velocity and acceleration of a particle measured in a **moving frame**, with the same quantities measured in a **fixed frame**. For the purposes of *kinematics*, the labels 'fixed' and 'moving' are arbitrary. Each frame is moving relative to the other and the labels could be reversed; we use them purely for convenience. In *dynamics* however the distinction between the fixed and moving frames is real. The 'fixed' frame is an inertial frame, in which Newton's laws

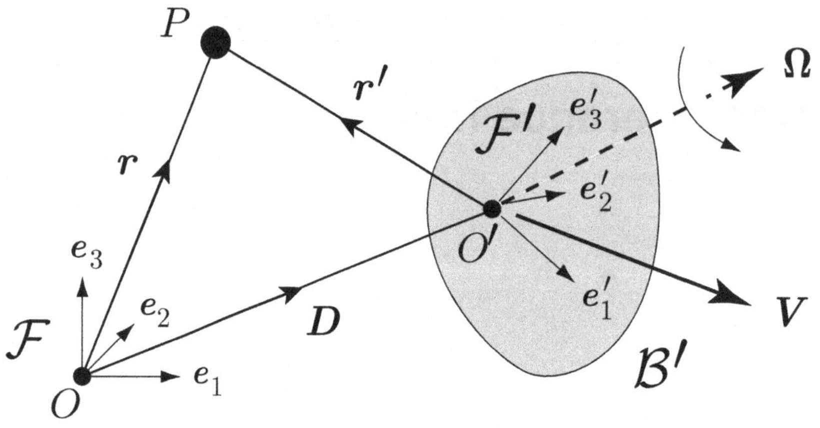

FIGURE 17.1 The moving frame $\mathcal{F}' \equiv \{O'; e_1', e_2', e_3'\}$ has translational velocity V and angular velocity $\boldsymbol{\Omega}$ relative to the fixed frame $\mathcal{F} \equiv \{O; e_1, e_2, e_3\}$.

apply, and the moving frame is a non-inertial frame, in which Newton's laws do not apply, at least in their standard form.

Let $\mathcal{F} \equiv \{O; e_1, e_2, e_3\}$ be the fixed frame and $\mathcal{F}' \equiv \{O'; e_1', e_2', e_3'\}$ be the moving frame, as shown in Figure 17.1. At time t, the frame $\mathcal{F}'$ has translational velocity* V and angular velocity $\boldsymbol{\Omega}$ relative to the frame $\mathcal{F}$. It is convenient to regard the moving reference frame $\mathcal{F}'$ as being embedded in a rigid body $\mathcal{B}'$ with reference particle O'. Then V and $\boldsymbol{\Omega}$ are the translational and angular velocities of the body $\mathcal{B}'$. The most important feature of reference frames in relative motion is this:

The rate of change of a ***vector*** *quantity measured in the frame $\mathcal{F}$ is generally not the same as the rate of change of the same quantity measured in the frame $\mathcal{F}'$.*

Question ***Why are there different rates of change?***

Suppose $u(t)$ is a vector quantity. Why should it have different rates of change when measured in the frames $\mathcal{F}$ and $\mathcal{F}'$?

Answer

Suppose the expression for u in terms of the basis set $\{e_1, e_2, e_3\}$ is

$$u = u_1 e_1 + u_2 e_2 + u_3 e_3, \tag{17.1}$$

and in terms of the basis set $\{e_1', e_2', e_3'\}$ is

$$u = u_1' e_1' + u_2' e_2' + u_3' e_3'. \tag{17.2}$$

In general, the components $\{u_1, u_2, u_3\}$ and $\{u_1', u_2', u_3'\}$ will be functions of the time t.

* This means that the origin O' has velocity V relative to $\mathcal{F}$. The angular velocity $\boldsymbol{\Omega}$ is independent of the choice of O'.

In the rate of change of $\boldsymbol{u}$ measured in $\mathcal{F}$, the **basis set** $\{\boldsymbol{e}_1, \boldsymbol{e}_2, \boldsymbol{e}_3\}$ is, by definition, **constant** so that

$$\left(\frac{d\boldsymbol{u}}{dt}\right)_{\mathcal{F}} = \dot{u}_1\,\boldsymbol{e}_1 + \dot{u}_2\,\boldsymbol{e}_2 + \dot{u}_1\,\boldsymbol{e}_3. \tag{17.3}$$

In contrast, in the rate of change of $\boldsymbol{u}$ measured in $\mathcal{F}'$, the **basis set** $\{\boldsymbol{e}'_1, \boldsymbol{e}'_2, \boldsymbol{e}'_3\}$ is, by definition, **constant** so that

$$\left(\frac{d\boldsymbol{u}}{dt}\right)_{\mathcal{F}'} = \dot{u}'_1\,\boldsymbol{e}'_1 + \dot{u}'_2\,\boldsymbol{e}'_2 + \dot{u}'_3\,\boldsymbol{e}'_3. \tag{17.4}$$

Note that the components $\{u_k\}$ and $\{u'_k\}$ are *scalar* functions of the time and so their rates of change are *independent of the reference frame*. This is why we do not need to label them as being observed in $\mathcal{F}$ or $\mathcal{F}'$. There is no reason why the two expressions (17.3) and (17.4) should be equal and, in general, they are *not* equal. Consider, for example, the case in which $\boldsymbol{u}$ is constant in $\mathcal{F}'$ so that $(d\boldsymbol{u}/dt)_{\mathcal{F}'} = \boldsymbol{0}$. However, the motion of $\mathcal{F}'$ relative to $\mathcal{F}$ means that $\boldsymbol{u}$ will not be constant in $\mathcal{F}$ and $(d\boldsymbol{u}/dt)_{\mathcal{F}}$ will not be zero. ∎

True and apparent values

In order to simplify the writing, we will, from now on, refer to the value of a quantity measured in the fixed frame $\mathcal{F}$ as its **true** value, and the value measured in the moving frame $\mathcal{F}'$ as its **apparent** value. For example, $(d\boldsymbol{u}/dt)_{\mathcal{F}}$ will be referred to the *true* value of $d\boldsymbol{u}/dt$, while $(d\boldsymbol{u}/dt)_{\mathcal{F}'}$ will be referred to as its *apparent* value.

Rates of change of the basis vectors $\{\boldsymbol{e}'_1, \boldsymbol{e}'_2, \boldsymbol{e}'_3\}$

Our first step is to find the rates of change of the fundamental basis vectors $\{\boldsymbol{e}'_1, \boldsymbol{e}'_2, \boldsymbol{e}'_3\}$ belonging to the frame $\mathcal{F}'$. Since these vectors are, by definition, constant in $\mathcal{F}'$, their *apparent* rates of change are zero. What we need to find are their *true* rates of change.

Let E be any particle of the body $\mathcal{B}'$ in which the frame $\mathcal{F}'$ is embedded, and let $\boldsymbol{e}$ and $\boldsymbol{e}'$ be the position vectors of E relative to O and O' respectively. Then, by the triangle law,

$$\boldsymbol{e} = \boldsymbol{D} + \boldsymbol{e}',$$

where $\boldsymbol{D}$ is the position vector of O' relative to O. Then

$$\left(\frac{d\boldsymbol{e}}{dt}\right)_{\mathcal{F}} = \left(\frac{d\boldsymbol{D}}{dt}\right)_{\mathcal{F}} + \left(\frac{d\boldsymbol{e}'}{dt}\right)_{\mathcal{F}} \tag{17.5}$$

$$= \boldsymbol{V} + \left(\frac{d\boldsymbol{e}'}{dt}\right)_{\mathcal{F}}, \tag{17.6}$$

where $\boldsymbol{V}$ is the true velocity of O' relative to O. But $(d\boldsymbol{e}/dt)_{\mathcal{F}}$ is, by definition, $\boldsymbol{v}^E$, the true velocity of the particle E, and, since E is a particle of the rigid body $\mathcal{B}'$, $\boldsymbol{v}^E$ is given by

$$\boldsymbol{v}^E = \boldsymbol{V} + \boldsymbol{\Omega}\times\boldsymbol{e}', \tag{17.7}$$

on using the kinematical formula (16.17). On comparing the equations (17.6) and (17.7), we see that

$$\left(\frac{d\boldsymbol{e}'}{dt}\right)_{\mathcal{F}} = \boldsymbol{\Omega}\times\boldsymbol{e}'.$$

This result applies to *any* vector $\boldsymbol{e}'$ that is the position vector (relative to O') of a particle of the rigid body $\mathcal{B}'$. In particular, since the basis vectors $\{\boldsymbol{e}'_1, \boldsymbol{e}'_2, \boldsymbol{e}'_3\}$ can be regarded as the position vectors of particles of $\mathcal{B}'$, we obtain the **fundamental relations**

$$\left(\frac{d\boldsymbol{e}'_j}{dt}\right)_{\mathcal{F}} = \boldsymbol{\Omega}\times\boldsymbol{e}'_j \qquad (1 \le j \le 3).$$

The notation $(d\boldsymbol{e}'_j/dt)_{\mathcal{F}}$ for the true rate of change of the vector $\boldsymbol{e}'_j$ is accurate but cumbersome. Since the apparent rate of change of these vectors is zero, there is little chance of confusion if, from now on, we replace $(d\boldsymbol{e}'_j/dt)_{\mathcal{F}}$ by the simple notation $\dot{\boldsymbol{e}}'_j$. Our result can then be expressed in the form:

True rates of change of the basis vectors $\{\boldsymbol{e}'_1, \boldsymbol{e}'_2, \boldsymbol{e}'_3\}$

$$\dot{\boldsymbol{e}}'_1 = \boldsymbol{\Omega}\times\boldsymbol{e}'_1 \qquad \dot{\boldsymbol{e}}'_2 = \boldsymbol{\Omega}\times\boldsymbol{e}'_2 \qquad \dot{\boldsymbol{e}}'_3 = \boldsymbol{\Omega}\times\boldsymbol{e}'_3 \qquad (17.8)$$

where $\boldsymbol{\Omega}$ is the angular velocity of the frame $\mathcal{F}'$ relative to the frame $\mathcal{F}$.

Relation between the true and apparent values of $d\boldsymbol{u}/dt$

Our next step is to find the relationship between the true and apparent values of $d\boldsymbol{u}/dt$, where $\boldsymbol{u}$ is any vector function of the time. To do this we differentiate the representation (17.2) with respect to t *while keeping the basis set* $\{\boldsymbol{e}_1, \boldsymbol{e}_2, \boldsymbol{e}_3\}$ *constant*. This gives

$$\begin{aligned}
\left(\frac{d\boldsymbol{u}}{dt}\right)_{\mathcal{F}} &= \left(\frac{d(u'_1\boldsymbol{e}'_1)}{dt}\right)_{\mathcal{F}} + \left(\frac{d(u'_2\boldsymbol{e}'_2)}{dt}\right)_{\mathcal{F}} + \left(\frac{d(u'_3\boldsymbol{e}'_3)}{dt}\right)_{\mathcal{F}}\\
&= \left(\dot{u}'_1\boldsymbol{e}'_1 + \dot{u}'_2\boldsymbol{e}'_2 + \dot{u}'_3\boldsymbol{e}'_3\right) + u'_1\dot{\boldsymbol{e}}'_1 + u'_2\dot{\boldsymbol{e}}'_2 + u'_3\dot{\boldsymbol{e}}'_3\\
&= \left(\frac{d\boldsymbol{u}}{dt}\right)_{\mathcal{F}'} + u'_1\left(\boldsymbol{\Omega}\times\boldsymbol{e}'_1\right) + u'_2\left(\boldsymbol{\Omega}\times\boldsymbol{e}'_2\right) + u'_3\left(\boldsymbol{\Omega}\times\boldsymbol{e}'_3\right)\\
&= \left(\frac{d\boldsymbol{u}}{dt}\right)_{\mathcal{F}'} + \boldsymbol{\Omega}\times\left(u'_1\boldsymbol{e}'_1 + u'_2\boldsymbol{e}'_2 + u'_3\boldsymbol{e}'_3\right)\\
&= \left(\frac{d\boldsymbol{u}}{dt}\right)_{\mathcal{F}'} + \boldsymbol{\Omega}\times\boldsymbol{u}.
\end{aligned}$$

The true and apparent values of $d\boldsymbol{u}/dt$ are therefore related by the formula:

Transformation formula for du/dt

$$\left(\frac{d\boldsymbol{u}}{dt}\right)_{\mathcal{F}} = \left(\frac{d\boldsymbol{u}}{dt}\right)_{\mathcal{F}'} + \boldsymbol{\Omega}\times\boldsymbol{u} \tag{17.9}$$

where $\boldsymbol{\Omega}$ is the angular velocity of the frame $\mathcal{F}'$ relative to the frame $\mathcal{F}$.

Velocity transformation formula

It is now easy to find how particle velocities transform between the two frames. Suppose a particle P has position vector $\boldsymbol{r}$ relative to $\mathcal{F}$ and position vector $\boldsymbol{r}'$ relative to $\mathcal{F}'$. It follows from the triangle law that

$$\boldsymbol{r} = \boldsymbol{D} + \boldsymbol{r}', \tag{17.10}$$

where $\boldsymbol{D}$ is the position vector of O' relative to O. Then

$$\left(\frac{d\boldsymbol{r}}{dt}\right)_{\mathcal{F}} = \left(\frac{d\boldsymbol{D}}{dt}\right)_{\mathcal{F}} + \left(\frac{d\boldsymbol{r}'}{dt}\right)_{\mathcal{F}},$$

that is,

$$\boldsymbol{v} = \boldsymbol{V} + \left(\frac{d\boldsymbol{r}'}{dt}\right)_{\mathcal{F}},$$

where $\boldsymbol{v}$ is the true velocity of P. From the transformation formula (17.9), it follows that

$$\begin{aligned}\left(\frac{d\boldsymbol{r}'}{dt}\right)_{\mathcal{F}} &= \left(\frac{d\boldsymbol{r}'}{dt}\right)_{\mathcal{F}'} + \boldsymbol{\Omega}\times\boldsymbol{r}'\\ &= \boldsymbol{v}' + \boldsymbol{\Omega}\times\boldsymbol{r}',\end{aligned} \tag{17.11}$$

where $\boldsymbol{v}'$ is the apparent velocity of P. On combining these results, we obtain:

Velocity transformation formula

$$\boldsymbol{v} = \boldsymbol{V} + \boldsymbol{\Omega}\times\boldsymbol{r}' + \boldsymbol{v}' \tag{17.12}$$

This is the **velocity transformation formula** connecting the true and apparent velocities of the particle P.

This formula has a simple interpretation. The true velocity $\boldsymbol{v}$ is the sum of (i) the apparent velocity $\boldsymbol{v}'$, and (ii) the velocity $\boldsymbol{V} + \boldsymbol{\Omega}\times\boldsymbol{r}'$ that is given to the particle by the motion of the frame $\mathcal{F}'$.

An immediate consequence of the velocity transformation formula is the corresponding transformation rule for **angular velocities.**

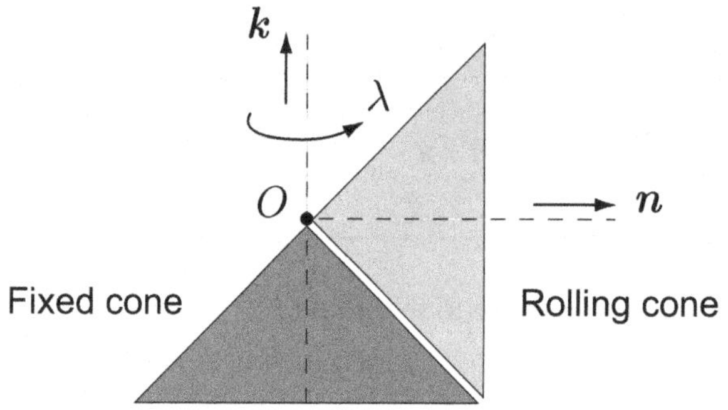

FIGURE 17.2 The dark shaded cone is fixed and the light shaded cone rolls around it. The vertex of each cone is at rest at O.

Theorem 17.1 ***Addition theorem for angular velocities*** *Suppose a rigid body in general motion has true angular velocity $\boldsymbol{\omega}$ and apparent angular velocity $\boldsymbol{\omega}'$. Then $\boldsymbol{\omega}$ and $\boldsymbol{\omega}'$ are related by the addition formula*

$$\boldsymbol{\omega} = \boldsymbol{\Omega} + \boldsymbol{\omega}', \tag{17.13}$$

where $\boldsymbol{\Omega}$ is the angular velocity of the frame $\mathcal{F}'$ relative to $\mathcal{F}$. [This result is the counterpart of the velocity transformation formula $\boldsymbol{v} = \boldsymbol{V} + \boldsymbol{v}'$ for *translating* frames.]

The proof of this theorem is the subject of Problem 17.2. We have already used this result in Chapter 16, Example 16.4, and we give a second example now.

Example 17.1 *Cone rolling on another cone*

A circular cone with semi-angle 45° is fixed with its axis of symmetry vertical and its vertex O upwards. An identical cone also has its vertex fixed at O and rolls on the first cone so that its axis of symmetry precesses around the upward vertical with angular speed λ as shown. Find the angular speed of the rolling cone.

Solution

Let $\mathcal{F}$ be a fixed frame and $\mathcal{F}'$ be a frame with origin at O and precessing with the rolling cone. Then $\mathcal{F}'$ has angular velocity $\lambda\,\boldsymbol{k}$ relative to $\mathcal{F}$, where the unit vector $\boldsymbol{k}$ points vertically upwards. In the frame $\mathcal{F}'$ the second cone has its axis of symmetry fixed and the first cone has angular velocity $-\lambda\,\boldsymbol{k}$. The rolling contact between the two cones means that the second cone must be rotating about its axis of symmetry $\{O, \boldsymbol{n}\}$ with angular speed λ in the negative sense. Its angular velocity is therefore $-\lambda\,\boldsymbol{n}$. In the notation of Theorem 17.1 we therefore have $\boldsymbol{\Omega} = \lambda\,\boldsymbol{k}$ and $\boldsymbol{\omega}' = -\lambda\,\boldsymbol{n}$. It follows that the **angular velocity** of the rolling cone is

$$\boldsymbol{\omega} = \boldsymbol{\Omega} + \boldsymbol{\omega}' = \lambda\,\boldsymbol{k} - \lambda\,\boldsymbol{n}.$$

The **angular speed** of the rolling cone is therefore $\sqrt{2}\,\lambda$. ■

Acceleration transformation formula

From the velocity transformation formula (17.12), it follows that

$$\left(\frac{d\boldsymbol{v}}{dt}\right)_{\mathcal{F}} = \left(\frac{d\boldsymbol{V}}{dt}\right)_{\mathcal{F}} + \left(\frac{d\boldsymbol{\Omega}}{dt}\right)_{\mathcal{F}} \times \boldsymbol{r}' + \boldsymbol{\Omega} \times \left(\frac{d\boldsymbol{r}'}{dt}\right)_{\mathcal{F}} + \left(\frac{d\boldsymbol{v}'}{dt}\right)_{\mathcal{F}},$$

that is,

$$\boldsymbol{a} = \boldsymbol{A} + \dot{\boldsymbol{\Omega}} \times \boldsymbol{r}' + \boldsymbol{\Omega} \times \left(\frac{d\boldsymbol{r}'}{dt}\right)_{\mathcal{F}} + \left(\frac{d\boldsymbol{v}'}{dt}\right)_{\mathcal{F}}, \tag{17.14}$$

where $\boldsymbol{a}$ is the true acceleration of the particle P, and $\boldsymbol{A}$, $\boldsymbol{\Omega}$ and $\dot{\boldsymbol{\Omega}}$ are the translational acceleration, the angular velocity, and the angular acceleration of the frame $\mathcal{F}'$, all relative to the frame $\mathcal{F}$.

The derivative $(d\boldsymbol{r}'/dt)_{\mathcal{F}}$ has already been evaluated in (17.11), and the derivative $(d\boldsymbol{v}'/dt)_{\mathcal{F}}$ is given by the transformation formula (17.9) to be

$$\begin{aligned}\left(\frac{d\boldsymbol{v}'}{dt}\right)_{\mathcal{F}} &= \left(\frac{d\boldsymbol{v}'}{dt}\right)_{\mathcal{F}'} + \boldsymbol{\Omega} \times \boldsymbol{v}', \\ &= \boldsymbol{a}' + \boldsymbol{\Omega} \times \boldsymbol{v}',\end{aligned} \tag{17.15}$$

where $\boldsymbol{a}'$ is the apparent acceleration of P. On combining these results, we obtain

$$\boldsymbol{a} = \boldsymbol{A} + \dot{\boldsymbol{\Omega}} \times \boldsymbol{r}' + \boldsymbol{\Omega} \times (\boldsymbol{v}' + \boldsymbol{\Omega} \times \boldsymbol{r}') + (\boldsymbol{a}' + \boldsymbol{\Omega} \times \boldsymbol{v}')$$

which, on simplification, gives the **acceleration transformation formula**:

Acceleration transformation formula

$$\boldsymbol{a} = \boldsymbol{A} + \dot{\boldsymbol{\Omega}} \times \boldsymbol{r}' + 2\,\boldsymbol{\Omega} \times \boldsymbol{v}' + \boldsymbol{\Omega} \times (\boldsymbol{\Omega} \times \boldsymbol{r}') + \boldsymbol{a}' \tag{17.16}$$

where $\boldsymbol{A}$ is the translational acceleration, $\boldsymbol{\Omega}$ is the angular velocity and $\dot{\boldsymbol{\Omega}}$ is the angular acceleration of the frame $\mathcal{F}'$, all relative to the frame $\mathcal{F}$.

The acceleration formula does not have a simple interpretation. The apparent acceleration is $\boldsymbol{a}'$ and the acceleration given to the particle by the motion of the frame $\mathcal{F}'$ is $\boldsymbol{A} + \dot{\boldsymbol{\Omega}} \times \boldsymbol{r}' + \boldsymbol{\Omega} \times (\boldsymbol{\Omega} \times \boldsymbol{r}')$. However, the true acceleration is *not* the sum of these two contributions! There is the additional term $2\,\boldsymbol{\Omega} \times \boldsymbol{v}'$ that appears *only when the particle is moving within the moving frame.* This **Coriolis*** term, as it is called, is hard to explain physically.

* After Gaspard-Gustave de Coriolis (1792–1843) whose paper on the subject was published in 1835. He also introduced the notion of 'work' with its present scientific meaning.

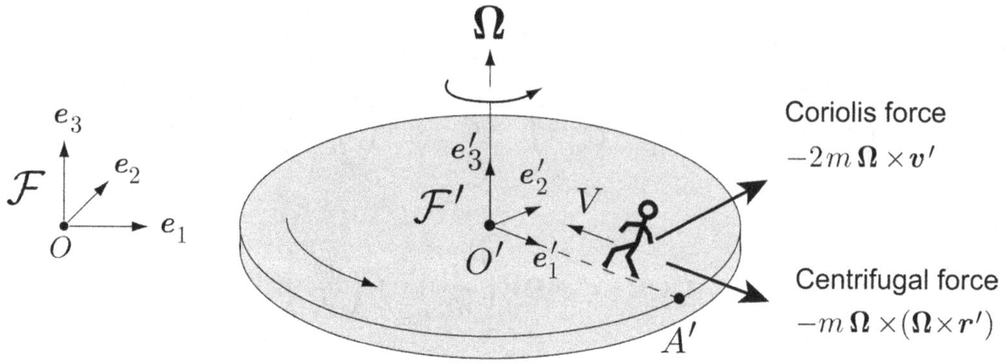

FIGURE 17.3 Man walking on a rotating roundabout.

17.2 PARTICLE DYNAMICS IN A NON-INERTIAL FRAME

Suppose now that $\mathcal{F}$ is an **inertial frame**. Then the equation of motion of a particle P of mass m moving under the force $\boldsymbol{F}$ is simply given by the Second Law

$$m\boldsymbol{a} = \boldsymbol{F}, \tag{17.17}$$

where $\boldsymbol{a}$ is the true acceleration of P. It follows from the acceleration transformation formula (17.16), that, when the *same motion* is observed from the frame $\mathcal{F}'$, equation (17.17) is replaced by the **transformed equation of motion**

Second Law in a general non-inertial frame

$$m\left[\boldsymbol{A} + \dot{\boldsymbol{\Omega}}\times\boldsymbol{r}' + 2\,\boldsymbol{\Omega}\times\boldsymbol{v}' + \boldsymbol{\Omega}\times(\boldsymbol{\Omega}\times\boldsymbol{r}') + \boldsymbol{a}'\right] = \boldsymbol{F}. \tag{17.18}$$

Here $\boldsymbol{r}'$, $\boldsymbol{v}'$, $\boldsymbol{a}'$ are the apparent position, velocity and acceleration of P, and $\boldsymbol{A}$, $\boldsymbol{\Omega}$ and $\dot{\boldsymbol{\Omega}}$ are the translational acceleration, angular velocity and angular acceleration of the frame $\mathcal{F}'$ relative to the frame $\mathcal{F}$.

This is the form taken by the Second Law in a general **non-inertial frame**. Note that there is *no new physics* in this equation. It is simply the Second Law written in terms of the apparent quantities observed in the non-inertial frame $\mathcal{F}'$.

Example 17.2 *Man walking on a rotating roundabout*

A fairground roundabout is made to rotate with constant angular speed Ω about its axis of symmetry which is fixed in a vertical position, the sense of the rotation being as shown in Figure 17.3. A man is on the roundabout and is walking grimly towards the centre O' along the moving radius $A'O'$ with constant speed V. What is the force that the roundabout exerts on the man?

Solution

Let the frame $\mathcal{F} \equiv \{O; \boldsymbol{e}_1, \boldsymbol{e}_2, \boldsymbol{e}_3\}$ be an inertial frame of reference attached to the ground and the frame $\mathcal{F}' \equiv \{O'; \boldsymbol{e}_1', \boldsymbol{e}_2', \boldsymbol{e}_3'\}$ be attached to the roundabout, as shown in Figure 17.3. Then $\mathcal{F}'$ has zero translational velocity and constant angular velocity $\boldsymbol{\Omega}\ (= \Omega\,\boldsymbol{e}_3)$ relative to $\mathcal{F}$. Hence $\boldsymbol{A} = \boldsymbol{0}$ and $\dot{\boldsymbol{\Omega}} = \boldsymbol{0}$ in this problem.

In the frame $\mathcal{F}'$, the motion of the man is uniform and rectilinear so that $\boldsymbol{r}' = x_1'\,\boldsymbol{e}_1'$, $\boldsymbol{v}' = -V\,\boldsymbol{e}_1'$ and $\boldsymbol{a}' = \boldsymbol{0}$. The transformed equation of motion (17.18) therefore reduces to

$$m\left[2\,(\Omega\,\boldsymbol{e}_3)\times(-V\,\boldsymbol{e}_1') + (\Omega\,\boldsymbol{e}_3)\times\big((\Omega\,\boldsymbol{e}_3)\times(x_1'\,\boldsymbol{e}_1')\big)\right] = -mg\,\boldsymbol{e}_3 + \boldsymbol{X},$$

where $\boldsymbol{X}$ is the force that the roundabout exerts on the man. Hence, $\boldsymbol{X}$ is given by

$$\boldsymbol{X} = (mg)\,\boldsymbol{e}_3 - (2m\Omega V)\,\boldsymbol{e}_2' - (m\Omega^2 x_1')\,\boldsymbol{e}_1'.$$

Because of the additional terms* in this expression, the man must lean forwards and to his left; otherwise he will fall over. ■

Fictitious forces

Equation (17.18) can be made to resemble the *standard* form of the Second Law by the simple device of transferring the four new terms on the left to the right side of the equation and regarding them as **fictitious forces** that act on P in addition to the real force $\boldsymbol{F}$. This gives

$$m\boldsymbol{a}' = \boldsymbol{F} + (-m\boldsymbol{A}) + (-m\,\dot{\boldsymbol{\Omega}}\times\boldsymbol{r}') + \underbrace{(-2m\,\boldsymbol{\Omega}\times\boldsymbol{v}')}_{\text{Coriolis force}} + \underbrace{(-m\,\boldsymbol{\Omega}\times(\boldsymbol{\Omega}\times\boldsymbol{r}'))}_{\text{centrifugal force}}, \tag{17.19}$$

which we will call the **fictitious force equation.**

In other words, the fact that the frame $\mathcal{F}'$ is non-inertial may be ignored provided that the four fictitious forces $-m\boldsymbol{A}$, $-m\dot{\boldsymbol{\Omega}}\times\boldsymbol{r}'$, $-2m\,\boldsymbol{\Omega}\times\boldsymbol{v}'$ and $-m\boldsymbol{\Omega}\times(\boldsymbol{\Omega}\times\boldsymbol{r}')$ are added to the real force $\boldsymbol{F}$. Two of these 'forces' have names. The 'force' $-2m\,\boldsymbol{\Omega}\times\boldsymbol{v}'$ is called the **Coriolis force** and the 'force' $-m\,\boldsymbol{\Omega}\times(\boldsymbol{\Omega}\times\boldsymbol{r}')$ is called the **centrifugal force.**† In the problem of the man on the rotating roundabout, only the Coriolis and centrifugal 'forces' are non-zero and their directions are shown in Figure 17.3.

The transformed equation of motion and the fictitious force equation are obviously equivalent and there is no logical need to introduce the notion of fictitious forces at all. However, the fictitious force approach is often useful when looking at a problem *qualitatively*. For example, in the problem of the man walking on the roundabout, the fictitious force description would be as follows:

---

* If the roundabout were at rest, then $\boldsymbol{X}$ would be simply $mg\,\boldsymbol{e}_3$.

† Although the name 'centrifugal force' for the expression $-m\,\boldsymbol{\Omega}\times(\boldsymbol{\Omega}\times\boldsymbol{r}')$ is standard in the literature, it is not really appropriate unless $O' = O$, that is, the two frames of reference have a common origin.

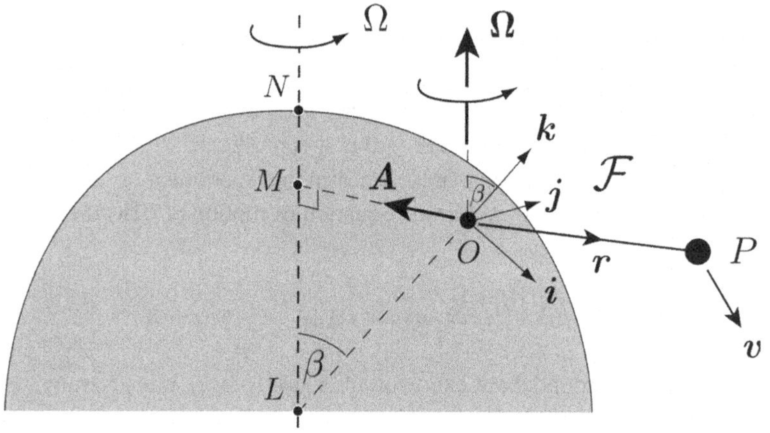

FIGURE 17.4 The non-inertial frame $\mathcal{F} \equiv \{O; \boldsymbol{i}, \boldsymbol{j}, \boldsymbol{k}\}$ is attached to the surface of the Earth.

'Forget that the roundabout is rotating, and introduce the Coriolis and centrifugal forces as shown in Figure 17.3. The man will then feel the Coriolis force pushing him to his right and the centrifugal force pushing him backwards. This is why he must lean forwards and to his left'.

Such arguments are fine at the qualitative level, where they provide a picture of what is going on. However, it is asking for trouble to have two parallel approaches available when solving problems. A choice must be made and, in this book, *we will always solve problems in rotating frame dynamics by using the transformed equation of motion* (17.18); *fictitious forces will only be mentioned in passing.*

17.3 MOTION RELATIVE TO THE EARTH

The most common non-inertial reference frame is the Earth. The effects of the Earth's motion are generally small and we have so far completely neglected them when solving problems in dynamics. However, there are circumstances in which the Earth's motion has a significant effect. In long range artillery, the Earth's rotation gives rise to an important correction,* and, in the hydrodynamics of the atmosphere and oceans, the Earth's rotation can have a dominant effect.

The Earth moves in its orbit around the Sun and also rotates on its axis. The effect of its orbital motion is *very* much smaller than the effect of its rotation and we will neglect

* Artillery shells are deflected to the *right* in the northern hemisphere and to the *left* in the southern. The following anecdote is due to the great English mathematician J.E. Littlewood. '*I heard an account of the battle of the Falkland Islands (early in the 1914 war) from an officer who was there. The German ships were destroyed at extreme range, but it took a long time and salvos were continually falling 100 yards to the left. The effect of the rotation of the Earth was incorporated into the gun-sights. But this involved the tacit assumption that Naval battles take place round about latitude 50°N, whereas the Falkland Islands are at about latitude 50°S. At extreme range, this double difference is of the order of 100 yards!*

the orbital motion of the Earth altogether. We will regard the Earth as an *axially symmetric rigid body** whose axis of symmetry is at rest in an inertial frame $\mathcal{I}$. Previously we have denoted the moving frame by $\mathcal{F}'$, but it would be insufferable to attach a dash to every quantity from now until the end of the chapter. We will therefore drop the dashes and, from now on, we will let $\mathcal{F} \equiv \{O; \boldsymbol{i}, \boldsymbol{j}, \boldsymbol{k}\}$ be a *non-inertial* frame attached to the surface of the Earth, as shown in Figure 17.4. The orientations of the basis vectors $\{\boldsymbol{i}, \boldsymbol{j}, \boldsymbol{k}\}$ will be described later. Because of the Earth's rotation, the frame $\mathcal{F}$ has an acceleration $\boldsymbol{A}$ and an angular velocity $\boldsymbol{\Omega}$ relative to the inertial frame $\mathcal{I}$. The acceleration $\boldsymbol{A}$ is caused by the circular motion of the origin O around the Earth's rotation axis and acts towards the axis. The angular velocity $\boldsymbol{\Omega}$ is the same as the angular velocity of the Earth itself; it is therefore parallel to the rotation axis and *constant*. If $\boldsymbol{n}$ is the unit vector in the direction $\overrightarrow{SN}$, where S and N are the north and south poles, then $\boldsymbol{\Omega} = +\Omega\boldsymbol{n}$, where Ω is 2π radians per day,† or about 0.000073 radians per second.

Let P be a particle of mass m and position vector $\boldsymbol{r}$ relative to O. Then the transformed equation of motion for P is

$$m\left[\boldsymbol{A} + 2\boldsymbol{\Omega}\times\boldsymbol{v} + \boldsymbol{\Omega}\times(\boldsymbol{\Omega}\times\boldsymbol{r}) + \boldsymbol{a}\right] = \boldsymbol{F}^E. \tag{17.20}$$

Here $\boldsymbol{r}$, $\boldsymbol{v}$ and $\boldsymbol{a}$ are the apparent position, velocity and acceleration of P, and $\boldsymbol{F}^E$ is the gravitational force exerted on P by the Earth. There may be other forces acting on P, but, for the moment, we suppose that the only force is the Earth's gravity.

Suppose first that P is *released from rest* (in $\mathcal{F}$) at the point with position vector $\boldsymbol{r}$. Then its initial acceleration $\boldsymbol{a}_0$ is given by equation (17.20) to be

$$\boldsymbol{a}_0 = m^{-1}\boldsymbol{F}^E - \boldsymbol{A} - \boldsymbol{\Omega}\times(\boldsymbol{\Omega}\times\boldsymbol{r}). \tag{17.21}$$

If the Earth were not rotating, this acceleration would be the acceleration due to the Earth's gravity at the point $\boldsymbol{r}$. With rotation present, it is still the observed acceleration of a body released from rest, but it is not caused by gravity alone. We call this the **apparent gravitational acceleration** (at the point $\boldsymbol{r}$). The direction opposite to $\boldsymbol{a}_0$ is the **apparent vertical** direction. Let $\boldsymbol{k}\ (= \boldsymbol{k}(\boldsymbol{r}))$ be the unit vector in the apparent vertical direction. Then $\boldsymbol{a}^I$ can be written in the form

$$\boldsymbol{a}_0 = -g(\boldsymbol{r})\,\boldsymbol{k}(\boldsymbol{r}), \tag{17.22}$$

where g is the magnitude of the **apparent gravitational acceleration.** These values of g and $\boldsymbol{k}$ differ slightly (by less than 1%) from the corresponding quantities due to Earth's gravity alone. In terms of the quantities g and $\boldsymbol{k}$, the transformed equation of motion for P simplifies to give

$$m\left[\frac{d\boldsymbol{v}}{dt} + 2\boldsymbol{\Omega}\times\boldsymbol{v}\right] = -m\,g(\boldsymbol{r})\,\boldsymbol{k}(\boldsymbol{r}).$$

* We are not restricted to regard the Earth as spherically symmetric.

† Actually, the Earth revolves not once every day, but every 23 h 56 m. Why?

As in the non-rotational theory, we will usually suppose that the extent of the motion of P is small compared to the Earth's radius so that the spatial variations of $g(\boldsymbol{r})$ and $\boldsymbol{k}(\boldsymbol{r})$ can be neglected. In this approximation, the transformed equation of motion for P becomes

Projectile equation on a rotating Earth

$$m\left[\frac{d\boldsymbol{v}}{dt} + 2\boldsymbol{\Omega}\times\boldsymbol{v}\right] = -mg\,\boldsymbol{k} \tag{17.23}$$

where g and $\boldsymbol{k}$ are now constants. This is the **projectile equation** for a **rotating Earth**. If any other forces act on P, they should be included on the right side of this equation. Note that, by introducing apparent gravity, this equation differs from the standard projectile equation only by the appearance of the Coriolis term.

We will now assign directions to the basis vectors $\{\boldsymbol{i}, \boldsymbol{j}, \boldsymbol{k}\}$ of our frame $\mathcal{F}$. The basis vector $\boldsymbol{k}$ is taken to be the same as the apparent upward vertical vector $\boldsymbol{k}$ (now assumed constant); the vector $\boldsymbol{i}$ is taken so that the Earth's rotation axis lies in the plane Oxz (see Figure 17.4); the vector $\boldsymbol{j}$ is then determined. In geographical language, $\boldsymbol{i}$ points south and $\boldsymbol{j}$ points east. With these choices, the Earth's angular velocity is then

$$\boldsymbol{\Omega} = \Omega(-\sin\beta\,\boldsymbol{i} + \cos\beta\,\boldsymbol{k}), \tag{17.24}$$

where β is the angle between the apparent vertical at O and the Earth's rotation axis. β is approximately $90° - \theta$, where θ is the geographical latitude of O. [Why only approximately?] Despite this small discrepancy, we will call the angle β the **co-latitude** of O.

Example 17.3 *Tower of Pisa experiment*

A lead ball is dropped from the top of the tower of Pisa. Show that its path deviates from that of a plumbline and find where it lands. [Neglect air resistance but include the effect of the Earth's rotation.]

Solution

Consider first the **plumbline**. The bob is subject to the apparent gravity force $-mg\boldsymbol{k}$ and the string tension force $\boldsymbol{T}$. When the plumbline is in equilibrium it follows from equation (17.23) that $\boldsymbol{0} = -mg\boldsymbol{k} + \boldsymbol{T}$, that is, $\boldsymbol{T} = mg\boldsymbol{k}$. Hence the tension in the plumbline is mg and the *string is parallel to the apparent vertical.*

Now consider the **ball**. Take the origin O of coordinates at the point where the ball is released and suppose that the apparent vertical through O meets the ground at the point $(0, 0, -h)$. The equation of motion for the ball is the **projectile equation** (17.23), together with the initial conditions $\boldsymbol{v} = \boldsymbol{0}$ and $\boldsymbol{r} = \boldsymbol{0}$ when $t = 0$. Since $\boldsymbol{\Omega}$ is constant, equation (17.23) can easily be integrated once with respect to t to give

$$\frac{d\boldsymbol{r}}{dt} + 2\boldsymbol{\Omega}\times\boldsymbol{r} = -gt\,\boldsymbol{k} + \boldsymbol{C},$$

where $\boldsymbol{C}$ is the constant of integration. The initial conditions then imply that $\boldsymbol{C} = \boldsymbol{0}$ so that the displacement of the ball satisfies the first order equation

$$\frac{d\boldsymbol{r}}{dt} + 2\,\boldsymbol{\Omega}\times\boldsymbol{r} = -gt\,\boldsymbol{k} \tag{17.25}$$

with the initial condition $\boldsymbol{r} = \boldsymbol{0}$ when $t = 0$. Since we expect the effect of the Earth's rotation to be a small correction, we will solve this equation approximately by an iterative method. On integrating equation (17.25) with respect to t and using the initial condition, we find that the unknown displacement $\boldsymbol{r}(t)$ satisfies the **integral equation***

$$\boldsymbol{r}(t) = -\tfrac{1}{2}gt^2\,\boldsymbol{k} - 2\boldsymbol{\Omega}\times\int_0^t \boldsymbol{r}(t')\,dt'. \tag{17.26}$$

(The variable of integration has been changed to the dummy variable t' because t appears as a limit of integration.) We now solve this equation approximately by iteration. The zeroth order approximation $\boldsymbol{r}^{(0)}$ corresponds to the case $\Omega = 0$, that is, the case of the non-rotating Earth. Thus the **zeroth order** approximation is

$$\boldsymbol{r}^{(0)} = -\tfrac{1}{2}gt^2\,\boldsymbol{k}, \tag{17.27}$$

which is just the elementary solution for vertical motion under uniform gravity. The first order approximation $\boldsymbol{r}^{(1)}$ is now obtained by substituting the zeroth order approximation into the integral term in equation (17.26). This gives

$$\begin{aligned}\boldsymbol{r}^{(1)} &= -\tfrac{1}{2}gt^2\,\boldsymbol{k} - 2\boldsymbol{\Omega}\times\int_0^t \boldsymbol{r}^{(0)}(t')\,dt'\\ &= -\tfrac{1}{2}gt^2\,\boldsymbol{k} - 2\boldsymbol{\Omega}\times\int_0^t \left(-\tfrac{1}{2}gt'^2\,\boldsymbol{k}\right)\,dt'\\ &= -\tfrac{1}{2}gt^2\,\boldsymbol{k} + \tfrac{1}{3}gt^3(\boldsymbol{\Omega}\times\boldsymbol{k}).\end{aligned}$$

Now the angular velocity $\boldsymbol{\Omega}$ is given by equation (17.24), where β is the co-latitude of the point O. It follows that $\boldsymbol{\Omega}\times\boldsymbol{k} = (\Omega\sin\beta)\,\boldsymbol{j}$ and so the **first order approximation** for the displacement $\boldsymbol{r}(t)$ is given by

$$\boldsymbol{r}^{(1)} = -\tfrac{1}{2}gt^2\,\boldsymbol{k} + \tfrac{1}{3}gt^3(\Omega\sin\beta)\,\boldsymbol{j}. \tag{17.28}$$

Higher approximations can be obtained in the same way, but they are progressively less important† (and more complicated). The first order approximation predicts that

---

* Any equation in which an unknown function appears under an integral sign is called an *integral equation*. There is an extensive theory of such equations. Integral equations like (17.26) have the impressive title of *Volterra integral equations of the second kind*. Their most important feature is that they can always be solved by the iteration process described.

† The iterative approximation scheme described above will converge rapidly when the dimensionless product $\Omega\tau$ is small, where τ is the total travel time. ($\Omega\tau$ is the angle through which the Earth rotates during the motion.) In the Tower of Pisa problem, the value of $\Omega\tau$ is about 0.0002, which explains why the correction is so small.

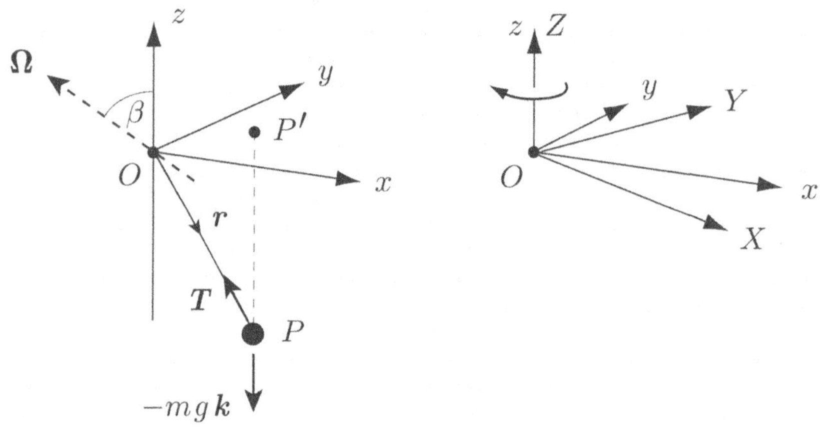

FIGURE 17.5 **Left**: The Foucault pendulum is an ordinary pendulum suspended from a point O fixed to the Earth. **Right**: The precessing frame $OXYZ$ rotates around the axis Oz with angular velocity $-(\Omega\cos\beta)\boldsymbol{k}$.

the ball will not travel down the apparent vertical, but will be displaced in the positive $\boldsymbol{j}$-direction, that is, to the east. It remains to determine this easterly displacement in terms of h, the height of the tower. For this we need the total travel time τ, but it is consistent with the first order approximation to determine τ from the *zero order* theory which gives $\tau = (2h/g)^{1/2}$. Hence, because of the Earth's rotation, the ball does not travel down the apparent vertical exactly but **drifts to the east** a distance

$$\left(\frac{8h^3}{9g}\right)^{1/2} \Omega \sin\beta.$$

For a ball dropped from the top of the **Tower of Pisa**, which is about 50 m high and is at latitude 44°N, this easterly drift is about 5.6 mm. ■

Example 17.4 *The Foucault pendulum*

How does the Earth's rotation affect the small oscillations of an ordinary pendulum?

Solution

Let the origin O be the suspension point of the pendulum as shown in Figure 17.5. The **transformed equation of motion** for the pendulum bob P is

$$m\left[\frac{d\boldsymbol{v}}{dt} + 2\boldsymbol{\Omega}\times\boldsymbol{v}\right] = -mg\,\boldsymbol{k} + \boldsymbol{T},$$

where $\boldsymbol{T}$ is the tension force exerted by the string. The magnitude T of the tension force is unknown but its direction is opposite to the position vector $\boldsymbol{r}$ so that

$$\boldsymbol{T} = -\left(\frac{T}{a}\right)\boldsymbol{r},$$

where a is the length of the string. The equation of motion for P can therefore be written

$$m\left[\frac{d\boldsymbol{v}}{dt}+2\boldsymbol{\Omega}\times\boldsymbol{v}\right]=-mg\,\boldsymbol{k}-\left(\frac{T}{a}\right)\boldsymbol{r}. \tag{17.29}$$

This non-linear equation, which holds for large oscillations of the pendulum, cannot be solved in closed form. We will therefore restrict attention to the case of **small oscillations** which allows us to replace (17.29) by a **linear approximation**. The zero order approximation corresponds to the 'motion' $\boldsymbol{r}=-a\boldsymbol{k}$, in which the pendulum hangs in equilibrium. Thus the zero order approximation to the tension is $T=mg$.

Furthermore, if we write $\boldsymbol{r}=x\boldsymbol{i}+y\boldsymbol{j}+z\boldsymbol{k}$, then, in any motion of the pendulum, $x^2+y^2+z^2=a^2$, from which it follows that

$$\begin{aligned} z &= \left(a^2-(x^2+y^2)\right)^{1/2} \\ &= a\left(1-\frac{x^2+y^2}{2a^2}+\cdots\right), \end{aligned}$$

which shows that the *change* in z is second order in the small quantities x and y. It can similarly be shown that v_z and dv_z/dt are also second order quantities. On using these approximations in equation (17.29), the **linearised equation** for the **horizontal motion** of P is found to be

$$\frac{d\boldsymbol{v}^H}{dt}+2\Omega\cos\beta(\boldsymbol{k}\times\boldsymbol{v}^H)+n^2\boldsymbol{r}^H=\boldsymbol{0}, \tag{17.30}$$

where $\boldsymbol{r}^H=x\boldsymbol{i}+y\boldsymbol{j}$, $\boldsymbol{v}^H=v_x\boldsymbol{i}+v_y\boldsymbol{j}$ and $n^2=g/a$. Note that n is the angular frequency the pendulum would have if the Earth were not rotating. Equation (17.30) is the linearised equation for the motion of the point P' shown in Figure 17.5.

The two-dimensional vector equation (17.30) is equivalent to a pair of homogeneous second order linear equations with constant coefficients and can therefore be solved by standard methods. However, this is a messy job and there is a much simpler way of generating the solutions. The term $2\Omega\cos\beta(\boldsymbol{k}\times\boldsymbol{v}^H)$ can be eliminated by viewing the motion from a frame* $OXYZ$ rotating with angular velocity $-(\Omega\cos\beta)\boldsymbol{k}$ relative to the frame $\mathcal{F}$ (see Figure 17.5). We will call this new frame the **precessing frame** and the horizontal displacement, velocity and acceleration of P in this frame will be denoted by $\boldsymbol{R}^H$, $\boldsymbol{V}^H$ and $\boldsymbol{A}^H$. The standard transformation formulae (17.10), (17.12), (17.16) can be used to express $\boldsymbol{r}^H$, $\boldsymbol{v}^H$, $\boldsymbol{a}^H$ in terms of $\boldsymbol{R}^H$, $\boldsymbol{V}^H$, $\boldsymbol{A}^H$, and if these formulae are substituted into equation (17.30), we find that the linearised equation of motion for P in the precessing frame can be written in the form

$$\frac{d^2\boldsymbol{R}^H}{dt^2}+\left(n^2+\Omega^2\cos^2\beta\right)\boldsymbol{R}^H=\boldsymbol{0}, \tag{17.31}$$

* Note that this operation does *not* simply return us to the underlying inertial frame.

which is the **two-dimensional SHM equation**. Its solutions are known to be ellipses with centre O and angular frequency $(n^2 + \Omega^2 \cos^2 \beta)^{1/2}$. The difference between this frequency and n is completely negligible.

Conclusion *In small oscillations, the period of the pendulum is unchanged by the Earth's rotation, but its motion now precesses* around the apparent vertical with angular velocity* $-(\Omega \cos \beta)\boldsymbol{k}$. *In particular, an oscillation that would take place in a fixed vertical plane through* O *when* $\Omega = 0$ *will now take place in a vertical plane through* O *that precesses slowly around the axis* Oz *with angular velocity* $-(\Omega \cos \beta)\boldsymbol{k}$. *This precession is in the negative sense in the northern hemisphere, and in the positive sense in the southern hemisphere.* ■

This practical demonstration of the Earth's rotation is called **Foucault's pendulum**† and is a favourite exhibit in science museums the world over. [But not in Singapore. Why?] From the duration of your museum visit, and the angle turned by the plane of motion of the pendulum while you were inside, you can work out your latitude!

Winds, ocean currents and bath water

Atmospheric winds and ocean currents are often strongly influenced by the Coriolis force induced by the Earth's rotation. We may analyse these fluid motions by analogy with particle dynamics.

By analogy with equation (17.23) for the motion of a particle over the Earth, the equation of motion for the steady flow of the atmosphere or the ocean is

$$\rho\, \boldsymbol{a} + 2\rho\, \boldsymbol{\Omega} \times \boldsymbol{v} = -\rho g\, \boldsymbol{k} - \operatorname{grad} p. \tag{17.32}$$

Here, $\boldsymbol{v}(\boldsymbol{r})$ and $\boldsymbol{a}(\boldsymbol{r})$ are the velocity and acceleration of the *fluid particle* with position vector $\boldsymbol{r}$, ρ is the fluid density and p is the pressure field in the fluid. The term $-\operatorname{grad} p$ is called the **pressure gradient** and represents the force exerted on a fluid particle by the fluid around it. The acceleration $\boldsymbol{a}$ can be written in terms of $\boldsymbol{v}$. If we follow the progress in time of a fluid particle, it follows from the chain rule that

$$\begin{aligned} \boldsymbol{a} &= \frac{\partial \boldsymbol{v}}{\partial x} \times \frac{dx}{dt} + \frac{\partial \boldsymbol{v}}{\partial y} \times \frac{dy}{dt} + \frac{\partial \boldsymbol{v}}{\partial z} \times \frac{dz}{dt} \\ &= v_x \frac{\partial \boldsymbol{v}}{\partial x} + v_y \frac{\partial \boldsymbol{v}}{\partial y} + v_z \frac{\partial \boldsymbol{v}}{\partial z}. \end{aligned} \tag{17.33}$$

Let us now compare the sizes of the two acceleration terms on the left of equation (17.32). Suppose that the flow has characteristic length scale L and velocity scale V. Then the Coriolis acceleration $2\,\boldsymbol{\Omega} \times \boldsymbol{v}$ is of order ΩV and, from equation (17.33), the acceleration $\boldsymbol{a}$ is of order V^2/L. The dimensionless ratio of these two magnitudes, given by

$$Ro = \frac{V}{\Omega L}, \tag{17.34}$$

is called the **Rossby number**‡ of the flow. *When the Rossby number is small, the Coriolis acceleration dominates over the real acceleration.* When this condition is satisfied, the equation of motion (17.32) can be approximated by

$$2\rho\, \boldsymbol{\Omega} \times \boldsymbol{v} = -\rho g\, \boldsymbol{k} - \operatorname{grad} p. \tag{17.35}$$

---

* A **precession** is a slow rotation superimposed on a fast motion.

† After Jean Bernard Leon Foucault (1819–1868), who first exhibited it at the 1851 Paris Exhibition. This was the first time the Earth's rotation had been demonstrated by an entirely terrestrial method.

‡ After the Swedish-American meteorologist Carl-Gustav Rossby (1898–1957).

which is called a **geostrophic flow**. If the flow velocity $\boldsymbol{v}$ is everywhere horizontal, then, on taking the scalar product of this equation with $\boldsymbol{v}$, we obtain

$$\boldsymbol{v} \cdot \operatorname{grad} p = 0. \tag{17.37}$$

Hence, in a horizontal geostrophic flow, the fluid velocity is perpendicular to the horizontal pressure gradient. In other words, the **fluid flows along the isobars.**

Weather maps often show winds circulating around areas of low pressure. In a large low pressure system on the Earth, $V \approx 10\,\mathrm{m\,s^{-1}}$, $L \approx 1000$ km and $\Omega \approx 10^{-4}$ radians per second. The corresponding Rossby number is $Ro \approx 0.1$. Such a flow is therefore quite accurately geostrophic. We would therefore expect the wind to follow the isobars in the sense predicted by equation (17.36) (anti-clockwise around a 'low' in the northern hemisphere) and qualitatively this is what is observed. On the other hand, in the wind flow near the centre of a hurricane, $V \approx 50\,\mathrm{m\,s^{-1}}$ and $L \approx 50$ km so that the Rossby number $Ro \approx 10$. Hence, contrary to common belief, the Coriolis force has an insignificant* effect on the wind in this part of a hurricane. Even less significant is the effect of the Earth's rotation on the swirling motion of water leaving a bath or a toilet bowl. The direction of the water in such flows is strongly influenced by plumbing but not by Coriolis force!

17.4 MULTI-PARTICLE SYSTEM IN A NON-INERTIAL FRAME

Some of the principles of **multi-particle mechanics** can be extended to non-inertial reference frames.

Linear momentum principle

The **linear momentum principle** (see Chapter 10) can easily be extended to the case in which the system is observed from a **non-inertial frame.**

In the notation for systems of particles used in Part II, the transformed equation of motion for the particle P_i of mass m_i is

$$m_i\left[\boldsymbol{A} + \dot{\boldsymbol{\Omega}}\times\boldsymbol{r}_i + 2\,\boldsymbol{\Omega}\times\boldsymbol{v}_i + \boldsymbol{\Omega}\times(\boldsymbol{\Omega}\times\boldsymbol{r}_i) + \frac{d\boldsymbol{v}_i}{dt}\right] = \boldsymbol{F}_i + \sum_{j=1}^{N}\boldsymbol{G}_{ij}$$

$(1 \le i \le N)$. Here, all quantities are observed in the moving frame $\mathcal{F}'$, the dashes being understood throughout. On summing these equations over all the particles we obtain

$$\left(\sum_{i=1}^{N} m_i\right)\boldsymbol{A} + \dot{\boldsymbol{\Omega}}\times\left(\sum_{i=1}^{N} m_i\boldsymbol{r}_i\right) + 2\,\boldsymbol{\Omega}\times\left(\sum_{i=1}^{N} m_i\boldsymbol{v}_i\right) + \boldsymbol{\Omega}\times\left[\boldsymbol{\Omega}\times\left(\sum_{i=1}^{N} m_i\boldsymbol{r}_i\right)\right] + \sum_{i=1}^{N} m_i\frac{d\boldsymbol{v}_i}{dt}$$

$$= \sum_{i=1}^{N}\boldsymbol{F}_i + \sum_{i=1}^{N}\left(\sum_{j=1}^{N}\boldsymbol{G}_{ij}\right).$$

* And yet the direction of such winds is always that predicted by the geostrophic theory! However, the flow *was* geostrophic in the early stages of the hurricane's formation, and this presumably determines the direction of flow thereafter.

As in the standard linear momentum principle, the double sum of the $\{G_{ij}\}$ is zero. In terms of centre of mass variables, this equation can be written

$$M\left[A + \dot{\Omega}\times R + 2\,\Omega\times V + \Omega\times(\Omega\times R) + \frac{dV}{dt}\right] = F, \tag{17.38}$$

where R and V are the position and velocity of the centre of mass, M is the total mass, and F is the total external force. This is the centre of mass form of the **linear momentum principle** in a **non-inertial frame**.

We have therefore proved that:

Linear momentum principle in a non-inertial frame

In *any* reference frame, the centre of mass of a system moves as if it were a particle of mass the total mass, and all the external forces acted upon it.

Hence, even in a non-inertial frame, the motion of the centre of mass of a *system* can still be calculated by *particle* mechanics.

Energy principle

There is also a form of **energy conservation** that applies in frames with a *fixed origin* and *constant angular velocity*. The proof is as follows:

In this case the acceleration A and the angular acceleration $\dot{\Omega}$ are zero and the transformed equation of motion for the particle P_i becomes

$$m_i\left[2\,\Omega\times v_i + \Omega\times(\Omega\times r_i) + \frac{dv_i}{dt}\right] = F_i + \sum_{j=1}^{N} G_{ij},$$

in the standard notation. If we now take the scalar poduct of this equation with v_i, the Coriolis term vanishes and we obtain

$$m_i\left[v_i\cdot\left[\Omega\times(\Omega\times r_i)\right] + v_i\cdot\frac{dv_i}{dt}\right] = F_i\cdot v_i + \sum_{j=1}^{N} G_{ij}\cdot v_i \tag{17.39}$$

($1 \le i \le N$). This is the same equation as for an inertial frame except for the centrifugal term. This term can be written

$$\begin{aligned} v_i\cdot\left[\Omega\times(\Omega\times r_i)\right] &= v_i\cdot\left[(\Omega\cdot r_i)\Omega - \Omega^2 r_i\right] \\ &= (\Omega\cdot v_i)(\Omega\cdot r_i) - \Omega^2 v_i\cdot r_i \\ &= \frac{d}{dt}\left[\tfrac{1}{2}(\Omega\cdot r_i)^2 - \tfrac{1}{2}\Omega^2 r_i\cdot r_i\right] \\ &= \frac{d}{dt}\left[-\tfrac{1}{2}\Omega^2 p_i^2\right], \end{aligned}$$

where p_i is the perpendicular distance of P from the rotation axis. On substituting this formula into equation (17.39) and summing over all the particles we obtain

$$\frac{d}{dt}\left[\sum_{i=1}^{N}\tfrac{1}{2}m_i v_i\cdot v_i - \tfrac{1}{2}\Omega^2\sum_{i=1}^{N} m_i p_i^2\right] = \sum_{i=1}^{N} F_i\cdot v_i + \sum_{i=1}^{N}\sum_{j=1}^{N} G_{ij}\cdot v_i,$$

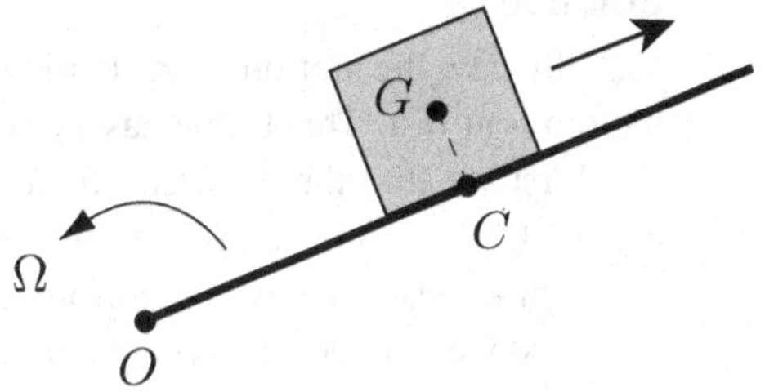

FIGURE 17.6 The straight rod rotates with constant angular speed Ω around O and the square lamina slides along the rod.

that is

$$\frac{d}{dt}\left(T - \tfrac{1}{2}I_{\mathcal{A}}\Omega^2\right) = \sum_{i=1}^{N} \boldsymbol{F}_i \cdot \boldsymbol{v}_i + \sum_{i=1}^{N}\sum_{j=1}^{N} \boldsymbol{G}_{ij} \cdot \boldsymbol{v}_i, \tag{17.40}$$

where T is the apparent total kinetic energy and $I_{\mathcal{A}}$ is the moment of inertia of the system about the rotation axis $\mathcal{A}$. The right side of equation (17.40) is the apparent total rate of working of all the forces. If the apparent total work done can be represented by the potential energy V, we will say that the system is **apparently conservative.** In this case, we have **conservation of energy** in the form

$$T + V - \tfrac{1}{2}I_{\mathcal{A}}\Omega^2 = E,$$

where E is a constant. The term $-\frac{1}{2}I_{\mathcal{A}}\Omega^2$ is sometimes called the **centrifugal potential energy** and we will call this equation the **transformed energy equation.** Our result is summarised as follows:

Energy conservation in a uniformly rotating frame

If a system is apparently conservative when viewed from a rotating reference frame with a fixed origin and constant angular velocity, **energy conservation** holds in the modified form

$$T + V - \tfrac{1}{2}I_{\mathcal{A}}\Omega^2 = E, \tag{17.41}$$

where T and V are the apparent kinetic and potential energies, Ω is the angular speed of the reference frame, and $I_{\mathcal{A}}$ is the moment of inertia of the system about the rotation axis.

For systems with one degree of freedom, this energy conservation principle is sufficient to determine the motion, as in the following example.

Example 17.5

A long straight smooth rod lies on a smooth horizontal table. One of its ends is fixed at point O on the table and the rod is made to rotate around O with constant angular speed Ω. A uniform square lamina of side $2a$ can slide freely on the table. One of its sides is in contact with the rod at all times and the lamina slides freely along the rod. Initially, the lamina is at rest (relative to the rod) with one corner at O. Find the subsequent displacement of the lamina as a function of the time.

Solution

We will view the motion of the lamina from a reference frame with origin O in which the rod is at rest. This frame has a *fixed* origin and *constant* angular velocity.

First we find the apparent kinetic energy. Let r be the distance OC shown in Figure 17.6. In the rotating frame, the motion of the lamina is rectilinear and its apparent **kinetic energy** is therefore given by $T = \frac{1}{2}M\dot{r}^2$.

Next we find the apparent potential energy. There are no specified external forces, but there is the constraint force exerted on the lamina by the rod, which, since the rod is smooth, acts *perpendicular* to the rod. In an inertial frame, this constraint force does work, but, in the frame rotating with the rod, the lamina simply moves *along* the stationary rod. The apparent rate of working of the constraint force is therefore zero. The internal forces within the lamina enforce rigidity and do no apparent work in total. The system is therefore **apparently conservative** with potential energy $V = 0$.

The transformed energy equation is therefore

$$\tfrac{1}{2}M\dot{r}^2 - \tfrac{1}{2}I_A\Omega^2 = E,$$

where I_A is the moment of inertia of the lamina about the vertical axis through O, and Ω is the angular speed of the rod. By the 'parallel axes' theorem (see the Appendix at the end of the book), I_A is given by

$$I_A = I_G + \left(a^2 + r^2\right),$$

so that the equation becomes

$$\tfrac{1}{2}M\dot{r}^2 - \tfrac{1}{2}\left(I_G + a^2 + r^2\right)\Omega^2 = E.$$

The **initial conditions** $r = a$ and $\dot{r} = 0$ when $t = 0$ give

$$E = -\tfrac{1}{2}\left(I_G + 2a^2\right)$$

so that the **transformed energy equation** for the lamina is finally given by

$$\dot{r}^2 = \Omega^2\left(r^2 - a^2\right).$$

This equation is sufficient to determine the motion. On taking square roots and integrating, we find that the **displacement** of the lamina at time t is

$$r = a\cosh\Omega t,$$

which is the same answer as for a particle sliding along a rotating wire. ■

Problems on Chapter 17

Answers and comments are at the end of the book.

Harder problems carry a star ($*$).

Kinematics in rotating frames

17.1 Use the velocity and acceleration transformation formulae to derive the standard expressions for the velocity and acceleration of a particle in plane polar coordinates.

17.2 ***Addition of angular velocities*** Prove the 'addition of angular velocities' theorem, Theorem 17.13. [*Hint*. For a general particle of the rigid body, find $\boldsymbol{v}$ and $\boldsymbol{v}'$ in terms of $\boldsymbol{\omega}$ and $\boldsymbol{\omega}'$ respectively. Then relate $\boldsymbol{v}$ and $\boldsymbol{v}'$ by the velocity transformation formula.]

17.3 A circular cone with semi-angle α is fixed with its axis of symmetry vertical and its vertex O upwards. A second circular cone has semi-vertical angle $(\pi/2) - \alpha$ and has its vertex fixed at O. The second cone rolls on the first cone so that its axis of symmetry precesses around the upward vertical with angular speed λ. Find the angular speed of the rolling cone.

Dynamics in rotating frames

17.4 A particle P of mass m can slide along a smooth rigid straight wire. The wire has one of its points fixed at the origin O, and is made to rotate in a plane through O with constant angular speed Ω. Show that r, the distance of P from O, satisfies the equation

$$\ddot{r} - \Omega^2 r = 0.$$

Initially, P is at rest (relative to the wire) at a distance a from O. Find r as a function of t in the subsequent motion.

17.5 ***Larmor precession*** A particle of mass m and charge e moves in the force field $\boldsymbol{F}(\boldsymbol{r})$ and the uniform magnetic field $B\boldsymbol{k}$, where $\boldsymbol{k}$ is a constant unit vector. Its equation of motion is then

$$m\frac{d\boldsymbol{v}}{dt} = \left(\frac{eB}{c}\right)\boldsymbol{v}\times\boldsymbol{k} + \boldsymbol{F}(\boldsymbol{r})$$

in cgs Gaussian units. Show that the term $(eB/c)\boldsymbol{v}\times\boldsymbol{k}$ can be removed from the equation by viewing the motion from an appropriate rotating frame.

For the special case in which $\boldsymbol{F}(\boldsymbol{r}) = -m\omega_0^2\boldsymbol{r}$, show that circular motions with two different frequencies are possible.

17.6 A bullet is fired vertically upwards with speed u from a point on the Earth with co-latitude β. Show that it returns to the ground west of the firing point by a distance $4\Omega u^3 \sin\beta/3g^2$.

17.7 An artillery shell is fired from a point on the Earth with co-latitude β. The direction of firing is due **south**, the muzzle speed of the shell is u and the angle of elevation of the barrel is

α. show that the effect of the Earth's rotation is to deflect the shell to the west by a distance

$$\frac{4\Omega u^3}{3g^2}\sin^2\alpha\,(3\cos\alpha\cos\beta+\sin\alpha\sin\beta)\,.$$

17.8 An artillery shell is fired from a point on the Earth with co-latitude β. The direction of firing is due **east**, the muzzle speed of the shell is u and the angle of elevation of the barrel is α. Show that the effect of the Earth's rotation is to deflect the shell to the south by a distance

$$\frac{4\Omega u^3}{3g^2}\sin^2\alpha\cos\alpha\cos\beta.$$

Energy conservation in rotating frames

17.9 Consider Problem 17.4 again. This time find the motion of the particle by using the transformed energy equation.

17.10 One end of a straight rod is fixed at a point O on a smooth horizontal table and the rod is made to rotate around O with constant angular speed Ω. A uniform circular disk of radius a lies flat on the table and can slide freely upon it. The disc remains in contact with the rod at all times and is constrained to *roll* along the rod. Initially, the disk is at rest (relative to the rod) with its point of contact at a distance a from O. Find the displacement of the disk as a function of the time.

17.11 A horizontal turntable is made to rotate about a fixed vertical axis with constant angular speed Ω. A *hollow* uniform circular cylinder of mass M and radius a can *roll* on the turntable. Initially the cylinder is at rest (relative to the turntable), with its centre of mass on the rotation axis, when it is slightly disturbed. Find the speed of the cylinder when it has rolled a distance x on the turntable.

∗ Find also an expression (in terms of x) for the force that the turntable exerts on the cylinder.

Hydrostatics in rotating frames

When a fluid is at rest in an **inertial frame**, the equation of hydrostatics is

$$\mathbf{0}=\boldsymbol{F}-\operatorname{grad}p,$$

where p is the pressure field in the fluid, and $\boldsymbol{F}$ is the body force (force per unit volume) acting on the fluid. This equation means that the body force is balanced by the pressure gradient. In the case of uniform gravity, $\boldsymbol{F}=-\rho g\boldsymbol{k}$ and, if ρ is constant, the equation integrates to give the well known 'hydrostatic pressure formula' $p=p_0-\rho g z$, where Oz points vertically upwards.

If a fluid is at rest in a **rotating frame** with a fixed origin and constant angular velocity $\boldsymbol{\Omega}$, then the equation of 'hydrostatics' in this frame is

$$\rho\,\boldsymbol{\Omega}\times(\boldsymbol{\Omega}\times\boldsymbol{r})=\boldsymbol{F}-\operatorname{grad}p,$$

where ρ is the fluid density. (This equation means that the body force, the pressure gradient and the centrifugal 'force' must balance.) This is all you need to know to answer the following questions.

17.12 ***Newton's bucket*** A bucket half full of water is made to rotate with angular speed Ω about its axis of symmetry, which is vertical. Find, to within a constant, the pressure field in

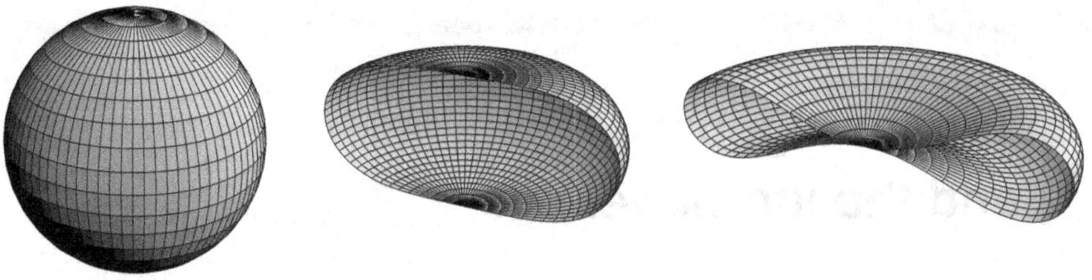

FIGURE 17.7 A mass of liquid in equilibrium (left) and rotating like a rigid body (centre and right, cut in half to show the cross sections). The right figure shows the mass rotating with the critical angular speed.

the fluid. By considering the isobars (surfaces of constant pressure) of this pressure field, find the shape of the free surface of the water.

What would the shape of the free surface be if the bucket were replaced by a cubical box?

17.13 A sealed circular can of radius a is three-quarters full of water of density ρ, the remainder being air at pressure p_0. The can is taken into gravity free space and then rotated about its axis of symmetry with constant angular speed Ω. Where will the water be when it comes to rest relative to the can? Find the water pressure at the wall of the can.

Computer assisted problem

The following problem is suitable for a **supervised project**.

17.14** If a mass of liquid in gravity free space is in equilibrium, then it has the form of a sphere, stabilised by its own surface tension (see Figure 17.7 (left)). Suppose that the mass now rotates like a rigid body with constant angular velocity. What will its equilibrium shape be now? Show that, as the angular speed increases, the mass becomes more oblate until its top and bottom surfaces become quite flat (Figure 17.7 (centre)). Show further that, at higher angular speeds, the top and bottom surfaces approach each other until they meet (Figure 17.7 (right)) and the mass breaks up. Find this critical angular speed for a sphere of mercury of radius 5 cm.

Chapter Eighteen

Tensor algebra and the inertia tensor

KEY FEATURES

The key features of this chapter are the **transformation formulae** for the components of tensors and **tensor algebra**; the **inertia tensor** and the calculation of the angular momentum and kinetic energy of a rigid body; and the **principal axes** and **principal moments of inertia** of a rigid body.

We have previously regarded a vector as a quantity that has magnitude and direction. Picturing a vector as a line segment with an arrow on it has helped us to understand essentially difficult concepts, such as the acceleration of a particle and the angular momentum of a rigid body. Neither of these quantities has any direct connection with line segments, but the picture is a valuable aid nonetheless. Useful though this notion of vectors is, it is not capable of generalisation and this is the main reason why we will now look at vectors from a different perspective. In reality, what we actually observe are the three *components* of a vector, and the values of these components will depend on the coordinate system in which they are measured. However, a general triple of real numbers $\{v_1, v_2, v_3\}$, defined in each coordinate system, does not necessarily constitute a vector. The reason is that the components of a vector in different coordinate systems are related to each other (in a way that we shall determine) and, unless the quantities $\{v_1, v_2, v_3\}$ satisfy this **transformation formula**, they are *not* the components of a vector. This leads us to a new definition of a vector as *a quantity that has three components in each coordinate system, whose values in different coordinate systems are related by the vector transformation formula.*

Defined in the above way, vectors can be regarded as part of a hierarchy of entities called **tensors** with a tensor of order n having 3^n components. A tensor of order zero, which has one component, is a **scalar**, and a tensor of order one, which has three components, is a **vector**. Tensors of higher order, which are entirely new objects, are defined to be quantities with 3^n components that obey the appropriate transformation formula. In this chapter we give a gentle introduction to the principles of **tensor algebra**, which should be useful, not only for its present application to dynamics, but also in other areas such as relativity. Any account of tensor algebra is mathematics, not mechanics, and some readers may find this indigestible. However, there are very few basic principles involved and, once these are mastered, the subject seems (and actually is) rather easy.

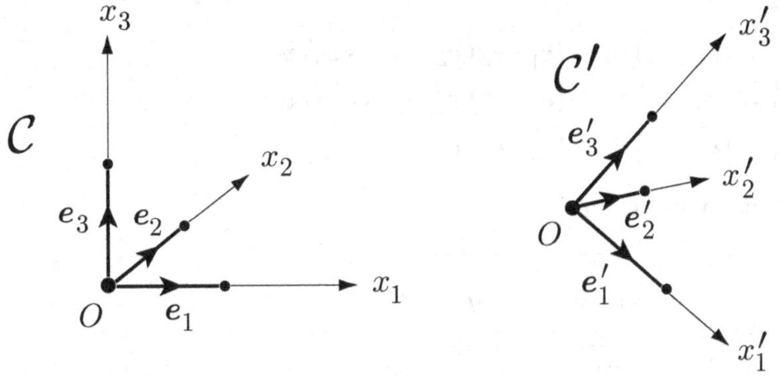

FIGURE 18.1 The Cartesian coordinate systems $\mathcal{C} \equiv Ox_1x_2x_3$ and $\mathcal{C}' \equiv Ox_1'x_2'x_3'$ have the associated unit vectors $\{\boldsymbol{e}_1, \boldsymbol{e}_2, \boldsymbol{e}_3\}$ and $\{\boldsymbol{e}_1', \boldsymbol{e}_2', \boldsymbol{e}_3'\}$ respectively. The two systems actually have a common origin but are drawn separately for clarity.

We then introduce the **inertia tensor**, a second order tensor that is used to calculate the angular momentum $\boldsymbol{L}$ and kinetic energy T of a generally rotating rigid body. It transpires that, by taking a special choice of axes, known as **principal axes**, the formulae for $\boldsymbol{L}$ and T are greatly simplified and reduce to non-tensorial forms. This makes it possible to solve most problems in rigid body dynamics without any knowledge of tensors at all.* The few results that one *really* needs in rigid body dynamics are given at the end of Section 18.6.

18.1 ORTHOGONAL TRANSFORMATIONS

Let $\mathcal{C} \equiv Ox_1x_2x_3$ and $\mathcal{C}' \equiv Ox_1'x_2'x_3'$ be two Cartesian coordinate systems having the common origin O and associated unit vectors $\{\boldsymbol{e}_1, \boldsymbol{e}_2, \boldsymbol{e}_3\}$ and $\{\boldsymbol{e}_1', \boldsymbol{e}_2', \boldsymbol{e}_3'\}$ respectively (see Figure 18.1). Then any vector $\boldsymbol{v}$ (understood in the elementary sense of Chapter 1) can be expanded in the form

$$\boldsymbol{v} = v_1\,\boldsymbol{e}_1 + v_2\,\boldsymbol{e}_2 + v_3\,\boldsymbol{e}_3, \tag{18.1}$$

where $\{v_1, v_2, v_3\}$ are the components of $\boldsymbol{v}$ in the coordinate system $\mathcal{C}$. Similarly $\boldsymbol{v}$ can be expanded in the form

$$\boldsymbol{v} = v_1'\,\boldsymbol{e}_1' + v_2'\,\boldsymbol{e}_2' + v_3'\,\boldsymbol{e}_3', \tag{18.2}$$

where $\{v_1', v_2', v_3'\}$ are the components of $\boldsymbol{v}$ in the coordinate system $\mathcal{C}'$. These two sets of components are related to each other. On taking the scalar product of the equality

$$v_1'\,\boldsymbol{e}_1' + v_2'\,\boldsymbol{e}_2' + v_3'\,\boldsymbol{e}_3' = v_1\,\boldsymbol{e}_1 + v_2\,\boldsymbol{e}_2 + v_3\,\boldsymbol{e}_3 \tag{18.3}$$

* Don't spread this information around!

successively with e'_1, e'_2 and e'_3 we find that

$$v'_1 = \left(e'_1 \cdot e_1\right) v_1 + \left(e'_1 \cdot e_2\right) v_2 + \left(e'_1 \cdot e_3\right) v_3,$$
$$v'_2 = \left(e'_2 \cdot e_1\right) v_1 + \left(e'_2 \cdot e_2\right) v_2 + \left(e'_2 \cdot e_3\right) v_3,$$
$$v'_3 = \left(e'_3 \cdot e_1\right) v_1 + \left(e'_3 \cdot e_2\right) v_2 + \left(e'_3 \cdot e_3\right) v_3,$$

which can be written in the matrix form

$$\boxed{\mathsf{v}' = \mathsf{A} \cdot \mathsf{v}} \tag{18.4}$$

where

$$\mathsf{v}' = \begin{pmatrix} v'_1 \\ v'_2 \\ v'_3 \end{pmatrix}, \qquad \mathsf{v} = \begin{pmatrix} v_1 \\ v_2 \\ v_3 \end{pmatrix}, \qquad \mathsf{A} = \begin{pmatrix} (e'_1 \cdot e_1) & (e'_1 \cdot e_2) & (e'_1 \cdot e_3) \\ (e'_2 \cdot e_1) & (e'_2 \cdot e_2) & (e'_2 \cdot e_3) \\ (e'_3 \cdot e_1) & (e'_3 \cdot e_2) & (e'_3 \cdot e_3) \end{pmatrix}, \tag{18.5}$$

and the '$\cdot$' means the *matrix product.* This is the required relationship between the components of $\boldsymbol{v}$ in the two coordinate systems. The elements of the **transformation matrix A** depend only on the *orientation* of the system C' relative to the system C.

The **inverse relationship** to (18.4) can be found by taking the scalar product of equation (18.3) successively with e_1, e_2 and e_3. The result is that

$$v'_1 = \left(e_1 \cdot e'_1\right) v_1 + \left(e_1 \cdot e'_2\right) v_2 + \left(e_1 \cdot e'_3\right) v_3,$$
$$v'_2 = \left(e_2 \cdot e'_1\right) v_1 + \left(e_2 \cdot e'_2\right) v_2 + \left(e_2 \cdot e'_3\right) v_3,$$
$$v'_3 = \left(e_3 \cdot e'_1\right) v_1 + \left(e_3 \cdot e'_2\right) v_2 + \left(e_3 \cdot e'_3\right) v_3,$$

which can be written in the matrix form

$$\boxed{\mathsf{v} = \mathsf{A}^T \cdot \mathsf{v}'} \tag{18.6}$$

where A^T is the **transpose** of A. If we now substitute equation (18.4) into equation (18.6) we obtain

$$\mathsf{v} = \mathsf{A}^T \cdot (\mathsf{A} \cdot \mathsf{v}) = \left(\mathsf{A}^T \cdot \mathsf{A}\right) \cdot \mathsf{v}.$$

Since this formula holds for every choice of v, it follows that $\mathsf{A}^T \cdot \mathsf{A} = \mathbf{1}$, where $\mathbf{1}$ is the identity matrix. This means that the transformation matrix A has the special property that $\mathsf{A}^T = \mathsf{A}^{-1}$. Such matrices are called **orthogonal**.

Definition 18.1 *Orthogonal matrix* *A matrix having the property that its transpose is also its inverse, that is,*

$$\mathsf{A}^T \cdot \mathsf{A} = \mathsf{A} \cdot \mathsf{A}^T = \mathbf{1}, \tag{18.7}$$

is called an ***orthogonal*** *matrix.*

Thus the transformation matrix $\mathbf{A}$, defined by (18.5), is an orthogonal matrix. For this reason, the transformation formula (18.4) is called an **orthogonal transformation.**

Example 18.1 ***Prove a matrix is orthogonal***

Prove that the matrix

$$\mathbf{B} = \frac{1}{9}\begin{pmatrix} 4 & 7 & -4 \\ 1 & 4 & 8 \\ 8 & -4 & 1 \end{pmatrix}$$

is orthogonal.

Solution

$$\mathbf{B}^T \cdot \mathbf{B} = \frac{1}{81}\begin{pmatrix} 4 & 1 & 8 \\ 7 & 4 & -4 \\ -4 & 8 & 1 \end{pmatrix}\begin{pmatrix} 4 & 7 & -4 \\ 1 & 4 & 8 \\ 8 & -4 & 1 \end{pmatrix} = \frac{1}{81}\begin{pmatrix} 81 & 0 & 0 \\ 0 & 81 & 0 \\ 0 & 0 & 81 \end{pmatrix} = \mathbf{1}.$$

There is no need to check that $\mathbf{B} \cdot \mathbf{B}^T = \mathbf{1}$ as this now follows automatically. ■

18.2 ROTATED AND REFLECTED COORDINATE SYSTEMS

In this section we work out the transformation matrices that correspond to specific **rotations** and/or **reflections** of the coordinate system C. Important though this is, it is not needed for the general understanding of tensors.

Suppose now that the coordinate system C' is *defined* by a specified rotation of C about an axis through O. This notion needs a little thought. It should be remembered that the coordinate system C does not actually move; the 'rotation' of C is hypothetical and is merely a way of describing the orientation of C' relative to C. What we are saying is that the orientation of C' is the same as C *would* have *if* it were rotated in the prescribed manner.

Consider first the **special case** in which C' is obtained by rotating C through an angle ψ about the axis Ox_3, as shown in Figure 18.2 (Left). In this case,

$$\begin{array}{lll} e_1' \cdot e_1 = \cos\psi & e_1' \cdot e_2 = \sin\psi & e_1' \cdot e_3 = 0 \\ e_2' \cdot e_1 = -\sin\psi & e_2' \cdot e_2 = \cos\psi & e_2' \cdot e_3 = 0 \\ e_3' \cdot e_1 = 0 & e_3' \cdot e_2 = 0 & e_3' \cdot e_3 = 1 \end{array}$$

and the **transformation matrix** between C and C' is therefore

$$\mathbf{A} = \begin{pmatrix} \cos\psi & \sin\psi & 0 \\ -\sin\psi & \cos\psi & 0 \\ 0 & 0 & 1 \end{pmatrix}. \tag{18.8}$$

Since this example is a special case of the preceding theory, the matrix $\mathbf{A}$ in (18.8) must be orthogonal. [You should check this directly.] A rotation of C about the Ox_1 or Ox_2 axis is treated in the same way.

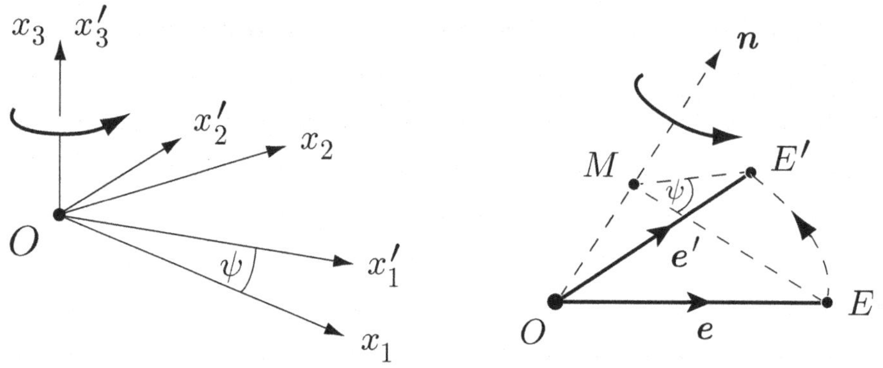

FIGURE 18.2 **Left**: The coordinate system $\mathcal{C}' \equiv Ox_1'x_2'x_3'$ is obtained by rotating the coordinate system $\mathcal{C} \equiv Ox_1x_2x_3$ through an angle ψ about the axis Ox_3. **Right**: The unit vector $\boldsymbol{e}'$ is obtained by rotating the unit vector $\boldsymbol{e}$ through an angle ψ about the general axis $\{O, \boldsymbol{n}\}$.

Example 18.2 *Rotation of $\mathcal{C}$ about a coordinate axis*

The coordinate system $\mathcal{C}'$ is obtained by rotating the coordinate system $\mathcal{C}$ through an angle of 30° about the axis Ox_1. Find the transformation matrix **A** between $\mathcal{C}$ and $\mathcal{C}'$.

A vector $\boldsymbol{v}$ has components $\{3, -1, 2\}$ in $\mathcal{C}$. What are its components in $\mathcal{C}'$?

Solution

The formula corresponding to (18.8) when $\mathcal{C}$ is rotated through an angle ψ about the axis Ox_1 is

$$\mathsf{A} = \begin{pmatrix} 1 & 0 & 0 \\ 0 & \cos\psi & \sin\psi \\ 0 & -\sin\psi & \cos\psi \end{pmatrix}.$$

Hence, when $\psi = 30°$, the **transformation matrix** is

$$\mathsf{A} = \begin{pmatrix} 1 & 0 & 0 \\ 0 & \sqrt{3}/2 & 1/2 \\ 0 & -1/2 & \sqrt{3}/2 \end{pmatrix}.$$

The components of the vector $\boldsymbol{v}$ in $\mathcal{C}'$ are then given by the elements of the column vector v', where

$$\mathsf{v}' = \mathsf{A}\cdot\begin{pmatrix} 3 \\ -4 \\ 2 \end{pmatrix} = \begin{pmatrix} 1 & 0 & 0 \\ 0 & \sqrt{3}/2 & 1/2 \\ 0 & -1/2 & \sqrt{3}/2 \end{pmatrix}\begin{pmatrix} 3 \\ -4 \\ 2 \end{pmatrix} = \begin{pmatrix} 3 \\ 1-2\sqrt{3} \\ 2+\sqrt{3} \end{pmatrix}.$$

The **components** of the vector $\boldsymbol{v}$ in $\mathcal{C}'$ are therefore $\{3,\ 1-2\sqrt{3},\ 2+\sqrt{3}\}$. ■

In the **general case**, the rotation of $\mathcal{C}$ can take place about an axis $\{O, \boldsymbol{n}\}$, where $\boldsymbol{n}$ is *any* unit vector. The argument is messier but we can still find an explicit formula for **A**.

Figure 18.2 (Right) shows how a typical unit vector $\boldsymbol{e}'$ of the coordinate system C' is obtained from the corresponding vector $\boldsymbol{e}$ belonging to C. Let $\boldsymbol{e}$ be resolved into parts parallel and perpendicular to the axis $\{O, \boldsymbol{n}\}$ in the form

$$\boldsymbol{e} = \boldsymbol{e}^{\parallel} + \boldsymbol{e}^{\perp},$$

where

$$\boldsymbol{e}^{\parallel} = \overrightarrow{OM} = (\boldsymbol{e} \cdot \boldsymbol{n})\,\boldsymbol{n}, \qquad \boldsymbol{e}^{\perp} = \overrightarrow{ME} = \boldsymbol{e} - (\boldsymbol{e} \cdot \boldsymbol{n})\,\boldsymbol{n}.$$

Then, if $\boldsymbol{e}'$ is similarly resolved as

$$\boldsymbol{e}' = \boldsymbol{e}'^{\parallel} + \boldsymbol{e}'^{\perp},$$

it follows that $\boldsymbol{e}'^{\parallel} = \boldsymbol{e}^{\parallel}$ and

$$\boldsymbol{e}'^{\perp} = \cos\psi\, \boldsymbol{e}^{\perp} + \sin\psi \left(\boldsymbol{n}\times\boldsymbol{e}^{\perp}\right) = \cos\psi\,(\boldsymbol{e} - (\boldsymbol{e}\cdot\boldsymbol{n})\,\boldsymbol{n}) + \sin\psi\,(\boldsymbol{n}\times\boldsymbol{e}).$$

Hence $\boldsymbol{e}'$ is related to $\boldsymbol{e}$ by the formula

$$\boldsymbol{e}' = \cos\psi\,\boldsymbol{e} + (1 - \cos\psi)\,(\boldsymbol{e}\cdot\boldsymbol{n})\,\boldsymbol{n} + \sin\psi\,(\boldsymbol{n}\times\boldsymbol{e}). \tag{18.9}$$

If we now take $\boldsymbol{e} = \boldsymbol{e}_1$, $\boldsymbol{e}' = \boldsymbol{e}_1'$ and take the scalar product of equation (18.9) successively with $\boldsymbol{e}_1$, $\boldsymbol{e}_2$ and $\boldsymbol{e}_3$, we obtain

$$\begin{aligned}
\boldsymbol{e}_1' \cdot \boldsymbol{e}_1 &= \cos\psi(\boldsymbol{e}_1\cdot\boldsymbol{e}_1) + (1-\cos\psi)(\boldsymbol{n}\cdot\boldsymbol{e}_1)(\boldsymbol{n}\cdot\boldsymbol{e}_1) + \sin\psi(\boldsymbol{n}\times\boldsymbol{e}_1)\cdot\boldsymbol{e}_1\\
&= \cos\psi + (1-\cos\psi)n_1^2 + 0,\\
\boldsymbol{e}_1' \cdot \boldsymbol{e}_2 &= \cos\psi(\boldsymbol{e}_1\cdot\boldsymbol{e}_2) + (1-\cos\psi)(\boldsymbol{n}\cdot\boldsymbol{e}_1)(\boldsymbol{n}\cdot\boldsymbol{e}_2) + \sin\psi(\boldsymbol{n}\times\boldsymbol{e}_1)\cdot\boldsymbol{e}_2\\
&= 0 + (1-\cos\psi)n_1n_2 + n_3\sin\psi,\\
\boldsymbol{e}_1' \cdot \boldsymbol{e}_3 &= \cos\psi(\boldsymbol{e}_1\cdot\boldsymbol{e}_3) + (1-\cos\psi)(\boldsymbol{n}\cdot\boldsymbol{e}_1)(\boldsymbol{n}\cdot\boldsymbol{e}_3) + \sin\psi(\boldsymbol{n}\times\boldsymbol{e}_1)\cdot\boldsymbol{e}_3\\
&= 0 + (1-\cos\psi)n_1n_3 - n_2\sin\psi,
\end{aligned}$$

where n_1, n_2, n_3 are the components of the vector $\boldsymbol{n}$ in the coordinate system C. This gives the elements of the first row of the transformation matrix A, and the elements of the second and third rows can be found in a similar manner. The full transformation matrix A is given by

$$\begin{pmatrix}
\cos\psi + (1-\cos\psi)n_1^2 & (1-\cos\psi)n_1n_2 + n_3\sin\psi & (1-\cos\psi)n_1n_3 - n_2\sin\psi\\
(1-\cos\psi)n_2n_1 - n_3\sin\psi & \cos\psi + (1-\cos\psi)n_2^2 & (1-\cos\psi)n_2n_3 + n_1\sin\psi\\
(1-\cos\psi)n_3n_1 + n_2\sin\psi & (1-\cos\psi)n_3n_2 - n_1\sin\psi & \cos\psi + (1-\cos\psi)n_3^2
\end{pmatrix} \tag{18.10}$$

This is the transformation matrix A when C' is obtained by rotating C through an angle ψ about the axis $\{O, \boldsymbol{n}\}$; n_1, n_2, n_3 are components of the unit vector $\boldsymbol{n}$ in the coordinate system C.

Example 18.3 *Rotation of C about a general axis*

The coordinate system C' is obtained by rotating the coordinate system C through an angle of $60°$ about the axis $\overrightarrow{OP}$, where P is the point with coordinates $(1, 1, -1)$ in C. Find the transformation matrix A between C and C'.

A vector $\boldsymbol{v}$ has components $\{3, -6, 9\}$ in C'. What are its components in C?

Solution

The unit vector $\boldsymbol{n}$ in the direction $\overrightarrow{OP}$ has components $\{1/\sqrt{3}, 1/\sqrt{3}, -1/\sqrt{3}\}$ in C, and $\psi = 60°$. It follows from the formula (18.10) that the transformation matrix from

$\mathcal{C}$ to $\mathcal{C}'$ is

$$\mathsf{A} = \frac{1}{3}\begin{pmatrix} 2 & -1 & -2 \\ 2 & 2 & 1 \\ 1 & -2 & 2 \end{pmatrix}.$$

The components of the vector $\boldsymbol{v}$ in $\mathcal{C}$ are then given by the elements of the column vector v', where

$$\mathsf{v}' = \mathsf{A}^T \cdot \begin{pmatrix} 3 \\ -6 \\ 9 \end{pmatrix} = \frac{1}{3}\begin{pmatrix} 2 & 2 & 1 \\ -1 & 2 & -2 \\ -2 & 1 & 2 \end{pmatrix}\begin{pmatrix} 3 \\ -6 \\ 9 \end{pmatrix} = \begin{pmatrix} 1 \\ -11 \\ 2 \end{pmatrix}.$$

The components of the vector $\boldsymbol{v}$ in $\mathcal{C}$ are therefore $\{1, -11, 2\}$. ■

Reflections

The coordinate system $\mathcal{C}'$ may also be defined by **reflecting** $\mathcal{C}$ in a plane through O. For example, if $\mathcal{C}$ is reflected in the coordinate plane Ox_1x_2, the transformation matrix is easily found to be

$$\mathsf{A} = \begin{pmatrix} 1 & 0 & 0 \\ 0 & 1 & 0 \\ 0 & 0 & -1 \end{pmatrix}.$$

In the general case, if $\mathcal{C}$ is reflected in the plane through O with unit normal vector $\boldsymbol{n}$, then the transformation matrix is

$$\mathsf{A} = \begin{pmatrix} 1-2n_1^2 & -2n_1n_2 & -2n_1n_3 \\ -2n_2n_1 & 1-2n_2^2 & -2n_2n_3 \\ -2n_3n_1 & -2n_3n_2 & 1-2n_3^2 \end{pmatrix}, \qquad (18.11)$$

where n_1, n_2, n_3 are the components of the vector $\boldsymbol{n}$ in $\mathcal{C}$. [The method of proof is similar to that for the general rotation.]

Example 18.4 *Reflection of axes*

The coordinate system $\mathcal{C}'$ is obtained by reflecting the coordinate system $\mathcal{C}$ in the plane $x_3 = 2x_1 + 2x_2$. Find the transformation matrix A between $\mathcal{C}$ and $\mathcal{C}'$.

Find also the transformation matrix when $\mathcal{C}'$ is obtained from $\mathcal{C}$ by performing the above reflection followed by a rotation of 90° about the *new* x_3-axis.

Solution

The equation $x_3 = 2x_1 + 2x_2$ can be written in the form

$$(x_1\boldsymbol{e}_1 + x_2\boldsymbol{e}_2 + x_3\boldsymbol{e}_3) \cdot (2\boldsymbol{e}_1 + 2\boldsymbol{e}_2 - \boldsymbol{e}_3)$$

and, by comparison with the standard vector equation $\boldsymbol{r} \cdot \boldsymbol{n} = p$ of a plane, we see that the unit vector

$$\boldsymbol{n} = \frac{2}{3}\boldsymbol{e}_1 + \frac{2}{3}\boldsymbol{e}_2 - \frac{1}{3}\boldsymbol{e}_3$$

is normal to the reflection plane. Hence $n_1 = 2/3$, $n_2 = 2/3$ and $n_3 = -1/3$. It follows from the formula (18.11) that the transformation matrix between C and C' is

$$\mathsf{A} = \frac{1}{9}\begin{pmatrix} 1 & -8 & 4 \\ -8 & 1 & 4 \\ 4 & 4 & 7 \end{pmatrix}.$$

If this transformation is followed by a rotation whose transformation matrix is

$$\begin{pmatrix} 0 & 1 & 0 \\ -1 & 0 & 0 \\ 0 & 0 & 1 \end{pmatrix}$$

then the overall transformation has matrix

$$\frac{1}{9}\begin{pmatrix} 0 & 1 & 0 \\ -1 & 0 & 0 \\ 0 & 0 & 1 \end{pmatrix}\begin{pmatrix} 1 & -8 & 4 \\ -8 & 1 & 4 \\ 4 & 4 & 7 \end{pmatrix} = \frac{1}{9}\begin{pmatrix} -8 & 1 & 4 \\ -1 & 8 & -4 \\ 4 & 4 & 7 \end{pmatrix}. \blacksquare$$

It can be proved that the most **general orthogonal transformation** corresponds either to (i) a rotation of coordinates, or (ii) a reflection followed by a rotation. Which of these classes a particular orthogonal transformation belongs to can be decided by examining the *sign* of the determinant det **A**.

If we take determinants of equation (18.7) we obtain

$$\det \mathbf{1} = \det\left(\mathsf{A}^T \cdot \mathsf{A}\right) = \det \mathsf{A}^T \times \det \mathsf{A} = \det \mathsf{A} \times \det \mathsf{A} = (\det \mathsf{A})^2,$$

from which it follows that det **A** can only take the values ± 1. Surprisingly, we can work out the determinant of the general **A** defined by equation (18.5) quite easily. The first row of **A** contains the three components of the vector e'_1 in the coordinate system C. Likewise, the second and third rows contain the components of the vectors e'_2 and e'_3. It follows from the vector formula (1.7) that, provided C is a **right-handed** coordinate system,

$$\det \mathsf{A} = \left(e'_1 \times e'_2\right) \cdot e'_3.$$

We can now see how the different signs for det **A** can occur. If C' is also a **right-handed** coordinate system, then the triple scalar product $\left(e'_1 \times e'_2\right) \cdot e'_3 = +1$, whereas if C' is a **left-handed** system then $\left(e'_1 \times e'_2\right) \cdot e'_3 = -1$. If two coordinate systems have the same handedness, then one can be made coincident with the other by a suitable **rotation** about an axis through O. If they have opposite handedness then a **reflection** in a plane through O is also needed. For this reason, orthogonal matrices with determinant $+1$ are sometimes called **rotation matrices.**

18.3 SCALARS, VECTORS AND TENSORS

Our view of a vector as a quantity that has magnitude and direction has served us well. By picturing a vector as a line segment with an arrow on it, we have been able to understand essentially difficult concepts, such as acceleration and angular momentum, far more easily. Neither of these quantities has any direct connection with line segments, but the picture is a valuable aid nonetheless. However, this notion of vectors cannot be generalised to tensors. This is the main reason we will now look at vectors (and even scalars) in a different light.

Scalars

We begin by giving the true definition of a scalar.

Definition 18.2 *Scalar* *Let ϕ be a real number defined in each coordinate system.* Then ϕ is said to be a **scalar** if it has the same value in every coordinate system. That is, for any pair of coordinate systems C and C',*

Definition of a scalar

$$\phi' = \phi. \tag{18.12}$$

*This can be alternatively expressed by saying that a scalar is **invariant** under change of coordinate system.*

It is this invariance under change of coordinate system that is the essential feature of a scalar. Thus the **mass** of a particle and the **length** of a line are scalars. However, the sum of the coordinates of a given point of space (which *is* a single real quantity defined in each coordinate system) is not a scalar since it is *not* invariant under change of coordinate system.

Vectors

We turn now to the definition of a vector in terms of its components. The individual components of a vector are certainly not invariants. They transform according to the transformation formula (18.4). It is this transformation formula that becomes our **new definition** of a vector.

Definition 18.3 *Vector* *Let $\{v_1, v_2, v_3\}$ be a set of three real numbers defined in each coordinate system. Then $\{v_1, v_2, v_3\}$ are said to be the components of a **vector** if their values in any pair of coordinate systems C and C' are related by the transformation formula*

Definition of a vector (matrix form)

$$\mathbf{v}' = \mathbf{A} \cdot \mathbf{v} \tag{18.13}$$

where $\mathbf{A}$ is the transformation matrix between C and C'.

* These are rectangular, Cartesian coordinate systems with common origin O.

This definition of a vector is clearly consistent with our previous point of view. Note that a set of three real numbers defined in each coordinate system does *not* in general constitute a vector; *the components must be related to each other by the transformation formula* (18.13). For example, the Cartesian coordinates (b_1, b_2, b_3) of a particle B are a set of three real numbers defined in each coordinate system and these numbers *are* related by the transformation formula (18.13). Thus $\{b_1, b_2, b_3\}$ is a **vector**. Suppose now that B is a moving particle with time dependent coordinates $(b_1(t), b_2(t), b_3(t))$. Then, at each time t, $\{\dot{b}_1(t), \dot{b}_2(t), \dot{b}_3(t)\}$ is also a set of three of real numbers defined in each coordinate system. The transformation formula (18.13) is again satisfied since the matrix **A** does not depend upon the time. Hence, at each time t, $\{\dot{b}_1(t), \dot{b}_2(t), \dot{b}_3(t)\}$ is a **vector**. This is, of course, the **velocity vector** of the particle B at time t, and a second differentiation yields the **acceleration vector** of B. In contrast, $\{2b_1, b_2, b_3\}$ is a set of three real numbers defined in each coordinate system, but the transformation formula is not satisfied. Therefore $\{2b_1, b_2, b_3\}$ is not a vector.

Tensors

In order to generalise our definition of a vector, it is necessary to write the vector transformation formula (18.13) in the following suffix form:

Definition of a vector (suffix form)

$$v_i' = \sum_{j=1}^{3} a_{ij}\, v_j \tag{18.14}$$

$(1 \le i \le 3)$, where a_{pq} is the element in the p-th row and q-th column of the matrix **A**. The matrix product in equation (18.13) is equivalent to the summation in equation (18.14).

Defined in the above way, vectors can be regarded as part of a hierarchy of entities called **tensors** with a tensor of order n having 3^n components. A tensor of order one, which has three components, is a vector; a tensor of order zero, which has one component, is a scalar. The general definition of tensors looks extremely daunting when first encountered. Below we give the definitions of second and third order tensors, from which the general case can be inferred.

Definition 18.4 ***Second order tensor*** *Let* $\{t_{ij}\}$ $(1 \le i, j \le 3)$ *be a set of* ***nine*** *real numbers defined in each coordinate system. Then the* $\{t_{ij}\}$ *are said to be the components of a* ***second order tensor*** *if their values in any pair of coordinate systems* C *and* C' *are related by the* ***transformation formula***

Definition of a second order tensor (suffix form)

$$t'_{ij} = \sum_{k=1}^{3}\sum_{l=1}^{3} a_{ik}\, a_{jl}\, t_{kl} \qquad (18.15)$$

$(1 \le i, j \le 3)$, where the $\{a_{pq}\}$ are the elements of the transformation matrix between C and C'.

Definition 18.5 ***Third order tensor*** *Let $\{t_{ijk}\}$ $(1 \le i, j, k \le 3)$ be a set of **twenty seven** real numbers defined in each coordinate system. Then the $\{t_{ijk}\}$ are said to be the components of a **third order tensor** if their values in any pair of coordinate systems C and C' are related by the **transformation formula***

Definition of a third order tensor (suffix form)

$$t'_{ijk} = \sum_{l=1}^{3}\sum_{m=1}^{3}\sum_{n=1}^{3} a_{il}\, a_{jm}\, a_{kn}\, t_{lmn} \qquad (18.16)$$

$(1 \le i, j, k \le 3)$, where the $\{a_{pq}\}$ are the elements of the transformation matrix between C and C'.

Notes on the tensor transformation formulae

The transformation formulae (18.14), (18.15), and (18.16) may seem incomprehensible but they do follow a pattern. In the definition of a vector there is only one summation and one appearance of a_{pq}; in the definition of a tensor of the second order, there are two summations and two appearances of a_{pq}, and so on. *Correct positioning of the suffices is vital.* The suffices of the tensor on the left (in order) must be the same as the first suffix of each of the a_{pq} (in order); and the suffices of the tensor on the right (in order) must be the same as the second suffix of each of the a_{pq} (in order).*

Power users of tensors write formulae such as those above *without* the summation signs. They adopt the **summation convention**† that any repeated suffix is deemed to be summed over its range. Although the summation convention is widely used in most applications of tensors, we do *not* use it here. This is for two reasons: (i) it is a good thing to see the summation signs written in when tensors are first encountered, and (ii) we have no need of the heavy tensor algebra for which the summation convention was designed.

---

* Be a hero. Write out the transformation formula for a fourth order tensor.

† Invented by Einstein to simplify the writing of the theory of general relativity.

Second order tensors in matrix form

In the general case it is not possible to give any simpler form for the tensor transformation formula. However, for the special case of tensors of the **second order** (which is what we are mainly concerned with), the transformation formula can be written in a more user-friendly form in terms of **matrix products**.

In each coordinate system, a second order tensor $\{t_{ij}\}$ has nine components, indexed by the integers i, j $(1 \le i, j \le 3)$. It is natural then to display these components as a 3×3 array, that is, as the elements of a 3×3 matrix. The tensor $\{t_{ij}\}$ can then be regarded as a 3×3 matrix

$$\mathbf{T} = \begin{pmatrix} t_{11} & t_{12} & t_{13} \\ t_{21} & t_{22} & t_{23} \\ t_{31} & t_{32} & t_{33} \end{pmatrix}.$$

defined in each coordinate system. Also, the tensor transformation rule (18.15) can be written in the form

$$\begin{aligned} t'_{ij} &= \sum_{k=1}^{3}\sum_{l=1}^{3} a_{ik}\, a_{jl}\, t_{kl} = \sum_{l=1}^{3}\left(\sum_{k=1}^{3} a_{ik}\, t_{kl}\right) a_{jl} \\ &= \sum_{l=1}^{3}\left(\sum_{k=1}^{3} a_{ik}\, t_{kl}\right) a^T_{lj}, \end{aligned}$$

where the $\{a^T_{pq}\}$ are the elements of the transposed matrix $\mathbf{A}^T$. The summations in this last expression are equivalent to matrix products and so the transformation rule for a second order tensor can be expressed in the form

$$\mathbf{T}' = \mathbf{A} \cdot \mathbf{T} \cdot \mathbf{A}^T.$$

This gives the following **alternative definition**:

Definition 18.6 ***Second order tensor (matrix form)*** *Let* $\mathbf{T}$ *be a* 3×3 *matrix defined in each coordinate system. Then* $\mathbf{T}$ *is said to represent a* ***second order tensor*** *if its values in any pair of coordinate systems* C *and* C' *are related by the* ***transformation formula***

Definition of a second order tensor (matrix form)

$$\mathbf{T}' = \mathbf{A} \cdot \mathbf{T} \cdot \mathbf{A}^T \qquad (18.17)$$

where the $\mathbf{A}$ *is the transformation matrix between* C *and* C'.

This definition has the virtue of being expressed in terms of the familiar operation of matrix multiplication. It only applies to second order tensors but that is exactly what we need. The **inertia tensor** is second order, as are many other important tensors of physics.

Example 18.5 *Transforming a second order tensor*

In the coordinate system $\mathcal{C}$, a second order tensor has the components $t_{11} = 1, t_{12} = 2, t_{13} = 3, t_{21} = 4, t_{22} = 5, t_{23} = 6, t_{31} = 7, t_{32} = 8, t_{33} = 9$. Find the components of the tensor in $\mathcal{C}'$ when

(i) $\mathcal{C}'$ is obtained from $\mathcal{C}$ by a rotation of 90° about the axis Ox_3, and

(ii) when $\mathcal{C}'$ is obtained from $\mathcal{C}$ by a reflection in the plane $x_3 = 0$.

Solution

In the coordinate system $\mathcal{C}$, the matrix of components is given to be

$$\mathsf{T} = \begin{pmatrix} 1 & 2 & 3 \\ 4 & 5 & 6 \\ 7 & 8 & 9 \end{pmatrix}.$$

Part (i). From the formula (18.8), the transformation matrix of the required rotation of coordinates is

$$\mathsf{A} = \begin{pmatrix} 0 & 1 & 0 \\ -1 & 0 & 0 \\ 0 & 0 & 1 \end{pmatrix}.$$

The transformation formula (18.17) then implies that, in the coordinate system $\mathcal{C}'$, the tensor is represented by the matrix

$$\mathsf{T}' = \mathsf{A} \cdot \mathsf{T} \cdot \mathsf{A}^T = \begin{pmatrix} 0 & 1 & 0 \\ -1 & 0 & 0 \\ 0 & 0 & 1 \end{pmatrix} \begin{pmatrix} 1 & 2 & 3 \\ 4 & 5 & 6 \\ 7 & 8 & 9 \end{pmatrix} \begin{pmatrix} 0 & -1 & 0 \\ 1 & 0 & 0 \\ 0 & 0 & 1 \end{pmatrix}$$

$$= \begin{pmatrix} 5 & -4 & 6 \\ -2 & 1 & -3 \\ 8 & -7 & 9 \end{pmatrix}.$$

Hence the **components** of the tensor in $\mathcal{C}'$ are $t'_{11} = 5, t'_{12} = -4, t'_{13} = 6, t'_{21} = -2, t'_{22} = 1, t'_{23} = -3, t'_{31} = 8, t'_{32} = -7, t'_{33} = 9$.

Part (ii). The transformation matrix of the required **reflection** of coordinates is

$$\mathsf{A} = \begin{pmatrix} 1 & 0 & 0 \\ 0 & 1 & 0 \\ 0 & 0 & -1 \end{pmatrix}.$$

The transformation formula (18.17) then implies that

$$\mathsf{T}' = \mathsf{A} \cdot \mathsf{T} \cdot \mathsf{A}^T = \begin{pmatrix} 1 & 0 & 0 \\ 0 & 1 & 0 \\ 0 & 0 & -1 \end{pmatrix} \begin{pmatrix} 1 & 2 & 3 \\ 4 & 5 & 6 \\ 7 & 8 & 9 \end{pmatrix} \begin{pmatrix} 1 & 0 & 0 \\ 0 & 1 & 0 \\ 0 & 0 & -1 \end{pmatrix}$$

$$= \begin{pmatrix} 1 & 2 & -3 \\ 4 & 5 & -6 \\ -7 & -8 & 9 \end{pmatrix}.$$

Hence the **components** of the tensor in C' are $t'_{11} = 1$, $t'_{12} = 2$, $t'_{13} = -3$, $t'_{21} = 4$, $t'_{22} = 5$, $t'_{23} = -6$, $t'_{31} = -7$, $t'_{32} = -8$, $t'_{33} = 9$. ■

The identity tensor

The simplest second order tensor is the **identity tensor**. This tensor has the components $\{\delta_{ij}\}$ in *every* coordinate system, where δ_{ij} is the Krönecker delta, defined by

$$\delta_{ij} = \begin{cases} 1 & (i = j) \\ 0 & (i \neq j) \end{cases}$$

In each coordinate system, the identity tensor is represented by the **identity matrix 1**. The transformation formula (18.17) is certainly satisfied since

$$\mathsf{A} \cdot \mathsf{1} \cdot \mathsf{A}^T = \mathsf{A} \cdot \mathsf{A}^T = \mathsf{1},$$

thus confirming that $\{\delta_{ij}\}$ is a tensor.

The identity tensor is one of a very restricted class of tensors that have the *same components in all coordinate systems*. Such tensors are called **isotropic** and they have an important physical rôle. All **scalars** are, by definition, isotropic; there are no (non-trivial) isotropic **vectors**; and the only isotropic second order **tensors** are scalar multiples of the identity tensor. It becomes increasingly difficult to identify the isotropic tensors of higher orders!

18.4 TENSOR ALGEBRA

From the linearity of the transformation formulae, it follows that a **scalar multiple** of a tensor is a tensor, and the **sum** of two tensors (of the same order) is a tensor. We will now look at two other ways in which new tensors can be created. This gives an indication how tensors can arise naturally.

The outer product of two tensors

Suppose, for example, that $\{u_{ij}\}$ is a **second order** tensor and $\{v_{ijk}\}$ a **third order** tensor. Then $\{t_{ijklm}\}$ defined* by

$$t_{ijklm} = u_{ij} v_{klm}$$

is a 243 component quantity, defined in each coordinate system. Although we will not prove this here, $\{t_{ijklm}\}$ is a **fifth order tensor** by virtue of satisfying the appropriate transformation formula. This is an example of the **outer product** of two tensors. In general, the *outer product of a tensor of order m and a tensor of order n is a new tensor of order $m + n$.*

* To see what this means, write out the definitions of a few elements of $\{t_{ijklm}\}$; for example, $t_{13231} = u_{13} v_{231}$. Note that, in an outer product, all the suffix *names* must be different.

The outer product is one way in which higher order tensors can be constructed from lower order ones. For example, suppose $\{u_i\}$ and $\{v_i\}$ are vectors (tensors of order one). Then the nine component quantity $\{u_i v_j\}$ is a **second order tensor** whose matrix form is

$$\begin{pmatrix} u_1v_1 & u_1v_2 & u_1v_3 \\ u_2v_1 & u_2v_2 & u_2v_3 \\ u_3v_1 & u_3v_2 & u_3v_3 \end{pmatrix}.$$

In the same way, if $\{u_i\}$, $\{v_i\}$ and $\{w_i\}$ are vectors, then the twenty seven component quantity $\{u_i v_j w_k\}$ is a **third order tensor**. Not all higher order tensors can be so constructed, but it is a common procedure. The inertia tensor makes use of this construction.

Contraction of a tensor

Suppose, for example, that $\{t_{ijkl}\}$ is a fourth order tensor. Now select **two** of its suffices (k and l say), set them **equal** (to m say)* and **sum** over the suffix m. The result is the nine component quantity

$$w_{ij} = \sum_{m=1}^{3} t_{ijmm},$$

defined in each coordinate system. Although we will not prove this here, $\{w_{ij}\}$ is a **second order tensor** by virtue of satisfying the appropriate transformation formula. The tensor $\{w_{ij}\}$ is called the **contraction** of $\{t_{ijkl}\}$ with respect to the suffix pair k, l. [There are six different contractions of $\{t_{ijkl}\}$. What are they?] In the general case, *contraction of a tensor produces a new tensor whose order is two less than that of the original tensor.*

Contraction of a second order tensor is the simplest case. If $\{t_{ij}\}$ is a second order tensor, then its contraction with respect to the suffix pair i, j, namely,

$$\sum_{i=1}^{3} t_{ii} = t_{11} + t_{22} + t_{33},$$

must be a tensor of order zero, that is, a **scalar invariant**. If $\{t_{ij}\}$ is represented by the matrix $\mathbf{T}$, then the contraction $t_{11} + t_{22} + t_{33}$ is the sum of the diagonal elements of $\mathbf{T}$. In linear algebra this is called the **trace** of $\mathbf{T}$. The fact that the *trace of a second order tensor is invariant*† is an important result. For example, suppose $\{u_i\}$ and $\{v_i\}$ are vectors. Then the outer product $\{u_i v_j\}$ is a second order tensor. It now follows that the trace of this tensor, namely $u_1v_1 + u_2v_2 + u_3v_3$ is an invariant. This is actually no surprise, since $u_1v_1 + u_2v_2 + u_3v_3$ is the scalar product of the vectors $\{u_i\}$ and $\{v_i\}$, which is known to be independent of the coordinate system in which the components are measured.

---

* The name of this repeated suffix can be any name not already in use. However, it is permissible to re-use either of the old suffix names that were set equal (k or l in this example).

† This result is by no means obvious. After all, the individual components of $\{t_{ij}\}$ are not invariants.

Another important application of contraction is as follows. Suppose that $\{t_{ij}\}$ is a second order tensor and that $\{u_i\}$ is a vector. Then the outer product $\{t_{ij}u_k\}$ is a third order tensor. If we now contract this third order tensor with respect to the suffix pair j, k, we obtain the vector $\{w_i\}$ given by

$$w_i = \sum_{j=1}^{3} t_{ij}u_j.$$

This formula looks nice in the matrix formulation. If $\{t_{ij}\}$ is represented by the matrix T and $\{u_i\}$ by the column vector u, then $\{v_i\}$ is represented by the column vector v given by

$$\mathsf{v} = \mathsf{T} \cdot \mathsf{u}.$$

In other words, *if the matrix* T *represents a second order tensor and the column vector* u *represents a vector, then the product* $\mathsf{T} \cdot \mathsf{u}$ *represents a vector.* Thus pre-multiplication by T transforms one vector into another vector. This is a common way in which tensors act in physics. For example, in crystalline materials, the electric vectors $\boldsymbol{E}$ and $\boldsymbol{D}$ are not parallel. In each coordinate system, they are related by the formula

$$\mathsf{D} = \mathsf{K} \cdot \mathsf{E},$$

where the 3×3 matrix K represents the **dielectric tensor**, the crystalline equivalent of the dielectric constant. This tensor relationship between $\boldsymbol{D}$ and $\boldsymbol{E}$ is the cause of double refraction in crystalline materials. The inertia tensor has a similar rôle, transforming the angular velocity vector into the angular momentum vector.

The angular velocity vector

We have already shown that the **velocity** of a particle is a vector in the sense of the definition (18.13), but we have so far said nothing about the transformation properties of the **angular velocity** of a rigid body. This is made more awkward to decide by the fact that $\boldsymbol{\omega}$ is defined *indirectly*. Suppose that one of the particles of a rigid body is fixed at the origin O. Then $\boldsymbol{\omega}$ is essentially defined to be that 'vector' that gives the velocities of the particles of the body by the formula $\boldsymbol{v} = \boldsymbol{\omega} \times \boldsymbol{r}$. In each coordinate system this formula can also be expressed by the *matrix* product

$$\mathsf{v} = \boldsymbol{\Omega} \cdot \mathsf{x},$$

where

$$\mathsf{v} = \begin{pmatrix} v_1 \\ v_2 \\ v_3 \end{pmatrix}, \qquad \boldsymbol{\Omega} = \begin{pmatrix} 0 & -\omega_3 & \omega_2 \\ \omega_3 & 0 & -\omega_1 \\ -\omega_2 & \omega_1 & 0 \end{pmatrix}, \qquad \mathsf{x} = \begin{pmatrix} x_1 \\ x_2 \\ x_3 \end{pmatrix}.$$

[Check this.] Now $\{x_i\}$ and $\{v_i\}$ are known to transform as vectors, but nothing is known about the transformation properties of the matrix $\boldsymbol{\Omega}$. However, it can be proved that if a matrix maps vectors into vectors by matrix multiplication, then that matrix represents a second order tensor. Hence $\mathbf{\Omega}$ transforms as a second order tensor. This must, in turn, imply a rule for transforming the $\{\omega_i\}$. It looks as if this rule will be something horrendous, but, amazingly, it turns out that the rule is

$$\begin{pmatrix} \omega_1' \\ \omega_2' \\ \omega_3' \end{pmatrix} = (\det \mathbf{A})\, \mathbf{A} \cdot \begin{pmatrix} \omega_1 \\ \omega_2 \\ \omega_3 \end{pmatrix}, \tag{18.18}$$

which is *not quite* the transformation formula for a vector because of the factor $\det \mathbf{A}$. If the transformation $\mathbf{A}$ acts between systems with the *same* handedness, then $\det \mathbf{A} = +1$ and (18.18) is the vector transformation formula. However, if the transformation acts between systems with opposite handedness, then $\det \mathbf{A} = -1$ and the quantity produced by (18.18) has the wrong sign. Three-component quantities that have this strange behaviour are known as **pseudovectors**. Pseudovectors are reasonably common. The cross product of any two (genuine) vectors is a pseudovector, not a vector. In particular, the moment of a force and the angular momentum of a body are pseudovectors. We do not wish to make an issue of this distinction. Instead, *we will restrict all our coordinate systems to be right-handed*. With this restriction, $\det \mathbf{A}$ is always $+1$ and $\{\omega_i\}$ (and all other pseudovectors) transform as vectors.* We will therefore have no need to distinguish between vectors and pseudovectors and we will call them all 'vectors'.

18.5 THE INERTIA TENSOR

Suppose the **rigid body** $\mathcal{B}$ has one of its particles held fixed at the origin O, but is in otherwise general motion. Then the angular momentum of $\mathcal{B}$ about O is defined by

$$\boldsymbol{L}_O = \sum_{i=1}^{N} \boldsymbol{r}_i \times (m_i \boldsymbol{v}_i) \tag{18.19}$$

in the standard notation, where the position vectors $\{\boldsymbol{r}_i\}$ are measured from the origin O. In Chapter 11 we found the formula for the angular momentum of a rigid body *about its own rotation axis*; this angular momentum is a scalar quantity. Now we will find the formula for the full vector value of $\boldsymbol{L}_O$.

In order that we may introduce component suffices without confusion, it is convenient to omit the suffix i in formula (18.19) and simply write

$$\boldsymbol{L}_O = \sum \boldsymbol{r} \times (m\boldsymbol{v}), \tag{18.20}$$

* It is an interesting fact that angular velocity exists, as a *vector*, only in a space of three dimensions. Had the universe been created with (say) four spatial dimensions, then $\mathbf{x}$ and $\mathbf{v}$ would be (four-dimensional) vectors and $\mathbf{v}$ would be given by the formula $\mathbf{v} = \mathbf{\Omega} \cdot \mathbf{x}$, where $\mathbf{\Omega}$ is a 4×4 anti-symmetric matrix representing a second order tensor. This matrix has *six* independent elements which cannot be fitted into a column of length *four*! The same applies to any spatial dimension other than three.

where the sum is deemed to be taken over all the particles of $\mathcal{B}$.

Since $\mathcal{B}$ is a rigid body and O is fixed, the velocity $\boldsymbol{v}$ is given by $\boldsymbol{v} = \boldsymbol{\omega}\times\boldsymbol{r}$, where $\boldsymbol{\omega}$ is the angular velocity of the body. Then

$$\begin{aligned}\boldsymbol{r}\times\boldsymbol{v} &= \boldsymbol{r}\times(\boldsymbol{\omega}\times\boldsymbol{r})\\ &= (\boldsymbol{r}\cdot\boldsymbol{r})\,\boldsymbol{\omega} - (\boldsymbol{r}\cdot\boldsymbol{\omega})\,\boldsymbol{r}.\end{aligned}$$

If the typical particle P has coordinates (x_1, x_2, x_3) and $\boldsymbol{\omega}$ has the components $\{\omega_1, \omega_2, \omega_3\}$, then the i-th component of $\boldsymbol{r}\times\boldsymbol{v}$ can be written

$$\begin{aligned}(\boldsymbol{r}\times\boldsymbol{v})_i &= \left(x_1^2 + x_2^2 + x_3^2\right)\omega_i - (x_1\omega_1 + x_2\omega_2 + x_3\omega_3)\,x_i\\ &= \left(\sum_{k=1}^{3} x_k x_k\right)\omega_i - \left(\sum_{j=1}^{3} x_j\,\omega_j\right)x_i.\end{aligned}$$

What we want to do now is to make ω_j a factor of this expression, which means that we would like ω_i to be replaced by ω_j. There is a standard trick for doing this, that is, to write

$$\omega_i = \sum_{j=1}^{3} \delta_{ij}\,\omega_j,$$

where δ_{ij} is the Krönecker delta. [This is equivalent to pre-multiplication of the column vector $\boldsymbol{\omega}$ by the identity matrix.] We then obtain

$$\begin{aligned}(\boldsymbol{r}\times\boldsymbol{v})_i &= \left(\sum_{k=1}^{3} x_k x_k\right)\left(\sum_{j=1}^{3} \delta_{ij}\,\omega_j\right) - \left(\sum_{j=1}^{3} x_j\,\omega_j\right)x_i\\ &= \sum_{j=1}^{3}\left(\left(\sum_{k=1}^{3} x_k x_k\right)\delta_{ij} - x_i\,x_j\right)\omega_j\end{aligned}$$

It follows that if $\{L_1, L_2, L_3\}$ are the components of the angular momentum $\boldsymbol{L}_O$, then

$$L_i = \sum_{j=1}^{3} I_{ij}\,\omega_j, \tag{18.21}$$

where the nine-component quantity $\{I_{ij}\}$ is defined by

$$I_{ij} = \sum m\left(\left(\sum_{k=1}^{3} x_k x_k\right)\delta_{ij} - x_i\,x_j\right).$$

It is easy to show directly that $\{I_{ij}\}$ is a second order **tensor**. For each particle of the body, the Cartesian coordinates $\{x_1, x_2, x_2\}$ transform as a vector. It follows that the outer product $\{x_i x_j\}$ is a tensor and its contraction $\sum_{k=1}^{3} x_k x_k$ is a scalar. It then follows by linearity that $\left(\sum_{k=1}^{3} x_k x_k\right)\delta_{ij} - x_i x_j$ is a tensor. Since the particle masses are certainly invariants, it once again follows by linearity that $\{I_{ij}\}$ must be a tensor.

Definition 18.7 ***Inertia tensor*** *The second order tensor $\{I_{ij}\}$ defined by*

Definition of the inertia tensor

$$I_{ij} = \sum m \left(\left(\sum_{k=1}^{3} x_k x_k \right) \delta_{ij} - x_i x_j \right) \tag{18.22}$$

is called the ***inertia tensor*** *of the body $\mathcal{B}$ at the point O. Note that the inertia tensor is* ***symmetric****, that is, $I_{ji} = I_{ij}$.*

We have therefore proved that, when a rigid body is rotating with angular velocity $\boldsymbol{\omega}$ about an axis through origin O, its angular momentum $\boldsymbol{L}_O$ about O is given by

$$\begin{pmatrix} L_1 \\ L_2 \\ L_3 \end{pmatrix} = \begin{pmatrix} I_{11} & I_{12} & I_{13} \\ I_{21} & I_{22} & I_{23} \\ I_{31} & I_{32} & I_{33} \end{pmatrix} \begin{pmatrix} \omega_1 \\ \omega_2 \\ \omega_3 \end{pmatrix}. \tag{18.23}$$

This can be written in matrix form as

Angular momentum formula

$$\mathsf{L}_O = \mathbf{I} \cdot \boldsymbol{\omega} \tag{18.24}$$

where the column vectors L_O and $\boldsymbol{\omega}$ are formed from the components of $\boldsymbol{L}_O$ and $\boldsymbol{\omega}$ respectively, and $\mathbf{I}$ is the matrix form of the inertia tensor $\{I_{ij}\}$. Because L_O is obtained from $\boldsymbol{\omega}$ by a *matrix* multiplication, it follows that $\boldsymbol{L}_O$ and $\boldsymbol{\omega}$ will generally lie in *different directions*. This fact (which does not show itself in planar mechanics) is what gives three-dimensional rigid body motion its special character.

The inertia tensor also appears in the corresponding expression for the **kinetic energy** of $\mathcal{B}$. By following a similar procedure to that used for angular momentum, it is found that T is given by

$$T = \tfrac{1}{2} \begin{pmatrix} \omega_1 & \omega_2 & \omega_3 \end{pmatrix} \begin{pmatrix} I_{11} & I_{12} & I_{13} \\ I_{21} & I_{22} & I_{23} \\ I_{31} & I_{32} & I_{33} \end{pmatrix} \begin{pmatrix} \omega_1 \\ \omega_2 \\ \omega_3 \end{pmatrix} \tag{18.25}$$

which can be written in the matrix form

Kinetic energy formula

$$T = \tfrac{1}{2} \boldsymbol{\omega}^T \cdot \mathbf{I} \cdot \boldsymbol{\omega} \tag{18.26}$$

The elements of the inertia tensor

The **diagonal** elements of **I** are actually familiar quantities. For example

$$\begin{aligned} I_{11} &= \sum m \left(\left(x_1^2 + x_2^2 + x_3^2 \right) \delta_{11} - x_1 x_1 \right) \\ &= \sum m \left(x_2^2 + x_3^2 \right) \\ &= \sum m p_1^2, \end{aligned}$$

where p_1 is the perpendicular distance of the typical particle P from the axis Ox_1. Hence I_{11} is just the *ordinary moment of inertia* of the body about the axis Ox_1. Similarly, I_{22} and I_{33} are the moments of inertia of the body about the axes Ox_2 and Ox_3 respectively.

The **off-diagonal** elements of **I** are given by

$$I_{12} = I_{21} = -\sum m x_1 x_2, \tag{18.27}$$

$$I_{23} = I_{32} = -\sum m x_2 x_3, \tag{18.28}$$

$$I_{31} = I_{13} = -\sum m x_3 x_1, \tag{18.29}$$

where, as usual, the sum is taken over the particles of the body. The quantities $\sum m x_1 x_2$, $\sum m x_2 x_3$, $\sum m x_3 x_1$ are known as **products of inertia**. Hence the off-diagonal elements of **I** are the *negatives* of their corresponding products of inertia.* Our results are summarised as follows:

Elements of the inertia tensor

- The **diagonal elements** of the inertia tensor, namely I_{11}, I_{22}, I_{33}, are the ordinary moments of inertia of the body about the coordinate axes Ox_1, Ox_2, Ox_3 respectively.
- The **off-diagonal elements** of the inertia tensor are the *negatives* of their corresponding products of inertia.

Products of inertia are new quantities and the reader will probably expect that we should now give several examples on their calculation. In fact, we will give just one easy one, which is a part of the next example. The reason is that, in almost all problems, the products of inertia can be inferred from symmetry and known results. Only a body with virtually no symmetry would *require* products of inertia to be calculated from first principles.

* Some authors include the negative sign in their definition of product of inertia, but this seems unnatural.

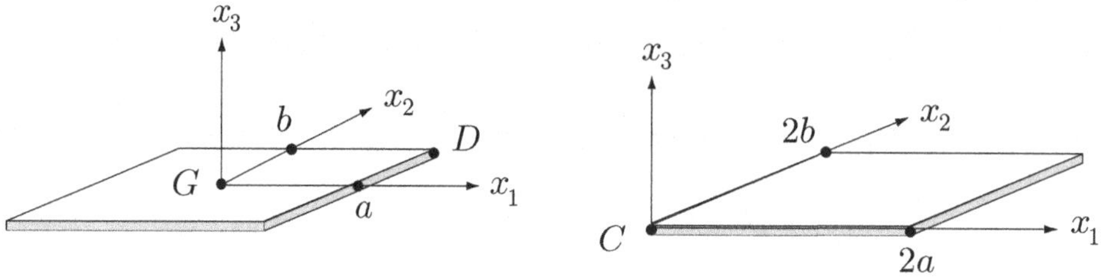

FIGURE 18.3 The uniform rectangular plate has mass M and sides $2a$ and $2b$

Example 18.6 *Calculating the inertia tensor*

A uniform rectangular plate has mass M and sides $2a$ and $2b$ as shown in Figure 18.3. Find the inertia tensor at the points G and C in the coordinate systems shown.

Solution

Consider first the **centre of mass** point G. Then I_{11} is the moment of inertia of the plate about the axis Gx_1 which is given in the Appendix (at the end of the book) to be $\frac{1}{3}Mb^2$. Similarly, $I_{22} = \frac{1}{3}Ma^2$ and, by the perpendicular axes theorem, $I_{33} = I_{11}+I_{22} = \frac{1}{3}M(a^2+b^2)$. The products of inertia are all zero. The products $\sum mx_2x_3$ and $\sum mx_3x_1$ are zero because all the mass lies in the plane $x_3 = 0$. The product $\sum mx_1x_2$ is zero because of symmetry; the plate is symmetrical about the axis $x_1 = 0$ but the terms of the sum are *odd* functions of x_1. The contributions to the sum from the right and left halves of the plate therefore cancel. Hence $I_{23} = I_{31} = I_{12} = 0$. Thus, in the given coordinate system, the **inertia tensor** at G is represented by the matrix

$$\mathbf{I}_G = \frac{1}{3}M\begin{pmatrix} b^2 & 0 & 0 \\ 0 & a^2 & 0 \\ 0 & 0 & a^2+b^2 \end{pmatrix}.$$

Now consider the **corner point** C. Then I_{11} is the moment of inertia of the plate about the axis Cx_1 which is given by the parallel axes theorem to be $\frac{1}{3}Mb^2 + Mb^2 = \frac{4}{3}Mb^2$. Similarly, $I_{22} = \frac{4}{3}Ma^2$ and, by the perpendicular axes theorem, $I_{33} = I_{11} + I_{22} = \frac{4}{3}M(a^2 + b^2)$. The elements I_{23} and I_{31} are zero for the same reason as before. However, the product $\sum mx_1x_2$ is not zero now because the plate is no longer symmetrically placed relative to the coordinate system based at C. We will therefore evaluate this product of inertia from first principles. Let the plate have uniform mass per unit area σ. Then

$$\sum mx_1x_2 = \int_{x_1=0}^{2a}\int_{x_2=0}^{2b} \sigma x_1x_2\, dx_1dx_2 = 4\sigma a^2b^2 = Mab,$$

since $M = 4\sigma ab$. Hence $I_{12} = Mab$. Thus, in the given coordinate system, the **inertia tensor** of the plate at C is represented by the matrix

$$\mathbf{I}_C = \frac{1}{3}M \begin{pmatrix} 4b^2 & -3ab & 0 \\ -3ab & 4a^2 & 0 \\ 0 & 0 & 4(a^2+b^2) \end{pmatrix}. \blacksquare$$

Example 18.7 *Transforming the inertia tensor*

Find $\mathbf{I}_G$ in the set of axes obtained by rotating the axes $Gx_1x_2x_3$ about Gx_3 so that the new axis Gx_1' lies along the diagonal GD.

Solution

In order to achieve the required position, the axes $Gx_1x_2x_3$ must be rotated through the acute angle $\alpha = \tan^{-1}(b/a)$. The corresponding transformation matrix is given by the formula (18.8) to be

$$\mathbf{A} = \begin{pmatrix} \cos\alpha & \sin\alpha & 0 \\ -\sin\alpha & \cos\alpha & 0 \\ 0 & 0 & 1 \end{pmatrix}.$$

The matrix $\mathbf{I}_G'$ representing the inertia tensor at G in the rotated coordinates is given by the transformation formula (18.17) to be

$$\begin{aligned}
\mathbf{I}_G' &= \mathbf{A}\cdot\mathbf{I}_G\cdot\mathbf{A}^T \\
&= \frac{1}{3}M \begin{pmatrix} \cos\alpha & \sin\alpha & 0 \\ -\sin\alpha & \cos\alpha & 0 \\ 0 & 0 & 1 \end{pmatrix} \begin{pmatrix} b^2 & 0 & 0 \\ 0 & a^2 & 0 \\ 0 & 0 & a^2+b^2 \end{pmatrix} \begin{pmatrix} \cos\alpha & -\sin\alpha & 0 \\ \sin\alpha & \cos\alpha & 0 \\ 0 & 0 & 1 \end{pmatrix} \\
&= \frac{1}{3}M \begin{pmatrix} a^2\sin^2\alpha + b^2\cos^2\alpha & (a^2-b^2)\sin\alpha\cos\alpha & 0 \\ (a^2-b^2)\sin\alpha\cos\alpha & a^2\cos^2\alpha + b^2\sin^2\alpha & 0 \\ 0 & 0 & a^2+b^2 \end{pmatrix} \\
&= \frac{M}{3(a^2+b^2)} \begin{pmatrix} 2a^2b^2 & ab(a^2-b^2) & 0 \\ ab(a^2-b^2) & a^4+b^4 & 0 \\ 0 & 0 & (a^2+b^2)^2 \end{pmatrix},
\end{aligned}$$

on inserting the values $\sin\alpha = b/(a^2+b^2)^{1/2}$ and $\cos\alpha = a/(a^2+b^2)^{1/2}$. This is the inertia tensor of the plate at G in the **rotated coordinate system.** ■

Example 18.8 *Finding kinetic energy using the inertia tensor*

Suppose that the plate in Figure 18.3 is made to rotate about one of its diameters with angular speed λ. Find its kinetic energy.

Solution

The kinetic energy can be evaluated by using either the original or the rotated coordinate system.

In the **original** coordinate system,

$$
\begin{aligned}
T &= \tfrac{1}{2}\boldsymbol{\omega}^T \cdot \mathbf{I}_G \cdot \boldsymbol{\omega} \\
&= \tfrac{1}{6}M \begin{pmatrix} \lambda\cos\alpha & \lambda\sin\alpha & 0 \end{pmatrix} \begin{pmatrix} b^2 & 0 & 0 \\ 0 & a^2 & 0 \\ 0 & 0 & a^2+b^2 \end{pmatrix} \begin{pmatrix} \lambda\cos\alpha \\ \lambda\sin\alpha \\ 0 \end{pmatrix} \\
&= \tfrac{1}{6}M\lambda^2 \left(b^2\cos^2\alpha + a^2\sin^2\alpha \right) \\
&= \frac{Ma^2b^2\lambda^2}{3(a^2+b^2)}.
\end{aligned}
$$

In the **rotated** coordinate system,

$$
\begin{aligned}
T &= \tfrac{1}{2}\boldsymbol{\omega}^T \cdot \mathbf{I}_G \cdot \boldsymbol{\omega} \\
&= \frac{M}{6(a^2+b^2)} \begin{pmatrix} \lambda & 0 & 0 \end{pmatrix} \begin{pmatrix} 2a^2b^2 & ab(a^2-b^2) & 0 \\ ab(a^2-b^2) & a^4+b^4 & 0 \\ 0 & 0 & (a^2+b^2)^2 \end{pmatrix} \begin{pmatrix} \lambda \\ 0 \\ 0 \end{pmatrix} \\
&= \frac{Ma^2b^2\lambda^2}{3(a^2+b^2)}.
\end{aligned}
$$

Starting from scratch, the first calculation is quicker. ■

18.6 PRINCIPAL AXES OF A SYMMETRIC TENSOR

The nine components of a second order tensor depend on the coordinate system, their values in different coordinate systems being related by the transformation formula (18.17). The question naturally arises as to whether we can simplify the representation of a tensor by a clever choice of coordinate system. If a tensor is represented by the matrix $\mathbf{T}$ in one coordinate system C, then, in another coordinate system C', it is represented by the matrix $\mathbf{T}'$ given by

$$
\mathbf{T}' = \mathbf{A} \cdot \mathbf{T} \cdot \mathbf{A}^T, \tag{18.30}
$$

where $\mathbf{A}$ is the orthogonal transformation matrix between C and C'. We therefore wish to find an orthogonal matrix $\mathbf{A}$ (with determinant $+1$) that makes the product $\mathbf{A} \cdot \mathbf{T} \cdot \mathbf{A}^T$ as 'simple' as possible, preferably a **diagonal** matrix. The question is therefore equivalent to a problem in linear algebra. There are a number of theorems available concerning the diagonalisation of matrices by transformations resembling (18.30) (see Anton [7]) and one of these results is exactly what we need. It is the **orthogonal diagonalisation** theorem:

Theorem 18.1 ***Orthogonal diagonalisation*** *Given any real* ***symmetric*** *matrix* $\mathbf{T}$, *there is an orthogonal matrix* $\mathbf{A}$ *such that the product* $\mathbf{A} \cdot \mathbf{T} \cdot \mathbf{A}^T$ *is a* ***diagonal*** *matrix.*

Since the **inertia tensor** is symmetric, the above theorem applies. Thus, at any point O of the body, one can find a set of axes in which $\mathbf{I}$ is a diagonal matrix. These are called **principal axes** of the body at O. The diagonal elements of $\mathbf{I}$ are simply the moments of inertia of the body about the three principal axes. They are called the **principal moments of inertia** at O and will be denoted by symbols A, B and C. This is summarised as follows:

Principal axes and principal moments of inertia

Relative to a set of **principal axes** $Ox_1x_2x_3$ at O, the inertia tensor $\mathbf{I}$ has the diagonal form

$$\mathbf{I} = \begin{pmatrix} A & 0 & 0 \\ 0 & B & 0 \\ 0 & 0 & C \end{pmatrix}, \qquad (18.31)$$

where the **principal moments of inertia** A, B, C are the moments of inertia of the body about the axes Ox_1, Ox_2, Ox_3 respectively.

Finding principal axes

There is a standard procedure for finding the orthogonal matrix A that reduces any symmetric matrix T to diagonal form in the above manner. The diagonal elements of the reduced matrix are the **eigenvalues** of the matrix T, and the rows of A contain the components of the corresponding normalised **eigenvectors** of T. However, for the purpose of solving problems in mechanics, we (almost) never need to carry through this procedure. The reason is that, in almost all problems, the orientation of the principal axes can be deduced by symmetry, as we will show in Section 18.7.

Expressions for L and T in principal axes

When principal axes are used, the expressions (18.24) for the angular momentum and (18.26) for the kinetic energy of a rigid body are much simplified. They reduce to the non-tensorial expressions

Expressions for L_O and T in principal axes at O

$$L_O = (A\,\omega_1)\,\boldsymbol{e}_1 + (B\,\omega_2)\,\boldsymbol{e}_2 + (C\,\omega_3)\,\boldsymbol{e}_3 \qquad (18.32)$$

$$T = \tfrac{1}{2}\left(A\,\omega_1^2 + B\,\omega_2^2 + C\,\omega_3^2\right)$$

where $\{\boldsymbol{e}_1, \boldsymbol{e}_2, \boldsymbol{e}_3\}$ are the unit vectors of the principal axes, and $\{\omega_1, \omega_2, \omega_3\}$ are the components of $\boldsymbol{\omega}$ in the principal axes, that is, $\boldsymbol{\omega} = \omega_1\boldsymbol{e}_1 + \omega_2\boldsymbol{e}_2 + \omega_3\boldsymbol{e}_3$.

These results apply to the case in which the body has a permanently fixed particle O (the vertex of a spinning top, for example). However, similar formulae apply to the case of completely **general motion**. In this case we have

Expressions for L_G and T in principal axes at G

$$L_G = (A\,\omega_1)\,e_1 + (B\,\omega_2)\,e_2 + (C\,\omega_3)\,e_3 \qquad (18.33)$$

$$T^G = \tfrac{1}{2}\left(A\,\omega_1^2 + B\,\omega_2^2 + C\,\omega_3^2\right)$$

where G is the centre of mass of the body. Here L_G is the angular momentum of the body about G, and T^G is the *rotational* kinetic energy of the body about G.

These are the results that you *really* need in order to solve problems in rigid body dynamics. Almost always, principal axes are used and only the equations (18.32) or (18.33) are required. Thus, it is possible to solve most problems in rigid body dynamics without any knowledge of tensors at all!

18.7 DYNAMICAL SYMMETRY

Finding principal axes from geometrical symmetries of the body

The directions of the principal axes of the inertia tensor of a body at a point O can often be inferred from the **geometrical symmetry** of the body about O. The following rules (which are consequences of the tensor transformation formula) are useful.

Rule 1: *If the body has **reflective symmetry** in a plane through O, then the line through O perpendicular to this plane is a principal axis.*

Rule 2: *If the body has any **rotational symmetry** about an axis through O, then this axis is a principal axis.*

Either of these rules is enough to show that, for a uniform **rectangular plate**, the axes shown in Figure 18.3 (left) are a set of principal axes at G. The plate has reflective symmetry in *each* of the coordinate planes, and also has rotational symmetry of order two about *each* of the coordinate axes. The set of parallel axes at C is not a principal set. For a **uniform cone** on a pentagonal base (see Figure 18.4 (left)), the axis of symmetry is a principal axis at G and there are *five* other principal axes through G. [What are they?] For a **spinning top** (see Figure 18.4 (centre)), the axis of symmetry is a principal axis at O and so is *any* axis through O that is perpendicular to it! We seem to be finding too many principal axes, but this will soon be explained.

Dynamical symmetry

Consider a rigid body $\mathcal{B}$ pivoted at the origin O. Then the motion of $\mathcal{B}$ under known forces is determined by the **angular momentum principle**, which, in view of equation (18.24), can be written

$$\frac{d}{dt}\left(\mathbf{I}_O \cdot \boldsymbol{\omega}\right) = \mathbf{K}_O$$

where $\boldsymbol{K}_O$ is the total moment about O of the external forces acting on $\mathcal{B}$. In principal axes, $\mathbf{I}_O$ contains only the three principal moments of inertia A, B and C and so the shape

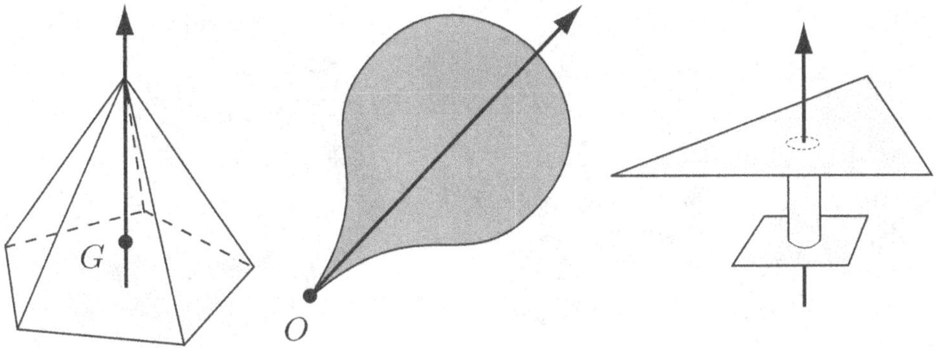

FIGURE 18.4 Three bodies with **dynamical axial symmetry**.

of the body enters into the equations of motion only by the values of these three numbers. Under the same set of forces and initial conditions, *different bodies that have the same principal moments of inertia have the same motion.*

The form of the motion, and the chance of solving problems, is very much influenced by the presence or absence of **dynamical symmetry**. There are three cases:

A, B, C all different

When $\{A, B, C\}$ are all different, the body is said to be **dynamically unsymmetric**, and, as might be expected, this is the most difficult case to treat analytically. For the rectangular plate shown in Figure 18.3, the principal moments of inertia at G are $A = \frac{1}{3}Mb^2$, $B = \frac{1}{3}Ma^2$ and $C = \frac{1}{3}M(a^2+b^2)$, which are all different. Thus, despite the apparently simple shape of the plate, it is no easier to treat than a lump of scrap metal. In the unsymmetrical case, the principal axes at O are *essentially unique.**

A and B equal, C different

Suppose now that two of the principal moments of inertia (A and B say) are equal, the third, C, being different. The body is then said to be **dynamically axially symmetric** about the axis Ox_3. This time, the axis of symmetry is a principal axis, and *the other two axes can be chosen arbitrarily*, provided that they form a right-handed orthogonal set. One might expect this to occur only when the body is a body of revolution about an axis through O, as in the case of a traditional spinning top. However, the following remarkable result shows that this is not necessary.

Rule 3: *If a body has a rotational symmetry about an axis through O, and the order*† *of this symmetry is three or more, then the body has* ***dynamical axial symmetry*** *about this axis.*

---

* The principal axes *are* unique in the sense that the three orthogonal unlabelled lines in space are unique. There are then 24 ways of labelling these lines to create a right-handed system of coordinates.

† If the mass distribution of a rigid body is left unchanged when the body is rotated through an angle $2\pi/n$, then the body is said to have **rotational symmetry of order** n about the rotation axis. Thus a uniform lamina having the shape of a regular pentagon has rotational symmetry of order five about the axis through G perpendicular to its plane.

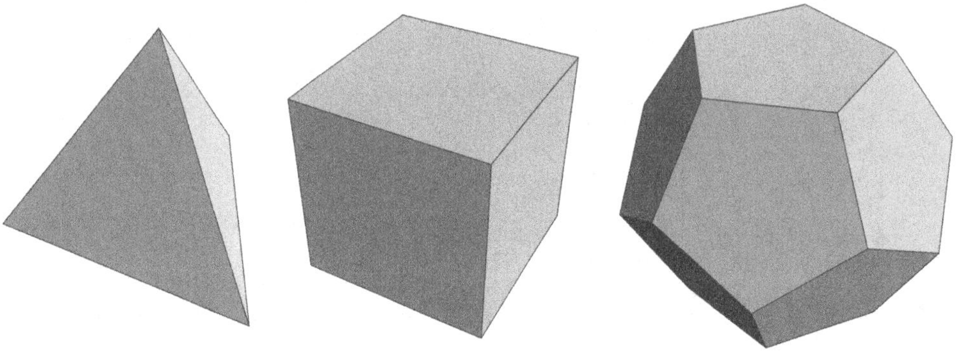

FIGURE 18.5 Three bodies with **dynamical spherical symmetry**.

The pentagonal cone in Figure 18.4 has rotational symmetry of order five and so Rule 3 applies. In particular, it follows that the cone has the same moment of inertia about *any* axis through G perpendicular to the symmetry axis!

Interestingly, it is possible for a body to have dynamical axial symmetry and yet have no rotational symmetry at all. The body in Figure 18.4 (right) has no rotational symmetry, but, since it is made up of three bodies that do have such symmetry, it must be dynamically axially symmetric!

A, B and C all equal

Suppose now that all three of the principal moments of inertia at O are equal. The body is then said to be **dynamically spherically symmetric** about O. This time, the inertia tensor is a multiple of the identity and *any* axis through O is a principal axis. The principal axes can therefore be chosen arbitrarily, provided that they form a right-handed orthogonal set. A uniform ball is obviously dynamically spherically symmetric, but there are many other examples. This follows from our last rule, Rule 4.

Rule 4: *If a body has two different axes of dynamical axial symmetry through O, then it must be* ***dynamically spherically symmetric*** *about O.*

Figure 18.5 shows three bodies that have **dynamical spherical symmetry** about their centres of mass. None of these bodies is a ball, but they move as if they were. For example, the uniform tetrahedron has *four* different axes of dynamical *axial* symmetry through G and so, by Rule 4, it must be dynamically *spherically* symmetric about G. As a consequence, it must have the same moment of inertia about *any* axis through G. [Don't try proving this any other way!]

Problems on Chapter 18

Answers and comments are at the end of the book.

Harder problems carry a star (∗).

Orthogonal transformations

18.1 Show that the matrix

$$\mathbf{A} = \frac{1}{7}\begin{pmatrix} 3 & 2 & 6 \\ -6 & 3 & 2 \\ 2 & 6 & -3 \end{pmatrix}$$

is orthogonal. If **A** is the transformation matrix between the coordinate systems $\mathcal{C}$ and $\mathcal{C}'$, do $\mathcal{C}$ and $\mathcal{C}'$ have the same, or opposite, handedness?

18.2 Find the transformation matrix between the coordinate systems $\mathcal{C}$ and $\mathcal{C}'$ when $\mathcal{C}'$ is obtained

(i) by rotating $\mathcal{C}$ through an angle of 45° about the axis Ox_2,
(ii) by reflecting $\mathcal{C}$ in the plane $x_2 = 0$,
(iii) by rotating $\mathcal{C}$ through a right angle about the axis $\overrightarrow{OB}$, where B is the point with coordinates (2, 2, 1),
(iv) by reflecting $\mathcal{C}$ in the plane $2x_1 - x_2 + 2x_3 = 0$.

In each case, find the new coordinates of the point D whose coordinates in $\mathcal{C}$ are (3, −3, 0).

18.3 Show that the matrix

$$\mathbf{A} = \frac{1}{3}\begin{pmatrix} 2 & -1 & -2 \\ 2 & 2 & 1 \\ 1 & -2 & 2 \end{pmatrix}$$

is orthogonal and has determinant +1. Find the column vectors **v** that satisfy the equation $\mathbf{A}\cdot\mathbf{v} = \mathbf{v}$. If **A** is the transformation matrix between the coordinate systems $\mathcal{C}$ and $\mathcal{C}'$, show that **A** represents a rotation of $\mathcal{C}$ about the axis $\overrightarrow{OE}$ where E is the point with coordinates (1, 1, −1) in $\mathcal{C}$.

∗ Find the rotation angle.

Tensor algebra

18.4 Write out the transformation formula for a fifth order tensor. [The main difficulty is finding enough suffix names!]

18.5 In the coordinate system $\mathcal{C}$, a certain second order tensor is represented by the matrix

$$\mathbf{T} = \begin{pmatrix} 1 & 0 & 1 \\ 0 & 1 & 0 \\ 1 & 0 & 1 \end{pmatrix}.$$

Find the matrix representing the tensor in the coordinate system $\mathcal{C}'$, where $\mathcal{C}'$ is obtained

(i) by rotating $\mathcal{C}$ through an angle of 45° about the axis Ox_1,
(ii) by reflecting $\mathcal{C}$ in the plane $x_3 = 0$.

18.6 The quantities t_{ijk} and u_{ijkl} are third and fourth order tensors respectively. Decide if each of the following quantities is a tensor and, if it is, state its order:

(i) $t_{ijk}u_{lmnp}$ (ii) $t_{ijk}t_{lmn}$ (iii) $\sum_{j=1}^{3} t_{ijj}$

(iv) $\sum_{j=1}^{3} t_{jij}$ (v) $\sum_{i=1}^{3} t_{iii}$ (vi) $\sum_{k=1}^{3} t_{ijk}u_{klmn}$

(vii) $\sum_{i=1}^{3}\sum_{j=1}^{3} u_{iijj}$ (viii) $\sum_{k=1}^{3} u_{kklmn}$ (ix) $\sum_{i=1}^{3}\sum_{j=1}^{3}\sum_{k=1}^{3} t_{ijk}t_{ijk}$.

18.7 Show that the sum of the squares of the elements of a tensor is an invariant. [First and second order tensors will suffice.]

18.8 If the matrix $\mathbf{T}$ represents a second order tensor, show that $\det \mathbf{T}$ is an invariant. [We have now found three invariant functions of a second order tensor: the sum of the diagonal elements, the sum of the squares of all the elements, and the determinant.]

18.9 In crystalline materials, the ordinary elastic moduli are replaced by c_{ijkl}, a fourth order tensor with eighty one elements. It appears that the most general material has eighty one elastic moduli, but this number is reduced because c_{ijkl} has the following symmetries:

(i) $c_{jikl} = c_{ijkl}$ (ii) $c_{ijlk} = c_{ijkl}$ (iii) $c_{klij} = c_{ijkl}$

How many elastic moduli does the most general material actually have?

Inertia tensor and principal axes

The following problems do *not* require moments or products of inertia to be evaluated by integration.

18.10 Show that $I_{\{O,\boldsymbol{n}\}}$, the moment of inertia of a body about an axis through O parallel to the unit vector $\boldsymbol{n}$, is given by

$$I_{\{O,\boldsymbol{n}\}} = \mathbf{n}^T \cdot \mathbf{I}_O \cdot \mathbf{n},$$

where $\mathbf{I}_O$ is the matrix representing the inertia tensor of the body at O (in some coordinate system), and $\mathbf{n}$ is the column vector that contains the components of $\boldsymbol{n}$ (in the same coordinate system).

Find the moment of inertia of a uniform rectangular plate with sides $2a$ and $2b$ about a diagonal.

18.11 Find the principal moments of inertia of a uniform circular disk of mass M and radius a (i) at its centre of mass, and (ii) at a point on the edge of the disk.

18.12 A uniform circular disk has mass M and radius a. A spinning top is made by fitting the disk with a light spindle AB which passes through the disk and is fixed along its axis of symmetry. The distance of the end A from the disk is equal to the disk radius a. Find the principal moments of inertia of the top at the end A of the spindle.

18.13 A uniform hemisphere has mass M and radius a. A spinning top is made by fitting the hemisphere with a light spindle AB which passes through the hemisphere and is fixed along its axis of symmetry with the curved surface of the hemisphere facing away from the end A. The distance of A from the point where the spindle enters the flat surface is equal to the radius a of the hemisphere. Find the principal moments of inertia of the top at the end A of the spindle.

18.14 Find the principal moments of inertia of a uniform cube of mass M and side $2a$ (i) at its centre of mass, (ii) at the centre of a face, and (iii) at a corner point.

Find the moment of inertia of the cube (i) about a space diagonal, (ii) about a face diagonal, and (iii) about an edge.

18.15 A uniform rectangular block has mass M and sides $2a$, $2b$ and $2c$. Find the principal moments of inertia of the block (i) at its centre of mass, (ii) at the centre of a face of area $4ab$. Find the moment of inertia of the block (i) about a space diagonal, (ii) about a diagonal of a face of area $4ab$.

18.16 Find the principal moments of inertia of a uniform cylinder of mass M, radius a and length $2b$ at its centre of mass G. Is it possible for the cylinder to have dynamical *spherical* symmetry about G?

18.17 Determine the dynamical symmetry (if any) of each the following bodies about their centres of mass:

(i) a frisbee,
(ii) a piece of window glass having the shape of an isosceles triangle,
(iii) a two bladed aircraft propeller,
(iv) a three-bladed ship propeller,
(v) an Allen screw (ignore the thread),
(vi) eight particles of equal mass forming a rigid cubical structure,
(vii) a cross-handled wheel nut wrench,
(viii) the great pyramid of Giza,
(ix) a molecule of carbon tetrachloride.

18.18* A uniform rectangular plate has mass M and sides $2a$ and $4a$. Find the principal axes and principal moments of inertia at a *corner* point of the plate. [Make use of the formula for $\mathbf{I}_C$ obtained in Example 18.6, with $b = 2a$.]

If you know how, do this question by finding the eigenvalues and eigenvectors of $\mathbf{I}_C$. If not, try the following homespun method: Starting from $\mathbf{I}_C$ in the coordinates used in Example 18.6, find $\mathbf{I}'_C$ in the coordinate system obtained by rotating through an angle α around the axis Cx_3. Then choose α to eliminate the off-diagonal elements.

Chapter Nineteen

Problems in rigid body dynamics

KEY FEATURES

The key features of this chapter are the use of the linear and angular momentum principles to generate the **equations of rigid body motion**; the importance of **body symmetry** in simplifying the problem; and the choice of solution method, **vectorial, Lagrangian** or **Eulerian**.

Readers who have reached this point have come a long way and deserve to be congratulated. With the linear and angular momentum principles and the inertia tensor behind us, we are now able to solve some of the most interesting and puzzling problems in mechanics. Rigid body motion is the pinnacle of achievement* in the 'problem solving' kind of mechanics but, given the background we now have, many problems are surprisingly easy. In this chapter, we classify problems by the dynamical symmetry of the body; first we consider **spherically symmetric** bodies, then **axially symmetric** bodies and finally **unsymmetric** bodies. Most problems can be done by either vectorial or analytical methods and we choose whatever is appropriate in each case.

19.1 EQUATIONS OF RIGID BODY DYNAMICS

Rigid body in general motion

Translational motion of G:

$$M\frac{d\boldsymbol{V}}{dt} = \boldsymbol{F} \tag{19.1}$$

Rotational motion about G:

$$\frac{d\boldsymbol{L}_G}{dt} = \boldsymbol{K}_G \tag{19.2}$$

* More advanced topics exist, but they are more concerned with 'grand general principles' than problem solving!

The governing equations of rigid body dynamics are the **linear** and **angular momentum principles** in their centre of mass form, as shown above; the notation is that used in Chapters 10 and 11. The linear momentum principle (19.1) determines the *translational* motion of the centre of mass G, and the angular momentum principle (19.2) determines the *rotational* motion of the body relative to G.

The above equations could be applied in all cases, but, when the body is **pivoted*** at a fixed point O, it is more convenient to drop the first equation and apply the angular momentum principle about the pivot point O instead of G. With this choice, the unknown reaction at the pivot is eliminated. We then have:

Rigid body pivoted at a fixed point O

Rotational motion about O:
$$\frac{d\boldsymbol{L}_O}{dt} = \boldsymbol{K}_O \qquad (19.3)$$

Calculating $\boldsymbol{L}_G$ and $\boldsymbol{L}_O$ in principal axes

In order to use the equation (19.2) (or (19.3)) we need to express $\boldsymbol{L}_G$ (or $\boldsymbol{L}_O$) in terms of $\boldsymbol{\omega}$, the angular velocity of the body. In general, this is a tensor relation, but, in principal axes, it reduces to the vector form obtained in Section 18.6. The results are as follows:

Suppose $Gx_1x_2x_3$ are principal axes at G with associated unit vectors $\{\boldsymbol{e}_1, \boldsymbol{e}_2, \boldsymbol{e}_3\}$. Then

$$\boxed{\boldsymbol{L}_G = A\omega_1\, \boldsymbol{e}_1 + B\omega_2\, \boldsymbol{e}_2 + C\omega_3\, \boldsymbol{e}_3} \qquad (19.4)$$

where A, B, C are the principal moments of inertia of the body about the axes Gx_1, Gx_2, Gx_3 respectively. Similarly, the angular momentum $\boldsymbol{L}_O$ can be expressed in terms of the angular velocity $\boldsymbol{\omega}$, relative to principal axes of the body at O, by the formula

$$\boxed{\boldsymbol{L}_O = A\omega_1\, \boldsymbol{e}_1 + B\omega_2\, \boldsymbol{e}_2 + C\omega_3\, \boldsymbol{e}_3} \qquad (19.5)$$

where A, B, C are now the principal moments of inertia of the body about the axes Ox_1, Ox_2, Ox_3 respectively. In each case, $\{\omega_1, \omega_2, \omega_3\}$ are the components of $\boldsymbol{\omega}$ in the principal axes, that is,

$$\boldsymbol{\omega} = \omega_1\, \boldsymbol{e}_1 + \omega_2\, \boldsymbol{e}_2 + \omega_3\, \boldsymbol{e}_3.$$

Dynamical symmetry of the body

The most important general feature of a moving rigid body is its **dynamical symmetry** (see Section 18.7). The more symmetry the body has, the simpler is its motion and the

* The spinning top with its vertex held at a fixed point is a typical example.

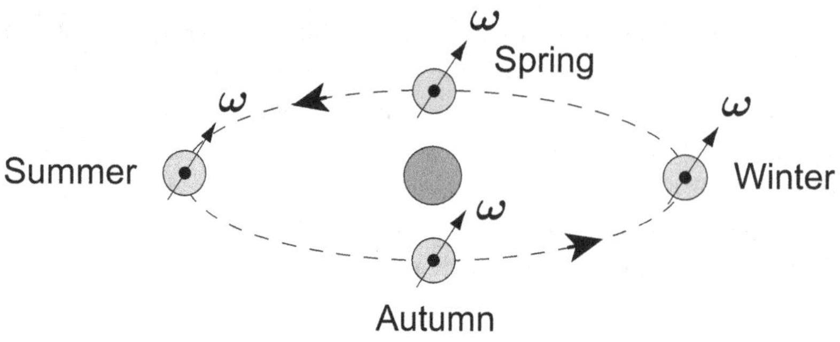

FIGURE 19.1 In the motion of the Earth, its angular velocity $\boldsymbol{\omega}$ is preserved and is *not* normal to the plane of the orbit. This gives rise to the seasons (shown here for the northern hemisphere).

easier it is to calculate it. We begin with the simplest case, which is when the body has dynamical **spherical symmetry**. Remember that, in order to have *dynamical* spherical symmetry, the body need not actually be spherical; indeed, one can find bodies that have no geometrical symmetry at all that are dynamically spherical.

19.2 MOTION OF 'SPHERES'

Suppose the body has dynamical **spherical symmetry** about its centre of mass G. The body could be a uniform sphere, but it does not have to be. All we require is that $A = B = C$ at G, in which case the body has the same moment of inertia A about every axis through G. Equation (19.4) then simplifies to give

$$\boldsymbol{L}_G = A\omega_1 \boldsymbol{e}_1 + A\omega_2 \boldsymbol{e}_2 + A\omega_3 \boldsymbol{e}_3 = A\left(\omega_1 \boldsymbol{e}_1 + \omega_2 \boldsymbol{e}_2 + \omega_3 \boldsymbol{e}_3\right) = A\,\boldsymbol{\omega},$$

so that $\boldsymbol{L}_G$ is simply proportional to $\boldsymbol{\omega}$. Equation (19.2) then becomes

$$A\,\dot{\boldsymbol{\omega}} = \boldsymbol{K}_G.$$

If the body is moving under **no forces**, or under **uniform gravity**, then the total moment $\boldsymbol{K}_G = \boldsymbol{0}$ and we obtain

$$\boldsymbol{\omega} = \boldsymbol{\omega}_0,$$

a constant. Thus, in either of these cases, the *angular velocity of the sphere is preserved* in the motion. Meanwhile, the motion of G, as determined by equation (19.1), is either straight line or parabolic motion respectively. Hence, when one throws a ball (or a cube!), G traces out the usual parabola and $\boldsymbol{\omega}$ retains the value that it was given initially. Note that $\boldsymbol{\omega}$ retains its *direction* as well as its magnitude, so that even though the direction of $\boldsymbol{V}$ changes, the rotation axis maintains a constant direction in space.

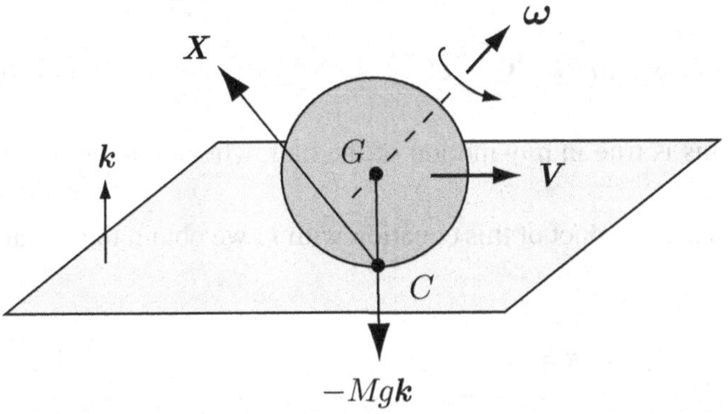

FIGURE 19.2 The snooker ball is a uniform sphere of mass M and radius b that moves in contact with a horizontal table.

The most important instance of this is the motion of the Earth in its orbit. If we assume that the Sun and the Earth are spherically symmetric bodies,* then the total moment of forces $\boldsymbol{K}_G$ exerted on the Earth by the Sun is zero and $\boldsymbol{\omega}$ is preserved. Since $\boldsymbol{\omega}$ is *not* normal to the plane of the orbit, the motion looks like Figure 19.1 (not to scale!), and this gives rise to the seasons.

19.3 THE SNOOKER BALL

Consider a snooker or pool ball, with mass M and radius b, that moves in contact with a horizontal table (see Figure 19.2). First we will consider **general motion** of the ball. In particular, we will allow skidding so that there is no special relation between $\boldsymbol{V}$, the velocity of G, and the angular velocity $\boldsymbol{\omega}$. The table is rough† so that the reaction $\boldsymbol{X}$ exerted by the table on the ball is not necessarily vertical. The governing equations (19.1) and (19.2) become

$$\begin{aligned} M\dot{\boldsymbol{V}} &= \boldsymbol{X} - Mg\,\boldsymbol{k} \\ \dot{\boldsymbol{L}}_G &= (-b\,\boldsymbol{k})\times\boldsymbol{X}, \end{aligned}$$

where $\boldsymbol{L}_G = A\,\boldsymbol{\omega}$. On eliminating the reaction $\boldsymbol{X}$ we find that

$$\begin{aligned} A\,\dot{\boldsymbol{\omega}} &= b\left(M\dot{\boldsymbol{V}} + Mg\,\boldsymbol{k}\right)\times\boldsymbol{k} \\ &= Mb\,\dot{\boldsymbol{V}}\times\boldsymbol{k}. \end{aligned}$$

* This is a very good approximation but (like everything else in astronomy) it is not exact. The Earth is slightly spheroidal and the gravitational fields of the Sun and Moon give rise to a small resultant moment about G. As a consequence, $\boldsymbol{\omega}$ actually precesses very slowly around the normal to the plane of the orbit (once every 25,730 years).

† But not so rough that the ball is *compelled* to roll.

On integrating with respect to t, we obtain the non-standard **conservation principle***

$$A\boldsymbol{\omega} + Mb\,\boldsymbol{k}\times\boldsymbol{V} = \boldsymbol{C}, \tag{19.6}$$

where $\boldsymbol{C}$ is a constant vector. This is true in *any* motion of the ball, whether skidding or rolling.

In particular, if we take the scalar product of this equation with $\boldsymbol{k}$, we obtain the scalar conservation principle

$$\boldsymbol{\omega}\cdot\boldsymbol{k} = n, \tag{19.7}$$

where n is a constant. Thus the *vertical component of* $\boldsymbol{\omega}$ *is conserved in* ***any*** *motion of the ball.*

We now examine the special case of **rolling**. In this case the contact particle C has zero velocity, so that the rolling condition is

$$\boldsymbol{V} + \boldsymbol{\omega}\times(-b\,\boldsymbol{k}) = \boldsymbol{0},$$

that is,

$$\boldsymbol{V} + b\,\boldsymbol{k}\times\boldsymbol{\omega} = \boldsymbol{0}. \tag{19.8}$$

Now, from the conservation principle (19.6), it follows that

$$\begin{aligned} A\,\boldsymbol{k}\times\boldsymbol{\omega} &= \boldsymbol{k}\times\boldsymbol{C} - Mb\,\boldsymbol{k}\times(\boldsymbol{k}\times\boldsymbol{V}) \\ &= \boldsymbol{k}\times\boldsymbol{C} - Mb\,((\boldsymbol{k}\cdot\boldsymbol{V})\,\boldsymbol{k} - (\boldsymbol{k}\cdot\boldsymbol{k})\,\boldsymbol{V}) \\ &= \boldsymbol{k}\times\boldsymbol{C} + Mb\,\boldsymbol{V}, \end{aligned}$$

since $\boldsymbol{k}$ is a unit vector and is perpendicular to $\boldsymbol{V}$. On substituting this result into the rolling condition (19.8) we obtain

$$\boldsymbol{V} + \left(\frac{Mb^2}{A}\right)\boldsymbol{V} = \left(\frac{b}{A}\right)\boldsymbol{C}\times\boldsymbol{k},$$

which shows that $\boldsymbol{V}$ must be constant in any rolling motion.

The corresponding value of $\boldsymbol{\omega}$ is

$$\boldsymbol{\omega} = \frac{1}{b}\,\boldsymbol{k}\times\boldsymbol{V} + n\,\boldsymbol{k}, \tag{19.9}$$

which is also constant. Hence:

---

* It is actually the angular momentum principle applied about the non-standard point C.

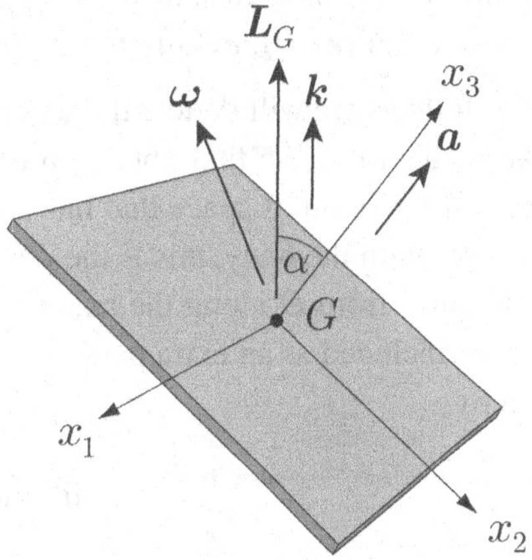

FIGURE 19.3 The free motion of a body with dynamic axial symmetry, depicted here as a square plate. The principal axis Gx_3 is the axis of symmetry and $\boldsymbol{a}$ $(= \boldsymbol{a}(t))$ is the axial unit vector.

The only rolling motions that are possible are straight line motions with constant $\boldsymbol{V}$ *and constant* $\boldsymbol{\omega}$*; this is despite the presence of the vertical angular velocity component* n.

This fact makes the games of pool and snooker possible. In practice, it is difficult to strike a ball and not give it some 'side' (a non-zero value of n). If rolling took place along curved lines when the ball had 'side', the game would be impossibly difficult. As it is, players often give the ball plenty of 'side' deliberately. The ball still rolls in a straight line, but the 'side' affects the bounce of the ball when it hits a cushion; good players use this fact to control the ball. As a corollary, it follows that, if a player actually wants to *swerve* a ball, then he must make the ball *skid* on the table.

19.4 FREE MOTION OF BODIES WITH AXIAL SYMMETRY

Bodies with dynamical **axial symmetry** are the most important objects whose motion we study. This is for three reasons: (i) They are much more common than spherically symmetric objects, (ii) they perform more interesting motions than spherically symmetric bodies, (iii) their motions can often be calculated in closed form.

Consider a rigid body with dynamical axial symmetry. It could be a baseball bat, a pencil, a top, a spacecraft, a freely-pivoted gyroscope, or the square plate shown in Figure 19.3. The centre of mass G of such a body must lie on the axis of symmetry and we will suppose that the principal moments of inertia of the body at G are $\{A, A, C\}$, with C being the moment of inertia about the *symmetry axis*. C may be greater or less than A. For example, if the body is a short fat cylinder, then $C > A$, but, if the body is a long thin cylinder, then $C < A$.

Consider such a body moving under either (i) **no forces**, or (ii) **uniform gravity**. Then the motion of G is given by particle mechanics and it only remains to calculate the rotational motion of the body relative to G. Since $\boldsymbol{K}_G = \boldsymbol{0}$ in each of these cases, the equation of rotational motion is $\dot{\boldsymbol{L}}_G = \boldsymbol{0}$, where, in the principal axes $Gx_1x_2x_3$,

$$\boldsymbol{L}_G = A\omega_1\,\boldsymbol{e}_1 + A\omega_2\,\boldsymbol{e}_2 + C\omega_3\,\boldsymbol{e}_3. \tag{19.10}$$

*It should be remembered that the principal axes $Gx_1x_2x_3$ move **with the body** and that the unit vectors $\{\boldsymbol{e}_1, \boldsymbol{e}_2, \boldsymbol{e}_3\}$ are therefore functions of the time.*

In what follows we will denote the **axial unit vector** $\boldsymbol{e}_3$ by $\boldsymbol{a}$ $(= \boldsymbol{a}(t))$; this is simply to improve readability. We first obtain an expression for the angular velocity $\boldsymbol{\omega}$ in terms of $\boldsymbol{a}$. Consider the point of space that has position vector $\boldsymbol{a}$ relative to G. Since the axial vector $\boldsymbol{a}$ moves with the body, this is the position vector of the particle A of the body that is at unit distance from O along the axis Gx_3. (If this point lies outside the body, then it can simply be included as an extra particle of zero mass.) It follows that $\dot{\boldsymbol{a}}$, the velocity of A, is given by

$$\dot{\boldsymbol{a}} = \boldsymbol{\omega} \times \boldsymbol{a}.$$

If we now take the vector product of this equation with $\boldsymbol{a}$, we obtain

$$\begin{aligned} \boldsymbol{a} \times \dot{\boldsymbol{a}} &= \boldsymbol{a} \times (\boldsymbol{\omega} \times \boldsymbol{a}) \\ &= (\boldsymbol{a} \cdot \boldsymbol{a})\,\boldsymbol{\omega} - (\boldsymbol{a} \cdot \boldsymbol{\omega})\,\boldsymbol{a} \\ &= \boldsymbol{\omega} - (\boldsymbol{a} \cdot \boldsymbol{\omega})\,\boldsymbol{a}. \end{aligned}$$

Hence $\boldsymbol{\omega}$ must have the form

$$\boldsymbol{\omega} = \boldsymbol{a} \times \dot{\boldsymbol{a}} + \lambda\, \boldsymbol{a}, \tag{19.11}$$

where λ $(= \boldsymbol{\omega} \cdot \boldsymbol{a})$ is a scalar function of the time.

The second step is to calculate the corresponding angular momentum $\boldsymbol{L}_G$. In the expression (19.11), the term $\lambda \boldsymbol{a}$ lies in the direction of the principal axis Gx_3 and the term $\boldsymbol{a} \times \dot{\boldsymbol{a}}$ is perpendicular to $\boldsymbol{a}$ and is therefore also in a principal direction. It follows from the formula (19.10) that the corresponding angular momentum $\boldsymbol{L}_G$ is given by

$$\boldsymbol{L}_G = A\, \boldsymbol{a} \times \dot{\boldsymbol{a}} + C \lambda\, \boldsymbol{a}. \tag{19.12}$$

The equation for the rotational motion of the body about G is therefore

$$\frac{d}{dt}\left(A\, \boldsymbol{a} \times \dot{\boldsymbol{a}} + C \lambda\, \boldsymbol{a}\right) = \boldsymbol{0}, \tag{19.13}$$

that is,

$$A\, \boldsymbol{a} \times \ddot{\boldsymbol{a}} + C\,(\dot{\lambda}\, \boldsymbol{a} + \lambda\, \dot{\boldsymbol{a}}) = \boldsymbol{0}.$$

If we now take the scalar product of this equation with $\boldsymbol{a}$, we obtain

$$A\, \boldsymbol{a} \cdot (\boldsymbol{a} \times \ddot{\boldsymbol{a}}) + C\,(\dot{\lambda}(\boldsymbol{a} \cdot \boldsymbol{a}) + \lambda(\boldsymbol{a} \cdot \dot{\boldsymbol{a}})) = 0.$$

Now the triple scalar product $\boldsymbol{a} \cdot (\boldsymbol{a} \times \ddot{\boldsymbol{a}})$ has two elements the same and is therefore zero; also, since $\boldsymbol{a}$ is a *unit* vector, $\boldsymbol{a} \cdot \boldsymbol{a} = 1$ and $\boldsymbol{a} \cdot \dot{\boldsymbol{a}} = 0$. It follows that $\dot{\lambda} = 0$, that is $\lambda = n$, a constant. Hence *the axial component of* $\boldsymbol{\omega}$ *is a constant.* We will call this axial component of $\boldsymbol{\omega}$ the **spin** of the body.

The **equation of motion** for the body is therefore

Equation of motion for a free axisymmetric body

$$A\,\boldsymbol{a}\times\dot{\boldsymbol{a}} + C n\,\boldsymbol{a} = \boldsymbol{L}_G \tag{19.14}$$

where the spin n $(= \boldsymbol{\omega}\cdot\boldsymbol{a})$ and the angular momentum $\boldsymbol{L}_G$ are constants determined by the initial conditions.

Surprisingly, this equation has a simple exact solution. First, let us write $\boldsymbol{L}_G = L\,\boldsymbol{k}$, where L is the magnitude of $\boldsymbol{L}_G$ and $\boldsymbol{k}$ is the unit vector in the same direction as $\boldsymbol{L}_G$, as shown in Figure 19.3. Then, on taking the *scalar* product of equation (19.14) with $\boldsymbol{a}$, we obtain

$$A\,\boldsymbol{a}\cdot(\boldsymbol{a}\times\dot{\boldsymbol{a}}) + C n\,\boldsymbol{a}\cdot\boldsymbol{a} = L\,(\boldsymbol{a}\cdot\boldsymbol{k}),$$

which simplifies to give

$$Cn = L(\boldsymbol{a}\cdot\boldsymbol{k}) = L\cos\alpha,$$

where α is the angle between $\boldsymbol{a}$ and $\boldsymbol{k}$.* It follows that α is constant and that n, the constant axial component of $\boldsymbol{\omega}$, is given by

$$n = \frac{L\cos\alpha}{C}. \tag{19.15}$$

Thus the *axis of symmetry of the body makes a constant angle with* **k** *and so sweeps out a cone around the axis* $\{G, \boldsymbol{k}\}$.

The progress of the axis of symmetry in time can be found by taking the *vector* product of equation (19.14) with $\boldsymbol{a}$. This gives

$$A\,\boldsymbol{a}\times(\boldsymbol{a}\times\dot{\boldsymbol{a}}) + C n\,\boldsymbol{a}\times\boldsymbol{a} = L\,(\boldsymbol{a}\times\boldsymbol{k}),$$

that is,

$$A\left((\boldsymbol{a}\cdot\dot{\boldsymbol{a}})\,\boldsymbol{a} - (\boldsymbol{a}\cdot\boldsymbol{a})\,\dot{\boldsymbol{a}}\right) + \boldsymbol{0} = L\,(\boldsymbol{a}\times\boldsymbol{k}).$$

Since $\boldsymbol{a}$ is a *unit* vector, $\boldsymbol{a}\cdot\boldsymbol{a} = 1$ and $\boldsymbol{a}\cdot\dot{\boldsymbol{a}} = 0$ and we obtain

$$\dot{\boldsymbol{a}} = \left(\frac{L}{A}\boldsymbol{k}\right)\times\boldsymbol{a}. \tag{19.16}$$

This equation shows that the *axis of symmetry of the body precesses around the axis* $\{G, \boldsymbol{k}\}$ *with constant angular speed* L/A.

---

* We will suppose that n is positive so that α is an acute angle.

Motion viewed from the precessing frame

It is instructive to view this motion from a rotating frame in which the axis of symmetry is at rest. The *true* angular velocity of the body is given by

$$\begin{aligned}\boldsymbol{\omega} &= \boldsymbol{a}\times\dot{\boldsymbol{a}} + n\,\boldsymbol{a}\\ &= \boldsymbol{a}\times\left(\frac{L}{A}(\boldsymbol{k}\times\boldsymbol{a})\right) + \left(\frac{L\cos\alpha}{C}\right)\boldsymbol{a}\\ &= \frac{L}{A}\,\boldsymbol{k} + L\cos\alpha\left(\frac{A-C}{AC}\right)\boldsymbol{a}.\end{aligned} \tag{19.17}$$

Suppose we now view the motion from a frame with origin at G that is rotating about the axis $\{G, \boldsymbol{k}\}$ with angular speed $(L/A)\boldsymbol{k}$. In this **precessing frame**, the axial vector $\boldsymbol{a}$ is at rest. Also, Theorem 17.1 (on the addition of angular velocities) tells us that, in the precessing frame, the *apparent* angular velocity of the body is

$$\boldsymbol{\omega}' = L\cos\alpha\left(\frac{A-C}{AC}\right)\boldsymbol{a},$$

that is, the body apparently rotates about its fixed axis of symmetry with angular velocity $L\cos\alpha(A-C)/AC$. The *true* motion is thus composed of (i) this constant axial rotation, and (ii) a precession around the axis $\{G, \boldsymbol{k}\}$ with constant angular speed L/A.

Our results are summarised as follows:

Free motion of an axisymmetric body

- The axis of symmetry of the body makes a constant angle α with the angular momentum vector $L\,\boldsymbol{k}$ and precesses around the axis $\{G, \boldsymbol{k}\}$ (in the positive sense) with constant angular speed L/A.
- In this motion, the axial spin n of the body has the constant value $L\cos\alpha/C$.
- In the precessing frame, the body apparently rotates about its fixed axis of symmetry with constant angular velocity $L\cos\alpha(A-C)/AC$.

The motion described above is very familiar, at least qualitatively. The precession of the symmetry axis looks like a 'wobble' superimposed on the spinning motion of the body about the symmetry axis.

Example 19.1 *Wobble on a frisbee*

A frisbee is spinning with constant angular speed Ω about its axis of symmetry when its motion is slightly disturbed. What is the angular frequency of the resulting wobble?

Solution

Suppose that the new angular momentum of the frisbee is $L\boldsymbol{k}$ with the axis of the frisbee making a small angle α with the fixed axis $\{G, \boldsymbol{k}\}$. Then the axis of the frisbee

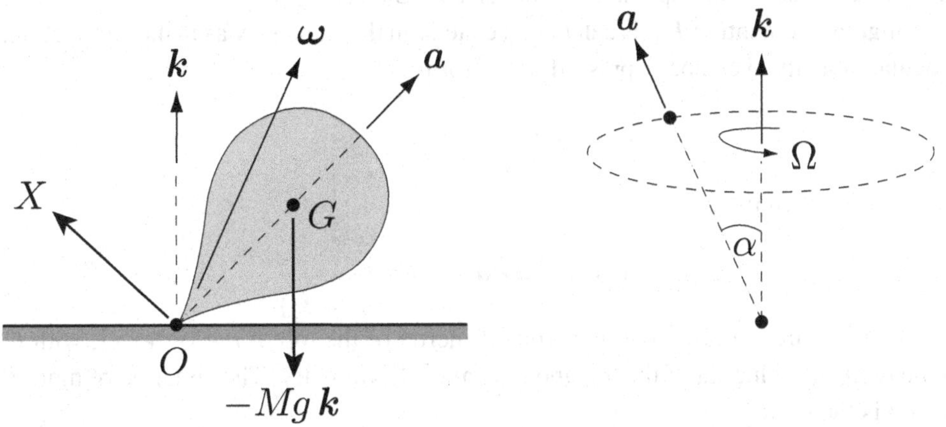

FIGURE 19.4 A symmetrical top with its vertex fixed at O. The top moves under uniform gravity and the reaction force $\boldsymbol{X}$ of the table.

precesses around the axis $\{G, \boldsymbol{k}\}$ with angular speed L/A. *It is this precession that is observed as a wobble.* The angular momentum magnitude L is not known exactly, but it differs only slightly from that in the undisturbed state, namely $C\Omega$. Hence the precession rate is $C\Omega/A$ approximately, and this is the angular frequency of the wobble. If we regard the frisbee as a circular disk, then $C = 2A$ and the **angular frequency of the wobble** is 2Ω, approximately twice the spin of the frisbee about its axis. ■

The above results can also be obtained by using Lagrangian mechanics. The method is described in Section 19.6.

19.5 THE SPINNING TOP

Consider the symmetrical spinning top* shown in Figure 19.4 (left), which has its vertex fixed at the origin O. One may imagine that the vertex of the top is lodged in a small smooth pit in the table from which it cannot escape. The top moves under uniform gravity and the unknown reaction force $\boldsymbol{X}$ exerted by the table. Since the top has a particle fixed at O, we take the equation of motion in the form

$$\dot{\boldsymbol{L}}_O = \boldsymbol{K}_O,$$

where $\boldsymbol{K}_O$ is the moment of the external forces about O. In the present case,

$$\begin{aligned}\boldsymbol{K}_O &= \boldsymbol{0}\times\boldsymbol{X} + (h\boldsymbol{a})\times(-Mg\,\boldsymbol{k})\\ &= -Mgh\,\boldsymbol{a}\times\boldsymbol{k},\end{aligned}$$

* A 'top' should be understood to mean *any* rigid body with dynamical axial symmetry that is pivoted at a point on its symmetry axis, and is moving under uniform gravity.

where M is the mass of the top and h is the distance OG.

The angular momentum $\boldsymbol{L}_O$ is calculated exactly in the same way as in the last section. The angular velocity $\boldsymbol{\omega}$ can be expressed in the form

$$\boldsymbol{\omega} = \boldsymbol{a}\times\dot{\boldsymbol{a}} + \lambda\,\boldsymbol{a},$$

and $\boldsymbol{L}_O$ is then given by

$$\boldsymbol{L}_O = A\,\boldsymbol{a}\times\dot{\boldsymbol{a}} + C\lambda\,\boldsymbol{a},$$

where $\{A, A, C\}$ are the principal moments of inertia of the top *at the vertex* O, with C being the moment of inertia of the top about its axis of symmetry. The equation of motion for the top is therefore

$$\frac{d}{dt}\left(A\,\boldsymbol{a}\times\dot{\boldsymbol{a}} + C\lambda\,\boldsymbol{a}\right) = -Mgh\,\boldsymbol{a}\times\boldsymbol{k},$$

that is

$$A\,\boldsymbol{a}\times\ddot{\boldsymbol{a}} + C\,(\dot{\lambda}\boldsymbol{a} + \lambda\dot{\boldsymbol{a}}) = -Mgh\,\boldsymbol{a}\times\boldsymbol{k}.$$

If we now take the scalar product of this equation with $\boldsymbol{a}$, we obtain

$$A\,\boldsymbol{a}\cdot(\boldsymbol{a}\times\ddot{\boldsymbol{a}}) + C\,(\dot{\lambda}(\boldsymbol{a}\cdot\boldsymbol{a}) + \lambda(\boldsymbol{a}\cdot\dot{\boldsymbol{a}})) = -Mgh\,\boldsymbol{a}\cdot(\boldsymbol{a}\times\boldsymbol{k}).$$

The triple scalar products are both zero since each has two elements the same. Also, since $\boldsymbol{a}$ is a *unit* vector, $\boldsymbol{a}\cdot\boldsymbol{a} = 1$ and $\boldsymbol{a}\cdot\dot{\boldsymbol{a}} = 0$. It follows that $\dot{\lambda} = 0$, so that $\lambda = n$, a constant. Hence *the axial component of* $\boldsymbol{\omega}$ *is a constant*. We call this axial component of $\boldsymbol{\omega}$ the **spin** of the top. The **equation of motion** for the top can therefore be written

Equation of motion for the top

$$\frac{d}{dt}\left(A\,\boldsymbol{a}\times\dot{\boldsymbol{a}} + Cn\,\boldsymbol{a}\right) = -Mgh\,\boldsymbol{a}\times\boldsymbol{k} \qquad (19.18)$$

where the spin $n\ (= \boldsymbol{\omega}\cdot\boldsymbol{a})$ is a constant determined by the initial conditions.

Steady precession

We will calculate the *general* motion of the top in the next section by using the Lagrangian method. Here we will investigate only the special, but important, motion called **steady precession**. In this motion, the axial vector $\boldsymbol{a}$ maintains a constant angle α with the vertical and precesses round the axis $\{O, \boldsymbol{k}\}$ with constant angular speed Ω, as shown in Figure 19.4 (right). In this steady precession the rate of change of $\boldsymbol{a}$ is given by

$$\dot{\boldsymbol{a}} = (\Omega\boldsymbol{k})\times\boldsymbol{a}. \qquad (19.19)$$

We will now investigate whether steady precession of the top can actually occur. From equation (19.16), it follows that, in steady precession,

$$\begin{aligned} \boldsymbol{a} \times \dot{\boldsymbol{a}} &= \boldsymbol{a} \times (\Omega \boldsymbol{k} \times \boldsymbol{a}) \\ &= \Omega (\boldsymbol{a} \cdot \boldsymbol{a}) \boldsymbol{k} - \Omega (\boldsymbol{a} \cdot \boldsymbol{k}) \boldsymbol{a} \\ &= \Omega (\boldsymbol{k} - \cos\alpha\, \boldsymbol{a}). \end{aligned}$$

Hence

$$\begin{aligned} \frac{d}{dt}(\boldsymbol{a} \times \dot{\boldsymbol{a}}) &= \Omega (\boldsymbol{0} - \cos\alpha\, \dot{\boldsymbol{a}}) \\ &= -\Omega \cos\alpha (\Omega \boldsymbol{k}) \times \boldsymbol{a} \\ &= \Omega^2 \cos\alpha (\boldsymbol{a} \times \boldsymbol{k}). \end{aligned} \tag{19.20}$$

On substituting equations (19.20) and (19.19) into the equation of motion (19.18), we obtain

$$\left(A \cos\alpha\, \Omega^2 - Cn\Omega + Mgh\right) (\boldsymbol{a} \times \boldsymbol{k}) = \boldsymbol{0}.$$

This is the condition for steady precession of the top. Since $\boldsymbol{a} \times \boldsymbol{k} \neq \boldsymbol{0}$, steady precession can only occur if there are values of α and Ω that satisfy the equation

$$A \cos\alpha\, \Omega^2 - Cn\Omega + Mgh = 0. \tag{19.21}$$

This quadratic equation will have *real roots* for the angular rate Ω if

$$C^2 n^2 \geq 4AMgh \cos\alpha. \tag{19.22}$$

If the angle α is obtuse, then this condition always holds, even when $n = 0$. However, the case in which the top executes a conical type of motion *below* the pivot O is not interesting. The interesting case is that in which α is an *acute* angle and the top precesses in an 'upright' position. Our result is as follows:

The top can undergo steady precession at an acute angle α to the upward vertical if n (the axial component of $\boldsymbol{\omega}$) is large enough to satisfy the condition (19.22). If this condition is satisfied there will generally be two different values of the precession rate for each choice of the angle α.

Fast and slow precession

The two solutions of equation (19.21) for the precession rate Ω are

$$\begin{aligned} \Omega^{F,S} &= \frac{Cn \pm \left(C^2 n^2 - 4AMgh \cos\alpha\right)^{1/2}}{2A \cos\alpha} \\ &= \frac{Cn}{2A \cos\alpha} \left[1 \pm \left(1 - \frac{4AMgh \cos\alpha}{C^2 n^2} \right)^{1/2} \right], \end{aligned} \tag{19.23}$$

where the fast (F) precession rate corresponds to the 'plus' choice and the slow precession rate (S) corresponds to the 'minus' choice. In practical circumstances the value of the spin n is often such that the dimensionless ratio

$$\frac{4AMgh}{C^2n^2}$$

is *small compared to unity*. In this case, Ω^F and Ω^S are given approximately by

$$\Omega^F \approx \frac{Cn}{A\cos\alpha}, \qquad \Omega^S \approx \frac{Mgh}{Cn}. \tag{19.24}$$

The **fast precession** rate is approximately *directly proportional* to n and is independent of the gravitational acceleration! In fact, it approximates the *force-free precession* found in the last section. This motion is almost impossible to observe in a real top. It has a precession rate of similar magnitude to n which would make the vertex of the top extremely difficult to secure. The fast precession may however explain the trembling motion sometimes seen when a top is spinning slowly with its axis almost vertical.

In contrast, the **slow precession** rate is approximately *inversely proportional to n and independent of the angle α*. This is the motion commonly observed; the faster the spin of the top, the slower the rate of precession.

Example 19.2 *A simple top*

A top is made by sticking a light pin of length 3 cm through the centre of a uniform circular disk of mass M and radius 8 cm. Find the rates of slow and fast precession of the top when it is given a spin of 10 revolutions per second and α is small. [Take $g = 10\ \mathrm{m\,s^{-2}}$.]

Solution

For a uniform disk of mass M and radius a, the principal moments of inertia at its centre of mass are $\{\frac{1}{4}Ma^2, \frac{1}{4}Ma^2, \frac{1}{2}Ma^2\}$ and so , by the theorem of parallel axes, the principal moments at the point of the pin are $A = B = \frac{1}{4}M(a^2 + 4h^2)$, $C = \frac{1}{2}Ma^2$, where h is the length of the pin. Hence, on putting in the given dimensions, $A = B = 25M/10000$ and $C = 32M/10000$, where M is the mass of the disk. We are also given that $n = 20\pi$ radians per second and that α is small. The quadratic equation for the precession rate Ω is therefore

$$\frac{25M}{10000}\Omega^2 - \frac{32M}{10000}(40\pi)\Omega + \frac{30M}{100} = 0.$$

The roots of this equation are $\Omega^S = 1.52$ and $\Omega^F = 78.9$ radians per second. The approximate formulae in equation (19.24) give 1.49 and 80.4 respectively. [In this case the dimensionless ratio $4AMgh/(C^2n^2) \approx 0.074$ so a reasonable approximation could be expected.] ■

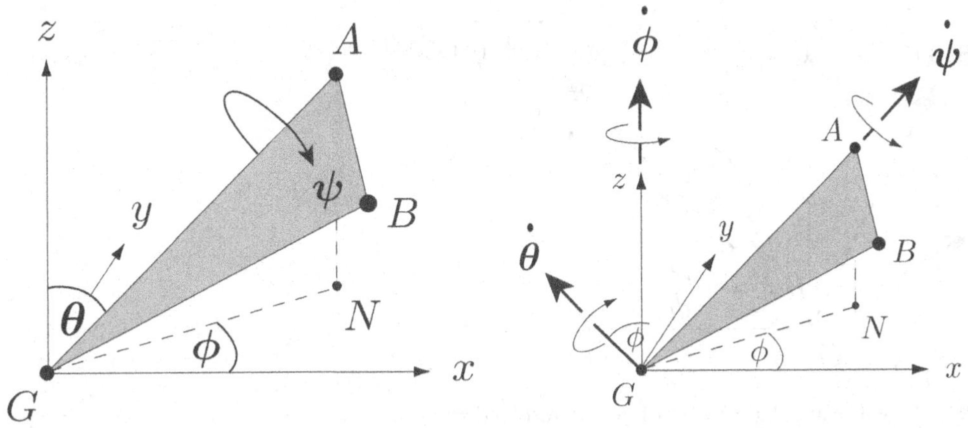

FIGURE 19.5 **Left:** The Euler angles θ, ϕ and ψ. **Right:** The corresponding 'velocity' diagram.

19.6 LAGRANGIAN DYNAMICS OF THE TOP

The Euler angles

In order to solve problems in rigid body dynamics by Lagrangian mechanics, we need a set of **generalised coordinates** for a generally moving rigid body. The position of the centre of mass G can be specified by its three Cartesian coordinates, but the position of the body *relative* to G still needs to be specified. The best practical coordinates for this purpose are the **Euler angles*** θ, ϕ and ψ shown in Figure 19.5 (left). In order to fix the position of the body, it is sufficient to specify the positions of *two more* of its particles, A and B say, chosen so that G, A, B do not lie in a straight line. The position of A is fixed by the 'polar angles' θ and ϕ, measured relative to the fixed Cartesian coordinate system $Gxyz$. This does not yet fix the position of B since the rigid triangle GAB can still rotate around GA. The position of B becomes fixed when we specify the angle ψ through which the triangle GAB has been rotated from some reference position.† Since the angles θ, ϕ, ψ are clearly independent variables, it follows that *the* ***Euler angles*** *θ, ϕ, ψ are a set of generalised coordinates for the* ***rotational*** *motion of a rigid body.*

The Euler angles are a particularly appropriate set of coordinates when the body has **axial symmetry**. In this case it is usual to take the A to be a particle *lying on the symmetry axis*. Then θ and ϕ determine the orientation of the symmetry axis and ψ is the rotation angle of the body around the symmetry axis.

We may now construct the **velocity diagram** corresponding to this choice of coordinates. When the coordinates θ, ϕ, ψ increase individually, each gives the body an *angular* velocity as shown in Figure 19.5 (right). The **angular velocity** of the body in general motion is then obtained by adding these three angular velocities vectorially.

---

* After Leonhard Euler (1707–1783), the greatest Swiss mathematician, and the most prolific of all time.

† The triangle GAB is in its standard reference position when it lies in the plane AGz with B 'below' the line GA. The configuration shown in Figure 19.5 thus has a ψ value of nearly 2π.

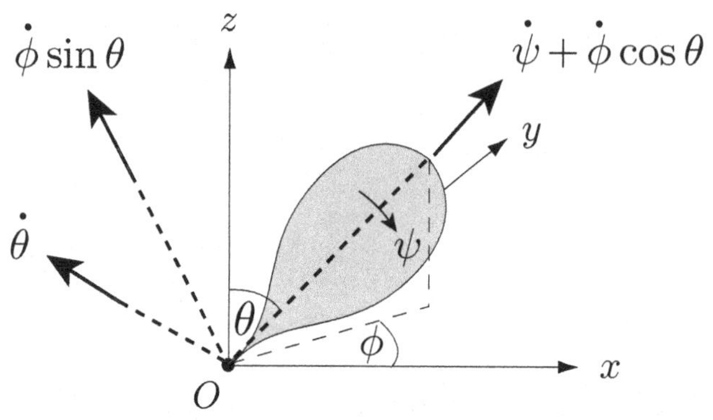

FIGURE 19.6 The angular velocity of the top resolved along three perpendicular principal directions.

Lagrangian for the top

For a top with its vertex fixed at the origin O, we take a fixed system of Cartesian coordinates $Oxyz$ with Oz vertically upwards and define the Euler angles as shown in Figure 19.5 (left) with G replaced by O. Since the top has axial symmetry we also take A to lie on the symmetry axis, which is a principal axis of the top at O. The top is thus a holonomic system and can be solved by Lagrangian method. To find the kinetic energy of the top in terms of the Euler angles, the angular velocities in Figure 19.5 need to be resolved along three perpendicular principal axes, as shown in Figure 19.6.* The **kinetic energy** of the top is therefore given by

$$T = \tfrac{1}{2}A\dot{\theta}^2 + \tfrac{1}{2}A\left(\dot{\phi}\sin\theta\right)^2 + \tfrac{1}{2}C\left(\dot{\psi} + \dot{\phi}\cos\theta\right)^2, \qquad (19.25)$$

where $\{A, A, C\}$ are the principal moments of inertia of the top at O. The gravitational **potential energy** of the top is simply

$$V = MgZ = Mgh\cos\theta,$$

where h is the distance OG. Hence the **Lagrangian** of the top, with the Euler angles as coordinates, is:

Lagrangian for the top

$$L = \tfrac{1}{2}A\dot{\theta}^2 + \tfrac{1}{2}A\left(\dot{\phi}\sin\theta\right)^2 + \tfrac{1}{2}C\left(\dot{\psi} + \dot{\phi}\cos\theta\right)^2 - Mgh\cos\theta \qquad (19.26)$$

Conserved quantities

Since ϕ and ψ appear in the Lagrangian only as $\dot{\phi}$ and $\dot{\psi}$, they are both **cyclic coordinates**. It follows that the generalised momenta p_ϕ and p_ψ are *constants of the motion*. On

* These principal axes are not 'embedded' in the top since they do not rotate as ψ increases.

evaluating p_ϕ $(= \partial L/\partial\dot{\phi})$ and p_ψ $(= \partial L/\partial\dot{\psi})$, these conservation principles become

$$A\dot{\phi}\sin^2\theta + C(\dot{\psi} + \dot{\phi}\cos\theta)\cos\theta = L_z, \tag{19.27}$$

$$C(\dot{\psi} + \dot{\phi}\cos\theta) = Cn, \tag{19.28}$$

where n and L_z are constants, determined by the initial conditions. In fact, *L_z is the z-component of the* ***angular momentum*** *$\mathbf{L}_O$, and n is the axial component of the angular velocity* $\boldsymbol{\omega}$, the quantity known as the **spin** of the top.* Without losing generality, we will assume that n is *positive*.

These two conservation principles, together with the **energy conservation** equation

$$\tfrac{1}{2}A\dot{\theta}^2 + \tfrac{1}{2}A(\dot{\phi}\sin\theta)^2 + \tfrac{1}{2}C(\dot{\psi} + \dot{\phi}\cos\theta)^2 + Mgh\cos\theta = E, \tag{19.29}$$

are sufficient to determine the motion of the top. The energy conservation equation is preferable to the Lagrange equation for the coordinate θ because it contains only *first* derivatives of the Euler angles.

The equation for the inclination angle θ

By making use of the conservation equations (19.27) and (19.28) in the energy equation (19.29), we can eliminate $\dot{\phi}$ and $\dot{\psi}$ to give

$$A\dot{\theta}^2 = 2E - Cn^2 - \frac{(L_z - Cn\cos\theta)^2}{A\sin^2\theta} - 2Mgh\cos\theta, \tag{19.30}$$

where the constants n, L_z, E are determined by the initial conditions. This ODE for the unknown function $\theta(t)$ has the form

$$\dot{\theta}^2 = \frac{F(\cos\theta)}{A^2\sin^2\theta},$$

where the function $F(u)$ is defined by

$$F(u) = A(2E - Cn^2)(1 - u^2) - (L_z - Cnu)^2 - 2AMghu(1 - u^2).$$

This equation is similar to the corresponding equation in the theory of the spherical pendulum (see Chapter 11) and we can make similar deductions about the behaviour of the coordinate θ. Because the left side of (19.30) is positive, it follows that the motion is restricted to those values of θ that make the function $F(\cos\theta)$ positive. Moreover, the maximum and minimum values of θ can only occur when $F(\theta) = 0$.

The equation $F(u) = 0$ is a cubic in the variable u and the positions of its roots can be deduced by considering the sign of F. We have the following information:

* Note that $n = \dot{\psi} + \dot{\phi}\cos\theta$, not $\dot{\psi}$.

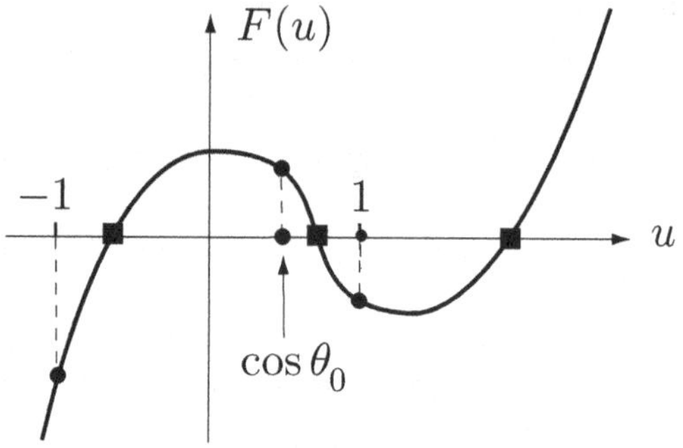

FIGURE 19.7 The form of the cubic function $F(u)$. The black squares are at its roots.

(i) As u tends to positive (or negative) infinity, $F(u)$ tends to positive (or negative) infinity.

(ii) $F(1) = -(L_z - Cn)^2$ and $F(-1) = -(L_z + Cn)^2$ which are both *negative*.†

(iii) Given that θ_0, the initial value of θ, lies in the range $0 < \theta_0 < \pi$ and that $\dot{\theta}$ is not zero initially, then $F(\cos\theta_0)$ must be *positive*.

The graph of $F(u)$ must therefore have the form shown in Figure 19.7. It follows that the equation $F(u) = 0$ must have a root u_{min} lying in $-1 < u < \cos\theta_0$, a root u_{max} lying in $\cos\theta_0 < u < 1$, and a third root lying in $u > 1$, as indicated by the black squares in Figure 19.7; this accounts for all three roots. The root greater than unity is not physically admissible since it lies outside the range of u when $u = \cos\theta$. It follows that, in the motion, u must oscillate in the range $u_{\text{min}} \le u \le u_{\text{max}}$. Hence, as in the case of the spherical pendulum, the inclination angle θ must perform **periodic oscillations** between two extreme values α and β. The difference here is that it is possible for both α and β to be *acute* angles (measured from the upward vertical); in other words, the top can stand 'upright'. This oscillation of the inclination angle θ is called **nutation** of the top.

Example 19.3 *Finding the range of the inclination angle*

Suppose that the top is released with its axis at rest and making an angle of $\pi/3$ with the upward vertical. Find the function $F(u)$. For the case in which $C^2n^2 = 4AMgh$, find the range of values of θ that occur in the subsequent motion.

Solution

On using the initial conditions $\theta = \pi/3$, $\dot{\theta} = 0$, $\dot{\phi} = 0$ when $t = 0$, we find that $L_z = \frac{1}{2}Cn$ and $E = \frac{1}{2}Cn^2 + \frac{1}{2}Mgh$. The function $F(u)$ therefore reduces to

$$F(u) = \tfrac{1}{4}(1-2u)\Big[4AMgh(1-u^2) - C^2n^2(1-2u)\Big],$$

and when $C^2n^2 = 4AMgh$, this becomes

$$F(u) = AMgh\,u(1-2u)(2-u).$$

† $F(\pm 1)$ is zero when $L_z = \pm Cn$ respectively. This corresponds to a special (and not very important) class of motions in which the axis of the top passes through the upward or downward vertical. We exclude these cases from consideration.

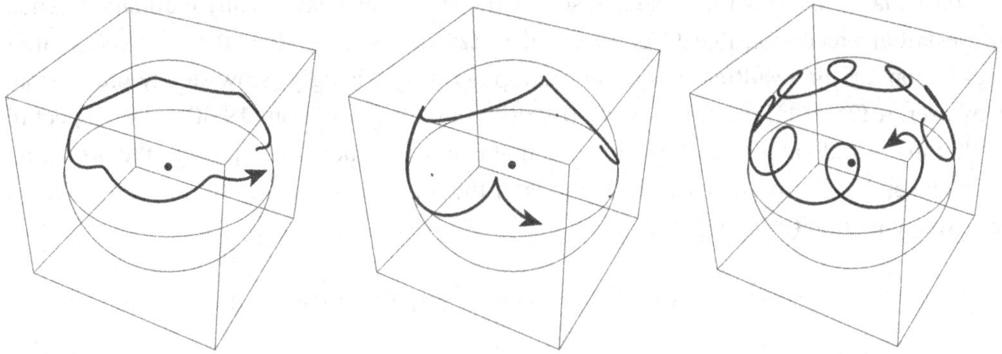

FIGURE 19.8 Three different forms for the motion of the axis of symmetry of the top. The paths show the movement of the point in which the axis of symmetry intersects a fixed sphere with centre O.

The physical roots of the equation $F(u) = 0$ are $u = 0$ and $u = \frac{1}{2}$ and so θ oscillates in the range $\pi/3 \le \theta \le \pi/2$. This is the motion shown in Figure 19.8 (centre). ∎

Question ***Finding the period of*** θ

Find the period of the angle θ in the same motion.

Answer

With the same set of initial conditions (and with $C^2n^2 = 4AMgh$), the ODE for θ reduces to

$$\dot{\theta}^2 = \frac{Mgh\cos\theta(1 - 2\cos\theta)(2 - \cos\theta)}{A\sin^2\theta}.$$

On taking square roots and integrating, the period τ of the function $\theta(t)$ is found to be

$$\tau = 2\left(\frac{A}{Mgh}\right)^{1/2}\int_{\pi/3}^{\pi/2}\frac{\sin\theta\, d\theta}{[\cos\theta(1 - 2\cos\theta)(2 - \cos\theta)]^{1/2}} \approx 3.37\left(\frac{A}{Mgh}\right)^{1/2}. \blacksquare$$

Precession of the axis

Once the function $\theta(t)$ has been determined (at least in principle), the precession angle $\phi(t)$ can be found from the angular momentum conservation equation

$$A\dot{\phi}\sin^2\theta + Cn\cos\theta = L_z. \tag{19.31}$$

Note that the precession rate $\dot{\phi}$ is *not constant* when the top is in general motion.

With the initial conditions $\theta = \alpha, \dot{\theta} = 0, \phi = 0, \dot{\phi} = \Omega$, the value of the constant L_z is $A\Omega\sin^2\alpha + Cn\cos\alpha$, and equation (19.31) becomes

$$A\dot{\phi}\sin^2\theta = A\Omega\sin^2\alpha + Cn(\cos\alpha - \cos\theta). \tag{19.32}$$

Suppose that $C^2n^2 \geq 4AMgh$ so that steady precession can exist at any inclination. Then if the initial precession rate Ω is set equal to Ω^S (the rate of slow steady precession at inclination α), the resulting motion of the top is (unsurprisingly) slow steady precession. However, if $\Omega < \Omega^S$, it can be shown (by differentiating equation (19.30) with respect to t) that the initial value of $\ddot{\theta}$ is positive so that θ initially increases and α is the *minimum* value taken by θ. (In other words, the axis of the top *falls* initially.) In this case, we can see from equation (19.32) that

$$A\Omega \sin^2\alpha \leq A\dot{\phi}\sin^2\theta \leq A\Omega\sin^2\alpha + Cn(\cos\alpha - \cos\beta),$$

where β is the *maximum* value taken by θ. It follows that

(i) If $\Omega > 0$, then the precession rate $\dot{\phi}$ is always *positive*. This is the case shown in Figure 19.8 (left). The critical case in which $\Omega = 0$ (when the axis is released from rest) is shown in Figure 19.8 (centre).

(ii) There is no reason why the top axis cannot be projected the 'wrong' way. If $\Omega < 0$ (but is not so negative that $A\Omega\sin^2\alpha + Cn(\cos\alpha - \cos\beta) < 0$), then $\dot{\phi}$ is *sometimes positive and sometimes negative*! In this case, the path of the axis crosses itself, as shown in Figure 19.8 (right).

(iii) If the value of Ω is so negative that $A\Omega\sin^2\alpha + Cn(\cos\alpha - \cos\beta) < 0$, then $\dot{\phi}$ is always *negative*. However, this requires a 'fast' precession rate that is unlikely to be observed.

In a similar way, one can show that when $\Omega^S < \Omega < \Omega^F$, the axis of the top *rises* initially and that $\dot{\phi}$ is always positive. The motion of the axis resembles that shown in Figure 19.8 (left).

Example 19.4 *The precession during a period of θ*

In the last example, find the angle through which the top precesses during one period of the inclination angle θ.

Solution

In the last example, the axis of the top is released from rest with $\theta = \pi/3$, and with $C^2n^2 = 4AMgh$. With $\alpha = \pi/3$ and $\Omega = 0$, the equation for the precession rate $\dot{\phi}$ becomes

$$\dot{\phi} = \frac{Cn(1 - 2\cos\theta)}{2A\sin^2\theta}.$$

It follows that Φ, the increase in ϕ during one period of the coordinate θ is given by

$$\Phi = \phi(\tau) - \phi(0) = \int_0^\tau \dot{\phi}\,dt$$
$$= \frac{Cn}{2A}\int_0^\tau \frac{1 - 2\cos\theta}{\sin^2\theta}\,dt = \left(\frac{Mgh}{A}\right)^{1/2} \times 2\int_{\pi/3}^{\pi/2}\left(\frac{1 - 2\cos\theta}{\sin^2\theta}\right)\frac{d\theta}{\dot{\theta}}.$$

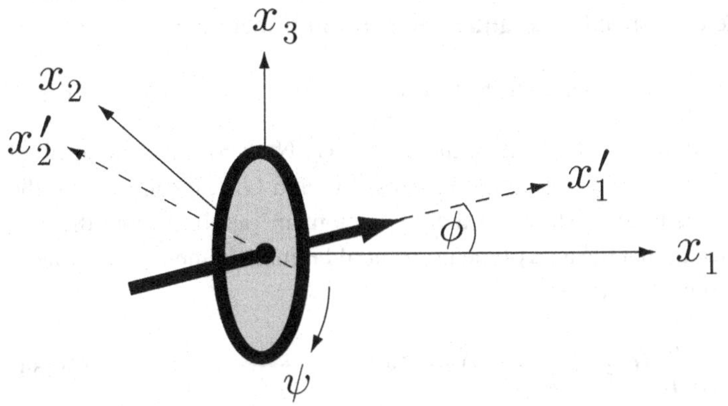

FIGURE 19.9 The gyrocompass is a gyroscope whose symmetry axis is constrained to remain in a horizontal plane.

On using the expression for $\dot{\theta}$ from the last example, we obtain

$$\Phi = 2\int_{\pi/3}^{\pi/2}\left[\frac{1-2\cos\theta}{\cos\theta(2-\cos\theta)}\right]^{1/2}\frac{d\theta}{\sin\theta} \approx 1.69,$$

which is about 97°. This motion is shown in Figure 19.8 (centre). ■

19.7 THE GYROCOMPASS

The gyrocompass is the most important navigational instrument used in the world today. Invented by Elmer Sperry* in 1908, it consists essentially of a gyroscope which is smoothly pivoted so that its centre of mass G remains fixed and its *axis of symmetry remains in the* ***horizontal*** *plane through* G; the gyroscope is however free to rotate about its axis of symmetry and about the vertical through G (see Figure 19.9). The constraint that the axis of the gyro should remain horizontal is vital to its operation.

Non-rotating Earth

Consider first the case in which the Earth has **no rotation**. Let $Gx_1x_2x_3$ be a frame fixed to the Earth, with the axis Gx_3 pointing vertically upwards and the axis of the gyro moving in the horizontal (x_1, x_2)-plane. Let $Gx_1'x_2'x_3$ be a second set of axes with Gx_1' pointing along the axis of symmetry of the gyro; these axes therefore *rotate with the gyro*. Since the Euler angle θ is constrained to be $\pi/2$ in this problem, the system has two degrees of freedom and the Euler angles ϕ and ψ (see Figure 19.9) form a set of generalised coordinates. In terms of these angles, the **angular velocity** of the gyroscope is

$$\boldsymbol{\omega} = \dot{\psi}\,\boldsymbol{e}_1' + \dot{\phi}\,\boldsymbol{e}_3, \tag{19.33}$$

* Elmer Ambrose Sperry (1860–1930), American inventor and entrepreneur.

where $\{e'_1, e'_2, e_3\}$ are the unit vectors of the coordinate system $Gx'_1x'_2x_3$; note that e'_1 and e'_2 are functions of the time. The corresponding **angular momentum** is therefore

$$L_G = C\dot{\psi}\, e'_1 + A\dot{\phi}\, e_3,$$

where $\{C, A, A\}$ are principal moments of inertia of the gyro at G. Now consider the **applied moment** $\boldsymbol{K}_G$. Since the gyro is free to rotate about the axes Gx'_1 and Gx_3, it follows that the applied moment about these axes is zero. However there is a moment* applied about the axis Gx'_2 which enforces the constraint that the axis of the gyro should remain in the (x_1, x_2)-plane. Hence the **equation of motion** of the gyro is

$$\frac{d}{dt}\left(C\dot{\psi}\, e'_1 + A\dot{\phi}\, e_3\right) = K\, e'_2, \tag{19.34}$$

where K is the unknown 'moment of constraint'. Now

$$\frac{d}{dt}\left(C\dot{\psi}\, e'_1 + A\dot{\phi}\, e_3\right) = C\ddot{\psi}\, e'_1 + C\dot{\psi}\,\frac{de'_1}{dt} + A\ddot{\phi}\, e_3,$$

and

$$\frac{de'_1}{dt} = \frac{de'_1}{d\phi} \times \frac{d\phi}{dt} = \dot{\phi}\, e'_2.$$

The last step follows since the unit vectors e'_1, e'_2 are analogous to the polar unit vectors $\widehat{\boldsymbol{r}}, \widehat{\boldsymbol{\theta}}$ and hence satisfy the relations

$$\frac{de'_1}{d\phi} = e'_2, \qquad \frac{de'_2}{d\phi} = -e'_1. \tag{19.35}$$

On combining these results, we obtain the three component equations

$$\ddot{\psi} = 0, \qquad C\dot{\phi}\dot{\psi} = K, \qquad \ddot{\phi} = 0.$$

Hence the **motion** of the gyro has the form

$$\dot{\psi} = n, \qquad \dot{\phi} = \Omega,$$

where n and Ω are constants determined by the initial conditions. Thus the *gyro has constant spin n and precesses at a constant rate Ω*. The 'moment of constraint' that keeps the gyro axis horizontal is $K = Cn\Omega$. This is the complete solution for the gyro on a non-rotating Earth.

Rotating Earth

A gyro that simply precesses at a constant rate does not seem much like a direction finding device, but everything changes when we introduce the effect of the **Earth's rotation**. Suppose now that the axes $Gx_1x_2x_3$ are fixed at some location on the Earth with Gx_3 in the direction of the *local* vertical; the axes are oriented so that Gx_1 points north and Gx_2 points west. Then $Gx_1x_2x_3$ is no longer an inertial frame and our equations of motion must be modified.

* This moment is applied by the gimbal ring that holds the gyro.

The angular velocity of the gyro *relative to the frame* $Gx_1x_2x_3$ is still given by equation (19.33) but now the frame $Gx_1x_2x_3$ is not inertial but rotates with the Earth's angular velocity $\boldsymbol{\Omega}^E$. This can be written in the form

$$\begin{aligned}\boldsymbol{\Omega}^E &= \Omega^E \sin\alpha\, \boldsymbol{e}_1 + \Omega^E \cos\alpha\, \boldsymbol{e}_3 \\ &= \Omega^E \sin\alpha\cos\phi\, \boldsymbol{e}'_1 - \Omega^E \sin\alpha\sin\phi\, \boldsymbol{e}'_2 + \Omega^E \cos\alpha\, \boldsymbol{e}_3 \end{aligned} \tag{19.36}$$

where the angle α is the co-latitude of the location where the gyro is situated. Hence, by the addition theorem for angular velocities, the true **angular velocity** of the gyro (relative to an *inertial* frame) is

$$\begin{aligned}\boldsymbol{\omega} &= (\dot\psi\, \boldsymbol{e}'_1 + \dot\phi\, \boldsymbol{e}_3) + \boldsymbol{\Omega}^E \\ &= \left(\dot\psi + \Omega^E \sin\alpha\cos\phi\right)\boldsymbol{e}'_1 - \left(\Omega^E \sin\alpha\sin\phi\right)\boldsymbol{e}'_2 + \left(\dot\phi + \Omega^E\cos\alpha\right)\boldsymbol{e}_3.\end{aligned}$$

It follows that the true **angular momentum** of the gyro is given by

$$\boldsymbol{L}_G = C\left(\dot\psi + \Omega^E \sin\alpha\cos\phi\right)\boldsymbol{e}'_1 - A\left(\Omega^E \sin\alpha\sin\phi\right)\boldsymbol{e}'_2 + A\left(\dot\phi + \Omega^E\cos\alpha\right)\boldsymbol{e}_3. \tag{19.37}$$

For the same reasons as before, the **applied moment** has the form $\boldsymbol{K}_G = K\boldsymbol{e}'_2$, where K is the unknown moment of constraint. The **equation of motion** of the gyro is therefore

$$\frac{d\boldsymbol{L}_G}{dt} = K\boldsymbol{e}'_2, \tag{19.38}$$

where $\boldsymbol{L}_G$ is given by equation (19.37). Now

$$\begin{aligned}\frac{d\boldsymbol{L}_G}{dt} &= C\frac{d}{dt}\left(\dot\psi + \Omega^E \sin\alpha\cos\phi\right)\boldsymbol{e}'_1 + C\left(\dot\psi + \Omega^E \sin\alpha\cos\phi\right)\frac{d\boldsymbol{e}'_1}{dt} \\ &- A\frac{d}{dt}\left(\Omega^E \sin\alpha\sin\phi\right)\boldsymbol{e}'_2 - A\left(\Omega^E \sin\alpha\sin\phi\right)\frac{d\boldsymbol{e}'_2}{dt} + A\ddot\phi\,\boldsymbol{e}_3 + A\left(\dot\phi + \Omega^E\cos\alpha\right)\frac{d\boldsymbol{e}_3}{dt}.\end{aligned}$$

The vectors $\{\boldsymbol{e}'_1, \boldsymbol{e}'_2, \boldsymbol{e}_3\}$ are the unit vectors of the frame $Gx'_1x'_2x_3$ and so their (true) rates of change are given by

$$\frac{d\boldsymbol{e}'_1}{dt} = \boldsymbol{\omega}^* \times \boldsymbol{e}'_1, \qquad \frac{d\boldsymbol{e}'_2}{dt} = \boldsymbol{\omega}^* \times \boldsymbol{e}'_2, \qquad \frac{d\boldsymbol{e}_3}{dt} = \boldsymbol{\omega}^* \times \boldsymbol{e}_3,$$

where $\boldsymbol{\omega}^*$ is the angular velocity of the frame $Gx'_1x'_2x_3$ relative to an inertial frame. By the addition theorem for angular velocities, this is given by

$$\begin{aligned}\boldsymbol{\omega}^* &= \dot\phi\,\boldsymbol{e}_3 + \boldsymbol{\Omega}^E \\ &= \left(\Omega^E \sin\alpha\cos\phi\right)\boldsymbol{e}'_1 - \left(\Omega^E \sin\alpha\sin\phi\right)\boldsymbol{e}'_2 + \left(\dot\phi + \Omega^E\cos\alpha\right)\boldsymbol{e}_3.\end{aligned}$$

On combining these results and simplifying, we obtain the three component equations of motion for the gyro. As before, the second of these equations serves to determine K and the first and third are

$$\frac{d}{dt}\left(\dot\psi + \Omega^E \sin\alpha\cos\phi\right) = 0,$$

and

$$A\ddot{\phi} + \Omega^E \sin\alpha \left[C\left(\dot{\psi} + \Omega^E \sin\alpha\cos\phi\right) - A\Omega^E \sin\alpha\cos\alpha \right] \sin\phi = 0.$$

It follows that

$$\dot{\psi} + \Omega^E \sin\alpha\cos\phi = n,$$

where n is a constant. Hence the **spin** $\boldsymbol{\omega} \cdot \boldsymbol{e}_1'$ of the gyro is a constant of the motion. The equation for the **precession angle** ϕ then becomes

$$A\ddot{\phi} + \Omega^E \sin\alpha \left[Cn - A\Omega^E \sin\alpha\cos\alpha \right] \sin\phi = 0. \qquad (19.39)$$

In practice, the ratio $A\Omega^E/Cn$ is *very* small so that $Cn - A\Omega^E \sin\alpha\cos\alpha$ can be closely approximated by Cn.

Equation (19.39) is the equation for the (large) oscillations of a pendulum about the direction $\phi = 0$, that is, about the axis Gx_1. Since this axis points north, it follows that *the axis of the gyro performs periodic oscillations about the northerly direction.* If these oscillations are damped (a feature not included in our model), then the axis of the gyro will eventually settle pointing north. The gyro does therefore 'home' to north and (despite statements in the literature to the contrary) does not have to be initially set pointing north.

Example 19.5 *The oscillation period of the gyrocompass*

Estimate the period of small oscillations of the gyro in a typical case.

Solution

In a typical case the ratio $\Omega^E/n < 10^{-6}$ so that the equation (19.39) is *very* accurately approximated by

$$A\ddot{\phi} + \left(Cn\Omega^E \sin\alpha\right) \sin\phi = 0,$$

which, for **small oscillations** is approximated by

$$A\ddot{\phi} + \left(Cn\Omega^E \sin\alpha\right) \phi = 0.$$

The period of these simple harmonic oscillations is

$$\tau = 2\pi \left(\frac{A}{Cn\Omega^E \sin\alpha} \right)^{1/2}.$$

With $\Omega^E = 0.000073$, $n = 50$ Hz, $C = 2A$ and $\alpha = 45°$ this gives a **period** of about 35 seconds. ■

19.8 EULER'S EQUATIONS

The simplest set of equations for the rotational motion of a rigid body are **Euler's equations**. They apply to any body, symmetrical or not, but they come with some deficiencies.

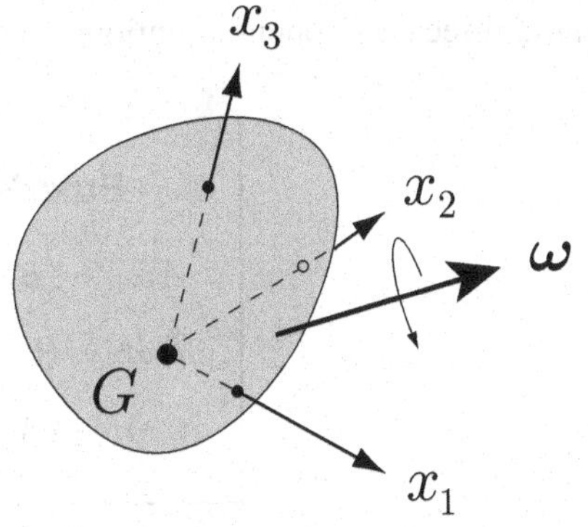

FIGURE 19.10 The principal axes $Gx_1x_2x_3$ are embedded in the body and have the same angular velocity as the body.

Consider a general rigid body with centre of mass G and principal axes $Gx_1x_2x_3$, as shown in Figure 19.10. If the body has an axis of dynamical symmetry through G, then there are infinitely many sets of principal axes at G and it is possible to find principal axes that are not rigidly attached to the body; indeed, this has been an essential feature when calculating the angular momentum $\boldsymbol{L}_G$ for a symmetrical body. However, we now wish to take a set of principal axes that is *rigidly fixed* within the body, even when this is not necessary. To emphasise the point, we will call these a set of **embedded axes** at G. The most important property of a reference frame embedded in a body is that its *angular velocity is the same as the angular velocity of the body in which it is embedded.*

Let $\boldsymbol{\omega}$, the **angular velocity** of the body, be expanded in the form

$$\boldsymbol{\omega} = \omega_1 \boldsymbol{e}_1 + \omega_2 \boldsymbol{e}_2 + \omega_3 \boldsymbol{e}_3, \tag{19.40}$$

where $\{\boldsymbol{e}_1, \boldsymbol{e}_2, \boldsymbol{e}_3\}$ are the unit vectors of the coordinate system $Gx_1x_2x_3$. Then the corresponding **angular momentum** is

$$\boldsymbol{L}_G = A\omega_1 \boldsymbol{e}_1 + B\omega_2 \boldsymbol{e}_2 + C\omega_3 \boldsymbol{e}_3, \tag{19.41}$$

where $\{A, B, C\}$ are the principal moments of inertia of the body. If the body is acted upon by the external moment $\boldsymbol{K}_G$, then the equation of rotational motion about G is

$$\frac{d}{dt}\big(A\omega_1 \boldsymbol{e}_1 + B\omega_2 \boldsymbol{e}_2 + C\omega_3 \boldsymbol{e}_3\big) = \boldsymbol{K}_G. \tag{19.42}$$

Now

$$\frac{d}{dt}(A\omega_1 \boldsymbol{e}_1 + B\omega_2 \boldsymbol{e}_2 + C\omega_3 \boldsymbol{e}_3) = A\dot{\omega}_1 \boldsymbol{e}_1 + B\dot{\omega}_2 \boldsymbol{e}_2 + C\dot{\omega}_3 \boldsymbol{e}_3 + \\ A\omega_1 \frac{d\boldsymbol{e}_1}{dt} + B\omega_2 \frac{d\boldsymbol{e}_2}{dt} + C\omega_3 \frac{d\boldsymbol{e}_3}{dt},$$

and, since the unit vectors $\{\boldsymbol{e}_1, \boldsymbol{e}_2, \boldsymbol{e}_3\}$ are embedded in the body, it follows that

$$\frac{d\boldsymbol{e}_1}{dt} = \boldsymbol{\omega}\times\boldsymbol{e}_1, \qquad \frac{d\boldsymbol{e}_2}{dt} = \boldsymbol{\omega}\times\boldsymbol{e}_2, \qquad \frac{d\boldsymbol{e}_3}{dt} = \boldsymbol{\omega}\times\boldsymbol{e}_3,$$

where $\boldsymbol{\omega}$ is given by equation (19.40). On combining these results and simplifying, we obtain the three component equations

Euler's equations

$$\begin{aligned} A\,\dot{\omega}_1 - (B - C)\,\omega_2\,\omega_3 &= K_1 \\ B\,\dot{\omega}_2 - (C - A)\,\omega_3\,\omega_1 &= K_2 \\ C\,\dot{\omega}_3 - (A - B)\,\omega_1\,\omega_2 &= K_3 \end{aligned} \qquad (19.43)$$

where $\{\omega_1, \omega_2, \omega_3\}$ and $\{K_1, K_2, K_3\}$ are the components of the angular velocity $\boldsymbol{\omega}$ and the applied moment $\boldsymbol{K}_G$ *in the embedded coordinate system* $Gx_1x_2x_3$. These are **Euler's equations of motion.**

As an example, we re-consider a problem solved earlier: the free motion of a body with axial symmetry.

Example 19.6 *Body with axial symmetry*

Solve Euler's equations for the free motion of a body with axial symmetry.

Solution

Let Gx_3 be the axis of symmetry so that $B = A$. In free motion, $K_1 = K_2 = K_3 = 0$ and the Euler equations reduce to

$$\begin{aligned} A\,\dot{\omega}_1 - (A - C)\,\omega_2\,\omega_3 &= 0, \\ A\,\dot{\omega}_2 - (C - A)\,\omega_3\,\omega_1 &= 0, \\ C\,\dot{\omega}_3 &= 0. \end{aligned}$$

The third equation gives

$$\omega_3 = n,$$

where n is a constant that we recognise as the spin $\boldsymbol{\omega} \cdot \boldsymbol{e}_3$. The first two equations then reduce to

$$\dot{\omega}_1 - \Omega\,\omega_2 = 0, \qquad \dot{\omega}_2 + \Omega\,\omega_1 = 0,$$

where $\Omega = (A - C)n/A$. On eliminating ω_2, we find that ω_1 satisfies the SHM equation

$$\ddot{\omega}_1 + \Omega^2\omega_1 = 0,$$

which has general solution

$$\omega_1 = P\sin(\Omega t + \gamma),$$

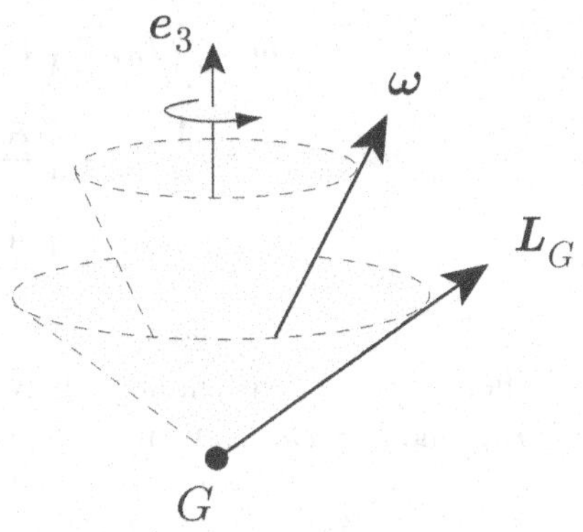

FIGURE 19.11 Relative to the embedded frame, $\boldsymbol{\omega}$ and $\boldsymbol{L}_G$ precess around the symmetry axis.

where P and γ are constants of integration; the corresponding solution for ω_2 is

$$\omega_2 = P\cos(\Omega t + \gamma).$$

The **general solution** of the Euler equations in this case is therefore

$$\omega_1 = P\sin(\Omega t + \gamma),$$
$$\omega_2 = P\cos(\Omega t + \gamma),$$
$$\omega_3 = n,$$

where P and γ are constants. The corresponding components of the **angular momentum** vector $\boldsymbol{L}_G$ are

$$L_1 = AP\sin(\Omega t + \gamma),$$
$$L_2 = AP\cos(\Omega t + \gamma),$$
$$L_3 = Cn.$$

The geometrical interpretation of these results is that *the vectors* $\boldsymbol{e}_3$*,* $\boldsymbol{\omega}$ *and* $\boldsymbol{L}_G$ *all lie in the same plane, where they make constant angles with each other, and that this plane rotates around the embedded axis* Gx_3 *at the constant rate* $(A - C)n/A$. This situation is shown in Figure 19.11 for the case in which $A > C$.

Motion of the body

What we have found is the time variation of $\boldsymbol{\omega}$ and $\boldsymbol{L}$ as seen from the *principal embedded frame* $Gx_1x_2x_3$. The true picture (relative to an inertial frame) is quite different. In this frame, $\boldsymbol{L}_G$ is constant and the time variation of $\boldsymbol{\omega}$ and the motion of the body have yet to be determined. In general, deducing the motion of the body is *more* difficult than solving the Euler equations. In the present case though, this can be done fairly easily. From the above expressions for $\boldsymbol{\omega}$ and $\boldsymbol{L}_G$, it follows that $\boldsymbol{\omega}$ can be written in the form

$$\begin{aligned}
\boldsymbol{\omega} &= \frac{1}{A}\boldsymbol{L}_G + \left(1 - \frac{C}{A}\right) n\,\boldsymbol{e}_3 \\
&= \frac{1}{A}\boldsymbol{L}_G + \left(\frac{A-C}{AC}\right) L_3\,\boldsymbol{e}_3 \\
&= \frac{1}{A}\boldsymbol{L}_G + \left(\frac{A-C}{AC}\right) (|\boldsymbol{L}_G| \cos\alpha)\,\boldsymbol{e}_3,
\end{aligned}$$

where α is the constant angle between $\boldsymbol{L}_G$ and the axis Gx_3. If we now write $L = |\boldsymbol{L}_G|$ and $\boldsymbol{k} = \boldsymbol{L}_G/L$, then

$$\boldsymbol{\omega} = \frac{L}{A}\boldsymbol{k} + L\cos\alpha\left(\frac{A-C}{AC}\right)\boldsymbol{e}_3,$$

which is precisely the solution found in section 19.4. ■

Deficiencies of Euler's equations

The above example illustrates the **first deficiency** in Euler's equations:

The solutions of Euler's equations yield the time variation of $\boldsymbol{\omega}$ *as seen from the* ***embedded reference frame****. The position of the body is still unknown.*

Thus Euler's equations do not tell us *where the body is* at time t, which is the most important unknown in the problem. In the example above, we were able to *deduce* the true motion of the body fairly simply. However, for an unsymmetrical body (which is the problem of principal interest that remains), there is no simple method and a further substantial calculation is required.

A **second** (but related) **deficiency** appears when the applied moment $\boldsymbol{K}_G$ is not zero:

When the applied moment $\boldsymbol{K}_G$ *is not zero, the components* K_1, K_2, K_3 *will not generally be known since the orientation of the body is unknown. In such a case, one cannot even begin the solution.*

For example, in the problem of the spinning top, the moment of the gravity force is known, but the *components* of this moment in the embedded reference frame are not known. The right sides of Euler's equations for the spinning top are therefore unknowns.

Dynamical balancing

The difficulties mentioned above disappear when the motion of the body is given, and we wish to find the moments that are being applied to the body to permit that motion. In such a case, Euler's equations give the answer immediately. Consider the case of an axially symmetric body which is smoothly constrained to rotate about a fixed axis through G but is inaccurately mounted so that the rotation axis is not quite coincident with the symmetry axis, as shown in Figure 19.12. What effect does this misalignment have on the pivots?

Let the symmetry axis be Gx_3 and take the embedded axis Gx_2 to be perpendicular to the *rotation axis* PQ; the direction of the embedded axis Gx_1 is then determined (see Figure 19.12). If the body is rotating about PQ at a constant rate λ, then the angular

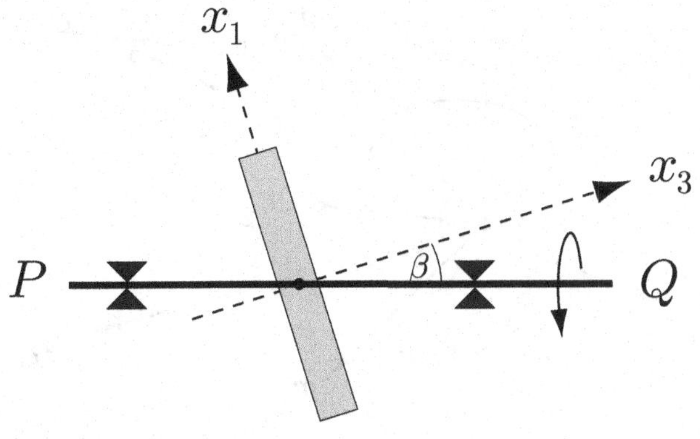

FIGURE 19.12 The axially symmetric body is constrained to rotate about the fixed axis PQ which makes an angle β with the axis of symmetry.

velocity components in the principal embedded frame $Gx_1x_2x_3$ are

$$\omega_1 = -\lambda \sin\beta$$
$$\omega_2 = 0$$
$$\omega_3 = \lambda \cos\beta.$$

To find the moment exerted by the pivots, we simply substitute these expressions into the left sides of Euler's equations. This gives

$$K_1 = 0, \qquad K_2 = (A - C)\lambda^2 \sin\beta\cos\beta, \qquad K_3 = 0.$$

Thus the applied moment is constant in the embedded frame. However, it is *not constant in an inertial frame*. The axis Gx_2 is perpendicular to PQ and rotates around it at a constant rate λ. *The pivots are therefore subjected to oscillating forces with angular frequency* λ *and magnitude proportional to* λ^2. In such a configuration, the body is said to be **dynamically unbalanced**. Since G lies on the rotation axis, the body can rest in equilibrium in any position, but, when rotating, it exerts oscillating forces on the pivots that may result in vibration of the mounting. This is why the wheels of motor cars must be re-balanced* when a new tyre is fitted. For a turbine with a mass of many tonnes rotating at 50 Hz say, extremely accurate alignment is needed, for otherwise the forces needed to restrain the turbine would be so large that it would break loose its mountings!

19.9 FREE MOTION OF AN UNSYMMETRICAL BODY

The free motion of unsymmetrical bodies is an important topic in space technology because of its application to the tumbling motion of spacecraft and space satellites.

* This is done by adding lead weights at determined positions around the rim of the wheel.

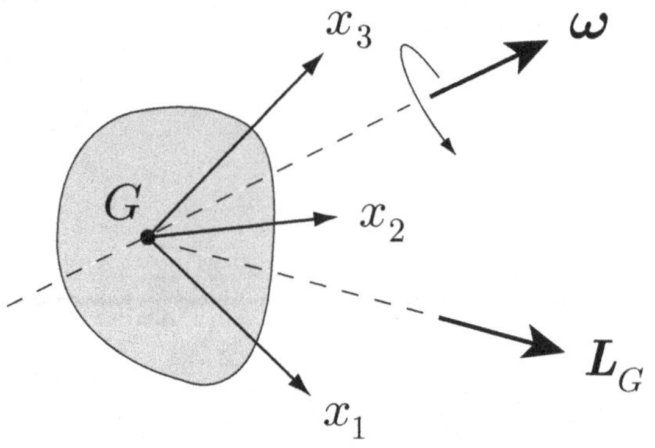

FIGURE 19.13 The unsymmetrical body moves under no forces with G at rest.

Such motions are considerably more difficult to analyse than for bodies with axial symmetry. The reader may be surprised then to learn that there is an exact **analytical solution** to the problem of an unsymmetrical body moving under no forces (see Landau & Lifshitz [6], Chapter VI). However, since this solution is expressed in terms of an impressive array of elliptic functions and theta functions, it gives one almost no idea of what the motion is actually like! Instead, we will use **Euler's equations** combined with **geometrical arguments**. This enables us to make interesting predictions about the motion without too much analysis.

Rotation of a free body about a fixed axis

Suppose a **free unsymmetrical body** $\mathcal{B}$ (an asteroid, for example) is observed from an inertial frame in which its centre of mass G is at rest. Is it possible for the body to rotate about a **fixed*** **axis**?

This is one of the few questions about unsymmetrical bodies that can be answered quite easily. If such an axis does exist, then it must pass through the fixed point G. Let $Gx_1x_2x_3$ be a set of principal axes of $\mathcal{B}$ at G. Since $\mathcal{B}$ is unsymmetrical, these principal axes are unique (apart from labelling) and are therefore necessarily *embedded* in $\mathcal{B}$; Euler's equations therefore apply.

The kinetic energy of $\mathcal{B}$ is given by $T = \frac{1}{2}I\,|\boldsymbol{\omega}|^2$, where I is the moment of inertia of $\mathcal{B}$ about the rotation axis and $\boldsymbol{\omega}$ is its angular velocity. Since $\mathcal{B}$ is rotating about a fixed axis, I is a constant. Also, since the motion takes place under no forces, the kinetic energy T is constant. It follows that the angular speed $|\boldsymbol{\omega}|$ must be constant and that the components $\{\omega_1, \omega_2, \omega_3\}$ of $\boldsymbol{\omega}$ in the embedded principal frame $Gx_1x_2x_3$ must be constants. Hence $\dot{\omega}_1 = \dot{\omega}_2 = \dot{\omega}_3 = 0$ and **Euler's equations** then imply that

$$(B - C)\omega_2\omega_3 = (C - A)\omega_3\omega_1 = (A - B)\omega_1\omega_2 = 0,$$

* This means an axis fixed in space and fixed in the body.

where $\{A, B, C\}$ are the principal moments of inertia about the axes Gx_1, Gx_2, Gx_3. Since $\mathcal{B}$ is an *unsymmetrical* body, A, B and C are all different and so

$$\omega_2\omega_3 = \omega_3\omega_1 = \omega_1\omega_2 = 0.$$

The only possible motions are therefore

$$\text{(i)}\quad \omega_1 = \text{constant}, \qquad \omega_2 = \omega_3 = 0,$$

$$\text{(ii)}\quad \omega_2 = \text{constant}, \qquad \omega_3 = \omega_1 = 0,$$

$$\text{(iii)}\quad \omega_3 = \text{constant}, \qquad \omega_1 = \omega_2 = 0.$$

The first solution corresponds to the body rotating with constant angular speed about the principal axis Gx_1. The other solutions correspond to rotations about the principal axes Gx_2 and Gx_3. Hence:

Free unsymmetrical body rotating about a fixed axis

A free unsymmetrical body can only rotate about a fixed axis if this axis is one of the three **principal axes** of the body at G. The rotation will then have constant angular speed.

For example, if it is required that a spacecraft should rotate about a fixed axis, this axis *must* be one of the three principal axes through its centre of mass. Later we will examine the **stability** of each of these steady motions.

General motion of a free unsymmetrical body

Once again we will suppose that the motion is viewed from an inertial frame in which G is at rest. It will be convenient to express the Euler equations in terms of the components of $\boldsymbol{L}$ in the frame $Gx_1x_2x_3$ (instead of the components of $\boldsymbol{\omega}$). Since $L_1 = A\omega_1$, $L_2 = B\omega_2$ and $L_3 = C\omega_3$, this gives the equivalent system of equations

$$\begin{aligned} BC\,\dot{L}_1 &= (B-C)L_2L_3,\\ AC\,\dot{L}_2 &= (C-A)L_3L_1,\\ AB\,\dot{L}_3 &= (A-B)L_1L_2. \end{aligned} \tag{19.44}$$

If we multiply the first equation by AL_1, the second by BL_2, the third by CL_3 and add, we obtain

$$ABC\left(L_1\dot{L}_1 + L_2\dot{L}_2 + L_3\dot{L}_3\right) = 0,$$

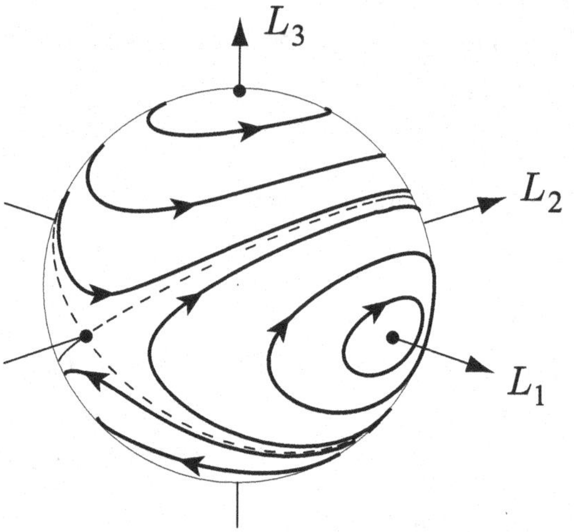

FIGURE 19.14 The L-sphere and the curves in which it meets different T-ellipsoids for the case $A < B < C$.

which integrates to give

$$L_1^2 + L_2^2 + L_3^2 = L^2, \tag{19.45}$$

where L is a constant. Similarly, if we multiply the first equation by L_1, the second by L_2, the third by L_3 and add, we obtain

$$BC\, L_1\dot{L}_1 + AC\, L_2\dot{L}_2 + AB\, L_3\dot{L}_3 = 0,$$

which integrates to give

$$\frac{L_1^2}{A} + \frac{L_2^2}{B} + \frac{L_3^2}{C} = 2T, \tag{19.46}$$

where T is a constant. The first of these integrals means that the **magnitude** of $\boldsymbol{L}_G$ is conserved* and the second means that the **kinetic energy** of the body is conserved.

The equations (19.45), (19.46) place geometric restrictions on the time variation of $\boldsymbol{L}_G$, *as seen from the embedded frame*. Equation (19.45) means that, if $\boldsymbol{L}_G$ is considered to be the position vector of a 'point' $\boldsymbol{L}$ in (L_1, L_2, L_3)-space, then that point must move on the surface of a sphere with centre O and radius L, which we will call L**-sphere**. In the same way, equation (19.46) means that the $\boldsymbol{L}$-point must also move on the surface of the ellipsoid† in (L_1, L_2, L_3)-space defined by equation (19.46). We will call this ellipsoid,

* Note that, in the embedded frame, the *individual components* L_1, L_2, L_3 *are not conserved*, but the magnitude $\left(L_1^2 + L_2^2 + L_3^2\right)^{1/2}$ is.

† An ellipsoid is the three-dimensional counterpart of an ellipse. The standard equation of an ellipsoid with semi-axes a, b, c is

$$\frac{x_1^2}{a^2} + \frac{x_2^2}{b^2} + \frac{x_3^2}{c^2} = 1.$$

which has the semi-axes $(2AT)^{1/2}$, $(2BT)^{1/2}$, $(2CT)^{1/2}$, the T**-ellipsoid**. It follows that *the path of the **L**-point must be a curve of intersection of the L-sphere and the T-ellipsoid.* Figure 19.14 shows the L-sphere for a fixed value of the parameter L. The paths on its surface are examples of the curves in which it is cut by the T-ellipsoid for different values of the parameter T.

Paths of the L-point

Suppose that $A < B < C$ and that L is fixed. For each value of L, values of T that are too small or too large are not admissible. Indeed we can see from equation (19.46) that

$$T_{\min} = \frac{L^2}{2C} \leq T \leq \frac{L^2}{2A} = T_{\max}.$$

If T lies outside this range, then the L-sphere and the T-ellipsoid do not intersect.

(i) When $T = T_{\min}$ the intersection consists of the two isolated points $(0, 0, \pm L)$. The constant solution $\boldsymbol{L} = (0, 0, L)$ corresponds to steady rotation in the positive sense about the principal axis Gx_3. The solution $\boldsymbol{L} = (0, 0, -L)$ corresponds to a steady rotation in the opposite sense.

(ii) Suppose now that T is slightly increased. The intersection point $(0, 0, L)$ then becomes a small closed curve* on the surface of the L-sphere that encircles the point $(0, 0, L)$ as shown in Figure 19.14. The $\boldsymbol{L}$-point must therefore move along this curve. The direction of motion of the $\boldsymbol{L}$-point can be deduced by examining the signs of $\dot{L}_1$ and $\dot{L}_2$ given by the first and second equations of (19.44). As T increases, the intersection curve becomes larger (and changes its shape) but it still encircles the point $(0, 0, L)$. This pattern continues until $T = L^2/2B$.

(iii) When $T = L^2/2B$, the L-sphere and the T-ellipsoid touch at the points $(0, \pm L, 0)$ and the intersection curves are the dashed lines in Figure 19.14. This is a transitional case between two different regimes of curves. It includes the case of steady rotation about the principal axis Gx_2.

(iv) When $T > L^2/2B$, the intersection curves are completely different. Now they encircle the point $(L, 0, 0)$ (or $(-L, 0, 0)$) and, as T approaches $T_{\max}$ they shrink to zero around these points. The case $T = T_{\max}$ corresponds to steady rotation about the principal axis Gx_1.

Stabilty of steady rotation about a principal axis

We can make some interesting deductions from the form of the above paths.

(i) In the general motion, the $\boldsymbol{L}$-point moves around a closed curve. It follows that the **time variation** of $\boldsymbol{L}$, as seen from the embedded frame, is **periodic**, a property not apparent from equations (19.44). This does not however mean that the motion of the *body* is periodic.

(ii) If the body is in steady rotation in the positive sense about the axis Gx_1 and is slightly disturbed, then the path of $\boldsymbol{L}$ remains close to the point $(L, 0, 0)$. This implies that,

* The projection of this curve on to the (x_1, x_2)-plane is a small ellipse with centre the origin.

in an inertial frame where $\boldsymbol{L}$ is constant, the axis Gx_1 remains close to $\boldsymbol{L}$. In other words, the original steady rotation about the axis Gx_1 is **stable** to small disturbances. The same applies to a steady rotation about the axis Gx_3. However, if a steady rotation about the axis Gx_2 is slightly disturbed, the paths of the $\boldsymbol{L}$-point lead far away from the original constant positions at $(0, \pm L, 0)$. This implies that a steady rotation about the axis Gx_2 is **unstable** to small disturbances. Hence:

The steady rotation of an unsymmetrical body about a principal axis is stable for the axes with the greatest and least moments of inertia, but unstable for the other axis.

Thus, it is useless to try to stabilise a satellite in steady rotation about the principal axis with the 'middle' moment of inertia.

(iii) The stability argument above is modified if the body has a means of **energy dissipation**. This cannot happen with rigid bodies, but real bodies, such as a satellite, can flex and slowly dissipate mechanical energy as heat. In this case, steady rotation of the body about the axis with the *least* moment of inertia (which corresponds to $T = T_{max}$) is no longer stable. The 'radius' of the path of $\boldsymbol{L}$ around the axis Gx_1 will slowly increase until $T = L^2/(2B)$ when it will switch to a large path encircling one of the points $(\pm 0, 0, \pm L)$. This path will then gradually shrink to zero. Thus, *in the presence of energy dissipation, the body will end up rotating about the axis with the largest moment of inertia*. In the early days of space exploration* this fact was learned by hard experience!

Motion of the body

Suppose that Euler's equations have been solved for the unknowns ω_1, ω_2, ω_3. What is the corresponding motion of the body? Let $\{\boldsymbol{e}_1, \boldsymbol{e}_2, \boldsymbol{e}_3\}$ be the unit vectors of the principal embedded frame $Gx_1x_2x_3$. Then the time variation of these vectors, as seen from an inertial frame, is given by

$$\dot{\boldsymbol{e}}_1 = \boldsymbol{\omega} \times \boldsymbol{e}_1, \qquad \dot{\boldsymbol{e}}_2 = \boldsymbol{\omega} \times \boldsymbol{e}_, \qquad \dot{\boldsymbol{e}}_3 = \boldsymbol{\omega} \times \boldsymbol{e}_3.$$

Unfortunately, these equations are not really uncoupled since Euler's equations give us $\boldsymbol{\omega}$ in the form

$$\boldsymbol{\omega} = \omega_1\, \boldsymbol{e}_1 + \omega_2\, \boldsymbol{e}_2 + \omega_3\, \boldsymbol{e}_3,$$

which also involves the unknown vectors $\{\boldsymbol{e}_1, \boldsymbol{e}_2, \boldsymbol{e}_3\}$. On combining these equations, we get the set of *coupled* ODEs

$$\begin{aligned} \dot{\boldsymbol{e}}_1 &= \omega_3\, \boldsymbol{e}_2 - \omega_2\, \boldsymbol{e}_3, \\ \dot{\boldsymbol{e}}_2 &= \omega_1\, \boldsymbol{e}_3 - \omega_3\, \boldsymbol{e}_1, \\ \dot{\boldsymbol{e}}_3 &= \omega_2\, \boldsymbol{e}_1 - \omega_1\, \boldsymbol{e}_2, \end{aligned} \tag{19.47}$$

* Kaplan [5] recounts the following incident: The first U.S. satellite, Explorer I, was set into orbit rotating about its longitudinal axis, which was the principal axis with the *least* moment of inertia. After only a few hours, radio signals indicated that a tumbling motion had developed and was increasing in amplitude in an unstable manner. It was concluded that the satellite's flexible antennae were dissipating energy and causing a transfer of body spin from the axis of minimum inertia to a transverse axis of maximum inertia.

for the unknown vectors $\{\boldsymbol{e}_1, \boldsymbol{e}_2, \boldsymbol{e}_3\}$. These equations cannot usually be solved explicitly, but they are suitable for numerical integration (see Problem 19.17) and for approximate solution by perturbation analysis. This is illustrated by the following example.

Example 19.7 *Motion of a principal axis in space*

An unsymmetrical body is in steady rotation about the principal axis with the *largest* moment of inertia. Find an approximation to the wobble of this principal axis if the body is slightly disturbed.

Solution

Suppose that $A < B < C$ and that, in the *initial motion*, $\omega_1 = \omega_2 = 0$ and $\omega_3 = \Lambda$, where Λ is a constant. In the *disturbed motion* we suppose that ω_1, ω_2 (and their time derivatives) remain small. If we now apply **Euler's equations** and then **linearise**, we obtain

$$\begin{aligned} A\,\dot{\omega}_1 - (B - C)\Lambda\,\omega_2 &= 0, \\ B\,\dot{\omega}_2 - (C - A)\Lambda\,\omega_1 &= 0, \\ C\,\dot{\omega}_3 &= 0, \end{aligned}$$

Hence, in the linear approximation to the disturbed motion, the coupled equations for ω_1, ω_2 are easily solved to give

$$\begin{aligned} \omega_1 &= \epsilon\Lambda\,((C - B)/A)^{1/2}\cos(\Omega t + \gamma), \\ \omega_2 &= \epsilon\Lambda\,((C - A)/B)^{1/2}\sin(\Omega t + \gamma), \end{aligned} \tag{19.48}$$

where ϵ, γ are dimensionless constants and

$$\Omega = \Lambda\left(\frac{(C - A)(C - B)}{AB}\right)^{1/2}.$$

The corresponding expressions for L_1, L_2 are

$$\begin{aligned} L_1 &= \epsilon\Lambda\,(A(C - B))^{1/2}\cos(\Omega t + \gamma), \\ L_2 &= \epsilon\Lambda\,(B(C - A))^{1/2}\sin(\Omega t + \gamma). \end{aligned} \tag{19.49}$$

Hence the projection of the path of the $\boldsymbol{L}$-point on to the (x_1, x_2)-plane is an ellipse executed in the anti-clockwise sense. This is entirely consistent with the paths shown in Figure 19.14 and our discussion of stability.

Now to find the actual time variation of the unit vector $\boldsymbol{e}_3$. In the initial motion we have

$$\begin{aligned} \boldsymbol{e}_1 &= \cos\Lambda t\,\boldsymbol{i} + \sin\Lambda t\,\boldsymbol{j}, \\ \boldsymbol{e}_2 &= -\sin\Lambda t\,\boldsymbol{i} + \cos\Lambda t\,\boldsymbol{j}, \\ \boldsymbol{e}_3 &= \boldsymbol{k}, \end{aligned}$$

where $\{\boldsymbol{i}, \boldsymbol{j}, \boldsymbol{k}\}$ is a *fixed* orthonormal set. Since ω_1, ω_2 are small in the disturbed motion, the vectors $\boldsymbol{e}_1$ and $\boldsymbol{e}_2$ that appear in the third equation of (19.47) can be

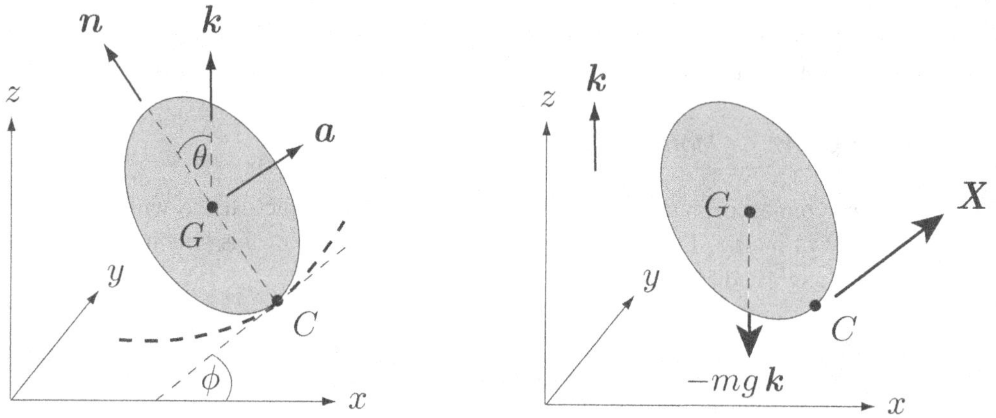

FIGURE 19.15 The wheel rolls on a rough horizontal plane. The dashed curve is the path of the point of contact C.

replaced by their steady (zero order) approximations to give

$$\dot{\boldsymbol{e}}_3 = \omega_2 (\cos \Lambda t\, \boldsymbol{i} + \sin \Lambda t\, \boldsymbol{j}) - \omega_1 (- \sin \Lambda t\, \boldsymbol{i} + \cos \Lambda t\, \boldsymbol{j})$$

where ω_1 and ω_2 are given by the expressions (19.48). This equation can now be integrated explicitly to give the first order approximation to the wobble of the axis vector $\boldsymbol{e}_3$.

We will not actually do this integration because the result is messy and the details of the formula are less interesting than its structure. By examining the form of the right side of the above equation one can see that $\boldsymbol{e}_3$ must have the form $\boldsymbol{e}_3 = \boldsymbol{k} +$ (small terms in the $\boldsymbol{i}$ and $\boldsymbol{j}$ directions), where these small terms contain oscillating functions with *two different frequencies* $\Omega + \Lambda$ and $\Omega - \Lambda$. In general the ratio of these frequencies will not be a rational number and so the **motion of the axis vector $\boldsymbol{e}_3$ is not periodic**. In fact, the projection of $\boldsymbol{e}_3$ onto the fixed plane through G perpendicular to $\boldsymbol{k}$ describes a Lissajous type path bounded by two ellipses. (See however Problem 19.14.) ■

19.10 THE ROLLING WHEEL

Suppose a wheel is free to *roll* on a rough horizontal floor. How does it move? This, our final problem, typifies problem solving in classical mechanics. Nothing could be easier to describe, but few could just sit down and write out the solution to this classic problem.

Since the rolling condition is a non-holonomic constraint, the rolling wheel cannot be solved by the Lagrangian methods presented in this book.* We therefore use vectorial mechanics. Figure 19.15 (left) shows a thin circular disk† of mass M and radius b rolling

* For an extension of the Lagrangian method to non-holonomic systems, see Goldstein [4], chapter 2.

† The disk need not be uniform, but it must have axial symmetry about its centre. For example, it could be a hoop.

on the (x, y)-plane. The thick dashed curve is the path of the point of contact C. The vector $\boldsymbol{a}$ $(= \boldsymbol{a}(t))$ is the axial unit vector and the vector $\boldsymbol{n}$ $(= \boldsymbol{n}(t))$ is the unit vector in the direction $\overrightarrow{CG}$. Since the disk has axisymmetry, the angular velocity of the disk can be expressed in the form

$$\boldsymbol{\omega} = \boldsymbol{a} \times \dot{\boldsymbol{a}} + \lambda \boldsymbol{a}, \tag{19.50}$$

where the spin λ $(= \boldsymbol{\omega} \cdot \boldsymbol{a})$ is an unknown scalar function of the time. The corresponding angular momentum $\boldsymbol{L}_G$ is then given by

$$\boldsymbol{L}_G = A\, \boldsymbol{a} \times \dot{\boldsymbol{a}} + C\lambda\, \boldsymbol{a}, \tag{19.51}$$

where $\{A, A, C\}$ are the principal moments of inertia of the disk at G.

The disk moves under the forces shown in Figure 19.15 (right). It follows that the equation of **translational motion** of the disk is

$$M\frac{d\boldsymbol{V}}{dt} = \boldsymbol{X} - Mg\,\boldsymbol{k}, \tag{19.52}$$

where $\boldsymbol{V}$ is the velocity of G, and that the equation of **rotational motion** is

$$\frac{d}{dt}\left(A\,\boldsymbol{a} \times \dot{\boldsymbol{a}} + C\lambda\, \boldsymbol{a}\right) = (-b\boldsymbol{n}) \times \boldsymbol{X}. \tag{19.53}$$

In addition, we have the **rolling condition** that the particle of the disk in contact with the floor has zero velocity, that is

$$\boldsymbol{V} + \boldsymbol{\omega} \times (-b\boldsymbol{n}) = \boldsymbol{0}. \tag{19.54}$$

Equations (19.52)–(19.54) are the **governing equations** of the motion. On eliminating $\boldsymbol{X}$ and $\boldsymbol{V}$, we obtain the equation

$$\frac{d}{dt}\left(A\,\boldsymbol{a} \times \dot{\boldsymbol{a}} + C\lambda\, \boldsymbol{a}\right) + Mb^2\left[\dot{\boldsymbol{\omega}} - (\dot{\boldsymbol{\omega}} \cdot \boldsymbol{n})\,\boldsymbol{n} - (\boldsymbol{\omega} \cdot \boldsymbol{n})\,\dot{\boldsymbol{n}}\right] + Mgb\,\boldsymbol{n} \times \boldsymbol{k} = \boldsymbol{0}. \tag{19.55}$$

where $\boldsymbol{\omega}$ is given by equation (19.50). The vectors $\boldsymbol{a}$ and $\boldsymbol{n}$ are not independent. In fact $\boldsymbol{n}$ can be expressed in terms of $\boldsymbol{a}$ and $\boldsymbol{k}$ in the form

$$\boldsymbol{n} = \frac{\boldsymbol{k} - (\boldsymbol{k} \cdot \boldsymbol{a})\,\boldsymbol{a}}{|\boldsymbol{k} - (\boldsymbol{k} \cdot \boldsymbol{a})\,\boldsymbol{a}|}. \tag{19.56}$$

Thus it is possible in principle to eliminate $\boldsymbol{n}$ from equation (19.55) and obtain a vector equation solely in terms of the unknown **axial vector** $\boldsymbol{a}$ and the **spin** λ. This equation is sufficient to determine $\boldsymbol{a}$ and λ.

All this begins to look easier when we take components of the vector equation (19.55) in three *well chosen* directions, namely, the (instantaneous) principal directions $\boldsymbol{a}$, $\boldsymbol{n}$ and

$\boldsymbol{a} \times \boldsymbol{n}$. On taking the scalar product of equation (19.55) with each of these vectors, we obtain the scalar equations

$$\begin{aligned}(C + Mb^2)\dot{\lambda} - Mb^2 \left[\boldsymbol{a}, \dot{\boldsymbol{a}}, \boldsymbol{n}\right] \left(\dot{\boldsymbol{n}} \cdot \boldsymbol{a}\right) &= 0,\\ A\left[\boldsymbol{a}, \ddot{\boldsymbol{a}}, \boldsymbol{n}\right] + C\lambda \left(\dot{\boldsymbol{a}} \cdot \boldsymbol{n}\right) &= 0,\\ (A + Mb^2)\left(\ddot{\boldsymbol{a}} \cdot \boldsymbol{n}\right) - (C + Mb^2)\left[\boldsymbol{a}, \dot{\boldsymbol{a}}, \boldsymbol{n}\right] - &\\ Mb^2\left[\boldsymbol{a}, \dot{\boldsymbol{a}}, \boldsymbol{n}\right]\left[\boldsymbol{a}, \boldsymbol{n}, \dot{\boldsymbol{n}}\right] - Mgb\left(\boldsymbol{a} \cdot \boldsymbol{k}\right) &= 0.\end{aligned} \tag{19.57}$$

Here $[\,\boldsymbol{p}, \boldsymbol{q}, \boldsymbol{r}\,]$ means the triple scalar product of the vectors $\boldsymbol{p}$, $\boldsymbol{q}$ and $\boldsymbol{r}$. These equations look better still when expressed in terms of scalar variables. Let θ be the angle between the plane of the disk and the vertical, and let ϕ be the angle between the tangent line to the disk at C and the axis Ox, as shown in Figure 19.15 (left). Then $\boldsymbol{a}$ and $\boldsymbol{n}$ take the form

$$\boldsymbol{a} = (\cos\theta\sin\phi)\,\boldsymbol{i} - (\cos\theta\cos\phi)\,\boldsymbol{j} + (\sin\theta)\,\boldsymbol{k}, \tag{19.58}$$

$$\boldsymbol{n} = -(\sin\theta\sin\phi)\,\boldsymbol{i} + (\sin\theta\cos\phi)\,\boldsymbol{j} + (\cos\theta)\,\boldsymbol{k}, \tag{19.59}$$

where $\{\boldsymbol{i}, \boldsymbol{j}, \boldsymbol{k}\}$ are the constant unit vectors of the inertial frame $Oxyz$. In terms of θ and ϕ, the equations (19.57) become

$$\begin{aligned}(C + Mb^2)\dot{\lambda} + Mb^2\cos\theta\,\dot{\theta}\,\dot{\phi} &= 0,\\ A\left(\cos\theta\,\ddot{\phi} - 2\sin\theta\,\dot{\theta}\dot{\phi}\right) + C\lambda\dot{\theta} &= 0,\\ (A + Mb^2)\ddot{\theta} + A\cos\theta\sin\theta\,\dot{\phi}^2 - (C + Mb^2)\lambda\cos\theta\,\dot{\phi} - Mgb\sin\theta &= 0.\end{aligned} \tag{19.60}$$

This is our final form of the **governing equations** for the rolling disk. Note that, unlike other problems involving an axially symmetric body, the *spin* λ *is not a constant of the motion*, but must be solved for, along with the angular coordinates θ, ϕ. The complexity of these coupled non-linear equations is such that explicit solutions are rare. However, they are suitable for numerical integration and (if they can be linearised) for finding approximate solutions. This is illustrated by the next example.

Example 19.8 *Stability of straight line rolling*

How fast must a wheel roll in a straight line so that it is stable to small disturbances?

Solution

In the **unperturbed motion** (along the x-axis say), we have $\theta = 0$, $\phi = 0$ and $\lambda = \Lambda$, where Λ is a positive constant. In the **perturbed motion**, we suppose that θ, ϕ and $\lambda - \Lambda$ (and their time derivatives) remain small so that we may **linearise** equations (19.60). This gives

$$\begin{aligned}\dot{\lambda} &= 0,\\ A\ddot{\phi} + C\Lambda\dot{\theta} &= 0,\\ (A + Mb^2)\ddot{\theta} - (C + Mb^2)\Lambda\dot{\phi} - Mgb\theta &= 0.\end{aligned}$$

Hence, in this linear approximation, $\lambda = \Lambda$, a constant. The second equation gives

$$A\dot{\phi} + C\Lambda\theta = \text{constant},$$

and, in order to be satisfied by the unperturbed state, the constant must be zero. On substituting this into the third equation, we obtain

$$A(A + Mb^2)\ddot{\theta} + \left[\Lambda^2 C(C + Mb^2) - AMgb\right]\theta = 0. \tag{19.61}$$

If the expression in the square brackets is *positive*, this is the SHM equation and θ will perform small oscillations about $\theta = 0$; the same is then true of ϕ. Hence, the **condition for stability** of straight line rolling is

$$\Lambda^2 C(C + Mb^2) - AMgb > 0,$$

that is

$$V^2 > \frac{AMgb^3}{C(C + Mb^2)},$$

where V is the speed of the disk. For a **hoop** with $A = \frac{1}{2}Mb^2$ and $C = Mb^2$, this condition becomes $V^2 > gb/4$. ■

Path of the point of contact

If $\boldsymbol{r}$ is the position vector of C, the point of contact, then

$$\boldsymbol{r} = \boldsymbol{R} - b\boldsymbol{n},$$

where $\boldsymbol{R}$ is the position vector of G. It follows that

$$\frac{d\boldsymbol{r}}{dt} = \boldsymbol{V} - b\dot{\boldsymbol{n}},$$

where $\boldsymbol{V}$ is the velocity of G. On using the rolling condition (19.54) and the expression (19.50) for $\boldsymbol{\omega}$, the differential equation for $\boldsymbol{r}$ becomes

$$\frac{d\boldsymbol{r}}{dt} = b\left[(\boldsymbol{a}\times\dot{\boldsymbol{a}})\times\boldsymbol{n} + \lambda\,\boldsymbol{a}\times\boldsymbol{n} - \dot{\boldsymbol{n}}\right]. \tag{19.62}$$

This is the **path equation** for the point of contact. Once the equations (19.60) have been solved for λ, θ and ϕ, the right side of (19.62) is known and the path of C can then be found by integration. Numerical integration shows that some possible paths can be quite exotic, as shown in Figure 19.16.

Stability of the bicycle

Our analysis of the rolling wheel gives some idea of the complexity of the mechanics of the bicycle. Indeed, the question of stability is not completely settled and rival sets of

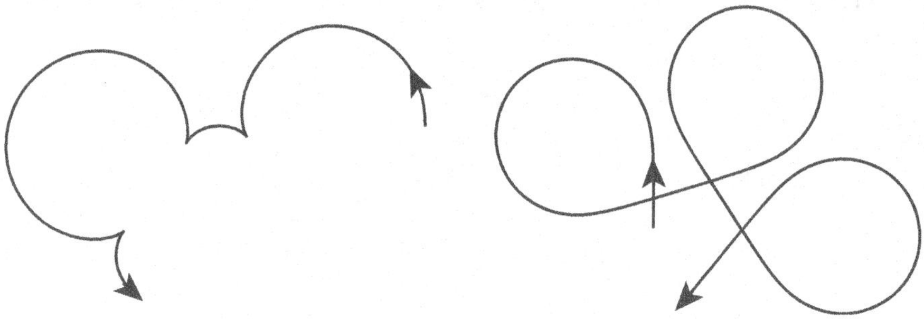

FIGURE 19.16 Two not-so-typical paths for the point of contact of a rolling wheel.

equations appear in the literature. Probably the most important paper on bicycle physics is that by Jones.* Jones constructed a non-standard bicycle in which the angular momentum of the wheels (the 'gyroscopic effect') was cancelled out by fitting extra wheels that spun in the opposite direction. It turned out that this bicycle was just as stable as a normal one and could be ridden 'no hands'. Despite Jones's findings, the myth that a bicycle is stabilised by the the gyroscopic effect of the spinning wheels still persists. However, the consensus of informed opinion is that the stability of a bicycle depends crucially on the geometry of the forks holding the front wheel. Jones constructed another non-standard bicycle in which the front wheel was four inches ahead of its normal position; it was almost unridable! The website `http://ruina.tam.cornell.edu/research/index.htm` has references to further interesting work on bicycle physics.

Problems on Chapter 19

Answers and comments are at the end of the book.

Harder problems carry a star (∗).

Rolling balls

19.1 ***Ball rolling on a slope*** A uniform ball can roll or skid on a rough plane inclined at an angle β to the horizontal. Show that, in *any* motion of the ball, the component of $\boldsymbol{\omega}$ perpendicular to the plane is conserved. If the ball *rolls* on the plane, show that the path of the ball must be a parabola.

19.2 ***Ball rolling on a rotating turntable*** ∗ A rough horizontal turntable is made to rotate about a fixed vertical axis through its centre O with *constant* angular velocity $\Omega\boldsymbol{k}$, where the unit vector $\boldsymbol{k}$ points vertically upwards. A uniform ball of radius a can roll or skid on the turntable. Show that, in *any* motion of the ball, the vertical spin $\boldsymbol{\omega}\cdot\boldsymbol{k}$ is conserved. If the ball *rolls* on the turntable, show that

$$\dot{\boldsymbol{V}} = \tfrac{2}{7}\Omega\,\boldsymbol{k}\times\boldsymbol{V},$$

* Jones, D.E.H., The stability of the bicycle, *Physics Today* **23** (1970) pp. 34–40.

where V is the velocity of the centre of the ball viewed from a *fixed* reference frame. Deduce the amazing result that the path of the rolling ball must be a circle.

Suppose the ball is held at rest (relative to the turntable), with its centre a distance b from the axis $\{O, \boldsymbol{k}\}$, and is then released. Given that the ball rolls, find the radius and the centre of the circular path on which it moves.

19.3 ***Ball rolling on a fixed sphere**** A uniform ball with radius a and centre C *rolls* on the rough outer surface of a fixed sphere of radius b and centre O. Show that the radial spin $\boldsymbol{\omega} \cdot \boldsymbol{c}$ is conserved, where $\boldsymbol{c}$ $(= \boldsymbol{c}(t))$ is the *unit* vector in the radial direction $\overrightarrow{OC}$. [Take care!] Show also that $\boldsymbol{c}$ satisfies the equation

$$7(a+b)\,\boldsymbol{c}\times\ddot{\boldsymbol{c}} + 2an\,\dot{\boldsymbol{c}} + 5g\,\boldsymbol{c}\times\boldsymbol{k} = \boldsymbol{0},$$

where n is the constant value of $\boldsymbol{\omega} \cdot \boldsymbol{c}$ and $\boldsymbol{k}$ is the unit vector pointing vertically upwards.

By comparing this equation with that for the spinning top, deduce the amazing result that the ball can roll on the spherical surface without ever falling off. Find the minimum value of n such that the ball is stable at the highest point of the sphere.

Axisymmetric bodies

19.4 Investigate the steady precession of a top for the case in which the axis of the top moves in the horizontal plane through O. Show that for any $n \neq 0$ there is just *one* rate of steady precession and find its value.

19.5 ***The sleeping top*** By performing a perturbation analysis, show that a top will be stable in the vertically upright position if

$$C^2n^2 > 4AMgh,$$

in the standard notation. [You can do this by either the vectorial or Lagrangian method.]

19.6 Estimate how large the spin n of a pencil would have to be for it to be stable in the vertically upright position, spinning on its point. [Take the pencil to be a uniform cylinder 15 cm long and 7 mm in diameter.]

19.7 A juggler is balancing a spinning ball of diameter 20 cm on the end of his finger. Estimate the spin required for stability (i) for a uniform solid ball, (ii) for a uniform thin hollow ball. Which do you suppose the juggler uses?

19.8 Solve the problem of the free motion of an axisymmetric body by the Lagrangian method. Compare your results with those in Section 19.4. [Surprisingly awkward. Take the axis Gz of coordinates $Gxyz$ to point in the direction of $\boldsymbol{L}_G$ and make use of the Lagrange equations for the Euler angles ϕ and ψ.]

19.9 ***Frisbee with resistance*** A (wobbling) frisbee moving through air is subject to a frictional couple equal to $-K\,\boldsymbol{\omega}$. Find the time variation of the axial spin λ $(= \boldsymbol{\omega} \cdot \boldsymbol{a})$, where $\boldsymbol{a}$ is the axial unit vector. Show also that $\boldsymbol{a}$ satisfies the equation

$$A\,\boldsymbol{a}\times\ddot{\boldsymbol{a}} + K\,\boldsymbol{a}\times\dot{\boldsymbol{a}} + C\lambda\,\dot{\boldsymbol{a}} = \boldsymbol{0}.$$

∗ By taking the cross product of this equation with $\dot{\boldsymbol{a}}$, find the time variation of $|\dot{\boldsymbol{a}}|$. Deduce that the angle between $\boldsymbol{\omega}$ and $\boldsymbol{a}$ decreases with time if $C > A$ (which it is for a normal frisbee). Thus, in the presence of linear resistance, the wobble dies away.

19.10 ***Spinning hoop on a smooth floor*** A uniform circular hoop of radius a rolls and slides on a *perfectly smooth* horizontal floor. Find its Lagrangian in terms of the Euler angles, and determine which of the generalised momenta are conserved. [Suppose that G has no *horizontal* motion.]

Investigate the existence of motions in which the angle between the hoop and the floor is a constant α. Show that Ω, the rate of steady precession, must satisfy the equation

$$\cos\alpha\,\Omega^2 - 2n\,\Omega - 2\frac{g}{a}\cot\alpha = 0,$$

where n is the constant axial spin. Deduce that, for $n \neq 0$, there are two possible rates of precession, a faster one going the 'same way' as n, and a slower one in the opposite direction. [These are interesting motions but one would need a *very* smooth floor to observe them.]

Euler's equations

19.11 ***Bicycle wheel*** A bicycle wheel (a hoop) of mass M and radius a is fitted with a smooth spindle lying along its symmetry axis. The wheel is spun with the spindle horizontal, and the spindle is then made to turn with angular speed Ω about a fixed vertical axis through the centre of the wheel. Show that n, the axial spin of the wheel, remains constant and find the moment that must be applied to the spindle to produce this motion.

19.12 ***Stability of steady rotation*** An unsymmetrical body is in steady rotation about a principal axis through G. By performing a perturbation analysis, investigate the stability of this motion for each of the three principal axes.

19.13 ***Frisbee with resistance*** Re-solve the problem of the frisbee with resistance (Problem 19.9) by using Euler's equations. [Find the time dependencies of ω_3 and $\omega_1^2 + \omega_2^2$.]

19.14 ***Wobble on spinning lamina*** An unsymmetrical lamina is in steady rotation about the axis through G perpendicular to its plane. Find an approximation to the wobble of this axis if the body is slightly disturbed. [This is a repeat of Example 19.7 for the special case in which the body is an unsymmetrical *lamina*; in this case $C = A+B$ and there is much simplification.]

19.15 ∗ ***Euler theory for the unsymmetrical lamina*** An unsymmetrical lamina has principal axes $Gx_1x_2x_3$ at G with the corresponding moments of inertia $\{A, B, A+B\}$, where $A < B$. Initially the lamina is rotating with angular velocity Ω about an axis through G that lies in the (x_1, x_2)-plane and makes an acute angle α with Gx_1. By using Euler's equations, show that, in the subsequent motion,

$$\begin{aligned}\omega_1^2 + \omega_2^2 &= \Omega^2,\\ (B-A)\,\omega_2^2 + (B+A)\,\omega_3^2 &= (B-A)\,\Omega^2\sin^2\alpha.\end{aligned}$$

Interpret these results in terms of the motion of the '$\boldsymbol{\omega}$-point' moving in $(\omega_1, \omega_2, \omega_3)$-space and deduce that $\boldsymbol{\omega}$ is periodic when viewed from the embedded frame.

Find an ODE satisfied by ω_2 alone and deduce that the lamina will once again be rotating about the same axis after a time

$$\frac{4}{\Omega}\left(\frac{B+A}{B-A}\right)^{1/2}\int_0^{\pi/2}\frac{d\theta}{(1-\sin^2\alpha\,\sin^2\theta)^{1/2}}.$$

Computer assisted problems

19.16 Solve Lagrange's equations for the top numerically and obtain the motions of the axis shown in Figure 19.8. Use Euler's angles as coordinates and use computer assistance to obtain the equations as well as solve them. [It seems easier not to use the conservation relations for p_ϕ and p_ψ in their integrated form, since this requires the initial conditions to be incorporated into the equations of motion.]

19.17 Obtain the paths of the $\boldsymbol{L}$-point for an unsymmetrical body, as shown in Figure 19.14. These were obtained by solving Euler's equations, together with the equations

$$\dot{\boldsymbol{e}}_1 = \omega_3\boldsymbol{e}_2 - \omega_2\boldsymbol{e}_3, \qquad \dot{\boldsymbol{e}}_2 = \omega_1\boldsymbol{e}_3 - \omega_3\boldsymbol{e}_1, \qquad \dot{\boldsymbol{e}}_3 = \omega_2\boldsymbol{e}_1 - \omega_1\boldsymbol{e}_2,$$

numerically. [*Mathematica* handled the solution of these twelve simultaneous equations with ease!]

Appendix

Centres of mass and moments of inertia

A.1 CENTRE OF MASS

Suppose we have a system $\mathcal{S}$ of particles $P_1, P_2, \ldots, P_N$ with masses $m_1, m_2, \ldots, m_N$, and position vectors $\boldsymbol{r}_1, \boldsymbol{r}_2, \ldots, \boldsymbol{r}_N$ respectively. Then the centre of mass of $\mathcal{S}$ is defined as follows:

Definition A.1 ***Centre of mass*** *The* ***centre of mass*** *of $\mathcal{S}$ is the* ***point of space*** *whose position vector* $\boldsymbol{R}$ *is defined by*

$$\boxed{\boldsymbol{R} = \frac{\sum_{i=1}^{N} m_i \boldsymbol{r}_i}{\sum_{i=1}^{N} m_i} = \frac{\sum_{i=1}^{N} m_i \boldsymbol{r}_i}{M},} \tag{A.1}$$

where M is the total mass of $\mathcal{S}$.

In Chapter 2, we gave some simple examples of centre of mass, but here we will suppose the system is a **rigid body** $\mathcal{B}$. Then G, the centre of mass of $\mathcal{B}$, moves as if it were a particle of the body and its position can be calculated once and for all.

Finding centres of mass by symmetry

For most bodies, the position of G can be found by symmetry arguments so that no summations (or integrations) are needed. Instead, one simply applies the following rules which follow from the definition (A.1) :

Rule 1: *If the body has* ***reflective symmetry*** *in a plane, then the centre of mass must lie in this plane.*

Rule 2: *If the body has any* ***rotational symmetry**** *about an axis, then the centre of mass must lie on this axis.*

* For a body to have a rotational symmetry, the mass distribution need not be preserved for *all* rotation angles about the axis. Just *one* angle (less than 2π) is enough.

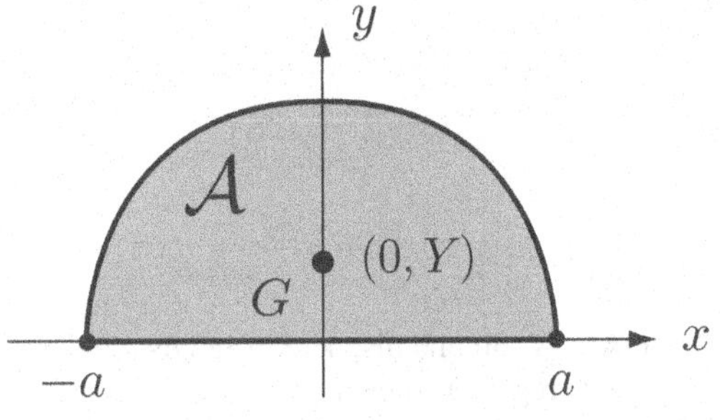

FIGURE A.1 A semi-circular lamina of radius a occupies the region $\mathcal{A}$ shown.

Suppose for example that the body $\mathcal{B}$ is a uniform **circular cylinder** occupying the region $x^2 + y^2 \le a^2$, $-b \le z \le b$. Then $\mathcal{B}$ has *reflective symmetry* in the plane $z = 0$ and so G lies in this plane. $\mathcal{B}$ also has full *rotational symmetry* about the z-axis and so G must lie on this axis. It follows that G must be at the origin. Similarly, if $\mathcal{B}$ is a uniform lamina in the shape of an **equilateral triangle**, $\mathcal{B}$ has *rotational symmetry* about each of its three medians. G must therefore lie at the intersection of the medians.

Not all bodies have enough symmetry to find G completely however. If $\mathcal{B}$ is a uniform **circular cone**, then it has full rotational symmetry about the axis connecting its vertex to the centre of its base. G must therefore lie somewhere on this axis but there is no other symmetry to determine exactly where. In such cases, a summation (or integration) becomes necessary. This process is illustrated by the following two examples.

Example A.1 *Centre of mass of a semi-circular lamina*

Find the position of the centre of mass of a uniform semi-circular lamina.

Solution

Suppose that the lamina has radius a and occupies the region $\mathcal{A}$ shown in Figure A.1. The lamina has *rotational symmetry* about the y-axis so that G must lie on the y-axis, as shown. It remains to find its y-coordinate Y which, from the definition (A.1), is given by

$$MY = \sum_{i=1}^{N} m_i y_i, \tag{A.2}$$

where M is the mass of the lamina. However, since the lamina is a *continuous distribution* of mass, this sum must now be interpreted as an integral.

Consider an element of area dA of the lamina; this has mass $M dA/(\frac{1}{2}\pi a^2)$. The contribution to the sum in equation (A.2) from this element is therefore

$$\left(M dA/(\tfrac{1}{2}\pi a^2)\right) y,$$

where y is the y-coordinate of the element. On 'summing' these contributions (and cancelling by M) we find that

$$Y = \frac{2}{\pi a^2} \int_{\mathcal{A}} y\, dA, \tag{A.3}$$

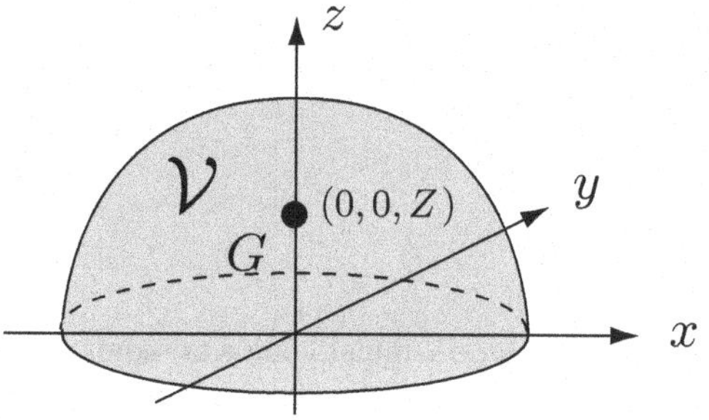

FIGURE A.2 A solid hemisphere of radius a occupies the region $\mathcal{V}$ shown.

where the integral is taken over the region $\mathcal{A}$ occupied by the lamina.

This double integral is most easily evaluated using standard polar coordinates. In these coordinates* $dA = (dr)(r\,d\theta) = r\,dr\,d\theta$, and $y = r\sin\theta$. The ranges of integration for r, θ are $0 \le r \le a$ and $0 \le \theta \le \pi$. We therefore obtain

$$\begin{aligned} Y &= \frac{2}{\pi a^2}\int_{r=0}^{r=a}\int_{\theta=0}^{\theta=\pi} r^2 \sin\theta\, dr\, d\theta \\ &= \frac{2}{\pi a^2}\left(\int_{r=0}^{r=a} r^2\, dr\right)\left(\int_{\theta=0}^{\theta=\pi} \sin\theta\, d\theta\right) \\ &= \frac{2}{\pi a^2}\left(\tfrac{1}{3}a^3\right)(2) = \frac{4a}{3\pi} \approx 0.424\, a. \end{aligned}$$

Hence the **centre of mass** of the semi-circular lamina is on the y-axis a distance $4a/3\pi$ from the origin. ■

Example A.2 *Centre of mass of a solid hemisphere*

Find the position of the centre of mass of a uniform solid hemisphere.

Solution

Suppose that the lamina has radius a and occupies the region $\mathcal{V}$ shown in Figure A.2. The hemisphere has full *rotational symmetry* about the z-axis so that G must lie on the z-axis, as shown. It remains to find its z-coordinate Z which, from the definition (A.1), is given by

$$MZ = \sum_{i=1}^{N} m_i z_i, \qquad \text{(A.4)}$$

where M is the mass of the hemisphere. However, since the hemisphere is a *continuous distribution* of mass, this sum must now be interpreted as an integral.

Consider a volume element dv of the hemisphere; this has mass $M dv/(\frac{2}{3}\pi a^2)$. The contribution to the sum in equation (A.4) from this element is therefore

$$\left(M dv/(\tfrac{2}{3}\pi a^2)\right) z,$$

* See Figure 3.6 (right) for a diagram of the element of area in plane polars.

where z is the z-coordinate of the volume element. On 'summing' these contributions (and cancelling by M) we find that

$$Z = \frac{3}{2\pi a^2}\int_{\mathcal{V}} z\, dv, \tag{A.5}$$

where the integral is taken over the region $\mathcal{V}$ occupied by the hemisphere.

This volume integral is most easily evaluated using standard spherical polar coordinates $r,\ \theta,\ \phi$. In these coordinates* $dv = (dr)(r\, d\theta)(r\sin\theta\, d\phi) = r^2\sin\theta\, dr\, d\theta\, d\phi$, and $z = r\cos\theta$. The ranges of integration for r, θ and ϕ are $0 \le r \le a$, $0 \le \theta \le \pi/2$ and $0 \le \phi \le 2\pi$. We therefore obtain

$$\begin{aligned} Z &= \frac{3}{2\pi a^2}\int_{r=0}^{r=a}\int_{\theta=0}^{\theta=\pi/2}\int_{\phi=0}^{\phi=2\pi} r^3 \sin\theta\cos\theta\, dr\, d\theta\, d\phi \\ &= \frac{3}{2\pi a^2}\left(\int_{r=0}^{r=a} r^3\, dr\right)\left(\int_{\theta=0}^{\theta=\pi/2} \sin\theta\cos\theta\, d\theta\right)\left(\int_{\phi=0}^{\phi=2\pi} d\phi\right) \\ &= \frac{3}{2\pi a^2}\left(\tfrac{1}{4}a^4\right)\left(\tfrac{1}{2}\right)(2\pi) = \frac{3a}{8}. \end{aligned}$$

Hence the **centre of mass** of the hemisphere is on the z-axis a distance $3a/8$ from the origin. ■

A.2 MOMENT OF INERTIA

Suppose that we have the same system $\mathcal{S}$ as before and that CD is a straight line. Then the moment of inertia of $\mathcal{S}$ about the axis† CD is defined as follows:

Definition A.2 ***Moment of inertia*** *The **moment of inertia** of the system $\mathcal{S}$ about the axis CD is defined by*

$$\boxed{I_{CD} = \sum_{i=1}^{N} m_i\, {p_i}^2} \tag{A.6}$$

where p_i is the perpendicular distance of the mass m_i from the axis CD.

Any system has a moment of inertia about any axis, but here we will suppose that the system is a **rigid body** $\mathcal{B}$. Then the value of I_{CD} about an *embedded* axis is a constant

* See Figure 3.7 (right) for a diagram of the volume element in spherical polars.

† It is traditional to call CD an *axis* whether or not $\mathcal{S}$ is a rigid body rotating about CD.

and can be calculated once and for all. Calculation of moments of inertia required us to evaluate the sum (or integral) appearing in the definition (A.6). This process is illustrated by the following examples.

Example A.3 *Uniform rod (about a perpendicular axis through G)*

Find the moment of inertia of a uniform rod of mass M and length $2a$ about an axis through its centre and perpendicular to its length.

Solution

Consider an element of length dx of the rod; this has mass $M dx/2a$. The contribution to the sum in (A.6) from this element is therefore $p^2(M dx/2a)$, where p is the distance of the element dx from the specified axis. On 'summing' these contributions over all the elements of length of the disk, we find that

$$I_{CD} = \frac{M}{2a}\int_{-a}^{a} p^2\,dx = \frac{M}{2a}\int_{-a}^{a} x^2\,dx,$$

since $p = |x|$ for the specified axis. Hence

$$I_{CD} = \frac{M}{2a}\left(\tfrac{2}{3}a^3\right) = \tfrac{1}{3}Ma^2.$$

Hence the **moment of inertia** of the rod about the specified axis is $\frac{1}{3}Ma^2$. ■

Example A.4 *Uniform circular disk (about its axis of symmetry)*

Find the moment of inertia of a uniform circular disk of mass M and radius a about its axis of symmetry.

Solution

Consider an element of area dA of the disk; this has mass $M dA/\pi a^2$. The contribution to the sum in (A.6) from this element is therefore

$$\left(M dA/\pi a^2\right)p^2,$$

where p is the distance of the element from the symmetry axis. On 'summing' these contributions over all the elements of area of the disk we find that

$$I_{CD} = \frac{M}{\pi a^2}\int_{\mathcal{A}} p^2\,dA, \qquad \text{(A.7)}$$

where the integral is taken over the region $\mathcal{A}$ occupied by the disk.

This double integral is most easily evaluated using standard plane polar coordinates r, θ. In these coordinates* $dA = (dr)(r\,d\theta) = r\,dr\,d\theta$, and $p = r$. The ranges

* See Figure 3.6 (right) for a diagram of the element of area in plane polars.

of integration for r, θ are $0 \le r \le a$ and $0 \le \theta \le 2\pi$. We therefore obtain

$$\begin{aligned} I_{CD} &= \frac{M}{\pi a^2} \int_{r=0}^{r=a} \int_{\theta=0}^{\theta=2\pi} r^3 \, dr \, d\theta \\ &= \frac{M}{\pi a^2} \left(\int_{r=0}^{r=a} r^3 \, dr \right) \left(\int_{\theta=0}^{\theta=2\pi} d\theta \right) \\ &= \frac{M}{\pi a^2} \left(\tfrac{1}{4} a^4 \right) (2\pi) = \tfrac{1}{2} M a^2. \end{aligned}$$

Hence the **moment of inertia** of the disk about its symmetry axis is $\frac{1}{2} M a^2$. ■

Example A.5 *Uniform solid sphere (about any axis through G)*

Find the moment of inertia of a uniform solid sphere of mass M and radius a about an axis through its centre.

Solution

Consider a volume element dv of the sphere; this has mass $M dv/(\frac{4}{3}\pi a^2)$. The contribution to the sum in equation (A.6) from this element is therefore

$$\left(M dv/(\tfrac{4}{3}\pi a^3) \right) p^2,$$

where p is the distance of the volume element dv from the axis CD. On 'summing' these contributions over all the volume elements of the sphere, we find that

$$I_{CD} = \frac{3M}{4\pi a^3} \int_{\mathcal{V}} p^2 \, dv, \tag{A.8}$$

where the integral is taken over the region $\mathcal{V}$ occupied by the sphere.

This volume integral is most easily evaluated using spherical polar coordinates r, θ, ϕ centred at the centre of the sphere, with the line $\theta = 0$ lying along the axis CD. In these coordinates* $dv = (dr)(r\, d\theta)(r \sin\theta \, d\phi) = r^2 \sin\theta \, dr \, d\theta \, d\phi$, and $p = r \sin\theta$. The ranges of integration for r, θ and ϕ are $0 \le r \le a$, $0 \le \theta \le \pi$ and $0 \le \phi \le 2\pi$. We therefore obtain

$$\begin{aligned} I_{CD} &= \frac{3M}{4\pi a^3} \int_{r=0}^{r=a} \int_{\theta=0}^{\theta=\pi} \int_{\phi=0}^{\phi=2\pi} r^4 \sin^3\theta \, dr \, d\theta \, d\phi \\ &= \frac{3M}{4\pi a^3} \left(\int_{r=0}^{r=a} r^4 \, dr \right) \left(\int_{\theta=0}^{\theta=\pi} \sin^3\theta \, d\theta \right) \left(\int_{\phi=0}^{\phi=2\pi} d\phi \right) \\ &= \frac{3M}{4\pi a^3} \left(\tfrac{1}{5} a^5 \right) \left(\tfrac{4}{3} \right) (2\pi) = \tfrac{2}{5} M a^2. \end{aligned}$$

Hence the **moment of inertia** of the sphere about an axis through its centre is $\frac{2}{5} M a^2$.
■

* See Figure 3.7 (right) for a diagram of the volume element in spherical polars.

TABLE OF MOMENTS OF INERTIA

Body	Axis	Moment of inertia
Thin rod mass M length $2a$	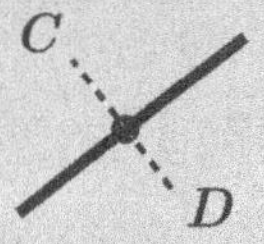	$I_{CD} = \frac{1}{3}Ma^2$
Circular hoop mass M radius a	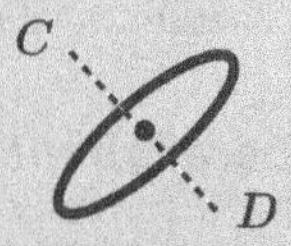	$I_{CD} = Ma^2$
Circular disk mass M radius a	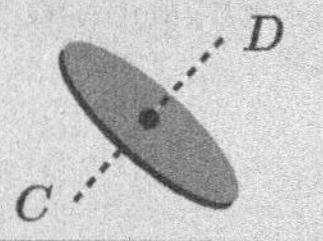	$I_{CD} = \frac{1}{2}Ma^2$
Solid sphere mass M radius a	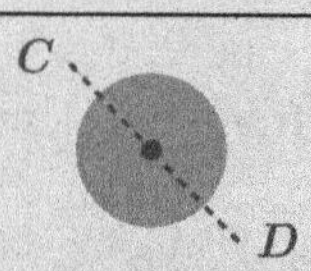	$I_{CD} = \frac{2}{5}Ma^2$
Spherical shell mass M radius a	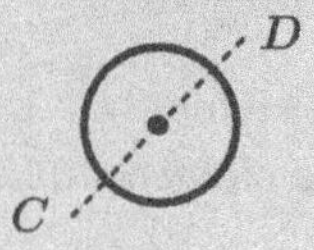	$I_{CD} = \frac{2}{3}Ma^2$
Circular cylinder mass M radius a length $2b$	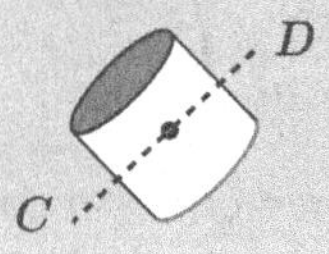	$I_{CD} = \frac{1}{4}Ma^2 + \frac{1}{3}Mb^2$

Table 3 Some useful moments of inertia. All the bodies are uniform and in each case the axis CD passes through the centre of mass.

Table of moments of inertia

We can now build up a table of the most useful moments of inertia. Table 3 consists of the three results derived earlier together with one that is obvious (the circular hoop), and two that are new* (the spherical shell and the circular cylinder).

It may seem that there are some obvious omissions. There is no mention of the circular cylinder about its axis of symmetry, nor the rectangular plate about *any* axis. However, the table has been deliberately kept short to make it more memorable. Any other moments of inertia that are likely to be required can be quickly deduced from those in the table by the methods given below. We begin with a couple of useful tricks.

* These two are less common than the others. They can be obtained by integration.

Two useful tricks

The **first trick** is based on the simple observation that the moment of inertia of a body is unaltered if its masses are moved in any manner *parallel* to the specified axis. In particular, the moment of inertia of a body is unaltered if it is squashed into a *lamina* perpendicular to the specified axis.

Suppose for example that the body is a uniform circular cylinder of radius a and length $2b$. Then its moment of inertia about its axis of symmetry is the same as that of a *uniform circular disk* of the same mass and radius about its own axis of symmetry. But this moment of inertia is known from Table 3 to be $\frac{1}{2}Ma^2$. Therefore the moment of inertia of the **circular cylinder** about its axis of symmetry is also $\frac{1}{2}Ma^2$. As a second example, suppose that the body is a uniform rectangular plate occupying the region $-a \le x \le a$, $-b \le y \le b$ of the (x, y)-plane. Then the moment of inertia of the plate about the x-axis is the same as that of a uniform rod of mass M and length $2b$ lying along the interval $-b \le y \le b$ of the y-axis. But this moment of inertia is known from the table to be $\frac{1}{3}Mb^2$. Therefore the moment of inertia of the **rectangular plate** about the x-axis is also $\frac{1}{3}Mb^2$. Similarly, the moment of inertia of the plate about the y-axis is $\frac{1}{3}Ma^2$. The moment of inertia of the plate about the z-axis will be deduced in the next section by using the perpendicular axes theorem.

To illustrate the **second trick**, suppose a uniform solid sphere of mass M and radius a is cut into two hemispheres. Then each of these hemispheres has mass $\frac{1}{2}M$ and (from the table) has moment of inertia $\frac{1}{2}\left(\frac{2}{5}Ma^2\right) = \frac{1}{5}Ma^2$ about its axis of symmetry, where M is the mass of the *whole* sphere. Hence, the moment of inertia of a **hemisphere** of mass M and radius a about its axis of symmetry must be $\frac{2}{5}Ma^2$, which is the same formula* as that for the complete sphere! *A similar argument applies whenever a body can be cut into a number of parts of equal mass that make equal contributions to the moment of inertia. The formula for the moment of inertia of each part is the same as that for the whole body.* Thus the moment of inertia of a segment of a *Terry's* chocolate orange about its straight edge is $\frac{2}{5}Ma^2$, where M is the mass of the segment and a is the radius of the orange.

A.3 PARALLEL AND PERPENDICULAR AXES

In this final section, we prove two important theorems that enable many more moments of inertia to be deduced. These are the **parallel axes** theorem and the **perpendicular axes** theorem.

Theorem A.1 ***Parallel axes theorem*** *Let I_G be the moment of inertia of a body about some axis through its centre of mass G, and let I be the moment of inertia of the body about a* **parallel** *axis. Then*

* It is the same formula, but remember that M is now the mass of the *hemisphere*.

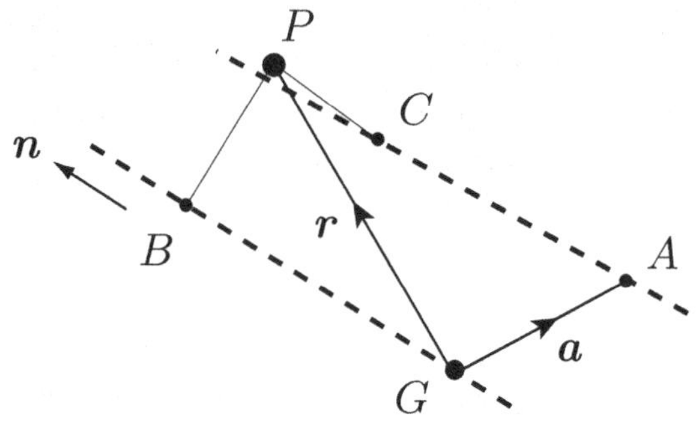

FIGURE A.3 The parallel axes theorem. The bold dashed lines are the two parallel axes.

$$\boxed{I = I_G + Ma^2} \tag{A.9}$$

where M is the mass of the body and a is the distance between the two parallel axes.

This is a powerful result. It means that, if you know the moment of inertia of a body about some axis, then you can easily calculate its moment of inertia about *any parallel axis*. The proof is as follows:

Proof. The configuration is shown in Figure A.3. The axis* $\{G, \boldsymbol{n}\}$ passes through G and $\{A, \boldsymbol{n}\}$ is the parallel axis. Here $\boldsymbol{n}$ is a unit vector parallel to the two axes, and A is chosen so that GA is perpendicular to the two axes. Then if $\overrightarrow{GA}$ represents the vector $\boldsymbol{a}$, the magnitude $|\boldsymbol{a}| = a$, the distance between the two axes.

Let P be a typical particle of the body with mass m and position vector $\boldsymbol{r}$ *relative to G*. Construct the perpendiculars PB, PC to the two axes, as shown. Then, by applications of Pythagoras,

$$PB^2 = GP^2 - GB^2 = |\boldsymbol{r}|^2 - (\boldsymbol{r} \cdot \boldsymbol{n})^2$$

and

$$\begin{aligned} PC^2 = AP^2 - AC^2 &= |\boldsymbol{r} - \boldsymbol{a}|^2 - \big((\boldsymbol{r} - \boldsymbol{a}) \cdot \boldsymbol{n}\big)^2 \\ &= (\boldsymbol{r} - \boldsymbol{a}) \cdot (\boldsymbol{r} - \boldsymbol{a}) - \big(\boldsymbol{r} \cdot \boldsymbol{n} - \boldsymbol{a} \cdot \boldsymbol{n}\big)^2 \\ &= |\boldsymbol{r}|^2 + |\boldsymbol{a}|^2 - 2\boldsymbol{r} \cdot \boldsymbol{a} - (\boldsymbol{r} \cdot \boldsymbol{n})^2 \\ &= PB^2 + a^2 - 2\boldsymbol{r} \cdot \boldsymbol{a}. \end{aligned}$$

If we now multiply this equality by m and sum over all the particles of the body, we obtain

$$\begin{aligned} I &= I_G + \left(\sum_{i=1}^{N} m_i\right) a^2 - 2\sum_{i=1}^{N} m_i \boldsymbol{r}_i \cdot \boldsymbol{a} = I_G + Ma^2 - 2\boldsymbol{a} \cdot \sum_{i=1}^{N} m_i \boldsymbol{r}_i \\ &= I_G + Ma^2 - 2\boldsymbol{a} \cdot (M\boldsymbol{R}), \end{aligned}$$

* The notation $\{G, \boldsymbol{n}\}$ means '*the axis through G parallel to the vector* $\boldsymbol{n}$'.

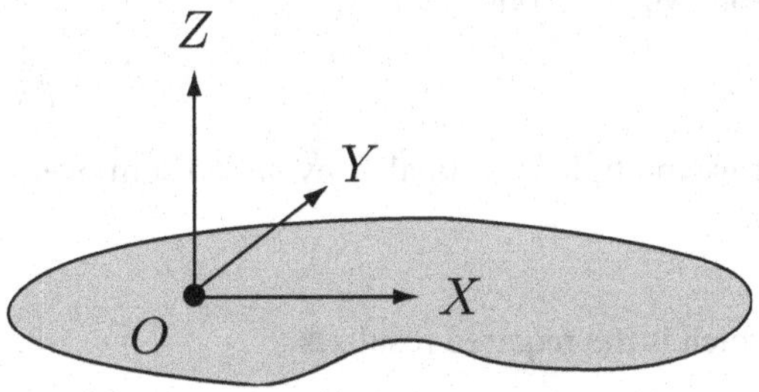

FIGURE A.4 The perpendicular axes theorem. The axes OX, OY lie in the plane of the lamina and OZ is perpendicular.

where $\boldsymbol{R}$ is the position vector of G. But since G is itself the origin of position vectors, $\boldsymbol{R} = \boldsymbol{0}$ and we obtain

$$I = I_G + Ma^2,$$

which is the required result. ■

Example A.6 *Rectangular plate (about an edge)*

A uniform rectangular plate has mass M and sides $2a$, $2b$. Find its moment of inertia about a side of length $2a$.

Solution

Suppose the plate occupies the region $-a \le x \le a$, $-b \le y \le b$ of the (x, y)-plane. It is already known that the moment of inertia of the plate about the x-axis is $\frac{1}{3}Mb^2$. The specified axis is parallel to the x-axis, the distance between the two axes being b. Hence the required **moment of inertia** is given by the parallel axes theorem to be $\frac{1}{3}Mb^2 + Mb^2 = \frac{4}{3}Mb^2$. ■

We will do more examples later, but first we will prove the perpendicular axes theorem.

Theorem A.2 ***Perpendicular axes theorem*** *Suppose the body is a* ***lamina*** *and that O is some point in its plane. Let $OXYZ$ be a set of Cartesian axes with OX, OY in the plane of the lamina and OZ perpendicular (see Figure A.4). Then the moments of inertia of the lamina about these* ***three perpendicular axes*** *are related by*

$$\boxed{I_{OZ} = I_{OX} + I_{OY}} \qquad \text{(A.10)}$$

that is, the moment of inertia of the lamina about the axis perpendicular to its plane is the ***sum*** *of the moments of inertia about the two in-plane axes.*

Proof. Let P be a typical particle of the lamina with mass m and coordinates $(x, y, 0)$ in the system $OXYZ$. Then p_X, p_Y, p_Z, the perpendicular distances of P from the axes OX, OY OZ are given by

$$p_X = |y|, \qquad p_Y = |x|, \qquad p_Z = \left(x^2 + y^2\right)^{1/2},$$

from which it follows that

$$p_Z^2 = p_X^2 + p_Y^2.$$

If we multiply this equality by m and sum over all the particles of the body, we obtain

$$I_{OZ} = I_{OX} + I_{OY},$$

which is the required result. ∎

Example A.7 *Rectangular plate (about the perpendicular axis through G)*

A uniform rectangular plate has mass M and sides $2a$, $2b$. Find its moment of inertia about the axis through its centre perpendicular to its plane.

Solution

Let the axes $OXYZ$ be placed so that O is at the centre of the plate with OZ perpendicular to its plane. Then, since the plate is a lamina, it follows from the perpendicular axes theorem that

$$I_{OZ} = I_{OX} + I_{OY}.$$

But it is already known that $I_{OX} = \frac{1}{3}Mb^2$ and $I_{OY} = \frac{1}{3}Ma^2$ so that $I_{OZ} = \frac{1}{3}Mb^2 + \frac{1}{3}Ma^2 = \frac{1}{3}M(a^2 + b^2)$. Hence the required **moment of inertia** is $\frac{1}{3}M(a^2 + b^2)$. ∎

Example A.8 *Circular disk (about a diameter)*

Find the moment of inertia of a uniform circular disk of mass M and radius a about a diameter.

Solution

Let the axes $OXYZ$ be placed so that O is at the centre of the disk with OZ perpendicular to its plane. Then, since the disk is a lamina, it follows from the perpendicular axes theorem that

$$I_{OZ} = I_{OX} + I_{OY}.$$

But I_{OZ} is known to be $\frac{1}{2}Ma^2$ and, by the rotational symmetry of the disk, $I_{OX} = I_{OY}$. It follows that $I_{OX} = I_{OY} = \frac{1}{4}Ma^2$. Hence the **moment of inertia** of the disk about any of its diameters is $\frac{1}{4}Ma^2$. ∎

Example A.9 *Rectangular block (about any of its edges)*

A uniform rectangular block of mass M has edges of lengths $2a$, $2b$, $2c$. Find its moment of inertia about any edge.

Solution

Suppose the block occupies the region $-a \le x \le a$, $-b \le y \le b$, $-c \le z \le c$. We will first find its moment of inertia about each of the coordinate axes. Suppose the block is squashed parallel to the x-axis into a lamina of mass M occupying the rectangular region $-b \leq y \leq b$, $-c \leq z \leq c$ of the (y, z)-plane. The moment of inertia of this 'plate' about the x-axis is already known to be $\frac{1}{3}M(b^2 + c^2)$ and the block must have the same moment of inertia about the x-axis. Hence the **moments of inertia** of the block about the x- y- and z-axes are $\frac{1}{3}M(b^2 + c^2)$, $\frac{1}{3}M(a^2 + c^2)$ and $\frac{1}{3}M(a^2 + b^2)$ respectively.

Consider now an axis lying along an edge of length $2a$. This axis is parallel to the x-axis, the distance between the two axes being $(b^2 + c^2)^{1/2}$. Hence the moment of inertia of the block about this edge is given by the theorem of parallel axes to be $\frac{1}{3}M(b^2 + c^2) + M(b^2 + c^2) = \frac{4}{3}M(b^2 + c^2)$. Hence the **moments of inertia** of the block about the edges of lengths $2a$, $2b$, $2c$ are $\frac{4}{3}M(b^2 + c^2)$, $\frac{4}{3}M(a^2 + c^2)$, $\frac{4}{3}M(a^2 + b^2)$ respectively. ■

Answers to the problems

Chapter 1 Vectors (page 21)

1.1 (i) $8\,\boldsymbol{i}+17\,\boldsymbol{j}-26\,\boldsymbol{k},\ 6.$ (ii) $3,\ 5,\ \cos^{-1}(14/15).$ (iii) $1/3,\ -9/5.$
(iv) $4\,\boldsymbol{i}+2\,\boldsymbol{j}+3\,\boldsymbol{k},\ -20\,\boldsymbol{i}-13\,\boldsymbol{j}-15\,\boldsymbol{k},\ 9\,\boldsymbol{i}-12\,\boldsymbol{k}.$
(v) $+3$. The set is *right*-handed. (vi) $-11\,\boldsymbol{i}+70\,\boldsymbol{j}-46\,\boldsymbol{k}.$

1.2 $\cos^{-1}(1/3)$

1.3 $\boldsymbol{c}=\boldsymbol{b}-\boldsymbol{a},\ \boldsymbol{d}=-\boldsymbol{a},\ \boldsymbol{e}=-\boldsymbol{b},\ \boldsymbol{f}=\boldsymbol{a}-\boldsymbol{b}.$

1.14 $6t\,\boldsymbol{i}+3t^2\,\boldsymbol{j}+\boldsymbol{k},\ 6\,\boldsymbol{i}+6t\,\boldsymbol{j},\ -6t(t+3)\,\boldsymbol{i}+6(t+3)\,\boldsymbol{j}+12t(t^2-2)\,\boldsymbol{k}.$

1.15 $2\,\boldsymbol{v}\cdot\dot{\boldsymbol{v}},\ (\dot{\boldsymbol{v}}\cdot\boldsymbol{k})\,\boldsymbol{v}+(\boldsymbol{v}\cdot\boldsymbol{k})\,\dot{\boldsymbol{v}},\ [\boldsymbol{v},\ddot{\boldsymbol{v}},\boldsymbol{k}].$

1.16 $\boldsymbol{t}=-\sin\theta\,\boldsymbol{i}+\cos\theta\,\boldsymbol{j},\ \boldsymbol{n}=-\cos\theta\,\boldsymbol{i}-\sin\theta\,\boldsymbol{j},\ \kappa^{-1}=a.$

1.17 $\boldsymbol{t}=(-a\sin\theta\,\boldsymbol{i}+a\cos\theta\,\boldsymbol{j}+b\,\boldsymbol{k})/(a^2+b^2)^{1/2},\ \boldsymbol{n}=(-\cos\theta\,\boldsymbol{i}-\sin\theta\,\boldsymbol{j}),\ \kappa^{-1}=(a^2+b^2)/a.$

1.18 $\boldsymbol{t}=(p\,\boldsymbol{i}+\boldsymbol{j})/(1+p^2)^{1/2},\ \boldsymbol{n}=(\boldsymbol{i}-p\,\boldsymbol{j})/(1+p^2)^{1/2},\ \kappa^{-1}=2a(1+p^2)^{3/2}.$

Chapter 2 Velocity, acceleration and scalar angular velocity (page 43)

2.1 $v=12t-3t^2\ \mathrm{m\,s^{-1}},\ a=12-6t\ \mathrm{m\,s^{-2}}$. P comes to rest when $t=0$ and $t=4$ s. At these times its displacement is 1 m and 33 m respectively.

2.2 $v=3t^2-4t-15\ \mathrm{m\,s^{-1}},\ x=t^3-2t^2-15t+20$ m. P comes to rest after 3 s. (The time $t=-5/3$ s is *before* the motion was set up.) At this time P has displacement -16 m.

2.3 If the acceleration were constant, then, from $v=u+at$, $a=14.9\ \mathrm{ft\,s^{-2}}$. (Use feet and seconds as units.) However, from $s=ut+\frac{1}{2}at^2$, $a=20.3\ \mathrm{ft\,s^{-2}}$. This indicates that the acceleration was *not* constant.

2.4 $|\boldsymbol{v}|=(\Omega^2b^2+c^2)^{1/2},\ |\boldsymbol{a}|=\Omega^2b.$

2.5 $0.034\ \mathrm{m\,s^{-2}}$ directed towards the centre of the Earth, $0.0060\ \mathrm{m\,s^{-2}}$ directed towards the Sun.

2.6 $\boldsymbol{v}=(\Omega b\,e^{\Omega t})\,\widehat{\boldsymbol{r}}+(\Omega b\,e^{\Omega t})\,\widehat{\boldsymbol{\theta}},\ \boldsymbol{a}=(2\Omega^2b\,e^{\Omega t})\,\widehat{\boldsymbol{\theta}}.$

2.7 $\cos^{-1}\left(b/(b^2+\alpha^2t^4)^{1/2}\right).$

2.8 $|\boldsymbol{a}|=v^2/(b\cos\alpha).$

2.9 $\boldsymbol{v}=2b\tau^{-2}(\tau-t)\,\widehat{\boldsymbol{r}}+b\tau^{-3}t(2\tau-t)\,\widehat{\boldsymbol{\theta}}$. Minimum $|\boldsymbol{v}|$ occurs when $t=\tau$. When $t=\tau$, $\boldsymbol{a}=-3b\tau^{-2}\,\widehat{\boldsymbol{r}}.$

2.13 Maximum speed relative to ground is 120 mph. $|\boldsymbol{a}|=121g.$

2.14 $|\boldsymbol{v}|=\Omega b(\cosh 2\Omega t)^{1/2},\ \boldsymbol{a}=(2\Omega^2b\sinh\Omega t)\,\widehat{\boldsymbol{\theta}}.$

2.15 Maximum speed is ωe, maximum acceleration is ω^2e.

2.16 $\omega^2b\left(1+(b/c)\right).$

2.17 Speed of C is $\frac{1}{2}(\omega_1 b_1 + \omega_2 b_2)$. Angular velocity of gear is $(\omega_2 b_2 - \omega_1 b_1)/(b_2 - b_1)$. Angular velocity of arm is $(\omega_1 b_1 + \omega_2 b_2)/(b_1 + b_2)$.

2.18 $\omega = \Omega b \cos \Omega t \left(a^2 - b^2 \sin^2 \Omega t\right)^{-1/2}$, speed of C is $\frac{1}{2}\Omega ab |\cos \Omega t| \left(a^2 - b^2 \sin^2 \Omega t\right)^{-1/2}$.

2.19 Bearing is 18.6° west of north, time taken is 4 h 46 m.

2.20 $aU \sin\alpha \left(U^2 + u^2 + 2Uu \cos\alpha\right)^{-1/2}$.

2.21 Maximum range is $R_0(v^2 - u^2)^{1/2}/v$, achieved when $\theta = \pm\pi/2$.

Chapter 3 Newton's laws and gravitation (page 71)

3.1 $\sqrt{6}m^2G/a^2$, $\sqrt{6}m^2G/4a^2$.

3.2 $\left[\sqrt{3} + \sqrt{3/2} + 1/3\right] m^2G/a^2$, $4\left[1 + 1/\sqrt{2} + 1/(3\sqrt{3})\right] m^2G/a^2$.

3.3 $mMG/(x'^2 - a^2)$, $(M^2G/4a^2)\ln\left(b^2/(b^2 - 4a^2)\right)$.

3.4 $(2MM'G/a^2 b)\left[a + b - (a^2 + b^2)^{1/2}\right]$.

3.6 $(mMG/a^3)r$.

3.7 $2\pi M\sigma G$.

3.8 $3M^2G/4a^2$.

Chapter 4 Problems in particle dynamics (page 98)

4.1 $\frac{1}{2}T_0$. $3T_0/2M$, $2T_0$.

4.2 $(m(b - z)/b + M)\, g$, $3\,(m(b - z)/b + M)\, g$.

4.4 $(\sin\alpha - \mu\cos\alpha)g$. If $\mu > \tan\alpha$, the motion will come to rest and remain at rest.

4.5 $20\,\text{m s}^{-1}$.

4.6 1445 m.

4.7 Terminal velocity is $(U^2 + V^2)^{1/2}$, inclined at an angle $\tan^{-1}(U/V)$ to the downward vertical.

4.8 $\tan^{-1}(ebE_0/mu^2)$.

4.9 $(2h - g\tau^2)^2/8g\tau^2$.

4.10 $r = \left[a^{3/2} + 3(MG/2)^{1/2}\, t\right]^{2/3}$, which tends to infinity as t tends to infinity.

4.11 $\sqrt{2}a^2/\gamma^{1/2}$.

4.12 65 days.

4.14 The accelerations of the three masses are $g/9$ *downwards*, $7g/9$ upwards and $5g/9$ downwards respectively.

4.15 $r = a\cosh\Omega t$, $N = 2ma\Omega^2 \sinh\Omega t$.

4.17 Maximum height is $(V^2/2g)\ln\left[1 + (u^2/V^2)\right]$, time taken is $(V/g)\tan^{-1}(u/V)$.

4.18 Calculated descent times are 2.32 s and 2.27 s, which could hardly have been distinguished.

4.19 Maximum height is $(V^2/2g)\tan^{-1}(u^2/V^2)$.

4.20 Charge is $6\pi a\mu(v_1 + v_2)/E_0$, where the droplet radius $a = 3(\mu v_1/2\rho g)^{1/2}$.

4.21 Ranges are 55 m and 57 m respectively.

4.22 Projection angle is 60°.

4.23 Man is not safe. At 60 m away he would need to be at least 25 m high.

4.24 Least projection speed to clear hill is $\left(g(a^2+h^2)^{1/2}-gh\right)^{1/2}$. Least muzzle speed needed to cover the hill is $20\,\mathrm{m\,s^{-1}}$, so this gun *can* do the job.

4.25 Landing point has displacement $(u^2/g)\,(\sin(2\alpha+\beta)-\sin\beta)\,/\cos^2\beta$ up the plane. $R^U = (u^2/g)(1-\sin\beta)/\cos^2\beta$, $R^D = (u^2/g)(1+\sin\beta)/\cos^2\beta$.

4.27 Path is an inclined straight line.

4.28 Estimated mass of the Earth is 6.06×10^{24} kg. Calculation supposes the Earth is fixed and so ignores the motion of the Earth induced by the Moon. Radius of geostationary orbit is 42,150 km.

4.30 Tension is $(mg/\sqrt{2})+(mu^2/a)$, normal reaction is $(mg/\sqrt{2})-(mu^2/a)$, maximum u is $(ag/\sqrt{2})^{1/2}$. If u exceeds the maximum, the particle loses contact with the sphere.

4.31 Angle turned by string is u/Kb. Tension is $(mu^2/b)e^{-2Kt}$.

4.32 Particle of mass m meets y-axis again at the point $y=-2mV/eB_0$.

Chapter 5 Linear oscillations and normal modes (page 126)

5.1 Amplitude is 2, time taken is $\pi/6$.

5.2 In parallel, $\omega^2=\Omega_1^2+\Omega_2^2$. In series, $\omega^2=\Omega_1^2\Omega_2^2/(\Omega_1^2+\Omega_2^2)$.

5.3 Distance travelled is $a\left(1+e^{-\pi k/n}\right)$.

5.4 Greater than 8.

5.5 In SI units, $\alpha=16.3$, $\beta=4.4$.

5.8 Lower block leaves floor after time $(a/g)^{1/2}\cos^{-1}(-5/6)$.

5.9 Angular frequencies less than 20 radians s^{-1} and greater than 40 radians s^{-1} are safe.

5.11 Amplitude is

$$\left(\frac{\Omega^4+4K^2p^2}{(\Omega^2-p^2)^2+4K^2p^2}\right)^{1/2} h_0.$$

5.12 Response is

$$x=2F_0\sum_{n=1}^{\infty}\left(\frac{(-1)^{n+1}}{n}\right)\left(\frac{(\Omega^2-n^2)\sin nt+2Kn\cos nt}{(\Omega^2-n^2)^2+4K^2n^2}\right).$$

5.13 Period is $2(\pi+4)/\Omega$.

5.14 Body comes to rest at $x=+F_0/(2\Omega^2)$. Time taken is $2\pi/\Omega$.

5.15 Period is $\pi(\Omega+\Omega_D)/(\Omega\Omega_D)$, where $\Omega_D=(\Omega^2-K^2)^{1/2}$. Ratio of successive maxima is $e^{-K\pi/\Omega_D}$.

5.16 Slow mode: $\omega_1=n/\sqrt{6}$, $A/B=2/3$, where $n=(\alpha/m)^{1/2}$. Fast mode: $\omega_2=n$, $A/B=-1$. General motion is

$$x=2\delta_1\cos(\omega_1 t-\gamma_1)+\delta_2\cos(\omega_2 t-\gamma_2),$$

$$y=3\delta_1\cos(\omega_1 t-\gamma_1)-\delta_2\cos(\omega_2 t-\gamma_2),$$

where $\delta_1,\delta_2,\gamma_1,\gamma_2$ are arbitrary constants.

5.17 Slow mode: $\omega_1 = n$, $A/B = 1$, where $n = (T/ma)^{1/2}$. Fast mode: $\omega_2 = \sqrt{3}\,n$, $A/B = -1$. General motion is

$$x = \delta_1 \cos(\omega_1 t - \gamma_1) + \delta_2 \cos(\omega_2 t - \gamma_2),$$
$$y = \delta_1 \cos(\omega_1 t - \gamma_1) - \delta_2 \cos(\omega_2 t - \gamma_2),$$

where $\delta_1, \delta_2, \gamma_1, \gamma_2$ are arbitrary constants. General motion is not periodic.

5.18 Slow mode: $\omega_1 = \sqrt{2/3}\,n$, $A/B = 1/2$, where $n = (g/a)^{1/2}$. Fast mode: $\omega_2 = \sqrt{2}\,n$, $A/B = -1/2$.

Chapter 6 Energy conservation (page 151)

6.1 Both are 16 J which illustrates the energy principle.

6.2 Man works at 500 W, which is quite a feat.

6.4 3.0×10^9 J.

6.5 Motion takes place in the interval $3 \le x \le 6$ and the period is $9\pi/\sqrt{2}$.

6.7 Particle penetrates a distance $\left(3mu^2/2K\right)^{1/3}$.

6.8 $V = -2mMG\left(a^2 + x^2\right)^{-1/2}$ and $v_{\max} = (4MG/5a)^{1/2}$.

6.9 $V = (2mMG/a^2)\left[z - (a^2 + z^2)^{1/2}\right]$ and speed on impact is $(8MG/3a)^{1/2}$.

6.10 Projection speed is $(\alpha a^2/4m)^{1/2}$.

6.11 Spring is compressed by $a/3$.

6.13 Minimum speed needed is $(g(a + 2h))^{1/2}$. When $h = \frac{1}{2}a$ the mortar should be placed a distance $a\left(\sqrt{3} - 1\right)/2$ from the wall of the building and inclined at 60° to the horizontal.

6.14 Escape speed is $\left(aq^2/m(b^2 - a^2)\right)^{1/2}$.

6.15 Escape speed is $\left(a^3q^2/mb^2(b^2 - a^2)\right)^{1/2}$.

6.16 (i) is stable while (ii) is unstable.

6.17 Arrival speed is $(4\pi bg)^{1/2}$, time taken is $\left(4\pi(a^2 + b^2)/gb\right)^{1/2}$.

6.18 When a/b is small, period is approximately

$$2\pi\left(\frac{b}{g}\right)^{1/2}\left[1 + \frac{a^2}{4b^2} + \cdots\right]$$

6.20 Time taken to hit post is $b^2/(2au)$.

6.21 String makes 60° with *upward* vertical when it first becomes slack.

6.22 Car was parked about 5 m from edge of building.

Chapter 7 Orbits in a central field (page 188)

7.1 Distance of closest approach is $(p^2V^2 + \gamma)^{1/2}/V$.

7.2 $a = c$, $b = c\sin\alpha$.

7.4 $a = \sqrt{2}c\cos(\frac{1}{2}\alpha)$, $b = \sqrt{2}c\sin(\frac{1}{2}\alpha)$.

7.5 120°.

7.6 Deflection angle is $2\tan^{-1}(\gamma/pV^2)$ and distance of closest approach is $\left[(\gamma^2 + p^2V^4)^{1/2} - \gamma\right] \Big/ V^2$, where $\gamma = M_{\odot}G$.

7.7 Path is $r = \frac{4}{15}p \big/ \sin\left(\frac{4}{15}\theta\right)$. Distance of closest approach is $\frac{4}{15}p$.

7.9 Path is $r = a\cos(\theta/3)$. Time taken to reach centre is $\pi a^2/(2\sqrt{2}\gamma)$.

7.10 Apsidal angle is $\pi/(1-\epsilon)^{1/2}$.

7.11 Advance of perihelion is $2\pi\epsilon$ per 'year'.

7.12 Advance of perihelion is $-4\pi^2\epsilon$ per 'year'.

7.13 Advance of perihelion is $6\pi MG/(ac^2)$ per 'year'.

7.14 Differential scattering cross section is $a^2/4$, a constant.

7.15 Differential scattering cross section is

$$\sigma(\theta) = \frac{\pi^2\gamma(\pi - \theta)}{V^2\theta^2(2\pi - \theta)^2 \sin\theta}.$$

7.16 Period was 89.6 minutes. Maximum speed was 7.84 km per second.

7.17 Apogee is 3910 km above the Earth. Apogee speed is 5.50 km per second. Period is 128 minutes.

7.18 Spacecraft will escape if $k \geq \sqrt{2}$. Eccentricity of escape orbit is $k^2 - 1$.

7.19 Spacecraft will go into orbit if

$$k^2 < \frac{2MG}{cV^2 + 2MG}.$$

7.20 Time average of kinetic energy is $\gamma/(2a)$.

7.22 $\Delta v^E = +3.13$ km per second, $\Delta v^M = +0.83$ km per second. Travel time = 119 hours.

7.23 The thrust should be applied in the direction of motion when the spacecraft is at its perigee.

7.24 The most fuel efficient one-impulse strategy is to use the first impulse of the Hohmann orbit. $\Delta v = -2.50$ km per second for Venus, and 2.95 km per second for Mars. Hence the Venus flyby uses less fuel. The travel time is 146 days to Venus, and 259 days to Mars.

7.25 The most fuel efficient one-impulse strategy is to apply the impulse parallel to the direction of motion, Δv being chosen to give an orbit with the correct period. For a synchronous orbit, $\Delta v = +2.77$ km per second. The apogee distance is 77,700 km.

7.26 In the given approximation, $r = (L_0^2/\gamma)e^{-2Kt}$ and $\dot{\theta} = (\gamma^2/L_0^3)e^{+3Kt}$. Hence $r\dot{\theta} = (\gamma/L_0)e^{+Kt}$, which *increases* as t increases.

7.27 In the given approximation, $r = (L_0/\Omega)^{1/2}e^{-Kt/2}$ and $\dot{\theta} = \Omega$. Hence $r\dot{\theta} = (\Omega L_0)^{1/2}e^{-Kt/2}$, which *decreases* as t increases.

Chapter 8 Non-linear oscillations and phase space (page 214)

8.1 When the oscillation has unit amplitude, $\omega = 1 - \frac{3}{8}\epsilon + O\left(\epsilon^2\right)$ and $x(t) = \cos s + \frac{1}{32}(\cos s - \cos 3s)\,\epsilon + O\left(\epsilon^2\right)$ where $s = \omega t$.

8.2 When the oscillation has unit amplitude, $\omega = 1 + \frac{5}{16}\epsilon + O\left(\epsilon^2\right)$.

8.3 When the maximum value achieved by $x(t)$ is unity, $x(t) = \cos s + \frac{1}{6}(-3 + 2\cos s + \cos 2s)\,\epsilon + O\left(\epsilon^2\right)$, where $s = \left(1 + O(\epsilon^2)\right)t$. The minimum value achieved by $x(t)$ is $-1 - \frac{2}{3}\epsilon + O\left(\epsilon^2\right)$.

8.4 At zero order, the limit cycle is a circle with centre the origin and radius 2. Correct to order ϵ^2, $\omega = 1 - \frac{1}{16}\epsilon^2 + O\left(\epsilon^3\right)$, and correct to order ϵ, $x(t) = 2\cos s + \frac{1}{12}(-3\sin s + \sin 3s) + O\left(\epsilon^2\right)$.

8.9 The period of the limit cycle is 2π.

8.13 Correct to order ϵ, the driven response is

$$x(t) = -\frac{\cos pt}{p^2 - 1} + \left(\frac{3p^3 \sin pt}{4(p^2-1)^4} - \frac{p^3 \sin 3pt}{4(p^2-1)^3(9p^2-1)}\right)\epsilon + O\left(\epsilon^2\right)$$

which is valid when $p \neq 1, 1/3, 1/5, \ldots$.

Chapter 9 The energy principle (page 241)

9.1 Equilibrium position is $\theta = \tan^{-1}(m/M)$.

9.2 The equilibrium positions are (i) vertically downwards (unstable), and (ii) inclined at 60° to the upward vertical (stable).

9.3 Angular frequency is $\left(4V_0/mb^2\right)^{1/2}$.

9.5 Acceleration of mass m up the plane, and the string tension, are

$$\left(\frac{M\sin\alpha - m\sin\beta}{M + m}\right) g \quad \text{and} \quad \frac{Mmg}{M+m}(\sin\alpha + \sin\beta).$$

9.6 Normal reactions on P and Q are

$$\tfrac{2}{3}mg\left(7\cos\theta + 2\sin\theta - 3\sqrt{2}\right) \quad \text{and} \quad \tfrac{1}{3}mg\left(4\cos\theta + 5\sin\theta - 3\sqrt{2}\right).$$

9.7 Speed of rope on leaving peg is $(ag)^{1/2}$

9.8 The velocity and acceleration of the free end are given by

$$v^2 = \left(\frac{x(2a-x)}{a-x}\right) g, \qquad \frac{dv}{dt} = \left(\frac{2a^2 - 2ax + x^2}{(a-x)^2}\right) g,$$

and the free end has acceleration $5g$ when $x = \frac{2}{3}a$.

9.9 Amplitude of oscillations is a, period is $4\pi(a/g)^{1/2}$.

9.10 Final speed of hoop is $\left(v^2 + gh\right)^{1/2}$.

9.11 Acceleration of ball is $\frac{5}{7}g\sin\alpha$.

9.12 Acceleration of yo-yo is $\frac{2}{3}g$.

9.13 Speed of roll when its radius is b is

$$\left(\frac{a^2V^2}{b^2} - \frac{4g(b^3 - a^3)}{3b^2}\right)^{1/2}.$$

9.14 Period of small oscillations is

$$2\pi\left(\frac{a^2+3b^2}{3gb}\right)^{1/2}.$$

9.15 The speed of G is $\left(\frac{10}{7}g(a+b)(1-\cos\theta)\right)^{1/2}$.

9.16 Period of small oscillations is $2\pi\,(7(b-a)/5g)^{1/2}$.

9.17 Period of small oscillations is $2\pi\,(b^2/3ga)^{1/2}$.

Chapter 10 The linear momentum principle (page 279)

10.1 The time average of the apparent weight is Mg.

10.2 Juggling the balls as he crosses the bridge is worse than carrying them. The *average* force his feet apply to the bridge is equal to the total weight of himself and the balls and, since this is certainly not a constant, there will be times when this force *exceeds* the average value.

10.4 The reaction exerted by the support is

$$R=\frac{MG}{4a}\left(\frac{2a^2+2ax-3x^2}{a-x}\right).$$

The support will give way when $x=\frac{2}{3}a$.

10.5 The chain first rises to height $\frac{1}{2}a$.

10.6 The speed of G is given by $v^2=\frac{10}{7}g(a+b)(1-\cos\theta)$.

10.8 The speed and height at burnout are $2100\,\mathrm{m\,s^{-1}}$ and $100\,\mathrm{km}$.

10.9 The maximum speed achieved is

$$\frac{u}{\epsilon}\left(1-\gamma^{-\epsilon}\right)\sim u\ln\gamma\left[1-\tfrac{1}{2}(\ln\gamma)\,\epsilon+O\left(\epsilon^2\right)\right]$$

when ϵ is small; $\gamma=M/m$.

10.12 The proportions of the total kinetic energy carried by the fragments are $\frac{1}{5},\frac{2}{5},\frac{2}{5}$.

10.14 Velocity of composite particle is $(m_1\boldsymbol{v}_1+m_2\boldsymbol{v}_2)/(m_1+m_2)$.

10.15 The proton lost 14% of its energy and the recoil angle of the helium nucleus was 62°.

10.16 The mystery nucleus has atomic mass 16 and is therefore oxygen.

10.18 Neutrons of energy $\frac{1}{4}E$ are found only at an angle of 60° to the incident beam.

10.20 The dark companion in Cygnus X-1 has a mass of about $16M_\odot$.

10.21 The emerging particles are scattered alpha particles and recoiling helium nuclei, which are identical. The angular distribution of emerging particles is

$$\frac{q^4}{E^2}\left(\frac{\cos\theta}{\sin^4\theta}+\frac{1}{\cos^3\theta}\right),$$

where θ is measured from the direction of the incident beam.

10.22 The scattering cross section for the incident neutrons is

$$\sigma^{TB}=\frac{A}{\pi}\cos\theta_1,$$

and the angular distribution of the recoiling protons is the same, that is, $(A/\pi)\cos\theta_2$.

10.24 The rod will hit the table after a time

$$\left(\frac{a}{3g}\right)^{1/2}\int_{\pi/3}^{\pi/2}\left(\frac{4-3\cos^2\theta}{1-2\cos\theta}\right)^{1/2} d\theta.$$

Chapter 11 The angular momentum principle (page 317)

11.2 Angular speed of target after impact is

$$\left(\frac{4mb}{Ma^2+4mb^2}\right)u.$$

11.3 Angular speed of cylinder after impact is

$$\left(\frac{2mb}{a^2(M+2m)}\right)u.$$

11.4 Angular speed of disk is $\frac{4}{5}(a^2/b^2)\Omega$ and increase in kinetic energy is

$$\frac{Ma^2}{25b^2}(4a^2-5b^2)\Omega^2.$$

11.5 Angle of the new conical motion is 84° approximately.

11.6 In the elastic case, the speed of the ball after impact is

$$\left(\frac{1-\beta}{1+\beta}\right)u, \quad \text{where} \quad \beta=\frac{m}{M}\left(1+\frac{b^2}{k^2}\right).$$

11.8 In Case A,

$$r=\left(a^2+u^2t^2\right)^{1/2}, \qquad \phi=\frac{\tan^{-1}(ut/a)}{\sin\alpha}.$$

In Case B, the required value of u is $\left(\frac{4}{3}ag\right)^{1/2}$.

11.9 After one lap by the bug, the hoop has rotated through the angle

$$\left(\frac{2m}{M+2m}\right)\pi.$$

11.10 Period of small oscillations is

$$2\pi\left(\frac{a^2+3b^2}{3gb}\right)^{1/2}.$$

11.11 Speed of ball after the onset of rolling is $\frac{5}{7}V$ and $\frac{2}{7}$ of the kinetic energy is lost in the process.

11.12 Ball will slide if $\tan\alpha > \frac{7}{2}\mu$ and will roll if $\tan\alpha < \frac{7}{2}\mu$. If ball slides, the acceleration is $g(\sin\alpha-\mu\cos\alpha)$. If ball rolls, the acceleration is $\frac{5}{7}g\sin\alpha$.

11.13 Bug will reach top of disk if

$$u^2 > \frac{8mag}{M+2m}.$$

11.14 Acceleration of yo-yo is $\frac{2}{3}g - \frac{1}{3}\ddot{Z}$ downwards.

11.15 Forward velocity of cylinder is

$$\left(\frac{k^2}{a^2+k^2}\right)V.$$

11.16

$$S = \tfrac{1}{4}\gamma(3\gamma-2)Mg\sin\theta, \qquad K = \tfrac{1}{2}\gamma^2(1-\gamma)Mga\sin\theta.$$

11.17 Reactions at B and C are $\frac{1}{6}mg\,\boldsymbol{i}$ and $\frac{1}{6}mg\,\boldsymbol{j}$. Reaction at the floor is $-\frac{1}{6}mg\,\boldsymbol{i} - \frac{1}{6}mg\,\boldsymbol{j} + mg\boldsymbol{k}$.

Chapter 12 Lagrange's equations and conservation principles (page 361)

12.1 Closed chain has N degrees of freedom. There are *four* conserved quantities (two horizontal components of linear momentum, the vertical component of angular momentum about G, and kinetic energy). Motion can be determined from conservation principles if $N \le 4$.

12.2 Upward acceleration of mass m is $2g/5$.

12.3 The accelerations of the three masses are (in order) $g/9$ *downwards*, $7g/9$ upwards and $5g/9$ downwards.

12.4 Period of small oscillations is $2\pi\,(4a/(3g\sin\alpha))^{1/2}$.

12.5 Normal reaction is $mg(7\cos\theta - 4)/3$. It is not realistic to assume that rolling persists until the normal reaction becomes zero. This would require an infinite coefficient of friction!

12.6 $p_x = (3/2)M\dot{x} + m(\dot{x} + b\dot{\theta}\cos\theta)$, which is *not* the horizontal linear momentum. Period of small oscillations is $2\pi(3Mb/(3M+2m)g)^{1/2}$.

12.7 Accelerations are (i) $2g\sin\alpha\cos\alpha/(7-2\cos^2\alpha)$, (ii) $5g\sin\alpha/(7-2\cos^2\alpha)$.

12.8 Reaction of the floor is $mg\left(4 - 6\cos\alpha\cos\theta + 3\cos^2\theta\right) / \left(1+3\sin^2\theta\right)^2$.

12.9 $p_\theta = mR^2\dot{\theta}$, which is the vertical component of angular momentum about O. Kinetic energy is not conserved because the force pulling the string down does work. Tension in the string is $m\left((L^2/R^3) - \ddot{R}\right)$.

12.10 Solution is $r = a\cosh\Omega t$. Energy function $h = \frac{1}{2}ma^2\left(\dot{r}^2 - \Omega^2 r^2\right) = -\frac{1}{2}ma^2\Omega^2$, which is constant.

12.11 Upward acceleration of yo-yo is $\frac{1}{3}(\ddot{Z} - 2g)$, so yo-yo can remain at same height (or move with constant velocity) if $\ddot{Z} = 2g$. The total energy at time t is $\frac{1}{6}m\left(\dot{Z}^2 + 2gZ\right)$.

12.12 Neither of E and h is conserved in general. (Consider, for example, the case in which $\dot{Z}$ is constant.)

12.13 Angle turned by hoop is $2\pi m/(M+2m)$.

12.14 $U = -f(t)x$.

12.15 The conserved momenta are $p_x = m\dot{x} + etz$ and $p_y = m\dot{y}$.

12.17 Conserved momentum $p_\theta = mr^2\dot{\theta}$, which is the component of angular momentum about O perpendicular to the plane of motion.

12.18 Conserved momentum $p_\phi = mr^2 \sin^2\alpha\, \dot{\phi}$, which is the angular momentum about the axis of the cone.

12.19 Conserved momenta are $p_\theta = mr^2\dot{\theta}$ and $p_z = m\dot{z} - (e\mu_0 I/2\pi)\ln r$.

12.20 In terms of obvious coordinates, the conserved quantities are (i) the vector $\boldsymbol{L}$ (ii) P_x, P_y and L_z (iii) L_z (iv) L_z (v) P_z and L_z.

12.21 Conserved quantity is $L_z + cP_z$.

Chapter 13 Calculus of variations and Hamilton's principle (page 388)

13.1 $x = t^4 + 2$.

13.2 $x = \sin t$.

13.4

$$z = \frac{1}{k}\ln\left(\frac{\cos kx}{\cos ka}\right).$$

13.5

$$\rho = \frac{a\cos\left(\frac{1}{2}\pi\sin\alpha\right)}{\cos\left(\theta\sin\alpha\right)}.$$

13.6 $x = t\,(2t + X - 8)/4$.

Chapter 14 Hamilton's equations and phase space (page 413)

14.1 $G = -v_1^2 - 3v_1v_2 - 2v_2^2 + 6wv_1 + 9wv_2 - 9w^2$.

14.2 The Hamiltonian is $H = p_\theta^2/(2m(a^2+b^2)) + mgb\,\theta$ and the equations are $\dot{\theta} = p_\theta/(m(a^2+b^2))$, $\dot{p}_\theta = -mgb$.

14.3 The Hamiltonian is $H = (p_x^2 + p_z^2)/2m + mgz$ and the equations are $\dot{x} = p_x/m$, $\dot{z} = p_z/m$, $\dot{p}_x = 0$, $\dot{p}_z = -mg$. The coordinate x is cyclic.

14.4 The Hamiltonian is

$$H = \frac{p_\theta^2}{2ma^2} + \frac{p_\phi^2}{2ma^2\sin^2\theta} - mga\cos\theta$$

and the equations are

$$\dot{\theta} = \frac{p_\theta}{ma^2}, \quad \dot{\phi} = \frac{p_\phi}{ma^2\sin^2\theta}, \quad \dot{p}_\theta = \frac{p_\phi^2\cos\theta}{ma^2\sin^3\theta} - mga\sin\theta, \quad \dot{p}_\phi = 0.$$

14.5 The Hamiltonian is

$$H = \frac{a^2p_x^2 - 2a\cos\theta\, p_x p_\theta + 2p_\theta^2}{2ma^2\left(2 - \cos^2\theta\right)} - mga\cos\theta.$$

14.6 The Hamiltonian is

$$H = \frac{p_\theta^2}{2m(a-Z)^2} - \tfrac{1}{2}m\dot{Z}^2 - mg(a-Z)\cos\theta$$

and the equations are $\dot{\theta} = p_\theta/m(a-Z)^2$, $\dot{p}_\theta = -mg(a-Z)\sin\theta$. Since H has an explicit time dependence through $Z(t)$, it is not conserved.

14.7 The Hamiltonian H is conserved when the fields $\{\boldsymbol{E}, \boldsymbol{B}\}$ are static.

Chapter 15 General theory of small oscillations (page 452)

15.1 $\omega_1^2 = \alpha/6m$, $\omega_2^2 = 3\alpha/4m$, $\boldsymbol{a}_1 = (2, 3)$, $\boldsymbol{a}_2 = (4, -1)$. In the subsequent motion,

$$x = \frac{2u}{14\omega_1\omega_2}(\omega_2 \sin\omega_1 t + 6\omega_1 \sin\omega_2 t), \qquad y = \frac{3u}{14\omega_1\omega_2}(\omega_2 \sin\omega_1 t - \omega_1 \sin\omega_2 t).$$

15.2 $\omega_1^2 = \alpha/6m$, $\omega_2^2 = \alpha/m$, $\boldsymbol{a}_1 = (2, 3)$, $\boldsymbol{a}_2 = (1, -1)$. The variables $\eta_1 = x + y$, $\eta_2 = 3x - 2y$ are a set of normal coordinates.

15.3 $\omega_1^2 = g/2a$, $\omega_2^2 = 15g/2a$, $\boldsymbol{a}_1 = (5, 6)$, $\boldsymbol{a}_2 = (3, -2)$. The ratio $\tau_1/\tau_2 = \sqrt{15}$, which is irrational; so general small motion is *not* periodic.

15.4 $\omega_1^2 = g/2a$, $\omega_2^2 = 3g/2a$, $\omega_3^2 = 3g/a$, $\boldsymbol{a}_1 = (1, 2, 3)$, $\boldsymbol{a}_2 = (1, 0, -3)$, $\boldsymbol{a}_3 = (1, -3, 3)$. The variables $\eta_1 = 3\theta + 2\phi + \psi$, $\eta_2 = 3\theta - \psi$, $\eta_3 = 3\theta - 3\phi + \psi$ are a set of normal coordinates.

15.5 The longitudinal modes have frequencies $\omega_1^2 = \alpha/m$ and $\omega_2^2 = 3\alpha/m$, and the transverse modes have frequencies $\omega_3^2 = T_0/ma$ and $\omega_4^2 = 3T_0/ma$.

15.6 $\omega_1^2 = 4g/9a$, $\omega_2^2 = g/a$, $\omega_3^2 = 4g/a$. The general small motion has period $6\pi(a/g)^{1/2}$.

15.7 The longitudinal mode has frequency $\omega_1^2 = 11g/48a$, and the transverse modes have frequencies $\omega_2^2 = g/6a$ and $\omega_3^2 = 3g/4a$.

15.8 The transverse modes have frequencies $\omega_2^2 = 12T_0/5Ma$ and $\omega_3^2 = 18T_0/Ma$.

15.9 The j-th normal frequency is

$$\omega_j = 2\left(\frac{\alpha}{m}\right)^{1/2} \sin\left(\frac{j\pi}{2(n+1)}\right) \qquad (1 \le j \le n).$$

15.10 The j-th normal frequency is

$$\omega_j = 2\left(\frac{T_0}{ma}\right)^{1/2} \sin\left(\frac{j\pi}{2(n+1)}\right) \qquad (1 \le j \le n).$$

15.11 Vibrational frequencies are $\omega_1^2 = \alpha/m$, $\omega_2^2 = 6\alpha/m$. Estimated vibrational frequencies of carbon oxysulphide are 2230 cm^{-1} and 880 cm^{-1}.

15.12 The antisymmetric mode has frequency $\omega_3^2 = 7k/4m$, and the symmetric modes have frequencies $\omega_1^2 = k/4m$ and $\omega_2^2 = 2k/m$.

Chapter 16 Vector angular velocity and rigid body kinematics (page 467)

16.1 Angular velocity is $2\boldsymbol{k}$ radians per second, particle velocity is $6\boldsymbol{i} + 8\boldsymbol{j}\ \text{m s}^{-1}$, speed is $10\ \text{m s}^{-1}$, and acceleration is $-16\boldsymbol{i} + 12\boldsymbol{j}\ \text{m s}^{-2}$.

16.2 Angular velocity is $2\boldsymbol{i} + 2\boldsymbol{j} + \boldsymbol{k}$ radians per second, particle velocity is $3\boldsymbol{i} - 6\boldsymbol{j} + 6\boldsymbol{k}\ \text{cm s}^{-1}$, speed is $9\ \text{cm s}^{-1}$, and acceleration is $18\boldsymbol{i} - 9\boldsymbol{j} - 18\boldsymbol{k}\ \text{cm s}^{-2}$.

16.4 Angular velocity of penny is $\dot{\theta}\,(\boldsymbol{k} - \cos\alpha\, \boldsymbol{a})$. Velocity of highest particle is zero.

16.5 Highest particle has the greatest speed which is $2h\cos\alpha\,|\dot{\theta}\,|$.

Chapter 17 Rotating reference frames (page 489)

17.3 Angular speed is $\lambda \sec\alpha$.

17.4 $r = a\cosh\Omega t$.

17.5 Take the angular velocity of the rotating frame to be $\boldsymbol{\Omega} = -(eB/2mc)\boldsymbol{k}$. The two frequencies are $\omega_0 \pm (eB/2mc)$ approximately.

17.10 $r = a\cosh\left(\sqrt{\tfrac{2}{3}}\,\Omega t\right)$.

17.11 Speed of cylinder is $\Omega x/\sqrt{2}$. Turntable exerts force $-\frac{1}{2}M\Omega^2 x\,\boldsymbol{i} + \sqrt{2}M\Omega^2 x\,\boldsymbol{j} + Mg\,\boldsymbol{k}$ on the cylinder, where $\boldsymbol{i}$ points in the direction of motion and $\boldsymbol{k}$ points vertically upwards.

17.12 In cylindrical coordinates with Oz along the rotation axis of the bucket, the isobars are the family of surfaces $z = (\Omega^2/2g)R^2 + c$, where c is a constant. The free surface of the water must be one of these, whatever the shape of the container.

17.13 In cylindrical coordinates, the water occupies the region $\frac{1}{2}a \le R \le a$. The pressure at the wall is $p_0 + \frac{3}{8}\rho\Omega^2 a^2$.

Chapter 18 Tensor algebra and the inertia tensor (page 519)

18.1 C and C' have opposite handedness.

18.2 The transformation matrices are

$$\text{(i)}\ \frac{1}{\sqrt{2}}\begin{pmatrix}1 & 0 & -1\\ 0 & \sqrt{2} & 0\\ 1 & 0 & 1\end{pmatrix}\quad \text{(ii)}\ \begin{pmatrix}1 & 0 & 0\\ 0 & -1 & 0\\ 0 & 0 & 1\end{pmatrix}\quad \text{(iii)}\ \frac{1}{9}\begin{pmatrix}4 & 7 & -4\\ 1 & 4 & 8\\ 8 & -4 & 1\end{pmatrix}\quad \text{(iv)}\ \frac{1}{9}\begin{pmatrix}1 & 4 & -8\\ 4 & 7 & 4\\ -8 & 4 & 1\end{pmatrix}$$

The new coordinates of the point D are $(3/\sqrt{2},\ -3,\ 3/\sqrt{2})$, $(3,\ 3,\ 0)$, $(-1, -1,\ 4)$, $(-1, -1, -4)$ respectively.

18.3 Solution for $\mathbf{v}$ is $(1, 1, -1)'$ or any multiple of it. The rotation angle about OE is $\pi/3$.

18.4 The transformation formula for a fifth order tensor is

$$t'_{ijklm} = \sum_{p=1}^{3}\sum_{q=1}^{3}\sum_{r=1}^{3}\sum_{s=1}^{3}\sum_{t=1}^{3} a_{ip}\,a_{jq}\,a_{kr}\,a_{ls}\,a_{mt}\,t_{pqrst}$$

18.5 The matrices representing the tensor are

$$\text{(i)}\ \frac{1}{\sqrt{2}}\begin{pmatrix}\sqrt{2} & 1 & 1\\ 1 & \sqrt{2} & 0\\ 1 & 0 & \sqrt{2}\end{pmatrix}\qquad \text{(ii)}\ \begin{pmatrix}1 & 0 & -1\\ 0 & 1 & 0\\ -1 & 0 & 1\end{pmatrix}.$$

18.6 (i) Yes, order 7, (ii) Yes, order 6, (iii) Yes, order 1 (a vector), (iv) Yes, order 1 (a vector), (v) No, (vi) Yes, order 5, (vii) Yes, order zero (a scalar), (viii) No, (ix) Yes, order zero (a scalar).

18.9 The most general material has 21 elastic moduli.

18.10 Moment of inertia of the plate about a diagonal is $\frac{2}{3}Ma^2b^2/(a^2+b^2)$.

18.11 Principal moments of inertia of the disk are (i) $\frac{1}{4}Ma^2$, $\frac{1}{4}Ma^2$, $\frac{1}{2}Ma^2$, (ii) $\frac{1}{4}Ma^2$, $\frac{5}{4}Ma^2$, $\frac{3}{2}Ma^2$.

18.12 Principal moments of inertia of the top at A are $\frac{5}{4}Ma^2$, $\frac{5}{4}Ma^2$, $\frac{1}{2}Ma^2$.

18.13 Principal moments of inertia of the top at A are $\frac{43}{20}Ma^2$, $\frac{43}{20}Ma^2$, $\frac{2}{5}Ma^2$.

18.14 Principal moments of inertia of the cube are (i) $\frac{2}{3}Ma^2$, $\frac{2}{3}Ma^2$, $\frac{2}{3}Ma^2$ (ii) $\frac{5}{3}Ma^2$, $\frac{5}{3}Ma^2$, $\frac{2}{3}Ma^2$ (iii) $\frac{11}{3}Ma^2$, $\frac{11}{3}Ma^2$, $\frac{2}{3}Ma^2$. The required moments of inertia are (i) $\frac{2}{3}Ma^2$, (ii) $\frac{5}{3}Ma^2$, (iii) $\frac{8}{3}Ma^2$.

18.15 Principal moments of inertia are (i) $\frac{1}{3}M(b^2+c^2)$, $\frac{1}{3}M(a^2+c^2)$, $\frac{1}{3}M(a^2+b^2)$ (ii) $\frac{1}{3}M(b^2+4c^2)$, $\frac{1}{3}M(a^2+4c^2)$, $\frac{1}{3}M(a^2+b^2)$. The required moments of inertia are

$$\text{(i)}\quad \frac{2M(a^2b^2+a^2c^2+b^2c^2)}{3(a^2+b^2+c^2)} \qquad \text{(ii)}\quad \frac{2M(a^2b^2+2a^2c^2+2b^2c^2)}{3(a^2+b^2)}$$

18.16 Principal moments of inertia at G are $\frac{1}{4}Ma^2+\frac{1}{3}Mb^2$, $\frac{1}{4}Ma^2+\frac{1}{3}Mb^2$, $\frac{1}{2}Ma^2$. Cylinder is dynamically spherical when $b=\sqrt{3}\,a/2$.

18.17 Symmetries are (i) axial, (ii) none, (iii) none, (iv) axial, (v) axial, (vi) spherical, (vii) axial, (viii) axial, (ix) spherical.

18.18 Principal moments of inertia at a corner point of the plate are $\frac{2}{3}(5+3\sqrt{2})Ma^2$, $\frac{2}{3}(5-3\sqrt{2})Ma^2$, $\frac{20}{3}Ma^2$. A set of principal axes is obtained by rotating the axes $Cx_1x_2x_3$ shown in Figure 18.3 through an angle of $-\pi/8$ about the axis Cx_3.

Chapter 19 Problems in rigid body dynamics (page 560)

19.2 Radius of circle is $\frac{7}{2}b$ and centre is a distance $\frac{5}{2}b$ from O.

19.3 Ball is stable at highest point if $n^2 > 35(a+b)g/a^2$.

19.4 Precession rate is Mgh/Cn.

19.6 For upright stability, the spin would have to be at least 3860 revolutions per second!

19.7 Spin needed for stability is (i) 9.3 and (ii) 6.2 revolutions per second. The juggler should therefore use a hollow ball. (It's lighter in any case!)

19.11 The required moment is $Ma^2\Omega n$ about the horizontal axis *perpendicular* to the symmetry axis of the wheel.

19.14 The motion of the axis is approximately periodic with period π/Λ.

Bibliography

This bibliography is not a survey of the mechanics literature. The books that appear below are included simply because they are referred to at some point in the text. Thus, no inference should be drawn regarding books that do not appear.

Classical mechanics

[1] Acheson, D., *From Calculus to Chaos: An Introduction to Dynamics*, Oxford University Press, 1997.
Not a mechanics textbook as such, but a selection of topics leading, as the title says, from calculus to chaos. Lots of interesting problems and photographs, including one of the Indian Rope Trick!

[2] Arnold, V.I., *Mathematical Methods of Classical Mechanics*, Springer-Verlag, 1978.
An elegant exposition of classical mechanics expressed in the language of differential topology. A classic, but you need to know classical mechanics *and* differential topology before reading it.

[3] French, A.P., *Newtonian Mechanics*, W.W. Norton & Co., 1971.
A leisurely introduction to classical mechanics from the Newtonian standpoint. Very good on the foundations and the background physics.

[4] Goldstein, H., Poole, C. & Safko, J., *Classical Mechanics*, third edition, Addison-Wesley, 2002.
Goldstein's classic graduate-level text, skilfully revised and extended by Poole & Safko. This is the first place to look for material not included in this book.

[5] Kaplan, M.H., *Modern Spacecraft Dynamics and Control*, John Wiley & Sons, 1976.
A good source for applications of classical mechanics to spacecraft orbits and manoeuvres.

[6] Landau, L.D. & Lifshitz, E.M., *Mechanics (Course of Theoretical Physics, Volume I)*, third edition, Butterworth-Heinemann, 1995.
Volume I of this famous *Course of Theoretical Physics* is a terse account of analytical mechanics. A classic, but not easy to follow.

Mathematics

[7] Anton, H., *Elementary Linear Algebra*, John Wiley & Sons, 2005.
An enduring introductory text on linear algebra. It has a good chapter on eigenvalues and eigenvectors, but does not *quite* prove the orthogonal diagonalisation theorem for symmetric matrices. For a full proof see Lang [10].

[8] Boyce, W. E. & DiPrima, R. C., *Elementary Differential Equations and Boundary Value Problems*, eighth edition, John Wiley & Sons, 2004.
One of several good textbooks on the *solution* of ODEs. It contains all the solution methods for ODEs that are used in this book.

[9] Coddington, E.A. & Levinson, N. *Theory of Ordinary Differential Equations*, McGraw-Hill, 1984.
The ultimate authority on the *theory* of ODEs. In particular, it has a correct proof of the Poincaré-Bendixson theorem in all its cases. The authors do everything as it should be done, but they take no prisoners.

[10] Lang, S., *Linear Algebra*, Springer-Verlag, 1993.
Contains a proof of the important result that *any* real symmetric matrix has a full set of eigenvectors (the orthogonal diagonalisation theorem).

[11] Schey, H.M., *Div, Grad, Curl and All That: An Informal Text on Vector Calculus*, W.W. Norton & Co., 1996.
A very readable book on vector calculus.

[12] Walters, P. *An Introduction to Ergodic Theory*, Springer-Verlag, 2000.
Despite its title, this is a graduate text on ergodic theory. In particular, it develops the theory of Poincaré recurrence further than we were able to do in this book.

Physics and chemistry

[13] Herzberg, G., *Molecular Spectra and Molecular Structure, Volume II: Infrared and Raman Spectra of Polyatomic Molecules*, Krieger Publishing Co., 1990.
This is the second part of Herzberg's epic trilogy on molecular spectra. It contains the classical theory of molecular vibrations, including group theoretic methods, and also has detailed analyses of the vibrational spectra of many common polyatomic molecules.

[14] Kaye, E.W.C., & Laby, T.H., *Tables of Physical and Chemical Constants*, Longman, 1986.
The title says it all.

Index

A page number printed in **bold** indicates that a definition or an important statement about the index entry appears on that page.

More information about this series at http://www.springer.com/series/7410

Liqun Chen · Ninghui Li ·
Kaitai Liang · Steve Schneider (Eds.)

Computer Security – ESORICS 2020

25th European Symposium on Research in Computer Security, ESORICS 2020
Guildford, UK, September 14–18, 2020
Proceedings, Part I

Springer

Editors
Liqun Chen
University of Surrey
Guildford, UK

Ninghui Li
Purdue University
West Lafayette, IN, USA

Kaitai Liang
Delft University of Technology
Delft, The Netherlands

Steve Schneider
University of Surrey
Guildford, UK

ISSN 0302-9743 ISSN 1611-3349 (electronic)
Lecture Notes in Computer Science
ISBN 978-3-030-58950-9 ISBN 978-3-030-58951-6 (eBook)
https://doi.org/10.1007/978-3-030-58951-6

LNCS Sublibrary: SL4 – Security and Cryptology

This Springer imprint is published by the registered company Springer Nature Switzerland AG
The registered company address is: Gewerbestrasse 11, 6330 Cham, Switzerland

Preface

The two volume set, LNCS 12308 and 12309, contain the papers that were selected for presentation and publication at the 25th European Symposium on Research in Computer Security (ESORICS 2020) which was held together with affiliated workshops during the week September 14–18, 2020. Due to the global COVID-19 pandemic, the conference and workshops ran virtually, hosted by the University of Surrey, UK. The aim of ESORICS is to further research in computer security and privacy by establishing a European forum, bringing together researchers in these areas by promoting the exchange of ideas with system developers and by encouraging links with researchers in related fields.

In response to the call for papers, 366 papers were submitted to the conference. These papers were evaluated on the basis of their significance, novelty, and technical quality. Except for a very small number of papers, each paper was carefully evaluated by three to five referees and then discussed among the Program Committee. The papers were reviewed in a single-blind manner. Finally, 72 papers were selected for presentation at the conference, yielding an acceptance rate of 19.7%. We were also delighted to welcome invited talks from Aggelos Kiayias, Vadim Lyubashevsky, and Rebecca Wright.

Following the reviews two papers were selected for Best Paper Awards and they share the 1,000 EUR prize generously provided by Springer: "Pine: Enabling privacy-preserving deep packet inspection on TLS with rule-hiding and fast connection establishment" by Jianting Ning, Xinyi Huang, Geong Sen Poh, Shengmin Xu, Jason Loh, Jian Weng, and Robert H. Deng; and "Automatic generation of source lemmas in Tamarin: towards automatic proofs of security protocols" by Véronique Cortier, Stéphanie Delaune, and Jannik Dreier.

The Program Committee consisted of 127 members across 25 countries. There were submissions from a total of 1,201 authors across 42 countries, with 24 countries represented among the accepted papers.

ESORICS 2020 would not have been possible without the contributions of the many volunteers who freely gave their time and expertise. We would like to thank the members of the Program Committee and the external reviewers for their substantial work in evaluating the papers. We would also like to thank the organization/department chair, Helen Treharne, the workshop chair, Mark Manulis, and all of the workshop co-chairs, the poster chair, Ioana Boureanu, and the ESORICS Steering Committee. We are also grateful to Huawei and IBM Research – Haifa, Israel for their sponsorship that enabled us to support this online event. Finally, we would like to express our thanks to the authors who submitted papers to ESORICS 2020. They, more than anyone else, are what made this conference possible.

We hope that you will find the proceedings stimulating and a source of inspiration for future research.

September 2020

Liqun Chen
Ninghui Li
Kaitai Liang
Steve Schneider

Organization

General Chair

Steve Schneider	University of Surrey, UK

Program Chairs

Liqun Chen	University of Surrey, UK
Ninghui Li	Purdue University, USA

Steering Committee

Sokratis Katsikas (Chair)
Michael Backes
Joachim Biskup
Frederic Cuppens
Sabrina De Capitani di Vimercati
Dieter Gollmann
Mirek Kutylowski
Javier Lopez
Jean-Jacques Quisquater
Peter Y. A. Ryan
Pierangela Samarati
Einar Snekkenes
Michael Waidner

Program Committee

Yousra Aafer	University of Waterloo, Canada
Mitsuaki Akiyama	NTT, Japan
Cristina Alcaraz	UMA, Spain
Frederik Armknecht	Universität Mannheim, Germany
Vijay Atluri	Rutgers University, USA
Erman Ayday	Bilkent University, Turkey
Antonio Bianchi	Purdue University, USA
Marina Blanton	University at Buffalo, USA
Carlo Blundo	Università degli Studi di Salerno, Italy
Alvaro Cardenas	The University of Texas at Dallas, USA
Berkay Celik	Purdue University, USA
Aldar C-F. Chan	BIS Innovation Hub Centre, Hong Kong, China
Sze Yiu Chau	Purdue University, USA

Rongmao Chen	National University of Defense Technology, China
Yu Chen	Shandong University, China
Sherman S. M. Chow	The Chinese University of Hong Kong, Hong Kong, China
Mauro Conti	University of Padua, Italy
Frédéric Cuppens	Polytechnique Montreal, Canada
Nora Cuppens-Boulahia	Polytechnique Montréal, Canada
Marc Dacier	Qatar Computing Research Institute (QCRI), Qatar
Sabrina De Capitani di Vimercati	Università degli Studi di Milano, Italy
Hervé Debar	Télécom SudParis, France
Stéphanie Delaune	University of Rennes, CNRS, IRISA, France
Roberto Di Pietro	Hamad Bin Khalifa University, Qatar
Tassos Dimitriou	Kuwait University, Kuwait
Josep Domingo-Ferrer	Universitat Rovira i Virgili, Spain
Changyu Dong	Newcastle University, UK
Wenliang Du	Syracuse University, Italy
Haixin Duan	Tsinghua University, China
François Dupressoir	University of Bristol, UK
Kassem Fawaz	University of Wisconsin-Madison, USA
Jose-Luis Ferrer-Gomila	University of the Balearic Islands, Spain
Sara Foresti	DI, Università degli Studi di Milano, Italy
David Galindo	University of Birmingham, UK
Debin Gao	Singapore Management University, Singapore
Joaquin Garcia-Alfaro	Télécom SudParis, France
Thanassis Giannetsos	Technical University of Denmark, Denmark
Dieter Gollmann	Hamburg University of Technology, Germany
Stefanos Gritzalis	University of the Aegean, Greece
Guofei Gu	Texas A&M University, USA
Zhongshu Gu	IBM Research, USA
Jinguang Han	Queen's University Belfast, UK
Feng Hao	University of Warwick, UK
Juan Hernández-Serrano	Universitat Politècnica de Catalunya, Spain
Xinyi Huang	Fujian Normal University, China
Syed Hussain	Purdue University, USA
Shouling Ji	Zhejiang University, China
Ghassan Karame	NEC Laboratories Europe, Germany
Sokratis Katsikas	Norwegian University of Science and Technology, Norway
Stefan Katzenbeisser	TU Darmstadt, Germany
Ryan Ko	The University of Queensland, Australia
Steve Kremer	Inria, France
Marina Krotofil	FireEye, USA
Yonghwi Kwon	University of Virginia, USA
Costas Lambrinoudakis	University of Piraeus, Greece
Kyu Hyung Lee	University of Georgia, USA

Shujun Li	University of Kent, UK
Yingjiu Li	Singapore Management University, Singapore
Kaitai Liang	Delft University of Technology, The Netherlands
Hoon Wei Lim	Trustwave, Singapore
Joseph Liu	Monash University, Australia
Rongxing Lu	University of New Brunswick, Canada
Xiapu Luo	The Hong Kong Polytechnic University, Hong Kong, China
Shiqing Ma	Rutgers University, USA
Leandros Maglaras	De Montfort University, UK
Mark Manulis	University of Surrey, UK
Konstantinos Markantonakis	Royal Holloway, University of London, UK
Fabio Martinelli	IIT-CNR, Italy
Ivan Martinovic	University of Oxford, UK
Sjouke Mauw	University of Luxembourg, Luxembourg
Catherine Meadows	NRL, USA
Weizhi Meng	Technical University of Denmark, Denmark
Chris Mitchell	Royal Holloway, University of London, UK
Tatsuya Mori	Waseda University, Japan
Haralambos Mouratidis	University of Brighton, UK
David Naccache	Ecole normale supérieur, France
Siaw-Lynn Ng	Royal Holloway, University of London, UK
Jianting Ning	Singapore Management University, Singapore
Satoshi Obana	Hosei University, Japan
Martín Ochoa	Universidad del Rosario, Colombia
Rolf Oppliger	eSECURITY Technologies, Switzerland
Manos Panousis	University of Greenwich, UK
Olivier Pereira	UCLouvain, Belgium
Günther Pernul	Universität Regensburg, Germany
Joachim Posegga	University of Passau, Germany
Indrajit Ray	Colorado State University, USA
Kui Ren	Zhejiang University, China
Giovanni Russello	The University of Auckland, New Zealand
Mark Ryan	University of Birmingham, UK
Reihaneh Safavi-Naini	University of Calgary, Canada
Brendan Saltaformaggio	Georgia Institute of Technology, USA
Pierangela Samarati	Università degli Studi di Milano, Italy
Damien Sauveron	XLIM, UMR University of Limoges, CNRS 7252, France
Einar Snekkenes	Norwegian University of Science and Technology, Norway
Yixin Sun	University of Virginia, USA
Willy Susilo	University of Wollongong, Australia

Pawel Szalachowski	SUTD, Singapore
Qiang Tang	Luxembourg Institute of Science and Technology, Luxembourg
Qiang Tang	New Jersey Institute of Technology, USA
Juan Tapiador	Universidad Carlos III de Madrid, Spain
Dave Jing Tian	Purdue University, USA
Nils Ole Tippenhauer	CISPA, Germany
Helen Treharne	University of Surrey, UK
Aggeliki Tsohou	Ionian University, Greece
Luca Viganò	King's College London, UK
Michael Waidner	Fraunhofer, Germany
Cong Wang	City University of Hong Kong, Hong Kong, China
Lingyu Wang	Concordia University, Canada
Weihang Wang	SUNY University at Buffalo, USA
Edgar Weippl	SBA Research, Austria
Christos Xenakis	University of Piraeus, Greece
Yang Xiang	Swinburne University of Technology, Australia
Guomin Yang	University of Wollongong, Australia
Kang Yang	State Key Laboratory of Cryptology, China
Xun Yi	RMIT University, Australia
Yu Yu	Shanghai Jiao Tong University, China
Tsz Hon Yuen	The University of Hong Kong, Hong Kong, China
Fengwei Zhang	SUSTech, China
Kehuan Zhang	The Chinese University of Hong Kong, Hong Kong, China
Yang Zhang	CISPA Helmholtz Center for Information Security, Germany
Yuan Zhang	Fudan University, China
Zhenfeng Zhang	Chinese Academy of Sciences, China
Yunlei Zhao	Fudan University, China
Jianying Zhou	Singapore University of Technology and Design, Singapore
Sencun Zhu	Penn State University, USA

Workshop Chair

Mark Manulis	University of Surrey, UK

Poster Chair

Ioana Boureanu	University of Surrey, UK

Organization/Department Chair

Helen Treharne	University of Surrey, UK

Organizing Chair and Publicity Chair

Kaitai Liang Delft University of Technology, The Netherlands

Additional Reviewers

Abbasi, Ali
Abu-Salma, Ruba
Ahlawat, Amit
Ahmed, Chuadhry Mujeeb
Ahmed, Shimaa
Alabdulatif, Abdulatif
Alhanahnah, Mohannad
Aliyu, Aliyu
Alrizah, Mshabab
Anceaume, Emmanuelle
Angelogianni, Anna
Anglés-Tafalla, Carles
Aparicio Navarro, Francisco Javier
Argyriou, Antonios
Asadujjaman, A. S. M.
Aschermann, Cornelius
Asghar, Muhammad Rizwan
Avizheh, Sepideh
Baccarini, Alessandro
Bacis, Enrico
Baek, Joonsang
Bai, Weihao
Bamiloshin, Michael
Barenghi, Alessandro
Barrère, Martín
Berger, Christian
Bhattacherjee, Sanjay
Blanco-Justicia, Alberto
Blazy, Olivier
Bolgouras, Vaios
Bountakas, Panagiotis
Brandt, Markus
Bursuc, Sergiu
Böhm, Fabian
Camacho, Philippe
Cardaioli, Matteo
Castelblanco, Alejandra
Castellanos, John Henry
Cecconello, Stefano
Chaidos, Pyrros
Chakra, Ranim
Chandrasekaran, Varun
Chen, Haixia
Chen, Long
Chen, Min
Chen, Zhao
Chen, Zhigang
Chengjun Lin
Ciampi, Michele
Cicala, Fabrizio
Costantino, Gianpiero
Cruz, Tiago
Cui, Shujie
Deng, Yi
Diamantopoulou, Vasiliki
Dietz, Marietheres
Divakaran, Dinil Mon
Dong, Naipeng
Dong, Shuaike
Dragan, Constantin Catalin
Du, Minxin
Dutta, Sabyasachi
Eichhammer, Philipp
Englbrecht, Ludwig
Etigowni, Sriharsha
Farao, Aristeidis
Faruq, Fatma
Fdhila, Walid
Feng, Hanwen
Feng, Qi
Fentham, Daniel
Ferreira Torres, Christof
Fila, Barbara
Fraser, Ashley
Fu, Hao
Galdi, Clemente
Gangwal, Ankit
Gao, Wei

Gardham, Daniel
Garms, Lydia
Ge, Chunpeng
Ge, Huangyi
Geneiatakis, Dimitris
Genés-Durán, Rafael
Georgiopoulou, Zafeiroula
Getahun Chekole, Eyasu
Ghosal, Amrita
Giamouridis, George
Giorgi, Giacomo
Guan, Qingxiao
Guo, Hui
Guo, Kaiwen
Guo, Yimin
Gusenbauer, Mathias
Haffar, Rami
Hahn, Florian
Han, Yufei
Hausmann, Christian
He, Shuangyu
He, Songlin
He, Ying
Heftrig, Elias
Hirschi, Lucca
Hu, Kexin
Huang, Qiong
Hurley-Smith, Darren
Iadarola, Giacomo
Jeitner, Philipp
Jia, Dingding
Jia, Yaoqi
Judmayer, Aljosha
Kalloniatis, Christos
Kantzavelou, Ioanna
Kasinathan, Prabhakaran
Kasra Kermanshahi, Shabnam
Kasra, Shabnam
Kelarev, Andrei
Khandpur Singh, Ashneet
Kim, Jongkil
Koay, Abigail
Kokolakis, Spyros
Kosmanos, Dimitrios
Kourai, Kenichi
Koutroumpouchos, Konstantinos
Koutroumpouchos, Nikolaos
Koutsos, Adrien
Kuchta, Veronika
Labani, Hasan
Lai, Jianchang
Laing, Thalia May
Lakshmanan, Sudershan
Lallemand, Joseph
Lan, Xiao
Lavranou, Rena
Lee, Jehyun
León, Olga
Li, Jie
Li, Juanru
Li, Shuaigang
Li, Wenjuan
Li, Xinyu
Li, Yannan
Li, Zengpeng
Li, Zheng
Li, Ziyi
Limniotis, Konstantinos
Lin, Chao
Lin, Yan
Liu, Jia
Liu, Jian
Liu, Weiran
Liu, Xiaoning
Liu, Xueqiao
Liu, Zhen
Lopez, Christian
Losiouk, Eleonora
Lu, Yuan
Luo, Junwei
Ma, Haoyu
Ma, Hui
Ma, Jack P. K.
Ma, Jinhua
Ma, Mimi
Ma, Xuecheng
Mai, Alexandra
Majumdar, Suryadipta
Manjón, Jesús A.
Marson, Giorgia Azzurra
Martinez, Sergio
Matousek, Petr

Mercaldo, Francesco
Michailidou, Christina
Mitropoulos, Dimitris
Mohammadi, Farnaz
Mohammady, Meisam
Mohammed, Ameer
Moreira, Jose
Muñoz, Jose L.
Mykoniati, Maria
Nassirzadeh, Behkish
Newton, Christopher
Ng, Lucien K. L.
Ntantogian, Christoforos
Önen, Melek
Onete, Cristina
Oqaily, Alaa
Oswald, David
Papaioannou, Thanos
Parkinson, Simon
Paspatis, Ioannis
Patsakis, Constantinos
Pelosi, Gerardo
Pfeffer, Katharina
Pitropakis, Nikolaos
Poettering, Bertram
Poh, Geong Sen
Polato, Mirko
Poostindouz, Alireza
Puchta, Alexander
Putz, Benedikt
Pöhls, Henrich C.
Qiu, Tian
Radomirovic, Sasa
Rakotonirina, Itsaka
Rebollo Monedero, David
Rivera, Esteban
Rizomiliotis, Panagiotis
Román-García, Fernando
Sachidananda, Vinay
Salazar, Luis
Salem, Ahmed
Salman, Ammar
Sanders, Olivier
Scarsbrook, Joshua
Schindler, Philipp
Schlette, Daniel
Schmidt, Carsten
Scotti, Fabio
Shahandashti, Siamak
Shahraki, Ahmad Salehi
Sharifian, Setareh
Sharma, Vishal
Sheikhalishahi, Mina
Shen, Siyu
Shrishak, Kris
Simo, Hervais
Siniscalchi, Luisa
Slamanig, Daniel
Smith, Zach
Solano, Jesús
Song, Yongcheng
Song, Zirui
Soriente, Claudio
Soumelidou, Katerina
Spielvogel, Korbinian
Stifter, Nicholas
Sun, Menghan
Sun, Yiwei
Sun, Yuanyi
Tabiban, Azadeh
Tang, Di
Tang, Guofeng
Taubmann, Benjamin
Tengana, Lizzy
Tian, Yangguang
Trujillo, Rolando
Turrin, Federico
Veroni, Eleni
Vielberth, Manfred
Vollmer, Marcel
Wang, Jiafan
Wang, Qin
Wang, Tianhao
Wang, Wei
Wang, Wenhao
Wang, Yangde
Wang, Yi
Wang, Yuling
Wang, Ziyuan
Weitkämper, Charlotte
Wesemeyer, Stephan
Whitefield, Jorden

Wiyaja, Dimaz
Wong, Donald P. H.
Wong, Harry W. H.
Wong, Jin-Mann
Wu, Chen
Wu, Ge
Wu, Lei
Wuest, Karl
Xie, Guoyang
Xinlei, He
Xu, Fenghao
Xu, Jia
Xu, Jiayun
Xu, Ke
Xu, Shengmin
Xu, Yanhong
Xue, Minhui
Yamada, Shota
Yang, Bohan
Yang, Lin
Yang, Rupeng
Yang, S. J.
Yang, Wenjie
Yang, Xu
Yang, Xuechao
Yang, Zhichao
Yevseyeva, Iryna
Yi, Ping
Yin, Lingyuan
Ying, Jason
Yu, Zuoxia
Yuan, Lun-Pin
Yuan, Xingliang
Zhang, Bingsheng
Zhang, Fan
Zhang, Ke
Zhang, Mengyuan
Zhang, Yanjun
Zhang, Zhikun
Zhang, Zongyang
Zhao, Yongjun
Zhong, Zhiqiang
Zhou, Yutong
Zhu, Fei
Ziaur, Rahman
Zobernig, Lukas
Zuo, Cong

Keynotes

Decentralising Information and Communications Technology: Paradigm Shift or Cypherpunk Reverie?

Aggelos Kiayias

University of Edinburgh and IOHK, UK

Abstract. In the last decade, decentralisation emerged as a much anticipated development in the greater space of information and communications technology. Venerated by some and disparaged by others, blockchain technology became a familiar term, springing up in a wide array of expected and some times unexpected contexts. With the peak of the hype behind us, in this talk I look back, distilling what have we learned about the science and engineering of building secure and reliable systems, then I overview the present state of the art and finally I delve into the future, appraising this technology in its potential to impact the way we design and deploy information and communications technology services.

Lattices and Zero-Knowledge

Vadim Lyubashevsky

IBM Research - Zurich, Switzerland

Abstract. Building cryptography based on the presumed hardness of lattice problems over polynomial rings is one of the most promising approaches for achieving security against quantum attackers. One of the reasons for the popularity of lattice-based encryption and signatures in the ongoing NIST standardization process is that they are significantly faster than all other post-quantum, and even many classical, schemes. This talk will discuss the progress in constructions of more advanced lattice-based cryptographic primitives. In particular, I will describe recent work on zero-knowledge proofs which leads to the most efficient post-quantum constructions for certain statements.

Accountability in Computing

Rebecca N. Wright

Barnard College, New York, USA

Abstract. Accountability is used often in describing computer-security mechanisms that complement preventive security, but it lacks a precise, agreed-upon definition. We argue for the need for accountability in computing in a variety of settings, and categorize some of the many ways in which this term is used. We identify a temporal spectrum onto which we may place different notions of accountability to facilitate their comparison, including prevention, detection, evidence, judgment, and punishment. We formalize our view in a utility-theoretic way and then use this to reason about accountability in computing systems. We also survey mechanisms providing various senses of accountability as well as other approaches to reasoning about accountability-related properties.

This is joint work with Joan Feigenbaum and Aaron Jaggard.

Contents – Part I

Software Security

Machine Learning Security

Network Security II

Privacy

Contents – Part II

Applied Cryptography III

Blockchain II

Database and Web Security

Pine: Enabling Privacy-Preserving Deep Packet Inspection on TLS with Rule-Hiding and Fast Connection Establishment

Jianting Ning[1,4], Xinyi Huang[1(✉)], Geong Sen Poh[2], Shengmin Xu[1], Jia-Chng Loh[2], Jian Weng[3], and Robert H. Deng[4]

[1] Fujian Provincial Key Laboratory of Network Security and Cryptology, College of Mathematics and Informatics, Fujian Normal University, Fuzhou, China
jtning88@gmail.com, xyhuang81@gmail.com, smxu1989@gmail.com

[2] NUS-Singtel Cyber Security Lab, Singapore, Singapore
pohgs@comp.nus.edu.sg, dcsljc@nus.edu.sg

[3] College of Information Science and Technology, Jinan University, Guangzhou, China
cryptjweng@gmail.com

[4] School of Information Systems, Singapore Management University, Singapore, Singapore
robertdeng@smu.edu.sg

Abstract. Transport Layer Security Inspection (TLSI) enables enterprises to decrypt, inspect and then re-encrypt users' traffic before it is routed to the destination. This breaks the end-to-end security guarantee of the TLS specification and implementation. It also raises privacy concerns since users' traffic is now known by the enterprises, and third-party middlebox providers providing the inspection services may additionally learn the inspection or attack rules, policies of the enterprises. Two recent works, BlindBox (SIGCOMM 2015) and PrivDPI (CCS 2019) propose privacy-preserving approaches that inspect encrypted traffic directly to address the privacy concern of users' traffic. However, BlindBox incurs high preprocessing overhead during TLS connection establishment, and while PrivDPI reduces the overhead substantially, it is still notable compared to that of TLSI. Furthermore, the underlying assumption in both approaches is that the middlebox knows the rule sets. Nevertheless, with the services increasingly migrating to third-party cloud-based setting, rule privacy should be preserved. Also, both approaches are static in nature in the sense that addition of any rules requires significant amount of preprocessing and re-instantiation of the protocols.

In this paper we propose Pine, a new Privacy-preserving inspection of encrypted traffic protocol that (1) simplifies the preprocessing step of PrivDPI thus further reduces the computation time and communication overhead of establishing the TLS connection between a user and a server; (2) supports *rule hiding*; and (3) enables dynamic rule addition without the need to re-execute the protocol from scratch. We demonstrate the

L. Chen et al. (Eds.): ESORICS 2020, LNCS 12308, pp. 3–22, 2020.
https://doi.org/10.1007/978-3-030-58951-6_1

superior performance of Pine when compared to PrivDPI through extensive experimentations. In particular, for a connection from a client to a server with 5,000 tokens and 6,000 rules, Pine is approximately 27% faster and saves approximately 92.3% communication cost.

Keywords: Network privacy · Traffic inspection · Encrypted traffic

1 Introduction

According to the recent Internet trends report [11], 87% of today's web traffic was encrypted, compared to 53% in 2016. Similarly, over 94% of web traffic across Google uses HTTPS encryption [7]. The increasing use of end-to-end encryption to secure web traffic has hampered the ability of existing middleboxes to detect malicious packets via deep packet inspection on the traffic. As a result, security service providers and enterprises deploy tools that perform Man-in-the-Middle (MitM) to decrypt, inspect and re-encrypt traffic before the traffic is sent to the designated server. Such approach is termed as Transport Layer Security Inspection (TLSI) by the National Security Agency (NSA), which recently issued an advisory on TLSI [12] citing potential security issues including insider threats. TLSI introduces additional risks whereby administrators may abuse their authorities to obtain sensitive information from the decrypted traffic. On the other hand, there exists growing privacy concern on the access to users' data by middleboxes as well as the enterprise gateways. According to a recent survey on TLSI in the US [16], more than 70% of the participants are concerned that middleboxes (or TLS proxies) performing TLSI can be exploited by hackers or used by governments, and close to 50% think it is an invasion to privacy. In general, participants are acceptable to the use of middleboxes by their employers or universities for security purposes but also want assurance that these would not be used by governments for surveillance or by exploited hackers.

To alleviate the above concerns on maintaining security of TLS while ensuring privacy of the encrypted traffic, Sherry *et al.* [20] introduced a solution called BlindBox to perform inspection on encrypted traffic directly. However, BlindBox needs a setup phase that is executed between the middlebox and the client. The setup phase performs two-party computation where the input of the middlebox are the rules, which means that the privacy of rules against the middlebox is not assured. In addition, this setup phase is built based on garbled circuit, and needs to be executed for every session. Due to the properties of garble circuit, such setup phase incurs significant computation and communication overheads. To overcome this limitation, Ning *et al.* [15] recently proposed PrivDPI with an improved setup phase. A new obfuscated rule generation technique was introduced, which enables the reuse of intermediate values generated during the first TLS session across subsequent sessions. This greatly reduces the computation and communication overheads over a series of sessions. However, there still exists considerable delay during the establishment of a TLS connection since each client is required to run a preprocessing protocol for each new connection. In addition, as we will show in Sect. 4.1, when the domain of the inspection or attack rules is small, the middlebox could perform brute force guessing for the rules in the setting of PrivDPI. This means that, as in BlindBox, PrivDPI does not provide privacy of rules against the middlebox. However, as noted in [20], most solution providers, such as McAfee, rely on the privacy of their rules in their business model. More so given the increasingly popular cloud-based middlebox services, the privacy of the rules should be preserved against the middleboxes.

Given the security and privacy concerns on TLSI, and the current status of the state-of-the-arts, we seek to introduce a new solution that addresses the following issues, in addition to maintaining the security and privacy provisions of BlindBox and PrivDPI: *(1) Fast TLS connection establishment without preprocessing in order to eliminate the session setup delay incurred in both BlindBox and PrivDPI; (2) Resisting brute force guessing of the rule sets even for small rule domains; (3) Supporting lightweight rule addition.*

Our Contributions. We propose Pine, a new protocol for privacy-preserving deep packet inspection on encrypted traffic, for a practical enterprise network setting, where clients connect to the Internet through an enterprise gateway. The main contributions are summarized as follows.

- **Identifying limitation of PrivDPI.** We revisit PrivDPI and demonstrate that in PrivDPI, when the rule domain is small, the middlebox could forge new encrypted rules that gives the middlebox the ability to detect the encrypted traffic with any encrypted rules it generates.
- **New solution with stronger privacy guarantee.** We propose Pine as the new solution for the problem of privacy-preserving deep packet inspection, where stronger privacy is guaranteed. First of all, the privacy of the traffic is protected unless there exists an attack in the traffic. Furthermore, privacy of rules is assured against the middlebox, we call this property *rule hiding*. This property ensures privacy of rules even when the rule domain is small (e.g. approximately 3000 rules as in existing Network Intrusion Detection (IDS) rules), which addresses the limitation of PrivDPI. In addition, privacy of rules is also assured against the enterprise gateway and the endpoints, we term this property *rule privacy*.
- **Amortized setup, fast connection establishment.** Pine enables the establishment of a TLS connection with low latency and without the need for an interactive preprocessing protocol as in PrivDPI and BlindBox. The latency-incurring preprocessing protocol is performed offline and is only executed once. Consequently, there is no per-user-connection overhead. Any client can setup a secure TLS connection with a remote server without preprocessing delay. In contrast, in PrivDPI and BlindBox, the more rules there are, the higher the per-connection setup cost is. The speed up of the connection is crucial for low-latency applications.
- **Lightweight rule addition.** Pine is a dynamic protocol in that it allows new rules being added on the fly without affecting the connection between a client and a server. The rule addition is seamless to the clients in the sense that the gateway can locally execute the rule addition phase with the middlebox

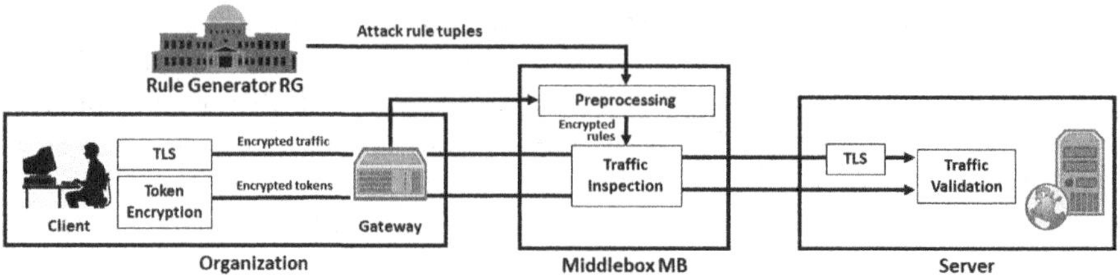

Fig. 1. Pine system architecture.

without any client involvement. This is beneficial as compared to BlindBox and PrivDPI, where the client would need to re-run the preprocessing protocol from scratch for every connection.

In addition to stronger privacy protection, we conduct extensive experiments to demonstrate the superior performance of Pine when compared to PrivDPI. For a connection from a client to a server with 5,000 tokens and a ruleset of 6,000, Pine is approximately 27% faster than PrivDPI, and saves approximately 92.3% communication cost. In particular, the communication cost of Pine is independent of the number of rules, while the communication cost of PrivDPI grows linear with the number of rules.

2 Protocol Overview

Pine shares a similar architecture with BlindBox and PrivDPI, as illustrated in Fig. 1. There are five entities in Pine: Client, Server, Gateway (**GW**), Rule Generator (**RG**) and Middlebox (**MB**). Client and server are the endpoints that send and receive network traffic protected by TLS. **GW** is a device located between a set of clients and servers that allows network traffic to flow from one endpoint to another endpoint. **RG** generates the attack rule tuples for **MB**. The attack rule tuples will be used by **MB** to detect attacks in the network traffic. Each attack rule describes an attack and contains one or more keywords to be matched in the network communication. Hereafter, we will use the terms "rule" and "attack rule" interchangeably. The role of **RG** can be performed by organization such as McAfee [18]. **MB** is a network device that inspects and filters network traffic using the attack rule tuples issued by **RG**.

System Requirements. The primary aim is to provide a privacy-preserving mechanism that can detect any suspicious traffic while at the same time ensure the privacy of endpoint's traffic. In particular, the system requirements include:

- *Traffic inspection*: Pine retains similar functionality of traditional IDS, i.e., to find a suspicious keyword in the packet.
- *Rule privacy*: The endpoints and **GW** should not learn the attack rules (i.e., the keywords). This is required especially for security solution providers that generate comprehensive and proprietary rule sets as their unique proposition that help to detect malicious traffic more effectively.

- *Traffic privacy*: On one hand, **MB** is not supposed to learn the plaintexts of the network traffic, except for the portions of the traffic that match the rules. On the other hand, **GW** is not allowed to read the content of the traffic.
- *Rule hiding*: **MB** is not supposed to learn the attack rules from the attack rule tuples issued by **RG** in a cloud-based setting where **MB** resides on a cloud platform. In such a case the cloud-based middlebox is not fully trusted. The security solution providers would want to protect the privacy of their unique rule sets, as was discussed previously in describing rule privacy.

Threat Model. There are three types of attackers described as follows.
- *Malicious endpoint.* The first type of attacker is the endpoint (i.e., the client or the server). Similar to BlindBox [20] and PrivDPI [15], at most one of the two endpoints is assumed to be malicious but not both. Such an attacker is the same as the attacker in the traditional IDS whose main goal is to evade detection. As in the traditional IDS [17], it is a fundamental requirement that at least one of the two endpoints is honest. This is because if two malicious endpoints agree on a private key and send the traffic encrypted by this particular key, detection of malicious traffic would be infeasible.
- *The attacker at the gateway.* As in conventional network setting, **GW** is assumed to be semi-honest. That is, **GW** honestly follows the protocol specification but may try to learn the plaintexts of the traffic. **GW** may also try to infer the rules from the messages it received.
- *The attacker at the middlebox.* **MB** is assumed to be semi-honest, which follows the protocol but may attempt to learn more than allowed from the messages it received. In particular, it may try to read the content of the traffic that passed through it. In addition, it may try to learn the underlying rules of the attack rule tuples issued by **RG**.

Protocol Flow. We present how each phase functions at a high level as follows.

- Initialization. **RG** initializes the system by setting the public parameters.
- Setup. **GW** subscribes the inspection service from **RG**, in which **RG** receives a shared secret from **GW**. **RG** issues the attack rule tuples to **MB**. The client and the server will derive some parameters from the key of the primary TLS handshake protocol and install a Pine HTTPS configuration, respectively.
- Preprocessing. In this phase, **GW** interacts with **MB** to generate a set of reusable randomized rules. In addition, **GW** generates and sends the initialization parameters to the clients within its domain.
- Preparation of Session Detection Rule. In this phase, the reusable randomized rules will be used to generate session detection rules.
- Token Encryption. In this phase, a client generates the encrypted token for each token in the payload. The encrypted tokens will be sent along with the traffic encrypted from the payload using regular TLS.
- Gateway Checking. For the first session, **GW** checks whether the attached parameters sent by the client is well-formed. This phase will be run when a client connects to a server for the first time.

- Traffic Inspection. **MB** generates a set of encrypted rules and performs inspection using these encrypted rules.
- Traffic Validation. One endpoint performs traffic validation in case the other endpoint is malicious.
- Rule Addition. A set of new attack rules will be added in this phase. **GW** interacts with **MB** to generate the reusable randomized rule set corresponding to these new attack rules.

3 Preliminaries

Complexity Assumption. The decision Diffie-Hellman (DDH) problem is stated as follows: given g, g^x, g^y, g^z, decide whether $z = xy$ (modulo the order of g), where $x, y, z \in \mathbb{Z}_p$. We say that a PPT algorithm $\mathcal{B}$ has advantage ϵ in solving the DDH problem if $|\Pr[\mathcal{B}^{(g,g^x,g^y,g^{xy})} = 1] - \Pr[\mathcal{B}^{(g,g^x,g^y,g^z)} = 1]| \geq \epsilon$, where the probability above is taken over the coins of $\mathcal{B}$, g, x, y, z.

Definition 1. *The DDH assumption holds if no PPT adversary has advantage at least ϵ in solving the DDH problem.*

Pseudorandom function. A pseudorandom function family PRF is a family of functions $\{\mathsf{PRF}_a : U \rightarrow V | a \in \mathcal{A}\}$ such that $\mathcal{A}$ could be efficiently samplable and all PRF, U, V, $\mathcal{A}$ are indexed by a security parameter λ. The security property of a PRF is: for any PPT algorithm $\mathcal{B}$ running in λ, it holds that $|\Pr[\mathcal{B}^{\mathsf{PRF}_a(\cdot)} = 1] - \Pr[\mathcal{B}^{R(\cdot)} = 1]| = \mathtt{negl}(\lambda)$, where $\mathtt{negl}$ is a negligible function of λ, a and R are uniform over $\mathcal{A}$ and $(U \rightarrow V)$ respectively. The probability above is taken over the coins of $\mathcal{B}$, a and R. For notational simplicity, we consider one version of the general pseudorandom function notion that is custom-made to fit our implementation. Specifically, the pseudorandom function PRF considered in this paper maps λ-bit strings to elements of $\mathbb{Z}_p$. Namely, $\mathsf{PRF}_a : \{0,1\}^\lambda \rightarrow \mathbb{Z}_p$, where $a \in G$.

Payload Tokenization. As in BlindBox and PrivDPI, we deploy window-based tokenization to tokenize keywords of a client's payload. Window-based tokenization follows a simple sliding window algorithm. We adopt 8 bytes per token when we implement the protocol. That is, given a payload "secret key", an endpoint will generate the tokens "secret k", "ecret ke" and "cret key".

4 Protocol

In this section, we first point out the limitation of PrivDPI. To address this problem and further reduce the connection delay, we then present our new protocol.

4.1 Limitation of PrivDPI

We show how PrivDPI fails when the domain of rule is small. We say that the domain of rule is small if one can launch brute force attack to guess the underlying rules given the public parameters. We first recall the setup phase of

PrivDPI. In the setup phase, a middlebox receives $(s_i, R_i, \mathsf{sig}(R_i))$ for rule r_i, where $R_i = g^{\alpha r_i + s_i}$ and $\mathsf{sig}(R_i)$ is the signature of R_i. With s_i and R_i, **MB** obtains the value $g^{\alpha r_i}$. Recall that in PrivDPI, the value $A = g^{\alpha}$ is included in the PrivDPI HTTPS configuration, **MB** could obtain this value via installing a PrivDPI HTTPS configuration. Since the domain of rule is small, with A and $g^{\alpha r_i}$, **MB** can launch brute force attack to obtain the value of r_i via trying every candidate value v by checking $A^v \stackrel{?}{=} g^{\alpha r_i}$ within the rule domain. In this way, **MB** could obtain the value r_i for R_i and r_j for R_j. After the completion of preprocessing protocol, **MB** obtains the reusable obfuscated rule $I_i = g^{k\alpha r_i + k^2}$ for rule r_i. Now, **MB** knows values r_i, r_j, $I_i = g^{k\alpha r_i + k^2}$, $I_j = g^{k\alpha r_j + k^2}$. It can then computes $(I_i/I_j)^{(r_i - r_j)^{-1}}$ to obtain a value $g^{k\alpha}$. With $g^{k\alpha}$, r_i and $I_i = g^{k\alpha r_i + k^2}$, it can compute $I_i/(g^{k\alpha})^{r_i} = g^{k^2}$. With g^{k^2} and $g^{k\alpha}$, **MB** could forge the reusable obfuscated rule successfully for any rule it chooses. With the forged (but valid) reusable obfuscated rule, **MB** could detect more than it is allowed, which violates the privacy requirement of the encrypted traffic.

4.2 Description of Our Protocol

Initialization. Let $\mathcal{R}$ be the domain of rules, PRF be a pseudorandom function, n be the number of rules and $[n]$ be the set $\{1, ..., n\}$. Let $\mathsf{AES}_a(\mathsf{salt})$ be the AES encryption with key a and message salt. Let $\mathsf{Enc}_a(\mathsf{salt}) = \mathsf{AES}_a(\mathsf{salt}) \mod \mathsf{R}$, where R is an integer used to reduce the ciphertext size [20]. The initialization phase takes in a security parameter λ and chooses a group G of prime order p. It then chooses a generator g of G, and sets the public parameters as (G, p, g).

Setup. **GW** chooses a key g^w for the pseudorandom function PRF, where $w \in \mathbb{Z}_p^*$. It subscribes the service from **RG** and sends w to **RG**. **RG** first computes $W = g^w$. For a rule set $\{r_i \in \mathcal{R}\}_{i \in [n]}$, for $i \in [n]$, **RG** chooses a randomness $k_i \in \mathbb{Z}_p$, calculates $r_{w,i} = \mathsf{PRF}_W(r_i)$ and $R_i = g^{r_{w,i} + k_i}$. **RG** chooses a signature scheme with sk as the secret key and pk as the public key. It then signs $\{R_i\}_{i \in [n]}$ with sk and generates the signature of R_i for $i \in [n]$, denote by σ_i. Finally, it sends the *attack rule tuples* $\{(R_i, \sigma_i, k_i)\}_{i \in [n]}$ to **MB**. Here, g^w is the key ingredient for ensuring the property of rule hiding. The key observation here is that since **MB** does not know g^w or w, it cannot guess the underlying r_i of R_i via brute forcing all the possible keywords it chooses. In particular, for a given attack rule tuple (R_i, σ_i, k_i), **MB** could obtain the value $g^{r_{w,i}}$ by computing R_i/g^{k_i}. Due to the property of pseudorandom function, $r_{w,i}$ is pseudorandom, and hence $g^{r_{w,i}}$ is pseudorandom. Without the knowledge of g^w or w, it is impossible to obtain r_i even if **MB** brute forces all possible keywords it chooses.

On the other hand, the client and the server install a Pine HTTPS configuration which contains a value R. Let k_{sk} be the key of the regular TLS handshake protocol established by a client and a server. With k_{sk}, the client (resp. the server) derives three keys k_T, c, k_s. Specifically, k_T is a standard TLS key, which is used to encrypt the traffic; c is a random value from $\mathbb{Z}_p$, which is used for generating session detection rules; k_s is a random value from $\mathbb{Z}_p$, which is used as a randomness to mask the parameters sent from the client to the server.

Preprocessing. In order to accelerate the network connection between a client and a server (compared to PrivDPI), we introduce a new approach that enables fast connection establishment without executing the preprocessing process per client as in PrivDPI. We start from the common networking scenario in an enterprise setting where there exists a gateway located between a set of clients and a server. The main idea is to let the gateway be the representative of the clients within its domain, who will run the preprocessing protocol with **MB** for only once. Both the clients and the gateway share the initialization parameters required for connection with the server. In this case, the connection between a client and a server can be established instantly without needing any preprocessing as in PrivDPI since the preprocessing is performed by the gateway and **MB** beforehand. In other words, we offload the operation of preprocessing to the gateway, which dramatically reduces the computation and communication overhead for the connection between a client and a server.

Specifically, in this phase, **GW** runs a preprocessing protocol with **MB** to generate a *reusable randomized rule* set as well as the initialization parameters for the clients within the domain of **GW**. The preprocessing protocol is run after the TLS handshake protocol, which is described in Fig. 2. Upon the completion of this phase, **MB** obtains a set of reusable randomized rules which enable **MB** to perform deep packet detection over the encrypted traffic across a series of sessions. The values I_0, I_1 and I_2 enable each client within the domain of **GW** to generate the encrypted tokens. Hence, for any network connection with a server, a client does not need to run the preprocessing phase with **MB** as compared to BlindBox and PrivDPI. This substantially reduces the delay and communication cost for the network connection between the client and the server, especially for large rule set. Furthermore, in case of adding new rules, a client does not need to re-run the preprocessing protocol as BlindBox and PrivDPI does. This means rule addition has no effect on the client side.

Preparation of Session Detection Rule. A set of session detection rules will be generated in this phase. These session detection rules are computed, tailored for every session, from the reusable randomized rules generated from the preprocessing protocol. The generated session detection rules are used as the inputs to generate the corresponding encrypted rules. The protocol is described in Fig. 3, and it is executed for every new session.

Token Encryption. Similar to BlindBox and PrivDPI, we adopt the window-based tokenization approach as described in Sect. 3. After the tokenization step, a client obtains a set of tokens corresponding to the payload. For the first time that a client connects with a server, the client derives a salt from c and stores the salt for future use, where c is the key derived from the key k_{sk} of the TLS handshake protocol. For each token t, a client runs the token encryption algorithm as described in Fig. 4. To prevent the count table T from growing too large, the client will clear T every Z sessions (e.g., $Z = 1,000$). In this case, the client will send a new salt to **MB**, where $\mathsf{salt} \leftarrow \mathsf{salt} + max_t \mathsf{count}_t + 1$.

In the above, we describe the token encryption when the endpoint is a client. When the endpoint is a server, the server will first run the same tokenization step, and encrypts the tokens as the step 1 and step 2 described in Fig. 4.

Gateway Checking. This phase will be executed when a client connects to a server for the first time. For the traffic sent from the client to a server for the first time, the client attaches $(\mathsf{salt}, C_{ks}, C_w, C_x, C_y)$. This enables the server to perform the validation of the encrypted traffic during the traffic validation phase. C_{ks} and k_s serve as the randomness to mask the values g^w, g^x and g^{xy}. The correctness of C_{ks} will be checked once the traffic reached the server. To ensure that g^w, g^x and g^{xy} are masked by C_{ks} correctly, **GW** simply checks whether the following equations hold: $C_w = (C_{ks})^w$, $C_x = (C_{ks})^x$ and $C_y = (C_{ks})^{xy}$.

Traffic Detection. During the traffic detection phase, **MB** performs the equality check between the encrypted tokens in the traffic and the encrypted rules it kept. The traffic detection algorithm is described as follows. **MB** first initializes a counter table CT_r to record the encrypted rule E_{r_i} for each rule r_i. The encrypted rule E_{r_i} for rule r_i is computed as $E_{r_i} = \mathsf{Enc}_{S_i}(\mathsf{salt} + \mathsf{count}_{r_i})$, where count_{r_i} is initialized to be 0. **MB** then generates a search tree that contains the encrypted rules. If a match is found, **MB** takes the corresponding action, deletes the old E_{r_i} corresponding to r_i, increases count_{r_i} by 1, computes and inserts a new E_{r_i} into the tree, where the new E_{r_i} is computes as $\mathsf{Enc}_{S_i}(\mathsf{salt} + \mathsf{count}_{r_i})$.

Traffic Validation. If it is the first session between a client and a server, upon receiving $(\mathsf{salt}, C_{ks}, C_w, C_x, C_y)$, the server checks whether the equation $C_{ks} = g^{k_s}$ holds, where k_s is derived (by the server) from the key k_{sk} of the regular TLS handshake protocol. If the equation holds, the server computes $(C_w)^{(k_s)^{-1}} = g^w$, $(C_x)^{(k_s)^{-1}} = g^x$, $(C_y)^{(k_s)^{-1}} = g^{xy}$. With the computed (g^w, g^x, g^{xy}), the server runs the same token encryption algorithm on the plaintext decrypted from the

Input: **MB** has inputs $\{(R_i, \sigma_i, k_i)\}_{i\in[n]}$, where $R_i = g^{r_{w,i}+k_i}$; **GW** has input pk.
The protocol is run between **GW** and **MB**:

1. **GW** chooses a random $x \in \mathbb{Z}_p^*$, computes $X = g^x$, and sends X to **MB**.
2. **MB** sends $\{(R_i, \sigma_i)\}_{i\in[n]}$ to **GW**.
3. Upon receiving $\{(R_i, \sigma_i)\}_{i\in[n]}$, **GW** does:
   (1) Check if σ_i is a valid signature on R_i using pk for $i \in [n]$; if not, halt and output $\perp$.
   (2) Choose a random $y \in \mathbb{Z}_p^*$ and compute $Y = g^y$. Compute $X_i = (R_i \cdot Y)^x = g^{xr_{w,i}+xk_i+xy}$ for $i \in [n]$, and return $\{X_i\}_{i\in[n]}$ to **MB**.
4. **MB** computes $K_i = X_i/(X)^{k_i} = g^{xr_{w,i}+xy}$ for $i \in [n]$ as the reusable randomized rule for rule r_i.
5. **GW** sets $I_0 = xy$, $I_1 = x$, $I_2 = g^w$ as the initialization parameters, and sends (I_0, I_1, I_2) to the clients within its domain.

Fig. 2. Preprocessing protocol

Input: The client (resp. the server) has input c. **MB** has input $\{K_i\}_{i\in[n]}$.
The protocol is run among a client, a server and **MB**:

1. The client computes $C = g^c$ and sends C to **MB** (through **GW**). Meanwhile, the server sets $C_s = c$ and sends C_s to **MB**.
2. **MB** checks whether C equals g^{C_s}. If yes, for $i \in [n]$, it calculates $S_i = (K_i{\cdot}C)^{C_s} = g^{c(xr_{w,i}+xy+c)}$ as the session detection rule for rule r_i.

Fig. 3. Session detection rule preparation protocol

encrypted TLS traffic as the client does. The server then checks whether the resulting encrypted tokens equal the encrypted tokens received from **MB**. If not, it indicates that the client is malicious. On the other hand, if it is the traffic sent from the server to the client, the client will do the same token encryption algorithm as the server does, and compares the resulting encrypted tokens with the received encrypted tokens from **MB** as well.

Rule Addition. In practice, new rules may be required to be added into the system. For a new rule $r'_i \in \mathcal{R}$ for $i \in [n']$, **RG** randomly chooses $k'_i \in \mathbb{Z}_p$, calculates $r'_{w,i} = \mathsf{PRF}_W(r'_i)$ and $R'_i = g^{r'_{w,i}+k'_i}$. It then signs the generated R'_i with sk to generate the signature σ'_i of R'_i. Finally, it sends the newly added attack rule tuples $\{R'_i, \sigma'_i, k'_i\}_{i\in[n']}$ to **MB**. For the newly added attack rule tuples, the rule addition protocol is described in Fig. 5, which is a simplified protocol of the preprocessing protocol.

Input: The client has inputs (I_0, I_1, I_2), a token t, the random keys k_s and c, the value R, a salt salt and a counter table T, where $I_0 = xy$, $I_1 = x$ and $I_2 = g^w$.
The algorithm is run by the client as follows:

1. Compute $I = I_0 + c = xy + c$.
2. For each token t:
   - If there exists no tuple corresponding to t in T: compute $t_w = \mathsf{PRF}_{I_2}(t)$, $T_t = g^{c(I_1 t_w + I)} = g^{c(xt_w+xy+c)}$, set $\mathsf{count}_t = 0$, compute the encryption of t as $E_t = \mathsf{Enc}_{T_t}(\mathsf{salt})$. Finally, insert tuple $(t, T_t, \mathsf{count}_t)$ into T.
   - If there exists a tuple $(t', T_{t'}, \mathsf{count}_{t'})$ in T where $t' = t$: update $\mathsf{count}_{t'} = \mathsf{count}_{t'} + 1$, and compute the encryption of t as $E_t = \mathsf{Enc}_{T_{t'}}(\mathsf{salt} + \mathsf{count}_{t'})$.
3. If it is the first session, compute $C_{ks} = g^{k_s}$, $C_w = (I_2)^{k_s} = g^{wk_s}$, $C_x = g^{I_1 k_s} = g^{xk_s}$ and $C_y = g^{I_0 k_s} = g^{xyk_s}$. The parameters $(\mathsf{salt}, C_{ks}, C_w, C_x, C_y)$ will be sent along with the encrypted token E_t for token t.

Fig. 4. Token encryption algorithm

Input: **MB** has newly added attack rule tuple set $\{(R'_i, \sigma'_i, k'_i)\}_{i\in[n']}$, where $R'_i = g^{r'_{w,i}+k'_i}$; **GW** has inputs Y, x.

The protocol is run between **GW** and **MB**:

1. **MB** sends $\{(R'_i, \sigma'_i)\}_{i\in[n']}$ to **GW**.
2. Upon receiving $\{(R'_i, \sigma'_i)\}_{i\in[n']}$, **GW** does: (1) Check if σ'_i is a valid signature on R'_i using pk for $i \in [n']$; if not, halt and output $\perp$. (2) Compute $X'_i = (R'_i \cdot Y)^x = g^{xr'_{w,i}+xk'_i+xy}$ for $i \in [n']$, and send $\{X'_i\}_{i\in[n']}$ to **MB**.
3. **MB** computes the reusable randomized rule $K'_i = X'_i/(X)^{k'_i}$ for $i \in [n']$.

Fig. 5. Rule addition protocol

5 Security

5.1 Middlebox Searchable Encryption

Definition. For a message space $\mathcal{M}$, a middlebox searchable encryption scheme consists of the following algorithms:

- $\mathsf{Setup}(\lambda)$: Takes a security parameter λ, outputs a key sk.
- $\mathsf{TokenEnc}(t_1, ..., t_n, \mathsf{sk})$: Takes a token set $\{t_i \in \mathcal{M}\}_{i\in[n]}$ and the key sk, outputs a set of ciphertexts $(\mathsf{c}_1, ..., \mathsf{c}_n)$ and a salt salt.
- $\mathsf{RuleEnc}(r, \mathsf{sk})$: Takes a rule $r \in \mathcal{M}$, the key sk, outputs an encrypted rule e_r.
- $\mathsf{Match}(\mathsf{e}_r, (\mathsf{c}_1, ..., \mathsf{c}_n), \mathsf{salt})$: Takes an encrypted rule e_r, ciphertexts $\{\mathsf{c}_i\}_{i\in[n]}$ and salt, outputs the set of indexes $\{\mathsf{ind}_i\}_{i\in[l]}$, where $\mathsf{ind}_i \in [n]$ for $i \in [l]$.

Correctness. We refer the reader to Appendix A for its definition.

Security. It is defined between a challenger $\mathcal{C}$ and an adversary $\mathcal{A}$.

- Setup. $\mathcal{C}$ runs $\mathsf{Setup}(\lambda)$ and obtains the key sk.
- $\mathsf{Challenge}$. $\mathcal{A}$ randomly chooses two sets of tokens $S_0 = \{t_{0,1}, ..., t_{0,n}\}$, $S_1 = \{t_{1,1}, ..., t_{1,n}\}$ from $\mathcal{M}$ and gives the two sets to $\mathcal{C}$. Upon receiving S_0 and S_1, $\mathcal{C}$ flips a random coin b, runs $\mathsf{TokenEnc}(t_{b,1}, ..., t_{b,n}, \mathsf{sk})$ to obtain a set of ciphertexts $(\mathsf{c}_1, ..., \mathsf{c}_n)$ and a salt salt. It then gives $(\mathsf{c}_1, ..., \mathsf{c}_n)$ and salt to $\mathcal{A}$.
- Query. $\mathcal{A}$ randomly chooses a set of rules $(r_1, ..., r_m)$ from $\mathcal{M}$ and gives the rules to $\mathcal{C}$. Upon receiving the set of rules, for $i \in [m]$, $\mathcal{C}$ runs $\mathsf{RuleEnc}(r_i, \mathsf{sk})$ to obtain encrypted rule e_{r_i}. $\mathcal{C}$ then gives the encrypted rules $\{\mathsf{e}_{r_i}\}_{i\in[m]}$ to $\mathcal{A}$.
- Guess. $\mathcal{A}$ outputs a guess b' of b.

Let $I_{0,i}$ be the index set that match r_i in S_0 and $I_{1,i}$ be the index set that match r_i in S_1. If $I_{0,i} = I_{1,i}$ and $b' = b$ for all i, we say that the adversary wins the above game. The advantage of the adversary in the game is defined as $\Pr[b' = b] - 1/2$.

Definition 2. *A middlebox searchable encryption scheme is secure if no PPT adversary has a non-negligible advantage in the game.*

Construction. The construction below captures the main structure from the security point of view.

- $\mathsf{Setup}(\lambda)$: Let PRF be a pseudorandom function. Generate $x, y, c, w \in \mathbb{Z}_p$, set (x, y, c, g^w) as the key.
- $\mathsf{TokenEnc}(t_1, ..., t_n, \mathsf{sk})$: Let salt be a random salt. For $i \in [n]$, do: (a) Let count be the number of times that token t_i repeats in the sequence t_1,...,t_{i-1}; (b) Calculate $t_{w,i} = \mathsf{PRF}_{g^w}(t_i)$, $T_{t_i} = g^{c(xt_{w,i}+xy+c)}$, $\mathsf{c}_i = H(T_{t_i}, \mathsf{salt}+\mathsf{count})$. Finally, the algorithm outputs $(\mathsf{c}_1, ..., \mathsf{c}_n)$ and salt.
- $\mathsf{RuleEnc}(r, \mathsf{sk})$: Compute $r_w = \mathsf{PRF}_{g^w}(r)$, $S = g^{c(xr_w+xy+c)}$, output $H(S)$.

Theorem 1. *Suppose H is a random oracle, the construction in Sect. 5.1 is a secure middlebox searchable encryption scheme.*

The proof of this theorem is provided in Appendix B.1.

5.2 Preprocessing Protocol

Definition. The preprocessing protocol is a two-party computation between **GW** and **MB**. Let $f : \{0,1\}^* \times \{0,1\}^* \to \{0,1\}^* \times \{0,1\}^*$ be the process of the computation, where for every inputs (a, b), the outputs are $(f_1(a, b), f_2(a, b))$. In our protocol, the input of **GW** is x and the input of **MB** is a derivation of r, and only **MB** receives the output.

Security. The security requirements include: (a) **GW** should not learn the value of each rule; (b) **MB** cannot forge any new reusable randomized rule that is different from the reusable randomized rules obtained during the preprocessing protocol. Intuitively, the second requirement is satisfied if **MB** cannot obtain the value x. Since both of **GW** and **MB** are assumed to be semi-honest, we adopt the security definition with static semi-honest adversaries as in [6]. Let π be the two-party protocol for computing f, View_i^π be the ith party's view during the execution of π, and Output^π be the joint output of **GW** and **MB** from the execution of π. For our protocol, since f is a deterministic functionality, we adopt the security definition for deterministic functionality as shown below.

Definition 3. *Let $f : \{0,1\}^* \times \{0,1\}^* \to \{0,1\}^* \times \{0,1\}^*$ be a deterministic functionality. We say that π securely computes f in the presence of static semi-honest adversaries if (a) Output^π equals $f(a, b)$; (b) there exist PPT algorithms $\mathcal{B}_1$ and $\mathcal{B}_2$ such that (1) $\{\mathcal{B}_1(a, f_1(a, b))\} \stackrel{c}{\equiv} \{\mathsf{View}_1^\pi(a, b)\}$, (2) $\{\mathcal{B}_2(b, f_2(a, b))\} \stackrel{c}{\equiv} \{\mathsf{View}_2^\pi(a, b)\}$, where $a, b \in \{0,1\}^*$ and $|a| = |b|$.*

Protocol. In Fig. 6, we provide a simplified protocol that outlines the main structure of the preprocessing protocol.

Lemma 1. *No computationally unbounded adversary can guess a rule r_i with probability greater than $1/|\mathcal{R}|$ with input R_i.*

The proof of this lemma is provided in Appendix B.2.

Theorem 2. *The preprocessing protocol securely computes f in the presence of static semi-honest adversaries assuming the DDH assumption holds.*

The proof of this theorem is provided in Appendix B.3.

Inputs: **GW** has inputs $x, y \in \mathbb{Z}_p$; **MB** has inputs $(\{R_i, k_i\}_{i\in[n]})$, where $R_i = g^{r_{w,i}+k_i}$.
The protocol is run between **GW** and **MB**:

1. **GW** computes $X = g^x$, and sends X to **MB**.
2. **MB** sends $\{R_i\}_{i\in[n]}$ to **GW**.
3. **GW** computes $X_i = (R_i \cdot g^y)^x$ for $i \in [n]$, and send $\{X_i\}_{i\in[n]}$ to **MB**.
4. **MB** computes $K_i = X_i/(X)^{k_i}$ as the reusable randomized rule for rule r_i.

Fig. 6. Simplified preprocessing protocol

5.3 Token Encryption

It captures the security requirement that **GW** cannot learn the underlying token when given an encrypted token.

Definition. For a message space $\mathcal{M}$, a token encryption scheme is as follows:

- $\mathsf{Setup}(\lambda)$: Takes as input a security parameter λ, outputs a secret key sk and the public parameters pk.
- $\mathsf{Enc}(\mathsf{pk}, \mathsf{sk}, t)$: Takes as input the public parameters pk, a secret key sk and a token $t \in \mathcal{M}$, outputs a ciphertext c.

Security. It is defined between a challenger $\mathcal{C}$ and an adversary $\mathcal{A}$.

- Setup: $\mathcal{C}$ runs $\mathsf{Setup}(\lambda)$ and sends the public parameters pk to $\mathcal{A}$.
- Challenge: $\mathcal{A}$ randomly chooses two tokens t_0, t_1 from $\mathcal{M}$ and sends them to $\mathcal{C}$. $\mathcal{C}$ flips a random coin $b \in \{0, 1\}$, runs $\mathsf{c} \leftarrow \mathsf{Enc}(\mathsf{pk}, \mathsf{sk}, t_b)$, and sends c to $\mathcal{A}$.
- Guess: $\mathcal{A}$ outputs a guess b' of b.

The advantage of an adversary is defined to be $\Pr[b' = b] - 1/2$.

Definition 4. *A token encryption scheme is secure if no PPT adversary has a non-negligible advantage in the security game.*

Construction. The construction presented below outlines the main structure from the security point of view.

- $\mathsf{Setup}(\lambda)$: Let PRF be a pseudorandom function. Choose random value $x, y, c, w \in \mathbb{Z}_p$, calculate $p_1 = g^c$, $p_2 = g^w$, $p_3 = x$ and $p_4 = y$. Finally, set c as sk and (p_1, p_2, p_3, p_4) as pk.
- $\mathsf{Enc}(\mathsf{pk}, \mathsf{sk}, t)$: Let salt be a random salt. Calculate $t_w = \mathsf{PRF}_{p_2}(t)$, $T_t = g^{c(xt_w+xy+c)}$, $\mathsf{c} = H(T_t, \mathsf{salt})$. Output c and salt.

Theorem 3. *Suppose H is a random oracle, the construction in Sect. 5.3 is a secure token encryption scheme.*

The proof of this theorem is provided in Appendix B.4.

5.4 Rule Hiding

It captures the security requirement that **MB** cannot learn the underlying rule when given an attack rule tuple (issued by **RG**).

Definition. For a message space $\mathcal{M}$, a rule hiding scheme is defined as follows:

- $\mathsf{Setup}(\lambda)$: Takes as input a security parameter λ, outputs a secret key sk and the public parameters pk.
- $\mathsf{RuleHide}(\mathsf{pk}, \mathsf{sk}, r)$: Takes as input the public parameters pk, a secret key sk and a rule $r \in \mathcal{M}$, outputs a hidden rule.

Security. The security definition for a rule hiding scheme is defined between a challenger $\mathcal{C}$ and an adversary $\mathcal{A}$ as follows.

- Setup: $\mathcal{C}$ runs $\mathsf{Setup}(\lambda)$ and gives the public parameters to $\mathcal{A}$.
- Challenge: $\mathcal{A}$ chooses two random rules r_0, r_1 from $\mathcal{M}$, and sends them to $\mathcal{C}$. Upon receiving r_0 and r_1, $\mathcal{C}$ flips a random coin b, runs $\mathsf{RuleHide}(\mathsf{pk}, \mathsf{sk}, r_b)$ and returns the resulting hidden rule to $\mathcal{A}$.
- Guess: $\mathcal{A}$ outputs a guess b' of b.

Construction.

- $\mathsf{Setup}(\lambda)$: Let PRF be a pseudorandom function. Choose random $k, w \in \mathbb{Z}_p$, set g^w as sk, k as pk.
- $\mathsf{RuleHide}(\mathsf{pk}, \mathsf{sk}, r)$: Calculate $r_w = \mathsf{PRF}_{\mathsf{sk}}(r)$, $R = g^{r_w + k}$, and output R.

Theorem 4. *Suppose* PRF *is a pseudorandom function, the construction in Sect. 5.4 is a secure rule hiding scheme.*

The proof of this theorem is provided in Appendix B.5.

6 Performance Evaluations

We investigate the performance of the network connection between a client and a server. Since PrivDPI perfoms better than BlindBox, we only present the comparison with PrivDPI. Let an *one-round connection* be a connection from the client to the server. The running time of a one-round connection reflects how fast a client can be connected to a server, and the communication cost captures the amount of overhead data need to be transferred for establishing this connection. Ideally, the running time for one-round connection should be as small as possible. The less running time it incurs, the faster a client can connect to a server. Similarly, it is desirable to minimize network communication overhead. We test the running time and the communication cost of one-round connection for our protocol and PrivDPI respectively. Our experiments are run on a Intel(R) Core i7-8700 CPU running at 3.20 Ghz with 8 GB RAM under 64bit Linux operating system. The CPU supports AES-NI instructions, where

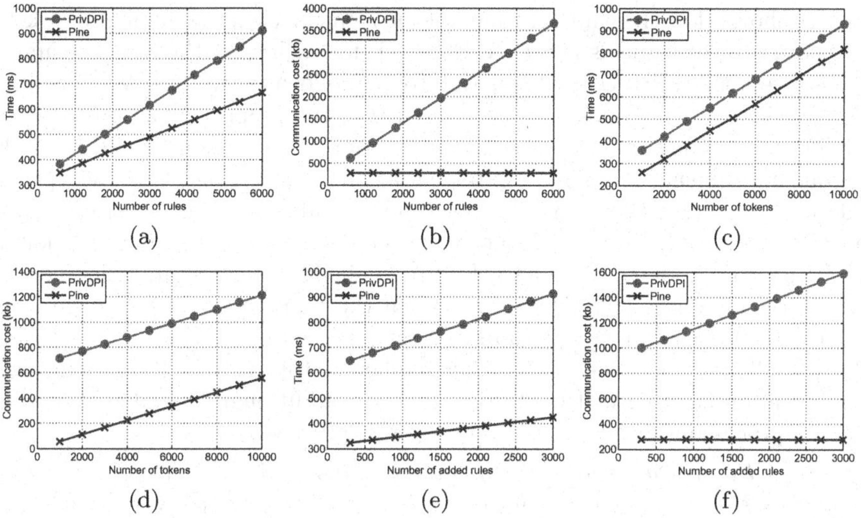

Fig. 7. Experimental performances

the encryption of token and the encryption of rule reflect this hardware support. The experiments are built on Charm-crypto [1], and is based on NIST Curve P-256. As stated in Sect. 3, both the rules and the tokens consist of 8 bytes. For simplicity, the payload that we test does not contain repeated tokens. We test each case for 20 times and takes the average.

How does the number of rules influence the one-round connection? Figure 7a illustrates the running time for one-round connection with 5,000 tokens when the number of rules range from 600 to 6,000. It is demonstrated that Pine takes less time than PrivDPI for each case, the more rules, the less time Pine takes compared to PrivDPI. This means that it takes less time for a client in Pine to connect to a server. In particular, for 5,000 tokens and 6,000 rules, it takes approximately 665 ms for Pine, while PrivDPI takes approximately 912 ms. That is, the delay for one-round connection of Pine is 27% less than PrivDPI; for 5,000 tokens and 3,000 rules, it takes approximately 488 ms for Pine, while PrivDPI takes approximately 616 ms. In other words, a client in Pine connects to a server with 20.7% faster speed than PrivDPI. Figure 7b shows the communication cost for one-round connection with 5,000 tokens when the number of rules range from 600 to 6,000. The communication cost of PrivDPI grows linearly with the number of rules, while for Pine it is constant. The more rules, the more communication cost PrivDPI incurs. This is because the client in PrivDPI needs to run the preprocessing protocol with **MB**, and the communication cost incurred by this preprocessing protocol is linear with the number of rules.

How does the number of tokens influence the one-round connection? We fix the number of rules to be 3,000, and test the running time and communication cost when the number of tokens range from 1,000 to 10,000.

Figure 7c shows that the running time of Pine is linear with the number of tokens in the payload, the same as PrivDPI. However, for each case, the time consumed of Pine is less than PrivDPI, this is due to the following two reasons. The first is that a client in Pine does not need to perform the preprocessing protocol for the 3,000 rules. The second is that, the encryption of a token in PrivDPI mainly takes one multiplication in G, one exponentiation in G, and one AES encryption. While in Pine, the encryption of a token mainly takes one hash operation, one exponentiation in G, and one AES encryption. That is, the token encryption of Pine is faster than that of PrivDPI. Figure 7d shows the communication cost of one-round connection with 3,000 rules when the number of tokens range from 1,000 to 10,000. Similar to the running time, the communication costs of Pine and PrivDPI are both linear with the number of tokens, but Pine incurs less communication than PrivDPI. This is due to the additional communication cost of the preprocessing protocol in PrivDPI for 3,000 rules.

How does the number of newly added rules influence the one-round connection? We test the running time and communication cost with 3,000 rules and 5,000 tokens when the number of newly added rules range from 300, to 3,000. Figure 7e shows that Pine takes less time than PrivDPI. For 3,000 newly added rules, Pine takes 424.96 ms, while PrivDPI takes 913.52 ms. That is, Pine is 53.48% faster than PrivDPI. Figure 7f shows that the communication cost of Pine is less than PrivDPI. In particular, the communication cost of Pine is independent of the number of newly added rules, while PrivDPI is linear with the number of newly added rules. This is because the client in Pine does not need to perform preprocessing protocol online.

7 Related Work

Our protocol is constructed based on BlindBox proposed by Sherry *et al.* [20] and PrivDPI proposed by Ning *et al.* [15], as was stated in the introduction. BlindBox introduces privacy-preserving deep packet inspection on encrypted traffic directly, while PrivDPI utilises an obfuscated rule generation mechanism with improved performance compared to BlindBox. Using the construction in BlindBox as the underlying component, Lan *et al.* [9] further proposed Embark that leverages on a trusted enterprise gateway to perform privacy-preserving detection in a cloud-based middlebox setting. In Embark, the enterprise gateway needs to be fully trusted and learns the content of the traffic and the detection rules, although in this case the client does not need to perform any operation as in our protocol. Our work focuses on the original setting of BlindBox and PrivDPI with further performance improvements, new properties and stronger privacy guarantee, while considering the practical enterprise gateway setting, in which the gateway needs not be fully trusted. Canard *et al.* [4] also proposed a protocol, BlindIDS, based on the concept of BlindBox, that has a better performance. The protocol consists of a token-matching mechanism that is based on pairing-based public key operation. Though practical, it is not compatible to TLS protocol.

Another related line of work focuses on accountability of the middlebox. This means the client and the server are aware of the middlebox that performs inspection on the encrypted traffic and are able to verify the authenticity of these middleboxes. Naylor *et al.* [14] first proposed such a scheme, termed mcTLS, where the existing TLS protocol is modified in order to achieve the accountability properties. However, Bhargavan *et al.* [3] showed that mcTLS can be tampered by an attacker to create confusion on the identity of the server that a middlebox is connected to, as well as the possibility for the attacker to inject its own data to the network. Due to this, a formal model on analyzing this type of protocols was proposed. Naylor *et al.* [13] further proposed a scheme, termed mbTLS, which does not modify the TLS protocol, thus allowing authentication of the middleboxes without needing to replace the existing TLS protocol. More recently, Lee *et al.* [10] proposed maTLS, a protocol that performs explicit authentication and verification of security parameters.

There are also proposals that analyse encrypted traffic without decrypting or inspecting the encrypted payloads. Machine learning models were utilised to detect anomalies based on the meta data of the encrypted traffic. Anderson *et al.* [2] proposed such techniques for malware detection on encrypted Traffic. Trusted hardware has also been deployed for privacy-preserving deep packet inspection. Most of the proposals utilize the secure enclave of Intel SGX. The main idea is to give the trusted hardware, resided in the middlebox, the session key. These include SGX-Box proposed by Han *et al.* [8], SafeBricks by Poddar *et al.* [19] and ShieldBox by Trach *et al.* [21] and LightBox by Duan *et al.* [5].

We note that our work can be combined with the accountability protocols, as well as the machine learning based works to provide comprehensive encrypted inspection that encompasses authentication and privacy.

8 Conclusion

In this paper, we proposed Pine, a protocol that allows inspection of encrypted traffic in a privacy-preserving manner. Pine builds upon the settings of BlindBox and techniques of PrivDPI in a practical setting, yet enables hiding of rule sets from the middleboxes with significantly improved performance compared to the two prior works. Furthermore, the protocol allows lightweight rules addition on the fly, which to the best of our knowledge has not been considered previously. Pine utilises the common practical enterprise setting where clients establish connections to Internet servers via an enterprise gateway, in such a way that the gateway assists in establishing the encrypted rule sets without learning the content of the client's traffic. At the same time, a middlebox inspects the encrypted traffic without learning both the underlying rules and content of the traffic. We demonstrated the improved performance of Pine over PrivDPI through extensive experiments. We believe Pine is a promising approach to detect malicious traffic amid growing privacy concerns for both corporate and individual users.

Acknowledgments. This work is supported in part by Singapore National Research Foundation (NRF2018NCR-NSOE004-0001) and AXA Research Fund, in part by the National Natural Science Foundation of China (61822202, 61872089, 61902070, 61972094, 61825203, U1736203, 61732021), Guangdong Provincial Special Funds for Applied Technology Research and Development and Transformation of Key Scientific and Technological Achievements (2016B010124009), and Science and Technology Program of Guangzhou of China (201802010061), and in part by the National Research Foundation, Prime Ministers Office, Singapore under its Corporate Laboratory@University Scheme, National University of Singapore, and Singapore Telecommunications Ltd.

A Correctness of Middlebox Searchable Encryption

On one hand, for every token that matches a rule r, the match should be detected with probability 1; on the other hand, for a token that does not match r, the probability of the match should be negligibly small. For every sufficiently large security parameter λ and any polynomial $n(\cdot)$ such that $n = n(\lambda)$, for all $t_1, ..., t_n \in \mathcal{M}^n$, for each rule $r \in \mathcal{M}$, for each index ind_i satisfying $r = t_{\mathsf{ind}_i}$ and for each index ind_j satisfying $r \neq t_{\mathsf{ind}_j}$, let $\mathsf{Exp}_1(\lambda)$ and $\mathsf{Exp}_2(\lambda)$ be experiments defined as follows:

Experiment $\mathsf{Exp}_1(\lambda)$:

$$\mathsf{sk} \leftarrow \mathsf{Setup}(\lambda); (\mathsf{c}_1, ..., \mathsf{c}_n), \mathsf{salt} \leftarrow \mathsf{TokenEnc}(t_1, ..., t_n, \mathsf{sk}); \mathsf{e}_\mathsf{r} \leftarrow \mathsf{RuleEnc}(r, \mathsf{sk});$$
$$\{\mathsf{ind}_k\}_{k\in[l]} \leftarrow \mathsf{Match}(\mathsf{e}_\mathsf{r}, (\mathsf{c}_1, ..., \mathsf{c}_n), \mathsf{salt}) : \mathsf{ind}_i \in \{\mathsf{ind}_k\}_{k\in[l]}$$

Experiment $\mathsf{Exp}_2(\lambda)$:

$$\mathsf{sk} \leftarrow \mathsf{Setup}(\lambda); (\mathsf{c}_1, ..., \mathsf{c}_n), \mathsf{salt} \leftarrow \mathsf{TokenEnc}(t_1, ..., t_n, \mathsf{sk}); \mathsf{e}_\mathsf{r} \leftarrow \mathsf{RuleEnc}(r, \mathsf{sk});$$
$$\{\mathsf{ind}_k\}_{k\in[l]} \leftarrow \mathsf{Match}(\mathsf{e}_\mathsf{r}, (\mathsf{c}_1, ..., \mathsf{c}_n), \mathsf{salt}) : \mathsf{ind}_j \notin \{\mathsf{ind}_k\}_{k\in[l]}$$

We have $\Pr[\mathsf{Exp}_1(\lambda)] = 1$, $\Pr[\mathsf{Exp}_2(\lambda)] = \mathsf{negl}(\lambda)$.

B Proofs

B.1 Proof of Theorem 1

The security is proved via one hybrid, which replaces the random oracle with deterministic random values. In particular, the algorithm $\mathsf{TokenEnc}$ now is modified as follows: $\mathsf{Hybrid.TokenEnc}(t_1, ..., t_n, \mathsf{sk})$: Let salt be a random salt. For $i \in [n]$, sample a random value T_i in the ciphertext space and set $\mathsf{c_i} = \mathsf{T}_i$. Finally, output $(\mathsf{c}_1, ..., \mathsf{c}_n)$ and salt. The algorithm $\mathsf{RuleEnc}(r, \mathsf{sk})$ is defined to output a random value R from the ciphertext space with the restriction that: (1) if r equals t_i for some t_i, R is set to be T_i; (2) for any future r' such that r equals r', the output is set to be R. We have that the outputs of algorithm $\mathsf{TokenEnc}$ and algorithm $\mathsf{RuleEnc}$ are random, while the the pattern of matching between tokens and rules are preserved. Clearly, the distributions for $S_0 = \{t_{0,1}, ..., t_{0,n}\}$, $S_1 = \{t_{1,1}, ..., t_{1,n}\}$ are the same. Hence, any PPT adversary has a change of distinguishing the two sets of exactly half.

B.2 Proof of Lemma 1

Fix a random $R = g^r$, where $r \in \mathbb{Z}_p$. We have that the probability for $R = R_i$ is the probability for $k_i = r - r_{w,i}$. Hence, for $\forall R \in G$, $\Pr[R = R_i] = 1/p$.

B.3 Proof of Theorem 2

We construct a simulator for each of the parties, $\mathcal{B}_1$ for **GW** and $\mathcal{B}_2$ for **MB**. For the case when **GW** is corrupted, $\mathcal{B}_1$ needs to generate the view of the incoming messages for **GW**. The message that **GW** received is R_i for $i \in [n]$. To simulate R_i for rule r_i, $\mathcal{B}_1$ chooses a random $u_i \in \mathbb{Z}_p$, calculate $U_i = g^{u_i}$ and sets U_i as the incoming message for rule r_i which simulates the incoming message from **MB** to **GW**. Following Lemma 1, the distribution of the simulated incoming message for **GW** (i.e., U_i) is indistinguishable from a real execution of the protocol. We next consider the case when **MB** is corrupted. The first and the third messages are the incoming message that **MB** received. For the first message, $\mathcal{B}_2$ randomly chooses a value $v \in \mathbb{Z}_p$, computes $V = g^v$, and sets V as the first incoming message for **MB**. For the third message, $\mathcal{B}_2$ randomly chooses a value $v_i \in \mathbb{Z}_p$ for $i \in [n]$, computes $V_i = g^{v_i}$ and sets V_i as the incoming message during the third step of the protocol. The view of **MB** for a rule r_i in an real execution of the protocol is $(k_i, R_i; X, X_i)$. The distributions of the real view and the simulated view are $(k_i, g^{r_{w,i}+k_i}; g^x, g^{xr_{w,i}+xk_i+xy})$ and $(k_i, g^{r_{w,i}+k_i}; g^v, g^{v_i})$. Clearly, a PPT adversary cannot distinguish these two distributions if DDH assumption holds.

B.4 Proof of Theorem 3

We prove the security by one hybrid, where we replace the random oracle with random values. In particular, the modified algorithm Enc is described as follows: $\mathsf{Enc}(\mathsf{pk}, \mathsf{sk}, t)$: Let salt be a random salt. Sample a random value c^* from the ciphertext space, and output c^* and salt. Now we have that the output of algorithm Enc is random. The distributions for challenge tokens t_0 and t_1 are the same. Hence, there exists no PPT adversary that has a chance of distinguishing the two tokes of exactly half.

B.5 Proof of Theorem 4

We proof the security via one hybrid, which replaces the output of PRF with a random value. In particular, during the challenge phase, the challenger chooses a random value v, computes $R^* = g^{v+k}$ (where k is publicly known to the adversary), returns R^* to the adversary. Clearly, if the adversary wins the security game, one can build a simulator that utilizes the ability of the adversary to break the pseudorandom property of PRF.

References

1. Akinyele, J.A., et al.: Charm: a framework for rapidly prototyping cryptosystems. J. Cryptographic Eng. **3**(2), 111–128 (2013). https://doi.org/10.1007/s13389-013-0057-3
2. Anderson, B., Paul, S., McGrew, D.A.: Deciphering malware's use of TLS (without decryption). J. Comput. Virol. Hacking Tech. **14**(3), 195–211 (2018). https://doi.org/10.1007/s11416-017-0306-6
3. Bhargavan, K., Boureanu, I., Delignat-Lavaud, A., Fouque, P.A., Onete, C.: A formal treatment of accountable proxying over TLS. In: S&P 2018, pp. 339–356. IEEE Computer Society (2018)
4. Canard, S., Diop, A., Kheir, N., Paindavoine, M., Sabt, M.: BlindIDS: market-compliant and privacy-friendly intrusion detection system over encrypted traffic. In: AsiaCCS 2017, pp. 561–574. ACM (2017)
5. Duan, H., Wang, C., Yuan, X., Zhou, Y., Wang, Q., Ren, K.: Lightbox: full-stack protected stateful middlebox at lightning speed. In: CCS 2019, pp. 2351–2367 (2019)
6. Goldreich, O.: Foundations of Cryptography: Volume 2, Basic Applications. Cambridge University Press (2009)
7. Google. HTTPS Encryption on the Web (2019). https://transparencyreport.google.com/https/overview?hl=en
8. Han, J., Kim, S., Ha, J., Han, D.: SGX-Box: enabling visibility on encrypted traffic using a secure middlebox module. In: APNet 2017, pp. 99–105. ACM (2017)
9. Lan, C., Sherry, J., Popa, R.A., Ratnasamy, S., Liu, Z.: Embark: securely outsourcing middleboxes to the cloud. In: NSDI 2016, pp. 255–273. USENIX Association (2016)
10. Lee, H., et al.: maTLS: how to make TLS middlebox-aware? In: NDSS 2019 (2019)
11. Meeker, M.: Internet trends (2019). https://www.bondcap.com/report/itr19/
12. National Security Agency. Managing Risk From Transport Layer Security Inspection (2019). https://www.us-cert.gov/ncas/current-activity/2019/11/19/nsa-releases-cyber-advisory-managing-risk-transport-layer-security
13. Naylor, D., Li, R., Gkantsidis, C., Karagiannis, T., Steenkiste, P.: And then there were more: secure communication for more than two parties. In: CoNEXT 2017, pp. 88–100. ACM (2017)
14. Naylor, D., et al.: Multi-context TLS (mcTLS): enabling secure in-network functionality in TLS. In: SIGCOMM 2015, pp. 199–212. ACM (2015)
15. Ning, J., Poh, G.S., Loh, J.C.N., Chia, J., Chang, E.C.: PrivDPI: privacy-preserving encrypted traffic inspection with reusable obfuscated rules. In: CCS 2019, pp. 1657–1670 (2019)
16. O'Neill, M., Ruoti, S., Seamons, K.E., Zappala, D.: TLS inspection: how often and who cares? IEEE Internet Comput. **21**(3), 22–29 (2017)
17. Paxson, V.: Bro: a system for detecting network intruders in real-time. Computer Netw. **31**(23–24), 2435–2463 (1999)
18. McAfee Network Security Platform (2019). http://www.mcafee.com/us/products/network-security-platform.aspx
19. Poddar, R., Lan, C., Popa, R.A., Ratnasamy, S.: SafeBricks: shielding network functions in the cloud. In: NSDI 2018, pp. 201–216. USENIX Association (2018)
20. Sherry, J., Lan, C., Popa, R.A., Ratnasamy, S.: BlindBox: deep packet inspection over encrypted traffic. In: SIGCOMM 2015, pp. 213–226 (2015)
21. Trach, B., Krohmer, A., Gregor, F., Arnautov, S., Bhatotia, P., Fetzer, C.: ShieldBox: secure middleboxes using shielded execution. In: SOSR 2018, pp. 2:1–2:14. ACM (2018)

Bulwark: Holistic and Verified Security Monitoring of Web Protocols

Lorenzo Veronese[1,2(✉)], Stefano Calzavara[1], and Luca Compagna[2]

[1] Università Ca' Foscari Venezia, Venezia, Italy
lorenzo.veronese@tuwien.ac.at
[2] SAP Labs France, Mougins, France

Abstract. Modern web applications often rely on third-party services to provide their functionality to users. The secure integration of these services is a non-trivial task, as shown by the large number of attacks against Single Sign On and Cashier-as-a-Service protocols. In this paper we present Bulwark, a new automatic tool which generates formally verified security monitors from applied pi-calculus specifications of web protocols. The security monitors generated by Bulwark offer holistic protection, since they can be readily deployed both at the client side and at the server side, thus ensuring full visibility of the attack surface against web protocols. We evaluate the effectiveness of Bulwark by testing it against a pool of vulnerable web applications that use the OAuth 2.0 protocol or integrate the PayPal payment system.

Keywords: Formal methods · Web security · Web protocols

1 Introduction

Modern web applications often rely on third-party services to provide their functionality to users. The trend of integrating an increasing number of these services has turned traditional web applications into *multi-party* web apps (MPWAs, for short) with at least three communicating actors. In a typical MPWA, a *Relying Party* (RP) integrates services provided by a *Trusted Third Party* (TTP). Users interact with the RP and the TTP through a *User Agent* (UA), which is normally a standard web browser executing a *web protocol*. For example, many RPs authenticate users through the Single Sign On (SSO) protocols offered by TTPs like Facebook, Google or Twitter, and use Cashier-as-a-Service (CaaS) protocols provided by payment gateway services such as PayPal and Stripe.

Unfortunately, previous research showed that the secure integration of third-party services is a non-trivial task [4,12,18,20–23]. Vulnerabilities might arise due to errors in the protocol specification [4,12], incorrect implementation practices at the RP [18,21,22] and subtle bugs in the integration APIs provided by the TTP [23]. To secure MPWAs, researchers proposed different approaches,

L. Veronese—Now at TU Wien.

L. Chen et al. (Eds.): ESORICS 2020, LNCS 12308, pp. 23–41, 2020.
https://doi.org/10.1007/978-3-030-58951-6_2

most notably based on *run time monitoring* [7,10,15,16,25]. The key idea of these proposals is to automatically generate security monitors allowing only the web protocol runs which comply with the expected, ideal run. Security monitors can block or try to automatically fix protocol runs which deviate from the expected outcome.

In this paper, we take a retrospective look at the design of previous proposals for the security monitoring of web protocols and identify important limitations in the current state of the art. In particular, we observe that:

1. existing proposals make strong assumptions about the placement of security monitors, by requiring them to be deployed either at the client [7,15] or at the RP [10,16,25]. In our work we show that both choices are sub-optimal, as they cannot prevent all the vulnerabilities identified so far in popular web protocols (see Sect. 4);
2. most existing proposals are not designed with formal verification in mind. They can ensure that web protocol runs are compliant with the expected run, e.g., derived from the network traces collected in an unattacked setting, however they do not provide any guarantee about the actual security properties supported by the expected run itself [10,16,25].

Based on these observations, we claim that none of the existing solutions provides a reliable and comprehensive framework for the security monitoring of web protocols in MPWAs.

Contributions. In this paper, we contribute as follows:

1. we perform a systematic, comprehensive design space analysis of previous work and we identify concrete shortcomings in all existing proposals by considering the popular OAuth 2.0 protocol and the PayPal payment system as running examples (Sect. 4);
2. we present Bulwark, a novel proposal exploring a different point of the design space. Bulwark generates formally verified security monitors from applied pi-calculus specifications of web protocols and lends itself to the appropriate placement of such monitors to have full visibility of the attack surface, while using modern web technologies to support an easy deployment. This way, Bulwark reconciles formal verification with practical security (Sect. 5);
3. we evaluate the effectiveness of Bulwark by testing it against a pool of vulnerable web applications that use the OAuth 2.0 protocol or integrate the PayPal payment system. Our analysis shows that Bulwark is able to successfully mitigate attacks on both the client and the server side (Sect. 6).

2 Motivating Example

As motivating example, shown in Fig. 1, we selected a widely used web protocol, namely OAuth 2.0 in explicit mode, which allows a RP to leverage a TTP for authenticating a user operating a UA.[1] The protocol starts (step 1) with the UA

[1] The protocol in the figure closely follows the Facebook implementation; details might slightly vary for different TTPs.

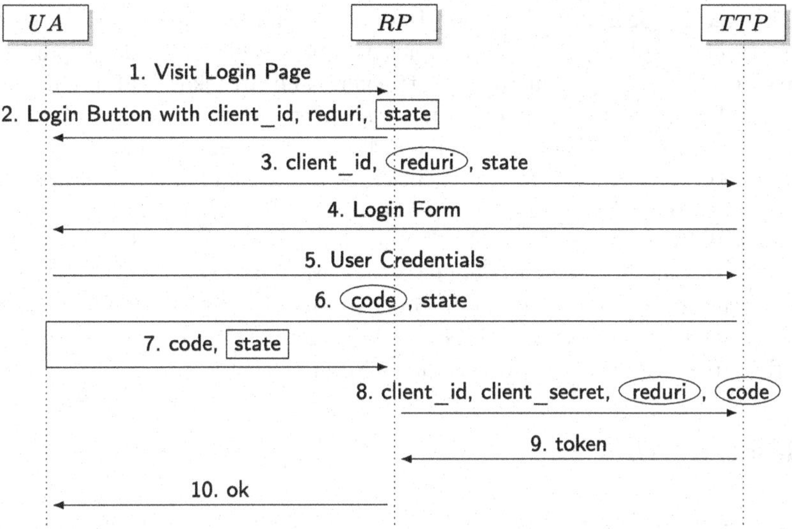

Fig. 1. Motivating example: Facebook OAuth 2.0 explicit mode

visiting the RP's login page. A login button is provided back that, when clicked, triggers a request to the TTP (steps 2–3). Such a request comprises: `client_id`, the identifier registered for the RP at the TTP; `reduri`, the URI at RP to which the TTP will redirect the UA after access has been granted; and `state`, a freshly generated value used by the RP to maintain session binding with the UA. The UA authenticates with the TTP (steps 4–5), which in turn redirects the UA to the `reduri` at RP with a freshly generated value `code` and the `state` value (steps 6–7). The RP verifies the validity of `code` in a back channel exchange with the TTP (steps 8–9): the TTP acknowledges the validity of `code` by sending back a freshly generated `token` indicating the UA has been authenticated. Finally, the RP confirms the successful authentication to the UA (step 10).

Securely implementing such a protocol is far from easy and many vulnerabilities have been reported in the past. We discuss below two representative attacks with severe security implications.

Session Swapping [21]**.** Session swapping exploits the lack of contextual binding between the login endpoint (step 2) and the callback endpoint (step 7). This is often the case in RPs that do not provide a `state` parameter or do not strictly validate it. The attack starts with the attacker signing in to the TTP and obtaining a valid `code` (step 6). The attacker then tricks an honest user, through CSRF, to send the attacker's `code` to the RP, which makes the victim's UA authenticate at the RP with the attacker's identity. From there on, the attacker can track the activity of the victim at the RP. The RP can prevent this attack by checking that the value of `state` at step 7 matches the one that was generated at step 2. The boxed shapes around `state` represent this invariant in Fig. 1.

Unauthorized Login by Code Redirection [4,14]. Code (and token) redirection attacks exploit the lack of strict validation of the `reduri` parameter and involve its manipulation by the attacker. The attacker crafts a malicious page which fools the victim into starting the protocol flow at step 3 with valid `client_id` and `state` from an honest RP, but with a `reduri` that points to the attacker's site. The victim then authenticates at the TTP and is redirected to the attacker's site with the `code` value. The attacker can then craft the request at step 7 with the victim's `code` to obtain the victim's `token` (step 9) and authenticate as her at the honest RP. The TTP can prevent this attack by (i) binding the `code` generated at step 6 to the `reduri` received at step 3, and (ii) checking, at step 8, that the received `code` is correctly bound to the supplied `reduri`. The rounded shapes represent this invariant in Fig. 1.

3 Related Work

We review here existing approaches to the security monitoring of web protocols, focusing in particular on their adoption in modern MPWAs. Each approach can be classified based on the placement of the proposed defense, i.e., we discriminate between client-side and server-side approaches.

We highlight here that none of the proposed approaches is able to protect the entire attack surface of web protocols. Moreover, none of the proposed solutions, with the notable exception of WPSE [7], is designed with formal verification in mind and provides clear, precise guarantees about the actual security properties satisfied by the enforced policy.

3.1 Client-Side Defenses

WPSE [7] extends the browser with a security monitor for web protocols that enforces the intended protocol flow, as well as the confidentiality and the integrity of messages. This monitor is able to mitigate many vulnerabilities found in the literature. The authors, however, acknowledge that some classes of attack cannot be prevented by WPSE. In particular, network attacks (like the HTTP variant of the IdP mix-up attack [12]), attacks that do not deviate from the intended protocol flow (like the automatic login CSRF from [4]) and purely server-side attacks are out of scope.

OAuthGuard [15] is a browser extension that aims to prevent five types of attacks on OAuth 2.0 and OpenID Connect, including CSRF and impersonation attacks. OAuthGuard essentially shares the same limitations of WPSE, due to the same partial visibility of the attack surface (the client side).

Recently Google has shown interest in extending its Chrome browser to monitor SSO protocols,[2] however their solution deals with a specific attack against their own implementation of SAML and is not a general approach designed to protect other protocols or TTPs.

[2] https://gsuiteupdates.googleblog.com/2018/04/more-secure-sign-in-chrome.html.

3.2 Server-Side Defenses

InteGuard [25] focuses on the server side of the RP, as its code appears to be more error-prone than that of the TTP. InteGuard is deployed as a proxy in front of the RP that checks invariants within the HTTP messages before they reach the web server. Different types of invariants are automatically generated from the network traces of SSO and CaaS protocols and enable the monitor to link together multiple HTTP sessions in transactions. Thanks to its placement, InteGuard can also monitor back channels (cf. steps 8–9 of Fig. 1). The authors explicitly mention that InteGuard does not offer protection on the TTP, expecting further efforts to be made in that direction. Unfortunately, several attacks can only be detected at the TTP, e.g., some variants of the unauthorized login by auth. code (or token) redirection attack from [4].

AEGIS [10] synthesizes run time monitors to enforce control-flow and data-flow integrity, authorization policies and constraints in web applications. The monitors are server-side proxies generated by extracting invariants from a set of input traces. AEGIS was designed for traditional two-party web applications, hence it does not offer comprehensive protection in the multi-party setting, e.g., due to its inability to monitor messages exchanged on the back channels. However, as mentioned by the authors, it can still mitigate those vulnerabilities which can be detected on the front channel of the RP (e.g., the shop-for-free TomatoCart example [10]). Similar considerations apply to BLOCK [16], a black-box approach for detecting *state violation* attacks, i.e., attacks which exploit logic flaws in the application to allow some functionality to be accessed at inappropriate states.

Guha et al. [13] apply a static control-flow analysis for JavaScript code to construct a *request-graph*, a model of a well-behaved client as seen by the server application. They then use this model to build a reverse proxy that blocks the requests that violate the expected control flow of the application, and are thus marked as potential attacks. Also this approach was designed for two-party web applications, hence does not offer holistic protection in the multi-party setting. Moreover, protection can only be enforced on web applications which are entirely developed in JavaScript.

4 Design Space Analysis

Starting from our analysis of related work, we analyze the design space of security monitors for web protocols, discussing pros and cons along different axes. Our take-away message is that solutions which assume a fixed placement of a single security monitor, which is the path taken by previous work, are inherently limited in their design for several reasons.

4.1 Methodology

We consider three possible deployment options for security monitors: the first two are taken from the literature, while the last one is a novel proposal we make. In particular, we focus on:

1. *browser extensions* [7,15]: a browser extension is a plugin module, typically written in JavaScript, that extends the web browser with custom code with powerful capabilities on the browser internals, e.g., arbitrarily accessing the cookie jar and monitoring network traffic;
2. *server-side proxies* [10,16,25]: a proxy server acts as an intermediary sitting between the web server hosting (part of) the web application and the clients that want to access it;
3. *service workers*: the Service Worker API[3] is a new browser functionality that enables websites to define JavaScript workers running in the background, decoupled from the web application logic. Service workers provide useful features normally not available to JavaScript, e.g., inspecting HTTP requests before they leave the browser, hence are an intriguing deployment choice for client-side security monitors.

We evaluate these options with respect to four axes, originally proposed as effective criteria for the analysis of web security solutions [8]:

1. *ease of deployment*: the practicality of a large-scale deployment of the defense mechanism, i.e., the price to pay for site operators to grant security benefits;
2. *usability*: the impact on the end-user experience, e.g., whether the user is forced to take actions to activate the protection mechanism;
3. *protection*: the effectiveness of the defense mechanism, i.e., the supported and unsupported security policies;
4. *compatibility*: the precision of the defense mechanism, i.e., potential false positives and breakages coming from security enforcement.

4.2 Ease of Deployment and Usability

Service workers are appealing, since they score best with respect to ease of deployment and usability. Specifically, the deployment of a service worker requires site operators to just add a JavaScript file to the web application: when a user visits the web application, the installation of the service worker is transparently performed with no user interaction.

Server-side proxies similarly have the advantage of ensuring transparent protection to end users. However, they are harder to deploy than service workers, as they require site operators to have control over the server networking. Even if site operators just needed to apply small modifications to the monitored application, they would have to reroute the inbound/outbound traffic to the proxy. This is typically easy for the TTP, which is usually a major company with full control over its deployment, but it can be impossible for some RPs. RPs are sometimes deployed on managed hosting platforms that may not allow any modification on the server itself, except for the application code. Note that site operators could implement the logic of the proxy directly in the application code, but this solution is impractical, since it would require a significant rewriting of the web

[3] https://developer.mozilla.org/en-US/docs/Web/API/Service_Worker_API.

application logic. This is also particularly complicated when the web application is built on top of multiple programming languages and frameworks.

Finally, browser extensions are certainly the worst solution with respect to usability and ease of deployment. Though installing a browser extension is straightforward, site operators cannot assume that every user will perform this manual installation step. In principle, site operators could require the installation of the browser extension to access their web application, but this would have a major impact on usability and could drive users away. Users' trust in browser extensions is another problem on its own: extensions, once installed and granted permissions, have very powerful capabilities on the browser internals. Moreover, the extension should be developed for the plethora of popular browsers which are used nowadays. Though major browsers now share the same extension architecture, many implementation details are different and would force developers to release multiple versions of their extensions, which complicates deployment. In the end, installing an extension is feasible for a single user, but relying on browser extensions for a large-scale security enforcement is unrealistic.

4.3 Protection and Compatibility

We study protection and compatibility together, since the *visibility* of the attack surface is the key enabler of both protection and compatibility. Indeed, the more the monitor has visibility of the protocol messages, the more it becomes able to avoid both false positives and false negatives in detecting potential attacks.

We use the notation $P_1 \leftrightarrow P_2$ to indicate the channel between two parties P_1 and P_2. If a monitor has visibility over a channel, then the monitor has visibility over all the messages exchanged on that channel.

Visibility. Browser extensions run on the UA and can have visibility of all the messages channeled through it. In particular, browser extensions can request *host permissions* upon installation to get access to the traffic exchanged with arbitrary hosts, which potentially enables them to inspect and edit any HTTP request and response relayed through the UA. In MPWAs, the UA can thus have visibility over both the channels UA $\leftrightarrow$ RP and UA $\leftrightarrow$ TTP (shortly indicated as UA $\leftrightarrow$ {RP, TTP}). However, the UA itself is not in the position to observe the entire attack surface against web protocols: for example, when messages are sent on the back channel between the RP and the TTP (RP $\leftrightarrow$ TTP) like in our motivating example (steps 8–9 in Fig. 1), an extension is unable to provide any protection, as the UA is not involved in the communication at all.

Server-side proxies can be categorized into *reverse proxies* and *forward proxies*, depending on whether they monitor incoming or outgoing HTTP requests respectively (plus their corresponding HTTP responses). Both approaches are useful and have been proposed in the literature, e.g., InteGuard [25] uses both a reverse proxy and a forward proxy at the RP to capture messages from the UA and to the TTP respectively. This way, InteGuard has full visibility of all the messages flowing through the RP, i.e., RP $\leftrightarrow$ {UA, TTP}. However, this is still not sufficient to fully monitor the attack surface. In particular, server-side proxies cannot inspect values that never leave the UA, like the *fragment identifier*, which instead plays an important role in the implicit flow of OAuth 2.0.

Finally, web applications can register service workers at the UA by means of JavaScript. Service workers can act as network proxies with access to all the traffic exchanged between the UA and the origin[4] which registered them. This way, service workers have the same visibility of a reverse-proxy sitting at the server; however, since they run on the UA, they also have access to values which never leave the client, like the fragment identifier. Despite this distinctive advantage, service workers are severely limited by the Same Origin Policy (SOP). In particular, they cannot monitor traffic exchanged between the UA and other origins, which makes them less powerful than browser extensions. For example, contrary to browser extensions like WPSE [7], a single service worker cannot monitor and defend both the RP and the TTP. This limitation can be mitigated by using multiple service workers and/or selectively relaxing SOP using CORS, which however requires collaboration between the RP and the TTP. Since service workers are more limited than browser extensions, they also share their inability to monitor back channels, hence they cannot be a substitute for forward proxies.

In the end, we conclude that browser extensions and server-side proxies are *complementary* in their ability to observe security-relevant protocol components, given their respective positioning, while service workers are strictly less powerful than their alternatives.

Cross-Site Scripting (XSS). XSS is a dangerous vulnerability which allows an attacker to inject malicious JavaScript code in a benign web application. Once a web application suffers from XSS, most confidentiality and integrity guarantees are lost, hence claiming security despite XSS is wishful thinking. Nevertheless, we discuss here how browser extensions can offer better mitigation than service workers in presence of XSS vulnerabilities. Specifically, since service workers can be removed by JavaScript, an attacker who was able to exploit an XSS vulnerability would also be able to void the protection offered by service workers. This can be mitigated by defensive programming practices, e.g., overriding the functions required for removing service workers, but it is difficult to assess both the correctness and the compatibility impact of this approach. For example, the deactivation of service workers might be part of the legitimate functionality of the web application or the XSS could be exploited before security-sensitive functions are overridden. Browser extensions, instead, cannot be removed by JavaScript and are potentially more robust against XSS. For example, WPSE [7] replaces secret values with random placeholders before they actually enter the DOM, so that secrets exchanged in the monitored protocol cannot be exfiltrated via XSS; the placeholders are then replaced with the intended values only when they leave the browser towards authorized parties.

[4] An origin is a triple including a scheme (HTTP, HTTPS, ...), a host (www.foo.com) and a port (80, 443, ...). Origins represent the standard web security boundary.

Table 1. Attacks on OAuth 2.0 and PayPal

	Attack	Channels to observe	UA	RP		TTP	
			ext	*sw*	*proxy*	*sw*	*proxy*
	OAuth 2.0						
1	307 Redirect attack [12]	UA ↔ TTP	✓	×	×	✓	✓
2	Access token eavesdropping [21]	UA ↔ RP	✓	✓	✓	×	×
3	Code/State Leakage via referer header [11,12]	UA ↔ RP	✓	✓	✓	×	×
4	Code/Token theft via XSS [21]	UA ↔ RP	✓	×	×	×	×
5	Cross Social-Network Request Forgery [4]	UA ↔ RP	✓	✓	✓	×	×
6	Facebook implicit AppId Spoofing [20,23]	UA ↔ TTP	×	×	×	✓	✓
7	Force/Automatic login CSRF [4,21]	UA ↔ RP	✓	✓	✓	×	×
8	IdP Mix-Up attack [12] (HTTP variant)	UA ↔ RP	×	×	✓	×	×
9	IdP Mix-Up attack [12] (HTTPS variant)	UA ↔ RP	✓	✓	✓	×	×
10	Naive session integrity attack [12]	UA ↔ RP	✓	✓	✓	×	×
11	Open Redirector in OAuth 2.0 [14,17]	UA ↔ {RP, TTP}	✓	✓	✓	✓	✓
12	Resource Theft by Code/Token Redirection [4,7]	UA ↔ TTP	✓	×	×	×	✓
13	Session swapping [14,21]	UA ↔ RP	✓	✓	✓	×	×
14	Social login CSRF on stateless clients [4,14]	UA ↔ RP	✓	✓	✓	×	×
15	Social login CSRF through TTP Login CSRF [4]	UA ↔ TTP	✓	×	×	✓	✓
16	Token replay implicit mode [14,20,24]	UA ↔ RP	✓	✓	×	×	×
17	Unauth. Login by Code Redirection [4,14]	UA ↔ TTP	✓	×	×	×	✓
	PayPal						
18	*NopCommerce* gross change in IPN callback [22]	RP ↔ {UA,TTP}	×	×	✓	×	×
19	*NopCommerce* gross change in PDT flow [22]	RP ↔ {UA,TTP}	×	×	✓	×	×
20	Shop for free by malicious *PayeeId* replay [18,20]	RP ↔ {UA,TTP}	×	×	✓	×	×
21	Shop for less by *Token* replay [18,20]	UA ↔ RP	×	×	✓	×	×

Tamper Resistance. Since both browser extensions and service workers are installed on the client, they can be tampered with or uninstalled by malicious users or software. This means that the defensive practices put in place by browser extensions and service workers are voided when the client cannot be trusted. This is particularly important for applications like CaaS, where malicious users might be willing to abuse the payment system, e.g., to shop for free. Conversely, server-side proxies are resilient by design to this kind of attacks, since they cannot be accessed at the client side, hence they are more appropriate for web applications where the client cannot be trusted to any extent.

Assessment on MPWAs. We now substantiate our general claims by means of a list of known attacks against the OAuth 2.0 protocol and the PayPal payment system, two popular protocols in MPWAs. Table 1 shows this list of attacks. For each attack, we show which channels need to be visible to detect the attack and we conclude whether the attack can be prevented by a browser extension (*ext*), a service worker (*sw*) or a server-side proxy (*proxy*) deployed on either the RP or the TTP.

In general we can see that, in the OAuth 2.0 setting, a browser extension is the most powerful tool, as it can already detect and block by itself most of the attacks (15 out of 17). The exceptions are the Facebook implicit AppId Spoofing attack [20,23], which can only be detected at the TTP, and the HTTP variant of the IdP Mix-Up attack [12], which is a network attack not observable at the client. Yet, remarkably, a comparable amount of protection can be achieved by using just service workers alone: in particular, the use of service workers at both the RP and the TTP can stop 13 out of 17 attacks. The only notable differences over the browser extension approach are that: (*i*) the code/token theft via XSS cannot be prevented, though we already discussed that even browser extensions can only partially mitigate the dangers of XSS, and (*ii*) the resource theft by code/token redirection and the unauthorized login by auth. code redirection cannot be stopped, because they involve a cross-origin redirect that service workers cannot observe by SOP. Remarkably, the combination of service workers and server-side proxies offers transparent protection that goes beyond browser extensions alone: 16 out of 17 attacks are blocked, with the only exception of code/token theft via XSS as explained.

The PayPal setting shows a very different trend with respect to OAuth 2.0. Although it is possible to detect the attacks on the client side, it is not safe to do so because both browser extensions and service workers can be uninstalled by malicious customers. For example, such client-side approaches cannot prevent the shop for free attack of [18], where a malicious user replaces the merchant id with her own account id. Moreover, it is worth noticing that PayPal deliberately makes heavy use of back channels (RP $\leftrightarrow$ TTP), since messages which are not relayed by the browser cannot be tampered with by malicious customers. This means that server-side proxies are the way to go to protect PayPal-like payment systems, as confirmed by the table.

4.4 Take-Away Messages

Here we highlight the main take-away messages of our design space analysis. In general, we claim that different web protocols require different protection mechanisms, hence every defensive solution which is bound to a specific placement of monitors does not generalize. More specifically:

- A clear total order emerges on the ease of deployment and usability axes. Service workers score best there, closely followed by server-side proxies, whose deployment is still feasible and transparent to end-users. Browser extensions are much more problematic, especially for large-scale security enforcement.
- With respect to the protection and compatibility axes, browser extensions are indeed a powerful tool, yet they can be replaced by a combination of service workers and server-side proxies to enforce transparent protection, extended to attacks which are not visible at the client alone.

In the end, we argue that a combination of service workers and server-side proxies has the potential to reconcile security, compatibility, ease of deployment and usability. In our approach, described in the next section, we thus pursue this research direction.

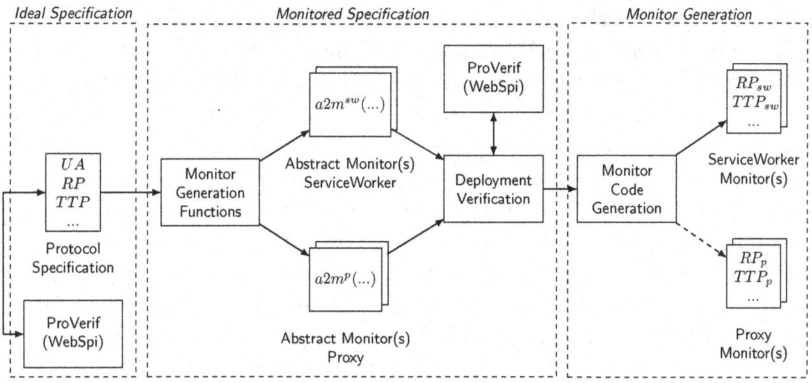

Fig. 2. Monitor generation pipeline

5 Proposed Approach: Bulwark

In this section we present Bulwark,[5] our formally verified approach to the holistic security monitoring of web protocols. For space reasons, we present an informal overview and we refer to the online technical report for additional details [2].

5.1 Overview

Bulwark builds on top of ProVerif, a state-of-the-art protocol verification tool [5]. ProVerif was originally designed for traditional cryptographic protocols, not for web protocols, but previous work showed how it can be extended to the web setting by using the WebSpi library [4]. In particular, WebSpi provides a ProVerif model of a standard web browser and includes support for important threats from the web security literature, e.g., *web attackers*, who lack traditional Dolev-Yao network capabilities and attack the protocol through a malicious website.

Bulwark starts from a ProVerif model of the web protocol to protect, called *ideal specification*, and generates formally verified security monitors deployed as service workers or server-side proxies. This process builds on an intermediate step called *monitored specification*. The workflow is summarized in Fig. 2.

To explain the intended use case of Bulwark, we focus on the typical setting of a multi-party web application including a TTP which offers integration to a set of RPs, yet the approach is general. The TTP starts by writing down its protocol in ProVerif, expressing the intended security properties by means of standard *correspondence assertions* (authentication) and *(syntactic) secrecy queries* supported by ProVerif. For example, the code/token redirection attack against OAuth 2.0 (cf. Sect. 2) can be discovered through the violation of a correspondence assertion [4]. The protocol can then be automatically verified for

[5] Bulwark is currently proprietary software at SAP: the tool could be made available upon request and an open-source license is under consideration.

security violations and the TTP can apply protocol fixes until ProVerif does not report any flaw. Since ProVerif is a sound verification tool [6], this process eventually leads to a security proof for an unbounded number of protocol sessions, up to the web model of WebSpi. The WebSpi model, although expressive, is not a complete model of the Web [4]. For example, it does not model advanced security headers such as `Content-Security-Policy`, frames and frame communication (`postMessage`). However, the library models enough components of the modern Web to be able to capture all the attacks of Table 1.

Once verification is done, the TTP can use Bulwark to automatically generate security monitors for its RPs from the ideal specification, e.g., to complement the traditional protocol SDK that the TTP normally offers anyway with protection for RPs, which are widely known to be the buggiest participants [25]. The TTP could also decide to use Bulwark to generate its own security monitors, so as to be protected even in the case of bugs in its own implementation.

5.2 Monitored Specification

In the *monitored specification* phase, Bulwark relaxes the ideal assumption that all protocol participants are implemented correctly. In particular, user-selected protocol participants are replaced by *inattentive* variants which comply with the protocol flow, but forget relevant security checks. Technically, this is done by replacing the ProVerif processes of the participants with new processes generated by automatically removing from the honest participants all the security checks (pattern matches, get/insert and conditionals) on the received messages, which include the invariants represented by the boxed checks in Fig. 1. This approximates the possible mistakes made by *honest-but-buggy* participants, obtaining processes that are interoperable with the other participants, but whose possible requests and responses are a super set of the original ones. Intuitively, an inattentive participant may be willing to install a monitor to prevent attackers from exploiting the lack of forgotten security checks. On the other hand, a deliberately malicious participant has no interest in doing so.

Then, Bulwark extracts from the ideal specification all the security invariant checks forgotten by the inattentive variants of the protocol participants and centralizes them within security monitors. This is done by applying two functions $a2m^p$ and $a2m^{sw}$, which derive from the participant specifications new ProVerif processes encoding security monitors deployed as a server-side proxy or a service worker respectively. The $a2m^p$ function is a modified version of the $a2m$ function of [19], which generates security monitors for cryptographic protocols. The proxy interposes and relays messages from/to the monitored inattentive participant, after performing the intended security checks. A subtle point here is that the monitor needs to keep track of the values that are already in its knowledge and those which are generated by the monitored participant and become known only after receiving them. A security check can only be executed when all the required values are part of the monitor knowledge.

The $a2m^{sw}$ function, instead, is defined on top of $a2m^p$ and the ideal UA process. This recalls that a service worker is a client-side defense that acts as a reverse proxy: a subset of the checks of both the server and the client side can be encoded into the final process running on the client. The function has three main responsibilities: *(i)* rewriting the proxy to be compatible with the service worker API; *(ii)* removing the channels and values that a service worker is not able to observe; *(iii)* plugging the security checks made by the ideal UA process into the service worker.

Example 1. Figure 3 illustrates how the RP of our OAuth 2.0 example from Sect. 2 is replaced by $I(RP)$, an inattentive variant in which the `state` parameter invariant is not checked. The right-hand side of the figure presents the RP as in Fig. 1 with a few more details on its internals, according to the ideal specification: (i) upon reception of message 1, the RP issues a new value for `state` and saves it together with the UA session cookie identifier (i.e., `state` is bound to the client session `sid(UA)`); and (ii) upon reception of message 7, the RP checks the `state` parameter and its binding to the client session. The inattentive version $I(RP)$, as generated by Bulwark, is shown on the left-hand side of Fig. 3: the `state` is neither saved by $I(RP)$ nor checked afterward. The left-hand side of the figure also shows the proxy $M(RP)$ generated by Bulwark as $a2m^p(RP)$ to enforce the `state` parameter invariant at RP. We can see that the saving and the checking of `state` are performed by $M(RP)$. It is worth noticing that the $M(RP)$ can only save `state` upon reception of message 2 from $I(RP)$. The service worker monitor $a2m^{sw}(RP)$ would look very similar to $M(RP)$.

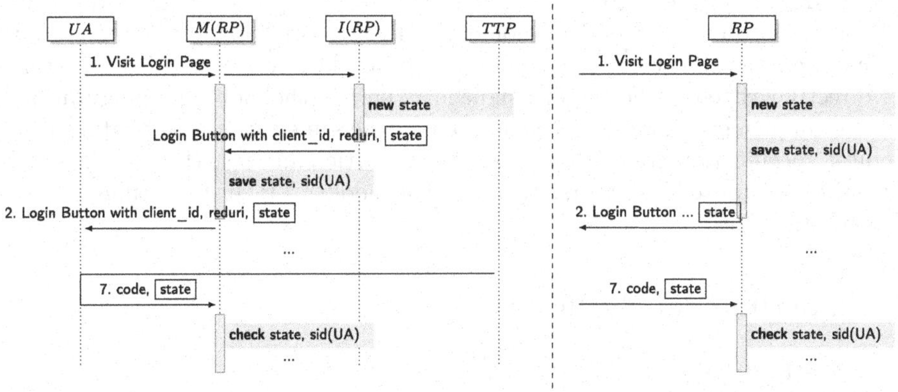

Fig. 3. Monitor invariant example

Finally, Bulwark produces a *monitored specification* where each inattentive protocol participant deploys a security monitor both at the client side (service worker) and at the server side (proxy). However, this might be overly conservative, e.g., a single service worker might already suffice for security. To optimize ease of deployment, Bulwark runs again ProVerif on the possible monitor deployment options, starting from the most convenient one, until it finds a setting which satisfies the security properties of the ideal specification. As an example, consider the system in which the only inattentive participant is the RP. There are three possible options, in decreasing order of ease of deployment:

1. $TTP \parallel I(RP) \parallel (a2m^{sw}(RP, UA) \parallel UA)$, where the monitor is deployed as a service worker registered by the RP at the UA;
2. $TTP \parallel (I(RP) \parallel a2m^{p}(RP)) \parallel UA$, where the monitor is a proxy at RP;
3. $TTP \parallel (I(RP) \parallel a2m^{p}(RP)) \parallel (a2m^{sw}(RP, UA) \parallel UA)$, with both.

The first option which passes the ProVerif verification is chosen by Bulwark.

5.3 Monitor Generation

Finally, Bulwark translates the ProVerif monitor processes into real service workers (written in JavaScript) or proxies (written in Python), depending on their placement in the monitored specification. This is a relatively direct one-to-one translation, whose key challenge is mapping the ProVerif messages to the real HTTP messages exchanged in the web protocol. Specifically, different RPs integrating the same TTP will host the protocol at different URLs and each TTP might use different names for the same HTTP parameters or rely on different message encodings (JSON, XML, etc.).

Bulwark deals with this problem by means of a configuration file, which drives the monitor generation process by defining the concrete values of the symbols and data constructors that are used by the ProVerif model. When the generated monitor needs to apply e.g., a data destructor on a name, it searches the configuration file for its definition and calls the corresponding function that deconstructs the object into its components. Since data constructors/destructors are directly written in the target language as part of the monitor configuration, different implementations can be generated for the same monitor, so that a single monitor specification created for a real-world participant e.g., the Google TTP, can be easily ported to others, e.g., the Facebook TTP, just by tuning their configuration files.

6 Experimental Evaluation

6.1 Methodology

To show Bulwark at work, we focus on the core MPWA scenarios discussed in Sect. 4.3. We first write ideal specifications of the OAuth 2.0 explicit protocol and the PayPal payment system in ProVerif + WebSpi. We also define appropriate correspondence assertions and secrecy queries which rule out all the attacks in Table 1 and we apply known fixes until ProVerif is able to prove security for the ideal specifications. Then, we setup a set of case studies representative of the key vulnerabilities plaguing these scenarios (see Table 2). In particular, we selected vulnerabilities from Table 1 so as to evaluate Bulwark on both the RP and TTP via a combination of proxy and service worker monitors. For each case study, we choose a set of inattentive participants and we collect network traces to define the Bulwark configuration files mapping ProVerif messages to actual protocol messages. Finally, we use Bulwark to generate appropriate security monitors and deploy them in our case studies. All our vulnerable case studies, their ideal specifications, and the executable monitors generated by Bulwark are provided as an open-source package to the community [1].

Case Studies. We consider a range of possibilities for OAuth 2.0. We start from an entirely artificial case study, where we develop both the RP and the TTP, introducing known vulnerabilities in both parties (CS1). We then consider integration scenarios with three major TTPs, i.e., Facebook, VK and Google, where we develop our own vulnerable RPs on top of public SDKs (CS2-CS4). Finally, we consider a case study where we have no control of any party, i.e., the integration between Overleaf and Google (CS5). We specifically choose this scenario, since the lack of the state parameter in the Overleaf implementation of OAuth 2.0 introduces known vulnerabilities.[6] To evaluate the CaaS scenario, we select legacy versions of three popular e-commerce platforms, suffering from known vulnerabilities in their integration with PayPal, in particular osCommerce 2.3.1 (CS6), NopCommerce 1.6 (CS7) and TomatoCart 1.1.15 (CS8).

Evaluation Criteria. We evaluate each case study in terms of four key aspects: (*i*) *security*: we experimentally confirm that the monitors stop the exploitation of the vulnerabilities; (*ii*) *compatibility*: we experimentally verify that the monitors do not break legitimate protocol runs; (*iii*) *portability*: we assess whether our ideal specifications can be used without significant changes across different case studies; and (*iv*) *performance*: we show that the time spent to verify the protocol and generate the monitors is acceptable for practical use.

Table 2. Test set of vulnerable applications

CS	RP	TTP	Protocol	Vuln. (Table 1)
1	*artificial RP 1*	*artificial IdP*	OAuth 2.0 explicit	#13 #17
2	*artificial RP 2*	`facebook.com`	OAuth 2.0 exp. (`graph-sdk 5.7`)	#13
3	*artificial RP 3*	`vk.com`	OAuth 2.0 exp. (`vk-php-sdk 5.100`)	#13
4	*artificial RP 4*	`google.com`	OAuth 2.0 exp. (`google/apiclient 2.4`)	#13
5	`overleaf.com`	`google.com`	OAuth 2.0 explicit	#13 #14
6	osCommerce 2.3.1	`paypal.com`	PayPal Standard	#18 #20
7	NopCommerce 1.6	`paypal.com`	PayPal Standard	#18
8	TomatoCart 1.1.15	`paypal.com`	PayPal Standard	#21

[6] We responsibly disclosed the issue to Overleaf and they fixed it before publication.

6.2 Experimental Results

The evaluation results are summarized in Table 3 and discussed below. In our case studies, we considered as inattentive participants all the possible sets of known-to-be vulnerable parties, leading to 10 experiments; when multiple experiments can be handled by a single run of Bulwark, their results are grouped together in the table, e.g., the experiments for CS2-CS4. Notice that for CS1 we considered three sets of inattentive participants: only TTP (vulnerability #17); only RP (vulnerability #13); and both RP and TTP (both vulnerabilities). Hence, we have 3 experiments for CS1, 3 experiments for CS2-CS4, 1 experiment for CS5 and 3 experiments for CS6-CS8.

Table 3. Generated monitors and run-time

CS	Ideal Spec.	Verification Time	Inattentive Parties	Monitor Verification time	#verif.	Gen. Monitors RP *sw*	RP *proxy*	TTP *sw*	TTP *proxy*	Prevented Vuln.
1	IS1	29m	TTP	41m	2	×	×	×	✓	#17
			RP	15m	1	✓	×	×	×	#13
			RP,TTP	54m	3	✓	×	×	✓	#13 #17
2 3 4	IS1	27m	RP	18m	1	✓	×	×	×	#13
5	IS1*	19m	RP	17m	1	✓	×	×	×	#13 #14
6 7 8	IS2	3m	RP	8m	1	×	✓	×	×	#18 #20 #21

Security and Compatibility. To assess security and compatibility, we created manual tests to exploit each vulnerability of our case studies and we ran them with and without the Bulwark generated monitors. In all the experiments, we confirmed that the known vulnerabilities were prevented only when the monitors were deployed (security) and that we were able to complete legitimate protocol runs successfully both with and without the monitors (compatibility). Based on Table 3, we observe that 5 experiments can be secured by a service worker alone, 4 experiments can be protected by a server-side proxy and only one experiment needed the deployment of two monitors. This heterogeneity confirms the need of holistic security solutions for web protocols.

Portability. We can see that the ideal specification IS1 created for our first case study CS1 is portable to CS2-CS4 without any change. This means that different TTPs supporting the OAuth 2.0 explicit protocol like Facebook, VK and Google can use Bulwark straightaway, by just tuning the configuration file to their settings. This would allow them to protect their integration scenarios with RPs that (like ours) make use of the state parameter. This is interesting, since different TTPs typically vary on a range of subtle details, which are all accounted for correctly by the Bulwark configuration files. However, the state parameter is not mandatory in the OAuth2 standard and thus TTPs tend to allow integration also with RPs that do not issue it. Case study CS5 captures this variant of OAuth 2.0: removing the state parameter from IS1 is sufficient to create a new ideal specification IS1*, which enables Bulwark towards these scenarios as well. As to PayPal, the ideal specification IS2 is portable to all the case studies CS6-CS8. Overall, our experience indicates that once an ideal specification is created for a protocol, then it is straightforward to reuse it on other integration scenarios based on the same protocol.

Performance. We report both the time spent to verify the ideal specification (Verification Time) as well as the time needed to verify the monitors (Monitor Verification). Both steps are performed offline and just once, hence the times in the table are perfectly fine for practical adoption. Verifying the ideal specification never takes more than 30 min, while verifying the monitors might take longer, but never more than one hour in our experiments. The time spent in the latter step depends on how many runs of ProVerif are required to reach a secure monitored specification (see the very end of Sect. 5.2). For example, the first experiment runs ProVerif twice (cf. #verif.) and requires 41 min, while the second experiment runs ProVerif just once and thus takes only 15 min.

7 Conclusion

In this paper we identified shortcomings in previous work on the security monitoring of web protocols and proposed Bulwark, the first holistic and formally verified defensive solution in this research area. Bulwark combines state-of-the-art protocol verification tools (ProVerif) with modern web technologies (service workers) to reconcile formal verification with practical security. We showed that Bulwark can generate effective security monitors on different case studies based on the OAuth 2.0 protocol and the PayPal payment system.

As future work, we plan to extend Bulwark to add an additional protection layer, i.e., on client-side communication based on JavaScript and the postMessage API. This is important to support modern SDKs making heavy use of these technologies, like the latest versions of the PayPal SDKs, yet challenging given the complexity of sandboxing JavaScript code [3]. On the formal side, we would like to strengthen our definition of "inattentive" participant to cover additional vulnerabilities besides missing invariant checks. For example, we plan to cover participants who forget to include relevant security headers and are supported by appropriately configured monitors in this delicate task. Finally, we would like to further engineer Bulwark to make it easier to use for people who have no experience with ProVerif, e.g., by including support for a graphical notation which is compiled into ProVerif processes, similarly to the approach in [9].

Acknowledgments. Lorenzo Veronese was partially supported by the European Research Council (ERC) under the European Unions Horizon 2020 research (grant agreement No. 771527-BROWSEC).

References

1. Bulwark case studies. https://github.com/secgroup/bulwark-experiments
2. Bulwark: holistic and verified security monitoring of web protocols (Technical report). https://secgroup.github.io/bulwark-experiments/report.pdf
3. Van Acker, S., Sabelfeld, A.: JavaScript sandboxing: isolating and restricting client-side JavaScript. In: Aldini, A., Lopez, J., Martinelli, F. (eds.) FOSAD 2015-2016. LNCS, vol. 9808, pp. 32–86. Springer, Cham (2016). https://doi.org/10.1007/978-3-319-43005-8_2
4. Bansal, C., Bhargavan, K., Maffeis, S.: Discovering concrete attacks on website authorization by formal analysis. In: CSF 2012. IEEE (2012)
5. Blanchet, B.: An efficient cryptographic protocol verifier based on prolog rules. In: CSFW 2001. IEEE (2001)
6. Blanchet, B.: Automatic verification of correspondences for security protocols. J. Comput. Secur. **17**(4), 363–434 (2009)
7. Calzavara, S., Focardi, R., Maffei, M., Schneidewind, C., Squarcina, M., Tempesta, M.: WPSE: fortifying web protocols via browser-side security monitoring. In: USENIX Security 18. USENIX Association (2018)
8. Calzavara, S., Focardi, R., Squarcina, M., Tempesta, M.: Surviving the web: a journey into web session security. ACM Comput. Surv. **50**(1), 1–34 (2017)
9. Carbone, R., Compagna, L., Panichella, A., Ponta, S.E.: Security threat identification and testing. In: ICST 2015. IEEE Computer Society (2015)
10. Compagna, L., dos Santos, D., Ponta, S., Ranise, S.: Aegis: automatic enforcement of security policies in workflow-driven web applications. In: CODASPY 2017. ACM (2017)
11. Fett, D., Küsters, R., Schmitz, G.: The web SSO standard OpenID connect: in-depth formal security analysis and security guidelines. In: CSF 2017. IEEE (2017)
12. Fett, D., Küsters, R., Schmitz, G.: A comprehensive formal security analysis of OAuth 2.0. CCS 2016. ACM (2016)
13. Guha, A., Krishnamurthi, S., Jim, T.: Using static analysis for ajax intrusion detection. In: WWW 2009. ACM (2009)
14. Hardt, D.: The OAuth 2.0 authorization framework. RFC 6749, October 2012
15. Li, W., Mitchell, C.J., Chen, T.: OAuthGuard: protecting user security and privacy with OAuth 2.0 and OpenID connect. In: SSR (2019)
16. Li, X., Xue, Y.: BLOCK: a black-bOx approach for detection of state violation attacks towards web applications. In: ACSAC 2011 (2011)
17. Lodderstedt, T., McGloin, M., Hunt, P.: OAuth 2.0 threat model and security considerations. RFC 6819, January 2013
18. Pellegrino, G., Balzarotti, D.: Toward black-box detection of logic flaws in web applications. In: NDSS (2014)
19. Pironti, A., Jürjens, J.: Formally-based black-box monitoring of security protocols. In: Massacci, F., Wallach, D., Zannone, N. (eds.) ESSoS 2010. LNCS, vol. 5965, pp. 79–95. Springer, Heidelberg (2010). https://doi.org/10.1007/978-3-642-11747-3_7
20. Sudhodanan, A., Armando, A., Carbone, R., Compagna, L.: Attack patterns for black-box security testing of multi-party web applications. In: NDSS (2016)
21. Sun, S.T., Beznosov, K.: The devil is in the (implementation) details: an empirical analysis of OAuth SSO systems. In: CCS 2012. ACM (2012)
22. Wang, R., Chen, S., Wang, X., Qadeer, S.: How to shop for free online - security analysis of cashier-as-a-service based web stores. In: S&P. IEEE (2011)

23. Wang, R., Chen, S., Wang, X.: Signing me onto your accounts through Facebook and Google: a traffic-guided security study of commercially deployed single-sign-on web services. In: S&P 2012. IEEE (2012)
24. Wang, R., Zhou, Y., Chen, S., Qadeer, S., Evans, D., Gurevich, Y.: Explicating SDKs: uncovering assumptions underlying secure authentication and authorization. In: USENIX (2013)
25. Xing, L., Chen, Y., Wang, X., Chen, S.: InteGuard: toward automatic protection of third-party web service integrations. In: NDSS 2013 (2013)

A Practical Model for Collaborative Databases: Securely Mixing, Searching and Computing

Shweta Agrawal[1], Rachit Garg[2(✉)], Nishant Kumar[3], and Manoj Prabhakaran[4]

[1] IIT Madras, Chennai, India
[2] UT Austin, Austin, USA
rachit0596@gmail.com
[3] Microsoft Research India, Bangalore, India
[4] IIT Bombay, Mumbai, India

Abstract. We introduce the notion of a *Functionally Encrypted Datastore* which collects data *anonymously* from *multiple data-owners*, stores it encrypted on an untrusted server, and allows untrusted clients to make *select-and-compute* queries on the collected data. Little coordination and no communication is required among the data-owners or the clients. Our notion is general enough to capture many real world scenarios that require controlled computation on encrypted data, such as is required for contact tracing in the wake of a pandemic. Our leakage and performance profile is similar to that of conventional *searchable encryption* systems, while the functionality we offer is significantly richer.

In more detail, the client specifies a query as a pair (Q, f) where Q is a filtering predicate which selects some subset of the dataset and f is a function on some computable values associated with the selected data. We provide efficient protocols for various functionalities of practical relevance. We demonstrate the utility, efficiency and scalability of our protocols via extensive experimentation. In particular, we evaluate the efficiency of our protocols in computations relevant to the *Genome Wide Association Studies* such as Minor Allele Frequency (MAF), Chi-square analysis and Hamming Distance.

1 Introduction

Many real world scenarios call for performing controlled computation on encrypted data belonging to multiple users. A case in point is that of contact tracing to control the COVID-19 pandemic, where cellphone users may periodically upload their (space, time) co-ordinates to enable tracing of infected persons, but desire the assurance that this data will not be used for any other purpose. Another example is Genome Wide Association Studies (GWAS), which look into entire genomes across different individuals to discover associations between genetic variants and particular diseases or traits [4].

R. Garg—Work done primarily when student at IIT Madras.

L. Chen et al. (Eds.): ESORICS 2020, LNCS 12308, pp. 42–63, 2020.
https://doi.org/10.1007/978-3-030-58951-6_3

More generally, enabling controlled computation on large-scale, multi-user, encrypted cloud storage is of much practical value in various privacy sensitive situations. Over the last several years, several tools have emerged that offer a variety of approaches towards this problem, offering different trade-offs among security, efficiency and generality. While theoretical schemes based on modern cryptographic tools like secure multi-party computation (MPC), fully homomorphic encryption (FHE) or functional encryption (FE) can provide strong security guarantees, their computational and communication requirements are often prohibitive for large-scale data (see Sect. 5.2, and Footnote 4). At the other end are efficient tools like CryptDB [25], Monomi [29], Seabed [22] and Arx [24], which add a lightweight encryption layer under the hood of conventional database queries, but offer only limited security guarantees and do not support collaborative databases (see Table 2). While there also exist tools which seek to strike a different balance by trading off some efficiency for more robust security guarantees and better support for collaboration – like Symmetric Searchable Encryption (SSE) [10,27], and Controlled Functional Encryption (CFE) [20] – they offer limited functionality.

Our Approach: We introduce *Functionally Encrypted Datastores* (FED), opening up new possibilities in secure cloud storage. FED is a secure cloud-based data store that can collect data *anonymously* from *multiple data-owners*, store it encrypted on untrusted servers, and allow untrusted clients to make *select-and-compute* queries on the collected data.

The "select and compute" functionality we support may be viewed as typical (relational) database operations on a single table. A query is specified as a pair (Q, f) where Q is a *filtering predicate* which selects some rows of the table, and f is a *function on the selected data.* Several real world scenarios involve such filtering – e.g., in contact tracing, records which have a (space, time) pair from a set of high-risk pairs are filtered; in GWAS, records are often filtered by presence of a disease (see [3] for more examples). A key feature we seek is that the computation overheads for a select-and-compute query *should not scale with the entire database size, but only with the number of selected records.*

1.1 Our Model

Before we describe our model, we outline our desiderata:

- *No central trusted authority.* The data-owners should not be required to trust any central server with their private data.
- *No coordination among offline data-owners.* The data-owners should not be required to trust, or even be aware of each other. Additionally, the data owners need not be online when queries arrive.
- *Untrusted clients, oblivious to the data-owners.* The data-owners should not be required to trust or even be aware of the clients who will query the datastore in the future.

- *Efficient on large scale data.* Enable handling of large databases (e.g. genomic data from hundreds of thousands of individuals, or census data) efficiently. This may be done by allowing some well-defined *leakage* to the servers, as is common in the searchable encryption literature.
- *Anonymity of data.* As we allow multiple data-owners, the data items referenced in the leakage above when obtained by a compromised server should not be traced back to the data-owners who contributed those items.

It is *impossible* to satisfy the first three requirements using a single (untrusted) server – it may internally play the role of clients and make any number of arbitrary queries to the database, violating security. Hence, we propose a solution with two servers: one with large storage that stores all the encrypted data, and an auxiliary server who stores only some key material. Either server by itself can be corrupt, but they will be assumed not to collude with each other. This model is well-accepted in the literature for privacy-preserving computation on genomic data [7,16], justified by the fact that in the real world, genomic data in the US [16] may be managed by distinct governmental organizations within the National Institutes of Health and the World Health Organization which are expected not to collude. More generally, this model has been used in multi-server Private Information Retrieval [8], CFE [20],[1] SSE [21], Secure Outsourced Computation [18] (including in the context of genomic data [28]), and even large-scale real-world systems like Riposte [9] and Splinter [31].

1.2 Our Results

As discussed above, our first contribution is the notion of a Functionally Encrypted Datastore (FED), which permits a data-owner to securely outsource its data to a storage server, such that, with the help of an auxiliary server, clients can carry out select-and-compute queries efficiently. We emphasize that our database is *anonymously collaborative* in the sense that it contains data belonging to multiple data owners but hides the association of the (encrypted) data items with their owners. In addition to this, our contributions include:

- A *general framework* for instantiating FED schemes. The framework is modular, and consists of two components, which may be of independent interest – namely, Searchably Encrypted Datastore (SED) and Computably Encrypted Datastore (CED). We present several constructions to instantiate this framework (see an overview in Sect. 1.3).
- We demonstrate the utility, efficiency and scalability of our protocols via *extensive experimentation.* In experiments that model *Genome Wide Association Studies* (GWAS), genomic records from 100,000 data owners (each contributing a single record) is securely procured in around 500 s, and a client query that filters up to 12,000 records is answered in less than 15 s with a maximum of 15 MB of total communication in the system. For standard functions in GWAS, like Minor Allele Frequency (MAF) and Chi Square Analysis,

[1] In CFE, the storage server is implicit as it carries out no computations.

our single data owner based FED protocol has an overhead of merely 1.5×, compared to SSE schemes (which offer no computing functionality).

To the best of our knowledge, no prior work achieves the features of collaborative databases, select and compute functionality and efficiency that we achieve in this work. See Table 2 for a comparison.

1.3 Overview of Constructions

We present several modular constructions of FED (and the single data-owner version sFED), which can be instantiated by plugging in different implementations of its components. Below we give a roadmap of how the two simpler primitives, Searchably Encrypted Datastore (SED) and Computably Encrypted Datastore (CED) can be securely dovetailed into a construction of FED.

The starting point of our constructions are single data-owner versions sSED and sCED. We show that these components can be implemented by leveraging constructions from the literature, namely, the Multi-Client SSE (MC-SSE) scheme due to Jarecki et al. [17] and CFE due to Naveed et al. [20]. The search query family supported by our sSED constructions are the same as in [17]. For sCED, we support a few specialized functions, as well as a general function family. The primitives sSED and sCED are of independent interest, and they can also be combined to yield a single data-owner version of FED (called sFED).

To upgrade sSED and sCED constructions into full-fledged (multi data-owner) SED and CED schemes, we require several new ideas. One challenge in this setting is to be able to hide the association of the (encrypted) data items with their data-owners. A simple approach, wherein each data-owner sends encryptions/secret shares of its data directly to the two servers, hence does not work. Our approach is to first securely merge the data from the different data-owners in a way that removes this association, and then use the single data-owner constructions on the merged data set. For this, both SED and CED constructions rely on *onion secret-sharing techniques*. Onion secret-sharing is a non-trivial generalization of the traditional mix-nets [6]. In a mix-net, a set of senders $\mathsf{D}_1, \cdots, \mathsf{D}_m$ want to send their messages $M_1, \cdots, M_m$ to a server S, with the help of an auxiliary server A (who does not collude with S), so that neither S nor A learns the association between the messages and the senders. We require the following generalization: each sender D_i wants to *secret-share* its message M_i between two servers S and A; that is, it sets $M_i = \sigma_i \oplus \rho_i$, and wants to send ρ_i to S and σ_i to A. While the senders want their messages to get randomly permuted in order to remove the association of data with themselves, the association between σ_i and ρ_i needs to be retained. As described in Sect. 4.1, onion secret-sharing provides a solution to this problem (and more).

In the case of CED, merging essentially consists of a multi-set union which can be solved using onion secret-sharing. But in the case of SED, merging entails merging "search indices" on individual data sets into a search index for the combined data set. A search index is a function that maps a keyword to a set of records; merging such indices requires that for each keyword, *the sets corresponding to it in the different indices* be identified and merged together. For this onion secret-sharing alone is not adequate. We propose two approaches to merge the indices – one in the random oracle model using an Oblivious Pseudorandom Function (OPRF) protocol, and another one with comparable efficiency in the standard model, relying on 2-party Secure Function Evaluation (SFE). Please see Sect. 4 for more details.

2 FED Framework

We use the notion of an *ideal functionality* [13] to specify our requirements from an FED system. FED is formulated as a two stage functionality, involving an *initialization* stage and a *query* stage. Figure 1 and Fig. 2 depict the FED functionality and protocol template schematically. The parties involved are:

- **Data-owners** D_i (for $i = 1, \cdots, m$, say) who are online only during the initialization phase. Each data-owner D_i has an input $Z_i \subseteq \mathcal{W} \times \mathcal{X}$, where for each $(w, x) \in Z_i$, w form the searchable attributes and x the computable values. $Z = \bigcup_i Z_i$ denotes the multi-set union of their inputs.
- **Storage Server** S, which is the only party with a large storage after the initialization stage.
- **Auxiliary Server** A, assumed not to collude with S.
- **Clients** which appear individually during the query phase. A client C has input a query of the form (Q, f) where $Q : \mathcal{W} \to \{0,1\}$ is a search query on the attributes, and f is a computation function on a multi-set of values. It receives in return $f(Q[Z])$ where $Q[Z]$ denotes the *multi-set* consisting of elements in $\mathcal{X}$, with the multiplicity of x being the total multiplicity of elements of the form (w, x) in Z, but restricted to w that are selected by Q; i.e., $\mu_{Q[Z]}(x) = \sum_{w \in \mathcal{W}:Q(w)=1} \mu_Z(w, x)$, where $\mu_R(y)$ denotes the multiplicity of an element y in multi-set R.

We instantiate our framework using protocols that provide security against active corruption of the clients and passive corruption of the other entities, allowing any subset not involving both the servers to collude. Our protocols provide different levels of (acceptable) leakage and are efficient on large scale data. We remark that these protocols indeed satisfy the desiderata outlined in Sect. 1: The data owners are not required to trust a single central authority or co-ordinate with each other, the clients are oblivious of data owners and of each other, and anonymity of data is maintained modulo the leakage to the storage server and auxiliary sever.

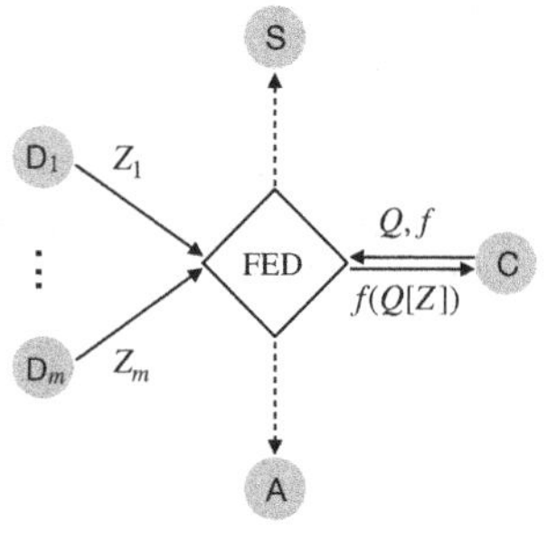

Fig. 1. FED functionality (dotted lines show leakage)

Keyword Search Queries: Following the searchable encryption literature we focus on "keyword queries." A keyword query is either a predicate about the presence of a single keyword in a record (document), or a boolean formula over such predicates. The searchable attribute for each record is a set of keywords, $w \subseteq \mathcal{K}$ where $\mathcal{K}$ is a given keyword space. In terms of the notation above, $\mathcal{W} = \mathcal{P}(\mathcal{K})$, the power set of $\mathcal{K}$. For instance, in a contact-tracing application, each keyword is a coarse-grained (space, time) pair, and a record is a set of such keywords. Risky co-ordinates are assumed to be explicitly enumerated in the query, and records matching any of them are returned for further computation. A basic search query could be a keyword occurrence query of the form Q_τ, for $\tau \in \mathcal{K}$, defined as $Q_\tau(w) = 1$ iff $\tau \in w$. A more complex search query can be specified as a boolean formula over several such keyword occurrence predicates.

Composite Queries: We shall sometimes allow Q and f to be more general than presented above. Specifically, we allow $Q = (Q_1, \cdots, Q_d)$, where each $Q_i : \mathcal{W} \to \{0,1\}$ and $f = (f_0, f_1, \cdots, f_d)$, where for $i > 0$, f_i are functions on multi-sets of values, and f_0 is a function on a d-tuple; we define $f(Q[Z]) := f_0(f_1(Q_1[Z]), \cdots, f_d(Q_d[Z]))$.

2.1 Protocol Template

All our protocols for FED use the template illustrated in Fig. 2. This is a fixed protocol (described below) that uses two ideal functionalities SED and CED, which in turn are implemented using secure protocols. Thus to obtain a full protocol for FED, we need to only specify the protocols for SED and CED.

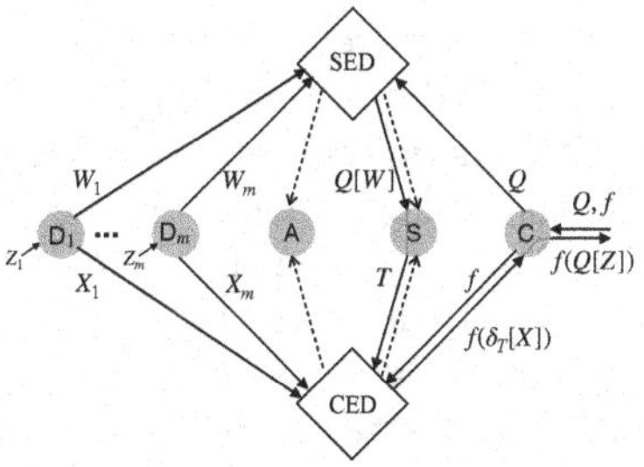

Fig. 2. FED protocol template

Searchably Encrypted Datastore. Recall that in an FED or sFED scheme, a query has two components – a *search query* Q and a computation function f. The Searchably Encrypted Datastore functionality (SED or sSED) has a similar structure, but supports only the search query; all the record identities that match the search query are revealed to the storage server S. This is referred as access pattern leakage in SSE literature and is leaked to S in most SSE schemes [10,17].

The functionality SED is depicted in Fig. 2: Each data-owner D_i maps its input Z_i to a pair (W_i, X_i), where $W_i \subseteq \mathcal{W} \times \mathcal{I}$ and $X_i \subseteq \mathcal{I} \times \mathcal{X}$ such that $(w, x) \mapsto ((w, \text{id}), (\text{id}, x))$ where id is randomly drawn from (a sufficiently large set) $\mathcal{I}$; the output that S receives when a client C inputs Q is the set of identities $Q[W] \subseteq \mathcal{I}$ where $W = \bigcup_i W_i$. Please see Sect. 3.1 and Sect. 4.2 for constructions of sSED and SED respectively.

Computably Encrypted Datastore. Our second functionality – CED, or its single data-owner variant sCED – helps us securely carry out a computation on an already filtered data set. The complexity of this computation will be related to the size of the filtered data rather than the entire contents of the data set.

In CED, as shown in Fig. 2, a data-owner D_i (who is online only during the setup phase) has an input in the form of $X_i \subseteq \mathcal{I} \times \mathcal{X}$. Later, during the query phase, clients can compute functions on a subset of data, $X = \bigcup_i X_i$. More precisely, a client C can specify a function f from a pre-determined function family, and the storage server S specifies a set $T \subseteq \mathcal{I}$, and C receives $f(\delta_T[X])$ where we define $\delta_T(\text{id}, x) = x$ if id $\in T$ and $\perp$ otherwise, and $\delta_T[X]$ is the multiset of x values obtained by applying $\delta_T(\text{id}, x)$ to all elements of X. Please see Sect. 3.2 and Sect. 4.2 for constructions of sCED and CED respectively.

Putting it Together. In the protocol for FED, during the initialization phase, each D locally maps its input Z_i to a pair (W_i, X_i), where $W_i \subseteq \mathcal{W} \times \mathcal{I}$ and $X_i \subseteq \mathcal{I} \times \mathcal{X}$ as follows: $(w, x) \mapsto ((w, \text{id}), (\text{id}, x))$ where id is randomly drawn from (a sufficiently large set) $\mathcal{I}$. Then, the parties D_i, S and A invoke the initialization phase of the SED functionality, and that of CED (possibly in parallel). During the query phase of FED, S, A and C first invoke the query phase of SED, so that S obtains $T = Q[W]$ as the output, where $W = \bigcup_i W_i$; then they invoke the query phase of CED (with the input of S being T), and C obtains $f(\delta_T[X])$. For correctness, i.e., $\delta_T[X] = Q[Z]$, we rely on there being no collisions when elements are drawn from $\mathcal{I}$.

Leakage: For every query (Q, f) from C, this protocol leaks the set $T = Q[W]$ to S, and in addition provides S and A with the leakage provided by the sSED and sCED functionalities. Note that D_i chooses ids at random to define W_i and X_i and the leakage functions of SED and CED are applied to these sets. Leaking T is what is known in the SSE literature as the *access pattern leakage*, i.e. it amounts to leaking the pattern of T over multiple queries. Specifically, since the ids are random, $T_1, \cdots, T_n$ contains only the information provided by the intersection *sizes* of all combinations of these sets. Formally, this leakage is given by $\text{pattern}(T_1, \cdots, T_n) := \{|\bigcap_{i \in R} T_i|\}_{R \subseteq [n]}$.

Notation Summary. For ease of reference, we provide a brief summary of our notation in the following Table 1.

Table 1. Notation index

Notation	Meaning	Notation	Meaning
$\mathcal{W}$	Universe of search-attributes in the data records	W	Input of D to SED functionality; $W \subseteq \mathcal{W} \times \mathcal{I}$
$\mathcal{I}$	Universe of identifiers used for the data records	Z	Input of D to FED functionality; $Z \subseteq \mathcal{W} \times \mathcal{X}$
$\mathcal{X}$	Universe of compute-attributes in the data records	X	Input of D to CED functionality; $X \subseteq \mathcal{I} \times \mathcal{X}$
(Q, f)	Input of C to FED;	T	Set of id's selected by a query; $T \subseteq \mathcal{I}$; $T = Q[W]$
$\mathcal{J}$	The set of unique identifiers for each record in the CED database; $\mathcal{J} \subseteq \mathcal{I}$; $\mathcal{J} = \{\text{id} \mid \exists x \text{ s.t. } (\text{id}, x) \in X\}$	DB(w)	For any $w \in \mathcal{K}$, the set of document indices that contain the keyword w
$\mathcal{L}$	Leakage function	$\mathcal{K}$	Universe of keywords

3 Single Data-Owner Protocols

Our protocols for the single data-owner setting follow the template from Sect. 2.1, but restricted to a single data owner D. We refer to the functionalities FED, SED and CED specialized to this setting as *s*FED, *s*SED and *s*CED respectively. As before, to instantiate a *s*FED protocol, we need only instantiate *s*SED and *s*CED protocols.

3.1 Instantiating *s*SED

We construct *s*SED by adapting constructions of symmetric searchable encryption (SSE) from the literature. Specifically, we use the multi-client extension of SSE by Jarecki et al. [17], where the data owner is separate from (and does not trust) the client. The main limitations of MC-SSE compared to *s*SED are that (1) in the former the data-owner D remains online throughout the protocol whereas in the latter D can be online only during the initialization phase, and (2) in the former the output is delivered to both S and C whereas in the latter it must be delivered only to S. We shall leverage the auxiliary server A to meet the additional requirements of *s*SED compared to MC-SSE. Our model also does not allow any leakage to the clients and handles active corruption of clients, in contrast to [17].

As a lightweight modification of the MC-OXT protocol of [17], our goals can be achieved by simply letting the auxiliary server A play the role of the data owner during the query phase, as well as the client in the MC-OXT protocol. This incurs some leakage to A, namely the search queries Q and the search outcome T. We provide some other alternatives with different security-efficiency tradeoffs in the full version [3] which maybe of independent interest. The leakage is summarized in Table 3. Note that since *s*SED is instantiated with random IDs, when used in the *s*FED protocol template (Fig. 2), only the *pattern* of the search outcomes, is revealed to A and S.

3.2 Instantiating *s*CED

We develop several *s*CED protocols for computing various functions on data filtered by a search query. The functionality, assumptions and leakage incurred by our protocols are summarized in Table 3. It will be convenient to define the set $\mathcal{J} \subseteq \mathcal{I}$ as $\mathcal{J} = \{\text{id} | \exists x \text{ s.t. } (\text{id}, x) \in X\}$. During each query, S has an input $T \subseteq \mathcal{J}$ (output from the SED functionality) and C has an input f from the computation function family.

- **Value Retrieval.** This is the functionality associated with standard SSE, where selected values are retrieved without any further computation on them. There is a single function in the corresponding computation function family, given by $f(\delta_T[X]) = \delta_T[X]$. Below we give a simple scheme which relies on A to support multiple clients who do not communicate directly with D.

Protocol sValRet

- *Initialization:* D picks a PRF key K, sets $\beta_{\text{id}} := x_{\text{id}} \oplus F_K(\text{id})$, where x_{id} is the value associated with id, and sends $\{(\text{id}, \beta_{\text{id}})\}_{\text{id} \in \mathcal{J}}$ to S and K to A.
- *Computation:* S picks a PRF key K_1, computes a random permutation of T and sends these to A. Additionally, S computes the same permutation of $\{\beta_{\text{id}} \oplus F_{K_1}(\text{id}) \mid \text{id} \in T\}$ and sends this to C. A sends $\{F_K(\text{id}) \oplus F_{K_1}(\text{id}) \mid \text{id} \in T\}$ to C. C outputs $\{a_i \oplus b_i\}_i$, where $\{a_i\}_i$ and $\{b\}_i$ are the messages it received from S and A.
- *Leakage,* $\mathcal{L}_{\texttt{sValRet}}$**:** On initialization, the ID-set $\mathcal{J}$ is leaked to S where $\mathcal{J} \subseteq \mathcal{I}, \mathcal{J} = \{\text{id} | \exists x \text{ s.t. } (\text{id}, x) \in X\}$. On each query, T is leaked to A.

- **Summation.** The family $\mathcal{F}_{\text{Sum}}$ consists of the single function f such that $f(S) = \sum_{x \in S} x$, where the summation is in a given abelian group which the domain of values is identified with. We provide a simple and efficient *s*CED protocol `sSum`, which is is similar to `sValRet`, but uses the homomorphic property of additive secret-sharing to aggregate the values while they remain shared. The value retrieval and summation protocols can also be extended to the setting where each value x is a vector $(x_1, \cdots, x_m)$. Please see the full version [3] for details.
- **General Functions.** We provide a *s*CED scheme for general functions. Our protocol for general functions, `sCktEval` makes use of garbled circuits and symmetric encryption, and can be seen as an adaptation of the CFE scheme from [20]. Despite the fact that garbled circuits have been optimised for performance in literature, garbled circuits incur high communication complexity and make this protocol less efficient than our other protocols. Due to space constraints, we refer the reader to [3] for the protocol description.
- **Composite Queries.** Recall that a composite query has $Q = (Q_1, \cdots, Q_d)$ and $f = (f_0, f_1, \cdots, f_d)$ such that $f(Q[Z]) := f_0(f_1(Q_1[Z]), \cdots, f_d(Q_d[Z]))$. We note that the *s*SED protocol for non-composite queries directly generalizes to composite queries, by simply running d instances of the original *s*SED functionality to let S learn $T_i = Q_i[W]$ for each i. But we need to adapt the *s*CED protocol to avoid revealing each $f_i(Q_i[Z])$ to C. Towards this, we observe a common feature in all our *s*CED protocols, namely that the last step has S and A sending secret shares for the output to C. We leverage this in our protocol `sComposite`, wherein, instead of S and A sending the final shares of $f_1, \ldots, f_d$ to C, they carry out a secure 2-party computation of f_0 with each input to f_0, namely $f_i(Q_i[Z])$ for $i \in [d]$ being secret-shared between S and A as described above. This can be implemented for a general f_0 using Yao's garbled circuits and oblivious transfer (OT) [32], or for special functions like linear combinations using standard and simpler protocols. The leakage includes the composed leakage of running d instances to compute $(f_1, \ldots, f_d)$. Also, the function f_0 is leaked to A and S.

4 Multiple Data-Owner Protocols

Following the template in Sect. 2.1, to instantiate a FED protocol, we instantiate SED and CED protocols in this section (see Table 3 for a quick summary). To do so, we first introduce the primitive of onion-secret sharing, which we will use in our constructions.

4.1 Onion Secret-Sharing

To illustrate onion secret-sharing, let us first consider the following simplified task: Each sender D_i wants to *share* its message M_i between two servers S and A; that is, it sets $M_i = \sigma_i \oplus \rho_i$, and wants to send σ_i to S and ρ_i to A. While the senders want their messages to get randomly permuted, removing the association of the data with themselves, the association between σ_i and ρ_i needs to be retained. Onion secret-sharing provides a solution to this problem, as follows. Below, let PK_{S} be the public-key of S for a semantically secure public-key encryption scheme (PK_{A} is defined analogously), and let $[\![M]\!]_{PK}$ denote encryption of M using a public-key PK. Now, each D_i sends $[\![(\rho_i, \zeta_i)]\!]_{PK_{\mathsf{S}}}$ to A, where ζ_i is of the form $[\![\sigma_i]\!]_{PK_{\mathsf{A}}}$. A mixes these ciphertexts and forwards them to S, who decrypts them to recover pairs of the form (ρ_i, ζ_i). Now, S reshuffles (or sorts) these pairs, stores ρ_i and sends ζ_i (in the new order); A recovers σ_i from ζ_i (in the same order as ρ_i are maintained by S).

This can be taken further to incorporate additional functionality. As an example of relevance to us, suppose A wants to add a short, private tag to the messages being secret-shared so that the tag persists even after random permutation. Among the messages which were assigned the same tag, A should not be able to link the shares it receives after the permutation to the ones it originally received; S should obtain no information about the tags. Looking ahead, this additional functionality will be required in one of our SED constructions. One solution is for A to add encrypted tags to the data items, and then while permuting the data items, S would *rerandomize* the ciphertexts holding the tags. We present an alternate approach, which does not require additional functionality from the public-key encryption scheme, but instead augments onion secret-sharing with extra functionality:

- D_i holds input M_i and shares it as $M_i = \sigma_i \oplus \rho_i$. It creates a 3-way additive secret sharing of 0 (the all 0's string), as $\alpha_i \oplus \beta_i \oplus \gamma_i = 0$, and sends $(\alpha_i, [\![\beta_i, \rho_i, [\![\gamma_i, \sigma_i]\!]_{PK_{\mathsf{A}}}]\!]_{PK_{\mathsf{S}}})$ to A.
- A assigns tags τ_i for each of them, and sends (in sorted order) $(\tau_i \oplus \alpha_i, [\![\beta_i, \rho_i, [\![\gamma_i, \sigma_i]\!]_{PK_{\mathsf{A}}}]\!]_{PK_{\mathsf{S}}})$ to S.
- S sends $(\tau_i \oplus \alpha_i \oplus \beta_i, [\![\gamma_i, \sigma_i]\!]_{PK_{\mathsf{A}}})$ to A, in sorted order; it stores ρ_i (in the same sorted order).
- A recovers $(\tau_i \oplus \alpha_i \oplus \beta_i \oplus \gamma_i, \sigma_i) = (\tau_i, \sigma_i)$.

This allows S and A to receive all the shares (in the same permuted order); S learns nothing about the tags; A cannot associate which shares originated from which D_i, except for what is revealed by the tag. (Even if S or A collude with some D_i, the unlinkability is retained for the remaining D_i). In Sect. 4.2, we use a variant of the above scheme.

4.2 Instantiating SED

We describe a general protocol template to realize the SED functionality, using access to the sSED functionality. The high-level plan is to let A create a merged database so that it can play the role of D for sSED. However, since we require privacy against A, the merged database should lack all information except statistics that we are willing to leak. Hence, during the initialization phase, we not only merge the databases, but also replace the keywords with pseudonyms and keep other associated data encrypted. We use pseudonyms for keywords (rather than encryptions) to support queries: during the query phase, the actual keywords will be mapped to these pseudonyms and revealed to A. These two tasks at the initialization and query phases are formulated as two sub-functionalities—$\mathsf{merge\text{-}map}$ and map—collectively referred to as the functionality mmap, and are explained below.

Functionality Pair $\mathsf{mmap} = (\mathsf{merge\text{-}map}, \mathsf{map})$

- Functionality $\mathsf{merge\text{-}map}$ takes as inputs W_i from each D_i; it generates a pair of "mapping keys" $K_{\mathsf{map}} = (K_{\mathsf{S}}, K_{\mathsf{A}})$ (independent of its inputs), and creates a "merged-and-mapped" input set, $\hat{W}$. Merging simply refers to computing the (multi-set) union $W = \bigcup_i W_i$; mapping computes the multi-set $\hat{W} = \{(\hat{w}, \hat{\mathrm{id}}) | (w, \mathrm{id}) \in W, \hat{w} = \mathfrak{M}_{K_{\mathsf{map}}}(w),$ and $\hat{\mathrm{id}} = \mathfrak{M}_{K_{\mathsf{map}}}(\mathrm{id})\}$, where the mapping function $\mathfrak{M}$ is to be specified when instantiating this template
- Functionality map takes K_{S} and K_{A} as inputs from S and A respectively, and a query Q from a client C; then it outputs a new query $\hat{Q}$ to C. We shall require that there is a decoding function D such that $Q[W] = \{D_{K_{\mathsf{S}}}(\hat{\mathrm{id}}) | \hat{\mathrm{id}} \in \hat{Q}(\hat{W})\}$, where $\hat{W}$ is as described above.

These functionalities may specify leakages to S and/or A, but not to the data-owners or clients. Note that since map gives an output $\hat{Q}$ to C, for security, we shall require that it can be simulated from Q.

We give two constructions, with different efficiency-security trade-offs. Recall that each data-owner D_i has an input $W_i \subseteq \mathcal{W} \times \mathcal{I}$. In both the solutions, D_i shall use a representation of W_i as a set $\widetilde{W_i} \subseteq \mathcal{K} \times \mathcal{I}$ such that $(w, \mathrm{id}) \in W_i$ iff $w = \{\tau | (\tau, \mathrm{id}) \in \widetilde{W_i}\}$ ($w \subseteq \mathcal{K}$ is the set of keywords (searchable attributes) and $\tau \in \mathcal{K}$ is a keyword). In the two solutions below, the mapping function $\mathfrak{M}$ maps the keywords differently; but in both the solutions, an identity id that occurs in $\widetilde{W_i}$ is mapped to $\zeta_i^{\mathrm{id}} \leftarrow [\![\mathrm{id}]\!]_{PK_{\mathsf{S}}}$ (an encryption of id under S's public-key). In one scheme, we use an oblivious pseudorandom function (OPRF) protocol to calculate $\mathfrak{M}$, while the other relies on a secure function evaluation for equality.

Protocol mmap-OPRF

merge-map:

- **Keys.** S generates a keypair for a CPA-secure PKE scheme, (PK_S, SK_S), and similarly A generates (PK_A, SK_A). The keys PK_S and PK_A are published. S also generates a PRF key K.
- **Oblivious Mapping.** Each data-owner, for each $\tau \in \mathcal{K}_i$, D_i engages in an *Oblivious PRF* (OPRF) evaluation protocol with S, in which D_i inputs τ, S inputs K and D_i receives as output $\hat{\tau} := F_K(\tau)$. D_i carries out $L_i - |\mathcal{K}_i|$ more OPRF executions (with an arbitrary $\tau \in \mathcal{K}_i$ as its input) with S.
- **Shuffling.** Each data-owner D_i computes $\zeta_i^{\text{id}} \leftarrow [\![\text{id}]\!]_{PK_S}$, for each id such that there is a pair of the form $(\tau, \text{id}) \in \widetilde{W}_i$. All the data-owners use the help of S to route the set of all pairs $(\hat{\tau}, \zeta_i^{\text{id}})$ to A, as follows:
  - Each D_i, for each (τ, id) computes $\xi_{i,\tau}^{\text{id}} \leftarrow [\![(\hat{\tau}, \zeta_i^{\text{id}})]\!]_{PK_A}$, and sends it to S. Further D_i computes $N_i - |\widetilde{W}_i|$ more ciphertexts of the form $[\![\bot]\!]_{PK_A}$.
  - S collects all such ciphertexts (N_i of them from D_i, for all i), and lexicographically sorts them (or randomly permutes them). The resulting list is sent to A.
  - A decrypts each item in the received list, and discards elements of the form $\bot$ to obtain a set consisting of pairs of the form $(\hat{\tau}, \hat{\text{id}})$, where $\hat{\text{id}} = \zeta_i^{\text{id}}$. $\hat{W}$ is defined as the set of pairs of the form $(\hat{w}, \hat{\text{id}})$ where $\hat{w}$ contains all $\hat{\tau}$ values for each $\hat{\text{id}}$.
- **Outputs.** A's outputs are $\hat{W}$ and an empty K_A; S's output is $K_S = (SK_S, K)$.

map: A client C has an input Q which is specified using one or more keywords. For each keyword τ appearing in Q, C engages in an OPRF execution with S, who inputs the key K that is part of K_S. The resulting query is output as $\hat{\tau}$.

Fig. 3. Protocol mmap-OPRF implementing functionality mmap.

Construction mmap-OPRF: The pair of protocols for the functionalities merge-map and map, collectively called mmap-OPRF, is shown in Fig. 3. In this solution, the mapping key K_A is empty, and K_S consists of a PRF key K, in addition to a secret-key for a public key encryption (PKE) scheme, SK_S. $\mathfrak{M}$ maps each keyword τ using a pseudorandom function with the PRF key K. This is implemented using an oblivious PRF execution with S, in both merge-map and map protocols. That is, for $w \subseteq \mathcal{K}$ we have $\mathfrak{M}_{K_{\mathsf{map}}}(w) = \{F_K(\tau) | \tau \in w\}$, where F is a PRF.

Note that mmap-OPRF uses two *a priori* bounds for each data-owner, N_i and L_i such that $|\widetilde{W}_i| \leq N_i$ and $|\mathcal{K}_i| \leq L_i$, where $\mathcal{K}_i = \{\tau | \exists \text{id s.t. } (\tau, \text{id}) \in \widetilde{W}_i\}$ is the set of keywords in D_i's data.

We point out the need for OPRF evaluations. If we allowed S to simply give the key K to every data-owner to carry out the PRF evaluations locally, then, if A colludes with even one data-owner D_i, it will be able to learn the actual keywords in the entire data-set.

Leakage, $\mathcal{L}^{\text{OPRF}}_{\text{mmap}}$**:** S learns the bounds N_i and L_i for each i, and A learns $\sum_i N_i$; from the map phase, S learns the number of keywords in the query.
We also include in $\mathcal{L}^{\text{OPRF}}_{\text{mmap}}$ what the output $\hat{W}$ to A reveals (being an output, this is not "leakage" for `mmap-OPRF`; but it manifests as leakage in the `SED` protocol that uses this functionality). $\hat{W}$ reveals an anonymized version of the keyword-identity incidence matrix of the merged data[2], where the actual labels of the rows and columns (i.e., the keywords and the ids) are absent. Assuming adaptive corruption of A after the setup phase (with reliable erasures), the adversary doesn't get this leakage. Please refer to the full version [3] for a discussion on implications of this leakage, and possible ways to avoid it.

Construction `mmap-SFE`: Our next protocol avoids OPRF evaluations. The idea here is to allow each data owner to send secret-shared keywords between S and A, and rely on secure function evaluation (SFE) to associate the keywords with pseudonymous handles, so that the same keyword (from different data owners) gets assigned the same handle (but beyond that, the handles are uninformative). Further, neither server should be able to link the keyword shares with the data owner from which it originated, necessitating a shuffle of the outputs. Due to the complexity of the above task, a standard application of SFE will be very expensive. Instead, we present a much more efficient protocol that relies on secure evaluation only for equality checks, with the more complex computations carried out locally by the servers; as a trade-off, we shall incur leakage similar to that of the OPRF-based protocol above. Below, we sketch the ideas behind the protocol, with the formal description in [3].

Consider two data owners who have shared keywords τ_1 and τ_2 among A and S, so that A has (α_1, α_2) and S has (β_1, β_2), where $\tau_1 = \alpha_1 \oplus \beta_1$ and $\tau_2 = \alpha_2 \oplus \beta_2$. Note that $\tau_1 = \tau_2 \Leftrightarrow \alpha_1 \oplus \beta_1 = \alpha_2 \oplus \beta_2 \Leftrightarrow \alpha_1 \oplus \alpha_2 = \beta_1 \oplus \beta_2$. So, for A to check if $\tau_1 = \tau_2$, A and S can locally compute $\alpha_1 \oplus \alpha_2$ and $\beta_1 \oplus \beta_2$ and use a secure evaluation of the equality function to compare them (with only A learning the result). By comparing all pairs of keywords in this manner, A can identify items with identical keywords and assign them pseudonyms (e.g., small integers), while holding only one share of the keyword. This can be done efficiently using *securely pre-computed correlations*. Let's say we wish to compute if strings $x = y$. We map x, y using a collision resistant hash H to 128 bits, and interpret them as elements in a field, say $\mathbb{F} = GF(2^{128})$. The protocol then uses the following pre-computed correlation: Alice holds $(a, p) \in \mathbb{F}^2$ and Bob holds $(b, q) \in \mathbb{F}^2$ such that $a + b = pq$ and p, q and one of a, b are uniformly random. Alice sends $u := H(x) + p$ to Bob, where x is her input. Bob replies with $v := q(u - H(y)) - b$, where y is his input. Alice concludes $x = y$ iff $a = v$.

The data shared by the data owners to the servers includes not just keywords, but also the records (document identifiers) associated with each keyword. To minimize the number of comparisons needed, each data owner packs all the records corresponding to a keyword τ into a single list that is secret-shared along with the keyword (rather than share separate (τ, id) pairs). However, after

[2] This is a 0–1 matrix with 1 in the position indexed by (τ, id) iff $(\tau, \text{id}) \in \bigcup_i \widetilde{W}_i$. The rows and columns may be considered to be ordered randomly.

handles have been assigned to the keywords, such lists should be unpacked and shuffled to erase the link between multiple records that came from a single list (and hence the same data owner). Each entry in each list can be tagged by A using the keyword handle assigned to the list, but then the entries need to be shuffled while retaining the tag itself. This is accomplished using onion secret-sharing – please see the full version [3] for details.

This construction uses $O((\sum_i L_i)^2)$ secure equality checks, where L_i is an upper bound on the total number of keywords in D_i's data. We describe an approach to improve this, the leakage and security analysis as well as a comparison with the OPRF protocol in the full version [3].

Instantiating CED: To upgrade a *s*CED protocol to a CED protocol we need additional tricks using onion secret-sharing. This results in a more involved security analysis and slower performance. Due to space constraints, we refer the reader to the full version [3] for the description.

5 Implementation and Experimental Results

We show the feasibility and performance of our protocols on realistically large-scale computations by running tasks representative of *Genome Wide Association Studies (GWAS)*, which is widely used to study associations between genetic variations and major diseases [4].

Implementation/Setup Details. The system was implemented in C++11 using open-source libraries The MC-OXT protocol of Jarecki et al. [17] was reimplemented for use in our SED protocols. Experiments were performed on a Linux system with 32 GB RAM and a 8-core 3.4 GHz Intel i7-6700 CPU. The network consisted of a simulated local area network (simulated using linux tc command), with an average bandwidth of 620 MBps and ping time of 0.1 ms.

Our Functionalities. The specific functions we choose as representative of GWAS were adapted from the *iDash competition* [1] for secure outsourcing of genomic data. Each record in a GWAS database corresponds to an individual, with fields corresponding to demographic and phenotypic attributes (like sex, race etc.), as well as genetic attributes. In a typical query, the demographic and phenotypic attributes are used for filtering, and statistics are computed over the genetic information. In our experiments, the following statistics were computed.

- **Minor Allele Frequency (MAF):** This can be described using the formula $f_0(f_1(Q[Z]), f_2(Q[Z]))$, where f_1 computes $N = |Q[Z]|$, f_2 computes the summation function, and f_0 computes the following formula: $f_0(x, y) = \frac{\min(y, 2x-y)}{2x}$. We implement f_2 using our `sSum` protocol, while f_0 is implemented using our `sComposite` protocol (using GC for f_0).

- **Chi-square analysis (ChiSq):** Using two search filters for queries Q and Q', χ^2 statistic can be abstractly described using the formula $f_0(f_1(Q[Z]), f_2(Q[Z]), f_3(Q'[Z]), f_4(Q'[Z]))$, where f_1, f_3 are summation functions, f_2 and f_4 compute $|Q[Z]|$ and $|Q'[Z]|$ respectively, and $f_0(a, b, c, d) = \frac{(b+d)(ad-bc)^2}{bd(a+c)((b+d)-(a+c))}$. As previously, f_1, f_3 are implemented using our `sSum` protocol and f_0 implements the above formula using our `sComposite` protocol.
- **Average Hamming Distance:** Hamming distance between 2 genome sequences, often used as a metric for genome comparison, is defined $f_{x^*}(Q[Z]) = \frac{\sum_{x \in Q[Z]} \Delta(x, x^*)}{|Q[Z]|}$, where Δ stands for the hamming distance of two strings. Here, we consider the entire genotype vector x for each individual (rather than the genotype at a given locus). We implement this using our general functions protocol `sCktEval`.
- **Genome retrieval:** This is a retrieval task, involving retrieval of genomic data at a locus for individuals selected by a search criteria. This is implemented using our `sValRet` protocol.

Dataset and Queries. We used synthetic data inspired by real-world applications [15], with 10,000–100,000 records and 50 SNPs, and with the number of filtered records ranging 2,000–12,000. For Hamming distance, which we implemented using our protocol for general functions (`sCktEval`) and costlier than our other protocols, we used 5,000–25,000 records with 600–3500 filtered records. The experimentation parameters were chosen to showcase the observations described below. Effects of changing the experimental setting (e.g., a WAN instead of a LAN model, bigger datasets and keyword spaces, etc.) are discussed in the full version.

Metrics. Our results are reported over two metrics – the total time taken and the total communication size, across all entities. These metrics are reported separately for the initialization and query phases. Also, the costs incurred by the SED component of the protocols are shown separately, as this could serve as the baseline search cost against which the FED cost can be compared. We performed experiments in both the single (sFED) and multiple data owner (FED) setting, and in the latter case with both the OPRF-based and SFE-based SED protocols.

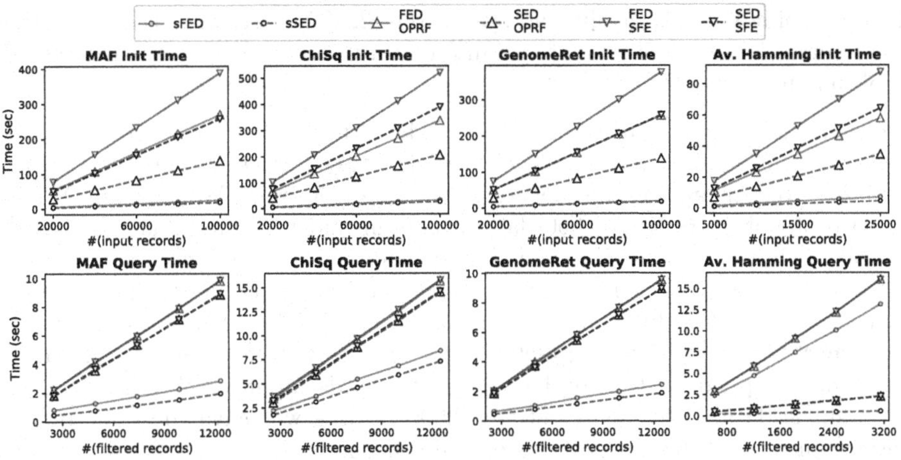

Fig. 4. Initialization and query-phase time plots for our four applications. (See Fig. 7 for communication costs.)

5.1 Observations

Linear in Data. In Fig. 4, we plot the initialization and query times (summed over all parties, with serial computation and LAN communication) for various functions, each one using our different protocols. The experiment runs on a single data-owner (even for FED) and the initialization times are plotted against the total number of records, while query times are plotted against the number of filtered records. As expected, all the times are linear in the number of records. Comparing the different versions of the protocols, we note that for initialization, the multiple data-owner protocols (FED-OPRF and FED-SFE) are slower that the single data-owner protocol (*s*FED) by a factor of about 10–12x and 15–18x respectively. The overhead in the query phase is a more modest 5x factor.

Efficiency of Query Phase. From our experimental results in Fig. 4, we observe that the query time for our CED and SED protocols is linearly proportional to the number of filtered records.

Comparison to SSE. We show a breakup of our *s*FED cost between search and compute components in Fig. 5. While the cost for Av. Hamming includes a dataset of 15,000 records and a filtered set of around 1800 records, other protocols involve a dataset of 20,000 records and a filtered set of around 2500 records. As can be observed, the overhead for supporting computation in addition to search ranges between 1.2–1.7x, except for Av. Hamming, which as a result of computing a large function using

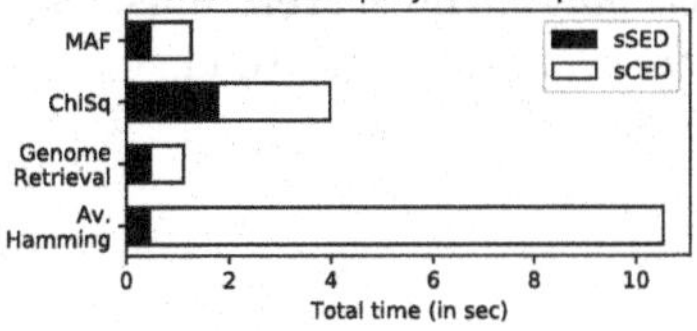

Fig. 5. Break-up of *s*FED cost.

the protocol sCktEval, involves an overhead of 22x. Nevertheless, the computation remains feasible: we evaluate Hamming distance between strings as large as 1.25k bits in 10 s.

Scalability with Number of Data-Owners. The setup phase of our protocols scale well with the number of data-owners participating in the system (see Fig. 6). Each data owner provides one record. As can be seen from the plot, our OPRF based protocol scales to 100,000 data owners in under 10 min. The SFE-based protocol (which scales quadratically, but avoids the use of random oracles) scales only up to 2000 owners in a similar time.[3] In our experiments data owners were serviced serially by S, but we estimate a drop of 100–200 s for our largest benchmarks if we exploited parallelism. Our protocols' performance in the query phase only depend on the total number of records input by the data-owners i.e. a single data owner contributing 10,000 records will have the same query performance as 10,000 data owners contributing 1 record.

Light-Weight/Efficient Clients. As is apparent from the complexity analysis of our protocols, our clients are extremely efficient and light-weight, typically only performing computation proportional to size of queries and outputs. Indeed, in all our experiments, client's computation time never exceeded 3 ms.

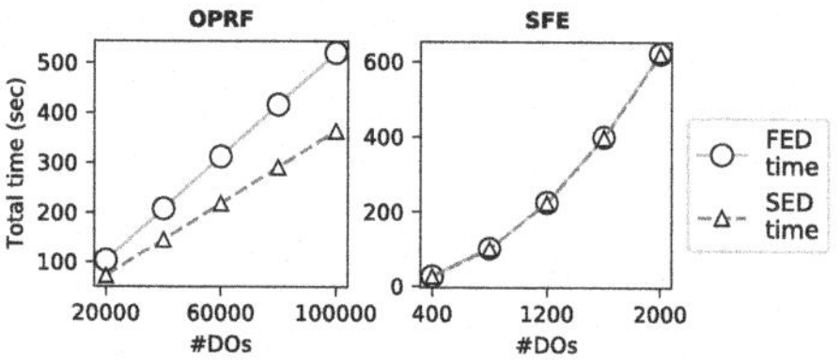

Fig. 6. Scaling of initialization time of FED/SED with increasing number of data-owners (each holding one record)

5.2 Comparison with Generic MPC Approaches

While MPC techniques cannot be directly deployed in our setting, as the data-owners cannot interact with each other during the query phase, they can be adapted to our 2-server architecture (assuming they do not collude with each other). The two servers could use generic semi-honest secure 2-party computation (2PC) on data that is kept secret-shared between them. However, as we note below, such a solution is significantly less efficient than our constructions.

We focus on the "Genome Retrieval" functionality for a single keyword match with a single data owner, as a minimalistic task for comparison. Note that *for each query*, the servers must engage in a joint computation proportional to the *entire database size*. To ensure a fair comparison, we allow the servers to obtain a

[3] The cost for generating correlated inputs to the server was not included, as this can be done ahead of time even before the setup phase in an efficient manner [19]. Here, the size of the field elements used for the secure equality check was aggressively set to 64 bits, which is adequate to avoid hash collisions for a modest number of data owners and distinct keywords.

similar leakage as our constructions do; this significantly simplifies the functions being securely computed, and reduces the costs.[4]

There are essentially 3 families of generic 2PC protocols which are relevant for the comparison: i) Garbled Circuits (GC) [11], ii) GMW [11] and iii) Fully Homomorphic Encryption (FHE) [2]. For the query phase, GC and GMW suffer heavily in communication cost, since communication between the servers is proportional to the circuit size. For instance, in the case of GC, we can lower bound the communication by restricting to the filtering step in the computation, and then calculating the size of the garbled circuit and oblivious transfers (OT) performed. Concretely, for 100,000 records, this implies a minimum communication size of 576 MB. For GMW, for simplicity, we omit the cost of generation of OT correlations (even though, in steady state, they need to be incurred as more queries come in). Then, considering merely 4 bits of communication per AND gate, for the same setting as above, we lower bound the communication during the query phase by 3.6 MB. On the other hand, for the same 100,000 records and a query for filtering around 12,000 records, our solution incurs only 2.8 MB of communication. For FHE based implementations, the bottleneck is homomorphic computation. A simple lower bound on execution time may be obtained by bounding the number of multiplications performed and considering the cost of a single homomorphic multiplication. For instance, for the smallest size parameters in the Microsoft SEAL homomorphic encryption library [2], the time for a single homomorphic multiplication is the order of milliseconds. Thus a simple estimate for execution time is at least 7 s, whereas for the same parameters, we achieve a time of 2.4 s. We note that our performance gains increase substantially with increase in the number of records in the database (indeed for 10 million records and a filtered set size of around 500, while we incur 112 KB of communication in $<$1.7 s, GC/GMW require 57.6 GB/360 MB and FHE is at least of the order of hundreds of seconds). Please see the full version [3] for more details including a discussion on the multiple data owner case.

[4] To avoid such leakage in this task, a *top-k sorting network* would be included in the computation circuit to collect selected items. A bitonic sorting network for selecting k items out of n has $\Theta(n \log^2 k)$ comparators [26]; selecting a thousand items from a database of a million items (each a few bytes long), would require *billions* of gates.

Table 2. Comparison of related works

	Functionality					Security				Query-phase efficiency**	
	Search-and-compute	Search supports Boolean formulas	Compute supports general functions	Multi data-owner	Multi client	Corruption level			Notes/additional assumptions	Client	Server
						Server (s)	Client (s)	Data-owner (s)			
CryptDB [25]	Y	Y	Y (SQL)	N	N	Passive	Passive	Passive	Trusted application server and passive DBMS server. No server-client collusion. Only snapshot attacker	$O(1)$	$O(t)$
Seabed [22]	Y	Y	Y (OLAP)	N	Y	Passive	Passive	Passive	No server-client collusion. Clients need a key given by the data owner. Only snapshot attacker	$O(1)$	$O(n)$
Arx [24]	Y	Y	N	N	N	Passive	Passive	Passive	Trusted application server and passive DBMS server. No server-client collusion. Only snapshot attacker	$O(1)$	$O(t \log n)$
OXT [5]	N	Y	NA	N	N	Passive	Passive	Passive	–	$O(t_1 k)$	$O(t_1 k)$
OSPIR [17]	N	Y	NA	N	Y	Passive	Malicious	Malicious	No server-client collusion. Also provides query privacy to clients assuming no server-data owner collusion	$O(t_1 k)$	$O(t_1 k)$
BlindSeer [12,23]	N	Y	NA	N	Y	Passive	Malicious	Passive	No server-client collusion. Also provides query privacy to clients assuming no server-data owner collusion	$O(t_1 \log n)$	$O(t_1 \log n)$
CFE [20]	N	NA	Y (circuits)	Y	Y	Passive	Malicious	Passive	No authority-client or authority-storage collusion. Can provide query privacy to clients	$O(n)$	$O(n)$
Multi-key/user SE [14,30]	N	N	N	Y	Y	Passive	Passive	Malicious	File sharing setting. Server allowed to collude with subset of data owners or clients. DOs aware of clients (fixed at start of protocol). Shared data can be traced back to original DO	$O(1)$	$O(n)$
This work	Y	Y	Y (circuits)	Y	Y	Passive	Malicious	Passive	Two non-colluding servers. Either server can collude with subset of clients/data owners. Can provide search/function query privacy to clients, anonymity to data owners. DOs oblivious of clients (can dynamically join system)	$O(1)$	$O(t_1 k)$

**Based on a query of the form SELECT SUM(COL) WHERE $\text{COL}_1 = \alpha_1$ AND $\text{COL}_2 = \alpha_2 \cdots \text{COL}_k = \alpha_k$, where COL and COL_i are column names. If computation is not supported, SUM is omitted; if search is not supported WHERE clause is omitted. n denotes the total number of records in the database and t denotes the number of records matching the search condition. WLOG, assume $\text{COL}_1 = \alpha_1$ filters the least number of records and t_1 denotes the number of records satisfying it. In each case, assume that all supported pre-processing has been done prior to the query.

Table 3. Summary of protocols

Protocol family	Instantiation	Reference	Functionality	Security assumptions	Leakage (S)	Leakage (A)
*s*SED	MD-MC-OXT	Section 3.1	Search	Non-collusion (A and S), Decision Diffie-Hellman, Random Oracle Model, Secure PRFs	Access pattern, Size Pattern, Equality Pattern, Conditional Intersection Pattern, Total Database Size	Access Pattern, Query
*s*CED	sValRet	Section 3.2	Value Retrieval	Non-collusion (A and S), Secure PRFs	Randomly Chosen IDset ($\mathcal{J}$)	Access Pattern
*s*CED	sSum	Section 3.2	Summation	Non-collusion (A and S), Secure PRFs	Randomly Chosen IDset ($\mathcal{J}$)	Access Pattern
*s*CED	sComposite	Section 3.2	$f_0(f_1(Q_1[Z]), \dots, f_d(Q_d[Z]))$	Non-collusion (A and S), Oblivious Transfer	Circuit for f_0 (only circuit structure if using Yao's GC) and respective leakages for $f_0, \dots, f_d$	Circuit for f_0 and respective leakages for $f_1, \dots, f_d$
*s*CED	sCktEval	[3]	Poly-time computable functions	Non-collusion (A and S), Secure PRFs	Randomly Chosen IDset ($\mathcal{J}$), Circuit Structure for f	Access Pattern, Circuit for f
SED	**mmap-OPRF, MD-MC-OXT	Section 4.2	Multiple D_i's + Search	**One More Diffie-Hellman, Public Key Encryption	**Upper bound for frequency of each keyword sent by D_i (called as N_i), Upper bound for number of keywords sent by D_i (called as L_i), Number of keywords in the query	**$\sum_i N_i$, $\hat{W}$ (erased after initialization is complete, can be removed if considering alternate SSE schemes)
SED	**mmap-SFE, MD-MC-OXT	Section 4.2	Multiple D_i's + Search	**Random Oracle Model*, Public Key Encryption	**Include leakage from above entry, Upper bound for total number of keywords (L)	**Include leakage from above entry, Number of keywords in the query
CED	**ValRet	[3]	Multiple D_i's + Value Retrieval	**Public Key Encryption	**Upper bound on computing data of D_i (formally $\lvert X_i \rvert$)	**Randomly Chosen IDset ($\mathcal{J}$)
CED	**Sum	[3]	Multiple D_i's + Summation	**Public Key Encryption	**Upper bound on computing data of D_i (formally $\lvert X_i \rvert$)	**Randomly Chosen IDset ($\mathcal{J}$)
CED	**CktEval	[3]	Multiple D_i's + Poly-time computable functions	**Public Key Encryption	**N/A	**Upper bound on computing data of D_i (formally $\lvert X_i \rvert$)

* Random Oracle needed for MC-OXT-mod, but not mmap-SFE.
** In addition to the assumption/leakage in the single data owner analogue.

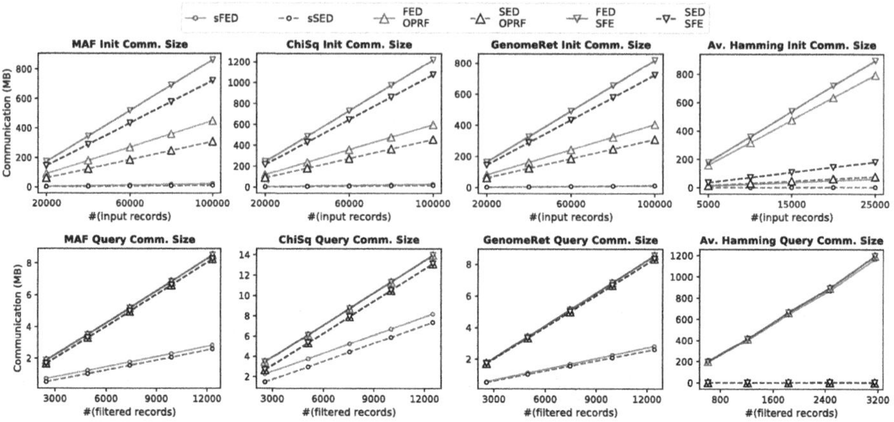

Fig. 7. Total communication across all entities in the initialization/query phases for single/multi data owner (contrasting two SED protocols in Sect. 4.2, OPRF and SFE) for our different applications.

References

1. IDASH Privacy & Security Workshop 2015 (2015). http://www.humangenomeprivacy.org/2015/competition-tasks.html
2. Microsoft SEAL (2019). https://github.com/Microsoft/SEAL
3. Agrawal, S., Garg, R., Kumar, N., Prabhakaran, M.: A practical model for collaborative databases: securely mixing, searching and computing. Cryptology ePrint Archive, Report 2019/1262 (2019). https://eprint.iacr.org/2019/1262
4. Bush, W.S., Moore, J.H.: Genome-wide association studies. PLoS Comput. Biol. (2012)
5. Cash, D., Jarecki, S., Jutla, C., Krawczyk, H., Roşu, M.-C., Steiner, M.: Highly-scalable searchable symmetric encryption with support for Boolean queries. In: Canetti, R., Garay, J.A. (eds.) CRYPTO 2013. LNCS, vol. 8042, pp. 353–373. Springer, Heidelberg (2013). https://doi.org/10.1007/978-3-642-40041-4_20
6. Chaum, D.L.: Untraceable electronic mail, return addresses, and digital pseudonyms. ACM Commun. **24**(2), 84–90 (1981)
7. Cho, H., Wu, D.J., Berger, B.: Secure genome-wide association analysis using multiparty computation. Nat. Biotechnol. **36**, 547–551 (2018)
8. Chor, B., Goldreich, O., Kushilevitz, E., Sudan, M.: Private information retrieval. In: FOCS (1995)
9. Corrigan-Gibbs, H., Boneh, D., Mazières, D.: Riposte: an anonymous messaging system handling millions of users. In: IEEE S&P (2015)
10. Curtmola, R., Garay, J., Kamara, S., Ostrovsky, R.: Searchable symmetric encryption: improved definitions and efficient constructions. In: ACM CCS (2006)
11. Demmler, D., Schneider, T., Zohner, M.: ABY - a framework for efficient mixed-protocol secure two-party computation. In: 22nd Annual Network and Distributed System Security Symposium, NDSS (2015)

12. Fisch, B.A., et al.: Malicious-client security in blind seer: a scalable private DBMS. In: IEEE S&P (2015)
13. Goldreich, O.: Foundations of Cryptography: Basic Applications. Cambridge University Press, Cambridge (2004)
14. Hamlin, A., Shelat, A., Weiss, M., Wichs, D.: Multi-key searchable encryption, revisited. In: Abdalla, M., Dahab, R. (eds.) PKC 2018. LNCS, vol. 10769, pp. 95–124. Springer, Cham (2018). https://doi.org/10.1007/978-3-319-76578-5_4
15. Hirokawa, M., et al.: A genome-wide association study identifies PLCL2 and AP3D1-DOT1L-SF3A2 as new susceptibility loci for myocardial infarction in Japanese. Eur. J. Hum. Genet. **23**(3), 374–380 (2015)
16. Jagadeesh, K.A., Wu, D.J., Birgmeier, J.A., Boneh, D., Bejerano, G.: Deriving genomic diagnoses without revealing patient genomes. Science **357**(6352), 692–695 (2017)
17. Jarecki, S., Jutla, C.S., Krawczyk, H., Rosu, M., Steiner, M.: Outsourced symmetric private information retrieval. In: ACM CCS (2013)
18. Kamara, S., Raykova, M.: Secure outsourced computation in a multi-tenant cloud. In: IBM Workshop on Cryptography and Security in Clouds (2011)
19. Keller, M., Orsini, E., Scholl, P.: MASCOT: faster malicious arithmetic secure computation with oblivious transfer. Cryptology ePrint Archive, Report 2016/505 (2016). https://eprint.iacr.org/2016/505
20. Naveed, M., et al.: Controlled functional encryption. In: CCS (2014)
21. Orencik, C., Selcuk, A., Savas, E., Kantarcioglu, M.: Multi-keyword search over encrypted data with scoring and search pattern obfuscation. Int. J. Inf. Secur. **15**, 251–269 (2016). https://doi.org/10.1007/s10207-015-0294-9
22. Papadimitriou, A., et al.: Big data analytics over encrypted datasets with seabed. In: OSDI (2016)
23. Pappas, V., et al.: Blind seer: a scalable private DBMS. In: IEEE S&P (2014)
24. Poddar, R., Boelter, T., Popa, R.A.: Arx: a strongly encrypted database system. IACR Cryptology ePrint Archive (2016)
25. Popa, R.A., Redfield, C., Zeldovich, N., Balakrishnan, H.: CryptDB: protecting confidentiality with encrypted query processing. In: Symposium on Operating Systems Principles (2011)
26. Shanbhag, A., Pirk, H., Madden, S.: Efficient top-k query processing on massively parallel hardware. In: Proceedings of the 2018 International Conference on Management of Data, pp. 1557–1570 (2018)
27. Song, D.X., Wagner, D., Perrig, A.: Practical techniques for searches on encrypted data. In: IEEE S&P (2000)
28. Tkachenko, O., Weinert, C., Schneider, T., Hamacher, K.: Large-scale privacy-preserving statistical computations for distributed genome-wide association studies. In: Asia CCS (2018)
29. Tu, S., Kaashoek, M.F., Madden, S., Zeldovich, N.: Processing analytical queries over encrypted data. In: PVLDB 2013 (2013)
30. Van Rompay, C., Molva, R., Önen, M.: Secure and scalable multi-user searchable encryption. IACR Cryptology ePrint Archive 2018 (2018)
31. Wang, F., Yun, C., Goldwasser, S., Vaikuntanathan, V., Zaharia, M.: Splinter: practical private queries on public data. In: NSDI 2017 (2017)
32. Yao, A.C.C.: How to generate and exchange secrets. In: FOCS (1986)

System Security I

Deduplication-Friendly Watermarking for Multimedia Data in Public Clouds

Weijing You[1], Bo Chen[2], Limin Liu[3(✉)], and Jiwu Jing[1]

[1] School of Computer Science and Technology, University of Chinese Academy of Sciences (UCAS), Beijing, China
[2] Department of Computer Science, Michigan Technological University, Houghton, MI, USA
[3] State Key Laboratory of Information Security, Institute of Information Engineering, Chinese Academy of Sciences (CAS), Beijing, China
liulimin@iie.ac.cn

Abstract. To store large volumes of cloud data, cloud storage providers (CSPs) use deduplication, by which if data from multiple owners are identical, only one unique copy will be stored. Deduplication can achieve significant storage saving, benefiting both CSPs and data owners. However, for ownership protection, data owners may choose to transform their outsourced multimedia data to "protected formats" (e.g., by watermarking) which disturbs deduplication since identical data may be transformed differently by different data owners.

In this work, we initiate research of resolving the fundamental conflict between deduplication and watermarking. We propose DEW, the first secure Deduplication-friEndly Watermarking scheme which neither requires any interaction among data owners beforehand nor requires any trusted third party. Our key idea is to introduce novel protocols which can ensure that identical data possessed by different data owners are watermarked to the same "protected format". Security analysis and experimental evaluation justify security and practicality of DEW.

Keywords: Public clouds · Multimedia data · Deduplication · Watermarking · Proofs of ownership

1 Introduction

Nowadays, outsourcing data to public cloud storage providers (CSPs) like Amazon S3 [1], iCloud [2], Microsoft Azure [3], has become an economical and convenient practice for data owners. A critical issue faced by the CSPs is how to manage the ever-surge volume of data [4]. This can be mitigated by deduplication [5], which ensures that when multiple identical copies are present in the CSPs, only one unique copy will be stored, eliminating unnecessary wasting of storage. According to recent reports [5], deduplication can save up to 87%

L. Chen et al. (Eds.): ESORICS 2020, LNCS 12308, pp. 67–87, 2020.
https://doi.org/10.1007/978-3-030-58951-6_4

storage. For multimedia data, deduplication also achieves significant storage saving [6].

Although cloud outsourcing can bring significant benefits to data owners, it also introduces significant security concerns. One concern is losing ownership of their valuable data. To protect ownership, data owners would choose to transform their data to "protected formats" before outsourcing them. One prevalent solution is through encryption. This, however, is cumbersome because both sharing and processing of encrypted data in clouds will be difficult[1]. Fortunately, for multimedia data (noise-tolerant signals such as images/audio/videos), an alternative solution is through digital watermarking [7]. The watermarked data are directly usable by other users, facilitating data sharing in the clouds with copyright protection.

The watermarking, however, creates significant obstacles for deduplication, because: identical data are usually transformed to different "formats" by different owners using their unique secrets, which cannot be deduplicated any more. To resolve this conflict, we initiate investigation of a deduplication-friendly watermarking design. The core idea of our design is to make sure that different data owners will always transform identical data to the same "watermarked format", such that deduplication will not get stuck. The challenge is, the data owners usually do not know each other and will not synchronize information (e.g., sharing secrets) among them; without information synchronization, it would be difficult for them to generate the same "watermarked format" for identical data possessed by them individually.

Message-locked encryption (MLE) [8] derives encryption key from message being encrypted, such that identical messages can always be encrypted into identical ciphertexts. Motivated by MLE, an immediate solution towards addressing the aforementioned challenge is to perform watermarking using secret derived by MLE. For identical data possessed by different data owners, the secret being derived will be identical, leading to identical watermark embedding and hence deduplication will not be disturbed. However, most the existing MLE instantiations are problematic, since they either rely on impractical assumptions or suffer from various attacks. The implementation like Dupless [9] requires an additional independent key server which is fully trusted, but such a trusted third party is rarely found in practice. Some MLE implementations [10,11] require cloud users to synchronize information initially which is also impractical. The only known MLE implementation, *convergent encryption* (CE) [12], does not require the trusted third party and the initial user synchronization, but is shown to be vulnerable to an offline brute-force attack [13], in which the attacker can try all possible content of a file since content space for a fixed-size file is limited.

In this work, for the first time, we adapt MLE to our setting by carefully considering both the nature of watermarking and deduplication. The resulting design, DEW, is the first secure and practical Deduplication-friEndly Watermarking scheme. Our key insights are threefold: 1) We separate secrecy of a

[1] Although the homomorphic encryption can be used to process data in public clouds, they are still far from practicality.

watermark into two parts: the watermark location (i.e., where the watermark is embedded) and the watermark bit stream (i.e., the content forming the watermark). CE is used to determine the location only, while the watermark bit stream is picked secretly by the first data owner, and securely propagated to other data owners during deduplication. 2) We design a new Proof of Ownership (PoW) scheme specifically for our purpose. Our new PoW protocol can allow the verifier to verify proofs of PoW without having access to the original file. 3) We design a novel technique which can securely and efficiently propagate the watermark bit stream to other data owners via the untrusted cloud server.
Contributions. We summarize our contributions in the following:

- We are the first to identify as well as resolve the conflict between deduplication and watermarking. The resulting design, DEW, is the first deduplication-friendly watermarking scheme without requiring interactions among data owners or any trusted third parties.
- We adapt MLE in the watermarking process to ensure that identical file data are always deduplicable after watermarking. Additionally, we propose a novel PoW scheme which can allow the cloud server to verify PoW proofs without having access to the original file.
- We analyze security of DEW. Additionally, we implement DEW and experimentally assess its practicality.

2 Background

Deduplication and Proofs of Ownership. Deduplication aims to eliminate unnecessary storage space by removing duplicate data among cloud users. Note that it is only interested in removing cross-user duplicates which are not used for data durability purpose [14–17]. Based on granularity of data being deduplicated, it can be categorized as *file-level* [12] (i.e., the entire file will be deduplicated) and *block-level* deduplication [18] (i.e. duplicate data will be detected by blocks or chunks). Based on where deduplication is performed, there are *server-side* deduplication, in which all files are uploaded and deduplicated by the cloud server transparently to clients, and *client-side* deduplication, where each client collaborates with the cloud server to perform deduplication. Specifically, the first client uploads the file normally; the following client checks if the file has been stored using a unique checksum of the file. If so, the file will not be uploaded again. Instead, the cloud server simply accepts the client as an owner of this file. Compared to the server-side deduplication, the client-side deduplication is more advantageous in saving both storage and bandwidth, and is thus used broadly in practice (e.g., Dropbox [19]).

However, the client-side deduplication faces a unique attack that, an adversary which obtains the checksum of a file (rather than the file itself), can claim ownership of it. Proofs of Ownership (PoW) [20], were explored to mitigate this attack. A PoW protocol allows the cloud server (i.e., the verifier) to check whether a client (i.e., the prover) indeed possesses the entire file. Halevi et al. [20] proposed the first PoW scheme based on Merkle-tree, where each leaf node in

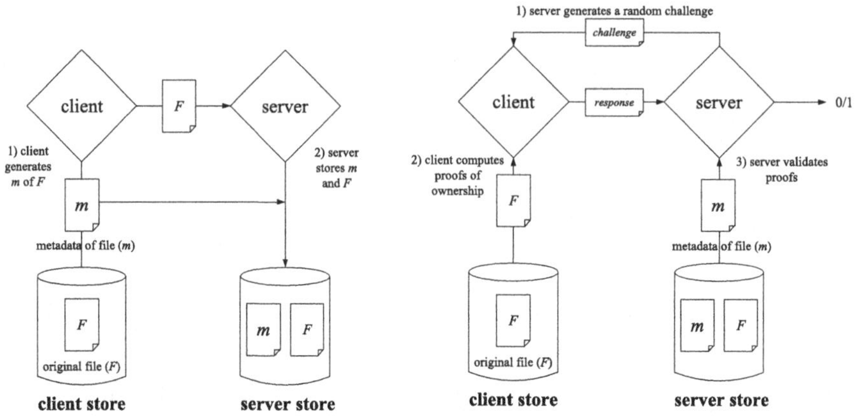

(a) Pre-processing and store (b) A Proofs of Ownership (PoWs)

Fig. 1. Client-side deduplication system

the tree is a hash value of a file block, and each non-leaf node is hash value of its child nodes. In their scheme (as shown in Fig. 1), the verifier is assumed to be fully trusted, and keeps a small amount of metadata for verification. When a prover comes, the verifier challenges the prover, and the prover returns a PoW proof. If the proof is validated by the verifier, the prover will passes the check.

Digital Watermarking. Digital watermarking is a kind of marker covertly embedded in a noise-tolerant signal (e.g. audios, videos, or images), and can be used to identify ownership of such signal (e.g., protecting authenticity or owners' copyright). Its core idea is to embed specific digital information in a carrier signal which is only perceptible under certain conditions. The watermark extraction process is the reverse operation of watermark embedding. The digital watermarking can be classified in terms of embedding domain (e.g., spatial/ frequency domain [21]), embedding method (e.g., spread-spectrum/ amplitude modulation/ quantization type), detection method (e.g, plaintext/ blind watermark), robustness (e.g., fragile/ robust watermark), information capacity (e.g., 1-bit/ n-bit watermark), and so on. In this paper, we choose n-bit watermarking in the spatial domain using amplitude modulation[2].

Message-Locked Encryption (MLE). Bellare et al. [8] proposed the message-lock encryption (MLE), which formalizes various cryptographic primitives that can be used to derive keys from messages being encrypted. By using MLE, different users owning an identical file are able to derive the same encryption key. A typical MLE scheme is convergent encryption (CE) [12], in which the

[2] A spread-spectrum watermark is modestly robust, but has a low information capacity; while a quantization watermark suffers from low robustness, though has a high information capacity. We thus choose amplitude modulation which can achieve a good balance between robustness and information capacity, and is particularly embedded in the spatial domain.

encryption key is the hash value of the message being encrypted. Due to the finite content space, CE is vulnerable to an offline brute-force attack [13].
Discrete Logarithm Problem (DLP). Let $\mathbb{G}$ be a finite cyclic group of order q, and g be a generator of $\mathbb{G}$. Given an element $y \in \mathbb{G}$, we want to find an integer x such that $g^x = y$ and $0 \leq x < q - 1$. This problem is computationally intractable, i.e., no algorithm can determine x in polynomial time [22].

3 System and Adversarial Model

System Model. We consider a cloud storage system consisting of three entities: the cloud server (S), the data owner (O), and the data user (U). S provides storage services and enables client-side deduplication. O outsources multimedia files to S but wants to retain their ownership, and therefore, O will watermark the files before outsourcing. Upon outsourcing a file, O will work with S to find out if this file has been stored by S. For a file which has been stored in S, O will not upload it, and S will simply add O to owner list of the file. U can use the outsourced files, e.g. viewing a watermarked video/picture. However, U is not supposed to have access to the original file data due to protection by watermark.
Adversarial Model. All data owners are assumed to be fully trusted. Otherwise, they can simply disclose the original file data, and using watermark to protect the original file will become useless. This assumption is reasonable, since a rational data owner will care about his/her ownership and has no motivation to disclose the original file data. We also assume that it is an authentic data owner which initializes upload of the original file. This assumption is also reasonable, since by uploading an arbitrary file, the adversary will need to pay for the storage service, and not gain any additional advantage on obtaining original data of any other files.

Both the cloud server and the data users are not trusted and may misbehave. The cloud server is honest-but-curious [23], which will honestly store the file, correctly execute all required operations, and timely respond to data owners/users as promised in the service-level agreement (SLA), but it is curious about the exact content of the original file, and may try all means to figure it out. The data users may be malicious. Their malicious behaviors include: removing the watermark to obtain the original file data, becoming owner of the file by fooling the cloud server, etc. However, we assume the cloud server and the data users *will not collude*. Otherwise, the cloud server can simply add the data users to the owner list of a file. This assumption is consistent with our "honest-but-curious" cloud model, considering that collusion is not an honest behavior.

We do not consider a special attacker, namely, a revoked data owner which still keeps the original file after being revoked, since this type of revoked data owner can simply use the preserved original file to pass the PoW check and become a data owner again. But we do consider an attacker which is a revoked data owner and has deleted the original file locally. Since a major concern of this work is ownership protection, attacks which are not directly related to ownership compromise are not considered here, like network attacks (e.g., DDoS), attacks that break robustness of watermarking (e.g., the adversary distorts the watermarked file arbitrarily).

All the communication channels among the cloud server, the data owner, and the cloud user are assumed to be secure, e.g., being protected using SSL/TLS.

4 A Deduplication-Friendly Watermarking Scheme

4.1 Design Rationale

To enable a secure deduplication-friendly watermarking design, we separate the watermark into two independent components: the watermark location and the watermark content. We handle each component as follows:

- The watermark location is derived from a secret key, which is generated using convergent encryption (CE). In this manner, if all the data owners possess an identical file, by applying CE on this file, they will be able to generate the same secret key, which can be used to derive the same watermark location.
- The watermark content is a random bit stream, which will be picked by the data owner who initially uploads the file, and will then be propagated secretly to other data owners during the client-side deduplication. In this way, all the data owners of the file can share the same random bit stream. Note that relying on the CE to derive both the watermark location and the watermark bit stream will be vulnerable to the offline brute-force attack [13].

Especially, in a client-side deduplication system, a data owner initially outsources a watermarked multimedia file to a cloud server. All the following data owners will check with the cloud server whether the file has been stored or not, and the cloud server will perform a PoW check on each following data owner; once the PoW check is passed, the corresponding data owner will be added to owner list of the file, and can delete the file locally without uploading it again. During each successful PoW process, the random bit stream will be propagated stealthily to each data owner. Three questions need to be answered further:

- How can each following data owner securely and efficiently find out whether a given file has been stored in the cloud server?
- How can the cloud server perform the PoW check without having access to the original file?
- How can the random bit stream (i.e., watermark content) be propagated to other data owners securely and efficiently via the untrusted cloud server?

We answer the *first* question. In traditional client-side deduplication (in a benign setting where the cloud server is fully trusted), the data owner can simply send a checksum (typically a hash value) of the original file to the cloud server before uploading it, using which the cloud server can check whether there is a duplicate file. However, in an adversarial setting, the checksum should not be computed over either the original file or the watermarked one. Simply computing a checksum over the original file is not secure, since the honest-but-curious cloud server may perform the offline brute-force attack by knowing this checksum due to the limit content space of a file. In addition, simply computing a checksum over the watermarked file is infeasible, because at this point, the watermarked file has not been generated yet. Fortunately, we observed that in a watermarked file, the file data excluding the watermark (we call this part of file data "public portion") are always known by the cloud server, and the checksum can be derived over the public portion of the file. In this manner, when performing the offline brute-force attack, the adversary can at most find out the public portion of the file, rather than the entire original file. In addition, since the watermark location is derived using CE, any data owners who possess the original file can easily derive the watermark location, exclude the watermark, and generate the checksum based on the public portion. Since the public portion is only part of the original file, matching of it does not provide 100% guarantee of matching of the file. In the case that two different files have the same public portion (the probability should be low), the mismatching will be detected later when the data owner proves ownership of the file. Therefore, to support existence detection of a particular file, we need to perform the following: 1) when outsourcing a watermarked file, the data owner should compute a checksum over its public portion, and upload to the server both the watermarked file and the checksum; and 2) during client-side deduplication, a data owner who wants to check whether a local file has been stored by the cloud server or not, will compute a checksum over the public portion of file, and send the checksum to the cloud server.

We now answer the *second* question. Suppose the watermarked file has been uploaded to the cloud server by a data owner (O_1) and another data owner (we call this data owner O_l, where $1 < l$) comes, who also possesses the same original file. After having found out the data owner may have a duplicate file (refer to our answer to the first question), the cloud server will run a PoW protocol to further verify whether O_l really possesses the original file. In the traditional PoW protocol [20] the cloud server is fully trusted and possesses the original file, which is different from our setting that the cloud server is assumed to be honest-but-curious and only stores the watermarked file. Therefore, the traditional PoW protocol cannot work and a new PoW protocol is designed as: the data owner generates a "miniature" of the original file to assist the cloud server to run the PoW protocol, under the condition that, the cloud server (i.e., PoW verifier) will not learn the original file from either the "miniature" or the PoW proofs, and the client (i.e., PoW prover) should not be able to pass the PoW verification without possessing the original file.

Specifically for security, the "miniature" should satisfy two properties: I) The "miniature" can be used to correctly verify a PoW proof, i.e., the "miniature" should be unique for the original file, and it is computationally infeasible to find two different files such that their miniatures are identical. II) The cloud server should not be able to derive anything else beyond the watermarked file from the "miniature". To create such a "miniature", an immediate solution is to use PoR [24]/PDP [25] tags which support public verifiability. This, however, turns out to be very expensive in tag generation [24,25]. An improvement of efficiency could be using the PoR tags supporting private verifiability [24] (note that a PDP tag supporting private verifiability is also expensive to be computed [25]). The privately verifiable PoR tag [24] is constructed as $\sum_{j=1}^{s} \alpha_j m_{ij} + F_\kappa(i)$, where s is the number of symbols in a file block m_i (with block index i, and $1 \leq i \leq n$ for a file with n blocks), $F_\kappa()$ is a pseudo-random function (PRF) with a key κ. Note that α_j (for $1 \leq j \leq s$) and κ are random numbers which should be kept private by the verifier. However, the privately verifiable PoR tag cannot immediately work here, since the verifier (i.e., the untrusted cloud sever here) needs to know α_j (for $1 \leq j \leq s$) and κ. By knowing α_j (for $1 \leq j \leq s$) and κ, the cloud server may be able to derive the original file content. To address this security issue, the design is tuned as follows: 1) The tag of each file block is computed as: $\sum_{j=1}^{s} \alpha_j m_{ij} + H(i||I_{wm})$, where H is a cryptographic hash function, and I_{wm} is the watermark bit stream of the file. 2) We disclose g^{α_j} (for $1 \leq j \leq s$) rather than α_j (for $1 \leq j \leq s$) to the untrusted cloud server, such that the cloud server can rely on g^{α_j} to check the PoW proofs without knowing α_j. 3) During the PoW process, we require the cloud server to stealthily distribute the watermark bit stream to the prover. Here "stealthily" means the cloud server is unable to figure out the watermark bit stream. The prover, if possesses the original file, will be able to extract the watermark bit stream I_{wm}, and to compute assisting information based on the I_{wm} together with its PoW proofs. This makes it possible that even if the verifier (i.e., the cloud server) does not know I_{wm}, it can still be able to check the PoW proofs returned by provers.

In addition, during each PoW process, the cloud server checks a random subset of file blocks from the original file (for a small file, the cloud server can simply check all the file blocks); the prover will return PoW proofs derived from file blocks being checked. A possible security issue here is that, the cloud server may learn sensitive information about the original file from the PoW proofs. Our solution is that, each time before returning a PoW proof, the prover will mask the proof with randomness; also, to facilitate the verification of the PoW proofs, assisting information relating to the randomness will be returned (can be combined together with the assisting information for I_{wm} mentioned above).

Here comes to the *third* question. Having observed that the watermark can be extracted through comparing the original file with the watermarked file, a straightforward solution is, the cloud server directly sends the watermarked file to the data owner during each PoW process. This solution suffers from a large communication overhead and, is insecure, since it discloses too much effective information of the original file to the prover before PoW checking. Having observed that, the watermark bit stream usually distributes sparsely in the watermarked file, we propose a novel approach to "compress" the watermarked file such that: 1) the size of the compressed watermarked file can be reduced significantly from $O(n)$ to $O(\sqrt{n})$, where n is the file size; and 2) without possessing the original file, one cannot extract the watermark bit stream from the compressed watermarked file (i.e., only a party which possesses the original file can extract the watermark bit stream). As a grayscale image[3] can be processed as a 2-dimension $N \times N$ pixel matrix, the key idea is to compress the matrix to a $d \times N$ row matrix and an $N \times d$ column matrix, where d is a compression parameter and $d \ll N$. The compression is performed in such a way that, compared to the original pixel matrix, the compressed matrices lose effective information significantly and are useless when used independently. Note that this special "compression" is completely different from the traditional compression which requires decompression, and therefore, it is also effective for image files in compressed format (e.g. jpeg).

4.2 Design Details

Our Deduplication-friEndly Watermarking scheme (DEW) consists of 5 phases: *Setup*, *Initial Upload*, *Client-side Deduplication*, *Retrieval*, and *Restoration*. For simple presentation, we assume there is only one file in the cloud server, which can be easily extended to a regular cloud storage system with multiple files. In addition, we focus on grayscale images which are commonly stored with 8 bits per sampled pixel.

Setup: Let β be a security parameter. The system selects a finite field $\mathbb{Z}_q$ of a prime order q, and initializes a multiplicative cyclic group $\mathbb{G}$ of the same order, and let $g \in \mathbb{G}$ be a generator. The system also selects three cryptographic hash functions, defined as H_1, H_2, and H_3 respectively: $\{0,1\}^* \rightarrow \{0,1\}^\beta$. There is also a pseudo-random permutation π: $\{0,1\}^\beta \times \{0,1\}^{log_2^n} \rightarrow \{0,1\}^{log_2^n}$, where n is file size in bytes. The compression parameter d for determining the size of compressed matrix is set to be a positive integer significantly less than n, i.e., $d \ll n$, and the watermark length is set as w, which usually is much smaller than n.

Initial Upload: In this phase, the data owner (O_1) processes the original file (f) into a deduplicable watermarked file (f_t), generates PoW tags for f, and stores PoW tags together with f_t to the cloud server (S) (Fig. 2(a)):

- Step 1: O_1 runs $K_{\mathsf{CE}} \leftarrow H_2(f)$ to derive a CE key from f. K_{CE} is used to derive the public portion, denoted as f_p. O_1 then computes a checksum (i.e. $checksum_f$) for f by running $checksum_f \leftarrow H_1(f_p)$. Lastly, O_1 sends a request with $checksum_f$ to S indicating that he/she wants to upload f.
- Step 2: S checks existence of f using $checksum_f$. If f has been stored, the *Initial Upload* phase will end, and *Client-side Deduplication* phase will be be invoked. Otherwise (denoted as $\times$), proceed to next Step 3.
- Step 3: O_1 generates a random bit stream (I_{wm}) with length w and uses Algorithm 1 (input: f, I_{wm}, w, K_{CE}) to generate a watermarked file f_t.
- Step 4: O_1 splits f into n blocks and each block contains s symbols in $\mathbb{Z}_q$. Each symbol is denoted as m_{ij}, where $1 \le i \le n$ and $1 \le j \le s$. Then O_1 selects s random numbers $\alpha_j \in \mathbb{Z}_q$ and computes g^{α_j}, for $1 \le j \le s$. The PoW tag for each file block i is computed as $\sigma_i = \sum_{j=1}^{s} \alpha_j m_{ij} + H_3(i||I_{wm})$, for $1 \le i \le n$.

[3] For simplicity, we focus on grayscale images in this paper.

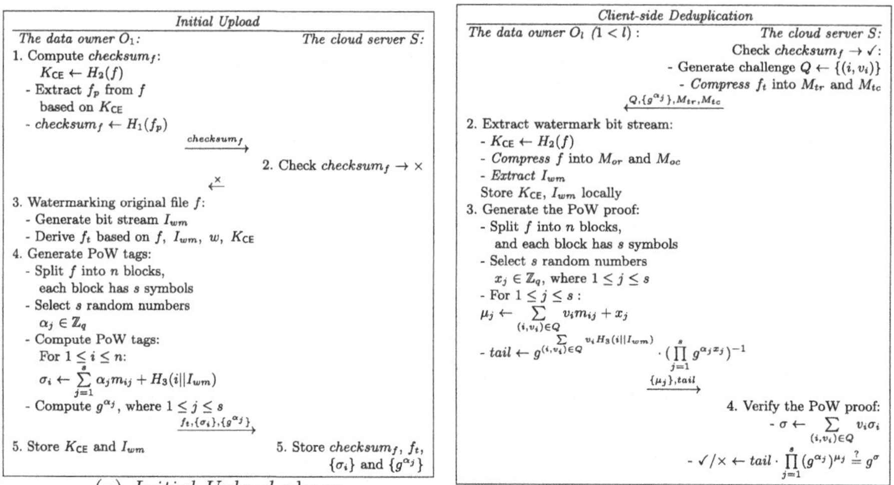

(a) *Initial Upload* phase (b) *Client-side Deduplication* phase

Fig. 2. Detailed steps in the *Initial Upload* and *Client-side Deduplication* phase

- Step 5: f_t is outsourced to S along with $\{g^{\alpha_j}|1 \le j \le s\}$ and $\{\sigma_i|1 \le i \le n\}$. Then O_1 deletes f and stores K_{CE} and I_{wm} locally.

Client-side Deduplication: If f has been stored by the cloud server, when a following data owner (O_l, where $1 < l$) sends the checksum (i.e., $checksum_f$), the cloud server will find a match. Then, the cloud server will ask O_l to prove ownership of f. The detailed steps (Fig. 2(b)) are elaborated as follows:

- Step 1: S generates a PoW challenge $Q = \{(i, v_i)|1 \le i \le c\}$, where i denotes a randomly selected block index, v_i denotes a random number picked from $\mathbb{Z}_q$, and c is the number of file blocks being challenged. S further compresses f_t using Algorithm 2 (input: f_t and d), generating a compressed row matrix M_{tr} and a compressed column matrix M_{tc}. S sends Q, M_{tr}, M_{tc}, $\{g^{\alpha_j}|1 \le j \le s\}$ to O_l.
- Step 2: O_l extracts I_{wm}: 1) O_l computes K_{CE} from f by running $K_{\mathsf{CE}} \leftarrow H_2(f)$. 2) O_l compresses the original file f following the Algorithm 2 (input: f and d), obtaining a compressed row matrix M_{or} and a compressed column matrix M_{oc}. 3) O_l runs Algorithm 3 (input: f, M_{or} and M_{oc}, M_{tr} and M_{tc}, d, w, K_{CE}) to obtain I_{wm}. K_{CE} and I_{wm} will be stored locally.
- Step 3: O_l generates the PoW proof according to Q: 1) O_l divides the file f into n blocks, and each block contains s symbols. Each symbol is m_{ij} where $1 \le i \le n$ and $1 \le j \le s$. 2) O_l selects s random numbers x_j from $\mathbb{Z}_q$, where $1 \le j \le s$ and computes: $\mu_j = \sum_{(i,v_i)\in Q} v_i m_{ij} + x_j$, for $1 \le j \le s$. 3) O_l computes $tail = g^{\sum_{(i,v_i)\in Q} v_i H_3(i||I_{wm})} (\prod_{j=1}^{s} g^{\alpha_j x_j})^{-1}$. O_l replies to S with the PoW proof: $\{\mu_j|1 \le j \le s\}$, $tail$.

Algorithm 1: Watermarking an image file (for grayscale images)

Input: An image file f, I_{wm}, watermark length w, K_{CE}
Output: A watermarked image file f_t
View the file f as an $N \times N$ pixel matrix M
for $i = 1 : w$ **do**
    $t = \pi_{K_{\mathsf{CE}}}(i)$;`// Derive watermark locations`
    $L[i] = ((t-1)/N + 1, (t-1) \bmod N + 1)$;`// Derive watermark index`
k=1;
for $i = 1 : N$ **do**
    for $j = 1 : N$ **do**
        if $L[k] == (i, j)$ **then**
            $M[i, j] = (M[i, j] + I_{wm}[k])(\text{mod } 256)$;`// embed a watermark bit`
            ++k;
The new matrix M will be output as a watermarked image file f_t.

– Step 4: After having received the PoW proof, S validates its correctness: 1) S computes $\sigma = \sum_{(i,v_i)\in Q} v_i\sigma_i$, and then computes g^σ. 2) S checks whether or not equation $g^\sigma = tail \cdot \prod_{j=1}^{s} (g^{\alpha_j})^{\mu_j}$ holds.

Algorithm 2: Compressing an image file (for grayscale images; $M_r[i, :]$ denotes all the N elements in row i of matrix M_r; $M_c[:, i]$ denotes all the N elements in column i of matrix M_c)

Input: An image file f, compression parameter d
Output: A compressed row matrix M_r, and a compressed column matrix M_c
View the file f as an $N \times N$ pixel matrix M
for $i = 1 : d$ **do**
    $M_r[i, :] = 0$;
    $M_c[:, i] = 0$;
    for $j = 0 : \frac{N}{d} - 1$ **do**
        $M_r[i, :] = (M[i + d \times j, :] + M_r[i, :])(\text{mod } 256)$;`// aggregated by row`
        $M_c[:, i] = (M[:, i + d \times j] + M_c[:, i])(\text{mod } 256)$;`// aggregated by column`

Retrieval: All valid data owners and data users are able to retrieve the f_t from the cloud server. Upon receiving a retrieval request, the cloud server authenticates the request (not the focus of this paper), and sends back f_t.
Restoration: Only data owners (O_l, where $1 \leq l$) are able to restore f from f_t, which is an inverse operation of the watermark embedding process in the *Initial*

Algorithm 3: Extracting watermark bit stream I_{wm}

```
Input: The original file f, M_or, M_oc, M_tr, M_tc, d, w, K_CE
Output: The extracted watermark bit stream I_wm
Initiate an N × N matrices M_1, and L ← ∅;
for i = 1 : w do
    t = π_{K_CE}(i);
    L = L ∪ {(t − 1)/N + 1, (t − 1) mod N + 1)};
for i = 1 : N do
    for j = 1 : N do
        if (i, j) ∈ L then M_1[i, j] = 2;// set 2 for watermark locations
        else
            M_1[i, j] = 3;// set 3 for other locations
for i = 1 : d do
    for j = 1 : N do
        M_r[i, j] = (M_tr[i, j] − M_or[i, j])(mod 256);
        M_c[j, i] = (M_tc[j, i] − M_oc[j, i])(mod 256);
for i = 1 : d do
    for j = 1 : N do
        t = 0;
        for k = 0 : N/d − 1 do
            if(M_1[k × d + i, j] == 2) then ++t;
        if t == 0 then continue;
        if M_r[i, j] == 0 then
            for k = 0 : N/d − 1 do
                if M_1[k × d + i, j] == 2 then M_1[k × d + i, j] = 0;
        else if M_r[i, j] == t then
            for k = 0 : N/d − 1 do
                if M_1[k × d + i, j] == 2 then
                    M_1[k × d + i, j] = 1;
                    M_c[k × d + i, (j − 1) mod d + 1]− = 1;
for i = 1 : N do
    for j = 1 : d do
        t = 0;
        for k = 0 : N/d − 1 do
            if (M_1[i, k × d + j] == 2) then ++t;
        if t == 0 then continue;
        if M_c[i, j] == 0 then
            for k = 0 : N/d − 1 do
                if M_1[i, k × d + j] == 2 then M_1[i, k × d + j] = 0;
        else if M_c[i, j] == t then
            for k = 0 : N/d − 1 do
                if M_1[i, k × d + j] == 2 then M_1[i, k × d + j] = 1;
For remaining elements with value 2 in M_1, retrieve corresponding elements from f_t,
determine them by comparing with f. Sequentially output elements with value 0 and 1
from matrix M_1 as I_wm.
```

Upload phase. Specifically, O_l downloads f_t from the cloud server and removes I_{wm} from watermark locations determined by K_{CE} to restore f.

5 Analysis and Discussion

5.1 Security Analysis

DEW aims to protect ownership of the outsourced multimedia files in a client-side deduplication-enabled cloud storage system, which implies: 1) any untrusted party, both the honest-but-curious cloud server and the malicious data users, should not obtain the original file; and 2) any party which does not possess the original file should not be able to pass the PoW check. Therefore, security of DEW is captured in Theorem 1, 2 and 3.

Theorem 1. *The cloud server cannot obtain the original file.*

Theorem 2. *The malicious data user cannot obtain the original file.*

Theorem 3. *The malicious data user cannot pass the PoW check.*

Due to space limitations, detailed proofs of Theorem 1, 2 and 3 are presented in our technical report [26].

5.2 Discussion

Regarding Watermark Extraction. During the Client-side Deduplication phase, to ensure watermark bit stream can be extracted from the compressed row and column matrix (i.e., M_{tr} and M_{tc}) sent by the cloud server, O_l may need to retrieve additional elements from the watermarked file f_t, if some of the elements in the watermarked locations cannot be determined (Algorithm 3). The number of additional elements needed to be retrieved is usually much smaller than the size of watermark bit stream according to our experimental evaluation (value r in Table 1), which only introduces a small extra communication overhead. In addition, this may disclose some of the watermark locations to the cloud server, but it will not help the cloud server to learn the original file, because only a small portion of watermark locations may be disclosed and most importantly, the watermark bit stream remains unknown to the cloud server. Note that to ensure Algorithm 3 can work correctly, $\frac{N}{d}$ should not reach 256 (i.e., to ensure that after aggregating all the potential watermark bits in a row/column, it will not be greater than or equal to 256, causing overflow).

Revoking Data Owners. A data owner may be revoked over time, e.g., after the owner deletes a multimedia file, he/she will be revoked. To prevent a revoked owner from restoring the file in the future by simply keeping the watermark (small in size), we should re-watermark the file so that after re-watermarking process is done, the revoked data owner will not be able to obtain the original file again by retrieving a watermarked file and subtracting the kept watermark. The re-watermarking process is usually triggered upon an owner is revoked. In the case that a file is shared by many owners which are revoked frequently, we may delay the re-watermarking process so that each such process will handle multiple revoked owners. After the re-watermarking process, the watermark bit stream may have been changed, and should be re-propagated to valid owners[4].

Side-Channel Attack Resistance. Traditional client-side deduplication is susceptible to a side-channel attack [27], where the checksum can be used to identify existence of a given file, and hence to learn file content. DEW can mitigate this attack since the checksum is computed from "public portion" of a file, i.e. the adversary cannot identify a file exactly, and at most learn "public portion" of the file.

Applications of DEW in Practice. DEW can be applied to a lot of real-world scenarios. For example, crowd-funding projects become popular nowadays, in which different users may collaborate for one project and produce multimedia files, which are later outsourced to clouds by individual users using watermarking for ownership protection. Due to COVID-19, online teaching becomes more and more prevalent, and individual teachers may store multimedia files (purchased from the same publisher) to clouds, using watermarking for ownership protection.

Extensions of DEW. Due to complication of multimedia data, we currently initiate DEW using an n-bit additive modulation watermarking scheme in a spatial domain with modest robustness, which is sufficient regarding ownership protection. However, other watermarking schemes, e.g., watermarking in frequency domain and quantization-based watermarking, may be supported by DEW, by slightly tuning the design to accommodate different types of watermarking process. This will be further investigated in our future work. Besides, DEW may be extended to other multimedia data. For example, 1) for a colored image (i.e. RGB image), it can be decomposed to three single-channel images, each can be processed like a grayscale image; 2) for a video file, it can be viewed as a collection of still images (either colored images or grayscale images).

6 Implementation and Evaluation

Both the cloud server and the client were run on a local machine (Intel Xeon E3-1225 v5 CPU 3.30GHz, 8.0 GB RAM, and Ubuntu 16.04 with kernel version 4.15.0-50-generic). The security parameter β was selected as 160. We worked on the type A pairings which are constructed on the curve $y^2 = x^3 + x$ [28] over the finite field $\mathbb{Z}_q$. The order (i.e., q) of the finite field $\mathbb{Z}_q$ and the cyclic group $\mathbb{G}$ is a 160-bit prime[5]. We used the pairing-based cryptographic library PBC-0.5.14 [28]. The hash function H_1, H_2 and H_3 were instantiated using SHA1, and the pseudo-random permutation π was implemented using $randomperm()$

[4] Using the old watermarked file as well as the old watermark, a valid owner can restore the original file, and by retrieving a compressed version of the new watermarked file, the owner is able to extract the new watermark bit stream. We will investigate a more efficient design in the future work.

[5] To ensure that each symbol is valid (i.e., not larger than the finite field), only 159 bits from a file were read each time.

in MATLAB-R2016a. d was selected as 8, and N were 128, 256, 512, 1024, and 2048, respectively. Note that $N = 2048$ is a corner case for satisfying the condition that $\frac{N}{d} < 256$ (Sect. 5.2), and we ensured that no overflow happened during the experiments. When the number of blocks in a file is less than or equal to 460, the cloud server will run the PoW protocol by checking the entire file; otherwise, the number of blocks being checked is 460, which can ensure that if 1% of the file has been corrupted, the server will detect the corruption with 99% probability [25]. We used 1000+ images for testing, the sizes of which range from 16KB to 10MB, under three test cases with different sizes of each file block and single symbol in a file block. Images were processed using MATLAB scripting language.

Evaluating the Initial Upload phase. In this phase, the data owner will generate the watermarked file, and compute PoW tags (from the original file) as well as $\{g^{\alpha_j}|1 \leq j \leq s\}$ for PoW verification.

Watermarking. The watermarking computation is shown in Fig. 3(a). We can observe that the watermarking computation is mainly determined by the watermark length rather than the file size. This is because, the major computation during watermarking is adding watermark bits to specific elements selected by the K_{CE}, which depends on the length of the watermark bit stream.

Preparing for the PoW. The computational cost for generating PoW tags and computing $\{g^{\alpha_j}|1 \leq j \leq s\}$ is shown in Fig. 3(b) and 3(c), respectively. Let a denotes the block size (in bytes) and b denotes the symbol size (in bytes). We can observe that: 1) for fixed block/symbol size, the computational cost for generating PoW tags increases linearly with the file size and decreases with the symbol size, while is independent to size of each file block. This is because, this cost is mainly determined by total numbers of symbol-wise operations, the computational complexity of which approximates $O(\frac{|F|}{b})$, i.e., determined mainly by the file size $|F|$ and the symbol size b; 2) for fixed block/symbol size, the time for computing $\{g^{\alpha_j}|1 \leq j \leq s\}$ is independent to the file size. This is because, it is mainly determined by $s = \frac{a}{b}$ rather than the file size. This also explains that, for fixed symbol size, if the block size is doubled, the computation time is also doubled; in addition, for fixed block size, if the symbol size is doubled, the computation time is approximately reduced by 50%.

Evaluating the Client-side Deduplication phase. In this phase, the data owner will compute the PoW proof, and the cloud server will validate correctness of the PoW proof. Besides, the watermark bit stream will be distributed to the data owner during the PoW process, which requires the cloud server to compress the watermarked file and the data owner to extract the watermark bit stream.

PoW proof generation. During the PoW process, the data owner will compute the PoW proof based on the cloud server's challenge and the compressed watermarked file. The computational cost for PoW proof generation is shown in Fig. 4(a). We have following observations: 1) The computational cost for PoW proof generation increases linearly with the file size before reaching a certain threshold. This is because, the cloud server will check all the file blocks when the file size is small (i.e., when the total number of file blocks is smaller than

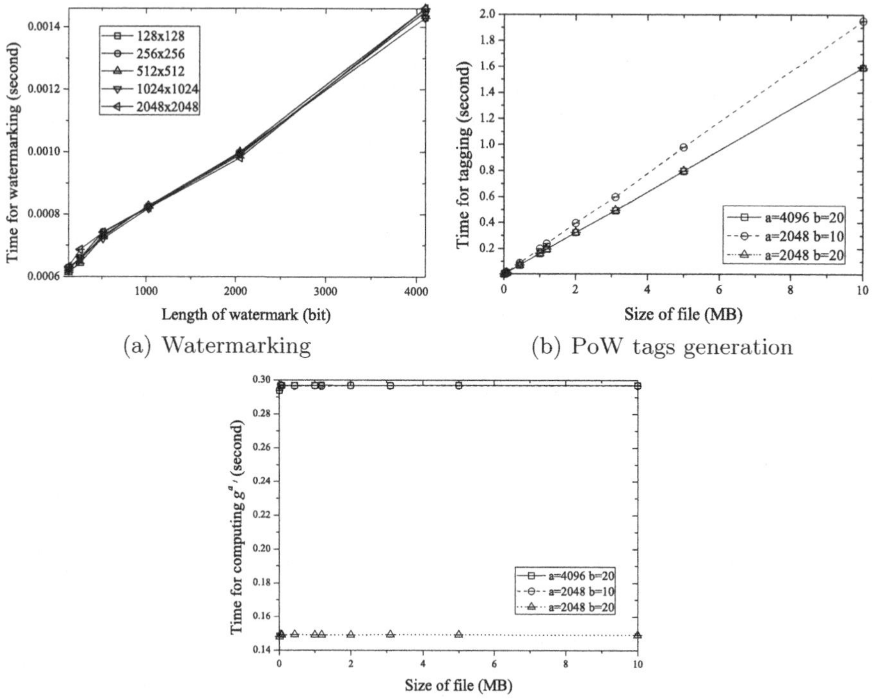

(a) Watermarking

(b) PoW tags generation

(c) Computing $\{g^{\alpha_j}\}$

Fig. 3. Computational cost of various components in Initial Upload phase

460), and the number of file blocks being checked depends on the file size for fixed block size; if the file size exceeds the threshold, it will always check 460 file blocks, and therefore, the computational cost turns constant. 2) After reaching the threshold, for a fixed symbol size, the computational cost for PoW proof generation will grow if the block size grows; for a fixed block size, this computational cost will grow if the symbol size decreases. The reason is, when the number of blocks being challenged is fixed, the computation for proof generation is mainly associated to $s = \frac{a}{b}$, where a is the block size and b is the symbol size.

PoW proof verification. After having received the PoW proof, the cloud server will verify its correctness and the computational cost is shown in Fig. 4(b). We can observe that: 1) for fixed block size and symbol size, the computational cost of verifying the proof is constant regardless of the file size; 2) for fixed symbol size, this cost will increase if the block size increases; 3) for fixed block size, this cost will increase if the symbol size decreases. The reason is similar to the aforementioned proof generation.

Distributing the watermark bit stream stealthily during PoW. In the PoW process, the cloud server will send two compressed matrices of the watermarked file to the data owner, and the data owner will extract the random watermark bit stream. The computational cost of compression and extracting watermark bit

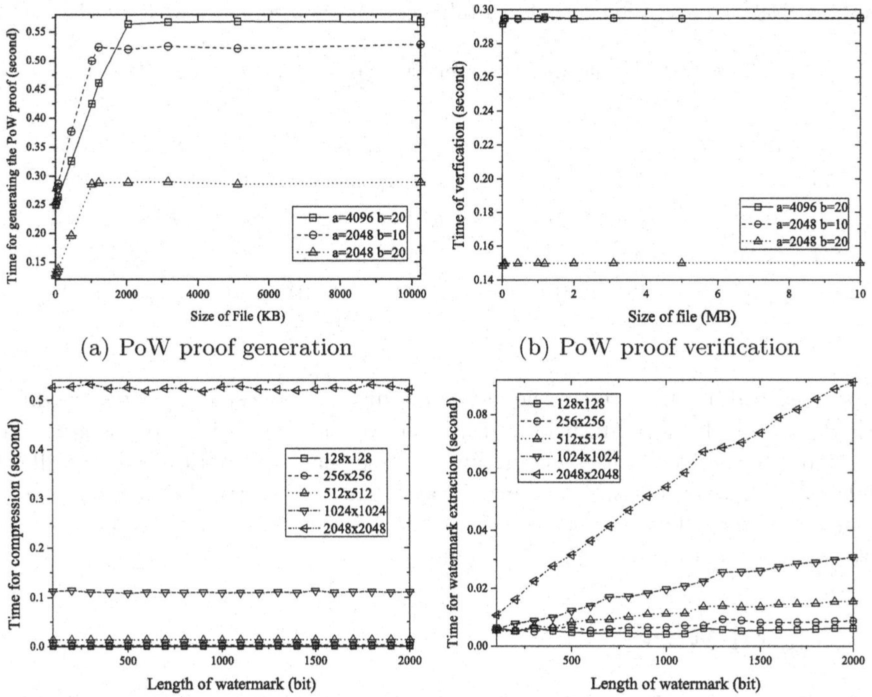

(a) PoW proof generation (b) PoW proof verification

(c) File compression (d) Watermark extraction

Fig. 4. Computational cost of various components in Client-side Deduplication

stream are shown in Fig. 4(c) and 4(d), respectively. We have following observations: 1) The compression time mainly depends on the file size, rather than the watermark length. This is because, the compression will iteratively add each d rows/columns of the image matrix together, which is determined by the size of the matrix, hence the size of the file. 2) When the file size is fixed, the computational cost of extracting watermark approximately grows linearly with the watermark length, which is the total amount of extracting operations; in addition, for fixed the watermark length, the computational cost grows with the file size, since the extracting process requires traversing the entire file.

We also assess effectiveness of our compression approach by calculating the compression rate $\rho = \frac{N^2-(2\times d\times N+r)}{N^2} \times 100\%$, where N^2 is the total number of elements in the original matrix, $d \times N$ is the number of elements in a compressed row/column matrix, and r is the number of elements which are sent additionally when uncertainty happen. The experiment was performed over files with different sizes (ranging from 128×128 to 2048×2048 in pixel) and different watermark lengths (ranging from 100 bits to 2,000 bits). The results are shown in Table 1, each is averaged over 100 runs. We can observe that, ρ is at least 84.42%, which justifies effectiveness of our compression approach. Additionally, ρ increases with file size but decreases with watermark length. This is because:

Table 1. Data compression rate. Each table cell shows ρ together with r in bracket.

File size (in pixel)	The length of watermark (bits)			
	100	500	1000	2000
128×128	87.5%(0.1)	87.36%(23.3)	86.73%(125.7)	84.42%(505)
256×256	93.75%(0.1)	93.73%(8.3)	93.67%(52.4)	93.34%(268.2)
512×512	96.88%(0.1)	96.87%(2.7)	96.86%(17.8)	96.83%(104.7)
1024×1024	98.43%(0)	98.43%(0.9)	98.42%(7.9)	98.41%(32.4)
2048×2048	99.22%(0)	99.22%(0.2)	99.21%(2.5)	99.21%(9.5)

1) for fixed watermark length, the larger file means more space to be embedded with watermark bit stream, leading to the more sparsely distributed watermark bits, hence less uncertainties occur during watermark extraction, i.e., r will be smaller; 2) for fixed file size, a shorter watermark bit stream leads to the more sparsely distributed watermark bits, and hence less uncertainties happen during watermark extraction, resulting in smaller r.

7 Related Work

Deduplication on Encrypted Data. To ensure the encrypted data remains deduplicable, a viable solution would be allowing each data owner generates the same encryption key for identical data. Douceur et al. [12] proposed convergent encryption (CE), in which the data owner derives the encryption key from the hash value of the message being encrypted. CE, however, suffers from the offline brute-force attack. To mitigate this attack, Bellare et al. proposed DupLESS [9], in which they introduced a trusted key server to further sign the CE key using its private key to enhance randomness of the CE key, and limited the number of requests for keys in a fixed time interval. Bellare et al.[8] formalized encryption schemes to a new cryptographic primitive, called Message-Locked Encryption (MLE) and presented complete security analysis. Liu et al. [11] removed the trusted key server by utilizing password-based authenticated key exchange protocol. Li et al. [29] and Lei et al. [23] further addressed the re-encryption problem for secure server-side deduplication and client-side deduplication, respectively. Once the encryption key is leaked, the encrypted data in a deduplication-based storage system will need to re-encrypted, but a challenge is how to re-encrypt the data efficiently. Using various techniques like all-or-nothing (AONT) transform [29] and delegated re-encryption [23], the re-encryption efficiency can be improved.

Proofs of Ownership (PoWs) and Cloud Storage Security. A PoW [20] protocol is a vital component for a secure client-side deduplication system, which can prevent an adversary from claiming ownership over a file without really possessing it. The first PoW scheme is based on Merkle-tree [20]. Pietro et al. [30] introduced a more efficient and secure PoW design, by asking the cloud server to pre-compute responses for a number of pre-computed PoW challenges. The two schemes assume the cloud server is fully trusted, but the client is untrusted. In parallel with PoWs, another direction of cloud storage security is to allow a client to verify whether the cloud server honestly stores the outsourced data, in which the client is the verifier while the cloud server is the prover. Typical protocols includes Provable Data Possession (PDP) [25,31,32] and Proofs of Retrievability (PoR) [33]. In this line of research, the client is assumed to be trusted while the server is untrusted, and new challenges include, the client is usually equipped with limited computation and storage resources, and hence needs to remain lightweight (e.g., the protocol should only require constant client storage, constant client computation).

8 Conclusion

In this work, we identify a fundamental conflict in deduplication-based cloud storage systems that, CSPs want to remove duplicate data (i.e., deduplication) to save storage space, while data owners want to protect ownership of their outsourced multimedia data by embedding watermarks which may hinder deduplication. We design a deduplication-friendly watermarking (DEW) scheme that can resolve the aforementioned conflict. A salient feature of our design is that, it does not require any interaction among data owners as well as any trusted third party. Security analysis and experimental evaluation show that DEW is secure at the cost of a modest additional overhead.

Acknowledgements. Weijing You, Limin Liu and Jiwu Jing were supported by National Key R&D Program of China (Grant No.2017YFB0802404).

References

1. "Amazon simple storage service" (2019). http://aws.amazon.com/cn/s3/
2. "Icloud" (2019). https://www.icloud.com/
3. "Microsoft azure" (2019). http://www.windowsazure.cn/?fb=002
4. "Cisco global cloud index: Forecast and methodology, 2016 2021 white paper" (2018). https://www.cisco.com/c/en/us/solutions/collateral/service-provider/global-cloud-index-gci/white-paper-c11-738085.html
5. Meyer, D.T., Bolosky, W.J.: A study of practical deduplication. ACM Trans. Storage **7**(4), 1–1 (2012)
6. Rashid, F., Miri, A.: Deduplication practices for multimedia data in the cloud. In: Srinivasan, S. (ed.) Guide to Big Data Applications. SBD, vol. 26, pp. 245–271. Springer, Cham (2018). https://doi.org/10.1007/978-3-319-53817-4_10
7. Tirkel, A.Z., Rankin, G., Van Schyndel, R., Ho, W., Mee, N., Osborne, C.F.: Electronic watermark. In: Digital Image Computing, Technology and Applications (DICTA 1993), pp. 666–673 (1993)
8. Bellare, M., Keelveedhi, S., Ristenpart, T.: Message-locked encryption and secure deduplication. In: Johansson, T., Nguyen, P.Q. (eds.) EUROCRYPT 2013. LNCS, vol. 7881, pp. 296–312. Springer, Heidelberg (2013). https://doi.org/10.1007/978-3-642-38348-9_18

9. Bellare, M., Keelveedhi, S., Ristenpart, T.: DupLESS: server-aided encryption for deduplicated storage. In: USENIX Conference on Security, pp. 179–194 (2013)
10. Shamir, A.: How to share a secret. Commun. ACM **22**(11), 612–613 (1979)
11. Liu, J., Asokan, N., Pinkas, B.: Secure deduplication of encrypted data without additional independent servers. In: Proceedings of the 22nd ACM SIGSAC Conference on Computer and Communications Security, pp. 874–885 (2015)
12. Douceur, J.R., Adya, A., Bolosky, W.J., Dan, S., Theimer, M.: Reclaiming space from duplicate files in a serverless distributed file system. In: International Conference on Distributed Computing Systems, pp. 617–624 (2002)
13. "Known attacks towards convergent encryption" (2013). https://tahoe-lafs.org/hacktahoelafs/drew_perttula.html
14. Chen, B., Curtmola, R., Ateniese, G., Burns, R.: Remote data checking for network coding-based distributed storage systems. In: Proceedings of the 2010 ACM workshop on Cloud computing security workshop, pp. 31–42. ACM (2010)
15. Chen, B., Curtmola, R.: Towards self-repairing replication-based storage systems using untrusted clouds. In: Proceedings of the third ACM conference on Data and application security and privacy, pp. 377–388. ACM (2013)
16. Chen, B., Ammula, A.K., Curtmola, R.: Towards server-side repair for erasure coding-based distributed storage systems. In: Proceedings of the 5th ACM Conference on Data and Application Security and Privacy, pp. 281–288. ACM (2015)
17. Chen, B., Curtmola, R.: Remote data integrity checking with server-side repair. J. Comput. Secur. **25**(6), 537–584 (2017)
18. Quinlan, S., Dorward, S.: Venti: a new approach to archival storage. FAST **2**, 89–101 (2002)
19. "Dropbox" (2019). https://www.dropbox.com/
20. Halevi, S., Harnik, D., Pinkas, B., Shulman-Peleg, A.: Proofs of ownership in remote storage systems. In: ACM Conference on Computer and Communications Security, pp. 491–500. ACM (2011)
21. Cox, I.J., Miller, M.L., Bloom, J.A., Honsinger, C.: Digital Watermarking, vol. 53. Springer, Heidelberg (2002)
22. Gordon, D.: Discrete Logarithm Problem. In: van Tilborg, H.C.A., Jajodia, S. (eds.) Encyclopedia of Cryptography and Security. Springer, Boston (2011). https://doi.org/10.1007/978-1-4419-5906-5_445
23. Lei, L., Cai, Q., Chen, B., Lin, J.: Towards efficient re-encryption for secure client-side deduplication in public clouds. In: Lam, K.-Y., Chi, C.-H., Qing, S. (eds.) ICICS 2016. LNCS, vol. 9977, pp. 71–84. Springer, Cham (2016). https://doi.org/10.1007/978-3-319-50011-9_6
24. Shacham, H., Waters, B.: Compact proofs of retrievability. In: Pieprzyk, J. (ed.) ASIACRYPT 2008. LNCS, vol. 5350, pp. 90–107. Springer, Heidelberg (2008). https://doi.org/10.1007/978-3-540-89255-7_7
25. Ateniese, G., et al.: Provable data possession at untrusted stores. In: Proceedings of the 14th ACM Conference on Computer and Communications Security, pp. 598–609, ACM (2007)
26. You, W., Chen, B., Liu, L., Jing, J.: Deduplication-friendly watermarking for multimedia data in public clouds. Technical report, Department of Computer Science, Michigan Tech, July 2020
27. Harnik, D., Pinkas, B., Shulman-Peleg, A.: Side channels in cloud services: deduplication in cloud storage. IEEE Secur. Priv. **8**(6), 40–47 (2010)
28. "Pairing based cryptographic library" (2019). https://crypto.stanford.edu/pbc/

29. Li, J., Qin, C., Lee, P.P.C., Li, J.: Rekeying for encrypted deduplication storage. In: IEEE/IFIP International Conference on Dependable Systems and Networks (2016)
30. Di Pietro, R., Sorniotti, R.: Boosting efficiency and security in proof of ownership for deduplication. In: Proceedings of the 7th ACM Symposium on Information, Computer and Communications Security, pp. 81–82. ACM (2012)
31. Erway, C.C., Küpçü, A., Papamanthou, C., Tamassia, R.: Dynamic provable data possession. ACM Trans. Inf. Syst. Secur. (TISSEC) **17**(4), 15 (2015)
32. Chen, B., Curtmola, R.: Robust dynamic provable data possession. In: 2012 32nd International Conference on Distributed Computing Systems Workshops, pp. 515–525. IEEE (2012)
33. Juels, A., Kaliski Jr, B.S.: PORs: proofs of retrievability for large files. In: Proceedings of the 14th ACM Conference on Computer and Communications Security, pp. 584–597. ACM (2007)

DANTE: A Framework for Mining and Monitoring Darknet Traffic

Dvir Cohen[1(✉)], Yisroel Mirsky[1,2], Manuel Kamp[3], Tobias Martin[3], Yuval Elovici[1], Rami Puzis[1], and Asaf Shabtai[1]

[1] Department of Software and Information Systems Engineering, Ben-Gurion University of the Negev, Beersheba, Israel
dvircohe@post.bgu.ac.il
[2] Georgia Institute of Technology, Atlanta, USA
[3] Deutsche Telekom Security GmbH, Bonn, Germany

Abstract. Trillions of network packets are sent over the Internet to destinations which do not exist. This 'darknet' traffic captures the activity of botnets and other malicious campaigns aiming to discover and compromise devices around the world. In this paper, we present DANTE: a framework and algorithm for mining darknet traffic. DANTE learns the meaning of targeted network ports by applying Word2Vec to observed port sequences. To detect recurring behaviors and new emerging threats, DANTE uses a novel and incremental time-series cluster tracking algorithm on the observed sequences. To evaluate the system, we ran DANTE on a full year of darknet traffic (over three Tera-Bytes) collected by the largest telecommunications provider in Europe, Deutsche Telekom and analyzed the results. DANTE discovered 1,177 new emerging threats and was able to track malicious campaigns over time.

Keywords: Darknet · Blackhole · Machine learning · Port embedding

1 Introduction

One way for Internet service providers (ISP) to obtain meaningful and actionable insights on malicious Internet campaigns, is to analyze traffic arriving at a subset of unassigned IP addresses. These IP addresses are sometimes referred to as a darknet [1,2]. In this paper, the term 'darknet' should not be confused with anonymous communication networks such as Tor. Darknet IP addresses are not associated with any registered host or services. Therefore, they are similar to honeypots [9,29] in that any incoming packets can be automatically considered unwanted and non-productive. Previous studies have shown that packets sent to darknet IP addresses are usually the result of network probing/scanning, worm propagation, and a DDoS attacks [34,41]. Therefore, darknet data can be used by an ISP's cyber emergency response team (CERT) to infer threat intelligence related to ongoing malicious activities or new emerging attacks [6] (see Fig. 1).

A great advantage to this approach is that darknet taps are easily deployed, inexpensive to implement, and can collect vast amounts of useful data. However,

L. Chen et al. (Eds.): ESORICS 2020, LNCS 12308, pp. 88–109, 2020.
https://doi.org/10.1007/978-3-030-58951-6_5

obtaining threat intelligence from darknet traffic is a challenging task for three reasons: (1) Darknet IPs are not assigned to actual hosts so the traffic only captures the initiation of a communication, and not the actual channel traffic (e.g., payloads). This is in contrast to honeypots which emulate real services (e.g., a Web application or SSH server). As a result, only metadata such as incoming packet's source IP (src IP), destination IP (dst IP), destination port (dst port), and packet size are available. (2) Darknet traffic is full of benign scanning activity from services/enterprises such as Amazon, Google, and Shodan which must be filtered out. (3) Attackers repackage old malwares (e.g., Mirai) and reuse known attack patterns which makes it difficult to identify novel threats. (4) Efficient algorithms are a necessity since terabytes of darknet data is generated every month.

Recent works proposed various methods for analyzing darknet traffic. Since the destination TCP or UDP port number provides a good indication of the sender's intentions (e.g., accessing port 23 may indicate an attempt to search for an accessible Telnet server), most of the previous research has focused on grouping ports into static clusters and detecting peaks or unusual trends in the volume of the clusters or individual ports [6,8,19,33]. However, attacks are becoming more sophisticated, automated, and noisy (in terms of port access). For example, attackers may perform multi-stage attacks or attempt to exploit multiple vulnerabilities to compromise a device [35]. Therefore, the methods proposed in previous works (1) cannot identify attacks which use more than one port in sequence, and (2) only perform one time clustering, and therefore do not provide any means for tracking on-going attacks, detecting a recurring/reused attack, or identifying novel emerging threats over time.

In order to detect attacks it is important for a security analyst to be able to analyze darknet data and provide insights on an hourly basis. However, this analysis is challenging as there are terabytes of darknet traffic data every month and this figure is expected to increase in the coming years. We believe the solution to this challenge will be based on utilizing the power of big data and using a distributed algorithm to provide hourly reports and alerts.

In this paper, we propose **Da**rk**n**et **T**raffic **E**mbedding (DANTE), a novel darknet analysis method for representing, monitoring, and detecting complex

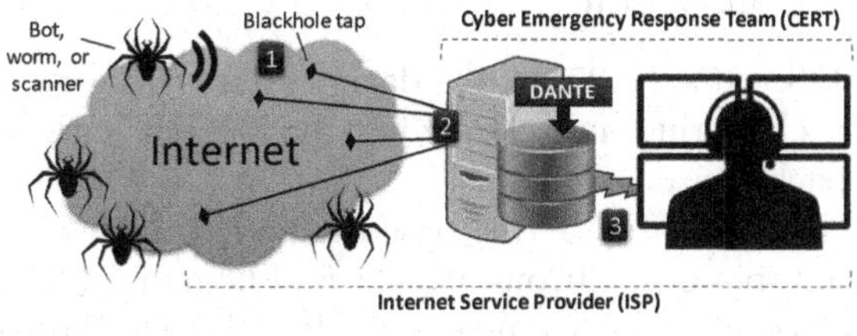

Fig. 1. Topological view of darknet analysis maintained by an ISP: (1) packets sent to non-existing IPs are captured at taps, then (2) logs are aggregated and sent to a cluster, and then (3) algorithms are used to generate emerging threat intelligence for the CERT.

emerging threats via darknet traffic. DANTE is designed to be scalable and work in a big data architecture. DANTE accomplishes this in two steps. First, for each time window, DANTE summarizes each hosts' activity as a feature vector which captures that host's behavior. The vector is made by (1) training a Word2vec [30] model on the targeted ports (words) found in observed sequences (sentences), and (2) summarizing newly observed port sequences from a host as the average of the ports' Word2Vec embeddings. The second step is to cluster the feature vectors (host activities) found in a time window, and use a novel technique to map these clusters to the previous time windows. Briefly, (1) the cluster labels from the previous time window are mapped to the current clusters using Jaccard similarity measures (*tracking*), (2) clusters which could not be mapped are then labeled using a collection of one-vs-all classifiers (*recurring concepts*), and then (3) the remaining unlabeled clusters are labeled by a members of the CERT and are given a new 1-vs-all classifiers (*novelty detection*).

To evaluate our method, we worked with the largest telecommunications provider in Europe, Deutsche Telekom AG. To support our research, they collected for us full year of darknet traffic (over three Tera-Bytes) in which DANTE discovered 1,177 new emerging threats and was able to track malicious campaigns over time. Deutsche Telekom is now using DANTE for threat intelligence in their CERT. We also evaluated the current best approach [4] on the same dataset and found that DANTE produced results were more concise and significantly more effective at capturing attack patterns.

In summary, this paper provides the following contributions: (1) A scalable and online framework for analyzing darknet traffic which can detect and track malicious campaigns and behaviors over time. (2) A method for representing a sequence of accessed ports, of variable length, as a numerical feature vector which captures the intent of an attacking host in a meaningful way. (3) A generic algorithm for performing temporal clustering which can (a) track cluster drift, detect novel/emerging clusters and reoccurring clusters, (b) run parallelized over a big data cluster with multiple data sources, and (c) can be used with any batch clustering algorithm according to the user's needs.

2 Related Work

2.1 Mining Darknet Traffic

In prior research [1–3,5,6,8,19,27,34,43], darknet data was used to detect botnet hosts, typically by clustering and classifying the src IPs with features such as the dst port and packet size.

In [27], the authors created a rule-based model to help categorize darknet records into several types of malicious attacks and benign activities, and showed how those categories evolved over ten years of data. The authors used attributes such as the number of source IPs and destination ports in order to categorize the data. However, they did not consider the sequence of destination ports coming from an IP. We found those sequences to be particularly informative in the detection of attack patterns as they can indicate the intention of the attacker.

Ban *et al.* [3–6] introduced a network incident analysis center for tactical emergency response (NICTER) that monitors around 300,000 darknet IPs in Japan. They used NICTER to find correlations between the malicious activities discovered on the darknet and activities extracted from different types of honeypots. In [3], and later in [4], Ban *et al.* used DT-growth, an association rule learning (ARL) algorithm, to port associations in the darknet data. Thonnard and Dacier [36] created a new clustering tool to detect groups of IPs that behave similarly. They used graph theory in order to find temporal correlation between port usage and thus created a way to group different IPs. Nevertheless, this work is different from ours, as Thonnard and Dacier [36] ignored the meaning and use of the ports in the sequence when clustering. In [15], the authors used DBSCAN to create clusters of packets and then used an algorithm from the field of topological data analysis in order to visualize the darknet and help an expert easily observe and analyze the data. To use DBSCAN, they treated the ports as integers by looking at the port number. In contrast to this work, we use an artificial neural network to learn the connections and relations between the ports to find an informative numeric representation.

In order to retrieve numeric information from network traffic packets, the majority of the research conducted extracted statistical features such as the number of destination IPs or unique ports [7,12,14,15,26,32,34,42]. Although these features proved to help in the detection and exploration of attacks, they are hand-picked, and it is difficult to choose the features that fit the task. In contrast to those methods, we used a neural network-based algorithm that automatically extracts meaningful representations of the packets.

Most of the aforementioned works apply their method on a static corpus of data. However, new data arrives continuously, and there is a need for an online system that can detect attacks in near real-time. To address this issue, our proposed method periodically analyzes the packets that have arrived from the sensor in the last L minutes and applies the detection mechanism in an online fashion by using a big data architectures.

2.2 Temporal Clustering

The well-known of clustering algorithms, such as k-means and DBSCAN [18], are batch algorithms. This means that they are applied once on the entire dataset and cannot track or monitor temporal trends. Although batch algorithms provide the best clustering quality, they are unsuitable as-is for processing data streams (unbounded sequences of observations).

To cluster data streams, one can use STREAM [21], Incremental DBSCAN [17], DenStream [10], CluStream, and many others [11]. However, these algorithms cannot perform novelty detection. In other words, they cannot differentiate between reoccurring clusters and novel clusters. This capability is required to detect emerging threats and monitor the re-use of known attack variants.

There are several algorithms for novelty detection in data streams [20], however these algorithms cannot be natively parallelized over a big data computing cluster and do not directly support the processing of multiple parallel sources.

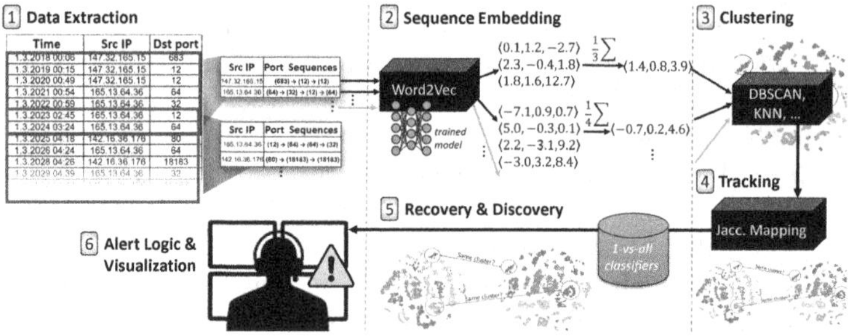

Fig. 2. An overview of the DANTE framework

Time	Src IP	Dst port
1.3.2018 00:06	147.32.165.15	683
1.3.2019 00:15	147.32.165.15	12
1.3.2020 00:49	147.32.165.15	12
1.3.2021 00:54	165.13.64.36	64
1.3.2022 00:59	165.13.64.36	32
1.3.2023 02:45	165.13.64.36	12
1.3.2024 03:24	165.13.64.36	64
1.3.2025 04:18	142.16.36.176	80
1.3.2026 04:24	165.13.64.36	64
1.3.2028 04:26	142.16.36.176	18183
1.3.2029 04:39	165.13.64.36	32
1.3.2030 04:48	142.16.36.176	18183

Time Window 1

Src IP	Port Sequences
147.32.165.15	(683) → (12) → (12)
165.13.64.36	(64) → (32) → (12) → (64)

Time Window 2

Src IP	Port Sequences
165.13.64.36	(12) → (64) → (64) → (32)
142.16.36.176	(80) → (18183) → (18183)

Fig. 3. Time window extraction.

Furthermore, each of these algorithms were designed to apply the principles of a particular kind of clustering algorithm over compressed data summaries. In addition to this being a lossy process, a user may need a different type of clustering algorithm to best fit the data (e.g., DBSCAN or K-means)). In contrast, the temporal clustering framework that we propose in this paper can be parallelized over a big data cluster while receiving data from multiple sources. In addition, the framework is flexible in terms of selecting a clustering algorithm. This enables the user to apply the most suitable batch algorithm in his/her arsenal.

3 The Darknet Analyzer Framework

In this study, we analyze the ongoing activities in the darknet by clustering common patterns/sequences of targeted (dst) ports. These patterns can be used to discover new attacks as well as explore the behavior of ongoing attacks and trends. The analysis process consists of the four stages described below.

1. **Sequence extraction** First, the data is split into sliding time windows, resulting in multiple windows with length L; for each time window, we group the destination port records of the same src IP into a port sequence. The final result is a table representing a time window with two attributes, the first being the src IP, and the second being the dst port's sequence, as shown in Fig. 3.

2. **Port sequence embeddings** By using a word embedding algorithm on the port sequences extracted from the previous stage and treating ports as words and port sequence as sentences, we are able to transform the port sequences into a meaningful numerical feature vectors. A description of how we expand on this method is presented in Sect. 4.
3. **Temporal clustering** With the feature vector obtained in the previous stage, we use a novel temporal clustering method to cluster the feature vectors over time. This step enables us to track attack patterns and distinguish between new (novel) and reoccurring patterns. This is explained further in Sect. 5.
4. **Alert logic and visualization** The current time window's clusters and labels are visualized for the CERT team (e.g., trends, or campaigns by country source). We can also use this information to create an alert rule regarding the reappearance of a cluster that an analyst classified as malicious in the past. These alerts can then be used in security information sharing platforms such as MISP [39].

Adversarial Attack Mitigation. We identify two groups of adversarial attacks: (1) where the attacker conceals him/herself by adding dummy port access as noise, and (2) where the attacker disguises the attack as a pattern that belongs to a known cluster. To detect these attacks one can issue an alert when (1) a novel cluster appears which is larger than a certain threshold, and (2) a cluster dramatically increases in size. Alternatively, one can recluster large clusters to find subpatterns within them.

4 Port Sequence Embedding

Threat agents (e.g., an attacker or bot) may send packets to unregistered IP addresses for several different reasons, such as to find a host with a vulnerability to exploit or, in the case of worms, to access a backdoor. We define a sequence S as the sequence of ports collected from a specific src IP to a specific dst IP.

A darknet sensor can observe these communications as a sequence of ports being accessed. For example, the sequence "42527, 80, 80" was observed in the wild. In this sequence, we can see that the attacker tried to access a high port (42527), and immediately send several packets to port 80 (HTTP); This sequence reflects a worm's attempt to detect if the target host has already been infected (via backdoor on port 42527), and then after failing, trying to compromise the host by exploiting two different vulnerabilities in the host's web server (port 80). From this, we can understand that the port targets in a given sequence reveal information regarding the intent of the attacker. From this example, it can be seen that the sequence of targeted ports not only reflects the attacker's goal and strategy, but also implicitly captures the type of threat agent. Moreover, threat agents such as bots, worms, and hackers in the same campaign act in unison enabling us to perform cluster analysis to detect new or recurring campaigns and other behaviors.

However, in order to cluster those sequences, we must find a representation which can summarize them as a numeric vector for the machine learning algorithm. This format is a fixed length vector $x \in \mathbb{R}^n$ where the euclidean distance between vectors x_i and x_j vectors measure similarity (closer vectors are more similar). Although TCP and UDP ports are numbers, the numerical relationship between ports is meaningless. For instance, port 21 is used for FTP, and port 22 is used for SSH, and there is no connection between the two. Therefore, in order to summarize the behavior of a scan, we first need to learn a numeric relationship between all of the ports.

To accomplish this we use Word2Vec. Word2vec, presented in [30] by Mikolov *et al.* , is a natural language processing (NLP) algorithm that aims to maximize the co-occurrence probability of words in the same sentence. Our method uses the same basic algorithm, but instead of looking at words in sentences, we use the port sequences where a sequence from a specific src IP corresponds to a sentence, and the port numbers correspond to the words in that sentence. Let p be a TCP/UDP port in the set $P = \{0, 1, \ldots 65, 536\}$ of all ports. Let $s = \{p_1, p_2, \ldots p_k\}$ be a sequence of ports sent from a specific src IP to a specific dst IP within some time interval L. We train a Word2Vec model on a corpus of example sequences S, such that a sequence $s \in S$ is treated as a sentence and $p \in P$ is treated as a word.

The trained Word2Vec model is able to convert a port p into a vector representation (embedding) $e_p \in \mathbb{R}^n$ such that e_p captures the meaning of p given its observed locality among other ports in sequences of S. An example of that property can be seen by looking at port 23 and port 2323, both of which are used for Telnet and hence are expected to appear in the scan data interchangeably. Therefore, they have remarkably similar embedding vectors. The same applies to arbitrary port numbers which are dynamic or not well documented. By using Word2vec, we do not need to consider the fact that multiple ports use the same service as the embedding process does this for us. Moreover, ports commonly found together in particular attack patterns will be associated as well, thus e_p captures the attack intent as well.

Unlike more straightforward methods, such as Bag of Words (BoW), Word2vec uses context windows in order to create the representation of each word. This method considers not only what other words are in the sentence, but also where they are, allowing for a better representation.

In order to build this port-to-embedding transformation model we need to supply it with a significant amount of scan data, which could be computationally heavy. Fortunately, this model does not have to be rebuilt in every time window, and it is possible to use a pretrained model for a long period of time. The intuition behind this rationale is that the uses of each port do not change often and a well-trained model should be sufficient for a considerable amount of time.

Additionally, there is no need to save the model itself once trained. Instead of keeping the entire neural network model, one can save a hash table where the key is the port number, and the value is the embedding for that port. This approach reduces the amount of data needed to be saved significantly, as the number of possible ports is limited by the number 65,536. Further more, by eliminating the need to execute a neural network, DANTE can run extremely fast in a big data framework.

After each port has an embedding vector of size d, we want to obtain an embedding vector with the same size, d, that represents an entire port sequence P that contains s number of ports. Although there are many methods for sentence embedding, recent research [40] discovered that the best way to do so is to average the embedding of each word in the sentence. In the port embeddings case, We summarize the overall behavior (intent) of s as a single vector $e_s \in \mathbb{R}$ by computing the average of that sequence's port embeddings. Concretely, we compute

$$e_s = \frac{1}{k}\sum_{i=1}^{k} e_{p_i} \tag{1}$$

The resulting feature vector can be used for any machine learning algorithm, such as a classifier or clustering algorithm.

5 Temporal Clustering

As described in Sect. 4, it is possible to summarize sequences of ports as their average embedding and analyze their behavior by performing cluster analysis. However, it is important to inspect the clusters over time. By doing so we can: (1) detect new attacks as they emerge (novelty detection), (2) track attack campaigns and how their strategies change, (3) follow the re-use of known attacks, e.g., variants of the Mirai botnet, and (4) analyze the trend of ongoing attacks, such as changes in volume, sources, and targets. However, darknet data is collected from X sources simultaneously. To manage this, the data it is typically stored in a big data cluster such as Hadoop. Therefore, we propose a temporal clustering framework which can be used with any batch clustering algorithm. The framework operates as follows.

5.1 Windowing

First, we sort and aggregate the most recent data into overlapping time windows. Let L be the width of the window in minutes, and let S be the step size in which we slide the window, where $S < L$. Following this process, let T_i be the i-th time window in our data, where T_{i+1} is the next sequential time window. Finally, let the ratio of observations shared between two neighboring windows be defined as

$$r_{i,i+1}\frac{|T_i| \cap |T_{i+1}|}{L} \tag{2}$$

The overlap between neighboring windows is necessary in order to track clusters. To ensure this, the parameter S should be small enough so that $0.2 \leq r_{i,i+1} \leq 0.8$.

5.2 Clustering

Next, we apply a clustering algorithm to the data of each time window to group the observations. We note that any batch clustering algorithm can be used. For example, K-means, Fuzzy C-means, Gaussian mixture models, hierarchical clustering, spectral clustering, and more. For our dataset, we found that the clustering algorithm, DBSCAN [17], worked best. The reason is because DBSCAN clusters data based on density. As a result, the number of clusters discovered is variable and does not need to be predefined (as in k-means). Another advantage is that DBSCAN can label outliers (points which are relatively far from the general distribution). This helps us analyze these cases separately without harming the quality of the clustering process.

5.3 Tracking

Between time window T_i and time window T_{i+1}, concept drift can occur. This means that the number of clusters and their shapes can change. Moreover, a cluster in T_{i+1} can be a *current* cluster (also found in T_i), an *old* cluster (found in T_j where $j < i$), or a *new* cluster (never seen before). Figure 4 illustrates this challenge. To annotate the clusters in T_{i+1}, we first find the current clusters by comparing T_i and T_{i+1}. A cluster in T_{i+1} is mapped to a cluster in T_i if there is a significant overlap of observations between them. We measure the overlap using the Jaccard similarity metric, defined as

$$\text{Jaccard}(A, B) = \frac{|A \cap B|}{|A \cup B|} = \frac{|A \cap B|}{|A| + |B| - |A \cap B|} \tag{3}$$

The Jaccard similarity metric measures the similarity between sets of items. This metric can be used in our case, because adjacent time windows overlap (by $L - S$). As a result, clusters which have a high Jaccard Similarity Score have a large number of overlapping observations and thus are considered to be the same pattern.

By using the distributed system, we simultaneously calculate the Jaccard similarity of all of the clusters in T_{i+1} with the clusters in T_i. If the Jaccard similarity is above a certain threshold for two clusters, then the cluster from T_{i+1} is considered to be the same as the cluster from T_i (i.e., *current* cluster). In cases in which the T_{i+1} cluster has no corresponding cluster from T_i, the cluster is considered new. The algorithm for mapping clusters between adjacent overlapping time windows is presented in Algorithm 1. Note that there is no need to use the embedding vector of each instance; only a key (the src IP in our case) is needed for the comparison.

5.4 Recovery and Discovery

The cluster mapping process presented in the previous Section enables us to align the clusters with the previous time window, but we also want to be able

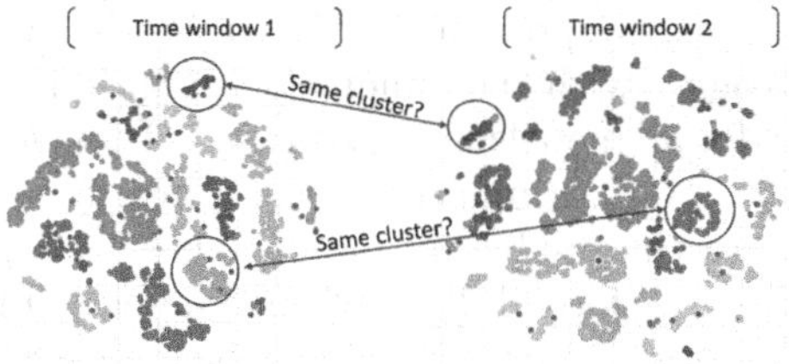

Fig. 4. An illustration of the challenge in mapping clusters found in different time windows (data here has been projected as 2D using T-SNE [28]).

to identify witch of the clusters are recurring concepts (to retrieve their annotations), as well as novel concepts. Because storing the entire past data is, in most cases, impractical, our approach is to build a classifier model for each of the observed clusters. Each model is a binary one-vs-all classifier trained on the time window where the said cluster was first seen. The instances that belong to the cluster get the label one, and the rest get the label zero. We found that Random Forest suits this problem well as this model, unlike classifiers such as K nearest neighbors, have no need to save the data points and only need to save the decision trees. We define the set of one-vs-all classifiers as M.

Let M be our database of our one-vs-all classifiers and let c be a cluster in T_{i+1} which we could not map to T_i. The probability that c belongs cluster c_j represented by $m_j \in M$, is computed as

$$p(c, c_j) = \frac{1}{|c|} \sum_{i=1}^{|c|} \text{predict}_{m_j}(e_{s_i}) \tag{4}$$

Algorithm 1. The cluster mapping algorithm for the current and previous time windows

Require: $T_i, T_{i+1}, Threshold$
Ensure: $ClusterToClusterMapping$
$ClusterToClusterMapping \leftarrow emptylist$
for all $p \in T_i$ **do:**
  for all $k \in T_{i+1}$ **do:**
    $Sim \leftarrow Jaccard(p, k)$
    if $Sim > Threshold$ **then**
      $ClusterToClusterMapping[p] \leftarrow k$;

Algorithm 2. The recovery & discovery algorithm for finding older appearances of a cluster

Require: $T_{I,c}, MS, Threshold$
Ensure: $TrackedCluster$
$N \leftarrow NumberOfInstances(T_{I,c})$
$TrackedCluster \leftarrow emptyString$
for all $M \in MS$ **do:**
  $C \leftarrow 0$
  for all $I \in T_{I,c}$ **do:**
    $P_I \leftarrow P(Y_i = 1|M)$
    $C \leftarrow C + P_I$
  $C_{normalized} \leftarrow C/N$
  if $C_{normalized} > Threshold$ **then**
    $TrackedCluster \leftarrow M.name$

where $\text{predict}_m(x)$ returns the probability that x belongs to the positive class using model m. We assign c with the annotation of cluster c_j if $p(c, c_j)$ obtains the highest probability for all models in M, and $p(c, c_j) > \beta$ for some user defined parameter β (we set $\beta = 0.7$). Similarly to the Jaccard similarity calculation, one can easily distribute the prediction part as those predictions can also be calculated simultaneously. A formal description is described in Algorithm 2.

In cases in which there is no match in any of the classifiers in M, we consider cluster c to be a *new* cluster. Once a new cluster is found we train a new classifier on this cluster's data as previously explained.

After some time, a concept drift may occur, and the patterns change slightly. To deal with this issue, in cases in which a known cluster appears in the data stream, we update and retrain the corresponding model.

6 Analysis of Darknet Traffic

Unlike other methods that used darknet data streams, e.g. [13], DANTE is not trying to find anomalies on specific ports, but rather find concepts and trends in the data. While methods that find correlations between ports exist [4], (1) they operate offline detecting patterns months after the fact, and (2) do not track the patterns over time. In contrast, DANTE finds new trends online (within minutes) and can detect recurring and novel patterns. To the best of our knowledge there are no online algorithms for detecting and tracking patterns in darknet traffic. Therefore, we demonstrate the usefulness of the proposed approach through a case study on over a year of data. The data was collected from a greynet [6,22], meaning that the unused IPs are from a network that is populated by both active and unused IP addresses. Because of the nature of greynet, it is harder for an attacker to detect if the target host is an active machine or a darknet sensor.

6.1 Configuration and Setup

Dataset. For the purpose of this research, the NSP (Network Service Provider) established 1,126 different unused IP addresses across 12 different subnets from the same organization. All traffic which was sent to these IPs were logged as darknet traffic. The traffic was collected in three batches; the first was recorded during a period of six weeks (44 days) from 10/25/2018 until 12/5/2018 (denoted by Batch 1), the second was recorded during a period of eight weeks (55 days) from 2/1/2019 until 3/26/2019 (denoted by Batch 2) and the third batch was recorded during a period of 37 weeks (257 days) from 1/21/2019 until 10/6/2019 (denoted by Batch 3). Note that Batch 2 and Batch 3 are overlapping, that is because the Batches 1 and 2 recorded in the research phase of the project, where Batch 3 recorded in the deployment stage, in real time. A deep analysis was performed on Batches 1 and 2. Batch 3 was used to demonstrate the system's long-term stability. In total, 7,918,787,884 packet headers from 4,887,568 different source IP addresses were recorded, resulting in over 3 terabytes of data. Figure 5 shows the number of packets and source IP addresses for every hour in the first two batches. Note that due to a technical problem, one hour at the end of October is missing. Because the missing time is insignificant, and the proposed method can deal with missing time windows, those missing values do not affect the overall results.

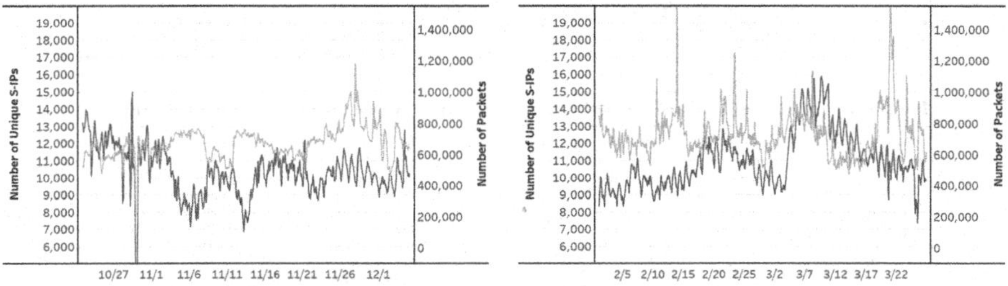

Fig. 5. The hourly number of different packets (orange) unique src IPs (blue) in the darknet after the data preprocessing stage. (Color figure online)

Configuration. We choose the step size S to be one hour and the window length L to be four hours, similar to the work of Ban *et al.* [4]. A one hour step size provides a sufficient amount of data while granting a security expert enough time to react to a detected attack. In addition, we choose the epsilon parameter of DB-SCAN to be 0.3, and the minPts parameter to be 30, as those parameters resulted in an average of four new clusters every day (agreed by the security experts to be a reasonable number of clusters to investigate each day). In addition, we do not want clusters with a small number of src IPs, as those clusters are too small to represent a significant trend in the network, and thus should be treated as noise.

Scalable Implementation. To scale to the number of sources and amount of data, DANTE was implemented and evaluated at the NSP's CERT using Spark on an Hadoop architecture. Those architectures utilized the power of big data to distribute the algorithm on several different machines. We tested the method on a Hadoop cluster consisting of 50 cores and 10 executors. The algorithm takes approximately 62 s to extract, embed, cluster, and map each four hours time window.

Data Preprocessing. IPs which sent less than three packets per window were discarded. The majority of these packets are likely miss-configurations (back scatter) and not a malicious attack. By removing those IPs, we reduced the number of packets by 33% percent.

6.2 Results

Because of time constrains, we use Batch 1 and 2 for a comprehensive analysis, and show that DANTE can find malicious attacks within the first few hours, long before any of the online reports. Batch 3 is used to prove that the algorithm can

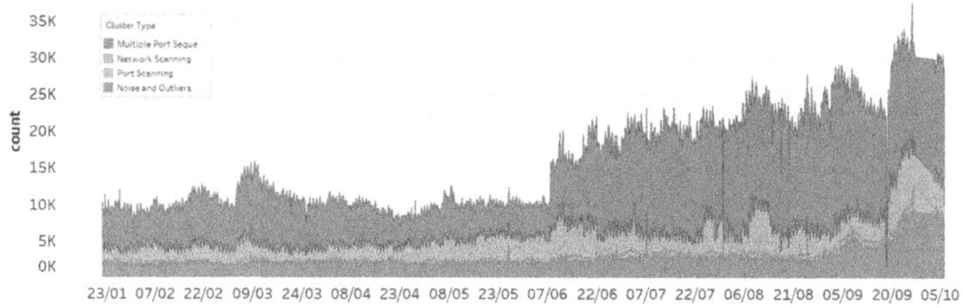

Fig. 6. Hourly number of sequences 10 month of data, by their types.

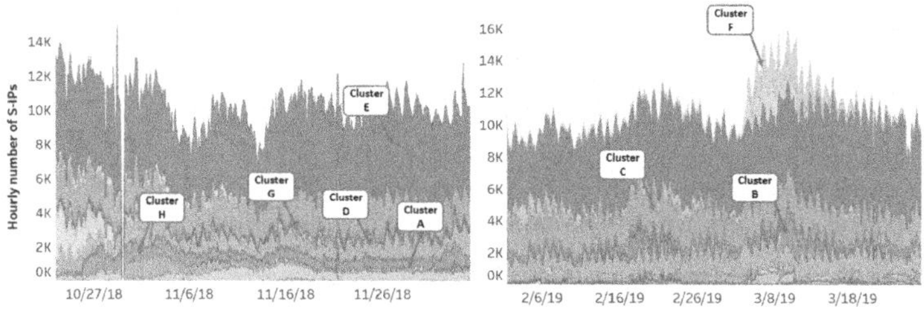

Fig. 7. An overview of the different clusters in the data during the data collection period, aggregated by hour. Each color represents a different cluster. (Color figure online)

work over a long period of time, and can detect new and reoccurring clusters even after almost a year. A visualisation of Batches 3 clusters over the 10 months can be seen in Fig. 6.

A total of 400 clusters (novel darknet behaviors) were discovered over Batches 1 and 2, and 1,141 clusters were discovered over Batch 3. As previously mentioned, the system discovered four new clusters each day, on average, as can be seen in Fig. 9. This number, determined by the set parameters, is small enough for an analyst to evaluate but large enough to detect new concepts. The vast majority of the src IPs belong to one of 16 large clusters (concepts) over the year. Some of these clusters reoccurred (e.g., botnet code reuse) and others always stayed present (e.g., worms looking for open telnet ports). Figure 7 plots the volume of the discovered clusters over time for Batches 1 and 2. The x-axis represents time, and the y-axis represents the total number of src IPs in the data for each hour. In the visualization component of DANTE (Sect. 3) this graph is updated periodically to help the analysts explore past and current trends.

The clusters discovered by DANTE can be roughly divided into four categories:

1. **Service Recon** Clusters whose sequences have six or more unique ports:*Either benign web scanners or agents performing reconnaissance on active services.*
2. **Basic Attacks & Host Recon** Clusters whose sequences consist of a single port: *Agents trying compromise a device via a single vulnerability, or performing reconnaissance to discover active network hosts.*

3. **Complex Attacks** Clusters whose sequences have two to five unique ports: *Agents which are attempting to exploit multiple vulnerabilities per device, are performing multi-step attacks, or are worms.*
4. **Noise and Outliers** A single cluster of benign sequences from misconfigurations, backscatter, or are too small to represent an ongoing trend.

In Table 1 we give statistics on these categories, and in Fig. 6 we present the volume of each category in Batch 3 over time. Table 2 provides concrete examples of discovered attack patterns, by cluster, as labeled in Fig. 7.

Table 1. The number of clusters, src IPs, and packets for each of the four cluster families.

Clusters Type	No. of clusters			No. of src IPs			No. of packets		
	Batch 1	Batch 2	Batch 3	Batch 1	Batch 2	Batch 3	Batch 1	Batch 2	Batch 3
Service Recon	24	27	326	16,909	78,644	101,316	9,603,305	39,111,889	162,499,127
Basic Attacks & Host Recon	77	71	202	246,864	313,457	1,216,077	50,937,141	56,881,565	380,244,017
Complex Attacks	**78**	**121**	**610**	**351,318**	**395,109**	**2,602,184**	**19,203,613**	**85,960,212**	**1,085,777,577**
Noise and Outliers	1	1	1	115,645	142,920	819,566	582,646,403	743,977,834	5,935,419,695

Service Recon. Service reconnaissance (port scanning) is performed by attackers to find and exploit backdoors or vulnerabilities in services [27]. In Batches 1 and 2, we identified 51 port scanning clusters with an average of 929 different ports scanned in each cluster. However, it is important to note that a cluster in this group is not necessarily malicious. For example, cluster A in Fig. 7 consists of src IPs of Censys, a security company which scans 40 ports to find and report vulnerable IoT devices. Interestingly, because of the embeddings, DANTE found that 2.2% of the IPs in cluster A do not belong Censys' subnet. Rather, we verified them to belong to a malicious actor copying Censys to stay under the radar.

In some cases, the service recon can consist of multiple ports that belong to the same service, in order to use an exploit on this service even if the host is using an alternative port. For example, Cluster B, occurred on 3/8/2019, consist only of ports that can be associated with HTTP. This cluster consists of 17 ports, such as 80, 8080, 8000, 8008, 8081 and 8181. In addition, most of the src IPs are located in Taiwan (18%), Iran (15%) and Vietnam (12%). Because DANTE assigns similar embedding to those ports, those port were group together and DANTE was able to detect this pattern and issue an alert. At the time of writing of this article, we were not able to find any information on this scan online. This lack of reports could be explained by the fact that there was no significant peak in any of the ports involved, which make it hard for conventional anomaly detectors to detect this pattern.

Basic Attacks and Host Recon. Many port sequences consist of a single port accessed repeatedly one or more times. This behavior captures agents which are trying to exploit a single vulnerability (e.g., how the Mirai malware compromises

Table 2. Summaries of the mentioned clusters.

Cluster name	Category	Number of src IPs	Number of packets	Sequence examples
A	Service Recon	141	873, 458	[2077, 2077, 8877, 7080, ..., 304, 3556]
B	Service Recon	20,982	2, 061, 655	[8000, 88, 80, 8000, 8081, 80, 80]
C	Basic Attacks & Host Recon	113, 407	7,061,042	[445, 445, 445]
D	Basic Attacks & Host Recon	895	68,065	[11390, 11390, ..., 11390, 11390]
E	Basic Attacks & Host Recon	285,651	42,225,387	[23, 23, 2323]
F	Complex Attacks	43,305	1,718,440	[9527, 9527, 9527, 5555, 5555, 5555]
G	Complex Attacks	535	7,270	[7550, 7550, 7547, 7547, 7547]
H	Complex Attacks	43,305	105,258	[7379, 7379, 5379, 5379, 6379, 6379]

devices via telnet). This behavior is also consistent with network scanning [27], where the attacker tries to detect live hosts to map out an organization. By following how campaigns (clusters) reoccur and grow in volume over time, analysts can use DANTE to find trends and discover new vulnerabilities. To visualize this, Fig. 8 contains a scatter plot of each port sequence cluster from Batch 1 and its corresponding port.

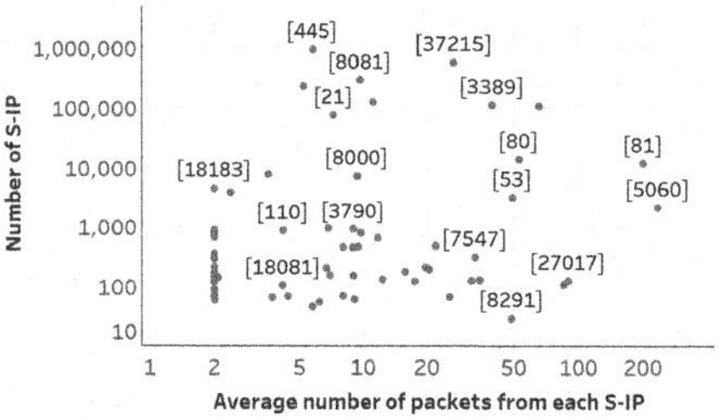

Fig. 8. A view of the Basic Attacks clusters from Batch 1. Each dot represents a cluster with the corresponding port shown above (x-axis and y-axis are logarithmic).

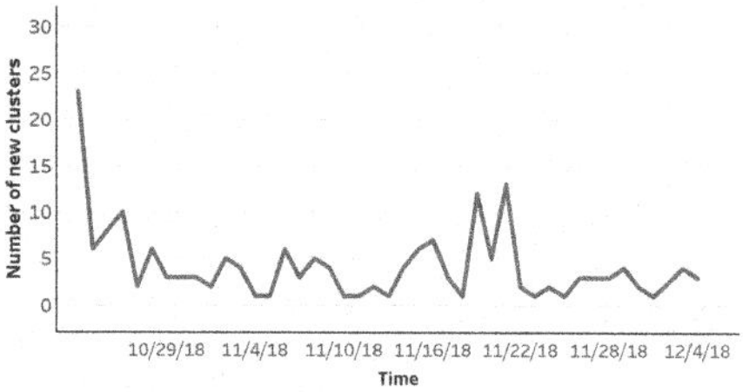

Fig. 9. Number of newly discovered clusters over time in the Batch 1 data (daily).

It is interesting to see that, for example, port 445 is accessed by many src IPs, each of which only accessed the port an average of 5.8 times. This type of cluster can be seen in the second largest cluster, C. This cluster consists of port 445 (SMB over IP: commonly used by worms such as Conficker [16,23,41] to compromise Windows machines) and was being attacked at a low rate, once every few hours. However, DANTE also detected that port 5060 (SIP: used to initiate and maintain voice/video calls) was trending, being attacked at a high rate from the USA.

Nonetheless, most of the clusters in this category are relatively small and only appear for a few hours. In another example, DANTE detected a network recon campaign between 1:00 and 6:00 AM GMT on 11/25/2018 (cluster D) on the unused port of 11390. The attack consisted of 895 different src IPs which mostly targeted one out of our 1,129 darknet IPs. This indicated which subnet the campaign was targeting and provided us with threat intelligence on a possible software vulnerability/backdoor on port 11390. By using Spark implementation system, DANTE was able to issue an alert to the CERT with that information about a minute after the data arrived.

Complex Attacks. These clusters capture patterns involving multiple attack steps or vulnerabilities. This category of clusters often captures the most interesting attack patterns, and the most difficult to detect because of their low volume of traffic which hides them under the noise floor. Without the proposed embedding approach, an attack which involves a sequence of ports would be clustered with the other Basic Attacks and thus the other ports in the attack would go unnoticed –making it impossible to distinguish from other attack campaigns. We note that cluster E contains 30% of all darknet sequences. This is because the cluster captures ports 23 and 2323 which are both used for Telnet. The Mirai botnet and its variants have been using these ports to search for vulnerable devices over last few years [23].

One example of this category occurred on 3/4/2019, where DANTE reported a new large cluster appeared (F in Fig. 7) consisting of two ports and originating from China and Brazil only. Here, 92% of the src IPs sent the sequence $\{9527, 9527, 9527\}$, and 8% of the src IPs sent $\{9527, 9527, 9527, 5555, 5555, 5555\}$ and in the reverse order. This is a clear indication of one or more campaigns launched at the same time aiming to exploit a new vulnerability. Four days *after* DANTE detected the attack, the attack was reported [31] and related to a vulnerability in an IP-Camera (CVE-2017-11632). Interestingly, none of the reports mentioned that the attackers were also targeting port 5555. This not only demonstrates how DANTE can provide threat intelligence, but how DANTE can detect ongoing attacks.

A third example of this category can be seen in a cluster, G, from 11/22/18, which was discovered by DANTE. This cluster contains a pattern of scanning two specific ports, 7547 and 7550, which occurs on 11/22/2018 from 5 to 9 PM GMT. According to the Internet Storm Center (ISC) port search [38], a free tool that monitors the level of malicious port activity, this is the most significant peak of activities for port 7550 in the past two years, however, to the best of our knowledge, there have been no reports of an attack that utilizes this port. The missing information in the reports could suggest a novel attack that utilize those two ports. In addition, port 7547 appears to have a large number of packets arriving each day at 10:00 AM GMT, that were assigned to a different cluster. Unlike cluster G, that cluster consists of port 7547 alone with no additional ports in the scan. 12/22/2018 is the only day when activity on this port peaks in a different hour (see Fig. 10). There are reports [37] of a known Mirai botnet variant that uses port 7547 to exploit routers. Based on this report we attribute cluster G to Mirai activity. Since DANTE detected that the two ports are used interchangeably by the same src IPs, we suspect that there is a new vulnerability tested on port 7550 of routers. At the time of writing of this article existence or absence of such vulnerability was not yet confirmed by any organization.

On 10/31/2018, DANTE generated an alert about a new attack (cluster H) one minute after it began. This attack continued everyday from 11/15/18 until the 11/18/18. In the attack, 1,789 different network hosts sent exactly four packets to two or three of the ports: 5379, 6379, and 7379. According to the ISC, this is the largest peak in the use of port 5379 and the third largest peak for port 7379, although these ports are considered unused. We could not find any report online which identified these ports being used together in an attack. Identifying the source of these IPs can lead a CERT team to uncovering an ongoing campaign.

Noise and Outliers. These are all of the patterns which DBScan considered outliers. For example, short sequences or those with too few occurrences. Darknet traffic which falls in this category is typically benign backscatter or misconfigured packets, but not a scan [22,27]. The size of this cluster is directly controlled by the minPts parameter in DBScan and can be changed at any time. As previously mentioned, we chose minPts to be 30 in order to create a reasonable number of clusters per day for a security expert to explore.

6.3 Comparative Discussion

In the previous sections, we demonstrated that DANTE can detect new and reoccurring attack patterns. Unlike methods such as [3,4,27,36] which detect port scanning and methods such as [24,25,27,36] which detect traffic spikes on individual port numbers, DANTE can detect both and distinguishing between them. Furthermore, because DANTE uses Word2vec, it can also detect complex patterns (e.g., CVE-2017-11632). Many well known and dangerous malware generate complex patterns captures. Table 3 exemplifies this claim by listing some of these malware along with the ports which they access during their lifetime. To the best of our knowledge, there are no other works which can detect these kinds of patterns as quickly and accurately as DANTE. Concretely, [15,24,25,27,36] observe traffic on individual ports. Although [3,4] finds the connection between ports, it ignores the meaning and the content of the sequences, preventing it from detecting behaviors such as cluster G (ports 7547 and 7550 together).

6.4 Comparison with Ban *et al.* [4]

To evaluate our contribution of using Word2vec as a means for mining meaningful patterns, we compare our embedding approach to Ban *et al.* [4]. Ban *et al.* used a frequent pattern mining (FPM) algorithm called FP-growth on darknet traffic to group similar ports together. By using FPM, Ban *et al.* was able to find subsets

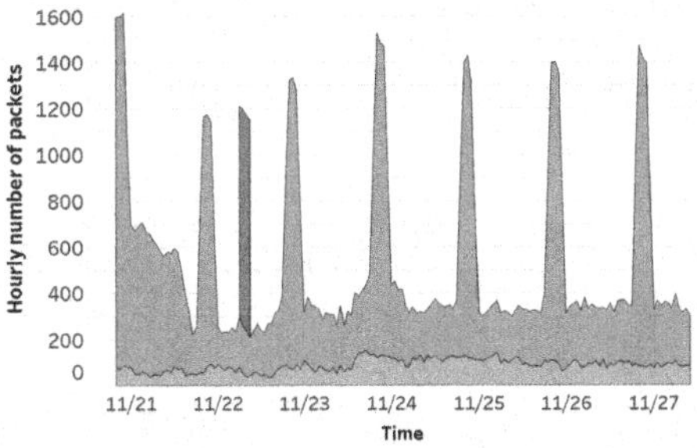

Fig. 10. Port 7547 during the week of 11/21/18. The orange cluster occurs each day at 10:00 AM GMT; the gray cluster is small and occurs continuously during the week. DANTE detected the attack (blue) which occurred at 5:00 PM 11/22/18. (Color figure online)

Table 3. Different malware and the ports they are using for infection and communication.

Botnet	Ports	Phase
Cryptojacking	6379, 2375, 2376	Infection
Wicked	8080, 8443, 80, 81	Infection
MIRAI	23, 2323	Infection
CONFICKER	445	Infection
EMOTET	80, 990, 8090, 50000, 8080, 7080, 443	Post-infection
CERBER	80, 6892	Post-infection
KOVTER	43, 8080	Post-infection
QAKBOT	443, 65400, 2222, 21, 41947	Post-infection
Cryptojacking	22, 7878, 2375, 2376	Post-infection
GH0ST	2011, 2013, 800	Post-infection
NANOCORE	3365	Post-infection

of ports that occur together in the data. To the best of our knowledge, this is the most recent work that mines patterns in darknet traffic to find new concepts. We applied their method using their parameters on Batch 1 using the same time windows. The top 20 largest subsets of both FPM and DANTE are presented in Table 4.

FPM cannot identify when different subsets are jointly used in novel attack patterns. This can be seen in cluster F, where DANTE clusters two different sets together, the set [9527] and the set [5555, 9527]. This cluster led DANTE to the discovery of an attack on Wireless IP Camera 360 devices (CVE-2017-11632). In addition, Ban *et al.*'s method omits all sequences with more than six different ports. However, these patterns are important to detect malicious service reconnaissance. For example, in cluster A, DANTE discovered a malicious actor who was copying Censys's port scanning behaviors.

By using the semantic embedding of ports, DANTE was able to relate similar ports (e.g., Telnet ports 23 and 2323 in cluster E). However, FPM created a set for every permutation although they all capture the same concept (see the three biggest FPM sets in Table 4).

Moreover, some ports typically appear in attack sequences and rarely appear alone. This means that there is semantic information which can be learned about these ports, such as the intent and how it is reflected on the other ports in the sequence. For example, in Table 4, port 80 (HTTP) is a single concept according to FPM due to the FP-growth process. In contrast, DANTE re-associates port 80 with other patterns. This can be seen in the top 20 largest concepts where port 80 appears in six different concepts but never appears by itself; in FPM, the singleton [80] is the fifth largest subset.

Lastly, FP-growth creates a large number of sets, even when setting the minimum support parameter to be a relatively large value (1,819 sets with minimum

Table 4. Top 20 biggest Batch 1 concepts according to DANTE and Ban et al.

Ban et al. [4]			DANTE						
Port 1	Port 2	Occur.	Port 1	Port 2	Port 3	Port 4	Port 5	Port 6	Occur.
23		5,783,748	23	2323					33,439,191
2323		1,637,119	37215						15,294,071
2323	23	1,577,142	445						4,404,525
445		982,528	80	8001	8080	8081	8088	and 10 other ports	3,699,342
80		886,540	80	8080	433				3,355,387
37215		672,897	22						2,697,434
8080		539,976	8080	8081					2,403,517
8081		383,209	3389	3398	9833	3839	8933		1,383,441
5555		322,826	5555						937,267
8080	80	277,633	1433	3413					850,124
37215	23	244,212	1235 different ports						798,760
81		187,815	80	8080					615,490
3389		178,766	3389	139	445	11211			519,732
8443		158,892	21						482,898
32764		132,361	81						463,355
1433		130,363	1252 different ports						344,761
9000		129,044	443						343,922
22		122,518	8080	8000	8333			and 44 other ports	237,758
80	23	106,023	123	47808	161	1883	8883	501	195,431
21		104181	5900	5901	5431				150,234

support of 1,000 for a period of six weeks). The resulting number of sets is impractical for a CERT analyst to investigate. An advantage of using DANTE is that only a reasonable number of concepts with high importance are identified per day (average of four a day, as shown in Fig. 9). At the same time, DANTE discovers small yet significant patterns that might represent dangerous attacks (such as cluster G).

7 Conclusion and Future Work

By mining darknet traffic, analysts can get frequent reports on-going and new merging threats facing their network. In this paper, we presented DANTE: a framework which enables network service providers to mine threat intelligence from massive darknet traffic streams. DANTE was able to detect hidden threats within real-world datasets and out performed the state of the art approach. Currently, DANTE is being deployed in Deutsche Telekom's networks to provide their CERT with better threat intelligence. We hope that the embedding and clustering techniques of this paper will assist researchers and the industry in better securing the Internet.

Acknowledgment. This research was partially supported by the CONCORDIA project that has received funding from the European Union's Horizon 2020 research and innovation programme under grant agreement number 830927. We would like to thank Nadav Maman for his help in implementing DANTE.

References

1. Bailey, M., Cooke, E., Jahanian, F., Myrick, A., Sinha, S.: Practical darknet measurement. In: 2006 40th Annual Conference on Information Sciences and Systems, pp. 1496–1501. IEEE (2006)
2. Bailey, M., Cooke, E., Jahanian, F., Nazario, J., Watson, D., et al.: The Internet motion sensor-a distributed blackhole monitoring system. In: NDSS (2005)
3. Ban, T., Eto, M., Guo, S., Inoue, D., Nakao, K., Huang, R.: A study on association rule mining of darknet big data. In: 2015 International Joint Conference on Neural Networks (IJCNN), pp. 1–7, July 2015
4. Ban, T., Pang, S., Eto, M., Inoue, D., Nakao, K., Huang, R.: Towards early detection of novel attack patterns through the lens of a large-scale darknet, pp. 341–349, July 2016
5. Ban, T., Zhu, L., Shimamura, J., Pang, S., Inoue, D., Nakao, K.: Detection of botnet activities through the lens of a large-scale darknet. In: Liu, D., Xie, S., Li, Y., Zhao, D., El-Alfy, E.-S.M. (eds.) ICONIP 2017. LNCS, vol. 10638, pp. 442–451. Springer, Cham (2017). https://doi.org/10.1007/978-3-319-70139-4_45
6. Ban, T., Zhu, L., Shimamura, J., Pang, S., Inoue, D., Nakao, K.: Behavior analysis of long-term cyber attacks in the darknet. In: Huang, T., Zeng, Z., Li, C., Leung, C.S. (eds.) ICONIP 2012. LNCS, vol. 7667, pp. 620–628. Springer, Heidelberg (2012). https://doi.org/10.1007/978-3-642-34500-5_73
7. Bartos, K., Sofka, M., Franc, V.: Optimized invariant representation of network traffic for detecting unseen malware variants (2016)
8. Bou-Harb, E., Debbabi, M., Assi, C.: A time series approach for inferring orchestrated probing campaigns by analyzing darknet traffic. In: 2015 10th International Conference on Availability, Reliability and Security, pp. 180–185. IEEE, August 2015
9. Bringer, M.L., Chelmecki, C.A., Fujinoki, H.: A survey: recent advances and future trends in honeypot research. Int. J. Comput. Netw. Inf. Secur. **4**(10), 63 (2012)
10. Cao, F., Estert, M., Qian, W., Zhou, A.: Density-based clustering over an evolving data stream with noise. In: Proceedings of the 2006 SIAM International Conference on Data Mining, pp. 328–339. SIAM (2006)
11. Carnein, M., Trautmann, H.: Optimizing data stream representation: an extensive survey on stream clustering algorithms. Bus. Inf. Syst. Eng. **61**(3), 277–297 (2019)
12. Casas, P., Mazel, J., Owezarski, P.: Unsupervised network intrusion detection systems: detecting the unknown without knowledge. Comput. Commun. **35**, 772–783 (2012)
13. Choi, S.S., Song, J., Kim, S., Kim, S.: A model of analyzing cyber threats trend and tracing potential attackers based on darknet traffic. Secur. Commun. Netw. **7**(10), n/a (2013)
14. Corchado, E., Herrero, Á.: Neural visualization of network traffic data for intrusion detection. Appl. Soft Comput. J. **11**, 2042–2056 (2010)
15. Coudriau, M., Lahmadi, A., François, J.: Topological analysis and visualisation of network monitoring data: darknet case study. In: 2016 IEEE International Workshop on Information Forensics and Security (WIFS), pp. 1–6 (2016)
16. Durumeric, Z., Bailey, M., Halderman, J.A.: An Internet-wide view of Internet-wide scanning. In: Proceedings of the 23rd USENIX Conference on Security Symposium, SEC 2014, pp. 65–78. USENIX Association (2014)

17. Ester, M., Wittmann, R.: Incremental generalization for mining in a data warehousing environment. In: Schek, H.-J., Alonso, G., Saltor, F., Ramos, I. (eds.) EDBT 1998. LNCS, vol. 1377, pp. 135–149. Springer, Heidelberg (1998). https://doi.org/10.1007/BFb0100982
18. Ester, M., Kriegel, H.P., Sander, J., Xu, X., et al.: A density-based algorithm for discovering clusters in large spatial databases with noise. In: KDD, vol. 96, pp. 226–231 (1996)
19. Fachkha, C., Bou-Harb, E., Debbabi, M.: Inferring distributed reflection denial of service attacks from darknet. Comput. Commun. **62**, 59–71 (2015)
20. Faria, E.R., Gonçalves, I.J.C.R., de Carvalho, A.C.P.L.F., Gama, J.: Novelty detection in data streams. Artif. Intell. Rev. **45**(2), 235–269 (2015). https://doi.org/10.1007/s10462-015-9444-8
21. Guha, S., Mishra, N., Motwani, R., O'Callaghan, L.: Clustering data streams. In: Proceedings of 41st Annual Symposium on Foundations of Computer Science, 2000, pp. 359–366. IEEE (2000)
22. Harrop, W., Armitage, G.: Defining and evaluating Greynets (sparse darknets). In: The IEEE Conference on Local Computer Networks 30th Anniversary (LCN 2005), vol. 1, pp. 344–350. IEEE (2005)
23. Heo, H., Shin, S.: Who is knocking on the Telnet port: a large-scale empirical study of network scanning. In: Proceedings of the 2018 on Asia Conference on Computer and Communications Security, pp. 625–636. ACM (2018)
24. Inoue, D., et al.: An incident analysis system NICTER and its analysis engines based on data mining techniques. In: Köppen, M., Kasabov, N., Coghill, G. (eds.) ICONIP 2008. LNCS, vol. 5506, pp. 579–586. Springer, Heidelberg (2009). https://doi.org/10.1007/978-3-642-02490-0_71
25. Kao, C.N., Chang, Y.C., Huang, N.F., Liao, I.J., Liu, R.T., Hung, H.W., et al.: A predictive zero-day network defense using long-term port-scan recording. In: 2015 IEEE Conference on Communications and Network Security (CNS), pp. 695–696. IEEE (2015)
26. Lagraa, S., Francois, J., Lahmadi, A., Miner, M., Hammerschmidt, C., State, R.: BotGM: unsupervised graph mining to detect botnets in traffic flows. In: 2017 1st Cyber Security in Networking Conference (CSNet), pp. 1–8. IEEE, October 2017
27. Liu, J., Fukuda, K.: Towards a taxonomy of darknet traffic. In: 2014 International Wireless Communications and Mobile Computing Conference (IWCMC), pp. 37–43. IEEE, August 2014
28. Maaten, L.V.D., Hinton, G.: Visualizing data using t-SNE. J. Mach. Learn. Res. **9**(Nov), 2579–2605 (2008)
29. Mairh, A., Barik, D., Verma, K., Jena, D.: Honeypot in network security: a survey. In: Proceedings of the 2011 International Conference on Communication, Computing and Security, pp. 600–605. ACM (2011)
30. Mikolov, T., Chen, K., Corrado, G., Dean, J.: Efficient estimation of word representations in vector space. arXiv preprint arXiv:1301.3781 (2013)
31. Nichols, S.: FBI warns of SIM-swap scams, IBM finds holes in visitor software, 13-year-old girl charged over Javascript prank (2019). https://www.theregister.co.uk/2019/03/09/security_roundup_080319
32. Owezarski, P.: A Near Real-Time Algorithm for Autonomous Identification and Characterization of Honeypot Attacks. Technical report (2015). https://hal.archives-ouvertes.fr/hal-01112926
33. Pa, Y.M.P., Suzuki, S., Yoshioka, K., Matsumoto, T., Kasama, T., Rossow, C.: IoTPOT: a novel honeypot for revealing current IoT threats. J. Inf. Process. **24**(3), 522–533 (2016)

34. Pang, S., et al.: Malicious events grouping via behavior based darknet traffic flow analysis. Wirel. Pers. Commun. **96**, 5335–5353 (2017)
35. Singhal, A., Ou, X.: Security risk analysis of enterprise networks using probabilistic attack graphs. Network Security Metrics, pp. 53–73. Springer, Cham (2017). https://doi.org/10.1007/978-3-319-66505-4_3
36. Thonnard, O., Dacier, M.: A framework for attack patterns' discovery in honeynet data. Digit. Invest. **5**, S128–S139 (2008)
37. Ullrich, J.: Port 7547 soap remote code execution attack against DSL modems (2016). https://isc.sans.edu/diary/Port+7547+SOAP+Remote+Code+Execution+Attack+Against+DSL+Modems/21759
38. Van Horenbeeck, M.: The sans Internet storm center. In: 2008 WOMBAT Workshop on Information Security Threats Data Collection and Sharing, pp. 17–23. IEEE (2008)
39. Wagner, C., Dulaunoy, A., Wagener, G., Iklody, A.: MISP: the design and implementation of a collaborative threat intelligence sharing platform. In: Proceedings of the 2016 ACM on Workshop on Information Sharing and Collaborative Security, pp. 49–56. ACM (2016)
40. Wieting, J., Bansal, M., Gimpel, K., Livescu, K.: Towards universal paraphrastic sentence embeddings. arXiv preprint arXiv:1511.08198 (2015)
41. Wustrow, E., Karir, M., Bailey, M., Jahanian, F., Huston, G.: Internet background radiation revisited. In: Proceedings of the 10th ACM SIGCOMM Conference on Internet Measurement, New York (2010)
42. Zhang, J., Tong, Y., Qin, T.: Traffic features extraction and clustering analysis for abnormal behavior detection. In: Proceedings of the 2016 International Conference on Intelligent Information Processing - ICIIP 2016, New York (2016)
43. Škrjanc, I., Ozawa, S., Dovžan, D., Tao, B., Nakazato, J., Shimamura, J.: Evolving cauchy possibilistic clustering and its application to large-scale cyberattack monitoring. In: 2017 IEEE Symposium Series on Computational Intelligence (SSCI), pp. 1–7, November 2017

Efficient Quantification of Profile Matching Risk in Social Networks Using Belief Propagation

Anisa Halimi[1] and Erman Ayday[1,2(✉)]

[1] Case Western Reserve University, Cleveland, OH, USA
{anisa.halimi,erman.ayday}@case.edu
[2] Bilkent University, Ankara, Turkey

Abstract. Many individuals share their opinions (e.g., on political issues) or sensitive information about them (e.g., health status) on the internet in an anonymous way to protect their privacy. However, anonymous data sharing has been becoming more challenging in today's interconnected digital world, especially for individuals that have both anonymous and identified online activities. The most prominent example of such data sharing platforms today are online social networks (OSNs). Many individuals have multiple profiles in different OSNs, including anonymous and identified ones (depending on the nature of the OSN). Here, the privacy threat is profile matching: if an attacker links anonymous profiles of individuals to their real identities, it can obtain privacy-sensitive information which may have serious consequences, such as discrimination or blackmailing. Therefore, it is very important to quantify and show to the OSN users the extent of this privacy risk. Existing attempts to model profile matching in OSNs are inadequate and computationally inefficient for real-time risk quantification. Thus, in this work, we develop algorithms to efficiently model and quantify profile matching attacks in OSNs as a step towards real-time privacy risk quantification. For this, we model the profile matching problem using a graph and develop a belief propagation (BP)-based algorithm to solve this problem in a significantly more efficient and accurate way compared to the state-of-the-art. We evaluate the proposed framework on three real-life datasets (including data from four different social networks) and show how users' profiles in different OSNs can be matched efficiently and with high probability. We show that the proposed model generation has linear complexity in terms of number of user pairs, which is significantly more efficient than the state-of-the-art (which has cubic complexity). Furthermore, it provides comparable accuracy, precision, and recall compared to state-of-the-art. Thanks to the algorithms that are developed in this work, individuals will be more conscious when sharing data on online platforms. We anticipate that this work will also drive the technology so that new privacy-centered products can be offered by the OSNs.

Keywords: Social networks · Profile matching · Deanonymization · Privacy risk quantification

L. Chen et al. (Eds.): ESORICS 2020, LNCS 12308, pp. 110–130, 2020.
https://doi.org/10.1007/978-3-030-58951-6_6

1 Introduction

Many individuals, to preserve their privacy and to protect themselves against potential damaging consequences, choose to share content anonymously in the digital space. For instance, people share their opinions about different topics or sensitive information about themselves (e.g., their health status) without sharing their real identities, hoping that they will remain anonymous. Unfortunately, this is non-trivial in today's interconnected world, in which different activities of individuals can be linked to each other. An attacker, by linking anonymous activities of individuals to their real identities (via other publicly available and identified information about them), can obtain privacy-sensitive information about them. Thus, individuals need tools that show them the scale of their vulnerability against such privacy risks when they share content. In this work, we tackle this problem by focusing on data sharing on online social networks (OSNs).

An OSN is a platform, in which, individuals share vast amount of information about themselves such as their social and professional life, hobbies, diseases, friends, and opinions. Via OSNs, people also get in touch with other people that share similar interests or that they already know in real-life [8]. With the widespread availability of the Internet, OSNs have been a part of our lives more than ever. Most individuals have multiple OSN profiles for different purposes. Furthermore, each OSN offers different services via different frameworks, leading individuals share different types of information [9]. Also, in some OSNs, users reveal their real identities (e.g., to find old friends), while in some OSNs, users prefer to remain anonymous (especially in OSNs in which users share anonymous opinions or sensitive information about themselves, such as their health status). Here, the privacy risk is the deanonymization of the anonymous OSN profile of a user using their other OSN profiles, in which the user is identified.

Such profile matching across OSNs (i.e., identifying profiles belonging to the same individuals) is a serious privacy threat, especially for individuals that have anonymous profiles in some OSNs and reveal their real identities in others. If an attacker can link anonymous profiles of individuals to their real identities, it can obtain privacy-sensitive information about individuals that is not intended to be linked to their real identities. Such sensitive information can then be used for discrimination or blackmailing. Thus, it is very important to quantify and show the risk of such profile matching attacks in an efficient and accurate way.

An OSN can be characterized by (i) its graphical structure (i.e., connections between its users) and (ii) the attributes of its users (i.e., types of information that is shared by its users). The graphical structures of most popular OSNs show strong resemblance to social connections of individuals in real-life (e.g., Facebook). Existing work shows that this fact can be utilized to link accounts of individuals from different OSNs [29]. However, without sufficient background information, just using graphical structure for profile matching becomes computationally infeasible. Furthermore, some OSNs or online platforms either do not have a graphical structure at all (e.g., forums) or their graphical structures do not resemble the real-life connections of the individuals (e.g., health-related OSNs such as PatientsLikeMe [3]). In these types of OSNs, an attacker can utilize the attributes of the users for profile matching. Thus, to show the scale of the profile matching threat, it is crucial to process both the graphical structure and the other attributes of the users in an efficient and accurate way.

In this work, we efficiently model the profile matching problem in OSNs by considering both the graphical structure and other attributes of the users, a step towards delivering real-time information to OSN users about their privacy risks for profile matching due to their sharings on online platforms. Designing efficient privacy risk quantification tools is non-trivial, especially considering the scale of the problem. To overcome this challenge, we develop a novel, graph-based model generation algorithm to solve the profile matching problem in a significantly more efficient and accurate way than the state-of-the-art.

We formulate the profile matching problem as finding the marginal probability distributions of random variables representing the possible matches between user profile pairs from the joint probability distribution of many variables. We factorize the joint probability distribution into simpler local functions to compute the marginal probability distributions efficiently. To do so, we formulate the model generation for profile matching by using a graph-based algorithm. That is, we formulate the problem on a factor graph and develop a novel belief propagation (BP)-based algorithm to generate the model efficiently and accurately (compared to the state-of-the-art). The outcome of the model generation will pave the way towards developing real-time risk quantification tools (i.e., inform users about their privacy loss and its consequences as they share new content).

Our results show that the proposed model generation algorithm can match user profiles with an accuracy of up to 90% (depending on the amount of information and attributes that users share). As more information is collected about the users profiles in social networks, the accuracy of the BP-based algorithm increases. Also, by analyzing the effect of social networks' size to obtained precision and recall values, we show the scalability of the proposed model generation algorithm. We also show that by controlling the structure of the proposed graphical model, we can simultaneously improve the efficiency of the proposed model generation algorithm and increase its accuracy.

The rest of the paper is organized as follows. In Sect. 2, we discuss the related work. In Sect. 3, we provide the threat model. In Sect. 4, we describe the proposed framework in detail. In Sect. 5, we implement and evaluate the proposed framework using real-life datasets belonging to various OSNs. In Sect. 6, we discuss how the proposed scheme can be used for real-time privacy risk quantification, potential mitigation techniques, and generalization of the proposed scheme for different OSNs. Finally, in Sect. 7, we conclude the paper.

2 Related Work

Several works in the literature have proposed profile matching schemes that leverage network structure, publicly available attributes of the users, or both of them. Profile matching based only on network (graph) structure is widely known as the deanonymization problem. Graph deanonymization (DA) attacks can be classified as (i) seed-based attacks [16,18,19,29,30], in which a set of seeds (users' accounts in two different networks which belong to the same individual) are known; and (ii) seed-free attacks [33,39], in which no seeds are used. Narayanan and Shmatikov were among the first that proposed a graph deanonymization algorithm [29]. Nilizadeh et al. [30] improved the attack proposed by Narayanan et al. by proposing a community-level deanonymization attack. Korula and Lattanzi [19] proposed a DA attack that by starting from a set of seeds, iteratively matches user pairs with the most number of neighboring mapped pairs. Ji et al. [16] quantified the deanonymizability of social networks from a theoretical perspective (i.e., focusing on social networks that follow a distribution model). Pedarsani et al. [33] proposed a Bayesian-based model to match users across social networks without using seeds. Their model uses node degrees and distances to other nodes. Sharad et al. [36] showed that users' re-identification (deanonymization) in anonymized social networks can be automated. Ji et al. [17] evaluated several anonymization techniques and deanonymization attacks and showed that all state-of-art anonymization techniques are vulnerable to modern deanonymization attacks. Recently, Zhou et al. [43] proposed DeepLink, a deep neural network based algorithm that leverages network structure for user linkage.

Another line of works [11,14,15,23,24,26,27,31,34,38,41,42] have leveraged public information in the users' profiles (such as user name, profile photo, description, location, and number of friends) for profile matching. Shu et al. [37] provided a broad review of the works that use public information for profile matching. Malhotra et al. [26] built classifiers on various attributes to determine whether two user profiles are matching or not. On the other hand, Zafarani et al. [42] explored user name by analyzing the behaviour patterns of the users, the language used, and the writing style to link users across social media sites. Goga et al. [11] showed that attributes that are hard to be controlled by users, such as location, activity, and writing style, may be sufficient for profile matching. Liu et al. [25] proposed a framework that mainly consists of three steps: behavior similarity modeling, structure consistency modeling, and multi-objective optimization. Goga et al. [12] conducted a detailed analysis of user profiles and their attributes identifying four properties: availability, consistency, non-impersonability, and discriminability. Andreou et al. [5] combined attribute and identity disclosure across social networks. Recently, Halimi et al. [13] proposed a more accurate profile matching framework based on machine learning techniques and optimization algorithms. One common thing about most of these aforementioned approaches is that they rely on training classifiers to determine whether a user pair is a match or not. We implemented some of these approaches and compared with the proposed framework in Sect. 5.

Contribution of this Paper: Previous works show that there exists a non-negligible risk of matching user profiles on offline datasets. Showing the risk on offline datasets is not effective since users need tools that guide them at the time of data sharing in digital world. However, building algorithms that will pave the way towards real-time privacy risk quantification is non-trivial considering the scale of the problem. In this paper, we develop a novel belief propagation

(BP)-based algorithm to generate the model efficiently and accurately (compared to the state-of-the-art). The proposed algorithm has linear complexity with respect to the number of user pairs (i.e., possible matches), while Hungarian algorithm [21], state-of-the-art that provides the highest accuracy (as shown in Sect. 5.4), has cubic complexity with respect to the number of users. We also show that the proposed algorithm achieves comparable accuracy with the Hungarian algorithm while providing this efficiency advantage.

3 Threat Model

We assume the attacker has access to user profiles in different OSNs. For simplicity we consider two OSNs: user profiles in OSN A (the auxiliary OSN) are linked to their identities, while in OSN T (the target OSN), the profiles of the individuals are anonymized. The attacker's goal is to match one or multiple user profiles from OSN T to the profiles in OSN A in order to determine the real identities of the users in OSN T. To do such profile matching, we assume that the attacker can only use the publicly available attributes of the users from OSNs A and T.

We study the extent of profile matching risk by means of two attacks: targeted attack and global attack. Targeted attack represents a scenario in which the attacker identifies the anonymous profile of a victim (or a set of victims) in OSN T and aims to find the corresponding unanonymized profile of the same victim in OSN A. Global attack represents the case in which the attacker aims to link all profiles in OSN T to their corresponding matches in OSN A.

4 Proposed Model Generation

Let A and T represent the auxiliary and the target OSNs, respectively, in which people publicly share attributes such as date of birth, gender, and location. We represent the profile of a user i in either A or T as U_i^k, where $k \in \{A, T\}$. We focus on the most common attributes that are shared in many OSNs and we categorize the profile of a user i as $U_i^k = \{n_i^k, \ell_i^k, g_i^k, p_i^k, f_i^k, a_i^k, t_i^k, s_i^k, r_i^k\}$, where n denotes the user name, ℓ denotes the location, g denotes the gender, p denotes the profile photo, f denotes the freetext provided by the user in the profile description, a denotes the activity patterns of the user (i.e., time instances at which the user posts), t denotes the interests of the user (based on the sharings of the user), s denotes the sentiment profile of the user, and r denotes the (graph) connectivity pattern of the user. As discussed, the main goal of the attacker is to link the profiles between two OSNs. The overview of the proposed framework is shown in Fig. 1. In the following, we describe the details of the proposed model generation algorithm.

4.1 Categorizing Attributes and Defining Similarity Metrics

Once the attributes of the users are extracted from their profiles, we first categorize them so that we can use them to compute the similarity values of attributes

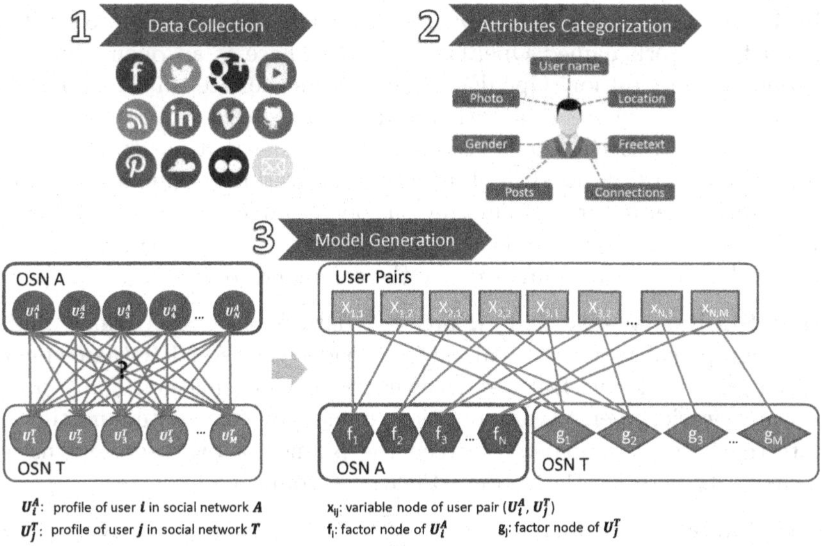

Fig. 1. Overview of the proposed framework. The proposed framework consists of 3 main steps: (1) data collection, (2) categorization of attributes and computation of attribute similarities, and (3) generation of the model.

between different users. In this section, we define the similarity metrics for each attribute between a user i in OSN A and user j in OSN T.

User Name Similarity - $S(n_i^A, n_j^T)$**:** We use Levenshtein distance [22] to compute the similarity between user names of profiles.

Location Similarity - $S(\ell_i^A, \ell_j^T)$**:** Location information collected from the users' profiles is usually text-based. We convert the textual information into coordinates via GoogleMaps API [1] and calculate the geographic distance between the corresponding coordinates.

Gender Similarity - $S(g_i^A, g_j^T)$**:** Availability of gender information is mostly problematic in OSNs. Some OSNs do not publicly share the gender information of their users. Furthermore, some OSNs do not even collect this information. In our model, if an OSN does not provide the gender information publicly (or does not have such information), we probabilistically infer the gender information by using a public name database. That is, we use the US social security name database[1] and look for a profile's name (or user name) to probabilistically infer the possible gender of the profile from the distribution of the corresponding name (among males and females) in the name database. We then use this probability as the $S(g_i^A, g_j^T)$ value between two profiles.

[1] US social security name database includes year of birth, gender, and the corresponding name for babies born in the United States.

Profile Photo Similarity - $S(p_i^A, p_j^T)$**:** We calculate the profile photo similarity through a framework named OpenFace [4]. OpenFace is an open source tool performing face recognition. OpenFace first detects the face (in the photo), and then preprocesses it to create a normalized and fixed-size input for the neural network. The features that characterize a person's face are extracted by the neural network and then used in classifiers or clustering techniques. OpenFace notably offers higher accuracy than previous open source frameworks. Given two profile photos p_i^A and p_j^T, OpenFace returns the photo similarity, $S(p_i^A, p_j^T)$, as a real value between 0 (meaning exactly the same photo) and 4.

Freetext Similarity - $S(f_i^A, f_j^T)$**:** Freetext data in an OSN profile could be a short biographical text or an "about me" page. We use NER (named-entity recognition) [10] to extract features from the freetext information. The extracted features are location, person, organization, money, percent, date, and time. To calculate the freetext similarity between the profiles of two users, we use the cosine similarity between the extracted features from each user.

Activity Pattern Similarity - $S(a_i^A, a_j^T)$**:** To compute the activity pattern similarity, we find the similarity between observed activity patterns of two profiles (e.g., likes or post). Let a_i^A represent a vector including the times of last $|a_i^A|$ activities of user i in OSN A. Similarly, a_j^T is a vector including the times of last $|a_j^T|$ activities of user j in OSN T. First, we compute the time difference between every entry in a_i^A and a_j^T and we determine $\min(|a_i^A|, |a_j^T|)$ pairs whose time difference is the smallest. Then, we compute the normalized distance between these $\min(|a_i^A|, |a_j^T|)$ pairs to compute the activity pattern similarity between two profiles.

Interest Similarity - $S(t_i^A, t_j^T)$**:** OSNs provide a platform in which users share their opinions via posts (e.g., tweets or tips), and this shared content is composed of different topics. In highlevel, first, we create a topic model using Latent Dirichlet Allocation (LDA) [7]. Then, by using the created model, we compute the topic distribution of each post generated by the users of the auxiliary and the target OSNs. Finally, we compute the interest similarity from the distance of the computed topic distributions.

Sentiment Similarity - $S(s_i^A, s_j^T)$**:** Users typically express their emotions when sharing their opinions about certain issues on OSNs. To determine whether the shared text (e.g., post or tweet) expresses positive or negative sentiment we use sentiment analysis through Python NLTK (natural language toolkit) Text Classification [2]. Given the text to analyze, the sentiment analysis tool returns the probability for positive and negative sentiment in the text. Users' moods are affected from different factors, so it is realistic to assume that they might change by time (e.g., daily). Thus, we compute the daily sentiment profile of each user, and daily sentiment similarity between the users. For this, first, we compute the normalized distribution of the positive and negative sentiments per day for each user, and then we find the normalized distance between these distributions for each user pair.

Graph Connectivity Similarity - $S(r_i^A, r_j^T)$: As in [36], for each user i, we define a feature vector $F_i = (c_0, c_1, ..., c_{n-1})$ of length n made up of components of size b. Each component contains the number of neighbors that have a *degree* in a particular range, e.g., c_k is the count of neighbors with a degree in range $[k \cdot b, (k+1) \cdot b]$. We use the feature vector length as 70 and bin size as 15 (as in [36]).

4.2 Generating the Model

We denote the set of profiles that are extracted for training from OSNs A and T as $\mathrm{A_t}$ and $\mathrm{T_t}$, respectively. Profiles are selected such that some profiles in $\mathrm{A_t}$ and $\mathrm{T_t}$ belong to the same individual. We let set G include pairs of profiles (U_i^A, U_j^T) from $\mathrm{A_t}$ and $\mathrm{T_t}$ that belong to the same individual (i.e., coupled profiles). Similarly, we let set I include pairs of profiles that belong to different individuals (i.e., uncoupled profiles). For each pair of users in sets G and I, we compute the attribute similarities based on the categorizations of the attributes (as discussed in Sect. 4.1). We label the pairs in sets G and I (as coupled and uncoupled) and add them to the training dataset. Then, to identify the weight (contribution) of each attribute, we use logistic regression.

Next, we select the profiles to be matched and construct the sets $\mathrm{A_e}$ (with size N) and $\mathrm{T_e}$ (with size M). Then, we compute the general similarity $S(U_i^A, U_j^T)$ between every user in $\mathrm{A_e}$ and $\mathrm{T_e}$ using the identified weights of the attributes to obtain the $N \times M$ similarity matrix R. Our goal is to obtain a one-to-one matching between the users in $\mathrm{A_e}$ and $\mathrm{T_e}$ that would also maximize the total similarity. One way of solving this problem is to formulate it as an optimization problem and use the Hungarian algorithm, a combinatorial optimization algorithm that solves the assignment problem in polynomial time [21]. It is also possible to formulate profile matching as a classification problem and solve it using machine learning algorithms. Thus, we evaluate and compare the solution of this problem by using both the Hungarian algorithm and other off-the-shelf machine learning algorithms including k-nearest neighbor (KNN), decision tree, random forest, and SVM.

Evaluations on different datasets (we will provide the details of the datasets later in Sect. 5.2) show us that Hungarian algorithm provides significantly better precision, recall, and accuracy compared to other machine learning techniques (we will provide the details of our evaluation in Sect. 5.4). However, assuming N users in set $\mathrm{A_e}$ and M users in set $\mathrm{T_e}$, the running time of the Hungarian algorithm for the above scenario is $O(\max\{N, M\}^3)$, and hence it is not scalable for large datasets. This raises the need for efficient, accurate, and scalable algorithms for model generation that will pave the way towards real-time privacy risk quantification.

4.3 Belief Propagation-Based Efficient Formulation of Model Generation

Inspired from the effective use of the message passing algorithms in information theory [35] and reputation management [6], in this research, for the first time, we formulate profile matching as an inference problem that infers the coupled profile pairs and develop an algorithm that relies on belief propagation (BP) on a graphical model. BP algorithm is based on a message-passing strategy for performing efficient inference using graphical models [32]. The problem we consider is different from [6,35] and so is the formulation. In this section we formalize our approach and present the different components that are needed to quantify the profile matching risk. Our goal is to obtain comparable precision, recall, and accuracy values as in the Hungarian algorithm with significantly better efficiency.

We represent the marginal probability distribution for a profile pair (i,j) to be a coupled pair as $p(x_{i,j})$, where $x_{i,j} = 1$ if profiles are matched as a result of the algorithm and $x_{i,j} = 0$, otherwise. Then, we formulate the profile matching (i.e., determining if a profile pair is coupled or uncoupled) as computing the marginal probability distributions of the variables in set $\mathrm{X} = \{x_{i,j} : i \in A, j \in T\}$, given the similarity values between the user pairs in the similarity matrix R. Since the number of users in OSNs is high, it is computationally infeasible to compute the marginal probability distributions from the joint probability distribution $p(\mathrm{X}|R)$. Thus, we propose to factorize $p(\mathrm{X}|R)$ into local functions using a factor graph and run the BP algorithm to compute the marginal probability distributions in linear time (with respect to the number of profile pairs).

A factor graph is a bipartite graph containing two sets of nodes (variable and factor nodes) and edges between these two sets. We form a factor graph by setting a variable node for each variable $x_{i,j}$ (i.e., each profile pair). Thus, each variable node represents the marginal probability distribution of that profile pair being coupled or uncoupled. We use two types of factor nodes: (i) "auxiliary" factor node (f_i), representing each user i in OSN A and (ii) "target" factor node (g_j), representing each user j in OSN T. Each factor node is connected to the variable nodes representing its potential matches. Factor nodes represent the statistical relationships between the user attributes and profile matching. Using the factor nodes, the joint probability distribution function can be factorized into products of several local functions, as follows:

$$p(X|R) = \frac{1}{Z}\left[\prod_{i=1}^{N} f_i(x_{\sigma f_i}, R) \prod_{j=1}^{M} g_j(x_{\sigma g_j}, R)\right], \tag{1}$$

where Z is a normalization constant, and σf_i (or σg_j) represents the indices of the variable nodes that are connected to factor node f_i (or g_j).

Figure 2 shows the factor graph representation of a toy example with 3 users from OSN A and 2 users from OSN T. Here, each user corresponds to a factor node in the graph (shown as a hexagon or rhombus, respectively). Each profile pair is represented by a variable node and shown as a rectangle. Each factor node is connected to the variable nodes it acts on. For example f_i is connected to all variable nodes (profile pairs) that contain U_i^A. The BP algorithm iteratively exchanges messages between the variable and the factor nodes, updating the beliefs on the values of the profile pairs (i.e., being a coupled or an uncoupled profile) at each iteration, until convergence.

Next, we introduce the messages between the variable and the factor nodes to compute the marginal distributions using BP. We denote the messages from the variable nodes to the factor nodes as μ. We also denote the messages from the auxiliary factor nodes to the variable nodes as λ and from the target factor nodes to the variable nodes as β. The message $\mu_{k\rightarrow i}^{(v+1)}\left(x_{i,j}^{(v)}\right)$ denotes the probability of $x_{i,j}^{v} = r$ ($r \in \{0,1\}$), at the vth iteration. Also, $\lambda_{i\rightarrow k}^{(v)}\left(x_{i,j}^{(v)}\right)$ denotes the probability that $x_{i,j}^{v} = r$ ($r \in \{0,1\}$) at the vth iteration given R (the messages β can be also expressed similarly). In the following, we describe the message exchange between the variable node $x_{1,4}$, the auxiliary factor node f_1, and the target factor node g_4 in Fig. 2. For clarity of presentation, we denote the variable and factor nodes $x_{1,4}$, f_1, and g_4 as i, k, z, respectively.

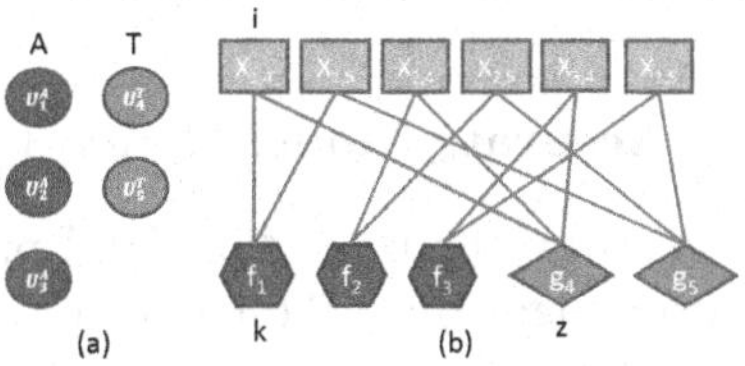

Fig. 2. Factor graph representation of 3 users from OSN A and 2 users from OSN T. (a) The users in both OSNs A and T. (b) Factor graph representation of all the possible profile pairs combinations between users in OSNs A and T.

Following the general rules of BP [20], the variable node i generates its message to auxiliary factor node k by multiplying all the messages it receives from its neighbors, excluding k. Note that each variable node has only two neighbors (one auxiliary factor node and one target factor node). Thus, the message from the variable node i to the auxiliary factor node k at the vth iteration is as follows:

$$\mu_{i\rightarrow k}^{(v)}\left(x_{1,4}^{(v)}\right) = \beta_{z\rightarrow i}^{(v-1)}\left(x_{1,4}^{(v-1)}\right). \tag{2}$$

This computation is done at every variable node. The message from the variable node i to the target factor node z is also constructed similarly.

Next, factor nodes generate their messages. The message from the auxiliary factor node k to the variable node i is given by:

$$\lambda_{k\rightarrow i}^{(v)}\left(x_{1,4}^{(v)}\right) = \frac{1}{Z} \times S(U_1^A, U_4^T) \times \prod_{d\in(\sim i)} f_d\left(\mu_{d\rightarrow k}^{(v)}\left(x_{1,4}^{(v)}\right)\right), \tag{3}$$

where $(\sim i)$ means all variable node neighbors of k, except i. We compute function f_d as:

$$f_d\left(\mu_{d\rightarrow k}^{(v)}\left(x_{1,4}^{(v)}\right)\right) = \left(1 - \mu_{d\rightarrow k}^{(v)}\left(x_{1,4}^{(v)}\right)\right). \tag{4}$$

The above computation must be performed for every neighbor of each auxiliary factor node. The message from the target factor node z to the variable node i is also computed similarly.

The next iteration is performed in the same way as the vth iteration. The algorithm starts at the variable nodes. In the first iteration (i.e., $v = 1$), all the variable nodes send to their neighboring factor nodes the same value $(\lambda_{i\rightarrow k}^{(1)}\left(x_{i,j}^{(1)}\right) = 1/N)$, where N is the total number of "auxiliary" factor nodes. The iterations stops when the probability distributions of all variables in X converge. The marginal probability distribution of each variable in X is computed by multiplying all the incoming messages at each variable node.

4.4 ϵ-Accurate Model Generation

We also study the limitations and properties of the proposed BP-based model generation algorithm. We particularly analyze if the proposed algorithm maintains any optimality in any sense. For this, we use the following definition:

Definition 1 ***ϵ-Accurate Model Generation.*** *We declare a model generation algorithm as ϵ-accurate if it can match at least ϵ% of the users accurately.*

Here, accuracy is the number of correctly matched coupled pairs by the proposed algorithm over the total number of coupled pairs. The above definition can also be made in terms of precision or recall (or both) of the proposed algorithm. Thus, for a fixed ϵ, we study the conditions for an ϵ-accurate algorithm. This also helps us understand the limits of profile matching in OSNs. To have an ϵ-accurate algorithm with a high ϵ value, it can be shown that, we require the BP-based algorithm to iteratively increase the accuracy until it converges. This brings about the following sufficient condition about ϵ-accuracy.

Definition 2 ***Sufficient Condition.*** *Accuracy of the model generation algorithm increases with each successive iteration (until convergence) if for all coupled profiles i and j, $Pr(x_{i,j}^{(2)} = 1) > Pr(x_{i,j}^{(1)} = 1)$ is satisfied.*

Depending on the fraction of the coupled profile pairs that meet the sufficient condition, ϵ-accuracy of the proposed algorithm can be obtained. In Sect. 5, we experimentally explore the cases in which this sufficient condition is satisfied with high probability.

5 Evaluation of the Proposed Mechanism

In this section, we evaluate the proposed BP-based algorithm by using real data from four OSNs. We also study the impact of various parameters to the ϵ-accuracy of the proposed algorithm.

5.1 Evaluation Metrics

To evaluate the proposed model, we mainly consider the global attack, in which the goal of the attacker is to match all profiles in A_e to all profiles in T_e. In other words, the goal is to deanonymize all anonymous users in the target OSN (who have accounts in the auxiliary OSN). For the evaluation metrics, we use precision, recall, and accuracy. Hungarian algorithm and the proposed BP-based algorithm provide a one-to-one match between all the users. However, we cannot expect that all anonymous users in the target OSN have profiles in the auxiliary OSN. Therefore, some of the provided matches are useless for us. Thus, we select a "similarity threshold" ("probability threshold" for machine learning techniques) for evaluation. Each matching scheme returns 1 (i.e., true match) if the similarity/probability of user pair is higher than the threshold, and 0 otherwise. So, we consider as true positives the pairs that are correctly matched by the algorithm and whose similarity/probability is greater than the threshold. We also compute accuracy as the number of correctly matched coupled pairs identified by the algorithm over the total number of coupled pairs.

5.2 Data Collection

To evaluate our proposed framework, we use three datasets: (i) Dataset 1 (D1): Google+ - Twitter [13], (ii) Dataset 2 (D2): Instagram - Twitter, and (iii) Dataset 3 (D3): Flickr social graph [40]. To collect the coupled profiles in D1 and D2, social links in Google+ profiles and about.me (a social network where users provide links to their OSN profiles) were used, respectively. In terms of dataset sizes, (i) D1 consists of 8000 users in each OSNs where 4000 of them are coupled profiles; (ii) D2 consists of more than 10000 coupled profiles (and more content about the OSN users compared to D1); and (iii) D3 consists of 50000 users. In D1, we use Twitter as our auxiliary OSN (A) and Google+ as our target OSN (T); in D2, we use Twitter as our auxiliary OSN (A) and Instagram as our target OSN (T); and in D3, we generate the auxiliary and the target OSN graphs as in [28,36] by using a vertex overlap of 1 and an edge overlap of 0.9.

5.3 Evaluation Settings

Since the model generation process is the same for all three datasets, in the rest of the paper, we hold the discussion over a target and auxiliary network. From each dataset, we select 3000 profile pairs (1500 coupled and 1500 uncoupled) for training. We also select 500 users from the auxiliary OSN and 500 users from the target OSN to construct sets A_e and T_e, respectively. Note that none of these users are involved in the training set. Among these profiles, we have 500 coupled pairs and 249500 uncoupled pairs, and hence the goal is to make sure that these 500 users are matched with high confidence in a global attack scenario. Note that we do not use cross-validation because we consider all the possible user combinations to test our model and it is time-consuming to compute all similarity metrics for all combinations. Considering all the combinations instead of randomly selecting some user pairs is a more realistic evaluation setting since one can never know which users pairs the attacker will have access to. In cases that there are missing attributes (that are not published by the users) in the dataset, we assign a value for the attribute similarity based on the distributions of the attribute similarity values between the coupled and uncoupled pairs.

5.4 Evaluation of BP-Based Model Generation

In Fig. 3, we show the comparison of the proposed BP-based model generation to [12,13,26,31,36] for each dataset (D1, D2, and D3). [12,26,31,36] use machine learning-based techniques (k-nearest neighbor (KNN), decision tree, random forest, and/or SVM), while [13] uses the Hungarian Algorithm. Our results show that the proposed scheme provides comparable precision and recall compared to the state-of-the-art Hungarian algorithm and it significantly outpowers machine leaning-based algorithms. For instance, the proposed algorithm provides a precision value of around 0.97 (for a similarity threshold of 0.5) in D1. This means, if our proposed algorithm returns a matched profile pair that has a similarity value above 0.5, the corresponding profiles belong to same individual with a high confidence. At the same time, the complexity of the proposed algorithm scales linearly with the number of user pairs, while the Hungarian algorithm suffers from cubic complexity. Note that precision and recall values obtained from D2 are higher compared to the ones from D1 as we collected more information about users in D2. We also compare the BP-based model generation to the deep neural network based algorithm (DeepLink) [43] in D3. For DeepLink, we use the same settings as in [43]. DeepLink achieves an accuracy of 84% in D3 which is slightly less than the one obtained by the proposed algorithm (90%). DeepLink achieves a precision of 0.84, and a recall of 1 while the BP-based algorithm achieves a precision of 0.93 and a recall of 0.9. DeepLink provides a match for each user even if that user does not have a match.

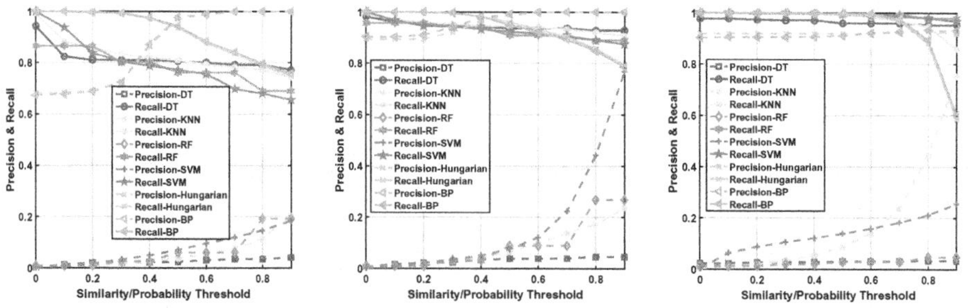

(a) D1 (b) D2 (c) D3

Fig. 3. Comparison of the proposed BP-based scheme with the Hungarian algorithm and machine learning techniques (decision tree - DT, KNN, random forest - RF, and SVM) in terms of precision and recall for D1, D2, and D3.

We also study the effect of the OSNs' size to precision and recall of the proposed algorithm. In Fig. 4, for each dataset, we show the precision/recall values of the BP-based algorithm when the number of users in the auxiliary OSN (OSN A) increases while the number of target users (i.e., users in OSN T) is fixed. We set the number of users in OSN T as 100 and increase the number of users in OSN A from 100 to 1000 in steps of 100. We observe that the precision/recall values of the proposed algorithm only slightly decrease with the increase of auxiliary OSN's size, which shows the scalability of our proposed algorithm. We achieve similar results for the other two scenarios: (i) when we fix the number of users in OSN A and vary the number of users in OSN T; and (ii) when we increase the number of users in both OSNs A and T. Due to the space constraints, we present the details of the results of scenarios (i) and (ii) in Fig. 7 and 8, respectively, in Appendix A.

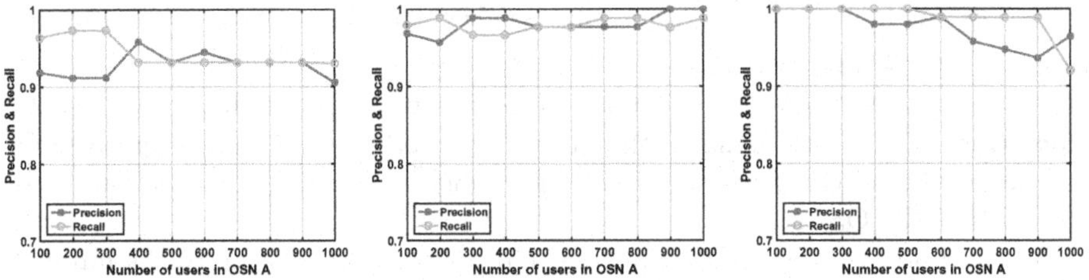

(a) D1 (b) D2 (c) D3

Fig. 4. The effect of auxiliary OSN's (OSN A) size to precision/recall when the size of target OSN (OSN T) is 100 in D1, D2, and D3.

Next, we evaluate the ϵ-accuracy of the proposed model generation algorithm (introduced in Sect. 4.4). There are many parameters to consider to analyze the ϵ-accuracy of the proposed algorithm, such as the average degree of factor nodes, total similarity of each user in the target OSN with the ones in the auxiliary OSN, and number of users in the target and auxiliary OSNs. Here, we experimentally analyze and show the ϵ-accuracy of the proposed algorithm considering such parameters. For evaluation, we use all datasets and pick 500 users from OSN T. For all the studied parameters, we observe that at least 97% of coupled profiles that can be correctly matched by the BP-based algorithm satisfy the sufficient condition (introduced in Sect. 4.4). We observe that ϵ value is inversely proportional to the average degree of the factor nodes. In D1, the ϵ-accuracy of the proposed algorithm is $\epsilon = 67$ and $\epsilon = 84$ when the average degrees of the factor nodes are 500 and 22, respectively (we discuss more about the results of this experiment in Sect. 5.5).

To study the impact of user pairs' similarity, for each user in OSN T, we compute the variance of the similarity values between that user and all users in OSN A. Then, we compute the accuracy of the proposed BP-based algorithm on users with varying variance values. For evaluation, we use D1 and D2. Our results show that ϵ-accuracy of the proposed algorithm is higher for users with higher variances (as shown in Fig. 5). For instance, in D1, we observe the ϵ-accuracy as $\epsilon = 42$ and $\epsilon = 90$, when we run the proposed algorithm only for the users with low variance (lower than 0.008) and high variance (higher than 0.012), respectively. These results show that the vulnerability of the users in an OSN (for the profile matching attack) can be identified by analyzing particular characteristics of the OSNs.

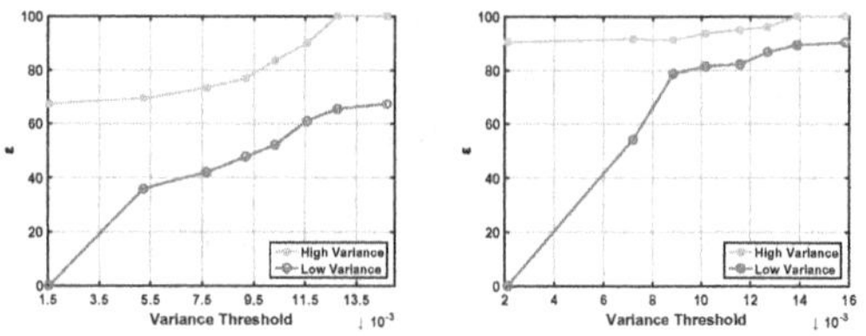

(a) D1 (b) D2

Fig. 5. The effect of variance threshold to ϵ-accuracy in D1 and D2. For "high variance", our goal is to match users in OSN T that have a variance (for the similarity values between a user in OSN T and all users in OSN A) greater than the variance threshold, while for "low variance", we consider users that have a variance value smaller than the variance threshold.

Furthermore, we observe that ϵ value is inversely proportional to the number of users in OSN A (as shown in Fig. 6). The proposed algorithm achieves an ϵ-accuracy of $\epsilon = 82$ and $\epsilon = 72$ in D1 when the number of users in A is 100 and 1000, respectively while the number of users in T is 100. This decrease in accuracy can be considered as low considering that the number of possible matches increases 10 times (from 10000 to 100000 user pairs). In D2, we observe a similar trend to D1. In D3, accuracy decreases faster with the increase in number of users in OSN A (compared to D1 and D2). This is because, in D3, we only use the graph connectivity attribute for profile matching. Thus, as the number of users in OSN A increases, the number of users with similar graph connectivity patterns also increases causing the decrease in accuracy.

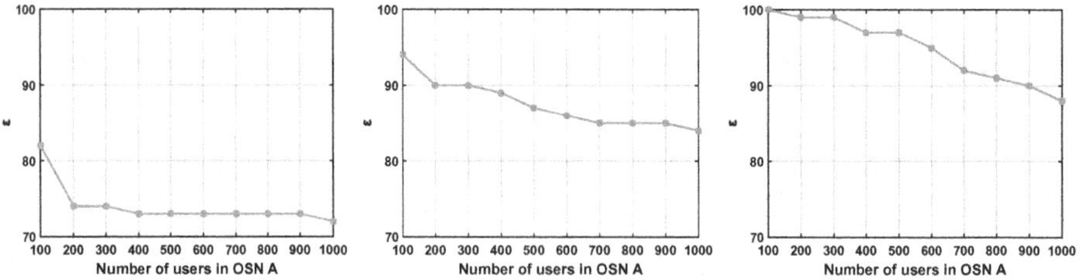

(a) D1 (b) D2 (c) D3

Fig. 6. The effect of auxiliary OSN's (OSN A) size to ϵ-accuracy in D1, D2, and D3. The size of target OSN (OSN T) is fixed to 100.

5.5 Complexity Analysis of the BP-Based Algorithm

The complexity of the BP-based algorithm is linear in the number of variable (or factor nodes). In the proposed BP-based algorithm, we generate a variable node for all potential matches between the target and the auxiliary OSNs. Assuming N users in both target and auxiliary OSNs, this results in N^2 variable nodes in the graph. To analyze the effect of number of variable nodes on the performance, we experimentally try to change the graph structure and limit the number of variable nodes for each user in the target OSN. We heuristically decrease the average degree of the factor nodes from N to $\sqrt{N}$ and $\log N$ by only leaving the variable nodes (user pairs) with the highest similarity values. For evaluation, we use all datasets (D1, D2, and D3). We pick 500 users from OSNs A and T to construct the test dataset (where there are 500 coupled and 249500 uncoupled profile pairs initially). In Table 1, we show the results with varying number of variable nodes. For instance, in D1, we obtain an accuracy of 67.4% when all the potential matches are considered (i.e., with N variable nodes for each user in T) and an accuracy of 96% when we use only $\log N$ variable nodes for each user. These results are important since they show that while reducing the complexity of the proposed BP-based algorithm, we can further improve its accuracy.

Table 1. Evaluation of the proposed BP-based algorithm with varying the number of variable nodes. N denotes the number of users in OSN T, and each value in the complexity column shows the number of variable nodes (i.e., the number of users pairs used for profile matching).

Dataset	Complexity	Precision	Recall	Accuracy
D1	N^2	0.978	0.944	67.4%
	$N\sqrt{N}$	0.955	0.939	84.6%
	$N \log N$	0.975	0.966	96%
D2	N^2	0.978	0.982	90.4%
	$N\sqrt{N}$	0.977	0.954	94.8%
	$N \log N$	0.934	0.946	90.6%
D3	N^2	0.925	0.976	90.4%
	$N\sqrt{N}$	0.92	0.973	91%
	$N \log N$	0.932	0.976	95%

6 Discussion

In this section we discuss how the proposed framework can be utilized for sensitive OSNs, and potential mitigation techniques against the identified profile matching risk.

6.1 Profile Matching on Sensitive OSNs

Note that in D1 and D2, users provide the links to their social networks publicly. It is quite hard to obtain coupled profiles from social networks where users share sensitive information such as PatientsLikeMe. We expect to obtain similar results as long as users share similar attributes across OSNs. Considering that these users are more privacy-cautious, mostly non-obvious attributes such as interest, activity, sentiment similarity, or writing style can be used.

6.2 Mitigation Techniques

We foresee that the OSN can provide recommendations to the users (about the content they share) to reduce their risk for profile matching attacks. Such recommendation may include (i) generalizing or distorting some shared content of the user (e.g., generalizing the shared location or posting a content at a later time); or (ii) choosing not to share some content (especially for attributes that are hard to generalize or distort, such as interest or sentiment). When generating such recommendations, there are two main objectives: (i) content shared by the user should not increase user's risk for profile matching and (ii) utility of the content shared by the user (or utility of user's profile) should not decrease due to the applied countermeasures. Using a utility metric for the user's profile, the proposed framework (in Sect. 4.3) can be used to formulate an optimization between the utility of the user's profile and privacy of the user. The solution of this optimization problem can provide recommendations to the user about how to (or whether to) share a new content on their profile.

7 Conclusion

In this work, we have proposed a novel message passing-based framework to model the profile matching risk in online social networks (OSNs). We have shown via simulations that the proposed framework provides comparable accuracy, precision, and recall compared to the state-of-the-art, while it is significantly more efficient in terms of its computational complexity. We have also shown that by controlling the structure of the proposed BP-algorithm we can further decrease the complexity of the algorithm while increasing its accuracy. We believe that the proposed framework will be instrumental for OSNs to educate their users about the consequences of their online sharings. It will also pave the way towards real-time privacy risk quantification in OSNs against profile matching attacks.

Acknowledgment. We would like to thank the anonymous reviewers and our shepherd Shujun Li for their constructive feedback which has helped us to improve this paper.

Appendix

A Scalability of the BP-Based Algorithm

We study the effect of the OSNs' size to precision and recall of the proposed algorithm. In Sect. 5.4, we provided the results when the number of users in OSN T is fixed. Here, we provide the results of the other two scenarios. In Fig. 7, for each dataset, we show the precision/recall values of the BP-based algorithm when the number of users in the target OSN (OSN T) increases while the number of auxiliary users (i.e., users in OSN A) is fixed. We set the number of users in

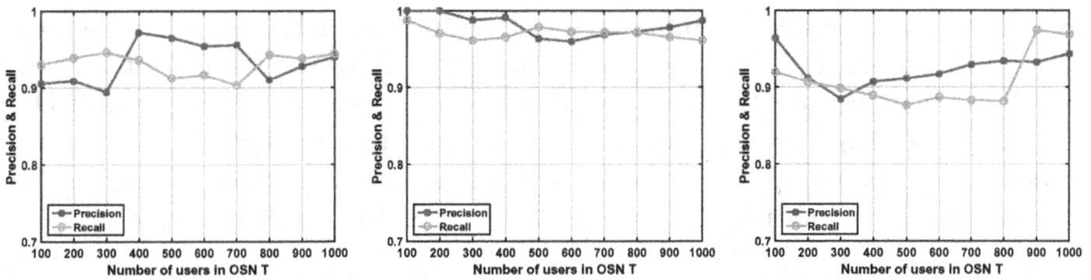

(a) D1 (b) D2 (c) D3

Fig. 7. The effect of target OSN's (OSN T) size to precision/recall when the size of auxiliary OSN (OSN A) is 1000 in D1, D2, and D3.

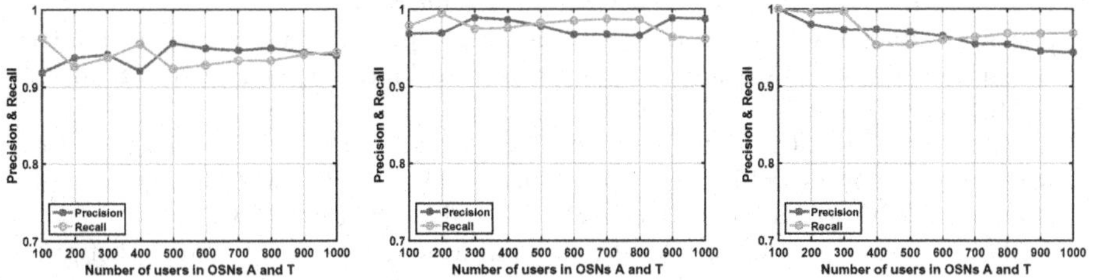

(a) D1 (b) D2 (c) D3

Fig. 8. The effect of auxiliary and target OSNs' (OSN A and T) size to precision/recall in D1, D2, and D3.

OSN A as 1000 and increase the number of users in OSN T from 100 to 1000 in steps of 100.

In Fig. 8, for each dataset, we show the precision/recall values of the BP-based algorithm when the number of users in both OSNs (i.e., OSN A and T) increases from 100 to 1000 in steps of 100. In both scenarios, we observe that the precision/recall values of the proposed algorithm only slightly decrease with the increase of the number of users in the target OSN, or the increase of the number of users in both OSNs, which shows the scalability of our proposed algorithm.

To further check the effect of the auxiliary OSN's size to precision and recall of the BP-based algorithm, we quantify the precision/recall values obtained by the proposed algorithm for larger scales in D3. We fix the number of users in the target OSN (i.e., OSN T) to 1000 while the number of users in the auxiliary OSN (i.e., OSN A) increases from 1000 to 8000 in steps of 1000 (in Fig. 4 the number of users in OSN T was fixed to 100 while the number of users in OSN A

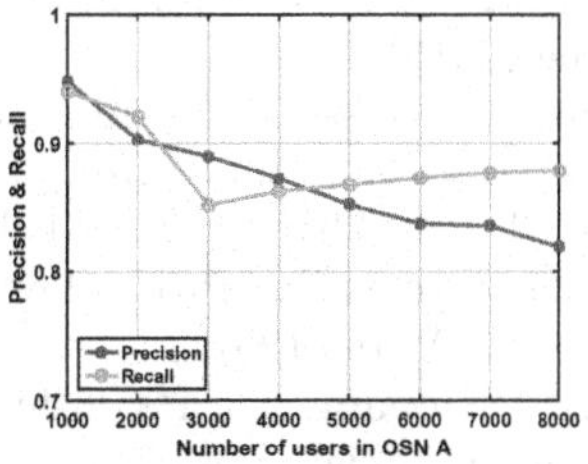

Fig. 9. The effect of auxiliary OSN's (OSN A) size to precision/recall when the size of target OSN (OSN T) is 1000 in D3.

was increasing from 100 to 1000). We show the results for D3 in Fig. 9. The precision/recall values slightly decrease with the increase of the number of users in OSN A, confirming the scalability of the proposed algorithm. Note that, in D3, we only use the graph connectivity attribute for profile matching. We expect that the decrease in precision/recall values will be smaller when both the graphical structure and other attributes of the users are used to generate the model.

References

1. Google maps API (2020). https://developers.google.com/maps/
2. Natural language toolkit (2020). http://www.nltk.org/
3. Patienslikeme (2020). https://www.patientslikeme.com/
4. Amos, B., Ludwiczuk, B., Satyanarayanan, M.: OpenFace: a general-purpose face recognition library with mobile applications. Technical report, CMU-CS-16-118, CMU School of Computer Science (2016)
5. Andreou, A., Goga, O., Loiseau, P.: Identity vs. attribute disclosure risks for users with multiple social profiles. In: Proceedings of the IEEE/ACM International Conference on Advances in Social Networks Analysis and Mining (2017)
6. Ayday, E., Fekri, F.: Iterative trust and reputation management using belief propagation. IEEE Trans. Dependable Secur. Comput. **9**(3), 375–386 (2012)
7. Blei, D.M., Ng, A.Y., Jordan, M.I.: Latent Dirichlet allocation. J. Mach. Learn. Res. **3**, 993–1022 (2003)
8. Boyd, D.M., Ellison, N.B.: Social network sites: definition, history, and scholarship. J. Comput.-Mediat. Commun. **13**(1), 210–230 (2007)
9. Debnath, S., Ganguly, N., Mitra, P.: Feature weighting in content based recommendation system using social network analysis. In: Proceedings of the International Conference on World Wide Web (2008)
10. Finkel, J.R., Grenager, T., Manning, C.: Incorporating non-local information into information extraction systems by gibbs sampling. In: Proceedings of the 43rd Annual Meeting of the Association for Computational Linguistics (2005)
11. Goga, O., Lei, H., Parthasarathi, S.H.K., Friedland, G., Sommer, R., Teixeira, R.: Exploiting innocuous activity for correlating users across sites. In: Proceedings of the 22nd International Conference on World Wide Web (2013)
12. Goga, O., Loiseau, P., Sommer, R., Teixeira, R., Gummadi, K.P.: On the reliability of profile matching across large online social networks. In: Proceedings of the ACM SIGKDD International Conference on Knowledge Discovery and Data Mining (2015)
13. Halimi, A., Ayday, E.: Profile matching across unstructured online social networks: threats and countermeasures. arXiv preprint arXiv:1711.01815 (2017)
14. Iofciu, T., Fankhauser, P., Abel, F., Bischoff, K.: Identifying users across social tagging systems. In: Proceedings of the International AAAI Conference on Web and Social Media (2011)
15. Jain, P., Kumaraguru, P., Joshi, A.: @i seek 'fb.me': identifying users across multiple online social network. In: Proceedings of the 22nd International Conference on World Wide Web (2013)
16. Ji, S., Li, W., Gong, N.Z., Mittal, P., Beyah, R.: On your social network deanonymizablity: quantification and large scale evaluation with seed knowledge. In: Proceedings of the Network and Distributed System Security Symposium (2015)

17. Ji, S., Li, W., Mittal, P., Hu, X., Beyah, R.: SecGraph: a uniform and open-source evaluation system for graph data anonymization and de-anonymization. In: Proceedings of the 24th USENIX Security Symposium (2015)
18. Ji, S., Li, W., Srivatsa, M., Beyah, R.: Structural data de-anonymization: quantification, practice, and implications. In: Proceedings of ACM SIGSAC Conference on Computer and Communications Security, pp. 1040–1053. ACM (2014)
19. Korula, N., Lattanzi, S.: An efficient reconciliation algorithm for social networks. Proc. VLDB Endowment **7**(5), 377–388 (2014)
20. Kschischang, F.R., Frey, B.J., Loeliger, H.A.: Factor graphs and the sum-product algorithm. IEEE Trans. Inf. Theory **47**(2), 498–519 (2001)
21. Kuhn, H.W.: The Hungarian method for the assignment problem. Naval Res. Logist. Q. **2**(1–2), 83–97 (1955)
22. Levenshtein, V.I.: Binary codes capable of correcting deletions, insertions, and reversals. Sov. Phys. Dokl. **10**(8), 707–710 (1966)
23. Liu, J., Zhang, F., Song, X., Song, Y.I., Lin, C.Y., Hon, H.W.: What's in the name?: an unsupervised approach to link users across communities. In: Proceedings of ACM International Conference on Web Search and Data Mining (2013)
24. Liu, S., Wang, S., Zhu, F.: Structured learning from heterogeneous behavior for social identity linkage. IEEE Trans. Knowl. Data Eng. **27**(7), 2005–2019 (2015)
25. Liu, S., Wang, S., Zhu, F., Zhang, J., Krishnan, R.: Hydra: large-scale social identity linkage via heterogeneous behavior modeling. In: Proceedings of the ACM SIGMOD International Conference on Management of Data (2014)
26. Malhotra, A., Totti, L., Meira Jr., W., Kumaraguru, P., Almeida, V.: Studying user footprints in different online social networks. In: Proceedings of the International Conference on Advances in Social Network Analysis and Mining (2012)
27. Motoyama, M., Varghese, G.: I seek you: searching and matching individuals in social networks. In: Proceedings of the 11th International Workshop on Web Information and Data Management (2009)
28. Narayanan, A., Shi, E., Rubinstein, B.I.P.: Link prediction by de-anonymization: how we won the kaggle social network challenge. In: Proceedings of the International Joint Conference on Neural Networks (2011)
29. Narayanan, A., Shmatikov, V.: De-anonymizing social networks. In: Proceedings of IEEE Symposium on Security and Privacy (2009)
30. Nilizadeh, S., Kapadia, A., Ahn, Y.Y.: Community-enhanced de-anonymization of online social networks. In: Proceedings of ACM Conference on Computer and Communications Security (2014)
31. Nunes, A., Calado, P., Martins, B.: Resolving user identities over social networks through supervised learning and rich similarity features. In: Proceedings of ACM Symposium on Applied Computing (2012)
32. Pearl, J.: Probabilistic Reasoning in Intelligent Systems: Networks of Plausible Inference (1988)
33. Pedarsani, P., Figueiredo, D.R., Grossglauser, M.: A Bayesian method for matching two similar graphs without seeds. In: Proceedings of the 51st Annual Allerton Conference on Communication, Control, and Computing (2013)
34. Perito, D., Castelluccia, C., Kaafar, M.A., Manils, P.: How unique and traceable are usernames? In: Fischer-Hübner, S., Hopper, N. (eds.) PETS 2011. LNCS, vol. 6794, pp. 1–17. Springer, Heidelberg (2011). https://doi.org/10.1007/978-3-642-22263-4_1
35. Pishro-Nik, H., Fekri, F.: Performance of low-density parity-check codes with linear minimum distance. IEEE Trans. Inf. Theory **52**(1), 292–300 (2005)

36. Sharad, K., Danezis, G.: An automated social graph de-anonymization technique. In: Proceedings of the 13th ACM Workshop on Privacy in the Electronic Society (2014)
37. Shu, K., Wang, S., Tang, J., Zafarani, R., Liu, H.: User identity linkage across online social networks: a review. ACM SIGKDD Explorations Newsletter **18**(2), 5–17 (2017)
38. Vosecky, J., Hong, D., Shen, V.Y.: User identification across multiple social networks. In: Proceedings of the International Conference on Networked Digital Technologies (2009)
39. Wondracek, G., Holz, T., Kirda, E., Kruegel, C.: A practical attack to de-anonymize social network users. In: Proceedings of IEEE Symposium on Security and Privacy (2010)
40. Zafarani, R., Liu, H.: Social computing data repository at ASU (2009). http://socialcomputing.asu.edu
41. Zafarani, R., Liu, H.: Connecting corresponding identities across communities. In: Proceedings of the 3rd International AAAI Conference on Web and Social Media (2009)
42. Zafarani, R., Liu, H.: Connecting users across social media sites: a behavioral-modeling approach. In: Proceedings of ACM SIDKDD Conference on Knowledge Discovery and Data Mining (2013)
43. Zhou, F., Liu, L., Zhang, K., Trajcevski, G., Wu, J., Zhong, T.: DeepLink: a deep learning approach for user identity linkage. In: Proceedings of IEEE International Conference on Computer Communications, pp. 1313–1321. IEEE (2018)

Network Security I

Anonymity Preserving Byzantine Vector Consensus

Christian Cachin[1], Daniel Collins[2,3(✉)], Tyler Crain[2], and Vincent Gramoli[2,3]

[1] University of Bern, Bern, Switzerland
cachin@inf.unibe.ch
[2] University of Sydney, Sydney, Australia
tcrainwork@gmail.com, vincent.gramoli@sydney.edu.au
[3] EPFL, Lausanne, Switzerland
daniel.collins@epfl.ch

Abstract. Collecting anonymous opinions has applications from whistleblowing to complex voting where participants rank candidates by order of preferences. Unfortunately, as far as we know there is no efficient distributed solution to this problem. Previous solutions either require trusted third parties, are inefficient or sacrifice anonymity.

In this paper, we propose a distributed solution called the *Anonymised Vector Consensus Protocol* (AVCP) that reduces the problem of agreeing on a set of anonymous votes to the binary Byzantine consensus problem. The key idea to preserve the anonymity of voters—despite some of them acting maliciously—is to detect double votes through traceable ring signatures. AVCP is resilient-optimal as it tolerates up to a third of Byzantine participants. We show that our algorithm is correct and that it preserves anonymity with at most a linear communication overhead and constant message overhead when compared to a recent consensus baseline. Finally, we demonstrate empirically that the protocol is practical by deploying it on 100 machines geo-distributed in three continents: America, Asia and Europe. Anonymous decisions are reached within 10 s with a conservative choice of traceable ring signatures.

Keywords: Anonymity · Byzantine agreement · Consensus · Vector consensus · Distributed computing

1 Introduction and Related Work

Consider a distributed survey where a group of mutually distrusting participants wish to exchange their opinions about some issue. For example, participants may wish to communicate over the Internet to rank candidates in order of preference to change the governance of a blockchain. Without making additional trust assumptions [1,18,38], one promising approach is to run a Byzantine consensus algorithm [40], or more generally a vector consensus algorithm [16,23,46] to allow for arbitrary votes. In vector consensus, a set of participants decide on a common vector of values, each value being proposed by one process. Unlike

L. Chen et al. (Eds.): ESORICS 2020, LNCS 12308, pp. 133–152, 2020.
https://doi.org/10.1007/978-3-030-58951-6_7

interactive consistency [40], a protocol solving vector consensus can be executed without fully synchronous communication channels, and as such is preferable for use over the Internet. Unfortunately, vector consensus protocols tie each participant's opinion to its identity to ensure one opinion is not overrepresented in the decision and to avoid double voting. There is thus an inherent difficulty in solving vector consensus while preserving anonymity.

In this paper, we introduce the *anonymity-preserving vector consensus* problem that prevents an adversary from discovering the identity of non-faulty participants that propose values in the consensus and we introduce a solution called Anonymised Vector Consensus Protocol (AVCP). To prevent the leader in some Byzantine consensus algorithms [12] from influencing the outcome of the vote by discarding proposals, AVCP reduces the problem to binary consensus that is executed without the need for a traditional leader [17].

We provide a mechanism to prevent Byzantine processes from double voting while also decoupling their ballots from their identity. In particular, we adopt *traceable ring signatures* [29,35], which enable participants to anonymously prove their membership in a set of participants, exposing a signer's identity if and only if they sign two different votes. This can disincentivise participants from proposing multiple votes to the consensus. Alternatively, we could use linkable ring signatures [41], but they cannot ensure that Byzantine processes are held accountable when double-signing. We could also have used blind signatures [11, 13], but this would have required an additional trusted authority.

We also identify interesting conditions to ensure anonymous communication. Importantly, participants must propagate their signatures anonymously to hide their identity. To this end, we could construct anonymous channels directly. However, these protocols either require additional trusted parties, are not robust or incur $O(n)$ message delays to terminate [14,32,33,37]. Thus, we assume access to anonymous channels, such as through publicly-deployed [22,56] networks which often require sustained network observation or large amounts of resources to deanonymise with high probability [31,44]. Anonymity is ensured then as processes do not reveal their identity via ring signatures and communicate over anonymous channels. If correlation-based attacks on certain anonymous networks are deemed viable [42] and for efficiency's sake, processes can anonymously broadcast their proposal and then continue the protocol execution over regular channels. However, anonymous channels alone cannot ensure anonymity when combined with regular channels as an adversary may infer identities through latency [2] and message transmission ordering [48]. For example, an adversary may relate late message arrivals from a single process over both channels to guess the identity of a slow participant. This is why, to ensure anonymity, the timing and order of messages exchanged over anonymous channels should be statistically independent (or computationally indistinguishable with a computational adversary) from that of those exchanged over regular channels. In practice, one may attempt to ensure that there is a low correlation by ensuring that message delays over anonymous and regular channels are each sufficiently randomised.

We construct our solution iteratively first by defining the *anonymity-preserving all-to-all reliable broadcast problem* that may be of independent interest. Here, anonymity and comparable properties to reliable broadcast [7] with n broadcasters are ensured. By constructing a solution to this problem with Bracha's classical reliable broadcast [5], AVCP can terminate after three regular and one anonymous message delay. With this approach, our experimental results are promising—with 100 geo-distributed nodes, AVCP generally terminates in less than ten seconds. We remark that, to ensure confidentiality of proposals until after termination, threshold encryption [20,52], in which a minimum set of participants must cooperate to decrypt any message, can be used at the cost of an additional message delay.

Related constructions. We consider techniques without additional trusted parties. Homomorphic tallying [18] encodes opinions as values that are summed in encrypted form prior to decryption. Such schemes that rely on a public bulletin board for posting votes [34] could use Byzantine consensus [40] and make timing assumptions on vote submission to perform an election. Unfortunately, homomorphic tallying is impractical when the pool of candidates is large and impossible when arbitrary. Nevertheless, using a multiplicative homomorphic scheme [25] requires exponential work in the amount of candidates at tallying time, and additive homomorphic encryption like Paillier's scheme [47] incurs large (RSA-size) parameters and costly operations. Fully homomorphic encryption [30] is suitable in theory, but at present is untenable for computing complex circuits efficiently. Self-tallying schemes [34], which use homomorphic tallying, are not appropriate as at some point all participants must be correct, which is untenable with arbitrary (Byzantine) behaviour. Constructions involving mix-nets [15] allow for arbitrary ballot structure. However, decryption mix-nets are not a priori robust to a single Byzantine or crash failure [15], and re-encryption mix-nets which use proofs of shuffle are generally slow in tallying [1]. With arbitrary encryptions, larger elections can take minutes to hours [39] to tally. Even with precomputation, $O(t)$ processes still need to perform proofs of shuffle [45] in sequence, which incurs untenable latency even without process faults. With faults and/or asynchrony, we could run $O(t)$ consensus instances in sequence but at an additional cost. DC-nets and subsequent variations [14,33] are not sufficiently robust and generally require $O(n)$ message delays for an all-to-all broadcast. On multi-party computation, general techniques like Yao's garbled-circuits [55] incur untenable overhead given enough complexity in the structure of ballots. Private-set intersection [27] can be efficient for elections that require unanimous agreement, but do not generalise arbitrarily.

Roadmap. The paper is structured as follows. Section 2 provides preliminary definitions and specifies the model. Section 3 presents protocols and definitions required for our consensus protocol. Section 4 presents our anonymity-preserving vector consensus solution. Section 5 benchmarks AVCP on up to 100 geo-distributed located on three continents. To conclude, Sect. 6 discusses the use of anonymous and regular channels in practice and future work.

2 Model

We assume the existence of a set of processes $P = \{p_1, \dots, p_n\}$ (where $|P| = n$, and the ith process is p_i), an adversary A who corrupts $t < \frac{n}{3}$ processes in P, and a trusted dealer D who generates the initial state of each process. For simplicity of exposition, we assume that cryptographic primitives are unbreakable. With concrete primitives that provide computational guarantees, each party could be modelled as being able to execute a number of instructions each message step bounded by a polynomial in a security parameter k [10]. In this case, the transformation of our proofs presented in the companion technical report [8] is straight forward given the hardness assumptions required by the underlying cryptographic schemes.

2.1 Network

We assume that P consists of asynchronous, sequential processes that communicate over reliable, point-to-point channels in an asynchronous network. An *asynchronous* process is one that executes instructions at its own pace. An *asynchronous* network is one where message delays are unbounded. A *reliable* network is such that any message sent by a correct process will eventually be delivered by the intended recipient. We assume that processes can also communicate using reliable one-way *anonymous* channels, which we soon describe.

Each process is equipped with the primitive "send M to p_j", which sends the message M (possibly a tuple) to process $p_j \in P$. For simplicity, we assume that a process can send a message to itself. A process receives a message M by invoking the primitive "receive M". Each process may invoke "broadcast M", which is short-hand for "**for each** $p_i \in P$ **do** send M to p_i **end for**". Analogously, processes may invoke "anon_send M to p_j" and "anon_broadcast M" over anonymous channels.

Since reaching consensus deterministically is impossible in failure-prone asynchronous message-passing systems [26], we assume that *partial synchrony* holds among processes in P in Sect. 4. That is, we assume there exists a point in time in protocol execution, the global stabilisation time (GST), unknown to all processes, after which the speed of each process and all message transfer delays are upper bounded by a finite constant [24].

2.2 Adversary

We assume that the adversary A schedules message delivery over the regular channels, restricted to the assumptions of our model (e.g., the reliability of the channels). For each send call made, A determines when the corresponding receive call is invoked. A portion of processes—exactly $t < \frac{n}{3}$ members of P—are initially corrupted by A and may exhibit Byzantine faults [40] over the lifetime of a protocol's execution. That is, they may deviate from the predefined protocol in an arbitrary way. We assume A can see all computations and messages sent and received by *corrupted* processes. A *non-faulty* process is one that is not corrupted by A and therefore follows the prescribed protocol. A can only observe anon_send and anon_receive calls made by corrupted processes. A cannot see the (local) computations that non-faulty processes perform. We do not restrict the amount or speed of computation that A can perform.

2.3 Anonymity Assumption

Consider the following experiment. Suppose that $p_i \in P$ is non-faulty and invokes "anon_send m to p_j", where $p_j \in P$ is possibly corrupted, and p_j invokes anon_receive with respect to m. No process can directly invoke send or invoke receive in response to a send call at any time. p_j is allowed to use anon_send to message corrupted processes if it is corrupted, and can invoke anon_recv with respect to messages sent by corrupted processes. Each process is unable to make oracle calls (described below), but is allowed to perform an arbitrary number of local computations. p_j then outputs a single guess, $g \in \{1, \ldots, n\}$ as to the identity of p_i. Then for any run of the experiment, $Pr(i = g) \leq \frac{1}{n-t}$.

As A can corrupt t processes, the anonymity set [19], i.e., the set of processes p_i is indistinguishable from, comprises $n - t$ non-faulty processes. Our definition captures the anonymity of the anonymous channels, but does not consider the effects of regular message passing and timing on anonymity. As such, we can use techniques to establish anonymous channels in practice with varying levels of anonymity with these factors considered.

2.4 Traceable Ring Signatures

Informally, a ring signature [29,50] proves that a signer has knowledge of a private key corresponding to one of many public keys of their choice without revealing the corresponding public key. Hereafter, we consider *traceable ring signatures* (or *TRSs*), which are ring signatures that expose the identity of a signer who signs two different messages. To increase flexibility, we can consider traceability with respect to a particular string called an *issue* [54], allowing signers to maintain anonymity if they sign multiple times, provided they do so each time with respect to a different issue.

We now present relevant definitions of the ring signatures which are analogous to those of Fujisaki and Suzuki [29]. Let $ID \in \{0,1\}^*$, which we denote as a *tag*. We assume that all processes may query an idealised distributed oracle, which implements the following *four* operations:

1. $\sigma \leftarrow \mathsf{Sign}(i, ID, m)$, which takes the integer $i \in \{1, \ldots, n\}$, tag $ID \in \{0,1\}^*$ and message $m \in \{0,1\}^*$, and outputs the signature $\sigma \in \{0,1\}^*$. We restrict Sign such that only process $p_i \in P$ may invoke Sign with first argument i.
2. $b \leftarrow \mathsf{VerifySig}(ID, m, \sigma)$, which takes the tag ID, message $m \in \{0,1\}^*$, and signature $\sigma \in \{0,1\}^*$, and outputs a bit $b \in \{0,1\}$. All parties may query VerifySig.
3. $out \leftarrow \mathsf{Trace}(ID, m, \sigma, m', \sigma')$, which takes the tag $ID \in \{0,1\}^*$, messages $m, m' \in \{0,1\}^*$ and signatures $\sigma, \sigma' \in \{0,1\}^*$, and outputs $out \in \{0,1\}^* \cup$

$\{1,\ldots,n\}$ (possibly corresponding to a process p_i). All parties may query Trace.

4. $x \leftarrow \mathsf{FindIndex}(ID, m, \sigma)$ takes a tag $ID \in \{0,1\}^*$, a message $m \in \{0,1\}^*$, and a signature $\sigma \in \{0,1\}^*$, and outputs a value $x \in \{1,\ldots,n\}$. FindIndex may not be called by any party, and exists only for protocol definitions.

The distributed oracle satisfies the following relations:

- $\mathsf{VerifySig}(ID, m, \sigma) = 1 \iff \exists\ p_i \in P$ which invoked $\sigma \leftarrow \mathsf{Sign}(i, ID, m)$.
- Trace is as below $\iff \sigma \leftarrow \mathsf{Sign}(i, ID, m)$ and $\sigma' \leftarrow \mathsf{Sign}(i', ID, m')$ where:

$$\mathsf{Trace}(ID, m, \sigma, m', \sigma') = \begin{cases} \text{“indep”} & \text{if } i \neq i', \\ \text{“linked”} & \text{else if } m = m', \\ i & \text{otherwise } (i = i' \wedge m \neq m'). \end{cases}$$

- If adversary D is given an arbitrary set of signatures S and must identify the signer p_i of a signature $\sigma \in S$ by guessing i, $Pr(i = g) \leq \frac{1}{n-t}$ for any D.

The concrete scheme proposed by Fujisaki and Suzuki [29] computationally satisfies these properties in the random oracle model [3] provided the Decisional Diffie-Hellman problem is intractable. In our protocols, where the ring comprises the n processes of P, the resulting signatures are of size $O(kn)$, where k is the security parameter. To simplify the presentation, we assume that its properties hold unconditionally in the following.

3 Communication Primitives

3.1 Traceable Broadcast

Suppose a process p wishes to anonymously send a message to a given set of processes P. By invoking anonymous communication channels, p can achieve this, but processes in $P \setminus \{p\}$ are unable to verify that p resides in P, and so cannot meaningfully participate in the protocol execution. By using (traceable) ring signatures, p can verify its membership in P over anonymous channels without revealing its identity. To this end, we outline a simple mechanism to replace the invocation of send and receive primitives (over regular channels) with calls to ring signature and anonymous messaging primitives (namely anon_send and anon_receive).

Let $(ID, TAG, \mathsf{label}, m)$ be a tuple for which $p_i \in P$ has invoked send with respect to. Instead of invoking send, p_i first calls $\sigma \leftarrow \mathsf{Sign}(i, T, m)$, where T is a tag uniquely defined by TAG and label. Then, p_i invokes anon_send M, where $M = (ID, TAG, \mathsf{label}, m, \sigma)$. Upon an anon_receive M call by $p_j \in P$, p_j verifies that $\mathsf{VerifySig}(T, m, \sigma) = 1$ and that they have not previously received (m', σ') such that $\mathsf{Trace}(T, m, \sigma, m', \sigma') \neq$ “indep”. Given this check passes, p_j invokes $\mathsf{receive}(ID, TAG, \mathsf{label}, m)$. In our protocols, processes always broadcast messages, so the transformation is used with respect to broadcast calls.

By the properties of the anonymous channels and the signatures, it follows that anonymity as defined in the previous section holds with additional adversarial access to the distributed oracles. We present the proof in the companion technical report [8]. Hereafter, we assume that calls to primitives send and receive are handled by the procedure presented in this subsection unless explicitly stated otherwise.

3.2 Binary Consensus

Broadly, the binary consensus problem involves a set of processes reaching an agreement on a binary value $b \in \{0, 1\}$. We first recall the definitions that define the binary Byzantine consensus (BBC) problem as stated in [17]. In the following, we assume that every non-faulty process proposes a value, and remark that only the values in the set $\{0, 1\}$ can be decided by a non-faulty process.

1. **BBC-Termination:** Every non-faulty process eventually decides on a value.
2. **BBC-Agreement:** No two non-faulty processes decide on different values.
3. **BBC-Validity:** If all non-faulty processes propose the same value, no other value can be decided.

For simplicity, we present the safe, non-terminating variant of the binary consensus routine from [17] in Algorithm 1. As assumed in the model (Sect. 2), the terminating variant relies on partial synchrony between processes in P. The protocols execute in asynchronous rounds.

State. A process keeps track of a binary value $est \in \{0, 1\}$, corresponding to a process' current estimate of the decided value, arrays $bin\_values[1..]$, in which each element is a set $S \subseteq \{0, 1\}$, a round number r (initialised to 0), an auxiliary binary value b, and lists of (binary) values $values_r$, $r = 1, 2, \ldots$, each of which are initially empty.

Messages. Messages of the form (EST, r, b) and (AUX, r, b), where $r \geq 1$ and $b \in \{0, 1\}$, are sent and processed by non-faulty processes. Note that we have omitted the dependency on a label *label* and identifier *ID* for simplicity of exposition.

BV-Broadcast. To exchange *EST* messages, the protocol relies on an all-to-all communication abstraction, BV-broadcast [43], which is presented in Algorithm 1. When a process adds a value $v \in \{0, 1\}$ to its array $bin\_values[r]$ for some $r \geq 1$, we say that v was BV-delivered.

Functions. Let $b \in \{0, 1\}$. In addition to BV-broadcast and the communication primitives in our model, a process can invoke bin_propose(b) to begin executing an instance of binary consensus with input b, and decide(b) to decide the value b. In a given instance of binary consensus, these two functions may be called exactly once. In addition, the function $list$.append(v) appends the value v to the list $list$.

Algorithm 1. Safe binary consensus routine

 1: bin_propose(v):
 2:     $est \leftarrow v; r \leftarrow 0;$
 3:     **repeat:**
 4:         $r \leftarrow r + 1;$
 5:         BV-broadcast(EST, r, est)
 6:         **wait until** ($bin\_values[r] \neq \emptyset$)
 7:         broadcast ($AUX, r, bin\_values[r]$)
 8:         **wait until** ($|values_r| \geq n - t) \wedge (val \in bin\_values[r]$ for all $val \in values_r$)
 9:         $b \leftarrow r \pmod 2$
10:         **if** $val = w$ for all $val \in values_r$ where $w \in \{0, 1\}$ **then**
11:             $est \leftarrow w;$
12:             **if** $w = b$ **then**
13:                 decide(v) if not yet invoked decide()
14:         **else**
15:             $est \leftarrow b$
16: **upon** intial receipt of (AUX, r, b) for some $b \in \{0, 1\}$ from process p_j
17:     $values_r$.append(b)

18: BV-broadcast(EST, r, v_i):
19:     broadcast (EST, r, v_i)

20: **upon** receipt of (EST, r, v)
21:     **if** (EST, r, v) received from ($t + 1$) processes and not yet broadcast **then**
22:         broadcast (EST, r, v)

23:     **if** (EST, r, v) received from ($2t + 1$) processes **then**
24:         $bin\_values[r] \leftarrow bin\_values[r] \cup \{v\}$

To summarise Algorithm 1, the BV-broadcast component ensures that only a value proposed by a correct process may be decided, and the auxiliary broadcast component ensures that enough processes have received a potentially decidable value to ensure agreement. The interested reader can verify the correctness of the protocol and read a thorough description of how it operates in [17], where the details of the corresponding terminating protocol also reside.

3.3 Anonymity-Preserving All-to-All Reliable Broadcast

To reach eventual agreement in the presence of Byzantine processes without revealing who proposes what, we introduce the *anonymity-preserving all-to-all reliable broadcast* problem that preserves the anonymity of each honest sender which is reliably broadcasting. In this primitive, all processes are assumed to (anonymously) broadcast a message, and all processes deliver messages over time. It ensures that all honest processes always receive the same message from one (possibly faulty) sender while hiding the identity of any non-faulty sender.

Let $ID \in \{0, 1\}^*$ be an *identifier*, a string that uniquely identifies an instance of anonymity-preserving all-to-all reliable broadcast, hereafter referred to as AARB-broadcast. Let m be a message, and σ be the output of the call $\mathsf{Sign}(i, T, m)$ for some $i \in \{1, \ldots, n\}$, where $T = f(ID, \mathsf{label})$ for some function f as in traceable broadcast. Each process is equipped with two operations, "AARBP" and "AARB-deliver". $\mathsf{AARBP}[ID](m)$ is invoked once with respect to ID and any message m, denoting the beginning of a process' execution of AARBP with respect to ID. $\mathsf{AARB\text{-}deliver}[ID](m, \sigma)$ is invoked between $n-t$ and n times over the protocol's execution. When a process invokes $\mathsf{AARB\text{-}deliver}[ID](m, \sigma)$, they are said to "AARB-deliver" (m, σ) with respect to ID. Then, given $t < \frac{n}{3}$, we define a protocol that implements AARB-broadcast with respect to ID as satisfying the following *six* properties:

1. **AARB-Signing:** If a non-faulty process p_i AARB-delivers a message with respect to ID, then it must be of the form (m, σ), where a process $p_i \in P$ invoked $\mathsf{Sign}(i, T, m)$ and obtained σ as output.
2. **AARB-Validity:** Suppose that a non-faulty process AARB-delivers (m, σ) with respect to ID. Let $i = \mathsf{FindIndex}(T, m, \sigma)$ denote the output of an idealised call to FindIndex. Then if p_i is non-faulty, p_i must have anonymously broadcast (m, σ).
3. **AARB-Unicity:** Consider any point in time in which a non-faulty process p has AARB-delivered more than one tuple with respect to ID. Let $delivered = \{(m_1, \sigma_1), \ldots, (m_l, \sigma_l)\}$, where $|delivered| = l$, denote the set of these tuples. For each $i \in \{1, \ldots, l\}$, let $out_i = \mathsf{FindIndex}(T, m_i, \sigma_i)$ denote the output of an idealised call to FindIndex. Then for all distinct pairs of tuples $\{(m_i, \sigma_i), (m_j, \sigma_j)\}$, $out_i \neq out_j$.
4. **AARB-Termination-1:** If a process p_i is non-faulty and invokes $\mathsf{AARBP}[ID](m)$, all the non-faulty processes eventually AARB-deliver (m, σ) with respect to ID, where σ is the output of the call $\mathsf{Sign}(i, T, m)$.
5. **AARB-Termination-2:** If a non-faulty process AARB-delivers (m, σ) with respect to ID, then all the non-faulty processes eventually AARB-deliver (m, σ) with respect to ID.

Firstly, we require AARB-Signing to ensure that the other properties are meaningful. Since messages are anonymously broadcast, properties refer to the index of the signing process determined by an idealised call to FindIndex. In spirit, AARB-Validity ensures if a non-faulty process AARB-delivers a message that was signed by a non-faulty process p_i, then p_i must have invoked AARBP. Similarly, AARB-Unicity ensures that a non-faulty process will AARB-deliver at most one message signed by each process. We note that AARB-Termination-1 is insufficient for consensus: without AARB-Termination-2, different processes may AARB-deliver different messages produced by the *same* process if it is faulty, as in the two-step algorithm implementing no-duplicity broadcast [6,49]. Finally, we state the anonymity property:

6. **AARB-Anonymity:** Suppose that non-faulty process p_i invokes $\mathsf{AARBP}[ID](m)$ for some m and a given ID, and has previously invoked an arbitrary number of $\mathsf{AARBP}[ID_j](m_j)$ calls where $ID \neq ID_j$ for all such j. Suppose that an adversary A is required to output a guess $g \in \{1, \ldots, n\}$, corresponding

to the identity of p_i after performing an arbitrary number of computations, allowed oracle calls and invocations of networking primitives. Then for any A and run, $Pr(i = g) \leq \frac{1}{n-t}$.

Informally, AARB-Anonymity guarantees that the source of an anonymously broadcast message by a non-faulty process is unknown to the adversary, in that it is indistinguishable from $n - t$ (non-faulty) processes. AARBP can be implemented by composing n instances of Bracha's reliable broadcast algorithm , which we describe and prove correct in the companion technical report [8].

4 Anonymity-Preserving Vector Consensus

In this section, we introduce the *anonymity-preserving vector consensus* problem and present and discuss the protocol Anonymised Vector Consensus Protocol (AVCP) that solves it. The anonymity-preserving vector consensus problem brings anonymity to the vector consensus problem [23] where non-faulty processes reach an agreement upon a vector containing at least $n - t$ proposed values. More precisely, anonymised vector consensus asserts that the identity of a process who proposes must be indistinguishable from that of all non-faulty processes. As in AARBP, instances of AVCP are identified uniquely by a given value ID. Each process is equipped with two operations. Firstly, "$\mathsf{AVCP}[ID](m)$" begins execution of an instance of AVCP with respect to ID and proposal m. Secondly, "$\mathsf{AVC\text{-}decide}[ID](V)$" signals the output of V from the instance of AVCP identified by ID, and is invoked exactly once per identifier. We define a protocol that solves anonymity-preserving vector consensus with respect to these operations as satisfying the following four properties. Firstly, we require an anonymity property, defined analogously to that of AARB-broadcast:

1. **AVC-Anonymity:** Suppose that non-faulty process p_i invokes $\mathsf{AVCP}[ID](m)$ for some m and a given ID, and has previously invoked an arbitrary number of $\mathsf{AVCP}[ID_j](m_j)$ calls where $ID \neq ID_j$ for all such j. Suppose that an adversary A is required to output a guess $g \in \{1, \ldots, n\}$, corresponding to the identity of p_i after performing an arbitrary number of computations, allowed oracle calls and invocations of networking primitives. Then for any A and run, $Pr(i = g) \leq \frac{1}{n-t}$.

It also requires the original agreement and termination properties of vector consensus:

2. **AVC-Agreement:** All non-faulty processes that invoke $\mathsf{AVC\text{-}decide}[ID](V)$ do so with respect to the same vector V for a given ID.
3. **AVC-Termination:** Every non-faulty process eventually invokes $\mathsf{AVC\text{-}decide}[ID](V)$ for some vector V and a given ID.

It also requires a validity property that depends on a pre-determined, deterministic validity predicate $\mathsf{valid}()$ [9,17] which we assume is common to all processes. We assume that all non-faulty processes propose a value that satisfies $\mathsf{valid}()$.

4. **AVC-Validity:** Consider each non-faulty process that invokes AVC-decide[ID](V) for some V and a given ID. Each value $v \in V$ must satisfy valid(), and $|V| \geq n - t$. Further, at least $|V| - t$ values correspond to the proposals of distinct non-faulty processes.

4.1 AVCP, the Anonymised Vector Consensus Protocol

We present a reduction to binary consensus which may converge in four message steps, as in the reduction to binary consensus of Democratic Byzantine Fault Tolerance (DBFT) [17], at least one of which must be performed over anonymous channels. We present the proof of correctness in the companion technical report [8]. We note that comparable problems [21], including agreement on a core set over an asynchronous network [4], rely on such a reduction but with a probabilistic solution. As in DBFT, we solve consensus deterministically by reliably broadcasting proposals which are then decided upon using n instances of binary consensus. The protocol is divided into two components. Firstly, the reduction component (Algorithm 2) reduces anonymity-preserving vector consensus to binary consensus. Here, one instance of AARBP and n instances of binary consensus are executed. But, since proposals are made anonymously, processes cannot associate proposals with binary consensus instances a priori. Consequently, processes start with n unlabelled binary consensus instances, and label them over time with the hash digest of proposals they deliver (of the form $h \in \{0,1\}^*$). To cope with messages sent and received in unlabelled binary consensus instances, we require a handler component (Algorithm 3) that replaces function calls made in binary consensus instances.

Functions. In addition to the communication primitives detailed in Sect. 2 and the two primitives "AVCP" and "AVC-decide", the following primitives may be called:

- "$inst$.bin_propose(v)", where $inst$ is an instance of binary consensus and $v \in \{0,1\}$, which begins execution of $inst$ with initial value v.
- "AARBP" and "AARB-deliver", as in Sect. 3.
- "valid()" as described above.
- "m.keys()" (resp. "m.values()"), which returns the keys (resp. values) of a map m.
- "$item$.key()", which returns the key of $item$ in a map m which is determined by context.
- "s.pop()", which removes and returns a value from set s.
- "$H(u,v)$", a cryptographic hash function which maps elements $u, v \in \{0,1\}^*$ to $h \in \{0,1\}^*$.

State. Each process tracks the following variables:

- $ID \in \{0,1\}^*$, a common identifier for a given instance of AVCP.

- *proposals*[], which maps labels of the form $l \in \{0,1\}^*$ to AARB-delivered messages of the form $(m, \sigma) \in (\{0,1\}^*, \{0,1\}^*)$ that may be decided, and is initially empty.
- *decision_count*, tracking the number of binary consensus instances for which a decision has been reached, initialised to 0.
- *decided_ones*, the set of proposals for which 1 was decided in the corresponding binary consensus instance, initialised to $\emptyset$.
- *labelled*[], which maps labels, which are the hash digest $h \in \{0,1\}^*$ of AARB-delivered proposals, to binary consensus instances, and is initially empty.
- *unlabelled*, a set of binary consensus instances with no current label, which is initially of cardinality n.
- *ones*[][], which maps two keys, EST and AUX, to maps with integer keys $r \geq 1$ which map to a set of labels, all of which are initially empty.
- *counts*[][], which maps two keys, EST and AUX, to maps with integer keys $r \geq 1$ which map to an integer $n \in \{0, \ldots, n\}$, all of which are initialised to 0.

Messages. In addition to messages propagated in AARBP, non-faulty processes process messages of the form $(ID, TAG, r, label, b)$, where $TAG \in \{EST, AUX\}$, $r \geq 1$, $label \in \{0,1\}^*$ and $b \in \{0,1\}$. A process buffers a message $(ID, TAG, r, label, b)$ until *label* labels an instance of binary consensus *inst*, at which point the message is considered receipt in *inst*. The handler, described below, ensures that all messages sent by non-faulty processes eventually correspond to a label in their set of labelled consensus instances (i.e., contained in *labelled*.keys()). Similarly, a non-faulty process can only broadcast such a message after labelling the corresponding instance of binary consensus. Processes also process messages of the form $(ID, TAG, r, ones)$, where $TAG \in \{EST\_ONES, AUX\_ONES\}$, $r \geq 1$, and *ones* is a set of strings corresponding to binary consensus instance labels.

Reduction. In the reduction, presented in Algorithm 2, n (initially unlabelled) instances of binary consensus are used, each corresponding to a value that one process in P may propose. Each (non-faulty) process invokes AARBP with respect to *ID* and their value m' (line 2), anonymously broadcasting (m', σ') inside the AARBP instance. On AARB-delivery of some message (m, σ), an unlabelled instance of binary consensus is deposited into *labelled*, whose key (label) is set to $H(m, \sigma)$ (line 10). Proposals that fulfil valid() are stored in *proposals* (line 12), and *inst*.bin_propose(1) is invoked with respect to the newly labelled instance $inst = labelled[H(m, \sigma)]$ if not yet done (line 13). Upon termination of each instance (line 14), provided 1 was decided, the corresponding proposal is added to *decided_ones* (line 16). For either decision value, *decision_count* is incremented (line 16). Once 1 has been decided in $n - t$ instances of binary consensus, processes will propose 0 in all instances that they have not yet proposed in (line 17). Note that upon AARB-delivery of valid messages after this point, bin_propose(1) is not invoked at line 13. Upon the termination of all n instances

Algorithm 2. AVCP (1 of 2): Reduction to binary consensus

```
1: AVCP[ID](m'):
2:    AARBP[ID](m')                        ▷ anonymised reliable broadcast of proposal
3:    wait until |decided_ones| ≥ n − t     ▷ wait until n − t instances terminate with 1
4:    for each inst ∈ unlabelled ∪ labelled.values() such that
5:    inst.bin_propose() not yet invoked do
6:       Invoke inst.bin_propose(0)      ▷ propose 0 in all binary consensus not yet invoked
7:    wait until decision_count = n  ▷ wait until all n instances of binary consensus terminate
8:    AVC-decide[ID](decided_ones)

9: upon invocation of AARB-deliver[ID](m, σ)
10:   labelled[H(m, σ)] ← unlabelled.pop()
11:   if valid(m, σ) then                    ▷ deterministic, common validity function
12:      proposals[H(m, σ)] ← (m, σ)
13:      Invoke labelled[H(m, σ)].bin_propose(1) if not yet invoked

14: upon inst deciding a value v ∈ {0, 1}, where inst ∈ labelled.values() ∪ unlabelled
15:   if v = 1 then ▷ store proposals for which 1 was decided in the corresponding binary consensus
16:      decided_ones ← decided_ones ∪ {proposals[inst.key()]}
17:   decision_count ← decision_count + 1
```

of binary consensus (after line 7), all non-faulty processes decide their set of values for which 1 was decided in the corresponding instance of binary consensus (line 8).

Handler. As proposals are anonymously broadcast, binary consensus instances cannot be associated with process identifiers a priori, and so are labelled by AARB-delivered messages. Thus, we require the handler, which overrides two of the three broadcast calls in the non-terminating variant of the binary consensus of [17] (Algorithm 1).

We now describe the handler (Algorithm 3). Let *inst* be an instance of binary consensus. On calling *inst*.bin_propose(b) ($b \in \{0, 1\}$) (and at the beginning of each round $r \geq 1$), processes invoke BV-broadcast (line 5 of Algorithm 1), immediately calling "broadcast (*ID*, *EST*, r, *label*, b)" (line 19 of Algorithm 1). If $b = 1$, (*ID*, *EST*, r, *label*, 1) is broadcast, and *label* is added to the set *ones*[*EST*][r] (line 21). Note that, given AARB-Termination-1, all messages sent by non-faulty processes of the form (*ID*, *EST*, r, *label*, 1) will be deposited in an instance *inst* labelled by *label*. Then, as the binary consensus routine terminates when all non-faulty processes propose the same value, all processes will decide the value 1 in $n - t$ instances of binary consensus (i.e., will pass line 3), after which they execute bin_propose(0) in the remaining instances of binary consensus.

Since these instances may not be labelled when a process wishes to broadcast a value of the form (*ID*, *EST*, r, *label*, 0), we defer their broadcast until "broadcast(*ID*, *EST*, r, *label*, b)" is called in all n instances of binary consensus. At this point (line 23), (*ID*, *EST_ONES*, r, *ones*[*EST*][r]) is broadcast (line 24).

Algorithm 3. AVCP (2 of 2): Handler of Algorithm 1

18: **upon** "broadcast $(ID, EST, r, label, b)$" in $inst \in labelled$.values() $\cup$ $unlabelled$
19: **if** b = 1 **then**
20: broadcast $(ID, EST, r, label, b)$
21: $ones[EST][r] \leftarrow ones[EST][r] \cup \{inst.\text{key}()\}$
22: $counts[EST][r] \leftarrow counts[EST][r] + 1$
23: **if** $counts[EST][r] = n \wedge |ones[EST][r]| < n$ **then**
24: broadcast $(ID, EST\_ONES, r, ones[EST][r])$
25: **upon** "broadcast $(ID, AUX, r, label, b)$" in $inst \in labelled$.values() $\cup$ $unlabelled$
26: **if** b = 1 **then**
27: broadcast $(ID, AUX, r, label, b)$
28: $ones[AUX][r] \leftarrow ones[AUX][r] \cup \{inst.\text{key}()\}$
29: $counts[AUX][r] \leftarrow counts[AUX][r] + 1$
30: **if** $counts[AUX][r] = n \wedge |ones[AUX][r]| < n$ **then**
31: broadcast $(ID, AUX\_ONES, r, ones[AUX][r])$
32: **upon** receipt of $(ID, TAG, r, ones)$ s.t. $TAG \in \{EST\_ONES, AUX\_ONES\}$
33: **wait until** $one \in labelled$.keys() $\forall one \in ones$
34: **if** $TAG = EST\_ONES$ **then**
35: $TEMP \leftarrow EST$
36: **else** $TEMP \leftarrow AUX$
37: **for each** $l \in labelled$.keys() such that $l \notin ones$ **do**
38: deliver $(ID, TEMP, r, l, 0)$ in $labelled[l]$
39: **for each** $inst \in unlabelled$ **do**
40: deliver $(ID, TEMP, r, \perp, 0)$ in $inst$

A message of the form $(ID, EST\_ONES, r, ones)$ is interpreted as the receipt of zeros in all instances not labelled by elements in $ones$ (at lines 38 and 40). This can only be done once all elements of $ones$ label instances of binary consensus (i.e., after line 33). Note that if $|ones[EST][r] = n|$, then there are no zeroes to be processed by receiving processes, and so the broadcast at line 24 can be skipped.

Handling "broadcast$(ID, AUX, r, label, b)$" calls (line 7 of Algorithm 1) is identical to the handling of initial " broadcast$(ID, EST, r, label, b)$" calls. Note that the third broadcast in the original algorithm, where $(ID, EST, r, label, b)$ is broadcast upon receipt from $t + 1$ processes if not yet done before (line 21 of Algorithm 1 (BV-Broadcast)), can only occur once the corresponding instance of binary consensus is labelled. Thus, it does not need to be handled. From here, we can see that messages in the handler are processed as if n instances of the original binary consensus algorithm were executing.

4.2 Complexity and Optimizations

Let k be a security parameter, S the size of a signature and c the size of a message digest. In Table 1, we compare the message and communication complexities of AVCP against DBFT [17], which, as written, can be easily altered to solve

Table 1. Comparing the complexity of AVCP and DBFT [17] after GST [24]

Complexity	AVCP	DBFT
Best-case message complexity	$O(n^3)$	$O(n^3)$
Worst-case message complexity	$O(tn^3)$	$O(tn^3)$
Best-case complexity	$O((S+c)n^3)$	$O(n^3)$
Worst-case bit complexity	$O((S+c)tn^3)$	$O(tn^3)$

vector consensus. We assume that AVCP is invoking the terminating variant of the binary consensus of [17]. When considering complexity, we only count messages in the binary consensus routines once the global stabilisation time (GST) has been reached [24]. Both best-free and worst-case message complexity are identical between the two protocols. We remark that there exist runs of AVCP where processes are faulty which has the best-case message complexity $O(n^3)$, such as when a process has crashed. AVCP mainly incurs greater communication complexity proportional to the size of the signatures, which can vary from size $O(k)$ [35,53] to $O(kn)$ [28]. If processes make a single anonymous broadcast per run, the best-case and worst-case bit complexities of AVCP are lowered to $O(Sn^2 + cn^3)$ and $O(Sn^2 + ctn^3)$.

As is done in DBFT [17], we can combine the anonymity-preserving all-to-all reliable broadcast of a message m and the proposal of the binary value 1 in the first round of a binary consensus instance. To this end, a process may skip the BV-broadcast step in round 1, which may allow AVCP to converge in four message steps, at least one of which must be anonymous. It may be useful to invoke "broadcast $TAG[r](b)$", where $TAG \in \{EST, AUX\}$ (lines 20 and 27) when the instance of binary consensus is labelled, rather than simply when $b = 1$ (i.e., the condition preceding these calls). Since it may take some time for all n instances of binary consensus to synchronise, doing this may speed up convergence in the "faster" instances.

5 Experiments

In order to evaluate the practicality of our solutions, we implemented our distributed protocols and deployed them on Amazon EC2 instances. We refer to each EC2 instance used as a *node*, corresponding to a 'process' as in the protocol descriptions. For each value of n (the number of nodes) chosen, we ran experiments with an equal number of nodes from *four* regions: Oregon (us-west-2), Ohio (us-east-2), Singapore (ap-southeast-1) and Frankfurt (eu-central-1). The type of instance chosen was c4.xlarge, which provide 7.5 GiB of memory, and 4 vCPUs, i.e., 4 cores of an Intel Xeon E5-2666 v3 processor. We performed between 50 and 60 iterations for each value of n and t we benchmarked. We varied n incrementally, and varied t both with respect to the maximum fault-tolerance (i.e., $t = \lfloor \frac{n-1}{3} \rfloor$), and also fixed $t = 6$ for values of $n = 20, 40, \ldots$ All networking code, and the application logic, was written in Python (2.7). As we have implemented our cryptosystems in golang, we call our libraries from Python using ctypes[1]. To simulate reliable channels, nodes communicate over TCP. Nodes begin timing once all connections have been established (i.e., after handshaking).

Our protocol, Anonymised Vector Consensus Protocol (AVCP), was implemented on top of the existing DBFT [17] codebase, as was the case with our implementation of AARB-broadcast, i.e., AARBP. We do not use the fast-path optimisation described in Sect. 4, but we hash messages during reliable broadcast to reduce bandwidth consumption. We use the most conservative choice of ring signatures, $O(kn)$-sized traceable ring signatures [29], which require $O(n)$ operations for each signing and verification call, and $O(n^2)$ work for tracing overall. Each process makes use of a single anonymous broadcast in each run of the algorithm. To simulate the increased latency afforded by using publicly-deployed anonymous networks, processes invoke a local timeout for 750 ms before invoking anon_broadcast, which is a regular broadcast in our experiments.

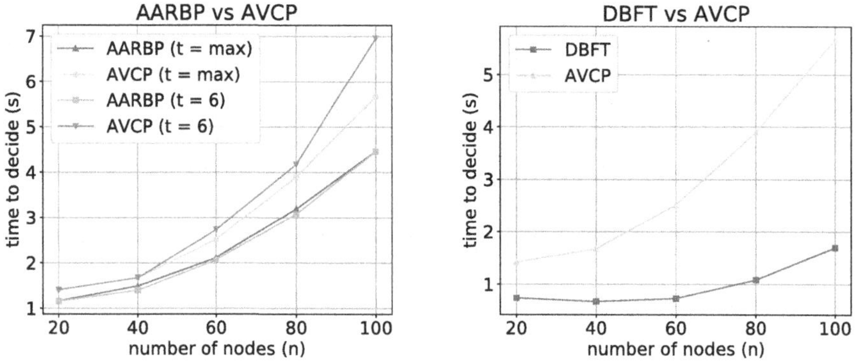

(a) Comparing AARBP and AVCP (b) Comparing DBFT and AVCP

Fig. 1. Evaluating the cost of the reliable anonymous broadcast (AARBP) in our solution (AVCP) and the performance of our solution (AVCP) compared to an efficient Byzantine consensus baseline (DBFT) without anonymity preservation

Figure 1a compares the performance of AARBP with that of AVCP. In general, convergence time for AVCP is higher as we need at least three more message steps for a process to decide. Given that the fast-path optimisation is used, requiring 1 additional message step over AARBP in the good case, the difference in performance between AVCP and AARBP would indeed be smaller.

Comparing AVCP with $t = max$ and $t = 6$, we see that when $t = 6$, convergence is slower. Indeed, AVC-Validity states at least $n - t$ values fulfilling valid() are included in a process' vector given that they decide. Consequently, as t is smaller, $n - t$ is larger, and so nodes will process and decide more values.

[1] https://docs.python.org/2/library/ctypes.html.

Although AARB-delivery may be faster for some messages, nodes generally have to perform more TRS verification/tracing operations. As nodes decide 1 in more instances of binary consensus, messages of the form $(ID, TAG, r, ones)$ are propagated where $|ones|$ is generally larger, slowing down decision time primarily due to the size of the message.

Figure 1b compares the performance of DBFT to solve vector consensus against AVCP. Indeed, the difference in performance between AVCP and DBFT when $n = 20$ and $n = 40$ is primarily due to AVCP's 750 ms timeout. As expected when scaling up n further, cryptographic operations result in increasingly slower performance for AVCP.

Overall, AVCP performs reasonably well, reaching convergence when $n = 100$ between 5 and 7 s depending on t, which is practically reasonable, particularly when used to perform elections which are typically occasional.

6 Discussion

It is clear that anonymity is preserved if processes only use anonymous channels to communicate, provided that processes do not double-sign with ring signatures for each message type. For performance and to prevent long-term correlation attacks on anonymous networks like Tor [42], it may be of interest to use anonymous message passing to propose a value, and then to use regular channels for the rest of a given protocol execution. In this setting, the adversary can de-anonymise a non-faulty process by observing message transmission time [2] and the order in which messages are sent and received [48]. For example, a single non-faulty process may be relatively slow, and so the adversary may deduce that messages it delivers late over anonymous channels were sent by that process.

Achieving anonymity in this setting in practice depends on the latency guarantees of the anonymous channels, the speed of each process, and the latency guarantees of the regular channels. One possible strategy could be to use public networks like Tor [22] where message transmission time through the network can be measured.[2] Then, based on the behaviour of the anonymous channels, processes can vary the timing of their own messages by introducing random message delays [42] to minimise the correlation between messages over the different channels. It may also be useful for processes to synchronise between protocol executions. This prevents a process from being de-anonymised when they, for example, invoke anon_send in some instance when all other processes are executing in a different instance.

In terms of future work, it is of interest to evaluate anonymity in different formal models [36,51] and with respect to various practical attack vectors [48]. It will be useful also to formalise anonymity under more practical assumptions so that the timing of anonymous and regular message passing do not correlate highly. In addition, a reduction to a randomized [10] binary consensus algorithm would remove the dependency on the weak coordinator used in each round of the binary consensus algorithm we rely on [17].

[2] https://metrics.torproject.org/.

Acknowledgment. This research is supported under Australian Research Council Discovery Projects funding scheme (project number 180104030) entitled "Taipan: A Blockchain with Democratic Consensus and Validated Contracts" and Australian Research Council Future Fellowship funding scheme (project number 180100496) entitled "The Red Belly Blockchain: A Scalable Blockchain for Internet of Things".

References

1. Adida, B.: Helios: web-based open-audit voting. In: USENIX Security, pp. 335–348 (2008)
2. Back, A., Möller, U., Stiglic, A.: Traffic analysis attacks and trade-offs in anonymity providing systems. In: International Workshop on Information Hiding, pp. 245–257 (2001)
3. Bellare, M., Rogaway, P.: Random oracles are practical: a paradigm for designing efficient protocols. In: CCS, pp. 62–73 (1993)
4. Ben-Or, M., Kelmer, B., Rabin, T.: Asynchronous secure computations with optimal resilience. In: PODC, pp. 183–192 (1994)
5. Bracha, G.: Asynchronous Byzantine agreement protocols. Inf. Comput. **75**(2), 130–143 (1987)
6. Bracha, G., Toueg, S.: Resilient consensus protocols. In: PODC, pp. 12–26 (1983)
7. Bracha, G., Toueg, S.: Asynchronous consensus and broadcast protocols. JACM **32**(4), 824–840 (1985)
8. Cachin, C., Collins, D., Crain, T., Gramoli, V.: Anonymity preserving Byzantine vector consensus. CoRR abs/1902.10010 (2020). http://arxiv.org/abs/1902.10010
9. Cachin, C., Kursawe, K., Petzold, F., Shoup, V.: Secure and efficient asynchronous broadcast protocols. In: CRYPTO, pp. 524–541 (2001)
10. Cachin, C., Kursawe, K., Shoup, V.: Random oracles in constantinople: practical asynchronous Byzantine agreement using cryptography. J. Cryptol. **18**(3), 219–246 (2005)
11. Camp, J., Harkavy, M., Tygar, J.D., Yee, B.: Anonymous atomic transactions. In: In Proceedings of the 2nd USENIX Workshop on Electronic Commerce (1996)
12. Castro, M., Liskov, B.: Practical Byzantine fault tolerance. In: OSDI, vol. 99, pp. 173–186 (1999)
13. Chaum, D.: Blind signatures for untraceable payments. In: Chaum, D., Rivest, R.L., Sherman, A.T. (eds.) Advances in Cryptology, pp. 199–203. Springer, Boston, MA (1983). https://doi.org/10.1007/978-1-4757-0602-4_18
14. Chaum, D.: The dining cryptographers problem: unconditional sender and recipient untraceability. J. Cryptol. **1**(1), 65–75 (1988)
15. Chaum, D.L.: Untraceable electronic mail, return addresses, and digital pseudonyms. CACM **24**(2), 84–90 (1981)
16. Correia, M., Neves, N.F., Veríssimo, P.: From consensus to atomic broadcast: time-free Byzantine-resistant protocols without signatures. Comput. J. **49**(1), 82–96 (2006)
17. Crain, T., Gramoli, V., Larrea, M., Raynal, M.: DBFT: efficient leaderless Byzantine consensus and its application to blockchains. In: NCA, pp. 1–8 (2018)
18. Cramer, R., Franklin, M., Schoenmakers, B., Yung, M.: Multi-authority secret-ballot elections with linear work. In: Eurocrypt, pp. 72–83 (1996)
19. Danezis, G., Diaz, C.: A survey of anonymous communication channels. Technical report, MSR-TR-2008-35, Microsoft Research (2008)

20. Desmedt, Y.: Threshold cryptosystems. In: Seberry, J., Zheng, Y. (eds.) AUSCRYPT 1992. LNCS, vol. 718, pp. 1–14. Springer, Heidelberg (1993). https://doi.org/10.1007/3-540-57220-1_47
21. Diamantopoulos, P., Maneas, S., Patsonakis, C., Chondros, N., Roussopoulos, M.: Interactive consistency in practical, mostly-asynchronous systems. In: ICPADS, pp. 752–759 (2015)
22. Dingledine, R., Mathewson, N., Syverson, P.: Tor: the second-generation onion router. In: USENIX Security, pp. 21–21 (2004)
23. Doudou, A., Schiper, A.: Muteness failure detectors for consensus with Byzantine processes. In: PODC, p. 315 (1997)
24. Dwork, C., Lynch, N., Stockmeyer, L.: Consensus in the presence of partial synchrony. JACM **35**(2), 288–323 (1988)
25. ElGamal, T.: A public key cryptosystem and a signature scheme based on discrete logarithms. IEEE Trans. Inf. Theory **31**(4), 469–472 (1985)
26. Fischer, M.J., Lynch, N.A., Paterson, M.S.: Impossibility of distributed consensus with one faulty process. JACM **32**(2), 374–382 (1985)
27. Freedman, M.J., Nissim, K., Pinkas, B.: Efficient private matching and set intersection. In: Cachin, C., Camenisch, J.L. (eds.) EUROCRYPT 2004. LNCS, vol. 3027, pp. 1–19. Springer, Heidelberg (2004). https://doi.org/10.1007/978-3-540-24676-3_1
28. Fujisaki, E.: Sub-linear size traceable ring signatures without random oracles. IEICE Trans. Fundam. Electron. Commun. Comput. Sci. **E95.A**, 393–415 (2011)
29. Fujisaki, E., Suzuki, K.: Traceable ring signature. In: Okamoto, T., Wang, X. (eds.) PKC 2007. LNCS, vol. 4450, pp. 181–200. Springer, Heidelberg (2007). https://doi.org/10.1007/978-3-540-71677-8_13
30. Gentry, C.: Fully homomorphic encryption using ideal lattices. In: STOC, pp. 169–178 (2009)
31. Gilad, Y., Herzberg, A.: Spying in the dark: TCP and Tor traffic analysis. In: Fischer-Hübner, S., Wright, M. (eds.) PETS 2012. LNCS, vol. 7384, pp. 100–119. Springer, Heidelberg (2012). https://doi.org/10.1007/978-3-642-31680-7_6
32. Golle, P., Jakobsson, M., Juels, A., Syverson, P.: Universal re-encryption for mixnets. In: Okamoto, T. (ed.) CT-RSA 2004. LNCS, vol. 2964, pp. 163–178. Springer, Heidelberg (2004). https://doi.org/10.1007/978-3-540-24660-2_14
33. Golle, P., Juels, A.: Dining cryptographers revisited. In: Cachin, C., Camenisch, J.L. (eds.) EUROCRYPT 2004. LNCS, vol. 3027, pp. 456–473. Springer, Heidelberg (2004). https://doi.org/10.1007/978-3-540-24676-3_27
34. Groth, J.: Efficient maximal privacy in boardroom voting and anonymous broadcast. In: Juels, A. (ed.) FC 2004. LNCS, vol. 3110, pp. 90–104. Springer, Heidelberg (2004). https://doi.org/10.1007/978-3-540-27809-2_10
35. Gu, K., Dong, X., Wang, L.: Efficient traceable ring signature scheme without pairings. Adv. Math. Commun. **14**(2), 207–232 (2019)
36. Halpern, J.Y., O'Neill, K.R.: Anonymity and information hiding in multiagent systems. J. Comput. Secur. **13**(3), 483–514 (2005)
37. Jakobsson, M.: A practical mix. In: Nyberg, K. (ed.) EUROCRYPT 1998. LNCS, vol. 1403, pp. 448–461. Springer, Heidelberg (1998). https://doi.org/10.1007/BFb0054145
38. Juels, A., Catalano, D., Jakobsson, M.: Coercion-resistant electronic elections. In: Chaum, D., Jakobsson, M., Rivest, R.L., Ryan, P.Y.A., Benaloh, J., Kutylowski, M., Adida, B. (eds.) Towards Trustworthy Elections. LNCS, vol. 6000, pp. 37–63. Springer, Heidelberg (2010). https://doi.org/10.1007/978-3-642-12980-3_2

39. Kulyk, O., Neumann, S., Volkamer, M., Feier, C., Koster, T.: Electronic voting with fully distributed trust and maximized flexibility regarding ballot design. In: EVOTE, pp. 1–10 (2014)
40. Lamport, L., Shostak, R., Pease, M.: The Byzantine generals problem. TOPLAS **4**(3), 382–401 (1982)
41. Liu, J.K., Wei, V.K., Wong, D.S.: Linkable spontaneous anonymous group signature for ad hoc groups. In: Wang, H., Pieprzyk, J., Varadharajan, V. (eds.) ACISP 2004. LNCS, vol. 3108, pp. 325–335. Springer, Heidelberg (2004). https://doi.org/10.1007/978-3-540-27800-9_28
42. Mathewson, N., Dingledine, R.: Practical traffic analysis: extending and resisting statistical disclosure. In: Martin, D., Serjantov, A. (eds.) PET 2004. LNCS, vol. 3424, pp. 17–34. Springer, Heidelberg (2005). https://doi.org/10.1007/11423409_2
43. Mostéfaoui, A., Moumen, H., Raynal, M.: Signature-free asynchronous binary Byzantine consensus with $t < n/3$, $O(n^2)$ messages, and $O(1)$ expected time. JACM **62**(4), 31 (2015)
44. Murdoch, S.J., Danezis, G.: Low-cost traffic analysis of Tor. In: S&P, pp. 183–195 (2005)
45. Neff, C.A.: A verifiable secret shuffle and its application to e-voting. In: CCS, pp. 116–125 (2001)
46. Neves, N.F., Correia, M., Verissimo, P.: Solving vector consensus with a wormhole. TPDS **16**(12), 1120–1131 (2005)
47. Paillier, P.: Public-key cryptosystems based on composite degree residuosity classes. In: Stern, J. (ed.) EUROCRYPT 1999. LNCS, vol. 1592, pp. 223–238. Springer, Heidelberg (1999). https://doi.org/10.1007/3-540-48910-X_16
48. Raymond, J.-F.: Traffic analysis: protocols, attacks, design issues, and open problems. In: Federrath, H. (ed.) Designing Privacy Enhancing Technologies. LNCS, vol. 2009, pp. 10–29. Springer, Heidelberg (2001). https://doi.org/10.1007/3-540-44702-4_2
49. Raynal, M.: Reliable broadcast in the presence of Byzantine processes. In: Fault-Tolerant Message-Passing Distributed Systems, pp. 61–73. Springer, Cham (2018). https://doi.org/10.1007/978-3-319-94141-7_4
50. Rivest, R.L., Shamir, A., Tauman, Y.: How to leak a secret. In: Boyd, C. (ed.) ASIACRYPT 2001. LNCS, vol. 2248, pp. 552–565. Springer, Heidelberg (2001). https://doi.org/10.1007/3-540-45682-1_32
51. Serjantov, A., Danezis, G.: Towards an information theoretic metric for anonymity. In: Dingledine, R., Syverson, P. (eds.) PET 2002. LNCS, vol. 2482, pp. 41–53. Springer, Heidelberg (2003). https://doi.org/10.1007/3-540-36467-6_4
52. Shoup, V., Gennaro, R.: Securing threshold cryptosystems against chosen ciphertext attack. J. Cryptol. **15**(2), 75–96 (2002)
53. Tsang, P.P., Wei, V.K.: Short linkable ring signatures for E-voting, E-cash and attestation. In: Deng, R.H., Bao, F., Pang, H.H., Zhou, J. (eds.) ISPEC 2005. LNCS, vol. 3439, pp. 48–60. Springer, Heidelberg (2005). https://doi.org/10.1007/978-3-540-31979-5_5
54. Tsang, P.P., Wei, V.K., Chan, T.K., Au, M.H., Liu, J.K., Wong, D.S.: Separable linkable threshold ring signatures. In: Canteaut, A., Viswanathan, K. (eds.) INDOCRYPT 2004. LNCS, vol. 3348, pp. 384–398. Springer, Heidelberg (2004). https://doi.org/10.1007/978-3-540-30556-9_30
55. Yao, A.C.C.: How to generate and exchange secrets. In: FOCS, pp. 162–167 (1986)
56. Zantout, B., Haraty, R.: I2P data communication system. In: ICN, pp. 401–409 (2011)

CANSentry: Securing CAN-Based Cyber-Physical Systems against Denial and Spoofing Attacks

Abdulmalik Humayed[1,3], Fengjun Li[1], Jingqiang Lin[2,4], and Bo Luo[1(✉)]

[1] The University of Kansas, Lawrence, KS, USA
{fli,bluo}@ku.edu

[2] School of Cyber Security, University of Science and Technology of China, Hefei, China

[3] Jazan University, Jazan, Saudi Arabia
ahumayed@jazanu.edu.sa

[4] Institute of Information Engineering, Chinese Academy of Sciences, Beijing, China
linjingqiang@iie.ac.cn

Abstract. The Controller Area Network (CAN) has been widely adopted as the *de facto* standard to support the communication between the ECUs and other computing components in automotive and industrial control systems. In its initial design, CAN only provided very limited security features, which is seriously behind today's standards for secure communication. The newly proposed security add-ons are still insufficient to defend against the majority of known breaches in the literature. In this paper, we first present a new stealthy denial of service (DoS) attack against targeted ECUs on CAN. The attack is hardly detectable since the actions are perfectly legitimate to the bus. To defend against this new DoS attack and other denial and spoofing attacks in the literature, we propose a CAN firewall, namely CANSentry, that prevents malicious nodes' misbehaviors such as injecting unauthorized commands or disabling targeted services. We implement CANSentry on a cost-effective and open-source device, to be deployed between any potentially malicious CAN node and the bus, without needing to modify CAN or existing ECUs. We evaluate CANSentry on a testing platform built with parts from a modern car. The results show that CANSentry successfully prevents attacks that have shown to lead to safety-critical implications.

1 Introduction

The *Controller Area Network* (*CAN*), also referred to as the *CAN bus*, has been widely adopted as the communication backbone of small and large vehicles, ships, planes, and industrial control systems. When CAN was originally developed, its nodes were not technically ready to be connected to the external world and thus assumed to be isolated and trusted. As a result, CAN was designed without basic security features such as encryption, authentication that are now considered

L. Chen et al. (Eds.): ESORICS 2020, LNCS 12308, pp. 153–173, 2020.
https://doi.org/10.1007/978-3-030-58951-6_8

essential to communication networks [14,19]. The protocol's broadcast nature also increases the likelihood of attacks exploiting these security vulnerabilities. Therefore, a malicious message injected by a compromised electronic control unit (ECU) on the bus, if it conforms to CAN specifications, will be treated the same as a legitimate message from a benign ECU and broadcasted over the bus. Moreover, wireless communication capabilities have been added to CAN nodes for expanded functionality (e.g., TPMS, navigation, entertainment systems) without carefully examining the potential security impacts [20,21,43,46]. This enlarges the attack vector for remote attacks by allowing some previously physically isolated units to connect to external entities through wireless connections. Various attacks have been reported in recent years, ranging from simple bus denial attacks [12,36] and spoofing attacks [27,32] to more sophisticated bus-off [23,42] and arbitration denial attacks [3,12,27,36]. In response to the vulnerabilities and attacks, significant research efforts have been devoted to automobile and CAN security. Solutions such as message authentication and intrusion detection, akin to those designed for Internet security, have been proposed to secure CAN [49,50]. However, these proposals suffer from major efficiency issues. For instance, the authentication-based schemes deploy cryptographic keys among ECUs to generate MACs, which inevitably incur non-negligible processing overhead and significantly impact CAN's transmission speed. Moreover, such defense mechanisms can be defeated if the adversary exploits compromised ECUs to nullify authentication or MAC frames [3]. The intrusion detection-based mechanisms leverage ECUs' behavioral features to detect abnormal frames on the bus, hence, they have to collect a sufficient number of frames by monitoring the bus and ECUs, which inevitably results in delays in the detection.

In this paper, we first present a new stealthy arbitration denial attack against CAN, which takes seemingly legitimate actions to prevent selected CAN nodes from sending messages to the bus. This attack bypasses CAN controllers to inject deliberately crafted messages without triggering packet-based detectors, which allows the adversary to extend the length of the attack and cause severe damages. For instance, the attack can block all steering and breaking messages from being sent to the bus without causing any bus error, while keeping other sub-systems such as the engine operating normally. We implement the attack using an STM32 Nucleo-144 board as the attacker and an instrument panel cluster from a 2014 passenger car as the victim and demonstrate a successful attack.

Moreover, we propose a new defense mechanism, namely CANSentry, against the family of general CAN denial and spoofing attacks. This system-level control mechanism addresses the fundamental CAN security problem caused by the protocol's broadcast nature and lack of authentication without demanding any modification to the CAN standard or the ECUs. CANSentry leverages the malicious behavior of the attacks, e.g., inconsistency between attack ECU state and the bus state, to filter frames from high-risk ECUs (i.e., a few *external-facing* ECUs) in a bit-by-bit fashion and block unauthorized frames from being sent to the bus. We implement a cost-efficient prototype of CANSentry using the STM32

Nucleo-144 development board and demonstrate its effectiveness against recent denial and spoofing attacks including the one we propose in this paper.

The contributions of this paper are three-fold: (1) we demonstrate a stealthy selective arbitration denial attack against CAN that prevents selected ECUs from transmitting to the CAN bus without triggering any error or anomaly. (2) More importantly, we design a proof-of-concept CAN firewall, CANSentry, which is deployed between high-risk CAN nodes and the bus to defend against bus/ECU denial attacks and ECU spoofing attacks. And (3) we implement the proposed CAN firewall using low-cost hardware and demonstrate its effectiveness on a platform consisting of components from modern passenger cars.

The rest of the paper is organized as follows: we introduce the CAN bus and CAN attacks in Sect. 2 and the threat model in Sect. 3. Then, we present the stealthy selective arbitration denial attack in Sect. 4, and CANSentry design and implementation in Sect. 5, followed by security analysis in Sect. 6. Finally, we conclude the paper in Sect. 7.

2 Background and Related Work

2.1 The Controller Area Network (CAN)

CAN has been widely used in many distributed real-time control systems . While components differ, the main concepts/mechanisms remain similar in all CAN applications. A CAN network consists of nodes interconnected by a differential bus. Each node is controlled by a microcontroller (MCU), a.k.a. electronic control unit (ECU) in automotive CAN networks. A node connects to the bus through a controller and a transceiver. The *controller* is a stand-alone circuit or an MCU module, which implements the protocol, e.g., encoding/decoding frames and error handling. The *transceiver* converts between logic data and physical bits.

Frame Prioritization. In CAN, frame prioritization is realized by the *arbitration ID* (a.k.a. *CAN ID*). Figure 1 (a) shows a simplified CAN frame. It starts with the 11-bit arbitration ID (or 29-bit in the extended format), which determines the frame's priority on the bus as well as the frame's relevance to receivers, based on which each receiver decides to accept or ignore the frame. When multiple nodes attempt to send to the bus simultaneously, the lowest ID indicates the highest priority and wins the *arbitration*. To support prioritization of frames, the CAN specification defines *dominant* and *recessive* bits, denoted by "0" and "1", respectively. Whenever a dominant and a recessive bit are sent at the same time by different nodes, the dominant bit will *dominate* the bus. This mechanism allows CAN to resolve real-time conflicts during arbitration and transmission.

CAN Error Handling. Error handling in CAN allows nodes to autonomously detect and resolve transmission errors without third-party intervention. It also supports fault confinement and the containment of defective nodes. There are five types of errors in CAN: bit, ACK, stuff, form, and CRC errors. Each ECU is responsible for detecting and keeping track of both transmission and receiving errors with two error counters, *transmit error counter* (TEC) and *receive error* *counter* (REC). Depending on the role of the ECU, one of the counters will increase when an error is detected or decrease after a successful transmission or reception. The ECU determines its error state according to the values of both counters, as shown in Fig. 1 (b). The transitions between error states allow nodes to treat temporary errors and permanent failures differently. In particular, when a node encounters persistent transmission errors that may affect other nodes, it switches to bus-off state, resulting in its complete isolation from the CAN bus.

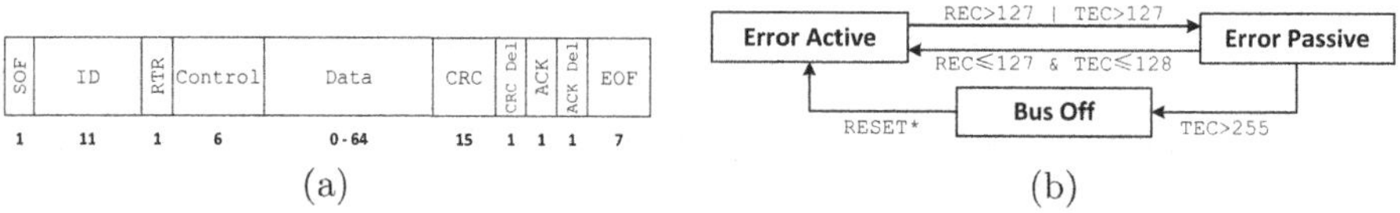

Fig. 1. The CAN Bus: (a) Format of a CAN data frame; (b) CAN error state transition. *: RESET or reception of 128 occurrences of 11 recessive bits.

2.2 Existing CAN Attacks

In this paper, we focus on *denial* and *spoofing* attacks against CAN. Denial attacks are further classified into bus, ECU, and arbitration denial.

Bus Denial. A naive approach is to flood the bus with a stream of 0×0 IDs so the bus is always occupied with this highest priority ID (*BD1*) [12,27]. Another approach is to occupy the bus by transmitting a stream of dominant bits and prevent any ECU from transmitting. However, this bit-stream does not conform to CAN protocol, hence, it cannot be dispatched by CAN controllers in regular mode. [12] exploited the *Test Mode* in some CAN controllers to flood the bus with dominant bits (*BD2*). [36] built a malicious ECU without the CAN controller to allow them to launch the dominant bit-stream attack (*BD3*).

ECU Denial. The attacker attempts to force an ECU to a bus-off state (TEC>255) and eliminates all the functionalities managed by the target ECU. Three types of ECU denial attacks have been proposed in the literature: (1) CAN Controller Abuse (*ED1*). [12] exploited the *ID Ready Interrupt* feature and the *Test Mode* on some ECUs to inject dominant bits when the target ID attempts to transmit. This would trigger transmission bit errors at the target ECU, as it detects the discrepancies between what it sends and what it sees on the bus. Repeating this attack would gradually increase the transmitter's TEC, and eventually force it to bus-off. (2) Malicious Frames (*ED2*). An adversarial ECU may send a frame with identical contents to the targeted ECU's, except replacing a recessive bit with a dominant one, to trigger a bit error at the transmitting ECU [3,12]. (3) Bypassed CAN Controller (*ED3*). Adversaries directly connect malicious/compromised ECUs to CAN transceivers to inject arbitrary bit streams to the bus. [23,36,42] implemented this attack to overwrite recessive bits from the target ECU by dominant ones to trigger bit errors at the transmitting ECU.

Arbitration Denial. The adversarial ECU attempts to inject frames with the lowest possible ID (0×0) to win the arbitration over any other CAN ID and prevent all non-0×0 IDs from sending to the bus, so that the bus becomes completely non-functional [12]. It could also target on a selected ID by transmitting an ID of higher priority whenever that ID starts transmission [22]. We categorize both cases as *AD1*. [36] monitored the bus with an MCU connected through a CAN transceiver (without CAN controller) to detect the target ID and then inject a dominant bit that replaces a recessive bit in the ID field (*AD2*). The target ID loses arbitration and is unaware of the attack, however, the attack results in an incomplete frame that causes a form error. In Sect. 4, we propose a stealthier version so that no error frame is generated (*AD3*).

Table 1. Summary of CAN security control mechanisms: features and effectiveness against attacks discussed in Sect. 2.2.

Control	Features				Effectiveness against attacks									
	Inj.	Aper.	RT	Cost	BD1	BD2	BD3	ED1	ED2	ED3	AD1	AD2	AD3	Spoof
C1	✗	✓	✗	✓	*D*	*D*	*D*	*D*	*D*	*D*	*D*	*D*	*D*	*D*
C2	✗	✓	✗	✗	*D*	–	–	–	*D*	–	*D*	–	–	*D*
C3	✗	✗	✗	✓	*D*	–	–	–	*D*	–	*D*	–	–	*D*
C4	✗	✓	✗	✓	–	–	–	*P*	*P*	*P*	*P*	*P*	*P*	*P*
C5	✗	✓	✓	✓	*P*	–	–	–	*P*	–	*P*	–	–	*P*
C6	✗	✓	✗	✗	*P*	–	–	–	*D*	–	*D*	–	–	*P*
C7	✗	✓	✗	✗	*P*	–	–	–	*D*	–	*P*	–	–	*P*
CANSentry	✓	✓	✓	✓	*P*	*P*	*P*	*P*	*P*	*P*	*P*	*P*	*P*	*P*

Features: Inj.: preventing injection of incomplete frames or random bits, **Aper.**: handling aperiodic attacks, **RT**: real-time defense; **Cost**: low cost.
Effectiveness: D: Detect, **P**: Prevent, **-**: No protection
Controls: C1: Anomaly-based IDS [13,19,37,38,51,52]; **C2**: Voltage-based IDS [4–7,10,35]; **C3**: Time-based IDS [4,15]; **C4**: CAN-ID Obfuscation [17,22,24,29,58,59]; **C5**: Counterattacking [8,28,30,48]; **C6**: Authentication [18,41,44,53,54]; **C7**: Application-level Firewalls [26,45].

Spoofing Attacks. Due to the lack of authentication in CAN and the broadcast nature of the bus, a compromised ECU could easily send CAN frames with any ID, including IDs that belong to other legitimate/critical ECUs. In [27, 32], attackers compromised an ECU through a remote channel, and sent CAN frames to unlock doors, stall the engine, or control the steering wheel. In the *masquerade attack*, the attacker first disables the target ECU and then transmits its frames [40]. In the *conquest attack*, the target ECU is fully compromised to transmit legitimate frames but with malicious intentions [40]. Lastly, [31] sent spoofed diagnostic frames to force a target ECU into diagnostic session, which would stop transmitting frames until the diagnostic session is terminated.

2.3 Existing Controls and Limitations

IDS for CAN. Due to the high predictability of CAN traffic, anomaly detection becomes a viable solution. For example, [13,19,37,38,51,52] monitor CAN traffic and detect frame injection attacks based on the analysis of traffic pattern and packet contents. Meanwhile, due to the lack of sender authentication in CAN, a number of fingerprinting approaches are proposed to identify senders based on physical layer properties. For example, ECUs could be profiled and identified with time/clock features [4,15] or electric/voltage features [5–7,10,25,35]. These approaches associate frames with their legitimate senders. When a mismatch is detected, an attack is assumed. They are effective against the spoofing attacks and some of the denial attacks, as shown in Table 1. The IDS mechanisms do not provide real-time detection except [10], which detects a malicious frame before it completes transmission. In addition, they all assume that packets conform to CAN protocol. Therefore, they cannot recognize many random bit or incomplete frame injection attacks discussed in Sect. 2.2.

CAN-ID Obfuscation. Many attacks discussed in Sect. 2.2 target specific CAN IDs. Several solutions have been proposed to obfuscate CAN IDs to confuse or deter the attackers, e.g., ID-Hopping [22,59], ID Randomization [17,24], ID Obfuscation [29], and ID Shuffling [58]. The effectiveness of these solutions rely on the secrecy of the obfuscation mechanism, i.e., the assumption that the obfuscated CAN IDs are known to the legitimate ECUs but not the attackers. This assumption may not be practical in real world applications.

Counterattacking. The counterattack mechanisms attempt to stop CAN intrusions by attacking the source. In [30], owners of CAN IDs (legitimate ECUs) would detect spoofed frames on the bus, and interrupt the sender by sending an error frame. [8] proposed a similar mechanism that induced collisions with the spoofed frames until the attacking ECU was forced into bus-off. [28] proposed a counterattack technique with a central monitoring node, which shared secret keys with legitimate ECUs for authentication and spoof detection. [48] launched a bus-off attack against the adversarial ECU of the attacks proposed in [3]. The counterattack approaches need to add detection and counterattacking capacity to all ECUs, so that they can protect their own CAN IDs. This is not cost-effective, and may be impractical for some mission-critical ECUs.

Authentication. Cryptography-based solutions have been proposed to enable node (ECU) authentication, and to support data confidentiality and integrity [34,54,55,60]. In addition, controls against attacks from external devices have been introduced [9,16,47,57]. They require encryption, key management, and collaboration between ECUs, which implies significant changes to the existing CAN infrastructure. Additional communication overhead and delays are also expected, which is undesired in real-time environments such as CAN.

Firewalls. The idea of an in-vehicle firewall was first suggested in [56] to enforce authentication and authorization of CAN nodes, but it did not articulate any technical details behind the concept. Application-level firewalls have been introduced in [26,45], which require intensive modification at the CAN node (in both software and hardware) and system-wide cryptographic capabilities to CAN. This is impractical in real world CAN systems. Finally, the attacks performed by the skipper model [36,42] cannot be prevented using such solutions.

In Table 1, we compare different CAN defense approaches with our CANSentry solution in terms of features and capabilities of detecting or preventing the attacks discussed in Sect. 2.2. We also like to note that existing countermeasures do not consider cases where the attacker abuses or bypasses the CAN controller to send arbitrary bit streams or incomplete frames, e.g. [12,23,36]. Meanwhile, controls that transmit frames to the bus, e.g., counterattacks, are also vulnerable to denial attacks that would nullify their effectiveness. Lastly, many existing solutions require overhead such as renovating the protocol, upgrading all or many ECUs, employing statistical learning for IDS, etc., which makes them less cost-efficient and impractical.

3 The Threat Model

Attackers' Objectives. In this paper, we study two attacks, *denial* and *spoofing*. In denial attacks, the attacker aims to nullify certain functionality of an ECU or the CAN bus, such as forcing an ECU into bus-off, stopping a CAN ID from sending to the bus, or occupying the bus. A *stealthy* denial is to achieve the goal without being noticed by the target ECU or the bus. In spoofing, the attacker aims to spoof a selected ID to send frames to the bus and deceive the receiver, which can trigger undesired consequences ranging from displaying incorrect information to disabling the brakes or the steering wheel [27,32,39].

Attacker's Capabilities and Limitations. Attackers with local or remote access have been demonstrated in the literature [2,11,27,32]. We consider two attacker models with different capabilities, based on which, different attacks could be realized that vary in impact and sophistication. In both cases, we assume that the attacker knows the design and specifications of the targeted system (automobile) model through open documentation or reverse engineering. For example, the attacker knows the IDs and functionalities of the ECUs.

1. CAN Abusers. The attacker has a local or remote access to an ECU. The attacker obtains full control of all software components of the ECU but cannot alter the hardware. The attacker is assumed to be able to sniff the bus and inject frames to abuse one or more of the CAN's basic functionalities: arbitration and error handling. Note that the integrity of the CAN controller is preserved, so that all the injected frames conform to the CAN protocol. When an attacker transmits a frame with a high priority ID, the frame wins arbitration over legitimate low priority IDs. On the other hand, the attacker can transmit an almost identical frame with a small difference to trigger error handling and eventually force a target ECU to bus-off state. The attacker may also abuse a feature provided by some CAN controllers, e.g., the "Test Mode", to inject dominant bits [12].

2. The Skippers. An attacker can sniff and inject arbitrary CAN traffic by skipping the CAN controller. This allows her to inject bits at any time without having to be limited by the rules enforced by CAN controllers. This could be achieved through direct or remote access: (1) Direct Access: the attacker could connect a malicious MCU and a CAN transceiver to the OBD-II port (e.g., [36,42]), or connect the MCU through an aftermarket OBD-II adapter (e.g. [11]). This approach requires brief physical access to the vehicle. (2) Remote Access: the attacker may first compromise an ECU (usually one with remote access capabilities such as WiFi or cellular), and then exploit certain hardware design vulnerabilities to directly connect to the bus (e.g. [32]). In particular, [12] showed that 78% of the MCUs deployed in vehicles have built-in CAN controllers, which connect to CAN transceivers through GPIO pins. A compromised MCU could change GPIO configurations to disconnect the CAN controller, and then directly connect itself to the transceiver. In both cases, the attacker "skips" the CAN controller, analyzes the traffic in a bit-by-bit fashion in a process known as "Bit Banging", and injects arbitrary bit-streams to the bus.

We assume that the attacker does not have unlimited and uncontrolled access to alter or to break the integrity of CAN hardware. The attacker could access the bus through any legitimate access point, or through existing hardware/software vulnerabilities, but she cannot alter the hardware to create new vulnerabilities or access points. That is, the attacker may have brief access to the car to connect a malicious ECU to the OBD-II port. However, she cannot disassemble the system to weld a new port or a new attacking device directly to the bus.

4 The Stealthy Selective Arbitration Denial Attack

In this section, we improve the design of the arbitration denial attack presented in [36] with two added features: stealthiness and inexpensive hardware implementation. We present the implementation of the new attack and evaluate it with an Instrument Panel Cluster (IPC) from a used passenger car.

Attack Objectives. The objective of the attack is to stealthily prevent frames with specific IDs from being sent to the bus. The attack is expected to have the following features: (1) Selective: only specific IDs are denied, while other ECUs/frames all function as expected. (2) Stealthy: the attack should conform to CAN standard and should not trigger any error on the bus. Hence, the attacker controls the damage and extends the length of the attack. And (3) Practical: the attack should not require expensive hardware or extended access to the bus.

Adversary Model. The attacker needs to bypass CAN controllers' restrictions to perform bit-by-bit analysis/manipulation on the bus. Therefore, the attacker connects directly to the CAN transceiver, i.e., the *Skipper* model in Sect. 3.

Challenges. First, most of existing CAN tools (e.g., CANoe, VehicleSpy, SocketCAN only work with full CAN frames. However, we need to monitor the bus at bit-level and inject bits at any arbitrary time. Next, the attack requires a high degree of precision to the bit-level timing. Last, since the skipper model operates without a CAN controller, any unexpected operation delay, premature injection, or malformed CAN frame will result in bus errors, which may render the attack unsuccessful or detectable.

The Attack. The proposed attack passively monitors the bus to detect a specific CAN ID in the arbitration phase. The attacker waits for the last recessive bit in the target ID, and overwrites it with a dominant one, to beat the targeted ECU in arbitration. The attacker completes the transmission with a fake frame, so that it would not trigger any error flag (as in [36]). Hence, the malfunctioning on the bus cannot be detected, and the attacking ECU cannot be identified.

Attack Implementation and Evaluation. Most of the current CAN research that require precise timing use automotive-grade micro-controllers or other expensive tools. We use an open-source tool, CANT [1], which facilitates the synchronization with the bus and bit-level analysis/manipulation on the bus. We connect the following to the CAN bus: (1) attacker: an STM32 Nucleo-144 board (connected through OBD-II) running CANT; (2) victim: the IPC of a used 2014 passenger car; (3) other nodes: simulated by BeagleBone Black (BBB) micro-controllers. In the experiments, when we launch the proposed attack against ID $0 \times 9A$ (turn signals), the turn signals become unresponsive regardless of the status of the turn signal lever, since their control messages are blocked.

We compare the stealthiness of the proposed attack with the other denial attacks discussed in Sect. 2.2: ECU denial [12,42] and arbitration denial[12,36]. We assign a BBB microcontroller as the victim ECU and capture its errors (TEC, REC and error state). The ECU denial attack (Fig. 2 (a)) causes a sharp increase of TEC at the victim ECU, to force it into bus-off state. It also increases the REC counters on other ECUs. The arbitration denial attack (Fig. 2 (b)) interrupts the victim ECU and causes a form error, which also drives all ECUs into error passive mode, since incomplete frames are detected on the bus. Finally, our attack (Fig. 2 (c)) is stealthy as it achieves the arbitration denial without causing any error on any ECU.

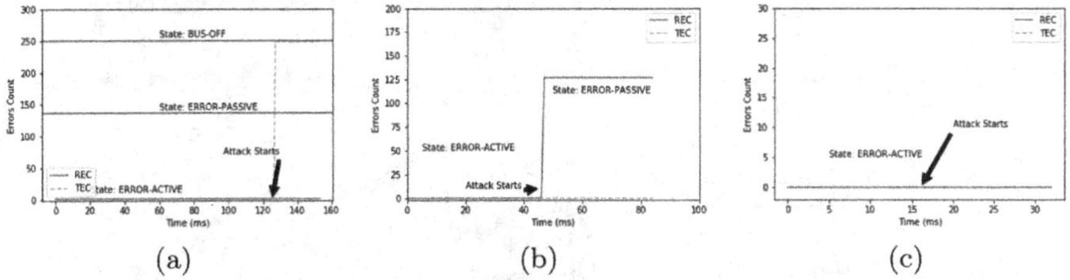

Fig. 2. Comparison of the stealthiness of CAN denial attacks: (a) ECU denial; (b) arbitration denial; (c) stealthy selective arbitration denial.

5 CANSentry: A Firewall for the CAN Bus

Next, we present CANSentry, an efficient and low-cost firewall for the CAN bus to defend against the attacks discussed in Sects. 2.2 and 4.

5.1 The Architecture of the CANSentry Firewall

The objective of CANSentry is to prevent an attacker node (either a CAN abuser or a skipper) from sending malicious frames onto the CAN bus without introducing any practical delay. This requires monitoring and filtering all the messages in real time, since we cannot block any ECU from accessing the bus before an abnormal activity is detected. To address this challenge, we propose a segmentation-based approach to separate high-risk CAN nodes, i.e., a few ECUs with interfaces to the external network, from the rest of the bus, using a firewall.

As shown in Fig. 3(a), CANSentry is deployed between each high-risk node and the main bus, which logically divides the original CAN bus into two segments, the *internal* bus CAN_{int} and the *external* bus CAN_{ext}. A high-risk ECU directly connects to CAN_{ext}, which further connects to CAN_{int} through the CANSentry firewall. Both segments send and receive messages following the original CAN specifications, where the bi-directional firewall functions mainly as a relay to transmit legitimate messages between CAN_{int} and CAN_{ext} without causing any collision. Therefore, CANSentry monitors the current transmission state of the main bus and decides to forward or block the messages from CAN_{ext}.

Introducing a firewall component for network segmentation enforces the necessary security control, which is missing in the current CAN standard, to regulate the activities of the high-risk, potentially vulnerable nodes. This defense mechanism could fundamentally overcome the vulnerabilities caused by the broadcast nature and lack of authentication/authorization capacities of the CAN bus by preventing the attacker node from broadcasting unauthorized frames. Note that CANSentry does not demand any modifications to the CAN standard, or require any changes to the existing ECUs, which makes it easily adoptable.

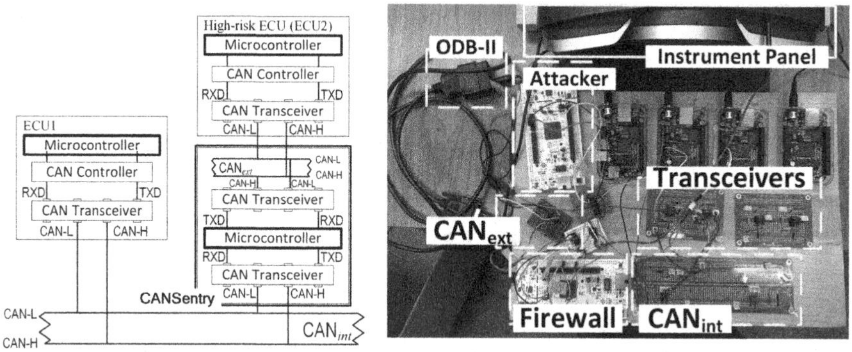

(a) (b)

Fig. 3. The CANSentry approach: (a) firewall architecture; (b) implementation.

5.2 Firewall Principle and Rules

The CANSentry firewall monitors the states of the internal and external buses. Similar to other network firewalls, it checks bi-directional traffic against a set of rules and enforces the forwarding or blocking actions. However, the design of the firewall rules is not straightforward. We have to leverage the features of CAN denial and spoofing attacks to derive appropriate firewall rules for prevention.

States and State Transitions of CAN Bus and Nodes. We examine the relationship between the states of a CAN node and the corresponding states of the CAN bus, as seen by the firewall. As shown in Fig. 4(a), each CAN node transits between four legitimate states. In particular, the RECEIVE state denotes the node is listening to the bus when another node is transmitting, while the IDLE state denotes the node is listening to the bus and waiting for other nodes to transmit. We combine them as the node takes the same action in both states. Moreover, the ARBITRATION state denotes the transmission of CAN ID bits and the TRANSMIT state denotes the transmission of all other data and control bits. Correspondingly, the CAN bus also transits between four legitimate states as shown in Fig. 4(b). For the internal bus, we further define TRANSMIT_{int} and TRANSMIT_{ext} states to distinguish the transmission due to an (directly connected) internal node or an (firewalled) external node, respectively.

From the aspect of bus state transitions, we can interpret the attacks discussed in Sect. 2.2 as exploiting malicious messages to compromise specific bus states. In particular, in the spoofing and arbitration denial attacks, the adversary node (e.g., a malicious abuser or skipper ECU) compromises the bus ARBITRATION state (denoted as "A1" in Fig. 4(b)) by injecting a partial or complete fake frame with a fraudulent CAN ID to falsely win the arbitration. In the ECU denial attacks, the attacker node compromises the bus TRANSMIT state by injecting messages with deliberately crafted data bits to falsely trigger the target node into transmission errors. Finally, the bus denial attack is a special case of compromising either ARBITRATION or TRANSMIT states with a stream of dominant bits to take over the entire bus.

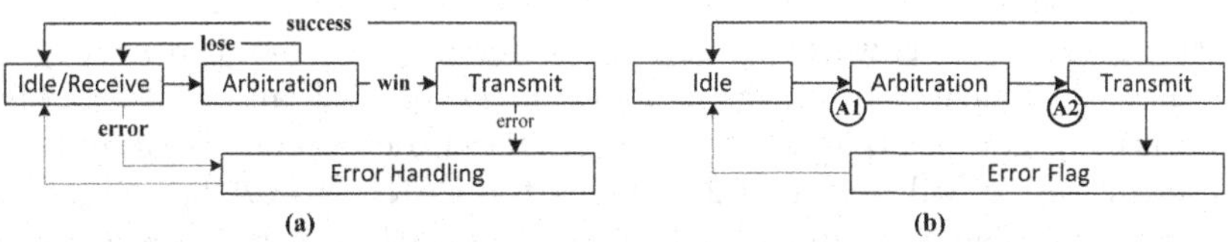

Fig. 4. State transitions of a CAN node and the CAN bus.

Firewall Rules. The fundamental *principle* of the firewall is to ensure that at any time high-risk nodes on the external bus operate in a state consistent with the state of the internal bus. So, for each bus state, we derive the corresponding consistent node states based on the CAN protocol. In particular, (i) for the bus

IDLE state, the consistent node states are IDLE/RECEIVE, ARBITRATION and TRANSMIT; (ii) for the bus ARBITRATION state, the consistent node state are IDLE/RECEIVE and ARBITRATION; and (iii) for the bus TRANSMIT or ERRORFLAG states, the consistent node state is only the IDLE/RECEIVE state. Finally, we derive a set of firewall rules following the above principle. Similar to network firewall enforcement, the rules will be executed in order such that the traffic blocked by an upper rule will not be evaluated by the lower rules.

$\mathbb{R}1$: When the internal bus is in either TRANSMIT_{int} or ERRORFLAG state, the firewall always *forwards* the traffic (bits or frames) from CAN_{int} to CAN_{ext} and *blocks* the traffic from external to internal, regardless of high-risk node's state.

$\mathbb{R}2$: When the internal bus is in either IDLE or ARBITRATION state, the firewall *forwards* all the traffic from CAN_{int} to CAN_{ext}. It also *forwards* traffic that has a CAN ID in the *arbitration whitelist* and conforms to CAN specifications from CAN_{ext} to CAN_{int}. It *blocks* all other traffic from external to internal.

$\mathbb{R}3$: When the internal bus is in the TRANSMIT_{ext}, the firewall *forwards* the traffic from CAN_{ext} to CAN_{int} that conforms CAN specifications, and *blocks* all traffic from CAN_{int} to CAN_{ext}, except for error flags.

Discussions. First, we assume that the traffic on the internal bus conforms to CAN specifications, because all the nodes behind the firewalls are considered low risk and more trustworthy than external nodes. Regulated by CAN controllers, their activities should not deviate from the CAN protocol. Meanwhile, in $\mathbb{R}1$, the firewall blocks all the traffic from CAN_{ext} to CAN_{int} so that it may block legitimate error flags generated by a benign external node. This may mistakenly block all error flags originated in CAN_{ext}, since we cannot distinguish if it comes from a benign or compromised external ECU. However, the impact of this blocking to both external and internal nodes is negligible: it does not affect the error counters of the external node, meantime, the internal node causing the error can still receive enough reliable error flags from internal nodes. Lastly, when the internal bus is in TRANSMIT_{ext} (i.e., an external node won the arbitration and is sending to the bus), no legitimate traffic should be sent by any other node on the internal bus. So, we block the CAN_{int}-to-CAN_{ext} circuit in $\mathbb{R}3$ as a protective measure. It is worth noting that error flags are handled as a special case – they are collected by an error buffer in the firewall, and sent to CAN_{ext} when identified. For example, when an internal node detects any error and sends out an error flag, it will trigger $\mathbb{R}2$ to block the CAN_{ext}-to-CAN_{int} circuit and forward the error flag to the transmission ECU on CAN_{ext}. Error flags will arrive at CAN_{ext} with a 5-bit delay, which would not cause any issue in error handling.

Rule Enforcement. Each firewall monitors both internal and external buses and detects the current internal bus state and external bus state (almost equivalent to the external node state since there is only one node on CAN_{ext}). Since $\mathbb{R}1$ and $\mathbb{R}3$ only involve two firewall actions, *forward* and *block*, the implementation is straightforward. However, $\mathbb{R}2$ requires the firewall to check the CAN ID originated from the external node against a pre-determined whitelist during the arbitration. The key challenge is to detect the malicious bit as soon as possible before a spoofed ID wins the arbitration. Therefore, we construct a *deterministic finite automaton* (DFA) to enforce this bit-by-bit ID matching.

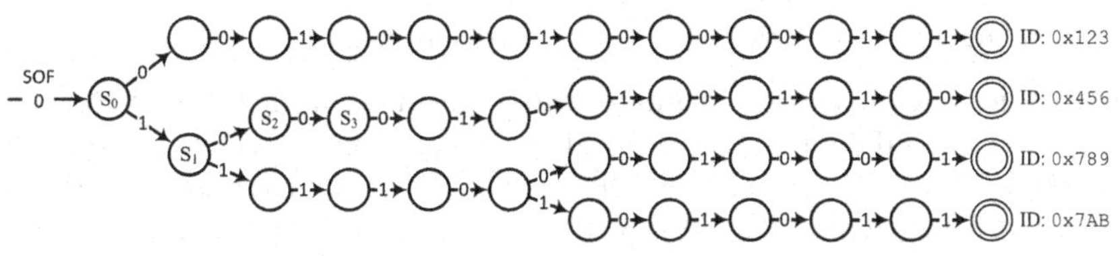

Fig. 5. An example DFA for CAN ID matching.

Formally, a DFA is defined as a 5-tuple $(Q, q_0, \Sigma, \delta, F)$, where Q is a finite set of states, $q_0 \in Q$ denotes the initial state, Σ is a finite set of input symbols, known as the input alphabet, $\delta : Q \times \Sigma \rightarrow Q$ is a transition function, and $F \subseteq Q$ denotes the accept states. To start the arbitration, a node always issues a Start-of-Frame (SOF) for hard synchronization, so the input of q_0 is always 0. $\Sigma = \{0, 1\}$ since the input is a bit stream. Since each firewall is in charge of one high-risk node on the external bus, it maintains a set of CAN IDs to which the external ECU is allowed to send messages. Based on the ID set, we derive the transition function δ and the accepted states $F = \{\text{CAN IDs}\}$. For example, Fig. 5 shows the DFA of a firewall whose whitelist contains four CAN IDs: "0×123", "0×456", "0×789" and "$0 \times 7AB$". For a spoofed ID "0×481", its fifth bit (i.e., "1") will be rejected by the state S_3 of the DFA, then the firewall will block all the remaining bits of this ID.

In practice, we cannot wait until an Arbitration ID reaches an accept state to disseminate it to the bus, since the arbitration phase is expected to be precisely synchronized bit-by-bit among all competing ECUs. To be precise, the proposed CAN firewall implements a Moore machine [33], which is a DFA that generates an output at each state. At each DFA state, an output bit will be disseminated to CAN_{int} instantly so that it competes with other nodes on the bus. Lastly, when a spoofed ID is rejected by the DFA, the prefix bits have already been sent to CAN_{int}. We will further discuss this and its impact to security in Sect. 6.

5.3 Implementation and Evaluation

To enable an efficient bit-by-bit monitoring and manipulation of bit streams transmitted over CAN, CANSentry is constructed with one MCU and two CAN transceivers, where one transceiver interfaces between the firewall and CAN_{int} and the other interfaces between the firewall and CAN_{ext}, as shown in Fig. 3(a). As the main component of the firewall, the MCU has a filter module, which implements firewall rules and the DFA for CAN ID matching, and then enforces the forwarding or blocking decisions for CAN traffic.

Hardware and Costs. We use the open-source tool CANT [1] on an STM32 Nucleo-144 development board to implement the proof-of-concept CANSentry. The MCU board is connected to CAN_{int} and CAN_{ext} through two designated transceivers. The hardware cost is about \$20 to us, which could be significantly lowered in mass production. Meanwhile, we only need to deploy firewalls to external-facing ECUs in a vehicle, which is very limited in number.

Transmission Delay. We evaluate the transmission delay introduced by CANSentry. Figure 6 depicts the received bit streams on the CAN_{int} and CAN_{ext} buses when a frame is relayed by the firewall from internal to external (i.e., Fig. 6(a)) and vice versa (i.e., Fig. 6(b)). In Fig. 6(b), the firewall processing includes DFA-based ID matching. Obviously, in both cases, there is no noticeable delay that could cause any bit error or synchronization error.

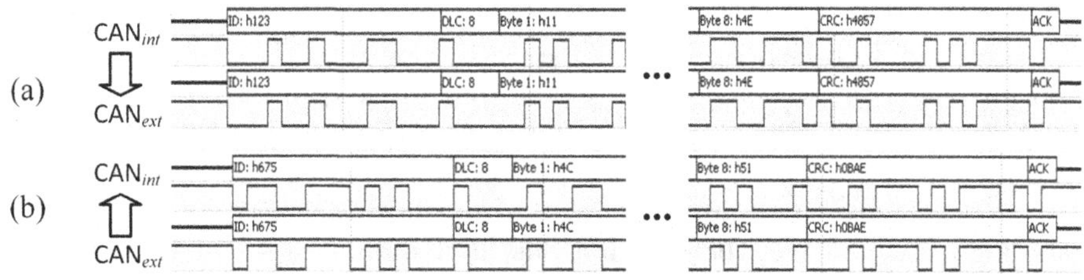

Fig. 6. Traffic between two buses: (a) CAN_{int} to CAN_{ext}; (b) CAN_{ext} to CAN_{int}.

Effectiveness Against Attacks. As shown in Fig. 3(b), CANSentry is connected to two CAN transceivers interfacing CAN_{ext} and CAN_{int}, respectively. We use another STM32 Nucleo-144 board to simulate the attacks, which is connected to CAN_{ext} through ODB-II. We further select an ECU from the Instrument Panel Cluster (IPC) from a used 2014 passenger car as the victim ECU. The IPC is connected to CAN_{int}. We simulate ten attacks discussed in Sect. 2.2 (Table 1) to evaluate the performance of CANSentry. In all the attacks, the attacker attempts to inject illegal bits that violate the CAN protocol (e.g., continuous dominant bits in bus-denial attack and arbitration-denial attack), a spoofed frame (e.g., spoofing attack and bus-denial attack), or a spoofed CAN ID (e.g., selective arbitration-denial attack). To evaluate the effectiveness of CANSentry, we monitor the bit stream on the `Tx` pin of the attacker ECU (i.e., the injected bits from the attacker), and the ones on the internal bus CAN_{int}, which shows the traffic on the main bus after the attacks, as shown in Fig. 7.

In the experiments, CANSentry was able to block all attack attempts from the adversary ECU, while not interfering with the normal operations on the internal bus CAN_{int}. Due to space limit, we only demonstrate the results of three attacks. Figure 7(a) shows a bus-off attack, when a skipper-type attacker injects arbitrary bits (denoted in the block with dashed-line) to trigger a receive bit error at the victim ECU, this attempt is blocked by CANSentry (i.e., no change is observed on the internal bus). Meanwhile, when the attacker injects a dominant bit "0" to win the arbitration and block the transmission attempt of the victim ECU (Fig. 7(b)), or even covers his trace with a complete frame (from a spoofed CAN ID) and a valid CRC (Fig. 7(c)), the malicious actions are prevented by CANSentry and thus have no effect on the internal bus. Finally, it is worth pointing out that although we evaluate CANSentry in the in-vehicle network setting, it can be adopted to secure any CAN-based system.

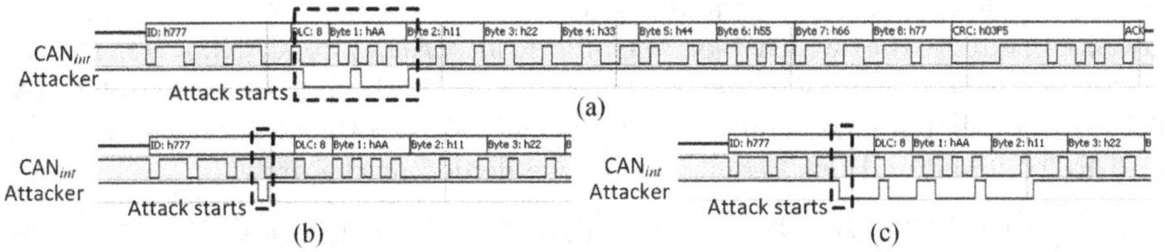

Fig. 7. Evaluation of firewall performance under: (a) bus-off attack; (b) arbitration denial attack; (c) the proposed stealthy selective arbitration denial attack.

6 Security Analysis and Discussions

Now we analyze the security guarantees of CANSentry and its effectiveness against threats introduced in Sect. 3, and discuss the remaining attack surfaces.

Security Analysis of Arbitration. A Deterministic Finite-state Automaton (DFA) is used to filter arbitration IDs (in binary strings).

Theorem 1. *Let $L(M)$ be the list of arbitration IDs on the whitelist, which is used to generate DFA M. When M is correctly generated, no CAN frames carrying an ID field that is not in $L(M)$ shall be disseminated to the bus.*

Theorem 1 roots on automata theory – a binary string I is accepted by M if and only if $I \in L(M)$. The firewall allows the transmission of the rest of the CAN frame only when I is accepted by M, and I wins arbitration. Theorem 1 ensures that CAN spoofing attacks are effectively blocked: when a malicious/compromised ECU tries to send CAN frames with IDs not in the whitelist, the attempt is rejected in the arbitration phase.

Theorem 2. *Let I_m be the lowest arbitration ID accepted by DFA M, any (partial) ID output from M cannot win arbitration against a target ID I_t that has higher priority than I_m (i.e., $I_t < I_m$).*

During the arbitration phase, the DFA generates an intermediate output bit at each state and transmits the output to CAN$_{int}$. It is possible that the prefix of a spoofed ID ($I_s \notin L(M)$) being sent to CAN$_{int}$. For instance, if the firewall only allows 0b1010110011, the ID 0b1010001011 will be denied at bit 5. However, its prefix 0b1010 will be sent to CAN_{int}. Theorem 2 shows that adversaries cannot exploit this mechanism to launch arbitration denial attacks against high-priority IDs. Please refer to Appendix A for the proof of this theorem.

Security Analysis of Data Frame Transmission. The external node is authorized to transmit its frame to the bus only when it wins arbitration on CAN_{int}. The firewall enforces **R3** to forward traffic from CAN_{ext} to CAN_{int}. For the simplicity of the design and the portability of the firewalls, we do not further audit the validity of the data frame. When CAN_{ext} loses arbitration, the firewall moves to enforce **R1**, which blocks traffic from CAN_{ext} to CAN_{int}. Therefore, an adversary node cannot inject dominate bit(s) to the bus to interrupt CAN frame from other nodes, which may eventually drive other nodes into bus-off. Such ECU denial and bus denial attacks are denied at the firewall.

Security of CANSentry. The physical and software security of CANSentry relies on the following factors: (1) CANSentry nodes are deployed in a physically isolated environment, i.e., inside the car. It is difficult for an adversary to bypass/alter it unless he has extended time to physically modify the in-vehicle components. (2) CANSentry only has two network interfaces: CAN_{ext} and CAN_{int}. The limited communication channel and the simplicity of CAN makes it impractical to compromise the operations of the firewall from CAN_{ext}. And (3) the simplicity of the firewall makes it unlikely to have significant software faults.

Updates. We do not consider *remote* updates to CANSentry in this paper, to keep the simplicity of CANSentry and to avoid adding a new attack vector. In case a CANSentry-protected ECU is *remotely* updated to send CAN packets with new IDs, such packets will be blocked. In practice, we are not aware of any existing remote update that adds new CAN IDs to the high-risk (e.g., remotely accessible) ECUs. Meanwhile, when the vehicle is updated in the shop, the corresponding CANSentry could be easily updated to add new CAN IDs to the whitelist.

Remaining Attack Surfaces. An adversary node may send to CAN_{int} using IDs on the whitelist, which may potentially block frames from higher CAN IDs.If the adversary node keeps sending at high frequency, it becomes arbitration denial attacks to nodes with *lower priority*. CANSentry cannot detect or block this attack. Fortunately, this may not be a severe problem, since: (1) the attack paths from known attacks (e.g. [27,32,39] are all attacking from low priority nodes (i.e., components with wireless interfaces and externally-facing vulnerabilities) to high priority nodes. Attacks from the other direction appear to be meaningless. (2) The *high-risk nodes*, i.e., nodes that have shown to be compromisable by remote adversaries, mostly belong to low priority nodes (entertainment units, navigation, etc), which are only allowed to send CAN frames with high arbitration IDs. (3) Arbitration denial attacks from such nodes are detectable with a simple IDS, which runs on the same chip as the firewall, monitors traffic from CAN_{int}, and informs the firewall to take actions when attacks are detected.

When a spoofing attempt is blocked by CANSentry, a partial frame, which only contains the first few bits of the arbitration ID, will be observed on the bus. An adversary node could then intentionally inject malformed frames to CAN_{int}. Other receiving nodes on the bus will detect *form errors* and raise error flags. This will increase RECs at the receiving nodes, and may drive them into error passive mode, in the worst case. This is not a severe problem since: (1) To force a node into error passive mode, REC>127 is needed. This requires continuous attack attempts from a adversary node (assuming that the node manages to avoid increasing its own TEC). (2) Such persistent attacks could be easily detected by a simple counter or IDS embedded in CANSentry. And (3) nodes in error passive mode still perform their functions (with a performance penalty), and they can recover from error passive mode when frames are correctly received.

In the attack model, internal ECUs are considered low risk. While we are unaware of any attack that compromises an internal ECU with only brief physical access, we cannot completely rule out this possibility. Lastly, a small number of units in modern automobiles, such as GM's OnStar, are both remotely compromisable and capable of sending high priority frames, e.g., stopping the vehicle by shutting down fuel injection. When such systems are compromised, they may take over control of the cars even with the existence of CANSentry. From the CAN bus perspective, they are fully authorized to send such frames. Defending against such attacks is outside of the scope of CANSentry.

7 Conclusion

In this paper, we first summarize existing DoS and spoofing attacks on the CAN bus, then implement and evaluate a stealthy selective arbitration denial attack. We present a novel CAN firewall, CANSentry, to be deployed between each high-risk ECU (external-facing ECUs with remote attack vectors) and the internal bus. It defends against attacks that violate CAN standards or abuse the error-handling mechanism to achieve malicious goals. To the best of our knowledge, CANSentry is the first mechanism to not only detect but also prevent denial attacks at data link layer that have not been addressed yet, in addition to preventing the traditional spoofing and denial attacks. Despite the simple yet powerful design, CANSentry effectively prevents the attacks with no noticeable overhead at a low cost with no need to modify CAN standard nor existing ECUs. With hardware implementation and evaluation, we demonstrate the effectiveness of the firewall against the bus/ECU denial attacks and ECU spoofing attacks.

Acknowledgements. We would like to thank the anonymous reviewers for their constructive comments. Fengjun Li and Bo Luo were sponsored in part by NSF CNS-1422206, DGE-1565570, NSA Science of Security Initiative H98230-18-D-0009, and the Ripple University Blockchain Research Initiative. Jingqiang Lin was partially supported by National Natural Science Foundation of China (No. 61772518) and Cyber Security Program of National Key RD Plan of China (2017YFB0802100).

A Proof of Theorem 2

Theorem 2. *Let I_m be the lowest arbitration ID accepted by DFA M, any (partial) ID output from M cannot win arbitration against a target ID I_t that has higher priority than I_m (i.e., $I_t < I_m$).*

Proof. Assume that the adversary attempts to spoof arbitration ID I_s ($I_s < I_t$) to win against I_t. Let $Bit(S, i)$ be the i-th bit of string S. If the longest common prefix of I_m, I_s, and I_t is an i-bit string P, then the first i bits of I_s would be accepted in M (since $Prefix(I_s, i) = Prefix(I_m, i)$) and sent to CAN accordingly. I_t and I_s would tie in the first i bits of arbitration (since $Prefix(I_s, i) = Prefix(I_t, i)$). At bit $i + 1$, we have the following three conditions:

- If $Bit(I_s, i + 1) < Bit(I_m, i + 1)$, I_s will be rejected by M, since: (1) I_s cannot be accepted by the DFA branch that contains I_m, since $Bit(I_s, i+1) \neq Bit(I_m, i+1)$; (2) if there exist another DFA branch (with ID I_n) that accepts $Bit(I_s, i+1)$, then we have $I_n < I_m$ (since they are identical in the first i bits and $I_n < I_m$ at bit $i+1$). This violates our assumption that I_m be the lowest arbitration ID accepted by M. Therefore, such I_n and the corresponding DFA branch does not exist. I_s will be rejected by M, and I_t wins arbitration against I_s.
- If $Bit(I_s, i+1) > Bit(I_m, i+1)$, then we have $I_s > I_m$, since they are identical in the first i bits and $I_s > I_m$ at bit $i + 1$. This violates our assumption that $I_s < I_t < I_m$.
- If $Bit(I_s, i + 1) = Bit(I_m, i + 1)$, then $Bit(I_s, i + 1)$ will be sent to the bus. Meanwhile, we need $Bit(I_s, i+1) \neq Bit(I_t, i+1)$, otherwise $Prefix(I_s, i+1)$ is a longer common prefix than P. In case $Bit(I_s, i + 1) > Bit(I_t, i + 1)$, I_s would lose arbitration against I_t. In case $Bit(I_s, i + 1) < Bit(I_t, i + 1)$, then we have $I_m = I_s < I_t$. This violates our assumption that $I_t < I_m$.

In summary, with the existence of M, any (partial) output generated by $I_s < I_m$ cannot win arbitration against a higher priority ID $I_t < I_m$. □

References

1. Grimm co. cant. https://github.com/bitbane/CANT
2. Checkoway, S., et al.: Comprehensive experimental analyses of automotive attack surfaces. In: USENIX Security Symposium (2011)
3. Cho, K.-T., Shin, K.G.: Error handling of in-vehicle networks makes them vulnerable. In: ACM CCS, pp. 1044–1055. ACM (2016)
4. Cho, K.-T., Shin, K.G.: Fingerprinting electronic control units for vehicle intrusion detection. In: USENIX Security Symposium (2016)
5. Cho, K.-T., Shin, K.G.: Viden: attacker identification on in-vehicle networks. In: ACM CCS (2017)
6. Choi, W., Jo, H.J., Woo, S., Chun, J.Y., Park, J., Lee, D.H.: Identifying ECUS using inimitable characteristics of signals in controller area networks. IEEE Trans. Veh. Tech. **67**(6), 4757–4770 (2018)

7. Choi, W., Joo, K., Jo, H.J., Park, M.C., Lee, D.H.: Voltageids: low-level communication characteristics for automotive intrusion detection system. IEEE TIFS **13**(8), 2114–2129 (2018)
8. Dagan, T., Wool, A.: Parrot, a software-only anti-spoofing defense system for the can bus. ESCAR EUROPE (2016)
9. Dardanelli, A., et al.: A security layer for smartphone-to-vehicle communication over bluetooth. IEEE Embed. Syst. Lett. **5**(3), 34–37 (2013)
10. Foruhandeh, M., Man, Y., Gerdes, R., Li, M., Chantem, T.: Simple: single-frame based physical layer identification for intrusion detection and prevention on in-vehicle networks. In: ACSAC, pp. 229–244 (2019)
11. Foster, I., Prudhomme, A., Koscher, K., Savage, S.: A story of telematic failures. In: USENIX WOOT, Fast and Vulnerable (2015)
12. Fröschle, S., Stühring, A.: Analyzing the capabilities of the CAN attacker. In: Foley, S.N., Gollmann, D., Snekkenes, E. (eds.) ESORICS 2017. LNCS, vol. 10492, pp. 464–482. Springer, Cham (2017). https://doi.org/10.1007/978-3-319-66402-6_27
13. Gmiden, M., Gmiden, M.H., Trabelsi, H.: An intrusion detection method for securing in-vehicle can bus. In: IEEE STA (2016)
14. Gupta, R.A., Chow, M.-Y.: Networked control system: overview and research trends. IEEE Trans. Ind. Electron. **57**(7), 2527–2535 (2010)
15. Halder, S., Conti, M., Das, S.K.: COIDS: a clock offset based intrusion detection system for controller area networks. In: ICDCN (2020)
16. Han, K., Potluri, S.D., Shin, K.G.: On authentication in a connected vehicle: secure integration of mobile devices with vehicular networks. In: ACM/IEEE ICCPS, pp. 160–169 (2013)
17. Han, K., Weimerskirch, A., Shin, K.G.: A practical solution to achieve real-time performance in the automotive network by randomizing frame identifier. In: Proceedings of Europe Embedded Security Cars (ESCAR), pp. 13–29 (2015)
18. Hartkopp, O., Schilling, R.M.: Message authenticated can. In: Escar Conference, Berlin, Germany (2012)
19. Hoppe, T., Kiltz, S., Dittmann, J.: Security threats to automotive CAN networks – practical examples and selected short-term countermeasures. In: Harrison, M.D., Sujan, M.-A. (eds.) SAFECOMP 2008. LNCS, vol. 5219, pp. 235–248. Springer, Heidelberg (2008). https://doi.org/10.1007/978-3-540-87698-4_21
20. Humayed, A., Lin, J., Li, F., Luo, B.: Cyber-physical systems security a survey. IEEE IoT J. **4**(6), 1802–1831 (2017)
21. Humayed, A., Luo, B.: Cyber-physical security for smart cars: taxonomy of vulnerabilities, threats, and attacks. In: ACM/IEEE ICCPS (2015)
22. Humayed, A., Luo, B.: Using ID-hopping to defend against targeted DOS on CAN. In: SCAV Workshop (2017)
23. Iehira, K., Inoue, H., Ishida, K.: Spoofing attack using bus-off attacks against a specific ECU of the can bus. In: IEEE CCNC (2018)
24. Karray, K., Danger, J.-L., Guilley, S., Elaabid, M.A.: Identifier randomization: an efficient protection against CAN-bus attacks. In: Koç, Ç.K. (ed.) Cyber-Physical Systems Security, pp. 219–254. Springer, Cham (2018). https://doi.org/10.1007/978-3-319-98935-8_11
25. Kneib, M., Huth, C.: Scission: signal characteristic-based sender identification and intrusion detection in automotive networks. In: ACM CCS (2018)
26. Kornaros, G., Tomoutzoglou, O., Coppola, M.: Hardware-assisted security in electronic control units: secure automotive communications by utilizing one-time-programmable network on chip and firewalls. IEEE Micro **38**(5), 63–74 (2018)

27. Koscher, K., et al.: Experimental security analysis of a modern automobile. In: IEEE S&P (2010)
28. Kurachi, R., Matsubara, Y., Takada, H., Adachi, N., Miyashita, Y., Horihata, S.: Cacan-centralized authentication system in can (controller area network). In: International Conference on ESCAR (2014)
29. Lukasiewycz, M., Mundhenk, P., Steinhorst, S.: Security-aware obfuscated priority assignment for automotive can platforms. ACM Trans. Des. Autom. Electron. Syst. (TODAES) **21**(2), 32 (2016)
30. Matsumoto, T., Hata, M., Tanabe, M., Yoshioka, K., Oishi, K.: A method of preventing unauthorized data transmission in controller area network. In: IEEE VTC (2012)
31. Miller, C., Valasek, C.: Adventures in automotive networks and control units. Def Con **21**, 260–264 (2013)
32. Miller, C., Valasek, C.: Remote exploitation of an unaltered passenger vehicle. Black Hat USA **2015**, 91 (2015)
33. Moore, E.F.: Gedanken-experiments on sequential machines. Automata Stud. **34**, 129–153 (1956)
34. Mundhenk, P., et al.: Security in automotive networks: lightweight authentication and authorization. ACM TODAES **22**(2), 1–27 (2017)
35. Murvay, P.-S., Groza, B.: Source identification using signal characteristics in controller area networks. IEEE Signal Process. Lett. **21**(4), 395–399 (2014)
36. Murvay, P.-S., Groza, B.: Dos attacks on controller area networks by fault injections from the software layer. In: ARES. ACM (2017)
37. Müter, M., Asaj, N.: Entropy-based anomaly detection for in-vehicle networks. In: IEEE Intelligent Vehicles Symposium (2011)
38. Narayanan, S.N., Mittal, S., Joshi, A.: Obd_securealert: an anomaly detection system for vehicles. In: IEEE SMARTCOMP (2016)
39. Nie, S., Liu, L., Yuefeng, D.: Free-fall: hacking tesla from wireless to can bus. Brief. Black Hat USA **25**, 1–16 (2017)
40. Nowdehi, N., Aoudi, W., Almgren, M., Olovsson, T.: CASAD: can-aware stealthy-attack detection for in-vehicle networks. arXiv:1909.08407 (2019)
41. Nürnberger, S., Rossow, C.: – vatiCAN – vetted, authenticated CAN bus. In: Gierlichs, B., Poschmann, A.Y. (eds.) CHES 2016. LNCS, vol. 9813, pp. 106–124. Springer, Heidelberg (2016). https://doi.org/10.1007/978-3-662-53140-2_6
42. Palanca, A., Evenchick, E., Maggi, F., Zanero, S.: A stealth, selective, link-layer denial-of-service attack against automotive networks. In: Polychronakis, M., Meier, M. (eds.) DIMVA 2017. LNCS, vol. 10327, pp. 185–206. Springer, Cham (2017). https://doi.org/10.1007/978-3-319-60876-1_9
43. Petit, J., Shladover, S.E.: Potential cyberattacks on automated vehicles. IEEE Trans. Intell. Transp. Syst. **16**(2), 546–556 (2015)
44. Radu, A.-I., Garcia, F.D.: LeiA: a lightweight authentication protocol for CAN. In: Askoxylakis, I., Ioannidis, S., Katsikas, S., Meadows, C. (eds.) ESORICS 2016. LNCS, vol. 9879, pp. 283–300. Springer, Cham (2016). https://doi.org/10.1007/978-3-319-45741-3_15
45. Rizvi, S., Willet, J., Perino, D., Marasco, S., Condo, C.: A threat to vehicular cyber security and the urgency for correction. Procedia Comput. Sci. **114**, 100–105 (2017)
46. Rouf, I., et al.: Security and privacy vulnerabilities of in-car wireless networks: a tire pressure monitoring system case study. In: USENIX Security Symposium (2010)
47. Sagstetter, F., et al.: Security challenges in automotive hardware/software architecture design. In: DATE. IEEE (2013)

48. Souma, D., Mori, A., Yamamoto, H., Hata, Y.: Counter attacks for bus-off attacks. In: Gallina, B., Skavhaug, A., Schoitsch, E., Bitsch, F. (eds.) SAFECOMP 2018. LNCS, vol. 11094, pp. 319–330. Springer, Cham (2018). https://doi.org/10.1007/978-3-319-99229-7_27
49. Studnia, I., Nicomette, V., Alata, E., Deswarte, Y., Kaaniche, M., Laarouchi, Y.: Survey on security threats and protection mechanisms in embedded automotive networks. In: IEEE/IFIP DSN (2013)
50. Taylor, A., Leblanc, S., Japkowicz, N.: Anomaly detection in automobile control network data with long short-term memory networks. In: IEEE DSAA (2016)
51. Theissler, A.: Detecting known and unknown faults in automotive systems using ensemble-based anomaly detection. Knowl.-Based Syst. **123**, 163–173 (2017)
52. Tian, D., et al.: an intrusion detection system based on machine learning for CAN-Bus. In: Chen, Y., Duong, T.Q. (eds.) INISCOM 2017. LNICST, vol. 221, pp. 285–294. Springer, Cham (2018). https://doi.org/10.1007/978-3-319-74176-5_25
53. Van Bulck, J., Mühlberg, J.T., Piessens, F.: Vulcan: efficient component authentication and software isolation for automotive control networks. In: ACSAC, pp. 225–237 (2017)
54. Van Herrewege, A., Singelee, D., Verbauwhede, I.: Canauth-a simple, backward compatible broadcast authentication protocol for can bus. In: ECRYPT Workshop on Lightweight Cryptography, vol. 2011 (2011)
55. Wang, Q., Sawhney, S.: Vecure: a practical security framework to protect the can bus of vehicles. In: IEEE International Conference on IOT (2014)
56. Wolf, M., Weimerskirch, A., Paar, C.: Security in automotive bus systems. In: Workshop on Embedded Security in Cars (2004)
57. Woo, S., Jo, H.J., Lee, D.H.: A practical wireless attack on the connected car and security protocol for in-vehicle can. IEEE Trans. Intell. Transp. Syst. **16**(2), 993–1006 (2014)
58. Woo, S., Moon, D., Youn, T.-Y., Lee, Y., Kim, Y.: Can ID shuffling technique (CIST): moving target defense strategy for protecting in-vehicle can. IEEE Access **7**, 15521–15536 (2019)
59. Wu, W., et al.: IDH-CAN: a hardware-based ID hopping can mechanism with enhanced security for automotive real-time applications. IEEE Access **6**, 54607–54623 (2018)
60. Ziermann, T., Wildermann, S., Teich, J.: Can+: a new backward-compatible controller area network (can) protocol with up to 16× higher data rates. In: DATE. IEEE (2009)

Distributed Detection of APTs: Consensus vs. Clustering

Juan E. Rubio, Cristina Alcaraz(✉), Ruben Rios, Rodrigo Roman, and Javier Lopez

Computer Science Department, University of Malaga,
Campus de Teatinos s/n, 29071 Malaga, Spain
{rubio,alcaraz,ruben,jlm}lcc.uma.es

Abstract. Advanced persistent threats (APTs) demand for sophisticated traceability solutions capable of providing deep insight into the movements of the attacker through the victim's network at all times. However, traditional intrusion detection systems (IDSs) cannot attain this level of sophistication and more advanced solutions are necessary to cope with these threats. A promising approach in this regard is Opinion Dynamics, which has proven to work effectively both theoretically and in realistic scenarios. On this basis, we revisit this consensus-based approach in an attempt to generalize a detection framework for the traceability of APTs under a realistic attacker model. Once the framework is defined, we use it to develop a distributed detection technique based on clustering, which contrasts with the consensus technique applied by Opinion Dynamics and interestingly returns comparable results.

Keywords: Clustering · Consensus · Opinion dynamics · Distributed detection · Traceability · Advanced persistent threat

1 Introduction

In recent years, there has been a growing interest for advanced event management systems in the industrial cyber-security community for two main reasons: (i) the integration of cutting-edge technologies (e.g., Big Data, Internet of Things) into traditionally isolated environments, which adds complexity to data collection and processing [1]; and (ii) the emergence of the new attack vectors as a result of the Industry 4.0 evolution, which have not been properly studied in context and may form part of an Advanced Persistent Threat (APT) [2].

APTs consist of sophisticated attacks perpetrated by resourceful adversaries which cost millions every year to diverse industrial sectors [3]. The main concern with these threats is that they are especially difficult to detect and trace. In this context, traditional Intrusion Detection Systems (IDS) only pose a first line of defense in an attempt to identify anomalous behaviours in very precise points of the infrastructure [4], and they are tailored to specific types of communication

L. Chen et al. (Eds.): ESORICS 2020, LNCS 12308, pp. 174–192, 2020.
https://doi.org/10.1007/978-3-030-58951-6_9

standards or types of data, which is not sufficient to track the wide range of attack vectors that might be used by an APT.

It is then necessary to fill this gap between classic security mechanisms and APTs. The premise is to find proper mechanisms capable of monitoring all the devices (whether physical or logical) that are interconnected within the organization, retrieve data about the production chain at all levels (e.g., alarms, network logs, raw traffic) and correlate events to trace the attack stages throughout its entire life-cycle. These measures would provide the ability to holistically detect and anticipate attacks as well as failures in a timely and autonomous way, so as to deter the attack propagation and minimize its impact.

To cope with this cyber-security scenario, novel candidate solutions such as the Opinion Dynamics approach emerge [5]. These alternatives propose to apply advanced correlation algorithms that analyze an industrial network from a holistic point of view, leveraging data mining and machine learning mechanisms in a distributed fashion. In this paper, we formalize a framework that enables the design and practical integration of such distributed mechanisms for the traceability of APTs, while also comparing the features of the aforementioned solutions according to the cyber-security needs of the industry nowadays, both qualitatively and experimentally. Altogether, we can summarize our contributions as:

- Characterization of the context in terms of security requirements and available solutions;
- Definition of a framework for developing solutions that enable the distributed correlation of APT events, based on these security needs and a new attacker model;
- Identification of effective techniques and algorithms for the traceability of APTs that satisfy the proposed framework;
- Qualitative and quantitative comparison of approaches in an Industry 4.0 scenario.

The remainder of the paper is organized as follows. Section 2 presents the state of the art of intrusion detection and anomaly correlation mechanisms, as well as the preliminary concepts involved in the studio. Then, Sect. 3 presents the security and detection requirements, whereas Sect. 4 defines the framework for developing solutions that fulfill them. Based on such framework, the studied solutions are addressed in Sect. 5, and experimentally analysed in Sect. 6. Finally, extracted conclusions are discussed in Sect. 7.

2 Background and Preliminaries

At present, there is a plethora of intrusion detection approaches tailored for traditional industrial scenarios (cf. [6]) and Industry 4.0 networks (cf. [7]). This includes specification-based IDSs, which compare the current state of the network with a model that describes its legitimate behaviour [8]; physics-based modeling systems, which simulates the effect of commands over the physical dynamics of the operations [9]; and other more traditional strategies such as signature and anomaly detection systems. Most of these detection approaches focus on the analysis of certain aspects of industrial control systems, such as the communication patterns, the behaviour of sensors and actuators, and others.

Still, industrial technologies are becoming more heterogeneous and attacks are extremely localized, which makes crucial to monitor all elements and evidences. Therefore, it is important for industrial ecosystems to set up more than one detection solution to ensure the maximum detection coverage [10]. Moreover, all solutions should coexist with advanced detection platforms that take the infrastructure from a holistic perspective, correlate all events and track all threats throughout their entire life-cycle [11]. This holistic perspective is even more necessary in light of the existence of APTs: sophisticated attacks comprised of several complex phases – from network infiltration and propagation to exfiltration and/or service disruption [3,12].

These advanced detection platforms have been explored in traditional Information Technology (IT) environments through forensic investigation solutions, using proactive (that analyse evidences as incidents occur) or reactive techniques (where evidences are processed once the events occur). Examples of these include flow-based analysis of traffic in real time [13] or the correlation of multiple IDS outputs to highlight and predict the movements of APTs, using information flow tracking [14] or machine learning [15]. Still, most of them are limited to a restricted set of attacks and are not applied to a real setup.

In turn, the progress in the Industry 4.0 has not been significant with respect to actual APT traceability solutions. In this sense, the Opinion Dynamics approach [5] paves the way for a new generation of solutions based on the deployment of distributed detection agents across the network. The anomalies reported by these agents are correlated to extract conclusions about the sequence of actions performed by the adversary, and also to identify the more affected areas of the infrastructure. Such assessment can be conducted in a centralized entity or using a distributed architecture of peers [16]. At the same time, it is open to integrate external IDS to examine anomalies in the vicinity of nodes, as well as the abstraction of diverse parameters such as the criticality of resources or the persistence of attacks.

Despite the many capabilities of this solution (explained in Sect. 5.1), it is necessary to define a more general detection model to lay the base for the precise application of more APT traceability solutions in the Industry 4.0 paradigm. The reason is that the Opinion Dynamics capabilities can be implemented modularly, they can be integrated into other correlation algorithms and each one has a different effect on many security, detection, deployment and efficiency constraints. These points will be addressed in the next section, where we define the security and detection requirements involved, to latter present the traceability framework.

3 Security and Detection Requirements

Based on the state of the art presented in Sect. 2, this section enumerates the requirements for the development of advanced solutions and systems that provide a holistic perspective on industrial ecosystems. According to [7], we should consider the following detection requirements:

(D1) **Coverage.** APTs make use of an extensive set of attack vectors that jeopardize organizations at all levels. Therefore, the system must be able to assimilate traffic and data from heterogeneous devices and sections of the network, while also incorporating the input of external detection systems.
(D2) **Holism.** In order to identify anomalous behaviors, the system must be able to process all the interactions between users, processes and outputs generated, as well as logs. This allows to generate anomaly and traceability reports at multiple levels (e.g., per application, device or portion of the network, as well as global health indicators).
(D3) **Intelligence.** Beyond merely detecting anomalous events within the network in a timely manner, the system must infer knowledge by correlating current events with past stages and anticipate future movements of the attacker. Similarly, it should provide mechanisms to integrate information from external sources – that is, cyber threat intelligence [17].
(D4) **Symbiosis.** The system should have the capability to offer its detection feedback to other Industry 4.0 services, by means of well-defined interfaces. This includes access control mechanisms (to adapt the authorization policies depending on the security state of the resources) or virtualization services (that permit to simulate response techniques under different scenarios without interfering the real setup), among others.

On the other hand, we can also establish the following security requirements with regards to the deployment of the detection solution over the network:

(S1) **Distributed data recollection.** It is necessary to find distributed mechanisms – such as local agents collaborating in a peer-to-peer fashion – that allow the collection and analysis of information as close as possible to field devices. The ultimate aim is to make the detection system completely autonomous and resistant to targeted attacks.
(S2) **Immutability.** The devised solution must be resistant to modifications of the detection data at all levels, including the reliability and veracity of data exchanged between agents (e.g., through trust levels that weigh the received security information), and the storage of such data (e.g., through unalterable storage mediums and data replication mechanisms such as immutable databases or distributed ledgers).
(S3) **Data confidentiality.** Apart from the protection against data modification, it is mandatory that the system provides authorization and cryptographic mechanisms to control the access to the information generated by the detection platform and all the interactions monitored.
(S4) **Survivability.** Not only the system must properly function even with the presence of accidental or deliberate faults in the industrial infrastructure, but also the system itself cannot be used as a point of attack. To achieve this, the detection mechanisms must be deployed in a separated network that can only retrieve information from the industrial infrastructure.

(S5) **Real-time performance.** The system must not introduce operational delays on the industrial infrastructure, and its algorithms should not impose a high complexity to ensure the generation of real-time detection information. Network segmentation procedures and separate computation nodes (e.g., Fog/Edge Computing nodes) can be used for this purpose.

4 APT Traceability Framework for the Industry 4.0

After defining the detection and security requirements that a conceptual APT traceability solution must fulfill, we now describe the guidelines for the design and construction of its deployment architecture, the algorithms to be used, and the attacker model under consideration.

4.1 Network Architecture and Information Acquisition

The industrial network topology is modelled with an acyclic graph $G(V, E)$, where V represents the devices and E is the set of communication links between them. This way, V can be assigned with parameters to represent, for instance, their criticality, vulnerability level or the degree of infection; whereas the elements in E can be associated with Quality of Service (QoS) parameters (e.g., bandwidth, delays), or compromise states that help to prioritize certain paths when running resilient routing algorithms.

For the interest of theoretical analysis, these networks are frequently generated using random distributions that model the architecture of real industrial systems. Also, the topology can be subdivided into multiple network segments with different distributions [5]. This is useful, for example, to study the effects of the attack and detection mechanisms over the corporate section (containing IT elements) and the operational section (OT, containing pure industrial assets), which can be connected by firewalls, so that $V = V_{IT} \cup V_{OT} \cup V_{FW}$.

Regardless of the topology configuration, the detection approach must acquire information from the whole set of nodes V to fulfill requirement D1 (Coverage, c.f. Sect. 3), by using agents that are in charge of monitoring such devices, complying with S1. Each of these agents can be either mapped to a individual device (following a 1:1 relationship), which would be ideal for S1, or aggregate the data from a set of physical devices belonging to the infrastructure. In either case, we can assume they are able to retrieve as much data as possible from their assigned devices, which encompasses **network-related parameters** (e.g., links with other nodes, number packets exchanged, delays), **host-based data** (e.g., storage, computational usage) or **communication information** (e.g., low-level commands issued by supervision protocols). These data items are aimed to feed the correlation algorithm with inputs in the form of an anomaly value for every device audited, which is formalized by vector x. This way, x_i represents the anomaly value sensed by the corresponding agent on device i, for all $i \in 1, 2, ..., |V|$. Such value can be calculated by each agent autonomously (e.g., applying some machine learning to determine deviations in every data item analyzed) or leveraging an external IDS that is configured to retrieve the raw data as input, thereby conforming to requirement D3.

From a deployment perspective, this leads to the question of where to locate the computation of anomalies and their subsequent correlation. As for the former, agents are implemented logically, since it is not always feasible to physically integrate monitoring devices into the industrial assets due to computational limitations. Consequently, these processes may have to run in separate computational nodes. However, we still want to achieve a close connection to field devices while avoiding a centralized implementation such as the one presented in [5]. The solution is then to introduce an intermediate approach based on the concept of *distributed data brokers*. These components collect the data from a set of individual devices via port-mirroring or network tapping, using data diodes to decouple agents from actual systems and ensure that data transmission is restricted to one direction, thereby shielding the industrial assets from outside access and complying with requirements S3 and S4.

These data brokers can also convey the detection reports (i.e., the anomalies sensed by its logical agents over the area where it is deployed) to other brokers in order to execute the correlation in a collaborative way. In consequence, they must be strategically deployed in a separate network such that there is at least one path between every two brokers.

Due to this distributed nature, the correlation algorithm can make use of two data models: *Replicated database*, which assumes that every agent has complete information of the whole network (e.g., through distributed ledgers), and *Distributed data endpoints*, where the information is fully compartmentalized and the cross-correlation is conducted at a local level. Both approaches have their advantages and disadvantages. The replicated database provides all agents with a vision of the network, although it imposes some overhead. As for the distributed data endpoints, they reduce the number of messages exchanged, yet the algorithm must deal with partial information coming from neighbour brokers.

Altogether, the ultimate election of the algorithm, data model and architectural design of the agents responds to performance and overhead restrictions. This composes the detection mechanism at a physical layer, while at an abstract level, it must also return a set of security insights that are based on the attacker model, described in the following.

4.2 Attacker Model

To analyze the parametrization of traceability algorithms, we need to characterize the chain of actions performed by an APT over the network. We will use the formalization described in [5] as a starting point. That paper reviews the most relevant APTs reported in the last decade and extracts a standard representation of an APT in the form of a finite succession of attacks stages: *initial intrusion*, *node compromise*, *lateral movement*, and *data exfiltration or destruction*. These stages cause different anomalies that are potentially inspected by the detection agents, according to an ordered set of probabilities [5].

Algorithm 1. Attacker model - anomaly calculation

```
input: attackSet_k, representing the chain of actions of APT k, 1 ≤ k ≤ numOfApts
local: Graph G(V, E) representing the network
output: x_i representing the anomaly value sensed by each agent i at the end of the APT
network, where x_i ∈ (0, 1)

x ← zeros(|V|) (initial opinion vector)
while attackSet_k ≠ ∅∀k ∈ {1, .., numOfApts} do
    for k ← 1 to numOfApts do
        if |attackSet_k| > 0 then
            {attack ← next attack from attackSet_k}
            if attack == initialIntrusion(IT, OT, FW) then
                attackedNode_k ← random v ∈ V_(IT,OT,FW)
                x(attackedNode_k) ← x_attackedNode_k + θ_3
            else if attack == compromise then
                x_attackedNode_k ← x_attackedNode_k + θ_2
                for neighbour in neighbours(attackedNode_k) do
                    x_attackedNode_k ← x_attackedNode_k + θ_5
                end for
            else if type(attack) == LateralMovement then
                previousAttackedNode ← attackedNode_k
                attackedNode_k ← SelectNextNode(G, attackedNode_k)
                x_previousAttackedNode_k ← x_previousAttackedNode_k + θ_5
                x_attackedNode_k ← x_attackedNode_k + θ_3,4
            else if attack == exfiltration then
                x_attackedNode_k ← x_attackedNode_k + θ_4
            else if attack == destruction then
                x_attackedNode_k ← x_attackedNode_k + θ_1
            end if

            x ← AttentauteOldAnomalies(x)
            attackSet_k ← attackSet_k \ attack
        end if
    end for
end while
```

Using these stages, APTs can be represented as a finite sequence of precise events. However, the original authors did not consider the possibility of parallel APT traces being executed simultaneously across the network, hence generating cross-related events. We refer to them as *concurrent routes* followed by the APT in the attack chain or different APTs taking place (that may eventually collaborate). We extend the aforementioned attacker model with this novel feature and formalize it in Algorithm 1.

In the algorithm, the effect of multiple APTs are implemented as a succession of updates on vector x, both in the concerned node (i.e., the anomaly sensed by its agent) and its neighbours when certain attack stages (e.g., a *compromise* stage) are involved. The routine continues until all attack stages have been executed for the entire set of APTs considered. On the other hand, the *theta* values refer to the ordered set of probabilities presented in [5]: the lower value the index of theta, the higher anomaly is reported by the agent after that phase. That paper also illustrates the attenuation function, which decreases the values of vector x over time based on the persistence of attacks, the criticality of the resources affected and their influence over posterior APT phases.

During the execution of these iterations, the correlation algorithm can be executed at any time to gain knowledge of the actual APT movements, by using the anomalies as input. The complete explanation of inputs and outputs for the traceability solution is further explained in next subsection.

4.3 Inputs and Outputs of the Traceability Solution

After introducing how the information from physical devices is collected by agents in practice and how the anomalies can be calculated in theoretical terms for our simulations, we summarize the set of inputs for traceability solutions as:

(I1) **Quantitative input.** expressed with vector x to assign every industrial asset with an anomaly value prior to conducting the correlation. As previously mentioned, it can be calculated by each associated agent or using external detection mechanisms integrated with the data broker by taking an extensive set of data inputs to comply with D1 and D2. In our simulations, this value is given by the attack phases executed on the network in a probabilistic way, without the detection mechanism having any knowledge about the actual stages.

(I2) **Qualitative input.** the previous values need to be enriched with information to correlate events in nearby devices and infer the presence of related attack stages, according to Sect. 4.2. At the same time, we also need to prioritize attacks that report a higher anomaly values. We assume that the resulting knowledge can be reflected in form of a weight w_{ij}, which is assigned by every agent i to each of its neighbours and represents the level of trust given to their anomaly indications when performing the correlation (fulfilling S2). This parameter can be subject to a threshold ε, which defines when two events should be correlated depending on the similarity of their anomalies. Further criteria could be introduced to associate anomalies from different agents.

With respect to the outputs of the traceability solutions, they should include, but are not limited to the following items:

(O1) Local result to determine whether the agent is generating an anomaly due to whether the actual infection of the associated node, as a result of a security threat in a neighbour device or a false positive.

(O2) Information at global level, to determine the degree of affection in the network and the nodes that have been previously taken over, filtered by zones.

(O3) Contextual information that permits to correlate past events and visualize the evolution of the threat, while anticipating the resources that are prone to be compromised (D3 & D4).

This comprehensive analysis of the requirements and techniques defines a framework for the development of distributed detection solutions for APTs in industrial scenarios, as depicted in Fig. 1. The following section presents some of the candidate solutions that implement them and hence achieve the APT traceability goals proposed so far.

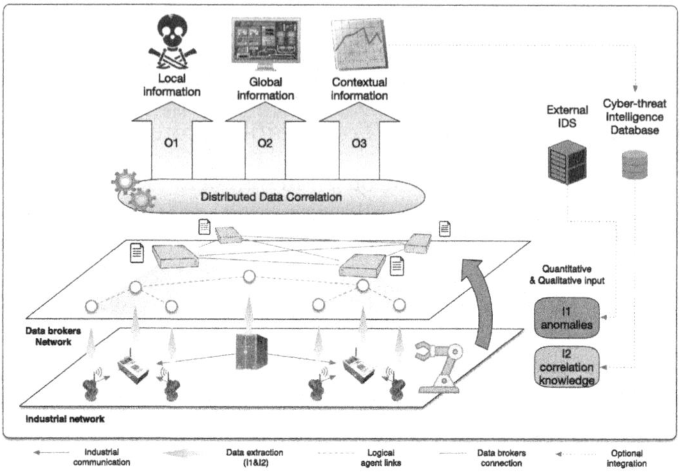

Fig. 1. Distributed detection framework

5 Distributed Traceability Solutions

After explaining the proposed framework, we look for feasible solutions that can effectively accommodate all of its statements. More specifically, we revisit the original Opinion Dynamics approach and compare it with two alternative mechanisms . The former is based on the concept of consensus and the two latter are based on clustering.

5.1 Opinion Dynamics

Despite its novelty, the Opinion Dynamics approach has faced a number of improvements since its inception in [18]. It was originally defined as a mechanism to address the anomalies caused by a theoretical APT, whose attacker model was better formalized in [19]. The event correlation and traceability capabilities were updated in [16], to latter show its implementation on an industrial testbed in [5]. Additionally, it has been studied its application to the Smart Grid [20] or the Industrial Internet of Things [21]. Compared to these, the aim of the framework is to ease the design of alternative solutions with results equal to or better than those of Opinion Dynamics.

The correlation approach of Opinion Dynamics is based on an iterative algorithm that takes the anomalies of individual agents as input, and generates their resulting 'opinions' based on this formula:

$$x_i(t+1) = \sum_{j=1}^{n} w_{ij} x_j(t)$$

Where $x_i(t)$ stands for the opinion of the agent ($i \in \{1, ..., |V|\}$) at iteration t, such that $x_i(0)$ contains the initial anomaly sensed in vector x prior to execute the correlation (I1), as stated in Sect. 4.3. As for w_{ij}, it represents the weight given by each agent i to the opinion of each neighbour j in $G(V, E)$, as to model the influence between them (I2). At this point, the original paper defined a $\varepsilon = 0.3$ threshold to hold the maximum difference in opinions between every pair of agents i and j, to associate a weight $w_{ij} > 0$. This way, the weight given by an agent i is equally divided and assigned to each other neighbour agent k that complies with ε (including itself), having $\sum_{k=1}^{n} w_{ik} = 1$. In [16], the authors provide a methodology to include further security factors and other metrics (e.g., QoS in communication links) when calculating these weight values.

Altogether, the correlation is performed by every agent as a weighted sum of the closest opinions, and such calculation can be performed by solely using the information from neighbouring agents, thereby adapting to the distributed architecture based on data brokers (either replicating data or not). When executing this algorithm with a high number of iterations, the outputs of all agents are distributed into different groups that expose the same anomaly value, which correspond to related attacks. As a result, the network is polarized based on their opinions, hence satisfying O1. From these values, it is also possible to study the degree of affection in different network zones and extract global security indicators (O2). Likewise, the evolution of a sequential APT can also be visualized if we account for the agents opinions over time (O3), as described in [5].

5.2 Distributed Anomaly Clustering Solutions

Opinion Dynamics belongs to a set of dynamic decision models in complex networks whose aim is to obtain a fragmentation of patterns within a group of interacting agents by means of *consensus*. This fragmentation process is locally regulated by the opinions and weights of the nodes, that altogether abstracts the APT dynamics and its effects on the underlying network. This ultimately enables to take snapshots of the current state of the network and highlight the most affected nodes, thereby tracing APT movements from anomaly events.

This rationale can also be applied to define different mechanisms with similar results. We propose to adapt clustering algorithms as an alternative for the correlation of events that fulfill the defined framework. These have been traditionally used as an unsupervised method for data analysis, where a set of instances are grouped according to some criteria of similarity. In our case, we have devices that are affected by correlated attacks (see Sect. 4.2) and show similar anomalies, which results in the devices being grouped together.

Classical clustering methods [22], such as K-means, partition a dataset by initially selecting k cluster centroids and assigning each element to its closest centroid. Centroids are repeatedly updated until the algorithm converges to a stable solution. In our case, the anomalies detected by the agents (denoted by the vector x) play the role of the data instances to be grouped into clusters. However, the parametrization of this kind of algorithms impose two main challenges to properly comply with the inputs and outputs of an APT traceability solution:

- **The election of k.** It is one classical drawback of the K-means, since that value has to be specified from the beginning and it is not usually known in advance, as in this case. Numerous works in the literature have proposed methods for selecting the number of clusters [23], including the use of statistical measures with assumptions about the underlying data distribution [24] or its determination by visualization [25]. It is also common to study the results of a set of values instead of a single k, which should be significantly smaller than the number of instances. The aim is to apply different evaluation criterion to find the optimal k, such as the Calinski and Harabasz score (also known as the Variance Ratio Criterion) [26], that minimizes the within-cluster dispersion and maximizes the between-cluster dispersion.
- **Representation of topological and security constraints.** By applying K-means, we assume the dataset consists of a set of multi-dimensional points. However, here we have an one-dimensional vector of anomalies in the range [0,1]. Also, the clusterization of these values is subject to the topology and the security correlation criteria which might determine that, for example, two data points should not be grouped in the same cluster despite having a similar anomaly value. Therefore, it becomes necessary to provide this knowledge to the algorithm and reflect these environmental conditions as inputs (I1 and I2) to the correlation. In this sense, some works have proposed a constrained K-means clustering [27], and specific schemes have been developed to divide a graph into clusters using Spanning Trees or highly connected components [28].

As for the first challenge, we can assume that the value of k is defined by the different classes of nodes within the network depending on their affection degree, which corresponds to the number of consensus between agents that Opinion Dynamics automatically finds. Here we can adopt two methodologies: (1) a *static* approach where we consider a fixed set of labels (e.g., 'low', 'medium', 'high' and 'critical' condition) to classify each agent; or (2) a *dynamic* approach where k is automatically determined based on the number and typology of attacks. In this case, we can study the Variance Ration Criterion in a range of k values (e.g., $k = \{1–5\}$) to extract the optimal value with the presence of an APT.

This procedure needs further improvements to make the solution fully distributed, so that each agent is in charge of locally deciding its own level of security based on the surrounding state, instead of adopting a global approach for all nodes. This bring us to the second challenge. A first naive solution would be to introduce additional dimensions to the data instances representing the coordinates of every node, together with the anomalies in vector x. We call this approach *location-based clustering*. However, this approach still needs to figure out an optimal value of k, and does not take into account the presence of actual links interconnecting nodes in $G(V, E)$.

To circumvent this issue while also adopting an automatic determination of the number of clusters, we propose an *accumulative anomaly clustering* scheme, which is formalized in Algorithm 2. This algorithm begins by selecting the most affected node within the network and subsequently applies the influence of their surrounding nodes. This is represented by adding an entire value to the anomalies

Algorithm 2. Accumulative anomaly clustering

```
input: x_i representing the initial anomaly value sensed by each agent i within the network,
where x_i ∈ (0, 1)
output: z_i representing the agents O1 output of each agent i after clustering
local: Graph G(V, E) representing the network, where V = V_IT ∪ V_OT ∪ V_FW

max ← |V|, k ← 0
y ← x, x' ← x sorted in descending order
for all i ∈ x' do
    anyNeighbourFound ← False
    for all j ∈ neighbours(i, G) do
        if y_j ≤ 1 AND |y_i − y_j| ≤ ϵ then
            y_j ← y_j + max * 10
            anyNeighbourFound ← True
        end if
    end for
    y_i = y_i + max * 10
    if anyNeighbourFound then
        k ← k + 1
    end if
    max ← max − 1
end for
clusters, centroid ← kmeans(y, k)
for all v_i ∈ V do
    c ← clusters(v_i)
    Z_i ← IntegerPart(centroid(c))
end for
```

of such agents (initially from 0 to 1), which is proportional to the anomaly of the influencing node (see *max* in the algorithm). This addition is performed as long as the difference between both anomalies (i.e., the influencing and influenced node) does not surpasses a defined threshold ε, similar to the Opinion Dynamics approach in order to comply with I2. Then, the algorithm continues by selecting the next one in the list of nodes inversely ordered by the anomaly value, until all nodes have been influenced or have influenced others. At that point, k is automatically assigned with the number of influencing nodes, and K-means is ready to be executed with the modified data instances. The resulting values of each agent corresponds to the decimal part of their associated centroid. This is comparable to the 'opinions' in the Opinion Dynamics approach.

The intuition behind this model of influence between anomalies (which can be enriched to include extra security factors to specify I2) assumes that successive attacks raise a similar anomaly value in closest agents, as Opinion Dynamics suggests. At the same time, it addresses the issue of selecting k and including topological information to the clusterization. It is validated from a theoretical point of view in Appendix A. In the following, the accuracy of these correlation approaches are compared under different attack and network configurations.

6 Experiments and Discussions

After presenting some alternative solutions to Opinion Dynamics that fulfill the distributed detection framework presented in Sect. 4, this section aims to put these approaches to the test. More specifically, we consider the attacker model explained in Sect. 4.2, which is applied against a network formalized by $G(V, E)$, following the structure introduced in Sect. 4.1. These theoretical APTs generate a set of anomalies that serve as input to compare the traceability capabilities of each correlation approach:

- **Location-based clustering:** as presented earlier, it consists of the K-means algorithm taking the anomalies and coordinates of each node as data instances. These are grouped in a number of clusters, k, which is selected in the range from 1 to 5 according to the Variance Ratio Criterion.
- **Accumulative clustering:** as previously presented, it allows to distributedly locate the infection while automatically determining the optimal k.
- **Opinion Dynamics:** is the approach that serves as inspiration for our framework and serves for comparison with the novel detection methods introduced above.

These traceability solutions are simulated under different network and attack configurations, as explained next. We start by running a brief attack test-case that illustrates the features of each approach in a simple network scenario. Based on Algorithm 1, Fig. 2 shows the detection outputs (O1 and O3) of the three approaches when correlating the anomalies of an APT perpetrated against a simple infrastructure. This network is modelled according to the concepts introduced in Sect. 4.1, to include an IT and OT section of nodes connected by a firewall. Concretely, the figure shows an snapshot of the detection state after the adversary has performed a lateral movement from IT node 2 to compromise the firewall. The numeric value assigned to each node represents O1, which will attenuate over time to highlight the most recent anomaly, according to O3.

As noted in the figure, location-based clustering fails to accurately determine where the threat is located and selects a wide affection area instead, which is composed by *IT1*, *IT2* and *FW1* nodes (i.e., grouped in the same cluster due to the average anomaly in such zone). On the other hand, the accumulative clustering and Opinion Dynamics show a similar result, and successfully identify both *IT2* and *FW1* as the affected nodes in this scenario. As for the rest of nodes, they agree on a subtle affection value due to the noise present in the network and the anomalies sensed in the vicinity of the attacked nodes. As previously stated, this is modelled in a probabilistic way [5].

We now execute these solutions with a more complex network and APT model in order to study their accuracy . In the context of cluster analysis, the 'purity' is an evaluation criteria of the cluster quality that is applicable in this particular scenario. It holds the percentage of the total number of data points that are classified correctly after executing the clustering algorithm, in the range [0,1]. It is calculated according to the following equation:

$$Purity = \frac{1}{N} \sum_{i=1}^{k} max|c_i \cap t_j| \tag{1}$$

where N is the number of nodes, k is the number of clusters, c_i is a cluster in C and t_j is the classification that has the max count for cluster c_i. In our case, by

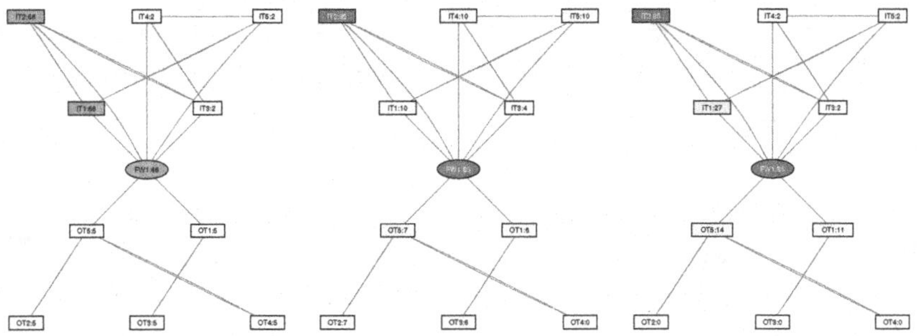

(a) Location-based clustering (b) Accumulative clustering (c) Opinion Dynamics

Fig. 2. Network topology used in the test case

'correct classification' we mean that a cluster c_i has identified a group of nodes that have actually been compromised, which is determined in the simulations (but not known by the traceability solutions). This value can be calculated after a single execution of these three approaches to study how the results of the initial test-case escalate to larger networks and more challenging APTs.

Specifically, we run 10 different APTs on randomly generated network topologies of 50, 100 and 150 nodes, respectively. For simplicity, we start by executing an individual instance of the Stuxnet APT [5] according to the attacker model established in Sect. 4.2. This attack can be formally defined by the following succession of stages:

$$attackSet_{Stuxnet} = \{initialIntrusion_{IT}, LateralMovement_{FW}, \\ LateralMovement_{OT}, destruction\}$$

At this point, it is worth mentioning that the lateral movement in the OT section is performed three times to model the real behavior of this APT and its successive anomalies, as explained in [5]. The purity value is then calculated after every attack stage of each of the ten APTs, to ultimately compute its average with respect to the number of nodes that have been successfully detected and grouped in the cluster with highest value of affection.

Figure 3(a) represents these average values in the form of boxplots, where each box represents the quartiles of each detection approach given the different network configurations. As it can be noted, the Opinion Dynamics stands out as the most accurate solution, closely followed by the accumulative clustering approach. The purity of the location-based clustering falls behind, and the three of them increase their value as the network grows in size due to the higher number of nodes that are successfully deemed as healthy and hence not mixed with those that are indeed affected by the APT.

Similar results are obtained when we execute two APT attacks in parallel over the same network configurations, as shown in Fig. 3(b). In this case, the former APT is coupled with another attack, which can be assumed to be part

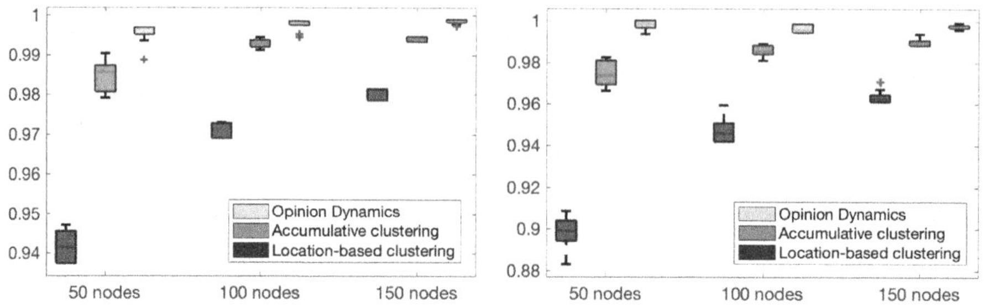

(a) Single APT

(b) Two APTs

Fig. 3. Purity average for the three test cases

of Stuxnet or a completely different attack trace within the network, composed by the following stages:

$$attackSet_{AnotherAPT} = \{initialIntrusion_{OT}, LateralMovement_{FW}, LateralMovement_{IT}, destruction\}$$

The second APT is located in a different area of the network so that it begins by sneaking into the OT section to subsequently propagate towards the IT portion of the infrastructure. This causes the spread of anomalies throughout the network hence putting location-based clustering to the test. Despite a subtle decline in the purity of the solutions (especially in the location approach due to the anomaly dispersion), they still output an appreciable accuracy.

On the other hand, the superiority of Opinion Dynamics and accumulative clustering over the first approach is also evident with the study of additional accuracy indicators, such as the *Rand Index*. It penalizes both false positive (FP) and false negative (FN) labeling of affected nodes during clustering, with respect to true positive (TP) and true negative (TN) decisions, according to the following formula:

$$Rand\,Index = \frac{TP + TN}{TP + FP + FN + TN} \tag{2}$$

Figure 4 shows the Rand Index value after each of the ten APTs in the previous experiment (each one composed of two parallel attack traces), for the largest network size (150 nodes). The plot clearly shows a steady accuracy of the two latter approaches (close to 1), contrasting with a lower value in the location-based approach, which faces a lack of precision when it comes to correctly locate the affection areas, for the same reasons discussed before. Despite the promising results of our clustering approach in terms of accuracy, we are currently in the process of assessing its performance (compliance with S5 requirement).

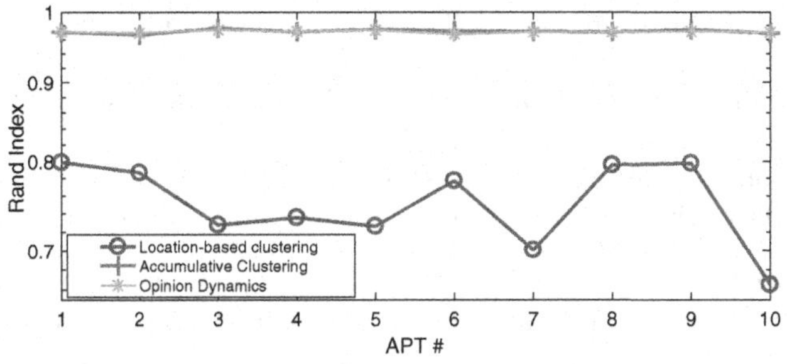

Fig. 4. Evolution of the Rand Index for 10 APTs and 150 nodes

7 Conclusions

The irruption of APTs is demanding for the development of innovative solutions capable of detecting, analyzing and protecting current and upcoming critical infrastructures. After an exhaustive analysis of the security and detection requirements for these solutions, we come up with a framework for developing APT traceability systems in Industry 4.0 scenarios, inspired by a promising approach called Opinion Dynamics. This framework considers various network architectures, types of attack and data acquisition models to later define the inputs and outputs that traceability solutions should include to support the aforementioned requirements. This lays the base for the development and comparison of novel solutions in this context. As a means to validate the proposed framework, we define two novel protection mechanisms based on clustering, which feature comparable results to the Opinion Dynamics (based on consensus). According to our experiments, the proposed clustering mechanism also presents optimal traceability of events in a distributed setting. The roadmap of this research now leads to further validation and possible extensions of the proposed framework, as well as to the application of these techniques to diverse real-world industrial scenarios.

Acknowledgments. This work has been partially supported by the EU H2020-SU-ICT-03-2018 Project No. 830929 CyberSec4Europe (cybersec4europe.eu), the EU H2020-MSCA-RISE-2017 Project No. 777996 (SealedGRID), and by a 2019 Leonardo Grant for Researchers and Cultural Creators of the BBVA Foundation. The first author has been partially financed by the Spanish Ministry of Education under the FPU program (FPU15/03213) and R. Rios by the 'Captacion del Talento Investigador' fellowship from the University of Malaga.

A Correctness Proof of the Clustering Detection Approach

This section presents the correctness proof of the consensus-based detection, both the location and accumulative approach. This problem is solved when these conditions are met:

1. The attacker is able to find an IT/OT device to compromise within the infrastructure.
2. The traceability solution is able to identify an affected node, thanks to the clustering mechanism and fulfilling O1.
3. The detection can continuously track the evolution of the APT and properly finish in a finite time (termination condition), complying with O2 and O3.

The first requirement is satisfied under the assumption that the attacker breaks into the network and then moves throughout the topology following a finite path, according to the model explained in Sect. 4.2. Thus, an APT is defined as at least one sequence of attack stages against the network defined by $G(V, E)$. If we study each of these traces independently, and based on the distribution of G, the attacker can either *compromise* the current node v_i in the chain (as well as performing a *data exfiltration or destruction*) or propagate to another $v_j \in V$, whose graph is connected by the means of firewalls, according to the interconnection methodology illustrated in [5] and summarized in Sect. 4.1.

As for the second requirement, it is met with the correlation of anomalies generated by agents in each attack phase. As presented with the attacker model, the value of these anomalies are determined in a probabilistic manner, depending on two possible causes: (1) the severity of the attack suffered and the criticality of the concerned resource; or (2) an indirect effect caused by another attack in the vicinity of the monitored node. Either way, the O1 correlation helps to actually determine whether the attack has been effectively perpetrated against that node, or it belongs to another APT stage in its surroundings. This information is deduced from the combination of I2 (the contextual information) together with these anomalies (i.e., I1), by using K-means to group these nodes and associate them with actual attacks.

We can easily demonstrate the third requirement (i.e., the termination of the approach) through induction. To do so, we specify the initial and final conditions as well as the base case:

Precondition: we assume the attacker models an APT against the network defined by graph $G(V, E)$ where $V \neq \oslash$, following the behaviour explained in Algorithm 1. On the other hand, the detection solution based on clustering can firstly sense the individual anomalies in every distributed agent, hence computing I1 and I2.

Postcondition: the attacker reaches at least one node in $G(V, E)$ and continues to execute all stages until $attackSet = \oslash$ in Algorithm 1. Over these steps, it is possible to visualize the threat evolution across the infrastructure, following the procedure described in Algorithm 2 in the case of accumulative clustering, and running K-means with both I1 and spatial information, in the case of location-based clustering.

Case 1: the adversary intrudes the network and takes control of the first node $v_i \in V$, and both clustering approaches cope with the scenario of grouping healthy nodes apart from the attacked node. This is calculated by the K-means algorithm within a finite time, by iteratively assigning data items to clusters and recomputing the centroids.

Case 2: the adversary propagates from a device node v_i to another v_j, so that there exist $(v_i, v_j) \in E$. In this case, the correlation with K-means aims to group both affected nodes within the same cluster, which can be visualized graphically. As explained before, this is influenced by the attack notoriety and the closeness in the anomalies sensed by their respective agents (i.e., the threshold ϵ in Algorithm 2), as well as extra information given by I2.

Induction: if we assume the presence of $k \geq 1$ APTs in the network, each one will consider Case 1 at the beginning and will separately consider Case 2 until $attackSet = \oslash$ for all k, ensuring the traceability of the threat and complying with the postcondition. Eventually, these APTs could affect the same subset of related nodes in G, which is addressed by the K-means to correlate the distribution of anomalies (again, attempting to distinguish between attacked nodes and devices that may sense side effects), in a finite time.

This way, we demonstrate the validity of the approach, since it finishes and it is able to trace the threats accordingly.

References

1. Khan, A., Turowski, K.: A survey of current challenges in manufacturing industry and preparation for industry 4.0. In: Proceedings of the First International Scientific Conference "Intelligent Information Technologies for Industry" (IITI 2016), pp. 15–26. Springer (2016). https://doi.org/10.1007/978-3-319-33609-1_2
2. Singh, S., Sharma, P.K., Moon, S.Y., Moon, D., Park, J.H.: A comprehensive study on APT attacks and countermeasures for future networks and communications: challenges and solutions. J. Supercomput. **75**(8), 4543–4574 (2016). https://doi.org/10.1007/s11227-016-1850-4
3. Lemay, A., Calvet, J., Menet, F., Fernandez, J.M.: Survey of publicly available reports on advanced persistent threat actors. Comput. Secur. **72**, 26–59 (2018)
4. Mitchell, R., Chen, I.-R.: A survey of intrusion detection techniques for cyber-physical systems. ACM Comput. Surv. (CSUR) **46**(4), 55 (2014)
5. Rubio, J.E., Roman, R., Alcaraz, C., Zhang, Y.: Tracking APTs in industrial ecosystems: a proof of concept. J. Comput. Secur. **27**(5), 521–546 (2019)
6. Zeng, P., Zhou, P.: Intrusion detection in SCADA system: a survey. In: Li, K., Fei, M., Du, D., Yang, Z., Yang, D. (eds.) ICSEE/IMIOT -2018. CCIS, vol. 924, pp. 342–351. Springer, Singapore (2018). https://doi.org/10.1007/978-981-13-2384-3_32
7. Rubio J.E., Roman R., Lopez J.: Analysis of cybersecurity threats in industry 4.0: the case of intrusion detection. In: The 12th International Conference on Critical Information Infrastructures Security, volume Lecture Notes in Computer Science, vol. 10707, pp. 119–130. Springer, August 2018. https://doi.org/10.1007/978-3-319-99843-5_11
8. Sekar, R., et al.: Specification-based anomaly detection: a new approach for detecting network intrusions. In: Proceedings of the 9th ACM Conference on Computer and Communications Security, pp. 265–274. ACM (2002)
9. Lin, H., Slagell, A., Kalbarczyk, Z., Sauer, P.W., Iyer, R.K.: Semantic security analysis of SCADA networks to detect malicious control commands in power grids. In: Proceedings of the First ACM Workshop on Smart Energy Grid Security, pp. 29–34. ACM (2013)

10. Rubio, J.E., Alcaraz, C., Roman, R., Lopez, J.: Current cyber-defense trends in industrial control systems. Comput. Secur. J. **87**, 101561 (2019)
11. Moustafa, N., Adi, E., Turnbull, B., Hu, J.: A new threat intelligence scheme for safeguarding industry 4.0 systems. IEEE Access **6**, 32910–32924 (2018)
12. Chhetri, S.R., Rashid, N., Faezi, S., Al Faruque, M.A.: Security trends and advances in manufacturing systems in the era of industry 4.0. In: 2017 IEEE/ACM International Conference on Computer-Aided Design (ICCAD), pp. 1039–1046. IEEE (2017)
13. Vance, A.: Flow based analysis of advanced persistent threats detecting targeted attacks in cloud computing. In: 2014 First International Scientific-Practical Conference Problems of Infocommunications Science and Technology, pp. 173–176. IEEE (2014)
14. Brogi, G., Tong, V.V.T.: Terminaptor: highlighting advanced persistent threats through information flow tracking. In: 2016 8th IFIP International Conference on New Technologies, Mobility and Security (NTMS), pp. 1–5. IEEE (2016)
15. Ghafir, I., et al.: Detection of advanced persistent threat using machine-learning correlation analysis. Future Gener. Comput. Syst. **89**, 349–359 (2018)
16. Rubio, J.E., Manulis, M., Alcaraz, C., Lopez, J.: Enhancing security and dependability of industrial networks with opinion dynamics. In: Sako, K., Schneider, S., Ryan, P.Y.A. (eds.) ESORICS 2019. LNCS, vol. 11736, pp. 263–280. Springer, Cham (2019). https://doi.org/10.1007/978-3-030-29962-0_13
17. Lee, S., Shon, T.: Open source intelligence base cyber threat inspection framework for critical infrastructures. In: 2016 Future Technologies Conference (FTC), pp. 1030–1033. IEEE (2016)
18. Rubio, J.E., Alcaraz, C., Lopez, J.: Preventing advanced persistent threats in complex control networks. In: Foley, S.N., Gollmann, D., Snekkenes, E. (eds.) ESORICS 2017. LNCS, vol. 10493, pp. 402–418. Springer, Cham (2017). https://doi.org/10.1007/978-3-319-66399-9_22
19. Rubio, J.E., Roman, R., Alcaraz, C., Zhang, Y.: Tracking advanced persistent threats in critical infrastructures through opinion dynamics. In: Lopez, J., Zhou, J., Soriano, M. (eds.) ESORICS 2018. LNCS, vol. 11098, pp. 555–574. Springer, Cham (2018). https://doi.org/10.1007/978-3-319-99073-6_27
20. Lopez, J., Rubio, J.E., Alcaraz, C.: A resilient architecture for the smart grid. IEEE Trans. Ind. Inform. **14**, 3745–3753 (2018)
21. Rubio, J.E., Roman, R., Lopez, J.: Integration of a threat traceability solution in the industrial Internet of Things. IEEE Trans. Ind. Inform. (2020). In Press
22. Rui, X., Wunsch, D.: Survey of clustering algorithms. IEEE Trans. Neural Netw. **16**(3), 645–678 (2005)
23. Pham, D.T., Dimov, S.S., Nguyen, C.D.: Selection of k in k-means clustering. Proc. Inst. Mech. Eng. Part C: J. Mech. Eng. Sci. **219**(1), 103–119 (2005)
24. Pelleg, D., Moore, A.W., et al.: X-means: extending k-means with efficient estimation of the number of clusters. In: Icml, vol. 1, pp. 727–734 (2000)
25. Bilmes, J., Vahdat, A., Hsu, W., Im, E.J.: Empirical observations of probabilistic heuristics for the clustering problem. Technical Report TR-97-018, International Computer Science Institute (1997)
26. Caliński, T., Harabasz, J.: A dendrite method for cluster analysis. Commun. Stat.-Theory Methods **3**(1), 1–27 (1974)
27. Wagstaff, K., Cardie, C., Rogers, S., Schrödl, S., et al.: Constrained k-means clustering with background knowledge. Icml **1**, 577–584 (2001)
28. Schaeffer, S.E.: Graph clustering. Comput. Sci. Rev. **1**(1), 27–64 (2007)

Designing Reverse Firewalls for the Real World

Angèle Bossuat[1(✉)], Xavier Bultel[4], Pierre-Alain Fouque[1], Cristina Onete[2], and Thyla van der Merwe[3]

[1] Univ Rennes, CNRS, IRISA, Rennes, France
angele.bossuat@irisa.fr
[2] University of Limoges/XLIM/CNRS, Limoges, France
[3] Mozilla, London, UK
[4] LIFO, INSA Centre Val de Loire, Université d'Orléans, Bourges, France

Abstract. Reverse firewalls (RFs) were introduced by Mironov and Stephens-Davidowitz to address algorithm-substitution attacks (ASAs) in which an adversary subverts the implementation of a provably-secure cryptographic primitive to make it insecure. This concept was applied by Dodis *et al.* in the context of secure key exchange (handshake phase), where the adversary wants to exfiltrate sensitive information by using a subverted client implementation. RFs are used as a means of "sanitizing" the client-side protocol in order to prevent this exfiltration. In this paper, we propose a new security model for both the handshake and record layers, a.k.a. secure channel. We present a signed, Diffie-Hellman based secure channel protocol, and show how to design a provably-secure reverse firewall for it. Our model is stronger since the adversary has a larger surface of attacks, which makes the construction challenging. Our construction uses classical and off-the-shelf cryptography.

1 Introduction

In 2013, Snowden revealed thousands of classified NSA documents indicating evidence of widespread mass-surveillance. In his essay on the moral character of cryptographic work [21], Rogaway suggests that the most important effect of the Snowden revelations was the realization of the existence of a new kind of adversary with much greater powers than ever imagined, aiming to enable mass-surveillance by eavesdropping on unsecured communications, and by negating the protection afforded by cryptography. It has massive computational resources at its disposal to mount conventional attacks against cryptography, and also tries to subvert the use of cryptographic primitives by introducing backdoors into cryptographic standards [6,7] or, equally insidiously, by using Algorithm Substitution Attacks (ASA). The latter class of attacks was formalised by Bellare *et al.* [1] in the context of symmetric encryption, but actually goes back to much

Full version available at eprint.iacr.org/2020/854.

L. Chen et al. (Eds.): ESORICS 2020, LNCS 12308, pp. 193–213, 2020.
https://doi.org/10.1007/978-3-030-58951-6_10

earlier work on *Kleptography* [22]. Bellare *et al.* [1] envisage a scenario where the adversary substitutes a legitimate and secure implementation of a protocol with one that leaks cryptographic keys to an adversary in an undetectable way. Sadly, they showed that all major secure communication protocols are vulnerable to ASAs because of their intrinsic reliance on randomness.

Reverse Firewalls (RF). Mironov and Stephens-Davidowitz [18] define 3 properties RF should satisfy: (1) *security preserving*: regardless of the client's behaviour, the RF will guarantee the same protocol security; (2) *functionality-maintaining*: if the user implementation is working correctly, the RF will not break the functionality of the underlying protocol; and (3) *exfiltration resistance*: regardless of the client's behaviour, the RF prevents the client from leaking information. In the key-exchange security game of RF, the adversary first corrupts the client by changing its code. Then, during the communication step, there is no more "direct" exchange, and the adversary has to exfiltrate information *from* the messages circulating between the RF and the server.

Dodis *et al.* [9] described an ASA allowing an adversary to break the security of many authenticated key exchanges (AKE). The attack is in the spirit of [1]: substitute an implementation of a (provably-secure) AKE protocol with a weaker one, in an undetected way for the servers, then exfiltrate information to a passively observing adversary. Next, they introduce a RF, used by either of the two endpoints, to prevent any exfiltration during the AKE protocol. The RF acts like a special entity in charge of enforcing the protocol's requirements (*e.g.*, checking signatures, adding randomness) to ensure the robustness of the exchanged keys even for a misbehaving client implementation. Delegating all sensitive steps to the RF would be simpler, but goes against the strong advocated model. RF should *preserve* security, not provide it, as security must hold even with a *malicious* RF. Moreover, for practical reasons, both endpoints should still be able to communicate even if the RF is not responding, *e.g.*, due to too many connections.

Dodis *et al.*'s protocol is a simple signed Diffie-Hellman key exchange modified to accommodate an RF. They prove that their new scheme is still a secure 2-party AKE protocol in the absence of an RF, and that it additionally provides exfiltration resistance when an RF is added even if the endpoints are corrupted. Furthermore, the RF learns no information about the session keys established by the endpoints. Finally, they show that the resulting AKE protocol can be composed with a secure messaging protocol and still provide a certain degree of exfiltration resistance for specific weak client implementations. To this end, they rerandomize the Diffie-Hellman inputs and perform verification tasks. Since the two parties no longer see the same tuple of DH elements, the signatures cannot be made on the transcripts directly: a bilinear pairing is cleverly used to sign a deterministic function of the transcripts.

There are many possibilities for a client implementation to become weak: (1) server authentication can be avoided and the adversary can play the role of a malicious server, (2) by using weak randomness or pre-determined "randoms", the adversary can predict the client's ephemeral DH secret and recover the common key exchange and (3) the client can skip the authenticated encryption (AEAD) security during the record phase. We point out that for channel security, because of weakness (3), we cannot just compose a key exchange protocol secure in this model with a secure channel without looking at the security of the channel. There is a subtle theoretic problem in the exfiltration resistance game: since the client knows the key, he can choose messages that depend on the key, which is not the case in traditional encryption security games.

Other Approaches. RFs have been extended to other contexts such as malleable smooth projective hash functions [8], and attribute-based encryption [17]. While these results have no real link to our work, they show that RFs are relatively versatile and a promising solution to ASAs. They also superficially resemble middleboxes, which have been studied in the context of TLS [12,19,20]. However, RFs fundamentally differ from middleboxes: the former are meant to *preserve* the confidentiality, integrity, and authenticity of the secure channel, while the latter break it in controlled ways.

Comparison Between Previous Attacker Model and Our Guarantees. Dodis *et al.* [9] consider three attack scenarios. The first is the security of the primitive in the absence of a firewall (but where the client and server honestly follow the protocol). The second entails that the primitive should still be secure even in the presence of a malicious firewall. The third, and most important security notion is *exfiltration resistance*, where a malicious implementation of the client tries to exfiltrate information to a network adversary able to monitor the channel between the (honest) RF and the server.

For this last definition, the adversary also controls the server. This has three consequences: (1) exfiltration resistance may at most be guaranteed for the handshake: if the RF is not present in the basic 2-party protocol, there is no security check at the record layer and the firewall cannot prevent a client from exfiltrating information to a malicious server at the record layer; (2) [9] have to restrict their malicious implementations to behave in a very particular way, called *functionality preserving*: in this model, the malicious client must follow protocol, although it may use weak parameters or bypass verification steps; (3) if, within a protocol, the client sends the first key-exchange element, then a commitment phase from the server must precede that message, essentially preventing the server from adaptively choosing a weak DH element for the key-exchange. A more in-depth comparison is given in Appendix A.

In contrast to [9] we consider security for both the key-agreement *and* the record-layer protocols. As mentioned above, this is a non-trivial extension with respect to *exfiltration resistance*, since the channel keys do not imply exfiltration resistance at the record layer. Our results are proven in the presence of a semi-honest server who follows the protocol specification. We argue that this restriction is necessary since nothing can prevent a malicious server – not monitored by a firewall – from forwarding all the data it receives to the adversary. We focus on guaranteeing exfiltration resistance (in the handshake *and* in the record layer) with respect to a MitM situated between the firewall and the server.

Adding RF with Exfiltration Resistance at the Record Layer. Protecting the record layer is more challenging than the key exchange, and means considering these two stages as a whole. Our key observation is that an RF cannot prevent exfiltration at this layer if it does not know the key used for the encryption: nothing prevents the adversary from choosing messages which, for the computed key, leak information to the adversary. Another difficulty is that, contrary to the key exchange which essentially involves public-key primitives (usually quite malleable), the record layer relies on symmetric key primitives, less tolerant to modifications. Our approach differs from [9] by introducing a new functionality preserving definition: we do not restrict the client's behaviour beyond requiring that a semi-honest server accepts the communication.

Our solution is to allow the RF to contribute to securing messages exchanged during the record layer, *without compromising the end-to-end security that the channel should provide.* As in [9], the main task of our RF is to rerandomize some elements and verify the validity of the signatures. The main difference we introduce is that this key exchange will now generate two keys: k_{cs} and $\mathsf{k}_{\mathsf{cfs}}$. The former is very similar to the one generated in [9] and will be only known to both endpoints. It will be used to encrypt messages at the record layer, ensuring security even against a malicious RF. The key $\mathsf{k}_{\mathsf{cfs}}$ will be known to the endpoints, but also to the RF, allowing it to preserve security by adding a second layer of encryption. To accomplish this, our RF will have a public key, which was not the case in [9]. This is non-trivial in practice: for transparency the RF must not modify the messages' format, and the endpoints must be oblivious to the RF's action. Indeed, if the RF was offline or not capable of protecting all of its clients, this must not be detectable by a corrupt implementation. Finally, we cannot rely on standard security properties for encryption to prove our exfiltration-resistance. This is related to the adversarial strategy of choosing messages whose ciphertexts have a distinctive pattern. We provide more details in Sect. 3.3 but, intuitively, the ciphertexts meant to be distinguished by the adversary are encryptions of messages chosen by a corrupt implementation that knows the channel key. Using key-dependent messages schemes [3], we design a very efficient solution based on hash functions and MAC schemes to prevent exfiltration at the record layer.

2 Security Model and Definitions

2.1 The Adversary Model

The adversary $\mathscr{A}$ is a MitM, interacting with honest parties (one client C, one server S, and sometimes an RF FW) via instances associated with unique labels. We assume that the client is always the initiator of each execution, and the server is the responder. The server and the firewall have an identity associated with a pair of private-public keys. Additionally, we require a setup phase for the firewall, run either by the challenger if the firewall is honest, or by the adversary if it is malicious. The $\mathsf{Setup}(1^\lambda)$ algorithms takes as input a security parameter 1^λ and generates secret parameters sparam (including the firewall's private credentials) and public parameters pparam (including FW).

To each instance label, there is an associated set of values: a *type* type in $\{C, S, FW\}$, the entity of which label is an instance, a *session identifier* sid, and two types of *session keys*, denoted $\mathsf{k_{cfs}}$ and $\mathsf{k_{cs}}$, respectively. Both keys are initially set to $\perp$. An *authentication-acceptance bit* accept, set to $\perp$ at the beginning of the instance, which may turn to either 1 (if the authentication of the partner succeeds) or to 0 otherwise. Only client and firewall instances have a non-$\perp$ accept bit (since only the server authenticates itself). A pair of *revealed bits* $\mathsf{revealed_{cfs}}$ and $\mathsf{revealed_{cs}}$, both initially set to 0. A *corrupt bit* corrupted initially set to 0, and a *test bit* b drawn at random at the initialization of each new instance label. We use the notation "label.attribute" to refer to a value attribute associated with the instance label.

Partnering. We define partnering in terms of type and session identifiers. Two instances, associated with labels label and label′, are *partnered* if and only if $\mathsf{label.sid} = \mathsf{label'.sid}$ and $\mathsf{label.type} \neq \mathsf{label'.type}$. As such, a client may be partnered with a server, or a server and a firewall.

The scenarios we consider are formally defined through security experiments, in which the adversary $\mathscr{A}$ has access to some (or all) of the following oracles.

- $\mathsf{NewInstance}(U)$: on input $U \in \{C, S, FW\}$, this oracle outputs an instance of party U labeled label. The adversary is given label.
- $\mathsf{Send}(\mathsf{label}, m)$: on input an existing client/RF/server instance label and a message m, this oracle simulates sending m to label, adds label to a list $\mathscr{L}_s$ (initialized as an empty list) and outputs its corresponding reply m'.
- $\mathsf{Reveal}(\mathsf{label}, \mathsf{kt})$: on input an existing label label and a string $\mathsf{kt} \in \{\mathsf{cs}, \mathsf{cfs}\}$ denoting the key type to reveal, this oracle adds label to a list $\mathscr{L}_{\mathsf{kt}}$ (initialized as an empty list), and outputs the values stored in $\mathsf{label.k_{kt}}$. The corresponding bit $\mathsf{label.revealed_{kt}}$ is also set to 1.
- $\mathsf{CorruptS}()$: this oracle yields the server's private long-term key. Upon corruption, all client and firewall instances with $\mathsf{label.accept} \neq 1$ change their corrupted values to 1. Moreover, any client or firewall instance generated after this CorruptS query will have corrupted initialized to 1.
- $\mathsf{Test_{kt}}(\mathsf{label})$: with index a string $\mathsf{kt} \in \{\mathsf{cs}, \mathsf{cfs}\}$, on input an instance label, this oracle first verifies that $\mathsf{label.k_{kt}} \neq \perp$ (otherwise, it outputs $\perp$). Then, depending on the test instance it responds as follows.
  - if label is a client instance, it outputs either the key stored in $\mathsf{label.k_{kt}}$ (if $\mathsf{label.b} = 1$), or a randomly-chosen key of the same length (if $\mathsf{label.b} = 0$);
  - if label is a server or firewall instance, it returns $\perp$ if it has no partnered instance label' of type C. Otherwise, it returns $\mathsf{Test_{kt}}(\mathsf{label'})$.
- $\mathsf{TestSend}(\mathsf{label}, m)$: on input an instance label and a message m, if $F[\mathsf{label}] = \perp$ (F is used to keep track of the firewalls), the oracle creates an RF instance labeled $F[\mathsf{label}]$:
  - if $\mathsf{label.b} = 1$, this oracle acts as $\mathsf{Send}(\mathsf{label}, m)$ except that it does not update $\mathscr{L}_s$;
  - else, this oracle acts as $\mathsf{Send}(F[\mathsf{label}], \mathsf{Send}(\mathsf{label}, \mathsf{Send}(F[\mathsf{label}], m)))$ except that it does not update $\mathscr{L}_s$ (without loss of generality, we assume that the firewall just forward the session initialization message);

This oracle adds label to a list $\mathscr{L}_*$ (initialized as an empty list).

As we only consider unilateral authentication, $\mathsf{Test}_{\mathsf{kt}}$ is defined asymmetrically, depending on which party the tested instance belongs to. Our last oracle TestSend either sends the unmodified message m, as in a Send query, or forces the RF to be active on this transmission. It will be used to show that the actions of our RF go unnoticed and so will only be used in the obliviousness and transparency experiments.

Forward Secrecy. Whenever CorruptS is queried, any ongoing instance label (*i.e.*, any instance such that label.accept $\neq 1$) has their corrupt bit set to 1. However, *completed, accepting* instances are excluded from this and can still be tested. This models forward secrecy. We will include the CorruptS query in all our security definitions, thus capturing forward secrecy by default.

Reveal and State Reveal. Krawczyk and Wee [16] also consider an oracle that reveals parts of the ephemeral state of a given party. In multi-stage AKE protocols, revealing state distinguishes inputs that the keys depend on from inputs whose knowledge does not affect the security of the channel. For simplicity, we omit state reveal queries in this paper, since we focus on exfiltration resistance and security in the presence of malicious firewalls, for which state reveals are less interesting.

Freshness. To rule out trivial attacks, we introduce the notion of *freshness*, which determines the instances that $\mathscr{A}$ is allowed to attack.

Definition 1 (Freshness). *Let* $\mathsf{kt} \in \{\mathsf{cs}, \mathsf{cfs}\}$. *An instance* label *is* k_{kt}-fresh *if:*

- label.accept $= 1$ *(for a client or firewall instance); and* label.corrupted $= 0$
- $\mathsf{label.revealed}_{\mathsf{kt}} = 0$, *and* $\mathsf{label'.revealed}_{\mathsf{kt}} = 0$ *for any partner* label$'$ *of* label*;*

Channel Security. Whenever the session keys are indistinguishable from random for the attacker, one can implicitly rely on existing work on constructing secure channels by composition, *e.g.* [4,14] to prove the security of the established channel. However, in the case of exfiltration resistance the hypothesis does not hold: the corrupt implementation of the client does have access to the keys.

2.2 The Security of k_{cs}

We look at the AKE security of the key k_{cs} when the firewall is malicious as it clearly implies security with an absent or an honest firewall. Our definition of AKE does not demand *explicit* server-authentication (see [2]). Since the adversary knows $\mathsf{k}_{\mathsf{cfs}}$, the default testing mode is $\mathsf{Test}_{\mathsf{cs}}(\cdot)$. The adversary first runs the Setup algorithm, outputting the public parameters to the challenger (since the server credentials are not included, the adversary does not learn them).

Definition 2 (CS-AKE security – k_{cs}). *A* cs-*AKE protocol* Π *is* secure *if for all PPT adversaries* $\mathscr{A}$, $\mathsf{Adv}_{\Pi,\mathscr{A}}^{\mathsf{CS\text{-}AKE}}(\lambda) = |2 \cdot \Pr[1 \leftarrow \mathsf{Exp}_{\Pi,\mathscr{A}}^{\mathsf{CS\text{-}AKE}}(\lambda)] - 1|$ *is negligible in the security parameter* λ, *where* $\mathsf{Exp}_{\Pi,\mathscr{A}}^{\mathsf{CS\text{-}AKE}}(\lambda)$ *is given in Fig. 1.*

$\mathsf{Exp}^{\mathsf{CS\text{-}AKE}}_{\Pi,\mathscr{A}}(\lambda)$:
1. $(\mathsf{sparam}, \mathsf{pparam}) \leftarrow \mathscr{A}^{\mathsf{Setup}(\cdot)}(1^\lambda)$
2. $\mathsf{Q} \leftarrow \{\mathsf{NewInstance}, \mathsf{Send}, \mathsf{Reveal}, \mathsf{CorruptS}, \mathsf{Test}_{\mathsf{cs}}(\cdot)\}$
3. $(\mathsf{label}, b_*) \leftarrow \mathscr{A}^{\mathsf{Q}}(\mathsf{sparam}, \mathsf{pparam})$
4. if (label is $\mathsf{k_{cs}}$-fresh) $\wedge$ ($\mathsf{label.b} = b_*$): return 1
5. else: return a random bit

$\mathsf{Exp}^{\mathsf{CFS\text{-}AKE}}_{\Pi,\mathscr{A}}(\lambda)$:
1. $(\mathsf{sparam}, \mathsf{pparam}) \leftarrow \mathsf{Setup}(\cdot)(1^\lambda)$
2. $\mathsf{Q} \leftarrow \{\mathsf{NewInstance}, \mathsf{Send}, \mathsf{Reveal}, \mathsf{CorruptS}, \mathsf{Test}_{\mathsf{cfs}}(\cdot)\}$
3. $(\mathsf{label}, b_*) \leftarrow \mathscr{A}^{\mathsf{Q}}(\mathsf{pparam})$
4. if (label is $\mathsf{k_{cfs}}$-fresh) $\wedge$ ($\mathsf{label.b} = b_*$): return 1
5. else: return a random bit

$\mathsf{Exp}^{\mathsf{Exf}}_{\Pi,\mathscr{A}}(\lambda)$:
1. $(\mathsf{sparam}, \mathsf{pparam}) \leftarrow \mathsf{Setup}(1^\lambda)$
2. $(\mathcal{P}^0, \mathcal{P}^1) \leftarrow \mathscr{A}(\mathsf{pparam})$
3. $\mathsf{Q} \leftarrow \{\mathsf{NewInstance}, \mathsf{Send}, \mathsf{Reveal}\}$
4. $(\mathsf{label}_{FW}, b_*) \leftarrow \mathscr{A}^{\mathsf{Q}}(\mathsf{pparam}, \mathcal{P}^0, \mathcal{P}^1)$
5. if (label_{FW} is exfiltration-fresh) $\wedge$ ($\mathsf{label}_{FW}.\mathsf{b} = b_*$): return 1
6. else: return a random bit

$\mathsf{Exp}^{\mathsf{OBL}}_{\Pi,\mathscr{A}}(\lambda)$:
1. $(\mathsf{sparam}, \mathsf{pparam}) \leftarrow \mathscr{A}^{\mathsf{Setup}(\cdot)}(1^\lambda)$
2. $\mathsf{Q} \leftarrow \{\mathsf{NewInstance}, \mathsf{Send}, \mathsf{Reveal}, \mathsf{CorruptS}, \mathsf{TestSend}\}$
3. $(\mathsf{label}, b_*) \leftarrow \mathscr{A}^{\mathsf{Q}}(\mathsf{sparam}, \mathsf{pparam})$
4. if (label is Send-fresh)$\wedge$ ($\mathsf{label.b} = b_*$): return 1
5. else: return a random bit

Fig. 1. Experiments for the CS-AKE, CFS-AKE, exfiltration and obliviousness games.

2.3 The Security of $\mathsf{k_{cfs}}$

The adversary can no longer control the RF (which is able to compute $\mathsf{k_{cfs}}$) and simply acts as a Man-in-the-Middle between the client and the firewall and/or the firewall and the server. The adversary first runs the Setup algorithm, outputting the parameters to the challenger. We focus on the first "layer" of keys, *i.e.*, on $\mathsf{k_{cfs}}$, thus, the default mode for the test oracle is $\mathsf{Test}_{\mathsf{cfs}}(\cdot)$.

Definition 3 (AKE security – $\mathsf{k_{cfs}}$). *A* cfs-*AKE protocol* Π *is* $\mathsf{k_{cfs}}$-secure *if for all PPT adversaries* $\mathscr{A}$, $\mathsf{Adv}^{\mathsf{CFS\text{-}AKE}}_{\Pi,\mathscr{A}}(\lambda) = |2 \cdot \Pr[1 \leftarrow \mathsf{Exp}^{\mathsf{CFS\text{-}AKE}}_{\Pi,\mathscr{A}}(\lambda)] - 1|$ *is negligible in* λ*, where* $\mathsf{Exp}^{\mathsf{CFS\text{-}AKE}}_{\Pi,\mathscr{A}}(\lambda)$ *is given in Fig. 1.*

2.4 The *Malicious Client* Scenario

The malicious client scenario is the core motivation behind this work. The firewall and server are both honest, but the adversary can tamper with the client-side implementation. The RF must guarantee the security of the session keys and ensure exfiltrate resistance. In our model, we also require this guarantee to hold *for record-layer transmissions.*

Client-Adversary Interaction. We consider a MitM adversary $\mathscr{A}$ situated between the firewall and the server[1]. The attacker has substituted the client-side implementation but this implementation has no direct or covert communication channel with the adversary.

The game starts with an honest Setup phase, then after being given the public parameters, the adversary $\mathscr{A}$ creates two client *programs* $\mathcal{P}^0, \mathcal{P}^1$ using

[1] Exfiltration resistance would obviously be unachievable elsewhere.

some suitable (unspecified) encoding, and outputs them to the challenger. The adversary will query $\mathsf{NewInstance}(FW)$ to create new firewall instance(s), which will (unbeknownst to $\mathscr{A}$) interact with one of the two client programs $\mathcal{P}^0$ or $\mathcal{P}^1$, provided by $\mathscr{A}$. In this case the firewall instance's test bit $\mathsf{b} \in \{0, 1\}$ indicates which of the two adversarial programs it interacts with.

In the malicious client scenario, the adversary interacts directly with the server and the firewall instances through the Send queries, but only indirectly (through the firewall) with the client programs. We note that $\mathscr{A}$'s interaction with the server and the firewall is still arbitrary, *i.e.*, $\mathscr{A}$ still actively controls the messages sent between them. The goal of the adversary is to guess which of the supplied client programs $\mathcal{P}^0$, $\mathcal{P}^1$ the firewall has been interacting with. Formally, $\mathscr{A}$ outputs a tuple consisting of a *firewall* instance label_{FW} and a bit d, representing its guess of $\mathsf{label}_{FW}.\mathsf{b}$.

Trivial Attacks. It is impossible to prove this "left-or-right" exfiltration resistance for *arbitrary* programs $\mathcal{P}^0, \mathcal{P}^1$: $\mathscr{A}$ could output a program $\mathcal{P}^0$ which produces illegal messages, whereas $\mathcal{P}^1$ emulates the protocol perfectly. The firewall will then abort for $\mathsf{label}_{FW}.\mathsf{b} = 0$, but not for $\mathsf{label}_{FW}.\mathsf{b} = 1$, allowing $\mathscr{A}$ to distinguish the two cases. Similarly, $\mathscr{A}$ can also trivially win if $\mathcal{P}^0$ and $\mathcal{P}^1$ output messages of different *lengths*.

Informally, we require that the *externally observable behavior* of the two programs $\mathcal{P}^0$, $\mathcal{P}^1$, *i.e.*, what the RF and server do in response to the programs' messages, should be similarly structured: the number of messages generated should be the same and these messages should pairwise have the same length. However, we do not restrict the semantics of these messages. This restriction applies only to the programs selected for the target instance; all other instances can behave in any way.

Formally, we consider a program $\mathcal{P}$, a firewall FW, and a server S. We denote by $\tau_{[\mathsf{label}_{FW}(\mathcal{P}) \leftrightarrow \mathsf{label}_S]}$ the ordered *transcript* of messages sent between the firewall (interacting with $\mathcal{P}$) and the server. Let $L = \left|\tau_{[\mathsf{label}_{FW}(\mathcal{P}) \leftrightarrow \mathsf{label}_S]}\right|$ denote the length of the transcript. For $i \in \{1, \ldots, L\}$, let $\tau_{[\mathsf{label}_S \leftrightarrow \mathsf{label}_{FW}(\mathcal{P})]}[i]$ denote the i-th message in the transcript, and let $\left|\tau_{[\mathsf{label}_S \leftrightarrow \mathsf{label}_{FW}(\mathcal{P})]}[i]\right|$ denote its length.

Definition 4 (Transcript equivalence). *Two programs $\mathcal{P}^0$ and $\mathcal{P}^1$ have* equivalent transcripts *in an execution with the firewall FW and the server S if the following conditions hold:*

1. $\left|\tau_{[\mathsf{label}_{FW}(\mathcal{P}^0) \leftrightarrow \mathsf{label}_S]}\right| = \left|\tau_{[\mathsf{label}_{FW}(\mathcal{P}^1) \leftrightarrow \mathsf{label}_S]}\right|$.
2. $\left|\tau_{[\mathsf{label}_{FW}(\mathcal{P}^0) \leftrightarrow \mathsf{label}_S]}[i]\right| = \left|\tau_{[\mathsf{label}_{FW}(\mathcal{P}^1) \leftrightarrow \mathsf{label}_S]}[i]\right|$ *for each* $i \in \{1, \ldots, L\}$.

Only allowing client programs that generate equivalent transcripts rules out the trivial attacks mentioned above. Interestingly, it is arguably much easier to decide whether two programs have equivalent transcripts (as it is only based on measurable quantities) than to determine if a corrupt program is "functionality-maintaining" in the sense of [9]. It allows realistic attacks, *e.g.*, selection of messages whose ciphertexts have very specific features, contrarily to [9].

Some of our requirements are fullfilled by using specific cryptographic tools, such as *length-hiding authenticated encryption* [13,15] which would, for example, make the second condition easily satisfied. Regarding the first condition, differences in the number of exchanged messages are likely to imply an unusual behaviour of the client (*e.g.*, a large number of aborted connections) and so can potentially be detected by an external security event management system.

Definition 5 (Exfiltration freshness). *A firewall* label_{FW} *is* exfiltration-fresh with respect to programs $\mathcal{P}^0, \mathcal{P}^1$ *if:*

- $\mathsf{label}_{FW}.\mathsf{accept} = 1$*; for any* label, $\mathsf{label.revealed_{cfs}} = 0$*;*
- *the programs* $\mathcal{P}^0$ *and* $\mathcal{P}^1$ *generate equivalent transcripts in their execution with the firewall instance* label_{FW} *and its partnering server instance* label_S.

Definition 6 (Exfiltration). *An AKE protocol* Π *is* exfiltration secure *if for all PPT adversaries* $\mathscr{A}$, $\mathsf{Adv}^{\mathsf{Exf}}_{\Pi,\mathscr{A}}(\lambda) = 2 \cdot |\Pr[1 \leftarrow \mathsf{Exp}^{\mathsf{Exf}}_{\Pi,\mathscr{A}}(\lambda)] - 1|$ *is negligible in* λ*, where* $\mathsf{Exp}^{\mathsf{Exf}}_{\Pi,\mathscr{A}}(\lambda)$ *is given in Fig. 1.*

2.5 Obliviousness

Dodis *et al.* take the *obliviousness* into account for their protocols: the client (resp. the server) cannot distinguish[2] whether it interacts with the firewall or the server (resp. the client). To discard trivial attacks, we define the *sending freshness.* A label is *sending-fresh* when it is never queried in Send and its related firewall $F[\mathsf{label}]$ is never an input of the Send and Reveal oracles.

Definition 7 (Sending freshness). *A firewall instance* label *is* send-fresh *if* $F[\mathsf{label}] \notin \mathscr{L}_s \cup \mathscr{L}_{\mathsf{cfs}}$ *and* $\mathsf{label} \notin \mathscr{L}_s$.

Definition 8 (Obliviousness). *An AKE protocol* Π *is* oblivious *if for all PPT adversaries* $\mathscr{A}$, $\mathsf{Adv}^{\mathsf{OBL}}_{\Pi}(\mathscr{A}) = |2\Pr[\mathsf{Exp}^{\mathsf{OBL}}_{\Pi,\mathscr{A}}(\lambda) \Rightarrow 1] - 1|$ *is negligible in* λ*, where* $\mathsf{Exp}^{\mathsf{OBL}}_{\Pi,\mathscr{A}}(\lambda)$ *is given in Fig. 1.*

3 Our Reverse Firewall

We now describe a message-transmission protocol compatible with a reverse firewall. We consider a setting of unilateral authentication, *i.e.*, only the server authenticates itself, using digital signatures. As in [9], we start with a simple DH key exchange protocol between a client and a server, modified to accommodate an RF. In particular, the goal of this new key exchange is for the client and the server to compute two keys: $\mathsf{k_{cs}}$ known only to them, and $\mathsf{k_{cfs}}$ which will also be known to the firewall. We manage to avoid the use of pairings and only use standard cryptographic tools (such as CPA-secure public key encryption) along with some tricks to ensure the obliviousness of our protocol.

[2] The related definition of *transparency* is given in Appendix B.

However, we must keep in mind that a shared key is just a tool to protect further communication, and we thus need to explain how the RF can proceed to preserve security of future messages sent by a corrupt client, while only having access to $\mathsf{k_{cfs}}$. This is particularly difficult as the record-later only involves symmetric key primitives, and moreover, other subtleties prevent us from using more "natural" solutions.

For the sake of clarity, we describe our protocol step by step, first presenting a key agreement with a passive firewall in Sect. 3.1 and then explaining in Sect. 3.2 how the latter can act to preserve security even with a corrupt client. Finally, we will consider in Sect. 3.3 the record-layer, describing how the keys generated during the key agreement should be used to prevent exfiltration at this stage.

3.1 Signed Diffie-Hellman with Passive/No Firewall

Setup. Before the protocol runs there is a one-time setup phase where the RF chooses a public/private key pair $(\mathsf{pk}_{FW}, \mathsf{sk}_{FW})$ for some public encryption scheme Enc, such as the one proposed by El Gamal [11], and sends pk_{FW} to the client. In case the client does not have an RF installed (and hence no pk_{FW} value), the client draws a new random element from the key space in every protocol run and uses this as a substitute for pk_{FW}.

Client $(\mathsf{pk}_{FW}, \mathsf{pk}_S)$		**Firewall** $(\mathsf{sk}_{FW}, \mathsf{pk}_{FW}, \mathsf{pk}_S)$		**Server** $(\mathsf{sk}_S, \mathsf{pk}_S)$
$x, c \xleftarrow{\$} \mathbb{Z}_p$				$y, d, \beta_1, \beta_2 \xleftarrow{\$} \mathbb{Z}_p$
$X = g^x, C = g^c$				$Y = g^y, D = g^d$
$e = \mathsf{Enc}_{\mathsf{pk}_{FW}}(c)$				
	$\xrightarrow{(X, C, e)}$	$c = \mathsf{Dec}_{\mathsf{sk}_{FW}}(e)$	$\xrightarrow{(X, C, e)}$	
				$\sigma = \mathsf{Sign}_{\mathsf{sk}_S}(Y, D, X^{\beta_1}, C^{\beta_2})$
If σ verifies:	$\xleftarrow{(\sigma, Y, D, \beta_1, \beta_2)}$		$\xleftarrow{(\sigma, Y, D, \beta_1, \beta_2)}$	
$\mathsf{k_{cs}} \leftarrow Y^{x \cdot \beta_1}$				$\mathsf{k_{cs}} \leftarrow X^{y \cdot \beta_1}$
$\mathsf{k_{cfs}} \leftarrow D^{c \cdot \beta_2}$		$\mathsf{k_{cfs}} \leftarrow D^{c \cdot \beta_2}$		$\mathsf{k_{cfs}} \leftarrow C^{d \cdot \beta_2}$
Else: abort.				

Fig. 2. A key agreement protocol between the client, a passive firewall, and the server.

The Protocol. Figure 2 depicts how the protocol runs in the presence of a passive firewall. The client begins by choosing two ephemeral DH shares $(X, C) \leftarrow (g^x, g^c)$ in some cyclic group $\mathbb{G}$ generated by g, encrypting c with the firewall public key pk_{FW} to get $e = \mathsf{Enc}_{\mathsf{pk}_{FW}}(c)$, and sending (X, C, e) to the server forwarded unmodified by the firewall. The latter decrypts e with sk_{FW} to get c.

The server in turn chooses two ephemeral DH shares $(Y, D) \leftarrow (g^y, g^d)$ and two random elements (β_1, β_2) from $\mathbb{Z}_p$, and sends these, together with a signature of $(Y, D, X^{\beta_1}, C^{\beta_2})$ to the client (the firewall just forwards this message). The server also computes the keys $\mathsf{k_{cs}} \leftarrow X^{y \cdot \beta_1}$ and $\mathsf{k_{cfs}} \leftarrow C^{d \cdot \beta_2}$. Provided the signature is valid, the client computes the keys $\mathsf{k_{cs}} \leftarrow Y^{x \cdot \beta_1}$ and $\mathsf{k_{cfs}} \leftarrow D^{c \cdot \beta_2}$. Finally, the RF computes $\mathsf{k_{cfs}} \leftarrow D^{c \cdot \beta_2}$. The scheme is correct since $X^y = Y^x = g^{x \cdot y}$ and $C^d = D^c = g^{c \cdot d}$.

Security. Intuitively, only the parties knowing x or y can recover $\mathsf{k_{cs}}$, and only the parties knowing d or c can recover $\mathsf{k_{cfs}}$ – which has no actual use in case the RF is passive or absent. The cs-AKE security of this protocol is implied by that of the following protocol, where we consider an active, potentially malicious, adversary.

Client $(\mathsf{pk}_{FW}, \mathsf{pk}_S)$		**Firewall** $(\mathsf{sk}_{FW}, \mathsf{pk}_{FW}, \mathsf{pk}_S)$		**Server** $(\mathsf{sk}_S, \mathsf{pk}_S)$
$x, c \xleftarrow{\$} \mathbb{Z}_p$		$\alpha_1, \alpha_2 \xleftarrow{\$} \mathbb{Z}_p$		$y, d, \beta_1, \beta_2 \xleftarrow{\$} \mathbb{Z}_p$
$X = g^x, C = g^c$				$Y = g^y, D = g^d$
$e = \mathsf{Enc}_{\mathsf{pk}_{FW}}(c)$				
	$\xrightarrow{(X, C, e)}$	$c = \mathsf{Dec}_{\mathsf{sk}_{FW}}(e)$		
		If $X \notin \mathbb{G}$ or $C \notin \mathbb{G}$ or $g^c \neq C$: abort		
		Else: $\tilde{e} = \mathsf{Enc}_{\mathsf{pk}_{FW}}(c \cdot \alpha_2)$		
		$\tilde{X} = X^{\alpha_1}, \tilde{C} = C^{\alpha_2}$	$\xrightarrow{(\tilde{X}, \tilde{C}, \tilde{e})}$	
				$\sigma = \mathsf{Sign}_{\mathsf{sk}_S}(Y, D, \tilde{X}^{\beta_1}, \tilde{C}^{\beta_2})$
			$\xleftarrow{(\sigma, Y, D, \beta_1, \beta_2)}$	
		If $(Y, D) \notin \mathbb{G}^2$ or invalid σ: abort.		
		Else: $\gamma_1 = \alpha_1 \cdot \beta_1, \gamma_2 = \alpha_2 \cdot \beta_2$		
If σ verifies:	$\xleftarrow{(\sigma, Y, D, \gamma_1, \gamma_2)}$			
$\mathsf{k_{cs}} \leftarrow Y^{x \cdot \gamma_1}$				$\mathsf{k_{cs}} \leftarrow \tilde{X}^{y \cdot \beta_1}$
$\mathsf{k_{cfs}} \leftarrow D^{c \cdot \gamma_2}$		$\mathsf{k_{cfs}} \leftarrow D^{c \cdot \gamma_2}$		$\mathsf{k_{cfs}} \leftarrow \tilde{C}^{d \cdot \beta_2}$
Else: abort.				

Fig. 3. An active reverse firewall for the protocol in Fig. 2, using the same notations.

3.2 Signed Diffie-Hellman with an Active Firewall

Next, we construct an active RF for the signed DH protocol described in Sect. 3.1. Note that the protocol in Fig. 2 is compatible with the rerandomization proposed by Dodis et al. [9]: the RF can rerandomize the keyshares without impacting the correctness of the resulting protocol.

In Fig. 3 we describe the active RF for the protocol of Fig. 2. The RF rerandomizes the two DH elements X and C with exponents α_1 and α_2, respectively, to obtain $\tilde{X}$ and $\tilde{C}$. It also decrypts e to obtain c, and encrypts $c \cdot \alpha_2$ (with its own public key[3] pk_{FW}) to get $\tilde{e}$, before forwarding $(\tilde{X}, \tilde{C}, \tilde{e})$ to the receiving endpoint. Upon reception, the server proceeds as before. The rerandomization made by the RF impacts the input to the server's signature; the firewall must therefore include its random values to the subsequent response to the client: the server sends $(\sigma, Y, D, \beta_1, \beta_2)$, and the firewall will forward $(\sigma, Y, D, \gamma_1 = \alpha_1 \cdot \beta_1, \gamma_2 = \alpha_2 \cdot \beta_2)$.

[3] For transparency, the message sent by the RF must have the same format as (X, C, e).

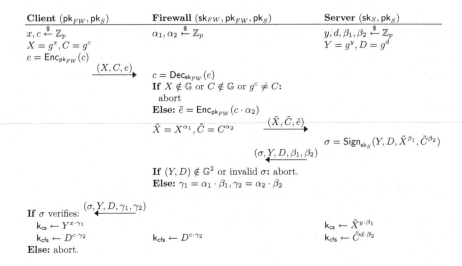

Apart from rerandomizing the transcript, the RF verifies that the received elements are from the correct groups and that they are not the neutral element of those groups. In addition, it checks the validity of the server's signature for the rerandomized transcript, *i.e.*, it checks that the server signed $(Y, D, \tilde{X}^{\beta_1}, \tilde{C}^{\beta_2})$.

This protocol ensures obliviousness, as stated in Theorem 4. This comes from the fact that the transcripts of the honestly-run protocol (in Fig. 2) and the ones from the protocol in Fig. 3 are identically distributed from the point of view of the endpoints, and a tampered client implementation cannot distinguish whether it is being monitored or not.

The security of this protocol (cs- and cfs-AKE security) is formally stated in Theorem 1 and Theorem 2, for which we give proof sketches in Sect. 4, and complete proofs in the full version.

Theorem 1. *The protocol given in Fig. 3 is* cfs-*AKE secure (Definition 3) under the DDH assumption, and assuming* Sig *is EUF-CMA and* Enc *is IND-CPA.*

$$\mathsf{Adv}_{\Pi}^{\mathsf{CFS\text{-}AKE}}(\lambda) \leq n_C \cdot n_S \cdot \left(\mathsf{Adv}_{Sig}^{\mathsf{EUF\text{-}CMA}}(\lambda) + \mathsf{Adv}_{\mathsf{Enc}}^{\mathsf{IND\text{-}CPA}}(\lambda) + \mathsf{Adv}_{\mathbb{G}}^{\mathsf{DDH}}(\lambda)\right)$$

Theorem 2. *The protocol given in Fig. 3 is* cs-*AKE secure (Definition 2) under the DDH assumption, and assuming* Sig *is EUF-CMA.*

$$\mathsf{Adv}_{\Pi}^{\mathsf{CS\text{-}AKE}}(\lambda) \leq n_C \cdot n_S \cdot \left(\mathsf{Adv}_{Sig}^{\mathsf{EUF\text{-}CMA}}(\lambda) + \mathsf{Adv}_{\mathbb{G}}^{\mathsf{DDH}}(\lambda)\right)$$

We discuss solutions to "stack" several RFs, and to adapt our RF to a real-world TLS-like protocol in the full version.

3.3 Record-Layer Firewall

Theorem 2 and Theorem 1 prove that our protocol is a secure AKE protocol, with or without an RF. What is still missing is a proof of exfiltration resistance, as per Definition 6. We do not prove this property directly for the protocol when viewed as an AKE, as our goal is to also cover exfiltration resistance when the AKE is combined with a subsequent encryption scheme, and we thus prove exfiltration resistance for the whole channel establishment protocol (AKE + encryption scheme). We explain how our RF can prevent exfiltration at the record-layer.

Double CPA Encryption Fails. When the client's implementation cannot be trusted, the security of the encryption algorithm used at the record layer becomes irrelevant. Even the RF cannot ensure the security of ciphertexts formed with $\mathsf{k_{cs}}$ because it does not have this key.

We cannot simply add an independent, external encryption layer, taking $\mathsf{k_{cfs}}$ as the key, known to the RF, hoping it could preserve security by first decrypting this external layer and then re-encrypting it; even if it seems that no matter how the client implementation behaves, the adversary would only see a valid ciphertext generated using $\mathsf{k_{cfs}}$, which it does not know. It might therefore be

Client $(\mathsf{k_{cfs}}, \mathsf{k_{cs}})$	Firewall $(\mathsf{k_{cfs}})$	Server $(\mathsf{k_{cfs}}, \mathsf{k_{cs}})$
$C \leftarrow \mathtt{AE.Enc}(\mathsf{k_{cs}}, \ell, \mathsf{he}, M, \mathsf{st_E})$		
$r \xleftarrow{\$} \{0,1\}^\lambda$, $k_1 \leftarrow H_1(r \Vert \mathsf{k_{cfs}})$	$\tilde{r} \xleftarrow{\$} \{0,1\}^\lambda$, $\tilde{k}_1 \leftarrow H_1(\tilde{r} \Vert \mathsf{k_{cfs}})$	
$k_2 \leftarrow H_2(r \Vert \mathsf{k_{cfs}})$,	$\tilde{k}_2 \leftarrow H_2(\tilde{r} \Vert \mathsf{k_{cfs}})$,	
$s \leftarrow k_1 \oplus C$, $t = \mathsf{MAC}_{k_2}(r \Vert s) \xrightarrow{(r,s,t)}$	**If** $\lvert s \rvert \neq \ell$: abort	
	$k_1 \leftarrow H_1(r \Vert \mathsf{k_{cfs}})$, $k_2 \leftarrow H_2(r \Vert \mathsf{k_{cfs}})$,	
	$\tilde{C} \leftarrow k_1 \oplus s$	
	$\tilde{s} \leftarrow \tilde{k}_1 \oplus \tilde{C}$, $\tilde{t} = \mathsf{MAC}_{\tilde{k}_2}(\tilde{r} \Vert \tilde{s}) \xrightarrow{(\tilde{r},\tilde{s},\tilde{t})}$	$\tilde{k}_1 \leftarrow H_1(\tilde{r} \Vert \mathsf{k_{cfs}})$
		$\tilde{k}_2 \leftarrow H_2(\tilde{r} \Vert \mathsf{k_{cfs}})$
		$\tilde{\tilde{C}} \leftarrow \tilde{k}_1 \oplus \tilde{s}$
		$\tilde{M} \leftarrow \mathtt{AE.Dec}(\mathsf{k_{cs}}, \mathsf{he}, \tilde{\tilde{C}}, \mathsf{st_D})$
		If $\tilde{M} = \bot$ or $\tilde{t}$ invalid: abort

Fig. 4. An active reverse firewall for the secure transmission of a message M, using an stLHAE scheme, where $H_1 : \{0,1\}^* \rightarrow \{0,1\}^\ell$ and $H_2\{0,1\}^* \rightarrow \mathscr{K}$ (where $\mathscr{K}$ is the key set of the MAC) are hash functions modelled as random oracles in the security analysis. The encryption/decryption states used for the inner layer is $(\mathsf{st_E}, \mathsf{st_D})$. A passive firewall simply forwards (r, s, t).

tempting to conclude that this protocol is exfiltration resistant, assuming IND-CPA security of the encryption scheme.

However, the IND-CPA security of an encryption scheme ensures that no adversary can decide if a given ciphertext encrypts a message m of its choice. Our adversary is much more powerful here: the client, controlled by the adversary, selects the messages to send while knowing the secret keys, which is not allowed in the IND-CPA security game. Moreover, the corrupt implementation could simply select messages whose ciphertexts contain some specific patterns that could be used as a distinguisher.

KDM Encryption Fails. Allowing the adversary to select messages while knowing the key is reminiscent of the key-dependent message model from [3]. Such an encryption scheme remains secure even if the adversary receives the encryption of some messages $m = f(\mathsf{sk})$, where f is chosen by the adversary and sk is the secret key. Unfortunately, this does not exactly match our scenario: here, we have two adversaries, one (the corrupt client) who selects the messages to be encrypted and knows the key, and one (the adversary between the RF and the server) who does not know the key and must distinguish the encryption of these messages. Our model is stronger than KDM, and we therefore cannot fully rely on this property.

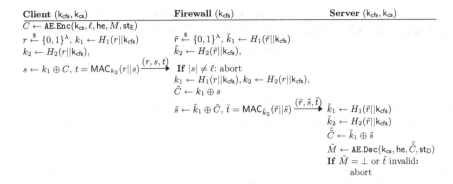

Our Solution. Nonetheless, the construction proposed in [3] *does* provide an answer to our problem. In our protocol, the client first encrypts the plaintext message into a ciphertext C using a length-hiding authenticated encryption and the secret key $\mathsf{k_{cs}}$. The client then picks r and computes $C' = H(r \Vert \mathsf{k_{cfs}}) \oplus C$, which is similar to the technique in [3]. It sends (r, C') to the firewall which decrypts C' and re-encrypts it by computing $\tilde{C}' = H(\tilde{r} \Vert \mathsf{k_{cfs}}) \oplus C$ using a fresh random $\tilde{r}$. We already showed that the key $\mathsf{k_{cfs}}$ is indistinguishable from random. Thus, as long as the adversary does not guess $\mathsf{k_{cfs}}$, the value $H(\tilde{r}||\mathsf{k_{cfs}})$ is indistinguishable from a random one (in the ROM) and acts as a one-time-pad on C. This means that $\tilde{C}'$ leaks no information to the adversary. Finally, since a one-time-pad does not prevent malleability, we need to additionally compute a MAC on $(\tilde{r}, \tilde{C}')$ by using a key derived from $\mathsf{k_{cfs}}$.

The resulting protocol is exfiltration resistant as stated in Theorem 3, and is also oblivious in both phases, as stated in Theorem 4. Proof sketches are given in Sect. 4 and Appendix C, respectively, with the complete proof of Theorem 3 given in the full version.

Theorem 3. *Protocol* Π *is exfiltration resistant (Definition 6) in the ROM under the CDH, and assuming that* Sig *is EUF-CMA and* Enc *is IND-CPA.*

$$\mathsf{Adv}_{\Pi}^{\mathsf{Exf}}(\lambda) \leq q_m^2/2^\lambda + \mathsf{Adv}_{\mathsf{Enc}}^{\mathsf{IND-CPA}}(\lambda) + n_s \cdot n_f \cdot (q_1 + q_2) \cdot \mathsf{Adv}_{\mathscr{G}}^{\mathsf{CDG}}(\lambda) + n_m \cdot q_s \cdot \mathsf{Adv}_{\mathsf{MAC}}^{\mathsf{EUF-CMA}}(\lambda)$$

Theorem 4. *Protocol* Π *is unconditionally oblivious (Definition 8).*

4 Security Proofs

We give a sketch of the proofs of Theorems 1 to 3, while the sketch of Theorems 4 is in Appendix C. Complete proofs are in the full version. For each security proof, we define the sid of an instance label (either at the firewall FW or the server S) to be its input to the signature scheme, i.e., referring to Fig. 3, we have $\mathsf{sid} = (Y, D, \tilde{X}^{\beta_1}, \tilde{C}^{\beta_2})$.

Proof of Theorem 1. Let $\mathscr{A}$ be an adversary against the CFS-AKE security of protocol Π. In the following sequence of games, let S_i denote the event that $\mathscr{A}$ is successful in Game i, and let $\epsilon_i = \Pr[S_i] - \frac{1}{2}$. Thus, S_i denotes the event that $\mathscr{A}$ correctly guesses the bit $\mathsf{label}_C.\mathsf{b}$ of a $\mathsf{k_{cfs}}$-fresh client instance label_C. We denote by label_T^i the ith label of type T created during the experiment.

Game 0. This game is the original experiment, thus: $\epsilon_0 = \mathsf{Adv}_{\Pi,\mathscr{A}}^{\mathsf{CFS\text{-}AKE}}(\lambda)$.

Game 1. Let n_C be the number of client labels created during the experiment. This game is the same as Game 0, except that the challenger picks $i \xleftarrow{\$} \{0, \ldots, n_C\}$ at the beginning of the game. If $\mathscr{A}$ does not return (label_C^i, d) for some bit d, the challenger returns a random bit. We have $\epsilon_0 \leq \epsilon_1 \cdot n_C$.

Game 2. Let n_S be the number of server labels created during the experiment. This game is the same as Game 1, except that the challenger picks $j \xleftarrow{\$} \{0, \ldots, n_S\}$ at the beginning of the game. If label_C^i and label_S^j are not partners, the challenger returns a random bit. We have $\epsilon_1 \leq \epsilon_2 \cdot n_S$.

Game 3. This is the same as Game 2 except that the challenger aborts, sets $\mathtt{abort}_3 = 1$ and returns a random bit if:

- $\mathscr{A}$ makes a $\mathsf{Send}(\mathsf{label}, (\sigma, Y, D, \gamma_1, \gamma_2))$ query before $\mathscr{A}$ makes any query to the oracle $\mathsf{CorruptS}$,

- $\mathsf{Send}(\mathsf{label}, \mathtt{init})$ previously output a triplet (X, C, e) such that σ is a valid signature on $(Y, D, X^{\gamma_1}, C^{\gamma_1})$, and
- the server did not previously output $(\sigma, Y, D, \beta_1, \beta_2)$ for some β_1 and β_2.

We claim that $|\Pr[S_2] - \Pr[S_3]| \leq \Pr[\mathtt{abort}_3] \leq \mathsf{Adv}^{\mathsf{EUF\text{-}CMA}}_{\mathsf{Sign}}(\lambda)$. We show that, using a PPT algorithm $\mathscr{A}$ that, with non-negligible probability, makes a $\mathsf{Send}(\mathsf{label}, (\sigma, Y, D, \gamma_1, \gamma_2))$ query, an adversary $\mathscr{B}$ can efficiently break the EUF-CMA experiment by outputting σ.

Game 4. This is the same as Game 3 except that when the challenger should encrypt a message c using the public key of the firewall pk_{FW}, the challenger picks a random value and encrypts it. Enc denotes the public key encryption scheme we use.

We claim that $|\Pr[S_3] - \Pr[S_4]| \leq \mathsf{Adv}^{\mathsf{IND\text{-}CPA}}_{\mathsf{Enc}}(\lambda)$. We show by reduction that distinguishing Games 3 and 4 is equivalent to distinguishing whether the ciphertexts c contain the valid messages or some random values, which breaks the IND-CPA security of Enc.

Game 5. This is the same as Game 4 except the challenger replaces $\mathsf{k_{cfs}}$ with random for the client label label^i_C.

We claim that $|\Pr[S_4] - \Pr[S_5]| \leq \mathsf{Adv}^{\mathsf{DDH}}_{\mathbb{G}}(\lambda)$. We prove this claim by reduction using an adversary $\mathscr{B}$ receiving a DDH challenge that it embeds into the game it simulates for $\mathscr{A}$, an adversary on Game 4. $\mathscr{B}$ can distinguish Game 4 from Game 5 depending on the behavior of $\mathscr{A}$.

By the changes in the games, we have that the adversary's $\mathsf{Test_{cfs}}$ query will always be answered with a random key, thus $\epsilon_5 = 0$, which concludes the proof.

Proof of Theorem 2. Let $\mathscr{A}$ be an adversary against the CS-AKE security of protocol Π. We use the same notation as for Theorem 1.

Game 0, 1, 2 and 3 Game 0 is the original $\mathsf{Exp}^{\mathsf{CS\text{-}AKE}}_{\Pi,\mathscr{A}}(\lambda)$ experiment. Games 1, 2 and 3 are defined as in the proof of Theorem 1, with similar reductions.

Game 4. This is the same as Game 3 except that the challenger replaces $\mathsf{k_{cs}}$ with random for the client label label^i_C.

We claim that $|\Pr[S_3]-\Pr[S_4]| \leq \mathsf{Adv}^{\mathsf{DDH}}_{\mathbb{G}}(\lambda)$. We prove this claim in a similar way as for Game 4 of Theorem 1, except that $\mathscr{B}$ sets $X = A$ for label^i_C, $Y = B$ for label^j_S, and Z for building the shared key $\mathsf{k_{cs}}$.

By the changes in the games, we have that the adversary's $\mathsf{Test_{cs}}$ query will always be answered with a random key, thus $\epsilon_4 = 0$, which concludes the proof.

Proof of Theorem 3. Recall that Π is the protocol that first runs the server-authenticated AKE protocol given in Fig. 3 to obtain the keys $\mathsf{k_{cs}}$ and $\mathsf{k_{cfs}}$, which are then used in the stLHAE protocol given in Fig. 4.

Let $\mathscr{A}$ be an adversary against the exfiltration resistance of protocol Π. Let $\mathcal{P}^0$ and $\mathcal{P}^1$ be the two programs output by $\mathscr{A}$ after the setup phase in experiment $\mathsf{Exp}^{\mathsf{Exf}}_{\Pi,\mathscr{A}}(\lambda)$. We use the same notations as in the proof of Theorem 2.

Game 0. This is the original experiment, hence $\epsilon_0 = \mathsf{Adv}^{\mathsf{Exf}}_{\Pi,\mathscr{A}}(\lambda)$.

Game 1. At the ith message sent by $\mathcal{P}^b$ to the firewall, the challenger simulates the firewall by picking a random element denoted $\tilde{r}_i \xleftarrow{\$} \{0,1\}^\lambda$. This game proceeds as in Game 0, except that if the challenger picks $\tilde{r}_i$ such that there exists $j < i$ with $\tilde{r}_i = \tilde{r}_j$, then the challenger aborts, sets $\texttt{abort}_1 = 1$, and returns a random bit. Let q_m be the number of messages sent by $\mathcal{P}^b$, we have $|\Pr[S_0] - \Pr[S_1]| \leq \Pr[\texttt{abort}_1] \leq q_m^2/2^\lambda$.

Game 2. This is the same as Game 1 except that the challenger aborts, sets $\texttt{abort}_2 = 1$, and returns a random bit if $\mathscr{A}$ makes a $\mathsf{Send}(\mathsf{label}_{FW}, (\sigma, Y, \beta_1, \beta_2))$ query such that σ is a valid signature on $(Y, X^{\beta_1}, C^{\beta_2})$, and a $(X, C, \cdot)$ was output by $\mathsf{Send}(\mathsf{label}_{FW}, \texttt{init})$, and the server did not previously output $(\sigma, Y, \beta_1, \beta_2)$. We claim that $|\Pr[S_1] - \Pr[S_2]| \leq \Pr[\texttt{abort}_2] \leq \mathsf{Adv}_{\mathsf{Sign}}^{\mathsf{EUF-CMA}}(\lambda)$. We prove this claim as in Game 3 in the proof of Theorem 1. *Game 3.* This is the same game as Game 2 except that each time the challenger should encrypt a message c using the public key of the firewall pk_{FW}, the challenger picks a random value and encrypts it. Enc denotes the public key encryption scheme used in our protocol. We claim that: $|\Pr[S_2] - \Pr[S_3]| \leq \mathsf{Adv}_{\mathsf{Enc}}^{\mathsf{IND-CPA}}(\lambda)$ We prove this claim as in the Game 3 in the proof of Theorem 1.

Game 4. We recall that in Game 3, each time $\mathcal{P}^b$ sends a message (r_i, s_i, t_i) to the RF, the RF picks $\tilde{r}_i$ and computes $\tilde{k}_{j,i} \leftarrow H_j(\tilde{r}_i \| \mathsf{k}_{\mathsf{cfs}})$ for $j \in \{0, 1\}$. We set $h_i = \tilde{r}_i \| \mathsf{k}_{\mathsf{cfs}}$. Game 4 proceeds as in Game 3, but now the challenger sets $\texttt{abort}_4 = 1$, aborts, and returns a random bit if the adversary sends one of the h_i to the random oracle that simulates H_1 or H_2.

We claim that $|\Pr[S_3] - \Pr[S_4]| \leq \Pr[\texttt{abort}_4] \leq n_s \cdot n_f \cdot (q_1 + q_2) \cdot \mathsf{Adv}_{\mathbb{G}}^{\mathsf{CDH}}(\lambda)$. where q_j is the number of queries sent to the random oracle H_j, and n_f (resp. n_s) is the number of firewall (resp. server) labels. To prove this, we show how to build an algorithm $\mathscr{B}$ that solves the CDH problem from an efficient algorithm $\mathscr{A}$ that triggers $\texttt{abort}_4$ with non-negligible probability.

We note that, at this step, k_2 can be viewed as a MAC key generated at random, independently from any other element of the protocol.

Game 5. This is the same as Game 4 except that the challenger aborts, sets $\texttt{abort}_5 = 1$ and returns a random bit if $\mathscr{A}$ makes a $\mathsf{Send}(\mathsf{label}_S, (r, s, t))$ query such that the server does not abort, and no $\mathsf{Send}(\mathsf{label}, \cdot)$ query received (r, s, t) as an answer. Let n_m be the number of messages sent by the firewall during the experiment, and q_s the number of queries sent to the sending oracle.

We claim that $|\Pr[S_4] - \Pr[S_5]| \leq \Pr[\texttt{abort}_5] \leq n_m \cdot q_s \cdot \mathsf{Adv}_{\mathsf{MAC}}^{\mathsf{EUF-CMA}}(\mathscr{B})$. To prove this claim, we show that, using an efficient algorithm $\mathscr{A}$ that makes such a $\mathsf{Send}(\mathsf{label}_S, (r, s, t))$ query with non-negligible probability, an adversary $\mathscr{B}$ can efficiently break the $\mathsf{EUF-CMA}$ experiment by outputting t.

No value output by the firewall or by the server depends on $\mathsf{label.b}$, which means that $\epsilon_5 = 0$, and concludes the proof.

5 Conclusion and Extension

Reverse firewalls aim to limit the damage done by subverted implementations. We revisited the original goals for this primitive in the AKE setting, as stated by Dodis *et al.* [9]. To thwart much larger classes of malicious implementations, we defined a new security model that is less restrictive, and significantly clearer than the one of [9], based on the notion of *functionality-preserving* adversaries.

Based on our model, we constructed a reverse firewall for communication protocols, taking into account both the key exchange and the record layer. Our construction resists complex strategies by only using simple cryptographic tools, such as hash functions or standard public key encryption. Our solution is thus efficient, only adding a reasonable overhead. Moreover, our RF is remarkably versatile, able to handle the elements of most recent secure communication protocols. It is therefore a truly practical solution, illustrating the benefits of reverse firewalls in the real world. To show this, we implement our reverse firewall in TLS1.3-like protocol, proving its security, compared with the Mint TLS1.3 implementation. The results are in the full version.

Acknowledgements. We would like to thank Håkon Jacobsen and Olivier Sanders for their contributions to the preliminary versions of this paper, as well as Kenny Paterson for the fruitful discussions on the subject. This work was supported in part by the French ANR, grants 16-CE39-0012 (SafeTLS) and 18-CE39-0019 (MobiS5).

A Our Security Model Compared to Dodis *et al.*'s

Reverse firewalls must *preserve*, not *create* security. If a secure-channel establishment protocol is run honestly, it should guarantee end-to-end security, no matter what is or isn't between the endpoints. This requirement is taken into account, both by Dodis *et al.*'s work [9] and by ours.

The use case for reverse firewalls (RFs) is one in which the adversary has tampered with the client implementation to *exfiltrate* information to a MitM adversary. It is expected that an honest[4] RF will preserve security, so the adversary should not be able to distinguish a transcript involving its corrupt implementation (protected by an RF) from one only involving honest parties. This property obviously cannot hold for *all* tampered implementation and it is thus necessary to place some restrictions on the adversarial behaviour. In [9], this is done by requiring *functionality-maintaining* implementations, whose definition we adapt and re-orient to fit our criteria.

Our Model. Our first goal is to define a model that places fewer restrictions on the adversarial behaviour, while making these restrictions easy to understand and quantify, via the notion of *transcript equivalence*, defined in Sect. 2.4.

As mentioned before, our second goal is to design an RF that preserves security for the key agreement *and* the record-layer protocols in this new model, which is a non-trivial extension compared to [9], as our model is much more permissive, allowing more realistic attacks, such as key-replacement in the record layer or choosing key-dependent messages. To achieve this, we consider a more general definition of the RF, allowing it to have a public key pk_{FW}. The existence of this public[5] key is an important difference compared to [9]: the client is aware

[4] This requirement is necessary: we cannot ensure any relevant security property when both the client and the RF are corrupt.

[5] Actually, only the client protected by the RF needs to know this *public* key, so we do not need a complex PKI infrastructure.

of the *existence* of the RF, which does not seem to be a significant restriction for most use-cases, such as the one of a company network. We stress that we can retain obliviousness and transparency for our protocol, meaning that the client cannot detect the status of the RF.

The firewall can, via its secret key, passively derive a shared secret $\mathsf{k}_{\mathsf{cfs}}$ (also known to both endpoints) from any key agreement involving one of the clients it protects. This key will be used to preserve the security of the record layer even if the client is corrupt. However, since we also want to ensure privacy for the client and the server *against* the RF (in case the latter is malicious), we allow the endpoints to derive *another* shared key k_{cs} (unknown to the RF), which is meant to provide end-to-end security (even with respect to the RF). In our protocol, one needs both keys to learn any information about the transmitted messages; yet the strength of only one key suffices to prevent exfiltration of information and provide security with respect to Man-in-the-Middle adversaries.

The RF is unable to break the security of the channel, yet tampered implementations cannot manage to exfiltrate information through the firewall – our RF performs its task *without being trusted*. In particular, our model considers malicious RFs and we prove that security still holds in this case, despite its additional knowledge of $\mathsf{k}_{\mathsf{cfs}}$. This additional key may be useful when considering more intricate key agreement protocols such as TLS 1.3, where parts of the handshake are encrypted.

Three Attack Scenarios. We first prove security in a context where everyone is honest, but we also need to separately study the case of k_{cs}, only known to the endpoints, and of $\mathsf{k}_{\mathsf{cfs}}$, known to the client, the firewall and the server. We also consider the possibility of a malicious client, which again is the main motivation of RFs. This leads to the three following scenarios:

1. *Security of* k_{cs}. The adversary controls a legitimate RF, situated between two honest endpoints, and its goal is to break the security of the channel, *i.e.*, get some information on k_{cs}. This scenario implies security without an RF, since the adversary can simply force the latter to be passive. It shows that the client server are still able to preserve their privacy even in the presence of a malicious RF.
2. *Security of* $\mathsf{k}_{\mathsf{cfs}}$. The protocol must ensure the security of the $\mathsf{k}_{\mathsf{cfs}}$ key established between three honest parties following the protocol. Our adversaries are stronger than those of Dodis *et al.*, since they can reveal all but the test-session keys[6] and since they may choose the target instance adaptively. In this scenario the adversary tries to get some information on $\mathsf{k}_{\mathsf{cfs}}$.
3. *Malicious client.* For exfiltration resistance, the adversary provides a malicious client implementation, which will then interact with the firewall and the server. The goal of the adversary is to obtain information about the data sent by the tampered implementation during the executions of the key agreement *and* secure messaging protocol. The server is honest.

[6] This is not the case for Dodis *et al.*, in particular with respect to their security against active adversaries.

For clarity, we choose to treat each case separately. Our modelling approach follows the models of Bellare and Rogaway [2] and Canetti and Krawczyk [5], although we also use ideas from the multi-stage model of Dowling *et al.* [10].

B Transparency

While obliviousness allows the adversary to obtain auxiliary information, such as revealed keys, transparency is solely based on the indistinguishability of messages coming from the endpoints, and those modified by the firewall. Therefore, it is entirely linked to the content of those messages. By definition, transparency is a sub-goal of obliviousness: it is necessary, but not sufficient. Being able to tell whether a message has been modified implies being able to tell whether a firewall is involved or not. If a protocol with a reverse firewall does not achieve transparency, *i.e.*, if there exists an PPT adversary who can distinguish the message originally issued by the client from a modified one with non-negligible probability, this protocol does not achieve obliviousness.

Definition 9 (Transparency). *An AKE protocol* Π *is* transparent *if for all PPT adversaries* $\mathscr{A}$, $\mathsf{Adv}_{\Pi}^{\mathsf{TRS}}(\mathscr{A}) = \Pr[\mathsf{Exp}_{\Pi,\mathscr{A}}^{\mathsf{TRS}}(\lambda) \Rightarrow 1] - \frac{1}{2}$ *is negligible in* λ*:*
$\mathsf{Exp}_{\Pi,\mathscr{A}}^{\mathsf{TRS}}(\lambda)$*:*
1. $(\mathsf{sparam}, \mathsf{pparam}) \leftarrow \mathscr{A}^{\mathsf{Setup}(\cdot)}(1^\lambda)$
2. $\mathsf{Q} \leftarrow \{\mathsf{NewInstance}, \mathsf{Send}, \mathsf{TestSend}\}$
3. $(\mathsf{label}, d) \leftarrow \mathscr{A}^{\mathsf{Q}}(\mathsf{sparam}, \mathsf{pparam})$
4. *if (*label *is Send-fresh)* $\wedge$ *(*$\mathsf{label.b} = d$*): return 1*
5. *else: return a random bit*

C Proof of Theorem 4

We will show that no adversary can distinguish a message sending by an entity (client or server) and a message randomized by the firewall:

Client key exchange initialization: the client sends a message $(g^x, g^c, e = \mathsf{Enc}_{\mathsf{pk}_f}(c))$ where x and c was picked at random. The firewall sends a message $(g^{x'}, g^{c'}, e = \mathsf{Enc}_{\mathsf{pk}_f}(c'))$ where where $x' = x \cdot \alpha_1$ and $c' = c \cdot \alpha_2$, such that α_1 and α_2 was picked at random. Since α_1 and α_2 perfectly randomizes x and y, $(g^x, g^c, e = \mathsf{Enc}_{\mathsf{pk}_f}(c))$ and $(g^{x'}, g^{c'}, e = \mathsf{Enc}_{\mathsf{pk}_f}(c'))$ follow the same distribution.

Server response: if the server receives $(g^x, g^c, e = \mathsf{Enc}_{\mathsf{pk}_f}(c))$ from the client, it sends a message $(\sigma, Y, D, \beta_1, \beta_2)$ where β_1 and β_2 are randomly chosen in $\mathbb{Z}_p$, and such that σ is a valid signature on $(Y, D, g^{x\cdot\beta_1}, g^{c\cdot\beta_2})$. if the server receives $(g^{x'}, g^{c'}, e = \mathsf{Enc}_{\mathsf{pk}_f}(c'))$ from the firewall, it sends a message $(\sigma, Y, D, \beta_1, \beta_2)$ where β_1 and β_2 are randomly chosen in $\mathbb{Z}_p$, and such that sigma is a valid signature on $(Y, D, g^{x'\cdot\beta_1}, g^{c'\cdot\beta_2})$. The firewall then returns $(\sigma, Y, D, \gamma_1, \gamma_2)$ where $\gamma_i = \alpha_i \cdot \beta_i$, so σ is a valid signature on $(Y, D, g^{x\cdot\gamma_1}, g^{c\cdot\gamma_2})$, and $(\sigma, Y, D, \beta_1, \beta_2)$ and $(\sigma, Y, D, \gamma_1, \gamma_2)$ follow the same distribution.

Transmission of a message: the client picks r at random an sends (r, s, t) where $s \leftarrow H_1(r \| \mathsf{k_{cfs}}) \oplus C$, and $t \leftarrow \mathsf{MAC}_{H_1(r\|\mathsf{k_{cfs}})}(r\|s)$. The firewall picks $\tilde{r}$

at random an sends the message $(\tilde{r}, \tilde{s}, \tilde{t})$ where $\tilde{s} \leftarrow H_1(\tilde{r}\|\mathsf{k_{cfs}}) \oplus C$, and $\tilde{t} \leftarrow \mathsf{MAC}_{H_1(\tilde{r}\|\mathsf{k_{cfs}})}(\tilde{r}\|\tilde{s})$, so (r, s, t) and $(\tilde{r}, \tilde{s}, \tilde{t})$ follow the same distribution.

References

1. Bellare, M., Paterson, K.G., Rogaway, P.: Security of symmetric encryption against mass surveillance. In: Garay, J.A., Gennaro, R. (eds.) CRYPTO 2014. LNCS, vol. 8616, pp. 1–19. Springer, Heidelberg (2014). https://doi.org/10.1007/978-3-662-44371-2_1
2. Bellare, M., Rogaway, P.: Entity authentication and key distribution. In: Stinson, D.R. (ed.) CRYPTO 1993. LNCS, vol. 773, pp. 232–249. Springer, Heidelberg (1994). https://doi.org/10.1007/3-540-48329-2_21
3. Black, J., Rogaway, P., Shrimpton, T.: Encryption-scheme security in the presence of key-dependent messages. In: Nyberg, K., Heys, H. (eds.) SAC 2002. LNCS, vol. 2595, pp. 62–75. Springer, Heidelberg (2003). https://doi.org/10.1007/3-540-36492-7_6
4. Brzuska, C., Jacobsen, H., Stebila, D.: Safely exporting keys from secure channels. In: Fischlin, M., Coron, J.-S. (eds.) EUROCRYPT 2016. LNCS, vol. 9665, pp. 670–698. Springer, Heidelberg (2016). https://doi.org/10.1007/978-3-662-49890-3_26
5. Canetti, R., Krawczyk, H.: Universally composable notions of key exchange and secure channels. In: Knudsen, L.R. (ed.) EUROCRYPT 2002. LNCS, vol. 2332, pp. 337–351. Springer, Heidelberg (2002). https://doi.org/10.1007/3-540-46035-7_22
6. Checkoway, S., et al.: A systematic analysis of the juniper dual EC incident. In: ACM SIGSAC 2016, pp. 468–479 (2016)
7. Checkoway, S., et al.: On the practical exploitability of dual EC in TLS implementations. In: USENIX 2014, pp. 319–335 (2014)
8. Chen, R., Mu, Y., Yang, G., Susilo, W., Guo, F., Zhang, M.: Cryptographic reverse firewall via malleable smooth projective hash functions. In: Cheon, J.H., Takagi, T. (eds.) ASIACRYPT 2016. LNCS, vol. 10031, pp. 844–876. Springer, Heidelberg (2016). https://doi.org/10.1007/978-3-662-53887-6_31
9. Dodis, Y., Mironov, I., Stephens-Davidowitz, N.: Message transmission with reverse firewalls—secure communication on corrupted machines. In: Robshaw, M., Katz, J. (eds.) CRYPTO 2016. LNCS, vol. 9814, pp. 341–372. Springer, Heidelberg (2016). https://doi.org/10.1007/978-3-662-53018-4_13
10. Dowling, B., Fischlin, M., Günther, F., Stebila, D.: A cryptographic analysis of the TLS 1.3 handshake protocol candidates. In: ACM CCS 2015, pp. 1197–1210 (2015)
11. Gamal, T.E.: A public key cryptosystem and a signature scheme based on discrete logarithms. In: CRYPTO 1984, pp. 10–18 (1984)
12. Grahm, R.: Extracting the SuperFish certificate (2015). http://blog.erratasec.com/2015/02/extracting-superfish-certificate.html
13. Jager, T., Kohlar, F., Schäge, S., Schwenk, J.: On the security of TLS-DHE in the standard model. In: Safavi-Naini, R., Canetti, R. (eds.) CRYPTO 2012. LNCS, vol. 7417, pp. 273–293. Springer, Heidelberg (2012). https://doi.org/10.1007/978-3-642-32009-5_17
14. Kohlweiss, M., Maurer, U., Onete, C., Tackmann, B., Venturi, D.: (De-)constructing TLS 1.3. In: Biryukov, A., Goyal, V. (eds.) INDOCRYPT 2015. LNCS, vol. 9462, pp. 85–102. Springer, Cham (2015). https://doi.org/10.1007/978-3-319-26617-6_5

15. Krawczyk, H., Paterson, K.G., Wee, H.: On the security of the TLS protocol: a systematic analysis. In: Canetti, R., Garay, J.A. (eds.) CRYPTO 2013. LNCS, vol. 8042, pp. 429–448. Springer, Heidelberg (2013). https://doi.org/10.1007/978-3-642-40041-4_24
16. Krawczyk, H., Wee, H.: The OPTLS protocol and TLS 1.3. In: Proceedings of Euro S&P, pp. 81–96 (2016)
17. Ma, H., Zhang, R., Yang, G., Song, Z., Sun, S., Xiao, Y.: Concessive online/offline attribute based encryption with cryptographic reverse firewalls—secure and efficient fine-grained access control on corrupted machines. In: Lopez, J., Zhou, J., Soriano, M. (eds.) ESORICS 2018. LNCS, vol. 11099, pp. 507–526. Springer, Cham (2018). https://doi.org/10.1007/978-3-319-98989-1_25
18. Mironov, I., Stephens-Davidowitz, N.: Cryptographic reverse firewalls. In: Oswald, E., Fischlin, M. (eds.) EUROCRYPT 2015. LNCS, vol. 9057, pp. 657–686. Springer, Heidelberg (2015). https://doi.org/10.1007/978-3-662-46803-6_22
19. Naylor, D., et al.: Multi-context TLS (mcTLS) enabling secure in-network functionality in TLS. In: SIGCOMM 2015, pp. 199–212 (2015)
20. O'Neill, M., Ruoti, S., Seamons, K., Zappala, D.: TLS proxies: friend or foe. In: IMC 2016, pp. 551–557 (2016)
21. Rogaway, P.: The moral character of cryptographic work (2015). http://web.cs.ucdavis.edu/rogaway/papers/moral-fn.pdf
22. Young, A., Yung, M.: Kleptography: using cryptography against cryptography. In: Fumy, W. (ed.) EUROCRYPT 1997. LNCS, vol. 1233, pp. 62–74. Springer, Heidelberg (1997). https://doi.org/10.1007/3-540-69053-0_6

Software Security

Follow the Blue Bird: A Study on Threat Data Published on Twitter

Fernando Alves[1(✉)], Ambrose Andongabo[2], Ilir Gashi[2], Pedro M. Ferreira[1], and Alysson Bessani[1]

[1] LASIGE, Faculdade de Ciências, Universidade de Lisboa, Lisbon, Portugal
fbalves@fc.ul.pt

[2] Centre for Software Reliability, City, University of London, London, UK

Abstract. Open Source Intelligence (OSINT) has taken the interest of cybersecurity practitioners due to its completeness and timeliness. In particular, Twitter has proven to be a discussion hub regarding the latest vulnerabilities and exploits. In this paper, we present a study comparing vulnerability databases between themselves and against Twitter. Although there is evidence of OSINT advantages, no methodological studies have addressed the quality and benefits of the sources available. We compare the publishing dates of more than nine-thousand vulnerabilities in the sources considered. We show that NVD is not the most timely or the most complete vulnerability database, that Twitter provides timely and impactful security alerts, that using diverse OSINT sources provides better completeness and timeliness of vulnerabilities, and provide insights on how to capture cybersecurity-relevant tweets.

Keywords: OSINT · Twitter · Vulnerabilities

1 Introduction

Cybersecurity has remained a hot research topic due to the increased number of vulnerabilities indexed and to the severe damage caused by recent attacks, from ransomware (*e.g.*, wannacry) to SCADA systems attacks (*e.g.*, the attacks on the Ukrainian power stations). A growing trend for obtaining cybersecurity news is to collect Open Source Intelligence (OSINT) from the Internet [57]. OSINT sources include vulnerability databases (*e.g.*, the National Vulnerability Database—NVD), online forums (*e.g.*, Reddit), social networks (*e.g.*, Twitter), and scientific literature. Although more technical, exploit databases (*e.g.*, ExploitDB) are a useful OSINT source providing code excerpts known as Proofs of Concept (PoC) that show how to exploit a vulnerability. PoCs can be analysed by a specialised audience capable of using the exploit's code to understand and counteract vulnerability exploitation, thereby removing the vulnerability.

The research community has shown many different uses for OSINT, from its collection and processing [26,29,36,37,40,43,48,54,59], vulnerability life cycle analysis [28,31,42,50,56], to evaluating vulnerability exploitability [24,31,32,

L. Chen et al. (Eds.): ESORICS 2020, LNCS 12308, pp. 217–236, 2020.
https://doi.org/10.1007/978-3-030-58951-6_11

38,45,51]. There are two predominant OSINT sources in the literature: NVD (*e.g.*, [24,31,45,51,55]), and Twitter (*e.g.*, [26,40,41,43,59]). The first provides curated vulnerability data, while the latter is more generic, concise, and covers more topics.

As Twitter's usage grew, various information sources began to link their content on Twitter to increase visibility and attract attention. Twitter's continued growth placed it among the most relevant communication tools used by the vast majority of companies who have a Twitter account to interact with the world. All this activity also caught the attention of the research community. The information flow and interaction graphs meant new research opportunities, such as detecting emerging topics [34,46], or finding events related to a specific topic, such as riots [25], patients experience with cancer treatment drugs [30], or earthquakes [52]. Twitter popularity instigated the development of tools to collect tweets (*e.g.*, Tweet Attacks Pro [20]), APIs for programming languages (*e.g.*, Tweepy [19]), and many OSINT-collecting tools developed specific plugins to collect tweets (*e.g.*, Elastic Stack [21]), including cybersecurity-oriented ones (*e.g.*, SpiderFoot [18]). The cybersecurity field also found opportunities in using Twitter (discussed in Sect. 2).

However, to the best of our knowledge, there is no evidence in the literature highlighting one security data source as advantageous over the others. For instance, the following questions are yet to be answered: Why use solely NVD when there are several reputable vulnerability databases? Is NVD the richest (in terms of number of vulnerabilities reported) and timeliest (does it contain the earliest reporting date of a vulnerability) vulnerability database? Why use Twitter to gather cybersecurity OSINT? Does Twitter provide any advantage over vulnerability databases? Is it useful for security practitioners?

In this paper, we present an extensive study on OSINT sources, comparing their timeliness and richness. We analysed the vulnerability OSINT sources indexed on vepRisk [27], which aggregates several vulnerability databases, advisory sites, and their relationships. We compared Twitter against these data sources to understand if there are any advantages in using it as a cybersecurity data source. To explore this topic, we formulated three research questions:

RQ1: Is NVD the richest and timeliest vulnerability database?
RQ2: Does Twitter provide a rich and timely vulnerability coverage?
RQ3: How are vulnerabilities discussed on Twitter?

Our findings show that: vulnerability databases complement one another in richness and timeliness (*i.e.*, no single source contains all the vulnerabilities; no single sources can be relied on as providing the earliest vulnerability reporting date); Twitter is a rich and timely vulnerability information source; and finally, Twitter complements other OSINT sources. In summary, our contributions are:

- A comparison between some of the most reputable and complete vulnerability databases in terms of timeliness and coverage;
- An analysis of the coverage and timeliness of Twitter with respect to vulnerability information;

- An analysis about "early alerts" on Twitter, *i.e.*, vulnerabilities disclosed or discussed on Twitter before their inclusion on vulnerability databases;
- An analysis on how vulnerabilities are discussed on Twitter;
- Insights on how to collect timely tweets;
- Insights regarding OSINT (and in particular Twitter's) usage for cybersecurity threat awareness.

2 Background and Related Work

The following sections present some vulnerability databases and previous research contributions related to this work.

2.1 Vulnerability Databases

MITRE Corporation [11] maintains the Common Vulnerabilities and Exposures list [3] (in short, CVE), a compilation of known vulnerabilities described in a standard format. A global index of known vulnerabilities simplifies complex analyses such as detecting advanced persistent threats. Therefore, indexing known vulnerabilities in CVE became standard practice for all kinds of security practitioners, including software vendors. Each CVE entry has an ID (CVE-ID), a short description, and the creation date.

NIST's National Vulnerability Database [12] mirrors and complements CVE entries on their database. Every hour, NVD contacts CVE to obtain newly disclosed vulnerabilities (we contacted NVD directly to get this information). Each vulnerability indexed in NVD undergoes a thorough analysis, including attributing an impact score based on the Common Vulnerability Scoring System (for both versions 2.0 [5] and 3.0 [4]), and links related to the vulnerability, such as advisory sites or technical discussions. NVD uses the CVE-ID in place of an ID of its own.

There is a significant difference between the dates of NVD and CVE entries. In CVE, it is the date when entries became *reserved*, but not yet *public*. NVD entries are always public, using the date when they were indexed, even prior to their analysis completion. Thus, in practice, a vulnerability has the same public disclosure date on both CVE and NVD, which is NVD creation date. Therefore, this study considers only the NVD vulnerability disclosure date.

Besides CVE and NVD, many online databases compile known vulnerabilities and provide unrestricted use of their contents, such as the Security Database [17] and PacketStorm [15]. The complementary information provided by each database differs, but in general, these provide a description, some analysis of the security issues raised by the vulnerabilities, known exploits, and possible fixes or mitigation actions.

2.2 Cybersecurity-Related OSINT Studies

To the best of our knowledge, Sauerwein *et al.* performed the most similar study to the one present on this paper [55]. For two years, the authors collected all tweets with a CVE-ID in its text. They show a comparison of the tweet publishing dates with the disclosure dates of those CVEs on NVD. The results show that 6232 vulnerabilities (25.7% of their dataset) were discussed on Twitter before their inclusion on NVD. However, this study falls short in some aspects. Firstly, the NVD is not always the first database to report new vulnerabilities, which changes the vulnerabilities first confirmed report date (see Sect. 4). Secondly, the authors search only for CVE-IDs on Twitter, which will not capture issues that have been disclosed to the public but not (yet) indexed on CVE or NVD. Finally, the analysis is focused solely on the vulnerabilities life cycle and Twitter appearance, overlooking vulnerability characterisation such as their impact.

There is some research work providing evidence that relevant and timely cybersecurity data is available on Twitter [33,41,44], *i.e.*, that some vulnerabilities were published on Twitter before their inclusion on vulnerability databases. However, these are case studies concerning a single vulnerability, and compare the tweets referring them solely with NVD. Other Twitter-based contributions include correlating security alerts from tweets with terms found in dark web sources [53], studying the propagation of vulnerabilities on Twitter [58], and finding that exploits are published on Twitter (on average) two days before the corresponding vulnerability is included in NVD [51].

In a similar research line, Rodriguez *et al.* [49] analysed vulnerability publishing delays on NVD when comparing to other OSINT sources: Security Focus, ExploitDB, Cisco, Wireshark, and Microsoft advisories. The authors report that NVD presents publishing delays (from 1 to more than 300 days) from 33% to 100% of the cases when comparing with those databases, *i.e.*, sometimes it publishes after these databases or it always publishes after these databases. However, the authors consider only the year of 2017. Similarly, the Recorded Future company reports that for 75% of the vulnerabilities NVD presents a 7-day disclosure delay [16]. However, the company does not reveal how it obtained these results.

The literature lacks a systematic and thorough analysis regarding the data published on Twitter and on vulnerability databases, including crucial aspects such as coverage, timeliness, and the actionability provided by such OSINT.

3 Methodology

The objective of this study is to compare some aspects of the information present on vulnerability databases with another OSINT source, namely Twitter. Instead of searching, collecting, and parsing a set of databases, we use the vepRisk database [27]. It contains several types of security-related public data, including all entries published on NVD, Security Database, Security Focus, and Packet-Storm databases, from their creation until the end of 2018.

We chose Twitter as an OSINT source, as it is a known aggregator of content posted by all kinds of users (hackers, security analysts, researchers, etc.), news sites, and blogs, among others who tweet about their content to increase visibility [9]. Thus, Twitter became an information hub for almost any kind of content. Unlike vulnerability databases—that contain only security data—Twitter includes discussions over a vast universe of topics. Since the results of this study are based on tweets mentioning indexed vulnerabilities, we decided to search for tweets mentioning the vulnerabilities indexed on NVD. Finally, to ensure the validity of our results, we opted to *manually* match tweets to vulnerabilities. These decisions raised two questions: 1) what part of the vulnerability description are we going to use as a search term? and 2) How to reduce the number of vulnerabilities to manually inspect?

The NVD description of some vulnerabilities includes a "colloquial" name for which the vulnerability became known. For example, CVE-2014-0160 is known as the "Heartbleed bug". These names fall mostly within two categories: a generic description of the vulnerability class (*e.g.*, "Microsoft Search Service Information Disclosure"), or some "creative" designation related to the vulnerability (*e.g.*, "Heartbleed" is a vulnerability on the "heartbeat" TLS packets which can be exploited to leak or "bleed" information). These colloquial names are easily recognisable since they always appear in the NVD vulnerability description after the "aka" acronym (for *also known as*). Therefore, to guide the search on Twitter, all vulnerabilities with a colloquial name were selected, and the names were used as query terms. This decision also reduced the number of vulnerabilities to analyse to 9,093, an amount of data manually processable. Additionally, vulnerabilities with colloquial names are more likely to be discussed on Twitter since most were "named" due to media attention. The IDs of the 9,093 vulnerabilities with a colloquial name that were used in this study are listed online [1].

We were unable to use the Twitter API to collect the tweets for the study as it only provides access to tweets published in the previous week. However, the Twitter web page allows searching for tweets published at any point in time. To automate the querying process, a library called `GetOldTweets` [7] was employed. It mimics a web browser performing queries on the Twitter page, enabling fast and programmatic retrieval of any number of tweets from any time.

Regarding matching tweets and vulnerabilities, we consider that a tweet t unequivocally refers a specific vulnerability v if and only if (1) t mentions in its text v's CVE-ID even if the vulnerability has not yet been disclosed on NVD, or (2) t contains a link mentioned in v's NVD description, even if the web page pointed by the link is currently down, or (3) t mentions a security advisory that is also referred by v's NVD links about that threat, or (4) t or t's links mention an ID associated with v. Two assumptions are made: 1) if an ID is present on a tweet, then the advisory has been published; and 2) a security analyst that receives a tweet containing a security advisory ID can search for this advisory, thus having the same result as publishing the advisory link on the tweet. If a vulnerability is mentioned by up to a thousand tweets, all tweets were manually inspected. The colloquial name of some vulnerabilities is also a word commonly used on tweets, such as "CRIME" (CVE-2012-4930) or "RESTLESS" (CVE-2018-12907). For those cases, where a search term can return more than 350,000 tweets, the manual inspection was done in two steps. First, the description is analysed to understand the vulnerability characteristics and related terminology. Then, a large set of informed searches were performed on the tweet set in search of tweets potentially referring the vulnerability. In total, *about a million tweets*

Table 1. The number of entries in each database (in bold) and the number of shared entries between database pairs.

	NVD	PS	SD	SF
NVD	**110,353**	–	–	–
PS	9,290	**129,130**	–	–
SD	110,353	9,344	**117,098**	–
SF	60,378	8,597	60,843	**98,445**

Table 2. The number of times one database or a group of databases were the first to disclose a vulnerability.

Database(s)	# Occurrences	%
NVD	0	0.00
PS	853	0.77
SD	0	0.00
SF	40,208	36.44
NVD, SD	51,238	46.43
PS, SF	1,265	1.15
NVD, SD, SF	16,580	15.02
NVD, PS, SD	85	0.08
NVD, PS, SD, SF	124	0.11

were manually inspected, and any links present in potential matches were also examined to confirm the matches. The data labelling was performed solely by a PhD student with a cybersecurity background, and it took roughly eight months to complete. All potential matches were triple checked to ensure their validity.

The time range considered in this study begins on March 2006 (Twitter's creation date) until the end of 2018. The tweets were collected between early 2017 and the end of 2019. The resulting dataset contains 3,461,098 tweets. The tweets publishing times were adjusted to the day time-scale to match the time granularity provided by the vulnerability databases. Therefore, all time comparisons performed in this study used the publishing day.

4 Vulnerability Database Comparison

As NVD is considered a standard for consulting vulnerability data, many research works use only it as their vulnerability database (*e.g.*, [42,47,50,55]). This is a natural choice since NVD includes multiple resources for further understanding of the issue at hand. However, other reputable vulnerability databases, with their own disclosure procedures and timings, provide useful information for security practitioners. Therefore, it is interesting to understand if there is evidence that supports using only NVD for practice or research work. To investigate this point, we collected data about two different aspects: the number of entries and their publishing date. The first measures the coverage of the database, while the second is related to its timeliness and practical usefulness.

Table 1 shows the number of entries in each of vepRisk's databases: NVD, PacketStorm (PS), Security Database (SD), and Security Focus (SF). It also shows the number of entries shared between each database pair. Tables 2 and 3 are related to timeliness. Table 2 is divided in two blocks. The first shows the number of occurrences where one database was the *first* to disclose a vulnerability ahead of other databases. The second block shows the number of occurrences where various groups of databases were *simultaneously first* to disclose a vulnerability. Table 3 complements the previous table by showing the percentage of time each database was one of the first to disclose a vulnerability.

There are five key takeaways obtained from analysing the tables: (1) NVD is not the most complete vulnerability database, with the Security Database and PacketStorm containing more entries; (2) NVD is not the most timely database. Alone, it was never the first to publish a vulnerability; (3) No database stands out as the most timely; (4) Security Database contains all of NVD's entries (this was manually verified); (5) With the exception of NVD, all databases publish different vulnerabilities. Therefore it is important to follow a set of data sources instead of relying solely on one.

Table 3. The percentage of times each database was one of the first to disclose a vulnerability.

Database	# Occurrences	%
NVD	68,027	61.64
Security Database	68,027	61.64
Security Focus	58,177	52.72
PacketStorm	2,327	2.11

5 Twitter Vulnerability Coverage and Timeliness

Coverage. A first validation on using Twitter for cybersecurity is verifying if vulnerability data reaches Twitter. We searched for tweets mentioning each of the CVE-IDs published on NVD after Twitter's creation. Of the 94,398 CVE-IDs searched, 71,850 (76.11%) were mentioned in tweets. However, by analysing Fig. 1 it is possible to observe that since the beginning of 2010, CVEs became regularly discussed on Twitter. In fact, from 2010 forward, the coverage became above 97.5%, validating the hypothesis that vulnerability data reaches Twitter.

The drastic increase in tweets mentioning CVEs in 2010 may be connected to the sudden growth Twitter underwent in that period [10]. Nevertheless, the turning point on cybersecurity threat awareness and on the importance of coordinated vulnerability disclosure mechanisms *may* have been in the beginning of 2010, when Google publicly disclosed that their infrastructure in China was targeted by an advanced persistent threat codenamed "Operation Aurora" [14]. Later on, it was discovered that other major companies were targeted, such as Adobe Systems, Rackspace, Yahoo, and Symantec. This event *could* have triggered two crucial social phenomena: that companies are attacked and should not be ashamed of it, and should disclose the details of these attacks in a coordinated effort to detect, understand, and prevent them; and that the users prefer transparency in cybersecurity events since when data breaches occur, typically it is the user data that is affected.

Timeliness. Regarding timeliness, we performed the analysis only for the 9,093 vulnerabilities that were manually analysed to ensure the correctness of the results. Figure 2 shows, for those vulnerabilities, which source discussed them first: either one of the vulnerability databases considered in this study, Twitter, or Twitter and at least one of the databases, simultaneously. There are also the cases where the vulnerability was not discussed on Twitter, which are predominant before 2010. Although we are not evaluating the whole databases, the figure shows a predominance of same day publishing cases (84.56% when considering 2006–2018, and 93.73% in 2010–2018). We consider that these results validate the hypothesis that Twitter is a timely source of vulnerability data. In the next section we present an in-depth study of the cases where the vulnerabilities were discussed on Twitter ahead of vulnerability databases.

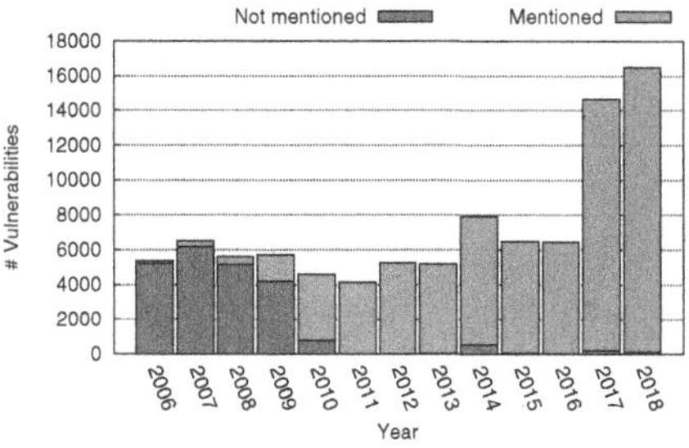

Fig. 1. Twitter's CVE coverage.

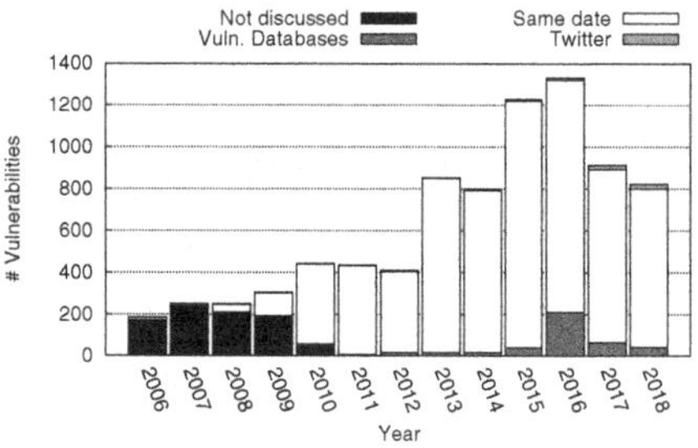

Fig. 2. A timeliness comparison between vulnerability databases and Twitter.

6 Early Vulnerability Alerts on Twitter

Of the 9,093 vulnerabilities analysed, 89 were referred by tweets before being published on at least one of the vulnerability databases considered. Even though these vulnerabilities represent a small percentage of the vulnerability sample under study (0.98%) we decided to characterize them to understand if searching for early alerts on Twitter is a worthy endeavour. The most mentioned vendors in the early alerts are the Ethereum blockchain (17 mentions), Microsoft products (5), Debian (5), Oracle (4), Linux (4), and Apple (4), while the most mentioned assets are Javascript (9), SSL/TLS (8), Xen Hypervisor (4), Safari Browser (3), Mercurial version control (3), and Cloud Foundry (3). These mentions provide evidence of the usefulness of these alerts, as both vendors and assets are some of the major players in their respective fields.

All vulnerabilities with Twitter early alerts can be found online [1], together with their publishing dates on the vulnerability databases and some extra notes. In the following sections these early alerts are further analysed on their impact and usefulness. We conclude the section with a discussion of the significance of these results.

6.1 Timeliness

Twitter Versus Vulnerability Databases. Figure 3 presents the distribution of early alerts over the years considered. The number of early alerts increased in the last two years, which matches the increase of vulnerabilities published on databases since the beginning of 2016.

Concerning publishing timing, the majority of early alerts were available up to thirty days ahead of vulnerability databases (78.65%–70 cases). Notably, four early alerts appeared between 31 and 50 days ahead, four between 51 and 100 days ahead, nine between 101 and 200 days ahead, and finally, the two cases with the highest antecedence were 371 and 528 days ahead. The number of days Twitter is ahead of vulnerability databases increased continuously since 2008, but only after 2016 we found more than 10 cases. No relevant patterns were discovered in the early alerts due to the small number of occurrences.

Twitter Versus Advisory Sites. Besides vulnerability databases and social media, advisory sites are an essential source of vulnerability information. Many companies use websites to announce software patches, along with which vulnerabilities are fixed. Therefore, we compare Twitter with advisory sites as these are specialized OSINT sources directly connected to the software vendors.

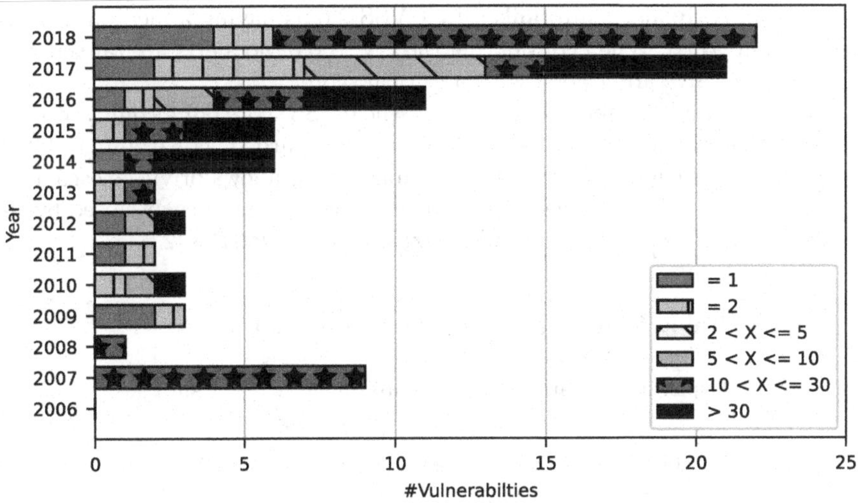

Fig. 3. The number of early alerts found per year.

We manually searched for advisory notices for each of the early alerts, obtaining only 33 advisories (approximately of early alerts). Table 4 presents the number of times either Twitter or the advisory was the first publisher, or when both published on the same day.

The majority of early alerts are not paired with an advisory, but the tweets referring them contain links that describe these vulnerabilities. This observation reinforces the idea that Twitter is a useful cybersecurity discussion hub by connecting various knowledge resources in a single place.

Table 4. The number of times Twitter, advisory site, or simultaneously both, were the first to publish an advisory notice.

1st publisher	#	%
Twitter	11	12.36
Same date	13	14.61
Advisory site	9	10.11
No Advisory	56	62.92

Table 5. The CVSS 2.0/3.0 impact of the early alert vulnerabilities.

(a) CVSS 2.0.

CVSS 2.0	#	%
Low	5	5.62
Medium	64	71.91
High	20	22.47

(b) CVSS 3.0.

CVSS 3.0	#	%
Low	0	0.00
Medium	16	17.98
High	37	41.57
Critical	5	5.62
N/A	31	34.83

6.2 Vulnerability Impact

Although the existence of early alerts is relevant by itself, it is essential to assess the impact of the vulnerabilities. Table 5 presents how many early alerts have low, medium, high, and critical (CVSS 3.0 only) CVSS scores according to the CVSS 2.0 and 3.0 scoring systems. As the CVSS 3.0 was released in 2015, 31 early alerts are ranked only according to CVSS 2.0 (the N/A line in Table 5b).

Almost all early alerts are ranked by CVSS 2.0 as having a medium or high impact (about 94%). When considering the CVSS 3.0, no alerts are ranked with low impact, and five are graded with a critical score.

Despite the small number of early alerts, the CVSS score points out that these are relevant vulnerabilities and should not be disregarded. For example, CVE-2016-7089 is a WatchGuard firewall vulnerability that allows privilege escalation via code injection. This vulnerability belongs to the set of issues disclosed by the "Shadow Brokers" [8], and has a public exploit on ExploitDB [23].

Table 6. The exploitation status of the early alerts.

Exploitation status	Twitter publishing		At disclosure	
	#	%	#	%
Exploited	21	23.60	22	24.72
PoC	11	12.36	11	12.36
No data	57	64.04	56	62.92

6.3 Vulnerabilities Exploited at Disclosure Time

A vulnerability only has an actual impact once it is exploited. Table 6 shows the exploitation status of the early alert vulnerabilities, both at the Twitter publishing and disclosure dates. The majority of vulnerabilities are not paired with observations of their exploits in the wild (64% or 57). A quarter of these cases (23.6% or 21) are known to be exploited. In a few cases (12% or 11), a PoC was referred by the vulnerability notice, describing how to exploit the vulnerability. As it is impossible to know if that PoC was used, we categorised these separately from the cases where the exploitation was confirmed.

We matched the early alerts with CVE-mentioning exploits present in ExploitDB [6] to complement the previous result. Only one case was found, published before the earliest vulnerability database and after the disclosing tweet. This information was used to update Table 6, adding the "At disclosure" column.

Considering vulnerabilities known to have been exploited and those with a PoC, the total amounts to about 34%. Current studies estimate that the percentage of vulnerabilities that are exploited in the wild is 5.5% [38], meaning that these early alerts include many appealing targets for hackers.

6.4 Actionability

Perhaps even more important than knowing the impact or exploitation status of a vulnerability, is to avoid exploitation. This can be achieved by applying patches or configurations to protect the vulnerable system. Table 7 shows which vulnerability mitigation measures can be reached by following the hyperlinks found in early alert tweets. In almost 40% of the cases, the tweet includes a link pointing to a patch that solves the vulnerability. For another 40% of vulnerabilities there is no patch available ("None"), or that information is not clear or the topic is not discussed ("No data"). The unreadable case is due to a page not written in English, where some parts of the text were not clear even after translation. The N/A entries are due to dead links, which blocked the analysis.

In the majority of cases (57%), the early alerts provided some information on how to protect the vulnerable systems from exploitation, either by patch or configuration. If the cases where we could not get more information (the "No data" cases) provided some solution, then the protection rate would increase to more than 70%. Therefore, we conclude that besides impact and exploitation relevance, early alerts are also useful due to the actionability they enable, as they inform security practitioners of possible actions to protect their systems.

Table 7. The actionability provided by the early alert tweets.

Action types	#	%
Patch	34	38.20
Configuration/patch	5	5.62
Configuration	12	13.48
None	23	25.84
No data	13	14.61
Unreadable	1	1.12
N/A	1	1.12

7 How Vulnerabilities are Discussed on Twitter

In this section we characterize some aspects of how vulnerabilities are discussed on Twitter. By identifying these aspects we provide guidelines for topic detection techniques oriented at capturing cybersecurity events. The following results are based on the analysis of the 9,093 vulnerabilities considered in this study.

7.1 Duration and Number of Tweets

Figure 4 (left) presents the discussion duration. We observed that half of the vulnerabilities were discussed during up to eight days. However, it is interesting to see that the other half is middling spread across to up to 2,000 days. In some cases, the discussion can continue to up to almost 3,800 days. Discussion periods are extensive on some vulnerabilities mainly due to three different reasons: being used as comparative examples when discussing new events (*e.g.*, CVE-2014-0160, the "Heartbleed" bug); being (partly) reused on new attacks or as a part of a campaign, (*e.g.*, CVE-2017-11882 [13]); as case studies, therefore being remembered by their impact, specificity, or technical details.

Figure 4 (middle) presents how many tweets discuss the vulnerabilities. From the graph we are omitting two outliers: one vulnerability discussed by 7,749 tweets, and another discussed by 15,733. Half of these vulnerabilities are discussed by two to thirteen tweets. These results are not surprising considering that most vulnerabilities are uneventful. The large majority of vulnerabilities are described, patched, and forgotten. Also, only a small percentage of vulnerabilities is exploited in the wild [38], which are the ones more likely to attract more attention. Only 351 vulnerabilities were discussed by more than 50 tweets, showing that although this content is posted on social media, relatively few vulnerabilities attract attention. However, taking a closer look at those 351 vulnerabilities, 14 of them have low severity rating according to CVSS 2.0, 124 have medium severity, and 213 have high severity. Although it is not implied that vulnerabilities with medium and high severity are going to be widely discussed, these results indicate that those referred by more tweets tend to have higher severity ratings.

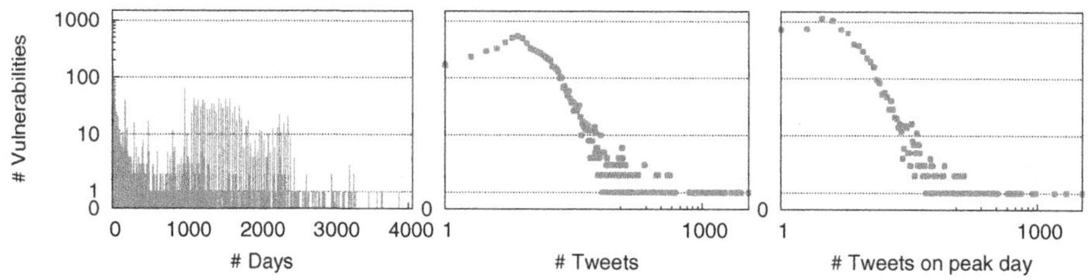

Fig. 4. Vulnerability discussion analysis: vulnerability discussion duration in days on Twitter (left), the total number of tweets discussing a vulnerability (middle), and the peak number of tweets in a single day (right).

Figure 4 (right) presents the daily peak discussion, *i.e.*, the maximum number of tweets discussing the vulnerability in a single day. Three-quarters of the vulnerabilities have daily peaks between one and ten tweets. This is an important factor for topic detection techniques, as most of these identify new trends based on detecting bursts of tweets discussing the same event [60,61].

7.2 Accounts

The tweets discussing security content used in this study were posted by 194,016 different accounts. We performed a quick analysis to understand if any account(s) stand out as sources of cybersecurity tweets. Out of the 194,016, only 5,863 of them published more that one relevant tweet, and only 228 posted more than 5. The highest tweet count for a single account is 73 tweets. Therefore, in this study, we did not find any best accounts to follow for cybersecurity content.

8 How to Find Timely Tweets

The results presented so far on this paper are a forensic analysis of the content posted on vulnerability databases and on Twitter. However, security practitioners are interested in capturing these posts live. Therefore, this section provides insights for data collection methods based on the aforementioned knowledge.

Systems that collect threat intelligence are designed to detect relevant news items while discarding non-relevant ones. The various systems proposed in the literature vary in terms of the complexity of the data selection approach, and as it was infeasible to test all of them, we selected three approaches to test against our data. The first is a simple heuristic-based approach. A tweet is considered relevant if it mentions a software element and a threat word from the VERIS [22] or ENISA [22] cybersecurity taxonomies. The second one is equal to the first but also detects the word "CVE". The third is a more sophisticated approach. We used a convolutional neural network-based approach (CNN) from a previous Twitter-based cyberthreat detection work developed by the authors [36,37]. The test simply measures if the approach correctly detects the target tweets.

Table 8 shows the results of these approaches. We distinguish pre- and post-2010 periods as the coverage differs significantly. The percentages on first four columns in the table are obtained against the labelled data used in Sects. 6 and 7, respectively. The last two columns are obtained by running the techniques mentioned above against Sect. 5. These are speculative results as the data is not labelled, but should transpose from the relatively large labelled data.

Table 8. Percentage of correctly detected tweets according to the various datasets and methods. The header row includes the dataset size between brackets.

	Early (89)	Early post 2010 (76)	Colloquial (9101)	Colloquial post 2010 (7923)	All CVES (94,398)	All CVES post 2010 (77409)
Heuristic	56.18%	63.16%	57.01%	63.98%	71.14%	70.99%
Heuristic + CVE	62.92%	71.05%	88.46%	99.24%	84.56%	93.73%
CNN	57.30%	45.68%	89.51%	87.76%	87.74%	87.59%

Using the simplest heuristic method is the worst method of the three except in one case. This means that the trivial approach works but lacks expressiveness regarding how vulnerabilities are discussed. Adding the word "CVE" to the detection mechanism enables it to detect tweets about already indexed vulnerabilities that do not follow the "software name and threat type" rule, drastically increasing the detection rates. Therefore, this is a suitable detection technique.

Finally, the CNN presents rather poor results regarding detecting early alerts but otherwise is consistently close to 90% accuracy. Although sometimes the CNN has a lower accuracy rate than the heuristics, this test does not cover false positive rates, where the CNN is expected to largely outperform heuristics.

We performed a follow-up analysis to the early alert tweets in an effort to understand the CNN results. We used BERT [35] to obtain semantic-rich feature vectors from the tweets, and cosine similarity [62] as a similarity measure. The tweets were grouped by similarity, where each tweet was grouped with its most similar peers, as long as each group had an average similarity rating above 0.8. In almost all cases, the CNN gave the same classification for all members of each group. By observing these groups we can assert that the CNN accurately classified as relevant tweets with a more cybersecurity-oriented speech ("*New "Lucky Thirteen" attack on TLS CBC...*" or "*Misfortune Cookie: The Hole in Your Internet Gateway...*"), while incorrectly classifying less structured tweets ("*Only in the IT world can you say things like "header smuggling" (...) or in regex "Did you escape the caret?"*" or "*The text for TBE-01-002 references TBE-01-004 - which does not seem to be included in the report. Is that intentional?*"). As our CNN was trained mostly using tweets directly discussing cybersecurity, any tweets not conforming to the pattern are likely to be discarded. Thereby we conclude that a diverse training set for neural networks is required towards a complete detection coverage.

9 Summary of Findings

In the following, we summarize the findings associated with each of the research questions formulated in this study.

RQ1: Is NVD the Richest and Timeliest Vulnerability Database? The NVD does not stand out as the most complete vulnerability database, as there are others that index more vulnerabilities. Also, NVD is not the most timely database. In fact, it was never the first database to publish a new vulnerability ahead of the others. However, NVD is known to have a strict publishing policy, allowing for consultation and comments from product vendors, which means that vulnerability publication may be delayed.

RQ2: Does Twitter Provide a Rich and Timely Vulnerability Coverage? Since the beginning of 2010 Twitter provides a timely and rich coverage of known vulnerabilities. Moreover, there is a small subset of vulnerabilities (less than 1% of those we inspected) that are discussed on Twitter before their inclusion on vulnerability databases. Although these are very few cases, our analysis shows that they are relevant, impactful, and in many cases provide useful security recommendations. Overall, we consider Twitter as a useful cybersecurity news feed that should be taken into account by security practitioners.

RQ3: How are Vulnerabilities Discussed on Twitter? Vulnerability discussion on Twitter is carried out mostly in small bursts of two to thirteen tweets. Most vulnerabilities stopped being discussed within eight days, although tweets about them can appear for several years. Vulnerabilities discussed by a higher-than-usual volume of tweets (more than 50) tend to have higher impacts.

10 Insights for Practical Usage

Beyond the comparative analysis presented in this paper, there are a set of insights that we gathered while analysing the tweets collected. Below we present practical takeaways related to OSINT usage and its advantages.

No Vulnerability Database Stands Out as the Best. NVD is an essential OSINT source, especially due to its thorough analysis and important link aggregation, but other reputable databases should be considered as a complement for four main reasons: *Timeliness*—NVD is not the most timely database (see Sect. 4); *Actionability*—NVD does not directly provide suggestions to mitigate or avoid vulnerabilities, unlike other databases (*e.g.*, PacketStorm, Security Focus); *Known exploits*—NVD does not collect information about known exploits, unlike other databases (*e.g.*, PacketStorm, Security Focus); *Completeness*—NVD is not the most complete database (see Table 1). Therefore security practitioners should use a database ensemble to collect security events.

Twitter is Relevant. OSINT is provided by many reputable sources and should be taken seriously. Besides the significant research efforts (*e.g.*, [26,40,41,43,59]), there are companies and tools dedicated to OSINT sharing and enrichment. Sects. 5 and 6 demonstrate that tweets provide timely, relevant, and useful cybersecurity news.

Twitter is a Natural Data Aggregator. Another clear advantage of using Twitter to gather information is its natural data aggregation capability. The 89 early alerts mention 73 different products from 59 different vendors. When considering the 9,093 vulnerabilities analysed, these numbers extend to 1,153 products from 346 vendors. Forty-two CVEs are not indexed in the Common Platform Enumeration—a database of standard machine-readable names of IT products and platforms [2] used by CVE.

Security Advisories May Not be Provided by Smaller Companies. The majority of vendors mentioned in the 89 early alerts do not provide an advisory site or news blog, while the vendors who provide advisories may not provide an API or a feed subscription. Since advisory sites link their content to Twitter at publication time, one can receive security updates by following the advisory accounts or by accessing Twitter's stream API and applying appropriate filters.

Twitter is Important but not Omniscient. We believe that a plausible trend for OSINT is to use Twitter as a front-end of the latest events. Since tweets have a relatively small size, messages tend to be concise, efficiently summarising the content of the associated links. This is one of Twitter's characteristics that made it so popular: reading a set of tweets is much faster than inspecting a collection of web sites. Therefore, Twitter naturally provides an almost standardised summary, quick and straightforward to process, which is very attractive for Security Operation Centres.

Tweets will not replace the current security publishing mechanisms in place. Once a security-related tweet is received, a visit to the associated site is practically mandatory to understand the issue at hand or to search for patches, among other relevant data. It is also arguable that a similar feed can be obtained by using an RSS feed. However, through Twitter, it is possible to monitor multiple accounts and to gather additional information not provided by RSS, such as timely breakthroughs or further discussion concerning the issue.

Collecting OSINT is a Continuous Process. Another takeaway for security practitioners is that it is essential to follow news about all layers of the software stack by including keywords related to network protocols (*e.g.*, SSL, HTTP) or purchased web services (*e.g.*, cloud services, issue tracking services). This may seem obvious to the reader given this paper's discussion. However, as part of a research work unrelated to this paper, when we asked security analysts of three industrial partners (nation-wide and global companies with dedicated Security Operations Centres) for keywords to describe their infrastructure (to guide our tweet collection), they did not include network protocols or hardware elements.

Moreover, beyond receiving updates about selected assets, it is vital to obtain trending security news. It is hard to describe all relevant elements of a large company thoroughly, and maybe not all software in use is indexed or known. By extending the collection elements with (for example) topic detection techniques (*e.g.*, [34,39,46,60]), one is more likely to cover all software in use. As Twitter can provide all these types of news and the research community has studied thoroughly topic detection on this platform, having trend detection might be mandatory for effective OSINT collection.

Diverse Sources Complement Each Other. Finally, and to complement the previous insight, it is important to follow a diverse set of accounts to observe the broad universe of software vendors. The early alerts were posted by 53 different accounts (for 89 alerts), demonstrating that diversity of sources is crucial for awareness. Moreover, during this study we collected tweets posted by about 194,000 accounts, reinforcing the idea that Twitter is a cybsersecurity discussion hub. It is also important to discuss critical cases like the exploitation of CVE-2017-0144, which became known as "wannacry". The vulnerability was published on CVE/NVD and Microsoft's security advisory, and patched a few months before the wannacry crisis. Therefore, by following Microsoft or CVE/NVD one would be aware of the issue and could avoid the ransomware.

Once the vulnerability started being exploited, several online discussions suggested a set of configurations that blocked the exploit. Therefore, those that did not patch their systems (and for the Windows versions that were not patched by Microsoft) could benefit from OSINT once the attacks began. The wannacry ransomware generated a massive discussion on Twitter: describing the issue, how to avoid it, and informing about the kill switch that eventually disabled it.

11 Conclusions and Limitations

In this paper we provide an analysis of the richness of coverage of vulnerabilities and timeliness (in terms of reporting dates of vulnerabilities) of some of the most important OSINT sources, namely Twitter and several vulnerability databases. Our key findings are the following: no source could be considered clearly better than others and therefore diverse OSINT sources should be used as they complement each other; when considering only confirmed vulnerabilities, NVD should not be the unique vulnerability database subscribed; since 2010, Twitter provides an almost perfect vulnerability coverage; Twitter discusses vulnerabilities ahead of databases for very few cases (about 1% for the vulnerabilities examined), and is as timely as the vulnerability databases for the remaining cases; and finally, most of the vulnerabilities reported early on Twitter have a high or critical impact, with the tweet leading to usable mitigation measures. Beyond the collected facts, we provide a set of insights for the security practitioner interested in using OSINT for cybersecurity. These insight are based on our experience of manual inspection of almost one million tweets, and analysing many thousands of vulnerabilities. We believe this knowledge should be valuable for security analysis and research both in industry and academia.

Limitations. The results presented in this paper are somewhat pessimistic in terms of the number of vulnerabilities that were found to have early alerts on Twitter. There could be more cases with media attention or early alerts that were not captured by our methodology since we cover a reduced amount of vulnerabilities. There is also the possibility of human error, as manual processing of tweets can lead to mistakes and missing some matches—all early alerts were triple checked to avoid false positives.

Another factor that we cannot control (since we are performing a forensic analysis) is that some early alert tweets could have been deleted before this study, and thus not captured. Many dead links also invalidated possible matches, especially when the tweet links used some shortening system, such as "dlvr.it", "hrbt.us", "url4.eu", "bit.ly", or "ow.ly".

Funding

This work was supported by the H2020 European Project DiSIEM (H2020-700692) and by the Fundação para a Ciência e a Tecnologia (FCT) through project ThreatAdapt (FCT-FNR/0002/2018) and the LASIGE Research Unit (UIDB/00408/2020 and UIDP/00408/2020).

References

1. Additional paper data. https://github.com/fernandoblalves/Follow-the-Blue-Bird-Paper-Additional-Data. Accessed 10 Jul 2020
2. Common platform enumeration. https://cpe.mitre.org/about/. Accessed 15 Apr 2020

3. Common vulnerabilities and exposures (cve). http://cve.mitre.org/. Accessed 15 Apr 2020
4. Common vulnerability scoring system version 3.0. https://www.first.org/cvss/v3-0/. Accessed 15 Apr 2020
5. Cvss v2 archive. https://www.first.org/cvss/v2/. Accessed 15 Apr 2020
6. Exploit database. www.exploit-db.com/. Accessed 15 Apr 2020
7. Get old tweets programatically. https://github.com/Jefferson-Henrique/GetOldTweets-java. Accessed 15 Apr 2020
8. Hackers say they hacked nsa-linked group, want 1 million bitcoins to share more. https://www.vice.com/en_us/article/ezpa9p/hackers-hack-nsa-linked-equation-group. Accessed 15 Apr 2020
9. How people use Twitter in general. https://www.americanpressinstitute.org/publications/reports/survey-research/how-people-use-twitter-in-general/. Accessed 15 Apr 2020
10. How Twitter evolved from 2006 to 2011. https://buffer.com/resources/how-twitter-evolved-from-2006-to-2011. Accessed 15 Apr 2020
11. The mitre corporation. https://www.mitre.org/. Accessed 15 Apr 2020
12. National vulnerability database. https://nvd.nist.gov/. Accessed 15 Apr 2020
13. New targeted attack in the middle east by APT34, a suspected iranian threat group, using cve-2017-11882 exploit. https://www.fireeye.com/blog/threat-research/2017/12/targeted-attack-in-middle-east-by-apt34.html. Accessed 15 Apr 2020
14. Operation aurora. https://en.wikipedia.org/wiki/Operation_Aurora. Accessed 15 Apr 2020
15. Packet storm. https://packetstormsecurity.com/. Accessed 15 Apr 2020
16. The race between security professionals and adversaries. https://www.recordedfuture.com/vulnerability-disclosure-delay/. Accessed 15 Apr 2020
17. Security database. https://www.security-database.com/. Accessed 15 Apr 2020
18. Spiderfoot. https://www.spiderfoot.net/documentation/. Accessed 15 Apr 2020
19. Tweepy. https://www.tweepy.org/. Accessed 15 Apr 2020
20. Tweet attacks pro. http://www.tweetattackspro.com/. Accessed 15 Apr 2020
21. Twitter input plugin. https://www.elastic.co/guide/en/logstash/current/plugins-inputs-twitter.html. Accessed 15 Apr 2020
22. Veris taxonomy. http://veriscommunity.net/enums.html#section-incident_desc. Accessed 13 Jun 2018
23. Watchguard firewalls - 'escalateplowman' ifconfig privilege escalation. https://www.exploit-db.com/exploits/40270. Accessed 15 Apr 2020
24. Almukaynizi, M., Nunes, E., Dharaiya, K., Senguttuvan, M., Shakarian, J., Shakarian, P.: Proactive identification of exploits in the wild through vulnerability mentions online. In: 2017 CyCon US (2017)
25. Alsaedi, N., Burnap, P., Rana, O.: Can we predict a riot? Disruptive event detection using Twitter. ACM TOIT **17**(2), 1–26 (2017)
26. Alves, F., Bettini, A., Ferreira, P.M., Bessani, A.: Processing tweets for cybersecurity threat awareness. Inf. Syst. **95**, 101586 (2020)
27. Andongabo, A., Gashi, I.: vepRisk - a web based analysis tool for public security data. In: 13th EDCC (2017)
28. Arora, A., Krishnan, R., Nandkumar, A., Telang, R., Yang, Y.: Impact of vulnerability disclosure and patch availability-an empirical analysis. In: Third Workshop on the Economics of Information Security (2004)

29. Behzadan, V., Aguirre, C., Bose, A., Hsu, W.: Corpus and deep learning classifier for collection of cyber threat indicators in Twitter stream. In: IEEE Big Data 2018 (2018)
30. Bian, J., Topaloglu, U., Yu, F.: Towards large-scale Twitter mining for drug-related adverse events. In: Proceedings of the SHB (2012)
31. Bozorgi, M., Saul, L.K., Savage, S., Voelker, G.M.: Beyond heuristics: learning to classify vulnerabilities and predict exploits. In: 16th ACM SIGKDD (2010)
32. Bullough, B.L., Yanchenko, A.K., Smith, C.L., Zipkin, J.R.: Predicting exploitation of disclosed software vulnerabilities using open-source data. In: 3rd ACM IWSPA (2017)
33. Campiolo, R., Santos, L.A.F., Batista, D.M., Gerosa, M.A.: Evaluating the utilization of Twitter messages as a source of security alerts. In: SAC 13 (2013)
34. Cataldi, M., Di Caro, L., Schifanella, C.: Emerging topic detection on Twitter based on temporal and social terms evaluation. In: 10th MDM/KDD (2010)
35. Devlin, J., Chang, M.W., Lee, K., Toutanova, K.: Bert: Pre-training of deep bidirectional transformers for language understanding (2018). arXiv:1810.04805
36. Dionísio, N., Alves, F., Ferreira, P.M., Bessani, A.: Cyberthreat detection from Twitter using deep neural networks. In: IJCNN 2019 (2019)
37. Dionísio, N., Alves, F., Ferreira, P.M., Bessani, A.: Towards end-to-end cyberthreat detection from Twitter using multi-task learning. In: IJCNN 2020 (2020)
38. Edkrantz, M., Truvé, S., Said, A.: Predicting vulnerability exploits in the wild. In: 2nd IEEE CSCloud (2015)
39. Fedoryszak, M., Frederick, B., Rajaram, V., Zhong, C.: Real-time event detection on social data streams. In: 25th ACM SIGKDD - KDD '9 (2019)
40. Le Sceller, Q., Karbab, E.B., Debbabi, M., Iqbal, F.: Sonar: automatic detection of cyber security events over the Twitter stream. In: 12th ARES (2017)
41. McNeil, N., Bridges, R.A., Iannacone, M.D., Czejdo, B., Perez, N., Goodall, J.R.: Pace: pattern accurate computationally efficient bootstrapping for timely discovery of cyber-security concepts. In: 12th ICMLA (2013)
42. McQueen, M.A., McQueen, T.A., Boyer, W.F., Chaffin, M.R.: Empirical estimates and observations of 0day vulnerabilities. In: 42nd HICSS (2009)
43. Mittal, S., Das, P.K., Mulwad, V., Joshi, A., Finin, T.: Cybertwitter: using Twitter to generate alerts for cybersecurity threats and vulnerabilities. In: 2016 ASONAM (2016)
44. Moholth, O.C., Juric, R., McClenaghan, K.M.: Detecting cyber security vulnerabilities through reactive programming. In: HICSS 2019 (2019)
45. Nayak, K., Marino, D., Efstathopoulos, P., Dumitraş, T.: Some vulnerabilities are different than others. In: 17th RAID (2014)
46. Petrović, S., Osborne, M., Lavrenko, V.: Streaming first story detection with application to Twitter. In: 11th NAACL HLT (2010)
47. Reinthal, A., Filippakis, E.L., Almgren, M.: Data modelling for predicting exploits. In: NordSec (2018)
48. Ritter, A., Wright, E., Casey, W., Mitchell, T.: Weakly supervised extraction of computer security events from Twitter. In: 24th WWW (2015)
49. Rodriguez, L.G.A., Trazzi, J.S., Fossaluza, V., Campiolo, R., Batista, D.M.: Analysis of vulnerability disclosure delays from the national vulnerability database. In: WSCDC-SBRC 2018 (2018)
50. Roumani, Y., Nwankpa, J.K., Roumani, Y.F.: Time series modeling of vulnerabilities. Comput. Secur. **51**, 32–40 (2015)

51. Sabottke, C., Suciu, O., Dumitraş, T.: Vulnerability disclosure in the age of social media: exploiting Twitter for predicting real-world exploits. In: 24th USENIX Security Symposium (2015)
52. Sakaki, T., Okazaki, M., Matsuo, Y.: Earthquake shakes Twitter users: real-time event detection by social sensors. In: 19th WWW (2010)
53. Sapienza, A., Bessi, A., Damodaran, S., Shakarian, P., Lerman, K., Ferrara, E.: Early warnings of cyber threats in online discussions. In: 2017 ICDMW (2017)
54. Sapienza, A., Ernala, S.K., Bessi, A., Lerman, K., Ferrara, E.: Discover: mining online chatter for emerging cyber threats. In: WWW'18 Companion (2018)
55. Sauerwein, C., Sillaber, C., Huber, M.M., Mussmann, A., Breu, R.: The tweet advantage: an empirical analysis of 0-day vulnerability information shared on Twitter. In: 33rd IFIP SEC (2018)
56. Shahzad, M., Shafiq, M.Z., Liu, A.X.: A large scale exploratory analysis of software vulnerability life cycles. In: 34th ICSE (2012)
57. Steele, R.D.: Open source intelligence: what is it? why is it important to the military. Am. Intell. J. **17**(1), 35–41 (1996)
58. Syed, R., Rahafrooz, M., Keisler, J.M.: What it takes to get retweeted: an analysis of software vulnerability messages. Comput. Hum. Behav. **80**, 207–215 (2018)
59. Trabelsi, S., Plate, H., Abida, A., Aoun, M.M.B., Zouaoui, A., et al.: Mining social networks for software vulnerabilities monitoring. In: 7th NTMS (2015)
60. Xie, W., Zhu, F., Jiang, J., Lim, E.P., Wang, K.: Topicsketch: real-time bursty topic detection from Twitter. IEEE TKDE **28**(8), 2216–2229 (2016)
61. Yan, X., Guo, J., Lan, Y., Xu, J., Cheng, X.: A probabilistic model for bursty topic discovery in microblogs. In: 29th AAAI (2015)
62. Zaki, M.J., et al.: Data Mining and Analysis: Fundamental Concepts and Algorithms. Cambridge University Press, Cambridge (2014)

Dynamic and Secure Memory Transformation in Userspace

Robert Lyerly(✉), Xiaoguang Wang(✉), and Binoy Ravindran(✉)

Virginia Tech, Blacksburg, VA, USA
{rlyerly,xiaoguang,binoy}@vt.edu

Abstract. Continuous code re-randomization has been proposed as a way to prevent advanced code reuse attacks. However, recent research shows the possibility of exploiting the runtime stack even when performing integrity checks or code re-randomization protections. Additionally, existing re-randomization frameworks do not achieve strong isolation, transparency and efficiency when securing the vulnerable application. In this paper we present Chameleon, a userspace framework for dynamic and secure application memory transformation. Chameleon is an out-of-band system, meaning it leverages standard userspace primitives to monitor and transform the target application memory from an entirely separate process. We present the design and implementation of Chameleon to dynamically re-randomize the application stack slot layout, defeating recent attacks on stack object exploitation. The evaluation shows Chameleon significantly raises the bar of stack object related attacks with only a 1.1% overhead when re-randomizing every 50 ms.

1 Introduction

Memory corruption is still one of the biggest threats to software security [43]. Attackers use memory corruption as a starting point to directly hijack program control flow [9,18,22], modify control data [24,25], or steal secrets in memory [20]. Recent works have shown that it is possible to exploit the stack even under new integrity protections designed to combat the latest attacks [23–25,31]. For example, position-independent return-oriented programming (PIROP) [23] leverages a user controlled sequence of function calls and un-erased stack memory left on the stack after returning from functions (e.g., return addresses, initialized local data) to construct a ROP payload. Data-oriented programming (DOP) also heavily relies on user-controlled stack objects to change the execution path in an attacker-intended way [24–26]. Both of these attacks defeat existing code re-randomization mechanisms, which continuously permute the locations of functions [50] or hide function locations to prevent memory disclosure vulnerabilities from constructing gadget chains [14].

In this work, we present Chameleon, a continuous stack randomization framework. Chameleon, like other continuous re-randomization frameworks, periodically permutes the application's memory layout in order to prevent attackers

L. Chen et al. (Eds.): ESORICS 2020, LNCS 12308, pp. 237–256, 2020.
https://doi.org/10.1007/978-3-030-58951-6_12

from using memory disclosure vulnerabilities to exfiltrate data or construct malicious payloads. Chameleon, however, focuses on randomizing the stack – it randomizes the layout of every function's stack frame so that attackers cannot rely on the locations of stack data for attacks. In order to correctly reference local variables in the randomized stack layout, Chameleon also rewrites every function's code, further disrupting code reuse attacks that expect certain instruction sequences (either aligned or unaligned). Chameleon periodically interrupts the application to rewrite the stack and inject new code. In this way, Chameleon can defeat attacks that rely on stack data locations such as PIROP or DOP.

Chameleon is also novel in how it implements re-randomization. Existing works build complex runtimes into the application's address space that add non-trivial performance overhead from code instrumentation [1,14,17,50]. Chameleon is instead an *out-of-band* framework that executes in userspace in an entirely separate process. Chameleon attaches to the application using standard OS interfaces for observation and re-randomization. This provides strong isolation between Chameleon and application – attackers cannot observe the re-randomization process (e.g., observe random number generator state, dump memory layout information) and Chameleon does not interact with any user-controlled input. Additionally, cleanly separating Chameleon from the application allows much of the re-randomization process to proceed in parallel. This design adds minimal overhead, as Chameleon only blocks the application when switching between randomized stack layouts. Chameleon can efficiently re-randomize an application's stack layout with randomization periods in the range of tens of milliseconds.

In this paper, we make the following contributions:

- We describe Chameleon, a system for continuously re-randomizing application stack layouts,
- We detail Chameleon's stack randomization process that relies on using compiler-generated function metadata and runtime binary reassembly,
- We describe how Chameleon uses the standard `ptrace` and `userfaultfd` OS interfaces to efficiently transform the application's stack and inject newly-rewritten code,
- We evaluate the security benefits of Chameleon and report its performance overhead when randomizing code on benchmarks from the SPEC CPU 2017 [42] and NPB [4] benchmark suites. Chameleon's out-of-band architecture allows it to randomize stack slot layout with only 1.1% overhead when changing the layout with a 50 millisecond period,
- We describe how Chameleon disrupts a real-world attack against the popular nginx webserver

The rest of this paper is organized as follows: in Sect. 2, we describe the background and the threat model. We then present the design and implementation of Chameleon in Sect. 3. We evaluate Chameleon security properties and performance in Sect. 4. We discuss related works in Sect. 5 and conclude the paper in Sect. 6.

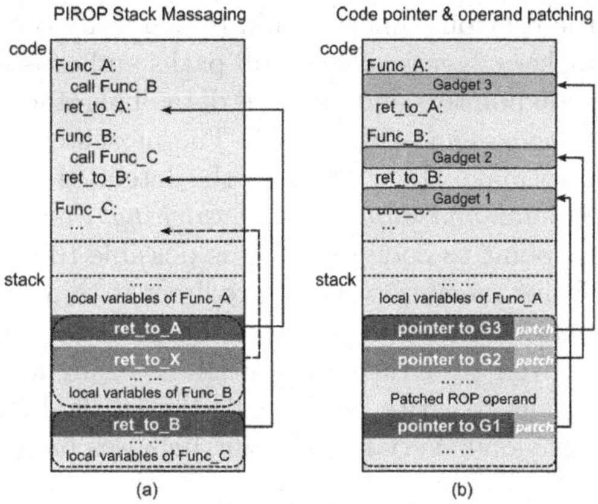

Fig. 1. Position Independent ROP. (a) *Stack massaging* uses code pointers that remain on the stack after the function returns. (b) *Code pointer and operand patching* write part of the massaged stack memory with relative memory writes to construct the payload.

2 Background and Threat Model

Before describing Chameleon, we first describe how stack object related attacks target vulnerable applications, including detailing a recent presented position-independent code reuse attack. We then define the threat model of these attacks.

2.1 Background

Traditional code reuse attacks rely on runtime application memory information to construct the malicious payload. Return-oriented programming (ROP) [36,40] chains and executes several short instruction sequences ending with `ret` instructions, called *gadgets*, to conduct Turing-complete computation. After carefully constructing the ROP payload of gadget pointers and data operands, the attacker then tricks the victim process into using the ROP payload as stack data. Once the ROP payload is triggered, gadget pointers are loaded into the program counter (which directs control flow to the gadgets) and the operand data is populated into registers to perform the intended operations (e.g., prepare parameters to issue an attacker-intended system call). Modern attacks, such as JIT-ROP [41], utilize a memory disclosure vulnerability to defeat coarse-grained randomization techniques such as ASLR [29] by dynamically discovering gadgets and constructing gadget payloads.

Position-independent ROP (PIROP) [23] proposes a novel way to reuse existing pointers on the stack (e.g., function addresses) as well as relative code offsets to construct the ROP payload. PIROP constructs the ROP payload agnostic to the code's absolute address. It leverages the fact that function call return addresses and local variables may remain on the stack even after the function returns, meaning the next function call may observe stack local variables and code pointers from the previous function call. By carefully controlling the application input, the attacker triggers specific call paths and constructs a stack with attacker-controlled code pointers and operand data. This stack construction procedure is called *stack massaging* (Fig. 1 (a)). The next step modifies some bits of the code pointers to make them point to the intended gadgets (Fig. 1 (b)). This is called *code pointer and data operand patching*. Since code pointers left from stack massaging point to code pages, it is possible to modify some bits of the pointer using relative memory writes to redirect it to a gadget on the same code page. Fundamentally, PIROP assumes function calls leave their stack slot contents on the stack even after the call returns. By using a temporal sequence of different function calls to write pointers to the stack, PIROP constructs a skeleton of the ROP payload. Very few existing defenses break this assumption.

2.2 Threat Model and Assumptions

The attacker communicates with the target application through typical I/O interfaces such as sockets, giving the attacker the ability to send arbitrary input data to the target. The attacker has the target application binary, thus they are aware of the relative addresses inside any 4 K memory page windows. The attacker can exploit a memory disclosure vulnerability to read arbitrary memory locations and can use PIROP to construct the ROP payload. The application is running using standard memory protection mechanisms such that no page has both write and execute permissions; this means the attacker cannot directly inject code but must instead rely on constructing gadget chains. However, the gadget chains crafted by the attacker can invoke system APIs such as `mprotect` to create such regions if needed. The attacker knows that the target is running under Chameleon's control and therefore knows of its randomization capabilities. We assume the system software infrastructure (compiler, kernel) is trusted and therefore the capabilities provided by these systems are correct and sound.

3 Design

Chameleon continuously re-randomizes the code section and stack layout of an application in order to harden it against temporal stack object reuse (i.e. PIROP), stack control data smashing and stack object disclosures. As a result of running under Chameleon, gadget addresses or stack object locations that are leaked by memory disclosures and that help facilitate other attacks (temporal stack object reuse, payload construction) are only useful until the next randomization, after which the attacker must re-discover the new layout and locations of sensitive data. Chameleon continuously randomizes the application (hereafter called the target or child) quickly enough so that it becomes probabilistically impossible for attackers to construct and execute attacks against the target.

Similarly to previous re-randomization frameworks [17,50], Chameleon is transparent – the target has no indication that it is being re-randomized. However, Chameleon's architecture is different from existing frameworks in that it executes outside the target application's address space and attaches to the target using standard OS interfaces. This avoids the need for bootstrapping and running randomization machinery inside an application, which adds complexity and high overheads. Chameleon runs all randomization machinery in a separate process, which allows generating the next set of randomization information in parallel with normal target execution. This also *strongly isolates* Chameleon from the target in order to make it extremely difficult for attackers to observe the randomization process itself. These benefits make Chameleon easier to use and less intrusive versus existing re-randomization systems.

3.1 Requirements

Chameleon needs a description of each function's stack layout, including location, size and alignment of each stack slot, so that it can randomize each stack slot's location. Ideally, Chameleon would be able to determine every stack slot's location, size and alignment by analyzing a function's machine code. In reality, however, it is impossible to tell from the machine code whether adjacent stack memory locations are separate stack slots (which can be relocated independently) or multiple parts of a single stack slot that must be relocated together (e.g., a struct with several fields)[1]. Therefore, Chameleon requires metadata from the compiler describing how it has laid out the stack.

While DWARF debugging information [21] can provide some of the required information, it is best-effort and does not capture a complete view of execution state needed for transformation (e.g., unnamed values created during optimization). Instead, Chameleon builds upon existing work [5] that extends LLVM's stack maps [30] to dump a complete view of function activations. The compiler instruments LLVM bitcode to track live values (stack objects, local variables) by adding stack maps at individual points inside the code. In the backend, stack maps force generation of a per-function record listing stack slot sizes, alignments and offsets. Stack maps also record locations of all live values at the location where the stack map was inserted. Chameleon uses each stack map to reconstruct the frame at that location. The modified LLVM extends stack maps to add extra semantic information for live values, particularly whether a live value is a pointer. This allows Chameleon to detect at runtime if the pointer references the stack, and if so, update the pointer to the stack slot's randomized location. The metadata also describes each function's location and size, which Chameleon uses to patch each function to match the randomized layout. All of the metadata is generated at compile time and is lowered into the binary.

Chameleon also needs to rewrite stack slot references in code to point to their new locations and must transform existing execution state, namely stack memory and registers, to adhere to the new randomized layout. To switch between different randomized stack layouts (named *randomization epochs*), Chameleon must be able to pause the target, observe current target execution state, rewrite

[1] This information could potentially be inferred heuristically, e.g., from a decompiler.

the existing state to match the new layout and inject code matching the new layout. Chameleon uses two kernel interfaces, `ptrace` and `userfaultfd`, to monitor and transform the target. `ptrace` [48] is widely used by debuggers to inspect and control the execution of *tracees*. `ptrace` allows *tracers* (e.g., Chameleon) to read and modify tracee state (per-thread register sets, signal masks, virtual memory), intercept events such as signals and system calls, and forcibly interrupt tracee threads. `userfaultfd` [27] is a Linux kernel mechanism that allows delegating handling of page faults for a memory region to user-space. When accesses to a region of memory attached to a `userfaultfd` file descriptor cause a page fault, the kernel sends a request to a process reading from the descriptor. The process can then respond with the data for the page by writing a response to the file descriptor. These two interfaces together give Chameleon powerful and flexible process control tools that add minimal overhead to the target.

3.2 Re-Randomization Architecture

Chameleon uses the mechanisms described in Sect. 3.1 to transparently observe the target's execution state and periodically interrupt the target to switch it to the next randomization epoch. In between randomization epochs, Chameleon executes in parallel with the target to generate the next set of randomized stack layouts and code. Figure 2 shows Chameleon's system architecture. Users launch the target application by passing the command line arguments to Chameleon. After reading the code and state transformation metadata from the target's binary, Chameleon forks the target application and attaches to it via `ptrace` and `userfaultfd`. From this point on, Chameleon enters a *re-randomization loop*. At the start of a new randomization cycle, a *scrambler* thread iterates through every function in the target's code, randomizing the stack layout as described below. At some point, a re-randomization timer fires, triggering a switch to the next randomization epoch. When the re-randomization event fires, the *event handler* thread interrupts the target and switches the target to the next randomization epoch by dropping the existing code pages and transforming the target's execution state (stack, registers) to the new randomized layout produced by the scrambler. After transformation, the event handler writes the execution state back into the target and resumes the child; it then blocks until the next re-randomization event. As the child begins executing, it triggers code page faults by fetching instructions from dropped code pages. A *fault handler* thread handles these page faults by serving the newly randomized code. In this way the entire re-randomization procedure is transparent to the target and incurs low overheads. We describe each part of the architecture in the following sections.
Randomizing stack layouts. Chameleon randomizes function stack layouts by logically permuting stack slot locations and adding padding between the slots. Chameleon also transforms stack memory references in code to point to their randomized locations. When patching the code, Chameleon must work within the space of the code emitted by the compiler. If, for example, Chameleon wanted to change the size of code by inserting arbitrary instructions or changing the operand encoding of existing instructions, Chameleon would need to update all

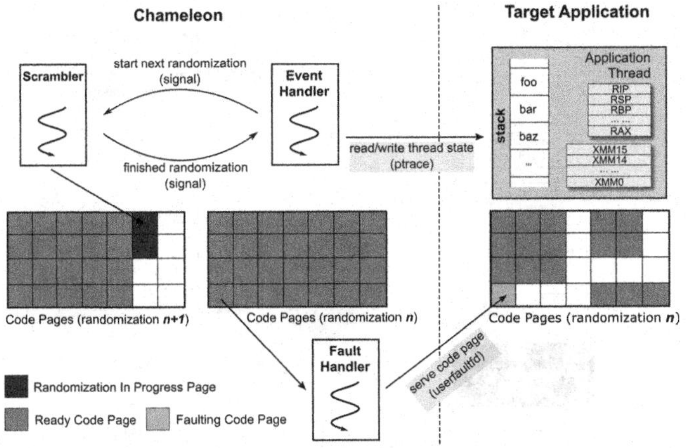

Fig. 2. Chameleon system architecture. An *event handler* thread waits for events in a target thread (e.g., signals), interrupts the thread, and reads/writes the thread's execution state (registers, stack) using `ptrace`. A *scrambler* thread concurrently prepares the next set of randomized code for the next re-randomization. A *fault handler* thread responds to page faults in the target by passing pages from the current code randomization to the kernel through `userfaultfd`.

code references affected by change in size (e.g., jumps between basic blocks, function calls/returns, etc.). Because finding and updating all code references is known to not be statically solvable [47], previous re-randomization works either leverage dynamic binary instrumentation (DBI) frameworks [7,17,32] or an indirection table [3,50] in order to allow arbitrary code instrumentation. There are problems with both approaches – the former often have large performance costs while the latter does not actually re-randomize the stack layout, instead opting to try and hide code pages from attackers. Chameleon instead applies stack layout randomization without changing the size of code to avoid these problems. In order to facilitate randomizing all elements of the stack, Chameleon modifies the compiler to *(1)* pad function prologues and epilogues with `nop` instructions that can be rewritten with other instructions and *(2)* force 4-byte immediate encodings for all memory operands[2].

Chameleon both permutes the ordering and adds random amounts of padding between stack slots; the latter is configurable so users can control how much memory is used versus how much randomness is added between slots. Figure 3 shows how Chameleon randomizes the following stack elements: *(1) Callee-saved registers*: the compiler saves and restores callee-saved registers through `push` and `pop` instructions. Chameleon uses the `nop` padding emitted by the compiler to rewrite them as `mov` instructions, allowing the scrambler to place callee-saved registers at arbitrary locations on the stack. Chameleon also randomizes the locations of the return address and saved frame base pointer by inserting `mov`

[2] x86-64 backends typically emit small immediate operands using a 1-byte encoding.

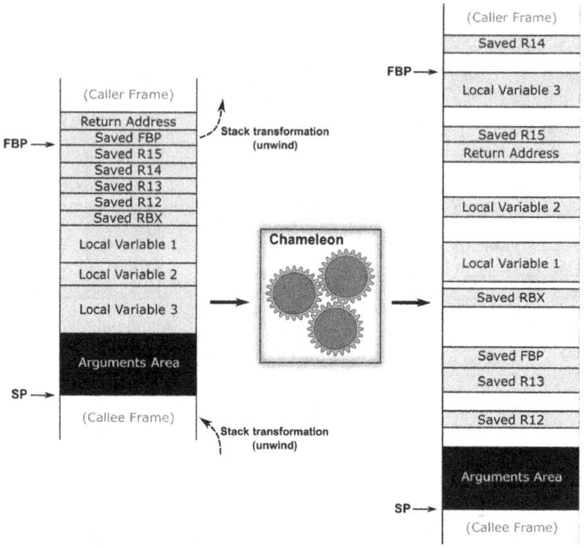

Fig. 3. Stack slot randomization. Chameleon permutes the ordering and adds random amounts of padding between slots.

intructions in the function's prologue and epilogue. *(2) Local variables*: compilers emit references to stack-allocated variables as offsets from the frame base pointer (FBP) or stack pointer (SP). Chameleon randomizes the locations of local variables by rewriting a variable's offset to point to the randomized location. Chameleon does not currently randomize the locations of stack arguments for called functions as the locations are dictated by the ABI and would require rewriting both the caller and callee with a new parameter passing convention. We plan these transformations as future work.

Serving code pages. Chameleon needs a mechanism to transparently and efficiently serve randomized code pages to the child. While Chameleon could use `ptrace` to directly write the randomized code into the address space of the child application, this would cause large delays when swapping between randomization epochs for applications with large code sections – Chameleon would have to bulk write the entire code section on every re-randomization. Instead Chameleon uses `userfaultfd` and page faults, which allows quicker switches between epochs by demand paging code into the target application's address space.

At startup, Chameleon attaches a `userfaultfd` file descriptor to the target's code memory virtual memory area (VMA). `userfaultfd` descriptors can only be opened by the process that owns the memory region for which faults should be handled. Chameleon cannot directly open a `userfaultfd` descriptor for the target but can induce the target application to create a descriptor and pass it to Chameleon over a socket before the target begins normal execution. Chameleon uses `compel` [15], a library that facilitates implanting *parasites* into applications controlled via `ptrace`. Parasites are small binary blobs which execute code in the context of the target application. For Chameleon, the parasite opens a `userfaultfd` descriptor and passes it to Chameleon through a socket. To execute the parasite, Chameleon takes a snapshot of the target process' main thread (registers, signal masks). Then, it finds an executable region of memory in the target and writes the parasite code into the target. Because `ptrace` allows writing a thread's registers, Chameleon redirects the target thread's program counter to the parasite and begins execution. The parasite opens a control socket, initializes a `userfaultfd` descriptor, passes the descriptor to Chameleon, and exits at well-known location. Chameleon intercepts the thread at the exit point, restores the thread's registers and signal mask to their original values and restores the code clobbered by the parasite.

After receiving the `userfaultfd` descriptor, Chameleon must prepare the target's code region for attaching (`userfaultfd` descriptors can only attach to anonymous VMAs [44]). Chameleon executes an `mmap` system call inside the target to remap the code section as anonymous and then registers the code section with the `userfaultfd` descriptor. The controller then starts the fault handler thread, which serves code pages through the `userfaultfd` descriptor from the scrambler thread's code buffer as the target accesses unmapped pages.

Switching Between Randomization Epochs. The event handler begins switching the target to the new set of randomized code when interrupted by the re-randomization alarm. The event handler interrupts the target, converts existing execution state (registers, stack memory) to the next randomization epoch, and drops existing code pages so the target can fetch fresh code pages on-demand.

The event handler issues a `ptrace` interrupt to grab control of the target. At this point Chameleon needs to transform the target's current stack to match the new stack layout. The compiler-emitted stackmaps only describe the complete stack layout at given points inside of a function, called *transformation points*. To switch to the next randomization epoch, Chameleon must advance the target to a transformation point. While the thread is interrupted, Chameleon uses `ptrace` to write trap instructions into the code at transformation points found during initial code disassembly and analysis[3]. Chameleon then resumes the target thread and waits for it to reach the trap. When it executes the trap, the kernel interrupts the thread and Chameleon regains control. Chameleon then restores the original instructions and begins state transformation.

Chameleon unwinds the stack using stackmaps similarly to other re-randomiz-ation systems [50]. During unwinding, however, Chameleon shuffles stack objects to their new randomized locations using information generated by the scrambler (Fig. 3). For each function, the scrambler creates a mapping between the original and randomized offsets of all stack slots. This mapping is used to move stack data from its current randomized location to the next randomized location. To access the target's stack, Chameleon reads the target's register values using `ptrace` and the target's stack using `/proc/<target`

[3] Chameleon uses the `int3` instruction.

`pid>/mem`[4]. After reaching a transformation point, Chameleon reads the thread's entire stack into a buffer, located using the target's stack pointer. Chameleon passes the stack pointer, register set and buffer containing stack data to a stack transformation library to transform it to the new randomized layout.

The final step in re-randomization is to map the new code into the target's address space using `userfaultfd`. Chameleon executes an `madvise` system call with the `MADV_DONTNEED` flag for the code section in the context of the target, which instructs the kernel to drop all existing code pages and cause fresh page faults upon subsequent execution. The fault handler begins serving page faults from the code buffer for the new randomization epoch and the target is released to continue normal execution. At this point the target is now executing under a new set of randomized code. The event handler thread signals the scrambler thread to begin generating the next randomization epoch. In this way, switching randomization epochs blocks the target only to transform the target thread's stack and drop the existing code pages. The most expensive work of generating newly randomized code happens in parallel with the target application's execution, highlighting one of the major benefits of cleanly separating re-randomization into a separate process from the target application.

Multi-process Applications. Chameleon supports multi-process applications such as web servers that fork children for handling requests. When the target forks a new child, the kernel informs Chameleon of a `fork` event. The new process inherits `ptrace` status from the parent, meaning the event handler also has tracing privileges for the new child. At this point, the controller instantiates a new scrambler, fault handler and event handler for the new child. Chameleon hands off tracer privileges from the parent to the child event handler thread so the new handler can control the new child. In order to do this, Chameleon first redirects the new child to a blocking read on a socket through code installed via parasite. The original event handler thread then detaches from the new child, allowing the new event handler thread to become the tracer for the new child while it is blocked. After attaching, the new event handler restores the new child to the fork location and removes the parasite. In this way, Chameleon always maintains complete control of applications even when they fork new processes.

3.3 A Prototype of Chameleon

Chameleon is implemented in 6092 lines of C++ code, which includes the event handler, scrambler and fault handler. Chameleon extends code from an open source stack transformation framework [5] to generate transformation metadata and transform the stack at runtime. Chameleon uses DynamoRIO [7] disassemble and re-assemble the target's machine code. Currently Chameleon supports x86-64. Chameleon's use of `ptrace` prevents attaching other `ptrace`-based applications such as GDB. However, it is unlikely users will want to use both, as GDB is most useful during development and testing. However, Chameleon could be

[4] This file allows tracers to seek to arbitrary addresses in the target's address space to read/write ranges of memory.

extended to dump randomization information when the target crashes to allow debugging core dumps in debuggers.

4 Evaluation

In this section we evaluate Chameleon's capabilities both in terms of security benefits and overheads:

- What kinds of security benefits does Chameleon provide? In particular, how much randomization does it inject into stack frame layouts? This includes describing a real-world case study of how Chameleon defeats a web server attack. (Sect. 4.1)
- How much overhead does Chameleon impose for these security guarantees, including how expensive are the individual components of Chameleon and how much overhead does it add to the total execution time? (Sect. 4.2)

Experimental Setup. Chameleon was evaluated on an x86-64 server containing an Intel Xeon 2620v4 with a clock speed of 2.1 GHz (max boost clock of 3.0 GHz). The Xeon 2620v4 contains 8 physical cores and has 2 hardware threads per core for a total of 16 hardware threads. The server contains 32 GB of DDR4 RAM. Chameleon is run on Debian 8.11 "Jessie" using Linux kernel 4.9. Chameleon was configured to add a maximum padding of 1024 bytes between stack slots. Chameleon was evaluated using benchmarks from the SNU C version of the NPB benchmarks [4,38] and SPEC CPU 2017 [42]. Benchmarks were compiled with all optimizations (`-O3`) using the previously described compiler, built on clang/LLVM v3.7.1. The single-threaded version of NPB was used.

4.1 Security Analysis

We first analyze both the quality of Chameleon runtime re-randomization in the target and describe the security of the Chameleon framework itself. Because Chameleon, like other approaches [17,45,50], relies on layout randomization to disrupt attackers, it cannot make any guarantees that attacks will not succeed. There is always the possibility that the attacker is lucky and guesses the exact randomization (both stack layout and randomized code) and is able to construct a payload to exploit the application and force it into a malicious execution. However, with sufficient randomization, the probability that such an attack will succeed is so low as to be practically impossible.

Target Randomization: Chameleon randomizes the target in two dimensions: *randomizing the layout of stack elements* and *rewriting the code* to match the randomized layout. We first evaluated how well *randomizing the stack disrupts attacks that utilize known locations of stack elements*. When quantifying the randomization quality of a given system, many works use *entropy* or the number of randomizable states as a measure of randomness. For Chameleon, entropy refers to the number of potential locations a stack element could be placed, i.e., the number of randomizable locations.

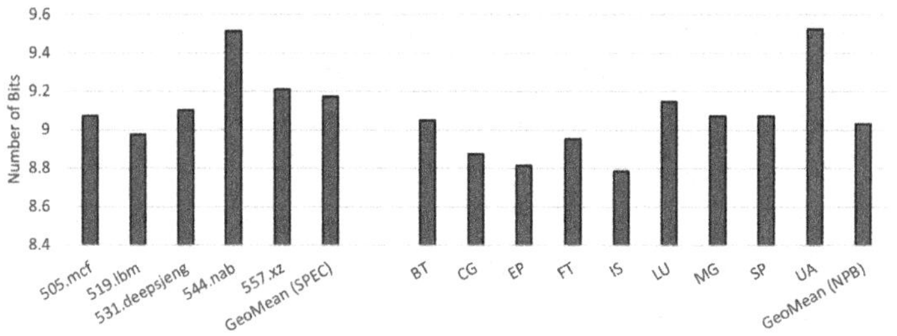

Fig. 4. Average number of bits of entropy for a stack element across all functions within each binary. Bits of entropy quantify in how many possible locations a stack element may be post-randomization – for example, 2 bits of entropy mean the stack element could be in $2^2 = 4$ possible locations with a $\frac{1}{4} = 25\%$ chance of guessing the location.

Figure 4 shows the average entropy created by Chameleon for each benchmark. For each application, the y-axis indicates the geometric mean of the number of bits of entropy across all stack slots in all functions. Chameleon provides a geometric mean 9.17 bits of entropy for SPEC and 9.03 bits for NPB. Functions with more stack elements have higher entropy as there are a larger number of permutations. SPEC's benchmarks tend to have higher entropy because they have more stack slots. While an attacker may be able to guess the location of a single stack element with 9 bits of entropy (probability of $\frac{1}{2^9} = 0.00195$), the attacker must chain together knowledge of multiple stack locations to make a successful attack. For an attack that must corrupt three stack slots, the attacker has a $0.00195^3 = 7.45 * 10^{-9}$ probability of correctly guessing the stack locations, therefore making successful attacks probabilistically impossible. It is also important to note that the amount of entropy can be increased arbitrarily by increasing the amount of padding between stack slots, which necessarily creates much larger stacks. We conclude that Chameleon makes it infeasible for attackers to guess stack locations needed in exploits.

Next, we evaluated how Chameleon's code patching disturbs gadget chains. Attackers construct malicious executions by chaining together gadgets that perform a very basic and low-level operation. Gadgets, and therefore gadget chains, are very frail – slight disruptions to a gadget's behavior can disrupt the entire intended functionality of the chain. As part of the re-randomization process, Chameleon rewrites the application's code to match the randomized stack layout. A side effect of this is that gadgets may be disrupted – Chameleon may overwrite part or all of a gadget, changing its functionality and disrupting the gadget chain. To analyze how many gadgets are disrupted, we searched for gadgets in the benchmark binaries and cross-referenced gadget addresses with instructions rewritten by Chameleon. We used Ropper[5], a gadget finder tool, to find all ROP gadgets (those that end in a return) and JOP gadgets (those that end in a call or jump) in the application binaries. We searched for gadgets of 6 instructions

[5] https://github.com/sashs/Ropper.

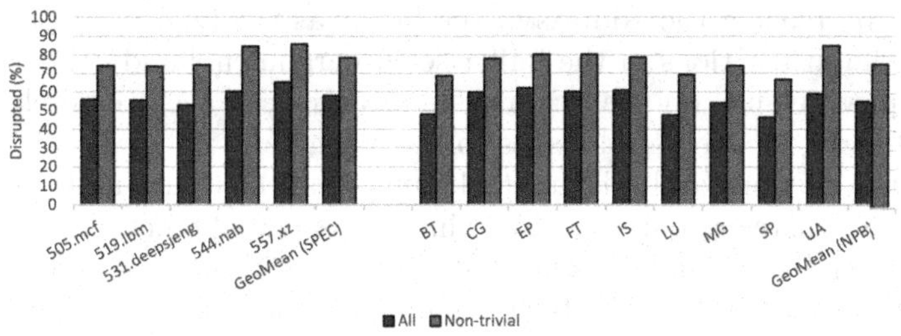

Fig. 5. Percent of gadgets disrupted by Chameleon's code randomization

or less, as longer gadgets become increasingly hard to use due to unintended gadget side effects (e.g., clobbering registers).

Figure 5 shows the percent of gadgets disrupted as part of Chameleon's stack randomization process. When searching the binary, Ropper may return single-instruction gadgets that only perform control flow. We term these "trivial" gadgets and provide results with and without trivial gadgets. Chameleon disrupted a geometric mean of 55.81% gadgets or 76.32% of non-trivial gadgets. While Chameleon did not disrupt all gadgets, it disrupted enough that attackers will have a hard time chaining together functionality without having to use one of the gadgets altered by Chameleon. To better understand the attacker's dilemma, previous work by Cheng et al. [12] mentions that the shortest useful ROP attack produced by the Q ROP compiler [37] consisted of 13 gadgets. Assuming gadgets are chosen with a uniform random possibility from the set of all available gadgets, attackers would have a probability of $(1-0.5582)^{13} = 2.44 \times 10^{-5}$ of being able to construct an unaltered gadget chain, or a $(1-0.7632)^{13} = 7.36 \times 10^{-9}$ probability if using non-trivial gadgets. Therefore, probabilistically speaking it is very unlikely that the attacker will be able to construct gadget chains that have not been altered by Chameleon.

Defeating Real Attacks. To better understand how re-randomization can help protect target applications from attackers, we used Chameleon to disrupt a flaw found in a real application. Nginx [35] is a lightweight and fast open-source webserver used by a large number of popular websites. CVE-2013-2028 [16] is a vulnerability affecting nginx v1.3.9/1.4.0 in which a carefully crafted set of requests can lead to a stack buffer overflow. When parsing an HTTP request, the Nginx worker process allocates a page-sized buffer on the stack and calls `recv` to read the request body. By using a "chunked" transfer encoding and triggering a certain sequence of HTTP parse operations through specifically-sized messages, the attacker can underflow the size variable used in the `recv` operation on the stack buffer and allow the attacker to send an arbitrarily large payload. VNSecurity published a proof-of-concept attack [46] that uses this buffer overflow to build a ROP gadget chain that remaps a piece of memory with read, write and execute permissions. After creating a buffer for injecting code, the ROP chain copies instruction bytes from the payload to the buffer and "returns" to the payload by placing the address of the buffer as the final return address on the stack. The instructions in the buffer set up arguments and call the `system` syscall to spawn a shell on the server. The attacker can then remotely connect to the shell and gain privileged access to the machine.

Chameleon randomizes both the stack buffer and the return address targeted by this attack. There are four stack slots in the associated function, meaning the vulnerable stack buffer can be in one of four locations in the final ordering. Using a maximum slot padding size of 1024, Chameleon will insert anywhere between 0 and 1024 bytes of padding between slots. The slot has an alignment restriction of 16, meaning there are $\frac{1024}{16} = 64$ possible amounts of padding that can be added between the vulnerable buffer and the preceding stack slot. Therefore, Chameleon can place the buffer at $4 * 64 = 256$ possible locations within the frame for 8 bits of entropy. Thus, an attacker has a probability of $\frac{1}{2^8} = 0.0039$ of guessing the correct buffer location. Additionally, the attacker must guess the location of the return address, which could be at $4 * \frac{1024}{8} = 512$ possible locations to initiate the attack, meaning the attacker will have probability of $7.62 * 10^{-6}$ of correctly placing data to start the attack.

Attacking Chameleon. We also analyzed how secure Chameleon is itself from attackers. Chameleon is most vulnerable when setting up the target as Chameleon communicates with the parasite over Unix domain sockets. However, these sockets are short lived, only available to local processes (not over the network) and only pass control flags and the `userfaultfd` file descriptor – Chameleon can easily validate the correctness of these messages. After the initial application setup, Chameleon only interacts with the outside world through `ptrace` and `ioctl` (for `userfaultfd`). The only avenue that attackers could potentially use to hijack Chameleon would be through corrupting state in the target binary/application which is then subsequently read during one of the re-randomization periods. Although it is conceivable that attackers could corrupt memory in such a way as to trigger a flaw in Chameleon, it is unlikely that they would be able to gain enough control to perform useful functionality; the most likely outcomes of such an attack are null pointer exceptions caused by Chameleon following erroneous pointers when transforming the target's stack. Additionally, because Chameleon is a small codebase, it could potentially be instrumented with safeguards and even formally verified. This is a large benefit of Chameleon's strong isolation – it is much simpler to verify its correctness. Thus, we argue that Chameleon's system architecture is safe for enhancing the security of target applications.

4.2 Performance

We next evaluated the performance of target applications executing under Chameleon's control. As mentioned in Sect. 3.2, Chameleon must perform a number of duties to continuously re-randomize applications. In particular, Chameleon runs a scrambler thread to generate a new set of randomized code, runs a fault handler to respond to code page faults with the current set of randomized code, and periodically switches the target application between randomization epochs.

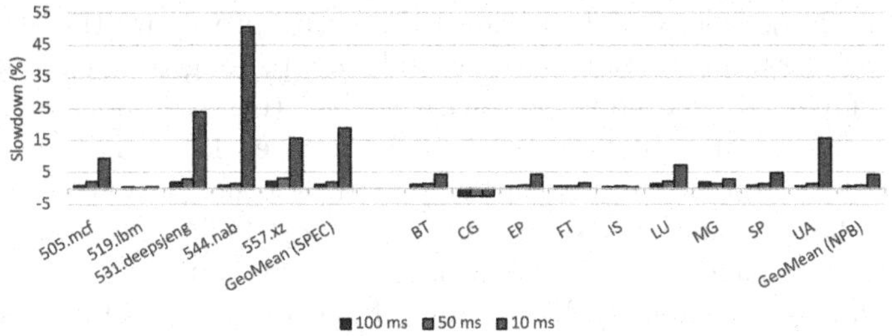

Fig. 6. Overhead when running applications under Chameleon versus execution without Chameleon. Overheads rise with smaller re-randomization periods, but are negligible in most cases.

Figure 6 shows the slowdown of each benchmark when re-randomizing the application every 100 ms, 50 ms and 10 ms versus execution without Chameleon. More frequent randomizations makes it harder for attackers to discover and exploit the current target application's layout at the cost of increased overhead. For SPEC, Chameleon re-randomizes target applications with a geometric mean 1.19% and 1.88% overhead with a 100 ms and 50 ms period, respectively. For NPB, the geometric means are 0.53% and 0.77%, respectively. Re-randomizing with a 10 ms period raises the overhead to 18.8% for SPEC and 4.18% for NPB. This is due to the time it takes the scrambler thread to randomize all stack layouts and rewrite the code to match – with a 10 ms period, the event handler thread must wait for the scrambler to finish generating the next randomization epoch before switching the target. With 100 ms and 50 ms periods, the scrambler's code randomization latency is completely hidden.

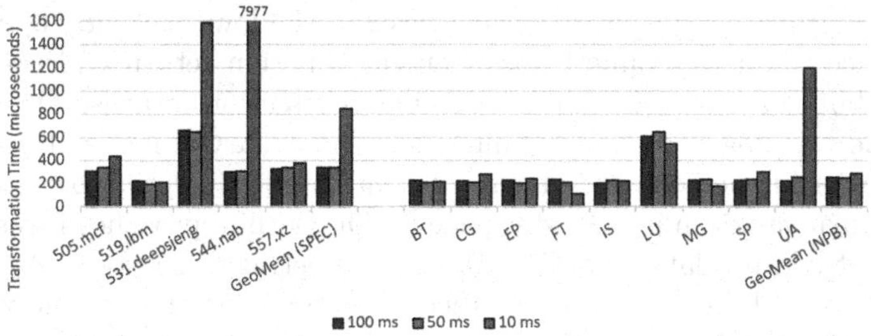

Fig. 7. Time to switch the target between randomization epochs, including advancing to a transformation point, transforming the stack and dropping existing code pages.

We also analyzed how long it took Chameleon to switch between randomization epochs as described in Sect. 3.2. Figure 7 shows the average switching cost for each benchmark. Switching between randomization epochs is an inexpensive process. For both 100 ms and 50 ms periods, it takes a geometric mean of 335 μs for SPEC and 250 μs for NPB to perform the entire procedure transformation. For these two re-randomization periods, only deepsjeng and LU take longer than 600 μs. This is due to large on-stack variables (e.g., LU allocates a 400 KB stack buffer) that must be copied between randomized locations. Nevertheless, as a percentage of the re-randomization period, transformations are inexpensive: 0.2% of the 100 ms re-randomization period and 0.5% of the 50 ms period. We also measured page fault overhead of 5.06 μs per fault. While Chameleon causes page faults throughout the lifetime of the target to bring in new randomized pages, we measured that this usually added less than 0.1% overhead to applications.

There are several performance outliers for the 10 ms transformation period. deepsjeng, nab and UA's overheads increase drastically due to code randomization overhead. When the event handler thread receives a signal to start a re-randomization, it advances the target to a transformation point and blocks until the scrambler thread signals it has finished re-randomizing the code. Because these applications have higher code randomization costs, the event handler thread is blocked waiting for a significant amount of time.

We conclude that Chameleon is able to inject significant amounts of entropy in target applications while adding minimal overheads.

5 Related Works

Stack object-based attacks were proposed a long time ago but are regaining popularity due to the recent data oriented attacks and position-independent code reuse attacks [23–26]. Traditional "stack smashing" attacks overflow the stack local buffer and modify the return address on the stack so that upon returning from the vulnerable function, the application jumps to the malicious payload [2]. There are a number of techniques proposed to prevent the return address from being corrupted, such as stack canaries and shadow stacks [8, 13, 49]. Stack canaries place a random value in between the function return address and the stack local buffer and re-checks the value before function returns. The program executes the warning code and terminates if the canary value is changed [13]. Shadow stacks further enforce backward control flow integrity by storing the function return values in a separate space [8, 49]. Both approaches focus on protecting direct control data on stack without protecting other stack objects.

Recent works have shown that stack objects other than function return addresses could also be used to generate exploits. Göktas et al. proposed using function return addresses and the initialized data that function calls left on the stack to construct position-independent ROP payloads. This way of legally using function calls to construct the malicious payload on the stack is named "stack massaging" [23]. Similarly, attackers can also manipulate other non-control data on the stack to fully control the target. Hu et al. proposed a general approach to automatically synthesize such data-oriented attacks, named data-oriented programming (DOP) [24, 25]. They used the fact that non-control data corruption could potentially be used to modify the program's control flow and implement memory loads and stores. By using a gadget dispatcher (normally a loop and a selector), the attacker could keep the program executing data-oriented gadgets. Note that both of these attacks leverage non-control data on the stack, bypassing existing control flow integrity checks.

Strict boundary checking could be a solution to preventing memory exploits. Such boundary checking could be either software-based [28,33,39] or hardware-based [19,34]. For example, Intel MPX introduces new bounds registers and an instruction set for boundary checking [34]. Besides the relatively large performance overhead introduced by strict boundary checks, the integrity-based approaches cannot defeat the stack object manipulation caused by temporal function calls [23]. StackArmor statically instruments the binary and randomly allocates discontinuous stack pages [10]. Although StackArmor can break the linear stack address space into discrete pages, the function call locality allows position-independent code reuse to succeed within a stack page size [23]. Timely code randomization breaks the constant locations used in the program layout, making it hard for attackers to reuse existing code to chain gadgets [6,11,50]. However, these approaches transform the code layout but not the stack slot layout, giving attackers the ability to exploit stack objects. Chameleon is designed to disrupt these kinds of attacks by continuously randomizing both the stack layout and code. By changing the stack layout, Chameleon makes it more difficult for attackers to corrupt specific stack elements.

6 Conclusion

We have presented the design, implementation and evaluation of Chameleon, a practical system for continuous stack re-randomization. Chameleon continually generates randomized stack layouts for all functions in the application, rewriting each function's code to match. Chameleon periodically interrupts the target to rewrite its existing execution state to a new randomized stack layout and injects matching code. Chameleon controls target applications from a separate address space using the widely available `ptrace` and `userfaultfd` kernel primitives, maintaining strong isolation between Chameleon and the target. The evaluation showed that Chameleon's lightweight user-level page fault handling and code transformation significantly raises the bar for stack exploitation with minimal overhead to target application.

The source code of Chameleon is publicly available as part of the Popcorn Linux project at http://popcornlinux.org.

Acknowledgments. This work is supported in part by the US Office of Naval Research (ONR) under grants N00014-18-1-2022 and N00014-16-1-2711, and by NAVSEA/NEEC under grant N00174-16-C-0018.

References

1. Aga, M.T., Austin, T.: Smokestack: thwarting DOP attacks with runtime stack layout randomization. In 2019 IEEE/ACM International Symposium on Code Generation and Optimization (CGO), pp. 26–36. IEEE (2019)

2. Aleph, O.: Smashing the stack for fun and profit (1996). http://www.shmoo.com/phrack/Phrack49/p49-14
3. Backes, M., Nürnberger, S.: Oxymoron: making fine-grained memory randomization practical by allowing code sharing. In: Proceedings of the 23rd USENIX Security Symposium, pp. 433–447 (2014)
4. Bailey, D.H., et al.: The NAS parallel benchmarks summary and preliminary results. In Supercomputing 1991: Proceedings of the 1991 ACM/IEEE Conference on Supercomputing, pp. 158–165. IEEE (1991)
5. Barbalace, A., et al.: Breaking the boundaries in heterogeneous-ISA datacenters. In: ACM SIGPLAN Notices, vol. 52, pp. 645–659. ACM (2017)
6. Bigelow, D., Hobson, T., Rudd, R., Streilein, W., Okhravi, H.: Timely rerandomization for mitigating memory disclosures. In: Proceedings of the 22nd ACM SIGSAC Conference on Computer and Communications Security, pp. 268–279. ACM (2015)
7. Bruening, D.: Efficient, transparent, and comprehensive runtime code manipulation. Ph.D thesis, Massachusetts Institute of Technology, September 2004
8. Burow, N., Zhang, X., Payer, M.: Shining light on shadow stacks (2018). arXiv preprint arXiv:1811.03165
9. Carlini, N., Barresi, A., Payer, M., Wagner, D., Gross, T.R.: Control-flow bending: on the effectiveness of control-flow integrity. In: 24th USENIX Security Symposium (USENIX Security 15), pp. 161–176 (2015)
10. Chen, X., Slowinska, A., Andriesse, D., Bos, H., Giuffrida, C.: Stackarmor: comprehensive protection from stack-based memory error vulnerabilities for binaries. In: NDSS. Citeseer (2015)
11. Chen, Y., Wang, Z., Whalley, D., Lu, L.: Remix: on-demand live randomization. In: Proceedings of the Sixth ACM Conference on Data and Application Security and Privacy, pp. 50–61. ACM (2016)
12. Cheng, Y., Zhou, Z., Miao, Y., Ding, X., Deng, R.H.: ROPecker: a generic and practical approach for defending against ROP attacks. In: Symposium on Network and Distributed System Security (NDSS) (2014)
13. Cowan, C., et al.: StackGuard: automatic adaptive detection and prevention of buffer-overflow attacks. In: Proceedings of the 7th USENIX Security Symposium, August 1998
14. Crane, S., et al.: Readactor: practical code randomization resilient to memory disclosure. In: 36th IEEE Symposium on Security and Privacy (Oakland), May 2015
15. CRIU. CRIU Compel. https://criu.org/Compel. Accessed 14 Apr 2019
16. CVE-2013-2028. https://cve.mitre.org/cgi-bin/cvename.cgi?name=CVE-2013-2028. Accessed 14 Apr 2019
17. Davi, L., Liebchen, C., Sadeghi, A.R., Snow, K.Z., Monrose, F.: Isomeron: code randomization resilient to (just-in-time) return-oriented programming. In: Proceedings of the 22nd Network and Distributed Systems Security Symposium (NDSS) (2015)
18. Davi, L., Sadeghi, A.R., Lehmann, D., Monrose, F.: Stitching the gadgets: on the ineffectiveness of coarse-grained control-flow integrity protection. In: Proceedings of the 23rd USENIX Conference on Security, SEC 2014 (2014)
19. Devietti, J., Blundell, C., Martin, M.M.K., Zdancewic, S.: Hardbound: architectural support for spatial safety of the C programming language. In: Proceedings of the 13th International Conference on Architectural Support for Programming Languages and Operating Systems (2008)
20. Durumeric, Z., et al.: The matter of heartbleed. In: Proceedings of the 2014 Conference on Internet Measurement Conference, pp. 475–488. ACM (2014)

21. DWARF Standards Committee. The DWARF Debugging Standard, February 2017
22. Göktas, E., Athanasopoulos, E., Bos, H., Portokalidis, G.: Out of control: overcoming control-flow integrity. In: Proceedings of the 2014 IEEE Symposium on Security and Privacy SP 2014 (2014)
23. Göktas, E., et al.: Position-independent code reuse: On the effectiveness of ASLR in the absence of information disclosure. In: 2018 IEEE European Symposium on Security and Privacy (EuroS&P), pp. 227–242. IEEE (2018)
24. Hu, H., Chua, Z.L., Adrian, S., Saxena, P., Liang, Z.: Automatic generation of data-oriented exploits. In: 24th USENIX Security Symposium (USENIX Security 15), pp. 177–192 (2015)
25. Hu, H., Shinde, S., Adrian, S., Chua, Z.L., Saxena, P., Liang, Z.: Data-oriented programming: on the expressiveness of non-control data attacks. In: 2016 IEEE Symposium on Security and Privacy (SP), pp. 969–986. IEEE (2016)
26. Ispoglou, K.K., AlBassam, B., Jaeger, T., Payer, M.: Block oriented programming: automating data-only attacks. In: Proceedings of the 2018 ACM SIGSAC Conference on Computer and Communications Security, pp. 1868–1882. ACM (2018)
27. kernel.org. Userfaultfd. https://www.kernel.org/doc/Documentation/vm/userfaultfd.txt. Accessed 14 Apr 2019
28. Kroes, T., Koning, K., van der Kouwe, E., Bos, H., Giuffrida, C.: Delta pointers: buffer overflow checks without the checks. In: Proceedings of the Thirteenth EuroSys Conference, p. 22. ACM (2018)
29. Linux Kernel Address Space Layout Randomization. http://lwn.net/Articles/569635/. Accessed 14 Apr 2019
30. LLVM Compiler Infrastructure. Stack maps and patch points in LLVM. https://llvm.org/docs/StackMaps.html. Accessed 14 Apr 2019
31. Lu, K., Walter, M.T., Pfaff, D., Nümberger, S., Lee, W., Backes, M.: Unleashing use-before-initialization vulnerabilities in the linux kernel using targeted stack spraying. In: NDSS (2017)
32. Luk, C.K., et al.: Pin: building customized program analysis tools with dynamic instrumentation. In: ACM SIGPLAN Notices, vol. 40, pp. 190–200. ACM (2005)
33. Nagarakatte, S., Zhao, J., Martin, M.M.K., Zdancewic, S.: SoftBound: highly compatible and complete spatial memory safety for C. In: Proceedings of the 2009 ACM SIGPLAN Conference on Programming Language Design and Implementation, PLDI 2009 (2009)
34. Oleksenko, O., Kuvaiskii, D., Bhatotia, P., Felber, P., Fetzer, C.: Intel MPX explained: a cross-layer analysis of the intel MPX system stack. Proc. ACM Measur. Anal. Comput. Syst. **2**(2), 28 (2018)
35. Reese, W.: Nginx: the high-performance web server and reverse proxy. Linux J. **2008**(173), 2 (2008)
36. Roemer, R., Buchanan, E., Shacham, H., Savage, S.: Return-oriented programming: systems, languages, and applications. ACM Trans. Inf. Syst. Secur. (TISSEC) **15**(1), 2 (2012)
37. Schwartz, E.J., Avgerinos, T., Brumley, D.: Q: exploit hardening made easy. In: USENIX Security Symposium, pp. 25–41 (2011)
38. Seo, S., Jo, G., Lee, J.: Performance characterization of the NAS parallel benchmarks in openCL. In: 2011 IEEE international symposium on workload characterization (IISWC), pp. 137–148. IEEE (2011)
39. Serebryany, K., Bruening, D., Potapenko, A., Vyukov, D.: AddressSanitizer: a fast address sanity checker. In: Presented as part of the 2012 USENIX Annual Technical Conference (USENIX ATC 12), pp. 309–318 (2012)

40. Shacham, H.: The geometry of innocent flesh on the bone: return-into-libc without function calls (on the x86). In: Proceedings of the 14th ACM Conference on Computer and Communications Security, October 2007
41. Snow, K.Z., Monrose, F., Davi, L., Dmitrienko, A., Liebchen, C., Sadeghi, A.R.: Just-in-time Code Reuse: on the effectiveness of fine-grained address space layout randomization. In: 2013 IEEE Symposium on Security and Privacy (SP), pp. 574–588. IEEE (2013)
42. Standard Performance Evaluation Corporation. SPEC CPU 2017. https://www.spec.org/cpu2017. Accessed 14 Apr 2019
43. Szekeres, L., Payer, M., Wei, T., Song, D.: Sok: eternal war in memory. In: 2013 IEEE Symposium on Security and Privacy (SP), pp. 48–62. IEEE (2013)
44. The Linux man-pages project. mmap(2) - Linux manual page, April 2020. http://man7.org/linux/man-pages/man2/mmap.2.html
45. Venkat, A., Shamasunder, S., Shacham, H., Tullsen, D.M.: Hipstr: heterogeneous-ISA program state relocation. In: ACM SIGARCH Computer Architecture News, vol. 44, pp. 727–741. ACM (2016)
46. Analysis of nginx 1.3.9/1.4.0 stack buffer overflow and x64 exploitation (CVE-2013-2028). https://www.vnsecurity.net/research/2013/05/21/analysis-of-nginx-cve-2013-2028.html. Accessed 14 Apr 2019
47. Wang, R., et al.: Ramblr: making reassembly great again. In: Proceedings of the 2017 Network and Distributed System Security Symposium (2017)
48. Wikipedia. Ptrace. http://en.wikipedia.org/wiki/Ptrace. Accessed 14 Apr 2019
49. Wikipedia. Shadow stack. https://en.wikipedia.org/wiki/Shadow_stack. Accessed 14 Apr 2019
50. Williams-King, D., et al.: Shuffler: fast and deployable continuous code re-randomization. In: OSDI, pp. 367–382 (2016)

Understanding the Security Risks of Docker Hub

Peiyu Liu[1], Shouling Ji[1(✉)], Lirong Fu[1], Kangjie Lu[2], Xuhong Zhang[1], Wei-Han Lee[3], Tao Lu[1], Wenzhi Chen[1(✉)], and Raheem Beyah[4]

[1] Zhejiang University, Hangzhou, China
{liupeiyu,sji,fulirong007,lutao,chenwz}@zju.edu.cn,
xuhongnever@gmail.com

[2] University of Minnesota Twin Cities, Minneapolis, USA
kjlu@umn.edu

[3] IBM Research, Yorktown Heights, USA
wei-han.lee1@ibm.com

[4] Georgia Institute of Technology, Atlanta, USA
rbeyah@ece.gatech.edu

Abstract. Docker has become increasingly popular because it provides efficient containers that are directly run by the host kernel. Docker Hub is one of the most popular Docker image repositories. Millions of images have been downloaded from Docker Hub billions of times. However, in the past several years, a number of high-profile attacks that exploit this key channel of image distribution have been reported. It is still unclear what security risks the new ecosystem brings. In this paper, we reveal, characterize, and understand the security issues with Docker Hub by performing the first large-scale analysis. First, we uncover multiple security-critical aspects of Docker images with an empirical but comprehensive analysis, covering sensitive parameters in run-commands, the executed programs in Docker images, and vulnerabilities in contained software. Second, we conduct a large-scale and in-depth security analysis against Docker images. We collect 2,227,244 Docker images and the associated meta-information from Docker Hub. This dataset enables us to discover many insightful findings. (1) run-commands with sensitive parameters expose disastrous harm to users and the host, such as the leakage of host files and display, and denial-of-service attacks to the host. (2) We uncover 42 malicious images that can cause attacks such as remote code execution and malicious cryptomining. (3) Vulnerability patching of software in Docker images is significantly delayed or even ignored. We believe that our measurement and analysis serves as an important first-step study on the security issues with Docker Hub, which calls for future efforts on the protection of the new Docker ecosystem.

1 Introduction

Docker has become more and more popular because it automates the deployment of applications inside containers by launching Docker images. Docker Hub, one

L. Chen et al. (Eds.): ESORICS 2020, LNCS 12308, pp. 257–276, 2020.
https://doi.org/10.1007/978-3-030-58951-6_13

of the most popular Docker image registries, provides a centralized market for users to obtain Docker images released by developers [4,31,35]. Docker has been widely used in many security-critical tasks. For instance, Solita uses Docker to handle the various applications and systems associated with their management of the Finnish National Railway Service [9]. Amazon ECS allows users to easily run applications on a managed cluster of Amazon EC2 instances in Docker containers [1]. In addition, millions of Docker images have been downloaded from Docker Hub for billion times by users for data management, website deployment, and other personal or business tasks.

The popularity of Docker Hub however brings many high-profile attacks. For instance, on June 13, 2018, a research institute reported that seventeen malicious Docker images on Docker Hub earned cryptomining criminals $90,000 in 30 days [8]. These images have been downloaded collectively for 5 million times in the past year. The report also explained the danger of utilizing unchecked images on Docker Hub: *"For ordinary users, just pulling a Docker image from the Docker Hub is like pulling arbitrary binary data from somewhere, executing it, and hoping for the best without really knowing what's in it"*. Therefore, a comprehensive and in-depth security study of Docker Hub is demanded to help users understand the potential security risks.

The study of Docker Hub differs from the ones of other ecosystems such as App store and virtual-machine image repositories [15,20,28,30,32] in the following aspects. (1) Docker images are started through run-commands. They are executed through special instructions called run-commands which are security-critical to the created containers. (2) The structures of Docker images are more complex than traditional applications. A single image may contain a large number of programs, environment variables, and configuration files; it is hard for a traditional analysis to scale to scan all images. (3) Docker images can bring new risks to not only the container itself but also the host because the lightweight virtualization technology leveraged in containers allows the sharing of the kernel. (4) The vulnerability-patching process of Docker images is significantly delayed because the programs are decoupled from the mainstream ones, and developers are less incentivized to update programs in Docker images. All the aforementioned differences require a new study for the security of Docker Hub.

The unique characteristics of Docker Hub call for an urgent study of its new security issues. However, a comprehensive and in-depth study entails overcoming multiple challenges. (1) It is not clear how to analyze the security impacts of various categories of information on Docker Hub. For example, run-commands are security-critical to Docker containers while a method for measuring the security impacts of run-commands is still missing. This requires significant empirical analysis and manual effort. (2) Obtaining and analyzing Docker images and the associated meta-information in a scalable manner is non-trivial. For example, it is difficult to perform a uniform analysis on Docker images, since a Docker image contains a large number of files in a broad range of types (e.g., ELF, JAR, and Shell Scripts).

In this paper, we perform the first security analysis against Docker Hub. Based on the unique characteristics of Docker Hub, we first empirically identify three major security risks, namely sensitive parameters in run-commands, malicious docker images, and unpatched vulnerabilities. We then conduct a large-scale and in-depth study against the three security risks. (1) Run-commands. We carefully analyze the parameters in run-commands to discover sensitive parameters that may pose threats to users. Moreover, we develop multiple new attacks (e.g., obtaining user files in the host and the host display) in Docker images to demonstrate the security risks of sensitive parameters in practice. We also conduct a user study to show that users, in general, are unaware of the risks from sensitive parameters. (2) Malicious executed programs. To study malicious Docker images efficiently, we narrow down our analysis to only the executed programs. We implement a framework to automatically locate, collect, and analyze executed programs. By leveraging this framework, we scan more than 20,000 Docker images to discover malicious executed programs. (3) CVE-assigned vulnerabilities. We provide a definition of the life cycle of vulnerability in Docker images and manually analyze the length of the time window of vulnerabilities in Docker images. To enable the analysis, we collect a large number of images and their meta-information from Docker Hub. Our collected dataset contains all the public information of 2,227,244 images from 975,858 repositories on Docker Hub.

The comprehensive analysis enables us to have multiple insightful findings. First of all, we find that the run-commands with sensitive parameters presented in Docker Hub may introduce serious security risks, including the suffering of denial-of-service attack and the leakage of user files in the host and the host display. Moreover, we observe that each recommended run-command in the repository description contains one sensitive parameter on average. Unfortunately, our user study reveals that users are not aware of the threats from sensitive parameters—they will directly execute run-commands specified by developers without checking and understanding them. Second, our analysis shows that malicious images are hidden among common ones on Docker Hub. Using our analysis framework, we have discovered 42 malicious images. The malicious behaviors include remote execution control and malicious cryptomining. Finally, we observe that the vulnerability patching for the software in Docker images is significantly delayed or even ignored. In particular, almost all the images on Docker Hub suffer from unpatched software vulnerabilities. In extreme cases, a single image may contain up to 7,500 vulnerabilities. In addition, vulnerabilities in the software of Docker images tend to have a much longer life cycle due to the lack of image updates. More critically, we find that the in-Docker vulnerabilities can even cause harms to the host machine through Docker, e.g., crashing the host.

Our analysis and findings reveal that the Docker ecosystem brings new security threats to users, contained software, and the host machine as well. To mitigate these threats, we suggest multiple potential solutions (see Sect. 7) such as automatically fixing vulnerabilities in images, detecting malicious images in Docker Hub, etc. We have reported all the security issues uncovered in this paper to Docker Hub and they are investigating to confirm these issues.

In summary, our work makes the following contributions.

- We empirically identify three major sources of security risks in Docker Hub, namely sensitive parameters in run-commands, malicious docker images, and unpatched vulnerabilities. We then conduct a large-scale and in-depth study against the three security risks based on all the public information of 2,227,244 images collected from 975,858 repositories on Docker Hub. We have open-sourced this dataset to support reproducibility and motivate future work in Docker security analysis [13].
- We uncover many new security risks on Docker Hub. 1) Sensitive parameters in run-commands can expose disastrous harm to users and the host, such as the leakage of host files and display, and denial-of-service attacks to the host. 2) We uncover 42 malicious images that can cause attacks such as remote code execution and malicious cryptomining. 3) Vulnerability patching of software in Docker images is significantly delayed by 422 days on average.
- Our analysis calls for attention to the security threats posed by the new Docker ecosystem. The threats should be addressed collectively by Docker image registry platforms, image developers, users, and researchers. Moreover, we have reported all the security issues uncovered in this paper to Docker Hub and suggest multiple mitigation approaches.

2 Background and Threat Model

In this section, we provide a brief introduction of Docker Hub and its critical risk resources. Then, we describe the threat model of our security analysis.

2.1 Critical Risk Sources in Docker Hub

Docker Hub is the world's largest registry of container images [5]. Images on Docker Hub are organized into repositories, which can be divided into official repositories and community repositories. For each image in a Docker Hub repository, besides the image itself, meta-information is also available to the users, such as repository description and history, Dockerfile [5], the number of stars and pulls of each repository, and developer information. To perform a risk analysis against Docker Hub, we first need to identify potential risk sources. We empirically identify risk sources based on *which major components control the behaviors of a Docker image.*

Run-Command and Sensitive Parameter. In order to run a Docker container, users need to execute an instruction called run-command. A run-command mainly specifies the image and parameters used to start a container. For instance, a developer may specify a recommended run-command on Docker Hub, such as *"Start container with: docker run --name flaviostutz-opencv2 --privileged -p 2222:22 flaviostutz/opencv-x86"*. For users who have never used the image before, the recommended run-commands can be helpful for deploying their containers. However, it is unclear to what extent users should trust the run-commands posted by the developers, who can publish run-commands without any obstruction because Docker Hub does not screen these content. In addition, a run-command may contain a variety of parameters that can affect the behavior of the container [6]. Some of these parameters are sensitive since they control the degree of isolation of networks, storage, or other underlying subsystems between a container and its host machine or other containers. For example, when users run an image with the parameter of `--privileged`, the container will get the root access to the host. Clearly, the misuse of run-commands containing sensitive parameters may lead to disastrous consequences on the container as well as the host (see Sect. 4).

Executed Programs. Previous work already shows that a large amount of software in Docker images is redundant [31]. Hence, when analyzing the content of a Docker image, we should focus on the executed programs that are bound up with the security of the image. Based on our empirical analysis, we find that the entry-file (an executable file in Docker images, specified by a configure file or run-commands) is always the first software triggered when a container starts. Besides, the entry-file can automatically trigger other files during execution. Therefore, the executed programs (the entry-file and subsequently triggered files) are key factors that directly affect the safety of a container. Furthermore, in general, it is less common for users to run software other than executed programs [4]. Therefore, in the current study, we choose to analyze executed programs to check malicious images for measurement purposes.

Vulnerabilities in Contained Software. A Docker image is composed of a large number of software packages, vulnerabilities in these software packages bring critical security risks for the following reasons. Vulnerabilities can be exploited by attackers to cause security impacts such as data leakage. Additionally, Docker software programs are often duplicated from original ones, and Docker developers lack incentives to timely fix vulnerabilities in the duplicated programs. As a result, the security risks with vulnerabilities are elevated in Docker because vulnerabilities take a much longer time to be fixed in Docker images.

2.2 Threat Model

As shown in Fig. 1, there are two different categories of threats that Docker Hub may face. (1) Vulnerable images. Developers upload their images and the associated meta-information to Docker Hub, which may contain vulnerabilities. If users download and run the vulnerable images, they are likely to

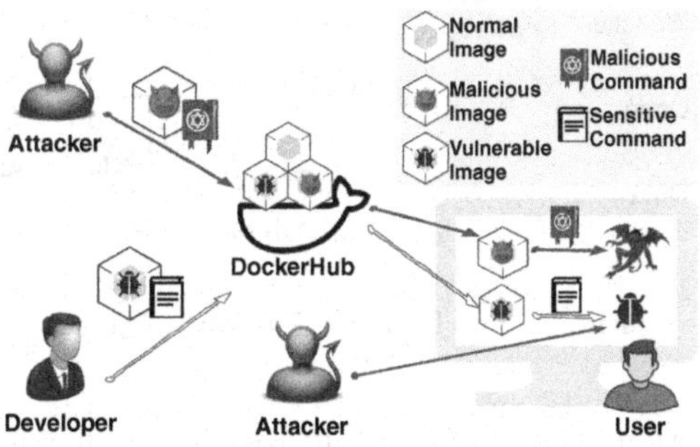

Fig. 1. The threat model of our security analysis.

become the targets of attackers who exploit vulnerabilities. Additionally, the run-commands announced by developers may contain sensitive parameters, which may bring more security concerns such as giving containers root access to the host (see Sect. 4). (2) Malicious images. Attackers may upload their malicious images to Docker Hub, and sometimes together with malicious run-commands in the description of their images. Due to the weak surveillance of Docker Hub, malicious images and metadata can easily hide themselves among benign ones (see Sect. 5) [8]. Once users download a malicious image and run it, they may suffer from attacks such as cryptomining. On the other hand, malicious run-commands can also lead to attacks such as host file leakage. In this study, since it is mainly for measurement purposes, we assume that attackers are not aware of the techniques we employ to analyze Docker images. Otherwise, they can hide the malicious code by bypassing the analysis, leading to an arms race.

3 Analysis Framework and Data Collection

In this section, we provide an overview of our analysis framework. Then we introduce our methods of data collection and provide a summary for the collected dataset.

3.1 Overview of Our Analysis Framework

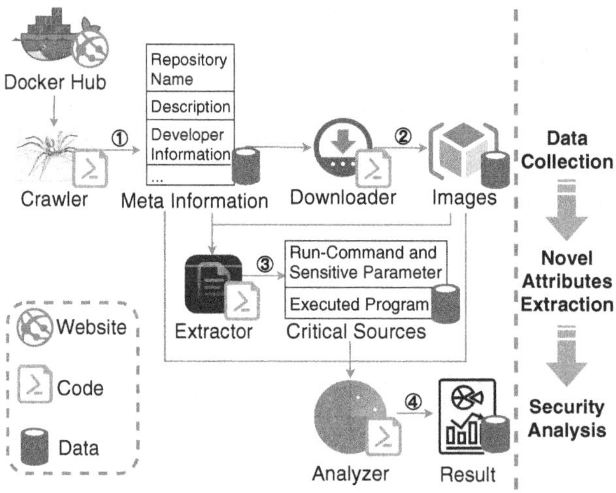

Fig. 2. The framework of our analysis.

In our security analysis, we focus on the key risk sources (run-commands, sensitive parameters, entry-files and vulnerabilities) in Docker Hub, as uncovered in Sect. 2.1. Our analysis framework for the study is outlined in Fig. 2. We first collect a large-scale dataset, including the images and all the public information of a Docker image such as image name, repository description, and developer information, from Docker Hub (①, ②). After data collection, the extractor utilizes a set of customized tools to obtain several previously-ignored essentials such as sensitive parameters and executed programs from the raw data (③). After obtaining the set of essentials, we perform systematic security analysis on Docker Hub from various aspects by leveraging `Anchore` [2], VirusTotal intelligence API [11], and a variety of customized tools (④).

3.2 Data Collection and Extraction

The abundant information (described in Sect. 2.1) of Docker images on Docker Hub is important in understanding the security of Docker Hub. However, most of this information has never been collected before, leading to incomprehensive analysis. Hence, we implement a customized web crawler that leverages Docker Hub API described in [3] to collect the Docker images and their associated meta-information from Docker Hub. All the information we obtain from Docker Hub is publicly available to anyone and it is legal to perform analysis on this dataset.

Summary of Our Collected Data. Our dataset, as shown in Table 1, contains all the public information of the top 975,858 repositories on Docker Hub. For each repository, the dataset contains the image files and the meta-information described in Sect. 2.1. Furthermore, to support in-depth analysis, we further extract the following data and code from the collected raw data.

Table 1. Data collected in our work.

	No. repositories	No. images	No. developers
Official	147	1,384	1
Community	975,711	2,225,860	349,860
Total	975,858	2,227,244	349,861

Collecting Run-Command and Sensitive Parameter. As discussed in Sect. 2.1, run-commands and sensitive parameters can make a great impact on the behavior of a container. In order to perform security analysis on them, we collect run-commands by extracting text contents that start with `''docker run''` from the repository descriptions and further obtain sensitive parameters from the run-commands through string matching.

Collecting Executed Program. As discussed in Sect. 2.1, the executed program is a key factor that directly affects the security of a container. Therefore, we develop an automatic parser to locate and extract the executed program. For each image, our parser first locates the entry-file according to the Dockerfile or the manifests file. Once obtaining the entry-file, the parser scans the entry-file to locate the files triggered by entry-file. Then, the parser scans the triggered file iteratively to extract all executed programs in the image. For now, our parser can analyze ELF files and shell scripts by leveraging `strings` [10] and a customized script interpreter, respectively.

4 Sensitive Parameters

In this section, we (1) identify sensitive parameters; (2) investigate the user awareness of sensitive parameters; (3) propose novel attacks exploiting sensitive parameters; (4) study the distribution of sensitive parameters on Docker Hub.

4.1 Identifying Sensitive Parameters

As described in Sect. 2.1, the parameters in run-commands can affect the behaviors of containers. However, among the over 100 parameters provided by Docker, it is unknown which parameters can cause security consequences. Therefore, we first obtain all the parameters and their corresponding descriptions from the documentation provided by Docker. Then, we manually identify the sensitive parameters by examining if they satisfy any of the following four proposed criteria: (1) Violate the isolation of file systems; (2) Violate the isolation of networking; (3) Break the separation of processes; (4) Escalate runtime privileges.

Next, we explain why we choose these criteria. (1) From the security perspective, each Docker container maintains its own file system isolated from the host file system. If this isolation is broken, a container can gain access to files on the host, which may lead to the leakage of host data. (2) By default, each Docker container has its own network stack and interfaces. If this isolation is broken, a container can have access to the host's network interfaces for sending/receiving network messages, which, for example, may cause denial of service attacks. (3) Generally, each Docker container has its own `process-tree` separated from the host. If this isolation is broken, a container can see and affect the processes on the host, which may allow containers to spy the defense mechanism on the host. (4) Most potentially dangerous Linux capabilities, such as loading kernel modules, are dropped in Docker containers. If a container obtains these capabilities, it may affect the host. For example, it is able to execute arbitrary hostile code on the host.

4.2 User Awareness of Sensitive Parameters

We find that nearly all the default container isolation and restrictions enforced by Docker can be broken by sensitive parameters in run-commands, such as `--privileged`, `-v`, `--pid`, and so on. We describe the impact of these sensitive parameters in Sect. 4.3. However, the real security impacts of these sensitive parameters on the users of Docker Hub in practice is not clear. Therefore, the first question we aim to answer is "are users aware of the sensitive parameters when the parameters in run-commands are visible to them?" To answer this question, we conduct a user study to characterize the behaviors of users of Docker Hub, which allows us to understand user preferences and the corresponding risks.

Specifically, we survey 106 users including 68 security researchers and 38 software engineers from both academia and industry fields. For all the 106 users, 97% of them only focus on the functionality of images and have never raised doubts about the descriptions, e.g., the developer identification, the run-commands, on Docker Hub. It is worth noting that, even for 68 users who have a background on security research, 95% of them trust the information provided on Docker Hub. 90% of security experts run an image by exactly following the directions provided by image developers. Only 10% of security experts indicate that they prefer to figure out what the run-commands would do. Indeed, the study can be biased due to issues such as the limited number of the investigated users, imbalanced gender and age distribution, and so on. However, our user study reveals that users, even the ones with security-research experience, do not realize the threats of sensitive parameters in general. Nearly 90% of users exactly execute run-commands specified by developers without checking and understanding them. Hence, we conclude that sensitive parameters are an overlooked risk source for Docker users. More details of the user study are deferred to Appendix A.1.

4.3 Novel Attacks Exploiting Sensitive Parameters

To demonstrate the security risks of sensitive parameters in practice, we develop a set of new attacks in Docker images that do not contain any malicious software packages. Our attacks rely on only run-commands with sensitive parameters to attack the host. Note that we successfully uploaded these images with our "malicious" run-commands to Docker Hub without any obstruction, confirming that Docker Hub does not carefully screen run-commands. However, to avoid harm to the community, we immediately removed the sensitive parameters in the run-commands after the uploading, and performed the attacks in our local lab machines only.

The Leakage of Host Files. As described in Fig. 3, we show how to leak user files in the host using sensitive parameters. Specifically, `--volume` or `-v` is used to mount a volume on the host to the container. If the operator uses parameter, `-v src:dest`, the container will gain access to `src` which is a volume on the host. Exploiting this parameter, attackers can maliciously upload user data saved in the host volume to their online repository, such as GitHub. It is important to note that this attack can be user-insensitive, i.e., the attacker can prepare a configure file in the malicious image to bypass the manual authentication.

```
git clone https://Attacker/data.git
cd data
git pull
cp -r usr-data .
git add .
cur_date=`date`
git commit -m "$cur_date"
git push
```

Fig. 3. Code example to implement the leakage of host files.

Other attacks are delayed to Appendix A.2. Overall, these novel attacks we proposed in this paper demonstrate that sensitive parameters can expose disastrous harm to the container as well as the host.

4.4 Distribution of Sensitive Parameters

Next, we study the distribution of sensitive parameters used in real images on Docker Hub. We observe that 86,204 (8.8%) repositories contain the recommended run-commands in their descriptions on Docker Hub. Moreover, as shown in Table 2, there are 81,294 sensitive parameters in these run-commands—on average, each run-command contains one sensitive parameter. Given the common usage of sensitive parameters and their critical security impacts, it is urgent to improve users' awareness of the potential security risks brought by sensitive parameters and propose effective vetting mechanisms to detect these risks.

Table 2. Distribution of sensitive parameters.

Criteria	No. sensitive parameters
Violate the isolation of file systems	33,951
Violate the isolation of networking	43,278
Break the separation of processes	56
Escalate runtime privileges	4,009
Total	81,294

5 Malicious Images

As reported in [8], high-profile attacks seriously damaged the profit of users. These attacks originate from the launching of malicious images such as electronic coin miners. In order to detect malicious images, manual security analysis on each software contained in a Docker image yields accurate results, which is however extremely slow and does not scale well for large and heterogeneous software packages. On the other hand, a dynamic analysis also becomes impractical since it is even more time-consuming to trace system calls, APIs, and network for millions of Docker images. Compared to the two methods above, a static analysis seems to be a potential solution to overcome these challenges. However, the number of software packages contained in each image varies from hundreds to thousands. It is still challenging to analyze billions of software packages using static analysis. Fortunately, we found that a majority of software packages in Docker images are redundant [31] which thus can be filtered out for efficient analysis. However, it is unclear which software deserves attention from security researchers. As stated in Sect. 2.1, we propose to focus on the executed programs. As long as a malicious executed program is detected, the corresponding image can be confirmed as malicious. Considering that our research goal is to characterize malware for measurement purposes instead of actually detecting them in practice, we focus on the executed programs only rather than other software for discovering malicious images in this study. We will discuss how to further improve the detection in Sect. 8.

5.1 Malicious Executed Programs

Detecting Malicious Executed Programs. By leveraging the parser proposed in Sect. 3.2, we can obtain all the executed programs in the tested images. We observe that the file types of the extracted executed programs could be JAR written by JAVA, ELF implemented by C++, Shell Script, etc. It is quite challenging to analyze many kinds of software at the same time, since generating and confirming fingerprints for malware are both difficult and time-consuming. Therefore, we turn to online malware analysis tools for help. In particular, VirusTotal [11] is a highly comprehensive malware detection platform that incorporates various signature-based and anomaly-based detection methods employed by over 50 anti-virus (AV) companies such as Kaspersky, Symantec. Therefore, it can detect various kinds of malware, including Trojan, Backdoor, and Bitcoin-Miner. As such, we employ VirusTotal to perform a primary screening. However, prior works have shown that VirusTotal may falsely label a benign program as malicious [14,22,29]. To migrate false positives, most of the prior works consider a program malicious if at least a threshold of AV companies detects it. In fact, there is no agreement in the threshold value. Existing detection has chosen two [14], four [22], or five [29] as the threshold. In this paper, to more precisely detect malicious programs, we consider a program as malicious only if at least five of the AV companies detect it. This procedure ensures that the tested programs are (almost) correctly split into benign and malicious ones.

The results of the primary screening only report the type of each malware provided by anti-virus companies. It is hard to demonstrate the accuracy of primary screening results, let along understanding the behavior of each malware. Therefore, after we obtain a list of potentially malicious files from the primary screening, a second screening is necessary to confirm the detection results and analyze the behavior of these files. Specifically, we dynamically run the potentially malicious files in a container and collect the logs of system call, network, and so on to expose the security violations of such files.

Finally, we implement a framework to finish the above pipeline. First, our framework utilizes our parser to locate and extract executed programs from the Docker images. Second, it leverages ViusTotal API to detect potentially malicious files. Third, we implement a container which contains a variety of tools such as `strace` and `tcpdump` for security analysis. Our framework leverages this container as a sandbox for automatically running and tracing the potentially malicious files to generate informative system logs. Since most benign images are filtered out by the primary screening, system logs are generated for only a few potentially malicious images. This framework greatly saves manual efforts and helps detect malicious images rapidly.

5.2 Distribution of Malicious Images

We first study the executed programs in the latest images in 147 official repositories, and we find that there are no malicious executed programs in these official images. Then, to facilitate in-depth analyses on community images, we extract the following subsets from the collected dataset for studying the executed programs. 1) The latest images in the top 10,000 community repositories ranked by popularity. 2) According to the popularity ranking, we divide the rest community repositories into 100 groups and randomly select 100 latest images from each group. In this way, we obtain 10,000 community images.

Results. On average, our parser proposed in Sect. 3.2 takes 5 and 0.15 seconds in analyzing one image and one file, respectively. The parser locates 693,757 executed programs from the tested images. After deduplication, we get 36,584 unique executed programs, in which there exist 13 malicious programs identified by our framework. The 13 malicious programs appear in 17 images. Moreover, we notice that all the malicious programs are entry-files in these malicious images. This observation indicates that it is common for malicious images to perform attacks by directly utilizing a malicious entry-files, instead of subsequently triggered files.

Intuitively, the developer of a malicious image may release other malicious images on Docker Hub. Therefore, we propose to check the images that are related to malicious images. By leveraging the metadata collected in Sect. 3.2, once we detect a malicious image, we can investigate two kinds of related images: (1) the latest 10 images in the same repository and (2) the latest images in the most popular 10 repositories created by the same developer. We obtain 48 and 84 of these two types of related images respectively, in which there are 27 images contain the same malicious file found in the previously-detected malicious images. After analyzing all the related images by leveraging the framework developed in Sect. 5.1, we further obtain 186 new executed programs, in which there are 20 new malicious programs in 25 images. This insightful finding indicates that heuristic approaches, such as analyzing related images proposed in this work, are helpful in discovering malicious images and programs effectively. We hope that the malicious images and insights discovered in this paper can serve as an indicator for future works. The case study of the detected malicious images is deferred to Appendix A.3.

6 CVE Vulnerabilities

In this section, we evaluate the vulnerabilities in Docker images which are identified through Common Vulnerabilities and Exposures (CVE) IDs because the information of CVEs is public, expansive, detailed, and well-formed [34]. First, we leverage `Anchore` [2] to perform vulnerability detection for each image and study the distribution of the discovered vulnerabilities in Docker images. The results demonstrate that both official and community images suffer from serious software vulnerabilities. More analysis about vulnerabilities in images are deferred to Appendix A.4. Then, we investigate the extra window of vulnerability in Docker images.

Defining of the Extra Window of Vulnerability. To understand the timeline of the life cycle of a vulnerability in Docker images, accurately determining the *discovery* and *patch* time of a vulnerability, the *release* and *update* time of an image is vital. However, several challenges exist in determining different times. For instance, a vulnerability may be patched multiple times. In addition, different vendors might release different *discovery* times. Hence, we propose to first define these times motivated by existing research [34].

- *Discovery-time* is the earliest reported date of a software vulnerability being discovered and recognized to pose a security risk to the public.
- *Patch-time* is the latest reported date that the vendor, or the originator of the software releases a fix, workaround, or a patch that provides protection against the exploitation of the vulnerability. If the vulnerability is fixed by

the upgrade of the software and the patch is not publicly available, we record the date of the upgrade of the software instead.
- For a vulnerability, *release-time* is the date that the developer releases an image that first brings in this vulnerability.
- For a vulnerability, *upgrade-time* is the date that the developer releases a new edition of the image, which fixes this vulnerability contained in the previous edition.

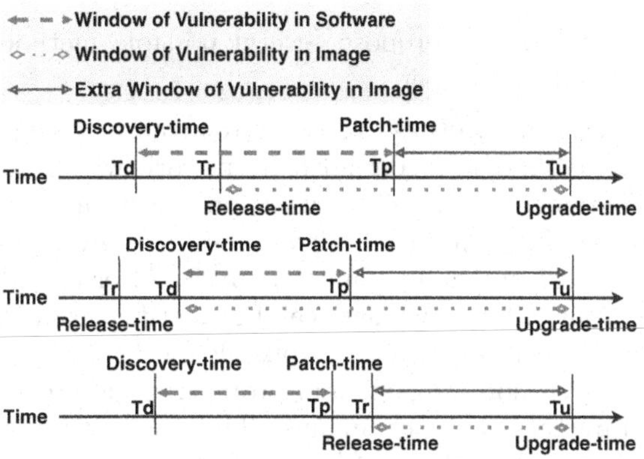

Fig. 4. Extra window of vulnerabilities in Docker images.

Suppose that a vulnerability of software S is discovered at T_d. Then after a period of time, the developer of S fixes this vulnerability at T_p. Image I first brings in this vulnerability at T_r. It takes extra time for the developer of I to fix the vulnerability and update this image at T_u. The developer of Image I is supposed to immediately fix the vulnerability once the patch is publicly available. However, it usually takes a long time before developers actually fix the vulnerability in an image. Therefore, we define W_e, the extra window of vulnerability in images, to measure how long it takes from the earliest time the vulnerability could be fixed to the time the vulnerability is actually fixed. Figure 4 presents three different cases of the extra window of vulnerability. In the first two cases, W_e is spreading from T_p to T_u. In the last case, even though the patch of a vulnerability is available, image I still brings in the vulnerability, so the W_e starts at T_r and ends at T_u. For all the cases, there is always an extra time window of vulnerability in image I before the developer updates I.

Obtaining Different Times. We choose the latest five editions of the 15 most popular images and randomly sample other 15 Docker images to investigate the extra window of vulnerabilities in images. By leveraging `Anchore` [2], we obtain 5,608 CVEs from these 30 images. After removing duplication, 3,023 CVEs remain, from which we randomly sample 1,000 CVEs for analysis. Then, we implement a tool to automatically collect *discovery-*, *patch*-time, and image *release-*, *update*-time for each CVE by leveraging the public information released on NVD Metrics [12]. We aim to obtain all the vital time of 1,000 CVEs from public information. However, the *discovery-* and *patch*-time of vulnerabilities are not always released in public. Therefore, we only obtain the complete information of vital time of 334 CVEs. It is worth noting that in some complex cases (e.g., there are multiple *discovery*-time for a CVE), we manually collect and confirm the complete information by reviewing the external references associated with CVEs.

Results. After analyzing the vital time of the 334 CVEs, we observe that it takes 181 days on average for common software to fix a vulnerability. However, the extra window of vulnerabilities in images is 422 days on average while the longest extra window of vulnerabilities could be up to 1,685 days, which allows sufficient time for attackers to craft corresponding exploitations of the vulnerabilities in Docker images.

7 Mitigating Docker Threats

In this section, we propose several possible methods to mitigate the threats uncovered in this paper.

Sensitive Parameters. To mitigate attacks abusing sensitive parameters, one possible method is to design a framework which automatically identifies sensitive parameters and alerts users on the webpages of repository descriptions on Docker Hub. First of all, it is necessary to maintain a comprehensive list of sensitive parameters by manual analysis. After that, sensitive parameters in the descriptions of Docker images can be identified easily by leveraging string matching. Docker Hub should be responsible for displaying the detection results in the image description webpage and prompting the users about the possible risks of the parameters in run-commands. The above framework can be implemented as a backend of the website of Docker Hub or a browser plug-in. Additionally, runtime alerts, as adopted by iOS and existing techniques [33,38], can warn users of potential risks before executing a run-command with sensitive parameters, which will be an effective mechanism to mitigate the abusing of sensitive parameters.

Malicious Images. To detect malicious images, traditional static and dynamic analysis, e.g., signature-based method, system call tracing can be certainly helpful. However, many challenges do exist. For example, if the redundant files of an image cannot be removed accurately, it will be extraordinarily time-consuming to analyze all the files in an image by traditional methods. The framework proposed in Sect. 5.1 can be utilized to solve this problem. Furthermore, heuristic approaches, such as analyzing related images proposed in Sect. 5.2, could also be beneficial in discovering malicious images.

Vulnerabilities. Motivated by previous research [18], we believe that automatic updating is an effective way to mitigate the security risks from vulnerable images. However, software update becomes challenging in Docker. Because the dependencies among a large quantity of software are complicated and arbitrary update may cause a broken image. We propose multiple possible solutions to automated updating for vulnerable software packages. First, various categories of existing tools such as `Anchore` [2] can be employed to obtain vulnerability description information including the CVE ID of the vulnerability, the edition of the corresponding vulnerable software packages that bring in the vulnerability, the edition of software packages that repairs this vulnerability. Second, package management tools, e.g., apt, yum, can be helpful to resolve the dependency relationship among software packages. After we obtain the above information, we may safely update vulnerable software and the related software.

8 Discussion

In this section, we discuss the limitations of our approach and propose several directions for future works.

Automatically Identifying Malicious Parameters. We perform the first analysis on sensitive parameters and show that these parameters can lead to disastrous security consequences in Sect. 4. However, the sensitive parameters we discuss in this paper are recognized by manual analysis. Obviously, there are other sensitive parameters in the field that still need to be discovered in the future. Additionally, even though we discover many sensitive parameters on Docker Hub, it is hard to identify which parameters are published for malicious attempts automatically. Hence, one future direction to improve our work is to automatically identify malicious parameters in repository descriptions. One possible method is gathering images with the similar functionalities and employ statistical analysis to detect the deviating uses of sensitive parameters—deviations are likely suspicious cases, given that most images are legitimate.

Improving Accuracy for Detecting Malicious Images. In Sect. 5, we narrow down the analysis of malicious images to the executed programs. Although this method achieves the goal of discovering security risks in Docker Hub, it still has shortcomings for detecting malicious images accurately. For example, malware detection is a well-known arms-race issue [17]. First of all, VirusTotal, the detector we used in the primary screening may miss malware. Additionally, we cannot detect malicious images that perform attacks with other files rather than their executed programs. For example, our approach cannot detect images that download malware during execution, since our parser performs static analysis. Furthermore, after we obtain the system log of potential malicious images, we conduct manual analysis on these log files. Motivated by [21], we plan to include more automatic techniques to parse and analyzing logs as future work. Moreover, we plan to analyze more images in the future.

Polymorphic Malware. Malware can change their behaviors according to different attack scenarios and environments [17,25]. Since our research goal is to characterize malware for measurement purposes instead of actually detecting them in practice, we employ existing techniques to find and analyze malware. However, it is valuable to understand the unique behaviors of Docker Malware. For instance, Docker has new fingerprints for malware to detect its running environments. How to emulate the Docker fingerprints to expose malicious behaviors can be an interesting future topic.

9 Related Work

Vulnerabilities in Docker Images. Many prior works focus on the vulnerabilities in Docker images [16,23,27,36,39]. For instance, Shu et al. proposed the *Docker Image Vulnerability Analysis (DIVA)* framework to automatically discover, download, and analyze Docker images for security vulnerabilities [36].

However, these studies only investigate the distribution of vulnerabilities in Docker images, while our work uniquely conducts in-depth analysis on new risks brought by image vulnerabilities, such as the extra window of vulnerability.

Security Reinforcement and Defense. Several security mechanisms have been proposed to ensure the safety of Docker containers [26,35,37]. For instance, Shalev et al. proposed `WatchIT`, a strategy that constrains IT personnel's view of the system and monitors the actions of containers for anomaly detection [35]. There also exist other Docker security works that focus on defenses for specific attacks [19,24]. For instance, Gao et al. discuss the root causes of the containers' information leakage and propose a two-stage defense approach [19]. However, these studies are limited to specific attack scenarios, which are not sufficient for a complete understanding of the security state of Docker ecosystems as studied in this work.

Registry Security. Researchers have conducted a variety of works on analyzing the code quality and security in third-party code store and application registries, such as GitHub and App Store [15,20,28,30,32]. For instance, Bugiel et al. introduced the security issues of VM image repository [15]. Duc et al. investigated Google Play and the relationship between the end-user reviews and the security changes in apps [30]. However, Docker image registries such as Docker Hub have not been fully investigated before. This work is the first attempt to fill the gap according to our best knowledge.

10 Conclusion

In this paper, we perform the first comprehensive study of Docker Hub ecosystem. We identified three major sources for the new security risks in Docker hub. We collected a large-scale dataset containing both images and the associated meta-information. This dataset allows us to discover novel security risks in Docker Hub, including the risks of sensitive parameters in repository descriptions, malicious images, and the failure of fixing vulnerabilities in time. We developed new attacks to demonstrate the security issues, such as leaking user files and the host display. As the first systematic investigation on this topic, the insights presented in this paper are of great significance to understanding the state of Docker Hub security. Furthermore, our results make a call for actions to improve the security of the Docker ecosystem in the future. We believe that the dataset and the findings of this paper can serve as a key enabler for improvements of the security of Docker Hub.

Acknowledgements. This work was partly supported by the Zhejiang Provincial Natural Science Foundation for Distinguished Young Scholars under No. LR19F020003, the National Key Research and Development Program of China under No. 2018YFB0804102, NSFC under No. 61772466, U1936215, and U1836202, the Zhejiang Provincial Key R&D Program under No. 2019C01055, and the Ant Financial Research Funding.

A Appendix

A.1 User Study on Sensitive Parameters

We design an online questionnaire that contains questions including "Do you try to fully understand every parameter of the run-commands provided on the Docker Hub website before running those commands?", "Do you make a security analysis of the compose.yml file before running the image?", etc. Our questionnaire was sent to our colleagues and classmates, and further spread by them. In order to ensure the authenticity and objectivity of the investigation results, we did not tell any respondents the purpose of this survey. We plan to conduct the user study in the official community of Docker Hub in the future.

Finally, we collected 106 feedback offered by 106 users from various cities in different countries. All of them have benefited from Docker Hub, i.e., they have experiences in using images from Docker Hub. Besides, they are from a broad range from both academia and industry fields, including students and researchers from various universities, software developers and DevOps engineers from different companies, etc.

As described in Sect. 4.2, the results of our user study show that 97% of users only care about if they can successfully run the image while ignoring how the images run, not to mention the sensitivity parameters in run-command and docker-compose.yml file. Even for 68 users who have a background in security research, only 10% of them indicate that they prefer to figure out the meaning of the parameters in run-commands.

A.2 Novel Attacks Exploiting Sensitive Parameters

Obtaining the Display of the Host. `--privileged` is one of the most powerful parameters provided by Docker, which may pose a serious threat to users. When the operator uses command `--privileged`, the container will gain access to all the devices on the host. Under this scenario, the container can do almost anything with no restriction, which is extremely dangerous to the security of users. More specifically, `--privileged` allows a container to mount a partition on the host. By taking a step further, the attacker can access all the user files stored on this partition. In addition to accessing user files, we design an attack to obtain the display of a user's desktop. In fact, with `--privileged`, a one-line code, `cp/dev/fb0 user_desktop.txt`, is sufficient for attackers to access user display data. Furthermore, by leveraging simple image processing software [7], attackers can see the user's desktop as if they were sitting in front of the user's monitor.

Spying the Process Information on the Host. `--pid` is a parameter related to namespaces. Providing `--pid=host` allows a container to share the host's PID namespace. In this case, if the container is under the control of an attacker, all the programs running on the user's host will become visible to the attacker inside the container. Then, the attacker can utilize these exposed information such as the `PID`, the `owner`, the path of the corresponding executable file and the execution parameters of the programs, to conduct effective attacks.

A.3 Case Study of Malicious Images

We manually conduct analysis on detected malicious images. For instance, the image `mitradomining/ffmpeg` on Docker Hub is detected as malicious by our framework. The entry-file of this image is `/opt/ffmpeg` [7]. According to the name and entry-file of the image, the functionality of this image should be image and video processing. However, our framework detects that the real functionality of the entry-file is mining Bit-coins. By leveraging the syscall log reported by our framework, we determine that the real identity of this image is a Bit-coin miner. Thus, once users run the image, their machines will become slaves for cryptomining.

A.4 Distribution of Vulnerabilities

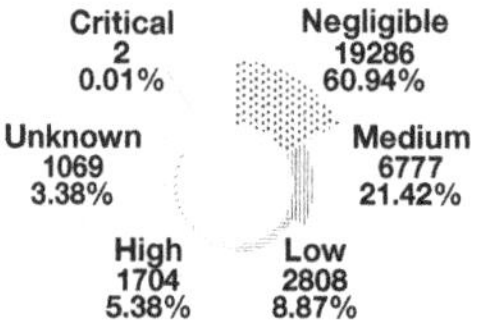

(a) Official images.

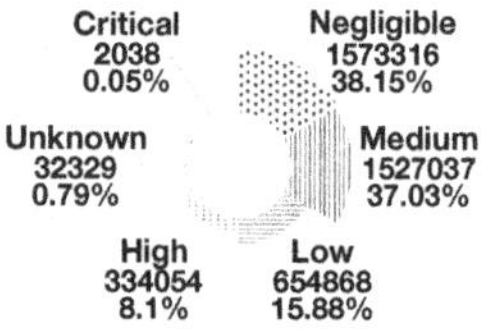

(b) Community images.

Fig. 5. Vulnerabilities existing in the latest images.

We investigate the distribution of vulnerabilities in the latest version of all official images. First, we observe that the latest official images contain 30,000 CVE vulnerabilities. Figure 5(a) categorizes these CVE vulnerabilities into 6 groups according to the severity levels assessed by the latest CVSSv3 scoring system [12]. Although only 6% of vulnerabilities are highly/critically severe, they exist in almost 30% of the latest official images. Furthermore, we conduct a similar analysis on the latest images in the 10,000 most popular community repositories. As shown in Fig. 5(b), the ratios of vulnerabilities with medium and high severity increase to over 37% and 8%, respectively, which are higher than those of official images. In addition, it is quite alarming that more than 64% of community images are affected by highly/critically severe vulnerabilities such as the denial of service and memory overflow. These results demonstrate that both official and community images suffer from serious software vulnerabilities. Additionally, community images contain more vulnerabilities with higher severity. Hence, we propose that software vulnerability is an urgent problem which seriously affects the security of Docker images.

References

1. Amazon Elastic Container Servicen, August 2019. https://aws.amazon.com/getting-started/tutorials/deploy-docker-containers
2. Anchore, August 2019., https://anchore.com/engine/
3. API to get Top Docker Hub images, August 2019. https://stackoverflow.com/questions/38070798/where-is-the-new-docker-hub-api-documentation
4. Docker, August 2019. https://www.docker.com/resources/what-container

5. Docker Hub Documents, August 2019. https://docs.docker.com/glossary/?term=Docker%20Hub
6. Docker Security Best-Practices, August 2019. https://dev.to/petermbenjamin/docker-security-best-practices-45ih
7. FFmpeg, August 2019. http://ffmpeg.org
8. Malicious Docker Containers Earn Cryptomining Criminals $90K, August 2019. https://kromtech.com/blog/security-center/cryptojacking-invades-cloud-how-modern-containerization-trend-is-exploited-by-attackers
9. Running Docker in Production, August 2019. https://ghost.kontena.io/docker-in-production-good-bad-ugly
10. strings(1) - Linux man page, August 2019. https://linux.die.net/man/1/strings
11. Virustotal Api, August 2019. https://pypi.org/project/virustotal-api/
12. Vulnerability Metrics, August 2019. https://nvd.nist.gov/vuln-metrics/cvss
13. Understanding the Security Risks of Docker Hub, July 2020. https://github.com/decentL/Understanding-the-Security-Risks-of-Docker-Hub
14. Arp, D., Spreitzenbarth, M., Hubner, M., Gascon, H., Rieck, K., Siemens, C.: DREBIN: effective and explainable detection of Android malware in your pocket. In: NDSS, vol. 14, pp. 23–26 (2014)
15. Bugiel, S., Nürnberger, S., Pöppelmann, T., Sadeghi, A.R., Schneider, T.: Amazonia: when elasticity snaps back. In: Proceedings of the 18th ACM Conference on Computer and Communications Security, pp. 389–400. ACM (2011)
16. Combe, T., Martin, A., Di Pietro, R.: To docker or not to docker: a security perspective. IEEE Cloud Comput. **3**(5), 54–62 (2016)
17. Cozzi, E., Graziano, M., Fratantonio, Y., Balzarotti, D.: Understanding Linux malware. In: 2018 IEEE Symposium on Security and Privacy (SP), pp. 161–175. IEEE (2018)
18. Duan, R., et al.: Automating patching of vulnerable open-source software versions in application binaries. In: NDSS (2019)
19. Gao, X., Gu, Z., Kayaalp, M., Pendarakis, D., Wang, H.: ContainerLeaks: emerging security threats of information leakages in container clouds. In: 2017 47th Annual IEEE/IFIP International Conference on Dependable Systems and Networks (DSN), pp. 237–248. IEEE (2017)
20. Gorla, A., Tavecchia, I., Gross, F., Zeller, A.: Checking app behavior against app descriptions. In: Proceedings of the 36th International Conference on Software Engineering, pp. 1025–1035. ACM (2014)
21. He, P., Zhu, J., He, S., Li, J., Lyu, M.R.: Towards automated log parsing for large-scale log data analysis. IEEE Trans. Dependable Secure Comput. **15**(6), 931–944 (2017)
22. Kotzias, P., Matic, S., Rivera, R., Caballero, J.: Certified PUP: abuse in authenticode code signing. In: Proceedings of the 22nd ACM SIGSAC Conference on Computer and Communications Security, pp. 465–478 (2015)
23. Tak, B., Kim, H., Suneja, S., Isci, C., Kudva, P.: Security analysis of container images using cloud analytics framework. In: Jin, H., Wang, Q., Zhang, L.-J. (eds.) ICWS 2018. LNCS, vol. 10966, pp. 116–133. Springer, Cham (2018). https://doi.org/10.1007/978-3-319-94289-6_8
24. Lin, X., Lei, L., Wang, Y., Jing, J., Sun, K., Zhou, Q.: A measurement study on Linux container security: attacks and countermeasures. In: Proceedings of the 34th Annual Computer Security Applications Conference, pp. 418–429. ACM (2018)
25. Liu, B., Zhou, W., Gao, L., Zhou, H., Luan, T.H., Wen, S.: Malware propagations in wireless ad hoc networks. IEEE Trans. Dependable Secure Comput. **15**(6), 1016–1026 (2016)

26. Loukidis-Andreou, F., Giannakopoulos, I., Doka, K., Koziris, N.: Docker-Sec: a fully automated container security enhancement mechanism. In: 2018 IEEE 38th International Conference on Distributed Computing Systems (ICDCS), pp. 1561–1564. IEEE (2018)
27. Martin, A., Raponi, S., Combe, T., Di Pietro, R.: Docker ecosystem-vulnerability analysis. Comput. Commun. **122**, 30–43 (2018)
28. Martin, W., Sarro, F., Yue, J., Zhang, Y., Harman, M.: A survey of app store analysis for software engineering. IEEE Trans. Softw. Eng. **43**(9), 817–847 (2017)
29. Miller, B., et al.: Reviewer integration and performance measurement for malware detection. In: Caballero, J., Zurutuza, U., Rodríguez, R.J. (eds.) DIMVA 2016. LNCS, vol. 9721, pp. 122–141. Springer, Cham (2016). https://doi.org/10.1007/978-3-319-40667-1_7
30. Nguyen, D., Derr, E., Backes, M., Bugiel, S.: Short text, large effect: measuring the impact of user reviews on Android app security and privacy. In: 2019 IEEE Symposium on Security and Privacy (SP), pp. 155–169. IEEE (2019)
31. Rastogi, V., Davidson, D., Carli, L.D., Jha, S., Mcdaniel, P.: Cimplifier: automatically debloating containers. In: Joint Meeting on Foundations of Software Engineering (2017)
32. Ray, B., Posnett, D., Filkov, V., Devanbu, P.: A large scale study of programming languages and code quality in GitHub. In: Proceedings of the 22nd ACM SIGSOFT International Symposium on Foundations of Software Engineering, pp. 155–165. ACM (2014)
33. Ringer, T., Grossman, D., Roesner, F.: Audacious: user-driven access control with unmodified operating systems. In: Proceedings of the 2016 ACM SIGSAC Conference on Computer and Communications Security, pp. 204–216. ACM (2016)
34. Shahzad, M., Shafiq, M.Z., Liu, A.X.: A large scale exploratory analysis of software vulnerability life cycles. In: 2012 34th International Conference on Software Engineering (ICSE), pp. 771–781. IEEE (2012)
35. Shalev, N., Keidar, I., Weinsberg, Y., Moatti, Y., Ben-Yehuda, E.: WatchIT: who watches your IT guy? In: Proceedings of the 26th Symposium on Operating Systems Principles, pp. 515–530. ACM (2017)
36. Shu, R., Gu, X., Enck, W.: A study of security vulnerabilities on docker hub. In: Proceedings of the Seventh ACM on Conference on Data and Application Security and Privacy, pp. 269–280. ACM (2017)
37. Sun, Y., Safford, D., Zohar, M., Pendarakis, D., Gu, Z., Jaeger, T.: Security namespace: making Linux security frameworks available to containers. In: 27th USENIX Security Symposium (USENIX Security 2018), pp. 1423–1439 (2018)
38. Wijesekera, P., et al.: The feasibility of dynamically granted permissions: aligning mobile privacy with user preferences. In: 2017 IEEE Symposium on Security and Privacy (SP), pp. 1077–1093. IEEE (2017)
39. Zerouali, A., Mens, T., Robles, G., Gonzalez-Barahona, J.M.: On the relation between outdated docker containers, severity vulnerabilities, and bugs, pp. 491–501 (2019)

DE-auth of the Blue! Transparent De-authentication Using Bluetooth Low Energy Beacon

Mauro Conti[1], Pier Paolo Tricomi[1(✉)], and Gene Tsudik[2]

[1] University of Padua, Padua, Italy
conti@math.unipd.it, tricomipierpaolo@gmail.com
[2] University of California, Irvine, USA
gene.tsudik@uci.edu

Abstract. While user authentication (e.g., via passwords and/or biometrics) is considered important, the need for de-authentication is often underestimated. The so-called "lunchtime attack", whereby a nearby attacker gains access to the casually departed user's active log-in session, is a serious security risk that stems from lack of proper de-authentication. Although there have been several proposals for automatic de-authentication, all of them have certain drawbacks, ranging from user burden to deployment costs and high rate of false positives.

In this paper we propose DE-auth of the Blue (DEB) – a cheap, unobtrusive, fast and reliable system based on the impact of the human body on wireless signal propagation. In DEB, the wireless signal emanates from a Bluetooth Low Energy Beacon, the only additional equipment needed. The user is not required to wear or to be continuously interacting with any device. DEB can be easily deployed at a very low cost. It uses physical properties of wireless signals that cannot be trivially manipulated by an attacker. DEB recognizes when the user physically steps away from the workstation, and transparently de-authenticates her in less than three seconds. We implemented DEB and conducted extensive experiments, showing a very high success rate, with a low risk of false positives when two beacons are used.

Keywords: De-authentication · Bluetooth beacon · Wireless signals · Information security

1 Introduction

In many environments, such as workplaces, schools and residences, computers are often used by multiple users, or multiple users' computers are co-located. To prevent unauthorized access, any secure system mandates authentication and de-authentication mechanisms. Most of the time, the user authenticates, e.g.,

The second author's work was done in part while visiting University of California, Irvine.

L. Chen et al. (Eds.): ESORICS 2020, LNCS 12308, pp. 277–294, 2020.
https://doi.org/10.1007/978-3-030-58951-6_14

by demonstrating the possession of secrets, once, at the beginning of the log-in session, and it is her own responsibility to explicitly log out when she leaves the session. Unfortunately, since users are often too lazy or careless, they neglect to terminate the log-in session by locking their workstation or logging out. Users who take a short (coffee or bathroom) break, often do not logout, since logging in again is annoying and time consuming. This behavior opens the door for so-called *Lunchtime Attacks* [9]. Such an attack occurs whenever an authenticated user, for whatever reason leaves the workplace, thus allowing the (typically insider) attacker to gain access to the current session.

The research community put a lot of effort into authentication techniques, producing several widely used and effective methods [1]. The most common technique, password-based authentication, remains popular despite predictions of its imminent demise. However, more and more alternative or supplemental techniques, (especially, biometric-based), are becoming popular. They reduce error rates and improve the overall security. Meanwhile, de-authentication is a relatively recent concern, arising mainly due to the threat of aforementioned lunchtime attacks. Many proposed methods are inaccurate, expensive, obtrusive and not transparent. Even continuous authentication (where the user is authenticated constantly after initial login) suffers from the same problems. It is clear that better de-authentication techniques are needed.

In this paper, we propose "DE-auth of the Blue" (DEB), a new de-authentication method that uses a Bluetooth Low Energy Beacon as the main tool to determine whether the user is still present. BLE Beacons are wireless sensors that continuously transmit Bluetooth Low Energy signals to allow mobile technologies to communicate in proximity. DEB takes advantage of the shadowing effect caused by the human body on high-frequency wireless signals. In particular, when a person stands on the Line-of-Sight (LoS) path between a transmitter and a receiver, the water in the human body affects signal strength, thus reducing the Received Signal Strength Indicator (RSSI) [22]. Therefore, by instrumenting a chair with a BLE Beacon, which spreads wireless signal, and monitoring the RSSI on the user's computer, we can infer the presence of a person, since it breaks the Line-of-Sight and lowers the RSSI. The power loss through human body is usually modeled as a time function due to random people's mobility [14,21]. However, in DEB the user is always sitting on a chair (i.e., not random), thus we can determine the shadowing loss by empirical measurements and use it as a threshold to determine whether there is human presence.

Contributions. Our contribution is twofold: (1) we present DEB (De-auth of the Blue), a new reliable, fast and inexpensive user de-authentication method, and (2) we give a comparative overview of the current de-authentication methods, discussing their benefits and flaws, w.r.t. DEB.

Organization: Sect. 2 overviews related work. Section 3 describes DEB and the adversary model. Section 4 discusses collected data, while experimental results and other de-authentication and continuous authentication techniques are treated in Sect. 5. Section 6 explores future work and we conclude in Sect. 7.

2 Related Work

Authentication and de-authentication techniques have been extensively studied in the last years. Meanwhile, usage of Bluetooth Low Energy (BLE) Beacons is quite recent: they are mainly used in the proximity context, for indoor localization and by retailers for managing the relationships with customers.

Faragher & Harle showed in [10] how to perform indoor localization with BLE fingerprinting, using 19 beacons in an area of ~600 m^2 to detect the position of a consumer device. This demonstrated the advantage of BLE over WiFi for positioning. In [16] Palimbo et al. used a stigmergic approach, a mechanism of indirect coordination, to create an on-line probability map identifying the user's position in an indoor environment. In [13], authors conducted experiments with indoor positioning using a BLE, achieving over 95% correct estimation rate and a fully accurate (100%) determination whether a device is in a given room. BLE beacons also make it possible to create smart offices and save energy, e.g., [6] Choi et al. created a smart office where BLE beacons and a mobile app determine whether a user enters or exits the office to adjust the power saving mode of the user's PCs, monitors, and lights.

De-authentication is required only after the initial authentication process. Classical authentication methods require one-time input of the user's secret(s). After, the user must either de-authenticate explicitly (e.g., by locking the screen or logging out) or be de-authenticated via some automated technique, in order to prevent lunchtime attacks. Recent techniques, which are mostly biometric, provide continuous authentication: the user is continuously monitored while interacting with the system. When this interactions stop, the user is automatically de-authenticated—with clear usability drawbacks, e.g., when the user is at her desk, just reading a document.

The most popular user authentication is based on passwords. They are intuitive, the user effort (burden) is low and no special equipment is needed. However, they also present some drawbacks. First, users often choose easy-to-remember passwords, which leads to the high probability of cracking [4]. Second, there is no way to automatically de-authenticate a user based on passwords: another task is required, such as clicking or pressing a button, which is often forgotten or deliberately ignored. Even if sophisticated techniques, such as fingerprint of iris recognition could address the first issue, passwords still are not amenable to automatic de-authentication. An inactivity time-out can address the second problem. However, this is ineffective, since interval length is difficult decide upon due to wide range of user (in)activity [19].

Technology is now moving towards continuous authentication based on biometrics. One popular method is keystroke dynamics [3], which creates a user model based on typing style, collecting features such as speed and strength. Though easy to deploy and not requiring any special equipment, it has two drawbacks. First, it is effective only against the *zero-effort* attack, whereby the attacker claims another user's identity without investing any effort. As shown in [20], a brief training is enough to learn how to reproduce a given user's typing style. Second, it does not allow for an inactivity period. The user can be continuously authenticated only while typing, whereas, in a common scenario where the user is present while not typing (e.g., reading emails or watching a video) the system is ineffective.

Kaczmarek et al. proposed *Assentication* [12], a new hybrid biometric based on user's seated posture pattern. The system captures the combination of physiological and behavioral traits, creating a unique model. It involves instrumenting an office chair with 16 tiny pressure sensors. However, though the system exhibits very low false positive and false negative, its cost is not negligible (∼150$ per chair), and it has low *permanence*, i.e., the biometric does not remain consistent over long term.

More complex systems such as gaze tracking [9] and pulse response [18] have been proposed. They are based on user's eye movements and human body impedance, respectively. Though highly accurate, both require expensive extra equipment which inhibits their large-scale adoption. Moreover, gaze tracking technique forces the user to always look in roughly the same direction. An inadvertent head movement can cause a false positive.

Proximity can be also as a basis for de-authentication. A typical technique of this sort required the user to carry a token, which communicates over a secure channel to a workstation, to continually authenticate the user [8]. The main drawback of this authentication is that the user must carry the token, and a serious risk can occur if it is stolen.

Another similar proposal is ZEBRA [15]. In it, a user wears a bracelet on the dominant wrist. The bracelet is equipped with a gyroscope, accelerometer and radio. When the user interacts with a computer keyboard, the bracelet records the wrist movement, processes it, and sends it to the terminal. After a comparison between received information and keyboard activity (made by the user who wears the bracelet), user's continued presence is confirmed only if they are correlated. There are two flaws here: (1) the user must wear a bracelet (thus, sacrificing unobtrusiveness), and (2) as shown in [11], attacker can, with reasonable probability, remain logged as the victim user in by opportunistically choosing when and how to interact.

There was one prior attempt to use attenuation of wireless signals caused by human body for de-authentication purposes. FADEWICH [7] involves an office setting instrumented with nine sensors that detect a user's position, measuring variations in signal strength. The system reaches 90% accuracy within four seconds, and 100% accuracy within six seconds. However, every office must be instrumented uniquely to determine where to place the sensors. Also, the deployment phase is not trivial and presence of other moving people nearby can cause false positives.

3 System and Adversary Model

We now present the system and adversary model we consider throughout this paper. In particular, Sect. 3.1 describes expected usage of the proposed system, while Sect. 3.2 discusses some attack scenarios.

3.1 System Model

We realized DEB using a BLE beacon and one antenna to capture the former's signals. The de-authentication process should be used in a common office, instrumented with chairs and workstations. DEB is completely unobtrusive. The user arrives to the office, sits down on the chair, logs into the computer (PC or a docked laptop) via usual authentication means (such as passwords), uses the computer and finally gets up, leaving her workspace. Right afterwards, the computer automatically locks her out, thus protecting the current session from unauthorized access. To accomplish this goal, every chair of the office is instrumented with a BLE beacon. A BLE antenna, usually integrated with laptops, collects beacon signals and sends them to our application. The application analyzes signals to decide whether to de-authenticate.

In contrast with other de-authentication or continuous authentication methods, DEB does not require the user to actually **use** the computer after initial login. As part of routine activities, a typical user might make or receive phonecalls, read a document or email, look at the smartphone, watch a video, etc. In all of these activities, the keyboard and mouse are not being used, yet the user remains present. In such cases, de-authentication is both unnecessary and annoying.

3.2 Adversary Model

The main anticipated adversary can be anyone with physical access to the office, such as a co-worker, a customer or a supervisor, with the goal of gaining access to the workstation and the active session of another the victim user. The adversary does not know the user's credentials. We assume that the adversary has access to the victim's inside office. There are two ways in which the adversary can try to subvert the system: (1) sit down on the victim's chair thus restoring a similar RSSI before DEB de-authenticates the user, or (2) alter the RSSI by attenuating the signal, e.g., putting an object in the line of sight, which yields a shadowing effect similar to the human body presence.

Another threat with a somewhat less impact is the user herself who might be annoyed by automatic de-authentication during a short break away from the workspace (e.g., to get a coffee, use the restroom). Re-logging in is generally viewed as tedious, thus the legitimate user might try to inhibit the system.

A selective attack on BLE advertisement has been demonstrated in [5], which makes packets unreadable. Other attacks annihilate the signal [17]. However, both the attacks cannot interfere with DEB. In fact, even if packets are unreadable, the RSSI is not changed, and if the receiver cannot detect the beacon, DEB automatically logs out the user, since it cannot provide security anymore. To succeed, the attacker must manipulate signals, changing the RSSI when the user leaves the workstation, reproducing the previous value. In this case, he should know this prior value, and it is very hard to alter the signal of one beacon when multiple beacons are located in the same room (e.g., in a common workplace scenario); thus this attack is easy to detect.

4 Proposed Method

We now present DEB – a transparent de-authentication method based on BLE beacons and the shadowing effect caused by the presence of the human body in Line of Sight. Section 4.1 overviews DEB. The correct positioning of the beacon is studied in Sect. 4.2. A two-beacon version of DEB is shown in Sect. 4.3.

4.1 DEB Overview

DEB is a low-cost and easy-to-deploy method for fast and trasparent user de-authentication. The only extra equipment is a BLE beacon placed on the underside of the user's chair. The computer continuously measures the RSSI between itself and the beacon. Since the body of the sitting user is directly in the LoS, it affects the signal power, causing the shadowing effect. The resulting RSSI is lower than the one in an unobstructed path. By monitoring the RSSI, DEB determines whether the user is present. DEB overview is shown in Fig. 1.

Fig. 1. Overview of the system.

The beacon located underneath the user's chair automatically spreads wireless signals at a fixed frequency. The user's PC has a BLE antenna which collects beacon's advertisements, measuring the RSSI. As mentioned in Sect. 3, laptops usually have a built-in BLE antenna, while a BLE USB dongle can be used for Desktops. In the deployment phase, it is sufficient to attach the beacon to the chair and give the ID of the beacon to the app running on the PC. This app is installed on the computer and runs in the background, continuously collecting BLE signals. At authentication time, the user is assumed to be sitting on the chair, and the measured RSSI becomes the reference value. When the user leaves, the RSSI increases, since human body impedance disappears. Noticing an RSSI increase that exceeds a pre-specified threshold (the expected impedance of the human body), DEB automatically de-authenticates the user. How to determine the optimal position for the beacon (in order to optimize the RSSI variation and the threshold value) is discussed below.

4.2 Beacon Positioning and Data Collection

The beacon and the receiver must be positioned to have LoS interrupted by the sitting user. Since we suppose that the computer is always on top of the desk, we need to decide where to attach the beacon. For this purpose, we considered 12 different positions: six on the back, and six on the seat of, a common office chair. The corresponding positions are shown in Fig. 2. We tried three different positions for the chair back: top, middle, bottom, as well as for the seat itself: behind, middle, front. Each position was tested on each side of the chair – behind and in front of the back, under and above the seat.

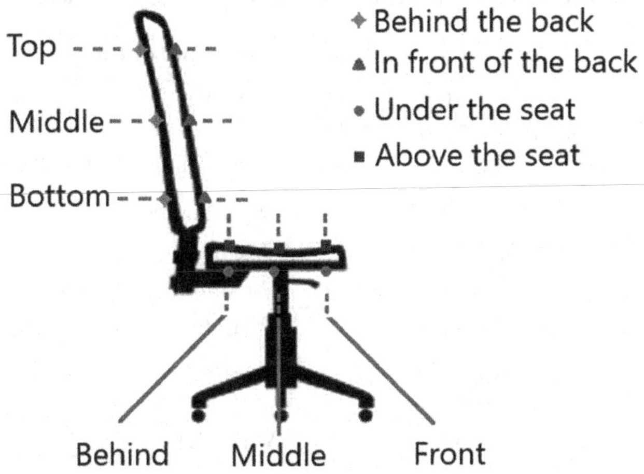

Fig. 2. Different beacon positions.

For the measurements, we used a RadBeacon, shown in Fig. 3. We chose this device due to its small dimensions (35 mm × 35 mm × 15 mm), coin cell power-supply, configurable transmission power and frequency, ranging from +3 dBm to −18 dBm, and one to ten times per second, respectively. After some tries, we set the transmission power at 0 dBm and the advertisement rate at three times per second, which we view as a good compromise between efficiency and energy savings. Due to these settings, transmission range is about five meters (the minimum achievable) and the battery is estimated to last about 220 days. Its small dimensions allow the beacon to be attached to many surfaces of the chair without being cumbersome or attracting attention.

To determine the best beacon position, we did a preliminary measurement using two average-sized people. We also considered the potential attack of putting an object into LoS, which could cause signal attenuation similar to that of a human body. Table 1 shows common objects with their level of interference, according to Apple documentation [2].

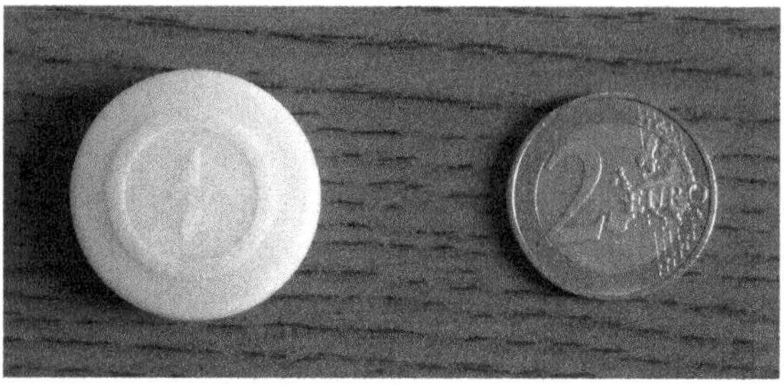

Fig. 3. Radbeacon used in measurements.

Table 1. Different materials and level of interference.

Type of barrier	Interference potential
Wood	Low
Synthetic material	Low
Glass	Low
Water	Medium
Bricks	Medium
Marble	Medium
Plaster	High
Concrete	High
Bulletproof glass	High
Metal	Very high

Since the human body attenuation is mainly introduced by water, that has a medium interference potential, to reproduce the same interference, intuitively, the attacker should use a medium-sized object made from medium-barrier material, or a small object with high interference. We believe that using a large object with low interference is infeasible. Since the object should be stealthily placed onto the chair, it would be hard to hide and manage. For the same reason, we believe that objects made from marble, plaster, concrete, bulletproof glass, and brick, would be noticed and are hard to use in limited time when the user is away. Thus, we experimented with two objects commonly found in a typical office: a bottle of water and a metal card with the form factor of a credit card.

We conducted RSSI measuring experiments from a laptop (Acer Aspire 7). In the initial phase, the user sat on the chair, breaking LoS between the beacon and the antenna. Eventually she (or the object) steps away from the workspace. An example graph of in Fig. 4 shows that – in the sitting period (from 0 to 5 seconds approximately) – the average RSSI is significantly lower than the one in free space propagation (around 25 dBm less from second 6 to 12 approximately). Also, changes occur almost instantaneously – within one second, the system decides whether the user left.

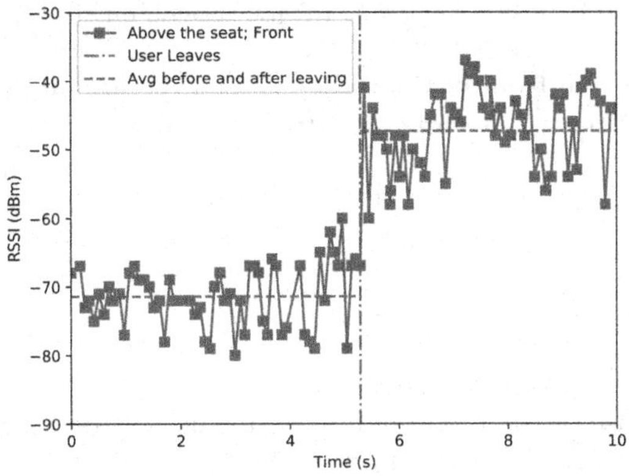

Fig. 4. Example of RSSI changes. Beacon located above the seat, in the front position.

Results of our study are shown in Tables 2 and 3. Table 2 refers to measurements conducted with the beacon under the seat, while in Table 3 the beacon on the seatback. The values represent the difference in RSSI when the user was absent and present. After the preliminary study, we chose the two best positions: *Under-Middle* for the seat and *Behind-Top* for the back. We chose these since they provide very good RSSI change and a noticeable difference between the human body and inanimate objects.

Table 2. RSSI changes in dBm - Beacon on the seat.

	Above the seat			**Under the seat**		
	Front	**Middle**	**Behind**	**Front**	**Middle**	**Behind**
Human body 1	25	25	15	10	7	11
Human body 2	25	25	10	5	5	10
Water bottle	22	15	20	0	0	6
Metal card	12	0	5	5	0	0

Obviously, since the chair is not in LoS between the user and the beacon, results of *above-the-seat* and *in-front-of-the-back* are better. However, we want to create system which is easy to deploy, and those positions require us to open the chair in order to install the beacon. This would be laborious and uncomfortable for the user. Therefore, those results are shown only to see that DEB could work even better if it were possible to place the beacon into the chair. Whereas, we decided to simply attach the beacon to the chair.

Table 3. RSSI changes in dBm - Beacon on the back.

	In front of the Back			Behind the Back		
	Top	Middle	Bottom	Top	Middle	Bottom
Human body 1	10	20	20	16	10	12
Human body 2	13	15	15	11	10	12
Water bottle	13	20	5	5	7	2
Metal card	8	7	0	3	7	5

In the final phase of the preliminar experiment, we used three average-sized people, the water bottle and the metal card. Tables 4 and 5 show results for *Under-the-seat-Middle* and *Behind-the-back-Top*, respectively. The average of measurements is reported for the human body. The experiment was conducted with the chair in front the desk (0°) and turning it in three different positions: 90° left, 90° right and 180°. Negative values indicate that the RSSI was higher than the initial one, measured with no one in the chair. This happens because the beacon is closer to the receiver in those positions.

4.3 Two-Beacon System

Efficacy of DEB depends on the quality of measurements. It can be affected by multiple factors, such as quality of transmitter and receiver, their synchronization and (obviously) presence of objects in LoS. Since the two devices are not paired, the advertising rate and the scanning frequency cannot be perfectly synchronized, which can lead to loss of some signals and lower performances. Also, if the user moves the chair, our assumptions might not hold any longer, thus causing some false positives. Consequently, in the later stage of the study, we decided to increase performances of DEB and decrease the false positive rate by introducing the second beacon. By using two different positions on the chair, we can gather more information and make more assumptions about the user's position, with a negligible difference in the deployment cost. Furthermore, fooling two beacons is more challenging for the attacker. The two-beacon version of DEB is discussed in Sect. 5.3.

Table 4. RSSI changes in dBm - Beacon under the seat in the middle.

	0°	90° Left	180°	90° Right
Human body	7	5	7	6
Water bottle	2	0	0	-3
Metal card	0	-2	2	0

Table 5. RSSI changes in dBm - Beacon behind the back on the top.

	0°	90° Left	180°	90° Right
Human body	18	15	-7	-4
Water bottle	2	6	11	13
Metal card	5	5	-6	10

5 Evaluations

After preliminary experiments, we conducted a user study to assess feasibility of DEB. In Sect. 5.1 we present the procedure of the user study and the algorithm for determining when the user leaves. In Sect. 5.2, measurements are performed using a single beacon, while Sect. 5.3 reports on the two-beacon version. Section 5.4 compares DEB with other de-authentication methods.

5.1 Data Collection Procedure and Algorithm

We collected data from 15 people (10 men and 5 women), with each person using a swivel chair and a normal (stationary) chair. For the latter, we asked participants to sit down for about five seconds and then stand up and leave. The beacon was positioned on the chair while the PC was on the desk. We tried two different positions, according to preliminary results: *Behind-the-back-Top* and *Under-the-seat-Middle*. For the swivel chair, since it can rotate, we asked subjects to sit down and then stand up and leave as before; in addition, we asked them to rotate the chair 90° left, 90° right and 180°. We decided to include rotational measurements to assess DEB resistance to false positives. More details are below.

To understand when the user leaves, we have to detect a significant increase of the RSSI. We compute the average of all measurements in a sliding window of three seconds, every second. The window of three seconds is the maximum amount of time we need to de-authenticate. Then, we compare the average of the current sliding window with those of its predecessors (in a limited range of five seconds to forget about history and remove the risk of false positives), to see whether the difference exceeds a fixed threshold. When this happens, we can infer the user left and, by using the sliding window approach, we can determine that in at most three seconds. The threshold was decided based on preliminary experiments, while a small variation of the algorithm was used in the two-beacon version (see below).

5.2 Single-Beacon System Results

Looking at preliminary results, we decided to use the position *Behind-the-back-Top* as the best for placing the beacon. While *Under-the-seat-Middle* is more stable, its value of 6–7 dBm is acceptable, though not very high. Also, the difference between the human body and the water bottle is not large. On the other hand, RSSI difference in *Behind-the-back-Top* is great, and it is very easy to distinguish between a human and a water bottle, or a metal card. Since the user is supposed to work in front of the desk most of the time, we could consider the 0° data only. In our particular case, with the swivel chair, the problems are present only in 90° right and 180°, since the beacon is closer to the receiver (i.e., in our PC the antenna was on the left side, so when the user rotates towards right the beacon goes towards left, closer to the receiver). However, in both cases, the value is higher than the one in the free LoS. This can be used to determine that the beacon is now closer; thus, the chair was rotated and the user might be sitting still. In 90° left the beacon is farther, so the RSSI would not be higher. A better solution to this problem is presented in Sect. 5.3.

Based on collected data, we fixed the threshold at ∼10 dBm. When the RSSI exceeds this threshold, the user is de-authenticated. With this threshold, DEB avoids false positives due to signal fluctuation, and none of the aforementioned objects mimic the same RSSI change as the one made by the human body. We believe that the attacker would not be able to interfere with DEB in any meaningful manner. With the frequency of three times per second, and the signal captured multiple times, we can compute an accurate average and notice the change within two seconds. To check for a false positive, we can give a grace period of one second, with a successful automatic de-authentication after three seconds. In this small amount of time, the attacker cannot sit down on the victim's chair or place an object in LoS without being seen. As mentioned in Sect. 3, another threat could emanate from the actual user. First, the user can place an object in LoS, while leaving. This is doable, however, the user should build a proper object to mimic the RSSI change, which is not a trivial task. Furthermore, re-entering a password is easier and less annoying than finding the right object and properly positioning it every time when leaves the workspace.

The user could also try to remove the beacon from the chair, and position it at another location to keep the constant RSSI. To do so, the user must find a position that mimics the RSSI reduced by the human presence. The user has two ways to succeed. One way is to place the beacon further than the normal position. However, if someone or something breaks LoS, the RSSI may become too low to be detected or to be credible and DEB will log out the user. Another way is to place the beacon nearby where nothing can interfere with LoS (i.e., on the desk) and reduce the RSSI by covering the beacon with an object. In this case, reproducing the desired RSSI is more difficult, and the fluctuation of the signal will be too regular to be produced by the human body. DEB would detect this attack.

Figure 5 shows sample RSSI changes and algorithm detection for one user. Dashed lines represent the averages of the sliding window. At second 6, the average is higher than its predecessors over the threshold; so, DEB detects that the user left. With all the 15 participants, we achieved 100% success with the beacon positioned *Behind-the-back-Top* and a threshold of 10 dBm, on the normal chair and on the swivel chair, without rotation. When the beacon was positioned *Under-the-seat-Middle*, we set the threshold to 6 dBm according to results from the preliminary study, again achieving 100% of success for both chairs.

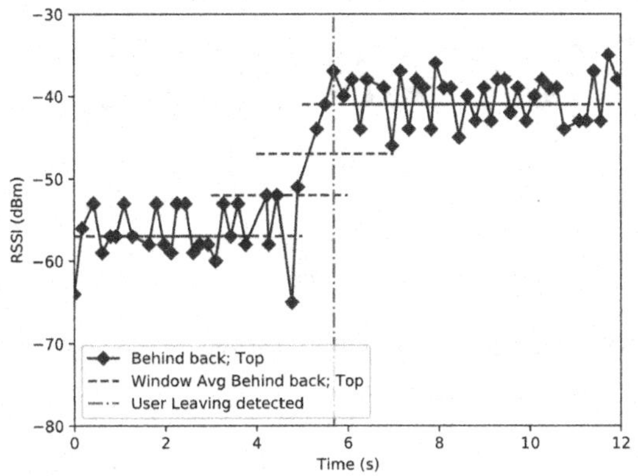

Fig. 5. RSSI changes: user leaves. Beacon in *Behind-the-back-Top* position.

5.3 Two-Beacon System Results

Positioning the beacon behind the back of the chair might cause a false positive, i.e., erroneous de-authentication, if the user rotates the chair. The RSSI might increase (if the beacon is closer to the receiver) even if the user is still sitting, which is clearly undesirable.

Previous results in Sect. 4 show that the RSSI of the signal spread by a beacon positioned *Under-the-seat-Middle* is almost constant, even if the chair is rotated. Based on this observation, we decided to use another beacon to improve DEB. Even if the RSSI change of a beacon located under the chair is not wide enough to distinguish between a human and another object, it is constant and high enough to be used as a double-check. The idea is to use the beacon *Behind-the-back-Top* as the first main means of deciding if the user is leaving. If the RSSI of its signal increases, we examine the signal of the second beacon to see whether this is a false positive. In the algorithm, we monitor averages using a sliding window, as before. When we detect a possible hit, we check the second beacon. If its average is also above its predecessor's more than the second threshold (appropriate for that position), the algorithm de-authenticates; otherwise, this is considered a rotation and no de-authentication takes place.

Two-beacon DEB is more difficult to attack, since the adversary needs to fool two beacons instead of one. Figure 6 shows a false positive, wherein the RSSI of the first beacon (light green diamonds - *Behind-the-back-Top*) is increasing, while the signal of the second (dark green circles - *Under-the-seat-Middle*) remains almost constant. Whereas, in Fig. 7, we see a true positive, where signals are increasing, since the user is leaving.

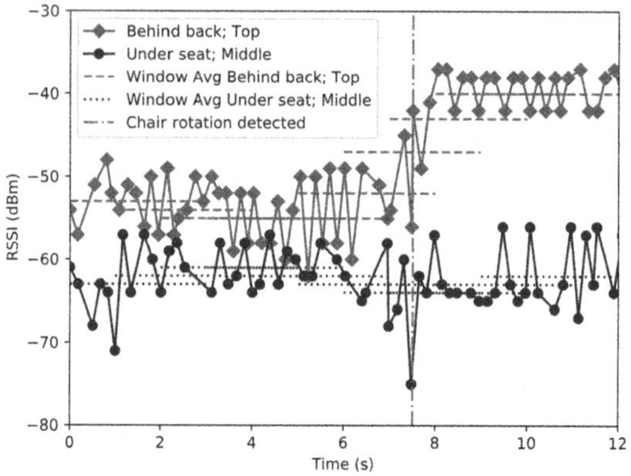

Fig. 6. RSSI changes: user chair rotation. (Color figure online)

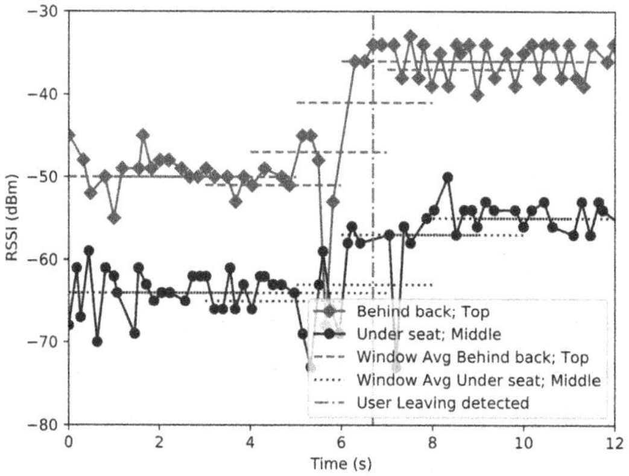

Fig. 7. RSSI changes: user leaves. (Color figure online)

From our user-study, all the 15 participants were correctly de-authenticated within three seconds. During the rotations, the false positive were probably to occur at 90° toward right and 180°, since the *Behind the back; Top* was closer to the PC's antenna. In the experiment, all the rotations were correctly classified. While at 90° toward left the signals RSSI were almost the same, in 90° toward right and 180° the *Behind the back; Top* showed possible de-authentication cases. However, the *Under the seat; Middle* never reached the threshold in these situations, avoiding false positives.

Table 6. Comparison between de-authentication techniques.

	Nothing to Carry	Non Biometric	Non-Cont. Auth.	Inactivity Allowed	Non Obtrusive	Difficult to Circumvent	Free from Enrollment	Price
Timeout	✓	✓	✓	✗	✓	✗	✓	Low (Free)
ZEBRA	✗	✓	✗	✗	✗	✗	✗	Medium ($100)
Gaze Track.	✓	✗	✗	✗	✓	✓	✗	High ($2-5k)
Popa	✓	✗	✗	✓	✓	✓	✓	Medium ($150)
Pulse resp.	✗	✗	✗	✓	✗	✓	✓	Unknown
Keystr. Dyn.	✓	✗	✗	✗	✓	✓	✗	Low (Free)
FADEWICH	✓	✓	✓	✓	✓	✗	✓	Unknown
BLE Beacon	✓	✓	✓	✓	✓	✓	✓	Low ($10)

5.4 Comparison with Other De-authentication Techniques

We now compare DEB with other de-authentication techniques. Table 6 shows key features of prominent current methods. First, a good method should not force the user to carry anything extra. In fact, it could be annoying, uncomfortable, unaesthetic and result in a serious problem if stolen or forgotten. Most of the techniques satisfy this objective, only ZEBRA with the bracelet and Pulse Response with electrodes violate it. The presented techniques can be divided into biometric and non-biometric. The former is desirable since it is more precise, and typically difficult to subvert. Also, they usually provide continuous authentication – de-authentication can be automatically triggered when the user can no longer verify identity. Even though biometric and continuous authentication methods are more precise, they usually have two issues. First, they require an enrollment phase which can be annoying and time-consuming. Second, biometric hardware can be very expensive. This is true especially for Gaze-Tracking, which costs \$2–5K per unit. Other methods,, such as ZEBRA and Popa, are cheaper than Gaze-Tracking. However, the cost per unit is still over \$100, which can be quite expensive if instrumenting a big office. Also, ZEBRA is not a purely biometric method; it has been shown to be less secure than others. Pulse Response and Keystroke Dynamics are both biometric-based and continually authenticate the user. Though the cost of Pulse Response is unclear, we believe that the hardware is not cheap. On the other hand, Keystroke Dynamics is completely free. DEB, similar to Timeout and FADEWICH, is not biometric-based and does not provide continuous authentication. Nonetheless, its effectiveness has been demonstrated. Furthermore, it requires no enrollment phase and its cost is very low.

Keystroke Dynamics might seem to be the best choice. However, as mentioned earlier, it has an important drawback in that it does not allow the user to be present-but-inactive. Another important property is user burden or unobtrusiveness. All considered methods are transparent, except for ZEBRA. DEB is completely unobtrusive since the user need not be aware and does not need to interact with it. Finally, difficulty of circumvention is also an important property. Inactivity time-outs is the least prone: a too-long time-out value can be too long to de-authenticate the user right after he leaves. ZEBRA was shown to be vulnerable, and FADEWICH can be circumvented if the office is crowded. We demonstrated that DEB is secure: its de-authentication is very fast, at under three seconds.

6 Future Work

Thus far, we tested DEB with 15 people in a communal workspace. In the future, a bigger study should be conducted to better evaluate DEB and discover new useful features, where machine learning could be used to improve efficacy. So far, we found a strong correlation between the RSSI and user actions, demonstrating that the human body attenuation does not vary a lot among different users, and can be used to de-authenticate. Moreover, we used a BLE beacon as the means to send signals. However, perhaps another more suitable device can be used. Since the beacon cannot be paired with the receiver, the RSSI value is affected by fluctuations that sometimes make results inaccurate. On the other hand, a connection between paired device is more stable. In this direction, we could try to find another cheap device that can work better than a BLE beacon. A stable signal will provide both more accuracy and more difficulties for an adversary. Finally, we imagine a similar system where the transmitter and the receiver are both on the user's chair. In this situation, no false positive are likely due to common actions, such as chair rotation.

7 Conclusion

This paper proposed a new system, DEB, to automatically de-authenticate users. DEB was implemented and tested with encouraging results. Although de-authentication is not a new topic, current methods have some limitations, such as extra equipment for users to carry, lack of inactivity permission, or imposition of expensive hardware to verify user presence. In contrast, for under \$10, DEB de-authenticate a user in under three seconds. This speed makes literally hard for the attacker to gain access to the victim's session. Though DEB is not biometric-based and is not a continuous authentication technique, it is completely unobtrusive, free from the tedious enrollment phase and is quite reliable. Finally, using a second beacon as an extra security factor, we can lower the rate of false positives.

References

1. Al Abdulwahid, A., Clarke, N., Stengel, I., Furnell, S., Reich, C.: A survey of continuous and transparent multibiometric authentication systems. In: European Conference on Cyber Warfare and Security, pp. 1–10 (2015)

2. Apple: Potential sources of wi-fi and bluetooth interference (2017). https://support.apple.com/en-us/HT201542. Accessed 5 July 2018
3. Banerjee, S.P., Woodard, D.L.: Biometric authentication and identification using keystroke dynamics: a survey. J. Pattern Recogn. Res. **7**(1), 116–139 (2012)
4. Bonneau, J.: The science of guessing: analyzing an anonymized corpus of 70 million passwords. In: 2012 IEEE Symposium on Security and Privacy, pp. 538–552. IEEE (2012)
5. Brauer, S., Zubow, A., Zehl, S., Roshandel, M., Mashhadi-Sohi, S.: On practical selective jamming of Bluetooth low energy advertising. In: 2016 IEEE Conference on Standards for Communications and Networking (CSCN), pp. 1–6. IEEE (2016)
6. Choi, M., Park, W.K., Lee, I.: Smart office energy management system using bluetooth low energy based beacons and a mobile app. In: 2015 IEEE International Conference on Consumer Electronics (ICCE), pp. 501–502. IEEE (2015)
7. Conti, M., Lovisotto, G., Martinovic, I., Tsudik, G.: Fadewich: fast deauthentication over the wireless channel. In: 2017 IEEE 37th International Conference on Distributed Computing Systems (ICDCS), pp. 2294–2301. IEEE (2017)
8. Corner, M.D., Noble, B.D.: Zero-interaction authentication. In: Proceedings of the 8th Annual International Conference on Mobile Computing and Networking, pp. 1–11. ACM (2002)
9. Eberz, S., Rasmussen, K., Lenders, V., Martinovic, I.: Preventing lunchtime attacks: fighting insider threats with eye movement biometrics (2015)
10. Faragher, R., Harle, R.: Location fingerprinting with Bluetooth low energy beacons. IEEE J. Sel. Areas Commun. **33**(11), 2418–2428 (2015)
11. Huhta, O., Shrestha, P., Udar, S., Juuti, M., Saxena, N., Asokan, N.: Pitfalls in designing zero-effort deauthentication: opportunistic human observation attacks. In: Network and Distributed System Security Symposium (NDSS), February 2016
12. Kaczmarek, T., Ozturk, E., Tsudik, G.: Assentication: user de-authentication and lunchtime attack mitigation with seated posture biometric. In: Preneel, B., Vercauteren, F. (eds.) ACNS 2018. LNCS, vol. 10892, pp. 616–633. Springer, Cham (2018). https://doi.org/10.1007/978-3-319-93387-0_32
13. Kajioka, S., Mori, T., Uchiya, T., Takumi, I., Matsuo, H.: Experiment of indoor position presumption based on RSSI of Bluetooth le beacon. In: 2014 IEEE 3rd Global Conference on Consumer Electronics (GCCE), pp. 337–339. IEEE (2014)
14. Kiliç, Y., Ali, A.J., Meijerink, A., Bentum, M.J., Scanlon, W.G.: The effect of human-body shadowing on indoor UWB TOA-based ranging systems. In: 2012 9th Workshop on Positioning Navigation and Communication (WPNC), pp. 126–130. IEEE (2012)
15. Mare, S., Markham, A.M., Cornelius, C., Peterson, R., Kotz, D.: Zebra: zero-effort bilateral recurring authentication. In: 2014 IEEE Symposium on Security and Privacy (SP), pp. 705–720. IEEE (2014)
16. Palumbo, F., Barsocchi, P., Chessa, S., Augusto, J.C.: A stigmergic approach to indoor localization using Bluetooth low energy beacons. In: 2015 12th IEEE International Conference on Advanced Video and Signal Based Surveillance (AVSS), pp. 1–6. IEEE (2015)
17. Pöpper, C., Tippenhauer, N.O., Danev, B., Capkun, S.: Investigation of signal and message manipulations on the wireless channel. In: Atluri, V., Diaz, C. (eds.) ESORICS 2011. LNCS, vol. 6879, pp. 40–59. Springer, Heidelberg (2011). https://doi.org/10.1007/978-3-642-23822-2_3
18. Rasmussen, K.B., Roeschlin, M., Martinovic, I., Tsudik, G.: Authentication using pulse- response biometrics. In: The Network and Distributed System Security Symposium (NDSS) (2014)

19. Sinclair, S., Smith, S.W.: Preventative directions for insider threat mitigation via access control. In: Stolfo, S.J., Bellovin, S.M., Keromytis, A.D., Hershkop, S., Smith, S.W., Sinclair, S. (eds.) Insider Attack and Cyber Security, pp. 165–194. Springer, Heidelberg (2008). https://doi.org/10.1007/978-0-387-77322-3_10
20. Tey, C.M., Gupta, P., Gao, D.: I can be you: questioning the use of keystroke dynamics as biometrics (2013)
21. Yoo, S.K., Cotton, S.L., Sofotasios, P.C., Freear, S.: Shadowed fading in indoor off-body communication channels: a statistical characterization using the (k-u)/Gamma composite fading model. IEEE Trans. Wireless Commun. **15**(8), 5231–5244 (2016)
22. Zhao, Y., Patwari, N., Phillips, J.M., Venkatasubramanian, S.: Radio tomographic imaging and tracking of stationary and moving people via kernel distance. In: 2013 ACM/IEEE International Conference on Information Processing in Sensor Networks (IPSN), pp. 229–240. IEEE (2013)

Similarity of Binaries Across Optimization Levels and Obfuscation

Jianguo Jiang[1,2], Gengwang Li[1,2], Min Yu[1,2(✉)], Gang Li[3], Chao Liu[1], Zhiqiang Lv[1], Bin Lv[1], and Weiqing Huang[1]

[1] Institute of Information Engineering, Chinese Academy of Sciences, Beijing, China
yumin@iie.ac.cn
[2] School of Cyber Security, University of Chinese Academy of Sciences, Beijing, China
[3] School of Information Technology, Deakin University, Melbourne, VIC, Australia

Abstract. Binary code similarity evaluation has been widely applied in security. Unfortunately, the compiler optimization and obfuscation techniques exert challenges that have not been well addressed by existing approaches. In this paper, we propose a prototype, ImOpt, for re-optimizing code to boost similarity evaluation. The key contribution is an immediate SSA (static single-assignment) transforming algorithm to provide a very fast pointer analysis for re-optimizing more thoroughly. The algorithm transforms variables and *even pointers* into SSA form on the fly, so that the information on *def-use* and reachability can be maintained promptly. By utilizing the immediate SSA transforming algorithm, ImOpt canonicalizes and eliminates junk code to alleviate the perturbation from optimization and obfuscation.

We illustrate that ImOpt can improve the accuracy of a state-of-the-art approach on similarity evaluation by 22.7%. Our experiment results demonstrate that the bottleneck part of our SSA transforming algorithm runs 15.7x faster than one of the best similar methods. Furthermore, we show that ImOpt is robust to many obfuscation techniques that based on data dependency.

Keywords: Binary code similarity · SSA transforming · Reverse engineering · Program analysis

1 Introduction

Binary similarity evaluation is a general technology with wide applications in security, including malware lineage inference [19,21,22,27], known bugs or vulnerabilities searching [11,12,24] in released binaries. However, the compiler optimization or obfuscation can remove or insert significant instructions respectively, which will cause much unpredictable damage to the similarity evaluation.

L. Chen et al. (Eds.): ESORICS 2020, LNCS 12308, pp. 295–315, 2020.
https://doi.org/10.1007/978-3-030-58951-6_15

```
int confound_compiler_opt() {
  int a, b, *c;
  if (&b > &a) c = &b - 1;
  else c = &b + 1;
  assert(c == &a);
  *c = 11;
  a = 10;
  return *c - a; // 1 if optimized
                 // else 0
}
```

(a) This code should output 0. But the compiler will simplify it as 1 mistakenly.

```
if (confound_compiler_opt()) {
    // obfuscation code that you
    // want to enable after the
    // optimization, this wouldn't
    // be executed in debug mode
} else {
    // the sensitive code which
    // would be eliminated after
    // the optimization
}
```

(b) An example for showing how to exploit the vulnerability.

Fig. 1. Compiler optimizer may change the outputs of the code. It has been tested with modern compilers `g++8` and `clang++9`.

Unfortunately, existing approaches to binary similarity evaluation cannot well address the perturbation brought by optimization and obfuscation. Static approaches usually count statistic features based on syntax sequence or instruction categories, which hardly reveal the semantic information. Besides, most of dynamic approaches tend to break the code into smaller fragments for overcoming the problems of path coverage and emulation. But once the junk code, which is inserted by obfuscation or should be eliminated but not, is broken into smaller fragments, these trivial fragments will be taken as a part of the function signature, which further affects the accuracy of similarity evaluation.

In this paper, we propose a prototype, ImOpt, to boost similarity evaluation. ImOpt tries to re-optimize lifted binary code to mitigate the impacts from optimization and obfuscation. Nevertheless, there are two challenges to do so. *Challenge 1: optimizing binary code demands a precise pointer analysis, which is lacked in compiler optimizers.* The irreversible compilation has erased variable information so that much routine processes of compiler optimizer are interrupted. The lack of pointer analysis also leads to the vulnerability that even compiler optimizers can be confounded by a few pointer dependencies, as shown in Fig. 1. *Challenge 2: traditional precise pointer analysis methods are costly.* A fast pointer analysis normally outputs an unsound result, while a precise pointer analysis always costs too much due to the iterative process [20]. Hasti et al. [15] accommodate to use SSA form for acceleration, while most SSA transforming algorithms only process variables but do not support pointer analysis.

To address *Challenge 1*, we build a framework by integrating a precise pointer analysis for optimizing the code lifted from binaries. The pointer analysis helps to resume the intercepted optimizations by pointers. Among the many optimizations, only those related with canonicalization and elimination are implemented because they influence the similarity evaluation most. Canonicalization tries to transform the logically equivalent expressions into a unified form. Besides, elimination clean out either useless or unreachable code according to the information on *def-use chain* and reachability. Thus, the framework is supposed to be resilient to the code transformation that based on data-dependency.

For *Challenge 2*, we propose an efficient algorithm to support a precise pointer analysis. This algorithm is named as immediate SSA transforming algorithm for satisfying the immediacy property (Sect. 2.2). The algorithm transforms variables and even pointers into SSA form on the fly to link each usage with its *dominating definition*. The processing order of code blocks is scheduled carefully to defer and reduce redundant re-computations so that only one pass of dense analysis is needed. Moreover, because of the immediacy property, the information on *dominating definition* can be maintained in $O(1)$ instead of $O(\sqrt{n})$ in the state-of-the-art [20], where n is the number of instructions in the code.

In summary, our contributions include:

- An algorithm called immediate SSA transforming algorithm to provide a fast and precise pointer analysis. By satisfying the immediacy property, it maintains the information on *dominating definition* in constant time (Sect. 3).
- A framework equipped with pointer analysis for re-optimizing binary code to boost binary similarity evaluation. It is supposed to be resilient to data-dependency-based code transformation (Sect. 4).
- An obfuscation method that confounds the compiler optimizer to eliminate any specified code blocks. This obfuscation illustrates it is vulnerable to re-optimize binary code with compiler optimizer directly (Sect. 5.2).
- We implement our designs in a prototype, ImOpt, and demonstrate that ImOpt can significantly improve the accuracy of a state-of-the-art approach [9] to similarity evaluation against optimization and obfuscation. We also show that the bottleneck module of ImOpt runs 15.7x faster than one of the best similar methods [20] (Sect. 6).

The rest of this paper is organized as follows. Section 2 introduces the terminologies. Section 3 details the immediate SSA transforming algorithm. In Sect. 4, the binary optimization framework is constructed. Experiment settings and results are provided in Sect. 5 and Sect. 6. Section 7 lists related works. Finally, Sect. 8 concludes the paper and envisages the future work.

2 Preliminaries

In this section, we present some terminologies related to pointer analysis and briefly recap the immediacy property.

2.1 Notations and Terminology

We assume readers have a basic familiarity with *SSA*, *dominating*, *dominator*, *dominance frontier* and *ϕ-function* from [5]. Here, we recap some key concepts together with their notations.

Symbols B, D, S, V are used to represent the set of blocks, definitions, statements and variables. The lowercase letters b, d, s and v denotes the elements in these sets respectively. The subscripts in b_i, d_i and s_i are used to stand for the number in the sequence while the subscript in v_k always refers to the SSA version number. A definition d for a variable v is *reachable* to a block b, denoted as $d \rightarrow b$, if and only if there is a path from the define-site to block b and that path contains no further definition for variable v. Such a definition is a *reachable definition* to the block b. If there is only one definition that reaches to the usage, then this definition is a *dominating definition* of that usage.

Moreover, three new concepts are introduced by us:

Backward/Forward block. An edge in control-flow graph (CFG) is a *back edge* if the source-end block is visited after the target-end block in reverse post-order. A block is a *backward block* if and only if there is a back edge pointing to that block. Otherwise, it is a *forward block.*

Backward/Forward definition. A definition d for variable v is a *forward definition* of block b, denoted as $d \overrightarrow{rc} b$, if there is such a path p from b' to b that no back edge or other definition of v on it. On the contrary a definition d is a *backward definition* of block b, denoted as $d \overleftarrow{rc} b$, if it can reach b but is not a forward definition.

Backward reachable graph. Given a CFG $G(B, E)$, where B is the set of all the basic blocks and E is the set of all the jump edges, we define the *backward reachable graph* (BRG) of a backward block b_i as the graph $G(B', E')$, where the nodes are $B' := \{\forall b_j \in B \mid b_j \overleftarrow{rc} b_i\}$ and the edges are $E' := \{\forall e' \in E \mid \forall b' \in B',\ e' \in p : b' \overleftarrow{rc} b\}$. The BRG of block b is denoted as $G_{BR}(b)$. Note that only the backward block has a BRG.

2.2 The Immediacy Property

The property of immediacy is the key to transform code into SSA form on the fly and to maintain the information on the dominating definition efficiently.

Property 1 (Immediacy). *An SSA transforming algorithm satisfies the immediacy property if it holds the following two invariants during processing:*

Global Invariant. *Before processing a block, for each free variable v in the block: a) if there is only one definition of it can reach the current block, then the version number of this definition should have been recorded and there is no ϕ-function for v in the front of the block; b) otherwise, a ϕ-function for v should be in the front of the block and the variable defined in the ϕ-function should have a different version number with other definitions of v.*

Local Invariant. *After processing a block, each variable v used in the block has only one definition can reach the usage and the used variable has been bound with the same version number as this definition.*

For a block under processing, the global invariant ensures every incoming definition is recorded and the necessary ϕ-functions are inserted before processing. After processing, the local invariant guarantees that each usage is linked with its dominating definition. These two invariants together preserve a correct SSA form. Besides, here we present the deductive condition that will be referred frequently in the following sections:

Condition 1 (Deductive Condition). *Suppose that traversing the CFG G in reverse post-order generates a block sequence, $\{b_n\}$. For the block b_i under processing, suppose our algorithm hold the invariants in Property 1 for every visited block b_j, where $j < i$.*

2.3 Useful Properties About Forward and Backward Blocks

The lemmas listed below describe some properties about forward and backward blocks. These properties inspire us on how to design an algorithm that satisfies Property 1 so that the information on dominating definition can be maintained in constant time. The proofs of these lemmas are presented in Appendix B.

Reverse post-order is a recommended alternative of the topological order for program analysis. It forms a good order for dominance:

Lemma 1. *The dominator block is ahead of the dominated one in reverse post-order traversal.*

Lemma 2. *A forward block has only forward definitions reaching itself if all the blocks except b_i on the path to the block are in SSA form.*

Lemma 3. *Given Condition 1, if block b_i is a forward block, we have:*

$$\forall b' \in B, \exists\, d \in b', d \rightarrow b_i \Rightarrow b' \text{ is ahead of } b_i \text{ in reverse post-order} \qquad (1)$$

The backward definition can reach the backward block along the iterative dominance frontier (IDF) chain of the block b' where that definition is defined:

Lemma 4. *Given Condition 1, if block b_i is a backward block, then:*

$$\forall b' \in B, \exists\, d \in b', d \overleftarrow{rc}\; b_i \Rightarrow b' \in G_{BR}(b_i) \wedge b_i \in DF^{+}(b') \qquad (2)$$

3 Immediate SSA Transforming Algorithm

In this section, we detail the immediate SSA transforming algorithm, which is proposed to transform code into SSA form on the fly for a prompt and thorough pointer analysis.

3.1 Main Function

The immediate SSA algorithm is implemented as the ImSsa procedure. Several components are maintained for the invariants during the processing:

- $DU \subseteq V \times B \times S$ is the set of *def-use edges*;
- $R \subseteq B \times B$ is the set of reachable edges;
- $E : V \rightarrow int$ traces the dominating definitions for each variable;
- $C : V \rightarrow int$ maintains each variable's *maximum SSA number*, which represents the unallocated SSA version number for every variable.

The ImSsa procedure traverses the CFG in reverse post-order. Due to the different properties as discussed in Sect. 2, the backward blocks are distinguished from forward ones to process separately.

1 ImSsa*(L) where L is the block list sorted in reverse post-order*
2 **for** $b_i \in L$ **do**
3 **if** b_i *is* `visited` **then continue**
4 **if** b_i *is forward block* **then**
5 ProcessForwardBlock(b_i, E_i)
6 **else** ProcessBackwardBlock(b_i, E_i)

3.2 Forward Block Analysis

Lemma 2 and Lemma 3 together reveal that for a forward block, all the definitions reaching it can be recorded before analyzing it. Therefore in theory, it is possible to transform the code into SSA form on the fly.

The ProcessForwardBlock procedure transforms the statements into SSA form on the fly (line 5).

1 ProcessForwardBlock$(b_i \in B)$
2 **if** *there is* **no** *reaching edge* $(\_, b_i)$ *in* R **then continue**
3 $V_{def}, E_i \leftarrow \{\}, E_{DOM(i)}$
4 **foreach** *statement* $s \in b_i$ **do**
5 ProcessStatement$(null, s, b_i, E_i, V_{def})$
6 Propagate(b_i, V_{def}, E_i)
7 PropagateReachability$(b_i, S_{invalid})$
8 Mark$(b_i,$ `visited`$)$

The table E_i, which records the dominating definitions, is inherited from the immediate dominator of b_i. As for the entry block, E is initially empty.

The ProcessStatement procedure links each variable usage with its dominating definition (line 2). When a new variable is found, it will be assigned with a unique SSA version number to distinguish with other definitions of the defined variable (line 6).

1 ProcessStatement$(v_k \in V, s \in S, b \in B, E, V_{def})$
2 SsaUse(s, E)
3 ClientAnalyse(s)
4 **if** $v \neq (u \leftarrow$ Lhs$(s))$ **then**
5 $v_k \leftarrow u_{E(u)}$
6 SsaDef(s, b, E, V_{def})
7 **return** v_k

3.3 Fetching Dominating Definitions in Constant Time

With Property 1, an algorithm can trace the dominating definitions in constant time. It is based on an observation of SSA-form code: the definition outgoing from

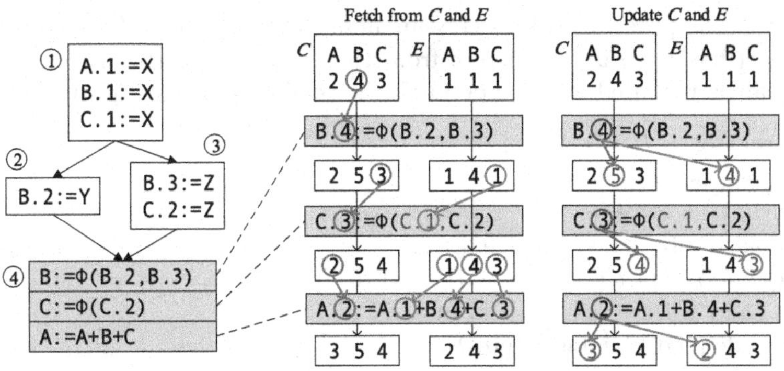

Fig. 2. Maintaining dominating definitions. Block 4 is under processing right now. The green(or blue) markers indicate the operations on C(or E). The middle sub-figure shows to number the usages(or definition) in statement's right(or left)-hand side by fetching from C(or E). And the right part shows to update E by copying from C and to update C by monotonic increment. (Color figure online)

the immediate dominator will be propagated to the current block eventually. If not, the statement that intercepts the definition will introduce a ϕ-function into the front of the current block. For the example in Fig. 2, the definition `A.1` outgoing from immediate dominator block 1 propagates into the current block 4. As for the definitions `B.1` and `C.3` that is intercepted by new definitions, the ϕ-functions for variable `B` and `C` are introduced into block 4.

Lemma 1 guarantees the dominator of a block can be visited in advance and Property 1 ensures the ϕ-functions are inserted before processing. Therefore, we can inherit the dominating definition table from immediate dominator to the current block and then correct that table by processing the ϕ-functions.

The SsaUse procedure transforms each used variable into SSA form:

1 SsaUse($s \in S, b_i \in B, E$)
2   **if** *s is ϕ-function* **then**
3   **foreach** *reaching edge* $(b_j, b_i) \in R$ **do**
4   **if** $b_j \notin s$ *and* $b_{DOM(i)}$ *dominates* b_j **then**
5   set $(b_j, \text{SSA}(v), true)$ in s
6   **else**
7   **foreach** *variable v used in* Rhs(s) **do**
8   replace usage of v by usage of Ssa(v)

What the Ssa(v) procedure returns actually is $v_{E(v)}$. But, if a definition coming from the immediate dominator to the current block is partially intercepted, a branch for it would be missing from the ϕ-function in the current block. Look at Fig. 2, definition `C.1` is partially intercepted by `C.2` and still can reach block 4. However, the ϕ-function in block 4 has no branch for `C.1` since block 4 is not block 1's IDF. To fix it, the missing branch is inserted (line 5).

The SSADEF procedure dispatches an unallocated number $C(v)$ to the defined variable v (line 4) to distinguish it from the other definitions. After that, the *def-use* pair is recorded and the definition is collected for later propagation.

1 SSADEF($s \in S, b \in B, E, V_{def}$)
2 $\quad v \leftarrow$ LHS(s)
3 $\quad E(v) \leftarrow C(v) + +$
4 $\quad$ replace v by SSA(v) as LHS(s)
5 $\quad$ **foreach** *variable v used in* RHS(s) **do**
6 $\qquad$ add def-use pair (v, b, s) into DU
7 $\quad$ add defined variable v into V_{def}

The invariants of C and E are still held after updating. $C(v)$ increases monotonously (line 3) hence it still records the maximum SSA number which is unallocated. Meantime, $E(v)$ is assigned with the new SSA number $C(v)$ thus $E(v)$ keeps the dominating definition for variable v.

3.4 Propagating Definition and Reachability

The PROPAGATE procedure introduces ϕ-functions into the direct dominance frontiers of block b for each outgoing definition (line 4). These ϕ-functions are tentatively marked $false$, which means they are not reachable yet. The PROPAGATEREACHABILITY procedure is responsible to enable these ϕ-functions.

1 PROPAGATE($b \in B, V_{def}, E$)
2 $\quad$ **foreach** *frontier edge* (b', b'') *of block* b **do**
3 $\qquad$ **foreach** $v \in V_{def}$ **do**
4 $\qquad\quad$ add branch $(b', \text{SSA}(v), false)$ into $\phi(v)$ in the front of b''

The goal of the PROPAGATEREACHABILITY procedure is to extend the reachability from the current block to its successors, to which the jump condition c is not always false. At this point, the disabled branches in the ϕ-functions that introduced by the PROPAGATE procedure are enabled.

1 PROPAGATEREACHABILITY($b \in B, E$)
2 $\quad$ **foreach** *jump edge* $(cond, b')$ *of block* b **do**
3 $\qquad$ **if** *cond is always false* **then continue**
4 $\qquad$ **foreach** $\phi(v) \in b'$ **do**
5 $\qquad\quad$ **if** $(b, v, false) \in \phi(v)$ **then**
6 $\qquad\qquad$ set branch $(b, \text{SSA}(v), true)$ in $\phi(v)$ in the front of b'
7 $\qquad$ add reaching edge (b, b') into R

Obviously, the ϕ-function is inserted when processing the dominator b. It is not enabled until the dominated block b' is processed. Here comes a question: *is it possible that the dominance frontier b'' is visited before its predecessor b' which causes the reachability information is missed?* That would not happen when b'' is a forward block since a forward block has no back edge. If b'' is a backward block, Sect. 3.5 shows how to treat the back edges as forward ones.

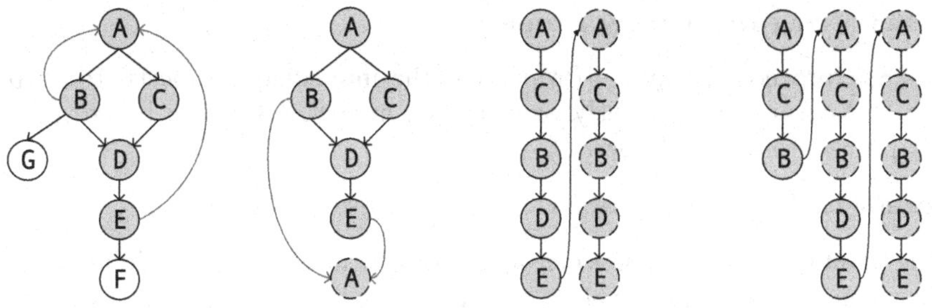

(a) Control-flow graph (b) (c) Process sequence 1 (d) Process sequence 2

Fig. 3. An example for showing how to handle the backward block and reduce the re-computations. (a) is a CFG starts with a backward block `A`, whose BRG is in blue. (b) flips the back edges by mirroring a fake block `A` (in dash circle) at the bottom of the graph so that only forward blocks remains. (c) is the process sequence of our method which does not re-compute until the end of graph. (d) is the sequence of **SAPR** [20], which re-computes once a processed node is invalidated (after `B` and `E`). (Color figure online)

3.5 Backward Block Analysis

Lemma 4 implies that for a backward block b_i, all the backward definitions that will reach it can be propagated back into itself along the IDF chain in $G_{BR}(b_i)$. Remember that for forward block, we propagate the definitions along its IDF chain. Hence, the basic idea to handle backward block is to convert its BRG into a graph which contains only forward blocks so that we can handle it with the procedures already presented in previous sections. Figure 3 gives an example.

Therefore, the ProcessBackwardBlock procedure processes the BRG of backward block b_i by taking b_i as a forward block (line 5).

1 ProcessBackwardBlock($b_i \in B, E_i, S_{invalid} = \{\}$)
2     $G' \leftarrow$ BackwardReachSubGraph(G_{cf}, b_i)
3     $L' \leftarrow$ ReversePostOrder(G')
4   **do**
5   ImSsaSparse($L', S_{invalid}, b_i$)
6   **while** $S_{invalid}(L')$ *is not empty*

The BackwardReachSubGraph procedure creates the BRG for the given backward block b_i. It scans the CFG in depth-first inversely by starting from the back edges of b_i. This process is very cheap and can be computed previously. The ImSsaSparse procedure is quite similar to the ImSsa procedure, except it only recomputes the invalid statements instead of the whole block. The implementation of ImSsaSparse can be found in Appendix A.

3.6 Satisfying Immediacy Property

The ImSsa procedure requires Property 1, the immediacy property, to output correct SSA-form code efficiently, as discussed in Sect. 3.3. It can be proved the ImSsa procedure meets Property 1.

Theorem 1. *The* ImSsa *procedure satisfies Property 1.*

The proof for this theorem is consigned to Appendix B.2. It should be emphasized that no algorithm can exactly hold Property 1 for backward block because it would require a topological sort for the cycling CFG, which is impossible. Here we say Property 1 can be held for backward block, we actually mean the property can be held at the last time of calling ImSsaSparse. Nevertheless, it is sufficient to preserve what we declaimed on the ImSsa procedure.

4 Binary Optimization Framework

The framework tries to canonicalize and eliminate the code for alleviating the impacts from optimization and obfuscation. Since the framework is driven by a pointer analysis, it is supposed to be robust to the code transformation that based on data-dependency.

4.1 Integration with Pointer Analysis

The framework is implemented in the ClientAnalyse procedure for utilizing the pointer analysis provided by the ImSsa procedure. Besides E and C, two more components are maintained: $D : V \rightarrow E$ is the table mapping variables into definition expressions and $A : E \rightarrow V$ is the table mapping address expressions into recovered variables.

Each variable usage, including the recovered one, is linked with its dominating definition (by Ssa) and then replaced by the corresponding expression (by Propagate). Thus, Canonicalize will have sufficient information to transform the expression correctly. After that, Recover recovers variables from the addresses of the memory-access instructions (such as STORE).

```
ClientAnalysis(e, E, C, D, A)
    match e with
        case VAR (v) do
            e_new ← Propagate(Ssa(v), D)
        case EXP (op, e ···) do
            for e' ∈ e ··· do
                ClientAnalysis(e', E, C, D, A)
            e_new ← Canonicalize(op, e ···)
            if op is STORE or LOAD then
                v ← Recover(e_new, E, C, A)
                e_new ← Propagate(Ssa(v), D)
    replace e by e_new
```

PROPAGATE substitutes the given SSA-form usage with the usage's definition expression. It complements the original expression with sufficient information to support more precise and thorough canonicalization. The validation of propagation can be guaranteed because the IMSSA procedure ensures each definition that reaches the usage is analyzed and recorded in advance. Besides, the expressions recorded in D consist of only free variables and constants. That is, after propagation, the statement is transformed into such a concrete form that no element in it can be propagated anymore. No doubt we can restrict the maximum expression size to avoid too costly reduction.

4.2 Canonicalization and Elimination

The canonicalization transforms the equivalent but dissimilar expressions into a unified form to preserve the similarity. It refines the pointer analysis in turn by providing normalized expressions. Moreover, it allows many cascade elimination techniques, such as common sub-expression elimination, sparse conditional constant elimination, etc. After that, the elimination runs to remove the junk code which can never be executed or has not effect the program results but will affect the similarity evaluation heavily.

For *Canonicalization*, we first collect about 44 basic algebraic reduction rules that applied in compilers and `BAP` [2]. Considering that most obfuscation of obfuscator `O-LLVM` works by complicating expressions or constants, such as substitution and bogus control-flow, obfuscation patterns are reversed into nearly 10 reduction rules. Though some obfuscation patterns may be overwritten or broken by the subsequent optimization, most of them still can be captured by patching a few rules. To match more patterns, we also enumerate a set of reduction rules with the size of 40 for the small size expressions. Some of these rules are like: $(x\&c) \oplus (x|c) \Leftrightarrow x \oplus c$, $(x\&c)|(x\&!c) \Leftrightarrow x$, etc. Once canonicalized, the unified code will reveal the reachability that depends on the obfuscated constant expression. That is the reason why the framework could be resilient to these obfuscations that based on data-dependency.

To *eliminate* the junk code, we follow Cytron et al.'s method [5] by substituting their flow analysis with ours. Initially, all the variables, except the ones assumed to affect program outputs, are tentatively marked *dead*. Next, the *live* marks propagate along the *def-use* chain and the reachability resulted from the pointer analysis phase. That is, the variables on which the *live* ones depend are also marked *live*. At last, the statements that define *dead* variables can be eliminated safely. It is clear that the correctness of the elimination relies on the validation of the information on *def-use* chain and reachability, which we have discussed in Sect. 3.3 and Sect. 3.4.

Here we only focus on two types of dead code, which are enough to show good results in our experiments. Nevertheless, our framework can be extended to handle other types by appending other kinds of initial *live* marks. We can also propagate the dependency across functions to deal with global dead code.

5 Implementation

Our designs are implemented on the analysis platform of BAP, because of BAP's rich suite of utilities and libraries for binary analysis.

We lift the compiled binaries into BIR (BAP's Intermediate Representation) code by using BAP. Since ImOpt will be compared to compiler LLVM's optimizer opt in Sect. 6.3, we also lift binaries to LLVM-IR code by virtue of the prominent framework Mcsema[1]. After optimized by opt, the LLVM-IR statements are then re-compiled into binaries with Mcsema for being converted into BIR code later. Our designs are language and platform-independent. In this paper, we implement our binary optimization framework into a prototype called ImOpt as a plugin on the analysis platform of BAP.

5.1 Evaluation Module

It is obvious ImOpt optimizes code but does not contain a evaluation module. ImOpt is supposed to be used as a preprocess module. To evaluate the effect of ImOpt, we reproduce *Asm2Vec* [9], a state-of-the-art of binary similarity evaluation, as the backend for the evaluation. *Asm2Vec* builds an NLP model to generate features for binary functions. Considering that *Asm2Vec* is not designed to process the BIR code, we decompose long BIR statements into a sequence of triples along the abstract syntax tree so that the model can digest.

5.2 Pointer-Based Obfuscation

We also construct a pointer-based obfuscation to demonstrate the vulnerability of optimizing binary code with opt. The obfuscation is similar to Fig. 1b but is implemented for LLVM-IR code. We implement it as an LLVM pass to obfuscate the query function. After obfuscation, the query function works just as before obfuscation. However, once someone wants to utilize opt to re-optimize the obfuscated procedure, this function will be processed such that only the junk code remains. opt's vulnerability shown in Fig. 1b is caused by early optimization, which runs before processing the IR code. The confound example will not happen when using opt to optimize the LLVM-IR code lifted by Mcsema. But, if we recover variables without pointer analysis to boost the performance of opt, that vulnerability will be reproduced.

6 Experiment

6.1 Experiment Settings

The experiments are performed in the system of Ubuntu 16.04 which is running on an Intel Core i7 @ 3.20 GHz CPU with 16G DDR4-RAM.

[1] https://www.trailofbits.com/research-and-development/mcsema/.

Table 1. Comparison across optimization. @K represents the Top-K accuracy.

Programs	IMOPT + Simulated *Asm2Vec*						Simulated *Asm2Vec*					
	O0 vs O3			O2 vs O3			O0 vs O3			O2 vs O3		
	@1	@5	@10	@1	@5	@10	@1	@5	@10	@1	@5	@10
busybox	.606	.703	.737	.854	.891	.905	.408	.505	.546	.816	.868	.882
sqlite	.496	.622	.662	.820	.875	.893	.111	.145	.166	.797	.861	.875
lua	.523	.623	.674	.800	.857	.890	.283	.345	.383	.809	.869	.902
putty	.542	.657	.709	.860	.896	.912	.389	.484	.522	.874	.913	.921
curl	.700	.812	.848	.927	.956	.964	.567	.658	.693	.948	.962	.967
openssl	.710	.787	.810	.975	.982	.985	.457	.539	.573	.926	.938	.942
Average	.596	.701	.740	.872	.909	.925	.369	.446	.480	.862	.902	.915

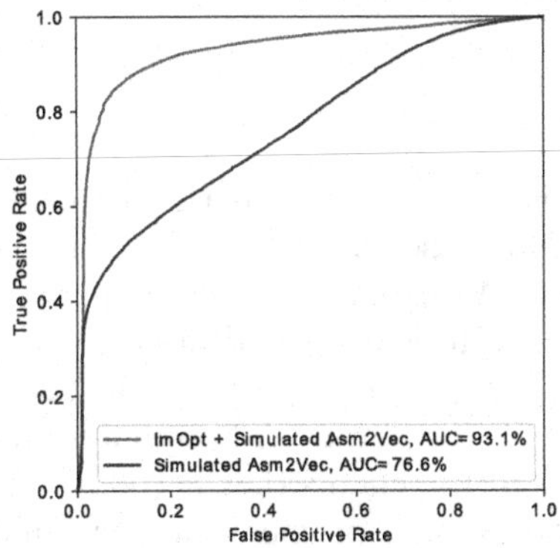

(a) O0 vs O3

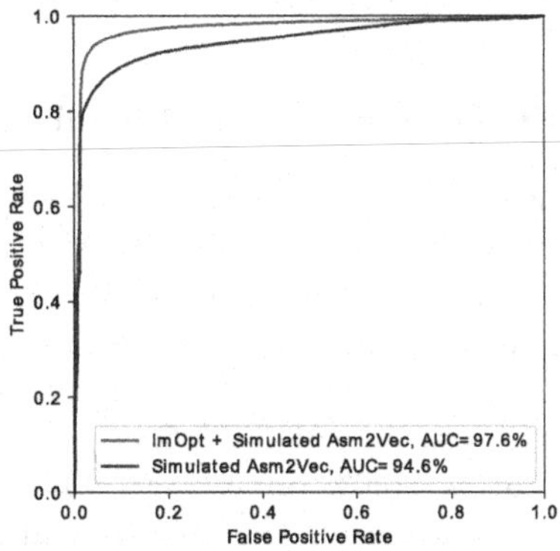

(b) O2 vs O3

Fig. 4. The ROC of the evaluation across optimization levels.

We build the dataset from six real-world open-source programs: *busybox*, *curl*, *lua*, *putty*, *openssl* and *sqlite*. Binary objects are built by compiler `GCC 5.4` from these open-source programs with different optimizations. To evaluate the robustness against the obfuscation, some of these programs are also compiled with the widely-used obfuscator `O-LLVM 4.0`[2]. Four kinds of obfuscation provided by `O-LLVM` are applied: *substitution* (`SUB`), *bogus control-flow* (`BCF`), *flatten* (`FLAT`) and *basic block split* (`BBS`). `BBS` therein is frequently used to improve `FLAT`, so we combine `FLAT` with `BBS` as one obfuscation. These obfuscation are configured to be run three times in our experiment.

Through the experiments, we are going to answer three research questions. We have argued these propositions, but they need to be supported empirically.

Q1: How can IMOPT boost evaluation against different optimization levels?
Q2: What is the performance of IMOPT against different obfuscation?
Q3: Is the SSA transforming of IMOPT more efficient than the state-of-the-art?

[2] https://github.com/obfuscator-llvm/obfuscator/tree/llvm-4.0.

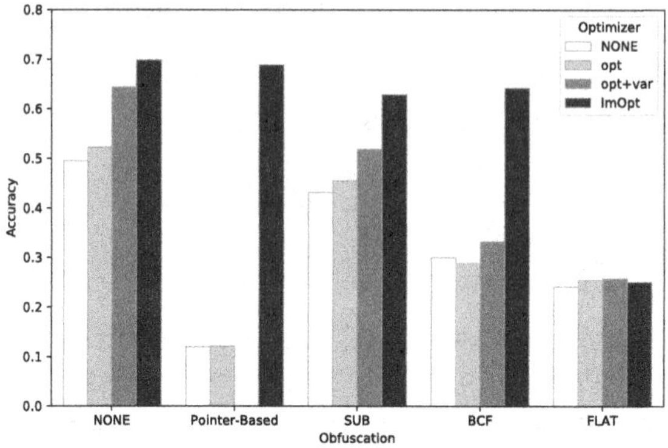

Fig. 5. Comparing ImOpt with optimizer `opt` against different obfuscation.

6.2 Evaluation Across Optimization Levels

To answer *Q1*, we simulate *Asm2Vec* [9] with no optimizer as the baseline and compare it to one equipped with ImOpt. Only case O0-vs-O3 and case O2-vs-O3 are focused in our experiment. The original SSA algorithm is not compared here because it is implied in optimizer `opt`, which will be evaluated in next section.

Table 1 shows that ImOpt improves the Top-1/5/10 accuracy in O0-vs-O3 case by no less than 13.3%/15.4%/15.5% for each program and promotes around 22.7%/25.5%/26.0% on average. Though *Asm2Vec* claims that it could capture semantic information, the re-optimization still helps to promote the performance. It is noted that the simulated *Asm2Vec* did not perform as well as the original model under case O0-vs-O3, and this is accordant with the findings in [16].

As for case O2-vs-O3, these two models are comparable in the Top-K accuracy. The reason why re-optimization has little influence in case O2-vs-O3 is probably because the optimization options enabled only by O3 are mostly about code structure, such as loop optimization, which is rarely related with the function of ImOpt. One might have noticed that ImOpt degraded in some programs, one big reason is the optimization magnified the impact of compilation-related constants for short functions.

Figure 4a shows a significant improvement brought by ImOpt in case O0-vs-O3 from another perspective. As for case O2-vs-O3, though simulated *Asm2Vec* has comparable Top-K accuracy to the one with ImOpt, the ROC in Fig. 4b says the latter will performs slightly better than the former after a very small false positive rate (less than 2.0%). Also, the good performance indicates that ImOpt has very little negative effect on similarity evaluation.

6.3 Comparison Against Obfuscation

To address *Q2*, we evaluate ImOpt on the O0-vs-O3 task against four kinds of obfuscation: `SUB`, `BCF`, `FLAT` (with `BBS`) and the pointer-based obfuscation

Table 2. Overhead on maintaining dominating definition. The % column indicates the proportion of the overhead in the cost of IMSSA.

Programs		*Overhead on Dominating Definition (ms)*					
Name	*Functions*	%	*SDA*		IMSSA		
			FETCH	UPDATE	FETCH	UPDATE	CLONE
lua	827	45.5%	658	87	46	22	24
putty	1,050	81.1%	4,795	231	117	50	176
curl	1,036	80.1%	20,159	576	207	95	771
sqlite	3,133	81.0%	17,608	1,177	594	259	586
busybox	4,279	64.3%	14,719	1,332	590	281	552
openssl	9,812	75.3%	64,021	2,201	1,175	508	2,053
Total	20,137		121,960	5,604	2,729	1,215	4,162

(Sect. 5.2). In this experiment, simulated *Asm2Vec* without optimizer continues to be used as the baseline. To explain the motivation for designing IMOPT, we further equip simulated *Asm2Vec* with three optimizers respectively: 1) the `LLVM` compiler's optimizer `opt`; 2) `opt` with variable recovery; and 3) IMOPT. The second control group which combines `opt` with variable recovery is constructed to live up to the potential of optimizer `opt`.

Figure 5 shows the evaluation results. IMOPT performs stably against `SUB`, `BCF` and pointer-based obfuscation, but has a bad performance against `FLAT`. As for `FLAT` which deals with structure instead of data, IMOPT has no special effect on it. The evaluation of IMOPT against `SUB` has a few more false positives. It is because compiler's early optimization may reform a few obfuscation rules so that the rules become too complex to be matched. IMOPT shows resilience to the obfuscation that based on data dependency.

Optimizing by optimizer `opt` without variable recovery has the similar performance as the `NONE` case. The main reason is that without variable recovery most optimizations of `opt` have been intercepted. By enabling variable recovery, optimizer `opt` archives comparable accuracy with IMOPT in the non-obfuscation case. Unfortunately, the performance of optimizer `opt` with variable recovery degraded into 0% when the pointer-based obfuscation is applied. This confirms that it is insufficient to optimize binary code by using optimizer `opt` directly.

6.4 Efficiency

Regarding *Q3*, we demonstrate the efficiency of the IMSSA procedure of IMOPT by comparing it to *SDA* [20], which is the most efficient approach that supports both pointer and reachability analysis. *SDA* leverages Quadtree to speed up the analysis. we mainly compare the overhead on dominating definition, since it takes up a large proportion of SSA transforming, as shown in Table 2. The experiment result shows that IMOPT costs only 8.1 s to maintain the information over 20,137 functions while *SDA* need 127.5 s.

The copies of table E between blocks may be the implicit cost of ImSsa. The Clone column in Table 2 shows that this extra overhead is very small and has a approximately linear relationship with the Update of ImOpt. For the most costly Fetch operation, ImSsa outperforms *SDA* by a factor of 14.3x to 97.4x as the average number of statements grows. All in all, ImSsa has a 15.7x speed-up on average for maintaining the information about dominating definition.

7 Related Works

Binary Code Similarity Evaluation. Most traditional static approaches deal with the statistic features based on syntax sequence or structure, such as the frequency of operators and operands [18], the categories of instructions [11,12], and the editing distance of code fragments [8]. Some methods [1,25] extract features based on the isomorphism of CFG. These statistic features rarely capture semantic information so that none of them can strip the junk code. Dynamic approaches based on the fact that the compilation does not change the execution result. To model the similarity, *Zsh* [6] and *BinHunt* [13] leverage theorem prover to verify the logical equivalence; and *Blanket* [10], *CACompare* [17], *BinGo* [3] analyse the input-output mappings of binary functions after emulating the executions. However, despite several trite questions existing in code emulation, they are not resilient to the elimination related optimizations. It is because most of them will split the binary function into smaller pieces (blocks [13], strands [6], tracelets [3], graphlets [28], blankets [10], etc.), where the junk code still cannot be distinguished. It is worth noting that *GitZ* [7] leverages `opt` to re-optimize `IR` code, but there are several defects for *GitZ*: 1) it only enables options `-early-cse` and `-instcombine`, which is not able to cover most of the optimization effect as ours; 2) it optimizes code strands; 3) as we have shown, compiler optimizer is bad at optimizing the reversed IR code.

Pointer Analysis. Flow-sensitive pointer analysis is demanded for a complete and precise data-flow analysis. Traditional iterative data-flow analysis [4] methods are too slow to be practical because of too much unnecessary re-analyzing. Pre-analysis based methods [14,23] sacrifice analysis precision for efficiency by constructing a sound approximation of the memory accessing addresses at the previous auxiliary analysis stage. To accelerate the analysis while retaining the precision, Tok et al. [26] and Madsen et al. [20] propose to compute the *def-use* information on the fly. But, the former is still an iterative method. And the latter costs $O\left(\sqrt{n}\right)$ to find dominating definitions, which only needs $O(1)$ with ours.

8 Conclusion

In this paper, we present a binary optimization framework to re-optimize lifted code for alleviating the affects by optimization. We have shown that optimizer `opt` is insufficient to re-optimize the lifted binary code. To support our framework with an efficient and precise pointer analysis, we proposed the immediate SSA transforming algorithm. The way we maintain the dominating definitions is faster than Madsen et al.'s method [20] by a factor of 15.7x on average. Furthermore, we implement our designs into a prototype called IMOPT. The experiments confirm that IMOPT can be used to boost the performance of a state-of-the-art [9] with an average accuracy improvement of 22.7%. It is also demonstrated that IMOPT is robust to data-dependency-based obfuscation, such as `O-LLVM` obfuscation and pointer-based obfuscation.

Of course, the proposed approach is not without limitations. The algebraic reduction rules equipped by IMOPT need to be manually constructed since the problem of algebraic reduction is NP-hard. That is, IMOPT cannot guarantee bug-free analysis though it is difficult to be defrauded. However, it is convenient to fix IMOPT by extending the reduction rule set.

Acknowledgments. This work is supported by National Natural Science Foundation of China (No. 61572469).

A Appendix 1: Sparse Analysis for Backward Block

The IMSSASPARSE procedure is quite similar to the IMSSA procedure, except it sparsely recomputes the visited but invalid block (line 4) to avoid unnecessary re-computation:

1 IMSSASPARSE$(L, G, S_{invalid}, b_{except})$
2   **for** $b_i \in L$ **do**
3   **if** b_i *is* `visited` **then**
4   SPARSERECOMPUTE$(b_i, E_i, S_{invalid})$
5   **else if** b_i *is forward block or* $b_i == b_{except}$ **then**
6   PROCESSFORWARDBLOCK(b_i, E_i)
7   **else** PROCESSBACKWARDBLOCK$(G, b_i, E_i, S_{invalid})$

The SPARSERECOMPUTE procedure is similar to the PROCESSSTATEMENT procedure, except the former only process invalid statements. Another difference is SPARSERECOMPUTE needs to propagate invalidation through the *def-use* chain.

B Appendix 2: Supplementary Proofs

B.1 Appended Proofs for Section 2.3

Proof (Proof of Lemma 2). Let us assume that there is a backward definition $d \in b'$ reaching the forward block b_i, and all the blocks except b_i on any path $p : b' \rightarrow b_i$ are in the SSA form. Then, there should be a back edge $e : b_j \rightarrow b_k$ on the path $p : b' \rightarrow b_i$. Note the block b_k can not be the block b_i since a forward block has no back edge pointing to itself. So, the block b_k should be in SSA form. However, in SSA-form code, a back edge will introduce ϕ-functions into the block b_k that is at the target end of the edge for every definition that passes through the block b_k. These ϕ-functions will prevent definition d reaching to the successor block b_i, which contradicts the assumption. Hence, a forward block has only forward definitions reaching itself in SSA-form code.

Proof (Proof of Lemma 3). According to Condition 1, the first $i-1$ blocks have already been in SSA form. Then following Lemma 2, only forward definitions can reach block b_i. Recall that if a forward definition d can reach a block b_i, then there is no back edge on the reaching path $p : d \overrightarrow{rc}\, b_i$, which means that the block b' which defines d is ahead of block b_i in reverse post-order.

Proof (Proof of Lemma 4). According to the definition of the BRG, it is trivial that $b' \in G_{BR}(b_i)$. If there is a path $p : b' \rightarrow b_i$, then b_i is either an IDF of block b' or dominated by an IDF. Besides, there should be a back edge on path p since $d \overleftarrow{rc}\, b_i$. If the back edge is in the middle of path p, then the ϕ-function introduced by it will intercept definition d. That contradicts with the fact that definition d can reach block b. Hence, the back edge can only be the last edge on path p. It means the backward block b cannot be dominated by any block on path p. Therefore, the backward block b is an IDF of block b'.

B.2 Appended Proofs for Section 3.6

To prove Lemma 1, we first show that given Condition 1 the property is satisfied for the i-th block, no matter the block is a forward or backward one.

Lemma 5. *Given Condition 1, if the i-th block b_i is a forward block, then the* ImSsa *procedure also satisfies the invariants in Property 1 for block b_i.*

Proof (Proof Sketch of Lemma 5). Based on Lemma 3, all the reachable definitions can be visited before visiting the current block. *Global invariant:* the variables with dominating definition are linked by $E_{DOM(i)}$ directly, whose validation is confirmed by Condition 1; the other variables, which have multiple reachable definitions, have the ϕ-functions being inserted in the front of the block when processing the blocks that define these definitions; Hence, the global invariant is held. *Local invariant:* E and C are well-defined at the beginning of local analysis since the global invariant is held; moreover, their invariants are preserved during the running of ProcessForwardBlock procedure. In conclusion, the local invariant is held.

Lemma 6. *If the CFG contains only forward block, the* ImSsa *procedure satisfies the invariants in Property 1 for all the blocks in that graph.*

Proof. This lemma simply follows Lemma 5.

Lemma 7. *Given Condition 1, if the i-th block b_i is a backward block, the* ImSsa *procedure also satisfies the invariants in Property 1. Furthermore, for each block in the BRG, the invariants in Property 1 are also satisfied.*

Proof (Proof Sketch of Lemma 7). Assume that all blocks except b_i in the BRG are forward blocks. According to Lemma 4 and Lemma 6, the ImSsa procedure can propagate all the backward definitions back into block b_i along with the IDF chain by processing $G_{BR}(b_i)$. Additionally, the following SparseRecompute procedure will reveal any new definitions caused by the re-computation and propagate them back, too. Hence, the *global invariant* is held. Recall that the invariants of E and C are also maintained during the sparse re-computations, therefore, the *local invariant* is also held. Furthermore, since the BRG $G_{BR}(b_i)$ contains only forward blocks, according to Lemma 6, these invariants are also satisfied for each block in the BRG. As for the case where the BRG contains backward blocks, it can be deductively reasoned from the previous case.

It follows that the ImSsa procedure satisfies Property 1:

Proof (Proof Sketch of Lemma 1). It is reasonable to assume a forward entry block since we can always insert an empty forward block at the beginning of the CFG. Therefore, the first block satisfies the global invariant because of no reachable definition. Meanwhile, the local invariant is also satisfied according to Lemma 5. That is, the ImSsa procedure satisfies the invariants for the first block. Then the conclusion can be deductively reasoned from Lemma 5 and Lemma 7.

References

1. Bourquin, M., King, A., Robbins, E.: BinSlayer: accurate comparison of binary executables. In: Proceedings of the 2nd ACM SIGPLAN Program Protection and Reverse Engineering Workshop (2013)
2. Brumley, D., Jager, I., Avgerinos, T., Schwartz, E.J.: BAP: a binary analysis platform. In: Gopalakrishnan, G., Qadeer, S. (eds.) CAV 2011. LNCS, vol. 6806, pp. 463–469. Springer, Heidelberg (2011). https://doi.org/10.1007/978-3-642-22110-1_37
3. Chandramohan, M., Xue, Y., Xu, Z., Liu, Y., Cho, C.Y., Tan, H.B.K.: Bingo: cross-architecture cross-OS binary search. In: International Symposium on Foundations of Software Engineering, pp. 678–689. ACM (2016)
4. Chase, D.R., Wegman, M., Zadeck, F.K.: Analysis of pointers and structures. ACM SIGPLAN Not. **25**(6), 296–310 (1990)
5. Cytron, R., Ferrante, J., Rosen, B.K., Wegman, M.N., Zadeck, F.K.: Efficiently computing static single assignment form and the control dependence graph. ACM Trans. Program. Lang. Syst. **13**(4), 451–490 (1991). https://doi.org/10.1145/115372.115320
6. David, Y., Partush, N., Yahav, E.: Statistical similarity of binaries. ACM SIGPLAN Not. **51**, 266–280 (2016)
7. David, Y., Partush, N., Yahav, E.: Similarity of binaries through re-optimization. ACM SIGPLAN Not. **52**, 79–94 (2017). https://doi.org/10.1145/3140587.3062387
8. David, Y., Yahav, E.: Tracelet-based code search in executables. ACM SIGPLAN Not. **49**, 349–360 (2014)
9. Ding, S.H., Fung, B.C., Charland, P.: Asm2Vec: boosting static representation robustness for binary clone search against code obfuscation and compiler optimization. In: 2019 IEEE Symposium on Security and Privacy (SP). IEEE (2019)

10. Egele, M., Woo, M., Chapman, P., Brumley, D.: Blanket execution: dynamic similarity testing for program binaries and components. In: USENIX Security Symposium, pp. 303–317 (2014)
11. Eschweiler, S., Yakdan, K., Gerhards-Padilla, E.: discovRE: efficient cross-architecture identification of bugs in binary code. In: NDSS (2016)
12. Feng, Q., Zhou, R., Xu, C., Cheng, Y., Testa, B., Yin, H.: Scalable graph-based bug search for firmware images. In: Proceedings of the 2016 ACM SIGSAC Conference on Computer and Communications Security, pp. 480–491 (2016)
13. Gao, D., Reiter, M.K., Song, D.: BinHunt: automatically finding semantic differences in binary programs. In: Chen, L., Ryan, M.D., Wang, G. (eds.) ICICS 2008. LNCS, vol. 5308, pp. 238–255. Springer, Heidelberg (2008). https://doi.org/10.1007/978-3-540-88625-9_16
14. Hardekopf, B., Lin, C.: Flow-sensitive pointer analysis for millions of lines of code. In: International Symposium on Code Generation and Optimization, pp. 289–298. IEEE (2011)
15. Hasti, R., Horwitz, S.: Using static single assignment form to improve flow-insensitive pointer analysis. ACM SIGPLAN Not. **33**(5), 97–105 (1998)
16. Hu, Y., Wang, H., Zhang, Y., Li, B., Gu, D.: A semantics-based hybrid approach on binary code similarity comparison. IEEE Trans. Softw. Eng. (2019)
17. Hu, Y., Zhang, Y., Li, J., Gu, D.: Binary code clone detection across architectures and compiling configurations. In: Proceedings of the 25th International Conference on Program Comprehension, pp. 88–98. IEEE Press (2017)
18. Jang, J., Woo, M., Brumley, D.: Towards automatic software lineage inference. In: Proceedings of the 22nd USENIX Conference on Security (2013)
19. Lindorfer, M., Di Federico, A., Maggi, F., Comparetti, P.M., Zanero, S.: Lines of malicious code: insights into the malicious software industry. In: Proceedings of the 28th Annual Computer Security Applications Conference, pp. 349–358 (2012)
20. Madsen, M., Møller, A.: Sparse dataflow analysis with pointers and reachability. In: Müller-Olm, M., Seidl, H. (eds.) SAS 2014. LNCS, vol. 8723, pp. 201–218. Springer, Cham (2014). https://doi.org/10.1007/978-3-319-10936-7_13
21. Ming, J., Xu, D., Jiang, Y., Wu, D.: BinSim: trace-based semantic binary diffing via system call sliced segment equivalence checking. In: USENIX Security Symposium, pp. 253–270 (2017)
22. Ming, J., Xu, D., Wu, D.: Memoized semantics-based binary diffing with application to malware lineage inference. In: Federrath, H., Gollmann, D. (eds.) SEC 2015. IAICT, vol. 455, pp. 416–430. Springer, Cham (2015). https://doi.org/10.1007/978-3-319-18467-8_28
23. Oh, H., Heo, K., Lee, W., Lee, W., Yi, K.: Design and implementation of sparse global analyses for C-like languages. **47**(6), 229–238 (2012)
24. Pewny, J., Schuster, F., Bernhard, L., Holz, T., Rossow, C.: Leveraging semantic signatures for bug search in binary programs. In: Proceedings of the 30th Annual Computer Security Applications Conference, pp. 406–415 (2014)
25. Shirani, P., Wang, L., Debbabi, M.: BinShape: scalable and robust binary library function identification using function shape. In: Polychronakis, M., Meier, M. (eds.) DIMVA 2017. LNCS, vol. 10327, pp. 301–324. Springer, Cham (2017). https://doi.org/10.1007/978-3-319-60876-1_14
26. Tok, T.B., Guyer, S.Z., Lin, C.: Efficient flow-sensitive interprocedural data-flow analysis in the presence of pointers. In: Mycroft, A., Zeller, A. (eds.) CC 2006. LNCS, vol. 3923, pp. 17–31. Springer, Heidelberg (2006). https://doi.org/10.1007/11688839_3

27. Walenstein, A., Lakhotia, A.: The software similarity problem in malware analysis. In: Dagstuhl Seminar Proceedings. Schloss Dagstuhl-Leibniz-Zentrum für Informatik (2007)
28. Wei, M.K., Mycroft, A., Anderson, R.: Rendezvous: a search engine for binary code. In: 2013 10th IEEE Working Conference on Mining Software Repositories (MSR) (2013)

HART: Hardware-Assisted Kernel Module Tracing on Arm

Yunlan Du[1], Zhenyu Ning[2], Jun Xu[3], Zhilong Wang[4], Yueh-Hsun Lin[5], Fengwei Zhang[2(✉)], Xinyu Xing[4], and Bing Mao[1]

[1] Nanjing University, Nanjing, China
[2] Southern University of Science and Technology, Shenzhen, China
zhangfw@sustech.edu.cn
[3] Stevens Institute of Technology, Hoboken, USA
[4] Pennsylvania State University, State College, USA
[5] JD Silicon Valley R&D Center, Mountain View, USA

Abstract. While the usage of kernel modules has become more prevalent from mobile to IoT devices, it poses an increased threat to computer systems since the modules enjoy high privileges as the main kernel but lack the matching robustness and security. In this work, we propose HART, a modular and dynamic tracing framework enabled by the Embedded Trace Macrocell (ETM) debugging feature in Arm processors. Powered by even the minimum supports of ETM, HART can trace binary-only modules without any modification to the main kernel efficiently, and plug and play on any module at any time. Besides, HART provides convenient interfaces for users to further build tracing-based security solutions, such as the modular AddressSanitizer HASAN we demonstrated. Our evaluation shows that HART and HASAN incur the average overhead of **5%** and **6%** on **6** widely-used benchmarks, and HASAN detects all vulnerabilities in various types, proving their efficiency and effectiveness.

Keywords: Kernel module · Dynamic tracing · ETM · Arm

1 Introduction

To enhance the extensibility and maintainability of the kernel, Linux allows third parties to design their own functions (*e.g.*, hardware drivers) to access kernel with loadable kernel modules. Unlike the main kernels developed by experienced experts with high code quality, the kernel modules developed by third parties lack the code correctness and testing rigorousness that happens with the core, resulting in more security flaws lying in the kernel modules. Some kernel vulnerability analyses [25,40] show that CVE patches to the Linux repository involving kernel drivers comprise roughly 19% of commits from 2005 to 2017. The conditions only worsen in complex ecosystems like Android and smart IoT devices. In

Y. Du and Z. Ning—These authors contributed equally to this work.

L. Chen et al. (Eds.): ESORICS 2020, LNCS 12308, pp. 316–337, 2020.
https://doi.org/10.1007/978-3-030-58951-6_16

2017, a study on vulnerabilities in the Android ecosystem [39] shows that 41% of 660 collected bugs came from kernel components, most of which were device drivers.

To mitigate the threats of vulnerable kernel modules, the literature brings three categories of solutions. As summarized in Table 1, all these solutions carry conditions that limit their use in practice. The first category of solution is to detect illegal memory access [5,6,11,16] at run-time, aiming to achieve memory protection in the kernel. As such solutions require to instrument memory accesses and perform execution-time validation, they need source code and incur high performance overhead. These properties limit solutions in this category for only debugging and testing. The second category of approaches explores to ensure the integrity of the kernel [23,26,53,60,64], including control flow, data flow and code integrity. However, these strategies introduce significant computation overhead and often require extensive modification to the main kernel. The last category of solutions isolates the modules from the core components [22,24,31, 48,51,54,58,61,65], so as to confine the potentially corrupted drivers running out of the kernel. By design, the isolation often requires source code and comes with significant instrumentation to the kernel and the driver.

Table 1. Comparison between existing kernel protection works and ours. *Non-intrusive* represents whether the tools have system-level modification of kernel, user space libraries, etc. (✓ = yes, ✗ = no, ✱ = partially supported).

Approach category	Binary-support	Non-intrusive	Low overhead	Representative works (in the order of time)
Memory Debugger	✗	✗	✗	Slub_debug [5], Kmemleak [16], Kmemcheck [11], KASAN [6]
Integrity Protection	✗	✗	✱	KOP [23], HyperSafe [60], HUKO [64], KCoFI [26], DFI for kernel [53]
Kernel Isolation	✗	✗	✱	Nooks [54], SUD [24], Livewire [22], SafeDrive [65], SecVisor [51]
Our method	✓	✓	✓	HASAN

To overcome these limitations, we propose HART, a high performance tracing framework on Arm. The motivation of HART is to dynamically monitor the execution of the third-party Linux kernel modules even without module source code. HART utilizes the ETM debugging component [9] for light-weight tracing of both control flow and data access. Combing hardware configurations and software hooks, HART further restricts the tracing to selectively work on specific module(s). With the support of the open interfaces provided by HART, various security solutions against module vulnerabilities can be established. We then demonstrate a modular AddressSanitizer named HASAN. Compared with the previous solutions, our approach is much less intrusive as it requires zero modification to other kernel components, and is fully compatible with any commercial kernel module since it requires no access to any source code. Besides, our empirical evaluation shows that HASAN can detect the occurrence of memory corruptions in binary-only modules with a negligible performance overhead.

The design of HART overcomes several barriers. First, for generality, we need to support ETM with the minimum configuration, particularly the size of Embedded Trace Buffer (ETB) that stores the trace. Take Freescale i.MX53 Quick Start Board as an example. Its ETB has a size of 4 KB and can get fully occupied very frequently. The second barrier is that many implementations of ETB raise no interrupt when the buffer is full, thereby the trace can be frequently overwritten. To tackle the above two challenges and retrieve the entire trace successfully, we leverage the Performance Monitoring Unit (PMU) to throw an interrupt after the CPU executes a certain number of instructions, ensuring that the trace size will never exceed the capacity of ETB. Finally, to optimize the performance of HART, we design an elastic scheduling to assure the decoding thread of rapid parsing while avoiding the waste of CPU cycles. Under this scheduling, the thread will occupy the CPU longer while the trace grows faster. Otherwise, the thread will yield the control of CPU more frequently.

Our main contributions are summarized as follows.

- We design HART, an ETM-assisted tracing framework for kernel modules. With the minimal hardware requirements of ETM, it achieves low intrusiveness, high efficiency, and binary-compatibility for kernel module protections.
- We implement HART with Freescale i.MX53 Quick Start Board running Linux 32-bit systems. Our evaluation with **6** widely used module benchmarks shows that HART incurs an average overhead of **5%**.
- We build HASAN on the top of HART to detect memory vulnerabilities triggered in the target module(s). We evaluate HASAN on **6** existing vulnerabilities ported from real-world CVE cases and the above benchmarks. All the vulnerabilities are detected with an average overhead of only **6%**.

2 Background and Problem Scope

In this section, we first introduce the background of kernel module and ETM, and then define our problem scope.

2.1 Loadable Kernel Module

In practice, the third-party modules are mostly developed as loadable kernel modules (LKM). An LKM is an object file that contains code to extend the main kernel, such as building support for new hardware, installing new file systems, etc. For the ease of practical use, an LKM can be loaded at anytime on-demand and unloaded once the functionality is no longer required. To support accesses

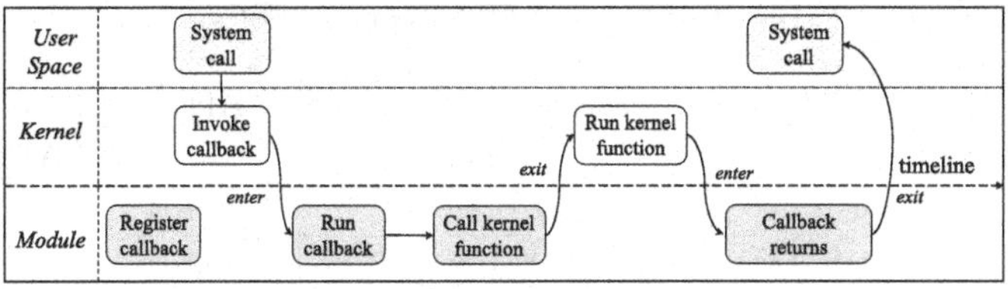

Fig. 1. Interactions between an LKM and the users space as well as kernel space.

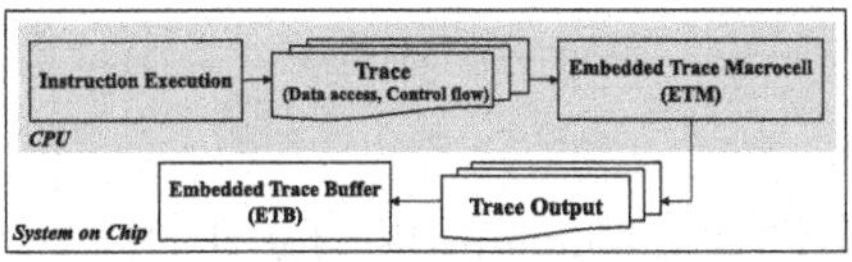

Fig. 2. A general model of ETM.

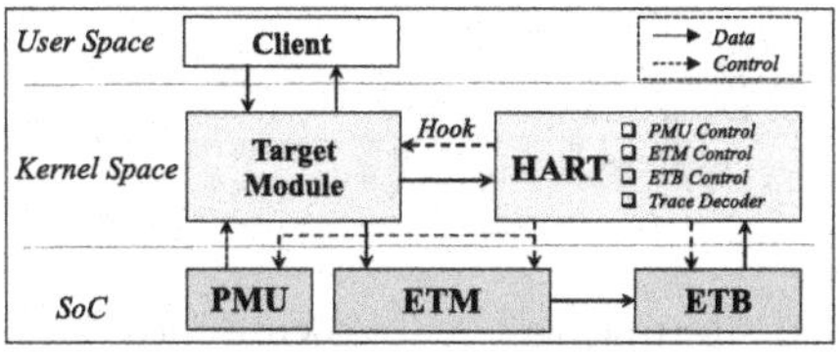

Fig. 3. Architecture of HART.

from the user space, LKM usually registers a group of its functions as callbacks to system calls in the kernel (*e.g.,* `ioctl`). In Fig. 1, a client (*e.g.,* user space application) can execute those callbacks via the system calls. Sitting in the kernel space, an LKM can also freely invoke functions exported by the main kernel.

2.2 Embedded Tracing Microcell

ETM is a hardware debugging feature in Arm processors since two decades ago. It works by tracing the control flow and memory accesses of a software's execution. Being aided by hardware, ETM incurs less than 0.1% performance overhead [43], barely decelerating the normal execution. While ETMs on different chips have varied properties, it generally follows the model in Fig. 2. When CPU executes instructions, ETM collects information about control flows and memory accesses. Such information can be compressed and stored into an on-chip memory called Embedded Trace Buffer (ETB), which is accessible through mapped I/O or debugging interfaces such as JTAG or Serial Wire interface.

Trace generated by ETM follows standard formats. A `Sync` packet is produced at the start of execution, containing information about the entry point and the process ID (`CID`); a `P-header` packet indicates the number of sequential instructions and not-taken instructions executed since the last packet; a `Data` packet records the memory address to be accessed by the instruction. We can fully reconstruct the control flow and memory accesses by decoding such packets.

2.3 Problem Scope

We consider the generality of our work from two aspects. On the one hand, the framework should have compatibility with the majority (if not all) of ETM. We investigate the ETM features on several early and low-end SoCs in Table 2,

Table 2. ETM and ETB on early SoCs.

SoC	Devices	ETB size	ETM feature	
			DAT[a]	ARF[b]
Qualcomm Snapdragon 200 [18]	Xperia E1, Moto E	≥4 KB	×	✓
Samsung Exynos 3110 [17]	Galaxy S, Nexus S	≥4 KB	✓	✓
Apple A4 [7]	iPhone 4	≥4 KB	✓	✓
Huawei Kirin 920 [36]	Mate 7, Honor 6	≥4 KB	×	✓
NXP i.MX53 [49]	iMX53 Quick Start Board	4 KB	✓	✓
Arm Juno [4]	Juno r0 Board	64 KB	×	✓

a: Data Address Trace
b: Address Range Filter

and summarize the following most preliminary ETM configuration for HART to work with: ① ETM supports tracing of both control flow and memory accesses. ② ETM supports filtering by address range. ③ ETB is available for trace storage, yet with a limited buffer size - 4 KB. ④ ETB raises no interrupts when it gets full.

On the other hand, the framework should not impose restrictions on the target modules. We then accordingly make a minimal group of assumptions about the target module: ① Source code of the target module is unavailable. ② The target module and other kernel components shall be intact without any modification. ③ The target module is not intentionally malicious.

3 HART Design and Implementation

In this section, we concentrate on HART, our selective tracing framework for kernel modules. We first give an overview of HART in Fig. 3. HART is designed as a standalone kernel driver. At the high-level, it manages ETM (and ETB) to trace the execution of target modules, and PMU to raise interrupts to address the problem of overwritten trace in ETB, and a Trace Decoder to parse the ETM trace. HART also hooks entrances/exits to/from the target modules, to coordinate with PMU and provide open interfaces for further usage. Following the workflow, we then delve into the details of our design and implementation.

3.1 HART Design

Initialization for Tracing. Before starting a module, the Linux kernel performs a set of initialization, including loading data and code from the module, relocating the references to external and internal symbols, and running the `init` function of the module. HART intercepts the initialization to prepare for tracing. Details are as follows.

First, HART builds a profile for the module, with some necessary information about the loading address, the `init` function, and the kernel data structure that manages this module, for later usage. Since such information will be available after the kernel has finished loading the module, we capture the initialization at this point of time with a callback registered through the trace-point in the `load_module` kernel function, without intrusion to the kernel.

Then, HART hooks the entrances to the module and the exits from the module, which is the base of PMU management and open interfaces for users of HART. As Fig. 1 depicted before, the modules are mainly entered and exited either through the callbacks registered to the kernel, or the external functions invoked by such callbacks. We deal with different cases as follows.

In the first case, those callbacks are typically functions within the module, waiting for the invoking from the kernel. They are usually registered through code pointers stored in the `.data` segment. In the course of module loading, the code pointers will be relocated to the run-time addresses during the initialization, so that we can intercept the relocation and alter these pointers to target the wrappers defined by HART. The comparison before and after the HART's alteration in the code pointers' relocation can be observed from Fig. 4a, presented with concrete examples. Thus, our wrappers are successfully registered as the callbacks to be invoked by the kernel, and can perform a series of HART-designed handlers, while ensuring the normal usage of original functions.

In the other case that callbacks invoke external functions, such external symbols (*e.g.,* `kmalloc`) also need a similar relocation to fix their references to the run-time addresses. Likewise, HART intercepts such relocation and directs the external symbols to the HART-designed wrappers, as is shown in Fig. 4b. So far, as we consider both conditions, all of the entrances and exits are under the control of HART via the adjusted relocation to our wrappers.

Finally, HART gets ETM and ETB online for tracing. To be specific, HART allocates a large continuous buffer `hart_buf`, for the sake of backing up ETB data. In our design, its buffer size is configured as `4 MB`. Meanwhile, HART spawns a child thread, which monitors data in `hart_buf` and does continuous decoding. We will cover more details about the decoding later in Sect. 3.1. Then, HART enables ETM with ETB, and configures ETM to only trace the range that stores the module. So far, HART finishes the initialization and starts the module by invoking the module's `init` function.

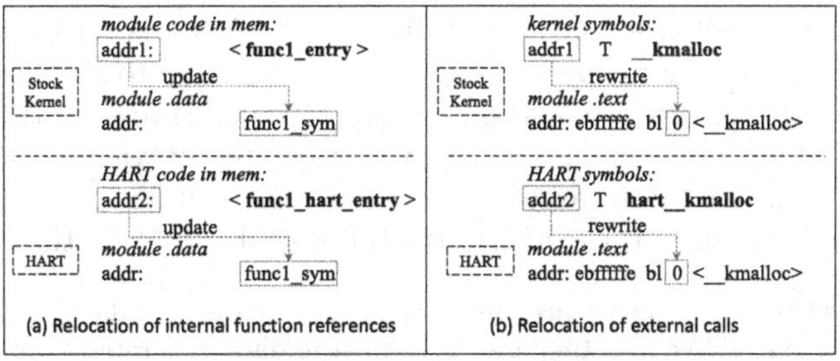

Fig. 4. Relocation of external calls and internal function references in modules.

Table 3. Wrapper for `func1` in Fig. 4a.

```
ENTRY(func1_hart):
    SAVE_CTX //save context
    MOV R0,LR //pass and save LR in HART
    BL resume_pmu //call resume_pmu
    MOV LR,R0 //return addr of ori. func1
    RESTORE_CTX //restore context
    BL LR //call original func1
    SAVE_CTX //save context
    BL stop_pmu //call stop_pmu
    MOV LR,R0 //return LR from HART
    RESTORE_CTX //restore context
    BL LR //return back
```

Table 4. Wrapper for `__kmalloc`.

```
void * hart__kmalloc(size_t size,
                     gfp_t flags){
   /*stop PMU*/
   cur_cnt = stop_pmu();
   /*instrumentation hooks*/
   pre_instrumention();
   /*original kmalloc*/
   addr = __kmalloc(size, flags);
   /*instrumentation hooks*/
   post_instrumentation();
   /*resume PMU*/
   reset_pmu(cur_cnt);
   return addr;
}
```

Continuous and Selective Tracing. After the above initialization, ETM will trace execution of the module and store its data in ETB. Under the assumption that the ETB hardware raises no interrupt to alert HART to a full ETB, the remaining challenge is to timely interrupt the module before ETB is filled up. In this work, HART leverages the instruction counter shipped with PMU.

Specifically, we start PMU on counting the instructions executed by the CPU. Once the number of instructions hits a threshold, PMU will raise an interrupt, during which HART can copy ETB data out to `hart_buf`. Generally, an Arm instruction can lead to at most `6` bytes of trace data[1]. This means a 4 KB ETB can support the execution of at least `680` instructions. Also considering that the PMU interrupts often come with a skid (less than 10 instructions [43]), we set up `680-10` as the threshold for PMU. In our evaluation with real-world benchmarks, we observe the threshold is consistently safe for avoiding overflow in ETB.

Though PMU aids HART in preventing trace loss, it causes an extra issue. PMU counts every instruction executed on the CPU, regardless of the context. This means PMU will interrupt not only the target module, but also the other kernel components, leading to high-frequency interruptions and tremendous down-gradation in the entire system. To restrict PMU to count inside the selective module(s), we rely on the hooks HART plants during initialization. When the kernel enters the module via a callback, the execution will first enter our wrapper function. In the wrapper, we resume PMU to count instructions, invoke the original callback, and stop PMU after the callback returns in order, as is shown in Table 3. Our wrapper avoids any modification to the stack and preserves all needed registers when stop/resume PMU, so as to keep the context to call the original `init` and maintain the context created by `init` back to the kernel. As aforementioned, the target module may also call for external functions, which are also hooked. In those hooks, HART first stops PMU, then invokes the

[1] Some instructions (*e.g.,* `LDM`) carry more than `6` bytes of trace data, but appear in a relatively few cases. We also take into account the data generated by data access in ETB. Actually, the information in the raw trace output is highly compressed, and a single byte in the trace output can represent up to 16 instructions in some cases.

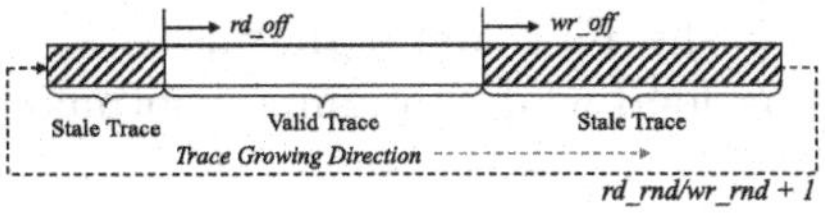

Fig. 5. Parallel trace data saving and decoding.

intended external function, and finally resumes PMU. Table 4 demonstrates the wrapper function of `__kmalloc` when it is called by the target module.

The last challenge derives from handling PMU interrupts. By intuition, we can simply register a handler to the kernel, which stops PMU, backs up ETB data, and then resumes PMU. However, this idea barely works in practice. Actually, on occurrence of an interrupt, the handling starts with a general interface `tzic_handle_irq`. This function retrieves the interrupt number, identifies the handler registered by users, and finally invokes that handler. After the user handler returns, `tzic_handle_irq` will check for further interrupts and handle them. Oftentimes, when we handle our PMU interrupts, new interrupts would come. If we resume PMU inside our handler, PMU would count the operations of `tzic_handle_irq` handling the newly arrived interrupts. Following the `tzic_handle_irq` handling process, it would trigger another PMU interrupt handler, eventually resulting in endless interrupt handling loop.

To address the above problem, we hook the `tzic_handle_irq` function with `hart_handle_irq`. And right before it exits to the target module, we resume PMU. When our PMU interrupts come, the handler `hart_pmu_handler` regis-

Table 5. Algorithm for backing up ETB data inside PMU interrupt handler and the parallel decoding thread for parsing trace with an elastic scheduling scheme.

```
void retrieve_trace_data(){
   /*get trace size*/
    trace_sz = bytes_in_etb();
    /*copy trace*/
    cpy_trace(trace_buf, etb, trace_sz);
    /*update write offset and rounds*/
    wr_off += trace_sz;
    if(wr_off >= trace_buf_sz){
        wr_off %= trace_buf_sz;
        wr_rnd ++;
    }
    /*overwrite is possible,
    pause the target module*/
    if(wr_rnd > rd_rnd &&
        rd_off - wr_off < 4KB)
        SIGSTP(target_pid);
}
```

(a) ETB data backup.

```
void parse_trace(){
   while(true){
      /*calculate valid trace size*/
      if(rd_rnd == wr_rnd) trace_sz =
         wr_off - rd_off;
      else trace_sz = trace_buf_sz -
         rd_off + wr_off;
      /*decode trace*/
      decode(trace_buf, rd_off, trace_sz);
      /*update rd_off and rd_rnd*/
      rd_off = wr_off;
      if(rd_rnd < wr_rnd)
      rd_rnd ++;
      if(target_paused(target_pid))
      SIGCONT(target_pid)
      /*yield CPU based on the
      workload of decoding*/
      msleep(100 - CONS *
         trace_sz/trace_buf_sz);
      yield();
   }
}
```

(b) Parallel decoding thread.

tered by us will back up the ETB data. In this way, we prevent the above endless interrupt handling loop while obtaining continuous and selective ETM trace.

Elastic Decoding. With the help of PMU, HART can timely interrupt the target module and back up the ETB data. To reduce time cost of backup, in the interrupt handler, HART simply copies the ETB data to the previously allocated `hart_buf`. Meanwhile, HART maintains `wr_off` and `wr_rnd`, respectively indicating the writing position to `hart_buf` and how many rounds HART has filled up this buffer. To decode the trace efficiently, HART spawns a decoding thread for each target module. This thread monitors and keeps decoding the remaining trace data, where `rd_off` and `rd_rnd` indicate the current parsing position and how many rounds it has read through the entire buffer. This design of parallel trace backup and decoding, as shown in Fig. 5, however, involves two issues. We describe the details and explain how we address them as follows.

First, an over-writing problem is likely to happen when the backup function runs beyond the decoder by more than one lap, leading to the error of losing valid data. A similar over-reading error comes due to a similar lack of synchronization. In order to maintain the correctness of the concurrency, we design algorithms for both the backup function and the decoder with checks of over-writing/reading possibilities. On the one hand, as the backup algorithm in Table 5a presents, if `wr_rnd` is larger than `rd_rnd` while `rd_off` is less than `4 KB` ahead of `wr_off`, a new operation of backing up `4 KB` data will potentially overwrite the previous unprocessed trace. In case of this, HART will pause the task that is using the target module, and restart via the decoder later when it is safe to continue. On the other hand, as the decoding algorithm in Table 5b depicts, if the decoder and the backup function arrive at the same location, the decoding thread will pause and yield the CPU due to no trace left to parse.

Second, since the generation of trace is oftentimes slow, a persistent query for any new trace will bring CPU with a waste of cycles. Such frequent occupation of CPU has a negative impact on other normal works, incurring heavy overhead in the performance and efficiency especially for the system equipped with fewer CPU cores. Thus, we introduce an elastic scheduling scheme to the decoding thread as a remedy for the waste, shown in Table 5. During a round of checking and decoding new trace, we calculate the size of the valid trace to represent the generation rate of the trace. Then, the sleep time for the decoding thread is set to inversely proportional to the computed growth rate. That is, the slower the trace grows, the longer the thread sleeps, and vice versa. In this way, HART leaves more resources for CPU while the decoding thread is not busy, but also assures the decoder of high efficiency when it owns a heavy workload.

Supports of Concurrency. We consider two types of concurrency, and support both conditions. First, multiple modules might be registered for protection under HART at the same time. In our design, the context of each module is maintained separately, and HART switches the context at module switch time.

The enter and exit events of a kernel module are captured by HART, and context of ETM tracing and PMU counting related to the module is saved at exiting and restored at reentering. Since traces pertaining to different modules are distinguishable based on the context-ID[2] packets, our design reuses the same `hart_buf` for different modules for the sake of memory efficiency.

Second, multiple user space clients may access the module(s) concurrently, which is common on multi-core SoCs. To handle this type of concurrency, we allocate multiple `hart_buf` on a device with many cores, with each of `hart_buf` uniquely serving one CPU core. This prevents race on trace storage among CPU cores. As both ETM and PMU are core-specific, their configurations and uses are independent across different cores. It is also worth noting that the ETB is shared by the cores. This does not interrupt our system. First, ETM traces from different cores are sequentially pushed to ETB, avoiding race on trace saving. Second, ETM attaches a context-ID whenever a task is being traced. This enables the decoder to separate the trace for individual tasks. Finally, we signal to pause the target module(s) on other cores while one core is doing ETB backup, so as to avoid losing trace data. Other than the trace buffer, we assign a separated decoding thread for each module to prevent confusion in the decoding process.

Open Interfaces. To facilitate the further establishment of various trace-based security applications, we provide users of HART with a comprehensive set of interfaces, including (1) **Initialization and Exiting interfaces** to help users register to the initialization and exiting process of HART, (2) **Decoding interfaces** to pass the parsed ETM trace data to users, and (3) **Instrumentation interfaces** to help hook external functions as shown in line `6` and `10` in Table 4.

3.2 HART Implementation

We implement the prototype of HART on Linux system with kernel 4.4.145-armv7-x16, running on a Freescale i.MX53 Quick Start Board [14] with ARMv7 Cortex-A8 processor [20]. As the processor only supports 32-bit mode, HART so far only runs with 32-bit systems. We believe there is no significant difference between the 32-bit and 64-bit implementations, and the only differences might lie in the driver for the involved hardware (*e.g.*, PMU, ETM, and ETB) and some configurations (*e.g.*, the instruction threshold for the trace buffer). In total, the HART system contains about `3,200` lines of `C` code.

To decode the trace data, we reuse an open-source decoder [57], which supports both ETM-V2 and ETM-V3. We optimize the decoding process by directly indexing the packet handler with the packet type. Thus HART avoids the cost of handler looking up and will achieve better efficiency.

After loading HART into the kernel as a standalone driver, we need to customize the native module utilities, `insmod` and `rmmod`, to inform HART of the installation and the removal of the target module to be protected. Before

[2] The context-ID of a task is identical to the process ID. With the context-ID, the processor and ETM can identify which task triggers the execution of the module.

installing the module, our `insmod` will send the notice to the virtual device registered by HART, which actually reminds HART of standby for the module protection. Our `rmmod` is customized similarly.

4 HASAN Design and Implementation

We also build an AddressSanitizer HASAN with HART's open interfaces to demonstrate the extended utilization of HART. The design of HASAN reuses the scheme of AddressSanitizer [50] and follows the implementation of the Kernel Address Sanitizer (KASAN) [15]. Instead of sanitizing the whole kernel, HASAN focuses on specific kernel modules. Without source code, we cannot locate local and global arrays, which prevents us from wrapping those objects with redzones. Thus, HASAN only focuses on memory objects on the heap. Note that if the module source code is available, HASAN can achieve the same level protection with KASAN. However, if the module source code is out of reach, HASAN can still achieve heap protection while KASAN will not work at all.

Table 6. Memory management interfaces HASAN hooked.

Category	Allocation	De-allocation
Kmem_cache	`kmem_cache_alloc kmem_cache_create`	`kmem_cache_free kmem_cache_destroy`
Kmalloc	`kmalloc krealloc kzalloc kcalloc`	`kfree`
Page operations	`alloc_pages __get_free_pages`	`free_pages __free_pages`

Design and Implementation: As HASAN aims to protect kernel memory, it only allocates shadow memory for the kernel space. As the kernel space ranges from `0xbf000000` to `0xffffffff` in our current system, HASAN allocates a 130 MB continuous virtual space for the shadow memory with each shadow memory bit mapping to a byte. Given an address `addr`, HASAN maps it to the shadow memory location using the following function, where `ADDR_OF_SHM` indicates the beginning of our virtual space.

```
void * addr_to_hasanaddr(addr){
    return ((u32)addr >> 3) +
        (addr - 0xbf000000) + ADDR_OF_SHM);
}
```

To wrap heap objects with redzones, HASAN hooks the `slab` interfaces for memory management[3]. As aforementioned, we achieve the hooking through the interfaces that HART plants to the wrappers of external calls. The interfaces that HASAN currently supports are listed in three categories in Table 6, with their handling methods explained as follows.

① When `kmem_cache_create` is invoked to creates a memory cache with *size* bytes, we first enlarge this *size* so as to reserve space for redzones around the

[3] We are yet to add supports for slub.

actual *object*. After `kmem_cache_alloc` has allocated the *object*, we mark the rest space as redzones. If `kmem_cache_alloc` requires memory from caches created by other kernel components (in many cases it will happen), HASAN will redirect the allocation to caches that are pre-configured in HASAN. These pre-configured caches will be destroyed when the target module is removed.

② In the second category, we first align the requested size by 8 bytes and then add another 16 bytes reserving for redzones. Then we call stock `kmalloc` and return to users with the address of the `9th` byte in the allocated buffer. The first 8 bytes together with the remaining space at the end will be poisoned.

De-allocations in both categories above follow similar strategies to KASAN. HASAN delays such operations (*e.g.,* `kfree`) and marks the freed regions as poisoned. These memories will be released until the size of delayed memory reaches a threshold or the module is unloaded.

③ In the third category, however, we only delay the corresponding de-allocation. For allocation, increasing the allocated *size* in those interfaces incurs a tremendous cost, as the *size* in their request indicates the order of pages to be allocated.

Table 7. Utility kernel functions for memory operations.

Category	Function name	Category	Function name
memory	`memcpy memccpy memmove memset`	print	`sprintf vsprintf`
string	`strcpy strncpy strcat strncat strdup`	kernel	`copy_from_user copy_to_user`

Since our design excludes the main kernel for tracing, it may miss to capture some vulnerable memory operations inside the kernel functions that listed in Table 7. To handle this issue, we make copies of those functions and load them as a new code section in HASAN. When a target module invokes those functions, we redirect the execution to our copies, enable ETM to trace on them and integrate this trace into ETB. To avoid switching hardware configurations on entering and leaving those functions, our implementation adds the memory holding our copies as an additional range to trace.

The detection of HASAN is straightforward and efficient. We register a call back to HART's decoder. On the arrival of a `normal_data` packet, the decoder will parse the memory address for HASAN. Then, HASAN maps the address to the shadow memory and checks the poison in it. To avoid unsynchronized cases such as decoding of valid accesses after memory de-allocation, we ensure that the decoding synchronizes with the execution in each interface as presented in Table 6.

Fine-Grained Synchronization: As our decoding proceeds after the execution, there is a delay from the occurrence of memory corruption to the detection of violations, which may be exploited to bypass HASAN. To reduce this attack surface, we force the decoder to synchronize with the execution whenever a callback is invoked. Specifically, our wrapper of that callback will pause the execution until the decoder consumes all the trace. Since a system call reaching the target module will invoke at least one callback, we ensure one or more synchronization(s) between successive system calls. This follows the existing research [21] that relies on system-call level synchronization to ensure security. Another rationale is that, even the state-of-art kernel exploits would require multiple system calls to compromise the execution [62,63].

5 Evaluation

In this section, to demonstrate the utility of our work, we sequentially test whether our work can achieve continuous tracing and measure their efficiency and effectiveness.

5.1 Setup

To understand the correctness and efficiency of our work, we run 6 widely-used kernel modules and come with standard benchmarks, detailing in Table 8. To further understand the non-intrusiveness in our work, we run the `lmbench` benchmarks. The evaluation is conducted on the aforementioned Freescale i.MX53 Quick Start Board [14] with 1 GB of RAM. As we aim to demonstrate the generality of HART with the minimal hardware requirements of the trace functionality of ETM and a small ETB, we consider i.MX53 QSB a perfect match. We do not measure the two types of concurrency described in Sect. 3.1, because the device has only one core, so concurrent tests would produce same results as running the tests sequentially. For comparison, we also test under the state-of-art solution `KASAN` [6]. Since the i.MX53 board we use can only support 32-bit systems while `KASAN` is only available on 64-bit systems, we run the `KASAN` based experiments in a `Raspberry Pi 3+` configured with the same computing power and memory.

To verify the effectiveness of detecting memory corruption, we collect **6** known memory vulnerabilities which cover different types and lie in different modules with various categories and complexities. Note we only cover heap-related buffer overflows because HASAN focuses on heap related memory errors.

5.2 Continuous Tracing

In Table 9, we summarize the frequency of HART backing up the data from the ETB, the size of data retrieved each time. Most importantly, the last column in Table 9 shows ETB has never been filled up. Besides, the maximal size of trace is uniformly less than 2K, not even reaching the middle of ETB. Both facts indicate that overflow in ETB never happened and all the trace data has been successfully backed-up and decoded. Thus, HART and HASAN achieve continuous tracing, which is the precondition for the correctness of the solution.

Table 8. Performance evaluation. The results are normalized using the tests on *Native img + Native module* as baseline 1. A larger number indicates a larger data transmission rate and a lower deceleration on performance.

Module		Benchmark		Result			
Type	*Name*	*Name*	*Setting*	*Native img +*		*KASAN img +*	
				HART *module*	HASAN *module*	*Native module*	*KASAN module*
Network	HSTCP [30]	iperf [29]	Local Comm.	1.00	1.00	0.29	0.28
	TCPW [2]	iperf [29]	Local Comm.	0.92	0.91	0.28	0.28
	H-TCP [12]	iperf [29]	Local Comm.	0.94	0.94	0.26	0.25
File system	HFS+ [13]	IOZONE [28]	Wr/fs=4048K/reclen=64	1.00	1.00	0.96	0.95
			Wr/fs=4048K/reclen=512	0.88	0.87	0.96	0.94
			Rd/fs=4048K/reclen=64	0.92	0.89	0.98	0.92
			Rd/fs=4048K/reclen=512	0.90	0.89	0.99	0.99
	UDF [19]	IOZONE [28]	Wr/fs=4048K/reclen=64	0.95	0.93	0.99	0.97
			Wr/fs=4048K/reclen=512	0.97	0.97	1.00	0.92
			Rd/fs=4048K/reclen=64	0.98	0.97	0.99	0.98
			Rd/fs=4048K/reclen=512	0.97	0.96	1.00	0.98
Driver	USB_STORAGE [8]	dd [45]	Wr/bs=1M/count=1024	1.00	1.00	1.00	0.43
			Wr/bs=4M/count=256	1.00	1.00	0.99	0.43
			Rd/bs=1M/count=1024	0.99	0.99	0.99	0.75
			Rd/bs=4M/count=256	1.00	1.00	1.00	0.76
Avg.	–	–	–	0.95	0.94	0.85	0.72

Table 9. Tracing evaluation of HART and HASAN.

Module		Retrieving times		Max size(Byte)		Min size(Byte)		Average size(Byte)		Full ETB	
Type	Name	HART	HASAN	HART	HASAN	HART	HASAN	HART	HASAN	HART	HASAN
Network	HSTCP	4243	3964	1100	1196	20	20	988	1056	0	0
	TCP-W	3728	3584	1460	1456	20	20	1128	1088	0	0
	H-TCP	3577	3595	1292	1304	20	20	1176	1168	0	0
File system	HFS+	30505	30278	1652	1756	20	20	144	148	0	0
	UDF	17360	20899	2424	2848	20	20	240	232	0	0
Driver	USB_STORAGE	9316	9325	1544	1692	20	20	448	448	0	0

We also observe that HART and HASAN have some similarities and differences in their tracing behaviours. Overall, HART and HASAN present the similar but non-identical results in the same set of module. The slight deviation mainly derives from the padding packets randomly inserted by ETM and the random skids of PMU. The differences lie across different families of modules. HART and HASAN back up ETB more frequently in the two file systems (nearly `2X` more than that in the driver module and `3-7X` more than the network modules), while they produce the least trace in each ETB backup with the file systems. This indicates that file systems execute more instructions yet they usually carry less information about control flow and data accesses. Due to such increase in instruction number, PMU has to raise more interrupts in the testing on file systems and contributes to the more frequent ETB backups. The statistics above are consistent with the following performance evaluation – a higher frequency of backing up incurs more trace management and thus, introduces higher performance overhead.

5.3 Performance Evaluation

For performance measurement, we record (1) the bandwidths of the server for network modules and (2) the read/write rates for file-systems as well as driver modules. The testing results are summarized in Table 8. Across all three types of drivers, HART and HASAN barely affects the performance. On average, HART introduces an overhead of **5%**, while HASAN introduces **6%**. In the worst-case scenario, HART brings a maximum of `12%` for (`reclen=64`) when testing on `HFS+`, and HASAN brings at most `13%` in the same case. These results well support that HART and HASAN are in general efficient. The most important reason for this efficiency, we believe, is caused by employing ETM to capture control flow and data flow with high efficiency. In general, the hardware-based tracing involves negligible performance overhead, and almost all the overhead introduced by HART is caused by decoding and analyzing the trace output.

On contrast, since `KASAN` cannot be deployed to the modules without `KASAN`-enabled kernel, we analyze the impacts of only enforcing `KASAN` through the differences between the tests with and without `KASAN`-enabled modules to some extent: ① The test performed on the network modules indicates `KASAN` introduces nearly `4X` slowdown. The reason is that, during a networking event (receiving and sending a message), the kernel performs large amounts of computations due to the complicated operations on the network stack, even though the IO through the network port is cheap as we are only doing local communication. As such, the overhead introduced by `KASAN` is not diluted by IO accesses. Furthermore, though the difference between two tests on `KASAN` is insignificant, it does *not* imply that `KASAN` will be efficient if we re-engineer it to protect the target module only. In a network activity, those drivers only take up a small fraction of computation in the kernel space. Overhead on such computation is covered up by the significant slowdown in the main kernel. ② In the two file systems, `KASAN` hardly affects the performance of the kernel or the target module. Again, this does *not* necessarily imply that `KASAN` is most efficient, because in a file system operation, most of the time is spent on disk access and the computation in the kernel is quite simple and quick. Therefore, the average overhead is insignificant. ③ In the test on `USB_STORAGE` driver, most of the computation in the kernel space is performed by the target module, and takes more time than the IO access. Thus, the cost of adding `KASAN` on the module is significant (about `2X`).

Besides reducing the overhead on the target module, we also verify our point of the non-intrusiveness of HART and HASAN. Running them without the target module, we find they introduce zero overhead to the main kernel, because

Table 10. Performance evaluation on KASAN with `lmbench`. HART and HASAN introduce *no* overhead to the main kernel, so the results are omitted here.

Func.	Setting	Native	KASAN	Overhead	Func.	Setting	Native	KASAN	Overhead
Processes (ms)	stat	3.08	16.4	5.3	File & VM system latency (ms)	0K File Create	44.0	136.1	3.1
	open clos	8.33	36.7	4.4		0K File Delete	35.2	227.1	6.5
	sig hndl	6.06	20.4	3.4		10K File Create	99.9	370.2	3.7
	fork proc	472	1940	4.1		10K File Delete	64.2	204.7	3.2
Local Comm. latency (ms)	Pipe	18.9	45.8	2.4		Mmap Latency	188000	385000	2.0
	AF UNIX	26.6	97.9	3.7		Prot Fault	0.5	0.5	1.0
	UDP	41.4	127.6	3.1		Page Fault	1.5	2.3	1.5
	TCP	53.4	176.4	3.3		100fd selct	6.6	13.7	2.1

they notice nothing to protect and are always sleeping. Differing from our works, `KASAN` places a significant burden on the main kernel (Table 10). This well supports the advantage of our works in terms of modification to the main kernel.

Table 11. Effectiveness evaluation on HASAN. For the result of `Detection`, "Y" means detected.

Vulnerability		Detection		
CVE-ID	Type	PoC	HASAN	KASAN
CVE-2016-0728	Use-after-free	REFCOUNT overflow [56]	Y	Y
CVE-2016-6187	Out-of-bound	Heap off-by-one [42]	Y	Y
CVE-2017-7184	Out-of-bound	xfrm_replay_verify_len [52]	Y	Y
CVE-2017-8824	Use-after-free	dccp_disconnect [33]	Y	Y
CVE-2017-2636	Double-free	n_hdlc [44]	Y	Y
CVE-2018-12929	Use-after-free	ntfs_read_locked_inode [46]	Y	Y

5.4 Effectiveness Evaluation

We present the results of our effectiveness evaluation in Table 11. As shown in this table, both HASAN and `KASAN` can detect all **6** cases. Since the selected cases cover different types of heap-related issues, including buffer overflow, off-by-one, use-after-free, and double free, we demonstrate that HASAN can cover a variety of vulnerabilities with a comparable security performance to the state-of-the-art solution in terms of the heap.

6 Discussion

Support of Other Architectures. As hardware tracing brings a significant improvement in the performance overhead during the debugging, it becomes a common feature in major architectures. While HART achieves tracing efficiency with the support of ETM on Arm platforms, we can easily extend our design to other architectures with such feature. For example, since Processor Tracing (PT) [3], supports instruction tracing and will be enhanced with `PTWRITE` [1] to support efficient data tracing, we can extend our design by replacing the hardware set-up and configuration (given a target instrumented with `PTWRITE`).

Support of Statically Linked Modules. HART mainly targets LKM, because most of the third-party modules are released as LKM for the ease of use. Our current design cannot work with statically linked modules, mainly because we require to hook the callbacks and calls to kernel functions, which can be achieved by adjusting the relocation. However, statically linked modules are compiled and built as port of the main kernel and hence, we have no access to place the hooks. To extend support for statically linked modules, we need the kernel builder to instrument the target modules and place the required hooks.

Breakage of Perf Tools. As perf tools have already included ETM tracing capabilities, enabling tracing functionality from the user-space, HART may lead to the breakage of perf tools. We think the breakage depends on the scenario. If HART and perf are tracing/profiling different tasks, we consider perf won't be affected since HART will save the context of the hardware at the entry and exit of the traced module. However, if HART and perf are tracing/profiling the same task, perf might be affected. One easy solution is to run HART and perf side by side instead of running them concurrently.

7 Related Work

7.1 Kernel Protection

Kernel Debugger. Kmemcheck [11] can dynamically check uninitialized memory usage in x86 kernel space, but its single-stepping debugging feature drags down the speed of program execution to a large extent. KASAN [6] is the kernel version of address sanitizer. However, the heavy overhead makes it impossible to be a real-time protection mechanism, while the low overhead of HASAN show us the potential. Furthermore, KASAN requires the source code for implementation. If a vulnerability is hidden in a close-source third-party kernel module, which is common, HASAN has the potential to detect it but KASAN would not. Ftrace [10] is an software-based tracer for the kernel. Comparing with Ftrace, HART is noninvasive since we neither modifies the target kernel module nor requires the source code of the kernel module. In regard to the trace granularity, Ftrace is used to monitor function calls and kernel events with predefined tracepoints, but HART can be used to trace executed instructions and data used in these instructions. Ftrace also introduces heavy performance overhead on some event tracing, and cannot be adopted as a real-time protection as well.

Kernel Integrity Protection. KCoFI [26] enforces the CFI policy by providing a complete implementation for event handling based on Secure Virtual

Architecture (SVA) [27]. However, all software including OS should be compiled to the virtual instruction set that SVA provides. The solution also incurs high computation overhead. With slight modification to the modern architectures, Song *et al.* [53] enforces DFI over both of control and non-control data that can affect security checks. Despite the aids from the re-modelled hardware, the DFI still introduces significant latency to various system calls.

Isolation and Introspection. BGI [22] enforces isolation through an additional interposition layer between kernel and the target module. However, it requires instrumentation over source code of the target modules. Similar ideas are followed by many other works [48,58,65]. Ninja [43] is a transparent debugging and tracing framework on Arm platform, and it can also be used to introspect the OS kernel from TrustZone. However, bridging the semantic gaps between Linux kernel and TrustZone would increase the performance overhead significantly.

7.2 Hardware Feature Assisted Security Solutions

Hardware Performance Counters. HDROP [66] and SIGDROP [59] leverage PMU to count for the abnormal increase of events and seek patterns of ROP at runtime. Morpheus [38] proposes a benchmarking tool to evaluate computational signatures based mobile malware detection, in which HPCs help to create runtime traces from applications for the ease of the signature comparison. Although these solutions [37,55] achieve significant overhead reduction with the assistance of HPCs, none of them applies this hardware feature to kernel module protection.

Hardware-Based Tracing. With the availability of Program Trace Interface, Kargos [41] monitors the execution and memory access events to detect code injection attacks. However, it brings extra performance overhead due to the modification to the kernel. Since the introduction in Intel's Broadwell microarchitecture, PT has been broadly applied in solutions for software security [32,34,35,47]. All the PT-based solutions are control flow specific, differing from HART that traces both data flow and control flow for security protection.

8 Conclusion

Kernel modules demand as much security protection as the main kernel. However, the current solutions are actually limited by the requirement of source code, significant intrusiveness and heavy overhead. We present HART as a general ETM-powered tracing framework specifically for kernel modules, with the most preliminary hardware support and the most compatible method. HART can trace the selective work on the binary-only module(s) continuously by combining hardware configurations and software hooks, and decode the trace efficiently following an elastic scheduling. Based on the framework, we then build a modular security solution, HASAN, to effectively detect memory corruptions without the aforementioned limitations. Testing on a set of benchmarks in different modules, both HART and HASAN perform significantly superior to the state-of-the-art KASAN, introducing the average overhead of **5%** and **6%**. Moreover, HASAN identifies all of the **6** vulnerabilities in different categories and modules, indicating its comparable effectiveness to the state-of-the-art solutions.

Acknowledgements. We sincerely thank our shepherd Prof. Dave Jing Tian and reviewers for their comments and feedback. This work was supported in part by grants from the Chinese National Natural Science Foundation (NSFC 61272078, NSFC 61073027).

References

1. PTWRITE - write data to a processor trace packet. https://hjlebbink.github.io/x86doc/html/PTWRITE.html
2. TCP westwood+ congestion control (2003). https://tools.ietf.org/html/rfc3649
3. Processor tracing (2013). https://software.intel.com/en-us/blogs/2013/09/18/processor-tracing
4. Juno ARM Development Platform SoC Technical Reference Manual (2014)
5. slub (2017). https://www.kernel.org/doc/Documentation/vm/slub.txt
6. Home Google/Kasan Wiki (2018). https://github.com/google/kasan/wiki
7. Apple A4 (2019). https://www.apple.com/newsroom/2010/06/07Apple-Presents-iPhone-4/
8. Config_usb_storage: USB mass storage support (2019). https://cateee.net/lkddb/web-lkddb/USB_STORAGE.html
9. Embedded trace macrocell architecture specification (2019). http://infocenter.arm.com/help/index.jsp?topic=/com.arm.doc.ihi0014q/index.html
10. ftrace - function tracer (2019). https://www.kernel.org/doc/Documentation/trace/ftrace.txt
11. Getting started with kmemcheck - the Linux kernel documentation (2019). https://www.kernel.org/doc/html/v4.14/dev-tools/kmemcheck.html
12. H-TCP - congestion control for high delay-bandwidth product networks (2019). http://www.hamilton.ie/net/htcp.htm
13. HFS plus (2019). https://www.forensicswiki.org/wiki/HFS%2B
14. i.MX53 quick start board—NXP (2019). https://www.nxp.com/products/power-management/pmics/power-management-for-i.mx-application-processors/i.mx53-quick-start-board:IMX53QSB
15. The kernel address sanitizer (Kasan) - the Linux kernel documentation (2019). https://www.kernel.org/doc/html/v4.14/dev-tools/kasan.html
16. Kmemleak (2019). https://www.kernel.org/doc/html/v4.14/dev-tools/kmemleak.html
17. Samsung Exynos 3110 (2019). https://www.samsung.com/semiconductor/minisite/exynos/products/mobileprocessor/exynos-3-single-3110/
18. Snapdragon 200 series (2019). https://www.qualcomm.com/snapdragon/processors/200
19. Universal disk format (2019). https://docs.oracle.com/cd/E19683-01/806-4073/fsoverview-8/index.html
20. ARM: Cortex-A8 Technical Reference Manual (2014)
21. Bigelow, D., Hobson, T., Rudd, R., Streilein, W., Okhravi, H.: Timely rerandomization for mitigating memory disclosures. In: Proceedings of the 22nd ACM SIGSAC Conference on Computer and Communications Security, pp. 268–279. ACM (2015)

22. Boyd-Wickizer, S., Zeldovich, N.: Tolerating malicious device drivers in Linux. In: USENIX Annual Technical Conference, Boston (2010)
23. Carbone, M., Cui, W., Lu, L., Lee, W., Peinado, M., Jiang, X.: Mapping kernel objects to enable systematic integrity checking. In: Proceedings of the 16th ACM Conference on Computer and Communications Security, pp. 555–565. ACM (2009)
24. Castro, M., et al.: Fast byte-granularity software fault isolation. In: Proceedings of the ACM SIGOPS 22nd Symposium on Operating Systems Principles, pp. 45–58. ACM (2009)
25. Chen, H., Mao, Y., Wang, X., Zhou, D., Zeldovich, N., Kaashoek, M.F.: Linux kernel vulnerabilities: state-of-the-art defenses and open problems. In: Proceedings of the Second Asia-Pacific Workshop on Systems, p. 5. ACM (2011)
26. Criswell, J., Dautenhahn, N., Adve, V.: KCoFI: complete control-flow integrity for commodity operating system kernels. In: 2014 IEEE Symposium on Security and Privacy (SP), pp. 292–307. IEEE (2014)
27. Criswell, J., Lenharth, A., Dhurjati, D., Adve, V.: Secure virtual architecture: a safe execution environment for commodity operating systems. In: Proceedings of 21st ACM SIGOPS Symposium on Operating Systems Principles, SOSP 2007, pp. 351–366. ACM (2007)
28. Don, C., Capps, C., Sawyer, D., Lohr, J., Dowding, G., et al.: IOzone filesystem benchmark (2016). http://www.iozone.org/
29. Dugan, J., Elliott, S., Mah, B.A., Poskanzer, J., Prabhu, K., et al.: iPerf - the ultimate speed test tool for TCP, UDP and SCTP (2018). https://iperf.fr/
30. Floyd, S.: Highspeed TCP for large congestion windows (2003). https://tools.ietf.org/html/rfc3649
31. Garfinkel, T., Rosenblum, M., et al.: A virtual machine introspection based architecture for intrusion detection. In: Proceedings of the Network and Distributed System Security Symposium, NDSS 2003, pp. 191–206 (2003)
32. Ge, X., Cui, W., Jaeger, T.: GRIFFIN: guarding control flows using intel processor trace. In: Proceedings of the 22nd ACM International Conference on Architectural Support for Programming Languages and Operating Systems (ASPLOS) (2017)
33. Ghannam, M.: CVE-2017-8824 Linux: use-after-free in DCCP code (2017). https://www.openwall.com/lists/oss-security/2017/12/05/1
34. Gu, Y., Zhao, Q., Zhang, Y., Lin, Z.: PT-CFI: transparent backward-edge control flow violation detection using intel processor trace. In: Proceedings of the 7th ACM International Conference on Data and Application Security and Privacy (CODASPY) (2017)
35. Hertz, J., Newsham, T.: Project triforce: run AFL on everything! (2016). https://www.nccgroup.trust/us/about-us/newsroom-and-events/blog/2016/june/project-triforce-run-afl-on-everything/
36. Hinum, K.: Hisilicon kirin 920 (2017). https://www.notebookcheck.net/HiSilicon-Kirin-920-SoC-Benchmarks-and-Specs.240088.0.html
37. Iyer, R.K.: An OS-level framework for providing application-aware reliability. In: 12th Pacific Rim International Symposium on Dependable Computing, PRDC 2006 (2007)
38. Kazdagli, M., Ling, H., Reddi, V., Tiwari, M.: Morpheus: benchmarking computational diversity in mobile malware. In: Proceedings of Hardware and Architectural Support for Security and Privacy (2014)
39. Linares-Vásquez, M., Bavota, G., Escobar-Velásquez, C.: An empirical study on android-related vulnerabilities. In: 2017 IEEE/ACM 14th International Conference on Mining Software Repositories (MSR), pp. 2–13. IEEE (2017)

40. Machiry, A., Spensky, C., Corina, J., Stephens, N., Kruegel, C., Vigna, G.: Dr. Checker: a soundy analysis for Linux kernel drivers. In: 26th USENIX Security Symposium (USENIX Security 2017), pp. 1007–1024. USENIX Association (2017)
41. Moon, H., Lee, J., Hwang, D., Jung, S., Seo, J., Paek, Y.: Architectural supports to protect OS kernels from code-injection attacks. In: Proceedings of Hardware and Architectural Support for Security and Privacy (2016)
42. Nikolenko, V.: Heap off-by-one POC (2016). http://cyseclabs.com/exploits/matreshka.c
43. Ning, Z., Zhang, F.: Ninja: towards transparent tracing and debugging on arm. In: 26th USENIX Security Symposium (USENIX Security 2017), pp. 33–49 (2017)
44. Popov, A.: CVE-2017-2636: exploit the race condition in the n_hdlc Linux kernel driver bypassing SMEP (2017). https://a13xp0p0v.github.io/2017/03/24/CVE-2017-2636.html
45. Rubin, P., MacKenzie, D., Kemp, S.: dd - convert and copy a file (2019). http://man7.org/linux/man-pages/man1/dd.1.html
46. Schumilo, S.: Multiple memory corruption issues in ntfs.ko (Linux 4.15.0-15.16) (2018). https://bugs.launchpad.net/ubuntu/+source/linux/+bug/1763403
47. Schumilo, S., Aschermann, C., Gawlik, R., Schinzel, S., Holz, T.: kAFL: hardware-assisted feedback fuzzing for OS kernels. In: Proceedings of the 26th Security Symposium (USENIX Security) (2017)
48. Sehr, D., et al.: Adapting software fault isolation to contemporary CPU architectures. In: 19th USENIX Security Symposium (USENIX Security 2010), pp. 1–12 (2010)
49. Freescale Semiconductor: i.MX53 Multimedia Applications Processor Reference Manual (2012)
50. Serebryany, K., Bruening, D., Potapenko, A., Vyukov, D.: AddressSanitizer: a fast address sanity checker. In: USENIX Annual Technical Conference, pp. 309–318 (2012)
51. Seshadri, A., Luk, M., Qu, N., Perrig, A.: SecVisor: a tiny hypervisor to provide lifetime kernel code integrity for commodity OSes. In: ACM SIGOPS Operating Systems Review, pp. 335–350. ACM (2007)
52. snorez: Exploit of CVE-2017-7184 (2017). https://raw.githubusercontent.com/snorez/exploits/master/cve-2017-7184/exp.c
53. Song, C., Lee, B., Lu, K., Harris, W., Kim, T., Lee, W.: Enforcing kernel security invariants with data flow integrity. In: NDSS (2016)
54. Swift, M.M., Martin, S., Levy, H.M., Eggers, S.J.: Nooks: an architecture for reliable device drivers. In: Proceedings of the 10th Workshop on ACM SIGOPS European Workshop, pp. 102–107. ACM (2002)
55. Tang, A., Sethumadhavan, S., Stolfo, S.J.: Unsupervised anomaly-based malware detection using hardware features. In: Stavrou, A., Bos, H., Portokalidis, G. (eds.) RAID 2014. LNCS, vol. 8688, pp. 109–129. Springer, Cham (2014). https://doi.org/10.1007/978-3-319-11379-1_6
56. Perception Point Team: Refcount overflow exploit (2017). https://github.com/SecWiki/linux-kernel-exploits/blob/master/2016/CVE-2016-0728/cve-2016-0728.c
57. virtuoso: virtuoso/etm2human: Arm's ETM v3 decoder (2009). https://github.com/virtuoso/etm2human
58. Wahbe, R., Lucco, S., Anderson, T.E., Graham, S.L.: Efficient software-based fault isolation. In: ACM SIGOPS Operating Systems Review, pp. 203–216. ACM (1994)
59. Wang, X., Backer, J.: SIGDROP: signature-based ROP detection using hardware performance counters. arXiv preprint arXiv:1609.02667 (2016)

60. Wang, Z., Jiang, X.: HyperSafe: a lightweight approach to provide lifetime hypervisor control-flow integrity. In: 2010 IEEE Symposium on Security and Privacy (SP), pp. 380–395. IEEE (2010)
61. Wang, Z., Jiang, X., Cui, W., Ning, P.: Countering kernel rootkits with lightweight hook protection. In: Proceedings of the 16th ACM Conference on Computer and Communications Security, pp. 545–554. ACM (2009)
62. Wu, W., Chen, Y., Xing, X., Zou, W.: Kepler: facilitating control-flow hijacking primitive evaluation for Linux kernel vulnerabilities. In: 28th USENIX Security Symposium (USENIX Security 2019), pp. 1187–1204 (2019)
63. Wu, W., Chen, Y., Xu, J., Xing, X., Gong, X., Zou, W.: Fuze: towards facilitating exploit generation for kernel use-after-free vulnerabilities. In: 27th USENIX Security Symposium (USENIX Security 2018), pp. 781–797 (2018)
64. Xiong, X., Tian, D., Liu, P., et al.: Practical protection of kernel integrity for commodity OS from untrusted extensions. In: NDSS, vol. 11 (2011)
65. Zhou, F., et al.: SafeDrive: safe and recoverable extensions using language-based techniques. In: Proceedings of the 7th Symposium on Operating Systems Design and Implementation, pp. 45–60. USENIX Association (2006)
66. Zhou, H.W., Wu, X., Shi, W.C., Yuan, J.H., Liang, B.: HDROP: detecting ROP attacks using performance monitoring counters. In: Huang, X., Zhou, J. (eds.) ISPEC 2014. LNCS, vol. 8434, pp. 172–186. Springer, Cham (2014). https://doi.org/10.1007/978-3-319-06320-1_14

Zipper Stack: Shadow Stacks Without Shadow

Jinfeng Li[1,2], Liwei Chen[1,2(✉)], Qizhen Xu[1,2], Linan Tian[1,2], Gang Shi[1,2], Kai Chen[1,2], and Dan Meng[1,2]

[1] Institute of Information Engineering, Chinese Academy of Sciences, Beijing, China
{lijinfeng,chenliwei,xuqizhen,tianlinan, shigang,chenkai,mengdan}@iie.ac.cn

[2] School of Cyber Security, University of Chinese Academy of Sciences, Beijing, China

Abstract. Return-Oriented Programming (ROP) is a typical attack technique that exploits return addresses to abuse existing code repeatedly. Most of the current return address protecting mechanisms (also known as the Backward-Edge Control-Flow Integrity) work only in limited threat models. For example, the attacker cannot break memory isolation, or the attacker has no knowledge of a secret key or random values.

This paper presents a novel, lightweight mechanism protecting return addresses, Zipper Stack, which authenticates all return addresses by a chain structure using cryptographic message authentication codes (MACs). This innovative design can defend against the most powerful attackers who have full control over the program's memory and even know the secret key of the MAC function. This threat model is stronger than the one used in related work. At the same time, it produces low-performance overhead. We implemented Zipper Stack by extending the RISC-V instruction set architecture, and the evaluation on FPGA shows that the performance overhead of Zipper Stack is only 1.86%. Thus, we think Zipper Stack is suitable for actual deployment.

Keywords: Intrusion detection · Control Flow Integrity

1 Introduction

In the exploitation of memory corruption bugs, the return address is one of the most widely exploited vulnerable points. On the one hand, code-reuse attacks (CRAs), such as ROP [5] and ret2libc [26], perform malicious behavior by chaining short sequences of instructions which end with a return via corrupted return

---

This work is partially supported by the National Natural Science Foundation of China (No. 61602469, U1836211), and the Fundamental theory and cutting edge technology Research Program of Institute of Information Engineering, CAS (Grant No. Y7Z0411105).

L. Chen et al. (Eds.): ESORICS 2020, LNCS 12308, pp. 338–358, 2020.
https://doi.org/10.1007/978-3-030-58951-6_17

addresses. These attacks require no code injection so they can bypass non-executable memory protection. On the other hand, the most widely exploited memory vulnerability, stack overflow, is also exploited by overwriting the return address. Both CRAs and stack smashing attacks are based on tampering with the return addresses.

In order to protect the return addresses, quite a few methods were presented, such as Stack Protector (also known as Stack Canary) [8,38], Address Space Layout Randomization (ASLR) [32], Shadow Stacks [6,9,23,29], Control Flow Integrity (CFI) [1,2], and Cryptography-based CFI [21,24]. However, they have encountered various problems in the actual deployment.

Stack Protector and ASLR rely on secret random values (cookies or memory layout). Both methods are widely deployed now. However, if there is a memory leak, both methods will fail [30]: the attacker can read the cookie and hold it unchanged while overwriting the stack to bypass the Stack Protector, and de-randomize ASLR to bypass ASLR. Some works have proved that they can be reliably bypassed in some circumstances [16,20,27], such as BROP [4]. Even if there is no information leaking, some approaches can still bypass ASLR and perform CRAs [25].

Shadow Stack is a direct mechanism that records all return addresses in a protected stack and checks them when returns occur. It has been implemented via both compiler-based and instrumentation-based approaches [9,10]. In recent years, commercial hardware support has also emerged [17]. But Shadow Stack relies heavily on the security of the memory isolation, which is difficult to guarantee in actual deployment. Some designs of Shadow Stack utilize ASLR to protect Shadow Stack. However, they cannot thwart the attacks contains any information disclosure [7,9]. Since most methods that bypass ASLR [20,27] are effective against this type of defense. Other designs use page attributes to protect Shadow Stack, for instance, CET [17]. However, defenses that rely on page attributes, such as NX (no-execute bit), have been bypassed by various technologies in actual deployment [18]: a single corrupted code pointer to the function in the library (via a JOP/COOP attack) may change the page attribute and disable the protection. In light of these findings, we think that mechanisms that do not rely on memory isolation are more reliable and imperative.

Cryptography-based protection mechanisms based on MAC authentication have also been proposed. These methods calculate the MACs of the return addresses and authenticate them before returns [21,24,40]. They do not rely on memory isolation because the attacker cannot generate new correct MACs without the secret key. But they face other problems: replay attacks (which reuse the existing code pointers and the MACs of them), high-performance overhead, and security risk of keys. A simple authentication of MAC is vulnerable facing replay attacks, so some designs introduce extra complexity to mitigate the problem (such as adding a nonce or adding stack pointer into the MAC input). Besides, these methods' performance overhead is also huge, since they use a cryptographic MAC and then save the result into memory. Both operations consume massive runtime overhead. The security of the secret key is also crucial. Theses works assume that there is no hardware attack or secret leak from the kernel. However, in the real world, the secret key can be leaked by a side-channel attack or an attack on context switching in the kernel. Therefore, we should go further over these assumptions, and consider a more powerful threat model.

In this paper, we propose a novel method to protect return addresses, Zipper Stack, which uses a chain of MACs to protect the return addresses and all the MACs. In the MAC chain, the newest MAC is generated from the latest return address and the previous MAC. The previous MAC is generated from the return address and MAC before the previous one, as Fig. 1 shows. So the newest MAC is calculated from all the former return addresses in the stack, although it is generated by computing the latest address and the previous MAC. Zipper Stack minimizes the amount of state requiring direct protection: only the newest MAC needs to be protected from tampering. Without tampering the newest MAC, an attacker cannot tamper with any return address because he cannot tamper the whole chain and keep the relation.

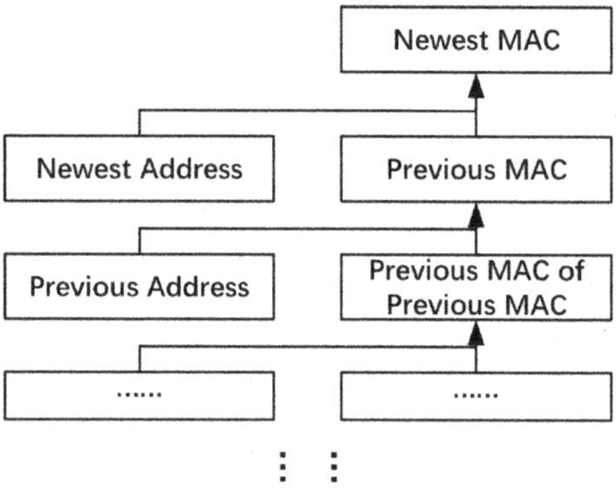

Fig. 1. Core of Zipper Stack

Zipper Stack avoids the problems of Shadow Stack and Cryptography-based protection mechanisms. Compared to Shadow Stack, it does not rely on memory isolation. Consequently, the attacks that modify both the shadow stack and the main stack cannot work in Zipper Stack. Compared with other cryptography-based mechanisms, Zipper Stack can resist against replay attacks itself and will not fail even if the secret key is leaked (see Sect. 4). In terms of efficiency, Zipper Stack performs even better. Our design is more suitable for parallel processing in the CPU pipeline, which avoids most performance overhead caused by the MAC calculation and memory access. The performance overhead of Zipper Stack with hardware support based on Rocket Core [35] (RISC-V CPU [36]) is only 1.86% based on our experiments.

The design of Zipper Stack solved three challenges: First, it avoids the significant runtime overhead that most cryptography-based mechanisms suffer since the newest MAC is updated in parallel. In our hardware implementation, most instructions that contain a MAC generation/authentication take only one cycle (See Sect. 6). Second, it utilized the LIFO order of return addresses to minimize the amount of state requiring direct protection. In general, a trust root authenticating all the data can help us defend against replay attacks or attacks containing secret key leaks (which most current return address protection method cannot). While in Zipper Stack, the authentication uses the newest MAC, at the same time, the MAC is a dynamic trust root itself. So it gets better security without extra overhead. Third, previous methods protect each return address separately, so any one been attacked may cause attacks. Zipper Stack, however, connect all the return addresses, leverage the prior information to increase the bar for attackers.

In order to demonstrate the design and evaluate the performance, we implement Zipper Stack in three deployments corresponding to three situations:

a) Hardware approach, which is suitable when hardware support of Zipper Stack is available. b) Customized compiler approach, which is suitable when hardware support is not available, but we can recompile the applications. c) Customized ISA approach, which is suitable when we cannot recompile the programs, but we can alter the function of CALL/RET instructions.

Ideally, the hardware approach is the best - it costs the lowest runtime overhead. The other two approaches, however, are suitable in some compromised situation. In the hardware approach, we instantiated Zipper Stack with a customized Rocket Core (a RISC-V CPU) on the Xilinx Zynq VC707 evaluation board (and hardware-based Shadow Stack as a comparison). In customized compiler approach, we implemented Zipper Stack in LLVM. In customized ISA approach, we use Qemu to simulate the modified ISA.

Contributions. In summary, this paper makes the following contributions:

1. **Design:** We present a novel, concise, efficient return address protection mechanism, called Zipper Stack, which protect return addresses against the attackers have full control of all the memory and know the secret keys, with no significant runtime overhead. Consequently, we analyze the security of our mechanism.
2. **Implementation:** To demonstrate the benefits of Zipper Stack, we implemented Zipper Stack on the FPGA board, and a hardware-based Shadow Stack as a comparison. To illustrate the potential of Zipper Stack to be further implemented, we also implemented it in LLVM and Qemu.
3. **Evaluation:** We quantitatively evaluated the runtime performance overhead of Zipper Stack, which is better than existing mechanisms.

2 Background and Related Work

2.1 ROP Attacks

Return Oriented Programming (ROP) [5,26] is the major form of code reuse attacks. ROP makes use of existing code snippets ending with return instructions called gadgets to perform malicious acts. In ROP attacks, the attackers link different gadgets by tampering with a series of return addresses. An ROP attack is usually made up of multiple gadgets. At the end of each gadget, a return instruction links the next gadget via the next address in the stack. The defenses against ROP mainly prevent return instructions from using corrupted return addresses or randomize the layout of the codes.

2.2 Shadow Stack and SafeStack

Shadow Stack is a typical technique to protect return addresses. Shadow Stack saves the return addresses in a separate memory area and checks the return addresses in the main stack when returns. It has been implemented in both compiler-based and instrumentation-based approaches [6,9,12,22,29]. SafeStack [19] is a similar way, which moves all the return addresses into a separated stack instead of backs up the return addresses. SafeStack is now implemented in LLVM as a component of CPI [33].

An isolated stack mainly brings about two problems: One is that memory isolation costs more memory overhead and implementation complexity. Another one is that security relies on the security of memory isolation, which is impractical. As the structure of the shadow stack is simply the copies of return addresses, it is fragile once the attacker can modify its memory area. For example, in Intel CET, the shadow stack's protection is provided by a new page attribute. But a similar approach in DEP is easily bypassed by a variety of methods modifying the page attribute in real-world attacks [18]. ASLR is also bypassed in real-world attacks that other implementations rely on. The previous work [7] constructed a variety of attacks on Shadow Stack.

2.3 CFI and Crypto-Based CFI

Control Flow Integrity (CFI), which first introduced by Abadi et al. [1,2], has been recognized as an important low-level security property. In CFI, runtime checks are added to enforce that jumps and calls/rets land only to valid locations that have been analyzed and determined ahead of execution [34,37,39]. The security and performance overhead of different implementations differ. Fine-grained CFI approaches will introduce significant overhead. However, coarse-grained CFI has lower performance overhead but enforces weaker restrictions, which is not secure enough [7]. Besides, Control-Flow Graphs (CFGs), which fine-grained CFI bases on, are constructed by analyzing either the disassembled binary or source code. CFG cannot be both sound and complete, so even if efficiency losses are not mainly considered, CFI is not a panacea for code reuse attacks [11]. Due to the above reasons, CFI is not widely deployed on real systems now.

To improve CFI, some implementations also introduce cryptography methods to solve problems such as inaccurate static analysis: CCFI [21], RAGuard [40], Pointer-Authentication [24]. Most of these methods are based on MAC: the MACs of protected key pointers, including return addresses, are generated, whereafter, authenticated before use. But all of these methods rely heavily on secret keys and cost tremendous performance overhead. Another problem of these mechanisms is replay attacks. The attackers can perform replay attacks by reusing the existing values in memory.

3 Threat Model

In this paper, we assume that a powerful attacker has the ability to read and overwrite arbitrary areas of memory. He tries to perform ROP (or ret2lib) attacks. This situation is widespread - for example, a controllable pointer out of bounds can help the attacker acquire the capability. Reasonably, the attacker cannot alter the value in the dedicated registers (called Top and Key registers in our design), since these registers cannot be accessed by general instructions.

The attacker in our assumption is more powerful than all previous works. Shadow Stacks assume that the attackers cannot locate or overwrite the shadow stack, which is part of the memory. In our work, we do not need that assumption, which means it can defend against more powerful attacks.

4 Design

In this section, we elaborate on the design of Zipper Stack in detail. Here, we take the hardware approach as an example. The design in other approaches is equivalent.

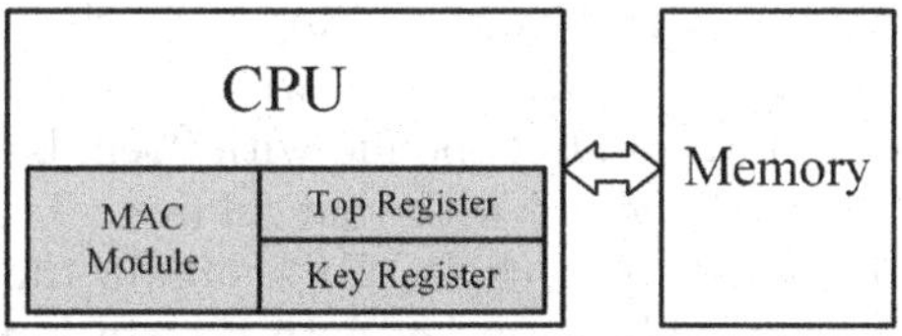

Fig. 2. Overview of Zipper Stack (hardware approach)

4.1 Overview

In the hardware approach, we need hardware support and the modification of memory layout. Figure 2 shows the overview of the hardware in Zipper Stack: Zipper Stack needs two dedicated registers and a MAC module in the CPU, but it requires no hardware modification of the memory. The registers include the Top register holding a MAC (Nm bits) and the Key register holding a secret key (Ns bits). Both the registers are initialized as random numbers at the beginning of a process, and they cannot be read nor rewritten by attackers. The secret key will not be altered in the same process. Therefore, we temporarily ignored this register for the sake of simplicity in the following. Assuming that the width of

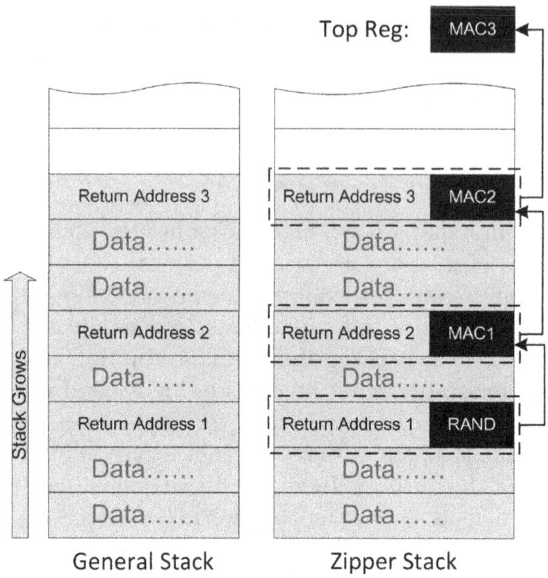

Fig. 3. Layout in Zipper Stack: The dotted rectangles in the figure indicate the input of the MAC function, and the solid lines indicate the storage location of the MAC.

return addresses is Na, the MAC module should perform a cryptographic MAC function with an input bit width of $Na + Nm$ and an output bit width of Nm.

We now turn to the memory layout of Zipper Stack. In Zipper Stack, all return addresses are bound to a MAC, as shown in Fig. 3. The novelty is, the MAC is not generated from the address bound with itself, but from the previous return address with the MAC bound with that address. This connection keeps all return addresses and MAC in a chain. To maintain the structure, the top one, namely the last return address pushed into the stack, is handled together with the previous MAC and the new MAC is saved into the Top register; while the bottom one, i.e., the first return address pushed into the stack is bound to a random number (exactly the initial value of Top register when the program begins).

4.2 Operations

Next, we describe how Zipper Stack works with return addresses in the runtime, i.e., how to handle the Call instructions and the Return instructions. As Fig. 4 shows.

Call: In general, the Call instructions perform two operations. First, push the address of the next instruction into the stack. Then, set the PC to the call destination. While in Zipper Stack, the Call instructions become slightly more complicated and need three steps:

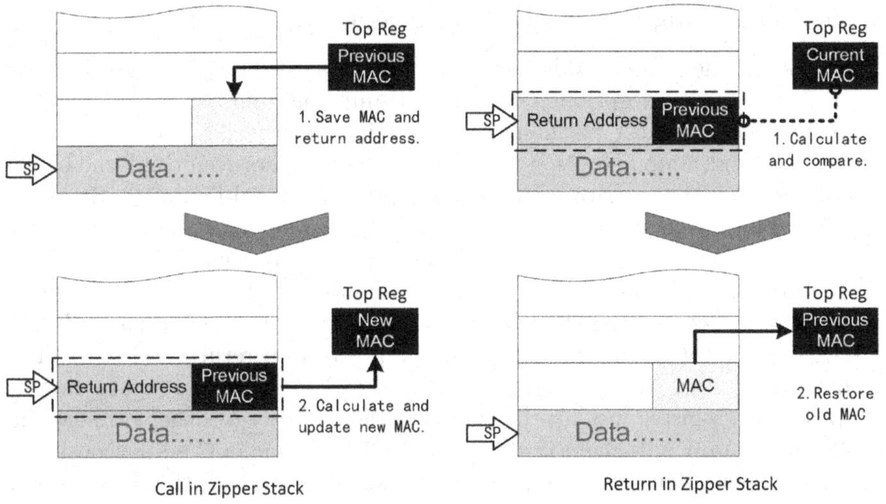

Fig. 4. The stack layout before/after a call/return. Previous MAC is generated from previous return address and the MAC with that return address. SP stands for stack pointer.

1. Push the Top register along with the return address into the main stack;
2. Calculate a new MAC of the Top register and the return address and save the new MAC into the Top register;
3. Set the PC to the call destination.

Return: In general, the Return instructions also perform two operations. First, pop the return address from the stack; second, set the PC to this address. Correspondingly, in Zipper Stack, Returns also become a little more complicated, including three steps.

1. Pop the return address and the previous MAC from the main stack and calculate the MAC of them. Then check whether it matches the Top register. If not, raise an exception (which means an attack).
2. Save the MAC poped from the stack into the Top register.
3. Set the PC to the return address.

Figure 4 shows the process of CALL and RET in Zipper Stack. We omit the normal operations about the PC and return addresses.

The core idea of Zipper Stack is to use a chain structure to link all return addresses together. Based on this structure, we only need to focus on the protection and verification of the top of the chain instead of protecting the entire structure. Just like a zipper, only the slider is active, and when the zipper is pulled up, the following structure automatically bites up. Obviously, protecting a MAC from tampering is much easier than protecting a series of MAC from tampering: Adding a special register in the CPU is enough, and there is no need to protect a special memory area.

4.3 Setjump/Longjump and C++ Exceptions

In most cases, the return addresses are used in a LIFO order (last in, first out). But there are exceptions, such as setjump/longjump and C++ exceptions. Consequently, most mechanisms protecting return addresses suffer from the Setjump/Longjump and C++ Exceptions, some papers even think the block-chaining like algorithm cannot work with exceptional control-flows [12]. However, Zipper Stack can accommodate both Setjump/Longjump and C++ Exceptions. Both Setjump/Longjump and C++ Exceptions mainly save and restore the context between different functions. The main task of them is stack unwind. The return addresses in the stack will not encounter any problem since Zipper Stack does not alter either the value nor the position of return addresses. The only problem is how to restore the Top register. The solution is quite simple: backup the Top register just like backup the stack pointer or other registers (in Setjump/Longjump save them in the jump buffer, similar in C++ Exceptions). When we need to Longjump or handle an exception, restore the Top register just as restoring other registers. The chain structure will remain tight.

However, the jump buffer is in memory, so this solution exposes the Top register (since the attacker can write on arbitrary areas of memory) and leaves an opportunity to overwrite the Top register. So additional protection is a must. In our implementation, we use a MAC to authenticate the jump buffer (or context record in C++ exception). In this way, we can reuse the MAC module.

5 Implementations

5.1 Hardware Approach

We introduce the hardware approach first. In hardware approach, we implemented a prototype of Zipper Stack by modifying the Rocket Chip Generator [35] and customized the RISC-V instruction set accordingly. We also added a MAC module, several registers, and several instructions into the core. Whereafter, we modified the toolchain, including the compiler and the library glibc. Besides, we implemented a similar Shadow Stack for comparison.

Core and New Instructions. In Rocket core, we added a Top register and a Key register, which correspond to those designed in the algorithm. These two registers cannot be loaded/stored via normal load/store instructions. At the beginning of a program, the Key and Top register are initialized by random values.

In RISC-V architecture, a CALL instruction will store the next PC, i.e., return address, to the *ra* register, and a RET instruction will read the address in the *ra* register and jump to the address. Consequently, two instructions were added in our prototype: ZIP (after call), UNZIP (before return). They will perform as a Zipper Stack's CALL/RET together with a normal CALL/RET.

For the sake of simplicity, we use a compressed structure. In RISC-V architecture, the return address in register *ra*, so we put the return address and the

MAC together into *ra*. In the current Rocket core, only lower 40 bits are used to store the address. Therefore, we use the upper 24 bits to hold the MAC. Correspondingly, our Top register is 24 bits. And our Key register is 64 bits.

Our new instructions will update a MAC (ZIP after a CALL) or check and restore a MAC (UNZIP before a RET). When a ZIP instruction is executed, the address in *ra* (only lower 40 bits) along with the old MAC in the Top register will be calculated into a new MAC. The new MAC is stored in the Top register. The old MAC is stored to the higher 24 bits in the *ra* register (the lower 40 bits remain unchanged). Correspondingly, when an UNZIP instruction is executed, the *ra* register (including the MAC and address) is calculated and compared with the Top register. If the values match, the MAC in *ra* (higher 24 bits) is restored into the Top register, and the higher 24 bits in *ra* is restored to zero. If the values do not match, an exception will be raised (which means an attack).

MAC Module. Next, we added a MAC module in the Rocket Core. Here, we use Keccak [3] (Secure Hash Algorithm 3 (SHA-3) in one special case of Keccak) as the MAC function. In our hardware implementation, the arguments of Keccak are as follows: $l = 4, r = 256, c = 144$ The main difference between our implementation and SHA-3 is that we use a smaller l: in SHA-3, $l = 6$. This module will take 20 cycles for one MAC calculation normally, and it costs 793 LUTs and 432 flip flops.

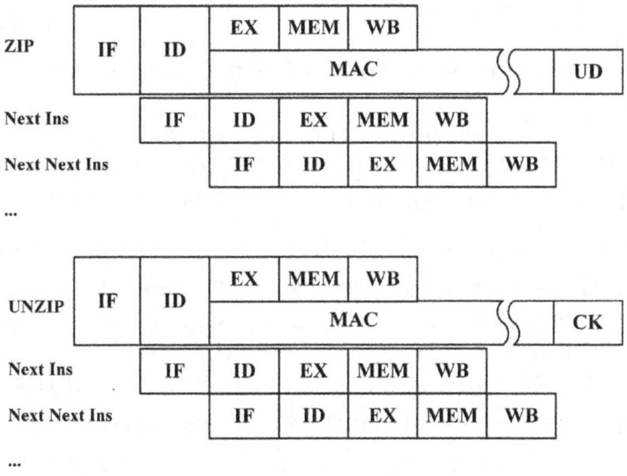

Fig. 5. Pipeline of ZIP and UNZIP

Pipeline. The pipeline in Rocket Core is a 5-stage single-issue in-order pipeline. To reduce performance overhead, the calculations are processed in parallel. If the next instruction that uses a MAC calculation arrives after the previous one finished, the pipeline will not stall. Figure 5 is a pipeline diagram of two instructions. In Fig. 5, IF, ID, EX, MEM, WB stand for fetch, decode, execute, memory, and write back stages. UD/CK stands for updating Top register/checking MACs. As the figure shows, if the next ZIP/UNZIP instruction arrives after the MAC calculation, only one extra cycle is added to the pipeline, which is equivalent to inserting a nop. It is worth noting that the WB stage in ZIP/UNZIP does not rely on the finish of the UD/CK stage: in the write back stage, it only writes the *ra* register, and the value does not rely on the current calculation. Fortunately, most functions require more cycles than 20. So in most cases, the ZIP/UNZIP instruction takes only one cycle.

Customized Compiler. To make use of the new instructions, we also customized the riscv-gcc. The modification on riscv-gcc is quite simple: Whenever we store a return address onto the stack, we add a ZIP instruction; whenever we pop a return address from the stack, we add a UNZIP instruction. It is noteworthy that, if a function will not call any function, *ra* register will not be spilled to the stack. So we only add the new instructions when the *ra* register is saved into/restored from the stack (rather than all calls and returns).

Setjmp/Longjmp Support. To support Setjmp/Longjmp, we also modified the glibc in the RISC-V tool chain. We have only modified two points:

1. Declaration of the Jump Buffer: Add additional space for the Top register and MAC.
2. Setjmp/Longjmp: Store/restore the Top register; Authenticate the data.

Our changes perfectly support Setjmp/Longjmp, which is verified in some benchmarks in SPEC2000, such as perlbmk. These benchmarks will not pass without Setjmp/Longjmp support.

Optimization. In order to further reduce the runtime overhead, we also optimized the MAC module. We added a small cache (with a size of 4) to cache the recent results. If a new request can be found in the cache, the calculation will take only one cycle. This optimization slightly increased the complexity of the hardware, but significantly reduced runtime overhead (by around 30%, see Sect. 6).

A Comparable Hardware Based Shadow Stack. In order to compare with Shadow Stacks, we also implemented a hardware supported Shadow Stack on Rocket Core. We tried to be consistent as much as possible: We added two instructions that back up or check the return address, and a pointer pointing the shadow stack. The compiler with Shadow Stacks inserts the instructions just like the way in Zipper Stack.

5.2 LLVM

In customized compiler approach, we implemented Zipper Stack algorithm based on LLVM 8.0. First we allocate some registers: we set the lower bits of XMM15 as the Top register, and several XMM registers as the Key register (we use different size of Key, corresponding to different numbers of rounds of AES-NI, see next part). We modified the backend of the LLVM so as to forbid these registers to be used by anything else. Therefore, the Key and Top register will not be leaked. At the beginning of a program, the Key and Top register are initialized by random numbers. We also modified the libraries so that the libraries will not use these registers.

MAC Function. Our implementation leverages the AES-NI instructions [13] on the Inter x86-64 architecture to minimize the performance impact of MAC calculation. We use the Key register as the round key of AES-NI, and use one 128-bit AES block as the input (64-bit address and 64-bit MAC). The 128-bit result is truncated into 64-bit in order to fit our design. We use different rounds of AES-NI to test the performance overhead: a) standard AES, performs 10 rounds of an AES encryption flow, we mark it as *full*; b) 5 rounds, marked as *half*; c) one round, marked as *single*. Obviously, the *full* MAC function is the most secure one, and the *single* is the fastest one.

Operations. In each function, we insert a prologue at the entry, and an epilogue before the return. In the prologue, the old Top register is saved onto stack, and updated to the new MAC of the current return address and the old Top register value. In the epilogue, the MAC in the stack and return address are authenticated using the Top register. If it doesn't match, an exception will be raised. Just as we introduced before.

5.3 Qemu

To figure out whether it is possible to run the existing binaries if we change the logic of calls and returns, we customized the x86-64 instruction set and simulated it with Qemu. All the simulation is in the User Mode of Qemu 2.7.0.

The modification is quite concise: we add two registers and change the logic of Call and Ret instructions. As the algorithm designed, the Call instruction will push the address and the Top register, update the Top register with a new MAC, while Ret instruction will pop the return address and check the MAC. Since Qemu uses lower 39 bits to address the memory, we use the upper 25 bits to store the MAC. Correspondingly, the width of the Top register is also 25 bits. The Key register here is 64 bits. We used SHA-3 as the MAC function in this implementation. Since we did not change the stack structure, this implementation has good binary compatibility. Therefore it can help us to evaluate the security with real x86-64 attacks.

Table 1. Security comparison

Adversary	Stack Protector	ASLR	Shadow Stack	MAC/Encryption	Zipper Stack
Stack Overflow	safe	safe	safe	safe	safe
Arbitrary Write	unsafe	safe	safe	safe	safe
Memory Leak & Arbitrary Write	unsafe	unsafe	unsafe	safe	safe
Secret Leak & Arbitrary Write	N/A	N/A	N/A	unsafe	safe
Replay Attack	unsafe	unsafe	N/A	unsafe	safe
Brute-force Attempts	N/A	N/A	N/A	2^{Ns-1}	$2^{Ns-1} + N * 2^{Nm-1}$ *

*: Under a probability of $1 - (1 - 1/e)^N$, the valid collision of a certain attack does not exist.
N: The number of gadgets in an attack; Nm: Bit width of MAC/ciphertext; Ns: Bit width of secret.
The Brute-force attack already contains the memory leak, so the ALSR is noneffective and not considered.

6 Evaluation and Analysis

We evaluated Zipper Stack in two aspects: Runtime Performance and Security. We evaluated the performance overhead with the SPEC CPU 2000 on the FPGA and SPEC CPU 2006 in LLVM approach. Besides, we also tested the compatibility of modified x86-64 ISA with existing applications.

6.1 Security

We first analyze the security of Zipper Stack, then show the attack test results, and finally, compare the security of different implementations.

Security Analysis. The challenge for the attacker is clear: how to tamper with the memory to make the fake return address be used and bypass our check? We list the defense effect of different methods of protecting return addresses in the face of different attackers in Table 1. The table shows that Zipper Stack has higher security than Shadow Stack and other cryptography-based protection mechanisms.

Direct Overwrite. First, we consider direct overwrite attacks. In the previous cryptography-based methods, the adversary cannot know the key and calculate the correct MAC, so it is secure. But we go further that the adversary may steal the key and calculate the correct MAC. As Fig. 6 shows, to tamper with any return address structure and bypass the check (let's say, Return Address N), the attacker must bypass the pre-use check. Even if the attacker has stolen the key, the attacker needs to tamper with the MAC, which is used to check the return address, i.e., the MAC stored beside Return Address N+1. Since the MAC and the return address are authenticated together, the attacker has to modify the MAC stored beside Return Address N+2 to tamper with the MAC bound to Return Address N+1. And so on, the attacker has to alter the MAC at the top, which is in the Top register. As we have assumed, the register is secure against tampering. As a result, an overwrite attack won't work.

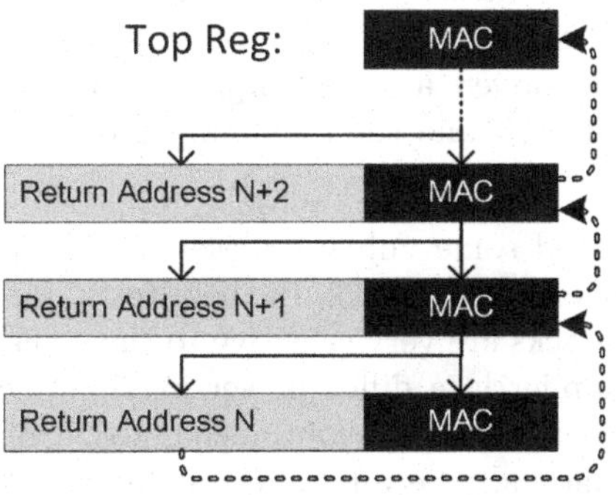

Fig. 6. Direct Overwrite Attack: The solid lines indicate the protect relation of the MACs, and the dotted lines show the order that the attacker should overwrite in.

Replay Attack. Next, we discuss replay attacks: since we have assumed that the attacker can read all the memory, is the attacker able to utilize the protected addresses and their MACs in memory to play a replay attack? In Zipper Stack, it is not feasible because once the call path has changed, the MAC will be updated immediately. The MACs in an old call path cannot work in a new call path, even if the address is the same. This is an advantage over previous cryptography-based methods: Zipper Stack resist replay attacks naturally.

Brute-Force Attack. Then we discuss the security of Zipper Stack in the face of brute force. Here we consider the attacks which read all related data in memory, guess the secret constantly, and finally construct the attack. In the cryptography-based approaches, security is closely related to the entropy of the secret. Here in Zipper Stack, the entropy of the secret is the bit width of Key register (Ns), which means the attacker needs to guess the correct Key register in a space of size 2^{Ns}. Once the attacker gets the correct key, the attack becomes equivalent to an attack with a secret leak.

Attack with Secret Leak. The difference between Zipper Stack and other cryptography-based approaches is, other approaches will fail to protect control flow once the attacker knows the secret, but Zipper Stack will not. Because even if the attacker knows the secret key, the Top register cannot be altered. If an attack contains N gadgets, the attacker needs to find N collisions whose input contains the ROP (or ret2lib) gadget addresses to use the gadgets and bypass the check[1]. Considering an ideal MAC function, one collision will take about 2^{Nm-1}

[1] The same gadget addresses won't share the same collision, because the MACs bound to them differ.

times of guesses on average[2]. So an attack with N gadgets will take $N * 2^{Nm-1}$ times of guesses on average even if the Key register is leaked. The total number of guesses is (guessing Key value and the collisions) $2^{Ns-1} + N * 2^{Nm-1}$. And more unfortunate for the attackers: under a certain probability $(1 - (1 - 1/e)^N$ for an attack contains N gadgets), the valid collision does not even exist. Take an attack that contains five gadgets as an example, the possibility that the collision does not exist is around 90%, and the possibility grows as the N grows.

Attack Tests. We tested some vulnerabilities and the corresponding attacks to evaluate the security of Zipper Stack. In these tests, we use Qemu implementation, because most attacks are very sensitive to the stack layout, a customized compiler (or just a compiler in a different version) may lead to failures. Using Qemu simulation can keep the stack layout unchanged, avoid the illusion that the defense works, which is actually because of the stack layout changes. We wrote a test suite contains 18 attacks and the corresponding vulnerable programs. Each attack contains at least one exploit on return addresses, including stack overflow, ROP gadget, or ret2lib gadget. All attacks are detected and stopped (all of them will alter the MAC and cannot pass the check during return). These tests show that Zipper Stack is reliable. Table 2 shows the difference between Zipper Stack and other defense methods.

Table 2. Attack test against defenses

Defence	# Applied	# Secured	# Bypassed
DEP	20	2	18
ASLR	8	2	6
Stack Canary	6	2	4
Coarse-Grained CFI	16	10	6
Shadow Stacks	2	0	2
Zipper Stack	20	20	0

Attacks Against Shadow Stacks. The above attacks can also be stopped by Shadow Stacks. To further prove the Zipper Stack's security advantages compared to Shadow Stacks, we also wrote two proof-of-concept attacks (corresponding to two common types of shadow stack) that can bypass the shadow stack. As introduced in previous work [7], Shadow Stack implementations have various

[2] The MAC function is a $(Nm + Na)$bit-to-Nmbit function, so choosing a gadget address will determine Na bits of input, which means there are 2^{Nm} optional values. On average, there is one collision in the values, since every $2^{Nm+Na}/2^{Nm} = 2^{Na}$ inputs share the same MAC value. Because of our special algorithm, this is not a birthday attack nor an ordinary second preimage attack, but a limited second preimage attack.

Table 3. Result of SPEC 2000 in hardware approach

Benchmark	Baseline (seconds)	Shadow Stack (seconds/slowdown)	Zipper Stack (seconds/slowdown)	Zipper Stack (optimized) (seconds/slowdown)
164.gzip	10923.10	10961.65 (0.35%)	10960.60 (0.34%)	10948.88 (0.24%)
175.vpr	7442.48	7528.06 (1.15%)	7490.49 (0.65%)	7485.40 (0.58%)
176.gcc	8227.93	8318.83 (1.10%)	8348.99 (1.47%)	8317.34 (1.09%)
181.mcf	11128.67	11153.01 (0.22%)	11183.31 (0.49%)	11168.93 (0.36%)
186.crafty	10574.27	10942.89 (3.49%)	10689.74 (1.09%)	10692.53 (1.12%)
197.parser	8318.16	8577.67 (3.12%)	8658.89 (4.10%)	8544.72 (2.72%)
252.eon	14467.81	15111.99 (4.45%)	15519.98 (7.27%)	15040.26 (3.96%)
253.perlbmk	7058.96	7310.78 (3.57%)	7388.20 (4.66%)	7342.20 (4.01%)
254.gap	7728.56	7850.32 (1.58%)	7926.10 (2.56%)	7817.52 (1.15%)
255.vortex	13753.47	14738.06 (7.16%)	14748.70 (7.24%)	14644.70 (6.48%)
256.bzip2	6829.01	6893.50 (0.94%)	6954.81 (1.84%)	6865.27 (0.53%)
300.twolf	11904.25	12044.16 (1.18%)	11974.22 (0.59%)	11917.37 (0.11%)
Average		2.36%	2.69%	1.86%

flaws and can be attacked via different vulnerabilities. We constructed similar attacks. In both PoC attacks, we corrupt the shadow stack before we overwrite the main stack and perform ROP attacks. In the first example, the shadow stack is parallel to the regular stack, as introduced in [9]. The layout of the shadow stack is easy to obtain because its offset to the main stack is fixed. In the second example, the shadow stack is compact (only stores the return address). The offset of this shadow stack to the main stack is not fixed, so we used a memory leak vulnerability to locate the shadow stack and the current offset. In both cases, Shadow Stack can not stop the attacks, but Zipper Stack can. We also added both attacks to Table 2.

Security of Different Approaches. The security of Zipper Stack is mainly affected by two aspects: the security of Top register and the length of MAC, corresponding to the risk of tamper of MAC and the risk of brute-force attack. From the Top register's perspective, the protection in the hardware and Qemu approach is complete. In both implementations, the program cannot access the Top register except the *call* and *return*, so there is no risk of tampering. In the LLVM approach, an XMM register is used as the Top register, so it faces the possibility of being accessed. We banned the use of XMM15 in the compiler and recompiled the libraries to prevent access to the Top register, but there are still risks. For example, there may be unintended instructions in the program that can access the XMM15 register. From the perspective of MAC length, MAC in LLVM approach is 64-bit wide and therefore has the highest security. MAC in hardware is 24-bit wide and Qemu is 25-bit wide, so they have comparable security but still lower than LLVM approach.

6.2 Performance Overhead

We evaluated the performance overhead of different approaches. The cost of the hardware approach is significantly lower than that of the LLVM approach. It is worth noting that QEMU is not designed to reflect the guest's actual performance, it only guarantees the correctness of logic[3]. Thus, we do not evaluate the performance overhead of the QEMU approach.

SPEC CINT 2000 in Hardware Approach. To evaluate the performance of Zipper Stack on RISC-V, we instantiated it on the Xilinx Zynq VC707 evaluation board and ran the SPEC CINT 2000 [14] benchmark suite (due to the limited computing power of the Rocket Chip on FPGA, we chose the SPEC 2000 instead of SPEC 2006). The OS kernel is Linux 4.15.0 with support for the RISC-V architecture. The hardware and GNU tool-chain are based on *freedom* (commit *cd9a525*) [28]. All the benchmarks are compiled with GCC version 7.2.0 and -O2 optimization level. We ran each benchmark for 3 times.

Table 3 shows the results of Zipper Stack and Shadow Stacks. The result shows that without optimization, Zipper Stack is slightly slower than Shadow Stacks (2.69% vs 2.36%); while with optimization (the cache), Zipper Stack is much faster than Shadow Stacks (1.86% vs 2.36%). To sum up, the runtime overhead of Zipper Stack is satisfactory (1.86%).

Table 4. Result of SPEC 2006 in LLVM approach

Benchmark	Baseline (seconds)	Full (seconds/slowdown)	Half (seconds/slowdown)	Single (seconds/slowdown)
401.bzip2	337.69	385.98 (14.30%)	358.32 (5.76%)	336.33 (−0.40%)
403.gcc	240.38	341.06 (41.89%)	305.94 (21.43%)	272.42 (13.33%)
445.gobmk	340.28	493.47 (45.02%)	427.90 (20.48%)	384.91 (13.12%)
456.hmmer	269.62	270.50 (0.33%)	268.81 (−0.30%)	268.74 (−0.33%)
458.sjeng	403.38	460.58 (14.18%)	435.94 (7.47%)	421.69 (4.54%)
462.libquantum	251.50	265.50 (5.57%)	261.17 (3.70%)	256.72 (2.07%)
464.h264ref	338.29	410.56 (21.37%)	381.11 (11.24%)	355.97 (5.23%)
473.astar	285.58	362.00 (50.11%)	333.91 (14.47%)	311.26 (8.99%)
Average		21.13%	8.96%	4.75%

Performance Overhead in LLVM Approach. To evaluate the performance of Zipper Stack on customized compiler, we run the SPEC CPU 2006 [15] compiled by our customized LLVM (with optimization -O2).

[3] For example, QEMU will optimize basic blocks. Some redundant instructions (e.g. a long nop slide) may be eliminated directly, which cannot reflect real execution performance.

Table 4 shows the performance overhead of Zipper Stack in LLVM approach[4]. Shadow Stack is reported to cost about 2.5–5% [9,31]. Our approach costs 4.75% ~21.13% accroding to the results, depending on how many rounds we perform in MAC function. It does not cost too much overhead if we use only one round of AES-NI, although it is still slower than hardware approach. Furthermore, it costs 21.13% when we perform a standard AES encryption, which is faster than CCFI [21], since we encrypt less pointers than CCFI.

6.3 Compatibility Test

Here, we test the binary compatibility of Zipper Stack. This test is only valid for Qemu implementation. In the other two implementations, due to we have modified the compiler, we can use Zipper Stack as long as we recompile the source code, so there is no compatibility issue. The purpose of this test is: If we only modify the Call and Ret instructions in the x86-64 ISA, and use the compression structure to maintain the stack layout, is it possible to maintain binary compatibility directly?

We randomly chose 50 programs in Ubuntu (under the path /usr/bin) to test the compatibility in Qemu. 42 out of 50 programs are compatible with our mechanism. Most failures are due to the Setjmp/Longjmp, which we have not supported in Qemu yet. So we think although some issues need to be solved (such as the setjmp/longjump), Zipper Stack can be used directly on most existing x86-64 binaries.

7 Conclusion

In this paper, we proposed Zipper Stack, a novel algorithm of return address protection, which authenticates all return addresses by a chain structure using MAC. It minimizes the amount of state requiring direct protection and costs low performance overhead.

Through our analysis, Zipper Stack is an ideal way to protect return addresses, and we think it is a better alternative to Shadow Stack. We discussed various possible attackers and attacks in detail, concluding that an attacker cannot bypass Zipper Stack and then counterfeit the return addresses. In most cases, Zipper Stack is more secure than existing methods. The simulation of attacks on Qemu also corroborates the security of Zipper Stack. Our experiment also evaluated the runtime performance of Zipper Stack, and the results have shown that the performance loss of Zipper Stack is very low. The performance overhead with hardware support based on Rocket Core is only 1.86% on average (versus a hardware-based Shadow Stack costs 2.36%). Thus, the proposed design is suitable for actual deployment.

[4] The performance gain is due to memory caching artifacts and fluctuations.

References

1. Abadi, M., Budiu, M., Erlingsson, U., Ligatti, J.: Control-flow integrity. In: Proceedings of the 12th ACM Conference on Computer and Communications Security, CCS 2005, pp. 340–353. ACM, New York (2005). https://doi.org/10.1145/1102120.1102165. http://doi.acm.org/10.1145/1102120.1102165
2. Abadi, M., Budiu, M., Erlingsson, U., Ligatti, J.: Control-flow integrity principles, implementations, and applications. ACM Trans. Inf. Syst. Secur. **13**(1), 4:1–4:40 (2009). https://doi.org/10.1145/1609956.1609960. http://doi.acm.org/10.1145/1609956.1609960
3. Bertoni, G., Daemen, J., Peeters, M., Van Assche, G.: Keccak sponge function family main document. Submission to NIST (Round 2) **3**(30) (2009)
4. Bittau, A., Belay, A., Mashtizadeh, A., Mazières, D., Boneh, D.: Hacking blind. In: IEEE Symposium on Security and Privacy (SP), pp. 227–242. IEEE (2014)
5. Buchanan, E., Roemer, R., Shacham, H., Savage, S.: When good instructions go bad: generalizing return-oriented programming to RISC. In: Proceedings of the 15th ACM Conference on Computer and Communications Security, CCS 2008, pp. 27–38. ACM, New York (2008). https://doi.org/10.1145/1455770.1455776. http://doi.acm.org/10.1145/1455770.1455776
6. Chiueh, T.C., Hsu, F.H.: RAD: a compile-time solution to buffer overflow attacks. In: Proceedings 21st International Conference on Distributed Computing Systems, pp. 409–417, April 2001. https://doi.org/10.1109/ICDSC.2001.918971
7. Conti, M., et al.: Losing control: on the effectiveness of control-flow integrity under stack attacks. In: Proceedings of the 22nd ACM SIGSAC Conference on Computer and Communications Security, pp. 952–963 (2015)
8. Cowan, C., et al.: StackGuard: automatic adaptive detection and prevention of buffer-overflow attacks. In: USENIX Security Symposium, San Antonio, TX, vol. 98, pp. 63–78 (1998)
9. Dang, T.H., Maniatis, P., Wagner, D.: The performance cost of shadow stacks and stack canaries. In: Proceedings of the 10th ACM Symposium on Information, Computer and Communications Security, ASIA CCS 2015, pp. 555–566. ACM, New York (2015). https://doi.org/10.1145/2714576.2714635. http://doi.acm.org/10.1145/2714576.2714635
10. Davi, L., Sadeghi, A.R., Winandy, M.: ROPdefender: a detection tool to defend against return-oriented programming attacks. In: Proceedings of the 6th ACM Symposium on Information, Computer and Communications Security, ASIACCS 2011, pp. 40–51. ACM, New York (2011). https://doi.org/10.1145/1966913.1966920. http://doi.acm.org/10.1145/1966913.1966920
11. Evans, I., et al.: Control jujutsu: on the weaknesses of fine-grained control flow integrity. In: Proceedings of the 22nd ACM SIGSAC Conference on Computer and Communications Security, CCS 2015, pp. 901–913. ACM, New York (2015). https://doi.org/10.1145/2810103.2813646. http://doi.acm.org/10.1145/2810103.2813646
12. Frantzen, M., Shuey, M.: StackGhost: hardware facilitated stack protection. In: USENIX Security Symposium, vol. 112 (2001)
13. Gueron, S.: Intel® advanced encryption standard (AES) new instructions set. Intel Corporation (2010)
14. Henning, J.L.: SPEC CPU2000: measuring CPU performance in the new millennium. Computer **33**(7), 28–35 (2000)

15. Henning, J.L.: SPEC CPU2006 benchmark descriptions. SIGARCH Comput. Archit. News **34**(4), 1–17 (2006). https://doi.org/10.1145/1186736.1186737. http://doi.acm.org/10.1145/1186736.1186737
16. Hund, R., Willems, C., Holz, T.: Practical timing side channel attacks against kernel space ASLR. In: 2013 IEEE Symposium on Security and Privacy (SP), pp. 191–205, May 2013. https://doi.org/10.1109/SP.2013.23. http://doi.ieeecomputersociety.org/10.1109/SP.2013.23
17. Intel: Control-flow enforcement technology preview (2016). https://software.intel.com/sites/default/files/managed/4d/2a/control-flow-enforcement-technology-preview.pdf
18. Katoch, V.: Whitepaper on bypassing ASLR/DEP (2011). https://www.exploit-db.com/docs/english/17914-bypassing-aslrdep.pdf
19. Kuznetsov, V., Szekeres, L., Payer, M., Candea, G., Sekar, R., Song, D.: Code-pointer integrity. In: OSDI, vol. 14 (2014)
20. Marco-Gisbert, H., Ripoll-Ripoll, I.: Exploiting Linux and PaX ASLR's weaknesses on 32-and 64-bit systems (2016)
21. Mashtizadeh, A.J., Bittau, A., Boneh, D., Mazières, D.: CCFI: cryptographically enforced control flow integrity. In: Proceedings of the 22nd ACM SIGSAC Conference on Computer and Communications Security, CCS 2015, pp. 941–951. ACM, New York (2015). https://doi.org/10.1145/2810103.2813676. http://doi.acm.org/10.1145/2810103.2813676
22. Ozdoganoglu, H., Vijaykumar, T., Brodley, C.E., Kuperman, B.A., Jalote, A.: SmashGuard: a hardware solution to prevent security attacks on the function return address. IEEE Trans. Comput. **55**(10), 1271–1285 (2006)
23. Prasad, M., Chiueh, T.C.: A binary rewriting defense against stack based buffer overflow attacks. In: USENIX Annual Technical Conference, General Track, pp. 211–224 (2003)
24. Qualcomm Technologies Inc.: Pointer authentication on armv8.3 (2017). https://www.qualcomm.com/media/documents/files/whitepaper-pointer-authentication-on-armv8-3.pdf
25. Seibert, J., Okhravi, H., Söderström, E.: Information leaks without memory disclosures: remote side channel attacks on diversified code. In: Proceedings of the 2014 ACM SIGSAC Conference on Computer and Communications Security, CCS 2014, pp. 54–65. ACM, New York (2014). https://doi.org/10.1145/2660267.2660309. http://doi.acm.org/10.1145/2660267.2660309
26. Shacham, H.: The geometry of innocent flesh on the bone: return-into-libc without function calls (on the x86). In: Proceedings of the 14th ACM Conference on Computer and Communications Security, CCS 2007, pp. 552–561. ACM, New York (2007). https://doi.org/10.1145/1315245.1315313. http://doi.acm.org/10.1145/1315245.1315313
27. Shacham, H., Page, M., Pfaff, B., Goh, E.J., Modadugu, N., Boneh, D.: On the effectiveness of address-space randomization. In: Proceedings of the 11th ACM Conference on Computer and Communications Security, pp. 298–307. ACM (2004)
28. SiFive: Sifive's freedom platforms (2015). https://github.com/sifive/freedom
29. Sinnadurai, S., Zhao, Q., Wong, W.-F.: Transparent runtime shadow stack: protection against malicious return address modifications (2008)
30. Strackx, R., Younan, Y., Philippaerts, P., Piessens, F., Lachmund, S., Walter, T.: Breaking the memory secrecy assumption. In: Proceedings of the Second European Workshop on System Security, EUROSEC 2009, pp. 1–8. ACM, New York (2009). https://doi.org/10.1145/1519144.1519145. http://doi.acm.org/10.1145/1519144.1519145

31. Szekeres, L., Payer, M., Wei, T., Song, D.: SoK: eternal war in memory. In: Proceedings of the IEEE Symposium on Security and Privacy (SP), pp. 48–62 (2013)
32. Pax Team: PaX address space layout randomization (ASLR) (2003)
33. The Clang Team: Clang 3.8 documentation safestack (2015). http://clang.llvm.org/docs/SafeStack.html
34. Tice, C., et al.: Enforcing forward-edge control-flow integrity in GCC & LLVM. In: USENIX Security, vol. 26, pp. 27–40 (2014)
35. UC Berkeley Architecture Research: Rocket chip generator (2012). https://github.com/freechipsproject/rocket-chip
36. UC Berkeley Architecture Research: The RISC-V instruction set architecture (2015). https://riscv.org/
37. van der Veen, V., et al.: A tough call: mitigating advanced code-reuse attacks at the binary level. In: 2016 IEEE Symposium on Security and Privacy (SP), pp. 934–953. IEEE (2016)
38. Wagle, P., Cowan, C., et al.: StackGuard: simple stack smash protection for GCC. In: Proceedings of the GCC Developers Summit, pp. 243–255. Citeseer (2003)
39. Zhang, C., et al.: Practical control flow integrity and randomization for binary executables. In: 2013 IEEE Symposium on Security and Privacy (SP), pp. 559–573. IEEE (2013)
40. Zhang, J., Hou, R., Fan, J., Liu, K., Zhang, L., McKee, S.A.: RAGuard: a hardware based mechanism for backward-edge control-flow integrity. In: Proceedings of the Computing Frontiers Conference, CF 2017, pp. 27–34. ACM, New York (2017). https://doi.org/10.1145/3075564.3075570. http://doi.acm.org/10.1145/3075564.3075570

Restructured Cloning Vulnerability Detection Based on Function Semantic Reserving and Reiteration Screening

Weipeng Jiang[1,2], Bin Wu[1,2(✉)], Xingxin Yu[1,2], Rui Xue[1,2], and Zhengmin Yu[1,2]

[1] State Key Laboratory of Information Security, Institute of Information Engineering, Chinese Academy of Sciences, Beijing, China
{jiangweipeng,wubin,yuxingxin,xuerui1,yuzhengmin}@iie.ac.cn
[2] School of Cyber Security, University of Chinese Academy of Sciences, Beijing, China

Abstract. Although code cloning may speed up the process of software development, it could be detrimental to the software security as undiscovered vulnerabilities can be easily propagated through code clones. Even worse, since developers tend not to simply clone the original code fragments, but also add variable and debug statements, detecting propagated vulnerable code clone is challenging. A few approaches have been proposed to detect such vulnerability- named as restructured cloning vulnerability; However, they usually cannot effectively obtain the vulnerability context and related semantic information. To address this limitation, we propose in this paper a novel approach, called RCVD++, for detecting restructured cloning vulnerabilities, which introduces a new feature extraction for vulnerable code based on program slicing and optimizes the code abstraction and detection granularity. Our approach further features reiteration screening to compensate for the lack of retroactive detection of fingerprint matching. Compared with our previous work RCVD, RCVD++ innovatively utilizes two granularities including line and function, allowing additional detection for exact and renamed clones. Besides, it retains more semantics by identifying library functions and reduces the false positives by screening the detection results. The experimental results on three different datasets indicate that RCVD++ performs better than other detection tools for restructured cloning vulnerability detection.

Keywords: Code vulnerability · Clone detection · Program slicing

1 Introduction

Given the diversity and accessibility of abundant open source code, code reuse occurs frequently during software development for the sake of convenience. Considering two similar software products P_A and P_B, if P_A's source code has already been open-accessed, then P_B's developers can quickly develop a similar function with the help of P_A's source code. At the same time, vulnerabilities in P_B's source code would also be introduced into P_B, which are the so called clone-caused vulnerabilities [1]. Although P_A's open source

L. Chen et al. (Eds.): ESORICS 2020, LNCS 12308, pp. 359–376, 2020.
https://doi.org/10.1007/978-3-030-58951-6_18

code may be patched after the vulnerability is disclosed, a large number of previous research work [2–9] shows that P_B's developers rarely upgrade and maintain the cloned codes, leaving a great number of security threats.

When cloning occurs, developers often modify the cloned code to meet specific functional requirements. Consider again the above two software P_A and P_B. After cloning, P_B's developers may add a bunch of assert statements for debugging or delete statements that are not related to its functional requirements. These modifications may or may not affect vulnerabilities in the code, making the vulnerability detection more difficult. For example, VUDDY [2], the most accurate cloning vulnerability detection tool, has been proven to be completely unable to detect such cloning vulnerabilities by experiments in literature [11].

If the cloned code still has vulnerabilities after the statement is modified, then this kind of vulnerability is called restructured cloning vulnerability. At present, detection tools for such cloning vulnerabilities include Deckard [10], ReDeBug [5], RCVD [11], etc., but they are less effective in retaining semantics of vulnerabilities, and the extracted vulnerability codes are not accurate enough. In addition, these tools also lack the retrospective filtering of the detection results, leading to incorrect identification of well-patched code.

In order to overcome the shortcomings of current detection tools in detecting restructured cloning vulnerability, we propose an approach that introduces semantic reserving function abstraction and reiteration screening. The proposed abstraction method can retain more semantic information, so it can obtain more accurate vulnerability fingerprints and improve the accuracy of detection. The reiteration screening method can effectively identify the patched codes in the detection results, and significantly reduce the detection false positives.

The main contributions of this paper are as follows:

Semantic Retention of C/C++ Library Functions. The semantic information related to functions is lost during tokenization and abstraction process of existing cloning vulnerability detection. Therefore, we propose a semantic-reserved code abstraction method, which could reduce semantic missing by retaining C/C++ library function names.

Screening of Detection Results Based on Retroactive Detection. The current detection methods cannot effectively identify similar codes before and after patching, which will result in a large number of patched codes still being mistakenly identified as vulnerabilities. Therefore, our method traces the detection results back to the hit fingerprints and looks for hash of official-patch fingerprint corresponding with hit fingerprints to furtherly eliminate this kind of false alarm results.

Accurate Cloning Vulnerabilities Detector. We implement our detection tool called RCVD++, which improves the abstraction method in fingerprint generation, and adds the features of retrospective filtering of the detection results, retaining more vulnerability semantics and reducing false positives. Compared to RCVD and ReDeBug, RCVD++ performs better. In comparison with Deckard and CloneWorks [12], the best restructured clone detection tools available today, RCVD++ outperforms their detection results.

The rest of the paper is organized as follows: Sect. 2 gives a brief introduction to clone detection and vulnerable code extraction. Then we propose the system architecture of our method in Sect. 3 and introduce the details of our method in Sect. 4 and Sect. 5. Moreover, we evaluate our method and compare it with other detection methods in Sect. 6, and introduce the related work about cloning vulnerable code detection in Sect. 7. At last, we conclude this paper in Sect. 8.

2 Background

In this section, we introduce some common concepts and describe the process of our proposed code clone detection, followed by the demonstration of program slicing.

2.1 Clone Detection

Clone Type. A code fragment is recognized as a cloned one if it satisfies several given definitions of similarity [13]. Currently, code clone mainly includes four types [2], which is widely accepted:

Type-1: Exact Clones. The code is copied directly without any modifications.
Type-2: Renamed Clones. These are syntactically identical clones except for the modification of identifiers, literals, types, whitespace, layout and comments.
Type-3: Restructured Clones. Based on the cloning of Type-2, copied code fragments are further modified such as added, deleted or modified statements. The proposed method can detect Type-1, Type-2 and Type-3 clones.
Type-4: Semantic Clones. The two code fragments implement the same function and have the same semantics, but their syntax is different.

Now, most of the research focuses on the first three types, while a few other efforts focus on detecting Type-4 cloning vulnerability. This is because vulnerabilities are sensitive to grammars; two pieces of code with the same functionality may not have the same vulnerabilities. Although Yamaguchi, et al. [14] propose a method to detect Type-4 clones, their method relies heavily on costly operations, and the detection accuracy is not precisely shown in their analysis.

Detection Granularity. Different detection methods apply different granularities, which influence the accuracy of detection result. The detection granularity is the smallest unit of comparison in the detection process. At present, the code clone detection is mainly composed of five different granularities [2].

Token: This is the smallest meaningful unit that makes up a program. For example, in the statement '`int x;`' three tokens exist: '`int`', 'x' and '`;`'.
Line: A sequence of tokens delimited by a new-line character.
Function: A collection of consecutive lines that perform a specific task.
File: A file contains a set of functions.
Program: A collection of files.

2.2 Vulnerable Code Extraction

When detecting the cloning vulnerable code, it is necessary to extract the corresponding vulnerable code from the function to characterize the vulnerability. The methods currently used mainly include program slicing, sliding windows and AST representation. The latter two methods are relatively simple, but the semantic information obtained is not as much as in the program slicing method, so we also adopt program slicing in our method.

Program Slicing

Program slicing is a technique for extracting code snippets which would affect specific data from a target program. The specific data may consist of variables, processes, objects or anything that users are interested in.

Typically, program slicing is based on a slicing criterion, which consists of a pair <*p*, *V*>, where *p* is a program point and *V* is a subset of program variables [17]. In addition, program slicing can be performed in a backward or forward fashion. A backward slice consists of all statements which could affect the computation of the criterion, while a forward slice contains all statements which could be affected by the criterion.

3 System Architecture

The main idea of RCVD++ is to use code patches and program slicing to obtain code fragments related to vulnerabilities, abstract the code fragments into precise fingerprint with semantic reserving function abstraction and retroactively screen the detection result to reduce false alarm.

3.1 Overall Structure

There are two stages in RCVD++ as shown in Fig. 1: fingerprint database generation and cloning vulnerability detection. The first stage includes preprocessing, Type-3 fingerprint and Type-1&Type-2 fingerprint generation. The second stage includes hash lookup, greedy-based fingerprint matching and reiteration screening.

3.2 System Flow

In the first stage, fingerprints of vulnerable code will be generated. We use two granularities, *Line* and *Function* to generate fingerprints, where *Line* granularity is used to generate Type-3 fingerprints, and Function granularity is used to generate Type-1&Type-2 fingerprints and official-patch fingerprints. The official-patch fingerprints will be linked with the Type-3 fingerprints for screening. Our method uses vulnerability-related keywords to retrieve vulnerable functions, and generates corresponding fingerprints. For Type-3 fingerprints, we first use program slicing to obtain discontinuous code fragments related to the vulnerability from the vulnerable function, and then apply the abstraction method of function semantic reservation to generate the corresponding fingerprints. In order to reduce the detection false positives caused by code patch, our method also proposes to retrospectively filter the detection results by linking Type-3 fingerprints with official-patch

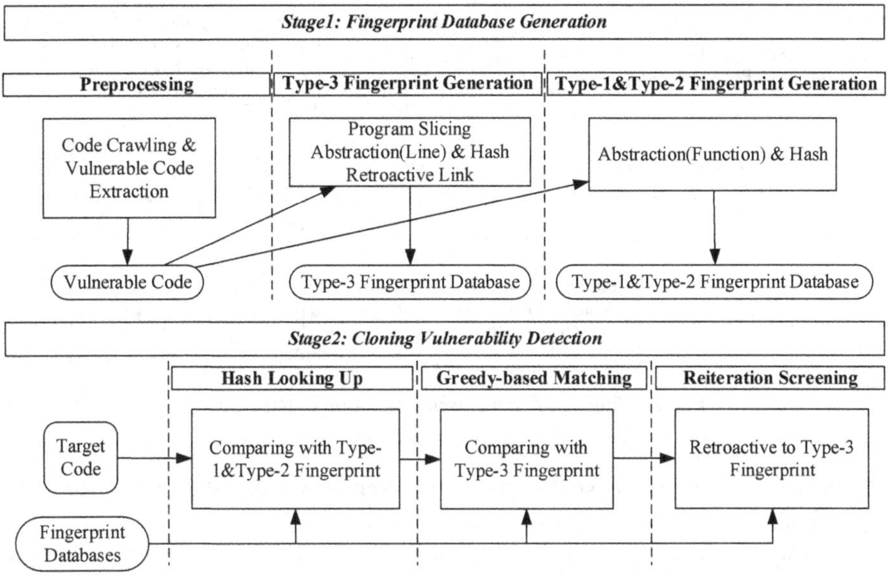

Fig. 1. Overall structure of RCVD++

fingerprints. For Type-1&Type-2 fingerprints, we take the entire vulnerable function as a whole and apply abstraction to generate the fingerprints.

In the second stage, the target code is converted into the same form as the fingerprint for detection. Specifically, there are three sub steps in the detection. First, we will use hash lookup to detect Type-1 and Type-2 clones; if the detection is successful, the detection process will end considering that if a clone is a Type-1 or Type-2 clone, then it cannot be a Type-3 clone. Otherwise, the greedy-based matching algorithm is used to compare the target code with Type-3 fingerprints. Once the detection is successful, we will further trace back to determine whether the code has been patched.

4 Establishment of Fingerprint Databases

In order to obtain the best detection result for different types of clones, our method utilizes two granularities to generate fingerprints including *Line* and *Function*. Type-3 fingerprints are generated by the granularity of *Line*. The official-patch fingerprints are generated by the granularity of *Function* and are linked to corresponding Type-3 fingerprints. Type-1&Type-2 fingerprints are generated by the granularity of *Function*. As shown in Fig. 2, the vulnerable code will be abstracted in the granularity of *Line* and *Function*.

4.1 Preprocessing

First, the source code database is formed by collecting open source projects. In order to retrieve the information about vulnerabilities more conveniently, the codes are crawled

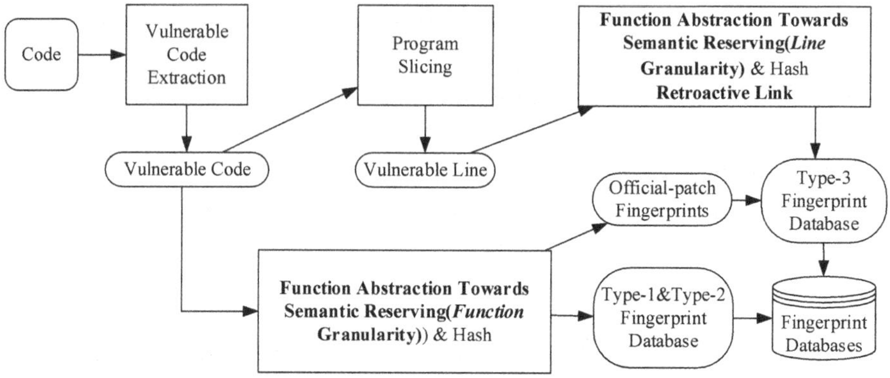

Fig. 2. Fingerprint databases generation in RCVD++

from open source projects hosted on GitHub, which also makes it easier to get patch information from the commit history.

To extract vulnerability related code, we leverage various information embedded within the commit history. Specifically, we look for vulnerability related keywords such as *use-after-free, double free, heap-based buffer overflow, stack-based buffer overflow, integer overflow, OOB, out-of-bounds read, out-of-bounds write* and `CVE` to retrieve vulnerable code. After that, the vulnerable codes, patches and the patched codes are saved in the vulnerable code database.

4.2 Program Slicing Based on Patches

Obviously, it is impossible to detect restructured clones with granularity of *Function* or higher. The granularity of Token will discard a lot of information about program, thus it is not suitable for detecting clone-caused vulnerabilities. Therefore, we take *Line* as granularity and use program slicing [18] to get the discontinuous code fragments, which would preserve the vulnerability related codes better. In order to detect the exact and renamed clones, we also take *Function* as the granularity to generate corresponding Type-1&Type-2 fingerprint. Besides, we also link the official-patch fingerprint to Type-3 fingerprint to implement the retroactive detection.

To determine the slicing criterion, the most intuitive idea is that the codes added or deleted in the patches are related to the vulnerability. However, the lines of added code in the patches do not exist in the original codes; we can only get the variables that exist in the original codes from added codes. If we use the statements in which these variables reside as vulnerable codes, we may introduce codes unrelated to vulnerability, which would greatly increase false positive. Therefore, we only take the deleted code as the slicing criterion. Note that other existing approaches such as VulDeePecker [1] and SySeVR [19] also adopt a similar strategy. Table 1 lists an example of program slicing based on patches. We take the deleted `free(a);` statement in the patch as slicing criterion to get forward slice and backward slice. These sliced codes contain the key statements that variable `a` is freed twice, and there are no statements unrelated to vulnerability.

This example demonstrates that patches-based program slicing can effectively extract the critical code lines related to vulnerability.

Table 1. Snippet of the code with double free, corresponding patch and the result of patch-based program slicing

Commit	Patch	Sliced Codes
char *a = malloc(100);		char *a = malloc(100);
int b;	@@ -14,5 +14,5 @@	scanf("%s",a);
char c;	}	if(strlen(a)>=100){
scanf("%s",a);	if(strlen(a)<100){	free(a);
if(strlen(a)>=100){	c=a[0];	}
printf("err");	+ free(a);	free(a);
free(a);	}	
}	- free(a);	
else{		
scanf("%d",&b);		
if (b == 0xffff){		
printf("useless");		
}		
}		
if(strlen(a)<100){		
c=a[0];		
}		
free(a);		

4.3 Function Abstraction Towards Semantic Reserving

In order to eliminate the impact of renamed clones, we need to abstract the code fragments [2]. First, the parameters from the arguments of function are all replaced with symbol `FPARAM`. Then all the local variables that appear in the body of a function are substituted with symbol `LVAR`. Next, the data types are all replaced with symbol `DTYPE`. Last, the names of called function are recognized. The C/C++ library function will not be replaced, while the others will be substituted with symbol `FUNCCALL`.

In the process of abstraction, the differences between identifiers will be eliminated. Although this can eliminate the detection impact of Type-2 cloning, it will also lead to the loss of relevant semantic information, especially the function names. In order to preserve the semantic information of fragile code as much as possible, we propose a code abstraction method for semantic reservation, which preserves more semantics during abstraction and obtains more accurate vulnerability fingerprints to reduce the number of detection errors.

The C/C++ code contains a large number of library function calls, such as I/O functions `scanf()` and `printf()`, memory operation functions `free()` and `malloc()`. The previous abstraction method in RCVD, VUDDY and other similar tools does not distinguish between functions. After abstraction, semantic information related to library functions is lost. Therefore, this paper proposes to retain the library function names during the abstraction process and retain more semantic information. Table 2 illustrates an example of function abstraction without and with semantic reserving. The first column shows abstracted function names, while the second one shows the actual function ones - clearly preserving more semantics.

Table 2. The result of function abstraction without and with semantic reserving

Function Abstraction without Semantic Reserving	Function Abstraction with Semantic Reserving
DTYPE * LVAR=FUNCCALL(100);	DTYPE * LVAR=malloc(100);
FUNCCALL("%s",LVAR);	scanf("%s",LVAR);
if(FUNCCALL(LVAR)>=100){	if(strlen(LVAR)>=100){
FUNCCALL(LVAR);	free(LVAR);
}	}
if(FUNCCALL(LVAR)<100){	if(strlen(LVAR)<100){
LVAR=LVAR[0];	LVAR=LVAR[0];
}	}
FUNCCALL(LVAR);	free(LVAR);

4.4 Fingerprint Generation

A direct comparison between two lines of code may be time-consuming, so it is necessary to convert the code to a shorter string. Besides, the slice results may only contain a few lines of code, which may lead to a large number of false positives. Therefore, two intuitive indicators are proposed for filtering: the lines of sliced code and the percentage of the lines of sliced code to the total lines of the function. For example, if a function has 10 lines and the sliced code contains 3 lines, then the percentage is 30%. In the process of fingerprint generation, those fingerprints that are too short and taking a small percentage will be ignored.

Type-3 Fingerprint Generation

Obviously, two different lines of code need to be completely different after conversion and the cost should be as little as possible. Since MD5 hash algorithm is naturally light-weight and strict enough for the requirement, it is employed in this process. We calculate the MD5 for each line of code to generate the fingerprint. An example for Type-3 fingerprint generation is shown in Table 3.

Type-1&Type-2 Fingerprint Generation

As for generating Type-1&Type-2 fingerprints, we take the granularity of *Function*. We do not have to compare the code line by line after abstraction because the exact and renamed clone will only modify the name of the identifier. Therefore, we further remove the '`\n`' in function after abstraction to get a string and use the length and the hash value of this string as the Type-1&Type-2 fingerprint.

Table 3. The result of abstraction and fingerprint generation for the sliced code in Table 1

Slice Code	Function Abstraction towards Semantic Reserving	Fingerprint
	DTYPE*LVAR=malloc(100);	b1cbec0b7cd55d6152648cfcd88f1bef
char *a =	scanf("%s",LVAR);	58fdbb28469ea169ae25a2fc5616005f
malloc(100);	if(strlen(LVAR)>=100){	e45e5f20daae179e6a5cbbf167294a86
scanf("%s",a);	free(LVAR);	e58a03ee48a74c4d50af5bee1ba18d08
if(strlen(a)>=100){	}	cbb184dd8e05c9709e5dcaedaa0495cf
free(a);	if(strlen(LVAR)<100){	19f40fd776b7634a317f6c67a558235f
}	LVAR=LVAR[0];	92fab8919eec2459041520cd398b3c6b
	}	cbb184dd8e05c9709e5dcaedaa0495cf
free(a);	free(LVAR);	e58a03ee48a74c4d50af5bee1ba18d08

5 Cloning Vulnerabilities Detection

In this stage, the target code will be transformed into a fingerprint, and then compared with different fingerprints to determine whether it is vulnerable. The entire process is shown in Fig. 3. RCVD++ will first pre-process the code to generate a fingerprint at *Function* granularity, and then use the Type-1&Type-2 fingerprint database to perform fingerprint matching. If the match is successful, a detection result will be generated directly; otherwise, the Type-3 clone detection will be performed. If the target code matches the Type-3 fingerprint successfully, RCVD++ will trace Type-3 fingerprint back to figure out whether the code is patched, and outputs the detection results.

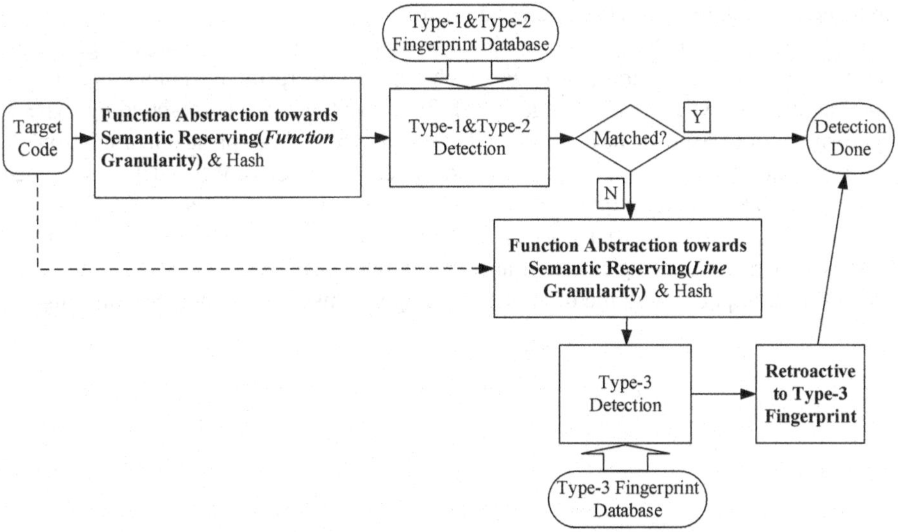

Fig. 3. Cloning vulnerability detection in RCVD++

5.1 Hash Lookup

In the matching process of Type-1&Type-2 fingerprints, we will first look up all the hash values corresponding to the length of target code string after abstraction, and then find out whether there is a same hash as the target code. According to the study of Kim et al. [2], the average time complexity of the this algorithm is O(1).

5.2 Greedy-Based Fingerprint Matching Algorithm

Normally, once the target code contains the fingerprints, it is not hard to conclude that the target code may contain a vulnerability. In other words if a fingerprint is the subsequence of target code, then the target code has the same vulnerability with this fingerprint. Therefore, fingerprint matching can be treated as a problem of the existence of subsequence. At present, greedy algorithm is usually used to solve such problems. We also propose a greedy-based matching algorithm to realize the process of matching. The detail of algorithm is shifted to Appendix.

5.3 Reiteration Screening

If the to-be-detected code has been patched by adding statements, we may still judge it as a vulnerable code. In order to eliminate such false positives, this paper proposes to utilize official patches to perform reiteration screening on the detection results. The main idea of filtering is to link corresponding Type-3 fingerprints with the official-patch fingerprints. For each function detected as Type-3 clone, RCVD++ will retroactively get the hash of linked official-patch fingerprint based o the hit fingerprint. If the target code is equivalent to the hash of official-patch fingerprint, it will be considered to be patched, and it should be deleted from the detection result.

The reiteration screening process above is based on the official-patch fingerprints generated at the *Function* granularity. For each Type-3 fingerprint, if a patched version is available, then we will link the official-patch fingerprint with the corresponding Type-3 fingerprint. Figure 4 shows the diagram of the linked Type-3 fingerprint and official-patch fingerprint data structure used in the reiteration screening process. When detecting Type-3 clones, RCVD++ obtains corresponding official-patch fingerprints based on the Type-3 fingerprint used during the detection, and then processes the target code. The process includes semantic reserving function abstraction and hash. Last, the MD5 value after hash will be compared with the official-patch fingerprints to implement the filtering.

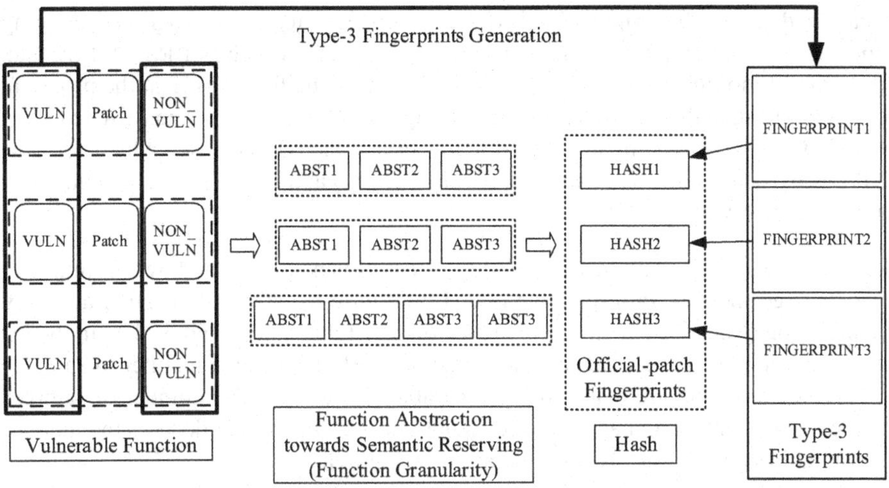

Fig. 4. Type-3 fingerprints and linked official-patch fingerprints

6 Experimental Evaluation

In order to evaluate the effectiveness of our method for detecting restructured clones, we implement RCVD++. In this section, we will compare RCVD++'s performance against RCVD, ReDeBug, Deckard and CloneWorks.

6.1 Experimental Setup

System Environment. The execution and detection performance of the above five tools are evaluated by conducting experiments on a machine running CentOS 7.2, with a 32-core Intel Xeon E5–2620 CPU operating at 2.10 GHz and 32 GB RAM.

Datasets. In this experiment, we use three datasets including NVD, NVD-CLONE and SARD. The NVD and SARD datasets are published by Li et al. in SySeVR [19]. NVD dataset collects code from 19 popular open source C/C++ projects, contains 874 vulnerable programs, and a total of 946 vulnerable functions. In addition, the dataset also contains the patched function code and the corresponding vulnerability patch (diff file). SARD contains production, synthetic and academic programs (also known as test cases), which are categorized as "good" (i.e., having no vulnerabilities), "bad" (i.e., having vulnerabilities), and "mixed" (i.e., having vulnerabilities whose patched versions are also available).

In addition, we also generate test codes of different clone types such as Type-1 to Type-3 based on the NVD dataset, which is called NVD-CLONE dataset. The cloned code is generated without affecting the vulnerability of the original code to verify the effectiveness of the detection tool. Specifically, when generating a Type-1 clone, the script will only add comments; when generating a Type-2 clone, the script will only modify the local variables; when generating a Type-3 clone, the script will only add a number of statements that output a single string, such as "`printf ("test\ n");`". In the case of deletion, the script will only delete the code related to the program debugging output, such as the output statement that contains keywords such as DEBUG, LOG, etc. In the case of modification, the script will only modify the fixed string in the program, such as the string in the output statement. Through the above steps, we constructed a test set NVD-CLONE based on the vulnerable function in the NVD dataset, which contains 1200 positive samples (i.e., containing a vulnerability) and 946 negative samples (i.e., not containing a vulnerability).

Metrics. The effectiveness of vulnerability detectors can be evaluated by the following 5 widely-used metrics: false-positive rate (FPR), false-negative rate (FNR), accuracy (A), precision (P), and F1-measure (F1) as shown in Table 4. TP denotes the number of vulnerable samples that are detected as vulnerable, FP denotes the number of samples are not vulnerable but are detected as vulnerable, TN denotes the number of samples that are not vulnerable and are detected as not vulnerable, and FN denotes the number of vulnerable samples that are detected as not vulnerable.

Table 4. Evaluation metrics

Metric	Formula	Meaning
FPR	$FPR = \frac{FP}{FP+TN}$	The proportion of false-positive samples in the total samples that are not vulnerable
FNR	$FNR = \frac{FN}{TP+FN}$	The proportion of false-positive samples in the total samples that are not vulnerable
A	$A = \frac{TP+TN}{TP+FP+TN+FN}$	The correctness of all detected samples
P	$P = \frac{TP}{TP+FP}$	The correctness of detected vulnerable samples
F1	$F1 = \frac{2P(1-FNR)}{P+(1-FNR)}$	The overall effectiveness considering both precision and false negative rate

6.2 Comparison with Other Cloning Vulnerability Detectors

First, we compare RCVD++ with cloning vulnerability detectors including RCVD and ReDeBug. All three tools use the vulnerable code in the NVD dataset and the vulnerable code in the Linux kernel to generate fingerprints. The parameter configuration of each tool is shown in Table 5, we use the default configuration of each tool.

The detection result is shown in Table 6 and Table 7. Note that all three detection tools could not detect any vulnerability in SARD dataset, because SARD does not contain any vulnerable code cloned from NVD or Linux kernel. It can be seen from the detection results that the F1-score of RCVD++ has reached 96.67% and 93.39%, which is significantly better than other tools. RCVD++ has a higher accuracy rate and precision, which means it can find more vulnerabilities without introducing too many false positives, reducing the cost of manual review. Hence, we can conclude that the

Table 5. Configuration for different tools

Tool	Configuration
RCVD++	Using NVD and Linux kernel for fingerprint generating
RCVD	Using NVD and Linux kernel for fingerprint generating
ReDeBug	n = 4; c = 10

abstraction method of semantic reservation proposed in this paper can effectively retain semantic information, leading to a more accurate generation of vulnerability fingerprints and subsequently higher detection accuracy. At the same time, RCVD++ has a lower FNR, which proves the effectiveness of the reiteration screening method proposed in this paper. After the screening, the false alarm rate is significantly reduced.

Table 6. Detection result in NVD dataset

Tool	FPR(%)	FNR(%)	P(%)	A(%)	F1(%)
RCVD++	2.7	3.93	97.28	96.68	96.67
RCVD	8	59.46	83.58	66.20	54.6
ReDeBug	4.47	38.99	93.21	78.23	73.75

Table 7. Detection result in NVD-CLONE dataset

Tool	FPR(%)	FNR(%)	P(%)	A(%)	F1(%)
RCVD ++	2.74	10.5	97.64	92.93	93.39
RCVD	8.11	40	90.344	74.08	72.11
ReDeBug	4.43	56.75	92.51	66.36	58.94

We further compare RCVD++ and RCVD in the NVD-CLONE dataset and set up four different experiments for comparison. The first experiment uses RCVD for detection. The second experiment is performed on RCVD++ with semantic reserving abstraction only. The third experiment adds Type-1&Type-2 fingerprints based on the second experiment, and the last experiment uses RCVD++ directly. The purpose of setting up experiments in this way is to verify the detection effect brought by the various methods proposed in this paper, while reducing unnecessary experiments. Table 8 shows the results achieved in the four experiments.

As shown, semantic reserving abstraction reduces the number of false positives by 2. Type-1&Type-2 fingerprint significantly improves the number of true positives by 354. Reiteration screening could effectively identify the patched code, resulting in a decline in the number of false positives by 56. The results fully prove that RCVD++ greatly improves the effectiveness of RCVD in detecting cloning vulnerabilities.

Table 8. Detection result of RCVD and RCVD++ with different optimization

Experiments	TP	FP
RCVD	720	76
RCVD++ (Abstraction)	720	74
RCVD++ (Type-1&Type-2 Fingerprint)	1074	82
RCVD++ (Reiteration Screening)	1074	26

6.3 Comparison with Other Clone Detectors

In order to further illustrate the effect of RCVD++ on clone detection, we also compared RCVD++ with clone detection tools including Deckard and CloneWorks. DECKARD depends on an algorithm for identifying similar subtrees to detect code clone. It uses a certain characterization of subtrees with numerical vectors, then clusters these vectors to judge which codes are similarity. DECKARD is language independent, and it is easy to scale to millions of lines of code with a good accuracy. CloneWorks provides a Type-3 clone detection tool which supports the detection of Java, C and C#. It can detect clone at the level of *Block*, *Function* and *File* granularity, and supports large-scale code detection with a good precision and recall. The configuration of Deckard and CloneWorks is listed in Table 9.

Table 9. Configuration of deckard and cloneworks

Tool	Configuration
Deckard	mint = 30; stride = 2; similarity = 0.95
CloneWorks	granularity = function; config = type3pattern

The experimental results are shown in Table 10. Compared with the clone detection tools, the indicators of RCVD++ are still superior to them. After analyzing the experimental results, we think this is mainly because the subtree matching algorithm used by Deckard is not sensitive to some Type-3 clones. CloneWorks does not pay much attention to code semantics, so the detection effect is poor.

Table 10. Detection results with clone detection tools in NVD-CLONE

Tool	FPR(%)	FNR(%)	P(%)	A(%)	F1(%)
RCVD++	2.74	10.5	97.64	92.93	93.39
Deckard	46.15	55.08	55.17	48.86	49.52
CloneWorks	41.94	51.5	59.39	52.72	53.39

7 Related Work

The earlier clone detecting tool is CCFinder [15] proposed in 2002. It measured the similarity of the sequence of tokens by a suffix-tree algorithm, which is computationally costly and consumes a large amount of memory. SourcererCC [16] uses a bag-of-tokens strategy to manage minor to specific changes in clones, which allows it to detect Type-3 clone. However, it is mainly designed to measure similarity and not suitable for vulnerable code clone detection. Li et al. proposed CP-Miner [20], the first vulnerable code clone detector. It parses a program and compares the token sequence with a heuristic algorithm. It can be seen that the early tools are still at the exploratory stage, with relatively high complexity.

CBCD [4] utilizes PDG to parse the vulnerable code, divides the graph into subgraphs with a small number of nodes, and finds vulnerability through the isomorphic matching algorithm of the graph. CLORIFI [6] uses n-token algorithm to process the input code and uses Bloom filter to find vulnerable code clone. It also verifies vulnerability with concolic test to reduce false positives. However, it cannot detect Type-2 clone either. CVdetector [7] traverses the grammar of vulnerable code fragments, constructs feature matrix and feature vector of key nodes, and detects various types of vulnerable code by applying clustering algorithm. This method implements a linear relationship between overhead and the amount of code. VulPecker [21] combines Token, AST, PDG and etc., extracts features of vulnerable code fragments according to its type, selects corresponding algorithms from similarity comparison algorithms, and detects the reuse of vulnerable code by using support vector machine. It improves the accuracy, but brings about a large computational overhead. VUDDY focuses on Type-1 and Type-2, and generates the fingerprint with the granularity of *Function*.

Deckard is a typical tool for detecting Type-3 clones, it builds AST for each file, then extracts characteristic vectors from the ASTs. After clustering vectors based on their Euclidean distance, vectors that lie in proximity to one another are identified as code clones. Such tree-based approaches require extensive execution time, as the subgraph isomorphism problem is time-consuming. The AST-based method it uses is not as good as the program slicing method in RCVD++, because it cannot obtain accurate vulnerable code. SecureSync [3] uses AST and directed graphs to describe the vulnerabilities caused by code reuse and API reuse. ReDeBug [5] utilizes a sliding window algorithm to obtain token streams and uses Bloom filter to find clones of vulnerable code. It supports detection on large amount of code with high efficiency, but it cannot cope with Type-2 clone. CloneWorks [12] provides users with customizable code fragments extraction methods, including conversion to tokens, extraction API, etc. However, these methods are still limited to the clone detection. RCVD [11] also utilizes program slicing, but user-defined functions and library functions are not distinguished during the abstraction process, which will lose semantic information. Besides, existing detection tools often fail to effectively identify patches present in the code, resulting in a large number of false positives. RCVD++ proposes the concept of backtracking and identifies false positives using official patches.

In addition, there are detection methods based on machine learning. Lin et al. [22] extract the features from AST and use LSTM to learn the representation. VulDeePecker [1] is the first to use deep learning for detection. It extracts features automatically but can only handle with the vulnerabilities about API. SySeVR [19] uses a variety of networks such as BGRU, DBN, and BLSTM for training, and achieves good detection accuracy in the test set. It can also find unknown vulnerabilities in real projects, but its detection metrics in real projects are not given. Besides, VulDeePecker and SySeVR are relatively inefficient, and cannot effectively distinguish similar codes before and after patching.

Through the introduction of vulnerable code cloning detection technology, we can infer that it is an inevitable trend to extract code fragments or features by program analysis. Compared with code clone detection, the vulnerable code detection needs more information about semantic or grammar and program analysis technology satisfies this requirement well.

8 Conclusion

In this paper, we propose a new method for cloning vulnerability detection. Our approach introduces a novel vulnerable code feature extraction based on program slicing, code abstraction and detection granularity. Our approach further features reiteration screening to compensate for the lack of retroactive detection of fingerprint matching. Compared to other detection tools, RCVD++ achieves a higher F1-score on three datasets, which fully proves the significance of our method.

So far, the work can also be extended in multiple directions. Firstly, dynamic analysis could be used to obtain more accurate vulnerable codes. In addition, we could further optimize the abstraction method and identify more functions to retain more semantic information. Moreover, we could consider combining machine learning algorithms to further improve the effectiveness of our method.

Acknowledgements. This research was supported by the National Nature Science Foundation of China under Grant No. U1936119, National Nature Science Foundation of China under Grant No. 61941116 and National Key R&D Program of China under Grant No. 2019QY(Y)0602.

Appendix

In this section, we provided the algorithms of greedy-based matching.

Greedy-Based Matching

Algorithm 1 introduces the pseudo code for matching algorithm. C is the target code and F is the fingerprint. The output R will be *True* if code C contains the fingerprint F, else it will be *False*. If the length of C is less than the length of F, it's impossible for C to match the F. If the nth element of C is the same as the mth element of F, the n and m will increase by one at the same time. Otherwise, only n will increase. If F is completely matched, then the fingerprint matching is considered as successful and returns *True*. From Algorithm 1, it's explicit that the time complexity is independent of fingerprint length and it has a linear relationship with the lines of code and the number of fingerprints.

Algorithm 1. Greedy-based matching algorithm

Input: C, F
Output: matching result R
1. L_c = length of C
2. L_f = length of F
3. R = False
4.   **if** $L_c < L_f$ **then**
5.   R = False
6.   **else**
7. $m = 0$
8.   **for** $n = 0, 1, …, Lc$ **do**
9.   **if** $C[n] == F[m]$ **then**
10. $m = m + 1$
11.   **end if**
12.   **if** $m == L_f$ **then**
13. R = True
14.   **break**
15.   **end if**
16.   **end for**
17. **end if**
18. **Output** R

References

1. Li, Z., et al.: Vuldeepecker: a deep learning-based system for vulnerability detection. In: Proceedings of the 25th Annual Network and Distributed System Security Symposium (NDSS) (2018)

2. Kim, S., Woo, S., Lee, H., Oh, H.: Vuddy: a scalable approach for vulnerable code clone discovery. In: 2017 IEEE Symposium on Security and Privacy (SP), pp. 595–614. IEEE, May 2017

3. Pham, N.H., Nguyen, T.T., Nguyen, H.A., Wang, X., Nguyen, A.T., Nguyen, T.N.: Detecting recurring and similar software vulnerabilities. In: Proceedings of the 32nd ACM/IEEE International Conference on Software Engineering, vol. 2, pp. 227–230. ACM, May 2010

4. Li, J., Ernst, M.D.: CBCD: cloned buggy code detector. In: Proceedings of the 34th International Conference on Software Engineering, pp. 310–320. IEEE Press, June 2012

5. Jang, J., Agrawal, A., Brumley, D.: ReDeBug: finding unpatched code clones in entire os distributions. In: 2012 IEEE Symposium on Security and Privacy, pp. 48–62. IEEE, May 2012

6. Li, H., Kwon, H., Kwon, J., Lee, H.: CLORIFI: software vulnerability discovery using code clone verification. Concurr. Comput. Pract. Exp. **28**(6), 1900–1917 (2016)

7. Gan, S., Qin, X., Chen, Z., Wang, L.: Software vulnerability code clone detection method based on characteristic metrics. J. Softw. **26**(2), 348–363 (2015)

8. Liu, Z., Wei, Q., Cao, Y.: Vfdetect: a vulnerable code clone detection system based on vulnerability fingerprint. In: 2017 IEEE 3rd Information Technology and Mechatronics Engineering Conference (ITOEC), pp. 548–553. IEEE, October 2017

9. Nishi, M.A., Damevski, K.: Scalable code clone detection and search based on adaptive prefix filtering. J. Syst. Softw. **137**, 130–142 (2018)

10. Jiang, L., Misherghi, G., Su, Z., Glondu, S.: Deckard: scalable and accurate tree-based detection of code clones. In: Proceedings of the 29th International Conference on Software Engineering, pp. 96–105. IEEE Computer Society, May 2007
11. Jiang, W., Wu, B., Jiang, Z., Yang, S.: Cloning vulnerability detection in driver layer of IoT devices. In: Zhou, J., Luo, X., Shen, Q., Xu, Z. (eds.) ICICS 2019. LNCS, vol. 11999, pp. 89–104. Springer, Cham (2020). https://doi.org/10.1007/978-3-030-41579-2_6
12. Svajlenko, J., Roy, C.K.: Cloneworks: a fast and flexible large-scale near-miss clone detection tool. In: 2017 IEEE/ACM 39th International Conference on Software Engineering Companion (ICSE-C), pp. 177–179. IEEE, May 2017
13. Roy, C.K., Cordy, J.R., Koschke, R.: Comparison and evaluation of code clone detection techniques and tools: a qualitative approach. Sci. Comput. Program. **74**(7), 470–495 (2009)
14. Yamaguchi, F., Lindner, F., Rieck, K.: Vulnerability extrapolation: assisted discovery of vulnerabilities using machine learning. In: Proceedings of the 5th USENIX Conference on Offensive Technologies, p. 13. USENIX Association, August 2011
15. Kamiya, T., Kusumoto, S., Inoue, K.: CCFinder: a multilinguistic token-based code clone detection system for large scale source code. IEEE Trans. Softw. Eng. **28**(7), 654–670 (2002)
16. Sajnani, H., Saini, V., Svajlenko, J., Roy, C.K., Lopes, C.V.: SourcererCC: scaling code clone detection to big-code. In: 2016 IEEE/ACM 38th International Conference on Software Engineering (ICSE), pp. 1157–1168. IEEE, May 2016
17. Xu, B., Qian, J., Zhang, X., Wu, Z., Chen, L.: A brief survey of program slicing. ACM Sigsoft Softw. Eng. Notes **30**(2), 1–36 (2005)
18. joern. https://joern.readthedocs.io
19. Li, Z., et al.: Sysevr: a framework for using deep learning to detect software vulnerabilities. arXiv preprint arXiv:1807.06756 (2018)
20. Li, Z., Lu, S., Myagmar, S., Zhou, Y.: CP-Miner: a tool for finding copy-paste and related bugs in operating system code. In: OSdi, vol. 4, no. 19, pp. 289–302, December 2004
21. Li, Z., et al.: VulPecker: an automated vulnerability detection system based on code similarity analysis. In: Proceedings of the 32nd Annual Conference on Computer Security Applications, pp. 201–213. ACM, December 2016
22. Lin, G., et al.: Cross-project transfer representation learning for vulnerable function discovery. IEEE Trans. Ind. Inform. **14**(7), 3289–3297 (2018)

LegIoT: Ledgered Trust Management Platform for IoT

Jens Neureither[1], Alexandra Dmitrienko[2(✉)], David Koisser[1], Ferdinand Brasser[1], and Ahmad-Reza Sadeghi[1]

[1] Technical University Darmstadt, Darmstadt, Germany
jens.neureither@gmail.com,
{david.koisser,ferdinand.brasser,ahmad.sadeghi}@trust.tu-darmstadt.de
[2] University of Würzburg, Würzburg, Germany
alexandra.dmitrienko@uni-wuerzburg.de

Abstract. We investigate and address the currently unsolved problem of trust establishment in large-scale Internet of Things (IoT) networks where heterogeneous devices and mutually mistrusting stakeholders are involved. We design, prototype and evaluate LegIoT, a novel, probabilistic trust management system that enables secure, dynamic and flexible (yet inexpensive) trust relationships in large IoT networks. The core component of LegIoT is a novel graph-based scheme that allows network devices (graph nodes) to re-use the already existing trust associations (graph edges) very efficiently; thus, significantly reducing the number of individually conducted trust assessments. Since no central trusted third party exists, LegIoT leverages *Distributed Ledger Technology* (DLT) to create and manage the trust relation graph in a decentralized manner. The trust assessment among devices can be instantiated by any appropriate assessment technique, for which we focus on remote attestation (integrity verification) in this paper. We prototyped LegIoT for Hyperledger Sawtooth and demonstrated through evaluation that the number of trust assessments in the network can be significantly reduced – e.g., by a factor of 20 for a network of 400 nodes and factor 5 for 1000 nodes.

Keywords: Trust management · Blockchain · Remote attestation

1 Introduction

The Internet of Things (IoT) is a technology trend that promises to bring our life standard to a revolutionary new level. The key aspect of IoT are smart "things" that connect physical objects, such as home appliances, vehicles, and smart city objects (parking lots, traffic lights, etc.) to the digital world through various sensors and communication interfaces. In this context, there is a growing importance of applications where IoT devices from various parties need to collaborate with each other. Examples can be found in smart energy grids [50], smart factories allowing for the automatic assembly of customized products [23], up to sensors and actors enhancing various cases, like smart locks [44].

L. Chen et al. (Eds.): ESORICS 2020, LNCS 12308, pp. 377–396, 2020.
https://doi.org/10.1007/978-3-030-58951-6_19

At the same time, the deployment of large IoT networks will lead to new threats. Malicious behavior and corrupted devices by one party can have devastating effects on the entire system. Attacks on smart meters [42] or smart locks [18] show how liability is already a concern in these systems. Worse, attacks on Industrial Control Systems (ICS) are increasing in number and sophistication in recent years [26] and have grave security implications, like *Stuxnet* [35] demonstrated. Thus, it is essential to have appropriate security mechanisms in place, which allow connected devices to establish trust relationships in the network. However, such trust relations are challenging to build and maintain, as we explore in the following.

Challenge 1: The large number of devices and their heterogeneity in terms of capabilities—due to different hardware and software architectures—complicate any attempts to use unified schemes for trust establishment.

Challenge 2: IoT networks interconnect devices produced, owned and controlled by many different parties, which are typically mutually mistrusting. Hence, solutions should not solely rely on a central entity, which needs to be trusted by all stakeholders and represents a single point of failure.

Challenge 3: Establishing trust in a scalable manner without a central authority is difficult. For example, methods based on pairwise trust links do not scale to large networks. Furthermore, if devices handle their individual trust establishment processes independently, the results within their limited point of view are a lot less significant compared to a scenario where information is shared throughout the network.

Trust Establishment. Existing trust establishment methods enable entities to gain trust into other entities in the system through either *behavioral monitoring* and detection of abnormal activities [14,25], or by means of a stronger type of trust indicator, such as cryptographic keys or *attestation evidence*—a primitive underlying Remote Attestation (RA), a key concept of Trusted Computing technology [41]. Behavioral monitoring can only detect malicious devices that deviate from the expected behavior, but are ineffective against malicious entities that play along established rules. Cryptographic keys help to tell apart malicious and benign devices; however, keys lose their benefits if the platform is compromised. In contrast, RA was specifically designed to allow an entity to remotely verify the state of another system and decide whether it is compromised in a process called *platform measurement* [3]. Typically, attestation assumes the existence of a trusted component, ranging from simple ROM to cryptographic co-processors.

However, RA is designed to work in a device-to-device fashion and does not scale well to large deployments. For instance, attempts to scale RA to large networks [10,30,33,34] have limitations, such as the inability to build trust relationships between individual devices dynamically when they are needed. Instead, groups of devices can only be attested statically in their entirety. In swarm attestation schemes [9,11,16], this leads to high workloads on individual devices and requires a trusted central entity.

Trust Management. Trust management schemes rely on underlying trust establishment methods and manage available trust information in the system. Existing solutions can be divided into two categories: Recommendation and evidence-based [36]. The first category relies on recommendations from intermediaries to establish the trust relationship between two strangers [5,51] and is well-suited for decentralized systems. Nevertheless, they cannot provide trust establishment on demand between arbitrary nodes, since the existence of intermediate trust links is required but cannot be guaranteed. The second category of evidence-based systems can be used for trust establishment on demand. However, they either rely on trusted third parties [22,27], or on out-of-band channels [21,47] to disseminate trustworthy information, and hence, are not suitable for large-scale networks of devices controlled by mutually mistrusted parties.

To the best of our knowledge, state-of-the-art distributed trust management systems cannot satisfy both the decentralized setting and on-demand trust establishment at once—a limitation we aim to address in this paper.

Contributions. In particular, we make the following contributions:

- We propose LegIoT, a trust management framework that can leverage various evidence-based methods for trust establishment to enable interoperability among heterogeneous devices (addressing challenge 1). It builds and maintains trust information in a system-wide fashion by managing dynamic (recommendation-like) trust chains across the network. As such, LegIoT combines recommendation and evidence-based approaches in one system.
- LegIoT enables on-demand trust establishment between mutually mistrusted devices, even if no intermediate trust relations pre-exist. In its heart, LegIoT has a novel graph enlargement algorithm, which adds new trust links to the system, while maximizing their re-usability to benefit the entire network. Our evaluation results demonstrate that the overall number of trust establishment processes can be significantly reduced (challenge 3), e.g., by as much as a factor of 20 for a network size of 400 and a factor of 5 for 1000 nodes.
- LegIoT utilizes Distributed Ledger Technology (DLT) to store, manage and process trust information, which enables mutually mistrusting parties to participate in the network (challenge 2). DLT is the previously missing piece of the puzzle that allows us to blend evidence-based and recommendation-based approaches to inherit advantages of both worlds: On-demand trust establishment and a decentralized setting. We instantiated LegIoT using Hyperledger Sawtooth and published the implementation as open source[1].

2 System Model

Our system model involves the following entities: (i) IoT devices $\mathcal{D}$ operating in a collaborative environment; (ii) device owners $\mathcal{O}$; (iii) device manufacturers $\mathcal{M}$, and (iv) distributed ledger $\mathcal{L}$. IoT devices $\mathcal{D}$ are manufactured by $\mathcal{M}$, and different devices $d_i, d_j \in \mathcal{D}$ can be manufactured by different $m \in \mathcal{M}$. We assume

[1] Available under the link: https://github.com/legiot/LegIoT.

that $d_i \in \mathcal{D}$ has the means to establish trust relations with another $d_j \in \mathcal{D}$, e.g., through known characteristics of its benign behavior or via remote attestation, and fulfills the requirements of the deployed trust mechanism. The distributed ledger $\mathcal{L}$ is a tamper-proof append-only data structure, which is maintained cooperatively by a distributed network of ledger validators. We assume that $\mathcal{L}$ has the ability to *store integrity-protected data* and *execute programs provably correct* through so-called smart contracts [15]. $\mathcal{L}$ is operated by a consortium of the stakeholders, which can include the owners $\mathcal{O}$, device manufacturers $\mathcal{M}$, as well as external entities like consumer advocacy organizations or government agencies. The parties maintaining $\mathcal{L}$ are mutually distrusting but we assume that only a minority of parties are malicious. New consortium members require a collective agreement by the consortium to participate, e.g., by vetting new members before admission, to prevent a party to gain disproportionate voting power.

Adversary Model. We assume that device manufacturers $\mathcal{M}$ are trusted to produce non-malicious hardware and firmware for IoT devices $\mathcal{D}$ and to provide correct information helping to verify that they are in the trusted state (e.g., benign software and hardware configurations). We also assume that devices $\mathcal{D}$ are initially trusted and are not infected prior deployment. After deployment, devices might be compromised. We denote an adversary by $\mathcal{A}$. At the network layer, we consider the *Dolev-Yao* model where $\mathcal{A}$ has unlimited access to the network [17,19]. $\mathcal{A}$ is unable to break cryptographic primitives but can perform both passive attacks such as *eavesdropping* and active attacks like *message spoofing* and *replaying*. Through these capabilities, the adversary can further run arbitrary malware on devices $\mathcal{D}$. Furthermore, we inherit any assumptions from the underlying trust establishment mechanisms, e.g., many attestation schemes assume a trust anchor on the device.

Moreover, we rule out Denial-of-Service (DoS) attacks, as it is a common assumption in the context of trust management literature. Additionally, we assume an attacker is not able to compromise and leave devices without traces instantaneously. We refer to this phenomenon as a *fast roaming adversary* and argue, that an attacker at least needs the time $\mathcal{T}$ to restore the uninfected state of a device and hide his prior presence.

3 LegIoT Design

LegIoT is a trust management framework that enables network participants to query trust information on demand. In a nutshell, our system builds and uses *indirect trust relationships* between devices via a trust relation graph. The intuition is, if the trustworthiness of a device d_i was successfully verified by d_j, and another device $d_k \in \mathcal{D}$ wants to assess the trustworthiness of d_j shortly after, d_k can gain indirect trust into d_i. This concept can be extended to consider more devices, and thus can build trust chains, which in turn can be combined into a trust graph. This graph is calculated and managed by LegIoT using Distributed

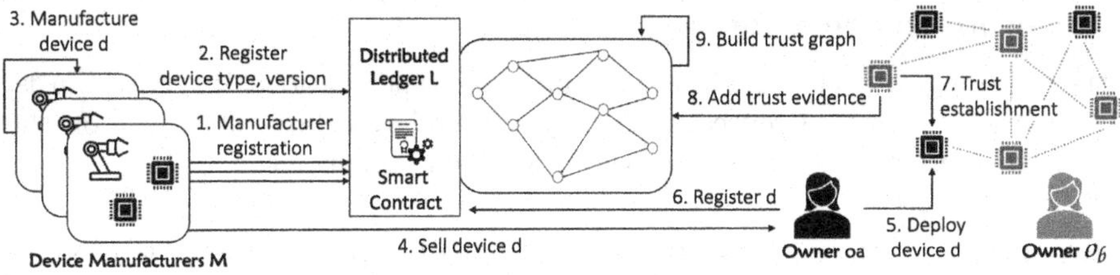

Fig. 1. Example of deploying a new device and integrating it into the trust graph in a collaborative environment.

Ledger Technology (DLT). This avoids a single point of failure and enables mutually mistrusting parties to reach consensus about the trust graph without relying on a central authority.

Figure 1 shows an exemplary deployment of a new device d and its integration into the trust graph. In Step 1, a device manufacturer $m \in \mathcal{M}$ registers itself on the distributed ledger $\mathcal{L}$. Additionally, m submits universal information about its devices required to publicly verify them by any party—namely, their trustworthy configurations or trustworthy behavioral characteristics (Step 2). After the owner o_a sets up and configures the device d acquired from m (Step 3, 4 and 5), o_a registers this individual deployment of d on $\mathcal{L}$ (Step 6), allowing any party to access the stored information to make their own trust assessments. The owners $\mathcal{O}$ are mutually mistrusting, so at some point there will be a request to establish trust in d by a device of another owner o_b in Step 7. After accessing all the necessary information from $\mathcal{L}$ to evaluate the trustworthiness of d, the result of this trust assessment is committed to $\mathcal{L}$ (Step 8). In Step 9, The ledger $\mathcal{L}$ then integrates this evidence into the trust graph via a smart contract.

Through the trust graph, $\mathcal{L}$ can serve queries regarding the trustworthiness of a device from any party. In such an instance we call the targeted device the prover $\mathcal{Prv}$ and the requesting party the verifier $\mathcal{Vrf}$. The smart contract handling the trust query then tries to find a continuous *trust path* to the $\mathcal{Prv}$, which may possibly consist of multiple hops. Whenever a prover's trust score is requested and no path exists, the $\mathcal{Vrf}$ will not perform a trust assessment of the device directly. Instead, it will follow suggestions of the trust management framework, which will find a $\mathcal{Prv}_{aux}$ as an optimal entry point into the trust graph. The trust link established from the suggested entry point will produce a valid trust path, while including as many devices as possible in between. Thus, a single trust evaluation process will result in gaining trust into several devices at once: the prover $\mathcal{Prv}$ at the end of the trust path and implicitly into all devices along the path from the prover node. We elaborate on the concept of indirect trust in Sect. 3.1. Hence, in the long-term, the number of individual trust establishment processes among devices can be largely reduced.

3.1 Theoretical Foundations

In the following, we define the notion of indirect trust relationships. Afterwards, we define a graph-based representation of trust relationships, elaborate on the idea of trust chains and propose methods for their valuations.

Indirect Trust. The notion of trust and indirect trust relationships was extensively studied in the context of social networks [31,53]. Generally, the concept of trust is not transferable via multiple entities, which makes it non-trivial to apply the notion of indirect trust to a system. For this purpose we refer to Jøsang et al., who introduced the notions of *functional* and *referral* trust [31]. Both referral and functional trust can be *direct* an *indirect*. In addition, each trust link is assigned a scope σ. Figure 2 depicts how referral trust can be combined with direct functional trust to derive indirect functional trust. Here, *Alice* wants to buy bread from the baker *Charlie* and relies on the recommendation of her friend *Bob* regarding the trust scope, bread quality (σ). To enable indirect trust chains with multiple hops, two important requirements must be fulfilled. First, the *Functional Trust Derivation Criterion* states that "referral trust requires that the last trust arc represents functional trust and all previous trust arcs represent referral trust" [31]. Second, the *Trust Scope Consistency Criterion* states that a "trust path requires that there exists a trust scope which is a common subset of all trust scores in the path. The derived trust scope then is the largest common subset." [31].

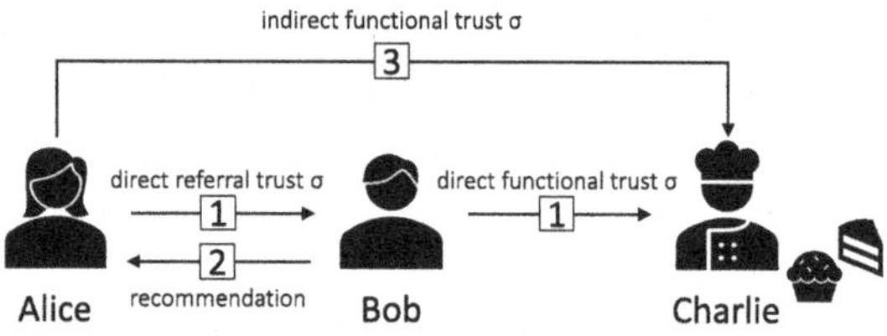

Fig. 2. Example regarding the process of gaining indirect functional trust.

Yu et al. define $T_i(j)^t$ as the "trust rating assigned by agent i to agent j at time t" [53]. We extend this definition in a way, that it also includes the trust scope σ between agent i and agent j: $T_i(j,\sigma)^t$. Further, we constrain:

$$0 \leq T_i(j,\sigma)^t \leq 1 \; and \; T_i(j,\sigma)^0 = 0. \tag{1}$$

Regarding trust propagation, the following rule holds for an indirect trust chain $x \implies y \implies z$ [53]:

$$(T_x(z,\sigma)^{t+1} \leq T_x(y,\sigma)^t) \wedge (T_x(z,\sigma)^{t+1} \leq T_y(z,\sigma)^t) \tag{2}$$

To satisfy this rule, multiplication can be used for determining indirect trust [53]:

$$T_x(z,\sigma)^{t+1} = T_x(y,\sigma)^t \; T_x(z,\sigma)^t \tag{3}$$

Trust Scope. The trust scope σ plays an important role in trust paths as it defines what the subject of a relationship actually is. Of particular interest is the question of whether evidences enable referral trust, functional trust or both. Devices may or may not provide referral trust, and the actual scope depends on the underlying scheme that is used for trust establishment. Generally, LegIoT can rely on various trust establishment methods.

Let us take remote attestation as an example, which can be classified into four groups. (i) software-based attestation schemes [3,45] aim to not rely on specific hardware components, while (ii) hybrid architectures [20,32], (iii) hardware-based architectures [41,43], and (iv) control flow attestation [2,4] leverage hardware trust anchors. Hence, they provide different levels of resilience against compromise and vary in trust scope. We provide a more thorough discussion on the various attestation schemes in Appendix A.

Trust Graph and Trust Chains. To represent the trust graph of a system, we consider a *directed and weighted graph* $(\mathcal{D}, \mathcal{E})$. $\mathcal{D}$ is the set of all devices participating in the network. $\mathcal{E}$ represents the set of all established trust links which create the edges in the graph[2]. The idea behind trust chains implies that devices do not necessarily need to be assessed directly. Instead, it suffices that a verifier evaluates any vertex $d \in \mathcal{D}$ where a path $p_{d \rightarrow \mathcal{P}rv}$ exists, if the following constraints apply:

1. *Path Scope:* $p_{d \rightarrow \mathcal{P}rv}$ fulfills the scope criteria σ (cf. Sect. 3.1 *Trust Scope*).
2. *Trust Scope:* d must be eligible for referral trust.
3. *Temporal Edge Validity:* All edges in the resulting path $p_{\mathcal{V}rf \rightarrow \mathcal{P}rv}$ must still be considered valid at the time of calculation.

The validity of edges is strongly bound to the temporal proximity τ of the trust evaluation process and the usage of the trust evidence $\tau = \mathcal{T}_{use} - \mathcal{T}_{eval}$. $\mathcal{T}_{use}$ is the time regarding the usage of the evidence and $\mathcal{T}_{eval}$ its inception). After a device was evaluated successfully, it might be compromised by adversaries. Thus, the *resilience* and *significance* of edges vanishes over time. We define a time threshold $\mathcal{T}_{min}$, assuming an adversary requires some time to progressively compromise devices in the network. As long as $\tau < \mathcal{T}_{min}$ the evidence is fully valid. Afterwards, the probability of an attack increases the longer the device was not checked. Therefore, the validity decreases according to a time function $v(\tau)$. We also define the expiration time $\mathcal{T}_{exp}$. After this time evidences are considered invalid.

Trust Chain Valuation. As a first step of a path valuation, the *weight* of a single edge is calculated. In addition to temporal validity, the valuation is also influenced by the resilience of an underlying trust evaluation scheme (cf. Sect. 3.1 *Trust Scope*). The reliability function r for a trust evaluation method type $\mathcal{E}val$

[2] An edge is equivalent to a direct trust rating $T_i(j)$ of two nodes; yet, we simply use $e \in \mathcal{E}$ for better readability.

produces a probabilistic result $r(\mathcal{Eval}) \in [0, 1]$. Together, the combined edge reliability $f(\tau, \mathcal{Eval})$ can be calculated as follows:

$$f(\tau, \mathcal{Eval}) = v(\tau) * r(\mathcal{Eval}) \tag{4}$$

The resulting value expresses the *probability* that the targeted node is in a *trustworthy condition* from the point of view of the verifying node. To chain the reliability of multiple sequential edges on a path together, we use the findings from Eq. (3). Based on this, the probability weights can be multiplied for each edge e in path p:

$$\phi(p) = \prod_{e \in p} f\Big(\tau(e), \mathcal{Eval}(e)\Big) \tag{5}$$

3.2 Protocols and Algorithms

We now proceed to describe protocols and algorithms based on the concepts described in the previous section.

System Initialization and Administration. Prior to deployment, multiple databases containing administrative data need to be initialized. In particular, every manufacturer $\mathcal{M}$ maintains a Policy DB on the ledger $\mathcal{L}$, which includes information about their devices as well as information to evaluate their state. Furthermore, an owner $\mathcal{O}$ needs to submit four databases to the ledger $\mathcal{L}$: (i) System Settings DB, (ii) Device DB, (iii) Verifier DB, and (iv) Trust Evaluation Methods DB.

System Settings DB includes system-wide settings, such as a security parameter denoting the maximum path distance between $\mathcal{Prv}$ and $\mathcal{Vrf}$. The Device DB maps a device identity (represented by public keys) to a device described in Policy DB. Verifier DB contains entities that are not included in Device DB; yet, are permitted to act as verifiers. These could be, for instance, external entities such as device owners. Trust Evaluation Methods DB stores parameters for each trust evaluation method, such as reliability score and timing characteristics, which denotes how fast the validity of an established trust evidence vanishes over time.

Trust Query. The trust query protocol offers an interface to query the state of another device. The protocol is depicted in Fig. 3. It is called by a verifier with an identity $\mathcal{ID}_{\mathcal{Vrf}}$ that requests to establish trust into a prover with $\mathcal{ID}_{\mathcal{Prv}}$. The query is represented by the transaction $trustQuery(\mathcal{ID}_{\mathcal{Vrf}}, \mathcal{ID}_{\mathcal{Prv}}, minReliability)$. With $minReliability$ the verifier can specify the lower bound of the final trust score the path is allowed to have. LegIoT continues with the following steps: (i) Verify that $\mathcal{ID}_{\mathcal{Prv}}$ and $\mathcal{ID}_{\mathcal{Vrf}}$ are found in Device DB or Verifier DB; (ii) run a graph search algorithm (cf. Sect. 3.2 *Graph Search Algorithm*) with parameters ($\mathcal{ID}_{\mathcal{Vrf}}$, $\mathcal{ID}_{\mathcal{Prv}}$, $minReliability$, $secParameter$) where the $secParameter$ denotes the maximum permitted path length; (iii) if a path exists with the desired $minReliability$, it is returned to the verifier (case A in Fig. 3).

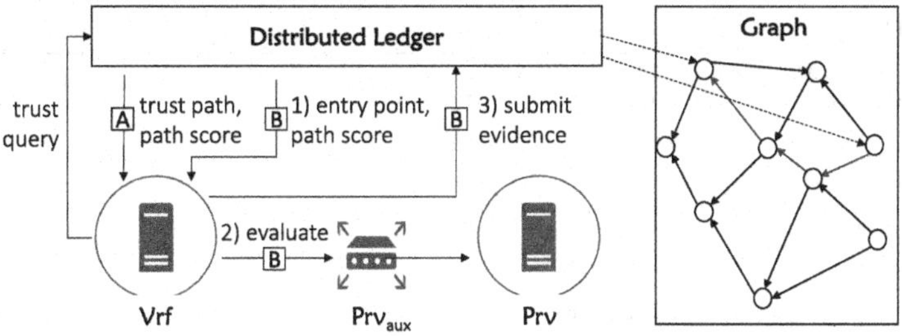

Fig. 3. Illustration of an exemplary trust query, sent by $\mathcal{Vrf}$.

If no path exists, an optimal entry point is returned according to the Graph Enlargement Policy (GEP) presented in Sect. 3.2 *Optimal Entry Point* (case B, Step 1 in Fig. 3). $\mathcal{Vrf}$ then proceeds to execute the trust evaluation process (Step 2) and submits the collected evidences (Step 3). Included in the submission is the obtained trust score $\mathcal{S}_{\mathcal{Vrf} \to \mathcal{Prv}}$, the trust evaluation method $\mathcal{Eval}$, and the prover's *device class* as well as its *version* regarding the Policy DB.

Graph Search Algorithm. To search the graph for an existing path in between $\mathcal{Prv}$ and $\mathcal{Vrf}$, LegIoT uses limited-depth breadth-first search with the parameters $\mathcal{ID}_{\mathcal{Vrf}}$, $\mathcal{ID}_{\mathcal{Prv}}$, *minReliability*, and *secParameter*. The functionality is described in Algorithm 1. The algorithm searches the graph for a direct path between $\mathcal{Vrf}$ and $\mathcal{Prv}$. The *expand* operation in line 10 denotes the listing of all predecessors for a given node that were not visited, yet. *Fringe* denotes the frontier of nodes that were already visited by a search algorithm. Nodes in the fringe directly border to nodes that were not yet visited and will be expanded next. Any valid path found (multiple may exist) will result in a positive trust assessment and the found path is returned. Otherwise, Algorithm 2 is called in line 22 to calculate the best entry point (described in Sect. 3.2 *Optimal Entry Point*).

Optimal Entry Point. If a trust query is issued and no path between $\mathcal{Prv}$ and the $\mathcal{Vrf}$ exists, the system needs to integrate $\mathcal{Vrf}$ to the trust graph in a way that it will be able to reach $\mathcal{Prv}$ through a valid path. We use a *GEP* to determine a node, which the verifier is supposed to evaluate. The idea behind this policy is to create and use the synergy effects of individual trust evaluation processes. The longer the path between verifier and prover is, the more likely it is that a path already exists for subsequent queries and no new trust evaluation is necessary. The algorithm for the entry point calculation is described in Algorithm 2. The task of the GEP is to find an entry point in the graph that meets the following conditions:

Algorithm 1. *BuildPath* function for establishing a path between $\mathcal{V}rf$ and $\mathcal{P}rv$ (shortened)

Input: $\mathcal{V}rf$, $\mathcal{P}rv$, $minReliability$, $secParameter$
Output: $pathFound$, $path$, $pathReliability$, $entrypoint$
1: $fringe\{\} \leftarrow \mathcal{P}rv$
2: $newFringe\{\} \leftarrow \emptyset$
3: $maxDepth \leftarrow (secParameter - 1)$
4: $visited[] \leftarrow \emptyset$
5: $currentDepth \leftarrow 0$
6: $path[] \leftarrow \emptyset$
7: $rating[] \leftarrow \emptyset$
8: **while** $currentDepth \leq maxDepth$ **do**
9:   | **for** $node$ in $fringe$ **do**
10:   | | $expanded \leftarrow$ **expand**$(node)$
11:   | | **for** $(newEdge, parent)$ in $expanded$ **do**
12:   | | | $edgeReliability \leftarrow$ **reliability**$(newEdge)$
13:   | | | $path[parent] \leftarrow path[node] \cup newEdge$
14:   | | | $rating[parent] \leftarrow rating[node] \cdot edgeReliability$
15:   | | | $visited \leftarrow visited \cup parent$
16:   | | | $newFringe \leftarrow newFringe \cup parent$
17:   | | **if** $\mathcal{V}rf \in expanded$ **then**
18:   | | | **return** $true, path[\mathcal{V}rf], rating[\mathcal{V}rf], \emptyset$
19:   | $currentDepth \leftarrow currentDepth + 1$
20:   | $fringe \leftarrow newFringe$
21:   | $newFringe \leftarrow \emptyset$
22: $entrypoint \leftarrow$ **CalculateEntryPoint**$(visited, minReliability)$
23: **return** $false, path[entrypoint], rating[entrypoint], entrypoint$

Algorithm 2. *CalculateEntryPoint* sub-algorithm to determine the best possible graph entry point

---
Input: $visited$, $minReliability$
Output: $entrypoint$
1: **for** $candidate$ in $visited$ **do**
2:   | **if** **reliability**$(candidate) < minReliability$ **then**
3:   | | **delete** $candidate$
4: **sortByReliability**$(visited, descending)$
5: **sortBySearchDepth**$(visited, descending)$
6: $entrypoint \leftarrow visited[0]$
7: **return** $entrypoint$

- *Validity:* The entry point enables at least one valid path to the prover.
- *Distance:* The entry point has the maximal hop distance to the prover that is permitted by the *secParameter*.
- *Optimality:* For multiple entry points at a given distance, the option with the highest resulting trust score is chosen.

4 Prototype Implementation

We implemented LegIoT based on Hyperledger Sawtooth, a framework for distributed ledgers [28]. The Hyperledger functionality is implemented in Python and consists of about 2900 LoC. It is included in the *Transaction Processors (TP)*—Sawtooth's equivalents of smart contracts in public blockchains. Three of them are provided by Sawtooth framework itself. Settings TP implements a voting-based administration of system settings among validators. Furthermore, Identity and Settings TPs manage permissioning of both validators and clients. Block-Info TP enables access to historic ledger-specific data such as *latest block number* and *current timestamp* in the context of transaction processing. Our implementation is included in the next two TPs described. Trust TP builds the core component of LegIoT and includes all application logic for handling submitted evidences and calculating paths for trust queries (Algorithms 1 and 2). Administration TP handles updates of the databases described in Sect. 3.2.

The Client side of LegIoT is implemented in two modules. (1) AdminClient serves administrative tasks, such as initialization of system parameters and populating databases (cf. Sect. 3.2 *System Initialization and Administration*). (2) IoTclient is used to execute the trust query protocol as described in Sect. 3.2 *Trust Query*. Both clients are written in Python and consist of 1200 and 900 LoC, respectively, and are available as open source (see footnote 1).

Additionally, we ported IoTclient to C to run it on the LPC55S69-EVK evaluation board from NXP. The board is based on ARM Cortex-M33 processor and includes the ARM TrustZone security architecture [8], which provides the necessary hardware features to support hybrid RA methods such as TrustLight [32] and SMART [20]. Furthermore, we implemented a middlebox to translate Message Queuing Telemetry Transport (MQTT)[3] requests issued by the board into HTTP requests processed by Hyperledger. The C version of the client contains some proprietary libraries, which prohibits us to open source it. We are willing to share it upon request for non-commercial use.

5 Evaluation

Within this section, we evaluate how LegIoT is able to enhance the *trust evaluation* performance compared to a scenario without trust management. If a trust path from $\mathcal{V}rf$ to $\mathcal{P}rv$ already exists, *zero* trust evaluation processes are needed to establish trust, which we call a *trust query hit.* If there is no valid path, *one* trust evaluation process must be carried out to the entry point, called a *trust query miss.* The goal is to maximize the hit percentage, which we define as:

$$HitPercentage = \frac{TrustQueryHits}{(TrustQueryHits + TrustQueryMisses)} * 100$$

Parameters. Several factors influence the hit percentage. Unless otherwise specified we use the following parameter values. The network size N is set to 200 and the *minReliability* to 0.80. The edge reliability function is set to:

$$f(\tau, \mathcal{E}val) = -0.0006666667 * x + 1.2$$

Further, we use $\mathcal{T}_{min} = 300$ s and $\mathcal{T}_{exp} = 600$ s. Attacks in the wild like on Industrial Control Systems (ICS) [26] last over weeks, with a stealthy adversary trying to

[3] http://mqtt.org/.

take over large parts of a network. Coupled with the fact that the adversary will not know when the next attestation towards compromised devices will take place, the system can certainly protect the network in its entirety with the chosen time parameters.

Note that LegIoT has an *initialization phase* before entering a normal *operation phase.* To ensure that the operation phase is reflected within the evaluation, *simulation duration* = 1200 s and *query rate* = $\frac{N}{2}$ are set accordingly. Additionally, simulations are repeated five times by default. Between simulation repetitions, the trust graph is not reset in order to reflect long-time operation.

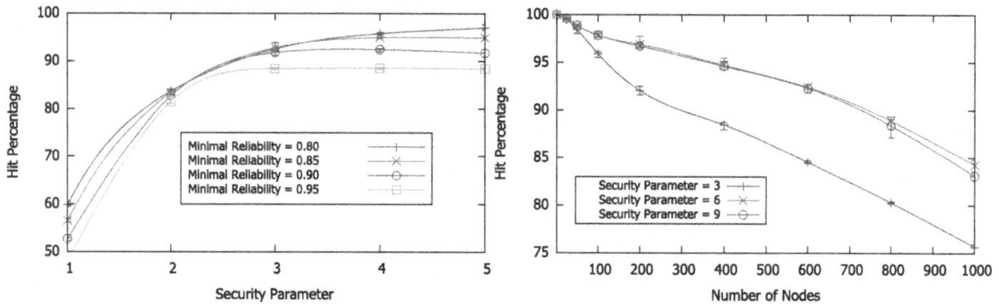

Fig. 4. Effects on the hit percentage for different minimal reliability parameters *(left)* and for increasingly larger networks with different security parameters *(right).*

Security Parameter Variation. The security parameter denotes the maximum permitted number of hops between $\mathcal{Vrf}$ and $\mathcal{Prv}$ on a trust path. In the following, we vary the security parameter from zero to five, where zero matches a case without using our system. Figure 4 on the left shows the influence of a larger security parameter. A *minReliability* = 0.80 and a security parameter of *three* already boosts the hit percentage to approximately 93%. This means that in 100 trust queries only 7 fresh trust assessments are needed. It is impossible for the hit percentage to reach 100% because evidences expire and need to be renewed. Thus, we state that a *horizontal asymptote* exists which the curve approaches. This asymptote changes depending on the other parameters.

Minimal Reliability Variation. In Fig. 4 on the left we display graphs for four different minimal reliabilities ranging from 0.80 to 0.95. Generally, lower required reliabilities result in a higher hit percentage. It can be observed that the higher the minimal reliability is, the faster the function approaches an asymptote. While the curve of '0.95' does not improve from security parameter 3 on, the curves '0.9' and '0.85' still profit from increasing the value. This phenomenon results from the fact, that the individual edge reliabilities are multiplied for the total reliability score. Thus, for a path that uses a higher security parameter the product of edge probabilities is lower than for a path that is only expanded regarding a lower security parameter.

Network Size Variation. To model different network sizes, we vary the number of nodes from $N = 25$ to $N = 1000$. Additionally, three different security parameters are tested with all network sizes. Results are depicted in Fig. 4 on the right. With the security parameter set to 6 and $N = 400$ a hit percentage of 95% is achieved (a reduction by a factor of 20) and 80% for $N = 1000$ (factor 5).

All curves have in common that a network enlargement lowers the hit percentage. This can be explained by the fact that if the graph grows, it is less probable that a random verifier and prover are located in the same area. To overcome the increasing distance, a higher security parameter has a positive impact. The curves with less-restrictive hop limitations (Security Parameter 6 and 9) descend more significantly at higher network sizes than the curve with Security Parameter 3. The curves for parameters 6 and 9 are nearly equal. This phenomenon is explained by the fact that the minimal reliability of '0.8' also limits search depth indirectly, due to multiplication of individual edge reliabilities. There is a high probability that the overall path reliability falls below '0.8' at a certain length. In order to make use of a greater security parameter, the minimal reliability must be lowered or a better suited path reliability calculation method must be used. While we stated that LegIoT does not profit from security parameter increases at a certain point in the context of minimal reliability limitations, it can still be useful to compensate for network size effects.

Edge Reliability Function Variation. Any arbitrary function can be used to describe desired time validity behavior. For exemplary purposes, we assume that the supported trust evaluation methods are different types of attestation schemes, which we described in Appendix A. They can be mapped to linear functions with different validity periods. Functions are noted in the format $[Timefunction, \mathcal{T}_{min}, \mathcal{T}_{exp}]$. All functions degrade linearly from '1' between $\mathcal{T}_{min}$ and $\mathcal{T}_{exp}$. They vary in their *gradient* as well as in *validity periods*. Short-lived functions like $[-0.003333333 * x + 1.2, 60, 120]$ reflect attestation types with weaker resilience such as *software-based attestation*. On the contrary schemes like *control-flow attestation* with stronger resilience against compromise might use functions with high validity for a long duration such as $[-0.000333333 * x + 1.2, 600, 1200]$.

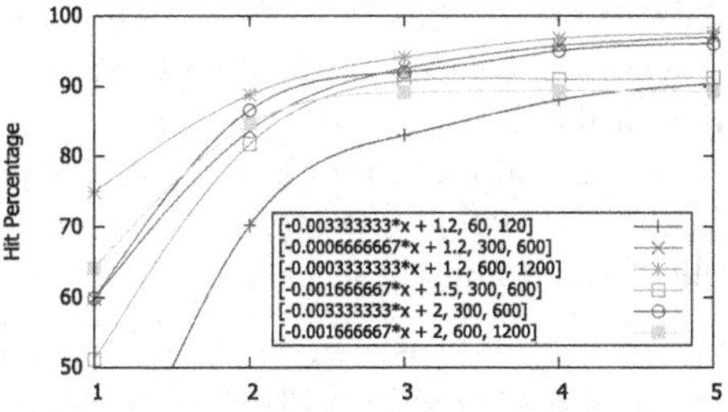

Fig. 5. Hit percentage effects regarding selected edge reliability functions.

In Fig. 5 we show how variations of these function parameters influence hit percentages with different security parameters. The first three functions that all degrade from '1' to '0.8' within their degeneration phase offer similar characteristics. As expected, the function with the longest validity achieves the highest hit percentages. It is noticeable, that the slowly descending functions profit from higher security parameters. Altogether, function differences in most cases have less impact the higher the security parameter is. Even though validity times are varied by the factor of ten, the hit percentage difference is less than 10%.

6 Security Analysis

In this section, we informally analyze possible attack vectors on LegIoT, explain their impact, and if needed, their mitigation. Our analysis focuses on LegIoT itself without taking into account potential security weaknesses of underlying technologies (e.g., distributed ledgers or crypto primitives).

Evidence Spoofing. The adversary may submit faulty evidence as he impersonates a verifier. Nevertheless, these edges are only used if another (honest) device evaluated the adversary's device to complete a trust path. As the device is infected, the trust evaluation method reports a lack of trust and no path is established. Thus, spoofed evidence is never used when considering the present attacker model that excludes *fast roaming adversaries.*

Roaming Adversaries. A roaming adversary is the one that infects the device to forge the trust evidence for a malicious node and leaves the device with no traces. If positively evaluated by another honest verifier while the forged evidence is still valid, trust is gained in the malicious node and in all other nodes connected to it. An attacker could use this strategy to create a great number of trust links to devices infected by him via one single device. Our system provides probabilistic protection against this attack vector. First, our attacker cannot leave devices without traces in no time (cf. Sect. 2). Second, he cannot know for sure when the next evaluation towards his device will take place, limiting his chances to succeed. Third, the estimated minimal compromise time $\mathcal{T}_{min}$ and the *temporal edge validity function* further restrict the motion range an adversary has.

Device Infection in Validity Period. A device can possibly be infected during the validity period of evidence created for it as a prover. Therefore, evidences towards the device might still exist even if it was corrupted in the meantime. The uncertainty of whether an infection happens in the validity period of an evidence is again reflected by $\mathcal{T}_{min}$, $\mathcal{T}_{exp}$ as well as the *temporal edge validity function.* Thus, the risk can be controlled by setting these parameters.

Timestamp Attacks. If timestamps can be forged or manipulated, system correctness is highly endangered. One important precaution is that timestamps are added by the Validator nodes that run the ledger. Using time from verifying devices as a timestamp would cause issues since this would require them to have a synchronized real-time clock, which is hard to implement in practice in the targeted setting. However, a malicious verifier could carry out a trust evaluation process for an honest prover and withhold this trust evidence. He might keep the evidence, infect the prover device and then submit the valid evidence. As a result, a valid evidence for a malicious device exists in the graph. Similar to *evidence spoofing*, the trust edge alone is not sufficient to convince other devices of the correctness of a prover. At the point where other verifiers try to build trust into the prover, the path leads via the corrupted verifier that uploaded the evidence. Another attack may be compromising a single Validator node. Here, in order to enforce deterministic transaction results, other Validators are unable to check the timestamps generated by the block creator since they never match exactly. To prevent this, Sawtooth offers rules that enforce timestamp monotonicity and a roughly synchronized network time [29].

Sneaky Hopping Attack. With *sneaky hopping*, an adversary tries to predict the next system actions to stay undetected. He always tries to compromise devices that will most likely not be attested in the near future. This also requires the adversary to be *roaming* but he has more time to infect, leave and hide traces (*slow roaming*). For LegIoT this is not applicable since it is unpredictable which evidence will be added or trust score queried next. Both *evidence submission* and *trust query* are initiated by entities whose decisions are out of range of the attacker.

7 Related Work

Trust Management. Trust Management is an important issue in distributed systems, which aims to track the trustworthiness of all individual devices in an up-to-date fashion. A key work in this field was proposed by Blaze et al. [13], which first tackled the problem of decentralized trust beyond the verification of certificates. The work of Abdul-Rahman and Hailes [1] then introduced the notion of a recommendation-based trust model, in which recommendations help to establish trust between two strangers. Another class of trust models derive trust based on evidence, such as cryptographic keys, and typically rely on trusted parties (e.g., for key distribution). Examples are the resurrecting duckling model [47], Pretty Good Privacy (PGP) [21] and the X.509 certificate [27] systems. Our approach is a combination of both recommendation- and evidence-based approaches. It relies on attestation evidences and indirect trust links built using recommendations, while enabling dynamic and on-demand trust establishment in networks between heterogeneous devices controlled by mutually mistrusting parties. Furthermore, it can leverage a stronger type of trust indicator, such as *attestation evidence*, which enables the detection of compromised devices. Our trust establishment is also directly applicable from a single request, in contrast to systems that need to converge over multiple recommendations, like bayesian trust models [49].

Another line of work in the area of trust management targets ad-hoc networks [14,37], which are dynamic and do not typically include trusted third parties. Such systems work by nodes disseminating trust information about other nodes through the network. Trust judgments are made by monitoring neighboring nodes and identifying suspicious behavior. In contrast, LegIoT relies on a stronger type of evidence that can detect compromised devices, even if they behave correctly. Furthermore, trust information is stored and managed on a distributed ledger, and thus nodes do not need to constantly disseminate information among peers.

DLT-Based Intrusion Detection. Research papers [7,25,38,46] deal with the question of whether distributed ledgers can be leveraged for detecting anomalies and isolating untrusted devices. All these systems rely on behavioral monitoring for trust establishment and maintain reputation scores for network nodes, while LegIoT can leverage stronger underlying trust establishment methods, such as *remote attestation*, and provides means to utilize indirect trust relationships.

DLT-Based Trust Management. Recently, DLT-based trust management systems were developed to support authentication systems [6,39]. Similarly to LegIoT, they use distributed ledgers to manage trust information in the system, but rely on evidence-based trust establishment methods based on cryptographic keys. In particular, Alexopoulos et al. [6] show how to enhance a public-key-to-id binding for PGP [21] and X.509 public key infrastructure (PKI) [27]. Their system can build and leverage indirect trust relationships; however, neither provides on-demand trust establishment nor maximizes benefits of pre-established trust connections. The solution by Moinet et al. [39] is intended to substitute PGP and X.509 PKI in Wireless Sensor Networks (WSN). Their approach to trust management is different, as it relies on the concept known as Human-like Knowledge based Trust (HKT), and does not include the notion of indirect trust.

DLT and Trusted Computing. TM-Coin [40] deals with "Trustworthy Management of TCB Measurements in IoT". The authors eliminate the redundancy of measurements by letting miners attest the Trusted Computing Base (TCB) of IoT devices. Afterwards, a remote verifier can query these results from the distributed ledger. In contrast, LegIoT does not involve miners into the attestation process—instead, it is carried out by other clients, which simplifies deployment. Furthermore, unlike TM-Coin, LegIoT utilizes attestation results submitted by other nodes to reduce the number of overall attestation processes in the system and can combine various attestation methods in one solution.

Banerjee et al. [12] propose a conceptual design for a blockchain-based security layer for identification, auditing, and isolation of misbehaving nodes in IoT networks, which relies on remote attestation to detect compromised entities. The attestation process in the system takes place between two clients and the result is recorded in a smart contract in a form of a *whitelist*. In contrast, LegIoT builds and manages a system trust graph and provides the means to reuse results of previously conducted trust evaluation processes. Furthermore, LegIoT supports various attestation schemes in one solution, while the system by Banerjee et al. is limited to only one method.

Last, Xu et al. [52] propose a security model for remote attestation for V2X applications with focus on anonymity of attested vehicles. In contrast to LegIoT, it does not focus on trust management and only supports *hardware-based* remote attestation that requires support of Trusted Platform Module (TPM). This limits heterogeneity and to a large extend, rules out IoT use-cases.

8 Conclusion

With LegIoT, we present a new approach to trust management in IoT networks, in which recommendation-like trust links provided by intermediate nodes are combined with evidence-based trust establishment methods. The design leverages distributed ledger technology to store, manage and distribute trust information in the network—the missing piece of the puzzle that made the combination of both approaches possible. The resulting solution inherits benefits of both worlds: On-demand trust management from evidence-based systems as well as the dynamics and trustless setting from reputation-based solutions. As a result, LegIoT can provide dynamic on-demand trust management in decentralized networks and can combine various underlying trust establishment methods to accommodate the various needs of heterogeneous IoT platforms. The novel algorithm for trust graph management in the heart of LegIoT maximizes the benefits of newly established trust links for the entire network, which improves the scalability of the system by multiple factors.

Acknowledgment. This research has been funded by the Federal Ministry of Education and Research of Germany (BMBF) in the framework KMU-innovativ-Verbundprojekt: Secure Internet of Things Management Platform - SIMPL (project number 16KIS0852), by BMBF within the project iBlockchain, by the European Space Operations Centre with the Networking/Partnering Initiative, and by the Intel Collaborative Research Institute for Collaborative Autonomous & Resilient Systems (ICRI-CARS).

A Attestation Schemes and Trust Scope

Remote attestation originally came to prominence as a feature of the TPM [41], the standard defined by the Trusted Computing Group (TCG) [48]. Many approaches to attestation have been developed, which differ in underlying requirements and security guarantees provided. Generally, they provide different levels of *resilience*, which refers to the general robustness of the underlying architecture against compromise. In the following we discuss attestation approaches of four categories, and suggest what resilience level and trust scope they can provide.

Hardware-Based Architectures. Include strong cryptographic co-processors like TPMs [41]. A different approach are Trusted Execution Environments (TEEs) that use an isolated processing environment [43]. Usually, they offer complex attestation mechanisms with arbitrary cryptographic functionality. Since cryptographic co-processors are well studied and strongly protected, hardware-based architectures generally have a *high* resilience. Thus, this architecture is able to attest other devices and create functional as well as referral trust.

Hybrid Architectures. Generally include minimal security features like Read Only Memory (ROM) and Memory Protection Unit (MPU) for secure storage [24]. Generally, hybrid schemes such as *SMART* [20] and *TrustLite* [32] attest a defined area of code only. Their limitations are less significant compared to software-based attestation schemes. Thus, their resilience is considered to be *medium*. If the attested code contains the segment that handles device functionality, functional trust is gained. In contrast, referral trust requires the *attestation component* of the prover to be attested.

Software-Based Attestation. Generally, secure co-processors are not available on low-end embedded devices due to minimal cost requirements. Thus, purely software-based approaches were developed [45]. They do not assume any secrets on the prover's device, since there is no secure storage available at the prover side. Instead, these schemes are based on using side-channel information to decide whether an attestation result is valid. However, this approach poses many assumptions on the network topology and adversarial capabilities. For instance, the verifier needs to have direct communication with the prover with no intermediate hops [3]. We consider *resilience* of this attestation type as *low* because the potential attack surface is comparatively high. As attestation statements made by such attestations about other parties cannot be trusted, they can only provide functional trust.

Control-Flow Attestation is a relatively recent development in the attestation landscape [2]. Static attestation, to which previously discussed attestation categories belong to, is not able to capture misbehavior of software during runtime. This is where runtime attestation comes into play by monitoring an application's control flow and detecting all deviations from the expected flow (documented in the security policy). This approach enables the highest trust guarantees of all attestation schemes. Runtime attestation schemes like *DIAT* [4] offer a *very high* resilience because they also protect against runtime adversaries, and thus can provide both referral and functional trust.

References

1. Abdul-Rahman, A., Hailes, S.: A distributed trust model. In: Proceedings of the 1997 Workshop on New Security Paradigms, pp. 48–60 (1998)
2. Abera, T., et al.: C-FLAT: control-flow attestation for embedded systems software. In: ACM SIGSAC CCS (2016)
3. Abera, T., et al.: Things, trouble, trust: on building trust in IoT systems. In: ACM DAC (2016)
4. Abera, T., Bahmani, R., Brasser, F., Ibrahim, A., Sadeghi, A.R., Schunter, M.: DIAT: data integrity attestation for resilient collaboration of autonomous systems. In: NDSS (2019)
5. Aberer, K., Despotovic, Z.: Managing trust in a peer-to-peer information system. In: ACM CIKM (2001)
6. Alexopoulos, N., Daubert, J., Mühlhäuser, M., Habib, S.M.: Beyond the hype: on using blockchains in trust management for authentication. In: IEEE Trustcom/BigDataSE/ICESS (2017)

7. Alexopoulos, N., Vasilomanolakis, E., Ivánkó, N.R., Mühlhäuser, M.: Towards blockchain-based collaborative intrusion detection systems. In: CRITIS (2017)
8. Alves, T., Felton, D.: TrustZone: integrated hardware and software security (2004)
9. Ambrosin, M., Conti, M., Ibrahim, A., Neven, G., Sadeghi, A.R., Schunter, M.: SANA: secure and scalable aggregate network attestation. In: ACM SIGSAC CCS (2016)
10. Ammar, M., Washha, M., Ramabhadran, G.S., Crispo, B.: Slimiot: scalable lightweight attestation protocol for the Internet of Things. In: IEEE DSC (2018)
11. Asokan, N., et al.: SEDA: scalable embedded device attestation. In: ACM SIGSAC CCS (2015)
12. Banerjee, M., Lee, J., Chen, Q., Choo, K.R.: Blockchain-based security layer for identification and isolation of malicious things in IoT: a conceptual design. In: ICCCN (2018)
13. Blaze, M., Feigenbaum, J., Lacy, J.: Decentralized trust management. In: IEEE S&P (1996)
14. Buchegger, S., Le Boudec, J.Y.: Performance analysis of the CONFIDANT protocol. In: ACM MOBIHOC (2002)
15. Buterin, V.: A next-generation smart contract and decentralized application platform. Whitepaper (2014). https://blockchainlab.com/pdf/Ethereum_white_paper-a_next_generation_smart_contract_and_decentralized_application_platform-vitalik-buterin.pdf
16. Carpent, X., ElDefrawy, K., Rattanavipanon, N., Tsudik, G.: Lightweight swarm attestation: a tale of two LISA-s. In: ACM AsiaCCS (2017)
17. Cervesato, I.: The Dolev-Yao intruder is the most powerful attacker. In: ACM/IEEE LICS (2001)
18. Dardaman, C.: Breaking & entering with Zipato SmartHubs (2019). https://blackmarble.sh/zipato-smart-hub/
19. Dolev, D., Yao, A.: On the security of public key protocols. IEEE Trans. Inf. Theory **29**, 198–208 (1983)
20. Eldefrawy, K., Tsudik, G., Francillon, A., Perito, D.: Smart: secure and minimal architecture for (establishing dynamic) root of trust. In: NDSS (2012)
21. Elkins, M., Torto, D.D., Levien, R., Roessler, T.: MIME Security with OpenPGP, IETF RFC 3156 (2001). www.ietf.org/rfc/rfc3156.txt
22. Eschenauer, L., Gligor, V., Baras, J.: On trust establishment in mobile ad-hoc networks. In: Security Protocols Workshop (2002)
23. WE Forum: This is how a smart factory actually works (2019). https://www.weforum.org/agenda/2019/06/connectivity-is-driving-a-revolution-in-manufacturing/
24. Francillon, A., Nguyen, Q., Rasmussen, K.B., Tsudik, G.: A minimalist approach to remote attestation. In: DATE (2014)
25. Golomb, T., Mirsky, Y., Elovici, Y.: CIoTA: collaborative IoT anomaly detection via blockchain. CoRR (2018)
26. Hemsley, K., Fisher, R.: History of industrial control system cyber incidents (2018)
27. Housley, R., Ford, W., Polk, W., Solo, D.: Internet X.509 Public Key Infrastructure, Certificate and CRL Profile, IETF RFC 2459 (1999). www.ietf.org/rfc/rfc2459.txt
28. Hyperledger: Hyperledger Sawtooth - a modular platform for building, deploying, and running distributed ledgers (2018). https://www.hyperledger.org/projects/sawtooth
29. Hyperledger: Hyperledger Sawtooth v1.1.4 documentation (2019). https://sawtooth.hyperledger.org/docs/core/releases/1.1.4/

30. Ibrahim, A., Sadeghi, A.R., Tsudik, G., Zeitouni, S.: DARPA: device attestation resilient to physical attacks. In: 9th ACM WiSec (2016)
31. Jøsang, A., Hayward, R., Pope, S.: Trust network analysis with subjective logic. In: Australasian Computer Science Conference (2006)
32. Koeberl, P., Schulz, S., Sadeghi, A.R., Varadharajan, V.: TrustLite: a security architecture for tiny embedded devices. In: EuroSys (2014)
33. Kohnhäuser, F., Büscher, N., Gabmeyer, S., Katzenbeisser, S.: Scapi: a scalable attestation protocol to detect software and physical attacks. In: ACM WiSec (2017)
34. Kohnhäuser, F., Büscher, N., Katzenbeisser, S.: Salad: secure and lightweight attestation of highly dynamic and disruptive networks. In: ACM AsiaCCS (2018)
35. Langner, R.: Stuxnet: dissecting a cyberwarfare weapon. IEEE S&P **9**, 49–51 (2011)
36. Li, H., Singhal, M.: Trust management in distributed systems. Computers **40**(2), 45–53 (2007)
37. Marti, S., Giuli, T.J., Lai, K., Baker, M.: Mitigating routing misbehavior in mobile ad hoc networks. In: MobiCom (2000)
38. Meng, W., Tischhauser, E.W., Wang, Q., Wang, Y., Han, J.: When intrusion detection meets blockchain technology: a review. IEEE Access **6**, 10179–10188 (2018)
39. Moinet, A., Darties, B., Baril, J.L.: Blockchain based trust & authentication for decentralized sensor networks. CoRR (2017)
40. Park, J., Kim, K.: TM-Coin: Trustworthy management of TCB measurements in IoT. In: PerCom Workshops. IEEE (2017)
41. Pearson, S., Balacheff, B.: Trusted Computing Platforms: TCPA Technology in Context. Prentice Hall Professional (2003)
42. Rayner, G.: Smart meters could leave British homes vulnerable to cyber attacks, experts have warned (2018). https://www.telegraph.co.uk/news/2018/02/18/smart-meters-could-leave-british-homes-vulnerable-cyber-attacks/
43. Sabt, M., Achemlal, M., Bouabdallah, A.: Trusted execution environment: what it is, and what it is not. In: IEEE Trustcom/BigDataSE/ISPA (2015)
44. Scout, S.L.: Guide on Airbnb smart locks (2019). https://www.postscapes.com/airbnb-smart-lock/
45. Seshadri, A., Perrig, A., van Doorn, L., Khosla, P.: Using software-based attestation for verifying embedded systems in cars. In: ESCAR Workshop (2004)
46. Signorini, M., Pontecorvi, M., Kanoun, W., Di Pietro, R.: Bad: blockchain anomaly detection. CoRR (2018)
47. Stajano, F., Anderson, R.: The resurrecting duckling: security issues for ad hoc wireless networks. In: Security Protocols Workshop (1999)
48. TCG: Trusted computing group. https://trustedcomputinggroup.org/
49. Wang, Y., Vassileva, J.: Bayesian network-based trust model. In: IEEE/WIC WI 2003, pp. 372–378. IEEE (2003)
50. World, T.: IoT in utilities market forecasted to grow to $53.8 billion by 2024 (2020). https://www.tdworld.com/grid-innovations/article/21120887/iot-in-utilities-market-worth-538-billion-by-2024
51. Xiong, L., Liu, L.: Building trust in decentralized peer-to-peer electronic communities. In: ICEC (2002)
52. Xu, C., Liu, H., Li, P., Wang, P.: A remote attestation security model based on privacy-preserving blockchain for v2x. IEEE Access **6**, 67809–67818 (2018)
53. Yu, B., Singh, M.P.: A social mechanism of reputation management in electronic communities. In: Klusch, M., Kerschberg, L. (eds.) CIA 2000. LNCS (LNAI), vol. 1860, pp. 154–165. Springer, Heidelberg (2000). https://doi.org/10.1007/978-3-540-45012-2_15

Machine Learning Security

PrivColl: Practical Privacy-Preserving Collaborative Machine Learning

Yanjun Zhang, Guangdong Bai(✉), Xue Li, Caitlin Curtis, Chen Chen, and Ryan K. L. Ko

The University of Queensland, St Lucia, Queensland, Australia
{yanjun.zhang,g.bai,c.curtis,chen.chen,ryan.ko}@uq.edu.au,
xueli@itee.uq.edu.au

Abstract. Collaborative learning enables two or more participants, each with their own training dataset, to collaboratively learn a joint model. It is desirable that the collaboration should not cause the disclosure of either the raw datasets of each individual owner or the local model parameters trained on them. This privacy-preservation requirement has been approached through differential privacy mechanisms, homomorphic encryption (HE) and secure multiparty computation (MPC), but existing attempts may either introduce the loss of model accuracy or imply significant computational and/or communicational overhead.

In this work, we address this problem with the lightweight additive secret sharing technique. We propose PrivColl, a framework for protecting local data and local models while ensuring the correctness of training processes. PrivColl employs secret sharing technique for securely evaluating addition operations in a multiparty computation environment, and achieves practicability by employing only the homomorphic addition operations. We formally prove that it guarantees privacy preservation even though the majority ($n-2$ out of n) of participants are corrupted. With experiments on real-world datasets, we further demonstrate that PrivColl retains high efficiency. It achieves a speedup of more than 45X over the state-of-the-art MPC-/HE-based schemes for training linear/logistic regression, and 216X faster for training neural network.

Keywords: Privacy · Machine learning · Collaborative learning

1 Introduction

The performance of machine learning largely relies on the availability of datasets. To take advantage of massive data owned by multiple entities, collaborative machine learning has been proposed to enable two or more data owners to construct a joint model. One typical scenario demanding collaborative learning is where the *features* of a same sample are held by multiple data owners. The collaboration among owners can improve the model accuracy by leveraging additional features from each other. A real-world example is that a recommender system

L. Chen et al. (Eds.): ESORICS 2020, LNCS 12308, pp. 399–418, 2020.
https://doi.org/10.1007/978-3-030-58951-6_20

can take use of the ratings of a same item among multiple online merchants to enhance its predictive power.

To address the privacy concerns arising from collaborative learning, many studies [7,12,40,44] have been proposed to provide *data locality* by distributing learning algorithms onto data owners such that the data can be confined within their owners. Despite this, their learning processes still entail sharing locally trained models, in order to synthesize the final models. However these local models are subject to information leakage. For example, model-inversion attacks [10,30] are able to restore training data from them. In addition, in the scenarios where the model itself represents intellectual property, e.g., in financial market systems, it is an essential requirement for the local models to be kept confidential [32].

To provide a supplementary, i.e., *privacy-preserving synthesis of the local models*, differential privacy mechanisms and cryptographic mechanisms may be employed. The former [1,8,19,23,37,38,48] usually entails adding noise on the model parameters, causing loss in the accuracy of the final models. The cryptographic mechanisms such as homomorphic encryption (HE) [7,29,36] and secure multiparty computation (MPC) [5,11,12,28,31] are able to yield identical models as those trained on plaintext data, but are known to be limited by the significant computational or communicational overheads.

This work focuses on the practicability of the cryptographic solutions. We propose a lightweight framework named PRIVCOLL for privacy-preserving collaborative learning in the distributed feature scenario. PRIVCOLL adopts the two-layer architecture commonly used in previous privacy-preserving collaborative learning frameworks [23,31,40]. It has a local node layer consisting of participating data owners, and an aggregation node (which can be untrustworthy). The main strategy of PRIVCOLL is to dispense the homomorphic *multiplication operations* and *non-linear functions* on ciphertext, as they are far more costly in computation and communication than the *addition operations* [3,13,15,42]. To this end, we redesign the workflow of collaborative learning, so that it employs only the homomorphic addition operations provided by *additive secret sharing scheme* [4] for synthesizing local models and intermediate outcomes. The computation that is carried out by the aggregation mode uses only the *sum* of the intermediate results that are generated by the local nodes (detailed in Sect. 3.3). As such, PRIVCOLL achieves significant cost savings, in comparison with state-of-the-art cryptographic solutions.

The redesigned workflow also ensures that both the raw data and local models are always kept with their owners. We formally prove PRIVCOLL preserves privacy in such a way that the *honest-but-curious* participants, who have access to the sum of the intermediate outcomes produced by the local nodes and additional knowledge learned from the training iterations, are unlikely (i.e., with a negligible probability ε) to reveal the raw training data or the local model parameters of other participants.

Notably, our new collaborative learning workflow in PRIVCOLL introduces no sacrifice to the accuracy of the models, and also supports a wide range of machine learning algorithms as previous work does [1,23,31]. Intuitively, our solution makes use of the chain rules in calculus to decompose gradient descent optimization into computational primitives, and to distribute them to the local nodes and the aggregation node respectively. When they collaborate together, these primitives can be recombined to achieve the correctness of learning. We prove that such correctness is guaranteed for any algorithm that uses gradient descent for optimization, including but not limited to linear regression, logistic regression, and a variety of neural networks.

Contributions. In general, our contributions can be summarized as follows.

- **A Novel Privacy-preserving Collaborative Framework.** We propose a novel framework PRIVCOLL for collaborative learning with distributed features. It preserves privacy while enabling a wide range of machine learning algorithms and achieving high computation efficiency. Not only does PRIVCOLL achieve the data locality as previous work does, but it also keeps the local models confidential.
- **Provable Privacy Preservation and Correctness Guarantee.** We prove the privacy preservation of PRIVCOLL, demonstrating a negligible probability of corrupted parties revealing either the original data or the trained parameters from other honest parties. We also prove that PRIVCOLL ensures the learned model is identical to that in the traditional non-distributed framework.
- **Experimental Evaluations.** We conduct experiments on real datasets, showing that PRIVCOLL achieves a significant improvement of efficiency over the state-of-the-art cryptographic solutions based on MPC and HE. For example, PRIVCOLL achieves around 22.5 min for a two-hidden-layer neural network to process all samples in the MNIST dataset [27], while it takes more than 81 h with a state-of-the-art MPC protocol SecureML [31].

2 Background

In this section, we introduce the background knowledge that is necessary to understand our framework.

2.1 Gradient Descent Optimization

Gradient descent is by far the most commonly used optimization strategy among various machine learning and deep learning algorithms. It is used to find the values of coefficients that minimize a cost function as far as possible. Given a defined cost function J, the coefficient matrix W is derived by the optimization $\arg\min_W J$, and is updated as:

$$W := W - \alpha \frac{\partial J}{\partial W} \tag{1}$$

Given a particular training dataset X and a label matrix y, the cost function J can be defined as $J(\sigma(XW), y, W)$, where σ is determined by the learning model. For example, in logistic regression, σ is usually a sigmoid function $1/(1+e^{-z})$, while in neural network, σ is a composite function that is known as forward propagation. According to the chain rule in calculus, the gradient with respect to W is computed as $\frac{\partial J}{\partial \sigma}\frac{\partial \sigma}{\partial XW}\frac{\partial XW}{\partial W} + \tau$, where τ is the gradient with respect to the regularization term, which is independent to X. Let Δ be $\frac{\partial J}{\partial \sigma}\frac{\partial \sigma}{\partial XW}$, and $\frac{\partial XW}{\partial W}$ equals to X, then the gradient with respect to W can be written as:

$$\frac{\partial J}{\partial W} = \Delta X + \tau. \tag{2}$$

As such, we can decompose the gradient descent optimization into Δ , X and τ, in which Δ is a function of XW. This provides an algorithmic foundation for PrivColl's distribution of learning algorithms.

2.2 Additive Secret Sharing Scheme

Secret sharing schemes aim to securely distribute secret values amongst a group of participants. PrivColl employs the secret sharing scheme proposed by [4], which uses *additive sharing* over $\mathbb{Z}_{2^{32}}$. In this scheme, a secret value srt is split to s shares $E^1_{srt}, ..., E^s_{srt} \in \mathbb{Z}_{2^{32}}$ such that

$$E^1_{srt} + E^2_{srt} + ... + E^s_{srt} \equiv srt \mod 2^{32}, \tag{3}$$

and any $s-1$ elements $E^{i_1}_{srt}, ..., E^{i_{s-1}}_{srt}$ are uniformly distributed. This prevents any participant who has part of the shares from deriving the value of srt, unless all participants join their shares.

In addition, the scheme has a homomorphic property that allows efficient and secure addition on a set of secret values $srt_1, ..., srt_s$ held by corresponding participants $S_1, ..., S_s$. To do this, each participant S_i executes a randomised sharing algorithm $Shr(srt_i, S)$ to split its secret srt_i into shares $E^1_{srt_i}, ..., E^s_{srt_i}$, and distributes each $E^j_{srt_i}$ to the participant S_j. Then, each S_i locally adds the shares it holds, $E^i_{srt_1}, ..., E^i_{srt_s}$, to produce $\sum_{j=1}^{s} E^i_{srt_j}$ (denoted by E^i for brevity). After that, a reconstruction algorithm $Rec(\{(E^i, S_i)\}_{S_i \in S})$, which takes E^i from each participant and add them together, can be executed by an aggregator to reconstruct the $\sum_{i=1}^{s} srt_i$ without revealing any secret addends srt_i.

3 Design of PrivColl

3.1 Scope and Threat Model

The involved parties in PrivColl are a set of local nodes (i.e., data owners) $S_1, ..., S_s$ and an aggregation node Agg. Each local node holds part of features of the training samples, denoted by X^l ($l \in \{1, ..., s\}$), and the corresponding local model W^l ($l \in \{1, ..., s\}$) trained on X^l. Each $X^l \in \mathbb{R}^{m \times d_l}$ is a $m \times d_l$ matrix representing m training samples with d_l features, and $W^l \in \mathbb{R}^{d_l \times k}$ is a matrix of coefficients, where k is the number of output classes. In PrivColl, m and k

are public and known by every party, and d_l is private and only known by the corresponding data owner S_l. We use X to denote the vertical concatenation of the local training datasets $X^1, ..., X^s$. Then we know that X is a $m \times n$ matrix where $n = \sum_{l=1}^{s} d_l$ (i.e., the total number of features in X). Since d_l is private, n is unknown unless all of the local nodes join their views.

PRIVCOLL aims to defend against an honest-but-curious adversary $\mathcal{A}$, who follows the collaboration protocols and training procedures, but is intending to obtain the datasets of other local nodes (i.e., X^l) and/or the model parameters trained out of them (i.e., W^l). The adversary may control Agg, and t out of s local nodes. Here, we conservatively assume $t < s-1$, which implies that at least two local nodes need to be out of the adversary's control, as $s-1$ comprised local nodes who has the sum of their shares, colluding with Agg who has the sum of all shares, will be able to obtain the share of the remaining node by a simple subtraction (detailed in Sect. 4).

3.2 Definitions of Privacy Preservation and Correctness

Keeping local data/model private and providing functional correctness are the main properties PRIVCOLL aims to achieve. Below we present the definitions of these two properties.

Definition 1 ***(ε-privacy).*** *A mechanism preserves ε-privacy if the probability for a probabilistic polynomial-time (PPT) adversary to derive X^l or W^l of any benign node S_l based on its knowledge is not greater than ε.*

Definition 2 ***(Correctness).*** *Given a function $\mathcal{F}$ that takes as input a training dataset X, and its distributed version $\mathcal{F}'$ that takes as inputs X's vertical partitions $X^1, ..., X^s$, we say $\mathcal{F}'$ is correct if $\mathcal{F}'(\{(S_l, X^l), Agg\}_{S_l \in S}) = \mathcal{F}(X)$.*

3.3 Workflow of PrivColl

Figure 1 illustrates the end-to-end workflow of PRIVCOLL, which is divided into the following steps.

Initialization: Each local node S_l holds its own training dataset $X^l \in \mathbb{R}^{m \times d_l}$, and randomly initializes its coefficient matrix $W^l \in \mathbb{R}^{d_l \times k}$.

Step 1: In each iteration of gradient descent, each S_l multiplies X^l by W^l locally, resulting in a $X^l W^l \in \mathbb{R}^{m \times k}$. The value of d_l, i.e., the number of features in X^l, is removed by such a matrix multiplication.

Step 2: Each S_l executes the sharing algorithm Shr to split $X^l W^l$ into s shares using the *additive secret sharing scheme.*

$$\{(S_i, E^i_{X^l W^l})\}_{S_i \in S} \leftarrow Shr(X^l W^l, S), \tag{4}$$

in which Shr takes as input a secret $X^l W^l$ and a set S of local nodes, and produces a set of shares $E^i_{X^l W^l} (i \in \{1, ..., s\})$, each of which is distributed to a different $S_i \in S$. Then, each S_l calculates the sum of all shares it receives, and gets $E^l = \sum_{j=1}^{s} E^l_{X^j W^j}$.

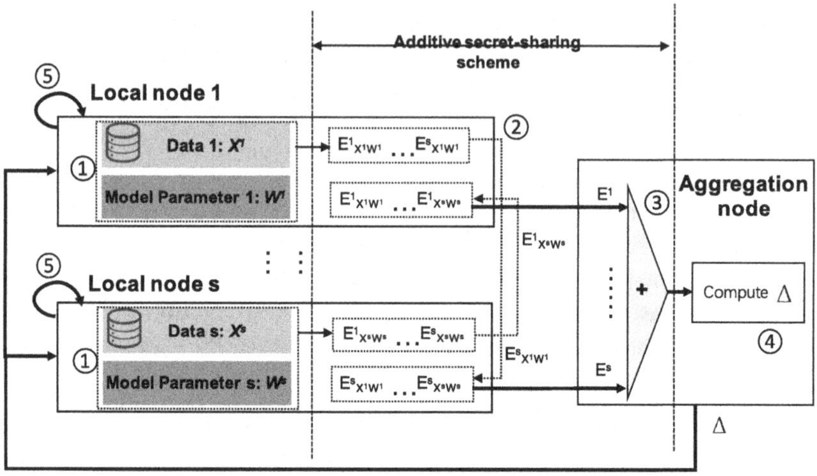

Fig. 1. The end-to-end workflow of the PrivColl

Step 3: *Agg* collects all E^l from local nodes and add them together. Since

$$\sum_{l=1}^{s} E^l = \sum_{l=1}^{s}\sum_{j=1}^{s} E^l_{X^jW^j} = \sum_{j=1}^{s}\sum_{l=1}^{s} E^l_{X^jW^j} = \sum_{j=1}^{s} X^jW^j, \tag{5}$$

this addition reconstructs the homomorphic addition result XW (equals $\sum_{l=1}^{s} X^lW^l$).

Step 4: *Agg* computes $\Delta = \frac{\partial J}{\partial \sigma}\frac{\partial \sigma}{\partial XW}$ (*c.f.*, Eq. 2), and sends Δ back to the local nodes.

Step 5: With the received Δ, each S_l updates its local coefficient matrix W^l.

$$W^l \leftarrow W^l - \alpha(\Delta X^l + \tau^l) \tag{6}$$

Step 1–5 are repeated for the next training iteration until convergence.

4 Privacy Preservation Analysis

In this section, we analyze the privacy preservation of PrivColl's learning process. To this end, we first investigate the overall knowledge that can be learned by an adversary from the training iterations (Sect. 4.2). Then we prove that the knowledge is limited such that the desired ε-privacy (Definition 1) is achieved with a negligible ε (Sect. 4.3).

4.1 Preliminaries

We start with the following lemma that is soon used in our proof.

Lemma 1. *Consider a positive (semi-)definite matrix A that is obtained as the product of a real number matrix B by its transpose B^T*

$$A = BB^T, \tag{7}$$

where B is of rank r. Without knowing the number of columns of B, the probability of solving the B given A, denoted as $P(B|A)$, is $\leq (r!)^{-1}$.

Proof. Since the matrix A is positive (semi-)definite, there exists an eigen-decomposition such that

$$A = U\Lambda U^T, \tag{8}$$

where U denotes a matrix of eigenvectors of A (each column of U is an eigenvector of A), and Λ denotes a diagonal matrix whose diagonal elements are the eigenvalues. For a positive (semi-)definite matrix, all the eigenvalues are non-negative. A different ordering of the eigenvector columns results in a different U and a corresponding Λ [16].

With the eigen-decomposition, B can be constructed by $B = U\Lambda^{\frac{1}{2}}$, where $\Lambda^{\frac{1}{2}}$ has the square roots of eigenvalues as its diagonal elements, and all its remaining values are zeros. Each eigen-decomposition leads to a different U and a corresponding Λ, resulting in a unique solution of B. A matrix B of rank r has r non-zero eigenvalues and thus there are $r!$ different possible orderings of eigenvector columns, implying $r!$ different Us and the corresponding Λs. Consequently, there are $r!$ different possible solutions of computing B, which gives the probability of solving B given A with eigen-decomposition $P_{eigen}(B|A) = (r!)^{-1}$.

In addition, without the knowledge of the number of columns in B, there are more than $r!$ possible solutions of solving B. This is because from the eigen-decomposition construction, B's columns are orthogonal. In general, the matrix B need not have orthogonal columns (it can be rectangular) [22]. Thus, $P(B|A) \leq (r!)^{-1}$.

4.2 Party Knowledge

We define the *party knowledge* as the overall knowledge that can be learned by adversary parties. It includes the parties' own inputs, and the additional knowledge that can be inferred from the training iterations. We prove that the party knowledge in PRIVCOLL is bounded within a certain range. In particular, we demonstrate that the overall *party knowledge* of adversary party I in PRIVCOLL is a set of $\{X^{l'} |_{l' \in I}, W^{l'} |_{l' \in I}, XW, \sum X^l W^l |_{l \in S \setminus I}, \sum X^l (X^l)^T |_{l \in S \setminus I}, XX^T\}$, where $\{X^{l'} |_{l' \in I}, W^{l'} |_{l' \in I}, XW\}$ are the adversary's own input in the workflow, and $\{\sum X^l W^l |_{l \in S \setminus I}, \sum X^l (X^l)^T |_{l \in S \setminus I}, XX^T\}$ are the additional information that can be inferred from the training iterations. The party knowledge we derive in this Section will be used in Sect. 4.3 to prove the privacy preservation of PRIVCOLL.

We use the simulation paradigm (also known as the *real/ideal model*) [15] to prove such a bound of party knowledge. The simulation paradigm compares what an adversary can do in a real protocol execution $REAL$ to what it can do in an ideal setting with a trusted functionality (simulation) SIM [15]. Formally, the protocol $\mathcal{P}$ securely computes a functionality $\mathcal{F}_{\mathcal{P}}$ if for every adversary in $REAL$, there exists an adversary in SIM, such that the view of the adversary from $REAL$ is indistinguishable from the view of the adversary from SIM. A perfect indistinguishability [6] between the view of $REAL$ and SIM guarantees that the adversary, without error probability, can learn nothing more than their own inputs and the information required by SIM for the simulation.

We introduce some notations used in our proof. We use $X^{S'} = \{X^l | S_l \in S'\}$ to indicate the inputs of any subset of local nodes $S' \subseteq S$. Given any subset $I_0 \subseteq S$ of the parties *without* the knowledge of XW, and subset $I_1 \subseteq S \cup \{Agg\}$ of the parties *with* the knowledge of XW, let $REAL(X^S, XW, S, t, \mathcal{P}, I)$ denote the combined views of all parties in $I = I_0 \cup I_1$ from the execution of a real protocol $\mathcal{P}$, where t is the adversary threshold (recall that $t < s - 1$). Let $SIM(X^I, Z, S, t, \mathcal{F}_{\mathcal{P}}, I)$ denote the views of I from an ideal execution, where Z is the information required by SIM for the simulation. In other words, the Z indicates the party knowledge that the adversary can and only can learn other than their own inputs.

Theorem 1 ***(Party Knowledge).*** *The simulator $SIM(X^I, Z, S, t, \mathcal{F}_{\mathcal{P}}, I)$ is perfectly indistinguishable from $REAL$ with respect to their outputs, namely*

$$REAL(X^S, XW, S, t, \mathcal{P}, I) \equiv SIM(X^I, Z, S, t, \mathcal{F}_{\mathcal{P}}, I)$$

if and only if

$$Z = \{z_1 = \sum X^l W^l \,|_{l \in S \setminus I}, z_2 = \sum X^l (X^l)^T \,|_{l \in S \setminus I}, z_3 = XX^T\}.$$

Proof. We define the simulator through each of the ith training iteration as:

- SIM_0: This is the simulator for the **Initialization**. In the step of initialization, the view of parties in $I_0 \cup I_1$ does not depend on the inputs of the parties not in $I_0 \cup I_1$. Therefore, instead of sending the actual $\{X^l W^{l(0)}\}_{l \in S \setminus \{I_0 \cup I_1\}}$ of the parties $S \setminus \{I_0 \cup I_1\}$ to the aggregation node, the simulator can produce a simulation by running the parties $S \setminus \{I_0 \cup I_1\}$ on a pseudorandom vector $\mu^{l(0)}$ in $\mathbb{R}^m$ as input, and then output the same pseudorandom vector to the aggregation node. Since the model parameter W^l is also randomized in the step of initialization, the pseudorandom vectors for the inputs of all honest parties $S \setminus \{I_0 \cup I_1\}$, and the joint view of parties in $\{I_0 \cup I_1\}$ will be identical to that in $REAL$

$$\{\mu^{l(0)} \,|_{l \in S \setminus \{I_0 \cup I_1\}}, X^{l'} W^{l'(0)} \,|_{l' \in \{I_0 \cup I_1\}}\} \equiv \{X^S W^{S(0)}\}.$$

- $SIM_i, i \geq 1$: This is the simulator for the ith training iteration ($i \geq 1$). The simulator computes $\Delta^{(i)} = f(\Sigma^{(i)})$, where the function f is determined by the learning model. For example, in the linear regression, $f(x) = x$, while in logistic regression, $f(x) = 1/(1 + e^{-x})$.
  We respectively consider the simulator for I_0, I_1. First, with respect to I_0, the simulator computes $\Sigma^{(i)}$ as

$$\Sigma^{(i)} = \sum \mu^{l(i)} \,|_{l \in S \setminus I_0} + \sum X^{l'} W^{l'(i)} \,|_{l' \in I_0} - y,$$

where $\mu^{l(i)}$ is computed by the result from the previous iteration as

$$\mu^{l(i)} = z_1^{l(i-1)} - \frac{\alpha}{m} z_2^l \Delta^{(i-1)} |_{l \in S \setminus I_0} .$$

Therefore,

$$\sum \mu^{l(i)} |_{l \in S \setminus I_0} = \sum z_1^{l(i-1)} - \frac{\alpha}{m} \sum z_2^l \Delta^{(i-1)} |_{l \in S \setminus I_0} .$$

Note $X^{l'} W^{l'(i)}$ is also computed by the result from the previous iteration, and

$$\sum X^{l'} W^{l'(i)} |_{l' \in I_0} = \sum X^{l'} W^{l'(i-1)} - \frac{\alpha}{m} \sum X^{l'} (X^{l'})^T \Delta^{(i-1)} |_{l' \in I_0} .$$

Then, the $\Sigma^{(i)}$ can be written as

$$\begin{aligned} \Sigma^{(i)} =& \sum z_1^{l(i-1)} |_{l \in S \setminus I_0} + \sum X^{l'} W^{l'(i-1)} |_{l' \in I_0} \\ & - \frac{\alpha}{m} (\sum z_2^l \Delta^{(i-1)} |_{l \in S \setminus I_0} + \sum X^{l'} (X^{l'})^T \Delta^{(i-1)} |_{l' \in I_0}) - y. \end{aligned}$$

Note that, in $REAL$, the output of $\Sigma^{(i)}$ by S is

$$\sum X^S W^{S(i)} - y = \sum X^S W^{S(i-1)} - \frac{\alpha}{m} \sum X^s (X^s)^T \Delta^{(i-1)} - y.$$

Thus, $\{\sum z_1^{l(i-1)} \quad |_{l \in S \setminus I_0}, \sum z_2^l \quad |_{l \in S \setminus I_0}, \sum X^{l'} W^{l'(i-1)} \quad |_{l' \in I_0}, \sum X^{l'} (X^{l'})^T \ |_{l' \in I_0}\}$ which is the joint view of all honest parties $S \setminus I_0$ and parties in I_0 will be perfectly indistinguishable to $\{\sum X^S W^{S(i-1)}, \sum X^s (X^s)^T\}$ which is the output in $REAL$.

Next, we consider the simulator for I_1. With respect to I_1, we let the simulator compute $\Sigma^{(i)}$ as

$$\Sigma^{(i)} = \mu^{(i)} - y,$$

where $\mu^{(i)}$ is computed by the result from the previous iteration as

$$\mu^{(i)} = XW^{(i-1)} - \frac{\alpha}{m} z_3 \Delta^{(i-1)}.$$

Then, the $\Sigma^{(i)}$ can be written as

$$\Sigma^{(i)} = XW^{(i-1)} - \frac{\alpha}{m} z_3 \Delta^{(i-1)} - y,$$

Note that, in $REAL$, the output of $\Sigma^{(i)}$ by S is

$$XW^{(i)} - y = XW^{(i-1)} - \frac{\alpha}{m} XX^T \Delta^{(i-1)} - y.$$

Thus, the joint view of all parties in SIM with knowledge of z_3 will be perfectly indistinguishable to $\{XW^{(i-1)}, XX^T\}$ which is the output in $REAL$.

All in all, the output of the simulator SIM of each training iteration is perfectly indistinguishable from the output of $REAL$. For the simulator with respect to $I = I_0 \cup I_1$, knowledge of $z_1 = \sum z_1^{l(i-1)} |_{l \in S \setminus I} = \sum X^l W^l |_{l \in S \setminus I}$, $z_2 = \sum z_2^l |_{l \in S \setminus I} = \sum X^l (X^l)^T |_{l \in S \setminus I}$ and $z_3 = XX^T$ is sufficient, completing the proof.

4.3 Privacy Preservation Guarantee

With lemma introduced in Sect. 4.1, and party knowledge discussed in Sect. 4.2, we give our theorem of privacy preservation guarantee.

Theorem 2. *Let I denote the adversary party with party knowledge $\{X^{l'} |_{l' \in I}, W^{l'} |_{l' \in I}, XW, \sum X^l W^l |_{l \in S \setminus I}, \sum X^l (X^l)^T |_{l \in S \setminus I}, XX^T\}$. Let X^H denote the vertical concatenation of $\{X^l |_{l \in S \setminus I}\}$, which is the concatenation of honest local nodes' training datasets. Let r denote the rank of X^H. PrivColl preserves ε-privacy against adversary party I on the training dataset $X \in \mathbb{R}^{m \times n}$, where $\varepsilon \leq (r!)^{-1}$.*

Proof. We give some sketches here.

We start with the party knowledge XX^T. In PrivColl, n (i.e. the number of column of X) is unknown given the adversary threshold $t < s - 1$. In addition, the rank of X, denoted as R, is $> r$. Invoking Lemma 1 gives that the probability of solving the X of rank R given XX^T is $\leq (R!)^{-1} < (r!)^{-1}$. Then we combine the party knowledge XW ($X \in \mathbb{R}^{m \times n}$, and $W \in \mathbb{R}^{n \times k}$). From Theorem 1, XX^T is the only information required by SIM with respect to I_1 with the knowledge of XW. In other words, combining the XW gives no more information other than XX^T. Therefore, given an unknown n, and the probability of solving the $X < (r!)^{-1}$, we have the probability of solving W is also $< (r!)^{-1}$.

Then we continue to combine the party knowledge of $\{\sum X^l W^l |_{l \in S \setminus I}, \sum X^l (X^l)^T |_{l \in S \setminus I}\}$. They are the sum of real number matrices and the sum of non-negative real number matrices respectively. Given the adversary threshold $t < s - 1$, which means the number of honest local nodes (i.e. $|l|_{l \in S \setminus I}$) is ≥ 2, we have a negligible probability to derive any $X^l W^l$ or $X^l (X^l)^T$ from their sum. In addition, d_l (the number of columns of X^l) is also unknown to I. Therefore, with the probability of solving the $X < (r!)^{-1}$, the probability of solving either X^l or W^l is also $< (r!)^{-1}$.

At last, we combine the party knowledge of $\{X^{l'} |_{l' \in I}, W^{l'} |_{l' \in I}\}$. First, they are the input of I, which are independent of $\{\sum X^l W^l |_{l \in S \setminus I}, \sum X^l (X^l)^T |_{l \in S \setminus I}\}$. Next, we combine them into $X(X)^T$, which will give the adversary the problem of solving $X^H (X^H)^T$. As each of $d_l |_{l \in S \setminus I}$ is unknown to I, we have the number of column of X^H is also unknown to I. Thus, with Lemma 1, we have the probability of solving the X^H of rank r is $\leq (r!)^{-1}$. Similarly, combing $\{X^{l'} |_{l' \in I}, W^{l'} |_{l' \in I}\}$ to XW will also give the probability of solving $W^H \leq (r!)^{-1}$.

Thus, PrivColl preserves ε-privacy against adversary parties I on X, and $\varepsilon \leq (r!)^{-1}$.

With Theorem 2, we demonstrate that, with a sufficient rank of the training dataset of honest parties, e.g. ≥ 35, which is common in real-world datasets, PrivColl achieves 10^{-40}-privacy.

5 Correctness Analysis and Case Study

In this section, we first prove the correctness of PRIVCOLL when distributing learning algorithms that are based on gradient descent optimization. Then we use a recurrent neural network as a case study to illustrate the collaborative learning process in PRIVCOLL.

5.1 Correctness of PRIVCOLL's Gradient Descent Optimization

The following theorem demonstrates that if a non-distributed gradient descent optimization algorithm taking X as input, denoted by $\mathcal{F}_{GD}(X)$, converges to a local/global minima η, then executing PRIVCOLL with the same hyper settings (such as cost function, step size, and model structure) on $X^1, ..., X^s$, denoted by $\mathcal{F}'_{GD}(\{(S_l, X^l), Agg\}_{S_l \in S})$, also converges to η.

Theorem 3. PRIVCOLL*'s distributed algorithm of solving gradient descent optimization* $\mathcal{F}'_{GD}(\{(S_l, X^l), Agg\}_{S_l \in S})$ *is correct.*

Proof. Let $\mathcal{F}_{GD}(X) = \eta$ denote the convergence of $\mathcal{F}_{GD}(X)$ to the local/global minima η. Let W_i denote the model parameters of $\mathcal{F}_{GD}$ at i^{th} training iteration. Let $W'_i = |\{W^l_i\}|_{l \in \{1,..,s\}}$ denote the vertical concatenation on $\{W^l_i\}_{l \in \{1,..,s\}}$, i.e., the model parameters of $\mathcal{F}'_{GD}$ at i^{th} training iteration.

In $\mathcal{F}_{GD}$, the ith $(i \geq 1)$ training iteration update W_i such that

$$\begin{aligned} W_i = W_{i-1} - \alpha \frac{\partial J}{\partial W} &= W_{i-1} - \alpha \frac{\partial J}{\partial (XW_{i-1})} \frac{\partial XW_{i-1}}{\partial W_{i-1}} \\ &= W_{i-1} - \alpha \frac{\partial J}{\partial (XW_{i-1})} X. \end{aligned} \tag{9}$$

In $\mathcal{F}'_{GD}$, each node S_l updates its local W^l_i in

$$\begin{aligned} W^l_i = W^l_{i-1} - \alpha \Delta X^l &= W^l_{i-1} - \alpha \frac{\partial J}{\partial (XW'_{i-1})} \frac{\partial X^l W^l_{i-1}}{\partial W^l_{i-1}} \\ &= W^l_{i-1} - \alpha \frac{\partial J}{\partial (XW'_{i-1})} X^l. \end{aligned}$$

Since $W'_i = |\{W^l_i\}|_{l \in \{1,..,s\}}$, and $X = |\{X^l\}|_{l \in \{1,..,s\}}$, we have in $\mathcal{F}'_{GD}$ that

$$W'_i = |\{(W^l_{i-1} - \alpha \frac{\partial J}{\partial (XW'_{i-1})} X^l)\}|_{l \in \{1,..,s\}} = W'_{i-1} - \alpha \frac{\partial J}{\partial (XW'_{i-1})} X. \tag{10}$$

Comparing Eqs. 9 and 10, we can find that W_i and W'_i are updated using the same equation. Therefore, with $\mathcal{F}_{GD}(X) = \eta$, the gradient descent guarantees $\mathcal{F}'_{GD}(\{(S_l, X^l), Agg\}_{S_l \in S})$ also converges to η.

Thus, $\mathcal{F'}_{GD}(\{(S_l, X^l), Agg\}_{S_l \in S}) = \mathcal{F}_{GD}(X)$.

5.2 Case Study

Figure 2 shows an example of a two-layer feed-forward recurrent neural network (RNN). Every neural layer is attached with a time subscript c. The weight matrix W maps the input vector $X^{(c)}$ to the hidden layer $h^{(c)}$. The weight matrix V propagates the hidden layer to the output layer $\hat{y}^{(c)}$. The weight matrix U maps the previous hidden layer to the current one.

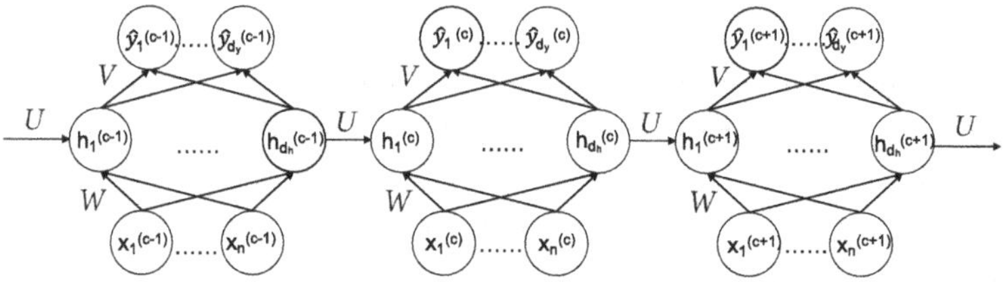

Fig. 2. An example of recurrent neural network

Original Algorithm. Recall that the original non-distributed version of the RNN is divided into the forward propagation and backward propagation through time. First, in the forward propagation, the output of the hidden layer propagated from the input layer is calculated as

$$\begin{aligned} Z_h^{(c)} &= X^{(c)}W + Uh^{(c-1)} + b_h \\ h^{(c)} &= \sigma_1(Z_h^{(c)}) \end{aligned} \tag{11}$$

The output of the output layer propagated from hidden layer is calculated as

$$\begin{aligned} Z_y^{(c)} &= Vh^{(c)} + b_y \\ \hat{y}^{(c)} &= \sigma_2(Z_y^{(c)}) \end{aligned} \tag{12}$$

Then, the cost function $\sum_{c=0}^{T} J(\hat{y}^{(c)}, y^{(c)})$, and the coefficient matrices W, U, V are updated using the backward propagation through time.

The gradients of $J^{(c)}$ with respect to V is calculated as $\frac{\partial J^{(c)}}{\partial V} = \frac{\partial J^{(c)}}{\partial \hat{y}^{(c)}} \frac{\partial \hat{y}^{(c)}}{\partial Z_y^{(c)}} \frac{\partial Z_y^{(c)}}{\partial V}$. We let $\frac{\partial J^{(c)}}{\partial \hat{y}^{(c)}} = \delta_{loss}^{(c)}$, $\delta_{\hat{y}}^{(c)} = \sigma_2'(Z_y^{(c)})$. The gradient of $J^{(c)}$ with respect to V can be written as:

$$\frac{\partial J^{(c)}}{\partial V} = [\delta_{loss}^{(c)} \circ \delta_{\hat{y}}^{(c)}](h^{(c)})^T. \tag{13}$$

The gradients of $J^{(c)}$ with respect to U is calculated as $\frac{\partial J^{(c)}}{\partial U} = \frac{\partial J^{(c)}}{\partial \hat{y}^{(c)}} \frac{\partial \hat{y}^{(c)}}{\partial h^{(c)}} \frac{\partial h^{(c)}}{\partial U}$, where $\frac{\partial J^{(c)}}{\partial \hat{y}^{(c)}} \frac{\partial \hat{y}^{(c)}}{\partial h^{(c)}} = V^T[\delta_{loss}^{(c)} \circ \delta_{\hat{y}}^{(c)}]$, and

$$\frac{\partial h^{(c)}}{\partial U} = \sum_{k=0}^{c} \frac{\partial h^{(c)}}{\partial h^{(k)}} \frac{\partial^+ h^{(k)}}{\partial U} = \sum_{k=0}^{c} (\prod_{i=k+1}^{c} \sigma_1'(Z_h^{(i)})U)\sigma_1'(Z_h^{(k)})h^{(k-1)}.$$

Let $\delta_h^{(c)} = \sigma_1'(Z_h^{(c)})$, then the gradient of $J^{(c)}$ with respect to U is

$$\frac{\partial J^{(c)}}{\partial U} = V^T \delta_{loss}^{(c)} \circ \delta_{\hat{y}}^{(c)} \left(\sum_{k=0}^{c} (\prod_{i=k+1}^{c} \delta_h^{(i)} U) \delta_h^{(k)} h^{(k-1)} \right). \tag{14}$$

Similarly, the gradients of $J^{(c)}$ with respect to W is calculated as:

$$\frac{\partial J^{(c)}}{\partial W} = V^T \delta_{loss}^{(c)} \circ \delta_{\hat{y}}^{(c)} \left(\sum_{k=0}^{c} (\prod_{i=k+1}^{c} \delta_h^{(i)} U) \delta_h^{(k)} X^{(k)} \right). \tag{15}$$

Algorithm 1. Privacy Preserving Collaborative Recurrent Neural Network

Input: Local training data $X^{l(c)}$ $(l = [1, ..., s], c = [0, ..., T])$,
    learning rate α
Output: Model parameters W^l $(l = [1, ..., s])$, U, V
Initialize: Randomize W^l $(l = [1, ..., s])$, U, V
repeat
    for all $S_l \in S$ **do in parallel**
        $S_l : X^{l(c)} W^l, c = [0, ..., T]$
        $S_l : srt^{(c)} \leftarrow X^{l(c)} W^l, c = [0, ..., T]$
        $\{(S_l, E_{srt}^{(c)})\}_{S_l \in S} \leftarrow Shr(srt^{(c)}, S)$
    end for
    $X^{(c)} W \leftarrow Rec(\{(S_l, E_{srt}^{(c)})\}_{S_l \in S})$
    $A : Z_h^{(c)} \leftarrow X^{(c)} W + U h^{(c-1)} + b_h$, $h^{(c)} \leftarrow \sigma_1(Z_h^{(c)})$
    $A : Z_y^{(c)} \leftarrow V h^{(c)} + b_y$, $\hat{y}^{(c)} = \sigma_2(Z_y^{(c)})$
    $A : \delta_{loss}^{(c)} \leftarrow \frac{\partial J^{(c)}}{\partial \hat{y}^{(c)}}, \delta_{\hat{y}}^{(c)} \leftarrow \sigma_2'(Z_y^{(c)}), \delta_h^{(c)} = \sigma_1'(Z_h^{(c)})$
    $A : \frac{\partial J^{(c)}}{\partial V} \leftarrow [\delta_{loss}^{(c)} \circ \delta_{\hat{y}}^{(c)}](h^{(c)})^T$
    $A : \frac{\partial J^{(c)}}{\partial U} \leftarrow V^T \delta_{loss}^{(c)} \circ \delta_{\hat{y}}^{(c)} \left(\sum_{k=0}^{c} (\prod_{i=k+1}^{c} \delta_h^{(i)} U) \delta_h^{(k)} h^{(k-1)} \right)$
    $A : V \leftarrow V - \alpha \sum_{c=0}^{T} \frac{\partial J^{(c)}}{\partial V}$
    $A : U \leftarrow U - \alpha \sum_{c=0}^{T} \frac{\partial J^{(c)}}{\partial U}$
    for all $S_l \in S$ **do in parallel**
        $\frac{\partial J^{(c)}}{\partial W^l} = V^T \delta_{loss}^{(c)} \circ \delta_{\hat{y}}^{(c)} \left(\sum_{k=0}^{c} (\prod_{i=k+1}^{c} \delta_h^{(i)} U) \delta_h^{(k)} X^{l(k)} \right)$
        $W^l \leftarrow W^l - \alpha \sum_{c=0}^{T} \frac{\partial J^{(c)}}{\partial W^l}$
    end for
until Convergence

PrivColl Algorithm. In PrivColl, each local node keeps $X^{l(c)}, c = [0, ..., T]$, and maintains the coefficient matrix W^l locally. The aggregation node maintains coefficient matrices U, V. Similar to its original non-distributed counterpart, the training process is divided into the forward propagation and backward propagation through time. Below we briefly outline these steps and the detailed algorithm is given by Algorithm 1.

In the forward propagation, local nodes compute $X^{l(c)}W^l, c = [0, ..., T]$ respectively (line 7) (line number in Algorithm 1), and $X^{(c)}W = \sum_{l=1}^{s} X^{l(c)}W^l$ is calculated using the secret-sharing scheme (line 8 to line 11). The aggregation node then computes $Z_h^{(c)}, h^{(c)}, Z_y^{(c)}, \hat{y}^{(c)}$ using Eqs. 11 and 12 (line 12 and line 13).

In the backward propagation through time, for each $J^{(c)}$ at time t, the aggregation node computes $\delta_{loss}^{(c)}, \delta_{\hat{y}}^{(c)}, \delta_h^{(k)}, k = [0, ..., t]$, and sends $\delta_{loss}^{(c)}, \delta_{\hat{y}}^{(c)}, \delta_h^{(k)}$ to local nodes (line 14). Then the aggregation node computes the gradients of $J^{(c)}$ with respect to V, U using Eqs. 13 and 14 (line 15 and line 16), and updates U, V (line 17 and line 18). The local nodes compute the gradients of $J^{(c)}$ with respect to W^l using Eq. 16, and update W^l respectively (line 19 to line 22).

$$\frac{\partial J^{(c)}}{\partial W^l} = V^T \delta_{loss}^{(c)} \circ \delta_{\hat{y}}^{(c)} \left(\sum_{k=0}^{c} (\prod_{i=k+1}^{c} \delta_h^{(i)} U) \delta_h^{(k)} X^{l(k)} \right). \tag{16}$$

6 Performance Evaluation

We implement PrivColl in C++. It uses the Eigen library [17] to handle matrix operations, and uses ZeroMQ library [21] to implement the distributed messaging. The experiments are executed on four Amazon EC2 c4.8xlarge machines with 60GB of RAM each, three of which act as local nodes and the other acts as the aggregation node. To simulate the real-world scenarios, we execute PrivColl on both LAN and WAN network settings. In the LAN setting, machines are hosted in a same region, and the average network bandwidth is 1GB/s. In the WAN setting, we host these machines in different continents. The average network latency (one-way) is 137.7 ms, and the average network throughput is 9.27 MB/s. We collect 10 runs for each data point in the results and report the average. We use the MNIST dataset [27], and duplicate its samples when its size is less than the sample size m ($m \geq 60,000$).

We take non-private machine learning which trains on the concatenated dataset as the baseline, and compare with MZ17 [31], which is the state-of-the-art cryptographic solution for privacy preserving machine learning. It is based on oblivious transfer (MZ17-OT) and linearly homomorphic encryption (MZ17-LHE). As shown in Fig. 3, PrivColl achieves significant efficiency improvement over MZ17, and due to parallelization in the computing of the local nodes, PrivColl also outperforms the non-private baselines in the LAN network setting.

Linear Regression and Logistic Regression. We use mini-batch stochastic gradient descent (SGD) for training the linear regression and logistic regression.

We set the batch size $|B| = 40$ with 4 sample sizes $(1,000\text{-}100,000)$ in the linear regression and logistic regression.

In the LAN setting, PrivColl achieves around 45x faster than MZ17-OT. It takes 13.11 s for linear regression (Fig. 3a) and 12.73 s for logistic regression (Fig. 3b) with sample size $m = 100,000$, while in MZ17-OT, 594.95 s and 605.95 s are reported respectively. PrivColl is also faster than the baseline, which takes 17.20 s and 17.42 s for linear/logistic regression respectively. In the WAN setting, PrivColl is around 9x faster than MZ17-LHE. It takes 1408.75 s for linear regression (Fig. 3c) and 1424.94 s for logistic regression (Fig. 3d) with sample size $m = 100,000$, while in MZ17-LHE, it takes 12841.2 s and 13441.2 s respectively with the same sample size. It is worth mentioning that in MZ17, an MPC-friendly alternative function is specifically designed to replace non-linear sigmoid functions for training logistic regression, while in our framework, the non-linear function is used as usual. To further break down the overhead to computation and communication, we summarize the results of linear regression and logistic regression on other sample sizes in Table 1.

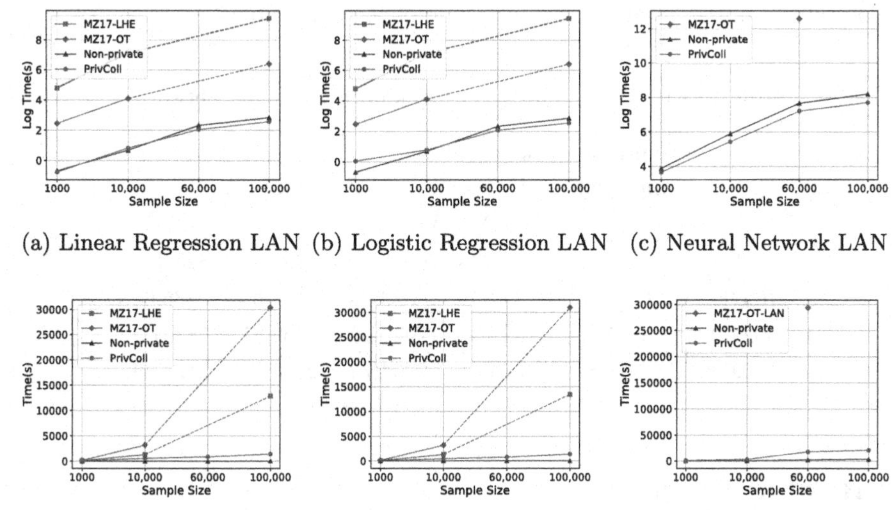

(a) Linear Regression LAN (b) Logistic Regression LAN (c) Neural Network LAN

(d) Linear Regression WAN (e) Logistic Regression WAN (f) Neural Network WAN

Fig. 3. Efficiency comparison. a-c) the natural logarithm of running time(s) as the sample size increases. d-f) running time(s) as the sample size increases.

Neural Network. We implement a fully connected neural network in PrivColl. It has two hidden layers with 128 neurons in each layer (same as MZ17) and takes a sigmoid function as the activation function. For training the neural network, we set the batch size $|B| = 150$ with 4 sample sizes $(1,000\text{–}60,000)$.

In the LAN network setting, PrivColl achieves 1352.87 s (around 22.5 min) (Fig. 3c) with sample size $m = 60,000$, while in MZ17, it takes 294,239.7 s (more

Table 1. PRIVCOLL's overhead breakdown for linear/logistic regression

	Linear regression					Logistic regression				
	Computation	Communication		Total		Computation	Communication		Total	
		LAN	WAN	LAN	WAN		LAN	WAN	LAN	WAN
m = 1,000	0.445 s	0.016 s	103.28 s	0.461 s	103.73 s	0.467 s	0.583 s	109.57 s	1.050 s	110.04 s
m = 10,000	1.293 s	0.978 s	561.53 s	2.271 s	562.83 s	1.368 s	0.812 s	506.64 s	2.180 s	508.01 s
m = 60,000	6.155 s	1.578 s	887.63 s	7.733 s	893.79 s	6.271 s	1.683 s	829.69 s	7.954 s	835.96 s
m = 100,000	10.07 s	3.037 s	1398.67 s	13.11 s	1408.75 s	10.29 s	2.434 s	1414.64 s	12.73 s	1424.94 s

than 81 h) with the same sample size. PRIVCOLL also outperforms the non-private baseline which takes 2,127.87 s. In the WAN setting, PRIVCOLL achieves 18,367.88 s (around 5.1 h) with sample size $m = 60,000$ (Fig. 3f), while in MZ17, it is not yet practical for training neural networks in WAN setting due to the high number of interactions and high communication. Note that, in Fig. 3f, we still plot the MZ17-OT-LAN result (294,239.7 s), showing that even when running our framework in the WAN setting, it is still much more efficient compared to the MPC solutions in MZ17 run in the LAN setting. The overhead breakdown on computation and communication is summarized in Table 2.

Table 2. PRIVCOLL's overhead breakdown for neural network

	Neural network				
	Computation	Communication		Total	
		LAN	WAN	LAN	WAN
m = 1,000	30.14 s	8.729 s	795.27 s	38.87 s	825.42 s
m = 10,000	223.08 s	4.683 s	2662.58 s	227.76 s	2885.66 s
m = 60,000	1320.77 s	32.10 s	17047.10 s	1352.87 s	18367.88 s
m = 100,000	2180.74 s	47.99 s	19364.73 s	2228.73 s	21545.47 s

7 Related Work

The studies most related to PRIVCOLL are [23,49]. Zheng et al. [49] employ the lightweight additive secret sharing scheme for secure outsourcing of the decision tree algorithm for classification. Hu et al. [23] propose FDML, which is a collaborative machine learning framework for distributed features, and the model parameters are protected by additive noise mechanism within the framework of differential privacy.

There also have been some previous research efforts which have explored collaborative learning without exposing their trained models [24,33,46,47]. For example, Papernot et al [33] make use of transfer learning in combination with differential privacy to learn an ensemble of *teacher* models on data partitions, and then use these models to train a private *student* model.

In addition, there are more works on generic privacy-preserving machine learning frameworks via HE/MPC solutions [9,14,20,25,26,34,35,41,43,45] or differential privacy mechanism [1,2,18,39]. Recent studies [11,12] propose a hybrid multi-party computation protocol for securely computing a linear regression model. In [28], an approach is proposed for transforming an existing neural network to an oblivious neural network supporting privacy-preserving predictions. In [43], a secure protocol is presented to calculate the delta function in the back-propagation training. In [31], a MPC-friendly alternative function is specifically designed to replace non-linear sigmoid and softmax functions, as the division and the exponentiation in these function are expensive to compute on shared values.

8 Conclusion

We have presented PrivColl, a practical privacy-preserving collaborative machine learning framework. PrivColl guarantees privacy preservation for both local training data and models trained on them, against an honest-but-curious adversary. It also ensures the correctness of a wide range of machine/deep learning algorithms, such as linear regression, logistic regression, and a variety of neural networks. Meanwhile, PrivColl achieves a practical applicability. It is much more efficient compared to other state-of-art solutions.

References

1. Abadi, M., et al.: Deep learning with differential privacy. In: Proceedings of the 2016 ACM SIGSAC Conference on Computer and Communications Security, pp. 308–318 (2016)
2. Abuadbba, S., et al.: Can we use split learning on 1D cnn models for privacy preserving training? arXiv preprint arXiv:2003.12365 (2020)
3. Albrecht, M., et al.: Homomorphic encryption security standard. Technical report, HomomorphicEncryption.org (2018)
4. Bogdanov, D., Laur, S., Willemson, J.: Sharemind: a framework for fast privacy-preserving computations. In: Jajodia, S., Lopez, J. (eds.) ESORICS 2008. LNCS, vol. 5283, pp. 192–206. Springer, Heidelberg (2008). https://doi.org/10.1007/978-3-540-88313-5_13
5. Bonawitz, K., et al.: Practical secure aggregation for privacy-preserving machine learning. In: Proceedings of the 2017 ACM SIGSAC Conference on Computer and Communications Security, pp. 1175–1191. ACM (2017)
6. Lindell, Y. (ed.): TCC 2014. LNCS, vol. 8349. Springer, Heidelberg (2014). https://doi.org/10.1007/978-3-642-54242-8
7. Chen, Y.R., Rezapour, A., Tzeng, W.G.: Privacy-preserving ridge regression on distributed data. Inf. Sci. **451**, 34–49 (2018)
8. Dwork, C., Roth, A., et al.: The algorithmic foundations of differential privacy. Found. Trends® Theor. Comput. Sci. **9**(3–4), 211–407 (2014)
9. Esposito, C., Su, X., Aljawarneh, S.A., Choi, C.: Securing collaborative deep learning in industrial applications within adversarial scenarios. IEEE Trans. Ind. Inf. **14**(11), 4972–4981 (2018)

10. Fredrikson, M., Jha, S., Ristenpart, T.: Model inversion attacks that exploit confidence information and basic countermeasures. In: Proceedings of the 22nd ACM SIGSAC Conference on Computer and Communications Security, pp. 1322–1333 (2015)
11. Gascón, A., et al.: Secure linear regression on vertically partitioned datasets. IACR Cryptology ePrint Archive 2016/892 (2016)
12. Gascón, A., et al.: Privacy-preserving distributed linear regression on high-dimensional data. Proc. Priv. Enhancing Technol. **2017**(4), 345–364 (2017)
13. Gentry, C.: Fully homomorphic encryption using ideal lattices. In: Proceedings of the Forty-First Annual ACM Symposium on Theory of Computing, pp. 169–178 (2009)
14. Gilad-Bachrach, R., Dowlin, N., Laine, K., Lauter, K., Naehrig, M., Wernsing, J.: CryptoNets: applying neural networks to encrypted data with high throughput and accuracy. In: International Conference on Machine Learning, pp. 201–210 (2016)
15. Goldreich, O., Micali, S., Wigderson, A.: How to play any mental game, or a completeness theorem for protocols with honest majority. In: Providing Sound Foundations for Cryptography: On the Work of Shafi Goldwasser and Silvio Micali, pp. 307–328 (2019)
16. Golub, G., Van Loan, C.: Matrix Computations, 3rd edn. The John Hopkins University Press, Baltimore (1996)
17. Guennebaud, G., Jacob, B., et al.: Eigen v3 (2010). http://eigen.tuxfamily.org
18. Gupta, O., Raskar, R.: Distributed learning of deep neural network over multiple agents. J. Netw. Comput. Appl. **116**, 1–8 (2018)
19. Hagestedt, I., et al.: Mbeacon: privacy-preserving beacons for DNA methylation data. In: NDSS (2019)
20. Hardy, S., et al.: Private federated learning on vertically partitioned data via entity resolution and additively homomorphic encryption. arXiv preprint arXiv:1711.10677 (2017)
21. Hintjens, P.: ZeroMQ: Messaging for Many Applications. O'Reilly Media, Inc. (2013)
22. Horn, R.A., Johnson, C.R.: Matrix Analysis. Cambridge University Press, Cambridge (2012)
23. Hu, Y., Niu, D., Yang, J., Zhou, S.: FDML: a collaborative machine learning framework for distributed features. In: Proceedings of the 25th ACM SIGKDD International Conference on Knowledge Discovery & Data Mining, pp. 2232–2240 (2019)
24. Jia, Q., Guo, L., Jin, Z., Fang, Y.: Privacy-preserving data classification and similarity evaluation for distributed systems. In: 2016 IEEE 36th International Conference on Distributed Computing Systems (ICDCS), pp. 690–699. IEEE (2016)
25. Ko, R.K.L., et al.: STRATUS: towards returning data control to cloud users. In: Wang, G., Zomaya, A., Perez, G.M., Li, K. (eds.) ICA3PP 2015. LNCS, vol. 9532, pp. 57–70. Springer, Cham (2015). https://doi.org/10.1007/978-3-319-27161-3_6
26. Kwabena, O.A., Qin, Z., Zhuang, T., Qin, Z.: MSCryptoNet: multi-scheme privacy-preserving deep learning in cloud computing. IEEE Access **7**, 29344–29354 (2019)
27. LeCun, Y., Cortes, C.: MNIST handwritten digit database (2010). http://yann.lecun.com/exdb/mnist/
28. Liu, J., Juuti, M., Lu, Y., Asokan, N.: Oblivious neural network predictions via MiniONN transformations. In: Proceedings of the 2017 ACM SIGSAC Conference on Computer and Communications Security, pp. 619–631 (2017)

29. Marc, T., Stopar, M., Hartman, J., Bizjak, M., Modic, J.: Privacy-enhanced machine learning with functional encryption. In: Sako, K., Schneider, S., Ryan, P.Y.A. (eds.) ESORICS 2019. LNCS, vol. 11735, pp. 3–21. Springer, Cham (2019). https://doi.org/10.1007/978-3-030-29959-0_1
30. Melis, L., Song, C., De Cristofaro, E., Shmatikov, V.: Exploiting unintended feature leakage in collaborative learning. In: 2019 IEEE Symposium on Security and Privacy (SP), pp. 691–706. IEEE (2019)
31. Mohassel, P., Zhang, Y.: SecureML: a system for scalable privacy-preserving machine learning. In: 2017 IEEE Symposium on Security and Privacy (SP), pp. 19–38. IEEE (2017)
32. Papernot, N., McDaniel, P., Sinha, A., Wellman, M.: Towards the science of security and privacy in machine learning. arXiv preprint arXiv:1611.03814 (2016)
33. Papernot, N., Song, S., Mironov, I., Raghunathan, A., Talwar, K., Erlingsson, Ú.: Scalable private learning with pate. arXiv preprint arXiv:1802.08908 (2018)
34. Ryffel, T., et al.: A generic framework for privacy preserving deep learning. arXiv preprint arXiv:1811.04017 (2018)
35. Sadat, M.N., Aziz, M.M.A., Mohammed, N., Chen, F., Wang, S., Jiang, X.: SAFETY: secure gwAs in federated environment through a hybrid solution with intel SGX and homomorphic encryption. arXiv preprint arXiv:1703.02577 (2017)
36. Sharma, S., Chen, K.: Confidential boosting with random linear classifiers for outsourced user-generated data. In: Sako, K., Schneider, S., Ryan, P.Y.A. (eds.) ESORICS 2019. LNCS, vol. 11735, pp. 41–65. Springer, Cham (2019). https://doi.org/10.1007/978-3-030-29959-0_3
37. Shokri, R., Shmatikov, V.: Privacy-preserving deep learning. In: Proceedings of the 22nd ACM SIGSAC Conference on Computer and Communications Security, pp. 1310–1321 (2015)
38. Song, S., Chaudhuri, K., Sarwate, A.D.: Stochastic gradient descent with differentially private updates. In: 2013 IEEE Global Conference on Signal and Information Processing, pp. 245–248. IEEE (2013)
39. Vepakomma, P., Gupta, O., Swedish, T., Raskar, R.: Split learning for health: distributed deep learning without sharing raw patient data. arXiv preprint arXiv:1812.00564 (2018)
40. Wang, S., Pi, A., Zhou, X.: Scalable distributed DL training: batching communication and computation. In: Proceedings of the AAAI Conference on Artificial Intelligence, vol. 33, pp. 5289–5296 (2019)
41. Will, M.A., Nicholson, B., Tiehuis, M., Ko, R.K.: Secure voting in the cloud using homomorphic encryption and mobile agents. In: 2015 International Conference on Cloud Computing Research and Innovation (ICCCRI), pp. 173–184. IEEE (2015)
42. Yao, A.C.C.: How to generate and exchange secrets. In: 27th Annual Symposium on Foundations of Computer Science (SFCS 1986), pp. 162–167. IEEE (1986)
43. Yuan, J., Yu, S.: Privacy preserving back-propagation neural network learning made practical with cloud computing. IEEE Trans. Parallel Distrib. Syst. **25**(1), 212–221 (2014)
44. Zhang, J., Chen, B., Yu, S., Deng, H.: PEFL: a privacy-enhanced federated learning scheme for big data analytics. In: 2019 IEEE Global Communications Conference (GLOBECOM), pp. 1–6. IEEE (2019)
45. Zhang, X., Ji, S., Wang, H., Wang, T.: Private, yet practical, multiparty deep learning. In: 2017 IEEE 37th International Conference on Distributed Computing Systems (ICDCS), pp. 1442–1452. IEEE (2017)
46. Zhang, Y., Bai, G., Zhong, M., Li, X., Ko, R.: Differentially private collaborative coupling learning for recommender systems. IEEE Intelligent Systems (2020)

47. Zhang, Y., Zhao, X., Li, X., Zhong, M., Curtis, C., Chen, C.: Enabling privacy-preserving sharing of genomic data for GWASs in decentralized networks. In: Proceedings of the Twelfth ACM International Conference on Web Search and Data Mining, pp. 204–212. ACM (2019)
48. Zheng, H., Ye, Q., Hu, H., Fang, C., Shi, J.: BDPL: a boundary differentially private layer against machine learning model extraction attacks. In: Sako, K., Schneider, S., Ryan, P.Y.A. (eds.) ESORICS 2019. LNCS, vol. 11735, pp. 66–83. Springer, Cham (2019). https://doi.org/10.1007/978-3-030-29959-0_4
49. Zheng, Y., Duan, H., Wang, C.: Towards secure and efficient outsourcing of machine learning classification. In: Sako, K., Schneider, S., Ryan, P.Y.A. (eds.) ESORICS 2019. LNCS, vol. 11735, pp. 22–40. Springer, Cham (2019). https://doi.org/10.1007/978-3-030-29959-0_2

An Efficient 3-Party Framework for Privacy-Preserving Neural Network Inference

Liyan Shen[1,2(✉)], Xiaojun Chen[1], Jinqiao Shi[3], Ye Dong[1,2], and Binxing Fang[4]

[1] Institute of Information Engineering, Chinese Academy of Sciences, Beijing, China
{shenliyan,chenxiaojun,dongye}@iie.ac.cn
[2] School of Cyber Security, University of Chinese Academy of Sciences, Beijing, China
[3] Beijing University of Posts and Telecommunications, Beijing, China
shijinqiao@bupt.edu.cn
[4] Institute of Electronic and Information Engineering of UESTC in Guangdong, Dongguan, China
fangbx@cae.cn

Abstract. In the era of big data, users pay more attention to data privacy issues in many application fields, such as healthcare, finance, and so on. However, in the current application scenarios of machine learning as a service, service providers require users' private inputs to complete neural network inference tasks. Previous works have shown that some cryptographic tools can be used to achieve the secure neural network inference, but the performance gap is still existed to make those techniques practical.

In this paper, we focus on the efficiency problem of privacy-preserving neural network inference and propose novel 3-party secure protocols to implement amounts of nonlinear activation functions such as ReLU and Sigmod, etc. Experiments on five popular neural network models demonstrate that our protocols achieve about 1.2×–11.8× and 1.08×–4.8× performance improvement than the state-of-the-art 3-party protocols (SecureNN [28]) in terms of computation and communication overhead. Furthermore, we are the first to implement the privacy-preserving inference of graph convolutional networks.

Keywords: Privacy-preserving computation · Neural network inference · Secret sharing

1 Introduction

Machine learning (ML) has been widely used in many applications such as medical diagnosis, credit risk assessment [4], facial recognition. With the rapid development of ML, Machine Learning as a Service (MLaaS) has been a prevalent business model. Many technology companies such as Google, Microsoft, and

L. Chen et al. (Eds.): ESORICS 2020, LNCS 12308, pp. 419–439, 2020.
https://doi.org/10.1007/978-3-030-58951-6_21

Amazon are providing MLaaS which helps clients benefit from ML without the cognate cost, time, and so on. The most common scenario of using MLaaS is neural network predictions that customers upload their input data to the service provider for inference or classification tasks. However, the data on which the inference is performed involves medical, genomic, financial and other types of sensitive information. For example, the patients would like to predict the probability of heart disease using Google's pre-trained disease diagnosis model. It will require patients to send personal electronic medical records to service providers for accessing inference services. More seriously, due to the restrictive laws and regulations [1,2], the MLaaS service would be prohibited. One naive solution is to let the consumers acquire the model and perform the inference task on locally trusted platforms. However, this solution is not feasible for two main reasons. First, these models were trained using massive amounts of private data, therefore these models have commercial value and the service providers should want them confidential to preserve companies' competitive advantage. Second, revealing the model may compromise the privacy of the training data containing sensitive information [22].

Recent advances in research provide many cryptographic tools like secure multiparty computation (MPC) and fully homomorphic encryption (FHE), to help us address some of these concerns. They can ensure that during the inference, neither the customers' data nor the service providers' models will be revealed by the others except themselves. The only information available for the customers is the inference or classification label, and for the service providers is nothing. Using these tools, privacy-preserving neural network inference could be achieved [15,18,22–24,27,28].

Concretely, the proposed privacy-preserving neural network computation methods mainly use cryptographic protocols to implement each layer of the neural network. Most of the works use mixed-protocols other than a single protocol for better performance. For the linear layers such as fully connected or convolution layers, they are usually represented as arithmetic circuits and evaluated using homomorphic encryption. And for the non-linear layers such as ReLU or Sigmoid functions, they are usually described as a boolean circuit and evaluated using Yao's garbled circuit [30]. There have been many secure computation protocols [6,9,13] used to solve the private computation problems under different circuit representations.

These protocols for the linear layers are practical enough, due to the use of lightweight cryptographic primitives, such as multiplication triples [8] generated by the three-parties [27,28] and homomorphic encryption with the SIMD batch processing technique [18,22]. However, for the non-linear layers, the protocols based on garbled circuits still have performance bottlenecks. Because garbled circuits incur a multiplicative communication overhead proportional to the security parameter. As noted by prior work [22], 2^{16} invocations of ReLU and Sigmoid could lead communication overhead about 10^3 MB and 10^4 MB respectively. This is not feasible in real applications with complex neural networks.

One solution to alleviate the overhead of non-linear functions is three-party secure protocols. SecureNN [28] constructs new and efficient protocols with the help of a third-party and results in a significant reduction in computation and communication overhead. They implement a secure protocol of derivative computation of ReLU function (DReLU) based on a series of sub-protocols such as private compare(PC), share convert(SC) and multiplication (MUL) etc. And then they use DReLU as a basic building block to implement the computation of non-linear layers such as ReLU and Maxpool. The main problem of DReLU protocol in SecureNN is the complexity and it has a long dependence chain of other sub-protocols.

Our work is motivated by SecureNN and attempt to implement a more efficient secure computation of non-linear functions with less complexity. Concretely, we propose a novel and more efficient three-party protocol for DReLU, which is several times faster than the version of SecureNN and only depends on the multiplication sub-protocol. And based on DReLU, we implement not only the activation functions ReLU and Maxpool that are done in SecureNN, but also the more complex activation function Sigmoid. In known secure protocols, Sigmoid is usually approximated by a set of piecewise continuous polynomials and implemented using garbled circuit [22]. Our Sigmoid implementation with a third-party can alleviate the overheads about 360×.

Besides, current works have been applied in the inference of many neural networks, such as linear regression, multi-layer perceptron (MLP), deep neural networks (DNN) and convolutional neural networks (CNN), we implement a privacy inference on graph convolutional networks (GCN) [29], which are popular when processing non-Euclidean data structures such as social network, finance transactions, etc. To the best of our knowledge, we are the first to implement the privacy-preserving inference on GCN.

1.1 Contribution and Roadmap

In this paper, we propose an efficient 3-party framework for privacy-preserving neural network inference. In detail, our contributions are described as follows:

- We propose a novel protocol of the DReLU which is much faster than that in [28]. Based on the protocol, our 3-party framework could efficiently implement the commonly used non-linear activation functions in neural networks.
- We are the first work to implement the privacy-preserving inference scheme of GCN on the MNIST dataset with an accuracy of more than 99%. And the GCN model is more complex than those in [28] in terms of the invocation numbers of linear and non-linear functions.
- We give a detailed proof of security and correctness. And we provide the same security as in SecureNN. Concretely, our protocols are secure against one single semi-honest corruption and against one malicious corruption under the privacy notion in Araki et al. [5].
- Experiments demonstrate that our novel protocol can obtain better performance in computation and communication overhead. Concretely, our proto-

cols outperform prior 3-party work SecureNN by 1.2×–11.8×. And our protocol for the secure computation of sigmoid is about 360× faster than that in MiniONN [22].

The remainder of this paper is organized as follows. In Sect. 2 we give the definition of symbols and primitives related to neural networks and cryptographic building blocks. An introduction of the general framework for the privacy-preserving inference of neural networks is given in Sect. 3. In Sect. 4 we propose our efficient 3-party protocol constructions. Then, we give detailed correctness and security analysis of our protocols in Sect. 5. In Sect. 6 we give the implementation of our protocols. The related work is presented in Sect. 7. Finally, we conclude this paper in Sect. 8.

2 Preliminary

2.1 Definitions

Firstly, we define the symbols used in the article in Table 1.

Table 1. Table for notations.

$\mathcal{S}$	Server
$\mathcal{C}$	Client
$\mathcal{P}$	Semi-honest Third Party
$\boldsymbol{a}$	Lowercase bold letter denotes vector
$\boldsymbol{A}$	Uppercase bold letter denotes matrix
$\boldsymbol{a}[i]$	Denotes the ith elements of $\boldsymbol{a}$
PRG	Pseudo Random Generator
$\in_R$	Uniformly random sampling from a distribution
$\in_R D$	Sampling according to a probability distribution D
s	The statistical security parameter

2.2 Neural Networks

The neural networks usually have similar structures with many layers stacked on top of each other, the output of the previous layer serves as the input of the next layer. And the layers are usually classified into linear layers and non-linear layers.

Linear Layers: The main operations in linear layers are matrix multiplications and additions. Concretely,

- The Fully Connected (FC) layer used in many neural networks can be formulated as $\boldsymbol{y} = \boldsymbol{W}\boldsymbol{x} + \boldsymbol{b}$, where $\boldsymbol{x}$ is the input vector, $\boldsymbol{y}$ is the output of FC layer, $\boldsymbol{W}$ is the weight matrix and $\boldsymbol{b}$ is the bias vector.
- Convolution layer used in CNN networks can be converted into matrix multiplication and addition as noted in [11] and can be formulated as $\boldsymbol{Y} = \boldsymbol{W}\boldsymbol{X} + \boldsymbol{B}$.
- The graph convolution layer used in GCN networks is essentially two consecutive matrix multiplication operations $\boldsymbol{Y} = \sum_{k=0}^{K-1} T_k(\tilde{\boldsymbol{L}})\boldsymbol{X}\boldsymbol{W}$ as we have described in Appendix A.
- The mean pooling operation used in CNN or GCN networks is the average of a region of the proceeding layer. It can be formulated as $y = mean(\boldsymbol{x})$.

Non-linear Layers: Operations in the non-linear layer are usually comparison, exponent, and so on. We identify three common categories:

- Piecewise linear functions [22]. e.g. Rectified Linear Units (ReLU): $f(\boldsymbol{y}) = [max(0, y_i)]$; the max pooling operation: $y = max(\boldsymbol{x})$.
- Smooth activation functions. e.g. Sigmoid: $f(\boldsymbol{y}) = [\frac{1}{1 + e^{-y_i}}]$.
- Softmax: $f(\boldsymbol{y}) = [\frac{e^{-y_i}}{\sum_j e^{-y_j}}]$. It is usually applied to the last layer and can be replaced by argmax operation. As in the prediction phase, it is order-preserving and will not change the prediction result.

2.3 Additive Secret Sharing

In our protocols, all values are secret-shared between the server and the client. Consider a 2-out-of-2 secret sharing protocol, a value is shared between two parties P_0 and P_1 such that the addition of two shares yields the true value. Suppose the value x is shared additively in the ring $\mathbb{Z}_{2^\ell}$ (integers modulo 2^ℓ), the two shares of x are denoted as $\langle x \rangle_0$ and $\langle x \rangle_1$.

- The algorithm `Share(x)` generates the two shares of x over $\mathbb{Z}_{2^\ell}$ by choosing $r \in_R \mathbb{Z}_{2^\ell}$ and sets $\langle x \rangle_0 = r$, $\langle x \rangle_1 = x - r \bmod 2^\ell$.
- The algorithm $\texttt{Reconst}(x_0, x_1)$ reconstructs the value $x = x_0 + x_1 \bmod 2^\ell$ using x_0 and x_1 as the two shares.

Suppose in two-party applications, both P_0 and P_1 share their inputs x,y with each other. Then, the two parties run secure computation protocols on the input shares.

- For addition operation, $\langle z \rangle = \langle x \rangle + \langle y \rangle$: P_i locally computes $\langle z \rangle_i = \langle x \rangle_i + \langle y \rangle_i$.
- For multiplication operation, $\langle z \rangle = \langle x \rangle \cdot \langle y \rangle$: it will be performed using pre-computed Multiplication Triples (MTs) [8] of the form $\langle c \rangle = \langle a \rangle \cdot \langle b \rangle$. Based on the shares of a MT, multiplication will be performed as follows:
  (1) P_i computes $\langle e \rangle_i = \langle x \rangle_i - \langle a \rangle_i$ and $\langle f \rangle_i = \langle y \rangle_i - \langle b \rangle_i$
  (2) P_i performs $\texttt{Reconst}(e_0, e_1)$ and $\texttt{Reconst}(f_0, f_1)$
  (3) P_i sets its output share of the multiplication as

$$\langle z \rangle_i = -i \times e \times f + f \times \langle x \rangle_i + e \times \langle y \rangle_i + \langle c \rangle_i$$

The secret sharing scheme could also be used for matrix $\boldsymbol{X}$ by sharing the elements of $\boldsymbol{X}$ component-wise.

2.4 Threat Model and Security

The parties involved in our protocols are service provider $\mathcal{S}$, the customer $\mathcal{C}$ and the third server $\mathcal{P}$. In our protocols, we consider an adversary who can corrupt only one of the three parties under semi-honest and malicious security.

Semi-honest Security: A semi-honest (passive) adversary is an adversary who corrupts parties but follows the protocol specification. That is, the corrupt parties run the protocol honestly but try to learn additional information from the received messages.

We follow the security proof method in the real-ideal paradigm [10]. Let π be a protocol running in the real interaction and $\mathcal{F}$ be the ideal functionality completed by a trusted party. The ideal-world adversary is referred to as a simulator Sim. We define the two interactions as follows:

- $\text{Real}_\pi(k, C; x_1, ..., x_n)$ run the protocol π with security parameter k, where each party P_i's input is x_i and the set of corrupted parties is C.
  Output $\{V_i, i \in C\}, (y_1, ..., y_n)$. The final view and final output of P_i are V_i and y_i respectively.
- $\text{Ideal}_{\mathcal{F},Sim}(k, C; x_1, ..., x_n)$ compute $(y_1, ..., y_n) \leftarrow \mathcal{F}(x_1, ..., x_n)$
  Output $Sim(C, \{(x_i, y_i), i \in C\}), (y_1, ..., y_n)$.

In the semi-honest model, a protocol π is secure, it must be possible that the ideal world adversary's view is indistinguishable from the real world adversary's view.

Privacy Against Malicious Adversary: A malicious (active) adversary may cause the corrupted parties to deviate from the protocol. Araki et al. [5] formalized the notion of privacy against malicious adversaries using an indistinguishability based argument, which is weaker than the full simulation-based malicious security. Nevertheless, it guarantees that privacy is not violated even if one of the parties behaves maliciously. The indistinguishability-based malicious security is that for any two inputs of the honest parties, the view of the adversary in the protocol is indistinguishable.

3 The General Framework for the Secure Inference of Neural Networks

The two-party mixed privacy-preserving inference protocol of neural networks is based on additive secret sharing. $\mathcal{S}$ and $\mathcal{C}$ run the interactive secure computation protocol for each layer on the input shares, during which they do not learn any intermediate information. And the outputs of each layer also be secretly shared among the two parties. Suppose the neural network is defined as: $\boldsymbol{y} = (\boldsymbol{W}^{L-1} \cdot f_{L-2}(...f_0(\boldsymbol{W}^0 \cdot \boldsymbol{X})...))$, the corresponding computation process of the model is presented in Fig. 1. Specifically, for each layer, $\mathcal{S}$ and $\mathcal{C}$ will each hold a share of $\boldsymbol{Y}^i$ such that `Reconst` of the shares are equal to the input/output to that layer which is performed in the version of plaintext computation of neural networks. The output values will be used as inputs for the next layer. Finally, $\mathcal{S}$ sends the output shares $\boldsymbol{y}_0$ to $\mathcal{C}$ who can reconstruct the output predictions.

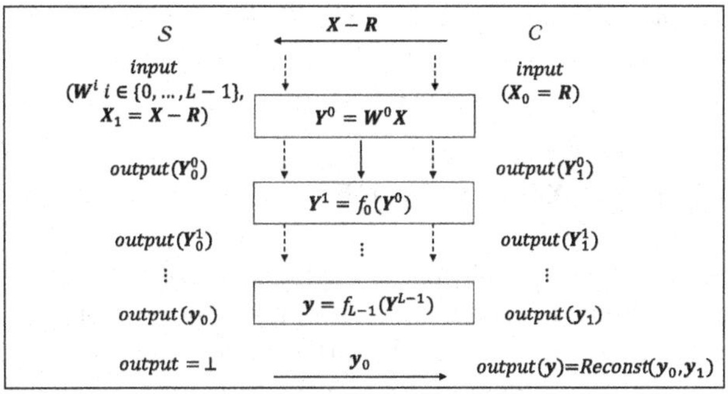

Fig. 1. The flow chart of the neural network computation process.

Most network weights and input/output features are represented as floating-point numbers, while in cryptographic protocols, they are typically encoded into integer form. In our paper, all values are secretly shared in the ring $\mathbb{Z}_L$, where $L = 2^{64}$. And we follow the work in [28] using fixed-point arithmetic to perform all computations with α and β bits (as in [28], we set $\beta = 13$) for integer and fraction parts respectively. Each value is represented in two's complement format, of which the most significant bit (MSB) is the sign bit. It works straightforwardly for the addition and subtraction of two fixed-point decimal numbers. While for multiplication, it needs to truncate the last β bits of the product.

4 Protocol Constructions

The idea of the 3-party protocol in our design is the same as the 2-party protocol as we described above, except that it needs a third-party $\mathcal{P}$ to assist in the computation for each layer protocol of the neural network. Concretely, for the linear layer, we utilize the protocol [27,28] based on the Beaver MTs which should be generated by the complex cryptographic protocol. We use $\mathcal{P}$ to provide the relevant randomness in the MTs, thereby it can significantly reduce the overhead. The corresponding ideal functionality $\mathcal{F}$ is $\boldsymbol{Z} = \boldsymbol{W}\boldsymbol{X}$, where $\boldsymbol{W} \in \mathbb{Z}_L^{m \times n}$ and $\boldsymbol{X} \in \mathbb{Z}_L^{n \times v}$. The detail of the protocol π_{Mul} is described in Algorithm 1. The parties generate the MTs in step 1. $\mathcal{S}$ and $\mathcal{C}$ compute the matrix multiplication in step 2–4, with each party obtain one share of $\boldsymbol{Z}$.

Algorithm 1. Matrix Multiplication. $\pi_{\text{Mul}}(\{\mathcal{S},\mathcal{C}\},\mathcal{P})$

Input:
$\mathcal{S}$: $\langle \boldsymbol{W}\rangle_0, \langle \boldsymbol{X}\rangle_0$, $\mathcal{C}$: $\langle \boldsymbol{W}\rangle_1, \langle \boldsymbol{X}\rangle_1$
Output:
$\mathcal{S}$: $\langle \boldsymbol{WX}\rangle_0$, $\mathcal{C}$: $\langle \boldsymbol{WX}\rangle_1$

1: $\mathcal{P}$ generates random matrices $\boldsymbol{A} \in_R \mathbb{Z}_L^{m\times n}$ and $\boldsymbol{B} \in_R \mathbb{Z}_L^{n\times v}$ using PRG, and sends $(\langle \boldsymbol{A}\rangle_0, \langle \boldsymbol{B}\rangle_0, \langle \boldsymbol{C}\rangle_0)$, $(\langle \boldsymbol{A}\rangle_1, \langle \boldsymbol{B}\rangle_1, \langle \boldsymbol{C}\rangle_1)$ to $\mathcal{S}$ and $\mathcal{C}$ respectively, where $\boldsymbol{C} = \boldsymbol{AB}$.
2: $\mathcal{S}$ computes $\langle \boldsymbol{E}\rangle_0 = \langle \boldsymbol{W}\rangle_0 - \langle \boldsymbol{A}\rangle_0$ and $\langle \boldsymbol{F}\rangle_0 = \langle \boldsymbol{X}\rangle_0 - \langle \boldsymbol{B}\rangle_0$,
$\mathcal{C}$ computes $\langle \boldsymbol{E}\rangle_1 = \langle \boldsymbol{W}\rangle_1 - \langle \boldsymbol{A}\rangle_1$ and $\langle \boldsymbol{F}\rangle_1 = \langle \boldsymbol{X}\rangle_1 - \langle \boldsymbol{B}\rangle_1$.
3: $\mathcal{S}$ and $\mathcal{C}$ run algorithm $\boldsymbol{E}$=$\texttt{Reconst}(\langle \boldsymbol{E}\rangle_0, \langle \boldsymbol{E}\rangle_1)$ and $\boldsymbol{F}$=$\texttt{Reconst}(\langle \boldsymbol{F}\rangle_0, \langle \boldsymbol{F}\rangle_1)$ by exchanging shares.
4: $\mathcal{S}$ outputs $\langle \boldsymbol{W}\rangle_0 \boldsymbol{F} + \boldsymbol{E}\langle \boldsymbol{X}\rangle_0 + \langle \boldsymbol{C}\rangle_0$,
$\mathcal{C}$ outputs $\langle \boldsymbol{W}\rangle_1 \boldsymbol{F} + \boldsymbol{E}\langle \boldsymbol{X}\rangle_1 + \langle \boldsymbol{C}\rangle_1 - \boldsymbol{EF}$.

For the non-linear layer, we design efficient protocol with the help of $\mathcal{P}$. The detailed description is as follows. We consider the non-linear functions in two categories. The first kind includes ReLU and Max-pooling operation. The second category includes some complex smooth activation functions such as Sigmoid. The secure computation protocols for the non-linear activation functions all are based on the computation of derivatives of ReLU.

Protocol for DReLU: The formula of DReLU is ReLU$'(x)$. It is 1 if $x \geq 0$ and 0 otherwise. It can be computed by the sign function of x, ReLU$'(x) = 1 - sign(x)$. Our protocol implements the computation of sign function with the help of $\mathcal{P}$. Concretely, for $\forall x$, $sign(x) = sign(x \cdot r)$ with $r > 0$. $\mathcal{S}$ and $\mathcal{C}$ need to generate random *positive* numbers $r_0 \in_R \mathbb{Z}_L$ and $r_1 \in_R \mathbb{Z}_L$ in advance respectively. It must satisfy that r=$\texttt{Reconst}(r_0, r_1)$ be positive number in $\mathbb{Z}_L$. Then $\mathcal{S}$ and $\mathcal{C}$ perform multiplication protocol based on Beaver MTs and the outputs of two parties are $\langle y\rangle_0 = \langle x \cdot r\rangle_0$ and $\langle y\rangle_1 = \langle x \cdot r\rangle_1$ respectively. Since r is randomly generated, the value of $x \cdot r$ is also a random number in $\mathbb{Z}_L$. $\mathcal{S}$ and $\mathcal{C}$ send $\langle y\rangle_0$ and $\langle y\rangle_1$ to $\mathcal{P}$ who performs $\texttt{Reconst}(\langle y\rangle_0, \langle y\rangle_1)$. $\mathcal{P}$ computes the sign function of the random number $z = sign(y)$, if $z = 1$, $\mathcal{P}$ generates secret shares of zero and sends each share to $\mathcal{S}$ and $\mathcal{C}$; if $z = 0$, $\mathcal{P}$ generates secret shares of one and sends to $\mathcal{S}$ and $\mathcal{C}$. We could also cut the communication of the last step to half by generating the output shares using a shared PRG key between $\mathcal{P}$ and one of the parties. The protocol is described in Algorithm 2.

It should be noted that after the multiplication protocol, it needs to truncate the last β bits of the product. Because for the multiplication of two fixed-point values, the rightmost 2β bits of the product now corresponds to the fraction part instead of β bits. While in this case, r is a random positive number in $\mathbb{Z}_L$ without scaling, the product of $x \cdot r$ does not need to truncate.

Algorithm 2. ReLU$'(x)$. $\pi_{\text{DReLU}}(\{\mathcal{S},\mathcal{C}\},\mathcal{P})$

Input:
$\mathcal{S}$: $\langle x\rangle_0$, $\mathcal{C}$: $\langle x\rangle_1$
Output:
$\mathcal{S}$: $\langle \text{ReLU}'(x)\rangle_0$, $\mathcal{C}$: $\langle \text{ReLU}'(x)\rangle_1$
1: $\mathcal{S}$ and $\mathcal{C}$ generate random positive numbers $r_0 \in_R \mathbb{Z}_L$ and $r_1 \in_R \mathbb{Z}_L$ resp., s.t. r=`Reconst`(r_0, r_1) be positive number in $\mathbb{Z}_L$.
$\mathcal{P}$ generates shares of zero $\langle u\rangle_i = \langle 0\rangle_i$ and shares of one $\langle v\rangle_i = \langle 1\rangle_i$, $i \in \{0, 1\}$.
2: $\mathcal{S}$, $\mathcal{C}$ and $\mathcal{P}$ perform $\pi_{\text{Mul}}(\{\mathcal{S},\mathcal{C}\},\mathcal{P})$ protocol with input $(\langle x\rangle_i, r_i)$ and output $\langle y\rangle_i$, $i \in \{0, 1\}$ of $\mathcal{S},\mathcal{C}$ resp.
3: $\mathcal{S}$ and $\mathcal{C}$ send the shares of $\langle y\rangle_i$ to $\mathcal{P}$ who performs `Reconst`$(\langle y\rangle_0, \langle y\rangle_1)$.
4: $\mathcal{P}$ computes the sign function of the random number $z = sign(y)$,
if $z = 1$, $\mathcal{P}$ sends $\langle u\rangle_0$, $\langle u\rangle_1$ to $\mathcal{S}$ and $\mathcal{C}$ resp, $\mathcal{S}$ outputs $\langle u\rangle_0$, $\mathcal{C}$ outputs $\langle u\rangle_1$;
if $z = 0$, $\mathcal{P}$ sends $\langle v\rangle_0$, $\langle v\rangle_1$ to $\mathcal{S}$ and $\mathcal{C}$ resp, $\mathcal{S}$ outputs $\langle v\rangle_0$, $\mathcal{C}$ outputs $\langle v\rangle_1$.

Protocol for ReLU: The formula of ReLU is ReLU$(x) = x\cdot$ReLU$'(x)$. It equals the multiplication of x and ReLU$'(x)$, the protocol is described in Algorithm 3. Similarly, the product of $x\cdot$ReLU$'(x)$ does not need to truncate.

Algorithm 3. ReLU. $\pi_{\text{ReLU}}(\{\mathcal{S},\mathcal{C}\},\mathcal{P})$

Input:
$\mathcal{S}$: $\langle x\rangle_0$, $\mathcal{C}$: $\langle x\rangle_1$
Output:
$\mathcal{S}$: $\langle \text{ReLU}(x)\rangle_0$, $\mathcal{C}$: $\langle \text{ReLU}(x)\rangle_1$
1: $\mathcal{S}$, $\mathcal{C}$ and $\mathcal{P}$ perform $\pi_{\text{DReLU}}(\{\mathcal{S},\mathcal{C}\},\mathcal{P})$ protocol with input $\langle x\rangle_i$ and output $\langle y\rangle_i$, $i \in \{0, 1\}$ of $\mathcal{S},\mathcal{C}$ resp.
2: $\mathcal{S}$, $\mathcal{C}$ and $\mathcal{P}$ perform $\pi_{\text{Mul}}(\{\mathcal{S},\mathcal{C}\},\mathcal{P})$ protocol with input $(\langle x\rangle_i, \langle y\rangle_i)$ and output $\langle c\rangle_i$, $i \in \{0, 1\}$ of $\mathcal{S},\mathcal{C}$ resp.

Protocol for Maxpool: The Maxpool protocol is the same as that in [28] except that we replace the DReLU module with our novel protocol π_{DReLU}. We give a brief description of the protocol in Appendix B due to space limits. And we point the reader to [28] for further details on the Maxpool protocol.

Protocol for Sigmoid: We adapt the method in [22] to approximate the smooth activation functions that can be efficiently computed and incurs negligible accuracy loss. The activation function $f()$ is split into $m+1$ intervals using m knots which are switchover positions for polynomials expressions. For simplicity, the author uses 1-degree polynomial to approximate sigmoid function:

$$\bar{f}(x) = \begin{cases} 0 & x < x_1 \\ a_1 x + b_1, & x_1 \le x < x_2 \\ \ldots & \\ a_{m-1}x + b_{m-1} & x_{m-1} \le x < x_m \\ 1 & x \ge x_m \end{cases} \quad (1)$$

However, the author uses garbled circuit to implement the approximate activation functions which results in a large amount of overhead. Instead, we propose a 3-party protocol for the approximate sigmoid function $\bar{f}(x)$. It is clear that $\bar{f}(x)$ should be public to all parties. According formula 2, $\mathcal{S}$ and $\mathcal{C}$ will be able to determine the range of x by the subtraction of x and knot x_i.

$$\bar{f}(x) = \begin{cases} 0 & \forall i, i \in \{1, ..., m\}, x - x_i < 0 \\ a_i x + b_i & \exists i, i \in \{1, ..., m-1\}, x - x_{i+1} < 0, x - x_i \geq 0 \\ 1 & \forall i, i \in \{1, ..., m\}, x - x_i \geq 0 \end{cases} \quad (2)$$

$\mathcal{S}$, $\mathcal{C}$ and $\mathcal{P}$ can perform a variation of the π_{DReLU} protocol to determine the range of x, which is described in Algorithm 4. $\mathcal{S}$ and $\mathcal{C}$ compute the subtraction of x and knot x_i in step 1. With the help of $\mathcal{P}$, the parties determine the range of x in step 2–5. The outputs of $\mathcal{S}$ and $\mathcal{C}$ are shares of a_i and b_i parameters in $\bar{f}()$. π_{DReLU} can be seen as a special form of π_{VoDReLU}, which only has one knot $x_1 = 0$.

Algorithm 4. VoDReLU. $\pi_{\text{VoDReLU}}(\{\mathcal{S}, \mathcal{C}\}, \mathcal{P})$

Input:

$\mathcal{S}$: $\langle x \rangle_0$, $\mathcal{C}$: $\langle x \rangle_1$

Output:

$\mathcal{S}$: $\langle a_i \rangle_0$, $\langle b_i \rangle_0$, $\mathcal{C}$: $\langle a_i \rangle_1$, $\langle b_i \rangle_1$

suppose x in the $(i+1)$th interval for $i \in \{0, ..., m\}$

1: $\mathcal{S}$ and $\mathcal{C}$ compute the subtraction of x and x_i locally, for $i \in \{0, ..., m-1\}$.
$\mathcal{S}$: $\langle \boldsymbol{y} \rangle_0 = \boldsymbol{x}$-$\hat{\boldsymbol{x}}$, with $\boldsymbol{x}[i] = \langle x \rangle_0$, $\hat{\boldsymbol{x}}[i] = x_{i+1}$.
$\mathcal{C}$: $\langle \boldsymbol{y} \rangle_1 = \boldsymbol{x}$, with $\boldsymbol{x}[i] = \langle x \rangle_1$. s.t. $\boldsymbol{y}[i] = x - x_{i+1}$.

2: $\mathcal{S}$ and $\mathcal{C}$ generate random positive numbers $\boldsymbol{r}_0 \in_R \mathbb{Z}_L^m$ and $\boldsymbol{r}_1 \in_R \mathbb{Z}_L^m$ resp, s.t. $\boldsymbol{r}$=`Reconst`$(\boldsymbol{r}_0, \boldsymbol{r}_1)$, $\boldsymbol{r}[i]$ be positive number in $\mathbb{Z}_L$.
$\mathcal{P}$ generates shares of a_i and b_i for $i \in \{0, ..., m\}$ and $a_0 = b_0 = a_m = 0$, $b_m = 1$.

3: $\mathcal{S}$, $\mathcal{C}$ and $\mathcal{P}$ perform $\pi_{\text{Mul}}(\{\mathcal{S}, \mathcal{C}\}, \mathcal{P})$ protocol element-wise with input $(\langle \boldsymbol{y} \rangle_j, \boldsymbol{r}_j)$ and output $\langle \boldsymbol{z} \rangle_j$, $j \in \{0, 1\}$ of $\mathcal{S}, \mathcal{C}$ resp.

4: $\mathcal{S}$ sends $\langle \boldsymbol{z} \rangle_0$ and $\mathcal{C}$ sends $\langle \boldsymbol{z} \rangle_1$ to $\mathcal{P}$ who performs `Reconst`$(\langle \boldsymbol{z} \rangle_0, \langle \boldsymbol{z} \rangle_1)$.

5: $\mathcal{P}$ computes the sign function of the random number $\boldsymbol{c} = sign(\boldsymbol{z})$
if $\exists i \in \{0, ..., m-2\}, \boldsymbol{c}[i] = 0$ and $\boldsymbol{c}[i+1] = 1$, $\mathcal{P}$ sends $\langle u \rangle_j = \langle a_i \rangle_j$, $\langle v \rangle_j = \langle b_i \rangle_j$ to $\mathcal{S}$ and $\mathcal{C}$ resp, for $j \in \{0, 1\}$;
if $\forall i \in \{0, ..., m-1\}, \boldsymbol{c}[i] = 1$, $\mathcal{P}$ sends $\langle u \rangle_j = \langle a_0 \rangle_j$, $\langle v \rangle_j = \langle b_0 \rangle_j$ to $\mathcal{S}$ and $\mathcal{C}$ resp, for $j \in \{0, 1\}$;
if $\forall i \in \{0, ..., m-1\}, \boldsymbol{c}[i] = 0$, $\mathcal{P}$ sends $\langle u \rangle_j = \langle a_m \rangle_j$, $\langle v \rangle_j = \langle b_m \rangle_j$ to $\mathcal{S}$ and $\mathcal{C}$ resp, for $j \in \{0, 1\}$;
$\mathcal{S}$ outputs $\langle u \rangle_0$, $\langle v \rangle_0$ and $\mathcal{C}$ outputs $\langle u \rangle_1$, $\langle v \rangle_1$;

The protocol of Sigmoid is described in Algorithm 5. In order to reduce the overhead, the outputs of π_{VoDReLU} could also be the value of i which indicates x in the $i+1$th interval. Then $\mathcal{S}$ and $\mathcal{C}$ could perform multiplication locally, that is $a_i x = a_i(\langle x \rangle_0 + \langle x \rangle_1)$, since all a_i, b_i are public.

The dependence of these protocols is presented in Fig. 2. It can be seen that the protocol for ReLU only depends on the DReLU and Mul subprotocol. Similarly, the protocol dependency of Sigmoid and Maxpool is simple.

Algorithm 5. Sigmoid. $\pi_{\text{Sigmoid}}(\{\mathcal{S},\mathcal{C}\},\mathcal{P})$

Input:
$\mathcal{S}$: $\langle x\rangle_0$, $\mathcal{C}$: $\langle x\rangle_1$

Output:
$\mathcal{S}$: $\langle \bar{f}(x)\rangle_0$, $\mathcal{C}$: $\langle \bar{f}(x)\rangle_1$

1: $\mathcal{S}$, $\mathcal{C}$ and $\mathcal{P}$ perform $\pi_{\text{VoDReLU}}(\{\mathcal{S},\mathcal{C}\},\mathcal{P})$ protocol with input $\langle x\rangle_i$ and output $\langle u\rangle_i$, $\langle v\rangle_i$ $i \in \{0,1\}$ of $\mathcal{S},\mathcal{C}$ resp.
2: $\mathcal{S}$, $\mathcal{C}$ and $\mathcal{P}$ perform $\pi_{\text{Mul}}(\{\mathcal{S},\mathcal{C}\},\mathcal{P})$ protocol with input $(\langle x\rangle_i, \langle u\rangle_i)$ and output $\langle c\rangle_i$, $i \in \{0,1\}$ of $\mathcal{S},\mathcal{C}$ resp.
3: $\mathcal{S}$ outputs $\langle c\rangle_0 + \langle v\rangle_0$, $\mathcal{C}$ outputs $\langle c\rangle_1 + \langle v\rangle_1$.

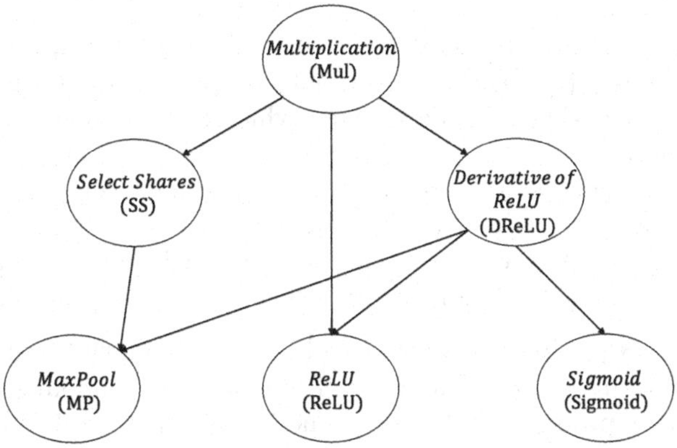

Fig. 2. Protocol dependence of the non-linear activation function.

5 Correctness and Security Analysis

5.1 Correctness Analysis

The protocols are correct as long as the core dependency module of the framework π_{DReLU} is correct. There are two conditions that must be satisfied.

1. As we described above, π_{DReLU} relies on the computation of sign function and it is obvious that $sign(x) = sign(x \cdot r)$ as long as $\forall r > 0$.
2. All the values in the protocol are in the ring $\mathbb{Z}_L$, in order to make sure the correctness of the protocol, all absolute value of any (intermediate) results will not exceed $\lfloor \frac{L}{2} \rfloor$.

Therefore, we let the bit length of random value $r \in_R \mathbb{Z}_L$ be relatively small, such as smaller than 32. Then r will be positive value in the ring and the probability that the result $x \cdot r$ exceeds $\lfloor \frac{L}{2} \rfloor$ will be ignored.

5.2 Security Analysis

Our protocols provide security against one single corrupted party under the semi-honest and malicious model. The universal composability framework [10] guarantees the security of arbitrary composition of different protocols. Therefore, we only need to prove the security of individual protocols. we first proof the security under semi-honest model in the real-ideal paradigm.

Semi-honest Security

Security for π_{DReLU}: The output of the simulator *Sim* which simulates for corrupted $\mathcal{S}$ is one share generated by the algorithm `Share` which is uniformly random chosen from $\mathbb{Z}_L$ and $\mathcal{S}$'s view in the real execution is also random share which is indistinguishable with that in ideal execution. It is the same for the simulation of $\mathcal{C}$. The output of the simulator *Sim* for $\mathcal{P}$ is a random value in $\mathbb{Z}_L$ that is indistinguishable with $x \cdot r$ generated in real-world execution.

Security for π_{ReLU}: $\mathcal{P}$ learns no information from the protocol, for both π_{DReLU} $(\{\mathcal{S},\mathcal{C}\},\mathcal{P})$ and $\pi_{\text{Mul}}(\{\mathcal{S},\mathcal{C}\},\mathcal{P})$ provide outputs only to $\mathcal{S}$ and $\mathcal{C}$. $\mathcal{S}$'s and $\mathcal{C}$'s view in the real execution is random share, which is indistinguishable with that in ideal execution.

Security for π_{Maxpool}: The Maxpool protocol in [28] has been proven secure, the only difference between us is that we have replaced the sub-protocol DReLU. π_{DReLU} has been proven secure as mentioned above. Due to the universal composability, the Maxpool protocol we use is secure under the semi-honest model.

Security for π_{VoDReLU}: The views of $\mathcal{S}$ and $\mathcal{C}$ are random shares in $\mathbb{Z}_L$, and the same as the views in π_{DReLU}. The view of $\mathcal{P}$ is a set of random shares $\boldsymbol{x} \circ \boldsymbol{r}$ ($\circ$ denotes hadamard product) rather than one in π_{DReLU}, and it is indistinguishable with that in ideal execution.

In order to reduce the overhead, we also provide an alternative solution. The outputs of $\mathcal{S}$ and $\mathcal{C}$ are plaintext values i indicating the interval information. Based on i, $\mathcal{S}$ and $\mathcal{C}$ could infer the range of input x and output $\bar{f}(x)$. However, the floating-point number corresponding to x in an interval is infinite. The only case that $\mathcal{S}$ could infer the classification result of $\mathcal{C}$ is that $\mathcal{S}$ knows exactly the output $\bar{f}(x)$ of hidden layer. We will prove that this situation doesn't exist.

Suppose the input dimension of the sigmoid function is p, where the number of input in range (x_1, x_m) is q, and the number of input in range $(-\infty, x_1), (x_m, \infty)$ is $p - q$. Then there will be $\alpha = (\eta \cdot 2^{\beta})^q$ possible values of $\bar{f}(\boldsymbol{x})$, where $\boldsymbol{x} \in \mathbb{Z}_L^p$, η is the interval size between two continuous knots. Our protocol could be secure when the statistical security parameter s is 40 by choosing appropriate value of α. For example, we set $\eta = 2$ (e.g. the sigmoid function is approximated in the set of knots $[-20, -18, ..., 18, 20]$) and $\beta = 13$, then as long as $q \geq 3$, statistical security can be satisfied.

In practical applications, the nodes in a hidden layer are usually high-dimensional vector, so our alternative scheme is secure under the semi-honest model.

Security for π_{Sigmoid}: $\mathcal{P}$ learns no information from the protocol, for both π_{VoDReLU} $(\{\mathcal{S},\mathcal{C}\},\mathcal{P})$ and $\pi_{\text{Mul}}(\{\mathcal{S},\mathcal{C}\},\mathcal{P})$ provide outputs only to $\mathcal{S}$ and $\mathcal{C}$. $\mathcal{S}$'s and $\mathcal{C}$'s view in the real execution is random share, which is indistinguishable with that in ideal execution.

Malicious Security
The same as [28], our protocols provide privacy against a malicious $\mathcal{S}$ or $\mathcal{P}$ in the indistinguishability paradigm [5]. This holds because all the incoming messages to $\mathcal{S}$ or $\mathcal{P}$ are random shares generated in $\mathbb{Z}_L$. For any two inputs of the honest $\mathcal{C}$, the view of the adversary is indistinguishable.

Collusion Analysis
Compared with SecureNN, our protocols for the non-linear functions can resist against collusion attack, as long as the underlying multiplication module is collusion resistant. For example, we can replace the multiplication protocol based on the 3-party with a traditional 2-party protocol based on HE or other techniques. Because the random number r_i owned by $\mathcal{S}$ and $\mathcal{C}$ is generated separately, it can be regarded as one-time-pad, thereby protecting the private information of each party. However, the protocols in SecureNN will reveal computation results at each layer, as long as the third-party collude with $\mathcal{S}$ or $\mathcal{C}$ even if the underlying multiplication protocol is collusion resistant.

6 Experimental Results

6.1 Experimental Enviroment

Our experiments are executed on a server with an Intel Xeon E5-2650 CPU(2.30 GHz) and 126GB RAM in two environments, respectively modeling LAN and WAN settings. The network bandwidth and latency are simulated using Linux Traffic Tools(tc) command. We consider the WAN setting with 40MB/s and 50 ms RTT (round-trip time). All our protocols and that in SecureNN has been implemented and executed more than 10 times on our server (The source code of SecureNN is obtained from Github[1] and we use their code directly in our comparison experiment). We take the average of experimental results for comparison and analysis.

6.2 Neural Networks

We evaluate the experiments in five different neural networks denoted by A, B, C, D and E over the MNIST dataset. More details about those neural networks can be found in Appendix C.
Network A is a 3-layer deep neural network comprising of three FC layers with a ReLU activation function next to every FC layer (Fig. 3).
Network B is a 3-layer convolutional neural network comprising of one convolutional layer and two full connection layers. Every layer is followed by a ReLU activation function (Fig. 4).
Network C is 4-layer convolutional neural networks, the first two layers of which are convolutional layers followed by ReLU and a 2×2 Maxpool. The last two layers are FC layers followed by the ReLU activation function (Fig. 5).

[1] https://github.com/snwagh/securenn-public.

Network D is 4-layer convolutional neural networks similar to **Network C**, and is a larger version with more output channels and more weights (Fig. 5). **Network E** is a graph convolutional neural network on MNIST datasets. The network consists of one graph convolutional layer with 32 output feature maps, followed by the ReLU activation layer and one FC layer (Fig. 6).

Network A, B, C and D were also evaluated in SecureNN, and graph convolutional neural network E is implemented for private inference for the first time. For network E, we construct an 8-nearest neighbor graph of the 2D grid with $|\mathcal{V}| = 784$ nodes. Each image x is transformed to a 784 dimension vector and the pixel value of an image x_i serves as the 1-dimensional feature on vertex v_i. We simplify the model in [12]. Network E just have one graph convolutional layer without pooling operation rather than the original model with two graph convolutional layer followed by a pooling operation separately, and the order of filter is set $K=5$ rather than $K=25$. The simplified network structure can faster the secure computation and also achieve an accuracy of more than 99% on MINST classification task.

6.3 Inference Results

Compared with previous two-party protocols, such as SecureML [24] and Minionn [22], the performance gains for 3-party protocols SecureNN come from using the third-party to generate Beaver MTs for the linear matrix multiplication and avoiding the use of garbled circuits for the non-linear activation functions. However, SecureNN still depends on a series of subprotocol for the secure computation of non-linear activation functions, such as DReLU protocol, it first needs to convert values that are shared over $\mathbb{Z}_L$ into shares over $\mathbb{Z}_{L-1}$ and then computes the MSB of this value based on private compare and multiplication subprotocol and finally convert results that are shared over $\mathbb{Z}_{L-1}$ into shares over $\mathbb{Z}_L$ for the computation of the next linear layer. Our protocol of the DReLU only depends on the third-party multiplication subprotocol which is obviously faster than SecureNN.

The experimental inference results of the five neural network are described in Table 2. We run the privacy-preserving inference for 1 prediction and batch of 128 predictions in both the LAN and WAN settings. It can be found that our protocol is about 1.2×–4.7× faster than SecureNN for 1 prediction, about 2.5×–11.8× faster for 128 predictions and the communication cost is about 1.08×–4.8× fewer than SecureNN. The neural network E is more complex than C or D, for they have 3211264, 1323520 and 1948160 invocations of ReLU respectively, and the invocations for multiplication module is also larger than that of network C and D. And the result of SecureNN for network E is performed based on their source code.

The experimental microbenchmark results for various protocols are described in Table 3. It is about 2×–40× faster than SecureNN and the communication cost of the DReLU, ReLU and Maxpool protocols are about 9.2×, 6.3× and 4.9× fewer than those in SecureNN respectively.

Table 2. Inference results for batch size 1 vs 128 on Networks A-E over MNIST.

Batch size		LAN (s)		WAN (s)		Comm (MB)	
		1	128	1	128	1	128
A	SecureNN	0.036	0.52	6.16	7.39	2.1	29
	Us	**0.02**	**0.11**	**3.09**	**3.29**	**1.94**	**8.28**
B	SecureNN	0.076	4.24	7.76	25.86	4.05	317.7
	Us	**0.03**	**0.68**	**3.87**	**7.43**	**2.28**	**90.45**
C	SecureNN	0.14	14.204	9.34	64.37	8.86	1066
	Us	**0.03**	**1.2**	**4.86**	**15.35**	**2.73**	**281.65**
D	SecureNN	0.24	19.96	10.3	89.05	18.94	1550
	Us	**0.07**	**2.25**	**5.11**	**19.84**	**9.92**	**395.21**
E	SecureNN	1.917	32.993	7.57	128.48	61.76	2398.83
	Us	**1.605**	**9.071**	**6.06**	**29.25**	**46.91**	**497.76**

Table 3. Microbenchmarks in LAN/WAN settings.

Protocol	Dimension	LAN (s)		WAN (s)		Comm (MB)	
		SecureNN	Us	SecureNN	Us	SecureNN	Us
DReLU	64×16	15.66	**1.61**	464.21	**190.79**	0.68	**0.074**
	128×128	204.54	**7.09**	1058.5	**200.146**	10.88	**1.18**
	576×20	134.05	**3.28**	874.36	**195.9**	7.65	**0.83**
ReLU	64×16	15.71	**1.65**	513.25	**233.98**	0.72	**0.115**
	128×128	210.2	**8.5**	1115.48	**281.543**	11.534	**1.835**
	576×20	135.8	**5.6**	921.76	**266.91**	8.11	**1.29**
Maxpool	$8 \times 8 \times 50$,4×4	64.9	**13.59**	7641.7	**3812.79**	2.23	**0.46**
	$24 \times 24 \times 16$,2×2	85.6	**5.79**	1724.7	**802.02**	5.14	**1.05**
	$24 \times 24 \times 20$,2×2	98.4	**6.82**	1763.9	**804.98**	6.43	**1.31**

Table 4. Overhead of secure computation of the sigmoid function.

	2^{16}		2^{20}	
	LAN (ms)	Comm (MB)	LAN (ms)	Comm (MB)
MiniONN	10^5	10^4	–	–
Us	**278.039**	**88.08**	**5470.12**	**1409.29**

We perform our sigmoid approximation protocol with the ranges as $[x_1, x_m] = [-11, 12]$ and 25 pieces, the experimental results are described in Table 4. Our protocol achieves about 360× improvement, and the communication cost is about 114× fewer than that in MiniONN when there are 2^{16} invocations of the sigmoid function. We also implement 2^{20} invocations of the sigmoid function with acceptable overhead.

7 Related Work

As a very active research field in the literature, privacy-preserving inference of neural networks has been extensively considered in recent years. And existing work mainly adopts different security protocols to realize the secure computation of neural networks. Early work in the area can be traced back to the work by Barni et al. [7,25], which is based on homomorphic encryption. Gilad-Bachrach et al. [15] proposed CryptoNets based on leveled homomorphic encryption (LHE) which supports additions and multiplications and only allows a limited number of two operations. Thus, only low-degree polynomial functions can be computed in a straightforward way. Due to the limitations LHE, the author proposed several alternatives to the activation functions and pooling operations used in the CNN layers, for example, they used square activation function instead of ReLU and mean pooling instead of max pooling. It will affect the accuracy of the model.

Instead of purely relying on HE, most of the work uses multiple secure protocols. SecureML presented protocols for privacy-preserving machine learning for linear regression, logistic regression and neural network training. The data owners distributed their data among two non-colluding servers to train various models using secure two-party computation (2PC). The author performed multiplications based on additive secret sharing and offline-generated multiplication triplets. And the non-linear activation function is approximated using a piecewise linear function, which is processed using garbled circuits.

MiniONN further optimized the matrix multiplication protocols. It performed leveled homomorphic encryption operations with the single instruction multiple data (SIMD) batch processing technique to complete the linear transformation in offline. The author used polynomial splines to approximate widely-used nonlinear functions (e.g. sigmoid and tanh) with negligible loss in accuracy. And the author used secret sharing and garbled circuits to complete the activation functions in online.

Chameleon [27] used the same technique as in [24] to complete the matrix multiplication operations. The difference is that [24] completed the multiplication triplets based on oblivious transfer [17,20] or HE, while Chameleon completed it with a third-party that can generate correlated randomness used in multiplication triplets. Besides, the oblivious transfer used in garbled circuits also completed based on the third-party. Almost all of the heavy cryptographic operations are precomputed in an offline phase which substantially reduces the computation and communication overhead. However, some information could be revealed if the third-party colluded with either party.

Gazelle [18] is state-of-the-art 2PC privacy-preserving inference framework of neural networks. It performed multiplications based on homomorphic encryption and used specialized packing schemes for the Brakerski-Fan-Vercauteren (BFV) [14] scheme. However, in order to achieve circuit privacy, the experiment result is about 3-3.6 times slow down for the HE component of Gazelle [26]. Similarly, it used garbled circuits for non-linear activations.

All prior works [22,24,27] use a secure computation protocol for Yao's garbled circuits to compute the non-linear activation functions. However, it has been noted [5] that garbled circuits incur a multiplicative overhead proportional to the security parameter in communication and are a major computational bottleneck. SecureNN constructed a novel protocol for non-linear functions such as ReLU and maxpool that completely avoid the use of garbled circuits. It is developed with the help of a third-party and is the state-of-the-art protocol in secure training and inference of CNNs.

Besides, there are some other works combine with quantization / binary neural networks [3,26] or some other constructions based on three-party protocols [23] (the three parties both have shares of the private input in the model).

8 Conclusions

In this paper, we propose novel and efficient 3-party protocols for privacy-preserving neural network inference. And we are the first work to implement the privacy-preserving inference scheme of the graph convolutional network. We conducted a detailed analysis of the correctness and security of this scheme. Finally, we give an implementation for the private prediction of five different neural networks, it has better performance than the previous 3-party work.

Acknowledgment. This work was supported by the Strategic Priority Research Program of Chinese Academy of Sciences, Grant (No.XDC02040400), DongGuan Innovative Research Team Program (Grant No.201636000100038).

A Graph Convolutional Network

For spectral-based GCN, the goal of these models is to learn a function of signals/features on a graph $\mathcal{G} = (\mathcal{V}, \mathcal{E}, \boldsymbol{A})$, where $\mathcal{V}$ is a finite set of $|\mathcal{V}| = \mathrm{n}$ vertices, $\mathcal{E}$ is a set of edges and $\boldsymbol{A} \in \mathbb{R}^{n \times n}$ is a representative description of the graph structure typically in the form of an adjacency matrix. It will be computed based on the training data by the server and the graph structure is identical for all signals [16]. $\boldsymbol{x} \in \mathbb{R}^n$ is a signal, each row x_v is a scalar for node v. An essential operator in spectral graph analysis is the graph Laplacian $\boldsymbol{L}$, and the normalized definition is $\boldsymbol{L} = \boldsymbol{I}_n - \boldsymbol{D}^{-1/2}\boldsymbol{A}\boldsymbol{D}^{-1/2}$, where $\boldsymbol{I}_n$ is the identity matrix and $\boldsymbol{D}$ is the diagonal node degree matrix of $\boldsymbol{A}$.

Defferrard et al. [12] proposed to approximate the graph filters by a truncated expansion in terms of Chebyshev polynomials. The Chebyshev polynomials are recursively defined as $T_k(x) = 2xT_{k-1}(x) - T_{k-2}(x)$, with $T_0(x) = 1$ and $T_1(x) = x$. And filtering of a signal $\boldsymbol{x}$ with a K-localized filter g_θ can be performed using: $g_\theta * \boldsymbol{x} = \sum_{k=0}^{K-1} \theta_k T_k(\tilde{\boldsymbol{L}})\boldsymbol{x}$, with $\tilde{\boldsymbol{L}} = \frac{2}{\lambda_{max}}\boldsymbol{L} - \boldsymbol{I}_n$. λ_{max} is the largest eigenvalue of $\boldsymbol{L}$. $\boldsymbol{\theta} \in \mathbb{R}^K$ is a vector of Chebyshev coefficients.

The definition generalized to a signal matrix $\boldsymbol{X} \in \mathbb{R}^{n \times c}$ with c dimensional feature vector for each node and f feature maps is as follows: $\boldsymbol{Y} = \sum_{k=0}^{K-1} T_k(\tilde{\boldsymbol{L}})\boldsymbol{X}\boldsymbol{\Theta}_k$ where $\boldsymbol{Y} \in \mathbb{R}^{n \times f}$, $\boldsymbol{\Theta}_k \in \mathbb{R}^{c \times f}$ and the total number of trainable parameters per layer is $c \times f \times K$ ($\boldsymbol{\Theta} \in \mathbb{R}^{K \times c \times f}$). Through graph convolution layer, GCN can capture feature vector information of all neighboring nodes. It preserves both network topology structure and node feature information.

However, the privacy-preserving inference of GCN is not suitable for node classification tasks, in which test nodes (without labels) are included in GCN training. It could not quickly generate embeddings and make predictions for unseen nodes [19].

B Maxpool Protocol

Algorithm 6. Maxpool. $\pi_{\mathrm{MP}}(\{\mathcal{S}, \mathcal{C}\}, \mathcal{P})$

Input:

$\mathcal{S}$: $\{\langle x_i \rangle_0\}_{i \in [n]}$, $\mathcal{C}$: $\{\langle x_i \rangle_1\}_{i \in [n]}$

Output:

$\mathcal{S}$: $\langle y \rangle_0$, $\mathcal{C}$: $\langle y \rangle_1$, s.t. $y = \mathrm{Max}(\{\mathrm{x_i}\}_{\mathrm{i} \in [\mathrm{n}]})$

1: $\mathcal{S}$ sets $\langle max_1 \rangle_0 = \langle x_1 \rangle_0$ and $\mathcal{C}$ sets $\langle max_1 \rangle_1 = \langle x_1 \rangle_1$.
2: for $i = \{2, ..., n\}$ do
3: $\mathcal{S}$ sets $\langle w_i \rangle_0 = \langle x_i \rangle_0 - \langle max_{i-1} \rangle_0$ and $\mathcal{C}$ sets $\langle w_i \rangle_1 = \langle x_i \rangle_1 - \langle max_{i-1} \rangle_1$.
4: $\mathcal{S}$, $\mathcal{C}$ and $\mathcal{P}$ perform $\pi_{\mathrm{DReLU}}(\{\mathcal{S}, \mathcal{C}\}, \mathcal{P})$ protocol with input $(\langle w_i \rangle_j)$ and output $\langle y_i \rangle_j$, $j \in \{0, 1\}$ of $\mathcal{S}, \mathcal{C}$ resp.
5: $\mathcal{S}$, $\mathcal{C}$ and $\mathcal{P}$ perform $\pi_{\mathrm{SS}}(\{\mathcal{S}, \mathcal{C}\}, \mathcal{P})$ protocol with input $(\langle y_i \rangle_j, \langle max_{i-1} \rangle_j, \langle x_i \rangle_j,)$ and output $\langle max_i \rangle_j$, $j \in \{0, 1\}$ of $\mathcal{S}, \mathcal{C}$ resp.
6: end for
7: $\mathcal{S}$ outputs $\langle max_n \rangle_0$ and $\mathcal{C}$ outputs $\langle max_n \rangle_1$.

Algorithm 7. SelectShare. $\pi_{\mathrm{SS}}(\{\mathcal{S}, \mathcal{C}\}, \mathcal{P})$

Input:

$\mathcal{S}$: $(\langle \alpha \rangle_0, \langle x \rangle_0, \langle y \rangle_0)$, $\mathcal{C}$: $(\langle \alpha \rangle_1, \langle x \rangle_1, \langle y \rangle_1)$

Output:

$\mathcal{S}$: $\langle z \rangle_0$, $\mathcal{C}$: $\langle z \rangle_1$, s.t. $z = (1 - \alpha)x + \alpha y$

1: $\mathcal{S}$ sets $\langle w \rangle_0 = \langle y \rangle_0 - \langle x \rangle_0$ and $\mathcal{C}$ sets $\langle w \rangle_1 = \langle y \rangle_1 - \langle x \rangle_1$.
2: $\mathcal{S}$, $\mathcal{C}$ and $\mathcal{P}$ perform $\pi_{\mathrm{Mul}}(\{\mathcal{S}, \mathcal{C}\}, \mathcal{P})$ protocol with input $(\langle \alpha \rangle_j, \langle w \rangle_j)$ and output $\langle c \rangle_j$, $j \in \{0, 1\}$ of $\mathcal{S}, \mathcal{C}$ resp.
3: $\mathcal{S}$ outputs $\langle z \rangle_0 = \langle x \rangle_0 + \langle c \rangle_0$ and $\mathcal{C}$ outputs $\langle z \rangle_1 = \langle x \rangle_1 + \langle c \rangle_1$.

C Neural Network Structure

(1) *FC*: input image 28 × 28, the output: $\mathbb{R}^{128\times1} \leftarrow \mathbb{R}^{128\times784} \cdot \mathbb{R}^{784\times1}$
(2) *ReLU activation*: calculates ReLU for each input.
(3) *FC*: input size 128, the output: $\mathbb{R}^{128\times1} \leftarrow \mathbb{R}^{128\times128} \cdot \mathbb{R}^{128\times1}$
(4) *ReLU activation*: calculates ReLU for each input.
(5) *FC*: input size 128, the output: $\mathbb{R}^{10\times1} \leftarrow \mathbb{R}^{10\times128} \cdot \mathbb{R}^{128\times1}$
(6) *ReLU activation*: calculates ReLU for each input.

Fig. 3. The neural network A presented in SecureML [24]

(1) *Convolution*: input image 28×28, window size 5×5, stride (2,2), output channels 5: $\mathbb{R}^{5\times196} \leftarrow \mathbb{R}^{5\times25} \cdot \mathbb{R}^{25\times196}$
(2) *ReLU activation*: calculates ReLU for each input.
(3) *FC*: input size 980, the output: $\mathbb{R}^{100\times1} \leftarrow \mathbb{R}^{100\times980} \cdot \mathbb{R}^{980\times1}$
(4) *ReLU activation*: calculates ReLU for each input.
(5) *FC*: input size 100, the output: $\mathbb{R}^{10\times1} \leftarrow \mathbb{R}^{10\times100} \cdot \mathbb{R}^{100\times1}$
(6) *ReLU activation*: calculates ReLU for each input.

Fig. 4. The neural network B presented in Chameleon [27]

(1) *Convolution*: input image 28×28, window size 5×5, stride (1,1), output channels 16: $\mathbb{R}^{16\times576} \leftarrow \mathbb{R}^{16\times25} \cdot \mathbb{R}^{25\times576}$ /output channels 20: $\mathbb{R}^{20\times576} \leftarrow \mathbb{R}^{20\times25} \cdot \mathbb{R}^{25\times576}$
(2) *ReLU activation*: calculates ReLU for each input.
(3) *Max Pooling*: window size 2 × 2, stride (2,2), the output: $\mathbb{R}^{16\times12\times12}/\mathbb{R}^{20\times12\times12}$
(4) *Convolution*: window size 5 × 5, stride (1,1), output channels 16: $\mathbb{R}^{16\times64} \leftarrow \mathbb{R}^{16\times400} \cdot \mathbb{R}^{400\times64}$/output channels 50: $\mathbb{R}^{50\times64} \leftarrow \mathbb{R}^{50\times400} \cdot \mathbb{R}^{400\times64}$
(5) *ReLU activation*: calculates ReLU for each input.
(6) *Max Pooling*: window size 2 × 2, stride (2,2), the output: $\mathbb{R}^{16\times4\times4}/\mathbb{R}^{50\times4\times4}$
(7) *FC*: input size 256, the output: $\mathbb{R}^{100\times1} \leftarrow \mathbb{R}^{100\times256} \cdot \mathbb{R}^{256\times1}$
/*FC*: input size 800, the output: $\mathbb{R}^{500\times1} \leftarrow \mathbb{R}^{500\times800} \cdot \mathbb{R}^{800\times1}$
(8) *ReLU activation*: calculates ReLU for each input.
(9) *FC*: input size 100, the output: $\mathbb{R}^{10\times1} \leftarrow \mathbb{R}^{10\times100} \cdot \mathbb{R}^{100\times1}$
/*FC*: input size 500, the output: $\mathbb{R}^{10\times1} \leftarrow \mathbb{R}^{10\times500} \cdot \mathbb{R}^{500\times1}$
(10) *ReLU activation*: calculates ReLU for each input.

Fig. 5. The neural network C/D presented in MiniONN [22] and [21] resp.

(1) *Graph convolution*: input image 28 × 28, with $T_k(\tilde{\boldsymbol{L}}) \in \mathbb{R}^{784\times784}$, $\boldsymbol{W}_k \in \mathbb{R}^{1\times32}$ output $\sum_{k=0}^{K-1} T_k(\tilde{\boldsymbol{L}})\boldsymbol{x}\boldsymbol{W}_k$, $\mathbb{R}^{784\times32} \leftarrow \mathbb{R}^{784\times784} \cdot \mathbb{R}^{784\times1} \cdot \mathbb{R}^{1\times32}$.
(2) *ReLU activation*: calculates ReLU for each input.
(3) *FC*: $\mathbb{R}^{10\times1} \leftarrow \mathbb{R}^{10\times(784\times32)} \cdot \mathbb{R}^{(784\times32)\times1}$

Fig. 6. Graph convolutional neural network trained from the MNIST dataset

References

1. The health insurance portability and accountability act of 1996 (hipaa). https://www.hhs.gov/hipaa/index.html
2. Regulation (eu) 2016/679 of the European parliament and of the council of 27 April 2016 on the protection of natural persons with regard to the processing of personal data and on the free movement of such data, and repealing directive 95/46/ec (gdpr). https://gdpr-info.eu/
3. Agrawal, N., Shahin Shamsabadi, A., Kusner, M.J., Gascón, A.: Quotient: two-party secure neural network training and prediction. In: Proceedings of the 2019 ACM SIGSAC Conference on Computer and Communications Security, pp. 1231–1247 (2019)
4. Angelini, E., di Tollo, G., Roli, A.: A neural network approach for credit risk evaluation. Q. Rev. Econ. Finan. **48**(4), 733–755 (2008)
5. Araki, T., Furukawa, J., Lindell, Y., Nof, A., Ohara, K.: High-throughput semi-honest secure three-party computation with an honest majority. In: Proceedings of the 2016 ACM SIGSAC Conference on Computer and Communications Security, pp. 805–817 (2016)
6. Barni, M., Failla, P., Kolesnikov, V., Lazzeretti, R., Sadeghi, A.-R., Schneider, T.: Secure evaluation of private linear branching programs with medical applications. In: Backes, M., Ning, P. (eds.) ESORICS 2009. LNCS, vol. 5789, pp. 424–439. Springer, Heidelberg (2009). https://doi.org/10.1007/978-3-642-04444-1_26
7. Barni, M., Orlandi, C., Piva, A.: A privacy-preserving protocol for neural-network-based computation. In: Proceedings of the 8th workshop on Multimedia and security, pp. 146–151 (2006)
8. Beaver, D.: Efficient multiparty protocols using circuit randomization. In: Feigenbaum, J. (ed.) CRYPTO 1991. LNCS, vol. 576, pp. 420–432. Springer, Heidelberg (1992). https://doi.org/10.1007/3-540-46766-1_34
9. Bogdanov, D., Laur, S., Willemson, J.: Sharemind: a framework for fast privacy-preserving computations. In: Jajodia, S., Lopez, J. (eds.) ESORICS 2008. LNCS, vol. 5283, pp. 192–206. Springer, Heidelberg (2008). https://doi.org/10.1007/978-3-540-88313-5_13
10. Canetti, R.: Universally composable security: a new paradigm for cryptographic protocols. In: Proceedings 42nd IEEE Symposium on Foundations of Computer Science, pp. 136–145. IEEE (2001)
11. Chellapilla, K., Puri, S., Simard, P.: High performance convolutional neural networks for document processing (2006)
12. Defferrard, M., Bresson, X., Vandergheynst, P.: Convolutional neural networks on graphs with fast localized spectral filtering. In: Advances in Neural Information Processing Systems, pp. 3844–3852 (2016)
13. Demmler, D., Schneider, T., Zohner, M.: Aby-a framework for efficient mixed-protocol secure two-party computation. In: NDSS (2015)

14. Fan, J., Vercauteren, F.: Somewhat practical fully homomorphic encryption. IACR Cryptology ePrint Archive 2012, 144 (2012)
15. Gilad-Bachrach, R., Dowlin, N., Laine, K., Lauter, K., Naehrig, M., Wernsing, J.: Cryptonets: Applying neural networks to encrypted data with high throughput and accuracy. In: International Conference on Machine Learning, pp. 201–210 (2016)
16. Henaff, M., Bruna, J., LeCun, Y.: Deep convolutional networks on graph-structured data. arXiv preprint arXiv:1506.05163 (2015)
17. Ishai, Y., Kilian, J., Nissim, K., Petrank, E.: Extending oblivious transfers efficiently. In: Boneh, D. (ed.) CRYPTO 2003. LNCS, vol. 2729, pp. 145–161. Springer, Heidelberg (2003). https://doi.org/10.1007/978-3-540-45146-4_9
18. Juvekar, C., Vaikuntanathan, V., Chandrakasan, A.: {GAZELLE}: a low latency framework for secure neural network inference. In: 27th {USENIX} Security Symposium ({USENIX} Security 18), pp. 1651–1669 (2018)
19. Kipf, T.N., Welling, M.: Semi-supervised classification with graph convolutional networks. arXiv preprint arXiv:1609.02907 (2016)
20. Kolesnikov, V., Kumaresan, R.: Improved OT extension for transferring short secrets. In: Canetti, R., Garay, J.A. (eds.) CRYPTO 2013. LNCS, vol. 8043, pp. 54–70. Springer, Heidelberg (2013). https://doi.org/10.1007/978-3-642-40084-1_4
21. LeCun, Y., Bottou, L., Bengio, Y., Haffner, P.: Gradient-based learning applied to document recognition. Proc. IEEE **86**(11), 2278–2324 (1998)
22. Liu, J., Juuti, M., Lu, Y., Asokan, N.: Oblivious neural network predictions via minionn transformations. In: Proceedings of the 2017 ACM SIGSAC Conference on Computer and Communications Security, pp. 619–631 (2017)
23. Mohassel, P., Rindal, P.: Aby3: a mixed protocol framework for machine learning. In: Proceedings of the 2018 ACM SIGSAC Conference on Computer and Communications Security, pp. 35–52 (2018)
24. Mohassel, P., Zhang, Y.: Secureml: a system for scalable privacy-preserving machine learning. In: 2017 IEEE Symposium on Security and Privacy (SP), pp. 19–38. IEEE (2017)
25. Orlandi, C., Piva, A., Barni, M.: Oblivious neural network computing via homomorphic encryption. EURASIP J. Inf. Secur. **2007**, 1–11 (2007). https://doi.org/10.1155/2007/37343
26. Riazi, M.S., Samragh, M., Chen, H., Laine, K., Lauter, K., Koushanfar, F.: {XONN}: Xnor-based oblivious deep neural network inference. In: 28th {USENIX} Security Symposium ({USENIX} Security 19), pp. 1501–1518 (2019)
27. Riazi, M.S., Weinert, C., Tkachenko, O., Songhori, E.M., Schneider, T., Koushanfar, F.: Chameleon: a hybrid secure computation framework for machine learning applications. In: Proceedings of the 2018 on Asia Conference on Computer and Communications Security, pp. 707–721 (2018)
28. Wagh, S., Gupta, D., Chandran, N.: Securenn: 3-party secure computation for neural network training. Proc. Priv. Enhanc. Technol. **2019**(3), 26–49 (2019)
29. Wu, Z., Pan, S., Chen, F., Long, G., Zhang, C., Yu, P.S.: A comprehensive survey on graph neural networks. arXiv preprint arXiv:1901.00596 (2019)
30. Yao, A.C.C.: How to generate and exchange secrets. In: 27th Annual Symposium on Foundations of Computer Science (sfcs 1986), pp. 162–167. IEEE (1986)

Deep Learning Side-Channel Analysis on Large-Scale Traces

A Case Study on a Polymorphic AES

Loïc Masure[1,3(✉)], Nicolas Belleville[2], Eleonora Cagli[1], Marie-Angela Cornélie[1], Damien Couroussé[2], Cécile Dumas[1], and Laurent Maingault[1]

[1] Univ. Grenoble Alpes, CEA, LETI, DSYS, CESTI, 38000 Grenoble, France
loic.masure@cea.fr
[2] Univ. Grenoble Alpes, CEA, List, 38000 Grenoble, France
[3] Sorbonne Université, UPMC Univ Paris 06, POLSYS, UMR 7606, LIP6, 75005 Paris, France

Abstract. *Code polymorphism* is an approach to efficiently address the challenge of automatically applying the *hiding* of sensitive information leakage, as a way to protect cryptographic primitives against side-channel attacks (SCA) involving layman adversaries. Yet, recent improvements in SCA, involving more powerful threat models, *e.g.*, using deep learning, emphasized the weaknesses of some hiding counter-measures. This raises two questions. On the one hand, the security of code polymorphism against more powerful attackers, which has never been addressed so far, might be affected. On the other hand, using deep learning SCA on code polymorphism would require to scale the state-of-the-art models to much larger traces than considered so far in the literature. Such a case typically occurs with code polymorphism due to the unknown precise location of the leakage from one execution to another. We tackle those questions through the evaluation of two polymorphic implementations of AES, similar to the ones used in a recent paper published in TACO 2019 [6]. We show on our analysis how to efficiently adapt deep learning models used in SCA to scale on traces 32 folds larger than what has been done so far in the literature. Our results show that the targeted polymorphic implementations are broken within 20 queries with the most powerful threat models involving deep learning, whereas 100,000 queries would not be sufficient to succeed the attacks previously investigated against code polymorphism. As a consequence, this paper pushes towards the search of new polymorphic implementations secured against state-of-the-art attacks, which currently remains to be found.

1 Introduction

1.1 Context

Side-channel analysis (SCA) is a class of attacks against cryptographic primitives that exploit weaknesses of their physical implementation. During the execution

L. Chen et al. (Eds.): ESORICS 2020, LNCS 12308, pp. 440–460, 2020.
https://doi.org/10.1007/978-3-030-58951-6_22

of the latter implementation, some *sensitive variables* are indeed processed that depend on both a piece of public data (*e.g.* a plain-text) and on some chunk of a secret value (*e.g.* a key). Hence, combining information about a sensitive variable with the knowledge of the public data enables an attacker to reduce the secret chunk search space. By repeating this attack several times, implementations of secure cryptographic algorithms such as the *Advanced Encryption Standard* (AES) [31] can then be defeated by recovering each byte of the secret key separately thanks to a *divide-and-conquer* strategy, thereby breaking the high complexity usually required to defeat such an algorithm. The information on sensitive variables is usually gathered by acquiring time series (a.k.a. *traces*) of physical measurements such as the power consumption or the electromagnetic emanations measured on the target device (*e.g.* a smart card). Nowadays, SCA are considered as one of the most effective threats against cryptographic implementations.

To protect against SCA, many counter-measures have been developed and have been shown to be practically effective so that their use in industrial implementations is today common. Their effects may be twofold. On the one hand it may force an attacker to require more traces to recover the secret data. In other words it may require more queries to the cryptographic primitive, possibly beyond the duration of a session key. On the other hand, it may increase the computational complexity of side-channel analysis, making the use of simple statistical tools and attacks harder. Countermeasures against SCA can be divided into two main families: *masking* and *hiding*. Masking, a.k.a. secret-sharing, consists in replacing any sensitive secret-dependent variable in the implementation by a $(d+1)-$*sharing*, so that each share subset is statistically independent of the sensitive variable. This counter-measure is known to be theoretically sound since it amplifies the noise exponentially with the number of shares d, though at the cost of a quadratic growth in the performance overheads [33,35]. This drawback, combined with the difficulty of properly implementing the counter-measure, makes masking still challenging for a non-expert developer. Hiding covers many techniques that aim at reducing the signal-to-noise ratio in the traces. Asynchronous logic [34] or dual rail logic [41] in hardware, shuffling independent computations [43], or injecting temporal de-synchronization in the traces [13] in software, are typical examples of hiding counter-measures that have been practically effective against SCA. Hence, in practice, masking and hiding are combined to ensure that a secured product meets security standards.

However, due to the skyrocketing production of IoT, there is a need for the automated application of protections to improve products' resistance against SCA while keeping the performance overhead sufficiently low. In this context, a recent work proposed a compiler toolchain to automatically apply a software hiding counter-measure called *code polymorphism* [6]. The working principle of the counter-measure relies on the execution of many variants of the machine code of the protected software component, produced by a runtime code generator. The successive execution of many variants aims at producing variable side-channel traces in order to increase the difficulty to realize SCA. We emphasize on the fact that, if code polymorphism is the only counter-measure applied to the target component, information leakage is still present in the side-channel traces. Yet, several works have shown the ability of code polymorphism and similar software mechanisms to be effective in practice against *vertical* SCA [2,14], *i.e.* up to the point that the *Test Vector Leakage Assessment* (TVLA) methodology [5], highly discriminating in the detection of side-channel leakage, would not be able to detect information leakage in the traces [3,6].

1.2 Problem Addressed in This Paper

Yet, though very promising, these results cannot draw an exhaustive guarantee concerning the security level against SCA, since other realistic scenarios have not been investigated.

Indeed, on the one hand the SCA literature proposes other ways to outperform vertical attacks when facing hiding counter-measures. *Re-synchronization* techniques might annihilate the misalignment effect occurred by code polymorphism, since it is successfully applied on hardware devices prone to jitter [16,29,44]. Likewise, *Convolutional Neural Networks* (CNN) can circumvent some software and hardware de-synchronization counter-measures, in a sense similar to code polymorphism [10,23]. It is therefore of great interest to use those techniques to assess the security provided by some code polymorphism configurations against more elaborated attackers.

On the other hand, until now the literature has only demonstrated the relevance of CNN attacks on restricted traces whose size does not exceed $5,000$ samples [7,10,23,42,45], which is small, *e.g.*, regarding the size of the raw traces in the public datasets of software AES implementations used in those papers [7,12,30]. This requires to restrict the acquired traces to a tight window where the attacker is confident that the relevant leakage occurs. Unfortunately, this is not possible since code polymorphism applies hiding in a systematic and pervasive way in the implementation. Likewise, other dimensionality reduction techniques like dedicated variants of *Principal Component Analysis* (PCA) [38] might be considered prior to the use of CNNs. Unfortunately, they do not theoretically provide any guarantee that relevant features will be extracted, especially for data prone to misalignment. As a consequence, attacking a polymorphic implementation necessarily requires to deal with large-scale traces. This generally spans serious issues for machine learning problems known under the name of *curse of dimensionality* [36]. That is why it currently remains an open question whether CNN attacks can scale on larger traces, or whether it represents a technical issue that some configurations of code polymorphism might benefit against these attacks. Hence, both problems, namely evaluating code polymorphism and addressing large-scale traces SCA, are closely connected.

1.3 Contribution

In this paper, we tackle the two problems presented so far by extending the security evaluation provided by Belleville *et al.* [6]. The evaluation aims to assess the security of the highest code polymorphism configuration they used, on same implementations, against stronger attackers.

Our evaluation considers a wide spectrum of threat models, ranging from automated attacks affordable by a layman attacker, to state-of-the-art techniques. The whole evaluation setup is detailed in Sect. 2. In particular, we propose to adapt the architectures used in the literature of CNN attacks, in order to handle the technical challenge of large scale traces. This is presented in Subsect. 2.6.

Our results show that vertical attacks fail due to the code polymorphism counter-measure, but that more elaborated scenarios lead to successful attacks. Compared to the ones conducted in [6], and depending on the different attack powers considered hereafter, the number of required queries is lowered by up to several orders of magnitude. In a worst case scenario, our trained CNNs are able to recover every secret key byte in less than 20 traces of dimensionality 160,000, which is 32 times higher than the traces used so far in CNN attacks. This therefore illustrates that large scale traces are not necessarily a technical challenge for deep-learning based SCA. Those results are presented in Sect. 3.

Thus, this study claims that though code polymorphism is a promising tool to increase the hardness of SCA against embedded devices, a sound polymorphic configuration, eventually coupled with other counter-measures, is yet to be found, in order to protect against state-of-the-art SCA. This is discussed in Sect. 4. However, the toolchain used by Belleville *et al.* [6] allows to explore many configurations beside the one considered here, the exploration of the securing capabilities of the toolchain is then beyond the scope of this paper, and left as an open question for further works.

2 Evaluation Methodology

This section presents all the settings necessary to proceed with the evaluation. Subsection 2.1 briefly presents the principle of the code polymorphism counter-measure, and precises the aim of the different code transformations included in the evaluated configuration. Subsection 2.2 presents the target device and the two AES implementations that have been used for this evaluation. Subsection 2.3 details the acquisition settings, while Subsect. 2.4 describes the shape of the acquired traces. Finally, Subsect. 2.5 presents the different threat models considered hereafter, and Subsect. 2.6 precises how the CNN attacks have been conducted.

2.1 Description of the Target Counter-Measure

We briefly describe the code polymorphism counter-measure applied by the toolchain used by Belleville *et al.* [6]. The compiler applies the counter-measure to selected critical parts of an unprotected source code: it inserts, in the target program, machine code generators, called *SGPC*s (Specialized Generators of Polymorphic Code), that can produce so-called *polymorphic instances*, i.e., many different but functionally-equivalent implementations of the protected components. At run-time, SGPCs are regularly executed to produce new machine code instances of the polymorphic components. Thus, the device will behave differently after each code generation but the results of the computations are not altered. The toolchain supports several polymorphic code transformations, which can be selected separately in the toolchain, and most of them offer a set of configuration parameters. A developer can then set the level and the nature of polymorphic transformations, hence the amount of behavioral variability.

Hereafter, we detail the specific code transformations that have been activated for the evaluation:

- **Register shuffling:** the index of the general purpose callee saved registers are randomly permuted.
- **Instruction shuffling:** the independent instructions are randomly permuted.
- **Semantic variants:** some instructions are randomly replaced by another semantically equivalent (sequence of) instruction(s). For example, variants of arithmetic instructions (*e.g.* `eor`, `sub`), remains arithmetically equivalent to the original instruction.
- **Noise instructions:** a random number of dummy instructions is added between the useful instructions in order to break the alignment of the leakage in the traces. Noise instructions are interleaved with useful instructions by the instruction shuffling transformation.

We emphasize on the fact that the sensitive variables (e.g., the AES key) are only manipulated by the polymorphic instances (i.e., the generated machine code), and not by the SGPCs themselves. SGPCs are specialized code generators, and their only input is a source of random data (a PRNG internal to the code generation runtime) that drives code generation. Hence, SGPCs only manipulate instruction and register encodings, and never manipulate secret data. Thus, performing a SCA on side-channel traces of executions of SGPCs cannot reveal a secret nor an information leakage. However, SGPCs manipulate data that are related to the contents of the buffer instances, *e.g.*, the structure of the generated code, the nature of the generated machine instructions (useful and noise instructions), etc. SCA performed on SGPC traces could possibly be helpful to reveal sensitive information about the code used by the polymorphic instances, but to the best of our knowledge, there is no such work in the literature. As such, this research question is out of the scope of this paper.

2.2 Target of Evaluation

In order to make a fair comparison with the results presented by Belleville *et al.* [6], we consider two out of the 15 implementations used in their benchmarks, namely the AES 8-bit and the mbedTLS that we briefly describe hereafter.

mbedTLS. This 32-bit implementation of AES from the ARM mbedTLS library [4] follows the so-called *T-table* technique [15]: the 16-byte state of AES is encoded into four `uint32_t` variables, each representing a column of the state. Each round of the AES is done by applying four different constant look-up tables, that are stored in flash memory.

AES 8-bit. This is a simple software unprotected implementation of AES written in C, and manipulating only variables of type `uint8_t`, similar to [1]. The SubBytes operation is computed byte-wise thanks to a look-up table, stored in RAM. This reduces information leakage on memory accesses, as compared to the use of the same look-up table stored in flash memory.

Target Device. We ran the different AES implementations on an evaluation board embedding an Arm Cortex-M4 32-bit core [40]. This device does not provide any hardware security mechanisms against side-channel attacks. This core originally operates at 72 MHz, but the core frequency was reduced to 8 MHz for the purpose of side-channel measurements. The target is similar to the one used by Belleville *et al.* [6], who considered a Cortex-M3 core running at 24 MHz. These two micro-controllers have an in-order pipeline architecture, but with a different pipeline organization. Thus, we cannot expect those two platforms to exhibit the same side-channel characteristics. However, our experience indicates that these two experimental setups would lead to similar conclusions with regards to the attacker models considered in our study. Similar findings on similar targets have also been reported by Heuser *et al.* [20]. Therefore, we assume in this study that the differences of side-channel characteristics between the two targets should not induce important differences in the results of such side-channel analysis.

Configuration of the Code Polymorphism Counter-Measure. For each evaluated implementation, the code polymorphism counter-measure is applied with a level corresponding to the configuration high described by Belleville *et al.* [6]: all the polymorphic code transformations are activated, the number of inserted noise instructions follows a probability distribution based on a truncated geometric law. The dynamic noise is activated and SGPCs produce a new polymorphic instance of the protected code for *each* execution (the regeneration period is set to 1).

2.3 Acquisition Setup

We measured side-channel traces corresponding to EM emanations with an EM probe RF-B 0.3-3 from Langer, equipped with a pre-amplifier, and a Rohde & Schwarz RTO 2044 oscilloscope with a 4 GHz bandwidth and a vertical resolution of 8 bits. We set the sampling rate to 200 MS/sec., with the acquisition mode *peak-detect* that collects the minimum and the maximum voltage values over each sampling period. We first verify that our acquisition setup is properly set. This is done by acquiring several traces where the code polymorphism was de-activated. Thus, we can verify that those traces are synchronized. Then, computing the *Welch's T-test* [28] enables to assess whether the probe is correctly positioned

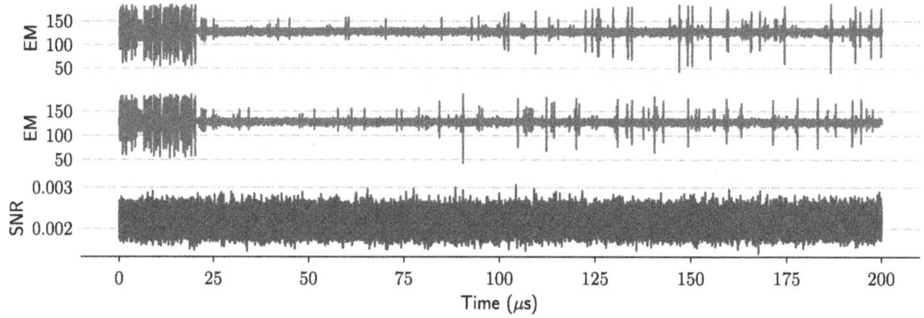

Fig. 1. Acquisitions on the mbedTLS implementation. Top: two traces containing the first AES round. Bottom: SNR computed on the 100,000 profiling traces.

and the sampling rate is high enough. Then, 100,000 profiling traces are acquired for each target implementation. Each acquisition campaign lasts about 12 h.

2.4 Preliminary Analysis of the Traces

Once the traces are acquired, and before we investigate the attacks that will be presented in Subsect. 2.5, we detail here a preliminary analysis of the component. The aim is to restrict as much as possible the target region acquired to a window covering the entire first AES round. Therefore there would not be any loss of informative leakage about the sensitive intermediate variable targeted in those experiments. In addition to that, a univariate leakage assessment, by computing the *Signal-to-Noise Ratio* (SNR) [27],[1] is provided hereafter in order to verify that there is no trivial leakage that could be exploited by a *weak, automatized* attacker (see Subsect. 2.5).

mbedTLS. We ran some preliminary acquisitions on 10^7 samples, in order to visualize all the execution. We could clearly distinguish the AES execution with sparse EM peaks, from the call to the SGPC with more frequent EM peaks. This enabled to focus on the first 10^6 samples of the traces corresponding to the AES execution.

Actually, the traces restricted to the AES execution seem to remain globally the same between each other, up to local elastic deformations along the time axis. This is in line with the effect of code polymorphism, since it involves transformations at the assembly level. Likewise, 10 patterns could be distinguished on each traces, which were clues to expect that it would correspond to the 10 rounds of AES. That is how we could restrict again our target window to the first round, up to comfortable margins because of the misalignment effect of code polymorphism. This represents 80,000 samples. An illustration of two traces restricted to the first round is given in Fig. 1 (top).

[1] A short description of this test is provided in Appendix B for the non-expert reader.

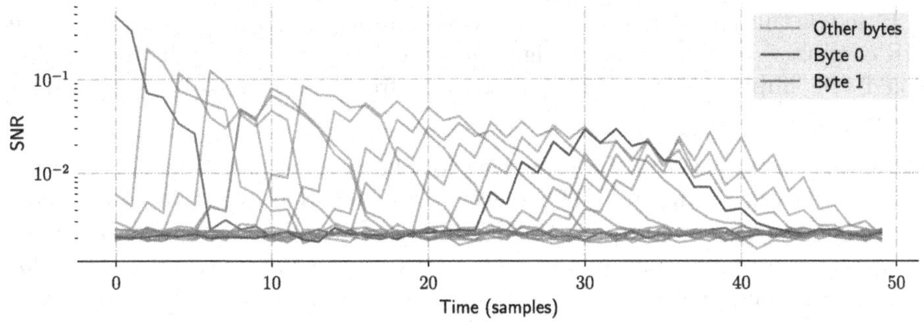

Fig. 2. The 16 SNR of the acquired traces from the mbedTLS implementation, one for each targeted byte, after re-aligned pattern extraction.

The SNR denoting here the potential univariate leakage of the first output byte of the SubBytes operation is computed based on the 100,000 acquired profiling traces, and is plotted in Fig. 1 (bottom). No distinguishing peak can be observed, which confirms the soundness of code polymorphism against vertical attacks on raw traces.

However, we observe in each trace approximately 16 EM peaks, which corresponds to the number of memory accesses to the look-up tables per encryption round. These memory accesses are known to carry sensitive information. This suggests that a trace re-alignment on EM peaks might be relevant to achieve successful vertical attacks. We proceed with such a re-alignment that we describe hereafter. For each 80,000-dimensional trace, the clock cycles corresponding to the region between two EM peaks are identified according to a thresholding on falling edges. Since the EM peak pattern delimiting two identified clock cycles may be spread over a different number of samples, from one pattern to another, we only keep the minimum and maximum points. Likewise, since the number of identified clock cycles can also differ from one trace to another, the extracted samples are eventually zero-padded to be of dimension $D = 50$.

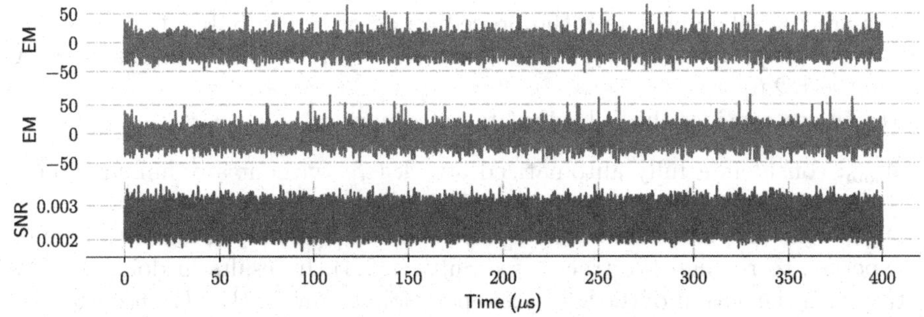

Fig. 3. Acquisitions on the AES 8-bit implementation. Top: two traces containing the first AES round. Bottom: SNR computed on the 100,000 profiling traces.

Based on this re-alignment, a new SNR is computed on Fig. 2. Contrary to the SNR computed on raw traces in Fig. 1, some leakages are clearly distinguishable though the amplitude of the SNR peaks vary from 5.10^{-2} to 5.10^{-1}, according to the targeted byte.

AES 8-bit. We proceed in the same way for the evaluation of AES 8-bit implementation. As with the mbedTLS traces, we could identify 10 successive patterns likely to correspond to the 10 AES rounds. Therefore we reduce the target window at the oscilloscope to the first AES round, which represents 160,000-dimensional traces plotted in Fig. 3 (top). This growth in the size of the traces is expected, since the naive AES 8-bit implementation is not optimized to be fast, contrary to the mbedTLS one.

Yet, here the peaks are hardly distinguishable from the level of noise. This is expected, due to the look-up tables being moved from the flash memory to the RAM, as mentioned in Subsect. 2.2. As a consequence, the memory accesses being less remarkable. That is why the re-alignment technique described for the mbedTLS implementation is not relevant here.

Finally, Fig. 3 (bottom) shows the SNR on the raw traces, ensuring once again that no trivial leakage can be exploited to recover the secret key.

2.5 Threat Models

We propose several threat models that we distinguish according to two powers that we precise hereafter.

First, the attacker may have the possibility to get a so-called *open sample*, which is a clone device behaving similarly to the actual target, in which he has a full access and control to the secret variables, *e.g.* the secret key. Having an open sample enables to run a *profiled* attack by building the exact leakage model of the target device. This is a necessary condition in order to evaluate the worst-case scenario from developer's point of view [21]. Though often seen as strong, this assumption can be considered realistic in our context: Here both the chip and the source codes used for the evaluation are publicly available.

Second, the attacker may eventually incorporate human expertise to improve attacks initially fully automatized. Here, this will concern either the preliminary task of trace re-alignment (see the procedure described in Subsect. 2.4), or the capacity to properly design a CNN architecture for the deep learning based SCA (see Subsect. 2.6).

Hence, we consider the following attack scenarios:

- $\mathcal{A}_{\mathsf{auto}}$: considers a fully automatized attack (*i.e.* without any human expertise), without access to an open sample.
- $\mathcal{A}_{\mathsf{CPA}}$: considers an attack without access to an open sample, but with human expertise to re-align the traces (see Subsect. 2.4). It results in doing a CPA targeting the output of the SubBytes operation, assuming the *Hamming weight* leakage model [8].
- $\mathcal{A}_{\mathsf{gT}}$: considers the same attack as $\mathcal{A}_{\mathsf{CPA}}$, *i.e.* targeting re-aligned traces, along with access to an open sample in addition. The profiling is done thanks to

Gaussian templates with *pooled* covariance matrices [11]. No dimensionality reduction technique is used here, beside the implicit reduction done through the re-alignment detailed in Subsect. 2.4.
- $\mathcal{A}_{\mathsf{CNN}}$: considers an attack with access to an open sample and human expertise to build a CNN for the profiled attack. This attack scenario is considered the most effective against de-synchronized traces with first-order leakage [7,10, 23]. Therefore, we do not assume $\mathcal{A}_{\mathsf{CNN}}$ to need to get access to re-aligned traces. In addition, no preliminary dimensionality reduction is done here.

We observed in the preliminary analysis conducted in Subsect. 2.4 that the SNR of the raw traces, computed on 100, 000 traces, did not emphasize any peak. Instead, the same SNR based on the re-aligned traces emphasized some peaks. Thanks to the works of Mangard *et al.* [26,27] and Oswald *et al.* [28], we can already draw the following conclusions: $\mathcal{A}_{\mathsf{auto}}$ will not succeed with less than $N_a^\star = 100,000$ queries, whereas $\mathcal{A}_{\mathsf{CPA}}$, $\mathcal{A}_{\mathsf{gT}}$ and $\mathcal{A}_{\mathsf{CNN}}$ are likely to succeed within the same amount of queries.

2.6 CNN-Based Profiling Attacks

As mentioned in Subsect. 2.5, CNN attacks may require some human expertise to properly set the model architecture. This section is devoted at describing the whole settings used to train the CNNs used in the attack scenario $\mathcal{A}_{\mathsf{CNN}}$, in order to tackle the challenge of large-scale traces. We quickly review the guidelines in the SCA literature, and argue why they are not suited to our traces. We then present the used architecture, and we detail the training parameters.

The Literature Guidelines. Though numerous papers have proposed CNN architectures [7,10,25], the state-of-the-art CNNs are currently given by Kim *et al.* [23] and Zaid *et al.* [45]. Their common point is to rely on the so-called *VGG*-like architecture [37]:

$$s \circ \lambda \circ [\sigma \circ \lambda]^{n_1} \circ [\delta_p \circ \sigma \circ \mu \circ \gamma_{w,k}]^{n_2} \circ \mu \ , \tag{1}$$

where $\gamma_{w,k}$ denotes a convolutional layer made of k filters of size w, μ denotes a batch-normalization layer [22], σ denotes an activation function *i.e.* a non-linear function, called *ReLU*, applied element-wise [18], δ_p denotes an *average* pooling layer of size p, λ denotes a dense layer, and s denotes the softmax layer.[2] Furthermore, the composition $[\delta_p \circ \sigma \circ \mu \circ \gamma_{w,k}]$ is denoted as a convolutional *block*. Likewise, $[\sigma \circ \lambda]$ denotes a dense block. We note n_1 (resp. n_2) the number of dense blocks (resp. convolutional blocks).

An intuitive approach would be to directly set the parameters or our architecture to the ones used by Kim *et al.* or by Zaid *et al.* Unfortunately, we argue in both case that such a transposition is not possible.

[2] The softmax outputs a normalized vector of positive scalars. Hence, it is similar to a discrete probability distribution that we want to fit with the actual leakage model.

Kim *et al.* propose particular guidelines to set the architecture [23]. They recommend to fix $w = 3$, $p = 2$, *i.e.*, the minimal possible values, and to set n_2 so that the time dimensionality at the output of the last block is reduced to one. Since each pooling divides the time dimensionality by p, $n_2 \leq \log_p(D)$.[3] Meanwhile they double the number of filters for each new block compared to the previous one, without exceeding 256. Unfortunately, using these guidelines is likely to increase n_2 from 10 in Kim *et al.*'s work to at least 17 in our context. As explained by He *et al.* [19], stacking such a number of layers is likely to make the training harder to tweak the learning parameters to optimal values. That is why an alternative architecture called *Resnet* has been introduced [19], and starts to be used in SCA as well [17,47]. This possibility will be discussed in Sect. 4. Second, due to the doubling number of filters at each new block, the number of learning parameters would be around 1.8 M, *i.e.* 10 folds more than the number of traces. In such a configuration, *over-fitting* is likely to happen [36], which degrades the performance of CNNs.

In order to improve the Kim *et al.*'s architecture, Zaid *et al.* [45] proposed thumb rules to set the filter size in the convolutional and the pooling layers depending on the maximum temporal amplitude of the de-synchronization. Unfortunately, it assumes to know the maximum amplitude of the de-synchronization, which is not possible here since it is hard to guess how many times those transformations are applied in the polymorphic instance.

Our Architecture. The drawbacks of Kim *et al.*'s and Zaid *et al.*'s guidelines in our particular context justify why we do not directly use them. Instead, we propose to take the Kim *et al.*'s architecture as a baseline, on which we modify some of the parameters as follows. First, we set the number of filters in this first block to $k_0 = 10$, we decrease the maximal number of filters from 256 to $k_{\max} = 100$, and we slightly change the way the number of filters is computed in the intermediate convolutional layers, according to Table 2 in Sect. A. Likewise, we remove the dense block (*i.e.* $n_1 = 0$). This limits the number of learning parameters, and thus avoids over-fitting. Second, we increase the pooling size to $p = 5$. This mechanically allows to decrease the minimal number of convolutional blocks n_2 from 17 to 6 for the mbedTLS traces and to 7 for the AES 8-bit ones. To be sure that this growth in p does not imply any loss of information in the pooling layers, we set the filter size to $w = 2p + 1 = 11$. Eventually, since with such numbers of convolutional blocks the output dimensionality is not equal to one yet, we add a *global* average pooling δ_G at the top of the convolutional blocks, which will force the reduction without adding any extra learning parameter [46]. Such an architecture would represent $177,500$ learning parameters, when targeting the mbedTLS implementation and $287,500$ for the AES 8-bit. As a comparison, we would expect at least 1.8M parameters if we apply the Kim *et al.*'s guidelines to our context.

[3] We recall that D denotes the dimensionality of the traces, *i.e.* $D = 80,000$ for the mbedTLS implementation ($D = 50$ for the re-aligned data) and $D = 160,000$ for the AES 8-bit.

First Convolutional Block. To decrease further the number of learning parameters, an attacker may even tweak the first convolutional block, by exploiting the properties of the input signal. Figure 4 sketches an EM trace chunk of about one clock cycle. We make the underlying assumption that the relevant information to extract from the traces is contained in the patterns that occur around each clock, mostly due to the change of states in the memory registers storing the sensitive variables [27].[4] Moreover, no additional relevant pattern is assumed to be contained in the trace until the next clock cycle, appearing $T = 50$ samples later.[5]

By carefully setting w_0 to the size of the EM patterns, and p_0 such that $w_0 \leq p_0 \leq T$, we optimally extract the relevant information from the patterns while avoiding the entanglement between two of them. In our experiments, we arbitrarily set $p_0 = 25$. This tweaked first convolutional block has then the same receptive field than one would have with two normal blocks of parameters $(w = 11, p = 5)$. Therefore, we spare one block (*i.e.* the last one), which decreases the number of learning parameters: our architecture now represents 84, 380 (resp. 177, 500) learning parameters for the model attacking the mbedTLS implementation (resp. AES 8-bit). Table 2 in Sect. A provides a synthesis of the description of our architecture, along with a comparison with the Kim *et al.* and Zaid *et al.*'s works.

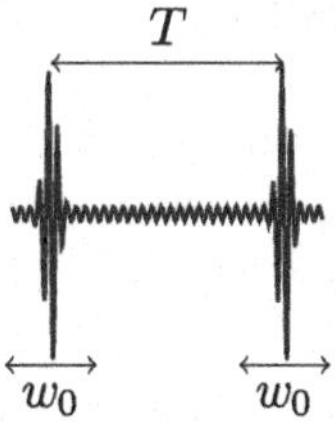

Fig. 4. Two EM patterns separated by one clock cycle.

Training Settings. The source code is implemented in Python thanks to the Pytorch [32] library and is run on a workstation with a Nvidia Quadro M4000 GP-GPU with 8 GB memory and 1664 cores. For each experiment, the whole data-set is split into a training and a test subsets, containing respectively 95, 000 and 5, 000 traces. The latter ones are used to simulate a key recovery based on the scores attributed to each hypothetical value of the sensitive target variable by the trained model. Moreover, the SH_{100} data augmentation method is applied to the training traces, following the description given in [10]: each trace is randomly shifted of maximum 100 points, which represents 2 clock cycles.[6]

The training is done by minimizing the *Negative Log Likelihood* (NLL) loss quantifying a dissimilarity between the output discrete probability distribution given by the softmax layer and the labels generated by the output of the SubBytes operation, with the *Adam* optimizer [24] during 200 epochs[7] which approx-

[4] This assumption has somewhat already been used for the pattern extraction re-alignment in Subsect. 2.4.

[5] We recall that despite the effect of code polymorphism, and in absence of hardware jitter, the duration of the clock period, in terms of samples, is roughly constant.

[6] This data augmentation is not applied on the attack traces.

[7] One epoch corresponds to the number of steps necessary to pass the whole training data-set through the optimization algorithm once.

imately represents a 16-hour long training for each targeted byte. The learning rate of the optimizer is always set to 10^{-5}.

2.7 Performance Metrics

This section explains how the performance metrics are computed in order to give a fair comparison between the different attack scenarios considered in this evaluation.

Based on an (eventually partially) trained model, we proceed a *key recovery*, by aggregating the output scores given by the softmax layer, computed from a given set of attack traces coming from the 5,000 test traces into a *maximum likelihood* distinguisher [21]. This outputs a scalar score for each key hypothesis. We evaluate the performance of a model during and after the training by computing its *Success Rate* (SR), namely the probability that the key recovery outputs the highest score to the right key hypothesis. In the following, if the success rate is at least $\beta = 90\%$, we will say that the attack was *successful*. Eventually, $N_a^\star$ will denote the required number of queries to the cryptographic primitive (*i.e.*, the number of traces) in order to obtain a successful attack. [8]

3 Results

Once the different threat models and their corresponding parameterization have been introduced in Sect. 2, we can now present the results of each attack, also summarized in Table 1.

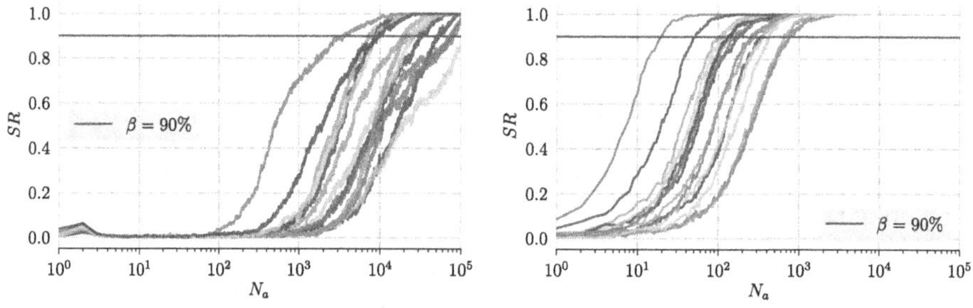

(a) $\mathcal{A}_{\mathsf{CPA}}$ on mbedTLS (b) $\mathcal{A}_{\mathsf{gT}}$ on mbedTLS

Fig. 5. Success Rate with respect to the number of attack traces. Vertical attacks on mbedTLS. The different colors denote the different targeted bytes.

As argued in Subsect. 2.4, we can directly conclude from the SNRs given by Fig. 1 (bottom) and Fig. 3 (bottom) that the fully automatized attack $\mathcal{A}_{\mathsf{auto}}$

[8] More details are given in Appendix C.

cannot succeed within the maximum amount of collected traces, *i.e.*, $N_a^\star > 10^5$, for both implementations.

Figure 5a depicts the performances of $\mathcal{A}_{\mathsf{CPA}}$ against the mbedTLS implementation, on each state byte at the output of the SubBytes operation. It can be seen that the re-alignment enables a first-order CPA to succeed within $N_a^\star \approx 10^3$ for the byte 1, and $N_a^\star \in [\![10^4, 10^5]\!]$ for the others.[9] Those results are in line with the rule of thumb stating that the higher the SNR on Fig. 2, the faster the success rate convergence towards 1 on Fig. 5a [27]. Since we argued in Subsect. 2.4 that the proposed re-alignment technique was not relevant on the AES 8-bit traces, we conclude that $\mathcal{A}_{\mathsf{CPA}}$ would require more that 10^5 queries on those traces.[10]

Figure 5b summarizes the outcomes of the attack $\mathcal{A}_{\mathsf{gT}}$. One can remark that the attack is successful for all the target bytes within 2,000 queries, which represents an improvement by one order of magnitude as compared to the scenario $\mathcal{A}_{\mathsf{CPA}}$. In other words, the access to an open sample provides a substantial advantage to $\mathcal{A}_{\mathsf{gT}}$ compared to $\mathcal{A}_{\mathsf{CPA}}$. As for the latter one, and for the same reasons, we conclude that the attack $\mathcal{A}_{\mathsf{gT}}$ would fail with 10^5 traces of the AES 8-bit implementation.

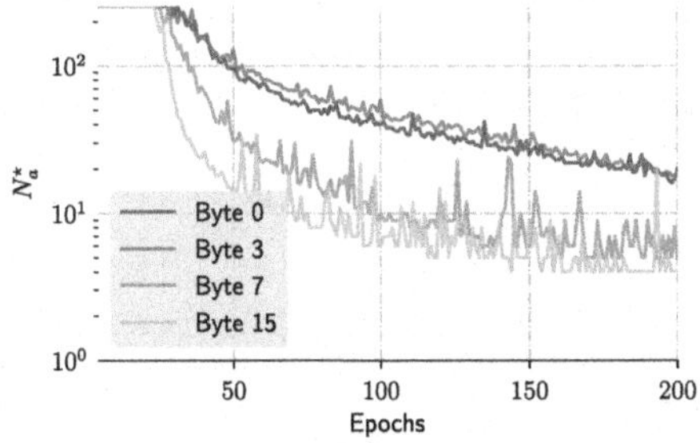

(a) mbedTLS

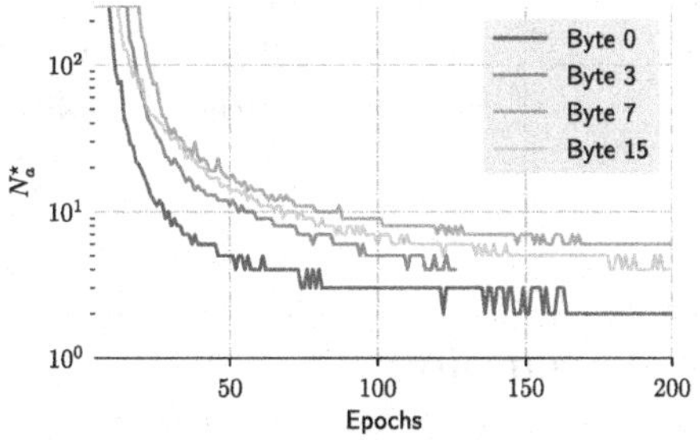

(b) AES 8-bit

Fig. 6. Evolution of $N_a^\star$ with respect to the number of training epochs during the open sample profiling by the CNN (attack $\mathcal{A}_{\mathsf{CNN}}$).

Figure 6 presents the results of the CNN attack $\mathcal{A}_{\mathsf{CNN}}$. In particular, Fig. 6a shows that training the CNN for 200 epochs allows to recover a secret byte in less than 20 traces in the case of the mbedTLS implementation. Likewise, Fig. 6b shows that a successful attack can be done in less than 10 traces on the AES 8-bit implementation. Moreover, both curves in Fig. 6 show that the latter observations can be generalized for each byte targeted in the attack $\mathcal{A}_{\mathsf{CNN}}$.[11] Finally, one can remark that training the CNNs during a lower number

[9] Targeting the output of the AddRoundKey operation instead of the output of the SubBytes operation has also been considered without giving better results.

[10] Section 4 discusses the possibility of relevant re-alignment techniques for the AES 8-bit implementation.

[11] Additional experiments realized on a setup close to $\mathcal{A}_{\mathsf{CNN}}$ confirm that the results can be generalized to any of the 16 state bytes.

of epochs (*e.g.*, 100 for mbedTLS, 50 for AES 8-bit), still leads to the same order of magnitude for $N_a^\star$.

Based on these observations, one can make the following interpretations. First, the attack $\mathcal{A}_{\mathsf{CNN}}$ leads to the best attack among the tested ones, by one or several orders of magnitude. Second, such attacks are reliable, since the results do not differ from one implementation to another, and from one targeted byte to another. Third, the training time for CNNs, currently set to roughly 16 hours for each byte (see Subsect. 2.6), can be halved or even quartered without requiring too much more queries to succeed the attack. Since in this scenario the profiling phase is here the *critical* (*i.e.* the longest) task, it might be interesting to find a trade-off between the training time and the resulting $N_a^\star$, depending on the attacker's abilities.

Table 1. Minimal number $N_a^\star$ of required queries to recover the target key bytes.

Scenario	mbedTLS	AES 8-bit
$\mathcal{A}_{\mathsf{auto}}$	$>10^5$	$>10^5$
$\mathcal{A}_{\mathsf{CPA}}$	$3.10^3 - 10^5$	$>10^5$
$\mathcal{A}_{\mathsf{gT}}$	$20 - 10^3$	$>10^5$
$\mathcal{A}_{\mathsf{CNN}}$	$\mathbf{>20}$	$\mathbf{>10}$

4 Discussion

So far Sect. 3 has presented and summarized the results of the attacks, depending on the threat models defined in Subsect. 2.5, against two implementations of a cryptographic primitive, protected by a given configuration of code polymorphism. This section proposes to discuss these results, the underlying assumptions behind the attacks, and eventually the consequences of them.

Choice of Our CNN Architecture. The small amount of queries to succeed the attack $\mathcal{A}_{\mathsf{CNN}}$, conducted on both implementations, shows the relevance of our choice of CNN architecture. This illustrates that an end-to-end attack with CNNs is possible when targeting large scale traces, without necessarily requiring very deep architectures. We emphasize that there may be other choices of parameters for the convolutional giving relevant results as well, if not better. Yet, we do not find necessary to further investigate this way here. The obtained minimal number of queries $N_a^\star$ was low enough so that any improvement in the CNN performances is not likely to change our interpretations of the vulnerability of the targets against $\mathcal{A}_{\mathsf{CNN}}$.

In particular, the advantage of Resnets [19] broadly used in image recognition typically relies on the necessity to use deep convolutional architectures in this field [37] and promising results have been obtained over the past few months with Resnets in SCA [17,47]. However, we demonstrate that we can take advantage of the distinctive features of side-channel traces to constrain the depth of our model and avoid typical issues related to deep architectures (*i.e.* vanishing gradient).

On the Re-alignment Technique. The re-alignment technique used in this work is based on the detection of leakage instants by thresholding. Other re-alignment techniques [16,29,44] may be used. Therefore, the results of $\mathcal{A}_{\mathsf{CPA}}$ and $\mathcal{A}_{\mathsf{gT}}$ might be improved. However, none of the re-alignment techniques in the literature provides strong theoretical guarantees of optimality, especially regarding the use of code polymorphism.

On the Security of Code Polymorphism. Our study exhibits the attacks $\mathcal{A}_{\mathsf{auto}}$ and $\mathcal{A}_{\mathsf{CPA}}$ with high $N_a^\star$ enough to enable a key refreshing period reasonably high, without compromising the confidentiality of the key. Unfortunately, this is not possible in presence of a stronger attacker that may have access to an open sample, as emphasized by attacks $\mathcal{A}_{\mathsf{gT}}$ and $\mathcal{A}_{\mathsf{CNN}}$, where a secret key can be recovered within the typical duration of a session key. This may be critical at first sight since massive IoT applications often rely on *Commercially available Of The Shelf* (COTS) devices, which implies that open samples may be easily accessible to any adversary.

More generally, this issue can be viewed from the perspective of the problem discussed by Bronchain *et al.* about the difficulty to prevent side-channel attacks in COTS devices, even with sophisticated counter-measures [9]. First, our experimental target is intrinsically highly vulnerable to SCA. Second, the use of software implementations of cryptographic primitives offers a large attack surface, which remains highly difficult to protect especially with a hiding countermeasure alone. This underlines the fact that a component may need to use hiding in combination with other counter-measures, *e.g.*, masking, to be secured against a strong side-channel attacker model.

5 Conclusion

So far, this paper answers two questions that may help both developers of secure implementations, and evaluators.

From a developer's point of view, this paper has studied the effect of two implementations of a code polymorphism counter-measure against several side channel attack scenarios, covering a wide range of potential attackers. In a nutshell, code polymorphism as an automated tool, is able to provide a strong protection against threat models considering automated and layman attackers, as the evaluated implementations were secure enough against our first attacker models. Yet, the implementations evaluated are not sound against stronger attacker models. The soundness of software hiding countermeasures, if used alone, remains to be demonstrated against state-of-the-art attacks, for example by using other configurations of the code polymorphism toolchain, or by proposing new code transformations. All in all, our results underline again, if need be, the necessity to combine the hiding and masking protection principles in a secured implementation.

From an evaluator's point of view, this paper illustrates how to leverage CNN architectures to tackle the problem of large-scale side-channel traces, thereby narrowing the gap between SCA literature and concrete evaluations of secure devices where pattern detection and re-alignment are not always possible. The idea lies in slight adaptations of the CNN architectures already used in SCA, eventually by exploiting the signal properties of the SCA traces. Surprisingly, our results emphasize that, though the use of more complex CNN architectures have been shown to be sound to succeed SCAs, their use might not be a necessary condition in a SCA context.

Acknowledgements. This work was partially funded thanks to the French national program "Programme d'Investissement d'Avenir IRT Nanoelec" ANR-10-AIRT-05. The authors would like to thank Emmanuel Prouff and Pierre-Alain Moëllic for their fruitful feedbacks and discussions on this work.

A Summary of the CNN Architecture Parameters

In the Zaid *et al.*'s methodology, T denotes the maximum assumed amount of random shift in the traces, and I denotes the assumed number of leakage temporal points in the traces.

Table 2. Our architecture and the recommendations from the literature.

	Kim *et al.* [23]	This paper	Zaid *et al.* [45]
n_1	1	0	2
n_2	$\log_p(D)$	$\log_p(D)$	3
p	2	$5(p_0 = 25)$	$2, \frac{T}{2}, \frac{D}{I}$
w	3	$11(w_0 = 10)$	$1, \frac{T}{2}, \frac{D}{I}$
k_n	$\min(k_0 \times 2^n, k_{\max})$	10, 20, 40, 40, 80, 100(100)	$k_0 \times 2^n$
k_0	8	10	8, 32
$k_{\max}$	256	100	–

B Signal-to-Noise Ratio (SNR)

To assess whether there is trivial leakage in the traces, one may compute a *Signal-to-Noise Ratio* (SNR) [27]. For each time sample t, it is estimated by the following statistics:

$$\mathrm{SNR}[t] \triangleq \frac{\underset{Z}{\mathbb{V}}\left(\mathbb{E}\left[\mathbf{X}[t] | Z = z\right]\right)}{\mathbb{V}\left(\mathbf{X}[t]\right)}, \tag{2}$$

where $\mathbf{X}[t]$ is the random continuous variable denoting the EM emanation measured at date t, Z is the random discrete variable denoting the sensitive target variable, $\mathbb{E}$ denotes the expected value and $\mathbb{V}$ denotes the variance. When the sample t does not carry any informative leakage, $\mathbf{X}[t]$ does not depend on Z so the numerator is zero, and inversely.

C Performance Metric Computation

Let $(p_i)_{i \leq N_a}$ the plaintexts of the N_a traces of an attack set, encrypted with varying keys $(k_i)_{i \leq N_a}$, and let $(\mathbf{y}_i)_{i \leq N_v}$ be the corresponding outputs of the CNN softmax. Let $\hat{k}$ be in the key chunk space. A *score* for the value $\hat{k}$ is defined by the maximum likelihood distinguisher [21] as: $\mathbf{d}_{N_a}[\hat{k}] = \sum_{i=1}^{N_a} \log(\mathbf{y}_i[z_i])$ where $z_i = \mathsf{SubBytes}[p_i \oplus k_i \oplus \hat{k}]$. The key is considered as recovered within N_a queries if the distinguisher $\mathbf{d}_{N_a}$ outputs the highest score for $\hat{k} = 0$.

To assess the difficulty of attacking a target device with profiling attacks (which is assumed to be the worst-case scenario for the attacked device), it has initially been suggested to measure or estimate the minimum number of traces required to get a successful key recovery [27]. Observing that many random factors may be involved during the attack, the latter measure has been refined to study the probability that the right key is ranked first according to the scores. This metric is called the ***Success Rate*** [39]: $\mathrm{SR}(N_a) = \Pr\left[\mathrm{argmax}_{\hat{k} \in \mathcal{K}} \mathbf{d}_{N_a}[\hat{k}] = 0\right]$.

In practice, to estimate $\mathrm{SR}(N_a)$, sampling many attack sets may be very prohibitive in an evaluation context, especially if we need to reproduce the estimations for many values of N_a until we find the smallest value $N_a^\star$ such that for $N_a \geq N_a^\star$ the success rate is higher than a given threshold β. One solution to circumvent this problem is, given a validation set of N_v traces, to sample some attack sets by permuting the order of the traces into the validation set (*e.g.* 500 times in our experiments). $\mathbf{d}_{N_a}$ can then be computed with a cumulative sum to get a score for each $N_a \in [\![1, N_v]\!]$. For each value of N_a, the success rate is estimated by the occurrence frequency of the event "$\mathrm{argmax}_{\hat{k} \in \mathcal{K}} \mathbf{d}_{N_a}[\hat{k}] = 0$". While this trick gives good estimations for $N_a \ll N_v$, one has to keep in mind that the estimates become biased when $N_a \to N_v$. Hopefully, in our experiments, the validation set size remains much higher than N_a afterwards: $N_v = 100,000$ for $\mathcal{A}_{\mathsf{auto}}$, $\mathcal{A}_{\mathsf{CPA}}$, $N_v = 20,000$ for $\mathcal{A}_{\mathsf{gT}}$, and $N_v = 5,000$ for $\mathcal{A}_{\mathsf{CNN}}$.

References

1. Small portable AES128 in C (2020). https://github.com/kokke/tiny-AES-c
2. Agosta, G., Barenghi, A., Pelosi, G.: A code morphing methodology to automate power analysis countermeasures. In: DAC (2012). https://doi.org/10.1145/2228360.2228376
3. Agosta, G., Barenghi, A., Pelosi, G., Scandale, M.: The MEET approach: securing cryptographic embedded software against side channel attacks. TCAD (2015). https://doi.org/10.1109/TCAD.2015.2430320

4. ARMmbed: 32-bit t-table implementation of aes for mbed tls (2019). https://github.com/ARMmbed/mbedtls/blob/master/library/aes.c
5. Becker, G., Cooper, J., De Mulder, E., Goodwill, G., Jaffe, J., Kenworthy, G., et al.: Test vector leakage assessment (TVLA) derived test requirements (DTR) with AES. In: International Cryptographic Module Conference (2013)
6. Belleville, N., Couroussé, D., Heydemann, K., Charles, H.: Automated software protection for the masses against side-channel attacks. TACO (2019). https://doi.org/10.1145/3281662
7. Benadjila, R., Prouff, E., Strullu, R., Cagli, E., Dumas, C.: Deep learning for side-channel analysis and introduction to ASCAD database. J. Cryptogr. Eng. **10**(2), 163–188 (2019). https://doi.org/10.1007/s13389-019-00220-8
8. Brier, E., Clavier, C., Olivier, F.: Correlation power analysis with a leakage model. In: Joye, M., Quisquater, J.-J. (eds.) CHES 2004. LNCS, vol. 3156, pp. 16–29. Springer, Heidelberg (2004). https://doi.org/10.1007/978-3-540-28632-5_2
9. Bronchain, O., Standaert, F.X.: Side-channel countermeasures' dissection and the limits of closed source security evaluations. IACR TCHES **2020**(2) (2020). https://doi.org/10.13154/tches.v2020.i2.1-25
10. Cagli, E., Dumas, C., Prouff, E.: Convolutional neural networks with data augmentation against jitter-based countermeasures. In: Fischer, W., Homma, N. (eds.) CHES 2017. LNCS, vol. 10529, pp. 45–68. Springer, Cham (2017). https://doi.org/10.1007/978-3-319-66787-4_3
11. Choudary, O., Kuhn, M.G.: Efficient template attacks. In: Francillon, A., Rohatgi, P. (eds.) CARDIS 2013. LNCS, vol. 8419, pp. 253–270. Springer, Cham (2014). https://doi.org/10.1007/978-3-319-08302-5_17
12. Coron, J.-S., Kizhvatov, I.: An efficient method for random delay generation in embedded software. In: Clavier, C., Gaj, K. (eds.) CHES 2009. LNCS, vol. 5747, pp. 156–170. Springer, Heidelberg (2009). https://doi.org/10.1007/978-3-642-04138-9_12
13. Coron, J.-S., Kizhvatov, I.: Analysis and improvement of the random delay countermeasure of CHES 2009. In: Mangard, S., Standaert, F.-X. (eds.) CHES 2010. LNCS, vol. 6225, pp. 95–109. Springer, Heidelberg (2010). https://doi.org/10.1007/978-3-642-15031-9_7
14. Couroussé, D., Barry, T., Robisson, B., Jaillon, P., Potin, O., Lanet, J.-L.: Runtime code polymorphism as a protection against side channel attacks. In: Foresti, S., Lopez, J. (eds.) WISTP 2016. LNCS, vol. 9895, pp. 136–152. Springer, Cham (2016). https://doi.org/10.1007/978-3-319-45931-8_9
15. Daemen, J., Rijmen, V.: AES and the wide trail design strategy. In: Knudsen, L.R. (ed.) EUROCRYPT 2002. LNCS, vol. 2332, pp. 108–109. Springer, Heidelberg (2002). https://doi.org/10.1007/3-540-46035-7_7
16. Durvaux, F., Renauld, M., Standaert, F.-X., van Oldeneel tot Oldenzeel, L., Veyrat-Charvillon, N.: Efficient removal of random delays from embedded software implementations using hidden Markov models. In: Mangard, S. (ed.) CARDIS 2012. LNCS, vol. 7771, pp. 123–140. Springer, Heidelberg (2013). https://doi.org/10.1007/978-3-642-37288-9_9
17. Gohr, A., Jacob, S., Schindler, W.: Efficient solutions of the CHES 2018 AES challenge using deep residual neural networks and knowledge distillation on adversarial examples. IACR Cryptology ePrint Archive 2020, 165 (2020). https://eprint.iacr.org/2020/165
18. Goodfellow, I.J., Bengio, Y., Courville, A.C.: Deep Learning. MIT Press, Cambridge (2016). http://www.deeplearningbook.org/

19. He, K., Zhang, X., Ren, S., Sun, J.: Deep residual learning for image recognition. In: CVPR (2016). https://doi.org/10.1109/CVPR.2016.90
20. Heuser, A., Genevey-Metat, C., Gerard, B.: Physical side-channel analysis on stm32f0, 1, 2, 3, 4 (2020). https://silm.inria.fr/silm-seminar
21. Heuser, A., Rioul, O., Guilley, S.: Good is not good enough - deriving optimal distinguishers from communication theory. In: Batina, L., Robshaw, M. (eds.) CHES 2014. LNCS, vol. 8731, pp. 55–74. Springer, Heidelberg (2014). https://doi.org/10.1007/978-3-662-44709-3_4
22. Ioffe, S., Szegedy, C.: Batch normalization: accelerating deep network training by reducing internal covariate shift. In: ICML (2015). http://jmlr.org/proceedings/papers/v37/ioffe15.html
23. Kim, J., Picek, S., Heuser, A., Bhasin, S., Hanjalic, A.: Make some noise. Unleashing the power of convolutional neural networks for profiled side-channel analysis. IACR TCHES **2019**(3) (2019). https://doi.org/10.13154/tches.v2019.i3.148-179
24. Kingma, D.P., Ba, J.: Adam: a method for stochastic optimization. In: ICLR (2015). http://arxiv.org/abs/1412.6980
25. Maghrebi, H., Portigliatti, T., Prouff, E.: Breaking cryptographic implementations using deep learning techniques. In: Carlet, C., Hasan, M.A., Saraswat, V. (eds.) SPACE 2016. LNCS, vol. 10076, pp. 3–26. Springer, Cham (2016). https://doi.org/10.1007/978-3-319-49445-6_1
26. Mangard, S.: Hardware countermeasures against DPA – a statistical analysis of their effectiveness. In: Okamoto, T. (ed.) CT-RSA 2004. LNCS, vol. 2964, pp. 222–235. Springer, Heidelberg (2004). https://doi.org/10.1007/978-3-540-24660-2_18
27. Mangard, S., Oswald, E., Popp, T.: Power Analysis Attacks - Revealing the Secrets of Smart Cards. Springer, Heidelberg (2007). https://doi.org/10.1007/978-0-387-38162-6
28. Mather, L., Oswald, E., Bandenburg, J., Wójcik, M.: Does my device leak information? An *a priori* statistical power analysis of leakage detection tests. In: Sako, K., Sarkar, P. (eds.) ASIACRYPT 2013. LNCS, vol. 8269, pp. 486–505. Springer, Heidelberg (2013). https://doi.org/10.1007/978-3-642-42033-7_25
29. Nagashima, S., Homma, N., Imai, Y., Aoki, T., Satoh, A.: DPA using phase-based waveform matching against random-delay countermeasure. In: ISCAS (2007). https://doi.org/10.1109/ISCAS.2007.378024
30. Nassar, M., Souissi, Y., Guilley, S., Danger, J.: RSM: a small and fast countermeasure for AES, secure against 1st and 2nd-order zero-offset SCAs. In: DATE (2012). https://doi.org/10.1109/DATE.2012.6176671
31. National Institute of Standards and Technology: Advanced encryption standard (AES). https://doi.org/10.6028/NIST.FIPS.197
32. Paszke, A., Gross, S., Massa, F., Lerer, A., Bradbury, J., Chanan, G., et al.: Pytorch: an imperative style, high-performance deep learning library. In: NeurIPS (2019)
33. Prouff, E., Rivain, M.: Masking against side-channel attacks: a formal security proof. In: Johansson, T., Nguyen, P.Q. (eds.) EUROCRYPT 2013. LNCS, vol. 7881, pp. 142–159. Springer, Heidelberg (2013). https://doi.org/10.1007/978-3-642-38348-9_9
34. Renaudin, M.: Asynchronous circuits and systems: a promising design alternative. Microelectron. Eng. **54**(1), 133–149 (2000)
35. Rivain, M., Prouff, E.: Provably secure higher-order masking of AES. In: Mangard, S., Standaert, F.-X. (eds.) CHES 2010. LNCS, vol. 6225, pp. 413–427. Springer, Heidelberg (2010). https://doi.org/10.1007/978-3-642-15031-9_28

36. Shalev-Shwartz, S., Ben-David, S.: Understanding Machine Learning: From Theory to Algorithms. Cambridge University Press, Cambridge (2014). https://doi.org/10.1017/CBO9781107298019
37. Simonyan, K., Zisserman, A.: Very deep convolutional networks for large-scale image recognition. In: ICLR (2015). http://arxiv.org/abs/1409.1556
38. Standaert, F.-X., Archambeau, C.: Using subspace-based template attacks to compare and combine power and electromagnetic information leakages. In: Oswald, E., Rohatgi, P. (eds.) CHES 2008. LNCS, vol. 5154, pp. 411–425. Springer, Heidelberg (2008). https://doi.org/10.1007/978-3-540-85053-3_26
39. Standaert, F.-X., Malkin, T.G., Yung, M.: A unified framework for the analysis of side-channel key recovery attacks. In: Joux, A. (ed.) EUROCRYPT 2009. LNCS, vol. 5479, pp. 443–461. Springer, Heidelberg (2009). https://doi.org/10.1007/978-3-642-01001-9_26
40. STMicroelectronics: NUCLEO-F303RE. https://www.st.com/content/st_com/en/products/evaluation-tools/product-evaluation-tools/mcu-mpu-eval-tools/stm32-mcu-mpu-eval-tools/stm32-nucleo-boards/nucleo-f303re.html
41. Suzuki, D., Saeki, M.: Security evaluation of DPA countermeasures using dual-rail pre-charge logic style. In: Goubin, L., Matsui, M. (eds.) CHES 2006. LNCS, vol. 4249, pp. 255–269. Springer, Heidelberg (2006). https://doi.org/10.1007/11894063_21
42. Timon, B.: Non-profiled deep learning-based side-channel attacks with sensitivity analysis. IACR TCHES **2019**(2) (2019). https://doi.org/10.13154/tches.v2019.i2.107-131
43. Veyrat-Charvillon, N., Medwed, M., Kerckhof, S., Standaert, F.-X.: Shuffling against side-channel attacks: a comprehensive study with cautionary note. In: Wang, X., Sako, K. (eds.) ASIACRYPT 2012. LNCS, vol. 7658, pp. 740–757. Springer, Heidelberg (2012). https://doi.org/10.1007/978-3-642-34961-4_44
44. van Woudenberg, J.G.J., Witteman, M.F., Bakker, B.: Improving differential power analysis by elastic alignment. In: Kiayias, A. (ed.) CT-RSA 2011. LNCS, vol. 6558, pp. 104–119. Springer, Heidelberg (2011). https://doi.org/10.1007/978-3-642-19074-2_8
45. Zaid, G., Bossuet, L., Habrard, A., Venelli, A.: Methodology for efficient CNN architectures in profiling attacks. IACR TCHES **2020**(1) (2019). https://doi.org/10.13154/tches.v2020.i1.1-36
46. Zhou, B., Khosla, A., Lapedriza, À., Oliva, A., Torralba, A.: Learning deep features for discriminative localization. In: CVPR (2016). https://doi.org/10.1109/CVPR.2016.319
47. Zhou, Y., Standaert, F.-X.: Deep learning mitigates but does not annihilate the need of aligned traces and a generalized ResNet model for side-channel attacks. J. Cryptogr. Eng. **10**(1), 85–95 (2019). https://doi.org/10.1007/s13389-019-00209-3

Towards Poisoning the Neural Collaborative Filtering-Based Recommender Systems

Yihe Zhang[1], Jiadong Lou[1], Li Chen[1], Xu Yuan[1(✉)], Jin Li[2], Tom Johnsten[3], and Nian-Feng Tzeng[1]

[1] University of Louisiana at Lafayette, Lafayette, LA 70503, USA
{yihe.zhang1,jiadong.lou1,li.chen,xu.yuan,nianfeng.tzeng}@louisiana.edu
[2] Guangzhou University, Guangzhou 510006, Guangdong, China
lijin@gzhu.edu.cn
[3] University of South Alabama, Mobile, AL 36688, USA
tjohnsten@southalabama.edu

Abstract. In this paper, we conduct a systematic study for the very first time on the poisoning attack to neural collaborative filtering-based recommender systems, exploring both availability and target attacks with their respective goals of distorting recommended results and promoting specific targets. The key challenge arises on how to perform effective poisoning attacks by an attacker with limited manipulations to reduce expense, while achieving the maximum attack objectives. With an extensive study for exploring the characteristics of neural collaborative filterings, we develop a rigorous model for specifying the constraints of attacks, and then define different objective functions to capture the essential goals for availability attack and target attack. Formulated into optimization problems which are in the complex forms of non-convex programming, these attack models are effectively solved by our delicately designed algorithms. Our proposed poisoning attack solutions are evaluated on datasets from different web platforms, *e.g.,* Amazon, Twitter, and MovieLens. Experimental results have demonstrated that both of them are effective, soundly outperforming the baseline methods.

1 Introduction

The recommender systems become prevalent in various E-commerce systems, social networks, and others, for promoting products or services to users of interest. The objective of a recommender system is to mine the intrinsic correlations of users' behavioral data so as to predict the relevant objects that may attract users' interest for promotion. Many traditional solutions leveraging the matrix factorization [16], association rule [6,8,22], and graph structure [10] techniques have been proposed by exploring such correlations to implement the recommender systems. Recently, the rapid advances in neural network (NN) techniques and their successful applications to diverse fields have inspired service

L. Chen et al. (Eds.): ESORICS 2020, LNCS 12308, pp. 461–479, 2020.
https://doi.org/10.1007/978-3-030-58951-6_23

providers to leverage such emerging solutions to deeply learn the intrinsic correlations of historical data for far more effective prediction, *i.e.*, recommendation, to improve users' experiences in using their web services [5, 7, 14, 23, 27]. A survey in [25] has summarized a series of NN-based recommender systems for objects recommendation. The neural collaborative filtering-based recommender system is among these emerging systems that have been proposed or envisioned for use in Youtube [7], Netflix [20], MovieLen [4], Airbnb [12], Amazon [18], and others.

However, existing studies have demonstrated that the traditional recommender systems are vulnerable to the poisoning attack [9, 19]. That is, an attacker can inject some fake users and operations on the network objects to disrupt the correlation relationships among objects. Such disrupted correlations can lead the recommender system to make wrong recommendation or promote attacker's specified objects, instead of original recommendation results that achieve its goal. For example, the poisoning attack solutions proposed in [19, 24], and [9] have been demonstrated with high efficiency in the matrix factorization-based, association rule-based, and graph-based recommender systems, respectively. To date, the poisoning attacks on the NN-based recommender systems have yet to be explored. It is still unsettled if this attack is effective to NN-based recommender systems.

Typically, such a recommender system takes all users' M latest operations to form a historical training dataset, which is used to train the weight matrices existing between two adjacent layers in neural networks. With the well trained weight matrices, the NN can recommend a set of objects that have the high correlations (similarity probabilities) within historical data. The goal of this paper is to perform poisoning attacks in neural collaborative filtering-based recommender systems and verify its effectiveness. The general idea of our poisoning attack approach is to inject a set of specified fake users and data. Once the recommender system takes the latest historical data for training, the injected data can be selected and act effectively to change the trained weight matrices, thus leading to wrong recommendation results. Specifically, two categories of attacks are studied, *i.e.*, availability attacks and target attacks. The first one aims to demote the recommended results by injecting poisoned data into the NN so as to change its output, *i.e.*, each object's recommendation probability. This category of attack refers to the scenario that some malicious users or service providers target to destroy the performance of other web applications' recommender systems. The second one aims to not only distort the recommended results but also promote a target set of objects to users. This category of attack connotes the application scenario that some users or business operators aim at promoting their target products, against the web servers' original recommendation.

In both attacks, we assume an attacker will inject as few operations as possible to minimize the expense of an attack while yielding the maximum attack outcomes. This objective is due to the fact that an attack's available resource may be limited and its detection should be avoided as best as possible. By defining the effective objective functions and modeling the resource constraints for each attack, we formulate each corresponding scenario into an optimization problem.

Even though the formulated problems are in the complex form of non-linear and non-convex programming, we design effective algorithms based on the gradient descent technique to solve them efficiently. To validate the effectiveness of our proposed two attack mechanisms, we take the real-world datasets from Amazon, Twitter, and MovieLens as the application inputs for evaluation. Experimental results show that our availability and target attacks both substantially outperform their baseline counterparts. Specifically, our availability attack mechanism lowers the accuracy by at least 60%, 17.7%, and 58%, in Amazon, Twitter, and MovieLens, respectively, given the malicious actions of only 5% users. Main contributions of this work are summarized as follows:

- We are the first to propose poisoning attack frameworks on the NN-based recommender systems. Through experiments, we have successfully demonstrated that the neural collaborative filtering-based recommender systems are also vulnerable to the poisoning attacks. This calls for the service providers to take into account the potential impact of poisoning attacks in the design of their NN-based recommender systems.
- We present two types of poisoning attacks, *i.e.*, availability attack and target attack, with the aim to demote recommendation results and promote the target objects, respectively. Through rigorous modeling and objective function creation, both types of attacks are formulated as the optimization problems with the goal of maximizing an attacker's profits while minimizing its involved operational cost (*i.e.*, the amount of injected data). Two effective algorithms are designed respectively to solve the two optimization problems.
- We implement our attack mechanisms in various real-world scenarios. Experimental results demonstrate that our poisoning attack mechanisms are effective in demoting recommended results and promoting target objects in the neural collaborative filtering-based recommender system.

2 Problem Statement

In this paper, we aim to design effective strategies for poisoning attacks on a general category of recommender systems based on neural collaborative filtering. Specifically, our goal is to achieve twofold distortions—*demoting* the recommended results and *promoting* the target objects—through *availability attack* and *target attack*, respectively.

2.1 Problem Setting

Given N users in the set $\mathcal{N}$ and D objects in the set $\mathcal{D}$, a recommender system is designed to recommend a small subset of objects to each user based on an estimate of the user's interest. In a neural collaborative-based recommender system, an NN is trained via historical data from each user to learn the probabilities of recommending objects to a given user, based on which the top-K objects will be selected for recommendation.

The historical operation behaviors (explicit [3] or implicit [15]) of a user can be learnt by a neural collaborative-based recommender system, such as marking, reviewing, clicking, watching, or purchasing. In practice, it typically truncates the M latest objects operated from each user n in the historical data, denoted as $\mathbf{u}_n = (u_n^1, u_n^2, ..., u_n^M)$, for learning. Let $\mathcal{U}$ denotes the entire dataset from all users, *i.e.*, $\mathcal{U} = \{\mathbf{u}_n | n \in \mathcal{N}\}$. The neural collaborative filtering model can be modeled as a function of $C^K(F(\mathcal{N}, \mathcal{D})) = \mathcal{P}^K$, which takes $\mathcal{N}$ and $\mathcal{D}$ as the input, and outputs the recommendation probability of each object j, i.e., $\mathbf{p}_n = \{p_n(j) | j \in \mathcal{D}\}$, for each user n. Here, F represents the neural collaborative filtering model and $p_n(j)$ satisfies $0 \leq p_n(j) \leq 1$ and $p_n(1) + p_n(2) + ... + p_n(D) = 1$ for $n \in \mathcal{N}$. $C^K(\cdot)$ function indicates the top-K objects selected according to the recommendation probabilities. For each user n, top-K recommended objects are denoted as $\mathbf{r}_n^K = (r_n^1, r_n^2, ..., r_n^K)$, and their corresponding scores (*i.e.*, probabilities) are represented as $\mathbf{p}_n^K$. The recommendation results of all users are denoted by $\mathcal{R} = \{\mathbf{r}_n^K | n \in \mathcal{N}\}$.

We perform poisoning attacks by injecting fake users into the network and enabling them to perform certain operations to distort the recommendation results for normal users. Assume there are $\hat{N}$ injected fake users with each performing at most M operations. Denote $\hat{\mathbf{u}}_n = (\hat{u}_n^1, \hat{u}_n^2, ..., \hat{u}_n^M)$ as the objects operated by a fake user n and $\hat{\mathcal{U}} = \{\hat{\mathbf{u}}_n | n \in \hat{\mathcal{N}}\}$ as the complete set of objects of $\hat{N}$ fake users. As the fake users disguise themselves over the normal users, the recommender system takes both normal users and fake users operations as the historical data for training, denoted by $\tilde{\mathcal{U}} = \mathcal{U} \cup \hat{\mathcal{U}}$, which naturally impacts the recommendation results. We denote $\tilde{\mathbf{r}}_n^K = (\tilde{r}_n^1, \tilde{r}_n^2, ..., \tilde{r}_n^K)$ as the top-K recommended objects for a normal user n after poisoning attacks and denote $\tilde{\mathcal{R}} = \{\tilde{\mathbf{r}}_n^K | n \in \mathcal{N}\}$ as the recommended results for all users. The attacker will control the amount of poisoned data operations (*i.e.*, fake users and their operations) to achieve its goal of demoting originally recommended objects or promoting target objects.

2.2 Attacks as Optimization Problems

We consider two types of poisoning attacks, *i.e.*, availability attack and target attack, with different profit requirements. In what follows, we will formulate them as optimization problems with the objective of optimizing their respective profits constrained by the available resources.

Availability Attack. The goal of availability attack is to demote the original recommendation results by poisoning data into the training dataset to change the NN output for the objects having top K highest probabilities. The profit of an attacker depends on the degree of distortion of the recommendation results. For its profit maximization, the attacker tries to achieve the largest discrepancy between recommendation results before and after the attack. We formulate an optimization problem in the following general form:

$$\begin{aligned} &\text{OPT-A:} \min S(\mathcal{R}, \tilde{\mathcal{R}}) \\ &s.t. \|\hat{\mathcal{U}}\|_0 \leq B,\ \hat{u}_n^m \in \{c_1, ..., c_d\}, \end{aligned} \tag{1}$$

where $S(\mathcal{R}, \tilde{\mathcal{R}})$ represents the *accuracy* metric of the recommendation results ($\tilde{\mathcal{R}}$) after the attack, as compared with those before the attack ($\mathcal{R}$). B represents the maximum limit of the poisoned data. With the $L0$-norm [21], the resource constraints address both limited fake users and their operations. The set of $\{c_1, ..., c_d\}$ represents a fake user's available operations. For example, in Ebay, this set is $\{0, 1\}$, representing whether a user clicks one product or not; In MovieLens, this set is $\{0, 1, 2, 3, 4, 5\}$, representing the rating score of one movie.

Target Attack. The goal of target attack is to not only distort the originally recommended results but also promote the target objects to users. Let $T = \{t^1, t^2, ..., t^K\}$ denote the K target objects that an attacker wishes to promote. We define a metric named the *successful score*, expressed as $H_T(\cdot)$, to measure the fraction of normal users whose top-K recommendation results include the target objects after the attack. To maximize the *successful score*, an attacker not only promotes target objects to as many users as possible but also promote as many target objects as possible for each user. This two-faceted consideration will be incorporated in the mathematical expression of $H_T(\cdot)$, to be discussed later. Such a problem is formulated in the following general form:

$$\begin{aligned} &\text{OPT-T: } \max H_T(\tilde{\mathcal{R}}) \\ &s.t. \|\hat{\mathcal{U}}\|_0 \leq B, \ \hat{u}_n^m \in \{c_1, ..., c_d\}. \end{aligned} \tag{2}$$

The difference between the problem formulations of the availability attack and the target attack (*i.e.*, OPT-A and OPT-T) lies in their objective functions. We hope to design a general attack strategy, which can flexibly achieve different goals by switching the objective function in a lightweight manner.

3 Availability Attack in Recommender System

In this section, we present our strategy for performing the availability attack in neural collaborative filtering-based recommender systems. Assume that an attacker can create or operate a set of fake users and inject bogus data into the recommender system and distort the recommended results. The structure of our attack model is illustrated in Fig. 1, where the flow with solid arrow lines illustrates the general framework of a neural collaborative filtering model and the flow with the dashed arrow lines represents our attack framework. The NN model takes a user vector $\mathbf{x}_n$ and an object vector $\bar{\mathbf{x}}_j$ as inputs to calculate the score $p_n(j)$, denoting the probability of recommending object $j \in \mathcal{D}$ to user $n \in \mathcal{U}$. To perform the attack, an attacker collects a set of recommended results and train an attack table A, used to serve as the guideline for the attacker to determine the number of fake users and their operations. Details of the attack process are elaborated in the sequence text.

3.1 Attack Model Design

In Fig. 1, $\mathbf{x}_n$ and $\bar{\mathbf{x}}_j$ indicates one-hot encoding vectors that keep 1 in their corresponding categorical entries and leave all other entries as 0. $\mathbf{W}_1$ and $\mathbf{W}_2$ are

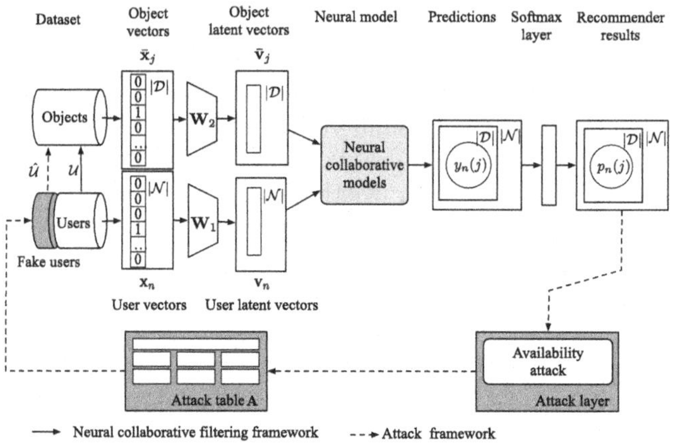

Fig. 1. Our attack framework.

two weight matrices that encode user vectors and object vectors into respective dense vectors, as follows:

$$\mathbf{v}_n = \mathbf{W}_1 \mathbf{x}_n,\ \bar{\mathbf{v}}_j = \mathbf{W}_2 \bar{\mathbf{x}}_j, \tag{3}$$

where $\mathbf{v}_n$ and $\bar{\mathbf{v}}_j$ indicate the user n's encoding vector and the object j's encoding vector, respectively. Here, both user and object encoding vectors are set to the same size. We assume the attacker knows the neural collaborative filtering-based recommender system used, but its detailed design is in black-box. According to characteristic of this system, we calculate the similarity of two encoding vectors from a user n and an object j by taking their *dot product* as follows: $y_n(j) = \bar{\mathbf{v}}_j \cdot \mathbf{v}_n^T$. Then, a Softmax layer is leveraged to calculate the probability of the object j by considering all the objects in the dataset:

$$p_n(j) = \frac{e^{\bar{\mathbf{v}}_j \mathbf{v}_n^T}}{\sum_{i=1}^{|\mathcal{D}|} e^{\bar{\mathbf{v}}_i \mathbf{v}_n^T}}. \tag{4}$$

To train this model to get W_1 and W_2, we employ the binary log likelihood [26] as follows:

$$G(\mathbf{v}_n, \bar{\mathbf{v}}_j) = L_{jn} \log(\sigma(\bar{\mathbf{v}}_j \mathbf{v}_n^T)) + (1 - L_{jn}) \log(\sigma(-\bar{\mathbf{v}}_j \mathbf{v}_n^T)), \tag{5}$$

where $L_{jn} = 1$ when a user n has operated on an object j in the training dataset (positive sample) and $L_{jn} = 0$, otherwise (negative sample) [11]. Sigmoid function σ [13] is leveraged to calculate the similarity of two encoded vectors and Log likelihood is employed here for easy gradient calculation.

Given the historical data $\mathcal{U}$, we can express the total likelihood function as:

$$G(\mathcal{U}) = \sum_{n \in \mathcal{N}} \sum_{\bar{\mathbf{v}}_j \in \bar{\mathbf{V}}_n} \left[L_{jn} \log(\sigma(\bar{\mathbf{v}}_j \mathbf{v}_n^T)) + (1 - L_{jn}) \log \sigma(-\bar{\mathbf{v}}_j \mathbf{v}_n^T)) \right], \tag{6}$$

where $\bar{\mathbf{V}}_n$ indicates the union of the positive and negative samples of a user n, which is sampled for efficiency consideration. To maximize $G(\mathcal{U})$, the stochastic gradient decent method is leveraged to solve it iteratively.

Based on our notations in Sect. 2, the number of normal users is N before the attack and the historical data of user operations is $\mathcal{U} = \{\mathbf{u}_n | 1 \le n \le N\}$. Assume the attacker injects a total of $\hat{N}$ users, with the whole set of operations being $\hat{U} = \{\mathbf{u}_n | 1 \le n \le \hat{N}\}$. With both historical and newly injected datasets, a recommender system will maximize the following objective in the training phase:

$$\begin{aligned} \tilde{G}(\tilde{U}) = &\sum_{n \in \mathcal{N}} \sum_{\bar{\mathbf{v}}_j \in \bar{\mathbf{V}}_n} \left[L_{jn} \log(\sigma(\bar{\mathbf{v}}_j \mathbf{v}_n^T)) + (1 - L_{jn}) \log \sigma(-\bar{\mathbf{v}}_j \mathbf{v}_n^T))\right] \\ &+ \sum_{n \in \hat{\mathcal{N}}} \sum_{\bar{\mathbf{v}}_j \in \bar{\mathbf{V}}_n} \left[L_{jn} \log(\sigma(\bar{\mathbf{v}}_j \mathbf{v}_n^T)) + (1 - L_{jn}) \log \sigma(-\bar{\mathbf{v}}_j \mathbf{v}_n^T))\right], \end{aligned} \tag{7}$$

where the first summation term deals with the historical data of normal users while the second summation term is for the injected bogus data.

To distort the recommended results through polluting NN training in Eq. (7), we first expand on the objective function of an availability attack presented in Sect. 2.2 and then model the constraints of injected fake users and their operations. With mathematical expressions of the goal and constraints, our availability attack is formulated and solved as an optimization problem, elaborated next.

3.2 Attack Objective Function

To distort the recommender system, the objective function of availability attack should be able to measure the discrepancy of recommended results before and after attack, i.e., $S(\mathcal{R}, \tilde{\mathcal{R}})$. As the recommender system typically recommends the top-K objects to a user, it is sufficient to consider only the top-K recommendations. Given the probabilities of all objects generated by the neural network, the recommended results $\mathcal{R}$ and $\tilde{\mathcal{R}}$ will store the objects that have the top K highest probabilities for each user before and after the attack, respectively. Before the attack, each object in the top-K recommendation list has a higher probability than any object after the K-th one. After the attack, if successful, at least one of the originally recommended objects, assuming object $r_n^k \in \mathcal{R}$ for a user n, will have a lower probability than the K-th object $\tilde{r}_n^K \in \tilde{\mathcal{R}}$ in the new ranking list, so that the object r_n^k will have a low chance to be recommended to the user. Thus, for each user n, we can define the following function to model the discrepancy of recommended results before and after attack:

$$\mathrm{sgn}[(p_n(r_n^k) - p_n(r_n^K)) \cdot (\tilde{p}_n(\tilde{r}_n^K) - \tilde{p}_n(r_n^k))] = \begin{cases} 1, & \tilde{p}_n(r_n^k) < \tilde{p}_n(\tilde{r}_n^K); \\ 0, & \tilde{p}_n(r_n^k) = \tilde{p}_n(\tilde{r}_n^K); \\ -1, & \tilde{p}_n(r_n^k) > \tilde{p}_n(\tilde{r}_n^K). \end{cases}$$

where $p_n(r_n^k)$ and $p_n(r_n^K)$ represent the probabilities of object r_n^k and the K-th object, respectively, in the recommendation list $\mathcal{R}$, while $\tilde{p}_n(r_n^k)$ and $\tilde{p}_n(\tilde{r}_n^K)$ stand for the probabilities of an object r_n^k and the Kth object in the recommendation results $\tilde{\mathcal{R}}$ after the attack. Note that $p_n(r_n^k)$ and $p_n(r_n^K)$ are known (*i.e.*, constant) while $\tilde{p}_n(r_n^k)$ and $\tilde{p}_n(\tilde{r}_n^K)$ are variables, resulted from the poisoning attack. The intuition of this equation is explained as follows. After the attack, if an object r_n^k that is previously in the recommendation list $\mathcal{R}$ has a lower probability than the K-th object $\tilde{r}_n^K \in \tilde{\mathcal{R}}$, r_n^k will not appear in the top-K recommendation list, and thus the function sgn(·) returns 1, indicating a successful attack. Otherwise, sgn(·) returns 0 or −1, meaning an unsuccessful attack.

Furthermore, considering all the users, the discrepancy of recommended results, i.e., $S(\mathcal{R}, \tilde{\mathcal{R}})$, is expressed as follows:

$$S(\mathcal{R}, \tilde{\mathcal{R}}) = \sum_{n=1}^{N} \sum_{k=1}^{K} \frac{1}{2}\{1 - \mathrm{sgn}[(p_n(r_n^k) - p_n(r_n^K)) \cdot (\tilde{p}_n(\tilde{r}_n^K) - \tilde{p}_n(r_n^k))]\}, \tag{8}$$

where $\frac{1}{2}(1 - \mathrm{sgn}(\cdot))$ is used to map the inner part over a value between 0 and 1. Minimizing $S(\mathcal{R}, \tilde{\mathcal{R}})$ implies minimizing the recommendation accuracy after poisoning attack, and it is equivalent to maximizing the discrepancy of recommended results before and after the attack. However, Eq. (8) is not continuous, unable to be solved directly. Based on the characteristics of $\frac{1}{2}(1-\mathrm{sgn}(x))$, we use $1 - \frac{1}{exp(-\theta x)}$ to approximate it by setting θ to an approximate value. After reformulating Eq. (8), we obtain the following objective function for $\min S(\mathcal{R}, \tilde{\mathcal{R}})$:

$$\min \sum_{n=1}^{N} \sum_{k=1}^{K} \{1 - \frac{1}{1 + \exp[-\theta_1(p_n(r_n^k) - p_n(r_n^K)) \cdot (\tilde{p}_n(\tilde{r}_n^K) - \tilde{p}_n(r_n^k))]}\}, \tag{9}$$

where θ_1 is a positive parameter with its value adjusted into a suitable range.

3.3 Attack Constraints

We present the resource constraints for the injected fake users and operations.

Budget Constraints. To meet the resource limitation, the amount of injected fake users and operations is constrained by a budget B, expressed as follows:

$$||\hat{\mathcal{U}}||_0 \leq B \ , \tag{10}$$

where $\hat{\mathcal{U}}$ denotes the set of fake users and their operations while $||\hat{\mathcal{U}}||_0$ represents the L_0-norm of their operations.

Operation Constraints. The operations of fake users are reflected in the attack table $\mathbf{A}$ as shown in Fig. 1, which is used by an attacker to determine the operations of each fake user on each object. Attack table A is defined with the dimensions of $\hat{N} \times D$, with rows and columns representing fake users and objects. Each entry δ_{nj} is a weight value in the range of $[0, 1]$, indicating the likelihood that a fake user n interacts with object j. By adjusting the weight values, the attacker can change the strategies of operating its fake users, thus bending the recommendation results. Each entry can be approximated as follows:

$$\delta_{nj} = \frac{1}{2}(\tanh(a_{nj}) + 1) \ , \tag{11}$$

where tanh(·) represents the *hyperbolic tangent function*.

We let each fake user n operate on at most M objects, *i.e.*, $\hat{\mathbf{u}}_n = (\hat{u}_n^1, \hat{u}_n^2, ..., \hat{u}_n^M)$. For a given fake user's dense vector $\mathbf{v}_n$ and an object's vector $\bar{\mathbf{v}}_j$, we can define our expected output of object j for a fake user n as a reference function $\tilde{y}_n(j)$:

$$\tilde{y}_n(j) = \delta_{nj} \cdot \sigma(\theta_p \bar{\mathbf{v}}_j \mathbf{v}_n^T) + (1 - \delta_{nj}) \cdot \sigma(-\theta_n \bar{\mathbf{v}}_j \mathbf{v}_n^T), \tag{12}$$

where θ_p and θ_n are shape parameters which adjust the sensitivity of user-object relationship, with the value of 0.5 as a threshold. If δ_{nj} is larger than this threshold, the attacker activates the operation. Since no attacker can manipulate normal users, δ_{nj} is a variable only related to fake users. For normal users, δ_{nj} is set to 1 if user i has relationship with object j, or set to 0, otherwise.

3.4 Solving the Optimization Problem

Based on our discussion, the availability attack can be reformulated as follows:

$$\begin{aligned} &\min\ S(\mathcal{R}, \tilde{\mathcal{R}}) \\ &s.t. \{\mathbf{W}_1, \mathbf{W}_2\} \leftarrow \operatorname{argmax} \tilde{G}(\tilde{\mathcal{U}}) \\ &\|\hat{\mathcal{U}}\|_0 \leq B,\ \hat{u}_n^m \in \{c_1, ..., c_d\}, \hat{u}_n^m \in \hat{\mathcal{U}}. \end{aligned} \tag{13}$$

We first solve the outer objective of $\min S(\mathcal{R}, \tilde{\mathcal{R}})$ and then solve the inner objective of $\operatorname{argmax} \tilde{G}(\tilde{\mathcal{U}})$.

We use $E_a(\mathbf{v}_n, \bar{\mathbf{v}}_k)$ to represent the inner part of Eq. (9), i.e.,

$$E_a(\mathbf{v}_n, \bar{\mathbf{v}}_k) = 1 - \frac{1}{1 + \exp[-\theta_1(p_n(r_n^k) - p_n(r_n^K)) \cdot (\tilde{p}_n(\tilde{r}_n^K) - \tilde{p}_n(r_n^k))]}. \tag{14}$$

Here, $p_n(r_n^k)$ and $p_n(r_n^K)$ are constants and can be obtained from Eq. (12) by setting δ_{nj} to 0 or 1. $\tilde{p}_n(r_n^k)$ and $\tilde{p}_n(\tilde{r}_n^K)$ are variables, derived via Eq. (12) by updating matrix $\mathbf{A}$. To reduce the computation cost, $\tilde{p}_n(\tilde{r}_n^K)$ will be updated only after we have finished one round of calculation on all objects and users. If our attack is successful, E will be close to 0.

To model the goal of taking as few operations as possible, the L_1 norm is added to the inner part of objective function Eq. (14), resulting in a new function, denoted as $loss_a(\mathbf{v}_n, \bar{\mathbf{v}}_k)$:

$$loss_a(\mathbf{v}_n, \bar{\mathbf{v}}_k) = E_a(\mathbf{v}_n, \bar{\mathbf{v}}_k) + c|\delta_{nk}|, \tag{15}$$

where c is a constant that adjusts the importance of two objective terms. Now, we use two phases to solve Eq. (15).

Phase I: This phase aims to update δ values in $\mathbf{A}$, by minimizing the loss function $loss_a$, with a two-step iterative procedure stated as follows:

Step 1: We initialize the random value for each a in Eq. (11) and employ the gradient descent method to update a iteratively, as follows:

$$a_{ik} := a_{ik} - \eta_a \nabla_a loss_a(\mathbf{v}_n, \mathbf{v}_k), \tag{16}$$

where η_a is the step size of updating a.

Step 2: We fix the attack table $\mathbf{A}$ and update the ranking of reference recommendation scores of all objects using Eq. (12). According to the new ranking, we can find the object $\tilde{r}_n^{K*}$.

The two steps repeat until the convergence criterion is satisfied, *i.e.*, the gradient difference between two consecutive iterations is less than a threshold.

Phase II: This phase is to add the fake users and update the vector representation of injected objects. Again, a two-step iterative procedure is used:

Step 3: Assume the attacker will control one fake user n and select S objects to perform the fake operations, we have: $S = \min\{\frac{B}{N'}, M\}$, where N' is the total number of fake users.

Step 4: After adding a fake user and its operations to the training dataset, we train new $\mathbf{W}_1$ and $\mathbf{W}_2$ by solving the inner optimization problem. Let $\mathbf{v}_n$ and $\bar{\mathbf{v}}_j$ denote the user vector and object vector, respectively. For an easy expression, we denote the terms inside the summation in Eq. (7) as $\mathbb{L}(\mathbf{v}_n, \bar{\mathbf{v}}_j)$ and then calculate the derivative of $\mathbb{L}(\mathbf{v}_n, \bar{\mathbf{v}}_j)$ with respect to $\mathbf{v}_n$ and $\bar{\mathbf{v}}_j$, i.e.,

$$\nabla_j \mathbb{L}(\mathbf{v}_n, \bar{\mathbf{v}}_j) = \frac{\partial \mathbb{L}(\mathbf{v}_n, \bar{\mathbf{v}}_j)}{\partial \bar{\mathbf{v}}_j} = (L_{jn} - \sigma(\bar{\mathbf{v}}_j \mathbf{v}_n^T))\mathbf{v}_n^T. \tag{17}$$

Then, we update $\mathbf{v}_j$ as follows: $\bar{\mathbf{v}}_j := \bar{\mathbf{v}}_j + \eta_j \cdot \nabla_j \mathbb{L}(\mathbf{v}_n, \bar{\mathbf{v}}_j)$, where η_j is the update step. With respect to $\mathbf{v}_n$, we have: $\nabla_n \mathbb{L}(\mathbf{v}_n, \bar{\mathbf{v}}_j) = \frac{\partial \mathbb{L}(\mathbf{v}_n, \bar{\mathbf{v}}_j)}{\partial \bar{\mathbf{v}}_j} = (L_{jn} - \sigma(\bar{\mathbf{v}}_j \mathbf{v}_n^T))\bar{\mathbf{v}}_j$. Then, for each $\mathbf{v}_i$ employed to aggregate $\mathbf{v}_n$, we have:

$$\mathbf{v}_n^T := \mathbf{v}_n^T + \eta_n \cdot \sum_j \nabla_n \mathbb{L}(\mathbf{v}_n, \bar{\mathbf{v}}_j)/(M-1) , \tag{18}$$

where η_n is the step size to update $\mathbf{v}_n$. Step 4 repeats until the convergence criterion is satisfied, i.e., the object vector expression result changes negligibly.

Phase 1 and **Phase 2** will repeat until the budget constraints are met or the attack purpose is achieved.

4 Target Attack

The goal of target attack is to promote specific objects to the normal users. The key challenge lies in designing an expression to measure the *successful score* of a targeted object. Denote $\mathcal{T} = \{t^1, t^2, ..., t^K\}$ as the target objects that an attacker wishes to promote to normal users. An attacker aims to promote the target objects in $\mathcal{T}$ to the top-K recommendation list by making them ranked higher than the highest recommended item r_n^1, formulated as follows:

$$\operatorname{sgn}[(p_n(r_n^t) - p_n(r_n^1)) \cdot (\tilde{p}_n(\hat{r}_n^t) - \tilde{p}_n(r_n^1))] = \begin{cases} -1, & \tilde{p}_n(r_n^t) > \tilde{p}_n(r_n^1); \\ 0, & \tilde{p}_n(r_n^t) = \tilde{p}_n(r_n^1); \\ 1, & \tilde{p}_n(r_n^t) < \tilde{p}_n(r_n^1). \end{cases} \tag{19}$$

In this function, r_n^t and r_n^1 are constant, which are known for a recommender system before attack. After performing poisoning attack, a success results in $\tilde{r}_n^t$ ranked higher than $\tilde{r}_n^1$ and sgn(x) returning -1; otherwise, the sgn(x) returns 1 or 0. Its corresponding *successful score* can be expressed by:

$$H_T(\tilde{\mathcal{R}}) = \sum_{n=1}^{N}\sum_{t=1}^{K}\frac{1}{2}\{1 - \text{sgn}[(p_n(r_n^t) - p_n(r_n^1)) \cdot (\tilde{p}_n(\hat{r}_n^t) - \tilde{p}_n(r_n^1))]\}. \quad (20)$$

To make this problem solvable, the *successful score* is approximated as follows:

$$\max\sum_{n=1}^{N}\sum_{t=1}^{K}\{1 - \frac{1}{1 + exp[-\theta_1(p_n(r_n^t) - p_n(r_n^1)) \cdot (\tilde{p}_n(\hat{r}_n^t) - \tilde{p}_n(r_n^1))]}\}. \quad (21)$$

The attack constraints illustrated in Sect. 3.3, i.e., budget and operation constraints, can also apply to the target attack. Thus, target attack can be formulated as follows:

$$\begin{aligned} &\max\ H_T(\tilde{\mathcal{R}}) \quad (22)\\ &s.t.\ \{\mathbf{W}_1, \mathbf{W}_2\} = \text{argmax}\,\tilde{G}(\tilde{U})\\ &\quad \|\hat{U}\|_0 \leq B,\ \hat{u}_n^m \in \{c_1, ..., c_d\}, \hat{u}_n^m \in \hat{U}. \end{aligned}$$

Like Eq. (14), we have:

$$loss_t(\mathbf{v}_n, \bar{\mathbf{v}}_t) = E_t(\mathbf{v}_n, \bar{\mathbf{v}}_t) + c|\delta_{nt}|, \quad (23)$$

where $E_t(\mathbf{v}_n, \bar{\mathbf{v}}_t)$ is the inner part of Eq. (21).

We can follow the similar procedure as in **Phases 1** and **2** of Sect. 3.4 to solve this problem iteratively.

5 Experiments

We implement the proposed attack frameworks and evaluate them by using three real-world datasets from Amazon, Twitter and MoiveLens, as outlined next.

5.1 Datasets

Amazon [2]**.** This dataset contains the item-to-item relationships, *e.g.*, a customer who bought one product X also purchased another product Y. The historical purchasing records are used by a recommender system to recommend product to a user. In our experiments, we take the data in the *Beauty* category as our dataset, which includes 1000 users and 5100 items.
Twitter [17]**.** The social closeness of users, represented by friend or follower relationships, is provided in this dataset. Specifically, we use 1000 users and 3457 friendships in our evaluation.
MoiveLens [1]**.** This dataset is from a non-commercial and personalized movie recommender system collected by GroupLens, consisting of 1000 users' ratings on 1700 movies. The rating scores range from 0 to 5, indicating the preference of users for movies.

5.2 Comparison

The state-of-the-art poisoning attacks target specific categories of recommender systems: [9,19,24] aiming to the matrix factorization based, association rule based, and graph based recommender systems, respectively. There is no straightforward method to apply them to the neural network-based recommender systems, making it infeasible to compare our solution with them. As far as we know, there is no existing poisoning attack solution for neural collaborative filtering-based recommender systems. Therefore, we propose to compare our solutions with the baseline methods, as described next, to show their performance gains.

Baseline Availability Attack. An attacker randomly selects M objects, with the number of injected fake users ranging from 1% to 30% of the users in the historical dataset. Each fake user operates on these M objects to poison the historical dataset. To be specific, in Amazon, each fake user purchases M selected products; in Twitter, each fake user follows M other users; in MovieLens, each user chooses M movies and rates them with random scores.

Baseline Target Attack. An attacker has a set of target objects $\mathcal{T}$. Each fake user randomly selects a target object from $\mathcal{T}$ and then selects another $M - 1$ popular objects in the network to operate on. The purpose of such a selection is to build close correlation of the target object and the popular objects so as to have the chance of being recommended. With respect to the selection of target objects in this attack, we consider two strategies: 1) selecting objects randomly and 2) selecting unpopular objects.

5.3 Performance of Availability Attack

We implement the neural collaborative filtering-based recommender systems for the Amazon, Twitter and MovieLens datasets and then perform both our poisoning attack solutions and the baseline availability attack to distort the recommended results. We set $K = 30$, i.e., recommending the top 30 objects.

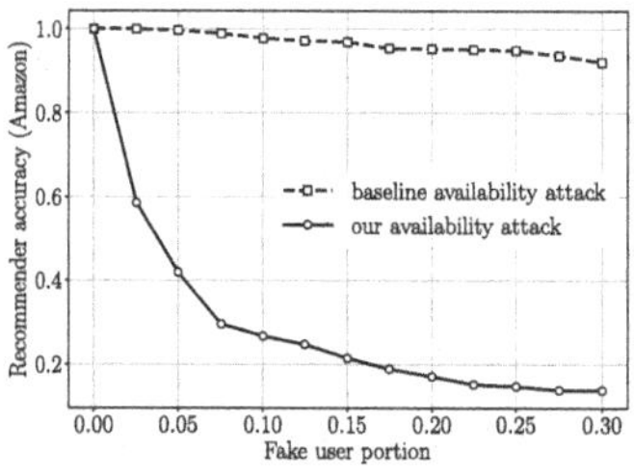

Fig. 2. Comparison of recommendation accuracy of our availability attack and the baseline method in Amazon with fake users portions varying from 0 to 30%.

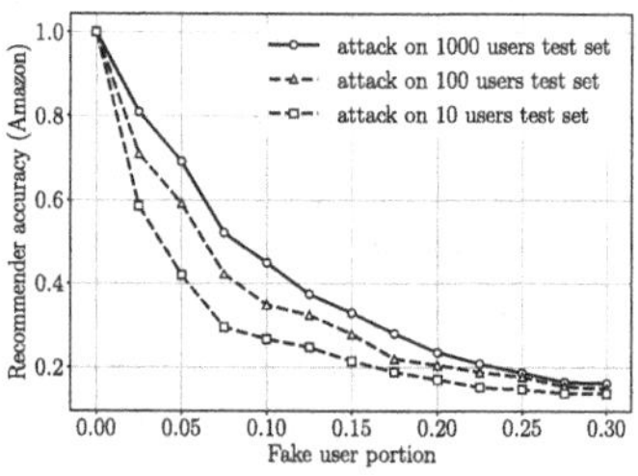

Fig. 3. The recommendation accuracy of our availability attack for the Amazon dataset under different amounts of users in test dataset.

Amazon. There are 1000 users selected, each with the latest $M = 8$ purchased products as the historical dataset $\mathcal{U}$. The neural collaborative filtering-based recommender system takes the historical dataset for training and then makes recommendations. We first randomly select 10 users for testing. Figure 2 shows the recommendation accuracy after performing our availability attack and baseline attack with the portion (the percentage of inject fake users over the total normal users in the historical data) of injected fake users increasing from 1% to 30%, *i.e.*, from 10 to 300 users. From this figure, it is obvious that our availability attack significantly outperforms the baseline method in terms of reducing the recommendation accuracy. Especially, the accuracy of recommendation drops by 60%, 80% and 90% when the percentages of fake users are 5%, 15% and 30%, respectively, in our availability attack. However, the recommendation accuracy drops only by 0.1%, 2% and 9%, respectively, in the baseline attack.

We now exam our solutions on various amounts of testing users, including 10, 100 and 1000. Figure 3 illustrates the recommendation accuracy outcomes, given different amounts of testing users when injecting the fake user count ranging from 1% to 30% in our attack. Intuitively, to achieve the same attack goal (in terms of the accuracy drop rate), more fake users need to be injected to change the correlation relationships of more testing users. However, our result is seen to be highly effective, i.e., recommendation accuracy drops less when the amount of testing users rises. Specifically, to achieve a 60% accuracy decrease, our solution only needs to inject 8% and 10% fake users respectively under 100 and 1000 users in the test set. This demonstrates that an attacker can achieve an excellent attack goal by resorting to only a small number of fake users.

Twitter. We select 1000 users and their corresponding relationships with other users sampled from [17] as the training dataset, while another 1000 users and their relationships are selected for testing. As it is hard to identify the latest friends in Twitter, we let each user randomly select 8 friends. 10 users are randomly selected from the test dataset as the targets to perform our proposed attack and the baseline availability attack. Figure 4 demonstrates the recommendation accuracy after injecting the fake user count in the range of 1% to 30% into the Twitter dataset. From this figure, the recommendation accuracy from our attack is found to drop by 17.7%, 45.2% and 61.8%, when the portions of fake users are 5%, 15% and 30%, respectively. Compared to the accuracy results attained by the baseline availability attack, which drops by only 0.07%, 1.5% and 5.7%, respectively, we conclude that our attack is starkly more effective.

To show the effectiveness of our attack on different types of users, we select 100 popular users that are followed by most users and 100 unpopular users that do not have any follower. Besides, we randomly select another 100 users as the third group of target users. Figure 5 shows the results after our attack on the three groups of users: the recommendation accuracy drops slower from the popular dataset than from both random and unpopular datasets, with the accuracy from the unpopular dataset dropping fastest. The reason is that the popular users have more knitted relationships with other users, thus requiring an attacker to invoke more operations to change such correlated friendships.

However, for an unpopular dataset, the friendships relationship is much lighter, thus making it far easier for our attack to change its correlation with others.

MovieLens. We select 1000 users and their rating scores in a total of 1700 movies sampled from [1] as the historical data and use other 10 users as the testing data. The portion of fake users increases from 0% to 5%, 10% and 30% when conducting our attack while K value increases from 1 to 30. Figure 6 depicts the recommendation accuracy of our attack with an increase in the K value under various portions of fake users. In this figure, we can see the recommendation accuracy drops less when K increases. The reason is that with a higher K value (more recommendation results), the fake users need to change more correlations among movies, thus lowering the successful rate. But with the portion of fake users rising, the recommendation accuracy can drop more as the attacker invokes more efforts to change the correlations among movies. Specifically, with 5% injected fake users, the recommendation accuracy drops by more than 50%. This demonstrates the superior effectiveness of our attack on the MovieLens.

Figure 7 shows the recommendation accuracy of baseline availability attack. It is seen the recommendation accuracy drops less than that from our attack (Fig. 6). Even when the portion of fake users rises to 30%, it drops less than 14%, which is still worse than our attack with only 1% fake users injected. Thus, the baseline solution is clearly outperformed by our attack.

5.4 Performance of Target Attack

We define a metric *hit ratio* to indicate the fraction of normal users whose top K recommendations contain the target items, after the attack.

Amazon. We use the same historical data in Amazon as in Sect. 5.3 and select 1000 other users for testing. 30 products outside the original top 30 recommendation list are randomly selected as our targets. Figure 8 compares hit ratios of our

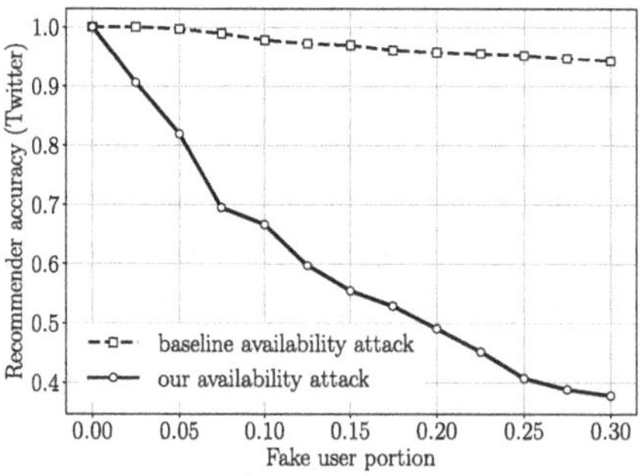

Fig. 4. Comparison of recommendation accuracy of our availability attack and the baseline method in Twitter under a range of fake users portions.

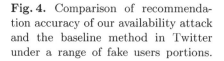

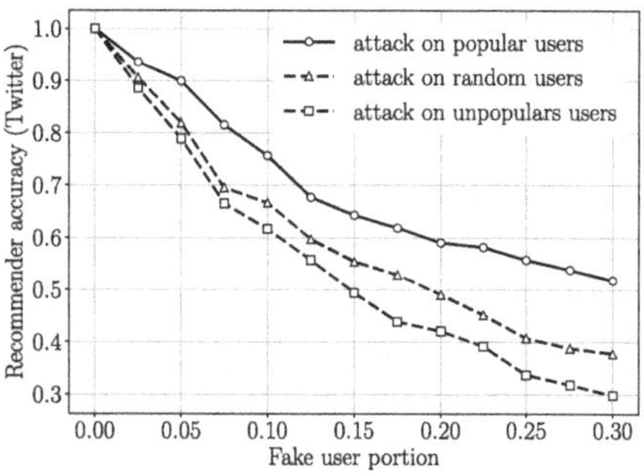

Fig. 5. Recommendation accuracy of availability attack in Twitter with respect to the popular, random and unpopular user datasets.

target attack and those of the baseline method for various amounts of injected fake users, ranging from 1% to 30%. It is seen that our target attack significantly outperforms the baseline approach. Specifically, our attack can achieve the hit ratios of 4.83%, 18.00% and 40.20%, respectively, under the injected fake user count of 5%, 15% and 30%. However, the baseline method can only attain the hit ratios of 0.00%, 0.7% and 5.36%, respectively.

We then fix the amount of fake users to 5% of all normal users in historical data (*i.e.*, 50 users) and change the amounts of both desired recommended results and target products. We randomly select the amount of target products from 1, 5, 10, to 30, while varying K from 1 to 5, 10 and 30. Table 1 lists the hit ratios from our attack and the baseline method. The hit ratios of both our attack and the baseline method are found to increase with a larger K or more target products. However, our attack always achieves a much higher hit ratio than the baseline counterpart. For example, when the amount of target products is 10, the hit ratios increase from 0.06% to 1.85% in our attack and from 0 to 0.09% in the baseline method, respectively, when K varies from 1 to 30.

Twitter. We use the same sampled training and test datasets from Sect. 5.3 to perform the target attack. The portion of fake users is fixed to 10%, *i.e.*, 100 users, and K ranges from 1 to 5, 10 and 30. The target users are randomly selected, with the user number varying from 1 to 5, 10 and 30. Table 2 provides the hit ratios of our attack and the baseline method. It is seen that the hit ratios increase with a larger K or more target users in both our attack and the baseline counterpart. But our attack significantly outperforms the baseline method. Especially, when $K = 30$, our attack can achieve the hit ratios of 0.39%, 1.02%, 2.21% and 4.62% with the number of target users varying from 1 to 5, 10 and 30, respectively, while the baseline method reaches the respective hit ratios of 0.05%, 0.08%, 0.09 and 0.18%.

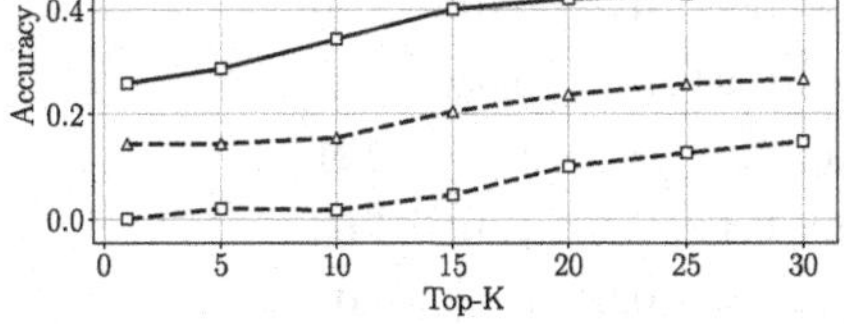

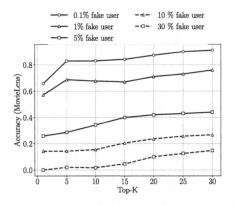

Fig. 6. Recommendation accuracy of availability attack in MovieLens with different portions of fake users and various K values.

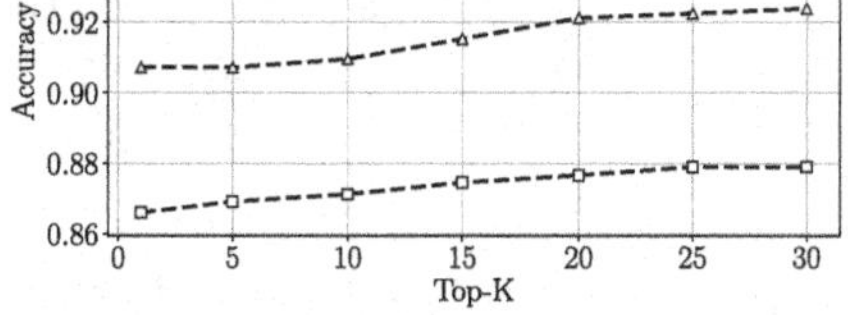

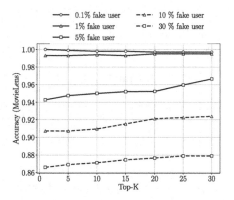

Fig. 7. Recommendation accuracy of baseline availability attack in MovieLens with different portions of fake users and various K values.

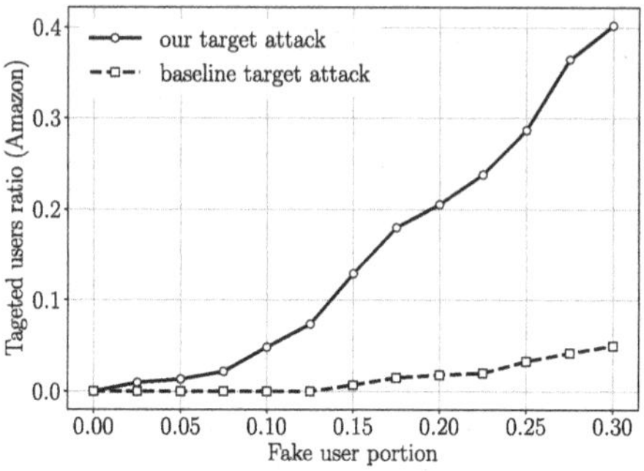

Fig. 8. Comparison of our target attack and the baseline method in Amazon under a range of fake users portions in the dataset.

Table 1. Hit ratios of our target attack and the baseline method in the Amazon dataset under a range of K values for a range of target products

Hit ratio (%)		K			
		1	5	10	30
# of target products (Our attack)	1	0.01	0.03	0.08	0.31
	5	0.03	0.10	0.54	1.02
	10	0.06	0.15	1.17	1.85
	30	0.08	0.27	1.32	2.17
# of target products (Baseline)	1	0.00	0.00	0.01	0.03
	5	0.00	0.01	0.02	0.04
	10	0.00	0.02	0.04	0.09
	30	0.00	0.03	0.05	0.11

Table 2. Hit ratios of our target attack and the baseline method in Twitter dataset under a range of K values for a range of target products

Hit ratio (%)		K			
		1	5	10	30
# of target users (Our attack)	1	0.02	0.03	0.011	0.39
	5	0.04	0.12	0.66	1.02
	10	0.07	0.20	1.72	2.21
	30	0.11	0.49	2.01	4.62
# of target users (Baseline)	1	0.01	0.01	0.03	0.05
	5	0.01	0.01	0.04	0.08
	10	0.01	0.02	0.05	0.09
	30	0.02	0.04	0.07	0.18

MovieLens. We use the same training dataset from Sect. 5.3 and sample three categories of test dataset: random, unpopular, and low-ranking movies. For each category, we randomly select 10 movies as the target set. For recommendation, we consider the top 30 ones. Figure 9(a) shows the hit ratios of our attack under the three categories of target movie set. From this figure, we can see the hit ratios of the unpopular and low ranking movie sets are lower than those from the random set. The reason is that the unpopular and low ranking targets have fewer correlations with others, thus making it much harder to promote them for recommendations when compared with the random target set. Still, our attack achieves 23% and 31% hit ratios by injecting 30% fake users. This demonstrates the advantages of our proposed attack on promoting products.

Figure 9(b) depicts the hit ratios of baseline target attack. Compared to Fig. 9(a), the hit ratios under baseline attack are much lower than those from our attack. Even in the random target movie set, the baseline has its hit ratio of only 4.8% with the injection of 30% fake users.

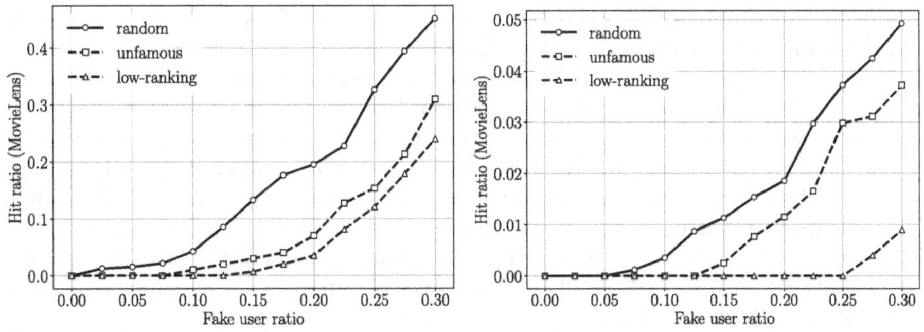

(a) Our Target Attack. (b) Baseline Target Attack.

Fig. 9. Comparisons of hit ratios of baseline and our target attack in MovieLens on the random, unpopular, low-ranking target sets for a range of fake users portions.

6 Conclusion

This paper proposes the first poisoning attack framework targeting the neural collaborative filtering-based recommender system. We have studied two types of poisoning attacks: the availability attack and the target attack, with the goals of demoting recommendation results and promoting specific target objects, respectively. By developing the mathematical models based on the resource constraints and establishing objective functions according to attacker's goals, we formulated these two types of attacks into optimization problems. Through algorithm designs, we solved the proposed optimization problems effectively. We have implemented the proposed solutions and conducted our attacks on the real-world datasets from Amazon, Twitter and MovieLens. Experimental results have demonstrated that the proposed availability attack and target attack are highly effective in demoting recommendation results and promoting specific targets, respectively, in the neural collaborative filtering-based recommender systems.

Acknowledgement. This work was supported in part by NSF under Grants 1948374, 1763620, and 1652107, and in part by Louisiana Board of Regents under Contract Numbers LEQSF(2018-21)-RD-A-24 and LEQSF(2019-22)-RD-A-21. Any opinion and findings expressed in the paper are those of the authors and do not necessarily reflect the view of funding agency.

References

1. grouplens.org (2019). https://grouplens.org/datasets/movielens/100k/
2. Mcauley, J.: (2019). http://jmcauley.ucsd.edu/data/amazon/
3. Amatriain, X., Pujol, J.M., Tintarev, N., Oliver, N.: Rate it again: increasing recommendation accuracy by user re-rating. In: Proceedings of the Third ACM Conference on Recommender Systems, pp. 173–180 (2009)
4. Bai, T., Wen, J.-R., Zhang, J., Zhao, W.X.: A neural collaborative filtering model with interaction-based neighborhood. In: Proceedings of the ACM Conference on Information and Knowledge Management, pp. 1979–1982 (2017)
5. Barkan, O., Koenigstein, N.: Item2vec: neural item embedding for collaborative filtering. In: Proceedings of the 26th International Workshop on Machine Learning for Signal Processing (MLSP), pp. 1–6 (2016)
6. Breese, J.S., Heckerman, D., Kadie, C.: Empirical analysis of predictive algorithms for collaborative filtering. In: Proceedings of the 14th Conference on Uncertainty in Artificial Intelligence, pp. 43–52 (1998)
7. Covington, P., Adams, J., Sargin, E.: Deep neural networks for YouTube recommendations. In: Proceedings of the 10th ACM Conference on Recommender Systems, pp. 191–198 (2016)
8. Davidson, J., et al.: The YouTube video recommendation system. In: Proceedings of the Fourth ACM Conference on Recommender Systems, pp. 293–296 (2010)
9. Fang, M., Yang, G., Gong, N.Z., Liu, J.: Poisoning attacks to graph-based recommender systems. In: Proceedings of the 34th Annual Computer Security Applications Conference (ACSAC), pp. 381–392 (2018)
10. Fouss, F., Pirotte, A., Renders, J.-M., Saerens, M.: Random-walk computation of similarities between nodes of a graph with application to collaborative recommendation. IEEE Trans. Knowl. Data Eng. **19**(3), 355–369 (2007)
11. Goldberg, Y., Levy, O.: Word2vec explained: deriving Mikolov et al'.s negative-sampling word-embedding method (2014)
12. Grbovic, M., Cheng, H.: Real-time personalization using embeddings for search ranking at airbnb. In: Proceedings of the 24th ACM SIGKDD International Conference on Knowledge Discovery & Data Mining, pp. 311–320 (2018)
13. Han, J., Moraga, C.: The influence of the sigmoid function parameters on the speed of backpropagation learning. In: Mira, J., Sandoval, F. (eds.) IWANN 1995. LNCS, vol. 930, pp. 195–201. Springer, Heidelberg (1995). https://doi.org/10.1007/3-540-59497-3_175
14. He, X., Liao, L., Zhang, H., Nie, L., Hu, X., Chua, T.-S.: Neural collaborative filtering. In: Proceedings of the 26th International Conference on World Wide Web (WWW), pp. 173–182 (2017)

15. Hu, Y., Koren, Y., Volinsky, C.: Collaborative filtering for implicit feedback datasets. In: Proceedings of the Eighth IEEE International Conference on Data Mining, pp. 263–272. IEEE (2008)
16. Koren, Y., Bell, R., Volinsky, C.: Matrix factorization techniques for recommender systems. IEEE Comput. **8**, 30–37 (2009)
17. Kwak, H., Lee, C., Park, H., Moon, S.: What is twitter, a social network or a news media? In: Proceedings of the 19th International Conference on World Wide Web (WWW), pp. 591–600 (2010)
18. Larry, H.: The history of amazon's recommendation algorithm. https://www.amazon.science/the-history-of-amazons-recommendation-algorithm (2020)
19. Li, B., Wang, Y., Singh, A., Vorobeychik, Y.: Data poisoning attacks on factorization-based collaborative filtering. In: Proceedings of the Advances in Neural Information Processing Systems (NIPS), pp. 1885–1893 (2016)
20. Liang, D., Krishnan, R.G., Hoffman, M.D., Jebara, T.: Variational autoencoders for collaborative filtering. In: Proceedings of the 27th World Wide Web Conference, pp. 689–698 (2018)
21. Mancera, L., Portilla, J.: L0-norm-based sparse representation through alternate projections. In: Proceedings of the IEEE International Conference on Image Processing, pp. 2089–2092 (2006)
22. Sarwar, B., Karypis, G., Konstan, J., Riedl, J.: Item-based collaborative filtering recommendation algorithms. In: Proceedings of the 10th International Conference on World Wide Web (WWW), pp. 285–295 (2001)
23. Wang, H., Wang, N., Yeung, D.-Y.: Collaborative deep learning for recommender systems. In: Proceedings of the 21th ACM SIGKDD International Conference on Knowledge Discovery and Data Mining, pp. 1235–1244 (2015)
24. Yang, G., Gong, N.Z., Cai, Y.: Fake co-visitation injection attacks to recommender systems. In: Proceedings of the 24th Annual Network and Distributed System Security Symposium (NDSS) (2017)
25. Zhang, S., Yao, L., Sun, A., Tay, Y.: Deep learning based recommender system: a survey and new perspectives. ACM Comput. Surv. (CSUR) **52**(1), 5 (2019)
26. Zhang, Z., Sabuncu, M.: Generalized cross entropy loss for training deep neural networks with noisy labels. In: Proceedings of the Advances in Neural Information Processing Systems (NIPS), pp. 8778–8788 (2018)
27. Zhou, C., et al.: Atrank: an attention-based user behavior modeling framework for recommendation. In: Proceedings of the 31th AAAI Conference on Artificial Intelligence (2018)

Data Poisoning Attacks Against Federated Learning Systems

Vale Tolpegin, Stacey Truex(✉), Mehmet Emre Gursoy, and Ling Liu

Georgia Institute of Technology, Atlanta, GA 30332, USA
{vtolpegin3,staceytruex,memregursoy}@gatech.edu, ling.liu@cc.gatech.edu

Abstract. Federated learning (FL) is an emerging paradigm for distributed training of large-scale deep neural networks in which participants' data remains on their own devices with only model updates being shared with a central server. However, the distributed nature of FL gives rise to new threats caused by potentially malicious participants. In this paper, we study targeted data poisoning attacks against FL systems in which a malicious subset of the participants aim to poison the global model by sending model updates derived from mislabeled data. We first demonstrate that such data poisoning attacks can cause substantial drops in classification accuracy and recall, even with a small percentage of malicious participants. We additionally show that the attacks can be targeted, i.e., they have a large negative impact only on classes that are under attack. We also study attack longevity in early/late round training, the impact of malicious participant availability, and the relationships between the two. Finally, we propose a defense strategy that can help identify malicious participants in FL to circumvent poisoning attacks, and demonstrate its effectiveness.

Keywords: Federated learning · Adversarial machine learning · Label flipping · Data poisoning · Deep learning

1 Introduction

Machine learning (ML) has become ubiquitous in today's society as a range of industries deploy predictive models into their daily workflows. This environment has not only put a premium on the ML model training and hosting technologies but also on the rich data that companies are collecting about their users to train and inform such models. Companies and users alike are consequently faced with 2 fundamental questions in this reality of ML: (1) How can privacy concerns around such pervasive data collection be moderated without sacrificing the efficacy of ML models? and (2) How can ML models be trusted as accurate predictors?

Federated ML has seen increased adoption in recent years [9,17,40] in response to the growing legislative demand to address user privacy [1,26,38]. Federated learning (FL) allows data to remain at the edge with only model parameters being shared with a central server. Specifically, there is no centralized data curator who collects and verifies an aggregate dataset. Instead, each

L. Chen et al. (Eds.): ESORICS 2020, LNCS 12308, pp. 480–501, 2020.
https://doi.org/10.1007/978-3-030-58951-6_24

data holder (participant) is responsible for conducting training on their local data. In regular intervals participants are then send model parameter values to a central parameter server or aggregator where a global model is created through aggregation of the individual updates. A global model can thus be trained over all participants' data without any individual participant needing to share their private raw data.

While FL systems allow participants to keep their raw data local, a significant vulnerability is introduced at the heart of question (2). Consider the scenario wherein a subset of participants are either malicious or have been compromised by some adversary. This can lead to these participants having mislabeled or poisonous samples in their local training data. With no central authority able to validate data, these malicious participants can consequently poison the trained global model. For example, consider Microsoft's AI chat bot Tay. Tay was released on Twitter with the underlying natural language processing model set to learn from the Twitter users it interacted with. Thanks to malicious users, Tay was quickly manipulated to learn offensive and racist language [41].

In this paper, we study the vulnerability of FL systems to malicious participants seeking to poison the globally trained model. We make minimal assumptions on the capability of a malicious FL participant – each can only manipulate the raw training data on their device. This allows for non-expert malicious participants to achieve poisoning with no knowledge of model type, parameters, and FL process. Under this set of assumptions, label flipping attacks become a feasible strategy to implement data poisoning, attacks which have been shown to be effective against traditional, centralized ML models [5,44,50,52]. We investigate their application to FL systems using complex deep neural network models.

We demonstrate our FL poisoning attacks using two popular image classification datasets: CIFAR-10 and Fashion-MNIST. Our results yield several interesting findings. First, we show that attack effectiveness (decrease in model utility) depends on the percentage of malicious users and the attack is effective even when this percentage is small. Second, we show that attacks can be targeted, i.e., they have large negative impact on the subset of classes that are under attack, but have little to no impact on remaining classes. This is desirable for adversaries who wish to poison a subset of classes while not completely corrupting the global model to avoid easy detection. Third, we evaluate the impact of attack timing (poisoning in early or late rounds of FL training) and the impact of malicious participant availability (whether malicious participants can increase their availability and selection rate to increase effectiveness). Motivated by our finding that the global model may still converge accurately after early-round poisoning stops, we conclude that largest poisoning impact can be achieved if malicious users participate in later rounds and with high availability.

Given the highly effective poisoning threat to FL systems, we then propose a defense strategy for the FL aggregator to identify malicious participants using their model updates. Our defense is based on the insight that updates sent from malicious participants have unique characteristics compared to honest participants' updates. Our defense extracts relevant parameters from the high-dimensional update vectors and applies PCA for dimensionality reduction. Results on CIFAR-10 and Fashion-MNIST across varying malicious participant rates (2–20%) show that the aggregator can obtain clear separation between malicious and honest participants' respective updates using our defense strategy. This enables the FL aggregator to identify and block malicious participants.

The rest of this paper is organized as follows. In Sect. 2, we introduce the FL setting, threat model, attack strategy, and attack evaluation metrics. In Sect. 3, we demonstrate the effectiveness of FL poisoning attacks and analyze their impact with respect to malicious participant percentage, choice of classes under attack, attack timing, and malicious participant availability. In Sect. 4, we describe and empirically demonstrate our defense strategy. We discuss related work in Sect. 5 and conclude in Sect. 6. Our source code is available[1].

2 Preliminaries and Attack Formulation

2.1 Federated Machine Learning

FL systems allow global model training without the sharing of raw private data. Instead, individual participants only share model parameter updates. Consider a deep neural network (DNN) model. DNNs consist of multiple layers of nodes where each node is a basic functional unit with a corresponding set of parameters. Nodes receive input from the immediately preceding layer and send output to the following layer; with the first layer nodes receiving input from the training data and the final layer nodes generating the predictive result.

In a traditional DNN learning scenario, there exists a training dataset $\mathcal{D} = (x_1, ..., x_n)$ and a loss function $\mathcal{L}$. Each $x_i \in \mathcal{D}$ is defined as a set of features $\mathbf{f}_i$ and a class label $c_i \in \mathcal{C}$ where $\mathcal{C}$ is the set of all possible class values. The final layer of a DNN architecture for such a dataset will consequently contain $|\mathcal{C}|$ nodes, each corresponding to a different class in $\mathcal{C}$. The loss of this DNN given parameters θ on $\mathcal{D}$ is denoted: $\mathcal{L} = \frac{1}{n}\sum_i^n \mathcal{L}(\theta, x_i)$.

When $\mathbf{f}_i$ is fed through the DNN with model parameters θ, the output is a set of predicted probabilities $\mathbf{p}_i$. Each value $p_{c,i} \in \mathbf{p}_i$ is the predicted probability that x_i has a class value c, and $\mathbf{p}_i$ contains a probability $p_{c,i}$ for each class value $c \in \mathcal{C}$. Each predicted probability $p_{c,i}$ is computed by a node n_c in the final layer of the DNN architecture using input received from the preceding layer and n_c's corresponding parameters in θ. The predicted class for instance x_i given a model M with parameters θ then becomes $M_\theta(x_i) = \text{argmax}_{c \in \mathcal{C}} p_{c,i}$. Given a cross entropy loss function, the loss on x_i can consequently can be calculated as $\mathcal{L}(\theta, x_i) = -\sum_{c \in \mathcal{C}} y_{c,i} \log(p_{c,i})$ where $y_{c,i} = 1$ if $c = c_i$ and 0 otherwise. The goal of training a DNN model then becomes to find the parameter values for θ which minimize the chosen loss function $\mathcal{L}$.

The process of minimizing this loss is typically done through an iterative process called stochastic gradient descent (SGD). At each step, the SGD algorithm

[1] https://github.com/git-disl/DataPoisoning_FL.

(1) selects a batch of samples $B \subseteq \mathcal{D}$, (2) computes the corresponding gradient $\mathbf{g}_B = \frac{1}{|B|}\sum_{x \in B} \nabla_\theta \mathcal{L}(\theta, x)$, and (3) then updates θ in the direction $-\mathbf{g}_B$. In practice, $\mathcal{D}$ is shuffled and then evenly divided into $|B|$ sized batches such that each sample occurs in exactly one batch. Applying SGD iteratively to each of the pre-determined batches is then referred to as one epoch.

In FL environments however, the training dataset $\mathcal{D}$ is not wholly available at the aggregator. Instead, N participants $\mathcal{P}$ each hold their own private training dataset $D_1, ..., D_N$. Rather than sharing their private raw data, participants instead execute the SGD training algorithm locally and then upload updated parameters to a centralized server (aggregator). Specifically, in the initialization phase (i.e., round 0), the aggregator generates a DNN architecture with parameters θ_0 which is advertised to all participants. At each global training round r, a subset $\mathcal{P}_r$ consisting of $k \leq N$ participants is selected based on availability. Each participant $P_i \in \mathcal{P}_r$ executes one epoch of SGD locally on D_i to obtain updated parameters $\theta_{r,i}$, which are sent to the aggregator. The aggregator sets the global parameters $\theta_r = \frac{1}{k}\sum_i \theta_{r,i}$ $\forall i$ where $P_i \in \mathcal{P}_r$. The global parameters θ_r are then advertised to all N participants. These global parameters at the end of round r are used in the next training round $r+1$. After R total global training rounds, the model M is finalized with parameters θ_R.

2.2 Threat and Adversary Model

Threat Model: We consider the scenario in which a subset of FL participants are malicious or are controlled by a malicious adversary. We denote the percentage of malicious participants among all participants $\mathcal{P}$ as $m\%$. Malicious participants may be injected to the system by adding adversary-controlled devices, compromising $m\%$ of the benign participants' devices, or incentivizing (bribing) $m\%$ of benign participants to poison the global model for a certain number of FL rounds. We consider the aggregator to be honest and not compromised.

Adversarial Goal: The goal of the adversary is to manipulate the learned parameters such that the final global model M has high errors for particular classes (a subset of $\mathcal{C}$). The adversary is thereby conducting a targeted poisoning attack. This differs from untargeted attacks which instead seek indiscriminate high global model errors across all classes [6,14,51]. Targeted attacks have the desirable property that they decrease the possibility of the poisoning attack being detected by minimizing influence on non-targeted classes.

Adversary Knowledge and Capability: We consider a realistic adversary model with the following constraints. Each malicious participant can manipulate the training data D_i on their own device, but cannot access or manipulate other participants' data or the model learning process, e.g., SGD implementation, loss function, or server aggregation process. The attack is not specific to the DNN architecture, loss function or optimization function being used. It requires training data to be corrupted, but the learning algorithm remains unaltered.

2.3 Label Flipping Attacks in Federated Learning

We use a label flipping attack to implement targeted data poisoning in FL. Given a source class c_{src} and a target class c_{target} from $\mathcal{C}$, each malicious participant P_i modifies their dataset D_i as follows: For all instances in D_i whose class is c_{src}, change their class to c_{target}. We denote this attack by $c_{src} \rightarrow c_{target}$. For example, in CIFAR-10 image classification, airplane → bird denotes that images whose original class labels are *airplane* will be poisoned by malicious participants by changing their class to *bird*. The goal of the attack is to make the final global model M more likely to misclassify airplane images as bird images at test time.

Label flipping is a well-known attack in centralized ML [43,44,50,52]. It is also suitable for the FL scenario given the adversarial goal and capabilities above. Unlike other types of poisoning attacks, label flipping does not require the adversary to know the global distribution of $\mathcal{D}$, the DNN architecture, loss function $\mathcal{L}$, etc. It is time and energy-efficient, an attractive feature considering FL is often executed on edge devices . It is also easy to carry out for non-experts and does not require modification or tampering with participant-side FL software.

Table 1. Notations used throughout the paper.

M, M_{NP}	Model, model trained with no poisoning
k	Number of FL participants in each round
R	Total number of rounds of FL training
$\mathcal{P}_r$	FL participants queried at round r, $r \in [1, R]$
$\theta_r, \theta_{r,i}$	Global model parameters after round r and local model parameters at participant P_i after round r
$m\%$	Percentage of malicious participants
c_{src}, c_{target}	Source and target class in label flipping attack
M^{acc}	Global model accuracy
c_i^{recall}	Class recall for class c_i
$m\_cnt_j^i$	Baseline misclassification count from class c_i to class c_j

Attack Evaluation Metrics: At the end of R rounds of FL, the model M is finalized with parameters θ_R. Let $\mathcal{D}_{test}$ denote the test dataset used in evaluating M, where $\mathcal{D}_{test} \cap D_i = \emptyset$ for all participant datasets D_i. In the next sections, we provide a thorough analysis of label flipping attacks in FL. To do so, we use a number of evaluation metrics.

Global Model Accuracy (M^{acc}): The global model accuracy is the percentage of instances $x \in \mathcal{D}_{test}$ where the global model M with final parameters θ_R predicts $M_{\theta_R}(x) = c_i$ and c_i is indeed the true class label of x.

Class Recall (c_i^{recall}): For any class $c_i \in \mathcal{C}$, its class recall is the percentage $\frac{TP_i}{TP_i+FN_i} \cdot 100\%$ where TP_i is the number of instances $x \in \mathcal{D}_{test}$ where $M_{\theta_R}(x) = c_i$ **and** c_i is the true class label of x; and FN_i is the number of instances $x \in \mathcal{D}_{test}$ where $M_{\theta_R}(x) \neq c_i$ and the true class label of x is c_i.

Baseline Misclassification Count ($m\_cnt_j^i$): Let M_{NP} be a global model trained for R rounds using FL without any malicious attack. For classes $c_i \neq c_j$, the baseline misclassification count from c_i to c_j, denoted $m\_cnt_j^i$, is defined as the number of instances $x \in \mathcal{D}_{test}$ where $M_{NP}(x) = c_j$ **and** the true class of x is c_i.

Table 1 provides a summary of the notation used in the rest of this paper.

3 Analysis of Label Flipping Attacks in FL

3.1 Experimental Setup

Datasets and DNN Architectures: We conduct our attacks using two popular image classification datasets: CIFAR-10 [22] and Fashion-MNIST [49]. CIFAR-10 consists of 60,000 color images in 10 object classes such as deer, airplane, and dog with 6,000 images included per class. The complete dataset is pre-divided into 50,000 training images and 10,000 test images. Fashion-MNIST consists of a training set of 60,000 images and a test set of 10,000 images. Each image in Fashion-MNIST is gray-scale and associated with one of 10 classes of clothing such as pullover, ankle boot, or bag. In experiments with CIFAR-10, we use a convolutional neural network with six convolutional layers, batch normalization, and two fully connected dense layers. This DNN architecture achieves a test accuracy of 79.90% in the centralized learning scenario, i.e. $N = 1$, without poisoning. In experiments with Fashion-MNIST, we use a two layer convolutional neural network with batch normalization, an architecture which achieves 91.75% test accuracy in the centralized scenario without poisoning. Further details of the datasets and DNN model architectures can be found in Appendix A.

Federated Learning Setup: We implement FL in Python using the PyTorch [35] library. By default, we have $N = 50$ participants, one central aggregator, and $k = 5$. We use an *independent and identically distributed* (*iid*) data distribution, i.e., we assume the total training dataset is uniformly randomly distributed among all participants with each participant receiving a unique subset of the training data. The testing data is used for model evaluation only and is therefore not included in any participant P_i's train dataset D_i. Observing that both DNN models converge after fewer than 200 training rounds, we set our FL experiments to run for $R = 200$ rounds total.

Label Flipping Process: In order to simulate the label flipping attack in a FL system with N participants of which $m\%$ are malicious, at the start of each experiment we randomly designate $N \times m\%$ of the participants from $\mathcal{P}$

as malicious. The rest are honest. To address the impact of random selection of malicious participants, by default we repeat each experiment 10 times and report the average results. Unless otherwise stated, we use $m = 10\%$.

For both datasets we consider three label flipping attack settings representing a diverse set of conditions in which to base adversarial attacks. These conditions include (1) a source class $\rightarrow$ target class pairing whose source class was very frequently misclassified as the target class in federated, non-poisoned training, (2) a pairing where the source class was very infrequently misclassified as the target class, and (3) a pairing between these two extremes. Specifically, for CIFAR-10 we test (1) 5: dog $\rightarrow$ 3: cat, (2) 0: airplane $\rightarrow$ 2: bird, and (3) 1: automobile $\rightarrow$ 9: truck. For Fashion-MNIST we experiment with (1) 6: shirt $\rightarrow$ 0: t-shirt/top, (2) 1: trouser $\rightarrow$ 3: dress, and (3) 4: coat $\rightarrow$ 6: shirt.

3.2 Label Flipping Attack Feasibility

We start by investigating the feasibility of poisoning FL systems using label flipping attacks. Figure 1 outlines the global model accuracy and source class recall in scenarios with malicious participant percentage m ranging from 2% to 50%. Results demonstrate that as the malicious participant percentage, increases the global model utility (test accuracy) decreases. Even with small m, we observe a decrease in model accuracy compared to a non-poisoned model (denoted by M_{NP} in the graphs), and there is an even larger decrease in source class recall. In experiments with CIFAR-10, once m reaches 40%, the recall of the source class decreases to 0% and the global model accuracy decreases from 78.3% in the non-poisoned setting to 74.4% in the poisoned setting. Experiments conducted on Fashion-MNIST show a similar pattern of utility loss. With $m = 4\%$ source class recall drops by $\sim 10\%$ and with $m = 10\%$ it drops by $\sim 20\%$. It is therefore clear that an adversary who controls even a minor proportion of the total participant population is capable of significantly impacting global model utility.

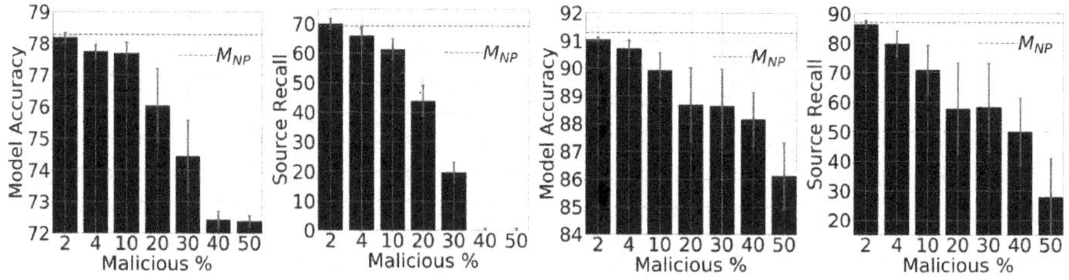

(a) CIFAR-10 M^{acc} (b) CIFAR-10 c_{src}^{recall} (c) F-MNIST M^{acc} (d) F-MNIST c_{src}^{recall}

Fig. 1. Evaluation of attack feasibility and impact of malicious participant percentage on attack effectiveness. CIFAR-10 experiments are for the $5 \rightarrow 3$ setting while Fashion-MNIST experiments are for the $4 \rightarrow 6$ setting. Results are averaged from 10 runs for each setting of $m\%$. The black bars are mean over the 10 runs and the green error bars denote standard deviation. (Color figure online)

While both datasets are vulnerable to label flipping attacks, the degree of vulnerability varies between datasets with CIFAR-10 demonstrating more vulnerability than Fashion-MNIST. For example, consider the 30% malicious scenario, Figure 1b shows the source class recall for the CIFAR-10 dataset drops to 19.7% while Fig. 1d shows a much lower decrease for the Fashion-MNIST dataset with 58.2% source class recall under the same experimental settings.

Table 2. Loss in source class recall for three source → target class settings with differing baseline misclassification counts in CIFAR-10 and Fashion-MNIST. Loss averaged from 10 runs. Highlighted bold entries are highest loss in each.

$c_{src} \rightarrow c_{target}$	$m\_cnt^{src}_{target}$	Percentage of Malicious Participants (m%)						
		2	4	10	20	30	40	50
CIFAR-10								
0 → 2	16	**1.42%**	2.93%	**10.2%**	14.1%	48.3%	**73%**	**70.5%**
1 → 9	56	0.69%	**3.75%**	6.04%	15%	36.3%	49.2%	54.7%
5 → 3	200	0%	3.21%	7.92%	**25.4%**	**49.5%**	69.2%	69.2%
Fashion-MNIST								
1 → 3	18	0.12%	0.42%	2.27%	2.41%	**40.3%**	**45.4%**	42%
4 → 6	51	**0.61%**	**7.16%**	**16%**	**29.2%**	28.7%	37.1%	**58.9%**
6 → 0	118	−1%	2.19%	7.34%	9.81%	19.9%	39%	43.4%

On the other hand, vulnerability variation based on source and target class settings is less clear. In Table 2, we report the results of three different combinations of source → target attacks for each dataset. Consider the two extreme settings for the CIFAR-10 dataset: on the low end the 0 → 2 setting has a baseline misclassification count of 16 while the high end count is 200 for the 5 → 3 setting. Because of the DNN's relative challenge in differentiating class 5 from class 3 in the non-poisoned setting, it could be anticipated that conducting a label flipping attack within the 5 → 3 setting would result in the greatest impact on source class recall. However, this was not the case. Table 2 shows that in only two out of the six experimental scenarios did 5 → 3 record the largest drop in source class recall. In fact, four scenarios' results show the 0 → 2 setting, the setting with the lowest baseline misclassification count, as the most effective option for the adversary. Experiments with Fashion-MNIST show a similar trend, with label flipping attacks conducted in the 4 → 6 setting being the most successful rather than the 6 → 0 setting which has more than 2× the number of baseline misclassifications. These results indicate that identifying the most vulnerable source and target class combination may be a non-trivial task for the adversary, and that there is not necessarily a correlation between non-poisoned misclassification performance and attack effectiveness.

We additionally study a desirable feature of the label flipping attack: they appear to be targeted. Specifically, Table 3 reports the following quantities for each source → target flipping scenario: loss in source class recall, loss in target

Table 3. Changes due to poisoning in source class recall, target class recall, and total recall for all remaining classes (non-source, non-target). Results are averaged from 10 runs in each setting. The maximum standard deviation observed was 1.45% in source class recall and 1.13% in target class recall.

$c_{src} \rightarrow c_{target}$	$\Delta\ c_{src}^{recall}$	$\Delta\ c_{target}^{recall}$	$\sum$ all other $\Delta\ c^{recall}$
CIFAR-10			
$0 \rightarrow 2$	−6.28%	1.58%	0.34%
$1 \rightarrow 9$	−6.22%	2.28%	0.16%
$5 \rightarrow 3$	−6.12%	3.00%	0.17%
Fashion-MNIST			
$1 \rightarrow 3$	−2.23%	0.25%	0.01%
$4 \rightarrow 6$	−9.96%	2.40%	0.09%
$6 \rightarrow 0$	−8.87%	2.59%	0.20%

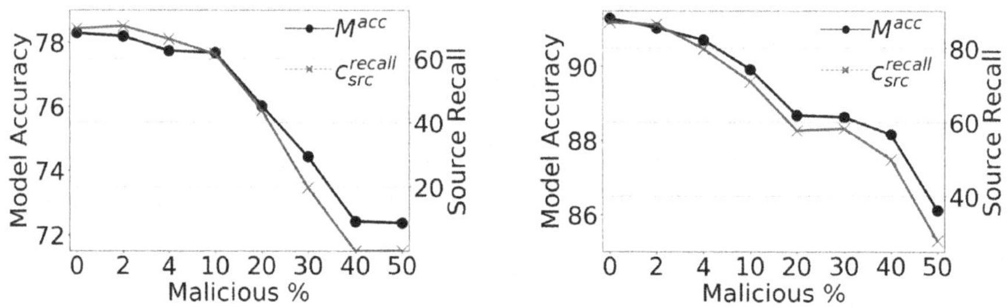

(a) CIFAR-10 (b) Fashion-MNIST

Fig. 2. Relationship between global model accuracy and source class recall across changing percentages of malicious participants for CIFAR-10 and Fashion-MNIST. As each dataset has 10 classes, the scale for M^{acc} vs c_{src}^{recall} is 1:10.

class recall, and loss in recall of all remaining classes. We observe that the attack causes substantial change in source class recall (> 6% drop in most cases) and target class recall. However, the attack impact on the recall of remaining classes is an order of magnitude smaller. CIFAR-10 experiments show a maximum of 0.34% change in class recalls attributable to non-source and non-target classes and Fashion-MNIST experiments similarly show a maximum change of 0.2% attributable to non-source and non-target classes, both of which are relatively minor compared to source and target classes. Thus, the attack is causing the global model to misclassify instances belonging to c_{src} as c_{target} at test time while other classes remain relatively unimpacted, demonstrating its targeted nature towards c_{src} and c_{target}. Considering the large impact of the attack on source class recall, changes in source class recall therefore make up the vast majority of the decreases in global model accuracy caused by label flipping attacks in FL systems. This observation can also be seen in Fig. 2 where the change in global model accuracy closely follows the change in source class recall.

The targeted nature of the label flipping attack allows for adversaries to remain under the radar in many FL systems. Consider systems where the data contain 100 classes or more, as is the case in CIFAR-100 [22] and ImageNet [13]. In such cases, targeted attacks become much more stealthy due to their limited impact to classes other than source and target.

3.3 Attack Timing in Label Flipping Attacks

While label flipping attacks can occur at any point in the learning process and last for arbitrary lengths, it is important to understand the capabilities of adversaries who are available for only part of the training process. For instance, Google's Gboard application of FL requires all participant devices be plugged into power and connected to the internet via WiFi [9]. Such requirements create cyclic conditions where many participants are not available during the day, when phones are not plugged in and are actively in use. Adversaries can take advantage of this design choice, making themselves available at times when honest participants are unable to.

We consider two scenarios in which the adversary is restricted in the time in which they are able to make malicious participants available: one in which the adversary makes malicious participants available only before the 75th training round, and one in which malicious participants are available only after the 75th training round. As the rate of global model accuracy improvement decreases with both datasets by training round 75, we choose this point to highlight how pre-established model stability may effect an adversary's ability to launch an effective label flipping attack. Results for the first scenario are given in Fig. 3 whereas the results for the second scenario are given in Fig. 4.

In Figure 3, we compare source class recall in a non-poisoned setting versus with poisoning only before round 75. Results on both CIFAR-10 and Fashion-MNIST show that while there are observable drops in source class recall during the rounds with poisoning (1–75), the global model is able to recover quickly after poisoning finishes (after round 75). Furthermore, the final convergence of the models (towards the end of training) are not impacted, given the models with and without poisoning are converge with roughly the same recall values. We do note that some CIFAR-10 experiments exhibited delayed convergence by an additional 50–100 training rounds, but these circumstances were rare and still eventually achieved the accuracy and recall levels of a non-poisoned model despite delayed convergence.

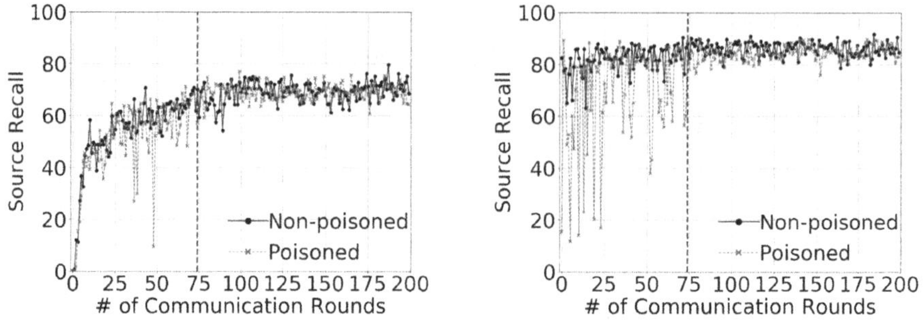

(a) CIFAR-10 (b) Fashion-MNIST

Fig. 3. Source class recall by round for experiments with "early round poisoning", i.e., malicious participation only in the first 75 rounds ($r < 75$). The blue line indicates the round at which malicious participation is no longer allowed.

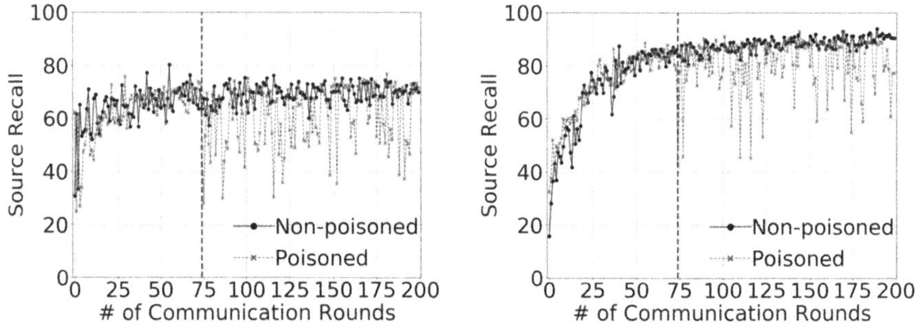

(a) CIFAR-10 (b) Fashion-MNIST

Fig. 4. Source class recall by round for experiments with "late round poisoning", i.e., malicious participation only after round 75 ($r \geq 75$). The blue line indicates the round at which malicious participation starts.

In Fig. 4, we compare source class recall in a non-poisoned setting versus with poisoning limited to the 75th and later training rounds. These results show the impact of such late poisoning demonstrating limited longevity; a phenomena which can be seen in the quick and dramatic changes in source class recall. Specifically, source class recall quickly returns to baseline levels once fewer malicious participants are selected in a training round even immediately

Table 4. Final source class recall when at least one malicious party participates in the final round R versus when all participants in round R are non-malicious. Results averaged for 10 runs for each experimental setting.

$c_{src} \rightarrow c_{target}$	Source Class Recall (c_{src}^{recall})	
	$m\% \in \mathcal{P}_R > 0$	$m\% \in \mathcal{P}_R = 0$
CIFAR-10		
$0 \rightarrow 2$	73.90%	82.45%
$1 \rightarrow 9$	77.30%	89.40%
$5 \rightarrow 3$	57.50%	73.10%
Fashion-MNIST		
$1 \rightarrow 3$	84.32%	96.25%
$4 \rightarrow 6$	51.50%	89.60%
$6 \rightarrow 0$	49.80%	73.15%

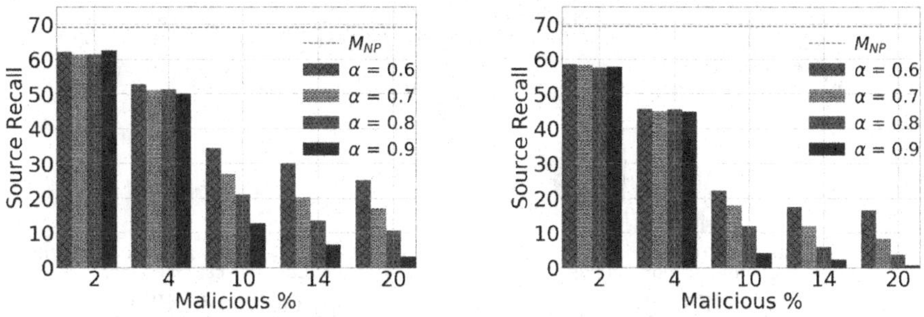

(a) CIFAR-10

(b) Fashion-MNIST

Fig. 5. Evaluation of impact from malicious participants' availability α on source class recall. Results are averaged from 3 runs for each setting.

following a round with a large number of malicious participants having caused a dramatic drop. However, the final poisoned model in the late-round poisoning scenario may show substantial difference in accuracy or recall compared to a non-poisoned model. This is evidenced by the CIFAR-10 experiment in Fig. 4, in which the source recall of the poisoned model is $\sim$10% lower compared to non-poisoned.

Furthermore, we observe that model convergence on both datasets is negatively impacted, as evidenced by the large variances in recall values between consecutive rounds. Consider Table 4 where results are compared when either (1) at least one malicious participant is selected for $\mathcal{P}_R$ or (2) $\mathcal{P}_R$ is made entirely of honest participants. When at least one malicious participant is selected, the final source class recall is, on average, 12.08% lower with the CIFAR-10 dataset and 24.46% lower with the Fashion-MNIST dataset. The utility impact from the label flipping attack is therefore predominantly tied to the number of malicious participants selected in the last few rounds of training.

3.4 Malicious Participant Availability

Given the impact of malicious participation in late training rounds on attack effectiveness, we now introduce a malicious participant availability parameter α. By varying α we can simulate the adversary's ability to control compromised participants' availability (i.e. ensuring connectivity or power access) at various points in training. Specifically, α represents malicious participants' availability and therefore likeliness to be selected relative to honest participants. For example, if $\alpha = 0.6$, when selecting each participant $P_i \in \mathcal{P}_r$ for round r, there is a 0.6 probability that P_i will be one of the malicious participants. Larger α implies higher likeliness of malicious participation. In cases where $k > N \times m\%$, the number of malicious participants in $\mathcal{P}_r$ is bounded by $N \times m\%$.

Figure 5 reports results for varying values of α in late round poisoning, i.e., malicious participation is limited to rounds $r \geq 75$. Specifically, we are interested in studying those scenarios where an adversary boosts the availability of

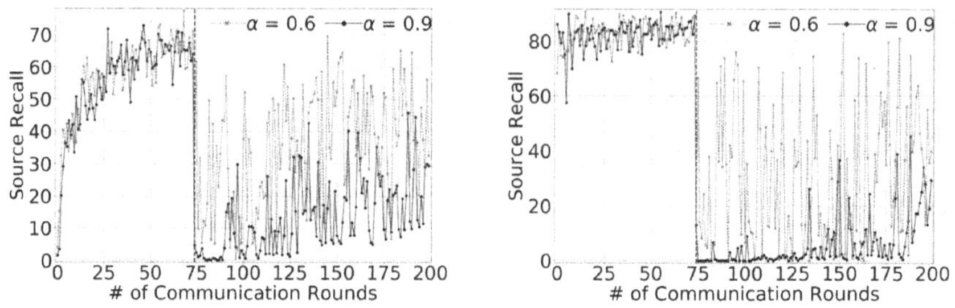

(a) CIFAR-10 (b) Fashion-MNIST

Fig. 6. Source class recall by round when malicious participants' availability is close to that of honest participants ($\alpha = 0.6$) vs significantly increased ($\alpha = 0.9$). The blue line indicates the round in which attack starts.

the malicious participants enough that their selection becomes more likely than the non-malicious participants, hence in Fig. 5 we use $\alpha \geq 0.6$. The reported source class recalls in Fig. 5 are averaged over the last 125 rounds (total 200 rounds minus first 75 rounds) to remove the impact of individual round variability; further, each experiment setting is repeated 3 times and results are averaged. The results show that, when the adversary maintains sufficient representation in the participant pool (i.e. $m \geq 10\%$), manipulating the availability of malicious participants can yield significantly higher impact on the global model utility with source class recall losses in excess of 20%. On both datasets with $m \geq 10\%$, the negative impact on source class recall is highest with $\alpha = 0.9$, which is followed by $\alpha = 0.8$, $\alpha = 0.7$ and $\alpha = 0.6$, i.e., in decreasing order of malicious participant availability. Thus, in order to mount an impactful attack, it is in the best interests of the adversary to perform the attack *with highest malicious participant availability in late rounds.* We note that when k is significantly larger than $N \times m\%$, increasing availability (α) will be insufficient for meaningfully increasing malicious participant selection in individual training rounds. Therefore, experiments where $m < 10\%$ show little variation despite changes in α.

To more acutely demonstrate the impact of α, Fig. 6 reports source class recall by round when $\alpha = 0.6$ and $\alpha = 0.9$ for both the CIFAR-10 and Fashion-MNIST datasets. In both datasets, when malicious participants are available more frequently, the source class recall is effectively shifted lower in the graph, i.e., source class recall values with $\alpha = 0.9$ are often much smaller than those with $\alpha = 0.6$. We note that the high round-by-round variance in both graphs is due to the probabilistic variability in number of malicious participants in individual training rounds. When fewer malicious participants are selected in one training round relative to the previous round, source recall increases. When more malicious participants are selected in an individual round relative to the previous round, source recall falls.

We further explore and illustrate our last remark with respect to the impact of malicious parties' participation in consecutive rounds in Fig. 7. In this figure, the x-axis represents the change in the number of malicious clients participating

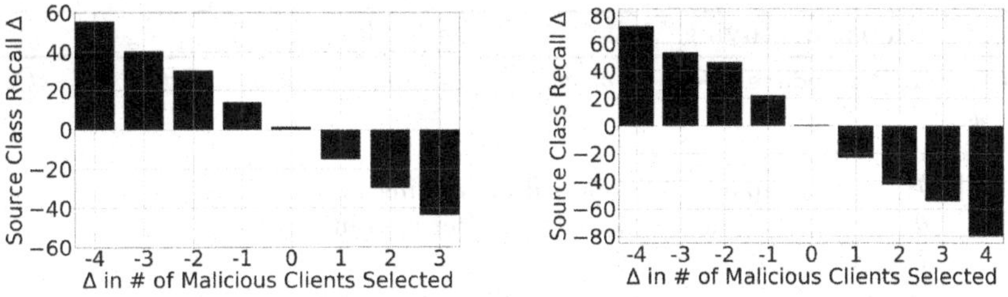

(a) CIFAR-10 (b) Fashion-MNIST

Fig. 7. Relationship between change in source class recall in consecutive rounds versus change in number of malicious participants in consecutive rounds. Specifically, $\forall r \geq 75$ the y-axis represents (c_{src}^{recall} @ round r) - (c_{src}^{recall} @ round $r-1$) while the x-axis represents (# of malicious $\in \mathcal{P}_r$) - (# of malicious $\in \mathcal{P}_{r-1}$).

in consecutive rounds, i.e., (# of malicious $\in \mathcal{P}_r$) − (# of malicious $\in \mathcal{P}_{r-1}$). The y-axis represents the change in source class recall between these consecutive rounds, i.e., (c_{src}^{recall} @ round r) − (c_{src}^{recall} @ round $r-1$). The reported results are then averaged across multiple runs of FL and all cases in which each participation difference was observed. The results confirm our intuition that, when $\mathcal{P}_r$ contains more malicious participants than $\mathcal{P}_{r-1}$, there is a substantial drop in source class recall. For large differences (such as +3 or +4), the drop could be as high as 40% or 60%. In contrast, when $\mathcal{P}_r$ contains fewer malicious participants than $\mathcal{P}_{r-1}$, there is a substantial increase in source class recall, which can be as high as 60% or 40% when the difference is −4 or −3. Altogether, this demonstrates the possibility that the DNN could recover significantly even in few rounds of FL training, if a large enough decrease in malicious participation could be achieved.

4 Defending Against Label Flipping Attacks

Given a highly effective adversary, how can a FL system defend against the label flipping attacks discussed thus far? To that end, we propose a defense which enables the aggregator to identify malicious participants.

After identifying malicious participants, the aggregator may blacklist them or ignore their updates $\theta_{r,i}$ in future rounds. We showed in Sects. 3.3 and 3.4 that high-utility model convergence can be eventually achieved after eliminating malicious participation. The feasibility of such a recovery from early round attacks supports use of the proposed identification approach as a defense strategy.

Our defense is based on the following insight: The parameter updates sent from malicious participants have unique characteristics compared to honest participants' updates for a subset of the parameter space. However, since DNNs have many parameters (i.e., $\theta_{r,i}$ is extremely high dimensional) it is non-trivial to analyze parameter updates by hand. Thus, we propose an automated strategy for identifying the relevant parameter subset and for studying participant updates using dimensionality reduction (PCA).

Algorithm 1: Identifying Malicious Model Updates in FL

def `evaluate_updates`($\mathcal{R}$: *set of vulnerable train rounds*, $\mathcal{P}$: *participant set*):

  $\mathcal{U} = \emptyset$

  for $r \in \mathcal{R}$ **do**

    $\mathcal{P}_r \leftarrow$ participants $\in \mathcal{P}$ queried in training round r

    $\theta_{r-1} \leftarrow$ global model parameters after training round $r-1$

    for $P_i \in \mathcal{P}_r$ **do**

      $\theta_{r,i} \leftarrow$ updated parameters after `train_DNN`(θ_{r-1}, D_i)

      $\theta_{\Delta,i} \leftarrow \theta_{r,i} - \theta_r$

      $\theta_{\Delta,i}^{src} \leftarrow$ parameters $\in \theta_{\Delta,i}$ connected to source class output node

      Add $\theta_{\Delta,i}^{src}$ to $\mathcal{U}$

  $\mathcal{U}' \leftarrow$ `standardize`($\mathcal{U}$)

  $\mathcal{U}'' \leftarrow$ `PCA`($\mathcal{U}'$, *components=2*)

  `plot`($\mathcal{U}''$)

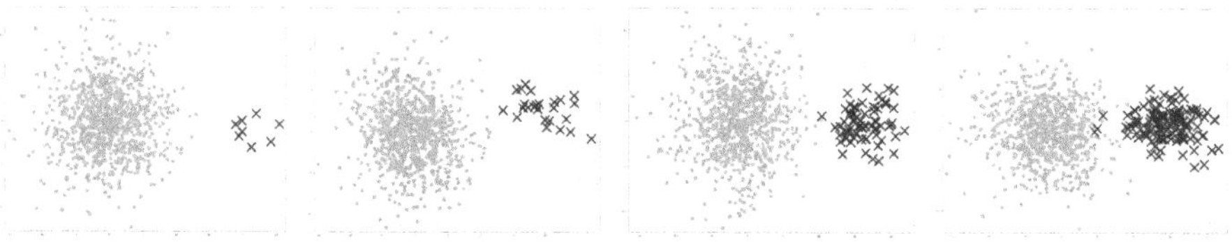

(a) CIFAR-10 m=2% (b) CIFAR-10 m=4% (c) CIFAR-10 m=10% (d) CIFAR-10 m=20%

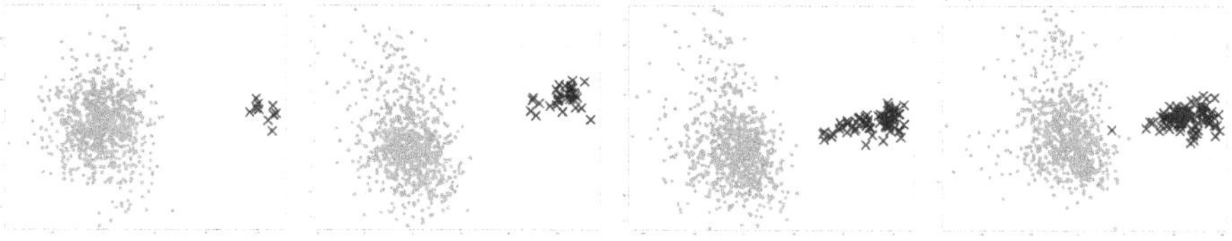

(e) F-MNIST m=2% (f) F-MNIST m=4% (g) F-MNIST m =10% (h) F-MNIST m=20%

Fig. 8. PCA plots with 2 components demonstrating the ability of Algorithm 1 to identify updates originating from a malicious versus honest participant. Plots represent relevant gradients collected from all training rounds $r > 10$. Blue Xs represent gradients from malicious participants while yellow Os represent gradients from honest participants. (Color figure online)

The description of our defense strategy is given in Algorithm 1. Let $\mathcal{R}$ denote the set of vulnerable FL training rounds and c_{src} be the class that is suspected to be the source class of a poisoning attack. We note that if c_{src} is unknown, the aggregator can defend against potential attacks such that $c_{src} = c \; \forall c \in \mathcal{C}$. We also note that for a given c_{src}, Algorithm 1 considers label flipping for all possible c_{target}. An aggregator therefore will conduct $|\mathcal{C}|$ independent iterations of Algorithm 1, which can be conducted in parallel. For each round $r \in \mathcal{R}$ and participant $P_i \in \mathcal{P}_r$, the aggregator computes the delta in participant's model update compared to the global model, i.e., $\theta_{\Delta,i} \leftarrow \theta_{r,i} - \theta_r$. Recall from Sect. 2.1 that a predicted probability for any given class c is computed by a specific node

n_c in the final layer DNN architecture. Given the aggregator's goal of defending against the label flipping attack from c_{src}, only the subset of the parameters in $\theta_{\Delta,i}$ corresponding to $n_{c_{src}}$ is extracted. The outcome of the extraction is denoted by $\theta^{src}_{\Delta,i}$ and added to a global list $\mathcal{U}$ built by the aggregator. After $\mathcal{U}$ is constructed across multiple rounds and participant deltas, it is standardized by removing the mean and scaling to unit variance. The standardized list $\mathcal{U}'$ is fed into Principal Component Analysis (PCA), which is a popular ML technique used for dimensionality reduction and pattern visualization. For ease of visualization, we use and plot results with two dimensions (two components).

In Fig. 8, we show the results of Algorithm 1 on CIFAR-10 and Fashion-MNIST across varying malicious participation rate m, with $\mathcal{R} = [10, 200]$. Even in scenarios with low m, as is shown in Figs. 8a and 8e, our defense is capable of differentiating between malicious and honest participants. In all graphs, the PCA outcome shows that malicious participants' updates belong to a visibly different cluster compared to honest participants' updates which form their own cluster. Another interesting observation is that our defense does not suffer from the "gradient drift" problem. Gradient drift is a potential challenge in designing a robust defense, since changes in model updates may be caused both by actual DNN learning and convergence (which is desirable) or malicious poisoning attempt (which our defense is trying to identify and prevent). Our results show that, even though the defense is tested with a long period of rounds (190 training rounds since $\mathcal{R} = [10, 200]$), it remains capable of separating malicious and honest participants, demonstrating its robustness to gradient drift.

A FL system aggregator can therefore effectively identify malicious participants, and consequently restrict their participation in mobile training, by conducting such gradient clustering prior to aggregating parameter updates at each round. Clustering model gradients for malicious participant identification presents a strong defense as it does not require access to any public validation dataset, as is required in [3], which is not necessarily possible to acquire.

5 Related Work

Poisoning attacks are highly relevant in domains such as spam filtering [10,32], malware and network anomaly detection [11,24,39], disease diagnosis [29], computer vision [34], and recommender systems [15,54]. Several poisoning attacks were developed for popular ML models including SVM [6,12,44,45,50,52], regression [19], dimensionality reduction [51], linear classifiers [12,23,57], unsupervised learning [7], and more recently, neural networks [12,30,42,45,53,58]. However, most of the existing work is concerned with poisoning ML models in the traditional setting where training data is first collected by a centralized party. In contrast, our work studies poisoning attacks in the context of FL. As a result, many of the poisoning attacks and defenses that were designed for traditional ML are not suitable to FL. For example, attacks that rely on crafting optimal poison instances by observing the training data distribution are inapplicable since the malicious FL participant may only access and modify the training data s/he holds. Similarly, server-side defenses that rely on filtering and eliminating poison instances through anomaly detection or k-NN [36,37] are inapplicable to FL since the server only observes parameter updates from FL participants, not their individual instances.

The rising popularity of FL has led to the investigation of different attacks in the context of FL, such as backdoor attacks [2,46], gradient leakage attacks [18,27,59] and membership inference attacks [31,47,48]. Most closely related to our work are poisoning attacks in FL. There are two types of poisoning attacks in FL: *data poisoning* and *model poisoning*. Our work falls under the data poisoning category. In data poisoning, a malicious FL participant manipulates their training data, e.g., by adding poison instances or adversarially changing existing instances [16,43]. The local learning process is otherwise not modified. In model poisoning, the malicious FL participant modifies its learning process in order to create adversarial gradients and parameter updates. [4] and [14] demonstrated the possibility of causing high model error rates through targeted and untargeted model poisoning attacks. While model poisoning is also effective, data poisoning may be preferable or more convenient in certain scenarios, since it does not require adversarial tampering of model learning software on participant devices, it is efficient, and it allows for non-expert poisoning participants.

Finally, FL poisoning attacks have connections to the concept of *Byzantine threats*, in which one or more participants in a distributed system fail or misbehave. In FL, Byzantine behavior was shown to lead to sub-optimal models or non-convergence [8,20]. This has spurred a line of work on Byzantine-resilient aggregation for distributed learning, such as Krum [8], Bulyan [28], trimmed mean, and coordinate-wise median [55]. While model poisoning may remain successful despite Byzantine-resilient aggregation [4,14,20], it is unclear whether optimal data poisoning attacks can be found to circumvent an individual Byzantine-resilient scheme, or whether one data poisoning attack may circumvent multiple Byzantine-resilient schemes. We plan to investigate these issues in future work.

6 Conclusion

In this paper we studied data poisoning attacks against FL systems. We demonstrated that FL systems are vulnerable to label flipping poisoning attacks and that these attacks can significantly negatively impact the global model. We also showed that the negative impact on the global model increases as the proportion of malicious participants increases, and that it is possible to achieve targeted poisoning impact. Further, we demonstrated that adversaries can enhance attack effectiveness by increasing the availability of malicious participants in later rounds. Finally, we proposed a defense which helps an FL aggregator separate malicious from honest participants. We showed that our defense is capable of identifying malicious participants and it is robust to gradient drift.

As poisoning attacks against FL systems continue to emerge as important research topics in the security and ML communities [4,14,21,33,56], we plan to continue our work in several ways. First, we will study the impacts of the attack and defense on diverse FL scenarios differing in terms of data size, distribution among FL participants (iid vs non-iid), data type, total number of instances available per class, etc. Second, we will study more complex adversarial behaviors such as each malicious participant changing the labels of only a small portion of source samples or using more sophisticated poisoning strategies to avoid being detected . Third, while we designed and tested our defense against the label flipping attack, we hypothesize the defense will be useful against model poisoning attacks since malicious participants' gradients are often dissimilar to those of honest participants. Since our defense identifies dissimilar or anomalous gradients, we expect the defense to be effective against other types of FL attacks that cause dissimilar or anomalous gradients. In future work, we will study the applicability of our defense against such other FL attacks including model poisoning, untargeted poisoning, and backdoor attacks.

Acknowledgements. This research is partially sponsored by NSF CISE SaTC 1564097 and 2038029. The second author acknowledges an IBM PhD Fellowship Award and the support from the Enterprise AI, Systems & Solutions division led by Sandeep Gopisetty at IBM Almaden Research Center. Any opinions, findings, and conclusions or recommendations expressed in this material are those of the author(s) and do not necessarily reflect the views of the National Science Foundation or other funding agencies and companies mentioned above.

A DNN Architectures and Configuration

All NNs were trained using PyTorch version 1.2.0 with random weight initialization. Training and testing was completed using a NVIDIA 980 Ti GPU-accelerator. When necessary, all CUDA tensors were mapped to CPU tensors before exporting to Numpy arrays. Default drivers provided by Ubuntu 19.04 and built-in GPU support in PyTorch was used to accelerate training. Details can be found in our repository: https://github.com/git-disl/DataPoisoning_FL.

Fashion-MNIST: We do not conduct data pre-processing. We use a Convolutional Neural Network with the architecture described in Table 6. In the table, Conv = Convolutional Layer, and Batch Norm = Batch Normalization.

CIFAR-10: We conduct data pre-processing prior to training. Data is normalized with mean [0.485, 0.456, 0.406] and standard deviation [0.229, 0.224, 0.225]. Values reflect mean and standard deviation of the ImageNet dataset [13] and are commonplace, even expected when using Torchvision [25] models. We additionally perform data augmentation with random horizontal flipping, random cropping with size 32, and default padding. Our CNN is detailed in Table 5.

Table 5. CIFAR-10 CNN.

Layer type	Size
Conv + ReLu + Batch Norm	$3 \times 3 \times 32$
Conv + ReLu + Batch Norm	$3 \times 32 \times 32$
Max Pooling	2×2
Conv + ReLu + Batch Norm	$3 \times 32 \times 64$
Conv + ReLu + Batch Norm	$3 \times 64 \times 64$
Max Pooling	2×2
Conv + ReLu + Batch Norm	$3 \times 64 \times 128$
Conv + ReLu + Batch Norm	$3 \times 128 \times 128$
Max Pooling	2×2
Fully Connected	2048
Fully Connected + Softmax	128/10

Table 6. Fashion-MNIST CNN.

Layer type	Size
Conv + ReLu + Batch Norm	$5 \times 1 \times 16$
Max Pooling	2×2
Conv + ReLu + Batch Norm	$5 \times 16 \times 32$
Max Pooling	2×2
Fully Connected	1568/10

References

1. An Act: Health insurance portability and accountability act of 1996. Public Law 104-191 (1996)
2. Bagdasaryan, E., Veit, A., Hua, Y., Estrin, D., Shmatikov, V.: How to backdoor federated learning. arXiv preprint arXiv:1807.00459 (2018)
3. Baracaldo, N., Chen, B., Ludwig, H., Safavi, J.A.: Mitigating poisoning attacks on machine learning models: a data provenance based approach. In: 10th ACM Workshop on Artificial Intelligence and Security, pp. 103–110 (2017)
4. Bhagoji, A.N., Chakraborty, S., Mittal, P., Calo, S.: Analyzing federated learning through an adversarial lens. In: International Conference on Machine Learning, pp. 634–643 (2019)
5. Biggio, B., Nelson, B., Laskov, P.: Support vector machines under adversarial label noise. In: Asian Conference on Machine Learning, pp. 97–112 (2011)
6. Biggio, B., Nelson, B., Laskov, P.: Poisoning attacks against support vector machines. In: Proceedings of the 29th International Conference on International Conference on Machine Learning, pp. 1467–1474 (2012)
7. Biggio, B., Pillai, I., Rota Bulò, S., Ariu, D., Pelillo, M., Roli, F.: Is data clustering in adversarial settings secure? In: Proceedings of the 2013 ACM Workshop on Artificial Intelligence and Security, pp. 87–98 (2013)
8. Blanchard, P., Guerraoui, R., Stainer, J., et al.: Machine learning with adversaries: byzantine tolerant gradient descent. In: NeurIPS, pp. 119–129 (2017)
9. Bonawitz, K., et al.: Towards federated learning at scale: System design. In: SysML 2019 (2019, to appear). https://arxiv.org/abs/1902.01046
10. Bursztein, E.: Attacks against machine learning - an overview (2018). https://elie.net/blog/ai/attacks-against-machine-learning-an-overview/
11. Chen, S., et al.: Automated poisoning attacks and defenses in malware detection systems: an adversarial machine learning approach. Comput. Secur. **73**, 326–344 (2018)
12. Demontis, A., et al.: Why do adversarial attacks transfer? Explaining transferability of evasion and poisoning attacks. In: 28th USENIX Security Symposium, pp. 321–338 (2019)
13. Deng, J., Dong, W., Socher, R., Li, L.J., Li, K., Fei-Fei, L.: ImageNet: a large-scale hierarchical image database. In: 2009 IEEE Conference on Computer Vision and Pattern Recognition, pp. 248–255. IEEE (2009)

14. Fang, M., Cao, X., Jia, J., Gong, N.Z.: Local model poisoning attacks to byzantine-robust federated learning. In: USENIX Security Symposium (2020, to appear)
15. Fang, M., Yang, G., Gong, N.Z., Liu, J.: Poisoning attacks to graph-based recommender systems. In: Proceedings of the 34th Annual Computer Security Applications Conference, pp. 381–392 (2018)
16. Fung, C., Yoon, C.J., Beschastnikh, I.: Mitigating sybils in federated learning poisoning. arXiv preprint arXiv:1808.04866 (2018)
17. Hard, A., et al.: Federated learning for mobile keyboard prediction. arXiv preprint arXiv:1811.03604 (2018)
18. Hitaj, B., Ateniese, G., Perez-Cruz, F.: Deep models under the GAN: information leakage from collaborative deep learning. In: Proceedings of the 2017 ACM SIGSAC Conference on Computer and Communications Security, pp. 603–618 (2017)
19. Jagielski, M., Oprea, A., Biggio, B., Liu, C., Nita-Rotaru, C., Li, B.: Manipulating machine learning: poisoning attacks and countermeasures for regression learning. In: 2018 IEEE Symposium on Security and Privacy (SP), pp. 19–35. IEEE (2018)
20. Kairouz, P., et al.: Advances and open problems in federated learning. arXiv preprint arXiv:1912.04977 (2019)
21. Khazbak, Y., Tan, T., Cao, G.: MLGuard: mitigating poisoning attacks in privacy preserving distributed collaborative learning (2020)
22. Krizhevsky, A., Hinton, G., et al.: Learning multiple layers of features from tiny images (2009)
23. Liu, C., Li, B., Vorobeychik, Y., Oprea, A.: Robust linear regression against training data poisoning. In: Proceedings of the 10th ACM Workshop on Artificial Intelligence and Security, pp. 91–102 (2017)
24. Maiorca, D., Biggio, B., Giacinto, G.: Towards adversarial malware detection: lessons learned from pdf-based attacks. ACM Comput. Surv. (CSUR) **52**(4), 1–36 (2019)
25. Marcel, S., Rodriguez, Y.: Torchvision the machine-vision package of torch. In: 18th ACM International Conference on Multimedia, pp. 1485–1488 (2010)
26. Mathews, K., Bowman, C.: The California consumer privacy act of 2018 (2018)
27. Melis, L., Song, C., De Cristofaro, E., Shmatikov, V.: Exploiting unintended feature leakage in collaborative learning. In: 2019 IEEE Symposium on Security and Privacy (SP), pp. 691–706. IEEE (2019)
28. Mhamdi, E.M.E., Guerraoui, R., Rouault, S.: The hidden vulnerability of distributed learning in byzantium. arXiv preprint arXiv:1802.07927 (2018)
29. Mozaffari-Kermani, M., Sur-Kolay, S., Raghunathan, A., Jha, N.K.: Systematic poisoning attacks on and defenses for machine learning in healthcare. IEEE J. Biomed. Health Inform. **19**(6), 1893–1905 (2014)
30. Muñoz-González, L., et al.: Towards poisoning of deep learning algorithms with back-gradient optimization. In: Proceedings of the 10th ACM Workshop on Artificial Intelligence and Security, pp. 27–38 (2017)
31. Nasr, M., Shokri, R., Houmansadr, A.: Comprehensive privacy analysis of deep learning: passive and active white-box inference attacks against centralized and federated learning. In: 2019 IEEE Symposium on Security and Privacy (SP), pp. 739–753. IEEE (2019)
32. Nelson, B., et al.: Exploiting machine learning to subvert your spam filter. LEET **8**, 1–9 (2008)
33. Nguyen, T.D., Rieger, P., Miettinen, M., Sadeghi, A.R.: Poisoning attacks on federated learning-based IoT intrusion detection system (2020)

34. Papernot, N., McDaniel, P., Sinha, A., Wellman, M.P.: SoK: security and privacy in machine learning. In: 2018 IEEE European Symposium on Security and Privacy (EuroS&P), pp. 399–414. IEEE (2018)
35. Paszke, A., et al.: PyTorch: an imperative style, high-performance deep learning library. In: NeurIPS, pp. 8024–8035 (2019)
36. Paudice, A., Muñoz-González, L., Gyorgy, A., Lupu, E.C.: Detection of adversarial training examples in poisoning attacks through anomaly detection. arXiv preprint arXiv:1802.03041 (2018)
37. Paudice, A., Muñoz-González, L., Lupu, E.C.: Label sanitization against label flipping poisoning attacks. In: Alzate, C., et al. (eds.) ECML PKDD 2018. LNCS (LNAI), vol. 11329, pp. 5–15. Springer, Cham (2019). https://doi.org/10.1007/978-3-030-13453-2_1
38. General Data Protection Regulation: Regulation (EU) 2016/679 of the European parliament and of the council of 27 April 2016 on the protection of natural persons with regard to the processing of personal data and on the free movement of such data, and repealing directive 95/46. Off. J. Eur. Union (OJ) **59**(1–88), 294 (2016)
39. Rubinstein, B.I., et al.: Antidote: understanding and defending against poisoning of anomaly detectors. In: Proceedings of the 9th ACM SIGCOMM Conference on Internet Measurement, pp. 1–14 (2009)
40. Ryffel, T., et al.: A generic framework for privacy preserving deep learning. arXiv preprint arXiv:1811.04017 (2018)
41. Schlesinger, A., O'Hara, K.P., Taylor, A.S.: Let's talk about race: identity, chatbots, and AI. In: Proceedings of the 2018 CHI Conference on Human Factors in Computing Systems, pp. 1–14 (2018)
42. Shafahi, A., et al.: Poison frogs! Targeted clean-label poisoning attacks on neural networks. In: Advances in Neural Information Processing Systems, pp. 6103–6113 (2018)
43. Shen, S., Tople, S., Saxena, P.: Auror: Defending against poisoning attacks in collaborative deep learning systems. In: Proceedings of the 32nd Annual Conference on Computer Security Applications, pp. 508–519 (2016)
44. Steinhardt, J., Koh, P.W.W., Liang, P.S.: Certified defenses for data poisoning attacks. In: NeurIPS, pp. 3517–3529 (2017)
45. Suciu, O., Marginean, R., Kaya, Y., Daume III, H., Dumitras, T.: When does machine learning fail? Generalized transferability for evasion and poisoning attacks. In: 27th USENIX Security Symposium, pp. 1299–1316 (2018)
46. Sun, Z., Kairouz, P., Suresh, A.T., McMahan, H.B.: Can you really backdoor federated learning? arXiv preprint arXiv:1911.07963 (2019)
47. Truex, S., Liu, L., Gursoy, M.E., Yu, L., Wei, W.: Towards demystifying membership inference attacks. arXiv preprint arXiv:1807.09173 (2018)
48. Truex, S., Liu, L., Gursoy, M.E., Yu, L., Wei, W.: Demystifying membership inference attacks in machine learning as a service. IEEE Trans. Serv. Comput. (2019)
49. Xiao, H., Rasul, K., Vollgraf, R.: Fashion-MNIST: a novel image dataset for benchmarking machine learning algorithms. arXiv preprint arXiv:1708.07747 (2017)
50. Xiao, H., Xiao, H., Eckert, C.: Adversarial label flips attack on support vector machines. In: ECAI, pp. 870–875 (2012)
51. Xiao, H., Biggio, B., Brown, G., Fumera, G., Eckert, C., Roli, F.: Is feature selection secure against training data poisoning? In: International Conference on Machine Learning, pp. 1689–1698 (2015)
52. Xiao, H., Biggio, B., Nelson, B., Xiao, H., Eckert, C., Roli, F.: Support vector machines under adversarial label contamination. Neurocomputing **160**, 53–62 (2015)

53. Yang, C., Wu, Q., Li, H., Chen, Y.: Generative poisoning attack method against neural networks. arXiv preprint arXiv:1703.01340 (2017)
54. Yang, G., Gong, N.Z., Cai, Y.: Fake co-visitation injection attacks to recommender systems. In: NDSS (2017)
55. Yin, D., Chen, Y., Kannan, R., Bartlett, P.: Byzantine-robust distributed learning: towards optimal statistical rates. In: International Conference on Machine Learning, pp. 5650–5659 (2018)
56. Zhao, L., et al.: Shielding collaborative learning: mitigating poisoning attacks through client-side detection. IEEE Trans. Dependable Secure Comput. (2020)
57. Zhao, M., An, B., Gao, W., Zhang, T.: Efficient label contamination attacks against black-box learning models. In: IJCAI, pp. 3945–3951 (2017)
58. Zhu, C., Huang, W.R., Li, H., Taylor, G., Studer, C., Goldstein, T.: Transferable clean-label poisoning attacks on deep neural nets. In: International Conference on Machine Learning, pp. 7614–7623 (2019)
59. Zhu, L., Liu, Z., Han, S.: Deep leakage from gradients. In: Advances in Neural Information Processing Systems, pp. 14747–14756 (2019)

Interpretable Probabilistic Password Strength Meters via Deep Learning

Dario Pasquini[1,2,3(✉)], Giuseppe Ateniese[1], and Massimo Bernaschi[3]

[1] Stevens Institute of Technology, Hoboken, USA
{dpasquin,gatenies}@stevens.edu
[2] Sapienza University of Rome, Rome, Italy
[3] Institute of Applied Computing, CNR, Rome, Italy
massimo.bernaschi@cnr.it

Abstract. Probabilistic password strength meters have been proved to be the most accurate tools to measure password strength. Unfortunately, by construction, they are limited to solely produce an opaque security estimation that fails to fully support the user during the password composition. In the present work, we move the first steps towards cracking the intelligibility barrier of this compelling class of meters. We show that probabilistic password meters inherently own the capability to describe the latent relation between password strength and password structure. In our approach, the security contribution of each character composing a password is disentangled and used to provide explicit fine-grained feedback for the user. Furthermore, unlike existing heuristic constructions, our method is free from any human bias, and, more importantly, its feedback has a clear probabilistic interpretation.

In our contribution: (1) we formulate the theoretical foundations of interpretable probabilistic password strength meters; (2) we describe how they can be implemented via an efficient and lightweight deep learning framework suitable for client-side operability.

Keywords: Password security · Strength meters · Deep learning

1 Introduction

Accurately measuring password strength is essential to guarantee the security of password-based authentication systems. Even more critical, however, is training users to select secure passwords in the first place. One common approach is to rely on password policies that list a series of requirements for a strong password. This approach is limited or even harmful [10]. Alternatively, Passwords Strength Meters (PSMs) have been shown to be useful and are witnessing increasing adoption in commercial solutions [15,26].

The first instantiations of PSMs were based on simple heuristic constructions. Password strength was estimated via either handcrafted features such as *LUDS* (which counts lower and uppercase letters, digits, and symbols) or heuristic

L. Chen et al. (Eds.): ESORICS 2020, LNCS 12308, pp. 502–522, 2020.
https://doi.org/10.1007/978-3-030-58951-6_25

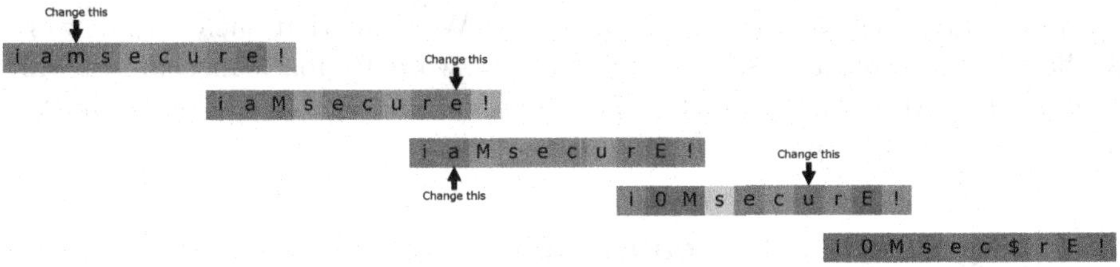

Fig. 1. Example of the character-level feedback mechanism and password composition process induced by our meter. In the figure, "*iamsecure!*" is the password initially chosen by the user. Colors indicate the estimated character security: red (insecure) → green (secure). (Color figure online)

entropy definitions. Unavoidably, given their heuristic nature, this class of PSMs failed to accurately measure password security [11,30].

More recently, thanks to an active academic interest, PSMs based on more sound constructions and rigorous security definitions have been proposed. In the last decade, indeed, a considerable research effort gave rise to more precise meters capable of accurately measuring password strength [9,20,28].

However, meters have also become proportionally more opaque and inherently hard to interpret due to the increasing complexity of the employed approaches. State-of-art solutions base their estimates on blackbox parametric probabilistic models [9,20] that leave no room for interpretation of the evaluated passwords; they do not provide any feedback to users on what is wrong with their password or how to improve it. We advocate for explainable approaches in password meters, where users receive additional insights and become cognizant of which parts of their passwords could straightforwardly improve. This makes the password selection process less painful since users can keep their passwords of choice mostly unchanged while ensuring they are secure.

In the present work, we show that the same rigorous probabilistic framework capable of accurately measuring password strength can also fundamentally describe the relation between password security and password structure. By rethinking the underlying mass estimation process, we create the first *interpretable probabilistic password strength* meter. Here, the password probability measured by our meter can be decomposed and used to estimate further the strength of every single character of the password. This explainable approach allows us to assign a security score to each atomic component of the password and determine its contribution to the overall security strength. This evaluation is, in turn, returned to the user who can tweak a few "weak" characters and consistently improve the password strength against guessing attacks. Figure 1 illustrates the selection process. In devising the proposed mass estimation process, we found it ideally suited for being implemented via a deep learning architecture. In the paper, we show how that can be cast as an efficient client-side meter employing deep convolutional neural networks. Our work's major contributions are: (i) We formulate a novel password probability estimation framework based on undirected probabilistic models. (ii) We show that such a framework can be used to build a precise and sound password feedback mechanism. (iii) We implement the proposed meter via an efficient and lightweight deep learning framework ideally suited for client-side operability.

2 Background and Preliminaries

In this section, we offer an overview of the fundamental concepts that are important to understand our contribution. Section 2.1 covers Probabilistic Password Strength Meters. Next, in Sect. 2.2, we cover structured probabilistic models that will be fundamental in the interpretation of our approach. Finally, Sect. 2.3 briefly discusses relevant previous works within the PSMs context.

2.1 Probabilistic Password Strength Meters (PPSMs)

Probabilistic password strength meters are PSMs that base their strength measure on an explicit estimate of password probability. In the process, they resort to probabilistic models to approximate the probability distribution behind a set of known passwords, typically, instances of a password leak. Having an approximation of the mass function, strength estimation is then derived by leveraging adversarial reasoning. Here, password robustness is estimated in consideration of an attacker who knows the underlying password distribution, and that aims at minimizing the guess entropy [19] of her/his guessing attack. To that purpose, the attacker performs an optimal guessing attack, where guesses are issued in decreasing probability order (i.e., high-probability passwords first). More formally, given a probability mass function $P(\mathbf{x})$ defined on the key-space $\mathbb{X}$, the attacker creates an ordering $\mathbb{X}_{P(\mathbf{x})}$ of $\mathbb{X}$ such that:

$$\mathbb{X}_{P(\mathbf{x})} = [x^0, x^1, \dots, x^n] \quad \text{where} \quad \forall_{i \in [0,n]} : P(x^i) \geq P(x^{i+1}). \tag{1}$$

During the attack, the adversary produces guesses by traversing the list $\mathbb{X}_{P(\mathbf{x})}$. Under this adversarial model, passwords with high probability are considered weak, as they will be quickly guessed. Low-probability passwords, instead, are assessed as secure, as they will be matched by the attacker only after a considerable, possibly not feasible, number of guesses.

2.2 Structured Probabilistic Models

Generally, the probabilistic models used by PPSMs are **probabilistic structured models** (even known as graphical models). These describe password distributions by leveraging a **graph notation** to illustrate the dependency properties among a set of random variables. Here, a random variable $\mathbf{x}_i$ is depicted as a vertex, and an edge between $\mathbf{x}_i$ and $\mathbf{x}_j$ exists whether $\mathbf{x}_i$ and $\mathbf{x}_j$ are statistically dependent. Structured probabilistic models are classified according to the orientation of edges. A direct acyclic graph (DAG) defines a **directed graphical**

model (or Bayesian Network). In this formalism, an edge asserts a cause-effect relationship between two variables; that is, the state assumed from the variable x_i is intended as a direct consequence of those assumed by its parents $par(\mathbf{x}_i)$ in the graph. Under such a description, a topological ordering among all the random variables can be asserted and used to factorize the joint probability distribution of the random variables effectively. On the other hand, an undirected graph defines an **undirected graphical model**, also known as Markov Random Field (MRF). In this description, the causality interpretation of edges is relaxed, and connected variables influence each other symmetrically. Undirected models permit the formalization of stochastic processes where causality among factors cannot be asserted. However, this comes at the cost of giving up to any simple form of factorization of the joint distribution.

2.3 Related Works

Here, we briefly review early approaches to the definition of PSMs. We focus on the most influential works as well as to the ones most related to ours.

Probabilistic PSMs: Originally thought for guessing attacks [21], Markov model approaches have found natural application in the password strength estimation context. Castelluccia et al. [9] use a stationary, finite-state Markov chain as a direct password mass estimator. Their model computes the joint probability by separately measuring the conditional probability of each pair of n-grams in the observed passwords. Melicher et al. [20] extended the Markov model approach by leveraging a character/token level Recurrent Neural Network (RNN) for modeling the probability of passwords. As discussed in the introduction, pure probabilistic approaches are not capable of any natural form of feedback. In order to partially cope with this shortcoming, a hybrid approach has been investigated in [24]. Here, the model of Melicher et al. [20] is aggregated with a series of 21 heuristic, hand-crafted feedback mechanisms such as detection of *leeting behaviors* or common tokens (e.g., keyboard walks).

Even if harnessing a consistently different form of feedback, our framework merges these solutions into a single and jointly learned model. Additionally, in contrast with [24], our feedback has a concrete probabilistic interpretation as well as complete freedom from any form of human bias. Interestingly enough, our model autonomously learns some of the heuristics hardwired in [24]. For instance, our model learned that capitalizing characters in the middle of the string could consistently improve password strength.

Token Look-Up PSMs: Another relevant class of meters is that based on the token look-up approach. Generally speaking, these are non-parametric solutions that base their strength estimation on collections of sorted lists of tokens like leaked passwords and word dictionaries. Here, a password is modeled as a combination of tokens, and the relative security score is derived from the ranking of the tokens in the known dictionaries. Unlike probabilistic solutions, token-based PSMs are able to return feedback to the user, such as an explanation for the

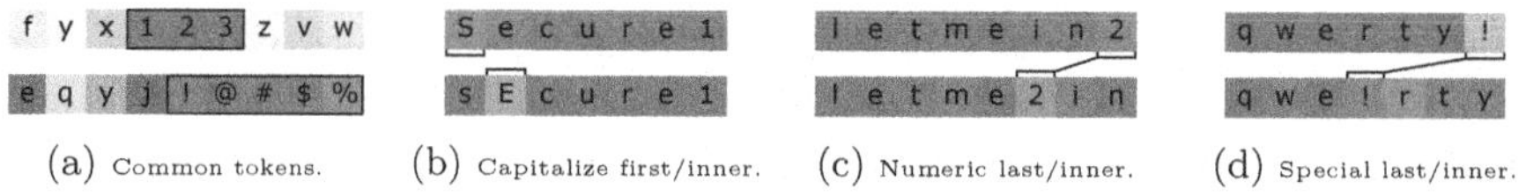
f y x 1 2 3 z v w
e q y j ! @ # $ %
S e c u r e 1
s E c u r e 1
l e t m e i n 2
l e t m e 2 i n
q w e r t y !
q w e ! r t y
(a) Common tokens.
(b) Capitalize first/inner.
(c) Numeric last/inner.
(d) Special last/inner.

Fig. 2. In Panel (a), the model automatically highlights the presence of weak substrings by assigning high probabilities to the characters composing them. Panels (b), (c), and (d) are examples of self-learned weak/strong password composition patterns. In panel (b), the model assigns a high probability to the capitalization of the first letter (a common practice), whereas it assigns low probability when the capitalization is performed on inner characters. Panel (c) and (d) report similar results for numeric and special characters. (Color figure online)

weakness of a password relying on the semantic attributed to the tokens composing the password. A leading member of token look-up meters is *zxcvbn* [31], which assumes a password as a combination of tokens such as *token, reversed, sequence repeat, keyboard, and date.* This meter scores passwords according to a heuristic characterization of the guess-number [19]. Such score is described as the number of combinations of tokens necessary to match the tested password by traversing the sorted tokens lists.

zxcvbn is capable of feedback. For instance, if one of the password components is identified as *"repeat"*, *zxcvbn* will recommend the user to avoid the use of repeated characters in the password. Naturally, this kind of feedback mechanism inherently lacks generality and addresses just a few human-chosen scenarios. As discussed by the authors themselves, *zxcvbn* suffers from various limitations. By assumption, it is unable to model the relationships among different patterns occurring in the same passwords. Additionally, like other token look-up based approaches, it fails to coherently model unobserved patterns and tokens.

3 Meter Foundations

In this section, we introduce the theoretical foundations of the proposed estimation process. First, in Sect. 3.1, we introduce and motivate the probabilistic character-level feedback mechanism. Later, in Sect. 3.2, we describe how that mechanism can be obtained using undirected probabilistic models.

3.1 Character-Level Strength Estimation via Probabilistic Reasoning

As introduced in Sect. 2.1, PPSMs employ probabilistic models to approximate the probability mass function of an observed password distribution, say $P(\mathbf{x})$. Estimating $P(\mathbf{x})$, however, could be particularly challenging, and suitable estimation techniques must be adopted to make the process feasible. In this direction, a general solution is to factorize the domain of the mass function (i.e., the key-space); that is, passwords are modeled as a concatenation of smaller factors, typically, decomposed at the character level. Afterward, password distribution is estimated by modeling stochastic interactions among these simpler components. More formally, every password is assumed as a realization $x = [x_1, \ldots, x_\ell]$ of a random vector of the kind $\mathbf{x} = [\mathbf{x}_1, \ldots, \mathbf{x}_\ell]$, where each disjoint random variable $\mathbf{x}_i$ represents the character at position i in the string. Then, $P(\mathbf{x})$ is described through **structured probabilistic models** that formalize the relations among those random variables, eventually defining a joint probability distribution. In the process, every random variable is associated with a **local conditional probability distribution** (here, referred to as Q) that describes the stochastic behavior of $\mathbf{x}_i$ in consideration of the conditional independence properties asserted from the underlying structured model i.e., $Q(\mathbf{x}_i) = P(\mathbf{x}_i \mid par(\mathbf{x}_i))$. Eventually, the joint measurement of probability is derived from the aggregation of the marginalized local conditional probability distributions, typically under the form $P(\mathbf{x}) = \prod_{i=1}^{\ell} Q(\mathbf{x}_i = x_i)$.

As introduced in Sect. 2.1, the joint probability can be employed as a good representative for password strength. However, such a global assessment unavoidably hides much fine-grained information that can be extremely valuable to a password meter. In particular, the joint probability offers us an atomic interpretation of the password strength, but it fails at disentangling the relation between password strength and password structure. That is, it does not clarify which factors of an evaluated password are making that password insecure. However, as widely demonstrated by non-probabilistic approaches [17,24,31], users benefit from the awareness of which part of the chosen password is easily predictable and which is not. In this direction, we argue that the **local conditional probabilities** that naturally appear in the estimation of the joint one, if correctly shaped, can offer detailed insights into the strength or the weakness of each factor of a password. **Such character-level probability assignments are an explicit interpretation of the relation between the structure of a password and its security.** The main intuition here is that: high values of $Q(x_i)$ tell us that x_i (i.e., the character at position i in the string) has a high impact on increasing the password probability and must be changed to make the password stronger. Instead, characters with low conditional probability are pushing the password to have low probability and must be maintained unchanged. Figure 2 reports some visual representations of such probabilistic reasoning. Each segment's background color renders the value of the local conditional probability of the character. Red describes high probability values, whereas green describes low probability assignments. Such a mechanism can naturally discover weak passwords components and explicitly guide the user to explore alternatives. For instance, local conditional probabilities can spot the presence of predictable tokens in the password without the explicit use of dictionaries (Fig. 2a). These measurements are able to automatically describe common password patterns like those manually modeled from other approaches [24], see Figs. 2b, 2c and 2d. More importantly, they can potentially describe latent composition patterns that have never been observed and modeled by human beings. In doing this, neither supervision nor human-reasoning is required.

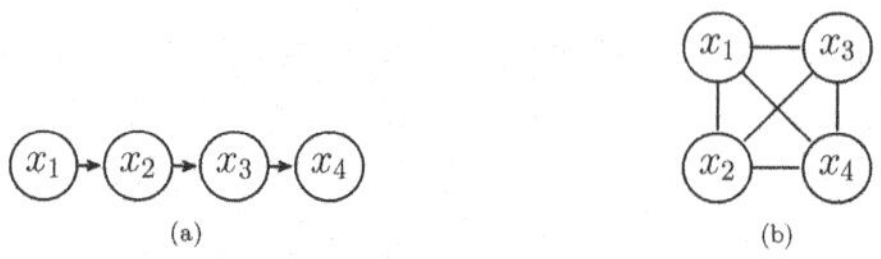

Fig. 3. Two graphical models describing different interpretations of the generative probability distribution for passwords of length four. Graph (a) represents a Bayesian network. Scheme (b) depicts a Markov Random Field.

Unfortunately, existing PPSMs, by construction, leverage arbitrary designed structured probabilistic models that fail to produce the required estimates. Those assume independence properties and causality relations among characters that are not strictly verified in practice. As a result, their conditional probability measurements fail to model correctly a coherent character-level estimation that can be used to provide the required feedback mechanism.

Hereafter, we show that relaxing these biases from the mass estimation process will allow us to implement the feedback mechanism described above. To that purpose, we have to build a new probabilistic estimation framework based on complete, undirected models.

3.2 An Undirected Description of Password Distribution

To simplify our method's understanding, we start with a description of the probabilistic reasoning of previous approaches. Then, we fully motivate our solution by comparison with them. In particular, we chose the state-of-the-art neural approach proposed in [20] (henceforth, referred to as FLA) as a representative instance, since it is the least biased as well as the most accurate among the existing PPSMs.

FLA uses a recurrent neural network (RNN) to estimate password mass function at the character level. That model assumes a stochastic process represented by a Bayesian network like the one depicted in Fig. 3a. As previously anticipated, such a density estimation process bears the burden of bold assumptions on its formalization. Namely, the description derived from a Bayesian Network implies a causality order among password characters. This assumption asserts that the causality flows in a single and specific direction in the generative process i.e., from the start of the string to its end. Therefore, characters influence their probabilities only asymmetrically; that is, the probability of $\mathbf{x}_{i+1}$ is conditioned by $\mathbf{x}_i$ but not *vice versa*. In practice, this implies that the observation of the value assumed from $\mathbf{x}_{i+1}$ does not affect our belief in the value expected from $\mathbf{x}_i$, yet the opposite does. Consequently, every character $\mathbf{x}_i$ is assumed **independent** from the characters that follow it in the string, i.e., $[\mathbf{x}_{i+1}, \ldots, \mathbf{x}_\ell]$. Assuming this property verified for the underlying stochastic process eventually simplifies the probability estimation; the local conditional probability of each character can be easily computed as $Q(\mathbf{x}_i) = P(\mathbf{x}_i \mid \mathbf{x}_1, \ldots, \mathbf{x}_{i-1})$, where $Q(\mathbf{x}_i)$ explicates that the i'th character solely depends on the characters that precede it in the string.

Just as easily, the joint probability factorizes in:
$P(\mathbf{x}) = \prod_{i=1}^{\ell} Q(\mathbf{x}_i) = P(\mathbf{x}_1) \prod_{i=2}^{\ell} P(\mathbf{x}_i \mid \mathbf{x}_1, \dots \mathbf{x}_{i-1})$ by chain rule.

Unfortunately, although those assumptions do simplify the estimation process, the conditional probability $Q(x_i)$, *per se*, does not provide a direct and coherent estimation of the security contribution of the single character x_i in the password. This is particularly true for characters in the first positions of the string, even more so for the first character x_1, which is assumed to be independent of any other symbol in the password; its probability is the same for any possible configuration of the remaining random variables $[x_2, \dots, x_\ell]$. Nevertheless, in the context of a **sound** character-level feedback mechanism, the symbol x_i must be defined as "weak" or "strong" according to the complete context defined by the entire string. For instance, given two passwords y = "*aaaaaaa*" and z = "*a######*", the probability $Q(\mathbf{x}_1 = \text{`}a\text{'})$ should be different if measured on $y$ or $z$. More precisely, we expect $Q(\mathbf{x}_1 = \text{`}a\text{'}|y)$ to be much higher than $Q(\mathbf{x}_1 = \text{`}a\text{'}|z)$, as observing $y_{2,7}$ = "*aaaaaa*" drastically changes our expectations about the possible values assumed from the first character in the string. On the other hand, observing $z_{2,7}$ = "*######*" tells us little about the event $\mathbf{x}_1 = \text{`}a\text{'}$. Yet, this interaction cannot be described through the Bayesian network reported in Fig. 3a, where $Q(\mathbf{x}_1 = \text{`}a\text{'}|y)$ eventually results equal to $Q(\mathbf{x}_1 = \text{`}a\text{'}|z)$. The same reasoning applies to trickier cases, as for the password x = "*(password)*". Here, arguably, the security contribution of the first character '(' strongly depends from the presence or absence of the last character[1], i.e., x_7 = ')'. The symbol x_1 = '(', indeed, can be either a good choice (as it introduces entropy in the password) or a poor one (as it implies a predictable template in the password), but this solely depends on the value assumed from another character in the string (the last one in this example). It should be apparent that the assumed structural independence prevents the resulting local conditional probabilities from being sound descriptors of the real character probability as well as of their security contribution. Consequently, such measures cannot be used to build the feedback mechanism suggested in Sect. 3.1. The same conclusion applies to other classes of PPSMs [9,29] which add even more structural biases on top of those illustrated by the model in Fig. 3a.

Under a broader view, directed models are intended to be used in contexts where the causality relationships among random variables can be fully understood. Unfortunately, even if passwords are physically written character after character by the users, it is impossible to assert neither independence nor cause-effect relations among the symbols that compose them. **Differently from plain dictionary words, passwords are built on top of much more complex structures and articulated interactions among characters that cannot be adequately described without relaxing many of the assumptions leveraged by existing PPSMs.** In the act of relaxing such assumptions, we base our estimation on an **undirected and complete[2] graphical model**,

[1] Even if not so common, strings enclosed among brackets or other special characters often appear in password leaks.

[2] "Complete" in the graph-theory sense.

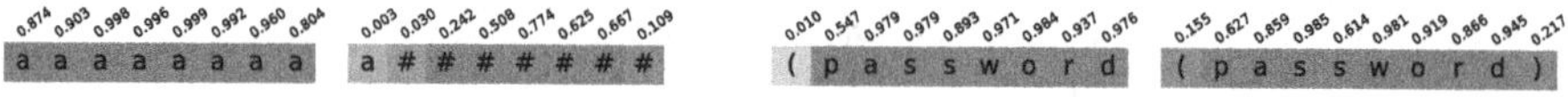

(a) (b)

Fig. 4. Estimated local conditional probabilities for two pairs of passwords. The numbers depicted above the strings report the $Q(x_i)$ value for each character (rounding applied).

as this represents the most general description of the password generative distribution. Figure 3b depicts the respective Markov Random Field (MRF) for passwords of length four. According to that description, the probability of the character x_i directly depends on any other character in the string, i.e., the full context. In other words, we model each variable $\mathbf{x}_i$ as a stochastic function of all the others. This intuition is better captured from the evaluation of local conditional probability (Eq. 2).

$$Q(\mathbf{x}_i) = \begin{cases} P(\mathbf{x}_i \mid \mathbf{x}_{i+1}, \dots \mathbf{x}_\ell) & i = 1 \\ P(\mathbf{x}_i \mid \mathbf{x}_1, \dots \mathbf{x}_{i-1}) & i = \ell \\ P(\mathbf{x}_i \mid \mathbf{x}_1, \dots, \mathbf{x}_{i-1}, \mathbf{x}_{i+1}, \dots \mathbf{x}_\ell) & 1 < i < \ell. \end{cases} \quad (2)$$

Henceforth, we use the notation $Q(\mathbf{x}_i)$ to refer to the local conditional distribution of the i'th character within the password x. When x is not clear from the context, we write $Q(\mathbf{x}_i \mid x)$ to make it explicit. The notation $Q(\mathbf{x}_i{=}s)$ or $Q(s)$, instead, refers to the marginalization of the distribution according to the symbol s.

Eventually, such undirected formalization intrinsically weeds out all the limitations observed for the previous estimation process (i.e., the Bayesian network in Fig. 3a). Now, every local measurement is computed within the context offered by any other symbol in the string. Therefore, relations that were previously assessed as impossible can now be naturally described. In the example, $y{=}$"*aaaaaaa*" / $z{=}$"*a######*", indeed, the local conditional probability of the first character can be now backward-influenced from the context offered from the subsequent part of the string. This is clearly observable from the output of an instance of our meter reported in Fig. 4a, where the value of $Q(\mathbf{x}_1{=}\text{'}a\text{'})$ drastically varies between the two cases, i.e., y and z. As expected, we have $Q(\mathbf{x}_1{=}\text{'}a\text{'}|y) \gg Q(\mathbf{x}_1{=}\text{'}a\text{'}|z)$ verified in the example. A similar intuitive result is reported in Fig. 4b, where the example x = "*(password)*" is considered. Here, the meter first scores the string x' = "*(password*", then it scores the complete password x = "*(password)*". In this case, we expect that the presence of the last character ')' would consistently influence the conditional measurement of the first bracket in the string. Such expectation is perfectly captured from the reported output, where appending at the end of the string the symbol ')' increases the probability of the first bracket of a factor ~15.

However, obtaining these improvements does not come for free. Indeed, under the MRF construction, the productory over the local conditional probabilities (better defined as potential functions or factors within this context) does not provide the exact joint probability distribution of $\mathbf{x}$. Instead, such product results in a unnormalized version of it: $P(\mathbf{x}) \propto \prod_{i=1}^{\ell} Q(\mathbf{x}_i) = \tilde{P}(\mathbf{x})$ with $P(x) = \frac{\tilde{P}(\mathbf{x})}{Z}$. In the equation, Z is the partition function. This result follows from the Hammersley–Clifford theorem [16]. Nevertheless, the unnormalized joint distribution preserves the core properties needed to the meter functionality. Most importantly, we have that:

$$\forall x, x' \; : \; P(x) \geq P(x') \Leftrightarrow \tilde{P}(x) \geq \tilde{P}(x'). \tag{3}$$

That is, if we sort a list of passwords according to the true joint $P(\mathbf{x})$ or according to the unnormalized version $\tilde{P}(\mathbf{x})$, we obtain the same identical ordering. Consequently, no deviation from the adversarial interpretation of PPSMs described in Sect. 2.1 is implied. Indeed, we have $\mathbb{X}_{P(\mathbf{x})} = \mathbb{X}_{\tilde{P}(\mathbf{x})}$ for every password distribution, key-space, and suitable sorting function. Furthermore, the joint probability distribution, if needed, can be approximated using suitable approximation methods, as discussed in Appendix B. Appendix A, instead, reports a more detailed description of the feedback mechanism.

It is important to highlight that, although we present our approach under the character-level description, our method can be directly applied to n-grams or tokens without any modification.

In summary, in this section, we presented and motivated an estimation process able to unravel the feedback mechanism described in Sect. 3.1. Maintaining a purely theoretical focus, no information about the implementation of such methodology has been offered to the reader. Next, in Sect. 4, we describe how such a meter can be shaped via an efficient deep learning framework.

4 Meter Implementation

In this section, we present a deep-learning-based implementation of the estimation process introduced in Sect. 3.2. Here, we describe the model and its training process. Then, we explain how the trained network can be used as a building block for the proposed password meter.

Model Training. From the discussion in Sect. 3.2, our procedure requires the parametrization of an exponentially large number of interactions among random variables. Thus, any tabular approach, such as the one used from Markov Chains or PCFG [29], is *a priori* excluded for any real-world case. To make such a meter feasible, we reformulate the underlying estimation process so that it can be approximated with a neural network. In our approach, we simulate the Markov Random Field described in Sect. 3.2 using a deep convolutional neural network trained to compute $Q(\mathbf{x}_i)$ (Eq. 2) for each possible configuration of the structured model. In doing so, we train our network to solve an *inpainting*-like task defined over the textual domain. Broadly speaking, inpainting is the task of reconstructing missing information from mangled inputs, mostly images with missing or damaged patches [32]. **Under the probabilistic perspective, the**

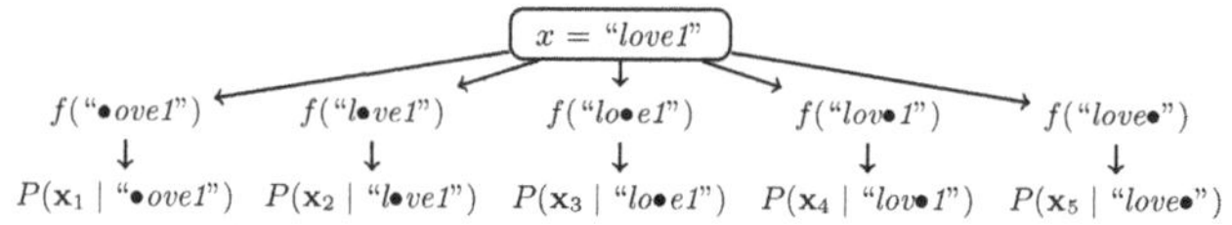

Fig. 5. Graphical depiction of the complete inference process for the password $x =$ *"love1"*. The function f refers to the trained autoencoder and the symbol '•' refers to the deleted character.

model is asked to return a probability distribution over all the unobserved elements of x, explicitly measuring the conditional probability of those concerning the observable context. Therefore, the network has to disentangle and model the semantic relation among all the factors describing the data (e.g., characters in a string) to reconstruct input instances correctly.

Generally, the architecture and the training process used for inpainting tasks resemble an auto-encoding structure [8]. In the general case, these models are trained to revert self-induced damage carried out on instances of a train-set $\mathbf{X}$. At each training step, an instance $x \in \mathbf{X}$ is artificially mangled with an information-destructive transformation to create a mangled variation $\tilde{x}$. Then, the network, receiving $\tilde{x}$ as input, is optimized to produce an output that most resembles the original x; that is, the network is trained to reconstruct x from $\tilde{x}$.

In our approach, we train a network to infer missing characters in a mangled password. In particular, we iterate over a password leak (i.e., our train-set) by creating mangled passwords and train the network to recover them. The mangling operation is performed by removing a randomly selected character from the string. For example, the train-set entry $x =$ *"iloveyou"* is transformed in $\tilde{x}$= "ilov•you" if the 5'th character is selected for deletion, where the symbol '•' represents the *"empty character"*. A compatible proxy-task has been previously used in [22] to learn a suitable password representation for guessing attacks.

We chose to model our network with a deep *residual* structure arranged to create an autoencoder. The network follows the same general Context Encoder [23] architecture defined in [22] with some modifications. To create an information bottleneck, the encoder connects with the decoder through a latent space junction obtained through two fully connected layers. We observed that enforcing a latent space, and a prior on that, consistently increases the meter effectiveness. For that reason, we maintained the same regularization proposed in [22]; a maximum mean discrepancy regularization that forces a standard normal distributed latent space. The final loss function of our model is reported in Eq. 4. In the equation, *Enc* and *Dec* refer to the encoder and decoder network respectively, s is the *softmax* function applied row-wise[3], the distance function d is the cross-entropy, and *mmd* refers to the *maximum mean discrepancy*.

$$\mathbb{E}_{x,\tilde{x}}[d(x,\ s(Dec(Enc(\tilde{x})))] + \alpha\mathbb{E}_{z\sim N(0,\mathbb{I})}[mmd(z, Enc(\tilde{x}))] \tag{4}$$

[3] The Decoder outputs ℓ estimations; one for each input character. Therefore, we apply the softmax function separately on each of those to create ℓ probability distributions.

Henceforth, we refer to the composition of the encoder and the decoder as $f(x) = s(Dec(Enc(x)))$. We train the model on the widely adopted *RockYou* leak [7] considering an 80/20 train-test split. From it, we filter passwords presenting fewer than 5 characters. We train different networks considering different maximum password lengths, namely, 16, 20, and 30. In our experiments, we report results obtained with the model trained on a maximum length equal to 16, as no substantial performance variation has been observed among the different networks. Eventually, we produce three neural nets with different architectures; a large network requiring 36 MB of disk space, a medium-size model requiring 18 MB, and a smaller version of the second that requires 6.6 MB. These models can be potentially further compressed using the same quantization and compression techniques harnessed in [20].[4]

Model Inference Process. Once the model is trained, we can use it to compute the conditional probability $Q(x_i)$ (Eq. 2) for each i and each possible configuration of the MRF. This is done by querying the network f using the same mangling trick performed during the training. The procedure used to compute $Q(x_i)$ for x is summarized in the following steps:

1. We substitute the i'th character of x with the *empty character* '•', obtaining a mangled password $\tilde{x}$.
2. Then, we feed $\tilde{x}$ to a network that outputs a probability distribution over Σ of the unobserved random variable $\mathbf{x}_i$ i.e., $Q(\mathbf{x}_i)$.
3. Given $Q(\mathbf{x}_i)$, we marginalize out x_i, obtaining the probability: $Q(x_i) = P(\mathbf{x}_i = x_i \mid \tilde{x})$.

For instance, if we want to compute the local conditional probability of the character '*e*' in the password $x =$ "*iloveyou*", we first create $\tilde{x} =$"ilov•you" and use it as input for the net, obtaining $Q(\mathbf{x}_5)$, then we marginalize that (i.e., $Q(\mathbf{x}_5 = \text{'}e\text{'})$) getting the probability $P(\mathbf{x}_5 = \text{'}e\text{'} \mid \tilde{x})$. From the probabilistic point of view, this process is equivalent to fixing the observable variables in the MRF and querying the model for an estimation of the single unobserved character.

At this point, to cast both the feedback mechanism defined in Sect. 3.1 and the unnormalized joint probability of the string, we have to measure $Q(x_i)$ for each character x_i of the tested password. This is easily achieved by repeating the inference operation described above for each character comprising the input string. A graphical representation of this process is depicted in Fig. 5. It is important to highlight that the ℓ required inferences are independent, and their evaluation can be performed in parallel (i.e., batch level parallelism), introducing almost negligible overhead over the single inference. Additionally, with the use of a feed-forward network, we avoid the sequential computation that is intrinsic in recurrent networks (e.g., the issue afflicting [20]), and that can be excessive for a reactive client-side implementation. Furthermore, the convolutional structure enables the construction of very deep neural nets with a limited memory footprint.

[4] The code, pre-trained models, and other materials related to our work are publicly available at: https://github.com/pasquini-dario/InterpretablePPSM.

In conclusion, leveraging the trained neural network, we can compute the potential of each factor/vertex in the Markov Random Field (defined as local conditional probabilities in our construction). As a consequence, we are now able to cast a PPSM featuring the character-level feedback mechanism discussed in Sect. 3.1. Finally, in Sect. 5, we empirically evaluate the soundness of the proposed meter.

5 Evaluation

In this section, we empirically validate the proposed estimation process as well as its deep learning implementation. First, in Sect. 5.1, we evaluate the capability of the meter of accurately assessing password strength at string-level. Next, in Sect. 5.2, we demonstrate the intrinsic ability of the local conditional probabilities of being sound descriptors of password strength at character-level.

5.1 Measuring Meter Accuracy

In this section, we evaluate the accuracy of the proposed meter at estimating password probabilities. To that purpose, following the adversarial reasoning introduced in Sect. 2.1, we compare the password ordering derived from the meter with the one from the ground-truth password distribution. In doing so, we rely on the guidelines defined in [15] for our evaluation. In particular, given a test-set (i.e., a password leak), we consider a weighted rank correlation coefficient between ground-truth ordering and that derived from the meter. The ground-truth ordering is obtained by sorting the unique entries of the test-set according to the frequency of the password observed in the leak. In the process, we compare our solution with other fully probabilistic meters. A detailed description of the evaluation process follows.

Test-Set. For modeling the ground-truth password distribution, we rely on the password leak discovered by 4iQ in the Dark Web [1] on 5th December 2017. It consists of the aggregation of ~250 leaks, consisting of 1.4 billion passwords in total. In the cleaning process, we collect passwords with length in the interval 5–16, obtaining a set of $\sim 4 \cdot 10^8$ unique passwords that we sort in decreasing frequency order. Following the approach of [15], we filter out all the passwords with a frequency lower than 10 from the test-set. Finally, we obtain a test-set composed of 10^7 unique passwords that we refer to as X_{BC}. Given both the large number of entries and the heterogeneity of sources composing it, we consider X_{BC} an accurate description of real-world passwords distribution.

Tested Meters. In the evaluation process, we compare our approach with other probabilistic meters. In particular:

- The Markov model [13] implemented in [3] (the same used in [15]). We investigate different n-grams configurations, namely, 2-g, 3-g and 4-g that we refer to as MM_2, MM_3 and MM_4, respectively. For their training, we employ the same train-set used for our meter.

- The neural approach of Melicher et al. [20]. We use the implementation available at [4] to train the main architecture advocated in [20], i.e., an RNN composed of three LSTM layers of 1000 cells each, and two fully connected layers. The training is carried out on the same train-set used for our meter. We refer to the model as FLA.

Metrics. We follow the guidelines defined by Golla and Dürmuth [15] for evaluating the meters. We use the **weighted** *Spearman* correlation coefficient (ws) to measure the accuracy of the orderings produced by the tested meters, as this has been demonstrated to be the most reliable correlation metrics within this context [15]. Unlike [15], given the large cardinality and diversity of this leak, we directly use the ranking derived from the password frequencies in X_{BC} as ground-truth. Here, passwords with the same frequency value have received the same rank in the computation of the correlation metric.

Results. Table 1 reports the measured correlation coefficient for each tested meter. In the table, we also report the required storage as auxiliary metric.

Our meters, even the smallest, achieve higher or comparable score than the most performant Markov Model, i.e., MM_4. On the other hand, our largest model cannot directly exceed the accuracy of the state-of-the-art estimator FLA, obtaining only comparable results. However, FLA requires more disk space than ours. Indeed, interestingly, our convolutional implementation permits the creation of remarkably lightweight meters. As a matter of fact, our smallest network shows a comparable result with MM_4 requiring more than a magnitude less disk space.

In conclusion, the results confirm that the probability estimation process defined in Sect. 3.2 is indeed sound and capable of accurately assessing password mass at string-level. The proposed meter shows comparable effectiveness with the state-of-the-art [20], whereas, in the *large* setup, it outperforms standard approaches such as Markov Chains. Nevertheless, we believe that even more accurate estimation can be achieved by investigating deeper architectures and/or by performing hyper-parameters tuning over the model.

5.2 Analysis of the Relation Between Local Conditional Probabilities and Password Strength

In this section, we test the capability of the proposed meter to correctly model the relation between password structure and password strength. In particular, we investigate the ability of the measured local conditional probabilities of determining the tested passwords' insecure components.

Our evaluation procedure follows three main steps. Starting from a set of weak passwords X:

(1) We perform a guessing attack on X in order to estimate the guess-number of each entry of the set.

Table 1. Rank correlation coefficient computed between X_{BC} and the tested meters.

	MM_2	MM_3	MM_4	FLA	ours (large)	ours (mid)	ours (small)
Weighted **Spearman** ↑	0.154	0.170	0.193	0.217	0.207	0.203	0.199
Required **Disk Space** ↓	1.1MB	94MB	8.8GB	60MB	36MB	18MB	6.6MB

Table 2. Strength improvement induced by different perturbations.

	$n = 1$	$n = 2$	$n = 3$
Baseline (PNP)	0.022	0.351	0.549
Semi-Meter (PNP)	0.036	0.501	0.674
Fully-Meter (PNP)	0.066	0.755	0.884
Baseline (**AGI**)	$3.0 \cdot 10^{10}$	$3.6 \cdot 10^{11}$	$5.6 \cdot 10^{11}$
Semi-Meter (**AGI**)	$4.6 \cdot 10^{10}$	$5.1 \cdot 10^{11}$	$6.8 \cdot 10^{11}$
Fully-Meter (**AGI**)	$8.2 \cdot 10^{10}$	$7.7 \cdot 10^{11}$	$8.9 \cdot 10^{11}$
Semi-Meter/Baseline (**AGI**)	**1.530**	**1.413**	**1.222**
Fully-Meter/Baseline (**AGI**)	**2.768**	**2.110**	**1.588**

(2) For each password $x \in X$, we substitute n characters of x according to the estimated local conditional probabilities (i.e., we substitute the characters with highest $Q(\mathbf{x}_i)$), producing a perturbed password $\tilde{x}$.
(3) We repeat the guessing attack on the set of perturbed passwords and measure the variation in the attributed guess-numbers. Hereafter, we provide a detailed description of the evaluation procedure.

Passwords Sets. The evaluation is carried out considering a set of weak passwords. In particular, we consider the first 10^4 most frequent passwords of the X_{BC} set.

Password Perturbations. In the evaluation, we consider three types of password perturbation:

(1) The first acts as a baseline and consists of the substitution of random positioned characters in the passwords with randomly selected symbols. Such a general strategy is used in [24] and [14] to improve the user's password at composition time. The perturbation is applied by randomly selecting n characters from x and substituting them with symbols sampled from a predefined character pool. In our simulations, the pool consists of the 25 most frequent symbols in X_{BC} (i.e., mainly lowercase letters and digits). Forcing this character-pool aims at preventing the tested perturbation procedures to create artificially complex passwords such as strings containing extremely uncommon *unicode* symbols. We refer to this perturbation procedure as **Baseline**.
(2) The second perturbation partially leverages the local conditional probabilities induced by our meter. Given a password x, we compute the conditional probability $Q(x_i)$ for each character in the string. Then, we select and substitute the character with maximum probability, i.e., $\arg\max_{x_i} Q(x_i)$. The symbol we use in the substitution is randomly selected from the same pool used for the baseline perturbation (i.e., top-25 frequent symbols). When n is

greater than one, the procedure is repeated sequentially using the perturbed password obtained from the previous iteration as input for the next step. We refer to this procedure as **Semi-Meter**.

(3) The third perturbation extends the second one by exploiting the local conditional distributions. Here, as in the Semi-Meter-based, we substitute the character in x with the highest probability. However, rather than choosing a substitute symbol in the pool at random, we select that according to the distribution $Q(\mathbf{x}_i)$, where i is the position of the character to be substituted. In particular, we choose the symbol the minimize $Q(\mathbf{x}_i)$, i.e., $\arg\min_{s \in \Sigma'} Q(\mathbf{x}_i = s)$, where Σ' is the allowed pool of symbols. We refer to this method as **Fully-Meter**.

Guessing Attack. We evaluate password strength using the *min-auto* strategy advocated in [27]. The ensemble we employ is composed of HashCat [2], PCFG [6,29] and the Markov chain approach implemented in [5,13]. We limit each tool to produce 10^{10} guesses. The total size of the generated guesses is $\sim$3 TB.

Metrics. In the evaluation, we are interested in measuring the increment of password strength caused by an applied perturbation. We estimate that value by considering the Average Guess-number Increment (henceforth, referred to as AGI); that is, the average delta between the guess-number of the original password and the guess-number of the perturbed password:

$$\text{AGI}(X) = \frac{1}{|X|} \sum_{i=0}^{|X|} [g(\tilde{x}^i) - g(x^i)]$$

where g is the guess-number, and $\tilde{x}^i$ refers to the perturbed version of the i'th password in the test set. During the computation of the guess-numbers, it is possible that we fail to guess a password. In such a case, we attribute an artificial guess-number equals to 10^{12} to the un-guessed passwords. Additionally, we consider the average number of un-guessed passwords as an ancillary metrics; we refer to it with the name of Percentage Non-Guessed Passwords (PNP) and compute it as:

$$\text{PNP(X)} = \frac{1}{|X|} |\{x^i \mid g(x^i) \neq \perp \wedge g(\tilde{x}^i) = \perp\}|,$$

where $g(x) = \perp$ when x is not guessed during the guessing attack.

Results. We perform the tests over three values of n (i.e., the number of perturbed characters), namely, 1, 2, and 3. Results are summarized in Table 2. The AGI caused by the two meter-based solutions is always greater than that produced by random perturbations. On average, that is twice more effective with respect to the Fully-Meter baseline and about 35% greater for the Semi-Meter. The largest relative benefit is observable when $n = 1$, i.e., a single character is modified. Focusing on the Fully-Meter approach, indeed, the guidance of the local conditional probabilities permits a guess-number increment 2.7 times bigger than the one caused by a random substitution in the string. This advantage drops to ∼1.5 when $n = 3$, since, after two perturbations, passwords tend to be already out of the dense zone of the distribution. Indeed, at $n = 3$ about 88% of the passwords perturbed with the Fully-Meter approach cannot be guessed during the guessing attack (i.e., PNP). This value is only ∼55% for the baseline. More interestingly, the results tell us that substituting two ($n = 2$) characters following the guide of the local conditional probabilities causes a guess-number increment greater than the one obtained from three ($n = 3$) random perturbations. In the end, these results confirm that the local conditional distributions are indeed sound descriptors of password security at the structural level.

Limitations. Since the goal of our evaluation was mainly to validate the soundness of the proposed estimation process, we did not perform user studies and we did not evaluate human-related factors such as password memorability although we recognize their importance.

6 Conclusion

In this paper, we showed that it is possible to construct interpretable probabilistic password meters by fundamentally rethinking the underlying password mass estimation. We presented an undirected probabilistic interpretation of the password generative process that can be used to build precise and sound password feedback mechanisms. Moreover, we demonstrated that such an estimation process could be instantiated via a lightweight deep learning implementation. We validated our undirected description and deep learning solution by showing that our meter achieves comparable accuracy with other existing approaches while introducing a unique character-level feedback mechanism that generalizes any heuristic construction.

Appendix

A Details on the Password Feedback Mechanism

Joint probability can be understood as a compatibility score assigned to a specific configuration of the MRF; it tells us the likelihood of observing a sequence of characters during the interaction with the password generative process. On a smaller scale, a local conditional probability measures the impact that a single character has in the final security score. Namely, it indicates how much the character contributes to the probability of observing a certain password x. Within this interpretation, low-probabilities characters push the joint probability of x to be closer to zero (secure), whereas high-probability characters (i.e., $Q(x_1) \lesssim 1$) make no significant contribution to lowering the password probability (insecure). Therefore, users can strengthen their candidate passwords by substituting high-probability characters with suitable lower-probability ones (e.g., Fig. 1).

Unfortunately, users' perception of password security has been shown to be generally erroneous [25], and, without explicit guidelines, it would be difficult for them to select suitable lower-probability substitutes. To address this limitation, one could adopt our approach based on local conditional distributions as an effective mechanism to help users select secure substitute symbols. Indeed, $\forall_i Q(\mathbf{x}_i)$ are able to clarify which symbol is a secure substitute and which is not for each character x_i of x. In particular, a distribution $Q(\mathbf{x}_i)$, defined on the whole alphabet Σ, assigns a probability to every symbol s that the character $\mathbf{x}_i$ can potentially assume. For a symbol $s \in \Sigma$, the probability $Q(\mathbf{x}_i{=}s)$ measures how much the event $\mathbf{x}_i{=}s$ is probable given all the observable characters in x. Under this interpretation, a candidate, secure substitution of x_i is a symbol with very low $Q(\mathbf{x}_i{=}s)$ (as this will lower the joint probability of x). In particular, every symbol s s.t. $Q(\mathbf{x}_i{=}s) < Q(\mathbf{x}_i{=}x_i)$ given x is a secure substitution for x_i. Table 3 better depicts this intuition. The Table reports the alphabet sorted by $Q(\mathbf{x}_i)$ for each x_i in the example password $x =$ "*PaSsW0rD!*". The bold symbols between parenthesis indicate x_i. Within this representation, all the symbols below the respective x_i for each $\mathbf{x}_i$ are suitable substitutions that improve password strength. This intuition is empirically proven in Sect. 5.2. It is important to note that the suggestion mechanism must be randomized to avoid any bias in the final password distribution.[5] To this end, one can provide the user with k random symbols among the pool of secure substitutions, i.e., $\{s \mid Q(\mathbf{x}_i{=}s) < Q(\mathbf{x}_i{=}x_i)\}$.

Table 3. First seven entries of the ordering imposed on Σ from the local conditional distribution for each character of the password $x =$ "*PaSsW0rD!*"

Rank	•aSsW0rD!	P•SsW0rD!	Pa•sW0rD!	PaS•W0rD!	PaSs•0rD!	PaSsW•rD!	PaSsW0•D!	PaSsW0r•!	PaSsW0rD•
0	**{P}**	A	s	S	w	O	R	d	1
1	S	**{a}**	**{S}**	**{s}**	**{W}**	o	**{r}**	**{D}**	S
2	p	@	c	A	#	**{0}**	N	t	2
3	B	3	n	T	f	I	0	m	s
4	C	4	t	E	k	i	L	l	3
5	M	I	d	H	1	#	D	k	**{!}**
6	1	1	r	O	F	A	n	e	5
7	c	5	x	$	3	@	X	r	9
8	s	0	$	I	0	)	S	f	4
⋮	⋮	⋮	⋮	⋮	⋮	⋮	⋮	⋮	⋮

B Estimating Guess-Numbers

Within the context of PPSMs, a common solution to approximate guess-numbers [19] is using the Monte Carlo method proposed in [12]. With a few adjustments, the same approach can be applied to our meter. In particular, we have to derive an approximation of the partition function Z. This can be done by leveraging the Monte Carlo method as follows:

$$Z \simeq N \cdot \mathbb{E}_x[P(x)] \tag{5}$$

[5] That is, if weak passwords are always perturbed in the same way, they will be easily guessed.

where N is the number of possible configurations of the MRF (i.e., the cardinality of the key-space), and x is a sample from the posterior distribution of the model. Samples from the model can be obtained in three ways: (1) sampling from the latent space of the autoencoder (as done in [22]), (2) performing Gibbs sampling from the autoencoder, or (3) using a dataset of passwords that follow the same distribution of the model. Once we have an approximation of Z, we can use it to normalize every joint probability, i.e., $P(x) = \frac{\tilde{P}(x)}{Z}$ and then apply the method in [12]. Alternatively, we could adopt a more articulate solution as in [18].

In any event, the estimation of the partition function Z is performed only once and can be done offline.

References

1. 1.4 Billion Clear Text Credentials Discovered. https://tinyurl.com/t8jp5h7
2. hashcat GitHub. https://github.com/hashcat
3. NEMO Markov Model GitHub. https://github.com/RUB-SysSec/NEMO
4. Neural Cracking GitHub. https://github.com/cupslab/neural_network_cracking
5. OMEN GitHub. https://github.com/RUB-SysSec/OMEN
6. PCFG GitHub. https://github.com/lakiw/pcfg_cracker
7. RockYou Leak. https://downloads.skullsecurity.org/passwords/rockyou.txt.bz2
8. Baldi, P.: Autoencoders, unsupervised learning, and deep architectures. In: Proceedings of ICML Workshop on Unsupervised and Transfer Learning, volume 27 of Proceedings of Machine Learning Research, pp. 37–49. PMLR, Bellevue, 02 July 2012
9. Castelluccia, C., Dürmuth, M., Perito, D.: Adaptive password-strength meters from Markov models. In: NDSS (2012)
10. Clair, L.S., Johansen, L., Enck, W., Pirretti, M., Traynor, P., McDaniel, P., Jaeger, T.: Password exhaustion: predicting the end of password usefulness. In: Bagchi, A., Atluri, V. (eds.) ICISS 2006. LNCS, vol. 4332, pp. 37–55. Springer, Heidelberg (2006). https://doi.org/10.1007/11961635_3
11. Dell' Amico, M., Michiardi, P., Roudier, Y.: Password strength: an empirical analysis. In: 2010 Proceedings IEEE INFOCOM, pp. 1–9, March 2010
12. Dell'Amico, M., Filippone, M.: Monte Carlo strength evaluation: fast and reliable password checking. In: Proceedings of the 22nd ACM SIGSAC Conference on Computer and Communications Security, CCS 2015, pp. 158–169. Association for Computing Machinery, New York (2015)
13. Dürmuth, M., Angelstorf, F., Castelluccia, C., Perito, D., Chaabane, A.: OMEN: faster password guessing using an ordered Markov enumerator. In: Piessens, F., Caballero, J., Bielova, N. (eds.) ESSoS 2015. LNCS, vol. 8978, pp. 119–132. Springer, Cham (2015). https://doi.org/10.1007/978-3-319-15618-7_10
14. Forget, A., Chiasson, S., van Oorschot, P.C., Biddle, R.: Improving text passwords through persuasion. In: Proceedings of the 4th Symposium on Usable Privacy and Security, SOUPS 2008, pp. 1–12. Association for Computing Machinery, New York (2008)
15. Golla, M., Dürmuth, M.: On the accuracy of password strength meters. In: Proceedings of the 2018 ACM SIGSAC Conference on Computer and Communications Security, CCS 2018, pp. 1567–1582. Association for Computing Machinery, New York (2018)

16. Koller, D., Friedman, N.: Probabilistic Graphical Models: Principles and Techniques - Adaptive Computation and Machine Learning. The MIT Press, Cambridge (2009)
17. Komanduri, S., Shay, R., Cranor, L.F., Herley, C., Schechter, S.: Telepathwords: preventing weak passwords by reading users' minds. In: 23rd USENIX Security Symposium (USENIX Security 2014), pp. 591–606. USENIX Association, San Diego, August 2014
18. Ma, J., Peng, J., Wang, S., Xu, J.: Estimating the partition function of graphical models using Langevin importance sampling. In: Proceedings of the Sixteenth International Conference on Artificial Intelligence and Statistics, volume 31 of Proceedings of Machine Learning Research, pp. 433–441. PMLR, Scottsdale, 29 April–01 May 2013
19. Massey, J.L.: Guessing and entropy. In: Proceedings of 1994 IEEE International Symposium on Information Theory, p. 204, June 1994
20. Melicher, W., et al.: Fast, lean, and accurate: modeling password guessability using neural networks. In: 25th USENIX Security Symposium (USENIX Security 2016), pp. 175–191. USENIX Association, Austin, August 2016
21. Narayanan, A., Shmatikov, V.: Fast dictionary attacks on passwords using time-space tradeoff. In: Proceedings of the 12th ACM Conference on Computer and Communications Security, CCS 2005, pp. 364–372. Association for Computing Machinery, New York (2005)
22. Pasquini, D., Gangwal, A., Ateniese, G., Bernaschi, M., Conti, M.: Improving password guessing via representation learning. In 2021 42th IEEE Symposium on Security and Privacy, May 2021
23. Pathak, D., Krähenbühl, P., Donahue, J., Darrell, T., Efros, A.: Context encoders: feature learning by inpainting. In: Computer Vision and Pattern Recognition (CVPR) (2016)
24. Ur, B., et al.: Design and evaluation of a data-driven password meter. In: CHI 2017 (2017)
25. Ur, B., Bees, J., Segreti, S.M., Bauer, L., Christin, N., Cranor, L.F.: Do users' perceptions of password security match reality? In: Proceedings of the 2016 CHI Conference on Human Factors in Computing Systems, CHI 2016, pp. 3748–3760. Association for Computing Machinery, New York (2016)
26. Ur, B., et al.: How does your password measure up? The effect of strength meters on password creation. In: Presented as part of the 21st USENIX Security Symposium (USENIX Security 2012), pp. 65–80. USENIX, Bellevue (2012)
27. Ur, B., et al.: Measuring real-world accuracies and biases in modeling password guessability. In: 24th USENIX Security Symposium (USENIX Security 2015), pp. 463–481. USENIX Association, Washington, D.C., August 2015
28. Wang, D., He, D., Cheng, H., Wang, P.: fuzzyPSM: a new password strength meter using fuzzy probabilistic context-free grammars. In: 2016 46th Annual IEEE/IFIP International Conference on Dependable Systems and Networks (DSN), pp. 595–606, June 2016
29. Weir, M., Aggarwal, S., Medeiros, B.D., Glodek, B.: Password cracking using probabilistic context-free grammars. In: 2009 30th IEEE Symposium on Security and Privacy, pp. 391–405, May 2009
30. Weir, M., Aggarwal, S., Collins, M., Stern, H.: Testing metrics for password creation policies by attacking large sets of revealed passwords. In Proceedings of the 17th ACM Conference on Computer and Communications Security, CCS 2010, pp. 162–175. Association for Computing Machinery, New York (2010)

31. Wheeler, D.L.: zxcvbn: low-budget password strength estimation. In: 25th USENIX Security Symposium (USENIX Security 16), pp. 157–173. USENIX Association, Austin, August 2016
32. Xie, J., Xu, L., Chen, E.: Image denoising and inpainting with deep neural networks. In: Advances in Neural Information Processing Systems 25, pp. 341–349. Curran Associates Inc. (2012)

Polisma - A Framework for Learning Attribute-Based Access Control Policies

Amani Abu Jabal[1(✉)], Elisa Bertino[1], Jorge Lobo[2], Mark Law[3], Alessandra Russo[3], Seraphin Calo[4], and Dinesh Verma[4]

[1] Purdue University, West Lafayette, IN, USA
{aabujaba,bertino}@purdue.edu
[2] ICREA - Universitat Pompeo Fabra, Barcelona, Spain
jorge.lobo@upf.edu
[3] Imperial College London, London, UK
{mark.law09,a.russo}@imperial.ac.uk
[4] IBM TJ Watson Research Center, Yorktown Heights, NY, USA
{scalo,dverma}@us.ibm.com

Abstract. Attribute-based access control (ABAC) is being widely adopted due to its flexibility and universality in capturing authorizations in terms of the properties (attributes) of users and resources. However, specifying ABAC policies is a complex task due to the variety of such attributes. Moreover, migrating an access control system adopting a low-level model to ABAC can be challenging. An approach for generating ABAC policies is to learn them from data, namely from logs of historical access requests and their corresponding decisions. This paper proposes a novel framework for learning ABAC policies from data. The framework, referred to as *Polisma*, combines data mining, statistical, and machine learning techniques, capitalizing on potential context information obtained from external sources (e.g., LDAP directories) to enhance the learning process. The approach is evaluated empirically using two datasets (real and synthetic). Experimental results show that *Polisma* is able to generate ABAC policies that accurately control access requests and outperforms existing approaches.

Keywords: Authorization rules · Policy mining · Policy generalization

1 Introduction

Most modern access control systems are based on the attribute-based access control model (ABAC) [3]. In ABAC, user requests to protected resources are granted or denied based on discretionary attributes of users, resources, and environmental conditions [7]. ABAC has several advantages. It allows one to specify access control policies in terms of domain-meaningful properties of users, resources, and environments. It also simplifies access control policy administration by allowing access decisions to change between requests by simply changing attribute values, without changing the user/resource relationships underlying the

L. Chen et al. (Eds.): ESORICS 2020, LNCS 12308, pp. 523–544, 2020.
https://doi.org/10.1007/978-3-030-58951-6_26

rule sets [7]. As a result, access control decisions automatically adapt to changes in environments, and in user and resource populations. Because of its relevancy for enterprise security, ABAC has been standardized by NIST and an XML-based specification, known as XACML, has been developed by OASIS [20]. There are several XACML enforcement engines, some of which are publicly available (e.g., AuthZForce [18] and Balana [19]). Recently a JSON profile for XACML has been proposed to address the verbosity of the XML notation. However, a major challenge in using an ABAC model is the manual specification of the ABAC policies that represent one of the inputs for the enforcement engine. Such a specification requires detailed knowledge about properties of users, resources, actions, and environments in the domain of interest [13,22]. One approach to address this challenge is to take advantage of the many data sources that are today available in organizations, such as user directories (e.g., LDAP directories), organizational charts, workflows, security tracking's logs (e.g., SIEM), and use machine learning techniques to automatically learn ABAC policies from data.

Suitable data for learning ABAC policies could be access requests and corresponding access control responses (i.e., *access control decisions*) that are often collected in different ways and forms, depending on the real-world situation. For example, an organization may log past decisions taken by human administrators [17], or may have automated access control mechanisms based on low-level models, e.g., models that do not support ABAC. If interested in adopting a richer access control model, such an organization could, in principle, use these data to automatically generate access control policies, e.g., logs of past decisions taken by the low-level mechanism could be used as labeled examples for a supervised machine learning algorithm[1].

Learned ABAC policies, however, must satisfy a few key requirements. They must be *correct* and *complete*. Informally, an ABAC policy set is correct if it is able to make the correct decision for any access request. It is complete if there are no access requests for which the policy set is not able to make a decision. Such a case may happen when the attributes provided with a request do not satisfy the conditions of any policy in the set.

To meet these requirements, the following issues need to be taken into account when learning ABAC policies:

- *Noisy examples.* The log of examples might contain decisions which are erroneous or inconsistent. The learning process needs to be robust to noise to avoid learning incorrect policies.
- *Overfitting.* This is a problem associated with machine learning [10] which happens when the learned outcomes are good only at explaining data given as examples. In this case, learned ABAC policies would be appropriate only

[1] Learning policies from logs of access requests does not necessarily mean that there is an existing access control system or that policy learning is aimed to reproduce existing policies or validate them. Such logs may consist of examples of access control decisions provided by a human expert, and learning may be used for example in a coalition environment where a coalition member can get logs from another coalition member to learn policies for similar missions.

for access requests observed during the learning process and fail to control any other potential access request, so causing the learned policy set to be incomplete. The learning process needs to generalize from past decisions.
- *Unsafe generalization.* Generalization is critical to address overfitting. But at the same time generalization should be *safe*, that is, it should not result in learning policies that may have unintended consequences, thus leading to learned policies that are unsound. The learning process has to balance the trade-off between overfitting and safe generalization.

This paper investigates the problem of learning ABAC policies, and proposes a learning framework that addresses the above issues. Our framework learns from logs of access requests and corresponding access control decisions and, when available, context information provided by external sources (e.g., LDAP directories). We refer to our framework as *Polisma* to indicate that our ABAC **po**licy **l**earner uses m**i**ning, **s**tatistical, and **ma**chine learning techniques. The use of multiple techniques enables extracting different types of knowledge that complement each other to improve the learning process. One technique captures data patterns by considering the frequency and another one exploits statistics and context information. Furthermore, another technique exploits data similarity. The assembly of these techniques in our framework enables better learning for ABAC policies compared to the other state-of-the-art approaches.

Polisma consists of four steps. In the first step, a data mining technique is used to infer associations between users and resources included in the set of decision examples and based on these associations a set of rules is generated. In the second step, each constructed rule is generalized based on statistically significant attributes and context information. In the third step, authorization domains for users and resources (e.g., which resources were accessed using which operations by a specific user) are considered in order to augment the set of generalized rules with "restriction rules" for restricting access of users to resources by taking into account their authorization domain. Policies learned by those three stages are safe generalizations with limited overfitting. To improve the completeness of the learned set, *Polisma* applies a machine learning (ML) classifier on requests not covered by the learned set of policies and uses the result of the classification to label these data and generate additional rules in an "ad-hoc" manner.

2 Background and Problem Description

In what follows, we introduce background definitions for ABAC policies and access request examples, and formulate the problem of learning ABAC policies.

2.1 ABAC Policies

Attribute-based access control (ABAC) policies are specified as Boolean combinations of conditions on attributes of users and protected resources. The following definition is adapted from Xu *et al.* [24]:

Definition 1 (cf. ABAC Model [24]). *An ABAC model consists of the following components:*

- $\mathcal{U}$, $\mathcal{R}$, $\mathcal{O}$, $\mathcal{P}$ *refer, respectively, to finite sets of users, resources, operations, and rules.*
- A_U *refers to a finite set of user attributes. The value of an attribute* $a \in A_U$ *for a user* $u \in \mathcal{U}$ *is represented by a function* $d_U(u, a)$*. The range of* d_U *for an attribute* $a \in A_U$ *is denoted by* $V_U(a)$*.*
- A_R *refers to a finite set of resource attributes. The value of an attribute* $a \in A_R$ *for a resource* $r \in \mathcal{R}$ *is represented by a function* $d_R(r, a)$*. The range of* d_R *for an attribute* $a \in A_R$ *is denote*
- *A user attribute expression* e_U *defines a function that maps every attribute* $a \in A_U$*, to a value in its range or* $\perp$*,* $e_U(a) \in V_U(a) \cup \{\perp\}$*. Specifically,* e_U *can be expressed as the set of attribute/value pairs* $e_U = \{\langle a_i, v_i\rangle \mid a_i \in A_U \wedge f(a_i) = v_i \in V_U(a_i)\}$*. A user* u_i *satisfies* e_U *(i.e., it belongs to the set defined by* e_U*) iff for every user attribute* a *not mapped to* $\perp$*,* $\langle a, d_U(u_i, a)\rangle \in e_U$*. Similarly, a resource* s_i *can be defined by a resource attribute expression* e_R*.*
- *A rule* $\rho \in \mathcal{P}$ *is a tuple* $\langle e_U, e_R, O, d\rangle$ *where* $\rho.e_U$ *is a user attribute expression,* e_R *is a resource attribute expression,* $O \subseteq \mathcal{O}$ *is a set of operations, and* d *is the decision of the rule* $(d \in \{permit, deny\})$[2]*.*

The original definition of an ABAC rule does not include the notion of "signed rules" (i.e., rules that specify positive or negative authorizations). By default, $d = permit$, and an access request $\langle u, r, o\rangle$ for which there exist at least a rule $\rho = \langle e_U, e_R, O, d\rangle$ in $\mathcal{P}$ such that u satisfies e_U (denoted by $u \models e_U$), r satisfies e_R (denoted by $r \models e_R$), and $o \in O$, is permitted. Otherwise, the access request is assumed to be denied. Even though negative authorizations are the default in access control lists, mixed rules are useful when dealing with large sets of resources organized according to hierarchies, and have been widely investigated [21]. They are used in commercial access control systems (e.g., the access control model of SQL Servers provides negative authorizations by means of the DENY authorization command), and they are part of the XACML standard.

2.2 Access Control Decision Examples

An access control decision example (referred to as a *decision example* (l)) is composed of an access request and its corresponding decision. Formally,

Definition 2 (Access Control Decisions and Examples (l)). *An access control decision is a tuple* $\langle u, r, o, d\rangle$ *where* u *is the user who initiated the access request,* r *is the resource target of the access request,* o *is the operation requested on the resource, and* d *is the decision taken for the access request. A decision example* l *is a tuple* $\langle u, e_U, r, e_R, o, d\rangle$ *where* e_U *and* e_R *are a user attribute expression, and a resource attribute expression such that* $u \models e_U$*, and* $r \models e_R$*, and the other arguments are interpreted as in an access control decision.*

[2] Throughout the paper, we will use the dot notation to refer to a component of an entity (e.g., $\rho.d$ refers to the decision of the rule ρ).

There exists an unknown set $\mathcal{F}$ of all access control decisions that should be allowed in the system, but for which we only have partial knowledge through examples. In an example l, the corresponding e_U and e_R can collect a few attributes (e.g., role, age, country of origin, manager, department, experience) depending on the available log information. Note that an access control decision is an example where e_U and e_R are the constant functions that assign to any attribute $\perp$. We say that an example $\langle u, e_U, r, e_R, o, d\rangle$ belongs to an access control decision set S iff $\langle u, r, o, d\rangle \in S$. Logs of past decisions are used to create an *access control decision example dataset* (see Definition 3). We say that a set of ABAC policies $\mathcal{P}$ *controls* an access control request $\langle u, r, o\rangle$ iff there is a rule $\rho = \langle e_U, e_R, O, d\rangle \in \mathcal{P}$ that satisfies the access request $\langle u, r, o\rangle$. Similarly, we say that the request $\langle u, r, o\rangle$ *is covered by* $\mathcal{F}$ iff $\langle u, r, o, d\rangle \in \mathcal{F}$, for some decision d. Therefore, $\mathcal{P}$ can be seen as defining the set of all access decisions for access requests controlled by $\mathcal{P}$ and, hence, be compared directly with $\mathcal{F}$ - and with some overloading in the notation, we want decisions in $\mathcal{P}$ ($\langle u, r, o, d\rangle \in \mathcal{P}$) to be decisions in $\mathcal{F}$ ($\langle u, r, o, d\rangle \in \mathcal{F}$).

Definition 3 (Access Control Decision Example Dataset ($\mathcal{D}$)). *An access control decision example dataset is a finite set of decision examples (i.e., $\mathcal{D} = \{l_1, \ldots, l_n\}$).*

$\mathcal{D}$ is expected to be mostly, but not necessarily, a subset of $\mathcal{F}$. $\mathcal{D}$ is considered to be *noisy* if $\mathcal{D} \nsubseteq \mathcal{F}$.

2.3 Problem Definition

Using a set of decision examples, extracted from a system history of access requests and their authorized/denied decisions, together with some context information (e.g., metadata obtained from LDAP directories), we want to learn ABAC policies (see Definition 4). The context information provides user and resource identifications, as well as user and resource attributes and their functions needed for an ABAC model. Additionally, it might also provide complementary semantic information about the system in the form of a collection of typed binary relations, $\mathcal{T} = \{t_1, \ldots, t_n\}$, such that each relation t_i relates pairs of attribute values, i.e., $t_i \subseteq V_X(a_1) \times V_Y(a_2)$, with $X, Y \in \{\mathcal{U}, \mathcal{R}\}$, and $a_1 \in A_X$, and $a_2 \in A_Y$. For example, one of these relations can represent the organizational chart (department names are related in an *is_member* or *is_part* hierarchical relation) of the enterprise or institution where the ABAC access control system is deployed.

Definition 4 (Learning ABAC Policies by Examples and Context (*LAPEC*)). *Given an access control decision example dataset $\mathcal{D}$, and context information comprising the sets $\mathcal{U}$, $\mathcal{R}$, $\mathcal{O}$, the sets A_U and A_R, one of which could be empty, the functions assigning values to the attributes of the users and resources, and a possibly empty set of typed binary relations, $\mathcal{T} = \{t_1, \ldots, t_n\}$, LAPEC aims at generating a set of ABAC rules (i.e., $\mathcal{P} = \{\rho_1 \ldots \rho_m\}$) that are able to control all access requests in $\mathcal{F}$.*

2.4 Policy Generation Assessment

ABAC policies generated by *LAPEC* are assessed by evaluation of two quality requirements: ***correctness***, which refers to the capability of the policies to assign a correct decision to any access request (see Definition 5), and ***completeness***, which refers to ensuring that all actions, executed in the domain controlled by an access control system, are covered by the policies (see Definition 6). This assessment can be done against example datasets as an approximation of $\mathcal{F}$. Since datasets can be noisy, it is possible that two different decision examples for the same request are in the set. Validation will be done only against consistent datasets as we assume that the available set of access control examples is noise-free.

Definition 5 (Correctness). *A set of ABAC policies $\mathcal{P}$ is correct with respect to a consistent set of access control decisions $\mathcal{D}$ if and only if for every request $\langle u, r, o\rangle$ covered by $\mathcal{D}$, $\langle u, r, o, d\rangle \in \mathcal{P} \rightarrow \langle u, r, o, d\rangle \in \mathcal{D}$.*

Definition 6 (Completeness). *A set of ABAC policies $\mathcal{P}$ is complete with respect to a consistent set of access control decisions $\mathcal{D}$ if and only if, for every request $\langle u, r, o\rangle$, $\langle u, r, o\rangle$ covered by $\mathcal{D} \rightarrow \langle u, r, o\rangle$ is controlled by $\mathcal{P}$.*

These definitions allow $\mathcal{P}$ to control requests outside $\mathcal{D}$. The aim is twofold. First, when we learn $\mathcal{P}$ from an example dataset $\mathcal{D}$, we want $\mathcal{P}$ to be correct and complete with respect to $\mathcal{D}$. Second, beyond $\mathcal{D}$, we want to minimize incorrect decisions while maximizing completeness with respect to $\mathcal{F}$. Outside $\mathcal{D}$, we evaluate correctness statistically through cross validation with example datasets which are subsets of $\mathcal{F}$, and calculating Precision, Recall, F1 score, and accuracy of the decisions made by $\mathcal{P}$. We quantify completeness by the Percentage of Controlled Requests (PCR) which is the ratio between the number of decisions made by $\mathcal{P}$ among the requests covered by an example dataset and the total number of requests covered by the dataset.

Definition 7 (Percentage of Controlled Requests (PCR)). *Given a subset of access control decision examples $\mathcal{N} \subseteq \mathcal{F}$, and a policy set $\mathcal{P}$, the percentage of controlled requests is defined as follows:*

$$PCR = \frac{|\{\langle u, r, o\rangle \mid \langle u, r, o\rangle \text{ is covered by } \mathcal{N} \text{ and } \langle u, r, o\rangle \text{ is controlled by } \mathcal{P}\}|}{|\{\langle u, r, o\rangle \mid \langle u, r, o\rangle \text{ is covered by } \mathcal{N}\}|}$$

3 The Learning Framework

Our framework comprises four steps, see Fig. 1. For a running example see Appendix B.

3.1 Rules Mining

$\mathcal{D}$ provides a source of examples of user accesses to the available resources in an organization. In this step, we use association rule mining (ARM) [1] to analyze the association between users and resources.

Thus, rules having high association metrics (i.e., support and confidence scores) are kept to generate access control rules. ARM is used for capturing the frequency and data patterns and formalize them in rules. *Polisma* uses Apriori [2] (one of the ARM algorithms) to generate rules that are *correct and complete* with respect to $\mathcal{D}$. This step uses only the examples in $\mathcal{D}$ to mine a first set of rules, referred to as *ground rules.* These ground rules potentially are overfitted (e.g., ρ_1 in Fig. 8 in Appendix B) or unsafely-generalized (e.g., ρ_2 in Fig. 8 in Appendix B). Overfitted rules impact completeness over $\mathcal{F}$ while unsafely-generalized rules impact correctness. Therefore, *Polisma* post-processes the ground rules in the next step.

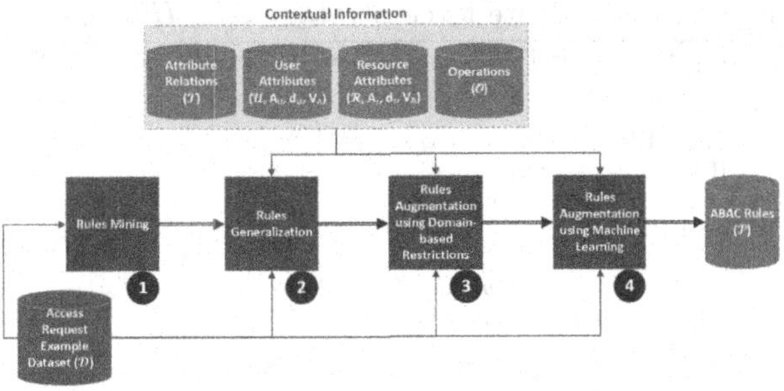

Fig. 1. The architecture of *Polisma*

3.2 Rules Generalization

To address overfitting and unsafe generalization associated with the ground rules, this step utilizes the set of user and resource attributes A_U, A_R provided by either $\mathcal{D}$, external context information sources, or both. Straightforwardly, e_U (i.e., user expression) or e_R (i.e., resource expression) of a rule can be adapted to expand or reduce the scope of each expression when a rule is overfitted or is unsafely generalized, respectively. In particular, each rule is post-processed based on its corresponding user and resource by analyzing A_U and A_R to statistically "choose" the most appropriate attribute expression that captures the subsets of users and resources having the same permission authorized by the rule according to $\mathcal{D}$. In this way, this step searches for an attribute expression that minimally generalizes the rules.

Polisma provides two strategies for post-processing ground rules. One strategy, referred to as Brute-force Strategy BS, uses only the attributes associated with users and resources appearing in $\mathcal{D}$. The other strategy assumes that attribute relationship metadata ($\mathcal{T}$) is also available. Hence, the second strategy, referred to as Structure-based Strategy SS, exploits both attributes and their relationships. In what follows, we describe each strategy.

Brute-Force Strategy (BS). To post-process a ground rule, this strategy searches heuristically for attribute expressions that cover the user and the resource of interest. Each attribute expression is weighted statistically to assess its "safety" level for capturing the authorization defined by the ground rule of interest.

The safety level of a user attribute expression e_U for a ground rule ρ is estimated by the proportion of the sizes of two subsets: a) the subset of decision

Algorithm 1. Brute-force Strategy (BS-U-S) for Generalizing Rules

Require: ρ_i: a ground rule, u_i: the user corresponding to ρ_i, $\mathcal{D}$, A_U.
Define w_{max} = -∞, $a_{selected}$= ⊥.
for $a_i \in A_U$ **do**
    $w_i = W_{uav}(u_i, a_i, d_U(u_i, a_i), \mathcal{D})$
    if $w_i > w_{max}$ **then**
        $a_{selected} = a_i$, $w_{max} = w_{a_i}$
    end if
end for
$e_U \leftarrow \langle\ a_{selected}, d_U(u_i, a_{selected})\ \rangle$
Create Rule $\rho_i' = \langle e_U, \rho_i.e_R, \rho_i.O, \rho_i.d\rangle$
return ρ_i'

Algorithm 2. Structure-based Strategy (SS) for Generalizing Rules

Require: ρ_i: a ground rule to be Generalized, u_i: the user corresponding to ρ_i, r_i: the resource corresponding to ρ_i, $\mathcal{T}$, A_U, A_R.
$g_i = G(u_i, r_i, A_U, A_R, \mathcal{T})$
$\langle x \in A_U, y \in A_R\rangle$ = First-Common-Attribute-Value(g_i)
$e_U \leftarrow \langle\ x, d_U(u_i, x)\ \rangle$; $e_R \leftarrow \langle\ y, d_R(r_i, y)\ \rangle$
Create Rule $\rho_i' = \langle e_U, e_R, \rho_i.O, \rho_i.d\rangle$
return ρ_i'

examples in $\mathcal{D}$ such that the access requests issued by users satisfy e_U while the remaining parts of the decision example (i.e., resource, operation, and decision) match the corresponding ones in ρ; and b) the subset of users who satisfy e_U. The safety level of a resource attribute expression is estimated similarly. Thereafter, the attribute expression with the highest safety level is selected to be used for post-processing the ground rule of interest. Formally:

Definition 8 (User Attribute Expression Weight $W_{uav}(u_i, a_j, d_U(u_i, a_j), \mathcal{D})$). *For a ground rule ρ and its corresponding user $u_i \in U$, the weight of a user attribute expression $\langle a_j, d_U(u_i, a_j)\rangle$ is defined by the formula: $W_{uav} = \frac{|U_D|}{|U_C|}$, where $U_D = \{l \in \mathcal{D} \mid d_U(u_i, a_j) = d_U(l.u, a_j) \wedge l.r \in \rho.e_R \wedge l.o \in \rho.O \wedge \rho.d = l.d\}$ and $U_C = \{u_k \in \mathcal{U} \mid d_U(u_i, a_j) = d_U(u_k, a_j)\}$.*

Different strategies for selecting attribute expressions are possible. They can be based on a single attribute or a combination of attributes, and they can consider either user attributes, resource attributes, or both. Strategies using only user attributes are referred to as BS-U-S "for a single attribute" and BS-U-C "for a combination of attributes". Similarly, BS-R-S and BS-R-C are strategies for a single or a combination of resources attributes. The strategies using the attributes of both users and resources are referred to as BS-UR-S and BS-UR-C. Due to space limitation, we show only the algorithm using the selection strategy BS-U-S (see Algorithm 1).

Structure-Based Strategy (SS). The BS strategy works without prior knowledge of $\mathcal{T}$. When $\mathcal{T}$ is available, it can be analyzed to gather information about "hidden" correlations between common attributes of a user and a resource. Such attributes can potentially enhance rule generalization. For example, users working in a department t_i can potentially access the resources owned by t_i. The binary relations in $\mathcal{T}$ are combined to form a directed graph, referred to as *Attribute-relationship Graph* G (see Definition 9). Traversing this graph starting from the lowest level in the hierarchy to the highest one allows one to find the common attributes values between users and resources. Among the common attribute values, the one with the least hierarchical level, referred to as *first common attribute value*, has heuristically the highest safety level for generalization because the generalization using such attribute value supports the least level of generalization. Subsequently, to post-process a ground rule, this strategy uses $\mathcal{T}$ along with A_U, A_R of both the user and resource of interest to build G. Thereafter, G is used to find the *first common attribute value* between the user and resource of that ground rule to be used for post-processing the ground rule of interest (as described in Algorithm 2).

Definition 9 (Attribute-relationship Graph G). *Given A_U, A_R, ρ and $\mathcal{T}$, the attribute-relationship graph (G) of ρ is a directed graph composed of vertices V and edges E where*

$$
\begin{aligned}
V = \; & \{v \mid \forall u_i \in \rho.e_U, \forall a_i \in A_U, v = d_U(u_i, a_i)\} \\
& \cup \{v \mid \forall r_i \in \rho.e_R, \forall a_i \in A_R, v = d_R(r_i, a_i)\}, \text{ and} \\
E = \; & \{(v_1, v_2) \mid \exists t_i \in \mathcal{T} \wedge (v_1, v_2) \in t_i \wedge v_1, v_2 \in V\}.
\end{aligned}
$$

Proposition 1. *Algorithms 1 and 2 output generalized rules that are correct and complete with respect to $\mathcal{D}$.*

As discussed earlier, the first step generates ground rules that are either overfitted or unsafely-generalized. This second step post-processes unsafely-generalized rules into safely-generalized ones; hence, improving correctness. It may also post-process overfitted rules into safely-generalized ones; hence, improving completeness. However, completeness can be further improved as described in the next subsections.

3.3 Rules Augmentation Using Domain-Based Restrictions

"Safe" generalization of ground rules is one approach towards fulfilling completeness. Another approach is to analyze the authorization domains of users and resources appearing in $\mathcal{D}$ to augment restriction rules, referred to as *domain-based restriction rules*[3]. Such restrictions potentially increase safety by avoiding erroneous accesses. Basically, the goal of these restriction rules is to augment negative authorization.

One straightforward approach for creating domain-based restriction rules is to analyze the authorization domain for each individual user and resource. However, such an approach leads to the creation of restrictions suffering from overfitting. Alternatively, the analysis can be based on groups of users or resources.

[3] This approach also implicitly improves correctness.

Identifying such groups requires pre-processing A_U and A_R. Therefore, this step searches heuristically for preferable attributes to use for partitioning the user and resource sets. The heuristic prediction for preferable attributes is based on selecting an attribute that partitions the corresponding set evenly. Hence, estimating the attribute distribution of users or resources allows one to measure the ability of the attribute of interest for creating even partitions[4]. One method for capturing the attribute distribution is to compute the attribute entropy as defined in Eq. 1. The attribute entropy is maximized when the users are distributed evenly among the user partitions using the attribute of interest (i.e., sizes of the subsets in $G_U^{a_i}$ are almost equal). Thereafter, the attribute with the highest entropy is the preferred one.

Given the preferred attributes, this step constructs user groups as described in Definition[5] 10. These groups can be analyzed with respect to $\mathcal{O}$ as described in Algorithm 3 Consequently, user groups $G_U^{a_i}$ and resource groups $G_R^{a_i}$ along with $\mathcal{O}$ comprise two types of authorization domains: operations performed by each user group, and users' accesses to each resource group. Subsequently, the algorithm augments restrictions based on these access domains as follows.

- It generates deny restrictions based on user groups (similar to the example in Fig. 11 in Appendix B). These restriction rules deny any access request from a specific user group using a specific operation to access any resource.
- It generates deny restrictions based on both groups of users and resources. These restriction rules deny any access request from a specific user group using a specific operation to access a specific resource group.

On the other hand, another strategy assumes prior knowledge about preferred attributes to use for partitioning the user and resource sets (referred to as Semantic Strategy (SS)).

Definition 10 (Attribute-based User Groups $G_U^{a_i}$). *Given $\mathcal{U}$ and $a_i \in A_U$, $\mathcal{U}$ is divided into is a set of user groups $G_U^{a_i}$ $\{g_1^{a_i}, \ldots, g_k^{a_i}\}$ where $(g_i^{a_i} = \{u_1, \ldots, u_m\} \mid \forall u', u'' \in g_i, d_U(u', a_i) = d_U(u'', a_i)\ , m \leq \mid \mathcal{U} \mid) \wedge (g_i \cap g_j = \phi \mid i \neq j) \wedge (k = \mid V_U(a_i) \mid)$.*

$$Entropy(G_U^{a_i}, a_i) = (\frac{-1}{\ln m} * \sum_{j=1, g_j \in G_U^{a_i}}^{m} p_j * \ln p_j), where\ m = |G_U^{a_i}|,\ p_j = \frac{|g_j|}{\sum_{l=1, g_l \in G_U^{a_i}}^{m} |g_l|} \quad (1)$$

Proposition 2. *Step 3 outputs generalized rules that are correct and complete with respect to $\mathcal{D}$.*

[4] Even distribution tends to generate smaller groups. Each group potentially has a similar set of permissions. A large group of an uneven partition potentially includes a diverse set of users or resource; hence hindering observing restricted permissions.

[5] The definition of constructing resource groups is analogous to that of user groups.

3.4 Rules Augmentation Using Machine Learning

The rules generated from the previous steps are generalized using domain knowledge and data statistics extracted from $\mathcal{D}$. However, these steps do not consider generalization based on the similarity of access requests. Thus, these rules cannot control a new request that is similar to an old one (i.e., a new request has a similar pattern to one of the old requests, but the attribute values of the new request do not match the old ones).

A possible prediction approach is to use an ML classifier that builds a model based on the attributes provided by $\mathcal{D}$ and context information. The generated model implicitly creates patterns of accesses and their decisions, and will be able to predict the decision for any access request based on its similarity to the ones controlled by the rules generated by the previous steps. Thus, this step creates new rules based on these predictions. Once these new rules are created, *Polisma* repeats Step 2 to safely generalize the ML-based augmented rules. Notice that ML classification algorithms might introduce some inaccuracy when performing the prediction. Hence, we utilize this step only for access requests that are not covered by the rules generated by the previous three steps. Thus, the inaccuracy effect associated with the ML classifier is minimized but the *correctness and completeness* are preserved.

Note that Polisma is used as a one-time learning. However, if changes arise in terms of regulations, security policies and/or organizational structure of the organization, the learning can be executed again (i.e., on-demand learning).

4 Evaluation

In this section, we summarize the experimental methodology and report the evaluation results of *Polisma*.

4.1 Experimental Methodology

Datasets. To evaluate *Polisma*, we conducted several experiments using two datasets: one is a synthetic dataset (referred to as *project management* (*PM*) [24]), and the other is a real one (referred to as *Amazon*[6]). The *PM* dataset has been generated by the Reliable Systems Laboratory at Stony Brook University based on a probability distribution in which the ratio is 25 for rules, 25 for resources, 3 for users, and 3 for operations. In addition to decision examples, *PM* is tagged with context information (e.g., attribute relationships and attribute semantics). We used such a synthetic dataset (i.e., *PM*) to show the impact of the availability of such semantic information on the learning process and the quality of the generated rules. Regarding the Amazon dataset, it is a historical dataset collected in 2010 and 2011. This dataset is an anonymized sample of access provisioned within the Amazon company. One characteristic of

[6] http://archive.ics.uci.edu/ml/datasets/Amazon+Access+Samples.

Table 1. Overview of datasets

	Datasets	
	Project Management (PM)	*Amazon (AZ)*
# of decision examples	550	1000
# of users	150	480
# of resources	292	271
# of operations	7	1
# of examples with a "Permit" decision	505	981
# of examples with a "Deny" decision	50	19
# of User Attributes	5	12
# of Resource Attributes	6	1

the Amazon dataset is that it is sparse (i.e., less than 10% of the attributes are used for each sample). Furthermore, since the Amazon dataset is large (around 700K decision examples), we selected a subset of the Amazon dataset (referred to as *AZ*). The subset was selected randomly based on a statistical analysis [11] of the size of the Amazon dataset where the size of the selected samples suffices a confidence score of 99% and a margin error score of 4%. Table 1 shows statistics about the *PM* and *AZ* datasets.

Comparative Evaluation. We compared *Polisma* with three approaches. We developed a *Naïve ML approach* which utilizes an ML classifier to generate rules. A classification model is trained using $\mathcal{D}$. Thereafter, the trained model is used to generate rules without using the generalization strategies which are used in *Polisma*. The rules generated from this approach are evaluated using another set of new decision examples (referred to as $\mathcal{N}$)[7]. Moreover, we compared *Polisma* with two recently published state-of-the-art approaches for learning ABAC policies from logs: a) Xu & Stoller Miner [24], and b) Rhapsody by Cotrini *et al.* [5].

Evaluation and Settings. On each dataset, *Polisma* is evaluated by splitting the dataset into training $\mathcal{D}$ (70%) and testing $\mathcal{N}$ (30%) subsets. We performed a 10-fold cross-validation on $\mathcal{D}$ for each dataset. As the ML classifiers used for the fourth step of *Polisma* and the naïve approach, we used a combined classification technique based on majority voting where the underlying classifiers are Random Forest and kNN[8]. All experiments were performed on a 3.6 GHz Intel Core i7 machine with 12 GB memory running on 64-bit Windows 7.

[7] Datasets are assumed to be noise-free, that is, $(\mathcal{N} \subset \mathcal{F}) \wedge (\mathcal{D} \subset \mathcal{F}) \wedge (\mathcal{N} \cap \mathcal{D} = \phi)$. Note that $\mathcal{F}$ is the complete set of control decisions which we will never have in a real system.

[8] Other algorithms can be used. We used Random Forest and kNN classifiers since they showed better results compared to SVM and Adaboost.

Evaluation Metrics. The correctness of the generated ABAC rules is evaluated using the standard metrics for the classification task in the field of machine learning. Basically, predicted decisions for a set of access requests are compared with their ground-truth decisions. Our problem is analogous to a two-class classification task because the decision in an access control system is either "permit" or "deny". Therefore, for each type of the decisions, a confusion matrix is prepared to enable calculating the values of accuracy, precision, recall, and F1 score as outlined in Eqs. 2–3[9].

$$Precision = \frac{TP}{TP + FP}, Recall = \frac{TP}{TP + FN} \tag{2}$$

$$F1\ Score = 2 \cdot \frac{Precision \cdot Recall}{Precision + Recall}, Accuracy = \frac{TP + TN}{TP + TN + FP + FN} \tag{3}$$

Regarding the completeness of the rules, they are measured using *PCR* (see Definition 7).

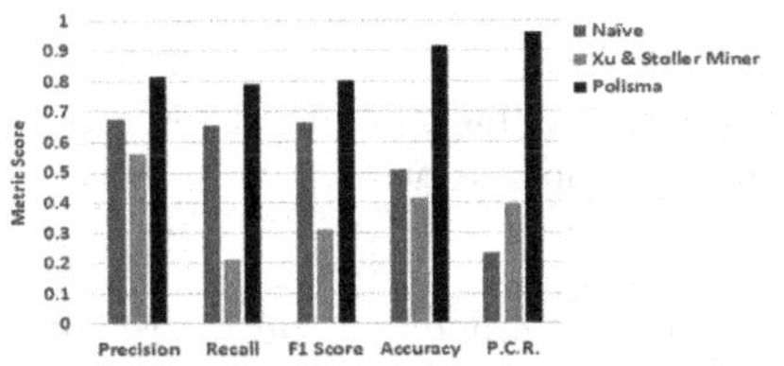

Fig. 2. Comparison of Naïve, Xu & Stoller Miner, and *Polisma* Using the *PM* Dataset

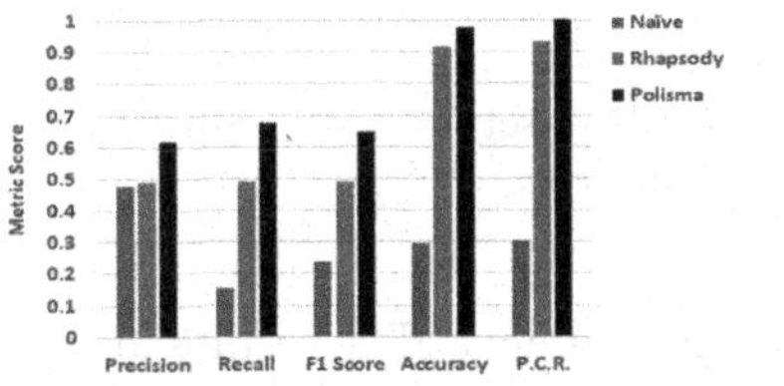

Fig. 3. Comparison of Naïve, Rhapsody, and *Polisma* Using the *AZ* Dataset

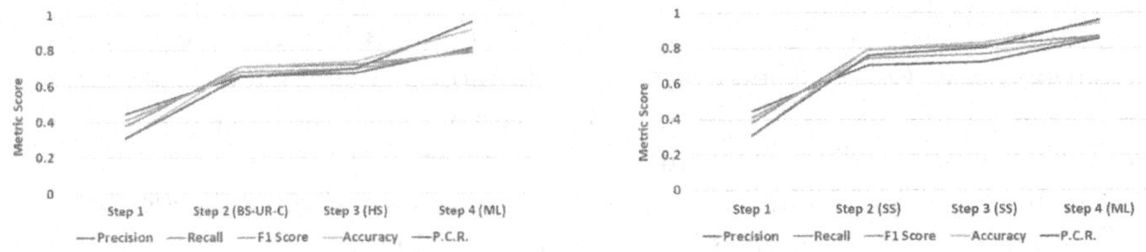

(a) *Polisma* (BS-HS-ML) (b) *Polisma* (SS-SS-ML)

Fig. 4. Evaluation of *Polisma* using the *PM* dataset.

4.2 Experimental Results

Polisma vs. Other Learning Approaches. Here, *Polisma* uses the default strategies in each step (i.e., the *BS-UR-C* strategy in the second step and the *HS* strategy in the third step[10]) and the impact of the other strategies in *Polisma*

[9] F1 Score is the harmonic mean of precision and recall.

[10] Since the *AZ* dataset does not contain resource attributes, *BS-R-C* (instead of *BS-UR-C*) is executed in the second step and the execution of the third step is skipped.

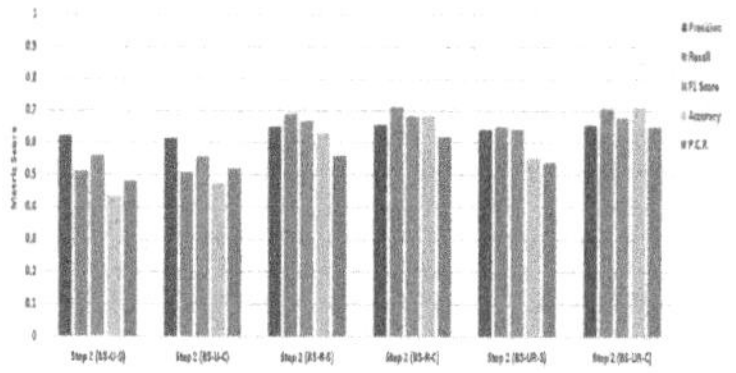

Fig. 5. Comparison between the variants of the brute-force strategy (Step 2) using the *PM* dataset.

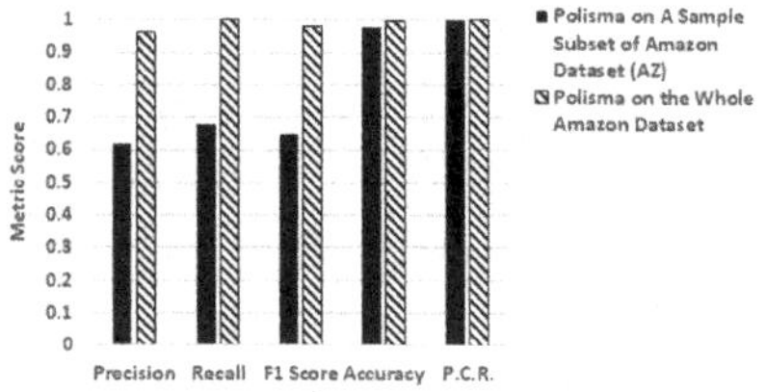

Fig. 6. *Polisma* Evaluation on the Amazon Dataset (a sample subset and the whole set).

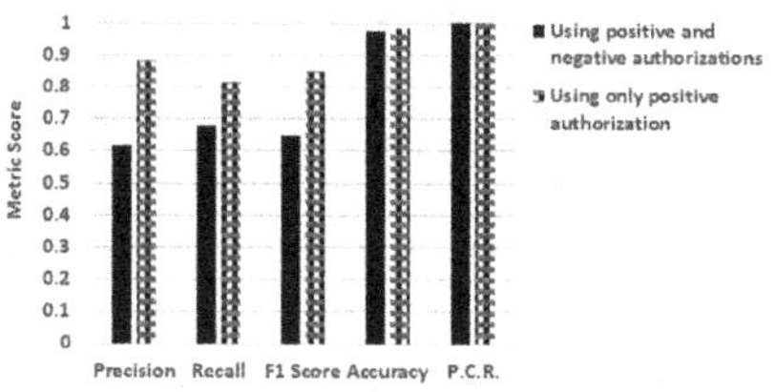

Fig. 7. *Polisma* Evaluation on a sample subset of Amazon Dataset for only positive authorizations.

is evaluated next. The results in Fig. 2 show that *Polisma* achieves better results compared to the naïve and Xu & Stoller Miner using the *PM* dataset. In this comparison, we did not include Rhapsody because the default parameters of Rhapsody leads to generating no rule. With respect to Xu & Stoller Miner, *Polisma*'s F1 score (*PCR*) improves by a factor of 2.6(2.4). This shows the importance of integrating multiple techniques (i.e., data mining, statistical, and ML techniques) in *Polisma* to enhance the policy learning outcome. Moreover, Xu & Stoller Miner requires prior knowledge about the structure of context information while *Polisma* benefits from such information when available. Xu & Stoller Miner [24] generates rules only for positive authorization (i.e., no negative authorization). This might increase the possibility of vulnerabilities and encountering incorrect denials or authorizations. Moreover, Xu & Stoller in their paper [24] used different metrics from the classical metrics for data mining that are used in our paper. In their work [24], they used a similarity measure between the generated policy and the real policy which given based on some distance function. However, such a metric does not necessarily reflect the correctness of the generated policy (i.e., two policies can be very similar, but their decisions are different). In summary, this experiment shows that the level of correctness (as indicated by the precision, recall and F1 score) and completeness (as indicated by the *PCR*) of the policies that are generated by *Polisma* is higher than for the policies generated by Xu & Stoller Miner and the naïve approach due to the reasons explained earlier (i.e., the lack of negative authorization rules), as well as generating generalized rules without considering the safety of generalization. Furthermore, as shown in Fig. 3, *Polisma* also outperforms the naïve approach and Rhapsody using the *AZ* dataset. Rhapsody [5] only considers positive authorization (similar to the problem discussed above about Xu & Stoller Miner [24]); hence increasing the chances of vulnerabilities. In this comparison, we have excluded Xu & Stoller Miner because of the shortage of available resource attributes information in the *AZ* dataset. Concerning the comparison with the naïve approach, on the *PM* dataset, the F1 score (*PCR*) of *Polisma* improves by a factor of 1.2 (4.1) compared to that of the naïve. On the *AZ* dataset, *Polisma*'s F1 score (*PCR*) improves by a factor of 2.7 (3.3) compared to the naïve. Both *Polisma* and the naïve approach use ML classifiers. However, *Polisma* uses an ML classifier for generating only some of the rules. This implies that among *Polisma* steps which improve the results of the generated rules (i.e., Steps 2 and Step 4), the rule generalization step is essential for enhancing the final outcome. In summary, the reported results show that the correctness level of the rules generated by *Polisma* is better than that of the ones generated by Rhapsody and the naïve approach (as indicated by the difference between the precision, recall, and F1 scores metrics). Meanwhile, the difference between the completeness level (indicated by PCR) of the generated rules using *Polisma* and that of Rhapsody is not large. The difference in terms of completeness and correctness is a result of Rhapsody missing the negative authorization rules.

Steps Evaluation. Figure 4 shows the evaluation results for the four steps of *Polisma* in terms of precision, recall, F1 score, accuracy, and *PCR*. In general, the quality of the rules improves gradually for the two datasets, even when using different strategies in each step. The two plots show that the strategies that exploit additional knowledge about user and resource attributes (e.g., $\mathcal{T}$) produce better results when compared to the ones that require less prior knowledge. Moreover, the quality of the generated rules is significantly enhanced after the second and the fourth steps. This shows the advantage of using safe generalization and similarity-based techniques. On the other hand, the third step shows only a marginal enhancement on the quality of rules. However, the rules generated from this step can be more beneficial when a large portion of the access requests in the logs are not allowed by the access control system.

We also conducted an evaluation on the whole Amazon dataset; the results are shown in Fig. 6. On the whole dataset, *Polisma* achieves significantly improved scores compared to those when using the *AZ* dataset because *Polisma* was able to utilize a larger set of decision examples. Nonetheless, given that the size of *AZ* was small compared to that of the whole Amazon dataset, *Polisma* outcomes using *AZ*[11] are promising. In summary, the increase of recall can be interpreted as reducing denials of authorized users (i.e., smaller number of false-negatives). A better precision value is interpreted as being more accurate with the decisions (i.e., smaller number of false-positives). Figures 4a–4b show that most of the reduction of incorrect denials is achieved in Steps 1 and 2, whereas the other steps improve precision which implies more correct decisions (i.e., decreas-

[11] We also experimented samples of different sizes (i.e., 2k–5k), the learning results using these sample sizes showed slight improvement of scores.

ing the number of false-positives and increasing the number of true-positives). As Fig. 7 shows, the results improve significantly when considering only positive authorizations. This is due to the fact that the negative examples are few and so they are not sufficient in improving the learning of negative authorizations. As shown in the figure, precision, recall, and F1 score increase indicating that the learned policies are correct. Precision is considered the most effective metric especially for only positive authorizations to indicate the correctness of the generated policies since higher precision implies less incorrect authorizations.

Variants of the Brute-Force Strategy. As the results in Fig. 5 show, using resource attributes for rules generalization is better than using the user attributes. The reason is that the domains of the resource attributes (i.e., $V_R(a_i) \mid a_i \in A_R$) is larger than that of the user attributes (i.e., $V_U(a_j) \mid a_j \in A_U$) in the *PM* dataset. Thus, the selection space of attribute expressions is expanded; hence, it potentially increases the possibility of finding better attribute expression for safe generalization. In particular, using resources attributes achieves 23% and 19% improvement F1 score and *PCR* when compared to that of using users attributes producing a better quality of the generated policies in terms of correctness (as indicated by the values of precision, recall, and F1 score) and completeness (as indicated by the *PCR* value). Similarly, performing the generalization using both user and resource attributes is better than that of using only user attribute or resource attributes because of the larger domains which can be exploited to find the best attribute expression for safe generalization. In particular, *BS-UR-C* shows a significant improvement compared to *BS-U-C* (22% for F1 score (which reflects a better quality of policies in terms of correctness), and 25% for *PCR* (which reflects a better quality of policies in terms of completeness)). In conclusion, the best variant of the *BS* strategy is the one that enables exploring the largest set of possible attribute values to choose the attribute expression which has the highest safety level.

5 Related Work

Policy mining has been widely investigated. However, the focus has been on RBAC role mining that aims at finding roles from existing permission data [16, 23]. More recent work has focused on extending such mining-based approaches to ABAC [5,8,14,15,24]. Xu and Stoller [24] proposed the first approach for mining ABAC policies from logs. Their approach iterates over a log of decision examples greedily and constructs candidate rules; it then generalizes these rules by utilizing merging and refinement techniques. Medvet *et al.* [14] proposed an evolutionary approach which incrementally learns a single rule per group of decision examples utilizing a divide-and-conquer algorithm. Cotrini *et al.* [5] proposed another rule mining approach, called Rhapsody, based on APRIORI-SD (a machine-learning algorithm for subgroup discovery) [9]. It is incomplete, only mining rules covering a significant number of decision examples. The mined rules are then filtered to remove any refinements. All of those approaches [5,14,24] mainly rely on decision logs assuming that they are sufficient for rule generalization. However, logs, typically being very sparse, might not contain sufficient information for mining rules of good quality. Thus, those approaches may not be always applicable. Moreover, neither of those approaches is able to generate negative authorization rules.

Mocanu *et al.* [15] proposed a deep learning model trained on logs to learn a Restricted Boltzmann Machine (RBM). Then, the learned model is used to generate candidate rules. Their proposed system is still under development and further testing is needed. Karimi and Joshi [8] proposed to apply clustering algorithms over the decision examples to predict rules, but they don't support rules generalization. To the best of our knowledge, *Polisma* is the first generic framework that incorporates context information, when available, along with decisions logs to increase accuracy. Another major difference of *Polisma* with respect to existing approaches is that *Polisma* can use ML techniques, such as statistical ML techniques, for policy mining while at the same time being able to generate ABAC rules expressed in propositional logics.

6 Conclusion and Future Work

In this paper, we have proposed *Polisma*, a framework for learning ABAC policies from examples and context information. *Polisma* comprises four steps using various techniques, namely data mining, statistical, and machine learning. Our evaluations, carried out on a real-world decision log and a synthetic log, show that *Polisma* is able to learn policies that are both complete and correct. The rules generated by *Polisma* using the synthetic dataset achieve an F1 score of 0.80 and *PCR* of 0.95; awhen using the real dataset, the generated rules achieve an F1 score of 0.98 and *PCR* of 1.0. Furthermore, by using the semantic information available with the synthetic dataset, *Polisma* improves the F1 score to reach 0.86. As part of future work, we plan to extend our framework by integrating an inductive learner [12] and a probabilistic learner [6]. We also plan to extend our framework to support policy transfer and policy learning explainability. In addition, we plan to extend *Polisma* with of conflict resolution techniques to deal with input noisy data. Another direction is to integrate in *Polisma* different ML algorithms, such as neural networks. Preliminary experiments about using neural networks for learning access control policies have been reported [4]. However, those results show that neural networks are not able to accurately learn negative authorizations and thus work is required to enhance them.

Acknowledgment. This research was sponsored by the U.S. Army Research Laboratory and the U.K. Ministry of Defence under Agreement Number W911NF-16-3-0001. The views and conclusions contained in this document are those of the authors and should not be interpreted as representing the official policies, either expressed or implied, of the U.S. Army Research Laboratory, the U.S. Government, the U.K. Ministry of Defence or the U.K. Government. The U.S. and U.K. Governments are authorized to reproduce and distribute reprints for Government purposes notwithstanding any copyright notation hereon. Jorge Lobo was also supported by the Spanish Ministry of Economy and Competitiveness under Grant Numbers TIN201681032P, MDM20150502, and the U.S. Army Research Office under Agreement Number W911NF1910432.

A Additional Algorithms

Algorithm 3 outlines the third step of *Polisma* for augmenting rules using domain-based restrictions.

Algorithm 3. Rules Augmentation using Domain-based Restrictions

Require: $\mathcal{D}$, $\mathcal{U}$, $\mathcal{R}$, x: a preferable attribute of users, y: a preferable attribute of resources, $\mathcal{O}$.

$G_U = \mathcal{G}_U^{a_i}(\mathcal{U}, x)$; $G_R = \mathcal{G}_R^{a_i}(\mathcal{R}, y)$

$\forall\, g_i \in G_U$, $O^u_{g_i} \rightarrow \{\}$; $\forall\, g_i \in G_R$, $O^r_{g_i} \rightarrow \{\}$; $\forall\, g_i \in G_R$, $U^r_{g_i} \rightarrow \{\}$

for $l_i \in \mathcal{D}$ **do**

    $g' \leftarrow g_i \in G_U \mid l_i.u \in g_i \wedge l_i.d = Permit$; $O^u_{g'} \rightarrow O^u_{g'} \cup l_i.o$

    $O^r_{g''} \rightarrow O^r_{g''} \cup l_i.o$; $U^r_{g''} \rightarrow U^r_{g''} \cup l_i.u$

end for

$\mathcal{P}' \rightarrow \{\}$

for $g_i \in G_U$ **do**

    $\rho_i = \langle\langle x, d_U(u_i, x)\rangle, *, o_i, Deny\rangle \mid u_i \in g_i \wedge o_i \in (\mathcal{O} \setminus O^u_{g_i})$; $\mathcal{P}' \rightarrow \mathcal{P}' \cup \rho_i$

end for

for $g_i \in G_R$ **do**

    $\rho_i = \langle\langle x, d_U(u_i, x)\rangle, \langle y, d_R(r_i, y)\rangle, *, Deny\rangle \mid r_i \in g_i \wedge u_i \in (\mathcal{U} \setminus U^r_{g_i})$; $\mathcal{P}' \rightarrow \mathcal{P}' \cup \rho_i$

end for

return $\mathcal{P}'$

B Running Example of Policy Learning Using *Polisma*

Consider a system including users and resources both associated with projects. User attributes, resource attributes, operations, and possible values for two selected attributes are shown in Table 2. Assume that a log of access control decision examples is given.

Table 2. Details about A Project Management System

User Attributes (A_U)	{id, role, department, project, technical area}
Resource Attributes (A_R)	{id, type, department, project, technical area}
Operations List ($\mathcal{O}$)	{request, read, write, setStatus, setSchedule, approve, setCost}
V_U(role)	{planner, contractor, auditor, accountant, department manager, project leader}
V_R(type)	{budget, schedule, and task}

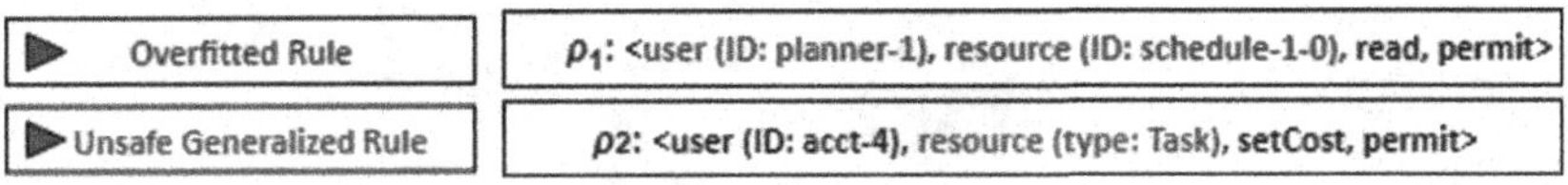

Fig. 8. Examples of ground rules generated from rule mining based on the specifications of the running example

B.1 Rules Generalization

Brute-Force Strategy (BS). *Assume that BS-UR-C is used to generalize ρ_2 defined in Fig. 8. A_U is {id, role, department, project, technical area} and A_R is {id, type, department, project, technical area}. Moreover, ρ_2 is able to control the access for the user whose ID is "acct-4' when accessing a resource whose type is task. The attribute values of the user and resources controlled by ρ_2 are analyzed. To generalize ρ_2 using BS-UR-C, each attribute value is weighted as shown in Fig. 9. For weighting each user/resource attribute value, the proportion of the sizes of two user/resource subsets is calculated according to Definition 8.*

In particular, for the value of the "department" attribute corresponding to the user referred by ρ_2 (i.e., "d1") (Fig. 9b), two user subsets are first found: a) the subset of the users belonging to department "d1"; and b) the subset of the users belonging to department "d1" and having a permission to perform the "setCost" operation on a resource of type "task" based on $\mathcal{D}$. Then, the ratio of the sizes of these subsets is considered as the weight for the attribute value "d1". The weights for the other user and resource attributes values are calculated similarly. Thereafter, the user attribute value and resource attribute value with the highest wights are chosen to perform the generalization. Assume that the value of the "department" user attribute is associated with the highest weight and the "project" resource attribute is associated with the highest weight. ρ_2 is generalized as:
$\rho_2' = \langle$user(department: d1), resource (type: task, project: d1-p1), setCost, permit$\rangle$

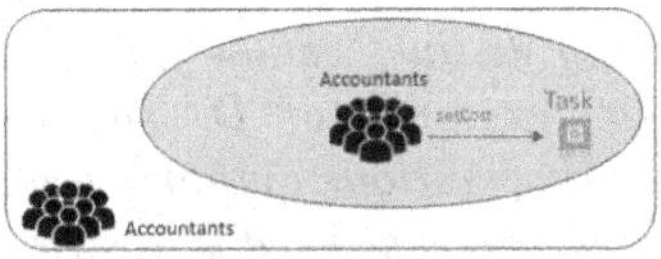

(a) w.r.t. $d_U(u_{id='acct-4'}$, "role")

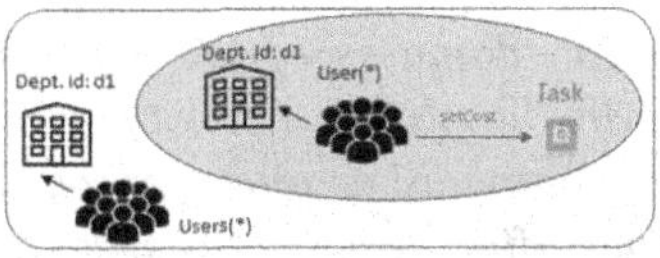

(b) w.r.t. $d_U(u_{id='acct-4'}$, "department")

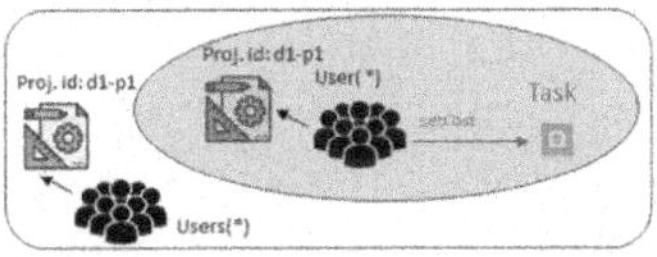

(c) w.r.t. $d_U(u_{id='acc-4'}$, "project")

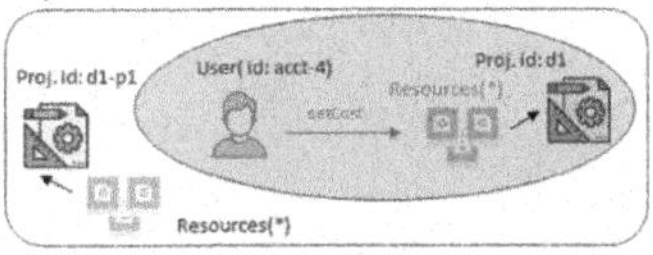

(d) w.r.t. $d_R(r_{type='task'}$, "project")

Fig. 9. Generalization of ρ_2 defined in Fig. 8 using the Brute Force Strategy (BS-UR-C)

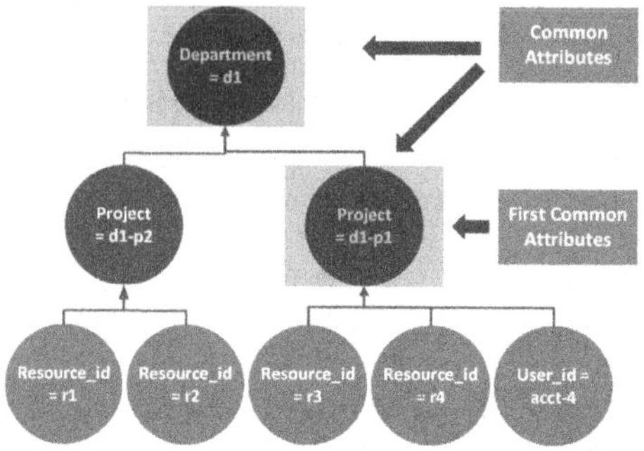

Fig. 10. Generalization of ρ_2 defined in Fig. 8 using Structure-based Strategy: An example of Attribute-relationship Graph

Structure-Based Strategy (SS). *Assume that SS is used to generalize ρ_2, defined in Fig. 8. Also, suppose that Polisma is given the following information:*

- *The subset of resources R' satisfying $\rho_2.e_R$ has two values for the project attribute (i.e., "d1-p1", "d1-p2").*
- *The user "acct-4" belongs to the project "d1-p1".*
- *R' and "acct-4" belong to the department "d1".*
- *$\mathcal{T}$ = {("d1-p1", "d1"), ("d1-p2", "d1")}.*

G is constructed as shown in Fig. 10. Using G, the two common attributes for "acct-4" and R' are "d1-p1" and "d1" and the first common attribute is "d1-p1". Therefore, ρ_2 is generalized as follows:
ρ_2'' = ⟨user(project: d1-p1), resource (type: task, project: d1-p1), setCost, permit⟩

B.2 Rules Augmentation Using Domain-Based Restrictions

Assume that we decide to analyze the authorization domain by grouping users based on the "role" user attribute. As shown in the top part of Fig. 11, the authorization domains of the user groups having distinct values for the "role" attribute are identified using the access requests examples of $\mathcal{D}$. These authorization domains allow one to recognize the set of operations authorized per user group. Thereafter, a set of negative authorizations are generated to restrict users having a specific role from performing specific operations on resources.

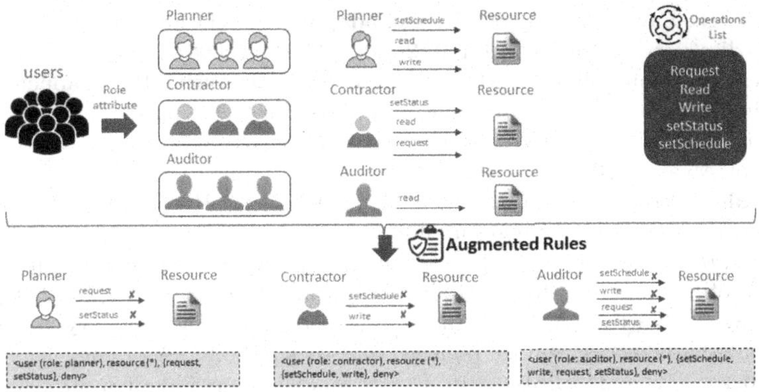

Fig. 11. Rules augmentation using domain-based restrictions

B.3 Rules Augmentation Using Machine Learning

Assume that $\mathcal{D}$ includes a decision example (l_i) for a user (id: "pl-1") accessing a resource (id: "sc-1") where both of them belong to a department "d1". Assuming Polisma generated a rule based on l_i as follows: ρ_i = ⟨ user(role: planner, department: d1), resource (type: schedule, department: d1), read, permit ⟩ Such a rule cannot control a new request by a user (id: "pl-5" for accessing a resource (id: "sc-5") where both of them belong to another department "d5"). Such a is similar to l_i. Therefore, a prediction-based approach is required to enable generating another rule.

References

1. Agrawal, R., Imieliński, T., Swami, A.: Mining association rules between sets of items in large databases. In: ACM SIGMOD Record, vol. 22, pp. 207–216. ACM (1993)
2. Agrawal, R., Srikant, R.: Fast algorithms for mining association rules. VLDB **1215**, 487–499 (1994)
3. Bertino, E., Ghinita, G., Kamra, A.: Access control for databases: concepts and systems. Found. Trends® Databases **3**(1–2), 1–148 (2011)
4. Cappelletti, L., Valtolina, S., Valentini, G., Mesiti, M., Bertino, E.: On the quality of classification models for inferring ABAC policies from access logs. In: Big Data, pp. 4000–4007. IEEE (2019)
5. Cotrini, C., Weghorn, T., Basin, D.: Mining ABAC rules from sparse logs. In: EuroS&P, pp. 31–46. IEEE (2018)
6. De Raedt, L., Dries, A., Thon, I., Van den Broeck, G., Verbeke, M.: Inducing probabilistic relational rules from probabilistic examples. In: IJCAI (2015)
7. Hu, V., et al.: Guide to attribute based access control (ABAC) definition and considerations (2017). https://nvlpubs.nist.gov/nistpubs/specialpublications/nist.sp.800-162.pdf
8. Karimi, L., Joshi, J.: An unsupervised learning based approach for mining attribute based access control policies. In: Big Data, pp. 1427–1436. IEEE (2018)

9. Kavšek, B., Lavrač, N.: APRIORI-SD: adapting association rule learning to subgroup discovery. Appl. Artif. Intell. **20**(7), 543–583 (2006)
10. Kohavi, R., Sommerfield, D.: Feature subset selection using the wrapper method: overfitting and dynamic search space topology. In: KDD, pp. 192–197 (1995)
11. Krejcie, R.V., Morgan, D.W.: Determining sample size for research activities. Educ. Psychol. Measur. **30**(3), 607–610 (1970)
12. Law, M., Russo, A., Elisa, B., Krysia, B., Jorge, L.: Representing and learning grammars in answer set programming. In: AAAI (2019)
13. Maxion, R.A., Reeder, R.W.: Improving user-interface dependability through mitigation of human error. Int. J. Hum.-Comput. Stud. **63**(1–2), 25–50 (2005)
14. Medvet, E., Bartoli, A., Carminati, B., Ferrari, E.: Evolutionary inference of attribute-based access control policies. In: Gaspar-Cunha, A., Henggeler Antunes, C., Coello, C.C. (eds.) EMO 2015. LNCS, vol. 9018, pp. 351–365. Springer, Cham (2015). https://doi.org/10.1007/978-3-319-15934-8_24
15. Mocanu, D., Turkmen, F., Liotta, A.: Towards ABAC policy mining from logs with deep learning. In: IS, pp. 124–128 (2015)
16. Molloy, I., et al.: Mining roles with semantic meanings. In: SACMAT, pp. 21–30. ACM (2008)
17. Ni, Q., Lobo, J., Calo, S., Rohatgi, P., Bertino, E.: Automating role-based provisioning by learning from examples. In: SACMAT, pp. 75–84. ACM (2009)
18. AuthZForce. https://authzforce.ow2.org/
19. Balana. https://github.com/wso2/balana
20. OASIS eXtensible Access Control Markup Language (XACML) TC. https://www.oasis-open.org/committees/tc_home.php?wg_abbrev=xacml
21. Rabitti, F., Bertino, E., Kim, W., Woelk, D.: A model of authorization for next-generation database systems. TODS **16**(1), 88–131 (1991)
22. Sadeh, N., et al.: Understanding and capturing people's privacy policies in a mobile social networking application. Pers. Ubiquitous Comput. **13**(6), 401–412 (2009)
23. Xu, Z., Stoller, S.D.: Algorithms for mining meaningful roles. In: SACMAT, pp. 57–66. ACM (2012)
24. Xu, Z., Stoller, S.D.: Mining attribute-based access control policies from logs. In: Atluri, V., Pernul, G. (eds.) DBSec 2014. LNCS, vol. 8566, pp. 276–291. Springer, Heidelberg (2014). https://doi.org/10.1007/978-3-662-43936-4_18

A Framework for Evaluating Client Privacy Leakages in Federated Learning

Wenqi Wei(✉), Ling Liu, Margaret Loper, Ka-Ho Chow, Mehmet Emre Gursoy, Stacey Truex, and Yanzhao Wu

Georgia Institute of Technology, Atlanta, GA 30332, USA
{wenqiwei,khchow,memregursoy,staceytruex,yanzhaowu}@gatech.edu, ling.liu@cc.gatech.edu, margaret.loper@gtri.gatech.edu

Abstract. Federated learning (FL) is an emerging distributed machine learning framework for collaborative model training with a network of clients (edge devices). FL offers default client privacy by allowing clients to keep their sensitive data on local devices and to only share local training parameter updates with the federated server. However, recent studies have shown that even sharing local parameter updates from a client to the federated server may be susceptible to gradient leakage attacks and intrude the client privacy regarding its training data. In this paper, we present a principled framework for evaluating and comparing different forms of client privacy leakage attacks. We first provide formal and experimental analysis to show how adversaries can reconstruct the private local training data by simply analyzing the shared parameter update from local training (e.g., local gradient or weight update vector). We then analyze how different hyperparameter configurations in federated learning and different settings of the attack algorithm may impact on both attack effectiveness and attack cost. Our framework also measures, evaluates, and analyzes the effectiveness of client privacy leakage attacks under different gradient compression ratios when using communication efficient FL protocols. Our experiments additionally include some preliminary mitigation strategies to highlight the importance of providing a systematic attack evaluation framework towards an in-depth understanding of the various forms of client privacy leakage threats in federated learning and developing theoretical foundations for attack mitigation.

Keywords: Privacy leakage attacks · Federated learning · Attack evaluation framework

1 Introduction

Federated learning enables the training of a high-quality ML model in a decentralized manner over a network of devices with unreliable and intermittent network connections [5,14,20,26,29,33]. In contrast to the scenario of prediction on edge devices, in which an ML model is first trained in a highly controlled Cloud environment and then downloaded to mobile devices for performing predictions,

L. Chen et al. (Eds.): ESORICS 2020, LNCS 12308, pp. 545–566, 2020.
https://doi.org/10.1007/978-3-030-58951-6_27

federated learning brings model training to the devices while supporting continuous learning on device. A unique feature of federated learning is to decouple the ability of conducting machine learning from the need of storing all training data in a centralized location [13].

Although federated learning by design provides the default privacy of allowing thousands of clients (e.g., mobile devices) to keep their original data on their own devices, while jointly learn a model by sharing only local training parameters with the server. Several recent research efforts have shown that the default privacy in FL is insufficient for protecting the underlaying training data from privacy leakage attacks by gradient-based reconstruction [9,32,34]. By intercepting the local gradient update shared by a client to the FL server before performing federated averaging [13,19,22], the adversary can reconstruct the local training data with high reconstruction accuracy, and hence intrudes the client privacy and deceives the FL system by sneaking into client confidential training data illegally and silently, making the FL system vulnerable to client privacy leakage attacks (see Sect. 2.2 on Threat Model for detail).

In this paper, we present a principled framework for evaluating and comparing different forms of client privacy leakage attacks. Through attack characterization, we present an in-depth understanding of different attack mechanisms and attack surfaces that an adversary may leverage to reconstruct the private local training data by simply analyzing the shared parameter updates (e.g., local gradient or weight update vector). Through introducing multi-dimensional evaluation metrics and developing evaluation testbed, we provide a measurement study and quantitative and qualitative analysis on how different configurations of federated learning and different settings of the attack algorithm may impact the success rate and the cost of the client privacy leakage attacks. Inspired by the attack effect analysis, we present some mitigation strategies with preliminary results to highlight the importance of providing a systematic evaluation framework for comprehensive analysis of the client privacy leakage threats and effective mitigation methods in federated learning.

2 Problem Formulation

2.1 Federated Learning

In federated learning, the machine learning task is decoupled from the centralized server to a set of N client nodes. Given the unstable client availability [21], for each round of federated learning, only a small subset of $K_t <$ clients out of all N participants will be chosen to participate in the joint learning.

Local Training at a Client: Upon notification of being selected at round t, a client will download the global state $w(t)$ from the server, perform a local training computation on its local dataset and the global state, i.e., $w_k(t+1) = w_k(t) - \eta \nabla w_k(t)$, where $w_k(t)$ is the local model parameter update at round t and ∇w is the gradient of the trainable network parameters. Clients can decide its training batch size B_t and the number of local iterations before sharing.

Update Aggregation at FL Server: Upon receiving the local updates from all K_t clients, the server incorporates these updates and update its global state, and initiates the next round of federated learning. Given that local updates can be in the form of either gradient or model weight update, thus two update aggregation implementations are the most representative. In *distributed SGD* [17,18,29,30], each client uploads the local gradients to the FL server at each round and the server iteratively aggregates the local gradients from all K_t clients into the global model: $w(t+1) = w(t) - \eta \sum_{k=1}^{K_t} \frac{n_t}{n} \nabla w_k(t)$, where η is the global learning rate and $\frac{n_t}{n}$ is the weight of client k. n_k is the number of data points at client k and n indicates the amount of total data from all participating clients at round t. In *federated averaging* [3,20], each client uploads the local training parameter update to the FL server and the server iteratively performs a weighted average of the received weight parameters to update the global model: $w(t+1) = \sum_{k=1}^{K_t} \frac{n_t}{n} w_k(t+1)$, where $\Delta w_k(t)$ denote the difference between the model parameter update before the iteration of training and the model parameter update after the training for client k.

2.2 Threat Model

In an FL system, clients are the most vulnerable attack surface for the client privacy leakage (CPL) attack. To focus on client gradient leakage vulnerability due to sharing its local gradient update, we assume that clients may be partially compromised: (i) an adversary may intercept the local parameter update prior to sending it via the client-to-server communication channel to the FL server and (ii) an adversary may gain access to the saved local model executable (checkpoint data) on the compromised client with no training data visible and launch white-box gradient leakage attacks to steal the private training data. Note that local data and model may not be compromised at the same time for many reasons, e.g., separated storage of data and model, or dynamic training data.

We also assume that the federated server is honest and will truthfully perform the aggregation of local parameter updates and manage the iteration rounds for jointly learning. However, the FL server may be curious and may analyze the periodic updates from certain clients to perform client privacy leakage attacks and gain access to the private training data of the victim clients. Given that the gradient-based aggregation and model weight-based aggregation are mathematically equivalent, one can obtain the local model parameter difference from the local gradient and the local training learning rate. In this paper, gradient-based aggregation is used without loss of generality.

2.3 The Client Privacy Leakage (CPL) Attack: An Overview

The client privacy leakage attack is a gradient-based feature reconstruction attack, in which the attacker can design a gradient-based reconstruction learning algorithm that will take the gradient update at round t, say $\nabla w_k(t)$, to be shared by the client, to reconstruct the private data used in the local training computation. For federated learning on images or video clips, the reconstruction algorithm will start by using a dummy image of the same resolution as its attack initialization seed, and run a test of this attack seed on the intermediate local model, to compute a gradient loss using a vector distance loss function between the gradient of this attack seed and the actual gradient from the client local training. The goal of this reconstruction attack is to iteratively add crafted small noises to the attack seed such that the generated gradient from this reconstructed attack seed will approximate the actual gradient computed on the local training data. The reconstruction attack terminates when the gradient of the attack seed reconstructed from the dummy initial data converges to the gradient of the training data. When the gradient-based reconstruction loss function is minimized, the reconstructed attack data will also converge to the training data with high reconstruction confidence.

Algorithm 1. Gradient-based Reconstruction Attack

1: **Inputs:**
$f(x; w(t))$: Differentiable learning model, $\nabla w_k(t)$: gradients produced by the local training on private training data $(x; y)$ at client k, $w(t)$, $w(t+1)$: model parameters before and after the current local training on $(x; y)$, η_k learning rate of local training
Attack configurations: $INIT(x.type)$: attack initialization method, $\mathbb{T}$: attack termination condition, η': attack optimization method, α: regularizer ratio.
2: **Output:** reconstructed training data $(x_{rec}; y_{rec})$
3: **Attack procedure**
4: **if** $w_k(t+1)$: **then**
5: $\quad \Delta w_k(t) \leftarrow w_k(t+1) - w(t)$
6: $\quad \nabla w_k(t) \leftarrow \frac{\Delta w_k(t)}{\eta_k}$
7: **end if**
8: $x^0_{rec} \leftarrow INIT(x.type)$
9: $y_{rec} \leftarrow \arg\min_i(\nabla_i w_k(t))$
10: **for** τ in $\mathbb{T}$ **do**
11: $\quad \nabla w^{\tau}_{att}(t) \leftarrow \frac{\partial loss(f(x^{\tau}_{rec}, w(t)), y_{rec})}{\partial w(t)}$
12: $\quad D^{\tau} \leftarrow ||\nabla w^{\tau}_{att}(t) - \nabla w_k(t)||^2 + \alpha||f(x^{\tau}_{rec}, w(t)) - y_{rec}||^2$
13: $\quad x^{\tau+1}_{rec} \leftarrow x^{\tau}_{rec} - \eta' \frac{\partial D^{\tau}}{\partial x^{\tau}_{rec}}$
14: **end for**

Algorithm 1 gives a sketch of the client privacy leakage attack method. Line 4–6 convert the weight update to gradient when the weight update is shared between the FL server and the client. The learning rate η_k for local training is assumed to be identical across all clients in our prototype system. Line 8 invokes the dummy attack seed initialization, which will be elaborated in Sect. 3.1. Line 9 is to get the label from the actual gradient shared from the local training. Since the local training update towards the ground-truth label of the training input data should be the most aggressive compared to other labels, the sign of gradient for the ground-truth label of the private training data will be different than other classes, and its absolute value is usually the largest. Line 10–14 presents the iterative reconstruction process that produces the reconstructed private training data based on the client gradient update. If the reconstruction algorithm converges, then the client privacy leakage attack is successful, or else the CPL attack is failed. Line 12–13 show that when the L_2 distance between the gradients of the attack reconstructed data and the actual gradient from the private training data is minimized, the reconstructed attack data from the dummy seed converges to the private local training data, leading to the client privacy leakage. In Line 12, a label-based regularizer is utilized to improve the stability of the attack optimization. An alternative way to reconstruct the label of the private training data is to initialize a dummy label and feed it into the iterative approximation algorithm for attack optimization [34], in a similar way as the content reconstruction optimization.

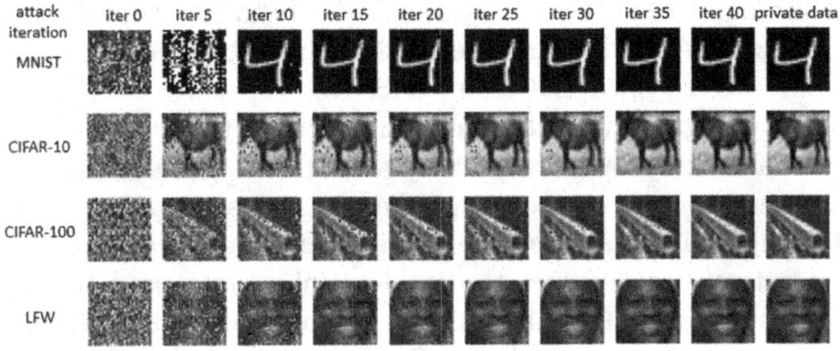

Fig. 1. Example illustration of the Client Privacy Leakage attack

Figure 1 provides a visualization of four illustrative example attacks over four datasets: MNIST [16], CIFAR10 [15], CIFAR100 [15], and LFW [12]. We make two interesting observations from Algorithm 1. *First*, multiple factors in the attack method could impact the attack success rate (ASR) of the client privacy leakage attack, such as the dummy attack seed data initialization method (Line 8), the attack iteration termination condition ($\mathbb{T}$), the selection of the gradient loss (distance) function (Line 12), the attack optimization method (Line 13). *Second*, the configuration of some hyperparameters in federated learning may also impact the effectiveness and cost of the CPL attack, including batch size, training data resolution, choice of activation function, and whether the gradient update is uploaded to the FL server using baseline communication protocol or a communication efficient method. In the next section, we present the design of our evaluation framework to further characterize the client privacy leakage attack of different forms and introduce cost-effect metrics to measure and analyze the adverse effect and cost of CPL attacks. By utilizing this framework, we provide a comprehensive study on how different attack parameter configurations and federated learning hyperparameter configurations may impact the effectiveness of the client privacy leakage attacks.

3 Evaluation Framework

3.1 Attack Parameter Configuration

Attack Initialization: We first study how different methods for generating the dummy attack seed data may influence the effectiveness of a CPL attack in terms of reconstruction quality or confidence as well as reconstruction time and convergence rate. A straightforward dummy data initialization method is to use a random distribution in the shape of dummy data type and we call this baseline the random initialization (CPL-random). Although random initialization is also used in [9,32,34], our work is, to the best of our knowledge, the first study on variations of attack initiation methods. To understand the role of random seed in the CPL attack, it is important to understand the difference of the attack reconstruction learning from the normal deep neural network (DNN) training. In a DNN training, it takes as the training input both the fixed data-label pairs and the initialization of the learnable model parameters, and iteratively learn the model parameters until the training converges, which minimizes the loss with respect to the ground truth labels. In contrast, the CPL attack performs reconstruction attack by taking a dummy attack seed input data, a fixed set of model parameters, such as the actual gradient updates of a client local training, and the gradient derived label as the reconstructed label y_{rec}, its attack algorithm will iteratively reconstruct the local training data used to generate the gradient, $\nabla w_k(t)$, by updating the dummy synthesized seed data, following the attack iteration termination condition $\mathbb{T}$, denoted by $\{x_{rec}^0, x_{rec}^1, ... x_s{}^{\mathbb{T}}\} \in \mathbb{R}^d$, such that the loss between the gradient of the reconstructed data x_{rec}^i and the actual gradient $\nabla w_k(t)$ is minimized. Here x_{rec}^0 denotes the initial dummy seed.

Theorem 1 ***(CPL Attack Convergence Theorem).*** *Let x_{rec}^* be the optimal synthesized data for $f(x)$ and attack iteration $t \in \{0, 1, 2, ...T\}$. Given the convexity and Lipschitz-smoothness assumption, the convergence of the gradient-based reconstruction attack is guaranteed with:*

$$f(x_{rec}^{\mathbb{T}}) - f(x_{rec}^*) \leq \frac{2L||x_{rec}^0 - x_{rec}^*||^2}{\mathbb{T}}. \tag{1}$$

The above CPL Attack Convergence theorem is derived from the well-established Convergence Theorem of Gradient Descent. Due to the limitation of space, the formal proof of Theorem 1 is provided in the Appendix. Note that the convexity assumption is generally true since the d-dimension trainable synthesized data can be seen as a one-hidden-layer network with no activation function. The fixed model parameters are considered as the input with optimization of the least square estimation problem as stated in Line 12 of Algorithm 1.

According to Theorem 1, the convergence of the CPL attack is closely related to the initialization of the dummy data x_{rec}^0. This motivates us to investigate different ways to generate dummy attack seed data. Concretely, we argue that different random seeds may have different impacts on both reconstruction learning efficiency (confidence) and reconstruction learning convergence (time or the number of iteration steps). Furthermore, using geometrical initialization as those introduced in [24] not only can speed up the convergence but also ensure attack stability. Consider a single layer of the neural network: $g(x) = \sigma(wx + b)$, a geometrical initialization considers the form $g(x) = \sigma(w_*(x - b_*)$ instead of directly initialing w and b with random distribution. For example, according to [31], the following partial derivative of the geometrical initialization.

$$\frac{\partial g}{\partial w_*} = \sigma'(w_*(x - b_*))(x - b_*), \tag{2}$$

is more independent of translation of the input space than $\frac{\partial g}{\partial w} = \sigma'(wx + b)x$, and is therefore more stable.

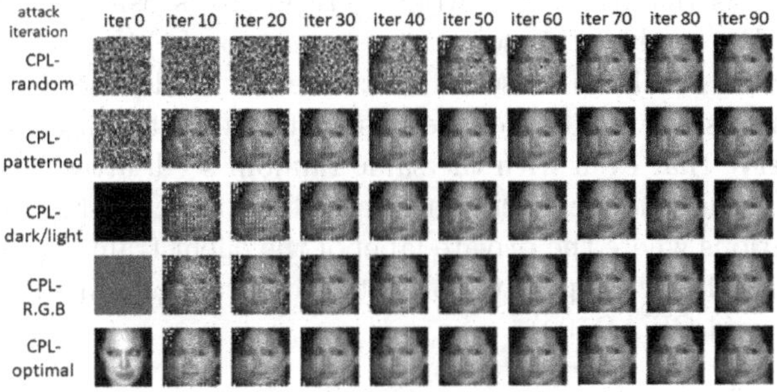

Fig. 2. Visualization of different initialization

Figure 2 provides a visualization of five different initialization methods and their impact on the CPL attack in terms of reconstruction quality and convergence (#iterations). In addition to CPL-random, CPL-patterned is a method that uses patterned random initialization. We initialize a small portion of the dummy data with a random seed and duplicate it to the entire feature space. An example of the portion can be 1/4 of the feature space. CPL-dark/light is to use a dark (or light) seed of the input type (size), whereas CPL-R.G.B. is to use red or green or blue solid color seed of the input type (size). CPL-optimal refers to the theoretical optimal initialization method, which uses an example from the same class as the private training data that the CPL attack aims to reconstruct. We observe from Fig. 2 that CPL-patterned, CPL-R.G.B., and CPL-dark/light can outperform CPL-random with faster attack convergence and more effective reconstruction confidence. We also include CPL-optimal to show that different CPL initializations can effectively approximate the optimal way of initialization in terms of reconstruction effectiveness.

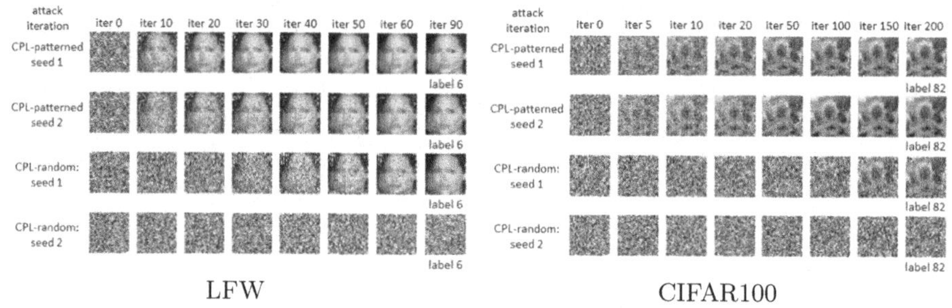

Fig. 3. Effect of different random seed

Figure 3 shows that the CPL attacks are highly sensitive to the choice of random seeds. We conduct this set of experiments on LFW and CIFAR100, and both confirm consistently our observations: different random seeds lead to diverse convergence processes with different reconstruction quality and confidence. From Fig. 3, we observe that even with the same random seed, attack with patterned initialization is much more efficient and stable than the CPL-random. Moreover, there are situations where the private label of the client training data is successfully reconstructed but the private content reconstruction fails (see the last row for both LFW and CIFAR100).

Attack Termination Condition: The effectiveness of a CPL attack is also sensitive to the setting of the attack termination condition. Recall Sect. 2.3 and Algorithm 1, there are two decision factors for termination. One is the maximum attack iteration [34] and the other is the L_2-distance threshold of the gradient loss [32], i.e., the difference between the gradient from the reconstructed data and the actual gradient from the local training using the private local data. Other gradient loss functions, e.g., cosine similarity, entropy, can be used to replace the L_2 function. Table 1 compares six different settings of attack iterations for configuring termination condition. A small attack iteration cap may cause the reconstruction attack to abort before it can succeed the attack, whereas a larger attack iteration cap may increase the cost of attack. The result in Table 1 confirms that choosing the cap for attack iterations may have an impact on both attack effectiveness and cost.

Table 1. Effect of termination condition

maximum attack iteration		10	20	30	50	100	300
LFW	CPL-patterned	0	0.34	0.98	1	1	1
	CPL-random	0	0	0	0.562	0.823	0.857
CIFAR10	CPL-patterned	0	0.47	0.93	0.973	0.973	0.973
	CPL-random	0	0	0	0	0.356	0.754
CIFAR100	CPL-patterned	0	0	0.12	0.85	0.981	0.981
	CPL-random	0	0	0	0	0.23	0.85

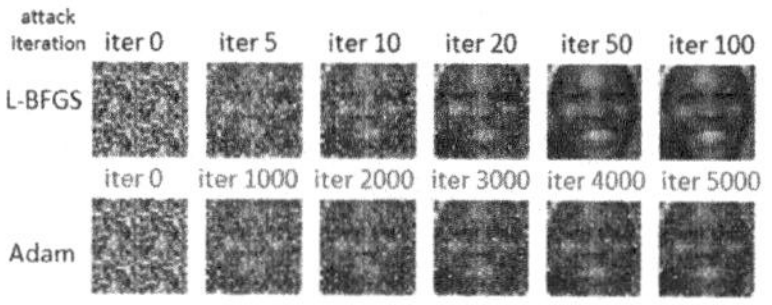

Fig. 4. Effect of attack optimization

Attack Optimization: Optimization methods, such as Stochastic Gradient descent, Adam, and Adagrad can be used to iteratively update the dummy data during the reconstruction of a CPL attack. While the first-order optimization techniques are easy to compute and less time consuming, the second-order techniques are better in escaping the slow convergence paths around the saddle points [4]. Figure 4 shows a comparison of L-BFGS [6] and Adam and their effects on the CPL-patterned attack for LFW dataset. It takes fewer attack iterations to get high reconstruction quality using the L-BFGS compared to using Adam as the optimizer in the reconstruction attack, confirming that choosing an appropriate optimizer may impact on attack effectiveness.

3.2 Hyperparameter Configurations in Federated Learning

Batch Size: Given that all forms of CPL attack methods are reconstruction learning algorithms that iteratively learn to reconstruct the private training data by inferencing over the actual gradient to perform iterative updates on the dummy attack seed data, it is obvious that a CPL attack is most effective when working with the gradient generated from the local training data of batch size 1. Furthermore, when the input data examples in a batch of size B belongs to only one class, which is often the case for mobile devices and the non-i.i.d distribution of the training data [33], the CPL attacks can effectively reconstruct the training data of the entire batch. This is especially true when the dataset has low inter-class variation, e.g., face and digit recognition. Figure 5 shows the visualization of performing a CPL-patterned attack on the LFW dataset with four different batch sizes. With a large batch size, the detail of a single input is being neutralized, making the attack reconstruction captures more of the shared features of the entire batch rather than specific to a single image.

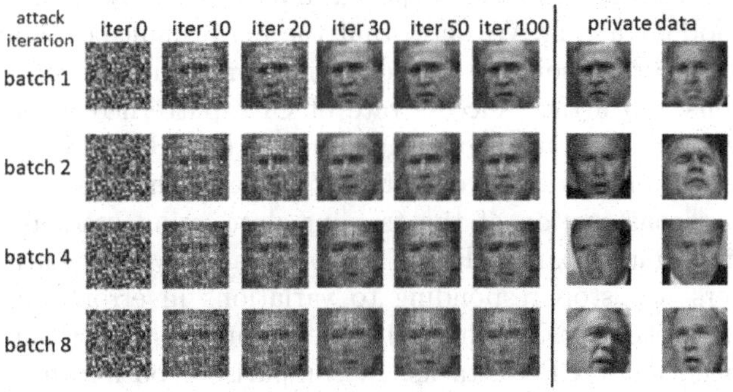

Fig. 5. Effect of batch size in CPL-patterned attacks on LFW

Training Data Resolution: In contrast to the early work [34] that fails to attack images of resolution higher than 64×64, we argue that the effectiveness of the CPL attack is mainly attributed to the model overfitting to the training data. In order to handle higher resolution training data, we double the number of filters in all convolutional layers to build a more overfitted model. Figure 6 shows the scaling results of CPL attack on the LFW dataset with input data size of 32×32, 64×64, and 128×128. CPL-random requires a much larger number of attack iterations to succeed the attack with high reconstruction performance. CPL-patterned is a significantly more effective attack for all three different resolutions with 3 to 4$\times$ reduction in the attack iterations compared to CPL-random. We also provide an example of attacking the 512×512 Indiana University Chest X-Rays image of very high resolution in Fig. 7.

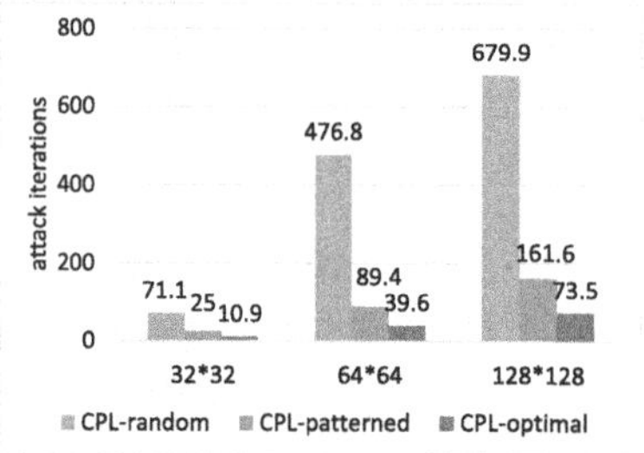

Fig. 6. Attack iteration of the training data scaling using LFW dataset

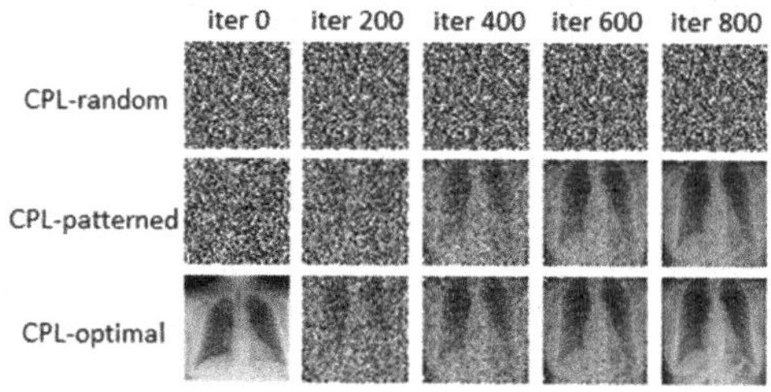

Fig. 7. Extreme scaling case of attacking 512×512 Chest X-ray image

Activation Function: The next hyperparameter of FL is the activation function used in model training. We show that the performance of the CPL attacks is highly related to the choice of the activation function. Figure 8 compares the attack iterations and attack success rate of CPL-patterned attack with three different activation functions: Sigmoid, Tanh, and LeakReLU. Due to space constraint, the results on MNIST is omitted. We observe that ReLU naturally prevents the full reconstruction of the training data using gradient because the gradient of the negative part of ReLU will be 0, namely, that part of the trainable parameters will stop responding to variations in error and will not get adjusted during optimization. This dying ReLU problem takes out the gradient information needed for CPL attacks. In comparison, both Sigmoid and Tanh are differentiable bijective and can pass the gradient from layer to layer in an almost lossless manner. LeakyReLU sets a slightly inclined line for the negative part of ReLU to mitigate the issue of dying ReLU and thus is vulnerable to CPL attacks.

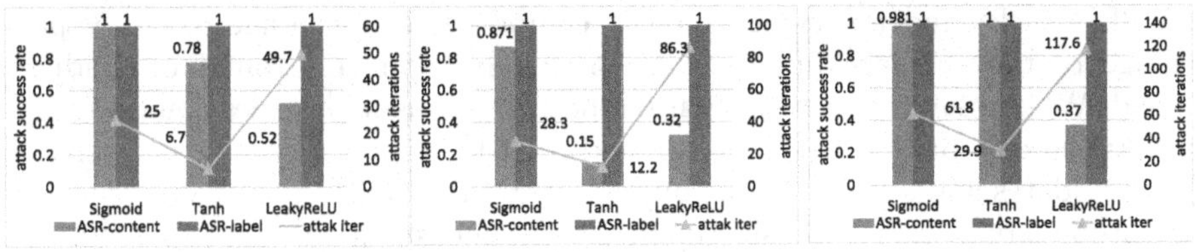

LFW CIFAR10 CIFAR100

Fig. 8. Effect of activation function on the CPL attack.

Motivated by the impact of activation function, we argue that any model components that discontinue the integrity and uniqueness of gradients can hamper CPL attacks. We observe from our experiments that an added dropout structure enables different gradient in every query, making $\nabla w_{att}^{\tau}(t)$ elusive and unable to converge to the uploaded gradients. By contrast, pooling cannot prevent CPL attacks since pooling layers do not have parameters.

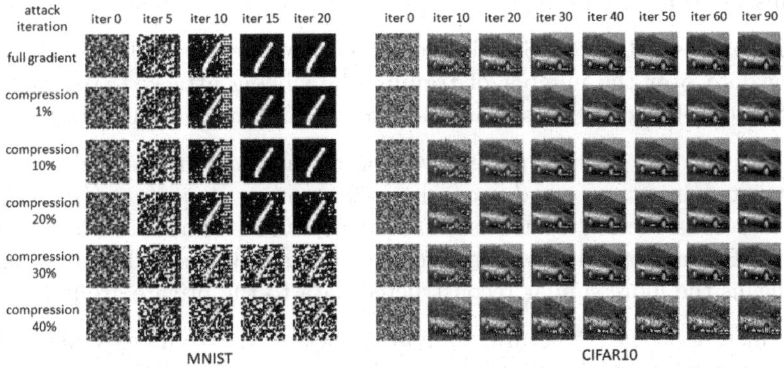

Fig. 9. Illustration of the CPL attack under communication-efficient update

Baseline Protocol v.s. Communication-Efficient Protocol: In the baseline communication protocol, the client sends a full vector of local training parameter update to the FL server in each round. For federated training of large models on complex data, this step is known to be the bottleneck of Federated Learning. Communication-efficient FL protocols were proposed [14,20] to improve the communication efficiency of parameter update and sharing by employing high precision vector compression mechanisms, such as structured updates and sketched updates. As more FL systems utilize a communication-efficient protocol to replace the baseline protocol, it is important to study the impact of using a communication efficient protocol on the performance of the CPL attacks, especially compared to the baseline client-to-server communication protocol. In this set of experiments, we measure the performance of CPL attacks under varying gradient compression percentage θ, i.e., θ percentage of the gradient update will be discarded in this round of gradient upload. We employ the compression method in [17] as it provides a good trade-off between communication-efficiency and model training accuracy. It leverages sparse updates and sends only the important gradients, i.e., the gradients whose magnitude larger than a threshold, and further measures are taken to avoid losing information. Locally, the client will accumulate small gradients and only send them when the accumulation is large enough. Figure 9 shows the visualization of the comparison on MNIST and CIFAR10. We observe that compared to the baseline protocol with full gradient upload, using the communication efficient protocol with θ up to 40%, the CPL attack remains to be effective for CIFAR10. Section 4.2 provides a more detailed experimental study on CPL attacks under communication-efficient FL protocol.

3.3 Attack Effect and Cost Metrics

Our framework evaluates the adverse effect and cost of CPL attacks using the following metrics. For data-specific metrics, we average the evaluation results over all successful reconstructions.

Attack Success Rate (ASR) is the percentage of successfully reconstructed training data over the number of training data being attacked. We use ASRc and ASRl to refer to the attack success rate on content and label respectively.

MSE uses the root mean square deviation to measure the similarity between reconstructed input x_{rec} and ground-truth input x: $\frac{1}{M}\sum_{i=1}^{M}(x(i) - x_{rec}(i))^2$ when the reconstruction is successful. M denotes total number of features in the input. MSE can be used on all data format, e.g. attributes and text. A smaller MSE means the more similar reconstructed data to the private ground truth.

SSIM measures the structural similarity between two images based on a perception-based model [28] that considers image degradation as perceived change.

Attack Iteration measures the number of attack iterations required for reconstruction learning to converge and thus succeed the attack, e.g., L_2 distance of the gradients between the reconstructed data and the local private training data is smaller than a pre-set threshold.

4 Experiments and Results

4.1 Experiment Setup

We evaluate CPL attacks on four benchmark datasets: MNIST, LFW, CIFAR10, CIFAR100. MNIST consists of 70000 grey-scale hand-written digits images of size 28×28. The 60000:10000 split is used for training and testing data. Labeled Faces in the Wild (LFW) people dataset has 13233 images from 5749 classes.

The original image size is 250 × 250 and we slice it to 32 × 32 and extract the 'interesting' part. Our experiments only consider 106 classes, each with more than 14 images. For a total number of 3735 eligible LFW data, a 3:1 train-test ratio is applied. CIFAR10 and CIFAR100 both have 60000 color images of size 32 × 32 with 10 and 100 classes respectively. The 50000:10000 split is used for training and testing. We perform CPL attacks with the following configurations as the default unless otherwise stated. The initialization method is patterned, the maximum attack iterations are 300, the optimization method is L-BFGS with attack learning rate 1. The attack is performed with full gradient communication. For each dataset, the attack is performed on 100 images with 10 different random seeds. For MNIST and LFW, we use a LeNet model with 0.9568 benign accuracy on MNIST and 0.695 on LFW. For CIFAR10 and CIFAR100, we apply a ResNet20 with benign accuracy of 0.863 on CIFAR10 and CIFAR100. We use 100 clients as the total client population and at each communication round, 10% of clients will be selected randomly to participate in federated learning.

4.2 Gradient Leakage Attack Evaluation

Comparison with Other Gradient Leakage Attacks. We first conduct a set of experiments to compare the CPL-patterned attack with two existing gradient leakage attacks: the deep gradient attack [34], and the gradient inverting attack [9], which replaces the L_2 distance function with cosine similarity and performs the optimization on the sign of the gradient. We measure the attack results on the four benchmark image datasets in Table 2. For all four datasets, the CPL attack is a much faster and more efficient attack with the highest attack success rate (ASR) and the lowest attack iterations on both content and label reconstruction. Meanwhile, the high SSIM and low MSE for CPL attack indicate the quality of the reconstructed data is almost identical to the private training data. We also observe that gradient inverting attack [9] can achieve a high ASR compared to deep gradient attack [34] but at a great cost of attack iterations. Note that the CPL attack offers slightly higher ASR compared to [9] but at much lower attack cost in terms of the required attack iteration.

Table 2. Comparison of different gradient leakage attacks

	CIFAR10			CIFAR100			LFW			MNIST		
	CPL	[34]	[9]	CPL	[34]	[9]	CPL	[34]	[9]	CPL	[34]	[9]
attack iter	**28.3**	114.5	6725	**61.8**	125	6813	**25**	69.2	4527	**11.5**	18.4	3265
ASRc	**0.973**	0.754	0.958	**0.981**	0.85	0.978	1	0.857	0.974	1	0.686	0.784
ASRl	1	0.965	1	1	0.94	1	1	0.951	1	1	0.951	1
SSIM	**0.9985**	0.9982	0.9984	**0.959**	0.953	0.958	**0.998**	0.997	0.9978	**0.99**	0.985	0.989
MSE	**2.2E-04**	2.5E-04	**2.2E-04**	**5.4E-04**	6.5E-04	**5.4E-04**	**2.2E-04**	2.9E-04	2.3E-04	**1.5E-05**	1.7E-05	1.6E-05

Table 3. Comparison of different geometrical initialization in CPL attacks

dataset \ initialization		baseline	patterned		dark/light		RGB			optimal
		random	2*2	4*4	dark	light	R	G	B	insider
CIFAR10	attack iter	91.14	28.3	**24.8**	34	52.3	35.9	77.5	79.1	23.2
	ASR	0.871	0.973	**0.976**	0.99	1	0.99	0.96	0.96	1
CIFAR100	attack iter	125	61.8	57.2	**57.5**	65.3	59.4	61.3	62.4	35.3
	ASR	1	0.981	0.995	**1**	1	1	0.88	1	1
LFW	attack iter	71.1	25	18.6	34	50.8	**20.3**	28	42.1	13.3
	ASR	0.86	1	0.997	1	1	**1**	1	1	1

Variation Study: Geometrical Initialization. This set of experiments measure and compare the four types of geometrical initialization methods: patterned random, dark/light, R.G.B., and optimal. For optimal initialization, we feed a piece of data that is randomly picked from the training set. This assumption is reasonable when different clients hold part of the information about one data item. Table 3 shows the result. We observe that the performance of all four geometrical initializations is always better than the random initialization (see the bold highlight in Table 3). Note that the optimal initialization is used in this experiment as a reference point as it assumes the background information about the data distribution. Furthermore, the performance of geometrical initializations is also dataset-dependent. CPL attack on CIFAR100 requires a longer time and more iterations to succeed than CPL on CIFAR10 and LFW.

Variation Study: Batch Size and Iterations. Motivated by the batch size visualization in Fig. 5, we study the impact of hyperparameters used in local training, such as batch size and the number of local training iterations, on the performance of CPL attack. Table 4a shows the results of the CPL attack on the LFW dataset with five different batch sizes. We observe that the ASR of CPL attack is decreased to 96%, 89%, 76%, and 13% as the batch size increases to 2, 4, 8, and 16. The CPL attacks at different batch sizes are successful at the attack iterations around 25 to 26 as we measure attack iterations only on successfully reconstructed instances. Table 4b shows the results of the CPL attack under five different settings of local iterations before sharing the gradient updates. We show that as more iterations are performed at local training before sharing the gradient update, the ASR of the CPL attack is decreasing with 97%, 85%, and 39% for iterations of 5, 7, and 9 respectively. This result confirms our analysis in Sect. 3.2 that a larger batch size for local training prior to sharing the local gradient update may help mitigate the CPL attack because the shared gradient data capture more shared features among the images in the batch instead of more specific to an individual image.

Table 4. Effect of local training hyperparameters on CPL attack (LFW)

batch size	1	2	4	8	16
ASR	1	0.96	0.89	0.76	0.13
attack iter	25	25.7	25.6	26.1	25.7
SSIM	0.998	0.635	0.525	0.456	0.401
MSE	2.2E-04	6.7E-03	8.0E-03	9.0E-03	1.0E-02

(a) Effect of Batch size on CPL

local iter	1	3	5	7	9
attack iter	25	42.5	94.2	95.6	97.9
ASR	1	1	0.97	0.85	0.39
SSIM	0.998	0.981	0.898	0.659	0.471
MSE	2.2E-04	6.8E-04	1.8E-03	3.8E-03	5.5E-03

(b) Effect of local training iterations

Variation Study: Leakage in Communication Efficient Sharing. This set of experiments measures and compares the gradient leakage in CPL under baseline protocol (full gradient sharing) and communication-efficient protocol (significant gradient sharing with low-rank filer). Table 5 shows the result. To illustrate the comparison results, we provide the accuracy of the baseline protocol and the communication-efficient protocol of varying compression percentages on all four benchmark datasets in Table 5(a). We make two interesting observations. (1) CPL attack can generate high confidence reconstructions (high ASR, high SSIM, low MSE) for MNIST and CIFAR10 at compression rate 40%, and for

Table 5. Effect of CPL attack under communication-efficient FL protocols

benign acc	0	1%	10%	20%	30%	40%	50%	70%	80%	90%
LFW	0.695	0.697	0.705	0.701	0.71	0.709	0.713	0.711	0.683	0.676
CIFAR100	0.67	0.673	0.679	0.685	0.687	0.695	0.689	0.694	0.676	0.668
CIFAR10	0.863	0.864	0.867	0.872	0.868	0.865	0.868	0.861	0.864	0.859
MNIST	0.9568	0.9567	0.9577	0.957	0.9571	0.9575	0.9572	0.9576	0.9573	0.9556

(a) Benign accuracy of four datasets with varying compression rates

	compression	original	1%	10%	20%	30%	40%	50%	70%	80%	90%
LFW	attack iter	25	25	24.9	24.9	25	24.8	25	24.6	24.5	300
	ASR	1	1	1	1	1	1	1	1	1	0
	SSIM	0.998	0.9996	0.9997	0.9978	0.9978	0.9975	0.998	0.9981	0.951	0.004
	MSE	2.2E-04	1.8E-04	1.7E-04	4.9E-04	4.8E-04	5.1E-04	4.5E-04	4.6E-04	1.6E-03	1.6E-01
CIFAR100	attack iter	61.8	61.8	61.8	61.7	61.7	61.5	61.8	60.1	59.8	300
	ASR	1	1	1	1	1	1	1	1	1	0
	SSIM	0.959	0.9994	0.9981	0.9981	0.998	0.9983	0.9982	0.9983	0.895	0.016
	MSE	5.4E-04	3.3E-04	3.7E-04	3.7E-04	3.8E-04	3.5E-04	3.6E-04	3.7E-04	1.5E-03	1.2E-01
CIFAR10	attack iter	28.3	28.3	28.1	26.5	25.8	25.3	300	300	300	300
	ASR	1	1	1	1	1	1	0	0	0	0
	SSIM	0.9985	0.9996	0.9996	0.9997	0.9992	0.87	0.523	0.0017	0.0019	0.0018
	MSE	2.2E-04	1.3E-04	1.2E-04	1.2E-04	2.1E-04	3.1E-03	9.6E-03	3.3E-01	3.3E-01	3.3E-01
MNIST	attack iter	11.5	11.5	11.2	10.7	7.2	300	300	300	300	300
	ASR	1	1	1	1	1	0	0	0	0	0
	SSIM	0.99	0.9899	0.9891	0.9563	0.9289	0.8889	0.8137	0.425	0.433	0.43
	MSE	2.4E-04	2.4E-04	2.2E-04	1.7E-03	8.8E-03	2.8E-02	5.8E-02	2.7E-01	2.7E-01	2.7E-01

(b) Attack performance of four datasets with varying compression rates

Table 6. Mitigation with Gaussian noise and Laplace noise

	CIFAR100				LFW			
Gaussian noise	original	G-10e-4	G-10e-3	G-10e-2	original	G-10e-4	G-10e-3	G-10e-2
benign acc	0.67	0.664	0.647	0.612	0.695	0.692	0.653	0.636
attack iter	61.8	61.8	61.8	300	25	25	25	300
ASR	1	1	1	0	1	1	1	0
SSIM	0.9995	0.9976	0.8612	0.019	0.998	0.9976	0.8645	0.013
MSE	5.4E-04	6.9E-04	4.1E-03	3.0E-01	2.2E-04	3.7E-04	3.0E-03	1.9E-01
Laplace noise	original	L-10e-4	L-10e-3	L-10e-2	original	L-10e-4	L-10e-3	L-10e-2
benign acc	0.67	0.651	0.609	0.578	0.695	0.683	0.632	0.597
attack iter	61.8	61.8	61.8	300	25	25	25	300
ASR	1	1	1	0	1	1	1	0
SSIM	0.9995	0.9956	0.7309	0.017	0.998	0.9965	0.803	0.009
MSE	5.4E-04	6.4E-04	6.4E-03	3.1E-01	2.2E-04	4.0E-04	3.9E-03	2.0E-01

CIFAR100 and LFW at the compression rate of 90%. Second, as the compression percentage increases, the number of attack iterations to succeed the CPL attack decreases. This is because a larger portion of the gradients are low significance and are set to 0 by compression. When the attack fails, it indicates that the reconstruction cannot be done even with the infinite(∞) attack iterations, but we measure SSIM and MSE of the failed attacks at the maximum attack iterations of 300. (2) CPL attacks are more severe with more training labels in the federated learning task. A possible explanation is that informative gradients are more concentrated when there are more classes.

4.3 Mitigation Strategies

Gradient Perturbation with Additive Noise. We consider Gaussian noise and Laplace noise with zero means and different magnitude of variance in this set of experiments. Table 6 provides the mitigation results on CIFAR100 and LFW. In both cases, the client privacy leakage attack is largely mitigated at some cost of accuracy if we add sufficient Gaussian noise (G-10e-2) or Laplace noise (L-10e-2). Figure 10 illustrates the effect of the additive noise using four examples. The large additive noise we use to obfuscate the gradient while performing the reconstruction, the small SSIM and the large MSE indicate the more dissimilar between the gradient of the reconstructed data and the gradient of the original sensitive input, leading to poor quality of the CPL attack.

Gradient Squeezing with Controlled Local Training Iterations. Instead of sharing the gradient from the local training computation at each round t, we schedule and control the sharing of the gradient only after M iterations of local training. Table 7 shows the results of varying M from 1 to 10 with step 1. It shows that as M increases, the ASR of CPL attack starts to decrease, with 97.1%, 83.5%, 50% and 29.2% for M = 3, 5, 8 and 10 respectively for CIFAR10, and with 100%, 97%, 78% and 7% for M = 3, 5, 8 and 10 respectively for LFW. An

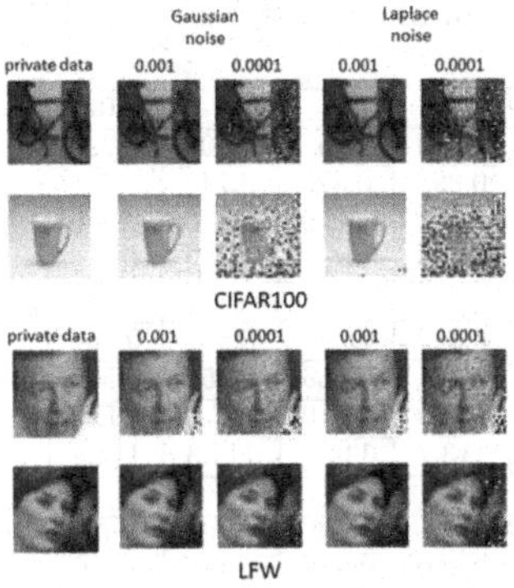

Fig. 10. Effect of additive noise

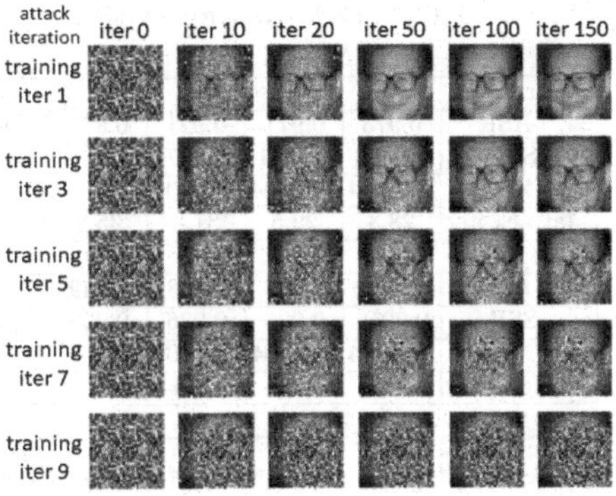

Fig. 11. Effect of local training

example of gradient squeezing with controlled local training iterations is provided in Fig. 11. This preliminary mitigation study shows that clients in federated learning may adopt some attack resilient optimizations when configuring their local training hyperparameters.

5 Related Work

Privacy in federated learning has been studied in two contexts: training-phase privacy attacks and prediction-phase privacy attacks. Gradient leakage attacks, formulated in CPL of different forms, or those in literature [9,32,34], are one type of privacy exploits in the training phase. In addition, Aono et al [1,2] proposed a privacy attack, which partially recovers private data based on the proportionality between the training data and the gradient updates in multi-layer perceptron models. However, their attack is not suitable for convolutional neural networks because the size of the features is far larger than the size of convolution weights. Hitaj et al [11] poisoned the shared model by introducing mislabeled samples into the training. In comparison, gradient leakage attacks are more aggressive since client privacy leakage attacks make no assumption on direct access to the training data as those in training poisoning attacks and yet can compromise the private training data by reconstruction attack based on only the local parameter updates to be shared with the federated server.

Privacy exploits at the prediction phase include model inversion, membership inference, and GAN-based reconstruction attack [10,11,27]. Fredrikson et al. [7] proposed the model inversion attack to exploit confidence values revealed along with predictions. Ganju et al. [8] inferred the global properties of the training data from a trained white-box fully connected neural network. Membership attacks [23,25] exploited the statistical differences between the model prediction on its training set and the prediction on unseen data to infer the membership of training data.

Table 7. Mitigation with controlled local training iterations

Dataset	local iter	1	2	3	4	5	6	7	8	9	10
CIFAR10	ASR	0.973	0.971	0.921	0.875	0.835	0.758	0.612	0.5	0.406	0.292
	SSIM	0.9985	0.9981	0.997	0.956	0.915	0.901	0.893	0.822	0.748	0.715
	MSE	2.2E-04	2.5E-04	2.9E-04	1.1E-03	2.4E-03	2.5E-03	2.7E-03	3.0E-03	4.5E-03	5.0E-03
	attack iter	28.3	29.5	31.6	35.2	42.5	71.5	115.3	116.3	117.2	117.5
CIFAR100	ASR	0.981	0.977	0.958	0.949	0.933	0.893	0.842	0.78	0.557	0.437
	SSIM	0.9959	0.996	0.996	0.959	0.907	0.803	0.771	0.666	0.557	0.505
	MSE	5.4E-04	5.8E-04	6.9E-04	1.1E-03	1.7E-03	2.2E-03	3.5E-03	4.2E-03	6.4E-03	6.9E-03
	attack iter	61.8	63.8	66.5	72.4	78.3	95.3	113.7	114.1	114.3	114.4
LFW	ASR	1	1	1	1	0.97	0.91	0.85	0.78	0.39	0.07
	SSIM	0.998	0.996	0.981	0.976	0.898	0.811	0.659	0.573	0.471	0.41
	MSE	2.2E-04	4.3E-04	6.8E-04	8.6E-04	1.8E-03	2.8E-03	3.8E-03	4.3E-03	5.5E-03	6.5E-03
	attack iter	25	34.7	42.5	68.3	94.2	95.5	95.6	98.3	97.9	98.1
MNIST	ASR	1	0.82	0.57	0.44	0.25	0.06	0	0	0	0
	SSIM	0.99	0.982	0.974	0.963	0.954	0.935	0.583	0.576	0.581	0.574
	MSE	1.5E-05	2.3E-04	2.8E-04	1.2E-03	1.5E-03	2.4E-03	1.7E-02	1.7E-02	1.7E-02	1.7E-02
	attack iter	11.5	34.7	93.2	96.7	97.1	96.5	300	300	300	300

6 Conclusion

We have presented a principled framework for evaluating and comparing different forms of client privacy leakage attacks. This framework showed that an effective mitigation method against client gradient leakage attacks should meet the two important criteria: (1) the defense can mitigate the gradient leakage attacks no matter how the attacker configures his reconstruction algorithm to launch the attack; and (2) the defense can mitigate the attacks no matter how the FL system is configured for joint training. Extensive experiments on four benchmark datasets highlighted the importance of providing a systematic evaluation framework for an in-depth understanding of the various forms of client privacy leakage threats in federated learning and for developing and evaluating different mitigation strategies.

Acknowledgements. The authors acknowledge the partial support from NSF CISE SaTC 1564097, NSF 2038029 and an IBM Faculty Award.

7 Appendices

7.1 Proof of Theorem 1

Assumption 1 ***(Convexity).*** *we say $f(x)$ is convex if*

$$f(\alpha x + (1-\alpha)x') \leq \alpha f(x) + (1-\alpha) f(x'), \tag{3}$$

where x, x' are data point in $\mathbb{R}^d$, and $\alpha \in [0,1]$.

Lemma 1. *If a convex $f(x)$ is differentiable, we have:*

$$f(x') - f(x) \geq \langle \nabla f(x), x' - x \rangle. \tag{4}$$

Proof. Equation 3 can be rewritten as: $\frac{f(x'+\alpha(x-x'))-f(x')}{\alpha} \leq f(x) - f(y)$. When $\alpha \to 0$, we complete the proof.

Assumption 2 ***(Lipschitz Smoothness).*** *With Lipschitz continuous on the differentiable function $f(x)$ and Lipschitz constant L, we have:*

$$||\nabla f(x) - \nabla f(x') \leq L||x - x'||, \tag{5}$$

Lemma 2. *If $f(x)$ is Lipschitz-smooth, we have:*

$$f(x^{t+1}) - f(x^t) \leq -\frac{1}{2L}||\nabla f(x^T)||_2^2 \tag{6}$$

Proof. Using the Taylor expansion of $f(x)$ and the uniform bound over Hessian matrix, we have

$$f(x') \leq f(x) + \langle \nabla f(x), x' - x \rangle + \frac{L}{2}||x' - x||_2^2. \tag{7}$$

By inserting $x' = x - \frac{1}{L}\nabla f(x)$ into Eq. 5 and Eq. 7, we have:

$$f(x - \frac{1}{L}\nabla f(x)) - f(x) \leq -\frac{1}{L}\langle \nabla f(x), \nabla f(x) \rangle + \frac{L}{2}||\frac{1}{L}\nabla f(x)||_2^2 = -\frac{1}{2L}||\nabla f(x)||_2^2$$

Lemma 3 ***(Co-coercivity).*** *A convex and Lipschitz-smooth $f(x)$ satisfies:*

$$\langle \nabla f(x') - \nabla f(x), x' - x \rangle \geq \frac{1}{L}||\nabla f(x') - \nabla f(x)|| \tag{8}$$

Proof. Due to Eq. 5,

$$\langle \nabla f(x') - \nabla f(x), x' - x \rangle \geq \langle \nabla f(x') - \nabla f(x), \frac{1}{L}(\nabla f(x') - \nabla f(x)) \rangle = \frac{1}{L}||\nabla f(x') - \nabla f(x)||$$

Then we can proof the attack convergence theorem: $f(x^T) - f(x^*) \leq \frac{2L||x^0 - x^*||^2}{T}$.

Proof. Let $f(x)$ be convex and Lipschitz-smooth. It follow that

$$\begin{aligned} ||x^{t+1} - x^*||_2^2 &= ||x^t - x^* - \frac{1}{L}\nabla f(x^t)||_2^2 \\ &= ||x^t - x^*||_2^2 - 2\frac{1}{L}\langle x^t - x^*, \nabla f(x^t) \rangle + \frac{1}{L^2}||\nabla f(x^t)||_2^2 \\ &\leq ||x^t - x^*||_2^2 - \frac{1}{L^2}||\nabla f(x^t)||_2^2 \end{aligned} \tag{9}$$

Equation 9 holds due to Eq. 8 in Lemma 3. Recall Eq. 6 in Lemma 2, we have:

$$f(x^{t+1}) - f(x^*) \leq f(x^t) - f(x^*) - \frac{1}{2L}||\nabla f(x^t)||_2^2. \tag{10}$$

By applying convexity,

$$\begin{aligned} f(x^t) - f(x^*) &\leq \langle \nabla f(x^t), x^t - x^* \rangle \\ &\leq ||\nabla f(x^t)||_2 ||x^t - x^*|| \\ &\leq ||\nabla f(x^t)||_2 ||x^1 - x^*||. \end{aligned} \tag{11}$$

Then we insert Eq. 11 into Eq. 10:

$$f(x^{t+1}) - f(x^*) \leq f(x^t) - f(x^*) - \frac{1}{2L}\frac{1}{||x^1 - x^*||^2}(f(x^t) - f(x^*))^2$$

$$\Rightarrow \frac{1}{f(x^t) - f(x^*)} \leq \frac{1}{f(x^{t+1}) - f(x^*)} - \beta\frac{f(x^t) - f(x^*)}{f(x^{t+1}) - f(x^*)} \tag{12}$$

$$\Rightarrow \frac{1}{f(x^t) - f(x^*)} \leq \frac{1}{f(x^{t+1}) - f(x^*)} - \beta \tag{13}$$

$$\Rightarrow \beta \leq \frac{1}{f(x^{t+1}) - f(x^*)} - \frac{1}{f(x^t) - f(x^*)}, \tag{14}$$

where $\beta = \frac{1}{2L}\frac{1}{||x^1 - x^*||^2}$. Equation 12 is done by divide both side with $(f(x^{t+1}) - f(x^*))(f(x^t) - f(x^*))$ and Eq. 13 utilizes $f(x^{t+1}) - f(x^*) \leq f(x^t) - f(x^*)$. Then, following by induction over $t = 0, 1, 2, ..T - 1$ and telescopic cancellation, we have

$$T\beta \leq \frac{1}{f(x^T) - f(x^*)} - \frac{1}{f(x^0) - f(x^*)} \leq \frac{1}{f(x^T) - f(x^*)}.$$

$$T\beta \leq \frac{1}{f(x^T) - f(x^*)} - \frac{1}{f(x^0) - f(x^*)} \leq \frac{1}{f(x^T) - f(x^*)} \tag{15}$$

$$\Rightarrow \frac{T}{2L}\frac{1}{||x^1 - x^*||^2} \leq \frac{1}{f(x^T) - f(x^*)} \tag{16}$$

$$\Rightarrow f(x^T) - f(x^*) \leq \frac{2L||x^0 - x^*||^2}{T}. \tag{17}$$

Thus complete the proof.

References

1. Phong, L.T., Aono, Y., Hayashi, T., Wang, L., Moriai, S.: Privacy-preserving deep learning: revisited and enhanced. In: Batten, L., Kim, D.S., Zhang, X., Li, G. (eds.) ATIS 2017. CCIS, vol. 719, pp. 100–110. Springer, Singapore (2017). https://doi.org/10.1007/978-981-10-5421-1_9
2. Aono, Y., Hayashi, T., Wang, L., Moriai, S., et al.: Privacy-preserving deep learning via additively homomorphic encryption. IEEE Trans. Inf. Forensics Secur. **13**(5), 1333–1345 (2017)

3. Bagdasaryan, E., Veit, A., Hua, Y., Estrin, D., Shmatikov, V.: How to backdoor federated learning. arXiv preprint arXiv:1807.00459 (2018)
4. Battiti, R.: First-and second-order methods for learning: between steepest descent and Newton's method. Neural Comput. **4**(2), 141–166 (1992)
5. Bonawitz, K., et al.: Towards federated learning at scale: System design. In: Proceedings of the 2nd SysML Conference, pp. 619–633 (2018)
6. Fletcher, R.: Practical Methods of Optimization. Wiley, Hoboken (2013)
7. Fredrikson, M., Jha, S., Ristenpart, T.: Model inversion attacks that exploit confidence information and basic countermeasures. In: Proceedings of the 22nd ACM SIGSAC Conference on Computer and Communications Security, pp. 1322–1333 (2015)
8. Ganju, K., Wang, Q., Yang, W., Gunter, C.A., Borisov, N.: Property inference attacks on fully connected neural networks using permutation invariant representations. In: Proceedings of the 2018 ACM SIGSAC Conference on Computer and Communications Security, pp. 619–633 (2018)
9. Geiping, J., Bauermeister, H., Dröge, H., Moeller, M.: Inverting gradients-how easy is it to break privacy in federated learning? arXiv preprint arXiv:2003.14053 (2020)
10. Hayes, J., Melis, L., Danezis, G., De Cristofaro, E.: Logan: evaluating privacy leakage of generative models using generative adversarial networks. arXiv preprint arXiv:1705.07663 (2017)
11. Hitaj, B., Ateniese, G., Perez-Cruz, F.: Deep models under the GAN: information leakage from collaborative deep learning. In: Proceedings of the 2017 ACM SIGSAC Conference on Computer and Communications Security, pp. 603–618 (2017)
12. Huang, G.B., Mattar, M., Berg, T., Learned-Miller, E.: Labeled faces in the wild: a database for studying face recognition in unconstrained environments. In: Technical report (2008)
13. Kamp, M., et al.: Efficient decentralized deep learning by dynamic model averaging. In: Berlingerio, M., Bonchi, F., Gärtner, T., Hurley, N., Ifrim, G. (eds.) ECML PKDD 2018. LNCS (LNAI), vol. 11051, pp. 393–409. Springer, Cham (2019). https://doi.org/10.1007/978-3-030-10925-7_24
14. Konečný, J., McMahan, H.B., Yu, F.X., Richtárik, P., Suresh, A.T., Bacon, D.: Federated learning: strategies for improving communication efficiency. In: NIPS Workshop on Private Multi-Party Machine Learning (2016)
15. Krizhevsky, A., Hinton, G., et al.: Learning multiple layers of features from tiny images. In: Technical report (2009)
16. LeCun, Y., Cortes, C., Burges, C.J.: The MNIST database of handwritten digits (1998). http://yann.lecun.com/exdb/mnist10, 34 (1998)
17. Lin, Y., Han, S., Mao, H., Wang, Y., Dally, W.J.: Deep gradient compression: reducing the communication bandwidth for distributed training. In: International Conference on Learning Representations (2018)
18. Liu, W., Chen, L., Chen, Y., Zhang, W.: Accelerating federated learning via momentum gradient descent. IEEE Trans. Parallel Distrib. Syst. (2020)
19. Ma, C., et al.: Adding vs. averaging in distributed primal-dual optimization. In: Proceedings of the 32nd International Conference on Machine Learning, vol. 37, pp. 1973–1982 (2015)
20. McMahan, B., Moore, E., Ramage, D., Hampson, S., Arcas, B.A.: Communication-efficient learning of deep networks from decentralized data. In: Artificial Intelligence and Statistics, pp. 1273–1282 (2017)
21. McMahan, B., Ramage, D.: Federated learning: Collaborative machine learning without centralized training data. Google Res. Blog **3** (2017)

22. McMahan, H.B., Moore, E., Ramage, D., Arcas, B.A.: Federated learning of deep networks using model averaging. corr abs/1602.05629 (2016). arXiv preprint arXiv:1602.05629 (2016)
23. Melis, L., Song, C., De Cristofaro, E., Shmatikov, V.: Exploiting unintended feature leakage in collaborative learning. In: 2019 IEEE Symposium on Security and Privacy (SP), pp. 691–706. IEEE (2019)
24. Rossi, F., Gégout, C.: Geometrical initialization, parametrization and control of multilayer perceptrons: application to function approximation. In: Proceedings of 1994 IEEE International Conference on Neural Networks (ICNN 1994), vol. 1, pp. 546–550. IEEE (1994)
25. Shokri, R., Stronati, M., Song, C., Shmatikov, V.: Membership inference attacks against machine learning models. In: 2017 IEEE Symposium on Security and Privacy (SP), pp. 3–18. IEEE (2017)
26. Vanhaesebrouck, P., Bellet, A., Tommasi, M.: Decentralized collaborative learning of personalized models over networks. In: Artificial Intelligence and Statistics (2017)
27. Wang, Z., Song, M., Zhang, Z., Song, Y., Wang, Q., Qi, H.: Beyond inferring class representatives: user-level privacy leakage from federated learning. In: IEEE INFOCOM 2019-IEEE Conference on Computer Communications, pp. 2512–2520. IEEE (2019)
28. Wang, Z., Bovik, A.C., Sheikh, H.R., Simoncelli, E.P.: Image quality assessment: from error visibility to structural similarity. IEEE Trans. Image Process. **13**(4), 600–612 (2004)
29. Yang, Q., Liu, Y., Chen, T., Tong, Y.: Federated machine learning: concept and applications. ACM Trans. Intell. Syst. Technol. (TIST) **10**(2), 1–19 (2019)
30. Yao, X., Huang, T., Zhang, R.X., Li, R., Sun, L.: Federated learning with unbiased gradient aggregation and controllable meta updating. arXiv preprint arXiv:1910.08234 (2019)
31. Zhang, Q., Benveniste, A.: Wavelet networks. IEEE Trans. Neural Netw. **3**(6), 889–898 (1992)
32. Zhao, B., Mopuri, K.R., Bilen, H.: iDLG: Improved deep leakage from gradients. arXiv preprint arXiv:2001.02610 (2020)
33. Zhao, Y., Li, M., Lai, L., Suda, N., Civin, D., Chandra, V.: Federated learning with non-IID data. arXiv preprint arXiv:1806.00582 (2018)
34. Zhu, L., Liu, Z., Han, S.: Deep leakage from gradients. In: Advances in Neural Information Processing Systems, pp. 14747–14756 (2019)

Network Security II

An Accountable Access Control Scheme for Hierarchical Content in Named Data Networks with Revocation

Nazatul Haque Sultan[1(✉)], Vijay Varadharajan[1], Seyit Camtepe[2], and Surya Nepal[2]

[1] University of Newcastle, Callaghan, Australia
{Nazatul.Sultan,Vijay.Varadharajan}@newcastle.edu.au
[2] CSIRO Data61, Marsfield, NSW 2122, Australia
{seyit.camtepe,surya.nepal}@data61.csiro.au

Abstract. This paper presents a novel encryption-based access control scheme to address the access control issues in Named Data Networking (NDN). Though there have been several recent works proposing access control schemes, they are not suitable for many large scale real-world applications where content is often organized in a hierarchical manner (such as movies in Netflix) for efficient service provision. This paper uses a cryptographic technique, referred to as Role-Based Encryption, to introduce inheritance property for achieving access control over hierarchical contents. The proposed scheme encrypts the hierarchical content in such a way that any consumer who pays a higher level of subscription and is able to access (decrypt) contents in the higher part of the hierarchy is also able to access (decrypt) the content in the lower part of the hierarchy using their decryption keys. Additionally, our scheme provides many essential features such as authentication of the consumers at the very beginning before forwarding their requests into the network, accountability of the Internet Service Provider, consumers' privilege revocations, etc. In addition, we present a formal security analysis of the proposed scheme showing that the scheme is provably secure against Chosen Plaintext Attack. Moreover, we describe the performance analysis showing that our scheme achieves better results than existing schemes in terms of functionality, computation, storage, and communication overhead. Our network simulations show that the main delay in our scheme is due to cryptographic operations, which are more efficient and hence our scheme is better than the existing schemes.

Keywords: Named Data Networking · Access control · Accountability · Revocation · Encryption · Authentication · Provable security

1 Introduction

Recent development in Named Data Networking (NDN) has given a new direction to build the future Internet. Unlike the traditional host-centric IP-based

L. Chen et al. (Eds.): ESORICS 2020, LNCS 12308, pp. 569–590, 2020.
https://doi.org/10.1007/978-3-030-58951-6_28

networking architecture, the emphasis in NDN is shifted from *where* the data is located to *what* the data is. NDN is built by keeping data (or content) as its pivotal point, which helps to cope with the bandwidth insufficiency issue of the traditional IP-based networking architecture [1]. Recently, Cisco predicted that 82% of the world business traffic will be constituted by Internet video by the year 2022 [2]. Cisco also predicted that there will be 5.3 billion total Internet users (66 percent of global population) by 2023 [3]. As such, the current *host-to-host* communication network (i.e., IP-based networking architecture) has become inadequate for such content and user intensive services [4]. On the other hand, the services of NDN such as in-network cache, support of mobility, multicast and in-built security make it an ideal architecture for today's ever growing content and user intensive services in the Internet [5].

In NDN, each content or data is represented using unique *content name*. A consumer (or user) sends *interest packet* containing the desired content name to the Content Provider. Once the interest packet reaches the Content Provider, it sends the requested content in a *data packet*. While forwarding the data packet, NDN routers keep a copy of the data packet in its cache, so that if any further interest packets arrive (either from the same consumer or from other consumers) for the same content, the routers can satisfy the requests by sending the data packets without contacting the actual Content Provider. This enables NDN to utilize efficiently the network resources such as network bandwidth, routers' cache space, and lower latency content delivery to consumers.

Although NDN provides the aforementioned benefits, it also brings new security challenges among which access control is an important one [5]. In NDN, as routers to keep copies of the data packets in their in-network caches, unauthorized consumers can easily obtain their desired content from the network without Content Providers' permission. This has a negative impact on Content Providers' business. Therefore, an effective access control mechanism is critical for the successful deployment of NDN. In the recent years, several research works have addressed the access control problem in data centric networks. In general, these solutions can be categorised into two groups, namely authentication-based solutions and encryption-based solutions. We will describe the existing solutions in the Related Work section.

The main rationale behind our proposed scheme is that most Content Providers arrange content in hierarchical structure providing better consumer choice as well as offering better flexibility for providing different subscription options for different categories based on consumers' preferences. Furthermore, hierarchical representation allows the inheritance property to be taken into account in system design. For instance, if a consumer is subscribing to a higher-level content (in the hierarchy) by paying a higher subscription fee, the consumer can also get access to the content at the lower level in the hierarchy. We have previously proposed a new encryption scheme called Role Based Encryption (RBE) [6–9], which takes into account of hierarchies in roles. Furthermore, RBE, *by design*, takes into account the inheritance property, which helps to simplify the key management problem. In this paper, we have used RBE to develop a novel encryption based approach for a secure access control scheme for hierarchically organized content in NDN. As far as we are aware, our proposed scheme is the first one that considers the design of a secure access control scheme for hierarchically organized encrypted content in NDNs. Note that our previous RBE schemes [6–9] addressed secure access control on encrypted data stored in the cloud. As there are fundamental differences between NDN and host-centric IP-based network cloud architectures, our work described in this paper is different from our previous works on RBE for cloud data security. Furthermore, certain mechanisms such as partial decryption computations done by cloud service providers on behalf of users are not suitable for the NDN environment.

Another important characteristic of our scheme is that of ISP (Internet Service Provider) accountability. Due to caches in routers, the ISP is able to satisfy the consumer requests for content. This makes it hard for the Content Provider to know the exact service amount that the ISP has provided. Our scheme enables the ISP to provide unforgeable evidence to the Content Provider on the number of consumers' requests serviced by the ISP. This makes it easier for the Content Provider to pay the ISP based on usage of content by its consumers. It can also be useful to have feedback information about consumers' preferences and content popularity to help the Content Provider improve its content services.

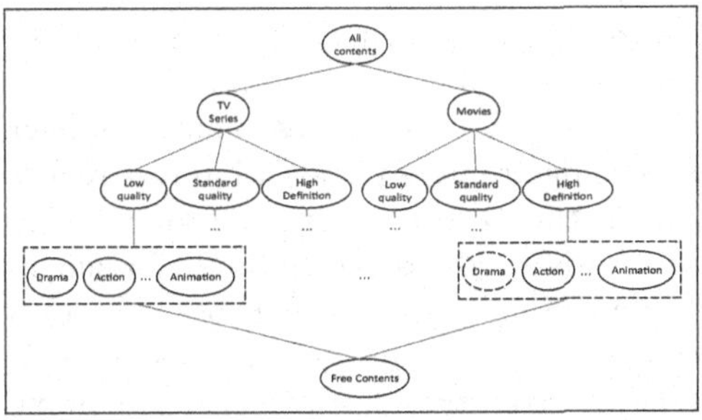

Fig. 1. Sample content hierarchy

1.1 Related Work

In the recent years, several research works have been proposed to address the access control problem in data centric networks, which are broadly categorised into two main groups, namely *authentication* and *encryption* [10]. The main idea of the authentication-based access control schemes is to verify authenticity of the consumers' content requests either at the routers (edge and internal routers) or Content Provider (CP) side before sending the contents. In [11], an access control scheme is proposed using a signature mechanism, where each router verifies integrity and authenticity of the encrypted contents before caching, and later send the cached contents only to the authorized consumers. In [12] and [13], CP requires to authenticate the consumers before sending any contents to them.

Moreover, [12] requires a secure-channel for sending the contents. In [14], the routers verify authenticity of the consumers using tags[1], which are assigned to them by the CP, before sending the contents to the consumers. In [15], an access control framework is proposed using the traditional *Kerberos*[2] mechanism. The scheme uses two separate online entities for authentication and authorization of the consumers. A shared key is established between the CP and the consumer for content encryption. In [16], CP verifies the authenticity of a consumer before establishing a session key for sending the encrypted contents. It is observed that all the existing authentication-based access control schemes: increases overhead at the routers [11,14], limits the utilization of NDN resources[3] [12–14,16] or needs online authority [12,13,15,16], which are not desirable for NDN [5].

On the hand, the main idea of the encryption-based access control schemes is to encrypt the contents that allows only the authorized consumers to decrypt. In the literature, different encryption techniques (e.g., Attribute-Based Encryption (ABE) [17], Broadcast Encryption (BE) [10,18–20], Identity-Based Encryption (IBE) [21], Proxy Re-encryption (PRE) [22], etc.) or a combination of those techniques have been used [23]. In [17], an ABE scheme for access control is proposed which allows any consumer having a qualified decryption key (i.e., a qualified set of attributes) to decrypt the ciphertexts. In [18], a BE scheme is proposed to share contents among a set of n consumers and can revoke t number of consumers from the system. In [19], another broadcast encryption technique is proposed, which integrates time tokens in the encrypted contents, and allows only the authorized consumers satisfying the time limitation to decrypt the ciphertexts. In [10] and [20], two BE based access control schemes have been proposed, which enables the edge routers to block interest requests of unauthorized consumers from entering the network using a signature-based authentication mechanism. [10] and [20] also address accountability issue of the ISP. In [21], a third-party entity ("Online Shop") uses an IBE based access control mechanism to share encrypted contents with the authorized consumers. In [23], another IBE based access control scheme is proposed, which also uses proxy re-encryption and broadcast encryption techniques, that embeds consumers' identities along with their subscription time in the encrypted content itself. This enables the consumers to access the contents within their respective subscription time. In [22], an in-device proxy re-encryption model is proposed that delegates the functionality of a proxy server to the consumers. However, all existing encryption-based works are not suitable for platforms like content services (such as Netflix (www.netflix.com) and Disney+ (www.disneyplus.com)), where contents are organized in hierarchies, such as genre, geographic region or video quality. Figure 1 shows an example of such a content hierarchy based video streaming service. To share the contents with the authorized consumers, the existing schemes would require issuing separate secret keys per category of content, which could be cumbersome.

[1] Each tag contains a signature, validity period, etc.
[2] www.tools.ietf.org/html/rfc4120.
[3] An encrypted content can only be shared with a particular consumer. As such, the other authorized consumers cannot take benefit of the cached contents.

Further, most of the schemes either do not support immediate revocation[4], e.g., [10,17,19], etc. (which is one of the essential requirements for NDN) or provides an inefficient revocation mechanism [18]. Moreover, authentication of the interest requests of the consumers before forwarding into the network at very beginning[5] is only provided by a few schemes such as [10,19,20]. However, [10,19,20] are vulnerable to collusion attack between a revoked consumer and ISP (i.e., edge routers).

1.2 Contributions

This paper proposes a novel encryption-based access control scheme for content organized in a hierarchical structure in NDN. A practical insight in this scheme is to reflect the real-world environment of the organization of content into hierarchical categories. Each category of content is then associated with a group of consumers who have subscribed to it and assigned appropriate unique decryption keys. The contents in a category are encrypted in such a way that any consumer associated with that category and ancestor categories of the content hierarchy can decrypt the content using their decryption key. This is where the idea from our previous work on RBE has been used. Content hierarchy is discussed further in Sect. 3.1 and 4.1. Our secure access scheme for NDN also achieves accountability enabling the ISP to provide unforgeable evidence (on the number of consumers' requests that it has serviced) to content providers. Furthermore, our scheme offers an efficient mechanism for revocation of privileges, which is a major stumbling block for the real-world deployment of NDN in content distribution.

The major contributions of this paper are as follows:

- Proposed a novel encryption based access control scheme for hierarchical content in NDN
- Designed an efficient signature mechanism, which enables the edge routers to authenticate the consumers' data requests before forwarding them to the Content Provider
- Proposed an accountability mechanism enabling the Content Provider to verify ISP's services to its consumers, thereby allowing the Content Provider to pay the correct amount to the ISP for its services
- Proposed a batch signature verification mechanism to verify correctly and efficiently the ISP's service
- Introduced revocation mechanisms, enabling the Content Provider to revoke the privileges of consumers from accessing its contents in an efficient manner. We introduce both expiration based and immediate revocation mechanisms. The former mechanism automatically revokes consumers after expiry of their subscription or validity periods. The latter mechanism enables the

[4] Privilege revocation of one or more consumers at any time.

[5] This is another essential requirement for NDN, which helps to prevent entering bogus interest requests into the network. This, in turn, prevents DoS attacks.

CP to revoke one or more consumers at anytime from accessing contents of a category.
- Provided a formal security analysis of the proposed scheme to demonstrate that the scheme is provably secure against Chosen Plaintext Attack (CPA)
- Demonstrated that the proposed scheme performs better than the existing works in terms of functionality, computation and communication overhead

This paper is organized as follows: Sect. 2 gives proposed model which includes system model, adversary model, and security model. Sect. 3 describes some preliminaries including content hierarchy, bilinear map, and complexity assumption, which are essential to understand our proposed scheme. Section 4 presents our proposed scheme in details. Analysis of the proposed scheme is provided in Sect. 5 and finally, Sect. 6 concludes this paper.

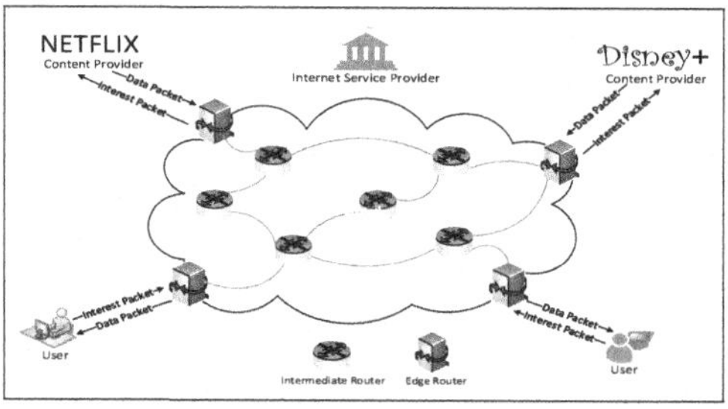

Fig. 2. System model

2 Proposed Model

2.1 System Model

Figure 2 shows NDN architecture with many Content Providers, an Internet Service Provider, and a large number of Consumers.

Content Providers (CPs) – They are the content producers like Netflix, Disney+ and Prime Video (www.primevideo.com). Each CP maintains a content hierarchy. A consumer may subscribe to one or more categories of content in the content hierarchy. For each subscription, CP issues access tokens in the form of secret keys to the consumers. Moreover, CP encrypts its contents before broadcasting them to the consumers.

Internet Service Provider (ISP) – It is responsible for managing the NDN network. It mainly consists of two types of routers, namely *Edge Routers* (ERs) and *Intermediate Routers* (IRs). Both ERs and IRs are responsible for forwarding consumers' interest packets to the CPs. The ERs and IRs are cache-enabled and hence can satisfy the consumers' interests without contacting the CPs if the content consumers' have the requested is in their caches. Additionally, ERs are responsible for authenticating the consumers before forwarding the interest packets into the network.

Consumers – They are the users of the content. Each consumer must register with a CP and subscribes for one or more categories of content.

2.2 Adversary Model

Our adversary model consists of two types of adversaries– ISP and consumers.

Honest-but-Curious and Greedy ISP – ISP is considered to be honest-but-curious. As a business, the ISP's aim is to increase its revenue from the CPs. As such, it honestly performs all the tasks assigned to it thereby providing incentives for the CPs to use its services. At the same time, it could be interested in learning as much as possible about the plaintext content. Moreover, ISPs can exaggerate or even cheat the CP, when it comes to the services provided to the consumers for gaining more financial benefit. In this sense, we regard the ISP to be greedy.

Malicious Consumers – The consumers are considered to be malicious in that they may collude among themselves or even with the ISP to gain access to content beyond what they are authorised to access. Note that, CPs are assumed as trusted entities, as they are the actual producers of the content and hence have a vested interest in protecting their content.

2.3 Security Model

The security model of the proposed scheme is defined by the *Semantic Security against Chosen Plaintext Attacks* (IND-CPA). IND-CPA is illustrated using a security game between a challenger $\mathcal{C}$ and an adversary $\mathcal{A}$, which is presented in Appendix A.

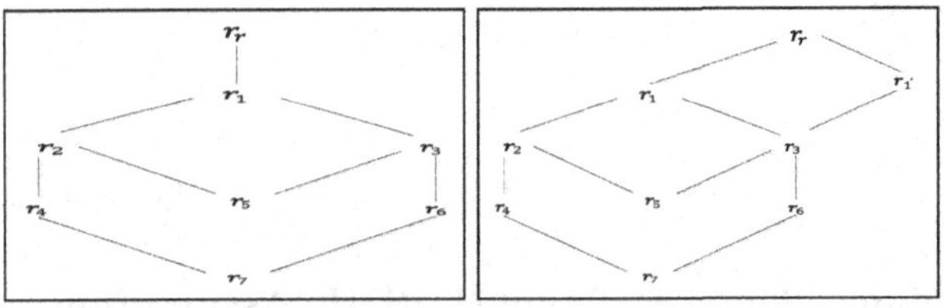

(a) Content Hierarchy 1. (b) Content Hierarchy 2.

Fig. 3. Sample content hierarchies.

3 Preliminaries

3.1 Content Hierarchy

In the proposed scheme, content is organized in a hierarchy, where consumers having access rights for ancestor category of content can inherit access rights of the descendant categories of content. For simplicity, from now onward, we may represent a category as a "role" and vise versa, as the content hierarchy is analogous to the role hierarchy. That is, consumers having access rights for the ancestor category or role inherits access rights to its descendent categories or roles. Figure 3a and Fig. 3b show two sample content hierarchies, namely Content Hierarchy 1 and Content Hierarchy 2 respectively. Let us consider the Content Hierarchy 1 as an example to define the notations that we will be using in a content hierarchy.

- r_r: root role of a content hierarchy. We assume that in any content hierarchy there can be only one root role.
- Ψ: set of all roles in the content hierarchy. For example, $\Psi = \{r_r, r_1, r_2, r_3, r_4, r_5, r_6, r_7\}$
- $\mathbb{A}_{r_i}$: ancestor set of the role r_i. For example, $\mathbb{A}_{r_1} = \{r_1, r_r\}$, $\mathbb{A}_{r_3} = \{r_r, r_1, r_3\}$ and $\mathbb{A}_{r_7} = \{r_r, r_1, r_2, r_4, r_6, r_7\}$.
- $\mathbb{D}_{r_i}$: descendent set of the role r_i. For example, $\mathbb{D}_{r_1} = \{r_2, r_3, r_4, r_5, r_6, r_7\}$, $\mathbb{D}_{r_3} = \{r_6, r_7\}$ and $\mathbb{D}_{r_7} = \emptyset$

3.2 Bilinear Map

Let $\mathbb{G}_1$ and $\mathbb{G}_T$ be two cyclic multiplicative groups. Let g be a generator of $\mathbb{G}_1$. The bilinear map $\hat{e} : \mathbb{G}_1 \times \mathbb{G}_1 \rightarrow \mathbb{G}_T$ should fulfil the following properties:

- Bilinear– $\hat{e}(g^a, g^b) = \hat{e}(g, g)^{ab}, \forall g \in \mathbb{G}_1, \forall (a, b) \in \mathbb{Z}_q^*$
- Non-degenerate– $\hat{e}(g, g) \neq 1$
- Computable– There exist an efficiently computable algorithm to compute $\hat{e}(g, g), \forall g \in \mathbb{G}_1$

3.3 Complexity Assumption

Our proposed scheme is based on the following complexity assumption.

Decisional Bilinear Diffie-Hellman (DBDH) Assumption. If $(a, b, c, z) \in \mathbb{Z}_q^*$ are chosen randomly, for any polynomial time adversary $\mathcal{A}$, then the ability for that adversary to distinguish the tuples $\langle g, g^a, g^b, g^c, Z = \hat{e}(g, g)^{a \cdot b \cdot c} \rangle$ and $\langle g, g^a, g^b, g^c, Z = \hat{e}(g, g)^z \rangle$ is negligible.

4 Proposed Scheme

This section first presents an overview of our proposed scheme followed by its main construction. In the overview section, we briefly describe main idea of our proposed encryption technique.

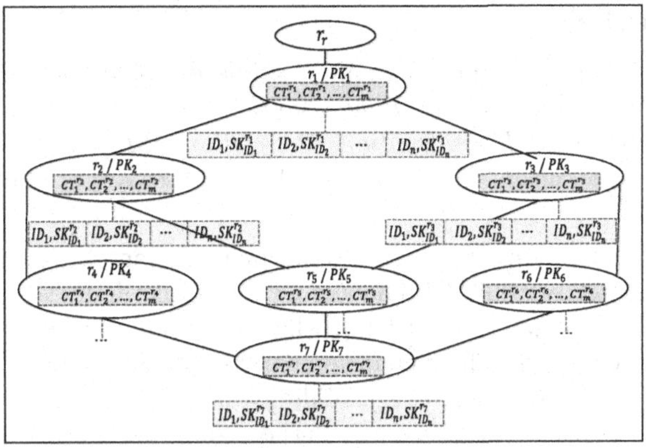

Fig. 4. A Sample Content Key Hierarchy (CKH)

4.1 Overview

In our proposed scheme, each CP maintains a content hierarchy CH, which consists of different categories of contents organized in a hierarchical structure. Each CP also maintains a Content Key Hierarchy (CKH) associated with the CH. A sample CKH is shown in Fig. 4. Each node in the CKH represents a category of the content. In the CKH, each node is associated with a public key, which is used to encrypt the content belonging to that category. Moreover, each node is also associated with a set of consumers who have subscribed to the associated category of content. For each subscription, the consumers are assigned unique secret keys (i.e., $\mathtt{SK}^{\mathtt{r_i}}_{\mathtt{ID_u}}$), so that they can access any content encrypted using the public key associated with that node (or category) as well as the content encrypted using the public keys associated with any descendent nodes (or sub-categories). For instance, if CP encrypts the content using the public key associated with r_2 category, then all consumers associated with r_1 and r_2 categories can decrypt the content using their respective secret keys. Similarly, all the ciphertexts associated with r_5 and r_7 can be decrypted by the consumers associated with r_1, r_2, r_3, r_5 and $r_1, r_2, r_3, r_4, r_5, r_6, r_7$ respectively. Note that, CP keeps the root node r_r as an unused category.

Our proposed scheme also provides a signature scheme that enables the edge routers to authenticate content requests (or interest packets) of the consumers before forwarding them to the networks. In the signature, each consumer includes his/her validity or subscription period, thereby enabling the edge router to prevent any consumer having expired subscription period from accessing the content and network resources. This achieves our time expiration-based revocation of consumers. Moreover, our proposed scheme introduces a novel immediate revocation scheme using the concept of bilinear-map accumulator. More details are given in Sect. 4.3.

4.2 Construction

Our proposed scheme comprises mainly 8 phases– *Content Provider Setup*, *Consumer Registration*, *Content Publication*, *Content Request and Consumer Authentication*, *Content Re-encryption*, *Decryption*, *ISP Service Accountability* and *Immediate Revocation*. The details of these phases are presented next.

Content Provider Setup. CP initiates this phase to generate system parameter $\mathtt{SP}$ and master secret key $\mathtt{MSK}$. CP chooses a content hierarchy Ψ, two cyclic multiplicative groups $\mathbb{G}_1$ and $\mathbb{G}_T$ of order q, where q is a large prime number. It also chooses a generator $g \in \mathbb{G}_1$, three collusion resistant hash functions $H : \{0,1\}^* \rightarrow \mathbb{G}_1, H_1 : \{0,1\}^* \rightarrow \mathbb{Z}_q^*, H_2 : \mathbb{G}_T \rightarrow \mathbb{Z}_q^*$, a bilinear map $\hat{e} : \mathbb{G}_1 \times \mathbb{G}_1 \rightarrow \mathbb{G}_T$, and random numbers $(\sigma, \delta, \rho, \{t_{r_i}\}_{\forall r_i \in \Psi}) \in \mathbb{Z}_q^*$. It then computes P, Q. It also computes public key $\mathbb{PK}_{r_i} = \langle \mathtt{PK}_{\mathtt{r_i}}, \mathbb{A}_{r_i}, r_i \rangle$, proxy re-encryption keys $\{\mathtt{PKey}_{\mathtt{r_i}}^{\mathtt{r_w}}\}_{\forall r_w \in \mathbb{A}_{r_i} \setminus \{r_i\}}$ and accumulator $\mathtt{Acc}_{\mathtt{r_i}}$ for each category $r_i \in (\Psi \setminus \{r_r\})$, where

$$P = \hat{e}(g,g)^{\sigma}; Q = \hat{e}(g,g)^{\delta}; \mathtt{PK}_{\mathtt{r_i}} = g^{\prod_{\forall r_j \in \mathbb{A}_{r_i} \setminus \{r_r\}} \left(\frac{t_{r_j}}{t_r}\right)}$$

$$\mathtt{PKey}_{\mathtt{r_i}}^{\mathtt{r_w}} = \prod_{\forall r_j \in \mathbb{A}_{r_i} \setminus \{r_w, r_r\}} \left(\frac{t_{r_j}}{t_r}\right); \mathtt{Acc}_{\mathtt{r_i}} = \hat{e}\left(H(r_i), g\right)^{\prod_{\forall \mathtt{ID_u} \in \mathbb{U}_{r_i}} \mathtt{sk}_{\mathtt{ID_u}}}$$

$\mathbb{U}_{r_i}$ represents the set of all the consumers who possess category r_i. The system parameter $\mathtt{SP}$ and master secret key $\mathtt{MSK}$ are $\mathtt{SP} = \Big\langle q, \mathbb{G}_1, \mathbb{G}_T, \hat{e}, g, H, H_1, H_2, P, Q, \{\mathbb{PK}_{r_i}\}_{\forall r_i \in \Psi} \Big\rangle$ and $\mathtt{MSK} = \Big\langle \sigma, \delta, \rho, \{t_{r_i}, \mathtt{Acc}_{\mathtt{r_i}}, \{\mathtt{PKey}_{\mathtt{r_i}}^{\mathtt{r_w}}\}_{\forall r_w \in \mathbb{A}_{r_i} \setminus \{r_i\}}\}_{\forall r_i \in \Psi \setminus \{r_r\}} \Big\rangle$ respectively. The system parameter $\mathtt{SP}$ is made public by keeping them in a public bulletin board, whereas the master secret key $\mathtt{MSK}$ is kept in a secure place. Note that, the proxy re-encryption keys are not generated for the root role. Further, CP securely sends the proxy re-encryption keys to the NDN routers when requested. We assume that CP uses a secure public key encryption technique such as the one in [24] for transmission of the proxy re-encryption keys.

Consumer Registration. This phase is initiated by the CP when a consumer wants to subscribe for its contents. Suppose a consumer, say $\mathtt{ID_u}$ ($\mathtt{ID_u}$ is the unique identity of the consumer), subscribes to the category r_x for a period of $\mathtt{VP}_{\mathtt{ID_u}}$. The CP issues a set of secret keys $\mathtt{SK}_{\mathtt{ID_u}}^{\mathtt{r_x}}$ and public keys $\mathtt{Pub}_{\mathtt{ID_u}}^{\mathtt{r_x}}$ to the subscribed consumer $\mathtt{ID_u}$. Moreover, the CP generates an updated accumulator $\mathtt{Acc}_{\mathtt{r_x}}^{\mathtt{ver_v}}$ associated with the category r_x. CP chooses a random number $\mathtt{sk}_{\mathtt{ID_u}} \in \mathbb{Z}_q^*$ (termed as private key) and computes $\mathtt{SK}_{\mathtt{ID_u}}^{\mathtt{r_x}} = \langle \mathtt{sk}_{\mathtt{ID_u}}, \mathtt{RK}_{\mathtt{ID_u,r_x}}^1, \mathtt{RK}_{\mathtt{ID_u,r_x}}^2 \rangle$, $\mathtt{Pub}_{\mathtt{ID_u}}^{\mathtt{r_x}} = \langle \mathtt{Pub1}_{\mathtt{ID_u}}^{\mathtt{r_x}}, \mathtt{Pub2}_{\mathtt{ID_u}}^{\mathtt{r_x}}, \mathtt{PubWit}_{\mathtt{ID_u}}^{\mathtt{r_x}} \rangle$, $\mathtt{Acc}_{\mathtt{r_x}}^{\mathtt{ver_v}}$, where

$$\mathtt{RK}_{\mathtt{ID_u,r_x}}^1 = g^{\frac{\sigma \cdot \mathtt{sk}_{\mathtt{ID_u}} + \delta}{t_{r_x}}}; \mathtt{RK}_{\mathtt{ID_u,r_x}}^2 = g^{\frac{\sigma \cdot \mathtt{sk}_{\mathtt{ID_u}} + \delta}{\prod_{\forall r_j \in (\mathbb{A}_{r_x} \setminus \{r_r\})} \left(\frac{t_{r_j}}{t_r}\right)}}$$

$$\mathtt{Pub1}_{\mathtt{ID_u}}^{\mathtt{r_x}} = H(r_x)^{\frac{1}{\mathtt{sk}_{\mathtt{ID_u}} + \rho \cdot H_1(\mathtt{VP}_{\mathtt{ID_u}})}}; \mathtt{Pub2}_{\mathtt{ID_u}}^{\mathtt{r_x}} = \hat{e}\left(H(r_x), g\right)^{\frac{\rho}{\mathtt{sk}_{\mathtt{ID_u}} + \rho \cdot H_1(\mathtt{VP}_{\mathtt{ID_u}})}}$$

$$\mathtt{PubWit}_{\mathtt{ID_u}}^{\mathtt{r_x}} = \hat{e}\left(H(r_x), g\right)^{\prod_{\forall \mathtt{ID_j} \in (\mathbb{U}_{r_x} \setminus \{\mathtt{ID_u}\})} \mathtt{sk}_{\mathtt{ID_j}}}; \mathtt{Acc}_{\mathtt{r_x}}^{\mathtt{ver_v}} = (\mathtt{Acc}_{\mathtt{r_x}}^{\mathtt{verv-1}})^{\mathtt{sk}_{\mathtt{ID_u}}}$$

$\mathtt{Acc}_{\mathtt{r_x}}^{\mathtt{ver_v}}$ and $\mathtt{Acc}_{\mathtt{r_x}}^{\mathtt{verv-1}}$ represent current and previous the accumulator associated with category r_x. The secret key $\mathtt{SK}_{\mathtt{ID_u}}$ is sent to the consumer using a secure public key encryption mechanism, as described in *Content Provider Setup* phase, and the public keys are kept in a public bulletin board. Note that, we shall refer to $\mathtt{Pub1}_{\mathtt{ID_u}}^{\mathtt{r_x}}, \mathtt{Pub2}_{\mathtt{ID_u}}^{\mathtt{r_x}}$ as public keys and $\mathtt{PubWit}_{\mathtt{ID_u}}^{\mathtt{r_x}}$ as witness public key. Also, the subscription or validity period $\mathtt{VP}_{\mathtt{ID_u}}$ can be of the format *dd– mm– yyyy*.

Content Publication. CP initiates this phase when it wants to broadcast a content to the authorized (subscribed) consumers. Suppose CP wants to broadcast a content $\mathbb{M}$ for the consumers having access rights for the category r_i. CP first encrypts the content using a random symmetric key $\mathtt{K} \in \mathbb{G}_T$ with a symmetric key encryption mechanism. It then encrypts the random symmetric key $\mathtt{K}$ using following encryption process. CP chooses two random numbers $(y_1, y_2) \in \mathtt{Z}_q^*$ and computes C_1, C_2, C_3, C_{r_i}, where

$$C_1 = \mathtt{K} \cdot P^{y_1} = \mathtt{K} \cdot \hat{e}(g,g)^{\sigma \cdot y_1}; C_2 = Q^{(y_1+y_2)} = \hat{e}(g,g)^{\delta \cdot (y_1+y_2)}$$

$$C_3 = P^{y_2} = \hat{e}(g,g)^{\sigma \cdot y_2}; C_{r_i} = (\mathrm{PK}_{\mathtt{r_i}})^{(y_1+y_2)} = g^{(y_1+y_2) \cdot \prod_{\forall r_j \in \mathbb{A}_{r_i} \setminus \{r_r\}} \left(\frac{t_{r_j}}{t_r}\right)}$$

Final, ciphertext is $\mathbb{CT} = \langle \mathbf{Enc}_{\mathtt{k}}(\mathbb{M}), C_1, C_2, C_3, C_{r_i} \rangle$, where $\mathbf{Enc}_{\mathtt{k}}(\mathbb{M})$ is the symmetric key encryption component of the actual content. Note that, the Advanced Encryption Standard (AES) scheme can be used for symmetric key encryption. Also note that, our proposed scheme uses a hierarchical content naming scheme similar to the one defined in [4]. For instance, a drama TV series name can be represented as "*/com/example/all-contents/tv-series/quality/drama/xyz.mp4/chunk_1*", where */com/example/* represents the CP's domain name (e.g., example.com), */all-contents/tv-series/quality/drama* represents category of the content in the CH (or CKH)[6], and */xyz.mp4* represents file name and */Chunk_1* specifies the contents in the file. This naming scheme helps the routers to identify the category of the requested content along with the ancestor categories in the CH.

Content Request and Consumer Authentication. This phase is divided into two sub-phases, namely *Content Request* and *Consumer Authentication.*

Content Request. A consumer initiates this sub-phase when he/she wants to access a content. Suppose, the consumer $\mathtt{ID_u}$, possessing access rights for the category r_x, wants to access a content of category r_i. The consumer $\mathtt{ID_u}$ chooses random number $v \in \mathbb{Z}_q^*$, timestamp ts, content name CN and computes a signature $\mathtt{Sig} = \langle S_1, S_2, S_3 \rangle$, where

$$S_1 = [\mathtt{sk}_{\mathtt{ID_u}} + H_1(ts||CN)] \cdot v; S_2 = \left(\mathtt{Pub2}_{\mathtt{ID_u}}^{\mathtt{r_x}}\right)^v = \hat{e}\left(H(r_x), g\right)^{\frac{v \cdot \rho}{\mathtt{sk}_{\mathtt{ID_u}} + \rho \cdot H_1(\mathtt{VP}_{\mathtt{ID_u}})}}; S_3 = g^v$$

The consumer $\mathtt{ID_u}$ sends the interest packet with the signature $\mathtt{Sig}$, current timestamp ts and validity period $\mathtt{VP}_{\mathtt{ID_u}}$ to the nearest ER.

[6] It also shows the ancestor categories in the CH.

Consumer Authentication. The ER initiates this sub-phase when it receives an interest packet from a consumer. In this phase, the ER verifies authenticity of the requested consumer before forwarding or satisfying his/her interest. ER first checks the timestamp ts and validity period $\mathtt{VP}_{\mathtt{ID_u}}$ of the consumer with the current time ts'. If they are invalid (i.e., $ts' - ts < \Delta t$ and $\mathtt{VP}_{\mathtt{ID_u}} < ts'$, where Δt is a predefined value), the ER ignores the request. Otherwise, it searches the database for an entry belonging to the consumer $\mathtt{ID_u}$. If there is no entry for the consumer, it sends a request to the CP asking for the public keys $\mathtt{Pub1}^{\mathtt{r_x}}_{\mathtt{ID_u}}, \mathtt{Pub2}^{\mathtt{r_x}}_{\mathtt{ID_u}}$. After receiving the public keys $\mathtt{Pub1}^{\mathtt{r_x}}_{\mathtt{ID_u}}, \mathtt{Pub2}^{\mathtt{r_x}}_{\mathtt{ID_u}}$, the ER verifies authenticity of the consumer $\mathtt{ID_u}$ possessing the access rights for the category r_x using the following process: ER computes V_1, V_2 and V_3,

$$\begin{aligned}
V_1 &= \hat{e}\left(\mathtt{Pub1}^{\mathtt{r_x}}_{\mathtt{ID_u}}, g\right)^{S_1} \cdot (S_2)^{H_1(\mathtt{VP}_{\mathtt{ID_u}})} \\
&= \hat{e}\left(H(r_x), g\right)^{\frac{\left[\mathtt{sk}_{\mathtt{ID_u}} + \rho \cdot H_1(\mathtt{VP}_{\mathtt{ID_u}})\right] \cdot v + H_1(ts||CN) \cdot v}{\mathtt{sk}_{\mathtt{ID_u}} + \rho \cdot H_1(\mathtt{VP}_{\mathtt{ID_u}})}} \\
&= \hat{e}\left(H(r_x), g\right)^{v} \cdot \hat{e}\left(H(r_x), g\right)^{\frac{H_1(ts||CN) \cdot v}{\mathtt{sk}_{\mathtt{ID_u}} + \rho \cdot H_1(\mathtt{VP}_{\mathtt{ID_u}})}} \\
V_2 &= \hat{e}\left(H(r_x), S_3\right) = \hat{e}\left(H(r_x), g\right)^{v} \\
V_3 &= \hat{e}\left(\left(\mathtt{Pub1}^{\mathtt{r_x}}_{\mathtt{ID_u}}\right)^{H_1(ts||CN)}, S_3\right) = \hat{e}\left(H(r_x), g\right)^{\frac{H_1(ts||CN) \cdot v}{\mathtt{sk}_{\mathtt{ID_u}} + \rho \cdot H_1(\mathtt{VP}_{\mathtt{ID_u}})}}
\end{aligned}$$

The ER compares V_1 and $V_2 \cdot V_3$. If $V_1 == V_2 \cdot V_3$, the ER successfully authenticates the consumer and it forwards the interest packet to the network if the data packets are not in its cache. Otherwise, the ER ignores the request. Note that, a consumer having expired subscription or validity period cannot pass through *Consumer Authentication* sub-phase. As such, this achieves our expiration-based consumer revocation.

Content Re-encryption. In this phase, a router (either ER or IR) re-encrypts a ciphertext. Suppose a router (either ER or IR) receives an interest request for r_i category of content from the consumer $\mathtt{ID_u}$ having access rights for the category r_x. Let, $\mathbb{CT}\langle \mathtt{Enc_K}(\mathbb{M}), C_1, C_2, C_3, C'_{r_i}\rangle$ be the requested ciphertext. If $r_x == r_i$, the router forwards the data packet to the consumer. Otherwise, if $r_x \neq r_i$ and $r_x \in \mathbb{A}_{r_i}$, the router performs a re-encryption operation over the ciphertext. The routers recomputes C_{r_i} component of the ciphertext and generates a re-encrypted ciphertext $\mathbb{CT}'$, which is sent to the consumer $\mathtt{ID_u}$. To recomputes C_{r_i}, the router requires the proxy re-encryption key $\mathtt{PKey}^{\mathtt{r_x}}_{\mathtt{r_i}}$. If the router does not have the proxy re-encryption key $\mathtt{PKey}^{\mathtt{r_x}}_{\mathtt{r_i}}$ in its database, it sends a request for it to the CP as described in *Content Provider Setup* phase. For the re-encryption, the router computes C'_{r_i}, where

$$C'_{r_i} = (C_{r_i})^{\frac{1}{\mathtt{PKey}^{\mathtt{r_x}}_{\mathtt{r_i}}}} = g^{y \cdot \prod_{\forall r_j \in (\mathbb{A}_{r_i} \setminus \{r_r\})} \left(\frac{t_{r_j}}{t_r}\right) \cdot \frac{1}{\prod_{\forall r_j \in \mathbb{A}_{r_i} \setminus \{r_x, r_r\}} \left(\frac{t_{r_j}}{t_r}\right)}} = g^{y \cdot t_{r_x}}$$

The final re-encrypted ciphertext is $\mathbb{CT}' = \langle \mathtt{Enc_K}(\mathbb{M}), C_1, C_2, C_3, C'_{r_i}\rangle$. Note that, after receiving the proxy re-encryption key, the router can keep it in its database for future use.

Content Decryption. A consumer initiates this phase to decrypt a ciphertext. The consumer $\mathtt{ID_u}$ first recovers the random symmetric key $\mathtt{K}$ from the ciphertext using his/her secret key $\mathtt{SK}^{\mathtt{r_x}}_{\mathtt{ID_u}} = \langle \mathtt{sk}_{\mathtt{ID_u}}, \mathtt{RK}^1_{\mathtt{ID_u},\mathtt{r_x}}, \mathtt{RK}^2_{\mathtt{ID_u},\mathtt{r_x}} \rangle$, and then decrypts the symmetric key encryption component of the ciphertext using $\mathtt{K}$ to get the plaintext content $\mathbb{M}$. If the consumer $\mathtt{ID_u}$ receives the ciphertext $\mathbb{CT} = \langle \mathtt{Enc_K}(\mathbb{M}), C_1, C_2, C_3, C_{r_i} \rangle$, i.e., $r_x = r_i$, then he/she computes D_1, where

$$\begin{aligned} D_1 &= \hat{e}\left(\mathtt{RK}^2_{\mathtt{ID_u},\mathtt{r_x}}, C_{r_i}\right) \\ &= \hat{e}\left(g^{\frac{\sigma \cdot \mathtt{sk}_{\mathtt{ID_u}} + \delta}{\prod_{\forall r_j \in (\mathbb{A}_{r_x} \setminus \{r_r\})} \left(\frac{t_{r_j}}{t_r}\right)}}, g^{(y_1+y_2) \cdot \prod_{\forall r_j \in (\mathbb{A}_{r_i} \setminus \{r_r\})} \left(\frac{t_{r_j}}{t_r}\right)} \right) \\ &= \hat{e}(g,g)^{[\sigma \cdot \mathtt{sk}_{\mathtt{ID_u}} + \delta] \cdot (y_1+y_2)} = \hat{e}(g,g)^{[\sigma \cdot \mathtt{sk}_{\mathtt{ID_u}}] \cdot (y_1+y_2)} \cdot \hat{e}(g,g)^{\delta \cdot (y_1+y_2)} \end{aligned}$$

If the consumer $\mathtt{ID_u}$ receives the re-encrypted ciphertext $\mathbb{CT}' = \langle \mathtt{Enc_K}(\mathbb{M}), C_1, C_2, C_3, C'_{r_i} \rangle$, i.e., $r_x \in \mathbb{A}_{r_i}$, then he/she computes D_1, where

$$\begin{aligned} D_1 &= \hat{e}\left(\mathtt{RK}^1_{\mathtt{ID_u},\mathtt{r_x}}, C'_{r_i}\right) \\ &= \hat{e}\left(g^{\frac{\sigma \cdot \mathtt{sk}_{\mathtt{ID_u}} + \delta}{t_{r_x}}}, g^{(y_1+y_2) \cdot t_{r_x}} \right) \\ &= \hat{e}(g,g)^{[\sigma \cdot \mathtt{sk}_{\mathtt{ID_u}} + \delta] \cdot (y_1+y_2)} = \hat{e}(g,g)^{[\sigma \cdot \mathtt{sk}_{\mathtt{ID_u}}] \cdot (y_1+y_2)} \cdot \hat{e}(g,g)^{\delta \cdot (y_1+y_2)} \end{aligned}$$

Then the consumer $\mathtt{ID_u}$ computes D_2, D_3 and recovers the secret key $\mathtt{K}$ as follows:

$$\begin{aligned} D_2 &= \frac{D_1}{C_2} = \hat{e}(g,g)^{\sigma \cdot \mathtt{sk}_{\mathtt{ID_u}} \cdot (y_1+y_2)} = \hat{e}(g,g)^{\sigma \cdot \mathtt{sk}_{\mathtt{ID_u}} \cdot y_1} \cdot \hat{e}(g,g)^{\sigma \cdot \mathtt{sk}_{\mathtt{ID_u}} \cdot y_2} \\ D_3 &= \frac{D_2}{(C_3)^{\mathtt{sk}_{\mathtt{ID_u}}}} = \frac{\hat{e}(g,g)^{\sigma \cdot \mathtt{sk}_{\mathtt{ID_u}} \cdot y_1} \cdot \hat{e}(g,g)^{\sigma \cdot \mathtt{sk}_{\mathtt{ID_u}} \cdot y_2}}{\hat{e}(g,g)^{\sigma \cdot \mathtt{sk}_{\mathtt{ID_u}} \cdot y_2}} = \hat{e}(g,g)^{\sigma \cdot \mathtt{sk}_{\mathtt{ID_u}} \cdot y_1} \\ \mathtt{K} &= \frac{C_1}{(D_3)^{\frac{1}{\mathtt{sk}_{\mathtt{ID_u}}}}} = \frac{\mathtt{K} \cdot \hat{e}(g,g)^{\sigma \cdot y_1}}{\hat{e}(g,g)^{\sigma \cdot y_1}} \end{aligned}$$

Finally, the consumer $\mathtt{ID_u}$ decrypts the symmetric key encryption component $\mathtt{Enc_K}(\mathbb{M})$ using $\mathtt{K}$ and gets the plaintext content $\mathbb{M}$.

ISP Service Accountability. This phase is initiated by the CP when it wants to evaluate the service information provided by the ISP (i.e., ERs). In our scheme, the ERs periodically send consumers' service information $\langle ts, CN, \mathtt{Sig}, \mathtt{ID_u} \rangle$ to the respective CP, where ts is the time at which the consumer $\mathtt{ID_u}$ sent the interest packet to the ER, CN is the requested content name, $\mathtt{Sig}$ is the signature of the consumer $\mathtt{ID_u}$. After receiving the service information, the CP first compares timestamp ts with the validity period $\mathtt{VP}_{\mathtt{ID_u}}$ of the consumer $\mathtt{ID_u}$. If $ts \leq \mathtt{VP}_{\mathtt{ID_u}}$, then only the CP needs to verify signature $\mathtt{Sig}$ of the consumer $\mathtt{ID_u}$. Otherwise, it ignores that service information of the consumer $\mathtt{ID_u}$, as the ISP has provided services to a consumer whose validity period has already expired. In this case, the CP can also take punitive actions against the ISP. Afterward, the CP employs a batch verification technique to verify the consumers' signatures. It computes V_1, V_2 and V_3 using the signatures of the consumers who had access rights for the same category of contents in the content hierarchy. Let's assume that the selected consumers' had access rights for the r_x category of contents in the content hierarchy and $\mathbb{S}_{r_x}$ is the set of those consumers. CP computes

$$\begin{aligned} V_1 &= \hat{e}\left(\prod_{\forall \mathrm{ID_u} \in \mathbb{S}_{r_x}} \left(\mathtt{Pub1}^{\mathtt{r_x}}_{\mathrm{ID_u}}\right)^{S_1}, g\right) \cdot \prod_{\forall \mathrm{ID_u} \in \mathbb{S}_{r_x}} (S_2)^{H_1(\mathtt{VP}_{\mathrm{ID_u}})} \\ &= \hat{e}\left(\prod_{\forall \mathrm{ID_u} \in \mathbb{S}_{r_x}} H(r_x)^{\frac{S_1}{\mathtt{sk}_{\mathrm{ID_u}} + \rho \cdot H_1(\mathtt{VP}_{\mathrm{ID_u}})}}, g\right) \cdot \prod_{\forall \mathrm{ID_u} \in \mathbb{S}_{r_x}} \hat{e}\left(H(r_x), g\right)^{\frac{v \cdot \rho \cdot H_1(\mathtt{VP}_{\mathrm{ID_u}})}{\mathtt{sk}_{\mathrm{ID_u}} + \rho \cdot H_1(\mathtt{VP}_{\mathrm{ID_u}})}} \\ &= \prod_{\forall \mathrm{ID_u} \in \mathbb{S}_{r_x}} \hat{e}\left(H(r_x), g\right)^{\frac{\mathtt{sk}_{\mathrm{ID_u}} + H_1(ts||CN) \cdot v + v \cdot \rho \cdot H_1(\mathtt{VP}_{\mathrm{ID_u}})}{\mathtt{sk}_{\mathrm{ID_u}} + \rho \cdot H_1(\mathtt{VP}_{\mathrm{ID_u}})}} \\ &= \hat{e}\left(H(r_x), g\right)^{\sum_{\forall \mathrm{ID_u} \in \mathbb{S}_{r_x}} v} \cdot \hat{e}\left(H(r_x), g\right)^{\sum_{\forall \mathrm{ID_u} \in \mathbb{S}_{r_x}} \frac{H_1(ts||CN) \cdot v}{\mathtt{sk}_{\mathrm{ID_u}} + \rho \cdot H_1(\mathtt{VP}_{\mathrm{ID_u}})}} \\ V_2 &= \hat{e}\left(H(r_x), \prod_{\forall \mathrm{ID_u} \in \mathbb{S}_{r_x}} S_3\right) = \hat{e}\left(H(r_x), g\right)^{\sum_{\forall \mathrm{ID_u} \in \mathbb{S}_{r_x}} v} \\ V_3 &= \hat{e}\left(H(r_x), \prod_{\forall \mathrm{ID_u} \in \mathbb{S}_{r_x}} (S_3)^{\frac{H_1(ts||CN)}{\mathtt{sk}_{\mathrm{ID_u}} + \rho \cdot H_1(\mathtt{VP}_{\mathrm{ID_u}})}}\right) = \hat{e}\left(H(r_x), g\right)^{\sum_{\forall \mathrm{ID_u} \in \mathbb{S}_{r_x}} \frac{H_1(ts||CN) \cdot v}{\mathtt{sk}_{\mathrm{ID_u}} + \rho \cdot H_1(\mathtt{VP}_{\mathrm{ID_u}})}} \end{aligned}$$

Note that the CP knows the private key $\mathtt{sk}_{\mathrm{ID_u}}$ and the validity period $\mathtt{VP}_{\mathrm{ID_u}}$ of each consumer in $\mathbb{S}_{r_x}$. It also knows ρ. As such, the CP can compute $\mathtt{sk}_{\mathrm{ID_u}} + \rho \cdot H_1(\mathtt{VP}_{\mathrm{ID_u}})$. If $V_1 = V_2 \cdot V_3$, then the verification process is successful, which demonstrates that the consumers' service information are legitimate.

4.3 Immediate Revocation

Our immediate revocation mechanism is designed using the concept of bilinear-map accumulator [25]. When the CP wants to prevent one or more consumers from accessing the contents of a category, this phase is initiated. We need a small modification in our proposed scheme to revoke consumers. Suppose, the CP wants to revoke a set of consumers having access rights for the category r_i. Let $\mathbb{R}_{r_i}$ be the set of revoked consumers possessing category r_i. The CP updates the system parameters such as public keys of the categories associated with $(\mathbb{D}_{r_i} \cup \{r_i\})$, public witness keys of all the non-revoked consumers and accumulator associated with r_i, where $\left\{\mathtt{PK}^{\mathtt{ver_v}}_{\mathtt{r_j}} = \left(\mathtt{PK}^{\mathtt{ver_{v-1}}}_{\mathtt{r_j}}\right)^{H_2(k^v)}\right\}_{\forall r_j \in\ (\mathbb{D}_{r_i} \cup \{r_i\})}$

$$\mathtt{PubWit}^{\mathtt{r_x},\mathtt{ver_v}}_{\mathrm{ID_u}} = \left(\mathtt{PubWit}^{\mathtt{r_x},\mathtt{ver_{v-1}}}_{\mathrm{ID_u}}\right)^{\frac{v}{\prod_{\forall \mathrm{ID_j} \in \mathbb{R}_{r_i}} \mathtt{sk}_{\mathrm{ID_j}}}} ; \mathtt{Acc}^{\mathtt{ver_v}}_{\mathtt{r_i}} = \left(\mathtt{Acc}^{\mathtt{ver_{v-1}}}_{\mathtt{r_i}}\right)^{\frac{v}{\prod_{\forall \mathrm{ID_j} \in \mathbb{R}_{r_i}} \mathtt{sk}_{\mathrm{ID_j}}}}$$

$k^v \in \mathbb{G}_T$, $v \in \mathbb{Z}_q^*$ are random numbers, and $\mathtt{ver_v}$ and $\mathtt{ver_{v-1}}$ represent current and previous version respectively.

Table 1. Functionality comparison

	Hierarchical Contents	Data Confidentiality	DoS Attack		Privilege Revocation			Accountability	Offline CP
			Authentication	Collusion Resistance	Expiration	Immediate	Enforcement		
[18]	No	Yes	No	N/A	No	Yes	CP	No	Yes
[10]	No	Yes	Yes	No	Yes	No	ER	Yes	Yes
[21]	No	Yes	No	N/A	No	No[1]	Online Authority	No	Yes
[11]	No	Yes	No	N/A	No	Yes	CP	No	Yes
[14]	No	Yes	Yes	No	Yes	No	ER, IR	No	Yes
[15]	No	Yes	No	N/A	Yes	No	Online Authority	No	No
[16]	No	Yes	No	N/A	No	No	N/A	No	No
[19]	No	Yes	Yes	No	Yes	No	ER	No	Yes
[22]	No	Yes	No	N/A	No	No	N/A	No	Yes
[23]	No	Yes	No	No	Yes	No	CP	No	Yes
[20]	No	Yes	Yes	No	Yes	No	ER	Yes	Yes
Our scheme	Yes	Yes	Yes	Yes	Yes	Yes	CP, ER	Yes	Yes

[1] [21] assumes that online authority, i.e., *Online Shop* will send contents only to the authorized consumers.
N/A– not applicable

CP also computes an additional component $\{Hdr_{r_j} = C_{h_{r_j}}^{\mathtt{ver_v}} = k^v \cdot \mathtt{Acc}_{\mathtt{r_j}}^{\mathtt{ver_v}}\}_{\forall r_j \in (\mathbb{D}_{r_i} \cup \{r_i\})}$, which is sent along with each newly encrypted ciphertexts. This additional component Hdr_{r_j} is mainly used to share the secret k^v with all the non-revoked authorized consumers of $r_j \in (\mathbb{D}_{r_i} \cup \{r_i\})$, so that they can use it to update their secret keys for decryption of the newly encrypted contents. Any non-revoked consumer having access rights for r_j can recover k^v from Hdr_{r_j} and updates his/her secret keys $(\mathtt{RK}^1_{\mathtt{ID_u,r_j}})^{H_2(k^v)}, (\mathtt{RK}^2_{\mathtt{ID_u,r_j}})^{H_2(k^v)}$, where $k^v = \frac{C_{h_{r_j}}^{\mathtt{ver_v}}}{(\mathtt{PubWit}_{\mathtt{ID_u}}^{\mathtt{r_j,ver_v}})^{\mathtt{sk_{ID_u}}}} = \frac{k^v \cdot \hat{e}(H(r_j),g)^{v \prod_{\forall \mathtt{ID_k} \in \mathbb{U}_{r_j}} \mathtt{sk_{ID_k}}}}{(\hat{e}(H(r_j),g)^{v \prod_{\forall \mathtt{ID_k} \in \mathbb{U}_{r_j} \setminus \{\mathtt{ID_u}\}} \mathtt{sk_{ID_k}}})^{\mathtt{sk_{ID_u}}}}$.

5 Analysis

5.1 Security Analysis

Chosen Plaintext Attack security of the proposed scheme can be defined by the following theorem and proof.

Theorem 1. *If a probabilistic polynomial time (PPT) adversary $\mathcal{A}$ can win the CPA security game defined in Sect. 2.3 with a non-negligible advantage ϵ, then a PPT simulator $\mathcal{B}$ can be constructed to break DBDH assumption with non-negligible advantage $\frac{\epsilon}{2}$.*

Proof. In this proof, we construct a PPT simulator $\mathcal{B}$ to break our proposed scheme with an advantage $\frac{\epsilon}{2}$ with the help of adversary $\mathcal{A}$. More details are given in Appendix B.

5.2 Performance Analysis

In this section, we evaluate the comprehensive performance of our proposed scheme in terms of functionality, computation, storage and communication overhead. We consider the same security level of all the studied schemes for the comparison purpose.

Table 2. Computation, communication and storage overhead comparison

		[10]	[20]	Ours
Signature	Generation	$(9T_{g_1} + 3T_{g_t} + 3T_p + T_h + 3T_{mul_{g_1}} + 3T_{mul_{g_t}})$	$T_{g_1} + T_h$	$(T_{g_1} + T_{g_t} + T_h)$
	Verification	$(8T_{g_1} + 4T_{g_t} + 8T_p + T_h + 4T_{mul_{g_1}} + 7T_{mul_{g_t}})$	$(4 + w')T_{g_1}$	$(T_{g_1} + 2T_{g_t} + 2T_h + 3T_p)$
Secret Key Generation		$(1 + 2w)T_{g_1}$	$(2w + w')T_{g_1}$	$(3T_{g_1} + 3T_{g_t} + T_h)$
Content Encryption		$2T_{g_1} + T_{g_t}$	$2T_{g_1} + T_{g_t}$	$(T_{g_1} + 3T_{g_t})$
Proxy Re-Encryption		N/A	N/A	T_{g_1}
Decryption		$2T_p$	$2T_p$	$(2T_{g_t} + T_p)$
Immediate Revocation	System Parameter Update	NA	NA	$(\lvert\mathbb{D}_{r_i}\rvert + \lvert\mathbb{NR}_{r_i}\rvert + 1)T_{g_1} + T_{g_t}$
	Ciphertext Component Generation			$T_{g_t} + T_{mul_{g_t}}$
	Secret Key Update			$2T_{g_1} + T_{g_t} + T_{mul_{g_t}} + T_h$
Batch Verification		$(10m+3)T_{g_1} + 2T_p + mT_h + (6m+3)T_{mul_{g_1}} + (1+m)T_{mul_{g_t}}$	N/A	$m(2T_{g_1} + T_{g_t} + T_h + 3T_{mul_{g_1}} + T_{mul_{g_t}}) + 3T_p$
Size	Ciphertext	$\lvert\mathbb{Enc}_K\rvert + 2\lvert\mathbb{G}_1\rvert$	$\lvert\mathbb{Enc}_K\rvert + 2\lvert\mathbb{G}_1\rvert + \lvert\mathbb{G}_T\rvert$	$(\lvert\mathbb{Enc}_K\rvert + \lvert\mathbb{G}_1\rvert + 3\lvert\mathbb{G}_T\rvert)$
	Signature	$3\lvert\mathbb{G}_1\rvert + \lvert\mathbb{G}_T\rvert + 6\lvert\mathbb{Z}_q^*\rvert$	$\lvert\mathbb{G}_1\rvert + \lvert\mathbb{Z}_q^*\rvert + l$	$(\lvert\mathbb{Z}_q^*\rvert + \lvert\mathbb{G}_1\rvert + \lvert\mathbb{G}_T\rvert)$
	Hdr_{r_i}	N/A	N/A	$\lvert\mathbb{G}_1\rvert$
	Secret Key	$(1 + 2w)\lvert\mathbb{G}_1\rvert + (1 + w)\mathbb{Z}_q^*$	$(1 + w')\lvert\mathbb{Z}_q^*\rvert + 2w\lvert\mathbb{G}_1\rvert$	$2\lvert\mathbb{G}_1\rvert + \lvert\mathbb{Z}_q^*\rvert$

T_h: computation time of one hash operation; $T_{mul_{g_1}}$: computation time of one group multiplication operation on $\mathbb{G}_1$; $T_{mul_{g_t}}$: computation time of one group multiplication operation on $\mathbb{G}_T$; $\lvert\mathbb{G}_1\rvert$, $\lvert\mathbb{G}_T\rvert$: size of an element in $\mathbb{G}_1$ and $\mathbb{G}_T$ respectively; $\lvert\mathbb{Z}_q^*\rvert$: size of an element in $\mathbb{Z}_q^*$; $\lvert\mathbb{D}_{r_i}\rvert$: number of descendent nodes or categories of r_i in the CKH; $\lvert\mathbb{NR}_{r_i}\rvert$: number of non-revoked consumers associated with r_i; $\lvert\mathbb{Enc}_K\rvert$: size of a symmetric key cipher

Comprehensive Performance Analysis. We have compared our scheme with other existing schemes in two parts, namely *Functionality Comparison* and *Computation, Communication and Storage Overhead Comparison*, which are presented next.

Functionality Comparison. Table 1 shows the functionality comparison of our proposed scheme with the following other notable works [10,11,14–16,18–23]. One can see that our proposed scheme supports several features that are essential, which the other existing schemes do not support. (i) Most notably, access control over hierarchical content is supported only by our proposed scheme, which has been the primary motivation behind our contribution. (ii) Our scheme also prevent unauthorized consumers from sending interest packets to the network through an authentication process, which helps to mitigate DoS attack. In addition, our scheme is resistant against collusion attack between the ISP (i.e., routers) and revoked consumers, which the existing schemes do not provide. (iii) Our scheme is the only existing scheme that supports both expiry-based (enforced at the ERs) and immediate consumer revocation (enforced by the CP). (iv) ISP's accountability is also supported by our scheme which enables the CP to compensate the ISP only for the actual services rendered to the consumers. (v) Moreover, as the consumers do not need the CP for accessing content, the CP can be offline for much of the time, which can give to rise to practical benefits.

Computation, Communication and Storage Overhead Comparison. In this comparison, we consider [10,20] as these two existing schemes support many of the features that our scheme has. Table 2 shows the computation, communication and storage overhead comparison between our scheme and [10,20]. The comparison is shown in asymptotic upper bound in the worst cases. We consider the most frequently operated phases in the comparison. The computation cost comparison is done in terms of number of cryptographic operations such as exponentiation (i.e., T_{g_1}, T_{g_t}), pairing (i.e., T_p), hash (i.e., T_h), group element multiplication (i.e., $T_{mul_{g_1}}, T_{mul_{g_t}}$) operations. In the comparison of storage and communication overhead, the sizes of ciphertext, signature, additional ciphertext component (i.e., Hdr_{r_i}), and secret keys of consumers have been considered, which are measured in terms of group element size (i.e., $|\mathbb{G}_1|, |\mathbb{G}_T|, |\mathbb{Z}_q^*|$) and size of a symmetric key cipher (i.e., $|\mathtt{Enc}_\mathtt{k}|$).

From Table 2, one can see that signature generation cost of our proposed scheme is less than that of [10], while it is comparable with [20]. On the other hand, our proposed scheme takes less cost to verify a signature compared with [10,20]. Our scheme also takes less cost than [10,20] to generate secret keys of the consumers; while content encryption and decryption costs are comparable with [10,20]. The additional proxy re-encryption operation in our scheme takes only one exponentiation operation to re-encrypt a ciphertext. As such, it only introduces a minimal overhead at the routers. Unlike [10,20], our proposed scheme supports immediate revocation, where most of the expensive operations are performed by the CP. A consumer needs to compute three exponentiation, one group multiplication, and one hash operations to update his/her secret keys. It can be also observed that our batch verification operations takes less cost than [10]. It is to be noted that [20] does not provide batch verification operation.

From Table 2, it can also be observed that ciphertext size our scheme is comparable with that of [10,20], while the size of the signature is considerably less than that of [10,20]. Hence our scheme incurs less communication burden. However, our scheme adds one additional $\mathbb{G}_T$ element per ciphertext during immediate revocation event, which increase only minimal overhead in the system. Moreover, a consumer in our scheme requires to maintain less number of secret keys compared to [10,20].

6 Concluding Remarks

In this paper, we have proposed a novel encryption-based access control scheme for NDN while organizing the contents into a hierarchy based on different categories. The proposed scheme employs the concept of Role-Based Encryption that introduces inheritance property, thereby enabling the consumers having access rights for ancestor categories of contents to access contents of the descendent categories. In addition, our proposed scheme enables the ERs to verify authenticity of a consumer before forwarding his/her interest request into the NDN network, which prevents from entering bogus requests into the network. Moreover, CP can revoke consumers any time without introducing heavy overhead into the system. Furthermore, our scheme provides ISP's accountability which is an important property as it enables the CPs to know reliably the amount of content sent to the consumers by the ISP. If a dispute arises between the ISP and the CP over the amount of service rendered to the consumers, then the information provided to the CP can be used in dispute resolution. We have also provided a formal security analysis of our scheme demonstrating that our scheme is provably secure against CPA. A comprehensive performance analysis also shows that our scheme outperforms other notable works in terms of functionality, storage, communication and computation overhead.

We have also been carrying out the simulation of our scheme using MiniNDN. The simulation involves the CP publishing regularly new content, and the consumers request either the new or old content. Our main objective has been to measure the average retrieval time for the consumer and the delay introduced by our scheme. In our experiments, we have varied the number of routers in the ISP as well as the bandwidth of each link and the cache in each of the routers. Our results show that the delay in our scheme between a user sending out an interest packet and receiving the corresponding data packet has been mainly due to the cryptographic operations (which we have shown to be superior to the existing schemes). We will be discussing in detail the simulation results in our technical report describing the implementation [26].

Appendix A Security Model

The security model of the proposed scheme is defined by the *Semantic Security against Chosen Plaintext Attacks* (IND-CPA) under *Selective-ID model*[7]. IND-CPA is illustrated using the following security game between a challenger $\mathcal{C}$ and an adversary $\mathcal{A}$.

- Init Adversary $\mathcal{A}$ submits a challenged category r_i and identity $\mathtt{ID_u}$ to the challenger $\mathcal{C}$.
- Setup Challenger $\mathcal{C}$ runs *Content Provider Setup* phase to generate system parameters and master secret. It also generates proxy re-encryption keys and a private key for the adversary $\mathcal{A}$. Challenger $\mathcal{C}$ sends the system parameters, proxy re-encryption keys and private key to the adversary $\mathcal{A}$.
- Phase 1 Adversary $\mathcal{A}$ sends an identity of a category $r_x^* \notin \mathbb{A}_{r_i}$ and a validity period $\mathtt{VP}^*_{\mathtt{ID_u}}$. Challenger $\mathcal{C}$ runs *Consumer Registration* phase and generates secret key $\mathtt{SK}^{\mathtt{r_x^*}}_{\mathtt{ID_u}}$ and public key $\mathtt{Pub}^{\mathtt{r_x^*}}_{\mathtt{ID_u}}$. Challenger $\mathcal{C}$ sends these keys to the adversary $\mathcal{A}$. Adversary $\mathcal{A}$ can ask for the secret and public keys for polynomially many times.
- Challenge After the Phase 1 is over, the adversary $\mathcal{A}$ sends two equal length messages $\mathtt{K_0}$ and $\mathtt{K_1}$ to the challenger $\mathcal{C}$. The challenger $\mathcal{C}$ flips a random coin $\mu \in \{0,1\}$ and encrypts $\mathtt{K}_\mu$ by initiating *Content Publication* phase. Challenger $\mathcal{C}$ sends the ciphertext of $\mathtt{K}_\mu$ to the adversary $\mathcal{A}$.
- Phase 2 Same as Phase 1.
- Guess Adversary $\mathcal{A}$ outputs a guess μ' of μ. The advantage of wining the game for the adversary $\mathcal{A}$ is $\mathtt{Adv}^{\mathtt{IND-CPA}} = |Pr[\mu' = \mu] - \frac{1}{2}|$.

[7] In the Selective-ID security model, the adversary must submit a challenged category before starting the security game. This is essential in our security proof to set up the system parameters.

Definition 1. *The proposed scheme is semantically secure against Chosen Plaintext Attack if* $\mathtt{Adv}^{\mathtt{IND-CPA}}$ *is negligible for any polynomial time adversary* $\mathcal{A}$.

Remark 1. In PHASE 1, the adversary $\mathcal{A}$ is also allowed to send queries for re-encryption of the ciphertexts and signature generation. In our security game, the simulator $\mathcal{B}$ gives all the proxy re-encryption keys to the adversary $\mathcal{A}$ in the SETUP phase. As such, the adversary $\mathcal{A}$ can answer the re-encryption queries. Similarly, as the private key is also given to the adversary $\mathcal{A}$, the adversary can also answer all the signature generation queries. Therefore, we do not include re-encryption and signature generation oracles in PHASE 1. We observe that the adversary $\mathcal{A}$ has at least the same capability as the NDN routers.

Appendix B Security Proof

Proof. In this proof, we construct a PPT simulator $\mathcal{B}$ to break our proposed scheme with an advantage $\frac{\epsilon}{2}$ with the help of an adversary $\mathcal{A}$.

The DBDH challenger $\mathcal{C}$ chooses random numbers $(a, b, c, z) \in \mathbb{Z}_q^*$ and computes $A = g^a, B = g^b, C = g^c$. It flips a random coin $l \in \{0, 1\}$. If $l = 0$, the challenger $\mathcal{C}$ computes $Z = \hat{e}(g, g)^{abc}$; otherwise it computes $Z = \hat{e}(g, g)^z$. The challenger $\mathcal{C}$ sends the tuple $\langle g, A, B, C, Z\rangle$ to a simulator $\mathcal{B}$ and asks the simulator $\mathcal{B}$ to output l. Now the simulator acts as the challenger in the rest of the security game.

INIT– Adversary $\mathcal{A}$ submits a challenged category r_i and identity $\mathtt{ID_u}$ to the simulator $\mathcal{B}$.

SETUP– Simulator $\mathcal{B}$ chooses random numbers $(x, \mathbb{sk}, \varrho, \{[\mathbb{sk}_i]_{\forall i \in \mathbb{U}_{r_j}}, \mathbb{t}_{r_j}\}_{\forall r_j \in \Psi}) \in \mathbb{Z}_q^*$. The simulator sets $\eta = x - ab \cdot \mathbb{sk}$ and computes the following parameters:

$$\hat{e}(g,g)^{ab} = \hat{e}(A,B); \hat{e}(g,g)^{\eta} = \hat{e}(g,g)^{x} \cdot \hat{e}(A,B)^{-\mathbb{sk}}; \left\{\mathtt{PK}_{\mathtt{r_j}} = C^{\prod_{\forall r_y \in \mathbb{A}_{r_j} \setminus \{r_r\}} \mathbb{t}_{r_y}}\right\}_{\forall r_j \in \mathbb{A}_{r_i}}$$

$$\left\{\mathtt{PK}_{\mathtt{r_j}} = g^{\prod_{\forall r_y \in \mathbb{A}_{r_j} \setminus \{r_r\}} \mathbb{t}_{r_y}}\right\}_{\forall r_j \in (\Psi \setminus \mathbb{A}_{r_i})}; \left\{\mathtt{PKey}_{\mathtt{r_j}}^{\mathtt{r_w}} = \prod_{\forall r_y \in \mathbb{A}_{r_j} \setminus \{r_w, r_r\}} \mathbb{t}_{r_y}\right\}_{\forall r_j \in \Psi}$$

Moreover, the simulator $\mathcal{B}$ computes $\left\{\mathtt{Acc}_{\mathtt{r_j}} = \hat{e}\left(H(r_x^*), g\right)^{\prod_{\forall i \in \mathbb{U}_{r_j}} \mathbb{sk}_i}\right\}_{\forall r_j \in \mathbb{A}_{r_i}}$ and $\left\{\mathtt{Acc}_{\mathtt{r_j}} = \hat{e}\left(H(r_x^*), g\right)^{\mathbb{sk} \prod_{\forall i \in \mathbb{U}_{r_j}} \mathbb{sk}_i}\right\}_{\forall r_j \in (\Psi \setminus \mathbb{A}_{r_i})}$. Simulator $\mathcal{B}$ sends the tuple $\left\langle q, \mathbb{G}, \mathbb{G}_T, \hat{e}, g, H, H_1, \hat{e}(g,g)^{ab}, \hat{e}(g,g)^{\eta}, \mathbb{PK}_{r_j} = \{\mathtt{PK}_{\mathtt{r_j}}, \mathbb{A}_{r_j}, r_j\}_{\forall r_j \in \Psi}\right\rangle$, proxy re-encryption keys $\{\mathtt{PKey}_{\mathtt{r_j}}^{\mathtt{r_w}}\}_{\forall r_j \in \Psi}$, and also $\mathbb{sk}$ (i.e., private key) to the adversary $\mathcal{A}$. It keeps other parameters by itself.

PHASE 1– Adversary $\mathcal{A}$ submits a category r_x^* such that $r_x^* \notin \mathbb{A}_{r_i}$ and validity period $\mathtt{VP}^*_{\mathtt{ID_u}}$ to the simulator $\mathcal{B}$. The simulator $\mathcal{B}$ computes a pair of secret keys $\mathtt{RK}^1_{\mathtt{ID_u}, \mathtt{r_x^*}}, \mathtt{RK}^2_{\mathtt{ID_u}, \mathtt{r_x^*}}$, another pair of public keys $\mathtt{Pub1}^{\mathtt{r_x^*}}_{\mathtt{ID_u}}, \mathtt{Pub2}^{\mathtt{r_x^*}}_{\mathtt{ID_u}}$ and a witness public key $\mathtt{PubWit}^{\mathtt{r_x^*}}_{\mathtt{ID_u}}$, where

$$\mathtt{RK}^1_{\mathtt{ID_u},\mathtt{r}_x^*} = g^{\frac{ab\cdot\mathtt{sk}+\eta}{\mathtt{t}_{\mathtt{r}_x^*}}} = g^{\frac{x}{\mathtt{t}_{\mathtt{r}_x^*}}}; \mathtt{RK}^2_{\mathtt{ID_u},\mathtt{r}_x^*} = g^{\frac{ab\cdot\mathtt{sk}+\eta}{\prod_{\forall r_j\in\mathbb{A}_{r_x^*}\setminus\{r_r\}}\frac{\mathtt{t}_{\mathtt{r_j}}}{\mathtt{t}_{\mathtt{r_r}}}}} = g^{\frac{x}{\prod_{\forall r_j\in\mathbb{A}_{r_x^*}\setminus\{r_r\}}\frac{\mathtt{t}_{\mathtt{r_j}}}{\mathtt{t}_{\mathtt{r_r}}}}}$$
$$\mathtt{Pub1}^{\mathtt{r}_x^*}_{\mathtt{ID_u}} = H(r_x^*)^{\frac{1}{\mathtt{sk}+\varrho\cdot H_1(\mathtt{VP}^*_{\mathtt{ID_u}})}}; \mathtt{Pub2}^{\mathtt{r}_x^*}_{\mathtt{ID_u}} = \hat{e}\left(H(r_x^*), g\right)^{\frac{\varrho}{\mathtt{sk}+\varrho\cdot H_1(\mathtt{VP}^*_{\mathtt{ID_u}})}}$$
$$\mathtt{PubWit}^{\mathtt{r}_x^*}_{\mathtt{ID_u}} = \hat{e}\left(H(r_x^*, g)\right)^{\prod_{\forall i\in\mathbb{U}_{r_x^*}}\mathtt{sk_i}}$$

Simulator $\mathcal{B}$ sends the pair of secret keys $\mathtt{RK}^1_{\mathtt{ID_u},\mathtt{r}_x^*}, \mathtt{RK}^2_{\mathtt{ID_u},\mathtt{r}_x^*}$, public keys $\mathtt{Pub1}^{\mathtt{r}_x^*}_{\mathtt{ID_u}}, \mathtt{Pub2}^{\mathtt{r}_x^*}_{\mathtt{ID_u}}$ and public witness key $\mathtt{PubWit}^{\mathtt{r}_x^*}_{\mathtt{ID_u}}$ to the adversary $\mathcal{A}$.

CHALLENGE– When the adversary $\mathcal{A}$ decides that PHASE 1 is over, it submits two equal length messages $\mathtt{K}_0$ and $\mathtt{K}_1$ to the simulator $\mathcal{B}$. The simulator chooses a random number $r \in \mathbb{Z}_q^*$ and flips a random binary coin μ. It then encrypts $\mathtt{K}_\mu$ using the challenged category r_i and generates a ciphertext $\mathbb{CT}_\mu = \langle C_1, C_2, C_3, C_{r_i}\rangle$, where

$$\begin{aligned}
C_1 &= \mathtt{K}_\mu \cdot Z\\
C_2 &= \hat{e}(g,g)^{\eta\cdot(c+r)}\\
&= \hat{e}(g,g)^{[x-ab\cdot\mathtt{sk}](c+r)}\\
&= \hat{e}(g,g)^{x(c+r)}\cdot\hat{e}(g,g)^{-ab(c+r)\cdot\mathtt{sk}} = \hat{e}(g,C)^x\cdot\hat{e}(g,g)^{x\cdot r}\cdot Z^{-\mathtt{sk}}\cdot\hat{e}(A,B)^{-r\cdot\mathtt{sk}}\\
C_3 &= \hat{e}(g,g)^{abr} = \hat{e}(A,B)^r; C_{r_i} = g^{(c+r)\cdot\prod_{\forall r_j\in\mathbb{A}_{r_i}\setminus\{r_r\}}\mathtt{t}_{r_j}} = (C\cdot g^r)^{\prod_{\forall r_j\in\mathbb{A}_{r_i}\setminus\{r_r\}}\mathtt{t}_{r_j}}
\end{aligned}$$

PHASE 2– Same as PHASE 1.

GUESS – The adversary $\mathcal{A}$ guesses a bit μ' and sends to the simulator $\mathcal{B}$. If $\mu' = \mu$ then the adversary $\mathcal{A}$ wins CPA game; otherwise it fails. If $\mu' = \mu$, simulator $\mathcal{B}$ answers "DBDH" in the game (i.e. outputs $l = 0$); otherwise $\mathcal{B}$ answers "random" (i.e. outputs $l = 1$).

If $Z = \hat{e}(g,g)^z$; then C_1 is completely random from the view of the adversary $\mathcal{A}$. So, the received ciphertext $\mathbb{CT}_\mu$ is not compliant to the game (i.e. invalid ciphertext). Therefore, the adversary $\mathcal{A}$ chooses μ' randomly. Hence, probability of the adversary $\mathcal{A}$ for outputting $\mu' = \mu$ is $\frac{1}{2}$.

If $Z = \hat{e}(g,g)^{abc}$, then adversary $\mathcal{A}$ receives a valid ciphertext. The adversary $\mathcal{A}$ wins the CPA game with non-negligible advantage ϵ (according to Theorem 1). So, the probability of outputting $\mu' = \mu$ for the adversary $\mathcal{A}$ is $\frac{1}{2} + \epsilon$, where probability ϵ is for guessing that the received ciphertext is valid and probability $\frac{1}{2}$ is for guessing whether the valid ciphertext $\mathbb{CT}_\mu$ is related to $\mathtt{K}_0$ or $\mathtt{K}_1$.

Therefore, the overall advantage $\mathtt{Adv}^{\mathtt{IND-CPA}}$ of the simulator $\mathcal{B}$ is $\frac{1}{2}(\frac{1}{2} + \epsilon + \frac{1}{2}) - \frac{1}{2} = \frac{\epsilon}{2}$.

References

1. Jacobson, V., et al.: Networking named content. In: Proceedings of the 5th International Conference on Emerging Networking Experiments and Technologies, CoNEXT 2009, pp. 1–12 (2009)
2. Cisco. 2020 Global Networking Trends Report. Accessed 5 July 2020

3. Cisco. Cisco Annual Internet Report (2018–2023), White paper. 2020. Accessed 5 July 2020
4. Zhang, L., et al.: Named data networking. SIGCOMM Comput. Commun. Rev. **44**(3), 66–73 (2014)
5. Tourani, R., Misra, S., Mick, T., Panwar, G.: Security, privacy, and access control in information-centric networking: a survey. IEEE Commun. Surv. Tutor. **20**(1), 566–600 (2018)
6. Zhou, L., Varadharajan, V., Hitchens, M.: Achieving secure role-based access control on encrypted data in cloud storage. IEEE Trans. Inf. Forensics Secur. **8**(12), 1947–1960 (2013)
7. Zhou, L., Varadharajan, V., Hitchens, M.: Trust enhanced cryptographic role-based access control for secure cloud data storage. IEEE Trans. Inf. Forensics Secur. **10**(11), 2381–2395 (2015)
8. Sultan, N.H., Varadharajan, V., Zhou, L., Barbhuiya, F.A.: A role-based encryption scheme for securing outsourced cloud data in a multi-organization context (2020). https://arxiv.org/abs/2004.05419
9. Sultan, N.H., Laurent, M., Varadharajan, V.: Securing organization's data: a role-based authorized keyword search scheme with efficient decryption (2020). https://arxiv.org/abs/2004.10952
10. Xue, K., et al.: A secure, efficient, and accountable edge-based access control framework for information centric networks. IEEE/ACM Trans. Netw. **27**(3), 1220–1233 (2019)
11. Li, Q., Zhang, X., Zheng, Q., Sandhu, R., Fu, X.: LIVE: lightweight integrity verification and content access control for named data networking. IEEE Trans. Inf. Forensics Secur. **10**(2), 308–320 (2015)
12. Fotiou, N., Polyzos, G.C.: Securing content sharing over ICN. In: Proceedings of the 3rd ACM Conference on Information-Centric Networking, ACM-ICN 2016, pp. 176–185 (2016)
13. AbdAllah, E.G., Zulkernine, M., Hassanein, H.S.: DACPI: a decentralized access control protocol for information centric networking. In: 2016 IEEE International Conference on Communications (ICC), pp. 1–6, May 2016
14. Tourani, R., Stubbs, R., Misra, S.: TACTIC: tag-based access control framework for the information-centric wireless edge networks. In 2018 IEEE 38th International Conference on Distributed Computing Systems (ICDCS), pp. 456–466, July 2018
15. Nunes, I.O., Tsudik, G.: KRB-CCN: lightweight authentication and access control for private content-centric networks. In: Preneel, B., Vercauteren, F. (eds.) ACNS 2018. LNCS, vol. 10892, pp. 598–615. Springer, Cham (2018). https://doi.org/10.1007/978-3-319-93387-0_31
16. Bilal, M., Pack, S.: Secure distribution of protected content in information-centric networking. IEEE Syst. J. **14**(2), 1–12 (2019)
17. Li, B., et al.: Attribute-based access control for ICN naming scheme. IEEE Trans. Dependable Secure Comput. **15**(2), 194–206 (2018)
18. Misra, S., et al.: AccConF: an access control framework for leveraging in-network cached data in the ICN-enabled wireless edge. IEEE Trans. Dependable Secure Comput. **16**(1), 5–17 (2019)
19. Xia, Q., et al.: TSLS: time sensitive, lightweight and secure access control for information centric networking. In: IEEE Global Communications Conference (GLOBECOM), pp. 1–6, December 2019
20. He, P., et al.: LASA: lightweight, auditable and secure access control in ICN with limitation of access times. In: IEEE International Conference on Communications (ICC), pp. 1–6, May 2018

21. Tseng, Y., Fan, C., Wu, C.: FGAC-NDN: fine-grained access control for named data networks. IEEE Trans. Netw. Serv. Manag. **16**(1), 143–152 (2019)
22. Suksomboon, K., et al.: In-device proxy re-encryption service for information-centric networking access control. In: IEEE 43rd Conference on Local Computer Networks (LCN), pp. 303–306, October 2018
23. Zhu, L., et al.: T-CAM: time-based content access control mechanism for ICN subscription systems. Future Gener. Comput. Syst. **106**, 607–621 (2020)
24. Boneh, D., Franklin, M.K.: Identity-based encryption from the Weil pairing. In: Proceedings of the 21st Annual International Cryptology Conference on Advances in Cryptology, CRYPTO 2001, pp. 213–229 (2001)
25. Au, M.H., Tsang, P.P., Susilo, W., Mu, Y.: Dynamic universal accumulators for DDH groups and their application to attribute-based anonymous credential systems. In: Fischlin, M. (ed.) CT-RSA 2009. LNCS, vol. 5473, pp. 295–308. Springer, Heidelberg (2009). https://doi.org/10.1007/978-3-642-00862-7_20
26. Sultan, N.H., Varadharajan, V.: On the design and implementation of a RBE based access control scheme for data centric model for content sharing. ACSRC Technical report, The University of Newcastle, Australia, April 2020

PGC: Decentralized Confidential Payment System with Auditability

Yu Chen[1,2,3,4], Xuecheng Ma[5,6], Cong Tang[7], and Man Ho Au[8](✉)

[1] School of Cyber Science and Technology, Shandong University, Qingdao 266237, China
[2] State Key Laboratory of Cryptology, P.O. Box 5159, Beijing 100878, China
[3] Key Laboratory of Cryptologic Technology and Information Security, Ministry of Education, Shandong University, Qingdao 266237, China
[4] Shandong Institute of Blockchain, Jinan, China
yuchen@sdu.edu.cn
[5] State Key Laboratory of Information Security, Institute of Information Engineering, Chinese Academy of Sciences, Beijing 100093, China
[6] School of Cyber Security, University of Chinese Academy of Sciences, Beijing 100049, China
maxuecheng@iie.ac.cn
[7] Beijing Temi Co., Ltd., pgc.info, Beijing, China
congtang.cn@gmail.com
[8] Department of Computer Science, The University of Hong Kong, Pok Fu Lam, China
allenau@cs.hku.hk

Abstract. Many existing cryptocurrencies fail to provide transaction anonymity and confidentiality. As the privacy concerns grow, a number of works have sought to enhance privacy by leveraging cryptographic tools. Though strong privacy is appealing, it might be abused in some cases. In decentralized payment systems, anonymity poses great challenges to system's auditability, which is a crucial property for scenarios that require regulatory compliance and dispute arbitration guarantee.

Aiming for a middle ground between privacy and auditability, we introduce the notion of *decentralized confidential payment* (DCP) system with auditability. In addition to offering confidentiality, DCP supports privacy-preserving audit in which an external party can specify a set of transactions and then request the participant to prove their compliance with a large class of policies. We present a generic construction of auditable DCP system from integrated signature and encryption scheme and non-interactive zero-knowledge proof systems. We then instantiate our generic construction by carefully designing the underlying building blocks, yielding a standalone cryptocurrency called PGC. In PGC, the setup is transparent, transactions are less than 1.3 KB and take under 38ms to generate and 15 ms to verify.

At the core of PGC is an additively homomorphic public-key encryption scheme that we newly introduce, twisted ElGamal, which is not only as secure as standard exponential ElGamal, but also friendly to

L. Chen et al. (Eds.): ESORICS 2020, LNCS 12308, pp. 591–610, 2020.
https://doi.org/10.1007/978-3-030-58951-6_29

Sigma protocols and Bulletproofs. This enables us to easily devise zero-knowledge proofs for basic correctness of transactions as well as various application-dependent policies in a modular fashion.

Keywords: Cryptocurrencies · Decentralized payment system · Confidential transactions · Auditable · Twisted ElGamal

1 Introduction

Cryptocurrencies such as Bitcoin [Nak08] and Ethereum [Woo14] realize decentralized peer-to-peer payment by maintaining an append-only public ledger known as blockchain, which is globally distributed and synchronized by consensus protocols. In blockchain-based payment systems, correctness of each transaction must be verified by the miners publicly before being packed into a block. To enable efficient validation, major cryptocurrencies such as Bitcoin and Ethereum simply expose all transaction details (sender, receiver and transfer amount) to the public. Following the terminology in [BBB+18], privacy for transactions consists of two aspects: (1) anonymity, hiding the identities of sender and receiver in a transaction and (2) confidentiality, hiding the transfer amount. While Bitcoin-like and Ethereum-like cryptocurrencies provide some weak anonymity through unlinkability of account addresses to real world identities, they lack confidentiality, which is of paramount importance.

1.1 Motivation

Auditability is a crucial property in all financial systems. Informally, it means that an auditor can not only check that if transactions satisfy some pre-fixed policies *before-the-fact*, but also request participants to prove that his/her transactions comply with some policies *after-the-fact*. In centralized payment system where there exists a trusted center, such as bank, Paypal or Alipay, auditability is a built-in property since the center knows details of all transactions and thus be able to conduct audit. However, it is challenging to provide the same level of auditability in decentralized payment systems with strong privacy guarantee.

In our point of view, strong privacy is a double-edged sword. While confidentiality is arguably the primary concern of privacy for any payment system, anonymity might be abused or even prohibitive for applications that require auditability, because anonymity provides plausible deniability [FMMO19], which allows participants to deny their involvements in given transactions. Particularly, it seems that anonymity denies the feasibility of *after-the-fact* auditing, due to an auditor is unable to determine who is involved in which transaction. We exemplify this dilemma via the following three typical cases.

<u>Case 1:</u> To prevent criminals from money laundering in a decentralized payment system, an auditor must be able to examine the details of suspicious transactions. However, if the system is anonymous, locating the participants of suspicious

transactions and then force them to reveal transaction details would be very challenging.

Case 2: Consider an employer pay salaries to employees via a decentralized payment system with strong privacy. If the employee did not receive the correct amount, how can he support his complaint? Likewise, how can the employer protect himself by demonstrating that he has indeed completed payment with the correct amount.

Case 3: Consider the employees also pay the income tax via the same payment system. How can he convince the tax office that he has paid the tax according to the correct tax rate.

Similar scenarios are ubiquitous in monetary systems that require after-the-fact auditing and in e-commerce systems that require dispute arbitration mechanism.

1.2 Our Contributions

The above discussions suggest that it might be impossible to construct a decentralized payment system offering the same level of auditability as centralized payment system while offering confidentiality and anonymity simultaneously without introducing some degree of centralization or trust assumption. In this work, we stick to confidentiality, but trade anonymity for auditability. We summarize our contributions as follows.

Decentralized Confidential Payment System with Auditability. We introduce the notion of *decentralized confidential payment* (DCP) system with auditability, and formalize a security model. For the sake of simplicity, we take an account-based approach and focus only on the *transaction layer*, and treat the network/consensus-level protocols as black-box and ignore attacks against them. Briefly, DCP should satisfy authenticity, confidentiality and soundness. The first security notion stipulates that only the owner of an account can spend his coin. The second security notion requires the account balance and transfer amount are hidden. The last security notion captures that no one is able to make validation nodes accept an incorrect transaction. In addition, we require the audit to be privacy-preserving, namely, the participant is unable to fool the auditor and the auditing results do not impact the confidentiality of other transactions.

We then present a generic construction of auditable DCP system from two building blocks, namely, integrated signature and encryption (ISE) schemes and non-interactive zero-knowledge (NIZK) proof systems. In our generic DCP construction, ISE plays an important role. First, it guarantees that each account can safely use a single keypair for both encryption and signing. This feature greatly simplifies the overall design from both conceptual and practical aspects. Second, the encryption component of ISE ensures that the resulting DCP system is *complete*, meaning that as soon as a transaction is recorded on the blockchain, the payment is finalized and takes effect immediately – receiver's balance increases with the same amount that sender's balance decreases, and the receiver can spend his coin on his will. This is in contrast to many commitment-based cryptocurrencies that require *out-of-band* transfer, which are thus not complete. NIZK not only enables a sender to prove transactions satisfy the most basic correctness policy, but also allows a user to prove that any set of confidential transactions he participated satisfy a large class of application-dependent policies, such as limit on total transfer amounts, paying tax rightly according the tax rate, and transaction opens to some exact value. In summary, our generic DCP construction is complete, and supports flexible audit.

PGC: A Simple and Efficient Instantiation. While the generic DCP construction is relatively simple and intuitive, an efficient instantiation is more technically involved. We realize our generic DCP construction by designing a new ISE and carefully devising suitable NIZK proof system. We refer to the resulting cryptocurrency as PGC (stands for Pretty Good Confidentiality). Notably, in addition to the advantages inherited from the generic construction, PGC admits transparent setup, and its security is based solely on the widely-used discrete logarithm assumption. To demonstrate the efficiency and usability of PGC, we implement it as a standalone cryptocurrency, and also deploy it as smart contracts. We report the experimental results in Sect. 5.

1.3 Technical Overview

We discuss our design choice in the generic construction of DCP, followed by techniques and tools towards a secure and efficient instantiation, namely, PGC.

1.3.1 Design Choice in Generic Construction of Auditable DCP

Pseudonymity vs. Anonymity. Early blockchain-based cryptocurrencies offers pseudonymity, that is, addresses are assumed to be unlinkable to their real world identities. However, a variety of de-anonymization attacks [RS13,BKP14] falsified this assumption. On the other hand, a number of cryptocurrencies such as Monero and Zcash sought to provide strong privacy guarantee, including both anonymity and confidentiality. In this work, we aim for finding a sweet balance between privacy and auditability. As indicated in [GGM16], identity is crucial to any regulatory system. Therefore, we choose to offer privacy in terms of confidentiality, and still stick to pseudonymous system. Interestingly, we view pseudonymity as a feature rather than a weakness, assuming that an auditor is able to link account addresses to real world identities. This opens the possibility to conduct after-the-fact audit. We believe this is the most promising avenue for real deployment of DCP that requires auditability.

PKE vs. Commitment. A common approach to achieve transaction confidentiality is to commit the balance and transfer amount using a global homomorphic commitment scheme (e.g., the Pedersen commitment [Ped91]), then derive a secret from blinding randomnesses to prove correctness of transaction and authorize transfer. The seminal DCP systems [Max, Poe] follow this approach.

Nevertheless, commitment-based approach suffers from several drawbacks. First, the resulting DCP systems are not *complete*. Due to lack of decryption capability, senders are required to honestly transmit the openings of outgoing commitments (includes randomness and amount) to receivers in an *out-of-band* manner. This issue makes the system much more complicated, as it must be assured that the out-of-band transfer is correct and secure. Second, users must be stateful since they have to keep track of the randomness and amount of each incoming commitment. Otherwise, failure to open a single incoming commitment will render an account totally unusable, due to either incapable of creating the NIZK proofs (lack of witness), or generating the signature (lack of signing key). This incurs extra security burden to the design of wallet (guarantee the openings must be kept in a safe and reliable way).

Observe that homomorphic PKE can be viewed as a computationally hiding and perfectly binding commitment, in which the secret key serves as a natural trapdoor to recover message. With these factors in mind, our design equips each user with a PKE keypair rather than making all users share a global commitment.

Integrated Signature and Encryption vs. SIG+PKE. Intuitively, to secure a DCP system, we need a PKE scheme to provide confidentiality, and a signature scheme to provide authenticity. If we follow the principle of *key separation*, i.e., use different keypairs for encryption and signing operations respectively, the overall design would be complicated. In that case, each account will be associated with two keypairs, and consequently deriving account address turns out to be very tricky. If we derive the address from one public key, which one should be chosen? If we derive the address from the two public keys, then additional mechanism is needed to link the two public keys together.

A better solution is to adopt *key reuse* strategy, i.e., use the same keypair for both encryption and signing. This will greatly simplify the design of overall system. However, reusing keypairs may create new security problems. As pointed out in [PSST11], the two uses may interact with one another badly, in such a way as to undermine the security of one or both of the primitives, e.g., the case of textbook RSA encryption and signature. In this work, we propose to use integrated signature and encryption (ISE) scheme with *joint security*, wherein a single keypair is used for both signature and encryption components in a secure manner, to replace the naive combination of signature and encryption. To the best of our knowledge, this is the first time that ISE is used in DCP system to ensure provable security. We remark that the existing proposal Zether [BAZB20] essentially adopts the key reuse strategy, employing a signature scheme and an encryption scheme with same keypair. Nevertheless, they do not explicitly identify key reuse strategy and formally address joint security.

1.3.2 Overview of Our Generic Auditable DCP

DCP from ISE and NIZK. We present a generic construction of DCP system from ISE and NIZK. We choose the account-based model for simplicity and usability. In our generic DCP construction, user creates an account by generating a keypair of ISE, in which the public key is used as account address and secret key is used to control the account. The state of an account consists of a serial number (a counter that increments with every outgoing transaction) and an encrypted balance (encryption of plaintext balance under account public key). State changes are triggered by transactions from one account to another. A blockchain tracks the state of every account.

Let $\tilde{C}_s$ and $\tilde{C}_r$ be the encrypted balances of two accounts controlled by Alice and Bob respectively. Suppose Alice wishes to transfer v coins to Bob. She constructs a confidential transaction via the following steps. First, she encrypts v under her public key pk_s and Bob's public key pk_r respectively to obtain C_s and C_r, and sets $\mathsf{memo} = (pk_s, pk_r, C_s, C_r)$. Then, she produces a NIZK proof π_{correct} for the most basic correctness policy of transaction: (i) C_s and C_r are two encryptions of the same transfer amount under pk_s and pk_r; (ii) the transfer amount lies in a right range; (iii) her remaining balance is still positive. Finally, she signs serial number sn together with memo and π_{correct} under her secret key, obtaining a signature σ. The entire transaction is of the form $(\mathsf{sn}, \mathsf{memo}, \pi_{\text{correct}}, \sigma)$. In this way, validity of a transaction is publicly verifiable by checking the signature and NIZK proof. If the transaction is valid, it will be recorded on the blockchain. Accordingly, Alice's balance (resp. Bob's balance) will be updated as $\tilde{C}_s = \tilde{C}_s - C_s$ (resp. $\tilde{C}_r = \tilde{C}_r + C_r$), and Alice's serial number increments. Such balance update operation implicitly requires that the underlying PKE scheme satisfies additive homomorphism.

In summary, the signature component is used to provide authenticity (proving ownership of an account), the encryption component is used to hide the balance and transfer amount, while zero-knowledge proofs are used to prove the correctness of transactions in a privacy-preserving manner.

Auditing Policies. A trivial solution to achieve auditability is to make the participants reveal their secret keys. However, this approach will expose all the related transactions to the auditor, which is not privacy-preserving. Note that a plaintext and its ciphertext can be expressed as an $\mathcal{NP}$ relation. Auditability can thus be easily achieved by leveraging NIZK. Moreover, note that the structure of memo is symmetric. This allows either the sender or the receiver can prove transactions comply with a variety of application-dependent policies. We provide examples of several useful policies as below:

- Anti-money laundering: the sum of a collection of outgoing/incoming transactions from/to a particular account is limited.
- Tax payment: a user pays the tax according to the tax rate.
- Selectively disclosure: the transfer amount of some transaction is indeed some value.

1.3.3 PGC: A Secure and Efficient Instantiation

A secure and efficient instantiation of the above DCP construction turns out to be more technical involved. Before proceeding, it is instructive to list the desirable features in mind:

1. transparent setup – do not require a trusted setup. This property is of utmost importance in the setting of cryptocurrencies.
2. efficient – only employ lightweight cryptographic schemes based on well-studied assumptions.
3. modular – build the whole scheme from reusable gadgets.

The above desirable features suggest us to devise efficient NIZK that admits transparent setup, rather than resorting to general-purpose zk-SNARKs, which are either heavyweight or require trusted setup. Besides, the encryption component of ISE should be zero-knowledge proofs friendly.

We begin with the instantiation of ISE. A common choice is to select Schnorr signature [Sch91] as the signature component and exponential ElGamal [CGS97] as the encryption component.[1] This choice brings us at least three benefits: (i) Schnorr signature and ElGamal PKE share the same discrete-logarithm (DL) keypair (i.e., $pk = g^{sk} \in \mathbb{G}, sk \in \mathbb{Z}_p$), thus we can simply use the common public key as account address. (ii) The signing operation of Schnorr's signature is largely independent to the decryption operation of ElGamal PKE, which indicates that the resulting ISE scheme could be jointly secure. (iii) ElGamal PKE is additively homomorphic. Efficient Sigma protocols can thus be employed to prove linear relations on algebraically encoded-values. For instance, proving plaintext equality: C_s and C_r encrypt the same message under pk_s and pk_r respectively.

It remains to design efficient NIZK proofs for the basic correctness policies and various application-dependent policies.

Obstacle of Working with Bulletproof. Besides using Sigma protocols to prove linear relations over encrypted values, we also need range proofs to prove the encrypted values lie in the right interval. In more details, we need to prove that the value v encrypted in C_s and the value encrypted in $\tilde{C}_s - C_s$ (the current balance subtracts v) lies in the right range. State-of-the-art range proof is Bulletproof [BBB+18], which enjoys efficient proof generation/verification, logarithmic proof size, and transparent setup. As per the desirable features of our instantiation, Bulletproof is an obvious choice. Recall that Bulletproof only accepts statements of the form of Pedersen commitment $g^r h^v$. To guarantee soundness, the DL relation between commitment key (g, h) must be unknown to the prover. Note that an ElGamal ciphertext C of v under pk is of the form $(g^r, pk^r g^v)$, in which the second part ciphertext can be viewed as a commitment of v under commitment key (pk, g). To prove v lies in the right range, it seems that we can simply run the Bulletproof on $pk^r g^v$. However, this usage is insecure since the prover owns an obvious trapdoor, say sk, of such commitment key.

There are two approaches to circumvent this obstacle. The first approach is to commit v with randomness r under commitment key (g, h), then prove (v, r) in $g^v h^r$ is consistent with that in the ciphertext $(g^r, pk^r g^v)$. Similar idea was used in Quisquis [FMMO19]. The drawback of this approach is that it brings extra overhead of proof size as well as proof generation/verification. The second

[1] In the remainder of this paper, we simply refer to the exponential ElGamal PKE as ElGamal PKE for ease of exposition.

approach is due to Bünz et al. [BAZB20] used in Zether. They extend Bulletproof to Σ-Bullets, which enables the interoperability between Sigma protocols and Bulletproof. Though Σ-Bullets is flexible, it requires custom design and analysis from scratch for each new Sigma protocols.

Twisted ElGamal - Our Customized Solution. We are motivated to directly use Bulletproof in a black-box manner, without introducing Pedersen commitment as a bridge or dissecting Bulletproof. Our idea is to twist standard ElGamal, yielding the twisted ElGamal. We sketch twisted ElGamal below. The setup algorithm picks two random generators (g, h) of $\mathbb{G}$ as global parameters, while the key generation algorithm is same as that of standard ElGamal. To encrypt a message $m \in \mathbb{Z}_p$ under pk, the encryption algorithm picks $r \xleftarrow{R} \mathbb{Z}_p$, then computes ciphertext as $C = (X = pk^r, Y = g^r h^m)$. The crucial difference to standard ElGamal is that the roles of key encapsulation and session key are switched and the message m is lifted on a new generator h.[2] Twisted ElGamal retains additive homomorphism, and is as efficient and secure as the standard exponential ElGamal. More importantly, it is zero-knowledge friendly. Note that the second part of twisted ElGamal ciphertext (even encrypted under different public keys) can be viewed as Pedersen commitment under the same commitment key (g, h), whose DL relation is unknown to all users. Such structure makes twisted ElGamal compatible with all zero-knowledge proofs whose statement is of the form Pedersen commitment. In particular, one can directly employ Bulletproof to generate range proofs for values encrypted by twisted ElGamal in a black-box manner, and these proofs can be easily aggregated. Next, we abstract two distinguished cases.

Prover Knows the Randomness. This case generalizes scenarios in which the prover is the producer of ciphertexts. Concretely, consider a twisted ElGamal ciphertext $C = (X = pk^r, Y = g^r h^v)$, to prove m lies in the right range, the prover first executes a Sigma protocol on C to prove the knowledge of (r, v), then invokes a Bulletproof on Y to prove v lies in the right range.

Prover Knows the Secret Key. This case generalizes scenarios where the prover is the recipient of ciphertexts. Due to lack of randomness as witness, the prover cannot directly invoke Bulletproof. We solve this problem by developing *ciphertext refreshing* approach. The prover first decrypts $\tilde{C}_s - C_s$ to v using sk, then generates a new ciphertext C_s^* of v under fresh randomness r^*, and proves that $\tilde{C}_s - C_s$ and C_s^* do encrypt the same message under his public key. Now, the prover is able to prove that v encrypted by C_s^* lies in the right range by composing a Sigma protocol and a Bulletproof, via the same way as the first case.

The above range proofs provide two specialized "proof gadgets" for proving encrypted values lie in the right range. We refer to them as Gadget-1 and Gadget-2 hereafter. As we will see, all the NIZK proofs used in this work can be built from these two gadgets and simple Sigma protocols. Such modular design helps

[2] As indicated in [CZ14], the essence of ElGamal is $F_{sk}(g^r) = pk^r$ forms a publicly evaluable pseudorandom functions over $\mathbb{G}$. The key insight of switching is that F_{sk} is in fact a permutation.

to reduce the footprint of overall cryptographic code, and have the potential to admit parallel proof generation/verification. We highlight the two "gadgets" are interesting on their own right as privacy-preserving tools, which may find applications in other domains as well, e.g., secure machine learning.

Enforcing more Auditing Policies. In the above, we have discussed how to employ Sigma protocols and Bulletproof to enforce the basic correctness policy for transactions. In fact, we are able to enforce a variety of polices that can be expressed as linear constraints over transfer amounts. In Sect. 4.2, we show how to enforce limit, rate and open policies using Sigma protocols and the two gadgets we have developed.

1.4 Related Work

The seminal cryptocurrencies such as Bitcoin and Ethereum do not provide sufficient level of privacy. In the past years privacy-enhancements have been developed along several lines of research. We provide a brief overview below.

The first direction aims to provide confidentiality. Maxwell [Max] initiates the study of *confidential transaction*. He proposes a DCP system by employing Pedersen commitment to hide transfer amount and using range proofs to prove correctness of transaction. Mimblewimble/Grin [Poe, Gri] further improve Maxwell's construction by reducing the cost of signatures. The second direction aims to enhance anonymity. A large body of works enhance anonymity via utilizing mixing mechanisms. For instance, CoinShuffle [RMK14], Dash [Das], and Mixcoin [BNM+14] for Bitcoin and Möbius [MM] for Ethereum. The third direction aims to attain both confidentiality and anonymity. Monero [Noe15] achieves confidentiality via similar techniques used in Maxwell's DCP system, and provides anonymity by employing linkable ring signature and stealth address. Zcash [ZCa] achieves strong privacy by leveraging key-private PKE and zk-SNARK.

Despite great progress in privacy-enhancement, the aforementioned schemes are not without their limitations. In terms of reliability, most of them require out-of-band transfer and thus are not complete. In terms of efficiency, some of them suffer from slow transaction generation due to the use of heavy advanced cryptographic tools. In terms of security, some of them are not proven secure based on well-studied assumption, or rely on trusted setup.

Another related work is zkLedger [NVV18], which offers strong privacy and privacy-preserving auditing. To attain anonymity, zkLedger uses a novel table-based ledger. Consequently, the size of transactions and audit efficiency are linear in the total number of participants in the system, and the scale of participants is fixed at the setup stage of system.

Concurrent and Independent Work. Fauzi et al. [FMMO19] put forward a new design of cryptocurrency with strong privacy called Quisquis in the UTXO model. They employ updatable public keys to achieve anonymity and a slight variant of standard ElGamal encryption to achieve confidentiality. Bünz et al. [BAZB20] propose a confidential payment system called Zether, which is compatible with Ethereum-like smart contract platforms. They use standard

ElGamal to hide the balance and transfer amount, and use signature derived from NIZK to authenticate transactions. They also sketch how to acquire anonymity for Zether via a similar approach as zkLedger [NVV18]. Both Quisquis and Zether design accompanying zero-knowledge proofs from Sigma protocols and Bulletproof, but take different approaches to tackle the incompatibility between ElGamal encryption and Bulletproof. Quisquis introduces ElGamal commitment to bridge ElGamal encryption, and uses Sigma protocol to prove consistency of bridging. Finally, Quisquis invokes Bulletproof on the second part of ElGamal commitment, which is exactly a Pedersen commitment. Zether develops a custom ZKP called Σ-Bullets, which is a dedicated integration of Bulletproof and Sigma protocol. Given an arithmetic circuit, a Σ-Bullets ensures that a public linear combination of the circuit's wires is equal to some witness of a Sigma protocol. This enhancement in turn enables proofs on algebraically-encoded values such as ElGamal encryptions or Pedersen commitments in different groups or using different commitment keys.

We highlight the following crucial differences of our work to Quisquis and Zether: (i) We focus on confidentiality, and trade anonymity for auditability. (ii) We use jointly secure ISE, rather than ad-hoc combination of signature and encryption, to build DCP system in a provably secure way. (iii) As for instantiation, PGC employs our newly introduced twisted ElGamal rather than standard ElGamal to hide balance and transfer amount. The nice structure of twisted ElGamal enables the sender to prove the transfer amount lies in the right range by directly invoking Bulletproof in a black-box manner, without any extra bridging cost as Quisquis. The final proof for correctness of transactions in PGC is obtained by assembling small "proof gadgets" together in a simple and modular fashion, which is flexible and reusable. This is opposed to Zether's approach, in which zero-knowledge proof is produced by a Σ-Bullets as a whole, while building case-tailored Σ-Bullets requires to dissect Bulletproof and skillfully design its interface to Sigma protocol.

Note. Due to space limit, we refer to the full version of this work [CMTA19] for definitions of standard cryptographic primitives and all security proofs.

2 Definition of DCP System

We formalize the notion of *decentralized confidential payment system* in account model, adapting the notion of *decentralized anonymous payment system* [ZCa].

2.1 Data Structures

We begin by describing the data structures used by a DCP system.

Blockchain. A DCP system operates on top of a blockchain B. The blockchain is publicly accessible, i.e., at any given time t, all users have access to B_t, the ledger at time t, which is a sequence of transactions. The blockchain is also append-only, i.e., $t < t'$ implies that B_t is a prefix of $B_{t'}$.

Public Parameters. A trusted party generate public parameters pp at the setup time of system, which is used by system's algorithms. We assume that pp always include an integer $v_{\max}$, which specifies the maximum possible number of coins that the system can handle. Any balance and transfer below must lie in the integer interval $\mathcal{V} = [0, v_{\max}]$.

Account. Each account is associated with a keypair (pk, sk), an encoded balance $\tilde{C}$ (which encodes plaintext balance $\tilde{m}$), as well as an incremental serial number sn (used to prevent replay attacks). Both sn, $\tilde{C}$, and pk are made public. The public key pk serves as account address, which is used to receive transactions from other accounts. The secret key sk is kept privately, which is used to direct transactions to other accounts and decodes encoded balance.

Confidential Transaction. A confidential transaction ctx consists of three parts, i.e., sn, memo and aux. Here, sn is the current serial number of sender account pk_s, $\mathsf{memo} = (pk_s, pk_r, C)$ records basic information of a transaction from sender account pk_s to receiver account pk_r, where C is the encoding of transfer amount, and aux denotes application-dependent auxiliary information.

2.2 Decentralized Confidential Payment System with Auditability

An auditable DCP system consists of the following polynomial-time algorithms:

- $\mathsf{Setup}(\lambda)$: on input a security parameter λ, output public parameters pp. A trusted party executes this algorithm once-for-all to setup the whole system.
- $\mathsf{CreateAccount}(\tilde{m}, \mathsf{sn})$: on input an initial balance $\tilde{m}$ and a serial number sn, output a keypair (pk, sk) and an encoded balance $\tilde{C}$. A user runs this algorithm to create an account.
- $\mathsf{RevealBalance}(sk, \tilde{C})$: on input a secret key sk and an encoded balance $\tilde{C}$, output the balance $\tilde{m}$. A user runs this algorithm to reveal the balance.
- $\mathsf{CreateCTx}(sk_s, pk_s, pk_r, v)$: on input a keypair (sk_s, pk_s) of sender account, a receiver account address pk_r, and a transfer amount v, output a confidential transaction ctx. A user runs this algorithm to transfer v coins from account pk_s to account pk_r.
- $\mathsf{VerifyCTx}(\mathsf{ctx})$: on input a confidential transaction ctx, output "0" denotes valid and "1" denotes invalid. Miners run this algorithm to check the validity of proposed confidential transaction ctx. If ctx is valid, it will be recorded on the blockchain B. Otherwise, it is discarded.
- $\mathsf{UpdateCTx}(\mathsf{ctx})$: for each fresh ctx on the blockchain B, sender and receiver update their encoded balances to reflect the change, i.e., the sender account decreases with v coins while the receiver account increases with v coins.
- $\mathsf{JustifyCTx}(pk, sk, \{\mathsf{ctx}\}, f)$: on input a user's keypair (pk, sk), a set of confidential transactions he participated and a policy f, output a proof π for $f(pk, \{\mathsf{ctx}\}) = 1$. A user runs this algorithm to generate a proof for auditing.
- $\mathsf{AuditCTx}(pk, \{\mathsf{ctx}\}, f, \pi)$: on input a user's public key, a set of confidential transactions he participated, a policy f and a proof, output "0" denotes accept and "1" denotes reject. An auditor runs this algorithm to check if $f(pk, \{\mathsf{ctx}\}) = 1$.

2.3 Correctness and Security Model

Correctness of basic DCP functionality requires that a valid ctx will always be accepted and recorded on the blockchain, and the states of associated accounts will be updated properly, i.e., the balance of sender account decreases the same amount as the balance of receiver account increases. Correctness of auditing functionality requires honestly generated auditing proofs for transactions complying with policies will always be accept.

As to security, we focus solely on the *transaction layer* of a cryptocurrency, and assume network-level or consensus-level attacks are out of scope.

Intuitively, a DCP system should provide *authenticity*, *confidentiality* and *soundness*. Authenticity requires that the sender can only be the owner of an account, nobody else (who does not know the secret key) is able to make a transfer from this account. Confidentiality requires that other than the sender and receiver (who does not know the secret keys of sender and receiver), no one can learn the value hidden in a confidential transaction. While the former two notions address security against outsider adversary, soundness addresses security against insider adversary (e.g. the sender himself). See the full version [CMTA19] for the details of security model.

3 A Generic Construction of Auditable DCP from ISE and NIZK

We present a generic construction of auditable DCP from ISE and NIZK. In a nutshell, we use homomorphic PKE to encode the balance and transfer amount, use NIZK to enforce senders to build confidential transactions honestly and make correctness publicly verifiable, and use digital signature to authenticate transactions. Let ISE = $(\mathsf{Setup}, \mathsf{KeyGen}, \mathsf{Sign}, \mathsf{Verify}, \mathsf{Enc}, \mathsf{Dec})$ be an ISE scheme whose PKE component is additively homomorphic on message space $\mathbb{Z}_p$. Let NIZK = $(\mathsf{Setup}, \mathsf{CRSGen}, \mathsf{Prove}, \mathsf{Verify})$[3] be a NIZK proof system for L_{correct} (which will be specified later). The construction is as below.

- $\mathsf{Setup}(1^\lambda)$: runs $pp_{\text{ise}} \leftarrow \text{ISE}.\mathsf{Setup}(1^\lambda)$, $pp_{\text{nizk}} \leftarrow \text{NIZK}.\mathsf{Setup}(1^\lambda)$, $crs \leftarrow \text{NIZK}.\mathsf{CRSGen}(pp_{\text{nizk}})$, outputs $pp = (pp_{\text{ise}}, pp_{\text{nizk}}, crs)$.
- $\mathsf{CreateAccount}(\tilde{m}, \mathsf{sn})$: on input an initial balance $\tilde{m} \in \mathbb{Z}_p$ and a serial number $\mathsf{sn} \in \{0,1\}^n$ (e.g., $n = 256$), runs ISE.$\mathsf{KeyGen}(pp_{\text{ise}})$ to generate a keypair (pk, sk), computes $\tilde{C} \leftarrow \text{ISE}.\mathsf{Enc}(pk, \tilde{m}; r)$ as the initial encrypted balance, sets sn as the initial serial number[4], outputs public key pk and secret key sk. Fix the public parameters, the KeyGen algorithm naturally induces an $\mathcal{NP}$ relation $\mathsf{R}_{\text{key}} = \{(pk, sk) : \exists r \text{ s.t. } (pk, sk) = \mathsf{KeyGen}(r)\}$.

[3] We describe our generic DCP construction using NIZK in the CRS model. The construction and security proof carries out naturally if using NIZK in the random oracle model instead.

[4] By default, $\tilde{m}$ and sn should be zero, r should be a fixed and publicly known randomness, say the zero string 0^λ. This settlement guarantees that the initial account state is publicly auditable. Here, we do not make it as an enforcement for flexibility.

- $\mathsf{RevealBalance}(sk, \tilde{C})$: on input secret key sk and encrypted balance $\tilde{C}$, outputs $\tilde{m} \leftarrow \text{ISE}.\mathsf{Dec}(sk, \tilde{C})$.
- $\mathsf{CreateCTx}(sk_s, pk_s, v, pk_r)$: on input sender's keypair (pk_s, sk_s), the transfer amount v, and receiver's public key pk_r, the algorithm first checks if $(\tilde{m}_s - v) \in \mathcal{V}$ and $v \in \mathcal{V}$ (here $\tilde{m}_s$ is the current balance of sender account pk_s). If not, returns $\perp$. Otherwise, it creates ctx via the following steps:
  1. compute $C_s \leftarrow \text{ISE}.\mathsf{Enc}(pk_s, v; r_1)$, $C_r \leftarrow \text{ISE}.\mathsf{Enc}(pk_r, v; r_2)$, set $\mathsf{memo} = (pk_s, pk_r, C_s, C_r)$, here (C_s, C_r) serve as the encoding of transfer amount;
  2. run $\text{NIZK}.\mathsf{Prove}$ with witness (sk_s, r_1, r_2, v) to generate a proof π_{correct} for $\mathsf{memo} = (pk_s, pk_r, C_s, C_r) \in L_{\text{correct}}$, where L_{correct} is defined as:

$$\{\exists sk_s, r_1, r_2, v \text{ s.t. } C_s = \mathsf{Enc}(pk_s, v; r_1) \wedge C_r = \mathsf{Enc}(pk_r, v; r_2) \\ \wedge \ v \in \mathcal{V} \ \wedge (pk_s, sk_s) \in \mathsf{R}_{\text{key}} \wedge \mathsf{Dec}(sk_s, \tilde{C}_s - C_s) \in \mathcal{V}\}$$

     L_{correct} can be decomposed as $L_{\text{equal}} \wedge L_{\text{right}} \wedge L_{\text{solvent}}$, where L_{equal} proves the consistency of two ciphertexts, L_{right} proves that the transfer amount lies in the right range, and L_{solvent} proves that the sender account is solvent.
  3. run $\sigma \leftarrow \text{ISE}.\mathsf{Sign}(sk_s, (\mathsf{sn}, \mathsf{memo}))$, sn is the serial number of pk_s;
  4. output $\mathsf{ctx} = (\mathsf{sn}, \mathsf{memo}, \mathsf{aux})$, where $\mathsf{aux} = (\pi_{\text{correct}}, \sigma)$.
- $\mathsf{VerifyCTx}(\mathsf{ctx})$: parses $\mathsf{ctx} = (\mathsf{sn}, \mathsf{memo}, \mathsf{aux})$, $\mathsf{memo} = (pk_s, pk_r, C_s, C_r)$, $\mathsf{aux} = (\pi_{\text{correct}}, \sigma)$, then checks its validity via the following steps:
  1. check if sn is a fresh serial number of pk_s;
  2. check if $\text{ISE}.\mathsf{Verify}(pk_s, (\mathsf{sn}, \mathsf{memo}), \sigma) = 1$;
  3. check if $\text{NIZK}.\mathsf{Verify}(crs, \mathsf{memo}, \pi_{\text{correct}}) = 1$.

  If all the above tests pass, outputs "1", miners confirm that ctx is valid and record it on the blockchain, sender updates his balance as $\tilde{C}_s = \tilde{C}_s - C_s$ and increments the serial number, and receiver updates his balance as $\tilde{C}_r = \tilde{C}_r + C_r$. Else, outputs "0" and miners discard ctx.

We further describe how to enforce a variety of useful policies.

- $\mathsf{JustifyCTx}(pk_s, sk_s, \{\mathsf{ctx}_i\}_{i=1}^n, f_{\text{limit}})$: on input pk_s, sk_s, $\{\mathsf{ctx}_i\}_{i=1}^n$ and f_{limit}, parses $\mathsf{ctx}_i = (\mathsf{sn}_i, \mathsf{memo}_i = (pk_s, pk_{r_i}, C_{s,i}, C_{r_i}), \mathsf{aux}_i)$, then runs $\text{NIZK}.\mathsf{Prove}$ with witness sk_s to generate π_{limit} for $(pk_s, \{C_{s,i}\}_{1 \le i \le n}, a_{\max}) \in L_{\text{limit}}$:

$$\{\exists sk \text{ s.t. } (pk, sk) \in \mathsf{R}_{\text{key}} \wedge v_i = \text{ISE}.\mathsf{Dec}(sk, C_i) \wedge \textstyle\sum_{i=1}^n v_i \le a_{\max}\}$$

  A user runs this algorithm to prove compliance with limit policy, i.e., the sum of v_i sent from account pk_s is less than $a_{\max}$. The same algorithm can be used to proving the sum of v_i sent to the same account is less than $a_{\max}$.
- $\mathsf{AuditCTx}(pk_s, \{\mathsf{ctx}_i\}_{i=1}^n, \pi_{\text{limit}}, f_{\text{limit}})$: on input pk_s, $\{\mathsf{ctx}_i\}_{i=1}^n$, π_{limit} and f_{limit}, first parses $\mathsf{ctx}_i = (\mathsf{sn}_i, \mathsf{memo}_i = (pk_s, pk_{r_i}, C_{s,i}, C_{r_i}), \mathsf{aux}_i)$, then outputs $\text{NIZK}.\mathsf{Verify}(crs, (pk_s, \{C_{s,i}\}_{1 \le i \le n}, a_{\max}), \pi_{\text{limit}})$. The auditor runs this algorithm to check compliance with limit policy.

- $\mathsf{JustifyCTx}(pk_u, sk_u, \{\mathsf{ctx}\}_{i=1}^2, f_{\text{rate}})$: on input pk_u, sk_u, $\{\mathsf{ctx}_i\}_{i=1}^2$ and f_{rate}, the algorithm parses $\mathsf{ctx}_1 = (\mathsf{sn}_1, \mathsf{memo}_1 = (pk_1, pk_u, C_1, C_{u,1}), \mathsf{aux}_1)$ and $\mathsf{ctx}_2 = (\mathsf{sn}_2, \mathsf{memo}_2 = (pk_u, pk_2, C_{u,2}, C_2), \mathsf{aux}_2)$, then runs NIZK.$\mathsf{Prove}$ with witness sk_u to generate π_{rate} for the statement $(pk_u, C_{u,1}, C_{u,2}, \rho) \in L_{\text{rate}}$:

$$\{\exists sk \text{ s.t. } (pk, sk) \in \mathsf{R}_{\text{key}} \wedge v_i = \text{ISE.}\mathsf{Dec}(sk, C_i) \wedge v_1/v_2 = \rho\}$$

  A user runs this algorithm to demonstrate compliance with tax rule, i.e., proving $v_1/v_2 = \rho$.
- $\mathsf{AuditCTx}(pk_u, \{\mathsf{ctx}_i\}_{i=1}^2, \pi_{\text{rate}}, f_{\text{rate}})$: on input pk_u, $\{\mathsf{ctx}_i\}_{i=1}^2$, π_{rate} and f_{rate}, parses $\mathsf{ctx}_1 = (\mathsf{sn}_1, \mathsf{memo}_1 = (pk_1, pk_u, C_1, C_{u,1}), \mathsf{aux}_1)$, $\mathsf{ctx}_2 = (\mathsf{sn}_2, \mathsf{memo}_2 = (pk_u, pk_2, C_{u,2}, C_2), \mathsf{aux}_2)$, outputs NIZK.$\mathsf{Verify}(crs, (pk_u, C_{u,1}, C_{u,2}, \rho), \pi_{\text{rate}})$. An auditor runs this algorithm to check compliance with rate policy.
- $\mathsf{JustifyCTx}(sk_u, \mathsf{ctx}, f_{\text{open}})$: on input pk_u, sk_u, ctx and f_{open}, parses $\mathsf{ctx} = (\mathsf{sn}, pk_s, pk_r, C_s, C_r, \mathsf{aux})$, then runs NIZK.$\mathsf{Prove}$ with witness sk_u (where the subscript u could be either s or r) to generate π_{open} for $(pk_u, C_u, v^*) \in L_{\text{open}}$:

$$\{\exists sk \text{ s.t. } (pk, sk) \in \mathsf{R}_{\text{key}} \wedge v^* = \text{ISE.}\mathsf{Dec}(sk, C)\}$$

  A user runs this algorithm to demonstrate compliance with open policy.
- $\mathsf{AuditCTx}(pk_u, \mathsf{ctx}, \pi_{\text{open}}, f_{\text{open}})$: on input pk_u, ctx, π_{open} and f_{open}, parses $\mathsf{ctx} = (\mathsf{sn}, pk_s, pk_r, C_s, C_r, \mathsf{aux})$, outputs NIZK.$\mathsf{Verify}(crs, (pk_u, C_u, v), \pi_{\text{open}})$, where the subscript u could be either s or r. An auditor runs this algorithm to check compliance with open policy.

4 PGC: An Efficient Instantiation

We now present an efficient realization of our generic DCP construction. We first instantiate ISE from our newly introduced twisted ElGamal PKE and Schnorr signature, then devise NIZK proofs from Sigma protocols and Bulletproof.

4.1 Instantiating ISE

We instantiate ISE from our newly introduced twisted ElGamal PKE and the classical Schnorr signature.

Twisted ElGamal. We propose twisted ElGamal encryption as the PKE component. Formally, twisted ElGamal consists of four algorithms as below:

- $\mathsf{Setup}(1^\lambda)$: run $(\mathbb{G}, g, p) \leftarrow \mathsf{GroupGen}(1^\lambda)$, pick $h \xleftarrow{\text{R}} \mathbb{G}^*$, set $pp = (\mathbb{G}, g, h, p)$ as global public parameters. The randomness and message spaces are $\mathbb{Z}_p$.
- $\mathsf{KeyGen}(pp)$: on input pp, choose $sk \xleftarrow{\text{R}} \mathbb{Z}_p$, set $pk = g^{sk}$.
- $\mathsf{Enc}(pk, m; r)$: compute $X = pk^r$, $Y = g^r h^m$, output $C = (X, Y)$.
- $\mathsf{Dec}(sk, C)$: parse $C = (X, Y)$, compute $h^m = Y/X^{sk^{-1}}$, recover m from h^m.

Remark 1. As with the standard exponential ElGamal, decryption can only be efficiently done when the message is small. However, it suffices to instantiate our generic DCP framework with small message space (say, 32-bits). In this setting, our implementation shows that decryption can be very efficient.

Correctness and additive homomorphism are obvious. The standard IND-CPA security can be proved in standard model based on the DDH assumption.

Theorem 1. *Twisted ElGamal is IND-CPA secure in the 1-plaintext/2-recipient setting based on the DDH assumption.*

ISE from Schnorr Signature and Twisted ElGamal Encryption. We choose Schnorr signature [Sch91] as the signature component. By merging the Setup and KeyGen algorithms of twisted ElGamal encryption and Schnorr signature, we obtain the ISE scheme, whose joint security is captured by the following theorem.

Theorem 2. *The obtained ISE scheme is jointly secure if the twisted ElGamal is IND-CPA secure (1-plaintext/2-recipient) and the Schnorr signature is EUF-CMA secure.*

4.2 Instantiating NIZK

Now, we design efficient NIZK proof systems for basic correctness policy (L_{correct}) and more extended policies (L_{limit}, L_{rate}, L_{open}).

As stated in Sect. 3, L_{correct} can be decomposed as $L_{\text{equal}} \wedge L_{\text{right}} \wedge L_{\text{solvent}}$. Let Π_{equal}, Π_{right}, Π_{solvent} be NIZK for L_{equal}, L_{right}, L_{solvent} respectively, and let $\Pi_{\text{correct}} := \Pi_{\text{equal}} \circ \Pi_{\text{right}} \circ \Pi_{\text{solvent}}$, where $\circ$ denotes sequential composition.[5] By the property of NIZK for conjunctive statements [Gol06], Π_{correct} is a NIZK proof system for L_{correct}. Now, the task breaks down to design Π_{equal}, Π_{right}, Π_{solvent}. We describe them one by one as below.

NIZK for L_{equal}. Recall that L_{equal} is defined as:

$$\{(pk_1, X_1, Y_1, pk_2, X_2, Y_2) \mid \exists r_1, r_2, v \text{ s.t. } X_i = pk_i^{r_i} \wedge Y_i = g^{r_i} h^v \text{ for } i = 1, 2\}.$$

For twisted ElGamal, randomness can be safely reused in the 1-plaintext/2-recipient setting. L_{equal} can thus be simplified to:

$$\{(pk_1, pk_2, X_1, X_2, Y) \mid \exists r, v \text{ s.t. } Y = g^r h^v \wedge X_i = pk_i^r \text{ for } i = 1, 2\}.$$

<u>Sigma protocol for L_{equal}.</u> To obtain a NIZK for L_{equal}, we first design a Sigma protocol $\Sigma_{\text{equal}} = (\mathsf{Setup}, P, V)$ for L_{equal}. The Setup algorithm of Σ_{equal} is same as that of the twisted ElGamal. On statement $(pk_1, pk_2, X_1, X_2, Y)$, P and V interact as below:

[5] In the non-interactive setting, there is no distinction between sequential and parallel composition.

1. P picks $a, b \xleftarrow{R} \mathbb{Z}_p$, sends $A_1 = pk_1^a$, $A_2 = pk_2^a$, $B = g^a h^b$ to V.
2. V picks $e \xleftarrow{R} \mathbb{Z}_p$ and sends it to P as the challenge.
3. P computes $z_1 = a + er$, $z_2 = b + ev$ using witness $w = (r, v)$, then sends (z_1, z_2) to V. V accepts iff the following three equations hold simultaneously:

$$pk_1^{z_1} = A_1 X_1^e \wedge pk_2^{z_1} = A_2 X_2^e \wedge g^{z_1} h^{z_2} = BY^e$$

Lemma 1. *Σ_{equal} is a public-coin SHVZK proof of knowledge for L_{equal}.*

Applying Fiat-Shamir transform to Σ_{equal}, we obtain Π_{equal}, which is actually a NIZKPoK for L_{equal}.

NIZK for L_{right}. Recall that L_{right} is defined as:

$$\{(pk, X, Y) \mid \exists r, v \text{ s.t. } X = pk^r \wedge Y = g^r h^v \wedge v \in \mathcal{V}\}.$$

For ease of analysis, we additionally define L_{enc} and L_{range} as below:

$$L_{\text{enc}} = \{(pk, X, Y) \mid \exists r, v \text{ s.t. } X = pk^r \wedge Y = g^r h^v\}$$
$$L_{\text{range}} = \{Y \mid \exists r, v \text{ s.t. } Y = g^r h^v \wedge v \in \mathcal{V}\}$$

It is straightforward to verify that $L_{\text{right}} \subset L_{\text{enc}} \wedge L_{\text{range}}$. Observing that each instance $(pk, X, Y) \in L_{\text{right}}$ has a unique witness, while the last component Y can be viewed as a Pedersen commitment of value v under commitment key (g, h), whose discrete logarithm $\log_g h$ is unknown to any users. To prove $(pk, X, Y) \in L_{\text{right}}$, we first prove $(pk, X, Y) \in L_{\text{enc}}$ with witness (r, v) via a Sigma protocol $\Sigma_{\text{enc}} = (\mathsf{Setup}, P_1, V_1)$, then prove $Y \in L_{\text{range}}$ with witness (r, v) via a Bulletproof $\Lambda_{\text{bullet}} = (\mathsf{Setup}, P_2, V_2)$.

Sigma protocol for L_{enc}. We begin with a Sigma protocol $\Sigma_{\text{enc}} = (\mathsf{Setup}, P, V)$ for L_{enc}. The Setup algorithm is same as that of twisted ElGamal. On statement $x = (pk, X, Y)$, P and V interact as below:

1. P picks $a, b \xleftarrow{R} \mathbb{Z}_p$, sends $A = pk^a$ and $B = g^a h^b$ to V.
2. V picks $e \xleftarrow{R} \mathbb{Z}_p$ and sends it to P as the challenge.
3. P computes $z_1 = a + er$, $z_2 = b + ev$ using witness $w = (r, v)$, then sends (z_1, z_2) to V. V accepts iff the following two equations hold simultaneously:

$$pk^{z_1} = AX^e \wedge g^{z_1} h^{z_2} = BY^e$$

Lemma 2. *Σ_{enc} is a public-coin SHVZK proof of knowledge for L_{enc}.*

Bulletproofs for L_{range}. We employ the logarithmic size Bulletproof $\Lambda_{\text{bullet}} = (\mathsf{Setup}, P, V)$ to prove L_{range}. To avoid repetition, we refer to [BBB+18, Section 4.2] for the details of the interaction between P and V.

Lemma 3 *[BBB+18, Theorem 3]***.** *Assuming the hardness of discrete logarithm problem, Λ_{bullet} is a public-coin SHVZK argument of knowledge for L_{range}.*

Sequential Composition. Let $\Gamma_{\text{right}} = \Sigma_{\text{enc}} \circ \Lambda_{\text{bullet}}$ be the sequential composition of Σ_{enc} and Λ_{bullet}. The Setup algorithm of Γ_{right} is a merge of that of Σ_{enc} and Λ_{bullet}. For range $\mathcal{V} = [0, 2^\ell - 1]$, it first generates a group $\mathbb{G}$ of prime order p together with two random generators g and h, then picks independent generators $\mathbf{g}, \mathbf{h} \in \mathbb{G}^\ell$. Let $P_1 = \Sigma_{\text{enc}}.P$, $V_1 = \Sigma_{\text{enc}}.V$, $P_2 = \Lambda_{\text{bullet}}.P$, $V_2 = \Lambda_{\text{bullet}}.V$. We have $\Gamma_{\text{right}}.P = (P_1, P_2)$, $\Gamma_{\text{right}}.V = (V_1, V_2)$.

Lemma 4. *Assuming the discrete logarithm assumption, $\Gamma_{\text{right}} = (\mathsf{Setup}, P, V)$ is a public-coin SHVZK argument of knowledge for L_{right}.*

Applying Fiat-Shamir transform to Γ_{right}, we obtain Π_{right}, which is actually a NIZKAoK for L_{right}.

NIZK for L_{Solvent}. Recall that L_{solvent} is defined as:

$$\{(pk, \tilde{C}, C) \mid \exists sk \text{ s.t. } (pk, sk) \in \mathsf{R}_{\text{key}} \wedge \text{ISE}.\mathsf{Dec}(sk, \tilde{C} - C) \in \mathcal{V}\}.$$

In the above, $\tilde{C} = (\tilde{X} = pk^{\tilde{r}}, \tilde{Y} = g^{\tilde{r}} h^{\tilde{m}})$ is the encryption of current balance $\tilde{m}$ of account pk under randomness $\tilde{r}$, $C = (X = pk^r, Y = g^r h^v)$ encrypts the transfer amount v under randomness r. Let $C' = (X' = pk^{r'}, Y' = g^{r'} h^{m'}) = \tilde{C} - C$, L_{solvent} can be rewritten as:

$$\{(pk, C') \mid \exists r', m' \text{ s.t. } C' = \text{ISE}.\mathsf{Enc}(pk, m'; r') \wedge m' \in \mathcal{V}\}.$$

We note that while the sender (playing the role of prover) learns $\tilde{m}$ (by decrypting $\tilde{C}$ with sk), v and r, it generally does not know the randomness $\tilde{r}$. This is because $\tilde{C}$ is the sum of all the incoming and outgoing transactions of pk, whereas the randomness behind incoming transactions is unknown. The consequence is that r' (the first part of witness) is unknown, which renders prover unable to directly invoke the Bulletproof on instance Y'.

Our trick is encrypting $m' = (\tilde{m} - v)$ under a fresh randomness r^* to obtain a new ciphertext $C^* = (X^*, Y^*)$, where $X^* = pk^{r^*}$, $Y^* = g^{r^*} h^{m'}$. C^* could be viewed as a refreshment of C'. Thus, we can express L_{solvent} as $L_{\text{equal}} \wedge L_{\text{right}}$, i.e., $(pk, C') \in L_{\text{solvent}} \iff (pk, C', C^*) \in L_{\text{equal}} \wedge (pk, C^*) \in L_{\text{right}}$.

To prove C' and C^* encrypting the same value under public key pk, we cannot simply use a Sigma protocol like Protocol for L_{equal}, in which the prover uses the message and randomness as witness, as the randomness r' behind C' is typically unknown. Luckily, we are able to prove this more efficiently by using the secret key as witness. Generally, L_{equal} can be written as:

$$\{(pk, C_1, C_2) \mid \exists sk \text{ s.t. } (pk, sk) \in \mathsf{R}_{\text{key}} \wedge \text{ISE}.\mathsf{Dec}(sk, C_1) = \text{ISE}.\mathsf{Dec}(sk, C_2)\}$$

When instantiated with twisted ElGamal, $C_1 = (X_1 = pk^{r_1}, Y_1 = g^{r_1} h^m)$ and $C_2 = (X_2 = pk^{r_2}, Y_2 = g^{r_2} h^m)$ are encrypted by the same public key pk, then proving membership of L_{equal} is equivalent to proving $\log_{Y_1/Y_2} X_1/X_2$ equals $\log_g pk$. This can be efficiently done by utilizing the Sigma protocol $\Sigma_{\text{ddh}} = (\mathsf{Setup}, P, V)$ for discrete logarithm equality due to Chaum and Pedersen [CP92].

Applying Fiat-Shamir transform to Σ_{ddh}, we obtain a NIZKPoK Π_{ddh} for L_{ddh}. We then prove $(pk, C^* = (X^*, Y^*)) \in L_{\text{right}}$ using the NIZKPoK Π_{right} as we described before. Let $\Pi_{\text{solvent}} = \Pi_{\text{ddh}} \circ \Pi_{\text{right}}$, we conclude that Π_{solvent} is a NIZKPoK for L_{solvent} by the properties of AND-proofs.

Putting all the sub-protocols described above, we obtain $\Pi_{\text{correct}} = \Pi_{\text{equal}} \circ \Pi_{\text{right}} \circ \Pi_{\text{solvent}}$.

Theorem 3. *Π_{correct} is a NIZKAoK for $L_{\text{correct}} = L_{\text{equal}} \wedge L_{\text{right}} \wedge L_{\text{solvent}}$.*

We then show the designs of NIZK for typical auditing policies.

NIZK for L_{limit}. Recall that $L_{\text{limit}} = \{pk, \{C_i\}_{1 \le i \le n}, a_{\max}\}$ is defined as:

$$\{\exists sk \text{ s.t. } (pk, sk) \in \mathsf{R}_{\text{key}} \wedge v_i = \text{ISE}.\mathsf{Dec}(sk, C_i) \wedge \sum_{i=1}^{n} v_i \le a_{\max}\}$$

Let $C_i = (X_i = pk^{r_i}, Y_i = g^{r_i} h^{v_i})$. By additive homomorphism of twisted ElGamal, the prover first computes $C = \sum_{i=1}^{n} C_i = (X = pk^r, Y = g^r h^v)$, where $r = \sum_{i=1}^{n} r_i$, $v = \sum_{i=1}^{n} v_i$. As aforementioned, in PGC users are not required to maintain history state, which means that users may forget the random coins when implementing after-the-fact auditing. Nevertheless, this is not a problem. It is equivalent to prove $(pk, C) \in L_{\text{solvent}}$, which can be done by using Gadget-2.

NIZK for L_{rate}. Recall that L_{rate} is defined as:

$$\{(pk, C_1, C_2, \rho) \mid \exists sk \text{ s.t. } (pk, sk) \in \mathsf{R}_{\text{key}} \wedge v_i = \text{ISE}.\mathsf{Dec}(sk, C_i) \wedge v_1/v_2 = \rho\}$$

Without much loss of generality, we assume $\rho = \alpha/\beta$, where α and β are two positive integers that are much smaller than p. Let $C_1 = (pk^{r_1}, g^{r_1} h^{v_1})$, $C_2 = (pk^{r_2}, g^{r_2} h^{v_2})$. By additive homomorphism of twisted ElGamal, we compute $C_1' = \beta \cdot C_1 = (X_1' = pk^{\beta r_1}, Y_1' = g^{\beta r_1} h^{\beta v_1})$, $C_2' = \alpha \cdot C_2 = (X_2' = pk^{\alpha r_2}, Y_2' = g^{\alpha r_2} h^{\alpha v_2})$. Note that $v_1/v_2 = \rho = \alpha/\beta$ if and only if $h^{\beta v_1} = h^{\alpha v_2}$,[6] L_{rate} is equivalent to $(Y_1'/Y_2', X_1'/X_2', g, pk) \in L_{\text{ddh}}$, which in turn can be efficiently proved via Π_{ddh} for discrete logarithm equality using sk as witness, as already described in Protocol for L_{solvent}.

NIZK for L_{open}. Recall that L_{open} is defined as:

$$\{(pk, C = (X, Y), v) \mid \exists sk \text{ s.t. } X = (Y/h^v)^{sk} \wedge pk = g^{sk}\}$$

The above language is equivalent to $(Y/h^v, X, g, pk) \in L_{\text{ddh}}$, which in turn can be efficiently proved via Π_{ddh} for discrete logarithm equality.

5 Performance

We first give a prototype implementation of PGC as a standalone cryptocurrency in C++ based on OpenSSL, and collect the benchmarks on a MacBook Pro with an Intel i7-4870HQ CPU (2.5 GHz) and 16 GB of RAM. The source code of PGC is publicly available at Github [lib]. For demo purpose only, we only focus on the transaction layer and do not explore any optimizations.[7] The experimental results are described in the table below.

[6] Since both v_1, v_2, α, β are much smaller than p, no overflow will happen.

[7] We expect at least 2× speedup after optimizations.

Table 1. The computation and communication complexity of PGC.

PGC	ctx size		Transaction cost (ms)	
	big-$\mathcal{O}$	Bytes	Generation	Verify
Confidential transaction	$(2\log_2(\ell)+20)\lvert\mathbb{G}\rvert+10\lvert\mathbb{Z}_p\rvert$	1310	40	14
Auditing policies	Proof size		Auditing cost (ms)	
	big-$\mathcal{O}$	Bytes	Generation	Verify
Limit policy	$(2\log_2(\ell)+4)\lvert\mathbb{G}\rvert+5\lvert\mathbb{Z}_p\rvert$	622	21.5	7.5
Rate policy	$2\lvert\mathbb{G}\rvert+1\lvert\mathbb{Z}_p\rvert$	98	0.55	0.69
Open policy	$2\lvert\mathbb{G}\rvert+1\lvert\mathbb{Z}_p\rvert$	98	0.26	0.42

Here we set the maximum number of coins as $v_{\max} = 2^\ell - 1$, where $\ell = 32$. PGC operates elliptic curve prime256v1, which has 128 bit security. The elliptic points are expressed in their compressed form. Each $\mathbb{G}$ element is stored as 33 bytes, and $\mathbb{Z}_p$ element is stored as 32 bytes.

Acknowledgments. We thank Benny Pinkas and Jonathan Bootle for clarifications on Sigma protocols and Bulletproofs in the early stages of this research. We particularly thank Shuai Han for many enlightening discussions. Yu Chen is supported by National Natural Science Foundation of China (Grant No. 61772522, No. 61932019). Man Ho Au is supported by National Natural Science Foundation of China (Grant No. 61972332).

References

[BAZB20] Bünz, B., Agrawal, S., Zamani, M., Boneh, D.: Zether: towards privacy in a smart contract world. In: Bonneau, J., Heninger, N. (eds.) FC 2020. LNCS, vol. 12059, pp. 423–443. Springer, Cham (2020). https://doi.org/10.1007/978-3-030-51280-4_23

[BBB+18] Bünz, B., Bootle, J., Boneh, D., Poelstra, A., Wuille, P., Maxwell, G.: Bulletproofs: short proofs for confidential transactions and more. In: 2018 IEEE Symposium on Security and Privacy, SP 2018, pp. 315–334 (2018)

[BKP14] Biryukov, A., Khovratovich, D., Pustogarov, I.: Deanonymisation of clients in bitcoin P2P network. In: Proceedings of the 2014 ACM SIGSAC Conference on Computer and Communications Security, CCS 2014, pp. 15–29 (2014)

[BNM+14] Bonneau, J., Narayanan, A., Miller, A., Clark, J., Kroll, J.A., Felten, E.W.: Mixcoin: anonymity for bitcoin with accountable mixes. In: Christin, N., Safavi-Naini, R. (eds.) FC 2014. LNCS, vol. 8437, pp. 486–504. Springer, Heidelberg (2014). https://doi.org/10.1007/978-3-662-45472-5_31

[CGS97] Cramer, R., Gennaro, R., Schoenmakers, B.: A secure and optimally efficient multi-authority election scheme. In: Fumy, W. (ed.) EUROCRYPT 1997. LNCS, vol. 1233, pp. 103–118. Springer, Heidelberg (1997). https://doi.org/10.1007/3-540-69053-0_9

[CMTA19] Chen, Y., Ma, X., Tang, C., Au, M.H.: PGC: pretty good confidential transaction system with auditability. Cryptology ePrint Archive, Report 2019/319 (2019). https://eprint.iacr.org/2019/319

[CP92] Chaum, D., Pedersen, T.P.: Wallet databases with observers. In: Brickell, E.F. (ed.) CRYPTO 1992. LNCS, vol. 740, pp. 89–105. Springer, Heidelberg (1993). https://doi.org/10.1007/3-540-48071-4_7

[CZ14] Chen, Yu., Zhang, Z.: Publicly evaluable pseudorandom functions and their applications. In: Abdalla, M., De Prisco, R. (eds.) SCN 2014. LNCS, vol. 8642, pp. 115–134. Springer, Cham (2014). https://doi.org/10.1007/978-3-319-10879-7_8

[Das] Dash. https://www.dash.org

[FMMO19] Fauzi, P., Meiklejohn, S., Mercer, R., Orlandi, C.: Quisquis: a new design for anonymous cryptocurrencies. In: Galbraith, S.D., Moriai, S. (eds.) ASIACRYPT 2019. LNCS, vol. 11921, pp. 649–678. Springer, Cham (2019). https://doi.org/10.1007/978-3-030-34578-5_23

[GGM16] Garman, C., Green, M., Miers, I.: Accountable privacy for decentralized anonymous payments. In: Grossklags, J., Preneel, B. (eds.) FC 2016. LNCS, vol. 9603, pp. 81–98. Springer, Heidelberg (2017). https://doi.org/10.1007/978-3-662-54970-4_5

[Gol06] Goldreich, O.: Foundations of Cryptography, vol. 1. Cambridge University Press, New York (2006)

[Gri] Grin. https://grin-tech.org/

[lib] libPGC. https://github.com/yuchen1024/libPGC

[Max] Maxwell, G.: Confidential transactions (2016). https://people.xiph.org/~greg/confidential_values.txt

[MM] Meiklejohn, S., Mercer, R.: Möbius: trustless tumbling for transaction privacy. PoPETs **2**, 105–121 (2018)

[Nak08] Nakamoto, S.: Bitcoin: a peer-to-peer electronic cash system (2008). https://bitcoin.org/bitcoin.pdf

[Noe15] Noether, S.: Ring signature confidential transactions for monero (2015). https://eprint.iacr.org/2015/1098

[NVV18] Narula, N., Vasquez, W., Virza, M.: zkLedger: privacy-preserving auditing for distributed ledgers. In: 15th USENIX Symposium on Networked Systems Design and Implementation, NSDI 2018, pp. 65–80 (2018)

[Ped91] Pedersen, T.P.: Non-interactive and information-theoretic secure verifiable secret sharing. In: Feigenbaum, J. (ed.) CRYPTO 1991. LNCS, vol. 576, pp. 129–140. Springer, Heidelberg (1992). https://doi.org/10.1007/3-540-46766-1_9

[Poe] Poelstra, A.: Mimblewimble. https://download.wpsoftware.net/bitcoin/wizardry/mimblewimble.pdf

[PSST11] Paterson, K.G., Schuldt, J.C.N., Stam, M., Thomson, S.: On the joint security of encryption and signature, revisited. In: Lee, D.H., Wang, X. (eds.) ASIACRYPT 2011. LNCS, vol. 7073, pp. 161–178. Springer, Heidelberg (2011). https://doi.org/10.1007/978-3-642-25385-0_9

[RMK14] Ruffing, T., Moreno-Sanchez, P., Kate, A.: CoinShuffle: practical decentralized coin mixing for bitcoin. In: Kutyłowski, M., Vaidya, J. (eds.) ESORICS 2014. LNCS, vol. 8713, pp. 345–364. Springer, Cham (2014). https://doi.org/10.1007/978-3-319-11212-1_20

[RS13] Ron, D., Shamir, A.: Quantitative analysis of the full bitcoin transaction graph. In: Sadeghi, A.-R. (ed.) FC 2013. LNCS, vol. 7859, pp. 6–24. Springer, Heidelberg (2013). https://doi.org/10.1007/978-3-642-39884-1_2

[Sch91] Schnorr, C.P.: Efficient signature generation by smart cards. J. Cryptol. **4**(3), 161–174 (1991). https://doi.org/10.1007/BF00196725

[Woo14] Wood, G.: Ethereum: a secure decentralized transaction ledger (2014). http://gavwood.com/paper.pdf, https://www.ethereum.org/

[ZCa] Zcash: privacy-protecting digital currency. https://z.cash/

Secure Cloud Auditing with Efficient Ownership Transfer

Jun Shen[1,2], Fuchun Guo[3], Xiaofeng Chen[1,2], and Willy Susilo[3(✉)]

[1] State Key Laboratory of Integrated Service Networks (ISN), Xidian University, Xi'an 710071, China
demon_sj@126.com, xfchen@xidian.edu.cn
[2] State Key Laboratory of Cryptology, P. O. Box 5159, Beijing 100878, China
[3] Institute of Cybersecurity and Cryptology, School of Computing and Information Technology, University of Wollongong, Wollongong, NSW 2522, Australia
{fuchun,wsusilo}@uow.edu.au

Abstract. Cloud auditing with ownership transfer is a provable data possession scheme meeting verifiability and transferability simultaneously. In particular, not only cloud data can be transferred to other cloud clients, but also tags for integrity verification can be transferred to new data owners. More concretely, it requires that tags belonging to the old owner can be transformed into that of the new owner by replacing the secret key for tag generation while verifiability still remains. We found that existing solutions are less efficient due to the huge communication overhead linear with the number of tags. In this paper, we propose a secure auditing protocol with efficient ownership transfer for cloud data. Specifically, we sharply reduce the communication overhead produced by ownership transfer to be independent of the number of tags, making it with a constant size. Meanwhile, the computational cost during this process on both transfer parties is constant as well.

Keywords: Cloud storage · Integrity auditing · Ownership transfer

1 Introduction

As cloud computing has developed rapidly, outsourcing data to cloud servers for remote storage has become an attractive trend [4,11,16]. However, when cloud clients store their data in the cloud, the integrity of cloud data would be threatened due to accidental corruptions or purposive attacks caused by a semi-trusted cloud server. Hence, cloud auditing was proposed as a significant security technology for integrity verification, and has been widely researched [12,17,18,20]. Concretely, a certain cloud client with secret and public key pair (sk_1, pk_1) generates tags $\sigma_i = T(m_i, sk_1)$ based on sk_1 for data blocks m_i, and uploads $\{(m_i, \sigma_i)\}$ to the cloud for remote storage. Once audited with $chal$, the cloud responds with a constant-size proof of possession P generated from $(m_i, \sigma_i, chal)$, rather than sending back the challenged cloud data for checking

L. Chen et al. (Eds.): ESORICS 2020, LNCS 12308, pp. 611–631, 2020.
https://doi.org/10.1007/978-3-030-58951-6_30

directly. The validity of P is checked via pk_1 and thereby the integrity of cloud data [5,6,17].

Ownership transfer requires that the owner of cloud data is changeable among cloud clients, instead of always being the original uploader. The ownership is usually indicated by signatures. When ownership transfer occurs, signatures of the old owner should be transformed to that of the new one with the old secret key contained in signatures replaced. Consequently, the new owner obtains the ownership and accesses these data without the approval from the old owner.

Cloud auditing with ownership transfer is a provable data possession scheme meeting verifiability and transferability simultaneously. In particular, not only data can be transferred to other cloud clients, but also tags for integrity verification can be transferred to new data owners. Specifically, when the ownership of some data is transferred to a new owner with key pair (sk_2, pk_2), verifiable tags σ_i should be $\sigma_i = T(m_i, sk_2)$ instead of $\sigma_i = T(m_i, sk_1)$. Thus, the transferred data belong to the new owner and are verifiable via pk_2.

A trivial solution to achieve cloud auditing with ownership transfer follows the "download-upload" mechanism. Specifically, the new owner downloads the cloud data approved and uploads new generated verifiable tags. It is obvious that such a way is not an advisable one for the huge computational overhead of tag generation and the communication cost even up to twice the size of transferred data. In order to relieve the computational burden, Wang *et al.* [25] considered to outsource computations of transfer for the first time, and proposed the concept of "Provable Data Possession with Outsourced Data Transfer" (DT-PDP). However, the communication overhead in their transfer protocol is still linear with the number of tags. To the best of our knowledge, there is no better way to further reduce the communication cost produced during ownership transfer.

1.1 Our Contribution

Inspired by the communication overhead, in this work, we are devoted to designing a secure cloud auditing protocol with efficient ownership transfer. Our contributions are summarized as follows.

- We focus on the efficiency of ownership transfer, reducing communications produced by transfer to be with a constant size and independent of the number of tags. Meanwhile, the majority of computational cost during ownership transfer is delegated to the computing server, leaving computations on both transfer parties constant as well.
- We analyze the security of the proposed protocol, demonstrating that it is provably secure under the k-CEIDH assumption in the random oracle model. The protocol achieves properties of correctness, soundness, unforgeability and detectability. In addition, when ownership transfer occurs, the protocol is secure against collusion attacks, making the data untransferred protected.

1.2 Related Work

Extensive researches have been conducted on the integrity verification for cloud storage. Such researches focus on various aspects, including but not limited to dynamic operations, privacy preservation, key-exposure resistance, etc.

In 2007, Ateniese *et al.* [1] proposed the first "Provable Data Possession" (PDP) scheme, making trusted third parties enabled to execute public verifications. The scheme employs random sampling and homomorphic authenticators to achieve the public auditing property. Almost at the same time, Juels and Kaliski [10] put forward the concept of "Proofs of Retrievability" (PoRs), which is slightly different from PDP but with the similar purpose of checking the remotely stored data. These data are encoded by error correcting codes to enable retrievability and with several sentinels inserted for data possession verification. Since such sentinels should be responded a few each time when audited, the number of challenge executions is extremely limited. In 2008, Shacham and Waters [19] first combined coding technology and PDP together. In this design, the data to be outsourced are divided twice as sectors for the first time, where sectors are equivalent to blocks in previous schemes and several sectors share one data tag, decreasing the size of the processed data storing in the cloud.

Subsequently, more researches were conducted on cloud auditing. In 2008, Ateniese *et al.* [2] first considered data auditing supporting dynamic operations and proposed a PDP scheme with partial dynamics, failing to support data insertion. To solve this problem, Erway *et al.* [7] designed the first PDP scheme with fully dynamic storage, while it suffers from high computational and communication overheads. Later, Wang *et al.* [28] employed the Merkle Hash Tree to establish another dynamic PDP scheme, which is much simpler. This topic is also researched in [9,22]. Besides, Wang *et al.* [24] focused on the study of privacy preservation in cloud auditing, combining homomorphic linear authenticators and random masking technology together. Later, Worku *et al.* [30] noted that the scheme in [24] leaks the identity privacy of data owners, and utilized ring signature to improve privacy preservation. What's more, key-exposure resistance is also considered in cloud auditing. Yu *et al.* [33] first considered the security problem caused by secret key exposure and employed the Merkle Hash Tree to give a solution, of which the efficiency is unsatisfying. Then, they released heavy computations in [32] and enhanced key-exposure resilience in [34].

Other aspects have also been studied these years. Auditing schemes for shared cloud data are researched in [8,21,31]. Data sharing with user revocation is achieved in [23] through using proxy re-signature to renew tags generated by the revoked user and in [15] via Shamir's Secret Sharing. To simplify certificate management, identity-based auditing schemes are proposed [14,26,36]. Besides, integrity auditing supporting data deduplication [13,27,35] has also been studied.

In 2019, it is Wang *et al.* [25] that considered the integrity checking for the remote purchased data with computations of ownership transfer outsourced for the first time, and proposed the concept of DT-PDP. In such a scheme, the computational and communication costs are reduced. Unfortunately, communications are still huge for its linearity with the number of tags. Besides, this scheme only achieves partial and one-time ownership transfer. In contrast, we explore how to achieve secure cloud auditing enabling thorough ownership transfer with high efficiency.

1.3 Organization

The rest of this paper is organized as follows. Section 2 introduces some preliminaries. Section 3 describes the system model and definition, as well as the security model. Section 4 presents details of the proposed secure cloud auditing protocol with efficient ownership transfer. Section 5 demonstrates the correctness and security of our design. Section 6 shows the performance analysis. Finally, conclusions are drawn in Sect. 7.

2 Preliminaries

To facilitate understandings, we present preliminaries including bilinear pairings, intractable problems in cyclic groups, and homomorphic authenticators.

2.1 Bilinear Pairings

Let $\mathbb{G}$ and $\mathbb{G}_T$ be cyclic groups of prime order p, and g is a generator of $\mathbb{G}$. The pairing $e : \mathbb{G} \times \mathbb{G} \rightarrow \mathbb{G}_T$ is a bilinear one iff the following conditions are satisfied:

- Bilinear: $e\left(u^a, v^b\right) = e\left(u, v\right)^{ab}$ holds for $\forall a, b \in \mathbb{Z}_p$ and $\forall u, v \in \mathbb{G}$;
- Non-degenerate: $e\left(g, g\right) \neq 1_{\mathbb{G}_T}$;
- Computable: $e\left(u, v\right)$ is efficiently computable for $\forall u, v \in \mathbb{G}$.

The following are two intractable problems in $\mathbb{G}$.

Definition 1 (Computational Diffie-Hellman problem). *The Computational Diffie-Hellman (CDH) problem in $\mathbb{G}$ is described as follows: given a tuple (g, g^x, g^y) for any $x, y \in_R \mathbb{Z}_p$ as input, output g^{xy}. Define that the CDH assumption holds in $\mathbb{G}$ if for any PPT adversary $\mathcal{A}$,*

$$\Pr\left[\mathcal{A}\left(1^\lambda, g, g^x, g^y\right) = g^{xy}\right] \leq \mathsf{negl}(\lambda)$$

holds for arbitrary security parameter λ, where $\mathsf{negl}(\cdot)$ is a negligible function.

Definition 2 (k-Computational Exponent Inverse Diffie-Hellman problem). *The k-Computational Exponent Inverse Diffie-Hellman problem (k-CEIDH problem) in $\mathbb{G}$ is defined as follows: given a $(2k+2)$-tuple $\left(e_1, e_2, \cdots, e_k, g, g^b, \; g^{\frac{1}{a+e_1}}, g^{\frac{1}{a+e_2}}, \cdots, g^{\frac{1}{a+e_k}}\right)$ as input, where k is a non-negative integer, g is a generator of $\mathbb{G}$ and $b, e_1, \cdots, e_k \in_R \mathbb{Z}_p$, output $\left(e_i, g^{\frac{b}{a+e_i}}\right)$ for any $e_i \in \{e_1, \cdots, e_k\}$. We define that the k-CEIDH assumption holds in $\mathbb{G}$ if for all PPT adversaries and arbitrary security parameter λ, there exists a negligible function $\mathsf{negl}(\cdot)$ s.t.*

$$\Pr\left[\mathcal{A}\left(1^\lambda, e_1, e_2, \cdots, e_k, g, g^b, g^{\frac{1}{a+e_1}}, g^{\frac{1}{a+e_2}}, \cdots, g^{\frac{1}{a+e_k}}\right) = g^{\frac{b}{a+e_i}}\right] \leq \mathsf{negl}(\lambda).$$

When $k = 1$, the k-CEIDH problem $\left(e_1, g, g^b, g^{\frac{1}{a+e_1}}\right)$ can be regarded as a CDH instance, which is computationally infeasible for $\mathcal{A}$ to give the solution $g^{\frac{b}{a+e_1}}$. When $k > 1$, the additional elements $\left(e_2, \cdots, e_k, g^{\frac{1}{a+e_2}}, \cdots, g^{\frac{1}{a+e_k}}\right)$ seem to give no assistance to $\mathcal{A}$ in solving the above problem. Thus, we can assume that the k-CEIDH problem is as difficult as the CDH problem.

2.2 Homomorphic Authenticators

The homomorphic authenticator is a homomorphic verifiable signature, allowing public verifiers to check the integrity of remote storage without specific data blocks, which is a significant building block employed in public cloud auditing mechanisms [3,13,19,20,22].

Given a bilinear pairing $e : \mathbb{G} \times \mathbb{G} \rightarrow \mathbb{G}_T$ and a data file $F = (m_1, m_2, \cdots, m_i, \cdots, m_n)$, where $m_i \in \mathbb{Z}_p$. Let the signer with key pair $(sk = a, pk = g^a)$ generate signatures $\sigma_i = (u^{m_i})^a$ on data blocks m_i for $i \in [1, n]$, where $a \in_R \mathbb{Z}_p$, g is a generator of $\mathbb{G}$ and $u \in_R \mathbb{G}$. The signature is a homomorphic authenticator iff the following properties are met:

- Blockless verifiability: the validity of σ_i is able to be batch authenticated via $\sum_{i \in I} m_i r_i \in \mathbb{Z}_p$ instead of blocks $\{m_i\}_{i \in I}$, where I is an integer set and $m_i r_i \in \mathbb{Z}_p$. Essentially, the verification equation could be written as

$$e\left(\prod_{i \in I} \sigma_i^{r_i}, g\right) \stackrel{?}{=} e\left(u^{\sum_{i \in I} m_i r_i}, pk\right).$$

- Non-malleability: given σ_i for m_i and σ_j for m_j, the signature on $m = am_i + bm_j$ can not be derived directly from the combination of σ_i and σ_j, where $am_i, bm_j \in \mathbb{Z}_p$.

3 System Models and Definitions

3.1 The System Model

Similar to but slightly different from the system model in [25], our model additionally involves a third party auditor since we achieve public verification. Specifically, such a model involves four entities: the cloud server (CS), the old data owner (PO), the new owner (NO) and the third party auditor (TPA).

- *The cloud server CS* is an entity with seemingly inexhaustible resources. It is responsible for data storage and computing delegations from cloud clients.
- *The old data owner PO* is the transferred party, being individuals or some organizations. It uploads data files to the cloud for remote storage to release local management burden.
- *The new data owner NO* is a cloud client being the target of data ownership transfer, i.e., the transferring party. Once ownership is transferred, *NO* inherits the jurisdiction on these data from *PO*.

- *The third party auditor TPA* is a trustworthy entity with professional knowledge of integrity verification. It is able to provide convincing results for auditing delegations from cloud clients.

We describe the system architecture in brief here. On the one hand, *PO* outsources its data to the *CS* and has on-demand access to these data. The integrity of such data is checked by the *TPA* with delegations from *PO*. Once receiving auditing challenges from the *TPA*, the *CS* generates corresponding proofs of intact storage. With such proofs, the *TPA* figures out and sends auditing results to *PO*. On the other hand, when *PO* decides to transfer the ownership of some data to *NO*, it generates and sends auxiliaries to *NO*. After processing these auxiliaries, *NO* completes the transfer with the computing service of the *CS*. Consequently, *NO* inherits the rights from *PO* and becomes the data owner currently.

3.2 Cloud Auditing with Ownership Transfer

Inspired by the DT-PDP scheme defined in [25], the formal definition for cloud auditing with ownership transfer is given as follows:

Definition 3. *The cloud auditing protocol with ownership transfer consists of five algorithms defined below.*

- SysGen $(1^\lambda) \rightarrow param$: *This algorithm is a probabilistic one run by the system. With the security parameter* λ *as input, it outputs the system parameter* $param$.
- KeyGen $(param) \rightarrow (SK, PK)$: *This algorithm is a probabilistic one run by cloud clients. Input param, the client generates secret and public key pair* (SK, PK)*, where* SK *is kept secret for tag generation and* PK *is distributed public.*
- TagGen $(F, n, m_i, SK, param) \rightarrow (\sigma_i, t)$: *This algorithm is a deterministic one run by the data owner with* SK*. Input the file abstract* F*, number of blocks* n*, data blocks* m_i*, and param, it outputs the file tag* t *and homomorphic verifiable tags* σ_i *corresponding to* m_i.
- Audit $(Q, t, m_i, \sigma_i, PK, param) \rightarrow \{0, 1\}$: *This algorithm is a probabilistic one run by the CS and the TPA in two steps: i) Input a challenge set* Q *from the TPA and file tag* t *along with data block and tag pairs* $\{(m_i, \sigma_i)\}_{i \in Q}$*, the CS sends the proof* P *to the TPA; ii) With* P*, the data owner's* PK *and param as inputs, the TPA verifies the validity of* P *and outputs "1" if succeed and "0" otherwise.*
- TagTrans $(F, SK_a, PK_a, SK_b, PK_b, \sigma_i, param) \rightarrow (\sigma_i', t')$: *This algorithm is a deterministic one run by the CS, PO and NO in three steps: i) Input the file abstract* F*, PO's* SK_a*, it sends auxiliaries Aux to NO; ii) With Aux, PO's* PK_a *and NO's* SK_b *as inputs, NO sends* Aux' *to the CS; iii) With* Aux'*, NO's* PK_b*, PO's tags* σ_i *and param as inputs, it outputs the renewed file tag* t' *and block tags* σ_i' *for NO.*

The correctness of the protocol is defined as follows:

Definition 4 (Correctness). *The cloud auditing with ownership transfer is correct iff the following conditions are satisfied:*

- *If PO and the CS execute honestly, then for any challenged data blocks* m_i*, there is always* Audit $(Q, t, m_i, \sigma_i, PK_a, param) \rightarrow 1$.
- *If PO, NO and the CS are honest, then* σ_i' *for NO generated by* Tag-Trans $(F, SK_a, PK_a, SK_b, PK_b, \sigma_i, param) \rightarrow (\sigma_i', t')$ *matches exactly with* m_i *uploaded by PO. Furthermore, for any auditing task from NO, the output of* Audit $(Q, t, m_i, \sigma_i', PK_b, param) \rightarrow \{0,1\}$ *is always "1".*

3.3 The Security Model

A secure cloud auditing protocol with efficient ownership transfer should satisfy properties of soundness, unforgeability, secure transferability and detectability, which are defined as follows.

Definition 5 (Soundness). *The cloud auditing with ownership transfer is sound if it is infeasible for the CS to provide a valid proof to pass the integrity verification, when the challenged data* m_i *is corrupted to be* $m_i' \neq m_i$*, i.e., for any security parameter* λ *and negligible function* $\mathsf{negl}(\cdot)$*, there is*

$$\Pr\Big[\mathcal{A}^{\mathcal{O}_{sign}(SK,\cdot)}(PK, param) \rightarrow (\sigma_i', t') \wedge m_i' \neq m_i \wedge (m_i', \sigma_i') \notin \{(m,\sigma)\}$$
$$\wedge \mathsf{Audit}(\{i\}, t', m_i', \sigma_i', PK, param) \rightarrow 1 :$$
$$(SK, PK) \leftarrow \mathsf{KeyGen}(param), param \leftarrow \mathsf{SysGen}(1^\lambda)\Big] \leq \mathsf{negl}(\lambda),$$

where $\mathcal{O}_{sign}(\cdot,\cdot)$ *is the oracle for signature queries, and* $\{(m,\sigma)\}$ *is the set of pairs that* $\mathcal{A}$ *had queried, the same below.*

Definition 6 (Unforgeability). *The cloud auditing with ownership transfer is unforgeable if the ownership of any block* m_s *is infeasible to be forged for all adversaries, i.e., for any security parameter* λ *and negligible function* $\mathsf{negl}(\cdot)$,

$$\Pr\Big[\mathcal{A}^{\mathcal{O}_{sign}(SK,\cdot)}(PK, param) \rightarrow (\sigma_s, t') \wedge (m_s, \sigma_s) \notin \{(m,\sigma)\}$$
$$\wedge \mathsf{Audit}(\{s\}, t', m_s, \sigma_s, PK, param) \rightarrow 1 :$$
$$(SK, PK) \leftarrow \mathsf{KeyGen}(param), param \leftarrow \mathsf{SysGen}(1^\lambda)\Big] \leq \mathsf{negl}(\lambda).$$

Definition 7 (Secure transferability). *The cloud auditing protocol is with secure ownership transfer iff the following conditions are satisfied:*

- *If PO is honest, it is resistant to the collusion attack launched by NO and the CS. In another word, such colluding adversaries can not produce any valid tags on PO's behalf. Suppose that the key pair of PO is* (SK_a, PK_a). *We define that for all PPT adversaries and arbitrary security parameter* λ,

$$\Pr\Big[\mathcal{A}^{\mathcal{O}_{sign}(SK_a,\cdot),\mathcal{O}_{aux}(\cdot)}(PK_a, SK_b, PK_b, param) \rightarrow (\sigma_s, t') \\ \wedge(m_s, \sigma_s) \notin \{(m,\sigma)\} \wedge \mathsf{Audit}(\{s\}, t', m_s, \sigma_s, PK_a, param) \rightarrow 1 : \\ (SK, PK) \leftarrow \mathsf{KeyGen}(param), param \leftarrow \mathsf{SysGen}(1^\lambda)\Big] \leq \mathsf{negl}(\lambda),$$

where $\mathsf{negl}(\cdot)$ *is a negligible function and* $\mathcal{O}_{aux}(\cdot)$ *is for auxiliary queries in ownership transfer.*

- *If NO is honest, it is protected from a colluding CS and PO. That is, even though with the combined ability of the CS and PO, tags belonging to NO cannot be produced. Suppose that the key pair of NO is* (SK_b, PK_b). *We define that for any PPT adversary and security parameter* λ,

$$\Pr\Big[\mathcal{A}^{\mathcal{O}_{sign}(SK_b,\cdot),\mathcal{O}_{aux}(\cdot)}(PK_b, SK_a, PK_a, param) \rightarrow (\sigma_s, t') \\ \wedge(m_s, \sigma_s) \notin \{(m,\sigma)\} \wedge \mathsf{Audit}(\{s\}, t', m_s, \sigma_s, PK_b, param) \rightarrow 1 : \\ (SK, PK) \leftarrow \mathsf{KeyGen}(param), param \leftarrow \mathsf{SysGen}(1^\lambda)\Big] \leq \mathsf{negl}(\lambda).$$

Definition 8 (Detectability). *The cloud auditing protocol with ownership transfer is* (ρ, δ)*-detectable, where* $0 \leq \rho, \delta \leq 1$, *if the probability that integrity corruptions of the transferred cloud data can be detected is no less than* δ *when there are a fraction* ρ *of corrupted data blocks.*

4 The Proposed Auditing Protocol

We begin by the overview of the proposed secure auditing protocol with efficient ownership transfer. Then we present it in more details.

4.1 Overview

To achieve cloud auditing with ownership transfer, we argue that it is impractical for the large costs produced, if the new owner downloads cloud data and uploads new generated tags for these data. On the other hand, if we delegate ownership transfer to a third party to reduce computations and improve efficiency, secrets of transfer parties may suffer from collusion attacks launched by the other transfer party and the third party employed. In such a delegation, the communication cost is still a primary concern. To simultaneously improve the security and efficiency, we construct a novel tag structure. With such a structure, ownership transfer is securely outsourced to relieve computations on transfer parties. Apart from that, auditing of the transferred data is consequently executed using the public key of the new owner. Last and most important, the communication cost produced by ownership transfer is constant and independent of the number of tags.

4.2 The Cloud Auditing with Efficient Ownership Transfer

The encoded file with abstract F to be uploaded by PO for remote storage is divided into n blocks, appearing as $\{m_1, m_2, \cdots, m_n\}$, where $m_i \in \mathbb{Z}_p$ for $i \in [1, n]$. Similar to [22,24,29], $\mathcal{S} = \langle \mathsf{Kgen}, \mathsf{Sig}, \mathsf{Vrf} \rangle$ is a signature scheme and $\mathcal{S}.\mathsf{Sig}(\cdot)_{ssk}$ is a signature under the secret signing key ssk, where $(ssk, spk) \leftarrow \mathcal{S}.\mathsf{Kgen}(1^\lambda)$.

The procedure of the protocol execution is as follows:

- $\mathsf{SysGen}(1^\lambda) \rightarrow param$. On input the security parameter λ, the system executes as follows:
  1. Select two cyclic multiplicative groups $\mathbb{G}$ and $\mathbb{G}_T$ with prime order p and a bilinear pairing $e: \mathbb{G} \times \mathbb{G} \rightarrow \mathbb{G}_T$.
  2. Pick two independent generators $g, u \in_R \mathbb{G}$.
  3. Choose two cryptographic hash functions $H_1: \{0,1\}^* \rightarrow \mathbb{G}$ and $H_2: \{0,1\}^* \rightarrow \mathbb{Z}_p$.
  4. Return the system parameter $param = \{\mathbb{G}, \mathbb{G}_T, p, g, u, e, H_1, H_2, \mathcal{S}\}$.
- $\mathsf{KeyGen}(param) \rightarrow (SK, PK)$. On input the system parameter $param$, PO and NO generate key pairs as follows:
  1. PO selects $sk_a = \alpha \in_R \mathbb{Z}_p$ and computes $pk_a = g^\alpha$.
  2. NO selects $sk_b = \beta \in_R \mathbb{Z}_p$ and computes $pk_b = g^\beta$.
  3. PO runs $\mathcal{S}.\mathsf{Kgen}$ to generate a signing key pair (ssk_a, spk_a).
  4. NO runs $\mathcal{S}.\mathsf{Kgen}$ to generate a signing key pair (ssk_b, spk_b).
  5. PO obtains key pair: $(SK_a, PK_a) = ((sk_a, ssk_a), (pk_a, spk_a))$, NO obtains key pair: $(SK_b, PK_b) = ((sk_b, ssk_b), (pk_b, spk_b))$.
- $\mathsf{TagGen}(F, n, m_i, SK, param) \rightarrow (\sigma_i, t)$. On input file abstract F, data blocks m_i and their amount n, and $param$, PO with SK_a executes as follows:
  1. Compute $r = H_2(\alpha \| F)$, $h = g^r$, and $h_1 = h^\alpha$.
  2. Calculate tags for $\{m_i\}_{i \in [1,n]}$ as
  $$\sigma_i = H_1(F \| i)^r \cdot (u^{m_i})^{\frac{1}{\alpha + H_2(F \| h)}}.$$
  3. Compute $t = (F \| n \| h \| h_1) \| \mathcal{S}.\mathsf{Sig}(F \| n \| h \| h_1)_{ssk_a}$ as the file tag.
  4. Upload $\{\{(m_i, \sigma_i)\}_{i \in [1,n]}, t\}$ to the cloud for remote storage and delete the local copy.
- $\mathsf{Audit}(Q, t, m_i, \sigma_i, PK, param) \rightarrow \{0, 1\}$. On input the file tag t, a c-element challenge set $Q = \{(i, \gamma_i \in_R \mathbb{Z}_p)\}$ chosen by the TPA, data blocks m_i along with tags σ_i, PK_a of PO, and $param$, the TPA interacts with the CS as follows:
  1. The CS calculates the data proof and tag proof as
  $$PM = \sum_{(i,\gamma_i) \in Q} m_i \gamma_i, \quad P\sigma = \prod_{(i,\gamma_i) \in Q} \sigma_i^{\gamma_i}.$$
  2. The CS sends the proof $P = (PM, P\sigma, t)$ to the TPA as the response.

3. The *TPA* obtains F, h, h_1 if $\mathcal{S}.\mathsf{Vrf}(F||n||h||h_1)_{spk_a} = 1$, and then checks the validity of P via the verification equation

$$e\left(P\sigma, pk_a g^{H_2(F||h)}\right) \stackrel{?}{=} e\left(\prod_{(i,\gamma_i)\in Q} H_1(F||i)^{\gamma_i}, h_1 h^{H_2(F||h)}\right) \cdot e\left(u^{PM}, g\right).$$

4. The *TPA* outputs "1" iff the equation holds, indicating that these data are correctly and completely stored. Otherwise, at least one of these challenged data blocks must have been corrupted in the cloud.

- $\mathsf{TagTrans}\ (F, SK_a, PK_a, SK_b, PK_b, \sigma_i, param) \rightarrow (\sigma_i', t')$. On input the file abstract F, data tags σ_i generated by *PO*, and *param*, *PO* with key pair (SK_a, PK_a) and *NO* with (SK_b, PK_b) interact with the *CS* as follows:
  1. *NO* computes $r' = H_2(\beta||F)$, $h' = g^{r'}$, and $h_1' = h'^{\beta}$.
  2. *PO* selects $x \in_R \mathbb{Z}_p$, computes auxiliaries

$$r = H_2(\alpha||F), \quad aux = -\frac{1}{\alpha + H_2(F||h)} - x, \quad v = u^x,$$

  and sends $Aux = (r||aux||v)\,||\mathcal{S}.\mathsf{Sig}(r||aux||v)_{ssk_a}$ to *NO*.
  3. *NO* parses Aux and recovers r, aux, v if $\mathcal{S}.\mathsf{Vrf}(r||aux||v)_{spk_a} = 1$; Otherwise, drops it and aborts.
  4. *NO* picks $x' \in_R \mathbb{Z}_p$, computes auxiliaries for tag recomputation:

$$R = r' - r, \quad aux' = \frac{1}{\beta + H_2(F||h')} - x' + aux, \quad V = vv' = vu^{x'},$$

  and sends $Aux' = (R||aux'||V)\,||\mathcal{S}.\mathsf{Sig}(R||aux'||V)_{ssk_b}$ along with the new file tag $t' = (F||n||h'||h_1')\,||\mathcal{S}.\mathsf{Sig}(F||n||h'||h_1')_{ssk_b}$ to the *CS*.
  5. The *CS* parses Aux' and obtains R, aux', V iff $\mathcal{S}.\mathsf{Vrf}(R||aux'||V)_{spk_b} = 1$, and stores t' iff $\mathcal{S}.\mathsf{Vrf}(F||n||h'||h_1')_{spk_b} = 1$.
  6. The *CS* computes the new tags belonging to *NO* as

$$\sigma_i' = \sigma_i \cdot H_1(F||i)^R \cdot (u^{m_i})^{aux'} \cdot V^{m_i} = H_1(F||i)^{r'} \cdot (u^{m_i})^{\frac{1}{\beta + H_2(F||h')}}.$$

This completes the description of the cloud auditing with efficient ownership transfer, where data tags before and after transfer are with the same structure. Such a fact enables ownership to be transferred to another cloud client by following algorithm $\mathsf{TagTrans}\ (F, SK_a, PK_a, SK_b, PK_b, \sigma_i, param) \rightarrow (\sigma_i', t')$. Note that Aux' for regenerating tags under the same file is with constant size and independent of the number of tags.

5 Correctness and Security Analysis

We give proofs of the following several theorems to demonstrate achievements of correctness, soundness, unforgeability, secure transferability and detectability defined.

Theorem 1. *The proposed protocol is correct. Concretely, if PO uploads its data honestly and the CS preserves them well, then the proof responded by the CS is valid with overwhelming probability.*

Proof. We demonstrate the correctness of the proposed protocol by proving the equality of the verification equation, since the equation only holds for valid proofs. The correctness of the equation is derived as follows:

$$
\begin{aligned}
& e\left(P\sigma, pk_a \cdot g^{H_2(F||h)}\right) \\
&= e\left(\prod_{(i,\gamma_i)\in Q} \sigma_i^{\gamma_i}, g^{\alpha} \cdot g^{H_2(F||h)}\right) \\
&= e\left(\prod_{(i,\gamma_i)\in Q} H_1(F||i)^{r\gamma_i} \cdot \prod_{(i,\gamma_i)\in Q} (u^{m_i})^{\frac{1}{\alpha+H_2(F||h)}\gamma_i}, g^{\alpha} g^{H_2(F||h)}\right) \\
&= e\left(\prod_{(i,\gamma_i)\in Q} H_1(F||i)^{\gamma_i}, h_1 \cdot h^{H_2(F||h)}\right) \cdot e\left(u^{\sum_{(i,\gamma_i)\in Q} m_i\gamma_i}, g\right) \\
&= e\left(\prod_{(i,\gamma_i)\in Q} H_1(F||i)^{\gamma_i}, h_1 \cdot h^{H_2(F||h)}\right) \cdot e\left(u^{PM}, g\right).
\end{aligned}
$$

If *PO* and *NO* calculate and output auxiliaries honestly and the *CS* stores data honestly as well as renews tags correctly, then the proof for the challenged data belonging to the new owner *NO* is valid. Similarly, the correctness is derivable from the equality of verification equation, of which the process is omitted here, since structures of verification equations are identical for the same tag structure before and after ownership transfer. □

Theorem 2. *The proposed protocol is sound. Concretely, if the k-CEIDH assumption holds, no adversary can cause the TPA to accept proofs generated from some corrupted data with non-negligible probability in the random oracle model.*

Proof. Suppose that there exists a PPT adversary $\mathcal{A}$ who can break the soundness of the protocol. We construct a simulator $\mathcal{B}$ to break the k-CEIDH assumption and collision resistance of hash functions by interacting with $\mathcal{A}$. Given as input a k-CEIDH problem instance $\left(e_1, e_2, \cdots, e_k, g, g^b, g^{\frac{1}{\alpha+e_1}}, g^{\frac{1}{\alpha+e_2}}, \cdots, g^{\frac{1}{\alpha+e_k}}\right)$, $\mathcal{B}$ controls random oracles and runs $\mathcal{A}$.

Let the file with abstract F_l and blocks $\{m_{l1}, m_{l2}, \cdots, m_{ln}\}$ along with tags $\{\sigma_{l1}, \sigma_{l2}, \cdots, \sigma_{ln}\}$ be challenged. The query set causing the challenger to abort is $Q = \{(i, \gamma_i)\}$ with $|Q| = c$, and the proof from $\mathcal{A}$ is $P_l^* = (PM_l^*, P\sigma_l^*, t_l^*)$.

Let the acceptable proof from the honest prover be

$$P_l = (PM_l = \sum_{(i,\gamma_i)\in Q} m_{li}\gamma_i,\ P\sigma_l = \prod_{(i,\gamma_i)\in Q} \sigma_{li}^{\gamma_i},\ t_l),$$

where $h_l = g^{r_l} = g^{H_2(\alpha||F_l)}$ and $h_{1l} = h_l^{\alpha}$ in t_l.

According to **Theorem** 1, it is required that the expected proof perfectly satisfies the verification equation, i.e.,

$$e\left(P\sigma_l, pk_a g^{H_2(F_l||h_l)}\right) = e\left(\prod_{(i,\gamma_i)\in Q} H_1(F_l||i)^{\gamma_i}, h_{1l} h_l^{H_2(F_l||h_l)}\right) e\left(u^{PM_l}, g\right).$$

Since $\mathcal{A}$ broke the soundness of the protocol, it is obvious that $P_l \neq P_l^*$ and that

$$e\left(P\sigma_l^*, pk_a g^{H_2(F_l||h_l^*)}\right) = e\left(\prod_{(i,\gamma_i)\in Q} H_1(F_l||i)^{\gamma_i}, h_{1l}^* h_l^{*H_2(F_l||h_l^*)}\right) e\left(u^{PM_l^*}, g\right).$$

First, we demonstrate that $h_l^* = h_l$ and $h_{1l}^* = h_{1l}$ if hash functions are collision-free. Since $h_l = g^{r_l} = g^{H_2(\alpha||F_l)}$ and $h_{1l} = h_l^\alpha = g^{r_l\alpha} = (g^\alpha)^{r_l} = {pk_a}^{r_l}$, where g, pk_a are public parameters, $h_l^* = h_l$ and $h_{1l}^* = h_{1l}$ indicate the equality of exponents r_l^* and r_l, implying a collision of H_2 occurring with negligible probability $1/p$.

Second, we show that $(PM_l^*, P\sigma_l^*) = (PM_l, P\sigma_l)$ if the assumption of k-CEIDH problem holds. The following are the details:

- *Setup.* Let $H_1 : \{0,1\}^* \to \mathbb{G}$ and $H_2 : \{0,1\}^* \to \mathbb{Z}_p$ be random oracles controlled by the simulator. $\mathcal{B}$ sets $u = g^b$, where $b \in_R \mathbb{Z}_p$.
- *H-query.* This phase is for hash queries of H_1 and H_2. Times of queries to $H_2(F_l||h_l)$ is q_{21}, to $H_1(F_l||i)$ is q_1 and to $H_2(\alpha||F_l)$ is q_{22}. Such query and response pairs are recorded in empty tables $T_{H_{21}}$, T_{H_1} and $T_{H_{22}}$ generated by $\mathcal{B}$. For queries of the ith block in file with abstract F_l, if they are searchable in tables, $\mathcal{B}$ returns these recorded values; otherwise, it executes as follows:
  - For query (F_l, h_l), $\mathcal{B}$ chooses $e_l \in_R \mathbb{Z}_p$, sets $H_2(F_l||h_l) = e_l$, records $(l, F_l||h_l, e_l, H_2(F_l||h_l), \mathcal{A})$ in $T_{H_{21}}$, and responds the hash query with $H_2(F_l||h_l)$.
  - For query (F_l, i), $\mathcal{B}$ chooses $x_{li} \in_R \mathbb{Z}_p$, sets $H_1(F_l||i) = g^{x_{li}}/u^{m_{li}}$, records $(l, i, F_l||i, x_{li}, H_1(F_l||i), \mathcal{A})$ in T_{H_1}, and responds the hash query with $H_1(F_l||i)$.
  - For query (α, F_l), $\mathcal{B}$ chooses $r_l' \in_R \mathbb{Z}_p$, sets $r_l = H_2(\alpha||F_l) = r_l' + \frac{1}{\alpha+e_l}$, records $(l, \alpha||F_l, r_l', H_2(\alpha||F_l), \mathcal{A})$ in $T_{H_{22}}$, and responds the hash query with $H_2(\alpha||F_l)$.
- *S-query.* This phase is for signature queries, which is conducted for q_s times. Let the ith block of file with abstract F_l be queried, we have

$$H_1(F_l||i) = \frac{g^{x_{li}}}{u^{m_{li}}}, \quad H_2(F_l||h_l) = e_l, \quad r_l = r_l' + \frac{1}{\alpha + e_l}.$$

  Then, $\mathcal{B}$ computes σ_{li} by

$$\sigma_{li} = \left(\frac{g^{x_{li}}}{u^{m_{li}}}\right)^{r_l' + \frac{1}{\alpha+e_l}} u^{\frac{m_{li}}{\alpha+e_l}} = \frac{g^{x_{li}r_l'} g^{x_{li}\frac{1}{\alpha+e_l}}}{u^{m_{li}r_l'} u^{m_{li}\frac{1}{\alpha+e_l}}} u^{\frac{m_{li}}{\alpha+e_l}} = \frac{g^{x_{li}r_l'} g^{\frac{1}{\alpha+e_l}x_{li}}}{u^{m_{li}r_l'}}.$$

  σ_{li} is the response to signature query on m_{li}.

- *Forgery.* Eventually, $\mathcal{A}$ forges a proof containing the corrupted i^{*th} block with $m^*_{li^*} \neq m_{li^*}$. The verification equation is rearranged as

$$e\left(P\sigma_l, g^{\alpha+e_l}\right) \stackrel{?}{=} e\left(\prod_{(i,\gamma_i)\in Q} H_1\left(F_l||i\right)^{\gamma_i}, h_l^{\alpha+e_l}\right) \cdot e\left(u^{PM_l}, g\right).$$

Divide the verification equation for $(P\sigma^*_l, PM^*_l)$ by that for $(P\sigma_l, PM_l)$, i.e.,

$$e\left(\left(P\sigma^*_l/P\sigma_l\right)^{\alpha+e_l}, g\right) = e\left(u^{PM^*_l}/u^{PM_l}, g\right).$$

Let $\Delta PM_l = PM^*_l - PM_l = \gamma_{i^*}\left(m^*_{li^*} - m_{li^*}\right)$, the division yields the solution to the k-CEIDH problem:

$$u^{\frac{1}{\alpha+e_l}} = \left(P\sigma^*_l/P\sigma_l\right)^{\frac{1}{\Delta PM_l}}.$$

This completes the simulation and solution. The correctness is shown below.

The simulation is indistinguishable from the real attack because randomnesses including b in setup, e_l, x_{li}, x_{li^*}, r'_l in hash responses and i^* in signature generation are randomly chosen and independent in the view of $\mathcal{A}$. Different randomnesses in signature and hash queries ensure the success of simulation. For the same file, such randomnesses vary in $H_1(F_l||i)$ merely with probability of $1 - \frac{q_1}{p}$, making the probability of successful simulation and useful attack be $\left(1 - \frac{q_1}{p}\right)^{q_s}$. Suppose $\mathcal{A}$ $(t, q_s, q_1, q_{21}, q_{22}, \epsilon)$-breaks the protocol. With $\mathcal{A}$'s ability, $\mathcal{B}$ solves the k-CEIDH problem with probability of $\left(1 - \frac{q_1}{p}\right)^{q_s} \cdot \epsilon \approx \epsilon$. The time cost of simulation is $T_S = O\left(q_1 + q_{21} + q_{22} + q_s\right)$. Therefore, $\mathcal{B}$ solves the k-CEIDH problem with $(t + T_S, \epsilon)$. □

Theorem 3. *The proposed protocol is unforgeable. Concretely, if the k-CEIDH assumption holds, it is computationally infeasible for all adversaries to forge provably valid tags for any data with non-negligible probability in the random oracle model.*

Proof. Suppose there is an adversary $\mathcal{A}$ who can break the unforgeability of the proposed protocol. We construct a simulator $\mathcal{B}$ to break the k-CEIDH assumption by interacting with $\mathcal{A}$. With the input of $\left(e_1, e_2, \cdots, e_k, g, g^b, g^{\frac{1}{\alpha+e_1}}, g^{\frac{1}{\alpha+e_2}}, \cdots, g^{\frac{1}{\alpha+e_k}}\right)$, the k-CEIDH adversary $\mathcal{B}$ simulates the security game for $\mathcal{A}$ below.

- *Setup.* Hash functions H_1 and H_2 are random oracles controlled by the simulator. $\mathcal{B}$ sets $u = g^b$, where $b \in_R \mathbb{Z}_p$.
- *H-query.* Hash queries are made in this phase, which are the same with that described in the proof of **Theorem** 2.
- *S-query.* Signature queries are made in this phase by $\mathcal{A}$ for q_s times. For a query on the ith block of file with abstract F_l, we have $H_1(F_l||i) = g^{x_{li}}/u^{m_{li}}$, $H_2(F_l||h_l) = e_l$, and $r_l = r'_l + \frac{1}{\alpha+e_l}$. $\mathcal{B}$ computes σ_{li} for m_{li} by

$$\sigma_{li} = \left(\frac{g^{x_{li}}}{u^{m_{li}}}\right)^{r'_l + \frac{1}{\alpha+e_l}} u^{\frac{m_{li}}{\alpha+e_l}} = \frac{g^{x_{li}r'_l} g^{\frac{1}{\alpha+e_l}x_{li}}}{u^{m_{li}r'_l}}.$$

- *Forgery.* In this phase, $\mathcal{A}$ aims to return a forged tag σ_{li^*} of the i^{*th} block m_{li^*} that has never been queried in file with abstract F_l. The corresponding hash responses from $\mathcal{B}$ are $H_1(F_l||i^*) = g^{x_{li^*}}/u^{m_{li^*}}$, $H_2(F_l||h_l) = e_l$, and $H_2(\alpha||F_l) = r_l$. Then there is

$$\sigma_{li^*} = H_1(F_l||i^*)^{r_l}(u^{m_{li^*}})^{\frac{1}{\alpha+e_l}} = \left(\frac{g^{x_{li^*}}}{u^{m_{li^*}}}\right)^{r_l}(u^{m_{li^*}})^{\frac{1}{\alpha+e_l}} = \frac{g^{r_l x_{li^*}}}{u^{r_l m_{li^*}}}u^{m_{li^*}\frac{1}{\alpha+e_l}}.$$

Now we have found the solution to the k-CEIDH problem:

$$u^{\frac{1}{\alpha+e_l}} = (\sigma_{li^*}u^{r_l m_{li^*}}/g^{r_l x_{li^*}})^{\frac{1}{m_{li^*}}}.$$

This completes the simulation and solution. The correctness is shown below.

The simulation is indistinguishable from the real attack because randomnesses including b in setup, e_l, x_{li}, x_{li^*}, r'_l in hash responses and i^* in signature generation are randomly chosen. For the same file, such randomnesses vary in $H_1(F_l||i)$ merely with probability of $1-\frac{q_1}{p}$, making the probability of successful simulation and useful attack be $\left(1-\frac{q_1}{p}\right)^{q_s}$. Suppose $\mathcal{A}$ $(t, q_s, q_1, q_{21}, q_{22}, \epsilon)$-breaks the protocol. With $\mathcal{A}$'s ability, $\mathcal{B}$ solves the k-CEIDH problem with probability of $\left(1-\frac{q_1}{p}\right)^{q_s}\cdot\epsilon \approx \epsilon$. The time cost of simulation is $T_S = O(q_1 + q_{21} + q_{22} + q_s)$. Therefore, $\mathcal{B}$ solves the k-CEIDH problem with $(t + T_S, \epsilon)$. □

Theorem 4. *The proposed protocol is with secure ownership transfer. Concretely, if the k-CEIDH assumption holds, it is computationally infeasible for any colluding adversary to forge provably valid tags on behalf of others with non-negligible probability in the random oracle model.*

Proof. First, we demonstrate that PO is secure against collusion attack launched by the CS and NO. Assume that there is a PPT adversary $\mathcal{A}$ who can break the transfer security of PO. Then, we construct a simulator $\mathcal{B}$ to break the k-CEIDH assumption by interacting with $\mathcal{A}$. Given as input a problem instance $\left(e_1, e_2, \cdots, e_k, g, g^b, g^{\frac{1}{\alpha+e_1}}, g^{\frac{1}{\alpha+e_2}}, \cdots, g^{\frac{1}{\alpha+e_k}}\right)$, $\mathcal{B}$ controls random oracles, runs $\mathcal{A}$ and works as follows.

- *Setup.* Since the CS and NO collude, $\mathcal{B}$ must be able to provide $\mathcal{A}$ with the secret key β of NO. Set $u = g^b$, where $b \in_R \mathbb{Z}_p$.
- *H-query.* Hash queries concerning data file belonging to PO are made as the same as that in **Theorem** 2. For queries of the ith block in file with abstract F_l, $\mathcal{B}$ sets $H_2(F_l||h_l) = e_l$, $H_1(F_l||i) = g^{x_{li}}/u^{m_{li}}$ and $r_l = H_2(\alpha||F_l) = r'_l + \frac{1}{\alpha+e_l}$, where $e_l, x_{li}, r'_l \in_R \mathbb{Z}_p$.
- *S-query.* Signature queries for tags belonging to PO are as the same as that in the proof of **Theorem** 2.
- *Aux-query.* Queries for auxiliaries to regenerate tags are made in this phase for q_a times. $r_l = r'_l + \frac{1}{\alpha+e_l}$ for PO is from H-query, and $aux_l = -\frac{1}{\alpha+e_l} - x_l$ and

$v_l = u^{x_l}$, where x_l is chosen by $\mathcal{B}$. NO computes $r_l^N = H_2(\beta||F_l)$, $h_l^N = g^{r_l^N}$, $h_{1l}^N = h_l^{N\beta}$, and picks $x_l^N \in_R \mathbb{Z}_p$, then there are:

$$R = r_l^N - r_l, \quad aux_l^N = \frac{1}{\beta + e_l^N} - x_l^N + aux_l, \quad V_l = v_l u^{x_l^N}.$$

- *Forgery.* Eventually, $\mathcal{A}$ forges a valid tag for a block m_{l^*i} in file with abstract F_{l^*} of PO that has not been queried. With hash responses $H_1(F_{l^*}||i) = g^{x_{l^*i}}/u^{m_{l^*i}}$ and $H_2(F_{l^*}||h_{l^*}) = e_{l^*}$, along with r_{l^*} from Aux-query, there is

$$\sigma_{l^*i} = H_1(F_{l^*}||i)^{r_{l^*}} (u^{m_{l^*i}})^{\frac{1}{\alpha+e_{l^*}}} = \frac{g^{r_{l^*}x_{l^*i}}}{u^{r_{l^*}m_{l^*i}}} u^{m_{l^*i}\frac{1}{\alpha+e_{l^*}}}.$$

Hence, we figure out the solution to the k-CEIDH problem:

$$u^{\frac{1}{\alpha+e_{l^*}}} = (\sigma_{l^*i} u^{r_{l^*}m_{l^*i}}/g^{r_{l^*}x_{l^*i}})^{\frac{1}{m_{l^*i}}}.$$

This completes the simulation and solution. The correctness is shown below.

The simulation is indistinguishable from the real attack because randomnesses including b in setup, e_l, e_{l^*},x_{li}, x_{l^*i}, r_l' in hash responses, x_l, r_{l^*} in aux queries and l^* in signature generation are randomly chosen. For different files, such randomnesses vary in $H_1(F_l||i)$ with $1-\frac{q_1}{p}$, $H_2(F_l||h_l)$ with $1-\frac{q_{21}}{p}$ and $H_2(\alpha||F_l)$ with $1-\frac{q_{22}}{p}$, making the probability of successful simulation and useful attack be $\left((1-\frac{q_1}{p})(1-\frac{q_{21}}{p})(1-\frac{q_{22}}{p})\right)^{q_s}$. Suppose $\mathcal{A}$ $(t, q_s, q_1, q_{21}, q_{22}, q_a, \epsilon)$-breaks the protocol. With $\mathcal{A}$'s ability, $\mathcal{B}$ solves the k-CEIDH problem with probability of $\left((1-\frac{q_1}{p})(1-\frac{q_{21}}{p})(1-\frac{q_{22}}{p})\right)^{q_s}\epsilon \approx \epsilon$. The time cost of simulation is $T_S = O(q_1 + q_{21} + q_{22} + q_s + q_a)$. Therefore, $\mathcal{B}$ solves the k-CEIDH problem with $(t + T_S, \epsilon)$.

Second, we show that NO is secure against collusion attack launched by the CS and PO. We assume that there is a PPT adversary $\mathcal{A}$ who can break the transfer security of NO. Then, we construct a simulator $\mathcal{B}$ to break the k-CEIDH assumption by interacting with $\mathcal{A}$. On input a k-CEIDH problem instance $\left(e_1, e_2, \cdots, e_k, g, g^b, g^{\frac{1}{\beta+e_1}}, g^{\frac{1}{\beta+e_2}}, \cdots, g^{\frac{1}{\beta+e_k}}\right)$, $\mathcal{B}$ controls random oracles, runs $\mathcal{A}$ and works as follows.

- *Setup.* Since the CS and PO collude, $\mathcal{B}$ must be able to provide $\mathcal{A}$ with the secret key α of PO. Set $u = g^b$, where $b \in_R \mathbb{Z}_p$.
- *H-query.* Hash queries are made similar to the above, only with the difference in that for query (β, F_l), $\mathcal{B}$ chooses $r_l' \in_R \mathbb{Z}_p$, sets $r_l = H_2(\beta||F_l) = r_l' + \frac{1}{\beta+e_l}$, and records $(l, \beta||F_l, r_l', H_2(\beta||F_l), \mathcal{A})$ in $T_{H_{22}}$.
- *S-query.* Signature queries for tags of NO are similar to that in the proof of **Theorem** 2, with the difference that the query is for tag belonging to NO. Hence, the tag response for m_{li} is:

$$\sigma_{li} = \left(\frac{g^{x_{li}}}{u^{m_{li}}}\right)^{r_l' + \frac{1}{\beta+e_l}} u^{\frac{m_{li}}{\beta+e_l}} = \frac{g^{x_{li}r_l'} g^{\frac{1}{\beta+e_l}x_{li}}}{u^{m_{li}r_l'}}.$$

- *Aux-query.* Queries for auxiliaries to regenerate tags are made in this phase. From NO, there is $Aux' = \left(r_l = r_l' + \frac{1}{\beta + e_l}, aux_l = -\frac{1}{\beta + e_l} - x_l, v_l = u^{x_l}\right)$, where r_l is from H-query and x_l is chosen by $\mathcal{B}$. PO computes $r_l^P = H_2(\alpha||F_l)$, $h_l^P = g^{r_l^P}$, $h_{1l}^P = h_l^{P\alpha}$, and picks $x_l^P \in_R \mathbb{Z}_p$, then there are:

$$R = r_l^P - r_l, \quad aux_l^P = \frac{1}{\alpha + e_l^N} - x_l^P + aux_l, \quad V_l = v_l u^{x_l^P}.$$

- *Forgery.* Eventually, $\mathcal{A}$ returns a valid tag for a block m_{l^*i} in file with abstract F_{l^*} belonging to NO that has not been queried. With $H_1(F_{l^*}||i) = g^{x_{l^*i}}/u^{m_{l^*i}}$, $H_2(F_{l^*}||h_{l^*}) = e_{l^*}$, and r_{l^*}, $\mathcal{B}$ computes

$$\sigma_{l^*i} = H_1(F_{l^*}||i)^{r_{l^*}}(u^{m_{l^*i}})^{\frac{1}{\beta + e_{l^*}}} = \frac{g^{r_{l^*}x_{l^*i}}}{u^{r_{l^*}m_{l^*i}}} u^{m_{l^*i}\frac{1}{\beta + e_{l^*}}}.$$

Then, the solution to k-CEIDH problem is:

$$u^{\frac{1}{\beta + e_{l^*}}} = (\sigma_{l^*i} u^{r_{l^*}m_{l^*i}} / g^{r_{l^*}x_{l^*i}})^{\frac{1}{m_{l^*i}}}.$$

This completes the simulation and solution. The correctness is analyzed similar to that in the first part, which is omitted here. Consequently, $\mathcal{B}$ solves the k-CEIDH problem with $(t + T_S, \epsilon)$. □

Theorem 5. *The proposed protocol is* $(f_c, 1 - (1 - f_c)^c)$*-detectable, where* f_c *is the ratio of corrupted data blocks and* c *is the number of challenged blocks.*

Proof. Given a fraction f_c of corrupted data after ownership transfer, the probability of corruption detectability is no less than $1 - (1 - f_c)^c$.

Table 1. Property and communication cost comparisons

Protocols		Protocol [25]	Our protocol
Outsourced computation of ownership transfer		Yes	Yes
Public verifiability		No	Yes
Thorough transferability		No	Yes
Unlimited times of ownership transfer		No	Yes
Data auditing	Challenges	$2\|q\| + log_2 n$	$c(log_2 n + \|p\|)$
	Responses	$4\|q\|$	$4\|p\|$
Ownership transfer	The transferred side	$n\|q\|$	$3\|p\|$
	The transferring side	$\|q\|$	$3\|p\|$
	The computing server side	$(n+1)\|q\|$	0

It is obvious that a block is regarded to be intact iff it is chosen from the fraction $1 - f_c$ of the whole. $(1 - f_c)^c$ is the probability that all the challenged c blocks are considered to be well-preserved, when the protocol is not detectable. Since the protocol is detectable, at least one of them should be in the f_c part, ruling out the case with the probability of $(1 - f_c)^c$. Hence, the probability that our protocol is detectable is at least $1 - (1 - f_c)^c$. □

6 Performance Analysis

We show the efficiency of our design through numerical analysis and experimental results compared with the state of the art. Following that, we discuss the performance of ownership transfer among multiple cloud clients.

6.1 Efficiency Analysis

Simulation experiments are conducted on the Ubuntu 12.04.5 (1 GB memory) VMware 10.0 in a laptop running Windows 10 with Intel(R) Core(TM) i5-8250U @ 1.6 GHZ and 8 GB RAM. The codes are written using C programming language with GMP Library (GMP-6.1.2) and PBC Library (pbc-0.5.14), of which the data results are used to draw figures in MATLAB 2019.

The property and communication cost comparisons are presented in Table 1, where n is the number of data blocks. Let the number of sectors in [25] set to be 1, and q equals to p appearing in our protocol, which is the prime order of $\mathbb{G}$. According to Table 1, both protocols enable computation of transfer outsourced, while we provide thorough transferability rather than only the partial one because the secret key of the old owner no longer remains in tags for new owners. In addition, we achieve public verifiability of data and unlimited times of ownership transfer among cloud clients. As for the communication overhead, though [25] costs less in data auditing, it trades for more computations in proof generation in turn. In ownership transfer, the communication cost in our protocol is with a smaller and constant size.

Table 2. Computational overhead comparison

Protocols	Protocol [25]	Our protocol
Tag generation	$(2n+2)E + 2nM$	$(2n+3)E + nM + I$
Proof generation	$cE + (4c-1)M$	$cE + (2c-1)M$
Proof verification	$4P + (c+1)E + 4cM$	$3P + (c+3)E + (c+1)M$
Aux generation on transferred side	$nE + nM + nI$	$E + I$
Aux' generation on transferring side	$E + I$	$E + M + I$
Tag recomputation on server side	$E + 2nM$	$4nE + 3nM$

The computational overhead analysis is shown in Table 2, where c is the number of challenged blocks. For the sake of simplicity, we denote modular exponentiation as E, point multiplication as M, bilinear pairing as P and inverse as I. We ignore hash functions in comparison, for the fact that its cost is negligible when compared with that of the pre-mentioned operations. According to Table 2, the computational cost of [25] in data auditing concerning the first three entries is larger than ours. The experimental results are shown in Fig. 1, where n ranges from 500 to 5000 in Fig. 1(a), and n is set to be 1000 in Fig. 1(b)(c). Costs in ownership transfer are described in Fig. 2 with n from 500 to 5000. According

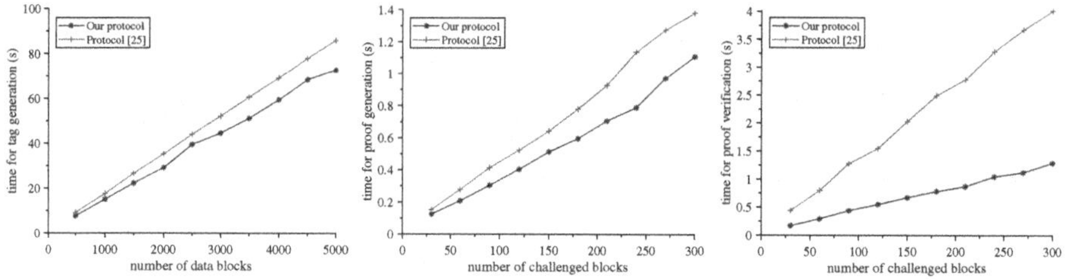

(a) tag generation (b) proof generation (c) proof verification

Fig. 1. Time cost of data outsourcing and auditing

to Fig. 2, though computation on the transferring side is more expensive in our protocol, the cost in [25] that could not be outsourced to the computing server is significantly higher than that of a constant size in ours. Besides, the total cost of ownership transfer is lower in our design. Note that, the more computational overhead concerning the last entry in our protocol seems like nothing, since the server is equipped with professional computing power.

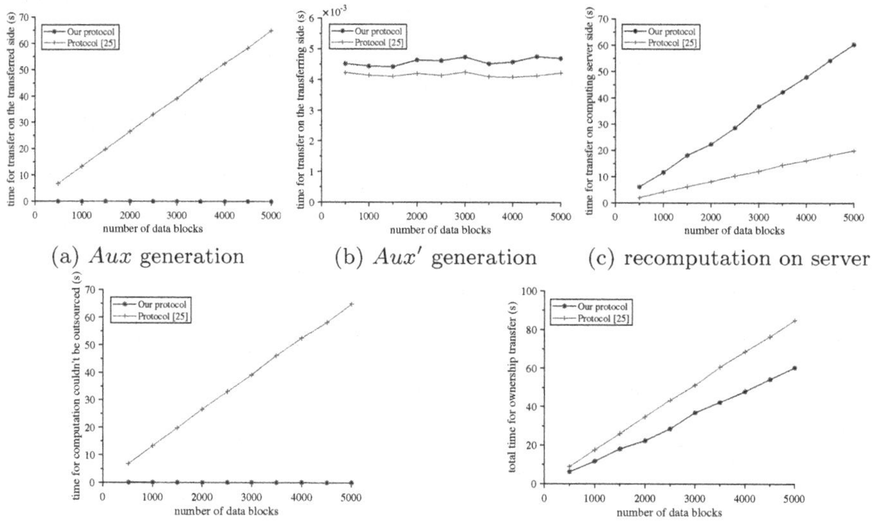

(a) Aux generation (b) Aux' generation (c) recomputation on server

(d) computations could not be outsourced (e) the total cost for transfer

Fig. 2. Time cost of data ownership transfer

6.2 Ownership Transfer Discussion

Recall that in the core part of Sect. 4, we have introduced the ownership transfer of cloud data from PO to NO. Due to the well-designed novel data tags in our protocol, the regenerated tag for NO is with the same structure with that for PO, except that the secret key embedded in now belongs to NO rather than PO. With such a tag structure, the transferred data looks just like the data originally possessed by NO and can be transferred continuously to other cloud clients. More generally speaking, our protocol supports ownership transfer of any piece of data among multiple clients for numerous times.

7 Conclusion

We propose a secure auditing protocol with efficient ownership transfer for cloud data, satisfying verifiability and transferability simultaneously. Compared with the state of the art, the communication cost produced by ownership transfer is constant rather than dependent of the number of tags, and the computational overhead during transfer on both transfer parties is constant as well. In addition, the protocol satisfies properties of correctness, soundness, unforgeability and detectability, which also protects the untransferred data from collusion attacks when ownership transfer occurs.

Acknowledgments. This work is supported by the National Nature Science Foundation of China (No. 61960206014) and the National Cryptography Development Fund (No. MMJJ20180110).

References

1. Ateniese, G., et al.: Provable data possession at untrusted stores. In: Proceedings of the 14th ACM Conference on Computer and Communications Security, pp. 598–609. ACM (2007)
2. Ateniese, G., Di Pietro, R., Mancini, L.V., Tsudik, G.: Scalable and efficient provable data possession. In: Proceedings of the 4th International Conference on Security and Privacy in Communication Networks, pp. 1–10. ACM (2008)
3. Chen, X., Lee, B., Kim, K.: Receipt-free electronic auction schemes using homomorphic encryption. In: Lim, J.-I., Lee, D.-H. (eds.) ICISC 2003. LNCS, vol. 2971, pp. 259–273. Springer, Heidelberg (2004). https://doi.org/10.1007/978-3-540-24691-6_20
4. Chen, X., Jin, L., Jian, W., Ma, J., Lou, W.: Verifiable computation over large database with incremental updates. In: Proceedings of European Symposium on Research in Computer Security, pp. 148–162 (2014)
5. Chen, X., Li, J., Huang, X., Ma, J., Lou, W.: New publicly verifiable databases with efficient updates. IEEE Trans. Dependable Secure Comput. **12**(5), 546–556 (2014)
6. Dillon, T., Wu, C., Chang, E.: Cloud computing: issues and challenges. In: Proceedings of the 24th IEEE International Conference on Advanced Information Networking and Applications, pp. 27–33. IEEE (2010)
7. Erway, C., Küpçü, A., Papamanthou, C., Tamassia, R.: Dynamic provable data possession. In: Proceedings of the 16th ACM Conference on Computer and Communications Security, pp. 213–222. ACM (2009)

8. Fu, A., Yu, S., Zhang, Y., Wang, H., Huang, C.: NPP: a new privacy-aware public auditing scheme for cloud data sharing with group users. IEEE Trans. Big Data (2017)
9. Jin, H., Jiang, H., Zhou, K.: Dynamic and public auditing with fair arbitration for cloud data. IEEE Trans. Cloud Comput. **6**(3), 680–693 (2016)
10. Juels, A., Kaliski Jr., B.S.: PORs: proofs of retrievability for large files. In: Proceedings of the 14th ACM Conference on Computer and Communications Security, pp. 584–597. ACM (2007)
11. Kamara, S., Lauter, K.: Cryptographic cloud storage. In: Sion, R., et al. (eds.) FC 2010. LNCS, vol. 6054, pp. 136–149. Springer, Heidelberg (2010). https://doi.org/10.1007/978-3-642-14992-4_13
12. Li, J., Tan, X., Chen, X., Wong, D.S., Xhafa, F.: OPoR: enabling proof of retrievability in cloud computing with resource-constrained devices. IEEE Trans. Cloud Comput. **3**(2), 195–205 (2014)
13. Li, J., Li, J., Xie, D., Cai, Z.: Secure auditing and deduplicating data in cloud. IEEE Trans. Comput. **65**(8), 2386–2396 (2015)
14. Li, Y., Yu, Y., Min, G., Susilo, W., Ni, J., Choo, K.K.R.: Fuzzy identity-based data integrity auditing for reliable cloud storage systems. IEEE Trans. Dependable Secure Comput. **16**(1), 72–83 (2017)
15. Luo, Y., Xu, M., Fu, S., Wang, D., Deng, J.: Efficient integrity auditing for shared data in the cloud with secure user revocation. In: Proceedings of 2015 IEEE Trustcom/BigDataSE/ISPA, vol. 1, pp. 434–442. IEEE (2015)
16. Mell, P., Grance, T., et al.: The NIST definition of cloud computing (2011)
17. Ramachandra, G., Iftikhar, M., Khan, F.A.: A comprehensive survey on security in cloud computing. Procedia Comput. Sci. **110**, 465–472 (2017)
18. Rittinghouse, J.W., Ransome, J.F.: Cloud Computing: Implementation, Management, and Security. CRC Press, Boca Raton (2017)
19. Shacham, H., Waters, B.: Compact proofs of retrievability. In: Pieprzyk, J. (ed.) ASIACRYPT 2008. LNCS, vol. 5350, pp. 90–107. Springer, Heidelberg (2008). https://doi.org/10.1007/978-3-540-89255-7_7
20. Shen, J., Shen, J., Chen, X., Huang, X., Susilo, W.: An efficient public auditing protocol with novel dynamic structure for cloud data. IEEE Trans. Inf. Forensics Secur. **12**(10), 2402–2415 (2017)
21. Shen, J., Zhou, T., Chen, X., Li, J., Susilo, W.: Anonymous and traceable group data sharing in cloud computing. IEEE Trans. Inf. Forensics Secur. **13**(4), 912–925 (2017)
22. Tian, H., et al.: Dynamic-hash-table based public auditing for secure cloud storage. IEEE Trans. Serv. Comput. **10**(5), 701–714 (2015)
23. Wang, B., Li, B., Li, H.: Panda: public auditing for shared data with efficient user revocation in the cloud. IEEE Trans. Serv. Comput. **8**(1), 92–106 (2013)
24. Wang, C., Chow, S.S., Wang, Q., Ren, K., Lou, W.: Privacy-preserving public auditing for secure cloud storage. IEEE Trans. Comput. **62**(2), 362–375 (2011)
25. Wang, H., He, D., Fu, A., Li, Q., Wang, Q.: Provable data possession with outsourced data transfer. IEEE Trans. Serv. Comput. (2019)
26. Wang, H., He, D., Tang, S.: Identity-based proxy-oriented data uploading and remote data integrity checking in public cloud. IEEE Trans. Inf. Forensics Secur. **11**(6), 1165–1176 (2016)
27. Wang, J., Chen, X., Li, J., Kluczniak, K., Kutylowski, M.: TrDup: enhancing secure data deduplication with user traceability in cloud computing. Int. J. Web Grid Serv. **13**(3), 270–289 (2017)

28. Wang, Q., Wang, C., Ren, K., Lou, W., Li, J.: Enabling public auditability and data dynamics for storage security in cloud computing. IEEE Trans. Parallel Distrib. Syst. **22**(5), 847–859 (2010)
29. Wang, Y., Wu, Q., Bo, Q., Shi, W., Deng, R.H., Hu, J.: Identity-based data outsourcing with comprehensive auditing in clouds. IEEE Trans. Inf. Forensics Secur. **12**(4), 940–952 (2017)
30. Worku, S.G., Xu, C., Zhao, J., He, X.: Secure and efficient privacy-preserving public auditing scheme for cloud storage. Comput. Electr. Eng. **40**(5), 1703–1713 (2014)
31. Yang, G., Yu, J., Shen, W., Su, Q., Fu, Z., Hao, R.: Enabling public auditing for shared data in cloud storage supporting identity privacy and traceability. J. Syst. Softw. **113**, 130–139 (2016)
32. Yu, J., Ren, K., Wang, C.: Enabling cloud storage auditing with verifiable outsourcing of key updates. IEEE Trans. Inf. Forensics Secur. **11**(6), 1362–1375 (2016)
33. Yu, J., Ren, K., Wang, C., Varadharajan, V.: Enabling cloud storage auditing with key-exposure resistance. IEEE Trans. Inf. Forensics Secur. **10**(6), 1167–1179 (2015)
34. Yu, J., Wang, H.: Strong key-exposure resilient auditing for secure cloud storage. IEEE Trans. Inf. Forensics Secur. **12**(8), 1931–1940 (2017)
35. Yuan, H., Chen, X., Jiang, T., Zhang, X., Yan, Z., Xiang, Y.: DedupDUM: secure and scalable data deduplication with dynamic user management. Inf. Sci. **456**, 159–173 (2018)
36. Zhang, Y., Yu, J., Hao, R., Wang, C., Ren, K.: Enabling efficient user revocation in identity-based cloud storage auditing for shared big data. IEEE Trans. Dependable Secure Comput. **17**(3), 608–619 (2020)

Privacy

Encrypt-to-Self: Securely Outsourcing Storage

Jeroen Pijnenburg[1(✉)] and Bertram Poettering[2]

[1] Royal Holloway, University of London, Egham, UK
jeroen.pijnenburg.2017@rhul.ac.uk
[2] IBM Research – Zurich, Rüschlikon, Switzerland
poe@zurich.ibm.com

Abstract. We put forward a symmetric encryption primitive tailored towards a specific application: outsourced storage. The setting assumes a memory-bounded computing device that inflates the amount of volatile or permanent memory available to it by letting other (untrusted) devices hold encryptions of information that they return on request. For instance, web servers typically hold for each of the client connections they manage a multitude of data, ranging from user preferences to technical information like database credentials. If the amount of data per session is considerable, busy servers sooner or later run out of memory. One admissible solution to this is to let the server *encrypt* the session data *to itself* and to let the client store the ciphertext, with the agreement that the client reproduce the ciphertext in each subsequent request (e.g., via a cookie) so that the session data can be recovered when required.

In this article we develop the cryptographic mechanism that should be used to achieve confidential and authentic data storage in the encrypt-to-self setting, i.e., where encryptor and decryptor coincide and constitute the only entity holding keys. We argue that standard authenticated encryption represents only a suboptimal solution for preserving confidentiality, as much as message authentication codes are suboptimal for preserving authenticity. The crucial observation is that such schemes instantaneously give up on *all* security promises in the moment the key is compromised. In contrast, data protected with our new primitive remains fully integrity protected and unmalleable. In the course of this paper we develop a formal model for encrypt-to-self systems, show that it solves the outsourced storage problem, propose surprisingly efficient provably secure constructions, and report on our implementations.

1 Introduction

We explore techniques that enable a computing device to securely outsource the storage of data. We start with motivating this area of research by describing three application scenarios where outsourcing storage might prove crucial.

The full version of this article is available at https://ia.cr/2020/847.

L. Chen et al. (Eds.): ESORICS 2020, LNCS 12308, pp. 635–654, 2020.
https://doi.org/10.1007/978-3-030-58951-6_31

WEB SERVER. We come back to the example considered in the abstract, giving more details. While it is difficult to make general statements about the setup of a web server back-end, it is fair to say that the processing of HTTP requests routinely also includes extracting a session identifier from the HTTP header and fetching basic session-related information (e.g., the user's password, the date and time of the last login, the number of failed login attempts, but also other kinds of data not related to security) from a possibly remote SQL database. To avoid the inherent bottleneck induced by the transmission and processing of the database query, such data can be cached on the web server, the limits of this depending only on the amount of available working memory (RAM). For some types of web applications and a large number of web sessions served simultaneously, these memory-imposed limits might represent a serious restriction to efficiency. This article scouts techniques that allow the web server to securely outsource the storage of session information to the (untrusted) web clients.

HARDWARE SECURITY MODULE. An HSM is a computing device that performs cryptographic and other security-related operations on behalf of the owning user. While such devices are internally built from off-the-shelf CPUs and memory chips, a key concept of HSMs is that they are specially encapsulated to protect them against physical attacks, including various kinds of side channel analysis. One consequence of this tamper-proof shielding is that the memory capacity of an HSM can never be physically extended—unlike it would be the case for desktop computers—so that the amount of available working memory might constitute a relevant obstacle when the HSM is deployed in applications with requirements that increase over time (e.g., due to a growing user base). This article scouts techniques that allow the HSM to securely outsource the storage of any kind of valuable information to the (untrusted) embedding host system.

SMARTCARD. A smartcard, most prominently recognized in the form of a payment card or a mobile phone security token, is effectively a tiny computing device. While fairly potent configurations exist (with 32-bit CPUs and a couple of 100 KBs of memory), as the costs associated with producing a smartcard scales roughly linearly with the amount of implemented physical memory, in order to be cost effective, mass-produced cards tend to come with only a small amount of memory. This article scouts techniques that allow smartcards to securely outsource the storage of valuable information to the infrastructure they connect to, e.g., a banking or mobile phone backbone, or a smartphone.

TRUSTED PLATFORM MODULE. A TPM is a discrete security chip that is embedded into virtually all laptops and desktop PCs produced in the past decade. A TPM supports its host system by offering trusted cryptographic services and is typically relied upon by boot loaders and operating systems. TPMs are located conceptually between HSMs and smartcards, and as much as these they benefit from a secure option to outsource storage.

Outsourced Storage Based on Symmetric Cryptography. If a computing device has access to some kind of external storage facility (a memory chip wired to it, a connected hard drive, cloud storage, etc.), then, intuitively, it can virtually extend the amount of memory available to it by outsourcing storage, i.e., by serializing data objects and communicating them to the storage facility which will reproduce them on request. In this article we focus on the case where neither the external storage facility nor the connection to it is considered trustworthy. More concretely, we assume that all infrastructure outside of the computing device itself is under control of an adversary that aims at reading or changing the data that is to be externally stored.[1] As a first approximation one might conclude that standard tools from the domain of symmetric encryption are sufficient to achieve security in this setting. Consider for instance the following approach based on authenticated encryption (AE, [13]): The computing device samples a fresh symmetric key; whenever it wants to store internal data on the outsourced storage, it encrypts and authenticates the data by invoking the AE encryption algorithm with its key and hands the resulting ciphertext over to the storage facility; to retrieve the data, it requests a copy of the ciphertext, and decrypts and verifies it. While this simple solution requires further tweaking to thwart replay attacks, as long as the AE key remains private it can be used to protect confidentiality and integrity as expected.

Our Contribution: Secure Outsourced Storage w/Key Leakage. While we confirm that standard cryptographic methods will securely solve the storage outsourcing problem if the used key material remains private, we argue that satisfactory solutions should go a step further by providing as much security as possible even if the latter assumption (that keys remain private) is not met. Indeed, different attacks against practical systems that lead to partial or full memory leakage continue to regularly emerge (including different types of side channel analysis against embedded systems, cold-boot attacks against memory chips, Meltdown/Spectre-like attacks against modern CPUs, etc.), and it is commonly understood that the corruption model considered for cryptographic primitives should always be as strong as possible and affordable. For two-party symmetric encryption (e.g., AE) this strongest model necessarily excludes any type of user corruption[2] as the keys of both parties are identical: Once any party is corrupted, any past or future ciphertext can be decrypted and ciphertexts can be forged for any message, i.e., no form of confidentiality or authenticity remains. We point out, however, that for outsourced storage a stronger corruption model is both feasible and preferable. Clearly, like in the AE case, if the adversary obtains a copy of the used key material then all confidentiality guarantees are lost (the adversary can decrypt what the device can decrypt, that is, everything), but a

[1] Certainly, the storage device can always decide to "fail" by not returning any data previously stored into it, leading to an attack on the *availability* of the computing device. We hence consider environments where this is either not a problem or where such an attack cannot be prevented anyway (independently of the storage technique). Note that this assumption holds for our three motivating scenarios.

[2] We use the terms 'key leakage', 'user corruption', and 'state corruption' synonymously.

similar reasoning with respect to integrity protection cannot be made. To see this, consider the encrypt-then-hash (EtH) solution where the computing device encrypts the outsourced data as described above, but in addition to having the ciphertext stored externally it internally registers a hash of it (computed with, say, SHA256). When the device decides to recover externally stored data, it requests a copy of the ciphertext, recomputes its hash value, and decrypts only if the hash value is consistent with the internally registered value. Note that even if the device is corrupted and its keys became public, all successfully decrypted ciphertexts are necessarily authentic.

The example just given shows that while no solution for secure storage outsourcing can do much about protecting data confidentiality against key leakage attacks, solutions can *fully* protect the integrity of the stored data in any case. Naive AE-based schemes do not provide this type of security, and the contribution of our work is to fill this gap and to explore corresponding constructions. Precisely, this article provides the following: (1) We identify the new *encrypt-to-self* (ETS) primitive as the right cryptographic tool to solve the outsourced storage problem and formalize its syntax and security properties. (2) We formalize notions more directly related to the outsourced storage problem and provably confirm that secure solutions based on ETS are indeed immediate. (3) We design provably secure constructions of ETS from established cryptographic primitives.[3] (4) We develop open-source implementations of our constructions that are optimized with respect to security and efficiency.

Related Work. While we are not aware of any former systematic treatment of the encrypt-to-self (ETS) primitive, a number of similar primitives or ad hoc constructions partially overlap with our results. We discuss these in the following, but emphasize that none of them provides general solutions to the ETS problem.

MEMORY ENCRYPTION IN MODERN CPUS. Recent desktop and server CPUs offer dedicated infrastructure for memory encryption, with the main applications in cloud computing and Trusted Execution Environments (TEEs). Prominent TEE examples include Intel SGX and ARM TrustZone in which every memory access of the processes that are executed within a TEE (aka 'enclave') is conducted through a memory encryption engine (MEE). This effectively implements outsourced data storage, but with quite different access rules and patterns than in the ETS case. While we consider the (stateless) encryption of a message to a ciphertext and then a decryption of a ciphertext back to a message, MEEs are stateful systems that consider the protected physical memory area a single ciphertext that is constantly locally modified with each write operation [8].

PASSWORD MANAGERS. A password manager can be seen as a database that stores security credentials in an encrypted form and requires e.g., a master password to be unlocked. Also this can be seen as an ETS instance, but the cryp-

[3] The above encrypt-then-hash (EtH) solution is secure in our models but requires two passes over the data. Our solutions are more efficient, getting along with just one pass.

tographic design of password managers has a different focus than general outsourced storage. More concretely, the central challenge solved by good password managers is the password-based key derivation, which typically involves invoking a time-expensive derivation function like PBKDF2 [9] or a memory-hard derivation function like ARGON2 [5]. Password-based key derivation is not considered in our treatment of the ETS primitive (we instead assume uniform keys).

ENCRYPTMENT. A symmetric encryption option that recently emerged as a proposal to protect messages in instant messaging is Encryptment [7]. Its features go beyond regular authenticated encryption in that the tags contained in ciphertexts act as (cryptographically strongly binding) commitments to the encoded messages. This committing feature was deemed helpful for the public resolution of cyber harassment cases by allowing affected parties to appeal to a judging authority by opening their ciphertexts by releasing their keys. On first sight this has nothing to do with our ETS setting (in which only one party holds a key, this key would never be deliberately shared, and a necessity of provably releasing message contents to anybody else is not considered). Interestingly, however, our constructions of ETS are very similar to those of [7]. The intuitive reason for this is that the ETS setting requires that ciphertexts remain unforgeable under key leakage, which somewhat aligns with the committing property of encryptment that is required to survive disclosing keys to a judge. Ultimately, however, the applications and thus security models of ETS and encryptment differ, and our constructions are actually more efficient than those in [7].[4]

Technical Approach. In addition to formalizing the security of the encrypt-to-self (ETS) primitive, in the course of this article we also propose efficient provably-secure constructions from standardized building blocks. As discussed above, the authenticity promises of ETS shall withstand adversaries that have knowledge of the key material. In this setting one cannot hope that standard secret-key authentication building blocks like MACs or universal hash functions will be of help, as generically they lose all security when the key is leaked. We instead employ, as they manifest *unkeyed* authentication primitives, cryptographic hash functions like SHA256. A first candidate construction, already hinted at above, would be the encrypt-then-hash (EtH) approach where the message is first encrypted (using any secret key scheme, e.g., AES-CTR) and the ciphertext is then hashed. Our constructions are more efficient than this by exploiting the structure of Merkle–Damgård (MD) hash functions and dual-use leveraging on the properties of their inner building block: the compression function (CF). Intuitively, for authentication we build on the collision resistance of the CF, and for confidentiality we build on a PRF-like property of the CF. More precisely, our message schedule for the CF is such that each invocation provides

[4] This is the case for at least two reasons: (1) The ETS primitive does not need to be committing to the key, which is the case for encryptment. (2) Our message padding is more sophisticated than that of [7] and does not require the processing of a length field.

both confidentiality *and* integrity for the processed block. This effectively halves the computational costs in comparison to the EtH approach.

We believe that a cryptographic analysis is not complete without also implementing the construction under consideration. This is because only implementing a scheme will enforce making conscious decisions about all its details and building blocks, and these decisions may crucially affect the obtained security and efficiency. We thus realized three ready-to-use instances of the ETS primitive, based on the CFs of the top performing hash functions SHA256, SHA512, and BLAKE2. In fact, observations from implementing the schemes led to considerable feedback to the theoretical design which was updated correspondingly. One example for this is connected to memory alignment: Computations on modern CPUs experience noticeable efficiency penalties if memory accesses are not aligned to specific boundaries. Our constructions reflect this at two different levels: at the register level and at the cache level (64 bit alignment for register-oriented operations, and 256 bit alignment for bulk memory transfers[5]).

2 Preliminaries

NOTATION. All algorithms considered in this article may be randomized. We let $\mathbb{N} = \{0, 1, \ldots\}$ and $\mathbb{N}^+ = \{1, 2, \ldots\}$. For the Boolean constants True and False we either write T and F, respectively, or 1 and 0, respectively, depending on the context. An alphabet Σ is any finite set of symbols or characters. We denote with Σ^n the set of strings of length n and with $\Sigma^{\leq n}$ the strings of length up to (and including) n. In the practical parts of this article we assume that $|\Sigma| = 256$, i.e., that all strings are byte strings. We denote string concatenation with $\|$. If *var* is a string variable and *exp* evaluates to a string, we write $var \overset{\|}{\leftarrow} exp$ shorthand for $var \leftarrow var \| exp$. Further, if *exp* evaluates to a string, we write $var \| var' \leftarrow_n exp$ to denote splitting *exp* such that we assign the first n characters from *exp* to *var* and assign the remainder to *var'*. When we do not need the remainder, we write $var \leftarrow_n exp$ shorthand for $var \| dummy \leftarrow_n exp$ and discard *dummy*. In pseudocode, if S is a finite set, expression $\$(S)$ stands for picking an element of S uniformly at random. Associative arrays implement the 'dictionary' data structure: Once the instruction $A[\cdot] \leftarrow exp$ initialized all items of array A to the default value *exp*, with $A[idx] \leftarrow exp$ and $var \leftarrow A[idx]$ individual items indexed by expression *idx* can be updated or extracted.

2.1 Security Games

Security games are parameterized by an adversary, and consist of a main game body plus zero or more oracle specifications. The execution of a game starts with the main game body and terminates when a 'Stop with *exp*' instruction is reached, where the value of expression *exp* is taken as the outcome of the game. The adversary can query all oracles specified by the game, in any order

[5] The value 256 stems from the size of the cache lines of 1st level cache.

and any number of times. If the outcome of a game G is Boolean, we write $\Pr[G(\mathcal{A})]$ for the probability that an execution of G with adversary $\mathcal{A}$ results in True, where the probability is over the random coins drawn by the game and the adversary. We define macros for specific combinations of game-ending instructions: We write 'Win' for 'Stop with T' and 'Lose' for 'Stop with F', and further 'Reward *cond*' for 'If *cond*: Win', 'Promise *cond*' for 'If $\neg$*cond*: Win', 'Require *cond*' for 'If $\neg$*cond*: Lose'. These macros emphasize the specific semantics of game termination conditions. For instance, a game may terminate with 'Reward *cond*' in cases where the adversary arranged for a situation—indicated by *cond* resolving to True—that should be awarded a win (e.g., the crafting of a forgery in an authenticity game).

2.2 Handling of Algorithm Failures

Regarding the algorithms of cryptographic schemes, we assume that *any* such algorithm can fail. Here, by failure we mean that an algorithm doesn't generate output according to its syntax specification, but instead outputs some kind of error indicator (e.g., an AE decryption algorithm that rejects an unauthentic ciphertext or a randomized signature algorithm that doesn't have sufficiently many random bits to its disposal). Instead of encoding this explicitly in syntactical constraints which would clutter the notation, we assume that if an algorithm invokes another algorithm as a subroutine, and the latter fails, then also the former immediately fails.[6] We assume the same for game oracles: If an invoked scheme algorithm fails, then the oracle immediately aborts as well. Further, we assume that the adversary learns about this failure, i.e., the oracle will return the error indicator when it aborts. Note that this implies that if a scheme's algorithms leak vital information through error messages, then the scheme will not be secure in our models. (That is, our models are particularly robust.) We believe that our way to handle errors implicitly rather than explicitly contributes to obtaining definitions with clean and clear semantics.

2.3 Memory Alignment

For n a power of 2, we say an address of computer memory is n-byte aligned if it is a multiple of n bytes. We further say that a piece of data is n-byte aligned if the address of its first byte is n-byte aligned. A modern CPU accesses a single (aligned) word in memory at a time. Therefore, the CPU performs reads and writes to memory most efficiently when the data is aligned. For example, on a 64-bit machine, 8 bytes of data can be read or written with a single memory

[6] This approach to handling algorithm failures is taken from [12] and borrows from how modern programming languages handle 'exceptions', where any algorithm can raise (or 'throw') an exception, and if the caller does not explicitly 'catch' it, the caller is terminated as well and the exception is passed on to the next level. See Wikipedia: Exception_handling_syntax for exception handling syntaxes of many different programming languages.

access if the first byte lies on an 8-byte boundary. However, if the data does not lie within one word in memory, the processor would need to access two memory words, which is considerably less efficient. Our scheme algorithms are designed such that when they need to move around data, they exclusively do this for aligned addresses. In practice, the preferred alignment value depends on the hardware used, so for generality in this article we refer to it abstractly as the memory alignment value mav. (A typical value would be mav = 8.)

2.4 Tweaking the Compression Functions of Hash Functions

The main NIST hash functions of the SHA2 family (FIPS 180-4, [10]) accomplish their task of hashing a message into a short string by strictly following the Merkle–Damgård framework: All inputs to their core building block—the compression function—are either directly taken from the message or from the chaining state. It has been recognized, however, that options to further contextualize or domain-separate the inputs of compression functions can be of advantage. Indeed, compression functions that are designed according to the alternative, more recent HAIFA framework [4] have a number of additional inputs, for instance an explicit salt input, that allow for weaving some extra bits of context information into the bulk hash operations. A concrete example for this is the compression function of the popular BLAKE2 hash function ([2,14], a HAIFA design), which takes as an additional input a Boolean finalization flag that is to be set specifically when processing the very last (padded) block of a hash computation. The idea behind making the last invocation "special" is that this effectively thwarts length extension attacks: While conducting extension attacks against the SHA2 hash functions, where the compression functions do not natively support any such marking mechanism, is quite immediate,[7] similar attacks against BLAKE2 are impossible [6]. We note that, generally speaking, an ad hoc way of augmenting the input of a compression function by an additional small number of bits is to XOR predefined constants into the hashing state (e.g., before or while the compression function is executed), with the choice of constants depending on the added bits. For instance, if the finalization flag is set, the BLAKE2 compression function will flip all bits of one of its inputs, but beyond that operate as normal.

While textbook SHA2 does not support contextualizing compression function invocations via additional inputs, we observe that NIST, in order to solve an emerging domain-separation problem in the definition of their FIPS 180-4 standard, employed ad hoc modifications of some SHA2 functions that can be seen as (implicitly) retrofitting a one-bit additional input into the compression function. Concretely, the SHA512/t functions [10], that intuitively represent plain SHA512 truncated to $0 < t < 512$ bits, are carefully designed such that for any $t_1 \neq t_2$ the functions SHA512/t_1 and SHA512/t_2 are independent of each

[7] For instance, an adversary who doesn't know a value x but instead the values $H(x)$ and y, can compute $H(x \parallel y)$ by just continuing the iterative MD computation from chain value $H(x)$ on. Note this does not require inverting the compression function.

other.[8] The separation of the individual SHA512/t versions works as follows [10, Sect. 5.3.6]: First compute the SHA512 hash value of the string `"SHA512/xxx"` (where placeholder `xxx` is replaced by the decimal encoding of t), then XOR the byte value `0xa5` (binary: `0b10100101`) into every byte of the resulting chain state, then continue with regular SHA512 steps from that state on, truncating the final hash value to t bits. While the XORing step is ad hoc, it arguably represents a fairly robust domain separation method for SHA2.

Our constructions of the encrypt-to-self primitive rely on compression functions that are *tweaked* with a single bit, that is, that support one bit as an additional input. When we implement this based on SHA2 compression functions, we employ precisely the mechanism scouted by NIST: When the additional tweak bit is set, we XOR constant `0xa5` into all state bytes and continue operation as normal. Our BLAKE2 based construction, on the other hand, uses the already existing finalization bit.

3 Foundations of Encrypt-to-Self

The overall goal of this article is to provide a secure solution for outsourced storage. We identified the novel encrypt-to-self (ETS) primitive, which provides one-time secure encryption with authenticity guarantees that hold beyond key compromise, as the right tool to construct outsourced storage.[9] In this section we first formalize and study ETS, then formalize outsourced storage, and finally show how the former immediately implies the latter. This allows us to leave the outsourced storage topic aside in the remaining part of the paper and lets us instead fully focus on constructing and implementing ETS.

3.1 Syntax and Security of ETS

ETS consists of an encryption and a decryption algorithm, where the former translates a message to a *binding tag* and a ciphertext, and the latter recovers the message from the tag-ciphertext pair. For versatility the two operations further support the processing of an associated-data input [13] which has to be identical for a successful decryption.

The task of the binding tag is to prevent forgery attacks: A user that holds an authentic copy of the binding tag will never accept any ciphertext they did not generate themselves, even if all their secrets become public. Note that while standard authenticated encryption (AE) does not provide this type of authentication, the encrypt-then-hash construction suggested in Sect. 1 does. In Sect. 4 we provide a considerably more efficient construction that uses a hash function's compression function as its core building block. Here, we define the generic syntax of ETS and formalize its security requirements.

[8] In particular, for instance, SHA512/128(`"a"`) is not a prefix of SHA512/192(`"a"`).

[9] While ETS is novel, note that prior work explored the quite similar Encryptment primitive [7]. Encryptment is stronger than ETS, and less efficient to construct.

Game SAFE$(ad, m, \mathcal{A})$	**Game** INT$(ad, m, \mathcal{A})$	**Game** IND$^b(ad, m^0, m^1, \mathcal{A})$
00 $k \leftarrow \$(\mathcal{K})$	09 $k \leftarrow \$(\mathcal{K})$	17 $k \leftarrow \$(\mathcal{K})$
01 $(bt, c) \leftarrow \mathrm{enc}(k, ad, m)$	10 $(bt, c) \leftarrow \mathrm{enc}(k, ad, m)$	18 Require $m^0 \equiv m^1$
02 Invoke $\mathcal{A}(k, ad, m, bt, c)$	11 Invoke $\mathcal{A}(k, ad, m, bt, c)$	19 $(bt, c) \leftarrow \mathrm{enc}(k, ad, m^b)$
03 Lose	12 Lose	20 $b' \leftarrow \mathcal{A}(ad, m^0, m^1, bt, c)$
		21 Stop with b'
Oracle Dec(ad', c')	**Oracle** Dec(ad', c')	**Oracle** Dec(ad', c')
04 $m' \leftarrow \mathrm{dec}(k, bt, ad', c')$	13 $m' \leftarrow \mathrm{dec}(k, bt, ad', c')$	22 $m' \leftarrow \mathrm{dec}(k, bt, ad', c')$
05 If $(ad', c') = (ad, c)$:	14 Reward $(ad', c') \neq (ad, c)$	23 If $(ad', c') = (ad, c)$:
06 Promise $m' = m$	15 $m' \leftarrow \perp$	24 $m' \leftarrow \perp$
07 $m' \leftarrow \perp$	16 Return m'	25 Return m'
08 Return m'		

Fig. 1. Games for ETS. For the values ad', c' provided by the adversary we require that $ad' \in \mathcal{AD}, c' \in \mathcal{C}$. Assuming $\perp \notin \mathcal{M}$, we encode suppressed messages with $\perp$. For the meaning of instructions Stop with, Lose, Promise, Reward, and Require see Sect. 2.1.

Definition 1. *Let $\mathcal{AD}$ be an associated data space and let $\mathcal{M}$ be a message space. An encrypt-to-self* (ETS) *scheme for $\mathcal{AD}$ and $\mathcal{M}$ consists of algorithms* enc, dec, *a key space $\mathcal{K}$, a binding-tag space $\mathcal{B}t$, and a ciphertext space $\mathcal{C}$. The encryption algorithm* enc *takes a key $k \in \mathcal{K}$, associated data $ad \in \mathcal{AD}$ and a message $m \in \mathcal{M}$, and returns a binding tag $bt \in \mathcal{B}t$ and a ciphertext $c \in \mathcal{C}$. The decryption algorithm* dec *takes a key $k \in \mathcal{K}$, a binding tag $bt \in \mathcal{B}t$, associated data $ad \in \mathcal{AD}$ and a ciphertext $c \in \mathcal{C}$, and returns a message $m \in \mathcal{M}$. A shortcut notation for this API is*

$$\mathcal{K} \times \mathcal{AD} \times \mathcal{M} \to \mathrm{enc} \to \mathcal{B}t \times \mathcal{C} \qquad\qquad \mathcal{K} \times \mathcal{B}t \times \mathcal{AD} \times \mathcal{C} \to \mathrm{dec} \to \mathcal{M}.$$

Correctness and Security. We require of an ETS scheme that if a message m is processed to a tag-ciphertext pair with associated data ad, and a message m' is recovered from this pair using the same associated data ad, then the messages m, m' shall be identical. This is formalized via the SAFE game in Fig. 1.[10] In particular, observe that if the adversary queries Dec(ad, c) (for the authentic ad and c that it receives in line 02) and the dec procedure produces output m', the game promises that $m' = m$ (lines 05,06). Recall from Sect. 2.1 that this means the game stops with output T if $m' \neq m$. Intuitively, the scheme is *safe* if we can rely on $m' = m$, that is, if the maximum advantage $\mathbf{Adv}^{\mathrm{safe}}(\mathcal{A}) := \max_{ad \in \mathcal{AD}, m \in \mathcal{M}} \Pr[\mathrm{SAFE}(ad, m, \mathcal{A})]$ that can be attained by realistic adversaries $\mathcal{A}$ is negligible. The scheme is perfectly safe if $\mathbf{Adv}^{\mathrm{safe}}(\mathcal{A}) = 0$ for all $\mathcal{A}$. We remark that the universal quantification over all pairs (ad, m) makes our advantage definition particularly robust.

[10] The SAFETY term borrows from the Distributed Computing community. SAFETY should not be confused with a notion of security. Informally, safety properties require that "bad things" will not happen. (In the case of encryption, it would be a bad thing if the decryption of an encryption yielded the wrong message.) For an initial overview we refer to Wikipedia: Safety_property and for the details to [1].

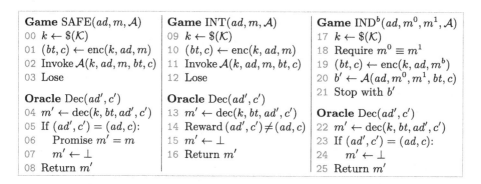

Our security notions demand that the integrity of ciphertexts be protected (INT-CTXT), and that encryptions be indistinguishable in the presence of chosen-ciphertext attacks (IND-CCA). The notions are formalized via the INT and $\text{IND}^0, \text{IND}^1$ games in Fig. 1, where the latter two depend on some equivalence relation $\equiv \subseteq \mathcal{M} \times \mathcal{M}$ on the message space.[11] For consistency, in lines 07, 15, 24 we suppress the message in all games if the adversary queries $\text{Dec}(ad, c)$. This is crucial in the IND^b games, as otherwise the adversary would trivially learn which message was encrypted, but does not harm in the other games as the adversary already knows m. Recall from Sect. 2.2 that all algorithms can fail, and if they do, then the oracles immediately abort. This property is crucial in the INT game where the dec algorithm must fail for unauthentic input such that the oracle immediately aborts. Otherwise, the game will reward the adversary, that is the game stops with T (line 14). We say that a scheme provides *integrity* if the maximum advantage $\mathbf{Adv}^{\text{int}}(\mathcal{A}) := \max_{ad \in \mathcal{AD}, m \in \mathcal{M}} \Pr[\text{INT}(ad, m, \mathcal{A})]$ that can be attained by realistic adversaries $\mathcal{A}$ is negligible, and that it provides *indistinguishability* if the same holds for the advantage $\mathbf{Adv}^{\text{ind}}(\mathcal{A}) := \max_{ad \in \mathcal{AD}, m^0, m^1 \in \mathcal{M}} |\Pr[\text{IND}^1(ad, m^0, m^1, \mathcal{A})] - \Pr[\text{IND}^0(ad, m^0, m^1, \mathcal{A})]|$.

3.2 Sufficiency of ETS for Outsourced Storage

We define the syntax of an outsourced storage scheme. We model such a scheme as a set of stateful algorithms, where algorithm write is invoked to store data and algorithm read is invoked to retrieve it. We indicate the statefulness of the algorithms by appending the term $\langle st \rangle$ to their names, where st is the state variable.

Definition 2. *Let $\mathcal{M}$ be a message space. A storage outsourcing scheme for $\mathcal{M}$ consists of algorithms* gen, write, read, *a state space $\mathcal{ST}$, and a ciphertext space $\mathcal{C}$. The state generation algorithm* gen *takes no input and outputs an (initial) state $st \in \mathcal{ST}$. The storage algorithm* write *takes a state $st \in \mathcal{ST}$ and a message $m \in \mathcal{M}$, and outputs an (updated) state $st \in \mathcal{ST}$ and a ciphertext $c \in \mathcal{C}$. The retrieval algorithm* read *takes a state $st \in \mathcal{ST}$ and a ciphertext $c \in \mathcal{C}$, and outputs an updated state $st \in \mathcal{ST}$ and a message $m \in \mathcal{M}$. A shortcut notation for this API is*

$$\text{gen} \to \mathcal{ST} \qquad \mathcal{M} \to \text{write}\langle \mathcal{ST} \rangle \to \mathcal{C} \qquad \mathcal{C} \to \text{read}\langle \mathcal{ST} \rangle \to \mathcal{M} \ .$$

Correctness and Security. We require of a storage outsourcing scheme that if a message m is processed to a ciphertext, and subsequently a message m' is recovered from this ciphertext, then the messages m, m' shall be identical. This is formalized via the SAFE game in Fig. 2. Observe boolean flag *is* ('in-sync') tracks

[11] We use relation $\equiv$ (in line 18 of IND^b) to deal with certain restrictions that practical ETS schemes may feature. Concretely, our construction does not take effort to hide the length of encrypted messages, implying that indistinguishability is necessarily limited to same-length messages. In our formalization such a technical restriction can be expressed by defining $\equiv$ such that $m^0 \equiv m^1 \Leftrightarrow |m^0| = |m^1|$.

whether the attack is active or passive. Initially $is = \mathtt{T}$, i.e., the attack is passive; however, once the adversary requests the reading of a ciphertext that is not the last created one, the game sets $is \leftarrow \mathtt{F}$ to flag the attack as active (line 11). For passive attacks the game promises that any m returned by the read procedure is the last one that was processed by the write procedure (line 13). Intuitively, the scheme is *safe* if the maximum advantage $\mathbf{Adv}^{\text{safe}}(\mathcal{A}) := \Pr[\text{SAFE}(\mathcal{A})]$ that can be attained by realistic adversaries $\mathcal{A}$ is negligible. The scheme is perfectly safe if $\mathbf{Adv}^{\text{safe}}(\mathcal{A}) = 0$ for all $\mathcal{A}$.

Our security notions demand that the integrity of ciphertexts be protected (INT-CTXT), and that encryptions be indistinguishable in the presence of chosen-ciphertext attacks (IND-CCA). The notions are formalized via the INT and $\text{IND}^0, \text{IND}^1$ games in Fig. 2, where the latter two depend on some equivalence relation $\equiv\, \subseteq \mathcal{M} \times \mathcal{M}$ on the message space (see also Footnote 13). Recall from Sect. 2.2 that all algorithms can fail, and if they do, the oracles immediately abort. This property is crucial in the INT game where the read algorithm must fail for unauthentic input such that the adversary is not rewarded in the subsequent line in the Read oracle. For consistency we suppress the message in the Read oracle for passive attacks in all games if the adversary queries $\text{Dec}(ad, c)$. This is crucial in the IND^b games, as otherwise the adversary would trivially learn which message was encrypted, but does not harm in the other games as the adversary already knows m for passive attacks. Furthermore, we remark the adversary is only allowed to query the Corrupt oracle if M contains at most 1 message, i.e., the ChWrite oracle was queried for $m^0 = m^1$. Otherwise, the adversary would be able to run the read procedure and trivially learn m. We say that a scheme provides *integrity* if the maximum advantage $\mathbf{Adv}^{\text{int}}(\mathcal{A}) := \Pr[\text{INT}(\mathcal{A})]$ that can be attained by realistic adversaries $\mathcal{A}$ is negligible, and that it provides *indistinguishability* if the same holds for the advantage $\mathbf{Adv}^{\text{ind}}(\mathcal{A}) := |\Pr[\text{IND}^1(\mathcal{A})] - \Pr[\text{IND}^0(\mathcal{A})]|$.

CONSTRUCTION FROM ETS. Constructing secure outsourced storage from ETS is immediate: The write procedure samples a uniformly random key and runs the enc procedure of ETS to obtain a binding tag and ciphertext. It stores the binding tag (and key) in the state and returns the ciphertext. The read procedure gets the key and binding tag from the state, runs the dec procedure of ETS and returns the message. The details of this construction are in Fig. 3. The security argument is obvious.

4 Construction of Encrypt-to-Self

We mentioned in Sect. 1 that a generic construction of ETS can be realized by combining standard symmetric encryption with a cryptographic hash function: one encrypts the message and computes the binding tag as the hash of the ciphertext. Here we provide a more efficient construction that builds on the compression function of a Merkle–Damgård hash function. To be more precise, our construction uses a tweakable compression function with tweak space $T =$

Game SAFE($\mathcal{A}$)	Game INT($\mathcal{A}$)	Game IND^b($\mathcal{A}$)
00 C, M ← ∅	16 C ← ∅	30 C, M ← ∅
01 is ← T	17 is ← T	31 is ← T
02 st ← gen	18 st ← gen	32 st ← gen
03 Invoke $\mathcal{A}$	19 Invoke $\mathcal{A}$	33 b' ← $\mathcal{A}$
04 Lose	20 Lose	34 Stop with b'
Oracle Write(m)	**Oracle** Write(m)	**Oracle** ChWrite(m^0, m^1)
05 c ← write⟨st⟩(m)	21 c ← write⟨st⟩(m)	35 Require $m^0 \equiv m^1$
06 If is:	22 If is: C ← {c}	36 c ← write⟨st⟩(m^b)
07 C ← {c}	23 Return c	37 If is:
08 M ← {m}		38 C ← {c}
09 Return c	**Oracle** Read(c)	39 M ← {m^0, m^1}
	24 m ← read⟨st⟩(c)	40 Return c
Oracle Read(c)	25 If $c \notin$ C: is ← F	
10 m ← read⟨st⟩(c)	26 Promise is	**Oracle** Read(c)
11 If $c \notin$ C: is ← F	27 If is: m ← ⊥	41 m ← read⟨st⟩(c)
12 If is:	28 Return m	42 If $c \notin$ C: is ← F
13 Promise $m \in$ M		43 If is: m ← ⊥
14 m ← ⊥	**Oracle** Corrupt	44 Return m
15 Return m	29 Return st	
		Oracle Corrupt
		45 Require |M| ≤ 1
		46 Return st

Fig. 2. Games for outsourced storage. For all values m, m^0, m^1, c provided by the adversary we require that $m, m^0, m^1 \in \mathcal{M}$ and $c \in \mathcal{C}$. Assuming $\perp \notin \mathcal{M}$, we encode suppressed messages with $\perp$. Boolean flag is ('in-sync') tracks whether the attack is active or passive. For the meaning of instructions Stop with, Lose, Promise, and Require see Sect. 2.1.

$\{0, 1\}$, i.e., the domain of the compression function is extended by one bit (see Sect. 2.4). We provide a general definition below.

Definition 3. *For Σ an alphabet, $c, d \in \mathbb{N}^+$ with $c \leq d$, and a tweak space T, we define a tweakable compression function to be a function $F: \Sigma^d \times T \times \Sigma^c \to \Sigma^c$ that takes as input a block $B \in \Sigma^d$ from the data domain, a tweak $t \in T$ from the tweak space, and a string $C \in \Sigma^c$ from the chain domain, and outputs a string $C' \in \Sigma^c$ in the chain domain.*

We will write $F_t(B, C)$ as shorthand notation for $F(B, t, C)$. For practical tweakable compression functions the memory alignment value mav (see Sect. 2.3) will divide both c and d. When constructing an ETS scheme from F, because the compression function only takes fixed-size input, we need to map the (ad, m) input to a series of block–tweak pairs (B, t). We will refer to this mapping as the input *encoding*. We take a modular approach by fixing the encoding independently of the encryption engine, and detail the former in Sect. 4.1 and the latter in Sect. 4.2. Together they form an efficient construction of ETS.

We first convey a rough overview of our ETS construction. In Fig. 4 we consider an example with block size d double the chaining value size c. We assume

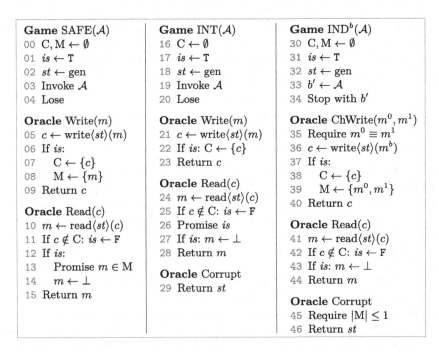

```
Proc gen
00 S ← ∅
01 st := S
02 Return st

Proc write⟨st⟩(m)
03 k ← $(K)
04 (bt, c) ← enc(k, ε, m)
05 S ← {(k, bt)}
06 Return c

Proc read⟨st⟩(c)
07 Require S ≠ ∅
08 {(k, bt)} ← S
09 m ← dec(k, bt, ε, c)
10 Return m
```

Fig. 3. Construction for outsourced storage from ETS. If in line 07 the condition is not met, the read algorithm aborts with some error indicator. Recall from Sect. 2.2 that the read algorithm also aborts if the dec invocation in line 09 fails.

that key k is padded to size d. The first block B_1 only contains associated data and we XOR B_1 with the key k before we feed it into the compression function. From the second block, we start processing message data. We fill the first half of the block with message data m_1 and the second half with associated data ad_3, and XOR with the key. We also XOR m_1 with the current chaining value C_1, to generate a partial ciphertext ct_1. The same happens in the third block and we append ct_2 to the ciphertext. If there is associated data left after processing all message data we can load the entire block with associated data, which occurs in the fourth block. Note, we no longer need to XOR the key into the block after we have processed all message data, because at this point the input to the compression function will already be independent of the message m. After processing all blocks, we XOR an offset $\omega \in \{\omega_0, \omega_1\}$ with the chaining value, where ω_0, ω_1 are two distinct constants. The binding tag will be (a truncation of) the last chaining value C^*. Therefore, it will be crucial to fix ω_0, ω_1 such that they remain distinct after truncation. Note that the task of the encoding is not only to partition ad and m into blocks $B_1, B_2, \ldots$ as described, but also to derive tweak values $t_1, t_2, \ldots$ and the choice of the final offset ω in such a way that the overall encoding is injective.

4.1 Message Block Encoding

We turn to the technical component of our ETS construction that encodes the (ad, m) input into a series of output pairs (B, t) and the final offset value ω. For

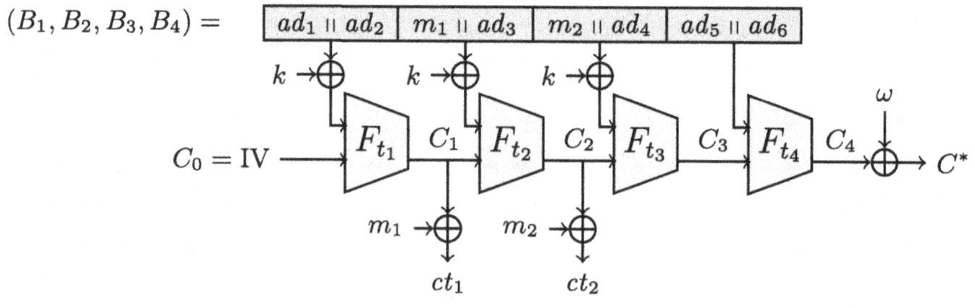

Fig. 4. Example for enc(k, ad, m) where $d = 2c$ and $ad = ad_1 \parallel \ldots \parallel ad_6$ and $m = m_1 \parallel m_2$ with $|ad| = 6c$ and $|m| = 2c$. For clarity we have made the blocks B_i, as they are output by the encoding function, explicit. Inspiration for this figure is drawn from https://www.iacr.org/authors/tikz/.

authenticity we require that the encoding is injective. For efficiency we require that the encoding is online (i.e., the input is read only once, left-to-right, and with small state), that the number of output pairs is as small as possible, and that the encoding preserves memory alignment (see Sect. 2.3). Syntactically, for the outputs we require that all $B \in \Sigma^d$, all $t \in T$, and $\omega \in \Omega$, where quantities c, d are those of the employed compression function, $T = \{0, 1\}$, and $\Omega \subseteq \Sigma^c$ is any two-element set. (Note that $|T| = 2$ allows us to use the tweaking approach from Sect. 2.4; further, in our implementations we use $\Omega = \{\omega_0, \omega_1\}$ where $\omega_0 = \texttt{0x00}^c$ and $\omega_1 = \texttt{0xa5}^c$.) Overall, the task we are facing is the following:

Task. Assume $|\Sigma| = 256$ and $\mathcal{AD} = \mathcal{M} = \Sigma^*$ and $T = \{0, 1\}$ and $|\Omega| = 2$. For $c, d \in \mathbb{N}^+$, $c < d$, find an injective encoding function encode: $\mathcal{AD} \times \mathcal{M} \to (\Sigma^d \times T)^* \times \Omega$ that takes as input two finite strings and outputs a finite sequence of pairs $(B, t) \in \Sigma^d \times T$ and an offset $\omega \in \Omega$.

A detailed specification of our encoding (and decoding) function can be found in Fig. 6, but we present it here in text. Our construction does not use the decoding function, but we provide it anyway to show that the encoding function is indeed injective. Roughly, we encode as follows. We fill the first block with *ad* and for any subsequent block we load the message in the first part of the block and *ad* in the second part of the block. When we have processed all the message data, we load the full block with *ad* again. Clearly, we need to pad *ad* if it runs out before we have processed all message data. We do this by appending a special termination symbol $\diamond \in \Sigma$ to *ad* and then appending null bytes as needed. Similarly, we need to pad the message if the message length is not a multiple of c. Naturally, one might want to pad the message to a multiple of c. However, this is suboptimal: Consider the scenario where there are $d - c + 1$ bytes remaining to be processed of associated data and 1 byte of message data. In principle, message and associated data would fit into a single block, but this would not be the case any longer if the message is padded to size c. On the other hand, for efficiency reasons we do not want to misalign all our remaining associated data. If we do not pad at all, when we process the next d bytes of associated data, we can only fit $d - 1$ bytes in the block and have to put 1 byte into the next block. Hence, we pad m up to a multiple of the memory alignment value mav. Therefore, we prepend a padded message with its length $|m|$; this will uniquely determine where m stops. We then fill up with null bytes until reaching a multiple of mav. This restricts us to $c \leq 256$ bytes such that $|m|$ always can be encoded into a single byte. As far as we are aware, any current practical compression function satisfies this requirement.

In Fig. 5, for the artificially small case with $c = 1$ and $d = 2$ we provide four examples of what the blocks would look like for different inputs. The top row shows the encoding of 8 bytes of associated data and an empty message. The second row shows the encoding of empty associated data and 3 bytes of message data. The third row shows the encoding of 6 bytes of associated data and 2 bytes of message data. The final row shows the encoding of 3 bytes of associated data and 3 bytes of message data.

$$(B_1, B_2, B_3, B_4) = \boxed{ad_1 \| ad_2} \boxed{ad_3 \| ad_4} \boxed{ad_5 \| ad_6} \boxed{ad_7 \| ad_8}$$

$$(B_1, B_2, B_3, B_4) = \boxed{\diamond \| \mathtt{0x00}} \boxed{m_1 \| \mathtt{0x00}} \boxed{m_2 \| \mathtt{0x00}} \boxed{m_3 \| \mathtt{0x00}}$$

$$(B_1, B_2, B_3, B_4) = \boxed{ad_1 \| ad_2} \boxed{m_1 \| ad_3} \boxed{m_2 \| ad_4} \boxed{ad_5 \| ad_6}$$

$$(B_1, B_2, B_3, B_4) = \boxed{ad_1 \| ad_2} \boxed{m_1 \| ad_3} \boxed{m_2 \| \diamond} \boxed{m_3 \| \mathtt{0x00}}$$

Fig. 5. Example encodings for the case $c = 1$ and $d = 2$.

We have two ambiguities remaining. (1) How to tell whether ad was padded or not? Consider the first row in Fig. 5. What distinguishes the case $ad = ad_1 \| \ldots \| ad_7$ from $ad = ad_1 \| \ldots \| ad_7 \| ad_8$ with $ad_8 = \diamond$? A similar question applies to the message. (2) How to tell whether a block contains message data or not? Compare e.g., the first row with the third row. This is where the tweaks come into play.

First of all, we tweak the first block if and only if the message is empty. This fully separates the authentication-only case from the case where we have message input.

Next, if the message is non-empty, we use the tweaks to indicate when we switch from processing message data to ad-only: we tweak when we have consumed all of m, but still have ad left. Note the first block never processes message data, so the earliest block this may tweak is the second block and hence this rule does not interfere with the first rule. Furthermore, observe this rule never tweaks the final block, as by definition of being the final block, we do not have any associated data left to process.

Next, we need to distinguish between the cases whether m is padded or not. In fact, as the empty message was already taken care of, we need to do this only if m is at least one byte in size. As in this case the final block does not coincide with the first block, we can exploit that its tweak is still unused; we correspondingly tweak the final block if and only if m is padded. Obviously, this does not interfere with the previous rules.

Finally, we need to decide whether ad was padded or not. We do not want to enforce a policy of 'always pad', as this could result in an extra block and hence an extra compression function invocation. Instead, we use our offset output. We set the offset ω to ω_1 if ad was padded; otherwise we set it to ω_0.

This completes our description of the encoding function. The decoding function is a technical exercise carefully unwinding the steps taken in the encoding function, which we perform in Fig. 6. We obtain that for all $m \in \mathcal{M}$, $ad \in \mathcal{AD}$ we have $\mathrm{decode}(\mathrm{encode}(ad, m)) = (ad, m)$. It immediately follows that our encoding function is injective. For readability we have implemented the core functionality of the encoding in a coroutine called nxt, rather than a subroutine. Instead of generating the entire sequence of (B, t) pairs and returning the result, it will 'Yield' one pair and suspend its execution. The next time it is called (e.g., the next step in a for loop), it will resume execution from where it called 'Yield',

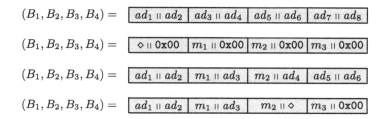

```
Proc encode(ad, m)
S[·] ← ·; i ← 0
For (B, t) ∈ nxt(ad, m):
    If B ≠ ε:
        i ← i + 1
        S[i] ← (B, t)
    Else: ω ← t
Return (S, ω)

Proc decode(S, ω)
ad ← ε; m ← ε
n ← |S|; j ← |S|
If n = 0:
    Return (ad, m)
For i ← 1 to n:
    (B_i, t_i) ← S[i]
For i ← 1 to n − 1:
    If t_i = 1: j ← i
ad ←‖ B_1
For i ← 2 to j − t_n:
    B_i ‖ B'_i ←_c B_i
    m ←‖ B_i
    ad ←‖ B'_i
If n > 1 ∧ t_n = 1:
    l ‖ B_j ←_1 B_j
    m' ‖ B_j ←_l B_j
    m ←‖ m'
    z ← −l mod mav
    _ ‖ ad' ←_{z−1} B_j
    ad ←‖ ad'
For i ← j + 1 to n:
    ad ←‖ B_i
If ω = ω_1:
    Split ad ‖ ◇ ‖ 0* ← ad
Return (ad, m)
```

```
Proc nxt(ad, m)
ad_main ← T
m_main ← T
ω ← ω_0; t̄ ← 0; n ← 0
While ad ≠ ε ∨ m ≠ ε:
    n ← n + 1
    (B_n, t_n) ← (ε, 0)
    If n > 1 ∧ m ≠ ε:
        If |m| < c:
            t̄ ← 1
            j ← |m| mod mav
            j ← mav − j − 1
            m ← |m| ‖ m ‖ 0^j
        l ← min(c, |m|)
        B_n ‖ m ←_l m
    d' ← d − |B_n|
    If |ad| < d':
        If ad_main:
            ω ← ω_1
            ad ←‖ ◇
            ad_main ← F
        j ← d' − |ad|
        ad ←‖ 0^j
    ad' ‖ ad ←_{d'} ad
    B_n ←‖ ad'
    If m_main ∧ m = ε:
        If n = 1 ∨ ad ≠ ε:
            t_n ← 1
            m_main ← F
    If ad = ε ∧ m = ε:
        If n > 1: t_n ← t̄
    Yield (B_n, t_n)
Yield (ε, ω)
```

```
Proc enc(k, ad, m)
ct ← ε; C ← IV; i ← 0
For (B, t) ∈ nxt(ad, m):
    If B = ε: Break
    i ← i + 1
    If i = 1 ∨ m ≠ ε:
        B ← B ⊕ k
    If i > 1 ∧ m ≠ ε:
        j ← min(c, |m|)
        m' ‖ m ←_j m
        C' ←_j C
        ct ←‖ m' ⊕ C'
    C ← F_t(B, C)
bt ←_taglen C ⊕ t
Return (bt, ct)

Proc dec(k, bt, ad, ct)
m ← ε; C ← IV; i ← 0
For (B, t) ∈ nxt(ad, ct):
    If B = ε: Break
    i ← i + 1
    If i = 1 ∨ ct ≠ ε:
        B ← B ⊕ k
    If i > 1 ∧ ct ≠ ε:
        If |ct| ≥ c:
            ct' ‖ ct ←_c ct
            m ←‖ ct' ⊕ C
            B ←⊕ C ‖ 0^{d−c}
        Else:
            C' ←_{|ct|} C
            m ←‖ ct ⊕ C'
            j ← d − |ct| − 1
            B ←⊕ 0 ‖ C' ‖ 0^j
    C ← F_t(B, C)
bt' ←_taglen C ⊕ t
If bt' ≠ bt: Fail
Return m
```

Fig. 6. ETS construction: encoder, decoder, encryptor, and decryptor. (Procedure nxt is a coroutine for encode, enc, and dec, see text.) Using global constants mav, c, d, taglen, and IV.

instead of at the beginning of the function, with all of its state intact. The encode procedure is a simple wrapper that runs the nxt procedure and collects its output, but our authenticated encryption engine described in Sect. 4.2 will call the nxt procedure directly.

4.2 Encryption Engine

We now turn our focus to the encryption engine. We assume that the associated data and message are present in encoded format, i.e., as a sequence of pairs (B, t), where $B \in \Sigma^d$ is a block and $t \in \{0, 1\}$ is a tweak, and additionally an offset $\omega \in \{\omega_0, \omega_1\}$. To be precise, we will use the nxt procedure that generates the next (B, t) pair on the fly.

We specify the encryption and decryption algorithms in Fig. 6 and assume they are provided with a key of length d. As illustrated in Fig. 4, the main idea is to XOR the key with all blocks that are involved with message processing. For the skeleton of the construction, we initialize the chaining value C to IV and loop through the sequence of pairs (B, t) output by the encoding function, each iteration updating the chaining value $C \leftarrow F_t(B, C)$. We now describe each iteration of the enc procedure in more detail, where numbers in brackets refer to line numbers in Fig. 6. If the block is empty [66], we are in the final iteration and do not do anything. Otherwise, we check if we are in the first iteration or if we have message data left [68]. In this case we XOR the key into the block [69]. This ensures we start with an unknown input block and that subsequent inputs are statistically independent of the message block. If we only have *ad* remaining we can use the block directly as input to the compression function. If we have message data left we will encrypt it starting from the second block [70]. To encrypt, we take a chunk of the message, XOR it with the chaining value of equal size and append the result to the ciphertext [71–74]. We only start encrypting from the second iteration as the first chaining value is public. Finally, we call the compression function F_t to update our chaining value [75]. Once we have finished the loop, the last pair (B, t) equals (ϵ, ω) by definition. So we XOR the offset ω with the chaining value C and truncate the result to obtain the binding tag [76]. We return the binding tag along with the ciphertext.

The dec procedure is similar to the enc procedure but needs to be slightly adapted. Informally, the nxt procedure now outputs a block $B = (ct \parallel ad)$ [79] instead of $B = (m \parallel ad)$ [65]. Hence, we XOR with the chaining variable [88, 93] such that the block becomes $B = (m \parallel ad)$ and the compression function call takes equal input compared to the enc procedure. The case distinction handles the slightly different positioning of ciphertext in the blocks. Finally, there obviously is a check if the computed binding tag is equal to the stored binding tag [96].

In order to prove security, we need further assumptions on our compression function than the standard assumption of preimage resistance and collision resistance. For example, we need F to be difference unpredictable. Roughly, this notion says it is hard to find a pair (x, y) such that $F(x) = F(y) \oplus z$ for a given difference z. Moreover, we truncate the binding tag, so actually it should be hard to find a tuple such that this equation holds for the first $|bt|$ bits. We note collision resistance of F does not imply collision resistance of a truncated version of F [3]. However, such assumptions can be justified if one views the compression function as a random function. Hence, instead of several ad hoc assumptions, we prove our construction secure directly in the random oracle model.

As described in Sect. 2.4 we tweak the SHA2 compression function by modifying the chaining value depending on the tweak. Let F be the tweakable compression function in Fig. 6, we denote with F' the (standard) compression function that will take as input the block and the (modified) chaining value. Let $H\colon \Sigma^d \times \Sigma^c \to \Sigma^c$ be a random oracle. We will substitute H for F' in our construction. We remark the BLAKE2b compression function is a tweakable compression function and it can be substituted for a random oracle directly. Hence, we focus on the security proof with a standard compression function, as the case with a tweakable compression function is a simplification of the proof with a standard compression function. In the random oracle model, our ETS construction from Fig. 6 provides integrity and indistinguishability (Theorem 1), assuming sufficiently large tag and key lengths. Here, we only state the theorem and we refer the reader to the full version [11] for the proof.

Theorem 1. *Let π be the construction given in Fig. 6, H a random oracle replacing the compression function, $\mathcal{A}$ an adversary, $\mathbf{Adv}_\pi^{\mathrm{int}}(\mathcal{A})$ the advantage that $\mathcal{A}$ has against π in the integrity game of Fig. 1, $\mathbf{Adv}_\pi^{\mathrm{ind}}(\mathcal{A})$ the advantage that $\mathcal{A}$ has against π in the indistinguishability games of Fig. 1, and q the number of random oracle queries (either directly or indirectly via* Dec*). We have,*

$$\mathbf{Adv}_\pi^{\mathrm{int}}(\mathcal{A}) \leq q \cdot 2^{-|bt|},$$

and

$$\mathbf{Adv}_\pi^{\mathrm{ind}}(\mathcal{A}) \leq q^2 \cdot 2^{-c} + q \cdot 2^{-|k|} + \mathbf{Adv}_\pi^{\mathrm{int}}(\mathcal{A}).$$

5 Implementation of Encrypt-to-Self

We implemented three versions of the ETS primitive. Precisely, we implemented the padding scheme and encryption engine from Fig. 6 in C, based on the compression functions of SHA256, SHA512, and BLAKE2 [10,14] and the tweaking approach described in Sect. 2.4. All three functions are particularly good performers in software, where specifically SHA512 and BLAKE2 excel on 64-bit platforms. Roughly, we measured that the BLAKE2 version is about 15% faster than the SHA512 version, which in turn is about 15% faster than the SHA256 version.[12] We note that our code is written in pure C and in principle would benefit from assembly optimizations. Fortunately, however, all three compression functions are ARX (add–rotate–xor) designs so that the penalty of not hand-optimizing is not too drastic. Further, many freely available assembly implementations of the SHA2 and BLAKE2 core functions exist, e.g., in the

[12] We do not provide cycle counts as we measured on outdated hardware (an Intel Core i3-2350M CPU @ 2.30 GHz) and our numbers would not allow for a meaningful efficiency estimation on current CPUs. Note that https://blake2.net/ reports a performance of raw BLAKE2(b) on Skylake of roughly 1 GB/s. Our ETS adds the message encryption step on top of this (a series of memory accesses and XOR operations per message block), so it seems fair to expect an overall performance of more than 800 MB/s.

OpenSSL package, and we made sure that our API abstractions are compatible with these, allowing for drop in replacements.

Acknowledgments. We thank the reviewers of ESORICS'20 for their comments and appreciate the feedback provided by Cristina Onete. The research of Pijnenburg was supported by the EPSRC and the UK government as part of the Centre for Doctoral Training in Cyber Security at Royal Holloway, University of London (EP/P009301/1). The research of Poettering was supported by the European Union's Horizon 2020 project FutureTPM (779391).

References

1. Alpern, B., Schneider, F.B.: Recognizing safety and liveness. Distrib. Comput. **2**(3), 117–126 (1987). https://doi.org/10.1007/BF01782772
2. Aumasson, J.-P., Neves, S., Wilcox-O'Hearn, Z., Winnerlein, C.: BLAKE2: simpler, smaller, fast as MD5. In: Jacobson, M., Locasto, M., Mohassel, P., Safavi-Naini, R. (eds.) ACNS 2013. LNCS, vol. 7954, pp. 119–135. Springer, Heidelberg (2013). https://doi.org/10.1007/978-3-642-38980-1_8
3. Biham, E., Chen, R.: Near-collisions of SHA-0. In: Franklin, M. (ed.) CRYPTO 2004. LNCS, vol. 3152, pp. 290–305. Springer, Heidelberg (2004). https://doi.org/10.1007/978-3-540-28628-8_18
4. Biham, E., Dunkelman, O.: A framework for iterative hash functions - HAIFA. Cryptology ePrint archive, report 2007/278 (2007). http://eprint.iacr.org/2007/278
5. Biryukov, A., Dinu, D., Khovratovich, D.: Argon2: new generation of memory-hard functions for password hashing and other applications. In: EuroS&P, pp. 292–302. IEEE (2016)
6. Chang, D., Nandi, M., Yung, M.: Indifferentiability of the hash algorithm BLAKE. Cryptology ePrint archive, report 2011/623 (2011). http://eprint.iacr.org/2011/623
7. Dodis, Y., Grubbs, P., Ristenpart, T., Woodage, J.: Fast message franking: from invisible salamanders to encryptment. In: Shacham, H., Boldyreva, A. (eds.) CRYPTO 2018. LNCS, vol. 10991, pp. 155–186. Springer, Cham (2018). https://doi.org/10.1007/978-3-319-96884-1_6
8. Gueron, S.: Memory encryption for general-purpose processors. IEEE Secur. Priv. **14**(6), 54–62 (2016)
9. Kaliski, B.: PKCS #5: password-based cryptography specification version 2.0. RFC 2898, September 2000. https://rfc-editor.org/rfc/rfc2898.txt
10. NIST: FIPS 180–4: Secure Hash Standard (SHS). Technical report, NIST (2015). https://doi.org/10.6028/NIST.FIPS.180-4
11. Pijnenburg, J., Poettering, B.: Encrypt-to-self: securely outsourcing storage. Cryptology ePrint archive, report 2020/847 (2020). https://eprint.iacr.org/2020/847
12. Pijnenburg, J., Poettering, B.: Key assignment schemes with authenticated encryption. IACR Trans. Symmetric Cryptol. **2020**(2), 40–67 (2020). Revisited
13. Rogaway, P.: Authenticated-encryption with associated-data. In: Atluri, V. (ed.) ACM CCS 2002, pp. 98–107. ACM Press, November 2002
14. Saarinen, M.O., Aumasson, J.: The BLAKE2 cryptographic hash and message authentication code (MAC). RFC 7693 (2015). https://rfc-editor.org/rfc/rfc7693.txt

PGLP: Customizable and Rigorous Location Privacy Through Policy Graph

Yang Cao[1(✉)], Yonghui Xiao[2], Shun Takagi[1], Li Xiong[2], Masatoshi Yoshikawa[1], Yilin Shen[3], Jinfei Liu[2], Hongxia Jin[3], and Xiaofeng Xu[2]

[1] Kyoto University, Kyoto, Japan
{yang,yoshikawa}@i.kyoto-u.ac.jp, s.takagi@db.soc.i.kyoto-u.ac.jp
[2] Emory University, Atlanta, Georgia
yohuxiao@gmail.com, {lxiong,jliu253}@emory.edu, xuxiaofeng1989@gmail.com
[3] Samsung Research America, Mountain View, USA
{yilin.shen,hongxia.jin}@samsung.com

Abstract. Location privacy has been extensively studied in the literature. However, existing location privacy models are either not rigorous or not customizable, which limits the trade-off between privacy and utility in many real-world applications. To address this issue, we propose a new location privacy notion called PGLP, i.e., *Policy Graph based Location Privacy*, providing a rich interface to release private locations with customizable and rigorous privacy guarantee. First, we design a rigorous privacy for PGLP by extending differential privacy. Specifically, we formalize location privacy requirements using a *location policy graph*, which is expressive and customizable. Second, we investigate how to satisfy an arbitrarily given location policy graph under realistic adversarial knowledge, which can be seen as constraints or public knowledge about user's mobility pattern. We find that a policy graph may not always be viable and may suffer *location exposure* when the attacker knows the user's mobility pattern. We propose efficient methods to detect location exposure and repair the policy graph with optimal utility. Third, we design an end-to-end location trace release framework that pipelines the detection of location exposure, policy graph repair, and private location release at each timestamp with customizable and rigorous location privacy. Finally, we conduct experiments on real-world datasets to verify the effectiveness and the efficiency of the proposed algorithms.

Keywords: Spatiotemporal data · Location privacy · Trajectory privacy · Differential privacy · Location-based services

---

This work is partially supported by JSPS KAKENHI Grant No. 17H06099, 18H04093, 19K20269, U.S. National Science Foundation (NSF) under CNS-2027783 and CNS-1618932, and Microsoft Research Asia (CORE16).
Yang and Yonghui contributed equally to this work.

L. Chen et al. (Eds.): ESORICS 2020, LNCS 12308, pp. 655–676, 2020.
https://doi.org/10.1007/978-3-030-58951-6_32

1 Introduction

As GPS-enabled devices such as smartphones or wearable gadgets are pervasively used and rapidly developed, location data have been continuously generated, collected, and analyzed. These personal location data connecting the online and offline worlds are precious, because they could be of great value for the society to enable ride sharing, traffic management, emergency planning, and disease outbreak control as in the current covid-19 pandemic via contact tracing, disease spread modeling, traffic and social distancing monitoring [4,19,26,30].

On the other hand, privacy concerns hinder the extensive use of big location data generated by users in the real world. Studies have shown that location data could reveal sensitive personal information such as home and workplace, religious and sexual inclinations [35]. According to a survey [18], 78% smartphone users among 180 participants believe that Apps accessing their location pose privacy threats. As a result, the study of *private location release* has drawn increasing research interest and many location privacy models have been proposed in the last decades (see survey [34]).

However, existing location privacy models for private location releases are either not rigorous or not customizable. Following the seminal paper [22], the early location privacy models were designed based on k-anonymity [37] and adapted to different scenarios such as mobile P2P environments [14], trajectory release [3] and personalized k-anonymity for location privacy [21]. The follow-up studies revealed that k-anonymity might not be rigorous because it syntactically defines privacy as a property of the final "anonymized" dataset [29] and thus suffers many realistic attacks when the adversary has background knowledge about the dataset [28,31]. To this end, the state-of-the-art location privacy models [1,12,38,39] were extended from differential privacy (DP) [15] to private location release since DP is considered a rigorous privacy notion which defines privacy as a property of the algorithm. Although these DP-based location privacy models are rigorously defined, they are not customizable for different scenarios with various requirements on privacy-utility trade-off. Taking an example of Geo-Indistinguishability [1], which is the first DP-based location privacy, the protection level is solely controlled by a parameter ϵ to achieve indistinguishability between any two possible locations (the indistinguishability is scaled to the Euclidean distance between any two possible locations).

This one-size-fits-all approach may not fit every application's requirement on utility-privacy trade-off. Different location-based services (LBS) may have different usage of the data and thus need different *location privacy policies* to strike the right balance between privacy and utility. For instance, a proper location privacy policy for weather apps could be "*allowing the app to access a user's city-level location but ensuring indistinguishability among locations in each city*", which guarantees both reasonable privacy and high usability for a city-level weather forecast. Similarly, for POI recommendation [2], trajectory mining [32] or crowd monitoring during the pandemic [26], a suitable location privacy policy could be "*allowing the app to access the semantic category (e.g., a restaurant or a shop) of a user's location but ensuring indistinguishability among locations with the*

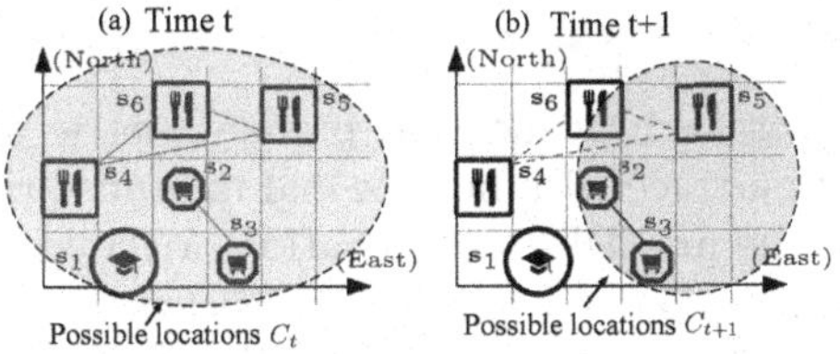

Fig. 1. An example of location policy graph and the constrained domains (i.e., possible locations) C_t and C_{t+1} at time t and $t+1$, respectively.

same category", so that the LBS provider may know the user is at a restaurant or a shop, but not sure which restaurant or which shop.

In this work, we study *how to release private location with customizable and rigorous privacy.* There are three significant challenges to achieve this goal. First, there is a lack of a rigorous and customizable location privacy metric and mechanisms. The closest work regarding customizable privacy is Blowfish privacy [25] for statistical data release, which uses a graph to represent a customizable privacy requirement, in which a node indicates a possible database instance to be protected, and an edge represents indistinguishability between the two possible databases. Blowfish privacy and its mechanisms are not applicable in our setting of private location release. It is because Blowfish privacy is defined on a statistical query over database with multiple users' data; whereas the input in the scenario of private location release is a single user's location.

The second challenge is how to satisfy an arbitrarily given location privacy policy under realistic adversarial knowledge, which is public knowledge about users' mobility pattern. In practice, as shown in [39,40], an adversary could take advantage of side information to rule out inadmissible locations[1] and reduce a user's possible locations into a small set, which we call *constrained domain.* We find that the location privacy policy may not be viable under a constrained domain and the user may suffer location exposure (we will elaborate how this could happen in Sect. 4.1).

The third challenge is how to release private locations continuously with high utility. We attempt to provide an end-to-end solution that takes the user's true location and a predefined location privacy policy as inputs, and outputs private location trace on the fly. We summarize the research questions below.

- How to design a rigorous location privacy metric with customizable location privacy policy? (Sect. 3)
- How to detect the problematic location privacy policy and repair it with high utility? (Sect. 4)
- How to design an end-to-end framework to release private location continuously? (Sect. 5)

[1] For example, it is impossible to move from Kyoto to London in a short time.

1.1 Contributions

In this work, we propose a new location privacy metric and mechanisms for releasing private location trace with flexible and rigorous privacy. To the best of our knowledge, this is the first DP-based location privacy notion with customizable privacy. Our contributions are summarized below.

First, we formalize Policy Graph based Location Privacy (PGLP), which is a rigorous privacy metric extending differential privacy with a customizable *location policy graph.* Inspired by the statistical privacy notion of Blowfish privacy [25], we design location policy graph to represent which information needs to be protected and which does not. In a location policy graph (such as the one shown in Fig. 1), the nodes are the user's possible locations, and the edges indicate the privacy requirements regarding the connected locations: an attacker should not be able to significantly distinguish which location is more probable to be the user's true location by observing the released location. PGLP is a general location privacy model compared with the prior art of DP-based location privacy notions, such as Geo-Indistinguishability [1] and Location Set Privacy [39]. We prove that they are two instances of PGLP under the specific configurations of the location policy graph. We also design mechanisms for PGLP by adapting the Laplace mechanism and Planar Isotropic Mechanism (PIM) (i.e., the optimal mechanism for Location Set Privacy [39]) w.r.t. a given location policy graph.

Second, we design algorithms that examine the feasibility of a given location policy graph under adversarial knowledge about the user's mobility pattern modeled by Markov Chain. We find that the policy graph may not always be viable. Specially, as shown in Fig. 1, some nodes (locations) in a policy graph may be *excluded* (e.g., s_4 and s_6 in Fig. 1 (b)) or *disconnected* (e.g., s_5 in Fig. 1 (b)) due to the limited set of the possible locations. Protecting the excluded nodes is a lost cause, but it is necessary to protect the disconnected nodes since it may lead to location exposure when the user is at such a location. Surprisingly, we find that a disconnected node may *not always* result in the location exposure, which also depends on the protection strength of the mechanism. Intuitively, this happens when a mechanism "overprotects" a location policy graph by implicitly guaranteeing indistinguishability that is not enforced by the policy. We design an algorithm to detect the disconnected nodes that suffer location exposure , which are named *isolated node.* We also design a *graph repair* algorithm to ensure no isolated node in a location policy graph by adding an optimal edge between the isolated node and another node with high utility.

Third, we propose an end-to-end private location trace release framework with PGLP that takes inputs of the user's true location at each time t and outputs private location continuously satisfying a pre-defined location policy graph. The framework pipelines the calculation of constrained domains, isolated node detection, policy graph repair, and private location release mechanism. We also reason about the overall privacy guarantee in multiple releases.

Finally, we implement and evaluate the proposed algorithms on real-world datasets, showing that privacy and utility can be better tuned with customizable location policy graphs.

2 Preliminaries

2.1 Location Data Model

Similar to [39,40], we employ two coordinate systems to represent locations for applicability for different application scenarios. A location can be represented by an index of *grid coordinates* or by a two-dimension vector of *longitude-latitude coordinate* to indicate any location on the map. Specifically, we partition the map into a grid such that each grid cell corresponds to an area (or a point of interest); any location represented by a longitude-latitude coordinate will also have a grid number or index on the grid coordinate. We denote the location domain as $\mathcal{S} = \{\mathbf{s}_1, \mathbf{s}_2, \cdots, \mathbf{s}_N\}$ where each $\mathbf{s}_i$ corresponds to a grid cell on the map, $1 \leq i \leq N$. We use $\mathbf{s}_t^*$ and $\mathbf{z}_t$ to denote the user's true location and perturbed location at time t. We also use t in $\mathbf{s}^*$ and $\mathbf{z}$ to refer the locations at a single time when it is clear from the context.

Location Query. For the ease of reasoning about privacy and utility, we use a location query $f : \mathcal{S} \rightarrow \mathbb{R}^2$ to represent the mapping from locations to the longitude and latitude of the center of the corresponding grid cell.

2.2 Problem Statement

Given a moving user on a map $\mathcal{S}$ in a time period $\{1, 2, \cdots, T\}$, our goal is to release the perturbed locations of the user to untrusted third parties at each timestamp under a pre-defined location privacy policy. We define ϵ-Indistinguishability as a building block for reasoning about our privacy goal.

Definition 1 (ϵ-Indistinguishability). *Two locations $\mathbf{s}_i$ and $\mathbf{s}_j$ are ϵ-indistin-guishable under a randomized mechanism $\mathcal{A}$ iff for any output $\mathbf{z} \subseteq Range(\mathcal{A})$, we have $\frac{\Pr(\mathcal{A}(s_i)=z)}{\Pr(\mathcal{A}(s_j)=z)} \leq e^{\epsilon}$, where $\epsilon \geq 0$.*

As we exemplified in the introduction, different LBS applications may have different metrics of utility. We aim at providing better utility-privacy trade-off by customizable ϵ-Indistinguishability between locations.

Adversarial Model. We assume that the attackers know the user's mobility pattern modeled by Markov chain, which is widely used for modeling user mobility profiles [10,20]. We use matrix $\mathbf{M} \in [0,1]^{N \times N}$ to denote the transition probabilities of Markov chain with m_{ij} being the probability of moving from location $\mathbf{s}_i$ to location $\mathbf{s}_j$. Another adversarial knowledge is the initial probability distribution of the user's location at $t = 1$. To generalize the notation, we denote probability distribution of the user's location at t by a vector $\mathbf{p}_t \in [0,1]^{1 \times N}$, and denote the ith element in $\mathbf{p}_t$ by $\mathbf{p}_t[i] = \Pr(\mathbf{s}_t^* = \mathbf{s}_i)$, where $\mathbf{s}_t^*$ is the user's true location at t and $\mathbf{s}_i \in \mathcal{S}$. Given the above knowledge, the attackers could infer the user's possible locations at time t, which is probably smaller than the location domain $\mathcal{S}$, and we call it a *constrained domain.*

Definition 2 (Constrained domain). *We denote* $\mathcal{C}_t = \{\mathbf{s}_i | \Pr(\mathbf{s}_t^* = \mathbf{s}_i) > 0, \mathbf{s}_i \in \mathcal{S}\}$ *as constrained domain, which indicates a set of possible locations at* t.

We note that the constrained domain can be explained as the requirement of LBS applications. For example, an App could only be used within a certain area, such as a university free shuttle tracker.

3 Policy Graph Based Location Privacy

In this section, we first formalize the privacy requirement using *location policy graph* in Sect. 3.1. We then design the privacy metric of PGLP in Sect. 3.2. Finally, we propose two mechanisms for PLGP in Sect. 3.3.

3.1 Location Policy Graph

Inspired by Blowfish privacy [25], we use an undirected graph to define what should be protected, i.e., privacy policies. The nodes are secrets, and the edges are the required indistinguishability, which indicates an attacker should not be able to distinguish the input secrets by observing the perturbed output. In our setting, we treat possible locations as nodes and the indistinguishability between the locations as edges.

Definition 3 (Location Policy Graph). *A location policy graph is an undirected graph* $\mathcal{G} = (\mathcal{S}, \mathcal{E})$ *where* $\mathcal{S}$ *denotes all the locations (nodes) and* $\mathcal{E}$ *represents indistinguishability (edges) between these locations.*

Definition 4 (Distance in Policy Graph). *We define the distance between two nodes* $\mathbf{s}_i$ *and* $\mathbf{s}_j$ *in a policy graph as the length of the shortest path (i.e., hops) between them, denoted by* $d_{\mathcal{G}}(\mathbf{s}_i, \mathbf{s}_j)$. *If* $\mathbf{s}_i$ *and* $\mathbf{s}_j$ *are disconnected,* $d_{\mathcal{G}}(\mathbf{s}_i, \mathbf{s}_j) = \infty$.

In DP, the two possible database instances with or without a user's data are called *neighboring databases.* In our location privacy setting, we define neighbors as two nodes with an edge in a policy graph.

Definition 5 (Neighbors). *The neighbors of location* $\mathbf{s}$, *denoted by* $\mathcal{N}(\mathbf{s})$, *is the set of nodes having an edge with* $\mathbf{s}$, *i.e.,* $\mathcal{N}(\mathbf{s}) = \{\mathbf{s}' | d_{\mathcal{G}}(\mathbf{s}, \mathbf{s}') = 1, \mathbf{s}' \in \mathcal{S}\}$.

We denote the nodes having a path with $\mathbf{s}$ by $\mathcal{N}^P(\mathbf{s})$, i.e., the nodes in the same connected component with $\mathbf{s}$. In our framework, we assume the policy graph is given and public. In practice, the location privacy policy can be defined application-wise and identical for all users using the same application.

3.2 Definition of PGLP

We now formalize Policy Graph based Location Privacy (PGLP), which guarantees ϵ-indistinguishability in Definition 1 for every pair of neighbors (i.e., for each edge) in a given location policy graph.

Definition 6 ($\{\epsilon, \mathcal{G}\}$-Location Privacy). *A randomized algorithm $\mathcal{A}$ satisfies $\{\epsilon, \mathcal{G}\}$-location privacy iff for all $z \subseteq Range(\mathcal{A})$ and for all pairs of neighbors s and s' in $\mathcal{G}$, we have $\frac{\Pr(\mathcal{A}(s)=z)}{\Pr(\mathcal{A}(s')=z)} \leq e^{\epsilon}$.*

In PGLP, privacy is rigorously guaranteed through ensuring indistinguishability between any two neighboring locations specified by a customizable location policy graph. The above definition implies the indistinguishability between two nodes that have a path in the policy graph.

Lemma 1. *An algorithm $\mathcal{A}$ satisfies $\{\epsilon, \mathcal{G}\}$-location privacy, iff any two nodes $s_i, s_j \in \mathcal{G}$ are $\epsilon \cdot d_{\mathcal{G}}(s_i, s_j)$-indistinguishable.*

Lemma 1 indicates that, if there is a path between two nodes $\mathbf{s}_i, \mathbf{s}_j$ in the policy graph, the corresponding indistinguishability is required at a certain degree; if two nodes are disconnected, the indistinguishability is not required (i.e., can be ∞) by the policy. As an extreme case, if a node is disconnected with any other nodes, it is allowed to be released without any perturbation.

Comparison with Other Location Privacy Models. We analyze the relation between PGLP and two well-known DP-based location privacy models, i.e., Geo-Indistinguishability (Geo-Ind) [1] and δ-Location Set Privacy [39]. We show that PGLP can represent them under proper configurations of policy graphs.

Geo-Ind [1] guarantees a level of indistinguishability between two locations $\mathbf{s}_i$ and $\mathbf{s}_j$ that is scaled with their Euclidean distance, i.e., $\epsilon \cdot d_E(\mathbf{s}_i, \mathbf{s}_j)$-indistinguishability, where $d_E(\cdot,\cdot)$ denotes Euclidean distance. Note that the unit length used in Geo-Ind scales the level of indistinguishability. We assume that, for any neighbors $\mathbf{s}$ and $\mathbf{s}'$, the unit length used in Geo-Ind makes $d_E(\mathbf{s}, \mathbf{s}') \geq 1$.

Let $\mathcal{G}_1$ be a location policy graph that every location has edges with its closest eight locations on the map, as shown in Fig. 2 (a). We can derive Theorem 1 by Lemma 1 with the fact of $d_{\mathcal{G}}(\mathbf{s}_i, \mathbf{s}_j) \leq d_E(\mathbf{s}_i, \mathbf{s}_j)$ for any $\mathbf{s}_i, \mathbf{s}_j \in \mathcal{G}_1$ (e.g., in Fig. 2(a), $d_{\mathcal{G}}(\mathbf{s}_1, \mathbf{s}_2) = 3$ and $d_E(\mathbf{s}_1, \mathbf{s}_2) = \sqrt{10}$).

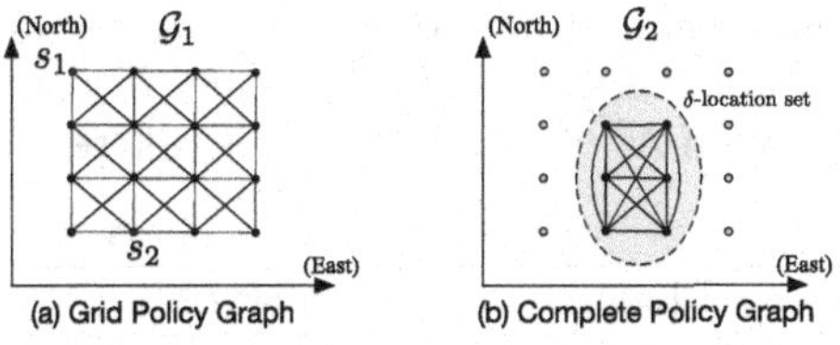

Fig. 2. Two examples of location policy graphs.

Theorem 1. *An algorithm satisfying $\{\epsilon, \mathcal{G}_1\}$-location privacy also achieves ϵ-Geo-Indistinguishability.*

δ-Location Set Privacy [39] extends differential privacy on a subset of possible locations, which is assumed as adversarial knowledge. We note that the constrained domain in Definition 2 can be considered a generalization of δ-location set, whereas we do not specify the calculation of this set in PGLP. δ-Location Set Privacy ensures indistinguishability among any two locations in the δ-location set. Let $\mathcal{G}_2$ be a location policy graph that is complete, i.e., fully connected among all locations in the δ-location set as shown in Fig. 2(b).

Theorem 2. *An algorithm satisfying $\{\epsilon, \mathcal{G}_2\}$-location privacy also achieves δ-Location Set privacy.*

We defer the proofs of the theorems to a full version because of space limitation.

3.3 Mechanisms for PGLP

In the following, we show how to transform existing DP mechanisms into one satisfying PGLP using *graph-calibrated sensitivity*. We temporarily assume the constrained domain $\mathcal{C} = \mathcal{S}$ and study the effect of $\mathcal{C}$ on policy $\mathcal{G}$ in Sect. 4.

As shown in Sect. 2.1, the problem of private location release can be seen as answering a location query $f : \mathcal{S} \to \mathbb{R}^2$ privately. Then we can adapt the existing DP mechanism for releasing private locations by adding random noises to longitude and latitude independently. We use this approach below to adapt the Laplace mechanism and Planar Isotropic Mechanism (PIM) (i.e., an optimal mechanism for Location Set Privacy [39]) to achieve PGLP.

Policy-Based Laplace Mechanism (P-LM). Laplace mechanism is built on the ℓ_1-norm sensitivity [16], defined as the maximum change of the query results due to the difference of neighboring databases. In our setting, we calibrate this sensitivity w.r.t. the neighbors specified in a location policy graph.

Definition 7 (Graph-calibrated ℓ_1-norm Sensitivity). *For a location $\mathbf{s}$ and a query $f(\mathbf{s})$: $\mathbf{s} \to \mathbb{R}^2$, its ℓ_1-norm sensitivity $S_f^{\mathcal{G}}$ is the maximum ℓ_1 norm of $\Delta f^{\mathcal{G}}$ where $\Delta f^{\mathcal{G}}$ is a set of points (i.e., two-dimension vectors) of $(f(\mathbf{s}_i) - f(\mathbf{s}_j))$ for $\mathbf{s}_i, \mathbf{s}_j \in \mathcal{N}^P(\mathbf{s})$ (i.e., the nodes with the same connected component of $\mathbf{s}$).*

We note that, for a true location $\mathbf{s}$, releasing $\mathcal{N}^P(\mathbf{s})$ does not violate the privacy defined by the policy graph. It is because, for any connected $\mathbf{s}$ and $\mathbf{s}'$, $\mathcal{N}^P(\mathbf{s})$ and $\mathcal{N}^P(\mathbf{s}')$ are the same; while, for any disconnected $\mathbf{s}$ and $\mathbf{s}'$, the indistinguishability between $\mathcal{N}^P(\mathbf{s})$ and $\mathcal{N}^P(\mathbf{s}')$ is not required by Definition 6.

Algorithm 1. Policy-based Laplace Mechanism (P-LM)

Require: ϵ, $\mathcal{G}$, the user's true location $\mathbf{s}$.
1: Calculate $S_f^{\mathcal{G}} = sup||(f(\mathbf{s}_i) - f(\mathbf{s}_j))||_1$ for all neighbors $\mathbf{s}_i, \mathbf{s}_j \in \mathcal{N}^P(\mathbf{s})$;
2: Perturb location $\mathbf{z}' = f(\mathbf{s}) + [Lap(S_f^{\mathcal{G}}/\epsilon), Lap(S_f^{\mathcal{G}}/\epsilon)]^T$;
3: **return** a location $\mathbf{z} \in \mathcal{S}$ that is closest to $\mathbf{z}'$ on the map.

Theorem 3. *P-LM satisfies* $\{\epsilon, \mathcal{G}\}$*-location privacy.*

Policy-Based Planar Isotropic Mechanism (P-PIM). PIM [39] achieves the low bound of differential privacy on two-dimension space for Location Set Privacy. It adds noises to longitude and latitude using K-norm mechanism [24] with *sensitivity hull* [39], which extends the convex hull of the sensitivity space in K-norm mechanism. We propose a *graph-calibrated sensitivity hull* for PGLP.

Definition 8 (Graph-calibrated Sensitivity Hull). *For a location* s *and a query* $f(s)$*:* $s \rightarrow \mathbb{R}^2$*, the graph-calibrated sensitivity hull* $K(\mathcal{G})$ *is the convex hull of* $\Delta f^{\mathcal{G}}$ *where* $\Delta f^{\mathcal{G}}$ *is a set of points (i.e., two-dimension vectors) of* $\big(f(s_i) - f(s_j)\big)$ *for any* $s_i, s_j \in \mathcal{N}^P(s)$ *and* s_i, s_j *are neighbors, i.e.,* $K(\mathcal{G}) = Conv(\Delta f^{\mathcal{G}})$*.*

We note that, in Definitions 7 and 8, the sensitivities are independent of the true location $\mathbf{s}$ and all the nodes in $\mathcal{N}(\mathbf{s})$ have the same sensitivity.

Definition 9. (K-norm Mechanism [24]). *Given any function* $f(s)$*:* $s \rightarrow \mathbb{R}^d$ *and its sensitivity hull* K*,* K*-norm mechanism outputs* z *with probability below.*

$$\Pr(z) = \frac{1}{\Gamma(d+1)\textsc{Vol}(K/\epsilon)} exp\left(-\epsilon||z - f(s)||_K\right) \tag{1}$$

where $\Gamma(\cdot)$ *is Gamma function and* $\textsc{Vol}(\cdot)$ *denotes volume.*

Algorithm 2. Policy-based Planar Isotropic Mechanism (P-PIM)

Require: ϵ, $\mathcal{G}$, the user's true location $\mathbf{s}$.
1: Calculate $K(\mathcal{G}) = Conv\big(f(\mathbf{s}_i) - f(\mathbf{s}_j)\big)$ for all neighbors $\mathbf{s}_i, \mathbf{s}_j \in \mathcal{N}^P(\mathbf{s})$;
2: $\mathbf{z}' = f(\mathbf{s}) + Y$ where Y is two-dimension noise drawn by Eq.(1) with sensitivity hull $K(\mathcal{G})$;
3: **return** a location $\mathbf{z} \in \mathcal{S}$ that is closest to $\mathbf{z}'$ on the map.

Theorem 4. *P-PIM satisfies* $\{\epsilon, \mathcal{G}\}$*-location privacy.*

We can prove Theorems 3 and 4 using Lemma 1. The sensitivity is scaled with the graph-based distance. We note that directly using Laplace mechanism or PIM can satisfy a fully connected policy graph over locations in the constrained domain as shown in Fig. 2(b).

Theorem 5. *Algorithm 2 has the time complexity* $O(|\mathcal{C}|\log(h) + h^2\log(h))$ *where* h *is number of vertices on the polygon of* $Conv(\Delta f^{\mathcal{G}})$*.*

4 Policy Graph Under Constrained Domain

In this section, we investigate and prevent the location exposure of a policy graph under constrained domain in Sect. 4.1 and 4.2, respectively; then we repair the policy graph in Sect. 4.3.

4.1 Location Exposure

As shown in Fig. 1 (right) and introduced in Sect. 1, a given policy graph may not be viable under adversarial knowledge of constrained domain (Definition 2). We illustrate the potential risks due to the constrained domain shown in Fig. 3.

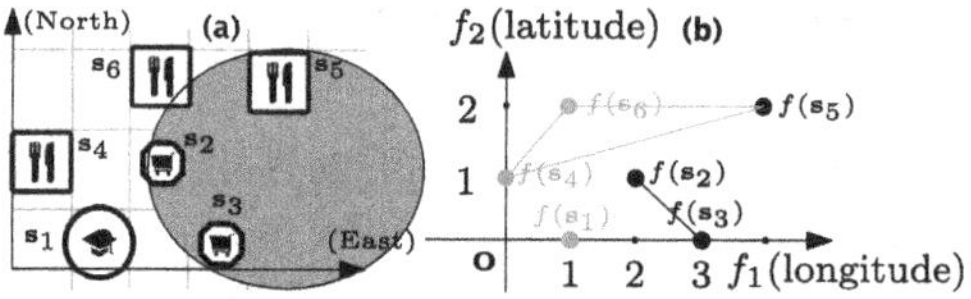

Fig. 3. **(a)** The constrained domain $\mathcal{C} = \{\mathbf{s}_2, \mathbf{s}_3, \mathbf{s}_5\}$; **(b)** The constrained policy graph.

We first examine the immediate consequences of the constrained domain to the policy graph by defining the excluded and disconnected nodes. We then show the disconnected node may lead to *location exposure*.

Definition 9 (Excluded node). *Given a location policy graph $\mathcal{G} = (\mathcal{S}, \mathcal{E})$ and a constrained domain $\mathcal{C} \subset \mathcal{S}$, if $s \in \mathcal{S}$ and $s \notin \mathcal{C}$, $\boldsymbol{s}$ is an excluded node.*

Definition 10 (Disconnected node). *Given a location policy graph $\mathcal{G} = (\mathcal{S}, \mathcal{E})$ and a constrained domain $\mathcal{C} \subset \mathcal{S}$, if a node $\boldsymbol{s} \in \mathcal{C}$, $\mathcal{N}(\boldsymbol{s}) \neq \emptyset$ and $\mathcal{N}(\boldsymbol{s}) \cap \mathcal{C} = \emptyset$, we call $\boldsymbol{s}$ a disconnected node.*

Intuitively, the *excluded* node is outside of the constrained domain $\mathcal{C}$, such as the gray nodes $\{\mathbf{s}_1, \mathbf{s}_4, \mathbf{s}_6\}$ in Fig. 3; whereas the *disconnected* node (e.g., $\mathbf{s}_5$ in Fig. 3) is inside of $\mathcal{C}$ and has neighbors, yet all its neighbors are outside of $\mathcal{C}$.

Next, we analyze the feasibility of a location policy graph under a constrained domain. The first problem is that, by the definition of excluded nodes, it is not possible to achieve indistinguishability between the excluded nodes and any other nodes. For example in Fig. 3, the indistinguishability indicated by the gray edges is not feasible because of $\Pr(\mathcal{A}(\mathbf{s}_4) = \mathbf{z}) = \Pr(\mathcal{A}(\mathbf{s}_6) = \mathbf{z}) = 0$ for any $\mathbf{z}$ given the adversarial knowledge of $\Pr(\mathbf{s}_4) = \Pr(\mathbf{s}_6) = 0$. Hence, one can only achieve a *constrained policy graph*, such as the one with nodes $\{\mathbf{s}_2, \mathbf{s}_3, \mathbf{s}_5\}$ in Fig. 3(b).

Definition 11 (Constrained Location Policy Graph). *A constrained location policy graph $\mathcal{G}^{\mathcal{C}}$ is a subgraph of the original location policy graph $\mathcal{G}$ under a constrained domain $\mathcal{C}$ that only includes the edges inside of $\mathcal{C}$. Formally, $\mathcal{G}^{\mathcal{C}} = (\mathcal{C}, \mathcal{E}^{\mathcal{C}})$ where $\mathcal{C} \subseteq \mathcal{S}$ and $\mathcal{E}^{\mathcal{C}} \subseteq \mathcal{E}$.*

Definition 12 (Location Exposure under constrained domain). *Given a policy graph $\mathcal{G}$, constrained domain $\mathcal{C}$ and an algorithm $\mathcal{A}$ satisfying $(\epsilon, \mathcal{G}^{\mathcal{C}})$-location privacy, for a disconnected node $\boldsymbol{s}$, if $\mathcal{A}$ does not guarantee ϵ-indistinguish-ability between $\boldsymbol{s}$ and any other nodes in $\mathcal{C}$, we call $\boldsymbol{s}$ an isolated node. The user suffers location exposure when she is at the location of the isolated node.*

4.2 Detecting Isolated Node

An interesting finding is that a disconnected node may not always lead to location exposure, which also depends on the algorithm for PGLP. Intuitively, the indistinguishability between a disconnected node and a node in the constrained domain could be guaranteed implicitly. We design Algorithm 3 to detect the isolated node in a constrained policy graph w.r.t. P-PIM. It could be extended to any other PGLP mechanism. For each disconnected node, we check whether it is indistinguishable with other nodes. The problem is equivalent to checking if there is any node inside the convex body $f(\mathbf{s}_i) + K(\mathcal{G}^{\mathcal{C}})$, which can be solved by the convexity property (if a point $\mathbf{s}_j$ is inside a convex hull K, then $\mathbf{s}_j$ can be expressed by the vertices of K with coefficients in $[0, 1]$) with complexity $O(m^3)$. We design a faster method with complexity $O(m^2 log(m))$ by exploiting the definition of convex hull: if $\mathbf{s}_j$ is inside $f(\mathbf{s}_i) + K(\mathcal{G}^{\mathcal{C}})$, then the new convex hull of the new graph by adding edge $\overline{\mathbf{s}_i\mathbf{s}_j}$ will be the same as $K(\mathcal{G}^{\mathcal{C}})$. We give an example of *disconnected but not isolated* node in Appendix A.

Algorithm 3. Finding Isolated Node

Require: $\mathcal{G}$, $\mathcal{C}$, disconnected node $\mathbf{s}_i \in \mathcal{C}$.
$\Delta f^{\mathcal{G}} = \bigvee_{\overline{\mathbf{s}_j\mathbf{s}_k} \in \mathcal{E}^{\mathcal{C}}} (f(\mathbf{s}_j) - f(\mathbf{s}_k))$; ▷ We use $\vee$ to denote Union operator.
$K(\mathcal{G}^{\mathcal{C}}) \leftarrow Conv(\Delta f^{\mathcal{G}})$;
for all $\mathbf{s}_j \in \mathcal{C}, \mathbf{s}_j \neq \mathbf{s}_i$ **do**
  if $Conv(\Delta f^{\mathcal{G}}, f(\mathbf{s}_j) - f(\mathbf{s}_i)) == K(\mathcal{G}^{\mathcal{C}})$ **then**
    return false ▷ not isolated
  end if
end for
return true ▷ isolated

4.3 Repairing Location Policy Graph

To prevent location exposure under the constrained domain, we need to make sure that there is no isolated node in a constrained policy graph. A simple way is to modify the policy graph to ensure the indistinguishability of the isolated node by adding an edge between it and another node in the constrained domain. The selection of such a node could depend on the requirement of the application. Without the loss of generality, a baseline method for repairing the policy graph could be choosing an *arbitrary* node from the constrained domain and adding an edge between it and the isolated node.

A natural question is how can we repair the policy graph with better utility. Since different ways of adding edges in the policy graph may lead to distinct graph-based sensitivity, which is propositional to the area of the sensitivity hull (i.e., a polygon on the map), the question is equivalent to adding an edge with the *minimum* area of sensitivity hull (thus the least noise). We design Algorithm 4 to find the minimum area of the new sensitivity hull, as shown in an example in Fig. 4. The analysis is shown in Appendix B.

We note that both Algorithms 3 and 4 are oblivious to the true location, so they do not consume the privacy budget. Additionally, the adversary may be able to "reverse" the process of graph repair and extract the information about the original location policy graph; however, this does not compromise our privacy notion since the location policy graph is public in our setting.

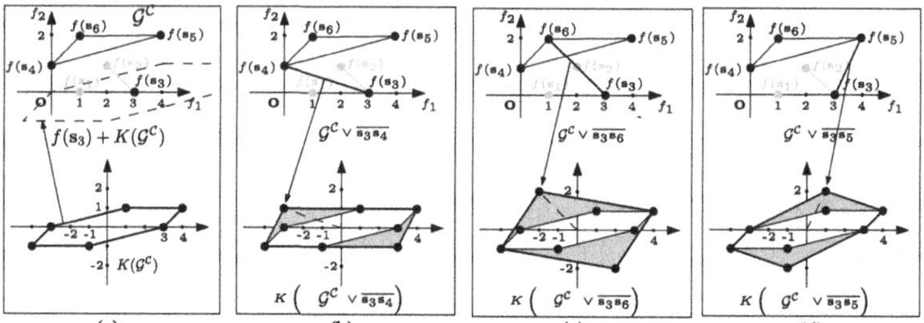

Fig. 4. An example of graph repair with high utility. (a): if $\mathcal{C} = \{\mathbf{s}_3, \mathbf{s}_4, \mathbf{s}_5, \mathbf{s}_6\}$, then $\mathbf{s}_3$ is isolated because $f(\mathbf{s}_3) + K(\mathcal{G}^{\mathcal{C}})$ only contains $\mathbf{s}_3$; to protect $\mathbf{s}_3$, we can re-connect $\mathbf{s}_3$ to one of the valid nodes $\{\mathbf{s}_4, \mathbf{s}_5, \mathbf{s}_6\}$. (b) shows the new sensitivity hull after adding $f(\mathbf{s}_4) - f(\mathbf{s}_3)$ to $K(\mathcal{G}^{\mathcal{C}})$; (c) shows the new sensitivity hull after adding $f(\mathbf{s}_6) - f(\mathbf{s}_3)$ to $K(\mathcal{G}^{\mathcal{C}})$; (d) shows the new sensitivity hull after adding $f(\mathbf{s}_5) - f(\mathbf{s}_3)$ to $K(\mathcal{G}^{\mathcal{C}})$. Because (b) has the smallest area of the sensitivity hull, $\mathbf{s}_3$ should be connected to $\mathbf{s}_4$.

Algorithm 4. Graph Repair with High Utility

Require: $\mathcal{G}$, $\mathcal{C}$, isolated node $\mathbf{s}_i$
$\mathcal{G}^{\mathcal{C}} \leftarrow \mathcal{G} \wedge \mathcal{C}$;
$K \leftarrow K(\mathcal{G}^{\mathcal{C}})$;
$\mathbf{s}_k \leftarrow \emptyset$;
$minArea \leftarrow \infty$;
for all $\mathbf{s}_j \in \mathcal{C}, \mathbf{s}_j \neq \mathbf{s}_i$ **do**
  $K \leftarrow K(\mathcal{G}^{\mathcal{C}} \vee \overline{\mathbf{s}_i\mathbf{s}_j})$; ▷ new sensitivity hull in $O(mlog(m))$
  $Area = \sum_{i=1,j=i+1}^{i=h} det(\mathbf{v}_i, \mathbf{v}_j)$ where $\mathbf{v}_{h+1} = \mathbf{v}_1$; ▷ $\Theta(h)$ time
  if $Area < minArea$ **then**
    $\mathbf{s}_k \leftarrow \mathbf{s}_j$;
    $minArea = Area$; ▷ find minimum area
  end if
end for
$\mathcal{G}^{\mathcal{C}} \leftarrow \mathcal{G}^{\mathcal{C}} \vee \overline{\mathbf{s}_i\mathbf{s}_k}$ ▷ add edge $\overline{\mathbf{s}_i\mathbf{s}_k}$ to the graph
return repaired policy graph $\mathcal{G}^{\mathcal{C}}$;

5 Location Trace Release with PGLP

5.1 Location Release via Hidden Markov Model

A remaining question for continuously releasing private location with PGLP is how to calculate the adversarial knowledge of constrained domain $\mathcal{C}_t$ at each time t. According to our adversary model described in Sect. 2.2, the attacker knows the user's mobility pattern modeled by the Markov chain and the initial probability distribution of the user's location. The attacker also knows the released mechanisms for PGLP. Hence, the problem of calculating the possible location domain (i.e., locations that $\Pr(\mathbf{s}_t^*) > 0$) can be modeled as an inference problem in Hidden Markov Model (HMM) in Fig. 5: the attacker attempts to infer the probability distribution of the true location $\mathbf{s}_t^*$, given the PGLP mechanism, the Markov model of $\mathbf{s}_t^*$, and the observation of $\mathbf{z}_1, \cdots, \mathbf{z}_t$ at the current time t. The constrained domain at each time is derived as the locations in the probability distribution of $\mathbf{s}_t^*$ with non-zero probability.

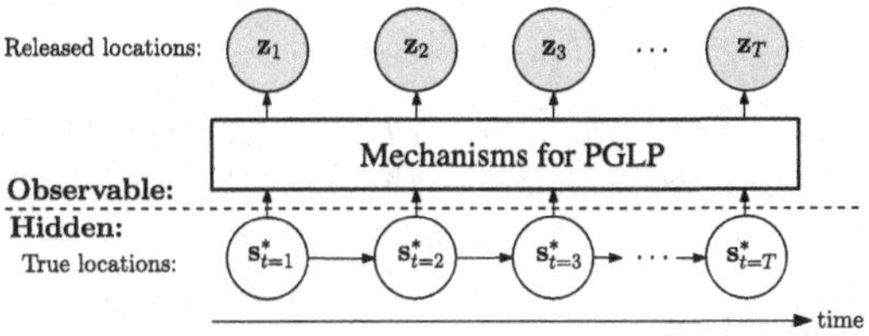

Fig. 5. Private location trace release via HMM.

We elaborate the calculation of the probability distribution of $\mathbf{s}_t^*$ as follows. The probability $\Pr(\mathbf{z}_t|\mathbf{s}_t^*)$ denotes the distribution of the released location $\mathbf{z}_t$ where $\mathbf{s}_t^*$ is the true location at any timestamp t. At timestamp t, we use $\mathbf{p}_t^-$ and $\mathbf{p}_t^+$ to denote the prior and posterior probabilities of an adversary about current state before and after observing $\mathbf{z}_t$ respectively. The prior probability can be derived by the (posterior) probability at previous timestamp $t-1$ and the Markov transition matrix as $\mathbf{p}_t^- = \mathbf{p}_{t-1}^+\mathbf{M}$. The posterior probability can be computed using Bayesian inference as follows. For each state $\mathbf{s}_i$:

$$\mathbf{p}_t^+[i] = \Pr(\mathbf{s}_t^* = \mathbf{s}_i|\mathbf{z}_t) = \frac{\Pr(\mathbf{z}_t|\mathbf{s}_t^* = \mathbf{s}_i)\mathbf{p}_t^-[i]}{\sum_j Pr(\mathbf{z}_t|\mathbf{s}_t^* = \mathbf{s}_j)\mathbf{p}_t^-[j]} \tag{2}$$

Algorithm 5 shows the location trace release algorithm. At each timestamp t, we compute the constrained domain (Line 2). For all disconnected nodes under the constrained domain, we check if they are isolated by Algorithm 3. If so, we derive a minimum protectable graph $\mathcal{G}_t$ by Algorithm 4. Next, we use the proposed PGLP mechanisms (i.e., P-LM or P-PIM) to release a perturbed location $\mathbf{z}_t$. Then the released $\mathbf{z}_t$ will also be used to update the posterior probability $\mathbf{p}_t^+$ (in the equation below) by Eq. (2), which subsequently will be used to compute the prior probability for the next time $t+1$. We note that, only Line 9 (invoking PGLP mechanisms) uses the true location $\mathbf{s}_t^*$. Algorithms 3 and 4 are independent of the true location, so they do not consume the privacy budget.

Algorithm 5. Location Trace Release Mechanism for PGLP

Require: ϵ, $\mathcal{G}$, $\mathbf{M}$, $\mathbf{p}_{t-1}^{+}$, $\mathbf{s}_t^*$

$\mathbf{p}_t^- \leftarrow \mathbf{p}_{t-1}^{+}\mathbf{M}$; ▷ `Markov transition`

$\mathcal{C}_t \leftarrow \{\mathbf{s}_i | \mathbf{p}_t^-[i] > 0\}$; ▷ `constraint`

$\mathcal{G}_t^{\mathcal{C}} \leftarrow \mathcal{G} \wedge \mathcal{C}_t$; ▷ `Definition 12`

for all disconnected node $\mathbf{s}_i$ in $\mathcal{G}_t^{\mathcal{C}}$ **do**

  if $\mathbf{s}_i$ is isolated **then** ▷ `isolated node detection by Algorithm 3`

    $\mathcal{G}_t^{\mathcal{C}} \leftarrow$ ALGORITHM 4($\mathcal{G}_t^{\mathcal{C}}$, $\mathcal{C}_t$, $\mathbf{s}_i$); ▷ `repair graph` $\mathcal{G}_t$ `by Algorithm 4`

  end if

end for

mechanisms for PGLP with parameters ϵ, $\mathbf{s}_t^*$, $\mathcal{G}_t$; ▷ `Algorithms 1 or 2`

Derive $\mathbf{p}_t^+$ by Equation (2); ▷ `inference go to next timestamp`

return ALGORITHM 5(ϵ, $\mathcal{G}_t^{\mathcal{C}}$, $\mathbf{M}$, $\mathbf{p}_t^+$, $\mathbf{s}_{t+1}^*$);

Theorem 6 (Complexity). *Algorithm 5 has complexity $O(dm^2log(m))$ where d is the number of disconnected nodes and m is the number of nodes in $\mathcal{G}_t^{\mathcal{C}}$.*

5.2 Privacy Composition

We analyze the composition of privacy for multiple location releases under PGLP. In Definition 6, we define $\{\epsilon, \mathcal{G}\}$-location privacy for single location release, where ϵ can be considered the privacy leakage w.r.t. the privacy policy $\mathcal{G}$. A natural question is what would be the privacy leakage of multiple releases at a single timestamp (i.e., for the same true location) or at multiple timestamps (i.e., for a trajectory). In either case, the privacy guarantee (or the upper bound of privacy leakage) in multiple releases depends on the achievable location policy graphs. Hence, the key is to study the composition of the policy graphs in multiple releases. Let $\mathcal{A}_1, \cdots, \mathcal{A}_T$ be T independent random algorithms that takes true locations $\mathbf{s}_1^*, \cdots, \mathbf{s}_T^*$ as inputs (note that it is possible $\mathbf{s}_1^* = \cdots = \mathbf{s}_T^*$) and outputs $\mathbf{z}_1, \cdots, \mathbf{z}_T$, respectively. When the viable policy graphs are the same at each release, we have Lemma 2 as below.

Lemma 2. *If all $\mathcal{A}_1, \cdots, \mathcal{A}_T$ satisfy $(\epsilon, \mathcal{G})$-location privacy, the combination of $\{\mathcal{A}_1, \cdots, \mathcal{A}_T\}$ satisfies $(T\epsilon, \mathcal{G})$-location privacy.*

As shown in Sect. 4, the feasibility of achieving a policy graph is affected by the constrained domain, which may change along with the released locations. We denote $\mathcal{G}_1, \cdots, \mathcal{G}_T$ as viable policy graphs at each release (for single location or for a trajectory), which could be obtained by algorithms in Sect. 4.2 and Sect. 4.3. We give a more general composition theorem for PGLP below.

Theorem 7. *If $\mathcal{A}_1, \cdots, \mathcal{A}_T$ satisfy $(\epsilon_1, \mathcal{G}_1), \cdots, (\epsilon_T, \mathcal{G}_T)$, -location privacy, respectively, the combination of $\{\mathcal{A}_1, \cdots, \mathcal{A}_T\}$ satisfies $\left(\sum_{i=1}^{T} \epsilon_i, \mathcal{G}_1 \wedge \cdots \wedge \mathcal{G}_T\right)$-location privacy, where $\wedge$ denotes the intersection between the edges of policy graphs.*

The above theorem provides a method to reason about the overall privacy in continuous releases using PGLP. We note that the privacy composition does not depend on the adversarial knowledge of Markov model, but replies on the soundness of the policy graph and PGLP mechanisms at each t. However, the resulting $\mathcal{G}_1 \wedge \cdots \wedge \mathcal{G}_T$ may not be the original policy graph. It is an interesting future work to study how to ensure a given policy graph across the timeline.

6 Experiments

6.1 Experimental Setting

We implement the algorithms use Python 3.7. The code is available in github[2]. We run the algorithms on a machine with Intel core i7 6770k CPU and 64 GB of memory running Ubuntu 15.10 OS.

Datasets. We evaluate the algorithms on three real-world datasets with similar configurations in [39] for comparison purpose. The Markov models were learned from the raw data. For each dataset, we randomly choose 20 users' location trace with 100 timestamps for testing.

- Geolife dataset [41] contains tuples with attributes of user ID, latitude, longitude and timestamp. We extracted all the trajectories within the Fourth Ring of Beijing to learn the Markov model, with the map partitioned into cells of $0.34 \times 0.34\ km^2$.
- Gowalla dataset [13] contains 6,442,890 check-in locations of 196,586 users over 20 months. We extracted all the check-ins in Los Angeles to train the Markov model, with the map partitioned into cells of $0.37 \times 0.37\ km^2$.
- Peopleflow dataset[3] includes 102,468 locations of 11,406 users with semantic labels of POI in Tokyo. We partitioned the map into cells of $0.27 \times 0.27\ km^2$.

Policy Graphs. We evaluate two types of location privacy policy graphs for different applications as introduced in Sect. 1. One is for the policy of "*allowing the app to access a user's location in which area but ensuring indistinguishability among locations in each area*", represented by G_{k9}, G_{k16}, G_{k25} below. The other is for the policy of "*allowing the app to access the semantic label (e.g., a restaurant or a shop) of a user's location but ensuring indistinguishability among locations with the same category*", represented by G_{poi} below.

- G_{k9} is a policy graph that all locations in each 3×3 region (i.e., 9 grid cells using *grid coordinates*) are fully connected with each other. Similarly, we have G_{k16} and G_{k25} for region size 4×4 and 5×5, respectively.
- G_{poi}: all locations with both the same category and the same 6×6 region are fully connected. We test the category of restaurant in Peopleflow dataset.

Utility Metrics. We evaluate three types of utility (error) for different applications. We run the mechanisms 200 times and average the results. Note that the lower value of the following metrics, the better utility.

[2] https://github.com/emory-aims/pglp.
[3] http://pflow.csis.u-tokyo.ac.jp/.

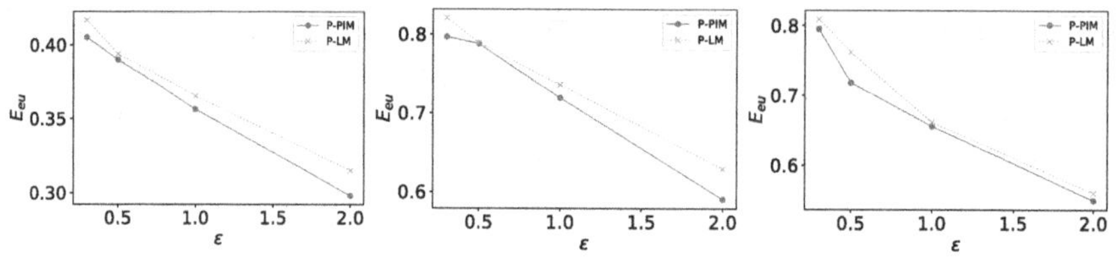

Fig. 6. Utility of P-LM vs. P-PIM with respect to G_{k9}, G_{k16} and G_{poi}.

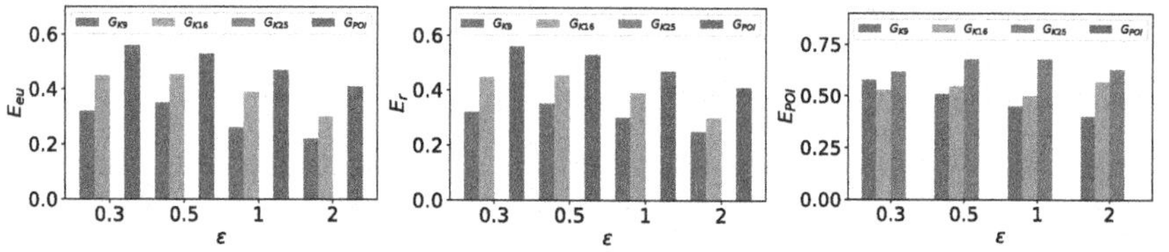

Fig. 7. Utility of different policy graphs.

- The general utility was measured by Euclidean distance (km), i.e., E_{eu}, between the released location and the true location as defined in Sect. 2.2.
- The utility for weather apps or road traffic monitoring, i.e., "whether the released location is in the same region with the true location". We measure it by $E_r = ||R(\mathbf{s}^*), R(\mathbf{z})||_0$ where $R(\cdot)$ is a region query that returns the index of the region. Here we define the region size as 5×5 grid cells.
- The utility for POI mining or crowd monitoring during the pandemic, i.e., "whether the released location is the same category with the true location". We measure it by $E_{poi} = ||C(\mathbf{s}^*), C(\mathbf{z})||_0$ where $C(\cdot)$ returns the category of the corresponding location. We evaluated the location category of "restaurant".

6.2 Results and Analysis

P-LM vs. P-PIM. Figure 6 compares the utility of two proposed mechanisms P-LM and P-PIM for PGLP under the policy graphs G_{k9}, G_{k16} , G_{k25} and G_{poi} on Peopleflow dataset. The utility of P-PIM outperforms P-LM for different policy graphs and different ϵ since the sensitivity hull could achieve lower error bound.

Utility Gain by Tuning Policy Graphs. Figure 7 demonstrates that the utility of different applications can be boosted with appropriate policy graphs. We evaluate the three types of utility metrics using different policy graphs on Peopleflow dataset. Figure 7 shows that, for utility metrics E_{eu}, E_r and E_{poi}, the policy graphs with the best utility are G_{k9}, G_{k25} and G_{poi}, respectively. G_{k9} has smallest E_{eu} because of the least sensitivity. When the query is 5×5 region query, G_{k25} has the full usability (E_r=0). When the query is POI query like the one mentioned above, G_{poi} leads to full utility (E_r=0) since G_{poi} allows to

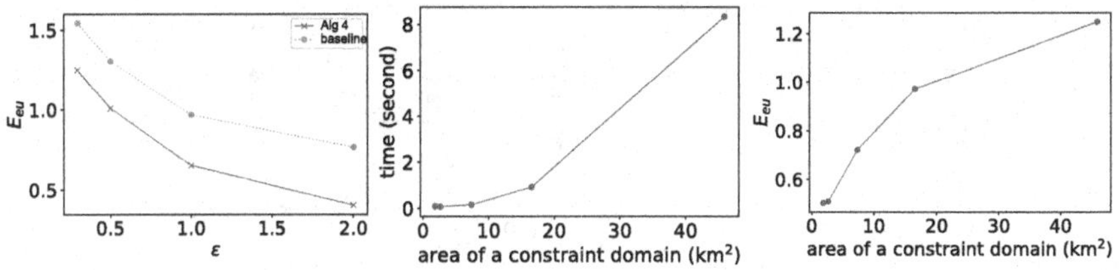

Fig. 8. Evaluation of graph repair.

disclose the semantic category of the true location while maintaining the indistinguishability among the set of locations with the same category. Note that E_{poi} is decreasing with larger ϵ for policy graph G_9 because the perturbed location has a higher probability to be the true location; while this effect is diminished in larger policy graphs such as G_{16} or G_{25} due to their larger sensitivities. We conclude that location policy graphs can be tailored flexibly for better utility-privacy trade-off.

Evaluation of Graph Repair. Figure 8 shows the results of graph repair algorithms. We compare the proposed Algorithm 4 with a baseline method that repairs the problematic policy graph by adding an edge between the isolated node with its nearest node in the constrained domain. It shows that the utility measured by E_{eu} of Algorithm 4 is always better than the baseline but at the cost of higher runtime. Notably, the utility is decreasing (i.e., higher E_{eu}) with larger constrained domains because of larger policy graph (thus higher sensitivity); a larger constrained domain also incurs higher runtime because more isolated nodes need to be processed.

Evaluation of Location Trace Release. We demonstrate the utility of private trajectory release with PGLP in Fig. 9 and Fig. 10. In Fig. 9, we show the results of P-LM and P-PIM on the Geolife Dataset. We test 20 users' trajectories with 100 timestamps and report average E_{eu} at each timestamp. We can see P-PIM has higher utility than P-PIM, which in line with the results for single location release. The error E_{eu} is increasing along with timestamps due to the enlarged constrained domain, which is in line with Fig. 8. The average of E_{eu} across 100 timestamps on different policy graphs, i.e., G_{k9}, G_{k16} and G_{k25} is also in accordance with the single location release in Fig. 7. G_{k9} has the least average error of E_{eu} due to the smallest sensitivity.

In Fig. 10, we show the utility of P-PIM with different policy graphs on two different datasets Geolife and Gowalla. The utility of G_{k9} is always the best over different timestamps for both datasets. In general, the Gowalla dataset has better utility than the Geolife dataset because the constraint domain of the Gowalla dataset is smaller. The reason is that the Gowalla dataset collects check-in locations that have an apparent mobility pattern, as shown in [13]. While Geolife dataset collects GPS trajectory with diverse transportation modes such as walk, bus, or train; thus, the trained Markov model is less accurate.

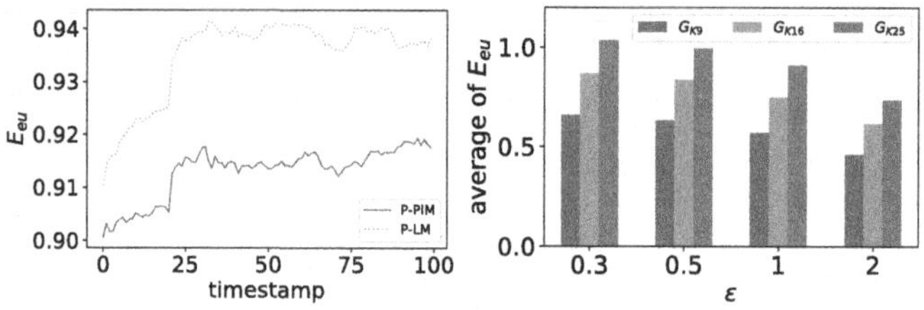

Fig. 9. Utility of private trajectory release with P-LM and P-PIM.

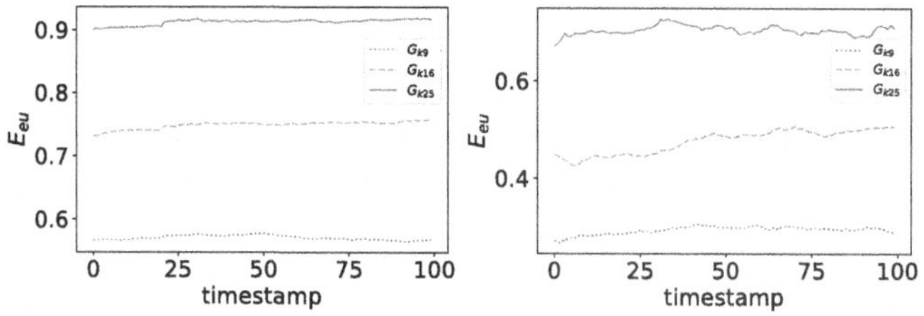

Fig. 10. Utility of private trajectory release with different policy graphs.

7 Conclusion

In this paper, we proposed a flexible and rigorous location privacy framework named PGLP, to release private location continuously under the real-world constraints with customized location policy graphs. We design an end-to-end private location trace release algorithm satisfying a pre-defined location privacy policy.

For future work, there are several promising directions. One is to study how to use the rich interface of PGLP for the utility-privacy tradeoff in the real-world location-based applications, such as carefully designing location privacy policies for COVID-19 contact tracing [4]. Another exciting direction is to design advanced mechanisms to achieve location privacy policies with less noise.

A An Example of Isolated Node

Intuition. We examine the privacy guarantee of P-PIM w.r.t. $\mathcal{G}^{\mathcal{C}}$ in Fig. 3(a). According to K-norm Mechanism [24] in Definition 9, P-PIM guarantees that, for any two neighbors $\mathbf{s}_i$ and $\mathbf{s}_j$, their difference is bounded in the convex body K, i.e. $f(\mathbf{s}_i) - f(\mathbf{s}_j) \in K$. Geometrically, for a location $\mathbf{s}$, all other locations in the convex body of $K + f(\mathbf{s})$ are ϵ-indistinguishable with $\mathbf{s}$.

Example 1 (Disconnected but Not Isolated Node). In Fig. 11, $\mathbf{s}_2$ is disconnected under constraint $\mathcal{C} = \{\mathbf{s}_2, \mathbf{s}_4, \mathbf{s}_5, \mathbf{s}_6\}$. However, $\mathbf{s}_2$ is not isolated because $f(\mathbf{s}_2) + K$ contains $f(\mathbf{s}_4)$ and $f(\mathbf{s}_5)$. Hence, $\mathbf{s}_2$ and other nodes in $\mathcal{C}$ are indistinguishable.

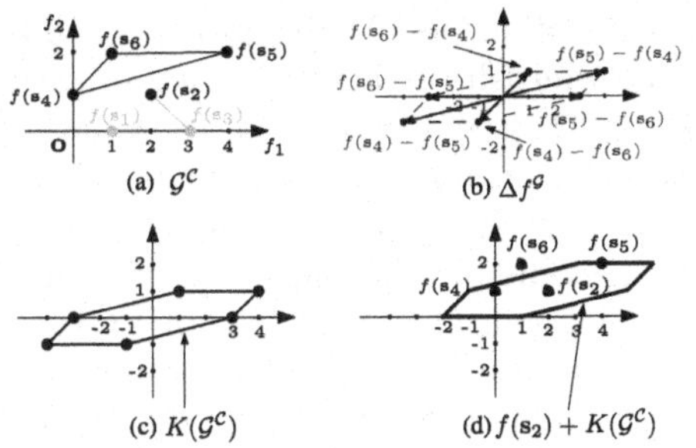

Fig. 11. (a) A policy graph under $\mathcal{C} = \{\mathbf{s}_2, \mathbf{s}_4, \mathbf{s}_5, \mathbf{s}_6\}$; (b) the $\Delta f^{\mathcal{G}}$ of vectors $f(\mathbf{s}_i) - f(\mathbf{s}_j)$; (c) the sensitivity hull $K(\mathcal{G}^{\mathcal{C}})$ covering the $\Delta f^{\mathcal{G}}$; (d) the shape $f(\mathbf{s}_2) + K(\mathcal{G}^{\mathcal{C}})$ containing $f(\mathbf{s}_4)$ and $f(\mathbf{s}_5)$. That is to say, $\mathbf{s}_2$ is indistinguishable with $\mathbf{s}_4$ and $\mathbf{s}_5$.

B Policy Graph Repair Algorithm

Figure 4(a) shows the graph under constraint $\mathcal{C} = \{\mathbf{s}_3, \mathbf{s}_4, \mathbf{s}_5, \mathbf{s}_6\}$. Then $\mathbf{s}_3$ is exposed because $f(\mathbf{s}_3) + K^{\mathcal{G}}$ contains no other node. To satisfy the PGLP without isolated nodes, we need to connect $\mathbf{s}_3$ to another node in $\mathcal{C}$, i.e. $\mathbf{s}_4$, $\mathbf{s}_5$ or $\mathbf{s}_6$.

If $\mathbf{s}_3$ is connected to $\mathbf{s}_4$, then Fig. 4(b) shows the new graph and its sensitivity hull. By adding two new edges $\{f(\mathbf{s}_3) - f(\mathbf{s}_4), f(\mathbf{s}_4) - f(\mathbf{s}_3)\}$ to $\Delta f^{\mathcal{G}}$, the shaded areas are attached to the sensitivity hull. Similarly, Figs. 4(c) and (d) show the new sensitivity hulls when $\mathbf{s}_3$ is connected to $\mathbf{s}_6$ and $\mathbf{s}_5$ respectively. Because the smallest area of $K^{\mathcal{G}}$ is in Fig. 4(b), the repaired graph is $\mathcal{G}^{\mathcal{C}} \vee \overline{\mathbf{s}_3\mathbf{s}_4}$, i.e., add edge $\overline{\mathbf{s}_3\mathbf{s}_4}$ to the graph $\mathcal{G}^{\mathcal{C}}$.

Theorem 8. *Algorithm 4 takes $O(m^2 log(m))$ time where m is the number of valid nodes (with edge) in the policy graph.*

Algorithm. Algorithm 4 derives the minimum protectable graph for location data when an isolated node is detected. We can connect the isolated node $\mathbf{s}_i$ to the rest (at most m) nodes, generating at most m convex hulls where $m = |\mathcal{V}|$ is the number of valid nodes. In two-dimensional space, it only takes $O(mlog(m))$ time to find a convex hull. To derive *Area* of a shape, we exploit the computation of determinant whose intrinsic meaning is the VOLUME of the column vectors. Therefore, we use $\sum_{i=1,j=i+1}^{i=h} det(\mathbf{v}_i, \mathbf{v}_j)$ to derive the Area of a convex hull with clockwise nodes $\mathbf{v}_1, \mathbf{v}_2, \cdots, \mathbf{v}_h$ where h is the number of vertices and $\mathbf{v}_{h+1} = \mathbf{v}_1$. By comparing the area of these convex hulls, we can find the smallest area in $O(m^2 log(m))$ time where m is the number of valid nodes. Note that Algorithms 3 and 4 can also be combined together to protect any disconnected nodes.

C Related Works

C.1 Differential Privacy

While differential privacy [15] has been accepted as a standard notion for privacy protection, the concept of standard differential privacy is not generally applicable for complicated data types. Many variants of differential privacy have been proposed, such as Pufferfish privacy [27], Geo-Indistinguishability [1] and Voice-Indistinguishability [23] (see Survey [33]). Blowfish privacy [25] is the first generic framework with customizable privacy policy. It defines sensitive information as secrets and known knowledge about the data as constraints. By constructing a policy graph, which should also be consistent with all constraints, Blowfish privacy can be formally defined. Our definition of PGLP is inspired by Blowfish framework. Notably, we find that the policy graph may not be viable under temporal correlations represented by Markov model, which was not considered in the previous work. This is also related to another line of works studying how to achieve differential privacy under temporal correlations [8–10,36].

C.2 Location Privacy

A close line of works focus on extending differential privacy to location setting. The first DP-based location privacy model is Geo-Indistinguishability [1], which scales the sensitivity of two locations to their Euclidean distance. Hence, it is suitable for proximity-based applications. Following by Geo-Indistinguishability, several location privacy notions [38,40] have been proposed based on differential privacy. A recent DP-based location privacy, spatiotemporal event privacy [5–7], proposed a new representation for secrets in spatiotemporal data called spatiotemoral events using Boolean expression. It is essentially different from this work since here we are considering the traditional representation of secrets, i.e., each single location or a sequence of locations.

Several works considered Markov models for improving utility of released location traces or web browsing activities [11,17], but did not consider the inference risks when an adversary has the knowledge of the Markov model. Xiao et al. [39] studied how to protect the true location if a user's movement follows Markov model. The technique can be viewed as a special instantiation of PGLP. In addition, PGLP uses a policy graph to tune the privacy and utility to meet diverse the requirement of location-based applications.

References

1. Andrés, M.E., Bordenabe, N.E., Chatzikokolakis, K., Palamidessi, C.: Geo-indistinguishability: differential privacy for location-based systems. In: CCS, pp. 901–914 (2013)
2. Bao, J., Zheng, Yu., Wilkie, D., Mokbel, M.: Recommendations in location-based social networks: a survey. GeoInformatica **19**(3), 525–565 (2015). https://doi.org/10.1007/s10707-014-0220-8
3. Bettini, C., Wang, X.S., Jajodia, S.: Protecting privacy against location-based personal identification. In: Jonker, W., Petković, M. (eds.) SDM 2005. LNCS, vol. 3674, pp. 185–199. Springer, Heidelberg (2005). https://doi.org/10.1007/11552338_13
4. Cao, Y., Takagi, S., Xiao, Y., Xiong, L., Yoshikawa, M.: PANDA: policy-aware location privacy for epidemic surveillance. In: VLDB Demonstration Track (2020, to appear)

5. Cao, Y., Xiao, Y., Xiong, L., Bai, L.: PriSTE: from location privacy to spatiotemporal event privacy. In: 2019 IEEE 35th International Conference on Data Engineering (ICDE), pp. 1606–1609 (2019)
6. Cao, Y., Xiao, Y., Xiong, L., Bai, L., Yoshikawa, M.: PriSTE: protecting spatiotemporal event privacy in continuous location-based services. Proc. VLDB Endow. **12**(12), 1866–1869 (2019)
7. Cao, Y., Xiao, Y., Xiong, L., Bai, L., Yoshikawa, M.: Protecting spatiotemporal event privacy in continuous location-based services. IEEE Trans. Knowl. Data Eng. (2019)
8. Cao, Y., Xiong, L., Yoshikawa, M., Xiao, Y., Zhang, S.: ConTPL: controlling temporal privacy leakage in differentially private continuous data release. VLDB Demonstration Track **11**(12), 2090–2093 (2018)
9. Cao, Y., Yoshikawa, M., Xiao, Y., Xiong, L.: Quantifying differential privacy under temporal correlations. In: 2017 IEEE 33rd International Conference on Data Engineering (ICDE), pp. 821–832 (2017)
10. Cao, Y., Yoshikawa, M., Xiao, Y., Xiong, L.: Quantifying differential privacy in continuous data release under temporal correlations. IEEE Trans. Knowl. Data Eng. **31**(7), 1281–1295 (2019)
11. Chatzikokolakis, K., Palamidessi, C., Stronati, M.: A predictive differentially-private mechanism for mobility traces. In: De Cristofaro, E., Murdoch, S.J. (eds.) PETS 2014. LNCS, vol. 8555, pp. 21–41. Springer, Cham (2014). https://doi.org/10.1007/978-3-319-08506-7_2
12. Chatzikokolakis, K., Palamidessi, C., Stronati, M.: Constructing elastic distinguishability metrics for location privacy. Proc. Priv. Enhancing Technol. **2015**(2), 156–170 (2015)
13. Cho, E., Myers, S.A., Leskovec, J.: Friendship and mobility: user movement in location-based social networks. In: KDD, pp. 1082–1090 (2011)
14. Chow, C.-Y., Mokbel, M.F., Liu, X.: Spatial cloaking for anonymous location-based services in mobile peer-to-peer environments. GeoInformatica **15**(2), 351–380 (2011)
15. Dwork, C.: Differential privacy. In: ICALP, pp. 1–12 (2006)
16. Dwork, C., McSherry, F., Nissim, K., Smith, A.: Calibrating noise to sensitivity in private data analysis. In: Halevi, S., Rabin, T. (eds.) TCC 2006. LNCS, vol. 3876, pp. 265–284. Springer, Heidelberg (2006). https://doi.org/10.1007/11681878_14
17. Fan, L., Bonomi, L., Xiong, L., Sunderam, V.: Monitoring web browsing behavior with differential privacy. In: WWW, pp. 177–188 (2014)
18. Fawaz, K., Shin, K.G.: Location privacy protection for smartphone users. In: CCS, pp. 239–250 (2014)
19. Furuhata, M., Dessouky, M., Ordóñez, F., Brunet, M.-E., Wang, X., Koenig, S.: Ridesharing: the state-of-the-art and future directions. Transp. Res. Part B: Methodol. **57**, 28–46 (2013)
20. Gambs, S., Killijian, M.-O., del Prado Cortez, M.N.: Next place prediction using mobility Markov chains. In: Proceedings of the First Workshop on Measurement, Privacy, and Mobility, pp. 1–6 (2012)
21. Gedik, B., Liu, L.: Protecting location privacy with personalized k-anonymity: Architecture and algorithms. IEEE Trans. Mob. Comput. **7**(1), 1–18 (2008)
22. Gruteser, M., Grunwald, D.: Anonymous usage of location-based services through spatial and temporal cloaking. In: MobiSys, pp. 31–42 (2003)
23. Han, Y., Li, S., Cao, Y., Ma, Q., Yoshikawa, M.: Voice-indistinguishability: protecting voiceprint in privacy-preserving speech data release. In: IEEE ICME (2020)

24. Hardt, M., Talwar, K.: On the geometry of differential privacy. In: STOC, pp. 705–714 (2010)
25. He, X., Machanavajjhala, A., Ding, B.: Blowfish privacy: tuning privacy-utility trade-offs using policies, pp. 1447–1458 (2014)
26. Ingle, M., et al.: Slowing the spread of infectious diseases using crowdsourced data. IEEE Data Eng. Bull. **12** (2020)
27. Kifer, D., Machanavajjhala, A.: A rigorous and customizable framework for privacy. In: PODS, pp. 77–88 (2012)
28. Li, N., Li, T., Venkatasubramanian, S.: t-closeness: privacy beyond k-anonymity and l-diversity. In: IEEE ICDE, pp. 106–115 (2007)
29. Li, N., Lyu, M., Su, D., Yang, W.: Differential privacy: from theory to practice (2016)
30. Luo, Y., Tang, N., Li, G., Li, W., Zhao, T., Yu, X.: DEEPEYE: a data science system for monitoring and exploring COVID-19 data. IEEE Data Eng. Bull. **12** (2020)
31. Machanavajjhala, A., Kifer, D., Gehrke, J., Venkitasubramaniam, M.: L-diversity: privacy beyond k-anonymity. In: IEEE ICDE, p. 24 (2006)
32. Parent, C., et al.: Semantic trajectories modeling and analysis. ACM Comput. Surv. **45**(4), 42:1–42:32 (2013)
33. Pejó, B., Desfontaines, D.: SoK: differential privacies. In: Proceedings on Privacy Enhancing Technologies Symposium (2020)
34. Primault, V., Boutet, A., Mokhtar, S.B., Brunie, L.: The long road to computational location privacy: a survey. IEEE Commun. Surv. Tutor. **21**, 2772–2793 (2018)
35. Recabarren, R., Carbunar, B.: What does the crowd say about you? Evaluating aggregation-based location privacy. WPES **2017**, 156–176 (2017)
36. Song, S., Wang, Y., Chaudhuri, K.: Pufferfish privacy mechanisms for correlated data. In: SIGMOD, pp. 1291–1306 (2017)
37. Sweeney, L.: K-anonymity: a model for protecting privacy. Int. J. Uncertain. Fuzziness Knowl.-Based Syst. **10**(5), 557–570 (2002)
38. Takagi, S., Cao, Y., Asano, Y., Yoshikawa, M.: Geo-graph-indistinguishability: protecting location privacy for LBS over road networks. In: Foley, S.N. (ed.) DBSec 2019. LNCS, vol. 11559, pp. 143–163. Springer, Cham (2019). https://doi.org/10.1007/978-3-030-22479-0_8
39. Xiao, Y., Xiong, L.: Protecting locations with differential privacy under temporal correlations. In: CCS, pp. 1298–1309 (2015)
40. Xiao, Y., Xiong, L., Zhang, S., Cao, Y.: LocLok: location cloaking with differential privacy via hidden Markov model. Proc. VLDB Endow. **10**(12), 1901–1904 (2017)
41. Zheng, Y., Chen, Y., Xie, X., Ma, W.-Y.: GeoLife2.0: a location-based social networking service. In: IEEE MDM, pp. 357–358 (2009)

Where Are You Bob? Privacy-Preserving Proximity Testing with a Napping Party

Ivan Oleynikov[1(✉)], Elena Pagnin[2], and Andrei Sabelfeld[1]

[1] Chalmers University, Gothenburg, Sweden
{ivanol,andrei}@chalmers.se
[2] Lund University, Lund, Sweden
elena.pagnin@eit.lth.se

Abstract. Location based services (LBS) extensively utilize proximity testing to help people discover nearby friends, devices, and services. Current practices rely on full trust to the service providers: users share their locations with the providers who perform proximity testing on behalf of the users. Unfortunately, location data has been often breached by LBS providers, raising privacy concerns over the current practices. To address these concerns previous research has suggested cryptographic protocols for privacy-preserving location proximity testing. Yet general and precise location proximity testing has been out of reach for the current research. A major roadblock has been the requirement by much of the previous work that for proximity testing between Alice and Bob both must be present online. This requirement is not problematic for one-to-one proximity testing but it does not generalize to one-to-many testing. Indeed, in settings like ridesharing, it is desirable to match against ride preferences of all users, not necessarily ones that are currently online.

This paper proposes a novel privacy-preserving proximity testing protocol where, after providing some data about its location, one party can go offline (nap) during the proximity testing execution, without undermining user privacy. We thus break away from the limitation of much of the previous work where the parties must be online and interact directly to each other to retain user privacy. Our basic protocol achieves privacy against semi-honest parties and can be upgraded to full security (against malicious parties) in a straight forward way using advanced cryptographic tools. Finally, we reduce the responding client overhead from quadratic (in the proximity radius parameter) to constant, compared to the previous research. Analysis and performance experiments with an implementation confirm our findings.

Keywords: Secure proximity testing · Privacy-preserving location based services · MPC

1 Introduction

As we digitalize our lives and increasingly rely on smart devices and services, *location based services (LBS)* are among the most widely employed ones. These

L. Chen et al. (Eds.): ESORICS 2020, LNCS 12308, pp. 677–697, 2020.
https://doi.org/10.1007/978-3-030-58951-6_33

range from simple automation like "switch my phone to silent mode if my location is office", to more advanced services that involve interaction with other parties, as in "find nearby coffee shops", "find nearby friends", or "find a ride".

Preserving Privacy for Location Based Services. Current practices rely on full trust to the LBS providers: users share their locations with the providers who manipulate location data on behalf of the users. For example, social apps Facebook and Tinder require access to user location in order to check if other users are nearby. Unfortunately, location and user data has been often breached by the LBS providers [27]. The ridesharing app Uber has been reported to violate location privacy of users by stalking journalists, VIPs, and ex-partners [22], as well as ex-filtrating user location information from its competitors [40]. This raises privacy concerns over the current practices.

Privacy-Preserving Proximity Testing. To address these concerns previous research has suggested cryptographic protocols for privacy-preserving location services. The focus of this paper is on the problem of *proximity testing*, the problem of determining if two parties are nearby without revealing any other information about their location. Proximity testing is a useful ingredient for many LBS. For example, ridesharing services are often based on determining the proximity of ride endpoints [18]. There is extensive literature (discussed in Sect. 6) on the problem of proximity testing [11,20,21,24,29,30,35,36,38,41,42,44].

Generality and Precision of Proximity Testing. Yet general and precise location proximity testing has been out of reach for the current research. A major roadblock has been the requirement that proximity testing between Alice and Bob is only possible in a pairwise fashion and both must be present online. As a consequence, Alice cannot have a single query answered with respect to multiple Bobs, and nor can she is able to check proximity with respect to Bob's preferences unless Bob is online.

The popular ridesharing service BlaBlaCar [4] (currently implemented as a full-trust service) is an excellent fit to illustrate our goals. This service targets intercity rides which users plan in advance. It is an important requirement that users might go off-line after submitting their preferences. The goal is to find rides that start and end at about the same location. Bob (there can be many Bobs) submits the endpoints of a desired ride to the service and goes offline (napping). At a later point Alice queries the service for proximity testing of her endpoints with Bob's. A key requirement is that Alice should be able to perform a one-to-many proximity query, against all Bobs, and compute answer even if Bob is offline. Unfortunately, the vast majority of the previous work [11,20,21, 24,30,35,36,38,41,42,44] fall short of addressing this requirement.

Another key requirement for our work is precision. A large body of prior approaches [11,26,29–31,41,42,44] resort to grid-based approximations where the proximity problem is reduced to the problem of checking whether the parties are located in the same cell on the grid. Unfortunately, grid-based proximity suffers from both false positives and negatives and can be exploited when crossing cell boundaries [9]. In contrast, our work targets precise proximity testing.

This paper addresses privacy-preserving proximity testing with respect to napping parties. Beyond the described offline functionality and precision we summarize the requirements for our solution as follows: (1) security, in the sense that Alice may not learn anything else about Bob's location other than the proximity; Bob should not learn anything about Alice's location; and the service provider should not learn anything about Alice's or Bob's locations; (2) generality, in the sense that the protocol should allow for one-to-many matching without demanding all users to be online; (3) precision, in the sense of a reliable matching method, not an approximate one; (2) lightweight client computation, in the sense of offloading the bulk of work to intermediate servers. We further articulate on these goals in Sect. 2.

Contributions. This paper proposes OLIC (OffLine Inner-Circle), a novel protocol for proximity testing (Sect. 4). We break away from the limitation of much of the previous work where the parties must be online. Drawing on Hallgren et al.'s two-party protocol InnerCircle [20] we propose a novel protocol for proximity testing that utilizes two non-colluding servers. One server is used to blind Bob's location in such a way that the other server can unblind it for any Alice. Once they have uploaded their locations users in our protocol can go offline and retrieve the match outcome the next time they are online.

In line with our goals, we guarantee security with respect to semi-honest parties, proving that the only location information leaked by the protocol is the result of the proximity test revealed to Alice (Sect. 4.2). We then show how to generically mitigate malicious (yet non-colluding) servers by means of zero knowledge proofs and multi-key homomorphic signatures (Sect. 4.3). Generality in the number of users follows from the fact that users do not need to be online in a pairwise fashion, as a single user can query proximity against the encrypted preferences of the other users. We leverage InnerCircle to preserve the precision, avoiding to approximate proximity information by grids or introducing noise. Finally, OLIC offloads the bulk of work from Bob to the servers, thus reducing Bob's computation and communication costs from quadratic (in the proximity radius parameter) to constant. We note, that while InnerCircle can also be trivially augmented with an extra server to offload Bob's computations to, this will add extra security assumptions and make InnerCircle less applicable in practice. OLIC, on the other hand, already requires the servers for Bob to submit his data to, and we get offloading for free. On Alice's side, the computation and communication costs stay unchanged. We develop a proof of concept implementation of OLIC and compare it with InnerCircle. Our performance experiments confirm the aforementioned gains (Sect. 5).

On the 2 Non-Colluding Servers Assumption. We consider this assumption to be realistic, in the sense that it significantly improves on the current practices of a single full-trust server as in BlaBlaCar, while at the same time being compatible with common assumptions in practical cryptographic privacy-preserving systems. For example, Sharemind [39] requires three non-colluding servers for its multi-party computation system based on 3-party additive secret sharing. To the best of our knowledge, OLIC represents the first 2-server solution to perform proximity testing against napping users in ridesharing scenarios, where privacy, precision and efficiency are all mandatory goals. Notably, achieving privacy using a single server is known to be impossible [17]. Indeed, if Bob was to submit his data to only one server, the latter could query itself on this data and learn Bob's location via trilateration attack.

2 Modeling Private Proximity Testing Using Two Servers

The goal of private proximity testing is to enable one entity, that we will call Alice, to find out whether another entity, Bob, is within a certain distance of her. Note that the functionality is asymmetric: only Alice learns whether Bob lies in her proximity. The proximity test should be performed without revealing any additional information regarding the precise locations to one another, or to any third party.

Our Setting. We consider a multi party computation protocol for four parties (Alice, Bob, Server_1, and Server_2) that computes the proximity of Alice's and Bob's inputs in a private way and satisfies the following three constraints:

(C-1) Alice does not need to know Bob's identity before starting the test, nor the parties need to share any common secret;
(C-2) Bob needs to be online only to update his location data. In particular, Bob can 'nap' during the actual protocol execution.
(C-3) The protocol is executed with the help of two servers.

In detail, constraint (C-1) ensures that Alice can look for a match in the database without necessarily targeting a specific user. This may be relevant in situations where one wants to check the availability of a ride 'near by', instead of searching if a specific cab is at reach and aligns with our generality goal towards one-to-many matching. Constraint (C-2) is rarely considered in the literature. The two most common settings in the literature are either to have Alice and Bob communicate directly to one another (which implies that either the two parties need to be online at the same time) [20], or to rely on a single server, which may lead either to 'hiccup' executions (lagging until the other party rejoins online) [30] or to the need for a trusted server. In order to ensure a smooth executions even with a napping Bob and to reduce the trust in one single server, we make use of two servers to store Bob's contribution to the proximity test, that is constraint (C-3). This aligns with our goal of lightweight client computation. We remark that, for a napping Bob, privacy is lost if we use a single server [17].

Finally, unlike [30] we do not require the existence of a social network among system users, to determine who trusts whom, nor do we rely on shared secret keys among users.

Formalizing 'Napping'. We formalize the requirement that 'Bob may be napping' during the protocol execution in the following way. It is possible to split the protocol in two sequential phases. In the first phase, Bob is required to be online and to upload data to the two servers. In the second phase Alice comes online and perform her proximity test query to one server, that we call Server_1. The servers communicate with one another to run the protocol execution, and finally Server_2 returns the result to Alice.

Ideal Functionality. We adopt an ideal functionality that is very close to the one in [20]: if Alice and Bob are within a distance r^2 of each other, the protocol outputs 1 (to Alice), otherwise it outputs 0 (to Alice). Figure 1 depicts this behavior. Alice and Bob are the only parties giving inputs to the protocol. Server_1 and Server_2 do not give any input nor receive any output. This approach aligns with our goal towards precision and a reliable matching method, and brakes away from approximate approaches. For simplicity, we assume the threshold value r^2 to be set a priori, but our protocol accommodates for Alice (or Bob) to choose this value.

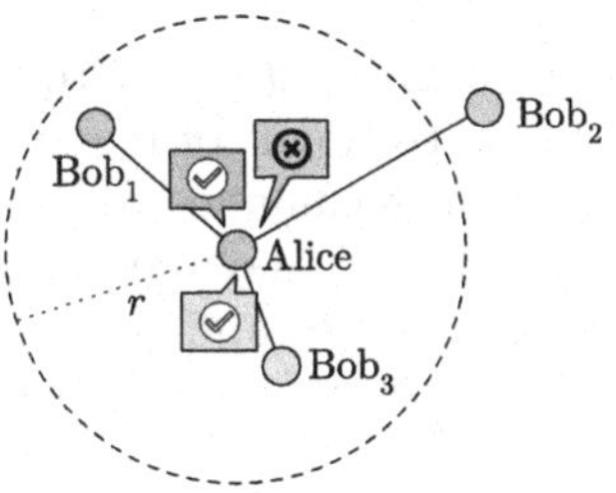

Fig. 1. Figurative representation of the functionality implemented by privacy-preserving location proximity.

Practical Efficiency. Finally, we are interested in solutions that are efficient and run in reasonable time on commodity devices (e.g., laptop computers, smartphones). Concretely, we aim at reducing the computational burden of the clients—Alice and Bob—so that their algorithms can run in just a few seconds, and can in the worst case match the performance of InnerCircle.

Attacker Model. We make the following assumptions:

(A-1) Server_1, Server_2 are not colluding;
(A-2) All parties (Alice, Bob, Server_1, and Server_2) are honest but curious, i.e., they meticulously follow the protocol but may log all messages and attempt to infer some further knowledge on the data they see.

In our model any party could be an attacker that tries to extract un-authorized information from the protocol execution. However, we mainly consider attackers that do so without deviating from their expected execution. In Sect. 4.3, we show that it is possible to relax assumption (A-2) and guarantee users' privacy against 2 malicious servers. In this case, the servers may misbehave and output a different result than what expected from the protocol execution. However, we do not let the two server collude and intentionally share information. While we can handle malicious servers, we cannot tolerate a malicious Alice or Bob. Detecting attacks such as Alice or Bob providing fake coordinates, or Alice trilaterating Bob is outside our scope and should be addressed by different means. We regard such attacks as orthogonal to our contribution and suggest to mitigate them by employing tamper-resistant location devices or location tags techniques [30].

Privacy. Our definition of privacy goes along the lines of [30]. Briefly, a protocol is private if it reveals no information other than the output of the computation, i.e., it has the same behavior as the ideal functionality. To show the privacy of a protocol we adopt the standard definition of simulation based security for deterministic functionalities [28]. Concretely, we will argue the indistinguishably between a real world and an ideal world execution of our protocol, assuming that the building blocks are secure, there is no collusion among any set of parties and all parties are honest but curious.

General Limitations. Proximity testing can be done in a private way only for a limited amount of requests [20,30] in order to avoid trilateration attacks [43]. Practical techniques to limit the number of requests to proximity services are available [34]. Asymmetric proximity testing is suitable for checking for nearby people, devices and services. It is known [44] that the asymmetric setting not directly generalize to mutual location proximity testing.

3 Recap of the InnerCircle Protocol

We begin by giving a high-level overview of the InnerCircle protocol by Hallgren et al. [20] and then describe its relation to our work.

The InnerCircle protocol in [20] allows two parties, Alice and Bob, to check whether the Euclidean distance between them is no greater than a chosen value r. The protocol is privacy preserving, i.e., Alice only learns a bit stating whether the Bob lies closer that r or not, while it does not reveal anything to Bob or to any external party.

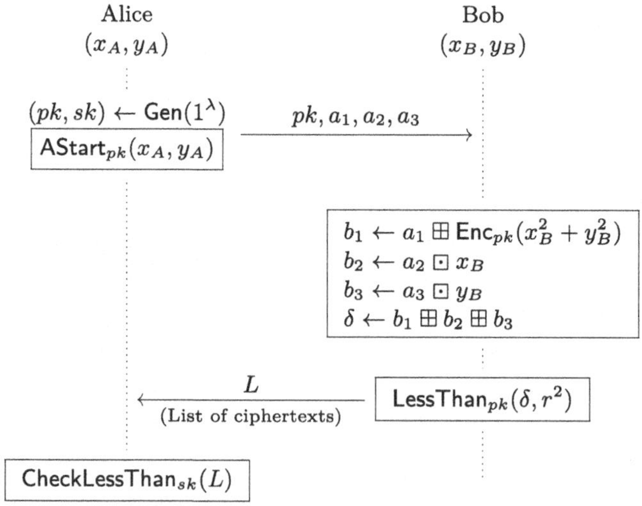

Fig. 2. Diagram of the InnerCircle protocol.

More formally, let (x_A, y_A) and (x_B, y_B) respectively denote Alice's and Bob's locations. Throughout the paper, we will use the shorthand $D = (x_A - x_B)^2 + (y_A - y_B)^2$ to denote the squared Euclidean distance between Alice and Bob and δ is an encryption of D. For for any fixed non-negative integer r, the functionality implemented by the InnerCircle protocol is defined by:

$$F_{\mathsf{InnerCircle}}((x_A, y_A), (x_B, y_B)) = (z, \varepsilon), \text{ where}$$
$$z = \begin{cases} 1, & \text{if } D \le r^2 \quad (D = (x_A - x_B)^2 + (y_A - y_B)^2) \\ 0, & \text{otherwise.} \end{cases}$$

InnerCircle is a single round protocol where the first message is from Alice to Bob and the second one is from Bob to Alice (see Fig. 2).

InnerCircle has three major steps (detailed in Fig. 3):

(1) Alice sends to Bob encryptions of her coordinates together with her public key pk; (2) Bob computes δ (an encryption of D) using Alice's data and his own coordinates, after that he runs LessThan to produce a list L of ciphertexts which encode whether $D \le r^2$ (without knowing D himself) and sends L to Alice; (3) Alice decrypts the list using CheckLessThan and learns whether $D \le r^2$ or not.

Figure 3 contains the definitions of the three procedures used in InnerCircle that will carry on to our OLIC protocol. Below we describe the major ones (LessThan and CheckLessThan) in more detail.

The procedure LessThan (Fig. 3b) produces a set of ciphertexts from which Alice can deduce whether $D \le r^2$ without disclosing the actual value D. The main idea behind this algorithm is that given $\delta = \mathsf{Enc}_{pk}(D)$ Bob can "encode" whether $D = i$ in the value $x \leftarrow \delta \oplus \mathsf{Enc}_{pk}(-i)$, and then "mask" it by multiplying it by a random element $l \leftarrow x \boxdot r$. Observe that if $D - i = 0$ then $\mathsf{Dec}_{sk}(l) = 0$; otherwise if $D - i$ is some invertible element of $\mathcal{M}$ then $\mathsf{Dec}_{sk}(l)$ is uniformly distributed on $\mathcal{M} \setminus \{0\}$. (When we instantiate our OLIC protocol for specific cryptosystem, we will ensure that $D - i$ is either 0 or an invertible element, hence we do not consider here the third case when neither of these holds.) The LessThan procedure terminates with a shuffling of the

$\mathsf{AStart}_{pk}(x_A, y_A)$

$a_1 \leftarrow \mathsf{Enc}_{pk}(x_A^2 + y_A^2)$
$a_2 \leftarrow \mathsf{Enc}_{pk}(2x_A)$
$a_3 \leftarrow \mathsf{Enc}_{pk}(2y_A)$
return (a_1, a_2, a_3)

(a) The AStart algorithm.

$\mathsf{LessThan}_{pk}(\delta, r^2)$

for $i \in \{0 \dots r^2 - 1\}$
  $x_i \leftarrow \delta \boxplus \mathsf{Enc}_{pk}(-i)$
  $t_i \leftarrow \mathcal{M} \setminus \{0\}$
  $l_i \leftarrow x_i \boxdot t_i$
$L \leftarrow [l_0, \dots l_{r^2-1}]$
return Shuffle(L)

(b) The LessThan algorithm.

$\mathsf{CheckLessThan}_{sk}(L)$

$[l_0, l_2 \dots l_{r^2-1}] \leftarrow L$
for $i \in \{0 \dots r^2 - 1\}$
  $v \leftarrow \mathsf{Dec}_{sk}(l_i)$
  if $v = 0$
    return 1
return 0

(c) The CheckLessThan algorithm.

Fig. 3. The core algorithms in the InnerCircle protocol.

list, i.e., the elements in L are reorganized in a random order so that the entry index no longer correspond to the value i.

The procedure CheckLessThan (called inProx in [20]) depicted in Fig. 3c. This procedure takes as input the list L output by LessThan and decrypts one by one all of its components. The algorithm returns 1 if and only if there is a list element that decrypts to 0. This tells Alice whether $D = i$ for some $i < r^2$. We remark that Alice cannot infer any additional information, in particular she cannot extract the exact value i for which the equality holds. In the LessThan procedure Bob computes such "encodings" l for all $i \in \{0 \dots r^2\}$ and accumulates them in a list. So if $D \in \{0 \dots r^2 - 1\}$ is the case, then one of the list elements will decrypt to zero and others will decrypt to uniformly random $\mathcal{M} \setminus \{0\}$ elements, otherwise all of the list elements will decrypt to random nonzero elements. If the cryptosystem allowed multiplying encrypted values, Bob could compute the product of the "encodings" instead of collecting them into a list and send only a single plaintext, but the cryptosystem used here is only additively homomorphic and does not allow multiplications.

4 Private Location Proximity Testing with Napping Bob

Notation. We denote an empty string by ε, and a list of values by $[\cdot]$ and the set of plaintexts (of an encryption scheme) by $\mathcal{M}$. We order parties alphabetically, protocols' inputs and outputs follow the order of the parties. Finally, we denote the computationally indistinguishablity of two distributions D_1, D_2 by $D_1 \stackrel{c}{\equiv} D_2$.

4.1 OLIC: Description of the Protocol

In what follows, we describe our protocol for privacy-preserving location proximity testing with a napping Bob. We name this protocol OLIC (for OffLine InnerCircle) to stress its connection with the InnerCircle protocol and the fact that it can run while Bob is offline. More specifically, instead of exchanging messages directly with one another (as in InnerCircle), in OLIC Alice and Bob communicate with (and through) two servers.

The Ideal Functionality of OLIC. At a high level, OLIC takes as input two locations (x_A, y_A) and (x_B, y_B), and a radius value r; and returns 1 if the Euclidean distance between the locations is less than or equal to r, and 0 otherwise. We perform this test with inputs provided by two parties, Alice and Bob, and the help of two servers, Server$_1$ and Server$_2$. Formally, our ideal functionality has the same input and output behavior as InnerCircle for Alice and Bob, but it additionally has two servers whose inputs and outputs are empty strings ε.

$$F_{\mathsf{OLIC}}((x_A, y_A), (x_B, y_B), \varepsilon, \varepsilon) = (\mathtt{res}, \varepsilon, \varepsilon, \varepsilon),$$
$$\text{where } \mathtt{res} = \begin{cases} 1, & \text{if } (x_A - x_B)^2 + (y_A - y_B)^2 \leq r^2 \\ 0, & \text{otherwise.} \end{cases} \tag{1}$$

In our protocol we restrict the input/output spaces of the ideal functionality of Eq. 1 to values that are meaningful to our primitives. In other words, we require that the values x_A, y_A, x_B, y_B are admissible plaintext (for the encryption scheme employed in OLIC), and that the following values are invertible (in the ring of plaintext) $x_B, y_B, D-i$ for $i \in \{0, \ldots, r^2-1\}$ and $D = (x_A - x_B)^2 + (y_A - y_B)^2$. We will denote the set of all suitable inputs as $\mathcal{S}_\lambda \subseteq \mathcal{M}$.

Figure 4 depicts the flow of the OLIC protocol. Figure 5 contains the detailed description of the procedures called in Fig. 4.

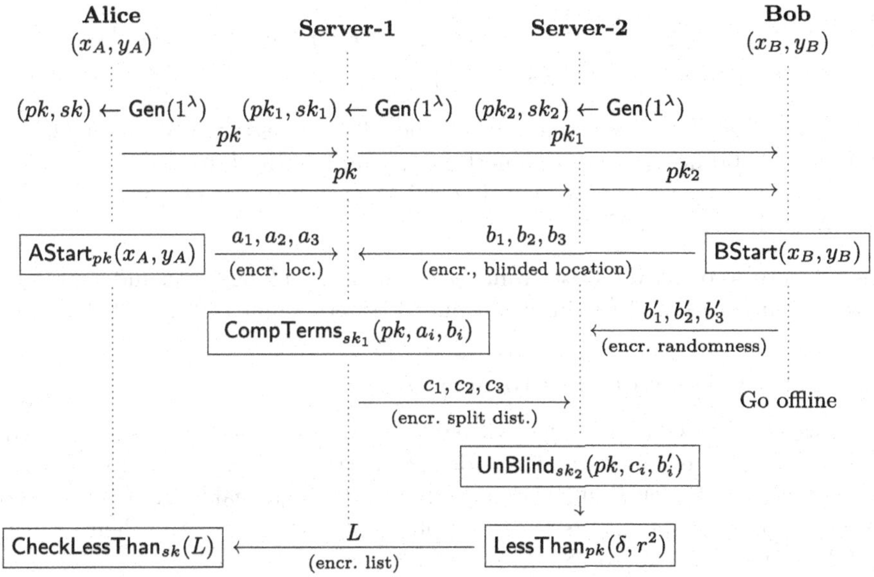

Fig. 4. Overview of the message flow of our OLIC protocol. The message exchanges are grouped to reduce vertical space; in a real execution of the protocol Bob may submit at any time before CompTerms is run.

Step-0: Alice, Server_1, and Server_2 independently generate their keys. Alice sends her pk to Server_1 and Server_2; Server_1 and Server_2 send their respective public keys pk_1, pk_2 to Bob.

Step-1: At any point in time, Bob encodes his location using the BStart algorithm (Fig. 5a), and sends its blinded coordinates to Server_1, and the corresponding blinding factors to Server_2.

Step-2: At any point in time Alice runs AStart (Fig. 3a) and sends her ciphertexts to Server_1.

Step-3: once Server_1 collects Alice's and Bob's data it can proceed with computing 3 addends useful to obtain the (squared) Euclidean distance between their locations. To do so, Server_1 runs $\mathsf{CompTerms}$ (Fig. 5b), and obtains:

$$c_1 = \mathsf{Enc}_{pk}(x_A^2 + y_A^2 + (x_B^2 + y_B^2 + r_1))$$
$$c_2 = \mathsf{Enc}_{pk}(2x_A x_B r_2)$$
$$c_3 = \mathsf{Enc}_{pk}(2y_A y_B r_3)$$

Server_1 sends the above three ciphertexts to Server_2.

Step-4: Server_2 runs $\mathsf{UnBlind}$ (Fig. 5c) to remove Bob's blinding factors from c_1, c_2, c_3 and obtains the encrypted (squared) Euclidean distance between Alice and Bob:

$$\delta = c_1 \boxplus \mathsf{Enc}_{pk}(-r_1) \boxplus c_2 \boxdot \mathsf{Enc}_{pk}(r_2^{-1}) \boxplus c_3 \boxdot \mathsf{Enc}_{pk}(r_3^{-1})$$
$$= \mathsf{Enc}_{pk}((x_A - x_B)^2 + (y_A - y_B)^2)$$

Then Server_2 uses δ and the radius value r to run $\mathsf{LessThan}$ (as in $\mathsf{InnerCircle}$, Fig. 3b), obtains the list of ciphertexts L and returns L to Alice.

Step-5: Alice runs $\mathsf{CheckLessThan}$ (Fig. 3c) to learn whether $D \le r^2$ in the same way as done in $\mathsf{InnerCircle}$.

The correctness of OLIC follows from the correctness of $\mathsf{InnerCircle}$ and a straightforward computation of the input-output of BStart, $\mathsf{CompTerms}$, and $\mathsf{UnBlind}$.

4.2 OLIC: Privacy of the Protocol

We prove that our OLIC provides privacy against semi-honest adversaries. We do so by showing an efficient simulator for each party involved in the protocol and then arguing that each simulator's output is indistinguishable from the view of respective party. More details on the cryptographic primitives and their security model can be found in [13,28].

Theorem 1. *If the homomorphic encryption scheme* $\mathsf{HE} = (\mathsf{Gen}, \mathsf{Enc}, \mathsf{Dec}, \mathsf{Eval})$ *used in* OLIC *is IND-CPA secure and function private then* OLIC *securely realizes the privacy-preserving location proximity testing functionality (in Eq. 1) against semi-honest adversaries.*

At a high level, to prove Theorem 1 we construct one simulator per party and prove that each simulator's output is computationally indistinguishable (by nonuniform algorithms) from the views of each single party in a real protocol execution. The simulators for Bob, Server_1, and Server_2 are trivial — all the messages in their views are either uniformly distributed or encrypted, both of which is easy to simulate (any encrypted value can be simulated as an encryption of 0). The simulator for Alice is slightly more involved, it needs to generate a list of ciphertexts L whose distribution is defined by the output bit `res`: when `res` $= 0$ the L is a list of encryptions of random nonzero elements of $\mathcal{M}$, otherwise one of its elements (the position of which is chosen uniformly at random) contains an encryption of zero. Given `res` this can also be done efficiently. Due to space restriction the detailed proof of Theorem 1 can be found in the extended version of this paper [32].

$\mathsf{BStart}(pk_1, pk_2, x_B, y_B)$

$r_1, \xleftarrow{\$} M$

$r_2, r_3 \xleftarrow{\$} M^*$

$b_1 \leftarrow \mathsf{Enc}_{pk_1}(x_B^2 + y_B^2 + r_1)$

$b_2 \leftarrow \mathsf{Enc}_{pk_1}(x_B r_2)$

$b_3 \leftarrow \mathsf{Enc}_{pk_1}(y_B r_3)$

$b_1' \leftarrow \mathsf{Enc}_{pk_2}(-r_1)$

$b_2' \leftarrow \mathsf{Enc}_{pk_2}(r_2^{-1})$

$b_3' \leftarrow \mathsf{Enc}_{pk_2}(r_3^{-1})$

$\textbf{return } (b_1, b_2, b_3, b_1', b_2', b_3')$

(a) The BStart algorithm.

$\mathsf{CompTerms}_{sk_1}(pk, a_1, a_2, a_3, b_1, b_2, b_3)$

$tmp_1 \leftarrow \mathsf{Dec}_{sk_1}(b_1)$

$tmp_2 \leftarrow \mathsf{Dec}_{sk_1}(b_2)$

$tmp_3 \leftarrow \mathsf{Dec}_{sk_1}(b_3)$

$c_1 \leftarrow a_1 \boxplus \mathsf{Enc}_{pk}(tmp_1)$

$c_2 \leftarrow a_2 \boxdot tmp_2$

$c_3 \leftarrow a_3 \boxdot tmp_3$

$\textbf{return } (c_1, c_2, c_3)$

(b) The CompTerms algorithm.

$\mathsf{UnBlind}_{sk_2}(pk, c_1, c_2, c_3, b_1', b_2', b_3')$

$tmp_1 \leftarrow \mathsf{Dec}_{sk_2}(b_1')$

$tmp_2 \leftarrow \mathsf{Dec}_{sk_2}(b_2')$

$tmp_3 \leftarrow \mathsf{Dec}_{sk_2}(b_3')$

$d_1 \leftarrow c_1 \boxplus \mathsf{Enc}_{pk}(tmp)_1$

$d_2 \leftarrow c_2 \boxdot tmp_2$

$d_3 \leftarrow c_3 \boxdot tmp_3$

$\textbf{return } \delta \leftarrow d_1 \boxplus d_2 \boxplus d_3$

(c) The UnBlind algorithm.

Fig. 5. The new subroutines in OLIC.

Privacy Remark on Bob. In case the two servers collude, Bob looses privacy, in the sense that Server_1 and Server_2 together can recover Bob's location (by unblinding the tmp_i in CompTerms, Fig. 5b). However Alice retains her privacy even in case of colluding servers (thanks to the security of the encryption scheme).

4.3 Security Against Malicious Servers

We describe a generic way to turn OLIC into a protocol secure against malicious, but non-colluding, servers. To do so we employ a suitable non-interactive zero knowledge proof system (NIZK) and a fairly novel cryptographic primitive called multi-key homomorphic signatures (MKHS) [1,10]. The proposed maliciously secure version of OLIC is currently more of a feasibility result rather than a concrete instantiation: to the best of our knowledge there is no combination of MKHS and NIZK that would fit our needs.

Homomorphic signatures [5,14] enable a signer to authenticate their data in such a way that any third party can homomorphically compute on it and obtain (1) the result of the computation, and (2) a signature vouching for the correctness of the latter result. In addition to what we just described, MKHS make it possible to authenticate computation on data signed by multiple sources. Notably, homomorphic signatures and MKHS can be used to efficiently verify that a computation has been carried out in the correct way on the desired data without need to access the original data [10]. This property is what we leverage to make OLIC secure against malicious servers.

At a high level, our proposed solution to mitigate malicious servers works as follows. Alice and Bob hold distinct secret signing keys, sk_A and sk_B respectively, and sign their ciphertexts before uploading them to the servers. In detail, using the notation in Fig. 4, Alice sends to Server_1 the three messages (ciphertexts) a_1, a_2, a_3 along with their respective signatures $\sigma_i^A \leftarrow \mathsf{MKHS.Sign}(\mathsf{sk}_A, a_i, \ell_i)$, where ℓ_i is an opportune label.[1] Bob acts similarly. Server_1 computes f_i (the circuit corresponding to the computation in CompTerms on the i-th input) on each ciphertext and each signature, i.e., $c_i \leftarrow f(a_i, b_i)$ and $\sigma_i' \leftarrow \mathsf{MKHS.Eval}(f, \{\mathsf{pk}_A, \mathsf{pk}_B\}, \sigma_i^A, \sigma_i^B)$. The unforgeability of MKHS ensures that each multi-key signature σ_i' acts as a proof that the respective ciphertext c_i' has been computed correctly (i.e., using f on the desired inputs[2]). Server_2 can be convinced of this fact by checking whether $\mathsf{MKHS.Verif}(f, \{\ell_j\}, \{\mathsf{pk_A}, \mathsf{pk}_B\}, c_i', \sigma_i')$ returns 1. If so, Server_2 proceeds and computes the value δ (and its multi-key signature σ) by evaluating the circuit g corresponding to the function UnBlind. As remarked in Sect. 4.2, privacy demands that the random coefficients, x_i:s, involved in the LessThan procedure are not leaked to Alice. However, without the x_i:s Alice cannot run the MKHS verification (as this randomness should be hardwired in the circuit h corresponding to the function LessThan). To overcome this obstacle we propose to employ a zero knowledge proof system. In this way Server_2 can state that it knows undisclosed values x_i:s such that the output data (the list $L \leftarrow [l_1, \ldots, l_{r^2-1}]$) passes the MKHS verification on the computations dependent on x_i:s. This can be achieved by interpreting the MKHS verification algorithm as a circuit v with inputs x_i (and l_i).

Security Considerations. To guarantee security we need the MKHS to be context hiding (e.g., [37], to prevent leakage of information between Server_1 and Server_2); unforgeable (in the homomorphic sense [10]); and the final proof to be zero knowledge. Implementing this solution would equip OLIC with a quite advanced and heavy machinery that enables Alice to detect malicious behaviors from the servers.

A Caveat on Labels. There is final caveat on data labels [10] needed for the MKHS schemes. We propose to set labels as a string containing the public information: day, time, identity of the user producing the location data, and data

[1] More details on labels at the end of the section.

[2] The 'desired' inputs are indicated by the labels, as we discuss momentarily.

type identifier (e.g., $(1, 3)$ to identify the ciphertext b_3, sent by Bob to Server_1, $(2, 1)$ to identify the ciphertext b'_1, sent by Bob to Server_2, and $(0, 2)$ to identify Alice's ciphertext a_2—Alice only sends data to Server_1). We remark that such label information would be retrievable by InnerCircle as well, as in that case Alice knows the identity of her interlocutor (Bob), and the moment (day and time) in which the protocol is run.

5 Evaluation

In this section we evaluate the performance of our proposal in three different ways: first we provide asymptotic bounds on time complexity of the algorithms in OLIC; second, we provide bounds on the total communication cost of the protocol; finally we develop a proof-of-concept implementation of OLIC to test its performance (running time and communication cost) and compare it against the InnerCircle protocol. Recall that InnerCircle and OLIC implement the same functionality (privacy-preserving proximity testing), however the former requires Bob to be online during the whole protocol execution while the latter does not.

Parameters. As described in Sect. 4.1, OLIC requires Alice to use an additive homomorphic encryption scheme. However, no special property is needed by the ciphertexts from Bob to the servers and between servers. Our implementation employs the ElGamal cryptosystem over a safe-prime order group for ciphertexts to and from Alice, while for the other messages it uses the Paillier cryptosystem. We refer to this implementation as **(non-EC)**, as it does not rely on any elliptic curive cryptograpy (ECC). In order to provide a fair comparison with its predecessor (InnerCircle), we additionally instantiate OLIC using only ECC cryptosystems **(EC)**, namely Elliptic Curve ElGamal.

We note that additive homomorphic ElGamal relies on the face that a plaintext value m is encoded into a group element as g^m (where g denotes the group generator). In this way, the multiplication of $g^{m_1} \cdot g^{m_2}$ returns an encoding of the addition of the corresponding plaintext values $m_1 + m_2$. In order to 'fully' decrypt a ciphertext, one would have to solve the discrete logarithm problem and recover m from g^m, which should be unfeasible, as this corresponds to the security assumption of the encryption scheme. Fortunately, this limitation is not crucial for our protocol. In OLIC, indeed, Alice only needs to check whether a ciphertext is an encryption of zero or not (in the CheckLessThan procedure), and this can be done efficiently without fully decrypting the ciphertext. However, we need to keep it into account when implementing OLIC with ECC. Indeed, in OLIC the servers need to actually fully decrypt Bob's ciphertexts in order to work with the corresponding (blinded) plaintexts. We do so by employing a non-homomorphic version of ElGamal based on the curve M383 in **(EC)** and Paillier cryptosystem in **(non-EC)**.

In our evaluation, we ignore the computational cost of initial key-generation as, in real-life scenarios, this process is a one-time set up and is not needed at every run of the protocol.

5.1 Asymptotic Complexity

Table 1 shows both the concrete numbers of cryptographic operations and our asymptotic bounds on the time complexity of the algorithms executed by each party involved in OLIC. We assume that ElGamal and Paillier encryption, decryption and arithmetic operations, are performed in time $\mathcal{O}(\lambda m)$ (assuming binary exponentiation), where λ here is the security parameter and m is the cost of λ-bit interger multiplication — depends on the specific multiplication algorithm used, e.g., $m = O(n \log n \log \log n)$ for Schönhage–Strassen algorithm. This applies to both **(EC)** and **(non-EC)**, although **(EC)** in practice is used with a lot smaller values of λ.

Table 1. Concrete number of operations and asymptotic time complexity for each party in OLIC (m is the cost of modular multiplication, λ the security parameter and r the radius). In our implementation Alice decryption is actually an (efficient) zero testing.

Party	Cryptosystem operations	Time bound
Alice	3 encryptions r^2 decryptions	$\mathcal{O}(r^2\lambda m)$
Bob	6 encryptions	$\mathcal{O}(\lambda m)$
Server_1	3 decryptions, 3 homomorphic operations,	$\mathcal{O}(\lambda m)$
Server_2	3 decryptions, $2r^2 + 5$ arithmetic operations, r^2 encryptions	$\mathcal{O}(r^2\lambda m)$

Communication Cost. The data transmitted among parties during a whole protocol execution amounts to $r^2 + 6$ ciphertexts between Alice and the servers and 6 ciphertexts between Bob and servers. Each ciphertext consists of 4λ bits in case of **(EC)**, and 2λ bits in case of **(non-EC)** (for both ElGamal and Paillier). Asymptotically, both cases require $\mathcal{O}(\lambda r^2)$ bit of communication—although **(EC)** is used with a lot smaller λ value in implementation.

5.2 Implementation

We developed a prototype implementation of OLIC in Python (available at [32]). The implementation contains all of the procedures shown in pseudocode in Figs. 3 and 5. To ensure a fair compare between OLIC and InnerCircle, we implemented latter in the same environment (following the nomenclature in Fig. 2).

Our benchmarks for the total communication cost of the OLIC protocol are reported in Fig. 6. We measured the total execution time of each procedure of both InnerCircle and OLIC. Figure 7 shows the outcome of our measurements for values of the proximity radius parameter r ranging from 0 to 100. Detailed values can be found in extended version of this paper [32].

Setup. We used cryptosystems from the cryptographic library `bettertimes` [19], which was used for benchmarking of the original InnerCircle protocol. The benchmarks were run on Intel Core i7-8700 CPU running at a frequency of 4.4 GHz. For **(non-EC)** we used 2048-bit keys for ElGamal (the same as in InnerCircle [20]), and 2148-bit keys for Paillier to accommodate a modulus larger than the ElGamal modulus (so that any possible message of Bob fits into the Paillier message space). For **(EC)** we used curves Curve25519 for additive homomorphic encryption and M383 for the ciphertexts exchanged among Bob and the servers from the `ecc-pycrypto` library [7]. We picked these curves because they were available in the library and also because M383 uses a larger modulus and allows us to encrypt the big values from the plaintexts field of Curve25519.

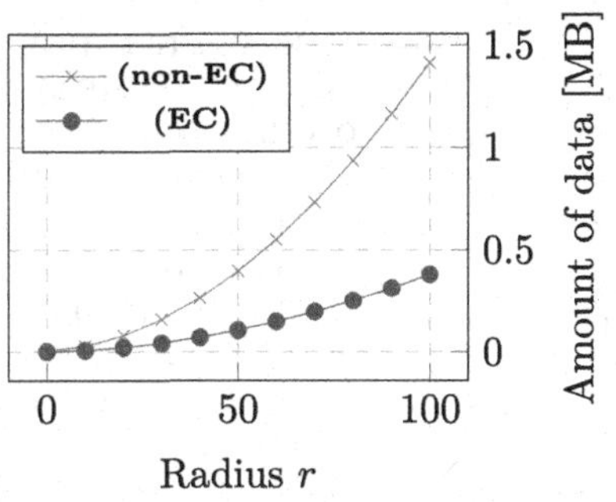

Fig. 6. Total communication cost of OLIC.

The plaintext ring for **(non-EC)** ElGamal-2048 has at least $|\mathcal{M}| \geq 2^{2047}$ elements, which allows Alice and Bob's points to lie on the grid $\{1, 2 \ldots 2^{1023}\}^2$ ensuring that the seqared distance between them never exceeds 2^{2047}. The corresponding plaintext ring size for **(EC)** is $|\mathcal{M}| \geq 2^{251}$ (the group size of Curve25519), and the grid is $\{1, 2 \ldots 2^{125}\}^2$. Since Earth equator is $\approx 2^{26}$ meters long, either of the two grids is more than enough to cover any location on Earth with 1 meter precision.

Optimizations. In InnerCircle [20], the authors perform three types of optimizations on the LessThan procedure:

(O-1) Iterating only through those values of $i \in \{0, \ldots, r^2 - 1\}$ which can be represented as a sum of two squares, i.e., such i that $\exists\, a, b : i = a^2 + b^2$.
(O-2) Precomputing the ElGamal ciphertexts $\mathsf{Enc}_{pk}(-i)$, and only for those i described in (O-1).
(O-3) Running the procedure in parallel, using 8 threads.

We adopt the optimizations (O-1) and (O-2) in our implementation as well. Note that (O-1) reduces the length of list L (Fig. 4), as well as the total communication cost. We disregard optimization (O-3) since we are interested in the total amount of computations a party needs to do (thus this optimization is not present in our implementations of OLIC and InnerCircle).

5.3 Performance Evaluation

Figure 7 shows a comparison of the total running time of each party in OLIC versus InnerCircle. One significant advantage of OLIC is that it offloads the execution of the LessThan procedure from Bob to the servers. This is reflected in

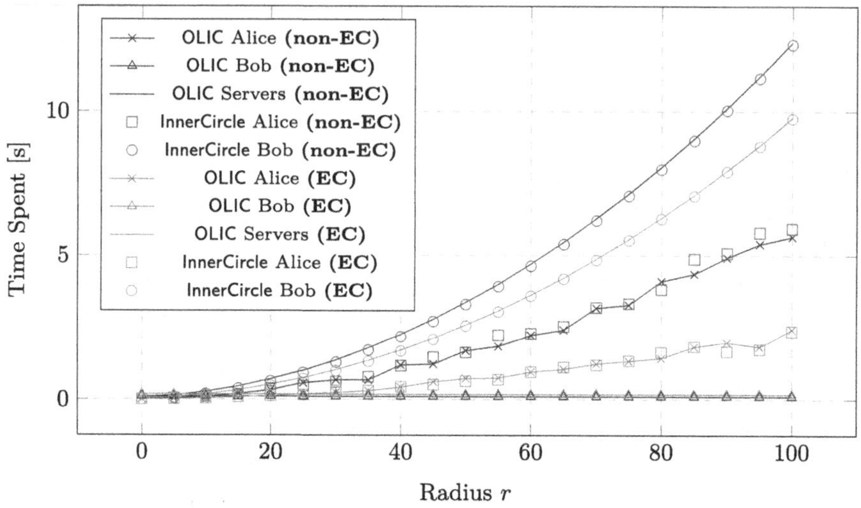

Fig. 7. Running times of each party in InnerCircle and OLIC for both **(non-EC)** and **(EC)** instantiations. (Reported times are obtained as average of 40 executions.)

Fig. 7, where the running time of Bob is flat in OLIC, in contrast to quadratic as in InnerCircle. Indeed, the combined running time of the two servers in OLIC almost matches the running time of Bob in InnerCircle. Note that the servers have to spend total 10–12 s on computations when $r = 100$, which is quite a reasonable amount of time, given that servers are usually not as resource-constrained as clients, and that the most time-consuming procedure LessThan consists of a single loop which can easily be executed in parallel to achieve even better speed. Finally, we remark that the amount of data being sent (Fig. 6) between the parties in OLIC is quite moderate.

In case Alice wants to be matched with multiple Bobs, say, with k of them, the amount of computations that she and the servers perform will grow linearly with k. The same applies to the communication cost. Therefore, one can obtain the counterparts of Figs. 7 and 6 for multiple Bobs by simply multiplying all the plots (except Bob's computations) by k.

6 Related Work

Zhong et al. [44] present the Louis, Lester and Pierre protocols for location proximity. The Louis protocol uses additively homomorphic encryption to compute the distance between Alice and Bob while it relies on a third party to perform the proximity test. Bob needs to be present online to perform the protocol. The Lester protocol does not use a third party but rather than performing proximity testing computes the actual distance between Alice and Bob. The Pierre protocol resorts to grids and leaks Bob's grid cell distance to Alice.

Hide&Crypt by Freni et al. [11] splits proximity in two steps. Filtering is done between a third party and the initiating principal. The two principals then execute computation to achieve finer granularity. In both steps, the granule which a principal is located is sent to the other party. C-Hide&Hash by Mascetti et al. [29] is a centralized protocol, where the principals do not need to communicate pairwise but otherwise share many aspects with Hide&Crypt. FriendLocator by Šikšnys et al. [42] presents a centralized protocol where clients map their position to different granularities, similarly to Hide&Crypt, but instead of refining via the second principal each iteration is done via the third party. VicinityLocator also by Šikšnys et al. [41] is an extension of FriendLocator, which allows the proximity of a principal to be represented not only in terms of any shape.

Narayanan et al. [30] present protocols for proximity testing. They cast the proximity testing problem as equality testing on a grid system of hexagons. One of the protocol utilizes an oblivious server. Parties in this protocol use symmetric encryption, which leads to better performance. However, this requires to have preshared keys among parties, which is less amenable to one-to-many proximity testing. Saldamli et al. [36] build on the protocol with the oblivious server and suggest optimizations based on properties from geometry and linear algebra. Nielsen et al. [31] and Kotzanikolaou et al. [26] also propose grid-based solutions.

Šeděnka and Gasti [38] homomorphically compute distances using the UTM projection, ECEF (Earth-Centered Earth-Fixed) coordinates, and the Haversine formula that make it possible to consider the curvature of the Earth. Hallgren et al. [20] introduce InnerCircle for parallizable decentralized proximity testing, using additively homomorphic encryption between two parties that must be online. The MaxPace [21] protocol builds on speed constraints of an InnerCircle-style protocol as to limit the effects of trilateration attacks. Polakis [34] study different distance and proximity disclosure strategies employed in the wild and experiment with practical effects of trilateration.

Sakib and Huang [35] explore proximity testing using elliptic curves. They require Alice and Bob to be online to be able to run the protocol. Järvinen et al. [24] design efficient schemes for Euclidean distance-based privacy-preserving location proximity. They demonstrate performance improvements over InnerCircle. Yet the requirement of the two parties being online applies to their setting as well. Hallgren et al. [18] how to leverage proximity testing for endpoint-based ridesharing, building on the InnerCircle protocol, and compare this method with a method of matching trajectories.

The computational bottle neck of privacy-preserving proximity testing is the input comparison process. Similarly to [20,33], we rely on homomorphic encryption to compare a private input (the distance between the submitted locations) with a public value (the threshold). Other possible approaches require the use of the arithmetic black box model [6], garbled circuits [25], generic two party computations [12], or oblivious transfer extensions [8].

To summarize, the vast majority [11,20,21,24,30,35,36,38,41,42,44] of the existing approaches to proximity testings require both parties to be online, thus not being suitable for one-to-many matching. A notable exception to the work above is the C-Hide&Hash protocol by Mascetti et al. [29], which allows one-to-many testing, yet at the price of not computing the precise proximity result but its grid-based approximation. Generally, a large number of approaches [11, 26, 29–31, 41, 42, 44] resort to grid-based approximations, thus loosing precision of proximity tests.

There is a number of existing works, which consider the problem of computing generic functions in the setting, where clients are not online during the whole execution. Hallevi et al. [17] consider a one-server scenario and show that the notion of security agains semi-honest adversary (which we prove for our protocol) is impossible to achive with one server. Additionally, the model from [17] lets all the parties know each other's public keys, i.e., the clients know all the other clients who supply inputs for the protocol—this does not allow one-to-many matching, which we achive in our work. Further works [2, 3, 15, 16, 23] also consider one-server scenarios.

7 Conclusions

We have presented OLIC, a protocol for privacy-preserving proximity testing with a napping party. In line with our goals, (1) we achieve privacy with respect to semi-honest parties; (2) we enable matching against offline users which is needed in scenarios like ridesharing; (3) we retain precision, not resorting to grid-based approximations, and (4) we reduce the responding client overhead from quadratic (in the proximity radius parameter) to constant.

Future work avenues include developing a fully-fledged ridesharing system based on our approach, experimenting with scalability, and examining the security and performance in the light of practical security risks for LBS services.

Acknowledgments. This work was partially supported by the Swedish Foundation for Strategic Research (SSF), the Swedish Research Council (VR) and the European Research Council (ERC) under the European Unions's Horizon 2020 research and innovation program under grant agreement No 669255 (MPCPRO). The authors would also like to thank Carlo Brunetta for insightful discussion during the initial development of the work.

References

1. Aranha, D.F., Pagnin, E.: The simplest multi-key linearly homomorphic signature scheme. In: Schwabe, P., Thériault, N. (eds.) LATINCRYPT 2019. LNCS, vol. 11774, pp. 280–300. Springer, Cham (2019). https://doi.org/10.1007/978-3-030-30530-7_14
2. Beimel, A., Gabizon, A., Ishai, Y., Kushilevitz, E., Meldgaard, S., Paskin-Cherniavsky, A.: Non-interactive secure multiparty computation. In: Garay, J.A., Gennaro, R. (eds.) CRYPTO 2014. LNCS, vol. 8617, pp. 387–404. Springer, Heidelberg (2014). https://doi.org/10.1007/978-3-662-44381-1_22

3. Benhamouda, F., Krawczyk, H., Rabin, T.: Robust non-interactive multiparty computation against constant-size collusion. In: Katz, J., Shacham, H. (eds.) CRYPTO 2017. LNCS, vol. 10401, pp. 391–419. Springer, Cham (2017). https://doi.org/10.1007/978-3-319-63688-7_13
4. BlaBlaCar - Trusted carpooling. https://www.blablacar.com/
5. Boneh, D., Freeman, D.M.: Homomorphic signatures for polynomial functions. In: Paterson, K.G. (ed.) EUROCRYPT 2011. LNCS, vol. 6632, pp. 149–168. Springer, Heidelberg (2011). https://doi.org/10.1007/978-3-642-20465-4_10
6. Catrina, O., de Hoogh, S.: Improved primitives for secure multiparty integer computation. In: Garay, J.A., De Prisco, R. (eds.) SCN 2010. LNCS, vol. 6280, pp. 182–199. Springer, Heidelberg (2010). https://doi.org/10.1007/978-3-642-15317-4_13
7. ChangLiu: Ecc-pycrypto Library (2019). https://github.com/lc6chang/ecc-pycrypto. Accessed 14 Apr 2020
8. Couteau, G.: New protocols for secure equality test and comparison. In: Preneel, B., Vercauteren, F. (eds.) ACNS 2018. LNCS, vol. 10892, pp. 303–320. Springer, Cham (2018). https://doi.org/10.1007/978-3-319-93387-0_16
9. Cuéllar, J., Ochoa, M., Rios, R.: Indistinguishable regions in geographic privacy. In: SAC, pp. 1463–1469 (2012)
10. Fiore, D., Mitrokotsa, A., Nizzardo, L., Pagnin, E.: Multi-key homomorphic authenticators. In: Cheon, J.H., Takagi, T. (eds.) ASIACRYPT 2016. LNCS, vol. 10032, pp. 499–530. Springer, Heidelberg (2016). https://doi.org/10.1007/978-3-662-53890-6_17
11. Freni, D., Vicente, C.R., Mascetti, S., Bettini, C., Jensen, C.S.: Preserving location and absence privacy in geo-social networks. In: CIKM, pp. 309–318 (2010)
12. Garay, J., Schoenmakers, B., Villegas, J.: Practical and secure solutions for integer comparison. In: Okamoto, T., Wang, X. (eds.) PKC 2007. LNCS, vol. 4450, pp. 330–342. Springer, Heidelberg (2007). https://doi.org/10.1007/978-3-540-71677-8_22
13. Gentry, C., Halevi, S., Vaikuntanathan, V.: i-hop homomorphic encryption and rerandomizable Yao circuits. In: Rabin, T. (ed.) CRYPTO 2010. LNCS, vol. 6223, pp. 155–172. Springer, Heidelberg (2010). https://doi.org/10.1007/978-3-642-14623-7_9
14. Gorbunov, S., Vaikuntanathan, V., Wichs, D.: Leveled fully homomorphic signatures from standard lattices. In: STOC, pp. 469–477. ACM (2015)
15. Gordon, S.D., Malkin, T., Rosulek, M., Wee, H.: Multi-party computation of polynomials and branching programs without simultaneous interaction. In: Johansson, T., Nguyen, P.Q. (eds.) EUROCRYPT 2013. LNCS, vol. 7881, pp. 575–591. Springer, Heidelberg (2013). https://doi.org/10.1007/978-3-642-38348-9_34
16. Halevi, S., Ishai, Y., Jain, A., Komargodski, I., Sahai, A., Yogev, E.: Non-interactive multiparty computation without correlated randomness. In: Takagi, T., Peyrin, T. (eds.) ASIACRYPT 2017. LNCS, vol. 10626, pp. 181–211. Springer, Cham (2017). https://doi.org/10.1007/978-3-319-70700-6_7
17. Halevi, S., Lindell, Y., Pinkas, B.: Secure computation on the web: computing without simultaneous interaction. In: Rogaway, P. (ed.) CRYPTO 2011. LNCS, vol. 6841, pp. 132–150. Springer, Heidelberg (2011). https://doi.org/10.1007/978-3-642-22792-9_8
18. Hallgren, P., Orlandi, C., Sabelfeld, A.: PrivatePool: privacy-preserving ridesharing. In: CSF, pp. 276–291 (2017)
19. Hallgren, P.: BetterTimes Python Library (2017). https://bitbucket.org/hallgrep/bettertimes/. Accessed 22 Jan 2020

20. Hallgren, P.A., Ochoa, M., Sabelfeld, A.: InnerCircle: a parallelizable decentralized privacy-preserving location proximity protocol. In: PST, pp. 1–6 (2015)
21. Hallgren, P.A., Ochoa, M., Sabelfeld, A.: MaxPace: speed-constrained location queries. In: CNS (2016)
22. Hern, A.: Uber employees 'spied on ex-partners, politicians and Beyoncé' (2016). https://www.theguardian.com/technology/2016/dec/13/uber-employees-spying-ex-partners-politicians-beyonce
23. Jarrous, A., Pinkas, B.: Canon-MPC, a system for casual non-interactive secure multi-party computation using native client. In: Sadeghi, A., Foresti, S. (eds.) WPES, pp. 155–166. ACM (2013)
24. Järvinen, K., Kiss, Á., Schneider, T., Tkachenko, O., Yang, Z.: Faster privacy-preserving location proximity schemes. In: Camenisch, J., Papadimitratos, P. (eds.) CANS 2018. LNCS, vol. 11124, pp. 3–22. Springer, Cham (2018). https://doi.org/10.1007/978-3-030-00434-7_1
25. Kolesnikov, V., Sadeghi, A.-R., Schneider, T.: Improved garbled circuit building blocks and applications to auctions and computing minima. In: Garay, J.A., Miyaji, A., Otsuka, A. (eds.) CANS 2009. LNCS, vol. 5888, pp. 1–20. Springer, Heidelberg (2009). https://doi.org/10.1007/978-3-642-10433-6_1
26. Kotzanikolaou, P., Patsakis, C., Magkos, E., Korakakis, M.: Lightweight private proximity testing for geospatial social networks. Comput. Commun. **73**, 263–270 (2016)
27. Lee, D.: Uber concealed huge data breach (2017). http://www.bbc.com/news/technology-42075306
28. Lindell, Y.: How to simulate it – a tutorial on the simulation proof technique. Tutorials on the Foundations of Cryptography. ISC, pp. 277–346. Springer, Cham (2017). https://doi.org/10.1007/978-3-319-57048-8_6
29. Mascetti, S., Freni, D., Bettini, C., Wang, X.S., Jajodia, S.: Privacy in geo-social networks: proximity notification with untrusted service providers and curious buddies. VLDB J. **20**(4), 541–566 (2011)
30. Narayanan, A., Thiagarajan, N., Lakhani, M., Hamburg, M., Boneh, D.: Location privacy via private proximity testing. In: NDSS (2011)
31. Nielsen, J.D., Pagter, J.I., Stausholm, M.B.: Location privacy via actively secure private proximity testing. In: PerCom Workshops, pp. 381–386. IEEE CS (2012)
32. Oleynikov, I., Pagnin, E., Sabelfeld, A.: Where are you Bob? Privacy-preserving proximity testing with a napping party (2020). https://www.cse.chalmers.se/research/group/security/olic/
33. Pagnin, E., Gunnarsson, G., Talebi, P., Orlandi, C., Sabelfeld, A.: TOPPool: time-aware optimized privacy-preserving ridesharing. PoPETs **2019**(4), 93–111 (2019)
34. Polakis, I., Argyros, G., Petsios, T., Sivakorn, S., Keromytis, A.D.: Where's wally?: Precise user discovery attacks in location proximity services. In: CCS (2015)
35. Sakib, M.N., Huang, C.: Privacy preserving proximity testing using elliptic curves. In: ITNAC, pp. 121–126. IEEE Computer Society (2016)
36. Saldamli, G., Chow, R., Jin, H., Knijnenburg, B.P.: Private proximity testing with an untrusted server. In: WISEC. pp. 113–118. ACM (2013)
37. Schabhüser, L., Butin, D., Buchmann, J.: Context hiding multi-key linearly homomorphic authenticators. In: Matsui, M. (ed.) CT-RSA 2019. LNCS, vol. 11405, pp. 493–513. Springer, Cham (2019). https://doi.org/10.1007/978-3-030-12612-4_25
38. Sedenka, J., Gasti, P.: Privacy-preserving distance computation and proximity testing on earth, done right. In: AsiaCCS, pp. 99–110 (2014)
39. Sharemind MPC Platform. https://sharemind.cyber.ee/sharemind-mpc/multi-party-computation/

40. Shu, C.: Uber reportedly tracked Lyft drivers using a secret software program named 'Hell' (2017). https://techcrunch.com/2017/04/12/hell-o-uber/
41. Siksnys, L., Thomsen, J.R., Saltenis, S., Yiu, M.L.: Private and flexible proximity detection in mobile social networks. In: MDM, pp. 75–84 (2010)
42. Siksnys, L., Thomsen, J.R., Saltenis, S., Yiu, M.L., Andersen, O.: A location privacy aware friend locator. In: SSTD, pp. 405–410 (2009)
43. Veytsman, M.: How I was able to track the location of any tinder user (2014). http://blog.includesecurity.com/
44. Zhong, G., Goldberg, I., Hengartner, U.: Louis, Lester and Pierre: three protocols for location privacy. In: PET, pp. 62–76 (2007)

Password and Policy

Distributed PCFG Password Cracking

Radek Hranický(✉), Lukáš Zobal(✉), Ondřej Ryšavý, Dušan Kolář, and Dávid Mikuš

Faculty of Information Technology, Brno University of Technology, Brno, Czech Republic
{ihranicky,izobal,rysavy,kolar}@fit.vutbr.cz
d.mikus@gmail.com

Abstract. In digital forensics, investigators frequently face cryptographic protection that prevents access to potentially significant evidence. Since users prefer passwords that are easy to remember, they often unwittingly follow a series of common password-creation patterns. A probabilistic context-free grammar is a mathematical model that can describe such patterns and provide a smart alternative for traditional brute-force and dictionary password guessing methods. Because more complex tasks require dividing the workload among multiple nodes, in the paper, we propose a technique for distributed cracking with probabilistic grammars. The idea is to distribute partially-generated sentential forms, which reduces the amount of data necessary to transfer through the network. By performing a series of practical experiments, we compare the technique with a naive solution and show that the proposed method is superior in many use-cases.

Keywords: Distributed · Password · Cracking · Forensics · Grammar

1 Introduction

With the complexity of today's algorithms, it is often impossible to crack a hash of a stronger password in an acceptable time. For instance, verifying a single password for a document created in MS Office 2013 or newer requires 100,000 iterations of SHA-512 algorithm. Even with the use of the popular hashcat[1] tool and a machine with 11 NVIDIA GTX 1080 Ti[2] units, brute-forcing an 8-character alphanumeric password may take over 48 years. Even the most critical pieces of forensic evidence lose value over such a time.

Over the years, the use of probability and statistics showed the potential for a rapid improvement of attacks against human-created passwords [12,13,19]. Various leaks of credentials from websites and services provide an essential source of knowledge about user password creation habits [2,20], including the use of existing words [4] or reusing the same credentials between multiple services [3].

[1] https://hashcat.net/.
[2] https://onlinehashcrack.com/tools-benchmark-hashcat-gtx-1080-ti-1070-ti.

L. Chen et al. (Eds.): ESORICS 2020, LNCS 12308, pp. 701–719, 2020.
https://doi.org/10.1007/978-3-030-58951-6_34

Therefore, the ever-present users' effort to simplify work is also their major weakness. People across the world unwittingly follow common password-creation patterns over and over.

Weir et al. showed how *probabilistic context-free grammars* (PCFG) [19] could describe such patterns. They proposed a technique for automated creation of probabilistic grammars from existing password datasets that serve as *training dictionaries*. A grammar is a mathematical model that allows representing the structure of a password as a composition of fragments. Each fragment is a finite sequence of letters, digits, or special characters. Then, by derivation using rewriting rules of the grammar, one can not only generate all passwords from the original dictionary but produce many new ones that still respect password-creation patterns learned from the dictionary.

The rewriting rules of PCFG have probability values assigned accordingly to the occurrence of fragments in the training dictionary. The probability of each possible password equals the product of probabilities of all rewriting rules used to generate it. Generating password guesses from PCFGs is deterministic and is performed in an order defined by their probabilities. Therefore, more probable passwords are generated first, which helps with a precise targeting of an attack.

Since today's challenges in password cracking often require distributed computing [11], it is necessary to find appropriate mechanisms for PCFG-based password generators. From the entire process, the most complex part that needs distribution is the calculation of password hashes. It is always possible to generate all password guesses on a single node and distribute them to others that perform the hash calculation. However, as we detected, the generating node and the interconnecting network may quickly become a bottleneck. Inspired by Weir's work with preterminal structures [19] and our previously published parallel PCFG cracker [10], we present a distributed solution that only distributes "partially-generated" passwords, and the computing nodes themselves generate the final guesses.

1.1 Contribution

We propose a mechanism for distributed password cracking using probabilistic context-free grammars. The concept uses preterminal structures as basic units for creating work chunks. In the paper, we demonstrate the idea by designing a proof-of-concept tool that also natively supports the deployment of the hashcat tool. Our solution uses adaptive work scheduling to reflect the performance of available computing nodes. We evaluate the technique in a series of experiments by cracking different hash algorithms using different PCFGs, network speeds, and numbers of computing nodes. By comparison with the naive solution, we illustrate the advantages of our concept.

1.2 Structure of the Paper

The paper is structured as follows. Section 2 provides a summary of related work. Section 3 describes the design of our distributed PCFG Manager. Section 4 shows experimental results of our work. Finally, Sect. 5 concludes the paper.

2 Background and Related Work

The use of probabilistic methods for computer-based password cracking dates back to 1980. Martin Hellman introduced a time-memory trade-off method for cracking DES cipher [6]. This chosen-plaintext attack used a precomputation of data stored in memory to reduce the time required to find the encryption key. With a method of distinguished points, Ron Rivest reduced the necessary amount of lookup operations [16]. Phillipe Oechslin improved the original concept and invented rainbow tables as a compromise between the brute-force attack and a simple lookup table. For the cost of space and precomputation, the rainbow table attack reduces the cracking time of non-salted hashes dramatically [14].

The origin of passwords provides another hint. Whereas a machine-generated password may be more or less random, human beings follow specific patterns we can describe mathematically. Markov chains are stochastic models frequently used in natural language processing [15]. Narayanan et al. showed the profit of using zero-order and first-order Markovian models based on the phonetical similarity of passwords to existing words [13]. Hashcat cracking tool utilizes this concept in the default configuration of a mask brute-force attack. The password generator uses a Markov model supplied in an external .hcstat file. Moreover, the authors provide a utility for the creation of new models by automated processing of character sequences from existing wordlists.

Weir et al. introduced password cracking using probabilistic context-free grammars (PCFG) [19]. The mathematical model is based on classic context-free grammars [5] with the only difference that each rewriting rule has a probability value assigned. Similarly to Markovian models, a grammar can be created automatically by training on an existing password dictionary. For generating passwords guesses from an existing grammar, Weir proposed the Next function together with a proof of its correctness. The idea profits from the existence of pre-terminal structures - sentential forms that produce password guesses with the same probability. By using PCFG on MySpace dataset (split to training and testing part), Weir et al. were able to crack 28% to 128% more passwords in comparison with the default ruleset from John the Ripper (JtR) tool[3] using the same number of guesses [19]. The original approach did not distinguish between lowercase and uppercase letter. Thus, Weir extended the original concept by adding special rules for capitalization of letter fragments. Due to high space complexity, the original Next function was later replaced by the Deadbeat dad algorithm [18].

[3] https://www.openwall.com/john/.

Through the following years, Weir's work inspired other researchers as well. Veras et al. proposed a semantic-based extension of PCFGs that makes the provision of the actual meaning of words that create passwords [17]. Ma et al. performed a large-scale evaluation of different probabilistic password models and proposed the use of normalization and smoothing to improve the success rate [12]. Houshmand et al. extended Weir's concept by adding an extra set of rules that respect the position of keys on keyboards to reflect frequent patterns that people use. The extension helped improve the success rate by up to 22%. Besides, they advised to use Laplace probability smoothing, and created guidelines for choosing appropriate attack dictionaries [8]. After that, Houshmand et al. also introduced targeted grammars that utilize information about a user who created the password [7]. Last but not least, Agarwall et al. published a well-arranged overview of new technologies in password cracking techniques, including the state-of-the-art improvements in the area of PCFGs and Markovian models [1].

In our previous research, we focused on the practical aspects of grammar-based attacks and identified factors that influence the time of making password guesses. We introduced parallel PCFG cracking and possibilities for additional filtering of an existing grammar. Especially, removing rules that rewrite the starting non-terminal to long base structures lead to a massive speedup of password guessing with nearly no impact on success rate [10]. In 2019, Weir released[4] a compiled PCFG password guesser written in pure C to achieve faster cracking and easier interconnection with existing tools like hashcat and JtR.

Making the password guessing faster, however, resolves only a part of the problem. Serious cracking tasks often require the use of distributed computing. But how to efficiently deliver the password guesses to different nodes? Weir et al. suggested the possible use of preterminal structures directly in a distributed password cracking trial [19]. To verify the idea, we decided to narrowly analyze the possibilities for distributed PCFG guessing, create a concrete design of intra-node communication mechanisms, and experimentally test its usability.

3 Distributed PCFG Manager

For distributed cracking, we assume a network consisting of a server and a set of clients, as illustrated in Fig. 1. The server is responsible for handling client requests and assigning work. Clients represent the cracking stations equipped with one or more OpenCL-compatible devices like GPU, hardware coprocessors, etc. In our proposal, we talk about a client-server architecture since the clients are actively asking for work, whereas the server is offering a "work assignment service."

In PCFG-based attacks, a probabilistic context-free grammar represents the source of all password guesses, also referred to as candidate passwords. Each guess represents a string generated by the grammar, also known as a *terminal structure* [18,19]. In a distributed environment, we need to deliver the passwords to the cracking nodes somehow. A naive solution is to generate all password

[4] https://github.com/lakiw/compiled-pcfg.

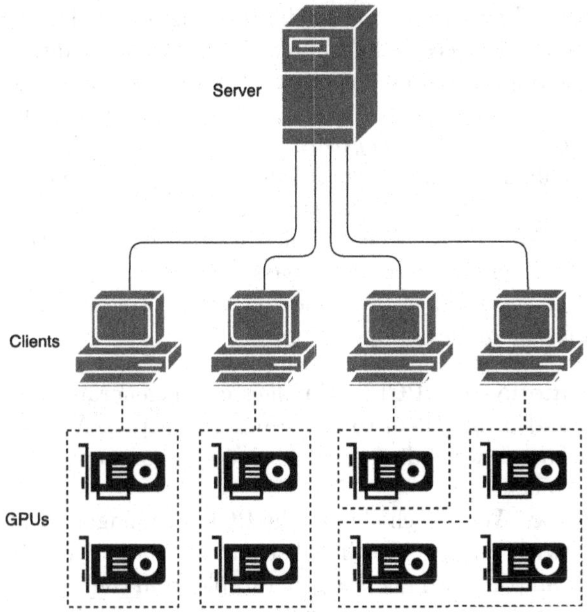

Fig. 1. An example of a cracking network

candidates on the server and distribute them to clients. However, such a method has high requirements on the network bandwidth, and as we detected in our previous research, also high memory requirements to the server [10]. Another drawback of the naive solution is limited scalability. Since all the passwords are generated on a single node, the server may easily become a bottleneck of the entire network.

And thus, we propose a new distributed solution that is inspired by our parallel one [10]. In the following paragraphs, we describe the design and communication protocol of our distributed PCFG Manager. To verify the usability of our concept, we created a proof-of-concept tool that is freely available[5] on GitHub. For implementation, we chose Go[6] programming language, because of its speed, simplicity, and compilation to machine language. The tool can run either as a server or as a client. The PCFG Manager server should be deployed on the server node. It is responsible for processing the input grammar and distribution of work. The PCFG Manager client running on the client nodes generates the actual password guesses. It either prints the passwords directly to the standard output or passes them to an existing hash cracker. The behavior depends on the mode of operation specified by the user. Details are explained in the following paragraphs.

The general idea is to divide the password generation across the computing nodes. The server only generates the preterminal structures (PT), while the

[5] https://github.com/nesfit/pcfg-manager.
[6] https://golang.org/.

terminal structures (T) are produced by the cracking nodes. The work is assigned progressively in smaller pieces called chunks. Each *chunk* produced by the server contains one or more preterminal structures, from which the clients generate the password guesses. To every created chunk, the server assigns a unique identifier called the *sequence number*. The *keyspace*, i.e., the number of possible passwords, of each chunk is calculated adaptively to fit the computational capabilities of a node that will be processing it. Besides that, our design allows direct cracking with hashcat tool. We chose hashcat as a cracking engine for the same reasons as we did for the Fitcrack distributed password cracking system [11], mainly because of its speed and range of supported hash formats. The proposed tool supports two different modes of operation:

- **Generating mode** - the PCFG Manager client generates all possible password guesses and prints them to the standard output. A user can choose to save them into a password dictionary for later use or to pass them to another process on the client-side.
- **Cracking mode** - With each chunk, the PCFG Manager client runs hashcat in stdin mode with the specified hashlist and hash algoritm. By using a pipe, it feeds it with all password guesses generatable from the chunk. Once hashcat processes all possible guesses, the PCFG Manager client returns a result of the cracking process, specifying which hashes were cracked within the chunk and what passwords were found.

3.1 Communication Protocol

The proposed solution uses remote procedure calls with the gRPC[7] framework. For describing the structure of transferred data and automated serialization of payload, we use the Protocol buffers[8] technology.

The server listens on a specified port and handles requests from client nodes. The behavior is similar to the function of Gouroutine M from the parallel version [10] - it generates PT and tailors workunits for client nodes. Each workunit, called chunk, contains one or more PTs. As shown in listing 1.1, the server provides clients an API consisting of four methods. Listing 1.2 shows an overview of input/output messages that are transferred with the calls of API methods.

```
service PCFG {
    rpc Connect (Empty) returns (ConnectResponse) {}
    rpc Disconnect(Empty) returns (Empty);
    rpc GetNextItems(Empty) returns (Items) {}
    rpc SendResult(CrackingResponse) returns (↩
   ResultResponse);
}
```

Listing 1.1. Server API

[7] https://grpc.io/.
[8] https://developers.google.com/protocol-buffers.

When a client node starts, it connects to the server using the `Connect()` method. The server responds with the `ConnectResponse` message containing a PCFG in a compact serialized form. If the desire is to perform an attack on a concrete list of hashes using hashcat, the `ConnectResponse` message also contains a *hashlist* (the list of hashes intended to crack) and a number that defines the *hash mode*[9], i.e., cryptographic algorithms used.

```
message ConnectResponse {
    Grammar grammar = 1;
    repeated string hashList = 2;
    string hashcatMode = 3;
}
message Items {
  repeated TreeItem preTerminals = 1;
}
message ResultResponse {
  bool end = 1;
}
message CrackingResponse {
  map<string, string> hashes = 1;
}
```

Listing 1.2. Messages transferred between the server and clients

Once connected, the client asks for a new chunk of preterminal structures using the `GetNextItems()` method. In response, the server assigns the client a chunk of 1 to N preterminal structures, represented by the `Items` message. After the client generates and processes all possible passwords from the chunk, using the `SendResult()` call, it submits the result in the `CrackingResponse` message. In cracking mode, the message contains a map (an associative array) of cracked hashes together with corresponding plaintext passwords. If no hash is cracked within the chunk or if the PCFG Manager runs in generating mode, the map is empty. With the `ResultResponse` message, the server then informs the client, if the cracking is over or if the client should ask for a new chunk by calling the `GetNextItems()` method.

The last message is `Disconnect()` that clients use to indicate the end of their participation, so that the server can react adequately. For instance, if a client had a chunk assigned, but disconnected without calling the `SendResult()` method, the server may reassign the chunk to a different client. The flow of messages between the server and a client is illustrated in Fig. 2.

3.2 Server

The server represents the controlling point of the computation network. It maintains the following essential data structures:

[9] https://hashcat.net/wiki/doku.php?id=example_hashes.

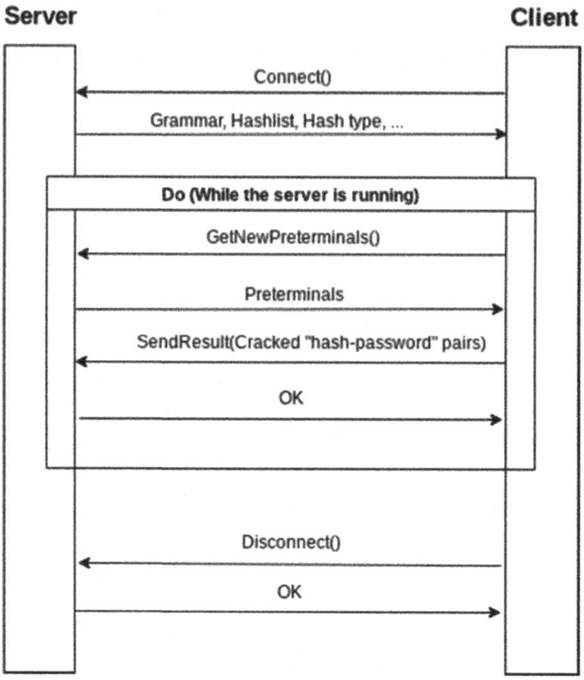

Fig. 2. The proposed communication protocol

- **Priority queue** - the queue is used by the *Deadbeat dad* algorithm [18] for generating preterminal structures.
- **Buffered channel** - the channel represents a memory buffer for storing already generated PTs from which the server creates chunks of work.
- **Client information** - for each connected client, the server maintains its IP address, current performance, the total number of password guesses performed by the client, and information about the last chunk that the client completed: its keyspace and timestamps describing when the processing started and ended. If the client has a chunk assigned, the server also stores its sequence number, PTs, and the keyspace of the chunk.
- **List of incomplete chunks** - the structure is essential for a failure-recovery mechanism we added to the server. If any client with a chunk assigned disconnects before reporting its result, the chunk is added to the list to be reassigned to a different client.
- **List of non-cracked hashes** (cracking mode only) - the list contains all input hashes that have not been cracked yet.
- **List of cracked hashes** (cracking mode only) - The list contains all hashes that have already been cracked, together with corresponding passwords.

Once started, the server loads an input grammar in the format used by Weir's PCFG Trainer[10]. Next, it checks the desired mode of operation and other configuration options – the complete description is available via the tool's help. In cracking mode, the server loads all input hashes to the list of non-cracked hashes. In generating mode, all hashlists remain empty. The server then allocates memory for the buffered channel, where the channel size can be specified by the user.

As soon as all necessary data structures get initialized, the server starts to generate PTs using the *Deadbeat dad* algorithm, and with each PT, it calculates and stores its keyspace. Generated PTs are sent to the buffered channel. The process continues as long as there is free space in the channel, and the grammar allows new PTs to be created. If the buffer infills, generating new PTs is suspended until the positions in the channel get free again.

When a client connects, the server adds a new record to the client information structure. In the `ConnectResponse` message, the client receives the grammar that should be used for generating passwords guesses. In cracking mode, the server also sends the hashlist and hash mode identifying the algorithms that should be used, as illustrated in Fig. 2.

Upon receiving the `GetNextItems()` call, the server pops one or more PTs from the buffered channel and sends them to the client as a new chunk. Besides, the server updates the client information structure to denote what chunk is currently assigned to the client. The number of PTs taken depends on their keyspace. Inspired by our adaptive scheduling algorithm that we Like in Fitcrack [9,11], the system schedules work adaptively to the performance of each client.

In our previous research, we introduced an algorithm for adaptive task scheduling. We integrated the algorithm into our proof-of-concept tool, Fitcrack[11] - a distributed password cracking system, and showed its benefits [9,11]. Therefore, we decided to use a similar technique and schedule work adaptively to each client's performance. In our distributed PCFG Manager, the performance of a client (p_c) in passwords per second is calculated from the keyspace (k_{last}) and computing time (Δt_{last}) of the last assigned chunk. The keyspace of a new chunk (k_{new}) assigned to the client depends on the client's performance and the `chunk_duration` parameter that the user can specify:

$$p_c = \frac{k_{last}}{\Delta t_{last}}, \tag{1}$$

$$k_{new} = p_c * chunk\_duration. \tag{2}$$

The server removes as many PTs from the channel as needed to make the total keyspace of the new chunk at least equal to k_{new}. If the client has not solved any chunk yet, we have no clue to find p_c. Therefore, for the very first chunk, the k_{new} is set to a pre-defined constant.

[10] https://github.com/lakiw/legacy-pcfg/blob/master/python_pcfg_cracker_version3/pcfg_trainer.py.

[11] https://fitcrack.fit.vutbr.cz.

An exception occurs if a client with a chunk assigned disconnects before reporting its result. In such a case, the server saves the assignment to the list of incomplete chunks. If the list is not empty, chunks in it have an absolute priority over the newly created ones. And thus, upon the following `GetNextItems()` call from any client, the server uses the previously stored chunk.

Once a client submits a result via the `SendResult()` call, the server updates the information about the last completed chunk inside the client information structure. For each cracked hash, the server removes it from the list of non-cracked hashes and adds it to the list of cracked hashes together with the resulting password. If all hashes are cracked, the server prints each of them with the correct password and ends. In generating mode, the server continues as long as new password guesses are possible. The same happens in the cracking mode in case there is a non-cracked hash.

Finally, if a client calls the `Disconnect()` method, the server removes its record from the client information structure. As described above, if the client had a chunk assigned, the server will save it for later use.

3.3 Client

After calling the `Connect()` method, a client receives the `ConnectResponse` message containing a grammar. In cracking mode, the message also include a hashlist and a hash mode. Then it calls the `GetNextItems()` method to obtain a chunk assigned by the server. Like in our parallel version, the client then subsequently takes one PT after another and uses the generating goroutines to create passwords from them [10]. In the generating mode, all password guesses are printed to the standard output.

For the cracking mode, it is necessary to have a compiled executable of hashcat on the client node. The user can define the path using the program parameters. The PCFG Manager then starts hashcat in the *dictionary attack* mode with the hashlist and hash mode parameters based on the information obtained from the `ConnectResponse` message. Since no dictionary is specified, hashcat automatically reads passwords from the standard input. And thus, the client creates a *pipe* with the end connected to the hashcat's input. All password guesses are sent to the pipe. From each password, hashcat calculates a cryptographic hash and compares it to the hashes in the hashlist. After generating all password guesses within the chunk, the client closes the pipe, waits for hashcat to end, and reads its return value. On success, the client loads the cracked hashes.

In the end, the client informs the server using the `SendResult()` call. If any hashes are cracked, the client adds them to the `CrackingResponse` message that is sent with the call. The architecture of the PCFG Manager client is displayed in Fig. 3.

4 Experimental Results

We conduct a number of experiments in order to validate several assumptions. First, we want to show the proposed solution results in a higher cracking per-

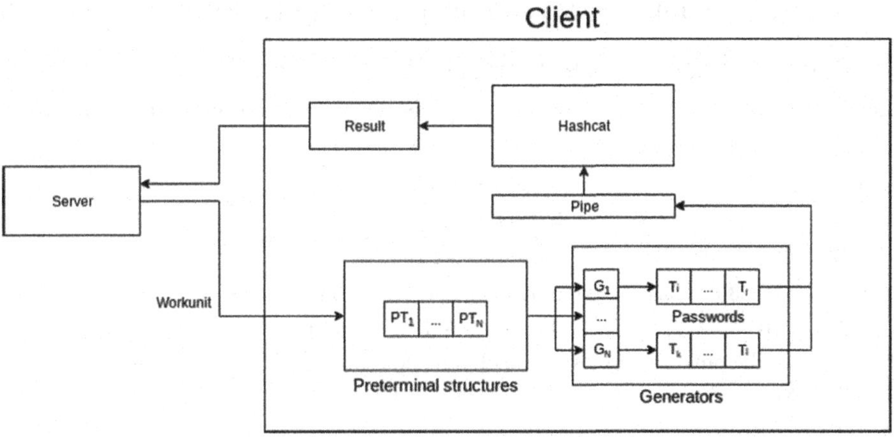

Fig. 3. The architecture of the client side

formance and lower network usage. We also show that while the naive terminal distribution quickly reaches the speed limit by filling the network bandwidth, our solution scales well across multiple nodes. We discuss the differences among different grammars and the impact of scrambling the chunks during the computation. In our experiments, we use up to 16 computing nodes for the cracking tasks and one server node distributing the chunks. All nodes have the following configuration:

- CentOS 7.7 operating system,
- NVIDIA GeForce GTX1050 Ti GPU,
- Intel(R) Core(TM) i5-3570K CPU, and 8 GB RAM.

The nodes are in a local area network connected with links of 10, 100, and 1000 Mbps bandwidth. During the experiments, we incrementally change the network speed to observe the changes. Furthermore, we limit the number of generated passwords to 1, 10, and 100 million.

With this setup, we conduct a number of cracking tasks on different hash types and grammars. As the hash cracking speed has a significant impact on results, we chose bcrypt with five iterations, a computationally difficult hash type, and SHA3-512, an easier, yet modern hash algorithm. Table 1 displays all chosen grammars with description. The columns cover statistics of the source dictionary: password count (pw-cnt) and average password length (avg-len), as well as statistics of the generated grammar: the number of possible passwords guesses (pw-cnt), the number of base structures (base-cnt), their average length (avg-base-len) and the maximum length of base structures (max-base-len), in nonterminals. One can also notice the enormous number of generated passwords, especially with the myspace grammar. Such a high number is caused only by few base structures with many nonterminals. We discussed the complexity added by long base structure in our previous research [10]. If not stated otherwise, the

Table 1. Grammars used in the experiments

Dictionary statistics			PCFG statistics			
name	**pw-cnt**	**avg-len**	**pw-cnt**	**base-cnt**	**avg-base-len**	**max-base-len**
myspace	37,145	8.59	6E+1874	1,788	4.50	600
cain	306,707	9.27	3.17E+15	167	2.59	8
john	3,108	6.06	1.32E+09	72	2.14	8
phpbb	184,390	7.54	2.84E+37	3,131	4.11	16
singles	12,235	7.74	6.67E+11	227	3.07	8
dw17[a]	10,000	7.26	2.92E+15	106	2.40	12

[a]https://github.com/danielmiessler/SecLists/blob/master/Passwords/darkweb2017-top10000.txt

grammars are generated from password lists found on SkullSecurity wiki page[12]. For each combination of described parameters, we run two experiments – first, with the naive terminal distribution (terminal), and second using our solution with the preterminal distribution (preterminals).

4.1 Computation Speedup and Scaling

The primary goal is to show that the proposed solution provides faster password cracking using PCFG. In Fig. 4, one can see the average cracking speed of SHA3-512 hash with *myspace* grammar, with different task sizes and network bandwidths.

Apart from the proposed solution being generally faster, we see a significant difference in speeds with the lower network bandwidths. This is well seen in the detailed graph in Fig. 5 which shows the cracking speed of SHA3-512 in a 10Mbps network with different task sizes. The impact of the network bandwidth limit is expected as the naive terminal distribution requires a significant amount of data in the form of a dictionary to be transmitted. In the naive solution, network links become the main bottleneck that prevents achieving higher cracking performance. In our solution, the preterminal distribution reduces data transfers dramatically, which removes the obstacle and allow for achieving higher cracking speeds.

Apart from SHA3-512, we conducted the same set of experiments with SHA1 and MD5 hashes. Results from cracking these hashes are not present in the paper as they are very similar to SHA3. The reason for this is the cracking performance of SHA1 and MD5 is very high, similar to the former.

Figure 6 illustrates the network activity using both solutions. The graph compares the data transfered in time in SHA3-512 cracking task with the lowest limit on network bandwidth. With the naive terminal distribution one can notice the transfer speed is limited for the entire experiment with small pauses for generating the terminals. As a result, the total amount of data transferred is more than

[12] https://wiki.skullsecurity.org/Passwords.

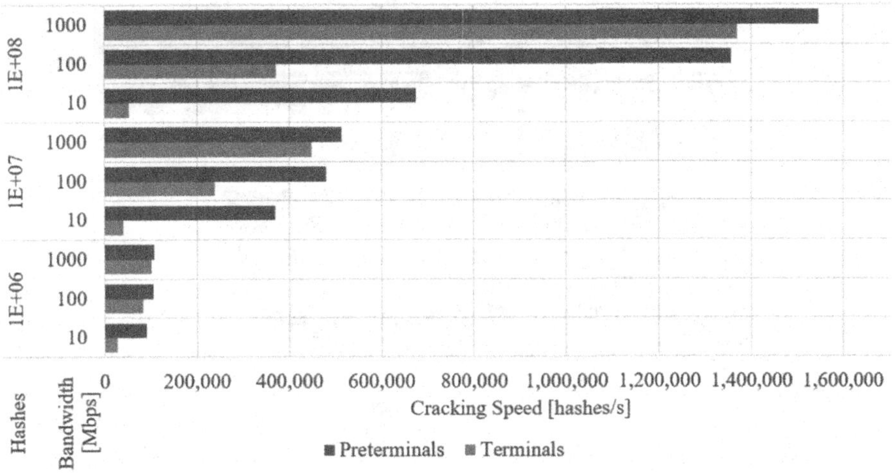

Fig. 4. Average cracking speed with different bandwidths and password count (SHA3-512/*myspace* grammar/4 nodes)

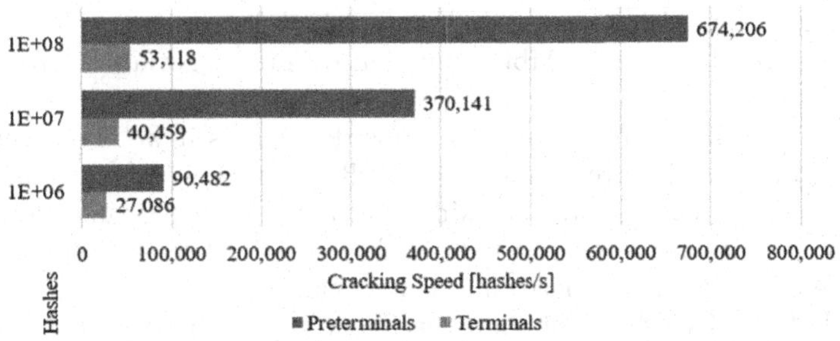

Fig. 5. Detail of 10Mbps network bandwidth experiment (SHA3-512/*myspace* grammar / 4 nodes)

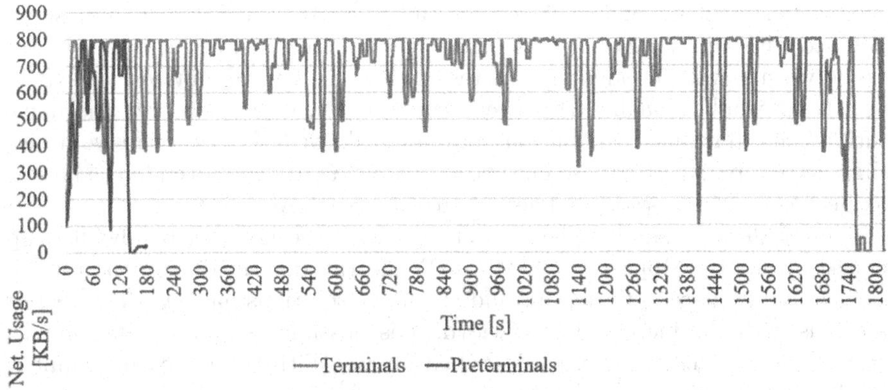

Fig. 6. Comparison of network activity (SHA3-512/*myspace* grammar/100 million hashes/10 Mbps network bandwidth/16 nodes)

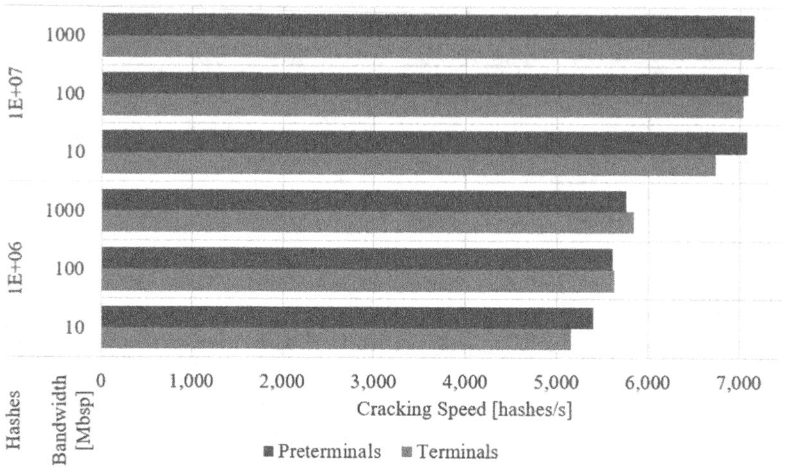

Fig. 7. Average cracking speed with different bandwidths and password count (bcrypt/*myspace* grammar)

15 times bigger than with the proposed preterminal distribution. The maximum speed has not reached the 10 Mbit limit due to auto-negotiation problems on the local network.

The difference between the two solutions disappears if we crack very complex hash algorithms. Figure 7 shows the results of cracking bcrypt hashes with *myspace* grammar. The average cracking speeds are multiple times lower than with SHA3. In this case, the two solutions do not differ because the transferred chunks have much lower keyspace since clients can not verify as many hashes as was possible for SHA3. Most of the experiment time is used by hashcat itself, cracking the hashes.

In the previous graphs, one could also notice the cracking speed increases with more hashes. This happens since smaller tasks cannot fully leverage the whole distributed network as the smallest task took only several seconds to crack.

Figure 8 shows that the average cracking speed is influenced by the number of connected cracking nodes. While for the smallest task, there is almost no difference with the increasing node count, for the largest task, the speed rises even between 8 and 16 nodes. As this task only takes several minutes, we expect larger tasks would visualize this even better. One can also observe the naive solution using terminal distribution does not scale well. Even though we notice a slight speedup up to 4 nodes. The speed is capped after that even in the largest task because of the network bottleneck mentioned above.

Figure 9 shows a similar picture but now with the network bandwidth cap increased from 100 Mbps to 1000 Mbps. While the described patterns remain visible, the difference between our and the naive solutions narrows. The reason for this is with the increased bandwidth, it is possible to send greater chunks of pre-generated dictionaries, as described above. With the increasing number of nodes and cracking task length, advantages of the proposed solution become more clear, as seen at the top of the described graph.

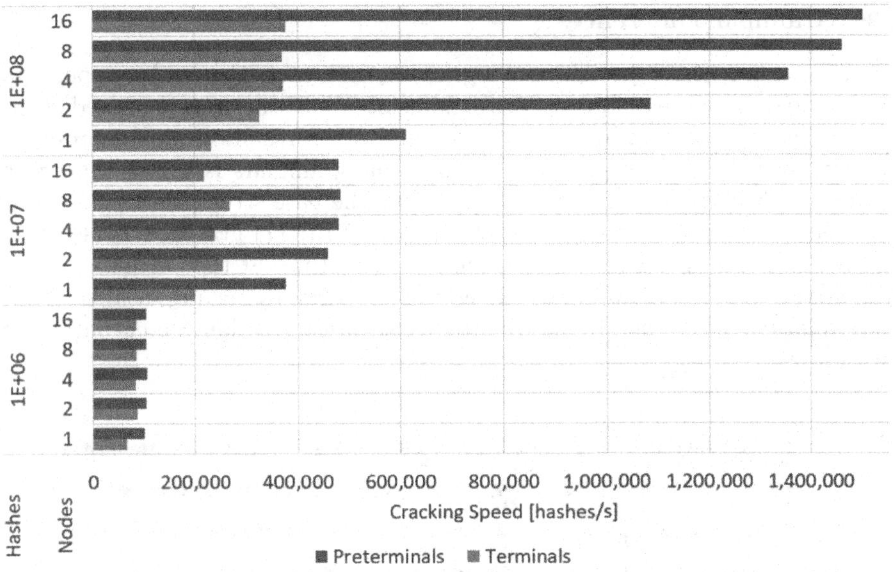

Fig. 8. Scaling across multiple cracking nodes (SHA3-512/*myspace* grammar/100 Mbps network bandwidth)

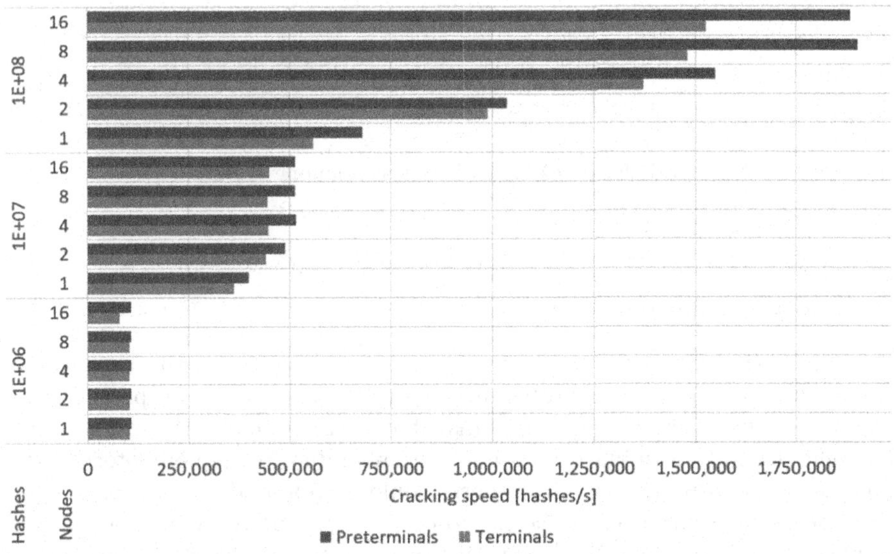

Fig. 9. Scaling across multiple cracking nodes (SHA3-512/*myspace* grammar/1000 Mbps network bandwidth)

4.2 Grammar Differences

Next, we measure how the choice of a grammar influences the cracking speed. In Fig. 10, we can see the differences are significant. While the cracking speed of the naive solution is capped by the network bandwidth, results from the proposed solution show generating passwords using some grammars is slower than with others – a phenomenon that is connected with the base structures lengths [10].

Generating passwords from the *Darkweb2017* (dw17) grammar is also very memory demanding because of the long base structures at the beginning of the grammar, and 8 GB RAM is not enough for the largest cracking task using the naive solution. With the proposed preterminal-based solution, we encounter no such problem.

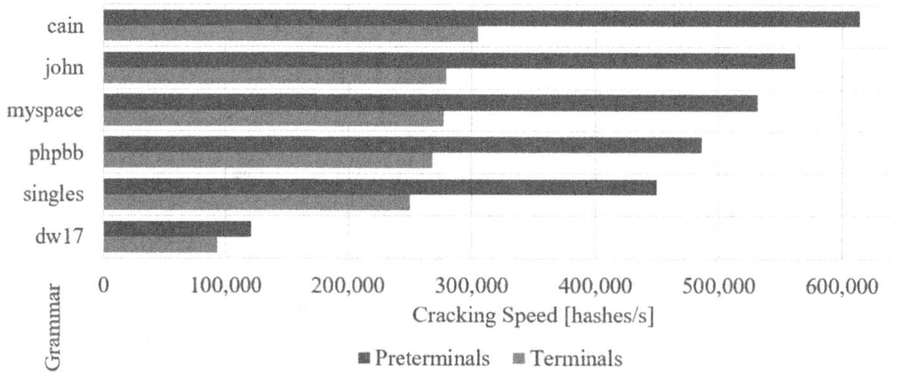

Fig. 10. Differences in cracking speed among grammars

4.3 Workunit Scrambling

The Deadbeat dad algorithm [18] ran on the server ensures the preterminal structures are generated in a probability order. The same holds for passwords, if generated sequentially. However, in a distributed environment, the order of outgoing chunks and incoming results may scramble due to the non-deterministic behavior of the network. Such a property could be removed by adding an extra intra-node synchronization to the proposed protocol. However, we do not consider that necessary if the goal is to verify all generated passwords in the shortest possible time.

Moreover, the scrambling does not affect the result as a whole. The goal of generating and verifying n most probable passwords is fulfilled. For example, for an assignment of generating and verifying 1 million most probable passwords from a PCFG, our solution generates and verifies 1 million most probable passwords, despite the incoming results might be received in a different order.

Though the scrambling has no impact on the final results, we study the extent of chunk scrambling in our setup. We observe the average difference between the expected and real order of chunks arriving at the server, calling it a *scramble factor* S_f. In the following equation, n is the number of chunks, and r_k is the index where k-th chunk was received:

$$S_f = \frac{1}{n} \sum_{k=1}^{n} |(k - r_k)|. \tag{3}$$

We identified two key features affecting the scramble factor – the number of the computing nodes in the system and the number of chunks distributed. In Figure 11, one can see that the scramble factor is increasing with the increasing number of chunks and computing nodes. This particular graph represents experiments with bcrypt algorithm, *myspace* grammar, capped at 10 million hashes. Other experiments resulted in a similar pattern. We conclude that the scrambling is relatively low in comparison with the number of chunks and nodes.

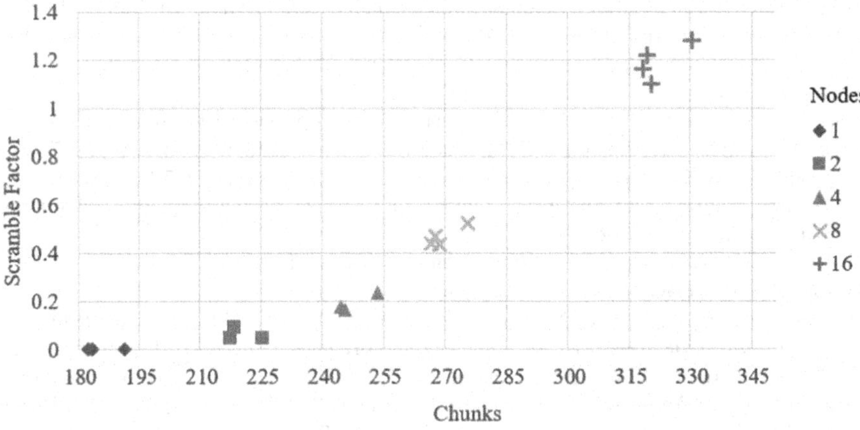

Fig. 11. Average scramble factor, bcrypt *myspace* grammar capped at 10 milion hashes

5 Conclusion

We proposed a method and a protocol for distributed password cracking using probabilistic context-free grammars. The design was experimentally verified using our proof-of-concept tool with a native hashcat support. We showed that distributing the preterminal structures instead of the final passwords reduces the bandwith requirements dramatically, and in many cases, brings a significant speedup. This fact confirms the hypothesis of Weir et al., who suggested preterminal distribution may be helpful in a distributed password cracking trial [19].

Moreover, we showed how different parameters, such as the hash algorithm, network bandwidth, or the choice of concrete grammar, affect the cracking process.

In the future, we would like to redesign the tool to become an envelope over Weir's compiled version of the password cracker in order to stay up to date with the new features that are added over time. The C-based version could figure as a module for our solution.

Acknowledgements. The research presented in this paper is supported by Ministry of Education, Youth and Sports of the Czech Republic from the National Programme of Sustainability (NPU II) project "IT4Innovations excellence in science" LQ1602.

References

1. Aggarwal, S., Houshmand, S., Weir, M.: New technologies in password cracking techniques. In: Lehto, M., Neittaanmäki, P. (eds.) Cyber Security: Power and Technology. ISCASE, vol. 93, pp. 179–198. Springer, Cham (2018). https://doi.org/10.1007/978-3-319-75307-2_11
2. Bonneau, J.: The science of guessing: analyzing an anonymized corpus of 70 million passwords. In: 2012 IEEE Symposium on Security and Privacy, pp. 538–552, May 2012. https://doi.org/10.1109/SP.2012.49
3. Das, A., Bonneau, J., Caesar, M., Borisov, N., Wang, X.: The tangled web of password reuse. NDSS **14**, 23–26 (2014)
4. Florencio, D., Herley, C.: A large-scale study of web password habits. In: Proceedings of the 16th International Conference on World Wide Web WWW 2007, pp. 657–666. ACM, New York (2007). https://doi.org/10.1145/1242572.1242661
5. Ginsburg, S.: The Mathematical Theory of Context Free Languages. McGraw-HillBook Company (1966)
6. Hellman, M.: A cryptanalytic time-memory trade-off. IEEE Trans. Inf. Theory **26**(4), 401–406 (1980)
7. Houshmand, S., Aggarwal, S.: Using personal information in targeted grammar-based probabilistic password attacks. DigitalForensics 2017. IAICT, vol. 511, pp. 285–303. Springer, Cham (2017). https://doi.org/10.1007/978-3-319-67208-3_16
8. Houshmand, S., Aggarwal, S., Flood, R.: Next gen PCFG password cracking. IEEE Trans. Inf. Forensics Secur. **10**(8), 1776–1791 (2015)
9. Hranický, R., Holkovič, M., Matoušek, P., Ryšavý, O.: On efficiency of distributed password recovery. J. Digit. Forensics Secur. Law **11**(2), 79–96 (2016). http://www.fit.vutbr.cz/research/viewpub.php.cs?id=11276
10. Hranický, R., Lištiak, F., Mikuš, D., Ryšavý, O.: On practical aspects of PCFG password cracking. In: Foley, S.N. (ed.) DBSec 2019. LNCS, vol. 11559, pp. 43–60. Springer, Cham (2019). https://doi.org/10.1007/978-3-030-22479-0_3
11. Hranický, R., Zobal, L., Ryšavý, O., Kolář, D.: Distributed password cracking with boinc and hashcat. Digit. Invest. **2019**(30), 161–172 (2019). https://doi.org/10.1016/j.diin.2019.08.001. https://www.fit.vut.cz/research/publication/11961
12. Ma, J., Yang, W., Luo, M., Li, N.: A study of probabilistic password models. In: 2014 IEEE Symposium on Security and Privacy, pp. 689–704 (May 2014). https://doi.org/10.1109/SP.2014.50
13. Narayanan, A., Shmatikov, V.: Fast dictionary attacks on passwords using time-space tradeoff. In: Proceedings of the 12th ACM Conference on Computer and Communications Security, pp. 364–372. CCS 2005. ACM, New York (2005). https://doi.org/10.1145/1102120.1102168

14. Oechslin, P.: Making a faster cryptanalytic time-memory trade-off. In: Boneh, D. (ed.) CRYPTO 2003. LNCS, vol. 2729, pp. 617–630. Springer, Heidelberg (2003). https://doi.org/10.1007/978-3-540-45146-4_36
15. Rabiner, L.R.: A tutorial on hidden markov models and selected applications in speech recognition. Proc. IEEE **77**(2), 257–286 (1989). https://doi.org/10.1109/5.18626
16. Robling Denning, D.E.: Cryptography and Data Security. Addison-Wesley Longman Publishing Co., Inc. (1982)
17. Veras, R., Collins, C., Thorpe, J.: On semantic patterns of passwords and their security impact. In: NDSS (2014)
18. Weir, C.M.: Using probabilistic techniques to aid in password cracking attacks. Ph.D. thesis, Florida State University (2010)
19. Weir, M., Aggarwal, S., De Medeiros, B., Glodek, B.: Password cracking using probabilistic context-free grammars. In: 2009 30th IEEE Symposium on Security and Privacy, pp. 391–405, May 2009. https://doi.org/10.1109/SP.2009.8
20. Weir, M., Aggarwal, S., Collins, M., Stern, H.: Testing metrics for password creation policies by attacking large sets of revealed passwords. In: Proceedings of the 17th ACM Conference on Computer and Communications Security, pp. 162–175 (2010)

Your PIN Sounds Good! Augmentation of PIN Guessing Strategies via Audio Leakage

Matteo Cardaioli[1,2(✉)], Mauro Conti[1,4], Kiran Balagani[3], and Paolo Gasti[3]

[1] University of Padua, Padua, Italy
matteo.cardaioli@phd.unipd.it
[2] GFT Italy, Milan, Italy
[3] New York Institute of Technology, Old Westbury, USA
[4] University of Washington, Seattle, USA

Abstract. Personal Identification Numbers (PINs) are widely used as the primary authentication method for Automated Teller Machines (ATMs) and Point of Sale (PoS). ATM and PoS typically mitigate attacks including shoulder-surfing by displaying dots on their screen rather than PIN digits, and by obstructing the view of the keypad. In this paper, we explore several sources of information leakage from common ATM and PoS installations that the adversary can leverage to reduce the number of attempts necessary to guess a PIN. Specifically, we evaluate how the adversary can leverage audio feedback generated by a standard ATM keypad to infer accurate inter-keystroke timing information, and how these timings can be used to improve attacks based on the observation of the user's typing behavior, partial PIN information, and attacks based on thermal cameras. Our results show that inter-keystroke timings can be extracted from audio feedback far more accurately than from previously explored sources (e.g., videos). In our experiments, this increase in accuracy translated to a meaningful increase in guessing performance. Further, various combinations of these sources of information allowed us to guess between 44% and 89% of the PINs within 5 attempts. Finally, we observed that based on the type of information available to the adversary, and contrary to common knowledge, uniform PIN selection is not necessarily the best strategy. We consider these results relevant and important, as they highlight a real threat to any authentication system that relies on PINs.

1 Introduction

Authentication via Personal Identification Numbers (PINs) dates back to the mid-sixties [5]. The first devices to use PINs were automatic dispensers and control systems at gas stations, while the first applications in the banking sector appeared in 1967 with cash machines [7]. PINs have found widespread use over the years in devices with numeric keypads rather than full keyboards [22].

L. Chen et al. (Eds.): ESORICS 2020, LNCS 12308, pp. 720–735, 2020.
https://doi.org/10.1007/978-3-030-58951-6_35

In the context of financial services, ISO 9564-1 [10] specifies basic security principles for PINs and PIN entry devices (e.g., PIN pads). For instance, to mitigate shoulder surfing attacks [12,13,17], ISO 9564-1 indicates that PIN digits must not be displayed on a screen, or identified using different sounds or sound duration for each key.

As a compromise between security and usability, PIN entry systems display a fixed symbol (e.g., a dot) to represent a key being pressed, and provide the same audio feedback (i.e., same tone, same duration) for all keys. While previous work has demonstrated that observing the dots as they appear on screen as a result of a key press reduces the search space for a PIN [4], to our knowledge no work has targeted the use of audio feedback to recover PINs.

In this paper, we evaluate how the adversary can reduce PIN search space using audio feedback, with (and without) using observable information such as PIN typing behavior (one- or two-handed), knowledge of one digit of the PIN, and knowledge of which keys have been pressed. We compare our attacks with an attack based on the knowledge of PIN distribution.

Exploiting audio feedback has several advantages compared to observing the user or the screen during PIN entry. First, sound is typically easier to collect. The adversary might not be able to observe the ATM's screen directly, and might risk being exposed when video-recording an ATM in a public space. In contrast, it is easy to record audio *covertly*, e.g., by casually holding a smartphone while pretending to stand in a line behind other ATM users. The sound emitted by ATMs is quite distinctive and can be easily isolated even in noisy environments. Second, sound enables higher time resolution compared to video. Conventional video cameras and smartphones record video between 24 and 120 frames per second. In contrast, audio can be recorded with a sampling rate between 44.1 kHz and 192 kHz, thus potentially allowing at least two orders of magnitude higher resolution.

Contributions. In this paper, we analyze several novel side channels associated with PIN entry. In particular:

1. We show that it is possible to retrieve accurate inter-keystroke timing information from audio feedback. In our experiments, we were able to correctly detect 98% of the keystroke feedback sounds with an average error of 1.8ms. Furthermore, 75% of inter-keystroke timings extracted by the software had absolute error under 15 ms. Our experiments also demonstrate that inter-keystroke timings extracted from audio can be more accurate than the same extracted from video recordings of PIN entry as done in [3,4].
2. We analyze how the behavior of the user affects the adversary's ability to guess PINs. Our results show that users who type PINs with one finger are more vulnerable to PIN guessing from inter-keystroke timings compared to users that enter their PIN using at least two fingers. In particular, the combining inter-keystroke timing with the knowledge that the user is a single-finger typist leads to 34-fold improvement over random guessing when the adversary is allowed to perform up to 5 guessing attempts.

3. We combine inter-keystroke timing information with knowledge of one key in the PIN (i.e., the adversary was able to see either the first or the last key pressed by the user), and with knowledge of *which* keys have been pressed by the user. The latter information is available, as shown in this paper as well as in recent work [1,11,16,24] when the adversary is able to capture a thermal image of the PIN pad after the user has typed her PIN. Our experiments show that inter-keystroke timing significantly improves performance for both attacks. For example, by combining inter-keystroke timing with a thermal attack, we were able to guess 15% of the PINs at the first attempt, reaching a four-fold improvement in performance. By combining multiple attacks, we were also able to drastically reduce the number of attempts required to guess a PIN. Specifically, we were able to guess 72% of the PINs within the first 3 attempts.
4. Finally, we show that uniform PIN selection might not be the best strategy against an adversary with access to one or more of the side-channel information discussed in this paper.

Organization. Section 2 reviews related work on password and PIN guessing. Section 3 presents our adversary model. We present our algorithms for inter-keystroke timing extraction in Sect. 4.1. In Sect. 4, we present the results of our experiments, while in Sect. 5 we analyze how different side-channels affect the guessing probability of individual PINs. We conclude in Sect. 6.

2 Related Work

Non-acoustic Side-channels. Vuagnoux and Pasini [20] demonstrated that it is possible to recover keystrokes by analyzing electromagnetic emanations from electronic components in wired and wireless keyboards. Marquardt et al. [15] showed that it is possible to recover key presses by recoding vibrations generated by a keyboard using an accelerometer. Other attacks focus on keystroke inference via motion detection from embedded sensors on wearable devices. For example, Sarkisyan et al. [18] and Wang et al. [21] infer smartphone PINs using movement data recorded by a smartwatch.

Those attacks require that the adversary is able to monitor the user's activity while the user is typing. However, there are attacks that allow the adversary to exploit information available several seconds after the user has typed her password. For instance, one such attack is based the observation that when a user presses a key, the heat from her finger is transferred to the keypad, and can be later be measured using a thermal camera [24]. Depending on the material of the keyboard, thermal residues have different dissipation rates [16], thus affecting the time window in which the attacks are effective. Abdelrahman et al. [1] evaluated how different PINs and unlock patterns on smartphones on can influence thermal attack performance. Kaczmarek et al. [11] demonstrated how a thermal attack can recover precise information about a password up to 30 s after it was typed, and partial information within 60 s.

Acoustic Side-channels. Asonov and Agrawal showed that each key on a keyboard emits a characteristic sound, and that this sound can be used to infer individual keys [2]. Subsequent work further demonstrated the effectiveness of sound emanation for text reconstruction. Berger et al. [6] combined keyboard acoustic emanation with a dictionary attack to reconstruct words, while Halevi and Saxena [9] analyzed keyboard acoustic emanations to eavesdrop over random password. Because ISO 9564-1 [10] specifications require that each key emits the same sound, those attacks do not apply to common keypads, including those on ATMs.

Another type of acoustic attack is based on time difference of arrivals (TDoA) [14,23,25]. These attacks rely on multiple microphones to triangulate the position of the keys pressed. Although this attacks typically result in good accuracies, they are difficult to instantiate in realistic environments.

Song et al. [19] presented an attack based on latency between key presses measured by snooping encrypted SSH traffic. Their experiments show that information about inter-keystroke timing can be used to narrow the password search space substantially. A similar approach was used by Balagani et al. [4], who reconstructed inter-keystroke timing from the time of appearance of the masking symbols (e.g., "dots") while a user types her password. Similarly, Balagani et al. [3] demonstrated that precise inter-keystroke timing information recovered from videos drastically reduces the number of attempts required to guess a PIN. The main limitation of [3,4] is that they require the adversary to video-record the ATM screen while the user is typing her PIN. Depending on the location and the ATM, this might not be feasible. Further, this reduces the set of vulnerable ATMs and payment systems to those that display on-screen feedback.

To our knowledge, this is the first paper to combine inter-keystroke timing information deduced from sound recording with observable information from other sources, and thereby drastically reduce the attempts to guess a PIN compared to prior work. Our attacks are applicable to a multitude of realistic scenarios. This poses an immediate and severe threat to current ATMs or PoS.

3 Adversary Model

In this section we evaluate four classes of information that the adversary can exploit to infer PINs. These classes are: (1) Key-stroke timing information extracted from audio recordings; (2) Knowledge of whether the user is a single- or multi-finger typist; (3) Information about the first or the last digit of the PIN; and (4) Information about which keys have been pressed, but not their order. Next, we briefly review how each of these classes of information can be collected by the adversary.

Class 1: Keystroke Timing. Keystroke timing measures the distance between consecutive keystroke events (e.g., the time between two key presses, or between the release of the key and the subsequent keypress). Collecting keystroke timing by compromising the software of an ATM located in a public space, or physically tampering with the ATM (e.g., by modifying the ATM's keyboard) is

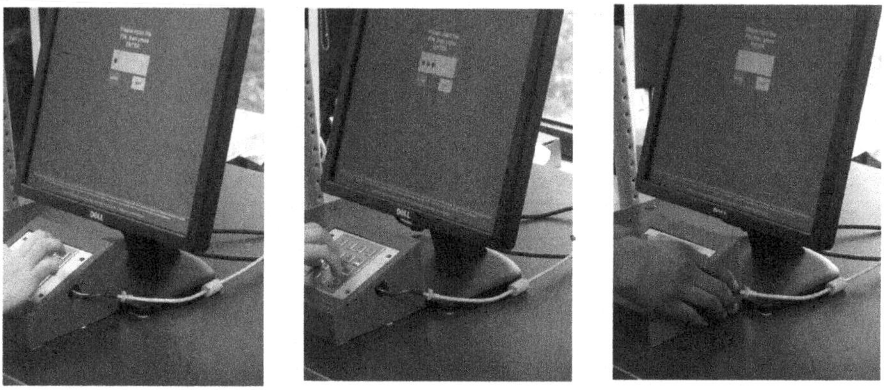

Fig. 1. Different typing strategies. Left: one finger; center: multiple fingers of one hand; right: multiple fingers of two hands.

not practical in most cases. However, as shown in [3], the adversary can infer keystroke timings without tampering with the ATM by using video recordings of the "dots" that appear on the screens when the user types her PIN. In this paper, we leverage audio signals to infer precise inter-keystroke timings.

Class 2: Single- or Multi-finger Typists. The adversary can typically directly observe whether the user is typing with one or more fingers. While the number of fingers used to enter a PIN does not reveal information about the PIN itself, it might be a useful constraint when evaluating other sources of information leakage. Figure 1 shows users typing using a different number of fingers.

Class 3: Information about the first or the last digit of the PIN. As users move their hands while typing their PIN, the adversary might briefly have visibility of the keypad, and might be able to see one of the keys as it is pressed (see Fig. 1). We model this information by disclosing either the first or the last digit of the PIN to the adversary.

Class 4: Which Keys Have Been Pressed. This information can be collected using various techniques. For instance, the adversary can use a thermal camera to determine which keys are warmer, thus learning which digits compose the PIN (see, e.g., Fig. 2). As an alternative, the adversary can place UV-sensitive powder on the keys before the user enters her PIN, and then check which keys had the powder removed by the users using a UV light.

While these attacks do not reveal the order in which the keys were pressed (except when the PIN is composed of one repeated digit), they significantly restrict the search space. Although this attack can be typically performed covertly, it requires specialized equipment.

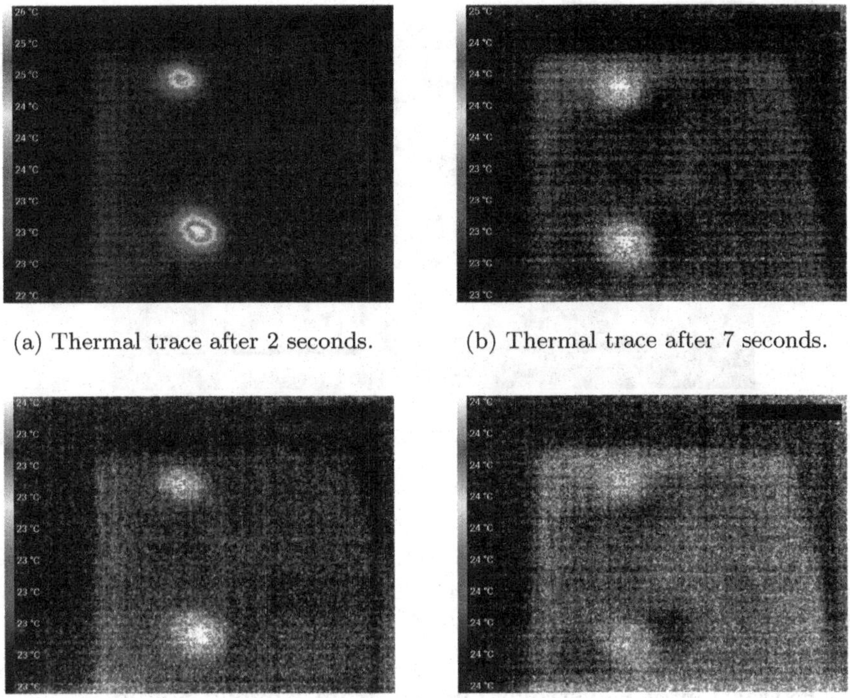

(a) Thermal trace after 2 seconds.

(b) Thermal trace after 7 seconds.

(c) Thermal trace after 10 seconds.

(d) Thermal trace after 15 seconds.

Fig. 2. Thermal image of a metallic PIN pad after applying a transparent plastic cover for PIN 2200.

4 Experiment Results

We extracted keystroke sounds using the dataset from [4]. This dataset was collected from 22 subjects, who typed several 4-digit PINS on a simulated ATM (see Fig. 3). Nineteen subjects completed three data collection sessions, while three subjects completed only one session.

In each session, subjects entered a total of 180 PINs as follows: each subject was shown a 4-digit PIN. The PIN remained on the screen for 10 s, during which the subject was encouraged to type the PIN multiple times. After 10 s, the PIN disappeared from the screen. At this point, the subject was asked to type the PIN 4 times from memory. In case of incorrect entry, the PIN was briefly displayed again on the screen, and the subject was allowed to re-enter it. This procedure was repeated in three batches of 15 PINs. As a result, each PIN was typed 12 times per session.

Each time a subject pressed a key, the ATM simulator emitted an audio feedback and logged the corresponding timestamp with millisecond resolution. Users were asked to type 44 different 4-digit PINs which represented all the Euclidean distances between keys on the keypad. Sessions were recorded in a

Fig. 3. Left: user typing a PIN using the ATM simulator. Right: close up view of the ATM simulator's keypad.

relatively noisy indoor public space (SNR −15 dB) using a Sony FDR-AX53 camera located approximately 1.5 m away from the PIN pad. The audio signal was recorded with a sampling frequency of 48 kHz.

4.1 Extraction of Keystroke Timings from Keypad Sound

To evaluate the accuracy of timing extraction from keystroke sounds, we first linearly normalized the audio recording amplitude in the interval $[-1, 1]$. We applied a 16-order Butterworth band-pass filter [8] centered at 5.6 kHz to isolate the characteristic frequency window of the keypad feedback sound. Finally, to isolate the signal from room noise, we processed the audio recording to "mute" all samples with an amplitude below a set threshold (0.01 in our experiments).

We then calculated the maximum amplitude across nearby values in a sliding window of 1,200 samples (consecutive windows had 1199 overlapping samples), corresponding to 25 ms of audio recording. We determined the length of the window by evaluating the distance between consecutive timestamps logged by the ATM simulator (ground truth), which was at least 100 ms for 99.9% of the keypairs. Figure 4 shows the result of this process.

We then extracted the timestamps of the peaks of the processed signal and compared them to the ground truth. Our results show that this algorithm can accurately estimate inter-keystroke timing information. We were able to correctly detect 98% of feedback sound with a mean error of 1.8 ms.

Extracting timings from audio led to more accurate time estimation than using video [4]. With the latter, 75% of the extracted keystroke timings had errors of up to 37 ms. In contrast, using audio we were able to extract 75% of the keystroke with errors below 15 ms. Similarly, using video, 50% of the estimated keystrokes timings had errors of up to 22 ms, compared to less than 7 ms with audio. Figure 5 shows the errors distribution for timings extracted from video and audio recordings.

4.2 PIN Inference from Keystroke Timing (Class 1)

This attack ranks PINs based on the estimated Euclidean distance between subsequent keys in each PIN. In particular, we calculated an inter-key Euclidean distance vector from a sequence of inter-keystroke timings inferred from audio feedback. As an example, the distance vector associated with PIN 5566 is $[0, 1, 0]$, where the first '0' is the distance between keys 5 and 5, '1' between keys 5 and 6, and '0' between 6 and 6. Any four-digit PIN is associated with one distance vector of size three. Each element of the distance vector can be 0, 1, 2, 3, diagonal distance 1 (e.g., 1–3), diagonal distance 2 (e.g., 3–7), short diagonal distance (e.g., 2–9), or long diagonal distance (e.g., 3–0). Different PINs might be associated with the same distance vector (e.g., 1234 and 4567). The goal of this attack is to reduce the search space by considering only PINs that match the estimated distance vector.

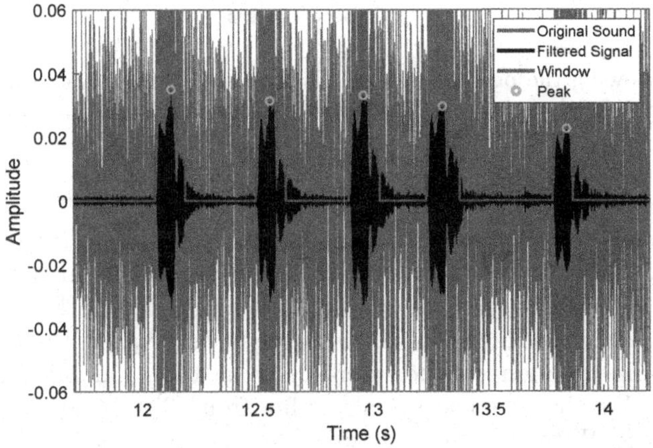

Fig. 4. Comparison between the original sound signal, filtered sound signal, windowed signal, and extracted peaks.

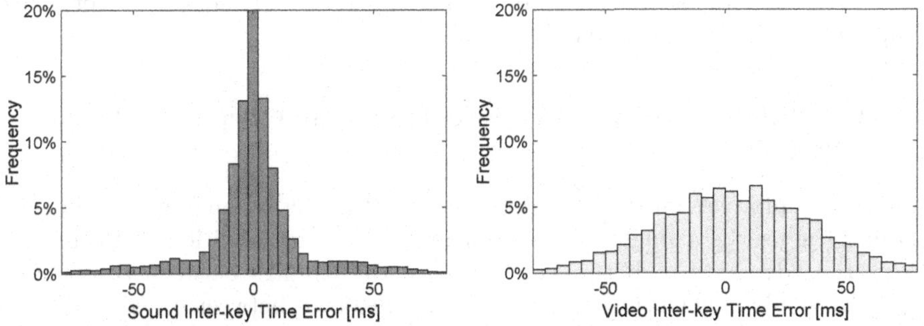

Fig. 5. Error distribution of estimated inter-keystroke timings. Left: timing errors from audio. Right: timing errors from video.

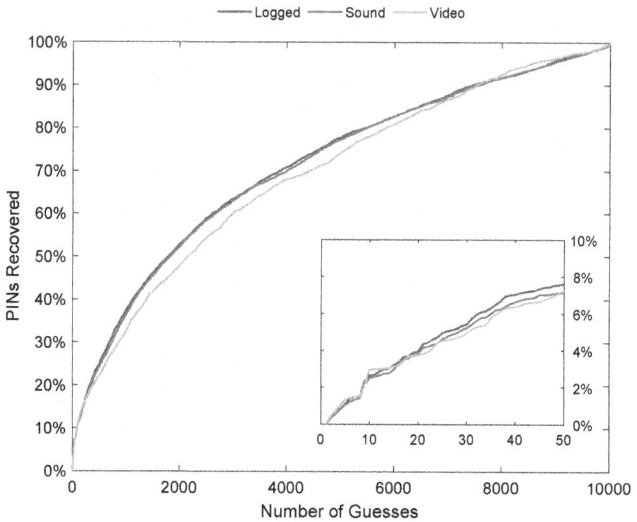

Fig. 6. CDF showing the percentage of PINs recovered using keystroke timing information derived from the ground truth (logged), sound feedback, and video.

For evaluation, we split our keystroke dataset into two sets. The first (training set) consists of 5195 PINs, typed by 11 subjects. The second (test set) consists of 5135 PINs, typed by a separate set of 11 subjects. This models the lack of knowledge of the adversary of the specific typing patterns of the victim user.

To estimate the Euclidean distances between consequent keys, we modeled a set of gamma function on the inter-keystroke timing distribution, one for each distance. We then applied the algorithm from [3] to infer PINs from estimated distances. With this strategy, we were able to guess 4% of PINs within 20 attempts—a 20-fold improvement compared to random guessing.

Figure 6 shows how timings extracted from audio and video feedback affect the number of PIN guessed by the algorithm compared to ground truth. Timings extracted from audio feedback exhibit a smaller decrease in guessing performance compared to timings extracted from video.

4.3 PIN Inference from Keystroke Timing and Typing Behavior (Class 2)

This attack improves on the keystroke timing attack by leveraging knowledge of whether the user is a single- or multi-finger typist. This additional information allows the adversary to better contextualize the timings between consecutive keys. For single-finger typists, the Euclidean distance between keys 1 and 0 is the largest (see Fig. 3), and therefore we expect the corresponding inter-keystroke timing to be the largest. However, if the user is a two-finger typist, then 1 might be typed with the right hand index finger, and 0 with the left hand index finger.

As a result, the inter-keystroke time might not be representative of the Euclidean distance between the two keys.

To systematically study typing behavior, we analyzed 61 videos from the 22 subjects. 70% of the subject were single-finger typists; 92% of them entered PINs using the index finger, and 8% with the thumb. We divided multi-finger typists into three subclasses: (1) PINs entered using fingers from two hands (38% of the PINs typed with more than one finger); (2) PINs entered with at least two fingers of the same hand (34% of the PINs typed with more than one finger); and (3) PINs that we were not able to classify with certainty due obfuscation of the PIN pad in the video recording (28% of the PINs typed with more than one finger).

In our experiments, subjects' typing behavior was quite consistent across PINs and sessions. Users that were predominantly single-finger typists entered 11% of their PINs using more than one finger, while multi-finger typists entered 23% of the PINs using one finger.

We evaluated guessing performance of timing information inferred from audio feedback on single-finger PINs and multi-finger PINs separately. We were able to guess a substantially higher number of PINs for each number of attempts for users single-finger typists (see Fig. 7) compared to multi-finger typists. In particular, the percentage of PINs recovered within 5 attempts was twice as high for PINs entered with one finger compared to PINs entered with multiple fingers. Further, the guessing rate within the first 5 attempts was 34 times higher compared to random guessing when using timing information on single-finger PINs. However, our ability to guess multi-finger PINs using timing information was only slightly better than random. This strongly suggests that the correlation

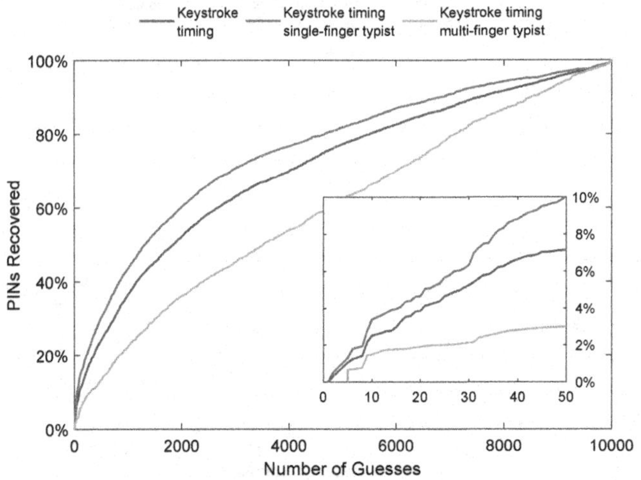

Fig. 7. CDF showing the percentage of PINs recovered using only keystroke timing information from audio feedback, compared to timing information for single- or multi-finger typists.

between inter-keystroke timing and Euclidean distance identified in [4] holds only quite strongly for PINs entered using a single finger, and only marginally for PINs entered with two or more fingers.

4.4 Knowledge of the First or the Last Digit of the PIN (Class 3)

In this section, we examine how information on the first or last digit of the PIN reduces the search space when combined with keystroke timings. Knowledge of one digit alone reduces the search space by a factor of 10 regardless of the digit's position, because the adversary needs to guess only the remaining three digits. (As a result, the expected number of attempts to guess a random PIN provided no additional information is 500.)

To determine how knowledge of the first or the last digit impacts PIN guessing based on keystroke timing, we applied the same procedure described in Sect. 4.2: for each PIN in the testing set, we associated a list of triplets of distances sorted by probability. We then pruned the set of PINs associated with those distance triplets to match the knowledge of the first or last PIN. For instance, given only the estimated distances 3, 0, and $\sqrt{2}$, the associated PINs are 0007, 0009, 2224, and 2226. If we know that the first digit of the correct PIN is 2, then our guesses are reduced to 2224 and 2226.

Information about the first or last digit of the PIN boosted the guessing performance of the keystroke-timing attack substantially, as shown in Fig. 8. In particular, guessing accuracy increased by 15–19 times within 3 attempts (4.36% guessing rate when the first digit was known, and 5.57% when the last digit was known), 7 times within 5 attempts, and about 4 times within 10 attempts,

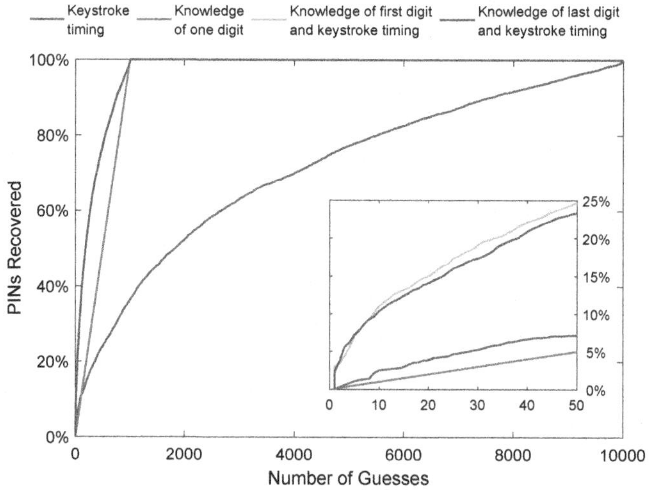

Fig. 8. CDF showing the percentage of PINs recovered using keystroke timings inferred from audio, random guessing over 3 digits of the PIN, and using inferred keystroke timings and knowledge of the first or the last digit of the PIN.

compared to timing information alone. In all three cases, timing information substantially outperformed knowledge of one of the digits in terms of guessing rate.

4.5 Knowledge of Which Keys Have Been Pressed (Class 4)

In this section, we evaluate how knowledge of *which digits* compose a PIN, but not *their order*, restricts the PIN search space, in conjunction with information about keystroke timings. The adversary can acquire this knowledge, for instance, by observing the keypad using a thermal camera shortly after the user has typed her PIN [11], or by placing UV-sensitive powder on the keys before the user enters her PIN, and then checking which keys were touched using a UV light.

Information on which digits compose a PIN can be divided as follows:

1. The user pressed only one key. In this case, the user must have entered the same digit 4 times. No additional information is required to recover the PIN.
2. The user pressed two distinct keys, and therefore each digit of the PIN might be repeated between one and three times, and might be in any position of the PIN. In this case, the number of possible PINs is $2^4 - 2 = 14$, i.e., the number of combinations of two values in four position, except for the combinations where only one of the two digits appears.
3. The user pressed three distinct keys. The number of possible PINs is equal to the combinations of three digits in four positions, i.e., $4 \cdot 3 \cdot 3 = 36$
4. The user pressed four distinct keys. The number of possible PINs is $4! = 24$.

We evaluated how many PINs the adversary could recover given keystroke timings and the set of keys pressed by the user while entering the PIN. Our

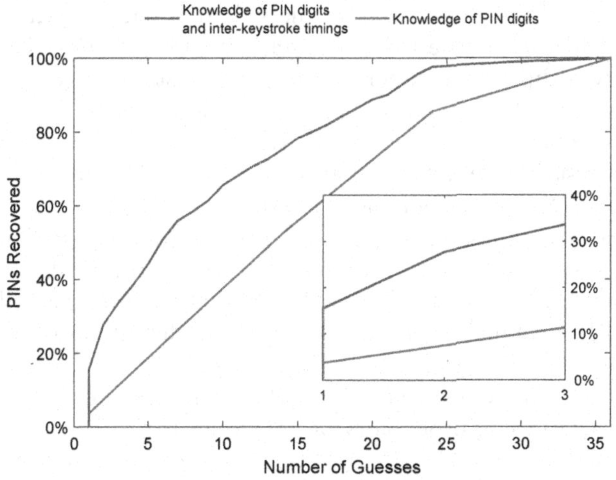

Fig. 9. CDF showing the percentage of PINs recovered with the knowledge of which keys have been pressed with and without inter-keystroke timing information.

results, presented in Fig. 9, show that combining these two sources of information leads to a high PIN recovery rate. Specifically, within the first three attempts, knowing only which keys were pressed led to the recovery of about 11% of the PINs. Adding timing information increased this value to over 33%.

4.6 Combining Multiple Classes of Information

In this section we examine how combining multiple classes of information leads to an improvement in the probability of correctly guessing a PIN.

First, we investigated how guessing probability increases when the adversary knows one of the digits of the PIN (first or last), the typing behavior (single-finger typist), and is able to infer inter-keystroke timing information from audio feedback. We used 3461 PINs typed by 11 subjects containing only single-finger PINs. In our experiments, we were able to guess 8.73% of the PINs within 5 attempts, compared to 6.97% with timing information and knowledge of one digit.

We then considered knowledge of the values composing the PIN, typing behavior, and inferred timing information. In this case, we successfully guessed 50.74% of the PINs within 5 attempts, and 71.39% within 10 attempts.

Finally, when we considered the values composing the PIN, one of the PIN's digits, and inferred timing information, we were able to guess 86.76% of the PINs in 5 attempts, and effectively all of them (98.99%) within 10 attempts.

All our results are summarized in Table 1.

5 PINs and Their Guessing Probability Distribution

In this section, we evaluate whether the classes of information identified in this paper make some of the PINs easier to guess than others, and thus intrinsically less secure. With respect to estimated inter-keystroke timings, different timing vectors identify a different number of PINs. For instance, vector [0,0,0] corresponds to 10 distinct PINs (`0000`, `1111`, ...), while vector [1,1,1] corresponds to 216 PINs (`0258`, `4569`, ...). This indicates that, against adversaries who are able to infer inter-keystroke timing information, choosing PINs uniformly at random from the entire PIN space is not the best strategy.

The adversary's knowledge of which digits compose the PIN has a similar effect of the guessing probability of individual PINs. In this case, PINs composed of three different digits are the hardest to guess, with a probability of 1/36, compared to PINs composed of a single digit, which can always be guessed at the first attempt.

The adversary's knowledge of one digit of the PIN and/or the typing behavior do not affect the guessing probability of individual PINs.

Table 1. Results from all combinations of attacks considered in this paper, sorting by guessing rate after 5 attempts. Because single finger reduces the PIN search space only in conjunction with inter-keystroke timings, we do not present results for single finger alone.

Information				PINs guessed within attempt				
Keystroke timing	Single finger	First digit	PIN digits	1	2	3	5	10
				0.01%	0.02%	0.03%	0.05%	0.10%
		o		0.10%	0.20%	0.30%	0.50%	1.00%
o				0.02%	0.31%	0.70%	1.05%	2.51%
o	o			0.03%	0.52%	0.91%	1.30%	3.38%
o		o		3.02%	3.72%	4.36%	6.97%	11.04%
o	o	o		3.73%	4.13%	5.43%	8.73%	14.01%
			o	3.76%	7.52%	11.28%	18.80%	37.60%
o			o	15.54%	27.79%	33.63%	44.25%	65.57%
o	o		o	19.04%	34.01%	40.60%	50.74%	71.31%
		o	o	13.27%	26.62%	39.88%	66.40%	92.80%
o		o	o	35.27%	53.46%	66.84%	86.76%	98.99%
o	o	o	o	40.86%	60.24%	71.77%	89.19%	99.28%

6 Conclusion

In this paper, we showed that inter-keystroke timing inferred from audio feedback emitted by a PIN pad compliant with ISO 9564-1 [10] can be effectively used to reduce the attempts to guess a PIN. Compared to prior sources of keystroke timing information, audio feedback is easier to collect, and leads to more accurate timing estimates (in our experiments, the average reconstruction error was 1.8 ms). Due to this increase in accuracy, we were able to reduce the number of attempts needed to guess a PIN compared to timing information extracted from videos.

We then analyzed how using inter-keystroke timing increases guessing performance of other sources of information readily available to the adversary. When the adversary was able to observe the first or the last digit of a PIN, inter-keystroke timings further increased the number of PINs guessed within 5 attempts by 14 times. If the adversaries was capable of observing which keys were pressed to enter a PIN (e.g., using a thermal camera), adding inter-keystroke timing information allowed the adversary to guess 15% of the PINs with a single attempt. This corresponds to a 4 times reduction in the number of attempts compared to knowing only which keys were pressed.

We evaluated how typing behavior affects guessing probabilities. Our results show that there is a strong correlation between Euclidean distance between keys and inter-keystroke timings when the user enters her PIN using one finger. However, this correlation was substantially weaker when users typed with more than one finger.

We then showed that the combination of multiple attacks can dramatically reduce attempts to guess the PIN. In particular, we were able to guess 72% of the

PINs within the first 3 attempts, and about 90% of the PINs within 5 attempts, by combining all the sources of information considered in this paper.

Finally, we observed that different adversaries require different PIN selection strategies. While normally PINs should be selected uniformly at random from the entire PIN space, this is not true when the adversary has access to inter-keystroke timings or thermal images. In this case, some classes of PINs (e.g., those composed of a single digit) are substantially easier to guess than other classes (e.g., those composed of three different digits). As a result, uniform selection from appropriate *subsets* of the entire PIN space leads to harder-to-guess PINs against those adversaries.

We believe that our results highlight a real threat to PIN authentication systems. The feasibility of these attacks and their immediate applicability in real scenarios poses a considerable security threat for ATMs, PoS-s, and similar devices.

References

1. Abdelrahman, Y., Khamis, M., Schneegass, S., Alt, F.: Stay cool! understanding thermal attacks on mobile-based user authentication. In: Proceedings of the 2017 CHI Conference on Human Factors in Computing Systems, pp. 3751–3763. ACM (2017)
2. Asonov, D., Agrawal, R.: Keyboard acoustic emanations. In: IEEE S&P (2004)
3. Balagani, K., et al.: Pilot: password and pin information leakage from obfuscated typing videos. J. Comput. Secur. **27**(4), 405–425 (2019)
4. Balagani, K.S., Conti, M., Gasti, P., Georgiev, M., Gurtler, T., Lain, D., Miller, C., Molas, K., Samarin, N., Saraci, E., Tsudik, G., Wu, L.: ***SILK-TV***: secret information leakage from keystroke timing videos. In: Lopez, J., Zhou, J., Soriano, M. (eds.) ESORICS 2018. LNCS, vol. 11098, pp. 263–280. Springer, Cham (2018). https://doi.org/10.1007/978-3-319-99073-6_13
5. Bátiz-Lazo, B., Reid, R.: The development of cash-dispensing technology in the UK. IEEE Ann. Hist. Comput. **33**(3), 32–45 (2011)
6. Berger, Y., Wool, A., Yeredor, A.: Dictionary attacks using keyboard acoustic emanations. In: Proceedings of the 13th ACM conference on Computer and communications security, pp. 245–254. ACM (2006)
7. Bonneau, J., Preibusch, S., Anderson, R.: A birthday present every eleven wallets? the security of customer-chosen banking PINs. In: Keromytis, A.D. (ed.) FC 2012. LNCS, vol. 7397, pp. 25–40. Springer, Heidelberg (2012). https://doi.org/10.1007/978-3-642-32946-3_3
8. Butterworth, S.: On the theory of filter amplifiers. Wireless Eng. **7**(6), 536–541 (1930)
9. Halevi, T., Saxena, N.: A closer look at keyboard acoustic emanations: random passwords, typing styles and decoding techniques. In: Proceedings of the 7th ACM Symposium on Information, Computer and Communications Security, pp. 89–90. ACM (2012)
10. ISO: Financial services - personal identification number (pin) management and security - part 1: Basic principles and requirements for pins in card-based systems (2017). https://www.iso.org/standard/68669.html

11. Kaczmarek, T., Ozturk, E., Tsudik, G.: Thermanator: thermal residue-based post factum attacks on keyboard password entry. arXiv preprint arXiv:1806.10189 (2018)
12. Kumar, M., Garfinkel, T., Boneh, D., Winograd, T.: Reducing shoulder-surfing by using gaze-based password entry. In: Proceedings of the 3rd symposium on Usable privacy and security, pp. 13–19. ACM (2007)
13. Kwon, T., Hong, J.: Analysis and improvement of a pin-entry method resilient to shoulder-surfing and recording attacks. IEEE Trans. Inf. Forensics Secur. **10**(2), 278–292 (2015)
14. Liu, J., Wang, Y., Kar, G., Chen, Y., Yang, J., Gruteser, M.: Snooping keystrokes with mm-level audio ranging on a single phone. In: Proceedings of the 21st Annual International Conference on Mobile Computing and Networking, pp. 142–154. ACM (2015)
15. Marquardt, P., Verma, A., Carter, H., Traynor, P.: (sp) iPhone: decoding vibrations from nearby keyboards using mobile phone accelerometers. In: Proceedings of the 18th ACM conference on Computer and communications security, pp. 551–562. ACM (2011)
16. Mowery, K., Meiklejohn, S., Savage, S.: Heat of the moment: characterizing the efficacy of thermal camera-based attacks. In: Proceedings of the 5th USENIX conference on Offensive technologies, p. 6. USENIX Association (2011)
17. Roth, V., Richter, K., Freidinger, R.: A pin-entry method resilient against shoulder surfing. In: Proceedings of the 11th ACM conference on Computer and communications security, pp. 236–245. ACM (2004)
18. Sarkisyan, A., Debbiny, R., Nahapetian, A.: Wristsnoop: smartphone pins prediction using smartwatch motion sensors. In: 2015 IEEE international workshop on information forensics and security (WIFS), pp. 1–6. IEEE (2015)
19. Song, D.X., Wagner, D., Tian, X.: Timing analysis of keystrokes and timing attacks on SSH. In: USENIX Security Symposium (2001)
20. Vuagnoux, M., Pasini, S.: Compromising electromagnetic emanations of wired and wireless keyboards. In: USENIX security symposium, pp. 1–16 (2009)
21. Wang, C., Guo, X., Wang, Y., Chen, Y., Liu, B.: Friend or foe?: your wearable devices reveal your personal pin. In: Proceedings of the 11th ACM on Asia Conference on Computer and Communications Security, pp. 189–200. ACM (2016)
22. Wang, D., Gu, Q., Huang, X., Wang, P.: Understanding human-chosen pins: characteristics, distribution and security. In: Proceedings of the 2017 ACM on Asia Conference on Computer and Communications Security, pp. 372–385. ACM (2017)
23. Wang, J., Zhao, K., Zhang, X., Peng, C.: Ubiquitous keyboard for small mobile devices: harnessing multipath fading for fine-grained keystroke localization. In: Proceedings of the 12th Annual International Conference on Mobile Systems, Applications, and Services, pp. 14–27. ACM (2014)
24. Zalewski, M.: Cracking safes with thermal imaging. ser (2005). http://lcamtuf.coredump.cx/tsafe
25. Zhu, T., Ma, Q., Zhang, S., Liu, Y.: Context-free attacks using keyboard acoustic emanations. In: ACM CCS (2014)

GDPR – Challenges for Reconciling Legal Rules with Technical Reality

Mirosław Kutyłowski[1(✉)], Anna Lauks-Dutka[1], and Moti Yung[2,3]

[1] Department of Fundamentals of Computer Science, Wrocław University of Science and Technology, Wrocław, Poland
{miroslaw.kutylowski,anna.lauks}@pwr.edu.pl

[2] Columbia University, New York, USA
motiyung@gmail.com

[3] Google LLC, New York City, NY, USA

Abstract. The main real impact of the GDPR regulation of the EU should be improving the protection of data concerning physical persons. The sharp GDPR rules have to create a controllable information environment, and to prevent misuse of personal data. The general legal norms of GDPR may, indeed, be regarded as justified and well motivated by the existing threats, however, substantial problems emerge when we attempt to implement GDPR in a real information processing systems setting.

This paper aims at bringing attention to some critical challenges related to the GDPR regulation from this technical implementation perspective. Our goal is to alert the community that due to incompatibility between the legal concepts (as understood by a layman) and the technical state-of-the-art, a literal implementation of the GDPR may, in fact, lead to a decrease in the attainable real security level, thus hurting privacy. Further, this situation may create barriers to information processing environments – including in critical evolving areas which are very important for citizens' security and safety. Demonstrating the problem, we provide a (possibly incomplete) list of concrete major clashes between the legal concepts of GDPR and security technologies. We also discuss possible solutions to these problems (from a technology perspective), and review related activities.

We hope that this work will encourage people to seek improvements and reforms of GDPR based on realistic privacy needs and computing goals, rather than the current situation where people involved in IT projects, merely attempt to only do things that are justified (and perhaps severely restricted) by GDPR.

Keywords: GDPR · Compliance · Privacy · Security

M. Yung—The opinions in this work are personal and do not represent the employers of the authors. The work of the first author has been initiated within the project 2014/15/B/ST6/02837 of Polish National Science Centre.

L. Chen et al. (Eds.): ESORICS 2020, LNCS 12308, pp. 736–755, 2020.
https://doi.org/10.1007/978-3-030-58951-6_36

1 Introduction

Since mid 2018, processing personal data in Europe as well as in some cases also outside Europe has to comply with the General Data Protection Regulation (GDPR) [19] of the European Union. In many other countries (including USA, P.R.C., Japan, Korea, Australia, ...) heavy personal data protection laws have been enforced – however there might be profound differences among these, both in terms of the overall concept and in term of the fine details.

In theory, harsh penalties for not adjusting to GDPR rules on the one hand, and an increased awareness of protection necessity as well as technological advances, on the other hand, should guarantee a swift transition to a world governed by the GDPR rules. However, the emerging reality is quite different: First, the GDPR may be regarded as a *paper tiger* that is not guarding against personal data acquisition (Edward Snowden, 2019), while on the other hand it has created substantial costs for running information systems. Moreover, even in Europe real adjustment to the fundamental principles is far from being the reality. In some areas the situation seems to be alarming (see Sect. 4.3 for some examples: – the situation of AI in Europe and epidemic information processing). Nevertheless, there is almost no discussion about GDPR itself and the necessity for corrections given the experience gathered. As the EU Commission is to present a review on GDPR in 2020, now it is, perhaps, the last moment to express concerns.

Paper Organization. Due to the lack of space we do not provide an introductory guide about the GDPR Regulation – there is an abundant literature on this topic. In Sect. 2 we present challenges for realization of the GPDR rules. These problems are not merely related to ambiguity of a legal text – most of them are due to overlooking the full complexity of issues in the information society. As the devil is in the details, we attempt to provide points that can be described as legal norms that would be appropriate from the technology point of view. In Sect. 3 we present a few general concepts de lege ferenda. In Sect. 4 we report some observed initiatives (or their lack) aimed at adjusting GDPR after two years of experience.

2 GDPR Challenges

2.1 Classification as "Personal Data"

The GDPR regulation says that data falls into the category of *personal data* if and only if it concerns an *identifiable person.* It does not specify the decision context. To make the problem harder, in many practical cases the decision must be automated. Of course, there are evident cases, where the data contain explicit identifiers of data subjects. The hard case is to guarantee that a data piece D does not fall into the category of *personal data.*

A procedure F classifying data as *personal data* or *non-personal data* may take into account the following context as its input:

Option 1: just the data D, by itself,
Option 2: D in the context of all other existing datasets,
Option 3 (hybrid solution): D and a piece of information about the other datasets available to the data controller.

Properties of Option 1: While it is easy to implement, classification as *not-personal data* may turn out to be obviously wrong taking into account the purpose and plausible interpretation of the GDPR. Adopting option 1 would enable easy circumvention of the GDPR principles.

Properties of Option 2: This option is infeasible in practice. Even if access to all other datasets is given, the analysis itself might be extremely complex and infeasible to automate. Definitely, it is very unlikely to get a decision in a real time. Last but not least, in general there is no free access to all other datasets. But even if the access is granted, the permissions are usually granted for a "fair use". The analysis being, in fact, a deanonymization attempt will not necessarily be treated as a "fair use". It can even be regarded as an offense against GDPR.

Properties of Option 3: This option is more realistic for implementation than option 2 (however it is still potentially hard and costly). Unfortunately, it may deliver misleading results from the point of view of GDPR goals.

As we see, every option leads to a kind of deadlock. The main problem is that the attribute *personal data* is regarded as a general property unrelated to the party processing this data. A different approach would be to determine this attribute in the context of the party processing the data. Thereby, the classification of the same dataset D might be different for different parties. Indeed, for instance, if data in D are encrypted homomorphically, then a party P_1 holding D but not the decryption key could perform nontrivial operations on ciphertexts from D without regarding them as *personal data.* At the same time a party P_2 holding the decryption key should regard D as a *personal data* (so even an unauthorized read operation executed by P_2 would be a GDPR violation). An immediate consequence of such an approach is that one cannot permanently classify a given data, the burden to classify data as personal or non-personal is on the party processing the data. (After all: Data is not information on papers which live in a static form forever, data in computations is processed within a context by different parties!)

Rule 1 (*personal data* as an attribute of data and processing party). *A data shall be considered a* personal data *by a party processing it, if and only if this party can identify a physical person related to these data.*

Let us note that Rule 1 has to be used with caution. While for internal processing Rule 1 can be easily applied, it does not automatically enable a data processor P to transfer a non-personal data D to other parties. In this case P has to make sure that D has still the attribute *non-personal* for the receiving party. If this is not the case, then the GDPR rules apply in full for the transfer. These precautions are necessary in particular if P aims to publish D. Then P has to make sure that by publishing D it will grant an unlawful access to personal data. We discuss these problems in detail in Sect. 2.2.

The proposed approach would have deep consequences for security practice. Assume for instance that party Y has technical capabilities to link the pseudonyms in an anonymous credentials system back to real identities of the users (e.g. thanks to a trapdoor information like the opening authority in group signature schemes which are used for multi-use anonymous credentials, or just thanks to extraordinary computational resources). The system of anonymous credentials can be used as usual, and the regular participants may safely assume the data processed do not fall into the category of personal data. On the other hand, party Y should keep hands off all data from the anonymous credential system, as any operation on these data would mean a violation of the GDPR principles. While Y would have technical possibility to access personal data, they would have no lawful way to take advantage of this knowledge.

In fact, the above situation would be uncomfortable for Y, as it would be necessary to collect a convincing evidence that it has not used its extra knowledge. Indeed, sooner or later it may be revealed that Y had such capabilities and Y might be accused of using it for own advantage. Therefore, for its own interest party Y should create a verifiable system guarding the use of the trapdoor information – e.g. via secret sharing and storing the shares in different locations controlled by different bodies, or an automatic audit of processing activities, and so on. A good (but still academic) example of a scheme that automatically guards against misusing knowledge advantage are fail-stop signatures: even if a powerful party Y can compute discrete logarithms, it cannot use this capability for forging signatures without creating a strong evidence of a forgery.

2.2 Consequences of Processing *Non-personal Data*

If data categorization depends on the processing party, the following problem arises:

Problem 2 (conversion to personal data). It may happen that a party X processes data D, which is *non-personal* from the point of view of X,and creates an output that converts $D$to *personal data* from the point of view of a party Z.

We propose the following rule:

Rule 3 (impact of processing data). A party X processing *non-personal data* is responsible for all consequences of that processing from the point of view of GDPR.

It can be argued that X can freely process data D, as literally taken these activities are not covered by GDPR. However, indirectly they may have a significant impact of giving access to personal data for other parties:

Example 1. A dataset D with no personal data from the point of view of party A is transmitted to party B in a country where personal data protection rules are not obligatory. This is not directly prohibited by GDPR as the regulation's scope are exclusively *personal data.* On the other hand, at the same time it may happen that for B or its partner C the data D falls into the category of personal data.

Assume now that B or C de-anonymizes the data D and publishes the resulting data D'. Neither B nor C can be accused of GDPR violations assuming that D' does not concern *offering goods or services* by B or C. Party A also may not be accused directly due to the fact that it has not performed any transfer operation of a *personal data.*

In order to avoid the problems one may adopt the following detailed interpretation of GDPR and the proposed Rule 3:

Rule 4 (Admissibility of data transfer). *A may send data D to B iff A can reasonably assume that either:*

- *D is not-personal data for B or any potential partner of B, or*
- *B complies with the GDPR obligation and has the right to keep this data.*

Note that according to Rule 4 it is legitimate to transfer data to party B that complies with the *Privacy Shield* framework.

2.3 Abandoned Data

The case so far ignored by GDPR is the situation when personal data is processed by a party P, but for some reason P becomes inactive (P may become inactive or even cease to exist). The data themselves may be hosted in a system maintained by a third party provider H. What should happen with these data? In an analogous case of a certification authority P, there are detailed rules how to manage data related to the qualified certificates if P terminates its services.

The host H has a dilemma of what to do with the personal data left by P: deleting the data is not allowed, as protection against destroying data is one of the main targets of GDPR. Erasure is allowed only if H is entitled by law or by the data subjects concerned. Keeping the data as they are, is also *processing personal data* in the sense of GDPR and requires appropriate legal grounds.

As long as the data cannot be classified as *personal data* by H, then the situation is slightly less problematic for H – there are no provisions in GDPR that would prohibit erasure of non-personal data. Therefore, in any kind of cloud services it is important for H to apply effective means of pseudonymization. The usual argument in favor of such approach is to guard against misuse of the private data by H. From the point of view of the provider H, the legal safety against supervisory authorities might be regarded as even more important.

Over time it may turn out that the used data protection mechanism becomes ineffective – e.g. due to advances in cryptanalysis. In this case H may be accused of GDPR violations when no countermeasures have been implemented. Note that H may be charged not only when key leakage is its direct fault – GDPR requires to implement appropriate safeguards against third party attacks as well.

2.4 Semantically Neutral Pseudonymization

GDPR attempts to be technologically neutral, however it frequently refers to techniques such as pseudonymization. One may have an impression of pseudonymization being the Holy Grail of personal data protection.

While pseudonymization might be extremely effective and useful in practice, we present an example showing that nontrivial semantic problems may arise. One may hope that pseudonymization or anonymization is a process that, apart from hiding the identity of the data subject, does not change the meaning of the data being processed. This is frequently the case, but not always:

Example 2. Consider in a medical dataset D containing the following note:

Alice suffers the same symptoms as her brother Bob.

There are the following options for pseudonymization of Bob's identity in the medical record of Alice (note that we do not aim to pseudonymize Alice, as she is a patient and her real ID must be available for her physician).

- Replacing *Bob* by a pseudonym X does not change the fact that X can be identified as Bob (provided that Alice has a single brother), so X becomes an *identifiable person* and the data record

  $$Alice\ suffers\ the\ same\ symptoms\ as\ her\ brother\ X$$

  still falls into the scope of *personal data* of data subject Bob. The original goal of hiding Bob's identity has not been achieved.
- In order to prevent immediate linking, we replace *her brother Bob* by X:

  $$Alice\ suffers\ the\ same\ symptoms\ as\ X$$

  This removes the link to Bob entirely, but semantically this is a different sentence. It merely says Alice is not the only person with these symptoms, while the original data record may indicate possibility of a genetic background.
- One can apply full pseudonymization and store the record

  $$Y suffers\ the\ same\ symptoms\ as\ her\ brother\ X.$$

  however this record is useless for the medical treatment of Alice.

Of course, the right for health protection of Alice may overwrite the right of Bob's privacy and this follows directly from GDPR. So, there is an excuse for not applying pseudonymization in medical records. Nevertheless, Example 2 has to show that replacing identifiers by pseudonyms cannot be regarded as a universal tool solving all problems of privacy protection:

Problem 5 (ineffective pseudonymization). There are data records for which pseudonymization either does not hide the identity of a data subject or changes semantic meaning of the data. In both cases pseudonymization fails its purpose.

2.5 Shared Personal Data

GDPR silently assumes that there is a link between a data record and at most one data subject. However, how should the rules apply, if a record D concerns more than one data subject?

Problem 6 (multiple data subjects problem). *Assume that a data record D concerns identifiable data subjects A and B. Then, a consent of which party is required according to GDPR to process D:*

- *a consent of a single party (either A or B), or*
- *a consent of both A and B?*

Example 3. A data record D stored by a controller M concerns data subjects A and B. Then A requests to store D, while B requests to remove D.

What should be the action of M in this situation? If a consent/request of one party suffices to process D, then M gets into troubles:

- if D is erased by M following the request of party B, then what about the right of A for protection of her data from erasure?
- if D is kept by M following the request of A, then what about the right-to-be-forgotten of B?

So whatever M decides to do, it can be accused of violating the obligations formulated in GDPR. If a consent of all data subjects is required, then again M is in a deadlock situation:

- M cannot continue to store D as the consent of all data subjects is missing,
- M cannot erase D as the consent of all data subjects is missing.

Problem 7 (data with multiple data subjects). *It is necessary to agree upon an interpretation or extension of GDPR so that the obligations of the data controller are defined explicitly in case of multiple data subjects of the data.*

An idea for solving this problem might be the following GDPR extension:

Rule 8 (Progressive/regressive data processing). Each data record should have a field or multiple fields *data subject.* The content of this field can be determined manually at the data creation time and can be changed during any editing operation.

The operations on personal data should be classified as progressive and regressive. A progressive operation is an operation that creates a new information contents. A regressive operation is an operation strictly limited to erasing information contents. For progressive operations a consent of all data subjects is necessary (or a corresponding legal reason replacing the consent). For regressive operations a request/withdrawal of the consent by just one data subject is enough to legitimize the operation.

If processing personal data complies with Rule 8, then the following invariant condition holds: if D is stored in a data set then we may assume that all data subjects of D gave their consent to store D in this form.

2.6 Linked Data and Local Categorization

GDPR silently assumes that data D storing personal data of a single party A can be published iff there is a consent of A or a legal reason for this. However, it may happen that there is no legal reason to publish D, A has not given a consent to publish it, but nevertheless the data gets effectively published:

Example 4. – Alice and Bob gave their consent to publish the following record D: "*Alice and Bob earn together x EUR*" in a public dataset M_1.
– Later Alice gives her consent to insert the record D': "*Alice earns y EUR*" in a public dataset M_2 – a cloud space provided by P_2.

Problem 9 (implicit personal data processing). *In the situation from Example 4, can the provider P_2 of M_2 store D' following the request of Alice?*

There are arguments that P_2 should follow the request as well as deny it:

- The data subject of D' is exclusively Alice, so according to GDPR (and a civil contract with Alice) P_2 may be obliged to store D'.
- On the other hand, publishing D' is equivalent to publication of a record D'': "*Bob earns x–y EUR.*" This happens without a consent of Bob. Moreover, the only data subject for record D'' is Bob.

Problem 10 (semantic analysis). *Is a data processor obliged to perform a semantical analysis of the request concerning data processing having in mind violations of personal data protection of people other than the explicit data subject?*

A positive answer to Problem 10 would raise the question what is the necessary scope of the analysis and how much is the data controller responsible for a misclassification? In practice any analysis may fail. For instance, the access to M_1 might be restricted (e.g. if this is a payed service or a service for a closed set of users). Moreover, P_2 may be even unaware of the existence of the data record D. Even if free access to all datasets that may contain relevant data records D is given, performing necessary analysis may enormously increase the cost and dramatically decrease efficiency of massive data processing. Indeed, an operation on a single data record would require parsing all datasets that may be related to that record (note that the records implying personal information may not be as simple as in the example, but can be, in fact, a complicated analysis of records about subset of people). The result would not be available in real time.

The only solution to this legal deadlock seems to be adoption of the following rule:

Rule 11 (extended context of a consent). *A consent of a party X concerning a data record D in a dataset M should be understood as the right to process D in M regardless of the context that may emerge outside M.*

In Example 4, given Rule 11, the right of Alice to publish D' should be unrestricted. Concurrently, when Bob gives his consent to publish D, he must keep in mind that Alice if free to disclose her personal data.

2.7 Data Aggregation

Assume that party P holds a dataset D containing personal data collected according to GDPR (e.g. the data necessary to run the contracts with the customers of P). Assume that P aggregates data from D: for instance, P may compute the average amount of money spent by the clients of P. Computing such characteristics might be commercially useful for P, but not always are the data processed for the original purpose (realization of a contract or fulfilling legal obligations).

Problem 12 (data aggregation). *Does the result of an aggregation operation fall into the category of personal data? Does aggregation fall into the category of processing personal data – based on the fact that its inputs are personal data?*

Answering the above on the grounds of GDPR is uneasy:

- The aggregated data does not concern a single person, but in the mathematical sense it brings some information about each data subject: e.g. the probability distribution for the salary of Alice might be different than the probability distribution of her salary given the median salary. Note that it may happen that the median belongs to Alice – and the exact amount is revealed. This might be a sensitive information. For instance, the income declared in the previous years is used as an authentication key for a current tax declarations in Poland. It follows that a large anonymity set is not enough to claim that the aggregation result can be revealed without violating cybersecurity conditions (regardless whether we regard it as a violation of GDPR).
- At which moment the aggregated data looses its attribute *personal data*? The legal system only considers a Boolean answer, while in reality there are plenty of possibilities in between.

There might be efforts to find simple shortcuts like: *data concerning a group of more than 10 persons is not personal data*, but it is so easy to misuse such rules in order to bypass the intended personal data protection. The concepts like differential privacy are attractive in theory, but using it in a standard practice could be a nightmare. For instance, what ϵ should be used in the context of ϵ-differential privacy? Or, since differential privacy techniques are based on probabilistic noise, what about cases where the exact sum of private data items has to be calculated for accounting or other commercial purposes?

There is the following legal dilemma:

Problem 13 (data processing classification). *In order to classify a process as processing personal data:*

- *is it enough to find that personal data are used as* ***input*** *in the process? or*
- *the personal data must appear (explicitly or implicitly) in the* ***output*** *of the process?*

A possible pragmatic solution to Problem 13 might be the following:

Rule 14 (narrow definition of data processing). *A data processing P where personal data are included in the input of P shall not be understood as processing of personal data, if the output of P (explicit and implicit) does not contain personal data.*

Rule 14 would be very useful for big data and AI computations like learning models based on private inputs which do not retain private properties. Adopting Rule 14 would also provide a positive answer to the following question:

Problem 15 (right to anonymize). *Is it legal to create a dataset Anon(D) by anonymization of all data records of D?*

A positive answer to Question 15 would solve a lot of problems concerning usage of data that are initially *contaminated* with personal data. This issue has been recognized – for this reason in the current version of GDPR there are exceptions from the general restrictions to process personal data. This touches, in particular, the issue of processing for research purposes (under the provision of respecting the fundamental rights of the data subjects). On the dark side, current exceptions from the general personal data protection rules make room for potential misuses and privacy violations. A rogue party may relatively easily masquerade personal data processing as a research activity. Adopting the right to anonymize data would eliminate the need for the GDPR exceptions in most of the cases occurring in practice. Thereby, one could eliminate the necessity of many exceptions without endangering the research targets.

Let us note that the current GDPR interpretation of European Data Protection Supervisor is not in line with Rule 14 (see Sect. 4.1).

2.8 Quorum Systems

The authors of GDPR seem to have in mind traditional data processing techniques originating from pre-electronic era: it has been silently assumed that each data record is stored in a single physical piece and there is a single party controlling the physical medium used. In such a situation, whenever a data record has to be modified, there is a corresponding physical operation executed on the corresponding physical memory location. GDPR allows a situation of more than one controller (*joint controllers*). In such a case the roles and responsibilities of the controllers have to be strictly defined.

In a quorum system this is not the case. Moreover, a quorum system can be run by a number of independent parties (in fact this is a preferred solution making a user independent from a particular service provider).

In a quorum system a data operation (read, write, erase) is initiated by separate requests to the servers of the system. The key property of the system is that some requests may result in a failure. A quorum system is immune to such failures, as long as the number of failures is limited.

According to GDPR, a data processor has strict obligations concerning data integrity. These obligations are not fulfilled by an individual member of a quorum system but by the system as a whole. On the other hand, a user has the right to exercise its own rights against any of joint controllers. Thereby, the only safe solution would be to run a quorum system by a single organization. However, this is just what we aim to avoid – a technical dependency on a single organization.

2.9 Secret Sharing

Assume that a personal data D is stored using a secret sharing scheme (e.g. a threshold system k-out-of-n) and that each share is kept by a different party. (This is a favored solution if we have to ensure that no single party can reconstruct the data).

Problem 16. How does the GDPR regulation apply to a party holding a share of a personal data according to a secret sharing scheme?

From the information theoretic point of view the data contained in a share might be purely random (e.g. for schemes based on Lagrange interpolation of polynomials), so one can argue that a share is not a data concerning a data subject. On the other hand, this data collectively with shares from other parties enables reconstruction of the personal data. So which obligations for a data processor apply in this case?

The discussion here is not purely academic – if we adopt the interpretation that a party holding only a share is not processing personal data, then among others the following problems may arise: First, a party keeping a share is no longer obliged by GDPR to protect the share from erasure or modification – indeed, these obligations concern only personal data. Second, after splitting a personal data into shares one could freely store them without a consent of a data subject – as long as each share is stored by a different party. Indeed, no consent is required by GDPR for processing non-personal data. Thereby, it would be very easy to circumvent the personal data protection requirements of GDPR. On the other hand, adopting the interpretation that storing a share falls into the scope of GDPR also creates problems: as long as at least k shares are available in a k-out-of-n threshold system, no complaints are likely. However, in case when less than k shares are intact, then who bears responsibility of the data loss? It is likely that the data subject would sue each share holder that has failed to keep it share. The legal ground would be violation of the GDPR requirement for protecting stored personal data.

2.10 Communication

A standard way of transmitting confidential data over public networks is to send the data encrypted with the key shared by the sender and the receiver (end-to-end encryption). In reality, the communication protocol may involve many operations that are not limited to forwarding the ciphertext to the destination point: the packets may be duplicated, dropped, sent over multiple paths, stored in the spooling systems, additional encryption layers may be imposed, etc. In case of a standard copyright law, these operations are not regarded as an exploitation of authors' copyright as long as they serve the original purpose of message delivery. Such an exempt does not exist in the case of GDPR: these operations are regarded as processing personal data, as long as the data themselves can be regarded as related to an identifiable person. In almost all communication protocols at least the destination address is explicitly given (the notable exception are the systems like TOR that aim to hide the identity of the communicating parties). In this case the data sent explicitly involves the receiving party. If this is a physical person, then the data fall into the category of *personal data.*

It can be argued that due to encryption the data are unreadable and therefore in some sense "erased." However, even in this case one can argue that encryption is a form of secret sharing (one share is a ciphertext or ciphertexts and one is a key or keys). Consequently, the same concerns apply. Last but not least, there is always some personal data leakage - the communication volume, which reveals the maximal Kolmogorov complexity of the plaintext data transmitted.

Just like in the case of secret sharing, creating a legal framework well capturing existing communication techniques and not creating obstacles for technical improvements is a challenging task. While there are no reports about supervisory authorities putting their hands on existing communication protocols, it cannot be excluded that in the future innovative technologies will be blocked due to such formal reasons. The supervisory authorities may for instance take a position that resilience against traffic analysis is one of GDPR requirements. Anyway, it is relatively likely that provisions of GDPR and the corresponding personal data protection acts can be used as an excuse for protectionist practices in the market of emerging communication networks (5G and beyond).

2.11 P2P Systems

Apart from other problems with distributed systems, there might be problems specific for P2P systems. One of the trouble sources is that the destination server for a given data is not known in advance – it is determined by P2P assignment of logical addresses to the servers. This, in turn, depends on the current state of the system. The owner of the data can be even unaware of the identity of the server storing his data. This prevents using a P2P system by a data controller even for storing encrypted records, as there must be a contract between the data controller and the data processor. For keeping the data by the data subject himself there are similar problems: the party processing the data has, according to GDPR, certain obligations to inform the data subject. However, in many P2P schemes the address of the data subject is not automatically available to the destination server.

Problem 17. *Certain rules of GDPR are hard or infeasible for P2P systems. The problems are caused, among others, by lack of an explicit linking between the party inserting the data and the party storing the data.*

2.12 Blockchain and Append-Only Data Systems

Recently, append-only data systems are gaining popularity, with the Blockchain as perhaps the most prominent example. The most important feature of such systems is that after a data entry is inserted into the system, it cannot be altered or erased. This is fundamental not only for cryptocurrencies, but also for achieving undeniability of transactions in the classical financial systems.

GDPR related problems start, if such a system admits storing personal data. The source of the troubles is the right-to-be-forgotten. No civil agreement may overwrite this right. A party P running the system may defend itself by storing only data that fall into the category of non-personal data. However, this does not guarantee that the problems will be avoided. For instance, a user submitting a signed data s may publish elsewhere a certificate with the public key for signature verification and own real ID. At this moment s becomes personal data, and consequently the user may exercise his right-to-be-forgotten and demand erasure of s. However, this should be technically infeasible and the system provider is trapped to violate GDPR rules.

3 De Lege Ferenda

Apart from solutions suggested in Sect. 2, we present here a few general concepts for privacy protection that would ease deploying pragmatic technical solutions while on the other hand imply high standards of personal data protection.

Processing Personal Data Versus Use of Personal Data. In the current legal situation there are strict rules for processing personal data. On the other hand, the processing itself might be unintentional and/or unconscious. For instance, as noted in Sect. 2.1 it might be non-trivial to determine whether processing of personal data takes place. In the gray area where it is unclear whether the regulation applies, a safe solution is to take precautionary steps and assume that the GDPR rules apply in full. However, such a strategy leads to severe limitations of data processing with profound negative consequences. For instance, it may hinder detection and prevention of financial criminality. Another important field of this kind is controlling the spread of infectious diseases, where adjustment to the rules of GDPR may decrease efficiency of identifying infection chains.

Most negative side effects would be avoided, if the regulation were concentrated on preventing negative consequences of a misconduct. The following limitation involves administrative fines which could serve this goal:

Rule 18. *Administrative fines may be imposed on a party P processing personal data without due diligence, if this results in:*

(a) a profit for P while the rights and freedoms of a data subject are violated, or
(b) a violation of the data subject's rights and freedoms by a third party.

Implementing Rule 18 would prevent any administrative fines when non-compliance with GDPR brings no profit to the data processor while the violations have strictly internal consequences.

The scope of Rule 18 should be understood broadly. It should apply even if the violation addressed by point (b) takes place outside the territorial scope of the GDPR. (Of course, the legal interpretation of our technical desire is in place).

Such an approach would solve some problems, e.g. in the area of data analysis. Even if big data analysis does not fully comply with the GDPR, there will be no sanctions as long as the data processor does not make profit based on a violation. Moreover, the data processor would be able to concentrate on making sure that the rights and freedoms of data subjects are respected, rather than on compliance issues. Last but not least, there might be the cases where the rights and freedoms of data subjects are overridden by other rights and higher values (the common good involving emergencies, is an example).

Responsibility. One of the sources of ineffectiveness of GDPR is focusing on administrative fines payed to the state supervisory authorities. This may result in a situation where the fines secure the state income and not the interests of the potential victims.

An alternative option would be a mechanism of an automatic compensation:

Rule 19. *On demand of a victim, a party getting a profit resulting from a violation of personal data protection rules has to pay a compensation proportional to this profit, independently of who is responsible for the original violation.*

For Rule 19, a default amount could be determined in order to ease risk analysis and eliminate legal disputes. The practice will determine the value of private information based on what a person loss is.

Due to Rule 19, a practical effect on deferring unlawful processing would be more predictable, while the penalties would lose its ad hoc character. In order to avoid penalties, any commercial activity should be supported by procedures that are transparent and provide a self-evident proof of lawful processing. A promising technique in this direction is the SPKI public key infrastructure [7]. What SPKI really provides is a user-centric framework of access control. The SPKI system of delegating access rights would provide a clear and efficient way of determining the right to process personal data.

4 State of Discussion and the Previous Work

While there are many activities regarding the GDPR regulation, there is not much focus on rethinking the general paradigms of GDPR. An implicit assumption is almost always that GDPR is not a subject for discussion and that with some (substantial) efforts one can comply with its requirements. Typically, only narrow aspects of personal data protection in selected IT systems are concerned.

The situation is somewhat surprising taking into account that GDPR itself did not declare itself sacred, and, rather, stipulates regular reviews on GDPR practice by European Commission, taking into account "the positions and findings of the European Parliament, of the Council, and of other relevant bodies or sources". So far, with a few exceptions mentioned below, there are not many such other sources, while the first review is scheduled for 2020. Below we review some ongoing and prior activities; our focus is technical though some related legal issues naturally enter into the discussion.

4.1 Activities of Authorities

European Data Protection Supervisor. Among others, the role of the EDPS is to evaluate regulatory initiatives and provide guidance to the interested parties. So one could expect that the problems of the GDPR implementation are reported in Annual Reports of EDPS. Indeed, there is such a case in the 2019 report [9]: *We received a complaint from a member of staff at an EU institution. He wanted access to his personal data relating to a harassment complaint, submitted against him by a colleague, which had been declared inadmissible.* The part easy to resolve was the invalid ground of data access refusal – protection of a potential victim while the complaint was found inadmissible. A non-trivial issue is that the accusation involved at the same time two other persons in the same institution. Finally, fulfilling the request of the complainant would include the personal data of all accused persons as well as the alleged victim. The position of the EDPS was: *We requested that the EU institution take all reasonable steps to ensure that the complainant's right of access to his personal data was granted.* Literally taken, this opinion indicates that the right to access the data by the complainant overrides the privacy rights of the victim and of the other accused persons. On the other hand, EDPS says that the EU institution should balance the rights and freedoms of the complainant and of the alleged victim. The problem is that such "balancing" is infeasible in a massive processing environment, and creates enormous costs when done manually.

Another interesting issue reported in [9] is *a temporary ban on the production of social media monitoring (SMM) reports* by EASO providing *news on the latest shifts in asylum and migration routes and smuggling offers, as well as an overview of conversations in the social media community relating to key issues.* In this case, the personal data were processed by EASO without a proper legal ground, but the result did not contain personal data, while EASO declared that they took *excess measures to ensure that no personal data was ever stored.* The ban shows that GDPR may block data processing even if the result is not violating privacy and freedoms of data subjects, while there is a clear public interest for processing. So, the current practice does not follow the proposed Rule 14.

EDPS attempts to provide guidance on interpretation of the privacy law. Quite useful are the guidelines proposing a framework for evaluation such aspects as *necessity* and *proportionality* of processing. However, they focus on privacy protection by the EU institutions and do not cover the problems arising outside the public administration.

ePrivacy. Closely related to GDPR is the ePrivacy regulation proposal [8] aiming at setting barriers for misusing personal data by providers of electronic communication. The general idea of ePrivacy is that a provider of electronic communication cannot use the data of the subscribers except for direct service maintaining. The current process of reaching an agreement upon ePrivacy is slow, with major disputes in the EU.

Quite interesting from the point of view of the problems discussed in Sect. 2.5 is admitting processing user's data by the provider *for the purpose of the provision of an explicitly requested service by an end-user for purely individual use if the requesting end-user has given consent and where such requested processing* ***does not adversely affect fundamental rights and interests of another person concerned*** *and does not exceed the duration necessary ....* This particular condition is a special case of the proposed Rule 8.

US CLOUD Act. Personal data protection is an area of conflict between the EU and the US authorities. The recent US CLOUD Act gives the US law enforcement authorities the right to access personal data processed by US providers overseas. On the other hand, the EU data protection authorities indicate that there is no common understanding regarding law enforcement and the scope of unlawful activities. This creates a very hard situation for some companies – it might be impossible to comply with data access rules of GDPR and CLOUD.

4.2 Academic Research

The academic IT community has devoted a substantial effort to ease implementation of GDPR. Usually, even if the authors point to certain difficulties, some kind of solution within the scope of the current regulation is proposed. Most of the papers present solutions addressed for very specific application areas.

GDPR is a legal concept focusing on essential legal principles to be achieved. Unfortunately, what is observed is that there is a *significant conceptual gap between legal and mathematical thinking around data privacy* (see e.g. [4]).

Compliance. Many researchers provide an evidence that implementation of GDPR creates high organizational costs. Just understanding the requirements by SME is already a substantial problem. Some works provide tools enabling translating the requirements into a form that can be addressed in an algorithmic way [11]. For bigger organizations achieving compliance becomes a very complex task due to its scale. There have been many efforts to ease this process by, say, semi-automated audits (see e.g. [1]).

Privacy Trade-offs. There are examples of substantial problems on the technical side. Among others, the authors point to metadata explosion and conflict with the previous efficiency goals in the area of database design [18], or log size explosion (a read operation has to be followed by a write operation) [16]. In case of technologies like distributed ledger, substantial problems must be solved in order to comply with the right-to-be-forgotten (RtbF) principle [10]. In some cases – like persuasive systems, where a user takes advice by looking at experience of others [17] – the GDPR requirements may block development.

It has also been observed that the procedures introduced due to GDPR may themselves create attack opportunities, despite the original intention. [12] shows such attacks based on abusing the right to access data by the data subject.

New Concepts. A interesting paradigm of the *right-not-to-be-deceived* has been proposed in [14]. It addresses the problem of user manipulation by personalization of the information contents provided to him. This could concern the cases when the user is actually willing to accept the biased information. Enforcing the *right-not-to-be-deceived* would dramatically change the current information processing landscape.

4.3 Industry and Independent Institutions

The controversies about GDPR have started already before it was adopted by the EU. In 2014, the decision by the Court of Justice of the European Union (CJEU) in the Google versus Spain case said that a search engine operator is responsible for processing personal information originating from web pages published by third parties. This decision has been later reflected by the right-to-be-forgotten and the right to data rectification – the most fundamental principles of GDPR. The controversy is that the decision creates an obligation to evaluate data and remove links not only by the parties holding the source data, but also by the parties processing already published data.

Feasibility of the right-to-be-forgotten (RtbF) has been in focus of, both, industry and the academic community [13]. One of the interesting legal concepts was *reputation bankruptcy* – a very controversial one, as the rights and freedoms of a bankrupt person are in conflict with the rights and conflicts of other persons. Balancing these rights or a selective application of RtbF in information systems is a procedural nightmare creating substantial legal risks.

RtbF creates also severe technical problems. It has been pointed out that physical removal of data may take a long time (like 180 days!). Fortunately, the GDPR regulation does not specify a concrete time limit, but refers to an *undue delay*. Nevertheless, the understanding of an *undue delay* may be different for a data controller and a supervision authority. Responding to the demand, some technical concepts have been developed. However, already an overview from 2012 by ENISA [6] has warned about discrepancy between the expectations and the technical reality. Thus far, the technological situation has not changed enough to withdraw this warning.

Another heavily discussed issue was article 22 and the right for a human review of an automated decision. While the regulation could sound reasonable for early IT systems, it ignores complexity of the modern state-of-the-art. For instance, if the decision is based on statistical correlation retrieved from big data, how one can review the outcome and explain the grounds for a data subject [15]?

The AI industry in Europe warns that GDPR creates severe problems for AI computing. Apparently, it has been overlooked by the legislators, as the immediate consequence is that the European industry is losing the competition with USA and China[1]. Article [20] points to the following critical issues:

1. *Requiring companies to manually review significant algorithmic decisions raises the overall cost of AI.*
2. *The right to explanation could reduce AI accuracy.*
3. *The right to erasure could damage AI systems.*
4. *The prohibition on re-purposing data will constrain AI innovation.*
5. *Vague rules could deter companies from using de-identified data.*
6. *The GDPR's complexity will raise the cost of using AI.*
7. *The GDPR increases regulatory risks for firms using AI.*
8. *Data-localization requirements raise AI costs.*

A deep reform of GDPR in the context of AI has been proposed in [2]. The goals proposed are *expanding authorized uses of AI in the public interest, allowing re-purposing of data that poses only minimal risk, not penalizing automated decision-making, permitting basic explanations of automated decisions, and making fines proportional to harm.* The problems for data processing has been also detected by the European authorities (see e.g. [5] for B2B communication problems). The data processing may involve very dynamic manipulation of data, while many GDPR concepts are focused on static data, which is not the emerging reality of modern information systems.

During the corona virus epidemics, it became evident that the current strict protection of personal data and lack of efficacy to make exempts from these rules for the vital public interests, makes GDPR one of the severe obstacles in the fight against corona virus [3]. Facing the catastrophic situation, Italy introduced extraordinary rules for collection and processing on medical data. However, [3] compares such rules to *sticking plaster on a wooden leg* and advocates for an international uniform framework. Definitely, the current situation proves that GDPR has been designed having in mind a standard situation. However, the most important case for evaluating efficacy of a security and safety framework are exceptional and critical situations. For example, while the Google-Apple mechanism for contact tracking carefully attempts to avoid any personal data processing, what about a request of an identifiable person (e.g. a well-known actress) to exercise the RtbF in relation to the BLE signals sent from her smart phone? Obviously, such a request when compared to the common good is a total absurd!

[1] Also in P.R.C. there are opinions pointing to the conflict between the recent cybersecurity law and its data protection chapter on one hand and feasibility of AI data processing.

5 Conclusions

We have presented and exemplified a number of areas and situation which demonstrate shortcomings in the current GDPR rules. Primarily, big data constraints, dynamic and evolving data manipulation and processing, cryptographic tools and techniques, evolving communication and distributed computing configurations, pose new ways by which private data is treated, and pause challenges to GDPR. We also proposed some suggestions (to be considered as initial attempts at remedy), and reviewed current related activities.

References

1. Arfelt, E., Basin, D., Debois, S.: Monitoring the GDPR. In: Sako, K., Schneider, S., Ryan, P.Y.A. (eds.) ESORICS 2019. LNCS, vol. 11735, pp. 681–699. Springer, Cham (2019). https://doi.org/10.1007/978-3-030-29959-0_33
2. Castro, D., Chivot, E.: The EU needs to reform the GDPR to remain competitive in the algorithmic economy. Center for Data Innovation (2019). https://www.datainnovation.org/2019/05/the-eu-needs-to-reform-the-gdpr-to-remain-competitive-in-the//-algorithmic-economy/
3. Chivot, E.: COVID-19 crisis shows limits of EU data protection rules and AI readiness. Center for Data Innovation (2020). https://www.datainnovation.org/2020/03/covid-19-crisis-shows-limits-of-eu-data-protection-rules-and//-ai-readiness/
4. Cohen, A., Nissim, K.: Towards formalizing the GDPR's notion of singling out. CoRR abs/1904.06009 (2019). http://arxiv.org/abs/1904.06009
5. Directorate-General for Communications Networks: Study on data sharing between companies in Europe. The European Commission (2018). https://publications.europa.eu/en/publication-detail/-/publication/8b8776ff-4834-11e8-be1d-01aa75ed71a1/language-en
6. Druschel, P., Backes, M., Tirtea, R.: The right to be forgotten - between expectations and practice. ENISA (2012). https://www.enisa.europa.eu/publications/the-right-to-be-forgotten/at_download/fullReport
7. Ellison, C.M.: SPKI requirements. RFC **2692**, 1–14 (1999). https://doi.org/10.17487/RFC2692
8. EU Presidency: Proposal for a Regulation of the European Parliament and of the Council concerning the respect for private life and the protection of personal data in electronic communications and repealing Directive 2002/58/EC (Regulation on Privacy and Electronic Communications) (amendments) (2020). https://privacyblogfullservice.huntonwilliamsblogs.com/wp-content/uploads/sites/28/2020/02/CONSIL_ST_5979_2020_INIT_EN_TXT.pdf
9. European Data Protection Supervisor: Annual report 2019 (2019). https://edps.europa.eu/sites/edp/files/publication/2020-03-17_annual_report_2020_en.pdf
10. Farshid, S., Reitz, A., Roßbach, P.: Design of a forgetting blockchain: A possible way to accomplish GDPR compatibility. In: Bui, T. (ed.) 52nd Hawaii International Conference on System Sciences, HICSS 2019, Grand Wailea, Maui, Hawaii, USA, 8–11 January 2019, pp. 1–9. ScholarSpace/AIS Electronic Library (AISeL) (2019). http://hdl.handle.net/10125/60145

11. Labadie, C., Legner, C.: Understanding data protection regulations from a data management perspective: a capability-based approach to EU-GDPR. In: Ludwig, T., Pipek, V. (eds.) Human Practice. Digital Ecologies. Our Future. 14. Internationale Tagung Wirtschaftsinformatik (WI 2019), 24–27 February 2019, Siegen, Germany, pp. 1292–1306. University of Siegen, Germany/AISeL (2019). https://aisel.aisnet.org/wi2019/track11/papers/3
12. Martino, M.D., Robyns, P., Weyts, W., Quax, P., Lamotte, W., Andries, K.: Personal information leakage by abusing the GDPR 'right of access'. In: Lipford, H.R. (ed.) Fifteenth Symposium on Usable Privacy and Security, SOUPS 2019, Santa Clara, CA, USA, 11–13 August 2019. USENIX Association (2019). https://www.usenix.org/conference/soups2019/presentation/dimartino
13. Politou, E.A., Alepis, E., Patsakis, C.: Forgetting personal data and revoking consent under the GDPR: challenges and proposed solutions. J. Cybersecur. **4**(1), 1–20 (2018). https://doi.org/10.1093/cybsec/tyy001
14. Reviglio, U.: Towards a right not to be deceived? An interdisciplinary analysis of media personalization in the light of the GDPR. In: Pappas, I.O., Mikalef, P., Dwivedi, Y.K., Jaccheri, L., Krogstie, J., Mäntymäki, M. (eds.) I3E 2019. IAICT, vol. 573, pp. 47–59. Springer, Cham (2020). https://doi.org/10.1007/978-3-030-39634-3_5
15. Roig, A.: Safeguards for the right not to be subject to a decision based solely on automated processing (article 22 GDPR). Eur. J. Law Technol. **8**(3) (2017). http://ejlt.org/article/view/570
16. Shah, A., Banakar, V., Shastri, S., Wasserman, M., Chidambaram, V.: Analyzing the impact of GDPR on storage systems. In: Peek, D., Yadgar, G. (eds.) 11th USENIX Workshop on Hot Topics in Storage and File Systems, HotStorage 2019, Renton, WA, USA, 8–9 July 2019. USENIX Association (2019). https://www.usenix.org/conference/hotstorage19/presentation/banakar
17. Shao, X., Oinas-Kukkonen, H.: How does GDPR (General Data Protection Regulation) affect persuasive system design: design requirements and cost implications. In: Oinas-Kukkonen, H., Win, K.T., Karapanos, E., Karppinen, P., Kyza, E. (eds.) PERSUASIVE 2019. LNCS, vol. 11433, pp. 168–173. Springer, Cham (2019). https://doi.org/10.1007/978-3-030-17287-9_14
18. Shastri, S., Banakar, V., Wasserman, M., Kumar, A., Chidambaram, V.: Understanding and benchmarking the impact of GDPR on database systems. PVLDB **13**(7), 1064–1077 (2020). http://www.vldb.org/pvldb/vol13/p1064-shastri.pdf
19. The European Parliament and the Council of the European Union: Regulation (EU) 2016/679 of the European Parliament and of the Council of 27 April 2016 on the protection of natural persons with regard to the processing of personal data and on the free movement of such data, and repealing Directive 95/46/ec (General Data Protection Regulation). Off. J. Eur. Union **119**(1) (2016)
20. Wallace, N., Castro, D.: The impact of the EU's new data protection regulation on AI. Center for Data Innovation (2018). http://www2.datainnovation.org/2018-impact-gdpr-ai.pdf

Author Index